Interactive LearningWare for General Chemistry CD and On-Line Learning Companion

The CD packaged with this text includes the features listed below, and the website password card included with this text allows students access to the On-Line Learning Companion described as well below.

Interactive LearningWare for General Chemistry CD A number of the chapter-ending exercises appear in the LearningWare as interactive exercises, intended to help students approach and solve these problems in an environment that patiently and carefully helps them through the problem. The exercises represent a wide range of core chemistry topics and complexity, and allow the student to reason through both qualitative and quantitative steps in solving the problems. Animations also help students visualize the chemistry involved in the problems. The text exercises converted into Interactive LearningWare exercises have been designated with a CD icon. Additional LearningWare exercises will be found at the website. The Interactive LearningWare for General Chemistry exercises were conceptualized and scripted by the authors James Brady and Joel Russell, and developed by Lenox SoftWorks.

Some of the exercises that appear as Interactive LearningWare exercises are:

- 1.75, 1.89, 3.32, 3.61, 3.72, 3.89, 3.95, 3.99, 3.103, 3.112, 3.117, 3.123, 4.53, 4.71, 5.40, 5.64, 6.52, 6.74, 6.80, 8.85, 8.101, 9.56, 10.47, 10.67, 10.77, 10.85, 16.41, 16.52, 16.82, 16.84

The On-Line Learning Companion The main goal of the website is to provide further conceptual underpinnings for the textbook's explanations of key concepts and problem solving approaches, and to allow students to test their understanding in an interactive on-line format.

For each chapter of the textbook, there will be a corresponding chapter on the website containing several features:

- *Overview.* This shows the main points of the chapter and the web-based multimedia assets developed for that chapter, so students can get a guided tour of the on-line materials.
- *Concept Visualizations.* These animations and videoclips show key figures from the chapters in an animated format, to allow students to see chemical processes.
- *Interactive Activities.* These exercises and simulations allow students to explore concepts in an interactive way; for example, an activity may ask students to predict outcomes of reactions or chemical systems, and give feedback on their predictions. There are 2-4 Interactive Activities per chapter.
- *Answers to "Thinking It Through" questions from the textbook.* These answers will have some animations to help students see the conceptual explanations.
- *Quizzing.* Using the WebTest engine, students will be able to take quizzes on-line, get limited feedback, and have the ability to work both on and off-line.

The On-Line Learning Companion has been conceptualized, scripted, and partially developed by Professor Gabriela Weaver, University of Colorado, Denver, and also developed by Lenox SoftWorks.

CHEMISTRY
Matter and Its Changes

3E

James E. Brady
St. John's University, New York

Joel W. Russell
Oakland University, Michigan

John R. Holum
Augsburg (Emeritus), Minnesota

John Wiley & Sons, Inc.

New York / Chichester / Weinheim / Brisbane / Singapore / Toronto

ACQUISITIONS EDITOR	Cliff Mills
DEVELOPMENT EDITOR	Beth Bortz
SENIOR PRODUCTION EDITOR	Elizabeth Swain
SENIOR MARKETING MANAGER	Charity Robey
DESIGNER	Dawn L. Stanley
ILLUSTRATION EDITORS	Sigmund Malinowski, Edward Starr
PHOTO EDITOR	Hilary Newman
PHOTO RESEARCHER	Ramon Rivera Moret
ELECTRONIC ILLUSTRATIONS	Precision Graphics
COVER ART	Norm Christensen
COVER PHOTOS	*Welder:* ©Lynn Johnson/AURORA
	Fireworks: ©Tony Wiles/Tony Stone Images/New York, Inc.
	Hot-air balloon: ©Donovan Reese/Tony Stone Images/New York, Inc.
	Lighthouse: ©John Lund/Tony Stone Images/New York, Inc.
	Diamond: ©Color Day Productions/The Image Bank

This book was set in 10/12 Times Ten by Progressive and printed and bound by Von Hoffmann. The cover was printed by Lehigh.

This book is printed on acid-free paper.

To order books or for customer service, call 1(800)-CALL-WILEY (225-5945).

Library of Congress Cataloging-in-Publication Data

Brady, James E., 1938–
    Chemistry : the study of matter and its changes / James E. Brady, Joel W. Russell,
John R. Holum.--3rd ed.
      p. cm.
    Includes bibliographical references and index.
    ISBN 0-471-18476-4 (acid-free paper)
      1. Chemistry. I. Title. II. Russell, Joel W. III. Holum, John R.

QD33.B82 2000
540 21--dc21
                                              99-042979

Printed in the United States of America

10 9 8 7 6 5 4

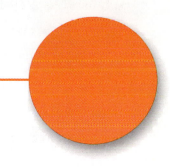

Preface

It is with a great deal of pleasure that we welcome Joel W. Russell to the author team. Such decisions are not made lightly. The potential co-author should be on top of developments in the field, be a leader in the discipline, and enjoy writing and teaching. Joel wonderfully fills these criteria. His work on this new edition greatly enhances the text.

As we approached the revision of this text, we renewed our commitment to provide a book that is appropriate for the mainstream general chemistry course for college science majors (e.g., chemistry, biology, pre-med). Our goal was to fashion a learning system that is capable of challenging the better student while providing a learning environment that will enable the less well prepared student to succeed. To accomplish this, we built on the strong features of previous editions, such as the writing style, student-friendly attitude, and clarity and thoroughness of our explanations of difficult concepts. We then modified the topic sequence somewhat, expanded the treatment of some topics and trimmed the treatment of others. We also strengthened the text's ability to foster a conceptual understanding of processes that take place at the molecular level. The result, we believe, is a textbook well suited to the needs of all students taking the course.

Overall Philosophy and Goals

The philosophy of the text continues to be based on our conviction that a general chemistry course serves a variety of goals in the education of a student. First, of course, it must provide a sound foundation in the basic facts and concepts of chemistry upon which the theoretical models of chemistry can be constructed. The general chemistry course should also give the student an appreciation of the central role that chemistry plays among the sciences as well as the importance of chemistry in society and day-to-day living. In addition, it should enable the student to develop skills in analytical thinking and problem solving.

Features That Serve the Philosophy and Goals

To fulfill the goals describe above, we've focused on a number of major areas, taking into account during the revision process comments and suggestions by reviewers, users of the text, and others.

Organization and Concept Development

From the outset our aim has been to provide a logical progression of topics that provide the maximum flexibility for the instructor in organizing his or her course. To improve upon this goal, we have made some significant changes in organization as well as revisions of individual chapters. The sequence of chapters provides an overview of the development of concepts in the text.

Chapters 1 through 6 develop a foundation in reaction chemistry, stoichiometry, and thermochemistry, with a basic introduction to the structure of matter and the periodic table.

To enable students to get an early start on understanding energy concepts, particularly the difference between heat and temperature, we have introduced some basic notions of kinetic theory in Chapter 1 when we first discuss the concept of energy. These are expanded upon further in Chapter 6 (Energy and Thermochemistry).

Because so many of the substances we encounter on a daily basis are organic compounds, we provide an early but *brief* overview of the kinds of compounds formed by carbon. Some simple hydrocarbons are introduced in Chapter 2 and the structural features of the most common functional groups are defined in Chapter 8 (Chemical Bonding: General Concepts). This overview will make it easier for students to understand later discussions of weak acids and bases as well as the physical properties of liquids, where organic compounds are often used as examples during discussions of basic principles.

In Chapter 3 we've extended the discussion of stoichiometry to include reactions in solution and molarity, which prepares students for meaningful experiments in the laboratory. We then apply these concepts to metathesis and redox reactions in Chapters 4 and 5.

Chapter 4 of the second edition has been divided into two chapters. Chapter 4 now deals with simple acid-base chemistry and metathesis reactions. Chapter 5, focusing on redox reactions, includes discussions of the activity series of metals and oxidation reactions involving oxygen. These two chapters provide students with a foundation in the basic descriptive chemistry of solution reactions and give students a knowledge base that serves as a foundation for theoretical concepts developed in Chapters 7–9, which deal with atomic structure and bonding.

Chapters 7 through 9 cover electronic structures of atoms and bonding in compounds. In Chapter 7 (Atomic and Electronic Structure), discussions of irregularities in periodic trends in ionization energy and electron affinity are covered as a special topic ("Facets of Chemistry"). This treatment enables teachers who do not wish to dwell on these details to easily omit them.

The discussion of bonding is divided between two chapters. Chapter 8 treats the topic at a relatively elementary level, describing the principal features of ionic and covalent bonds using Lewis structures. Chapter 9 deals with molecular structure (VSEPR theory) and the valence bond and molecular orbital theories. Lewis acids and bases have been moved to a later chapter.

Chapters 10 through 12 focus on the physical properties of the states of matter and solutions. In Chapter 11, we have expanded the discussion of factors that affect the strengths of London forces.

Chapters 13 and 14 examine rates of chemical reactions and chemical equilibrium. These are followed by a new Chapter 15 (Acids and Bases: A Second Look), which brings together the various views of acids and bases, including Brønsted and Lewis acid-base concepts. The pH concept is introduced in Chapter 15 and applied to solutions of strong acids and bases.

Chapters 16 and 17 cover equilibria in aqueous solutions. Equilibria involving weak acids and bases, including polyprotic acids and their salts, are now all discussed in the same chapter (Chapter 16). As in the last edition, treatment of problems requiring the quadratic formula or the method of successive approximations are placed in a separate section. Chapter 17 now deals with solubility and complex ion equilibria.

The discussion of thermodynamics has been moved to Chapter 18. By placing this subject after the chapters on equilibria, we were able to incorporate the treatment of thermodynamic equilibrium constants. In this edition, we have expanded the treatment of the second law.

Chapter 19 (Electrochemistry) ties together concepts of thermodynamics and equilibrium as well as practical applications of electrolysis and galvanic cells. We have added a discussion of the use of standard reduction potentials in predicting electrolysis reactions. Our treatment of concepts in this chapter has been highly rated; we have strived to improve the clarity of discussions even further in order to avoid student conceptual errors.

Following the chapter on electrochemistry, most texts include several chapters on descriptive inorganic chemistry. Here we take a somewhat radical approach. Recognizing the fact that most instructors do not have time to teach much of this material, we have placed most of the classical descriptive chemistry in a supplement entitled Descriptive Chemistry of the Elements. Within the text itself we provide two chapters (one of them new to this edition) that deal with descriptive chemistry. With our unique approach we look at trends in properties that can be explained using principles of bonding, thermodynamics and kinetics developed in earlier chapters. We also examine trends in properties and structure that extend across periods and down groups in the periodic table.

Chapter 20 examines metals and what kinds of reactions are required to recover them from their compounds. We also study covalent character of metal compounds and the factors that affect their colors. This chapter also includes a discussion of structures, nomenclature and bonding involving complex ions, particularly those of the transition metals. Chapter 21 looks at nonmetals and includes discussions of the structures of the free elements, preparation of hydrogen and oxygen compounds of the nonmentals, and the structural relationships among their oxides, oxoacids, and polymeric oxoacids, oxoanions, and oxides.

Chapter 22 presents an overview of nuclear reactions and the role they play in chemistry and society. This chapter, which could actually be presented earlier in the course should the instructor elect to do so, has been revised to keep it up to date.

Chapter 22, the final chapter of the text, serves as an introduction to organic and biochemistry, and includes some elements of polymer chemistry.

Level

We have developed discussions carefully to provide students with clear, easily understood explanations of difficult topics while maintaining a light and student-friendly writing style. As in the previous edition, students are not assumed to have had a previous course in chemistry and are expected to have mastered only basic algebra. The level of the discussions in this book is probably best judged by the topic coverage in Chapters 13–18, those that deal with the driving forces for reactions, with chemical equilibria, and with exercises calling for a combination of thinking and computational skills. We believe that the level is right for the mainstream general college chemistry course.

Learning Features

Aware of the importance of problem solving and analytical thinking, we have adopted a unique approach to developing thinking skills. We distinguish three types of learning aids: those that further comprehension and learning; those that enhance problem-solving skills; and those that extend the breadth and knowledge of the student.

Features That Further Comprehension and Learning

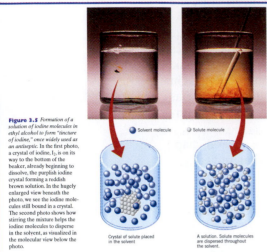

Figure 3.5 *Formation of a solution of iodine molecules in ethyl alcohol to form "tincture of iodine," once widely used as an antiseptic.* In the first photo, a crystal of iodine, I_2, is on its way to the bottom of the beaker, already beginning to dissolve, the purplish iodine crystal forming a reddish brown solution. In the hugely enlarged view beneath the photo, we see the iodine molecules still bound in a crystal. The second photo shows how stirring the mixture helps the iodine molecules to disperse in the solvent, as visualized in the molecular view below the photo.

● Solvent molecule   ○ Solute molecule

Crystal of solute placed in the solvent

A solution. Solute molecules are dispersed throughout the solvent.

Macro-to-Micro Illustrations. To help students make the connection between the macroscopic world we see and events that take place at the molecular level, we have added many illustrations that combine both views. A photograph, for example, will show a chemical reaction as well as an artist's rendition of the chemical interpretation of what is taking place between the atoms, molecules, or ions involved. The goal is to show how models of nature enable chemists to better understand their observations and to get students to visualize events at the molecular level.

Margin comments make it easy to enrich a discussion, without carrying the aura of being essential. Some margin comments jog the student's memory concerning a definition of a term.

Periodic Table Correlations. One of our goals was to call particular attention to the usefulness of the periodic table in correlating chemical and physical properties of the elements.

Boldface terms alert the student to "must-learn" items. Especially important equations are highlighted with a yellow background.

Problem Analysis at a Glance. Where appropriate, figures contain flow-charts that summarize the relationships involved in solving problems, the approach used to analyze the method of attack on problems, or the approach to applying rules of chemical nomenclature.

Chapter summaries use the boldface terms to show how the terms fit into statements that summarize concepts.

Features that Enhance Problem-Solving and Critical Thinking Skills

Many students entering college today lack experience in analytical thinking. Because problem solving in chemistry operates on two levels, a course in chemistry should provide an ideal opportunity to help students sharpen their reasoning skills. In addition to mathematics many problems also involve the application of theoretical concepts. Students have difficulty at both levels, and one of the goals of this text has been to develop a unified approach that addresses each level.

Chemical Tools Approach to Problem Analysis. In studying chemistry, students learn

Tools *you have learned*

The table below lists the tools that you have learned in this chapter. You will need them to answer chemistry questions. Review them if necessary, and refer to the tools when working on the Thinking-It-Through problems and the Review Problems that follow.

Tool	Function
Subscripts in a formula *(page 50)*	Specify the number of atoms of each element in a formula unit of a substance. You must be able to interpret the formula in terms of the numbers of atoms of each element expressed by the formula.
Rules for writing formulas of ionic compounds *(page 73)*	Permit us to write correct chemical formulas for ionic compounds. You will need to learn to use the periodic table to remember the charges on the cations and anions of the representative metals and nonmetals. You also should learn the ions formed by the transition and post-transition metals in Table 2.5, and it is essential that you learn the names and formulas of the polyatomic ions in Table 2.6.
Rules for naming compounds *(pages 81, 83, 84)*	Allow us to write a formula from a name, and a name from a formula.

many simple skills, such as finding the number of grams in a mole of a substance or writing the Lewis structure of a molecule. Problem solving involves
bringing together a sequence of such simple tasks. Therefore, if we are to teach
problem solving, we must teach students how to seek out the necessary relationships required to obtain solutions to problems.

Our innovative approach to problem solving makes use of an analogy between the abstract tools of chemistry and the concrete tools of a mechanic. Students are taught to think of simple skills as tools that can be used to solve more
complex problems. When faced with a new problem, the student is encouraged
to examine the tools that have been taught and to select those that bear on the
problem at hand.

To foster this approach, we present a comprehensive program of reinforcement and review:

Tools Icon. The Tools icon in the margin calls attention to each chemical tool
when it is first introduced and is accompanied by a brief statement that identifies the tool. Following the Summary at the end of the chapter, the tools are reviewed under the heading **Tools You Have Learned,** preparing students for the
exercises that follow.

Worked Examples generally follow a three-step process. They begin with an
Analysis that describes the thought processes involved in selecting the tools to
solve the problem. This is followed by the *Solution* in which the problem is
solved following the plan developed in the *Analysis.* Examples usually conclude
with a section titled *Is the Answer Reasonable?* in which the answer is studied to
see whether it "makes sense."

Practice Exercises follow the worked examples to enable the student to apply
what has just been studied to a similar problem.

EXAMPLE 3.3
Converting Moles
to Grams

Gold nugget and bars.

How many grams of gold are in 0.150 mol of Au?

Analysis: The question can be restated as follows.

$$0.150 \text{ mol Au} \Leftrightarrow ? \text{ g Au}$$

We'll round the atomic mass of gold from 196.96655 to 197.0, so the following is
true about gold.

$$1 \text{ mol Au} \Leftrightarrow 197.0 \text{ g Au}$$

The conversion factor that we need to convert the given number of moles of Au to
the number of grams of Au comes from this equivalency.

Solution: The given is 0.150 mol Au, so we multiply it by the ratio of grams of Au
to moles of Au obtained from the atomic mass of gold.

$$0.150 \text{ mol Au} \times \frac{197.0 \text{ g Au}}{1 \text{ mol Au}} = 29.6 \text{ g Au}$$

Thus, we can write 0.150 mol Au $\Leftrightarrow$ 29.6 g Au.

Is the Answer Reasonable?
We have between 1/10 and 2/10 of a mole of gold. One-tenth would weigh 19.7 g
(about 20 g) and two-tenths of a mole would weigh approximately 40 g. Our answer of about 30 g seems just right.

Practice Exercise 3

How many grams of silver are in 0.263 mol of Ag? ◆

Thinking it Through is a selection of end-of-chapter problems that focuses on the information and method needed to solve the problem, rather than the answer itself. The intent of the Thinking It Through problems is to develop the student's ability to solve conceptual problems and to further accustom the student to analyzing a problem before trying to carry out any computations. The problems are divided into two categories. Students who have developed reasonably good thinking skills can skip the Level 1 problems and proceed directly to the more challenging problems in Level 2.

THINKING IT THROUGH

The goal in the following problems is *not* to find the answer itself; instead, you are asked to assemble the available information needed to obtain the answer, state what additional data (if any) are needed, and describe *how* you would use the data to answer the question. Where problems can be solved by the factor-label method, set up the solution to the problem by arranging the conversion factors so the units cancel correctly to give the desired units of the answer.

The problems are divided into two groups. Those in Level 2 are more challenging than those in Level 1 and provide an opportunity to really hone your problem solving skills. Detailed answers to these problems can be found at our Web site: http://www.wiley.com/college/brady

Level 1 Problems

1. How many times heavier than an atom of ^{12}C is the average atom of iron? (Explain how you can obtain the answer.)

2. Suppose the atomic mass unit had been defined as $\frac{1}{10}$th of the mass of an atom of phosphorus. What would the atomic mass of carbon be on this scale?

3. How do we find the chemical symbol for a particle that has 43 electrons outside a nucleus which contains 45 protons and 58 neutrons?

Level 2 Problems

9. Carbon forms a compound with element X that has the formula CX_4. A sample of this compound was analyzed and found to contain 1.50 g of C and 39.95 g of X. Explain how you would calculate the atomic mass of X.

10. In an experiment it was found that 3.50 g of phosphorus combines with 12.0 g of chlorine to form the compound PCl_3. How many grams of chlorine would combine with 8.50 g of phosphorus to form this same compound? (Set up the calculation.)

End-of-Chapter Exercises have been divided into three sections. **Review Questions,** classified according to topic, enable students to gauge their progress in learning concepts presented in a chapter. **Review Problems,** also classified according to topic, are presented in pairs of similar problems, with the first member of each pair having its answer in Appendix D. These problems provide routine practice in the use of basic tools as well as opportunities to incorporate these tools in the solution of more complex problems. **Additional Exercises** at the ends of the problem sets are unclassified. Many problems are cumulative, requiring two or more concepts, and several in later chapters require skills learned in earlier chapters. In this edition we have added a number of more difficult problems to both the **Review Problems** and **Additional Exercises.** These serve to increase the range of problem difficulty available to the instructor when assigning homework.

Features That Extend the Breadth of Knowledge of the Student

Facets of Chemistry. The topics of these brief boxes range from historical background to chemistry and the environment and other applications of chemistry. A full list is given on p. xvii.

Facets of Chemistry 4.1

Boiler Scale and Hard Water

Precipitation reactions occur around us all the time and we hardly ever take notice until they cause a problem. One common problem is caused by **hard water**—groundwater that contains the "hardness ions," Ca^{2+}, Mg^{2+}, Fe^{2+}, or Fe^{3+}, in concentrations high enough to form precipitates with ordinary soap. Soap normally consists of the

Like most gases, carbon dioxide becomes less soluble as the temperature is raised, so CO_2 is driven from the hot solution and the HCO_3^- is gradually converted to CO_3^{2-}. As the carbonate ions form, they are able to precipitate the hardness ions. This precipitate, which sticks to the inner walls of pipes and hot water boilers, is called *boiler scale*. In

Chemicals in Our World. These are two-page, illustrated between-chapter essays following odd-numbered chapters. The essays offer students the opportunity to learn about real-world applications of chemistry in industry, medicine, and the environment. Essay topics include "Xenon—*Star Wars* Propulsion 'Fuel' Today," "Greenhouse Gases and Global Warming," "The Carbonate Buffer and Mountaineering," "Synthetic Diamonds and Diamond Coatings," and "DNA Biochips."

Chemistry in Practice. These very brief descriptions of practical applications of chemistry are interspersed throughout the text.

> ▶**Chemistry in Practice**◀ Engineers who design vehicle engines or furnaces, both of which produce carbon dioxide, CO_2, always want to supply enough air so that oxygen is in sufficient supply to make CO_2, and not the highly poisonous carbon monoxide, CO. In a poor supply of oxygen, much CO forms. ◆

Illustrations. Most of the illustrations, which are distributed liberally throughout the text, have been redrawn using modern computer techniques to provide accurate, eye-appealing complements to discussions. Color is used constructively rather than for its own sake. For example, a consistent set of colors is used to identify atoms of the elements in drawings that illustrate molecular structure. These are shown in the margin.

Photographs. The large number of striking photographs in the book serve two purposes. One is to provide a sense of reality and color to the chemical and physical phenomena described in the text. Toward this end, many photographs of chemicals and chemical reactions are included. The other purpose of the photographs is to illustrate how chemistry relates to the world outside the laboratory. The chapter-opening photos, for example, call the students' attention to the relationship between the chapter's content and common (and often not-so-common) things. Similar photos within the chapters illustrate practical examples and applications of chemical reactions and physical phenomena.

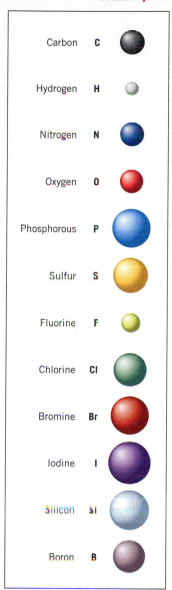

Carbon	**C**
Hydrogen	**H**
Nitrogen	**N**
Oxygen	**O**
Phosphorous	**P**
Sulfur	**S**
Fluorine	**F**
Chlorine	**Cl**
Bromine	**Br**
Iodine	**I**
Silicon	**Si**
Boron	**B**

Supplements

A comprehensive package of supplements has been created to assist both the instructor and the student and includes the following:

Study Guide by James E. Brady and John R. Holum. This Guide has been written to further enhance understanding of concepts. It is an invaluable tool for students and contains chapter overviews, additional worked-out problems, and alternative problem-solving approaches as well as extensive review exercises.

Chemistry LearningWare. A selection of the chapter-ending exercises have been converted into Chemistry LearningWare interactive exercises, intended to help students solve these problems in a patient and highly interactive tutorial format. The exercises represent a wide range of core chemistry topics and complexity and allow the student to reason through both conceptual and quantitative steps in solving the posed problems. Animations also help students visualize the chemistry involved in the problems. The text exercises converted into LearningWare exercises have been designated with a CD icon. Additional LearningWare exercises

will be found at the website. The Chemistry LearningWare exercises were conceptualized and scripted by the authors James Brady and Joel Russell, and were developed by Lenox Softworks.

Chemistry Website. The Website to accompany the text provides chapter
overviews, animations of key topics, answers to the "Thinking It Through" questions, and additional self-quizzing and assessment. The site was conceptualized
and developed by Professor Gabriela Weaver of the University of Colorado,
Denver, and co-developed by Lenox Softworks.

Solutions Manual by Michael Kenney of Crabtree & Company and Cynthia K.
Anderson of Robert E. Lee High School. The Manual contains worked-out solutions for text problems whose answers appear in Appendix D.

Laboratory Manual for Principles of General Chemistry, Sixth Edition, by Jo
Beran, of Texas A & M University, Kingsville. This comprehensive laboratory
manual is for use in the general chemistry course. Beran has been popular for
five editions because of its broad selection of topics and experiments, and for its
clear layout and design. Containing enough material for two or three terms, this
lab manual emphasizes techniques, helping students learn the time and situation
for their correct use. The accompanying Instructor's Manual presents the details
of each experiment, including overviews, an instructor's lecture outline, and
teaching hints. The IM also contains answers to the prelab assignment and laboratory questions.

Take Note! A notebook for students that contains black and white versions of
key figures from the text and allows students to easily take notes directly on the
pages during lecture.

Instructor's Manual by Mark A. Benvenuto of University of Detroit—Mercy.
In addition to lecture outlines, alternate syllabi, and chapter overviews, this
Manual contains suggestions for small group active-learning projects, class discussions, and short writing projects. Conversion notes are included allowing the
instructor to use class notes from other texts.

Test Bank by Raymond X. Williams of Howard University. The Text Bank contains over 1,800 test questions, including multiple-choice, true-false, short answer
questions, and critical thinking problems.

Computerized Test Bank. IBM and Macintosh versions of the entire Test Bank
are available with full editing features to help the instructor customize tests.

Four-Color Overhead Transparencies. Over 125 four-color illustrations from
the text are provided in a form suitable for projections in the classroom.

Instructor's Resource CD-ROM. This CD-ROM, which contains four color illustrations and photos from the text, allows instructors to create and save slide
shows, add their own text/illustration slides to presentations, and create four-
color transparencies.

JAMES E. BRADY
JOEL W. RUSSELL
JOHN R. HOLUM

Acknowledgments

We begin with fond words of thanks to June Brady and Susan Russell for their constant support and encouragement, and we remember with love and gratitude the support that Mary Holum gave to all of the work involved with the several versions and editions of this text. We express thanks to our children Liz, Ann, and Kathryn Holum, Mark and Karen Brady, and Andrea, Natalie, Ethan, Theresa, and Joel Russell for their love and inspiration. Our appreciation is also extended to our colleague Sandra Olmsted of Augsburg College, to Ernest Birnbaum, Neil Jespersen, and William Pasfield of St. John's University, and to Mike Sevilla and Mark Severson of Oakland University for their helpful discussions and stimulating ideas. It is with particular pleasure that we thank the staff at Wiley for their careful work, encouragement, and sense of humor, particularly our editor Cliff Mills, our developmental editor Beth Bortz, our picture editor Hilary Newman, our designer Dawn Stanley, our illustrator Sigmund Malinowski, and our production supervisor Elizabeth Swain. We also extend our most sincere appreciation to Connie Parks for her diligence with the copy editing and proofreading, and to Krisztina Varga of St. John's University and Michael Kenney of Crabtree & Co. for their assistance with the development of the problem sets and answers.

We express gratitude especially to the colleagues listed below, whose careful reviews, helpful suggestions, and thoughtful criticisms of the manuscript have been of such great value to us in developing this book. We also appreciate the suggestions for end-of-chapter Exercises by some reviewers of the text and we have cited their specific contributions.

Reviewers of the Third Edition

Roma Advani
Prairie State University

Hugh Akers
Lamar University

Robert D. Allendoerfer
State University of New York, Buffalo

George C. Bandik
University of Pittsburgh

Mark A. Benvenuto
University of Detroit-Mercy

Steven W. Buckner
Columbus State University

Jerry Burns
Pellissippi State Technical College

Deborah Carey
Stark Learning Center

Jefferson D. Cavalieri
Dutchess Community College

Michael Chetcuti
University of Notre Dame

Ronald J. Clark
Florida State University

Wendy Clevenger
University of Tennessee at Chattanooga

Paul S. Cohen
The College of New Jersey

John E. Davidson
Eastern Kentucky University

William B. Euler
University of Rhode Island

John H. Forsberg
St. Louis University

Ronald A. Garber
California State University, Long Beach

John I. Gelder
Oklahoma State University

David B. Green
Pepperdine University

Michael Guttman
Miami-Dade Community College

Daniel T. Haworth
Marquette University

Carl A. Hoeger
University of California, San Diego

Paul A. Horton
Indian River Community College

Thomas Huang
East Tennessee State University

Peter Iyere
Tennessee State University

Denley Jacobson
Purdue University

Louis J. Kirschenbaum
University of Rhode Island

Barbara McGoldrick
Union County College

William A. Meena
Rock Valley College

Brian H. Nordstrom
Embry-Riddle Aeronautical University

Robert H. Paine
Rochester Institute of Technology

Naresh Pandya
Kapiolani Community College

Jerry L. Sarquis
Miami University

Paula Secondo
Western Connecticut State University

Venkatesh Shanbhag
Mississippi State University

S. Paul Steed
Sacramento City College

David White
State University of New York, New Paltz

S.D. Worley
Auburn University

David Young
Ohio University

E. Peter Zurbach
St. Joseph's University

Reviewers of the Second Edition

Wesley Bentz
Alfred University

Mark Benvenuto
University of Detroit-Mercy

C. Eugene Burchill
University of Manitoba

Kathleen Crago
Loyola University

David Dobberpuhl
Creighton University

Joseph Dreisbach
University of Scranton

Barbara Drescher
Middlesex County College

Thomas Greenbowe
Iowa State University

Peter Hambright
Howard University

Henry Harris
Armstrong State College

Ernest Kho
University of Hawaii-Hilo

Ken Loach
SUNY-Plattsburgh

Glen Loppnow
University of Alberta

David Marten
Westmont College

Jeanette Medina
SUNY-Geneseo

Robert Nakon
West Virginia University

Brian Nordstrom
Embry-Riddle Aeronautical University

Sabrina Godfrey Novick
Hofstra University

Les Pesterfield
Western Kentucky University

Ronald See
St. Louis University

Anton Shurpik
U.S. Merchant Marine Academy

Reuben Simoyi
West Virginia University

Agnes Tenney
University of Portland

Roselin Wagner
Hofstra University

Reviewers of the First Edition

Mark Amman
Alfred State College

Dale Arrington
South Dakota School of Mines and Technology

Keith O. Berry
Oklahoma State University

William Bitner
Alvin Community College

Simon Bott
University of North Texas

Donald Brandvold
New Mexico Institute of Mining and Technology

Robert F. Bryan
University of Virginia

Barbara A. Burke
California State Polytechnic University-Pomona

Judith Burstyn
University of Wisconsin

Henry Daley
Bridgewater State College

Diana Daniel
General Motors Institute

William Davies
Emporia State University

William Deese
Louisiana Tech University

David F. Dever
Macon College

Wendy Elcesser
Indiana University of Pennsylvania

James Farrar
University of Rochester

David Frank
Ferris State University

DonnaJean A. Fredeen
Southern Connecticut State University

Donna Friedman
Florrisant Valley Community College

Paul Gaus
College of Wooster

Andrew Jorgensen
University of Toledo

Wendy L. Keeney-Kennicutt
Texas A & M University

Janice Kelland
Memorial University of Newfoundland

Henry C. Kelly
Texas Christian University

Reynold Kero
Saddleback College

Nina Klein
Montana Tech

Larry Krannich
University of Alabama-Birmingham

Robert M. Kren
University of Michigan-Flint

Russell D. Larsen
Texas Technical University

Patricia Moyer
Phoenix College

James Niewahner
Northern Kentucky University

Karl Seff
University of Hawaii

Edward Senkbeil
Salisbury State University

Ralph W. Sheets
Southwest Missouri State University

Mary Sohn
Florida Institute of Technology

Darel Straub
University of Pittsburgh

Warren Yeakel
Henry Ford Community College

To The Student

You are about to begin what could be one of the most exciting courses that you will undertake in college. It offers you the opportunity to learn what makes our world "tick" and to gain insight into the roles that natural and synthetic chemicals play in nature and in our society. This knowledge will not come without some effort, however, and this book has been carefully designed and written with an awareness of the kinds of difficulties you may face. Therefore, before you begin, let us outline some of the key features of the book that will aid you in your studies.

Chemical Vocabulary.   Every specialty has its own unique vocabulary, and chemistry is certainly no exception. In this text you will find new terms set in bold type when you first encounter them. *We have not assumed that you've had a previous course in chemistry;* thus these terms will not be used until they have been clearly defined. New terms are also set in bold type in the Summary that accompanies each chapter, and they are included in a Glossary at the end of the book, so that you can look up their meanings at a later time if necessary.

Problem Solving.   Students often equate problem solving with doing mathematical calculations. It is really much more than that. Solving a problem involves *thinking* and *analyzing information*. This is true whether the problem relates to chemistry, physics, psychology, or to some aspect of your private life. Learning to solve problems is learning *how* to think and *how* to incorporate information into the solution.

When first faced with a genuine problem, many people have no idea where to begin. One of the goals of this text is to provide you with a way to think about problem solving, so that you can be successful. You may find it difficult and slow going at first, but you can succeed if you're willing to put in the effort. And it is an effort that will pay off, not only in your chemistry course, but in other endeavors as well.

The Chemical Toolkit Approach to Problem Solving.   Imagine for a moment what it would be like to be an auto mechanic. Among other things, you would need to be familiar with an assortment of tools. Then, when faced with a repair job, you would study the problem and select the proper tools for the work.

Solving chemistry problems is not much different from repairing a car; we simply use different tools. When faced with a problem, we study it to see which

xv

of the tools we have learned fit the job. After we've selected the necessary tools, we use them to obtain the solution.

This is such a useful approach to solving problems that we have organized much of the book around it. In each chapter, you will be introduced to certain important concepts that will form the basis of problem solving tools. Sometimes they will be mathematical in nature and other times they will involve theoretical concepts. The Tool icon that appears in the margin will be used to highlight the tools as we go along, and at the end of each chapter there is a summary of the tools you have been taught.

Within the text you will encounter worked-out Examples that illustrate how the "chemical tools" are applied. Often, the examples will begin with an *Analysis* section that describes the thought processes involved in solving the problem, and, at the end, a *Check* to see if the answer is reasonable. Following worked-out examples are Practice Exercises that give you an immediate opportunity to apply what you've just learned.

At the end of a chapter you will find two sets of exercises. The first set, titled *Thinking It Through,* asks you to describe *how* you would solve various problems. The goal is not the answer itself, but rather a description of the tools and method needed to obtain the answer. Some of these problems are simple, while others require a lot of thinking. Be sure to work on them all, however, because once you've learned *how* to solve a problem, obtaining the answer is simple.

Following the Thinking It Through exercises is a comprehensive set of *Review Questions* and *Review Problems* that test your knowledge of all the topics discussed in the chapter. This is also another opportunity for you to practice your problem solving skills.

LearningWare. To further assist you in developing problem solving skills, we provide easy-to-use computer software, which we call ***LearningWare,*** on a CD provided with the book. This tutorial software, compatible with both PC and Macintosh, has been carefully designed to guide you through the solutions to a variety of typical chemistry problems selected from the *Review Problems* in the textbook. (The problems are identified by the CD icon shown in the margin.). We urge you to use the software as an additional study aid.

Facets of Chemistry

Brief Contents

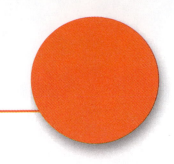

Table of Contents

Were it not for chemical research, most of the materials from which consumer products and medicines are made would not exist. One of the goals of this book is to help you see the many ways by which chemistry enters our lives and to give you an appreciation of the pivotal place that chemistry holds among the sciences.

Fundamentals of Chemical Change

This Chapter in Context The goal of this chapter is to provide you with a foundation of basic concepts and tools that will be critical to your success in later chapters. You will learn about the nature of chemistry and how science in general works. We will introduce you to the tools and units of measurement and explain how to properly apply them in calculations.

If you have had a prior course in chemistry, perhaps in high school, you may already be familiar with many of the topics that we cover in this introductory chapter. Nevertheless, it is important to be sure you have a mastery of these subjects, because if you don't start this course with a firm understanding of the basics, you may find yourself in trouble later on.

1.1 Chemistry and Its Place among the Sciences

A good part of the quality of life that we and others in the world enjoy today can be traced to the success of chemistry as a science. Almost everything we touch or eat or drink or take into our bodies for nutrition or to cure illness bears the marks of chemical research. Synthetic fibers make up much of the fabric of the clothes we wear or form the threads that hold more "natural" fabrics together. Many food products contain preservatives that retard the growth of bacteria or mold, artificial flavors that enhance taste, or substances that improve texture and color. Low calorie beverages contain artificial sweeteners such as saccharin or aspartame. And, of course, drugs developed by pharmaceutical companies cure illnesses today that would have been fatal only a short time ago.

You are about to embark on a study of chemistry—a science that seeks to answer fundamental questions about the makeup of all the materials that compose our world. More specifically, **chemistry**[1] seeks to determine what substances are composed of and how the properties of substances are influenced by their compositions and interactions with each other. As part of this quest, chemistry deals with the way substances change, often dramatically, when they interact with

In our discussions, we do not assume that you have had a prior course in chemistry. However, we do urge you to study this chapter thoroughly, because the concepts developed here will be used in later chapters.

---

[1]Important terms will be set in bold type to call them to your attention. Be sure you learn their meanings.

each other in **chemical reactions.** And permeating all of this is a search for knowledge about the basic underlying structure of matter and the forces that determine the properties we are able to observe through our senses. From such knowledge has come the ability to create materials never before found on Earth, materials with especially desirable properties that fulfill the needs of society.

Chemistry holds a unique place among the sciences because all material things are composed of chemicals. A cricket, for example, is made up of a complex set of chemicals that possess the unique quality we call life. Chemists are interested in these chemicals for the way they behave toward each other. A biologist, on the other hand, might wish to study how the cricket metabolizes nutrients and how it derives energy from the process. In doing so, the biologist is likely to draw on knowledge gained by chemists. And a physicist might be interested in the mechanics of motion of the limbs of the cricket, which allow it to jump. Thus, chemists, biologists, and physicists may all have interests in the cricket and study the creature from slightly different points of view. Today, in fact, the lines between traditional disciplines such as chemistry and biology have become blurred, so there is little difference, for example, between a molecular biologist and a biochemist.

> ►**Chemistry in Practice**◄ If you plan to major in chemistry, it is clear why you are in this course. But even if you are not a chemistry major, as a science student you will surely encounter chemical substances as part of your studies. This is why you are being asked to take this course—in pursuing your major you are likely to find it helpful, and perhaps even necessary, to view the world with the eye of a chemist. As the authors of your textbook, we hope you will seek to understand the value of chemistry as it pertains to your chosen career. By doing so, you will be better able to appreciate both your chemistry course and the future that awaits you. ◆

1.2 The Scientific Method: Building Models of Nature

As a science, chemistry is a dynamic subject, constantly changing as new discoveries are made by scientists who work in university, industrial, and government laboratories. The general approach that these scientists bring to their work is called the *scientific method.* It is, quite simply, a commonsense approach to developing an understanding of natural phenomena.

A scientific study normally begins with some observation that fires our curiosity and raises some questions about the behavior of nature. Usually, we begin to search for answers in the work of others who have published in scientific journals. As our knowledge grows, we begin to plan experiments that permit us to make our own observations. Generally, these experiments are performed in a laboratory under controlled conditions so the observations are *reproducible.* In fact, the ability to obtain the same results when experiments are repeated is what separates a true science from a pseudoscience such as astrology.

Empirical Facts and Scientific Laws

The observations we make in the course of performing our experiments provide us with **empirical facts**—so named because we learn them by *observing* some physical, chemical, or biological system. These facts are referred to as our **data.** For example, if we study the behavior of gases, such as the air we breathe, we soon discover that the volume of a gas depends on a number of factors, includ-

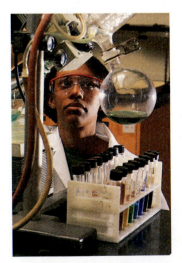

A scientist working in a chemical research laboratory.

Webster's defines *empirical* as "pertaining to, or founded upon, experiment or experience."

ing the mass of the gas, its temperature, and its pressure. The bits of information we record relating these factors are our data.

One of the goals of science is to organize facts so that relationships or generalizations among the data can be established. For instance, one generalization we would make from our observations is that when the temperature of a gas rises, the gas tends to expand and occupy a larger volume. If we were to repeat our experiments many times with numerous different gases, we would find that this generalization is uniformly applicable to them all. Such a broad generalization, based on the results of many experiments, is called a **law** or **scientific law.**

Hypotheses and Theories; Models of Nature

As useful as they may be in summarizing the results of experiments, laws can only state what happens. They do not explain *why* substances behave the way they do. *Why,* for example, do gases expand when they're heated? More specifically, *what must gases be like at the most basic, elementary level for them to behave as they do?* Answering such questions when they first arise is no simple task and requires much speculation. But gradually, scientists build mental pictures, called **theoretical models,** that enable them to explain observed laws.

In the development of a theoretical model, tentative explanations, called **hypotheses,** are formed. As these explanations are tested, usually by performing additional experiments that test predictions derived from the model, the model is refined and adjusted to account for new data. Eventually, if the model survives repeated testing, it gradually achieves the status of a theory. A **theory** *is a tested explanation of the behavior of nature.* Most useful theories are broad, with many far-reaching and subtle implications. It is impossible to perform every test that might show such a theory to be wrong, so we can never be *absolutely* sure the theory is correct.

The sequence of steps just described—observation, explanation through the creation of a theoretical model, and the testing of the model by additional experiments—constitute the **scientific method.** Despite its name, this method is not used only by those who call themselves scientists. An auto mechanic follows essentially the same steps when fixing your car. First, tests are performed (*observations* are made) that enable the mechanic to suggest the probable cause of the problem (*a hypothesis*). Then parts are replaced and the car is checked to see whether the problem has been solved (*testing the hypothesis by experiment*). In short, we all use the scientific method as much by instinct as by design.

From the preceding discussion you may get the impression that scientific progress always proceeds in a dull, orderly, and stepwise fashion. This isn't true; science is exciting and provides a rewarding outlet for cleverness and creativity. Luck, too, sometimes plays an important role. For example, in 1828 Frederick Wöhler, a German chemist, was heating a substance called ammonium cyanate in an attempt to add support to one of his hypotheses. His experiment, however, produced an unexpected substance, which out of curiosity he analyzed and found to be urea (a constituent of urine). This was an exciting discovery, because it was the first time anyone had knowingly ever made a substance produced only by living creatures from a chemical not having a life origin. The fact that this could be done led to the beginning of a whole branch of chemistry called *organic chemistry.* Yet, had it not been for Wöhler's curiosity and his application of the scientific method to his unexpected results, the significance of his experiment might have gone unnoticed.

As an additional note, it is significant that the most spectacular and dramatic changes in science occur when major theories are proved to be wrong. Although this happens only rarely, when it does occur, scientists are sent scrambling to develop new theories, and exciting new frontiers are opened.

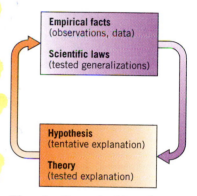

The scientific method is cyclical. Observations suggest explanations, which suggest new experiments, which suggest new explanations, and so on.

Many breakthrough discoveries in science have come about by accident.

Figure 1.1 *Atoms combine to form molecules.* Illustrated here are molecules of water, each of which consists of one atom of oxygen and two atoms of hydrogen. (*a*) Colored spheres are used to represent individual atoms, gray for hydrogen and red for oxygen. (*b*) A drawing that illustrates the shape of a water molecule. (*c*) A glass of liquid water contains an enormous number of tiny water molecules jiggling about.

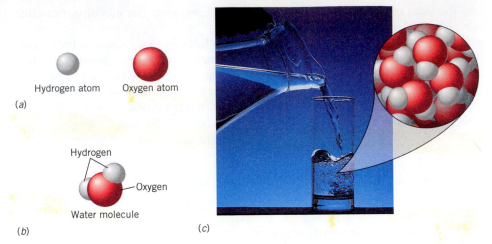

The Atomic Theory as a Model of Matter

Virtually every scientist would agree that the most significant theoretical model of nature ever formulated is the atomic theory. According to this theory, all chemical substances are composed of tiny particles that we call **atoms.** Individual atoms combine in diverse ways to form more complex particles called **molecules.** Molecules of water are illustrated in Figure 1.1, where we see that each is composed of the same atoms (hydrogen and oxygen) in the same proportions. In a similar fashion, each of the different chemicals we find around us is composed of atoms of various kinds in specific combinations.

The concept of atoms and molecules enables us to visualize what takes place when atoms combine in various ways. As you will see, it also enables us to understand, explain, and sometimes predict a wide range of observations concerning the behavior of nature. To help us, we will frequently use drawings to illustrate molecules, and Figure 1.2 shows some of the ways molecules can be represented.

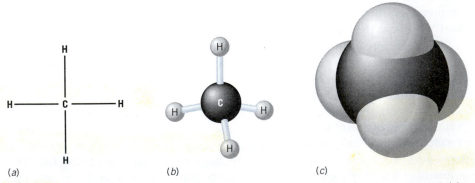

Figure 1.2 *Some of the different ways that the structures of molecules are represented.* (*a*) A structure using chemical symbols to stand for atoms and dashes to indicate how the atoms are connected to each other. The molecule is methane, the substance present in natural gas that fuels stoves and Bunsen burners. A methane molecule is composed of one atom of carbon (C) and four atoms of hydrogen (H). (*b*) A *ball-and-stick model* of methane. The dark gray ball is the carbon atom and the light gray balls are hydrogen atoms. (*c*) A *space-filling model* of methane that shows the relative sizes of the C and H atoms. Ball-and-stick and space-filling models are used to illustrate the three-dimensional shapes of molecules.

We will expand on the atomic model of matter as we proceed in our discussions, and we will frequently make the connection between what we physically observe in our large, macroscopic world, and what we believe takes place in the tiny submicroscopic world of atoms and molecules. Learning to recognize these connections should be one of your major goals in studying this course.

We use the term macroscopic to mean the world we observe with our senses, whether it be in the laboratory or the world we encounter in our day-to-day living.

1.3 Matter and Energy; Chemical Change

The two terms *matter* and *energy* basically define the focus of chemistry, and as we shall see, they are intimately related. Therefore, let's begin our study of chemistry by exploring the meanings of these terms and how they apply to the study of our subject.

Matter is defined as *anything that occupies space and has mass.* It is the stuff our universe is made of, and all of the chemicals that make up tangible things, from rocks to pizza to people, are examples of matter.

Notice that in setting out this definition, we have used the term *mass* rather than *weight.* The words mass and weight are often used interchangeably even though they refer to different things. **Mass** *refers to how much matter there is in a given object, whereas* **weight** *refers to the force with which the object is attracted by gravity.* For example, a golf ball contains a certain amount of matter and has a certain mass, which is the same regardless of the golf ball's location. However, the weight of the golf ball can vary. On Earth its weight is approximately six times larger than the weight it would have on the moon because the gravitational attraction of Earth is six times that of the moon. Because mass does not vary from place to place, we use mass rather than weight when we specify the amount of matter in an object.

Mass is measured by a procedure called **weighing,** even though our goal is not to measure the sample's weight. As we will discuss in the next section, mass is measured with an instrument called a balance.

When we observe a sample of matter (a chemical), it can be in one or more of three different **physical states:** solid, liquid, or gas. Water, for instance, can exist as solid ice, as liquid water, or as gaseous steam, depending on the temperature. Even though the physical appearances of these three states are quite different, the chemical makeup of water is the same in all of them.

Mass is also related to an object's inertia, or resistance to a change in its motion. An object of large mass has a large inertia and is difficult to move.

Solid, liquid, and gas are referred to as *states of matter.*

Chemical Changes

Chemistry is especially concerned with learning and understanding the *changes* that take place between chemicals, changes that we call chemical reactions. In a **chemical reaction,** chemicals interact with each other to form entirely *different* substances with different properties. Sometimes these changes can be quite dramatic, as illustrated by the reaction of sodium with chlorine.

Sodium, shown in Figure 1.3*a*, is a metal. Like other metals, it is shiny and conducts electricity well. It is a very soft metal and is easily cut with a knife. Notice that the outside of the bar of sodium is coated with a white film. This was formed by the reaction of sodium with oxygen and moisture in the air. The tendency of sodium to react rapidly with oxygen and water makes it a dangerous chemical with which to work. Sodium reacts violently in water, producing a lot of heat and liberating the flammable gas hydrogen. In this same reaction sodium

The iceberg and the sea that surrounds it are both composed of the same substance, water. They represent the solid and liquid states of this substance.

(*a*)

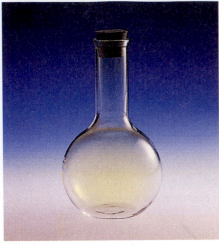

(*b*)

(*c*)

Figure 1.3 *Sodium reacts with chlorine to form sodium chloride.* (*a*) A freshly cut piece of sodium reveals a shiny metallic surface. The metal reacts with oxygen and moisture, so it cannot be touched with bare fingers. (*b*) Chlorine is a pale green gas. (*c*) When a small piece of sodium is melted in a metal spoon and thrust into the flask of chlorine, it burns brightly as the two elements react to form sodium chloride. The smoke coming from the flask is composed of fine crystals of salt.

also forms a substance called sodium hydroxide, commonly known as lye, which is very corrosive toward flesh. Contact of sodium with your skin can cause severe chemical burns.

Chlorine, shown in the flask in Figure 1.3*b*, is different from sodium in many ways. It is a pale, yellow-green gas. If you've ever smelled a liquid laundry bleach such as Clorox, you have smelled chlorine, which escapes from the bleach in small amounts. In concentrated form chlorine is especially dangerous to inhale, causing severe lung damage that can easily lead to death. In fact, chlorine gas has been used as a weapon of war.

When metallic sodium and gaseous chlorine come together, they react violently as shown in Figure 1.3*c*. The substance formed in this reaction is a white powder, quite different in appearance from either sodium or chlorine. Its chemical name is sodium chloride, although it's known to most people by its common name, *table salt* (or simply, *salt*).

If you think about this reaction for a moment, perhaps it will strike you as really quite amazing, and even a bit like magic. Here we have two chemicals, sodium and chlorine, whose ingestion can produce severe medical problems or even death. Yet, when they react with each other they form a substance our bodies cannot do without! Such startling events help make chemistry fascinating. As you study this course, perhaps you will discover for yourself that same sense of magic that caught the imagination of others and produced generations of chemists.

> ▶**Chemistry in Practice**◀ Most chemical reactions are not quite as spectacular as the one between sodium and chlorine. Nevertheless, they take place in us and around us all the time. We metabolize the foods we consume while photosynthesis in plants creates their replacements. Chemical reactions in the air lead to smog while in the upper atmosphere sunlight interacting with gases released from aerosol cans and damaged air conditioners degrades the natural ozone layer that shields Earth from harmful ultraviolet radiation. Understanding these reactions and seeking to control the ones that can be harmful to us are important roles for chemistry in modern science and in our society. ◆

Energy

When chemical reactions occur, *energy* is almost always absorbed or given off (as heat or light, for instance). For example, the vigorous reaction between sodium and chlorine shown in Figure 1.3 produces light and heat. Similarly, the combustion reaction between gasoline and oxygen liberates energy in the form of heat that we harness to power vehicles. In fact, one of the practical uses of chemical reactions is to satisfy the energy needs of society.

The concept of energy is more difficult to grasp than that of matter because energy is intangible; you can't hold it in your hand to study it and you can't put it in a bottle. **Energy** *is something an object has if the object is able to do work.* It can be possessed by the object in two different ways, as kinetic energy and as potential energy.

Kinetic energy *is the energy an object has when it is moving.* It depends on the object's mass and velocity; the larger its mass and the greater its velocity, the more kinetic energy it has. A simple equation relates kinetic energy (KE) to these quantities[2]:

$$KE = \frac{1}{2}mv^2 \qquad (1.1)$$

where m is the mass and v is the velocity.

Potential energy *is energy an object has that can be changed to kinetic energy; it can be thought of as* ***stored energy.*** For example, when you wind an alarm clock, you transfer energy to a spring. The spring holds this stored energy (potential energy) and gradually releases it, in the form of kinetic energy, to make the clock's mechanism work. Water, high in the mountains, also has potential energy because of its attraction by gravity. When the water falls to a lower altitude, we can harness the energy that's released to turn turbines that generate electricity or operate machinery. Chemicals also possess potential energy, which is sometimes called **chemical energy.** This is energy stored in chemicals that can be liberated during chemical reactions. For example, the foods we eat have potential energy that is changed to kinetic energy by the process we call metabolism. This kinetic energy ultimately manifests itself as movements of muscles and as body heat.

One of the most important facts about energy is that *it cannot be created or destroyed; it can only be changed from one form to another.* This fact was established as the result of many experiments and observations, and is known today as the **law of conservation of energy.** (When we say in science that something is "conserved," we mean that it is unchanged or remains constant.) You observe this law whenever you toss something—a ball, for instance—into the air. You give the ball some initial amount of kinetic energy when you throw it. As it rises, its potential energy increases. Because energy cannot come from nothing, the rise in potential energy comes at the expense of the ball's kinetic energy, so the ball slows down. When all the kinetic energy has changed to potential energy the ball can go no higher; it has stopped moving and its potential energy is at a maximum. The ball then begins to fall, and its potential energy is changed back to kinetic energy.

Heat and Temperature

In any object, the atoms and molecules are not stationary; they are constantly moving and colliding with each other. Because of these motions, the particles have varying amounts of kinetic energy, and the terms heat and temperature are

Because chemical reactions involve an absorption or release of energy, the study of energy is an integral part of the study of chemistry.

Work is done by an object when it causes something to move. A moving car has energy because it can move another car in a collision.

Unlike kinetic energy, there is no single, simple equation that can be used to calculate the amount of potential energy an object has.

Because the ball's mass can't change, the decrease in kinetic energy must be accompanied by a decrease in the ball's velocity.

[2]Many equations will be numbered to make it easy to refer to them later on in the book or in class discussions.

The difference between heat and temperature. Both fires could be at the same temperature if the *average* kinetic energies of the atoms in the flames are the same. The forest fire has many more atoms with high kinetic energies, so its *total* kinetic energy is greater than that of the match flame. Therefore, the forest fire has more heat in it.

related to this energy. Specifically, the **temperature** of an object is proportional to the *average* kinetic energy of its particles—the higher the average kinetic energy, the higher the temperature. What this means is that when the temperature of an object is increased, the molecules move faster. (Recall that $KE = \frac{1}{2}mv^2$. Increasing the average kinetic energy doesn't increase the masses of the atoms, so it must increase their speeds.)

Heat is energy that is transferred between objects caused by a difference in their temperatures, and, as you know, heat always passes spontaneously from the object with the higher temperature to the one that is colder. This energy transfer continues until both objects come to the same temperature.

Energy can also be transferred as light, which we will discuss later in the book.

The way we interpret the transfer of heat is as a transfer of kinetic energy between the atoms of the two objects. This takes place in a manner that's similar to the transfer of kinetic energy between balls on a billiard table. Here we may see a fast-moving ball strike a slow one, causing the slow one to speed up and the fast one to slow down. In this collision, kinetic energy is transferred from the ball that slows to the ball that goes faster. Similarly, when a hot object is placed in contact with a cold one, the faster atoms of the hot object collide with and lose kinetic energy to the slower atoms of the cold object. This decreases the average kinetic energy of the particles in the hot object, causing its temperature to drop. Likewise, the average kinetic energy of the particles in the cold object is raised, causing the temperature of the object to rise. Eventually, the average kinetic energies of the atoms in both objects become the same and the objects reach the same temperature.

Energy and Its Relationship to Chemical Change

The concepts discussed in the preceding paragraphs form the foundation of one of science's most important theories, which we will discuss in much greater detail in Chapter 6. For now, however, let's see how these concepts allow us to explain how changes in chemical energy that accompany chemical reactions make themselves known.

Recall that chemical energy is the potential energy stored in chemicals.

In a gasoline engine, the combustion of gasoline produces hot gases from which heat is extracted to perform work. What is the origin of the heat that appears in this reaction?

When gasoline burns, it reacts with oxygen to form the gases carbon dioxide and water. These gases have a much higher temperature than the starting materials, which means that the average kinetic energy of the atoms has increased during the reaction. But where has this increase in kinetic energy come from? Because kinetic energy cannot come from nothing, it must have come from the potential energy (chemical energy) of the gasoline and oxygen. Therefore, we are able to conclude that during this reaction there is a drop in potential energy and that the carbon dioxide and water formed in the reaction must have a lower potential energy than the gasoline and oxygen.

As you can see, the analysis of temperature changes in chemical reactions can tell us about the potential energy changes that occur. And as we shall see, the changes in potential energy that accompany reactions also reveal a great deal about the chemicals themselves.

> The rise in temperature occurs because there is a lowering of the chemical energy during the course of the reaction.

1.4 Tools of Chemistry: Measured Quantities and Their Units

The Importance of Measurements

In our description of the reaction of sodium with chlorine, we mentioned that sodium is a metal with the ability to conduct electricity and that chlorine is a pale-green gas. Observations of this kind are said to be **qualitative** because they do not involve numbers.

However interesting, qualitative observations are of limited usefulness in any science because they simply do not provide enough information. To make significant progress, every science must resort to **quantitative observations,** and these involve making numerical measurements.

Units of Measurement

A necessary aspect of all measurements is a uniform and generally accepted system of units. In the United States most people are accustomed to using the *English system* of units in which we measure distances in inches, feet, and miles; volume in ounces, quarts, and gallons; and mass in ounces and pounds. However, in the sciences and increasingly for consumer products (Figure 1.4) a metric based system is used. Its chief advantage is that larger and smaller units are related to each other by multiples of ten.

> Any measurement consists of two parts, a numerical value and a unit of measurement.

The International System of Units

In 1960, a modification of the original metric system was adopted by the General Conference on Weights and Measures (an international body). It is called the **International System of Units,** abbreviated **SI** from the French name, *Le Système International d'Unités.*

For the most part, scientists have willingly adopted the SI units, although there is still some usage of the older metric units. The primary emphasis throughout this book will be on the SI, but there will be times where we will use some older units because of their convenience in dealing with certain laboratory measurements.

SI Base Units

The SI has as its foundation a set of **base units** (Table 1.1) for seven basic measured quantities. These quantities were chosen because all other units of measurement can be derived from them. The size of each base unit is very precisely defined. For example, the international standard for the SI unit of mass, the **kilogram,** is a carefully preserved platinum–iridium alloy block stored at the

355ml 454g 1 liter
454g

Figure 1.4 *Metric units are becoming commonplace on many consumer products.*

Table 1.1 The SI Base Units

Measurement	Unit	Symbol
Length	meter	m
Mass	kilogram	kg
Time	second	s
Electric current	ampere	A
Temperature	kelvin	K
Amount of substance	mole	mol
Luminous intensity	candela	cd

Although not all of these units will have meaning for you now, we will encounter most of them later in this book, and their meanings will be made clear when they are needed.

Scientists are working on a method of accurately counting atoms whose masses are accurately known. Their goal is to develop a new definition of the kilogram that doesn't rely on an object that can be stolen, lost, or destroyed.

The units we attach to numbers are sometimes referred to as *dimensions.*

International Bureau of Weights and Measures in France (Figure 1.5). This metal block is *defined* as having a mass of *exactly* one kilogram. Most countries keep a carefully calibrated replica of the international standard kilogram in some central location. For example, the National Institute of Standards and Technology located near Washington, D.C., has its own standard kilogram which is, as near as possible, a duplicate of the one in France. This U.S. kilogram serves indirectly as the calibrating standard for all "weights" used for scales and balances in the United States. Thus, the masses you measure on the balances in your general chemistry laboratory can be traced to the U.S. standard, and ultimately to the international standard.

Objects created by human hands, of course, can be damaged or lost and so are not the most desirable choices for standards. In the SI, only the kilogram is defined by using an object. The other base units are established in terms of reproducible physical phenomena. For instance, the meter is defined as exactly the distance light travels in a vacuum in 1/299,792,458 of a second. Everyone has access to this standard because light and a vacuum are available to all.

Derived Units

The base units, which form the foundation of the SI, are used to define additional **derived units.** For example, there is no SI base unit for area, but we know that to calculate the area of a rectangular room we multiply its length by its width. Therefore, the *unit* for area is derived by multiplying the *unit* for length by the *unit* for width.

$$\text{length} \times \text{width} = \text{area}$$
$$(\text{meter}) \times (\text{meter}) = (\text{meter})^2$$
$$\text{m} \quad \times \quad \text{m} \quad = \quad \text{m}^2$$

The SI unit for area is therefore m^2 (read as *meter squared,* or *square meter*). In obtaining this unit we employ a very important concept that we will use repeatedly throughout this book when we perform calculations: *Units undergo the same kinds of mathematical operations that numbers do.* This is a concept we will discuss at greater length in Section 1.6.

Figure 1.5 *The international standard kilogram.* The SI standard for mass is this cylinder of a platinum–iridium alloy (an alloy chosen because it will not corrode in air). It is kept at the International Bureau of Weights and Measures in France. Other nations such as the United States maintain their own standard masses that have been carefully calibrated against this international standard.

Decimal Multipliers

When making measurements, we sometimes find that the basic units are of an awkward size. For instance, the meter (slightly larger than a yard) is not convenient for expressing the sizes of very small objects such as bacteria. Nor is it convenient for expressing the very large distances between planets or stars. In the

SI, we form larger or smaller units that more closely suit our needs by modifying the basic units with **decimal multipliers.** These modifiers and the prefixes we use to identify them are given in Table 1.2. Those we shall use most frequently are given in bold colored type. Be sure to learn them.

When the name of a unit is preceded by one of these prefixes, the size of the unit is modified by the corresponding decimal multiplier. For instance, the prefix *kilo* indicates a multiplying factor of 10^3, or 1000. Therefore, a kilometer is a unit of length equal to 1000 meters. The symbol for kilometer (km) is formed by applying the symbol meaning kilo (k) as a prefix to the symbol for meter (m). Thus 1 km = 1000 m (or alternatively, 1 km = 10^3 m). Similarly a decimeter (dm) is $\frac{1}{10}$ of a meter, so 1 dm = 0.1 m (1 dm = 10^{-1} m). The following Practice Exercise will give you some experience applying these decimal multipliers. Refer to Table 1.2 if necessary.

Tools

SI Prefixes

Exponential notation and powers of ten are reviewed in Appendix A.

Practice Exercise 1

Give the abbreviation for (a) milligram, (b) micrometer, (c) picosecond.
How many meters are in (a) 1 nm, (b) 1 cm, (c) 1 pm?
What symbol is used to represent (a) 10^{-2} g, (b) 10^6 m, (c) 10^{-9} s? ◆

Units for Laboratory Measurements

The most common measurements you will make in the laboratory will be those of length, volume, mass, and temperature. Although you will nearly always use SI units, it is wise to be aware of how they relate in size to the perhaps more familiar English units. Some common conversions between the English system and the SI are given in Table 1.3 and inside the rear cover of the book.[3]

Tools

Units in
laboratory
measurements

Length

The SI base unit for length, the **meter (m),** is too large for most laboratory purposes. More convenient units are the **centimeter (cm)** and the **millimeter (mm).** They are related to the meter as follows.

$$1 \text{ cm} = 10^{-2} \text{ m} = 0.01 \text{ m}$$

$$1 \text{ mm} = 10^{-3} \text{ m} = 0.001 \text{ m}$$

It is also useful to know the relationships

$$1 \text{ m} = 100 \text{ cm} = 1000 \text{ mm}$$

$$1 \text{ cm} = 10 \text{ mm}$$

Many 12-inch rulers have one side marked in centimeters and millimeters (Figure 1.6). It is helpful to remember that 1 in. equals approximately 2.5 cm or about 25 mm (1 in. = 2.54 cm, *exactly*).

Volume

Volume has the dimensions of (distance)^3. For example, the volume of a room is calculated by multiplying its length times its width times its height. With these

Table 1.2 Decimal Multipliers That Serve as SI Prefixes

Prefix	Symbol	Multiplication factor
exa	E	10^{18}
peta	P	10^{15}
tera	T	10^{12}
giga	g	10^{9}
mega	**M**	10^{6}
kilo	**k**	10^{3}
hecto	h	10^{2}
deka	da	10^{1}
deci	**d**	10^{-1}
centi	**c**	10^{-2}
milli	**m**	10^{-3}
micro	**μ**	10^{-6}
nano	**n**	10^{-9}
pico	**p**	10^{-12}
femto	f	10^{-15}
atto	a	10^{-18}

---

[3]Originally, these conversions were established by measurement. For example, if a metric ruler is used to measure the length of an inch, it is found that 1 in. equals 2.54 cm. Later, to avoid confusion about the accuracy of such measurements, it was agreed that these relationships would be taken to be exact. For instance, 1 in. is now defined as *exactly* 2.54 cm. Exact relationships also exist for the other quantities, but for simplicity many have been rounded off. For example, 1 lb = 453.59237 g, *exactly.*

Figure 1.6 *A comparison between English and SI units. Centimeters and millimeters are conveniently sized units for most laboratory measurements of length. Here is a common ruler that is marked in both English and SI units.*

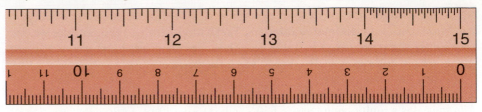

dimensions expressed in meters, the derived SI unit for volume is the **cubic meter, m³**.

In chemistry, measurements of volume usually arise when we measure amounts of liquids. The traditional metric unit of volume used for this is the **liter (L).** In SI terms, a liter is defined as exactly 1 cubic decimeter.

$$1 \text{ L} = 1 \text{ dm}^3$$

However, even the liter is too large to conveniently express most volumes measured in the lab. The glassware we normally use, such as that illustrated in Figure 1.7, is marked in **milliliters (mL).**[4]

$$1 \text{ L} = 1000 \text{ mL}$$

Because 1 dm = 10 cm, then 1 dm³ = 1000 cm³. Therefore, 1 mL is exactly the same as 1 cm³.

$$1 \text{ cm}^3 = 1 \text{ mL}$$

$$1 \text{ L} = 1000 \text{ cm}^3 = 1000 \text{ mL}$$

Sometimes you may see cm³ abbreviated cc (especially in medical applications), although the SI frowns on this symbol.

Mass

In the SI, the base unit for the measurement of mass is the **kilogram (kg),** although the **gram (g)** is a more conveniently sized unit for most laboratory measurements. One gram, of course, is $\frac{1}{1000}$ of a kilogram (1 kilogram = 1000 g, so 1 g must equal 0.001 kg).

Mass is measured by comparing the weight of a sample with the weights of known standard masses. The apparatus used is called a **balance** (some examples

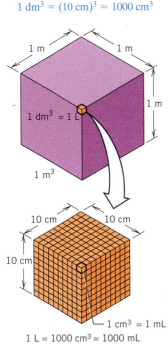

$1 \text{ dm}^3 = (10 \text{ cm})^3 = 1000 \text{ cm}^3$

1 m
1 m
1 m
1 dm³ = 1 L
1 m³

10 cm
10 cm
10 cm
1 cm³ = 1 mL
1 L = 1000 cm³ = 1000 mL

Comparison among the cubic meter, the liter (cubic decimeter), and the milliliter (cubic centimeter).

One cubic meter is somewhat larger than one cubic yard, which is much too large to conveniently express most volumes measured in the laboratory.

The difference between mass and weight was discussed earlier.

Table 1.3 **Some Useful Conversions**

Measurement	English to Metric	Metric to English
Length	1 in. = 2.54 cm	1 m = 39.37 in.
	1 yd = 0.9144 m	1 km = 0.6215 mile
	1 mile = 1.609 km	
Mass	1 lb = 453.6 g	1 kg = 2.205 lb
	1 oz = 28.35 g	
Volume	1 gal = 3.785 L	1 L = 1.057 qt
	1 qt = 946.4 mL	
	1 oz (fluid) = 29.6 mL	

[4]Use of the abbreviations L for liter and mL for milliliter is rather recent. Confusion between the printed letter l and the number 1 prompted the change from l for liter and ml for milliliter. You may encounter the abbreviation ml in other books or on older laboratory glassware.

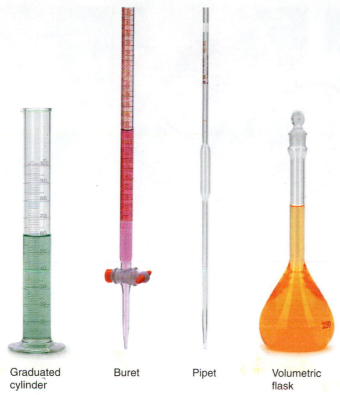

Graduated cylinder   Buret   Pipet   Volumetric flask

Figure 1.7 *Apparatus for measuring volumes of liquids.* Common laboratory glassware used for measuring volumes includes a graduated cylinder, a buret, a pipet, and a volumetric flask.

are shown in Figure 1.8). For the balance in Figure 1.8*a*, we would place our sample on the left pan and then add standard masses to the other. When the weight of the sample and the total weight of the standards are in balance (when they match), their masses are then equal.

Notice that two of the balances in Figure 1.8 have only one pan. This is true of most modern balances. In some, the standard masses supplied by the manufacturer are located within the balance case where they are protected from dust (and the probing fingers of curious people). In many modern balances, such as

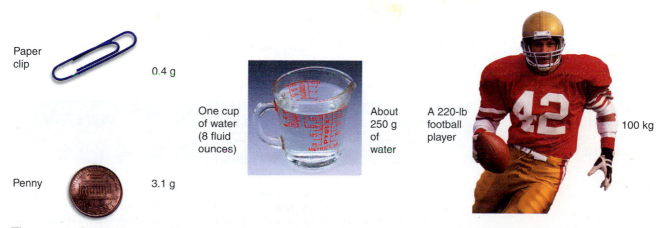

Paper clip   0.4 g

Penny   3.1 g

One cup of water (8 fluid ounces)   About 250 g of water

A 220-lb football player   100 kg

The masses of some common things expressed in SI units.

(a)

(b)

(c)

Figure 1.8 *Typical laboratory balances.* (*a*) A traditional two-pan analytical balance capable of measurements to the nearest 0.0001 g. (*b*) A modern top-loading balance capable of mass measurements to the nearest 0.001 g (fitted with a cover to reduce the effects of air currents and thereby improve precision). (*c*) A modern analytical balance capable of measurements to the nearest 0.0001 g.

the one shown in Figure 1.8*b*, there are no standard masses at all. These electronic balances work on an altogether different principle which relies on electrically generated magnetic forces to balance the weight of an object placed on the balance pan.

Temperature

To measure temperature we use a thermometer, which consists of a long glass tube with a very thin bore connected to a reservoir containing a liquid, usually mercury (Figure 1.9) or alcohol. As the temperature rises, the liquid in the reservoir expands and its height in the column increases.

Thermometers are graduated in *degrees* according to one of two temperature scales. Both scales use as reference points the temperature at which water freezes[5] and the temperature at which it boils. On the **Fahrenheit scale** water freezes at 32 °F and boils at 212 °F. If you've been raised in the United States, this is probably the scale you're most familiar with. In recent times, however, you have probably noticed an increased use of the Celsius scale, especially in weather broadcasts. This is the scale we use most often in the sciences. On the **Celsius scale** water freezes at 0 °C and boils at 100 °C. (See Figure 1.10.)

As you can see in Figure 1.10, on the Celsius scale there are 100 degree units between the freezing and boiling points of water, while on the Fahrenheit scale this same temperature range is spanned by 180 degree units. Therefore, each

Biologists who study microorganisms often carry out their experiments at 37 °C because that is normal human body temperature.

Notice that the name of the temperature scale, the Kelvin scale, is capitalized, but the name of the unit, the kelvin, is not. However, the symbol for the kelvin is the capital letter K.

Figure 1.9 *A typical laboratory thermometer.* This thermometer contains mercury as the fluid that expands when heated.

[5]Water freezes and ice melts at the same temperature, and a mixture of ice and water will maintain a constant temperature of 32 °F or 0 °C. If heat is added, some ice melts; if heat is removed, some liquid water freezes, but the temperature doesn't change. This constancy of temperature is what makes the "ice point" convenient for calibrating thermometers.

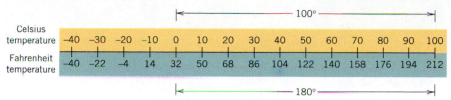

Figure 1.10 *Comparison of the Celsius and Fahrenheit temperature scales.*

Celsius degree is nearly twice as large as a Fahrenheit degree (actually, 5 Celsius degrees is the same as 9 Fahrenheit degrees). If it is necessary to convert between these temperature scales, we can use the equation

$$t_F = \left(\frac{9\ °F}{5\ °C}\right) t_C + 32\ °F \qquad (1.2)$$

Temperature conversions

where t_F is the Fahrenheit temperature and t_C is the Celsius temperature.

The SI unit of temperature is the **kelvin (K),** which is the degree unit on the **Kelvin scale.** Notice that the temperature unit is K, not °K (the degree symbol, °, is omitted). Kelvin temperatures must be used in many equations in which the temperature enters directly into the calculations. We will come across this situation many times throughout the book.

Figure 1.11 shows how the Kelvin, Celsius, and Fahrenheit temperature scales relate to each other. Notice that the kelvin is *exactly* the same size as the Celsius degree. *The only difference between these two temperature scales is the zero point.* The zero point on the Kelvin scale is called **absolute zero** and corresponds to nature's coldest temperature. It is 273.15 degree units below the zero point on the Celsius scale, which means that 0 °C equals 273.15 K, and 0 K equals −273.15 °C. Thermometers are never marked with the Kelvin scale, so when we need to express a temperature in kelvins, we have to do some arithmetic. The relationship between the Kelvin temperature, T_K, and the Celsius temperature, t_C, is

$$T_K = t_C + 273.15 \qquad (1.3)$$

In our calculations, we will nearly always round this to the nearest whole degree and use the equation

$$T_K = t_C + 273 \qquad (1.4)$$

We will use a capital T to stand for the Kelvin temperature and a lowercase t (as in t_C) to stand for the Celsius temperature. If we wished to be very precise about the cancellation of units, we could express Equation 1.3 as

$$T_K = \left(\frac{1\ K}{1\ °C}\right) t_C + 273.15\ K$$

Temperature conversions

EXAMPLE 1.1

Converting among Temperature Scales

Thermal pollution, the release of large amounts of heat into rivers and other bodies of water, is a serious problem near power plants and can affect the survival of some species of fish. For example, trout will die if the temperature of the water rises above approximately 25 °C. (a) What is this temperature in °F? (b) What is this temperature in kelvins?

Analysis: This is your first opportunity in this text to study the principles of problem solving. If you have not already done so, we suggest you read the section titled "To the Student" located at the beginning of the book where we examine in a general way the strategy we will use to tackle problems of all sorts. Here the problem is relatively simple, but we will use the same approach in future examples.

Our first job in solving a problem is determining the kinds of tools required to do the work. Both parts of the problem here deal with temperature conversions. Therefore, we ask ourselves "What relationships do we have that relate temperature scales to each other?" Let's write them:

Figure 1.11 *Comparison among Kelvin, Celsius, and Fahrenheit temperature scales. Normal human body temperature is indicated on each of the scales.*

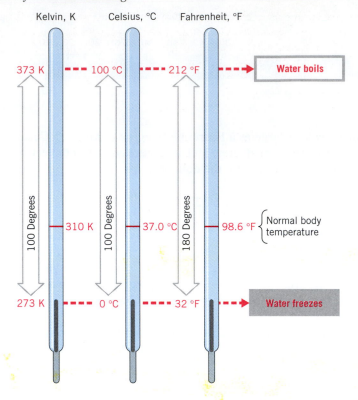

Equation 1.2

$$t_F = \left(\frac{9\ °F}{5\ °C}\right) t_C + 32\ °F$$

Equation 1.4

$$T_K = t_C + 273$$

Equation 1.2 relates Fahrenheit temperatures to Celsius temperatures, so this is the relationship we need to answer part (a). Equation 1.4 relates Kelvin temperatures to Celsius temperatures, which is what we need for part (b). Now that we have what we need, the rest follows.

Solution to (a): We substitute the value of the Celsius temperature (25 °C) for t_C.

$$t_F = \left(\frac{9\ °F}{5\ °C}\right)(25\ °C) + 32\ °F$$

$$= 77\ °F$$

Therefore, 25 °C = 77 °F. (Notice that we have canceled the unit °C in the equation above. As noted earlier, units behave the same as numbers do in calculations.)

Solution to (b): Once again, we have a simple substitution. Since $t_C = 25$ °C, the Kelvin temperature is

$$T_K = (25) + 273$$

$$= 298\ K$$

Thus, 25 °C = 298 K.

Is the Answer Reasonable?
Before leaving a problem, it is always wise to examine the answer to see whether it makes sense. Ask yourself "Is the answer too large, or too small?" Judging the an-

swers to such questions serves as a check on the arithmetic as well as on the method of obtaining the answer and can help you find obvious errors. Let's look at how we might estimate the magnitude of the answers to both parts (a) and (b).

For part (a), we know that a Fahrenheit degree is about half the size of a Celsius degree, so 25 Celsius degrees should be about 50 Fahrenheit degrees. The positive value for the Celsius temperature tells us we have a temperature *above* the freezing point of water. Since water freezes at 32 °F, the Fahrenheit temperature should be approximately 32 °F + 50 °F = 82 °F. The answer of 77 °F is quite close.

For part (b), we recall that 0 °C = 273 K. A temperature above 0 °C must be higher than 273 K. Our calculation, therefore, appears to be correct.

Practice Exercise 2

What Celsius temperature corresponds to 50 °F? What Kelvin temperature corresponds to 68 °F (expressed to the nearest whole Kelvin unit)? ◆

1.5 Accuracy and Precision; Significant Figures

In the preceding section, we discussed the importance of obtaining measurements in science. In making measurements, scientists are concerned with two critical questions. The first is "How close is the measured value to the true or correct value?" The second question is "How reproducible is the measurement?" (In other words, if we were to make repeated measurements, how close would they be to each other?) These same questions could also be phrased "How *accurate* is the measurement and how *precise* is it?" A practical example of how accuracy and precision differ is illustrated in Figure 1.12.

For measurements to be **accurate,** the measuring device must be carefully calibrated. This means that it must be adjusted to give correct values when some standard reference is used with the device. For example, to calibrate an electronic balance, a known reference mass is placed on the balance and a calibration routine within the balance is initiated. Once calibrated, the balance will give correct readings, the accuracy of which is determined by the accuracy of the standard mass used. Standard reference masses (also called "weights") can be purchased from scientific supply companies.

When we use the term **precision,** we refer to how closely repeated measurements of the same quantity come to each other. This is determined often by our ability to read the measuring scale on the instrument. Consider, for example, the following illustration.

Suppose you asked two different people to measure the width, in meters, of a large room. The first person reports that the room is 11.2 m wide. The second tells you that he measured the room at the same place and obtained 11.13 m. Obviously, the room can have only one true width at any given place. What, then, do these two numbers tell us?

The first number, 11.2 m, implies that the width of the room was measured with a tape measure that required the person using it to estimate the tenths place. Figure 1.13*a* illustrates how a tape measure such as this would be marked. Because the tenths place must be *estimated,* different people measuring the width of the room might report distances that differ by ±0.1 m. In other words, we might expect another person's estimate of the distance to be 0.1 m larger or smaller than the reported value of 11.2 m. Therefore, we view this measurement as being *uncertain* by ±0.1 m.

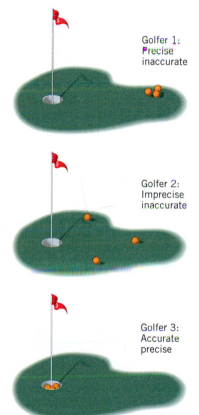

Golfer 1: Precise inaccurate

Golfer 2: Imprecise inaccurate

Golfer 3: Accurate precise

Figure 1.12 *The difference between accuracy and precision in the game of golf.* Golfer 1 hits shots that are precise (because they are tightly grouped), but the accuracy is poor because the balls are not near the target. Golfer 2 needs help. His shots are neither precise *nor* accurate. Golfer 3 wins the prize with shots that are precise (tightly grouped) and accurate (in the hole).

Figure 1.13 *Measurements with tape measures graduated differently.* (*a*) Measurement of length with a tape measure marked only every 1 m. An estimate must be made of the tenths place. (*b*) A portion of the scale of another tape measure that is marked every 0.1 m. This scale permits estimation of the hundredths place.

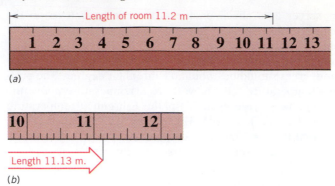

The second measurement, 11.13 m, suggests that the tape measure used to obtain it was more finely divided, as shown in Figure 1.13*b*. The markings on this scale allow the user to be certain of the tenths place, but require that the hundredths place be estimated. Therefore, we can expect the second measurement to be uncertain by about ±0.01 m. The second measurement is also said to be more precise because 11.13 ± 0.01 m has a smaller uncertainty than 11.2 ± 0.1 m.

We would certainly expect the second measurement to be more reliable than the first because it has more digits and a smaller amount of uncertainty. *The reliability of a piece of data is indicated by the number of digits used to represent it.* This is so important we have a special terminology to describe numbers that come from measurement.

> *Digits that result from measurement such that only the digit farthest to the right is not known with certainty are called* **significant figures.**

The number of significant figures in a measured value is equal to the number of digits known for sure *plus* one that is uncertain.

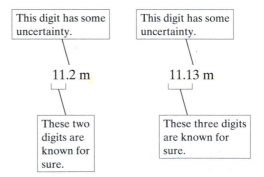

The first measurement of 11.2 m has three significant figures; the second, more precise value of 11.13 m has four significant figures. In general, for a given measurement, *the greater the number of significant figures, the greater is the degree of precision.*

We usually assume that a very precise measurement is also of high accuracy. We can be wrong, however, if our instruments are improperly calibrated. For example, the improperly marked ruler in Figure 1.14 might yield measurements with a precision of ±0.01 cm, but all the measurements would be too large by 1 cm, a case of good precision but poor accuracy.

How accurate would measurements be with this ruler?

Figure 1.14 *An improperly marked ruler.* This ruler will yield measurements that are each wrong by one whole unit. The measurements might be precise, but the accuracy would be very poor.

Counting Significant Figures

Usually, it is simple to determine the number of significant figures in a number; we just count the digits. Thus 3.25 has three significant figures and 56.205 has five of them. When zeros come at the beginning or the end of a number, however, they sometimes cause confusion.

When zeros appear at the end of a number **and** *to the right of the decimal point, they are always counted as significant figures.* Thus, 4.500 m has four significant figures because the zeros would not be written unless those digits were known to be zeros.

Any zeros that precede the first nonzero digit are never counted as significant figures. For instance, a length of 2.3 mm is the same as 0.0023 m. Since we are dealing with the same measured value, its number of significant figures cannot change when we change the units. Therefore, both quantities have two significant figures and in 0.0023 m we don't count the zeros that come before the 2. They are needed, however, to locate the position of the decimal point.

With large numbers, zeros are also needed often to locate the decimal point, and this can lead to questions as to the number of significant figures in a reported value. For instance, suppose you were told that at a football game there were 45,000 fans. If this was just a rough estimate, it might be uncertain by as much as several thousand, in which case the value 45,000 represents just two significant figures and none of the zeros count as significant figures. On the other hand, suppose the official count of tickets collected was reported to be 45,000 "give or take about 10." In this case, the value represents 45,000 ± 10 fans and contains four significant figures, two of which are zeros. We can see, therefore, that a simple statement such as "there were 45,000 fans at the game" is ambiguous. We can't tell whether any of the zeros should count as significant figures unless the statement is accompanied by a description of how uncertain the value is.

To avoid this type of confusion, so that we can indicate the proper number of significant figures as well as the location of the decimal, scientific notation comes in handy. Thus, we can write the rough estimate of 45,000 as 4.5×10^4. The 4.5 shows the number of significant figures and the 10^4 tells us the location of the decimal. The value obtained from the ticket count, on the other hand, can be expressed as 4.500×10^4. This time the 4.500 shows four significant figures and an uncertainty of ±10 people.

$$4.5 \times 10^4 \qquad \text{two significant figures}$$
$$4.500 \times 10^4 \qquad \text{four significant figures}$$

Tools

Rules for counting significant figures

For the rough estimate, the number of fans present is 45,000 ± 1000.

Scientific notation is reviewed in Appendix A.

Combining Numbers in Calculations

When several numbers are obtained in an experiment they are usually combined in some way to calculate a desired quantity. For example, to determine the area of a rectangular carpet we require two measurements, length and width, which are then multiplied to give the answer we want. If one of these measurements is very precise and the other is not, we can't expect too much precision in the calculated area; the large uncertainty in the less precise measurement carries through to give a large uncertainty in the area. Therefore, to get some idea of how precise the area really is, we need a way to take into account the precision of the various values used in the calculation. To make sure this happens, we follow certain rules according to the kinds of arithmetic being performed.

Multiplication and Division

Tools

Rules for
arithmetic and
significant figures

For multiplication and division, the number of significant figures in the answer should not be greater than the number of significant figures in the least precise measurement. Let's look at a typical problem involving some measured quantities.

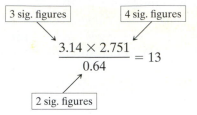

The answer to this calculation displayed on a calculator[6] is 13.497093. However, the rule says that the answer should have only as many significant figures as the least precise factor. Because the least precise factor, 0.64, has only two significant figures, the answer should have only two. The correct answer, 13, is then obtained by rounding off the calculator answer.[7]

Addition and Subtraction

Tools

Rules for
arithmetic and
significant figures

For addition and subtraction, the answer should have the same number of decimal places as the quantity with the fewest number of decimal places. As an example, consider the following addition of measured quantities.

$$
\begin{array}{r}
3.247 \\
41.36 \\
+\ 125.2 \quad\leftarrow\ \text{This number has only 1 decimal place.}\\
\hline
169.8 \quad\leftarrow\ \text{The answer has been rounded to 1 decimal place.}
\end{array}
$$

In this calculation, the digits beneath the 6 and the 7 are unknown; they could be anything. (They're not necessarily zeros because if we *knew* they were zeros, then zeros would have been written there.) Adding an unknown digit to the 6 or 7 will give an answer that's also unknown, so for this sum we are not justified in writing digits in the second and third places after the decimal point. Therefore, we round the answer to the nearest tenth.

Exact Numbers

Not all the numbers we use come from measurement. Those that come from definitions, such as 12 in. = 1 ft, and those that come from a direct count, such as the number of people in a room, have no uncertainty and are called **exact numbers.** When exact numbers are used in a calculation, we assume them to have an infinite number of significant figures and we don't take them into account when we apply the rules described above.

---

[6]Calculators usually give too many significant figures. An exception is when the answer has zeros at the right that are significant figures. For example, an answer of 1.200 would be displayed on most calculators as 1.2. If the zeros belong in the answer, be sure to write them down.

[7]When we wish to round off a number at a certain point, we simply drop the digits that follow if the first of them is less than 5. Thus, 8.1634 rounds to 8.16, if we wish to have only two decimal places. If the first digit after the point of round off is larger than 5, or if it is 5 followed by other nonzero digits, then we add 1 to the preceding digit. Thus 8.167 and 8.1653 both round to 8.17. Finally, when the digit after the point of round off is a 5 and no other digits follow the 5, then we drop the 5 if the preceding digit is even and add 1 if it is odd. Thus, 8.165 rounds to 8.16 and 8.175 rounds to 8.18.

1.6 Computations and the Factor-Label Method

As you saw in Example 1.1, solving a mathematical problem involves two steps. The first is assembling the necessary information and the second is using the information correctly to obtain the answer. Our goal is to teach you both, but our principal focus in this section is on a system called the **factor-label method** (also called **dimensional analysis**), which scientists use to help them perform the correct arithmetic to solve a problem.

In the factor-label method we treat a numerical problem as one involving a conversion of units from one kind to another. To do this we use one or more *conversion factors* to change the units of the given quantity to the units of the answer.

(given quantity) × (conversion factor) = (desired quantity)

A **conversion factor** *is a fraction formed from a <u>valid</u> relationship or equality between units and is used to switch from one system of measurement and units to another.* To illustrate, suppose we want to express a person's height of 72.0 inches in centimeters. To do this we need the relationship between the inch and the centimeter. We can obtain this from Table 1.3.

2.54 cm = 1 in. (exactly)   (1.5)

If we divide both sides of this equation by 1 in., we obtain a conversion factor.

$$\frac{2.54 \text{ cm}}{1 \text{ in.}} = \frac{\cancel{1 \text{ in.}}}{\cancel{1 \text{ in.}}} = 1$$

Notice that we have canceled the units from both the numerator and denominator of the center fraction. *Units behave just as numbers do in mathematical operations,* which is a key part of the factor-label method. This leaves the first fraction equaling 1. Let's see what happens if we multiply 72.0 inches, the height that we mentioned, by this fraction.

$$72.0 \text{ }\cancel{\text{in.}} \times \frac{2.54 \text{ cm}}{1 \text{ }\cancel{\text{in.}}} = 183 \text{ cm}$$

$$\left(\begin{array}{c}\text{given}\\\text{quantity}\end{array}\right) \times \left(\begin{array}{c}\text{conversion}\\\text{factor}\end{array}\right) = \left(\begin{array}{c}\text{desired}\\\text{quantity}\end{array}\right)$$

Because we have multiplied 72.0 in. by something that is equal to 1, we know we haven't changed the magnitude of the person's height. We have, however, changed the units. Notice that we have canceled the unit inches. The only unit left is centimeters, which is the unit we want for the answer. The result, therefore, is the person's height in centimeters.

One of the benefits of the factor-label method is that it lets you know when you have done the *wrong* arithmetic. From the relationship in Equation 1.5, we can actually construct two conversion factors:

$\dfrac{2.54 \text{ cm}}{1 \text{ in.}}$ and $\dfrac{1 \text{ in.}}{2.54 \text{ cm}}$

We used the first one correctly, but what would have happened if we had used the second by mistake?

$$72.0 \text{ in.} \times \frac{1 \text{ in.}}{2.54 \text{ cm.}} = 28.3 \text{ in.}^2/\text{cm}$$

In this case, none of the units cancel. We get units of in.^{2}/cm because in. × in. = in.^{2}. Even though our calculator may be very good at arithmetic, we've got the

Tools

Factor-label method

To solve a problem, you first have to collect all the information you need. The factor-label method will then help you set up the arithmetic correctly.

To construct a valid conversion factor, the relationship between the units must be true. For example, the statement

3 ft = 41 in.

is false. Although you might make a conversion factor out of it, any answers you would calculate are sure to be incorrect. *Correct answers require correct relationships between units.*

The relationship between the inch and the centimeter is exact, so the numbers in 1 in. = 2.54 cm have an infinite number of significant figures.

Most chemists and physicists use the factor-label method. There must be a good reason why.

wrong answer. *The factor-label method lets us know we have the wrong answer because the units are wrong!*

We will use the factor-label method extensively throughout this book to aid us in setting up the proper arithmetic in problems. In fact, we will see that it also helps us assemble the information we need to solve the problem. The following examples illustrate the method.

EXAMPLE 1.2

Applying the Factor-Label Method

Convert 3.25 m to millimeters (mm).

Analysis: A good way to begin a problem is to restate the question in the form of an equation. We will do this by writing the given quantity (with its units) on the left and the *units* of the desired answer on the right.

$$3.25 \text{ m} = ? \text{ mm}$$

To solve this by the factor-label method we need a conversion factor that can be used to change the unit meter to the unit millimeter, and this requires a relationship between the two units. Reviewing what has been covered so far we realize that the tool we need to solve this problem comes from the table of decimal multipliers. The prefix "milli" means "$\times 10^{-3}$," so we can write

$$1 \text{ mm} = 10^{-3} \text{ m}$$

Notice that this relationship connects the units given to the units desired.

Now we have the information we need to solve the problem, so let's proceed with the solution.

Solution: From the relationship above, we can make two conversion factors.

$$\frac{1 \text{ mm}}{10^{-3} \text{ m}} \quad \text{and} \quad \frac{10^{-3} \text{ m}}{1 \text{ mm}}$$

We know we have to cancel the unit meter, so we need to multiply by a conversion factor with this unit in the denominator. Therefore, we select the one on the left. This gives

$$3.25 \text{ m} \times \frac{1 \text{ mm}}{10^{-3} \text{ m}} = 3.25 \times 10^3 \text{ mm}$$

Notice we have expressed the answer to three significant figures because that is how many there are in the given quantity, 3.25 m.

Is the Answer Reasonable?
We know that millimeters are much smaller than meters, so 3.25 m must represent a lot of millimeters. Our answer, therefore, makes sense.

EXAMPLE 1.3

Applying the Factor-Label Method

A liter, which is slightly larger than a quart, is defined as 1 cubic decimeter (1 dm^3). How many liters are there in 1 cubic meter (1 m^3)?

Analysis: Let's begin once again by stating the problem in equation form.

$$1 \text{ m}^3 = ? \text{ L}$$

Next, we assemble the tools. What relationships do we know that relate these various units? We are given the relationship between liters and cubic decimeters,

$$1 \text{ L} = 1 \text{ dm}^3 \tag{1.6}$$

From the table of decimal multipliers, we also know the relationship between decimeters and meters,

$$1 \text{ dm} = 0.1 \text{ m}$$

but we need a relationship between cubic units. Since units undergo the same kinds of operations numbers do, we simply cube each side of this equation (being careful to cube *both* the numbers and the units).

$$(1 \text{ dm})^3 = (0.1 \text{ m})^3$$

$$1 \text{ dm}^3 = 0.001 \text{ m}^3 \tag{1.7}$$

Notice how Equations 1.6 and 1.7 provide a path from the given units to those we seek. Such a path is always a necessary condition when we apply the factor-label method.

$$m^3 \xrightarrow{\text{Equation 1.7}} dm^3 \xrightarrow{\text{Equation 1.6}} L$$

Now we are ready to solve the problem.

Solution: The first step is to eliminate the units m^3. We use Equation 1.7.

$$1 \text{ m}^3 \times \frac{1 \text{ dm}^3}{0.001 \text{ m}^3} = 1000 \text{ dm}^3$$

Then we use Equation 1.6 to take us from dm^3 to L.

$$1000 \text{ dm}^3 \times \frac{1 \text{ L}}{1 \text{ dm}^3} = 1000 \text{ L}$$

Thus, 1 m^3 = 1000 L.

Usually, when a problem involves the use of two or more conversion factors, they can be "strung together" to avoid having to compute intermediate results. For example, this problem can be set up as follows.

$$1 \text{ m}^3 \times \frac{1 \text{ dm}^3}{0.001 \text{ m}^3} \times \frac{1 \text{ L}}{1 \text{ dm}^3} = 1000 \text{ L}$$

Is the Answer Reasonable?
One liter is about a quart. A cubic meter is about a cubic yard. Therefore, we expect a large number of liters in a cubic meter, so our answer seems reasonable. (Notice here that in our analysis we have approximated the quantities in the calculation in units of quarts and cubic yards, which are more familiar than liters and m^3 if you've been raised in the United States. We get a feel for the approximate magnitude of the answer using our familiar units and then relate this to the actual units of the problem.)

EXAMPLE 1.4

Applying the Factor-Label Method

In 1998, Jani Soininen of Finland won the Bronze Medal for the ski jump at the Nagano Winter Olympics with a jump of 136 meters. What is this distance expressed in feet?

Analysis: The problem can be stated as

$$136 \text{ m} = ? \text{ ft}$$

One of several sets of relationships we can use is

$$1 \text{ cm} = 10^{-2} \text{ m} \quad \text{(from Table 1.2)}$$

Jani Soininen

$$1 \text{ in.} = 2.54 \text{ cm} \qquad \text{(from Table 1.3)}$$

$$1 \text{ ft} = 12 \text{ in.}$$

Notice how they take us from meters to centimeters to inches to feet.

Solution: Now we apply the factor-label method by eliminating unwanted units to bring us to the units of the answer.

$$136 \text{ m} \times \frac{1 \text{ cm}}{10^{-2} \text{ m}} \times \frac{1 \text{ in.}}{2.54 \text{ cm}} \times \frac{1 \text{ ft}}{12 \text{ in.}} = 446 \text{ ft}$$

m to cm
m to in.
m to ft

Notice that if we were to stop after the first conversion factor, the units of the answer would be centimeters; if we stop after the second, the units would be inches, and after the third we get feet—the units we want. This time the answer has been rounded to three significant figures because that's how many there were in the measured distance. Notice that the numbers 12 and 2.54 do not affect the number of significant figures in the answer because they are exact numbers derived from definitions.

This is not the only way we could have solved this problem. Other sets of conversion factors could have been chosen. For example, we could have used

$$1 \text{ yd} = 0.9144 \text{ m}$$

$$3 \text{ ft} = 1 \text{ yd}$$

Then the problem would have been set up as

$$136 \text{ m} \times \frac{1 \text{ yd}}{0.9144 \text{ m}} \times \frac{3 \text{ ft}}{1 \text{ yd}} = 446 \text{ ft}$$

Many problems that you meet, just like this one, have more than one path to the answer. There isn't necessarily any *one* correct way to set up the solution. *The important thing is for you to be able to reason your way through a problem and find some set of relationships that can take you from the given information to the answer.* The factor-label method can help you search for these relationships if you keep in mind the units that must be eliminated by cancellation.

Is the Answer Reasonable?
A meter is slightly longer than a yard. In 136 m, there are slightly more than 3 × 136 = 408 ft, so the answer of 446 ft seems to be reasonable.

Practice Exercise 3

Use the factor-label method to perform the following conversions: (a) 3.00 yd to inches; (b) 1.25 km to centimeters; (c) 3.27 mm to feet; (d) 20.2 miles/gallon to kilometers/liter. ◆

1.7 Properties of Substances; Density and Specific Gravity

Properties of Matter

If you were searching through a pile of books for your chemistry book, you would no doubt recognize it by its size, color, and the printing on the cover. These are the characteristics that books have that help you distinguish among them. Similarly, in chemistry we use the characteristics, or *properties,* of matter to distinguish one kind from another. To help organize our thinking we classify these properties into different types.

Physical and Chemical Properties

A **physical property** *is one that can be observed without changing the chemical makeup of a substance.* For example, you know that when you stir sugar in water the solid sugar disappears as it dissolves. This ability of sugar to dissolve in water is a physical property because the act of dissolving the sugar doesn't alter its chemical composition. Figure 1.15 illustrates that the molecules of sugar and water simply intermingle when the sugar dissolves. By evaporating the water we can show that the chemical makeup of the sugar has not been affected; the sugar is recovered chemically unchanged. Because the sugar isn't altered chemically as it dissolves, forming the solution is said to be a **physical change.**

A **chemical property** *describes a chemical change (chemical reaction) that a substance undergoes.* If we melt sugar in a pan and heat it to a high temperature, the color of the sugar darkens as it begins to decompose into carbon and water. The decomposition of sugar at high temperatures is a chemical property. The only way we can observe this property is by having the chemical reaction occur, a reaction that leads to a permanent change in chemical composition that is not reversed by cooling the sugar to room temperature.

When we observe a chemical property of a substance, the substance undergoes a reaction and changes into something else. When we observe a physical property, the substance is not changed chemically.

Intensive and Extensive Properties

Some properties that we observe depend on the size of the sample under study and others do not. For example, the masses and volumes of two different pieces of gold can be different, but both have the same characteristic shiny yellow color

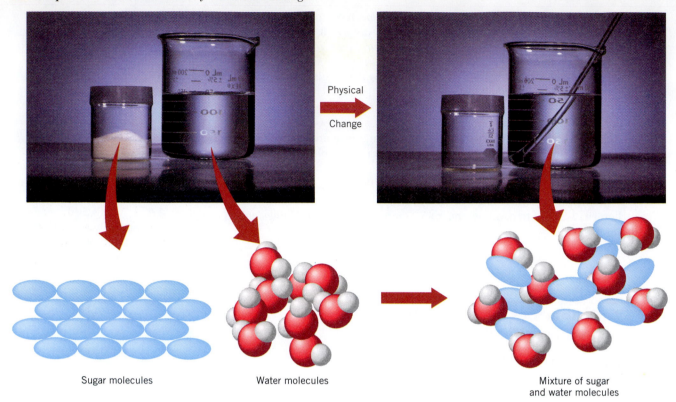

Sugar molecules   Water molecules   Mixture of sugar
                                                   and water molecules

Figure 1.15 *Formation of a solution of sugar in water.* As the sugar mixes with the water, sugar molecules (shown here in a simplified way) mingle with water molecules. The chemical makeups of the individual sugar and water molecules do not change, however.

and both conduct electricity. Mass and volume are **extensive properties,** *or properties that depend on sample size.* Color and electrical conductivity are **intensive properties,** *or properties that are independent of sample size.*[8]

Density and Specific Gravity

One of the interesting things about extensive properties is that if you take the ratio of two of them, the resulting quantity is usually independent of sample size. In effect, the sample size cancels out and the calculated quantity becomes an intensive property. One useful property obtained this way is **density,** *which is defined as the ratio of an object's mass to its volume.* Using the symbols d for density, m for mass, and V for volume, we can express this mathematically as

Tools →

Density

$$d = \frac{m}{V} \qquad (1.8)$$

Notice that to determine an object's density we make two measurements, mass and volume.

Each pure substance has its own characteristic density (Table 1.4). Gold, for instance, is much more dense than iron. Each cubic centimeter of gold has a

[8]Color can often be a useful property for identification, but there are instances where you can be fooled, particularly when particle size is very small. For example, silver is a white metal with a high luster; however, in a very finely divided state, as in the image on black-and-white photographic film or paper, metallic silver appears black.

EXAMPLE 1.5
Calculating Density

A student measured the volume of an iron nail to be 0.880 cm^3. She found that its mass was 6.92 g. What is the density of iron?

Analysis: This is a very straightforward calculation that requires that you remember the definition of density given by Equation 1.8. Don't neglect to learn such definitions, because they are tools you will need in situations like this one.

Solution: To determine the density we simply take the ratio of mass to volume.

$$\text{density} = \frac{6.92 \text{ g}}{0.880 \text{ cm}^3}$$

$$= 7.86 \text{ g/cm}^3$$

This could also be written as

$$\text{density} = 7.86 \text{ g/mL}$$

because $1 \text{ cm}^3 = 1 \text{ mL}$.

Is the Answer Reasonable?
In calculating the density, we are dividing 6.92 by a number slightly smaller than 1. The answer should therefore be slightly larger than 6.92. Our answer of 7.86 seems to be okay.

Table 1.4 Densities of Some Common Substances in g/cm³ at Room Temperature

Water	1.00
Aluminum	2.70
Iron	7.86
Silver	10.5
Gold	19.3
Glass	2.2
Air	0.0012

mass of 19.3 g, so its density is 19.3 g/cm^3. By comparison, the density of water is 1.00 g/cm^3 and the density of air at room temperature is about 0.0012 g/cm^3.

Most substances, such as the mercury in the bulb of a thermometer, expand slightly when they are heated. The same amount of matter occupies a larger volume at a higher temperature, so the amount of matter packed into each cubic centimeter is less. Therefore, density usually decreases slightly with increasing temperature.[9] For solids and liquids the size of this change is small, as you can see from the data for water in Table 1.5. Except when very precise calculations are involved we can generally ignore the variation of density with temperature.

There is more mass in 1 cm^3 of gold than in 1 cm^3 of iron.

Although the density of water varies slightly with temperature, it is useful to remember the value 1.00 g/cm^3. It can be used if the water is near room temperature and only three (or fewer) significant figures are required.

►**Chemistry in Practice**◄ The densities of metals are one of the important physical properties considered when designing parts for aircraft and spacecraft such as the space shuttle. The metals aluminum and titanium are used extensively because they provide significant strength with minimal weight. Titanium is particularly useful because it is as strong as steel but about 40% lighter (for a given volume of metal). Titanium is 60% heavier than aluminum but twice as strong, and its melting point is approximately 1000 °C higher. ◆

Table 1.5 Density of Water as a Function of Temperature

Temperature (°C)	Density (g/cm³)
10	0.999700
15	0.999099
20	0.998203
25	0.997044
30	0.995646

Using Density

One reason why density is a useful property is that it provides a way to convert between the mass and volume of a substance. In Example 1.5, we found that iron has a density of 7.86 g/cm^3. This density defines a relationship, which we will call

[9]Liquid water behaves oddly. Its maximum density is at 4 °C, so when water at 0 °C is warmed, its density increases until the temperature reaches 4 °C. As the temperature is increased further the density of water gradually decreases.

an **equivalence,** between the amount of mass and its volume. In words, we could express this by saying that 7.86 g of iron are equivalent to 1.00 cm^3 of iron, or alternatively that 1.00 cm^3 of iron is equivalent to 7.86 g of iron. We express this relationship symbolically as

$$7.86 \text{ g iron} \Leftrightarrow 1.00 \text{ cm}^3 \text{ iron}$$

where we have used the symbol $\Leftrightarrow$ to mean "is equivalent to." (We can't really use an equals sign in this expression because grams can't *equal* cubic centimeters; one is a unit of mass and the other is a unit of volume.)

In setting up calculations by the factor-label method, an equivalence can be used to construct conversion factors just as equalities can. From the equivalence we have just written we can form two conversion factors:

$$\frac{7.86 \text{ g iron}}{1.00 \text{ cm}^3 \text{ iron}} \quad \text{and} \quad \frac{1.00 \text{ cm}^3 \text{ iron}}{7.86 \text{ g iron}}$$

The following example illustrates how we use density in calculations.

EXAMPLE 1.6

Calculations Using Density

A sample of vegetable oil has a density of 0.916 g/mL. (a) What is the mass of 225 mL of the oil? (b) How many milliliters are occupied by 45.0 g of the oil?

Analysis: What we need here is a tool that lets us convert between the mass and volume of a substance, and that is precisely what the density is. For this oil the density tells us that

$$1.00 \text{ mL oil} \Leftrightarrow 0.916 \text{ g oil}$$

From this relationship we can construct two conversion factors.

$$\frac{1.00 \text{ mL oil}}{0.916 \text{ g oil}} \quad \text{and} \quad \frac{0.916 \text{ g oil}}{1.00 \text{ mL oil}}$$

Solution to (a): The question can be restated as: 225 mL oil = ? g oil. We need to eliminate the unit *mL oil,* so we choose the conversion factor on the right.

$$225 \text{ mL oil} \times \frac{0.916 \text{ g oil}}{1.00 \text{ mL oil}} = 206 \text{ g oil}$$

Thus, 225 mL of the oil has a mass of 206 g.

Solution to (b): The question is: 45.0 g oil = ? mL oil. This time we need to eliminate the unit *g oil,* so we use the conversion factor on the left.

$$45.0 \text{ g oil} \times \frac{1.00 \text{ mL oil}}{0.916 \text{ g oil}} = 49.1 \text{ mL oil}$$

Thus, 45.0 g of the oil has a volume of 49.1 mL.

Are the Answers Reasonable?
Notice that the density tells us that 1 mL of oil has a mass of slightly less than 1 g. Therefore, for part (a) we might expect that 225 mL of oil should have a mass slightly less than 225 g. Our answer, 206 g, is reasonable. For part (b), 45 g of oil should have a volume not too far from 45 mL, so our answer of 49.1 mL is also "in the right ballpark."

Practice Exercise 4

A bar of aluminum has a volume of 1.45 mL. Its mass is 3.92 g. What is its density? ◆

Practice Exercise 5

The density of silver is 10.5 g/cm^3. (a) What volume would 2.86 g of silver occupy? (b) What is the mass of 16.3 cm^3 of silver? ◆

Specific Gravity

The numerical value for the density of a substance depends on the units used for mass and volume. For example, if we express mass in grams and volume in milliliters, the density of water is 1.00 g/mL. However, if mass is given in pounds and volume in gallons, the density of water is 8.34 lb/gal; and if mass is in pounds and volume is in cubic feet, water's density is 62.4 lb/ft^3. In a similar way, the densities of other substances have different numerical values for different units. One way to avoid having to tabulate densities in all sorts of different units is to tabulate specific gravities instead. *The **specific gravity** of a substance is defined as the ratio of the density of the substance to the density of water.*

$$\text{sp.gr.} = \frac{d_{substance}}{d_{water}} \qquad (1.9)$$

Tools ←

Specific gravity

The specific gravity tells us how much denser than water a substance is. For instance, if the specific gravity of a substance is 2.00, then it is twice as dense as water; if its specific gravity is 0.50, then it is only half as dense as water. This means that if we know the density of water in a particular set of units, we can multiply it by a substance's specific gravity to obtain the substance's density in these same units. Rearranging Equation 1.9 gives

$$d_{substance} = (\text{sp.gr.})_{substance} \times d_{water} \qquad (1.10)$$

Let's look at an example.

EXAMPLE 1.7
Using Specific Gravity

Methanol, a liquid fuel that can be made from coal, has a specific gravity of 0.792. What is the density of methanol in units of g/mL, lb/gal, and lb/ft^3?

Analysis: The question asks us to relate specific gravity to density, so Equation 1.9 (or its equivalent, Equation 1.10) is the tool we need to get the answer. We also need some data—the density of water in each of the requested units. Ordinarily we would look for them in a table of densities of water, although this time the necessary data were given in the preceding discussion.

$$d_{water} = 1.00 \text{ g/mL}$$

$$= 8.34 \text{ lb/gal}$$

$$= 62.4 \text{ lb/ft}^3$$

Solution: Now that we have all the necessary tools and data, we bring them together to obtain the answers.

$$d_{methanol} = (\text{sp. gr.})_{methanol} \times d_{water}$$

$$= 0.792 \times 1.00 \text{ g/mL}$$

$$= 0.792 \text{ g/mL}$$

Similarly, for the other units,

$$d_{methanol} = 0.792 \times 8.34 \text{ lb/gal}$$
$$= 6.61 \text{ lb/gal}$$

and

$$d_{methanol} = 0.792 \times 62.4 \text{ lb/ft}^3$$
$$= 49.4 \text{ lb/ft}^3$$

Notice that when the density is expressed in g/mL, it is numerically the same as the specific gravity.

Are the Answers Reasonable?

If the specific gravity is less than 1.00, the density of the substance is less than that of water. With a specific gravity of 0.792, therefore, the density of the methanol must be smaller than that of water. Notice that in each case, we have obtained an answer that is smaller than the density of water. Our answers, therefore, seem to be correct.

Practice Exercise 6

The density of aluminum is 2.70 g/mL. What is its specific gravity? What is the density of aluminum in units of lb/ft³? ◆

Practice Exercise 7

Ethyl acetate is a clear colorless solvent having a fruity odor that is often used in the manufacture of plastics. Its specific gravity is 0.902. What is its density in both g/mL and lb/gal? ◆

Identification of Substances by Their Properties

A job chemists are often called upon to perform is chemical analysis. They're asked "What is a particular sample composed of?" To answer such a question, the chemist relies on the properties of the chemicals that make up the sample.

For identification purposes, intensive properties are more useful than extensive ones because every sample of a given substance exhibits the same set of intensive properties. For instance, if you were asked to decide whether a particular liquid sample was water, finding either its mass or volume wouldn't help in your identification because mass and volume are not unvarying properties of a substance. By taking appropriate amounts, it is possible for *any* liquid to have a given mass or a given volume. However, if we measure both the mass *and* volume of the sample, we can calculate the density of the liquid, and density is an intensive property. Since water has a density of 1.00 g/mL, we would know for sure that the liquid is not water if its density differs from this value. If the measured density is 1.00 g/mL, there is a reasonable possibility the liquid is water, so we might determine some additional properties such as melting point and boiling point. If after doing our experiments we find that the liquid in question is colorless, has a density of 1.00 g/mL, a freezing point of 0 °C, and a boiling point of 100 °C, we can feel quite certain that it *is* water, because these are properties that all samples of water have in common. It is also unlikely that another liquid would have these same identical properties.

(a)

(b)

Figure 1.16 *Identification of substances by their properties.* Gold and the mineral pyrite (also called iron pyrite) have similar colors, which is why some miners mistook pyrite for gold. (*a*) A nugget of pure gold. (*b*) A sample of iron pyrite. The color of the mineral accounts for its nickname, "fool's gold."

Color, density, freezing point, and boiling point are examples of physical properties that can help us identify substances. Chemical properties are also intensive properties and also can be used for identification. For example, gold miners were able to distinguish between real gold and fool's gold (a mineral also called pyrite, Figure 1.16) by heating the material in a flame. Nothing happens to the gold, but the pyrite sputters, smokes, and releases bad smelling fumes because of its ability when heated to react chemically with oxygen in the air.

The Importance of Accuracy and Precision of Measurements

If we are to rely on properties such as density for identification of substances, it is very important that our measurements be reliable. The terms accuracy and precision address this question.

The importance of accuracy is obvious. If we have no confidence that our measured values are close to the true values, we certainly cannot trust the results of our experiments.

Precision of measurements can be equally important. For example, suppose we had two samples of a substance and believed them to be of the same mass. We could check our suspicion by weighing them. Suppose we selected an inexpensive balance capable of measurements to the nearest ±0.1 g and obtained values of 12.3 g and 12.4 g. Are the masses really different? With an uncertainty of ±0.1 g, we cannot tell whether the difference between the values is the result of an actual difference in mass or just a manifestation of the uncertainty in the measurements.

Suppose we now measure the masses with a higher quality balance capable of measurements to the nearest ±0.0001 g and obtain values of 12.3115 g and 12.3667 g. These values differ by 0.0552 g. This difference is much larger than the uncertainty in the measured masses (±0.0001 g), so we can be sure there is a real difference in the masses of the two samples.

The point of the preceding discussion is that to trust the results of our experiments, we must be sure our measurements are accurate and that they are of sufficient precision to be meaningful. This has a lot to do with how we plan experiments in the laboratory.

SUMMARY

Chemistry and Chemical Reactions. Scientists from all fields must be concerned with chemistry because the material things they study are composed of chemicals. Today, as in the past, chemistry plays a crucial role in fulfilling the needs of society. **Chemistry** is a **natural science** that studies the properties and composition of matter and the way substances interact with each other in chemical reactions. When substances undergo a **chemical reaction** (or **chemical change**) they change into new substances that have different properties from the starting materials. Before the reaction begins we observe the properties of the starting materials; after the reaction is over we observe the properties of the new substances that are formed.

Scientific Method. This is the general procedure by which science advances. **Observations** are made, and the **empirical facts** or **data** that are collected from many experiments are often summarized in **scientific laws.** Scientists formulate mental images or models of nature to explain observed behavior. Models begin as **hypotheses,** and those that survive repeated testing become known as **theories.** The **atomic theory** proposes that matter is made of tiny particles (atoms) that combine to form more complex substances. Molecules are particles that contain two or more atoms. Although the general pattern in the development of science is the cycle of observation–explanation–observation–explanation . . . , many discoveries are made accidentally by people who have learned to be observant through scientific training.

Matter and Energy. **Matter** is anything that has mass and occupies space. **Mass** is proportional to the amount of matter in a substance, and it differs from **weight,** which is determined by how mass is attracted to Earth by gravity. **Energy** is the capacity to do work. An object can have energy as **kinetic energy** (because of its motion) or as **potential energy** (stored energy). All forms of energy can be converted to heat. Heat and temperature are different: **heat** is energy that's transferred between objects because of a difference in their temperatures. **Temperature** is a measure of the *average* kinetic energy of the atoms in an object.

Units of Measurement. **Qualitative observations** lack numerical information, whereas **quantitative observations** require numerical measurements. The units used for measurements are based on the set of seven **SI base units** which can be combined to give various **derived units.** These all can be scaled to various sizes by applying **decimal multiplying factors.** In the laboratory we routinely measure length, volume, mass, and temperature. Convenient units for these purposes are, respectively, **centimeters** or **millimeters, liters** or **milliliters, grams,** and **kelvins** or **degrees Celsius.**

Significant Figures. The **precision** of a measured quantity is revealed by the number of **significant figures** that it contains, which equals the number of digits known with certainty plus the first one that possesses some uncertainty. Measured values are **precise** if they contain many significant figures and therefore differ from each other by small amounts. A measurement is **accurate** if its value lies very close to the true value. When measurements are combined by multiplication or division, the answer should not contain more significant figures than the least precise factor. When addition or subtraction is used, the answer is rounded to the same number of decimal places as the quantity having the fewest number of decimal places. **Exact numbers** do not enter into determining the number of significant figures in a calculated quantity because they have no uncertainty. **Scientific notation** is useful for writing large or small numbers in compact form and for expressing unambiguously the number of significant figures in a number.

Factor-Label Method. The **factor-label method** is based on the ability of units to undergo the same mathematical operations as numbers. **Conversion factors** are constructed from *valid relationships* between units. These relationships can be either equalities or equivalencies between units. Unit cancellation serves as a guide to the use of conversion factors and aids us in correctly setting up the arithmetic for a problem.

Properties of Matter. When studying matter we are concerned with its physical and chemical properties. **Physical properties** can be measured without changing the chemical makeup of the sample. **Chemical properties** relate to how substances change into other substances in chemical reactions. **Intensive properties** are independent of sample size; **extensive properties** depend on sample size.

Density and Specific Gravity. **Density** is a useful intensive property that is defined as the ratio of a sample's mass to its volume. Besides serving as a means for identifying substances, density is a conversion factor that relates mass to volume. **Specific gravity** is the ratio of a sample's density to that of water. The numerical values of specific gravity and density are the same if density is expressed in the units g/mL.

Identification of Substances by Their Properties. Intensive properties are more useful than extensive properties for identification of substances. Accuracy and precision of measurements affect our ability to make positive identifications.

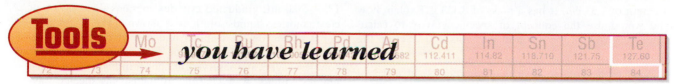

Tools *you have learned*

The table below lists the tools that you have learned in this chapter that are applicable to problem solving. Review them if necessary, and refer to them when working on the Thinking-It-Through problems and the Review Exercises that follow.

Tool	Function
SI prefixes (*Table 1.2, page 11*)	To create larger and smaller units. To form conversion factors for converting between differently sized units.
Units in laboratory measurements (*page 11*)	To convert between commonly used laboratory units of measurement.
Temperature conversions (*page 15*)	To convert among temperature units. Be especially sure you can convert between Celsius and Kelvin temperatures, because that will be needed most in this course.
Rules for counting significant figures (*page 19*)	To determine the number of significant figures in a number.
Rules for arithmetic and significant figures (*page 20*)	To round answers to the correct number of significant figures.
Factor-label method (*page 21*)	To set up the arithmetic in numerical problems by assembling necessary relationships into conversion factors and applying them to obtain the desired units of the answer.
Density (*page 26*)	To calculate the density of a substance from measurements of mass and volume. To convert between mass and volume for a substance.
Specific gravity (*page 29*)	To convert density from one set of units to a different set of units.

THINKING IT THROUGH

The goal for each of the following problems is to give you practice in *thinking* your way through problems. The goal is *not* to find the answer itself; instead, you are just asked to assemble the available information needed to obtain the answer, state what additional data (if any) are needed, and describe how you would use the data to answer the question. For problems involving unit conversions, list the relationships among the units that are needed to carry out the conversions. Construct the conversion factors that can be formed from these relationships.

Then set up the solution to the problem by arranging the conversion factors so the units cancel correctly to give the desired units of the answer.

The problems are divided into two groups. Those in Level 2 are more challenging than those in Level 1 and provide an opportunity to really hone your problem solving skills. Detailed answers to these problems can be found at our Web site: http://www.wiley.com/college/brady.

Level 1 Problems

1. The following ask you to perform unit conversions. Show how you would accomplish these conversions, using appropriate conversion factors and abbreviations for the units involved.
(a) How many nanometers are there in 14.6 cm?
(b) Suppose a projectile is fired at a velocity of 1450 m/s. What is this speed expressed in kilometers per hour?
(c) A substance has a density of 6.85 g/cm^3. What is this density expressed in units of $\mu g/\mu m^3$?

(d) How many cubic millimeters are there in 2.20 cm^3?
(e) A person is 5 ft 9 in. tall. What is this height expressed in centimeters?

2. Normal human body temperature is 98.6 °F. What is this temperature expressed in kelvins?

3. In one year (365 days), light travels a distance of 9.46×10^{12} km. What is the speed of light expressed in *miles per hour?*

4. A cylindrical metal bar has a diameter of 0.753 cm and a length of 2.33 cm. It has a mass of 8.423 g. Explain how you would use the concept of specific gravity to calculate the density of the metal in the units lb/ft³. Set up the calculation.

5. What is the volume (in cubic centimeters) of a bar of lead that has a mass of 256.4 g? The density of lead is 11.34 g/cm³.

6. For the preceding question, explain how you determine the number of significant figures that should be reported in the answer.

7. A liquid has a specific gravity of 1.12. Show how you would calculate the mass in pounds of 3.45 gallons of the liquid.

Level 2 Problems

8. A liquid has a specific gravity of 0.824. What is the mass in pounds of 146 in.³ of it?

9. Because of the serious consequences of lead poisoning, the Federal Centers for Disease Control in Atlanta has set a threshold of concern for lead levels in children's blood. This threshold was based on a study that suggested that lead levels in blood as low as *10 micrograms of lead per deciliter of blood* can result in subtle effects of lead toxicity. Suppose a child had a lead level in her blood of 2.5×10^{-4} grams of lead per liter of blood. Is this person in danger of exhibiting the effects of lead poisoning?

10. Gold has a density of 19.31 g/cm³. How many grams of gold are required to provide a gold coating 0.50 mm thick on a ball bearing having a diameter of 0.500 in.?

11. Methanol has a specific gravity of 0.792. What is its density in units of pounds per cubic meter?

12. It is possible to separate molecules based on size using a specially prepared membrane made of polycarbonate, a synthetic polymer. By chemically depositing gold from solution onto the polymer, it is possible to shrink the size of the pores in a controlled fashion and thus to design filters for molecules of various sizes. Suppose a polycarbonate membrane has pores 30 nm in diameter. How would you calculate the weight of gold required to reduce the size of a pore from 30 nm to 0.6 nm? What other information would you need to carry out the calculation?

13. A 10.0 mL graduated cylinder contains three layers of clear colorless liquids/solutions as shown in Fig. 1A. A stopper is placed on the cylinder and it is vigorously shaken. After shaking and standing for a few minutes two clear colorless layers are observed, as shown in Fig. 1B. Rank the liquid/solutions (A–E) in order of increas-ing specific gravity. Explain how you obtained your ranking.

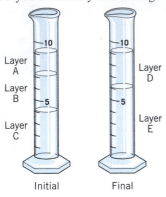

Figure 1A   *Figure 1B*

14. An irregularly shaped figure with mass 122.6 g claimed to be composed of pure gold is added to the graduated cylinder containing a liquid shown in Fig. 2A. The result is shown in Fig. 2B. Is the figure pure gold? What concept discussed in this chapter did you use to answer this problem? What additional information did you need to look up to answer the problem?

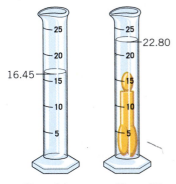

Figure 2A   *Figure 2B*

REVIEW QUESTIONS

Introduction

1.1 After some thought, give two reasons why a course in chemistry will benefit *you* in the pursuit of your particular major.

1.2 What does the science of chemistry seek to study?

1.3 Look around and list ten items you see that are made of synthetic materials, that is, materials not found in nature.

1.4 In answering Question 1.3, what have you learned about the contribution of chemistry to modern civilization?

1.5 What is a chemical reaction?

1.6 Lye is a common name for a substance called sodium hydroxide. Muriatic acid is the common name for hydrochloric acid. Either of these chemicals can cause severe burns if left in contact with the skin, but when water solutions of them are mixed in just the right proportions, the resulting solution contains only sodium chloride. From this description, how do you know that a chemical reaction occurs between sodium hydroxide and hydrochloric acid?

1.7 If you swallow a water solution of baking soda, a gas (carbon dioxide) forms in your stomach, which causes you to burp. Has a chemical reaction occurred? Explain your answer.

Scientific Method

1.8 What steps are involved in the scientific method?

1.9 What is the function of a laboratory?

1.10 Define (a) data, (b) hypothesis, (c) law, and (d) theoretical model.

1.11 What role does luck play in the advancement of science?

Matter and Energy

1.12 Define *matter* and *energy*. Which of the following are examples of matter? (a) air, (b) a pencil, (c) a cheese sandwich, (d) a squirrel, (e) your mother.

1.13 What are the three states of matter?

1.14 How do mass and weight differ?

1.15 Define (a) *kinetic energy* and (b) *potential energy*. What two things determine how much kinetic energy an object possesses?

1.16 State the equation used to calculate an object's kinetic energy. Define the symbols used in the equation.

1.17 How do *temperature* and *heat* differ?

1.18 State the law of conservation of energy. When a ball rolls down a hill, what happens to its potential energy? What happens to its kinetic energy? Are these observations consistent with the law of conservation of energy? Explain.

1.19 When gasoline burns, it reacts with oxygen in the air and forms carbon dioxide and water vapor. How does the potential energy of the gasoline and oxygen compare with the potential energy of the carbon dioxide and water vapor?

1.20 What is *mechanical energy?*

SI Units

1.21 Several commercial products that use metric units are pictured in Figure 1.4. Can you find any others among the items with which you come in contact each day?

1.22 What does the abbreviation SI stand for?

1.23 Which SI base unit is defined in terms of a physical object?

1.24 On page 7 we saw that we could calculate kinetic energy by using the equation $KE = \frac{1}{2}mv^2$. The SI unit for mass is the kilogram (kg), and the derived unit for velocity or speed is meter/second (m/s). What is the SI derived unit for energy? (The unit is called the joule, abbreviated J.)

1.25 What is the meaning of each of the following prefixes? (a) centi, (b) milli, (c) kilo, (d) micro, (e) nano, (f) pico, (g) mega.

1.26 What abbreviation is used for each of the prefixes named in Question 1.25?

1.27 What units are most useful in the laboratory for measuring (a) length, (b) volume, and (c) mass?

1.28 How is mass measured? What is the difference between a balance and a spring scale found in a market?

1.29 What reference points do we use in calibrating the scale of a thermometer? What temperature on the Celsius scale do we assign to each of these reference points?

1.30 In each pair, which is larger: (a) A Fahrenheit degree or a Celsius degree? (b) A Celsius degree or a kelvin? (c) A Fahrenheit degree or a kelvin?

Significant Figures; the Factor-Label Method

1.31 Define the term *significant figures*.

1.32 What is *accuracy?* What is *precision?*

1.33 Suppose someone suggested using the fraction $\dfrac{1 \text{ yd}}{2 \text{ ft}}$ as a conversion factor to change a length expressed in feet to its equivalent in yards. What is wrong with this conversion factor? Can we construct a valid conversion factor relating centimeters to meters from the equation 1 cm = 1000 m? Explain your answer.

1.34 In 1 hour there are 3600 seconds. By what conversion factor would you multiply 250 seconds to convert it to hours? By what conversion factor would you multiply 3.84 hours to convert it to seconds?

1.35 If you were to convert the measured length 4.165 ft to yards by multiplying by the conversion factor (1 yd/3 ft), how many significant figures should the answer contain? Why?

Properties of Substances

1.36 What is a *physical property?* What is a *chemical property?* What is the chief distinction between physical and chemical properties? Define the terms *intensive property* and *extensive property*. Give two examples of each.

1.37 How does a physical change differ from a chemical change? When a substance changes from one state of matter to another (for example, from solid to liquid), is the change a physical or chemical change?

1.38 "A 2.50 g sample of calcium (an electrically conducting white metal that is shiny, relatively soft, has a density of 1.54 g/cm^3, melts at 850 °C, and boils at 1440 °C) was placed into 250 mL of liquid water that was at 25 °C. The calcium reacted slowly with the water to give bubbles of gaseous hydrogen and a solution of the substance calcium hydroxide." In this description, what physical properties and what chemical properties are described?

1.39 In the statement in the preceding question, which of the properties are intensive and which are extensive?

1.40 Suppose you were told that behind a screen there were two samples of liquid, one of them water and the other gasoline. You are told that Sample 1 occupies 3 fluid ounces and Sample 2 occupies 7 fluid ounces.
(a) What kind of property (intensive/extensive) is volume?
(b) Can you use the information given to you in this ques-

tion to determine which sample is water and which is gasoline? Explain.

1.41 Name two intensive properties that you *could* use to distinguish between water and gasoline. Give one chemical property you could use.

1.42 Many reference books list physical properties that can aid in the identification of substances. One such book is the *Handbook of Chemistry and Physics,* which no doubt is available in your school library.
(a) Use the Table of Physical Constants of Inorganic Compounds in the *Handbook of Chemistry and Physics* to tabulate the melting points, boiling points, and colors of cadmium iodide, lead iodide, and bismuth tribromide.
(b) Which of these tabulated properties could you use to distinguish among these three substances?
(c) A chemist who was asked to analyze a sample was able to isolate a yellow solid from an experiment. This substance was found to melt at 402 °C and boil at 954 °C. Which of the chemicals described in part (a) of this question could the chemist have obtained from the experiment?

REVIEW PROBLEMS

Answers to problems whose numbers are printed in color are given in Appendix D.
More challenging problems are marked with asterisks.

SI Prefixes

1.43 What number should replace the question mark in each of the following?
(a) 1 cm = ? m   (c) 1 m = ? pm   (e) 1 g = ? kg
(b) 1 km = ? m   (d) 1 dm = ? m   (f) 1 cg = ? g

1.44 What numbers should replace the question marks below?
(a) 1 nm = ? m   (c) 1 kg = ? g   (e) 1 mg = ? g
(b) 1 μg = ? g   (d) 1 Mg = ? g   (f) 1 dg = ? g

Temperature Conversions

1.45 Perform the following conversions.
(a) 60 °C to °F   (c) 45.5 °F to °C   (e) 40 °C to K
(b) 20 °C to °F   (d) 59 °F to °C   (f) −20 °C to K

1.46 Perform the following conversions.
(a) 86 °F to °C   (c) −25 °C to °F   (e) 298 K to °C
(b) −4 °F to °C   (d) 373 K to °C   (f) 30 °C to K

1.47 A clinical thermometer registers a patient's temperature to be 37.46 °C. What is this temperature in °F? Is the patient a well person?

1.48 The coldest permanently inhabited place on Earth is the Siberian village of Oymyakon in Russia. In 1964 the temperature reached a shivering −96 °F! What is this temperature in °C?

1.49 Helium has the lowest boiling point of any liquid. It boils at 4 K. What is its boiling point in °C and °F?

1.50 Natural gas is mostly methane, a substance that boils at a temperature of 111 K. What is its boiling point in °C and °F?

1.51 Liquid nitrogen is used as a commercial refrigerant to flash freeze foods. Nitrogen boils at −196 °C. What is this temperature on the Kelvin temperature scale?

1.52 Liquid oxygen is used as an oxidizer in the launch of the space shuttle. It boils at −183 °C. What is this temperature on the Kelvin temperature scale?

Significant Figures

1.53 How many significant figures do the following measured quantities have?
(a) 27.53 cm   (c) 102.0 g   (e) 0.06080 m
(b) 39.240 cm   (d) 0.00021 kg   (f) 0.0002 L

1.54 How many significant figures do the following measured quantities have?
(a) 0.0240 g   (c) 0.008 kg   (e) 1.00050 L
(b) 101.303 m   (d) 615.00 mg   (f) 3.5105 mm

1.55 Perform the following arithmetic and round off the answers to the correct number of significant figures. Assume that all of the numbers were obtained by measurement.
(a) 0.0022×315
(b) $83.25 - 0.01075$
(c) $(84.45 \times 0.02)/(31.2 \times 9.8)$
(d) $(22.4 + 102.7 + 0.005)/(5.478)$
(e) $(315.44 - 208.1) \times 7.2234$

1.56 Perform the following arithmetic and round off the answers to the correct number of significant figures. Assume that all of the numbers were obtained by measurement.
(a) $3.58/1.739$
(b) $4.02 + 0.001$
(c) $(22.4 \times 8.3)/(1.142 \times 0.002)$
(d) $(1.345 + 0.022)/(13.36 \times 8.4115)$
(e) $(74.335 - 74.332)/(4.75 \times 1.114)$

Review of Scientific Notation (See Appendix A.1)

1.57 Express the following numbers in scientific notation. Assume three significant figures in each number.
(a) 2,340   (c) 0.000287   (e) 0.00000400
(b) 31,000,000   (d) 45,000   (f) 324,000

1.58 Express the following numbers in scientific notation. Assume, in this problem, that only the nonzero digits are significant figures.
(a) 389   (c) 81,300   (e) 2.33
(b) 0.00075   (d) 0.00225   (f) 28,320

1.59 Write the following numbers in standard, nonexponential form.
(a) 2.1×10^5 (c) 3.8×10^3 (e) 34.6×10^{-7}
(b) 3.35×10^{-6} (d) 4.6×10^{-12} (f) 8.5×10^8

1.60 Write the following numbers in standard, nonexponential form:
(a) 4.27×10^{-8} (c) 33.5×10^{-9} (e) 5.0000×10^7
(b) 7.11×10^5 (d) 2.85×10^{-2} (f) 17.2×10^2

1.61 Perform the following arithmetic and express the answers in scientific notation.

(a) $\dfrac{(1.0 \times 10^7) \times (4.0 \times 10^5)}{(2.0 \times 10^8)}$

(b) $\dfrac{(4.0 \times 10^{-5}) \times (6.0 \times 10^{10})}{(3.0 \times 10^{-2})}$

(c) $\dfrac{(5.0 \times 10^{-4}) \times (2.0 \times 10^{-6})}{(1.0 \times 10^{-12})}$

(d) $(3.0 \times 10^4) + (2.1 \times 10^5)$
(e) $(8.0 \times 10^{12}) \div (2.0 \times 10^{-3})^2$

1.62 Perform the following arithmetic and express the answers in scientific notation.

(a) $\dfrac{(8.0 \times 10^6)}{(2.0 \times 10^5) \times (1.0 \times 10^3)}$

(b) $\dfrac{(1.6 \times 10^{15}) \times (1.0 \times 10^{-5})}{(8.0 \times 10^4)}$

(c) $\dfrac{(4.5 \times 10^{28})}{(3.0 \times 10^{-6})^2}$

(d) $(1.4 \times 10^5) - (3.0 \times 10^4)$
(e) $(3.3 \times 10^{-4}) + (2.52 \times 10^{-2})$

Unit Conversions by the Factor-Label Method
1.63 Perform the following conversions.
(a) 32.0 dm to km   (c) 75.3 mg to kg   (e) 0.025 L to mL
(b) 8.2 mg to μg   (d) 137.5 mL to L   (f) 342 pm to dm

1.64 Perform the following conversions; express your answers in scientific notation.
(a) 183 nm to cm   (c) 6.22 km to nm   (e) 0.55 dm to km
(b) 3.55 g to dg   (d) 33 dm to mm   (f) 53.8 ng to pg

1.65 Perform the following conversions. If necessary, refer to Tables 1.2 and 1.3.
(a) 36 in. to cm   (d) 1 cup (8 oz) to mL
(b) 5.0 lb to kg   (e) 55 mi/hr to km/hr
(c) 3.0 qt to mL   (f) 50.0 mi to km

1.66 Perform the following conversions. If necessary, refer to Tables 1.2 and 1.3.
(a) 250 mL to qt   (d) 1.75 L to fluid oz
(b) 3.0 ft to m   (e) 35 km/hr to mi/hr
(c) 1.62 kg to lb   (f) 80.0 km to mi

1.67 Cola is often sold in 12-oz cans. How many milliliters are in each can?

1.68 How many fluid ounces are in a 2.00-L bottle of cola?

1.69 A metric ton is 1000 kg. How many pounds is this?

1.70 In the United States, a "long ton" is 2240 lb. What is this mass expressed in metric tons (1 metric ton = 1000 kg)?

1.71 If a person is 6 ft 2 in. tall, what is the person's height in centimeters?

1.72 If a person weighs 81.6 kg, what is the person's weight in pounds?

1.73 Perform the following conversions.
(a) 6.2 yd² to m²   (b) 4.8 in.² to mm²   (c) 3.7 ft³ to L

1.74 Perform the following conversions.
(a) 9.5×10^2 mm² to in.²   (c) 288 in.² to m²
(b) 7.3 m³ to ft³

1.75 A bullet is fired at a speed of 2235 ft/s. What is this speed expressed in kilometers per hour?

1.76 In the United States, the speed limit on some highways is 65 miles per hour. What is this speed expressed in kilometers per hour?

Density and Specific Gravity
1.77 A sample of kerosene weighs 36.4 g. Its volume was measured to be 45.6 mL. What is the density of the kerosene?

1.78 A block of magnesium has a mass of 14.3 g and a volume of 8.46 cm³. What is the density of magnesium in g/cm³?

1.79 Acetone, the solvent in nail polish remover, has a density of 0.791 g/mL. What is the volume of 25.0 g of acetone?

1.80 A glass apparatus contains 26.223 g of water when filled at 25 °C. At this temperature, water has a density of 0.99704 g/mL. What is the volume of this apparatus?

1.81 Chloroform, a chemical once used as an anesthetic, has a density of 1.492 g/mL. What is the mass in grams of 185 mL of chloroform?

1.82 Gasoline has a density of about 0.65 g/mL. How much does 34 L (approximately 18 gallons) weigh in kilograms? In pounds?

1.83 A graduated cylinder was filled with water to the 15.0-mL mark and weighed on a balance. Its mass was 27.35 g. An object made of silver was placed in the cylinder and completely submerged in the water. The water level rose to 18.3 mL. When reweighed, the cylinder, water, and silver object had a total mass of 62.00 g. Calculate the density of silver.

1.84 Titanium is a metal used to make golf clubs. A rectangular bar of this metal measuring 1.84 cm × 2.24 cm × 2.44 cm was found to have a mass of 45.7 g. What is the density of titanium?

1.85 Ethyl ether has been used as an anesthetic. Its density is 0.715 g/mL. What is its specific gravity?

1.86 Propylene glycol is used as a food additive and in cosmetics. Its density is 8.65 lb/gal. Calculate the specific gravity of propylene glycol.

1.87 Trichloroethylene is a common dry-cleaning solvent. Its specific gravity is 1.47 measured at 20°C. What is the mass in grams of a liter of this solvent?

1.88 The space shuttle uses liquid hydrogen as its fuel. The external fuel tank used during takeoff carries 227,641 lb of hydrogen with a volume of 385,265 gallons. Use the concept of specific gravity to calculate the density of liquid hydrogen in units of g/mL. (Express your answer to three significant figures.)

1.89 Gold has a specific gravity of 19.3. Calculate the mass, in pounds, of one cubic foot of gold.

1.90 From the data in Problem 1.88, you can calculate that liquid hydrogen has a specific gravity of 0.0708. How many cubic feet of liquid hydrogen weigh 565 pounds?

ADDITIONAL EXERCISES

1.91 You are the science reporter for a daily newspaper and your editor asks you to write a story based on a report in the scientific literature. The report states that analysis of the sediments in Hausberg Tarn (elevation 4350 m) on the side of Mount Kenya (elevation 5200 m) shows that the average temperature of the water rose by 4 °C between 350 BC and AD 450. Your editor tells you that she wants all the data expressed in the English system of units. Make the appropriate conversions.

1.92 Distances between stars are often expressed in units of *light-years*. A light-year is the distance that light travels during 1 year. Use the factor-label method to calculate the number of miles in one light-year, given that light travels at a speed of 3.0×10^8 meters per second. [*Hint:* Begin with *1 year* (365 days) and then use conversion factors to change that to a distance traveled by light.]

1.93 A pycnometer is a glass apparatus used for accurately determining the density of a liquid. When dry and empty, a certain pycnometer had a mass of 27.314 g. When filled with distilled water at 25.0 °C, it weighed 36.842 g. When filled with chloroform (a liquid once used as an anesthetic before its toxic properties were known), the apparatus weighed 41.428 g. At 25.0 °C, the density of water is 0.99704 g/mL. (a) What is the volume of the pycnometer? (b) What is the density of chloroform?

1.94 Radio waves travel at the speed of light, 3.0×10^8 meters per second. If you were to broadcast a question to an astronaut on the moon, which is 239,000 miles from Earth, what is the minimum time that you would have to wait to receive a reply?

1.95 The strange Muloh tribe of a former French colony thrives on cabbage and canned potatoes! Cabbage costs 42 francs per head and potatoes cost 26 francs per can. The average Muloh consumes 3 cans of potatoes per head of cabbage. If an average Muloh spends 186 francs on cabbage, how much money will be spent on potatoes?

1.96 Suppose you have a job in which you earn $4.50 for each 30 minutes that you work.
(a) Express this information in the form of an equivalence between dollars and minutes worked.

(b) Use the equivalence defined in (a) to calculate the number of dollars earned in 1 hr 45 min.
(c) Use the equivalence defined in (a) to calculate the number of minutes you would have to work to earn $17.35.

1.97 When an object floats in water, it displaces a volume of water that has a weight equal to the weight of the object. If a ship has a weight of 4255 tons, how many cubic feet of seawater will it displace? Seawater has a specific gravity of 1.025, and 1 ton = 2000 lb.

1.98 Carbon tetrachloride and water do not dissolve in each other. When mixed, they form two separate layers. Carbon tetrachloride has a density of 13.3 lb/gal. Which liquid will float on top of the other when the two are mixed?

1.99 If you were given only the masses of samples of water and methanol, or if you were given only the volume of each liquid, you could not tell them apart. However, if you were given both their masses *and* their volumes, you could determine which was the methanol and which was the water. How?

1.100 A liquid known to be either ethanol (ethyl alcohol) or methanol (methyl alcohol) was found to have a density of 0.798 ± 0.001 g/mL. Consult the *Handbook of Chemistry and Physics* to determine which liquid it is. What other measurements could help to confirm the identity of the liquid?

1.101 An unknown liquid was found to have a density of 69.22 lb/ft^3. The density of ethylene glycol (the liquid used in antifreeze) is 1.1088 g/mL. Is the unknown liquid ethylene glycol?

1.102 The density of propylene glycol, a substance used as a nontoxic antifreeze, is 1.040 g/mL at 20 °C. What is the density of this substance at 20 °C expressed in units of (a) grams per liter and (b) kilograms per cubic meter?

1.103 The density of isopropyl alcohol, used in rubbing alcohol, is 785 kg/m^3 at 20 °C. What is the density of this alcohol at 20 °C expressed in units of (a) pounds per gallon and (b) grams per liter?

*1.104 A standard LP phonograph record has a diameter of 12 in. and rotates at $33\frac{1}{3}$ rpm (revolutions per minute) while it is being played. How fast is a point on the edge of the record moving, expressed in miles per hour?

1.105 When an object is heated to a high temperature, it glows and gives off light. The color balance of this light depends on the temperature of the glowing object. Photographic lighting is described, in terms of its color balance, as a temperature in kelvins. For example, a certain electronic flash gives a color balance (called color temperature) rated at 5800 K. What is this temperature expressed in °C?

1.106 Suppose that when a chemical reaction occurs in a solution in an insulated container, the solution becomes cool.
(a) What has happened to the average kinetic energy of the atoms in the liquid?
(b) What has happened to the total kinetic energy of the atoms in the liquid?
(c) What has happened to the total potential energy of the substances in the solution?

* **1.107** There exists a single temperature at which the value reported in °F is numerically the same as the value reported in °C. What is this temperature?

* **1.108** In the text, the Kelvin scale of temperature is defined as an absolute scale in which one Kelvin degree unit is the same size as one Celsius degree unit. A second absolute temperature scale exists called the Rankine scale. On this scale, one Rankine degree unit (°R) is the same size as one Fahrenheit degree unit. (a) What is the only temperature at which the Kelvin and Rankine scales possess the same numerical value? Explain your answer. (b) What is the boiling point of water expressed in °R?

Ascorbic Acid— Limes, Scurvy, and the Scientific Method

Before the 1500s, sailors who went on long voyages without being resupplied with fresh vegetables and fruit came down with scurvy, a disease characterized by bleeding gums, weakened and spongy gum tissue, and the loss of teeth. The victims also bruised easily and became susceptible to fatal diseases. A high percentage died. "What was going on here?" The answer could not be given yet, but its unfolding illustrates how science works through observations, accidental discoveries, testable hypotheses, and the constant seeking after solutions to puzzles and questions.

Does Citrus Juice Kill Germs?

As early as the mid-1500s, the Dutch observed that their sailors were healthier when their diets included citrus fruits. Two centuries later, James Lind (1716–1794), a Scottish physician, discovered that the health effect of citrus juice was specifically the prevention or cure of scurvy. The cure was very startling, like magic, and Lind published a treatise on it in 1753.

Scurvy had long been a scourge of the British navy. Of the 1955 sailors who participated in a naval voyage around the world, lasting from 1740 to 1744, 1051 died, chiefly of this disease. The British Admiralty finally issued formal orders in 1795 to include daily issues of lemon juice to all sailors. Lemons were called limes then, and British sailors were soon popularly called "limeys."

Still, a question persisted. "What's going on here with lemon juice?" What's in lemon juice (or any other citrus fruit) that prevents scurvy? By the early 1800s, chemistry was advancing, improving the methods for isolating, purifying, and identifying chemicals. The antiscurvy substance was eventually isolated, and it turned out to be ascorbic acid (vitamin C). But this advance still begged the question, "How does ascorbic acid manage to do what it does?" Perhaps germs cause scurvy

and ascorbic acid kills them. At one time this would have been a reasonable hypothesis, *one that could be tested.* Or does vitamin C work in another way? Having a *testable* hypothesis was (and always is) critical in science.

Nature Bats Last

You can see that the use of the scientific method begins with a puzzle, "How does something work?" Or "What is behind this or that observation?" Then a possible explanation, a hypothesis, is put forward, one that is reasonable in the light of the observations and the knowledge existing at the time. The hypothesis, if well made, suggests an approach, perhaps an experiment, that would be a test for it. It was reasonable to suppose that scurvy might be caused by germs, like so many other diseases, but no scurvy germs could be found. So the hypothesis got nowhere. Despite how reasonable it once might have been, the hypothesis was wrong, which neatly illustrates that the test of reason isn't alone sufficient in science. A hypothesis must be aligned with nature as it really is. Scientists always know that "nature bats last."

By the early twentieth century, partly as the result of the research of a Polish chemist, Casimir Funk, we came to know that not all diseases are caused by germs. Many arise from the absence of specific chemicals in the diet, now called *vitamins,* chemicals that are not carbohydrates, proteins, or edible fats and oils. Ascorbic acid is a vitamin, and scurvy is a *vitamin-deficiency disease,* not a germ-caused condition.

Ascorbic Acid and Collagen

We still have the persistent, altogether human question, "How does ascorbic acid do what it does?" It wasn't until several decades into the twentieth century that ascorbic acid was shown to be essential to the proper formation of a protein called collagen. A huge amount

chemicals in our world

of biochemical research into proteins had to take place before the linkage between ascorbic acid and collagen could develop.

Collagen functions something like strong flexible reinforcing rods in bones, teeth, cartilage, tendons, skin, blood vessels, and certain ligaments. Collagen is required for healthy gum tissue, whose deterioration is simply an early sign that collagen is not being made properly.

But we're not finished yet. "Just exactly how does ascorbic acid work in the manufacture of collagen?" The "sound bite" answer is that ascorbic acid is needed to make an enzyme without which proper collagen cannot be made. This answer lands us finally at the *molecular* basis of scurvy.

From Puzzle to Solution

Exactly how ascorbic acid molecules help to form the molecules that make up strong collagen is at one of the many frontiers of chemistry. The scurvy story illustrates much of what is fun, beautiful, exciting, and useful in chemistry. There's a problem, a really serious one: scurvy. Then there is a wonderful accidental discovery: citrus fruit juice prevents it. Which leads to a big puzzle: how does it do this? The answer awaits advances in a number of fields of chemistry, but there are always a few chemists who do not let go of the puzzle until the *molecular* basis is the big part of the answer.

All of the processes of life and disease, and all of the strengths and weaknesses of materials, have a basis in the properties of atoms and molecules, which the study of chemistry is mainly about. Somewhere along the way you'll pick your own favorite frontier. On reaching it, you'll apply the lessons of how science works to solve the puzzles that interest you most. And you'll be amazed that people actually pay you to have all the fun of discovery and to enjoy all its beauty.

Figure 1a The juice of lemons or limes can prevent scurvy, as mariners from the 16th century on discovered.

We've all been told how important milk is in the diet, especially for growing children, because the calcium in milk is important for growing bones. One of the hazards of nuclear testing, however, is the release of a radioactive form of the element strontium, which has chemical properties that are very similar to calcium. If this form of strontium finds its way into milk, it will be incorporated into growing bones as well, which could lead to serious health problems at a later time. In this chapter you will learn that we can use the periodic table to find similarities in chemical properties, like those that exist between calcium and strontium.

The Periodic Table and Some Properties of the Elements

This Chapter in Context One of the goals of chemistry is to organize information about chemical substances so that similarities and differences are easy to spot. Organization is the theme of this chapter, and we begin here with a discussion of three principal classes of matter—elements, compounds, and mixtures. We will also study in this chapter the basic structures of atoms as well as the periodic table, the chemist's chief tool for organization. You will learn how to use the periodic table as an aid in writing the formulas and names of compounds. By doing so you begin to build your store of factual knowledge, a process that will continue in Chapters 4 and 5.

2.1 Elements, Compounds, and Mixtures

Elements

If a chemical reaction changes one substance into two or more others, a **decomposition** has occurred. For example, if we pass electricity through molten (melted) sodium chloride (salt), the silvery metal sodium and the pale green gas chlorine are formed. In this example, we have decomposed sodium chloride into two simpler substances. No matter how we try, however, sodium and chlorine cannot be decomposed further by chemical reactions into still simpler substances that can be stored and studied.

In chemistry, *substances that cannot be decomposed into simpler materials by chemical reactions are called* **elements.** Sodium and chlorine are two examples. Others you may be familiar with include iron, chromium, lead, copper, aluminum, sulfur, and carbon (as in charcoal). Some elements are gases at room temperature, including chlorine, oxygen, hydrogen, nitrogen, and helium. Elements are the simplest forms of matter that chemists work with directly. All more complex substances are composed of elements in various combinations.

So far, scientists have discovered 90 existing elements in nature and have made 25 more, for a total of 115. Each element is assigned a unique **chemical**

Table 2.1 **Elements That Have Symbols Derived from Their Latin Names**

Element	Symbol	Latin Name
Sodium	Na	Natrium
Potassium	K	Kalium
Iron	Fe	Ferrum
Copper	Cu	Cuprum
Silver	Ag	Argentum
Gold	Au	Aurum
Mercury	Hg	Hydrargyrum
Antimony	Sb	Stibium
Tin	Sn	Stannum
Lead	Pb	Plumbum

symbol. In most cases, the symbol is formed from one or two letters of the English name for the element. For instance, the symbol for carbon is C, for bromine it is Br, and for silicon it is Si. For some elements, the symbols are derived from the Latin names given to those elements long ago. Table 2.1 contains a list of elements whose symbols come to us in this way.[1]

Latin was the universal language of science in the early days of chemistry.

Regardless of the origin of the symbol, the first letter is always capitalized and the second letter, if there is one, is always written lowercase. Thus, the symbol for copper is Cu, not CU. Be careful to follow this rule so you can avoid confusion between such symbols as Co (cobalt) and CO (carbon monoxide). The names and chemical symbols of the elements are given on the inside front cover of the book. Although the list may seem long, many elements are rare, and we will be most interested in only a relatively small number of them.

Compounds

By means of chemical reactions, elements combine in various *specific proportions* to give all the more complex substances in nature. Thus, hydrogen and oxygen combine to form water, and sodium and chlorine combine to form sodium chloride. Water and sodium chloride are examples of compounds. **A compound** *is a substance formed from two or more different elements in which the elements are always combined in the same fixed (i.e., constant) proportions by mass.* For example, if any sample of pure water is decomposed, the mass of oxygen obtained is *always* eight times the mass of hydrogen. Similarly, when hydrogen and oxygen react to form water, the mass of oxygen consumed is always eight times the mass of hydrogen, never more and never less.

Water and sodium chloride must be more complex than elements because they can be decomposed to elements.

In our description of the makeup of a compound, there is a subtle but significant point to be understood. For example, we say that the compound carbon dioxide, a substance formed in the combustion of many fuels, is *composed of* carbon and oxygen. However, in this compound these two elements are not present in the same form as we find them in their pure states. We can see this by comparing the properties of the compound with those of the elements from which it is formed. Ordinarily, carbon is found as a black solid; it is found in coal

[1]The symbol for tungsten is W, from the German name *wolfram*. This is the only element whose symbol is neither related to its English name nor derived from its Latin name.

and charcoal. But at room temperature carbon dioxide isn't a solid, it's a color-less gas. Oxygen is also a gas, but it supports combustion, whereas carbon dioxide is used to extinguish fires. Thus, the properties of carbon dioxide are much different from those of both carbon and oxygen. When carbon and oxygen combine to form carbon dioxide, the properties of carbon and oxygen disappear and in their place we find the unique properties of the compound carbon dioxide.

Mixtures

Elements and compounds are examples of **pure substances.**[2] The composition of a pure substance is always the same, regardless of its source. All samples of table salt, for example, contain sodium and chlorine combined in the same proportions by mass. Similarly, all samples of water contain the same proportions by mass of hydrogen and oxygen. Pure substances are rare, however. Usually we encounter mixtures of compounds or elements. Unlike elements and compounds, **mixtures *can have variable compositions.*** Carbon dioxide (CO_2) and water (H_2O) are examples of compounds because each consists of elements chemically combined in definite proportions. But, when we dissolve carbon di-oxide in water to give a carbonated beverage such as seltzer, we can vary the amounts of the two compounds. As a result, we can vary the amount of "fizz" in the beverage.

Orange juice, Coca-Cola, and pancake syrup are mixtures that contain sugar. The amount of sugar varies from one to another because mixtures can have variable compositions.

Mixtures can be either homogeneous or heterogeneous. A **homogeneous mixture *has the same properties throughout the sample.*** An example is a thoroughly stirred mixture of sugar in water. We call such a homogeneous mixture a **solution.** Solutions need not be liquids, just homogeneous. Brass, for example, is a solid solution of copper and zinc, and clean air is a gaseous solution of oxygen, nitrogen, and a number of other gases.

Solid solutions of metals, such as brass and stainless steel, are called *alloys*.

[2]We have used the term substance rather loosely until now. Strictly speaking, **substance** really means *pure substance.* Each unique chemical element and compound is a *substance;* a mixture consists of two or more substances.

A **heterogeneous mixture** *consists of two or more regions called* **phases** *that differ in properties.* A mixture of olive oil and vinegar in a salad dressing, for example, is a two-phase mixture in which the oil floats on the vinegar as a separate layer (Figure 2.1). The phases in a mixture don't have to be chemically different substances like oil and vinegar, however. A mixture of ice and liquid water is a two-phase heterogeneous mixture in which the phases have the same chemical composition but occur in different physical states.

An important way that mixtures differ from compounds is in the changes that occur when they form. Consider, for example, the elements iron and sulfur, which are pictured in powdered form in Figure 2.2a. We can make a mixture simply by dumping them together and stirring them. In the mixture (Figure 2.2b), both elements retain their original properties. The process we use to create this mixture involves a **physical change,** rather than a chemical change, because no new chemical substances form. To separate the mixture, we could similarly use just physical changes. For example, we could remove the iron by stirring the mixture with a magnet—a physical operation. The iron powder sticks to the magnet as we pull it out, leaving the sulfur behind (Figure 2.3). The mixture also could be separated by treating it with a liquid called carbon disulfide, which is able to dissolve the sulfur but not the iron. Filtering the sulfur solution from the solid iron, followed by evaporation of the liquid carbon disulfide from the sulfur solution, gives the original components, iron and sulfur, separated from each other.

As we noted earlier, formation of a compound from its elements involves a chemical reaction. Iron and sulfur, for example, combine to form a compound often called "fool's gold" because of its appearance (Figure 2.4). In this compound the elements no longer have the same properties they had before they were combined, and they cannot be separated by physical means. Just as the formation of a compound involves a chemical reaction, so also does its decomposition. The decomposition of fool's gold into iron and sulfur is therefore a chemical change.

The relationships among elements, compounds, and mixtures are shown in Figure 2.5 on page 48.

Figure 2.1 *A heterogeneous mixture.* The salad dressing shown here contains vinegar and vegetable oil (plus assorted other flavorings). Vinegar and oil do not dissolve in each other and form two layers. The mixture is heterogeneous because each of the separate phases (oil, vinegar, and other solids) has its own set of properties that differ from the properties of the other phases.

2.2 Dalton's Atomic Theory

In modern science, we have come to take for granted the existence of atoms, and we have referred to atoms in our discussion of heat. Even in nontechnical circles we hear talk of *atomic energy* and *nuclear energy.* However, scientific evidence for the existence of atoms is relatively recent, and chemistry did not progress very far until that evidence was found. Therefore, let's take a brief look at how atomic theory evolved.

The concept of atoms began nearly 2500 years ago when certain Greek philosophers expressed the belief that matter is ultimately composed of tiny indivisible particles, and it is from the Greek word *atomos,* meaning "not cut," that the word atom is derived. The philosophers' conclusions, however, were not supported by any evidence; they were derived simply from philosophical reasoning.

Laws of Chemical Combination

The concept of atoms remained a philosophical belief, having limited scientific usefulness, until the discovery of two quantitative laws of chemical combination—the *law of conservation of mass* and the *law of definite proportions*. The evidence that led to the discovery of these laws came from the work of many scientists in the eighteenth and early nineteenth centuries.

(*a*)   (*b*)

Figure 2.2 *Formation of a mixture of iron and sulfur.* (*a*) Samples of powdered sulfur and powdered iron. (*b*) A mixture of sulfur and iron is made by stirring the two powders together.

Figure 2.3 *Formation of a mixture is a physical change.* Here we see that forming a mixture has not changed the iron and sulfur into a compound of these two elements. The mixture can be separated by pulling the iron out with a magnet.

Law of Conservation of Mass
No detectable gain or loss of mass occurs in chemical reactions. Mass is *conserved.*

Law of Definite Proportions
In a given chemical compound, the elements are always combined in the same proportions by mass.

The **law of conservation of mass** means that if we place chemical reactants in a sealed vessel that permits no matter to enter or escape, the mass of the vessel and its contents after a reaction will be identical to its mass before. Although this may seem quite obvious to us now, it wasn't quite so clear in the early history of modern chemistry. Not until scientists made sure that *all reactants* and *all products,* including any that were gases, were included when masses were measured could the law of conservation of mass be truly tested.

The **law of definite proportions** is illustrated by the compound zinc sulfide. This compound is formed in a violent reaction when a mixture of powdered zinc and sulfur is ignited, as shown in Figure 2.6 on page 47. Whenever samples of zinc sulfide are decomposed, 1.000 g of Zn is obtained for each 0.490 g of S. In this compound, therefore, the zinc-to-sulfur ratio is 1.000 g Zn/0.490 g S. Similarly, it is *always* found that when 1.000 g of Zn combines with sulfur, exactly 0.490 g of S reacts. If a mixture of 1.000 g of Zn and 1.000 g of S is ignited, only 0.490 g of the sulfur reacts and the rest, 0.510 g of S, is left over unreacted. It is impossible to cause 1.000 g of zinc to combine with more than 0.490 g of S to form zinc sulfide.

This constancy of composition of compounds is one of the most useful fundamental laws of chemistry. In fact, on page 44 we *defined* a compound as a substance in which two or more elements are chemically combined in a *definite fixed proportion by mass.*

At one time gases weren't even recognized as substances.

Figure 2.4 *A compound of iron and sulfur.* Iron pyrite, commonly called "fool's gold," has an appearance that is much different from either iron or sulfur, and it cannot be separated into its elements by physical means.

The Atomic Theory

The law of definite proportions suggests a question also raised by the law of conservation of mass: "What must be true about the nature of matter, given the truth of these laws?" In other words, what is matter made of?

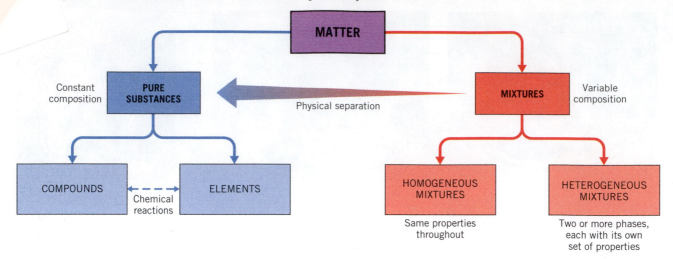

Figure 2.5 *Classification of matter.*

At the beginning of the nineteenth century, John Dalton (1766–1844), an English scientist, used the Greek concept of atoms to make sense out of the laws of conservation of mass and definite proportions. Dalton reasoned that if atoms really exist, they must have certain properties to account for these laws. He described such properties, and the list constitutes what we now call **Dalton's atomic theory.**

Dalton's theory was not *perfectly* correct. For example, we now know that atoms can be broken into smaller pieces and that most elements occur as mixtures of two or more isotopes, which are atoms of an element with slightly different masses. However, neither of these facts affected the ability of the theory to explain the laws of chemical combination.

Dalton's Atomic Theory

1. *Matter consists of tiny particles called atoms.*
2. *Atoms are indestructible. In chemical reactions, the atoms rearrange but they do not themselves break apart.*
3. *The atoms of one particular element are all identical in mass and other properties.*
4. *The atoms of different elements differ in mass and other properties.*
5. *When atoms of different elements combine to form compounds, new and more complex particles form. However, in a given compound the constituent atoms are always present in the same fixed **numerical** ratio.*

Dalton's theory easily explained the law of conservation of mass. According to the theory, a chemical reaction is simply a reordering of atoms from one combination to another. If no atoms are gained or lost and if the masses of the atoms can't change, then the mass after the reaction must be the same as the mass before. This explanation of the law of conservation of mass works so well that it serves as the reason for balancing chemical equations, which we will discuss in the next section.

As mentioned in Chapter 1, we use the term **molecule** to mean a distinct particle composed of two or more atoms.

The law of definite proportions is also easy to explain. According to the theory a given compound always has atoms of the same elements in the same numerical ratio. Suppose, for example, that two elements, *A* and *B*, combine to form molecules in which one atom of *A* is combined with one of *B*. If the mass of a *B* atom is twice that of an *A* atom, then every time we encounter a molecule of this compound, the mass ratio (*A* to *B*) is 1 to 2. This same mass ratio would exist regardless of how many molecules we had in our sample, so in samples of this compound the elements *A* and *B* are always present in the same proportion by mass.

Law of Multiple Proportions

Strong support for Dalton's theory came when Dalton and other scientists studied elements that are able to combine to give two (or more) compounds. For example, sulfur and oxygen form two different compounds which we call sulfur dioxide and sulfur trioxide. If we decompose a 2.00 g sample of sulfur dioxide, we find it contains 1.00 g of S and 1.00 g of O. If we decompose a 2.50 g sample of sulfur trioxide, we find it also contains 1.00 g of S, but this time the mass of O is 1.50 g. This is summarized in the following table.

Compound	Sample Size	Mass of Sulfur	Mass of Oxygen
Sulfur dioxide	2.00 g	1.00 g	1.00 g
Sulfur trioxide	2.50 g	1.00 g	1.50 g

First, notice that sample sizes aren't the same; they were chosen so that each has the *same mass of sulfur*. Second, the ratio of the masses of oxygen in the two samples is one of small whole numbers.

$$\frac{\text{mass of oxygen in sulfur trioxide}}{\text{mass of oxygen in sulfur dioxide}} = \frac{1.50 \text{ g}}{1.00 \text{ g}} = \frac{3}{2}$$

Similar observations are made when we study other elements that form more than one compound with each other, and these observations form the basis of the **law of multiple proportions.**

> **Law of Multiple Proportions**
> Whenever two elements form more than one compound, the different masses of one element that combine with the same mass of the other element are in the ratio of small whole numbers.

Dalton's theory explains the law of multiple proportions in a very simple way. Suppose a molecule of sulfur trioxide contains one sulfur and three oxygen atoms and a molecule of sulfur dioxide contains one sulfur and two oxygen atoms (Figure 2.7). If we had just one molecule of each, then our samples each would have one sulfur atom and therefore the same mass of sulfur. Then, comparing the oxygen atoms, we find they are in a numerical ratio of 3 to 2. But because oxygen atoms all have the same mass, the mass ratio must also be 3 to 2.

The law of multiple proportions was not known before Dalton presented his theory. It was discovered because the theory suggested its existence. Although this law is not of much practical use in chemistry today, for many years it was one of the strongest arguments in favor of the existence of atoms.

2.3 Symbols, Formulas, and Equations

Earlier we noted that each element has been assigned a chemical symbol, which we can use to stand for the name of the element. Often, we also use the symbol to stand for an atom of the element, which is the smallest bit of the element that retains its chemical identity. Thus, Co can be used in place of the name cobalt and also to mean one atom of cobalt.

Only a small number of elements exist in nature as isolated atoms. An example is helium, a gas used to inflate balloons and the Goodyear blimp. Some elements occur as **diatomic molecules** (molecules composed of two atoms each).

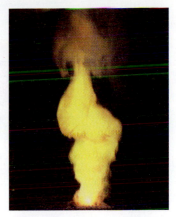

Figure 2.6 *The law of definite proportions is illustrated by the reaction of zinc with sulfur to form zinc sulfide. In this violent reaction, zinc and sulfur combine in the ratio of 1.000 g of Zn for each 0.490 g of S.*

The discovery of the law of multiple proportions is a fine illustration of the scientific method at work.

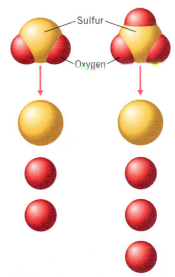

Figure 2.7 *Oxygen compounds of sulfur demonstrate the law of multiple proportions. Illustrated here are molecules of sulfur trioxide, SO_3, and sulfur dioxide, SO_2. Each has one sulfur atom, and therefore the same mass of sulfur. The oxygen ratio is 3 to 2, both by atoms and by mass.*

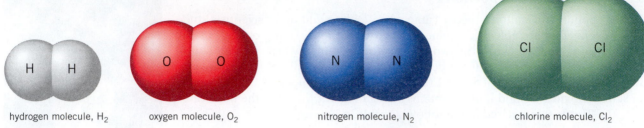

hydrogen molecule, H_2 oxygen molecule, O_2 nitrogen molecule, N_2 chlorine molecule, Cl_2

Figure 2.8 *Models that depict the diatomic molecules of hydrogen, oxygen, nitrogen, and chlorine. Each contains two atoms per molecule; their different sizes reflect differences* in the sizes of the atoms that make up the molecules. The atoms are shaded by color to indicate the element (hydrogen, white; oxygen, red; nitrogen, blue; and chlorine, green).

Elements That Occur Naturally as Diatomic Molecules

Hydrogen	H_2	Fluorine	F_2
Nitrogen	N_2	Chlorine	Cl_2
Oxygen	O_2	Bromine	Br_2
		Iodine	I_2

A "space-filling" model of H_2O.

Subscripts in a formula

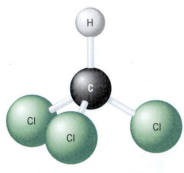

A ball-and-stick model of the chloroform molecule, $CHCl_3$. A standard color scheme is used to represent the different elements: carbon, black; hydrogen, white; and chlorine, green.

Among them are the gases hydrogen, oxygen, nitrogen, and chlorine. We represent them with **chemical formulas** in which subscripts indicate the number of atoms in the molecule. Thus, the formula for molecular hydrogen is H_2, and those for oxygen, nitrogen, and chlorine are O_2, N_2, and Cl_2. (See Figure 2.8.) The elements that occur as diatomic molecules are listed in the table in the margin. Be sure to learn them; you will encounter them often throughout the course. Other elements have their atoms arranged in even more complex combinations, but we won't discuss them further at this point.

Formulas of Compounds

Chemical formulas are also used to represent compounds. In one sense a chemical formula is a shorthand way of writing the name for a compound, just as chemical symbols are shorthand notations for the names of elements. The most important characteristic of a formula, however, is that it specifies the composition of a substance.

In the formula of a compound, each element present is identified by its chemical symbol. When more than one atom of an element is present, the number of atoms is given by a subscript. For example, the iron oxide in rust has the formula Fe_2O_3, which tells us that the compound is composed of iron and oxygen and that in this compound there are two atoms of iron for every three atoms of oxygen. When no subscript is written, we assume it to be 1. Thus, the formula H_2O tells us that in water there are two H atoms for every one O atom. Similarly, the formula for chloroform, $CHCl_3$, indicates that one atom of carbon, one atom of hydrogen, and three atoms of chlorine have combined.

For more complicated compounds, we sometimes find formulas containing parentheses. An example is the formula for urea, $CO(NH_2)_2$, which tells us that the group of atoms within the parentheses, NH_2, occurs twice. (The formula for urea also could be written as CON_2H_4, but there are good reasons for writing certain formulas with parentheses, as you will see later.)

Hydrates

Certain compounds form crystals that contain water molecules. An example is ordinary plaster—the material often used to coat the interior walls of buildings. Plaster consists of crystals of calcium sulfate, $CaSO_4$, that contain two molecules of water for each $CaSO_4$. These water molecules are not held very tightly and can be driven off by heating the crystals. The dried crystals absorb water again if exposed to moisture, and the amount of water absorbed always gives crystals in which the H_2O-to-$CaSO_4$ ratio is 2 to 1. Compounds whose crystals contain water molecules in fixed ratios are quite common and are called **hydrates.** The formula for this hydrate of calcium sulfate is written $CaSO_4 \cdot 2H_2O$ to show that there are two molecules of water per $CaSO_4$. The raised dot is used to indicate that the water molecules are not bound too tightly in the crystal and can be removed.

Sometimes the *dehydration (removal of water) of hydrate crystals produces changes in color.* An example is copper sulfate, which is sometimes used as an agricultural fungicide. Copper sulfate forms blue crystals with the formula $CuSO_4 \cdot 5H_2O$ in which there are five water molecules for each $CuSO_4$. When these blue crystals are heated, most of the water is driven off and the solid that remains, now nearly pure $CuSO_4$, is almost white (Figure 2.9). If left exposed to the air, the $CuSO_4$ will absorb moisture and form blue $CuSO_4 \cdot 5H_2O$ again.

When all the water is removed, the solid is said to be **anhydrous,** meaning *without water.*

▶**Chemistry in Practice**◀ The tendency of certain compounds to form hydrates has practical applications. For example, calcium chloride, $CaCl_2$, forms a number of hydrates including $CaCl_2 \cdot 6H_2O$, and the anhydrous compound will absorb water from the air to form hydrates. Products such as the one pictured in the margin are available in hardware stores and contain porous bags of $CaCl_2$. When placed in a damp basement or closet, the $CaCl_2$ absorbs moisture and reduces the humidity, slowing the growth of mildew. ◆

Counting Atoms in Formulas

Counting the number of atoms of the elements in a chemical formula is an operation you will have to perform many times, so let's look at an example.

EXAMPLE 2.1

Counting Atoms in Formulas

How many atoms of each element are in the formulas (a) $Al_2(SO_4)_3$ and (b) $CoCl_3 \cdot 6H_2O$?

Solution: (a) Here we must recognize that all the atoms within the parentheses occur three times. Therefore, the formula $Al_2(SO_4)_3$ shows

2 Al 3 S 12 O

(b) This is a formula for a hydrate, as indicated by the raised dot. It contains six water molecules, each with two H and one O, for every $CoCl_3$. Therefore, the formula $CoCl_3 \cdot 6H_2O$ represents

1 Co 3 Cl 12 H 6 O

Practice Exercise 1

How many atoms of each element are expressed by the formulas: (a) $NiCl_2$, (b) $FeSO_4$, (c) $Ca_3(PO_4)_2$, (d) $Co(NO_3)_2 \cdot 6H_2O$? ◆

(a)

(b)

Figure 2.9 *Water can be driven from hydrates by heating.* Shown here are blue crystals of copper sulfate hydrate ($CuSO_4 \cdot 5H_2O$), which readily lose water when they are heated. The white solid that forms is pure $CuSO_4$.

Chemical Equations

A **chemical equation** *describes what happens when a chemical reaction occurs.* It uses chemical formulas to provide a before-and-after picture of the chemical substances involved. Consider, for example, the reaction that occurs when a mixture of powdered zinc (Zn) and sulfur (S) is ignited. This reaction, shown in Figure 2.6 on page 49, produces the compound zinc sulfide, which has the formula ZnS. The reaction can be represented by the chemical equation

$$Zn + S \longrightarrow ZnS$$

The two substances that appear to the left of the arrow, Zn and S, are called the **reactants.** These are the substances that exist before the reaction occurs. To the right of the arrow is the formula for zinc sulfide, ZnS, which is the **product** of the reaction. In this example, only one substance is formed in the reaction, so there is only one product. As we will see, however, in most chemical reactions there is more than one product. The products are the substances that are formed and that exist after the reaction is over. The arrow means "reacts to yield." Thus, this equation tells us that *zinc and sulfur react to yield zinc sulfide.*

Coefficients

Most chemical equations are more complex than the one we've just examined. For instance, consider the reaction between the gases H_2 and O_2 to form H_2O illustrated in Figure 2.10. We see that to use up both oxygen atoms of the O_2 molecule, we need two hydrogen molecules and we obtain two molecules of H_2O. The equation for the reaction is written to reflect this:

$$2H_2 + O_2 \longrightarrow 2H_2O$$

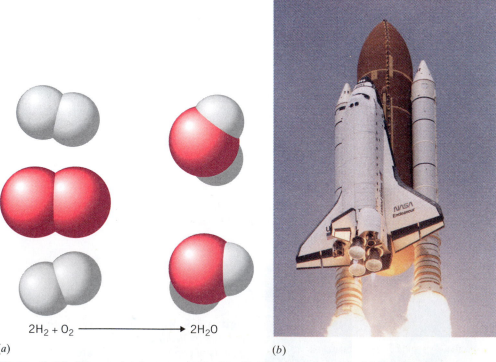

$$2H_2 + O_2 \longrightarrow 2H_2O$$

(*a*)   (*b*)

Figure 2.10 *The reaction between molecules of hydrogen and oxygen.* (*a*) The reaction between two molecules of hydrogen and one molecule of oxygen gives two molecules of water. (*b*) The space shuttle uses this reaction to provide thrust with its main rocket engines.

The numbers in front of the formulas are called **coefficients,** and they indicate the number of molecules of each kind among the reactants and products. Thus, $2H_2$ means two molecules of H_2, and $2H_2O$ means two molecules of H_2O. When no number is written, the coefficient is assumed to be 1 (so the coefficient of O_2 equals 1).

Coefficients are required to make an equation conform to the law of conservation of mass, with an equal number of atoms of each element on both sides of the arrow. When this condition is met, we say the equation is **balanced.** The following is another example of a balanced equation—the equation for the combustion of butane, C_4H_{10}, the fluid in disposable cigarette lighters (Figure 2.11). The chemical equation for this reaction is

$$2C_4H_{10} + 13O_2 \longrightarrow 8CO_2 + 10H_2O$$

The 2 before the C_4H_{10} tells us that two molecules of butane react. This involves a total of 8 carbon atoms and 20 hydrogen atoms. (Notice we have multiplied the numbers of atoms of C and H in one molecule of C_4H_{10} by the coefficient 2.) On the right we find 8 molecules of CO_2, which contain a total of 8 carbon atoms. Similarly, 10 water molecules contain 20 hydrogen atoms. Finally, we can count 26 oxygen atoms on both sides of the equation. You will learn to balance equations such as this in Chapter 3.

In a chemical equation we sometimes specify the physical states of the reactants and products, that is, whether they are solids, liquids, or gases. This is done by writing s, l, or g in parentheses after the chemical formulas. For example, the equation for the combustion of the carbon in a charcoal briquette can be written as

$$C(s) + O_2(g) \longrightarrow CO_2(g)$$

At times we will also find it useful to indicate that a particular substance is dissolved in water. We do this by writing aq, meaning aqueous solution, in parentheses after the formula. For instance, the reaction between stomach acid (HCl) and the active ingredient in Tums, $CaCO_3$, is

$$2HCl(aq) + CaCO_3(s) \longrightarrow CaCl_2(aq) + H_2O(l) + CO_2(g)$$

Figure 2.11 *The combustion of butane, C_4H_{10}. The products are carbon dioxide and water vapor.*

s = solid
l = liquid
g = gas

Aqueous means dissolved in water.

Practice Exercise 2

How many atoms of each element appear on each side of the arrow in the following equation?

$$Mg(OH)_2 + 2HCl \longrightarrow MgCl_2 + 2H_2O \quad \blacklozenge$$

Practice Exercise 3

Rewrite the equation in Practice Exercise 2 to show that $Mg(OH)_2$ is a solid, HCl and $MgCl_2$ are dissolved in water, and H_2O is a liquid. $\blacklozenge$

2.4 Atomic Masses

One of the most useful concepts to come from Dalton's atomic theory is that atoms of an element have a constant, characteristic **atomic mass** (or **atomic weight**). This concept opened the door to the determination of chemical formulas and ultimately to one of the most useful devices chemists have for organizing chemical information, the periodic table of the elements. But how can the masses of atoms be measured?

Individual atoms are much too small to weigh on a balance, so their masses in grams can't be measured directly. However, the *relative masses* of the atoms of elements can be determined *provided that the formula of a compound containing them is known.* Let's look at an example to see how this could work.

Hydrogen (H) combines with the element fluorine (F) to form the compound hydrogen fluoride. The formula of this compound is HF, which means that in *any* sample of this substance the fluorine-to-hydrogen *atom ratio* is always 1 to 1. It is found that when a sample of HF is decomposed, the mass of fluorine obtained is always 19.0 times larger than the mass of hydrogen, so the fluorine-to-hydrogen *mass ratio* is always 19.0 to 1.00.

F-to-H atom ratio: 1 to 1

F-to-H mass ratio: 19.0 to 1.00

How could a 1-to-1 atom ratio give a 19.0-to-1.00 mass ratio? *Only if each fluorine atom is 19.0 times heavier than each H atom.*

Notice that even though we haven't found the actual masses of F and H atoms, we do now know how their masses compare (i.e., we know their *relative masses*). Similar procedures with other elements in other compounds are able to establish relative mass relationships among the other elements as well. What we need next is a way to place all these masses on the same mass scale, but before we study this, let's take a closer look at one of Dalton's hypotheses.

> To establish the first table of relative atomic masses, the formulas for compounds needed to be known. At the time Dalton presented his atomic theory no method was available to determine what the formula of a compound actually was. This difficulty was not overcome for nearly 50 years, so a reliable set of relative atomic masses was not established until the mid-1800s.

Isotopes

The cornerstone of Dalton's theory was the idea that all of the atoms of an element have identical masses. Actually, most elements occur in nature as uniform mixtures of two or more kinds of atoms that have slightly different masses, which we call **isotopes.** An iron nail, for example, is made up of a mixture of four isotopes, and the chlorine in salt is a mixture of two isotopes.

The existence of isotopes did not affect the development of Dalton's theory for two reasons. First, all the isotopes of a given element have virtually identical *chemical* properties—all give the same kinds of chemical reactions. Second, the relative proportions of its different isotopes are essentially constant, regardless of where on Earth or in the atmosphere the element is found. As a result, every sample of a particular element has the same isotopic composition, and the *average* mass per atom is the same from sample to sample. Therefore, in the laboratory elements *behave* as though their atoms have masses equal to the average.

> Even the smallest laboratory sample of an element has so many atoms that the relative proportions of the isotopes is constant from one sample to another.

The Carbon-12 Atomic Mass Scale

To establish a uniform mass scale for atoms it is necessary to define a standard against which the relative masses can be compared. Currently, the agreed upon reference is the most abundant isotope of carbon, called carbon-12 and symbolized ^{12}C. An atom of this isotope is assigned *exactly* 12 units of mass. These are called **atomic mass units,** symbolized in the SI by the letter **u.** Notice that this assignment establishes the size of the atomic mass unit to be 1/12th of the mass of a single carbon-12 atom, or stated another way:

> The older abbreviation for the atomic mass unit is **amu.**

$$1 \text{ atom } {}^{12}C \Leftrightarrow 12 \text{ u (exactly)}$$

In modern terms, then, the atomic mass of an element is the average mass of the element's atoms (as they occur in nature) relative to an atom of carbon-12, which is assigned a mass of 12 units. Thus, if an average atom of an element has a mass twice that of a ^{12}C atom, its atomic mass would be 24 u.

The definition of the size of the atomic mass unit is really quite arbitrary. It could just as easily have been selected to be 1/24th of the mass of a carbon atom, or 1/10th of the mass of an iron atom, or any other value. Why 1/12th of the mass of a ^{12}C atom? First, carbon is a very common element, available to any scientist. Second, and most important, by choosing the atomic mass unit of this size, the atomic masses of nearly all the other elements are almost whole numbers, with the lightest atom (hydrogen) having a mass of approximately 1 u.

Chemists generally work with whatever *mixture* of isotopes comes with a given element as it occurs naturally. Because the composition of this isotopic mixture is very nearly constant regardless of the source of the element, we can speak of an *average atom* of the element — average in terms of mass. For example, naturally occurring hydrogen is a mixture of two isotopes in the relative proportions given in the margin. The "average atom" of the element hydrogen, as it occurs in nature, has a mass that is 0.083992 times that of a ^{12}C atom. Since 0.083992×12.000 u $= 1.0079$ u, the average atomic mass of hydrogen is 1.0079 u. Notice that this average value is just a little larger than the atomic mass of ^{1}H because naturally occurring hydrogen also contains a little ^{2}H.

The atomic mass unit is sometimes called a **dalton (D).**

$$1 \text{ u} = 1 \text{ dalton}$$

Hydrogen Isotope	Mass	Percentage Abundance
^{1}H	1.007825 u	99.985
^{2}H	2.0140 u	0.015

Average Atomic Masses from Isotopic Abundances

Originally, the relative atomic masses of the elements were determined in a way similar to that described for hydrogen and fluorine in our earlier discussion. A sample of a compound was analyzed and from the formula of the substance the relative atomic masses were calculated. These were then adjusted to place them on the unified atomic mass scale. In modern times, methods have been developed to measure very precisely both the relative abundances of the isotopes of the elements and their atomic masses. This kind of information has permitted the calculation of more precise values of the average atomic masses, which are found in the table on the inside front cover of the book. Example 2.2 illustrates how this calculation is done.

EXAMPLE 2.2

Calculating Average Atomic Masses from Isotopic Abundances

Naturally occurring chlorine is a mixture of two isotopes. In every sample of this element 75.77% of the atoms are ^{35}Cl and 24.23% are atoms of ^{37}Cl. The accurately measured atomic mass of ^{35}Cl is 34.9689 u and that of ^{37}Cl is 36.9659 u. From these data, calculate the average atomic mass of chlorine.

Analysis: In a sample containing many atoms of chlorine, 75.77% of the mass is contributed by atoms of ^{35}Cl and 24.23% is contributed by atoms of ^{37}Cl. This means that when we calculate the mass of the "average atom," we have to weight it according to both the masses of the isotopes and their relative abundances. To do this it is convenient to imagine an "average atom" to be composed of 75.77% of ^{35}Cl and 24.23% of ^{37}Cl. (Of course, such an atom doesn't really exist, but this is a simple way to see how we can calculate the average atomic mass of this element.)

Solution: We will calculate 75.77% of the mass of an atom of ^{35}Cl, which is the contribution of this isotope to the "average atom." Then we will calculate 24.23% of the mass of an atom of ^{37}Cl, which is the contribution of this isotope. Adding these contributions gives the total mass of the "average atom."

$$0.7577 \times 34.9689 \text{ u} = 26.50 \text{ u} \qquad \text{(for } ^{35}Cl)$$

$$\underline{0.2423 \times 36.9659 \text{ u} = 8.957 \text{ u}} \qquad \text{(for } ^{37}Cl)$$

Total mass of average atom $= 35.46$ u (rounded)

The average atomic mass of chlorine is therefore 35.46 u.

Is the Answer Reasonable?
Naturally occurring chlorine will have an average atomic mass somewhere between approximately 35 and 37. If the abundances of the two isotopes were equal, the average would be nearly 36. But there is more ^{35}Cl than ^{37}Cl, so a value between 35 and 36 seems reasonable.

Practice Exercise 4

Aluminum atoms have a mass that is 2.24845 times that of an atom of ^{12}C. What is the atomic mass of aluminum? ◆

Practice Exercise 5

How much heavier than an atom of ^{12}C is the average atom of naturally occurring copper? Refer to the table inside the front cover of the book for the necessary data. ◆

Practice Exercise 6

Naturally occurring boron is composed of 19.8% of ^{10}B and 80.2% of ^{11}B. Atoms of ^{10}B have a mass of 10.0129 u and those of ^{11}B have a mass of 11.0093 u. Calculate the average atomic mass of boron. ◆

A knowledge of the internal structure of atoms is important to understanding the chemical and physical properties of the elements.

2.5 The Structure of Matter: Atoms and Subatomic Particles

As you probably know, atoms are not quite as indestructible as Dalton had thought, and Facets of Chemistry 2.1 describes experiments that were performed in the late 1800s and early 1900s that demonstrated that atoms are composed instead of simpler **subatomic particles.** These experiments led to the current theoretical model of atomic structure which we will examine in general terms in this chapter. A more detailed discussion of atomic structure will follow in Chapter 7.

Protons are in *all* nuclei. Except for ordinary hydrogen, all nuclei also contain neutrons.

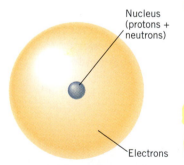

Nucleus (protons + neutrons)

Electrons

Figure 2.12 *Internal structure of an atom.* An atom is composed of a tiny nucleus that holds all the protons and neutrons, plus electrons that fill the space outside the nucleus.

Physicists have found a large number of subatomic particles, but protons, neutrons, and electrons are the only ones that will concern us at this time.

Subatomic Particles: Protons, Neutrons, and Electrons

Experiments have shown that atoms are composed of three principal kinds of subatomic particles: **protons, neutrons,** and **electrons.** Experiments also revealed that at the center of an atom there exists a very tiny, extremely dense core called the **nucleus,** which is where an atom's protons and neutrons are found. Because they are found in nuclei, protons and neutrons are sometimes called **nucleons.** The electrons in an atom surround the nucleus and fill the remaining volume of the atom. (*How* the electrons are distributed around the nucleus is the subject of Chapter 7.) The properties of the subatomic particles are summarized in Table 2.2, and the general structure of the atom is illustrated in Figure 2.12.

Table 2.2 Properties of Subatomic Particles

Particle	Mass (g)	Mass (u)	Electrical Charge	Symbol
Electron	$9.1093897 \times 10^{-28}$	0.000548579903	1−	$^{0}_{-1}e$
Proton	$1.6726231 \times 10^{-24}$	1.007276470	1+	$^{1}_{1}H^{+}, ^{1}_{1}p$
Neutron	$1.6749286 \times 10^{-24}$	1.008664904	0	$^{1}_{0}n$

Facets of Chemistry 2.1

The Discovery of Subatomic Particles

Our current knowledge of atomic structure was pieced together from facts obtained from experiments by scientists that began in the nineteenth century. In 1834, Michael Faraday discovered that the passage of electricity through aqueous solutions could cause chemical changes, which was the first hint that matter was electrical in nature. Later in that century scientists began to experiment with *gas discharge tubes* in which a high-voltage electric current was passed through a gas at low pressure in a glass tube (Figure 1). Such a tube is fitted with a pair of metal *electrodes,* and when the electricity begins to flow between them the gas in the tube glows. This flow of electricity is called an electric discharge, which is how the tubes got their names. (Modern neon signs work this way.)

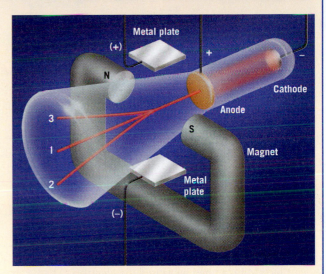

Figure 2 *Thomson's cathode ray tube.* The apparatus which was used to measure the charge-to-mass ratio for the electron.

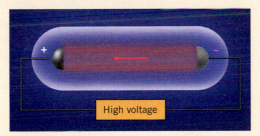

Figure 1 *A gas discharge tube.* Cathode rays flow from the negatively charged cathode to the positively charged anode.

The physicists who first studied this phenomenon did not know what caused the tube to glow, but tests soon revealed that negatively charged particles were moving from the negative electrode (the *cathode*) to the positive electrode (the *anode*). The physicists called these emissions *rays,* and because the rays came from the cathode, they were called *cathode rays.*

In 1897 the British physicist J. J. Thomson constructed a special gas discharge tube to make quantitative measurements of the properties of cathode rays. In some ways, the *cathode ray tube* he used was similar to a television picture tube, as Figure 2 shows. In Thomson's tube, a beam of cathode rays was focused on a glass surface coated with a phosphor that glows when the cathode rays strike it (point 1). The cathode ray beam passed between the poles of a magnet and between a pair of metal electrodes that could be given electrical charges. The magnetic field tends to bend the beam in one direction (to point 2) while the charged electrodes bend the beam in the opposite direction (to point 3). By adjusting the charge on the electrodes, the two effects can be made to cancel, and from the amount of charge on the electrodes required to balance the effect of the magnetic field, Thomson was able to calculate the first bit of quantitative information about a cathode ray particle—the ratio of its charge to its mass (often expressed as e/m, where e stands for charge and m stands for mass). The charge-to-mass ratio has a value of -1.76×10^8 coulombs/gram, where the coulomb (C) is the SI unit of electrical charge, and the negative sign reflects the negative charge on the particle.

Many experiments were performed using the cathode ray tube, and they demonstrated that cathode ray particles are in all matter. They are, in fact, *electrons.*

Measuring the Charge and Mass of the Electron

In 1909 a researcher at the University of Chicago, Robert Millikan, designed a clever experiment that enabled him to measure the electron's charge (Figure 3). During an experiment he would spray a fine mist of oil droplets above a pair of parallel metal plates, the top one of which had a small hole in it. As the oil drops settled, some would pass through this hole into the space between the plates, where he would irradiate them briefly with X rays. The X rays knocked electrons off molecules in the air, and the electrons became attached to the oil drops, which thereby were given an electrical charge. By observing the rate of fall of the charged drops both when the metal plates were electrically charged and when they were not, Millikan was

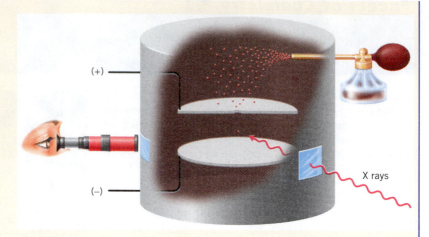

Figure 3 *Millikan's oil drop experiment.* Electrons, which are ejected from air molecules by the X rays, are picked up by very small drops of oil falling through the tiny hole in the upper metal plate. By observing the rate of fall of the charged oil drops, with and without electrical charges on the metal plates, Millikan was able to calculate the charge carried by an electron.

(+)

(−)

X rays

able to calculate the amount of charge carried by each drop. When he examined his results, he found that all the values he obtained were whole-number multiples of -1.60×10^{-19} C. He reasoned that since a drop could only pick up whole numbers of electrons, this value must be the charge carried by each individual electron.

Once Millikan had measured the electron's charge, its mass could then be calculated from Thomson's charge-to-mass ratio. This mass was calculated to be 9.09×10^{-28} g. More precise measurements have since been made, and the mass of the electron is currently reported to be $9.1093897 \times 10^{-28}$ g.

Discovery of the Proton

The removal of electrons from an atom gives a positively charged particle (called an *ion*). To study these, a modification was made in the construction of the cathode ray tube to give a new device called a *mass spectrometer*. This apparatus is described in Facets of Chemistry 2.2 and was used to measure the charge-to-mass ratios of positive ions. These

ratios were found to vary, depending on the chemical nature of the gas in the discharge tube, showing that their masses also varied. The lightest positive particle observed was produced when hydrogen was in the tube, and its mass was about 1800 times as heavy as an electron. When other gases were used, their masses always seemed to be whole-number multiples of the mass observed for hydrogen atoms. This suggested the possibility that clusters of the positively charged particles made from hydrogen atoms made up the positively charged particles of other gases. The hydrogen atom, minus an electron, thus seemed to be a fundamental particle in all matter and was named the *proton*, after the Greek word *proteios*, meaning "of first importance."

Discovery of the Atomic Nucleus

Early in the twentieth century, Hans Geiger and Ernest Marsden, working under Ernest Rutherford at Great Britain's Manchester University, studied what happened when *alpha rays* hit thin metal foils. Alpha rays are composed of particles having masses four times those of the

Notice that two of the subatomic particles carry electrical charges. Protons carry a single unit of **positive charge,** which is one type of electrical charge. Electrons carry one unit of the opposite charge, a **negative charge.** These electrical charges have the property that like charges repel each other and opposite charges attract, so negatively charged electrons are attracted to positively charged protons. In fact, it is this attraction that holds the electrons around the nucleus. Neutrons have no charge and are said to be electrically neutral (hence the name *neut*ron).

Because of their identical charges, electrons repel each other. The repulsions between the electrons keep them spread out throughout the volume of the atom, and it is the *balance* between the attractions the electrons feel toward the nucleus and the repulsions they feel toward each other that controls the sizes of atoms.

The *strong nuclear force* is discussed in Chapter 22.

Protons also repel each other, but they are able to stay together in the small volume of the nucleus because their repulsions are apparently offset by powerful nuclear forces that involve other subatomic particles we will not study.

Matter as we generally find it in nature appears to be electrically neutral,

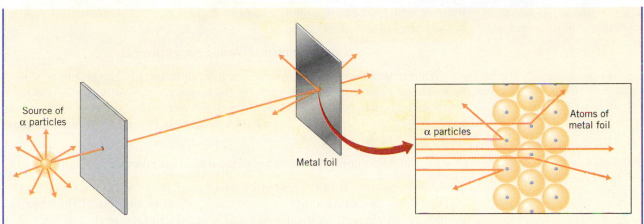

Figure 4 *Scattering of alpha particles by a metal foil.* Alpha particles are scattered in all directions by a thin metal foil. Some hit something very massive head-on and are deflected backward. Many sail through. Some, making near misses with the massive "cores" (nuclei), are still deflected, because alpha particles have the same kind of charge (+) as these cores.

proton and bearing two positive charges; they are emitted by certain unstable atoms in a phenomenon called radioactivity. Most of the alpha particles sailed right on through as if the foils were virtually empty space (Figure 4). A significant number of alpha particles, however, were deflected at very large angles. Some were even deflected backward, as if they had hit stone walls. Rutherford was so astounded that he compared the effect to that of firing a 15-in. artillery shell at a piece of tissue paper and having it come back and hit the gunner! From studying the angles of deflection of the particles, Rutherford reasoned that only something extraordinarily massive and positively charged could cause such an occurrence. Since most of the alpha particles went straight through, he further reasoned that the metal atoms in the foils must be mostly empty space. Rutherford's ultimate conclusion was that virtually all of the mass of an atom must be concentrated in a particle having a very small volume located in the center of the atom. He called this massive particle the atom's *nucleus.*

Discovery of the Neutron

From the way alpha particles were scattered by a metal foil, Rutherford and his students were able to estimate the number of positive charges on the nucleus of an atom of the metal. This had to be equal to the number of protons in the nucleus, of course. But when they computed the nuclear mass based on this number of protons, the value always fell short of the actual mass. In fact, Rutherford found that only about half of the nuclear mass could be accounted for by protons. This led him to suggest that there were other particles in the nucleus that had a mass close to or equal to that of a proton, but with no electrical charge. This suggestion initiated a search that finally ended in 1932 with the discovery of the *neutron* by Sir James Chadwick, a British physicist.

which means that it contains equal numbers of positive and negative charges. Therefore, *in a neutral atom, the number of electrons must equal the number of protons.*

The proton and neutron are much more massive than the electron, so in any atom almost all of the atomic mass is contributed by the particles that are found in the nucleus. It is also interesting to note, however, that the diameter of the atom is approximately 10,000 times the diameter of its nucleus, so almost all the *volume* of an atom is occupied by its electrons, which fill the space around the nucleus. (To place this on a more meaningful scale, if the nucleus was 1 ft in diameter, it would lie at the center of an atom with a diameter of approximately 1.9 miles!)

The electron has a mass that is only about 1/1830 that of a proton or neutron.

Atomic Numbers, Mass Numbers, and Isotopes

What distinguishes one element from another is the number of protons in the nuclei of its atoms, because *all the atoms of a particular element have an identical number of protons.* This means that each element has associated with it a unique

It is the atomic number of an atom that identifies which element it is.

number, which we call its **atomic number** (Z), that equals the number of protons in the nuclei of any of its atoms.

$$\text{atomic number, } Z = \text{number of protons}$$

We will use the symbol Z to stand for atomic number in later discussions.

What makes isotopes of the same element different are the numbers of neutrons in their nuclei. *The* **isotopes** *of a given element have atoms with the same number of protons but different numbers of neutrons.* The numerical sum of the protons and neutrons in the atoms of a particular isotope is called the **mass number** (A) of the isotope.

The mass number is simply the number of nucleons and has no units.

$$\text{mass number, } A = \text{number of protons} + \text{number of neutrons}$$

Thus, every isotope is fully defined by two numbers, its atomic number and its mass number. Sometimes these numbers are added to the left of the chemical symbol of an element as a subscript and a superscript, respectively. Thus, if X stands for the chemical symbol for the element, an isotope of X is represented as

$$^{A}_{Z}X$$

The isotope of uranium used in nuclear reactors, for example, can be symbolized as follows.

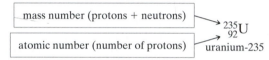

It is useful to remember that for a neutral atom, the atomic number equals both the number of protons and the number of electrons.

As indicated, the name of this isotope is uranium-235. Each neutral atom contains 92 protons and $(235 - 92) = 143$ neutrons as well as 92 electrons. In writing the symbol for the isotope, the atomic number is often omitted because it is really redundant. Every atom of uranium has 92 protons, and every atom that has 92 protons is an atom of uranium. Therefore, this uranium isotope can be represented simply as ^{235}U.

One could define an element as all atoms that have the same atomic number. Thus, all atoms that have atomic number 6, regardless of the number of neutrons, are atoms of carbon.

In naturally occurring uranium, a more abundant isotope is ^{238}U. Atoms of this isotope also have 92 protons, but the number of neutrons is 146. Thus, atoms of ^{235}U and ^{238}U have the identical number of protons, but differ in the numbers of neutrons.

In general, the mass number of an isotope differs slightly from the atomic mass of the isotope. For instance, the isotope ^{35}Cl has an atomic mass of 34.968852 u. In fact, the *only* isotope that has an atomic mass equal to its mass number is ^{12}C; *by definition* the mass of this atom is exactly 12 u.

Practice Exercise 7

Write the symbol for the isotope of plutonium (Pu) that contains 146 neutrons. ◆

Practice Exercise 8

How many protons, neutrons, and electrons are in each atom of $^{35}_{17}Cl$? ◆

2.6 The Periodic Table

When we study different kinds of substances, we find that some are elements and others are compounds. Among compounds, some are composed of discrete molecules. Others, as you will learn, are made up of atoms that have acquired

Facets of Chemistry 2.2

The Mass Spectrometer and the Measurement of Atomic Masses

When a spark is passed through a gas, electrons are knocked off the gas molecules. Because electrons are negatively charged, the particles left behind carry positive charges; they are called positive ions. These positive ions have different masses, depending on the masses of the molecules from which they are formed. Thus, some molecules have large masses and give heavy ions, while some have small masses and give light ions.

The device that is used to study the positive ions produced from gas molecules is called a *mass spectrometer* (illustrated in the figure at the right). In a mass spectrometer, positive ions are created by passing an electrical spark (called an electric discharge) through a sample of the particular gas being studied. As the positive ions are formed, they are attracted to a negatively charged metal plate that has a small hole in its center. Some of the positive ions pass through this hole and travel onward through a tube that passes between the poles of a powerful magnet.

One of the properties of charged particles, both positive and negative, is that their paths become curved as they pass through a magnetic field. This is exactly what happens to the positive ions in the mass spectrometer as they pass between the poles of the magnet. However, the extent to which their paths are bent depends on the masses of the ions. This is because the path of a heavy ion, like that of a speeding cement truck, is difficult to change, but the path of a light ion, like that of a motorcycle, is influenced more easily. As a result, heavy ions emerge from between the magnet's poles along different lines than the lighter ions. In effect, an entering beam containing ions of

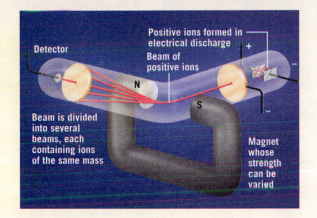

Detector
Positive ions formed in electrical discharge
Beam of positive ions
N
S
Beam is divided into several beams, each containing ions of the same mass
Magnet whose strength can be varied

different masses is sorted by the magnet into a number of beams, each containing ions of the same mass. This spreading out of the ion beam thus produces an array of different beams called a *mass spectrum*.

In practice, the strength of the magnetic field is gradually changed, which sweeps the beams of ions across a detector located at the end of the tube. As a beam of ions strikes the detector, its intensity is measured and the masses of the particles in the beam are computed based on the strength of the magnetic field and the geometry of the apparatus.

Among the benefits derived from measurements using the mass spectrometer are very accurate isotopic masses and relative isotopic abundances. These serve as the basis for the very precise values of the atomic masses that you find in the periodic table.

electrical charges. For elements such as sodium and chlorine, we mentioned metallic and nonmetallic properties. If we were to continue on this way, without attempting to build our subject around some central organizing structure, it would not be long before we became buried beneath a mountain of information in the form of seemingly unconnected facts.

The need for organization was recognized by many early chemists, and there were numerous attempts to discover relationships among the chemical and physical properties of the elements. A number of different sequences of elements were tried in the search for some sort of order or pattern. A few of these arrangements came quite close, at least in some respects, to our current periodic table, but either they were flawed in some way or they were presented to the scientific community in a manner that did not lead to their acceptance.

Mendeleev's Periodic Table

The periodic table we use today is based primarily on the efforts of a Russian chemist, Dmitri Ivanovich Mendeleev (1834–1907), and a German physicist, Julius Lothar Meyer (1830–1895). Working independently, these scientists

Dmitri Mendeleev

Periodic refers to the recurrence of properties at regular intervals.

developed similar periodic tables only a few months apart in 1869. Mendeleev is usually given the credit, however, because he had the good fortune to publish first.

Mendeleev was preparing a chemistry textbook for his students at the University of St. Petersburg. He found that when he arranged the elements in order of increasing atomic mass, similar chemical properties occurred over and over again at regular intervals. For instance, the elements lithium (Li), sodium (Na), potassium (K), rubidium (Rb), and cesium (Cs) have similar chemical properties. Each forms a water-soluble chlorine compound with the general formula *M*Cl: LiCl, NaCl, KCl, RbCl, and CsCl. Moreover, the elements that immediately follow each of these elements also constitute a set with similar chemical properties. Beryllium (Be) follows lithium, magnesium (Mg) follows sodium, calcium (Ca) follows potassium, strontium (Sr) follows rubidium, and barium (Ba) follows cesium. All of these elements form a water-soluble chlorine compound with the general formula MCl_2: $BeCl_2$, $MgCl_2$, $CaCl_2$, $SrCl_2$, and $BaCl_2$. On the basis of extensive observations of this type, Mendeleev devised the original form of the **periodic law.** It stated that *the chemical and physical properties of the elements vary in a periodic way with their atomic weights.* Mendeleev used this law to construct his periodic table, which is illustrated in Figure 2.13.

The elements in Mendeleev's table are arranged in rows, called **periods,** in order of increasing atomic mass. When the rows are broken at the right places and stacked, the elements fall naturally into columns, called **groups,** in which the elements of a given group have similar chemical properties.

Mendeleev's genius rested on his placing elements with similar properties in the same group even when this left occasional gaps in the table. For example, he

		Group I	Group II	Group III	Group IV	Group V	Group VI	Group VII	Group VIII
	1	H 1							
	2	Li 7	Be 9.4	B 11	C 12	N 14	O 16	F 19	
	3	Na 23	Mg 24	Al 27.3	Si 28	P 31	S 32	Cl 35.5	
	4	K 39	Ca 40	—44	Ti 48	V 51	Cr 52	Mn 55	Fe 56, Co 59 Ni 59, Cu 63
Periods	5	(Cu 63)	Zn 65	—68	—72	As 75	Se 78	Br 80	
	6	Rb 85	Sr 87	?Yt 88	Zr 90	Nb 94	Mo 96	—100	Ru 104, Rh 104 Pd 105, Ag 108
	7	(Ag 108)	Cd 112	In 113	Sn 118	Sb 122	Te 128	I 127	
	8	Cs 133	Ba 137	?Di 138	?Ce 140	—	—	—	———
	9	—	—	—	—	—	—	—	
	10	—	—	?Er 178	?La 180	Ta 182	W 184	—	Os 195, Ir 197 Pt 198, Au 199
	11	(Au 199)	Hg 200	Tl 204	Pb 207	Bi 208	—		
	12	—	—	—	Th 231	—	U 240	—	——— ———

Figure 2.13 The first periodic table. Mendeleev's periodic table roughly as it appeared in 1871. The numbers next to the symbols are atomic masses.

Table 2.3 Comparison of Some Predicted Properties of Eka-silicon and Observed Properties of Germanium

Property	Observed for Silicon (Si)	Predicted for Eka-silicon (Es)	Observed for Tin (Sn)	Found for Germanium (Ge)
Atomic mass	28	72	118	72.59
Melting point (°C)	1410	high	232	947
Density (g/cm^3)	2.33	5.5	7.28	5.35
Formula of oxide	SiO_2	EsO_2	SnO_2	GeO_2
Density of oxide (g/cm^3)	2.66	4.7	6.95	4.23
Formula of chloride[a]	$SiCl_4$	$EsCl_4$	$SnCl_4$	$GeCl_4$
Boiling point of chloride (°C)	57.6	100	114	83

[a]Tin and germanium also form chlorides with formulas of $SnCl_2$ and $GeCl_2$.

placed arsenic (As) in Group V under phosphorus because its chemical properties were similar to those of phosphorus, even though this left gaps in Groups III and IV. Mendeleev reasoned, correctly, that the elements that belonged in these gaps had simply not yet been discovered. In fact, on the basis of the location of these gaps Mendeleev was able to predict with remarkable accuracy the properties of these yet-to-be-found substances. His predictions helped serve as a guide in the search for the missing elements. Table 2.3 compares Mendeleev's predicted properties of "eka-silicon" and some of its compounds with the actual properties measured for the element germanium after it was discovered.

Two elements, tellurium (Te) and iodine (I), caused Mendeleev some problems. According to the best estimates at that time, the atomic mass of tellurium was greater than that of iodine. Yet if these elements were placed in the table according to their atomic masses, they would not fall into the proper groups required by their properties. Therefore, Mendeleev switched their order and in so doing violated his own periodic law. (Actually, he believed that the atomic mass of tellurium had been incorrectly measured, but this wasn't so.)

The table that Mendeleev developed is in many ways similar to the one we use today. One of the main differences, though, is that Mendeleev's table lacks the column containing the elements helium (He) through radon (Rn). In Mendeleev's time, none of these elements had yet been found because they are relatively rare and because they have virtually no tendency to undergo chemical reactions. When these elements were finally discovered, beginning in 1894, another problem arose. Two more elements, argon (Ar) and potassium (K), did not fall into the groups required by their properties if they were placed in the table in the order required by their atomic masses. Another switch was necessary and another exception to Mendeleev's periodic law had been found. Apparently, then, atomic mass was not the true basis for the periodic repetition of the properties of the elements. To determine what the true basis was, however, scientists had to await the discoveries of the atomic nucleus, the proton, and atomic numbers.

The Modern Periodic Table

Once atomic numbers had been discovered, it was soon realized that the elements in Mendeleev's table are arranged in precisely the order of increasing atomic number. In other words, if we take atomic numbers as the basis for arranging the elements in sequence, no annoying switches are required and the elements Te and I or Ar and K are no longer a problem. This leads to our present statement of the **periodic law.**

> **The Periodic Law**
> The chemical and physical properties of the elements vary in a periodic way with their atomic *numbers*.

The fact that it is the atomic number—the number of protons in the nucleus of an atom—that determines the order of elements in the table is very significant. We will see later that this has important implications with regard to the relationship between the number of electrons in an atom and the atom's chemical properties.

The modern periodic table is shown in Figure 2.14 and also appears on the inside front cover of the book. We will refer to the table frequently, so it is important for you to become familiar with it and with some of the terminology applied to it.

Terminology Associated with the Periodic Table

Recall that the symbol Z stands for atomic number.

As in Mendeleev's table, the elements are arranged in rows called **periods,** but here they are arranged in order of increasing atomic number. For identification purposes the periods are numbered. We will find these numbers useful later on. Below the main body of the table are two long rows of 14 elements each. These actually belong in the main body of the table following La ($Z = 57$) and

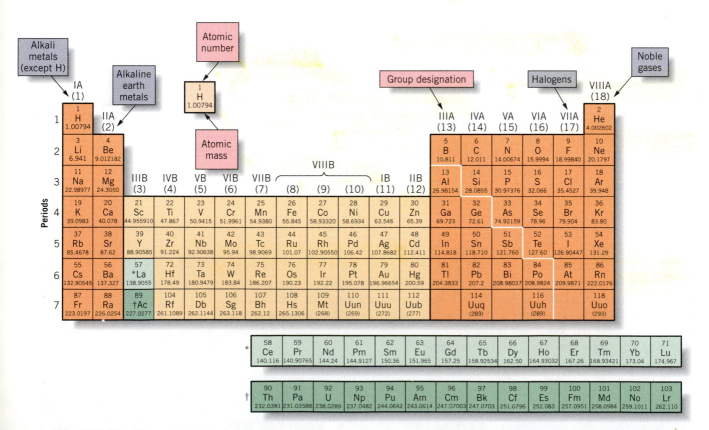

Figure 2.14 *The modern periodic table.*

1 H																															2 He
3 Li	4 Be																											5 B	6 C	7 N	8 O	9 F	10 Ne
11 Na	12 Mg																											13 Al	14 Si	15 P	16 S	17 Cl	18 Ar
19 K	20 Ca	21 Sc																22 Ti	23 V	24 Cr	25 Mn	26 Fe	27 Co	28 Ni	29 Cu	30 Zn	31 Ga	32 Ge	33 As	34 Se	35 Br	36 Kr
37 Rb	38 Sr	39 Y																40 Zr	41 Nb	42 Mo	43 Tc	44 Ru	45 Rh	46 Pd	47 Ag	48 Cd	49 In	50 Sn	51 Sb	52 Te	53 I	54 Xe
55 Cs	56 Ba	57 La	58 Ce	59 Pr	60 Nd	61 Pm	62 Sm	63 Eu	64 Gd	65 Tb	66 Dy	67 Ho	68 Er	69 Tm	70 Yb	71 Lu	72 Hf	73 Ta	74 W	75 Re	76 Os	77 Ir	78 Pt	79 Au	80 Hg	81 Tl	82 Pb	83 Bi	84 Po	85 At	86 Rn
87 Fr	88 Ra	89 Ac	90 Th	91 Pa	92 U	93 Np	94 Pu	95 Am	96 Cm	97 Bk	98 Cf	99 Es	100 Fm	101 Md	102 No	103 Lr	104 Rf	105 Db	106 Sg	107 Bh	108 Hs	109 Mt	110 Uun	111 Uuu	112 Uub		114 Uuq		116 Uuh		118 Uuo

Figure 2.15 *Extended form of the periodic table.* The two long rows of elements below the main body of the table in Figure 2.14 are placed in their proper places in this table.

Ac ($Z = 89$), as shown in Figure 2.15. They are almost always placed below the table simply to conserve space. If the fully spread-out table is printed on one page, the type is so small that it's difficult to read. Notice that in the fully extended form of the table, with all the elements arranged in their proper locations, there is a great deal of empty space. An important requirement of a detailed atomic theory, which we will get to in Chapter 7, is that it must explain not only the repetition of properties, but also why there is so much empty space in the table.

Again, as in Mendeleev's table, the vertical columns are called **groups.** However, there is not uniform agreement among chemists on how they should be numbered. In the past, the groups were labeled with Roman numerals and divided into A groups and B groups, separated by the three short columns headed by Fe, Co, and Ni (Group VIIIB), as indicated in Figure 2.14. This corresponds quite closely to Mendeleev's original designations, and is preferred by many chemists in the United States. However, another version of the table, popular in Europe, has the first seven groups from left to right labeled A, followed by the three short columns headed by Fe, Co, and Ni, and then the next seven groups labeled B. In an attempt to standardize the table, the International Union of Pure and Applied Chemistry (the IUPAC), an international body of scientists responsible for setting standards in chemistry, officially adopted a third system in which the groups are simply numbered sequentially from left to right using Arabic numerals. Thus, Group IA in the old system is Group 1 in the IUPAC table, and Group VIIA in the old system is Group 17 in the IUPAC table. In Figure 2.14 and on the inside front cover of the book, we have used both the older U.S. labels as well as those preferred by the IUPAC. Because of the lack of uniform agreement among chemists on how the groups should be specified, and because many chemists still prefer the more traditional A-group/B-group designations in Figure 2.14, we will use just the latter when we wish to specify a particular group.

As we have already noted, the elements in a given group bear similarities to each other. Because of such similarities, groups are sometimes referred to as **families of elements.** The elements in the longer columns (the A groups) are known as the **representative elements** or **main group elements.** Those that fall into the B groups in the center of the table are called **transition elements.** The elements in the two long rows below the main body of the table are the **inner transition elements,** and each row is named after the element that it follows in the main body of the table. Thus, elements 58–71 are called the **lanthanide elements** because they follow lanthanum ($Z = 57$) and elements 90–103 are called the **actinide elements** because they follow actinium ($Z = 89$).

Some of the groups have acquired common names. For example, except for hydrogen, the Group IA elements are metals. They form compounds with

If you continue in chemistry you will certainly need to know the customary U.S. system because so many references use it.

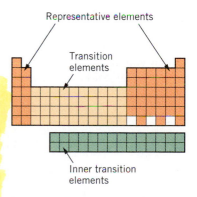

Representative elements

Transition elements

Inner transition elements

The IUPAC scheme also recognizes the groupings called the **lanthanide** and **actinide series.**

oxygen that dissolve in water to give solutions that are strongly alkaline, or caustic. As a result, they are called the **alkali metals,** or simply the **alkalis.** The Group IIA elements are also metals. Their oxygen compounds are alkaline, too, but many compounds of the Group IIA elements are unable to dissolve in water and are found in deposits in the ground. Because of their properties and where they occur in nature, the Group IIA elements became known as the **alkaline earth metals.**

On the right side of the table, in Group VIIIA, are the **noble gases.** They used to be called the inert gases until it was discovered that the heavier members of the group show a small degree of reactivity. The term *noble* is used when we wish to suggest a very limited degree of chemical reactivity. Gold, for instance, is often referred to as a noble metal because so few chemicals are capable of reacting with it.

Finally, the elements of Group VIIA are called the **halogens,** derived from the Greek word meaning sea or salt. Chlorine (Cl), for example, is found in familiar table salt, NaCl, a compound that accounts in large measure for the salty taste of seawater.

Practice Exercise 9

Circle the correct choices.

(a) Representative elements are: K, Cr, Pr, Ar, Al.
(b) A halogen is: Na, Fe, O, Cl, Cu.
(c) An alkaline earth metal is: Rb, Ba, La, As, Kr.
(d) A noble gas is: H, Ne, F, S, N.
(e) An alkali metal is: Zn, Ag, Br, Ca, Li.
(f) An inner transition element is: Ce, Pb, Ru, Xe, Mg. ◆

2.7 Metals, Nonmetals, and Metalloids

The periodic table organizes all sorts of chemical and physical information about the elements and their compounds. It allows us to study systematically the way properties vary with an element's position within the table and, in turn, makes the similarities and differences among the elements easier to understand and remember.

Even a casual inspection of samples of the elements reveals that some are familiar metals and that others, equally familiar, are not metals. Most of us are already familiar with metals such as lead, iron, or gold and nonmetals such as oxygen or nitrogen. A closer look at the nonmetallic elements, though, reveals that some of them, silicon and arsenic to name two, have properties that lie between those of true metals and true nonmetals. These elements are called **metalloids.** Division of the elements into the categories of metals, nonmetals, and metalloids is not even, however (see Figure 2.16). Most elements are metals, slightly over a dozen are nonmetals, and only a handful are metalloids.

Notice in Figure 2.16 that the metalloids are grouped around the bold stair-step line that is drawn diagonally from boron (B) to astatine (At).

Metals

You probably know a metal when you see one. Metals tend to have a shine so unique that it's called a *metallic luster*. For example, the silvery sheen of the freshly exposed surface of sodium in Figure 1.3*a* (page 6) would most likely lead you to identify sodium as a metal even if you had never seen or heard of it before. We also know that metals conduct electricity. Few of us would hold an iron

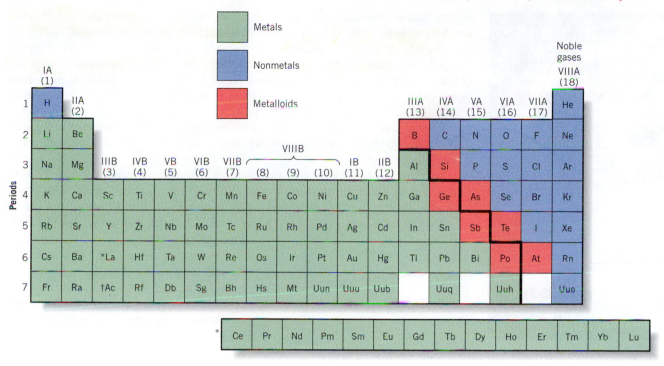

Figure 2.16 *Distribution of metals, nonmetals, and metalloids among the elements in the periodic table.*

nail in our hand and poke it into an electrical outlet. In addition, we know that metals conduct heat very well. On a cool day, metals always feel colder to the touch than do neighboring nonmetallic objects because metals conduct heat away from your hand very rapidly. Nonmetals seem less cold because they can't conduct heat away as quickly and therefore their surfaces warm up faster.

Other properties that metals possess, to varying degrees, are **malleability**—the ability to be hammered or rolled into thin sheets—and **ductility**—the ability to be drawn into wire. The production of sheet steel for automobiles and household appliances as well as the ability of a blacksmith to fashion horseshoes from a bar of iron (Figure 2.17) depend on the malleability of iron and steel, and the manufacture of electrical wire (Figure 2.18) is based on the ductility of copper.

▶**Chemistry in Practice**◀ One of the most malleable of metals is gold. It can be hammered into sheets so thin that they are only about 1/280,000 of an inch thick, which corresponds to a thickness of only several hundred atoms. (The sheets are so thin that some light can pass through.) Thin gold sheets are sold as "gold leaf" and are used for applying gold in a variety of decorative ways. For example, gold lettering is applied to doors by first painting the letters using a varnish and then, while the varnish is still tacky, applying the gold leaf. Gold leaf is also used as a decorative finish on various objects, such as the statue of Prometheus that overlooks the skating rink at Rockefeller Center in New York City. ◆

The statue of Prometheus that stands above the skating rink in Rockefeller Center in New York City is covered by a thin, decorative layer of gold leaf.

Another important physical property that we usually think of for metals is hardness. Some metals, such as chromium or iron, are indeed quite hard; but others, like copper and lead, are rather soft. The alkali metals are so soft they can be cut with a knife, but they are also so chemically reactive that we rarely get to see them as free elements.

All the metallic elements, except mercury, are solids at room temperature (Figure 2.19). Mercury's low freezing point (-39 °C) and fairly high boiling point (357 °C) make it useful as a fluid in thermometers. Most of the other metals have much higher melting points, and some are used primarily because of this. Tungsten, for example, has the highest melting point of any metal (3400 °C, or 6150 °F), which explains its use as filaments that glow white hot in electric lightbulbs.

The chemical properties of metals vary tremendously. Some, such as gold and platinum, are very unreactive toward almost all chemical agents. This property, plus their natural beauty and rarity, makes them highly prized for use in jewelry. Other metals, however, are so reactive that few people except chemists and chemistry students ever get to see them in their "free" states. For instance, in Chapter 1 you learned that the metal sodium reacts very quickly with oxygen or moisture in the air, and its bright metallic surface tarnishes almost immediately. In contrast, compounds of sodium are quite stable and very common. Examples are table salt (NaCl), baking soda ($NaHCO_3$), lye (NaOH), and bleach (NaOCl). We will have more to say about the chemical properties of metals in Section 2.8.

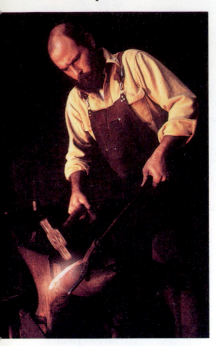

Figure 2.17 *Malleability of iron.* A blacksmith uses the malleability of hot iron to fashion horse shoes from an iron bar.

Nonmetals

Substances such as plastics, wood, and glass that lack the properties of metals are said to be *nonmetallic,* and an element that has nonmetallic properties is called a **nonmetal.** Most often, we encounter the nonmetals in the form of compounds or mixtures of compounds. There are some nonmetals, however, that are very important to us in their elemental forms. The air we breathe, for instance, contains mostly nitrogen, N_2, and oxygen, O_2. Both are gaseous, colorless, and odorless nonmetals. Since we can't see, taste, or smell them, however, it's difficult to experience their existence. (Although if you step into an atmosphere without oxygen, your body will very quickly tell you that something is missing!) Probably the most commonly *observed* nonmetallic element is carbon. We find it as the graphite in pencils, as coal, and as the charcoal used for barbecues. It also occurs in a more valuable form as diamond. Although diamond and graphite differ in appearance, each is a form of elemental carbon.

Figure 2.18 *The ductility of copper.* This property allows the metal to be drawn into wire. Here copper wire passes through one die after another as it is drawn into thinner and thinner wire.

Photographs of some of the nonmetallic elements appear in Figure 2.20. Their properties are almost completely opposite those of metals. Each of these elements lacks the characteristic appearance of a metal. They are poor conductors of heat and, with the exception of the graphite form of carbon, are also poor conductors of electricity. The electrical conductivity of graphite appears to be an accident of molecular structure, since the structures of metals and graphite are completely different.

Many of the nonmetals are solids at room temperature and atmospheric pressure, while many others are gases. All of the Group VIIIA elements are gases in which the particles consist of single atoms. The other gaseous elements—hydrogen, oxygen, nitrogen, fluorine, and chlorine—are composed of diatomic molecules. Their formulas are H_2, O_2, N_2, and Cl_2. Bromine and iodine are also diatomic, but bromine is a liquid and iodine is a solid at room temperature. (The formulas of the diatomic nonmetals were mentioned on page 50.)

Mercury and bromine are the only two liquid elements at room temperature and pressure.

The nonmetallic elements lack the malleability and ductility of metals. A lump of sulfur crumbles when hammered and breaks apart when pulled on. Diamond cutters rely on the brittle nature of carbon when they split a gem-quality stone by carefully striking a quick blow with a sharp blade.

As with metals, nonmetals exhibit a broad range of chemical reactivity. Fluorine, for instance, is extremely reactive. It reacts readily with almost all the other elements. At the other extreme is helium, the gas used to inflate children's balloons and the Goodyear blimp. This element does not react with anything, a fact that chemists find useful when they want to provide a totally *inert* (unreactive) atmosphere inside some apparatus.

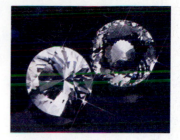

Diamonds such as these are simply another form of the element carbon.

Metalloids

The properties of metalloids lie between those of metals and nonmetals. This shouldn't surprise us since the metalloids are located between the metals and the nonmetals in the periodic table. In most respects, metalloids behave as nonmetals, both chemically and physically. However, in their most important physical property, electrical conductivity, they somewhat resemble metals. Metalloids tend to be **semiconductors;** they conduct electricity, but not nearly so well as metals. This property, particularly as found in silicon and germanium, is responsible for the remarkable progress made during the last four decades in the field of solid-state electronics. The operation of every computer, hi-fi stereo system, TV receiver, VCR, CD player, and AM-FM radio relies on transistors made from semiconductors. Perhaps the most amazing advance of all has been the fantastic reduction in the size of electronic components that semiconductors have allowed (Figure 2.21). To it, we owe the development of small and versatile TV cameras, compact disc players, hand-held calculators, and microcomputers. The heart of these devices is a microcircuit embedded in the surface of a tiny silicon chip.

Figure 2.19 *Mercury.* The metal mercury (once known as quicksilver) is a liquid at room temperature, unlike other metals, which are solids.

Trends in the Periodic Table

The occurrence of the metalloids between the metals and the nonmetals is our first example of trends in properties within the periodic table. We will frequently see that as we move from position to position across a period or down a group, chemical and physical properties change in a more or less regular fashion. There

Figure 2.20 *Some nonmetallic elements.* In the bottle on the left is dark-red liquid bromine, which vaporizes easily to give a deeply colored orange vapor. Pale green chlorine fills the round flask in the center. Solid iodine lines the bottom of the flask on the right and gives off a violet vapor. Powdered red phosphorus occupies the dish in front of the flask of chlorine, and black powdered graphite is in the watch glass. Also shown are lumps of yellow sulfur.

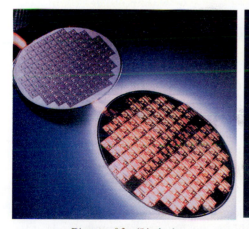

Diameter: 0.2m (8 inches)
Population: 10 billion components

Diameter: 13×10^8 m (8×10^3 miles)
Population: 6 billion people

Figure 2.21 *Uses of semiconductors.* Modern electronic circuits rely on the semiconductor properties of silicon. The silicon wafer shown here contains more electronic components (10 billion) than there are people on our entire planet (about 6 billion)!

are few abrupt changes in the characteristics of the elements as we scan across a period or down a group. The location of the metalloids can be seen, then, as an example of the gradual transition between metallic and nonmetallic properties. From left to right across period 3, we go from aluminum, an element that has every appearance of a metal; to silicon, a semiconductor; to phosphorus, an element with clearly nonmetallic properties. A similar gradual change is seen going down Group IVA. Carbon is certainly a nonmetal, silicon and germanium are metalloids, and tin and lead are metals. Trends such as these are useful to spot, because they help us remember properties.

2.8 Reactions of the Elements: Formation of Molecular and Ionic Compounds

A property possessed by nearly every element is the ability to combine with other elements to form compounds, although not all combinations appear to be possible. For example, sodium reacts vigorously with chlorine to form sodium chloride, but no compound is formed between sodium and iron. When we examine the chemical properties of the elements, though, we do find that certain generalizations are possible. It is worth looking at these now for a number of reasons. First, they provide a general framework within which we can study chemical behavior in greater detail later. Second, they illustrate how the periodic table can make it easier to organize and remember chemical properties. And finally, they show us that to *understand* the chemical behavior of the elements, we will need a theoretical model that explains what actually occurs between atoms when they react.

Molecular and Ionic Compounds

Atoms combine with each other in two broad general ways, either by the sharing of electrons between atoms or by the transfer of one or more electrons from one atom to another. Electron sharing gives discrete electrically neutral particles that we call **molecules**, whereas electron transfer produces electrically charged particles that we call **ions.** Compounds composed of molecules are called *molecular compounds* and those composed of ions are called *ionic compounds.*

Formation of Molecular Compounds

Molecular compounds are formed when nonmetallic elements combine. For example, you learned earlier that H_2 and O_2 combine to form molecules of water. Similarly, carbon and oxygen combine to form either carbon monoxide, CO, or carbon dioxide, CO_2. (Both are gases that are formed in various amounts as products in the combustion of fuels such as gasoline and charcoal.) Although molecular compounds can be formed by the direct combination of elements, often they are also the products of reactions between compounds. You will encounter such reactions frequently later in this book.

Molecules vary in size from small to very large. Some contain as few as two atoms; we call them diatomic molecules. Most molecules are more complex, however, and contain more atoms. Molecules of water, for example, have the formula H_2O and those of ordinary table sugar have the formula $C_{12}H_{22}O_{11}$. There also are molecules that are very large, such as those that occur in plastics and in living organisms, some of which contain millions of atoms.

The attractions that hold atoms to each other in molecules, called **chemical bonds,** are electrical in nature. As mentioned, they arise from the sharing of

Carbon monoxide is a poisonous gas found in the exhaust of automobiles.

electrons between one atom and another. We will discuss such bonds at considerable length in Chapters 8 and 9. What is important to know about molecules now is that the group of atoms that make up a molecule move about together and behave as a single particle, just as the various parts that make up a car move about as one unit. The chemical formulas that we write to describe the compositions of molecules are called **molecular formulas.**

Formation of Ionic Compounds

Ionic compounds are generally formed when metals react with nonmetals. Under appropriate conditions, the atoms transfer electrons between them when they react. This is what happens, for example, when the metal sodium combines with the nonmetal chlorine. When these elements react, each sodium atom gives up one electron to a chlorine atom, which thereby gains one. We can diagram the changes in equation form by using the symbol e^- to stand for an electron.

$$\overset{e^-}{\overset{\frown}{Na + Cl}} \longrightarrow Na^+ + Cl^-$$

The electrically charged particles formed in this reaction are called **ions,** specifically a sodium ion (Na^+) and a chloride ion (Cl^-). The sodium ion has a positive 1+ charge, indicated by the superscript plus sign, because it now has one more proton in its nucleus than there are electrons outside. Similarly, by gaining one electron the chlorine atom has added one more negative charge, so the chloride ion has a single negative charge indicated by the minus sign. Solid sodium chloride is composed of these charged sodium and chloride ions and is said to be an **ionic compound.**

Some atoms gain or lose more than one electron. For example, when calcium atoms react, they lose two electrons to form Ca^{2+} ions and when oxygen atoms form ions they each gain two electrons to give O^{2-} ions. (We will have to wait until a later chapter to study the reasons why certain atoms gain or lose one electron each, whereas other atoms gain or lose two or more electrons.)

Figure 2.22 illustrates the structures of water and sodium chloride and demonstrates an important difference between molecular and ionic compounds.

Here we are concentrating on what happens to the individual atoms, so we have not shown chlorine as diatomic Cl_2 molecules.

Na = sodium atom

Na^+ = sodium ion

A neutral sodium atom has 11 protons and 11 electrons; a sodium ion has 11 protons and 10 electrons, so it carries a unit positive charge. A neutral chlorine atom has 17 protons and 17 electrons; a chloride ion has 17 protons and 18 electrons, so it carries a unit negative charge.

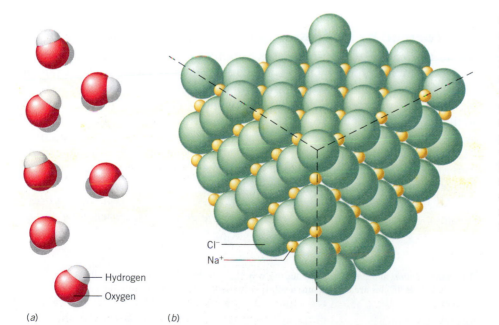

Cl⁻
Na⁺

— Hydrogen
— Oxygen

(a)   (b)

Figure 2.22 *Molecular and ionic compounds.* (*a*) In water there are discrete molecules that each consist of one atom of oxygen and two atoms of hydrogen. Each particle has the formula H_2O. (*b*) In sodium chloride, ions are packed in the most efficient way. Each Na^+ is surrounded by six Cl^-, and each Cl^- is surrounded by six Na^+. Because individual molecules do not exist, we simply specify the ratio of ions as NaCl.

In water it is safe to say that two hydrogen atoms "belong" to each oxygen atom in a particle having the formula H_2O. However, in NaCl it is impossible to say that a particular Na^+ ion belongs to a particular Cl^- ion. The ions in a crystal of NaCl are simply packed in the most efficient way, so that positive ions and negative ions can be as close to each other as possible. In this way, the attractions between oppositely charged ions, which are responsible for holding the compound together, can be as strong as possible.

Because molecules don't exist in ionic compounds, the subscripts in their formulas are always chosen to specify the smallest whole-number ratio of the ions. This is why the formula of sodium chloride is given as NaCl rather than Na_2Cl_2 or Na_3Cl_3. Although the smallest unit of an ionic compound can't be called a molecule, the idea of "smallest unit" is still quite often useful. Therefore, we take the smallest unit of an ionic compound to be whatever is represented in its formula and call this unit a **formula unit.** Thus, one formula unit of NaCl consists of one Na^+ and one Cl^-, whereas one formula unit of the ionic compound $CaCl_2$ consists of one Ca^{2+} and two Cl^- ions. (In a broader sense, we can use the term *formula unit* to refer to whatever is represented by a formula. Sometimes the formula specifies a set of ions, as in NaCl; sometimes it is a molecule, as in O_2 or H_2O; sometimes it can be just an ion, as in Cl^- or Ca^{2+}; and sometimes it might be just an atom, as in Na.)

Notice that the charges on the ions are omitted in writing the formula.

2.9 Ionic Compounds and Their Properties

In the preceding section we noted that metals combine with nonmetals to form ionic compounds. In such reactions, metal atoms lose one or more electrons to become positively charged ions and nonmetal atoms gain one or more electrons to become negatively charged ions. In referring to these particles, we will frequently call a positively charged ion a **cation** (pronounced *CAT-i-on*) and a negatively charged ion an **anion** (pronounced *AN-i-on*).[3] Thus, solid NaCl is composed of sodium cations and chloride anions.

Ions Formed by Representative Metals and Nonmetals

One of the reasons the periodic table is useful is that it can help us remember the kinds of ions formed by many of the representative elements. For example, except for hydrogen, the neutral atoms of the Group IA elements always lose one electron each when they react, thereby becoming ions with a charge of 1+. Similarly, atoms of the Group IIA elements always lose two electrons when they react; so these elements always form ions with a charge of 2+. In Group IIIA, the only important positive ion we need consider now is that of aluminum, Al^{3+}; an aluminum atom loses three electrons when it reacts to form this ion.

Positive ions are formed by metals.

All these ions are listed in Table 2.4. *Notice that the number of positive charges on each of the cations is the same as the group number when we use the*

---

[3]The names *cation* and *anion* come from the way the ions behave when electrically charged metal plates called electrodes are dipped into a solution that contains them. We will discuss this in detail in Chapter 19.

Table 2.4 Some Ions Formed from the Representative Elements

			Group Number			
IA	IIA	IIIA	IVA	VA	VIA	VIIA
Li^+	Be^{2+}		C^{4-}	N^{3-}	O^{2-}	F^-
Na^+	Mg^{2+}	Al^{3+}	Si^{4-}	P^{3-}	S^{2-}	Cl^-
K^+	Ca^{2+}				Se^{2-}	Br^-
Rb^+	Sr^{2+}				Te^{2-}	I^-
Cs^+	Ba^{2+}					

traditional numbering of groups in the periodic table. Thus, sodium is in Group IA and forms an ion with a 1+ charge, barium (Ba) is in Group IIA and forms an ion with a 2+ charge, and aluminum is in Group IIIA and forms an ion with a 3+ charge. Although this generalization doesn't work for all the metallic elements (it doesn't work for the transition elements, for instance), it does help us remember what happens to the metallic elements of Groups IA and IIA and aluminum when they react.

Among the nonmetals on the right side of the periodic table we also find some useful generalizations. For example, when they combine with metals, the halogens (Group VIIA) form ions with a 1− charge and the nonmetals in Group VIA form ions with a 2− charge. Notice that *the number of negative charges on the anion is equal to the number of spaces to the right that we have to move in the periodic table to get to a noble gas.*

Negative ions are formed by the nonmetals when they combine with metals.

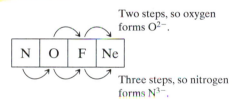

Two steps, so oxygen forms O^{2-}.

Three steps, so nitrogen forms N^{3-}.

Writing Formulas for Ionic Compounds

Because all chemical compounds are electrically neutral, the ions in an ionic compound always occur in a ratio such that the total positive charge is equal to the total negative charge. This is why the formula for sodium chloride is NaCl; the 1-to-1 ratio of Na^+ to Cl^- gives electrical neutrality. In addition, as we've already mentioned, discrete molecules do not exist in ionic compounds, so we always use the smallest set of subscripts that specify the correct ratio of the ions. The following, therefore, are the rules we use in writing the formulas of ionic compounds.

A substance is electrically neutral, with a net charge of zero, if the total positive charge equals the total negative charge.

Tools

Rules for writing formulas of ionic compounds

Rules for Writing Formulas of Ionic Compounds

1. The positive ion is given first in the formula. (This isn't required by nature, but it is a custom we always follow.)
2. The subscripts in the formula must produce an electrically neutral formula unit. (Nature does require electrical neutrality.)
3. The subscripts should be the smallest set of whole numbers possible.

EXAMPLE 2.3
Writing Formulas for Ionic Compounds

Write the formulas for the ionic compounds formed from (a) Ba and S, (b) Al and Cl, and (c) Al and O.

Solution: In each case, the ions must be combined in a ratio that produces an electrically neutral formula unit with the smallest set of whole-number subscripts. (a) The ions here are Ba^{2+} and S^{2-}. Since the charges are equal but opposite, a 1-to-1 ratio will give a neutral formula unit. Therefore, putting the positive ion first, the formula is BaS.
(b) For these elements the ions are Al^{3+} and Cl^-. We can obtain a neutral formula unit by combining one Al^{3+} with three Cl^-. (The charge on Cl is $1-$; the 1 is understood.)

$$1(3+) + 3(1-) = 0$$

The formula is $AlCl_3$.
(c) For these elements, the ions are Al^{3+} and O^{2-}. In the formula we seek there must be the same number of positive charges as negative charges. This number must be a whole-number multiple of both 3 and 2. The smallest number that satisfies this condition is 6, so there must be two Al^{3+} and three O^{2-} in the formula.

$$2(3+) = 6+$$
$$\underline{3(2-) = 6-}$$
$$\text{sum} = 0$$

The formula is Al_2O_3.

_____ **Practice Exercise 10** _____

Write formulas for ionic compounds formed from (a) Na and F, (b) Na and O, (c) Mg and F, and (d) Al and C. ◆

There is another rather simple way to obtain the formulas of the compounds in Example 2.3. The procedure is to make the subscript for one ion equal to the number of charges on the other. For example, for Al^{3+} and Cl^-, we can write

$$Al\,③^+ \quad \diagdown\!\!\!\diagup \quad Cl\,①^-$$

which gives Al_1Cl_3 or simply $AlCl_3$.
For the ions Al^{3+} and O^{2-} we write

$$Al\,③^+ \quad \diagdown\!\!\!\diagup \quad O\,②^-$$

This gives the formula Al_2O_3.
If you are not careful, you can be fooled in following this procedure. For example, if you apply this method to the compound formed from the ions Ba^{2+} and S^{2-}, it gives the formula Ba_2S_2. However, by convention we always choose the smallest whole-number ratio of ions (Rule 3 above). Notice in Ba_2S_2 that both subscripts are divisible by 2. Therefore, to obtain the correct formula we reduce the subscripts to the smallest set of whole numbers, which gives BaS. Exercising appropriate care, you might go back and try this method on Practice Exercise 10.

Many of our most important chemicals are ionic compounds. We have mentioned NaCl, common table salt, and $CaCl_2$, which is a substance often used to melt ice on walkways in the winter. Other examples are sodium fluoride, NaF, used by dentists to give fluoride treatments to teeth, and calcium oxide, CaO, an important ingredient in cement.

Transition Metals and Post-transition Metals

The transition elements are located in the center of the periodic table, from Group IIIB on the left to Group IIB on the right. All of them lie to the left of the metalloids, and they all are metals. Included here are some of our most familiar metals, including iron, chromium, copper, silver, and gold.

Most of the transition metals are much less reactive than the metals of Groups IA and IIA, but when they react they also transfer electrons to nonmetal atoms to form ionic compounds. However, the charges on the ions of the transition metals do not follow as straightforward a pattern as do those of the alkali and alkaline earth metals. One of the characteristic features of the transition metals is the ability of many of them to form more than one positive ion. Iron, for example, can form two different ions, Fe^{2+} and Fe^{3+}. This means that iron can form more than one compound with a given nonmetal. For example, with chloride ion, Cl^-, iron forms two compounds, with the formulas $FeCl_2$ and $FeCl_3$. With oxygen, we find the compounds FeO and Fe_2O_3. As usual, we see that the formulas contain the ions in a ratio that gives electrical neutrality. Some of the most common ions of the transition metals are given in Table 2.5.

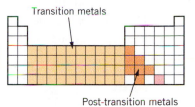

Distribution of transition and post-transition metals in the periodic table.

The prefix *post* means "after."

Practice Exercise 11

Write formulas for the chlorides and oxides formed by (a) chromium and (b) copper. ◆

The **post-transition metals** are those metals that occur in the periodic table immediately following a row of transition metals. The two most common and important ones are tin (Sn) and lead (Pb). These post-transition metals are quite different from the metals that precede the transition metals. One of the most significant differences is their ability to form two different ions, and therefore two different compounds with a given nonmetal. For example, tin forms two oxides, SnO and SnO_2. Lead also forms two oxides that have similar formulas (PbO and PbO_2). The ions that these metals form are also included in Table 2.5.

Compounds Containing Ions Composed of More than One Element

The metal compounds that we have discussed so far have been **binary compounds**—compounds formed between *two different elements*. There are many other ionic compounds that contain more than two elements. These substances usually contain **polyatomic ions,** which are ions that are themselves composed of two or more atoms linked by the same kinds of bonds that hold molecules together. Polyatomic ions differ from molecules, however, in that they contain either too many or too few electrons to make them electrically neutral. Table 2.6 lists some important polyatomic ions. The formulas of ionic compounds formed from them are determined in the same way as are those of binary ionic compounds; the ratio of the ions must be such that the formula unit is electrically neutral, and the smallest set of whole-number subscripts is used.

Table 2.5 Ions of Some Transition Metals and Post-transition Metals

Transition Metals	
Chromium	Cr^{2+}, Cr^{3+}
Manganese	Mn^{2+}, Mn^{3+}
Iron	Fe^{2+}, Fe^{3+}
Cobalt	Co^{2+}, Co^{3+}
Nickel	Ni^{2+}
Copper	Cu^+, Cu^{2+}
Zinc	Zn^{2+}
Silver	Ag^+
Cadmium	Cd^{2+}
Gold	Au^+, Au^{3+}
Mercury	Hg_2^{2+}, Hg^{2+}
Post-transition Metals	
Tin	Sn^{2+}, Sn^{4+}
Lead	Pb^{2+}, Pb^{4+}
Bismuth	Bi^{3+}

A substance is **diatomic** if it is composed of molecules that contain only two atoms. It is a **binary compound** if it contains two different elements, regardless of the number of each. Thus, BrCl is a binary compound and is also diatomic; CH_4 is a binary compound but is not diatomic.

Table 2.6 Formulas and Names of Some Polyatomic Ions

Ion	Name (Alternate Name in Parentheses)	Ion	Name (Alternate Name in Parentheses)
NH_4^+	ammonium ion	CO_3^{2-}	carbonate ion
H_3O^+	hydronium ion[a]	HCO_3^-	hydrogen carbonate ion (bicarbonate ion)[b]
OH^-	hydroxide ion	SO_3^{2-}	sulfite ion
CN^-	cyanide ion	HSO_3^-	hydrogen sulfite ion (bisulfite ion)[b]
NO_2^-	nitrite ion	SO_4^{2-}	sulfate ion
NO_3^-	nitrate ion	HSO_4^-	hydrogen sulfate ion (bisulfate ion)[b]
ClO^-	hypochlorite ion (often written OCl^-)	SCN^-	thiocyanate ion
ClO_2^-	chlorite ion	$S_2O_3^{2-}$	thiosulfate ion
ClO_3^-	chlorate ion	CrO_4^{2-}	chromate ion
ClO_4^-	perchlorate ion	$Cr_2O_7^{2-}$	dichromate ion
MnO_4^-	permanganate ion	PO_4^{3-}	phosphate ion
$C_2H_3O_2^-$	acetate ion	HPO_4^{2-}	monohydrogen phosphate ion
$C_2O_4^{2-}$	oxalate ion	$H_2PO_4^-$	dihydrogen phosphate ion

[a]You will only encounter this ion in aqueous solutions.

[b]You will often see and hear the alternate names for these ions.

EXAMPLE 2.4

Formulas That Contain Polyatomic Ions

One of the minerals responsible for the strength of bones is the ionic compound calcium phosphate, which is formed from Ca^{2+} and PO_4^{3-}. Write the formula for this compound.

Solution: As before, we write the positive ion first and then make the number of charges on one ion equal to the subscript for the other.

$$Ca^{②+} \qquad PO_4^{③-}$$

The formula is written with parentheses to show that the PO_4^{3-} ion occurs two times in the formula unit.

$$Ca_3(PO_4)_2$$

Practice Exercise 12

Write formulas for the ionic compound formed from (a) Na^+ and CO_3^{2-}, (b) NH_4^+ and SO_4^{2-}, (c) potassium ion and acetate ion, (d) strontium ion and nitrate ion, and (e) Fe^{3+} and acetate ion. ◆

In general, polyatomic ions are not formed in the direct combination of elements. They are the products of reactions between compounds.

Polyatomic ions are found in a large number of very important compounds. Some common ones are $CaSO_4$ (calcium sulfate, in plaster of Paris), $NaHCO_3$ (sodium bicarbonate, also called baking soda), $NaOCl$ (sodium hypochlorite, in liquid laundry bleach), $NaNO_2$ (sodium nitrite, a meat preservative), $MgSO_4$ (magnesium sulfate, also known as Epsom salts), and $NH_4H_2PO_4$ (ammonium dihydrogen phosphate, a fertilizer).

Properties of Ionic Compounds

The properties of ionic compounds reflect the way ions interact with each other. The attractive force between ions of opposite charge is very large, as is the repelling force between ions of like charge. Therefore, in an ionic compound the ions arrange themselves so that the attractions between oppositely charged ions are at a maximum and the repulsions between like-charged ions are at a minimum.

In a crystal at room temperature, the ions jiggle about within a cage of other ions. They are held too tightly to move away. When heat is added to raise the temperature of the crystal, the ions have more kinetic energy and neighboring ions bounce off each other more violently. Eventually a temperature is reached at which the violent motions overcome the attractions between the ions and the crystal collapses—the compound melts. However, because the net attractions between the ions are so large, the temperature required is very high. This is why all ionic compounds are solids at room temperature and tend to have high melting points.

You have probably never seen an ionic substance melt. Most of them melt well above room temperature. For example, ordinary table salt, NaCl, melts at 801 °C. Some ionic substances melt only at extremely high temperatures. For instance, aluminum oxide, Al_2O_3, melts at about 2000 °C, and for this reason it is made into special bricks that are used to line the inside walls of furnaces.

Another property of ionic compounds is that their solids are generally relatively hard and quite brittle. Molecular substances, such as paraffin wax or car wax, tend to be soft and easily crushed, but a crystal of rock salt is much harder. When struck by a hammer, however, the salt crystal shatters. The slight movement of a layer of ions within an ionic crystal suddenly places ions of the *same* charge next to each other, and for that instant there are large repulsive forces that split the solid, as illustrated in Figure 2.23.

Notice how the notion of temperature being related to the average kinetic energy of the particles within a substance helps us explain the high melting points of ionic compounds.

Electrical Properties

In the solid state, ionic compounds do not conduct electricity. This is because electrical conductivity requires the movement of electrical charges, and in the solid the attractive forces prevent the movement of ions through the crystal. When the solid is melted, however, the ions become free to move about and the liquid conducts electricity well.

An apparatus like that in Figure 2.24, which has a pair of electrodes that can be dipped into a container, can be used to test the electrical conductivity of a substance. One of the electrodes is wired to an electric lightbulb that will glow if electricity is able to flow. When this apparatus is used to test the electrical

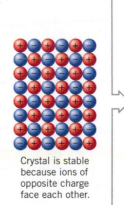

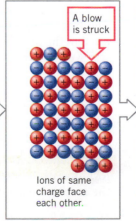

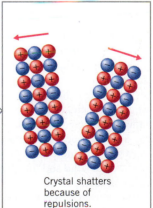

Crystal is stable because ions of opposite charge face each other.

Ions of same charge face each other.

A blow is struck

Crystal shatters because of repulsions.

Figure 2.23 *An ionic crystal shatters when struck.* In the photo on the left, we see what happens when a crystal of sodium chloride is struck with a hammer. In the micro-view, we see that striking the crystal causes some of the layers to shift. This can bring ions of like charge face-to-face. The repulsions between the ions can then force parts of the crystal apart, causing the crystal to shatter.

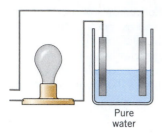

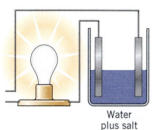

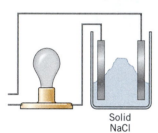

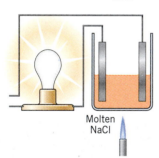

Figure 2.24 *An apparatus to test for electrical conductivity.* The electrodes are dipped into the substance to be tested. If the lightbulb glows when electricity is applied, the sample is an electrical conductor. Here we see that neither pure water (a molecular substance) nor solid sodium chloride conduct. Salt does conduct, however, if it is melted or dissolved in water.

conductivity of solid salt crystals, the bulb fails to light. However, when a flame is applied, the bulb lights brightly as soon as the salt melts.

The apparatus in Figure 2.24 can also be used to show that water solutions of ionic compounds conduct electricity. If the electrodes are dipped into pure distilled water, the bulb remains dark; pure water is not a conductor of electricity. If salt crystals are then added to the water and the mixture stirred, the solution that's formed conducts electricity well. The reason it conducts is that the ions become separated when the salt dissolves, and they are therefore free to move about. This freedom of movement of the charged ions permits the conduction of electricity[4] and has a profound influence on the reactions of ionic compounds.

> ▶**Chemistry in Practice**◀ The ability of electricity to make the ions in solutions of ionic substances move is used in one step of getting a DNA "fingerprint." The step uses a technique that causes ions with different ratios of mass to charge to move at different speeds. As a result, ions with different mass-to-charge ratios become separated, like the runners in a race. Later steps in the procedure find out how far the ions have traveled, and the results are displayed like a store's bar code. Each individual produces his or her own unique "bar code" that represents that person's DNA "fingerprint." ◆

2.10 Molecular Compounds and Their Properties

You've learned that nonmetals form ionic compounds when they combine with metals. However, nonmetals are also able to form compounds with other nonmetals, and such combinations yield molecular substances.

Although there are relatively few nonmetals, the number of molecular substances formed by them is huge. This is because of the variety of ways in which they combine as well as the varying degrees of complexity of their molecules. This reaches a maximum with compounds in which carbon is combined with a handful of other elements such as hydrogen, oxygen, and nitrogen. There are so many of these compounds, in fact, that their study encompasses the chemical specialties called organic chemistry and biochemistry.

Even in their elemental states, most nonmetals exist as molecules. The only ones that do not are the noble gases, which occur as simple uncombined atoms. The formulas of the nonmetals in Groups IVA through VIIA are shown in the table in the margin on page 79.

At this early stage we can only begin to look for signs of order among the many nonmetal–nonmetal compounds, so we will look only briefly at some simple compounds that the nonmetals form with hydrogen, as well as some simple compounds of carbon.

Compounds of Nonmetals with Hydrogen

Compounds that elements form with hydrogen are often called *hydrides.* The formulas of the simple hydrides of the nonmetals are given in Table 2.7.[5] As with the anions formed by the nonmetals, we can use the periodic table to help us re-

---

[4]This kind of conduction takes place by a different means than the conduction in metals, and is described in more detail in Chapter 19.

[5]Table 2.7 shows how the formulas are normally written. The order in which the hydrogens appear in the formula is not of concern to us now. Instead, we are interested in the *number* of hydrogens that combine with a given nonmetal.

Table 2.7 Simple Hydrogen Compounds of the Nonmetallic Elements

	Group			
Period	IVA	VA	VIA	VIIA
2	CH_4	NH_3	H_2O	HF
3	SiH_4	PH_3	H_2S	HCl
4	GeH_4	AsH_3	H_2Se	HBr
5		SbH_3	H_2Te	HI

Molecular Formulas of the Nonmetals

Group Number			
IVA	VA	VIA	VIIA
C^a	N_2	O_2	F_2
	P_4	S_8	Cl_2
	As_4	Se_8	Br_2
			I_2

^{a} Elemental carbon can be found in three forms: graphite, in which the crystals consist of flat sheets of huge numbers of carbon atoms linked in a two-dimensional network; diamond, whose crystals are formed of an interlocking three-dimensional network of carbon atoms; and fullerenes, whose large molecules, with formulas like C_{60}, resemble spherical cages of graphite.

member formulas of the simple hydrides: The number of hydrogens combined with the nonmetal atom equals the number of negative charges carried by the simple anion of the nonmetal. Thus, the anion of oxygen is O^{2-}, with *two* negative charges. The simple hydride of oxygen is water, H_2O, in which *two* hydrogens are combined with the oxygen.

Notice that the formulas of the simple hydrides are similar for nonmetals within a given group of the periodic table. If you know the formula for the hydride of the top member of the group, then you know the formulas of all of them in that group.

Many of the nonmetals form more complex compounds with hydrogen, but we will not discuss them here.

Compounds of Carbon

Among all the elements, carbon is unique in the variety of compounds it forms with elements such as hydrogen, oxygen, and nitrogen. As a consequence, the number and complexity of such compounds is enormous, and their study constitutes the major specialty called **organic chemistry.** The term "organic" here comes from an early belief that these compounds could only be made by living organisms. We now know this isn't true, but the name organic chemistry persists nonetheless.

The study of organic chemistry begins with **hydrocarbons** (compounds of carbon and hydrogen). The simplest hydrocarbon is methane, CH_4, which is a member of a series of hydrocarbons with the general formula C_nH_{2n+2}, where n is a whole number. The first six members of this series, called the **alkane** series, are given in Table 2.8 along with their boiling points. Molecules of methane, ethane, and propane are illustrated in Figure 2.25.

The alkanes are common compounds. They are the principal constituents of petroleum, and most of our useful fuels are produced from petroleum. Methane itself is the major component of natural gas. Gas-fired barbecues use propane as a fuel, and butane is the fuel in inexpensive cigarette lighters. Hydrocarbons with higher boiling points are found in gasoline, kerosene, paint thinners, diesel fuel, and even candle wax.

Table 2.8 Hydrocarbons Belonging to the Alkane Series

Compound	Name	Boiling Point (°C)
CH_4	Methane^a	-161.5
C_2H_6	Ethane^a	-88.6
C_3H_8	Propane^a	-42.1
C_4H_{10}	Butane^a	-0.5
C_5H_{12}	Pentane	36.1
C_6H_{14}	Hexane	68.7

^{a}Gases at room temperature (25 °C) and atmospheric pressure.

Propane is the fuel that cooks hamburgers and hot dogs on this gas-fired barbecue grill.

methane, CH$_4$ ethane, C$_2$H$_6$ propane, C$_3$H$_8$

Figure 2.25 *The first three members of the alkane series of hydrocarbons.* White atoms represent hydrogen and black atoms represent carbon in these space-filling models that demonstrate the shapes of the molecules.

The chemically correct names for ethylene and acetylene are ethene and ethyne, respectively. Naming organic compounds is discussed in Chapter 23.

Methanol is also known as wood alcohol. It is quite poisonous. Ethanol in high doses is also a poison.

Alkanes are not the only class of hydrocarbons. For example, there are three two-carbon hydrocarbons. In addition to ethane, C$_2$H$_6$, there are also ethylene, C$_2$H$_4$ (from which polyethylene is made), and acetylene, C$_2$H$_2$ (the fuel used in *acetylene* torches).

The hydrocarbons serve as the foundation for organic chemistry. Derived from hydrocarbons are various other classes of organic compounds. An example is the class of compounds called alcohols, in which the atoms OH replace a hydrogen in the hydrocarbon. Thus, *methanol*, CH$_3$OH (also called *methyl alcohol*), is related to methane, CH$_4$, by removing one H and replacing it with OH. Methanol is used as a fuel and as a raw material for making other organic chemicals. Another familiar alcohol is *ethanol* (also called *ethyl alcohol*), C$_2$H$_5$OH. Ethanol, also known as grain alcohol because it is obtained from the fermentation of grains, is in alcoholic beverages. It is also mixed with gasoline to give a fuel mixture known as gasohol.

Alcohols constitute just one kind of compound derived from hydrocarbons. We will discuss some others when you have learned more about how atoms bond to each other and about the structures of molecules.

Properties of Molecular Compounds

The properties of molecular substances usually differ markedly from those of ionic compounds. Within the individual molecules the atoms are held to each other very strongly, but between neighboring molecules the attractions are very weak. These weak attractions are responsible for many of the properties of molecular substances, just as the strong attractions between ions are responsible for many of the properties of ionic compounds. For example, molecular compounds such as water and candle wax tend to have low melting points. The molecules of these substances don't have to bounce around inside their crystals as violently as do the ions in an ionic crystal to overcome the attractive forces and become a liquid. Crystals of molecular compounds are also usually soft because the molecules easily slide past each other.

Molecular substances differ from ionic compounds in their electrical characteristics, too. Molecules are uncharged particles, so they do not conduct electricity in the solid state or when melted. Pure water, for example, is a poor conductor of electricity. In addition, most molecular substances will not conduct electricity when dissolved in water. For example, if the electrodes of the conductivity apparatus shown in Figure 2.24 are dipped into a solution of sugar in water, the lightbulb won't glow. Sugar molecules carry no electrical charge, so there are no electrical charges in the solution to provide conduction. As we will see later, however, there are certain kinds of molecules that *react* with water to give ions, and their solutions do conduct electricity.

So then why do you get electrocuted when in water?

2.11 Inorganic Chemical Nomenclature

In conversation, chemists rarely use formulas to describe compounds. Instead, names are used. For example, you already know that water is the name for the compound having the formula H_2O and that sodium chloride is the name of NaCl.

At one time there was no uniform procedure for assigning names to compounds, and those who discovered compounds used whatever method they wished. Without some sort of system, however, remembering names for the rapidly increasing number of compounds soon became impossible. The search for a solution led chemists around the world to agree on a systematic method for naming substances. By using this method we are able to write the correct formula for any particular compound given the correct name, and vice versa.

In this section we discuss the **nomenclature** (naming) of simple **inorganic compounds.** In general, these are substances that would *not* be considered to be derived from hydrocarbons such as methane (CH_4), ethane (C_2H_6), and other carbon–hydrogen compounds.

Binary Compounds Containing a Metal and a Nonmetal

Tools

Rules for naming binary compounds of metals and nonmetals

For ionic compounds formed from a metal and a nonmetal, the name of the cation is given first, followed by the name of the anion formed from the nonmetal. The latter is created by adding the suffix *-ide* to the stem of the name for the nonmetal. A familiar example is NaCl, sodium chloride. Table 2.9 lists some common *monatomic* (one-atom) negative ions and their names. Other examples of compounds formed from them are

CaO	calcium oxide
ZnS	zinc sulfide
Mg_3N_2	magnesium nitride

The *-ide* suffix is usually used only for monatomic ions, although there are two common exceptions—*hydroxide ion* (OH^-) and *cyanide ion* (CN^-).

EXAMPLE 2.5

Naming Compounds and Writing Formulas

(a) What is the name of $SrBr_2$? (b) What is the formula for aluminum selenide?

Solution: (a) The compound is composed of the ions Sr^{2+} and Br^-. The cation simply takes the name of the metal, which is strontium. The anion's name is derived from bromine by replacing *-ine* with *-ide;* it is the bromide ion. The name of the compound is strontium bromide.
(b) The name tells us that the cation is the aluminum ion, Al^{3+} and the anion is the selenide ion, which is formed from selenium. This ion is Se^{2-}. The correct formula must represent an electrically neutral formula unit, so the formula is Al_2Se_3.

Table 2.9 Monatomic Negative Ions

H^-	hydride	N^{3-}	nitride	O^{2-}	oxide	F^-	fluoride
C^{4-}	carbide	P^{3-}	phosphide	S^{2-}	sulfide	Cl^-	chloride
Si^{4-}	silicide	As^{3-}	arsenide	Se^{2-}	selenide	Br^-	bromide
				Te^{2-}	telluride	I^-	iodide

If a metal forms only one positive ion, the cation is simply specified by naming the metal, as illustrated in the preceding Example. This applies to the metals in Groups IA and IIA plus aluminum.

Many of the transition metals and post-transition metals are able to form more than one positive ion. Iron, a typical example, forms ions with either a 2+ or a 3+ charge (Fe^{2+} or Fe^{3+}). Compounds that contain these different iron ions have different formulas, so in their names it is necessary to specify which iron ion is present. There are two ways of doing this. In the old system, the suffix *-ous* is used to specify the ion with the lower charge and the suffix *-ic* is used to specify the ion with the higher charge. With this method, we use the Latin stem for elements whose symbols are derived from their Latin names. Some examples are

Fe^{2+}	ferrous ion	$FeCl_2$	ferrous chloride
Fe^{3+}	ferric ion	$FeCl_3$	ferric chloride
Cu^+	cuprous ion	$CuCl$	cuprous chloride
Cu^{2+}	cupric ion	$CuCl_2$	cupric chloride

Additional examples are given in Table 2.10. Notice that mercury is an exception; we use the English stem when naming its ions.

The currently preferred method for naming ions of metals that can have more than one charge in compounds is called the **Stock system.** Here we use the English name followed, *without a space,* by the numerical value of the charge written as a Roman numeral in parentheses.

Alfred Stock (1876–1946), a German inorganic chemist, was one of the first scientists to warn the public of the dangers of mercury poisoning.

Fe^{2+}	iron(II)	$FeCl_2$	iron(II) chloride
Fe^{3+}	iron(III)	$FeCl_3$	iron(III) chloride
Cr^{2+}	chromium(II)	CrS	chromium(II) sulfide
Cr^{3+}	chromium(III)	Cr_2S_3	chromium(III) sulfide

Copper(I) sulfate is Cu_2SO_4.
Copper(II) sulfate is $CuSO_4$.

Remember that the Roman numeral is the positive charge on the metal ion; it is not necessarily a subscript in the formula. You must figure out the formula from the ionic charges, as discussed in Section 2.9 and illustrated in the preceding example. Even though the Stock system is now preferred, some chemical companies still label bottles of chemicals using the old system. These old names also appear in the older scientific literature, which still holds much excellent data. Unfortunately, this means that you must know both systems.

Practice Exercise 13

Name the compounds K_2S, Mg_3P_2, $NiCl_2$, and Fe_2O_3. Use the Stock system where appropriate. ◆

Practice Exercise 14

Write formulas for (a) aluminum sulfide, (b) strontium fluoride, (c) titanium(IV) oxide, and (d) chromous bromide. ◆

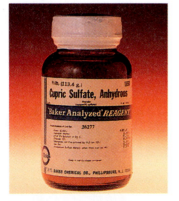

The older system of nomenclature is still found on the labels of many laboratory chemicals. This bottle contains copper(II) sulfate, which according to the older system is called cupric sulfate.

Table 2.10 Metals That Form More than One Ion

Cr^{2+}	chromous	Mn^{2+}	manganous	Fe^{2+}	ferrous	Cu^+	cuprous
Cr^{3+}	chromic	Mn^{3+}	manganic	Fe^{3+}	ferric	Cu^{2+}	cupric
Hg_2^{2+}	mercurous	Sn^{2+}	stannous	Pb^{2+}	plumbous	Co^{2+}	cobaltous
Hg^{2+}	mercuric	Sn^{4+}	stannic	Pb^{4+}	plumbic	Co^{3+}	cobaltic

Binary Compounds between Two Nonmetals

In naming binary compounds containing two nonmetals, we usually use a method that specifies the actual numbers of atoms in a molecule. This system makes use of the following Greek prefixes.

Rules for
naming binary
compounds of
two nonmetals

mono-	= 1 (often omitted)	hexa-	= 6
di-	= 2	hepta-	= 7
tri-	= 3	octa-	= 8
tetra-	= 4	nona-	= 9
penta-	= 5	deca-	= 10

For example, NO_2 is nitrogen *dioxide* and N_2O_4 is *dinitrogen tetraoxide*. (Sometimes we drop the *a* before an *o* for ease of pronunciation. N_2O_4 would then be named dinitrogen tetroxide.) Some other examples are

HCl	hydrogen chloride (mono- omitted)	$AsCl_3$	arsenic trichloride
CO	carbon monoxide	SF_6	sulfur hexafluoride

Practice Exercise 15

Name the following compounds using Greek prefixes when needed: PCl_3, SO_2, Cl_2O_7. ◆

Binary Acids and Their Salts

Rules for
naming binary
acids

Acids are substances that react with water to yield hydronium ions, H_3O^+, and anions. An example is hydrogen chloride, HCl. In the pure state, this compound is a gas and consists of molecules. When it dissolves in water, however, it reacts to yield ions according to the equation

$$HCl(g) + H_2O \longrightarrow H_3O^+(aq) + Cl^-(aq)$$

Thus, the acid transfers an H^+ ion to a water molecule to form H_3O^+.

The binary compounds of hydrogen with many of the nonmetals are acidic, and in their water solutions they are referred to as **binary acids.** Some other examples are HBr and H_2S. In naming these substances as acids, we add the prefix *hydro-* and the suffix *-ic* to the stem of the nonmetal name, followed by the word *acid*. For example, water solutions of hydrogen chloride and hydrogen sulfide are named as follows:

Name of the molecular compound		*Name of the binary acid*	
$HCl(g)$	hydrogen chloride	$HCl(aq)$	*hydro*chloric acid
$H_2S(g)$	hydrogen sulfide	$H_2S(aq)$	*hydro*sulfuric acid

Notice that the gaseous molecular substances are named in the usual way as binary compounds. *It is their aqueous solutions that are named as acids.*

One of the important reactions that acids undergo is called **neutralization.** This occurs when an acid reacts with a substance called a **base,** often a metal hydroxide such as NaOH. For example, aqueous HCl reacts with aqueous NaOH as follows:

$$HCl(aq) + NaOH(aq) \longrightarrow NaCl(aq) + H_2O$$

(a neutralization reaction)

In general, the reaction of an acid with a base produces water and an ionic compound, in this case, sodium chloride, or salt. This reaction is so general, in fact, that the term salt has taken on a broader meaning than just NaCl. We will use the word **salt** to mean any ionic compound that doesn't contain either OH^- or O^{2-} ion. (Compounds that contain these ions are bases.)

When acids are neutralized, the salt that is produced contains the anion formed by removing a hydrogen ion, H^+, from the acid molecule. Thus HCl yields salts containing the chloride ion, Cl^-. Similarly, HBr gives salts containing the bromide ion, Br^-.

Practice Exercise 16

Name the water solutions of the following acids: HF, HBr. Name the sodium salts formed by neutralizing these acids with NaOH. ◆

Tools ▶

Rules for naming oxoacids

Oxoacids and Their Salts

Acids that contain hydrogen, oxygen, plus another element are called **oxoacids.** Examples are H_2SO_4 and HNO_3. These acids do not take the prefix *hydro-*. Many nonmetals form two or more oxoacids that differ in the number of oxygen atoms in their formulas, and they are named according to the number of oxygens. When there are two oxoacids, the one with the larger number of oxygens takes the suffix *-ic* and the one with the fewer number of oxygens takes the suffix *-ous.*

H_2SO_4	sulfur*ic acid*	HNO_3	nitr*ic acid*
H_2SO_3	sulfur*ous acid*	HNO_2	nitr*ous acid*

The halogens can occur in as many as four different oxoacids. The oxoacid with the most oxygens has the prefix *per-*, and the one with the least has the prefix *hypo-*.

HClO	*hypo*chlor*ous acid* (usually written HOCl)	$HClO_3$	chlor*ic acid*
$HClO_2$	chlor*ous acid*	$HClO_4$	*per*chlor*ic acid*

The neutralization of oxoacids produces negative polyatomic ions. There is a very simple relationship between the name of the polyatomic ion and that of its parent acid.

(1) *-ic* acids give *-ate* anions: HNO_3 (nitr*ic acid*) $\longrightarrow NO_3^-$ (nitr*ate* ion)
(2) *-ous* acids give *-ite* anions: H_2SO_3 (sulfur*ous acid*) $\longrightarrow SO_3^{2-}$ (sulf*ite* ion)

In naming polyatomic anions, the prefixes *per-* and *hypo-* carry over from the name of the parent acid. Thus perchloric acid, $HClO_4$, gives perchlorate ion, ClO_4^-, and hypochlorous acid, HClO, gives hypochlorite ion, ClO^-.

Practice Exercise 17

The formula for arsenic acid is H_3AsO_4. What is the name of the salt Na_3AsO_4? ◆

Tools ▶

Rules for naming acid salts

Acid Salts

The hydrogen ion (H^+) given up by an acid is just a proton. A **proton** is all that's left after the single electron of hydrogen is removed from a neutral hydrogen atom.

In the formula of an acid such as HCl or H_2SO_4 the hydrogens that can be removed by neutralization are almost always written first. When the acid has just one hydrogen that can be neutralized, it is called a **monoprotic acid,** but if it contains more than one such hydrogen it is called a **polyprotic acid.** Neutralization of polyprotic acids occurs stepwise and can be halted before all the hydrogens have been removed. For example, if sulfuric acid is combined with sodium hydroxide in a ratio of one formula unit of acid to two formula units of base, then complete neutralization of the acid occurs and the SO_4^{2-} ion is formed.

$$H_2SO_4 + 2NaOH \longrightarrow Na_2SO_4 + 2H_2O$$

However, if the acid and base are combined in a one-to-one ratio, only half of the available hydrogens are neutralized.

$$H_2SO_4 + NaOH \longrightarrow NaHSO_4 + H_2O$$

The salt $NaHSO_4$, which can be isolated as crystals by evaporating the reaction mixture, is referred to as an **acid salt** because it contains the HSO_4^- ion which is still capable of furnishing additional H^+.

In naming acid salts formed from ions such as HSO_4^-, we specify the number of hydrogens that can still be neutralized if the salt were to be treated with additional base. For example,

$NaHSO_4$	sodium hydrogen sulfate
NaH_2PO_4	sodium dihydrogen phosphate

For acid salts of **diprotic acids** (acids capable of releasing two H^+), the prefix *bi-* is still often used.

$NaHCO_3$	sodium bicarbonate
	or
	sodium hydrogen carbonate

Notice that the prefix bi- does *not* mean "two"; it means that there is an acidic hydrogen in the compound.

Practice Exercise 18

What is the formula for sodium bisulfite? What is the chemically correct name for this compound? ◆

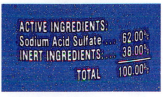

Many acid salts have useful applications. As its active ingredient, this familiar product contains sodium hydrogen sulfate (sodium bisulfate), which the manufacturer calls "sodium acid sulfate."

Common Names

Not every compound is named according to the systematic procedure described above. Many familiar substances were discovered long before a systematic method for naming them had been developed, and they acquired common names that are so well known that no attempt has been made to rename them. For example, following the scheme described above we might expect that H_2O would have the name dihydrogen oxide. Although this isn't wrong, the common name water is so well known that it is always used. Another example is ammonia, NH_3, whose odor you have no doubt experienced while using household ammonia solutions or the glass cleaner Windex.

Another class of compounds for which common names are often used is very complex substances. A common example is sucrose, which is the chemical name for table sugar, $C_{12}H_{22}O_{11}$. The structure of this compound is pretty complex, and its name assigned following the systematic method is equally complex. It is much easier to say the simple name sucrose, and be understood, than to struggle with the cumbersome systematic name for this common compound.

How to Name a Compound

In the preceding pages, we've described a number of sets of rules for naming different classes of compounds. When given a compound to name, therefore, your first task is to determine which of the sets of rules to apply. Figure 2.26 provides a systematic approach that will lead you to use the correct rules, and the example below illustrates how to use it. When working on the Review Questions, you may want to refer to Figure 2.26 until you are able to develop the skills that will enable you to work without it.

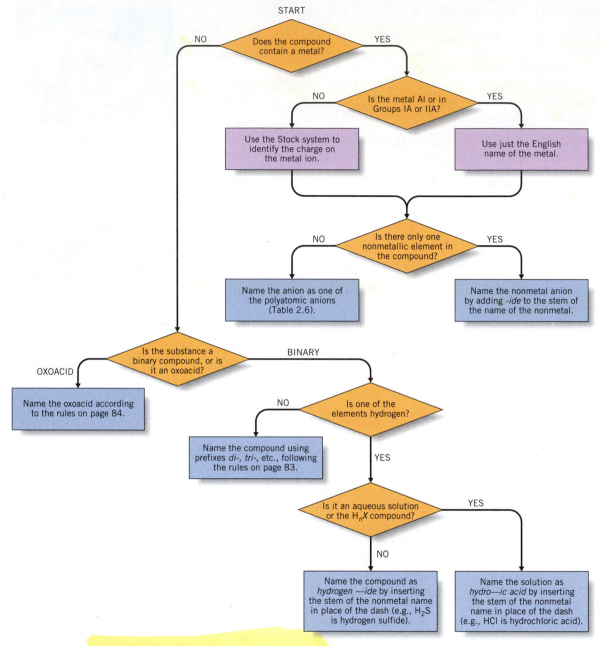

Figure 2.26 *Flowchart for naming inorganic compounds.*

EXAMPLE 2.4

Formulas That Contain Polyatomic Ions

What is the name of (a) $CrCl_3$ and (b) P_4S_3?

Analysis: For each compound, we will begin at the top of Figure 2.26 and follow the decision processes to arrive at the way to name the compound.

Solution: (a) Starting at the top of Figure 2.26, we first determine that the compound contains a metal. Next, we see that the metal is not a member of Group IA or IIA, so we have to apply the Stock method. The anion is Cl^-, so the metal ion must be Cr^{3+}, so we name the metal as *chromium(III)*. Next, we see that there is only one nonmetallic element in the compound, Cl, so the anion ends in -ide; it's the *chloride* ion. The compound is therefore named *chromium(III) chloride*.

(b) Once again, we start at the top. First we determine that the compound doesn't contain a metal. Next, we see that it is a binary compound (i.e., it contains only two elements); it's not an oxoacid. In the next step we determine that neither of the elements is hydrogen, so we are led to the decision that we must use Greek prefixes to specify the numbers of atoms of each element. Applying the procedure on page 83, the name of the compound is *tetraphosphorus trisulfide*.

Practice Exercise 19

What is the name of (a) $Zn(NO_3)_2$ and (b) AsF_3? ◆

SUMMARY

Elements, Compounds, and Mixtures. An **element** is a substance that cannot be decomposed into something simpler by a chemical reaction. In more modern terms, it is a substance whose atoms all have the same number of protons in their nuclei. Elements combine in fixed proportions to form **compounds.** Elements and compounds are **pure substances** that may be combined in *varying* proportions to give **mixtures.** A one-phase mixture is called a **solution** and is **homogeneous.** If a mixture consists of two or more phases it is **heterogeneous.** Formation or separation of a mixture into its components can be accomplished by a **physical change** in which the chemical properties of the components do not change. Formation or decomposition of a compound takes place by a **chemical change.**

Laws of Chemical Combination. When accurate masses of all the reactants and products in a reaction are measured and compared, no observable changes in mass accompany chemical reactions (the **law of conservation of mass**). The mass ratios of the elements in any compound are constant regardless of the source of the compound or how it is prepared (the **law of definite proportions**). Whenever two elements form more than one compound, then the different masses of one element that combine with a fixed mass of the other are in a ratio of small whole numbers (the **law of multiple proportions,** discovered after Dalton had proposed his theory)

Dalton's Atomic Theory. Dalton explained the laws of chemical combination by proposing that matter consists of indestructible atoms with masses that do not change during chemical reactions. During a chemical reaction, atoms may change partners, but they are neither created nor destroyed.

Symbols, Formulas, and Equations. Each element has an internationally agreed upon **chemical symbol.** Symbols are used in place of the name of an element and also to represent an atom of the element. Chemical symbols are used to write **formulas** for chemical compounds. Subscripts are used to specify how many atoms of each element are present. Some compounds form crystals, called **hydrates,** that contain water molecules in definite proportions. Heating a hydrate usually drives off the water. **Chemical equa-** tions present before-and-after descriptions of chemical reactions. When **balanced,** an equation contains **coefficients** that make the number of atoms of each kind the same among the **reactants** and the **products.**

Atomic Masses. An element's **atomic mass (atomic weight)** is the relative mass of its atoms on a scale in which atoms of carbon-12 have a mass of exactly 12 u (**atomic mass units**). Most elements consist of a small number of **isotopes** whose masses differ slightly. However, all isotopes of an element have very nearly identical chemical properties and the percentages of the isotopes that make up an element are generally so constant throughout the world that we can say that the average mass of their atoms is a constant.

Atomic Structure. Atoms can be split into **subatomic particles** such as **electrons, protons,** and **neutrons. Nucleons** are particles that make up the atomic **nucleus** and include the protons (charge = $1+$) and neutrons (no charge). The number of protons is called the **atomic number (Z)** of the element. Each element has a different atomic number. The electrons (each with a charge of $1-$) are found outside the nucleus; their number equals the atomic number in a neutral atom. Isotopes of an element have identical atomic numbers but different numbers of neutrons.

The Periodic Table. The search for similarities and differences among the properties of the elements led Mendeleev to discover that when the elements are placed in (approximate) order of increasing atomic mass, similar properties recur at regular, repeating intervals. In the modern **periodic table** the elements are arranged in rows, called **periods,** in order of increasing atomic number. The rows are stacked so that elements in the columns, called **groups** or **families,** have similar chemical and physical properties. The A-group elements (IUPAC Groups 1, 2, and 13–18) are called **representative elements;** the B-group elements (IUPAC Groups 3–12) are called **transition elements.** The two long rows of **inner transition elements** located below the main body of the table consist of the **lanthanides,** which follow La ($Z = 57$), and the **actinides,** which follow Ac ($Z = 89$). Certain groups are given family names: Group IA (Group 1), except for hydrogen, are the **alkali**

metals (the alkalis); Group IIA (Group 2), the **alkaline earth metals;** Group VIIA (Group 17), the **halogens;** Group VIIIA (Group 18), the **noble gases.**

Metals, Nonmetals, and Metalloids. Most elements are **metals;** they occupy the lower left-hand region of the periodic table (to the left of a line drawn approximately from boron, B, to astatine, At). **Nonmetals** are found in the upper right-hand region of the table. **Metalloids** occupy a narrow band between the metals and nonmetals.

Metals exhibit a **metallic luster,** tend to be **ductile** and **malleable,** and conduct electricity. They react with nonmetals to form ionic compounds called **salts.** Nonmetals tend to be brittle, lack metallic luster, and are nonconductors of electricity. Many nonmetals are gases. Besides combining with metals, nonmetals combine with each other to form molecules without undergoing electron transfer. Bromine (a nonmetal) and mercury (a metal) are the two elements that are liquids at ordinary room tem-perature. **Metalloids** have properties intermediate between those of metals and nonmetals and are **semiconductors** of electricity.

Molecular and Ionic Compounds. When atoms chemically combine to form compounds, the reactions produce either ions of opposite charge or neutral molecules, depending on the elements involved. When molecules form, the atoms are linked by the sharing of electrons. Molecules in **molecular compounds** carry no electrical charge. **Organic compounds** are **hydrocarbons** or compounds considered to be derived from hydrocarbons by replacing H atoms with other atoms.

When ionic **binary compounds** are formed, electrons are transferred from a metal to a nonmetal. The metal atom becomes a positive ion (a **cation**); the nonmetal atom becomes a negative ion (an **anion**). The formulas of ionic compounds are controlled by the requirement that the compound must be electrically neutral. Many ionic compounds also contain **polyatomic ions**—ions that are composed of two or more atoms. Ionic compounds tend to be brittle, high-melting, nonconducting solids. When melted or dissolved in water, however, they do conduct electricity. Most molecular compounds tend to be soft and melt at relatively low temperatures. Molecular compounds do not conduct electricity in either the solid state or when melted.

Naming Compounds. International agreement between chemists provides a system that allows us to write a single formula from a compound's name. For **salts** (ionic compounds not containing OH^- or O^{2-}), the **Stock system** is preferred, but the older system must also be learned. Compounds between nonmetals use Greek prefixes to specify number. **Acids** are substances that react with water to give H_3O^+ (hydronium ions). They react with **bases** such as NaOH in **neutralization** reactions that yield water and salts. Binary acids are *hydro . . . ic acids* and give *-ide* anions. **Oxoacids** and their anions are related: *-ic acid* produces *-ate* ion; *-ous acid* produces *-ite* ion. **Acid salts,** formed by the partial neutralization of **polyprotic acids,** contain acidic hydrogens; when named, they use the prefix *bi-* or contain the name, hydrogen.

The table below lists the tools that you have learned in this chapter. You will need them to answer chemistry questions. Review them if necessary, and refer to the tools when working on the Thinking-It-Through problems and the Review Questions that follow.

Tool	Function
Subscripts in a formula *(page 50)*	Specify the number of atoms of each element in a formula unit of a substance. You must be able to interpret the formula in terms of the numbers of atoms of each element expressed by the formula.
Rules for writing formulas of ionic compounds *(page 73)*	Permit us to write correct chemical formulas for ionic compounds. You will need to learn to use the periodic table to remember the charges on the cations and anions of the representative metals and nonmetals. You also should learn the ions formed by the transition and post-transition metals in Table 2.5, and it is essential that you learn the names and formulas of the polyatomic ions in Table 2.6.
Rules for naming compounds *(pages 81, 83, 84)*	Allow us to write a formula from a name, and a name from a formula.

THINKING IT THROUGH

The goal in the following problems is *not* to find the answer itself; instead, you are asked to assemble the available information needed to obtain the answer, state what additional data (if any) are needed, and describe *how* you would use the data to answer the question. Where problems can be solved by the factor-label method, set up the solution to the problem by arranging the conversion factors so the units cancel correctly to give the desired units of the answer.

The problems are divided into two groups. Those in Level 2 are more challenging than those in Level 1 and provide an opportunity to really hone your problem solving skills. Detailed answers to these problems can be found at our Web site: http://www.wiley.com/college/brady

Level 1 Problems

1. How many times heavier than an atom of ^{12}C is the average atom of iron? (Explain how you can obtain the answer.)

2. Suppose the atomic mass unit had been defined as $\frac{1}{10}$th of the mass of an atom of phosphorus. What would the atomic mass of carbon be on this scale?

3. How do we find the chemical symbol for a particle that has 43 electrons outside a nucleus which contains 45 protons and 58 neutrons?

4. How can you tell whether the element that has a nucleus which contains 49 protons is a metal, a nonmetal, or a metalloid?

5. Describe how you would determine the chemical formula for the compound formed between the element with atomic number 56 and the element with atomic number 35.

6. A certain compound formed by two elements is a gas at room temperature. When cooled to a very low temperature, it liquefies and then freezes to form a solid that is soft and easily crushed. How can you tell whether this compound is more likely to be molecular or ionic?

7. How do we know that the name *hydroselenic acid* refers to a binary acid?

8. What reasoning is involved in determining the formula for sodium bioxalate? Oxalic acid is $H_2C_2O_4$.

Level 2 Problems

9. Carbon forms a compound with element X that has the formula CX_4. A sample of this compound was analyzed and found to contain 1.50 g of C and 39.95 g of X. Explain how you would calculate the atomic mass of X.

10. In an experiment it was found that 3.50 g of phosphorus combines with 12.0 g of chlorine to form the compound PCl_3. How many grams of chlorine would combine with 8.50 g of phosphorus to form this same compound? (Set up the calculation.)

11. Aluminum combines with oxygen to form aluminum oxide. In an experiment a chemist found that 15.0 g of aluminum combined with 13.3 g of oxygen. If this experiment were repeated with 25.0 g of aluminum, how many grams of aluminum oxide would be formed? (Set up the calculations.)

12. When the element phosphorus burns in air it is found that 2.40 g of phosphorus combines with 2.89 g of oxygen to form the oxide P_2O_5. If this experiment were repeated with a mixture of 5.60 g of phosphorus and 8.00 g of oxygen, how many grams of P_2O_5 would be formed and how many grams of oxygen would remain unreacted? (Set up the calculations.)

13. Manganese forms several compounds with oxygen. In one of them it was found that 1.45 g of Mn was combined with 0.845 g of oxygen. In a sample of a different compound it was found that 3.26 g of Mn was combined with 1.42 g of oxygen. Set up the calculations that would show that these two compounds obey the law of multiple proportions.

14. Arsenic forms two compounds with sulfur. In 6.00 g of one of these compounds there was 3.65 g of As. In 8.00 g of the other compound there was 3.86 g of As. Explain in detail how you would show that these compounds exhibit the law of multiple proportions.

15. Shown below are six formulas representing possible compounds used in Problems 13 and 14. Find the answers to Problems 13 and 14 and then identify all possible pairs of compounds that are consistent with the results. In these formulas, A represents manganese or arsenic and B represents oxygen or sulfur. What concepts discussed in this chapter did you use to answer this problem?

AB_2	AB_3	AB_4	A_2B_5	A_2B_4	A_2B_5
(a)	(b)	(c)	(d)	(e)	(f)

16. Two compounds have formulas A_2B_6 and CD_3. One of these compounds is an ionic compound and one a molecular one. How can you tell from the formulas which is ionic? For the ionic compound, if the negative ion has a 1− charge, what is the charge of the positive ion? Suggest two ions that might form this compound and draw a sketch to represent one formula unit of this compound. How would a sketch representing the molecular compound differ from the one for the ionic compound?

REVIEW QUESTIONS

Elements, Compounds, and Mixtures

2.1 Define (a) element, (b) compound, (c) mixture, (d) homogeneous, (e) heterogeneous, (f) phase, (g) solution, and (h) physical change.

2.2 What kind of change is needed to separate a compound into its elements?

2.3 In places like Saudi Arabia, freshwater is scarce and is recovered from seawater. When seawater is boiled, the water evaporates and the steam can be condensed to give pure water that people can drink. If all the water is evaporated, solid salt is left behind. Are the changes described here chemical or physical?

Laws of Chemical Combination and Dalton's Theory

2.4 Name and state the three laws of chemical combination. Which law was not discovered until after Dalton had proposed his theory?

2.5 Which postulate of Dalton's theory is based on the law of conservation of mass? Which is based on the law of definite proportions?

2.6 Why didn't the existence of isotopes affect the apparent validity of the atomic theory?

2.7 In your own words, describe how Dalton's theory explains the law of conservation of mass and the law of definite proportions.

2.8 Which of the laws of chemical combination is used to define the term *compound*?

Chemical Symbols

2.9 What is the chemical symbol for each of the following elements? (a) chlorine, (b) sulfur, (c) iron, (d) silver, (e) sodium, (f) phosphorus, (g) iodine, (h) copper, (i) mercury, (j) calcium.

2.10 What is the name of each of the following elements? (a) K, (b) Zn, (c) Si, (d) Sn, (e) Mn, (f) Mg, (g) Ni, (h) Al, (i) C, (j) N.

2.11 What are two ways to interpret a chemical symbol?

Chemical Formulas

2.12 What is the difference between an atom and a molecule?

2.13 How many atoms of each element are represented in each of the following formulas? (a) $K_2C_2O_4$, (b) H_2SO_3, (c) $C_{12}H_{26}$, (d) $HC_2H_3O_2$, (e) $(NH_4)_2HPO_4$, (f) $(CH_3)_3COH$.

2.14 How many atoms of each kind are represented in the following formulas? (a) Na_3PO_4, (b) $Ca(H_2PO_4)_2$, (c) C_4H_{10}, (d) $Fe_3(AsO_4)_2$, (e) $Cu(NO_3)_2$, (f) $MgSO_4 \cdot 7H_2O$.

2.15 How many atoms of each kind are represented in the following formulas? (a) $Ni(ClO_4)_2$, (b) $CuCO_3$, (c) $K_2Cr_2O_7$, (d) CH_3CO_2H, (e) $(NH_4)_2HPO_4$, (f) $C_3H_5(OH)_3$.

2.16 Asbestos, known to cause cancer, has as a typical formula, $Ca_3Mg_5(Si_4O_{11})_2(OH)_2$. How many atoms of each element are given in this formula?

2.17 Write the formulas and names of the elements that exist in nature as diatomic molecules.

Chemical Equations

2.18 The combustion of a thin wire of magnesium metal (Mg) in an atmosphere of pure oxygen produces the brilliant light of a flashbulb. After the reaction, a thin film of magnesium oxide is seen on the inside of the bulb. The equation for the reaction is

$$2Mg + O_2 \longrightarrow 2MgO$$

(a) State in words how this equation is read.
(b) Give the formula(s) of the reactants.
(c) Give the formula(s) of the products.
(d) Rewrite the equation to show that Mg and MgO are solids and O_2 is a gas.

2.19 What are the numbers called that are written in front of the formulas in a balanced chemical equation?

2.20 How many atoms of each element are represented in each of the following expressions?
(a) $3N_2O$ (c) $4NaHCO_3$ (e) $2CuSO_4 \cdot 5H_2O$
(b) $7CH_3CO_2H$ (d) $2(NH_2)_2CO$ (f) $5K_2Cr_2O_7$

2.21 Consider the balanced equation

$$2Fe(NO_3)_3 + 3Na_2CO_3 \longrightarrow Fe_2(CO_3)_3 + 6NaNO_3$$

(a) How many atoms of Na are on each side of the equation?
(b) How many atoms of C are on each side of the equation?
(c) How many atoms of O are on each side of the equation?

2.22 Consider the balanced equation for the combustion of octane, a component of gasoline:

$$2C_8H_{18} + 25O_2 \longrightarrow 16CO_2 + 18H_2O$$

(a) How many atoms of C are on each side of the equation?
(b) How many atoms of H are on each side of the equation?
(c) How many atoms of O are on each side of the equation?

2.23 Rewrite the chemical equation in the preceding question so that it specifies gasoline and water as liquids and oxygen and carbon dioxide (CO_2) as gases.

Atomic Masses and Atomic Structure

2.24 Write the symbol for the isotope that forms the basis of the atomic mass scale. What is the mass of this atom expressed in atomic mass units?

2.25 What are the names, symbols, electrical charges, and masses (expressed in u) of the three subatomic particles introduced in this chapter?

2.26 Where is nearly all of the mass of an atom located? Explain your answer in terms of the particles that contribute to this mass.

2.27 What is a *nucleon?* Which ones have we studied?

2.28 Define the terms *atomic number* and *mass number.*

2.29 Which is better related to the chemistry of an element, its mass number or its atomic number? Give a brief explanation in terms of the basis for the periodic table.

2.30 In terms of the structures of atoms, how are isotopes of the same element alike? How do they differ?

2.31 Is it possible for atoms of two different elements to have the same mass number? Explain.

2.32 Consider the symbol $_b^a X_d^c$, where X stands for the chemical symbol for an element. What information is given in locations (a) a, (b) b, (c) c, and (d) d?

2.33 Write the symbols of the isotopes that contain the following. (Use the table of atomic masses and numbers printed inside the front cover for additional information, as needed.)
(a) An isotope of iodine whose atoms have 78 neutrons.
(b) An isotope of strontium whose atoms have 52 neutrons.
(c) An isotope of cesium whose atoms have 82 neutrons.
(d) An isotope of fluorine whose atoms have 9 neutrons.

The Periodic Table

2.34 What is the general formula for the compounds formed by Li, Na, K, Rb, and Cs with chlorine? What is the general formula for the compounds formed by Be, Mg, Ca, Sr, and Ba with chlorine? How did this kind of information lead Mendeleev to develop his periodic table?

2.35 On what basis did Mendeleev construct his periodic table? On what basis are the elements arranged in the modern periodic table?

2.36 In the periodic table, what is a "period"? What is a "group"?

2.37 Why were there gaps in Mendeleev's periodic table?

2.38 In the text, we identified two places in the modern periodic table where the atomic mass order was reversed. Using the table on the inside front cover, locate two other places where this occurs.

2.39 Below are some data for the elements sulfur and tellurium.

	Sulfur	Tellurium
Atomic mass	32.06	127.60
Melting point (°C)	112.8	449.5
Boiling point (°C)	445	990
Formula of oxide	SO_2	TeO_2
Melting point of oxide (°C)	−72.7	733
Density (g/cm³)	2.07	6.25

(a) Had the element selenium not been known in the time of Mendeleev, he would have called it eka-sulfur. Estimate its properties as an average of those of sulfur and tellurium.
(b) In the *Handbook of Chemistry and Physics* (located in your school library), look up the values for the properties of selenium and its oxide and compare them with the values you obtained in (a).

2.40 On the basis of their positions in the periodic table, why is it not surprising that strontium-90, a dangerous radioactive isotope of strontium, replaces calcium in newly formed bones?

2.41 In the refining of copper, sizable amounts of silver and gold are recovered. Why is this not surprising?

2.42 Why would you reasonably expect cadmium to be a contaminant in zinc but not in silver?

2.43 Make a rough sketch of the periodic table and mark off those areas where you would find (a) the representative elements, (b) the transition elements, and (c) the inner transition elements.

2.44 What group numbers are used to designate the representative elements following (a) the traditional U.S. system for designating groups and (b) the IUPAC system?

2.45 Supply the IUPAC group numbers that correspond to the following customary U.S. designations: (a) Group IA, (b) Group VIIA, (c) Group IIIB, (d) Group IB, (e) Group IVA.

2.46 Based on discussions in this chapter, explain why it is unlikely that scientists will discover a new element, never before observed, having an atomic weight of approximately 73.

2.47 Which of the following is
(a) an alkali metal? Ca, Cu, In, Li, S.
(b) a halogen? Ce, Hg, Si, O, I.
(c) a transition element? Pb, W, Ca, Cs, P.
(d) a noble gas? Xe, Se, H, Sr, Zr.
(e) a lanthanide element? Th, Sm, Ba, F, Sb.
(f) an actinide element? Ho, Mn, Pu, At, Na.
(g) an alkaline earth metal? Mg, Fe, K, Cl, Ni.

Physical Properties of Metals, Nonmetals, and Metalloids

2.48 Give five physical properties that we usually observe for metals.

2.49 Why is mercury used in thermometers? Why is tungsten used in lightbulbs?

2.50 What property of metals allows them to be drawn into wire?

2.51 Gold can be hammered into sheets so thin that some light can pass through them. What property of gold allows such thin sheets to be made?

2.52 Only two metals are colored (the rest are "white," like iron or lead). You have surely seen both of them. Which metals are they?

2.53 Which nonmetals occur as monatomic gases (gases whose particles consist of single atoms)?

2.54 Which nonmetals occur in nature as diatomic molecules? Which are gases?

2.55 Which two elements exist as liquids at room temperature and pressure?

2.56 Which physical property of metalloids distinguishes them from metals and nonmetals?

2.57 Sketch the shape of the periodic table and mark off those areas where we find (a) metals, (b) nonmetals, and (c) metalloids.

2.58 Which metals can you think of that are commonly used to make jewelry? Why isn't iron used to make jewelry?

Ionic Compounds

2.59 Describe what kind of event must occur (involving electrons) if the atoms of two different elements are to react to form (a) an ionic compound or (b) a molecular compound.

2.60 With what kind of elements do metals react?

2.61 In what major way do compounds formed between two nonmetals differ from those formed between a metal and a nonmetal?

2.62 Why are nonmetals found in more compounds than are metals, even though there are fewer nonmetals than metals?

2.63 What is an ion? How does it differ from an atom or a molecule?

2.64 Consider the sodium atom and the sodium ion.
(a) Write the chemical symbol of each.
(b) Do these particles have the same number of nuclei?
(c) Do they have the same number of protons?
(d) Could they have different numbers of neutrons?
(e) Do they have the same number of electrons?

2.65 What holds an ionic compound together? Can we identify individual molecules in an ionic compound?

2.66 Why do we use the term *formula unit* for ionic compounds instead of the term *molecule*?

2.67 Define *cation* and *anion*.

2.68 When calcium reacts with chlorine, each calcium atom loses two electrons and each chlorine atom gains one electron. What are the formulas of the ions created in this reaction?

2.69 How many electrons has a titanium atom lost if it has formed the ion Ti^{4+}? What are the total numbers of protons and electrons in a Ti^{4+} ion?

2.70 If an atom gains an electron to become an ion, what kind of electrical charge does the ion have?

2.71 How many electrons has a nitrogen atom gained if it has formed the ion N^{3-}? How many protons and electrons are in an N^{3-} ion?

2.72 What three rules do we apply when we write the formulas for ionic compounds?

2.73 What is wrong with the formula $RbCl_3$? What is wrong with the formula SNa_2?

2.74 A student wrote the formula for an ionic compound of titanium as Ti_2O_4. What is wrong with this formula? What should the formula be?

2.75 The formula for a compound is correctly given as C_4H_{10}. How do we know that this is a molecular compound and not an ionic compound?

2.76 Use the periodic table, but not Table 2.4, to write the symbols for the ions of (a) K, (b) Br, (c) Mg, (d) S, and (e) Al.

2.77 Use the periodic table, but not Table 2.4, to write the symbols for ions of (a) barium, (b) oxygen, (c) fluorine, (d) strontium, and (e) rubidium.

2.78 Which are the post-transition metals? Write their symbols.

2.79 What are the formulas of the ions formed by (a) iron, (b) cobalt, (c) mercury, (d) chromium, (e) tin, (f) lead, and (g) copper?

2.80 Write formulas for ionic compounds formed between (a) Na and Br, (b) K and I, (c) Ba and O, (d) Mg and Br, and (e) Ba and F.

2.81 Which of the following formulas are incorrect? (a) NaO_2, (b) $RbCl$, (c) K_2S, (d) Al_2Cl_3, (e) MgO_2.

2.82 What are the formulas for (a) cyanide ion, (b) ammonium ion, (c) nitrate ion, (d) sulfite ion, and (e) chlorate ion?

2.83 What are the formulas for (a) hypochlorite ion, (b) bisulfate ion, (c) phosphate ion, (d) dihydrogen phosphate ion, and (e) permanganate ion?

2.84 What are the names of the following ions? (a) $Cr_2O_7^{2-}$, (b) OH^-, (c) $C_2H_3O_2^-$, (d) CO_3^{2-}, (e) CN^-, and (f) ClO_4^-.

2.85 Write formulas for the ionic compounds formed from (a) K^+ and nitrate ion, (b) Ca^{2+} and acetate ion, (c) ammonium ion and Cl^-, (d) Fe^{3+} and carbonate ion, and (e) Mg^{2+} and phosphate ion.

2.86 Write formulas for the ionic compounds formed from (a) Zn^{2+} and hydroxide ion, (b) Ag^+ and chromate ion, (c) Ba^{2+} and sulfite ion, (d) Rb^+ and sulfate ion, and (e) Li^+ and bicarbonate ion.

2.87 Write formulas for two compounds formed between O^{2-} and (a) lead, (b) tin, (c) manganese, (d) iron, and (e) copper.

2.88 Write formulas for the ionic compounds formed from Cl^- and (a) cadmium ion, (b) silver ion, (c) zinc ion, and (d) nickel ion.

2.89 From what you have learned in Section 2.9, write correct balanced equations for the reactions between (a) calcium and chlorine, (b) magnesium and oxygen, (c) aluminum and oxygen, and (d) sodium and sulfur.

Molecular Compounds of Nonmetals

2.90 Which are the only elements that exist as free, individual atoms when not chemically combined with other elements?

2.91 What are the chemical formulas for the molecules formed by the following nonmetals in their elemental states? (a) chlorine (b) sulfur, (c) phosphorus, (d) nitrogen, (e) oxygen, (f) hydrogen.

2.92 Without referring to Table 2.7, but using the periodic table, write chemical formulas for the simplest hydrogen compounds of (a) carbon, (b) nitrogen, (c) tellurium, and (d) iodine.

2.93 Astatine forms a compound with hydrogen. Predict its chemical formula.

2.94 Under appropriate conditions, tin can be made to form a simple molecular compound with hydrogen. Predict its formula.

2.95 What are the formulas of (a) methanol and (b) ethanol?

2.96 What is the formula for the alkane that has 10 carbon atoms?

2.97 Candle wax is a mixture of hydrocarbons, one of which is an alkane with 23 carbon atoms. What is the formula for this hydrocarbon?

Properties of Ionic and Molecular Compounds

2.98 Compare the properties of ionic and molecular compounds.

2.99 Why does molten KCl conduct electricity even though solid KCl does not?

2.100 What happens when an ionic compound dissolves in water that allows its solution to conduct electricity?

2.101 Why are ionic compounds so brittle?

2.102 Why do most ionic compounds have high melting points? Naphthalene (moth flakes) consists of soft crystals that melt at a relatively low temperature of 80 °C. Does this substance have characteristics of an ionic or a molecular compound?

Naming Compounds

2.103 What is the difference between a binary compound and one that is diatomic? Give examples that illustrate this difference.

2.104 Name the following.
(a) CaS   (c) $AlBr_3$ (e) Na_3P (g) Ba_3As_2
(b) NaF   (d) Mg_2C (f) Li_3N (h) Al_2O_3

2.105 Name the following, using both the old nomenclature system and the Stock system.
(a) $CrCl_3$ (c) CuO   (e) SnO_2 (g) $CoCl_2$
(b) Mn_2O_3 (d) Hg_2Cl_2 (f) PbS   (h) FeS

2.106 Name the following.
(a) SiO_2 (c) XeF_4 (e) P_4O_{10} (g) Cl_2O_7
(b) ClF_3 (d) S_2Cl_2 (f) N_2O_5 (h) $AsCl_5$

2.107 Define *acid, base, neutralization, polyprotic acid, monoprotic acid,* and *salt.* What is the name of the ion H_3O^+?

2.108 Iodine, like chlorine, forms several acids. What are the names of the following?
(a) HIO_4 (c) HIO_2 (e) HI
(b) HIO_3 (d) HIO

2.109 For the acids in the preceding question, (a) write the formulas and (b) name the ions formed by removing a hydrogen ion (H^+) from each acid.

2.110 Name the following. If necessary, refer to Table 2.6 on page 76.
(a) $NaNO_2$ (d) $NH_4C_2H_3O_2$ (g) KSCN
(b) K_3PO_4 (e) $BaSO_4$ (h) $Na_2S_2O_3$
(c) $KMnO_4$ (f) $Fe_2(CO_3)_3$

2.111 Write the formula for (a) chromic acid, (b) carbonic acid, and (c) oxalic acid. (*Hint:* Check the table of polyatomic ions.)

2.112 Name the following:
(a) $CrCl_2$ (h) $H_2Se(g)$ (n) $GeBr_4$
(b) NH_4NO_2 (i) $H_2Se(aq)$ (o) K_2CrO_4
(c) KIO_3 (j) $V(NO_3)_3$ (p) $Fe(OH)_2$
(d) $HClO_2$ (k) $Co(C_2H_3O_2)_2$ (q) I_2O_4
(e) $CaSO_3$ (l) Au_2S_3 (r) I_4O_9
(f) AgCN   (m) Au_2S (s) P_4Se_3
(g) $ZnBr_2$

2.113 Write formulas for the following.
(a) sodium monohydrogen phosphate
(b) lithium selenide
(c) sodium hydride
(d) chromic acetate
(e) nickel(II) cyanide
(f) iron(III) oxide
(g) stannic sulfide
(h) antimony pentafluoride
(i) dialuminum hexachloride
(j) tetraarsenic decaoxide
(k) magnesium hydroxide
(l) cupric bisulfate
(m) ammonium thiocyanate
(n) potassium thiosulfate

2.114 Write formulas for the following.
(a) ammonium sulfide
(b) chromium(III) sulfate
(e) iron(III) oxide
(f) calcium bromate
(g) mercury(II) acetate
(h) barium bisulfite
(i) silicon tetrafluoride
(j) boron trichloride
(c) molybdenum(IV) sulfide
(d) tin(IV) chloride
(k) stannous sulfide
(l) calcium phosphide
(m) magnesium dihydrogen phosphate
(n) calcium oxalate

2.115 Write formulas for the following:
(a) gold(III) nitrate
(b) copper(II) sulfate
(c) ammonium bromate
(d) nickel(II) iodate
(e) lead(IV) oxide
(f) antimony(III) sulfide
(g) platinum(II) chloride
(h) cadmium sulfide
(i) barium acetate
(j) mercury(I) chloride
(k) mercury(II) chloride
(l) strontium phosphate
(m) barium arsenide
(n) cobaltous hydroxide
(o) aluminum sulfite
(p) ammonium dichromate
(q) diiodine pentaoxide
(r) tetraphosphorus heptasulfide
(s) disulfur decafluoride

2.116 Name the following acid salts:
(a) $NaHCO_3$ (b) KH_2PO_4 (c) $(NH_4)_2HPO_4$

REVIEW PROBLEMS

Answers to problems whose numbers are printed in color are given in Appendix D.
More challenging questions are marked with asterisks.

Laws of Chemical Combination

2.117 Laughing gas is a compound formed from nitrogen and oxygen in which there are 1.75 g of nitrogen for every 1.00 g of oxygen. Below are given the compositions of several nitrogen–oxygen compounds. Which of these is laughing gas?
(a) 6.35 g nitrogen, 7.26 g oxygen
(b) 4.63 g nitrogen, 10.58 g oxygen
(c) 8.84 g nitrogen, 5.05 g oxygen
(d) 9.62 g nitrogen, 16.5 g oxygen
(e) 14.3 g nitrogen, 40.9 g oxygen

2.118 One of the substances used to melt ice on sidewalks and roadways in cold climates is calcium chloride. In this compound calcium and chlorine are combined in a ratio of 1.00 g of calcium to 1.77 g of chlorine. Which of the following calcium–chlorine mixtures will give calcium chloride with no calcium or chlorine left over after the reaction is complete?
(a) 3.65 g calcium, 4.13 g chlorine
(b) 0.856 g calcium, 1.56 g chlorine
(c) 2.45 g calcium, 4.57 g chlorine
(d) 1.35 g calcium, 2.39 g chlorine
(e) 5.64 g calcium, 9.12 g chlorine

2.119 Ammonia is composed of hydrogen and nitrogen in a ratio of 9.33 g of nitrogen to 2.00 g of hydrogen. If a sample of ammonia contains 6.28 g of hydrogen, how many grams of nitrogen does it contain?

2.120 A compound of phosphorus and chlorine used in the manufacture of a flame-retardant treatment for fabrics contains 1.20 grams of phosphorus for every 4.12 g of chlorine. Suppose a sample of this compound contains 6.22 g of chlorine. How many grams of phosphorus does it contain?

2.121 Refer to the data about ammonia in Problem 2.119. If 4.56 g of nitrogen combined completely with hydrogen to form ammonia, how many grams of ammonia would be formed?

2.122 Refer to the data about the phosphorus–chlorine compound in Problem 2.120. If 12.5 g of phosphorus combined completely with chlorine to form this compound, how many grams of the compound would be formed?

2.123 A compound of nitrogen and oxygen has the formula NO. In this compound there are 1.143 g of oxygen for each 1.000 g of nitrogen. A different compound of nitrogen and oxygen has the formula NO_2. How many grams of oxygen would be combined with each 1.000 g of nitrogen in NO_2?

2.124 Tin forms two compounds with chlorine, $SnCl_2$ and $SnCl_4$. When combined with the same mass of tin, what would be the ratio of the masses of chlorine in the two compounds? In the compound $SnCl_2$, 0.597 g of chlorine is combined with each 1.000 g of tin. In $SnCl_4$, how many grams of chlorine would be combined with 1.000 g of tin?

Atomic Masses and Isotopes

2.125 The mass in grams of one atomic mass unit is $1.6605402 \times 10^{-24}$ g. Using this value, calculate the mass in grams of one atom of carbon-12.

2.126 Use the mass in grams of the atomic mass unit given in the preceding problem to calculate the average mass of one atom of sodium.

2.127 In the compound CH_4, 0.33597 g of hydrogen is combined with 1.0000 g of carbon-12. Use this information to calculate the atomic mass of the element hydrogen.

2.128 A certain element X forms a compound with oxygen that has the formula X_2O_3. In this compound, 1.125 g of X are combined with 1.000 g of oxygen. Use the average atomic mass of oxygen to calculate the average atomic mass of X. Identify the element.

2.129 If an atom of carbon-12 had been assigned a relative mass of 24.0000 u, what would be the average atomic mass of hydrogen relative to this mass?

2.130 One atom of ^{109}Ag has a mass that is 9.0754 times that of a ^{12}C atom. What is the atomic mass of this isotope of silver expressed in atomic mass units?

2.131 Naturally occurring copper is composed of 69.17% of ^{63}Cu, with an atomic mass of 62.9396 u, and 30.83% of ^{65}Cu, with an atomic mass of 64.9278 u. Use these data to calculate the average atomic mass of copper.

2.132 Naturally occurring magnesium is composed of 78.99% of ^{24}Mg (atomic mass, 23.9850 u), 10.00% of ^{25}Mg (atomic mass, 24.9858 u), and 11.01% of ^{26}Mg (atomic mass, 25.9826 u). Use these data to calculate the average atomic mass of magnesium.

2.133 Give the numbers of neutrons, protons, and electrons in the atoms of each of the following isotopes. (Use the table of atomic masses and numbers printed inside the front cover for additional information, as needed.)
(a) radium-226   (b) carbon-14   (c) $^{206}_{82}$Pb   (d) $^{23}_{11}$Na

2.134 Give the numbers of electrons, protons, and neutrons in the atoms of each of the following isotopes. (As necessary, consult the table of atomic masses and numbers printed inside the front cover.)
(a) cesium-137   (c) $^{238}_{92}$U
(b) iodine-131   (d) $^{197}_{79}$Au

2.135 How many electrons, protons, and neutrons are in each of the following particles? (a) $^{81}_{35}Br^-$ (b) $^{58}_{26}Fe^{3+}$ (c) $^{63}_{29}Cu^{2+}$ (d) $^{87}_{37}Rb^+$

2.136 How many electrons, protons, and neutrons are in each of the following particles? (a) $^{39}_{20}Ca^{2+}$ (b) $^{7}_{3}Li^+$ (c) $^{32}_{16}S^{2-}$ (d) $^{112}_{47}Ag^+$

ADDITIONAL EXERCISES

2.137 An element has 28 protons in its nucleus. When this element reacts with chlorine, is the compound formed likely to be ionic or molecular? Justify your answer.

*⁎**2.138** Elements X and Y form the compound XY_4. When these elements react, it is found that 1.00 g of X combines with 5.07 g of Y. When X combines with oxygen, it forms the compound XO_2 in which 1.00 g of X combines with 1.14 g of O. What is the atomic mass of Y?

2.139 An iron nail is composed of four isotopes with the percentage abundances and atomic masses given in the table below. Calculate the average atomic mass of iron.

Isotope	Percentage Abundance	Atomic Mass (u)
^{54}Fe	5.80	53.9396
^{56}Fe	91.72	55.9349
^{57}Fe	2.20	56.9354
^{58}Fe	0.28	57.9333

*⁎**2.140** Naturally occurring bromine is composed of two isotopes, ^{79}Br with a mass of 78.9183 u and ^{81}Br with a mass of 80.9163 u. Use this information and the average atomic mass of bromine given in the table on the inside front cover of the book to calculate the percentage abundances of these two isotopes.

2.141 Write balanced chemical equations for the complete neutralization of H_2SO_4 by (a) $Al(OH)_3$ and (b) $Ca(OH)_2$.

2.142 Write the formulas for all the acid salts that could be formed from the reaction of NaOH with the acid H_3PO_4.

2.143 Name the following oxoacids and give the names and formulas of the salts formed from them by neutralization with NaOH: (a) HOCl, (b) HIO_2, (c) $HBrO_3$, (d) $HClO_4$.

2.144 Why is the formula for mercury(I) chloride written as Hg_2Cl_2 instead of HgCl?

2.145 One atomic mass unit has a mass of $1.6605402 \times 10^{-24}$ g. Calculate the mass, in grams, of one atom of magnesium. What is the mass of one atom of iron, expressed in grams? Use these two answers to determine how many atoms of Mg are in 24.305 g of magnesium and how many atoms of Fe are in 55.847 g of iron. Compare your answers. How many atoms do you think would be in 40.078 g of calcium? What conclusions can you draw from the results of these calculations?

*⁎**2.146** The main propulsion system that lifts the space shuttle during takeoff uses the reaction between hydrogen and oxygen to provide thrust. During this reaction, the mass of oxygen consumed equals eight times the mass of hydrogen consumed. The external fuel tank of the space shuttle holds 385,200 gallons of liquid hydrogen. What must be the minimum capacity of the liquid oxygen tank, expressed in gallons, if it were to carry enough liquid oxygen to completely react with all of the liquid hydrogen during takeoff? (The density of liquid oxygen is 1.139 g/mL and the density of liquid hydrogen is 0.0708 g/mL.)

*⁎**2.147** The compound Fe_2O_3 contains iron and oxygen in the ratio of 2.325 g Fe to 1.000 g O. Another compound of iron and oxygen contains these elements in the ratio of 2.616 g Fe to 1.000 g O. What is the formula for this other iron–oxygen compound?

Reactants come together rapidly when they have been dissolved in a solvent before they are combined. We'll learn in this chapter how to obtain precise amounts of dissolved reactants simply by measuring volumes of solution.

Stoichiometry: Quantitative Chemical Relationships

This Chapter in Context We studied enough concepts in sufficient detail in the first two chapters to let us now venture into the real world where chemical reactions are planned and carried out in the laboratory. Two particularly important relationships among reacting substances are used in lab work. One is the ratio *by atoms* of the elements in a particular compound. The other is the ratio *by lab-sized reacting units* of the substances in reactions. The field dealing with these ratios is called *stoichiometry,* (stoy-kee-ah-meh-tree, from the Greek *stoicheion* for "element" and *metron* for "measure"). Stoichiometry is dominated by the concept of the *mole.*

3.1 The Mole Concept

The ratios of atoms in compounds are generally expressed by *whole-number* subscripts in chemical formulas. Fractional numbers are not used because fractions of atoms are not found in compounds. For example, in water, H_2O, the combining ratio is 2 whole atoms of H to 1 whole atom of O, the subscript "1" being understood. Thus, to make only *one molecule* of H_2O with no atoms of H or O left over, nature gives us no options. We are bound by the formula of water to take *two* atoms of hydrogen and *one* of oxygen.

$$2 \text{ atoms H} + 1 \text{ atom O} \longrightarrow 1 \text{ molecule } H_2O$$

To make a dozen (12) molecules of H_2O, we must take 2 dozen atoms of H and 1 dozen atoms of O, still a *ratio* of 2:1.

$$2 \text{ doz H atoms} + 1 \text{ doz O atoms} \longrightarrow 1 \text{ doz } H_2O \text{ molecules}$$

Thus, whether we express it by atoms or by dozens of atoms, the important ratio in H_2O is the same, 2 H:1 O. This ratio is the stoichiometric "law" for water. *Every pure substance has a stoichiometric "law" disclosed by its formula.*

►**Chemistry in Practice**◄ Engineers who design vehicle engines or furnaces, both of which produce carbon dioxide, CO_2, always want to supply enough air so that oxygen is in sufficient supply to make CO_2, and not the highly poisonous carbon monoxide, CO. In a poor supply of oxygen, much CO forms. ◆

Avogadro's Number and the Mole

The extreme smallness of atoms and molecules—we need far more than dozens of them to have samples that we can manipulate—has led chemists to define a very large "standard sample," a quantity called the **mole** (abbreviated **mol**). It's the SI base unit for *amount of chemical substance.*

One **mole** of any element or compound has the same number of formula units as there are atoms in exactly 12 g of carbon-12.

Amedeo Avogadro (1776–1856) was an Italian scientist who did important early work in stoichiometry.

The number of formula units[1] in a mole is called **Avogadro's number** in honor of Amedeo Avogadro. Its best experimental value is 6.0221367×10^{23}, a number so huge that it is beyond easy comprehension. A mole of dollars, for example, would be enough to guarantee every person in the world, roughly 6 billion people, an income of over $3 million *per second* for a hundred years. Yet, because atoms really are small, Avogadro's number of carbon-12 atoms has a mass of only 12 g.

When we need Avogadro's number in calculations, we usually round it to three or four significant figures, namely, to 6.02×10^{23} or 6.022×10^{23}. Thus, to four significant figures, we can write

Tools
Avogadro's number

$$1 \text{ mol things} = 6.022 \times 10^{23} \text{ things}$$

The "things" of importance in chemistry, of course, are atoms, ions, molecules, and formula units. Thus, 1 mol of carbon has Avogadro's number of C atoms, 1 mol of water has the same number of H_2O molecules, and 1 mol of NaCl has Avogadro's number of formula units. Thus, *equal numbers of moles contain equal numbers of units, be they atoms, molecules, or formula units.* This simple idea lies at the heart of all stoichiometry and is the foundation of all quantitative reasoning in chemistry.

When we refer to a mole of something, it is always important to identify clearly what that "something" is. For example, the expression "one mole of oxygen" is ambiguous; it could be taken to mean either one mole of oxygen atoms or one mole of oxygen molecules. To avoid such confusion, we usually associate a chemical formula with the unit *mol.* There is no ambiguity, therefore, if we write "1 mol O" to mean "one mole of oxygen atoms" and "1 mol O_2" to mean "one mole of oxygen molecules."

We can now use the mole concept to re-express the ratios of the atoms in molecules of water, H_2O.

Ratio by *atoms*	Ratio by *dozens* of atoms	Ratio by *moles*
2 H atoms	2 doz H atoms	2 mol H atoms
1 O atom	1 doz O atoms	1 mol O atoms

In all, the key stoichiometric ratio is 2:1, and one of the most important yet easy ideas in all of chemistry is that *moles of atoms always combine in the same ratio as the individual atoms themselves.*

[1]Recall that we use the term "formula unit" instead of "molecule" for ionic compounds. One formula unit consists of whatever is represented by the ionic compound's formula (see Section 2.9).

Stoichiometric Equivalencies

The **stoichiometric equivalence** of two elements in a chemical formula is the mole ratio in which they occur in the formula. For example, the subscripts in P_4O_{10} give us the ratio of P to O in this substance, both by atoms and by moles of atoms as follows.

By atoms: 4 P atoms to 10 O atoms

By moles: 4 mol P to 10 mol O

Using the symbol $\Leftrightarrow$ to stand for the phrase "is *stoichiometrically* equivalent to," we can write the following in terms of numbers of moles.

$$4 \text{ mol P} \Leftrightarrow 10 \text{ mol O}$$

Notice how this relationship is immediately available from the formula P_4O_{10}. Thus, a chemical formula becomes a tool for figuring out a stoichiometric equivalency that we might need in working a problem, whether it's one involving a planned experiment or one in this book.

Stoichiometric equivalencies let us prepare conversion factors as we need them. Thus, the equivalency given by $4 \text{ mol P} \Leftrightarrow 10 \text{ mol O}$ easily translates into the following factors.

$$\frac{4 \text{ mol P}}{10 \text{ mol O}} \quad \text{or} \quad \frac{10 \text{ mol O}}{4 \text{ mol P}}$$

The formula P_4O_{10} also implies other equivalencies, each with its two associated conversion factors.

$$1 \text{ mol } P_4O_{10} \Leftrightarrow 4 \text{ mol P} \quad \text{or} \quad \frac{1 \text{ mol } P_4O_{10}}{4 \text{ mol P}} \text{ and } \frac{4 \text{ mol P}}{1 \text{ mol } P_4O_{10}}$$

$$1 \text{ mol } P_4O_{10} \Leftrightarrow 10 \text{ mol O} \quad \text{or} \quad \frac{1 \text{ mol } P_4O_{10}}{10 \text{ mol O}} \text{ and } \frac{10 \text{ mol O}}{1 \text{ mol } P_4O_{10}}$$

Chemical formula

The reaction of phosphorus with oxygen that gives P_4O_{10} produces a brilliant light, often used in fireworks displays.

EXAMPLE 3.1
Stoichiometry of Chemical Formulas

How many moles of oxygen atoms are combined with 4.20 moles of chlorine atoms in Cl_2O_7?

Analysis: Note the words *How many moles . . . combined with . . . moles*. These words are the clue that we have a stoichiometric equivalency question. We can restate the problem as follows.

$$4.20 \text{ mol Cl} \Leftrightarrow ? \text{ mol O}$$

(Or, in words, "4.20 mol of Cl is equivalent to how many moles of O in Cl_2O_7?") What we need therefore is a moles-to-moles conversion factor. The subscripts in Cl_2O_7 supply it, because the formula, Cl_2O_7, actually means

$$2 \text{ mol Cl} \Leftrightarrow 7 \text{ mol O}$$

So we have the following conversion factors.

$$\frac{2 \text{ mol Cl}}{7 \text{ mol O}} \quad \text{or} \quad \frac{7 \text{ mol O}}{2 \text{ mol Cl}}$$

Cl_2O_7 is an oily and very explosive liquid.

Solution: To find the moles of O atoms stoichiometrically equivalent to 4.20 moles of Cl atoms in Cl_2O_7, we multiply 4.20 mol Cl by the second factor. Notice how the units *mol Cl* (not just *mol*) cancel properly. (Draw in the cancel lines yourself.)

$$4.20 \text{ mol Cl} \times \frac{7 \text{ mol O}}{2 \text{ mol Cl}} = 14.7 \text{ mol O}$$

Thus 14.7 mol of O is combined with 4.20 mol of Cl in Cl_2O_7. We leave three significant figures in the answer because the least precise number, 4.20, has three significant figures. (The 7 and the 2 in the conversion factor, which come from the formula's subscripts, are *exact* numbers of moles, and so they have an infinite number of significant figures.)

Is the Answer Reasonable?
Let's check to be sure the *size* of the answer makes sense. The 7-to-2 ratio tells us that the amount of oxygen, by moles, is a little less than four times the amount of chlorine. The answer, 14.7, is somewhat less than 4×4.20, so the answer "makes sense."

Practice Exercise 1

How many moles of nitrogen atoms are combined with 8.60 mol of oxygen atoms in dinitrogen pentoxide, N_2O_5? ◆

3.2 Measuring Moles of Elements and Compounds

Diamond is one form of pure carbon. A 12 g diamond would be a 60 carat stone.

The SI definition of the mole says that exactly 1 mol of carbon-12 has a mass of exactly 12 g. However, naturally occurring carbon is not pure carbon-12 but is a mixture of isotopes, consisting everywhere in nature of 98.90% carbon-12 and 1.10% carbon-13. The "average" mass of the carbon atoms in the mixture is 12.011 u, which is the atomic mass of carbon given in tables. Avogadro's number of these "average atoms" therefore has a mass of 12.011 g. In other words, for naturally occurring carbon, 1 mol C equals 12.011 g C. The extension to all the elements is worth emphasizing.[2]

> One mole of any element has a mass in grams numerically equal to the element's atomic mass.

The sizes *in grams* of the one-mole samples of four elements shown in Figure 3.1 depend on the individual atomic masses, but each sample contains the *same number* of atoms (Avogadro's number).

Being able to use an element's atomic mass to determine the mass of one mole of the element provides us with a convenient way, in the laboratory, of dispensing elements in any mole ratio we want. Our tool is the balance. For example, suppose we needed one mole of magnesium atoms for an experiment. The atomic mass of magnesium is 24.305. Therefore, 1 mol Mg has a mass of 24.305 g. Similarly, from the table of atomic masses we know that 1 mol S equals 32.066 g.

Atomic mass

Rounding Atomic Masses

We will often obtain data from the Table of Atomic Numbers and Masses or from the periodic table, where many atomic masses are given to six or more significant figures. Most of the numerical problems in this and later chapters involve data with only three or four significant figures. As a rule, we will round

---

[2]From here on, when we use the term "atomic mass" it will mean the *average* atomic mass of the element's isotopes as they occur in nature.

Figure 3.1 *Moles of elements.* Each sample of these elements contains the same number of atoms, Avogadro's number.

atomic masses so they have one more significant figure than required by the data in the problem. This will ensure that any uncertainty in the answers reflect uncertainty in the data provided in the problem.

Now let's consider one of the most common kinds of chemical calculations, namely, finding the number of *moles* in a given *mass*.

EXAMPLE 3.2
Converting Grams to Moles

How many moles of silicon are in 4.60 g of Si?

Analysis: It's often helpful to restate the problem either as an incomplete equation or an equivalency. Thus, for this problem we write

$$4.60 \text{ g Si} \Leftrightarrow ? \text{ mol Si}$$

In other words, 4.60 g of silicon is equivalent to how many moles of silicon? The tool for converting a given mass of an element into number of moles is a conversion factor obtained from the atomic mass. For silicon, this is 28.09 (rounded from 28.0855). We first write the atomic mass as an equivalency.

$$28.09 \text{ g Si} \Leftrightarrow 1 \text{ mol Si}$$

Now two possible conversion factors become clear; one is the tool we need for the grams-to-moles calculation.

$$\frac{28.09 \text{ g Si}}{1 \text{ mol Si}} \quad \text{and} \quad \frac{1 \text{ mol Si}}{28.09 \text{ g Si}}$$

If we multiply the given, 4.60 g Si, by the second conversion factor, the units will cancel properly to give the answer.

Solution: Draw in the cancel lines yourself.

$$4.60 \text{ g Si} \times \frac{1 \text{ mol Si}}{28.09 \text{ g Si}} = 0.164 \text{ mol Si}$$

In other words,

$$4.60 \text{ g Si} \Leftrightarrow 0.164 \text{ mol Si}$$

If you are writing in your own cancel lines, *be sure to show them canceling the full unit,* "g Si," not just "g."

Circular silicon wafers are being loaded into a high temperature furnace during one step in the manufacture of silicon chips.

Is the Answer Reasonable?

A simple way to *estimate* the answer is to round off the given data and then do some simple mental arithmetic. Let's say that 28.09 is about 30, and 4.60 is about 5. If the atomic mass of silicon were 30, then 5 grams of silicon would be about 1/6th of a mole. One-sixth equals 0.167, so our answer looks good.

Practice Exercise 2

How many moles of sulfur are present in 35.6 g of sulfur? ◆

EXAMPLE 3.3
Converting Moles to Grams

Gold nugget and bars.

How many grams of gold are in 0.150 mol of Au?

Analysis: The question can be restated as follows.

$$0.150 \text{ mol Au} \Leftrightarrow ? \text{ g Au}$$

We'll round the atomic mass of gold from 196.96655 to 197.0, so the following is true about gold.

$$1 \text{ mol Au} \Leftrightarrow 197.0 \text{ g Au}$$

The conversion factor that we need to convert the given number of moles of Au to the number of grams of Au comes from this equivalency.

Solution: The given is 0.150 mol Au, so we multiply it by the ratio of grams of Au to moles of Au obtained from the atomic mass of gold.

$$0.150 \ \overline{\text{mol Au}} \times \frac{197.0 \text{ g Au}}{1 \ \overline{\text{mol Au}}} = 29.6 \text{ g Au}$$

Thus, we can write 0.150 mol Au ⇔ 29.6 g Au.

Is the Answer Reasonable?

We have between 1/10 and 2/10 of a mole of gold. One-tenth would weigh 19.7 g (about 20 g) and two-tenths of a mole would weigh approximately 40 g. Our answer of about 30 g seems just right.

Practice Exercise 3

How many grams of silver are in 0.263 mol of Ag? ◆

Using Avogadro's Number in Calculations

Avogadro's number is not actually used in lab work. It's employed, for example, when we'd like to know the mass in grams of one or more atoms of an element or when we want to calculate the number of atoms in a weighed sample of an element. The following example illustrates how we carry out such a calculation.

EXAMPLE 3.4
Converting Mass to Number of Atoms

How many atoms are in a sample of uranium with a mass of 1.00×10^{-6} g (1.00 μg)?

Analysis: The "How many atoms" phrase in the question tells us that we need Avogadro's number. But this number tells us the number of atoms in one *mole.* We're given *grams* of uranium, however, not moles. So we must use the atomic mass of uranium, our tool for finding the number of moles in a given mass. Thus, the calculation must "flow" as follows, and you can see how the units will properly cancel until "atoms U" remains. (Draw in the cancel lines yourself.)

$$\text{g U} \times \frac{\text{mol U}}{\text{g U}} \times \frac{\text{atoms U}}{\text{mol U}} = \text{atoms U}$$

| given mass of U | from atomic mass of U | from Avogadro's Number |

The first half gives us "mol U," but let's not pause to do this calculation; we can easily go on. The last half converts "mol U" to "atoms U." (Recall that you learned to use such chain calculations in Chapter 1.)

Solution: From the atomic mass given in the table inside the front cover of the book we can write

$$1 \text{ mol U} = 238.0 \text{ g U}$$

The first conversion factor whose need was identified in the preceding equation, mol U/g U, is therefore

$$\frac{1 \text{ mol U}}{238.0 \text{ g U}}$$

The second conversion factor is simply restating Avogadro's number as it applies to uranium.

$$\frac{6.022 \times 10^{23} \text{ atoms U}}{1 \text{ mol U}}$$

So the solution is calculated as follows.

$$1.00 \times 10^{-6} \text{ g U} \times \frac{1 \text{ mol U}}{238.0 \text{ g U}} \times \frac{6.022 \times 10^{23} \text{ atoms U}}{\text{mol U}} = 2.53 \times 10^{15} \text{ atoms U}$$

Thus a 1-microgram sample of uranium, a tiny speck, contains over 2.5 million billion ($10^6 \times 10^9$) atoms.

Is the Answer Reasonable?
Let's check to see that the magnitude of the answer is reasonable. To do this, let's round off the numbers and express them in exponential notation. The arithmetic involved is therefore approximated as follows (we've approximated 238 by 200):

$$\frac{(1 \times 10^{-6})(6 \times 10^{23})}{(2 \times 10^2)} = \frac{6}{2} \times \frac{10^{-6} \times 10^{23}}{10^2} = 3 \times 10^{15}$$

This agrees pretty well with our answer.

The check we've just outlined may seem to require quite a bit of work, but with some practice, you will find it is not so difficult at all. Learning to approximate answers this way is well worth the effort.

Practice Exercise 4

How many atoms are in 1.00×10^{-9} g of lead? ◆

Figure 3.2 *Moles of compounds.* One mole of four different compounds. Each sample contains the identical number of formula units or molecules, Avogadro's number.

Formula Masses and Molecular Masses

The mole concept works with compounds exactly as it does with elements. Just as each element has an atomic mass, each compound has a **formula mass,** the sum of the atomic masses of the atoms shown in the formula. When a substance is known to be molecular, not ionic, we refer to its **molecular mass,** reserving *formula mass* for ionic compounds. (Many chemists use the terms "formula weight" and "molecular weight" for formula mass and molecular mass.)

Figure 3.2 shows 1.00 mol samples of four compounds. Their sizes in *grams* vary with their molecular or formula masses, but they have identical numbers of molecules or formula units. *For every chemical substance, its molecular or formula mass in grams equals one mole of it.* This makes the calculation of a molecular or formula mass from a formula a common operation. Let's see how it's done.

EXAMPLE 3.5

Calculating the Formula Mass of a Compound

Calcium phosphate is a major source of phosphate fertilizer. Those in agriculture call it "white gold," it's so valuable.

What is the formula mass of calcium phosphate, $Ca_3(PO_4)_2$, expressed to the second decimal place?

Analysis: We simply add up the atomic masses, remembering that a subscript *outside* the parentheses is a multiplier for all atoms inside.

Solution: Using atomic masses of 40.08 for Ca, 30.97 for P, and 16.00 for O,

$$Ca_3(PO_4)_2 = 3\,Ca + 2\,P + 8\,O$$
$$\text{Formula mass of } Ca_3(PO_4)_2 = (3 \times 40.08) + (2 \times 30.97) + (8 \times 16.00)$$
$$= 120.24 + 61.94 + 128.00$$
$$= 310.18$$

To have one mole of calcium phosphate, the sample's mass must be 310.18 g. In other words, 1 mol $Ca_3(PO_4)_2 \Leftrightarrow 310.18$ g of $Ca_3(PO_4)_2$.

Practice Exercise 5

What is the mass in grams of 1 mol Na_2CO_3, expressed to the second decimal place? ◆

We calculate grams from moles or moles from grams for compounds, using their molecular or formula masses, just as we do for elements, using atomic masses.

Formula mass;
molecular mass

EXAMPLE 3.6

Calculating Grams from Moles for Compounds

A student has worked out the details for an experiment that will use 0.115 mol of $Ca_3(PO_4)_2$, calcium phosphate, as a starting material. How many grams of $Ca_3(PO_4)_2$ should be weighed out?

Analysis: The problem can be restated as follows.

$$0.115 \text{ mol } Ca_3(PO_4)_2 \Leftrightarrow ? \text{ g } Ca_3(PO_4)_2$$

The tool we need is the moles-to-grams tool implied by the formula mass of $Ca_3(PO_4)_2$, which we found in Example 3.5 to be 310.18. Thus,

$$1 \text{ mol } Ca_3(PO_4)_2 \Leftrightarrow 310.18 \text{ g } Ca_3(PO_4)_2$$

Solution: The given is 0.115 mol $Ca_3(PO_4)_2$, so we multiply it by the ratio of grams to moles obtained from the formula mass of $Ca_3(PO_4)_2$.

$$0.115 \text{ mol } Ca_3(PO_4)_2 \times \frac{310.18 \text{ g } Ca_3(PO_4)_2}{1 \text{ mol } Ca_3(PO_4)_2} = 35.7 \text{ g } Ca_3(PO_4)_2$$

In other words,

$$0.115 \text{ mol } Ca_3(PO_4)_2 \Leftrightarrow 35.7 \text{ g } Ca_3(PO_4)_2$$

Is the Answer Reasonable?
Does the *size* of the answer make sense? Yes. The student requires a little over one-tenth of a mole of $Ca_3(PO_4)_2$, so a bit over one-tenth of 310 g is needed.

Practice Exercise 6

How many grams of sodium carbonate, Na_2CO_3, must be measured to obtain 0.125 mol of Na_2CO_3? ◆

Practice Exercise 7

A sample of 45.8 g of H_2SO_4 contains how many moles of H_2SO_4? ◆

Applications of Stoichiometry

One of the most common uses of stoichiometry in the lab occurs when we must relate the mass of one substance needed to make a compound to the *mass* of another that is also required.

EXAMPLE 3.7

Reaction Stoichiometry

How many grams of Cl are needed to combine with 24.4 g of Si to make silicon tetrachloride, $SiCl_4$?

Analysis: Let's begin, as usual, by restating the problem as follows.

$$24.4 \text{ g Si} \Leftrightarrow ? \text{ g Cl} \qquad \text{(for the compound } SiCl_4)$$

We have not studied a *direct* way to go from the number of *grams* of Si to the number of *grams* of Cl, so to solve this problem we will need to combine more than one step. Where do we begin? Let's start by examining what we *do* know to attempt to find the tools that will take us to the answer.

$SiCl_4$ is used to make transistor-grade elemental silicon.

We do know a formula, $SiCl_4$, and it is the key to the problem's solution. This formula is a tool that permits us to write the following mole equivalency.

$$1 \text{ mol Si} \Leftrightarrow 4 \text{ mol Cl}$$

We also know the mass of Si. We must go from 24.4 g of Si to *moles* of Si. Then we can use the number of moles of Si and the previous equivalency (1 mol Si $\Leftrightarrow$ 4 mol Cl) to calculate the number of *moles* of Cl. Finally, we can use the number of moles of Cl to calculate the corresponding number of grams of Cl. Our successive moves can be diagrammed as follows. The starting point is at 24.4 g Si. The tools we need are by the arrows.

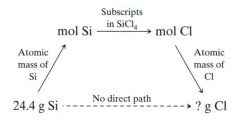

The top level, the mole level, is where the stoichiometric information in the formula $SiCl_4$ comes into play. *In general, the key calculation in a stoichiometry problem occurs at the mole level.*

Solution: We'll do this both stepwise and by a chain calculation.

Stepwise, we first move from "g Si" to "mol Si" using the atomic mass of Si.

$$1 \text{ mol Si} \Leftrightarrow 28.09 \text{ g Si}$$

So,

$$24.4 \text{ g Si} \times \frac{1 \text{ mol Si}}{28.09 \text{ g Si}} = 0.869 \text{ mol Si}$$

Next we use the stoichiometric equivalency we earlier found in the formula $SiCl_4$ to find the number of moles of Cl that are equivalent to 0.869 mol of Si.

$$0.869 \text{ mol Si} \times \frac{4 \text{ mol Cl}}{1 \text{ mol Si}} = 3.48 \text{ mol Cl}$$

Thus, 3.48 mol Cl $\Leftrightarrow$ 0.869 mol Si in $SiCl_4$.

Finally, we change 3.48 mol of Cl to the number of grams of Cl. Now we use chlorine's atomic mass.

$$1 \text{ mol Cl} \Leftrightarrow 35.45 \text{ g Cl}$$

So,

$$3.48 \text{ mol Cl} \times \frac{35.45 \text{ g Cl}}{1 \text{ mol Cl}} = 123 \text{ g Cl}$$

The answer is that 123 g of Cl combines with 24.4 g of Si to make $SiCl_4$.

A more efficient way to solve this problem, as you no doubt now realize, is to string together or "chain" the conversion factors. Then it isn't necessary to write down the intermediate answers. (Insert the cancel lines yourself to show that the answer is in the correct units. Notice how the calculation flows smoothly from "grams Si" to "grams Cl.")

$$24.4 \text{ g Si} \times \frac{1 \text{ mol Si}}{28.09 \text{ g Si}} \times \frac{4 \text{ mol Cl}}{1 \text{ mol Si}} \times \frac{35.45 \text{ g Cl}}{1 \text{ mol Cl}} = 123 \text{ g Cl}$$

grams Si $\longrightarrow$ mole Si $\longrightarrow$ mole Cl $\longrightarrow$ grams Cl

Is the Answer Reasonable?
It seems like a great deal of Cl, 123 g, for the 24.4 g of Si, particularly since the atomic masses of Si and Cl are not too different, 28.09 versus 35.45. But notice that, *in moles,* the formula $SiCl_4$ demands four times as much Cl as Si, and 4 times 24.4 is approaching the numerical part of the answer.

Practice Exercise 8

How many grams of iron are needed to combine with 25.6 g of O to make Fe_2O_3? ◆

3.3 Percentage Composition

Among the most exciting experimental activities in chemistry are making an entirely new compound or isolating a new compound from a natural source. Immediate questions are then, What is it? What is its formula? To find out, the research begins with **qualitative analysis,** a series of procedures to identify all of the elements making up the substance. Next come procedures for **quantitative analysis,** which determine the mass of each element in a weighed sample of the substance.

The usual form for describing the relative masses of the elements in a compound is a list of *percentages by mass* called the compound's **percentage composition.** The **percentage by mass** of an element is the number of grams of the element present in 100 g of the compound. In general, a percentage by mass is found by using the following equation.

$$\% \text{ by mass of element} = \frac{\text{mass of element}}{\text{mass of whole sample}} \times 100\% \qquad (3.1)$$

Tools

Percentage composition

EXAMPLE 3.8

Calculating a Percentage Composition from Chemical Analysis

A sample of a liquid with a mass of 8.657 g was decomposed into its elements and gave 5.217 g of carbon, 0.9620 g of hydrogen, and 2.478 g of oxygen. What is the percentage composition of this compound?

Analysis: We must apply Equation 3.1 for each element. The "mass of whole sample" here is 8.657 g, so we take each element in turn and perform the calculations. The "check" is that the percentages must add up to 100%, allowing for small differences caused by rounding.

Solution:

$$\text{For C:} \quad \frac{5.217 \text{ g}}{8.657 \text{ g}} \times 100\% = 60.26\% \text{ C}$$

$$\text{For H:} \quad \frac{0.9620 \text{ g}}{8.657 \text{ g}} \times 100\% = 11.11\% \text{ H}$$

$$\text{For O:} \quad \frac{2.478 \text{ g}}{8.657 \text{ g}} \times 100\% = \underline{28.62\%} \text{ O}$$

Sum of percentages: 99.99%

The results tell us that in 100.00 g of the liquid there are 60.26 g of carbon, 11.11 g of hydrogen, and 28.62 g of oxygen.

From 0.5462 g of a compound there was isolated 0.1417 g of nitrogen and 0.4045 g of oxygen. What is the percentage composition of this compound? Are any other elements present? ◆

Theoretical Percentage Composition

Another kind of calculation is to find the *theoretical percentage composition, that which is calculated from the formula.* This might be needed, for example, when the same elements can combine in two or more ways. Nitrogen and oxygen, for example, form all of the following compounds: N_2O, NO, NO_2, N_2O_3, N_2O_4, and N_2O_5. By comparing the percentage composition found by experiment for some unknown compound of nitrogen and oxygen to the theoretical percentages for each possible formula, a choice can be made among the candidates. Which formula, for example, fits the percentage composition calculated in Practice Exercise 9? We will work this as an example.

NO_2 causes the red-brown color of smog.

EXAMPLE 3.9

Calculating a Theoretical Percentage Composition from a Chemical Formula

Do the mass percentages of 25.94% N and 74.06% O match the formula N_2O_5?

Analysis: To calculate the theoretical percentages by mass of N and O in N_2O_5, we need the masses of N and O in a specific sample of N_2O_5. *If we choose 1 mol of the given compound to be this sample, calculating the rest of the data will be simpler.*

Solution: We know that 1 mol of N_2O_5 must contain 2 mol N and 5 mol O. The corresponding number of grams of N and O are found as follows.

$$2 \text{ N:} \qquad 2 \text{ mol N} \times \frac{14.01 \text{ g N}}{1 \text{ mol N}} = 28.02 \text{ g N}$$

$$5 \text{ O:} \qquad 5 \text{ mol O} \times \frac{16.00 \text{ g O}}{1 \text{ mol O}} = 80.00 \text{ g O}$$

$$\overline{\qquad 1 \text{ mol } N_2O_5 = 108.02 \text{ g } N_2O_5}$$

Now we can calculate the percentages.

$$\text{For \% N:} \qquad \frac{28.02 \text{ g}}{108.02 \text{ g}} \times 100\% = 25.94\% \text{ N in } N_2O_5$$

$$\text{For \% O:} \qquad \frac{80.00 \text{ g}}{108.02 \text{ g}} \times 100\% = 74.06\% \text{ O in } N_2O_5$$

Thus the experimental values do match the theoretical percentages for the formula N_2O_5.

Calculate the theoretical percentage composition of N_2O_3. ◆

A calculation similar to the one studied in Example 3.9 lets us calculate the mass of an element present in a given mass of one of its compounds.

> ►**Chemistry in Practice**◄ Engineers working with metals have to know how well they're doing when they isolate a metal from one of its natural compounds. The next example illustrates a calculation that might be done when copper metal is obtained from a sulfide ore. It answers the question, "What's the most we can rightfully expect." ◆

EXAMPLE 3.10

Calculating the Mass of an Element in a Sample of a Compound

One of the forms in which copper (Cu) occurs in nature is copper(I) sulfide, Cu_2S. How many grams of copper metal can theoretically be obtained from 10.0 grams of this compound?

Analysis: The question can be restated as follows.

$$10.0 \text{ g } Cu_2S \Leftrightarrow ? \text{ g } Cu$$

The easiest way to get at this, however, is to take advantage of what we do when we calculate the formula mass of Cu_2S. We'll use the atomic masses of Cu and S and translate them into grams per mole for each element. Thus, in 1 mol of Cu_2S, we have

2 mol Cu: $\quad 2 \text{ mol Cu} \times \dfrac{63.55 \text{ g Cu}}{\text{mol Cu}} = 127.1 \text{ g Cu}$

1 mol S: $\quad 1 \text{ mol S} \times \dfrac{32.07 \text{ g S}}{\text{mol S}} = 32.07 \text{ g } S$

$$\overline{\rule{4cm}{0pt}}$$

$$1 \text{ mol of } Cu_2S = 159.2 \text{ g } Cu_2S$$

Thus, in 159.2 g of Cu_2S there is 127.1 g of Cu. We can express this as a mass equivalency for Cu_2S, namely,

$$159.2 \text{ g } Cu_2S \Leftrightarrow 127.1 \text{ g } Cu$$

From it we can fashion a conversion factor to calculate how many grams of Cu are in only 10.0 g Cu_2S.

Solution: The needed conversion factor is

$$\dfrac{127.1 \text{ g Cu}}{159.2 \text{ g } Cu_2S}$$

To calculate the mass of Cu in 10.00 g Cu_2S, we simply do the following.

$$10.0 \text{ g } Cu_2S \times \dfrac{127.1 \text{ g Cu}}{159.2 \text{ g } Cu_2S} = 7.98 \text{ g Cu}$$

Thus, in 10.0 g Cu_2S there is 7.98 g Cu.

Is the Answer Reasonable?
Notice that because of the high atomic mass of Cu and the low atomic mass of S, most of a Cu_2S sample is copper, and 7.98 g is "most of" 10.0 g.

Practice Exercise 11

How many grams of iron are in a 15.0 g sample of iron(III) oxide, Fe_2O_3, an important iron ore called hematite? ◆

3.4 Empirical and Molecular Formulas

The compound that forms when phosphorus burns in oxygen consists of molecules with the formula P_4O_{10}. When a formula gives the composition of one *molecule*, it is called a **molecular formula.** Notice, however, that both the subscripts 4 and 10 are divisible by 2, so the *smallest* numbers that tell us the *ratio* of P to O are 2 and 5. A simpler (but less informative) formula that expresses this ratio is P_2O_5. This is sometimes called the *simplest formula* for the compound. It is also called the **empirical formula** because it can be obtained from an experimental analysis of the compound.

To obtain an empirical formula experimentally, we need to determine the number of grams of each element in a sample of the compound. Grams are then converted to moles, from which we obtain the mole ratios of the elements. Because the ratio by moles is the same as the ratio by atoms, we are able to construct the empirical formula.

The next four examples illustrate how we can calculate empirical formulas. We will then look at what additional data are required to obtain a compound's molecular formula.

EXAMPLE 3.11

Calculating an Empirical Formula from Mass Data

A sample of a tin and chlorine compound with a mass of 2.57 g was found to contain 1.17 g of tin. What is the compound's empirical formula?

Analysis: The empirical formula must specify the ratio *by atoms* of Sn and Cl, but how do we obtain this from the *mass* data given? The answer lies in the mole concept. Recall that for any compound, *moles of atoms always combine in the same ratio as the individual atoms themselves.* Thus if we can find the *mole* ratio of Sn to Cl, then we know it will be identical to the atom ratio and we will have the empirical formula. The first step, therefore, is to convert the numbers of grams of Sn and Cl to the numbers of moles of Sn and Cl. Then we convert these numbers into their simplest *whole-number* ratio.

You no doubt noticed that the problem did not give the mass of chlorine in the 2.57 g sample. But there are only two elements present in the compound, tin and chlorine. We know the mass of one of them (Sn) and we know the total mass. Therefore, the mass of Cl is simply the difference between 2.57 g and the mass of Sn present in the sample, namely, 1.17 g of Sn.

Solution: First, we find the mass of Cl in 2.57 g of compound:

$$\text{mass of Cl} = 2.57 \text{ g compound} - 1.17 \text{ g Sn} = 1.40 \text{ g Cl}$$

Now we use the atomic masses to convert the mass data for tin and chlorine into moles.

$$1.17 \ \cancel{\text{g Sn}} \times \frac{1 \text{ mol Sn}}{118.7 \ \cancel{\text{g Sn}}} = 0.00986 \text{ mol Sn}$$

$$1.40 \ \cancel{\text{g Cl}} \times \frac{1 \text{ mol Cl}}{35.45 \ \cancel{\text{g Cl}}} = 0.0395 \text{ mol Cl}$$

We could now write a formula: $Sn_{0.00986}Cl_{0.0395}$, which does express the mole ratio, but we want whole numbers as subscripts so the formula can also express the proper atom ratio. To change the subscripts in $Sn_{0.00986}Cl_{0.0395}$ into a set of whole numbers *that are in the same ratio,* we divide each by a common divisor. The easiest common divisor is the smaller number in the set. *This is always the way to begin the search for whole-number subscripts; pick the smallest number of the set as the divisor.* It's guaranteed to make at least one subscript a whole number, namely, 1. Here, we divide both numbers by 0.00986.

$$Sn_{\frac{0.00986}{0.00986}} Cl_{\frac{0.0395}{0.00986}} = Sn_{1.00}Cl_{4.01}$$

As a rule, if a calculated subscript, expressed to the proper number of significant figures, differs from a whole number by less than a few units in the last significant figure, we can safely round to the nearest whole number. We may round 4.01 to 4, so the empirical formula is $SnCl_4$.

Is the Answer Reasonable?
For problems of this kind, there is no simple check other than the fact that we obtained a formula with whole-number subscripts.

Practice Exercise 12

A 1.525 g sample of a compound between nitrogen and oxygen contains 0.712 g of nitrogen. Calculate its empirical formula. (Write the symbol of nitrogen first.) ◆

Sometimes our strategy of using the lowest common divisor does not give whole numbers. Let's see how to handle such a situation.

EXAMPLE 3.12

Calculating an Empirical Formula from Mass Composition

One of the compounds of iron and oxygen, "black iron oxide," occurs naturally in the mineral magnetite. When a 2.448 g sample was analyzed it was found to have 1.771 g of Fe. Calculate the empirical formula of this compound.

Analysis: When we calculate the mass of O in the sample by difference, we'll have the masses in grams of both elements, but we need the numbers of *moles* for a formula. So we'll use the atomic masses of Fe and O as tools for converting their respective masses to numbers of moles. Then we'll write a trial formula and see whether we can adjust the coefficients to their smallest whole numbers by the strategy learned in Example 3.11.

Solution: The mass of O is, as we said, found by difference.

$$2.448 \text{ g compound} - 1.771 \text{ g Fe} = 0.677 \text{ g O}$$

The moles of Fe and O in the sample can now be calculated.

$$1.771 \text{ g Fe} \times \frac{1 \text{ mol Fe}}{55.845 \text{ g Fe}} = 0.03171 \text{ mol Fe}$$

$$0.677 \text{ g O} \times \frac{1 \text{ mol O}}{16.00 \text{ g O}} = 0.0423 \text{ mol O}$$

The mineral magnetite, like any magnet, is able to affect the orientation of a compass needle.

These results let us write the formula as $Fe_{0.03171}O_{0.0423}$.

Our first effort to change the ratio of 0.03171 to 0.0423 into whole numbers is to divide both by the smaller, 0.03171.

$$Fe_{\frac{0.03171}{0.03171}} O_{\frac{0.0423}{0.03171}} = Fe_{1.000}O_{1.33}$$

This time the procedure seems to fail. Given the fact that our number of significant figures is three, we can be sure that the digits 1.3 in 1.33 are known with certainty. Therefore the subscript for O, 1.33, is much too far from a whole number to round off and retain the precision allowed. In a *mole* sense, the ratio of 1 to 1.33 is correct; we just have not yet found a way to state the ratio in whole numbers. Let's look at a strategy that will give us whole-number subscripts.

Let's multiply each subscript in $Fe_{1.000}O_{1.33}$ by a whole number, 2. *This does not change the ratio;* it changes only the size of the numbers used to state it.

The numbers before the last place are known with certainty, according to our definition of significant figures.

$$Fe_{(1.000 \times 2)}O_{(1.33 \times 2)} = Fe_{2.000}O_{2.66}$$

This didn't work either; 2.66 is also too far from a whole number (based on the allowed precision) to be rounded off. Let's try using 3 instead of 2 *on the first ratio of 1.000 to 1.33.*

$$Fe_{(1.000 \times 3)}O_{(1.33 \times 3)} = Fe_{3.000}O_{3.99}$$

We are now justified in rounding; 3.99 is acceptably close to 4. The empirical formula of the oxide of iron is Fe_3O_4.

Practice Exercise 13

A 2.012 g sample of a compound of nitrogen and oxygen has 0.522 g of nitrogen. Calculate its empirical formula. ◆

Empirical Formulas from Percentage Compositions

Only rarely is it possible to obtain the masses of every element in a compound by the use of just one weighed sample. Two or more analyses carried out on different samples are needed. For example, suppose an analyst is given a compound known to consist exclusively of calcium, chlorine, and oxygen. The mass of calcium in one weighed sample and the mass of chlorine in another sample would be determined in separate experiments. Then the mass data for calcium and chlorine would be converted to percentages by mass *so that the data from different samples relate to the same sample size, namely, 100 g of the compound.* The percentage of oxygen would be calculated by difference because %Ca + %Cl + %O = 100%. Each mass percentage represents a certain number of grams of the element, which is next converted into the corresponding number of moles of the element. The mole proportions are converted to whole numbers in the way we just studied, giving us the subscripts for the empirical formula. Let's see how this works.

EXAMPLE 3.13

Calculating an Empirical Formula from Percentage Composition

A white powder used in paints, enamels, and ceramics has the following percentage composition: Ba, 69.6%; C, 6.09%; and O, 24.3%. What is its empirical formula?

Analysis: The percentages are *numerically* equal to masses if we choose a 100 g sample of the compound. So we have to convert masses to moles using the atomic masses of the elements. Then we proceed as before.

Solution: A 100 g sample of the compound would contain 69.6 g Ba, 6.09 g C, and 24.3 g O. Let's convert these to moles.

Ba: $\quad 69.6 \text{ g Ba} \times \dfrac{1 \text{ mol Ba}}{137.3 \text{ g Ba}} = 0.507 \text{ mol Ba}$

C: $\quad 6.09 \text{ g C} \times \dfrac{1 \text{ mol C}}{12.01 \text{ g C}} = 0.507 \text{ mol C}$

O: $\quad 24.3 \text{ g O} \times \dfrac{1 \text{ mol O}}{16.00 \text{ g O}} = 1.52 \text{ mol O}$

Our preliminary empirical formula is then

$$Ba_{0.507}C_{0.507}O_{1.52}$$

We next divide each subscript by the smallest, 0.507.

$$Ba_{\frac{0.507}{0.507}} C_{\frac{0.507}{0.507}} O_{\frac{1.52}{0.507}} = Ba_{1.00}C_{1.00}O_{3.00}$$

The subscripts are whole numbers, so the empirical formula is $BaCO_3$, representing barium carbonate.

Practice Exercise 14

A white solid used to whiten paper has the following percentage composition: Na, 32.4%; S, 22.6%. The unanalyzed element is oxygen. What is the compound's empirical formula? ◆

Indirect Analysis

A compound is seldom broken down completely to its *elements* in a quantitative analysis, as our examples may lead you to think. Instead, the compound is changed into other *compounds*. The reactions separate the elements by capturing each one entirely (quantitatively) in a *separate* compound *whose formula is known.*

In the following example, we illustrate an indirect analysis of a compound made entirely of carbon, hydrogen, and oxygen. Such compounds burn completely in pure oxygen—the reaction is called *combustion*—and the sole products are carbon dioxide and water. (This particular kind of indirect analysis is sometimes called a *combustion analysis.*) The complete combustion of methyl alcohol (CH_3OH), for example, occurs according to the following equation.

$$2CH_3OH + 3O_2 \longrightarrow 2CO_2 + 4H_2O$$

The carbon dioxide and water can be separated and are individually weighed. Notice that all of the carbon atoms in the original compound end up among the CO_2 molecules, and all of the hydrogen atoms are in H_2O molecules. In this way at least two of the original elements, C and H, are entirely separated.

We will calculate the mass of carbon in the CO_2 collected, which equals the mass of carbon in the original sample. Similarly, we will calculate the mass of hydrogen in the H_2O collected, which equals the mass of hydrogen in the original sample. When added together, the mass of C and mass of H are less than the total mass of the sample because part of the sample is composed of oxygen. By subtracting the sum of the C and H masses from the original sample weight we can obtain the mass of oxygen in the sample of the compound.

EXAMPLE 3.14

Calculating an Empirical Formula from Indirect Analysis

A 0.5438 g sample of a liquid consisting of only C, H, and O was burned in pure oxygen, and 1.039 g of CO_2 and 0.6369 g of H_2O were obtained. What is the empirical formula of the compound?

Analysis: There are two broad parts to this problem. For the first part, we will find the number of grams of C in the CO_2 and the number of grams of H in the H_2O. (This kind of calculation was illustrated in Example 3.10.) These values represent the number of grams of C and H in the original sample. Adding them together and subtracting the sum from the mass of the original sample will give us the mass of oxygen in the sample. In short, we have the following series of calculations.

$$\text{grams } CO_2 \longrightarrow \text{grams } C$$

$$\text{grams } H_2O \longrightarrow \text{grams } H$$

We find the mass of oxygen by difference.

$$0.5438 \text{ g sample} - (\text{g C} + \text{g H}) = \text{g O}$$

In the second half of the solution, we use the masses of C, H, and O to calculate the empirical formula as in Example 3.13.

Solution: First we find the number of grams of C in the CO_2 and of H in the H_2O. In one mole of CO_2 (44.009 g) there are 12.011 g of C. Therefore, in 1.039 g of CO_2 we have

$$1.039 \text{ g } CO_2 \times \frac{12.011 \text{ g C}}{44.009 \text{ g } CO_2} = 0.2836 \text{ g C}$$

In one mole of H_2O (18.015 g) there are 2.0158 g of H. For the number of grams of H in 0.6369 g of H_2O,

$$0.6369 \text{ g } H_2O \times \frac{2.0158 \text{ g H}}{18.015 \text{ g } H_2O} = 0.07127 \text{ g H}$$

The total mass of C and H is therefore the sum of these two quantities.

$$\text{Total mass of C and H} = 0.2836 \text{ g C} + 0.07127 \text{ g H} = 0.3549 \text{ g}$$

The difference between this total and the 0.5438 g in the original sample is the mass of oxygen (the only other element).

$$\text{mass of O} = 0.5438 \text{ g} - 0.3549 \text{ g} = 0.1889 \text{ g O}$$

Now we can convert the masses of the elements to an empirical formula.

For C:
$$0.2836 \text{ g C} \times \frac{1 \text{ mol C}}{12.011 \text{ g C}} = 0.02361 \text{ mol C}$$

For H:
$$0.07127 \text{ g H} \times \frac{1 \text{ mol H}}{1.0079 \text{ g H}} = 0.07071 \text{ mol H}$$

For O:
$$0.1889 \text{ g O} \times \frac{1 \text{ mol O}}{15.999 \text{ g O}} = 0.01181 \text{ mol O}$$

Our preliminary empirical formula is thus $C_{0.02361}H_{0.07071}O_{0.01181}$. We divide all of these subscripts by the smallest number, 0.01181.

$$C_{\frac{0.02361}{0.01181}}H_{\frac{0.07071}{0.01181}}O_{\frac{0.01181}{0.01181}} = C_{1.999}H_{5.987}O_{1.000}$$

The results are acceptably close to C_2H_6O, the answer.

Practice Exercise 15

The combustion of a 5.048 g sample of a compound of C, H, and O gave 7.406 g CO_2 and 3.027 g H_2O. Calculate the empirical formula of the compound. ◆

As we said earlier, more than one sample of a substance must be analyzed whenever more than one reaction is necessary to separate the elements. This is also true in indirect analyses. For example, if a compound contains C, H, N, and O, combustion converts the C and H to CO_2 and H_2O, which are separated by special techniques and weighed. The mass of C in the CO_2 sample and the mass of H in the H_2O sample are then calculated in the usual way. A second reaction with a different sample of the compound can be used to obtain the nitrogen, either as N_2 or as NH_3. Because different size samples are used, the masses of C and H from one sample and the mass of N from another cannot be used to calculate the empirical formula. So we convert the masses of the elements found in their respective samples into percentages of the element by mass in the compound. We then add up the percentages, subtract from 100 to get the percentage of O, and then calculate the empirical formula from the percentage composition as described earlier.

$\%C + \%H + \%N + \%O$

$= 100\%$

Molecular Formulas from Empirical Formulas and Molecular Masses

What we obtain by using mass data to calculate a formula is an *empirical* formula, as we have said, and the empirical formula is all that we can get for ionic compounds. For molecular compounds, however, chemists prefer *molecular* formulas because they contain more information, such as the exact composition of one molecule.

Sometimes an empirical formula and a molecular formula are the same. Two examples are H_2O and NH_3. Usually, however, the subscripts of a molecular formula are whole-number multiples of those in the empirical formula. The subscripts of the molecular formula P_4O_{10}, for example, are each two times those in the empirical formula, P_2O_5, as you saw earlier. The molecular mass of P_4O_{10} is likewise two times the formula mass of P_2O_5. This observation provides us with a way to find out the molecular formula for a compound provided we have a way of determining experimentally the molecular mass of the compound. (Later in this book you will learn how molecular masses can be measured.) If the experimental molecular mass *equals* the calculated empirical formula mass, the empirical formula itself is also a molecular formula. Otherwise, the experimental molecular mass will be some whole-number multiple of the value calculated from the empirical formula. Whatever the whole number is, it's a common multiplier for the subscripts of the empirical formula.

When the empirical and molecular formulas are identical, the term of choice is "molecular formula."

EXAMPLE 3.15
Calculating a Molecular Formula

Styrene, the raw material for polystyrene foam plastics, has an empirical formula of CH. Its molecular mass is 104. What is its molecular formula?

Analysis: The molecular mass of styrene, 104, is some simple multiple of the formula mass calculated for its empirical formula, CH. So we have to compute the latter, and then see how many times larger 104 is.

Solution: For CH, the formula mass is

$$12.01 + 1.008 = 13.02$$

To find how much larger 104 is than 13.02, we divide.

$$\frac{104}{13.02} = 7.99$$

Rounding this to 8, we see that 104 is 8 times larger than 13.02, so the correct molecular formula of styrene must have subscripts 8 times those in CH. Styrene, therefore, is C_8H_8.

Practice Exercise 16

The empirical formula of hydrazine is NH_2, and its molecular mass is 32.0. What is its molecular formula? ◆

3.5 Writing and Balancing Chemical Equations

In a chemistry lab, much of our time is spent dealing with chemical reactions. It is natural, therefore, to work with chemical equations. We use them to describe what happens and, for quantitative work, we use them to plan experiments.

Chemical Equations

Coefficient
↓
$3CO_2$
↑
Subscript

We learned in Chapter 2 that a *chemical equation* is a shorthand, quantitative description of a chemical reaction. An equation is *balanced* when all atoms present among the reactants are also somewhere among the products. As we learned, coefficients, the numbers in front of formulas, are multiplier numbers for their respective formulas, and the values of the coefficients determine whether an equation is balanced.

Balancing Equations

Always approach the balancing of an equation as a two-step process.

Step 1. *Write the unbalanced "equation."* Organize the formulas in the pattern of an equation with plus signs and an arrow. Use *correct* formulas. (You learned to write many of them in Chapter 2, but until we have studied more chemistry, you will usually be given formulas.)

Step 2. *Adjust the coefficients to get equal numbers of each kind of atom on both sides of the arrow.*

When doing step 2, make no changes in the formulas, either in the atomic symbols or their subscripts. If you do, the equation will involve different substances from those intended. You may still be able to balance it, but the equation will not be for the reaction you want.

We'll begin with simple equations that can be balanced easily by inspection. An example is the reaction of zinc metal with hydrochloric acid (see margin photo). First, we need the correct formulas, and this time we'll include the physical states because they are different. The reactants are zinc, $Zn(s)$, and hydrochloric acid, an aqueous solution of the gas hydrogen chloride, HCl, and so symbolized as $HCl(aq)$. We also need formulas for the products. Zn changes to a water-soluble compound, zinc chloride, $ZnCl_2(aq)$, and hydrogen gas, $H_2(g)$, bubbles out as the other product. (Recall that hydrogen occurs naturally as a *diatomic molecule,* not as atoms.)

The reactants could be displayed as $HCl(aq) + Zn(s)$. The products could be shown as $H_2(g) + ZnCl_2(aq)$. The orders, in other words, are unimportant.

Step 1. Write an unbalanced equation.

$$Zn(s) + HCl(aq) \longrightarrow ZnCl_2(aq) + H_2(g) \qquad \text{(unbalanced)}$$

Step 2. Adjust the coefficients to get equal numbers of each kind of atom on both sides of the arrow.

There is no simple set of rules for adjusting coefficients. Experience is the greatest help, and experience has taught chemists that the following guidelines often get to the solution most directly when they are applied in the order given.

Some Guidelines for Balancing Equations

1. Balance elements other than H and O first.

2. Balance as a group those polyatomic ions that appear unchanged on both sides of the arrow.

3. Balance separately those elements that appear somewhere by themselves (e.g., Zn in our example).

Using the guidelines given here, we'll look at Cl first in our example. Because there are two Cl to the right of the arrow but only one to the left, we put a 2 in front of the HCl on the left side. The result is

$$Zn(s) + 2HCl(aq) \longrightarrow ZnCl_2(aq) + H_2(g)$$

Everything is now balanced. On each side we find 1 Zn, 2 H, and 2 Cl. Equations that can be balanced by inspection do not get much harder than this.

One complication is that an infinite number of *balanced* equations can be written for any given reaction! We might, for example, have adjusted the coefficients so that our equation came out as follows.

$$2Zn(s) + 4HCl(aq) \longrightarrow 2ZnCl_2(aq) + 2H_2(g)$$

This equation is also balanced. So we add one further convention to the rules about balanced equations, and it is only a *convention*, not a law of nature. The *smallest* whole-number coefficients are generally used to write balanced equations. Occasionally we ignore this convention, but only for a special reason.

Zinc metal reacts with hydrochloric acid.

EXAMPLE 3.16

Writing a Balanced Equation

Sodium hydroxide, NaOH, and phosphoric acid, H_3PO_4, react as aqueous solutions to give sodium phosphate, Na_3PO_4, and water. The sodium phosphate remains in solution. Write the balanced equation for this reaction.

Solution: We first write an unbalanced equation that includes all of the formulas arranged according to the conventions for an equation. We include the designation (*aq*) for all substances dissolved in water (except H_2O itself; we'll not give it any designation when it is in its liquid state).

$$NaOH(aq) + H_3PO_4(aq) \longrightarrow Na_3PO_4(aq) + H_2O \quad \text{(unbalanced)}$$

There are several things not in balance, but our guidelines suggest that we work with Na first rather than with O, H, or PO_4. There are 3 Na on the right side, so we put a 3 in front of NaOH on the left, as a trial.

$$3NaOH(aq) + H_3PO_4(aq) \longrightarrow Na_3PO_4(aq) + H_2O \quad \text{(unbalanced)}$$

Now the Na are in balance. The unit of PO_4 is balanced also. Not counting the PO_4, we have on the left 3 O and 3 H in 3NaOH plus 3 H in H_3PO_4, for a net of 3 O and 6 H on the left. On the right, in H_2O, we have 1 O and 2 H. The ratio of 3 O to 6 H on the left is equivalent to the ratio of 1 O to 2 H on the right, so we write the multiplier (coefficient) 3 in front of H_2O.

$$3NaOH(aq) + H_3PO_4(aq) \longrightarrow Na_3PO_4(aq) + 3H_2O \quad \text{(balanced)}$$

We now have a balanced equation. On each side we have 3 Na, 1 PO_4, 6 H, and 3 O besides those in PO_4, and it is obvious that our coefficients cannot be reduced to smaller whole numbers.

In deciding where to start, try leaving the balancing of H and O atoms (when they are involved) to the last.

"PO_4" is actually the phosphate ion, $PO_4{}^{3-}$.

Practice Exercise 17

When aqueous solutions of calcium chloride, $CaCl_2$, and potassium phosphate, K_3PO_4, are mixed, a reaction occurs in which solid calcium phosphate, $Ca_3(PO_4)_2$, separates from the solution. The other product is $KCl(aq)$. Write the balanced equation. ◆

The strategy in Example 3.16 of balancing whole units of atoms, like PO_4, unchanged is extremely useful. These are usually polyatomic ions that go through reactions unaffected. If we leave them alone, we have less atom counting to do.

Another suggestion for picking trial coefficients for balancing an equation is to use subscripts to suggest them. The reaction of aluminum with oxygen illustrates how this works.

EXAMPLE 3.17

Balancing Chemical Equations

Although they are bright and shiny, aluminum objects are actually covered with a tight, invisible coating of solid aluminum oxide, Al_2O_3. This coating forms instantly when fresh aluminum metal is exposed to oxygen in the air. Write the balanced equation for the formation of this oxide. Remember that oxygen occurs in air as the diatomic molecule, $O_2(g)$.

Solution: The unbalanced equation comes first.

$$Al(s) + O_2(g) \longrightarrow Al_2O_3(s)$$

It doesn't matter where we start, but starting with the *least simple* adjustment can actually work best overall. So let's begin with oxygen. More O is on the right than on the left, so we will try its *subscript* in Al_2O_3 as the *coefficient* of O_2 on the left. We write 3 in front of O_2.

Subscript of O suggests
coefficient of O_2

$$Al(s) + 3O_2(g) \longrightarrow Al_2O_3(s)$$

Now we have (3×2) O on the left yet only 3 O on the right, but now the *subscript* in O_2 suggests a *coefficient* for Al_2O_3. So we place a 2 before Al_2O_3.

Subscript of O_2 suggests
coefficient of Al_2O_3

$$Al(s) + 3O_2(g) \longrightarrow 2Al_2O_3(s)$$

This leaves (2×2) Al on the right, so we write a 4 in front of Al on the left, and the equation is then balanced.

$$4Al(s) + 3O_2(g) \longrightarrow 2Al_2O_3(s)$$

Practice Exercise 18

Iron reacts with chlorine, a gaseous, diatomic element, to form the solid, $FeCl_3$. Write the balanced equation. ◆

3.6 Using Chemical Equations in Calculations

A balanced equation serves as a tool for relating moles of substances that react or are produced.

For quantitative work the most significant thing about an equation is that the coefficients give us the *ratios by moles* of the reactants and products. In the following reaction, for example,

$$3NaOH(aq) + H_3PO_4(aq) \longrightarrow Na_3PO_4(aq) + 3H_2O$$

the coefficients tell us that to make 1 mol of Na_3PO_4 from 1 mol of H_3PO_4, we must also use 3 mol of NaOH. We're not compelled, of course, to carry out the reaction with these actual numbers of moles. The coefficients tell us only that we must take whatever quantities we do choose in the *proportions* set by the coefficients. *The chemist makes the decision about the actual quantities, the magnitude, or the scale of the reaction. The natural world tells us what must be the combining proportions in moles, whatever the scale may be.* Regardless of the scale of a reaction, the coefficients of a chemical equation give the ratio in which the *moles* of one substance react with or produce *moles* of another. When you fully understand this fact, you will have mastered the heart of the stoichiometry of chemical reactions.

It is useful to look upon a balanced equation as a calculating tool, because its coefficients give us *stoichiometric equivalencies* between the substances involved. We have the following equivalencies, for example, in the reaction shown above between sodium hydroxide and phosphoric acid.

Use of a balanced equation

$$3 \text{ mol NaOH} \Leftrightarrow 1 \text{ mol H}_3\text{PO}_4$$

$$3 \text{ mol NaOH} \Leftrightarrow 1 \text{ mol Na}_3\text{PO}_4$$

$$3 \text{ mol NaOH} \Leftrightarrow 3 \text{ mol H}_2\text{O}$$

$$1 \text{ mol H}_3\text{PO}_4 \Leftrightarrow 1 \text{ mol Na}_3\text{PO}_4$$

$$1 \text{ mol H}_3\text{PO}_4 \Leftrightarrow 3 \text{ mol H}_2\text{O}$$

Any of these can be used to construct conversion factors for stoichiometric calculations.

EXAMPLE 3.18

Stoichiometry of Chemical Reactions

How many moles of sodium phosphate, Na_3PO_4, can be made from 0.240 mol of NaOH by the following reaction?

$$3\text{NaOH}(aq) + \text{H}_3\text{PO}_4(aq) \longrightarrow \text{Na}_3\text{PO}_4(aq) + 3\text{H}_2\text{O}$$

Analysis: The question asks about a *moles-to-moles relationship* within a given reaction. We can restate the question as follows.

$$0.240 \text{ mol NaOH} \Leftrightarrow ? \text{ mol Na}_3\text{PO}_4$$

The balanced equation is our tool because it gives the coefficients we need. From the coefficients, we know the following.

$$3 \text{ mol NaOH} \Leftrightarrow 1 \text{ mol Na}_3\text{PO}_4$$

This enables us to prepare the conversion factor that we need.

Solution: We convert 0.240 mol NaOH to the moles of Na_3PO_4 equivalent to it in the reaction as follows.

$$0.240 \text{ mol NaOH} \times \frac{1 \text{ mol Na}_3\text{PO}_4}{3 \text{ mol NaOH}} = 0.0800 \text{ mol Na}_3\text{PO}_4$$

Thus, in this reaction,

$$0.240 \text{ mol NaOH} \Leftrightarrow 0.0800 \text{ mol Na}_3\text{PO}_4$$

Thus we can make 0.0800 mol Na_3PO_4 from 0.240 mol NaOH.

Is the Answer Reasonable?
Don't forget to ask, "Does the *size* of the answer make sense?" Yes. The equation told us that 3 mol NaOH $\Leftrightarrow$ 1 mol Na_3PO_4, so the actual number of moles of Na_3PO_4 (0.0800 mol) should be one-third the actual number of moles of NaOH (0.240 mol), and it is.

Practice Exercise 19

How many moles of sulfuric acid, H_2SO_4, are needed to react with 0.366 mol of NaOH by the following reaction?

$$2NaOH(aq) + H_2SO_4(aq) \longrightarrow Na_2SO_4(aq) + 2H_2O \; \blacklozenge$$

EXAMPLE 3.19

Reaction Stoichiometry

Amateur gold seekers, unaware of the extremely poisonous nature of sodium cyanide, sometimes died using it carelessly.

Gold metal is obtained from $NaAu(CN)_2$ by a reaction with zinc metal.

$Zn(s) + 2NaAu(CN)_2(aq) \longrightarrow$

$2Au(s) + Na_2Zn(CN)_4(aq)$

Gold particles can be separated from crushed rock by passing air and an aqueous solution of sodium cyanide, NaCN, through the mixture. The gold dissolves by the following reaction in which oxygen in the air is also a reactant.

$$2Au(s) + 4NaCN(aq) + O_2(g) + 2H_2O \longrightarrow$$
$$2NaAu(CN)_2(aq) + H_2O_2(aq) + 2NaOH(aq)$$

How many moles of NaCN are required to dissolve 0.136 mol of Au?

Analysis: The question can be restated as follows.

$$0.136 \text{ mol Au} \Leftrightarrow ? \text{ mol NaCN}$$

The coefficients tell us the following.

$$2 \text{ mol Au} \Leftrightarrow 4 \text{ mol NaCN}$$

We will use this equivalency to make a conversion factor.

Solution: We multiply the given, 0.136 mol Au, by the appropriate conversion factor.

$$0.136 \text{ mol Au} \times \frac{4 \text{ mol NaCN}}{2 \text{ mol Au}} = 0.272 \text{ mol NaCN}$$

Thus the reaction requires 0.272 mol NaCN.

Is the Answer Reasonable?
Does the *size* of the answer make sense? Of course. The number of moles of NaCN has to be twice the number of moles of Au because the ratio of the coefficients in 4NaCN and 2Au is 4:2 or 2:1.

Practice Exercise 20

If 0.575 mol of CO_2 is produced by the combustion of propane, C_3H_8, how many moles of oxygen are consumed? The balanced equation is

$$C_3H_8 + 5O_2 \longrightarrow 3CO_2 + 4H_2O \; \blacklozenge$$

Before continuing, please notice a point of major importance. *Stoichiometric equivalencies obtained from an equation's coefficients apply only to the given reaction.* It is possible in the lab, for example, to arrange a reaction between sodium hydroxide and phosphoric acid according to any of the following equations.

$$NaOH(aq) + H_3PO_4(aq) \longrightarrow NaH_2PO_4(aq) + H_2O$$

where 1 mol NaOH $\Leftrightarrow$ 1 mol H_3PO_4

$$2NaOH(aq) + H_3PO_4(aq) \longrightarrow Na_2HPO_4(aq) + 2H_2O$$

where 2 mol NaOH $\Leftrightarrow$ 1 mol H_3PO_4

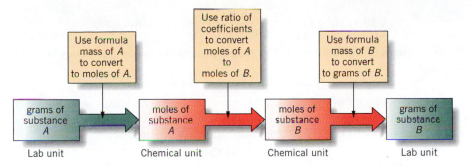

Figure 3.3 *The flow of the calculation for stoichiometry problems.* The flow applies to the calculation of the required or the expected mass of *any* reactant or product.

$$3NaOH(aq) + H_3PO_4(aq) \longrightarrow Na_3PO_4(aq) + 3H_2O$$

$$\text{where } 3 \text{ mol NaOH} \Leftrightarrow 1 \text{ mol H}_3PO_4$$

The stoichiometric equivalencies between NaOH and H_3PO_4 are different in all three reactions. *Stoichiometric equivalencies between substances are therefore always a function of the particular reaction being studied.*

Stoichiometric Mass Calculations

In practical work, a chemist is often confronted by a question such as the following. "If I start with so many grams of reactant *A*, how many grams of reactant *B* ought I use, and how many grams of a particular product should be produced?" Notice that the question concerns *grams,* not moles, for the practical reason that masses in *grams* are delivered by laboratory balances. The coefficients of the desired reaction, however, say nothing about grams, only about relative numbers of moles. This is why the solutions to such questions flow from a given mass to the corresponding moles, then through stoichiometric equivalencies to the moles of something else, and finally to the equivalent in mass of the latter. Figure 3.3 outlines this flow. We emphasize again that when we know two facts, namely, the *balanced equation* and the *mass* of any substance in it, we can calculate the required or expected mass of *any* other substance in the equation. Example 3.20 illustrates how it works.

EXAMPLE 3.20

Stoichiometric Mass Calculations

Portland cement is a mixture of the oxides of calcium, aluminum, and silicon. The raw material for its calcium oxide is calcium carbonate, which occurs as the chief component of a natural rock, limestone. When calcium carbonate is strongly heated it decomposes by the following reaction. One product, CO_2, is driven off to leave the desired CaO as the only other product.

$$CaCO_3(s) \xrightarrow{\text{heat}} CaO(s) + CO_2(g)$$

A chemistry student is to prepare 1.50×10^2 g of CaO in order to test a particular "recipe" for portland cement. How many grams of $CaCO_3$ should be used, assuming that all will be converted?

Analysis: We're given the number of *grams* of one substance, but we know that the fundamental chemical reasoning must be done in terms of *moles.* So we must first translate the desired mass of product into moles. Then we can employ the balanced equation as a tool, using its coefficients and the calculated number of moles of CaO to find the equivalent number of moles of the reactant, $CaCO_3$. Finally, we convert the calculated moles of the reactant into grams.

Solution: Making a conversion factor out of the formula mass of CaO, 56.08, we translate 1.50×10^2 g CaO into moles as follows. Draw in all appropriate cancel lines.

Portland cement gets its name from the color of stone formations near Portland, England.

CaO is commonly called "quicklime" and it is an ingredient in portland cement.

$$1.50 \times 10^2 \text{ g CaO} \times \frac{1 \text{ mol CaO}}{56.08 \text{ g CaO}} = 2.67 \text{ mol CaO}$$

The coefficients of the equation tell us that the stoichiometric equivalency between CaO and $CaCO_3$ is

$$1 \text{ mol CaCO}_3 \Leftrightarrow 1 \text{ mol CaO}$$

Concrete is a hardened mixture of cement and sand, gravel, or rock.

So the 2.67 mol of CaO requires 2.67 mol of $CaCO_3$. To weigh out this much $CaCO_3$, the student must convert the number of moles of $CaCO_3$ into grams. The formula mass of $CaCO_3$ is 100.09, so we do the following calculation.

$$2.67 \text{ mol CaCO}_3 \times \frac{100.09 \text{ g CaCO}_3}{1 \text{ mol CaCO}_3} = 2.67 \times 10^2 \text{ g CaCO}_3$$

Thus to prepare 1.50×10^2 g of CaO, 2.67×10^2 g of $CaCO_3$ has to be used.

Chain Calculation Route: Had we solved the problem by a chain calculation, not leaving out the obvious stoichiometric step of the 1:1 mole ratio of mol $CaCO_3$ to mol CaO, the setup would have looked like

$$150 \text{ g CaO} \times \frac{1 \text{ mol CaO}}{56.08 \text{ g CaO}} \times \frac{1 \text{ mol CaCO}_3}{1 \text{ mol CaO}} \times \frac{100.09 \text{ g CaCO}_3}{1 \text{ mol CaCO}_3} = 268 \text{ g CaCO}_3$$

grams CaO $\longrightarrow$ moles CaO $\longrightarrow$ moles $CaCO_3$ $\longrightarrow$ grams $CaCO_3$

The difference between the first answer, 267 g, and the second, 268 g, is caused by rounding. Notice how the calculation flows from grams of CaO to moles of CaO, then to moles of $CaCO_3$ (using the equation), and finally to grams of $CaCO_3$. We cannot emphasize too much that *the key step in all calculations of reaction stoichiometry is the use of the balanced equation.*

Is the Answer Reasonable?
Does the *size* of the answer, which is numerically larger than the grams of CaO to be made, make sense? It does, because the formula mass of $CaCO_3$ is larger than that of CaO. ◆

EXAMPLE 3.21

Stoichiometric Mass Calculations

One of the most spectacular reactions of aluminum, the thermite reaction, is with iron oxide, Fe_2O_3, by which metallic iron is made. So much heat is generated that the iron forms in the liquid state (Figure 3.4). The equation is

$$2Al(s) + Fe_2O_3(s) \longrightarrow Al_2O_3(s) + 2Fe(l)$$

A certain welding operation, used over and over, requires that each time at least 86.0 g of Fe be produced. What is the minimum mass in grams of Fe_2O_3 that must be used for each operation? (Do a chain calculation.) Calculate also how many grams of aluminum are needed.

Analysis: Remember that all problems in reaction stoichiometry must be solved at the mole level because an equation's coefficients disclose *mole* ratios, not mass ratios. So we have to convert the number of grams of Fe to moles. Then we can use the stoichiometric equivalency given by the coefficients in the balanced equation,

$$1 \text{ mol Fe}_2O_3 \Leftrightarrow 2 \text{ mol Fe}$$

to see how many *moles* of Fe_2O_3 are needed. We finally convert this answer into grams of Fe_2O_3. The other calculations follow the same pattern.

Solution: We'll set up the first calculation as a chain. Draw all cancel lines.

$$86.0 \text{ g Fe} \times \frac{1 \text{ mol Fe}}{55.85 \text{ g Fe}} \times \frac{1 \text{ mol Fe}_2O_3}{2 \text{ mol Fe}} \times \frac{159.70 \text{ g Fe}_2O_3}{1 \text{ mol Fe}_2O_3} = 123 \text{ g Fe}_2O_3$$

grams Fe $\longrightarrow$ moles Fe $\longrightarrow$ moles Fe_2O_3 $\longrightarrow$ grams Fe_2O_3

A minimum of 123 g of Fe_2O_3 is required to make 86.0 g of Fe.

Next, we calculate the number of grams of Al needed, but we know that we must first find the number of *moles* of Al required. Only from this can the grams of Al be calculated. The relevant stoichiometric equivalency, again using the balanced equation, is

$$2 \text{ mol Al} \Leftrightarrow 2 \text{ mol Fe}$$

Expressing this in the smallest whole numbers, we have

$$1 \text{ mol Al} \Leftrightarrow 1 \text{ mol Fe}$$

Employing another chain calculation to find the mass of Al needed to make 86.0 g of Fe, we have (using 26.98 as the atomic mass of Al)

$$86.0 \text{ g Fe} \times \frac{1 \text{ mol Fe}}{55.85 \text{ g Fe}} \times \frac{1 \text{ mol Al}}{1 \text{ mol Fe}} \times \frac{26.98 \text{ g Al}}{1 \text{ mol Al}} = 41.5 \text{ g Al}$$

$$\text{grams Fe} \longrightarrow \text{moles Fe} \longrightarrow \text{moles Al} \longrightarrow \text{grams Al}$$

Is the Answer Reasonable?

Think about the size of the answer, 41.5 g of Al, in relationship to the following: the mass of Fe needed, the stoichiometric equivalency in this reaction of Fe and Al (1 mol : 1 mol), and the relative atomic masses of Fe (55.85) and Al (26.98). Does the answer make sense? (Should it take *less* mass of Al than the mass of Fe made?)

Practice Exercise 21

How many grams of aluminum oxide are also produced by the reaction described in Example 3.21? ◆

Figure 3.4 *The thermite reaction.* Pictured here is a device for making white hot iron by the reaction of aluminum with iron oxide and letting the molten iron run down into a mold between the ends of two steel railroad rails. The rails are thereby welded together.

3.7 Limiting Reactant Calculations

In many situations a chemist deliberately mixes reactants in a mole ratio that does not agree with the coefficients of the equation. Some reactions proceed better when one reactant is in stoichiometric excess, for example. One such situation is the preparation of ammonia, NH_3, from its elements. The equation is

$$N_2(g) + 3H_2(g) \longrightarrow 2NH_3(g)$$

Suppose that a chemist mixed 1.00 mol of N_2 with 5.00 mol of H_2. What is the maximum number of moles of product that could form? We first note that the coefficients tell us that 1 mol of N_2 consumes 3 mol of H_2.

$$1 \text{ mol } N_2 \Leftrightarrow 3 \text{ mol } H_2$$

But 5 mol of H_2 was used, not 3, so there will be 2 mol of H_2 left over. Once the 1 mol of N_2 taken is consumed, no additional NH_3 can form. In this sense, the reactant that is *completely* consumed limits the amount of product that forms, so it is called the **limiting reactant.** *We must always base the calculation of the maximum yield of product on the stoichiometric equivalency between it and the limiting reactant.* Because N_2 is the limiting reactant in our example, the equivalency we must use is

$$1 \text{ mol } N_2 \Leftrightarrow 2 \text{ mol } NH_3$$

This tells us that the 1.00 mol of N_2 can be converted to a maximum of 2.00 mol of NH_3.

The identity of the limiting reactant depends on the actual composition of the mixture of starting materials. Suppose, for example, that we had taken a mixture of 2.00 mol N_2 and 5.00 mol H_2. We can see (from 1 mol $N_2 \Leftrightarrow 3$ mol H_2) that 2.00 mol of N_2 will require 6.00 mol of H_2, but we took only 5.00 mol of H_2.

The industrial synthesis of ammonia is a major operation worldwide, because ammonia is applied to fields as a source of nitrogen that crops can use.

Therefore all of the H_2 will be used up, and so now H_2 is the limiting reactant. The amount of product that forms (based on 3 mol $H_2 \Leftrightarrow$ 2 mol NH_3) is found as follows.

$$5.00 \ \overline{\text{mol } H_2} \times \frac{2.00 \text{ mol } NH_3}{3.00 \ \overline{\text{mol } H_2}} = 3.33 \text{ mol } NH_3$$

In Example 3.22 we'll see how to solve limiting reactant problems when the amounts of the reactants are given in mass units.

EXAMPLE 3.22

Limiting Reactant

The death mask of ancient Egyptian King Tutankhamun is made of gold, lapis lazuli, carnelian, turquoise, and other materials.

Gold(III) hydroxide, $Au(OH)_3$, is used for electroplating gold onto other metals. It can be made by the following reaction.

$$2KAuCl_4(aq) + 3Na_2CO_3(aq) + 3H_2O \longrightarrow$$

$$2Au(OH)_3(aq) + 6NaCl(aq) + 2KCl(aq) + 3CO_2(g)$$

To prepare a fresh supply of $Au(OH)_3$, a chemist at an electroplating plant has mixed 20.00 g of $KAuCl_4$ with 25.00 g of Na_2CO_3 (both dissolved in a large excess of water). What is the maximum number of grams of $Au(OH)_3$ that can form?

Analysis: The clue that tells us to *begin* the solution by figuring out what is the limiting reactant is that *the quantities of two reactants are given.* Once we determine which reactant limits the product, we use its mass in our calculation.

To find the limiting reactant, we arbitrarily pick one of the candidates ($KAuCl_4$ or Na_2CO_3)—there is no general rule for making the choice—and calculate whether it would all be used up. If so, we've found the limiting reactant. If not, the other reactant limits. (We were told that water is in excess, so we know that it does not limit the reaction.)

Solution: We'll arbitrarily work with the first candidate, $KAuCl_4$, and calculate how many grams of the second, Na_2CO_3, *should* be provided to react with 20.00 g of $KAuCl_4$. The formula masses are 377.88 for $KAuCl_4$, and 105.99 for Na_2CO_3. We'll set up a chain calculation using the following stoichiometric equivalency. Set all of the cancel lines yourself.

$$2 \text{ mol } KAuCl_4 \Leftrightarrow 3 \text{ mol } Na_2CO_3$$

$$20.00 \text{ g } KAuCl_4 \times \frac{1 \text{ mol } KAuCl_4}{377.88 \text{ g } KAuCl_4} \times \frac{3 \text{ mol } Na_2CO_3}{2 \text{ mol } KAuCl_4} \times \frac{105.99 \text{ g } Na_2CO_3}{1 \text{ mol } Na_2CO_3} = 8.415 \text{ g } Na_2CO_3$$

grams $KAuCl_4$ $\longrightarrow$ moles $KAuCl_4$ $\longrightarrow$ moles Na_2CO_3 $\longrightarrow$ grams Na_2CO_3

We find that 20.00 g of $KAuCl_4$ needs 8.415 g of Na_2CO_3. The 25.00 g of Na_2CO_3 taken is therefore more than enough to let the $KAuCl_4$ react completely. $KAuCl_4$ is limiting.

Had we tested Na_2CO_3 instead of $KAuCl_4$ as the limiting reactant we would have calculated how many grams of $KAuCl_4$ are required by 25.00 g of Na_2CO_3.

$$25.00 \text{ g } Na_2CO_3 \times \frac{1 \text{ mol } Na_2CO_3}{105.99 \text{ g } Na_2CO_3} \times \frac{2 \text{ mol } KAuCl_4}{3 \text{ mol } Na_2CO_3} \times \frac{377.88 \text{ g } KAuCl_4}{1 \text{ mol } KAuCl_4} = 59.42 \text{ g } KAuCl_4$$

grams Na_2CO_3 $\longrightarrow$ moles Na_2CO_3 $\longrightarrow$ moles $KAuCl_4$ $\longrightarrow$ grams $KAuCl_4$

The given mass of Na_2CO_3 would require much more $KAuCl_4$ than provided, so we confirm that $KAuCl_4$ is the limiting reactant.

From here on, we have a routine calculation going from mass of reactant, $KAuCl_4$, to mass of product, $Au(OH)_3$. We know from the equation's coefficients that

$$1 \text{ mol } KAuCl_4 \Leftrightarrow 1 \text{ mol } Au(OH)_3$$

Using this, plus formula masses, we do the following chain calculation.

$$20.00 \text{ g KAuCl}_4 \times \frac{1 \text{ mol KAuCl}_4}{377.88 \text{ g KAuCl}_4} \times \frac{1 \text{ mol Au(OH)}_3}{1 \text{ mol KAuCl}_4} \times \frac{247.99 \text{ g Au(OH)}_3}{1 \text{ mol Au(OH)}_3} = 13.13 \text{ g Au(OH)}_3$$

grams KAuCl$_4$ $\longrightarrow$ moles KAuCl$_4$ $\longrightarrow$ moles Au(OH)$_3$ $\longrightarrow$ grams Au(OH)$_3$

Thus from 20.00 g of KAuCl$_4$ we can make a maximum of 13.13 g of Au(OH)$_3$.

In this synthesis, some of the initial 25.00 g of Na$_2$CO$_3$ is left over; 20.00 g of KAuCl$_4$ requires only 8.415 g of Na$_2$CO$_3$ out of 25.00 g Na$_2$CO$_3$. The difference, (25.00 g − 8.415 g) = 16.58 g of Na$_2$CO$_3$, remains unreacted. It is possible that the chemist used an excess to ensure that every last bit of the very expensive KAuCl$_4$ would be changed to Au(OH)$_3$.

Are the Answers Reasonable?
The "head check" has to be done at each step. We'll illustrate just the first. Is it reasonable that around 8 g of Na$_2$CO$_3$ would be equivalent to 20 g of KAuCl$_4$ in the reaction whose equation is given? Yes. Notice that the formula mass of Na$_2$CO$_3$ is between $\frac{1}{3}$ and $\frac{1}{4}$ the size of the formula mass of KAuCl$_4$. So on this score alone, it's reasonable that the calculated mass of Na$_2$CO$_3$ be quite a bit less than the mass of KAuCl$_4$ taken. Of course, the equation calls for 3 mol of Na$_2$CO$_3$ to 2 mol of KAuCl$_4$, so this ratio calls for more Na$_2$CO$_3$ than thus far suggested in our "head check." To correct for this, we multiply $\frac{1}{4}$ by $\frac{3}{2}$ to give $\frac{3}{8}$ and we multiply $\frac{3}{8}$ by 20 g (of KAuCl$_4$) to give us a little less than the more precisely calculated mass of Na$_2$CO$_3$. So, the answer is reasonable.

Practice Exercise 22

In an industrial process for making nitric acid, the first step is the reaction of ammonia with oxygen at high temperature in the presence of a platinum gauze. Nitrogen monoxide forms as follows.

$$4NH_3 + 5O_2 \longrightarrow 4NO + 6H_2O$$

How many grams of nitrogen monoxide can form if a mixture of 30.00 g of NH$_3$ and 40.00 g of O$_2$ is taken initially? ◆

3.8 Theoretical Yield and Percentage Yield

In most experiments designed for chemical synthesis, the amount of a product actually isolated falls short of the calculated maximum amount. Losses occur for several reasons. Some are mechanical, such as materials sticking to glassware. In some reactions, losses occur by the evaporation of a *volatile* product. In others, a product is a solid that precipitates from the solution as it forms because it is largely insoluble. The solid is removed by filtration. What stays in solution, although relatively small, contributes to some loss of product.

One of the common causes of obtaining less than the stoichiometric amount of a product is the occurrence of a **competing reaction.** It produces a **by-product,** a substance made by a reaction that competes with the **main reaction.** The synthesis of phosphorus trichloride, for example, gives some phosphorus pentachloride as well, because PCl$_3$ can react further with Cl$_2$.

Main reaction: $\quad 2P(s) + 3Cl_2(g) \longrightarrow 2PCl_3(l)$

Competing reaction: $\quad PCl_3(l) + Cl_2(g) \longrightarrow PCl_5(s)$

The competition is between newly formed PCl$_3$ and still unreacted phosphorus for still unchanged chlorine.

The **actual yield** of desired product is simply how much is isolated, stated in mass units or moles. The **theoretical yield** of the product is what must be obtained if no losses occur. When less than the theoretical yield of product is ob-

A **volatile** liquid is one with a low boiling point that evaporates quickly, like acetone (bp 56 °C), which is used in nail polish remover.

A competing reaction is often referred to as a **side reaction.**

Theoretical yield is a calculated quantity. The actual yield is a measured quantity; in general, it cannot be calculated.

tained, chemists generally calculate the *percentage yield* of product to describe how well the preparation went. The **percentage yield** is the actual yield calculated as a percentage of the theoretical yield.

Tools →

Theoretical, actual, and percentage yields

$$\text{Percentage yield} = \frac{\text{actual yield}}{\text{theoretical yield}} \times 100\%$$

Both the actual and theoretical yields must be in the same units, of course. Let's now work an example that combines the determination of the limiting reactant with a calculation of percentage yield.

EXAMPLE 3.23

Calculating a Percentage Yield

A chemist set up a synthesis of phosphorus trichloride by mixing 12.0 g P with 35.0 g Cl_2 and obtained 42.4 g of PCl_3. Calculate the percentage yield of this compound. The equation for the main reaction is, as we noted earlier,

$$2P(s) + 3Cl_2(g) \longrightarrow 2PCl_3(l)$$

Analysis: Notice that the masses of *both* reactants are given, so we know at the outset that we have a limiting reactant problem. We must first figure out which reactant, P or Cl_2, is the limiting reactant, because we have to base the calculation of the theoretical yield of PCl_3 on it.

Solution: Recall that in any limiting reactant problem, we arbitrarily pick one reactant and do a calculation to see whether it can be entirely used up. We'll choose phosphorus and see whether there is enough to react with 35.0 g of chlorine. The following chain calculation gives us the answer. Put in the cancel lines.

$$12.0 \text{ g P} \times \frac{1 \text{ mol P}}{30.97 \text{ g P}} \times \frac{3 \text{ mol Cl}_2}{2 \text{ mol P}} \times \frac{70.90 \text{ g Cl}_2}{1 \text{ mol Cl}_2} = 41.2 \text{ g Cl}_2$$

Thus, with 35.0 g of Cl_2 provided but 41.2 g of Cl_2 needed, there is not enough Cl_2 for the 12.0 g of P. The Cl_2 can be all used up, so Cl_2 is the limiting reactant. We therefore base the calculation of the theoretical yield of PCl_3 on Cl_2.

To find the *theoretical yield* of PCl_3, we calculate how many grams of PCl_3 could be made from 35.0 g of Cl_2 if everything went perfectly according to the equation given.

$$35.0 \text{ g Cl}_2 \times \frac{1 \text{ mol Cl}_2}{70.90 \text{ g Cl}_2} \times \frac{2 \text{ mol PCl}_3}{3 \text{ mol Cl}_2} \times \frac{137.32 \text{ g PCl}_3}{1 \text{ mol PCl}_3} = 45.2 \text{ g PCl}_3$$

grams Cl_2 ⟶ moles Cl_2 ⟶ moles PCl_3 ⟶ grams PCl_3

The actual yield was 42.4 g of PCl_3, not 45.2 g, so the percentage yield is calculated as follows.

$$\text{Percentage yield} = \frac{42.4 \text{ g PCl}_3}{45.2 \text{ g PCl}_3} \times 100\% = 93.8\%$$

Thus 93.8% of the theoretical yield of PCl_3 was obtained.

Is the Answer Reasonable?
The obvious check is that the calculated or theoretical yield can never be *less* than the actual yield. So on this score, our answer is reasonable. But doing a "head calculation" beyond the obvious in a problem like this is difficult for most people, even when data are vigorously rounded. Doing chain calculations in which both numbers and their full units are given (for example, 1 mol Cl_2, not just 1 mol), so that full units can be canceled, should help you avoid the most common errors.

Practice Exercise 23

Ethyl alcohol, C_2H_6O, can be converted to acetic acid (the acid in vinegar), $HC_2H_3O_2$, by the action of sodium dichromate in aqueous sulfuric acid according to the following equation.

$$3C_2H_6O(aq) + 2Na_2Cr_2O_7(aq) + 8H_2SO_4(aq) \longrightarrow$$
$$3HC_2H_3O_2(aq) + 2Cr_2(SO_4)_3(aq) + 2Na_2SO_4(aq) + 11H_2O$$

In one experiment, 24.0 g of C_2H_6O, 90.0 g of $Na_2Cr_2O_7$, and a known excess of sulfuric acid were mixed, and 26.6 g of acetic acid ($HC_2H_3O_2$) was isolated. Calculate the percentage yield of $HC_2H_3O_2$. ◆

3.9 Reactions in Solution

When two or more reactants are involved in a chemical reaction, their particles—atoms, ions, or molecules—must make physical contact. In other words, the particles need freedom of motion, which exists in gases and liquids but not in solids. Thus, whenever possible, reactions are carried out with all of the reactants in one fluid phase, liquid or gas. When possible, a liquid is used to dissolve solid reactants so their particles can move about.

> ▶**Chemistry in Practice**◀ Medications that fizz when water is added, like Alka-Seltzer, can be stored a long time if they are kept completely dry. Without the water, two key fizz-causing reactants in Alka-Seltzer, citric acid and sodium bicarbonate, are both in the solid state. (Aspirin is another ingredient.) When added to water, the solids dissolve and so gain the freedom to find each other, react, and liberate gaseous carbon dioxide. Representing citric acid by H_3Cit, the equation is
>
> $$2H_3Cit(aq) + 6NaHCO_3(aq) \longrightarrow 6CO_2(g) + 2Na_3Cit(aq) + 6H_2O \quad ◆$$

Carbon dioxide forms when solid sodium bicarbonate and citric acid, two ingredients of Alka-Seltzer, dissolve in water.

We now need to define some terms to continue our study.

Some Definitions

A **solution** is a homogeneous mixture in which the molecules or ions of the components are fully intermingled (see Figure 3.5). When a solution forms, at least two substances are involved. One is called the *solvent* and all of the others are called *solutes*. The **solvent** is the medium into which the solutes are mixed or dissolved. Water is a typical and very common solvent, but the solvent can actually be in any physical state, a solid, liquid, or gas. Unless stated otherwise, we will assume that any solutions we mention are *aqueous solutions*, so that liquid water is understood to be the solvent.

A **solute** is any substance dissolved in the solvent. It might be a gas, like the carbon dioxide dissolved in carbonated beverages. Some solutes are liquids, like ethylene glycol dissolved in water to protect a vehicle's radiator against freezing. Solids, of course, can be solutes, like the sugar dissolved in lemonade.

To describe the composition of a solution, we often give a ratio called a **concentration,** the *ratio* of the quantity of solute either to the quantity of solvent or to the quantity of solution. The quantities may be in any units we desire, but grams are often used. For example, we might have a solution of salt in water in which the concentration is 2 g salt/100 g of water, in other words, a

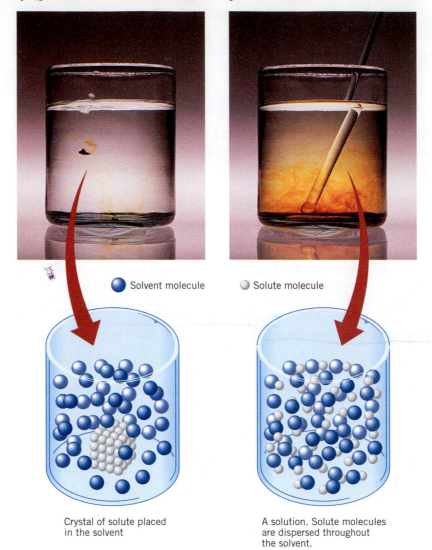

Figure 3.5 *Formation of a solution of iodine molecules in ethyl alcohol to form "tincture of iodine," once widely used as an antiseptic.* In the first photo, a crystal of iodine, I_2, is on its way to the bottom of the beaker, already beginning to dissolve, the purplish iodine crystal forming a reddish brown solution. In the hugely enlarged view beneath the photo, we see the iodine molecules still bound in a crystal. The second photo shows how stirring the mixture helps the iodine molecules to disperse in the solvent, as visualized in the molecular view below the photo.

● Solvent molecule   ○ Solute molecule

Crystal of solute placed in the solvent

A solution. Solute molecules are dispersed throughout the solvent.

solute-to-*solvent* ratio. Often, the concentration of salt in seawater is given as 3 g salt/100 g seawater, which is a solute-to-*solution* ratio. When the ratio is like this, namely, number of grams of solute per 100 g of *solution,* we call the ratio the solution's **percentage concentration.**

The *relative* amounts of solute and solvent are often loosely given without specifying actual quantities. In a **dilute solution,** for example, the ratio of solute to solvent is small, sometimes very small. A few crystals of salt in a glass of water make a very dilute salt solution. In a **concentrated solution,** the ratio of solute to solvent is large. Syrup, for example, is a very concentrated solution of sugar in water.

Concentrated and dilute are relative terms. For example, a solution of 100 g of sugar in 100 mL of water is concentrated compared to one with just 10 g of sugar in 100 mL of water, but the latter solution is more concentrated than one that has 1 g of sugar in 100 mL of water.

Usually there is a limit to the amount of a solute that can dissolve in a given amount of solvent. For example, at 20 °C only 36.0 g of sodium chloride is able to dissolve in 100 g of water. If more solute is added, it simply rests at the bottom of the solution. We say this is a **saturated solution** because at its present temperature it cannot dissolve any more solute. To make sure that a solution is

Table 3.1 Solubilities of Some Common Compounds in Water

Substance	Formula	Solubility (g/100 g water)[a]
Ammonium chloride	NH_4Cl	29.7 (0 °C)
Boric acid	H_3BO_3	6.35 (30 °C)
Calcium carbonate	$CaCO_3$	0.0015 (25 °C)
Calcium chloride	$CaCl_2$	74.5 (20 °C)
Copper sulfide	CuS	3.3×10^{-5} (18 °C)
Lead sulfate	$PbSO_4$	4.3×10^{-3} (25 °C)
Sodium hydroxide	$NaOH$	42 (0 °C)
		347 (100 °C)
Sodium chloride	$NaCl$	35.7 (0 °C)
		39.12 (100 °C)

[a]At the temperature given in parentheses.

saturated, extra solid solute is sometimes put into its container. The presence of this solid at the bottom of the container over a period of time is visible evidence that the solution is saturated.

The **solubility** of a solute is usually described by the number of grams that dissolve in 100 g of *solvent* at a given temperature to make a saturated solution. The temperature must be specified because solubilities vary with temperature. Table 3.1 shows that some saturated aqueous solutions, like sodium hydroxide, are concentrated, but others, like copper sulfide, are extremely dilute.

A solution that contains less solute than required for saturation is called an **unsaturated solution.** It is able to dissolve more solute.

Usually, the solubility of a solute increases as the temperature of the solution increases. In such cases, more solute can be dissolved by heating the mixture. If the temperature of such a warmer, saturated solution is subsequently lowered, the additional solute should separate from the solution, and indeed this tends to happen spontaneously. However, sometimes the solute doesn't separate, leaving us with a **supersaturated solution,** a solution that actually contains more solute than required for saturation at a given temperature.

Supersaturated solutions are unstable. They can only be prepared if there are no traces of undissolved solute left in contact with the solution. If even a tiny crystal of the solute is present or is added, the extra solute crystallizes. Figure 3.6

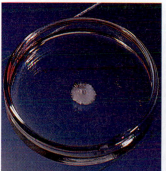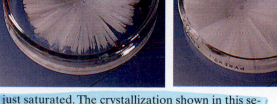

Figure 3.6 *Crystallization.* When a small seed crystal of sodium acetate is added to a supersaturated solution of the salt, excess solute crystallizes rapidly until the solution is just saturated. The crystallization shown in this sequence took less than 10 seconds.

shows what happens when a tiny crystal of sodium acetate is added to a supersaturated solution of this compound in water.

Often, when a reaction is carried out in a solution, one of the products that forms has a low solubility in the solvent. As this substance forms, it separates from the solution as a solid, which we call a **precipitate.** A reaction that produces a precipitate is called a **precipitation reaction.** An example is the reaction of lead nitrate and potassium iodide which is described below.

Solution Stoichiometry

A reaction between two substances might depend altogether on their being in solution, but the stoichiometric calculations are basically the same whether they are or not. Crystals of lead nitrate, $Pb(NO_3)_2$, for example, can be mixed with crystals of potassium iodide, KI, but the particles in the crystals are not free to seek each other out and react. So you don't see anything happen. But if an aqueous *solution* of potassium iodide, $KI(aq)$, is poured into an aqueous solution of lead nitrate, $Pb(NO_3)_2(aq)$, an almost instantaneous reaction occurs, and a bright yellow solid, lead iodide, $PbI_2(s)$, forms as a precipitate (see Figure 3.7). The equation is

$$Pb(NO_3)_2(aq) + 2KI(aq) \longrightarrow PbI_2(s) + 2KNO_3(aq)$$

The reactants are related by the following mole equivalency.

$$1 \text{ mol } Pb(NO_3)_2 \Leftrightarrow 2 \text{ mol } KI$$

So if we want to use up, say, 0.10 mol of $Pb(NO_3)_2$, we'd need 0.20 mol of KI. The formula mass of $Pb(NO_3)_2$ is 331 and that of KI is 166, so we'd want to take 33.1 g of $Pb(NO_3)_2$ and 16.6×2 or 33.2 g of KI. We would dissolve the substances in water, in separate beakers, and then pour one into the other.

Sometimes only one reactant has to be in solution as we learned earlier in this chapter for the reaction between solid pieces of zinc metal with hydrochloric acid, $HCl(aq)$.

Figure 3.7 *Precipitation of lead iodide.* A solution of potassium iodide being added to a solution of lead nitrate produces solid lead iodide (yellow), which has a very low solubility in water.

3.10 Molar Concentration

A very useful way of working quantitatively with the stoichiometry of reactions involving solutions involves a special method of describing concentrations, namely, by **molar concentration,** or **molarity** (abbreviated M). The molarity of a solution is the number of moles of solute per liter of solution. We can write this as a ratio of the moles of solute to the volume of the solution expressed in liters.

Tools

Molar concentration

$$\text{Molarity } (M) = \frac{\text{moles of solute}}{\text{liters of solution}} \quad (3.2)$$

Thus, a solution that contains 0.100 mol of NaCl in 1.00 L has a molarity of 0.100 M, and we would refer to this solution as 0.100 *molar* NaCl or as 0.100 M NaCl. The same concentration would result if we dissolved 0.0100 mol of NaCl in 0.100 L (100 mL) of solution, because the *ratio* of moles of solute to volume of solution is the same.

$$\frac{0.100 \text{ mol NaCl}}{1.00 \text{ L NaCl soln}} = \frac{0.0100 \text{ mol NaCl}}{0.100 \text{ L NaCl soln}} = 0.100 \, M \text{ NaCl}$$

Molarity is particularly useful because it lets us obtain a given number of moles of a substance simply by measuring a volume of a previously prepared solution, and this measurement is quick and easy to do in the lab. If we had a stock supply of 0.100 M NaCl solution, for example, and we needed 0.100 mol of NaCl for a reaction, we would simply measure out 1.00 L of the solution because in this volume there is 0.100 mol of NaCl.

We need not (and seldom do) work with whole liters of solutions. It's the *ratio,* not the total volume, that matters, that is, the ratio of the number of moles of solute to the number of liters of solution. If we dissolved, say, 0.0200 mol of sodium chromate, Na_2CrO_4, in 0.250 L (250 mL) of final solution, the molarity of the solution would be found by the following simple calculation, using Equation 3.2.

$$\text{Molarity} = \frac{0.0200 \text{ mol } Na_2CrO_4}{0.250 \text{ L } Na_2CrO_4 \text{ soln}} = 0.0800 \text{ } M \text{ } Na_2CrO_4$$

A *volumetric flask* is a narrow-necked bottle having an etched mark high on its neck. When filled to the mark, the flask contains the volume given by the flask's label. Figure 3.8 shows how to use a 250-mL volumetric flask to prepare a solution of known molarity.

Many sizes of volumetric flasks are available from scientific supply houses.

Molarity as a Conversion Factor

A solution's molarity is a source of two conversion factors that relate moles of solute to volume of solution. When a solution is 0.100 M NaCl, for example, we can construct the following conversion factors.

$$\frac{0.100 \text{ mol NaCl}}{1.00 \text{ L NaCl soln}} \quad \text{and} \quad \frac{1.00 \text{ L NaCl soln}}{0.100 \text{ mol NaCl}}$$

Generally, the volumes used in experiments are better expressed in milliliters rather than liters. Many students find it easier to translate the "1.00 L" part of these factors into the equivalent 1000 mL here rather than convert between liters and milliliters at some other stage of the calculation. Factors such as the two above, therefore, can be rewritten as follows whenever it is convenient. (Remember that "1000" in the following is regarded as having an infinite number of significant figures because, standing as it does for 1 L, it is part of the definition of molarity and is an exact number.)

$$\frac{0.100 \text{ mol NaCl}}{1000 \text{ mL NaCl soln}} \quad \text{and} \quad \frac{1000 \text{ mL NaCl soln}}{0.100 \text{ mol NaCl}}$$

EXAMPLE 3.24
Calculating the Molarity of a Solution

To study the effect of dissolved salt on the rusting of an iron sample, a student prepared a solution of NaCl by dissolving 1.461 g of NaCl in a 250.0-mL volumetric flask. What is the molarity of this solution?

Analysis: To work with molarity, it is imperative that you know its definition; molarity is the ratio of *moles of solute* to *liters of solution.* To calculate this ratio from the given data, we first have to convert 1.461 g of NaCl into moles of NaCl, and we have to convert 250.0 mL into liters. Then we simply divide the number of moles by the number of liters to find the molarity.

Solution: The number of moles of NaCl are found using the formula mass of NaCl, 58.443.

(a)   *(b)*   *(c)*

(d)   *(e)*

Figure 3.8 *The preparation of a solution having a known molarity.* (*a*) A 250-mL volumetric flask, one of a number of sizes available for preparing solutions. When filled to the line etched around its neck, this flask contains exactly 250 mL of solution. The flask here already contains a weighed amount of solute. (*b*) Water is being added. (*c*) The solute is brought completely into solution before the level is brought up to the narrow neck of the flask. (*d*) More water is added to bring the level of the solution to the etched line. (*e*) The flask is stoppered and then inverted several times to mix its contents thoroughly.

$$1.461 \ \overline{\text{g NaCl}} \times \frac{1 \ \text{mol NaCl}}{58.443 \ \overline{\text{g NaCl}}} = 0.02500 \ \text{mol NaCl}$$

The volume of the solution in liters equivalent to 250.0 mL is 0.2500 L. The ratio of moles to liters, therefore, is

$$\frac{0.02500 \ \text{mol NaCl}}{0.2500 \ \text{L}} = 0.1000 \ M \ \text{NaCl}$$

Is the Answer Reasonable?
The mass of NaCl taken, 1.461 g, is a small fraction of 1 mol NaCl (58.443 g), but the given volume is by no means so small a fraction of 1 L. Therefore, the ratio of the two has to be less than 1 M, which it is.

The salt in sea water accelerated the rusting of this anchor.

_____ **Practice Exercise 24** _____

As part of a study to determine if just the chloride ion in NaCl accelerates the rusting of iron, a student decided to see if iron rusted as rapidly when exposed to a solution of Na_2SO_4. This solution was made by dissolving 3.550 g of Na_2SO_4 in water using a 100.0-mL volumetric flask to adjust the final volume. What is the molarity of this solution? ◆

EXAMPLE 3.25

Using Molar Concentrations

How many milliliters of 0.250 M NaCl solution must be measured to obtain 0.100 mol of NaCl?

Analysis: Once again we use the given molarity as a tool to construct a conversion factor. From the molarity we have the following options.

$$\frac{0.250 \text{ mol NaCl}}{1.00 \text{ L NaCl soln}} \quad \text{or} \quad \frac{1.00 \text{ L NaCl soln}}{0.250 \text{ mol NaCl}}$$

Solution: We operate with the second factor on 0.100 mol NaCl.

$$0.100 \text{ mol NaCl} \times \frac{1.00 \text{ L NaCl soln}}{0.250 \text{ mol NaCl}} = 0.400 \text{ L of } 0.250 \text{ } M \text{ M NaCl}$$

Because 0.400 L corresponds to 400 mL, 400 mL of 0.250 NaCl provides 0.100 mol of NaCl.

Is the Answer Reasonable?
A whole liter provides 0.250 mol, so we need a fraction of a liter (1000 mL) to provide just 0.100 mol. The answer, 400 mL, is reasonable.

_____ **Practice Exercise 25** _____

To continue the experiment on rusting (Practice Exercise 24), a student decided to see how rapidly iron rusted when chloride ion but not sodium ion was present. How many milliliters of 0.150 M KCl solution must be taken to provide 0.0500 mol of KCl? ◆

In the laboratory, we often must prepare a solution with a specific molarity. Example 3.25 illustrates the kind of calculation required, and also demonstrates a useful relationship. *Whenever you know **both** the volume and the molarity of a solution, you can calculate the number of moles of solute in that volume.* This is because the product of molarity and volume (expressed in liters) equals the moles of solute.

Tools

molarity × volume (L) = moles of solute   (3.3)

Molarity × volume (L) = moles

$$\frac{\text{mol solute}}{\text{L soln}} \times \text{L soln} = \text{mol solute}$$

This equation is a tool for solving problems in solution stoichiometry.

EXAMPLE 3.26
Preparing a Solution with a Known Molarity

A student needs to prepare 250 mL of 0.100 M $Cd(NO_3)_2$ solution. How many grams of cadmium nitrate are required?

Analysis: Because we know both the molarity and the volume of the solution, we can calculate the number of moles of $Cd(NO_3)_2$ that will be in solution after it's prepared. Once we know the number of moles of $Cd(NO_3)_2$, we can calculate the mass using the formula mass of the salt.

Solution: The stated molarity, 0.100 M $Cd(NO_3)_2$, provides the following conversion factors.

$$\frac{0.100 \text{ mol } Cd(NO_3)_2}{1.00 \text{ L } Cd(NO_3)_2 \text{ soln}} \quad \text{or} \quad \frac{1.00 \text{ L } Cd(NO_3)_2 \text{ soln}}{0.100 \text{ mol } Cd(NO_3)_2}$$

We have 250 mL of solution, or 0.250 L $Cd(NO_3)_2$ solution. To find moles of $Cd(NO_3)_2$ we multiply by the first factor.

$$0.250 \text{ L } Cd(NO_3)_2 \text{ soln} \times \frac{0.100 \text{ mol } Cd(NO_3)_2}{1.00 \text{ L } Cd(NO_3)_2 \text{ soln}} = 0.0250 \text{ mol } Cd(NO_3)_2$$

We next convert from moles to grams using the formula mass of $Cd(NO_3)_2$, which is 236.43.

$$0.0250 \text{ mol } Cd(NO_3)_2 \times \frac{236.43 \text{ g } Cd(NO_3)_2}{1 \text{ mol } Cd(NO_3)_2} = 5.91 \text{ g } Cd(NO_3)_2$$

Thus, to prepare the solution, 5.91 g of $Cd(NO_3)_2$ would be weighed out, placed in a 250.0-mL volumetric flask, and then dissolved in water. More water would be added to make the solution reach the flask's etched mark.

We could have set this up as a chain calculation as follows.

$$0.250 \text{ L } Cd(NO_3)_2 \text{ soln} \times \frac{0.100 \text{ mol } Cd(NO_3)_2}{1.00 \text{ L } Cd(NO_3)_2 \text{ soln}} \times \frac{236.43 \text{ g } Cd(NO_3)_2}{1 \text{ mol } Cd(NO_3)_2}$$
$$= 5.91 \text{ g } Cd(NO_3)_2$$

Is the Answer Reasonable?
If we were working with a full liter of this solution, it would contain 0.1 mol of $Cd(NO_3)_2$. The formula mass of the salt is 236.43, so 0.1 mol is about 24 g. However, we are working with just a quarter of a liter (250 mL), so the amount of $Cd(NO_3)_2$ needed is about a quarter of 24 g, or 6 g. The answer, 5.91 g, is very close to this, so it makes sense. (The more you try to work these rough calculations in your head, the quicker and surer you will become at understanding and solving stoichiometry problems.)

Practice Exercise 26

How many grams of $CaCl_2$ are needed to prepare 250 mL of 0.125 M $CaCl_2$ solution? ◆

Figure 3.9 *Volumetric glassware.* Shown here is a selection of volumetric pipets, each marked for a particular volume and each with an etched line to designate where the filling of the pipet should stop.

It is not always necessary to begin with a solute in pure form to prepare a solution with a known molarity. The solute can already be dissolved in a solution of relatively high concentration, and this can be *diluted* to make a solution of the desired lower concentration. The choice of apparatus used depends on the precision required. If high precision is necessary, pipets (Figure 3.9) and volumetric flasks are used. If we can do with less precision, we might use graduated cylinders instead.

▶**Chemistry in Practice**◀ Solutions of medications are, for economy, sometimes prepared, shipped, and stored at concentrations much too high for giving to patients. The physician prescribes the correct concentration, and the pharmacist or nurse must then make the proper dilution. Mistakes in preparing a more dilute solution could literally be deadly. ◆

The process of diluting a solution involves distributing a given amount of solute throughout a larger volume of solution. The amount of solute remains constant, however. Since the product of molarity and volume equals the number of moles of solute, it must therefore be true that

$$\left(\begin{array}{c}\text{volume of}\\ \text{dilute solution}\\ \text{to be prepared}\end{array}\right) \times M_{\text{dilute}} = \left(\begin{array}{c}\text{volume of}\\ \text{concentrated solution}\\ \text{to be used}\end{array}\right) \times M_{\text{concd}}$$

Or,

$$V_{\text{dil}} \cdot M_{\text{dil}} = V_{\text{concd}} \cdot M_{\text{concd}} \qquad (3.4)$$

Tools

Diluting
solutions of
known
molarity

Any units can be used for volume in Equation 3.4 provided that they are the same volume units on both sides of the equation. We thus normally solve dilution problems using *milliliters* directly in Equation 3.4.

EXAMPLE 3.27

Preparing a Solution of Known Molarity by Dilution

How can we prepare 100 mL of 0.0400 M $K_2Cr_2O_7$ from 0.200 M $K_2Cr_2O_7$?

Analysis: This is the way such a question comes up in the lab, but what it is really asking is, "How many milliliters of 0.200 M $K_2Cr_2O_7$ (the more concentrated solution) must be diluted to give a solution with a final volume of 100 mL and a final molarity of 0.0400 M?" Once we see the question this way, we see that Equation 3.4 applies.

Solution: It's a good idea to assemble the data first, noting what is missing (and therefore what has to be calculated).

$$V_{\text{dil}} = 100 \text{ mL} \qquad M_{\text{dil}} = 0.0400 \ M$$

$$V_{\text{concd}} = ? \qquad M_{\text{concd}} = 0.200 \ M$$

Next, we use Equation 3.4.

$$100 \text{ mL} \times 0.0400 \ M = V_{\text{concd}} \times 0.200 \ M$$

Solving for V_{concd} gives

$$V_{\text{concd}} = \frac{100 \text{ mL} \times 0.0400 \ M}{0.200 \ M}$$

$$= 20.0 \text{ mL}$$

This answer means that we would withdraw 20.0 mL of 0.200 M $K_2Cr_2O_7$, place it in a 100-mL volumetric flask, and then add water until the final volume is 100 mL (see Figure 3.10).

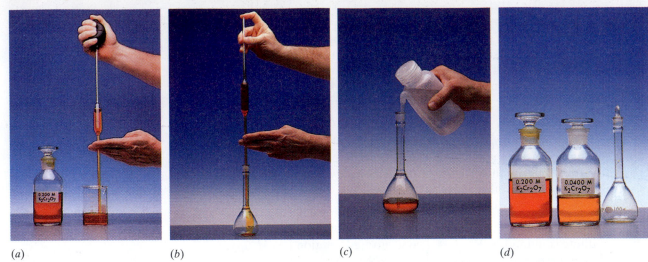

(a)   (b)   (c)   (d)

Figure 3.10 *Preparing a solution by dilution.* (*a*) The calculated volume of the more concentrated solution is withdrawn from the stock solution by means of a volumetric pipet. (*b*) The solution is allowed to drain entirely from the pipet into the volumetric flask. (*c*) Water is added to the flask, the contents are mixed, and the final volume is brought up to the etch mark on the narrow neck of the flask. (*d*) The new solution is put into a labeled container.

Is the Answer Reasonable?
Think about the *magnitude* of the dilution, from 0.2 M to 0.04 M. Notice that the concentrated solution is 5 times as concentrated as the dilute solution ($5 \times 0.04 = 0.2$). So we should make a 1:5 dilution, and we see that 100 mL is 5 times 20 mL.

Practice Exercise 27

How could 100 mL of 0.125 M H_2SO_4 solution be made from 0.500 M H_2SO_4 solution? ◆

3.11 Stoichiometry of Reactions in Solution

In this section we will work an example that illustrates how to do stoichiometry problems involving solutions of known molar concentrations.

EXAMPLE 3.28

Stoichiometry Involving Reactions in Solution

One of the solids present in photographic film is silver bromide, AgBr. This compound is quite insoluble in water, and one way to prepare it is to mix solutions of two water-soluble compounds, silver nitrate and calcium bromide. Suppose we wished to prepare AgBr by the following precipitation reaction.

$$2AgNO_3(aq) + CaBr_2(aq) \longrightarrow 2AgBr(s) + Ca(NO_3)_2(aq)$$

How many milliliters of 0.125 M $CaBr_2$ solution must be used to react with the solute in 50.0 mL of 0.115 M $AgNO_3$?

Analysis: The key equivalency is seen in the coefficients of the equation.

$$2 \text{ mol } AgNO_3 \Leftrightarrow 1 \text{ mol } CaBr_2$$

We can calculate the moles of $AgNO_3$ from the data given, using the volume and the molarity of the $AgNO_3$ solution. We then use the coefficients in the equation to translate to moles of $CaBr_2$. Finally, we use the molarity of the $CaBr_2$ solution as a conversion factor to find the volume of the solution needed. The calculation flow will look like the following.

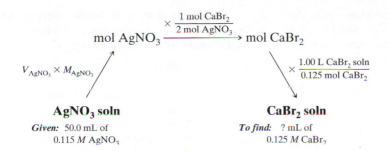

$$\text{mol AgNO}_3 \xrightarrow{\times \frac{1 \text{ mol CaBr}_2}{2 \text{ mol AgNO}_3}} \text{mol CaBr}_2$$

$$V_{\text{AgNO}_3} \times M_{\text{AgNO}_3}$$

$$\times \frac{1.00 \text{ L CaBr}_2 \text{ soln}}{0.125 \text{ mol CaBr}_2}$$

AgNO₃ soln
Given: 50.0 mL of
0.115 *M* AgNO₃

CaBr₂ soln
To find: ? mL of
0.125 *M* CaBr₂

Silver bromide (AgBr) precipitates (separates as a solid) when solutions of calcium bromide and silver nitrate are mixed.

Solution: First, we find the moles of $AgNO_3$ taken. Changing 50.0 mL to 0.050 L,

$$0.0500 \text{ L AgNO}_3 \text{ soln} \times \frac{0.115 \text{ mol AgNO}_3}{1.00 \text{ L AgNO}_3 \text{ soln}} = 5.75 \times 10^{-3} \text{ mol AgNO}_3$$

Next, we use the coefficients of the equation to calculate the amount of $CaBr_2$ required.

$$5.75 \times 10^{-3} \text{ mol AgNO}_3 \times \frac{1 \text{ mol CaBr}_2}{2 \text{ mol AgNO}_3} = 2.88 \times 10^{-3} \text{ mol CaBr}_2$$

Finally we calculate the volume (mL) of 0.125 *M* $CaBr_2$ that contains this many moles of $CaBr_2$.

$$2.88 \times 10^{-3} \text{ mol CaBr}_2 \times \frac{1.00 \text{ L CaBr}_2 \text{ soln}}{0.125 \text{ mol CaBr}_2} = 0.0230 \text{ L CaBr}_2 \text{ soln}$$

Thus 0.0230 L, or 23.0 mL, of 0.125 *M* $CaBr_2$ has enough solute to combine with the $AgNO_3$ in 50.0 mL of 0.115 *M* $AgNO_3$.

Is the Answer Reasonable?

The molarities of the two solutions are about the same, but only 1 mole of $CaBr_2$ is needed for each 2 mol $AgNO_3$. Therefore, the volume of $CaBr_2$ solution needed (23.0 mL) should be about half the volume of $AgNO_3$ solution taken (50.0 mL), which it is.

Practice Exercise 28

How many milliliters of 0.124 *M* NaOH contain enough NaOH to react with the H_2SO_4 in 15.4 mL of 0.108 *M* H_2SO_4 according to the following equation?

$$2NaOH(aq) + H_2SO_4(aq) \longrightarrow Na_2SO_4(aq) + 2H_2O \blacklozenge$$

SUMMARY

Stoichiometric Equivalencies. The ratios by *particles* in chemical compounds and in reactions are given by the subscripts within formulas and by the coefficients within **balanced equations.** These ratios describe the **stoichiometric equivalencies** within compounds or between substances in reactions.

Mole Concept and Formula Mass or Molecular Mass. In the SI *definition,* one **mole** of any substance is an amount with the same number, **Avogadro's number** (6.022 $\times 10^{23}$), of atoms, molecules, or formula units as there are atoms in 12 g (exactly) of carbon-12. The sum of the atomic

masses of all of the atoms appearing in a chemical formula gives the **formula mass** or the **molecular mass.**

Formula (or molecular) mass (g) ⇔ 1 mol of substance

Like an atomic mass, a formula or molecular mass is a tool for grams-to-moles or moles-to-grams conversions.

Chemical Formulas. The actual composition of a molecule is given by its **molecular formula.** An **empirical formula** gives the ratio of atoms, but in the smallest whole numbers, and it is the *only* formula we write for ionic compounds. In the case of a molecular compound, the molecular mass is equal either to that calculated from the compound's empirical formula or to a simple multiple of it. When the two calculated masses are the same, then the molecular formula is the same as the empirical formula.

An empirical formula can be found from the masses of the elements obtained by the quantitative analysis of a known sample of the compound or it can be calculated from the **percentage composition.** The percentage of an element in a compound is the same as the number of grams of the element in 100 g of the compound. If there is $A\%$ of X in X_xZ_z, then 100.00 g of X_xZ_z contains A g of X. From the grams of X and Z, the moles of X and Z can be calculated. When these are adjusted to their corresponding whole-number ratios, the subscripts in the empirical formula are obtained.

Formula Stoichiometry. A chemical formula is a tool for stoichiometric calculations, because its subscripts tell us the mole ratios in which the various elements are combined. In X_xZ_z, the stoichiometric equivalency between X and Z is simply

$$z \text{ mol } Z \Leftrightarrow x \text{ mol } X$$

The conversion factors available from this equivalency are

$$\frac{z \text{ mol } Z}{x \text{ mol } X} \quad \text{and} \quad \frac{x \text{ mol } X}{z \text{ mol } Z}$$

Balanced Equations and Reaction Stoichiometry. A balanced equation is a tool for reaction stoichiometry because its coefficients disclose the stoichiometric equivalencies. When balancing an equation, only the coefficients can be adjusted, never the subscripts. In a generalized reaction,

$$xA + mW \longrightarrow yB + nZ$$

we know that

$$x \text{ mol } A \Leftrightarrow m \text{ mol } W$$

$$x \text{ mol } A \Leftrightarrow y \text{ mol } B$$

$$x \text{ mol } A \Leftrightarrow n \text{ mol } Z$$

and so on.

All problems of reaction stoichiometry must be solved at the mole level of the substances involved. If grams are given, they must first be changed to moles.

Yields of Products. A reactant taken in a quantity less than required by another reactant, as determined by the reaction's stoichiometry, is called the **limiting reactant.** The **theoretical yield** of a product can be no more than permitted by the limiting reactant. Sometimes **competing reactions** (side reactions) producing by-products reduce the **actual yield.** The ratio of the actual to the theoretical yields, expressed as a percentage, is the **percentage yield.**

Solution Vocabulary. A considerable vocabulary has developed for describing **solutions.** Some terms are qualitative—**dilute, concentrated, saturated, unsaturated,** and **supersaturated,** for example. The **solubility** of a substance is often stated in grams of **solute** per 100 g of **solvent.** An insoluble substance that forms in a reaction is called a **precipitate.**

Solution Stoichiometry. Molar concentration (molarity) equals the number of moles of solute per liter of solution. This concentration unit makes available two conversion factors,

$$\frac{\text{mol solute}}{1 \text{ L soln}} \quad \text{and} \quad \frac{1 \text{ L soln}}{\text{mol solute}}$$

Solutions of known molarity are made in two ways. One is to dissolve a known number of moles of the solute into a solvent until the final volume of the solution reaches a known value. The other is to take a specific volume of a relatively concentrated solution and dilute it with the solvent until the final volume is reached. For dilution problems, remember the relationship

$$V_{dil} \times M_{dil} = V_{concd} \times M_{concd}$$

Figure 3.11 gives an overview of how to use stoichiometric equivalencies to move from moles to the units commonly used in measurements—grams, or volumes of solutions of known molarities.

Figure 3.11 *Paths for working stoichiometry problems involving solutions.* The central feature is the conversion between moles of one reactant and moles of another, which makes use of the coefficients of the balanced chemical equation. Moles are thus the central fundamental chemical units used to construct chemical equivalencies. We can obtain moles from more than one laboratory unit, namely, from grams or from volumes of solutions of known molarities.

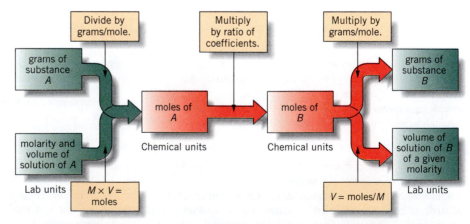

Tools *you have learned*

The table below lists the tools that you have learned in this chapter that are applicable to problem solving. Review them if necessary, and refer to them when working on the Thinking-It-Through problems and the Review Problems that follow.

Tool	Function
Avogadro's number (*page 98*)	To relate macroscopic lab-sized quantities (e.g., moles) to numbers of individual atomic-sized particles such as atoms, molecules, or ions.
Chemical formula (*page 99*)	To allow use of subscripts in a formula to establish atom ratios and mole ratios between the elements in the substance.
Atomic mass (*page 100*)	To form a conversion factor to calculate mass from moles of an element, or moles from the mass of an element.
Formula mass; molecular mass (*page 105*)	To form a conversion factor to calculate mass from moles of a compound, or moles from the mass of a compound.
Percentage composition (*page 107*)	To represent the composition of a compound and be the basis for computing the empirical formula. Comparing experimental and theoretical percentage compositions can help establish the identity of a compound.
Balanced chemical equation (*page 119*)	To allow use of the coefficients to establish stoichiometric equivalencies that relate moles of one substance to moles of another in a chemical reaction. The coefficients also establish the ratios by formula units among the reactants and products.
Theoretical, actual, and percentage yields (*page 126*)	To estimate the efficiency of a reaction. Remember that the theoretical yield is calculated from the limiting reactant using the balanced chemical equation.
Molar concentration (*page 130*)	To relate the amount of solute in a solution to the volume of the solution. Molarity is used to construct conversion factors based on the relationship mol solute ⇔ liters of solution

THINKING IT THROUGH

Remember, the goal in the following problems is *not* to find the answer itself; instead, assemble the needed information and describe *how* you would use the data to answer the question. (For some problems, you may find your explanations easier if you actually perform some of the intermediate calculations.) Where problems can be solved by the factor-label method, be sure to set up the solution to the problem by arranging the conversion factors so the units cancel correctly to give the desired units of the answer. Obtain data for atomic masses from the inside front cover of the book.

The problems are divided into two groups. Those in Level 2 are more challenging than those in Level 1 and provide an opportunity to really hone your problem solving skills. Detailed answers to these problems can be found at our Web site: http://www.wiley.com/college/brady.

Level 1 Problems

1. Show how you would answer the following questions: How many moles of carbon are combined with 0.60 mol of hydrogen atoms in the rocket fuel dimethylhydrazine, $(CH_3)_2N_2H_2$? How many moles of H_2 gas could be made from all the hydrogen in 0.56 mol of $(CH_3)_2N_2H_2$?

2. Sulfur and oxygen atoms are combined in sulfur trioxide, SO_3. How many atoms of S react with 0.400 mol of O_2 molecules to make molecules of SO_3? Set up the calculation.

3. The following information applies to a certain compound of carbon and hydrogen: In 4 mol of the compound there are 12 mol of C. In 6 molecules of the compound, there are 48 atoms of H. Describe how you would use this information to obtain the empirical formula for the compound.

4. Chloroform, a chemical once used as an anesthetic, has the formula $CHCl_3$. Set up the solutions to the following questions: (a) How many molecules of $CHCl_3$ would have to be decomposed to give 18 atoms of Cl? (b) How many moles of $CHCl_3$ would have to be decomposed to give enough chlorine atoms to make 0.36 mol of Cl_2?

5. A 2.05 g sample of a certain compound of tin and chlorine was found to contain 1.11 g of chlorine. Explain how you would use this information to obtain the empirical formula for the compound.

6. Concerning a certain compound of sodium, sulfur, and oxygen, you are given the following stoichiometric equivalencies:

$$4 \text{ mol Na} \Leftrightarrow 10 \text{ mol O}$$

$$5 \text{ mol O} \Leftrightarrow 2 \text{ mol S}$$

Explain how you would determine the empirical formula. Set up the calculation to determine the number of moles of sulfur combined with 2.7 mol Na.

7. Explain how you would calculate the atomic mass of an element M if 36.0 g of M combines with 24.0 g of oxygen to give a compound with the formula MO.

8. A 1.015 g sample of a binary compound of calcium and bromine was dissolved in water and treated with $Na_2C_2O_4$, which caused CaC_2O_4 to form. The CaC_2O_4 was recovered from the reaction mixture and was found to weigh 0.650 g. Outline the steps needed to calculate the percentage of calcium in the original calcium–bromine compound. Explain how you would use the data given here to calculate the empirical formula for the calcium–bromine compound. Show relevant conversion factors.

9. A sample of the hydrate $CuSO_4 \cdot 5H_2O$ has a mass of 12.5 g. If it is completely dehydrated, how many grams of $CuSO_4$ would remain? How many moles of water would be released during the dehydration process? Show how you would set up the calculations.

10. Calcium hydroxide and hydrobromic acid react as follows:

$$Ca(OH)_2(s) + 2HBr(aq) \longrightarrow CaBr_2(aq) + 2H_2O$$

Describe in detail how you would answer the following:
(a) How many grams of HBr are needed to produce 0.125 mol of $CaBr_2$?
(b) If the reaction is carried out by combining 1.24 g of $Ca(OH)_2$ with 2.65 g of HBr, what is the theoretical yield of $CaBr_2$ expressed in grams?
(c) For part (b), how many grams of which reactant remain after the reaction is complete?

11. How many milliliters of 0.150 M $Al_2(SO_4)_3$ are needed to react with 25.0 mL of 0.250 M $BaCl_2$ according to the following equation?

$$Al_2(SO_4)_3(aq) + 3BaCl_2(aq) \rightarrow 2AlCl_3(aq) + 3BaSO_4(s)$$

Set up the solution following the factor-label method. Be sure to show cancellation of units.

Level 2 Problems
12. A sample of iron ore in which the iron occurs in the mineral Fe_3O_4 is found to contain 12.5% (by mass) of iron. Describe how you would calculate the percentage by mass of Fe_3O_4 in the ore.

13. A white powder was known to be either Na_3PO_4, Na_2HPO_4, or NaH_2PO_4. A sample of it was analyzed and found to contain 21.82% phosphorus by mass. Explain how you would obtain the percentage by mass of sodium in the sample.

14. If compound X and compound Y react with each other in a 1-to-1 ratio by moles, and if 25.0 g of X combines exactly with 29.0 g of Y, which compound has the larger formula mass? Explain your answer.

15. Al_2O_3 reacts with HI as follows:

$$Al_2O_3(s) + 6HI(aq) \longrightarrow 2AlI_3(aq) + 3H_2O$$

A mixture containing 16.0 g of Al_2O_3 and 140 g of HI produces 127 g of AlI_3. Explain in detail how you would determine the number of grams of which reactant were present in excess in the original reaction mixture.

16. A 1.000 g sample of a mixture of NaCl and $CaCl_2$ was treated in solution with an excess of $AgNO_3$, which yielded a precipitate of AgCl. The AgCl weighed 2.53 g. Describe in detail how the percentage by mass of NaCl in the sample can be calculated.

17. A 1.24 g portion of $H_2C_2O_4$ was dissolved in exactly 250 mL of water. Next, a 20.0-mL portion of this solution was removed and placed into another container, to which was also added 50.0 mL of 0.100 M NaOH solution. NaOH reacts with $H_2C_2O_4$ following the equation

$$2NaOH + H_2C_2O_4 \longrightarrow Na_2C_2O_4 + 2H_2O$$

When the reaction is over, how many moles of NaOH remain unreacted? Explain how you would obtain the answer.

18. Concentrated sulfuric acid contains 96% H_2SO_4 by mass and has a specific gravity at room temperature of 1.839. You are assigned the preparation of 250 mL of 0.10 M H_2SO_4. How would you proceed to prepare the more dilute solution in the volume specified? Describe any necessary calculations.

19. How many distinct empirical formulas are shown by the following models for compounds formed between elements A and B? Explain. (Element A is represented by a black sphere and element B by a light gray sphere.)

20. Molecules containing *A* and *B* react to form *AB* as shown below. Based on the equations and the contents of the boxes labeled "Initial," sketch for each reaction the molecular models of what is present after the reaction is over. (In both cases, the species *B* exists as B_2. In reaction 1, *A* is monatomic; in reaction 2, *A* exists as diatomic molecules A_2.)

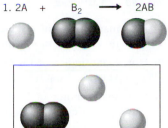

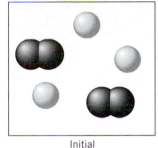

Initial   Initial

REVIEW QUESTIONS

3.1 What is the definition of the mole?

3.2 Write the stoichiometric equivalence, expressed in terms of atoms, that exists between the elements in each of the following compounds.
(a) SO_2 (b) As_2O_3 (c) PbN_6 (d) Al_2Cl_6

3.3 Write the stoichiometric equivalence, expressed in moles, that exists between the elements in each of the following compounds.
(a) Mn_3O_4 (b) Sb_2S_5 (c) P_4O_6 (d) Hg_2Cl_2

3.4 Why is the expression "1.0 mol of oxygen" ambiguous? Why doesn't a similar ambiguity exist in the expression "64 g of oxygen?"

3.5 The atomic mass of aluminum is 26.98. What specific conversion factors does this value make available for relating a mass of aluminum (in grams) and a quantity of aluminum given in moles?

3.6 In general, what fundamental information, obtained from experimental measurements, is required to calculate the empirical formula of a compound?

3.7 Why is the changing of subscripts not allowed when balancing a chemical equation?

3.8 When given the *unbalanced* equation

$$Na(s) + Cl_2(g) \longrightarrow NaCl(s)$$

and asked to balance it, student *A* wrote

$$Na(s) + Cl_2(g) \longrightarrow NaCl_2(s)$$

and student *B* wrote

$$2Na(s) + Cl_2(g) \longrightarrow 2NaCl(s)$$

(a) Both equations are balanced, but which student is correct?
(b) Explain why the other student's answer is incorrect.

3.9 If two substances react completely in a 1-to-1 ratio *both* by mass and by moles, what must be true about these substances?

3.10 A mixture of 0.020 mol of Mg and 0.020 mol of Cl_2 reacted completely to form $MgCl_2$ according to the equation

$$Mg + Cl_2 \longrightarrow MgCl_2$$

(a) What information describes the *stoichiometry* of this reaction?
(b) What information gives the *scale* of the reaction?

3.11 In a report to a supervisor, a chemist described an experiment in the following way: "0.0800 mol of H_2O_2 decomposed into 0.0800 mol of H_2O and 0.0400 mol of O_2." Express the chemistry and stoichiometry of this reaction by a conventional chemical equation.

3.12 Define the following: (a) solvent, (b) solute, (c) concentration.

3.13 Define the following: (a) concentrated, (b) dilute, (c) saturated, (d) unsaturated, (e) supersaturated.

3.14 What is meant by the term solubility?

3.15 Why are chemical reactions often carried out using solutions?

3.16 Define *percentage concentration* and *molar concentration*.

REVIEW PROBLEMS

Answers to problems whose numbers are printed in color are given in Appendix D.
More challenging problems are marked with asterisks.

The Mole Concept and Stoichiometric Equivalencies

3.17 In what smallest whole-number ratio must N and O atoms combine to make dinitrogen tetroxide, N_2O_4? What is the mole ratio of the elements in this compound?

3.18 In what atom ratio are the elements present in methane, CH_4 (the chief component of natural gas)? In what mole ratio are the atoms of the elements present in this compound?

3.19 How many moles of sodium atoms correspond to 1.56×10^{21} atoms of sodium?

3.20 How many moles of nitrogen molecules correspond to 1.80×10^{24} molecules of N_2?

3.21 Sucrose (table sugar) has the formula $C_{12}H_{22}O_{11}$. In this compound, what is the
(a) atom ratio of C to H?   (c) atom ratio of H to O?
(b) mole ratio of C to O?   (d) mole ratio of H to O?

3.22 Nail polish remover is usually the volatile liquid acetone, $(CH_3)_2CO$. In this compound, what is the
(a) atom ratio of C to O?   (c) atom ratio of C to H?
(b) mole ratio of C to O?   (d) mole ratio of C to H?

3.23 How many moles of Al atoms are needed to combine with 1.58 mol of O atoms to make aluminum oxide, Al_2O_3?

3.24 How many moles of vanadium atoms, V, are needed to combine with 0.565 mol of O atoms to make vanadium pentoxide, V_2O_5?

3.25 How many moles of Al are in 2.16 mol of Al_2O_3?

3.26 How many moles of O atoms are in 3.25 mol of calcium carbonate, $CaCO_3$, the chief constituent of seashells?

3.27 Aluminum sulfate, $Al_2(SO_4)_3$, is a compound used in sewage treatment plants.
(a) Construct a pair of conversion factors that relate moles of aluminum to moles of sulfur for this compound.
(b) Construct a pair of conversion factors that relate moles of sulfur to moles of $Al_2(SO_4)_3$.
(c) How many moles of Al are in a sample of this compound if the sample also contains 0.900 mol S?
(d) How many moles of S are in 1.16 mol $Al_2(SO_4)_3$?

3.28 Magnetite is a magnetic iron ore. Its formula is Fe_3O_4.
(a) Construct a pair of conversion factors that relate moles of Fe to moles of Fe_3O_4.
(b) Construct a pair of conversion factors that relate moles of Fe to moles of O in Fe_3O_4.
(c) How many moles of Fe are in 2.75 mol of Fe_3O_4?
(d) If this compound could be prepared from Fe_2O_3 and O_2, how many moles of Fe_2O_3 would be needed to prepare 4.50 mol Fe_3O_4?

3.29 How many moles of H_2 and N_2 can be formed by the decomposition of 0.145 mol of ammonia, NH_3?

3.30 How many moles of O_2 are needed to combine with 0.225 mol Al to give Al_2O_3?

3.31 How many moles of UF_6 would have to be decomposed to provide enough fluorine to prepare 1.25 mol of CF_4? (Assume sufficient carbon is available.)

3.32 How many moles of Fe_3O_4 are required to supply enough iron to prepare 0.260 mol Fe_2O_3? (Assume sufficient oxygen is available.)

3.33 How many atoms of carbon are combined with 3.13 mol H in a sample of the compound propane, C_3H_8? (Propane is used as the fuel in gas barbecues.)

3.34 How many atoms of hydrogen are found in 2.31 mol of propane, C_3H_8?

3.35 What is the total number of atoms in 0.260 mol of glucose, $C_6H_{12}O_6$?

3.36 What is the total number of atoms in 0.356 mol of ammonium nitrate, NH_4NO_3, an important fertilizer?

Measuring Moles of Elements and Compounds

3.37 How many atoms are in 6.00 g of carbon-12?

3.38 How many atoms are in 1.50 mol of carbon-12? How many grams does this much carbon-12 weigh?

3.39 What is the mass of 1.00 mol of each of the following elements?
(a) sodium   (b) sulfur   (c) chlorine

3.40 What is the mass of 1.00 mol of each of the following elements?
(a) nitrogen   (b) iron   (c) magnesium

3.41 Determine the mass in grams of each of the following:
(a) 1.35 mol Fe   (b) 24.5 mol O   (c) 0.876 mol Ca

3.42 Determine the mass in grams of each of the following:
(a) 0.536 mol S   (b) 4.29 mol N   (c) 8.11 mol Al

3.43 How many moles of nickel are in 18.7 g of Ni?

3.44 How many moles of chromium are in 65.7 g of Cr?

3.45 Calculate the formula mass of each of the following and round your answer to the nearest 0.1 u.
(a) $NaHCO_3$ (c) $(NH_4)_2CO_3$ (e) $CuSO_4 \cdot 5H_2O$
(b) $K_2Cr_2O_7$ (d) $Al_2(SO_4)_3$

3.46 Calculate the formula mass of each of the following and round your answer to the nearest 0.1 u.
(a) $Ca(NO_3)_2$ (c) $Mg_3(PO_4)_2$ (e) $Na_2SO_4 \cdot 10H_2O$
(b) $Pb(C_2H_5)_4$ (d) $Fe_4[Fe(CN)_6]_3$

3.47 Calculate the mass in grams of the following.
(a) 1.25 mol $Ca_3(PO_4)_2$ (c) 0.600 mol C_4H_{10}
(b) 0.625 mol $Fe(NO_3)_3$ (d) 1.45 mol $(NH_4)_2CO_3$

3.48 What is the mass in grams of the following?
(a) 0.754 mol $ZnCl_2$ (c) 0.322 mol $POCl_3$
(b) 0.194 mol KIO_3 (d) 4.31×10^{-3} mol $(NH_4)_2HPO_4$

3.49 Calculate the number of moles of each compound in the following samples.
(a) 21.5 g $CaCO_3$ (c) 16.8 g $Sr(NO_3)_2$
(b) 1.56 g NH_3 (d) 6.98 µg Na_2CrO_4

3.50 Calculate the number of moles of each compound in the following samples.
(a) 9.36 g $Ca(OH)_2$ (c) 4.29 g H_2O_2
(b) 38.2 g $PbSO_4$ (d) 4.65 mg $NaAuCl_4$

3.51 One sample of CaC_2 contains 0.150 mol of carbon. How many moles and how many grams of calcium are also in the sample? [Calcium carbide, CaC_2, was once used to make signal flares for ships. Water dripped onto CaC_2 reacts to give acetylene (C_2H_2) which burns brightly.]

3.52 How many moles of iodine are in 0.500 mol of $Ca(IO_3)_2$? How many grams of calcium iodate are needed to supply this much iodine? [Iodized salt contains a trace amount of calcium iodate, $Ca(IO_3)_2$, to help prevent a thyroid condition called goiter.]

3.53 How many moles of nitrogen, N, are in 0.650 mol of $(NH_4)_2CO_3$? How many grams of this compound supply this much nitrogen?

3.54 How many moles of nitrogen, N, are in 0.556 mol of NH_4NO_3? How many grams of NH_4NO_3 supply this much nitrogen?

Percentage Composition

3.55 Calculate the percentage composition by mass for each of the following: (a) NaH_2PO_4, (b) $NH_4H_2PO_4$, (c) $(CH_3)_2CO$, (d) $CaSO_4$, (e) $CaSO_4 \cdot 2H_2O$.

3.56 Calculate the percentage composition by mass of each of the following: (a) $(CH_3)_2N_2H_2$, (b) $CaCO_3$, (c) $Fe(NO_3)_3$, (d) C_3H_8, (e) $Al_2(SO_4)_3$.

3.57 It was found that 2.35 g of a compound of phosphorus and chlorine contained 0.539 g of phosphorus. What are the percentages by mass of phosphorus and chlorine in this compound?

3.58 An analysis revealed that 5.67 g of a compound of nitrogen and oxygen contained 1.47 g of nitrogen. What are the percentages by mass of nitrogen and oxygen in this compound?

3.59 Phencyclidine ("angel dust") is $C_{17}H_{25}N$. A sample suspected of being this illicit drug was found to have a percentage composition of 83.71% C, 10.42% H, and 5.61% N. Do these data acceptably match the theoretical data for phencyclidine?

3.60 The hallucinogenic drug LSD has the molecular formula $C_{20}H_{25}N_3O$. One suspected sample contained 74.07% C, 7.95% H, and 9.99% N.

(a) What is the percentage O in the sample?
(b) Are these data consistent for LSD?

3.61 How many grams of O are combined with 7.14×10^{21} atoms of N in the compound N_2O_5?

3.62 How many grams of C are combined with 4.25×10^{23} atoms of H in the compound C_5H_{12}?

Empirical Formulas

3.63 Write empirical formulas for the following compounds.
(a) S_2Cl_2 (c) NH_3 (e) H_2O_2
(b) $C_6H_{12}O_6$ (d) As_2O_6

3.64 What are the empirical formulas of the following compounds?
(a) $C_2H_4(OH)_2$ (c) C_4H_{10} (e) C_2H_5OH
(b) $H_2S_2O_8$ (d) B_2H_6

3.65 Quantitative analysis of a sample of sodium pertechnetate with a mass of 0.896 g found 0.111 g of sodium and 0.477 g of technetium. The remainder was oxygen. Calculate the empirical formula of sodium pertechnetate, arranging the atomic symbols in the formula in the order of NaTcO. (Radioactive sodium pertechnetate is used as a brain-scanning agent in medicine.)

3.66 A sample of Freon, a gas once used as a propellant in aerosol cans, was found to contain 0.423 g of C, 2.50 g Cl, and 1.34 g F. What is the empirical formula of this compound?

3.67 A dry-cleaning fluid composed of only carbon and chlorine was found to be composed of 14.5% C and 85.5% Cl (by mass). What is the empirical formula of this compound?

3.68 One compound of mercury with a formula mass of 519 contains 77.26% Hg, 9.25% C, and 1.17% H (with the balance being O). Calculate the empirical and molecular formulas, arranging the symbols in the order HgCHO.

3.69 A substance was found to be composed of 22.9% Na, 21.5% B, and 55.7% O. What is the empirical formula of this compound?

3.70 Vanillin, a compound used as a flavoring agent in food products, has the following percentage composition: 63.2% C, 5.26% H, and 31.6% O. What is the empirical formula of vanillin?

3.71 When 0.684 g of an organic compound containing only carbon, hydrogen, and oxygen was burned in oxygen, 1.312 g CO_2 and 0.805 g H_2O were obtained. What is the empirical formula of the compound?

3.72 Methyl ethyl ketone (often abbreviated MEK) is a powerful solvent with many commercial uses. A sample of this compound (which contains only C, H, and O) weighing 0.822 g was burned in oxygen to give 2.01 g CO_2 and 0.827 g H_2O. What is the empirical formula for MEK?

3.73 When 6.853 mg of a sex hormone was burned in a combustion analysis, 19.73 mg of CO_2 and 6.391 mg of H_2O were obtained. What is the empirical formula of the compound?

3.74 When a sample of a compound in the vitamin D family was burned in a combustion analysis, 5.983 mg of the compound gave 18.490 mg of CO_2 and 6.232 mg of H_2O. What is the empirical formula of the compound?

Molecular Formulas

3.75 The following are empirical formulas and the masses per mole for three compounds. What are their molecular formulas?
(a) NaS_2O_3; 270.4 g/mol   (c) C_2HCl; 181.4 g/mol
(b) C_3H_2Cl; 147.0 g/mol

3.76 The following are empirical formulas and the masses per mole for three compounds. What are their molecular formulas?
(a) Na_2SiO_3; 732.6 g/mol   (c) CH_3O; 62.1 g/mol
(b) $NaPO_3$; 305.9 g/mol

3.77 The compound described in Problem 3.73 was found to have a molecular mass of 290. What is its molecular formula?

3.78 The compound described in Problem 3.74 was found to have a molecular mass of 399. What is the molecular formula of this compound?

3.79 A sample of a compound of mercury and bromine with a mass of 0.389 g was found to contain 0.111 g bromine. Its molecular mass was found to be 561. What are its empirical and molecular formulas?

3.80 A 0.6662 g sample of "antimonal saffron" was found to contain 0.4017 g of antimony. The remainder was sulfur. The formula mass of this compound is 404. What are the empirical and molecular formulas of this pigment, arranging their atomic symbols in the order SbS? (This compound is a red pigment used in painting.)

3.81 A sample of a compound of C, H, N, and O, with a mass of 0.6216 g, was found to contain 0.1735 g C, 0.01455 g H, and 0.2024 g N. Its formula mass is 129. Calculate its empirical and molecular formulas, arranging their symbols in alphabetical order.

3.82 Strychnine, a deadly poison, has a formula mass of 334 and a percentage composition of 75.42% C, 6.63% H, 8.38% N, and the balance oxygen. Calculate the empirical and molecular formulas of strychnine (arranging the symbols alphabetically).

Balancing Chemical Equations

3.83 How many moles of iron are part of the expression $2Fe_4[Fe(CN)_6]_3$, taken from a balanced equation?

3.84 How many moles of oxygen are part of the expression $3Fe(H_2PO_2)_3$, taken from a balanced equation?

3.85 Write the equation that expresses in acceptable chemical shorthand the following statement: "Iron can be made to react with molecular oxygen (O_2) to give iron oxide with the formula Fe_2O_3."

3.86 The conversion of one air pollutant, NO, produced in vehicle engines, into another, NO_2, occurs when NO reacts with molecular oxygen in the air. Write the balanced equation for this reaction.

3.87 Balance the following equations.
(a) $Ca(OH)_2 + HCl \longrightarrow CaCl_2 + H_2O$
(b) $AgNO_3 + CaCl_2 \longrightarrow Ca(NO_3)_2 + AgCl$
(c) $Fe_2O_3 + C \longrightarrow Fe + CO_2$
(d) $NaHCO_3 + H_2SO_4 \longrightarrow Na_2SO_4 + H_2O + CO_2$
(e) $C_4H_{10} + O_2 \longrightarrow CO_2 + H_2O$

3.88 Balance the following equations.
(a) $SO_2 + O_2 \longrightarrow SO_3$
(b) $P_4O_{10} + H_2O \longrightarrow H_3PO_4$
(c) $Pb(NO_3)_2 + Na_2SO_4 \longrightarrow PbSO_4 + NaNO_3$
(d) $Fe_2O_3 + H_2 \longrightarrow Fe + H_2O$
(e) $Al + H_2SO_4 \longrightarrow Al_2(SO_4)_3 + H_2$

3.89 Balance the following equations.
(a) $Mg(OH)_2 + HBr \longrightarrow MgBr_2 + H_2O$
(b) $HCl + Ca(OH)_2 \longrightarrow CaCl_2 + H_2O$
(c) $Al_2O_3 + H_2SO_4 \longrightarrow Al_2(SO_4)_3 + H_2O$
(d) $KHCO_3 + H_3PO_4 \longrightarrow K_2HPO_4 + H_2O + CO_2$
(e) $C_9H_{20} + O_2 \longrightarrow CO_2 + H_2O$

3.90 Balance the following equations.
(a) $CaO + HNO_3 \longrightarrow Ca(NO_3)_2 + H_2O$
(b) $Na_2CO_3 + Mg(NO_3)_2 \longrightarrow MgCO_3 + NaNO_3$
(c) $(NH_4)_3PO_4 + NaOH \longrightarrow Na_3PO_4 + NH_3 + H_2O$
(d) $LiHCO_3 + H_2SO_4 \longrightarrow Li_2SO_4 + H_2O + CO_2$
(e) $C_4H_{10}O + O_2 \longrightarrow CO_2 + H_2O$

Stoichiometry Based on Chemical Equations

3.91 Chlorine is used by textile manufacturers to bleach cloth. Excess chlorine is destroyed by its reaction with sodium thiosulfate, $Na_2S_2O_3$, as follows.

$$Na_2S_2O_3(aq) + 4Cl_2(g) + 5H_2O \longrightarrow$$

$$2NaHSO_4(aq) + 8HCl(aq)$$

(a) How many moles of $Na_2S_2O_3$ are needed to react with 0.12 mol of Cl_2?
(b) How many moles of HCl can form from 0.12 mol of Cl_2?
(c) How many moles of H_2O are required for the reaction of 0.12 mol of Cl_2?
(d) How many moles of H_2O react if 0.24 mol HCl is formed?

3.92 The octane in gasoline burns according to the following equation.

$$2C_8H_{18} + 25O_2 \longrightarrow 16CO_2 + 18H_2O$$

(a) How many moles of O_2 are needed to react fully with 6 mol of octane?
(b) How many moles of CO_2 can form from 0.5 mol of octane?

(c) How many moles of water are produced by the combustion of 8 mol of octane?

(d) If this reaction is used to synthesize 6.00 mol of CO_2, how many moles of oxygen are needed? How many moles of octane?

3.93 The incandescent white of a fireworks display is caused by the reaction of phosphorus with O_2 to give P_4O_{10}.
(a) Write the balanced chemical equation for the reaction.
(b) How many grams of O_2 are needed to combine with 6.85 g of P?
(c) How many grams of P_4O_{10} can be made from 8.00 g of O_2?
(d) How many grams of P are needed to make 7.46 g of P_4O_{10}?

3.94 The combustion of butane, C_4H_{10}, produces carbon dioxide and water. When one sample of C_4H_{10} was burned, 3.46 g of water was formed.
(a) Write the balanced chemical equation for the reaction.
(b) How many grams of butane were burned?
(c) How many grams of O_2 were consumed?
(d) How many grams of CO_2 were formed?

3.95 In *dilute* nitric acid, HNO_3, copper metal dissolves according to the following equation.

$$3Cu(s) + 8HNO_3(aq) \longrightarrow$$
$$3Cu(NO_3)_2(aq) + 2NO(g) + 4H_2O$$

How many grams of HNO_3 are needed to dissolve 11.45 g of Cu according to this equation?

3.96 The reaction of hydrazine, N_2H_4, with hydrogen peroxide, H_2O_2, has been used in rocket engines. One way these compounds react is described by the equation

$$N_2H_4 + 7H_2O_2 \longrightarrow 2HNO_3 + 8H_2O$$

According to this equation, how many grams of H_2O_2 are needed to react completely with 852 g of N_2H_4?

Limiting Reactant Calculations
3.97 The reaction of powdered aluminum and iron(III) oxide,

$$2Al + Fe_2O_3 \longrightarrow Al_2O_3 + 2Fe$$

produces so much heat the iron that forms is molten. Because of this, railroads use the reaction to provide molten steel to weld steel rails together when laying track. Suppose that in one batch of reactants 4.20 mol of Al was mixed with 1.75 mol of Fe_2O_3.
(a) Which reactant, if either, was the limiting reactant?
(b) Calculate the number of grams of iron that can be formed from this mixture of reactants.

3.98 Phosphorus pentachloride reacts with water to give phosphoric acid and hydrogen chloride according to the following equation.

$$PCl_5 + 4H_2O \longrightarrow H_3PO_4 + 5HCl$$

In one experiment, 0.360 mol of PCl_5 was slowly added to 2.88 mol of water.
(a) Which reactant, if either, was the limiting reactant?
(b) How many grams of HCl were formed in the reaction?

3.99 Silver nitrate, $AgNO_3$, reacts with iron(III) chloride, $FeCl_3$, to give silver chloride, AgCl, and iron(III) nitrate, $Fe(NO_3)_3$. A solution containing 18.0 g of $AgNO_3$ was mixed with a solution containing 32.4 g of $FeCl_3$. How many grams of which reactant remains after the reaction is over?

3.100 Chlorine dioxide, ClO_2, has been used as a disinfectant in air conditioning systems. It reacts with water according to the equation

$$6ClO_2 + 3H_2O \longrightarrow 5HClO_3 + HCl$$

If 142.0 g of ClO_2 is mixed with 38.0 g of H_2O, how many grams of which reactant remain if the reaction is complete?

Theoretical Yield and Percentage Yield
3.101 Barium sulfate, $BaSO_4$, is made by the following reaction.

$$Ba(NO_3)_2(aq) + Na_2SO_4(aq) \rightarrow BaSO_4(s) + 2NaNO_3(aq)$$

An experiment was begun with 75.00 g of $Ba(NO_3)_2$ and an excess of Na_2SO_4. After collecting and drying the product, 63.45 g of $BaSO_4$ was obtained. Calculate the theoretical yield and percentage yield of $BaSO_4$.

3.102 The Solvay process for the manufacture of sodium carbonate begins by passing ammonia and carbon dioxide through a solution of sodium chloride to make sodium bicarbonate and ammonium chloride. The equation for the overall reaction is

$$H_2O + NaCl + NH_3 + CO_2 \longrightarrow NH_4Cl + NaHCO_3$$

In the next step, sodium bicarbonate is heated to give sodium carbonate and two gases, carbon dioxide and steam.

$$2NaHCO_3 \longrightarrow Na_2CO_3 + CO_2 + H_2O$$

What is the theoretical yield of sodium carbonate, expressed in grams, if 120 g NaCl was used in the first reaction? If 85.4 g of Na_2CO_3 was obtained, what was the percentage yield?

3.103 Aluminum sulfate can be made by the following reaction.

$$2AlCl_3(aq) + 3H_2SO_4(aq) \rightarrow Al_2(SO_4)_3(aq) + 6HCl(aq)$$

It is quite soluble in water, so to isolate it the solution has to be evaporated to dryness. This drives off the volatile HCl, but the residual solid has to be heated to a little over 200 °C to drive off all of the water. In one experiment, 25.0 g of $AlCl_3$ was mixed with 30.0 g of H_2SO_4. Eventually, 28.46 g of pure $Al_2(SO_4)_3$ was isolated. Calculate the percentage yield.

3.104 The combustion of methyl alcohol in an abundant excess of oxygen follows the equation

$$2CH_3OH + 3O_2 \longrightarrow 2CO_2 + 4H_2O$$

When 6.40 g of CH_3OH was mixed with 10.2 g of O_2 and ignited, 6.12 g of CO_2 was obtained. What was the percentage yield of CO_2?

3.105 The potassium salt of benzoic acid, potassium benzoate ($KC_7H_5O_2$), can be made by the action of potassium permanganate on toluene (C_7H_8) as follows.

$$C_7H_8 + 2KMnO_4 \rightarrow KC_7H_5O_2 + 2MnO_2 + KOH + H_2O$$

If the yield of potassium benzoate cannot realistically be expected to be more than 71%, what is the minimum number of grams of toluene needed to achieve this yield while producing 11.5 g of potassium benzoate?

3.106 Manganese trifluoride, MnF_3, can be prepared by the following reaction.

$$2MnI_2(s) + 13F_2(g) \longrightarrow 2MnF_3(s) + 4IF_5(l)$$

What is the minimum number of grams of F_2 that must be used to react with 12.0 g of MnI_2 if the overall yield of MnF_3 is no more than 75% of theory?

Molar Concentration
3.107 A solution is labeled 0.25 *M* HCl. Construct two conversion factors that relate moles of HCl to the volume of solution expressed in liters.

3.108 A solution is labeled 0.75 *M* $CuSO_4$. Construct two conversion factors that relate moles of $CuSO_4$ to the volume of solution, expressed in liters.

3.109 Calculate the molarity of a solution prepared by dissolving
(a) 4.00 g of NaOH in 100.0 mL of solution.
(b) 16.0 g of $CaCl_2$ in 250.0 mL of solution.
(c) 14.0 g of KOH in 75.0 mL of solution.
(d) 6.75 g of $H_2C_2O_4$ in 500 mL of solution.

3.110 Calculate the molarity of a solution that contains
(a) 3.60 g of H_2SO_4 in 450 mL of solution.
(b) 2.0×10^{-3} mol $Fe(NO_3)_2$ in 12.0 mL of solution.
(c) 1.65 mol HCl in 2.16 L of solution.
(d) 18.0 g HCl in 0.375 L of solution.

3.111 Calculate the number of grams of each solute that has to be taken to make each of the following solutions.
(a) 125 mL of 0.200 *M* NaCl
(b) 250 mL of 0.360 *M* $C_6H_{12}O_6$ (glucose)
(c) 250 mL of 0.250 *M* H_2SO_4

3.112 How many grams of solute are needed to make each of the following solutions?

(a) 250 mL of 0.100 *M* K_2SO_4 (c) 500 mL of 0.400 *M*
(b) 100 mL of 0.250 *M* K_2CO_3 KOH

Dilution of Solutions
3.113 If 25.0 mL of 0.56 *M* H_2SO_4 is diluted to a volume of 125 mL, what is the molarity of the resulting solution?

3.114 A 150 mL sample of 0.45 *M* HNO_3 is diluted to 450 mL. What is the molarity of the resulting solution?

3.115 To what volume must 25.0 mL of 18.0 *M* H_2SO_4 be diluted to produce 1.50 *M* H_2SO_4?

3.116 To what volume must 50.0 mL of 1.50 *M* HCl be diluted to produce 0.200 *M* HCl?

3.117 How many milliliters of water must be added to 150 mL of 2.5 *M* KOH to give a 1.0 *M* solution? (Assume volumes are additive.)

3.118 How many milliliters of water must be added to 120 mL of 1.50 *M* HCl to give 1.00 *M* HCl?

Solution Stoichiometry
3.119 What is the molarity of an aqueous solution of potassium hydroxide if 21.34 mL is exactly neutralized by 20.78 mL of 0.116 *M* HCl by the following reaction?

$$KOH(aq) + HCl(aq) \longrightarrow KCl(aq) + H_2O$$

3.120 What is the molarity of an aqueous sulfuric acid solution if 12.88 mL is neutralized by 26.04 mL of 0.1024 *M* NaOH? The reaction is

$$2NaOH(aq) + H_2SO_4(aq) \longrightarrow Na_2SO_4(aq) + 2H_2O$$

3.121 How many milliliters of 0.25 *M* $NiCl_2$ solution are needed to react completely with 20.0 mL of 0.15 *M* Na_2CO_3 solution? How many grams of $NiCO_3$ will be formed? The reaction is

$$Na_2CO_3(aq) + NiCl_2(aq) \longrightarrow NiCO_3(s) + 2NaCl(aq)$$

3.122 Epsom salts is $MgSO_4 \cdot 7H_2O$. How many grams of this compound are needed to react completely with 50.0 mL of 0.125 *M* $BaCl_2$ solution? These solutes react according to the equation

$$MgSO_4(aq) + BaCl_2(aq) \longrightarrow MgCl_2(aq) + BaSO_4(s)$$

3.123 How many milliliters of 0.100 *M* NaOH are needed to completely neutralize 25.0 mL of 0.250 *M* H_3PO_4? The reaction is

$$3NaOH(aq) + H_3PO_4(aq) \longrightarrow Na_3PO_4(aq) + 3H_2O$$

3.124 How many grams of baking soda, $NaHCO_3$, are needed to react with 162 mL of stomach acid having an HCl concentration of 0.052 *M*? The reaction is

$$NaHCO_3(aq) + HCl(aq) \longrightarrow NaCl(aq) + CO_2(g) + H_2O$$

ADDITIONAL EXERCISES

3.125 Suppose you had one mole of pennies and that you were going to spend 500 million dollars (5.00×10^8 dollars) each and every second until you spent your entire fortune. How many years would it take you to spend all this cash? (Assume 1 year = 365 days.)

3.126 A 0.1246 g sample of a compound of chromium and chlorine was dissolved in water. All of the chloride ion was then captured by silver ion in the form of AgCl. A mass of 0.3383 g of AgCl was obtained. Calculate the empirical formula of the compound of Cr and Cl.

* **3.127** A compound of Ca, C, N, and S was subjected to quantitative analysis and formula mass determination, and the following data were obtained. A 0.250 g sample was mixed with Na_2CO_3 to convert all of the Ca to 0.160 g of $CaCO_3$. A 0.115 g sample of the compound was carried through a series of reactions until all of its S was changed to 0.344 g of $BaSO_4$. A 0.712 g sample was processed to liberate all of its N as NH_3, and 0.155 g NH_3 was obtained. The formula mass was found to be 156. Determine the empirical and molecular formulas of this compound.

3.128 Ammonium nitrate will detonate if ignited in the presence of certain impurities. The equation for this reaction at a high temperature is

$$2NH_4NO_3(s) \xrightarrow{>300\,°C} 2N_2(g) + O_2(g) + 4H_2O(g)$$

Notice that all of the products are gases and so must occupy a vastly greater volume than the solid reactant.
(a) How many moles of *all* gases are produced from 1 mol of NH_4NO_3?
(b) If 1.00 ton of NH_4NO_3 exploded according to this equation, how many moles of *all* gases would be produced? (1 ton = 2000 lb.)
(In April 1947 a ship cargo of ammonium nitrate, NH_4NO_3, blew up in the harbor of Texas City, Texas, with a loss of nearly 600 lives.)

3.129 A lawn fertilizer is rated as 6.00% nitrogen, meaning 6.00 g of N in 100 g of fertilizer. The nitrogen is present in the form of urea, $(NH_2)_2CO$. How many grams of urea are present in 100 g of the fertilizer to supply the rated amount of nitrogen?

3.130 Based solely on the amount of available carbon, how many grams of sodium oxalate, $Na_2C_2O_4$, could be obtained from 125 g of C_6H_6? (Assume that no loss of carbon occurs in any of the reactions needed to produce the $Na_2C_2O_4$.)

3.131 According to NASA, the space shuttle's external fuel tank for the main propulsion system carries 1,361,936 lb of liquid oxygen and 227,641 lb of liquid hydrogen. During takeoff, these chemicals are consumed as they react to form water. If the reaction is continued until all of one reactant is gone, how many pounds of which reactant are left over?

* **3.132** For a research project, a student decided to test the effect of the lead(II) ion (Pb^{2+}) on the ability of salmon eggs to hatch. This ion was obtainable from the water-soluble salt, lead(II) nitrate, $Pb(NO_3)_2$, which the student decided to make by the following reaction. (The desired product was to be isolated by the slow evaporation of the water.)

$$PbO(s) + 2HNO_3(aq) \longrightarrow Pb(NO_3)_2(aq) + H_2O$$

Losses of product for various reasons were expected, and a yield of 86.0% was expected. In order to have 5.00 g of product at this yield, how many grams of PbO should be taken? (Assume that sufficient nitric acid, HNO_3, would be used.)

* **3.133** How many milliliters of 0.10 *M* HCl must be added to 50.0 mL of 0.40 *M* HCl to give a final solution that has a molarity of 0.25 *M*?

* **3.134** A mixture of sodium carbonate and sodium chloride with a mass of 0.326 g was dissolved in water and 0.116 *M* HCl was added until no more CO_2 could be removed from the solution. By then, 43.6 mL of the standard HCl had been added. Calculate the percentage by mass of the sodium carbonate in the mixture. The reaction of sodium carbonate with hydrochloric acid is

$$Na_2CO_3(aq) + 2HCl(aq) \rightarrow 2NaCl(aq) + CO_2(g) + H_2O$$

Silicon—An Element for the Twenty-first Century

2

The potential for *new* uses of silicon in the twenty-first century may be even greater than its current uses in transistors and computer chips. New on the rapidly approaching horizon are *MEMS* and *biochips*. MEMS stands for micro-electromechanical systems. They're tiny motors and gears *no larger than specks of dust* made out of and embedded in wafers of silicon. Some are microscopic laboratories no larger than credit cards. Biochips are silicon wafers "loaded" not with the ones and zeros of binary code but with a kind of molecular quaternary code that enables physicians to diagnose diseases and genetic defects.

MEMS

The relatively low cost of silicon, its chemical stability, and our ability to machine it or etch it to various shapes have widened the applications of this element immensely. The twenty-first century will see an explosion of uses of silicon-based MEMS. Figure 2a is a close-up view of an example, a section of the safety lock system developed by Sandia Laboratories (New Mexico) for nuclear missiles. MEMS are thus microscopic machines. They are now used in airbag systems where a special MEM device converts sharp mechanical jolts into voltage changes that instantly trigger the deployment of the airbag. Propulsion systems for small space satellites use MEMS.

Of greatest interest to chemists is the possibility of carving the pipes, tubes, and flasks of a laboratory into a silicon chip, enabling reactions of chemical analyses to be carried out very rapidly and on very small scales.

Why Silicon?

Silicon is a shiny, blue-gray metal that can be machined and buffed. It is exceptionally stable to heat, not melting until 1420 °C. It is relatively inexpensive, being the second most abundant element (after oxygen) in the

Figure 2a A micro-electromechanical system or MEMS. This is a portion of the 50-micrometer gears in Sandia's safety lock for nuclear missiles. (Courtesy of Sandia National Laboratories' Intelligent Micromachine Initiative; www.mdl.sandia.gov/Micromachine).

earth's crust and present in silicate minerals and sand or silicon dioxide (SiO_2).

Another advantage of silicon is its exceptional chemical stability at or near room temperature. Although solid silicon does react with oxygen, the ultrathin oxide coating that forms shields the remaining silicon from almost all other substances. (If the silicon thus protected is melted, then the coating is broken up and the silicon reacts rapidly with almost anything.) At room temperature only the most reactive element, fluorine, attacks silicon, and fluorine is never present in its free state in the environment.

Silicon is obtained by heating sand and carbon in an electric arc furnace.

$$SiO_2 + 2C \longrightarrow Si + 2CO$$

The silicon so obtained is only about 96% pure, however, not nearly pure enough for use in computer chips where the level of impurities must be less than 10^{-9} to

chemicals in our world

10^{-10} percent. One way to make raw silicon purer is to convert it by hot chlorine to silicon tetrachloride, $SiCl_4$, which is a liquid that boils at only 59 °C and is far easier to make exceptionally pure. When $SiCl_4$ is treated with either very pure zinc or magnesium, silicon is released.

$$SiCl_4 + 2Mg \longrightarrow Si + 2MgCl_2$$

The chief advantage of silicon for its use in computer chips is that it is a semiconductor. We'll have to defer a discussion of what "semiconductor" means to Chapter 9.

Biochips

The "bio" of "biochip" refers to a biochemical substance, DNA, which, as you probably learned in eighth or ninth grade, is the chemical of genes. In its native state in cells, DNA consists of two extremely long molecules that intertwine as a coil, the famous DNA double helix. A single strand of DNA, one of the two pieces of a double helix, is like a long charm bracelet made up of four different kinds of "charms" suspended from a common molecular "chain." The charms have simple symbols, namely, A, T, G, and C. (The C does not stand for carbon here.) A very, very short segment of a single strand of DNA might be made up to have the "charms" strung out in the following sequence on a chain.

. . . ATTCGGGCCAATAGCCCAATTAGGC. . .

Biochips incorporate molecules of preselected lengths and preselected sequences of A, T, G, and C. The molecules are actually affixed to the chip. So this is why a biochip can be described as carrying information not in a binary code of ones and zeros, but in a molecular code of four DNA "letters." We'll defer to *Chemicals in Our World 12,* following Chapter 23, to see how biochips work.

Additional Reading

I. Amato, "Fomenting a Revolution, in Miniature," *Science,* October 16, 1998, page 402.

Questions

1. What are some of the properties of silicon that make it so useful as a material for making MEMS?

2. Carbon is even cheaper and more inert at room temperature than silicon, but it's not useful for making MEMS. Why do you suppose this is so?

The beautiful pink sands of Bermuda were formed by small marine animals (coral) that extract calcium ions and carbonate ions from seawater to produce insoluble calcium carbonate for their shells. In this chapter we will explore a variety of reactions between ions in aqueous solution, including those that form insoluble products.

4

Reactions between Ions in Aqueous Solutions

This Chapter in Context You learned in Chapter 2 that ionic compounds constitute an important class of substances and that they contain either simple monatomic ions such as Na$^+$ and Cl$^-$, or polyatomic ions such as NO$_3^-$ and SO$_4^{2-}$. Many ionic compounds are soluble in water, and in this chapter we will discuss what happens to them when they dissolve. Often, aqueous solutions of ionic substances undergo chemical reactions when they are combined, and we will explore in some depth the nature of these reactions and the products that form. We will also extend the principles of stoichiometry to deal with solutions of ionic substances.

4.1 Electrolytes and Nonelectrolytes

In Chapter 2 you learned that pure water is a very poor conductor of electricity. This is because water consists of uncharged molecules that are incapable of transporting electrical charge. However, when an ionic compound such as CuSO$_4$ or NaCl is dissolved in the water, an electrically conducting solution is formed. This electrical conductivity is demonstrated in Figure 4.1*a* for a solution of copper sulfate.

Solutes such as CuSO$_4$ or NaCl, which yield electrically conducting aqueous solutions, are called **electrolytes.** Their ability to conduct electricity suggests the presence of electrically charged particles that are able to move through the solution. The generally accepted reason is that when an ionic compound dissolves in water, the ions separate from each other and enter the solution as more or less independent particles that are surrounded by molecules of the solvent. This change is called the **dissociation** of the ionic compound, and is illustrated in Figure 4.2. In general, *we will assume that the dissociation of an ionic compound is complete* and that the solution contains no undissociated formula units of the salt. Thus, an aqueous solution of NaCl is really a solution that contains Na$^+$ and Cl$^-$ ions, with no undissociated NaCl "molecules" in the solution. Because these solutions contain so many ions, they are strong conductors of electricity, and salts are said to be **strong electrolytes.**

Keep in mind that a strong electrolyte is 100% dissociated in aqueous solution.

151

Figure 4.1 *Electrical conductivity of solutions of electrolytes versus nonelectrolytes.* (*a*) The copper sulfate solution is a strong conductor, and $CuSO_4$ is a strong electrolyte. (*b*) Neither sugar nor water is an electrolyte, and this sugar solution is a nonconductor.

(*a*)

(*b*)

Many ionic compounds have low solubilities in water. An example is AgBr, the light-sensitive compound in most photographic film. Although only a minute amount of this compound dissolves in water, all of it that does dissolve is completely dissociated. However, because of the extremely low solubility, the number of ions in the solution is extremely small and the solution doesn't conduct electricity well. Nevertheless, it is still convenient to think of AgBr as a strong electrolyte because it serves to remind us that salts are completely dissociated in aqueous solution.

Aqueous solutions of most molecular compounds do not conduct electricity, and such solutes are called **nonelectrolytes.** Examples are sugar and ethylene glycol (the solute in antifreeze solutions). Both of these solutes consist of mole-

Figure 4.2 *Dissociation of an ionic compound as it dissolves in water.* Ions separate from the solid and become surrounded by molecules of water. The ions are said to be hydrated. In the solution, the ions are able to move freely, and the solution is able to conduct electricity.

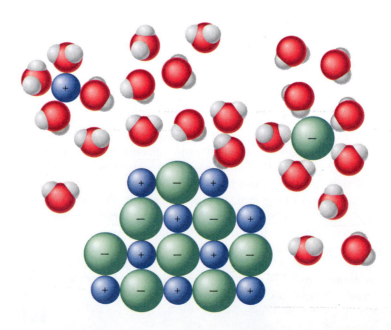

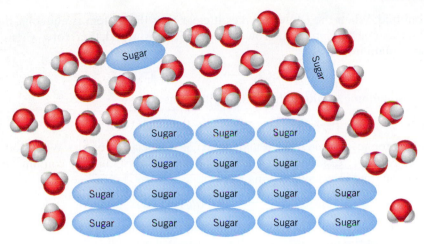

Figure 4.3 *Formation of an aqueous solution of a nonelectrolyte.* When a nonelectrolyte dissolves in water, the molecules of solute separate from each other and mingle with the water molecules. The solute molecules stay intact and do not dissociate into smaller particles.

cules that stay intact when they dissolve. They simply intermingle with water molecules when their solutions form (Figure 4.3). The solution contains no electrically charged particles, so it is unable to conduct electricity.

Ethylene glycol, $C_2H_4(OH)_2$, is a type of alcohol. Other alcohols, such as ethanol and methanol, are also nonelectrolytes.

▶**Chemistry in Practice**◀ Electroencephalograms (EEGs) and electrocardiograms (EKGs) are diagnostic procedures for evaluating the conditions of the brain and the heart, respectively. They depend on the presence of strong electrolytes in all body fluids, because these solutes must carry weak electrical currents used in the tests. An electrolyte paste is applied to those skin spots to which electrodes will be taped to provide a conducting connection into and through the skin. ◆

Equations for Dissociation Reactions

A convenient way to describe the dissociation of an ionic compound is with a chemical equation.

$$NaCl(s) \longrightarrow Na^+(aq) + Cl^-(aq)$$

We use the symbol *aq* after a charged particle to mean that it is dissolved in water and surrounded by water molecules. When a solute particle is surrounded by water molecules, we say it is **hydrated.** By writing the formulas of the ions separately, we mean that they are essentially independent of each other in the solution.

We can write a similar equation for the dissociation of calcium chloride, $CaCl_2$.

$$CaCl_2(s) \longrightarrow Ca^{2+}(aq) + 2Cl^-(aq)$$

This shows that each formula unit of $CaCl_2(s)$ releases three ions, one $Ca^{2+}(aq)$ and two $Cl^-(aq)$. The symbols (*s*) and (*aq*) once again indicate that the change is the dissociation of the ionic compound. Quite often, however, these symbols are omitted. When the context makes it clear that the system is aqueous, these symbols can be "understood." You should not be disturbed, therefore, when you see an equation such as

$$CaCl_2 \longrightarrow Ca^{2+} + 2Cl^-$$

Unless something is said to the contrary, it means

$$CaCl_2(s) \longrightarrow Ca^{2+}(aq) + 2Cl^-(aq)$$

Polyatomic ions generally remain intact as dissociation occurs. When sodium sulfate, Na_2SO_4, dissolves in water, for example, the solution contains both sodium ions and intact sulfate ions, released as follows.

$$Na_2SO_4(s) \longrightarrow 2Na^+(aq) + SO_4^{2-}(aq)$$

To write equations such as this correctly, you must know both the formulas and the charges of the polyatomic ions. If necessary, refer to Table 2.6 (page 76) for a review of the formulas of polyatomic ions.

Practice Exercise 1

Write equations that show what happens when the following solid ionic compounds dissolve in water: (a) $MgCl_2$, (b) $Al(NO_3)_3$, (c) Na_2CO_3, (d) $(NH_4)_2SO_4$. ◆

4.2 Equations for Ionic Reactions

In Figure 3.7 on page 130 we illustrated the reaction between aqueous solutions of lead nitrate, $Pb(NO_3)_2$, and potassium iodide, KI, to give a bright yellow precipitate of lead iodide, PbI_2. The equation we gave for the reaction is

$$Pb(NO_3)_2(aq) + 2KI(aq) \longrightarrow PbI_2(s) + 2KNO_3(aq) \tag{4.1}$$

As you learned in Chapter 3, this equation is perfectly satisfactory for performing stoichiometric calculations. However, let's take a closer look at the reaction to examine other ways that we might write the chemical equation.

Lead nitrate and potassium iodide are both ionic compounds. When they dissolve in water, they undergo complete dissociation, so their aqueous solutions actually contain ions, not "molecules" of $Pb(NO_3)_2$ and KI (Figure 4.4a). When the two solutions are mixed, the ions mingle. If no reaction were to take place, we simply would have a mixture of all four ions as shown in Figure 4.4b. However, it turns out that the salt PbI_2 has a very low solubility in water, so the large concentrations of Pb^{2+} and I^- ions in the mixture are exactly what would be present in a highly *supersaturated* solution of PbI_2. As you learned in Chapter 3, supersaturated solutions are unstable and tend to deposit crystals of the dissolved solute, and this is exactly what happens here; lead iodide crystals form and we observe the formation of a precipitate. This is illustrated in Figure 4.4c. Notice that after the precipitation of PbI_2, the solution contains only the ions of the other product, KNO_3. The products of the reaction can be separated from each other by filtering the solid from the rest of the mixture, as pictured in Figure 4.5.

We now see that the reaction between $Pb(NO_3)_2$ and KI is really a *reaction between ions,* and can be appropriately referred to as an **ionic reaction.** We can also see that the equation given above (Equation 4.1) does not describe what actually takes place in the reaction mixture. (Chemists refer to an equation such

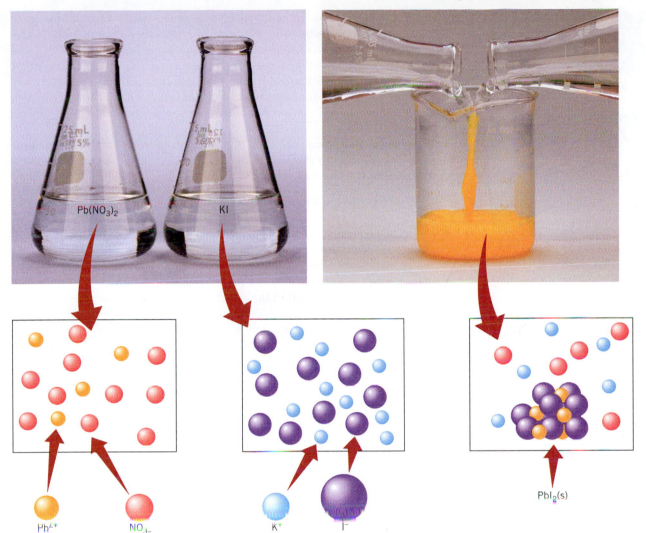

Figure 4.4 *The reaction of Pb(NO₃)₂ with KI.* On the left are flasks containing solutions of lead nitrate and potassium iodide. These solutes exist as separated ions in their respective solutions. On the right, we observe that when the solutions of the ions are combined, there is an immediate reaction as the Pb²⁺ ions join with the I⁻ ions to give a precipitate of small crystals of solid, yellow PbI₂. The reaction is so rapid that the yellow color develops where the two streams of liquid come together. If the Pb(NO₃)₂ and KI are combined in a 1-to-2 mole ratio, the solution now contains only K⁺ and NO₃⁻ ions (the ions of KNO₃).

as 4.1 as a **molecular equation** because all the formulas are written with the ions together, as if the substances in solution consist of neutral molecules.) A more accurate representation of the reaction is given by the **ionic equation,** *in which all soluble strong electrolytes are written in "dissociated" form.* To write the ionic equation for the reaction of Pb(NO₃)₂ with KI, we write the formulas of both reactants and the soluble product, KNO₃, in dissociated form. This gives

$$Pb^{2+}(aq) + 2NO_3^-(aq) + 2K^+(aq) + 2I^-(aq) \longrightarrow$$

$$PbI_2(s) + 2K^+(aq) + 2NO_3^-(aq)$$

Notice that we have *not* separated PbI₂ into its ions in this equation. This is because after the reaction is over, the Pb²⁺ and I⁻ ions are no longer able to move independently. They are trapped in the insoluble product, PbI₂.

Figure 4.5 *Separating a precipitate from a solution by filtration.* The reaction mixture is passed through filter paper held in a funnel. The precipitate is caught by the filter paper while the clear solution passes through.

The ionic equation for this reaction gives a much better representation of what actually takes place when the solutions of reactants are combined. It also lets us see that the reaction in solution really takes place between the Pb^{2+} and I^- ions; these are the ions that come together to form the product. The other ions, K^+ and NO_3^-, are unchanged by the reaction; they are present before the reaction begins and are present afterward. *Ions that do not actually take part in a reaction are sometimes called* **spectator ions;** *in a sense, they just "stand by and watch the action."*

To emphasize the actual reaction that occurs, we can write the **net ionic equation,** *which is obtained by eliminating spectator ions from the ionic equation.* Let's cross out the K^+ and NO_3^- ions.

$$Pb^{2+}(aq) + \cancel{2NO_3^-(aq)} + \cancel{2K^+(aq)} + 2I^-(aq) \longrightarrow$$
$$PbI_2(s) + \cancel{2K^+(aq)} + \cancel{2NO_3^-(aq)}$$

What remains is the net ionic equation,

$$Pb^{2+}(aq) + 2I^-(aq) \longrightarrow PbI_2(s)$$

Notice how it calls our attention to the ions that are actually participating in the reaction as well as the change that occurs.

The net ionic equation is especially useful because it permits us to *generalize.* It tells us that if we combine *any* solution that contains Pb^{2+} with *any* other solution that contains I^-, we ought to expect a precipitate of PbI_2. And this is exactly what happens! For example, another soluble lead salt is lead acetate, $Pb(C_2H_3O_2)_2$, and another soluble iodide is sodium iodide, NaI. If aqueous solutions of these salts are prepared and then mixed, a yellow precipitate of PbI_2 forms immediately. Example 4.1 below demonstrates how we write the molecular, ionic, and net ionic equations for the reaction.

Criteria for a Balanced Ionic or Net Ionic Equation

In the ionic and net ionic equations we've written, not only are the atoms in balance, but so is the net electrical charge, which is the same on both sides of the equation. Thus, in the ionic equation above, the sum of the charges of the ions on the left (Pb^{2+}, $2NO_3^-$, $2K^+$, and $2I^-$) is zero, which matches the sum of the charges on all of the formulas of the products (PbI_2, $2K^+$, and $2NO_3^-$). In the net ionic equation the charges on both sides are also the same: on the left we have Pb^{2+} and $2I^-$, with a net charge of zero, and on the right we have PbI_2, also with a charge of zero. We now have an additional requirement for an ionic equation or net ionic equation to be balanced: *the net electrical charge on both sides of the equation must be the same.*

Balancing ionic equations

> **Criteria for Balanced Ionic and Net Ionic Equations**
> **1. Material balance.** There must be the same number of atoms of each kind on both sides of the arrow.
> **2. Electrical balance.** The net electrical charge on the left must equal the net electrical charge on the right (although this charge does not necessarily have to be zero).

EXAMPLE 4.1
Writing Molecular, Ionic, and Net Ionic Equations

Write the molecular, ionic, and net ionic equations for the reaction of aqueous solutions of lead acetate and sodium iodide, which yields a precipitate of lead iodide and leaves the salt sodium acetate in solution.

Solution: We begin with the molecular equation, which we form by writing the complete formulas for the reactants and products. Since only the names of the reactants and products are given, they must be translated into chemical formulas, which requires a knowledge of chemical nomenclature. Let's begin by writing the names and chemical formulas of the reactants and products.

If necessary, review Section 2.11, which discusses naming compounds.

Reactants		Products	
lead acetate	$Pb(C_2H_3O_2)_2$	lead iodide	PbI_2
sodium iodide	NaI	sodium acetate	$NaC_2H_3O_2$

The Molecular Equation

Now that we have the chemical formulas, we assemble them into the molecular equation.

$$Pb(C_2H_3O_2)_2(aq) + NaI(aq) \longrightarrow PbI_2(s) + NaC_2H_3O_2(aq)$$

Notice that we've indicated which are in solution and which is a precipitate. Next, we balance the equation.

$$Pb(C_2H_3O_2)_2(aq) + 2NaI(aq) \longrightarrow PbI_2(s) + 2NaC_2H_3O_2(aq)$$

This is the balanced molecular equation.

Ionic Equation

To write the ionic equation, we keep in mind that all soluble salts are completely dissociated in water and that the formulas of precipitates are written in "molecular" form. This gives

$$Pb^{2+}(aq) + 2C_2H_3O_2^-(aq) + 2Na^+(aq) + 2I^-(aq) \longrightarrow$$
$$PbI_2(s) + 2Na^+(aq) + 2C_2H_3O_2^-(aq)$$

This is the ionic equation. Notice that to properly write the ionic equation it is necessary to know both the formulas and charges of the ions.

Net Ionic Equation

We obtain this from the ionic equation by eliminating spectator ions. The spectator ions in this reaction are Na^+ and $C_2H_3O_2^-$. Removing them gives

$$Pb^{2+}(aq) + 2I^-(aq) \longrightarrow PbI_2(s)$$

(Notice this is the same net ionic equation as in the reaction of lead nitrate with potassium iodide. Notice, also, that the net charges are the same on both sides of the arrow in the ionic and net ionic equations.)

The net ionic equation tells us that any soluble lead salt will react with any soluble iodide salt to give lead iodide. This prediction is borne out here as a precipitate of lead iodide is formed when a solution of sodium iodide is added to a solution of lead acetate.

Practice Exercise 2

Write molecular, ionic, and net ionic equations for the reaction of aqueous solutions of cadmium chloride and sodium sulfide to give a precipitate of cadmium sulfide and a solution of sodium chloride. ◆

In this section we've examined three ways to write an equation for a reaction between solutions of electrolytes. Is one better than another? Not really. The molecular equation is often best for planning an experiment and performing stoichiometric calculations. The ionic equation is useful when we want to provide a better description of what actually takes place in the solution during the reaction. And finally, the net ionic equation is useful because it focuses our attention on the chemical changes that take place during the reaction and it helps us generalize the reaction so that we can select other sets of reactants that give the same net chemical change.

4.3 Predicting Reactions That Produce Precipitates

The reaction between $Pb(NO_3)_2$ and KI, which we studied in the preceding section, is just one example of a large class of ionic reactions in which cations and anions change partners. The technical term we use to describe these reactions is **metathesis,** but they are also sometimes called **double replacement reactions.** (In the formation of the products, PbI_2 and KNO_3, the I^- replaces NO_3^- in the lead compound and NO_3^- replaces I^- in the potassium compound.) Metathesis reactions in which a precipitate forms are sometimes called **precipitation reactions.**

> ▶**Chemistry in Practice**◀ As you know, when someone picks up an object with their bare hands, they leave fingerprints on it. Crime investigators call them *latent* fingerprints, which means they cannot be seen unless developed in some way. (*Latent* means *hidden.*) There's a number of ways latent fingerprints can be made visible, including dusting with a fine powder, as seen on TV crime shows. Another method makes use of an ionic reaction of the kind discussed in this section.
>
> When you touch something, among the substances present in the fingerprint you leave behind is NaCl. This comes from the small amount of sweat present on your fingers. To develop a fingerprint, a dilute solution of silver nitrate is sprayed on the object of interest. As the sodium chloride in the fingerprint dissolves in the solution, the silver nitrate reacts with it to give a precipitate of silver chloride, AgCl.
>
> $$NaCl(aq) + AgNO_3(aq) \longrightarrow AgCl(s) + NaNO_3(aq)$$
>
> Silver chloride is white, but when exposed to sunlight it decomposes to give a black deposit of metallic silver, which reveals the lines and swirls of the fingerprint. ◆

For precipitation reactions, if we know which products are insoluble, we can predict the correct net ionic equation. Fortunately, there is a set of rules that we can use to determine, in many cases, whether an ionic compound is soluble or insoluble. These **solubility rules** are given in Table 4.1. First, let's look at some examples to see how we apply them, and then we can see how the rules can be used to predict precipitation reactions.

Applying the Solubility Rules

To make them easier to remember, the rules given in Table 4.1 are divided into two categories. The first includes compounds that are soluble, with some exceptions. The second describes compounds that are generally insoluble, with some exceptions. Some examples will help clarify their use.

Rule 1 states that all compounds of the alkali metals are soluble in water. This means that you can expect *any* salt containing Na^+ or K^+, or any of the Group IA metal ions, *regardless of the anion,* to be soluble. If one of the reactants in a metathesis is Na_3PO_4, you now know from this rule that it is soluble. Therefore, you would write it in *dissociated* form in the ionic equation. Similarly,

Table 4.1 Solubility Rules for Ionic Compounds in Water

Solubility rules

Soluble Compounds

1. All compounds of the alkali metals (Group IA) are soluble.
2. All salts containing NH_4^+, NO_3^-, ClO_4^-, ClO_3^-, and $C_2H_3O_2^-$ are soluble.
3. All chlorides, bromides, and iodides (salts containing Cl^-, Br^-, or I^-) are soluble *except* when combined with Ag^+, Pb^{2+}, and Hg_2^{2+} (note the subscript "2").
4. All sulfates (salts containing SO_4^{2-}) are soluble *except* those of Pb^{2+}, Ca^{2+}, Sr^{2+}, Hg_2^{2+}, and Ba^{2+}.

Insoluble Compounds

5. All metal hydroxides (ionic compounds containing OH^-) and all metal oxides (ionic compounds containing O^{2-}) are insoluble *except* those of Group IA and of Ca^{2+}, Sr^{2+}, and Ba^{2+}.

   When metal oxides do dissolve, they react with water to form hydroxides. The oxide ion, O^{2-}, does not exist in water. For example,

   $$Na_2O(s) + H_2O \rightarrow 2NaOH(aq)$$

6. All salts that contain PO_4^{3-}, CO_3^{2-}, SO_3^{2-}, and S^{2-} are insoluble, *except* those of Group IA and NH_4^+.

Rule 6 states, in part, that all carbonate salts are *insoluble* except those of the alkali metals and the ammonium ion. Thus, if one of the products in a metathesis reaction is $CaCO_3$, you'd expect it to be insoluble, because the cation is not an alkali metal or NH_4^+. Therefore, you would write its formula in its undissociated form when you construct the ionic equation. (Facets of Chemistry 4.1 describes how the insolubility of $CaCO_3$ bears on the softening of "hard water.")

▶**Chemistry in Practice**◀ The solubility rules tell us that iron(II) sulfide is insoluble. It's also a greenish compound that's responsible for the green-yellow color of the yoke of a hard-boiled egg that has not been properly prepared. Cooking an egg releases tiny amounts of hydrogen sulfide from proteins of the albumen (egg white). Iron(II) ions are in the yoke waiting to be made into hemoglobin for a fetal chicken's blood. Proper cooking ensures that the hydrogen sulfide escapes. If it doesn't, cooling of the freshly cooked egg draws this gas back toward the iron(II) ions where FeS can then precipitate. ◆

Predicting Precipitation Reactions

Now let's look at some examples that illustrate how we use the solubility rules to predict reactions in which a precipitate forms.

EXAMPLE 4.2

Predicting Reactions and Writing Their Equations

Predict the reaction that will occur when aqueous solutions of $Pb(NO_3)_2$ and $Fe_2(SO_4)_3$ are mixed. Write molecular, ionic, and net ionic equations for it.

Analysis: We first have to determine the makeup of the products. We begin, therefore, by predicting what a double replacement (metathesis) might produce. Then

Facets of Chemistry 4.1

Boiler Scale and Hard Water

Precipitation reactions occur around us all the time and we hardly ever take notice until they cause a problem. One common problem is caused by **hard water**—groundwater that contains the "hardness ions," Ca^{2+}, Mg^{2+}, Fe^{2+}, or Fe^{3+}, in concentrations high enough to form precipitates with ordinary soap. Soap normally consists of the sodium salts of organic acids derived from animal fats or oils (so-called *fatty acids*). An example is sodium stearate, $NaC_{18}H_{35}O_2$. The negative ion of the soap forms an insoluble "scum" with hardness ions, which reduces the effectiveness of the soap for removing dirt and grease.

Hardness ions can be removed from water in a number of ways. One way is to add hydrated sodium carbonate, $Na_2CO_3 \cdot 10H_2O$, often called washing soda, to the water. The carbonate ion forms insoluble precipitates with the hardness ions; an example is $CaCO_3$.

$$Ca^{2+}(aq) + CO_3^{2-}(aq) \longrightarrow CaCO_3(s)$$

Once precipitated, the hardness ions are not available to interfere with the soap.

Another problem when the hard water of a particular locality is rich in bicarbonate ion is the precipitation of insoluble carbonates on the inner walls of hot water pipes. When solutions containing HCO_3^- are heated, the ion decomposes as follows:

$$2HCO_3^-(aq) \longrightarrow H_2O + CO_2(g) + CO_3^{2-}(aq)$$

Like most gases, carbon dioxide becomes less soluble as the temperature is raised, so CO_2 is driven from the hot solution and the HCO_3^- is gradually converted to CO_3^{2-}. As the carbonate ions form, they are able to precipitate the hardness ions. This precipitate, which sticks to the inner walls of pipes and hot water boilers, is called *boiler scale*. In locations that have high concentrations of Ca^{2+} and HCO_3^- in the water supply, boiler scale is a very serious problem, as illustrated in the accompanying photograph.

Boiler scale built up on the inside of a water pipe.

we proceed to expand the molecular equation into an ionic equation, and finally we drop spectator ions to obtain the net ionic equation. To do this we need to know solubilities, and here we apply the solubility rules.

Always write equations in two steps: First write correct formulas for the reactants and products, then adjust the coefficients to balance the equation.

Solution: Our reactants are $Pb(NO_3)_2$ and $Fe_2(SO_4)_3$, which contain the ions Pb^{2+} and NO_3^-, and Fe^{3+} and SO_4^{2-}, respectively. To write the formulas of the products, we interchange anions. We combine Pb^{2+} with SO_4^{2-}, and for electrical neutrality, we must use one ion of each. Therefore, we write $PbSO_4$ as one possible product. For the other product, we combine Fe^{3+} with NO_3^-. Electrical neutrality now demands that we use *three* NO_3^- to *one* Fe^{3+} to make $Fe(NO_3)_3$. The correct formulas of the products,[1] then, are $PbSO_4$ and $Fe(NO_3)_3$.

Next, we need to determine solubilities. The reactants are ionic compounds and we are told that they are in solution, so we know they are water soluble. Solubility Rules 2 and 4 tell us this also. For the products, we find that Rule 2 says that all ni-

[1] Some students might be tempted (without thinking) to write $Pb(SO_4)_2$ and $Fe_2(NO_3)_3$, or even $Pb(SO_4)_3$ and $Fe_2(NO_3)_2$. This is a common error. Always be careful to figure out the charges of the ions that must be combined in the formula. Then take the ions in a ratio that gives a neutral formula unit.

trates are soluble, so $Fe(NO_3)_3$ is soluble; Rule 4 tells us that the sulfate of Pb^{2+} is *insoluble*. This means that a precipitate of $PbSO_4$ will form. Let's now write the *unbalanced* molecular equation:

$$Fe_2(SO_4)_3(aq) + Pb(NO_3)_2(aq) \longrightarrow Fe(NO_3)_3(aq) + PbSO_4(s) \quad \text{(unbalanced)}$$

When it is balanced, we obtain the *molecular equation*.

$$Fe_2(SO_4)_3(aq) + 3Pb(NO_3)_2(aq) \longrightarrow 2Fe(NO_3)_3(aq) + 3PbSO_4(s)$$

Next, we expand this to give the ionic equation in which soluble compounds are written in dissociated (separated) form as ions, and insoluble compounds are written in "molecular" form. We can place the following labels beneath the formulas of the molecular equation.

$$\underset{\substack{\text{soluble}\\\text{(Rule 4)}}}{Fe_2(SO_4)_3(aq)} + \underset{\substack{\text{soluble}\\\text{(Rule 2)}}}{3Pb(NO_3)_2(aq)} \longrightarrow \underset{\substack{\text{soluble}\\\text{(Rule 2)}}}{2Fe(NO_3)_3(aq)} + \underset{\substack{\text{insoluble}\\\text{(Rule 4)}}}{3PbSO_4(s)}$$

We now separate the ions of the water-soluble species to obtain the *ionic equation*.

$$2Fe^{3+}(aq) + 3SO_4^{2-}(aq) + 3Pb^{2+}(aq) + 6NO_3^-(aq) \longrightarrow$$
$$2Fe^{3+}(aq) + 6NO_3^-(aq) + 3PbSO_4(s)$$

By removing spectator ions (Fe^{3+} and NO_3^-), we obtain

$$3Pb^{2+}(aq) + 3SO_4^{2-}(aq) \longrightarrow 3PbSO_4(s)$$

Finally, we reduce the coefficients to give us the correct *net ionic equation*.

$$Pb^{2+}(aq) + SO_4^{2-}(aq) \longrightarrow PbSO_4(s)$$

Practice Exercise 3

Predict the reaction that occurs on mixing the following solutions. Write molecular, ionic, and net ionic equations for the reactions that take place. (a) $AgNO_3$ and NH_4Cl, (b) Na_2S and $Pb(C_2H_3O_2)_2$. ◆

4.4 Acids and Bases as Electrolytes

Some of our most familiar chemicals as well as many important laboratory reagents are acids and bases. The vinegar in a salad dressing, the sour juice of a lemon, the gastric juice that aids digestion in the stomach, vitamin C, and the liquid in the battery of an automobile are similar in at least one respect—they all contain acids. The white crystals of lye in certain drain cleaners, the white substance that makes milk of magnesia opaque, and household ammonia are all bases.

Usually, we use the term **reagent** to mean a chemical commonly kept on hand to be used in chemical reactions.

Common Properties of Acids and Bases

There are some general properties that are common to aqueous solutions of acids and bases. For example, **acids** generally have a tart (sour) taste. You know this from having tasted vinegar and lemon juice. However, taste is *never* used as a laboratory test for acids; some acids are poisonous and others, such as sulfuric acid in battery fluid, are extremely corrosive to animal tissue. (In general, never taste chemicals in the laboratory!)

Some common acids.

Some common bases.

Acids (and bases) also affect the colors of certain natural dye substances called acid–base indicators. An example is litmus, which has a red color in an acidic solution and a blue color in a basic solution.[2]

Substances classified as acids also corrode many metals. (Perhaps you have seen this effect following the spill of battery acid somewhere on a metal surface under the hood of a car.)

If you spill an acid or base on yourself in the lab, be sure to rinse it off with a lot of water as you ask someone to get your instructor.

Although some acids can be dangerous, many acids are essential to our well-being. Ascorbic acid, for example, is also known as vitamin C and is a necessary part of our diet. (See *Chemicals in Our World 1.*) And, of course, acids such as citric acid and vinegar flavor our foods and make them more enjoyable.

Bases also have some properties in common. For example, aqueous solutions of **bases** have a somewhat bitter taste, and they turn litmus blue. They also have a soapy "feel." (Actually, bases change oils and fat in your skin into soap, which is why they feel slippery.) Like acids, bases pose varying risks. The base in milk of magnesia, $Mg(OH)_2$, can be taken internally as a laxative or as something to soothe the stomach when too much gastric acid (hydrochloric acid) is present. But another base, sodium hydroxide (lye), is *extremely* hazardous, particularly to the eyes but even to the skin. (The hazards posed by acids and bases are good reasons for wearing eye protective devices whenever you observe or do laboratory work.)

Litmus paper, a strip of paper impregnated with the dye litmus, becomes blue in aqueous ammonia (a base) and red in lemon juice (which contains citric acid).

Arrhenius Definition of Acids and Bases

The first comprehensive theory concerning acids, bases, and electrically conducting solutions appeared in 1884 in the Ph.D. thesis of a Swedish chemist, Svante Arrhenius. He proposed that ions form directly when salts dissolve in water, which was a radical notion at the time. This is why *all* aqueous solutions of salts conduct electricity, according to Arrhenius.

Arrhenius also theorized that all acids release hydrogen ions, H^+, in water, and all bases release hydroxide ions, OH^-. This explains why acids have many common properties and why bases behave similarly.

[2] Litmus paper, commonly found among the items in a locker in the general chemistry lab, consists of strips of absorbent paper that have been soaked in a solution of litmus and dried. Red litmus paper is used to test if a solution is basic. A basic solution turns red litmus blue. To test if the solution is acidic, blue litmus paper is used. Acidic solutions turn blue litmus red.

One of the properties of acids and bases is their reaction with each other. For example, if solutions of hydrochloric acid, HCl(aq), and sodium hydroxide, NaOH(aq), are mixed in a 1-to-1 ratio by moles, the resulting solution has no effect on litmus paper because the following reaction occurs.

$$HCl(aq) + NaOH(aq) \longrightarrow NaCl(aq) + H_2O$$

The acidic and basic properties of the solutes disappear; the solution of the products is neither acidic nor basic. We say that an *acid–base neutralization* has occurred. According to Arrhenius, **acid–base neutralization** is simply the combination of a hydrogen ion with a hydroxide ion to produce a water molecule, thus making H^+ ions and OH^- ions disappear.

Arrhenius was nearly failed by his examining committee for proposing such a bizarre idea that electrically charged particles exist in solution. In 1903, this view won him the Nobel Prize.

Ionization Reactions for Acids and Bases in Water

Today we know that hydrogen ions, H^+, can only exist in water when they are attached to something else. When they attach themselves to water molecules they form hydronium ions, H_3O^+. We view H_3O^+ as a species that carries H^+ in water. However, for the sake of convenience, we often use the term *hydrogen ion* as a substitute for *hydronium ion,* and in many equations, we use $H^+(aq)$ to stand for $H_3O^+(aq)$. In fact, whenever you see the symbol $H^+(aq)$, you should realize that we are actually referring to $H_3O^+(aq)$.

For most purposes, we find that the following modified versions of Arrhenius' definitions work satisfactorily when we deal with aqueous solutions.

Even the formula H_3O^+ is something of a simplification. In water the H^+ ion is associated with more than one molecule of water, but we use the formula H_3O^+ as a simple representation.

> **Arrhenius Definition of Acids and Bases**
> An **acid** is a substance that reacts with water to produce hydronium ion, H_3O^+.
> A **base** is a substance that produces hydroxide ion in water, or is able to react with hydronium ion.

Substances That Are Acids

In general, **acids** are molecular substances that react with water to produce ions, one of which is the hydronium ion. For example, pure hydrogen chloride is a gas and is molecular, not ionic. As it dissolves in water, however, the following chemical reaction occurs, which produces ions that did not pre-exist in HCl(g).

$$HCl(g) + H_2O \longrightarrow H_3O^+(aq) + Cl^-(aq)$$

A reaction like this in which ions form where none existed before is called an **ionization reaction.** One result of such a reaction in water is a solution that conducts electricity, so acids can be classified as electrolytes.

If gaseous HCl is cooled to about $-85\ °C$, it condenses to a liquid that doesn't conduct electricity. No ions are present in pure liquid HCl.

Acids are electrolytes that yield ions by reaction with water.

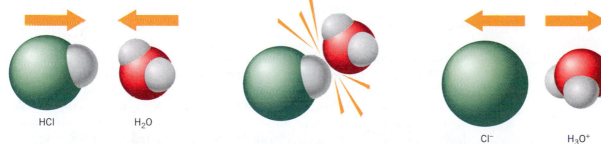

Ionization of HCl in water. Collisions between HCl molecules and water molecules lead to a transfer of H^+ from the HCl to the H_2O, giving Cl^- and H_3O^+ as products.

Similar ionization reactions occur for other acids as well. For example, nitric acid reacts with water according to the equation

$$HNO_3(l) + H_2O \longrightarrow H_3O^+(aq) + NO_3^-(aq)$$

Notice that in the ionization reactions of HCl and HNO_3 described above, the hydronium ion is formed by the transfer of an H^+ ion from the acid molecule to the water molecule. The particle left after loss of the H^+ is an anion. We might represent this in general terms by the equation

$$\text{Acid molecule} + H_2O \longrightarrow H_3O^+ + \text{anion}$$

If we represent the acid molecule by the general formula HA, this becomes

$$HA + H_2O \longrightarrow H_3O^+ + A^-$$

As noted earlier, the active ingredient in the hydronium ion is H^+, which is why $H^+(aq)$ is often used in place of $H_3O^+(aq)$ in equations. Using this simplification, the ionization of HCl and HNO_3 in water can be represented as

$$HCl(g) \xrightarrow{H_2O} H^+(aq) + Cl^-(aq)$$

and

$$HNO_3(l) \xrightarrow{H_2O} H^+(aq) + NO_3^-(aq)$$

Sometimes an acid also contains hydrogen atoms that are not able to transfer to water molecules. An example is acetic acid, $HC_2H_3O_2$, the acid that gives vinegar its sour taste. This acid reacts with water as follows.

$$HC_2H_3O_2(l) + H_2O \longrightarrow H_3O^+(aq) + C_2H_3O_2^-(aq)$$

Notice that only the hydrogen written first in the formula is able to transfer to H_2O to give hydronium ions. (We will discuss why the hydrogens are different at a later time.)

The molecules HCl, HNO_3, and $HC_2H_3O_2$ are said to be **monoprotic acids** because they are capable of furnishing only one H^+ per molecule of acid. **Polyprotic acids** can furnish more than one H^+ per molecule of acid and undergo reactions similar to those of HCl and HNO_3, except that the loss of H^+ by the acid occurs in two or more steps. Thus, the ionization of sulfuric acid, a **diprotic acid,** takes place by two successive steps.

$$H_2SO_4(aq) + H_2O \longrightarrow H_3O^+(aq) + HSO_4^-(aq)$$

$$HSO_4^-(aq) + H_2O \longrightarrow H_3O^+(aq) + SO_4^{2-}(aq)$$

Hydrogen ions that are able to be transferred to water molecules to form hydronium ions are usually written first in the formula for the acid.

Practice Exercise 4

Write equations for (a) the ionization of $HCHO_2$ (formic acid) in water and (b) the stepwise ionization of the **triprotic acid** H_3PO_4 in water. ◆

Nonmetal Oxides as Acids. The acids we've discussed so far have been molecules that contain hydrogen atoms that are able to be transferred to water molecules. Another class of compounds that fits our description of acids is *nonmetal oxides.* Here we have oxides such as SO_3, CO_2, and N_2O_5 whose aqueous solutions contain H_3O^+ and turn litmus red. These oxides are called **acidic anhydrides,** where *anhydride* means "without water." They react with water to form molecular acids containing hydrogen, which are then able to undergo reaction with water to yield H_3O^+.

$$SO_3(g) + H_2O \longrightarrow H_2SO_4(aq) \qquad \text{sulfuric acid}$$

$$N_2O_5(g) + H_2O \longrightarrow 2HNO_3(aq) \qquad \text{nitric acid}$$

$$CO_2(g) + H_2O \longrightarrow H_2CO_3(aq) \qquad \text{carbonic acid}$$

Although carbonic acid is too unstable to be isolated as a pure compound, its solutions in water are quite common. Carbon dioxide from the atmosphere dissolves in rainwater and the waters of lakes and streams, for example, where it exists partly as carbonic acid and its ions (HCO_3^- and CO_3^{2-}). This makes these waters naturally slightly acidic. Carbonic acid is also present in carbonated beverages.

Not all nonmetal oxides are acidic anhydrides, only those that are able to react with water. For example, carbon monoxide doesn't react with water, so its solutions in water are not acidic; carbon monoxide, therefore, is not classified as an acidic anhydride.

Substances That Are Bases

Bases fall into two categories: ionic compounds that contain OH^- or O^{2-}, and molecular compounds that react with water to give hydroxide ions. Because solutions of bases contain ions, they conduct electricity, so bases are electrolytes.

Ionic bases include metal hydroxides, such as NaOH and $Ca(OH)_2$. When dissolved in water, they dissociate just like other soluble ionic compounds.

$$NaOH(s) \longrightarrow Na^+(aq) + OH^-(aq)$$

$$Ca(OH)_2(s) \longrightarrow Ca^{2+}(aq) + 2OH^-(aq)$$

$Ca(OH)_2$ is not highly soluble in water. It could be called slightly soluble or partially soluble.

Soluble metal oxides are **basic anhydrides** because they react with water to form the hydroxide ion as one of the products. Calcium oxide is typical.

$$CaO(s) + H_2O \longrightarrow Ca(OH)_2(aq)$$

This reaction occurs when water is added to dry cement or concrete because calcium oxide or "quicklime" is an ingredient in these materials. In this case it is the oxide ion, O_2^-, that actually forms OH^-.

$$O^{2-} + H_2O \longrightarrow 2OH^-$$

Continual contact of your hands with fresh portland cement can lead to irritation because the mixture is quite basic.

Molecular Bases. The most common molecular base is the gas ammonia, NH_3, which dissolves in water and reacts to give a basic solution.

$$NH_3(aq) + H_2O \longrightarrow NH_4^+(aq) + OH^-(aq)$$

See the figure at the top of page 166.

This is also an ionization reaction because ions have been formed where none previously existed. Notice that when a molecular base reacts with water an H^+ is lost by the water molecule and gained by the base. One product is a cation that has one more H and one more positive charge than the reactant base. Loss of H^+ by the water gives the other product, the OH^- ion, which is why the solution is basic. We might represent this by the general equation

$$Base + H_2O \longrightarrow BaseH^+ + OH^-$$

If we represent the base by the symbol B, this becomes

$$B + H_2O \longrightarrow BH^+ + OH^-$$

Since bases yield solutions that contain ions, they are also electrolytes.

Ionization of ammonia in water. Collisions between NH_3 molecules and water molecules lead to a transfer of H^+ from H_2O to NH_3, giving NH_4^+ and OH^- ions.

4.5 Strong and Weak Acids and Bases

Sodium chloride and calcium chloride are both examples of **strong electrolytes**—electrolytes that break up essentially 100% into ions in water. No "molecules" of either NaCl or $CaCl_2$ are detectable in their aqueous solutions; they exist entirely as ions. This behavior is typical of all ionic compounds in dilute solutions.

Hydrochloric acid is also a strong electrolyte. Its ionization in water is essentially complete, and no molecules of HCl can be detected in its solutions. As a result, solutions of hydrochloric acid are highly acidic, and this acid is called a strong acid. In general, acids that are strong electrolytes are called **strong acids.** Actually, there are very few strong acids. The list below gives the most common ones. You will find it useful to memorize this list.

All strong acids are strong electrolytes.

Strong acids

Strong Acids

$HClO_4(aq)$	Perchloric acid	$HI(aq)$	Hydriodic acid
$HClO_3(aq)$	Chloric acid	$HNO_3(aq)$	Nitric acid
$HCl(aq)$	Hydrochloric acid	$H_2SO_4(aq)$	Sulfuric acid
$HBr(aq)$	Hydrobromic acid		

Metal hydroxides are ionic compounds, so they are also strong electrolytes. Those that are soluble are the hydroxides of Group IA and the hydroxides of calcium, strontium, and barium of Group IIA. Solutions of these compounds are strongly basic, so these substances are considered to be **strong bases.** (See below.)

Strong bases are ionic metal hydroxides, such as NaOH. Weak bases are molecular bases, such as NH_3.

Strong Bases (Soluble Metal Hydroxides)

	Group IA		Group IIA
LiOH	lithium hydroxide		
NaOH	sodium hydroxide		
KOH	potassium hydroxide	$Ca(OH)_2$	calcium hydroxide
RbOH	rubidium hydroxide	$Sr(OH)_2$	strontium hydroxide
CsOH	cesium hydroxide	$Ba(OH)_2$	barium hydroxide

The hydroxides of other metals have very low solubilities in water. They are strong electrolytes in the sense that the small amounts of them that dissolve in solution are completely dissociated. However, because of their low solubility in water, their solutions are very weakly basic, so many chemists would not consider them to be strong bases.

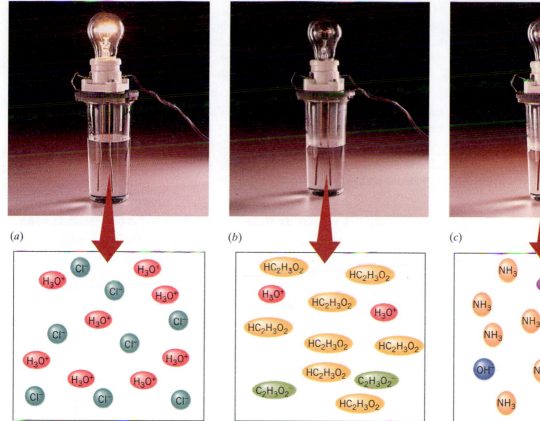

(a)

(b)

(c)

All the HCl is ionized in the solution, so there are many ions present.

Only a small fraction of the acetic acid is ionized, so there are few ions to conduct electricity. Most of the acetic acid is present as neutral molecules of $HC_2H_3O_2$.

Only a small fraction of the ammonia is ionized, so few ions are present to conduct electricity. Most of the ammonia is present as neutral molecules of NH_3.

Figure 4.6 *Electrical conductivity of solutions of strong and weak acids and bases.* (*a*) 1 *M* HCl is 100% ionized and is a strong conductor, enabling the light to glow brightly, (*b*) 1 *M* $HC_2H_3O_2$ is a weaker conductor than 1 *M* HCl because the extent of its ionization is far less, so the light is dimmer. (*c*) 1 *M* NH_3 also is a weaker conductor than 1 *M* HCl because the extent of its ionization is low, and the light remains dim.

Weak Acids and Bases

Most acids are not completely ionized in water. For instance, a solution of acetic acid, $HC_2H_3O_2$, is a relatively poor conductor of electricity compared to a solution of HCl with the same concentration (Figure 4.6), so acetic acid is classified as a **weak electrolyte.** This is because in the acetic acid solution only a small fraction of the molecules of the acid actually exist as H_3O^+ and $C_2H_3O_2^-$ ions. (Actually, less than 0.5% of the acid is ionized in a 1 *M* solution.) The rest of the acetic acid is present as molecules of $HC_2H_3O_2$. For now, let's represent this weak ionization as follows.

$$HC_2H_3O_2(aq) + H_2O \xrightarrow{\text{small percentage}} H_3O^+(aq) + C_2H_3O_2^-(aq)$$

Solutions of acetic acid do not have high concentrations of H_3O^+ and are not strongly acidic. Acetic acid is therefore said to be a **weak acid.** All weak acids are weak electrolytes. Other examples are carbonic acid, H_2CO_3, and nitrous acid, HNO_2. In fact, *if an acid is not one of the strong acids listed above, you can assume it to be a weak acid.*

Molecular bases, such as ammonia, are also weak electrolytes and have a low percentage ionization. They are classified as **weak bases.** See Figure 4.6c. In a solution of ammonia, only a small fraction of the solute is ionized to give NH_4^+ and OH^-. Most of the base is present as ammonia molecules.

$$NH_3(aq) + H_2O \xrightarrow{\text{small percentage}} NH_4^+(aq) + OH^-(aq)$$

It is interesting to note that *virtually all molecular bases are weak bases.*

Let's briefly summarize the results of our discussion.

> Weak acids and bases are weak electrolytes.
> Strong acids and bases are strong electrolytes.

Dynamic Equilibria in Solutions of Weak Acids and Bases

A question that might have occurred to you during the discussion of weak acids and bases is, "If some of the solute molecules in aqueous ammonia or aqueous acetic acid can react with water, why can't all of them?" Actually, all do have the potential to react, but we meet here examples of a very important chemical phenomenon—*dynamic equilibrium.* Let's mentally go inside an acetic acid solution to see what's happening.

As soon as the acetic acid is mixed with water, solute and solvent molecules start to collide with each other. Because of the chemical nature of acetic acid, however, only a small percentage of such collisions produces ions. As a result, the rate (or speed) at which the ions form is quite small compared with the rate of collision. There is a *low* probability of the following reaction occurring. When it happens, we call it the *forward reaction* in the acetic acid–water system.

$$HC_2H_3O_2(aq) + H_2O \xrightarrow[\text{low probability}]{\text{small percentage}} H_3O^+(aq) + C_2H_3O_2^-(aq)$$

Once H_3O^+ and $C_2H_3O_2^-$ ions form, they wander around in the solution and experience collisions with solvent molecules. Occasionally, of course, an H_3O^+ ion and a $C_2H_3O_2^-$ ion meet and collide. Because of the chemical nature of these ions, there is a *high* probability that such a collision transfers H^+ from H_3O^+ back to $C_2H_3O_2^-$. We also represent this reaction by an equation, but one in which the arrow goes from right to left. We call it the *reverse reaction* because it tends to undo the forward reaction.

$$HC_2H_3O_2(aq) + H_2O \xleftarrow[\text{high probability}]{\text{high percentage}} H_3O^+(aq) + C_2H_3O_2^-(aq)$$

Let us go over the dynamics of this again. Just after adding $HC_2H_3O_2$ to water, the concentrations of H_3O^+ and $C_2H_3O_2^-$ build up because of the ionization reaction, the *forward reaction.* As their concentrations increase, however, the ions naturally encounter each other more frequently. So the *reverse reaction* occurs with increasing frequency. Eventually the concentrations of the ions become large enough *to make the rate of the reverse reaction equal the rate of the forward reaction.* At this point, ions disappear as rapidly as they form, so their concentrations remain constant from this moment on. The entire chemical system consisting of H_2O, $HC_2H_3O_2$, H_3O^+, and $C_2H_3O_2^-$ has reached a state of balance called **chemical equilibrium.** It is also said to be a **dynamic equilibrium** because the forward and reverse reactions don't cease; they continue to occur even after the concentrations have stopped changing.

We indicate a dynamic equilibrium by using double arrows ($\rightleftharpoons$) in the chemical equation. The ionization of acetic acid is thus represented as follows.

For now, consider *rate* to mean the number of events per unit of volume per unit of time, like collisions per cm^3 per second.

The probability of ions forming or disappearing as a result of collisions varies from solute to solute.

For this juggler, balls are going up at the same rate as they are coming down, so the number of balls in the air stays constant. A somewhat similar situation exists in a dynamic equilibrium, where products form at the same rate as they disappear and the amount of products stays constant.

$$HC_2H_3O_2(aq) + H_2O \rightleftharpoons H_3O^+(aq) + C_2H_3O_2^-(aq)$$

In describing an equilibrium such as this, we will often talk about the *position of equilibrium*. By this we mean the extent to which the forward reaction proceeds toward completion. If very little of the products are present at equilibrium, the forward reaction has not gone far toward completion and we say "the position of equilibrium lies to the left," toward the reactants. On the other hand, if large amounts of the products are present at equilibrium, we say "the position of equilibrium lies to the right."

For any weak electrolyte, only a small percentage of the solute is actually ionized at any instant after equilibrium is reached, so the position of equilibrium lies to the left. To call acetic acid a *weak* acid, for example, is just another way of saying that the forward reaction in this equilibrium is far from completion.

With molecular compounds that are strong electrolytes, the tendency of the forward ionization reaction to occur is very high, while the tendency of the reverse reaction to occur is extremely small. In aqueous HCl, for example, there is little tendency during a collision between Cl^- and H_3O^+ for neutral molecules of HCl and H_2O to form. As a result, the reverse reaction has essentially no tendency to occur, so in a very brief time all of the HCl molecules dissolved in water are converted to ions—the acid becomes 100% ionized. For this reason, *we do not use double arrows in describing what happens when HCl(g) or any other strong electrolyte undergoes ionization or dissociation.*

The reaction read from left to right is the forward reaction, and the one read in the opposite direction is the reverse reaction.

Practice Exercise 5

Nitrous acid, HNO_2, is a weak acid. Write the chemical equation that represents the equilibrium for this reaction. ◆

4.6 Acid–Base Neutralization

Acid–base neutralizations generally produce water and a **salt** (which we defined in Chapter 2 as any ionic compound that does not involve H^+, OH^-, or O^{2-} ions). The salt is composed of the anion of the acid and the cation of the base. An example is the reaction of nitric acid with potassium hydroxide.

$$HNO_3(aq) + KOH(aq) \longrightarrow KNO_3(aq) + H_2O$$

The products are water and a salt, KNO_3, which is made of K^+ and NO_3^- ions.

Acid Salts

Polyprotic acids are able to provide two or more H^+ ions per molecule of the acid. For instance, H_2SO_4 furnishes two H^+, and during neutralization either one or both of these H^+ ions can react with base, depending on the mole ratio of base to acid. Thus, if two moles of a base such as NaOH are added to one mole of H_2SO_4, complete neutralization results, as illustrated by the equation

$$H_2SO_4(aq) + 2NaOH(aq) \longrightarrow Na_2SO_4(aq) + 2H_2O$$

However, if only one mole of NaOH is added to one mole of H_2SO_4, partial neutralization occurs.

$$H_2SO_4(aq) + NaOH(aq) \longrightarrow NaHSO_4(aq) + H_2O$$

The salt produced in this case, sodium hydrogen sulfate, $NaHSO_4$, is called an *acid salt.* It contains the HSO_4^- ion which is capable of furnishing another H^+ in the neutralization of a base.

$$NaHSO_4(aq) + NaOH(aq) \longrightarrow Na_2SO_4(aq) + H_2O$$

Neutralization of a Strong Acid by a Strong Base

Nitric acid and potassium hydroxide are examples of a strong acid and a strong base. The molecular equation for their reaction was given above. Let's rewrite it here.

$$HNO_3(aq) + KOH(aq) \longrightarrow KNO_3(aq) + H_2O$$

Both nitric acid and potassium hydroxide are strong electrolytes, which means that when we write the ionic equation for the reaction, we write their formulas in dissociated form. On the right side of the equation only one product, KNO_3, is a strong electrolyte. The other product, water, is present in solution as molecules, so when we write the ionic equation only the KNO_3 is written in dissociated form. This gives the ionic equation,

$$H^+(aq) + NO_3^-(aq) + K^+(aq) + OH^-(aq) \longrightarrow K^+(aq) + NO_3^-(aq) + H_2O$$

The spectator ions are K^+ and NO_3^-. Removing them from the ionic equation gives the net ionic equation,

$$H^+(aq) + OH^-(aq) \longrightarrow H_2O$$

This net ionic equation applies to any reaction of a strong acid with a strong base, provided the salt formed in the reaction is water soluble.

Practice Exercise 6

Write the molecular, ionic, and net ionic equations for the neutralization of $HCl(aq)$ by $Ca(OH)_2(aq)$. ◆

Neutralization When One Reactant Is a Weak Acid or Base

Let's look at the ionic and net ionic equations when acetic acid ($HC_2H_3O_2$), a weak acid, and sodium hydroxide, a strong base, react to give a salt, sodium acetate ($NaC_2H_3O_2$), and water. The molecular equation is

$$HC_2H_3O_2(aq) + NaOH(aq) \longrightarrow NaC_2H_3O_2(aq) + H_2O$$

In this reaction, only two substances are strong electrolytes, NaOH and $NaC_2H_3O_2$. Acetic acid is a weak electrolyte and is present in solution mostly in the form of molecules. Therefore, when we write the ionic equation, we do not express the acid in ionic form. We simply write the formula for molecular acetic acid. The ionic equation for this reaction is therefore

In writing an ionic equation, we always write the formulas of weak electrolytes in "molecular form."

$$HC_2H_3O_2(aq) + Na^+(aq) + OH^-(aq) \longrightarrow Na^+(aq) + C_2H_3O_2^-(aq) + H_2O$$

This time there is only one spectator ion, Na^+, and when we eliminate it from the equation we obtain the net ionic equation,

$$HC_2H_3O_2(aq) + OH^-(aq) \longrightarrow C_2H_3O_2^-(aq) + H_2O$$

The net ionic equation tells us this time that the reaction in the solution is actually one between hydroxide ions and molecules of acetic acid.

A slightly different situation exists in the reaction of hydrochloric acid with aqueous ammonia, a weak base, to give the salt ammonium chloride (NH_4Cl). The molecular equation for the reaction is

$$NH_3(aq) + HCl(aq) \longrightarrow NH_4Cl(aq)$$

Notice that no water is formed in this reaction. This is because the molecular base does not contain hydroxide ions. A solution of ammonia is basic because of the equilibrium

$$NH_3(aq) + H_2O \rightleftharpoons NH_4^+(aq) + OH^-(aq)$$

but the reaction occurs to a very small extent (less than 1% in a 1 M solution); thus, in a solution of ammonia nearly all of the base is present as molecules. In the neutralization reaction, therefore, ammonia is the principal reactant. The net ionic equation for the neutralization is

$$NH_3(aq) + H^+(aq) \longrightarrow NH_4^+(aq)$$

Practice Exercise 7

Write molecular, ionic, and net ionic equations for the reaction of (a) HCl with KOH, (b) $HCHO_2$ with LiOH, and (c) N_2H_4 with HCl. ◆

Reactions of Acids with Insoluble Hydroxides and Oxides

The tendency for water to be formed in neutralization reactions is so strong that it serves to drive reactions of acids (both strong and weak) with insoluble hydroxides and oxides. An example is the reaction of magnesium hydroxide with hydrochloric acid, pictured in Figure 4.7. This is the reaction that occurs when milk of magnesia neutralizes stomach acid.

Magnesium hydroxide has a very low solubility in water and is the solid that makes milk of magnesia white. The molecular equation for its reaction with $HCl(aq)$ is

$$Mg(OH)_2(s) + 2HCl(aq) \longrightarrow MgCl_2(aq) + 2H_2O$$

When we write the ionic equation, we do not write the $Mg(OH)_2$ in dissociated form because it is insoluble. Magnesium chloride, on the other hand, is water soluble and is fully dissociated. (Notice in Figure 4.7 that the reaction mixture is not cloudy after the reaction is over, indicating that all the solid has dissolved. This tells us that the $MgCl_2$ is soluble in water.) The ionic equation is

$$Mg(OH)_2(s) + 2H^+(aq) + 2Cl^-(aq) \longrightarrow Mg^{2+}(aq) + 2Cl^-(aq) + 2H_2O$$

The net ionic equation is obtained by eliminating the Cl^- spectator ions.

$$Mg(OH)_2(s) + 2H^+(aq) \longrightarrow Mg^{2+}(aq) + 2H_2O$$

Many metal oxides are similarly dissolved by acids. For example, before steel can be given a protective coating of zinc (a process called *galvanizing*), rust must first be stripped from the steel. This is usually done by dissolving the rust, mostly Fe_2O_3, in hydrochloric acid. The molecular, ionic, and net ionic equations are

$$Fe_2O_3(s) + 6HCl(aq) \longrightarrow 2FeCl_3(aq) + 3H_2O$$

$$Fe_2O_3(s) + 6H^+(aq) + 6Cl^-(aq) \longrightarrow 2Fe^{3+}(aq) + 6Cl^-(aq) + 3H_2O$$

$$Fe_2O_3(s) + 6H^+(aq) \longrightarrow 2Fe^{3+}(aq) + 3H_2O$$

Practice Exercise 8

Write molecular, ionic, and net ionic equations for the reaction of $Al(OH)_3$ with HCl. (Aluminum hydroxide is an ingredient in the antacid Digel.) ◆

4.7 Ionic Reactions That Produce Gases

Sometimes a product of a metathesis reaction is a substance that normally is a gas at room temperature and is not very soluble in water. An example is hydrogen sulfide, H_2S, the compound that gives rotten eggs their foul odor. It is a weak electrolyte and is formed when a strong acid such as HCl is added to a metal sulfide like sodium sulfide, Na_2S. Hydrogen sulfide has a low solubility in water, so it tends to bubble out of the solution as it forms and thereby escape as

Figure 4.7 *Hydrochloric acid is neutralized by milk of magnesia. A solution of hydrochloric acid is added to a beaker containing milk of magnesia. The thick white solid in milk of magnesia is magnesium hydroxide,* $Mg(OH)_2$, *which is able to neutralize the acid. The mixture is clear where the solid* $Mg(OH)_2$ *has reacted and dissolved.*

The solubility rules tell us that $Mg(OH)_2$ is insoluble and that $MgCl_2$ is soluble.

H_2S can be detected by odor when its concentration is only 0.15 ppb (parts per billion) or 21 μg H_2S/m^3 air.

Recall that a gas is indicated in a chemical equation by placing (g) after its formula.

a gas. Once it has escaped, there is no possible way for it to participate in any sort of reverse reaction, so its departure drives the reaction to completion. (The departure of H_2S removes ions, $2H^+$ and S^{2-}.) The molecular, ionic, and net ionic equations for the reaction are as follows.

Molecular Equation

$$2HCl(aq) + Na_2S(aq) \longrightarrow 2NaCl(aq) + H_2S(g)$$

Ionic Equation

$$2H^+(aq) + 2Cl^-(aq) + 2Na^+(aq) + S^{2-}(aq) \longrightarrow 2Na^+(aq) + 2Cl^-(aq) + H_2S(g)$$

Net Ionic Equation

$$2H^+(aq) + S^{2-}(aq) \longrightarrow H_2S(g)$$

Reactions of Carbonates and Bicarbonates with Acids

Carbonic acid (H_2CO_3) forms when an acid reacts with either a bicarbonate or a carbonate. For example, consider the reaction of sodium bicarbonate ($NaHCO_3$) with hydrochloric acid, which is pictured in Figure 4.8. As the sodium bicarbonate solution is added to the hydrochloric acid, bubbles of carbon dioxide are released. This is the same reaction that occurs if you take sodium bicarbonate to soothe an upset stomach. Stomach acid is HCl, and its reaction with the $NaHCO_3$ both neutralizes the acid and produces CO_2 gas (burp!). The molecular equation for the reaction is

Figure 4.8 *The reaction of sodium bicarbonate with hydrochloric acid.* The bubbles contain the gas carbon dioxide.

$$HCl(aq) + NaHCO_3(aq) \longrightarrow NaCl(aq) + H_2CO_3(aq)$$

Earlier it was mentioned that carbonic acid is too unstable to be isolated in pure form. When it forms in appreciable amounts as a product in a metathesis reaction, it decomposes into its anhydride, the gas CO_2, and water. Carbon dioxide is only slightly soluble in water, so most of the CO_2 bubbles out of the solution. The decomposition reaction is

$$H_2CO_3(aq) \longrightarrow H_2O + CO_2(g)$$

Therefore, the overall molecular equation for the reaction is

$$HCl(aq) + NaHCO_3(aq) \longrightarrow NaCl(aq) + H_2O + CO_2(g)$$

The ionic equation is

$$H^+(aq) + Cl^-(aq) + Na^+(aq) + HCO_3^-(aq) \longrightarrow$$
$$Na^+(aq) + Cl^-(aq) + H_2O + CO_2(g)$$

and the net ionic equation is

$$H^+(aq) + HCO_3^-(aq) \longrightarrow H_2O + CO_2(g)$$

Similar results are obtained if we begin with a carbonate instead of a bicarbonate. Consider the reaction of sodium carbonate with hydrochloric acid. Writing the equation as a metathesis reaction yields

$$Na_2CO_3(aq) + 2HCl(aq) \longrightarrow 2NaCl(aq) + H_2CO_3(aq)$$

where we have just exchanged cations and anions. Recognizing that H_2CO_3 decomposes into H_2O and CO_2 gives the following molecular equation.

$$Na_2CO_3(aq) + 2HCl(aq) \longrightarrow 2NaCl(aq) + CO_2(g) + H_2O$$

Carbonic acid formed and then decomposed when sodium carbonate, pumped by a highway snowblower, was used to neutralize concentrated nitric acid spilling from a ruptured tank car in a switching yard in Denver, Colorado (April, 1983).

In the net ionic equation derived from this, we first have the carbonate ion reacting with $H^+(aq)$ to give carbonic acid, followed by its decomposition to CO_2 and H_2O.

$$2H^+(aq) + CO_3^{2-}(aq) \longrightarrow H_2CO_3(aq)$$
$$H_2CO_3(aq) \longrightarrow H_2O + CO_2(g)$$

Overall, the net reaction is

$$2H^+(aq) + CO_3^{2-}(aq) \longrightarrow H_2O + CO_2(g)$$

Reactions of Insoluble Carbonates with Acid

The release of CO_2 by the reaction of a carbonate with an acid is such a strong driving force for reaction that it enables insoluble carbonates to dissolve in acids (strong and weak). The reaction of limestone, $CaCO_3$, with hydrochloric acid is shown in Figure 4.9. The molecular, ionic, and net ionic equations for the reaction are as follows.

Figure 4.9 *Limestone reacts with acid.* Bubbles of CO_2 are formed in the reaction of limestone ($CaCO_3$) with hydrochloric acid.

$$CaCO_3(s) + 2HCl(aq) \longrightarrow CaCl_2(aq) + CO_2(g) + H_2O$$
$$CaCO_3(s) + 2H^+(aq) + 2Cl^-(aq) \longrightarrow Ca^{2+}(aq) + 2Cl^-(aq) + CO_2(g) + H_2O$$
$$CaCO_3(s) + 2H^+(aq) \longrightarrow Ca^{2+}(aq) + CO_2(g) + H_2O$$

> ▶**Chemistry in Practice**◀ The ability of calcium carbonate to dissolve in acid is responsible for one of nature's most marvelous wonders, limestone caverns. This same ability is also responsible for the destructive effects of "acid rain" on marble statuary and building materials. Marble is a form of calcium carbonate. (See *Chemicals in Our World 7.*) ◆

Reactions of Acids with Sulfites

Another unstable weak acid that decomposes when formed in large amounts is sulfurous acid, H_2SO_3. The reaction is

$$H_2SO_3(aq) \longrightarrow H_2O + SO_2(g)$$

H_2SO_3 is so unstable that it could be written $SO_2 \cdot H_2O$.

Whenever a solution contains the ions necessary to form H_2SO_3, we can expect a release of the gas SO_2. For example, the net ionic equations for the reaction of bisulfite and sulfite ions with a strong acid are

$$HSO_3^-(aq) + H^+(aq) \longrightarrow H_2O + SO_2(g)$$
$$SO_3^{2-}(aq) + 2H^+(aq) \longrightarrow H_2O + SO_2(g)$$

Reaction of Bases with Ammonium Salts

In writing an equation for a metathesis reaction between NH_4Cl and $NaOH$, one of the products we are tempted to write is NH_4OH—so-called "ammonium hydroxide."

$$NaOH + NH_4Cl \longrightarrow NaCl + NH_4OH \quad \text{(by exchanging cations among anions)}$$

However, NH_4OH has never been detected, either in pure form or in an aqueous solution. Instead, "ammonium hydroxide" is simply a solution of NH_3 in water. (Note the equivalence in terms of atoms: $NH_4OH \Leftrightarrow NH_3 + H_2O$.) Therefore, in place of NH_4OH we write $NH_3 + H_2O$.

$$NaOH(aq) + NH_4Cl(aq) \longrightarrow NaCl(aq) + NH_3(aq) + H_2O$$

The net ionic equation, after eliminating Na^+ and Cl^-, is

$$NH_4^+(aq) + OH^-(aq) \longrightarrow NH_3(g) + H_2O$$

Table 4.2 Gases Formed in Metathesis Reactions

Tools

Gases from metathesis reactions

Gas	Formed by Reaction of Acids with:	Equation for Formation[a]
H_2S	Sulfides	$2H^+ + S^{2-} \rightarrow H_2S$
CO_2	Bicarbonates (hydrogen carbonates)	$H^+ + HCO_3^- \rightarrow (H_2CO_3) \rightarrow H_2O + CO_2$
	Carbonates	$2H^+ + CO_3^{2-} \rightarrow (H_2CO_3) \rightarrow H_2O + CO_2$
SO_2	Bisulfites (hydrogen sulfites)	$H^+ + HSO_3^- \rightarrow (H_2SO_3) \rightarrow H_2O + SO_2$
	Sulfites	$2H^+ + SO_3^{2-} \rightarrow (H_2SO_3) \rightarrow H_2O + SO_2$
HCN	Cyanides	$H^+ + CN^- \rightarrow HCN$

Gas	Formed by Reaction of Bases with:	Equation for Formation
NH_3	Ammonium salts[b]	$NH_4^+ + OH^- \rightarrow NH_3 + H_2O$

[a]Formulas in parentheses are of unstable compounds that break down according to the continuation of the sequence.

[b]In writing a metathesis reaction, you may be tempted sometimes to write NH_4OH as a formula for "ammonium hydroxide." This compound does not exist. In water, it is nothing more than a solution of NH_3.

This equation shows that *any* ammonium salt will react with a strong base to yield ammonia. Even though ammonia is very soluble in water, some escapes as a gas and its odor can be easily detected over the reaction mixture.

Table 4.2 contains a more comprehensive list of substances that are released as gases in metathesis reactions. You should learn these reactions so that you can use them in writing ionic and net ionic equations.

4.8 Predicting When Ionic Reactions Occur

Recall that a metathesis reaction occurs between two compounds in solution by an exchange of two cations between two anions. An example is the reaction of silver nitrate with sodium chloride to yield insoluble silver chloride and the soluble compound sodium nitrate. The molecular, ionic, and net ionic equations for the reaction are as follows.

Molecular equation

$$AgNO_3(aq) + NaCl(aq) \longrightarrow AgCl(s) + NaNO_3(aq)$$

Ionic equation

$$Ag^+(aq) + NO_3^-(aq) + Na^+(aq) + Cl^-(aq) \longrightarrow AgCl(s) + Na^+(aq) + NO_3^-(aq)$$

Net ionic equation

$$Ag^+(aq) + Cl^-(aq) \longrightarrow AgCl(s)$$

The reason this reaction occurs is because a precipitate of silver chloride forms. In a sense, forming the precipitate serves as the "driving force" for the reaction, and we can tell that a reaction occurs because when we drop the spectator ions from the ionic equation, a net ionic equation remains.

In contrast, let's look at what happens when we mix solutions of sodium chloride and potassium nitrate. Anticipating a reaction, we might assemble a molecular equation by exchanging cations between the anions.

$$NaCl + KNO_3 \longrightarrow KCl + NaNO_3$$

The solubility rules tell us that both reactants and both products are soluble, so in the ionic equation we write all of them in dissociated form.

$$Na^+(aq) + Cl^-(aq) + K^+(aq) + NO_3^-(aq) \longrightarrow$$
$$K^+(aq) + Cl^-(aq) + Na^+(aq) + NO_3^-(aq)$$

Notice that both sides are the same except for the order in which the ions are written. If we cancel spectator ions, there is nothing left. There is no net ionic equation, which means there is no net reaction. The two solutions combined simply give a mixture of all four ions.

We now see that the existence or nonexistence of a net ionic equation reveals whether or not an ionic reaction occurs between a pair of reactants. With this in mind, we can review those factors that we know will yield a net ionic equation.

Formation of a Precipitate from Soluble Reactants

We've just discussed this. A net ionic equation will exist when a precipitate is formed by mixing two soluble salts. If both products are soluble strong electrolytes, a net ionic equation will not exist and no reaction occurs.

Formation of Water or a Weak Electrolyte

Earlier in this chapter we discussed acid–base neutralization in which acids and bases react to form water and a salt. In these reactions it is the formation of water that is the driving force. In fact, formation of water in acid–base neutralization is such a strong driving force for reaction that strong and weak acids will react with insoluble oxides and hydroxides.

The formation of water in a reaction is a special case of a more general driving force for ionic reactions—the formation of a weak electrolyte from reactants that are strong electrolytes. Consider the reaction between solutions of hydrochloric acid, HCl, and sodium formate, $NaCHO_2$. The molecular equation, written as a metathesis reaction, is

$$HCl(aq) + NaCHO_2(aq) \longrightarrow HCHO_2(aq) + NaCl(aq)$$

The reactants are both strong electrolytes; HCl is a strong acid and $NaCHO_2$ is a salt. Among the products, we recognize NaCl as a salt, which is a strong electrolyte. The acid $HCHO_2$ is not on our list of strong acids, so we can expect that it is a weak acid, and therefore a weak electrolyte. The ionic equation for the reaction, then, is

$$H^+(aq) + Cl^-(aq) + Na^+(aq) + CHO_2^-(aq) \longrightarrow HCHO_2(aq) + Na^+(aq) + Cl^-(aq)$$

Removing spectator ions gives the net ionic equation,

$$H^+(aq) + CHO_2^-(aq) \longrightarrow HCHO_2(aq)$$

Formation of a Gas

When a gas is formed in a reaction, it leaves the solution, and by doing so serves as the driving force for the reaction. This permits even insoluble compounds such as calcium carbonate to dissolve in acids. For example, the net ionic equation for the reaction of $CaCO_3$ with HCl is

$$CaCO_3(s) + 2H^+(aq) \longrightarrow Ca^{2+}(aq) + CO_2(g) + H_2O$$

Notice, once again, the existence of a net ionic equation for the reaction.

Summary

The factors we've discussed above can be summarized as follows:

A net ionic equation will exist when:
1. A precipitate forms from a solution of soluble reactants.
2. Water forms in the reaction of an acid and a base.
3. A weak electrolyte forms from a solution of strong electrolytes.
4. A gas forms that escapes from the reaction mixture.

To predict whether an ionic reaction will occur between a pair of reactants, we proceed as follows:

- We begin by writing a molecular equation in the form of a metathesis reaction, exchanging anions between the two cations.

- We next translate the molecular equation into an ionic equation.

- We cancel any spectator ions.

If a net ionic equation remains after removing the spectator ions, we conclude that a net reaction *does* occur. On the other hand, if there is nothing left after canceling spectator ions, *there is no net reaction.*

Let's look at several more examples that illustrate how we use these criteria to establish whether a reaction occurs between potential reactants.

EXAMPLE 4.3
Predicting a Reaction

What reaction, if any, occurs between $Ba(OH)_2$ and $(NH_4)_2SO_4$ in an aqueous system?

Analysis: We will consider the possible products of a metathesis reaction and apply the criteria established above to determine what actually happens.

Solution: Exchanging cations between the two reactants gives as possible products $BaSO_4$ and NH_4OH. However, earlier you learned that NH_4OH doesn't really exist and that we should write instead $NH_3(g) + H_2O$. Let's assemble this into a balanced molecular equation and apply the solubility rules next.

$$Ba(OH)_2 + (NH_4)_2SO_4 \longrightarrow BaSO_4 + 2NH_3(g) + 2H_2O$$

soluble	soluble	insoluble
(Rule 5)	(Rule 2)	(Rule 4)

Now we write the ionic equation, expressing soluble ionic compounds in dissociated form.

$$Ba^{2+}(aq) + 2OH^-(aq) + 2NH_4^+(aq) + SO_4^{2-}(aq) \longrightarrow$$
$$BaSO_4(s) + 2NH_3(g) + 2H_2O$$

Notice that there are no ions that appear the same on both sides, so there are no spectator ions to eliminate. Therefore, the ionic equation is also the net ionic equation. ◆

EXAMPLE 4.4
Predicting Reactions and Writing Their Equations

What reaction (if any) occurs when solutions of ammonium carbonate, $(NH_4)_2CO_3$, and propionic acid, $HC_3H_5O_2$, are mixed?

Analysis: As before, we write a potential metathesis equation in molecular form. Then we determine which of the reactants and products are soluble strong electrolytes, so we can expand them to dissociated form in the ionic equation. We will also look for the possibility of forming a precipitate or a gas.

Solution: We begin by constructing a molecular equation, treating the reaction as a metathesis. For the acid, we take the cation to be H^+ and the anion to be $C_3H_5O_2^-$. Therefore, exchanging cations between the two anions we can obtain the following balanced molecular equation.

$$(NH_4)_2CO_3 + 2HC_3H_5O_2 \longrightarrow 2NH_4C_3H_5O_2 + H_2CO_3$$

Here we should recognize that H_2CO_3 decomposes into $CO_2(g)$ and H_2O, so we know a net reaction will occur. Let's rewrite the molecular equation taking this into account.

$$(NH_4)_2CO_3 + 2HC_3H_5O_2 \longrightarrow 2NH_4C_3H_5O_2 + CO_2(g) + H_2O$$

Next, we need to determine which of the substances are soluble strong electrolytes. The solubility rules tell us that both ammonium salts are soluble, and we know that all salts are strong electrolytes. Therefore, we will write $(NH_4)_2CO_3$ and $NH_4C_3H_5O_2$ in dissociated form. In the statement of the problem we are told that we are working with a solution of $HC_3H_5O_2$, so we know it's soluble. Also, it is not on the list of strong acids, so we expect that it is a weak acid; we will write it in molecular form in the ionic equation. Now we are ready to expand the molecular equation into the ionic equation.

$$2NH_4^+(aq) + CO_3^{2-}(aq) + 2HC_3H_5O_2(aq) \longrightarrow$$
$$2NH_4^+(aq) + 2C_3H_5O_2^-(aq) + CO_2(g) + H_2O$$

The only spectator ion is NH_4^+. Dropping this gives the net ionic equation.

$$CO_3^{2-}(aq) + 2HC_3H_5O_2(aq) \longrightarrow 2C_3H_5O_2^-(aq) + CO_2(g) + H_2O \quad \blacklozenge$$

EXAMPLE 4.5

Predicting Reactions and Writing Their Equations

What reaction (if any) occurs in water between KNO_3 and NH_4Cl?

Analysis: Let's proceed by writing molecular, ionic, and net ionic equations as in the preceding example.

Solution: Solubility Rule 2 tells us that both KNO_3 and NH_4Cl are soluble. So at the instant of mixing solutions of these compounds, we'll have the following ions intermingling: K^+, NO_3^-, NH_4^+, and Cl^-. By exchange of partners, the new combinations link K^+ with Cl^- as KCl, and NH_4^+ with NO_3^- as NH_4NO_3. By solubility Rules 1 and 2, both products are also soluble in water. The anticipated molecular equation is therefore

$$\underset{\text{soluble}}{KNO_3(aq)} + \underset{\text{soluble}}{NH_4Cl(aq)} \longrightarrow \underset{\text{soluble}}{KCl(aq)} + \underset{\text{soluble}}{NH_4NO_3(aq)}$$

and the ionic equation is

$$K^+(aq) + NO_3^-(aq) + NH_4^+(aq) + Cl^-(aq) \longrightarrow$$
$$K^+(aq) + Cl^-(aq) + NH_4^+(aq) + NO_3^-(aq)$$

Notice that the right side of the equation is the same as the left side except for the order in which the ions are written. When we eliminate spectator ions, everything goes. There is no net ionic equation, which means there is no net reaction. (In fact, when we mix solutions of salts whose potential products in a metathesis reaction are all soluble ionic compounds, we can anticipate in advance that no reaction will occur. It really isn't necessary to actually write the chemical equations.)

Predict whether a reaction will occur between the following pairs of substances. If a reaction does occur, derive the net ionic equation. (a) $KCHO_2$ and HCl, (b) $CuCO_3$ and $HC_2H_3O_2$, (c) $Ca(C_2H_3O_2)_2$ and $AgNO_3$, and (d) $NaOH$ and $NiCl_2$. ◆

4.9 Stoichiometry of Ionic Reactions

In dealing quantitatively with reactions in solution, it is necessary to describe the concentrations of the reactants and products. As you learned in Chapter 3, for the purpose of stoichiometric calculations, molar concentration is the most useful concentration unit.

Calculating Concentrations of Ions in Solutions of Electrolytes

To work problems involving solutions of electrolytes, it is often necessary to know the concentrations of the ions in a solution. This information is easy to calculate from the molar concentration of the electrolyte solute. For example, suppose we are working with a solution labeled "0.10 M $CaCl_2$." In 1.0 L of this solution there is dissolved 0.10 mol of $CaCl_2$, which is fully dissociated into Ca^{2+} and Cl^- ions.

$$CaCl_2 \longrightarrow Ca^{2+} + 2Cl^-$$

The solution doesn't actually contain any $CaCl_2$, even though this is the solute used to prepare the solution. Instead, the solution contains Ca^{2+} and Cl^- ions.

From the stoichiometry of the dissociation, we see that 1 mol Ca^{2+} and 2 mol Cl^- are formed from 1 mol $CaCl_2$. Therefore, 0.10 mol $CaCl_2$ will yield 0.10 mol Ca^{2+} and 0.20 mol Cl^-. In 0.10 M $CaCl_2$, then, the concentration of Ca^{2+} is 0.10 M and the concentration of Cl^- is 0.20 M. *Notice that the concentration of a particular ion equals the concentration of the salt multiplied by the number of ions of that kind in one formula unit of the salt.*

EXAMPLE 4.6

Calculating the Concentrations of Ions in a Solution

What are the molar concentrations of the ions in 0.20 M $Al_2(SO_4)_3$?

Analysis: The concentrations of the ions are determined by the stoichiometry of the salt. Therefore, we determine the number of ions of each kind formed from one formula unit of $Al_2(SO_4)_3$. These values are then used along with the given concentration of the salt to calculate the ion concentrations.

Solution: Each formula unit of $Al_2(SO_4)_3$ yields two Al^{3+} ions and three SO_4^{2-} ions, so 0.20 mol $Al_2(SO_4)_3$ yields 0.40 mol Al^{3+} and 0.60 mol SO_4^{2-}. Therefore, the solution contains 0.40 M Al^{3+} and 0.60 M SO_4^{2-}.

The answers here have been obtained by simple mole reasoning. We could have found the answers in a more formal manner using the factor-label method. We can start with the given concentration and proceed as follows:

$$\frac{0.20 \text{ mol } Al_2(SO_4)_3}{1.0 \text{ L soln}} \times \frac{2 \text{ mol } Al^{3+}}{1 \text{ mol } Al_2(SO_4)_3} = \frac{0.40 \text{ mol } Al^{3+}}{1.0 \text{ L soln}} = 0.40 \text{ } M \text{ } Al^{3+}$$

$$\frac{0.20 \text{ mol } Al_2(SO_4)_3}{1.0 \text{ L soln}} \times \frac{3 \text{ mol } SO_4^{2-}}{1 \text{ mol } Al_2(SO_4)_3} = \frac{0.60 \text{ mol } SO_4^{2-}}{1.0 \text{ L soln}} = 0.60 \text{ } M \text{ } SO_4^{2-}$$

Study both methods. With just a little practice, you will have little difficulty with the reasoning approach that we used first.

A student found that the sulfate ion concentration in a solution of $Al_2(SO_4)_3$ was 0.90 M. What was the concentration of $Al_2(SO_4)_3$ in the solution?

Analysis: Once again, we use the formula of the salt to determine the number of ions released when it dissociates. This time we use the information to work backward to find the salt concentration.

Solution: Let's set up the problem using the factor-label method to be sure of our procedure. We will use the fact that 1 mol $Al_2(SO_4)_3$ yields 3 mol SO_4^{2-} in solution.

$$1 \text{ mol } Al_2(SO_4)_3 \Leftrightarrow 3 \text{ mol } SO_4^{2-}$$

Therefore,

$$\frac{0.90 \text{ mol } SO_4^{2-}}{1.0 \text{ L soln}} \times \frac{1 \text{ mol } Al_2(SO_4)_3}{3 \text{ mol } SO_4^{2-}} = \frac{0.30 \text{ mol } Al_2(SO_4)_3}{1.0 \text{ L soln}} = 0.30 \text{ M } Al_2(SO_4)_3$$

The reasoning approach here would have worked as follows. We know that 1 mol $Al_2(SO_4)_3$ yields 3 mol SO_4^{2-} in solution. Therefore, the number of moles of $Al_2(SO_4)_3$ is only one-third the number of moles of SO_4^{2-}. So the concentration of $Al_2(SO_4)_3$ must be one-third of 0.90 M, or 0.30 M.

Practice Exercise 10

What are the molar concentrations of the ions in 0.40 M $FeCl_3$? ◆

Practice Exercise 11

In a solution of Na_3PO_4, the PO_4^{3-} concentration was determined to be 0.250 M. What was the sodium ion concentration in the solution? ◆

EXAMPLE 4.7

Calculating the Concentration of a Salt from the Concentration of One of Its Ions

Using Net Ionic Equations in Solution Stoichiometry Calculations

You have seen that a net ionic equation is convenient for focusing on the net chemical change in an ionic reaction. Let's study some examples that illustrate how such equations can be used in stoichiometric calculations.

How many milliliters of 0.100 M $AgNO_3$ solution are needed to react completely with 25.0 mL of 0.400 M $CaCl_2$ solution? The net ionic equation for the reaction is

$$Ag^+(aq) + Cl^-(aq) \longrightarrow AgCl(s)$$

Analysis: The problem describes solutions of salts with formulas expressed in "molecular form," but the equation relates a reaction between the ions of the salts. Therefore, the first thing we will do is calculate the concentrations of the ions in the solutions being mixed. Then, using the volume and molarity of the Cl^- solution, we calculate the moles of Cl^- available. This value equals the moles of Ag^+ that react, because Ag^+ and Cl^- combine in a 1-to-1 mole ratio. Once we have the number of moles of Ag^+ that react, we use the molarity of the Ag^+ solution to determine the volume of the 0.100 M $AgNO_3$ solution needed.

Solution: In 0.100 M $AgNO_3$, the Ag^+ concentration is 0.100 M. In 0.400 M $CaCl_2$, the Cl^- concentration is 0.800 M. Having these values, we can now restate the problem: How many milliliters of 0.100 M Ag^+ solution are needed to react completely with 25.0 mL of 0.800 M Cl^- solution?

We have the volume and molarity of the Cl^- solution, so we calculate the number of moles of Cl^- available for reaction.

EXAMPLE 4.8

Stoichiometric Calculations Using a Net Ionic Equation

A solution of $AgNO_3$ is added to a solution of $CaCl_2$, producing a precipitate of AgCl.

$$0.0250 \ \cancel{L \ Cl^- \ soln} \times \frac{0.800 \ mol \ Cl^-}{1.00 \ \cancel{L \ Cl^- \ soln}} = 0.0200 \ mol \ Cl^-$$

(Notice we have been careful to express the volume of the chloride solution in liters so the units cancel correctly.)

Since 0.0200 mol Cl^- is available, the amount of Ag^+ required for reaction is also 0.0200 mol. Now we calculate the volume of the Ag^+ solution using its molarity as a conversion factor.

$$0.0200 \ \cancel{mol \ Ag^+} \times \frac{1.00 \ L \ Ag^+ \ soln}{0.100 \ \cancel{mol \ Ag^+}} = 0.200 \ L \ Ag^+ \ soln$$

Our calculations tell us that we must use 0.200 L or 200 mL of the $AgNO_3$ solution. We could have used the following chain calculation, of course.

$$0.0250 \ \cancel{L \ Cl^- \ soln} \times \frac{0.800 \ \cancel{mol \ Cl^-}}{1.00 \ \cancel{L \ Cl^- \ soln}} \times \frac{1 \ \cancel{mol \ Ag^+}}{1 \ \cancel{mol \ Cl^-}} \times \frac{1.00 \ L \ Ag^+ \ soln}{0.100 \ \cancel{mol \ Ag^+}}$$
$$= 0.200 \ L \ Ag^+ \ soln$$

Is the Answer Reasonable?
The silver ion concentration is one-eighth as large as the chloride ion concentration. Since the ions react one-for-one, we will need eight times as much silver ion solution as chloride solution. Eight times the amount of chloride solution, 25 mL, is 200 mL, which is the answer we obtained. Therefore, the answer appears to be correct.

<hr>

Practice Exercise 12

How many milliliters of 0.500 *M* KOH are needed to react completely with 60.0 mL of 0.250 *M* $FeCl_2$ solution to precipitate $Fe(OH)_2$? ◆

EXAMPLE 4.9

Calculation Involving the Stoichiometry of an Ionic Reaction

Milk of magnesia is a remedy for acid indigestion.

Actually, the reaction occurs as the solutions are being poured together. For purposes of the calculations, however, we may pretend that the solutions can be mixed first and then allowed to react. The initial and final states will still be the same.

Milk of magnesia is a suspension of $Mg(OH)_2$ in water. It can be made by adding a base to a solution containing Mg^{2+}. Suppose that 40.0 mL of 0.200 *M* NaOH solution is added to 25.0 mL of 0.300 *M* $MgCl_2$ solution. What mass of $Mg(OH)_2$ will be formed, and what will be the concentrations of the ions in the solution after the reaction is complete? The net ionic equation for the reaction is

$$Mg^{2+}(aq) + 2OH^-(aq) \longrightarrow Mg(OH)_2(s)$$

Analysis: The first thing we should notice here is that we've been given the volume and molarity for *both* solutions. By now, you should realize that volume and molarity give us moles, so in effect, we have been given the number of moles of each reactant. This means we have a limiting reactant problem (similar to the kind we studied on page 123).

To answer all the parts of this question, the best approach is to begin by tabulating the number of *moles* of each ion supplied by their respective solutions. Then we'll use the ionic equation to see which ions are removed and calculate which, if any, is limiting. Then we can go after the concentrations of the remaining ions and the mass of $Mg(OH)_2$ that forms.

Solution: Let's begin by determining the number of moles of NaOH and $MgCl_2$ supplied by the volumes of their solutions used. The conversion factors are taken from their molarities—0.200 *M* NaOH and 0.300 *M* $MgCl_2$.

$$0.0400 \ \cancel{L \ NaOH \ soln} \times \frac{0.200 \ mol \ NaOH}{1.00 \ \cancel{L \ NaOH \ soln}} = 8.00 \times 10^{-3} \ mol \ NaOH$$

$$0.0250 \ \cancel{L \ MgCl_2 \ soln} \times \frac{0.300 \ mol \ MgCl_2}{1.00 \ \cancel{L \ MgCl_2 \ soln}} = 7.50 \times 10^{-3} \ mol \ MgCl_2$$

From this information, we obtain the number of moles of each ion present *before* any reaction occurs. In doing this, notice we take into account that 1 mol of $MgCl_2$ gives 2 mol Cl^-.

Moles of Ions before Reaction

Mg^{2+} 7.50×10^{-3} mol Cl^- 15.0×10^{-3} mol
Na^+ 8.00×10^{-3} mol OH^- 8.00×10^{-3} mol

The number of moles of Cl^- equals $2 \times (7.50 \times 10^{-3})$.

Now we refer to the net ionic equation, where we see that only Mg^{2+} and OH^- react. From the coefficients of the equation,

$$1 \text{ mol } Mg^{2+} \Leftrightarrow 2 \text{ mol } OH^-$$

This means that 7.50×10^{-3} mol Mg^{2+} would require 15.0×10^{-3} mol OH^-, but we have only 8.00×10^{-3} mol OH^-. Insufficient OH^- is available to react with all of the Mg^{2+}, so OH^- must be the limiting reactant. Therefore, all of the OH^- will be used up and some Mg^{2+} will be unreacted. The amount of Mg^{2+} that *does* react to form $Mg(OH)_2$ can be found as follows.

$$8.00 \times 10^{-3} \text{ mol OH}^- \times \frac{1 \text{ mol } Mg^{2+}}{2 \text{ mol OH}^-} = 4.00 \times 10^{-3} \text{ mol } Mg^{2+}$$

(This amount reacts.)

Now we can tabulate the number of moles of each ion left in the mixture *after* the reaction is complete. For Mg^{2+}, the amount remaining equals the initial amount minus the amount that reacts.

Moles of Ions after Reaction

Mg^{2+} $(7.50 \times 10^{-3} \text{ mol}) - (4.00 \times 10^{-3} \text{ mol}) = 3.50 \times 10^{-3} \text{ mol}$
Cl^- 15.0×10^{-3} mol (no change)
Na^+ 8.00×10^{-3} mol (no change)
OH^- 0.00 mol (all used up)

Let's assemble all these numbers into one table to make them easier to understand.

Ion	Moles Present before Reaction	Moles That React	Moles Present after Reaction
Mg^{2+}	7.50×10^{-3} mol	4.00×10^{-3} mol	3.50×10^{-3} mol
Cl^-	15.0×10^{-3} mol	0.00 mol	15.0×10^{-3} mol
Na^+	8.00×10^{-3} mol	0.00 mol	8.00×10^{-3} mol
OH^-	8.00×10^{-3} mol	8.00×10^{-3} mol	0.00 mol

The question asks for the concentrations of the ions in the final reaction mixture, so we must now divide each of the number of moles in the last column by the *total volume of the final solution* (40.0 mL + 25.0 mL = 65.0 mL). This volume must be expressed in liters (0.0650 L). For example, for Mg^{2+}, its concentration is

$$\frac{3.50 \times 10^{-3} \text{ mol } Mg^{2+}}{0.0650 \text{ L soln}} = 0.0538 \, M \, Mg^{2+}$$

Performing similar calculations for the other ions gives the following.

Concentrations of Ions after Reaction

Mg^{2+} 0.0538 *M* Cl^- 0.231 *M*
Na^+ 0.123 *M* OH^- 0.00 *M*

Now we can turn our attention to the mass of $Mg(OH)_2$ that's formed. If 4.00×10^{-3} mol Mg^{2+} reacts, then 4.00×10^{-3} mol $Mg(OH)_2$ must be produced. The

Stoichiometrically,
$1 \text{ mol } Mg^{2+} \Leftrightarrow 1 \text{ mol } Mg(OH)_2$

formula mass of $Mg(OH)_2$ is 58.32 g mol^{-1}, so the mass of $Mg(OH)_2$ formed is

$$4.00 \times 10^{-3} \overline{\text{mol Mg(OH)}_2} \times \frac{58.32 \text{ g Mg(OH)}_2}{1 \overline{\text{mol Mg(OH)}_2}} = 0.233 \text{ g Mg(OH)}_2$$

The amount of $Mg(OH)_2$ formed is 0.233 g.

Is the Answer Reasonable?
This was a fairly complex problem, and there are no simple checks we can make to verify our answers.

Practice Exercise 13

How many moles of $BaSO_4$ will form if 20.0 mL of 0.600 M $BaCl_2$ is mixed with 30.0 mL of 0.500 M $MgSO_4$? What will the concentrations of each ion be in the final reaction mixture? ◆

Chemical Analyses

Chemical analyses fall into two categories. In a **qualitative analysis** we simply determine which substances are present in a sample without measuring their amounts. In a **quantitative analysis,** our goal is to measure the amounts of the various substances in a sample.

When chemical reactions are used in a quantitative analysis, a useful strategy is to capture all of a desired chemical species in a compound with a known formula. From the amount of this compound obtained, we can determine how much of the desired chemical species was present in the original sample.

EXAMPLE 4.10

Calculation Involving a Quantitative Analysis

A certain insecticide is a compound known to contain carbon, hydrogen, and chlorine. Reactions were carried out on a 1.000 g sample of the compound that converted all of its chlorine to chloride ion dissolved in water. This aqueous solution was treated with an excess amount of $AgNO_3$ solution, and the AgCl precipitate was collected and weighed. Its mass was 2.022 g. What was the percentage by mass of Cl in the original insecticide sample?

Analysis: The strategy is to determine the mass of Cl in 2.022 g of AgCl. We then assume that all this chlorine was in the original 1.000 g sample, and calculate the percentage Cl.

Solution: The formula mass of AgCl is 143.32, and the atomic mass of Cl is 35.453. We also know that in one mole of AgCl there must be one mole of Cl, so in 143.32 g AgCl there must be 35.453 g Cl. With this information, we can now calculate the mass of Cl in 2.022 g AgCl.

$$2.022 \text{ g AgCl} \times \frac{1 \text{ mol AgCl}}{143.32 \text{ g AgCl}} \times \frac{1 \text{ mol Cl}}{1 \text{ mol AgCl}} \times \frac{35.453 \text{ g Cl}}{1 \text{ mol Cl}} = 0.5002 \text{ g Cl}$$

The percentage Cl in the sample can be calculated as

$$\% \text{ Cl} = \frac{\text{mass of Cl}}{\text{mass of sample}} \times 100\%$$

$$= \frac{0.5002 \text{ g Cl}}{1.000 \text{ g sample}} \times 100\%$$

$$= 50.02\%$$

The insecticide was 50.02% Cl.

Is the Answer Reasonable?

Let's look at the formula masses of AgCl and Cl (143.32 and 35.453, respectively). If we round these off to 140 and 35, we can say that AgCl is approximately 25% Cl, by mass ($35 \div 140 = 0.25$). Approximately 2 g of AgCl was obtained, and 25% of 2 g equals 0.5 g of Cl. The original sample weighed 1 g, so the compound must be approximately 50% Cl by mass. (Notice how closely our approximation comes to the value we obtained by the precise calculation!)

Practice Exercise 14

A sample of a mixture containing $CaCl_2$ and $MgCl_2$ weighed 2.000 g. The sample was dissolved in water and H_2SO_4 was added until the precipitation of $CaSO_4$ was complete. The $CaSO_4$ was filtered, dried completely, and weighed. A total of 0.736 g of $CaSO_4$ was obtained.

(a) How many moles of Ca^{2+} were in the $CaSO_4$?
(b) How many moles of Ca^{2+} were in the original 2.000 g sample?
(c) How many moles of $CaCl_2$ were in the 2.000 g sample?
(d) How many grams of $CaCl_2$ were in the 2.000 g sample?
(e) What was the percentage of $CaCl_2$ in the mixture? ◆

Titrations

Titration is an important laboratory procedure used in performing chemical analyses. The apparatus is shown in Figure 4.10. The long tube is called a **buret,** and it is marked for volumes, usually in increments of 0.10 mL. In a typical titration, a solution containing one reactant is placed in the receiving flask. Carefully

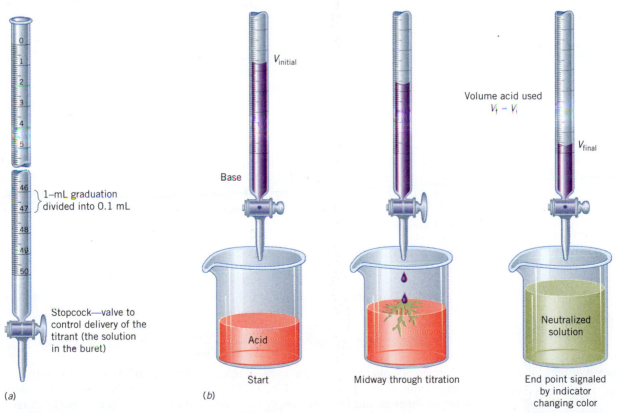

1–mL graduation divided into 0.1 mL

Stopcock—valve to control delivery of the titrant (the solution in the buret)

(a)

$V_{initial}$

Base

Start

V_{final}

Volume acid used $V_f - V_i$

Acid

Midway through titration

Neutralized solution

End point signaled by indicator changing color

(b)

Figure 4.10 *Titration.* (*a*) A buret. (*b*) The titration of an acid by a base in which an acid–base indicator is used to signal the end point—the point at which all of the acid has been neutralized and addition of the base is halted.

measured volumes of a solution of the other reactant are then added from the buret. (One of the two solutions is of a precisely known concentration and is called a **standard solution**). This addition is continued until some visual effect, like a color change, signals that the two reactants have been combined in just the right ratio to give a complete reaction.

The valve at the bottom of the buret is called a **stopcock,** and it permits the analyst to control the amount of **titrant** (the solution in the buret) that is delivered to the receiving flask. This permits the addition of the titrant to be stopped once the reaction is complete.

Acid–Base Titrations

The titration procedure is used for many kinds of reactions.

Titrations are often used for acid–base reactions. Prior to the start of an acid–base titration, a drop or two of an *indicator* solution is added to the solution in the receiving flask. An **acid–base indicator** is a dye that has one color in an acidic solution and a different color in a basic solution. We have already mentioned litmus. Another, phenolphthalein, is more commonly used in acid–base titrations. It is colorless in acid and pink in base. (The theory of acid–base indicators is discussed in Chapter 16.)

During an acid–base titration, the analyst adds titrant slowly, watching for a change in the indicator color, which would signal that the solution has changed from acidic to basic (or basic to acidic, depending on the nature of the reactants in the buret and flask). Usually this color change is very abrupt, and occurs with the addition of only one final drop of the titrant just as the end of the reaction is reached. When the color change is observed, the **end point** has been reached. At this time the addition of titrant is stopped and the total volume of the titrant that's been added to the receiving flask is recorded.

EXAMPLE 4.11

Calculation Involving Acid–Base Titration

A student prepares a solution of hydrochloric acid that is approximately 0.1 *M* and wishes to determine its precise concentration. A 25.00-mL portion of the HCl solution is transferred to a flask, and after a few drops of indicator are added, the HCl solution is titrated with 0.0775 *M* NaOH solution. The titration requires exactly 37.46 mL of the standard NaOH solution to reach the end point. What is the molarity of the HCl solution?

Analysis: We'll need the balanced equation to learn the stoichiometric equivalency between HCl and NaOH. Then we calculate the number of moles of NaOH used (from the volume and molarity of the NaOH solution) to find the equivalent moles of HCl and use this and the volume of the HCl solution to calculate the desired molarity.

Solution: The chemical equation for the neutralization reaction is

$$HCl(aq) + NaOH(aq) \longrightarrow NaCl(aq) + H_2O$$

From the molarity and volume of the NaOH solution, we calculate the number of moles of NaOH consumed in the titration.

$$0.03746 \ \cancel{\text{L NaOH soln}} \times \frac{0.0775 \ \text{mol NaOH}}{1.00 \ \cancel{\text{L NaOH soln}}} = 2.90 \times 10^{-3} \ \text{mol NaOH}$$

Stoichiometrically,

$1 \ \text{mol NaOH} \Leftrightarrow 1 \ \text{mol HCl}$

The coefficients in the equation tell us that NaOH and HCl react in a one-to-one mole ratio, so in this titration, 2.90×10^{-3} mol HCl was in the flask. To calculate the molarity of the HCl, we simply take the ratio of the number of moles of HCl that reacted (2.90×10^{-3} mol HCl) to the volume (in liters) of the HCl solution used (25.00 mL, or 0.02500 L).

$$\text{Molarity of HCl soln} = \frac{2.90 \times 10^{-3} \text{ mol HCl}}{0.02500 \text{ L HCl soln}}$$

$$= 0.116 \ M \text{ HCl}$$

The molarity of the hydrochloric acid is 0.116 M.

Is the Answer Reasonable?

The volume of NaOH solution used is larger than the volume of HCl solution. Because the reactants combine in a one-to-one ratio, this must mean that the HCl solution is more concentrated than the NaOH solution. The value we obtained, 0.116 M, is larger than 0.0775 M, so our answer seems reasonable.

Practice Exercise 15

In a titration, a sample of H_2SO_4 solution having a volume of 15.00 mL required 36.42 mL of 0.147 M NaOH solution for *complete* neutralization. What is the molarity of the H_2SO_4 solution? ◆

Practice Exercise 16

"Stomach acid" is hydrochloric acid. A sample of gastric juice having a volume of 5.00 mL required 11.00 mL of 0.0100 M KOH solution for neutralization in a titration. What was the molar concentration of HCl in this fluid? If we assume a density of 1.00 g mL^{-1} for the fluid, what was the percentage by weight of HCl? ◆

A solution of HCl is titrated with a solution of NaOH using phenolphthalein as the acid–base indicator. The pink color that phenolphthalein has in a basic solution can be seen where a drop of the NaOH solution has entered the HCl solution, to which a few drops of the indicator had been added.

SUMMARY

Electrolytes. Substances that **dissociate** or **ionize** in water to produce cations and anions are **electrolytes;** those that do not are called **nonelectrolytes.** Electrolytes include salts and metal hydroxides as well as molecular acids and bases that ionize by reaction with water. In water, ionic compounds are completely dissociated into ions and are **strong electrolytes.**

Metathesis Reactions. Metathesis or **double replacement** reactions occur between ions and are called **ionic reactions.** Solutions of soluble strong electrolytes often yield an insoluble product which appears as a **precipitate.** Equations for these reactions can be written in three different ways. In a **molecular equation,** complete formulas for all reactants and products are used. In an **ionic equation,** soluble strong electrolytes are written in dissociated (ionized) form; "molecular" formulas are used for solids and weak electrolytes. A **net ionic equation** is obtained by eliminating **spectator ions** from the ionic equation, and such an equation allows us to identify other combinations of reactants that give the same net reaction. An ionic or net ionic equation is balanced only if both atoms *and* net charge are balanced.

Acids and Bases as Electrolytes. The modern version of the **Arrhenius definition of acids and bases** is that an **acid** is a substance that produces hydronium ions, H_3O^+, when dissolved in water, and a **base** produces hydroxide ions, OH^-, when dissolved in water.

The oxides of nonmetals are generally **acidic anhydrides** and react with water to give acids. Metal oxides are usually **basic anhydrides** because they tend to react with water to give metal hydroxides or bases.

Strong acids and bases are also strong electrolytes. **Weak acids and bases** are **weak electrolytes,** which are incompletely ionized in water. In a solution of a weak electrolyte there is a **chemical equilibrium (dynamic equilibrium)** between the non-ionized molecules of the solute and the ions formed by the reaction of the solute with water.

Acid–Base Neutralization and Ionic Reactions. Acids react with bases in neutralization reactions to produce a salt and water. Acids react with insoluble oxides and hydroxides to form water and the corresponding salt. Most acid–base neutralization reactions can be viewed as a type of metathesis reaction in which one product is water.

Reactions That Produce Gases. Some metathesis reactions produce gaseous products with low solubility in water or compounds that are unstable and decompose to give slightly soluble gases. In particular, acids react with soluble and insoluble carbonates and with bicarbonates to yield carbon dioxide, a gaseous decomposition product of carbonic acid, H_2CO_3.

Predicting Metathesis Reactions. A metathesis reaction will occur if there is a net ionic equation. Factors that lead to the existence of a net ionic equation are formation of a precipitate from soluble products, formation of water in an acid–base neutralization, formation of a gas, and formation of a weak electrolyte from soluble strong electrolytes. You should learn the **solubility rules** (Table 4.1), and remember that all salts are strong electrolytes. Remember that all strong acids and bases are strong electrolytes, too. Everything else is a weak electrolyte or a nonelectrolyte. Be sure to learn the reactions that produce gases in metathesis reactions, which are found in Table 4.2.

Solution Stoichiometry. Molar concentration (molarity) is a useful concentration unit for working problems in the stoichiometry of ionic reactions. The concentrations of ions in the solution of a salt can be derived from the molar con-

centration of the salt, taking into account the number of ions formed per formula unit of the salt.

Titration is a technique used to make quantitative measurements of the amounts of solutions needed to obtain a complete reaction. In an acid–base titration, the **end point** is normally detected visually using an **acid–base indicator.** A color change indicates complete reaction, at which time addition of **titrant** is stopped and the volume added is recorded.

Tools *you have learned*

The table below lists the tools that you have learned in this chapter. You will need them to answer chemistry questions. Review them if necessary, and refer to the tools when working on the

Thinking-It-Through problems and the Review Questions and Review Problems that follow.

Tool	Function
Criteria for balanced ionic equations (*page 156*)	To obtain balanced ionic and net ionic equations.
Solubility rules (*page 159*)	To tell if a given salt is soluble in water and dissociates fully. To enable the prediction of metathesis reactions.
List of strong acids (*page 166*)	To enable the recognition of any other acids as weak. To decide whether to write the formula of an acid in ionic or molecular form in an ionic equation.
Types of substances that react with either acids or bases to give gases (*page 174*)	To predict when to expect CO_2, SO_2, HCN, H_2S, or NH_3 to form from a reaction.

THINKING IT THROUGH

The goal in the following problems is *not* to find the answer itself; instead, you are asked to assemble the available information needed to obtain the answer, state what additional data (if any) are needed, and describe *how* you would use the data to answer the question. Where problems can be solved by the factor-label method, set up the solution to the problem by arranging the conversion factors so the units cancel correctly to give the desired units of the answer.

The problems are divided into two groups. Those in Level 2 are more challenging than those in Level 1 and provide an opportunity to really hone your problem solving skills. Detailed answers to these problems can be found at our Web site: http://www.wiley.com/college/brady

Level 1 Problems

1. You are planning an experiment that requires the presence of the hydroxide ion at a concentration of at least 0.2 *M*. Which of the following could you use as the solute: NaOH, KOH, or $Mg(OH)_2$? Describe the information you need to know to answer the question.

2. As the first step in answering this question, write both a molecular and a net ionic equation that describes the reaction of solid $MgCO_3$ with an aqueous solution of hydrobromic acid. Suppose that you mix 100.0 mL of 0.1250 *M* HBr with a 10.54 g sample of solid $MgCO_3$ in a large beaker. Explain the calculations required to answer the fol-

lowing questions. How many grams of the $MgCO_3$ dissolve, and how much remains after the reaction is over? What are the concentrations of the ions dissolved in the solution after the reaction is over?

3. An unknown compound was either NaCl, NaBr, or NaI. When 0.564 g of the compound was dissolved in water, it took 22.1 mL of 0.248 *M* $AgNO_3$ to combine with all of the halide ion in the solution. Describe how you would use this information to determine which compound the unknown was.

4. Consider two solutions, one of them containing $Fe(NO_3)_2$ and the other containing $Fe_2(SO_4)_3$. In the first solution, the NO_3^- concentration is 0.240 *M*; in the other the SO_4^{2-} concentration is 0.210 *M*. How would you determine which solution has the higher concentration of iron?

Level 2 Problems

5. In an experiment, 20.0 mL of 0.25 *M* NaCl will be added to 35.0 mL of 0.12 *M* AgNO₃. Explain how you would determine the mass of AgCl that will be formed in this reaction mixture.

6. A sample of $Cr_2(SO_4)_3$ weighing 0.500 g was dissolved in water and treated with an excess of NaOH solution, causing a precipitate to form. The precipitate was collected by filtration. Describe the calculations required to determine the number of milliliters of 0.400 *M* HNO₃ needed to dissolve this precipitate.

7. A 0.100 g sample of an acid with the molecular formula $C_6H_8O_6$ was titrated with 0.150 *M* NaOH. A volume of 7.58 mL of the base was required to completely neutralize the acid. Is the acid a monoprotic acid or a diprotic acid? (Describe the calculation involved.)

8. A mixture of two monoprotic acids—lactic acid, $HC_3H_5O_3$ (found in sour milk), and caproic acid, $HC_6H_{11}O_2$ (a foul-smelling substance found in excretions from male goats)—was titrated with 0.0500 *M* NaOH. A 0.1000 g sample of the mixture required 20.4 mL of the base. Explain in detail how you would calculate the mass in grams of each acid in the sample.

9. Lead chloride, $PbCl_2$, is a white solid that is slightly soluble in water at 25 °C. If a sample of $PbCl_2$ is added to a beaker of water, which of the following diagrams best represents the results? (Black spheres represent Pb and gray spheres Cl. Explain what is wrong with the other four diagrams.

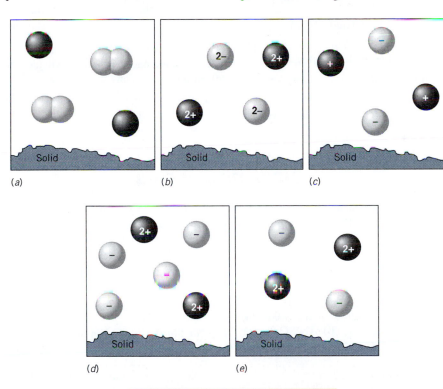

(a)   (b)   (c)

(d)   (e)

REVIEW QUESTIONS

Electrolytes

4.1 What is an electrolyte? What is a nonelectrolyte?

4.2 Define dissociation as it applies to ionic compounds that dissolve in water.

4.3 Experimentally, what is the observable difference between a strong electrolyte and a weak electrolyte?

4.4 If a substance is a weak electrolyte, what does this mean in terms of the tendency of the ions to react to re-form the molecular compound? How does this compare with strong electrolytes?

Ionic Reactions

4.5 What are the differences among molecular, ionic, and net ionic equations? What are spectator ions?

4.6 What two conditions must be fulfilled by a balanced ionic equation?

4.7 The following equation is not balanced. What is wrong with it?

$$NO_2 + H_2O \longrightarrow NO_3^- + 2H^+$$

Metathesis Reactions

4.8 What is another name for *metathesis reaction?* What is a precipitate?

4.9 Write equations for the dissociation of the following in water: (a) $CaCl_2$, (b) $(NH_4)_2SO_4$, (c) $NaC_2H_3O_2$.

4.10 If a solution of trisodium phosphate, Na_3PO_4, is poured into seawater, precipitates of calcium phosphate and magnesium phosphate are formed. (Magnesium and calcium ions are among the principal ions found in seawater.) Write net ionic equations for these reactions.

4.11 How would the electrical conductivity of a solution of $Ba(OH)_2$ change as a solution of H_2SO_4 is added slowly to it? Use a net ionic equation to justify your answer.

4.12 Washing soda is $Na_2CO_3 \cdot 10H_2O$. Explain, using chemical equations, how this substance is able to remove Ca^{2+} ions from "hard water."

Acids and Bases

4.13 Give two general properties of an acid. Give two general properties of a base.

4.14 How did Arrhenius define an acid and a base?

4.15 What is the difference between *dissociation* and *ionization* as applied to the discussion of electrolytes?

4.16 Write the formulas for the strong acids discussed in Section 4.5. (If you must peek, they are listed on page 166.)

4.17 Why don't we use double arrows in the equation for the reaction of a strong acid with water?

4.18 Which of the following would yield an acidic solution when they react with water? (a) P_4O_{10}, (b) K_2O, (c) SeO_3, (d) Cl_2O_7

Reactions That Produce Gases

4.19 Suppose you suspected that a certain solution contained ammonium ions. What simple chemical test could you perform that would tell you whether your suspicion was correct?

4.20 What gas is formed if HCl is added to (a) $NaHCO_3$, (b) Na_2S, (c) K_2SO_3?

Predicting Ionic Reactions

4.21 What factors lead to the existence of a net ionic equation in a reaction between ions?

4.22 Complete the following equations and then determine the "driving force" in each reaction.

(a) $CuCl_2 + Na_2CO_3 \longrightarrow$ (c) $NaClO + H_2SO_4 \longrightarrow$
(b) $K_2SO_3 + H_2SO_4 \longrightarrow$ (d) $Ba(OH)_2 + HNO_3 \longrightarrow$

Chemical Analyses and Titrations

4.23 Describe each of the following: (a) buret, (b) titration, (c) titrant, and (d) end point.

4.24 What is the function of an indicator in a titration? What color is phenolphthalein in (a) an acidic solution and (b) a basic solution?

REVIEW PROBLEMS

Answers to problems whose numbers are printed in color are given in Appendix D.
Challenging problems are marked with asterisks.

Electrolytes

4.25 Write equations for the dissociation of the following ionic compounds in water: (a) LiCl, (b) $BaCl_2$, (c) $Al(C_2H_3O_2)_3$, (d) $(NH_4)_2CO_3$, (e) $FeCl_3$.

4.26 Write equations for the dissociation of the following ionic compounds in water: (a) $CuSO_4$, (b) $Al_2(SO_4)_3$, (c) $CrCl_3$, (d) $(NH_4)_2HPO_4$, (e) $KMnO_4$.

Metathesis Reactions That Produce Precipitates

4.27 Write ionic and net ionic equations for these reactions.
(a) $(NH_4)_2CO_3(aq) + MgCl_2(aq) \longrightarrow$

$$2NH_4Cl(aq) + MgCO_3(s)$$

(b) $CuCl_2(aq) + 2NaOH(aq) \longrightarrow$

$$Cu(OH)_2(s) + 2NaCl(aq)$$

(c) $3FeSO_4(aq) + 2Na_3PO_4(aq) \longrightarrow$

$$Fe_3(PO_4)_2(s) + 3Na_2SO_4(aq)$$

(d) $2AgC_2H_3O_2(aq) + NiCl_2(aq) \longrightarrow$

$$2AgCl(s) + Ni(C_2H_3O_2)_2(aq)$$

4.28 Write *balanced* ionic and net ionic equations for these reactions.
(a) $CuSO_4(aq) + BaCl_2(aq) \longrightarrow CuCl_2(aq) + BaSO_4(s)$

(b) $Fe(NO_3)_3(aq) + LiOH(aq) \longrightarrow$

$$LiNO_3(aq) + Fe(OH)_3(s)$$

(c) $Na_3PO_4(aq) + CaCl_2(aq) \longrightarrow$

$$Ca_3(PO_4)_2(s) + NaCl(aq)$$

(d) $Na_2S(aq) + AgC_2H_3O_2(aq) \longrightarrow$

$$NaC_2H_3O_2(aq) + Ag_2S(s)$$

4.29 Aqueous solutions of sodium sulfide, Na_2S, and copper nitrate, $Cu(NO_3)_2$, are mixed. A precipitate of copper sulfide, CuS, forms at once. Left behind is a solution of sodium nitrate, $NaNO_3$. Write the net ionic equation for this reaction.

4.30 Silver bromide is "insoluble." What does this mean about the concentrations of Ag^+ and Br^- in a saturated solution of AgBr? Explain why a precipitate of AgBr forms when solutions of the soluble salts $AgNO_3$ and NaBr are mixed.

4.31 Silver bromide (mentioned in the preceding problem) is the chief light-sensitive substance used in the manufacture of photographic film. It can be prepared by mixing solutions of $AgNO_3$ and NaBr. Taking into account that $NaNO_3$ is soluble, write molecular, ionic, and net ionic equations for this reaction.

4.32 If a solution of Na_3PO_4 is added to one containing a calcium salt such as $CaCl_2$, a precipitate of calcium phosphate is formed. Write molecular, ionic, and net ionic equations for the reaction.

Acids and Bases as Electrolytes

4.33 Pure $HClO_4$ is molecular. In water it is a strong acid. Write an equation for its ionization in water.

4.34 HBr is a molecular substance that is a strong acid in water. Write an equation for its ionization in water.

4.35 Hydrazine is a toxic substance that can form when household ammonia is mixed with a bleach such as Clorox. Its formula is N_2H_4, and it is a weak base. Write a chemical equation showing its reaction with water.

4.36 Methylamine, CH_3NH_2, is a weak base with a fishy odor. Write a chemical equation showing its reaction with water.

4.37 Nitrous acid, HNO_2, is a weak acid. Write an equation showing its reaction with water.

4.38 Formic acid, $HCHO_2$, is a weak acid. Write an equation showing its reaction with water.

4.39 Carbonic acid, H_2CO_3, is a weak diprotic acid. Write chemical equations for the equilibria involved in the stepwise ionization of this acid in water.

4.40 Phosphoric acid, H_3PO_4, is a weak acid that undergoes ionization in three steps. Write chemical equations for the equilibria involved in each of these reactions.

Acid–Base Neutralization

4.41 Complete and balance the following equations. For each, write the molecular, ionic, and net ionic equations. (All the products are soluble in water.)
(a) $Ca(OH)_2(aq) + HNO_3(aq) \longrightarrow$
(b) $Al_2O_3(s) + HCl(aq) \longrightarrow$
(c) $Zn(OH)_2(s) + H_2SO_4(aq) \longrightarrow$

4.42 Complete and balance the following equations. For each, write the molecular, ionic, and net ionic equations. (All the products are soluble in water.)
(a) $HC_2H_3O_2(aq) + Mg(OH)_2(s) \longrightarrow$
(b) $HClO_4(aq) + NH_3(aq) \longrightarrow$
(c) $H_2CO_3(aq) + NH_3(aq) \longrightarrow$

Ionic Reactions That Produce Gases

4.43 Write balanced net ionic equations for the reaction between:
(a) $HNO_3(aq) + K_2CO_3(aq)$
(b) $Ca(OH)_2(aq) + NH_4NO_3(aq)$

4.44 Write balanced net ionic equations for the reaction between:
(a) $H_2SO_4(aq) + NaHSO_3(aq)$
(b) $HNO_3(aq) + (NH_4)_2CO_3(aq)$

Predicting Metathesis Reactions

4.45 Explain why the following reactions take place.
(a) $CrCl_3 + 3NaOH \longrightarrow Cr(OH)_3 + 3NaCl$
(b) $ZnO + 2HBr \longrightarrow ZnBr_2 + H_2O$

4.46 Explain why the following reactions take place.
(a) $MnCO_3 + H_2SO_4 \longrightarrow MnSO_4 + H_2O + CO_2$
(b) $Na_2C_2O_4 + 2HNO_3 \longrightarrow 2NaNO_3 + H_2C_2O_4$

4.47 Use the solubility rules to decide which of the following compounds are *soluble* in water.
(a) $Ca(NO_3)_2$ (c) $Ni(OH)_2$ (e) $BaSO_4$
(b) $FeCl_2$ (d) $AgNO_3$ (f) $CuCO_3$

4.48 Predict which of the following compounds are *soluble* in water.
(a) $HgBr_2$ (c) Hg_2Br_2 (e) PbI_2
(b) $Sr(NO_3)_2$ (d) $(NH_4)_3PO_4$ (f) $Pb(C_2H_3O_2)_2$

4.49 Use the solubility rules to decide which of the following compounds are *insoluble* in water.
(a) $AgCl$ (c) $(NH_4)_2CO_3$ (e) $Al(C_2H_3O_2)_3$
(b) $Cr_2(SO_4)_3$ (d) $Ca_3(PO_4)_2$ (f) ZnO

4.50 Predict which of the following compounds are *insoluble* in water.
(a) $Ca(OH)_2$ (d) $CsOH$ (g) K_3PO_4
(b) NiS (e) $MgSO_3$ (h) $Mg(ClO_4)_2$
(c) $PbCl_2$ (f) Fe_2O_3 (i) $CaCO_3$

4.51 Complete and balance the molecular, ionic, and net ionic equations for the following reactions.
(a) $HNO_3 + Cr(OH)_3 \longrightarrow$
(b) $HClO_4 + NaOH \longrightarrow$
(c) $Cu(OH)_2 + HC_2H_3O_2 \longrightarrow$
(d) $ZnO + H_2SO_4 \longrightarrow$

4.52 Complete and balance molecular, ionic, and net ionic equations for the following reactions.
(a) $NaHSO_3 + HBr \longrightarrow$
(b) $(NH_4)_2CO_3 + NaOH \longrightarrow$
(c) $(NH_4)_2CO_3 + Ba(OH)_2 \longrightarrow$
(d) $FeS + HCl \longrightarrow$

4.53 Complete and balance the following reactions, and then write the ionic and net ionic equations. If all ions cancel, indicate that no reaction (N.R.) takes place.
(a) $Na_2SO_3 + Ba(NO_3)_2 \longrightarrow$
(b) $HCHO_2 + K_2CO_3 \longrightarrow$
(c) $NH_4Br + Pb(C_2H_3O_2)_2 \longrightarrow$
(d) $NH_4ClO_4 + Cu(NO_3)_2 \longrightarrow$

4.54 Complete and balance the following reactions, and then write the ionic and net ionic equations. If all ions cancel, indicate that no reaction (N.R.) takes place.
(a) $(NH_4)_2S + NaCl \longrightarrow$
(b) $Cr_2(SO_4)_3 + K_2CO_3 \longrightarrow$
(c) $Sr(OH)_2 + MgCl_2 \longrightarrow$
(d) $Ba(NO_3)_2 + Na_2SO_3 \longrightarrow$

4.55 Choose reactants that would yield the following net ionic equations. Write molecular equations for each.
(a) $HCO_3^- + H^+ \longrightarrow H_2O + CO_2(g)$
(b) $Fe^{2+} + 2OH^- \longrightarrow Fe(OH)_2(s)$
(c) $Ba^{2+} + SO_3^{2-} \longrightarrow BaSO_3(s)$
(d) $2Ag^+ + S^{2-} \longrightarrow Ag_2S(s)$
(e) $ZnO(s) + 2H^+ \longrightarrow Zn^{2+} + H_2O$

**4.56 Suppose that you wished to prepare copper(II) carbonate by a precipitation reaction involving Cu^{2+} and CO_3^{2-}. Which of the following pairs of reactants could you use as solutes?
(a) $Cu(OH)_2 + Na_2CO_3$ (d) $CuCl_2 + K_2CO_3$
(b) $CuSO_4 + (NH_4)_2CO_3$ (e) $CuS + NiCO_3$
(c) $Cu(NO_3)_2 + CaCO_3$

Stoichiometry of Ionic Reactions

4.57 Calculate the number of moles of each of the ions in the following solutions.
(a) 35.0 mL of 1.25 M KOH
(b) 32.3 mL of 0.45 M $CaCl_2$
(c) 50.0 mL of 0.40 M $AlCl_3$

4.58 Calculate the number of moles of each of the ions in the following solutions.
(a) 18.5 mL of 0.40 M $(NH_4)_2CO_3$
(b) 30.0 mL of 0.35 M $Al_2(SO_4)_3$
(c) 60.0 mL of 0.12 M $Zn(C_2H_3O_2)_2$

4.59 Calculate the concentrations of each of the ions in (a) 0.25 M $Cr(NO_3)_2$, (b) 0.10 M $CuSO_4$, (c) 0.16 M Na_3PO_4, (d) 0.075 M $Al_2(SO_4)_3$.

4.60 Calculate the concentrations of each of the ions in (a) 0.060 M $Ca(OH)_2$, (b) 0.15 M $FeCl_3$, (c) 0.22 M $Cr_2(SO_4)_3$, (d) 0.60 M $(NH_4)_2SO_4$.

4.61 A solution contains only the solute Na_3PO_4. What is the molar concentration of this salt if the sodium concentration is 0.21 M?

4.62 A solution contains $Fe_2(SO_4)_3$ as the only solute. The Fe^{3+} concentration in the solution is 0.48 M. What is the molar concentration of the sulfate ion?

4.63 In a solution of $Al_2(SO_4)_3$ the Al^{3+} concentration is 0.12 M. How many grams of $Al_2(SO_4)_3$ are in 50.0 mL of this solution?

4.64 In a solution of $NiCl_2$, the Cl^- concentration is 0.055 M. How many grams of $NiCl_2$ are in 250 mL of this solution?

Solution Stoichiometry

4.65 What is the molarity of an aqueous solution of barium hydroxide if 21.34 mL is exactly neutralized by 20.78 mL of 0.116 M HCl?

4.66 What is the molarity of an aqueous sulfuric acid solution if 12.88 mL is neutralized completely by 26.04 mL of 0.1024 M NaOH?

4.67 How many milliliters of 0.25 M $NiCl_2$ solution are needed to react completely with 20.0 mL of 0.15 M $AgNO_3$ solution? How many grams of AgCl will be formed? The net ionic equation for the reaction is

$$Ag^+(aq) + Cl^-(aq) \longrightarrow AgCl(s)$$

4.68 Epsom salts is $MgSO_4 \cdot 7H_2O$. How many grams of this compound are needed to react completely with 50.0 mL of 0.125 M KOH solution? The net ionic equation for the reaction is

$$Mg^{2+}(aq) + 2OH^-(aq) \longrightarrow Mg(OH)_2(s)$$

4.69 How many milliliters of 0.100 M NaOH are needed to completely neutralize 25.0 mL of 0.250 M H_3PO_4?

4.70 How many grams of baking soda, $NaHCO_3$, are needed to react with 162 mL of stomach acid having an HCl concentration of 0.052 M?

4.71 Consider the reaction of aluminum chloride with silver acetate. How many milliliters of 0.250 M $AlCl_3$ would be needed to react completely with 20.0 mL of 0.500 M $AgC_2H_3O_2$ solution? The net ionic equation for the reaction is

$$Ag^+(aq) + Cl^-(aq) \longrightarrow AgCl(s)$$

4.72 How many milliliters of ammonium sulfate solution having a concentration of 0.250 M are needed to react completely with 50.0 mL of 1.00 M NaOH solution? The net ionic equation for the reaction is

$$NH_4^+(aq) + OH^-(aq) \longrightarrow NH_3(g) + H_2O$$

4.73 Suppose that 4.00 g of solid Fe_2O_3 is added to 25.0 mL of 0.500 M HCl solution. What will the concentration of the Fe^{3+} be when all the HCl has reacted? What mass of Fe_2O_3 will not have reacted?

4.74 Suppose 3.50 g of solid $Mg(OH)_2$ is added to 30.0 mL of 0.500 M H_2SO_4 solution. What will the concentration of Mg^{2+} be when all of the acid has been neutralized? How many grams of $Mg(OH)_2$ will not have dissolved?

4.75 Suppose that 25.0 mL of 0.440 M NaCl is added to 25.0 mL of 0.320 M $AgNO_3$.
(a) How many moles of AgCl would precipitate?
(b) What would be the concentrations of each of the ions in the reaction mixture after the reaction?

4.76 A mixture is prepared by adding 25.0 mL of 0.185 M Na_3PO_4 to 34.0 mL of 0.140 M $Ca(NO_3)_2$.
(a) What mass of $Ca_3(PO_4)_2$ will be formed?
(b) What will be the concentrations of each of the ions in the mixture after reaction?

Titrations and Chemical Analyses

4.77 A certain lead ore contains the compound $PbCO_3$. A sample of the ore weighing 1.526 g was treated with nitric acid, which dissolved the $PbCO_3$. The resulting solution was filtered from undissolved rock and treated with Na_2SO_4. This gave a precipitate of $PbSO_4$ which was dried and found to weigh 1.081 g. What is the percentage by mass of lead in the ore?

4.78 An ore of barium contains $BaCO_3$. A 1.542 g sample of the ore was treated with HCl to dissolve the $BaCO_3$. The resulting solution was filtered to remove insoluble material and then treated with H_2SO_4 to precipitate $BaSO_4$. The precipitate was filtered, dried, and found to weigh 1.159 g. What is the percentage by mass of barium in the ore?

4.79 In a titration, 23.25 mL of 0.105 M NaOH was needed to react with 21.45 mL of HCl solution. What is the molarity of the acid?

4.80 A 12.5-mL sample of vinegar, containing acetic acid, $HC_2H_3O_2$, was titrated using 0.504 M NaOH solution. The titration required 20.65 mL of the base.
(a) What was the molar concentration of acetic acid in the vinegar?
(b) Assuming the density of the vinegar to be 1.0 g mL^{-1}, what was the percent (by mass) of acetic acid in the vinegar?

4.81 Lactic acid, $HC_3H_5O_3$, is a monoprotic acid that forms when milk sours. An 18.5 mL sample of a solution of lactic acid required 17.25 mL of 0.155 M NaOH to reach an end point in a titration. How many moles of lactic acid were in the sample?

4.82 Ascorbic acid (vitamin C) is a diprotic acid having the formula $H_2C_6H_6O_6$. A sample of a vitamin supplement was analyzed by titrating a 0.1000 g sample dissolved in water with 0.0200 M NaOH. A volume of 15.20 mL of the base was required to completely neutralize the ascorbic acid. What was the percentage by mass of ascorbic acid in the sample?

4.83 Magnesium sulfate forms a hydrate known as *Epsom salts*. A student dissolved 1.24 g of this hydrate in water and added a $BaCl_2$ solution until the precipitation of $BaSO_4$ was complete. The precipitate was filtered, dried, and found to weigh 1.174 g. Determine the formula for Epsom salts.

4.84 A sample of iron chloride weighing 0.300 g was dissolved in water, and the solution was treated with $AgNO_3$ solution to precipitate the chloride as AgCl. After precipitation was complete, the AgCl was filtered, dried, and found to weigh 0.678 g. Determine the empirical formula of the iron chloride.

4.85 To a mixture of NaCl and Na_2CO_3 with a mass of 1.243 g was added 50.00 mL of 0.240 M HCl (an excess of HCl). The mixture was warmed to expel all of the CO_2 and then the unreacted HCl was titrated with 0.100 M NaOH. The titration required 22.90 mL of the NaOH solution. What was the percentage by mass of NaCl in the original mixture of NaCl and Na_2CO_3?

4.86 A mixture was known to contain both KNO_3 and K_2SO_3. To 0.486 g of the mixture, dissolved in enough water to give 50.00 mL of solution, was added 50.00 mL of 0.150 M HCl (an excess of HCl). The reaction mixture was heated to drive off all of the SO_2, and then 25.00 mL of the reaction mixture was titrated with 0.100 M KOH. The titration required 13.11 mL of the KOH solution to reach an end point. What was the percentage by mass of K_2SO_3 in the original mixture of KNO_3 and K_2SO_3?

ADDITIONAL EXERCISES

4.87 Classify each of the following as a strong electrolyte, weak electrolyte, or nonelectrolyte.
(a) KCl
(b) glycerin
(c) NaOH
(d) $C_{12}H_{22}O_{11}$ (sucrose, or table sugar)
(e) $HC_2H_3O_2$ (acetic acid)
(f) CH_3OH (methyl alcohol)
(g) H_2SO_4
(h) NH_3

4.88 Write the net ionic equations for the following reactions. Include the designations for the states, solid, liquid, gas, or aqueous solution. (For any substance soluble in water, assume that it is used as its aqueous solution.)
(a) $CaCO_3 + HNO_3 \longrightarrow Ca(NO_3)_2 + CO_2 + H_2O$
(b) $CaCO_3 + H_2SO_4 \longrightarrow CaSO_4 + CO_2 + H_2O$
(c) $FeS + HBr \longrightarrow H_2S + FeBr_2$
(d) $KOH + SnCl_2 \longrightarrow KCl + Sn(OH)_2$

4.89 Examine each pair of substances, determine if they will react together, and then write the molecular and net ionic equations for any reactions that you predict.
(a) $Na_2S(aq)$ and $H_2SO_4(aq)$
(b) $LiHCO_3(s)$ and $HNO_3(aq)$
(c) $(NH_4)_3PO_4(aq)$ and $KOH(aq)$
(d) $K_2SO_3(aq)$ and $HCl(aq)$
(e) $BaCO_3(s)$ and $HBr(aq)$
(f) $CaCl_2(aq)$ and $HNO_3(aq)$

*4.90 A 0.113 g sample of a mixture of sodium chloride and sodium nitrate was dissolved in water, and enough silver nitrate solution was added to precipitate all of the chloride ion. The mass of silver chloride obtained was 0.277 g. What mass percentage of the sample was sodium chloride?

*4.91 Aspirin is a monoprotic acid called acetylsalicylic acid. Its formula is $HC_9H_7O_4$. A certain pain reliever was analyzed for aspirin by dissolving 0.250 g of it in water and titrating it with 0.0300 M KOH solution. The titration required 29.40 mL of base. What is the percentage by weight of aspirin in the drug?

*4.92 In an experiment, 40.0 mL of 0.270 M $Ba(OH)_2$ was mixed with 25.0 mL of 0.330 M $Al_2(SO_4)_3$.
(a) Write the net ionic equation for the reaction that takes place.
(b) What is the total mass of precipitate that forms?
(c) What are the molar concentrations of the ions that remain in the solution after the reaction is complete?

*4.93 Qualitative analysis of an unknown acid found only carbon, hydrogen, and oxygen. Quantitative analysis gave the following data. A 10.46 mg sample, when burned in oxygen, gave 22.17 mg CO_2 and 3.40 mg H_2O. Its formula weight was determined to be 166. When a 0.1680 g sample of the acid was titrated with 0.1250 M NaOH, the end point was reached after 16.18 mL of the base had been added.
(a) Calculate the percentage composition of the acid.
(b) What is its empirical formula?
(c) What is its molecular formula?
(d) Is the acid mono-, di-, or triprotic?

In this chapter we discuss a variety of chemical reactions that occur by the transfer of electrons between one substance and another. These reactions have many practical applications, including serving as the source of power in batteries that permit us to operate portable laptop computers.

Oxidation–Reduction Reactions

This Chapter in Context In the preceding chapter you learned about some important reactions that take place in aqueous solutions. This chapter expands on that knowledge with a discussion of a class of reactions that can be viewed as involving the transfer of one or more electrons from one reactant to another. This is a very broad class of reactions with many practical examples that range from combustion to batteries to the metabolism of foods by our bodies. We will introduce you to some of these reactions in this chapter, but we will have to wait until a later time to discuss some of the others.

5.1 Oxidation–Reduction Reactions

Among the first reactions studied by early scientists were those that involved oxygen. The combustion of fuels and the reactions of metals with oxygen to give oxides were described by the word *oxidation*. The removal of oxygen from metal oxides to give the metals in their elemental forms was described as *reduction*.

As time went by, scientists realized that reactions involving oxygen were actually special cases of a much more general phenomenon, one in which electrons are transferred from one substance to another. Collectively, electron transfer reactions came to be called **oxidation–reduction reactions,** or simply **redox reactions.** The term **oxidation** was applied to the loss of electrons by one reactant, and **reduction** to the gain of electrons by another. For example, the reaction between sodium and chlorine involves a loss of electrons by sodium (*oxidation* of sodium) and a gain of electrons by chlorine (*reduction* of chlorine). We can write these changes in equation form including the electrons, which we represent by the symbol e^-.

$$Na \longrightarrow Na^+ + e^- \qquad \text{(oxidation)}$$

$$Cl_2 + 2e^- \longrightarrow 2Cl^- \qquad \text{(reduction)}$$

We say that sodium is oxidized and chlorine is reduced.

Oxidation and reduction always occur together. No substance is ever oxidized unless something else is reduced. Otherwise, electrons would appear as a product of the reaction, and this is never observed. During a redox reaction, then, some substance must accept the electrons that another substance loses. This electron-accepting substance is called the **oxidizing agent** because it helps something else to be oxidized. The substance that supplies the electrons is called the **reducing agent** because it helps something else to be reduced. Sodium is a reducing agent, for example, when it supplies electrons to chlorine. In the process, sodium is oxidized. Chlorine is an oxidizing agent when it accepts electrons from the sodium, and when that happens, chlorine is reduced to chloride ion. One way to remember this is by the following summary:

The *oxidizing agent* causes oxidation to occur by accepting electrons (it gets reduced). The *reducing agent* causes reduction by supplying electrons (it gets oxidized).

> The reducing agent is the substance that is oxidized.
> The oxidizing agent is the substance that is reduced.

Redox reactions are very common. They occur in batteries, which are built so that the electrons transferred can pass through some external circuit where they are able to light a flashlight or power a pocket calculator. The metabolism of foods, which supplies our bodies with energy, also occurs by a series of redox reactions. And ordinary household bleach works by oxidizing substances that stain fabrics, making them colorless or easier to remove from the fabric. (See Facets of Chemistry 5.1.)

Liquid bleach contains the hypochlorite ion, OCl^-, as the oxidizing agent.

EXAMPLE 5.1
Identifying Oxidation–Reduction

Fireworks display over Washington, D.C.

The bright light produced by the reaction between magnesium and oxygen often is used in fireworks displays. The product of the reaction is magnesium oxide, MgO, an ionic compound.

$$2Mg + O_2 \longrightarrow 2MgO$$

Which element is oxidized and which is reduced? What are the oxidizing and reducing agents?

Analysis: To answer this question, we have to determine how electrons are transferred when magnesium and oxygen react. Then we apply the definitions.

Solution: When magnesium reacts with oxygen, its atoms change to Mg^{2+} ions, which means that the magnesium atoms must lose electrons.

$$Mg \longrightarrow Mg^{2+} + 2e^-$$

Because Mg loses electrons, it is oxidized. Since Mg is oxidized, it must be the reducing agent.

When oxygen reacts with magnesium, it forms O^{2-} ions, which means that the atoms in O_2 must gain electrons.

$$O_2 + 4e^- \longrightarrow 2O^{2-}$$

Because O_2 gains electrons, it is reduced and must be the oxidizing agent.

Practice Exercise 1

Identify the substances oxidized and reduced and the oxidizing and reducing agents in the reaction of aluminum and chlorine to form aluminum chloride. ◆

Facets of Chemistry 5.1

Chlorine and Laundry Bleach

Although the combustion of fuels is probably the best known example of a practical application of redox reactions, there are other lesser-known reactions that are almost as important. Each year huge amounts of chlorine, manufactured from salt, are used in purifying municipal drinking water supplies and treating wastewater. Chlorine is a powerful oxidizing agent, and its function in water purification is to kill bacteria. It does this by oxidizing the bacteria and causing their death.

Another use for chlorine and the chlorine-containing anion ClO^- is as a bleach. Ordinary chlorine bleaches such as Clorox consist of a dilute solution of sodium hypochlorite, $NaClO$. The hypochlorite ion is a strong oxidizing agent and performs its bleaching action by destroying dyes and other colored matter. In the paper and cotton processing industries, this bleaching function has been accomplished by using chlorine directly.

Today, industry is moving away from the use of chlorine as a bleach because of environmental concerns. It is feared that the products of the reaction of chlorine with organic substances may be toxic, and in fact, many well-known insecticides are chlorine-containing compounds. Even more disturbing are reports that chlorine-containing organic compounds mimic the behavior of the hormone estrogen.

"Chlorine" bleach, $NaClO(aq)$, destroys dyes by oxidizing them to colorless products.

Oxidation Numbers

The reaction of magnesium with oxygen in the preceding example is clearly a redox reaction. However, not all reactions with oxygen produce ionic products. For example, sulfur reacts with oxygen to give sulfur dioxide, SO_2, which is molecular. Nevertheless, it is convenient to also view this as a redox reaction, but to do so we must make some changes in the way we define oxidation and reduction. To do this, chemists have developed a bookkeeping system called **oxidation numbers** which provides a way to keep tabs on electron transfers.

The oxidation number of an element in a particular compound is assigned according to a set of rules, which are described below. For simple, monatomic ions in a compound such as NaCl, the oxidation numbers are the same as the charges on the ions, so in NaCl the oxidation number of Na^+ is +1 and the oxidation number of Cl^- is −1. The real value of oxidation numbers is that they can also be assigned to atoms in molecular compounds. In such cases, it is important to realize that the oxidation number does not actually equal a charge on an atom. To be sure not to confuse oxidation numbers with actual electrical charges, we will specify the sign *before* the number when writing oxidation numbers, and *after* the number when writing electrical charges. Thus, a sodium ion has a charge of 1+ and an oxidation number of +1.

A term that is frequently used interchangeably with *oxidation number* is **oxidation state.** In NaCl, for example, sodium has an oxidation number of +1 and is said to be "in the +1 oxidation state." Similarly, the chlorine in NaCl is said to be in the −1 oxidation state. There are times when it is convenient to specify the oxidation state of an element when its name is written out. This is done by writing the oxidation number as a Roman numeral in parentheses after the name of the element. For example, "iron(III)" means iron in the +3 oxidation state.

We can use these new terms now to redefine a redox reaction.

> A **redox reaction** is a chemical reaction in which changes in oxidation numbers occur.

We will find it easy to follow redox reactions by taking note of the changes in oxidation numbers. To do this, however, we must be able to assign oxidation numbers to atoms in a quick and simple way.

Assignment of Oxidation Numbers

We can use some basic knowledge learned earlier plus the set of rules below to determine the oxidation numbers of the atoms in almost any compound.

Assigning oxidation numbers

Rule 1 means that O in O_2, P in P_4, and S in S_8 all have oxidation numbers of zero.

> **Rules for Assigning Oxidation Numbers**
> 1. The oxidation number of any free element (an element not combined chemically with a different element) is zero, regardless of how complex its molecules might be.
> 2. The oxidation number for any simple, monatomic ion (e.g., Na^+ or Cl^-) is equal to the charge on the ion.
> 3. The sum of all the oxidation numbers of the atoms in a molecule or polyatomic ion must equal the charge on the particle.
> 4. In its compounds, fluorine has an oxidation number of −1.
> 5. In its compounds, hydrogen has an oxidation number of +1.
> 6. In its compounds, oxygen has an oxidation number of −2.

As you will see shortly, occasionally there is a conflict between two of these rules. When this happens, *we apply the rule with the lower number and ignore the conflicting rule.*

In addition to these basic rules, there is some other chemical knowledge you will need to remember. Recall that we can use the periodic table to help us remember the charges on certain ions of the elements. For instance, all the metals in Group IA form ions with a 1+ charge, and all those in Group IIA form ions with a 2+ charge. This means that when we find sodium in a compound, we can assign it an oxidation number of +1 because its simple ion, Na^+, has a charge of 1+. (We have applied Rule 2.) Similarly, calcium in a compound exists as Ca^{2+} and has an oxidation number of +2.

In binary ionic compounds with metals, the nonmetals have oxidation numbers equal to the charges on their anions (Table 2.4 on page 73). For example, the compound Fe_2O_3 contains the oxide ion, O^{2-}, which is assigned an oxidation number of −2. Similarly, Mg_3P_2 contains the phosphide ion, P^{3-}, which has an oxidation number of −3.

The numbered rules given above usually come into play when an element is capable of having more than one oxidation state. For example, you learned in Chapter 2 that transition metals can form more than one ion. Iron, for example, forms Fe^{2+} and Fe^{3+} ions, so in an iron compound we have to use the rules to figure out which iron ion is present. Similarly, when nonmetals are combined with hydrogen and oxygen in compounds or polyatomic ions, their oxidation numbers can vary and must be calculated using the rules.

With this as background, let's look at some examples that illustrate how we apply the rules, especially when they don't explicitly cover all the atoms in a formula.

Molybdenum disulfide, MoS_2, has a structure that allows it to be used as a dry lubricant, much like graphite. What are the oxidation numbers of the atoms in MoS_2?

Solution: This is a binary compound of a metal and a nonmetal, so for the purposes of assigning oxidation numbers we assume it to contain ions. Molybdenum is a transition metal, so there is no simple rule to tell what its ions are. However, sulfur is a nonmetal in Group VIA, and its ion is S^{2-}. (If necessary, you might review Table 2.4 on page 73, which summarizes the ions formed by the representative elements.) Once we know the oxidation number of sulfur, we can use Rule 3 (the summation rule) to determine the oxidation number of molybdenum.

$$\begin{array}{llll} S & (2 \text{ atoms}) \times (-2) = -4 & (\text{Rule 2}) \\ \underline{Mo} & \underline{(1 \text{ atom}) \times (x) = \quad x} \\ & \quad\quad\quad\quad\quad Sum = \quad 0 & (\text{Rule 3}) \end{array}$$

The value of x must be $+4$ for the sum to be zero. Therefore, the oxidation numbers are

$$Mo = +4 \quad \text{and} \quad S = -2$$

EXAMPLE 5.2

Assigning Oxidation Numbers

Molybdenum disulfide lubricant.

Chlorite ion, ClO_2^-, has been shown to be a potent disinfectant, and solutions of it are sometimes used to disinfect air conditioning systems in cars. What are the oxidation numbers of chlorine and oxygen in this ion?

Solution: Examining our rules, we see we have one for oxygen, so let's use it and the summation rule to figure out the oxidation number of chlorine.

$$\begin{array}{llll} O & (2 \text{ atoms}) \times (-2) = -4 & (\text{Rule 6}) \\ \underline{Cl} & \underline{(1 \text{ atom}) \times (x) = \quad x} \\ & \quad\quad\quad\quad\quad Sum = -1 & (\text{Rule 3}) \end{array}$$

For the sum to be -1, the value of x must be $+3$, which is the oxidation number of the chlorine.

$$Cl = +3 \quad \text{and} \quad O = -2$$

EXAMPLE 5.3

Assigning Oxidation Numbers

Determine the oxidation numbers of the atoms in hydrogen peroxide, H_2O_2, a common antiseptic purchased in pharmacies.

Solution: This time we are going to have a conflict between two of the rules. Rule 5 tells us to assign an oxidation number of $+1$ to hydrogen, and Rule 6 tells us to give oxygen an oxidation number of -2. Both of these can't be correct because the sum must be zero (Rule 3).

$$\begin{array}{llll} H & (2 \text{ atoms}) \times (+1) = +2 & (\text{Rule 5}) \\ \underline{O} & \underline{(2 \text{ atoms}) \times (-2) = -4} & (\text{Rule 6}) \\ & \quad\quad\quad\quad\quad Sum \neq \quad 0 \end{array}$$

As mentioned above, when there is a conflict between the rules, we ignore the one with the higher number that causes the conflict. Rule 6 is the one with the higher number, so we will ignore it this time and just apply Rules 3 and 5.

EXAMPLE 5.4

Assigning Oxidation Numbers

Hydrogen peroxide destroys bacteria by oxidizing them.

$$H \qquad (\text{2 atoms}) \times (+1) = +2 \qquad (\text{Rule 5})$$

$$O \qquad \underline{(\text{2 atoms}) \times (x) = 2x }$$

$$\text{Sum} = 0 \qquad (\text{Rule 3})$$

For the sum to be zero, $2x = -2$, so $x = -1$. Therefore, in this compound, oxygen has an oxidation number of -1.

$$H = +1 \qquad O = -1$$

EXAMPLE 5.5
Assigning Oxidation Numbers

Calcium reacts with hydrogen to form the compound CaH_2. What are the oxidation numbers of the atoms in this compound?

Solution: Calcium is a metal in Group IIA, so we expect it to exist as Ca^{2+} in its compounds. According to Rule 2, then, the oxidation number of Ca must be $+2$. Rule 3 demands that the sum of the oxidation numbers be equal to the charge on the formula, which is zero. We also have a rule for hydrogen (Rule 5), but we can't apply it here because it would require that hydrogen be $+1$. If the oxidation number of hydrogen were equal to $+1$, Rule 3 would be violated.

$$Ca \qquad (\text{1 atom}) \times (+2) = +2 \qquad (\text{Rule 2})$$

$$H \qquad \underline{(\text{2 atoms}) \times (+1) = +2 }$$

$$\text{Sum} \neq 0 \qquad (\text{violation of Rule 3})$$

Rule 3 is higher up than Rule 5, so Rule 3 takes precedence and we will ignore Rule 5. This gives

$$Ca \qquad (\text{1 atom}) \times (+2) = +2 \qquad (\text{Rule 2})$$

$$H \qquad \underline{(\text{2 atoms}) \times (x) = 2x }$$

$$\text{Sum} = 0 \qquad (\text{Rule 3})$$

For the sum to be zero, $2x$ must equal -2, so x must equal -1. The oxidation number of hydrogen is -1 in this compound.

Sometimes, oxidation numbers calculated by the rules have fractional values, as illustrated in the next example.

EXAMPLE 5.6
Assigning Oxidation Numbers

The air bags used as safety devices in modern autos are inflated by the very rapid decomposition of the ionic compound sodium azide, NaN_3. The reaction gives elemental sodium and gaseous nitrogen. What is the average oxidation number of the nitrogen in sodium azide?

Solution: In ionic compounds, sodium exists as the ion Na^+. Therefore, azide ion (N_3^-) must have a charge of $1-$ (because NaN_3 is neutral).

$$Na \qquad (\text{1 atom}) \times (+1) = +1 \qquad (\text{Rule 5})$$

$$N \qquad \underline{(\text{3 atoms}) \times (x) = 3x }$$

$$\text{Sum} = 0 \qquad (\text{Rule 3})$$

The sum of the oxidation numbers of the three nitrogen atoms in this ion must add up to -1, so each nitrogen must have an oxidation number of $-\frac{1}{3}$.

Many compounds contain familiar polyatomic ions such as ammonium (NH_4^+), sulfate (SO_4^{2-}), nitrate (NO_3^-), and acetate ($C_2H_3O_2^-$). The charges on these ions can be considered to represent the *net oxidation number* of the atoms that make up the ion. We can use this sometimes to help assign oxidation numbers, as shown in Example 5.7.

The charge on the polyatomic ion equals the sum of the oxidation numbers of its atoms.

EXAMPLE 5.7

Assigning Oxidation Numbers

What is the oxidation number of chromium in the compound $Cr(NO_3)_3$?

Solution: The nitrate ion has a charge of $1-$, which we can take to be its net oxidation number. Then we apply the summation rule.

$$
\begin{array}{lll}
Cr & (1\text{ atom}) \times (x) = & x \\
NO_3^- & (3\text{ ions}) \times (-1) = & -3 \\
\hline
& \text{Sum} = & 0 \quad \text{(Rule 3)}
\end{array}
$$

Obviously, the oxidation number of chromium is $+3$.

Practice Exercise 2

Assign oxidation numbers to each atom in (a) $NiCl_2$, (b) Mg_2TiO_4, (c) $K_2Cr_2O_7$, (d) HPO_4^{2-}, (e) $V(C_2H_3O_2)_3$. ◆

Practice Exercise 3

Iron forms an oxide with the formula Fe_3O_4 that contains both Fe^{2+} and Fe^{3+} ions. What is the *average* oxidation number of iron in this oxide? ◆

Oxidation Numbers and Redox Reactions

Oxidation numbers can be used in several ways. One is to define oxidation and reduction in the most comprehensive manner, as follows.

Oxidation is an increase in oxidation number.
Reduction is a decrease in oxidation number.

Let's see how these definitions apply to the reaction of hydrogen with chlorine. To avoid ever confusing oxidation numbers with actual electrical charges, we will write oxidation numbers directly above the chemical symbols of the elements.

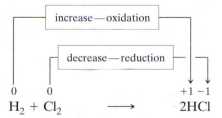

Notice we have assigned the atoms in H_2 and Cl_2 oxidation numbers of zero, in accord with Rule 1. The changes in oxidation number tell us that hydrogen is oxidized and chlorine is reduced.

EXAMPLE 5.8

Using Oxidation Numbers to Follow Redox Reactions

Identify the substance oxidized and the substance reduced as well as the oxidizing and reducing agents in the reaction

$$2KCl + MnO_2 + 2H_2SO_4 \longrightarrow K_2SO_4 + MnSO_4 + Cl_2 + 2H_2O$$

Solution: To identify redox changes using oxidation numbers, we first must assign oxidation numbers to each atom on both sides of the equation. Following the rules, we get

$$\overset{+1-1}{2KCl} + \overset{+4\ -2}{MnO_2} + \overset{+1+6-2}{2H_2SO_4} \longrightarrow \overset{+1+6-2}{K_2SO_4} + \overset{+2+6-2}{MnSO_4} + \overset{0}{Cl_2} + \overset{+1-2}{2H_2O}$$

Next we look for changes, keeping in mind that an increase in oxidation number is oxidation and a decrease is reduction.

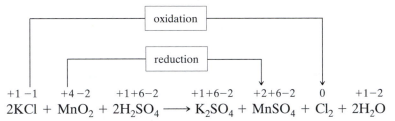

$$\overset{+1\ -1}{2KCl} + \overset{+4-2}{MnO_2} + \overset{+1+6-2}{2H_2SO_4} \longrightarrow \overset{+1+6-2}{K_2SO_4} + \overset{+2+6-2}{MnSO_4} + \overset{0}{Cl_2} + \overset{+1-2}{2H_2O}$$

Thus the Cl in KCl is oxidized and the Mn in MnO_2 is reduced. The reducing agent is KCl and the oxidizing agent is MnO_2.

5.2 Balancing Redox Equations by the Ion–Electron Method

Many redox reactions take place in aqueous solution and many of these involve ions; they are ionic reactions. An example is the reaction of laundry bleach with substances in the wash water. The active ingredient in the bleach is hypochlorite ion, OCl^-, which is the oxidizing agent in these reactions. To study redox reactions, it is often helpful to write ionic and net ionic equations, just as we did in our analysis of metathesis reactions in Chapter 4. Balancing net ionic equations for redox reactions is especially easy if we follow a procedure called the ion–electron method.

Basic Principles of the Ion–Electron Method

In the **ion–electron method,** we divide the oxidation and reduction processes into individual equations called **half-reactions** that are balanced separately. Then we combine the balanced half-reactions to obtain the fully balanced net ionic equation.

To illustrate the method, let's balance the net ionic equation for the reaction of iron(III) chloride, $FeCl_3$, with tin(II) chloride, $SnCl_2$, which changes the iron to Fe^{2+} and the tin to Sn^{4+}. In the reaction the chloride ion is unaffected—it's a spectator ion.

Both $FeCl_3$ and $SnCl_2$ are soluble salts.

We begin by writing a **skeleton equation,** which shows only the ions (or sometimes molecules, too) involved in the reaction. The reactants are Fe^{3+} and Sn^{2+}, and the products are Fe^{2+} and Sn^{4+}. The skeleton equation is therefore

$$Fe^{3+} + Sn^{2+} \longrightarrow Fe^{2+} + Sn^{4+}$$

The next step is to divide the equation into two half-reactions. In this case, we have one for tin and one for iron. We choose one of the reactants, let's say Sn^{2+}, and write it at the left of an arrow. On the right, we write what Sn^{2+} changes to, which is Sn^{4+}. This gives us the beginnings of one half-reaction. For the second half-reaction, we write the other reactant, Fe^{3+}, on the left and the other product, Fe^{2+}, on the right.

$$Sn^{2+} \longrightarrow Sn^{4+}$$

$$Fe^{3+} \longrightarrow Fe^{2+}$$

Next, we balance the half-reactions so that each obeys *both* criteria for a balanced ionic equation: *both atoms and charge have to balance.* Obviously, the atoms are already balanced in each equation. The charge, however, is not. *To bring the charge into balance, we add electrons.* For the first half-reaction, 2 electrons are added to the right, so the net charge on both sides will be 2+. In the second half-reaction, one electron is added to the left, which makes the net charge on both sides equal to 2+.

$$Sn^{2+} \longrightarrow Sn^{4+} + 2e^-$$

$$Fe^{3+} + e^- \longrightarrow Fe^{2+}$$

In any redox reaction, the number of electrons gained always equals the number lost. For the electron gain to equal the electron loss, the second reaction has to occur twice each time the first reaction occurs once. We make this so by multiplying each of the coefficients in the second equation by 2.

$$Sn^{2+} \longrightarrow Sn^{4+} + 2e^-$$

$$2(Fe^{3+} + e^- \longrightarrow Fe^{2+})$$

This gives

$$Sn^{2+} \longrightarrow Sn^{4+} + 2e^-$$

$$2Fe^{3+} + 2e^- \longrightarrow 2Fe^{2+}$$

Finally, we combine the balanced half-reactions by adding them together. The result is

$$Sn^{2+} + 2Fe^{3+} + 2e^- \longrightarrow Sn^{4+} + 2Fe^{2+} + 2e^-$$

Dropping $2e^-$ from both sides gives us the final balanced equation.

$$Sn^{2+} + 2Fe^{3+} \longrightarrow Sn^{4+} + 2Fe^{2+}$$

Notice that the charge *and* the number of atoms are now balanced.

Practice Exercise 4

Balance the following equation by the ion–electron method.

$$SnCl_3^-(aq) + HgCl_2(aq) \longrightarrow SnCl_6^{2-}(aq) + Hg_2Cl_2(s)$$

(*Hint:* In this equation, you will have to use Cl^- to help balance the half-reactions.) ◆

Reactions Involving H^+ and OH^-

In many redox reactions in aqueous solutions, H^+ or OH^- ions play an important role, as do water molecules. For example, when solutions of $K_2Cr_2O_7$ and $FeSO_4$ are mixed, the acidity of the mixture decreases as dichromate ion, $Cr_2O_7^{2-}$, oxidizes Fe^{2+}. This is because the reaction uses up H^+ as a reactant and produces H_2O as a product. In other reactions, OH^- is consumed, while in still others H_2O is a reactant. Another fact is that in many cases the products (or even the reactants) of a redox reaction will differ depending on the acidity of the solution. For example, in an acidic solution MnO_4^- is reduced to Mn^{2+} ion,

Each half-reaction must have the same elements on both sides. If Sn is on one side, it must also be on the other side. Keep this in mind when dividing a skeleton equation into half-reactions.

Electrons must be added to whichever side of the half-reaction is more positive (or less negative).

In an oxidation the electrons appear as a product in the half-reaction, and in reduction they appear as a reactant. That's the way we identify which is oxidation and which is reduction when we balance an equation this way. In this example, Sn^{2+} is oxidized and Fe^{3+} is reduced.

A solution of $K_2Cr_2O_7$ being added to an acidic solution containing Fe^{2+}.

but in a neutral or slightly basic solution, the reduction product is insoluble MnO_2.

Because of these factors, redox reactions are generally carried out in solutions containing a substantial excess of either acid or base, so before we can apply the ion–electron method, we have to know whether the reaction occurs in an acidic or a basic solution. (This information will always be given to you in this book.)

▶**Chemistry in Practice**◀ A Breathalyzer test for alcohol makes use of the oxidation of alcohol by dichromate ion. The dichromate ion is orange, but its reduction product is green Cr^{3+} ion. The breath of a person who is intoxicated contains alcohol vapor, which passes through an acidic solution containing $Cr_2O_7^{2-}$ when the person blows through the Breathalyzer device. Any alcohol in the exhaled air is oxidized, which is signaled by the appearance of green Cr^{3+}. The more alcohol the person has consumed, the greater the intensity of the green color. ◆

Balancing Redox Equations for Acidic Solutions

As you just learned, $Cr_2O_7^{2-}$ reacts with Fe^{2+} in an acidic solution to give Cr^{3+} and Fe^{3+} as products. This information gives the skeleton equation,

$$Cr_2O_7^{2-} + Fe^{2+} \longrightarrow Cr^{3+} + Fe^{3+}$$

We then proceed through the following steps to find the balanced equation.

As we balance the equation, the ion–electron method will tell us how H^+ and H_2O are involved in the reaction. We don't need to know this information in advance.

Step 1. *Divide the skeleton equation into half-reactions.*
We create the beginnings of two half-reactions. Except for hydrogen and oxygen, the same elements must appear on both sides of a given half-reaction.

$$Cr_2O_7^{2-} \longrightarrow Cr^{3+}$$
$$Fe^{2+} \longrightarrow Fe^{3+}$$

Step 2. *Balance atoms other than H and O.*
There are two Cr atoms on the left and only one on the right, so we place a coefficient of 2 in front of Cr^{3+}. The second half-reaction is already balanced in terms of atoms, so nothing need be done to it.

$$Cr_2O_7^{2-} \longrightarrow 2Cr^{3+}$$
$$Fe^{2+} \longrightarrow Fe^{3+}$$

Many students tend to forget this step. If you do, you may end up in trouble later on.

Step 3. *Balance oxygen by adding H_2O to the side that needs O.*
There are seven oxygen atoms on the left of the first half-reaction and none on the right. Therefore, we add $7H_2O$ to the right side of the first half-reaction to balance the oxygens. (There is no oxygen imbalance in the second half-reaction, so there's nothing to do there.)

We use H_2O, not O or O_2, to balance oxygen atoms because H_2O is what is actually present in the solution.

$$Cr_2O_7^{2-} \longrightarrow 2Cr^{3+} + 7H_2O$$
$$Fe^{2+} \longrightarrow Fe^{3+}$$

Step 4. *Balance hydrogen by adding H^+ to the side that needs H.*
After adding the water, we see that the first half-reaction has 14 hydrogens on the right and none on the left. To balance these, we add $14H^+$ to the left side of this equation. When you do this step (or others) *be careful to write the charges*

By balancing oxygen with H_2O, we have created an imbalance in H. We are allowed to correct this by adding H^+ because the solution is acidic.

on the ions. If they are omitted, you will not obtain a balanced equation in the end.

$$14H^+ + Cr_2O_7^{2-} \longrightarrow 2Cr^{3+} + 7H_2O$$

$$Fe^{2+} \longrightarrow Fe^{3+}$$

Now each half-reaction is balanced in terms of atoms. Next we will balance the charge.

Step 5. *Balance the charge by adding electrons.*
First we compute the net electrical charge on each side. For the first half-reaction we have

$$\underbrace{14H^+ + Cr_2O_7^{2-}}_{\text{Net charge} = (14+) + (2-) = 12+} \longrightarrow \underbrace{2Cr^{3+} + 7H_2O}_{\text{Net charge} = 2(3+) + 0 = 6+}$$

This is a critical step. Be sure you understand how to perform it.

The algebraic difference between the net charges on the two sides equals the number of electrons that must be added to the more positive (less negative) side. In this instance, we must add $6e^-$ to the left side of the equation.

$$6e^- + 14H^+ + Cr_2O_7^{2-} \longrightarrow 2Cr^{3+} + 7H_2O$$

Check the net charge after adding e^-; it must be the same on both sides.

This half-reaction is now complete; it is balanced in terms of both atoms and charge.

To balance the other half-reaction, we add one electron to the right.

$$Fe^{2+} \longrightarrow Fe^{3+} + e^-$$

Step 6. *Make the number of electrons gained equal to the number lost and then add the two half-reactions.*
At this point we have the two balanced half-reactions

$$6e^- + 14H^+ + Cr_2O_7^{2-} \longrightarrow 2Cr^{3+} + 7H_2O$$

$$Fe^{2+} \longrightarrow Fe^{3+} + e^-$$

Six electrons are gained in the first, but only one is lost in the second. Therefore, before combining the two equations we multiply all of the coefficients of the second one by 6.

$$6e^- + 14H^+ + Cr_2O_7^{2-} \longrightarrow 2Cr^{3+} + 7H_2O$$

$$6(Fe^{2+} \longrightarrow Fe^{3+} + e^-)$$

(Sum) $\quad 6e^- + 14H^+ + Cr_2O_7^{2-} + 6Fe^{2+} \longrightarrow$

$$2Cr^{3+} + 7H_2O + 6Fe^{3+} + 6e^-$$

Because we know the electrons will cancel, we really don't have to carry them down into the combined equation. We've done so here just for emphasis.

Step 7. *Cancel anything that is the same on both sides.*
This is the final step. Six electrons cancel from both sides to give the final balanced equation.

$$14H^+ + Cr_2O_7^{2-} + 6Fe^{2+} \longrightarrow 2Cr^{3+} + 7H_2O + 6Fe^{3+}$$

Notice that both the charge and the atoms balance.

In some reactions, after adding the two half-reactions you may have H_2O or H^+ on both sides—for example, $6H_2O$ on the left and $2H_2O$ on the right. Cancel as many as you can. Thus,

$$\ldots + 6H_2O \ldots \longrightarrow \ldots + 2H_2O \ldots$$

gives

$$\ldots + 4H_2O \ldots \longrightarrow \ldots$$

The following is a summary of the steps we've followed for balancing an equation for a redox reaction in an acidic solution.

Tools

Ion–electron method (acidic solution)

If you don't skip any steps and you perform them in the order given, you will always obtain a properly balanced equation.

> **Ion–Electron Method—Acidic Solution**
> **Step 1.** Divide the equation into two half-reactions.
> **Step 2.** Balance atoms other than H and O.
> **Step 3.** Balance O by adding H_2O.
> **Step 4.** Balance H by adding H^+.
> **Step 5.** Balance net charge by adding e^-.
> **Step 6.** Make e^- gain equal e^- loss; then add half-reactions.
> **Step 7.** Cancel anything that's the same on both sides.

EXAMPLE 5.9

Using the Ion–Electron Method

Balance the following equation. The reaction occurs in an acidic solution.

$$MnO_4^- + H_2SO_3 \longrightarrow SO_4^{2-} + Mn^{2+}$$

Solution: We follow the steps given above.

Step 1. Divide into half-reactions

$$MnO_4^- \longrightarrow Mn^{2+}$$
$$H_2SO_3 \longrightarrow SO_4^{2-}$$

Step 2. There is nothing to do for this step. All the atoms except H and O are already in balance.

Step 3. Add H_2O to balance oxygens.

$$MnO_4^- \longrightarrow Mn^{2+} + 4H_2O$$
$$H_2O + H_2SO_3 \longrightarrow SO_4^{2-}$$

Step 4. Add H^+ to balance H.

$$8H^+ + MnO_4^- \longrightarrow Mn^{2+} + 4H_2O$$
$$H_2O + H_2SO_3 \longrightarrow SO_4^{2-} + 4H^+$$

Step 5. Balance charge by adding electrons to the more positive side.

$$5e^- + 8H^+ + MnO_4^- \longrightarrow Mn^{2+} + 4H_2O$$
$$H_2O + H_2SO_3 \longrightarrow SO_4^{2-} + 4H^+ + 2e^-$$

Step 6. Make electron loss equal to electron gain, then add half-reactions.

$$2(5e^- + 8H^+ + MnO_4^- \longrightarrow Mn^{2+} + 4H_2O)$$
$$5(H_2O + H_2SO_3 \longrightarrow SO_4^{2-} + 4H^+ + 2e^-)$$

$$10e^- + 16H^+ + 2MnO_4^- + 5H_2O + 5H_2SO_3 \longrightarrow$$
$$2Mn^{2+} + 8H_2O + 5SO_4^{2-} + 20H^+ + 10e^-$$

Step 7. Cancel $10e^-, 16H^+$, and $5H_2O$ from both sides. The final equation is

$$2MnO_4^- + 5H_2SO_3 \longrightarrow 2Mn^{2+} + 3H_2O + 5SO_4^{2-} + 4H^+$$

Once again, notice that each side of the equation has the same number of atoms of each element and the same net charge. This is what makes it a balanced equation.

Practice Exercise 5

What is the balanced equation for the following reaction in an acidic solution?

$$Cu + NO_3^- \longrightarrow Cu^{2+} + N_2O \; \blacklozenge$$

Balancing Redox Equations for Basic Solutions

In basic solutions, the concentration of H^+ is very small; the dominant species are H_2O and OH^-. Strictly speaking, these should be used to balance the half-reactions. However, the simplest way to obtain a balanced equation for a basic solution is to first *pretend* that the solution is acidic. We balance the equation using the seven steps just described, and then we use a simple three-step procedure described below to convert the equation to the correct form for a basic solution. The conversion uses the fact that H^+ and OH^- react in a 1-to-1 ratio to give H_2O.

Tools

Ion–electron
method
(basic solution)

Additional Steps in the Ion–Electron Method for Basic Solutions

Step 8. Add to <u>both</u> sides of the equation the same number of OH^- as there are H^+.

Step 9. Combine H^+ and OH^- to form H_2O.

Step 10. Cancel any H_2O that you can.

As an example, suppose we wanted to balance the following equation for a basic solution.

$$SO_3^{2-} + MnO_4^- \longrightarrow SO_4^{2-} + MnO_2$$

Following Steps 1 through 7 for acidic solutions gives

$$2H^+ + 3SO_3^{2-} + 2MnO_4^- \longrightarrow 3SO_4^{2-} + 2MnO_2 + H_2O$$

Conversion of this equation to one appropriate for a basic solution proceeds as follows

Step 8. *Add to <u>both</u> sides of the equation the same number of OH^- as there are H^+.*

The equation for acidic solution has $2H^+$ on the left, so we add $2OH^-$ to *each* side. This gives

$$2OH^- + 2H^+ + 3SO_3^{2-} + 2MnO_4^- \longrightarrow 3SO_4^{2-} + 2MnO_2 + H_2O + 2OH^-$$

Step 9. *Combine H^+ and OH^- to form H_2O.*

The left side has $2OH^-$ and $2H^+$, which become $2H_2O$. So in place of $2OH^- + 2H^+$ we write $2H_2O$.

$$2H_2O + 3SO_3^{2-} + 2MnO_4^- \longrightarrow 3SO_4^{2-} + 2MnO_2 + H_2O + 2OH^-$$

The result of Steps 8 and 9 is to replace H^+ with an equivalent number of H_2O, and the addition of that number of OH^- to the other side of the equation.

Step 10. *Cancel any H_2O that you can.*

In this equation, one H_2O can be canceled from both sides. The final equation, balanced for basic solution, is

$$H_2O + 3SO_3^{2-} + 2MnO_4^- \longrightarrow 3SO_4^{2-} + 2MnO_2 + 2OH^-$$

Practice Exercise 6

Balance the following equation for a basic solution.

$$MnO_4^- + C_2O_4^{2-} \longrightarrow MnO_2 + CO_3^{2-} \; \blacklozenge$$

Figure 5.1 *Zinc reacts with sulfuric acid.* Bubbles of hydrogen are formed when a solution of sulfuric acid comes in contact with metallic zinc. The same reaction takes place if sulfuric acid is spilled on galvanized (zinc coated) steel.

5.3 Reactions of Metals with Acids

In Chapter 4 we began our discussion of acids by examining the way they ionize in water to produce solutions that contain the hydronium ion, H_3O^+. Typical reactions of acids that we studied were neutralizations, which involve reactions of the H_3O^+ ion with OH^- or O^{2-}. In this section we will look at another way that acids are able to react in redox reactions.

An important property of acids is their tendency to react with certain metals. Battery acid (which is sulfuric acid, H_2SO_4) will begin to dissolve unprotected iron or steel parts of an automobile if it's spilled and not flushed away with water. A similar reaction occurs even faster, accompanied by vigorous bubbling, if the acid is spilled on the zinc surface of galvanized steel (see Figure 5.1).

In general, the reaction of an acid with a metal is a redox reaction in which the metal is oxidized and the acid is reduced. But in these reactions, the part of the acid that is reduced depends on the composition of the acid itself as well as on the metal.

When sulfuric acid reacts with the zinc coating on a galvanized garbage pail or a galvanized nail, the reaction is

$$Zn(s) + H_2SO_4(aq) \longrightarrow ZnSO_4(aq) + H_2(g)$$

The observed bubbling is caused by the release of hydrogen gas. If we assign oxidation numbers, we can analyze the oxidation–reduction changes that occur. In the diagram below, we use the symbol Δ (Greek letter delta) to stand for the change in oxidation number (positive, as we said, for an increase in oxidation number, negative for a decrease).

$$\Delta = (-1 \text{ per H}) \times (2\text{H}) = -2 \text{ units}$$

$$\underset{0}{Zn(s)} + \underset{+1 +6 -2}{H_2SO_4(aq)} \longrightarrow \underset{+2 +6 -2}{ZnSO_4(aq)} + \underset{0}{H_2(g)}$$

$$\Delta = +2 \text{ units per Zn}$$

The oxidation number of zinc increases from 0 to $+2$, so zinc is oxidized. The oxidation number of hydrogen decreases from $+1$ to 0, so the hydrogen of the acid is reduced.

An even clearer picture of what occurs is seen if we write the net ionic equation. Sulfuric acid, you recall, is a strong acid, so in aqueous solution we can write it in ionized form. Similarly, zinc sulfate is a salt and is therefore a strong electrolyte; we write it in fully dissociated form as well. The ionic equation for the reaction is therefore

$$Zn(s) + 2H^+(aq) + SO_4^{2-}(aq) \longrightarrow Zn^{2+}(aq) + SO_4^{2-}(aq) + H_2(g)$$

After canceling spectator ions, the net ionic equation for the reaction is

$$Zn(s) + 2H^+(aq) \longrightarrow Zn^{2+}(aq) + H_2(g)$$

Once again, assigning oxidation numbers reveals that the zinc is oxidized and the H^+ of the acid is reduced.

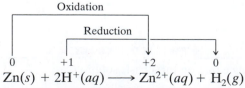

Stated another way, the H^+ of the acid is the oxidizing agent.

In many cases,

Metal + acid $\longrightarrow H_2$ + salt

Many metals react with acids just as zinc does—by being oxidized by hydrogen ions. In these reactions a metal salt and gaseous hydrogen are the products.

Practice Exercise 7

Write balanced molecular, ionic, and net ionic equations for the reaction of hydrochloric acid with (a) magnesium and (b) aluminum. (Both are oxidized by hydrogen ions.) ◆

The ease with which metals lose electrons varies considerably from one metal to another. Some metals are easily oxidized and others are not. For many metals, like iron and zinc, the hydrogen ion is a sufficiently strong oxidizing agent to do the job, so when such metals are placed in a solution of an acid they are oxidized while the hydrogen ions are reduced. Such metals are said to be *more active* than hydrogen (H_2), and they dissolve in acids like HCl and H_2SO_4 to give hydrogen gas and a salt that contains the anion of the acid. For other metals, however, hydrogen ions are not powerful enough to cause their oxidation. Copper, for example, is significantly less reactive than zinc or iron, and H^+ cannot oxidize it. If a copper penny is dropped into sulfuric or hydrochloric acid, it just sits there. No reaction occurs. Copper is an example of a metal that is *less active* than H_2.

Oxidizing Power of Acids

Hydrochloric acid contains H_3O^+ ions (which we abbreviate as H^+) and Cl^- ions. The hydrogen ion in hydrochloric acid is able to be an oxidizing agent by being reduced to H_2. However, the Cl^- ion in the solution has no tendency at all to be an oxidizing agent, because it would have to become a Cl^{2-} ion to gain an electron. This is just not feasible, so in a solution of HCl, the only oxidizing agent is H^+.

In an aqueous solution of sulfuric acid, we find a similar situation. In water this acid ionizes to give hydrogen ions and sulfate ions. Although the sulfate ion can be reduced (to SO_3^{2-}, for example), hydrogen ion is more easily reduced. Therefore, when we add a metal such as zinc to sulfuric acid, it is H^+, rather than SO_4^{2-}, that removes the electrons from Zn.

Compared with many other chemicals that we will study, the hydrogen ion in water is really a rather poor oxidizing agent, so hydrochloric acid and sulfuric acid have rather poor oxidizing abilities. For this reason, they are often called **nonoxidizing acids,** even though their hydrogen ions are able to oxidize certain metals. Actually, when we call something a nonoxidizing acid we are saying that the *anion* of the acid is a weaker oxidizing agent than H^+ (which is equivalent to saying that the anion of the acid is more difficult to reduce than H^+).

There are also **oxidizing acids,** acids whose anions are stronger oxidizing agents than H^+. (See Table 5.1.) An example is nitric acid, HNO_3. When dissolved in water, nitric acid ionizes to give H^+ and NO_3^- ions. However, in this solution the nitrate ion is a more powerful oxidizing agent than the hydrogen ion. In the competition for electrons, therefore, it is the nitrate ion that wins, and when nitric acid reacts with a metal, it is the nitrate ion that is reduced.

Because the NO_3^- ion is a stronger oxidizing agent than H^+, it is able to oxidize metals that H^+ cannot. For example, if a copper penny is dropped into concentrated nitric acid, it reacts violently (see Figure 5.2). The reddish brown gas is nitrogen dioxide, NO_2, which is formed by the reduction of the NO_3^- ion. The molecular equation for the reaction is

$$Cu(s) + 4HNO_3(aq) \longrightarrow Cu(NO_3)_2(aq) + 2NO_2(g) + 2H_2O$$

The strongest oxidizing agent in a solution of a "nonoxidizing" acid is H^+.

Tools

Oxidizing and nonoxidizing acids

Table 5.1 **Nonoxidizing and Oxidizing Acids**

Nonoxidizing Acids	
$HCl(aq)$	
$H_2SO_4(aq)$[a]	
$H_3PO_4(aq)$	
Most organic acids (e.g., $HC_2H_3O_2$)	
Oxidizing Acids	**Reduction Reaction**
HNO_3	(conc.) $NO_3^- + 2H^+ + e^- \rightarrow NO_2(g) + H_2O$ (dilute) $NO_3^- + 4H^+ + 3e^- \rightarrow NO(g) + 2H_2O$ (very dilute, with strong reducing agent) $NO_3^- + 10H^+ + 8e^- \rightarrow NH_4^+ + 3H_2O$
H_2SO_4	(hot, conc.) $SO_4^{2-} + 4H^+ + 2e^- \rightarrow SO_2(g) + 2H_2O$ (hot conc., with strong reducing agent) $SO_4^{2-} + 10H^+ + 8e^- \rightarrow H_2S(g) + 4H_2O$

[a]H_2SO_4 is a nonoxidizing acid when cold and dilute.

If we change this to a net ionic equation and then assign oxidation numbers, we can see easily which is the oxidizing agent and which is the reducing agent.

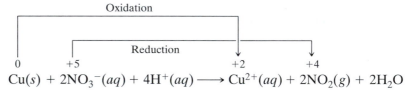

Oxidation (0 → +2); Reduction (+5 → +4)

$$\overset{0}{Cu(s)} + 2\overset{+5}{N}O_3^-(aq) + 4H^+(aq) \longrightarrow \overset{+2}{Cu^{2+}}(aq) + 2\overset{+4}{N}O_2(g) + 2H_2O$$

Notice that even though it is the nitrogen whose oxidation number is decreasing, we identify the substance that contains this nitrogen, the NO_3^- ion, as the oxidizing agent.

Nitrate ion is reduced and copper is oxidized. Therefore, nitrate ion is the oxidizing agent and copper is the reducing agent. Notice that in this reaction, *no hydrogen gas is formed.* The H^+ ions of the HNO_3 are an essential part of the reaction, but they just become part of water molecules without a change in oxidation number.

$$Cu + 4HNO_3 \longrightarrow Cu(NO_3)_2 + 2NO_2 + 2H_2O$$

Figure 5.2 *The reaction of copper with concentrated nitric acid.* A copper penny reacts vigorously with concentrated nitric acid, as this sequence of photographs shows. The dark red-brown vapors are nitrogen dioxide, the same gas that gives smog its characteristic color.

When oxidizing acids react with metals, it is often difficult to predict the products. Reduction of the nitrate ion, for example, can produce all sorts of compounds having different oxidation states for the nitrogen, depending on the reducing power of the metal and the concentration of the acid. When *concentrated* nitric acid reacts with a metal, it often produces nitrogen dioxide, NO_2, as the reduction product. When *dilute* nitric acid is used to dissolve a metal, the product is often nitric oxide, NO, instead. Copper, for example, can display either behavior toward nitric acid. The molecular and net ionic equations for the reactions are as follows:

The oxidation number of N in NO_2 is $+4$. In NO, the oxidation number is $+2$.

Concentrated HNO$_3$

$$Cu(s) + 4HNO_3(aq) \longrightarrow Cu(NO_3)_2(aq) + 2NO_2(g) + 2H_2O$$

$$Cu(s) + 4H^+(aq) + 2NO_3^-(aq) \longrightarrow Cu^{2+}(aq) + 2NO_2(g) + 2H_2O$$

Dilute HNO$_3$

$$3Cu(s) + 8HNO_3(aq) \longrightarrow 3Cu(NO_3)_2(aq) + 2NO(g) + 4H_2O$$

$$3Cu(s) + 8H^+(aq) + 2NO_3^-(aq) \longrightarrow 3Cu^{2+}(aq) + 2NO(g) + 4H_2O$$

If very dilute nitric acid reacts with a metal that is a particularly strong reducing agent, like zinc, the nitrogen can be reduced all the way down to the -3 oxidation state that it has in NH_4^+ (or NH_3). The net ionic equation is

$$4Zn(s) + 10H^+(aq) + NO_3^-(aq) \longrightarrow 4Zn^{2+}(aq) + NH_4^+(aq) + 3H_2O$$

The nitrate ion in the presence of hydrogen ions makes nitric acid quite a powerful oxidizing acid. All metals except the very unreactive ones, such as platinum and gold, are attacked by it. Nitric acid also does a good job of oxidizing organic substances, so it is wise to be especially careful when working with this acid in the laboratory. Very serious accidents have occurred when inexperienced people have used concentrated nitric acid around organic substances.

Nitric acid causes severe skin burns, so be careful when you work with it in the laboratory. If you spill any on your skin, wash it off immediately and seek the help of your lab teacher.

5.4 Displacement of One Metal by Another from Compounds

The evolution of hydrogen in the reaction of a metal with an acid is a special case of a more general phenomenon—one element displacing (pushing out) another element from a compound by means of a redox reaction. In the case of a metal–acid reaction, it is the metal that displaces hydrogen from the acid, changing $2H^+$ to H_2.

Another reaction of this same general type occurs when one metal displaces another metal from its compounds, and is illustrated by the experiment shown in Figure 5.3. Here we see a brightly polished strip of metallic zinc that is dipped into a solution of copper sulfate. After the zinc is in the solution for a while, a reddish brown deposit of metallic copper forms on the zinc, and if the solution were analyzed, we would find that it now contains zinc ions, as well as some remaining unreacted copper ions.

The results of this experiment can be summarized by the equation

$$Zn(s) + CuSO_4(aq) \longrightarrow Cu(s) + ZnSO_4(aq)$$

A reaction such as this, in which one element replaces another in a compound, is sometimes called a **single replacement reaction.**

In a *double replacement* reaction such as

$$CdCl_2(aq) + Na_2S(aq) \longrightarrow$$
$$CdS(s) + 2NaCl(aq)$$

two anions (Cl^- and S^{2-}) acquire different cations. In a *single replacement* reaction, *one* anion acquires a different cation. Single replacement reactions are redox reactions; double replacement reactions are not.

Figure 5.3 *The reaction of zinc with copper ion. (Left)* A piece of shiny zinc next to a beaker containing a copper sulfate solution. *(Center)* When the zinc is placed in the solution, copper ions are reduced to the free metal while the zinc dissolves. *(Right)* After a while the zinc becomes coated with a red-brown layer of copper. Notice that the solution is a lighter blue than before, showing that some of the copper ions have left the solution.

The redox changes in the reaction above become clear if we write the net ionic equation. Both copper sulfate and zinc sulfate are soluble salts, so they are completely dissociated. Sulfate ion is a spectator ion, and the net ionic equation is

$$Zn(s) + Cu^{2+}(aq) \longrightarrow Cu(s) + Zn^{2+}(aq)$$

We see that zinc has reduced the copper ion to metallic copper, and the copper ion has oxidized metallic zinc to the zinc ion. In the process, zinc ions have taken the place of the copper ions, so a solution of copper sulfate is changed to a solution of zinc sulfate.

The reaction of zinc with copper ion is quite similar to the reaction of zinc with sulfuric acid. The more "active" zinc replaces another less "active" element in a compound. Furthermore, it is easy to show that the reverse reaction doesn't occur. Nothing happens if a brightly polished piece of copper is dipped into a solution of zinc sulfate (see Figure 5.4). No matter how long the copper is in the solution, its surface remains untarnished. This means that the reaction of copper with zinc sulfate doesn't occur.

$$Cu(s) + ZnSO_4(aq) \longrightarrow \text{no reaction}$$

In other words, the less "active" copper is unable to displace the more "active" zinc from the solution.

The Activity Series

Throughout the discussion in the preceding paragraph we used the word *active* to mean "easily oxidized." In other words, an element that is more easily oxidized will displace one that is less easily oxidized from its compounds. The relative ease of oxidation of two metals can be established in experiments just as simple as the ones pictured in Figures 5.3 and 5.4. After such comparisons are made for many pairs, the metals can be arranged in order of their ease of oxidation to give what is often called an **activity series** (see Table 5.2). According to the way the metals have been arranged in Table 5.2, those at the bottom are more easily oxidized

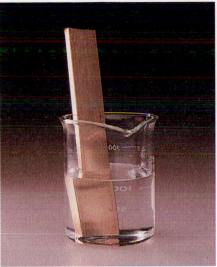

Figure 5.4 *The inability of copper to react with zinc ion.* Although metallic zinc will displace copper from a solution containing Cu^{2+} ion, metallic copper will not displace Zn^{2+} from its solutions. Here we see that the copper bar is unaffected by being dipped into a solution of zinc sulfate.

Table 5.2 **Activity Series for Some Metals (and Hydrogen)**

Activity series

	Element	Oxidation Product
Least Active	Gold	Au^{3+}
	Mercury	Hg^{2+}
	Silver	Ag^+
	Copper	Cu^{2+}
	HYDROGEN	H^+
	Lead	Pb^{2+}
	Tin	Sn^{2+}
	Cobalt	Co^{2+}
	Cadmium	Cd^{2+}
	Iron	Fe^{2+}
	Chromium	Cr^{3+}
	Zinc	Zn^{2+}
	Manganese	Mn^{2+}
	Aluminum	Al^{3+}
	Magnesium	Mg^{2+}
	Sodium	Na^+
	Calcium	Ca^{2+}
	Strontium	Sr^{2+}
	Barium	Ba^{2+}
	Potassium	K^+
	Rubidium	Rb^+
Most Active	Cesium	Cs^+

Increasing Ease of Oxidation of the Metal

Figure 5.5 *Metallic sodium reacts violently with water.* The heat of the reaction ignites the sodium metal, which can be seen burning and sending sparks from the surface of the water. In the reaction, sodium is oxidized to Na^+ and water molecules are reduced to give hydrogen gas and hydroxide ions. When the reaction is over, the solution contains sodium hydroxide.

If a metal is above hydrogen in the activity series (Table 5.2), it will not react with nonoxidizing acids.

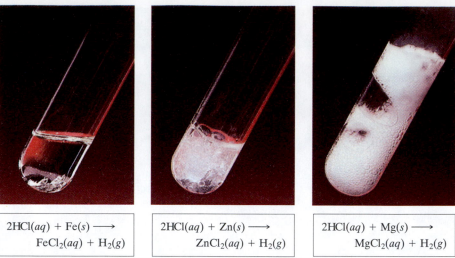

$$2HCl(aq) + Fe(s) \longrightarrow FeCl_2(aq) + H_2(g)$$

$$2HCl(aq) + Zn(s) \longrightarrow ZnCl_2(aq) + H_2(g)$$

$$2HCl(aq) + Mg(s) \longrightarrow MgCl_2(aq) + H_2(g)$$

Figure 5.6 *The relative ease of oxidation of metals parallels their rates of reaction with hydrogen ions of an acid.* The products are hydrogen gas and the metal ion in solution. All three test tubes contain HCl(aq) at the same concentration. The first also contains pieces of iron, the second, pieces of zinc, and the third, pieces of magnesium. Among these three metals, the ease of oxidation increases from iron to zinc to magnesium.

(are more active) than those at the top. *This means that a given element will be displaced from its compounds by any metal below it in the table.*

Notice that we have included hydrogen in the activity series. Metals below hydrogen in the series can displace hydrogen from solutions containing H^+. These are the metals that are capable of reacting with the nonoxidizing acids discussed in the previous section. On the other hand, metals above hydrogen in the table do not react with acids having H^+ as the strongest oxidizing agent.

Metals at the very bottom of the table are very easily oxidized and are extremely strong reducing agents. They are so reactive, in fact, that they are able to reduce the hydrogen in water. Sodium, for example, reacts vigorously with water to liberate H_2 according to the following equation (see Figure 5.5).

$$2Na(s) + 2H_2O \longrightarrow H_2(g) + 2NaOH(aq)$$

For metals below hydrogen in the activity series, an interesting parallel exists between the ease of oxidation of the metal and the speed with which it reacts with H^+. For example, in Figure 5.6, we see samples of iron, zinc, and magnesium reacting with solutions of hydrochloric acid. In each test tube the initial HCl concentration is the same, but we see that the magnesium reacts more rapidly than zinc, which reacts more rapidly than iron. You can see that the order of reactivity in Table 5.2 is the same; magnesium is more easily oxidized than zinc, which is more easily oxidized than iron.

Using the Activity Series

The activity series in Table 5.2 permits us to make predictions of the outcome of single replacement redox reactions, as illustrated in the following examples.

EXAMPLE 5.10
Using the Activity Series

What will happen if an iron nail is dipped into a solution containing copper sulfate? If a reaction occurs, write its chemical equation.

Analysis: Because of the discussion above, it is obvious we will need to refer to the activity series. However, if this question had been asked in another setting, how would you know what to do?

Reading the question, we have to ask, what *could* happen? If a chemical reaction were to occur, iron would have to react with the copper sulfate. A *metal* possibly reacting with the *salt of another metal* should suggest the possibility of a single replacement reaction. In turn, you should recognize that such reactions can be predicted from the locations of the potential reactants in the activity series. According to the activity series in Table 5.2, iron is more easily oxidized than copper, so the iron will displace the copper from the copper sulfate. A reaction will occur. To write an equation for the reaction, we have to know the final oxidation state of the iron. In the table, this is indicated as $+2$, so the Fe atoms change to Fe^{2+} ions and pair with SO_4^{2-} ions to give $FeSO_4$. Copper(II) ions are reduced to copper atoms.

Solution: The analysis told us that a reaction will occur and gave us the products, so the equation is

$$Fe(s) + CuSO_4(aq) \longrightarrow Cu(s) + FeSO_4(aq)$$

Check: The equation is balanced as it stands; the question is answered.

EXAMPLE 5.11
Using the Activity Series

What happens if an iron nail is dipped into a solution of aluminum sulfate?

Analysis: Scanning the activity series, we see that aluminum metal is *more* easily oxidized than iron metal. This means that aluminum atoms would be able to displace iron ions from an iron compound. But it also means that iron atoms cannot displace aluminum ions from its compounds, and iron *atoms* plus aluminum *ions* are what we're given.

Solution: If nature does not permit electrons to transfer from iron atoms to aluminum ions, we must conclude that no reaction can occur.

$$Fe(s) + Al_2(SO_4)_3(aq) \longrightarrow \text{no reaction}$$

Practice Exercise 8

Write a chemical equation for the reaction that will occur, if any, when (a) aluminum metal is added to a solution of copper chloride and (b) silver metal is added to a solution of magnesium sulfate. If no reaction will occur, write "no reaction" in place of the products. ◆

5.5 Molecular Oxygen as an Oxidizing Agent

In 1789, the French chemist Antoine Lavoisier discovered that *combustion* involves the reaction of chemicals in various fuels, like wood and coal, not just with air but specifically with the oxygen in air. Oxygen is a plentiful chemical; it's in the air and available to anyone who wants to use it, chemist or not. Furthermore, O_2 is a very reactive oxidizing agent, so its reactions have been well stud-

Antoine Laurent Lavoisier (1743–1794), a French chemist, is deservedly called the "father of modern chemistry." He was the first to insist on *quantitative* data from chemical research; he systematized chemical nomenclature; and his text *Traité élémentaire de chimie* was to modern chemistry what Newton's *Principia* was to the development of physics.

ied. When substances combine with oxygen, the products are generally oxides, *molecular oxides* when the oxygen reacts with nonmetals and *ionic oxides* when the oxygen reacts with metals.

Combustion of Organic Compounds

Combustion is normally taken to mean a particularly *rapid* reaction of a substance with oxygen in which both heat and light are given off. If you had to build a fire to keep warm, you no doubt would look for combustible materials like twigs, logs, or other pieces of wood to use as fuel. You know that wood burns. Experience has also taught you that certain other substances burn. When you drive a car, for example, it is probably powered by the combustion of gasoline. Wood and gasoline are examples of substances or mixtures of substances that chemists call *organic compounds*—compounds whose structures are determined primarily by the linking together of carbon atoms. When organic compounds burn, the products of the reactions are usually easy to predict.

Combustion of Hydrocarbons

The general, unbalanced equation is

$$\text{Hydrocarbon} + O_2 \longrightarrow$$
$$CO_2 + H_2O$$

Fuels such as natural gas, gasoline, kerosene, heating oil, and diesel fuel are examples of *hydrocarbons*—compounds containing only the elements carbon and hydrogen. Natural gas is composed principally of methane, CH_4. Gasoline is a mixture of hydrocarbons, the most familiar of which is octane, C_8H_{18}. Kerosene, heating oil, and diesel fuel are mixtures of hydrocarbons in which the molecules contain even more atoms of carbon and hydrogen.

When hydrocarbons burn in a *plentiful* supply of oxygen, the products of combustion are always carbon dioxide and water. Thus, methane and octane combine with oxygen according to the equations

$$CH_4 + 2O_2 \longrightarrow CO_2 + 2H_2O$$

$$2C_8H_{18} + 25O_2 \longrightarrow 16CO_2 + 18H_2O$$

Many people don't realize that water is one of the products of the combustion of hydrocarbons, even though they have seen evidence for it. Perhaps you have seen clouds of condensed water vapor coming from the exhaust pipes of automobiles on cold winter days, or you may have noticed that shortly after you first start a car, drops of water fall from the exhaust pipe. This is water that has been formed during the combustion of the gasoline.

Clouds of condensed water vapor coming from the stacks of an oil-fired electric generating plant during the winter.

Practice Exercise 9

Write a balanced equation for the combustion of butane, C_4H_{10}, in an abundant supply of oxygen. Butane is the fuel used in disposable cigarette lighters. ◆

When there is less than an abundant supply of oxygen during the combustion of a hydrocarbon, not all of the carbon is converted to carbon dioxide. Instead, some of it forms carbon monoxide. Its formation is a pollution problem associated with the use of gasoline engines, as you may know.

$$2CH_4 + 3O_2 \longrightarrow 2CO + 4H_2O \qquad \text{(in a limited oxygen supply)}$$

When the oxygen supply is extremely limited, only the hydrogen of a hydrocarbon mixture is converted to the oxide (water). The carbon atoms emerge as elemental carbon. For example, the combustion of methane in a very limited oxygen supply follows the equation

$$CH_4 + O_2 \longrightarrow C + 2H_2O \qquad \text{(in a very limited oxygen supply)}$$

The carbon that forms is very finely divided and would be called *soot* by almost anyone observing the reaction. Nevertheless, such soot has considerable commercial value when collected and marketed under the name *lampblack*. This sooty form of carbon is used to manufacture inks, and much of it is used in the production of rubber tires, where it serves as a binder and a filler. When soot from incomplete combustion is released into air, its tiny particles constitute one component of air pollution, namely, *particulates*, which contribute to the haziness of smog.

Combustion of Organic Compounds That Contain Oxygen

Earlier we mentioned that you might choose wood to build a fire. The chief combustible ingredient in wood is cellulose, a fibrous material that gives plants their structural strength. Cellulose is composed of the elements carbon, hydrogen, and oxygen. Each cellulose molecule consists of many small, identical groups of atoms that are linked together to form a very long molecule, although the lengths of the molecules differ. For this reason we cannot specify a molecular formula for cellulose. Instead, we use the empirical formula of cellulose, $C_6H_{10}O_5$, which represents the small, repeating "building block" units in large cellulose molecules. When cellulose burns, the products are also carbon dioxide and water. The only difference between its reaction and the reaction of a hydrocarbon with oxygen is that some of the oxygen in the products comes from the cellulose.

$$C_6H_{10}O_5 + 6O_2 \longrightarrow 6CO_2 + 5H_2O$$

The complete combustion of all other organic compounds containing only carbon, hydrogen, and oxygen produces the same products, CO_2 and H_2O, and follows similar equations.

Practice Exercise 10

Ethanol, C_2H_5OH, is now mixed with gasoline, and the mixture is sold under the name *gasohol*. Write a chemical equation for the combustion of ethanol. ◆

Combustion of Organic Compounds That Contain Sulfur

A major pollution problem in industrialized countries is caused by the release into the atmosphere of sulfur dioxide formed by the combustion of fuels that contain sulfur or its compounds. The products of the combustion of organic compounds of sulfur are carbon dioxide, water, and sulfur dioxide. A typical reaction is

$$2C_2H_5SH + 9O_2 \longrightarrow 4CO_2 + 6H_2O + 2SO_2$$

A solution of sulfur dioxide in water is acidic, and dissolved SO_2 is one solute that converts rainwater into "acid rain" (see *Chemicals in Our World 7*).

Reactions of Metals with Oxygen

We don't often think of metals as undergoing combustion, but have you ever seen an old-fashioned flashbulb fired to take a photograph? The source of light is the reaction of the metal magnesium with oxygen (see Figure 5.7). A close look at a fresh flashbulb reveals a fine web of thin magnesium wire within the glass envelope. The wire is surrounded by an atmosphere of oxygen, a clear colorless gas. When the flashbulb is used, a small electric current surges through the

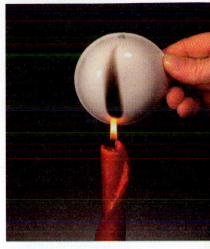

The bright yellow color of a candle flame is caused by glowing particles of elemental carbon. Here we see that a black deposit of carbon is formed when the flame contacts a cold porcelain surface.

The formula for cellulose can be expressed as $(C_6H_{10}O_5)_n$, which indicates that the molecule contains the $C_6H_{10}O_5$ unit repeated some large number n times.

Figure 5.7 *An old-fashioned flashbulb, before and after firing.* Fine magnesium wire in an atmosphere of oxygen fills the flashbulb at the left. After being used (right), the interior of the bulb is coated with a white film of magnesium oxide.

Figure 5.8 *Cutting steel with an oxyacetylene torch.* An oxygen–acetylene flame is used to heat steel until it glows red-hot. Then the acetylene is turned off and the steel is cut by a stream of pure oxygen whose reaction with the white-hot metal produces enough heat to melt the steel and send a shower of burning steel sparks flying.

Nitrogen is one nonmetal that doesn't react readily with oxygen, especially at normal temperatures. But nitrogen oxides formed at high temperatures in automobile engines are a serious source of air pollution.

Most scientists are convinced that SO_2 released by coal-burning power plants in the Midwest is responsible for acid rain in the northeastern United States and Canada.

The label on a bag of charcoal shows a warning about carbon monoxide.

thin wire, causing it to become hot enough to ignite, and it burns rapidly in the oxygen atmosphere. The equation for the reaction is

$$2Mg + O_2 \longrightarrow 2MgO$$

Most metals react directly with oxygen, although not so spectacularly, and usually we refer to the reaction as **corrosion** or tarnishing because the oxidation products dull the shiny metal surface. Iron, for example, is oxidized fairly easily, especially in the presence of moisture. As you know, under these conditions the iron corrodes—it rusts. Rust is a form of iron(III) oxide, Fe_2O_3, that also contains an appreciable amount of absorbed water. The formula for rust is therefore normally given as $Fe_2O_3 \cdot xH_2O$ to indicate its somewhat variable composition. Although the rusting of iron is a slow reaction, the combination of iron with oxygen can be speeded up if the metal is heated to a very high temperature (see Figure 5.8).

An aluminum surface, unlike that of iron, is not noticeably dulled by the reaction of aluminum with oxygen. Aluminum is a common metal found around the home in uses ranging from aluminum foil to aluminum window frames, and it surely appears shiny. Yet, aluminum is a rather easily oxidized metal, as can be seen from its position in the activity series (Table 5.2). A *freshly* exposed surface of the metal does react very quickly with oxygen and becomes coated with a very thin film of aluminum oxide, Al_2O_3, so thin that it doesn't obscure the shininess of the metal beneath. Fortunately, the oxide coating adheres very tightly to the surface of the metal and makes it very difficult for additional oxygen to combine with the aluminum. Therefore, further oxidation of aluminum occurs very slowly.

Practice Exercise 11

The oxide formed in the reaction shown in Figure 5.8 is Fe_2O_3. Write a balanced chemical equation for the reaction. ◆

Reactions of Nonmetals with Oxygen

Most nonmetals combine as readily with oxygen as do the metals, and their reactions usually occur rapidly enough to be described as combustion. To most people, the most important nonmetal combustion is that of carbon, because the reaction is a source of heat. Coal and charcoal, for example, are common carbon fuels. Coal is used worldwide in large amounts to generate electricity, and charcoal is a popular barbecue fuel for broiling hamburgers. If plenty of oxygen is available, the combustion of carbon gives CO_2, but when the supply of oxygen is limited, some CO also forms. Manufacturers that package charcoal briquettes, therefore, print a warning on the bag that the charcoal shouldn't be used indoors for cooking or heating.

Sulfur is another nonmetal that readily burns in oxygen. In the manufacture of sulfuric acid, the first step is the combustion of sulfur to produce sulfur dioxide. As mentioned earlier, sulfur dioxide also forms when sulfur compounds burn, and the presence of both sulfur and sulfur compounds as impurities in coal and petroleum is a major source of air pollution. Combustion of these fuels releases their sulfur content in the form of SO_2, which enters the atmosphere, where it drifts on the wind until it finally dissolves in rainwater. Then, as a dilute solution of sulfurous acid, it falls to Earth as one component of acid rain.

Practice Exercise 12

Elemental phosphorus can exist in a white, waxy form that consists of molecules having the formula P_4. When phosphorus burns in an abundant supply of oxygen, it forms an oxide composed of molecules of P_4O_{10}. Write an equation for the reaction. ◆

5.6 Stoichiometry and Redox Reactions

In general, redox reactions are somewhat more complex than metathesis reactions. You have witnessed this in constructing balanced net ionic equations for both types of reactions. Nevertheless, stoichiometry problems involving redox reactions are approached in the same manner as those we've discussed earlier, as illustrated by the following example.

EXAMPLE 5.12
Stoichiometry of a Redox Reaction

How many grams of sodium sulfite are needed to react completely with 12.4 g of potassium dichromate in an acidic solution? The products of the reaction include sulfate ion and chromium(III) ion.

Analysis: To solve the problem we will need a balanced chemical equation. We will use the ion–electron method to construct the net ionic equation. Then we can proceed as with other stoichiometry problems.

Solution: To write the balanced equation, we need to first construct the skeleton equation, which we can then balance by the ion–electron method. On the product side of the skeleton equation we will write sulfate ion (SO_4^{2-}) and chromium(III) ion (Cr^{3+}). This suggests that the reactants will be sulfite ion (SO_3^{2-}) and dichromate ion ($Cr_2O_7^{2-}$).

$$SO_3^{2-} + Cr_2O_7^{2-} \longrightarrow SO_4^{2-} + Cr^{3+}$$

Balancing the equation by the procedure described in Section 5.2 for an acidic solution gives

$$8H^+ + 3SO_3^{2-} + Cr_2O_7^{2-} \longrightarrow 3SO_4^{2-} + 2Cr^{3+} + 4H_2O$$

The stoichiometric equivalence between sulfite and dichromate ions is therefore

$$3 \text{ mol } SO_3^{2-} \Leftrightarrow 1 \text{ mol } Cr_2O_7^{2-}$$

The complete formulas of the reactants are Na_2SO_3 and $K_2Cr_2O_7$, so we can represent this equivalence as

$$3 \text{ mol } Na_2SO_3 \Leftrightarrow 1 \text{ mol } K_2Cr_2O_7$$

Using the factor-label method, the solution to the problem is

$$12.4 \text{ g } K_2Cr_2O_7 \times \frac{1 \text{ mol } K_2Cr_2O_7}{294.2 \text{ g } K_2Cr_2O_7} \times \frac{3 \text{ mol } Na_2SO_3}{1 \text{ mol } K_2Cr_2O_7} \times \frac{126.1 \text{ g } Na_2SO_3}{1 \text{ mol } Na_2SO_3}$$
$$= 15.9 \text{ g } Na_2SO_3$$

The amount of sodium sulfite required is 15.9 g.

Is the Answer Reasonable?
If the formula masses of the two compounds were the same, the amount of Na_2SO_3 needed would be three times the amount of $K_2Cr_2O_7$, or approximately 36 g of Na_2SO_3. However, the formula mass of Na_2SO_3 is less than half that of $K_2Cr_2O_7$, so the amount of Na_2SO_3 needed will be somewhat less than half of 36 g, or somewhat less than 18 g. Therefore, our answer, 15.9 g, seems reasonable.

Practice Exercise 13

A researcher planned to use chlorine gas in an experiment and wished to trap excess chlorine to prevent it from escaping into the atmosphere. To accomplish this, the reaction of sodium thiosulfate ($Na_2S_2O_3$) with chlorine gas in an acidic aqueous solution to give sulfate ion and chloride ion would be used. How many grams of $Na_2S_2O_3$ are needed to trap 4.25 g of chlorine? ◆

Figure 5.9 *Reduction of MnO_4^- by Fe^{2+}.* A solution of $KMnO_4$ is added to a stirred acidic solution containing Fe^{2+}. The reaction oxidizes the pale blue-green Fe^{2+} to Fe^{3+} while the MnO_4^- is reduced to the almost colorless Mn^{2+} ion. The purple color of the permanganate ion will continue to be destroyed until all of the Fe^{2+} has reacted. Only then will the iron-containing solution take on a pink or purple color. This ability of MnO_4^- to signal the completion of the reaction makes it especially useful in redox titrations.

Using Redox Reactions in the Laboratory

Because so many reactions involve oxidation and reduction, it should not be surprising that they have useful applications in the laboratory. An example of such an application is described in the preceding practice exercise. Some redox reactions are especially useful in chemical analyses, particularly in titrations.

Redox Titrations

Unlike in acid–base titrations, there are no simple indicators that can be used to conveniently detect the end points in redox titrations. One of the most useful reactants for redox titrations is potassium permanganate, $KMnO_4$, especially when the reaction can be carried out in an acidic solution. Permanganate ion is a powerful oxidizing agent, so it oxidizes most substances that are capable of being oxidized. That's one reason why it is used. Especially important, though, is the fact that the MnO_4^- ion has a deep purple color and its reduction product in acidic solution is the almost colorless Mn^{2+} ion. Therefore, when a solution of $KMnO_4$ is added from a buret to a solution of a reducing agent, the chemical reaction that occurs forms a nearly colorless product. As the $KMnO_4$ solution is added, the purple color continues to be destroyed as long as there is any reducing agent left (Figure 5.9). However, after the last trace of the reducing agent has been consumed, the MnO_4^- ion in the next drop of titrant has nothing to react with, so it colors the solution pink. This signals the end of the titration. In this way, permanganate ion serves as its own indicator in redox titrations. Example 5.13 illustrates a typical analysis using $KMnO_4$ in a redox titration.

In concentrated solutions, MnO_4^- is purple, but dilute solutions of the ion appear pink.

EXAMPLE 5.13

Redox Titrations in Chemical Analysis

All the iron in a 2.000 g sample of an iron ore was dissolved in an acidic solution and converted to Fe^{2+}, which was then titrated with 0.1000 M $KMnO_4$ solution. In the titration the iron was oxidized to Fe^{3+}. The titration required 27.45 mL of the $KMnO_4$ solution to reach the end point.

(a) How many grams of iron were in the sample?
(b) What was the percentage iron in the sample?
(c) If the iron was present in the sample as Fe_2O_3, what was the percentage by mass of Fe_2O_3 in the sample?

Analysis: The first step, of course, will be to write a balanced equation. Let's analyze the problem after that point. From the volume of the $KMnO_4$ solution and its concentration we can determine the number of moles of titrant used. The coefficients of the equation then allow us to compute the number of moles of iron(II) that reacted. Since all the iron had been previously changed to iron(II), this is the number of moles of iron in the sample. Converting moles of iron to grams of iron gives the mass of iron in the sample, from which the percentage of iron can be calculated easily. Let's tackle the first two parts of the problem now, and then return to the last part afterward.

Solution: The skeleton equation for the reaction is

$$Fe^{2+} + MnO_4^- \longrightarrow Fe^{3+} + Mn^{2+}$$

Balancing it by the ion–electron method for acidic solutions gives

$$5Fe^{2+} + MnO_4^- + 8H^+ \longrightarrow 5Fe^{3+} + Mn^{2+} + 4H_2O$$

The number of moles of $KMnO_4$ consumed is calculated from the volume of the solution used in the titration and its concentration.

$$0.02745 \text{ L } KMnO_4 \times \frac{0.1000 \text{ mol } KMnO_4}{1.000 \text{ L } KMnO_4} = 0.002745 \text{ mol } KMnO_4$$

The chemical equation tells us five moles of iron react per mole of permanganate consumed. The number of moles of iron that reacted is

$$0.002745 \text{ mol } KMnO_4 \times \frac{5 \text{ mol Fe}}{1 \text{ mol } KMnO_4} = 0.01372 \text{ mol Fe}$$

and the mass of iron in the sample is

$$0.01372 \text{ mol Fe} \times \frac{55.845 \text{ g Fe}}{1 \text{ mol Fe}} = 0.7662 \text{ g Fe}$$

This is the answer to part (a) of the problem. Next we calculate the percentage of iron in the sample, which is the mass of iron divided by the mass of the sample, all multiplied by 100%.

$$\% \text{ Fe} = \frac{\text{mass of Fe}}{\text{mass of sample}} \times 100\%$$

Substituting gives

$$\% \text{ Fe} = \frac{0.7662 \text{ g Fe}}{2.000 \text{ g sample}} \times 100\% = 38.31\% \text{ Fe}$$

Is the Answer Reasonable?

We've used approximately 30 mL, or 0.030 L, of the $KMnO_4$ solution, which is 0.1 M. Multiplying these tells us we've used approximately 0.003 mol of $KMnO_4$. From the coefficients of the equation, five times as many moles of Fe^{2+} react, so the amount of Fe in the sample is approximately $5 \times 0.003 = 0.015$ mol. The atomic mass of Fe is about 55 g/mol, so the mass of Fe in the sample is approximately $0.015 \times 55 = 0.8$ g. Our answer (0.7662 g) is reasonable.

Analysis Continued: Now we can work on the last part of the question. Earlier in the problem we determined the number of moles of iron that reacted, 0.01372 mol Fe. How many moles of Fe_2O_3 would have contained this number of moles of iron? That's the critical question we have to answer. Once we know this, we can calculate the mass of the Fe_2O_3 and the percentage of Fe_2O_3 in the original sample.

Solution Continued: The chemical formula for the iron oxide gives us

$$1 \text{ mol } Fe_2O_3 \Leftrightarrow 2 \text{ mol Fe}$$

This provides the conversion factor we need to determine how many moles of Fe_2O_3 were present in the sample. Working with the number of moles of Fe,

$$0.01372 \text{ mol Fe} \times \frac{1 \text{ mol } Fe_2O_3}{2 \text{ mol Fe}} = 0.006860 \text{ mol } Fe_2O_3$$

This is the number of moles of Fe_2O_3 in the sample. The formula mass of Fe_2O_3 is 159.69 g/mol, so the mass of Fe_2O_3 in the sample was

$$0.006860 \ \overline{\text{mol Fe}_2\text{O}_3} \times \frac{159.69 \text{ g Fe}_2\text{O}_3}{1 \ \overline{\text{mol Fe}_2\text{O}_3}} = 1.095 \text{ g Fe}_2\text{O}_3$$

Finally, the percentage of Fe_2O_3 in the sample was

$$\% \ \text{Fe}_2\text{O}_3 = \frac{1.095 \text{ g Fe}_2\text{O}_3}{2.000 \text{ g sample}} \times 100\% = 54.75\% \ \text{Fe}_2\text{O}_3$$

The ore sample contained 54.75% Fe_2O_3.

Is the Answer Reasonable?

We've noted that the amount of Fe in the sample is approximately 0.015 mol. The amount of Fe_2O_3 that contains this much Fe is 0.0075 mol. The formula mass of Fe_2O_3 is about 160, so the mass of Fe_2O_3 in the sample was approximately $0.0075 \times 160 = 1.2$ g, which isn't too far from the mass we obtained (1.095 g). Since 1.095 g is about half of the total sample mass of 2.000 g, the sample was approximately 50% Fe_2O_3, in agreement with the answer we obtained.

Practice Exercise 14

A sample of a tin ore weighing 0.3000 g was dissolved in an acid solution and all the tin in the sample was changed to tin(II). The solution was titrated with 8.08 mL of 0.0500 M $KMnO_4$ solution, which oxidized the tin(II) to tin(IV).

(a) What is the balanced equation for the reaction in the titration?
(b) How many grams of tin were in the sample?
(c) What was the percentage by weight of tin in the sample?
(d) If the tin in the sample had been present in the compound SnO_2, what would have been the percentage by weight of SnO_2 in the sample? ◆

SUMMARY

Oxidation–Reduction. Oxidation is the loss of electrons or an increase in oxidation number; **reduction** is the gain of electrons or a decrease in oxidation number. Both always occur together in **redox reactions.** The substance oxidized is the **reducing agent;** the substance reduced is the **oxidizing agent. Oxidation numbers** are a bookkeeping device that we use to follow changes in redox reactions. They are assigned according to the rules on page 196. The term **oxidation state** is equivalent to oxidation number. General definitions of oxidation and reduction are the following: **oxidation** is an algebraic increase in oxidation number; **reduction** is an algebraic decrease in oxidation number.

Ion–Electron Method. In a balanced redox equation, the number of electrons gained by one substance is always equal to the number lost by another substance. This fact forms the basis for the **ion–electron method** (pages 204 and 205), which provides a systematic method for deriving a net ionic equation for a redox reaction in aqueous solution. According to this method the *skeleton* net ionic equation is divided into two **half-reactions,** which are balanced separately before being recombined to give the final bal-

anced net ionic equation. For reactions in basic solution, the equation is balanced as if it occurred in an acidic solution, and then the balanced equation is converted to its proper form for basic solution by adding an appropriate number of OH^-.

Metal–Acid Reactions. In **nonoxidizing acids,** the strongest oxidizing agent is H^+ (Table 5.1). The reaction of a metal with a nonoxidizing acid gives hydrogen gas and a salt of the acid. Only metals more active than hydrogen react this way. These are metals that are located below hydrogen in the activity series (Table 5.2). **Oxidizing acids,** like HNO_3, contain an anion that is a stronger oxidizing agent than H^+, and they are able to oxidize many metals that nonoxidizing acids cannot.

Metal-Displacement Reactions. If one metal is more easily oxidized than another, it can displace the other metal from its compounds by a redox reaction. Such reactions are sometimes called **single replacement reactions.** Atoms of the more active metal become ions; ions of the less active metal generally become atoms. In this manner, any metal in the activity series can displace any of the others above it in the series from their compounds.

Oxidations by Molecular Oxygen. **Combustion,** the rapid reaction of a substance with oxygen, is accompanied by the evolution of heat and light. Combustion of a hydrocarbon in the presence of excess oxygen gives CO_2 and H_2O, two molecular oxides. When the supply of oxygen is limited, some CO also forms, and in a very limited supply of oxygen the products are H_2O and very finely divided, elemental carbon (as soot or lampblack). The combustion of organic compounds containing only carbon, hydrogen, and

oxygen also gives the same products, CO_2 and H_2O. Sulfur burns to give SO_2, which also forms when sulfur-containing fuels burn. Most nonmetals also burn in oxygen to give molecular oxides.

Many metals combine with oxygen, but only sometimes is the reaction rapid enough to be considered combustion. The products are ionic metal oxides.

Redox Titrations. Potassium permanganate is often used in redox titrations because it is a powerful oxidizing agent and serves as its own indicator. In acidic solutions, the purple MnO_4^- ion is reduced to the nearly colorless Mn^{2+} ion.

Tools *you have learned*

The table below lists the tools that you have learned in this chapter. You will need them to solve chemistry problems. Review them if necessary, and refer to the tools when working on the

Thinking-It-Through problems and the Review Problems that follow.

Tool	Function
Rules for assigning oxidation numbers (*page 196*)	To assign oxidation numbers to the atoms in a chemical formula.
	To use changes in oxidation numbers to enable you to identify oxidation and reduction.
Ion–electron method (*Follow the steps outlined on page 204 for acidic solutions, and on page 205 for basic solutions.*)	To enable you to write balanced net ionic equations for redox reactions.
Table of oxidizing and nonoxidizing acids (*page 208*)	To enable you to determine the products of reactions of metals with acids.
Activity series of metals (*page 211*)	To allow you to predict the outcome of single displacement reactions.

THINKING IT THROUGH

The goal in the following problems is *not* to find the answer itself; instead, you are asked to assemble the available information needed to obtain the answer, state what additional data (if any) are needed, and describe *how* you would use the data to answer the question. Where problems can be solved by the factor-label method, set up the solution to the problem by arranging the conversion factors so the

units cancel correctly to give the desired units of the answer.

The problems are divided into two groups. Those in Level 2 are more challenging than those in Level 1 and provide an opportunity to really hone your problem solving skills. Detailed answers to these problems can be found at our Web site: http://www.wiley.com/college/brady.

Level 1 Problems

1. Is the reaction $HIO_4 + 2H_2O \rightarrow H_5IO_6$ a redox reaction? (Describe how you would go about answering this question.)

2. Will cadmium metal react spontaneously with Ni^{2+} ion to give Cd^{2+} and metallic nickel? (How do you determine the answer to this question?)

3. A solution contains Au^{3+} and Fe^{2+} ions. Into the solution are placed strips of metallic Au and Fe. What reaction will occur? (Explain how you would answer this question.)

4. A few minutes after you first start your car, drops of a colorless liquid can be seen falling from the exhaust pipe. After a while, this stops. Explain these observations in terms of the chemistry involved in the running of the engine and the physical changes that occur.

5. Iodate ion, IO_3^-, oxidizes $NO_2(g)$ to give iodide ion and nitrate ion as products. How many milliliters of 0.0200 M IO_3^- solution will react with 0.230 g of NO_2? (Set up the solution to the problem.)

6. A certain reaction is expected to evolve Cl_2 gas as a product, and it is desired to keep it from escaping into the lab. Therefore, the gases formed in the reaction will be trapped and passed through a solution of $Na_2S_2O_3$. The Cl_2 reacts with $S_2O_3^{2-}$ to give Cl^- and SO_4^{2-}. If 0.020 mol of Cl_2 is expected to be formed, what is the minimum volume (in milliliters) of 0.500 M $Na_2S_2O_3$ that will be required to react with all the Cl_2? (Set up the solution to the problem.)

7. How many milliliters of 0.200 M $Na_2S_2O_3$ solution are required to react completely with 0.020 mol of I_3^-? In the reaction, I_3^- is reduced to I^- and $S_2O_3^{2-}$ is oxidized to $S_4O_6^{2-}$. (Set up the solution to the problem.)

Level 2 Problems

8. A mixture is made by combining 300 mL of 0.0200 M $Na_2Cr_2O_7$ with 400 mL of 0.0600 M $Fe(NO_3)_2$. Initially, the H^+ concentration in the mixture is 0.400 M. Dichromate ion oxidizes Fe^{2+} to Fe^{3+} and is reduced to Cr^{3+}. After the reaction in the mixture has ceased, how many milliliters of 0.0100 M NaOH will be required to neutralize the remaining H^+? (Set up the solution to the problem.)

9. A solution contained a mixture of SO_3^{2-} and $S_2O_3^{2-}$. A 100.0 mL portion of the solution was found to react with 80.00 mL of 0.0500 M CrO_4^{2-} in a basic solution to give CrO_2^-. The only sulfur-containing product was SO_4^{2-}. After the reaction, the solution was treated with excess $BaCl_2$ solution, which precipitated $BaSO_4$. This solid was filtered from the solution, dried, and found to weigh 0.9336 g. Explain in detail how you can determine the molar concentrations of SO_3^{2-} and $S_2O_3^{2-}$ in the original solution. (Set up the solution to the problem.)

10. An organic compound contains carbon, hydrogen, and sulfur. A sample of it with a mass of 1.045 g was burned in oxygen to give gaseous CO_2, H_2O, and SO_2. These gases were passed through 500 mL of an acidified 0.0200 M $KMnO_4$ solution, which caused the SO_2 to be oxidized to

SO_4^{2-}. Only part of the available $KMnO_4$ was reduced to Mn^{2+}. Next, 50.00 mL of 0.0300 M $SnCl_2$ was added to a 50.00 mL portion of this solution, which still contained unreduced $KMnO_4$. There was more than enough added $SnCl_2$ to cause all of the remaining MnO_4^- in the 50 mL portion to be reduced to Mn^{2+}. The excess Sn^{2+} that still remained after the reaction was then allowed to react with 0.0100 M $KMnO_4$, using up 27.28 mL of the $KMnO_4$ solution. Show how you would calculate the percentage of sulfur in the original sample of the organic compound that had been burned.

11. A bar of copper weighing 32.00 g was dipped into 50.0 mL of 0.250 M $AgNO_3$ solution. If all the silver that deposits adheres to the copper bar, how much will the bar weigh after the reaction is complete? (Describe the calculations necessary to solve the problem. Write and balance any necessary chemical equations.)

12. A compound AY_2 reacts with another compound BW in an oxidation reduction reaction. Y is an element in AY_2 with an oxidation number of +3. W is an element of BW with an oxidation number of −2. In one of the products, CY, Y has an oxidation number of +1. The other product is DW. A, B, C, and D are the remaining atoms in these compounds, none of which undergo oxidation or reduction.

$$AY_2 + BW \longrightarrow CY + DW$$

(a) What are possible oxidation numbers for W in DW?
(b) If the coefficients of AY_2 and BW in the balance equation are both equal to one, what is the oxidation number of W in DW?
(c) What are the smallest possible coefficients of AY_2 and BW in the balanced equation if the oxidation number of W in DW is +1?

13. You are given strips of four pure metals (A, B, C, and D) and four bottles containing 0.10 M solutions of the nitrate salts of these same four metals. Describe experiments you could perform in order to arrange these four metals in order of increasing activity or ease of oxidation. The only other items available to you are a balance and as many beakers as you need. You have as many strips of each metal as you need along with emery paper to clean the metal surfaces.

REVIEW QUESTIONS

Oxidation–Reduction

5.1 Define oxidation and reduction (a) in terms of electron transfer and (b) in terms of oxidation numbers.

5.2 In the reaction $2Mg + O_2 \rightarrow 2MgO$, which substance is the oxidizing agent and which is the reducing agent? Which substance is oxidized and which is reduced?

5.3 Why must both oxidation and reduction occur simultaneously during a redox reaction? What is an oxidizing

agent? What happens to it in a redox reaction? What is a reducing agent? What happens to it in a redox reaction?

5.4 In the compound As_4O_6, arsenic has an oxidation number of +3. What is the oxidation state of the arsenic in this compound?

5.5 Is the following a redox reaction? Explain.

$$2NO_2 \rightarrow N_2O_4$$

5.6 Is the following a redox reaction? Explain.

$$2CrO_4^{2-} + 2H^+ \rightarrow Cr_2O_7^{2-} + H_2O$$

5.7 If the oxidation number of nitrogen in a certain molecule changes from +3 to -2 during a reaction, is the nitrogen oxidized or reduced? How many electrons are gained (or lost) by each nitrogen atom?

Ion–Electron Method

5.8 The following equation is not balanced. Why? Use the ion–electron method to balance it.

$$Ag + Fe^{2+} \rightarrow Ag^+ + Fe$$

5.9 Use the ion–electron method to balance the following equation.

$$Cr^{3+} + Zn \rightarrow Cr + Zn^{2+}$$

5.10 What are the net charges on the left and right sides of the following equations? How many electrons must be added to which side of each half-reaction?
(a) $NO_3^- + 10H^+ \rightarrow NH_4^+ + 3H_2O$
(b) $Cl_2 + 4H_2O \rightarrow 2ClO_2^- + 8H^+$

5.11 In the preceding question, which half-reaction represents oxidation? Which represents reduction?

Reactions of Metals with Acids and the Activity Series

5.12 What is a *single replacement reaction?*

5.13 What is a nonoxidizing acid? Give two examples. What is the oxidizing agent in a nonoxidizing acid?

5.14 What is the strongest oxidizing agent in an aqueous solution of nitric acid?

5.15 Where in the activity series do we find the best reducing agents? Where do we find the best oxidizing agents?

5.16 Which metals in Table 5.2 will not react with nonoxidizing acids?

5.17 Which metals in Table 5.2 will react with water? Write chemical equations for each of these reactions.

5.18 When manganese reacts with silver ion, is manganese oxidized or reduced? Is it an oxidizing agent or a reducing agent?

Oxygen as an Oxidizing Agent

5.19 Define combustion.

5.20 Why is "loss of electrons" described as oxidation?

5.21 What products are produced in the combustion of $C_{10}H_{22}$ (a) if there is an excess of oxygen available? (b) If there is a slightly limited oxygen supply? (c) If there is a very limited supply of oxygen?

5.22 If one of the impurities in diesel fuel has the formula C_2H_6S, what products will be formed when it burns? Write a balanced chemical equation for the reaction.

5.23 Burning ammonia in an atmosphere of oxygen produces stable N_2 molecules as one of the products. What is the other product? Write the balanced equation for the reaction.

REVIEW PROBLEMS

Answers to problems whose numbers are printed in color are given in Appendix D.
More challenging problems are marked with asterisks.

Oxidation–Reduction; Oxidation Numbers

5.24 For the following reactions, identify the substance oxidized, the substance reduced, the oxidizing agent, and the reducing agent.
(a) $2HNO_3 + 3H_3AsO_3 \rightarrow 2NO + 3H_3AsO_4 + H_2O$
(b) $NaI + 3HOCl \rightarrow NaIO_3 + 3HCl$
(c) $2KMnO_4 + 5H_2C_2O_4 + 3H_2SO_4 \rightarrow$
$\qquad 10CO_2 + K_2SO_4 + 2MnSO_4 + 8H_2O$
(d) $6H_2SO_4 + 2Al \rightarrow Al_2(SO_4)_3 + 3SO_2 + 6H_2O$

5.25 For the following reactions, identify the substance oxidized, the substance reduced, the oxidizing agent, and the reducing agent.
(a) $Cu + 2H_2SO_4 \rightarrow CuSO_4 + SO_2 + 2H_2O$
(b) $3SO_2 + 2HNO_3 + 2H_2O \rightarrow 3H_2SO_4 + 2NO$
(c) $5H_2SO_4 + 4Zn \rightarrow 4ZnSO_4 + H_2S + 4H_2O$
(d) $I_2 + 10HNO_3 \rightarrow 2HIO_3 + 10NO_2 + 4H_2O$

5.26 Assign oxidation numbers to the atoms indicated by boldface type: (a) $\mathbf{S}^{2-}$, (b) $\mathbf{S}O_2$, (c) $\mathbf{P}_4$, (d) $\mathbf{P}H_3$.

5.27 Assign oxidation numbers to the atoms indicated by boldface type: (a) $\mathbf{Cl}O_4^-$, (b) $\mathbf{Cr}Cl_3$, (c) $\mathbf{Sn}S_2$, (d) $\mathbf{Au}(NO_3)_3$.

5.28 Assign oxidation numbers to all of the atoms in the following compounds: (a) Na_2HPO_4, (b) $BaMnO_4$, (c) $Na_2S_4O_6$, (d) ClF_3.

5.29 Assign oxidation numbers to all the atoms in the following ions: (a) NO_3^-, (b) SO_3^{2-}, (c) NO^+, (d) $Cr_2O_7^{2-}$.

5.30 Assign oxidation numbers to nitrogen in the following.
(a) NO   (c) NH_2OH (e) N_2H_4
(b) N_2O_5 (d) NO_2

5.31 Assign oxidation numbers to nitrogen in the following.
(a) NH_3 (c) N_2 (e) HNO_2
(b) N_2O_3 (d) NaN_3

5.32 Assign oxidation numbers to each atom in the following: (a) $NaOCl$, (b) $NaClO_2$, (c) $NaClO_3$, (d) $NaClO_4$.

5.33 Assign oxidation numbers to the elements in the following. (a) $Ca(VO_3)_2$, (b) $SnCl_4$, (c) MnO_4^{2-}, (d) MnO_2

5.34 Assign oxidation numbers to the elements in the following. (a) PbS, (b) $TiCl_4$, (c) $Sr(IO_3)_2$, (d) Cr_2S_3

5.35 Assign oxidation numbers to the elements in the following. (a) OF_2, (b) HOF, (c) CsO_2, (d) O_2F_2

Ion–Electron Method

5.36 Balance the following half-reactions occurring in an acidic solution. Indicate whether each is an oxidation or a reduction.
(a) $BiO_3^- \rightarrow Bi^{3+}$ (b) $Pb^{2+} \rightarrow PbO_2$

5.37 Balance the following half-reactions occurring in an acidic solution. Indicate whether each is an oxidation or a reduction.
(a) $NO_3^- \rightarrow NH_4^+$ (b) $Cl_2 \rightarrow ClO_3^-$

5.38 Balance the following half-reactions occurring in a basic solution. Indicate whether each is an oxidation or a reduction.
(a) $Fe \rightarrow Fe(OH)_2$ (b) $SO_2Cl_2 \rightarrow SO_3^{2-} + Cl^-$

5.39 Balance the following half-reactions occurring in a basic solution. Indicate whether each is an oxidation or a reduction.
(a) $Mn(OH)_2 \rightarrow MnO_4^{2-}$ (b) $H_4IO_6^- \rightarrow I_2$

5.40 Balance the following equations for reactions occurring in an acidic solution.
(a) $S_2O_3^{2-} + OCl^- \rightarrow Cl^- + S_4O_6^{2-}$
(b) $NO_3^- + Cu \rightarrow NO_2 + Cu^{2+}$
(c) $IO_3^- + AsO_3^{3-} \rightarrow I^- + AsO_4^{3-}$
(d) $SO_4^{2-} + Zn \rightarrow Zn^{2+} + SO_2$
(e) $NO_3^- + Zn \rightarrow NH_4^+ + Zn^{2+}$
(f) $Cr^{3+} + BiO_3^- \rightarrow Cr_2O_7^{2-} + Bi^{3+}$
(g) $I_2 + OCl^- \rightarrow IO_3^- + Cl^-$
(h) $Mn^{2+} + BiO_3^- \rightarrow MnO_4^- + Bi^{3+}$
(i) $H_3AsO_3 + Cr_2O_7^{2-} \rightarrow H_3AsO_4 + Cr^{3+}$
(j) $I^- + HSO_4^- \rightarrow I_2 + SO_2$

5.41 Balance these equations for reactions occurring in an acidic solution.
(a) $Sn + NO_3^- \rightarrow SnO_2 + NO$
(b) $PbO_2 + Cl^- \rightarrow PbCl_2 + Cl_2$
(c) $Ag + NO_3^- \rightarrow NO_2 + Ag^+$
(d) $Fe^{3+} + NH_3OH^+ \rightarrow Fe^{2+} + N_2O$
(e) $HNO_2 + I^- \rightarrow I_2 + NO$
(f) $C_2O_4^{2-} + HNO_2 \rightarrow CO_2 + NO$
(g) $HNO_2 + MnO_4^- \rightarrow Mn^{2+} + NO_3^-$
(h) $H_3PO_2 + Cr_2O_7^{2-} \rightarrow H_3PO_4 + Cr^{3+}$
(i) $VO_2^+ + Sn^{2+} \rightarrow VO^{2+} + Sn^{4+}$
(j) $XeF_2 + Cl^- \rightarrow Xe + F^- + Cl_2$

5.42 Balance equations for these reactions occurring in a basic solution.
(a) $CrO_4^{2-} + S^{2-} \rightarrow S + CrO_2^-$
(b) $MnO_4^- + C_2O_4^{2-} \rightarrow CO_2 + MnO_2$
(c) $ClO_3^- + N_2H_4 \rightarrow NO + Cl^-$
(d) $NiO_2 + Mn(OH)_2 \rightarrow Mn_2O_3 + Ni(OH)_2$
(e) $SO_3^{2-} + MnO_4^- \rightarrow SO_4^{2-} + MnO_2$

5.43 Balance equations for these reactions occurring in a basic solution.
(a) $CrO_2^- + S_2O_8^{2-} \rightarrow CrO_4^{2-} + SO_4^{2-}$
(b) $SO_3^{2-} + CrO_4^{2-} \rightarrow SO_4^{2-} + CrO_2^-$
(c) $O_2 + N_2H_4 \rightarrow H_2O_2 + N_2$
(d) $Fe(OH)_2 + O_2 \rightarrow Fe(OH)_3 + OH^-$
(e) $Au + CN^- + O_2 \rightarrow Au(CN)_4^- + OH^-$

5.44 Write a balanced net ionic equation for the reaction of NaOCl with $Na_2S_2O_3$. The OCl^- is reduced to chloride ion and the $S_2O_3^{2-}$ is oxidized to sulfate ion.

5.45 Write a balanced net ionic equation for the oxidation of $H_2C_2O_4$ by $K_2Cr_2O_7$ in an acidic solution. The reaction yields Cr^{3+} and CO_2 among the products.

Reactions of Metals with Acids

5.46 Write balanced molecular, ionic, and net ionic equations for the reactions of the following metals with hydrochloric acid to give hydrogen plus the metal ion in solution.
(a) Manganese (gives Mn^{2+})   (d) Nickel (gives Ni^{2+})
(b) Cadmium (gives Cd^{2+})   (e) Chromium (gives Cr^{3+})
(c) Tin (gives Sn^{2+})

5.47 Write balanced molecular, ionic, and net ionic equations for the reaction of each metal in the previous problem with dilute sulfuric acid.

5.48 On the basis of the discussions in this chapter, suggest chemical equations for the oxidation of metallic silver to silver(I) ion with (a) dilute HNO_3 and (b) concentrated HNO_3.

5.49 When hot and concentrated, sulfuric acid is a fairly strong oxidizing agent. Write a balanced net ionic equation for the oxidation of metallic copper to copper(II) ion by hot concentrated H_2SO_4, in which the sulfur is reduced to SO_2. Write a balanced molecular equation for the reaction.

Displacement Reactions and the Activity Series

5.50 Use Table 5.2 to predict the outcome of the following reactions. If no reaction occurs, write N.R. If a reaction occurs, write a balanced equation for it.
(a) $Fe + Mg^{2+} \rightarrow$ (c) $Ag^+ + Fe \rightarrow$
(b) $Cr + Pb^{2+} \rightarrow$ (d) $Ag + Au^{3+} \rightarrow$

5.51 Use Table 5.2 to predict the outcome of the following displacement reactions. If no reaction occurs, write N.R. If a reaction occurs, write a balanced chemical equation for it.
(a) $Mn + Fe^{2+} \rightarrow$ (c) $Mg + Co^{2+} \rightarrow$
(b) $Cd + Zn^{2+} \rightarrow$ (d) $Cr + Sn^{2+} \rightarrow$

Reactions of Oxygen

5.52 Write balanced chemical equations for the complete combustion (in the presence of excess oxygen) of the following:
(a) C_6H_6 (benzene, an important industrial chemical and solvent)
(b) C_3H_8 (propane, a gaseous fuel used in many stoves)
(c) $C_{21}H_{44}$ (a component of paraffin wax)

5.53 Write balanced chemical equations for the complete combustion (in the presence of excess oxygen) of the following:
(a) $C_{12}H_{26}$ (a component of kerosene)
(b) $C_{18}H_{36}$ (a component of diesel fuel)
(c) C_7H_8 (toluene, a raw material in the production of TNT)

5.54 Write balanced equations for the combustion of the hydrocarbons in Problem 5.52 in (a) a slightly limited supply of oxygen and (b) a very limited supply of oxygen.

5.55 Write balanced equations for the combustion of the hydrocarbons in Problem 5.53 in (a) a slightly limited supply of oxygen and (b) a very limited supply of oxygen.

5.56 Methanol, CH_3OH, has been suggested as an alternative to gasoline as an automotive fuel. Write a balanced chemical equation for its complete combustion.

5.57 Metabolism of carbohydrates such as glucose, $C_6H_{12}O_6$, produces the same products as complete combustion. Write a chemical equation representing the metabolism (combustion) of glucose.

Redox Reactions and Stoichiometry

5.58 Manganese(II) ion is oxidized to permanganate ion by bismuthate ion, BiO_3^-, in an acidic solution. In the reaction, BiO_3^- is reduced to Bi^{3+}.
(a) Write a balanced net ionic equation for the reaction.
(b) How many grams of $NaBiO_3$ are needed to oxidize the manganese in 18.5 g of $Mn(NO_3)_2$?

5.59 Iodate ion, IO_3^-, reacts with sulfite ion to give sulfate ion and iodide ion.
(a) Write a balanced net ionic equation for the reaction.
(b) How many grams of Na_2SO_3 are needed to react with 5.00 g of $NaIO_3$?

5.60 How many grams of copper must react to displace 12.0 g of silver from a solution of silver nitrate?

5.61 How many grams of aluminum must react to displace all the silver from 25.0 g of silver nitrate? The reaction occurs in aqueous solution.

5.62 In an acidic solution, MnO_4^- reacts with Sn^{2+} to give Mn^{2+} and Sn^{4+}.
(a) Write a balanced net ionic equation for the reaction.
(b) How many milliliters of 0.230 M $KMnO_4$ solution are needed to react completely with 40.0 mL of 0.250 M $SnCl_2$ solution?

5.63 In an acidic solution, HSO_3^- ion reacts with ClO_3^- ion to give SO_4^{2-} ion and Cl^- ion.
(a) Write a balanced net ionic equation for the reaction.
(b) How many milliliters of 0.150 M $NaClO_3$ solution are needed to react completely with 30.0 mL of 0.450 M $NaHSO_3$ solution?

5.64 A sample of a copper ore with a mass of 0.4225 g was dissolved in acid. A solution of potassium iodide was added, which caused the reaction

$$2Cu^{2+}(aq) + 5I^-(aq) \rightarrow I_3^-(aq) + 2CuI(s)$$

The I_3^- that formed reacted quantitatively with exactly 29.96 mL of 0.02100 M $Na_2S_2O_3$ according to the following equation.

$$I_3^-(aq) + 2S_2O_3^{2-}(aq) \rightarrow 3I^-(aq) + S_4O_6^{2-}(aq)$$

(a) What was the percentage by mass of copper in the ore?
(b) If the ore contained $CuCO_3$, what was the percentage by mass of $CuCO_3$ in the ore?

5.65 A 1.362 g sample of an iron ore that contained Fe_3O_4 was dissolved in acid and all the iron was reduced to Fe^{2+}. The solution was then acidified with H_2SO_4 and titrated with 39.42 mL of 0.0281 M $KMnO_4$, which oxidized the iron to Fe^{3+}. The net ionic equation for the reaction is

$$5Fe^{2+} + MnO_4^- + 8H^+ \rightarrow 5Fe^{3+} + Mn^{2+} + 4H_2O$$

(a) What was the percentage by mass of iron in the ore?
(b) What was the percentage by mass of Fe_3O_4 in the ore?

5.66 Hydrogen peroxide (H_2O_2) can be purchased in drug stores for use as an antiseptic. A sample of such a solution weighing 1.000 g was acidified with H_2SO_4 and titrated with a 0.02000 M solution of $KMnO_4$. The net ionic equation for the reaction is

$$6H^+ + 5H_2O_2 + 2MnO_4^- \rightarrow 5O_2 + 2Mn^{2+} + 8H_2O$$

The titration required 17.60 mL of $KMnO_4$ solution.
(a) How many grams of H_2O_2 reacted?
(b) What is the percentage by mass of the H_2O_2 in the original antiseptic solution?

5.67 Sodium nitrite, $NaNO_2$, is used as a preservative in meat products such as frankfurters and bologna. In an acidic solution, nitrite ion is converted to nitrous acid, HNO_2, which reacts with permanganate ion according to the equation

$$H^+ + 5HNO_2 + 2MnO_4^- \rightarrow 5NO_3^- + 2Mn^{2+} + 3H_2O$$

A 1.000 g sample of a water-soluble solid containing $NaNO_2$ was dissolved in dilute H_2SO_4 and titrated with 0.01000 M $KMnO_4$ solution. The titration required 12.15 mL of the $KMnO_4$ solution. What was the percentage by mass of $NaNO_2$ in the original 1.000 g sample?

5.68 A sample of a chromium-containing alloy weighing 3.450 g was dissolved in acid, and all the chromium in the sample was oxidized to CrO_4^{2-}. It was then found that 3.18 g of Na_2SO_3 was required to reduce the CrO_4^{2-} to CrO_2^- in a basic solution, with the SO_3^{2-} being oxidized to SO_4^{2-}.
(a) Write a balanced equation for the reaction of CrO_4^{2-} with SO_3^{2-} in a basic solution.
(b) How many grams of chromium were in the alloy sample?
(c) What was the percentage by mass of chromium in the alloy?

5.69 Solder is an alloy containing the metals tin and lead. A particular sample of this alloy weighing 1.50 g was dissolved in acid. All the tin was then converted to the $+2$ oxidation state. Next, it was found that 0.368 g of $Na_2Cr_2O_7$ was required to oxidize the Sn^{2+} to Sn^{4+} in an acidic solution. In the reaction the chromium was reduced to Cr^{3+} ion.

(a) Write a balanced net ionic equation for the reaction between Sn^{2+} and $Cr_2O_7^{2-}$ in an acidic solution.

(b) Calculate the number of grams of tin that were in the sample of solder.

(c) What was the percentage by mass of tin in the solder?

5.70 Both calcium chloride, $CaCl_2$, and sodium chloride are used to melt ice and snow on roads in the winter. A certain company was marketing a mixture of these two compounds for this purpose. A chemist, wishing to analyze the mixture, dissolved 2.463 g of it in water and precipitated the calcium by adding sodium oxalate, $Na_2C_2O_4$.

$$Ca^{2+} + C_2O_4^{2-} \longrightarrow CaC_2O_4(s)$$

The calcium oxalate was then carefully filtered from the solution, dissolved in sulfuric acid, and titrated with 0.1000 M $KMnO_4$ solution. The reaction that occurred was

$$6H^+ + 5H_2C_2O_4 + 2MnO_4^- \rightarrow 10CO_2 + 2Mn^{2+} + 8H_2O$$

The titration required 21.62 mL of the $KMnO_4$ solution.

(a) How many moles of $C_2O_4^{2-}$ were present in the CaC_2O_4 precipitate?

(b) How many grams of $CaCl_2$ were in the original 2.463 g sample?

(c) What was the percentage by mass of $CaCl_2$ in the sample?

5.71 A way to analyze a sample for nitrite ion is to acidify a solution containing NO_2^- and then allow the HNO_2 that is formed to react with iodide ion in the presence of excess I^-. The reaction is

$$2HNO_2 + 2H^+ + 3I^- \longrightarrow 2NO + 2H_2O + I_3^-$$

Then the I_3^- is titrated with $Na_2S_2O_3$ solution.

$$I_3^- + 2S_2O_3^{2-} \longrightarrow 3I^- + S_4O_6^{2-}$$

In a typical analysis, a 1.104 g sample that was known to contain $NaNO_2$ was treated as described above. The titration required 29.25 mL of 0.3000 M $Na_2S_2O_3$ solution to reach the end point.

(a) How many moles of I_3^- had been produced in the first reaction?

(b) How many moles of NO_2^- had been in the original 1.104 g sample?

(c) What was the percentage by mass of $NaNO_2$ in the original sample?

ADDITIONAL EXERCISES

5.72 What is the average oxidation number of carbon in (a) C_2H_5OH (grain alcohol), (b) $C_{12}H_{22}O_{11}$ (sucrose—table sugar), (c) $CaCO_3$ (limestone), (d) $NaHCO_3$ (baking soda)?

5.73 The following chemical reactions are *observed to occur* in aqueous solution.

$$2Al + 3Cu^{2+} \longrightarrow 2Al^{3+} + 3Cu$$

$$2Al + 3Fe^{2+} \longrightarrow 3Fe + 2Al^{3+}$$

$$Pb^{2+} + Fe \longrightarrow Pb + Fe^{2+}$$

$$Fe + Cu^{2+} \longrightarrow Fe^{2+} + Cu$$

$$2Al + 3Pb^{2+} \longrightarrow 3Pb + 2Al^{3+}$$

$$Pb + Cu^{2+} \longrightarrow Pb^{2+} + Cu$$

Arrange the metals Al, Pb, Fe, and Cu in order of increasing ease of oxidation.

5.74 In the preceding problem, were all the experiments described actually necessary to establish the order?

5.75 According to the activity series in Table 5.2, which of the following metals react with nonoxidizing acids? (a) silver, (b) gold, (c) zinc, (d) magnesium

5.76 In each pair below, choose the metal that would most likely react more rapidly with a nonoxidizing acid such as HCl. (a) aluminum or iron, (b) zinc or nickel, (c) cadmium or magnesium

5.77 Use Table 5.2 to predict whether the following displacement reactions should occur. If no reaction occurs, write N.R. If a reaction does occur, write a balanced chemical equation for it.

(a) $Zn + Sn^{2+} \rightarrow$ (d) $Mn + Pb^{2+} \rightarrow$
(b) $Cr + H^+ \rightarrow$ (e) $Zn + Co^{2+} \rightarrow$
(c) $Pb + Cd^{2+} \rightarrow$

5.78 Sucrose, $C_{12}H_{22}O_{11}$, is ordinary table sugar. Write a balanced chemical equation representing the metabolism of sucrose. (See Review Problem 5.57.)

5.79 Write chemical equations for the reaction of oxygen with (a) zinc, (b) aluminum, (c) magnesium, (d) iron, and (e) calcium.

*5.80 Balance the following equations by the ion–electron method.

(a) $NBr_3 \rightarrow N_2 + Br^- + HOBr$ (basic solution)
(b) $Cl_2 \rightarrow Cl^- + ClO_3^-$ (basic solution)
(c) $H_2SeO_3 + H_2S \rightarrow S + Se$ (acidic solution)
(d) $MnO_2 + SO_3^{2-} \rightarrow Mn^{2+} + S_2O_6^{2-}$ (acidic solution)
(e) $XeO_3 + I^- \rightarrow Xe + I_2$ (acidic solution)
(f) $(CN)_2 \rightarrow CN^- + OCN^-$ (basic solution)

5.81 Lead(IV) oxide reacts with hydrochloric acid to give chlorine. The equation for the reaction is

$$PbO_2 + 4Cl^- + 4H^+ \longrightarrow PbCl_2 + 2H_2O + Cl_2$$

How many grams of PbO_2 must react to give 15.0 g of Cl_2?

*5.82 A solution contains $Ce(SO_4)_3{}^{2-}$ at a concentration of 0.0150 M. It was found that in a titration, 25.00 mL of this solution reacted completely with 23.44 mL of 0.032 M $FeSO_4$ solution. The reaction gave Fe^{3+} as a product in the solution. In this reaction, what is the final oxidation state of the Ce?

*5.83 A copper bar with a mass of 12.340 g is dipped into 255 mL of 0.125 M $AgNO_3$ solution. When the reaction that occurs has finally ceased, what will be the mass of unreacted copper in the bar? If all the silver that forms adheres to the copper bar, what will be the total mass of the bar after the reaction?

5.84 A solution containing 0.1244 g of $K_2C_2O_4$ was acidified, changing the $C_2O_4{}^{2-}$ ions to $H_2C_2O_4$. The solution was then titrated with 13.93 mL of a $KMnO_4$ solution to reach a faint pink end point. In the reaction, $H_2C_2O_4$ was oxidized to CO_2 and $MnO_4{}^-$ was reduced to Mn^{2+}. What was the molarity of the $KMnO_4$ solution used in the titration?

*5.85 It was found that a 20.0 mL portion of a solution of oxalic acid, $H_2C_2O_4$, requires 6.25 g of 0.200 M $K_2Cr_2O_7$ for complete reaction in an acidic solution. In the reaction, the oxidation product is CO_2 and the reduction product is Cr^{3+}. How many milliliters of 0.450 M NaOH are required to completely neutralize the $H_2C_2O_4$ in a separate 20.00 mL sample of the same oxalic acid solution?

Xenon—
Star Wars Propulsion "Fuel" Today

3

What was once just fantasy in the *Star Wars* saga has now become a novel use of ion propulsion to propel primary spacecraft. The concept is simple: use solar energy to convert the heavy atoms of a gas (in reality, the element xenon) into positively charged ions, and let their natural repulsions provide thrust for the spacecraft. NASA engineers deployed such a propulsion system as the *primary* engine for *Deep Space 1,* an unmanned spacecraft launched in October, 1998. It's purpose was to rendezvous with and study an orbiting asteroid, called 1992 KD (see Figure 3*a*). If all went well, *Deep Space 1* was to fly on to study the death throes of the Wilson-Harrington comet.

Solar-Electric Propulsion

An ion propulsion engine, unlike conventional spacecraft engines, produces thrust over the life of the craft, not just at its start. Visualize the force exerted by a single sheet of typing paper resting on the palm of your hand. This tiny, almost unnoticeable force is roughly how much thrusting force is steadily generated by the ion engine of *Deep Space 1*. Yet, in the near vacuum of space, such a force is able to increase the speed of the spacecraft by about 9.1 m/sec/day (30 ft/sec/day). That is all that is needed to make the final speed over the life of the mission roughly 13×10^3 km/hr (8.0×10^3 miles per hour). Here's how the system works.

Figure 3*b* shows a cutaway drawing of the ion engine. Both xenon and electrons enter on the left, xenon from a "tank" initially holding a little over 400 kg of this Group VIIIA gaseous element. Solar energy is gathered from large solar panels attached to the engine and funneled into a cathode ray tube where electrons are generated. The electrons are accelerated toward positively charged surfaces on the chamber walls. In transit, collisions between these electrons and xenon atoms generate positively charged xenon ions. The xenon ions leave the chamber through its rear, open end where electrically charged grids focus the ions into a high speed beam, which causes the propelling thrust. Another cathode directs an electron beam on the exiting xenon ions, changing them back to xenon atoms so that there is no buildup of negative charge inside the engine.

Why Xenon?

Xenon (atomic number 54) is a Group VIIIA element, the least reactive of all elements. So part of the answer to the question, "Why xenon?" is that xenon will not react at low temperature and pressure with anything.[1] For a long space voyage, you wouldn't want the fuel to corrode the engine or the tank.

So now we know that it makes sense to use a Group VIIIA element. But out of 100 moles of all of the gases in dry air, only 0.087×10^{-6} moles are xenon, the lowest concentration of all of air's components. Yet air, for all practical purposes, is the only source of this element that we have. So now we must expand the question and ask, "Why xenon instead of helium or argon?" Argon is over 107,000 times as concentrated as xenon in air and helium is 60 times as concentrated. It would seem to be cheaper to use the more abundant and more easily isolated elements.

Remember that xenon is not really a "fuel" until it's changed to xenon cations. Electrons generated by solar energy do this vital work, so *Deep Space 1* is ultimately powered by the sun. To keep the demand for solar energy to a minimum, you would want to use an element with as low an ionization energy as possible. So part of the answer to "Why xenon instead of helium or argon?" is that xenon requires the least energy to be changed to ions (has the lowest ionization energy) of all of the nonradioactive Group VIIIA elements (see page 309).

Xenon atoms are also the most massive of the atoms of the nonradioactive Group VIIIA elements. Therefore, when xenon cations are focused and accelerated, they produce more thrust than could the cations of the lighter Group VIIIA elements.

[1] Under the right conditions xenon, unlike helium, neon, and argon, can be made to form fluorides and oxides. But you can be sure that neither fluorine nor oxygen is present in an ion propulsion engine.

chemicals in our world

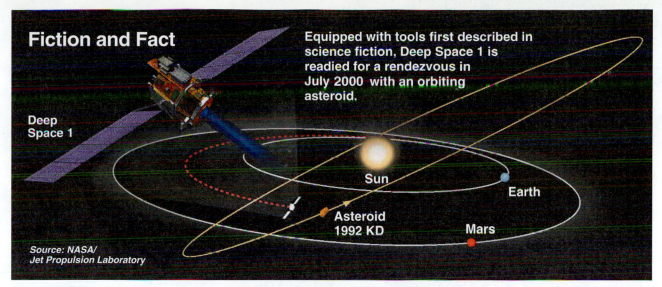

Figure 3a The route of *Deep Space 1* to asteroid 1992 KD.

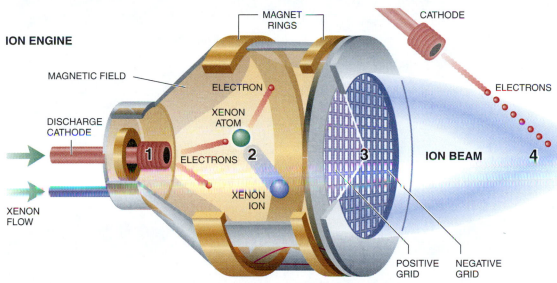

Figure 3b Ion propulsion engine of *Deep Space 1*. (1) Electrons boil off a cathode using solar energy. Xenon atoms enter here, too. (2) Electrons, attracted to and accelerated by magnetic rings, strike xenon atoms, creating xenon cations. (3) Xenon cations are focused by electrically charged grids and stream out at high velocity into space, thus providing the engine's thrust. (4) Electron beam neutralizes xenon cations, preventing the engine from developing a negative charge.

Suggested Readings

T. Beardsley, "The Way to Go in Space," *Scientific American*, February 1999, page 81. See also W. E. Leary, *The New York Times*, October 6, 1998, page B11.

Questions

1. Why is xenon, one of the least available of all of the gaseous elements, used as the "fuel" for the ion propulsion engine?

2. Solar energy, not xenon, is the ultimate source of the energy for propelling *Deep Space 1*. Explain.

In huge "pours" of concrete, such as needed to build the Grand Coulee Dam on the Columbia River in eastern Washington, so much heat evolves that it must be carried away by cold water circulating in networks of pipes embedded in the structure. Thermochemistry deals with heats of reaction.

Energy and Thermochemistry

This Chapter in Context Of the two great topics studied by chemists, matter and energy, we have so far focused mostly on the first. We turn now to the second. The study of the energies given off or absorbed by chemical reactions makes up a field called *thermochemistry*. Out of it has come many useful applications as well as important clues about the nature of matter itself. In Chapter 7, for example, you will learn that atoms can be made to emit and absorb light, and that the amounts of energy associated with such light helped scientists figure out the internal structures of atoms.

▶**Chemistry in Practice**◀ We carry out many reactions not to make compounds but to obtain energy. Every time we burn fuels in a furnace, for example, we are running reactions to produce heat. The explosive energy from chemical reactions in vehicle engines causes gases to expand that impel pistons or rockets, and so makes things move. The energy from the burning of coal, oil, or natural gas at power stations makes the wiring of turbines turn in a magnetic field and so generates electricity. Batteries give us portable electricity because the energies of chemical reactions can be directed to make electrons move in wires without the need for turbines. ◆

6.1 Kinetic and Potential Energy Revisited

In Chapter 1 we defined energy as something an object has if the object is able to do work. There are basically two forms of energy, namely, kinetic and potential energy. Let's briefly review them before we move into thermochemistry itself.

Kinetic and Potential Energy

As you learned in Section 1.3, energy is not a *physical* thing to be weighed and bottled. Rather it's an *ability* possessed by things, the ability to do work. You learned that objects can have energy in two fundamental ways, kinetic and

Kinetic is from the Greek *kinetikos,* meaning "of motion."

Work is defined more precisely on page 240.

potential. *Kinetic energy* or KE, the energy of motion associated with *mechanical work,* is calculated from a moving object's mass (m) and velocity (v) by the equation $KE = \frac{1}{2}mv^2$.

Potential energy, or PE, is energy "in storage," existing *because there are in the universe natural attractions or repulsions that objects or their parts experience for each other.* When you lift a book off your desk, for example, you convert kinetic energy (associated with your motions) into stored or potential energy. The book and Earth have a natural force of attraction for each other called the gravitational force, so you have to use some of your own kinetic energy to increase the distance between Earth and the book. We say that you *do work* on the book to move it upward, against the gravitational attraction. Your work energy, now transformed into potential energy possessed by the book, can be recovered as kinetic energy by dropping the book. The higher you move the book, the more you work, but then the more is the potential energy that you give the book, and the more is the kinetic energy released when you drop it. Thus, *one very important property of potential energy is that it can be converted into kinetic energy.* Likewise, kinetic energy can be converted into potential energy.

The work of squeezing or compressing a spring is another example of converting kinetic energy to potential energy, that of the compressed spring. When the spring is released, it might push on something and thus transform its potential energy back into kinetic energy and mechanical work.

These two examples illustrate the two basic mechanisms for increasing the supply of potential energy in something.

Winding a watch also puts energy into storage in the mainspring. It then changes to the kinetic energy of the moving hands.

How Potential Energy Increases

1. Potential energy increases whenever objects that attract each other are pulled apart, as in lifting a book to a higher level above Earth.

2. Potential energy increases whenever objects that naturally experience a repelling force are forced together, as in compressing a spring.

Paralleling these, of course, are two mechanisms by which potential energy can decrease.

How Potential Energy Decreases

1. Potential energy decreases when objects that attract each other come closer together, as when a book falls toward Earth.

2. Potential energy decreases when things that repel are allowed to move farther apart, as when a compressed spring pushes objects apart.

Another important aspect of energy introduced in Section 1.3 was the *law of conservation of energy.* Energy *cannot* be created nor destroyed, but only changed from one form to another.[1] An implication of this concept is that *the total energy of the universe is constant.*

$$\text{Energy} = \text{a constant in the universe}$$

Furthermore, regardless of the kinds of changes and transformations of KE into PE or PE into KE, their sum total remains unchanged.

The energy used to compress a spring remains as the spring's potential energy.

[1]The famous physicist Albert Einstein showed that matter and energy are interconvertible, and in some circumstances matter can actually be changed to energy and vice versa. This happens, for example, in nuclear reactions, which are discussed in Chapter 22. The law of conservation of energy, therefore, should really be stated as the *law of conservation of mass–energy.* The total mass and energy of the universe is a constant. However, for chemical reactions under normal conditions the amount of such interconversion, mass to energy or energy to mass, is entirely negligible.

Units of Energy

The SI unit of energy is called the **joule** (abbreviated **J**) and corresponds to the amount of kinetic energy possessed by a 2 kilogram object moving at a speed of 1 meter per second (approximately the energy of a 4.4 lb object traveling at 2.25 mi/hr).

$$1 \text{ J} = \frac{1}{2} (2 \text{ kg}) \left(\frac{1 \text{ m}}{1 \text{ s}} \right)^2$$

$$1 \text{ J} = 1 \text{ kg m}^2 \text{ s}^{-2}$$

The joule is actually a rather small amount of energy, and in most cases we will use the larger unit, the **kilojoule (kJ).**

$$1 \text{ kJ} = 10^3 \text{ J}$$

Another energy unit you may be familiar with is called the **calorie (cal).** Originally, it was defined as the energy needed to raise the temperature of 1 gram of water by 1 degree Celsius. With the introduction of the SI, the calorie has been redefined as follows:

$$1 \text{ cal} = 4.184 \text{ J} \qquad \text{(exactly)}$$

The larger unit **kilocalorie (kcal),** which equals 1000 calories, can also be related to the kilojoule.

$$1 \text{ kcal} = 10^3 \text{ cal}$$

$$1 \text{ kcal} = 4.184 \text{ kJ} \qquad \text{(exactly)}$$

In much of the older scientific literature (and even in some current journals), energies are expressed in calories and kilocalories. Today, most scientists have adopted the SI units of joules and kilojoules, and those are the energy units required by most current scientific journals. However, because the literature contains data in both sets of units, you may someday find it necessary to apply these conversions.

The joule is named in honor of James Prescott Joule (1818–1889), an English physicist who determined the amount of work needed to produce one unit of heat.

The nutritional Calorie (note the capital), Cal, is actually one kilocalorie.

$$1 \text{ Cal} = 1 \text{ kcal} = 4.184 \text{ kJ}$$

Chemical Bonds and Chemical Energy

We have emphasized the relationship of potential energy to attractions and repulsions because of its importance to the study of matter and its changes. Matter, after all, is made of things that have natural attractions and repulsions for each other, namely, *nuclei* and *electrons* (see Section 2.5). Nuclei are positively charged because they carry an atom's protons; electrons, located outside the nucleus, are negatively charged. Therefore, compounds made up of two or more nuclei and electrons have built into them both attractive and repulsive forces. The electrons and nuclei within molecules, being oppositely charged, *attract* each other. Electrons, however, *repel* other electrons, and atomic nuclei *repel* other atomic nuclei because like charged particles repel. Therefore, *net attractive forces must be at work in compounds to bind their atomic nuclei and electrons together.* The term *chemical bond* is simply the name we give to such a net attractive force that exists between a pair of atomic nuclei.

We'll have much more to say about chemical bonds in Chapters 8 and 9, but remember that attractive and repulsive forces are related to energy in one form or another. What is important, therefore, about chemical bonds for the present chapter is simply that they give rise to a compound's potential energy. The potential energy that resides in chemical bonds that can be changed to kinetic energy is called **chemical energy.** Thus, the chemical energy in things like firecrackers or gasoline is associated with the chemical bonds within them. The

Unlike charges attract; like charges repel.

chemical energy of the gasoline in the fuel tank contributes much to the overall potential energy of a parked car.

6.2 Kinetic Theory of Matter

The concept introduced in Section 1.3 that *heat* is actually a form of kinetic energy that is transferred between objects at different temperatures is part of a major understanding of matter called the **kinetic theory of matter.** Its central idea is that atoms and molecules are in constant motion of some sort, either moving at random from one place to another or, in solids, simply jiggling and vibrating in place. We'll use the term **molecular kinetic energy** for the energy associated with such motions. Each particle has a certain value of molecular kinetic energy at any given moment.

Evidence for the motions of molecules comes from several sources, some of which you have experienced yourself. The fragrance of a drop of perfume or cologne, when placed at one corner of a room, even in still air, will eventually be noticed by someone at the opposite corner. The only way that the molecules responsible for the fragrance can make this trip from corner to corner in a room without drafts is by actually being in motion themselves. Because of collisions between these moving molecules, either with themselves or with air molecules, the motions are quite *random,* going in every conceivable direction. A given molecule does not go far before it hits another and careens off on a different course. So it may take a given molecule a long time to make much net progress in one direction. With time, however, the random motions and collisions of the perfume's molecules result in their being uniformly mixed with the air in the room.

> Between collisions, molecules in air at room temperature move with an average speed of about 400 m s^{-1} (900 mph)!

> Not all collisions are head-on collisions. It's like a demolition derby, or bumper cars at an amusement park.

Kinetic Energy Distribution in a Sample of Matter

In a large collection of molecules, say, those of a sample of air, there is a wide distribution of molecular kinetic energies because there is a wide distribution of molecular velocities. At some particular moment, for example, an extremely small number of *individual* molecules will have zero kinetic energy, because balanced collisions happen to cause them to stop for an instant and so have zero velocity. (When v in $\frac{1}{2}mv^2$ goes to zero, KE becomes zero.) Similarly, a small number of the molecules will have very high molecular kinetic energies because unbalanced collisions have, by chance, given them very high velocities. Between these extremes of kinetic energy there will be many molecules with intermediate amounts of kinetic energy.

In Chapter 1 you learned that the **temperature** of a sample is related to the *average kinetic energy of its atoms and molecules*. This relationship is described in detail by the kinetic theory and is one that we will call on frequently to explain a variety of physical and chemical phenomena later in the book.

Figure 6.1 shows graphs that describe the kinetic energy distributions among molecules in a sample at two different temperatures. The vertical axis represents the *fraction* of molecules with a given kinetic energy (i.e., the number of molecules with a given KE divided by the total number of molecules in the sample). Each curve in Figure 6.1 therefore describes how the fraction of molecules with a given KE varies with KE.

Notice that each curve starts out with a fraction equal to zero for zero KE. This is because the fraction of molecules that are motionless (and therefore have zero KE) at any given moment is essentially zero, regardless of the temperature. Moving to higher values of KE, we find greater fractions of molecules. For

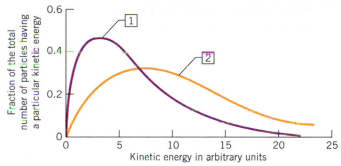

Figure 6.1 *The distribution of kinetic energies among gas particles.* The distribution of individual kinetic energies changes in going from a lower temperature, curve 1, to a higher temperature, curve 2. The highest point on each curve is the most probable value of kinetic energy for that temperature. It's the value of the molecular kinetic energy that we would most frequently find, if we could get inside the system and observe and measure the kinetic energy of each molecule. The most probable value of molecular kinetic energy is less at the lower temperature. At the higher temperature, more molecules have high speeds and fewer molecules have low speeds, so the maximum shifts to the right and the curve flattens.

very large values of KE, the fraction drops off again because fewer molecules have very large velocities.

Each curve in Figure 6.1 has a characteristic peak or maximum, each maximum corresponding to the most frequently experienced values for molecular KE. Because the curves are not symmetrical, the *average* values of molecular KE lie slightly to the right of the maxima. *An important part of the kinetic theory is that the Kelvin temperature of the sample is directly proportional to the average molecular kinetic energy.* When the cooler sample (curve 1 in Figure 6.1) is heated, the curve flattens and the maximum shifts to a higher value of average molecular KE (see curve 2). The reason the curve flattens is because the area under each curve corresponds to the sum of all the fractions, and when all the fractions are added, they must equal one. (No matter how you cut up a pie, if you add all the fractions, you must end up with one pie!)

Temperature Changes and What They Tell Us

In a very cold object, the *average* molecular kinetic energy is small. As the temperature increases, so does the average molecular kinetic energy. For a given object, therefore, we can interpret a change in its temperature as *both* a change in the *average* molecular kinetic energy of the particles that make up the object and a change in the *total* amount of kinetic energy (heat energy) the object has. Consequently, measurements of temperature are essential procedures in experimental thermochemistry, particularly the determination of the initial and final temperatures of a mixture undergoing a chemical reaction.

▶**Chemistry in Practice**◀ Civil engineers need some knowledge of thermochemistry. The setting of concrete, for example, involves reactions that give off heat. This is no problem when the size of the "pour" is small; the heat escapes well enough. But in huge "pours," like those made during the building of large dams, the buildup of unescaped heat can cause the setting concrete to expand and crack. So engineers put pipes in place to carry cooling water that removes the heat. The Grand Coulee Dam, completed in 1942 over the Columbia River in eastern Washington State, for example, is laced with elaborate networks of cooling pipes now no longer needed. ◆

Kinetic Theory and Liquids and Solids

Random motions of molecules occur in liquids, too. When you add sugar without stirring to hot coffee, the sugar molecules eventually become equally distributed among the other molecules in the cup as they move randomly about. The reason that the whole cup of coffee does not rattle around on the table is because the effects of the random molecular motions within the coffee cancel each other out entirely. On the average, for every molecule moving upward there are some moving in every other possible direction. Similarly, the particles in solids do not move from place to place, but they do vibrate about their positions within the solid.

It is significant that the distributions of kinetic energy among the molecules in a liquid, solid, or gas are identical. This means that if we have a gas, a liquid, and a solid all at the same temperature, the molecules in each phase have the same average kinetic energy. Their potential energies, however, are quite different, as you will learn later in the book.

Kinetic Theory and the Transfer of Heat

Without any reference to molecules in motion, we would still have a way of knowing that an object possesses heat energy, namely, by the object's ability to change the temperature of another, *colder* object when the two are brought into some kind of contact.

Transfers of Heat—A Molecular Explanation

The concept of molecular kinetic energy lets us better understand a common way by which heat transfers. When you pour boiling water into a cool cup, for example, the cup will soon be too hot to hold (except by its handle). Figure 6.2 shows how the heat transfer occurs at the molecular level. Regard the lengths of the arrows in this figure as proportional to average molecular kinetic energies (in other words, to temperature). The directions of the arrows are meant to be at random. Let's now imagine what happens.

High energy molecules in the boiling water strike the lower energy particles in the cup's material. These collisions give some of the molecular kinetic energy of the water molecules to the cup's molecules, which vibrate more energetically. The water molecules slow down as a result, and have less average molecular ki-

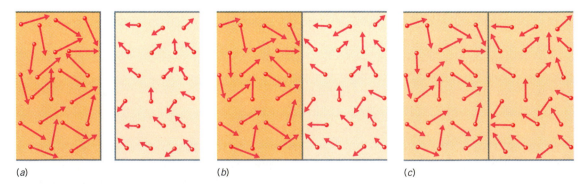

(a)   (b)   (c)

Figure 6.2 *Energy transfer from a warmer to a cooler object.* (*a*) The longer arrows on the left denote higher average kinetic energies and so something warmer, like hot water just before it's poured into a cooler coffee cup. (*b*) The hot water and the cup's inner surface are now in thermal contact. Energetic water molecules are slowed down by collisions with molecules in the cup's material. Thus energy transfers from the water to the material of the cup. (*c*) Thermal equilibrium is established; the average kinetic energies of the molecules of both materials are now the same, so their temperatures are now equal.

netic energy. By this mechanism, some of the molecular kinetic energy in the hot water transfers to the material of the cup. The temperature of the water decreases and the temperature of the cup itself increases.

6.3 Energy Changes in Chemical Reactions

The Energy Consequences of Bond Making and Breaking

Chemical reactions generally involve *both* the breaking and making of chemical bonds (Section 2.8). In most reactions, when bonds *form,* things that attract each other move closer together, which tends to decrease the potential energy of the reacting system. When bonds *break,* on the other hand, things that normally attract each other are forced apart, which increases the potential energy of the reacting system. Each reaction, therefore, has a certain net overall potential energy change, a net balance between the "costs" of breaking bonds and the "profits" from making them.

In many reactions, the products have *less* potential energy (chemical energy) than the reactants. When the methane gas burns in a Bunsen burner, for example, molecules of methane and oxygen with large amounts of chemical energy but relatively low amounts of molecular kinetic energy change into products having less chemical energy but much more molecular kinetic energy, molecules of carbon dioxide and water. Thus some chemical energy changes into molecular *kinetic* energy that can be released to the surroundings as heat, so we'll write heat as a product.

$$CH_4(g) + 2O_2(g) \longrightarrow CO_2(g) + 2H_2O(g) + heat$$

Any reaction in which heat is a product is said to be **exothermic.**

In some reactions the products have *more* chemical energy than the reactants. For example, plants make energy-rich molecules of glucose ($C_6H_{12}O_6$) and oxygen from carbon dioxide and water by a multistep process called *photosynthesis.* It requires a continuous supply of energy, which is supplied by portions of the sun's energy. The plant's green pigment, chlorophyll, is the solar energy absorber for photosynthesis.

$$6CO_2 + 6H_2O + solar\ energy \xrightarrow[\text{(several steps)}]{\text{chlorophyll}} C_6H_{12}O_6 + 6O_2$$

Reactions that consume energy are called **endothermic.**[2]

> ▶**Chemistry in Practice**◀ Thermochemistry helps us understand what is at stake in photosynthesis. The overall equation for this multistep process shows that we depend altogether on the sun *and chlorophyll containing systems,* like the trees of rain forests and the microscopic organisms (phytoplankton) in the ocean, to remove carbon dioxide from the atmosphere, construct plant materials, *and regenerate oxygen.* The carbon dioxide used by photosynthesis can no longer contribute to global warming. The solar energy becomes lodged in the high energy chemical bonds of glucose (and things made of glucose). Our metabolism harvests this energy when we eat plants or animals that feed on plants. ◆

When we study energy changes in chemical reactions, our concern is with changes in *potential energy* (chemical energy), not kinetic energy.

Photosynthesis converts solar energy, trapped by chlorophyll, into plant compounds that are rich in chemical energy.

Global warming is discussed in *Chemicals in Our World 4,* following chapter 7.

[2]Strictly speaking, when the energy absorbed (or given off) is not heat energy but some other form of energy, we should describe the event as *endergonic* (or *exergonic*), which are terms based on the name of an old unit of energy, the *erg,* but the distinctions are not always honored.

6.4 First Law of Thermodynamics: Heat and Work

By mentioning exothermic and endothermic events, we have moved into a major field shared by both physics and chemistry, namely, **thermodynamics,** the study of the *flow* of energy. The connection to thermochemistry is that thermochemical measurements provide some of the basic data for thermodynamics. We'll study thermodynamics in greater depth in Chapter 18, but some of its details are now needed for our study of thermochemistry.

We particularly need to define three frequently used terms, namely, *system, surroundings,* and *boundary*. By **system** we mean the particular part of the universe that we have decided to study, such as a mixture undergoing a chemical reaction in a beaker. Literally everything else in the universe besides the system constitutes the system's **surroundings.** Separating the system from its surroundings is a **boundary,** which might be visible (like the walls of a beaker) or invisible (like the boundary that separates the atmosphere's troposphere and stratosphere). An **insulated system** is one that permits no energy to cross the system's boundary in any direction. An **open system** is one open to the atmosphere; a **closed system** is not. All insulated systems are closed systems, but not all closed systems are insulated. A burning lightbulb is an example of a noninsulated, closed system. A corked Thermos bottle approximates a closed system that is insulated. When no heat can flow between a system and its surroundings, the change occurring in the system is called an **adiabatic change.**

> The troposphere is the zone of the atmosphere nearest Earth's surface.

> From the Greek *a + diabatos,* not passible.

The total energy of an insulated system, the sum of its molecular kinetic and potential energies, is constant (law of conservation of energy), but transformations of one energy form into the other can occur. When a chemical reaction, like the combustion of gasoline, occurs in an *insulated* system, the system's potential energy decreases, so its molecular kinetic energy must increase. Otherwise, the total energy would not be constant. Increasing the average molecular kinetic energy means increasing the system's temperature. Hence when a spontaneous reaction in an *insulated* system produces an increase in the average molecular kinetic energy by a lowering of the potential energy, the temperature of the system increases. *The only way that an insulated system's temperature can increase spontaneously is by a decrease in the system's potential energy and an increase in its molecular kinetic energy.*

> *Spontaneous* means happening without sustained human intervention.

When the same reaction is carried out in an *uninsulated* container, the buildup of molecular kinetic energy (heat) can be offset by a flow of heat *out of the system* into the cooler surroundings. Because of the *direction* of this heat flow, the reaction is said to be *exothermic;* heat is flowing *out* of the system.

Similarly, systems in *insulated* vessels that undergo reactions leading to an increase in potential energy at the expense of molecular kinetic energy become cooler. *The only way that a closed or insulated system's temperature can decrease spontaneously is by an increase in the system's potential energy and a decrease in its molecular kinetic energy.* In an open or *uninsulated* vessel where cooling is occurring, heat will tend to flow from the warmer surroundings into the container. Such reactions are *endothermic;* heat is flowing *into* the system. Let's summarize these important relationships.

Exothermic and Endothermic Reactions and Heat Flow

> *exo* means "out"
> *endo* means "in"
> *therm* means "heat"

1. During an **exothermic reaction,** the system becomes warmer and a net decrease of the potential energy of the substances occurs. If uninsulated, the system gives off heat to the surroundings.

2. During an **endothermic reaction,** the system becomes cooler and a net increase in the potential energy of the substances occurs. If uninsulated, the system receives heat from the surroundings.

Usually, chemical reactions are open systems and so are not insulated from their surroundings. Heat can flow into or out of then. Heat flows out if the temperature of a reaction mixture increases above that of the surroundings. Heat flows in when a reaction causes the temperature to become lower than that of the surroundings. If such heat exchanges allow the system to maintain a constant temperature, the change is an **isothermal change.**

> Heat always flows *spontaneously* from a region of higher temperature to one of lower temperature. Always.

State Functions and the State of a System

One of the interesting facts about the amount of energy in a system is that it does not depend on how the system received it nor on what might become of the energy at some future time. In other words, a system's energy is a function only of what we will now call the *state of the system*. Any property that does not depend on a system's history or future is called a **state function.** By "history" we mean all the details about how the system reached its present state, details that make up what we call the *path* to the current state from some previous state.

Chemists define the **state of a system** as being its chemical composition (substances and numbers of moles) plus its pressure, temperature, and volume. Pressure, temperature, and volume are examples of *state functions* and it's easy to understand why. A system's current temperature, for example, does not depend on what it was yesterday. Nor does it matter *how* the system acquired it, that is, the path to its current value. If it's now 25 °C, for example, we know all we can or need to know about its temperature. We do not have to specify how it got there or where it's going. Also, if the temperature were to increase, say, to 35 °C, the change in temperature, Δt, is simply the difference between the final and the initial temperatures.

> "State," as used in thermochemistry, doesn't have the same meaning as when it's used in terms such as "liquid state."

$$\Delta t = t_{final} - t_{initial}$$

> The symbol Δ, pronounced "delta," denotes a *change between some initial and final state.*

To make the calculation of Δt, we do not have to know what caused the temperature change—exposure to sunlight, heating by an open flame, or any other mechanism. All we need are the initial and final values. *This independence from the method or mechanism by which a change occurs is the important feature of all state functions.* As we'll see later, the advantage of recognizing that some property is a *state* function is that many calculations are then much, much easier.

Internal Energy

The *total energy* of a system is called its **internal energy** and is given the symbol E or E_{system}. Like its temperature, *a system's internal energy is a state function*. It is the sum of all the kinetic energies (KE) and potential energies (PE) of the system's particles.

$$E_{system} = (KE)_{system} + (PE)_{system}$$

Despite the importance of this definition, we can never actually know the value of E_{system}. Let's see why.

Think for a moment about a beaker just loaded with reactants ready to react. The beaker and its contents do have a *total* energy, namely, the sum of their combined kinetic and potential energy. But how can we ever know what, for example, the total kinetic energy actually is? It's certainly not zero simply because the system is *resting* at one location in the lab. The beaker is on spinning planet Earth and so has velocity and kinetic energy. And Earth is orbiting the sun, and

> Earth orbits the sun at about 6.7×10^4 miles per second. The solar system moves through the Milky Way at about 2×10^2 miles per second. The Milky Way rushes through space at about 6×10^2 miles per second.

both are members of a galaxy also in motion. We really have no way of knowing all of the associated kinetic energies of the beaker. However, as long as the galactic and planetary motions remain steady, we do not need to know the beaker's total kinetic energy, or even its total potential energy. Our interest is solely in the *change* in energy brought about in the reacting system itself. It's only from such a *change* in energy that we might get something useful from the system. (The reaction might be one that makes a battery work, for example.) Thus, even though we can *imagine* total values of initial and final energy, $E_{initial}$ and E_{final}, we can obtain by experiment only a *change* in energy, ΔE (read "delta E"), which is defined as follows.

$$\Delta E = E_{final} - E_{initial}$$

Both ΔE and E are state functions.

For a chemical reaction, E_{final} corresponds to the internal energy of the products, so we'll write it as $E_{product}$. Similarly, we'll use the symbol $E_{reactants}$ for $E_{initial}$. So for a chemical reaction the change in internal energy is given by

$$\Delta E = E_{products} - E_{reactants}$$

Notice carefully an important convention illustrated by this equation. Changes in something like temperature (Δt) or in internal energy (ΔE) are always figured by taking "final minus initial" or "products minus reactants." This means that if a system *absorbs* energy from its surroundings during a change, its final energy is greater than its initial energy and ΔE is positive. This is what happens, for example, when photosynthesis occurs or when something in the surroundings supplies energy to charge a battery. As the system (the battery) absorbs the energy, its internal energy increases and is then available for later use elsewhere.

Heat and Work

Thermodynamics considers two ways by which a system can exchange energy with its surroundings, *heat* and *work*. When a system undergoes a change, it might absorb or lose heat. When heat is absorbed, the system's energy increases, and when heat is lost, the system's energy decreases.

Work is exchanged when a pushing force moves something through a distance. For example, the system does work on the surroundings if it pushes back an opposing force, as when the hot expanding gases produced by fuel combustion in the cylinder of an engine push a piston. As the piston is pushed, the system of gases loses some of its energy. On the other hand, a system might have some work done on it, as in the compression of a system's gas, and thereby increase its amount of stored energy.

The **first law of thermodynamics** relates the change in the internal energy of a system to the exchanges of heat and work by the equation

ΔE = heat input + work input

$$\Delta E = q + w \tag{6.1}$$

where q represents the amount of heat *absorbed by* the system, and w is the amount of work *done on* the system. In effect, Equation 6.1 states that the change in the internal energy of a system is the sum of the energy it gains as heat plus the energy it gains from having work done on it.

The values of q or w can be either positive or negative, of course. Equation 6.1 actually represents them in such ways as to establish their algebraic signs under exothermic or endothermic circumstances. Thus, q has a positive sign if heat flows into the system and a negative sign if heat flows out. Similarly, w has a positive sign if the system gains energy by having work done on it, as in compressing a spring. And w has a negative sign if the system performs work so that its energy reserves are depleted somewhat (i.e., the system loses energy). These sign conventions are summarized as follows.

q is $(+)$	Heat is absorbed by the system.
q is $(-)$	Heat is released by the system.
w is $(+)$	Work is done on the system.
w is $(-)$	Work is done by the system.

When energy (heat or work) exits the system, the sign is negative (because the system loses internal energy); when heat or work enters the system, the sign is positive. Now let's look more closely at q and w and why, significantly, they are *not* state functions even though their sum (ΔE) is.

Heat, Work, and State Functions

Although ΔE is independent of how a change takes place, the values of q and w are not. Thus, neither q nor w is a state function. Their values depend on the *path* of the change. For example, consider the discharge of an automobile battery by two different paths (see Figure 6.3). Both paths take us between the same two states, one being the fully charged state and the other the fully discharged state. Because ΔE is a state function and because the same initial and final states are involved, *ΔE must be the same for both paths*. But how about q and w?

In path 1, we simply short the battery by placing a heavy wrench across the terminals. If you have ever done this, even by accident, you know how violent the result can be. Sparks fly and the wrench becomes very hot as the battery quickly discharges. Lots of heat is given off, but *the system does no work ($w = 0$)*. All of the ΔE appears as heat.

In path 2, we discharge the battery more slowly by using it to operate a motor. Along this path, much of the energy represented by ΔE appears as work

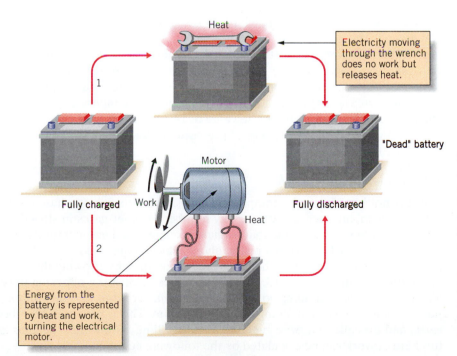

Heat

Electricity moving through the wrench does no work but releases heat.

1

"Dead" battery

Motor

Fully charged   Work

Heat

Fully discharged

2

Energy from the battery is represented by heat and work, turning the electrical motor.

Figure 6.3 *Energy, heat, and work.* The complete discharge of a battery along two different paths yields the same total amount of energy, ΔE. However, if the battery is simply shorted with a heavy wrench, as shown in path 1, this energy appears entirely as heat. Path 2 gives part of the total energy as heat, but much of the energy appears as work done by the motor.

(running the motor) and only a relatively small amount appears as heat (from the friction within the motor and the electrical resistance of the wires).

There are two vital lessons to be learned from this example. The first is that neither q nor w is a state function. Their values depend *entirely* on the path between the initial and final states. Yet, their sum, ΔE, as we said, *is* a state function.

Work Related to Gas Expansion or Compression

Most of the reactions that we carry out in the lab do not produce electricity, and therefore are not able to do electrical work. But there is another important kind of work in which these systems can be involved that is related to expansion or contraction of a gas under the influence of an outside pressure, like atmospheric pressure.

Pressure

If you have ever blown up a balloon or worked with an auto or bicycle tire, you have already learned something about *pressure.* You know that when a tire holds air at a relatively high pressure, it is hard to dent or push in its sides because "something" is pushing back. That something is the air pressure inside the tire.

Pressure, in general, is defined as the amount of *force* acting on a unit of area; it's the ratio of force to area.

$$\text{Pressure} = \frac{\text{force}}{\text{area}}$$

You can increase the pressure in a tire by pushing or forcing air into it. Its pressure, when you quit, is whatever pressure you have added *over and above the initial pressure,* namely, the atmospheric pressure. We describe the air in the tire as being *compressed,* meaning that it's at a higher pressure than the atmospheric pressure.

Atmospheric pressure is the pressure exerted by Earth's air simply by virtue of its weight. Imagine a one square inch area of Earth's surface at sea level. Above it is a column of air reaching to outer space and becoming thinner as you ascend. The column has a total weight of about 14.7 lbs, depending on the temperature. This weight of the air in the column is the force with which the air pushes on Earth, and we call it the *atmospheric pressure.* In the units we have used, the atmospheric pressure, the ratio of force to area, is approximately 14.7 lb in.$^{-2}$ at sea level. The ratio does vary a bit with temperature and weather, and it varies a lot with altitude. The value of 14.7 lb in.$^{-2}$ is very close to what we now call the **standard atmosphere of pressure** or **atm,** which will be discussed in Chapter 10.

Pressure–Volume Work

Any gas, not just air, can be compressed. Thus, scientists can purchase pure oxygen, hydrogen, nitrogen, helium, or any number of other gases in strong-walled steel tanks. Let's now imagine that we have a compressed gas in a special device, namely, a cylinder fitted with a piston above which is the pressure of the atmosphere (see Figure 6.4). When the pin is pulled out, thereby allowing the piston to move, the pressurized gas pushes the piston back against the opposing pressure of the atmosphere. In doing so, *the gas does some work on the surroundings.* It pushes some of the surrounding air out of the way. This is properly described as *work,* and it's called *pressure–volume work.* We'll next show that the amount of this kind of work can be calculated by the following equation.

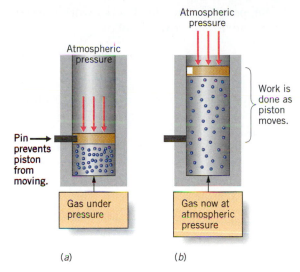

Figure 6.4 *Pressure–volume work.* (*a*) A gas is confined under pressure in a cylinder fitted with a piston that is held in place by a sliding pin. (*b*) When the piston is released, the gas inside the cylinder expands and pushes the piston upward against the opposing pressure of the atmosphere. As it does so, the gas does some pressure–volume work on the surroundings.

$$w = -P\Delta V \qquad (6.2)$$

P is the *opposing pressure* against which the piston pushes, and ΔV is the change in the volume of the system (the gas) during the expansion, i.e., $V_{final} - V_{initial}$. Because V_{final} is greater than $V_{initial}$, ΔV must be positive. Therefore, the negative sign in Equation 6.2 is required by the convention that work done *by the system* on the surroundings be negative.

If a gas is *compressed* by an external pressure, w is positive because work is done on the gas in the system.

Let's now establish that the product in Equation 6.2, namely, $P\Delta V$, actually is an energy term, which it must be if we're to refer to it as work. Work, as physicists define it, is accomplished when an opposing force, F, is pushed through some distance or length, L. The amount of work done is equal to the strength of the opposing force multiplied by the distance the force is moved.

$$\text{Work} = F \times L$$

Because pressure is force (F) per unit area, and area is simply length squared, L^2, we can write the following equation for pressure.

$$P = \frac{F}{L^2}$$

Volume (or a volume change) has dimensions of length cubed, L^3, so pressure times volume change is

$$P\Delta V = \frac{F}{L^2} \times L^3$$

$$= F \times L$$

But, from above, $F \times L$ also equals work. Thus, pressure times volume-change, or **pressure–volume work,** really is basically the same as any other kind of work. As we've noted, a common example of $P–V$ work is the work done by the expanding gases in a cylinder of a car engine as they move a piston. Another example is the work done by expanding gases to lift a rocket from the ground.

Heats of Reaction at Constant Volume

The amount of heat absorbed or released in a chemical reaction is called the **heat of reaction.** Usually it is measured under conditions of either constant volume or constant pressure.

Suppose that the only kind of work a system can do is $P–V$ work. Then we can modify Equation 6.1 as follows.

$$\Delta E = q + w = q + (-P\Delta V)$$
$$= q - P\Delta V$$

Let's apply this equation to a chemical reaction occurring under laboratory controlled conditions of constant volume. Energy changes in chemical reactions are measured using an apparatus called a **calorimeter** (calorie or heat "meter"). Figure 6.5 shows one kind, namely, a *bomb calorimeter,* so called because the vessel holding the reaction itself resembles a small bomb. The "bomb" has rigid walls, so the volume of the reaction mixture cannot change. Hence, ΔV must be equal to zero when the reaction occurs. This means, of course, that $P\Delta V$ must also be zero. Therefore, under conditions of constant volume,

$$\Delta E = q - 0 = q$$

Thus, the heat of reaction measured in a bomb calorimeter is the **heat of reaction at constant volume,** symbolized as q_v, and it corresponds to ΔE.

$$\Delta E = q_v$$

Enthalpy

Most reactions that are of interest to us do not occur at constant volume. Instead, they are carried out in open containers where they experience the constant pressure of the atmosphere. Now some pressure–volume work might be involved besides heat. Because of this, scientists have defined a quantity similar to E called **enthalpy,** or H. By definition,

From the Greek en + thalpein, to heat or to warm.

$$H = E + PV$$

Like E, H is a state function.

An **enthalpy change, ΔH,** is defined by the equation

$$\Delta H = H_{\text{final}} - H_{\text{initial}}$$

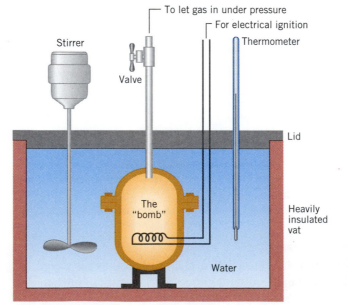

Figure 6.5 *A bomb calorimeter.* The water bath is usually equipped with devices for adding or removing heat from the water, thus keeping its temperature constant up to the moment when the reaction occurs in the bomb. The reaction chamber is of fixed volume, so $P\Delta V$ must equal zero for reactions in this apparatus.

Like H, ΔH *is a state function.* For a chemical reaction, the equation for ΔH may be rewritten as follows.

$$\Delta H = H_{\text{products}} - H_{\text{reactants}} \qquad (6.3)$$

Equation 6.3 serves to define the meaning of positive and negative values of ΔH in the same way by which we defined such meanings for ΔE. When a system absorbs energy from its surroundings, the change is endothermic, so H_{products} has to be larger than $H_{\text{reactants}}$. Therefore, *in endothermic changes ΔH is positive.*

On the other hand, if a change is exothermic, the system loses energy so H_{products} must be less than $H_{\text{reactants}}$. Therefore, *in exothermic changes ΔH is negative.*

The meanings of $+$ and $-$ values of ΔH are the same as for ΔE.

Enthalpy Changes and Heats of Reaction at Constant Pressure

Let's look again at the equation for enthalpy.

$$H = E + PV$$

At constant pressure, as for a reaction in an open container, the value of P is constant, but the other three values are not necessarily so. When we allow the volume to change, but not the pressure, then

$$\Delta H = \Delta E + P\Delta V$$

According to Equation 6.2, the work involved, w, is simply $-P\Delta V$. Because ΔE equals $(q + w)$, we can substitute the expression $-P\Delta V$ for w and write for situations involving only pressure–volume work

$$\Delta E = q + (-P\Delta V) = q - P\Delta V$$

Substituting this expression for ΔE into the preceding equation for ΔH, we get

$$\Delta H = q - P\Delta V + P\Delta V$$

which reduces to

$$\Delta H = q_p$$

The subscript p indicates that the heat involved this time is the **heat of reaction at constant pressure, q_p.** It's the value of ΔH when all of the energy change of a reaction at constant pressure involves heat energy. The sign of q_p, as you can see, is the sign of ΔH.

> For exothermic changes, q_p is negative (because ΔH is negative).

> For endothermic changes, q_p is positive (because ΔH is positive).

The Difference between ΔE and ΔH

We saw above that under constant pressure,

$$\Delta H = \Delta E + P\Delta V$$

So,

$$\Delta H - \Delta E = P\Delta V$$

Thus ΔH and ΔE differ by $P\Delta V$. Now let's see what this means, physically.

Suppose we have a reaction for which ΔE is negative (i.e., an exothermic reaction), and also suppose that during this reaction a gas is formed. Let's consider first what happens if we carry out the reaction in a bomb calorimeter, where the

The pressure inside the reaction vessel increases, but the gas can't do any work because the volume can't change.

volume can't change. When the reaction occurs, a certain total amount of energy is given off. Since the volume is unable to change, ΔV must be zero, so the system can't do any $P-V$ work. This means that *all* the energy released must appear as heat, and the amount of heat that we measure, q_v, is equal to ΔE.

Now, consider what happens if we carry out the same reaction at constant pressure (still supposing that it produces a gas). We could do this by letting the reaction take place inside a cylinder with a movable piston. As the gas forms inside the cylinder, the piston is pushed back against the opposing pressure of the atmosphere and the pressure within the system remains constant. But, by moving the piston, the expanding gas system does some work on the surroundings, and the energy for this work comes from the total energy given off by the reaction. Therefore, the amount of energy left over to be given off as the heat of reaction at constant pressure (ΔH) is equal to the total amount of energy released in the reaction (ΔE) minus the amount that was lost in pushing back the piston. Thus, the difference between ΔE and ΔH is the $P\Delta V$ work that the system does as it expands against the opposing pressure of the atmosphere.

Whenever a reaction system expands under the constant pressure of the atmosphere, the measured ΔH is a little less negative than ΔE. This is a small energy penalty that we pay for the convenience of carrying out the reaction at constant pressure. On the other hand, if there is a decrease in the volume of the reacting system at constant pressure, then the value of ΔH is actually a bit more negative than ΔE, and we get a bonus. In either case, however, the $P\Delta V$ term is generally small in relation to the values of ΔE and ΔH, so differences between ΔE and ΔH are usually small. The difference vanishes, of course, and ΔE equals ΔH for reactions having no changes in volume, typically reactions in which no gases are generated or consumed.

The values of ΔV for reactions involving only solids and liquids are very tiny, so ΔE and ΔH for these reactions are nearly identical.

6.5 Measuring Energy Changes: Calorimetry

The Special Place of Heat as a Form of Energy

One of the important facts about our world is that *all forms of energy can be converted quantitatively into heat.* The mechanical energy of a moving car, for example, is converted into heat by the frictional action of the brakes, and the brake shoes and drums can become very hot. When a current of electrons is forced into something with high electrical resistance, like the heating element in a toaster, the electrical energy of the current changes entirely into heat. Any and all forms of energy can be changed entirely into heat, and so *any form of energy can be completely determined by letting it change quantitatively into heat.*[3]

One way to measure the heat generated by an exothermic reaction is to let the heat flow into a known mass of cooler water and measure the increase in the temperature of the water. To calculate the amount of heat that transfers by using the temperature change of the water, we need to know something about the capacity of water to absorb heat. This capacity is one of the *thermal properties* of water. **Thermal properties** are those that describe a system's ability to absorb or

---

[3]The opposite is not true; heat cannot be converted *quantitatively* into another form of energy. The heat generated, for example, by burning coal at an electrical power plant cannot *even in principle* be changed 100% into electrical energy. This is a major limitation, set by the inherent nature of things, on what we can do, and we'll leave further discussion of it to Chapter 18.

release heat without a chemical change. They include *specific heat, heat capacity,* and *molar heat capacity.*

Specific Heat

The quantity of heat needed to raise the temperature of 1 gram of a substance by 1 °C is called its **specific heat.** Historically, the need for a standard unit of heat was met by *defining* the unit, the **calorie,** abbreviated **cal,** as the heat that increases the temperature of a 1 g sample of water by 1 °C, specifically the degree between 14.5 and 15.5 °C. In other words, the *specific heat* of water is 1 cal g^{-1} °C^{-1} (read, one calorie per gram per degree Celsius change).[4] Today, specific heats are generally given in joules. Because 1 cal equals 4.184 J, the specific heat of water is 4.184 J g^{-1} °C^{-1}.

Relative to water, metals have low specific heats (see Table 6.1). Notice that the specific heat of water is roughly nine times that of iron, which explains why 1 calorie added to 1 g of iron raises its temperature 9 times as much as when 1 calorie is added to 1 g of water. To put it another way, the 1 cal of heat that increases the temperature of only 1 g of water by 1 °C will increase the temperature of 9 g of iron that much.

Heat Capacity

Samples are almost always much different than 1 g, of course. How, then, do we treat the abilities of real samples to absorb or release heat? No doubt you have figured out that a 2 g sample of water would take 2 cal to make its temperature increase by 1 degree. Or, if it received only 1 cal, its temperature would increase by only half a degree. In other words, an object's ability to absorb or release heat, an ability called its *heat capacity,* depends on both its specific heat and mass. The **heat capacity** of a given object is the product of its specific heat and mass.

$$\text{Heat capacity} = \text{specific heat} \times \text{mass} \qquad (6.4)$$

Or, putting in the units most often used,

$$\text{Heat capacity} = \frac{J}{g\ °C} \times g \qquad (6.5)$$

Thus, the usual units of heat capacity are J/°C (joules per degree Celsius change). We should not, of course, limit our discussion to temperature *increases* caused by heat absorption. The heat capacity of an object also tells us how much heat it must lose for a given decrease in temperature to occur.

Specific heat is an *intensive property;* it's an inherent thermal property and is unique for each substance (see Table 6.1). At 25 °C, iron's specific heat, for example, is 0.45 J g^{-1} °C^{-1}, much lower than that of water. Silver's specific heat, 0.235 J g^{-1} °C^{-1}, is still lower.

Heat capacity, on the other hand, is an *extensive property* because it depends on the size of the sample. We could not, for example, speak of the heat capacity of water but only of the heat capacity of a specific quantity of water, like a glassful or a lakeful.

Table 6.1 Specific Heats

Substance	Specific Heat, J g^{-1} °C^{-1} (25 °C)
Carbon (graphite)	0.711
Copper	0.387
Ethyl alcohol	2.45
Gold	0.129
Granite	0.803
Iron	0.4498
Lead	0.128
Olive oil	2.0
Silver	0.235
Water (liquid)	4.1796

[4]The value of specific heat varies somewhat with the initial temperature we choose, but not by much. At 25 °C, the value for water is 4.1796 J g^{-1} °C^{-1}, which is very close to 4.184 J g^{-1} °C^{-1}. To two significant figures, liquid water's specific heat is 4.2 J g^{-1} °C^{-1} over the whole range between its melting and boiling points.

Lake Superior is a huge reservoir that traps heat in summer and releases it in winter, making Duluth, MN, seen here with its famous lift bridge, cooler in summer and warmer in winter than places many miles inland.

▶**Chemistry in Practice**◀ Because the heat capacity of a lake is much greater than that of a glassful of water, cities on large lakes or the ocean tend to have milder temperatures, year-round, than cities even a few miles inland. Because of its huge heat capacity, a lake must absorb a large amount of heat before its temperature increases even one degree. In the spring, during a lake's warm-up, some of the needed heat comes from the sun, but some is also delivered from the land near the lakeshore by air movements and by heat radiation. Because the lake soaks up heat, lakeside land tends to have a cooler temperature even far into the summer than it would if it were miles away from the lake. Visitors to and residents of cities like Duluth, Minnesota (on Lake Superior, the most westerly Great Lake), or Buffalo, New York (on Lake Erie, another Great Lake) have experienced this effect.

By the time a huge lake has warmed up during the summer, fall begins and temperatures start to drop. But because of the long summer warm-up, the lake now holds a great quantity of heat, and this is slowly drawn into the surrounding air and land as winter advances. Hence, cities near such lakes tend to have milder winter temperatures than places several miles inland. ◆

The units of heat capacity, J °C^{-1}, tell us that the heat capacity of some object corresponds to the amount of heat needed to cause one degree Celsius of temperature change. In other words, the *heat capacity* of any object, like the bomb calorimeter of Figure 6.5 (including all objects immersed in the water plus the walls of the water bath), is *the amount of heat required to raise its temperature one degree Celsius or the amount of heat it must lose for its temperature to decrease by one degree Celsius.* Heat capacity is thus the ratio of the energy gained or lost by the object to the degrees of the associated temperature change, Δt.

$$\text{Heat capacity} = \frac{\text{heat exchanged}}{\Delta t} \qquad (6.6)$$

You can see that if we have determined in a separate experiment the heat capacity of the bomb calorimeter of Figure 6.5, and if we can measure Δt caused in the surrounding water by a reaction within the bomb, we can calculate the heat of the reaction in the bomb. The calculation merely involves a rearranged form of Equation 6.6.

> The sign of Δt will determine the sign of the heat exchanged.

$$\text{Heat exchanged} = \text{heat capacity} \times \Delta t \qquad (6.7)$$

Equation 6.4 gives us an expression for heat capacity to put into Equation 6.7, namely, the product of specific heat and mass. Thus when we know the specific heat of a substance, its mass, and a change in temperature that it has undergone, we can calculate the heat exchange as follows.

Heat exchanged

$$\text{Heat exchanged} = (\text{specific heat} \times \text{mass}) \times \Delta t$$

Or, in the units that we have been using,

$$\text{Heat exchanged} = \frac{\text{J}}{\text{g °C}} \times \text{g} \times \Delta t \, (\text{in °C}) \qquad (6.8)$$

If a gold ring with a mass of 5.5 g changes in temperature from 25.0 to 28.0 °C, how much energy (in joules) has it absorbed?

Analysis: The question actually asks for the *change* in heat energy that occurs in the gold ring as a result of the temperature change, Δt. Equation 6.8 gives the relationship among these factors, and to use it we must have the specific heat of gold. This wasn't given in the problem, so we must look it up. We will use the value of the specific heat of gold given for 25 °C in Table 6.1, namely, 0.129 J g^{-1} °C^{-1}.

Solution: The temperature increases from 25.0 to 28.0 °C, so Δt is 3.0 °C. The mass of the ring is 5.5 g. We substitute our values of specific heat, temperature change, and grams into Equation 6.8 and solve for the *energy change*.

$$\text{Heat exchanged} = \frac{J}{g \, °C} \times g \times °C \qquad \text{(Equation 6.8)}$$

$$= 0.129 \frac{J}{g \, °C} \times 5.5 \text{ g} \times 3.0 \, °C$$

$$= 2.1 \text{ J}$$

Thus, only 2.1 J raises the temperature of 5.5 g of gold by 3.0 °C. Because Δt is positive, so is the heat, 2.1 J. Thus the *sign* of the energy change signifies that the ring *absorbs* heat.

Is the Answer Reasonable?
If the ring had a mass of only 1 g and its temperature increased by 1 °C, we'd know from the specific heat of gold (let's round it to 0.13 J g^{-1} °C^{-1}) that the ring would absorb 0.13 J. For three degrees of increase, the answer would be three times as much, or 0.39 J, nearly 0.4 J. For a ring a little heavier than 5 g, the heat absorbed would be five times as much, or about 2.0 J. So our answer (2.1 J) is clearly reasonable.

EXAMPLE 6.1
Calculating an Energy Change from a Temperature Change, Mass, and Specific Heat

Δt is positive because ($t_{final} - t_{initial}$) is positive.

<hr>

Practice Exercise I

The temperature of 250 g of water is changed from 25.0 to 30.0 °C. How much energy was transferred into the water? Calculate your answer in joules, kilojoules, calories, and kilocalories. ◆

Molar Heat Capacity

When dealing with the thermal properties of a pure substance, it is sometimes more relevant to work with one mole instead of one gram. We therefore define the **molar heat capacity** of a substance as the heat needed to raise the temperature of one mole of it by one degree Celsius. The usual units of molar heat capacity are J mol^{-1} °C^{-1}.

Because the molar mass of water is 18.0 g mol^{-1}, the molar heat capacity of water is 18.0 times its specific heat, namely, 75.2 J mol^{-1} °C^{-1} or, in the old units, 18.0 cal mol^{-1} °C^{-1}.

The Unusual Thermal Properties of Water

Water, as we see from the data in Table 6.1, has an unusually high specific heat. This means that it is relatively difficult to make the temperature of a large mass of water to swing widely. As we've noted, a lake heats up or cools down only slowly as the seasons change.

Facets of Chemistry 6.1

Hypothermia—When the Body's "Thermal Cushion" Is Overtaxed

All of the reactions in the body, collectively called *metabolism,* have rates sensitive to temperature changes. Over the range of body temperatures, metabolic reactions generally are faster at higher temperatures and slower at lower temperatures. We will discuss here a consequence of overusing the body's thermal cushion.

Some of the critical functions sustained by the energy of metabolism are the operation of the central nervous system, including the brain, and the working of the heart. *Hypothermia* is a decrease in the temperature of the inner core of the body. As the inner core cools, rates of metabolism slow down and the symptoms given in the table progressively occur. When a newspaper states that someone died of "exposure," nearly always it is a case of hypothermia.

The onset of hypothermia is a serious medical emergency requiring prompt action by others. The victim may either not recognize the problem or be confused by it. First aid consists of getting the victim out of the wind, dried off, wrapped in warm, dry blankets, and fed warm fluids or sweet food. Contrary to the legend about the St.

Bernard dogs, brandy should never be given because, once in the bloodstream, the alcohol in brandy causes the blood capillaries to dilate. When capillaries near the cold skin suddenly release extra loads of chilled blood toward the body's core, the core cooling accelerates.

Core Temperature (°C)	Symptoms
37	Normal body temperature
35–36	Intense, uncontrollable shivering
33–35	Amnesia, confusion
30–32	Muscular rigidity, reduced shivering
26–30	Unconsciousness, erratic heartbeat
Below 26	Death by pulmonary edema and heart failure

A Reference

Doris R. Kimbrough, "Heat Capacity, Body Temperature, and Hypothermia," *Journal of Chemical Education,* January 1998, page 48.

An infant's body is about 80% water.

▶**Chemistry in Practice**◀ The adult body is about 60% water by mass, so it has a high heat capacity. This makes it relatively easy for the human body to maintain a steady temperature of 37 °C, which is vital to survival. In other words, the body can exchange considerable energy with the environment but experience only a small change in temperature. With a substantial thermal "cushion," the body adjusts to large and sudden changes in outside temperature while experiencing very small fluctuations of its core temperature. *Hypothermia* can still happen, of course; it's an emergency brought on when the body cannot prevent its core temperature from decreasing (see Facets of Chemistry 6.1). ◆

Calorimetry

Most heats of reaction are determined by measuring the change in temperature caused by the heat exchanged between the reacting system and the contents of a surrounding calorimeter, mostly water. Calorimeter design is not standard, varying according to the kind of reaction and the precision desired. The science of using a calorimeter for determining heats of reaction is called **calorimetry.**

There is no instrument that directly measures energy. A calorimeter does not do this; its thermometer is the only part to provide raw data, the temperature change. We calculate the energy change.

Calorimeters can be very sophisticated, like the bomb calorimeter of Figure 6.5, or very simple, like the "coffee cup calorimeter," which is a simple kind of *solution calorimeter,* that you might use in the lab. The bomb calorimeter involves constant volume calorimetry and yields data for calculating q_v, the heat of reaction at constant volume. The solution calorimeter involves constant pressure calorimetry and gives us data for calculating q_p, the heat of reaction at constant pressure.

Coffee Cup Calorimeter, a Solution Calorimeter

A simple solution calorimeter, dubbed the coffee cup calorimeter, is made of two nested and capped cups made of Styrofoam, a very good insulator. When a reaction occurs in such a calorimeter, little heat is lost to the surroundings (or gained from it), particularly if the reaction is rapid. Thus, a rapid and easily measured temperature change will be observed. From the predetermined heat capacity of the calorimeter and its contents before the reaction, the heat absorbed or evolved by the reaction is found by Equation 6.7. Only a negligible quantity of heat is absorbed by the Styrofoam and the thermometer, and we'll ignore it in our calculations.

In research, calorimeters of much greater precision than the coffee cup calorimeter would be used.

EXAMPLE 6.2

Solution Calorimetry

The reaction of hydrochloric acid and sodium hydroxide is very rapid and exothermic. The equation is

$$HCl(aq) + NaOH(aq) \longrightarrow NaCl(aq) + H_2O$$

In one experiment, a student placed 50.0 mL of 1.00 M HCl at 25.5 °C in a coffee cup calorimeter. To this was added 50.0 mL of 1.00 M NaOH solution also at 25.5 °C. The mixture was stirred, and the temperature quickly increased to a maximum of 32.2 °C. What is the energy evolved in joules per mole of HCl? Because the solutions are relatively dilute, we can assume that their specific heats are close to that of water, 4.18 J g^{-1} °C^{-1}. The density of 1.00 M HCl is 1.02 g mL^{-1} and that of 1.00 M NaOH is 1.04 g mL^{-1}. (We will neglect the heat lost to the Styrofoam itself, to the thermometer, or to the surrounding air.)

Analysis: Before we can use Equation 6.7 to find the number of joules evolved, we have to calculate two things, namely, the system's heat capacity and the temperature change. To calculate the former, using the given specific heats of 4.18 J g^{-1} °C^{-1}, we need Equation 6.4.

$$\text{Heat capacity} = \text{mass (g)} \times \text{specific heat}$$

The mass here refers to the *total* grams of the combined solutions, but we were given volumes. So we have to use their densities to calculate their masses, which we learned how to do in Section 1.7.

$$\text{Mass} = \text{density} \times \text{volume}$$

For the HCl solution,

$$\text{Mass (HCl)} = 1.02 \text{ g mL}^{-1} \times 50.0 \text{ mL} = 51.0 \text{ g}$$

For the NaOH solution,

$$\text{Mass (NaOH)} = 1.04 \text{ g mL}^{-1} \times 50.0 \text{ mL} = 52.0 \text{ g}$$

The mass of the final solution is thus the sum, 103.0 g.

Solution: We can now calculate the heat capacity of the final solution at the instant of mixing and immediately before any reaction occurs to increase its temperature. It was given that the solution's specific heat is 4.18 J g^{-1} °C^{-1}, so

$$\text{Heat capacity} = \text{mass} \times \text{specific heat} \qquad \text{(Equation 6.4)}$$
$$= 103.0 \text{ g} \times 4.18 \text{ J g}^{-1} °C^{-1}$$
$$= 431 \text{ J } °C^{-1}$$

Thus the heat capacity of the solution in the calorimeter is 431 J °C^{-1}.

The reaction changes the system's temperature by ($t_{final} - t_{initial}$), so

$$\Delta t = 32.2 °C - 25.5 °C = 6.7 °C$$

The energy evolved by the reaction can now be calculated from Equation 6.7.

$$\text{Heat (evolved)} = \text{heat capacity} \times \Delta t$$

Or

$$\text{Heat evolved} = 431 \text{ J} \,^{\circ}\text{C}^{-1} \times 6.7 \,^{\circ}\text{C}$$

$$= 2.9 \times 10^3 \text{ J}$$

This is the heat evolved for the specific mixture prepared, but the problem calls for joules *per mole* of HCl. We calculate the number of moles of HCl used with the molarity-to-moles tool given by the definition of molarity. The molarity of the acid is 1.00 *M*, so in 50.0 mL of HCl solution (0.0500 L) we have 0.0500 mol HCl.

$$\text{Molarity} = M = \frac{\text{moles solute}}{1 \text{ L soln}}$$

$$= \frac{\text{moles solute}}{1000 \text{ mL soln}}$$

$$0.0500 \text{ L HCl soln} \times \frac{1.00 \text{ mol HCl}}{1.00 \text{ L HCl soln}} = 0.0500 \text{ mol HCl}$$

The neutralization of 0.0500 mol of acid released 2.9×10^3 J. So, properly put, we should express the heat released as the following ratio.

$$\text{Energy evolved per mole of HCl} = \frac{2.9 \times 10^3 \text{ J}}{0.0500 \text{ mol HCl}} = 58 \times 10^3 \text{ J mol}^{-1}$$

$$= 58 \text{ kJ mol}^{-1}$$

Thus the energy of neutralizing HCl by NaOH is 58 kJ mol^{-1}.

Is the Answer Reasonable?

Let's first review the logic of the steps we used. *Notice how the logic is driven by definitions, which carry specific units.* Working backward, knowing that we want units of joules per mole in the answer, we must calculate separately the number of moles of acid neutralized and the number of joules that evolved. The latter will emerge when we multiply the solution's heat capacity (units of J °C^{-1}) by the degrees of temperature increase (°C). But we don't know the heat capacity directly; we have to calculate it. That's why we need the total mass of the sample as well as the sample's specific heat (which is given); mass times specific heat gives us heat capacity.

The step in the calculation that involves the mass of the final sample is clearly easy. Because the densities are close to 1 g mL^{-1}, the masses mixed are about 50 g per solution or a total of 100 g.

The next step, figuring out the heat capacity of the final solution, is also easy to check. Its specific heat is around 4.2 J g^{-1} °C^{-1}, which means that if the final solution weighed 1 g, its heat capacity would be 4.2 J °C^{-1}. But if the mass is 100 g, then the heat capacity is 100 times as much, or 420 J °C^{-1}.

The temperature increases about 7 °C, so when we multiply 420 J °C^{-1} by 7, we get the heat evolved, namely, about 2940 J, or about 2.9×10^3 J; this is the same as 2.9 kJ.

This much heat is associated with neutralizing 0.05 mol HCl, so the heat per mole is 2.9 kJ divided by 0.05 mol, or 58 kJ mol^{-1}.

Practice Exercise 2

When pure sulfuric acid dissolves in water, much heat is given off. To measure it, 175 g of water was placed in a coffee cup calorimeter and chilled to 10.0 °C. Then 4.90 g of sulfuric acid (H_2SO_4), also at 10.0 °C, was added, and the mixture was quickly stirred with a thermometer. The temperature rose rapidly to 14.9 °C. Assume that the value of the specific heat of the solution is 4.18 J g^{-1} °C^{-1}, and that all of the heat evolved is absorbed by the solution. Calculate the heat evolved in kilojoules by the formation of this solution. (Remember to use the *total* mass of the solution, the water plus the solute.) Calculate also the heat evolved *per mole* of sulfuric acid. ◆

6.6 Enthalpy Changes in Chemical Reactions

Although the total energy of the universe is assumed to be constant, the energy possessed by any given *system* within the universe can vary widely. The energy in the *surroundings* simply adjusts as transfers of energy occur into or out of systems. Our attention in what follows is on reacting *systems*, not their surroundings, and on the associated enthalpy changes.

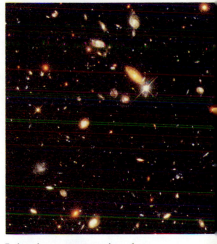

Standard Conditions for Enthalpy Changes

To facilitate comparisons of enthalpy changes among different reacting systems, chemists have agreed to a set of standard conditions. The standard reference temperature is 25 °C, and the standard reference pressure is the standard atmosphere, the atm. Although the atmospheric pressure fluctuates a little from day to day, the *standard atmosphere* has a nonfluctuating value.

It is a huge assumption that the energy of the *universe* is a constant, but it works every time it is applied here on Earth.

Standard Heats of Reaction

The **standard heat of reaction** is the value of ΔH for a reaction occurring under standard conditions and involving the actual numbers of *moles* specified by the coefficients of the equation. To show that ΔH is for *standard* conditions, we add a superscript to ΔH, the degree sign, to make $\Delta H°$. The units of $\Delta H°$ are normally kilojoules.

Unless we specify otherwise, whenever we write ΔH we mean ΔH for the *system*, not the surroundings.

To illustrate clearly what we mean by $\Delta H°$, let us use the reaction between gaseous nitrogen and hydrogen that produces gaseous ammonia.

$$N_2(g) + 3H_2(g) \longrightarrow 2NH_3(g)$$

When specifically 1.000 mol of N_2 and 3.000 mol of H_2 react to form 2.000 mol of NH_3 at 25 °C and 1 atm, the reaction releases 92.38 kJ. Hence, for the reaction *as given by the above equation,* $\Delta H° = -92.38$ kJ. Often the enthalpy change is given immediately after the equation; for example,

The values of $\Delta H°$ are taken from available reference sources.

$$N_2(g) + 3H_2(g) \longrightarrow 2NH_3(g) \qquad \Delta H° = -92.38 \text{ kJ}$$

An equation that also shows the value of $\Delta H°$ is called a **thermochemical equation.** It always gives the physical states of the reactants and products, and *its $\Delta H°$ value is true only when the coefficients of the reactants and products are taken to mean moles of the corresponding substances.* The above equation, for example, shows a release of 92.38 kJ if *two* moles of NH_3 form. If we were to make twice as much or 4.000 mol of NH_3 (from 2.000 mol of N_2 and 6.000 mol of H_2), then twice as much heat (184.8 kJ) would be released. On the other hand, if only 0.5000 mol of N_2 and 1.500 mol of H_2 were to react to form only 1.000 mole of NH_3, then only half as much heat (46.19 kJ) would be released. For the various sizes of the reactions just described, for example, we would have the following thermochemical equations.

Tools

Thermochemical equations

$$N_2(g) + 3H_2(g) \longrightarrow 2NH_3(g) \qquad \Delta H° = -92.38 \text{ kJ}$$
$$2N_2(g) + 6H_2(g) \longrightarrow 4NH_3(g) \qquad \Delta H° = -184.8 \text{ kJ}$$
$$\tfrac{1}{2}N_2(g) + \tfrac{3}{2}H_2(g) \longrightarrow NH_3(g) \qquad \Delta H° = -46.19 \text{ kJ}$$

Because the coefficients of a thermochemical equation always mean *moles,* not molecules, we may use fractional coefficients. Normally we don't because we cannot have fractions of *molecules,* but we can have fractions of moles.

We emphasize again that physical states must be specified in thermochemical equations. The combustion of 1 mol of methane, for example, has different values of $\Delta H°$ if the water produced is in its liquid or its gaseous state.

$$CH_4(g) + 2O_2(g) \longrightarrow CO_2(g) + 2H_2O(l) \qquad \Delta H° = -890.5 \text{ kJ}$$
$$CH_4(g) + 2O_2(g) \longrightarrow CO_2(g) + 2H_2O(g) \qquad \Delta H° = -802.3 \text{ kJ}$$

(The difference in $\Delta H°$ values for these two reactions is the quantity of energy that would be released by the physical change of 2 mol of water vapor at 25 °C to 2 mol of liquid water at 25 °C.)

EXAMPLE 6.3

Writing a Thermochemical Equation

The following thermochemical equation is for the exothermic reaction of hydrogen and oxygen that produces water.

$$2H_2(g) + O_2(g) \longrightarrow 2H_2O(g) \qquad \Delta H° = -517.8 \text{ kJ}$$

What is the thermochemical equation for this reaction when it is conducted to produce 1.000 mol H_2O?

Analysis: The given equation is for 2.000 mol of H_2O, and any changes in the coefficient for water must be made identically to all other coefficients, *as well as to the value of* $\Delta H°$.

Solution: We divide everything by 2, to obtain

$$H_2(g) + \tfrac{1}{2}O_2(g) \longrightarrow H_2O(g) \qquad \Delta H° = -258.9 \text{ kJ}$$

Is the Answer Reasonable?
Compare the equation just found with the initial equation to see that the coefficients and the value of $\Delta H°$ are all divided by 2.

Practice Exercise 3

What is the thermochemical equation for the formation of 2.500 mol of H_2O? ◆

The great dirigible, *Hindenburg,* was filled with hydrogen, and its explosive reaction with oxygen destroyed the airship in minutes on May 6, 1937, at Lakehurst, New Jersey.

6.7 Combining Thermochemical Equations: Hess's Law

Once we have the thermochemical equation for a given reaction, we can write the equation for the reverse reaction, regardless of how hard it might actually be to make it happen. For example, the thermochemical equation for the combustion of carbon in oxygen to give carbon dioxide is

$$C(s) + O_2(g) \longrightarrow CO_2(g) \qquad \Delta H° = -393.5 \text{ kJ}$$

The reverse reaction, which is extremely difficult to carry out, would be the decomposition of carbon dioxide to carbon and oxygen.

$$CO_2(g) \longrightarrow C(s) + O_2(g) \qquad \Delta H° = ?$$

Despite how hard it would be to cause this reaction, we can still know what its $\Delta H°$ *must* be. It must be $+393.5$ kJ, that is, equal in size but opposite in sign to $\Delta H°$ for the reaction written in the opposite direction. *The law of conservation of energy requires this remarkable result.* If the values of $\Delta H°$ for the forward and the reverse reactions were not equal but opposite in sign, then perpetual motion machines would be possible. But they are not, not even in theory (see Facets of Chemistry 6.2). Thus, regardless of the difficulty of directly decomposing CO_2 into its elements, we can still write a thermochemical equation for it.

$$CO_2(g) \longrightarrow C(s) + O_2(g) \qquad \Delta H° = +393.5 \text{ kJ}$$

Facets of Chemistry 6.2

Perpetual Motion Machines—Impossible

A perpetual motion machine is one that creates more energy than it uses and so theoretically could run forever. The U.S. Patent Office, wearied by claims for these impossibilities, now requires that all patent applications be accompanied by "reductions to practice," that is, *working models*. Because working models of perpetual motion machines are impossible, the Patent Office is no longer bothered with such applications.

Here's how such a machine would work in principle. Suppose that it takes only 200 kJ to break 1 mol of CO_2 back to C plus O_2. If we get 394 kJ by burning a mole of carbon to CO_2 but only 200 kJ is required to recycle the

system back to C plus O_2, our profit is 194 kJ per mole. Over and over, we could let the elements burn, get 394 kJ per mole, save 200 kJ out of this, and get 194 kJ left over. We could use this energy to swat flies, bake cookies, repair worn out parts of the machinery, or run tractors while we took it easy. Very nice—but impossible. We haven't explained *why* it's impossible; we don't know why, except that this is the way of nature. All we have done is *discover* that it is impossible. Out of this discovery, often repeated by would-be inventors of perpetual motion machines, came the law of conservation of energy.

To repeat, if we know $\Delta H°$ for a given reaction, then we also know $\Delta H°$ for the reverse reaction; it has the same numerical value, but the opposite algebraic sign. This extremely useful fact makes thermochemical data available that would otherwise be impossible to measure.

Multiple versus Single Path Routes and Enthalpy Changes

What we're leading up to is a method for combining known thermochemical equations in a way that will allow us to calculate an unknown $\Delta H°$ for some other reaction. This requires experience in other kinds of manipulations of equations. Let's revisit the combustion of carbon to see how this works and, as a bonus, see a neat demonstration that $\Delta H°$ is a state function.

We can imagine two paths leading from 1 mole each of carbon and oxygen to 1 mol of carbon dioxide.

One-Step Path

Let C and O_2 react to give CO_2 directly.

$$C(s) + O_2(g) \longrightarrow CO_2(g) \qquad \Delta H° = -393.5 \text{ kJ}$$

Two-Step Path

Let C and O_2 react to give CO, and then let CO react with O_2 to give CO_2.

Step 1: $\qquad C(s) + \frac{1}{2}O_2(g) \longrightarrow CO(g) \qquad \Delta H° = -110.5 \text{ kJ}$

Step 2: $\qquad CO(g) + \frac{1}{2}O_2(g) \longrightarrow CO_2(g) \qquad \Delta H° = -283.0 \text{ kJ}$

Overall, the two-step path consumes 1 mol each of C and O_2 to make 1 mol of CO_2, just like the one-step path. The initial and final states for the two routes to CO_2, in other words, are identical.

If $\Delta H°$ is a state function dependent only on the initial and final states and is independent of path, then the values of $\Delta H°$ for both routes should be identical. We can see that this is exactly true simply by adding the equations for the two-step path and comparing the result with the equation for the single step.

Step 1: $\qquad C(s) + \frac{1}{2}O_2(g) \longrightarrow CO(g) \qquad\qquad \Delta H° = -110.5 \text{ kJ}$

Step 2: $\qquad CO(g) + \frac{1}{2}O_2(g) \longrightarrow CO_2(g) \qquad\qquad \Delta H° = -283.0 \text{ kJ}$

$$CO(g) + C(s) + O_2(g) \longrightarrow CO_2(g) + CO(g) \qquad \Delta H° = -110.5 \text{ kJ}$$
$$+ (-283.0 \text{ kJ})$$
$$\Delta H° = -393.5 \text{ kJ}$$

The equation resulting from adding Steps 1 and 2 has "$CO(g)$" appearing *identically* on opposite sides of the arrow. The two may therefore be canceled to obtain the net equation. Such a cancellation is permitted only when both the formula and the physical state of a species are identically given on opposite sides of the arrow. The net thermochemical equation for the two-step process, therefore, is

$$C(s) + O_2(g) \longrightarrow CO_2(g) \qquad \Delta H° = -393.5 \text{ kJ}$$

The results, chemically and thermochemically, are thus identical for both routes to CO_2, demonstrating that $\Delta H°$ is a state function.

Enthalpy Diagrams

Tools

Enthalpy diagrams

The energy relationships among the alternative pathways for the same overall reaction are clearly seen using a graphical construction called an *enthalpy diagram*. One is illustrated in Figure 6.6 for the formation of CO_2 from C and O_2. Horizontal lines correspond to different *absolute* values of enthalpy, H, which we cannot know. However, an upper line represents a larger value of H than a lower line. *Changes* in enthalpy, $\Delta H°$, which we can know, are represented by the vertical distances between these lines. Arrows pointing upward represent increases in enthalpy and have positive values of ΔH. Arrows pointing downward represent decreases in enthalpy and have negative values of ΔH.

For an exothermic reaction, the reactants are always on a higher line than the products, because the reactants have more potential energy and so more enthalpy than the products. Notice in Figure 6.6 that the line for the absolute enthalpy of $C(s) + O_2(g)$, taken as the sum, is above the line for the final product,

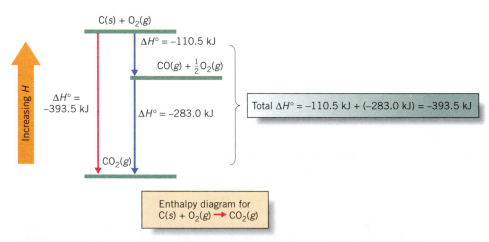

Figure 6.6 *An enthalpy diagram for the formation of $CO_2(g)$ from its elements by two different paths.* On the left is path 1, the direct conversion of $C(s)$ and $O_2(g)$ to $CO_2(g)$. On the right, path 2 shows two shorter, downward pointing arrows. The first step of path 2 takes the elements to $CO(g)$, and the second step takes $CO(g)$ to $CO_2(g)$. The overall enthalpy change is identical for both paths, as it must be, because enthalpy is a state function.

CO_2. Thus we think of a horizontal line as representing the *sum* of the enthalpies of all of the substances on the line *in the physical states specified*.

Near the left side of Figure 6.6, we see a long downward arrow that connects the enthalpy level for the reactants, $C(s) + O_2(g)$, to that of the final product, $CO_2(g)$. This is for the one-step path, the direct path. On the right we have the two-step path. Here, the overall change stops at an intermediate enthalpy level corresponding to the intermediate products, $CO(g) + \frac{1}{2}O_2(g)$, which are those made or left over from the first step. Then Step 2 occurs to give the final product. What the enthalpy diagram shows (indeed, *must* show) is that the total decrease in enthalpy is the same regardless of the path.

EXAMPLE 6.4

Preparing an Enthalpy Diagram

Hydrogen peroxide, H_2O_2, decomposes into water and oxygen by the following equation.

$$H_2O_2(l) \longrightarrow H_2O(l) + \tfrac{1}{2}O_2(g)$$

Construct an enthalpy diagram for the following two reactions of hydrogen and oxygen, and use the diagram to determine the value of $\Delta H°$ for the decomposition of hydrogen peroxide.

$$H_2(g) + O_2(g) \longrightarrow H_2O_2(l) \qquad \Delta H° = -188 \text{ kJ}$$
$$H_2(g) + \tfrac{1}{2}O_2(g) \longrightarrow H_2O(l) \qquad \Delta H° = -286 \text{ kJ}$$

Analysis: The two given reactions are exothermic, so their values of $\Delta H°$ will be associated with downward pointing arrows. The highest enthalpy level, therefore, must be for the elements themselves. The lowest level must be for the product of the reaction with the largest negative $\Delta H°$, the formation of water by the second equation. The enthalpy level for the H_2O_2, formed by a reaction with the less negative $\Delta H°$, must be in between.

Solution: We diagram the two reactions having known values of $\Delta H°$ and then think about how the answer to the problem might be determined.

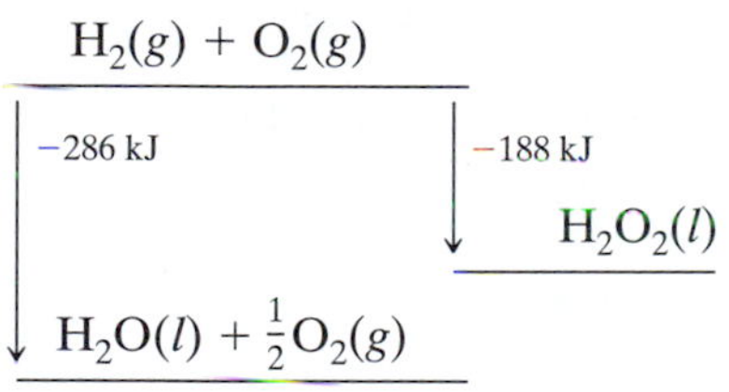

Notice that the gap on the right side corresponds exactly to the change of $H_2O_2(l)$ into $H_2O(l) + \frac{1}{2}O_2(g)$, the reaction for which $\Delta H°$ is sought. This enthalpy separation corresponds to the difference between the enthalpy levels, -286 kJ and -188 kJ. Thus for the decomposition of H_2O_2 by the equation given,

$$\Delta H° = [-286 \text{ kJ} - (-188 \text{ kJ})] = -98 \text{ kJ}$$

Our completed enthalpy diagram looks like this, showing that $\Delta H°$ for the decomposition of $H_2O_2(l)$ is -98 kJ.

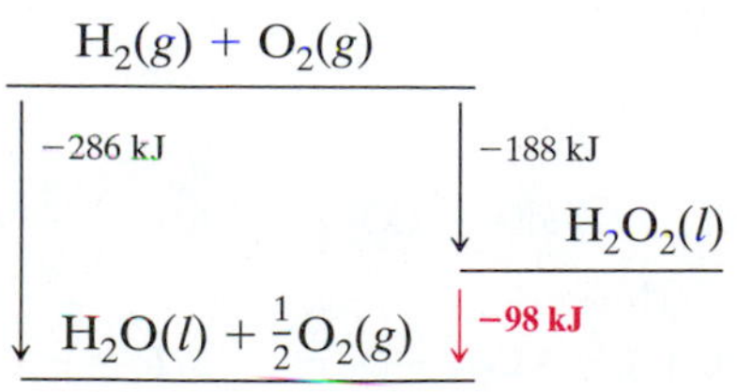

Is the Answer Reasonable?
First be sure that the chemical formulas are arrayed properly on the horizontal lines in the enthalpy diagram. The arrows pointing downward should be associated with negative values of $\Delta H°$, and these arrows must be consistent *in direction* with the given equations (in other words, they must point from reactants to products). Finally, notice that the total amount of energy associated with the two-step process, namely, (-188 kJ) plus (-98 kJ) equals -286 kJ, the energy for the one-step process.

Practice Exercise 4

Two oxides of copper can be made from copper by the following reactions.

$$2Cu(s) + O_2(g) \longrightarrow 2CuO(s) \qquad \Delta H° = -310 \text{ kJ}$$

$$2Cu(s) + \tfrac{1}{2}O_2(g) \longrightarrow Cu_2O(s) \qquad \Delta H° = -169 \text{ kJ}$$

Using these data, construct an enthalpy diagram that can be used to find $\Delta H°$ for the reaction, $Cu_2O(s) + \tfrac{1}{2}O_2(g) \rightarrow 2CuO(s)$. ◆

Hess's Law

Enthalpy diagrams, while instructive, are not necessary to calculate $\Delta H°$ for a reaction from known thermochemical equations. Using the tools for manipulating equations, we ought to be able to calculate $\Delta H°$ values simply by algebraic summing. G. H. Hess was the first to realize this, so the associated law is called **Hess's law of heat summation** or, simply, **Hess's law.**

Germain Henri Hess (1802–1850) anticipated the law of conservation of energy in the law named after him.

> **Hess's Law**
> The value of $\Delta H°$ for any reaction that can be written in steps equals the sum of the values of $\Delta H°$ of each of the individual steps.

The chief use of Hess's law is to calculate the enthalpy change for a reaction for which such data cannot be determined experimentally or are otherwise unavailable. Because this requires that we manipulate equations, let's recapitulate the few rules that govern these operations.

Rules for Manipulating Thermochemical Equations

We reverse an equation by interchanging reactants and products. This leaves the arrow still pointing from left to right.

1. When an equation is reversed—written in the opposite direction—the sign of $\Delta H°$ must also be reversed.[5]
2. Formulas canceled from both sides of an equation must be for the substance in identical physical states.
3. If all the coefficients of an equation are multiplied or divided by the same factor, the value of $\Delta H°$ must likewise be multiplied or divided by that factor.

[5]To illustrate, the reverse of the equation

$$C(s) + O_2(g) \longrightarrow CO_2(g) \qquad \Delta H° = -394 \text{ kJ}$$

is the following equation.

$$CO_2(g) \longrightarrow C(s) + O_2(g) \qquad \Delta H° = +394 \text{ kJ}$$

EXAMPLE 6.5

Using Hess's Law

Carbon monoxide is often used in metallurgy to remove oxygen from metal oxides and thereby give the free metal. The thermochemical equation for the reaction of CO with iron(III) oxide, Fe_2O_3, is

$$Fe_2O_3(s) + 3CO(g) \longrightarrow 2Fe(s) + 3CO_2(g) \qquad \Delta H° = -26.7 \text{ kJ}$$

Use this equation and the equation for the combustion of CO,

$$CO(g) + \tfrac{1}{2}O_2(g) \longrightarrow CO_2(g) \qquad \Delta H° = -283.0 \text{ kJ}$$

to calculate the value of $\Delta H°$ for the following reaction.

$$2Fe(s) + \tfrac{3}{2}O_2(g) \longrightarrow Fe_2O_3(s)$$

Analysis: We cannot simply add the two given equations, because this will not produce the equation we want. We first have to manipulate these equations so that when we add them we will get the target equation.

Solution: We can manipulate the two given equations as follows.

Step 1. We begin by trying to get the iron atoms to come out right. The target equation must have 2Fe on the *left,* but the first equation above has 2Fe to the *right* of the arrow. To move it to the left, we must reverse the *entire* equation, remembering also to reverse the sign of $\Delta H°$. We also are pleased to note that this manipulation places Fe_2O_3 to the right of the arrow, which is where it has to be after we add our adjusted equations. After these manipulations, and reversing the sign of $\Delta H°$, we have

$$2Fe(s) + 3CO_2(g) \longrightarrow Fe_2O_3(s) + 3CO(g) \qquad \Delta H° = +26.7 \text{ kJ}$$

Step 2. There must be $\tfrac{3}{2}O_2$ on the left, and we must be able to cancel *three* CO and *three* CO_2 when the equations are added. If we multiply the second of the equations given above by 3, we will obtain the necessary coefficients. We must also multiply the value of $\Delta H°$ of this equation by 3, because three times as much chemicals are now involved in the reaction. When we have done this, we have

$$3CO(g) + \tfrac{3}{2}O_2(g) \longrightarrow 3CO_2(g) \qquad \Delta H° = 3 \times (-283.0 \text{ kJ}) = -849.0 \text{ kJ}$$

Let's now put our two equations together and find the answer.

$$2Fe(s) + 3CO_2(g) \longrightarrow Fe_2O_3(s) + 3CO(g) \qquad \Delta H° = +26.7 \text{ kJ}$$
$$3CO(g) + \tfrac{3}{2}O_2(g) \longrightarrow 3CO_2(g) \qquad \Delta H° = -849.0 \text{ kJ}$$

Sum: $\quad 2Fe(s) + \tfrac{3}{2}O_2(g) \longrightarrow Fe_2O_3(s) \qquad \Delta H° = -822.3 \text{ kJ}$

Thus the value of $\Delta H°$ for the oxidation of 2 mol Fe(s) to 1 mol $Fe_2O_3(s)$ is -822.3 kJ. (The reaction is *very* exothermic.)

Is the Answer Reasonable?
There is no quick "head check." But for each step, double-check that you have heeded the rules for manipulating thermochemical equations.

Practice Exercise 5

Ethanol, C_2H_5OH, is made industrially by the reaction of water with ethylene, C_2H_4. Calculate the value of $\Delta H°$ for the reaction

$$C_2H_4(g) + H_2O(l) \longrightarrow C_2H_5OH(l)$$

given the following thermochemical equations.

$$C_2H_4(g) + 3O_2(g) \longrightarrow 2CO_2(g) + 2H_2O(l) \qquad \Delta H° = -1411.1 \text{ kJ}$$
$$C_2H_5OH(l) + 3O_2(g) \longrightarrow 2CO_2(g) + 3H_2O(l) \qquad \Delta H° = -1367.1 \text{ kJ} \quad \blacklozenge$$

6.8 Standard Heats of Formation and Hess's Law

The **standard enthalpy of formation, ΔH_f°,** of a substance, also called its **standard heat of formation,** is the amount of heat absorbed or evolved when specifically *one mole* of the substance is formed at 25 °C and 1 atm from its elements in their *standard states*. An element is in its **standard state** when it is at 25 °C and 1 atm and in its most stable form and physical state (solid, liquid, or gas). Oxygen, for example, is in its standard state only as a gas at 25 °C and 1 atm and only as O_2 molecules, not as O atoms or O_3 (ozone) molecules. Carbon must be

Table 6.2 Standard Enthalpies of Formation of Typical Substances

Substance	ΔH_f° (kJ mol^{-1})	Substance	ΔH_f° (kJ mol^{-1})
$Ag(s)$	0	$H_2O_2(l)$	-187.6
$AgBr(s)$	-100.4	$HBr(g)$	-36
$AgCl(s)$	-127.0	$HCl(g)$	-92.30
$Al(s)$	0	$HI(g)$	26.6
$Al_2O_3(s)$	-1669.8	$HNO_3(l)$	-173.2
$C(s, \text{graphite})$	0	$H_2SO_4(l)$	-811.32
$CO(g)$	-110.5	$HC_2H_3O_2(l)$	-487.0
$CO_2(g)$	-393.5	$Hg(l)$	0
$CH_4(g)$	-74.848	$Hg(g)$	60.84
$CH_3Cl(g)$	-82.0	$I_2(s)$	0
$CH_3I(g)$	14.2	$K(s)$	0
$CH_3OH(l)$	-238.6	$KCl(s)$	-435.89
$CO(NH_2)_2(s)$ (urea)	-333.19	$K_2SO_4(s)$	-1433.7
$CO(NH_2)_2(aq)$	-391.2	$N_2(g)$	0
$C_2H_2(g)$	226.75	$NH_3(g)$	-46.19
$C_2H_4(g)$	52.284	$NH_4Cl(s)$	-315.4
$C_2H_6(g)$	-84.667	$NO(g)$	90.37
$C_2H_5OH(l)$	-277.63	$NO_2(g)$	33.8
$Ca(s)$	0	$N_2O(g)$	81.57
$CaBr_2(s)$	-682.8	$N_2O_4(g)$	9.67
$CaCO_3(s)$	-1207	$N_2O_5(g)$	11
$CaCl_2(s)$	-795.0	$Na(s)$	0
$CaO(s)$	-635.5	$NaHCO_3(s)$	-947.7
$Ca(OH)_2(s)$	-986.59	$Na_2CO_3(s)$	-1131
$CaSO_4(s)$	-1432.7	$NaCl(s)$	-411.0
$CaSO_4 \cdot \frac{1}{2}H_2O(s)$	-1575.2	$NaOH(s)$	-426.8
$CaSO_4 \cdot 2H_2O(s)$	-2021.1	$Na_2SO_4(s)$	-1384.5
$Cl_2(g)$	0	$O_2(g)$	0
$Fe(s)$	0	$Pb(s)$	0
$Fe_2O_3(s)$	-822.2	$PbO(s)$	-219.2
$H_2(g)$	0	$S(s)$	0
$H_2O(g)$	-241.8	$SO_2(g)$	-296.9
$H_2O(l)$	-285.9	$SO_3(g)$	-395.2

in the form of graphite, not diamond, to be in its standard state, because the graphite form of carbon is the more stable form under standard conditions.

Standard enthalpies of formation for a variety of substances are given in Table 6.2, and a more extensive table of standard heats of formation can be found in Appendix E. Notice in particular that all values of ΔH_f° for the elements in their standard states are zero. (Forming an element *from itself,* of course, would yield no change in enthalpy.) In most tables, values of ΔH_f° for the elements are not included for this reason.

It is important to keep in mind the significance of the subscript f in the symbol ΔH_f°. It is applied to a value of ΔH° only when *one mole* of the substance is formed *from its elements in their standard states.* Consider, for example, the following four thermochemical equations and their corresponding values of ΔH°.

$$H_2(g) + \tfrac{1}{2}O_2(g) \longrightarrow H_2O(l) \qquad \Delta H_f^\circ = -285.9 \text{ kJ/mol}$$

$$2H_2(g) + O_2(g) \longrightarrow 2H_2O(l) \qquad \Delta H^\circ = -571.8 \text{ kJ}$$

$$CO(g) + \tfrac{1}{2}O_2(g) \longrightarrow CO_2(g) \qquad \Delta H^\circ = -283.0 \text{ kJ}$$

$$2H(g) + O(g) \longrightarrow H_2O(l) \qquad \Delta H^\circ = -971.1 \text{ kJ}$$

Only in the first equation is ΔH° given the subscript f. It is the only reaction that satisfies both of the conditions specified above for standard enthalpies of formation. The second equation shows the formation of *two* moles of water, not one. The third involves a *compound* as one of the reactants. The fourth involves the elements as *atoms,* which are *not standard states* for these elements. Also notice that the units of ΔH_f° are kilojoules *per mole,* not just kilojoules, because the value is for the formation of 1 mol of the compound (from its elements). We can obtain the enthalpy of formation of two moles of water (ΔH° for the second equation) simply by multiplying the ΔH_f° value for 1 mol of H_2O by the factor, 2 mol $H_2O(l)$.

$$\left(\frac{-285.9 \text{ kJ}}{\text{mol } H_2O(l)} \right) \times 2 \text{ mol } H_2O(l) = -571.8 \text{ kJ}$$

EXAMPLE 6.6

Writing an Equation for a Standard Heat of Formation

What equation must be used to represent the formation of nitric acid, $HNO_3(l)$, when we want to include its value of ΔH_f°?

Analysis: Because we are allowed to show only *one mole* of the product, we begin with its formula and take whatever fractions of moles of the elements needed to make it. We also remember to include the physical states. Table 6.2 gives the value of ΔH_f° for $HNO_3(l)$, -173.2 kJ mol^{-1}.

Solution: The three elements, H, N, and O, all occur as diatomic molecules in the gaseous state, so the following fractions of moles supply exactly enough to make one mole of HNO_3.

$$\tfrac{1}{2}H_2(g) + \tfrac{1}{2}N_2(g) + \tfrac{3}{2}O_2(g) \longrightarrow HNO_3(l) \qquad \Delta H_f^\circ = -173.2 \text{ kJ mol}^{-1}$$

Is the Answer Reasonable?
The answer correctly shows only 1 mol of HNO_3, and this governs the coefficients for the reactants. So simply check if everything is balanced.

Practice Exercise 6

Write the thermochemical equation that would be used to represent the standard heat of formation of sodium bicarbonate, $NaHCO_3(s)$. ◆

Standard enthalpies of formation are useful because they provide a convenient method for applying Hess's law without having to manipulate thermochemical equations. This is possible because, as we will demonstrate, the net $\Delta H^\circ_{\text{reaction}}$ equals the sum of the heats of formation of the products minus the sum of the heats of formation of the reactants, each ΔH°_f value multiplied by the appropriate coefficient given by the thermochemical equation. In other words, we can express Hess's law in the form of the following, known as the **Hess's law equation.**

Tools

Hess's law equation

$$\Delta H^\circ_{\text{reaction}} = \left(\begin{array}{c} \text{sum of } \Delta H^\circ_f \text{ of all} \\ \text{of the products} \end{array} \right) - \left(\begin{array}{c} \text{sum of } \Delta H^\circ_f \text{ of all} \\ \text{of the reactants} \end{array} \right) \qquad (6.9)$$

Let's now demonstrate that Equation 6.9 works. Consider the reaction given by the following equation.

$$SO_3(g) \longrightarrow SO_2(g) + \tfrac{1}{2}O_2(g) \qquad \Delta H^\circ = ?$$

We wish to calculate the heat of reaction using standard enthalpies of formation.

If we use the first method we learned, namely, the manipulation of thermochemical equations, we would need to imagine a path from the reactant to the products that involves first decomposing SO_3 into its elements in their standard states and then recombining the elements to form the products. This path is shown in Figure 6.7. The first step, whose enthalpy change is indicated as ΔH°_1, corresponds to the decomposition of SO_3 into sulfur and oxygen. This is just the *reverse* of the formation of SO_3, so we can write

$$\Delta H^\circ_1 = -\Delta H^\circ_{f\,SO_3(g)}$$

The minus sign is written because when we reverse a process, we change the sign of its ΔH.

The second step in Figure 6.7, whose enthalpy change is indicated as ΔH°_2, is for the formation of SO_2 plus half a mole of O_2 from sulfur and oxygen. Therefore, we can write

$$\Delta H^\circ_2 = \Delta H^\circ_{f\,SO_2(g)} + \tfrac{1}{2}\,\Delta H^\circ_{f\,O_2(g)}$$

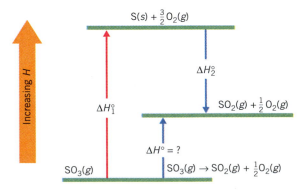

Figure 6.7 *Enthalpy diagram.* The reaction diagrammed is

$$SO_3(g) \longrightarrow SO_2(g) + \tfrac{1}{2}O_2(g)$$

The path of the reaction in this diagram involves the elements in their standard states as the intermediate state. Here we see the reactant being decomposed into its elements (longer upward pointing arrow); then the elements are recombined to form the products (short downward pointing arrow). The difference in the lengths of these two arrows is proportional to the net enthalpy change (shorter upward pointing arrow).

The sum of these two steps gives the net change we want, so the sum of ΔH_1° and ΔH_2° must equal the desired ΔH°.

$$\Delta H^\circ = \Delta H_1^\circ + \Delta H_2^\circ$$

By substitution

$$\Delta H^\circ = [-\Delta H_{fSO_3(g)}^\circ] + [\Delta H_{fSO_2(g)}^\circ + \tfrac{1}{2}\Delta H_{fO_2(g)}^\circ]$$

This can be rewritten as

$$\Delta H^\circ = [\Delta H_{fSO_2(g)}^\circ + \tfrac{1}{2}\Delta H_{fO_2(g)}^\circ] + [-\Delta H_{fSO_3(g)}^\circ]$$

The change of sign gives

$$\Delta H^\circ = [\Delta H_{fSO_2(g)}^\circ + \tfrac{1}{2}\Delta H_{fO_2(g)}^\circ] - [\Delta H_{fSO_3(g)}^\circ]$$

Notice carefully the net result. $\Delta H_{reaction}^\circ$ equals the sum of the heats of formation of the products minus the sum of the heats of formation of the reactant(s), each multiplied by the appropriate coefficient. So we could have obtained the identical result by using the Hess's law equation, Equation 6.9, directly, instead of by the more laborious method of manipulating thermochemical equations.

Some chefs keep baking soda, $NaHCO_3$, handy to put out grease fires. When thrown on the fire, baking soda partly smothers the fire, and the heat decomposes it to give CO_2, which further smothers the flame. The equation for the decomposition of $NaHCO_3$ is

$$2NaHCO_3(s) \longrightarrow Na_2CO_3(s) + H_2O(l) + CO_2(g)$$

Use the data in Table 6.2 to calculate the ΔH° for this reaction in kilojoules.

Analysis: Hess's law equation (Equation 6.9) is now our basic tool for computing values of ΔH°. So we calculate the values of ΔH° for the products, taken as a set, and the values of ΔH° for the reactants, also taken as a set. Then we subtract the latter from the former to calculate ΔH°. We compute values of ΔH° using ΔH_f° data from Table 6.2 and the coefficients of the chemical equation.

Solution:

$$\Delta H^\circ = [1\ \text{mol}\ Na_2CO_3(s) \times \Delta H_{fNa_2CO_3(s)}^\circ + 1\ \text{mol}\ H_2O(l) \times \Delta H_{fH_2O(l)}^\circ$$
$$+ 1\ \text{mol}\ CO_2(g) \times \Delta H_{fCO_2(g)}^\circ] - [2\ \text{mol}\ NaHCO_3(s) \times \Delta H_{fNaHCO_3(s)}^\circ]$$

We now use Table 6.2 to find the values of ΔH_f° for each substance in its proper physical state.

$$\Delta H^\circ = \left[1\ \text{mol}\ Na_2CO_3 \times \frac{-1131\ \text{kJ}}{\text{mol}\ Na_2CO_3} + 1\ \text{mol}\ H_2O \times \frac{-285.9\ \text{kJ}}{\text{mol}\ H_2O} \right.$$
$$\left. + 1\ \text{mol}\ CO_2 \times \frac{-393.5\ \text{kJ}}{\text{mol}\ CO_2} \right]$$
$$- \left[2\ \text{mol}\ NaHCO_3 \times \frac{-947.7\ \text{kJ}}{\text{mol}\ NaHCO_3} \right]$$

$$\Delta H^\circ = (-1810\ \text{kJ}) - (-1895\ \text{kJ})$$
$$= +85\ \text{kJ}$$

Thus, under standard conditions, the reaction is endothermic by 85 kJ. (Notice that we did not have to manipulate any equations.)

EXAMPLE 6.7

Using Hess's Law and Standard Enthalpies of Formation

Is the Answer Reasonable?
Double-check that all of the coefficients found in the given equation are correctly applied as multipliers in the solution. Be sure that the *signs* of the values of ΔH_f° have all been carefully used. Have you taken the order of subtracting specified in Equation 6.9? Keeping track of minus signs requires particular care.

Practice Exercise 7

Calculate ΔH° for the following reactions.

(a) $2NO(g) + O_2(g) \longrightarrow 2NO_2(g)$

(b) $NaOH(s) + HCl(g) \longrightarrow NaCl(s) + H_2O(l)$ ◆

SUMMARY

Kinetic and Potential Energy. Because electrical attractions and repulsions occur within atoms, molecules, and ions, substances have **chemical energy,** a form of **potential energy.** The **total energy** of a system, E, is the sum of its kinetic and potential energies.

The particles that make up matter, namely, atoms, molecules, or ions, are in constant motion, so they also possess kinetic energy, specifically, **molecular kinetic energy.** In gases and liquids the motions take the particles from one location to another; in solids the particles jiggle about fixed positions. **Heat** is transferred as molecular kinetic energy from an object at a higher temperature to one at a lower temperature. The Kelvin temperature of the sample is proportional to the *average* value of the molecular kinetic energies.

Energy Changes in Reactions. **Thermochemistry** is the study of energy changes in chemical systems. A **system** is anything we choose it to be, such as a reaction occurring in an open container. Everything else in the universe is the system's **surroundings.** A **boundary,** real or imaginary, exists that separates the system from its surroundings. The values of E before the reaction ($E_{reactants}$) or after the reaction ($E_{products}$) cannot be known. However, the difference between the two, the *change* in internal energy or ΔE, is knowable.

$$\Delta E = E_{products} - E_{reactants}$$

Chemical reactions involve the breaking of bonds and the forming of different bonds. The net energy effect is the net of the bond breaking "costs" and the bond forming "profits." When some combinations of substances react, potential energy is converted into molecular kinetic energy or heat, called the **heat of reaction,** q. If the system is **adiabatic** (no heat leaves it), the internal temperature increases. Otherwise, the heat has a tendency to leave the system. Either way, the reaction is said to be **exothermic.**

In **endothermic** reactions, molecular kinetic energy of the reactants is converted into potential energy of the products. If the system is adiabatic, the reduction of the molecular kinetic energy causes the system's temperature to decrease. Otherwise, the system tends to absorb energy from the surroundings (heat "flows" in). One way of stating the **law of conservation of energy** is that any energy which leaves a system is identical in magnitude to the energy that enters its surroundings.

First Law of Thermodynamics. **Thermodynamics** is the study of the flow of energy between a system and its surroundings. Energy can flow into or out of a system either as heat, q, or as work, w. Electrical work is one kind of work. Another kind is called pressure–volume work. **Pressure** is the ratio of the force applied to the area over which it is applied. Atmospheric pressure is the pressure that air exerts because it has mass. The **standard atmosphere** of pressure or 1 atm is approximately what the atmospheric pressure is at sea level. When a gas evolves and pushes against the atmosphere or causes a piston to move in a cylinder, pressure–volume work is being done. When the volume change, ΔV, occurs at constant opposing pressure, P, the associated pressure–volume work is given by $w = -P\Delta V$.

The **first law of thermodynamics** says that no matter how the change in energy accompanying a reaction may be allocated between q and w, their sum is the same, ΔE.

$$\Delta E = q + w$$

The algebraic sign for q and w is negative when the system gives off heat to or does work on the surroundings. The sign is positive when the system absorbs heat or receives work energy done to it.

The **state of the system** is known when we know the system's pressure, temperature, volume, and composition (in terms of the numbers of moles of the system's substances and their identities). Pressure, temperature, and volume are **state functions** because changes in their values are independent of the path or mechanism taken to attain them. ΔE is another state function, but neither q nor w is; the values of q and w depend on the path taken from the initial to the final state.

When a system has rigid walls, no pressure–volume work can be done to or by the atmosphere, Consequently, w is zero, and the heat of reaction, now symbolized as q_v, is the change in the internal energy of the system, ΔE. Values of q_v are determined using a bomb calorimeter or a similar device.

When the system is under constant pressure (e.g., open to the atmosphere), pressure–volume work is now possible, so in this circumstance, the energy of the system is called its **enthalpy, H.**

$$H = E + PV$$

Like E, values of H cannot be known in an absolute sense, but a difference in enthalpy between reactants and products can be determined. The heat of reaction is now called the **enthalpy change, ΔH.** Under constant pressure,

$$\Delta H = \Delta E + P\Delta V$$

ΔH is a state function. Its value differs from that of ΔE by the work involved in interacting with the atmosphere when the change occurs at constant atmospheric pressure. In general, the difference between ΔE and ΔH is quite small.

Exothermic reactions have negative values of ΔH; endothermic changes have positive values. The heat exchanged with the surroundings by a reaction run at constant pressure is symbolized by q_p. Its sign depends on whether the reaction is exothermic (negative sign) or endothermic (positive sign).

Various techniques of **calorimetry** are used to determine heat evolved or absorbed. For constant volume work, a bomb calorimeter is used if gases are involved. Otherwise, solution calorimeters are used. The "coffee cup" calorimeter works well when low precision is acceptable.

Specific Heat and Heat Capacity. Two thermal properties that all substances have are *specific heat* and *heat capacity*. **Specific heat** is the heat needed to change the temperature of 1 g of a substance by 1 °C. **Heat capacity** is the heat needed to change the temperature of an entire sample by one degree Celsius. Thus, specific heat can also be thought of as heat capacity per gram. Heat capacity is spe-

cific heat times mass. Heat capacity calculated on a per mole basis is called **molar heat capacity.** Water has an unusually high specific heat.

Standard Heats of Reaction. The reference conditions for thermochemistry, called the **standard conditions,** are 25 °C and 1 atm of pressure. An enthalpy change measured under these conditions is called the **standard enthalpy of reaction** or the **standard heat of reaction.** Its symbol is $\Delta H°$ where the degree sign signifies that standard conditions are involved. The value of $\Delta H°$ is a function of the amounts of substances involved in the reaction. If the number of moles of reactants is doubled, the standard heat of the reaction is twice as large. The units for $\Delta H°$ are generally joules or kilojoules.

When the enthalpy change is for the formation of *one* mole of a substance under standard conditions from its *elements in their standard states*, it is called the **standard heat of formation** of the compound, symbolized as $\Delta H_f°$, and it is generally given in units of kilojoules per mole (kJ mol^{-1}).

A balanced chemical equation that includes both the enthalpy change and the physical states of the substances is called a **thermochemical equation.** These can be added, reversed (reversing also the sign of the enthalpy change), or multiplied by a constant multiplier (doing the same to the enthalpy change). If formulas are canceled or added, they must be of substances in identical physical states.

Hess's Law. **Hess's law of heat summation** is possible because enthalpy is a state function. The value of $\Delta H_f°$ for a reaction can be calculated as the sum of the $\Delta H_f°$ values of the products minus the sum of the $\Delta H_f°$ values of the reactants. The coefficients in the equation and the values of $\Delta H_f°$ from tables are used to calculate values of $\Delta H_f°$ for products and reactants. Otherwise, values of $\Delta H°$ can be determined by the manipulation of any combination of thermochemical equations that add up to the final net equation.

The table below lists the tools that you have learned in this chapter that are applicable to problem solving. Review them if necessary, and refer to them when working on the Thinking-It-Through problems and the Review Questions and Review Problems that follow.

Tool	Function
Heat capacity and specific heat *(page 248)*	To use a heat capacity and a temperature change to calculate the heat of a reaction or to use specific heat data and a mass together with a temperature change to calculate the heat of a reaction.
Thermochemical equations *(page 253)*	To use thermochemical equations for one set of reactions to write a thermochemical equation for some net reaction.
Enthalpy diagrams *(page 256)*	To visualize enthalpy changes in multistep reactions.
Hess's law equation *(page 262)*	To use standard heats of formation to calculate the enthalpy of a reaction.

THINKING IT THROUGH

The goal for each of the following problems is to give you practice in *thinking* your way through problems. The goal is *not* to find the answer itself; instead, you are just asked to assemble the available information needed to obtain the answer, state what additional data (if any) are needed, and describe how you would use the data to answer the question. For problems involving unit conversions, list the relationships among the units that are needed to carry out the conversions. Construct the conversion factors that can be formed from these relationships. Then set up the solution to the problem by arranging the conversion factors so the units cancel correctly to give the desired units of the answer.

The problems are divided into two groups. Those in Level 2 are more challenging than those in Level 1 and provide an opportunity to really hone your problem solving skills. Detailed answers to these problems can be found at our Web site: http://www.wiley.com/college/brady

Level 1 Problems

1. Which vehicle possesses more kinetic energy and why, vehicle X with a mass of 2000 kg and a velocity of 50 km hr^{-1} or vehicle Y with a mass of 1000 kg and a velocity of 100 km hr^{-1}?

2. How many grams of water can be heated from 25.0 °C to 35.0 °C by the heat released from 85.0 g of iron that cools from 85.0 °C to 30.0 °C? (Set up the calculation.)

3. If you know the specific heat of an object, what else besides the temperature change has to be known before one can calculate the energy flow caused by a temperature change?

4. The heat capacity of a 4.50 kg bar of a metal is 1.44 kJ °C^{-1}. What is the specific heat of the metal expressed in the units cal g^{-1} °C^{-1}? (Set up the calculation.)

5. If you want to calculate how much energy is absorbed by a bar of iron when it has been kept in a fireplace with a blazing log fire, which of the following information is needed: the initial temperature of the iron before it went into the fire, the length of time it is in the fire, the final temperature of the iron, the mass of the iron bar, the volume of the iron bar, the temperature of the burning material, the specific heat of iron?

6. Consider the reaction,

$$C_{12}H_{22}O_{11}(s) + 12O_2(g) \longrightarrow 12CO_2(g) + 11H_2O(l)$$

How do the values of ΔE and ΔH compare?

7. Consider the thermochemical equation

$$H_2(g) + \tfrac{1}{2}O_2(g) \longrightarrow H_2O(l)$$

What is the value of $\Delta H°$ for the following reaction? (Describe the calculation involved.)

$$3H_2O(l) \longrightarrow 3H_2(g) + \tfrac{3}{2}O_2(g)$$

8. How many kilojoules of heat are evolved in the combustion of 35.5 g of methyl alcohol? The *unbalanced equation* for the reaction is

$$CH_3OH(l) + O_2(g) \longrightarrow CO_2(g) + H_2O(g)$$

(Set up the calculations. Include any data required.)

Level 2 Problems

9. A rock with a mass of 20 kg falls 30 ft to the ground and strikes a lever that tosses a 10 kg object into the air. How will the speed of the 10 kg object leaving the ground compare with the speed of the 20 kg object just as it strikes the lever? (Assume all the KE of the 20 kg object is transferred to the 10 kg object.)

10. A 2.00 kg piece of granite with a specific heat of 0.803 J g^{-1} °C^{-1} and a temperature of 95.0 °C is placed into 2.00 L of water at 22.0 °C. When the granite and water come to the same temperature, what will that temperature be? (Set up the calculation.)

11. A piece of metallic lead with a mass of 51.36 g was heated to 100.0 °C in boiling water. It was quickly dried and immersed in 10.0 g of liquid water that was at a temperature of 24.6 °C. The system (water plus lead) came to a temperature of 35.2 °C. What is the specific heat of lead? (Describe the calculations involved.)

12. Suppose a truck with a mass of 14.0 tons (1 ton = 2000 lb) is traveling at a speed of 45.0 mi/hr. If the truck driver slams on the brakes, the kinetic energy of the truck is changed to heat as the brakes slow the truck to a stop. How much would the temperature of 5.00 gallons of water increase if all this heat could be absorbed by the water? (Explain in detail all the calculations involved. Apply the factor-label method where appropriate.)

13. Consider the thermochemical equation

$$C_2H_2(g) + H_2(g) \longrightarrow C_2H_4(g) \qquad \Delta H° = -175.1 \text{ kJ}$$
$$\text{acetylene} \qquad\qquad\qquad \text{ethylene}$$

Without referring to Table 6.2 or performing any calculations, determine which gas, acetylene or ethylene, has the more positive standard heat of formation. Explain your reasoning.

14. Two 1 liter boxes contain identical amounts of argon gas, represented in the figure below as blue spheres. The insulating partition in the center can be lifted to expose the thermal conducting partition.
(a) If you were to make an animated movie showing how the argon atoms move inside the two boxes, what differences would you show in the motions of atoms in the left and right boxes? Explain.
(b) The insulating partition in the center is lifted to allow atoms on both sides to strike the conducting partition. Explain how you would change your movies over time to represent what happens to the argon atoms in the left and right boxes.

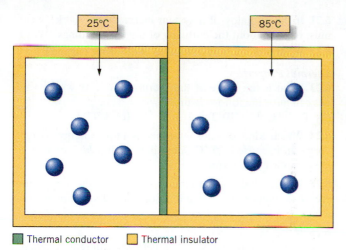

▮ Thermal conductor   ▯ Thermal insulator

(c) What would the two thermometers show over time after the insulating partition had been lifted? Would they eventually stop changing values and, if so, what values would each show?

15. In an experiment, 100 mL of a clear, colorless, aqueous solution of a compound *A* is added to a beaker containing 100 mL of a clear blue aqueous solution of a compound *B*. Prior to mixing, both solutions were at 25 °C. After mixing the re-sulting solution was clear and colorless. The mixed solutions were stirred for several seconds with a thermometer. The temperature was observed to rise rapidly to a maximum value of 53 °C and then to fall slowly over a period of 30 minutes back to 25 °C. Explain the origins of the changes in temperature.

16. The left beaker in the figure contains a 200.0 g piece of pure copper in 100 g of boiling water. The right beaker contains 75 g of water at 10 °C. If the cold water is poured rapidly into the left beaker and the temperature of the water monitored, what will be the final temperature? Assume no heat loss to the surroundings. List the data you will need to look up to work this problem and set up the equation necessary for its solution.

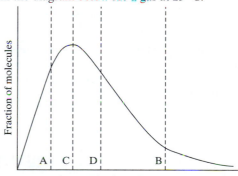

100°C   10°C

REVIEW QUESTIONS

Kinetic and Potential Energy

6.1 How are the total kinetic energy and total potential energy of atomic-sized particles related to each other in an isolated system?

6.2 What is meant by the term *chemical energy?*

6.3 In a certain chemical reaction, there is a decrease in the potential energy (chemical energy) as the reaction proceeds. (a) How does the total kinetic energy of the particles change? (b) How does the temperature of the reaction mixture change?

6.4 How does the potential energy change (increase, decrease, or no change) for each of the following?
(a) Two electrons come closer together.
(b) An electron and a proton become farther apart.
(c) Two atomic nuclei approach each other.
(d) A ball rolls downhill.

6.5 Describe how the potential energy of the system of atomic-sized particles changes when each of the following events occurs.
(a) The wax of a candle burns in air, giving a yellow flame. (The system consists of the wax and O_2 in the air.)
(b) Ammonium nitrate dissolves in water to produce a cooling effect.

Energy Units

6.6 How is the joule defined?

6.7 How is the calorie defined according to the SI? How was it originally defined?

6.8 How many joules are in one nutritional Calorie?

Kinetic Theory of Matter

6.9 Consider the distribution of molecular kinetic energies shown in the diagram below for a gas at 25 °C.

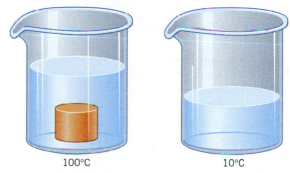

(a) Which point corresponds to the most frequently occurring (also called the most probable) molecular kinetic energy?
(b) Which point corresponds to the average molecular kinetic energy?
(c) If the temperature of the gas is raised to 50 °C, how will the height of the curve at B change?
(d) If the temperature of the gas is raised to 50 °C, how will the height of the curve at A change? How will the maximum height of the curve change?
(e) If the Kelvin temperature of the gas is doubled, which of the kinetic energy values, A, B, C, or D, will exactly double?

6.10 Suppose the temperature of an object is raised from 100 °C to 200 °C by heating it with a Bunsen burner. Which of the following will be true?
(a) The average kinetic energy will increase.
(b) The average kinetic energy will double.
(c) The total kinetic energy of all the molecules will increase.
(d) The number of fast-moving molecules will increase.
(e) The number of slow-moving molecules will increase.
(f) The chemical potential energy will decrease.

Energy Changes in Chemical Reactions

6.11 What term do we use to describe a reaction that liberates heat to its surroundings? How does the chemical energy change during such a reaction?

6.12 What term is used to describe a reaction that absorbs heat from the surroundings? How does the chemical energy change during such a reaction?

Internal Energy and Enthalpy

6.13 What are some facts about a system that we should specify to describe the *state* of the system?

6.14 What is a state function? Give two examples.

6.15 What does the internal energy, E, of a system represent in terms of the kinetic and potential energies of its particles?

6.16 Write the equation that states the first law of thermodynamics. In your own words, what does this statement mean in terms of energy exchanges between a system and its surroundings?

6.17 Which state function is given by the heat of reaction at constant volume? Which state function is given by the heat of reaction at constant pressure? Under what conditions are these heats of reaction equal?

6.18 If a system containing gases expands and pushes back a piston against a constant opposing pressure, what equation describes the work done on the system?

6.19 How is enthalpy defined?

6.20 What equation defines ΔH in general terms? How is this expressed when the system involves a chemical reaction? In general, why are chemists and biologists more interested in values of ΔH than ΔE?

6.21 If the enthalpy of a system increases by 100 kJ, what must be true about the enthalpy of the surroundings? Why?

6.22 What is the *sign* of ΔH for an exothermic change?

Thermal Properties

6.23 What is the name of the thermal property whose values can have the following units?
(a) $J\,g^{-1}\,°C^{-1}$ (b) $J\,mol^{-1}\,°C^{-1}$ (c) $J\,°C^{-1}$

6.24 Which kind of substance needs more energy to undergo an increase of 5 °C, something with a *high* or with a *low* specific heat? Explain.

6.25 Which kind of substance experiences the larger increase in temperature when it absorbs 100 J, something with a *high* or a *low* specific heat?

6.26 If the specific heat values in Table 6.1 were in units of kJ $kg^{-1}\,K^{-1}$, would the values be *numerically* different? Explain.

Enthalpy and Heats of Reaction

6.27 Why do standard reference values for temperature and pressure have to be selected when we consider and compare heats of reaction for various reactions? What are the values for the standard temperature and standard pressure?

6.28 What distinguishes a *thermochemical* equation from an ordinary chemical equation?

6.29 In a thermochemical equation, what do the coefficients represent in terms of the amounts of the reactants and products?

6.30 Why are fractional coefficients permitted in a balanced thermochemical equation? If a thermochemical equation has a coefficient of $\frac{1}{2}$ for a formula, what does it signify?

Hess's Law

6.31 What fundamental fact about ΔH makes Hess's law possible?

6.32 What *two* conditions must be met by a thermochemical equation so that its standard enthalpy change can be given the symbol ΔH_f°?

6.33 What is Hess's law of heat summation expressed in terms of standard heats of formation?

REVIEW PROBLEMS

Answers to problems whose numbers are printed in color are given in Appendix D.
More challenging problems are marked with asterisks.

Energy Units and Conversions

6.34 What is the kinetic energy, in kilojoules, of a 2150 kg automobile traveling at a speed of 80.0 km/hr (approximately 50 mph)?

6.35 What is the kinetic energy, in kilojoules, of a cement truck with a mass of 2.04×10^4 kg traveling at 80.0 km/hr?

6.36 Perform the following conversions: (a) 457 kJ to kcal, (b) 127 kcal to kJ.

6.37 Perform the following conversions: (a) 187 J to cal, (b) 568 cal to J.

First Law of Thermodynamics

6.38 If a system does 45 J of work and receives 28 J of heat, what is the value of ΔE for this change?

6.39 If a system absorbs 48 J of heat and does 22 J of work, what is the value of ΔE for this change?

Thermal Properties, Measuring Energy Changes

6.40 How much heat in kilojoules must be removed from 175 g of water to lower its temperature from 25.0 to 15.0 °C (which would be like cooling a glass of lemonade)?

6.41 How much heat in kilojoules is needed to bring 1.0 kg of water from 25 to 99 °C (comparable to making four cups of coffee)?

6.42 How many joules are needed to increase the temperature of 15.0 g of Fe from 20.0 to 40.0 °C?

6.43 The addition of 250 J to 30.0 g of copper initially at 22 °C will change its temperature to what final value?

6.44 A 50.0 g mass of a metal was heated to 100 °C and then plunged into 100 g of water at 24.0 °C. The temperature of the resulting mixture became 28.0 °C.
(a) How many joules did the water absorb?
(b) How many joules did the metal lose?
(c) What is the heat capacity of the metal sample?
(d) What is the specific heat of the metal?

6.45 A sample of copper was heated to 120 °C and then thrust into 200 g of water at 25.00 °C. The temperature of the mixture became 26.50 °C.
(a) How much heat in joules was absorbed by the water?
(b) The copper sample lost how many joules?
(c) What was the mass in grams of the copper sample?

6.46 Fat tissue is 85% fat and 15% water. The complete breakdown of the fat itself converts it to CO_2 and H_2O, which releases 9.0 kcal per gram of fat in the tissue.
(a) How many kilocalories are released by the loss of 1.0 lb of fat *tissue* in a weight-reduction program?
(b) Running 8.0 miles hr^{-1} expends about 5.0×10^2 kcal hr^{-1} of extra energy. How far does a person have to run to burn off 1.0 lb of fat *tissue* by this means alone?

6.47 A well-nourished person adds about 0.50 lb of fat tissue for each 3.5×10^2 kcal of food energy taken in over and above that needed. Suppose you decided to reduce your weight simply by omitting oils and butter but keeping every other aspect of your diet and your activities the same. How many days would be needed to lose 1.0 kg of fat *tissue* by this strategy alone? The caloric content of dietary fats and oils is 9.0 kcal g^{-1}. Suppose that you have been eating 0.25 lb of salad oil and butter per day.

6.48 Calculate the molar heat capacity of iron in J mol^{-1} $°C^{-1}$. Its specific heat is 0.4498 J g^{-1} $°C^{-1}$.

6.49 What is the molar heat capacity of ethyl alcohol, C_2H_5OH, in units of J mol^{-1} $°C^{-1}$, if its specific heat is 0.586 cal g^{-1} $°C^{-1}$?

Calorimetry

6.50 A vat of 4.54 kg of water underwent a decrease in temperature from 60.25 to 58.65 °C. How much energy in kilojoules left the water? (For this range of temperature, use a value of 4.18 J g^{-1} $°C^{-1}$ for the specific heat of water.)

6.51 A container filled with 2.46 kg of water underwent a temperature change from 25.24 °C to 27.31 °C. How much heat, measured in kilojoules, did the water absorb?

6.52 Nitric acid, HNO_3, reacts with potassium hydroxide, KOH, as follows.

$$HNO_3(aq) + KOH(aq) \longrightarrow KNO_3(aq) + H_2O(l)$$

A student placed 55.0 mL of 1.3 M HNO_3 in a coffee cup calorimeter, noted that the temperature was 23.5 °C, and added 55.0 mL of 1.3 M KOH, also at 23.5 °C. The mixture was stirred quickly with a thermometer, and its temperature rose to 31.8 °C. Calculate the heat of reaction in joules. Assume that the specific heats of all solutions are 4.18 J g^{-1} $°C^{-1}$ and that all densities are 1.00 g mL^{-1}. Calculate the heat of reaction per mole of acid (in units of kJ mol^{-1}).

6.53 A dilute solution of hydrochloric acid with a mass of 610.29 g and containing 0.33183 mol of HCl was exactly neutralized in a calorimeter by the sodium hydroxide in 615.31 g of a comparably dilute solution. The temperature increased from 16.784 to 20.610 °C. The specific heat of the HCl solution was 4.031 J g^{-1} $°C^{-1}$; that of the NaOH solution was 4.046 J g^{-1} $°C^{-1}$. The heat capacity of the calorimeter was 77.99 J $°C^{-1}$. Use these data to calculate the heat evolved by the following reaction.

$$HCl(aq) + NaOH(aq) \longrightarrow NaCl(aq) + H_2O(l)$$

What is the heat of neutralization per mole of HCl? Assume that the original solutions made independent contributions to the total heat capacity of the system following their mixing.

6.54 A 1.00 mol sample of propane, a gas used for cooking in many rural areas, was placed in a bomb calorimeter with excess oxygen and ignited. The reaction was

$$C_3H_8(g) + 5O_2(g) \longrightarrow 3CO_2(g) + 4H_2O(l)$$

The initial temperature of the calorimeter was 25.000 °C and its total heat capacity was 97.1 kJ $°C^{-1}$. The reaction raised the temperature of the calorimeter to 27.282 °C.
(a) How many joules were liberated in this reaction?
(b) What is the heat of reaction of propane with oxygen expressed in kilojoules per mole of C_3H_8 burned?

6.55 Toluene, C_7H_8, is used in the manufacture of explosives such as TNT (trinitrotoluene). A 1.500 g sample of liquid toluene was placed in a bomb calorimeter along with excess oxygen. When the combustion of the toluene was initiated, the temperature of the calorimeter rose from 25.000 °C to 26.413 °C. The products of the combustion are $CO_2(g)$ and $H_2O(l)$, and the heat capacity of the calorimeter was 45.06 kJ $°C^{-1}$. The reaction was

$$C_7H_8(l) + 9O_2(g) \longrightarrow 7CO_2(g) + 4H_2O(l)$$

(a) How many joules were liberated by the reaction?
(b) How many joules would be liberated under similar conditions if 1.00 mol of toluene were burned?

Enthalpy Changes and Heats of Reaction

6.56 One thermochemical equation for the reaction of carbon monoxide with oxygen is

$$3CO(g) + \tfrac{3}{2}O_2(g) \longrightarrow 3CO_2(g) \qquad \Delta H° = -849 \text{ kJ}$$

(a) Write the thermochemical equation for the reaction using 2 mol of CO.

(b) What is $\Delta H°$ for the formation of 1 mol of CO_2 by this reaction?

6.57 Ammonia reacts with oxygen as follows.

$$4NH_3(g) + 7O_2(g) \longrightarrow 4NO_2(g) + 6H_2O(g)$$
$$\Delta H° = -1132 \text{ kJ}$$

(a) Calculate the enthalpy change for the combustion of 1 mol of NH_3.

(b) Write the thermochemical equation for the reaction in which one mole of H_2O is formed.

6.58 Aluminum and iron(III) oxide, Fe_2O_3, react to form aluminum oxide, Al_2O_3, and iron. For each mole of aluminum used, 426.9 kJ of energy is released under standard conditions. Write the thermochemical equation that shows the consumption of 4 mol of Al. (All of the substances are solids.)

6.59 Liquid benzene, C_6H_6, burns in oxygen to give carbon dioxide gas and liquid water (when all products are returned to 25 °C and 1 atm pressure). The combustion of 1.00 mol of benzene liberates 3271 kJ. Write the thermochemical equation for the combustion of 3.00 mol of liquid benzene.

6.60 Magnesium burns in air to produce a bright light and is often used in fireworks displays. The combustion of magnesium follows the thermochemical equation

$$2Mg(s) + O_2(g) \longrightarrow 2MgO(s) \quad \Delta H° = -1203 \text{ kJ}$$

How much heat (in kilojoules) is liberated by the combustion of 6.54 g of magnesium?

6.61 Methanol is the fuel in "canned heat" containers (e.g., Sterno) that are used to heat foods at cocktail parties. The combustion of methanol follows the thermochemical equation

$$2CH_3OH(l) + 3O_2(g) \longrightarrow 2CO_2(g) + 4H_2O(g)$$
$$\Delta H° = -1199 \text{ kJ}$$

How many kilojoules are liberated by the combustion of 46.0 g of methanol?

Hess's Law

6.62 Construct an enthalpy diagram that shows the enthalpy changes for a one-step conversion of germanium, $Ge(s)$, into $GeO_2(s)$, the dioxide. On the same diagram, show the two-step process, first to the monoxide, $GeO(s)$, and then its conversion to the dioxide. The relevant thermochemical equations are the following.

$$Ge(s) + \tfrac{1}{2}O_2(g) \longrightarrow GeO(s) \quad \Delta H° = -255 \text{ kJ}$$

$$Ge(s) + O_2(g) \longrightarrow GeO_2(s) \quad \Delta H° = -534.7 \text{ kJ}$$

Using this diagram, determine $\Delta H°$ for the following reaction.

$$GeO(s) + \tfrac{1}{2}O_2(g) \longrightarrow GeO_2(s)$$

6.63 Construct an enthalpy diagram for the formation of $NO_2(g)$ from its elements by two pathways: first, from its elements and, second, by a two-step process, also from the elements. The relevant thermochemical equations are

$$\tfrac{1}{2}N_2(g) + O_2(g) \longrightarrow NO_2(g) \quad \Delta H° = +33.8 \text{ kJ}$$
$$\tfrac{1}{2}N_2(g) + \tfrac{1}{2}O_2(g) \longrightarrow NO(g) \quad \Delta H° = +90.37 \text{ kJ}$$
$$NO(g) + \tfrac{1}{2}O_2(g) \longrightarrow NO_2(g) \quad \Delta H° = ?$$

Be sure to note the signs of the values of $\Delta H°$ associated with arrows pointing up or down. Using the diagram, determine the value of $\Delta H°$ for the third equation.

6.64 Show how the equations

$$N_2O_4(g) \longrightarrow 2NO_2(g) \quad \Delta H° = +57.93 \text{ kJ}$$
$$2NO(g) + O_2(g) \longrightarrow 2NO_2(g) \quad \Delta H° = -113.14 \text{ kJ}$$

can be manipulated to give $\Delta H°$ for the following reaction.

$$2NO(g) + O_2(g) \longrightarrow N_2O_4(g)$$

6.65 We can generate hydrogen chloride by heating a mixture of sulfuric acid and potassium chloride according to the equation

$$2KCl(s) + H_2SO_4(l) \longrightarrow 2HCl(g) + K_2SO_4(s)$$

Calculate $\Delta H°$ in kilojoules for this reaction from the following thermochemical equations.

$$HCl(g) + KOH(s) \longrightarrow KCl(s) + H_2O(l)$$
$$\Delta H° = -203.6 \text{ kJ}$$

$$H_2SO_4(l) + 2KOH(s) \longrightarrow K_2SO_4(s) + 2H_2O(l)$$
$$\Delta H° = -342.4 \text{ kJ}$$

6.66 Calculate $\Delta H°$ in kilojoules for the following reaction, the preparation of the unstable acid nitrous acid, HNO_2.

$$HCl(g) + NaNO_2(s) \longrightarrow HNO_2(l) + NaCl(s)$$

Use the following thermochemical equations.

$$2NaCl(s) + H_2O(l) \longrightarrow 2HCl(g) + Na_2O(s)$$
$$\Delta H° = +507.31 \text{ kJ}$$

$$NO(g) + NO_2(g) + Na_2O(s) \longrightarrow 2NaNO_2(s)$$
$$\Delta H° = -427.14 \text{ kJ}$$

$$NO(g) + NO_2(g) \longrightarrow N_2O(g) + O_2(g)$$
$$\Delta H° = -42.68 \text{ kJ}$$

$$2HNO_2(l) \longrightarrow N_2O(g) + O_2(g) + H_2O(l)$$
$$\Delta H° = +34.35 \text{ kJ}$$

6.67 Barium oxide reacts with sulfuric acid as follows.

$$BaO(s) + H_2SO_4(l) \longrightarrow BaSO_4(s) + H_2O(l)$$

Calculate $\Delta H°$ in kilojoules for this reaction. The following thermochemical equations can be used.

$$SO_3(g) + H_2O(l) \longrightarrow H_2SO_4(l) \qquad \Delta H° = -78.2 \text{ kJ}$$
$$BaO(s) + SO_3(g) \longrightarrow BaSO_4(s) \qquad \Delta H° = -213 \text{ kJ}$$

6.68 Copper metal can be obtained by heating copper oxide, CuO, in the presence of carbon monoxide, CO, according to the following reaction.

$$CuO(s) + CO(g) \longrightarrow Cu(s) + CO_2(g)$$

Calculate $\Delta H°$ in kilojoules, using following thermochemical equations.

$$2CO(g) + O_2(g) \longrightarrow 2CO_2(g) \qquad \Delta H° = -566.1 \text{ kJ}$$
$$2Cu(s) + O_2(g) \longrightarrow 2CuO(s) \qquad \Delta H° = -310.5 \text{ kJ}$$

6.69 Calcium hydroxide reacts with hydrochloric acid by the following equation.

$$Ca(OH)_2(aq) + 2HCl(aq) \longrightarrow CaCl_2(aq) + 2H_2O(l)$$

Calculate $\Delta H°$ in kilojoules for this reaction, using the following equations as needed.

$$CaO(s) + 2HCl(aq) \longrightarrow CaCl_2(aq) + H_2O(l)$$
$$\Delta H° = -186 \text{ kJ}$$

$$CaO(s) + H_2O(l) \longrightarrow Ca(OH)_2(s) \qquad \Delta H° = -65.1 \text{ kJ}$$

$$Ca(OH)_2(s) \xrightarrow[\text{in water}]{\text{dissolving}} Ca(OH)_2(aq) \qquad \Delta H° = -12.6 \text{ kJ}$$

6.70 Given the following thermochemical equations,

$$CaO(s) + Cl_2(g) \longrightarrow CaOCl_2(s) \qquad \Delta H° = -110.9 \text{ kJ}$$
$$H_2O(l) + CaOCl_2(s) + 2NaBr(s) \longrightarrow$$
$$2NaCl(s) + Ca(OH)_2(s) + Br_2(l) \qquad \Delta H° = -60.2 \text{ kJ}$$
$$Ca(OH)_2(s) \longrightarrow CaO(s) + H_2O(l) \qquad \Delta H° = +65.1 \text{ kJ}$$

calculate the value of $\Delta H°$ (in kilojoules) for the reaction

$$\tfrac{1}{2}Cl_2(g) + NaBr(s) \longrightarrow NaCl(s) + \tfrac{1}{2}Br_2(l)$$

6.71 Given the following thermochemical equations,

$$2Cu(s) + S(s) \longrightarrow Cu_2S(s) \qquad \Delta H° = -79.5 \text{ kJ}$$
$$S(s) + O_2(g) \longrightarrow SO_2(g) \qquad \Delta H° = -297 \text{ kJ}$$
$$Cu_2S(s) + 2O_2(g) \longrightarrow 2CuO(s) + SO_2(g)$$
$$\Delta H° = -527.5 \text{ kJ}$$

calculate the standard enthalpy of formation (in kilojoules per mole) of $CuO(s)$.

6.72 Given the following thermochemical equations,

$$4NH_3(g) + 7O_2(g) \longrightarrow 4NO_2(g) + 6H_2O(g)$$
$$\Delta H° = -1132 \text{ kJ}$$

$$6NO_2(g) + 8NH_3(g) \longrightarrow 7N_2(g) + 12H_2O(g)$$
$$\Delta H° = -2740 \text{ kJ}$$

calculate the value of $\Delta H°$ (in kilojoules) for the reaction

$$4NH_3(g) + 3O_2(g) \longrightarrow 2N_2(g) + 6H_2O(g)$$

6.73 Given the following thermochemical equations,

$$3Mg(s) + 2NH_3(g) \longrightarrow Mg_3N_2(s) + 3H_2(g)$$
$$\Delta H° = -371 \text{ kJ}$$

$$\tfrac{1}{2}N_2(g) + \tfrac{3}{2}H_2(g) \longrightarrow NH_3(g) \qquad \Delta H° = -46 \text{ kJ}$$

calculate $\Delta H°$ (in kilojoules) for the following reaction.

$$3Mg(s) + N_2(g) \longrightarrow Mg_3N_2(s)$$

6.74 Use the following thermochemical equations,

$$8Mg(s) + Mg(NO_3)_2(s) \longrightarrow Mg_3N_2(s) + 6MgO(s)$$
$$\Delta H° = -3884 \text{ kJ}$$

$$Mg_3N_2(s) \longrightarrow 3Mg(s) + N_2(g) \quad \Delta H° = +463 \text{ kJ}$$

$$2MgO(s) \longrightarrow 2Mg(s) + O_2(g) \qquad \Delta H° = +1203 \text{ kJ}$$

to calculate the standard heat of formation (in kilojoules per mole) of $Mg(NO_3)_2(s)$.

6.75 Given the following thermochemical equations,

$$2H_2(g) + O_2(g) \longrightarrow 2H_2O(l) \qquad \Delta H° = -571.5 \text{ kJ}$$

$$N_2O_5(g) + H_2O(l) \longrightarrow 2HNO_3(l) \qquad \Delta H° = -76.6 \text{ kJ}$$

$$\tfrac{1}{2}N_2(g) + \tfrac{3}{2}O_2(g) + \tfrac{1}{2}H_2(g) \longrightarrow HNO_3(l)$$
$$\Delta H° = -174 \text{ kJ}$$

calculate $\Delta H°$ for the reaction

$$2N_2(g) + 5O_2(g) \longrightarrow 2N_2O_5(g)$$

Hess's Law and Standard Heats of Formation

6.76 Which of the following equations has a value of $\Delta H°$ that would properly be labeled as $\Delta H_f°$?
(a) $CaCO_3(s) \longrightarrow CaO(s) + CO_2(g)$
(b) $Ca(s) + \tfrac{1}{2}O_2(g) \longrightarrow CaO(s)$
(c) $2Cu(s) + O_2(g) \longrightarrow 2CuO(s)$

6.77 Which of the following equations has a value of $\Delta H°$ that would properly be labeled as $\Delta H_f°$?
(a) $2Fe(s) + O_2(g) \longrightarrow 2FeO(s)$
(b) $SO_2(g) + \tfrac{1}{2}O_2(g) \longrightarrow SO_3(g)$
(c) $N_2(g) + \tfrac{5}{2}O_2(g) \longrightarrow N_2O_5(g)$

6.78 Write the thermochemical equations, including values of $\Delta H_f°$ in kilojoules per mole (from Table 6.2), for the formation of each of the following compounds from their elements, everything in standard states.
(a) $HC_2H_3O_2(l)$, acetic acid
(b) $NaHCO_3(s)$, sodium bicarbonate
(c) $CaSO_4 \cdot 2H_2O(s)$, gypsum

6.79 Write the thermochemical equations, including values of $\Delta H_f°$ in kilojoules per mole (from Table 6.2), for the formation of each of the following compounds from their elements, everything in standard states.
(a) $CO(NH_2)_2(s)$, urea
(b) $CaSO_4 \cdot \tfrac{1}{2}H_2O(s)$, plaster of Paris
(c) $CH_3OH(l)$, methyl alcohol

6.80 Using data in Table 6.2, calculate $\Delta H°$ in kilojoules for the following reactions.
(a) $2H_2O_2(l) \longrightarrow 2H_2O(l) + O_2(g)$
(b) $HCl(g) + NaOH(s) \longrightarrow NaCl(s) + H_2O(l)$

6.81 Using data in Table 6.2, calculate $\Delta H°$ in kilojoules for the following reactions.
(a) $CH_4(g) + Cl_2(g) \longrightarrow CH_3Cl(g) + HCl(g)$
(b) $2NH_3(g) + CO_2(g) \longrightarrow CO(NH_2)_2(s) + H_2O(l)$

6.82 Write the thermochemical equations that would be associated with the standard heats of formation of the following compounds. (a) $HCl(g)$ (b) $NH_4Cl(s)$

6.83 Write the thermochemical equations that would be associated with the standard heats of formation of the following compounds. (a) $C_2H_5OH(l)$ (b) $Na_2CO_3(s)$

6.84 The enthalpy change for the combustion of *one* mole of a compound under standard conditions is called the standard heat of combustion, and its symbol is $\Delta H°_{combustion}$. The value for sucrose, $C_{12}H_{22}O_{11}$, is -5.65×10^3 kJ mol^{-1}. Write the thermochemical equation for the combustion of 1 mol of sucrose and calculate the value of $\Delta H°_f$ for this compound. The sole products of combustion are $CO_2(g)$ and $H_2O(l)$.

6.85 The thermochemical equation for the combustion of acetylene gas, $C_2H_2(g)$, is

$$2C_2H_2(g) + 5O_2(g) \longrightarrow 4CO_2(g) + 2H_2O(l)$$

$$\Delta H° = -2599.3 \text{ kJ}$$

Using data in Table 6.2, determine the value of $\Delta H°_f$ for acetylene gas.

ADDITIONAL EXERCISES

6.86 A body of water with a mass of 750 g changed in temperature from 25.50 to 19.50 °C.
(a) What would have to be done to cause such a change?
(b) How much energy (in kJ) is involved in this change?

* **6.87** A bar of hot iron with a mass of 1.000 kg and a temperature of 100.00 °C was plunged into an insulated vat of water. The mass of the water was 2.000 kg, and its initial temperature was 25.00 °C. What was the temperature of the resulting system when it stabilized?

6.88 Sulfur trioxide reacts with water to produce sulfuric acid according to the following equation. Calculate $\Delta H°$ for the reaction.

$$SO_3(g) + H_2O(l) \longrightarrow H_2SO_4(l)$$

6.89 Iron metal can be obtained from iron ore, which we can assume for this problem to be Fe_2O_3, by its reaction with hot carbon.

$$2Fe_2O_3(s) + 3C(s) \longrightarrow 4Fe(s) + 3CO_2(g)$$

What is $\Delta H°$ (in kJ) for this reaction? Is the reaction exothermic or endothermic?

6.90 In the recovery of iron from iron ore, the reduction of the ore is actually accomplished by reactions involving carbon monoxide. Use the following thermochemical equations,

$$Fe_2O_3(s) + 3CO(g) \rightarrow 2Fe(s) + 3CO_2(g) \quad \Delta H° = -28 \text{ kJ}$$

$$3Fe_2O_3(s) + CO(g) \rightarrow 2Fe_3O_4(s) + CO_2(g)$$

$$\Delta H° = -59 \text{ kJ}$$

$$Fe_3O_4(s) + CO(g) \rightarrow 3FeO(s) + CO_2(g) \quad \Delta H° = +38 \text{ kJ}$$

to calculate $\Delta H°$ for the reaction

$$FeO(s) + CO(g) \longrightarrow Fe(s) + CO_2(g)$$

6.91 Use the results of Problem 6.90 and data in Table 6.2 to calculate the value of $\Delta H°_f$ for FeO. Express the answer in units of kilojoules per mole.

6.92 Phosphorus burns in air to give tetraphosphorus decaoxide.

$$4P(s) + 5O_2(g) \longrightarrow P_4O_{10}(s) \quad \Delta H° = -3062 \text{ kJ}$$

The product combines with water to give phosphoric acid, H_3PO_4.

$$P_4O_{10}(s) + 6H_2O(l) \longrightarrow 4H_3PO_4(l) \quad \Delta H° = -257.2 \text{ kJ}$$

Using these equations (and any others in the chapter, as needed), write the thermochemical equation for the formation of 1 mol of $H_3PO_4(l)$ from the elements. Include $\Delta H°_f$.

6.93 The amino acid glycine, $C_2H_5NO_2$, is one of the compounds used by the body to make proteins. The equation for its combustion is

$$4C_2H_5NO_2(s) + 9O_2(g) \rightarrow 8CO_2(g) + 10H_2O(l) + 2N_2(g)$$

For each mole of glycine that burns, 973.49 kJ of heat is liberated. Use this information plus values of $\Delta H°_f$ for the products of combustion to calculate $\Delta H°_f$ for glycine.

6.94 Exothermic reactions occur spontaneously more frequently than do endothermic reactions. In which direction will the following system most likely react, left to right or right to left? Calculate $\Delta H°$ for the reaction and then answer the question.

$$2Ag(s) + Zn(NO_3)_2(aq) \longrightarrow Zn(s) + 2AgNO_3(aq)$$

The following thermochemical equations can be used.

$$Cu(NO_3)_2(aq) + Zn(s) \longrightarrow Zn(NO_3)_2(aq) + Cu(s)$$

$$\Delta H° = -258 \text{ kJ}$$

$$2AgNO_3(aq) + Cu(s) \longrightarrow Cu(NO_3)_2(aq) + 2Ag(s)$$
$$\Delta H° = -106 \text{ kJ}$$

6.95 Calculate the $\Delta H°$ for the following reaction using the given thermochemical equations and others, as needed, that use data tabulated in this chapter or in Appendix E.

$$LiOH(aq) + HCl(aq) \longrightarrow LiCl(aq) + H_2O(l)$$

$$Li(s) + \tfrac{1}{2}O_2(g) + \tfrac{1}{2}H_2(g) \longrightarrow LiOH(s) \quad \Delta H° = -487.0 \text{ kJ}$$

$$2Li(s) + Cl_2(g) \longrightarrow 2LiCl(s) \quad \Delta H° = -815.0 \text{ kJ}$$

$$LiOH(s) \xrightarrow[\text{dissolving in water}]{} LiOH(aq) \quad \Delta H° = -19.2 \text{ kJ}$$

$$HCl(g) \xrightarrow[\text{dissolving in water}]{} HCl(aq) \quad \Delta H° = -77.0 \text{ kJ}$$

$$LiCl(s) \xrightarrow[\text{dissolving in water}]{} LiCl(aq) \quad \Delta H° = -36.0 \text{ kJ}$$

6.96 The value of $\Delta H_f°$ for $HBr(g)$ was first evaluated using the following standard enthalpy values obtained experimentally. Use these data to calculate the value of $\Delta H_f°$ for $HBr(g)$.

$$Cl_2(g) + 2KBr(aq) \longrightarrow Br_2(aq) + 2KCl(aq)$$
$$\Delta H° = -96.2 \text{ kJ}$$

$$H_2(g) + Cl_2(g) \longrightarrow 2HCl(g) \quad \Delta H° = -184 \text{ kJ}$$

$$HCl(aq) + KOH(aq) \longrightarrow KCl(aq) + H_2O(l)$$
$$\Delta H° = -57.3 \text{ kJ}$$

$$HBr(aq) + KOH(aq) \longrightarrow KBr(aq) + H_2O(l)$$
$$\Delta H° = -57.3 \text{ kJ}$$

$$HCl(g) \xrightarrow[\text{dissolving in water}]{} HCl(aq) \quad \Delta H° = -77.0 \text{ kJ}$$

$$Br_2(g) \xrightarrow[\text{dissolving in water}]{} Br_2(aq) \quad \Delta H° = -4.2 \text{ kJ}$$

$$HBr(g) \xrightarrow[\text{dissolving in water}]{} HBr(aq) \quad \Delta H° = -79.9 \text{ kJ}$$

*6.97 Acetylene, C_2H_2, is a gas commonly used in welding. It is formed in the reaction of calcium carbide, CaC_2, with water. Given the thermochemical equations below, calculate the value of $\Delta H_f°$ for acetylene in units of kilojoules per mole.

$$CaO(s) + H_2O(l) \longrightarrow Ca(OH)_2(s) \quad \Delta H° = -65.3 \text{ kJ}$$

$$CaO(s) + 3C(s) \longrightarrow CaC_2(s) + CO(g) \quad \Delta H° = +462.3 \text{ kJ}$$

$$CaCO_3(s) \longrightarrow CaO(s) + CO_2(g) \quad \Delta H° = +178 \text{ kJ}$$

$$CaC_2(s) + 2H_2O(l) \longrightarrow Ca(OH)_2(s) + C_2H_2(g)$$
$$\Delta H° = -126 \text{ kJ}$$

$$2C(s) + O_2(g) \longrightarrow 2CO(g) \quad \Delta H° = -220 \text{ kJ}$$

$$2H_2O(l) \longrightarrow 2H_2(g) + O_2(g) \quad \Delta H° = +572 \text{ kJ}$$

6.98 The reaction for the metabolism of sucrose, $C_{12}H_{22}O_{11}$, is the same as for its combustion in oxygen to yield $CO_2(g)$ and $H_2O(l)$. The standard heat of formation of sucrose is -2230 kJ mol^{-1}. Use data in Table 6.2 to compute the amount of energy (in kJ) released by metabolizing 1 oz (28.4 g) of sucrose.

*6.99 For ethanol, C_2H_5OH, which is mixed with gasoline to make the fuel gasohol, $\Delta H_f° = -277.63$ kJ/mol. Calculate the number of kilojoules released by burning completely 1 gallon of ethanol. The density of ethanol is 0.787 g cm^{-3}. Use data in Table 6.2 to help in the computation.

6.100 Consider the following thermochemical equations:

$$(1) \quad CH_3OH(l) + O_2(g) \longrightarrow HCHO_2(l) + H_2O(l)$$
$$\Delta H° = -411 \text{ kJ}$$

$$(2) \quad CO(g) + 2H_2(g) \longrightarrow CH_3OH(l) \quad \Delta H° = -128 \text{ kJ}$$

$$(3) \quad HCHO_2(l) \longrightarrow CO(g) + H_2O(l) \quad \Delta H° = -33 \text{ kJ}$$

Suppose Equation (1) is reversed and divided by 2, Equations (2) and (3) are multiplied by $\tfrac{1}{2}$, and then the three adjusted equations are added. What is the net reaction, and what is the value of $\Delta H°$ for the net reaction?

6.101 Chlorofluoromethanes (CFMs) are carbon compounds of chlorine and fluorine and are also known as Freons. Examples are Freon-11 ($CFCl_3$) and Freon-12 (CF_2Cl_2), which have been used as aerosol propellants. Freons have also been used in refrigeration and air-conditioning systems. In the stratosphere CFMs absorb high-energy radiation from the sun and split off chlorine atoms that hasten the decomposition of ozone, O_3. Possible reactions are

$$(1) \quad O_3(g) + Cl(g) \longrightarrow O_2(g) + ClO(g) \quad \Delta H° = -126 \text{ kJ}$$

$$(2) \quad ClO(g) + O(g) \longrightarrow Cl(g) + O_2(g) \quad \Delta H° = -268 \text{ kJ}$$

$$(3) \quad O_3(g) + O(g) \longrightarrow 2O_2(g)$$

The O atoms in Equation 2 come from the breaking apart of O_2 molecules caused by radiation from the sun. Use Equations 1 and 2 to calculate the value of $\Delta H°$ (in kilojoules) for Equation 3, the net reaction for the removal of O_3 from the atmosphere.

6.102 Given the following thermochemical equations,

$$2Cu(s) + O_2(g) \longrightarrow 2CuO(s) \quad \Delta H° = -155 \text{ kJ}$$

$$Cu(s) + S(s) \longrightarrow CuS(s) \quad \Delta H° = -53.1 \text{ kJ}$$

$$S(s) + O_2(g) \longrightarrow SO_2(g) \quad \Delta H° = -297 \text{ kJ}$$

$$4CuS(s) + 2CuO(s) \longrightarrow 3Cu_2S(s) + SO_2(g)$$
$$\Delta H° = -13.1 \text{ kJ}$$

calculate $\Delta H°$ (in kilojoules) for the reaction

$$CuS(s) + Cu(s) \longrightarrow Cu_2S(s)$$

One of nature's marvels is a rainbow following a rain shower. In bygone days, many myths grew around these elusive colors in the sky, including the "pot of gold" to be found at the base of a rainbow. We now know that a rainbow is caused by the refraction of sun light by tiny water droplets in the air, dividing white light into its myriad of colors. In this chapter we will study the properties of light and how, by observing the colors of light emitted by atoms when they've become excited, scientists were able to determine the way electrons distribute themselves around atomic nuclei. Such knowledge was the first step in understanding the forces that hold chemical substances together.

Atomic and Electronic Structure

7

This Chapter in Context In previous chapters you learned about the basic structures of atoms and how the elements can be arranged in the periodic table. You learned that some elements are metals and others are nonmetals, and you learned how metallic and nonmetallic properties vary in a systematic way across the rows and down the columns of the periodic table. You were also introduced to some of the kinds of compounds formed by the elements, and you learned to use the periodic table to write the formulas for simple ions of metals and nonmetals. As you learned these chemical facts, perhaps you wondered *what* makes an element behave as a metal or a nonmetal and *why* the metals form positive ions and the nonmetals negative ones. Perhaps you wondered *why* the periodic table is able to correlate so many properties of the elements and *why* there are no elements that fill the spaces between Be and B and between Mg and Al. Are these elements to be discovered?

The search for answers to such questions has given rise to theories that attempt to describe the inner structures of atoms in greater detail and to *explain* how atoms combine to form compounds. The subject of this chapter is the modern theory of atomic structure, which describes the atom's *electronic structure*—the way the electrons are arranged around the nucleus of an atom. We will study how atoms combine in the next two chapters.

7.1 Electromagnetic Radiation

In Chapter 2 you learned that an atom consists of a small dense nucleus, which contains all of the atom's neutrons and protons, surrounded by electrons that fill the remaining volume of the atom. The nucleus, by way of the number of protons it contains, serves to identify which element the atom is, and it determines the number of electrons the atom must have to be electrically neutral. However, when two or more atoms join to form a compound, the nuclei of the atoms stay relatively far apart. Only the atoms' outer regions—the regions inhabited by

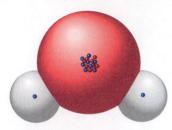

Atoms, not drawn to scale, as they are joined in a molecule of water. The nuclei stay far apart and only the outer parts of the atoms touch.

electrons—come in close contact. For this reason, the chemical properties of the elements, including their similarities and differences, are related to the way the electrons are distributed around the various nuclei.

The basic clue to the electronic structures of atoms comes from the study of the light emitted when atoms of the elements are *excited,* or energized. To learn about this, however, we must first learn a little about light itself.

The Nature of Electromagnetic Radiation

You've learned that objects can have energy in only two ways, as kinetic energy and as potential energy. You also learned that energy can be transferred between things, and in Chapter 6 our principal focus was on the transfer of heat (actually the transfer of molecular kinetic energy through collisions between atomic-sized particles). You may recall, however, that energy can also be transferred between things as light energy (more properly called *electromagnetic energy*). This is a very important form of energy in chemistry. For example, many chemical systems emit visible light as they react (see Figure 7.1).

Electromagnetic energy is energy carried through space or matter by means of wavelike oscillations. These oscillations are systematic fluctuations in the intensities of very tiny electrical and magnetic forces. Each changes rhythmically with time, and the successive series of these oscillations that travel through space is called **electromagnetic radiation** (popularly, a light wave).

Figure 7.2 shows how the **amplitude** or intensity of the wave varies with time and with distance as the wave travels through space. In Figure 7.2*a*, we see two complete oscillations or *cycles* of the wave during a one second interval. The number of cycles per second is called the **frequency** of the electromagnetic radiation, and its symbol is ν (the Greek letter *nu*, pronounced "new").

The concept of frequency extends beyond just electromagnetic radiation to other events that recur at regular intervals. For example, you go to school 5 days *per week,* or you pay your bills once *each month.* Each of these statements de-

Figure 7.1 *Light is given off in a variety of chemical reactions. (a) Combustion. (b) Cyalume light sticks. (c) A lightning bug.*

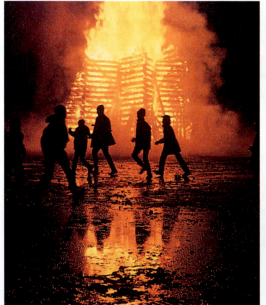

(*a*)

(*b*)

(*c*)

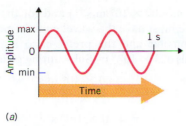

(a)

(b)

Figure 7.2 *Two views of electromagnetic radiation.* (a) The frequency, ν, of a light wave is the number of complete oscillations each second. Here two cycles span a one second time interval, so the frequency is 2 cycles per second, or 2 Hz. (b) Electromagnetic radiation frozen in time. This curve shows how the amplitude varies along the direction of travel. The distance between two peaks is the wavelength, λ, of the electromagnetic radiation.

scribes how frequently an event occurs, and they have in common the notion of *per unit of time,* or $\frac{1}{\text{time}}$. In the SI, the unit of time is the second (s), so frequency is given the unit "per second," which is $\frac{1}{\text{second}}$, or $(\text{second})^{-1}$. This unit is given the special name **hertz (Hz).**

$$1 \text{ Hz} = 1 \text{ s}^{-1}$$

> The SI symbol for the second is s.
>
> $$s^{-1} = \frac{1}{s}$$

As electromagnetic radiation moves away from its source, the positions of maximum and minimum amplitude are regularly spaced. The distance separating maximum values (or minimum values) is called the radiation's **wavelength,** symbolized by $\boldsymbol{\lambda}$ (the Greek letter, *lambda*). See Figure 7.2b. Because wavelength is a distance, it has distance units (for example, meters).

If we multiply the wavelength by frequency, the result is the speed of the wave. We can see this if we analyze the units.

$$\text{meters} \times \frac{1}{\text{second}} = \frac{\text{meters}}{\text{second}} = \text{speed}$$

$$\text{m} \times \frac{1}{\text{s}} = \frac{\text{m}}{\text{s}} = \text{m s}^{-1}$$

The speed of electromagnetic radiation in a vacuum is a constant and is commonly called the *speed of light.* Its value to three significant figures is 3.00×10^8 m/s (or m s^{-1}). This important physical constant is given the symbol c.

$$c = 3.00 \times 10^8 \text{ m s}^{-1}$$

> The speed of light is one of our most carefully measured constants, and a more precise value is given in a table on the inside rear cover of this book.

From the preceding discussion we obtain a very important relationship that allows us to convert between λ and ν.

$$\lambda \times \nu = c = 3.00 \times 10^8 \text{ m s}^{-1} \qquad (7.1)$$

Wavelength–frequency relationship

EXAMPLE 7.1

Calculating Frequency from Wavelength

What is the frequency in hertz of yellow light that has a wavelength of 625 nm?

Analysis: To convert between wavelength and frequency we use Equation 7.1. However, we must be careful about the units.

Solution: To calculate the frequency, we solve Equation 7.1 for ν.

$$\nu = \frac{c}{\lambda}$$

Next, we substitute for c (3.00×10^8 m s^{-1}) and for the wavelength. However, to cancel units correctly, we must have the wavelength in meters. Recall from Chapter 1 that nm means nanometer and the prefix nano implies the factor "$\times 10^{-9}$." Therefore, 625 nm equals 625×10^{-9} m. Substituting gives

$$\nu = \frac{3.00 \times 10^8 \text{ m s}^{-1}}{625 \times 10^{-9} \text{ m}}$$

$$= 4.80 \times 10^{14} \text{ s}^{-1}$$

$$= 4.80 \times 10^{14} \text{ Hz}$$

EXAMPLE 7.2

Calculating Wavelength from Frequency

Radio station WGBB on Long Island, New York, broadcasts its AM signal, a form of electromagnetic radiation, at a frequency of 1240 kHz. What is the wavelength of these radio waves expressed in meters?

Solution: This time we solve Equation 7.1 for the wavelength.

$$\lambda = \frac{c}{\nu}$$

Now we must cancel the unit s^{-1} in 3.00×10^8 m s^{-1}. The prefix "k" in kHz means kilo and stands for "$\times 10^3$" and Hz means s^{-1}. The frequency is therefore 1240×10^3 s^{-1}. Substituting gives

$$\lambda = \frac{3.00 \times 10^8 \text{ m s}^{-1}}{1240 \times 10^3 \text{ s}^{-1}}$$

$$= 242 \text{ m}$$

Practice Exercise 1

A certain shade of green light has a wavelength of 550 nm. What is the frequency of this light in hertz? ◆

Practice Exercise 2

An FM radio station in West Palm Beach, Florida, broadcasts electromagnetic radiation at a frequency of 104.3 MHz (megahertz). What is the wavelength of these radio waves, expressed in meters? ◆

The Electromagnetic Spectrum

Electromagnetic radiation comes in a broad range of frequencies called the **electromagnetic spectrum,** illustrated in Figure 7.3. Some portions of the spectrum have popular names. For example, radio waves are electromagnetic radiations having very low frequencies (and therefore very long wavelengths). Microwaves, which also have low frequencies, are emitted by radar instruments such as those the police use to monitor the speeds of cars. In microwave ovens, similar radiation heats the water in foods, causing the food to cook quickly. Infrared radiation is emitted by hot objects and consists of the range of frequencies that can make molecules of most substances vibrate internally. You can't see infrared radiation, but you can feel how it is absorbed by your body by holding your hand near a hot radiator; the absorbed radiation makes your hand warm. Gamma rays are at the high-frequency end of the electromagnetic spectrum. They are produced by certain elements that are radioactive. X rays are very much like gamma rays, but they are usually made by special equipment. Both X rays and gamma rays penetrate living things easily.

Remember that there is an inverse relationship between wavelength and frequency. The lower the frequency, the longer the wavelength.

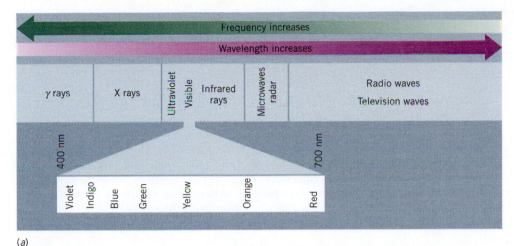

(a)

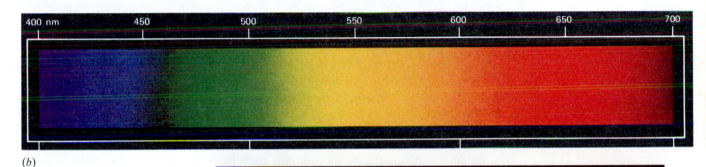

(b)

Figure 7.3 *The electromagnetic spectrum.* (*a*) The electromagnetic spectrum is divided into regions according to the wavelengths of the radiation. (*b*) The visible spectrum is composed of wavelengths that range from about 400 to 700 nm. (*c*) The production of a visible spectrum by splitting white light into its rainbow of colors.

(c)

Most of the time, you are bombarded with electromagnetic radiations from all portions of the electromagnetic spectrum. Radio and TV signals pass through you; you feel infrared radiation when you sense the warmth of a radiator; X rays and gamma rays fall on you from space; and light from a lamp reflects into your eyes from the page you're reading. Of all these radiations, your eyes are able to sense only a very narrow band of wavelengths ranging from about 400 to 700 nm. This band is called the **visible spectrum** and consists of all the colors you can see, from red through orange, yellow, green, blue, and violet. White light is composed of all these colors in roughly equal amounts, and it can be separated

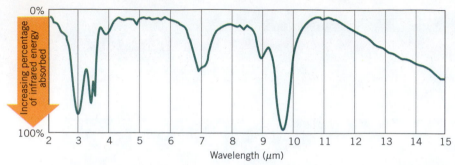

Figure 7.4 *Infrared absorption spectrum of methyl alcohol.* Methyl alcohol, also called wood alcohol, is the fuel in "canned heat" products such as Sterno. In an infrared spectrum, the usual practice is to show the amount of light absorbed increasing from top to bottom in the graph. Thus, there is a peak in the percentage of light absorbed at about 3 μm. (Spectrum courtesy Sadtler Research Laboratories, Inc., Philadelphia.)

into them by focusing a beam of white light through a prism, which spreads the various wavelengths apart. This is illustrated in Figure 7.3*b*. A photograph showing the production of a visible spectrum is given in Figure 7.3*c*.

The way substances absorb electromagnetic radiation often can help us characterize them. For example, each substance absorbs a uniquely different set of infrared frequencies. A plot of the wavelengths absorbed versus the intensities of absorption is called an infrared absorption spectrum. It can be used to identify a compound, because each infrared spectrum is as unique as a set of fingerprints. (See Figure 7.4.) Many substances absorb visible and ultraviolet radiations in unique ways, too, and they have visible and ultraviolet spectra (Figure 7.5).

(*a*)

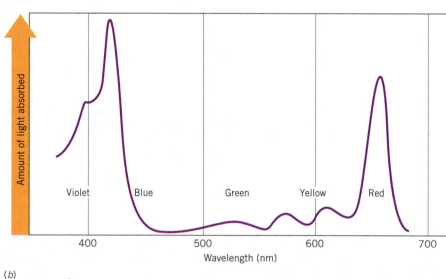

(*b*)

Figure 7.5 *Absorption of light by chlorophyll.* (*a*) Chlorophyll is the green pigment plants use to harvest solar energy for photosynthesis. (*b*) In this visible absorption spectrum of chlorophyll, the percentage of light absorbed increases from bottom to top. Thus, there is a peak in the light absorbed at about 420 nm and another at about 660 nm. This means the pigment strongly absorbs blue-violet and red light. The green color we *see* is the light that's *not* absorbed. It's composed of the wavelengths of visible light that are reflected. (Our eyes are most sensitive to green, so we don't notice the yellow components of the reflected light.)

Facets of Chemistry 7.1

Photoelectricity and Its Applications

One of the earliest clues to the relationship between the frequency of light and its energy was the discovery of the photoelectric effect. In the latter part of the nineteenth century, it was found that certain metals acquired a positive charge when they were illuminated by light. Apparently, light is capable of kicking electrons out of the surface of the metal.

When this phenomenon was studied in detail, it was discovered that electrons could only be made to leave a metal's surface if the frequency of the incident radiation was above some minimum value, which was named the threshold frequency. This threshold frequency differs for different metals, depending on how tightly the metal atom holds onto electrons. Above the threshold frequency, the kinetic energy of the emitted electron increases with increasing frequency of the light. Interestingly, however, its kinetic energy does not depend on the intensity of the light. In fact, if the frequency of the light is below the minimum frequency, no electrons are observed at all, no matter how bright the light is. To physicists of that time, this was very perplexing because they believed the energy of light was related to its brightness. The explanation of the phenomenon was finally given by Albert Einstein in the form of a very simple equation,

$$KE = h\nu - w$$

where KE is the kinetic energy of the electron that is emitted, $h\nu$ is the energy of the photon of frequency ν, and w is the minimum energy needed to eject the electron from the metal's surface. Stated another way, part of the energy of the photon is needed just to get the electron off the surface of the metal. This amount is w. Any energy left over $(h\nu - w)$ appears as the electron's kinetic energy.

Besides its important theoretical implications, the photoelectric effect has many practical applications. For example, automatic "electric eye" door openers use this phenomenon by sensing the interruption of a light beam caused by the person wishing to use the door. The phenomenon is also responsible for photoconduction by certain substances that are used in light meters in cameras and other devices. The production of sound in motion pictures was first made possible by incorporating a strip along the edge of the film (called the sound track) that causes the light passing through it to fluctuate in intensity according to the frequency of the sound that's been recorded. A photocell converts this light to a varying electric current that is amplified and played through speakers in the theater. Even the sensitivity of photographic film to light is related to the release of photoelectrons within tiny grains of silver bromide that are suspended in a coating on the surface of the film.

The Energy of Electromagnetic Radiation

In 1900 a German physicist named Max Planck (1858–1947) launched one of the greatest upheavals in the history of science when he proposed that electromagnetic radiation is emitted only in tiny packets or **quanta** of energy that were later called **photons.** Each photon pulses with a frequency, ν, and travels with the speed of light. Planck proposed, and Albert Einstein (1879–1955) confirmed, that *the energy of a photon of electromagnetic radiation is proportional to its frequency,* not to its intensity or brightness as had been believed up to that time. (See Facets of Chemistry 7.1.)

The energy of one photon is called one **quantum** of energy.

$$\text{Energy of a photon} = E = h\nu \qquad (7.2)$$

In this expression, **h** is a proportionality constant that we now call **Planck's constant.**

Planck's and Einstein's discovery was really quite surprising. If a particular event requiring energy, such as photosynthesis in green plants, is initiated by the absorption of light, it is the frequency of the light that is important, not its intensity or brightness. An analogy would be a group of pole vaulters trying to get over a wall. If they have long enough poles, each can clear the wall. If the poles are too short, however, they can batter the wall with as much intensity as they

Energy of a photon

The value of Planck's constant is 6.63×10^{-34} J s. It has units of energy (joules) multiplied by time (seconds).

Figure 7.6 *Production and observation of an atomic spectrum.* Light emitted by excited atoms is formed into a narrow beam and passed through a prism, which divides the light into a relatively few narrow beams with frequencies that are characteristic of the particular element that's emitting the light. When these beams fall on a screen, a series of lines is observed, which is why the spectrum is also called a *line spectrum*.

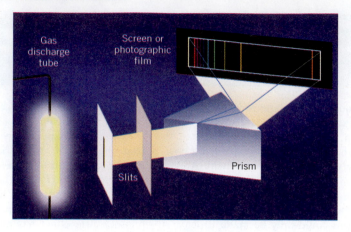

want, but they'll never clear the top. Not even doubling or tripling the number of vaulters with short poles will get anyone across.

7.2 Atomic Spectra and the Bohr Model of the Hydrogen Atom

The spectrum described in Figure 7.3 is called a **continuous spectrum** because it contains a continuous unbroken distribution of light of *all* colors. It is formed when the light from the sun, or any other object that's been heated to a very high temperature (such as the filament in an electric light bulb), is split by a prism and displayed on a screen. A rainbow after a summer shower is a continuous spectrum that most people have seen. In this case, the colors contained in sunlight are spread out by tiny water droplets in the air.

Atoms of an element can also be excited by adding them to the flame of a Bunsen burner.

A rather different kind of spectrum is observed if we examine the light that is given off when an *electric discharge,* or spark, passes through a gas such as hydrogen. The electric discharge is an electric current that *excites,* or energizes, the atoms of the gas. More specifically, the electric current excites and gives energy to the electrons in the atom. The atoms then emit the absorbed energy in the form of light as the electrons return to a lower energy state. When a narrow beam of this light is passed through a prism, as shown in Figure 7.6, we do *not* see a continuous spectrum. Instead, only a few colors are observed, displayed as a series of individual lines. This series of lines is called the element's **atomic**

An emission spectrum is also called a *line spectrum* because the light corresponding to the individual emissions appear as lines on the screen.

spectrum or **emission spectrum.** Figure 7.7 shows the visible portions of the atomic spectra of two common elements, sodium and hydrogen, and how they compare with a continuous spectrum. Notice that the spectra of these elements are quite different. In fact, each element has its own unique atomic spectrum that is as characteristic as a fingerprint.

▶**Chemistry in Practice**◀ At one time, most of the street lighting in towns and cities was provided by incandescent lamps, in which a tungsten filament is heated white-hot by an electric current. Unfortunately, much of the light from this kind of lamp is infrared radiation, which we cannot see. As a result, only a relatively small fraction of the electrical energy used to operate the lamp actually results in visible light. Modern streetlights consist of high intensity sodium or mercury vapor lamps in which the light is produced by passing an electric discharge through the vapors of these metals. The electric current excites the

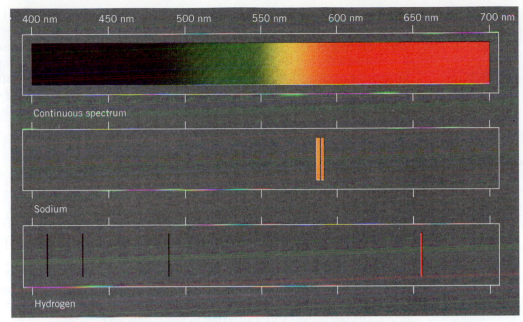

Figure 7.7 *Continuous and atomic emission spectra.* (*a*) The continuous visible spectrum produced by the sun or an incandescent lamp. (*b*) The atomic emission spectrum (line spectrum) produced by sodium. The emission spectrum of sodium actually contains more than 90 lines in the visible region. The two brightest lines are shown here. All the others are less than 1% as bright as these. (*c*) The atomic spectrum produced by hydrogen. There are only four lines in this visible spectrum. They vary in brightness by only a factor of five, so they are all shown.

atoms, which then emit their characteristic atomic spectra. In these lamps, most of the electrical energy is converted to light in the visible region of the spectrum, so they are much more energy efficient (and cost efficient) than incandescent lamps. Sodium emits intense light at a wavelength of 589 nm, which is yellow. The golden glow of streetlights in scenes like that shown in the photograph in the margin is from this emission line of sodium, which is produced in high-pressure sodium vapor lamps. There is another, much more yellow light emitted by low-pressure sodium lamps that you may also have seen. Both kinds of sodium lamps are very efficient, and almost all communities now use this kind of lighting to save money on the costs of electricity. In some places, the somewhat less efficient mercury vapor lamps are still used. These lamps give a bluish-white light.

Fluorescent lamps also depend on an atomic emission spectrum as the primary source of light. In these lamps an electric discharge is passed through mercury vapor which emits some of its light in the ultraviolet region of the spectrum. The inner wall of the fluorescent tube is coated with a phosphor which is caused to glow white when struck by the UV light. These lamps are also highly efficient and convert most of the electrical energy they consume into visible light. ◆

Park Avenue in New York City is brightly lit by sodium vapor lamps in this photo taken during a recent Christmas season.

The Atomic Spectrum of Hydrogen

The first success in explaining atomic spectra quantitatively came with the study of the spectrum of hydrogen. This is the simplest element, since its atoms have only one electron, and it produces the simplest spectrum with the fewest lines.

The atomic spectrum of hydrogen actually consists of several series of lines. One series is in the visible region of the electromagnetic spectrum and is shown in Figure 7.7. Another series is in the ultraviolet region, and the rest are in the infrared. In 1885, J. J. Balmer found an equation that was able to give the wavelengths of the lines in the visible portion of the spectrum. This was soon extended to a more general equation, called the **Rydberg equation,** that could be used to calculate the wavelengths of *all* the spectral lines of hydrogen.

It's really rather amazing that an equation as simple as this, involving only a constant and a set of integers, can be used to calculate the wavelengths of all the spectral lines of hydrogen.

Rydberg equation:

$$\frac{1}{\lambda} = R_H\left(\frac{1}{n_1^2} - \frac{1}{n_2^2}\right)$$

The symbol λ stands for the wavelength, R_H is a constant ($109{,}678 \text{ cm}^{-1}$), and n_1 and n_2 are variables whose values are whole numbers that range from 1 to ∞. The only restriction is that the value of n_2 must be larger than n_1. (This assures that the calculated wavelength has a positive value.) Thus, if $n_1 = 1$, acceptable values of n_2 are $2, 3, 4, \ldots, \infty$. The Rydberg constant, R_H, is an *empirical constant,* which means its value was chosen so that the equation gives values for λ that match the ones determined experimentally. The use of the Rydberg equation is straightforward, as illustrated in the following example.

EXAMPLE 7.3

Calculating the Wavelength of a Line in the Hydrogen Spectrum

The lines in the visible portion of the hydrogen spectrum are called the Balmer series, for which $n_1 = 2$ in the Rydberg equation. Calculate, to four significant figures, the wavelength in nanometers of the spectral line in this series for which $n_2 = 4$.

Solution: To solve this problem, we substitute values into the Rydberg equation, which will give us $1/\lambda$. Taking the reciprocal will then give the wavelength. As usual, we must be careful with the units.

Substituting $n_1 = 2$ and $n_2 = 4$ into the Rydberg equation gives

$$\frac{1}{\lambda} = 109{,}678 \text{ cm}^{-1}\left(\frac{1}{2^2} - \frac{1}{4^2}\right)$$

$$= 109{,}678 \text{ cm}^{-1}\left(\frac{1}{4} - \frac{1}{16}\right)$$

$$= 109{,}678 \text{ cm}^{-1}\,(0.2500 - 0.0625)$$

$$= 109{,}678 \text{ cm}^{-1}\,(0.1875)$$

$$= 2.056 \times 10^4 \text{ cm}^{-1} \qquad \text{(rounded)}$$

Taking the reciprocal gives the wavelength in centimeters.

$$\lambda = \frac{1}{2.056 \times 10^4 \text{ cm}^{-1}}$$

$$= 4.864 \times 10^{-5} \text{ cm}$$

Finally, we convert to nanometers.

$$\lambda = 4.864 \times 10^{-5} \text{ cm} \times \frac{10^{-2} \text{ m}}{1 \text{ cm}} \times \frac{1 \text{ nm}}{10^{-9} \text{ m}}$$

$$= 486.4 \text{ nm}$$

This is the green line in the hydrogen spectrum in Figure 7.7.

Practice Exercise 3

Calculate the wavelength in nanometers of the spectral line in the visible spectrum of hydrogen for which $n_1 = 2$ and $n_2 = 3$. What color is this line? ◆

The discovery of the Rydberg equation was both exciting and perplexing. The fact that the wavelength of any line in the hydrogen spectrum can be calculated by a simple equation involving just one constant and the reciprocals of the squares of two whole numbers is remarkable. What is there about the behavior of the electron in the atom that could account for such simplicity?

The Significance of Atomic Spectra

Earlier you saw that there is a simple relationship between the frequency of light and its energy, $E = h\nu$. Because excited atoms emit light of only certain specific frequencies, it must be true that only certain characteristic energy changes are able to take place within the atoms. For instance, in the spectrum of hydrogen there is a red line (see Figure 7.7) that has a wavelength of 656.4 nm and a frequency of 4.567×10^{14} Hz. As shown in the margin, the energy of a photon of this light is 3.026×10^{-19} J. Whenever a hydrogen atom emits red light, the frequency of the light is always precisely 4.567×10^{14} Hz and the energy of the atom decreases by *exactly* 3.026×10^{-19} J, never more and never less. Atomic spectra, then, tell us that *when an excited atom loses energy, not just any arbitrary amount can be lost.*

How is it that atoms of a given element always undergo exactly the same specific energy changes? The answer seems to be that in an atom an electron can have only certain definite amounts of energy and no others. In the words of science, we say that the electron is restricted to certain **energy levels,** and that the energy of the electron is **quantized.**

The energy of an electron in an atom might be compared to the potential energy of a ball on a staircase (see Figure 7.8). The ball can only come to rest on a step, so at rest it can only have certain specific amounts of potential energy as determined by the "energy levels" of the various steps of the staircase. If the ball is raised to a higher step, its potential energy is increased. When it drops to a lower step, its potential energy decreases. But each time the ball stops, it stops on one of the steps, never in between. Therefore, the energy changes for the ball are restricted to the differences in potential energy between the steps.

So it is with an electron in an atom. The electron can only have energies corresponding to the set of electron energy levels in the atom. When the atom is supplied with energy (by an electric discharge, for example), an electron is raised from a low-energy level to a higher one. When the electron drops back, energy equal to the difference between the two levels is released and emitted as a photon. Because only certain energy jumps can occur, only certain frequencies of light can appear in the spectrum.

The existence of specific energy levels in atoms, as implied by atomic spectra, forms the foundation of all theories about electronic structure. Any model of the atom that attempts to describe the positions or motions of electrons must also account for atomic spectra.

The Bohr Model of the Hydrogen Atom

The first theoretical model of the hydrogen atom that successfully accounted for the Rydberg equation was proposed in 1913 by Niels Bohr (1885–1962), a Danish physicist. In his model, Bohr likened the electron moving around the

$E = h\nu$

$h = 6.626 \times 10^{-34}$ J s

$E = (6.626 \times 10^{-34}\text{ J s})$

$\quad \times (4.567 \times 10^{14}\text{ s}^{-1})$

$\quad = 3.026 \times 10^{-19}$ J

The potential energy of the ball at rest is quantized.

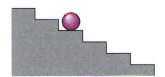

Figure 7.8 *A ball on a stair-case.* The ball can have only certain amounts of potential energy when at rest. Similarly, the energy of the electron is restricted to certain values, which correspond to the various energy levels in an atom.

Niels Bohr won the 1922 Nobel Prize in physics for his work on atomic structure.

nucleus to a planet circling the sun. He suggested that the electron moves around the nucleus along fixed paths, or orbits. His model broke with the classical laws of physics by placing restrictions on the sizes of the orbits and the energy that the electron could have in a given orbit. This ultimately led Bohr to an equation that described the energy of the electron in the atom. The equation includes a number of physical constants such as the mass of the electron, its charge, and Planck's constant. It also contains an integer, n, that Bohr called a **quantum number.** Each of the orbits is identified by its value of n. When all the constants are combined, Bohr's equation becomes

$$E = \frac{-b}{n^2} \tag{7.3}$$

Classical physical laws, such as those discovered by Issac Newton, place no restrictions on the sizes or energies of orbits.

Bohr's equation for the energy actually is

$$E = -\frac{2\pi^2 m e^4}{n^2 h^2}$$

where m is the mass of the electron, e is the charge on the electron, n is the quantum number, and h is Planck's constant. Therefore, in Equation 7.3,

$$b = \frac{2\pi^2 m e^4}{h^2}.$$

where E is the energy of the electron and b is the combined constant (its value is 2.18×10^{-18} J). The allowed values of n are whole numbers that range from 1 to ∞ (i.e., n could equal 1, 2, 3, 4, . . . , ∞). From this equation the energy of the electron in any particular orbit could be calculated.

Because of the negative sign in Equation 7.3, the lowest (most negative) energy value occurs when $n = 1$, which corresponds to the *first Bohr orbit.* In general, the lowest energy state of an atom is the most stable one and is called the **ground state.** For hydrogen, the ground state occurs when its electron has $n = 1$. According to Bohr's theory, this orbit brings the electron closest to the nucleus.

In general, we will use the term *ground state* to describe the lowest energy state for any given atom, molecule, or ion.

When a hydrogen atom absorbs energy, as it does when an electric discharge passes through it, the electron is raised from the orbit having $n = 1$ to a higher orbit, to $n = 2$ or $n = 3$ or even higher. These higher orbits are less stable than the lower ones, so the electron quickly drops to a lower orbit. When this happens, energy is emitted in the form of light (see Figure 7.9). Since the energy of the electron in a given orbit is fixed, a drop from one particular orbit to another, say, from $n = 2$ to $n = 1$, always releases the same amount of energy, and the frequency of the light emitted because of this change is always precisely the same.

The success of Bohr's theory was in its ability to account for the Rydberg equation. When the atom emits a photon, an electron drops from a higher initial energy E_h to a lower final energy E_l. If the initial quantum number of the electron is n_h and the final quantum number is n_l, then the energy change, calculated as a positive quantity, is

$$\Delta E = E_h - E_l$$

$$= \left(\frac{-b}{n_h^2} \right) - \left(\frac{-b}{n_l^2} \right)$$

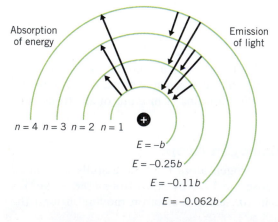

Absorption of energy

Emission of light

$n = 4$ $n = 3$ $n = 2$ $n = 1$

$E = -b$
$E = -0.25b$
$E = -0.11b$
$E = -0.062b$

Figure 7.9 *Absorption of energy and emission of light by the hydrogen atom.* When the atom absorbs energy, the electron is raised to a higher energy level. When the electron falls to a lower energy level, light of a particular energy and frequency is emitted.

This can be rearranged to give

$$\Delta E = b \left(\frac{1}{n_l^2} - \frac{1}{n_h^2} \right) \quad \text{with } n_h > n_l$$

By combining Equations 7.1 and 7.2, the relationship between the energy "ΔE" of a photon and its wavelength λ is

$$\Delta E = \frac{hc}{\lambda} = hc \left(\frac{1}{\lambda} \right)$$

Substituting and solving for $1/\lambda$ give

$$\frac{1}{\lambda} = \frac{b}{hc} \left(\frac{1}{n_l^2} - \frac{1}{n_h^2} \right) \quad \text{with } n_h > n_l$$

Notice how closely this equation derived from Bohr's theory matches the Rydberg equation, which was obtained solely from the experimentally measured atomic spectrum of hydrogen. Equally satisfying is that the combination of constants, b/hc, has a value of 109,730 cm^{-1}, which differs by only 0.05% from the experimentally derived value of R_H in the Rydberg equation.

Why the Bohr Theory Had to Be Abandoned

Bohr's model of the atom was both a success and a failure. By calculating the energy changes that occur between energy levels, Bohr was able to account for the Rydberg equation and, therefore, for the atomic spectrum of hydrogen. However, the theory was not able to explain quantitatively the spectra of atoms more complex than hydrogen, and all attempts to modify the theory to make it work met with failure. Gradually, it became clear that Bohr's picture of the atom was flawed and that another theory would have to be found. Nevertheless, the concepts of quantum numbers and fixed energy levels were important steps forward.

7.3 Wave Properties of Matter and Wave Mechanics

Bohr's efforts to develop a theory of electronic structure were doomed from the very beginning because the classical laws of physics—those known in his day— simply do not apply to particles as tiny as the electron. Classical physics fails for atomic particles because matter is not really as our physical senses perceive it. Under appropriate circumstances, small particles such as an electron behave not like solid particles, but instead like waves. This idea was proposed in 1924 by a young French graduate student, Louis de Broglie.

In Section 7.1 you learned that light waves are characterized by their wavelengths and their frequencies. The same is true of matter waves. De Broglie suggested that the wavelength of a matter wave, λ, is given by the equation

$$\lambda = \frac{h}{mv} \tag{7.4}$$

where h is Planck's constant, m is the particle's mass, and v is its velocity.

When first encountered, the concept of a particle of matter behaving as a wave rather than as a solid object is difficult to comprehend. This book certainly seems solid enough, especially if you drop it on your toe! The reason for the book's apparent solidity is that in de Broglie's equation (Equation 7.4) the mass appears in the denominator. This means that heavy objects have extremely short wavelengths. The peaks of the matter waves for heavy objects are so close to-

All the objects that had been studied by scientists until the time of Bohr were large and massive in comparison with the electron, so no one had detected the limits of classical physics.

De Broglie was awarded a Nobel Prize in 1929.

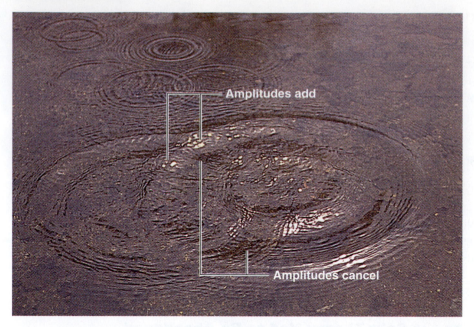

Figure 7.10 *Diffraction of water waves on the surface of a pond.* As the waves cross, the amplitudes increase where the waves are in phase and cancel where they are out of phase.

gether that the wave properties go unnoticed and can't even be measured experimentally. But tiny particles with very small masses have much longer wavelengths, so their wave properties become an important part of their overall behavior.

Perhaps by now you've begun to wonder if there is any way to *prove* that matter has wave properties. Actually, these properties can be demonstrated by a phenomenon that you have probably witnessed. When raindrops fall on a quiet pond, ripples spread out from where the drops strike the water, as shown in Figure 7.10. When two sets of ripples cross, there are places where the waves are *in phase,* which means that the peak of one wave coincides with the peak of the other. At these points the intensities of the waves add and the height of the water is equal to the sum of the heights of the two crossing waves. At other places the crossing waves are *out of phase,* which means the peak of one wave occurs at the trough of the other. In these places the intensities of the waves cancel. This reinforcement and cancellation of wave intensities, referred to, respectively, as *constructive* and *destructive interference,* is a phenomenon called **diffraction.** It is examined more closely in Figure 7.11.

Gigantic waves, called rogue waves, with heights up to 100 ft have been observed in the ocean and are believed to be formed when a number of wave sets moving across the sea become in phase simultaneously.

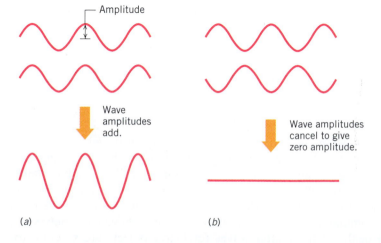

Figure 7.11 *Constructive and destructive interference.* (*a*) Waves in phase produce constructive interference and an increase in intensity. (*b*) Waves out of phase produce destructive interference and yield cancellation of intensity.

If you've ever noticed the rainbow of colors that shine from the surface of a compact disk (Figure 7.12), you have seen the diffraction of light waves. When white light that contains radiations of all wavelengths is reflected from the closely spaced "grooves" on the CD, it is divided into many individual light beams. The light waves in these beams experience interference with each other, and for a given angle between the incoming and reflected light, all colors cancel except for one, which reinforces. Because we only see colors that reinforce, as the angle changes, so do the colors of the light we see.

Diffraction is a phenomenon that can only be explained as a property of waves, and we have seen how it can be demonstrated with water waves and light waves. Experiments can also be done to show that electrons, protons, and neutrons experience diffraction, which demonstrates their wave nature. In fact, diffraction is the principle on which the electron microscope is based.

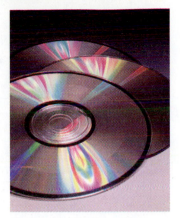

Figure 7.12 *Diffraction of light waves.* A rainbow of colors is produced by the diffraction of reflected light from the closely spaced grooves on the surface of a compact disk.

Properties of Waves

Before we can discuss how electron waves behave in atoms, we need to know a little more about waves in general. There are basically two kinds of waves, *traveling waves* and *standing waves*. On a lake or ocean the wind produces waves whose crests and troughs move across the water's surface, as shown in Figure 7.13. The water moves up and down while the crests and troughs travel horizontally in the direction of the wind. These are examples of **traveling waves.**

A more important kind of wave for us is the standing wave. An example is the vibrating string of a guitar. When the string is plucked, its center vibrates up and down while the ends, of course, remain fixed. The crest, or point of maximum amplitude of the wave, occurs at one position. At the ends of the string are points of zero amplitude, called **nodes,** and their positions are also fixed. A **standing wave,** then, is one in which the crests and nodes do not change position. One of the interesting things about standing waves is that they lead naturally to "quantum numbers." Let's see how this works using the guitar as an example.

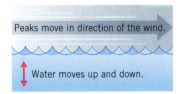

Peaks move in direction of the wind.

Water moves up and down.

Figure 7.13 *Traveling waves.*

As you know, many notes can be played on a guitar string by shortening its effective length with a finger placed at frets along the neck of the instrument. But even without shortening the string, we can play a variety of notes. For instance, if the string is touched momentarily at its midpoint at the same time it is plucked, the string vibrates as shown in Figure 7.14 and produces a tone an octave higher. The wave that produces this higher tone has a wavelength exactly half of that formed when the untouched string is plucked. In Figure 7.14 we see that other wavelengths are possible, too, and each gives a different note.

If you examine Figure 7.14, you will see that there are some restrictions on the wavelengths that can exist. Not just any wavelength is possible because the nodes at either end of the string are in fixed positions. The only waves that can occur are those for which a half-wavelength is repeated *exactly* a whole number of times. Expressed another way, the length of the string is a whole-number multiple of half-wavelengths. In a mathematical form we could write this as

$$L = n\left(\frac{\lambda}{2}\right)$$

where L is the length of the string, λ is a wavelength (therefore, $\lambda/2$ is half the wavelength), and n is an integer. Rearranging this to solve for the wavelength gives

$$\lambda = \frac{2L}{n}$$

Notes played on a guitar rely on standing waves. The ends of the strings correspond to nodes of the standing waves. Different notes can be played by shortening the effective lengths of the strings with fingers placed along the neck of the instrument.

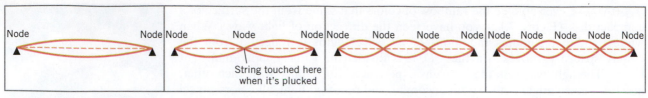

Figure 7.14 *Standing waves on a guitar string.*

We see that the waves that are possible are determined quite naturally by a set of whole numbers (similar to quantum numbers).

Electron Waves in Atoms

The wave properties of matter form the foundation for a theory called **wave mechanics,** which serves as the basis of all current theories of electronic structure. The term **quantum mechanics** is also used because wave mechanics predicts quantized energy levels. In 1926 Erwin Schrödinger (1887–1961), an Austrian physicist, became the first scientist to successfully apply the concept of the wave nature of matter to an explanation of electronic structure. His work and the theory that developed from it are highly mathematical. Fortunately, we need only a qualitative understanding of electronic structure, and the main points of the theory can be understood without all the math.

Schrödinger won a Nobel Prize in 1933 for his work.

Standing Electron Waves in Atoms

Having discussed standing waves, we can now look at matter waves, concentrating on electron waves in particular. To study this topic, you may find it best to read through the entire discussion rather quickly to get an overview of the subject, and then read it again more slowly. All of the information fits together in the end, somewhat like the pieces of a puzzle.

The theory of quantum mechanics tells us that in the atom, electron waves are standing waves. Similar to guitar strings, electrons can have many different waveforms or wave patterns. Each of these waveforms, which are called **orbitals,** is described in quantum mechanics by a mathematical expression called a **wave function,** usually represented by the symbol ψ (Greek letter *psi*). The wave function can be used to describe the shape of the electron wave and its energy. Not all of the energies of the waves are different, but most are. *Energy changes within an atom are simply the result of an electron changing from a wave pattern with one energy to a wave pattern with a different energy.*

Electron waves are described by the term *orbital* to differentiate them from the notion of *orbits,* which was part of the Bohr model of the atom.

We will be interested in two properties of orbitals, their energies and their shapes. Their energies are important because when an atom is in its most stable state (its **ground state**), the atom's electrons have waveforms with the lowest possible energies. The shapes of the wave patterns (i.e., where their amplitudes are large and where they are small) are important because the theory tells us that the amplitude of a wave at any particular place is related to the likelihood of finding the electron there. This will be important when we study how and why atoms form chemical bonds to each other.

"Most stable" almost always means "lowest energy."

In much the same way that the characteristics of a wave on a one-dimensional guitar string can be related to a single integer, wave mechanics tells us that the three-dimensional electron waves (orbitals) can be characterized by a set of *three* integer quantum numbers, n, ℓ, and m_ℓ. In discussing the energies of the orbitals, it is usually most convenient to sort the orbitals into groups according to these quantum numbers.

The Principal Quantum Number, n

The quantum number *n* is called the **principal quantum number,** and all orbitals that have the same value of *n* are said to be in the same **shell.** The values of *n* can range from *n* = 1 to *n* = ∞. The shell with *n* = 1 is called the *first shell,* the shell with *n* = 2 is the *second shell,* and so forth. The various shells are also sometimes identified by letters, beginning (for no significant reason) with K for the first shell (*n* = 1).

The principal quantum number is related to the size of the electron wave (i.e., how far the wave effectively extends from the nucleus). The higher the value of *n*, the larger is the electron's average distance from the nucleus. This quantum number is also related to the energy of the orbital. As *n* increases, the energies of the orbitals also increase.

Bohr's theory took into account only the principal quantum number *n*. His theory worked fine for hydrogen because hydrogen just happens to be the one element in which all orbitals having the same value of *n* also have the same energy. Bohr's theory failed for atoms other than hydrogen, however, because when the atom has more than one electron, different orbitals with the same value of *n* can have different energies.

> The term shell comes from an early notion that atoms could be thought of as similar to onions, with the electrons being arranged in layers around the nucleus.

> Bohr was fortunate to have used the element hydrogen to develop his model of the atom. If he had chosen a different element, his model would not have worked.

The Secondary Quantum Number, ℓ

The **secondary quantum number, ℓ,** divides the shells into smaller groups of orbitals called **subshells.** The value of *n* determines which values of ℓ are allowed. For a given *n*, ℓ can range from ℓ = 0 to ℓ = (*n* − 1). Thus, when *n* = 1, (*n* − 1) = 0, so the only value of ℓ that's allowed is zero. This means that when *n* = 1, there is only one subshell (the shell and subshell are really identical). When *n* = 2, ℓ can have values of 0 or 1. (The maximum value of ℓ = *n* − 1 = 2 − 1 = 1.) This means that when *n* = 2, there are two subshells. One has *n* = 2 and ℓ = 0, and the other has *n* = 2 and ℓ = 1. The relationship between *n* and the allowed values of ℓ are summarized in the table in the margin.

Subshells could be identified by their value of ℓ. However, to avoid confusing numerical values of *n* with those of ℓ, a letter code is normally used to specify the value of ℓ.

> ℓ is also called the *azimuthal quantum number.*

Relationship between *n* and ℓ

Value of *n*	Values of ℓ
1	0
2	0, 1
3	0, 1, 2
4	0, 1, 2, 3
5	0, 1, 2, 3, 4
n	0, 1, 2, . . ., (*n* − 1)

Value of ℓ	0	1	2	3	4	5	. . .
Letter designation	*s*	*p*	*d*	*f*	*g*	*h*	. . .

To designate a particular subshell, we write the value of its principal quantum number followed by the letter code for the subshell. For example, the subshell with *n* = 2 and ℓ = 1 is the 2*p* subshell; the subshell with *n* = 4 and ℓ = 0 is the 4*s* subshell. Notice that because of the relationship between *n* and ℓ, every shell has an *s* subshell (1*s*, 2*s*, 3*s*, etc.); all the shells except the first have a *p* subshell (2*p*, 3*p*, 4*p*, etc.); all but the first and second shells have a *d* subshell (3*d*, 4*d*, etc.); and so forth.

The secondary quantum number determines the shape of the orbital, which we will examine more closely later. Except for the special case of hydrogen, which has only one electron, the value of ℓ also affects the energy. This means that in atoms with two or more electrons, the subshells within a given shell differ slightly in energy, with the energy of the subshell increasing with increasing ℓ. Therefore, within a given shell, the *s* subshell is lowest in energy, *p* is the next lowest, followed by *d*, then *f*, and so on. For example,

> The number of subshells in a given shell equals the value of *n* for that shell. For example, when *n* = 3, there are three subshells.

$$4s < 4p < 4d < 4f$$
—increasing energy →

Practice Exercise 4

What subshells would be found in the shells with $n = 3$ and $n = 4$? ◆

The Magnetic Quantum Number, m_ℓ

Spectroscopists used m_ℓ to explain additional lines that appear in atomic spectra when atoms emit light while in a magnetic field, which explains how this quantum number got its name.

The third quantum number, m_ℓ, is known as the **magnetic quantum number.** It divides the subshells into individual orbitals, and its values are related to the way the individual orbitals are oriented relative to each other in space. As with ℓ, there are restrictions as to the possible values of m_ℓ, which can range from $+\ell$ to $-\ell$. When $\ell = 0$, m_ℓ can have only the value 0 because $+0$ and -0 are the same. An s subshell, then, has just a single orbital. When $\ell = 1$, the possible values of m_ℓ are $+1$, 0, and -1. A p subshell therefore has three orbitals: one with $\ell = 1$ and $m_\ell = 1$, another with $\ell = 1$ and $m_\ell = 0$, and a third with $\ell = 1$ and $m_\ell = -1$. Similarly, we find that a d subshell has five orbitals and an f subshell has seven orbitals. The numbers of orbitals in the subshells are easy to remember because they follow a simple arithmetic progression.

$$
\begin{array}{ccccc}
s & p & d & f & \cdots \\
1 & 3 & 5 & 7 & \cdots
\end{array}
$$

Practice Exercise 5

How many orbitals are there in a g subshell? ◆

The Whole Picture

The relationships among all three quantum numbers are summarized in Table 7.1. In addition, the relative energies of the subshells in an atom containing two or more electrons are depicted in Figure 7.15. Several important features should be noted. First, observe that each orbital on this energy diagram is indicated by a separate circle—one for an s subshell, three for a p subshell, and so forth. Second, notice that all the orbitals of a given subshell have the *same* energy. Third,

Table 7.1 Summary of Relationships among the Quantum Numbers n, ℓ, and m_ℓ

Value of n	Value of ℓ	Values of m_ℓ	Subshell	Number of Orbitals
1	0	0	$1s$	1
2	0	0	$2s$	1
	1	$-1, 0, 1$	$2p$	3
3	0	0	$3s$	1
	1	$-1, 0, 1$	$3p$	3
	2	$-2, -1, 0, 1, 2$	$3d$	5
4	0	0	$4s$	1
	1	$-1, 0, 1$	$4p$	3
	2	$-2, -1, 0, 1, 2$	$4d$	5
	3	$-3, -2, -1, 0, 1, 2, 3$	$4f$	7

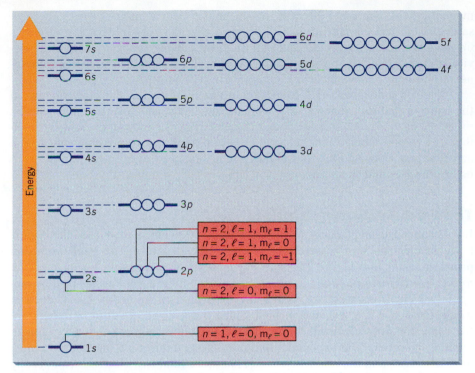

Figure 7.15 *Approximate energy level diagram for atoms with two or more electrons.* The quantum numbers associated with the orbitals in the first two shells are also shown.

In an atom, an electron can take on many different energies and waveshapes, each of which is called an orbital that is identified by a set of values for n, ℓ, and m_ℓ. When the electron wave possesses a given set of n, ℓ, and m_ℓ, we say the electron "occupies the orbital" with that set of quantum numbers.

note that, in going upward on the energy scale, the spacing between successive shells decreases as the number of subshells increases. This leads to the overlapping of shells having different values of n. For instance, the $4s$ subshell is lower in energy than the $3d$ subshell, $5s$ is lower than $4d$, and $6s$ is lower than $5d$. In addition, the $4f$ subshell is below the $5d$ subshell and $5f$ is below $6d$.

We will see shortly that Figure 7.15 is very useful for predicting the electronic structures of atoms. Before discussing this, however, we must study another very important property of the electron, a property called spin.

We really don't know whether an electron actually spins, but if it does spin, that would account for the electron's magnetic properties.

7.4 Electron Spin and the Pauli Exclusion Principle

Earlier it was stated that an atom is in its most stable state (its ground state) when its electrons have the lowest possible energies. This occurs when the electrons "occupy" the lowest energy orbitals that are available. But what determines how the electrons "fill" these orbitals? Fortunately, there are some simple rules that can help. These govern both the maximum number of electrons that can be in a particular orbital and how orbitals with the same energy become filled. One important factor that influences the distribution of electrons is the phenomenon known as *electron spin*.

Magnetic Properties of the Electron

The concept of **electron spin** is based on the fact that electrons behave as tiny magnets. This can be explained by imagining that an electron spins around its axis, like a toy top. The revolving electrical charge of the electron creates its own

Electrons moving in curved paths through wires in the large electromagnet suspended from a crane produce a magnetic field that is strong enough to pick up large portions of scrap steel.

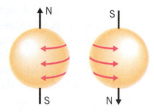

The electron can spin in either of two directions.

Pauli received the 1945 Nobel Prize in physics for his discovery of the exclusion principle.

magnetic field. The same effect is used to make electric motors work. The passage of electrical charge through the curved windings of an electric motor sets up magnetic forces that push and pull, thereby causing the rotor within the device to turn. Electricity passing through wire coils also enables large electromagnets to lift heavy steel objects, as illustrated in the margin.

Electron spin gives us a fourth quantum number for the electron, called the **spin quantum number, m_s,** which can take on two possible values: $m_s = +\frac{1}{2}$ or $m_s = -\frac{1}{2}$. It is as though the electron, like a toy top, can spin in either of two directions, giving the spin quantum number two possible values. The actual values of m_s and the reason they are not integers aren't very important to us, but the fact that there are *only* two values is very significant.

The Pauli Exclusion Principle

In 1925 an Austrian physicist, Wolfgang Pauli (1900–1958), expressed the importance of electron spin in determining electronic structure. The **Pauli exclusion principle** states that *no two electrons in the same atom can have identical values for all four of their quantum numbers.* To understand the significance of this, suppose two electrons were to occupy the 1s orbital of an atom. Each electron would have $n = 1$, $\ell = 0$, and $m_\ell = 0$. Since these three quantum numbers are the same for both electrons, the exclusion principle requires that their fourth quantum numbers (their spin quantum numbers) be different; one electron must have $m_s = +\frac{1}{2}$ and the other, $m_s = -\frac{1}{2}$. No more than two electrons can occupy the 1s orbital of the atom simultaneously because there are only two possible values of m_s. Thus the Pauli exclusion principle is really telling us that *the maximum number of electrons in any orbital is two,* and that *when two electrons are in the same orbital, they must have opposite spins.*

The limit of two electrons per orbital also limits the maximum electron populations of the shells and subshells. For the subshells we have

Subshell	Number of Orbitals	Maximum Number of Electrons
s	1	2
p	3	6
d	5	10
f	7	14

The maximum electron population per shell is shown below.

Shell	Subshells	Maximum Shell Population	
1	1s	2	
2	2s 2p	8	(2 + 6)
3	3s 3p 3d	18	(2 + 6 + 10)
4	4s 4p 4d 4f	32	(2 + 6 + 10 + 14)

In general, the maximum electron population of a shell is $2n^2$.

Magnetic Properties of Atoms

We have seen that when two electrons occupy the same orbital they must have different values of m_s. When this occurs, we say that the spins of the electrons are *paired,* or simply that the electrons are *paired.* Such pairing leads to the cancellation of the magnetic effects of the electrons because the north pole of one electron magnet is opposite the south pole of the other. Atoms with more electrons that spin in one direction than in the other are said to contain *unpaired* electrons. For these atoms, the magnetic effects do not cancel and the atoms themselves become tiny magnets which can be attracted to an external magnetic field. This weak attraction to a magnet of a substance containing unpaired electrons is called **paramagnetism.** Substances in which all the electrons are paired are not attracted to a magnet and are said to be **diamagnetic.**

Paramagnetism and diamagnetism are measurable properties that provide experimental verification of the presence or absence of unpaired electrons in substances. In addition, the quantitative measurement of the strength of the attraction of a paramagnetic substance toward a magnetic field permits the calculation of the number of unpaired electrons in its atoms, molecules, or ions.

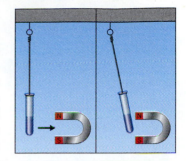

A paramagnetic substance is attracted to a magnetic field.

Diamagnetic substances are actually weakly repelled by a magnetic field.

7.5 Electronic Structures of Multielectron Atoms

The distribution of electrons among the orbitals of an atom is called the atom's **electronic structure** or **electron configuration.** This is something very useful to know about an element because the arrangement of electrons in the outer parts of an atom, which is determined by its electron configuration, controls the chemical properties of the element.

We shall be interested in the ground state electron configurations of the elements. This is the configuration that yields the lowest energy for an atom and can be predicted for many of the elements by the use of the energy level diagram in Figure 7.15 and application of the Pauli exclusion principle. To see how we go about this, let's begin with the simplest atom of all, hydrogen.

Hydrogen has an atomic number, Z, equal to 1, so a neutral hydrogen atom has one electron. In its ground state this electron occupies the lowest energy orbital that's available, which is the $1s$ orbital. To indicate symbolically the electron configuration we list the subshells that contain electrons and indicate their electron populations by appropriate superscripts. Thus the electron configuration of hydrogen is written as

$$H \qquad 1s^1$$

Another way of expressing electron configurations that we will sometimes find useful is the **orbital diagram.** In it, each orbital will be represented by a circle and arrows will be used to indicate the individual electrons, head up for spin in one direction and head down for spin in the other. The orbital diagram for hydrogen is simply

$$H \qquad \uparrow$$
$$1s$$

It doesn't matter whether the arrow points up or down. The energy of the electron is the same whether it has one spin or the other. For consistency, when an orbital is half-filled, we will show the electron with an arrow that points up.

The Aufbau Principle

To arrive at the electron configuration of an atom of another element, we imagine that we begin with a hydrogen atom and then add one proton after another (plus whatever neutrons are also needed) until we obtain the nucleus of the atom of interest. As we proceed, we also add electrons, one at a time to the lowest available orbital, until we have added enough electrons to give the neutral atom of the element in question. This imaginary process for obtaining the electronic structure of an atom is known as the **aufbau principle,** the word *aufbau* being German for *building up.* Let's look at the way this works for helium, which has $Z = 2$. This atom has two electrons, both of which are permitted to occupy the $1s$ orbital. The electron configuration of helium can therefore be written as

$$\text{He} \qquad 1s^2 \qquad \text{or} \qquad \text{He} \quad ⊕ \atop 1s$$

Notice that the orbital diagram shows that both electrons in the $1s$ orbital are paired.

We can proceed in the same fashion to predict successfully the electron configurations of most of the elements in the periodic table. For example, the next two elements in the table are lithium, Li ($Z = 3$), and beryllium, Be ($Z = 4$), which have three and four electrons, respectively. For each of these, the first two electrons enter the $1s$ orbital with their spins paired. The Pauli exclusion principle tells us that the $1s$ subshell is filled with two electrons, and Figure 7.15 shows that the orbital of next lowest energy is the $2s$, which can also hold up to two electrons. Therefore, the third electron of lithium and the third and fourth electrons of beryllium enter the $2s$. We can represent the electronic structures of lithium and beryllium as

$$\text{Li} \qquad 1s^2 2s^1 \qquad \text{or} \qquad \text{Li} \quad ⊕ \atop 1s \quad ↑ \atop 2s$$

$$\text{Be} \qquad 1s^2 2s^2 \qquad \text{or} \qquad \text{Be} \quad ⊕ \atop 1s \quad ⊕ \atop 2s$$

After beryllium comes boron, B ($Z = 5$). Referring to Figure 7.15, we see that the first four electrons of this atom complete the $1s$ and $2s$ subshells, so the fifth electron must be placed into the $2p$ subshell.

$$\text{B} \qquad 1s^2 2s^2 2p^1$$

In the orbital diagram for boron, the fifth electron can be put into any one of the $2p$ orbitals—which one doesn't matter because they are all of equal energy.

$$\text{B} \qquad ⊕ \atop 1s \quad ⊕ \atop 2s \quad ↑◯◯ \atop 2p$$

Notice, however, that when we give this orbital diagram we show *all* of the orbitals of the $2p$ subshell even though two of them are empty.

Next we come to carbon, which has six electrons. As before, the first four electrons complete the $1s$ and $2s$ orbitals. The remaining two electrons go in the $2p$ subshell to give

$$\text{C} \qquad 1s^2 2s^2 2p^2$$

Now, however, to write the orbital diagram we have to make a decision as to where to put the two p electrons. (At this point you may have an unprintable suggestion! But try to bear up. It's really not all that bad.) To make this decision, we apply **Hund's rule,** which states that *when electrons are placed in a set of orbitals of equal energy, they are spread out as much as possible to give as few paired electrons as possible.* Both theory and experiment have shown that if we follow this rule, we obtain the electron configuration with the lowest energy. For carbon, it means that the two p electrons are in separate orbitals and their spins are in the same direction.[1]

C   
1s   2s   2p

It doesn't matter which two orbitals are shown as occupied. Any of these are okay for the ground state of carbon.

2p

Applying the Pauli exclusion principle and Hund's rule, we can now complete the electron configurations and orbital diagrams for the rest of the elements of the second period.

		1s	2s	2p
N	$1s^2 2s^2 2p^3$	⇅	⇅	↑ ↑ ↑
O	$1s^2 2s^2 2p^4$	⇅	⇅	⇅ ↑ ↑
F	$1s^2 2s^2 2p^5$	⇅	⇅	⇅ ⇅ ↑
Ne	$1s^2 2s^2 2p^6$	⇅	⇅	⇅ ⇅ ⇅

We can continue to predict electron configurations in this way, using Figure 7.15 as a guide to tell us which subshells become occupied and in what order. For instance, after completing the $2p$ subshell at neon, Figure 7.15 predicts that the next two electrons enter the $3s$, followed by the filling of the $3p$. Then we find the $4s$ lower in energy than the $3d$, so it is filled first. Next, the $3d$ is completed before we go on to fill the $4p$, and so forth.

On the basis of this discussion, it would seem that Figure 7.15 must be available to us whenever we wished to write the electron configuration of an element. However, as you will see in the next section, all the information contained in this figure is also contained in the periodic table.

7.6 Electron Configurations and the Periodic Table

In Chapter 2 you learned that when Mendeleev constructed his periodic table, elements with similar chemical properties were arranged in vertical columns called groups. Later work led to the expanded version of the periodic table we use today. The basic structure of this table is one of the strongest empirical supports for the quantum theory, which we have been using to predict electron configurations, and it also permits us to use the periodic table itself as a device for predicting electron configurations.

Consider, for example, the way the table is laid out (Figure 7.16). On the left there is a block of *two* columns of elements, on the right there is a block of *six* columns, in the center there is a block of *ten* columns, and below the table there are two rows consisting of *fourteen* elements each. These numbers—2, 6, 10, and 14—are *precisely* the numbers of electrons that the quantum theory tells us can

[1]Hund's rule gives us the *lowest* energy (ground state) distribution of electrons among the orbitals. However, configurations such as

C   ⇅   ⇅   ↑ ↓ ○
C   ⇅   ⇅   ⇅ ○ ○
    1s   2s   2p

are not impossible, it is just that neither of them corresponds to the lowest energy distribution of electrons in the carbon atom.

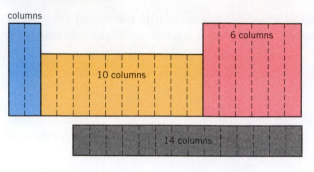

Figure 7.16 *The overall structure of the periodic table.* The table is naturally divided into regions of 2, 6, 10, and 14 columns, which are the numbers of electrons that can occupy *s*, *p*, *d*, and *f* subshells.

This would be an amazing coincidence if the theory were wrong.

occupy *s*, *p*, *d*, and *f* subshells, respectively. In fact, *we can use the structure of the periodic table to predict the filling order of the subshells when we write the electron configuration of an element.*

To use the periodic table to predict electron configurations, we follow the aufbau principle as before. We start with hydrogen and then move through the table row after row until we reach the element of interest, noting as we go along which regions of the table we pass through. For example, consider the element calcium. Using Figure 7.15, we obtain the electron configuration

$$\text{Ca} \qquad 1s^2 2s^2 2p^6 3s^2 3p^6 4s^2$$

To see how this configuration relates to the periodic table, refer to the inside front cover of the book. Notice that the first period has just two elements, H and He. Starting with hydrogen and passing through this period, two electrons are added to the atom. These enter the 1*s* subshell. Next we move across the second period, where the first two elements, Li and Be, are in the block of two columns. As you saw above, the last electron to enter each of these atoms goes into the 2*s* subshell. Then we enter the block of six columns, and as we move across this region in the second row we gradually fill the 2*p* subshell. Now we go to the third period where we first pass through the block of two columns, filling the 3*s* subshell, and then through the block of six columns, filling the 3*p*. Finally, we move to the fourth period. To get to calcium we step through two elements in the two-column block, filling the 4*s* subshell.

Now let's see what we can learn from this. Notice that every time we move across the block of two columns we add electrons to an *s* subshell that has a principal quantum number equal to the period number (second period, 2*s*; third period, 3*s*; fourth period, 4*s*). Also note that every time we move across the block of six columns we add electrons to a *p* subshell that has a principal quantum number equal to the period number (second period, 2*p*; third period, 3*p*). This is summarized in Figure 7.17.

Periodic table and electron configurations

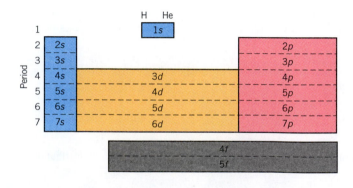

Figure 7.17 *Subshells that become filled as we cross periods.*

Figure 7.17 also includes the subshells that are filled when we cross through rows of transition elements and through rows of inner transition elements. For the transition elements, we simply subtract 1 from the period number to find the principal quantum number of the d subshell that's filled. For instance, as we move through the transition elements of *period 4,* we fill the $3d$ subshell, through those in *period 5* we fill the $4d$, and through those in *period 6* we fill the $5d$. For the inner transition elements, we fill an f subshell that has a principal quantum number that is two less than the period number of the element. Thus, as we cross the lanthanides in period 6, we add electrons to the $4f$ subshell.

EXAMPLE 7.4

Predicting Electron Configurations

What is the electron configuration of (a) Mn and (b) Bi?

Analysis: We use the periodic table as a tool to tell us which subshells become filled. It is best if you can do this without referring to Figure 7.17.

Solution: (a) To get to manganese, we cross the following regions of the table, with the results indicated.

Period 1   Fill the $1s$ subshell
Period 2   Fill the $2s$ and $2p$ subshells
Period 3   Fill the $3s$ and $3p$ subshells
Period 4   Fill the $4s$ and then move 5 places in the $3d$ region

The electron configuration of Mn is therefore

$$\text{Mn} \qquad 1s^2 2s^2 2p^6 3s^2 3p^6 4s^2 3d^5$$

Often it is desirable to group all subshells of the same shell together. For manganese, this gives

$$\text{Mn} \qquad 1s^2 2s^2 2p^6 3s^2 3p^6 3d^5 4s^2$$

(b) To get to bismuth, we fill the following subshells:

Period 1   Fill the $1s$
Period 2   Fill the $2s$ and $2p$
Period 3   Fill the $3s$ and $3p$
Period 4   Fill the $4s$, $3d$, and $4p$
Period 5   Fill the $5s$, $4d$, and $5p$
Period 6   Fill the $6s$, $4f$, $5d$, and then add 3 electrons to the $6p$

This gives

$$\text{Bi} \qquad 1s^2 2s^2 2p^6 3s^2 3p^6 4s^2 3d^{10} 4p^6 5s^2 4d^{10} 5p^6 6s^2 4f^{14} 5d^{10} 6p^3$$

Grouping subshells with the same value of n gives

$$\text{Bi} \qquad 1s^2 2s^2 2p^6 3s^2 3p^6 3d^{10} 4s^2 4p^6 4d^{10} 4f^{14} 5s^2 5p^6 5d^{10} 6s^2 6p^3$$

Practice Exercise 6

Use the periodic table to predict the electron configurations of (a) Mg, (b) Ge, (c) Cd, and (d) Gd. Group subshells of the same shell together. ◆

Practice Exercise 7

Draw orbital diagrams for (a) Na, (b) S, and (c) Fe. ◆

Electronic Basis for the Periodic Recurrence of Properties

You learned that Mendeleev constructed the periodic table by placing elements with similar properties in the same group. We are now ready to understand the reason for these similarities in terms of the electronic structures of atoms.

When we consider the chemical reactions of atoms, our attention is usually focused on the distribution of electrons in the **outer shell** of the atom (the occupied shell with the largest value of *n*). This is because the **outer electrons** (those in the outer shell) are the ones that are exposed to other atoms when the atoms react. The inner electrons of an atom, called the **core electrons,** are buried deep within the atom and normally do not play a role when chemical bonds are formed. It seems reasonable, therefore, that elements with similar properties should have similar outer shell electron configurations. This is, in fact, exactly what we observe. For example, let's look at the alkali metals of Group IA. Going by our rules, we obtain the following configurations.

$$\text{Li} \qquad 1s^2\mathbf{2s^1}$$

$$\text{Na} \qquad 1s^22s^22p^6\mathbf{3s^1}$$

$$\text{K} \qquad 1s^22s^22p^63s^23p^6\mathbf{4s^1}$$

$$\text{Rb} \qquad 1s^22s^22p^63s^23p^63d^{10}4s^24p^6\mathbf{5s^1}$$

$$\text{Cs} \qquad 1s^22s^22p^63s^23p^63d^{10}4s^24p^64d^{10}5s^25p^6\mathbf{6s^1}$$

Each of these elements has just one outer shell electron, which is in an *s* subshell. We know that when they react, the alkali metals each lose one electron to form ions with a charge of 1+. For each, the electron that is lost is this outer *s* electron, and the electron configuration of the ion that is formed is the same as that of the preceding noble gas.

Li^+	$1s^2$		He	$1s^2$
Na^+	$1s^22s^22p^6$		Ne	$1s^22s^22p^6$
K^+	$1s^22s^22p^63s^23p^6$		Ar	$1s^22s^22p^63s^23p^6$

etc.

If you write the electron configurations of the members of any of the groups in the periodic table, you will find the same kind of similarity among the configurations of the outer shell electrons. The differences are in the value of the principal quantum number of these outer electrons.

Abbreviated Electron Configurations

Because we are interested primarily in the electrons of the outer shell, we often write electron configurations in an abbreviated, or shorthand, form. To illustrate, let's consider the elements sodium and magnesium. Their full electron configurations are

$$\text{Na} \qquad 1s^22s^22p^63s^1$$

$$\text{Mg} \qquad 1s^22s^22p^63s^2$$

The outer electrons of both atoms are in the 3*s* subshell, and each has the core configuration $1s^22s^22p^6$, which is the same as that of the noble gas neon.

To write the *shorthand configuration* for an element we indicate what the core is by placing in brackets the symbol of the noble gas whose electron configuration is the same as the core configuration. This is followed by the configuration of the outer electrons for the particular element. The noble gas used is almost always the one that occurs at the end of the period preceding the period containing the element whose configuration we wish to represent. Thus, for sodium and magnesium we write

$$\text{Na} \quad \text{[Ne]} \, 3s^1$$

$$\text{Mg} \quad \text{[Ne]} \, 3s^2$$

Even with the transition elements, we need only concern ourselves with the outermost shell and the d subshell just below. For example, iron has the configuration

$$\text{Fe} \quad 1s^2 2s^2 2p^6 3s^2 3p^6 3d^6 4s^2$$

Only the $4s$ and $3d$ electrons play a role in the chemistry of iron. In general, the electrons below the outer s and d subshells of a transition element—the *core* electrons—are relatively unimportant. In every case these core electrons have the electron configuration of a noble gas. Therefore, for iron the core is $1s^2 2s^2 2p^6 3s^2 3p^6$, which is the same as the electron configuration of argon. The abbreviated configuration of iron is therefore written as

$$\text{Fe} \quad \text{[Ar]} \, 3d^6 4s^2$$

EXAMPLE 7.5

Writing Shorthand Electron Configurations

What is the shorthand electron configuration of manganese? Draw the orbital diagram for manganese using this shorthand configuration.

Solution: Manganese is in period 4. The preceding noble gas is argon, Ar, so to write the abbreviated configuration we write the symbol for argon in brackets followed by the electron configuration that exists beyond the argon core. We can obtain this by noting that to get to Mn in period 4, we first cross the "s region" by adding two electrons to the $4s$ subshell, and then go five steps into the "d region" where we add five electrons to the $3d$ subshell. Therefore, the shorthand electron configuration for Mn is

$$\text{Mn} \quad \text{[Ar]} \, 4s^2 3d^5$$

Placing the electrons that are in the highest shell farthest to the right gives

$$\text{Mn} \quad \text{[Ar]} \, 3d^5 4s^2$$

To draw the orbital diagram, we simply distribute the electrons in the $3d$ and $4s$ orbitals following Hund's rule. This gives

$$\text{Mn} \quad \text{[Ar]} \quad \uparrow \uparrow \uparrow \uparrow \uparrow \qquad \uparrow\downarrow$$
$$\phantom{\text{Mn} \quad \text{[Ar]} \quad} 3d \quad 4s$$

Notice that each of the $3d$ orbitals is half-filled. The atom contains five unpaired electrons and is paramagnetic.

Practice Exercise 8

Write shorthand configurations and abbreviated orbital diagrams for (a) P and (b) Sn. Where appropriate, place the electrons that are in the highest shell farthest to the right. How many unpaired electrons does each of these atoms have? ◆

Valence Shell Electron Configurations

For the representative elements (those in the longer columns), the only electrons that are normally important in controlling chemical properties are the ones in the outer shell. This outer shell is known as the **valence shell,** and it is

"Valence" stems from the Latin *valere,* "to be strong," suggesting "bonding strength."

always the occupied shell with the largest value of *n*. The electrons in the valence shell are called **valence electrons.** (The term *valence* comes from the study of chemical bonding and relates to the combining capacity of an element, but that's not important here.)

For the representative elements it is very easy to determine the electron configuration of the valence shell by using the periodic table. *The valence shell always consists of just the s and p subshells that we encounter crossing the period that contains the element in question.* Thus, to determine the valence shell configuration of sulfur, a period 3 element, we note that to reach sulfur in period 3 we place two electrons into the 3*s* and four electrons into the 3*p*. The valence shell configuration of sulfur is therefore

$$S \qquad 3s^2 3p^4$$

EXAMPLE 7.6

Writing Valence Shell Configurations

Predict the electron configuration of the valence shell of arsenic ($Z = 33$).

Solution: To reach arsenic in period 4, we add electrons to the 4*s*, 3*d*, and 4*p* subshells. But the 3*d* is not part of the fourth shell and therefore not part of the valence shell, so all we need be concerned with are the electrons in the 4*s* and 4*p* subshells. This gives us the valence shell configuration of arsenic,

$$As \qquad 4s^2 4p^3$$

Practice Exercise 9

What is the valence shell electron configuration of (a) Se, (b) Sn, and (c) I? ◆

7.7 Some Unexpected Electron Configurations

The rules you've learned for predicting electron configurations work most of the time, but not always. Appendix B gives the electron configurations of all of the elements as determined experimentally. Close examination reveals that there are quite a few exceptions to the rules. Some of these exceptions are important to us because they occur with common elements.

Two important exceptions are for chromium and copper. Following the rules, we would expect the configurations to be

$$Cr \qquad [Ar]\, 3d^4 4s^2$$

$$Cu \qquad [Ar]\, 3d^9 4s^2$$

However, the actual electron configurations, determined experimentally, are

$$Cr \qquad [Ar]\, 3d^5 4s^1$$

$$Cu \qquad [Ar]\, 3d^{10} 4s^1$$

The corresponding orbital diagrams are

Cr   [Ar] ⦵⦵⦵⦵⦵ ⦵
Cu   [Ar] ⦶⦶⦶⦶⦶ ⦵
                  3*d*   4*s*

Notice that for chromium, an electron is "borrowed" from the 4*s* subshell to give a 3*d* subshell that is exactly half-filled. For copper the 4*s* electron is borrowed to give a completely filled 3*d* subshell. A similar thing happens with silver and gold, which have filled 4*d* and 5*d* subshells, respectively.

$$\text{Ag} \qquad [\text{Kr}] \, 4d^{10}5s^1$$

$$\text{Au} \qquad [\text{Xe}] \, 4f^{14}5d^{10}6s^1$$

Apparently, half-filled and filled subshells (particularly the latter) have some special stability that makes such borrowing energetically favorable. This subtle but nevertheless important phenomenon affects not only the ground state configurations of atoms but also the relative stabilities of some of the ions formed by the transition elements.

7.8 Shapes of Atomic Orbitals

The same theory that tells us of the energies of atomic orbitals also describes their shapes. How do we know these shapes are correct? We don't for sure, but many of the predictions that have been made using the theory seem to be borne out by experiments. This gives the theoretical explanations strength and support. For example, the fact that the results of wave mechanics account for the shape of the periodic table so very well gives the theory a good deal of credibility.

The difficulty in describing where electrons are in an atom stems from the basic problem that we face when we attempt to picture a particle as a wave. There is nothing in our worldly experience that is comparable. The way we get around this perplexing conceptual problem, so that we can still think of the electron as a particle in the usual sense, is to speak in terms of the statistical probability of the electron being found at a particular place.

Describing the electron's position in terms of statistical probability is based on more than simple convenience. The German physicist Werner Heisenberg showed mathematically that it is impossible to measure with complete precision both a particle's velocity and position at the same instant, and that such measurements are subject to a minimum uncertainty. This was Heisenberg's famous **uncertainty principle.** The theoretical limitations on measuring speed and position are not significant for large objects. However, for small particles such as the electron, these limitations prevent us from ever knowing or predicting where in an atom an electron will be at a particular instant, so we speak of probabilities instead.

Wave mechanics views the probability of finding an electron at a given point in space as equal to the square of the amplitude of the electron wave (given by the square of the wave function, ψ^2) at that point. It seems quite reasonable to relate probability to amplitude, or intensity, because where a wave is intense its presence is strongly felt. The amplitude is squared because, mathematically, the amplitude can be either positive or negative, but probability only makes sense if it is positive. Squaring the amplitude assures us that the probabilities will be positive. We need not be very concerned about this point, however.

The notion of electron probability leads to two very important and frequently used concepts. One is that an electron behaves as if it were spread out around the nucleus in a sort of **electron cloud.** Figure 7.18*a* is a *dot-density diagram* that illustrates the way the probability of finding the electron varies in space for a 1*s* orbital. In those places where the dot density is large (i.e., where there are large numbers of dots per unit volume), the amplitude of the wave is large and the probability of finding the electron is also large. Figure 7.18*b* shows

Heisenberg won the 1932 Nobel Prize in physics for his work.

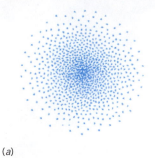

(a)

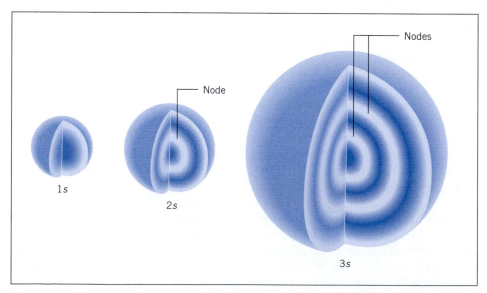

(b)

Figure 7.18 *Electron distribution in a 1s orbital.* (a) A dot-density diagram that illustrates the electron probability distribution for a 1s electron. (b) A graph that shows how the probability of finding the electron around a given point, ψ^2, decreases as the distance from the nucleus increases.

how the electron probability for a 1s orbital varies as we move away from the nucleus. As you might expect, the probability of finding the electron close to the nucleus is large and decreases with increasing distance from the nucleus.

The other important concept that stems from the notion that the electron probability varies from place to place is **electron density,** which relates to how much of the electron's charge is packed into a given volume. In regions of high probability there is a high concentration of electrical charge (and mass) and the electron density is large; in regions of low probability, the electron density is small.

On first thought, it is difficult to imagine an object such as an electron as being spread throughout some relatively large volume. To help you with this picture, imagine an ice cube. The frozen water in the cube behaves as though it's a single object. But the same amount of water that's in the ice cube can also be in a cloud floating in the sky. Here the water is spread out as a mist through a large volume. In some places the cloud is dense (meaning there is a lot of water mist per unit volume) and we can't see through it; in others the mist might be thinner and semitransparent. Thus, the "water mist density" can vary from place to place in the cloud just as the electron density can vary from place to place around the nucleus. This is a useful picture to keep in mind as you try to visualize the shapes of atomic orbitals.

Shapes and Sizes of *s* and *p* Orbitals

In looking at the way the electron density distributes itself in atomic orbitals, we are interested in three things—the *shape* of the orbital, its *size,* and its *orientation* in space relative to other orbitals.

The electron density in an orbital doesn't end abruptly at some particular distance from the nucleus. It gradually fades away. Therefore, to define the size and shape of an orbital, it is useful to picture some imaginary surface enclosing, say, 90% of the electron density of the orbital, and on which the probability of finding the electron is everywhere the same. For the 1s orbital in Figure 7.18, we find that if we go out a given distance from the nucleus in *any* direction, the

Figure 7.19 *Size variations among s orbitals.* The orbitals become larger as the principal quantum number, *n*, becomes larger.

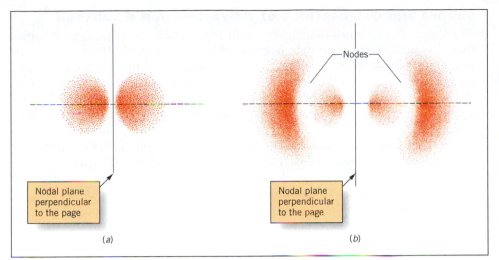

Figure 7.20 *Distribution of electron density in p orbitals.* (*a*) Dot-density diagram that represents a cross section of the probability distribution in a 2*p* orbital. There is a nodal plane between the two lobes of the orbital. (*b*) Cross section of a 3*p* orbital. Note the nodes in the electron density that are in addition to the nodal plane passing through the nucleus.

probability of finding the electron is the same. This means that all the points of equal probability lie on the surface of a sphere, so we can say that the shape of the orbital is spherical. In fact, all *s* orbitals are spherical. As suggested earlier, their sizes increase with increasing *n*. This is illustrated in Figure 7.19. Notice that beginning with the 2*s* orbital, there are certain places where the electron density drops to zero. These are the nodes of the electron wave. It is interesting that electron waves have nodes just like the waves on a guitar string. For electron waves, however, the nodes consist of imaginary *surfaces* on which the electron density is zero.

The *p* orbitals are quite different from *s* orbitals, as shown in Figure 7.20. Notice that the electron density is equally distributed in two regions on opposite sides of the nucleus. Figure 7.20*a* illustrates the two "lobes" of *one* 2*p* orbital. The electron density of each lobe is concentrated around an imaginary line that passes through the nucleus. At the nucleus and perpendicular to the line passing through the centers of the two lobes there is a *nodal plane*—an imaginary flat surface on which every point has an electron density of zero. The size of the *p* orbitals also increase with increasing *n* as illustrated by the cross section of a 3*p* orbital in Figure 7.20*b*. The 3*p* and higher *p* orbitals have additional nodes besides the nodal plane that passes through the nucleus.

Figure 7.21*a* illustrates the shape of a surface of constant probability for a 2*p* orbital. Often chemists will simplify this shape by drawing two "balloons" connected at the nucleus and pointing in opposite directions as shown in Figure 7.21*b*. Both representations emphasize the point that a *p* orbital has two equal-sized lobes that extend in opposite directions along a line that passes through the nucleus.

Orientations of Orbitals in a *p* Subshell

As you've learned, a *p* subshell consists of three orbitals of equal energy. Wave mechanics tells us that the lines along which the orbitals have their maximum electron densities are oriented at 90° angles to each other, corresponding to the axes of an imaginary *xyz* coordinate system (Figure 7.22). For convenience in referring to the individual *p* orbitals they are often labeled according to the axis along which they lie. The *p* orbital concentrated along the *x* axis is labeled p_x, and so forth.

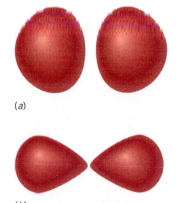

(a)

(b)

Figure 7.21 *Representations of the shapes of p orbitals.* (*a*) Shape of a surface of constant probability for a 2*p* orbital. (*b*) A simplified representation of a *p* orbital that emphasizes the directional nature of the orbital.

Shapes and Orientations of *d* Orbitals in a *d* Subshell

The shapes of the *d* orbitals, illustrated in Figure 7.23, are a bit more complex than are those of the *p* orbitals. Because of this, and because there are five orbitals in a *d* subshell, we haven't attempted to draw all of them at the same time on the same set of coordinate axes. Notice that four of the five *d* orbitals have the same shape and consist of four lobes of electron density. These orbitals differ only in their orientations around the nucleus (their labels come from the mathematics of wave mechanics). The fifth *d* orbital, labeled d_{z^2}, has two lobes that point in opposite directions along the *z* axis plus a doughnut-shaped ring of electron density around the center that lies in the $x-y$ plane. We will see that the *d* orbitals are important in the formation of chemical bonds in certain molecules, and that their shapes and orientations are important in understanding the properties of the transition metals, which will be discussed in Chapter 20.

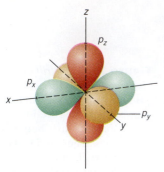

Figure 7.22 *The orientations of the three p orbitals in a p subshell.* Because the directions of maximum electron density lie along lines that are mutually perpendicular, like the axes of an *xyz* coordinate system, it is convenient to label the orbitals $p_x, p_y,$ and p_z.

The *f* orbitals are even more complex than the *d* orbitals, but we will have no need to discuss their shapes.

7.9 Variation of Atomic Properties with Electronic Structure

There are many chemical and physical properties that vary in a more or less systematic way according to an element's position in the periodic table. For example, in Chapter 2 we noted that the metallic character of the elements increases from top to bottom in a group and decreases from left to right across a period. In this section we discuss several physical properties of the elements that have an important influence on chemical properties. We will see how these properties correlate with an atom's electron configuration, and because electron configuration is also related to the location of an element in the periodic table, we will study their periodic variations as well.

Effective Nuclear Charge

Many of an atom's properties are determined by the amount of positive charge felt by the atom's outer electrons. Except for hydrogen, this positive charge is always *less* than the full nuclear charge, because the negative charge of the electrons in inner shells partially offsets, or "neutralizes," the positive charge of the nucleus.

The inner electrons partially shield the outer electrons from the nucleus, so the outer electrons "feel" only a fraction of the full nuclear charge.

To gain a better understanding of this, consider the element lithium, which has the electron configuration $1s^2 2s^1$. The core electrons $(1s^2)$, which lie beneath the valence shell $(2s^1)$, are tightly packed around the nucleus and for the most part lie between the nucleus and the electron in the outer shell. This core has

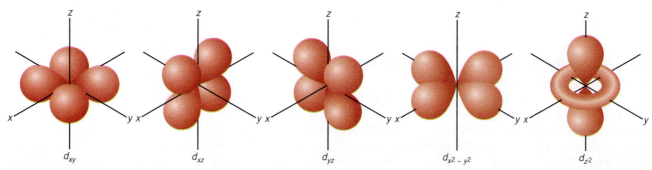

Figure 7.23 *The shapes and directional properties of the five d orbitals of a d subshell.*

a charge of 2− and it surrounds a nucleus that has a charge of 3+. When the outer 2s electron "looks toward" the center of the atom, it "sees" the 3+ charge of the nucleus reduced to only about 1+ because of the intervening 2− charge of the core. In other words, the 2− charge of the core effectively neutralizes two of the positive charges of the nucleus, so the net charge that the outer electron feels, which we call the **effective nuclear charge,** is only about 1+. This is illustrated in an overly simplified way in Figure 7.24.

Although electrons in inner shells shield the electrons in outer shells quite effectively from the nuclear charge, electrons in the *same* shell are much less effective at shielding each other. For example, in the element beryllium ($1s^2 2s^2$) each of the electrons in the outer 2s orbital is shielded quite well from the nuclear charge by the inner $1s^2$ core, but one 2s electron doesn't shield the other 2s electron very well at all. This is because electrons in the same shell are at about the same average distance from the nucleus, and in attempting to stay away from each other they only spend a very small amount of time one below the other, which is what's needed to provide shielding. Since electrons in the same shell hardly shield each other at all from the nuclear charge, *the effective nuclear charge felt by the outer electrons is determined primarily by the difference between the charge on the nucleus and the charge on the core.* With this as background, let's examine some properties controlled by the effective nuclear charge.

Sizes of Atoms and Ions

The wave nature of the electron makes it difficult to define exactly what we mean by the "size" of an atom or ion. As we've seen, the electron cloud doesn't simply stop at some particular distance from the nucleus; instead it gradually fades away. Nevertheless, atoms and ions do behave in many ways as though they have characteristic sizes. For example, in a whole host of hydrocarbons, ranging from methane (CH_4, natural gas) to octane (C_8H_{18}, in gasoline) to many others, the distance between the nuclei of carbon and hydrogen atoms is virtually the same. This would suggest that carbon and hydrogen have the same relative sizes in each of these compounds.

Experimental measurements reveal that the diameters of atoms range from about 1.4×10^{-10} to 5.7×10^{-10} m. Their radii, which is the usual way that size is specified, range from about 7.0×10^{-11} to 2.9×10^{-10} m. Such small numbers are difficult to comprehend. A million carbon atoms placed side by side in a line would extend a little less than 0.2 mm, or about the diameter of the period at the end of this sentence.

The sizes of atoms and ions are rarely expressed in meters because the numbers are so cumbersome. Instead, a unit is chosen that makes the values easier to comprehend. A unit that scientists have traditionally used is called the **angstrom** (symbolized **Å**), which is defined as

$$1 \text{ Å} = 1 \times 10^{-10} \text{ m}$$

However, the angstrom is not an SI unit, and in many current scientific journals, atomic dimensions are given in picometers, or sometimes in nanometers ($1 \text{ pm} = 10^{-12}$ m and $1 \text{ nm} = 10^{-9}$ m). In this book, we will normally express atomic dimensions in picometers, but because much of the scientific literature has these quantities in angstroms, you may someday find it useful to remember the conversions:

$$1 \text{ Å} = 100 \text{ pm}$$

$$1 \text{ Å} = 0.1 \text{ nm}$$

An electron spends very little time between the nucleus and another electron in the same shell, so it shields that other electron poorly.

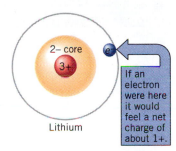

Figure 7.24 *Effective nuclear charge of the valence electron in lithium.* If the 2− charge of the $1s^2$ core of lithium were 100% effective at shielding the 2s electron from the nucleus, the valence electron would feel an effective nuclear charge of only about 1+.

The C—H distance in most hydrocarbons is about 110 pm (110×10^{-12} m).

The angstrom is named after Anders Jonas Ångström (1814–1874), a Swedish physicist who was the first to measure the wavelengths of the four most prominent lines of the hydrogen spectrum.

Periodic Variations in Atomic Size

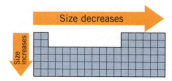

Size decreases

Size increases

The variations in atomic radii within the periodic table are illustrated in Figure 7.25. Here we see that atoms generally become larger going from top to bottom in a group, and they become smaller going from left to right across a period. To understand these variations we must consider two factors. One is the value of the principal quantum number of the valence electrons, and the other is the effective nuclear charge felt by the valence electrons.

Going from top to bottom within a group, the effective nuclear charge felt by the outer electrons remains nearly constant, while the principal quantum number of the valence shell increases. For example, consider the elements of Group IA. For lithium, the valence shell configuration is $2s^1$; for sodium, it is $3s^1$; for potassium, it is $4s^1$; and so forth. For each of these elements, the core has a negative charge that is one less than the nuclear charge, so the valence electron of each experiences a nearly constant effective nuclear charge of about 1+. However, as we descend the group, the value of n for the valence shell increases, and as you learned earlier, the larger the value of n, the larger is the orbital. Therefore, the atoms become larger as we go down a group simply because the orbitals containing the valence electrons become larger. This same argument applies whether the valence shell orbitals are s or p.

Moving from left to right across a period, electrons are added to the same shell. The orbitals holding the valence electrons all have the *same* value of n. In this case we have to examine the variation in the effective nuclear charge felt by the valence electrons.

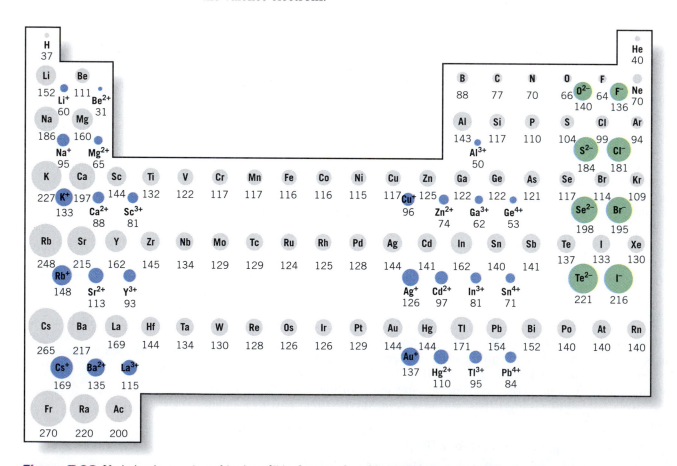

Figure 7.25 *Variation in atomic and ionic radii in the periodic table.* Values are in picometers.

As we move from left to right across a period, the nuclear charge increases, the outer shells of the atoms become more populated, but the inner core remains the same. For example, from lithium to fluorine the nuclear charge increases from 3+ to 9+. The core ($1s^2$) stays the same, however. As a result, the outer electrons feel an increase in positive charge (i.e., effective nuclear charge), which causes them to be drawn inward, and thereby causes the sizes of the atoms to decrease.

Across a row of transition elements or inner transition elements, the size variations are less pronounced than among the representative elements. This is because the outer shell configuration remains essentially the same while an inner shell is filled. From atomic numbers 21 to 30, for example, the outer electrons occupy the $4s$ subshell while the underlying $3d$ subshell is gradually completed. The amount of shielding provided by the addition of electrons to this inner $3d$ level is greater than the amount of shielding that would occur if the electrons were added to the outer shell, so the effective nuclear charge felt by the outer electrons increases more gradually. As a result, the decrease in size with increasing atomic number is also more gradual.

Trends in the Sizes of Ions

Figure 7.25 also illustrates how sizes of the ions compare with those of the neutral atoms. As you can see, when atoms gain or lose electrons to form ions, rather significant size changes take place. The reasons are easy to understand and remember.

When electrons are added to an atom, the mutual repulsions between them increase. This causes the electrons to push apart and occupy a larger volume. Therefore, *negative ions are always larger than the atoms from which they are formed* (Figure 7.26).

When electrons are removed from an atom, the electron–electron repulsions decrease, which allows the remaining electrons to be pulled closer together around the nucleus. Therefore, *positive ions are always smaller than the atoms from which they are formed*. This is also illustrated in Figure 7.26 for the elements lithium and iron. For lithium, removal of the outer $2s$ electron completely empties the valence shell and exposes the smaller $1s^2$ core. When a metal is able to form more than one positive ion, the sizes of the ions decrease as the amount of positive charge on the ion increases. To form the Fe^{2+} ion, an iron atom loses its outer $4s$ electrons. To form the Fe^{3+} ion, an additional electron is lost from the $3d$ subshell that lies beneath the $4s$. Comparing sizes, we see that the radius of an iron atom is 116 pm whereas the radius of the Fe^{2+} ion is 76 pm. Removing yet another electron to give Fe^{3+} decreases electron–electron repulsions in the d subshell and gives the Fe^{3+} ion a radius of 64 pm.

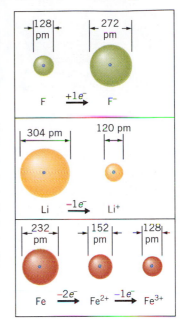

Figure 7.26 *Changes in size when atoms gain or lose electrons to form ions.* Adding electrons leads to an increase in the size of the particle, as illustrated for fluorine. Removing electrons leads to a decrease in the size of the particle, as shown for lithium and iron.

Adding electrons creates an ion that is larger than the neutral atom; removing electrons produces an ion that is smaller than the neutral atom

Li	$1s^2\,2s^1$
Li^+	$1s^2$
Fe	$[Ar]\,3d^6 4s^2$
Fe^{2+}	$[Ar]\,3d^6$
Fe^{3+}	$[Ar]\,3d^5$

Practice Exercise 10

Use the periodic table to choose the largest atom or ion in each set.
(a) Ge, Te, Se, Sn   (b) C, F, Br, Ga   (c) Cr, Cr^{2+}, Cr^{3+} (d) O, O^{2-}, S, S^{2-} ◆

Ionization Energy

The **ionization energy** (abbreviated **IE**) *is the energy required to remove an electron from an isolated, gaseous atom or ion in its ground state.* For an atom of an element X, it is the increase in potential energy associated with the change

$$X(g) \longrightarrow X^+(g) + e^-$$

In effect, the ionization energy is a measure of how much work is required to pull an electron from an atom, so it reflects how tightly the electron is held by the atom. Usually, the ionization energy is expressed in units of kilojoules per mole (kJ/mol), so we can also view it as the energy needed to remove one mole of electrons from one mole of gaseous atoms.

Table 7.2 gives the ionization energies of the first 12 elements. As you can see, atoms with more than one electron have more than one ionization energy. These correspond to the stepwise removal of electrons, one after the other. Lithium, for example, has three ionization energies because it has three electrons. Removing the outer 2s electrons from one mole of isolated lithium atoms to give one mole of gaseous lithium ions, Li^+, requires 520 kJ; so the *first ionization energy* of lithium is 520 kJ/mol. The second IE of lithium is 7297 kJ/mol, and corresponds to the process

$$Li^+(g) \longrightarrow Li^{2+}(g) + e^-$$

This involves the removal of an electron from the now-exposed 1s core of lithium. Removal of the third (and last) electron requires the third IE, which is 11,810 kJ/mol. In general, successive ionization energies always increase because each subsequent electron is being pulled away from an increasingly more positive ion, and that requires more energy.

Periodic Trends in Ionization Energies

Within the periodic table there are trends in the way IE varies that are useful to know and to which we will refer in later discussions. We can see these by examining a graph that shows how the first ionization energy varies with an element's position in the table, which is shown in Figure 7.27. Notice that the elements with the largest ionization energies are the nonmetals in the upper right of the periodic table, and that those with the smallest ionization energies are the metals in the lower left of the table. In general, then, the following trends are observed.

It is often helpful to remember that the trends in IE are just the opposite of the trends in atomic size within the periodic table; when size increases, IE decreases.

Tools

Periodic trends in ionization energy

> Ionization energy generally increases from bottom to top within a group and increases from left to right within a period.

Table 7.2 Successive Ionization Energies in kJ/mol for Hydrogen through Magnesium

	1st	2nd	3rd	4th	5th	6th	7th	8th
H	1312							
He	2372	5250						
Li	520	7297	11,810					
Be	899	1757	14,845	21,000				
B	800	2426	3659	25,020	32,820			
C	1086	2352	4619	6221	37,820	47,260		
N	1402	2855	4576	7473	9442	53,250	64,340	
O	1314	3388	5296	7467	10,987	13,320	71,320	84,070
F	1680	3375	6045	8408	11,020	15,160	17,860	92,010
Ne	2080	3963	6130	9361	12,180	15,240	—	—
Na	496	4563	6913	9541	13,350	16,600	20,113	25,666
Mg	737	1450	7731	10,545	13,627	17,995	21,700	25,662

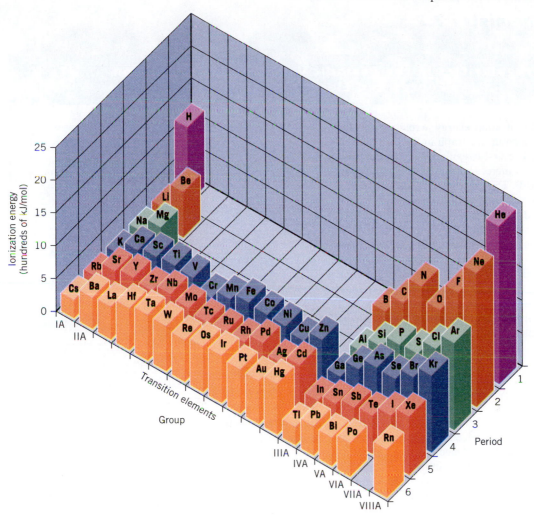

Figure 7.27 *Variation in first ionization energy with location in the periodic table.* Elements with the largest ionization energies are in the upper right of the periodic table. Those with the smallest ionization energies are at the lower left.

The same factors that affect atomic size also affect ionization energy. As the value of *n* increases going down a group, the orbitals become larger and the outer electrons are farther from the nucleus. Electrons farther from the nucleus are bound less tightly, so IE decreases from top to bottom. Of course, this is just the same as saying that it increases from bottom to top.

As you can see, there is a gradual overall increase in IE as we move from left to right across a period, although the horizontal variation of IE is somewhat irregular (see Facets of Chemistry 7.2). The reason for the overall trend is the increase in effective nuclear charge felt by the valence electrons as we move across a period. As we've seen, this draws the valence electrons closer to the nucleus and leads to a decrease in atomic size as we move from left to right. But the increasing effective nuclear charge also causes the valence electrons to be held more tightly, which makes it more difficult to remove them.

The results of these trends place elements with the largest IE in the upper right-hand corner of the periodic table. It is very difficult to cause these atoms to lose electrons. In the lower left-hand corner of the table are elements that have

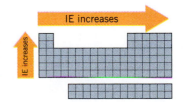

Facets of Chemistry 7.2

Irregularities in the Periodic Variations in Ionization Energy and Electron Affinity

The variation in first ionization energy across a period is not a smooth one, as seen in the graph in Figure 1 for the elements in period 2. The first irregularity occurs between Be and B, where the IE increases from Li to Be but then decreases from Be to B. This happens because there is a change in the nature of the subshell from which the electron is being removed. For Li and Be, the electron is removed from the $2s$ subshell, but at boron the first electron comes from the higher energy $2p$ subshell where it is not bound so tightly.

Another irregularity occurs between nitrogen and oxygen. For nitrogen, the electron that's removed comes from a singly occupied orbital. For oxygen, the electron is taken from an orbital that already contains an electron. We can diagram this as follows:

Figure 1 Variation in IE for the period 2 elements Li through Ne.

For oxygen, repulsions between the two electrons in the p orbital that's about to lose an electron help the electron leave. This "help" is absent for the electron that's about to leave the p orbital of nitrogen. As a result, it is not as difficult to remove one electron from an oxygen atom as it is to remove one electron from a nitrogen atom.

As with ionization energy, there are irregularities in the periodic trends for electron affinity. For example, the Group IIA elements have little tendency to acquire electrons because their outer shell s orbitals are filled. The incoming electron must enter a higher energy p orbital. We also see that the EA for elements in Group VA are either endothermic or only slightly exothermic. This is because the incoming electron must enter an orbital already occupied by an electron.

One of the most interesting irregularities occurs between periods 2 and 3 among the nonmetals. In any group, the element in period 2 has a less exothermic electron affinity than the element below it. The reason seems to be the small size of the nonmetal atoms of period 2, which are among the smallest elements in the periodic table. Repulsions between the many electrons in the small valence shells of these atoms leads to a lower than expected attraction for an incoming electron and a less exothermic electron affinity than the element below in period 3.

loosely held valence electrons. These elements form positive ions relatively easily, as you learned in Chapter 2.

Stability of the Noble Gas Configuration

Table 7.2 shows that, for a given element, successive ionization energies increase gradually until the valence shell is emptied. Then a very much larger increase in IE occurs as the core is broken into. This is illustrated graphically in Figure 7.28 for the period 2 elements lithium through fluorine. For lithium, we see that the first electron (the $2s$ electron) is removed rather easily, but the second and third electrons, which come from the $1s$ core, are much more difficult to dislodge. For beryllium, the large jump in IE occurs after two electrons (the two $2s$ electrons)

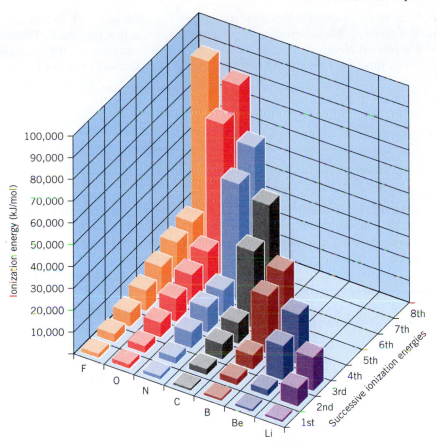

Figure 7.28 *Variations in successive ionization energies for the elements lithium through fluorine.*

are removed. In fact, for all of these elements, the big jump in IE happens when the core is broken into.

The data displayed in Figure 7.28 suggest that although it may be moderately difficult to empty the valence shell of an atom, it is *extremely* difficult to break into the noble gas configuration of the core electrons. As you will learn, this is one of the factors that influences the number of positive charges on ions formed by the representative metals.

<hr>

Practice Exercise 11

Use the periodic table to select the atom with the largest IE: (a) Na, Sr, Be, Rb; (b) B, Al, C, Si. ◆

Electron Affinity

The **electron affinity** (abbreviated **EA**) *is the potential energy change associated with the addition of an electron to a gaseous atom or ion in its ground state.* For an element X, it is the change in potential energy associated with the process

$$X(g) + e^- \longrightarrow X^-(g)$$

As with ionization energy, electron affinities are usually expressed in units of kilojoules per mole, so we can also view the IE as the energy change associated with adding one mole of electrons to one mole of gaseous atoms or ions.

For nearly all the elements, the addition of one electron to the neutral atom is exothermic, and the EA is given as a negative value. This is because the

Notice that the sign convention used here agrees with the one that we followed in determining the sign of ΔH in Chapter 6.

incoming electron experiences an attraction to the nucleus which causes the potential energy to be lowered as the electron approaches the atom. However, when a second electron must be added, as in the formation of the oxide ion, O^{2-}, work must be done to force the electron into an already negative ion.

Change	EA (kJ/mol)
$O(g) + e^- \longrightarrow O^-(g)$	-141
$O^-(g) + e^- \longrightarrow O^{2-}(g)$	$+844$
$O(g) + 2e^- \longrightarrow O^{2-}(g)$	$+703$ (net)

Notice that more energy is absorbed adding an electron to the O^- ion than is released by adding an electron to the O atom. Overall, the formation of an isolated oxide ion leads to a net increase in potential energy (so we say its formation is *endothermic*). The same applies to the formation of *any* negative ion with a charge larger than $1-$.

The electron affinities of the representative elements are given in Table 7.3, and we see that periodic trends in electron affinity roughly parallel those for ionization energy. (See Facets of Chemistry 7.4 for a discussion of some of the irregularities in the trends.)

Periodic trends in electron affinity

Although there are some irregularities, overall, the electron affinities of the elements become more *exothermic* going from left to right across a period and from bottom to top in a group (Figure 7.29).

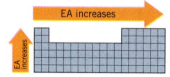

Figure 7.29 *Variation of electron affinity (as an exothermic quantity) within the periodic table.*

This shouldn't be surprising, because a valence shell that loses electrons easily (low IE) will have little attraction for additional electrons (small EA). On the other hand, a valence shell that holds its electrons tightly will also tend to bind an additional electron tightly.

Table 7.3 Electron Affinities of the Representative Elements (kJ/mol)

IA	IIA	IIIA	IVA	VA	VIA	VIIA
H						
-73						
Li	Be	B	C	N	O	F
-60	$+238$	-27	-122	$\sim +9$	-141	-328
Na	Mg	Al	Si	P	S	Cl
-53	$+230$	-44	-134	-72	-200	-348
K	Ca	Ga	Ge	As	Se	Br
-48	$+155$	-30	-120	-77	-195	-325
Rb	Sr	In	Sn	Sb	Te	I
-47	$+167$	-30	-121	-101	-190	-295
Cs	Ba	Tl	Pb	Bi	Po	At
-45	$+50$	-30	-110	-110	-183	-270

SUMMARY

Electromagnetic Energy. Electromagnetic energy, or light energy, travels through space at a constant speed of 3.00×10^8 m s^{-1} in the form of waves. The **wavelength, λ,** and **frequency, ν,** of the wave are related by the equation $\lambda\nu = c$, where c is the **speed of light.** The SI unit for frequency is the **hertz (Hz).** Light also behaves as if it consists of small packets of energy called **photons** or **quanta.** The energy delivered by a photon is proportional to the frequency of the light, and is given by the equation $E = h\nu$ where h is **Planck's constant.** White light is composed of all the frequencies visible to the eye and can be split into a **continuous spectrum.** Visible light represents only a small portion of the entire **electromagnetic spectrum,** which also includes **X rays, ultraviolet, infrared, microwaves,** and **radio** and **TV** waves.

Atomic Spectra. The occurrence of **line spectra** tells us that atoms can emit energy only in discrete amounts and suggests that the energy of the electron is **quantized;** that is, the electron is restricted to certain specific **energy levels** in an atom. Niels Bohr recognized this and, although his theory was later shown to be incorrect, he was the first to propose a model that was able to account for the **Rydberg equation.** Bohr was the first to introduce the idea of **quantum numbers.**

Matter Waves. The wave behavior of electrons and other tiny particles, which can be demonstrated by **diffraction** experiments, was suggested by de Broglie. Schrödinger applied wave theory to the atom and launched the theory we call **wave mechanics** or **quantum mechanics.** This theory tells us that electron waves in atoms are **standing waves** whose crests and **nodes** are stationary. Each standing wave, or **orbital,** is characterized by three quantum numbers, n, ℓ, m_ℓ (**principal, secondary,** and **magnetic quantum numbers,** respectively). **Shells** are designated by n (which can range from 1 to ∞), **subshells** by ℓ (which can range from 0 to $n - 1$), and orbitals within subshells by m_ℓ (which can range from $-\ell$ to $+\ell$).

Electron Configurations. The electron has magnetic properties that are explained in terms of spin. The **spin quantum number, m_s,** can have values of $+\frac{1}{2}$ or $-\frac{1}{2}$. The **Pauli exclusion principle** limits orbitals to a maximum population of two electrons with **paired spins.** Substances with unpaired electrons are **paramagnetic** and are attracted weakly to a magnetic field. Substances with only paired electrons are **diamagnetic** and are slightly repelled by a magnetic field. The **electron configuration** of an element in its **ground state** is obtained by filling orbitals beginning with the $1s$ subshell and following the Pauli exclusion principle and **Hund's rule** (which states that electrons spread out as much as possible in orbitals of equal energy). The periodic table serves as a guide in predicting electron configurations. **Abbreviated configurations** show subshell populations outside a noble gas **core. Valence shell configurations** show the populations of subshells in the **outer shell** of an atom of the representative elements. Sometimes we represent electron configurations using **orbital diagrams.** Unexpected configurations occur for chromium and copper because of the extra stability of half-filled and filled subshells.

Orbital Shapes. The **Heisenberg uncertainty principle** says we cannot know exactly the position and velocity of an electron both at the same instant. Consequently, wave mechanics describes the probable locations of electrons in atoms. In each orbital the electron is conveniently viewed as an **electron cloud** with a varying **electron density.** All s orbitals are spherical; each p orbital consists of two lobes with a **nodal plane** between them. A p subshell has three p orbitals whose axes are mutually perpendicular and point along the x, y, and z axes of an imaginary coordinate system centered at the nucleus. Four of the five d orbitals in a d subshell have the same shape, with four lobes of electron density each. The fifth has two lobes of electron density pointing in opposite directions along the z axis and a ring of electron density in the $x-y$ plane.

Atomic Properties. The amount of positive charge felt by the valence electrons of an atom is the **effective nuclear charge.** This is less than the actual nuclear charge because core electrons partially shield the valence electrons from the full positive charge of the nucleus. **Atomic radii** depend on the value of n of the valence shell orbitals and the effective nuclear charge experienced by the valence electrons. These radii are expressed in units of picometers or nanometers, or an older unit called the **angstrom (Å):** 1 Å = 100 pm = 0.1 nm. Atomic radii decrease from left to right in a period and from bottom to top in a group in the periodic table. Negative ions are larger than the atoms from which they are formed; positive ions are smaller than the atoms from which they are formed.

Ionization energy (IE) is the energy needed to remove an electron from an isolated gaseous atom, molecule, or ion in its ground state; it is endothermic. The first ionization energies of the elements increase from left to right in a group and from bottom to top in a period. (Irregularities occur in a period when the nature of the orbital from which the electron is removed changes and when the electron removed is first taken from a doubly occupied p orbital.) Successive ionization energies become larger, but there is a very large jump when the next electron must come from the noble gas core beneath the valence shell.

Electron affinity (EA) is the potential energy change associated with the addition of an electron to a gaseous atom or ion in its ground state. For atoms, the first EA is usually exothermic. When more than one electron is added to an atom, the overall potential energy change is endothermic. In general, electron affinity becomes more exothermic from left to right in a period and from bottom to top in a group. (However, the EA of second period nonmetals is less exothermic than for the nonmetals of the third period. Irregularities across a period occur when the electron being added must enter the next higher energy subshell and when it must enter a half-filled *p* subshell.)

Tools *you have learned*

The table below lists the tools that you have learned in this chapter. Notice that only two of them are related to numerical calculations. The others are conceptual tools that we use in analyzing properties of substances in terms of the underlying structure of matter. Review all these tools and refer to them, if necessary, when working on the Thinking-It-Through problems and the Review Problems that follow.

Tool	Function
Wavelength–frequency relationship *(page 277)*	To convert between wavelength and frequency.
Energy of a photon *(page 281)*	To calculate the energy carried by a photon of frequency ν. Also, ν can be calculated if E is known.
Periodic table *(page 298)*	We will use the periodic table as a tool for many purposes. In this chapter you learned to use the periodic table as an aid in writing electron configurations of the elements and as a tool to correlate an element's location in the table to properties such as atomic radius, ionization energy, and electron affinity.
Periodic trends in atomic and ionic size *(page 308)*	To compare the sizes of atoms and ions.
Periodic trends in ionization energy *(page 310)*	To compare the ease with which atoms of the elements lose electrons.
Periodic trends in electron affinity *(page 314)*	To compare the tendency of atoms or ions to gain electrons.

THINKING IT THROUGH

For the following problems, the goal is *not* to find the answer itself; instead, you are just asked to assemble the available information needed to obtain the answer, state what additional data (if any) are needed, and describe how you would use the data to answer the question. For problems involving unit conversions, list the relationships among the units that are needed to carry out the conversions. Construct the conversion factors that can be formed from these relationships. Then set up the solution to the problem by arranging the conversion factors so the units cancel correctly to give the desired units of the answer.

The problems are divided into two groups. Those in Level 2 are more challenging than those in Level 1. Detailed answers to these problems can be found at our Web site: http://www.wiley.com/college/brady

Level 1 Problems

1. An FM radio station broadcasts at a frequency of 101.9 MHz. What is the wavelength of these radio waves expressed in meters? How many peaks of these waves pass a given point each second? (Set up the calculation.)

2. Ultraviolet (UV) radiation can cause sunburn and has been implicated as a causative factor in the formation of certain skin cancers. A typical photon of UV light has a wavelength of 150 nm. How can you calculate the energy of a photon of this light? How many times larger is this than the energy of a photon emitted by a high-voltage wire carrying 60 Hz alternating current?

3. For a hydrogen atom, explain how to calculate the frequency of the photon that is emitted when the electron drops from an energy level that has $n = 6$ to $n = 2$. Describe how you would calculate the wavelength of this light. How can you determine what color the light is?

4. Atoms of which of the following elements are paramagnetic? (a) Mg, (b) Zn, (c) As, (d) Ar, (e) Ag

5. The atoms of which element have valence electrons that experience the greatest effective nuclear charge, silicon or sulfur? Explain the reasoning required to answer this question.

6. State the specific relationship that allows you to determine
(a) which element has the larger atoms, Ge or S.
(b) which element is likely to have the larger first ionization energy, phosphorus or sulfur.
(c) which element is likely to have the more exothermic electron affinity, phosphorus or sulfur.

Level 2 Problems

7. Explain in detail how to calculate the ionization energy of the hydrogen atom from information provided in this chapter.

8. Magnesium forms the ion Mg^{2+} when a magnesium atom loses two electrons. Show how you would calculate the energy required to remove two electrons from a mole of magnesium atoms. How much water (in grams) could have its temperature raised from room temperature (25 °C) to the boiling point (100 °C) by the energy needed to remove two electrons from a mole of magnesium atoms? Set up the calculation.

9. Explain how one can easily determine the ionization energy of the fluoride ion, F^-. Is this ionization energy endothermic or exothermic?

10. How can you determine the electron affinity of an Al^{3+} ion?

11. How can you show that $Na(g) + Cl(g)$ is more stable (of lower energy) than $Na^+(g) + Cl^-(g)$, assuming infinite distance of separation between atoms or ions? The potential energy of the Na^+ and Cl^- ions is inversely proportional to the distance between the ions and is given by the equation

$$E = \frac{z^+ z^- e^2}{4\pi d \epsilon_0^2}$$

where z^+ and z^- are the number of charges on the cation and anion, respectively, e is the electronic charge in units of coulombs, d is the distance between the ions in meters, and ϵ_0 is a constant which equals 8.85×10^{-12} $C^2 m^{-1} J^{-1}$. Describe how you would use this equation to determine the distance below which the energy of the Na^+ and Cl^- ions is lower than the energy of the neutral atoms.

12. What wavelength of light would be required to eject an electron from an atom of sodium? Set up the calculation.

13. Using the concepts of effective nuclear charge and sizes of electron shells in atoms, arrange the following atoms in order of increasing first ionization energies and explain your order: Al, Ba, Ca, Cs, and P.

14. Using effective nuclear charge and sizes of electron shells, explain why you can identify the larger atom in some of the following pairs but not in others.

   Sn and Se   S and Br   K and Sr   K and Mg

Use Figure 7.25 to confirm your predictions. For the pairs for which you could not predict the larger atom, use this figure to determine which of these two effects is dominant.

15. The five atomic orbitals shown below have three different principle quantum numbers which are consecutive integers. One of the principle quantum numbers is the lowest allowed one for that type of orbital. Identify the five atomic orbitals using the notation $1s$, $2s$, etc.

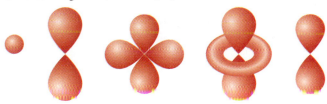

16. A certain paramagnetic element commonly forms two ions. The ion with the lower charge is diamagnetic while the one with the higher charge is paramagnetic. Which of the following elements and its ions is consistent with these facts: Ba, Cu, Fe, or S?

REVIEW QUESTIONS

Electromagnetic Radiation

7.1 In general terms, why do we call light *electromagnetic radiation*?

7.2 In general, what does the term *frequency* imply? What is meant by the term *frequency of light*? What symbol is used for it, and what is the SI unit (and symbol) for frequency?

7.3 What is meant by the term *wavelength* of light? What symbol is used for it?

7.4 Sketch a picture of a wave and label its wavelength and its amplitude.

7.5 Arrange the following regions of the electromagnetic spectrum in order of increasing wavelength (i.e., shortest wavelength → longest wavelength): microwave, TV, X-ray, ultraviolet, visible, infrared, gamma rays.

7.6 What wavelength range is covered by the *visible spectrum*?

7.7 Arrange the following colors of visible light in order of increasing wavelength: orange, green, blue, yellow, violet, red.

7.8 Write the equation that relates the wavelength and frequency of a light wave.

7.9 How is the frequency of a particular type of radiation related to the energy associated with it? (Give an equation, defining all symbols.)

7.10 What is a photon?

7.11 Show that the energy of a photon is given by the equation $E = hc/\lambda$.

7.12 Examine each of the following pairs and state which of the two has the higher *energy*.
(a) microwaves and infrared
(b) visible light and infrared
(c) ultraviolet light and X rays
(d) visible light and ultraviolet light

7.13 What is a quantum of energy?

Atomic Spectra
7.14 What is an atomic spectrum? How does it differ from a continuous spectrum?

7.15 What fundamental fact is implied by the existence of atomic spectra?

Bohr Atom and the Hydrogen Spectrum
7.16 Describe Niels Bohr's model of the structure of the hydrogen atom.

7.17 In qualitative terms, how did Bohr's model account for the atomic spectrum of hydrogen?

7.18 What is the term used to describe the lowest energy state of an atom?

7.19 In what way was Bohr's theory a success? How was it a failure?

Wave Nature of Matter
7.20 How does the behavior of very small particles differ from that of the larger, more massive objects that we meet in everyday life? Why don't we notice this same behavior for the larger, more massive objects?

7.21 Describe the phenomenon called diffraction. How can this be used to demonstrate that de Broglie's theory was correct?

7.22 What is the difference between a *traveling wave* and a *standing wave*?

Electron Waves in Atoms
7.23 What are the names used to refer to the theories that apply the matter–wave concept to electrons in atoms?

7.24 What is the term used to describe a particular waveform of a standing wave for an electron?

7.25 What are the two properties of orbitals in which we are most interested? Why?

Quantum Numbers
7.26 What are the allowed values of the principal quantum number?

7.27 What is the value for n for (a) the K shell and (b) the M shell?

7.28 Why does every shell contain an s subshell?

7.29 How many orbitals are found in (a) an s subshell, (b) a p subshell, (c) a d subshell, and (d) an f subshell?

7.30 If the value of m_ℓ for an electron in an atom is 2, could another electron in the same subshell have $m_\ell = -3$?

7.31 Suppose an electron in an atom has the following set of quantum numbers: $n = 2$, $\ell = 1$, $m_\ell = 1$, $m_s = +\frac{1}{2}$. What set of quantum numbers is impossible for another electron in this same atom?

Electron Spin
7.32 What physical property of electrons leads us to propose that they spin like a toy top?

7.33 What is the name of the magnetic property exhibited by atoms that contain unpaired electrons?

7.34 What is the Pauli exclusion principle? What effect does it have on the populating of orbitals by electrons?

7.35 What are the possible values of the spin quantum number?

Electron Configurations of Atoms
7.36 What do we mean by the term *electronic structure*?

7.37 Within any given shell, how do the energies of the s, p, d, and f subshells compare?

7.38 What fact about the energies of subshells was responsible for the apparent success of Bohr's theory about electronic structure?

7.39 How do the energies of the orbitals belonging to a given subshell compare?

7.40 Give the electron configurations of the elements in period 2 of the periodic table.

7.41 Give the correct electron configurations of (a) Cr and (b) Cu.

7.42 What is the correct electron configuration of silver?

7.43 How are the electron configurations of the elements in a given group similar? Illustrate your answer by writing shorthand configurations for the elements in Group VIA.

7.44 Define the terms *valence shell* and *valence electrons*.

Shapes of Atomic Orbitals
7.45 Why do we use probabilities when we discuss the position of an electron in the space surrounding the nucleus of an atom?

7.46 Sketch the approximate shape of (a) a 1s orbital and (b) a 2p orbital.

7.47 How does the size of a given type of orbital vary with n?

7.48 How are the p orbitals of a given p subshell oriented relative to each other?

7.49 What is a *nodal plane?*

7.50 On appropriate coordinate axes, sketch the shape of the following d orbitals: (a) d_{xy}, (b) $d_{x^2-y^2}$, (c) d_{z^2}.

Atomic and Ionic Size

7.51 What is the meaning of *effective nuclear charge?* How does the effective nuclear charge felt by the outer electrons vary going down a group? How does it change as we go from left to right across a period?

7.52 In what region of the periodic table are the largest atoms found? Where are the smallest atoms found?

7.53 Going from left to right in the periodic table, why are the size changes among the transition elements more gradual than those among the representative elements?

Ionization Energy

7.54 Define ionization energy. Why are ionization energies of atoms and positive ions endothermic quantities?

7.55 For oxygen, write an equation for the change associated with (a) its first ionization energy and (b) its third ionization energy.

7.56 Explain why ionization energy increases from left to right in a period and decreases from top to bottom in a group.

7.57 Why is an atom's second ionization energy always larger than its first ionization energy?

7.58 Why is the fifth ionization energy of carbon so much larger than its fourth?

7.59 Why is the first ionization energy of aluminum less than the first ionization energy of magnesium?

7.60 Why does phosphorus have a larger first ionization energy than sulfur?

Electron Affinity

7.61 Define *electron affinity.*

7.62 For sulfur, write an equation for the change associated with (a) its first electron affinity and (b) its second electron affinity. How should they compare?

7.63 Why does Cl have a more exothermic electron affinity than F? Why does Br have a less exothermic electron affinity than Cl?

7.64 Why is the second electron affinity of an atom always endothermic?

7.65 How is electron affinity related to effective nuclear charge? On this basis, explain the relative magnitudes of the electron affinities of oxygen and fluorine.

REVIEW PROBLEMS

Answers to problems whose numbers are printed in color are given in Appendix D.
More challenging problems are marked with asterisks.

Electromagnetic Radiation

7.66 What is the frequency in hertz of blue light having a wavelength of 430 nm?

7.67 A certain substance strongly absorbs infrared light having a wavelength of 6.85 μm. What is the frequency of this light in hertz?

7.68 Ozone protects Earth's inhabitants from the harmful effects of ultraviolet light arriving from the sun. This shielding is a maximum for UV light having a wavelength of 295 nm. What is the frequency in hertz of this light?

7.69 Radar signals are electromagnetic radiations in the microwave region of the spectrum. A typical radar signal has a wavelength of 3.19 cm. What is its frequency in hertz?

7.70 In New York City, radio station WCBS broadcasts its FM signal at a frequency of 101.1 megahertz (MHz). What is the wavelength of this signal in meters?

7.71 Sodium vapor lamps are often used in residential street lighting. They give off a yellow light having a frequency of 5.09×10^{14} Hz. What is the wavelength of this light in nanometers?

7.72 There has been some concern in recent times about possible hazards to people who live very close to high-volt-age electric power lines. The electricity in these wires oscillates at a frequency of 60 Hz, which is the frequency of any electromagnetic radiation that they emit. What is the wavelength of this radiation in meters? What is it in kilometers?

7.73 An X-ray beam has a frequency of 1.50×10^{18} Hz. What is the wavelength of this light in nanometers and in picometers?

7.74 Calculate the energy in joules of a photon of red light having a frequency of 4.0×10^{14} Hz. What is the energy of one mole of these photons?

7.75 Calculate the energy in joules of a photon of green light having a wavelength of 560 nm.

Atomic Spectra

7.76 In the spectrum of hydrogen, there is a line with a wavelength of 410.3 nm. (a) What color is this line? (b) What is its frequency? (c) What is the energy of each of its photons?

7.77 In the spectrum of sodium, there is a line with a wavelength of 589 nm. (a) What color is this line? (b) What is its frequency? (c) What is the energy of each of its photons?

7.78 Use the Rydberg equation to calculate the wavelength in nanometers of the spectral line of hydrogen for which

$n_2 = 6$ and $n_1 = 3$. Would we be expected to see the light corresponding to this spectral line? Explain your answer.

7.79 Use the Rydberg equation to calculate the wavelength in nanometers of the spectral line of hydrogen for which $n_2 = 5$ and $n_1 = 2$. Would we be expected to see the light corresponding to this spectral line? Explain your answer.

7.80 Calculate the wavelength of the spectral line produced in the hydrogen spectrum when an electron falls from the tenth Bohr orbit to the fourth. In which region of the electromagnetic spectrum (UV, visible, or infrared) is the line?

7.81 Calculate the energy in joules and the wavelength in nanometers of the spectral line produced in the hydrogen spectrum when an electron falls from the fourth Bohr orbit to the first. In which region of the electromagnetic spectrum (UV, visible, or infrared) is the line?

Quantum Numbers

7.82 What is the letter code for a subshell with (a) $\ell = 1$ and (b) $\ell = 3$?

7.83 What is the value of ℓ for (a) an f orbital and (b) a d orbital?

7.84 Give the values of n and ℓ for the following subshells: (a) $3s$, (b) $5d$.

7.85 Give the values of n and ℓ for the following subshells: (a) $4p$, (b) $6f$.

7.86 For the shell with $n = 6$, what are the possible values of ℓ?

7.87 In a particular shell, the largest value of ℓ is 7. What is the value of n for this shell?

7.88 What are the possible values of m_ℓ for a subshell with (a) $\ell = 1$ and (b) $\ell = 3$?

7.89 If the value of ℓ for an electron in an atom is 5, what are the possible values of m_ℓ that this electron could have?

7.90 If the value of m_ℓ for an electron in an atom is -4, what is the smallest value of ℓ that the electron could have? What is the smallest value of n that the electron could have?

7.91 How many orbitals are there in an h subshell ($\ell = 5$)? What are their values of m_ℓ?

7.92 Give the complete set of quantum numbers for all of the electrons that could populate the $2p$ subshell of an atom.

7.93 Give the complete set of quantum numbers for all of the electrons that could populate the $3d$ subshell of an atom.

7.94 In an antimony atom, how many electrons have $\ell = 1$? How many electrons have $\ell = 2$ in an antimony atom?

7.95 In an atom of barium, how many electrons have (a) $\ell = 0$ and (b) $m_\ell = 1$?

Electron Configurations of Atoms

7.96 Predict the electron configurations of (a) S, (b) K, (c) Ti, and (d) Sn.

7.97 Predict the electron configurations of (a) As, (b) Cl, (c) Ni, and (d) Si.

7.98 Which of the following atoms in their ground states are paramagnetic: (a) Mn, (b) As, (c) S, (d) Sr, (e) Ar?

7.99 Which of the following atoms in their ground states are diamagnetic: (a) Ba, (b) Se, (c) Zn, (d) Si?

7.100 How many unpaired electrons would be found in the ground state of (a) Mg, (b) P, and (c) V?

7.101 How many unpaired electrons would be found in the ground state of (a) Cs, (b) S, and (c) Ni?

7.102 Write the shorthand electron configurations for (a) Ni, (b) Cs, (c) Ge, (d) Br, and (e) Bi.

7.103 Write the shorthand electron configurations for (a) Al, (b) Se, (c) Ba, (d) Sb, and (e) Gd.

7.104 Draw complete orbital diagrams for (a) Mg and (b) Ti.

7.105 Draw complete orbital diagrams for (a) As and (b) Ni.

7.106 Draw orbital diagrams for the shorthand configurations of (a) Ni, (b) Cs, (c) Ge, and (d) Br.

7.107 Draw orbital diagrams for the shorthand configurations of (a) Al, (b) Se, (c) Ba, and (d) Sb.

7.108 What is the value of n for the valence shells of (a) Sn, (b) K, (c) Br, and (d) Bi?

7.109 What is the value of n for the valence shells of (a) Al, (b) Se, (c) Ba, and (d) Sb?

7.110 Give the configuration of the valence shell for (a) Na, (b) Al, (c) Ge, and (d) P.

7.111 Give the configuration of the valence shell for (a) Mg, (b) Br, (c) Ga, and (d) Pb.

7.112 Draw the orbital diagram for the valence shell for (a) Na, (b) Al, (c) Ge, and (d) P.

7.113 Draw the orbital diagram for the valence shell of (a) Mg, (b) Br, (c) Ga, and (d) Pb.

Atomic Properties

7.114 If the core electrons were 100% effective at shielding the valence electrons from the nuclear charge and the valence electrons provided no shielding for each other, what would be the effective nuclear charge felt by a valence electron in (a) Na, (b) S, (c) Cl?

7.115 If the core electrons were 100% effective at shielding the valence electrons from the nuclear charge and the valence electrons provided no shielding for each other, what would be the effective nuclear charge felt by a valence electron in (a) Mg, (b) Si, (c) Br?

7.116 Choose the larger atom in each pair: (a) Na or Si; (b) P or Sb.

7.117 Choose the larger atom in each pair: (a) Al or Cl; (b) Al or In.

7.118 Choose the largest atom among the following: Ge, As, Sn, Sb.

7.119 Place the following in order of increasing size: N^{3-}, Mg^{2+}, Na^+, Ne, F^-, O^{2-}.

7.120 Choose the larger particle in each pair: (a) Na or Na^+; (b) Co^{3+} or Co^{2+}; (c) Cl or Cl^-.

7.121 Choose the larger particle in each pair: (a) S or S^{2-}; (b) Al^{3+} or Al; (c) Au^+ or Au^{3+}.

7.122 Choose the atom with the larger ionization energy in each pair: (a) B or C; (b) O or S; (c) Cl or As.

7.123 Choose the atom with the larger ionization energy in each pair: (a) Li or Cs ; (b) Al or Si; (c) F or O.

7.124 Choose the atom with the more exothermic electron affinity in each pair: (a) Cl or Br; (b) Se or Br.

7.125 Choose the atom with the more exothermic electron affinity in each pair: (a) P or As; (b) Si or Ga.

7.126 Use the periodic table to select the element in the following list for which there is the largest difference between the second and third ionization energies: Na, Mg, Al, Si, P, Se, Cl.

7.127 Use the periodic table to select the element in the following list for which there is the largest difference between the fourth and fifth ionization energies: Na, Mg, Al, Si, P, Se, Cl.

ADDITIONAL EXERCISES

7.128 Microwaves are used to heat food in microwave ovens. The microwave radiation is absorbed by moisture in the food. This heats the water, and as the water becomes hot, so does the food. How many photons having a wavelength of 3.00 mm would have to be absorbed by 1.00 g of water to raise its temperature by 1.00 °C?

7.129 In the spectrum of hydrogen, there is a line with a wavelength of 410.3 nm. Use the Rydberg equation to calculate the value of n for the higher energy Bohr orbit involved in the emission of this light. Assume the value of n for the lower energy orbit equals 2.

7.130 Calculate the wavelength in nanometers of the shortest wavelength of light emitted by a hydrogen atom.

7.131 Which of the following electronic transitions could lead to the emission of light from an atom?

$$1s \longrightarrow 4p \longrightarrow 3d \longrightarrow 5f \longrightarrow 4d \longrightarrow 2p$$

*7.132 A neon sign is a gas discharge tube in which electrons traveling from the cathode to the anode collide with neon atoms in the tube and knock electrons off of them. As electrons return to the neon ions and drop to lower energy levels, light is given off. How fast would an electron have to be moving to eject an electron from an atom of neon, which has a first ionization energy equal to 2080 kJ mol^{-1}?

*7.133 How many grams of water could have its temperature raised by 5.0 °C by a mole of photons that have a wavelength of (a) 600 nm and (b) 300 nm?

*7.134 It has been found that when the chemical bond between chlorine atoms in Cl_2 is formed, 328 kJ is released per mole of Cl_2 formed. What is the wavelength of light that would be required to break chemical bonds between chlorine atoms?

7.135 Calculate the wavelengths of the lines in the spectrum of hydrogen that result when an electron falls from a Bohr orbit with (a) $n = 5$ to $n = 1$, (b) $n = 4$ to $n = 2$, and (c) $n = 6$ to $n = 4$. In which regions of the electromagnetic spectrum are these lines?

7.136 What, if anything, is wrong with the following electron configurations for atoms in their ground states?
(a) $1s^2 2s^1 2p^3$ (c) $1s^2 2s^2 2p^4$
(b) $[Kr] 3d^7 4s^2$ (d) $[Xe] 4f^{14} 5d^8 6s^1$

7.137 Suppose students gave the following orbital diagrams for the $2s$ and $2p$ subshells in the ground state of an atom. What, if anything, is wrong with them? Are any of these electron distributions impossible?

(a) ⃝ ⇅⇅⇅   (c) ↑ ↑↑↑

(b) ↑ ↓⃝⃝   (d) ⇈ ↑↑⃝

7.138 How many electrons are in p orbitals in an atom of germanium?

7.139 What are the quantum numbers of the electrons that are lost by an atom of iron when it forms the ion Fe^{2+}?

*7.140 The removal of an electron from the hydrogen atom corresponds to raising the electron to the Bohr orbit that has $n = \infty$. On the basis of this statement, calculate the ionization energy of hydrogen in units of (a) joules per atom and (b) kilojoules per mole.

7.141 Use orbital diagrams to illustrate what happens when an oxygen atom gains two electrons. On the basis of what you have learned about electron affinities and electron configurations, why is it extremely difficult to place a third electron on the oxygen atom?

7.142 From the data available in this chapter, determine the ionization energy of (a) F^-, (b) O^-, and (c) O^{2-}. Are any of these energies exothermic?

7.143 For an oxygen atom, which requires more energy, the addition of two electrons or the removal of one electron?

Greenhouse Gases and Global Warming

4

Why Earth Stays Relatively Warm

Earth constantly receives energy, mostly from the sun, but some is from radioactive processes within Earth. It also radiates energy to outer space. It must do so at exactly the same average rate as it receives energy or Earth's average temperature would change drastically.

What modulates Earth's temperature is the atmosphere. It acts somewhat as the glass of a greenhouse, an analogy first proposed in 1822 by a French mathematician, Jean Fourier.[1] Without the atmosphere, Earth would be as cold and lifeless as Mars appears to be today. On the other hand, with too much of certain constituents in the atmosphere, Earth would heat up to a point at which life would also be impossible.

The insulating effect of the atmosphere is called the **greenhouse effect.** It's caused not by oxygen and nitrogen, which are diatomics and air's chief components, but by trace polyatomic substances called **greenhouse gases,** chiefly water vapor, carbon dioxide, methane, dinitrogen monoxide (N_2O), and some synthetic chemicals. The greenhouse gases absorb outgoing energy and reradiate some of it back to Earth.[2] This reradiation causes the insulating effect, making Earth's surface an estimated 33 °C warmer than it would otherwise be (see Figure 4a). Although the greenhouse gases are largely of natural origin and perform an essential service, the concern to-day is about their concentrations becoming steadily higher as a result of human activities.

Carbon Dioxide as a Greenhouse Gas

The atmosphere annually receives an estimated 9×10^{15} mol of CO_2, largely from three sources: 93% from the respiration of plants and animals, 2% from forest fires and the burning of other vegetable matter, and 5% from the burning of fossil fuels: oil, coal, and natural gas.

Two natural "sinks," each contributing about equally, remove over 95% of the released CO_2. One is *photosynthesis,* the conversion of CO_2 and H_2O in plants to plant materials, using solar energy and chlorophyll (the green pigment in leaves). The other sink is the uptake of CO_2 by calcium and magnesium ions in the oceans, forming carbonate deposits. These activities still leave 5% of the CO_2 in the air, a small but important net amount of greenhouse gas.

The increased rate of burning of fossil fuels throughout the twentieth century is chiefly responsible for the extra CO_2. From about 1900, the average atmospheric CO_2 level has risen from 280 to 345 ppm. (One part per million, or ppm, is the same as 1 mg/L.) Some scientists see the level increasing to 600 ppm in another 50 to 75 years, maybe in less time. Such a change could alone increase the average global temperature by 0.5 to 1.5 °C.

The Greenhouse Effect and Climate

Small changes in the *average* Earth temperature could cause major changes in world climate patterns, placing millions of people at risk of hunger, water shortages, and coastal flooding. Tropical and subtropical regions would expand, opening larger populations to malaria, yellow fever, encephalitis, and dengue fever. Tropical storms and hurricanes would intensify and perhaps destroy the coral reefs that protect coastlines. Huge ecosystems, like the Arctic Ocean and nearby Arctic tundra, as well as mangrove swamps and other coastal wetlands, would be under severe pressure. If global warming caused the entire volume of the West Antarctic ice sheet (3.8×10^6 km^3) to be released to the ocean, the global mean rise in sea level would be 4 to 6 m, enough to flood all port facilities now existing, most of Florida and Louisiana, and many of the world's largest cities.

---

[1]The analogy is not too good. The glass of a real greenhouse primarily prevents the wind from moving cold air in and warm air out. Still, the glass does absorb some outward bound heat and reradiate it back inside.

[2]Both methane and dinitrogen oxide arise by the action of bacteria on materials in the soil, including fertilizers. Together they make up an estimated 20% of greenhouse gases. Dinitrogen oxide also forms in mufflers equipped with catalytic converters. Synthetic chemicals include methyl bromide (a pesticide) and chlorofluorocarbons, although they are being phased out.

chemicals in our world

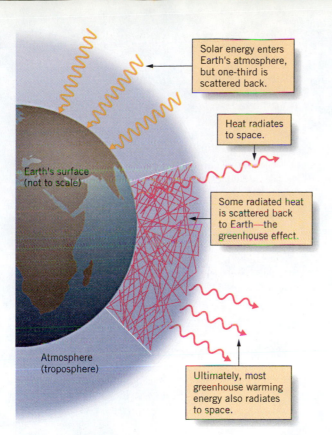

Figure 4a Greenhouse gases insulate Earth. (Adapted with permission from D. B. Botkin and E. A. Keller, *Environmental Science*, 1995, John Wiley & Sons, Inc., New York. Developed by M. S. Manalis and E. A. Keller.)

It isn't that Earth has never before undergone large changes in overall climate with successful adaptations by most living species. But an accelerated greenhouse warming would entail an unprecedented *rate* of change with which vulnerable species could not cope.

Atmospheric Haze and the Greenhouse Effect

Some of the possible greenhouse warming is being canceled by haze particles that reflect incoming solar radiation back to space and so work to cool the planet. Haze particles are largely colloidal aerosols of moisture containing sulfuric acid. This forms from sulfur oxides made by the combustion of sulfur-bearing fuels and vegeta-

tion as well as from several huge volcanic eruptions of recent years.

Where Are We Now?

What complicates forecasts of global temperatures are their normal fluctuations brought about by El Niño surges and particles thrown up by volcanoes and forest fires. Human influence on global climate is real, however. In a 1995 report, the United Nations Intergovernmental Panel on Climate Change (IPCC) predicted an upward trend of 0.1 °C per decade, which may be compared to a rate of 0.04 °C per decade in the early 1900s. The World Meteorological Organization reported that the average global temperature in the year 1998 was 15 °C (58 °F), or an amazing one full degree warmer than the average annual temperature of 1991 to 1998 and a quarter of a degree warmer than 1997. Of the ten warmest years on record, eight have occurred between 1990 and 1999.

The Importance of Carbon Dioxide. CO_2 is responsible for about 55% of the total increase in the concentration of the greenhouse gases—more than all the others combined. A worldwide, phased reduction in CO_2 emissions was, therefore, the subject of international negotiations in the 1990s. A United Nations conference in Kyoto, Japan, in 1997, called for a 5.2% reduction in greenhouse gas emissions, relative to 1990 levels, by the year 2010.

Some scientists say that it's later than we think. They began urging in the late 1990s that we take steps now not just to reduce CO_2 emissions but also to get ready for and to adapt to the coming changes in world climate. Even heroic reductions in CO_2 emissions will have little positive effect on populations at risk of hunger, tropical diseases, water shortages, and coastal flooding.

Questions

1. Name the greenhouse gases and describe their principal origins.

2. Explain briefly how the greenhouse effect works. What happens that tends to promote global warming?

3. What "sinks" are at work to reduce the CO_2 level of the atmosphere?

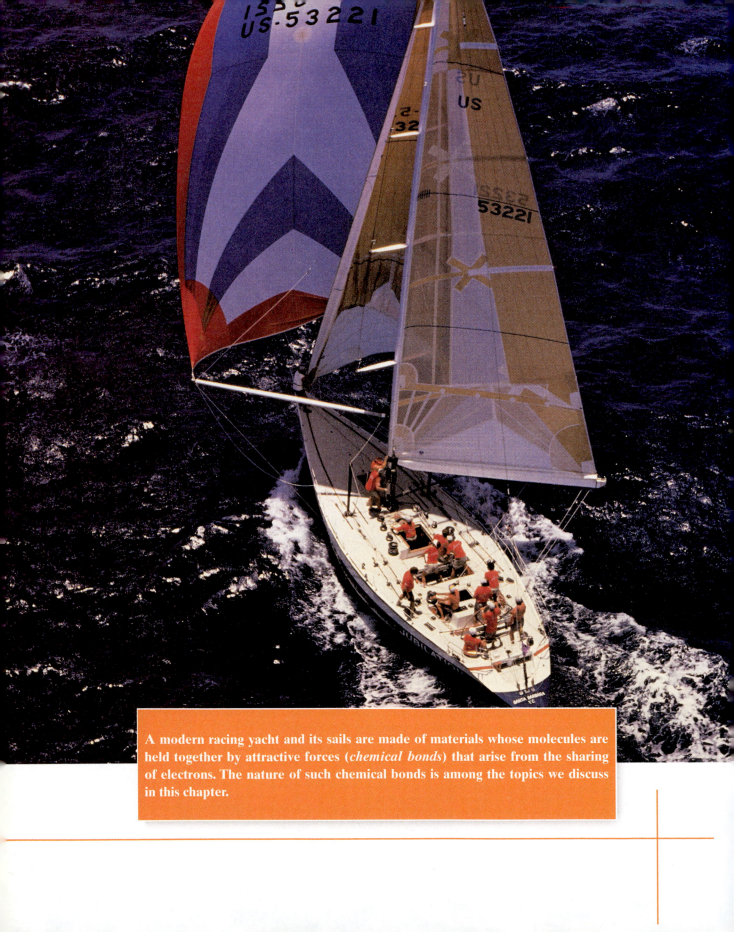

A modern racing yacht and its sails are made of materials whose molecules are held together by attractive forces (*chemical bonds*) that arise from the sharing of electrons. The nature of such chemical bonds is among the topics we discuss in this chapter.

Chemical Bonding: General Concepts

This Chapter in Context In Chapter 7 you learned that our understanding of the electronic structures of atoms evolved. For example, the Bohr model of the atom was the first to successfully provide an explanation for the atomic spectrum of hydrogen. However, the failure of the model to explain the spectra of more complex atoms led to its ultimate replacement by another theory called wave mechanics. By treating the electron as a wave, scientists were able to devise an entirely new description of matter, one that gives us the electron configurations of atoms and an ability to explain the periodic variations of atomic properties such as size, ionization energy, and electron affinity.

With this understanding of the electronic structures of atoms, we can now proceed to learn how atoms combine to form compounds through the formation of *chemical bonds*—the attractions that bind atoms to each other. As with electronic structure, models of chemical bonding have also evolved, and in this chapter we will begin with relatively simple theories. Although more complicated theories exist (some of which we will explore in Chapter 9), the basic concepts you will study in this chapter still find many useful applications in modern chemical thought.

8.1 Electron Transfer and the Formation of Ionic Compounds

There are two principal classes of bonding attractions. One involves the transfer of electrons between atoms; this produces ions and is called *ionic bonding*. The other occurs in molecules and involves the sharing of electrons; it's called *covalent bonding*. Ionic bonding is simpler to understand, so we will discuss it first.

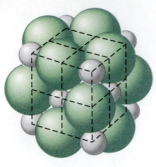

Packing of ions in NaCl.

Keep in mind the relationship between potential energy changes and endothermic and exothermic processes:

endothermic ⇔ increase in PE

exothermic ⇔ decrease in PE

Remember, enthalpy is a state function, so its change is determined only by the initial and final states, not by the path between them.

The name lattice energy comes from the word *lattice,* which is used to describe the regular pattern of ions or atoms in a crystal.

Remember, when things that attract one another come closer together, their potential energy decreases.

Conditions That Favor the Formation of Ionic Compounds

In Chapter 2 you learned that ionic compounds are formed when metals react with nonmetals. Among the examples discussed was sodium chloride, table salt. You learned that when this compound is formed from its elements, each sodium atom loses one electron to form a sodium ion, Na^+, and each chlorine atom gains one electron to become a chloride ion, Cl^-. Once formed, these ions become tightly packed together, as illustrated in the margin, because their opposite charges attract. *This attraction between positive and negative ions in an ionic compound is what we call an* **ionic bond.**

The reason Na^+ and Cl^- ions attract each other is easy to understand. But *why* are electrons transferred between these and other atoms? *Why* does sodium form Na^+ and not Na^- or Na^{2+}? And *why* does chlorine form Cl^- instead of Cl^+ or Cl^{2-}? To answer these questions we must consider a number of factors that are related to the potential energy of the system of reactants and products. This is because *for any stable compound to form from its elements, there must be a net lowering of the potential energy.* In other words, the reaction must be exothermic.

Importance of the Lattice Energy

To appreciate the importance of the lattice energy, we call on the law of conservation of energy and Hess's law. This will allow us to divide the overall equation for the reaction of the elements to give the compound into a number of simple steps. Then, if the sum of the steps yields the overall reaction, the sum of the energy changes equals the overall energy change.

Figure 8.1 is an enthalpy diagram that illustrates two paths that take us from the elements, $Na(s)$ and $\frac{1}{2}Cl_2(g)$, to the product, $NaCl(s)$. Because both paths take us from the same initial state (the elements) to the same final state (solid NaCl), the law of conservation of energy demands that the net enthalpy change must be the same along each path.

The direct path at the bottom has as its enthalpy change the heat of formation of NaCl, ΔH_f°. The alternative path is divided into a number of steps. The first two steps, both of which are endothermic, change $Na(s)$ and $Cl_2(g)$ into gaseous atoms, $Na(g)$ and $Cl(g)$. The next two steps change these atoms to ions, first by the endothermic ionization energy (IE) of Na followed by the exothermic electron affinity (EA) of Cl. Notice that at this point, if we add all the energy changes, the ions are at a considerably higher energy than the reactants. If these were the only energy terms involved in the formation of NaCl, the heat of formation would be endothermic and the compound would be unstable; it could not be formed by direct combination of the elements.

The reason ionic compounds such as NaCl are stable is because of the final energy term in the alternative path. It is called the **lattice energy,** defined as *the amount that the potential energy of the system decreases when the ions in one mole of the compound are brought from a gaseous state to the positions the ions occupy in a crystal of the compound.* The potential energy is lowered because the ions have a net attraction for each other. As they are being brought closer together, from isolated gaseous ions to the tightly packed arrangement found in the solid, the potential energy drops. You learned earlier that when the potential energy decreases, the change is exothermic. This is why the lattice energy appears with a negative sign in Figure 8.1.

As you can see in Figure 8.1, the lattice energy is the largest of all the contributing energy factors in the formation of NaCl. It is able to overcome the net increase in potential energy associated with the formation of the isolated ions

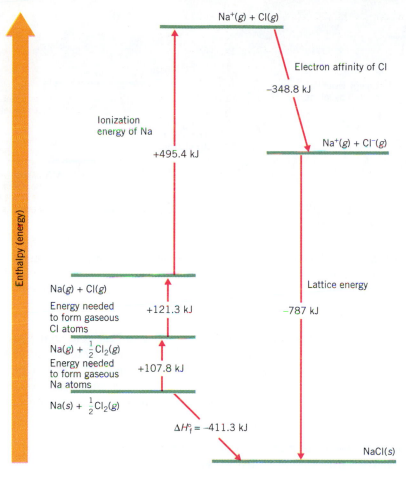

Figure 8.1 *Energy changes in the formation of NaCl, presented in the form of an enthalpy diagram.* The path labeled ΔH_f° leads directly to NaCl. The upper path involves the formation of gaseous atoms from the elements, formation of ions from the gaseous atoms, and finally the condensation of the gaseous ions to give solid NaCl. Both paths yield the same net energy change.

and causes the formation of the compound to be exothermic overall, which it must be for the compound to be stable.

What we have seen here for the formation of NaCl applies to other ionic compounds as well. Formation of the gaseous ions from the elements involves a *net increase* in potential energy; condensing the ions to give the ionic solid yields a *decrease* in potential energy corresponding to the lattice energy. If the decrease in PE is larger than the increase, the ionic compound can form from the elements. However, if the opposite is true, the ionic compound is unstable. It can't be made from the elements, and if made by some other reaction, it will tend to decompose into the elements. Figure 8.2 summarizes this in a somewhat simplified way.

Why Metals Form Cations and Nonmetals Form Anions

Having discussed the energy factors involved in the formation of an ionic compound, we can now understand why metals tend to form positive ions and nonmetals tend to form negative ions. At the left of the periodic table are the metals—elements with small ionization energies and electron affinities. Relatively little energy is needed to remove electrons from them to produce positive ions. At the upper right of the periodic table are the nonmetals—elements with large ionization energies and generally exothermic electron affinities. It is quite difficult to remove electrons from these elements, but sizable amounts of energy are released when they gain electrons. On an energy basis, therefore, it is least "expensive"

For an ionic compound to be formed from its elements, the exothermic lattice energy must be larger than the endothermic combination of factors involved in the formation of the ions themselves, which primarily involve the IE of the metal and the EA of the nonmetal.

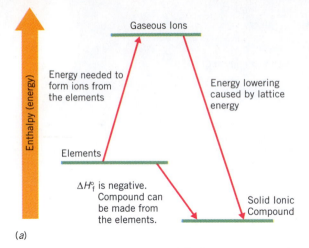

(a)

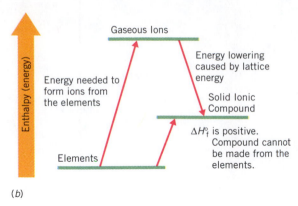

(b)

Figure 8.2 *The importance of the lattice energy in the formation of ionic compounds.* (*a*) The energy lowering of the lattice energy is larger than the energy needed to form the ions, so the ionic compound is stable and can be made from the elements. (*b*) The energy lowering of the lattice energy is not large enough to compensate for the energy needed to form the ions, so the ionic compound is unstable.

to form a cation from a metal and an anion from a nonmetal, so it is relatively easy for the energy-lowering effect of the lattice energy to exceed the net energy-raising effect of the ionization energy and electron affinity. In fact, metals combine with nonmetals to form ionic compounds simply because ionic bonding is favored energetically over other types whenever atoms with small ionization energies combine with atoms that have large exothermic electron affinities.

Changes in Electron Configurations when Ions Form: The Octet Rule

Let's look now at how the electronic structures of the elements affect the kinds of ions they form. We begin by examining what happens when sodium loses an electron. The electron configuration of Na is

$$\text{Na} \quad 1s^2 2s^2 2p^6 3s^1$$

The electron that is lost is the one least tightly held, which is the single outer $3s$ electron. The electronic structure of the Na^+ ion, then, is

$$\text{Na}^+ \quad 1s^2 2s^2 2p^6$$

Notice that this is identical to the electron configuration of the noble gas neon. We say the Na^+ ion has achieved a *noble gas configuration.*

The removal of the first electron from Na does not require much energy because the first ionization energy of sodium is small. Therefore, an input of energy equal to the first ionization energy can be easily recovered by the exothermic lattice energy of ionic compounds that contain the Na^+ ion. However, removal of a second electron from sodium is *very* difficult because it involves breaking into the $2s^2 2p^6$ core. The second ionization energy of Na is enormous, as indicated in the margin. As a result, the amount of energy required to create a Na^{2+} ion is *much* larger than the amount of energy that can be recovered by the lattice energy, so overall the formation of a compound containing Na^{2+} is endothermic and, therefore, energetically unfavorable. This is why we never observe compounds that contain Na^{2+}, and why sodium stops losing electrons once it has achieved a noble gas configuration.

For sodium:

1st IE = 496 kJ/mol

2nd IE = 4563 kJ/mol

A somewhat similar situation exists for other metals, too. Consider calcium, for example. You learned in Chapter 2 that this metal forms ions with a 2+ charge. This means that when a calcium atom reacts, it loses its two outermost electrons (giving it the same electron configuration as argon).

$$\text{Ca} \qquad 1s^2 2s^2 2p^6 3s^2 3p^6 4s^2$$

$$\text{Ca}^{2+} \quad 1s^2 2s^2 2p^6 3s^2 3p^6$$

The two $4s$ electrons of Ca are not held too tightly, so the total amount of energy that must be invested to remove them (the sum of the first and second ionization energies) can be recovered easily by the large lattice energy of a Ca^{2+} compound. However, the removal of yet another electron from calcium to form Ca^{3+} requires breaking into the noble gas core. As in the case of sodium, a tremendous amount of energy is needed to accomplish this, much more than can be regained by the lattice energy of a Ca^{3+} compound. Therefore, a calcium atom loses just two electrons when it reacts.

For calcium:

1st IE = 590 kJ/mol

2nd IE = 1146 kJ/mol

3rd IE = 4940 kJ/mol

In the case of sodium and calcium, we find that the stability of the noble gas core that lies below the outer shell of electrons of these metals effectively limits the number of electrons that they lose, so the ions that are formed have noble gas electron configurations.

Nonmetals also tend to have noble gas configurations when they form anions. For example, when a chlorine atom reacts, it gains one electron. For the chlorine atom we have

$$\text{Cl} \quad 1s^2 2s^2 2p^6 3s^2 3p^5$$

and when an electron is gained, its configuration becomes

$$\text{Cl}^- \quad 1s^2 2s^2 2p^6 3s^2 3p^6$$

At this point, electron gain ceases, because if another electron were to be added, it would have to enter an orbital in the next higher shell. With oxygen, a similar situation exists. The formation of the oxide ion, O^{2-}, is endothermic, as you learned in the previous chapter.

$$\text{O} \, (1s^2 2s^2 2p^4) + 2e^- \longrightarrow O^{2-} \, (1s^2 2s^2 2p^6) \qquad EA(\text{net}) = +703 \text{ kJ/mol}$$

However, this energy input is modest and can be overcome without much difficulty by the large lattice energies of ionic metal oxides. But we never observe the formation of O^{3-} because, once again, the third electron would have to enter an orbital in the next higher shell, and this is *very* energetically unfavorable.

Summary

We see here that a balance of energy factors causes many atoms to form ions that have a noble gas electron configuration. Historically, this is expressed in the form of a generalization: *When they form ions, atoms of most of the representative elements tend to gain or lose electrons until they have obtained a configuration identical to that of the nearest noble gas.* Because all the noble gases except helium have outer shells with eight electrons, this rule has become known as the **octet rule,** which can be stated as follows: *Atoms tend to gain or lose electrons until they have achieved an outer shell that contains an* **octet of electrons** *(eight electrons).* Sodium and calcium achieve an octet by emptying their valence shells; chlorine and oxygen achieve an octet by gaining enough electrons to reach a noble gas configuration.

Tools

Electron configurations of ions of the representative elements

Hydrogen cannot obey the octet rule because its outer shell can contain just two electrons.

Exceptions to the Octet Rule

The octet rule, as applied to ionic compounds, really works well only for the cations of the Group IA and IIA metals and for the anions of the nonmetals. It does not work so well for the transition metals and post-transition metals (the metals that follow a row of transition metals).

To obtain the correct electron configuration of the cations of these metals, we apply the following rules:

Electron configurations of ions of transition and post-transition metals

Applying these rules to the metals of Groups IA and IIA also gives the correct electron configurations.

1. The first electrons to be lost by an atom or ion are *always* those from the shell with the largest value of n (i.e., the outer shell).

2. As electrons are removed from a given shell, they come from the highest energy occupied subshell first, before any are removed from a lower energy subshell. Within a given shell, the energies of the subshells vary as follows: $s < p < d < f$. This means that f is emptied before d, which is emptied before p, which is emptied before s.

Let's look at two examples that illustrate these rules.

Tin (a post-transition metal) forms two ions, Sn^{2+} and Sn^{4+}. The electron configurations are

$$Sn \quad [Kr]\,4d^{10}5s^25p^2$$

$$Sn^{2+} \quad [Kr]\,4d^{10}5s^2$$

$$Sn^{4+} \quad [Kr]\,4d^{10}$$

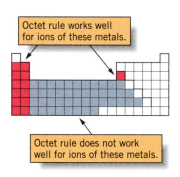

Octet rule works well for ions of these metals.

Octet rule does not work well for ions of these metals.

Notice that the Sn^{2+} ion is formed by the loss of the higher energy $5p$ electrons first. Then, further loss of the two $5s$ electrons gives the Sn^{4+} ion. However, neither of these ions has a noble gas configuration.

For the transition elements, the first electrons lost are the s electrons of the outer shell. Then, if additional electrons are lost, they come from the underlying d subshell. An example is iron, which forms the ions Fe^{2+} and Fe^{3+}. The element iron has the electron configuration

$$Fe \quad [Ar]\,3d^64s^2$$

Iron loses its $4s$ electrons fairly easily to give Fe^{2+}, which has the electron configuration

$$Fe^{2+} \quad [Ar]\,3d^6$$

The Fe^{3+} ion forms when another electron is removed, this time from the $3d$ subshell.

$$Fe^{3+} \quad [Ar]\,3d^5$$

Iron is able to form Fe^{3+} because the $3d$ subshell is close in energy to the $4s$, so it is not very difficult to remove the third electron. Notice once again that the first electrons to be removed come from the shell with the largest value of n (the $4s$ subshell). Then, after this shell is emptied, the next electrons are removed from the shell below.

Because so many of the transition elements are able to form ions in a way similar to that of iron, the ability to form more than one positive ion is usually cited as one of the characteristic properties of the transition elements. Frequently, one of the ions formed has a 2+ charge, which arises from the loss of the two outer s electrons. Ions with larger positive charges result when additional d electrons are lost. Unfortunately, it is not easy to predict exactly which ions can form for a given transition metal, nor is it simple to predict their relative stabilities.

How do the electron configurations change (a) when a nitrogen atom forms the N^{3-} ion and (b) when an antimony atom forms the Sb^{3+} ion?

Analysis: For the nonmetals, you've learned that the octet rule does work, so the ion that is formed by nitrogen will have a noble gas configuration.

When a cation of a representative element is formed, electrons are removed from the outer shell of the atom (the shell with the largest value of the principal quantum number, n). Within a given shell, electrons are always removed first from the subshell highest in energy.

Solution: (a) The electron configuration for nitrogen is

$$N \quad [He]\, 2s^2 2p^3$$

To form N^{3-}, three electrons are gained. These enter the $2p$ subshell because it is the lowest available energy level. The configuration for the ion is therefore

$$N^{3-} \quad [He]\, 2s^2 2p^6$$

(b) Let's begin with the ground state electron configuration for antimony.

$$Sb \quad [Kr]\, 4d^{10} 5s^2 5p^3$$

To form the Sb^{3+} ion, three electrons must be removed. These will come from the outer shell, which has $n = 5$. Recall that within this shell, the energies of the sub-shells increase in the order $s < p < d < f$. Therefore, the $5p$ subshell is higher in energy than the $5s$, so all three electrons are removed from the $5p$. This gives

$$Sb^{3+} \quad [Kr]\, 4d^{10} 5s^2$$

EXAMPLE 8.1

Writing Electron Configurations of Ions

What is the electron configuration of the V^{3+} ion? Give the orbital diagram for the ion.

Analysis: We will begin with the electron configuration of the neutral atom and then remove three electrons to obtain the electron configuration of the ion.

Solution: The abbreviated electron configuration of vanadium is

$$V \quad [Ar]\, 3d^3 4s^2$$

Notice that we've written the configuration showing the outer shell $4s$ electrons farthest to the right. To form the V^{3+} cation, three electrons must be removed from the neutral atom. The first two come from the $4s$ subshell and the third comes from the $3d$. This gives

$$V^{3+} \quad [Ar]\, 3d^2$$

For the orbital diagram, we have

$$V^{3+} \quad [Ar]\; \uparrow\uparrow\bigcirc\bigcirc\bigcirc$$
$$3d$$

Note that we've shown all five orbitals of the $3d$ subshell, even though only two of the orbitals are occupied.

EXAMPLE 8.2

Writing Electron Configurations of Ions

Practice Exercise 1

How do the electron configurations change when a chromium atom forms the following ions: (a) Cr^{2+}, (b) Cr^{3+}, (c) Cr^{6+}? ◆

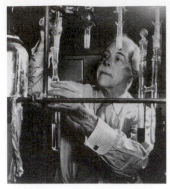

Gilbert N. Lewis, chemistry professor at the University of California, helped develop theories of chemical bonding. In 1916, he proposed that atoms form bonds by sharing pairs of electrons between them.

Lewis symbols

This is one of the advantages of the U.S. system for numbering groups in the periodic table.

8.2 Electron Bookkeeping: Lewis Symbols

In the last section you saw how the valence shells of atoms change when electrons are transferred during the formation of ions. We will soon see that many atoms share their valence electrons with each other when they form covalent bonds. In these discussions of bonding it is useful to be able to keep track of valence electrons. To help us do this, we use a simple bookkeeping device called Lewis symbols, named after their inventor, a famous American chemist, G. N. Lewis (1875–1946).

To draw the **Lewis symbol** for an element, we write its chemical symbol surrounded by a number of dots (or some other similar mark), which represent the atom's valence electrons. For example, the element lithium, which has one valence electron in its $2s$ subshell, has the Lewis symbol

$$\text{Li}\cdot$$

In fact, each element in Group IA has a similar Lewis symbol, because each has only one valence electron. The Lewis symbols for all of the Group IA metals are

$$\text{Li}\cdot \quad \text{Na}\cdot \quad \text{K}\cdot \quad \text{Rb}\cdot \quad \text{Cs}\cdot$$

The Lewis symbols for the eight A-group elements of period 2 are[1]

Group	IA	IIA	IIIA	IVA	VA	VIA	VIIA	VIIIA
Symbol	Li·	·Be·	·Ḃ·	·Ċ·	·N̈·	·Ö:	·F̈:	:N̈e:

The elements below each of these in their respective groups have identical Lewis symbols except, of course, for the chemical symbol of the element. Notice that when an atom has more than four valence electrons, the additional electrons are shown to be paired with others. Also notice that *for the representative elements, the group number is equal to the number of valence electrons* when the U.S. convention for numbering groups in the periodic table is followed.

What is the Lewis symbol for arsenic, As?

Solution: Arsenic is in Group VA and therefore has five valence electrons. The first four are placed around the symbol for arsenic as follows:

$$\cdot \overset{\textstyle\cdot}{\underset{\textstyle\cdot}{\text{As}}} \cdot$$

The fifth electron is paired with one of the first four. This gives

$$\cdot \overset{\textstyle\cdot}{\underset{\textstyle\cdot}{\text{As}}} :$$

The location of the fifth electron doesn't really matter, so equally valid Lewis symbols are

$$\cdot \overset{\textstyle\cdot\cdot}{\text{As}} \cdot \quad \text{or} \quad : \overset{\textstyle\cdot}{\text{As}} \cdot \quad \text{or} \quad \cdot \underset{\textstyle\cdot}{\text{As}} \cdot$$

Practice Exercise 2

Write Lewis symbols for (a) Se, (b) I, and (c) Ca. ◆

[1] For beryllium, boron, and carbon, the number of unpaired electrons in the Lewis symbol doesn't agree with the number predicted from the atom's electron configuration. Boron, for example, has two electrons paired in its $2s$ orbital and a third electron in one of its $2p$ orbitals; therefore, there is actually only one unpaired electron in a boron atom. The Lewis symbols are drawn as shown, however, because when beryllium, boron, and carbon form bonds, they *behave* as if they have two, three, and four unpaired electrons, respectively.

Although we will use Lewis symbols mostly to follow the fate of valence electrons in covalent bonds, they can also be used to describe what happens during the formation of ions. For example, when a sodium atom reacts with a chlorine atom, the electron transfer can be depicted as

$$\text{Na} \cdot + \cdot \ddot{\underset{..}{\text{Cl}}}: \longrightarrow \text{Na}^+ + \left[:\ddot{\underset{..}{\text{Cl}}}: \right]^-$$

The valence shell of the sodium atom is emptied, so no dots remain. The outer shell of chlorine, which formerly had seven electrons, gains one to give a total of eight. The brackets are drawn around the chloride ion to show that all eight electrons are the exclusive property of the Cl^- ion.

We can diagram a similar reaction between calcium and chlorine atoms.

$$:\ddot{\underset{..}{\text{Cl}}} \cdot \quad \cdot \text{Ca} \cdot \quad \cdot \ddot{\underset{..}{\text{Cl}}}: \longrightarrow \text{Ca}^{2+} + 2\left[:\ddot{\underset{..}{\text{Cl}}}: \right]^-$$

EXAMPLE 8.4
Using Lewis Symbols

Use Lewis symbols to diagram the reaction that occurs between sodium and oxygen atoms to give Na^+ and O^{2-} ions.

Solution: First let's draw the Lewis symbols for Na and O.

$$\text{Na} \cdot \qquad \cdot \ddot{\text{O}}:$$

It takes two electrons to complete the octet around oxygen. Each Na can supply only one, so we need two Na atoms. Therefore,

$$\text{Na} \cdot \quad \cdot \ddot{\underset{..}{\text{O}}}: \quad \cdot \text{Na} \longrightarrow 2\text{Na}^+ + \left[:\ddot{\underset{..}{\text{O}}}: \right]^{2-}$$

Notice that we have put brackets around the oxide ion.

Practice Exercise 3

Diagram the reaction between magnesium and oxygen atoms to give Mg^{2+} and O^{2-} ions. ◆

8.3 Electron Sharing: The Formation of Covalent Bonds

Most of the substances we encounter in our daily lives are not ionic. Rather than existing as collections of electrically charged particles (ions), they occur as electrically neutral combinations of atoms called **molecules.** Water, as you already know, consists of molecules made from two hydrogen atoms and one oxygen atom, and the formula for one particle of this compound is H_2O. Most substances consist of much larger molecules. For example, you've learned that the formula for table sugar is $C_{12}H_{22}O_{11}$.

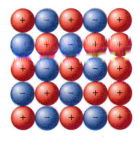

An ionic
substance

Energy Changes in the Formation of a Covalent Bond

Earlier we saw that for ionic bonding to occur, the energy-lowering effect of the lattice energy must be greater than the combined energy-raising effects of the ionization energy (IE) and electron affinity (EA). Many times this is not possible, particularly when the ionization energies of all the atoms involved are large. This happens, for example, when nonmetals combine with each other. In such cases, nature uses a different way to lower the energy—electron sharing.

A molecular
substance

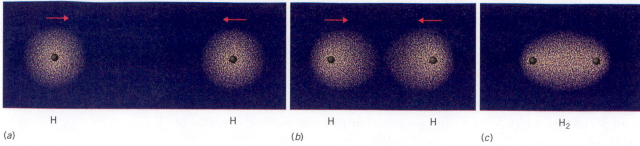

H   H   H   H H_2

(a)   (b)   (c)

Figure 8.3 *Formation of a bond between two hydrogen atoms.* (*a*) Two H atoms separated by a large distance. (*b*) As the atoms approach each other, their electron densities begin to shift to the region between the two nuclei. (*c*) The electron density becomes concentrated between the nuclei.

Let's look at what happens when two hydrogen atoms join to form an H_2 molecule (Figure 8.3). As the two atoms approach each other, the electron of each atom begins to feel the attraction of both nuclei. This causes the electron density around each nucleus to shift toward the region between the two atoms. Therefore, as the distance between the nuclei decreases, there is an increase in the probability of finding either electron near either nucleus. In effect, as the molecule is formed, each of the hydrogen atoms in the H_2 molecule acquires a share of two electrons.

When the electron density shifts to the region between the two hydrogen atoms, it attracts both nuclei and pulls them together. Being of the same charge, however, the two nuclei also repel each other, as do the two electrons. In the molecule that forms, therefore, the atoms are held at a distance at which all these attractions and repulsions are balanced. Overall, the nuclei are kept from separating, and the net force of attraction produced by the sharing of the pair of electrons is called a **covalent bond.**

> As the distance between the two nuclei and the electron cloud that lies between them decreases, the potential energy decreases.

Every covalent bond is characterized by two quantities: the average distance between the nuclei held together by the bond, and the amount of energy needed to separate the two atoms to produce neutral atoms again. In the hydrogen molecule, the attractive forces pull the nuclei to a distance of 75 pm, and this distance is called the **bond length** (or sometimes, the **bond distance**). Because a covalent bond holds atoms together, work must be done (energy must be supplied) to separate them. When the bond is *formed,* an equivalent amount of energy is released as the potential energies of the atoms are lowered. The amount of energy released when the bond is formed (or the amount of energy needed to "break" the bond) is called the **bond energy.**

Figure 8.4 shows how the potential energy changes when two hydrogen atoms come together to form H_2. We see that the minimum potential energy occurs at a bond distance of 75 pm, and that 1 mol of hydrogen molecules is more stable than 2 mol of hydrogen atoms by 435 kJ. In other words, the bond energy of H_2 is 435 kJ/mol.

Pairing of Electrons

> In Chapter 7 you learned that when two electrons occupy the same orbital, and therefore share the same space, their spins must be paired. The pairing of electrons is an important part of the formation of a covalent bond.

Before joining, each of the separate hydrogen atoms has one electron in its $1s$ orbital. When these electrons are shared, the $1s$ orbital of each atom is, in a sense, filled. Because the electrons now share the same space, they become paired as required by the Pauli exclusion principle; that is, m_s is $+\frac{1}{2}$ for one of the electrons and $-\frac{1}{2}$ for the other. In general, we almost always find that the electrons involved become paired when atoms form covalent bonds. In fact, a covalent bond is sometimes referred to as an **electron pair bond.**

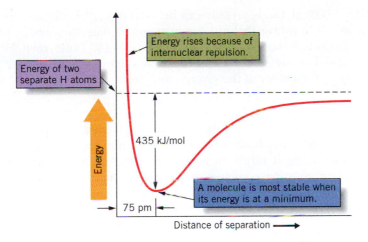

Figure 8.4 *Changes in the potential energy of two hydrogen atoms as they approach each other to form the H_2 molecule.*

Lewis symbols are often used to keep track of electrons in covalent bonds. The electrons that are shared between two atoms are shown as a pair of dots placed between the symbols for the bonded atoms. The formation of H_2 from hydrogen atoms, for example, can be depicted as

$$H\cdot + H\cdot \longrightarrow H\!:\!H$$

Because the electrons are shared, each H atom is considered to have two electrons.

(Two electrons can be counted around each of the H atoms.)

For simplicity, the electron pair in the covalent bond is usually represented as a dash. Thus, the hydrogen molecule is represented as

$$H\!-\!H$$

One dash stands for two electrons.

A formula such as this, which is drawn with Lewis symbols, is called a **Lewis formula** or **Lewis structure**. It is also called a **structural formula** because it shows which atoms are present in the molecule *and* how they are attached to each other.

Covalent Bonding and the Octet Rule

You have seen that when a nonmetal atom forms an anion, electrons are gained until the *s* and *p* subshells of its valence shell are completed. This tendency to finish with a completed valence shell, usually consisting of eight electrons, also influences the number of electrons an atom tends to acquire by sharing, and it thereby controls the number of covalent bonds that an atom forms.

Hydrogen, with just one electron in its $1s$ orbital, can complete its valence shell by obtaining a share of just one electron from another atom. When this other atom is hydrogen, the H_2 molecule is formed. Since hydrogen obtains a stable valence shell configuration when it shares just one pair of electrons with another atom, a hydrogen atom only forms one covalent bond.

Many atoms form covalent bonds by sharing enough electrons to give them complete *s* and *p* subshells in their outer shells. This is the noble gas configuration mentioned earlier and is the basis of the octet rule described in Section 8.1. As applied to covalent bonding, the **octet rule** can be stated as follows: *When atoms form covalent bonds, they tend to share sufficient electrons so as to achieve an outer shell having eight electrons.*

As you will see, it is useful to remember that hydrogen atoms only form one covalent bond.

Often, the octet rule can be used to explain the number of covalent bonds an atom forms. This number normally equals the number of electrons the atom needs to have a total of eight (an octet) in its outer shell. For example, the halogens (Group VIIA) all have seven valence electrons. The Lewis symbol for a typical member of this group, chlorine, is

$$\cdot \ddot{\underset{\cdot\cdot}{Cl}} :$$

We can see that only one electron is needed to complete an octet. Of course, chlorine can actually gain this electron and become a chloride ion. This is what it does when it forms an ionic compound such as sodium chloride (NaCl). But when chlorine combines with another nonmetal, the complete transfer of an electron is not energetically favorable. Therefore, in forming such compounds as HCl or Cl_2, chlorine gets the one electron it needs by forming a covalent bond.

$$H\cdot + \cdot\ddot{\underset{\cdot\cdot}{Cl}}: \longrightarrow H:\ddot{\underset{\cdot\cdot}{Cl}}:$$

$$:\ddot{\underset{\cdot\cdot}{Cl}}\cdot + \cdot\ddot{\underset{\cdot\cdot}{Cl}}: \longrightarrow :\ddot{\underset{\cdot\cdot}{Cl}}:\ddot{\underset{\cdot\cdot}{Cl}}:$$

The HCl and Cl_2 molecules can also be represented using dashes for the bonds.

$$H\!-\!\ddot{\underset{\cdot\cdot}{Cl}}: \quad \text{and} \quad :\ddot{\underset{\cdot\cdot}{Cl}}\!-\!\ddot{\underset{\cdot\cdot}{Cl}}:$$

There are many nonmetals that form more than one covalent bond. For example, the three most important elements in biochemical systems are carbon, nitrogen, and oxygen.

$$\cdot\dot{\underset{\cdot}{C}}\cdot \qquad \cdot\dot{\underset{\cdot\cdot}{N}}\cdot \qquad \cdot\dot{\underset{\cdot\cdot}{O}}:$$

The simplest hydrogen compounds of these elements are methane, CH_4, ammonia, NH_3, and water, H_2O. Their Lewis structures are

H	H	H
H:C̈:H	H:N̈:H	H:Ö:
Ḧ		

or	or	or

H	H	H
|	|	|
H—C—H	H—N—H	H—Ö:
|		
H		
methane	ammonia	water

In most of the compounds in which they occur, carbon forms four covalent bonds, nitrogen forms three, and oxygen forms two.

Multiple Bonds

The molecules CH_4, NH_3, and H_2O have only single bonds.

The bond produced by the sharing of one pair of electrons between two atoms is called a **single bond.** So far, these have been the only kind we have discussed. There are, however, many molecules in which more than one pair of electrons is shared between two atoms. For example, we can diagram the formation of the bonds in CO_2 as follows.

$$:\ddot{O}\cdot \leftrightarrow \cdot\dot{\underset{\cdot}{C}}\cdot \leftrightarrow \cdot\ddot{O}: \longrightarrow :O::C::\ddot{O}:$$

The central carbon atom shares two of its electrons with each of the oxygen atoms, and each oxygen shares two electrons with carbon. The result is the formation of two **double bonds.** Notice that in the Lewis formula for the molecule, the two shared pairs of electrons of a double bond are placed between the two atoms that are joined by the bond. Once again, if we circle the valence shell electrons that "belong" to each atom, we see that each has an octet.

$$:O::C::O:$$

8 electrons

The Lewis structure for CO_2, using dashes, is

$$:O=C=O:$$

Sometimes three pairs of electrons are shared between two atoms. The most abundant gas in the atmosphere, nitrogen, occurs in the form of diatomic molecules, N_2. As we've just seen, the Lewis symbol for nitrogen is

$$\cdot N:$$

and each nitrogen atom needs three electrons to complete its octet. When the N_2 molecule is formed, each of the nitrogen atoms shares three electrons with the other.

$$:N\cdot \leftrightarrow \cdot N: \longrightarrow :N:::N:$$

The result is called a **triple bond.** Again, notice that we place all three electron pairs of the bond between the two atoms. We count all of these electrons as though they belong to both of the atoms. Each nitrogen therefore has an octet.

8 electrons   8 electrons

$$:N:::N:$$

The triple bond is usually represented by three dashes, so the bonding in the N_2 molecule is normally shown as

$$:N\equiv N:$$

> The locations of the unshared pairs of electrons around the oxygen are unimportant. Two equally valid Lewis structures for CO_2 are
>
> $$:\ddot{O}=C=\ddot{O}: \quad \text{and} \quad \ddot{O}=C=\ddot{O}$$

8.4 Some Important Compounds of Carbon

Carbon, hydrogen, oxygen, and nitrogen are the elements most commonly found in **organic compounds.** In Chapter 2 you learned that these compounds are substances that can be considered to be derived from hydrocarbons—compounds in which the basic molecular "backbones" are composed of carbon atoms linked to one another in a chainlike fashion. (Hydrocarbons themselves are the principal constituents of petroleum.)

One of the chief features of organic compounds is the tendency of carbon to complete its octet by forming four covalent bonds. It can accomplish this by bonding to atoms of other nonmetallic elements or to other carbon atoms. For example, the structures of methane, ethane, and propane are

$$\begin{array}{ccc}
\text{H} & \text{H } \text{H} & \text{H } \text{H } \text{H} \\
| & | \quad | & | \quad | \quad | \\
\text{H—C—H} & \text{H—C—C—H} & \text{H—C—C—C—H} \\
| & | \quad | & | \quad | \quad | \\
\text{H} & \text{H } \text{H} & \text{H } \text{H } \text{H} \\
\text{methane} & \text{ethane} & \text{propane}
\end{array}$$

When more than four carbon atoms are present, matters become more complex because there is more than one way to arrange the atoms. For example, butane

> A more comprehensive discussion of organic compounds is found in Chapter 23. In this chapter we will look at some simple ways carbon atoms combine with other atoms to form certain important classes of organic substances that we encounter frequently.
>
> The shapes of these molecules are illustrated in Figure 2.25 on page 80.
>
> These structures can be written in a condensed form as
>
> $$CH_4$$
>
> $$CH_3CH_3$$
>
> $$CH_3CH_2CH_3$$

has the formula C_4H_{10}, but there are two ways to arrange the carbon atoms. They occur in compounds commonly called butane and isobutane.

The phenomenon seen here is one of the reasons there are so many compounds of carbon. For example, the formula $C_{20}H_{42}$ is found for 366,319 distinct compounds that differ in the way the carbon atoms are attached to each other.

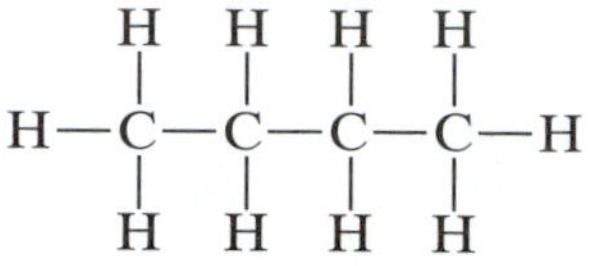

butane
C_4H_{10}
bp = -0.5 °C

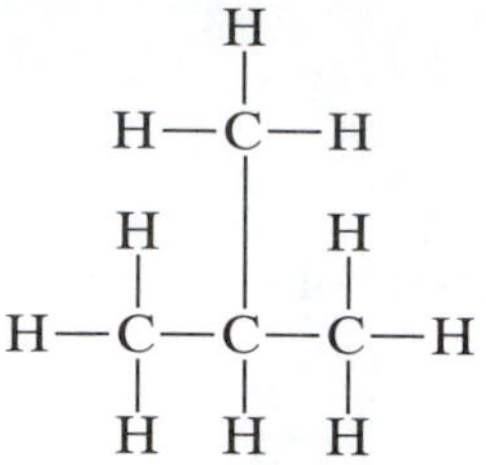

isobutane
C_4H_{10}
bp = -11.7 °C

Even though they have the same molecular formula, these are actually different compounds with different properties, as you can see from the boiling points listed below their structures. The ability of atoms to arrange themselves in more than one way to give different compounds that have the same molecular formula is called *isomerism,* and is discussed more fully in Chapters 20 and 23.

Butane and isobutane are said to be *isomers* of each other.

Carbon can complete its octet by forming four single bonds, or it can form double or triple bonds. The structures of ethylene, C_2H_4, and acetylene, C_2H_2, are

Ethylene is the raw material used to make polyethylene.

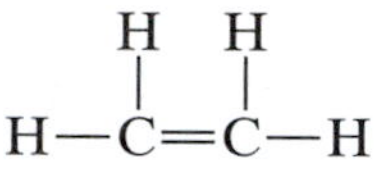

ethylene 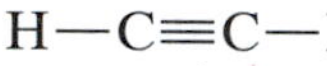   acetylene

Compounds That Also Contain Oxygen and Nitrogen

Most organic compounds contain elements in addition to carbon and hydrogen. For organizational purposes, it is convenient to consider such compounds to be derived from hydrocarbons by replacing one or more hydrogens by other groups of atoms. Such compounds can be divided into various families according to the nature of the groups attached to the parent hydrocarbon fragment. Some such families are summarized in Table 8.1.

In Chapter 2 it was noted that alcohols are organic compounds in which one of the hydrogen atoms of a hydrocarbon is replaced by OH. The family name for these compounds is *alcohol.* Examples are methyl alcohol and ethyl alcohol, which have the structures

Ethyl alcohol is added to gasoline to produce a motor fuel known as gasohol.

These also can be written as CH_3OH and CH_3CH_2OH.

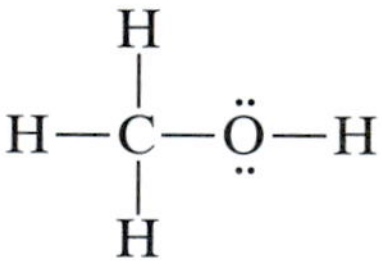

methyl alcohol 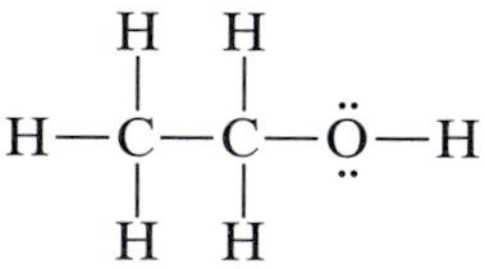   ethyl alcohol

Notice that the oxygen forms two bonds to complete its octet, just as in water. In alcohols these are single bonds, but oxygen can also form double bonds, as you saw for CO_2. One family of compounds in which a doubly bonded oxygen replaces a pair of hydrogen atoms is called *ketones.* The simplest example is acetone, the solvent in nail polish remover.

Ketones are important solvents.

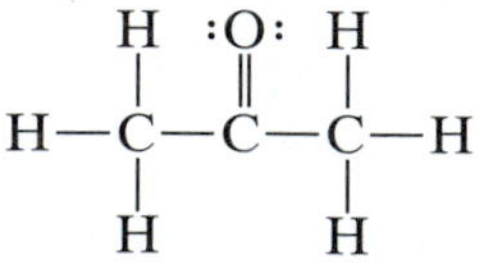

Acetone can also be represented by the formula

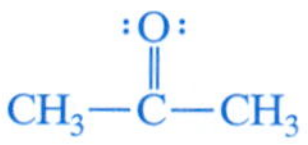

CH_3—C—CH_3 (with $:O:$ above C)

acetone

Table 8.1 Some Families of Oxygen- and Nitrogen-Containing Organic Compounds

Family Name	General Formula[a]	Example
Alcohols	R—OH	CH$_3$—OH methyl alcohol
Aldehydes	R—C—H (with =O above C)	CH$_3$—C—H (with =O above C) acetaldehyde
Ketones	R—C—R (with =O above C)	CH$_3$—C—CH$_3$ (with =O above C) acetone
Acids	R—C—OH (with =O above C)	CH$_3$—C—OH (with =O above C) acetic acid
Amines	R—NH$_2$ R—NH—R R—N—R ⎮ R	CH$_3$—NH$_2$ methylamine

[a]R stands for a hydrocarbon fragment such as CH$_3$—, or CH$_3$CH$_2$—.

Methyl alcohol is the fuel in this can of Sterno.

Notice that the carbon bonded to the oxygen is also attached to *two* other carbon atoms. If at least one of the atoms attached to the C=O group (called a *carbonyl group*, pronounced *car-bon-EEL*) is a hydrogen, a different family of compounds is formed called *aldehydes*. An example is formaldehyde, which is used to preserve biological specimens, for embalming, and to make plastics.

$$H-\overset{\displaystyle :O:}{\underset{}{\overset{\|}{C}}}-H$$
formaldehyde

Another important family of oxygen-containing organic compounds are the organic acids, also called *carboxylic acids*. An example is acetic acid.

$$H-\overset{\displaystyle H}{\underset{\displaystyle H}{\overset{|}{C}}}-\overset{\displaystyle :O:}{\overset{\|}{C}}-\overset{..}{\underset{..}{O}}-H$$
acetic acid

Notice that the acid has both a doubly bonded oxygen and an OH group attached to the end carbon atom. In general, molecules that contain the

$$-\overset{\displaystyle O}{\overset{\|}{C}}-O-H$$ group (called a *carboxyl group*) are acids. When they react with water, the hydrogen that's transferred to the water to give H$_3$O$^+$ is the one bonded to the oxygen.

Acetaldehyde, shown below, is used in the manufacture of perfumes, dyes, plastics, and other products.

$$H-\overset{\displaystyle H}{\underset{\displaystyle H}{\overset{|}{C}}}-\overset{\displaystyle :O:}{\overset{\|}{C}}-H$$
acetaldehyde

In general, compounds with the carboxyl group are weak acids.

$$H-\overset{\displaystyle H}{\underset{\displaystyle H}{\overset{|}{C}}}-\overset{\displaystyle :O:}{\overset{\|}{C}}-\overset{..}{\underset{..}{O}}-H + H_2O \longrightarrow H_3O^+ + \left[H-\overset{\displaystyle H}{\underset{\displaystyle H}{\overset{|}{C}}}-\overset{\displaystyle :O:}{\overset{\|}{C}}-\overset{..}{\underset{..}{O}}:\right]^-$$
acetic acid   acetate ion

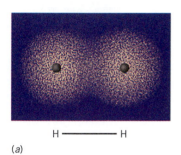

ammonia

methylamine

Amino acids combine in long chains to form the proteins in our bodies.

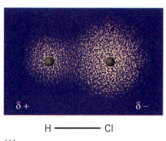

(a)
H ———————— H

(b)
δ+   δ−
H ———————— Cl

Figure 8.5 *Nonpolar and polar covalent bonds.* (a) The electron density of the pair of bonding electrons is distributed evenly in the bond between two H atoms. This gives a nonpolar covalent bond. (b) The electron density of the bonding electron pair between H and Cl is distributed unevenly in HCl, with more of the charge around the chlorine end of the bond. This causes the bond to be polar.

Experimental measurement of dipole moments is one way to check our theoretical models.

Nitrogen atoms need three electrons to complete an octet and in most of its compounds, nitrogen forms three bonds. The common nitrogen-containing organic compounds can be imagined as being derived from ammonia by replacing one or more of the hydrogens of NH_3 with hydrocarbon groups. They're called *amines,* and an example is methylamine, CH_3NH_2, as shown in the margin. Amines are strong-smelling compounds and often have a "fishy" odor. Like ammonia, they're basic.

$$CH_3NH_2(aq) + H_2O \rightleftharpoons CH_3NH_3^+(aq) + OH^-(aq)$$

Some molecules can incorporate atom groups that belong to two or more classes of compounds. An example from biology is a series of compounds called amino acids, which contain the amine group ($-NH_2$) as well as a carboxyl group ($-CO_2H$). The simplest of these is the amino acid glycine,

$$:NH_2 - CH_2 - \overset{\overset{\displaystyle :O:}{\|}}{C} - \ddot{O}H$$

glycine

We have just touched the surface of the subject of organic chemistry here, but we have introduced you to some of the kinds of organic compounds you will encounter in chapters ahead.

8.5 Electronegativity and the Polarity of Bonds

When two identical atoms form a covalent bond, as in H_2 or Cl_2, each has an equal share of the electron pair in the bond. The electron density at both ends of the bond is the same, because the electrons are equally attracted to both nuclei. However, when different kinds of atoms combine, as in HCl, one nucleus usually attracts the electrons in the bond more strongly than the other.

The result of unequal attractions for the bonding electrons is an unbalanced distribution of electron density within the bond. For example, it has been found that a chlorine atom attracts electrons more strongly than does a hydrogen atom. In the HCl molecule, therefore, the electron cloud is pulled more tightly around the Cl, and that end of the molecule experiences a slight buildup of negative charge. The electron density that shifts toward the chlorine is removed from the hydrogen, which causes the hydrogen end to acquire a slight positive charge.

In HCl, electron transfer is incomplete. The electrons are still shared, but unequally. The charges on either end of the molecule are less than full 1+ and 1− charges. In HCl, for example, the hydrogen carries a charge of +0.17 and the chlorine a charge of −0.17. These are called **partial charges** and are usually indicated by the lowercase Greek letter delta, δ (see Figure 8.5). Partial charges can also be indicated on Lewis structures. For example,

$$H - \underset{\delta+}{\overset{\displaystyle \ddot{}}{C}}\underset{\delta-}{\ddot{l}}:$$

A bond that carries partial positive and negative charges on opposite ends is called a **polar covalent bond,** or often simply a **polar bond** (the word *covalent* is understood). The term *polar* comes from the notion of *poles* of opposite charge at either end of the bond. Because there are *two poles* of charge involved, the bond is said to be a **dipole.**

The polar bond in HCl causes the molecule as a whole to have opposite charges on either end, so we say that HCl is a **polar molecule.** The HCl molecule as a whole is also a dipole, and the extent of its polarity is expressed quantitatively through its **dipole moment,** which is found by multiplying the amount of charge on either end by the distance between the charges.

The degree to which a covalent bond is polar depends on the difference in the abilities of the bonded atoms to attract electrons. The greater the difference, the more polar is the bond and the more the electron density is shifted toward the atom that attracts electrons more.

The term that we use to describe the relative attraction of an atom for the electrons in a bond is called the **electronegativity** of the atom. In HCl, for example, chlorine is *more electronegative* than hydrogen. The electron pair of the covalent bond spends more of its time around the more electronegative atom, which is why that end of the bond acquires a partial negative charge.

Numerical values for electronegativity have been assigned for each element, as shown in Figure 8.6. This information is useful because the *difference* in electronegativity provides an estimate of the degree of polarity of a bond. In addition, the relative magnitudes of the electronegativities indicate which end of the bond carries the partial negative charge. For instance, fluorine is even more electronegative than chlorine. Therefore, we expect an HF molecule to be more polar than an HCl molecule. In addition, hydrogen is less electronegative than either fluorine or chlorine, so in both of these molecules the hydrogen bears the partial positive charge.

The noble gases are assigned electronegativities of zero and are omitted from the table.

$$H-\ddot{\underset{\cdot\cdot}{F}}\colon \qquad\qquad H-\ddot{\underset{\cdot\cdot}{Cl}}\colon$$
$$\delta+ \quad \delta- \qquad\qquad\qquad \delta+ \quad \delta-$$

Practice Exercise 4

For each of the following bonds, choose the atom that carries the partial negative charge: (a) P—Br, (b) Si—Cl, (c) S—Cl. ◆

Examination of electronegativity values and their differences reveals that there is no sharp dividing line between ionic and covalent bonding. Ionic bonding and *nonpolar covalent bonding* simply represent the two extremes. A bond is mostly ionic when the difference in electronegativity between two atoms is very large; the more electronegative atom acquires essentially complete control of the bonding electrons. In a **nonpolar covalent bond,** there is no difference in electronegativity, so the pair of bonding electrons is shared equally.

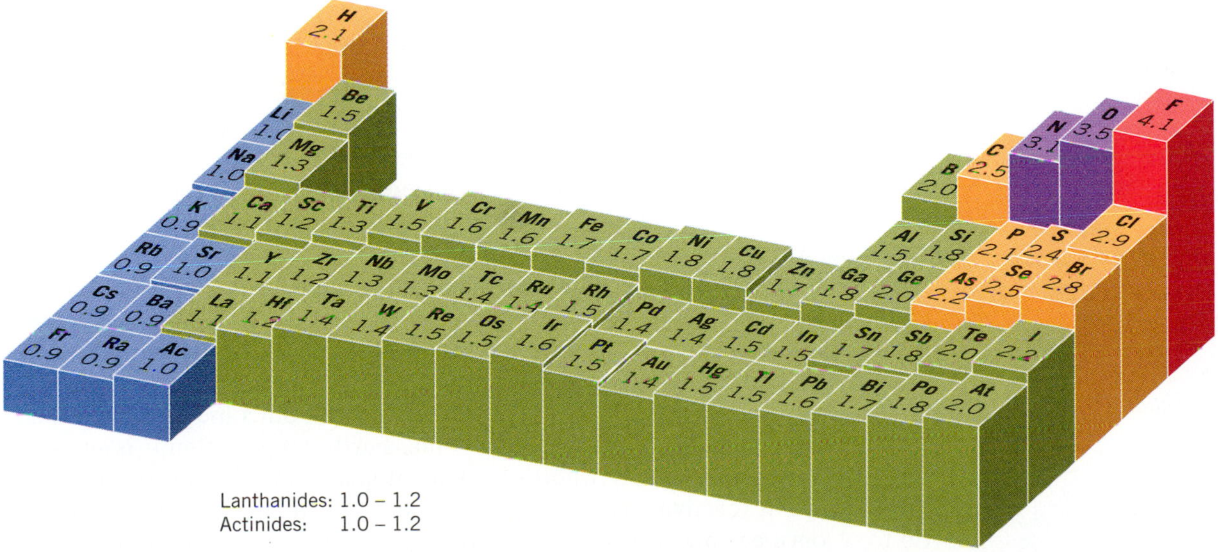

Lantanides: 1.0 – 1.2
Actinides: 1.0 – 1.2

Figure 8.6 *The electronegativities of the elements.*

$$Cs^+ \left[:\ddot{\underset{..}{F}}: \right]^-$$

"bonding pair" held
exclusively by fluorine

$$:\ddot{\underset{..}{F}}:\ddot{\underset{..}{F}}:$$

bonding pair
shared equally

The degree to which the bond is polar, which we might think of as the amount of **ionic character** of the bond, varies in a continuous way with changes in the electronegativity difference (Figure 8.7). The bond becomes more than 50% ionic when the electronegativity difference exceeds approximately 1.7.

Trends in Electronegativity in the Periodic Table

Figure 8.6 also illustrates the trends in electronegativity within the periodic table; *electronegativity increases from bottom to top in a group, and from left to right in a period.* Notice that the trends follow those for ionization energy. This is because an atom that has a small IE will give away an electron more easily than an atom with a large IE, just as an atom with a small electronegativity will lose its share of an electron pair more readily than an atom with a large electronegativity.

Elements located in the same region of the table (for example, the nonmetals) have similar electronegativities, which means that if they form bonds with each other, the electronegativity differences will be small and the bonds will be more covalent than ionic. On the other hand, if elements from widely separated regions of the table combine, large electronegativity differences occur and the bonds will be predominantly ionic. This is what happens, for example, when an element from Group IA or Group IIA reacts with a nonmetal from the upper right-hand corner of the periodic table.

Tools

Periodic trends
in
electronegativity

It is found that the electronegativity is proportional to the average of the ionization energy and the electron affinity of an element.

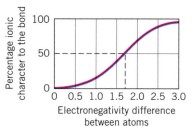

Figure 8.7 *Variation in the percentage ionic character of a bond with electronegativity difference.* The bond becomes approximately 50% ionic when the electronegativity difference equals 1.7, which means each atom in the bond carries a partial charge of approximately 0.5 unit.

8.6 Electronegativity and the Reactivities of Metals and Nonmetals

There are parallels between an element's electronegativity and its reactivity—its tendency to undergo reactions. In this section we will examine some of these trends.

Reactivities of Metals

When we raise questions about the reactivity of a metal, we are concerned with how easily the metal is oxidized. This is because in nearly every compound containing a metal, the metal exists in a positive oxidation state, and when a free metal reacts to form a compound, it is oxidized. As a result, a metal like sodium, which is very easily oxidized, is said to be very reactive, whereas a metal like platinum, which is very difficult to oxidize, is said to be unreactive.

There are several ways to compare how easily metals are oxidized. In Chapter 5 we saw that by comparing the abilities of metals to displace each other from compounds we are able to establish their relative ease of oxidation. This was the basis for the activity series (Table 5.2).

As useful as the activity series is in predicting the outcome of certain redox reactions, it is difficult to remember in detail. Furthermore, often it is sufficient just to know approximately where an element stands in relation to others in a broad range of reactivity. This is where the periodic table can be especially useful to us once again, because there are trends and variations in reactivity within the periodic table that are simple to identify and remember.

Reactivity refers in general to the tendency of a substance to react with something. The *reactivity of a metal* refers to its tendency specifically to undergo *oxidation*.

Figure 8.8 illustrates how the ease of oxidation (the reactivity) of metals varies in the periodic table. In general, these trends roughly follow the variations in electronegativity, with the metal being less easily oxidized as its electronegativity increases. You might expect this, since electronegativity is a measure of how strongly the atom of an element attracts electrons when combining with an atom of a different element. The more strongly the atom attracts electrons, the more difficult it is to oxidize. This relationship between reactivity and electronegativity is only approximate, however, because many other factors affect the stability of the compounds that are formed.

By examining Figure 8.8, we see that the metals that are most easily oxidized are found at the far left in the periodic table. The metals in Group IA, for example, are so easily oxidized that all of them react with water to liberate hydrogen. Because of their reactivity toward moisture and oxygen, they have no useful applications that require exposure to the atmosphere, so we rarely encounter them as free metals. The same is true of the heavier metals in Group IIA, calcium through barium. These elements also react with water to liberate hydrogen.

You've learned how the electronegativities of the elements vary within the periodic table, increasing as we move from left to right in a period. This places metals with the lowest electronegativities at the left of the table—in other words, in Groups IA and IIA. It should be no surprise, therefore, that these elements are the easiest to oxidize. You also learned that *electronegativity decreases as we go down a group*. This explains why the heavier elements in Group IIA are easier to oxidize than those at the top of the group.

In Figure 8.8 we can also locate the metals that are the most difficult to oxidize. They occur for the most part among the heavier transition elements in the center of the periodic table, where we find the very unreactive elements platinum and gold—metals used to make fine jewelry. Their bright luster and lack of any tendency to corrode in air or water combine to make them particularly attractive for this purpose. This same lack of reactivity also is responsible for their industrial uses. Gold, for example, is used to coat the electrical contacts in low-voltage circuits found in microcomputers, because even small amounts of corrosion on more reactive metals would be sufficient to impede the flow of electricity so much as to make the devices unreliable.

Study Figure 8.8 to identify where the very reactive metals are found and where the very unreactive ones are found.

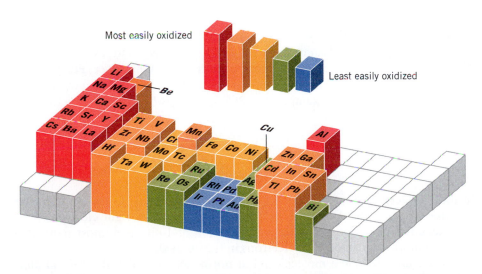

Figure 8.8 *The variation of the ease of oxidation of metals with position in the periodic table.*

Oxidizing Power of Nonmetals

The reactivity of a metal is determined by its ease of oxidation, and therefore its ability to serve as a reducing agent. *For a nonmetal, reactivity is usually gauged by its ability to serve as an oxidizing agent.* This ability also varies according to the nonmetal's electronegativity. Nonmetals with high electronegativities have strong tendencies to acquire electrons and are therefore strong oxidizing agents. In parallel with changes in electronegativities in the periodic table, *the oxidizing abilities of nonmetals increase from left to right across a period and from bottom to top in a group.* Thus, the most powerful oxidizing agent is fluorine, followed closely by oxygen, both in the upper right-hand corner of the periodic table.

Single replacement reactions occur among the nonmetals, just as with the metals. For example, heating a metal sulfide in oxygen causes the sulfur to be replaced by oxygen. The replaced sulfur then combines with oxygen to give sulfur dioxide. The equation for a typical reaction is

$$CuS(s) + \tfrac{3}{2}O_2(g) \longrightarrow CuO(s) + SO_2(g)$$

Replacement reactions are especially evident among the halogens, where a particular halogen, as an element, will oxidize the *anion* of any halogen below it in Group VIIA. Thus, F_2 will oxidize Cl^-, Br^-, and I^-. However, Cl_2 will only oxidize Br^- and I^-, and Br_2 will only oxidize I^-. Thus, the reactions in the margin are observed.

> You studied single replacement reactions of metals in Chapter 5.
>
> *Fluorine:*
>
> $F_2 + 2Cl^- \longrightarrow 2F^- + Cl_2$
>
> $F_2 + 2Br^- \longrightarrow 2F^- + Br_2$
>
> $F_2 + 2I^- \longrightarrow 2F^- + I_2$
>
> *Chlorine:*
>
> $Cl_2 + 2Br^- \longrightarrow 2Cl^- + Br_2$
>
> $Cl_2 + 2I^- \longrightarrow 2Cl^- + I_2$
>
> *Bromine:*
>
> $Br_2 + 2I^- \longrightarrow 2Br^- + I_2$

▶**Chemistry in Practice**◀ Chlorine is used to kill bacteria in drinking water and swimming pools. Because chlorine is such a strong oxidizing agent, bacteria that might enter the water are quickly killed. A variety of solid chemicals are used to provide the chlorine in swimming pools. For municipal water treatment, chlorine is added to water directly as the gas. ◆

8.7 Drawing Lewis Structures

Lewis structures are very useful in chemistry because they give us a relatively simple way to describe the structures of molecules. As a result, much chemical reasoning is based on them. Also, as you will learn in the next chapter, we can use the Lewis structure to make a reasonably accurate prediction about the shape of a molecule.

In Sections 8.3 and 8.4 you saw Lewis structures for a variety of molecules that obey the octet rule. Examples included CO_2, Cl_2, N_2, and a variety of compounds of carbon. The octet rule is not always obeyed, however. For instance, there are some molecules in which one or more atoms must have more than an octet in the valence shell. Examples are PCl_5 and SF_6, whose Lewis structures are

In these molecules the formation of more than four bonds to the central atom requires that the central atom have a share of more than eight electrons. With the exception of period 2 elements like carbon and nitrogen, most nonmetals can have more than an octet of electrons in the outer shell.

There are also some molecules (but not many) in which the central atom behaves as though it has less than an octet. The most common examples involve compounds of beryllium and boron.

> Lewis structures are not meant to depict the shapes of molecules. They just describe which atoms are bonded to each other and the kinds of bonds involved. Thus, the Lewis structure for water can be drawn as
>
> $H—\ddot{O}—H$
>
> but it does not mean the water molecule is linear, with all the atoms in a straight line. The actual shape of a water molecule is depicted below.
>
>
>
> When an atom forms more than four bonds, it must have more than four electron pairs in its valence shell.

$$\cdot Be \cdot + 2 \cdot \overset{\displaystyle \cdot \cdot}{\underset{\displaystyle \cdot \cdot}{Cl}} : \longrightarrow \quad : \overset{\displaystyle \cdot \cdot}{\underset{\displaystyle \cdot \cdot}{Cl}} - Be - \overset{\displaystyle \cdot \cdot}{\underset{\displaystyle \cdot \cdot}{Cl}} :$$

four electrons around Be

$$: \overset{\displaystyle \cdot \cdot}{\underset{\displaystyle \cdot \cdot}{Cl}} :$$
$$\cdot \dot{B} \cdot + 3 \cdot \overset{\displaystyle \cdot \cdot}{\underset{\displaystyle \cdot \cdot}{Cl}} : \longrightarrow \quad : \overset{\displaystyle \cdot \cdot}{\underset{\displaystyle \cdot \cdot}{Cl}} - \overset{|}{B} - \overset{\displaystyle \cdot \cdot}{\underset{\displaystyle \cdot \cdot}{Cl}} :$$

six electrons around B

Although Be and B sometimes have less than an octet, the elements in period 2 never exceed an octet. The reason is because their valence shells, having $n = 2$, can hold a maximum of only 8 electrons. (This explains why the octet rule works so well for atoms of carbon, nitrogen, and oxygen.) However, elements in periods below period 2, such as phosphorus and sulfur, sometimes do exceed an octet, because their valence shells can hold more than 8 electrons. For example, the valence shell for elements in period 3, for which $n = 3$, can hold a maximum of 18 electrons, and the valence shell for period 4 elements, which have $s, p, d,$ and f subshells, can hold as many as 32 electrons.

A Method for Drawing Lewis Structures

We have used the Lewis structures of a variety of compounds in our discussions of covalent bonds. Such structures are also drawn for polyatomic ions, which are held together by covalent bonds, too. We now turn our attention to how these structures can be obtained in a systematic way.

The method of writing Lewis structures can be broken down into a number of steps, which are summarized in Figure 8.9. The first is to decide which atoms are bonded to each other, so that we know where to put the dots or dashes. This is not always a simple matter. Many times the formula suggests the way the atoms are arranged because the central atom, which is usually the least electronegative one, is usually written first. Examples are CO_2 and ClO_4^-, which have the following *skeletal structures* (i.e., arrangements of atoms):

Sometimes, obtaining the skeletal structure is not quite so simple, especially when more than two elements are present. There are some generalizations possible, however. For example, the skeletal structure of nitric acid, HNO_3, is

O
H   O   N   O   (correct)

rather than one of the following.

O
O N O   or   H O O N O   (incorrect)
H

Nitric acid is an oxoacid (Section 2.11), and it happens that the hydrogen atoms that can be released from molecules of oxoacids are always bonded to oxygen atoms, which are in turn bonded to the third nonmetal atom. Therefore, recognizing HNO_3 as the formula of an oxoacid allows us to predict that the three oxygen atoms are bonded to the nitrogen, and the hydrogen is bonded to one of the oxygens. (It is also useful to remember that hydrogen forms only one bond, so we would not choose it to be a central atom.)

Method for drawing Lewis structures

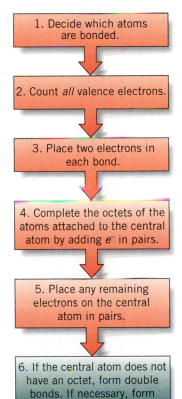

1. Decide which atoms are bonded.

2. Count *all* valence electrons.

3. Place two electrons in each bond.

4. Complete the octets of the atoms attached to the central atom by adding e^- in pairs.

5. Place any remaining electrons on the central atom in pairs.

6. If the central atom does not have an octet, form double bonds. If necessary, form triple bonds.

Figure 8.9 *Summary of steps in the writing of Lewis structures.* These rules lead to Lewis structures in which the octet rule is obeyed by the maximum number of atoms.

There are times when no reasonable basis can be found for choosing a particular skeletal structure. If you must make a guess, choose the most symmetrical arrangement of atoms, because it has the greatest chance of being correct.

Practice Exercise 5

Predict reasonable skeletal structures for SO_2, NO_3^-, $HClO_3$, and H_3PO_4. ◆

After you've decided on the skeletal structure, the next step is to count all of the *valence electrons* to find out how many dots must appear in the final formula. Using the periodic table, locate the groups in which the elements in the formula occur to determine the number of valence electrons contributed by each atom. If the structure you wish to draw is that of an ion, *add one additional valence electron for each negative charge or remove a valence electron for each positive charge*. Some examples are

SO_3	Sulfur (Group VIA) contributes $6e^-$.	$1 \times 6 = 6$
	Each oxygen (Group VIA) contributes $6e^-$.	$3 \times 6 = 18$
		Total $\quad 24e^-$

ClO_4^-	Chlorine (Group VIIA) contributes $7e^-$.	$1 \times 7 = 7$
	Each oxygen (Group VIA) contributes $6e^-$.	$4 \times 6 = 24$
	Add $1e^-$ for the $1-$ charge.	$+1$
		Total $\quad 32e^-$

NH_4^+	Nitrogen (Group VA) contributes $5e^-$.	$1 \times 5 = 5$
	Each hydrogen (Group IA) contributes $1e^-$.	$4 \times 1 = 4$
	Subtract $1\,e^-$ for the $1+$ charge.	-1
		Total $\quad 8e^-$

Practice Exercise 6

How many valence electrons should appear in the Lewis structures of SO_2, PO_4^{3-}, and NO^+? ◆

After we have determined the number of valence electrons, we place them into the skeletal structure in pairs following the steps outlined in Figure 8.9. Let's look at some examples of how we go about this.

EXAMPLE 8.5
Drawing Lewis Structures

What is the Lewis structure of the chloric acid molecule, $HClO_3$?

Solution: First, we select a reasonable skeletal structure. Since the substance is an oxoacid, we expect the hydrogen to be bonded to an oxygen, which in turn is bonded to the chlorine. The other two oxygens would also be bonded to the chlorine. This gives

$$O$$
$$H \quad O \quad Cl \quad O$$

The total number of valence electrons is 26 ($1e^-$ from H, $6e^-$ from each O, and $7e^-$ from Cl). Continuing with the steps in Figure 8.9, we place a pair of electrons in each bond, because we know that there must be at least one pair of electrons between each pair of atoms.

$$\overset{\text{O}}{\text{H}\colon\!\text{O}\colon\!\ddot{\text{C}}\text{l}\colon\!\text{O}}$$

This has used $8e^-$, so we still have $18e^-$ to go. Next, we work on the atoms surrounding the chlorine (which is the central atom in this structure). No additional electrons are needed around the H, because $2e^-$ are all that can occupy its valence shell. Therefore, we next complete the octets of the oxygens.

$$\overset{\cdot\ddot{\text{O}}\cdot}{\text{H}\colon\!\ddot{\text{O}}\colon\!\ddot{\text{C}}\text{l}\colon\!\ddot{\text{O}}\colon}$$

We have now used a total of $24e^-$, so there are two electrons left. These are placed on the Cl (the central atom) to give

$$\overset{\colon\ddot{\text{O}}\colon}{\text{H}\colon\!\ddot{\text{O}}\colon\!\ddot{\text{C}}\text{l}\colon\!\ddot{\text{O}}\colon}$$

which we can also write as follows, using dashes for the electron pairs in the bonds.

$$\overset{\colon\ddot{\text{O}}\colon}{\overset{|}{\text{H}-\ddot{\text{O}}-\text{Cl}-\ddot{\text{O}}\colon}}$$

The chlorine and the three oxygens have octets, and the hydrogen is complete with $2e^-$, so we are finished.

Write the Lewis structure for the SO_3 molecule.

Solution: We expect the skeletal structure to be

$$\overset{\text{O}}{\text{O} \quad \text{S} \quad \text{O}}$$

The total number of electrons in the formula is 24 ($6e^-$ from the sulfur, plus $6e^-$ from each oxygen). We begin to distribute the electrons by placing a pair in each bond. This gives

$$\overset{\text{O}}{\text{O}\colon\!\dot{\text{S}}\colon\!\text{O}}$$

We have used $6e^-$, so there are $18e^-$ left. We next complete the octets around the oxygens, which uses the remaining electrons.

$$\overset{\colon\ddot{\text{O}}\colon}{\colon\ddot{\text{O}}\colon\!\ddot{\text{S}}\colon\!\ddot{\text{O}}\colon}$$

At this point all the electrons have been placed into the structure, but we see that the sulfur still lacks an octet. We cannot simply add more dots because the total must be 24. Therefore, according to the last step of the procedure in Figure 8.9, we have to create a multiple bond. To do this we move a pair of electrons that we have shown to belong solely to an oxygen into a sulfur–oxygen bond so that it can be counted as belonging to both the oxygen *and* the sulfur. In other words, we place a double bond between sulfur and one of the oxygens. It doesn't matter which oxygen we choose for this honor.

$$\overset{\colon\ddot{\text{O}}\colon}{\colon\ddot{\text{O}}\colon\!\ddot{\text{S}}\colon\!\ddot{\text{O}}\colon} \quad\text{gives}\quad \overset{\colon\ddot{\text{O}}\colon}{\colon\ddot{\text{O}}\colon\colon\!\ddot{\text{S}}\colon\!\ddot{\text{O}}\colon} \quad\text{or}\quad \overset{\colon\ddot{\text{O}}\colon}{\overset{|}{\colon\ddot{\text{O}}=\text{S}-\ddot{\text{O}}\colon}}$$

Notice that each atom has an octet.

EXAMPLE 8.6

Drawing Lewis Structures

EXAMPLE 8.7

Drawing Lewis Structures

Carbon monoxide detector. Carbon monoxide is a colorless, odorless, and very poisonous gas that's sometimes produced by faulty heating equipment. Devices such as the one shown here can be used in the home to detect CO and sound a warning.

What is the Lewis structure of CO?

Solution: The skeletal structure is simply

$$C \qquad O$$

The total number of valence electrons is 10, and we begin distributing them by placing a pair into the bond.

$$C:O$$

Now we try to complete the octets with the remaining 8 electrons. To do this, we arbitrarily select one atom, say carbon, to be the "central atom" and complete the valence shell of the other. The two electrons that remain are then placed on the carbon to give

$$:C:\ddot{O}:$$

Carbon has only four electrons at this point, so we must move two unshared pairs of electrons that are on the oxygen into the bond. This gives a triple bond.

 gives :C:::O: or :C≡O:

EXAMPLE 8.8

Drawing Lewis Structures

What is the Lewis structure for SF_4?

Solution: The skeletal structure is

$$\begin{array}{ccc} & F & \\ F & S & F \\ & F & \end{array}$$

and there must be $6 + 28 = 34$ electrons in the structure. First we place $2e^-$ into each bond, and then we complete the octets of the fluorine atoms. This uses 32 electrons.

$$\begin{array}{cc} :\ddot{F}: & :\ddot{F}: \\ :\ddot{F}:\ddot{S}:\ddot{F}: & \text{or} \quad :\ddot{F}{-}S{-}\ddot{F}: \\ :\ddot{F}: & :\ddot{F}: \end{array}$$

There are two electrons left, and according to step 5 in Figure 8.9 they are placed on the central atom as a *pair* of electrons. We will redraw the structure to make room for them.

$$\begin{array}{c} :\ddot{F} \\ \diagdown \\ \ddot{.F.} \quad S{-}\ddot{F}: \\ | \\ :\ddot{F}: \end{array}$$

We are now finished. Each fluorine atom has an octet and the central atom does not have *less* than an octet. In fact, in this molecule, the sulfur violates the octet rule by having five pairs of electrons in its valence shell.

Practice Exercise 7

Draw Lewis structures for OF_2, NH_4^+, SO_2, NO_3^-, ClF_3, and $HClO_4$. ◆

8.8 Formal Charge and the Selection of Lewis Structures

Lewis structures are meant to describe how atoms share electrons in chemical bonds. Such descriptions are theoretical explanations or predictions that relate to the forces that hold molecules and polyatomic ions together. But, as you learned in Chapter 1, a theory is only as good as the observations on which it is based, so to have confidence in a theory about chemical bonding, we need to have a way to check it. We need experimental observations that relate to the description of bonding.

Two properties that are related to the number of electron pairs shared between two atoms are **bond length,** the distance between the nuclei of the bonded atoms, and **bond energy,** the energy required to separate the bonded atoms to give neutral particles. For example, we mentioned in Section 8.3 that measurements have shown the H_2 molecule has a bond length of 75 pm and a bond energy of 435 kJ/mol, which means that it takes 435 kJ to break the bonds of 1 mol of H_2 molecules to give 2 mol of hydrogen atoms.

For bonds between the same elements, the bond length and bond energy depend on the **bond order,** which is defined as *the number of pairs of electrons shared between two atoms.* The bond order is a measure of the amount of electron density in the bond, and the greater the electron density, the more tightly the nuclei are held and the more closely they are drawn together. This is illustrated by the data in Table 8.2, which gives typical bond lengths and bond energies for single, double, and triple bonds between carbon atoms. In summary:

> As the bond order increases, the bond length decreases and the bond energy increases, provided we are comparing bonds between the same elements.

With this as background, let's examine the Lewis structure of sulfuric acid, drawn following the rules given in the preceding section.

$$H—\ddot{O}—S—\ddot{O}—H \qquad \text{(Structure 1)}$$

It obeys the octet rule, and there doesn't seem to be any need to attempt to write any other structures for it. But a problem arises if we compare the predicted bond lengths with those found experimentally. In our Lewis structure, all four sulfur–oxygen bonds are single bonds, which means they should have about the same bond lengths. However, experimentally it has been found that the bonds are not of equal length, as illustrated in Figure 8.10. The S—O bonds are shorter than the S—OH bonds, which means they must have a larger bond order. Therefore, we need to modify our Lewis structure to make it conform to reality.

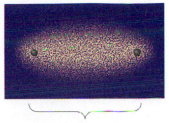

Bond length = 75 pm

The hydrogen molecule has a bond length of 75 pm.

A single bond has a bond order of 1, a double bond a bond order of 2, and a triple bond a bond order of 3.

Correlation between bond properties and bond order

Table 8.2 Average Bond Lengths and Bond Energies Measured for Carbon–Carbon Bonds

Bond	Bond Length (pm)	Bond Energy (kJ/mol)
C—C	154	348
C=C	134	615
C≡C	120	812

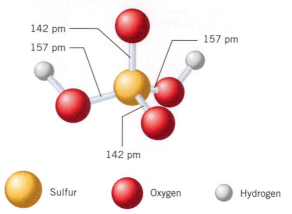

142 pm

157 pm

157 pm

142 pm

Sulfur   Oxygen   Hydrogen

Figure 8.10 *The structure of sulfuric acid in the vapor state.* Notice the differences in the sulfur–oxygen bond lengths.

Because sulfur is in period 3, its valence shell has $3s$, $3p$, and $3d$ subshells, which together can accommodate more than eight electrons. Therefore, we are able to have more than four bonds to sulfur and we are able to increase the bond order in the S—O bonds by moving electron pairs to create sulfur–oxygen double bonds as shown below.

$$H—\overset{..}{\underset{..}{O}}—\overset{:\overset{..}{O}⊖}{\underset{:\overset{..}{O}⊖}{S}}—\overset{..}{\underset{..}{O}}—H \qquad \text{gives} \qquad H—\overset{..}{\underset{..}{O}}—\overset{:\overset{..}{O}}{\underset{:\overset{..}{O}}{S}}—\overset{..}{\underset{..}{O}}—H \qquad \text{(Structure II)}$$

Now we have a Lewis structure that better fits experimental observations because the sulfur–oxygen double bonds are expected to be shorter than the sulfur–oxygen single bonds. Because this second Lewis structure agrees better with the actual structure of the molecule, it is the *preferred* Lewis structure, even though it violates the octet rule.

Formal Charges

Are there any criteria that we could have applied that would have allowed us to predict that the second Lewis structure for H_2SO_4 is better than the one with only single bonds, even though it seems to violate the octet rule unnecessarily? To answer this question, let's take a closer look at the two Lewis structures we've drawn.

In Structure I, there are only single bonds between the sulfur and oxygen atoms. If the electrons in the bonds are shared equally by S and O, then each atom "owns" half of the electron pair, or the equivalent of one electron. In other words, the four single bonds place the equivalent of four electrons in the valence shell of the sulfur. A single atom of sulfur by itself, however, has six valence electrons. This means that in this Lewis structure, the sulfur has two electrons *less* than it does as just an isolated atom. Thus, at least in a bookkeeping sense, it would appear that if sulfur obeyed the octet rule in H_2SO_4, it would have a charge of $2+$. This *apparent charge* on the sulfur atom is called its **formal charge.**

The formal charge on any atom in a Lewis structure can be calculated as follows:

> Each unshared electron contributes one $1e^-$ to the total an atom has in its valence shell.

Tools

Formal charges

$$\text{Formal charge} = \left(\begin{array}{c}\text{number of } e^- \\ \text{in valence shell of} \\ \text{the isolated atom}\end{array}\right) - \left(\begin{array}{c}\text{number of bonds} \\ \text{to the atom}\end{array} + \begin{array}{c}\text{number of} \\ \text{unshared } e^-\end{array}\right)$$

(8.1)

For example, for the sulfur in Structure I, we get

$$\text{Formal charge on S} = 6 - (4 + 0) = +2$$

Let's also calculate the formal charges on the hydrogen and oxygen atoms in Structure I. An isolated H atom has one electron. In Structure I each H has one bond and no unshared electrons. Therefore,

$$\text{Formal charge on H} = 1 - (1 + 0) = 0$$

The hydrogens in this structure have no formal charge.

We add up the number of bonds and the number of unshared electrons and then subtract this total from the number of valence electrons possessed by a neutral atom of the element.

$$6 - (4 \text{ bonds} + 0 \text{ unshared}) = +2 \qquad 1 - (1 \text{ bond} + 0 \text{ unshared}) = 0$$

$$\text{H}-\overset{..}{\underset{..}{\text{O}}}-\underset{\underset{\overset{..}{\underset{..}{\text{O}}}}{|}}{\overset{\overset{..}{\text{O}}}{|}{\text{S}}}-\overset{..}{\underset{..}{\text{O}}}-\text{H} \qquad \text{(Structure I)}$$

In Structure I, we also see that there are two kinds of oxygens to consider. An isolated oxygen atom has six electrons, so we have, for the oxygens also bonded to hydrogen,

$$\text{Formal charge} = 6 - (2 + 4) = 0$$

and, for the oxygens not bonded to hydrogen,

$$\text{Formal charge} = 6 - (1 + 6) = -1$$

$$\text{H}-\overset{..}{\underset{..}{\text{O}}}-\underset{\underset{\overset{..}{\text{O}}:}{|}}{\overset{\overset{:\text{O}:}{|}}{\text{S}}}-\overset{..}{\underset{..}{\text{O}}}-\text{H} \qquad \text{(Structure I)}$$

$$6 - (2 \text{ bonds} + 4 \text{ unshared}) = 0 \qquad 6 - (1 \text{ bond} + 6 \text{ unshared}) = -1$$

Nonzero formal charges are indicated in a Lewis structure by placing them in circles alongside the atoms, as shown below.

$$\text{H}-\overset{..}{\underset{..}{\text{O}}}-\underset{\underset{:\overset{..}{\text{O}}:_{\ominus}}{|}}{\overset{\overset{:\overset{..}{\text{O}}:^{\ominus}}{|}}{\text{S}^{\text{(2+)}}}}-\overset{..}{\underset{..}{\text{O}}}-\text{H}$$

Notice that the sum of the formal charges in the molecule adds up to zero. It is useful to remember that, in general, *the sum of the formal charges in any Lewis structure adds up to the charge on the species.*

Now let's look at the formal charges in Structure II. For sulfur we have

$$\text{Formal charge on S} = 6 - (6 + 0) = 0$$

so the sulfur has no formal charge. The hydrogens and the oxygens that are also bonded to H are the same in this structure as before, so they have no formal charges. And finally, the oxygens that are not bonded to hydrogen have

$$\text{Formal charge} = 6 - (2 + 4) = 0$$

These oxygens also have no formal charges.

Now let's compare the two structures side by side.

$$\text{H}-\overset{..}{\underset{..}{\text{O}}}-\underset{\underset{\overset{..}{\underset{..}{\text{O}}}}{\|}}{\overset{\overset{\overset{..}{\text{O}}}{\|}}{\text{S}}}-\overset{..}{\underset{..}{\text{O}}}-\text{H} \qquad \text{H}-\overset{..}{\underset{..}{\text{O}}}-\underset{\underset{:\overset{..}{\text{O}}:_{\ominus}}{|}}{\overset{\overset{:\overset{..}{\text{O}}:^{\ominus}}{|}}{\text{S}^{\text{(2+)}}}}-\overset{..}{\underset{..}{\text{O}}}-\text{H}$$

Imagine changing the one with the double bonds to the one with the single bonds. This involves creating two pairs of positive–negative charge from something electrically neutral. Stated another way, it involves separating negative charges from positive charges, and this requires an increase in the potential energy. Our conclusion is that the singly bonded structure on the right has a higher potential energy than the one with the double bonds. In general, the lower the potential energy of a molecule, the more stable it is. Therefore, the lower energy structure with the double bonds is, in principle, the more stable structure, so it is preferred over the one with only single bonds. This now gives us a rule that we can use in selecting the best Lewis structures for a molecule or ion:

> When several Lewis structures are possible, those with the smallest formal charges are the most stable and are preferred.

EXAMPLE 8.9

Selecting Lewis Structures Based on Formal Charges

A student drew three Lewis structures for the nitric acid molecule:

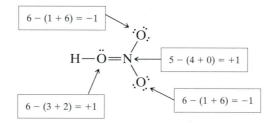

(I)   (II)   (III)

Which one is preferred?

Solution: Structure III can be eliminated at the very beginning. This structure is not acceptable because nitrogen has 10 electrons in its valence shell. Period 2 elements *never* exceed an octet because their valence shells have only *s* and *p* subshells and can accommodate a maximum of eight electrons. Let's now calculate formal charges on the atoms in each of the other two structures.

Structure I:

Formal charge on N = $5 - (4 + 0) = +1$

Formal charge on O attached to H = $6 - (3 + 2) = +1$

Formal charge on O not attached to H = $6 - (1 + 6) = -1$

Structure II:

Formal charge on N = $+1$

Formal charge on O attached to H = $6 - (2 + 4) = 0$

Formal charge on singly bonded O not attached to H = -1

Formal charge on doubly bonded O = $6 - (2 + 4) = 0$

$$H-\overset{..}{\underset{..}{O}}-\overset{\overset{\displaystyle \overset{..}{O}.}{|}}{\underset{\displaystyle \underset{..}{\overset{..}{O}.}}{N}} \quad \boxed{5-(4+0)=+1}$$

$$\boxed{6-(1+6)=-1}$$

Next, we place these formal charges on the atoms in the structures.

$$H-\overset{..}{\underset{\oplus}{O}}=\overset{\displaystyle \overset{\ominus}{\overset{..}{O}:}}{\underset{\displaystyle \underset{\ominus}{:\overset{..}{O}.}}{N\oplus}} \qquad\qquad H-\overset{..}{\underset{..}{O}}-\overset{\overset{\displaystyle \overset{..}{O}.}{\parallel}}{\underset{\displaystyle \underset{\ominus}{:\overset{..}{O}.}}{N\oplus}}$$

(I)   (II)

Because Structure II has fewer formal charges than Structure I, it is the lower energy, preferred Lewis structure for HNO_3.

In Structure III, the formal charges are zero on each of the atoms, but this cannot be the "preferred structure" because the nitrogen atom has too many electrons in its valence shell.

EXAMPLE 8.10

Selecting Lewis Structures Based on Formal Charges

What is the preferred Lewis structure for the molecule $POCl_3$, in which phosphorus is bonded to one oxygen and three chlorine atoms?

Analysis: First we must draw a Lewis structure following the rules in the previous section. Then we can assign formal charges. If there is a way to reduce these formal charges by moving electrons, we will do so to obtain a better Lewis structure.

Solution: By following the rules from the previous section, we obtain the following Lewis structure.

$$\overset{\displaystyle :\overset{..}{Cl}:}{\underset{\displaystyle :\overset{..}{Cl}:}{:\overset{..}{\underset{..}{Cl}}-P-\overset{..}{O}:}}$$

Now we calculate the formal charges.

	Phosphorus	Oxygen	Chlorine
Formal charge	$5-(4+0)=+1$	$6-(1+6)=-1$	$7-(1+6)=0$

Adding formal charges to the Lewis structure gives

$$\overset{\displaystyle :\overset{..}{Cl}:}{\underset{\displaystyle :\overset{..}{Cl}:}{:\overset{..}{\underset{..}{Cl}}-\overset{\oplus}{P}-\overset{..}{\underset{..}{O}}\underset{\ominus}{:}}}$$

We can reduce the formal charges if we can place another electron on phosphorus and remove one from the oxygen. This can be accomplished if we make one of the unshared pairs of electrons on the oxygen into a bond between P and O (which we are permitted to do because phosphorus is a period 3 element; its valence shell can hold more than eight electrons).

$$\overset{\displaystyle :\overset{..}{Cl}:}{\underset{\displaystyle :\overset{..}{Cl}:}{:\overset{..}{\underset{..}{Cl}}-P\overset{\curvearrowleft}{\overset{..}{O}:}}} \quad \text{gives} \quad \overset{\displaystyle :\overset{..}{Cl}:}{\underset{\displaystyle :\overset{..}{Cl}:}{:\overset{..}{\underset{..}{Cl}}-P=\overset{..}{O}:}}$$

Since this structure contains no formal charges, it is preferred over the previous structure.

When an atom with a negative formal charge is bonded to an atom with a positive formal charge, the atom with the negative formal charge can donate a pair of electrons to form a double bond.

EXAMPLE 8.11

Selecting Lewis Structures Based on Formal Charges

Two structures can be drawn for BCl_3, as shown below

$$:\ddot{Cl}: \qquad :\ddot{Cl}:$$
$$\overset{|}{:\ddot{Cl}-B-\ddot{Cl}:} \qquad :\ddot{Cl}=B-\ddot{Cl}:$$
$$\text{(I)} \qquad\qquad \text{(II)}$$

Why is the one that violates the octet rule preferred?

Solution: Let's assign formal charges.

$$\text{(I)} \quad :\ddot{Cl}: \qquad \text{(II)} \quad :\ddot{Cl}:$$
$$:\ddot{Cl}-B-\ddot{Cl}: \qquad :\overset{\oplus}{Cl}=\underset{\ominus}{B}-\ddot{Cl}:$$

In Structure I all the formal charges are zero. In Structure II, two of the atoms have formal charges, so this alone would argue in favor of Structure I. There is another argument in favor as well. Notice that the formal charges in Structure II place the positive charge on the more electronegative chlorine atom and the negative charge on the less electronegative boron. If charges could form in this molecule, they certainly would not be expected to form in this way. Therefore, there are two factors that make the structure with the double bond unfavorable, so we usually write the Lewis structure for BCl_3 as shown in (I).

Practice Exercise 8

Assign formal charges to the atoms in the following Lewis structures:

(a) $:\ddot{N}-N\equiv O:$ (b) $\left[:\ddot{S}=C=\ddot{N}:\right]^{-}$ ◆

Practice Exercise 9

Select the preferred Lewis structures for (a) SO_2, (b) $HClO_3$, and (c) H_3PO_4. ◆

Formic acid is an organic acid. It has the structure

$$:\overset{\ddot{O}}{\underset{\|}{}}$$
$$H-C-\ddot{O}-H$$

The name formic acid comes from *formica,* the Latin word for ant. The one shown here is a fire ant.

8.9 Resonance: When a Single Lewis Structure Fails

There are some molecules and ions for which we cannot write Lewis structures that agree with experimental measurements of bond length and bond energy. One example is the formate ion, CHO_2^-, which is produced by neutralizing formic acid, $HCHO_2$ (the substance that causes the stinging sensation in bites from fire ants). The skeletal structure for this ion is

$$\begin{array}{cc} & O \\ H & C \\ & O \end{array}$$

and, following the usual steps, we would write its Lewis structure as

$$\left[H-C \begin{array}{c} \ddot{O}: \\ \diagup \\ \diagdown \\ :\ddot{O}: \end{array} \right]^{-}$$

This structure suggests that one carbon–oxygen bond should be longer than the other, but experiment shows that they are identical. In fact, the C—O bond lengths are about halfway between the expected values for a single bond and a double bond. The Lewis structure doesn't match the experimental evidence, and there's no way to write one that does. It would require showing all of the electrons in pairs and, at the same time, showing 1.5 pairs of electrons per bond.

The way we get around problems like this is through the use of a concept called **resonance.** We view the actual structure of the molecule or ion, which we cannot draw satisfactorily, as a composite, or average, of a number of Lewis structures that we can draw. For example, for formate we write

$$\left[\ H-C{\overset{\displaystyle \cdot\cdot O\cdot}{\underset{\displaystyle \cdot O\cdot}{\Big\langle}}} \ \right]^{-} \longleftrightarrow \left[\ H-C{\overset{\displaystyle \cdot\cdot O\cdot}{\underset{\displaystyle \cdot\cdot O\cdot}{\Big\langle}}} \ \right]^{-}$$

No atoms have been moved; the electrons have just been redistributed.

where we have simply shifted electrons around in going from one structure to the other. The bond between the carbon and a particular oxygen is depicted as a single bond in one structure and as a double bond in the other. The average of these is 1.5 bonds (halfway between a single and a double bond), which is in agreement with the experimental bond lengths. These two Lewis structures are called **resonance structures** or **contributing structures,** and the actual structure of the ion is said to be a **resonance hybrid** of the two resonance structures that we've drawn. The double-ended arrow is used to show that we are drawing resonance structures and implies that the true hybrid structure is a composite of the two resonance structures.

The term *resonance* is often misleading to the beginning student. The word itself suggests that the actual structure flip-flops back and forth between the two structures shown. This is *not* the case! A mule, which is the *hybrid* offspring of a donkey and a horse, isn't a donkey one minute and a horse the next! Although it may have characteristics of both parents, a mule is a mule. A *resonance hybrid* also has characteristics of its "parents," but it never has the exact structure of any of them.

There is a simple way to determine when resonance should be applied to Lewis structures. If you find that you must move electrons to create one or more double bonds while following the procedures developed in the previous sections, the number of resonance structures is equal to the number of equivalent choices for the locations of the double bonds. For example, in drawing the Lewis structure for the NO_3^- ion, we reach the stage

A double bond must be created to give the nitrogen an octet. Since it can be placed in any one of three locations, there are three resonance structures for this ion.

$$\left[{\overset{\displaystyle \cdot\cdot O \cdot\cdot}{\underset{\displaystyle \cdot O \ \ N \ \ O \cdot}{|}}} \right]^{-} \longleftrightarrow \left[{\overset{\displaystyle \cdot\cdot O \cdot\cdot}{\underset{\displaystyle \cdot\cdot O \ \ N \ \ O \cdot}{|}}} \right]^{-} \longleftrightarrow \left[{\overset{\displaystyle \cdot O \cdot}{\underset{\displaystyle \cdot\cdot O \ \ N \ \ O \cdot\cdot}{|}}} \right]^{-}$$

The three oxygens in NO_3^- are said to be equivalent; that is, they are all alike in their chemical environment. The only thing that each oxygen is bonded to is the nitrogen atom.

Notice that each structure is the same, except for the location of the double bond.

In the nitrate ion, the extra bond that moves around from one structure to another is divided among all three bond locations. Therefore, the average bond order in the N—O bonds is expected to be $1\frac{1}{3}$, or 1.33.

EXAMPLE 8.12
Drawing Resonance Structures

Draw resonance structures for the sulfite ion, SO_3^{2-}.

Solution: Following our usual procedure we obtain the Lewis structure

$$\begin{bmatrix} \overset{\cdot\cdot}{\underset{\cdot\cdot}{O}} \\ | \\ :\overset{\cdot\cdot}{O}-S-\overset{\cdot\cdot}{O}: \\ \cdot\cdot \quad\quad \cdot\cdot \end{bmatrix}^{2-}$$

and it doesn't seem that we need the concept of resonance. However, if we assign formal charges, we get

$$\begin{bmatrix} \overset{\ominus}{:}\overset{\cdot\cdot}{\underset{}{O}}: \\ | \\ :\overset{\ominus}{\underset{\cdot\cdot}{O}}-S-\overset{\cdot\cdot}{O}:^{\ominus} \\ \underset{\oplus}{} \end{bmatrix}^{2-}$$

We can obtain a better Lewis structure by reducing the number of formal charges, which can be accomplished by forming a double bond between one of the oxygen atoms and the sulfur. Since there are three choices for the location of this double bond, there are three resonance structures.

$$\begin{bmatrix} :\overset{\cdot\cdot}{O}: \\ | \\ \overset{\cdot\cdot}{O}=S-\overset{\cdot\cdot}{O}: \\ \cdot\cdot \end{bmatrix}^{2-} \longleftrightarrow \begin{bmatrix} :O: \\ || \\ :\overset{\cdot\cdot}{O}-S-\overset{\cdot\cdot}{O}: \\ \cdot\cdot \quad\quad \cdot\cdot \end{bmatrix}^{2-} \longleftrightarrow \begin{bmatrix} :\overset{\cdot\cdot}{O}: \\ | \\ :\overset{\cdot\cdot}{O}-S=\overset{\cdot\cdot}{O} \\ \cdot\cdot \end{bmatrix}^{2-}$$

As with the nitrate ion, we expect an average bond order of 1.33.

Practice Exercise 10

Draw the resonance structures for (a) NO_2^- and (b) PO_4^{3-}. ◆

Effect of Resonance on the Stability of Molecules and Ions

One of the benefits that a molecule or ion derives from existing as a resonance hybrid is that its total energy is lower than any one of its resonance structures. A particularly important example of this occurs with the compound benzene, C_6H_6. This is a ring-shaped molecule with a basic structure that appears in many important organic molecules, ranging from plastics to amino acids.

Two resonance structures are drawn for benzene.

Polystyrene plastic. This common polymer contains benzene rings joined to every other carbon atom in a long hydrocarbon chain. When a gas is blown into melted polystyrene, the lightweight product called styrofoam is formed, which is used to make wide variety of packaging products for consumer goods.

These are usually represented as hexagons with dashes showing the locations of the double bonds. It is assumed that at each apex of the hexagon there is a carbon bonded to a hydrogen as well as to the adjacent carbon atoms.

Usually, the actual structure of benzene—that of its resonance hybrid—is represented as a hexagon with a circle in the center. This is intended to show that the electron density of the three extra bonds is evenly distributed around the ring.

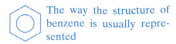

Although the individual resonance structures for benzene show double bonds, the molecule does not react like other organic molecules that have true carbon–carbon double bonds. The reason appears to be that the resonance hybrid is considerably more stable than either of these resonance forms. In fact, it has been calculated that the actual structure of the benzene molecule is more stable than either of its resonance structures by approximately 146 kJ/mol. This extra stability achieved through resonance is called the **resonance energy.**

8.10 Coordinate Covalent Bonds

When an ammonia molecule is formed from a nitrogen atom and three hydrogen atoms, each hydrogen shares an electron with one of the electrons of the nitrogen atom. By this process, three bonds are formed by nitrogen and an octet of electrons is achieved around the central atom.

$$3H\cdot + \cdot\ddot{N}\!: \longrightarrow H\!-\!\underset{\underset{H}{|}}{\overset{\overset{H}{|}}{N}}\!:$$

One might expect that at this point, no further bonds to nitrogen can be made. However, when NH_3 is placed into an acidic solution, it picks up a hydrogen ion, H^+, and becomes NH_4^+ in which there are four bonds between nitrogen and hydrogen. To understand how this can happen, let's keep tabs on all of the electrons by using red dots for the hydrogen electrons and black dots for nitrogen electrons.

$$H\!:\!\underset{\underset{H}{..}}{\overset{..}{N}}\!: + H^+ \longrightarrow \left[H\!:\!\underset{\underset{H}{..}}{\overset{H}{N}}\!:\!H\right]^+$$

All electrons are alike, of course. We are using different colors for them so we can see where the electrons in the bond came from.

Three of the N—H bonds are formed in the usual way, with each atom contributing one electron toward the pair that's shared. Notice, however, that the fourth bond is created differently, with both electrons in the bond being contributed by the nitrogen.

The reason this fourth bond can form is because the H^+ ion has an empty $1s$ orbital in its valence shell which can accommodate both of the electrons from the nitrogen. The result is that when the H^+ becomes bonded to the nitrogen of NH_3, the nitrogen donates both of the electrons to the bond. *This type of bond, in which both electrons of the shared pair come from just one of the two atoms, is called a* **coordinate covalent bond.**

It is important to understand that even though we make a distinction as to the *origin* of the electrons, once the bond is formed it is really the same as any other covalent bond. In other words, we can't tell where the electrons in the bond came from after the bond has been formed. In NH_4^+, for instance, all four of the N—H bonds are equivalent once they've been formed, and no distinction is made between them. We usually write the structure of NH_4^+ as

The notion of coordinate covalent bonds is used to explain how bonds are formed, rather than what the bonds are like after they have been formed.

$$\left[H\!-\!\underset{\underset{H}{|}}{\overset{\overset{H}{|}}{N}}\!-\!H\right]^+$$

The concept of a coordinate covalent bond is sometimes useful when we are attempting to understand what happens to atoms in a chemical reaction. For example, it is known that if ammonia is added to boron trichloride, an exothermic reaction takes place and the compound NH_3BCl_3 is formed in which there is a boron–nitrogen bond. Using Lewis structures, we can diagram this reaction as follows.

Compounds like BCl_3NH_3, which are formed by simply joining two smaller molecules, are sometimes called **addition compounds.**

In the reaction, a boron atom with less than an octet achieves a complete valence shell by sharing a pair of electrons that originates on the nitrogen atom. We might say that "the boron forms a coordinate covalent bond with the nitrogen of the ammonia molecule."

An arrow sometimes is used to represent the donated pair of electrons in the coordinate covalent bond. The direction of the arrow indicates the direction in which the electron pair is donated, in this case from the nitrogen to the boron.

We will find the notion of coordinate covalent bonds useful later in the book when we describe certain kinds of acid–base reactions as well as certain compounds of transition metal ions.

Practice Exercise 11

Use Lewis structures to explain how the reaction between hydroxide ion and hydrogen ion involves the formation of a coordinate covalent bond. ◆

SUMMARY

Ionic Bonding. The forces of attraction between positive and negative ions in an ionic compound are called **ionic bonds.** The formation of ionic compounds by electron transfer is favored when atoms of low ionization energy react with atoms of high electron affinity. The chief stabilizing influence in the formation of ionic compounds is the **lattice energy**—the lowering of the potential energy that occurs when gaseous ions are brought together to form the crystal of the ionic compound. When atoms of the elements in Groups IA and IIA as well as the nonmetals form ions, they usually gain or lose enough electrons to achieve a noble gas electron configuration. Transition elements lose their outer *s* electrons first, followed by loss of *d* electrons from the shell below the outer shell. Post-transition metals lose electrons from their outer *p* subshell first, followed by loss of electrons from the *s* subshell in the outer shell.

Covalent Bonding. Electron sharing between atoms occurs when electron transfer is energetically too "expen-

sive." Shared electrons attract the positive nuclei, and this leads to a lowering of the potential energy of the atoms as a covalent bond forms. Electrons generally become paired when they are shared. An atom tends to share enough electrons to complete its valence shell. Except for hydrogen, the valence shell usually holds eight electrons, which forms the basis of the octet rule. The **octet rule** states that atoms of the representative elements tend to acquire eight electrons in their outer shells when they form bonds. **Single, double,** and **triple bonds** involve the sharing of one, two, and three pairs of electrons, respectively, between two atoms. Compounds of B and Be often have less than an octet around the central atom. Atoms of the elements of period 2 cannot have more than an octet because their outer shells can hold only eight electrons. Elements in period 3 and below can exceed an octet if they form more than four bonds.

Bond energy (the energy needed to separate the bonded atoms) and **bond length** (the distance between the nuclei of the atoms connected by the bond) are two experimentally measurable quantities that can be related to the number of pairs of electrons in the bond. For bonds between atoms of the same elements, bond energy increases and bond length decreases as the **bond order** increases.

Organic Compounds. Carbon normally forms four covalent bonds, either to atoms of other elements or to other carbon atoms. Oxygen-containing organic compounds include **alcohols** (which have an O—H group bonded to carbon), **ketones** (which have a doubly bonded oxygen atom attached to a carbon that is also bonded to two other carbon atoms), **aldehydes** (in which the C=O group is attached to a hydrogen), and **carboxylic acids** (organic acids), which contain the carboxyl group. Amines are nitrogen-containing compounds derived from ammonia in which a hydrogen of NH_3 is replaced by a hydrocarbon group. **Amino acids** contain both a carboxyl group and an amine group.

Electronegativity. The attraction an atom has for the electrons in a bond is called the atom's **electronegativity.** When atoms of different electronegativities form a bond, the electrons are shared unequally and the bond is **polar,** with **partial positive** and **partial negative** charges at opposite ends. When the two atoms have the same electronegativity, the bond is **nonpolar.** The extent of polarity of the bond depends on the electronegativity difference between the two bonded atoms. When the electronegativity difference is very large, ionic bonding results. A bond is approximately 50% ionic when the electronegativity difference is 1.7. In the periodic table, electronegativity increases from left to right across a period and from bottom to top in a group.

Reactivity and Electronegativity. The reactivity of metals is related to the ease with which they are oxidized (lose electrons); for nonmetals it is related to the ease with which they are reduced (gain electrons). Metals with low electronegativities lose electrons easily, are good reducing agents, and tend to be very reactive. The most reactive metals are located in Groups IA and IIA, and their ease of oxidation increases going down the group. For nonmetals, the higher the electronegativity, the stronger is their ability to serve as oxidizing agents. The strongest oxidizing agent is fluorine. Among the halogens, oxidizing strength decreases from fluorine to iodine.

Lewis Symbols and Lewis Structures. Lewis symbols are a bookkeeping device used to keep track of valence electrons in ionic and covalent bonds. The **Lewis symbol** of an element consists of the element's chemical symbol surrounded by a number of dots equal to the number of valence electrons. In the **Lewis structure** for an ionic compound, the Lewis symbol for the anion is enclosed in

brackets (with the charge written outside) to show that all the electrons belong entirely to the ion. The Lewis structure for a molecule or polyatomic ion uses pairs of dots between chemical symbols to represent shared pairs of electrons. The electron pairs in covalent bonds usually are represented by dashes; one dash equals two electrons. The following procedure is used to draw the Lewis structure: (1) decide on the skeletal structure (remember that the least electronegative atom is usually the central atom and is usually first in the formula); (2) count all the valence electrons, taking into account the charge, if any; (3) place a pair of electrons in each bond; (4) complete the octets of atoms other than the central atom (but remember that hydrogen can only have two electrons); (5) place any leftover electrons on the central atom in pairs; (6) if the central atom still has *less than* an octet, move electron pairs to make double or triple bonds.

Formal Charges. The **formal charge** assigned to an atom in a Lewis structure is the difference between the number of valence electrons of an isolated atom of the element and the number of electrons that "belong" to the atom because of its bonds to other atoms and its unshared valence electrons. The sum of the formal charges always equals the net charge on the molecule or ion. The most stable (lowest energy) Lewis structure for a molecule or ion is the one with the fewest formal charges. This is usually the preferred Lewis structure for the particle.

Resonance. Two or more atoms in a molecule or polyatomic ion are chemically equivalent if they are attached to the same kinds of atoms or groups of atoms. Bonds to chemically equivalent atoms must be the same; they must have the same bond length and the same bond energy, which means they must involve the sharing of the same number of electron pairs. Sometimes the Lewis structures we draw suggest that the bonds to chemically equivalent atoms are not the same. Typically, this occurs when it is necessary to form multiple bonds during the drawing of a Lewis structure. When alternatives exist for the location of this multiple bond among two or more equivalent atoms, then each possible Lewis structure is actually a **resonance structure** or **contributing structure,** and we draw them all. In drawing resonance structures, the relative locations of the nuclei must be identical in all. Remember that none of the resonance structures corresponds to a real molecule, but their composite—the **resonance hybrid**—does approximate the actual structure of the molecule or ion.

Coordinate Covalent Bonding. For bookkeeping purposes, we sometimes single out a covalent bond whose electron pair originated from one of the two bonded atoms. An arrow is sometimes used to indicate the donated pair of electrons. Once formed, a coordinate covalent bond is no different than any other covalent bond.

Tools → *you have learned*

The table below lists the tools that you have learned in this chapter. You will need them to answer chemistry questions. Review them if necessary, and refer to the tools when working on the Thinking-It-Through problems and the Review Problems that follow.

Tool	Function
Rules for electron configurations of ions *(pages 329 and 330)*	To be able to write correct electron configurations for cations and anions.
Lewis symbols *(page 332)*	A bookkeeping device for keeping track of valence electrons in atoms and ions. Learn to use the periodic table to construct the Lewis symbol.
Periodic trends in electronegativity *(page 342)*	To compare the relative polarities of chemical bonds and to identify the most (or least) electronegative element in a compound.
Method for drawing Lewis structures *(page 345)*	A systematic method for constructing the Lewis structure for a molecule or polyatomic ion.
Correlation between bond properties and bond order *(page 349)*	To compare experimental properties that are related to covalent bonds with those predicted by theory.
Formal charges *(page 350)*	To select the best Lewis structure for a molecule or polyatomic ion.
Method for determining resonance structures *(page 355)*	Resonance structures provide a way of describing the structures of molecules for which a single satisfactory Lewis structure cannot be drawn.

THINKING IT THROUGH

By now you know that the goal here is *not* to find the answer itself; instead, you are asked to assemble the available information needed to obtain the answer, state what additional data (if any) are needed, and describe *how* you would use the data to answer the question. For problems involving unit conversions, list the relationships among the units that are needed to carry out the conversions. Construct the conversion factors that can be formed from these relationships. Then set up the solution to the problem by arranging the conversion factors so the units cancel correctly to give the desired units of the answer.

The problems are divided into two groups. Those in Level 2 are more challenging than those in Level 1 and provide an opportunity to really hone your problem solving skills. Detailed answers to these problems can be found at our Web site: http://www.wiley.com/college/brady

Level 1 Problems

1. How can you determine how much energy would be absorbed or given off in the formation of the gaseous ions that could eventually come together to form one mole of Na_2O?

2. Describe how we could use Lewis symbols to follow the formation of K_2S from potassium and sulfur atoms.

3. Explain how you would figure out the changes in electron configuration that accompany the following:

$$Pb \longrightarrow Pb^{2+} \longrightarrow Pb^{4+}$$

4. Why is K_2S ionic but H_2S is molecular? What factors must you consider in answering this question?

5. How would you determine which molecule is likely to have the more polar bonds, PCl_3 or SCl_2? In these molecules, how can you tell which atoms carry the partial positive and partial negative charges?

Level 2 Problems

6. How many double bonds are in the Lewis structure for $HBrO_3$? Explain in detail how you would answer this question.

7. What is the preferred Lewis structure for H_3AsO_4? In detail, explain what knowledge you must have to answer this question.

8. Are the following Lewis structures considered to be resonance structures? Explain. How can you predict which is the more likely structure for $POCl_3$?

9. How can you make predictions about the relative S—O bond lengths in the SO_3^{2-} and SO_4^{2-} ions?

10. What wavelength of light, if absorbed by a hydrogen molecule, would cause the molecule to split into two atoms of hydrogen? Set up the calculation. (The data required are available in this and previous chapters.)

11. What wavelength of light, if absorbed by a hydrogen molecule, could cause the molecule to split into the ions H^+ and H^-? Set up the calculations. (The data required are available in this and previous chapters.)

12. The lattice energy is defined as the change in potential energy when the ions in one mole of an ionic compound are brought from the gaseous state to their positions in the crystalline lattice. Lattice energies are always exothermic because the ions have a net attraction for each other. Explain what the phrase *net attraction* means. Examine the figure in the margin of page 326. Note that every ion in the crystal shows an attractive Coulombic force with all other ions of the opposite charge and a repulsive Coulombic force with all other ions of the same charge.

13. The formation of $NaCl(s)$ from $Na(s)$ and $Cl_2(g)$ can be viewed as occurring by a direct path (Step 1 in the figure below) or by a multistep path (Steps 2–6 in the figure).

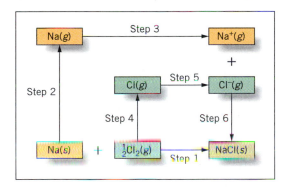

(a) For each of Steps 2–6 in the multistep path, describe the nature of the physical change that occurs and explain whether the change is exothermic or endothermic.
(b) Look up the standard heat of formation of $NaCl(s)$ and discuss how this value relates to the sum of the ΔHs for Steps 2–6.

14. By ranking the strengths of the attractive forces between the cations and anions in the compounds below, arrange them in order of increasing lattice energies. Note: Consider both the charges of the ions and their relative sizes.

NaCl   MgO   KBr   CaO

15. Use Figures 8.6 and 8.7 to discuss the bonding in the following compounds (some of which are hypothetical): CO_2, CS_2, CSe_2, CTe_2, MgO, MgS, MgSe, and MgTe. Is it reasonable to claim that binary compounds between metals and nonmetals are always ionic while those between two nonmetals are always covalent? Explain.

16. Figure 8.7 shows that, as a function of electronegativity differences between atoms forming a chemical bond, the percent ionic character of the bond can vary from zero to over 90 percent. Is this figure consistent with the statement that the bonding pair for CsF is held *exclusively* by fluorine? Explain what the figure should show if this statement were precisely true. Is the extension of Figure 8.7 to a 3.2 electronegativity difference likely to give this result?

REVIEW QUESTIONS

Ionic Bonding

8.1 What must be true about the change in the total potential energy of a collection of atoms for a stable compound to be formed from the elements?

8.2 What is an *ionic bond*?

8.3 How is the tendency to form ionic bonds related to the IE and EA of the atoms involved?

8.4 Define the term *lattice energy*. In what ways does the lattice energy contribute to the stability of ionic compounds?

8.5 Magnesium forms the ion Mg^{2+}, but not the ion Mg^{3+}. Why?

8.6 Why doesn't chlorine form the ion Cl^{2-}?

8.7 Why do many of the transition elements in period 4 form ions with a 2+ charge?

Lewis Symbols

8.8 The Lewis symbol for an atom only accounts for electrons in the valence shell of the atom. Why?

8.9 What relationship exists between the number of dots in a Lewis symbol and the location of the element in the periodic table?

Electron Sharing

8.10 Describe what happens to the electron density around two hydrogen atoms as they come together to form an H_2 molecule.

8.11 In terms of the potential energy change, why doesn't ionic bonding occur when two nonmetals react with each other?

8.12 Is the formation of a covalent bond from separated atoms endothermic or exothermic?

8.13 What happens to the energy of two hydrogen atoms as they approach each other? What happens to the spins of the electrons?

8.14 Why is a covalent bond sometimes called an *electron pair bond*?

8.15 What holds the two nuclei together in a covalent bond?

8.16 What factors control the bond length in a covalent bond?

Covalent Bonding and the Octet Rule

8.17 What is the *octet rule?* What is responsible for it?

8.18 How many covalent bonds are normally formed by (a) hydrogen, (b) carbon, (c) oxygen, (d) nitrogen, and (e) chlorine?

8.19 Why do period 2 elements never form more than four covalent bonds? Why are period 3 elements able to exceed an octet when they form bonds?

8.20 Define (a) *single bond,* (b) *double bond,* and (c) *triple bond.*

8.21 What is a structural formula?

8.22 $H—C\equiv N:$ is the Lewis structure for hydrogen cyanide. Draw circles enclosing electrons to show that carbon and nitrogen obey the octet rule.

8.23 Why doesn't hydrogen obey the octet rule? How many covalent bonds does a hydrogen atom form?

Compounds of Carbon

8.24 What special "talent" does carbon have that allows it to form so many compounds?

8.25 Sketch the Lewis structures for (a) methane, (b) ethane, and (c) propane.

8.26 Draw the structure for a hydrocarbon that has a chain of five carbon atoms linked by single bonds. How many hydrogen atoms does the molecule have? What is the molecular formula for the compound?

8.27 How many *different* molecules have the formula C_5H_{12}? Sketch their structures.

8.28 Match the compounds on the left with the family types on the right.

$$CH_3—CH_2—\overset{\displaystyle O}{\overset{\displaystyle \|}{C}}—OH \qquad \text{hydrocarbon}$$

$$CH_3—\overset{\displaystyle O}{\overset{\displaystyle \|}{C}}—CH_2—CH_3 \qquad \text{aldehyde}$$

$$CH_3—CH_2—\overset{\displaystyle OH}{\overset{\displaystyle |}{C}}H—CH_3 \qquad \text{amine}$$

$$CH_3—CH_2—\overset{\displaystyle O}{\overset{\displaystyle \|}{C}}—H \qquad \text{acid}$$

$$CH_3—\overset{\displaystyle H}{\overset{\displaystyle |}{N}}—CH_2—CH_3 \qquad \text{alcohol}$$

$$CH_3—CH_2—\overset{\displaystyle |}{C}H—CH_3 \qquad \text{ketone}$$
$$\overset{\displaystyle |}{CH_3}$$

8.29 Write a chemical equation for the ionization in water of the acid in Question 8.28. Is the acid strong or weak?

8.30 Write a chemical equation for the reaction of the amine in Question 8.28 with water.

8.31 Write the net ionic equation for the reaction of the acid and amine in Question 8.28 with each other.

8.32 Draw the structures of (a) ethylene and (b) acetylene.

Polar Bonds and Electronegativity

8.33 What is a polar covalent bond? Define *dipole moment.*

8.34 Define *electronegativity.*

8.35 Which element has the highest electronegativity? Which is the second most electronegative element?

8.36 Which elements are assigned electronegativities of zero? Why?

8.37 Among the following bonds, which are more ionic than covalent? (a) Si—O, (b) Ba—O, (c) Se—Cl, (d) K—Br

8.38 If an element has a low electronegativity, is it likely to be a metal or a nonmetal? Explain your answer.

Electronegativity and the Reactivities of the Elements

8.39 When we say that aluminum is more reactive than iron, which property of these elements are we concerned with?

8.40 In what groups in the periodic table are the most reactive metals found? Where do we find the least reactive metals?

8.41 How is the electronegativity of a metal related to its reactivity?

8.42 Arrange the following metals in their approximate order of reactivity (most reactive first, least reactive last) based on their locations in the periodic table: (a) iridium, (b) silver, (c) calcium, (d) iron.

8.43 Complete and balance equations for the following. If no reaction occurs, write "N.R."
(a) $KCl + Br_2 \rightarrow$ (d) $CaBr_2 + Cl_2 \rightarrow$
(b) $NaI + Cl_2 \rightarrow$ (e) $AlBr_3 + F_2 \rightarrow$
(c) $KCl + F_2 \rightarrow$ (f) $ZnBr_2 + I_2 \rightarrow$

8.44 In each pair, choose the better oxidizing agent.
(a) O_2 or F_2 (c) Br_2 or I_2 (e) Se_8 or Cl_2
(b) As_4 or P_4 (d) P_4 or S_8 (f) As_4 or S_8

Failure of the Octet Rule

8.45 How many electrons are in the valence shells of (a) Be in $BeCl_2$, (b) B in BCl_3, and (c) H in H_2O?

8.46 What is the minimum number of electrons that would be expected to be in the valence shell of As in $AsCl_5$?

Bond Length and Bond Energy

8.47 Define *bond length* and *bond energy.*

8.48 Define bond order. How are bond energy and bond length related to bond order? Why are there these relationships?

8.49 The energy required to break the H—Cl bond to give

H^+ and Cl^- ions would not be called the H—Cl bond energy. Why?

Formal Charge

8.50 What is the definition of formal charge?

8.51 How are formal charges used to select the best Lewis structure for a molecule? What is the basis for this method of selection?

Resonance

8.52 Why is the concept of resonance needed?

8.53 What is a resonance hybrid? How does it differ from the resonance structures drawn for a molecule?

Coordinate Covalent Bonds

8.54 What is a *coordinate covalent bond*?

8.55 Once formed, how (if at all) does a coordinate covalent bond differ from an ordinary covalent bond?

8.56 BCl_3 has an incomplete valence shell. Explain how it could form a coordinate covalent bond with a water molecule.

REVIEW PROBLEMS

Answers to problems whose numbers are printed in color are given in Appendix D.
More challenging problems are marked with asterisks.

Electron Configurations of Ions

8.57 Explain what happens to the electron configurations of Mg and Br when they react to form the compound $MgBr_2$.

8.58 Describe what happens to the electron configurations of lithium and nitrogen when they react to form the compound Li_3N.

8.59 What are the electron configurations of the Pb^{2+} and Pb^{4+} ions?

8.60 What are the electron configurations of the Bi^{3+} and Bi^{5+} ions?

8.61 Write the abbreviated electron configuration of the Mn^{3+} ion. How many unpaired electrons does the ion contain?

8.62 Write the abbreviated electron configuration of the Co^{3+} ion. How many unpaired electrons does the ion contain?

Lewis Symbols

8.63 Write Lewis symbols for the following atoms: (a) Si, (b) Sb, (c) Ba, (d) Al, (e) S.

8.64 Write Lewis symbols for the following atoms: (a) K, (b) Ge, (c) As, (d) Br, (e) Sc.

8.65 Write Lewis symbols for the following ions: (a) K^+, (b) Al^{3+}, (c) S^{2-}, (d) Si^{4-}, (e) Mg^{2+}.

8.66 Write Lewis symbols for the following ions: (a) Br^-, (b) Se^{2-}, (c) Li^+, (d) C^{4-}, (e) As^{3-}.

8.67 Use Lewis symbols to diagram the reactions between (a) Ca and Br, (b) Al and O, and (c) K and S.

8.68 Use Lewis symbols to diagram the reactions between (a) Mg and S, (b) Mg and Cl, and (c) Mg and N.

Bond Energy

8.69 How many grams of water could have its temperature raised from 25 °C (room temperature) to 100 °C (the boiling point of water) by the amount of energy released in the formation of 1 mol of H_2 from hydrogen atoms? The bond energy of H_2 is 435 kJ/mol.

8.70 How much energy, in joules, is required to break the bond in *one* chlorine molecule? The bond energy of Cl_2 is 242.6 kJ/mol.

8.71 A carbon–carbon single bond has a bond energy of approximately 348 kJ per mole. What wavelength of light is required to provide sufficient energy to break the C—C bond? In which region of the electromagnetic spectrum is this wavelength located?

8.72 A mixture of H_2 and Cl_2 is stable, but a bright flash of light passing through it can cause the mixture to explode. The light causes Cl_2 molecules to split into Cl atoms, which are highly reactive. What wavelength of light is necessary to cause the Cl_2 molecules to split? The bond energy of Cl_2 is 242.6 kJ per mole.

Covalent Bonds and the Octet Rule

8.73 Use Lewis structures to diagram the formation of (a) Br_2, (b) H_2O, and (c) NH_3 from neutral atoms.

8.74 Chlorine tends to form only one covalent bond because it needs just one electron to complete its octet. What are the Lewis structures for the simplest compound formed by chlorine with (a) nitrogen, (b) carbon, (c) sulfur, and (d) bromine?

8.75 Use the octet rule to predict the formula of the simplest compound formed from hydrogen and (a) Se, (b) As, and (c) Si. (Remember, however, that the valence shell of hydrogen can hold only two electrons.)

8.76 What would be the formula for the simplest compound formed from (a) P and Cl, (b) C and F, and (c) I and Cl?

Electronegativity

8.77 Use Figure 8.6 to choose the atom in each of the following bonds that carries the partial positive charge: (a) N—S, (b) Si—I, (c) N—Br, (d) C—Cl.

8.78 Use Figure 8.6 to choose the atom in each of the following bonds that carries the partial negative charge: (a) Hg—I, (b) P—I, (c) Si—F, (d) Mg—N.

8.79 Which of the bonds in Problem 8.77 is the most polar?

8.80 Which of the bonds in Problem 8.78 is the least polar?

Drawing Lewis Structures

8.81 What are the expected skeletal structures for (a) $SiCl_4$, (b) PF_3, (c) PH_3, and (d) SCl_2?

8.82 What are the expected skeletal structures for (a) HIO_3, (b) H_2CO_3, (c) HCO_3^-, and (d) PCl_4^+?

8.83 How many dots must appear in the Lewis structures of (a) $SiCl_4$, (b) PF_3, (c) PH_3, and (d) SCl_2?

8.84 How many dots must appear in the Lewis structures of (a) HIO_3, (b) H_2CO_3, (c) HCO_3^-, and (d) PCl_4^+?

8.85 Draw Lewis structures for (a) $AsCl_4^+$, (b) ClO_2^-, (c) HNO_2, and (d) XeF_2.

8.86 Draw Lewis structures for (a) TeF_4, (b) ClF_5, (c) PF_6^-, and (d) XeF_4.

8.87 Draw Lewis structures for (a) $SiCl_4$, (b) PF_3, (c) PH_3, and (d) SCl_2.

8.88 Draw Lewis structures for (a) HIO_3, (b) H_2CO_3, (c) HCO_3^-, and (d) PCl_4^+.

8.89 Draw Lewis structures for (a) CS_2 and (b) CN^-.

8.90 Draw Lewis structures for (a) SeO_3 and (b) SeO_2.

8.91 Draw Lewis structures for (a) AsH_3, (b) $HClO_2$, (c) H_2SeO_3, and (d) H_3AsO_4.

8.92 Draw Lewis structures for (a) NO^+, (b) NO_2^-, (c) $SbCl_6^-$, and (d) IO_3^-.

8.93 Draw the Lewis structure for (a) CH_2O (the central atom is carbon, which is attached to two hydrogens and an oxygen), and (b) $SOCl_2$ (the central atom is sulfur, which is attached to an oxygen and two chlorines).

8.94 Draw Lewis structures for (a) $GeCl_4$, (b) CO_3^{2-}, (c) PO_4^{3-}, and (d) O_2^{2-}.

Formal Charge

8.95 Assign formal charges to each atom in the following structures:

(a) $H—\ddot{\underset{..}{O}}—\ddot{\underset{..}{Cl}}—\ddot{\underset{..}{O}}:$ (c) $\ddot{\underset{..}{O}}=\ddot{S}—\ddot{\underset{..}{O}}:$

(b) $\ddot{\underset{..}{O}}=\overset{\overset{:\ddot{O}:}{|}}{S}—\ddot{\underset{..}{O}}:$

8.96 Assign formal charges to each atom in the following structures:

(a) $:\ddot{\underset{..}{Cl}}—\overset{\overset{:O:}{\|}}{N}—\ddot{\underset{..}{O}}:$ (c) $:\ddot{\underset{..}{F}}—\overset{\overset{:\ddot{O}:}{|}}{\underset{\underset{:\ddot{O}:}{|}}{S}}—\ddot{\underset{..}{F}}:$

(b) $:\ddot{\underset{..}{F}}—\overset{\overset{:\ddot{F}:}{|}}{\underset{\underset{:\ddot{F}:}{|}}{N}}—\ddot{\underset{..}{O}}:$

8.97 Draw the Lewis structure for $HClO_4$ according to the rules in Section 8.7. Assign formal charges to each atom in

the formula. Determine the preferred Lewis structure for this compound.

8.98 Draw the Lewis structure for $SOCl_2$ (sulfur bonded to two Cl and one O). Assign formal charges to each atom. Determine the preferred Lewis structure for this molecule.

8.99 Below are two structures for $BeCl_2$. Give two reasons why the one on the left is the preferred structure.

$$:\ddot{\underset{..}{Cl}}—Be—\ddot{\underset{..}{Cl}}: \qquad :\ddot{\underset{..}{Cl}}—Be=\ddot{Cl}:$$

8.100 The following are two Lewis structures that can be drawn for phosgene, a substance that has been used as a war gas.

$$:\ddot{\underset{..}{Cl}}—\overset{\overset{:\ddot{O}:}{|}}{C}=\ddot{Cl}: \qquad :\ddot{\underset{..}{Cl}}—\overset{\overset{\ddot{O}:}{\|}}{C}—\ddot{\underset{..}{Cl}}:$$

Which is the better Lewis structure? Why?

Resonance

8.101 Draw the resonance structures for CO_3^{2-}. Calculate the average C—O bond order.

8.102 Draw all the resonance structures for the N_2O_4 molecule and determine the average N—O bond order. The skeletal structure of the molecule is

$$\begin{matrix} O & & & O \\ & N & N & \\ O & & & O \end{matrix}$$

8.103 How should the N—O bond lengths compare in the NO_3^- and NO_2^- ions?

8.104 Arrange the following in order of increasing C—O bond length: CO, CO_3^{2-}, CO_2, HCO_2^- (formate ion, page 354).

8.105 The Lewis structure of CO_2 was given as

$$\ddot{\underset{..}{O}}=C=\ddot{\underset{..}{O}}$$

but two other resonance structures can also be drawn for it. What are they? On the basis of formal charges, why are they not preferred structures?

8.106 Write Lewis structures for the nitrate ion, NO_3^-, and for nitric acid, HNO_3. How many equivalent oxygens are there in each structure? Compare the number of resonance structures for each.

8.107 Following the rules in Figure 8.9, draw the Lewis structure for the sulfate ion. Assign formal charges to each atom in this structure. Reduce formal charges to obtain the best Lewis structure for the ion. Draw the resonance structures for the ion. What is the average sulfur–oxygen bond order in the ion?

8.108 Use formal charges to establish the preferred Lewis structures for the ClO_3^- and ClO_4^- ions. Draw resonance structures for both ions and determine the average Cl—O bond order in each. Which of these ions would be expected to have the shorter Cl—O bond length?

Coordinate Covalent Bonds

8.109 Use Lewis structures to show that the hydronium ion, H_3O^+, can be considered to be formed by the creation of a coordinate covalent bond between H_2O and H^+.

8.110 Use Lewis structures to show that the reaction

$$BF_3 + F^- \longrightarrow BF_4^-$$

involves the formation of a coordinate covalent bond.

ADDITIONAL EXERCISES

8.111 Use data from the tables of ionization energies and electron affinities on pages 310 and 314 to calculate the energy changes for the following reactions.

$$Na(g) + Cl(g) \longrightarrow Na^+(g) + Cl^-(g)$$

$$Na(g) + 2Cl(g) \longrightarrow Na^{2+}(g) + 2Cl^-(g)$$

Approximately how many times larger would the lattice energy of $NaCl_2$ have to be compared to the lattice energy of $NaCl$ for $NaCl_2$ to be more stable than $NaCl$?

8.112 Changing 1 mol of $Mg(s)$ and 1/2 mol of $O_2(g)$ to gaseous atoms requires a total of approximately 150 kJ of energy. The first and second ionization energies of magnesium are 737 and 1450 kJ/mol; the first and second electron affinities of oxygen are -141 and $+844$ kJ/mol; and the standard heat of formation of $MgO(s)$ is -602 kJ/mol. Construct an enthalpy diagram similar to the one in Figure 8.1 and use it to calculate the lattice energy of magnesium oxide. How does the lattice energy of MgO compare with that of NaCl? What might account for the difference?

*__8.113__ Use an enthalpy diagram to calculate the lattice energy of $CaCl_2$ from the following information. Energy needed to vaporize one mole of $Ca(s)$ is 192 kJ. For calcium, 1st IE = 589.5 kJ mol^{-1}, 2nd IE = 1146 kJ mol^{-1}. Electron affinity of Cl is 348 kJ mol^{-1}. Bond energy of Cl_2 is 242.6 kJ per mole of Cl—Cl bonds. Standard heat of formation of $CaCl_2$ is -795 kJ mol^{-1}.

8.114 Use an enthalpy diagram and the following data to calculate the electron affinity of bromine. Standard heat of formation of NaBr is -360 kJ mol^{-1}. Energy needed to vaporize one mole of $Br_2(l)$ to give $Br_2(g)$ is 31 kJ mol^{-1}. The first ionization energy of Na is 495.4 kJ mol^{-1}. The bond energy of Br_2 is 192 kJ per mole of Br—Br bonds. The lattice energy of NaBr is -743.3 kJ mol^{-1}.

8.115 In many ways, tin(IV) chloride behaves more like a covalent molecular species than as a typical ionic chloride. Draw the Lewis structure for the tin(IV) chloride molecule.

8.116 In each pair, choose the one with the more polar bonds. (Use the periodic table to answer the question.)
(a) PCl_3 or $AsCl_3$ (c) $SiCl_4$ or SCl_2
(b) SF_2 or GeF_4 (d) SrO or SnO

8.117 How many electrons are in the outer shell of the Zn^{2+} ion?

8.118 The Lewis structure for carbonic acid (formed when CO_2 dissolves in water) is usually given as

What is wrong with the following structures?

(a)

(b)

(c)

8.119 Assign formal charges to all the atoms in the following Lewis structure of hydrazoic acid, HN_3.

$$H-N\equiv N-\ddot{N}:$$

Suggest a lower energy resonance structure for this molecule.

8.120 Assign formal charges to all the atoms in this Lewis structure. Suggest a lower energy Lewis structure for this molecule.

8.121 The inflation of an "air bag" when a car experiences a collision occurs by the explosive decomposition of sodium azide, NaN_3, which yields nitrogen gas that inflates the bag. The following resonance structures can be drawn for the azide ion, N_3^-. Identify the best and worst of them.

8.122 How should the sulfur–oxygen bond lengths compare for the species SO_3, SO_2, SO_3^{2-}, and SO_4^{2-}?

8.123 What is the most reasonable Lewis structure for S_2Cl_2?

*__8.124__ There are two acids that have the formula HCNO. Which of the following skeletal structures are most likely for them? Justify your answer.

H C O N   H N O C   H O C N

H C N O   H N C O   H O N C

These rock climbers trust their lives to safety ropes that keep them from falling to their deaths. Indirectly, they also trust their lives to the covalent bonds that link atoms together in the long chain-like molecules that make up the rope fibers. In this chapter we will study covalent bonds in depth and learn about theories that enable us to explain why covalent bonds form and why molecules have the shapes they do.

Chemical Bonding and Molecular Structure

This Chapter in Context The structure of an ionic compound, such as NaCl, is controlled primarily by the sizes of the ions and their charges. The attractions between the ions have no preferred directions, so if an ionic compound is melted, this structure is lost and the array of ions collapses into a jumbled liquid state. Molecular substances are quite different, however. Molecules have three-dimensional shapes that are determined by the relative orientations of their covalent bonds, and this structure is maintained regardless of whether the substance is a solid, liquid, or a gas.

Many of the properties of a molecule depend on the three-dimensional arrangement of its atoms. For example, the functioning of enzymes, which are substances that affect how fast biochemical reactions occur, requires that there be a very precise fit between one molecule and another. Even slight alterations in molecular geometry can destroy this fit and deactivate the enzyme, which in turn prevents the biochemical reaction involved from occurring.

In this chapter we will explore molecular geometry and study theoretical models that allow us to explain, and in some cases predict, the shapes of molecules. We will also examine theories that explain, in terms of wave mechanics and the electronic structures of atoms, *why* covalent bonds form and *why* they are so highly directional in nature. You will find the knowledge gained here helpful in later discussions of physical properties of substances such as melting point and boiling point.

Describing the three-dimensional structure of a molecule is a tool you will use later in analyzing the physical properties of substances. It will also be useful to you in other chemistry-related courses you might take.

9.1 Some Common Molecular Structures

Molecular shape only becomes a question when there are at least three atoms present. If there are only two, there is no doubt as to how they are arranged; one is just alongside the other. But when there are three or more atoms in a molecule we find that its shape can usually be considered to be derived from one or another of five basic geometrical structures, which are described on the pages ahead. As you study these structures, you should try hard to visualize them in three dimensions. You should also learn how to sketch them in a way that conveys the three-dimensional information.

Basic molecular shapes

Linear Molecules

In a **linear molecule** the atoms lie in a straight line. All diatomic molecules could therefore be considered to be linear. When there are three atoms present the angle formed by the covalent bonds, which we call the **bond angle,** equals 180° as illustrated below.

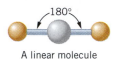

A linear molecule

Planar Triangular Molecules

A **planar triangular molecule** is one in which three atoms are located at the corners of a triangle and are bonded to a fourth atom that lies in the center of the triangle.

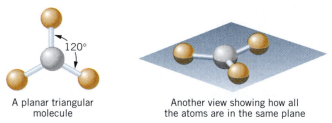

A planar triangular molecule   Another view showing how all the atoms are in the same plane

In this molecule, all four atoms lie in the same plane and the bond angles are all equal to 120°.

Tetrahedral Molecules

A **tetrahedron** is a four-sided geometric figure shaped like a pyramid with triangular faces. A **tetrahedral molecule** is one in which four atoms, located at the vertices of a tetrahedron, are bonded to a fifth atom in the center of the structure.

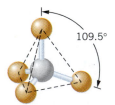

A tetrahedron   A tetrahedral molecule

All the bond angles in a tetrahedral molecule are the same and are equal to 109.5°.

This is the first molecular shape that you've encountered that can't be set down entirely on a two-dimensional surface, so you may need some practice in drawing the structure. It is really quite simple, as illustrated in Figure 9.1.

Drawing molecular shapes is a tool you will find useful in your study of chemistry, so learning to do so is worth the time spent.

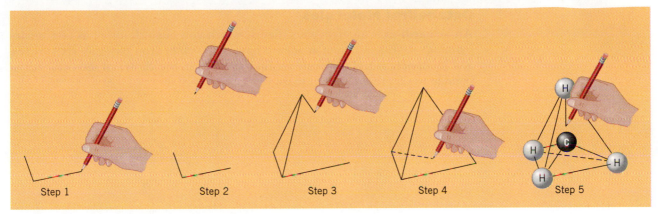

Figure 9.1 *Steps in drawing a tetrahedral molecule.*

Trigonal Bipyramidal Molecules

A **trigonal bipyramid** consists of two *trigonal pyramids* (pyramids with triangular faces) that share a common base.

A trigonal bipyramid

In a **trigonal bipyramidal molecule,** the central atom is located in the middle of the triangular plane shared by the upper and lower trigonal pyramids and is bonded to five atoms that are at the vertices of the figure.

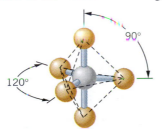

A trigonal bipyramidal molecule

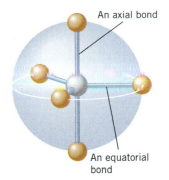

Axial and equatorial bonds in a trigonal bipyramidal molecule.

In this structure, not all the bonds are equivalent. If we imagine the trigonal bipyramid centered inside a sphere similar to Earth, as illustrated in the margin, the atoms in the triangular plane are located around the equator. The bonds to these atoms are called **equatorial bonds.** The angle between two equatorial bonds is 120°. The two vertical bonds pointing along the north and south axis of the sphere are 180° apart and are called **axial bonds.** The bond angle between an axial bond and an equatorial bond is 90°.

A simplified representation of a trigonal bipyramid is illustrated in the margin. The equatorial triangular plane is sketched as it would look tilted backward, so we're looking at it from its edge. The axial bonds are represented as lines pointing up and down. To add more three-dimensional character, notice that the bond pointing down appears to be partially hidden by the triangular plane in the center.

Simplified representation of a trigonal bipyramid.

Octahedral Molecules

An **octahedron** is an eight-sided figure, which you might think of as two *square pyramids* sharing a common square base. The octahedron has only six vertices, and in an **octahedral molecule** we find an atom in the center of the octahedron bonded to six other atoms at these vertices.

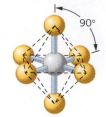

An octahedron   An octahedral molecule

Simplified representation of an octahedron.

All the bonds in an octahedral molecule are equivalent, with angles between adjacent bonds equal to 90°.

A simplified representation of an octahedron is shown in the margin. The square plane in the center of the octahedron, when drawn in perspective and viewed from its edge, looks like a parallelogram. The bonds to the top and bottom of the octahedron are shown as vertical lines. Once again, the bond pointing down is drawn so it appears to be partially hidden by the square plane in the center.

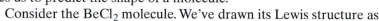

9.2 Predicting the Shapes of Molecules: VSEPR Theory

For a theoretical model to be useful, it should explain known facts and it should be capable of making accurate predictions. A theory that is remarkably successful at both of these, and that is also conceptually simple, has come to be called the **valence shell electron pair repulsion theory,** or **VSEPR theory,** for short. The theory is based on the notion that *valence shell electron pairs, being negatively charged, stay as far apart as possible so that the repulsions between them are at a minimum.* Let's look at an example that illustrates how this simple concept enables us to predict the shape of a molecule.

Consider the $BeCl_2$ molecule. We've drawn its Lewis structure as

$$:\ddot{C}l—Be—\ddot{C}l:$$

But how are these atoms arranged? Is $BeCl_2$ linear or is it nonlinear; that is, do the atoms lie in a straight line, or do they form some angle less than 180°?

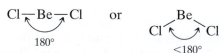

According to VSEPR theory, the shape of a molecule is determined by how the electron pairs in the valence shell of the central atom minimize their mutual repulsions. For $BeCl_2$, there are two pairs of electrons around the beryllium atom. To minimize their repulsions the two electron pairs will seek positions as far apart as possible, which occurs when the electron pairs are on opposite sides of the nucleus. We can represent this as

to suggest the approximate locations of the electron clouds of the valence shell electron pairs. In order for the electrons to be in the Be—Cl bonds, the Cl atoms must be placed where the electrons are; the result is that we predict that a $BeCl_2$ molecule should be linear.

$$Cl-Be-Cl$$

In fact, this is the shape of $BeCl_2$ molecules in the vapor state.

Let's look at another example, the BCl_3 molecule. Its Lewis structure is

$$:\ddot{C}l:$$
$$|$$
$$:\ddot{C}l-B-\ddot{C}l:$$

Here the central atom has three electron pairs. What arrangement will lead to minimum repulsions? As you may have guessed, the electron pairs will be as far apart as possible when they are located at the corners of a triangle with the boron in the center. This places the three electron pairs and the boron nucleus all in the same plane.

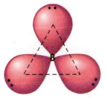

When we attach the Cl atoms, we obtain a planar triangular molecule.

$$Cl$$
$$|$$
$$B$$
$$Cl \quad Cl$$

Experimentally, it has been proved that this is the shape of BCl_3.

Figure 9.2 shows the orientations expected for different numbers of electron pairs around the central atom. It can be shown that these provide the minimum repulsions between the electron pairs. Notice that they yield the same five shapes that we discussed in Section 9.1.

Using Lewis Structures to Predict Molecular Shapes

To predict the shape of a molecule or ion, we need to know how many sets of electron pairs surround the central atom. We can obtain this information by drawing the Lewis structure. For this purpose, we just need the Lewis structure constructed following the rules in Figure 8.9. It is not necessary to assign formal charges and go through the procedure of trying to select "preferred" Lewis structures. Nor is it necessary to draw multiple resonance structures when they might otherwise be appropriate. Just one simple Lewis structure is all we need.

EXAMPLE 9.1

Predicting Molecular Shapes

Carbon tetrachloride was once used as a cleaning fluid until it was discovered that it causes liver damage if absorbed by the body. What is the shape of the CCl_4 molecule?

Analysis: We will begin by drawing a Lewis structure for CCl_4. Then we will count the number of electron pairs in the valence shell of the central atom. From this, we will decide the structure of the molecule.

Number of Electron Pairs	Shape		Example	
2		Linear	$BeCl_2$	180°
3		Planar triangular	BCl_3	120°
4		Tetrahedral (A tetrahedron is pyramid shaped. It has four triangular faces and four corners.)	CH_4	109.5°
5		Trigonal bipyramidal (This figure consists of two three-sided pyramids joined by sharing a common face—the triangular plane through the center.)	PCl_5	
6		Octahedral (An octahedron is an eight-sided figure with *six* corners. It consist of two square pyramids that share a common square base.)	SF_6	

Figure 9.2 *Shapes expected for different numbers of electron pairs around a central atom, M.*

Solution: Following the rules, the Lewis structure of CCl_4 is

$$:\ddot{C}l - C - \ddot{C}l:$$

There are four electron pairs in the valence shell of carbon. These would have minimum repulsions when arranged tetrahedrally, so the molecule is expected to be tetrahedral. (This is, in fact, its structure.)

What shape is expected for the SbCl₅ molecule? ◆

Molecular Shapes when Some Electron Pairs Are Not in Bonds

In the molecules that we've considered so far, all the electron pairs around the central atom have been in bonds. This isn't always the case, though. Some molecules have a central atom with one or more unshared electron pairs. Still, because they are in the valence shell, these unshared electron pairs—also called **lone pairs**—affect the geometry of the molecule. An example is $SnCl_2$.

$$:\ddot{C}l\!-\!\ddot{S}n\!-\!\ddot{C}l:$$

There are three electron pairs around the tin atom, two in bonds plus the lone pair. As in BCl_3, mutual repulsions will direct them toward the corners of a triangle.

$SnCl_2$ is another molecule that behaves as though it has less than an octet around the central atom.

Placing the two chlorine atoms where two of the electron pairs are gives

We can't describe this molecule as triangular, even though that is how the electron pairs are arranged. *Molecular shape describes the arrangement of atoms, not the arrangement of electron pairs.* Therefore, we describe the shape of the $SnCl_2$ molecule as being **nonlinear** or **bent** or **V-shaped.**

Notice, now, that when there are three electron pairs around the central atom, *two* different molecular shapes are possible. If all three electron pairs are in bonds, as in BCl_3, a molecule with a planar triangular shape is formed. If one of the electron pairs is a lone pair, as in $SnCl_2$, the arrangement of the atoms in the molecule is said to be nonlinear. The predicted shapes of both, however, are *derived* by first noting the triangular arrangement of electron pairs around the central atom and *then* adding the necessary number of atoms.

We use the orientations of the electron pairs in the valence shell of the central atom to determine how the atoms in the structure are arranged. The arrangement of the atoms is what we mean by the *shape of the molecule.*

Molecules with Four Electron Pairs in the Valence Shell

The most common type of molecule or ion has four electron pairs (an octet) in the valence shell of the central atom. When these electron pairs are used to form four bonds, as in methane (CH_4), the resulting molecule is tetrahedral, as we've seen. There are many examples, however, where some of the pairs are lone pairs. For example,

$$H\!-\!\ddot{N}\!-\!H \qquad H\!-\!\ddot{O}\!-\!H$$
$$\quad\ \ |$$
$$\quad\ \ H$$

one lone pair two lone pairs

Figure 9.3 shows how the lone pairs affect the shapes of molecules of this type.

Number of Pairs in Bonds	Number of Lone Pairs	Structure	
4	0	109.5°	**Tetrahedral** (example, CH_4) All bond angles are 109.5°.
3	1		**Trigonal pyramidal** (pyramid shaped) (example, NH_3)
2	2		**Nonlinear, bent** (example, H_2O)

Figure 9.3 *Molecular shapes with four electron pairs around the central atom.* The molecules MX_4, MX_3, and MX_2 shown here all have four electron pairs arranged tetrahedrally around the central atom. The shapes of the molecules and their descriptions are derived from the way the X atoms are arranged around M, ignoring the lone pairs.

With one lone pair, the central atom is at the top of a pyramid with three atoms at the corners of the triangular base. The resulting structure is said to be **trigonal pyramidal.** Notice once again that in specifying the shape of the molecule, we ignore the lone pair. Even though the *electron pairs* are arranged in a tetrahedron, the *atoms* are arranged in a trigonal pyramid.

When there are two lone pairs in the tetrahedron, we see that the three atoms (the central atom plus the two atoms bonded to it) do not lie in a straight line, so the structure is **nonlinear.**

If a molecule contains three atoms, they either lie in a straight line and the molecule is linear, *or they are not in a straight line and the molecule is* nonlinear.

Molecules with Five Electron Pairs in the Valence Shell

When five electron pairs are present around the central atom, they are directed toward the vertices of a trigonal bipyramid. Molecules such as PCl_5 have this geometry, as we saw in Figure 9.2, and are said to be **trigonal bipyramidal.**

When there are lone pairs in the valence shell of the central atom, their locations are determined by which arrangement gives the minimum overall repulsions. To determine this, we must compare the sizes of lone pairs and bonding pairs. A bond pair (BP) is attracted to the positive nuclei of two atoms (one at each end of the bond), so its effective size is smaller than a lone pair (LP), which is attracted to the positive nucleus of only one atom. Because of their greater size, lone pairs repel others more strongly than do bond pairs, so we can say that LP–LP repulsions are greater than LP–BP repulsions, which in turn are greater than BP–BP repulsions.

In the trigonal bipyramid, not all the electron pairs have the same nearest neighbor environments. Those in axial positions each have three close neighbors at angles of 90° and one at a much larger angle of 180°. Those in the equatorial plane (the triangular plane through the center of the molecule) each have two neighbors at 90° and two at a much larger angle of 120°. The repulsions that most affect the geometry of molecules are the nearest neighbor interactions, so

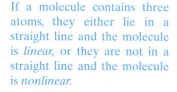

Bonding electron pair

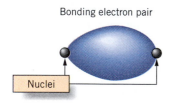

Nuclei

Lone electron pair

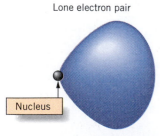

Nucleus

Relative sizes of bonding pairs and lone pairs of electrons.

when lone pairs occur in this structure, they *always* are located in the equatorial plane, because this yields the fewest strong nearest neighbor repulsions.

Figure 9.4 shows the kinds of geometries that we find for different numbers of lone pairs in the trigonal bipyramid. When there is only one lone pair, as in SF_4, the structure is described as a **distorted tetrahedron** in which the central atom lies along one edge of a four-sided figure. With two lone pairs in the equatorial plane, the molecule is **T-shaped** (shaped like the letter T), and when there are three lone pairs in the equatorial plane, the molecule is **linear.**

Molecules with Six Electron Pairs in the Valence Shell

Finally, we come to those molecules or ions that have six electron pairs around the central atom. When all are in bonds, as in SF_6, the molecule is octahedral. When one lone pair is present the molecule or ion has the shape of a **square pyramid,** and when two lone pairs are present they take positions on opposite sides of the nucleus and the molecule or ion has a **square planar** structure. These shapes are shown in Figure 9.5.

If the lone pair is in an axial position, there are three strong BP–LP repulsions. If it is an equatorial position, there are only two strong BP–LP repulsions. Therefore, the favored position for the lone pair is in the equatorial triangle.

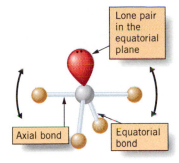

Lone pair in the equatorial plane

Axial bond

Equatorial bond

The distorted tetrahedron is sometimes described as a **seesaw structure,** after the playground apparatus of the same name. The origin of this description can be seen if we tip the structure over so it stands on the two atoms in the equatorial plane, with the lone pair pointing up. The structure can rock back and forth like a seesaw, as indicated by the arrows.

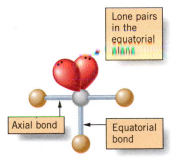

Lone pairs in the equatorial plane

Axial bond

Equatorial bond

A molecule with two lone pairs in the equatorial plane of the trigonal bipyramid. When tipped over, the molecule has the shape of the letter T, so it's described as T-shaped.

Figure 9.4 *Molecular shapes with five electron pairs around the central atom.* Four different molecular structures are possible, depending on the number of lone pairs in the valence shell of the central atom *M.*

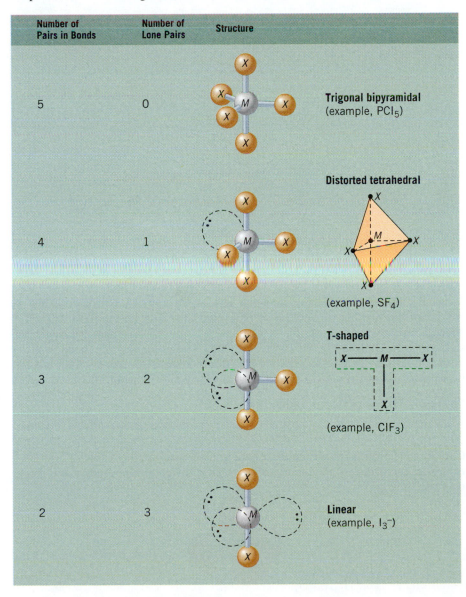

Number of Pairs in Bonds	Number of Lone Pairs	Structure	
5	0		Trigonal bipyramidal (example, PCl_5)
4	1		Distorted tetrahedral (example, SF_4)
3	2		T-shaped (example, ClF_3)
2	3		Linear (example, I_3^-)

Number of Pairs in Bonds	Number of Lone Pairs	Structure	
6	0	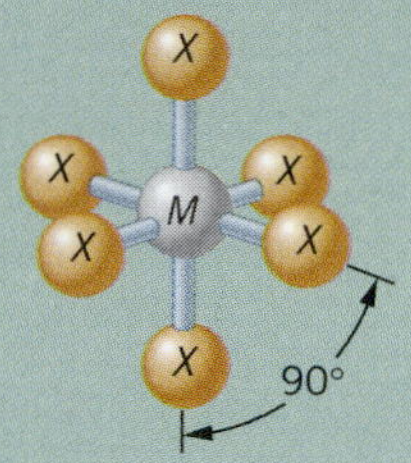	**Octahedral** (example, SF_6) all bond angles are 90°
5	1	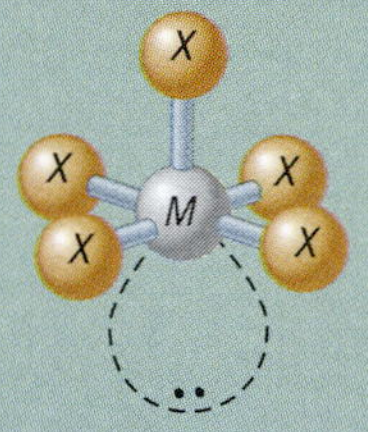	**Square pyramidal** 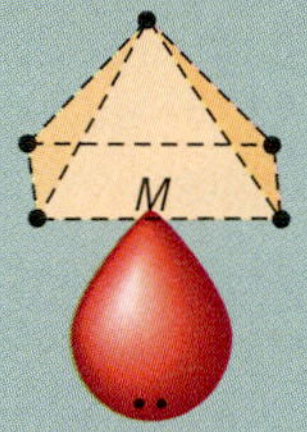 (example, BrF_5)
4	2	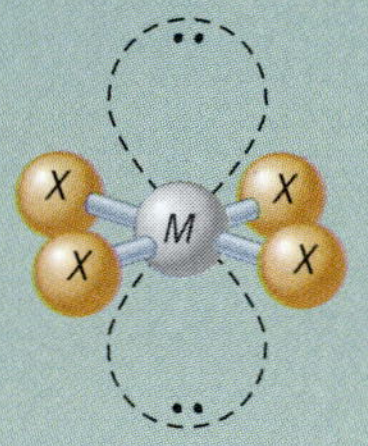	**Square planar** (example, XeF_4)

Figure 9.5 *Molecular shapes with six electron pairs around the central atom.* Although more are theoretically possible, only three different molecular shapes are observed, depending on the number of lone pairs around the central atom.

EXAMPLE 9.2

Predicting the Shapes of Molecules and Ions

Do we expect the ClO_2^- ion to be linear?

Analysis: To apply VSEPR theory, we first need the Lewis structure for the ClO_2^- ion, so drawing the Lewis structure is the first step in solving the problem. Then we will count the number of electron pairs around the central atom. This will tell us which of the basic geometries on which to base the molecular shape. We sketch this shape and then attach the two oxygens. Finally, we ignore any lone pairs to arrive at the description of the shape of the ion. In other words, in this last step we note how the *atoms* are arranged, not how the electron pairs are arranged.

Solution: Following our procedure, we obtain

$$\left[\ddot{\text{O}}\!-\!\ddot{\text{Cl}}\!-\!\ddot{\text{O}}\!:\right]^-$$

Counting electron pairs, we see that there are four around the chlorine. Four electron pairs (according to the theory) are always arranged tetrahedrally. For the electron pairs, this gives

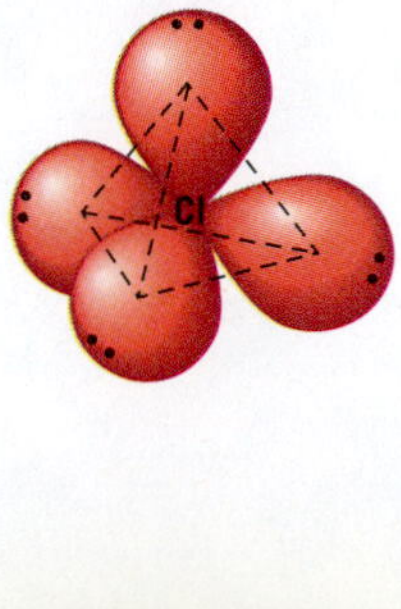

Now we add the two oxygens. It doesn't matter which locations in the tetrahedron we choose because all the bond angles are equal.

We see that the O—Cl—O angle is less than 180°, so the ion is expected to be nonlinear.

EXAMPLE 9.3

Predicting the Shapes of Molecules and Ions

Xenon is one of the noble gases, and is generally quite unreactive. In fact, it was long believed that all the noble gases were totally unable to form compounds. It came as quite a surprise, therefore, when it was discovered that some compounds could be made. One of these is xenon difluoride, XeF_2. What would you expect the geometry of XeF_2 to be, linear or nonlinear?

Analysis: To apply VSEPR theory, we need a Lewis structure. Then we count electron pairs around the central atom to determine which of the basic geometries forms the basis for the structure. We sketch the structure and attach the fluorine atoms. Finally, we ignore the electron pairs and observe how the atoms are arranged to describe the shape of the molecule.

Solution: The outer shell of xenon, of course, has a noble gas configuration, which contains 8 electrons. Each fluorine has 7 valence electrons, so the total number of electrons that must be placed in the structure is 22. These are distributed following our usual procedure, taking Xe as the central atom. The result is

$$:\ddot{F}—\overset{..}{\underset{..}{Xe}}—\ddot{F}:$$

Next we count electron pairs around xenon; there are five of them. When there are five electron pairs, they are arranged in a trigonal bipyramid.

Now we must add the fluorine atoms. In a trigonal bipyramid, the lone pairs always occur in the equatorial plane through the center, so the fluorines go on the top and bottom. This gives

The three atoms, F—Xe—F, are arranged in a straight line, so the molecule is linear.

EXAMPLE 9.4

Predicting the Shapes of Molecules and Ions

What is the probable geometry of the $XeOF_4$ molecule, which contains an oxygen atom and four fluorine atoms each bonded to xenon?

Solution: By now you should know the routine. First we draw the Lewis structure.

$$\ddot{\underset{\displaystyle :\ddot{O}}{\overset{\displaystyle :\ddot{F}:}{\underset{\displaystyle}{\overset{\displaystyle |}{Xe}}}}}$$

There are six electron pairs around xenon, and they must be arranged octahedrally.

Next we attach the oxygen and four fluorines. The most symmetrical structure is

which we would describe as a square pyramid. Although the oxygen might, in principle, be placed in one of the positions at the corners of the square base (in place of one of the fluorines), the structure shown is the actual structure of the molecule.

Shapes of Molecules and Ions with Double or Triple Bonds

Our discussion to this point has dealt only with the shapes of molecules or ions having single bonds. Luckily, the presence of double or triple bonds does not complicate matters at all. In a double bond, both electron pairs must stay together between the two atoms; they can't wander off to different locations in the valence shell. This is also true for the three pairs of electrons in a triple bond. *For the purposes of predicting molecular geometry, then, we can treat double and triple bonds just as we do single bonds.* For example, the Lewis formula for CO_2 is

$$:\ddot{O}=C=\ddot{O}:$$

The carbon atom has no lone pairs. Therefore, the two groups of electron pairs that make up the double bonds are located on opposite sides of the nucleus and a linear molecule is formed. Similarly, we would predict the following shapes for NO_2^- and NO_3^-.

We only need to consider one of the resonance structures to reach a conclusion about molecular structure.

$$\left[\ddot{\underset{:\ddot{O}}{\overset{:}{N}}}_{\displaystyle\ddot{O}:}\right]^{-}\qquad\left[\ddot{\underset{:\ddot{O}}{\overset{:\ddot{O}:}{N}}}_{\displaystyle\ddot{O}:}\right]^{-}$$

nonlinear (bent)   planar triangular

EXAMPLE 9.5

Predicting the Shapes of Molecules and Ions

The Lewis structure for the very poisonous gas hydrogen cyanide, HCN, is

$$H—C≡N:$$

Is the HCN molecule linear or nonlinear?

Solution: The triple bond behaves like a single bond for the purposes of predicting the overall geometry of a molecule. Therefore, we expect the triple and single bonds to locate themselves 180° apart, yielding a linear HCN molecule.

Practice Exercise 2

Predict the shape of SO_3^{2-}, XeO_4, and OF_2. ◆

Practice Exercise 3

Predict the geometry of CO_3^{2-}. ◆

9.3 Molecular Shape and Molecular Polarity

One of the main reasons we are concerned about whether a molecule is polar or not is because many physical properties, such as melting point and boiling point, are affected by it. This is because polar molecules attract each other. The positive end of one polar molecule attracts the negative end of another, as illustrated in Figure 9.6. The strength of the attraction depends both on the amount of charge on either end of the molecule and on the distance between the charges; in other words, it depends on the molecule's dipole moment.

The dipole moment of a molecule is a property that can be determined experimentally, and when this is done, an interesting observation is made. There are many molecules that have no dipole moment even though they contain bonds that are polar. Stated differently, they are nonpolar molecules, even though they have polar bonds. The reason for this can be seen if we examine the key role that molecular structure plays in determining molecular polarity.

Let's begin with the H—Cl molecule, which has only two atoms and therefore only one bond. As you've learned, this bond is polar because the two atoms differ in electronegativity. This means that the atoms at opposite ends of the bond carry partial charges of opposite sign. A molecule with equal but opposite charges on opposite ends is polar, so HCl is a polar molecule. In fact, any molecule composed of just two atoms that differ in electronegativity must be polar.

For molecules that contain more than two atoms, we have to consider the combined effects of all the polar bonds. Sometimes, when all the atoms attached to the central atom are the same, the effects of the individual polar bonds cancel and the molecule as a whole is nonpolar. Some examples are shown in Figure 9.7. In this figure the dipoles associated with the bonds themselves—the **bond dipoles**—are shown as arrows crossed at one end, ↦. The arrowhead indicates the negative end of the bond dipole and the crossed end corresponds to the positive end.

In CO_2, both bonds are identical, so each bond dipole is of the same magnitude. Because CO_2 is a linear molecule, these bond dipoles point in opposite directions and work against each other. The net result is that their effects cancel,

———— Attractions
———— Repulsions

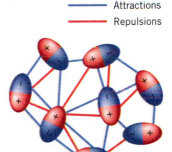

Figure 9.6 *Attractions between dipoles.* Molecules tend to orient themselves so that the positive end of one molecular dipole is near the negative end of another.

The polar HCl molecule has partial positive and negative charges on opposite ends, with the hydrogen carrying the positive charge.

$$H—Cl$$
$$\delta+ \quad \delta-$$

Tools

Molecular shape and molecular polarity

Bond dipoles can be treated as vectors, and the polarity of a molecule is predicted by taking the *vector sum* of the bond dipoles. If you've studied vectors before, this may help your understanding.

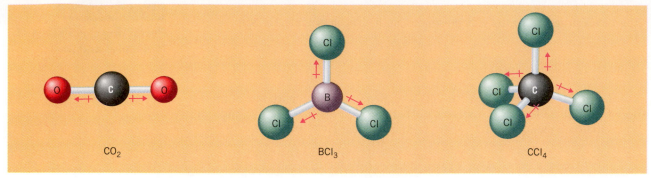

Figure 9.7 *Cancellation of bond dipoles.* In symmetric molecules such as these, the bond dipoles cancel to give nonpolar molecules.

and CO_2 is a nonpolar molecule. Although it isn't so easy to visualize, the same thing also happens in BCl_3 and CCl_4. In each of these molecules, the influence of one bond dipole is canceled by the effects of the others.

Perhaps you've noticed that the structures of the molecules in Figure 9.7 correspond to three of the basic shapes that we used in the VSEPR theory to derive the shapes of molecules. Molecules with the remaining two structures, trigonal bipyramidal and octahedral, also are nonpolar if all the atoms attached to the central atom are the same. Examples are shown in Figure 9.8. The trigonal bipyramidal structure can be viewed as a planar triangular set of atoms (shown in blue) plus a pair of atoms arranged linearly (shown in red). All the bond dipoles in the planar triangle cancel, as do the two dipoles of the bonds arranged linearly, so the molecule is nonpolar overall. Similarly, we can look at the octahedral molecule as consisting of three linear sets of bond dipoles. Cancellation of bond dipoles occurs in each set, so overall the octahedral molecule is also nonpolar.

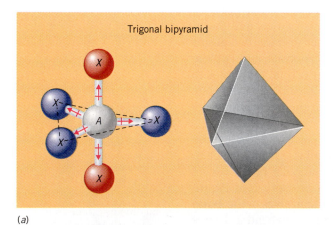

(a)

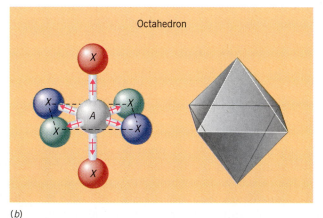

(b)

Figure 9.8 *Cancellation of bond dipoles in trigonal bipyramidal and octahedral molecules.* (a) A trigonal bipyramidal molecule, AX_5, in which the central atom A is bonded to five identical atoms X. The set of three bond dipoles in the triangular plane in the center (in blue) cancel, as do the linear set of dipoles (red). Overall, the molecule is nonpolar. (b) An octahedral molecule AX_6 in which the central atom is bonded to six identical atoms. This molecule contains three linear sets of bond dipoles. Cancellation occurs for each set, so the molecule is nonpolar overall.

Polar Molecules

If all the atoms attached to the central atom are not the same, or if there are lone pairs in the valence shell of the central atom, the molecule is usually polar. For example, in $CHCl_3$, one of the atoms in the tetrahedral structure is different from the others. The C—H bond is less polar than the C—Cl bonds, and the bond dipoles do not cancel (Figure 9.9).

Two familiar molecules that have lone pairs in the valence shells of their central atoms are shown in Figure 9.10. Here the bond dipoles are oriented in such a way that their effects do not cancel. In water, for example, each bond dipole points partially in the same direction, toward the oxygen atom. As a result, the bond dipoles partially add to give a net dipole moment for the molecule. The same thing happens in ammonia where three bond dipoles point partially in the same direction and add to give a polar NH_3 molecule.

Not every structure that contains lone pairs on the central atom produces polar molecules. The following are two exceptions.

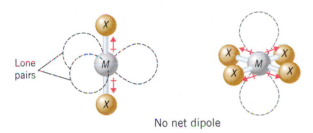

No net dipole

In the first case, we have a pair of bond dipoles arranged linearly, just as in CO_2. In the second, the bonded atoms lie at the corners of a square, which can be viewed as two linear sets of bond dipoles. If the atoms attached to the central atom are the same, cancellation of bond dipoles is bound to occur and produce nonpolar molecules. This means that molecules such as linear XeF_2 and square planar XeF_4 are nonpolar.

> **In Summary:** A molecule will be nonpolar if (a) the bonds are nonpolar or (b) there are no lone pairs in the valence shell of the central atom and all the atoms attached to the central atom are the same.
>
> A molecule in which the central atom has lone pairs of electrons will be usually be polar, with the two exceptions described above.

On the basis of the preceding discussions, let's see how we can use VSEPR theory to determine whether molecules are expected to be polar or nonpolar.

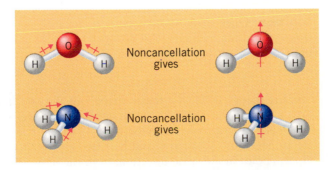

Noncancellation gives

Noncancellation gives

The lone pairs also influence the polarity of a molecule, but we will not explore this any further here.

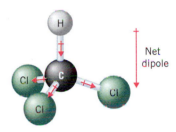

Net dipole

Figure 9.9 *Bond dipoles in the chloroform molecule, $CHCl_3$. Because C is slightly more electronegative than H, the C—H bond dipole points toward the carbon. The small C—H bond dipole actually adds to the effects of the C—Cl bond dipoles. All the bond dipoles are additive, and this causes $CHCl_3$ to be a polar molecule.*

Figure 9.10 *Noncancellation of bond dipoles. When lone pairs occur on the central atom, the bond dipoles usually do not cancel, and polar molecules result.*

EXAMPLE 9.6

Predicting Molecular Polarity

Do we expect the PCl₃ molecule to be polar or nonpolar?

Analysis: First, we need to know whether the bonds in the molecule are polar. If not, the molecule will be nonpolar regardless of its structure. If the bonds are polar, then we need to determine its structure. On the basis of the structure, we can then decide whether the bond dipoles cancel.

Solution: The electronegativities of the atoms (P = 2.1, Cl = 2.9) tell us that the individual P—Cl bonds will be polar. Therefore, to predict whether the molecule is polar, we need to know its shape. First we draw the Lewis structure following our usual procedure.

$$:\ddot{C}l:$$
$$|$$
$$:\ddot{C}l-P-\ddot{C}l:$$

There are four electron pairs around the phosphorus, so according to VSEPR theory, they should be arranged tetrahedrally. This means that the PCl₃ molecule should have a trigonal pyramidal shape, as shown below.

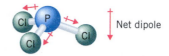

Net dipole

Because of the structure, the bond dipoles do not cancel, and we expect the molecule to be polar.

EXAMPLE 9.7

Predicting Molecular Polarity

Would you expect the molecule SO₃ to be polar or nonpolar?

Solution: Oxygen is more electronegative than sulfur, so we expect the S—O bonds to be polar. Let's look at the molecular shape to see if the bond dipoles cancel. The simple Lewis structure for SO₃ is

$$:\ddot{O}:$$
$$|$$
$$:\ddot{O}=S-\ddot{O}:$$

VSEPR theory tells us that the molecule should have a planar triangular shape. We saw in Figure 9.7 that such a molecule is nonpolar when all the attached atoms are the same, because the bond dipoles cancel.

EXAMPLE 9.8

Predicting Molecular Polarity

Would you expect the molecule HCN to be polar or nonpolar?

Solution: Once again we have polar bonds because carbon is slightly more electronegative than hydrogen and nitrogen is slightly more electronegative than carbon. The Lewis structure of HCN is

$$H-C\equiv N:$$

VSEPR theory predicts a linear shape, but the two bond dipoles do not cancel. One reason is that they are not of equal magnitude, which we know because the difference in electronegativity between C and H is 0.4, whereas the difference in electronegativity between C and N is 0.6. The other reason is because both bond dipoles point in the same direction, from the atom of low electronegativity to the one of high electronegativity. This is illustrated on the next page.

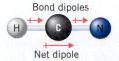

Notice that the bond dipoles add to give a net dipole moment for the molecule.

9.4 Wave Mechanics and Covalent Bonding: Valence Bond Theory

So far, we have taken a very simple view of covalent bonding based primarily on the use of Lewis structures for electron bookkeeping. Lewis structures, however, tell us nothing about *why* covalent bonds are formed or *how* electrons manage to be shared between atoms. Nor does the VSEPR theory, as useful and accurate as it generally is at predicting molecular geometry, explain *how* the electron pairs in the valence shell of an atom manage to avoid each other. Thus, as helpful as these simple models are, we must look beyond them if we wish to understand more fully the covalent bond and the factors that determine molecular geometry.

Modern theories of bonding are based on the principles of wave mechanics. This is the theory, you recall, that gives us the electron configurations of atoms and the description of the shapes of atomic orbitals. When wave mechanics is applied to molecules, it considers how the orbitals of the atoms that come together to form a covalent bond interact with each other, and in so doing it attempts to explain in detail how atoms share electrons.

There are fundamentally two theories of covalent bonding that have evolved, and in many ways they complement one another. They are called the **valence bond theory (or VB theory,** for short) and the **molecular orbital theory (MO theory).** They differ principally in the way they construct a theoretical model of the bonding in a molecule. The valence bond theory imagines individual atoms, each with its own orbitals and electrons, coming together to form the covalent bonds of the molecule. On the other hand, the molecular orbital theory doesn't concern itself with *how* the molecule is formed. It just views a molecule as a collection of positively charged nuclei surrounded in some way by electrons that occupy a set of *molecular orbitals,* in much the same way that the electrons in an atom occupy *atomic orbitals.* (In a sense, this theory would look at an atom as if it were a special case — a molecule having only one positive center, instead of many.)

In their simplest forms, the VB and MO theories appear to be rather different. However, it has been found that both theories can be extended and refined to give the same results. This is as it should be, of course, since they both attempt to explain the same set of facts — the structures and shapes of molecules, and the strengths of chemical bonds. We will examine both theories in an elementary way, but the greater emphasis will be on the VB theory because it is easier to understand.

Basic Postulates of the Valence Bond Theory

According to VB theory, *a bond between two atoms is formed when a **pair of electrons** with their **spins paired** is shared by two **overlapping** atomic orbitals, one orbital from each of the atoms joined by the bond.* By **overlap of orbitals**

Remember, just *two electrons* with *paired spins* can be shared between two overlapping atomic orbitals.

we mean that portions of two atomic orbitals from different atoms share the same space.

An important part of the theory, as suggested by the bold italic type above, is that only *one* pair of electrons, with paired spins, can be shared by two overlapping orbitals. This electron pair becomes concentrated in the region of overlap and helps "cement" the nuclei together, so the amount that the potential energy is lowered when the bond is formed is determined in part by the extent to which the orbitals overlap. Therefore, *atoms tend to position themselves so that the maximum amount of orbital overlap occurs because this yields the minimum potential energy and therefore the strongest bonds.* As we will see, this is one of the major factors that control the shapes of molecules.

The way VB theory views the formation of a hydrogen molecule is shown in Figure 9.11. As the two atoms approach each other, their $1s$ orbitals overlap and the electron pair spreads out over both orbitals, thereby giving the H—H bond. The description of the bond in this molecule provided by VB theory is essentially the same as that discussed in Section 8.3.

Now let's look at the HF molecule, which is a bit more complex than H_2. Following the usual rules we can write its Lewis structure as

$$H—\ddot{\underset{\cdot\cdot}{F}}:$$

and we can diagram the formation of the bond as

$$H\cdot + \cdot\ddot{\underset{\cdot\cdot}{F}}: \longrightarrow H—\ddot{\underset{\cdot\cdot}{F}}:$$

Our Lewis symbols suggest that the H—F bond is formed by the pairing of electrons, one from hydrogen and one from fluorine. To explain this according to VB theory, we must have two half-filled orbitals, one from each atom, that can be joined by overlap. (They must be half-filled, because we can't place more than two electrons into the bond.) To see clearly what must happen, it is best to look at the orbital diagrams of the valence shells of hydrogen and fluorine.

The requirements for bond formation are met by overlapping the half-filled $1s$ orbital of hydrogen with the half-filled $2p$ orbital of fluorine; there are then two orbitals plus two electrons whose spins can adjust so they are paired. The formation of the bond is illustrated in Figure 9.12.

The overlap of orbitals provides a means for sharing electrons, thereby allowing each atom to complete its valence shell. It is sometimes convenient to indicate this using orbital diagrams. The diagram on the next page shows how the fluorine atom completes its $2p$ subshell by acquiring a share of an electron from hydrogen.

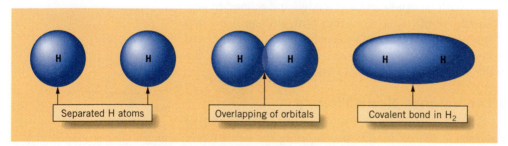

Figure 9.11 *Formation of the hydrogen molecule according to valence bond theory.*

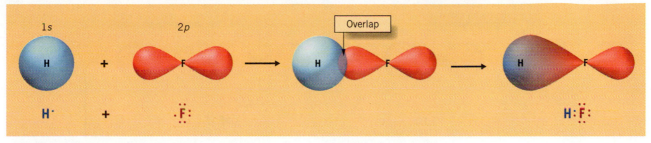

Figure 9.12 *Formation of the hydrogen fluoride molecule according to valence bond theory.* Only the half-filled 2p orbital of fluorine is shown.

F (in HF)   ⇅ ⇅⇅⇵   (Colored arrow is the H electron.)
              2s   2p

Notice that both the Lewis and VB descriptions of the formation of the H—F bond account for the completion of the atoms' valence shells. Therefore, *a Lewis structure can be viewed, in a very qualitative sense, as a shorthand notation for the valence bond description of a molecule.*

Let's look now at a still more complex molecule, hydrogen sulfide, H_2S. Experiments have shown that this is a nonlinear molecule in which the H—S—H bond angle is about 92°.

H_2S is the compound that gives rotten eggs their foul odor.

$$\overset{\displaystyle S}{\underset{92°}{H \quad H}}$$

Valence bond theory explains this structure as follows.

The orbital diagram for the valence shell of sulfur is

S   ⇅ ⇅↑↑
     3s   3p

Sulfur has two 3p orbitals that each contain only one electron. Each of these can overlap with the 1s orbital of a hydrogen atom, as shown in Figure 9.13. This overlap completes the 3p subshell of sulfur because each hydrogen provides one electron.

S (in H_2S)   ⇅ ⇅⇵⇵   (Colored arrows are H electrons.)
                 3s   3p

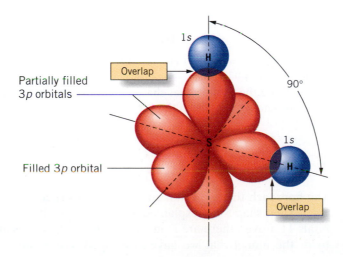

Partially filled 3p orbitals

Overlap

Filled 3p orbital

1s
H
Overlap
90°
1s
H
Overlap

Figure 9.13 *Bonding in H_2S.* We expect the hydrogen 1s orbitals to position themselves so that they can best overlap with the two partially filled 3p orbitals of sulfur, which gives a predicted bond angle of 90°. The experimentally measured bond angle of 92° is very close to the predicted angle.

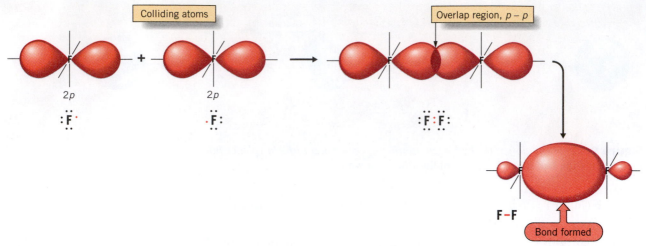

Figure 9.14 *Bonding in the fluorine molecule according to valence bond theory.* The two completely filled *p* orbitals on each fluorine atom are omitted, for clarity.

Two orbitals from different atoms never overlap simultaneously with opposite ends of the same *p* orbital.

In Figure 9.13, notice that when the 1*s* orbital of a hydrogen atom overlaps with a *p* orbital of sulfur, the best overlap occurs when the hydrogen atom lies along the axis of the *p* orbital. Because *p* orbitals are oriented at 90° angles to each other, the H—S bonds are expected to be at this angle, too. Therefore, the predicted bond angle is 90°. This is very close to the actual bond angle of 92° found by experiment. Thus, the VB theory requirement for maximum overlap quite nicely accounts for the geometry of the hydrogen sulfide molecule.

The overlap of *p* orbitals with each other is also possible. For example, according to VB theory the bonding in the fluorine molecule, F_2, occurs by the overlap of two 2*p* orbitals, as shown in Figure 9.14. The formation of the other diatomic molecules of the halogens, all of which are held together by single bonds, could be similarly described.

Practice Exercise 5

Use the principles of VB theory to explain the bonding in HCl. Give the orbital diagram for chlorine in the HCl molecule and indicate the orbital that shares the electron with one from hydrogen. Sketch the orbital overlap that gives rise to the H—Cl bond. ◆

Practice Exercise 6

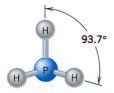

The phosphine molecule, PH_3, has a trigonal pyramidal shape with H—P—H bond angles equal to 93.7°. Give the orbital diagram for phosphorus in the PH_3 molecule and indicate the orbitals that share electrons with those from hydrogen. On a set of *xyz* coordinate axes, sketch the orbital overlaps that give rise to the P—H bonds. ◆

9.5 Hybrid Orbitals

The approach that we've taken so far has worked well with some simple molecules. Their shapes are explained quite nicely by the overlap of simple atomic orbitals. However, there are many molecules with shapes and bond angles that fail to fit the model that we have developed. For example, methane, CH_4, has a shape that the VSEPR theory predicts (correctly) to be tetrahedral. The H—C—H bond angles in this molecule are 109.5°. But no simple atomic orbitals are oriented at this angle with respect to each other. Therefore, to explain the bonds in molecules such as CH_4 we must study the way atomic orbitals *of the same atom* can interact with each other when bonds are formed.

Formation of *sp*, *sp²*, and *sp³* Hybrid Orbitals

The theory of wave mechanics suggests that when atoms form bonds, their simple atomic orbitals often mix to form new orbitals that we call **hybrid atomic orbitals.** These new orbitals have new shapes and new directional properties. The reason this mixing occurs can be understood if we look at the shapes of hybrid orbitals.

One kind of hybrid atomic orbital is formed by mixing an *s* orbital and a *p* orbital. This creates *two* new orbitals called *sp* **hybrid orbitals.** The label *sp* identifies the kinds of pure atomic orbitals from which the hybrids are formed, in this case, one *s* and one *p* orbital. The shapes and directional properties of the *sp* hybrid orbitals are illustrated in Figure 9.15. Notice that each of the hybrid orbitals has the same shape; each has one large lobe and another much smaller one. The large lobe extends farther from the nucleus than either the *s* or *p* orbital from which the hybrid was formed. This allows the hybrid orbital to overlap more effectively with an orbital on another atom when a bond is formed. Therefore, hybrid orbitals form stronger, more stable bonds than would be possible if just simple atomic orbitals were used.

Another point to notice in Figure 9.15 is that the large lobes of the two *sp* hybrid orbitals point in opposite directions; that is, they are 180° apart. If bonds are formed by the overlap of these hybrids with orbitals of other atoms, the other atoms will occupy positions on opposite sides of this central atom. Let's look at a specific example, the linear beryllium hydride molecule, BeH_2, as it would be formed in the gas phase.[1]

In general, the greater the overlap of two orbitals, the stronger is the bond. At a given internuclear distance, the greater "reach" of a hybrid orbital gives better overlap than either an *s* or *p* orbital.

The mathematics of wave mechanics predicts this 180° angle between *sp* hybrid orbitals.

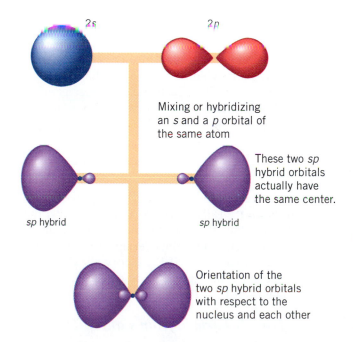

Mixing or hybridizing an *s* and a *p* orbital of the same atom

These two *sp* hybrid orbitals actually have the same center.

sp hybrid   *sp* hybrid

Orientation of the two *sp* hybrid orbitals with respect to the nucleus and each other

Figure 9.15 *Formation of sp hybrid orbitals.* Mixing of the 2*s* and 2*p* atomic orbitals produces a pair of *sp* hybrid orbitals. The large lobes of these orbitals point in opposite directions.

[1] In the solid state, BeH_2 has a complex structure not consisting of simple BeH_2 molecules.

The orbital diagram for the valence shell of beryllium is

Be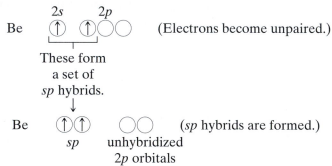

Notice that the 2*s* orbital is filled and the three 2*p* orbitals are empty. For bonds to form at a 180° angle between beryllium and the two hydrogen atoms, two conditions must be met: (1) the two orbitals that beryllium uses to form the Be—H bonds must point in opposite directions, and (2) each of the beryllium orbitals must contain only one electron. The reason for the first requirement is obvious; the overlap of Be and H orbitals must give the correct experimentally measured bond angle. The reason for the second is that each bond must contain just two electrons. Since a hydrogen 1*s* orbital supplies one electron, the beryllium orbital with which it overlaps must also contain one electron. A filled Be orbital won't do, because overlap of a half-filled hydrogen 1*s* with a filled orbital on Be would produce a "bond" with three electrons in it—a situation that's forbidden by VB theory. An empty Be orbital won't do either, because overlap with a hydrogen 1*s* orbital would give a bond with only one electron in it. Although such a bond isn't forbidden, it would be much weaker than a bond with two electrons, so electron pair bonds are definitely preferred. The net effect of all this is that when the Be—H bonds form, the electrons of the beryllium atom become unpaired, and the resulting half-filled *s* and *p* atomic orbitals become hybridized.

Now the 1*s* orbitals of the hydrogen atoms can overlap with the *sp* hybrids of beryllium as shown in Figure 9.16. Because the two *sp* hybrid orbitals of beryllium are identical in shape, the two Be—H bonds are identical except for the directions in which they point, and we say that the bonds are *equivalent*. Since the bonds point in opposite directions, the linear geometry of the molecule is also explained. The orbital diagram for beryllium in this molecule is

These sp hybrid orbitals are said to be equivalent because they have the same size, shape, and energy.

Even if we had not known the shape of the BeH_2 molecule, we could have obtained the same bonding picture by applying the VSEPR theory first. The Lewis structure for BeH_2 is H:Be:H, and VSEPR theory predicts that the molecule is linear. Once the shape is known, we can apply the VB theory to explain the bonding in terms of orbital overlaps. Thus, the VB and VSEPR theories complement each other well. VSEPR theory allows us to predict geometry in a simple way, and once the geometry is known, it is relatively easy to analyze the bonding in terms of VB theory.

Other Hybrids Formed from s and p Orbitals

Hybrid orbitals can also be formed by combining an *s* orbital with two or three *p* orbitals. When two *p* orbitals are mixed with an *s* orbital, a set of *three sp²*

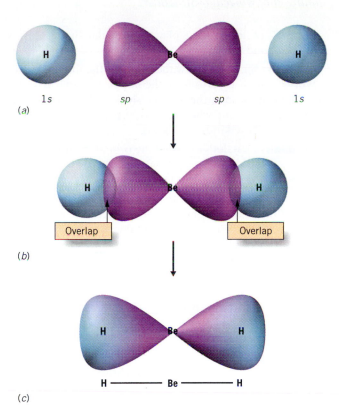

(a)

(b)

(c)

Figure 9.16 *Bonding in BeH$_2$ according to valence bond theory.* Only the larger lobe of each *sp* hybrid orbital is shown. (*a*) The two hydrogen 1*s* orbitals approach the pair of *sp* hybrid orbitals of beryllium. (*b*) Overlap of the hydrogen 1*s* orbitals with the *sp* hybrid orbitals. (*c*) A representation of the distribution of electron density in the two Be—H bonds after they have been formed.

hybrid orbitals is formed. When three *p* orbitals are mixed with an *s* orbital, a set of *four sp*^{3} **hybrid orbitals** is formed. The superscript 2 or 3 indicates the number of *p* orbitals in the mix. Notice that the number of hybrid orbitals in a set equals the number of atomic orbitals used to form them.

The three *sp*^{2} hybrids lie in the same plane and point to the corners of a triangle. The four *sp*^{3} hybrids point to the corners of a tetrahedron. Their directional properties are summarized in Figure 9.17. The following examples illustrate how they can be used to explain the bonding in molecules.

There is a conservation of orbitals during hybrid orbital formation. The number of hybrid orbitals of a given kind is always equal to the number of atomic orbitals that are mixed.

Tools

Orientations of hybrid orbitals

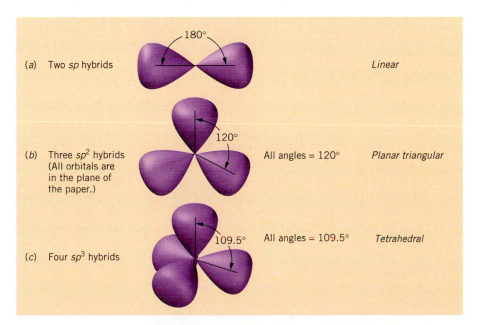

(a) Two *sp* hybrids — 180° — *Linear*

(b) Three *sp*^{2} hybrids (All orbitals are in the plane of the paper.) — 120° — All angles = 120° — *Planar triangular*

(c) Four *sp*^{3} hybrids — 109.5° — All angles = 109.5° — *Tetrahedral*

Figure 9.17 *Directional properties of hybrid orbitals formed from s and p atomic orbitals.* (*a*) *sp* hybrid orbitals oriented at 180° to each other. (*b*) *sp*^{2} hybrid orbitals formed from an *s* orbital and two *p* orbitals. The angle between them is 120°. (*c*) *sp*^{3} hybrid orbitals formed from an *s* orbital and three *p* orbitals. The angle between them is 109.5°.

EXAMPLE 9.9

Explaining Bonding with Hybrid Orbitals

The BCl_3 molecule has a planar triangular shape. How is this explained in terms of valence bond theory?

Analysis: Since we know the shape of the molecule, we can anticipate the kinds of hybrid orbitals used to form the bonds. A planar triangular shape is consistent with sp^2 hybridization in the valence shell of the central atom, so we proceed from that point to determine whether the electronic structures of the atoms involved fit with the use of these kinds of orbitals.

Solution: First, let's examine the orbital diagram for the valence shell of boron.

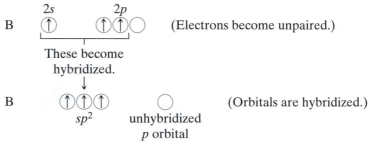

To form the three bonds to chlorine, boron needs three half-filled orbitals. These can be obtained by unpairing the electrons in the $2s$ orbital and placing one of them in the $2p$. Then the $2s$ orbital and two of the $2p$ orbitals can be combined to give the set of sp^2 hybrids.

B $2s$ $2p$ (Electrons become unpaired.)

These become hybridized.

B sp^2 unhybridized p orbital   (Orbitals are hybridized.)

Now let's look at the valence shell of chlorine.

Cl $3s$ $3p$

We see that the half-filled $3p$ orbital of a chlorine atom can overlap with a hybrid sp^2 orbital of boron to give a B—Cl bond.

B (in BCl_3) sp^2 unhybridized p orbital   (Colored arrows are Cl electrons.)

    Figure 9.18 illustrates the overlap of the orbitals to give the bonding in the molecule. Notice that the arrangement of atoms yields a planar triangular molecule, which is in agreement with the structure predicted by the VSEPR theory.

Figure 9.18 *Valence bond description of the bonding in BCl_3.* Each B—Cl bond is formed by the overlap of a half-filled p orbital of chlorine with an sp^2 hybrid orbital of boron. (Only the half-filled p orbital of each chlorine is shown.)

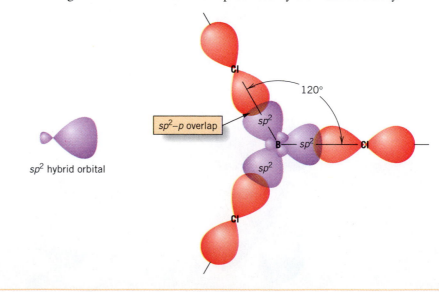

EXAMPLE 9.10

Explaining Bonding with Hybrid Orbitals

Methane, CH_4, is a tetrahedral molecule. How is this explained in terms of valence bond theory?

Solution: The tetrahedral structure of the molecule suggests that sp^3 hybrid orbitals are involved in bonding. Let's examine the valence shell of carbon.

$$C \quad \boxed{\uparrow\downarrow} \quad \boxed{\uparrow}\boxed{\uparrow}\boxed{}$$
$$2s \qquad 2p$$

To form four C—H bonds, we need four half-filled orbitals. Unpairing the electrons in the $2s$ and moving one to the vacant $2p$ orbital satisfies this requirement. Then we can hybridize all the orbitals to give the desired sp^3 set.

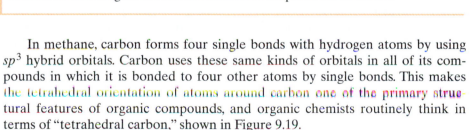

$$
\begin{array}{cc}
2s & 2p \\
\end{array}
$$
$$C \quad \boxed{\uparrow} \quad \boxed{\uparrow}\boxed{\uparrow}\boxed{\uparrow} \qquad \text{(Electrons become unpaired.)}$$

These become
hybridized.
↓

$$C \quad \boxed{\uparrow}\boxed{\uparrow}\boxed{\uparrow}\boxed{\uparrow}$$
$$sp^3 \text{ hybrids}$$

Then we form the four bonds to hydrogen $1s$ orbitals.

$$C \text{ (in } CH_4) \quad \boxed{\uparrow\downarrow}\boxed{\uparrow\downarrow}\boxed{\uparrow\downarrow}\boxed{\uparrow\downarrow} \qquad \text{(Colored arrows are H electrons)}$$
$$sp^3$$

This is illustrated in the margin. Notice that the positions of the hydrogen atoms around the carbon give the correct tetrahedral shape for the molecule.

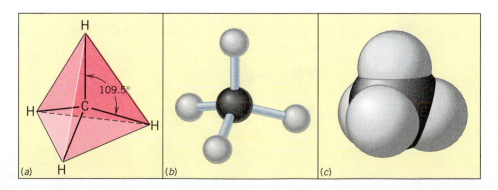

Region of overlap

In methane, carbon forms four single bonds with hydrogen atoms by using sp^3 hybrid orbitals. Carbon uses these same kinds of orbitals in all of its compounds in which it is bonded to four other atoms by single bonds. This makes the tetrahedral orientation of atoms around carbon one of the primary structural features of organic compounds, and organic chemists routinely think in terms of "tetrahedral carbon," shown in Figure 9.19.

In the alkane hydrocarbons, carbon atoms are bonded to other carbon atoms. An example is ethane, C_2H_6.

$$
\begin{array}{ccc}
& H \quad H & \\
& | \quad\; | & \\
H- & C-C & -H \\
& | \quad\; | & \\
& H \quad H &
\end{array}
$$
ethane

Figure 9.19 The "tetrahedral carbon." (a) The heavy lines are the axes of the bonds. (b) A ball-and-stick model of the CH_4 molecule. (c) A space filling model of CH_4 that indicates the relative volumes occupied by the electron clouds.

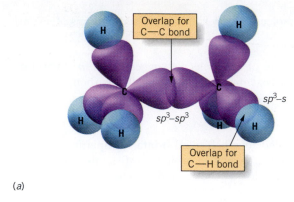

(a)

Figure 9.20 *The bonds in the ethane molecule. (a) Overlap of orbitals. (b) The degree of overlap of the sp^3 orbitals in the carbon–carbon bond is not appreciably affected by the rotation of the two CH_3 groups relative to each other around the bond.*

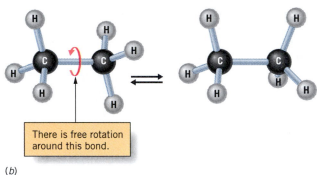

(b)

In this molecule, the carbons are bonded together by the overlap of sp^3 hybrid orbitals (Figure 9.20). One of the most important characteristics of this bond is that the overlap of the orbitals in the C—C bond is hardly affected at all if one portion of the molecule rotates relative to the other around the bond axis. Such rotation, therefore, is said to occur freely and permits different possible relative orientations of the atoms in the molecule. These different relative orientations are called **conformations.** With complex molecules, the number of possible conformations is enormous. For example, Figure 9.21 illustrates three of the enor-

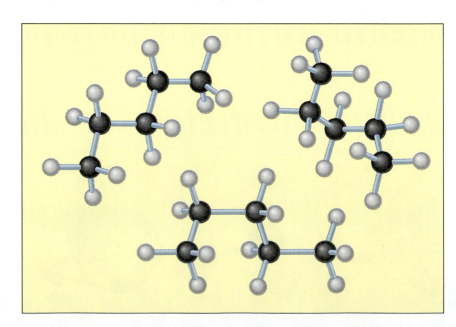

Figure 9.21 *Three of the many conformations of the atoms in the pentane molecule, C_5H_{12}. Free rotation around single bonds makes these different conformations possible.*

Tools

Orientations of hybrid orbitals

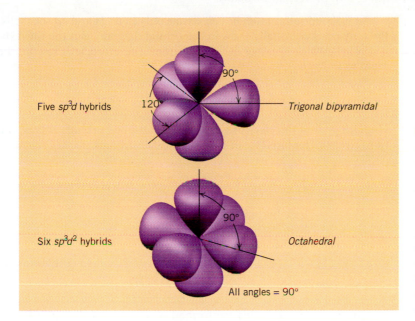

Figure 9.22 *Orientations of hybrid orbitals that involve d atomic orbitals.* (a) sp^3d *hybrid orbitals, formed by mixing an s orbital, three p orbitals, and a d orbital. The orbitals point to the corners of a trigonal bipyramid.* (b) sp^3d^2 *hybrid orbitals, formed by mixing an s orbital, three p orbitals, and two d orbitals. The orbitals point to the corners of an octahedron.*

mous number of possible conformations of the pentane molecule, C_5H_{12}, one of the low molecular weight organic compounds in gasoline.

pentane

Hybridization when the Central Atom Has More than an Octet

Earlier we saw that certain molecules have atoms that must violate the octet rule because they form more than four bonds. In these cases, the atom must reach beyond its s and p valence shell orbitals to form sufficient half-filled orbitals for bonding. This is because the s and p orbitals can be mixed to form a maximum of only four hybrid orbitals. When five or more hybrid orbitals are needed, d orbitals are brought into the mix. The two most common kinds of hybrid orbitals involving d orbitals are **sp^3d** and **sp^3d^2 hybrid orbitals.** Their directional properties are illustrated in Figure 9.22. Notice that the sp^3d hybrids point toward the corners of a trigonal bipyramid and the sp^3d^2 hybrids point toward the corners of an octahedron.

EXAMPLE 9.11

Explaining Bonding with Hybrid Orbitals

The sulfur hexafluoride molecule has an octahedral shape. Describe the bonding in this molecule in terms of valence bond theory.

Solution: As before, let's examine the valence shell of sulfur.

To form six bonds to fluorine atoms we need six half-filled orbitals, but we see only four orbitals altogether. The structure of the molecule, however, gives us a clue to where we can obtain the others. An octahedral structure suggests the use of sp^3d^2 hybrid orbitals, so we need to find some d orbitals to form the hybrids.

An isolated sulfur atom has electrons only in its $3s$ and $3p$ subshells, so these are the only ones we usually show in the orbital diagram. But the third shell also has a d subshell, which is empty in a sulfur atom. Therefore, let's rewrite the orbital diagram to show the vacant $3d$ subshell.

S $3s$ $3p$ $3d$

Unpairing all of the electrons to give six half-filled orbitals, followed by hybridization, gives the required set of half-filled sp^3d^2 orbitals.

$$S \quad \overset{3s}{\uparrow} \quad \overset{3p}{\uparrow\,\uparrow\,\uparrow} \quad \overset{3d}{\uparrow\,\uparrow} \bigcirc\bigcirc\bigcirc \qquad \text{(Electrons become unpaired.)}$$

These become hybridized.

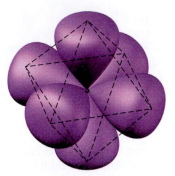

sp^3d^2 hybrid orbitals of sulfur in SF_6.

$$S \quad \underset{sp^3d^2}{\uparrow\,\uparrow\,\uparrow\,\uparrow\,\uparrow\,\uparrow} \qquad \underset{\substack{\text{unhybridized} \\ 3d \text{ orbitals}}}{\bigcirc\bigcirc\bigcirc} \qquad (sp^3d^2 \text{ hybrids are formed.})$$

Finally, the six S—F bonds are formed by overlap of the half-filled $2p$ orbitals of fluorine with these half-filled sp^3d^2 hybrids.

$$S \text{ (in } SF_6) \quad \underset{sp^3d^2}{\updownarrow\,\updownarrow\,\updownarrow\,\updownarrow\,\updownarrow\,\updownarrow} \qquad \underset{\substack{\text{unhybridized} \\ 3d \text{ orbitals}}}{\bigcirc\bigcirc\bigcirc} \qquad \begin{array}{l} \text{(Colored arrows are} \\ \text{F electrons.)} \end{array}$$

Practice Exercise 7

Use valence bond theory to describe the bonding in the molecule $AsCl_5$, which has a trigonal bipyramidal shape. ◆

Using VSEPR Theory to Predict Hybridization

We have seen that if we know the structure of a molecule, we can make a reasonable guess as to the kind of hybrid orbitals that the central atom uses to form its bonds. Because the VSEPR theory works so well in predicting geometry, we can use it to help us obtain VB descriptions of bonding. For example, the Lewis structures of CH_4 and SF_6 are

In CH_4, there are four electron pairs around carbon. The VSEPR model tells us that they should be arranged tetrahedrally. The only hybrid orbitals that are tetrahedral are sp^3 hybrids, and we have seen that they explain the structure of this molecule well. Similarly, VSEPR theory tells us that the six electron pairs around sulfur should be arranged octahedrally. The only octahedrally oriented hybrids are sp^3d^2, so the sulfur in the SF_6 molecule must use these hybrids.

Practice Exercise 8

What kind of hybrid orbitals are expected to be used by the central atom in (a) SiH_4 and (b) PCl_5? ◆

Hybridization in Molecules That Have Lone Pairs of Electrons

Methane is a tetrahedral molecule with sp^3 hybridization of the orbitals of carbon and H—C—H bond angles that are each equal to 109.5°. In ammonia, NH_3, the H—N—H bond angles are 107°, and in water the H—O—H bond angle is 104.5°. Both NH_3 and H_2O have H—X—H bond angles that are close to the bond angles expected for a molecule whose central atom has sp^3 hybrids. The use of sp^3 hybrids by oxygen and nitrogen, therefore, is often used to explain the geometry of H_2O and NH_3.

O   2s   2p

Hybridization and bond formation occur.

O (in H_2O)   (Colored arrows are H electrons.)
sp^3

N   2s   2p

Hybridization and bond formation occur.

N (in NH_3)   (Colored arrows are H electrons.)
sp^3

According to these descriptions, not all of the hybrid orbitals of the central atom must be used for bonding. Lone pairs of electrons can be accommodated in them too. In fact, there is evidence to suggest that the lone pair on the nitrogen in ammonia does in fact reside in an sp^3 hybrid orbital. (The case is less convincing for water, however, as noted in Facets of Chemistry 9.1.)

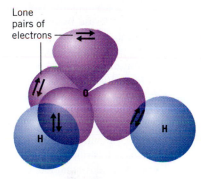

Lone pairs of electrons

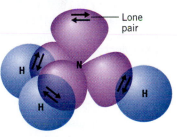

Lone pair

EXAMPLE 9.12

Explaining Bonding with Hybrid Orbitals

Use valence bond theory to explain the bonding in the SF_4 molecule.

Solution: Let's begin by constructing the Lewis formula for the molecule. Following the usual procedure, we obtain

The VSEPR theory predicts that the electron pairs around the sulfur should be in a trigonal bipyramidal arrangement, and the only hybrids that fit this geometry are sp^3d. To see how they are formed, we look at the valence shell of sulfur, including the vacant $3d$ subshell.

S
3s   3p   3d

To form the four bonds to fluorine atoms, we need four half-filled orbitals, so we unpair the electrons in one of the filled orbitals. This gives

3s   3p   3d
S

Next, we form the hybrid orbitals. In doing this, we use all the valence shell orbitals that have electrons in them.

Facets of Chemistry 9.1

What's the Truth about Water?

Over the years there's been quite a bit of controversy about the nature of the bonding in the water molecule. Specifically, the question has been: What kind of orbitals does oxygen really use in forming its bonds to the hydrogen atoms?

In our discussion of this molecule, we've given what has become a standard explanation for the H—O—H bond angle—the use by oxygen of sp^3 hybrid orbitals to overlap with the $1s$ orbitals of the hydrogen atoms. On the basis of bond angle data alone, this is plausible because the angles between these kinds of hybrids is quite close to the bond angle in H_2O. It also fits with the general rule of thumb that we've followed in the remainder of the discussion of VB theory—that atoms generally use hybrid orbitals to form bonds to neighboring atoms. By following this rule, we obtain reasonable explanations of molecular structure. Nevertheless, as useful as this rule may be for other molecules, there is considerable evidence that it doesn't really work well for water.

In our discussion of hybridization, we've ignored the fact that forming hybrid orbitals requires energy, which is usually more than paid back by the formation of strong covalent bonds. In oxygen and other period 2 elements, however, it is especially "expensive" to form the hybrids, because the energy of the $2s$ orbital is so much lower than the energy of the $2p$ orbitals. This makes it less attractive for these elements to use hybrid orbitals if unhybridized orbitals can also do the job, and this does appear to be the case with water, based on detailed calculations and some experimental evidence that are beyond the scope of this discussion. But, if oxygen uses its pure $2p$ orbitals to form the O—H bonds, how come the bond angle isn't close to 90°, as it is in H_2S (page 385) where sulfur uses its $3p$ orbitals to form the S—H bonds?

To explain the spreading of the H—O—H angle in water from 90° to 104.5°, we have to consider the small size of the oxygen atom. As you can see in the figure below, if the bond angle were 90°, the two hydrogen atoms would interpenetrate each other, which would cause their completed valence shells to overlap. This can't happen, however, because two pairs of electrons can never occupy the same space, so the angle spreads to alleviate this problem. In H_2S, the bond angle doesn't have to increase appreciably from 90° because the sulfur atom is larger and the hydrogens don't interfere with each other.

The point of this discussion is that often there are alternative theoretical explanations of the same phenomenon (in this instance, the bonding in water). Some are simple and others are more complex. Which one we use depends on how precise we must be. The VSEPR theory, for example, is very simple, but it doesn't attempt to explain bonding in terms of the atomic orbitals involved. The model we've provided in this chapter using hybrid orbitals also works well most of the time, so we find it useful. But even this model is not completely accurate all the time, so when it really matters, many factors have to be weighed in deciding what is as close to the "truth" as possible.

(*Left*) If the bond angle in H_2O were 90°, the H atoms would overlap severely. (*Center*) At a bond angle of 104°, the hydrogens don't interfere with each other very much. (*Right*) Because S is larger than O, the bond angle in H_2S can be much closer to 90°.

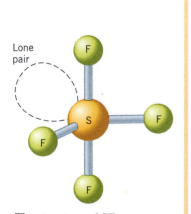

Lone pair

The structure of SF_4.

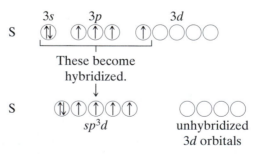

$$3s \quad\quad 3p \quad\quad\quad 3d$$

S   [↑↓] [↑][↑][↑] [↑][ ][ ][ ][ ]

These become hybridized.

↓

S   [↑↓][↑][↑][↑][↑]   [ ][ ][ ][ ]
         sp^3d unhybridized
                              $3d$ orbitals

Now, four S—F bonds can be formed by overlap of half-filled $2p$ orbitals of fluorine with the sp^3d hybrid orbitals of sulfur.

S (in SF$_4$)   [ ][ ][ ][ ]   (Colored arrows
                    sp^3d unhybridized   are F electrons.)
                                          $3d$ orbitals

Practice Exercise 9

What kind of hybrid orbitals would we expect the central atom to use for bonding in (a) PCl_3 and (b) ClF_3? ◆

Coordinate Covalent Bonds and Hybrid Orbitals

In Section 8.10 we defined a coordinate covalent bond as one in which both of the shared electrons are provided by just one of the joined atoms. For example, boron trifluoride, BF_3, can combine with an additional fluoride ion to form the tetrafluoroborate ion, BF_4^-, according to the equation

$$BF_3 + F^- \longrightarrow BF_4^-$$
tetrafluoroborate ion

We can diagram this reaction as follows:

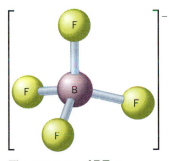

The structure of BF_4^-.

As we mentioned previously, the coordinate covalent bond is really no different from any other covalent bond once it has been formed. The distinction between them is made *only* for bookkeeping purposes. One place where such bookkeeping is useful is in keeping track of the orbitals and electrons used when atoms bond together.

The VB theory requirements for bond formation—two overlapping orbitals sharing two paired electrons—can be satisfied in two ways. One, as we have seen, is by the overlapping of two half-filled orbitals. This gives an "ordinary" covalent bond. The other is the overlapping of one filled orbital with one empty orbital. The shared pair of electrons is donated by the atom with the filled orbital and a coordinate covalent bond is formed.

The structure of the BF_4^- ion can therefore be explained as follows. First we examine the orbital diagram for boron.

To form four bonds, we need four hybrid orbitals. Since VSEPR theory predicts that the ion will have a tetrahedral shape, the boron will use sp^3 hybrids. Notice we spread the electrons out over the hybrid orbitals as much as possible.

Boron forms three ordinary covalent bonds with fluorine atoms plus one coordinate covalent bond with a fluoride ion.

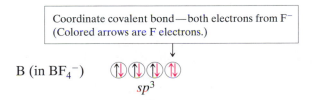

Practice Exercise 10

What hybrid orbitals are used by phosphorus in PCl_6^-? Draw the orbital diagram for phosphorus in PCl_6^-. What is the shape of the PCl_6^- ion? ◆

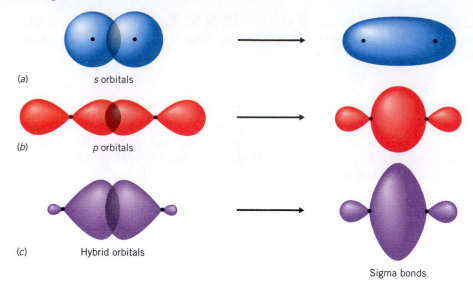

(a) s orbitals

(b) p orbitals

(c) Hybrid orbitals

Sigma bonds

Figure 9.23 *Formation of sigma bonds.* Sigma bonds concentrate electron density along the line between the two atoms joined by the bond. (*a*) From the overlap of *s* orbitals. (*b*) From the end-to-end overlap of *p* orbitals. (*c*) From the overlap of hybrid orbitals.

9.6 Double and Triple Bonds

The types of orbital overlap that we have described so far produce bonds in which the electron density is concentrated most heavily between the nuclei of the two atoms along an imaginary line that joins their centers. Any bond of this kind, whether formed from the overlap of *s* orbitals, *p* orbitals, or hybrid orbitals (Figure 9.23), is called a **sigma bond** (or **σ bond**).

Another way that *p* orbitals can overlap is shown in Figure 9.24. This produces a bond in which the electron density is divided between two separate regions that lie on opposite sides of an imaginary line joining the two nuclei. This kind of bond is called a **pi bond** (or **π bond**). Notice that a π bond, like a *p* orbital, consists of two parts, and each part makes up just half of the π bond; it takes *both* of them to equal *one* π bond.

The formation of π bonds allows atoms to form double and triple bonds. To see how this occurs, let's look at the bonding in the compound ethene, C_2H_4, which has the Lewis structure

Ethene is also called ethylene. Polyethylene, a common plastic, is made from C_2H_4.

$$\begin{array}{c} \text{H} \qquad\qquad \text{H} \\ \diagdown \qquad\quad \diagup \\ \text{C}=\text{C} \\ \diagup \qquad\quad \diagdown \\ \text{H} \qquad\qquad \text{H} \end{array}$$

ethene

The molecule is planar and each carbon atom lies in the center of a triangle surrounded by three other atoms (two H and one C atom). As we've seen, this

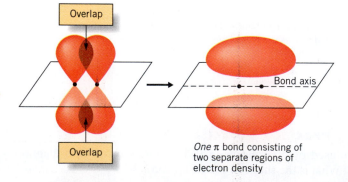

Overlap

Overlap

Bond axis

One π bond consisting of two separate regions of electron density

Figure 9.24 *Formation of a π bond.* Two *p* orbitals overlap sideways instead of end to end. The electron density is concentrated in two regions on opposite sides of the bond axis.

structure suggests that carbon uses sp^2 hybrid orbitals to form its bonds. There-fore, let's look at the distribution of electrons among the orbitals that carbon has available in its valence shell, assuming sp^2 hybridization.

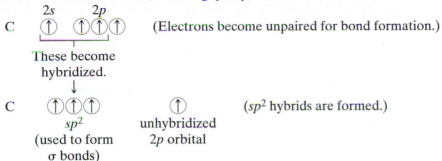

(Electrons become unpaired for bond formation.)

These become hybridized.

↓

C   ⬆⬆⬆   ⬆   (sp^2 hybrids are formed.)

sp^2 unhybridized
(used to form   2p orbital
σ bonds)

Notice that the carbon atom has an unpaired electron in an unhybridized $2p$ orbital. This p orbital is oriented perpendicular to the triangular plane of the sp^2 hybrid or-bitals, as shown in Figure 9.25. Now we can see how the molecule goes together.

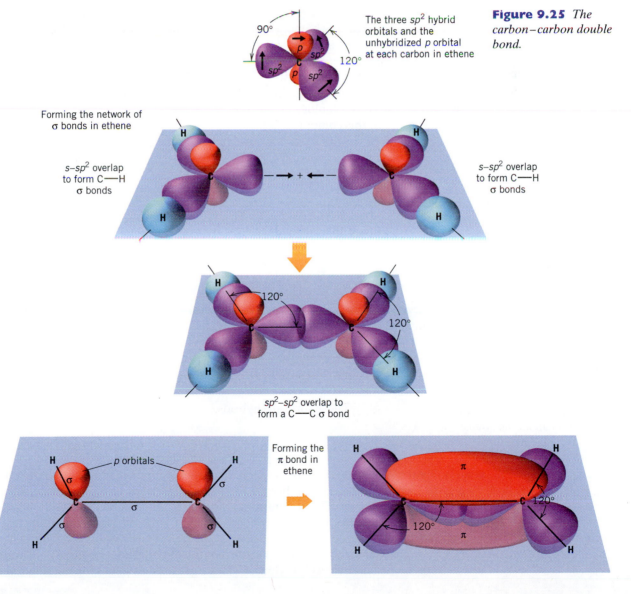

The three sp^2 hybrid orbitals and the unhybridized p orbital at each carbon in ethene

Forming the network of σ bonds in ethene

s–sp^2 overlap to form C—H σ bonds

s–sp^2 overlap to form C—H σ bonds

sp^2–sp^2 overlap to form a C—C σ bond

p orbitals

Forming the π bond in ethene

Figure 9.25 *The carbon–carbon double bond.*

The basic framework of the molecule is determined by the formation of σ bonds. Each carbon uses two of its sp^2 hybrids to form σ bonds to hydrogen atoms. The third sp^2 hybrid on each carbon is used to form a σ bond between the two carbon atoms, thereby accounting for one of the two bonds of the double bond. Finally, the remaining unhybridized $2p$ orbitals, one from each carbon atom, overlap to produce a π bond, which accounts for the second bond of the double bond.

Notice how well this description of bonding fits with the observed (and predicted) structure of the molecule. The use of sp^2 hybrids by carbon makes available the unpaired electrons in the unhybridized $2p$ orbitals, which in turn makes it possible for the carbon atoms to form the extra bond between them.

This explanation also accounts for one of the most important properties of double bonds: rotation of one portion of the molecule relative to the rest around the axis of the double bond occurs only with great difficulty. The reason for this is illustrated in Figure 9.26. We see that as one CH_2 group is rotated relative to the other around the carbon–carbon bond, the unhybridized p orbitals become misaligned and can no longer overlap effectively. This destroys the π bond. In effect, then, rotation around a double bond involves bond breaking, which requires more energy than is normally available to molecules at room temperature. As a result, rotation around the axis of a double bond usually doesn't take place.

In almost every instance, a double bond consists of a σ bond and a π bond. Another example is the compound formaldehyde (the substance used as a preservative for biological specimens and as an embalming fluid). The Lewis structure of this compound is

$$\begin{array}{c} H \\ \diagdown \\ C{=}\ddot{\underset{\cdot\cdot}{O}} \\ \diagup \\ H \end{array}$$

formaldehyde

Restricted rotation around double bonds has many important consequences in organic chemistry.

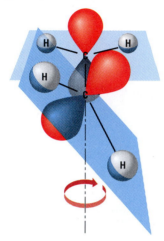

Figure 9.26 *Restricted rotation around a double bond.* If the CH_2 group closest to us were to rotate relative to the one at the rear, the unhybridized p orbitals would become misaligned, as shown here. This would destroy the overlap and break the π bond. Bond breaking requires a lot of energy, more than is available to the molecule through the normal bending and stretching of its bonds at room temperature. Because of this, rotation about the double bond axis is hindered or "restricted."

The two filled sp^2 hybrids on the oxygen become lone pairs on the oxygen atom in the molecule.

As with ethene, the carbon forms sp^2 hybrids, leaving an unpaired electron in an unhybridized p orbital.

$$C \quad \boxed{\uparrow}\boxed{\uparrow}\boxed{\uparrow} \quad \boxed{\uparrow}$$
$$sp^2 \qquad \text{unhybridized } 2p \text{ orbital}$$

The oxygen can also form sp^2 hybrids, with electron pairs in two of them and an unpaired electron in the third. This means that the remaining unhybridized p orbital also has an unpaired electron.

$$\begin{array}{cccc} & 2s & 2p & \\ O & \boxed{\uparrow\downarrow} & \boxed{\uparrow\downarrow}\boxed{\uparrow}\boxed{\uparrow} & \quad \text{(ground state of oxygen)} \end{array}$$

These form sp^2 hybrids.
$$\downarrow$$
$$O \quad \boxed{\uparrow\downarrow}\boxed{\uparrow\downarrow}\boxed{\uparrow} \quad \boxed{\uparrow}$$
$$sp^2 \qquad \text{unhybridized } 2p \text{ orbital}$$

Figure 9.27 shows how the carbon, hydrogen, and oxygen atoms come together to form the molecule. As before, the basic framework of the molecule is formed by the σ bonds. These determine the molecular shape. The carbon–oxygen double bond also contains a π bond formed by the overlap of the unhybridized p orbitals.

Now let's look at a molecule containing a triple bond. An example is ethyne, also known as acetylene, C_2H_2 (a gas used as a fuel for welding torches).

$$H-C\equiv C-H$$

ethyne
(acetylene)

In the linear acetylene molecule, each carbon needs two hybrid orbitals to form two σ bonds—one to a hydrogen atom and one to the other carbon atom. These can be provided by mixing the $2s$ and one of the $2p$ orbitals to form sp hybrids. To help us visualize the bonding, we will imagine that there is an xyz coordinate system centered at each carbon atom and that it is the $2p_z$ orbital that becomes mixed in the hybrid orbitals.

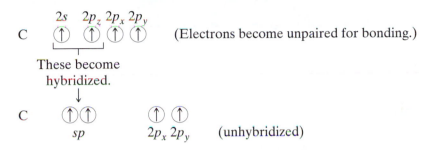

(Electrons become unpaired for bonding.)

We label the orbitals p_x, p_y, and p_z just for convenience; they are really all equivalent.

These become hybridized.

C sp $2p_x$ $2p_y$ (unhybridized)

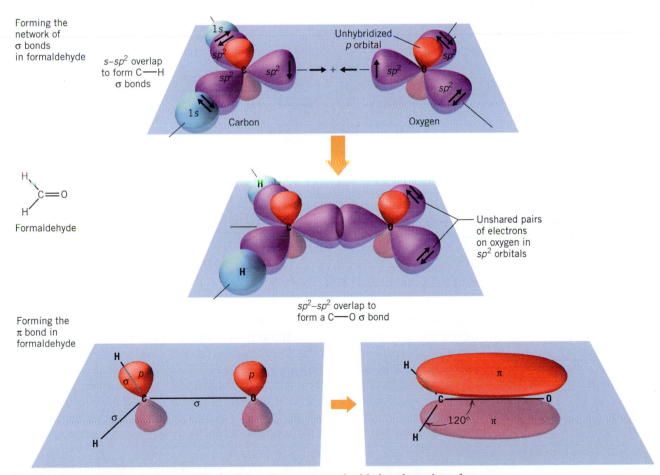

Forming the network of σ bonds in formaldehyde

s–sp^2 overlap to form C—H σ bonds

1s

Unhybridized p orbital

sp^2

sp^2

sp^2

sp^2

sp^2

1s

Carbon

Oxygen

Formaldehyde

sp^2–sp^2 overlap to form a C—O σ bond

Unshared pairs of electrons on oxygen in sp^2 orbitals

Forming the π bond in formaldehyde

H

σ

p

p

σ

σ

H

H

π

120°

π

H

Figure 9.27 *Bonding in formaldehyde.* The carbon–oxygen double bond consists of a σ bond and a π bond.

H—C≡C—H
acetylene

Forming the
σ-bond
network in
acetylene

(a)

(b)

(c)

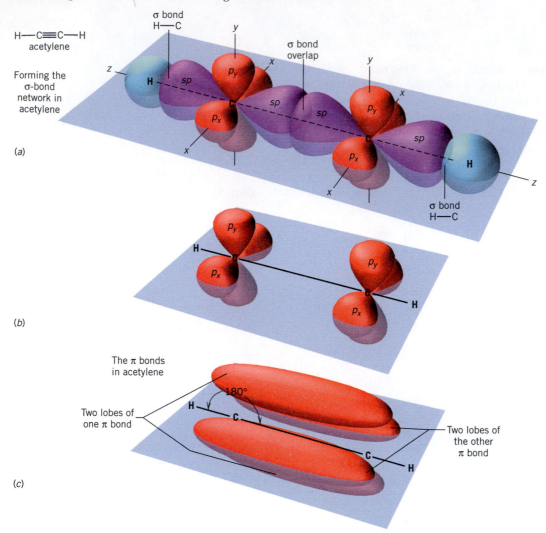

Figure 9.28 *The carbon–carbon triple bond in acetylene.* (*a*) The *sp* hybrid orbitals on each carbon atom are used to form the sigma bonds. (*b*) Sideways overlap of unhybridized $2p_x$ and $2p_y$ orbitals of the carbon atoms produces two π bonds. (*c*) The two π bonds in acetylene after they've formed.

Figure 9.28 shows how the molecule is formed. The *sp* orbitals point in opposite directions and are used to form the σ bonds. The unhybridized $2p_x$ and $2p_y$ orbitals are perpendicular to the C—C bond axis and overlap sideways to form two separate π bonds that surround the C—C σ bond. Notice that we now have three pairs of electrons in three bonds—one σ bond and two π bonds—whose electron densities are concentrated in different places. The three electron pairs therefore manage to avoid each other as much as possible. Also notice that the use of *sp* hybrid orbitals for the σ bonds allows us to explain the linear arrangement of atoms in the molecule.

Similar descriptions can be used to explain the bonding in other molecules that have triple bonds. Figure 9.29, for example, shows how the nitrogen mole-

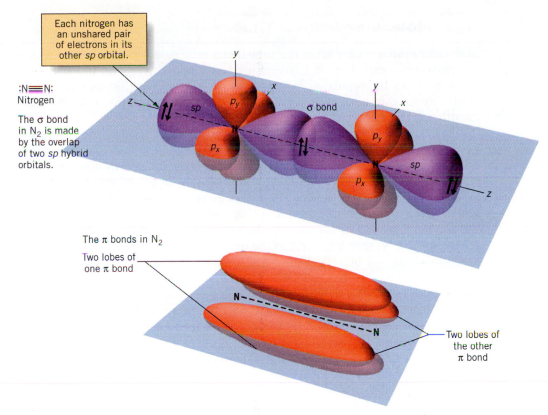

Each nitrogen has an unshared pair of electrons in its other *sp* orbital.

:N≡N:
Nitrogen

The σ bond in N_2 is made by the overlap of two *sp* hybrid orbitals.

The π bonds in N_2

Two lobes of one π bond

Two lobes of the other π bond

Figure 9.29 *Bonding in nitrogen.* The triple bond in N_2 is formed like the triple bond in acetylene. A sigma bond is formed by overlap of *sp* hybrid orbitals. The two unhybridized $2p$ orbitals on each nitrogen atom overlap to give the two π bonds. On each nitrogen, there is a lone pair of electrons in the *sp* hybrid orbital that's not used to form the sigma bond.

cule, N_2, is formed. In it, too, the triple bond is composed of one σ bond and two π bonds.

A Brief Summary

On the basis of the preceding discussion, we can make some observations that are helpful in applying the valence bond theory to a variety of molecules.

1. The basic molecular framework of a molecule is determined by the arrangement of its σ bonds.

2. Hybrid orbitals are used by an atom to form its σ bonds and to hold lone pairs of electrons.

3. The number of hybrid orbitals needed by an atom in a structure equals the number of atoms to which it is bonded *plus* the number of lone pairs of electrons in its valence shell.

4. When there is a double bond in a molecule, it consists of one σ bond and one π bond.

5. When there is a triple bond in a molecule, it consists of one σ bond and two π bonds.

9.7 Molecular Orbital Theory

Molecular orbital theory takes the view that a molecule is similar to an atom in one important respect. Both have energy levels that correspond to various orbitals that can be populated by electrons. In atoms, these orbitals are called atomic orbitals; in molecules, they are called **molecular orbitals.** (We shall frequently call them **MOs.**)

In most cases, the actual shapes and energies of molecular orbitals cannot be determined exactly. Nevertheless, theoreticians have found that reasonably good estimates of their shapes and energies can be obtained by combining the electron waves corresponding to the atomic orbitals of the atoms that make up the molecule. In forming molecular orbitals, these waves interact by constructive and destructive interference just like other waves that we've seen. Their intensities are either added or subtracted when the atomic orbitals overlap. The way this occurs can be seen if we look at the overlap of a pair of 1s orbitals from two atoms in a molecule like H_2 (Figure 9.30). The *two 1s* orbitals combine when the molecule is formed to give *two* MOs. In one MO, the intensities of the electron waves add together between the nuclei, which gives a buildup of electron density that helps hold the nuclei near each other. Such an MO is said to be a **bonding molecular orbital.** *Electrons in bonding MOs tend to stabilize a molecule.*

> The number of MOs formed is always equal to the number of atomic orbitals that are combined.

In the other MO, cancellation of the electron waves reduces the electron density between the nuclei. The reduced amount of negative charge between the nuclei allows the nuclei to repel each other strongly, so this MO is called an **antibonding molecular orbital.** *Antibonding MOs tend to destabilize a molecule when occupied by electrons.*

Both the bonding and antibonding MOs formed by the overlap of *s* orbitals have their maximum electron density on an imaginary line that passes through the two nuclei. Earlier, we called bonds that have this property sigma bonds. Molecular orbitals like this are also designated as sigma (σ). An asterisk is used to indicate the MOs that are antibonding, and a subscript is written to report which atomic orbitals make up the MO. For example, the bonding MO formed by the overlap of 1s orbitals is symbolized as σ_{1s} and the antibonding MO is written as σ_{1s}^*.

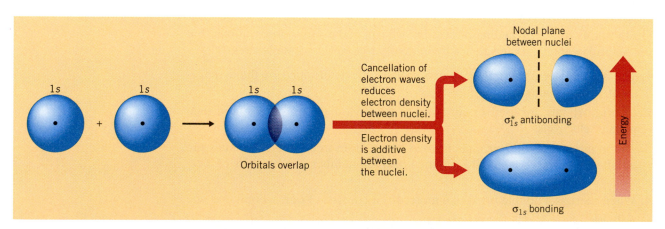

Figure 9.30 *Interaction of 1s atomic orbitals to produce bonding and antibonding molecular orbitals.* These are σ-type orbitals because the electron density is concentrated along the imaginary line that passes through both nuclei.

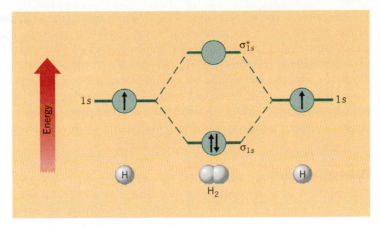

Figure 9.31 *Molecular orbital energy level diagram for H_2.*

Bonding MOs are lower in energy than antibonding MOs formed from the same atomic orbitals. This is also depicted in Figure 9.30. When electrons populate molecular orbitals, they fill the lower energy, bonding MOs first. The rules that apply to filling MOs are the same as those for filling atomic orbitals: *electrons spread out over orbitals of equal energy (Hund's rule) and two electrons can only occupy the same orbital if their spins are paired.*

Why Some Molecules Exist and Others Do Not

Let's see how molecular orbital theory can be used to account for the existence of certain molecules, as well as the nonexistence of others. Figure 9.31 is an MO energy level diagram for H_2. The energies of the separate $1s$ atomic orbitals are indicated at the left and right; those of the molecular orbitals are shown in the center. The H_2 molecule has two electrons, and both can be placed in the σ_{1s} orbital. The shape of this bonding orbital, shown in Figure 9.30, should be familiar. It's the same as the shape of the electron cloud that we described using the valence bond theory.

Next, let's consider what happens when two helium atoms come together. Why can't a stable molecule of He_2 be formed? Figure 9.32 is the energy diagram for He_2. Notice that both bonding and antibonding orbitals are filled. In situations such as this there is a net destabilization because the antibonding MO is raised in energy more than the bonding MO is lowered, relative to the orbitals of the separated atoms. This means the total energy of He_2 is larger than that of two separate He atoms, so the "molecule" is unstable and immediately comes apart.

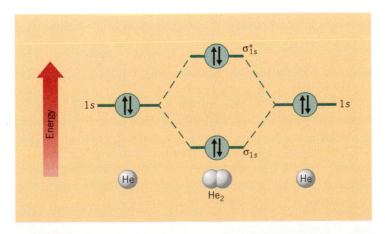

Figure 9.32 *Molecular orbital energy level diagram for He_2.*

In general, the effects of *antibonding electrons* (those in antibonding MOs) cancel the effects of an equal number of bonding electrons, and molecules with equal numbers of bonding and antibonding electrons are unstable. If we remove an antibonding electron from He_2 to give He_2^+, there is a net excess of bonding electrons, and the ion should be capable of existence. In fact, the emission spectrum of He_2^+ can be observed when an electric discharge is passed through a helium-filled tube, which shows that He_2^+ is present during the electric discharge. However, the ion is not very stable and cannot be isolated.

He₂⁺ has two bonding electrons and one antibonding electron.

Bond Orders

The concept of bond order was introduced in Chapter 8. Recall that it was defined as the number of pairs of electrons shared between two atoms. Thus, the sharing of one pair gives a single bond with a bond order of 1, two pairs give a double bond and a bond order of 2, and three pairs give a triple bond with a bond order of 3. To translate the MO description into these terms, we compute the bond order as follows:

$$\text{Bond order} = \frac{(\text{number of bonding } e^-) - (\text{number of antibonding } e^-)}{2}$$

For the H_2 molecule, we have

$$\text{Bond order} = \frac{2 - 0}{2} = 1$$

A bond order of 1 corresponds to a single bond. For He_2 we have

$$\text{Bond order} = \frac{2 - 2}{2} = 0$$

A bond order of zero means no bond exists, so the He_2 molecule is unable to exist. However, the He_2^+ ion does form, and its calculated bond order is

$$\text{Bond order} = \frac{2 - 1}{2} = 0.5$$

Notice that the bond order does not have to be a whole number.

Bonding in Diatomic Molecules of Period 2

The outer shell of a period 2 element (Li through Ne) consists of $2s$ and $2p$ subshells. When atoms of this period bond to each other, the atomic orbitals of these subshells interact strongly to produce molecular orbitals. The $2s$ orbitals, for example, overlap to form σ_{2s} and σ_{2s}^* molecular orbitals having essentially the same shapes as the σ_{1s} and σ_{1s}^* MOs, respectively. Figure 9.33 shows the shapes of the bonding and antibonding MOs produced when the $2p$ orbitals overlap. If we label those that point toward each other as $2p_z$, a set of bonding and antibonding MOs are formed that we can label as σ_{2p_z} and $\sigma_{2p_z}^*$. The $2p_x$ and $2p_y$ orbitals, which are perpendicular to the $2p_z$ orbitals, overlap sideways to give π-type molecular orbitals. They are labeled π_{2p_x} and $\pi_{2p_x}^*$, and π_{2p_y} and $\pi_{2p_y}^*$.

Tools

Molecular orbital energy diagram

The approximate relative energies of the MOs formed from the second shell atomic orbitals are shown in Figure 9.34. Notice that from Li to N, the energies of the π_{2p_x} and π_{2p_y} orbitals are lower than the energy of the σ_{2p_z}. Then from O to Ne, the energies of the two levels are reversed.

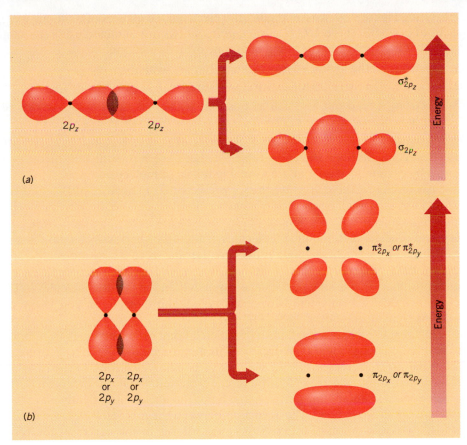

Figure 9.33 *Formation of molecular orbitals by the overlap of p orbitals.* (a) Two p_z orbitals that point at each other give bonding and antibonding σ-type MOs. (b) Perpendicular to the $2p_z$ orbitals are $2p_x$ and $2p_y$ orbitals that overlap to give two sets of bonding and antibonding π-type MOs.

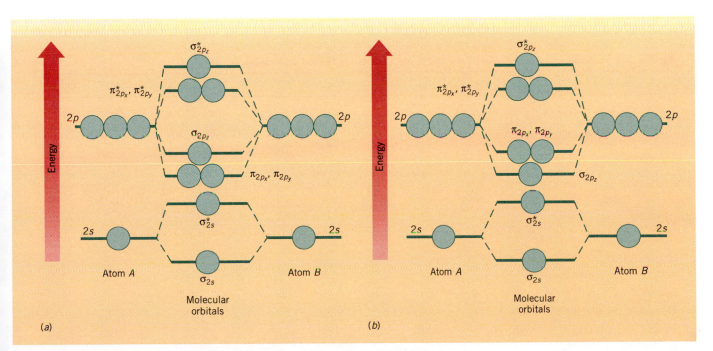

Figure 9.34 *Approximate relative energies of molecular orbitals in second period diatomic molecules.* (a) Li_2 through N_2. (b) O_2 through Ne_2.

Using Figure 9.34, we can predict the electronic structures of diatomic molecules of period 2. These *MO electron configurations* are obtained using the same rules that are applied to the filling of atomic orbitals in atoms.

1. Electrons fill the lowest energy orbitals that are available.
2. No more than two electrons, with spins paired, can occupy any orbital.
3. Electrons spread out as much as possible, with spins unpaired, over orbitals that have the same energy.

Applying these rules to the valence electrons of period 2 atoms gives the MO electron configurations shown in Table 9.1. Let's see how well MO theory performs by examining data that are available for these molecules.

According to Table 9.1, MO theory predicts that molecules of Be_2 and Ne_2 should not exist at all because they have bond orders of zero. In beryllium vapor and in gaseous neon, no evidence of Be_2 or Ne_2 has ever been found. MO theory also predicts that diatomic molecules of the other period 2 elements should exist because they all have bond orders greater than zero. These molecules have, in fact, been observed. Although lithium, boron, and carbon are complex solids under ordinary conditions, they can be vaporized. In the vapor, molecules of Li_2, B_2, and C_2 can be detected. Nitrogen, oxygen, and fluorine, as you know, are gaseous elements that exist as N_2, O_2, and F_2.

In Table 9.1, we also see that the predicted bond order increases from boron to carbon to nitrogen and then decreases from nitrogen to oxygen to fluorine. As the bond order increases, the *net* number of bonding electrons increases, which means that more electron density is concentrated between the nuclei. This greater concentration of negative charge binds the nuclei more tightly and therefore gives a stronger bond. The attraction between the increased electron density and the positive nuclei also draws the nuclei closer to the center of the bond, thereby decreasing the bond length. The *experimentally measured* bond energies and bond lengths given in Table 9.1 follow these predictions quite nicely.

Molecular orbital theory is particularly successful in explaining the electronic structure of the oxygen molecule. Experiments show that O_2 is paramagnetic; the molecule contains two unpaired electrons. In addition, the bond length

Molecular oxygen is attracted weakly by a magnet.

Table 9.1 Molecular Orbital Populations and Bond Orders for Period 2 Diatomic Molecules[a]

	Li₂	Be₂	B₂	C₂	N₂		O₂	F₂	Ne₂
$\sigma^*_{2p_z}$	○	○	○	○	○	$\sigma^*_{2p_z}$	○	○	⇅
$\pi^*_{2p_x}, \pi^*_{2p_y}$	○○	○○	○○	○○	○○	$\pi^*_{2p_x}, \pi^*_{2p_y}$	↑ ↑	⇅ ⇅	⇅ ⇅
σ_{2p_z}	○	○	○	○	⇅	π_{2p_x}, π_{2p_y}	⇅ ⇅	⇅ ⇅	⇅ ⇅
π_{2p_x}, π_{2p_y}	○○	○○	↑ ↑	⇅ ⇅	⇅ ⇅	σ_{2p_z}	⇅	⇅	⇅
σ^*_{2s}	○	⇅	⇅	⇅	⇅	σ^*_{2s}	⇅	⇅	⇅
σ_{2s}	⇅	⇅	⇅	⇅	⇅	σ_{2s}	⇅	⇅	⇅
Number of Bonding Electrons	2	2	4	6	8		8	8	8
Number of Antibonding Electrons	0	2	2	2	2		4	6	8
Bond Order	1	0	1	2	3		2	1	0
Bond Energy (kJ/mol)	110	—	300	612	953		501	129	—
Bond Length (pm)	267	—	158	124	109		121	144	—

The vertical axis on both the left and right sides of the table is labeled **Energy**.

[a]Although the order of the energy levels corresponding to the σ_{2p_z} and the π bonding MOs become reversed at oxygen, either sequence would yield the same result—a triple bond for N_2, a double bond for O_2, and a single bond for F_2.

in O_2 is about what is expected for an oxygen–oxygen double bond. These data are not explained by valence bond theory. For example, if we write a Lewis structure for O_2 that shows a double bond and also obeys the octet rule, all the electrons appear in pairs.

$:\ddot{O}::\ddot{O}:$ (not acceptable based on experimental
evidence because all electrons are paired)

On the other hand, if we show the unpaired electrons, the structure has only a single bond and doesn't obey the octet rule.

$:\ddot{O}:\ddot{O}:$ (not acceptable based on experimental
evidence because of the O—O single bond)

With MO theory, we don't have any of these difficulties. By applying Hund's rule, the two electrons in the π^* orbitals of O_2 spread out over these orbitals with their spins unpaired because both orbitals have the same energy. The electrons in the two antibonding π^* orbitals cancel the effects of two electrons in the two bonding π orbitals, so the net bond order is 2 and the bond is effectively a double bond.

Although MO theory handles easily the things that VB theory has trouble with, MO theory loses the simplicity of VB theory. For even quite simple molecules, MO theory is too complicated to make predictions without extensive calculations.

Practice Exercise 11

The MO energy level diagram for the nitric oxide molecule, NO, is essentially the same as that shown in Table 9.1 for O_2. Indicate which MOs are populated in NO and calculate the bond order for the molecule. ◆

9.8 Delocalized Molecular Orbitals

One of the least satisfying aspects of the way valence bond theory explains chemical bonding is the need to write resonance structures for certain molecules and ions. In Chapter 8 we discussed this in terms of Lewis structures, which we have since learned can be equated to shorthand notations for valence bond descriptions of molecules. As an example, let's look at the formate ion, CHO_2^-, which we described in Chapter 8 as a resonance hybrid of the following two structures.

$\left[H-C {\overset{\displaystyle :\ddot{O}:}{\underset{\displaystyle .\ddot{O}:}{\Big\langle}}} \right]^- \longleftrightarrow \left[H-C {\overset{\displaystyle :O:}{\underset{\displaystyle .\ddot{O}:}{\Big\langle}}} \right]^-$

In terms of the overlap of orbitals, VB theory would picture the formate ion as shown in Figure 9.35. The σ-bond framework is formed by overlap of the sp^2 hybrid orbitals of carbon with the $1s$ orbital of hydrogen and the p orbitals of

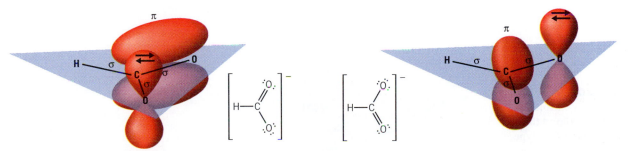

Figure 9.35 *Valence bond descriptions of the resonance structures of the formate ion,* CHO_2^-. *Only the p orbitals of oxygen that can form π bonds to carbon are shown. The ion is planar because carbon uses sp^2 hybrid orbitals to form the σ-bond framework.*

the oxygen atoms. As you can see in Figure 9.35, the π bond can be formed with either oxygen atom, which is how we obtain the two resonance structures.

Molecular orbital theory avoids the problem of resonance by recognizing that electron pairs can sometimes be shared among overlapping orbitals from three or more atoms. For the CHO_2^- ion, it allows *three p* orbitals (one from carbon and one each from the two oxygen atoms) to overlap simultaneously to form one large π-type molecular orbital that spreads over all three nuclei, as shown in Figure 9.36. Since the π electrons in this MO are not required to stay "localized" between just two nuclei, we say that the bond is **delocalized.** Delocalized π-type molecular orbitals permit a single description of the electronic structure of molecules and ions.

> A **localized bond** is one in which the electrons spend all their time shared between just two atoms. The electrons in a delocalized bond become shared among more than two atoms.

The delocalized nature of π bonds that extend over three or more nuclei can be indicated with dotted lines rather than dashes when Lewis-type structural formulas are drawn. The formate ion, for example, can be drawn as

$$\left[H-C \begin{array}{c} O \\ \\ O \end{array} \right]^-$$

> The bonding in each of these ions or molecules is explained by just one structure in the MO theory.

Delocalized π bonds are quite common in many kinds of molecules and ions. Some other examples are

$$\left[O=C \begin{array}{c} O \\ \\ O \end{array} \right]^{2-} \qquad \left[O=N \begin{array}{c} O \\ \\ O \end{array} \right]^- \qquad \left[O=\ddot{N}=O \right]^-$$

$$CO_3^{2-} \qquad\qquad NO_3^- \qquad\qquad NO_2^-$$

An important example from the realm of organic chemistry is benzene, C_6H_6. As you learned in Chapter 8, this molecule has a ring structure whose resonance structures can be written as

> Benzene is an important industrial solvent, but it is also toxic.

In benzene, the sigma-bond framework requires that the carbon atoms use sp^2 hybrid orbitals because each carbon forms three σ bonds. This leaves each carbon atom with a half-filled unhybridized *p* orbital perpendicular to the plane of the ring. These *p* orbitals overlap to give a delocalized π-electron cloud that looks something like two doughnuts with the sigma-bond framework sandwiched between them (Figure 9.37). The delocalized nature of the pi electrons is the reason we usually represent the structure of benzene as

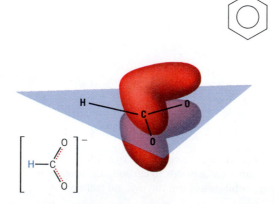

Figure 9.36 *Delocalized π bond in CHO_2^-.* Molecular orbital theory allows the electron pair in the π-type bond to be spread out, or delocalized, over all three atoms of the $—CO_2^-$ unit in CHO_2^-.

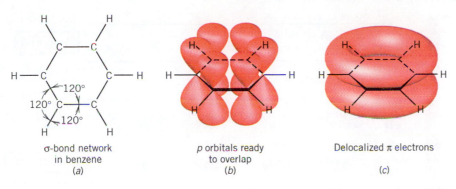

σ-bond network
in benzene
(a)

p orbitals ready
to overlap
(b)

Delocalized π electrons
(c)

Figure 9.37 *Benzene.* (*a*) The σ-bond framework. All atoms lie in the same plane. (*b*) The unhybridized *p* orbitals at each carbon prior to side-to-side overlap. (*c*) The double doughnut-shaped electron cloud formed by the delocalized π electrons.

One of the special characteristics of delocalized bonds, like those found in CO_3^{2-}, NO_3^-, NO_2^-, and C_6H_6, is that they make a molecule or ion more stable than it would be if it had localized bonds. In Chapter 8 this was described in terms of resonance energy. In the molecular orbital theory, we no longer speak of resonance; instead, we refer to the electrons as being delocalized. The extra stability that is associated with this delocalization is therefore described, in the language of MO theory, as the **delocalization energy.**

Functionally, the terms *resonance energy* and *delocalization energy* are the same; they just come from different approaches to bonding theory.

9.9 Bonding in Solids

In Chapter 2 we described the division of the elements into three broad classes: metals, nonmetals, and metalloids. One of the significant properties of these substances is their differing abilities to conduct electricity. To adequately explain these differences, a theory of bonding in solids was developed called the **band theory of solids.**

In a solid, an **energy band** is composed of a very large number of closely spaced energy levels that are formed by combining atomic orbitals of similar energy from each of the atoms within the substance. For example, in sodium the $1s$ atomic orbitals, one from each atom, combine to form a single $1s$ band. The number of energy levels in the band equals the number of $1s$ orbitals supplied by the entire collection of sodium atoms. The same thing occurs with the $2s$, $2p$, etc., orbitals, so that we also have $2s$, $2p$, etc., bands within the solid.

Figure 9.38 illustrates the energy bands in solid sodium. Notice that the electron density in the $1s$, $2s$, and $2p$ bands does not extend far from each individual nucleus, so these bands produce effectively localized energy levels in the solid. However, the $3s$ band is delocalized and extends continuously through the solid. The same applies to energy bands formed by higher energy orbitals.

Sodium atoms have filled $1s$, $2s$, and $2p$ orbitals, so the corresponding bands in the solid are also filled. The $3s$ orbital of sodium, however, is only half-filled, which leads to a half-filled $3s$ band. The $3p$ and higher energy bands in sodium are completely empty. When a voltage is applied across sodium, electrons in filled bands cannot move through the solid because orbitals in the same band on neighboring atoms are already filled and cannot accept an additional electron. In the $3s$ band, which is half-filled, an electron can hop from atom to atom with ease, and this allows sodium to conduct electricity well.

We refer to the band containing the outer shell (valence shell) electrons as the **valence band.** Any band that is either vacant or partially filled and uninterrupted throughout the lattice is called a **conduction band,** because electrons in it are able to move through the solid and thereby serve to carry electricity.

In metallic sodium the valence band and conduction band are the same, so sodium is a good conductor. In magnesium the $3s$ valence band is filled and

The formation of delocalized energy bands in the solid is like the formation of the delocalized bonds discussed in the preceding section.

Figure 9.38 *Energy bands in solid sodium.* The 1s, 2s, and 2p bands do not extend far in any direction from the nucleus, so the electrons in them are localized around each nucleus and can't move through the crystal. The delocalized 3s valence band extends throughout the entire solid, as do bands such as the 3p band formed from higher energy orbitals of the sodium atoms.

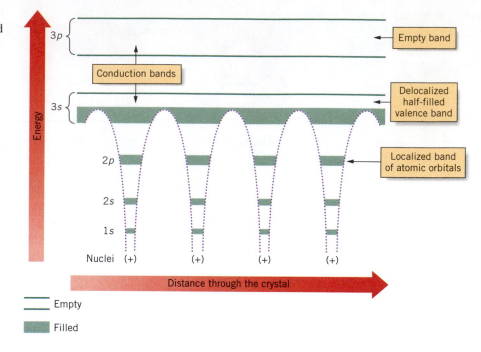

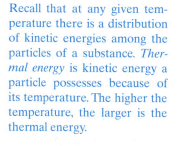

therefore cannot be used to transport electrons. However, the vacant 3p conduction band actually overlaps the valence band and can easily be populated by electrons when voltage is applied (Figure 9.39a). This permits magnesium to be a conductor.

In an insulator such as glass, diamond, or rubber, all the valence electrons are used to form covalent bonds, so all the orbitals of the valence band are filled and cannot contribute to electrical conductivity. In addition, the energy separation, or **band gap,** between the filled valence band and the nearest conduction band (empty band) is large. As a result, electrons cannot populate the conduction band, so these substances are unable to conduct electricity (Figure 9.39b).

In a semiconductor such as silicon or germanium, the valence band is also filled, but the band gap between the filled valence band and the nearest conduction band is small (Figure 9.39c). At room temperature, thermal energy pos-

Recall that at any given temperature there is a distribution of kinetic energies among the particles of a substance. *Thermal energy* is kinetic energy a particle possesses because of its temperature. The higher the temperature, the larger is the thermal energy.

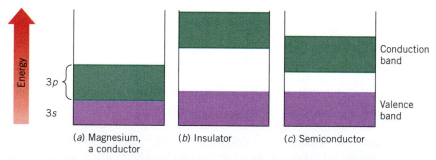

Figure 9.39 *Energy bands in different types of solids.* (*a*) In magnesium, a good electrical conductor, the empty 3p conduction band overlaps the filled 3s valence band and provides a way for this metal to conduct electricity. (*b*) In an insulator, the energy gap between the filled valence band and the empty conduction band prevents electrons from populating the conduction band. (*c*) In a semiconductor, there is a small band gap, and thermal energy can promote some electrons from the filled valence band to the empty conduction band. This enables the solid to conduct electricity weakly.

sessed by the electrons is sufficient to promote some electrons to the conduction band, and a small degree of electrical conductivity is observed. One of the interesting properties of semiconductors is that their electrical conductivity increases with increasing temperature. This is because as the temperature rises, the number of electrons with enough energy to populate the conduction band also increases.

Transistors and Other Electronic Devices

One of the most significant discoveries in the twentieth century was the way that the electrical characteristics of semiconductors can be modified by the controlled introduction of carefully selected impurities. This led to the discovery of transistors which made possible all the marvelous electronic devices we now take for granted, such as portable TVs, CD players, radios, calculators, and microcomputers. In fact, almost everything we do today is influenced in some way by electronic circuits etched into tiny silicon chips.

In a semiconductor such as silicon, all the valence electrons are used to form covalent bonds to other atoms. If a small amount of a Group IIIA element such as boron is added to silicon, it can replace silicon atoms in the structure of the solid. (We say the silicon has been **doped** with boron.) However, for each boron added, one of the covalent bonds in the structure will be deficient in electrons because boron has only three valence electrons instead of the four that silicon has. Under an applied voltage, an electron from a neighboring atom can move to fill this deficiency and thereby leave a positive "hole" behind. When this "hole" becomes filled, another is created elsewhere, and electrical conduction results from the migration of this positive "hole" through the solid. Because of the positive nature the moving charge carrier, the substance is said to be a **p-type semiconductor.**

If the impurity added to the silicon is a Group VA element such as arsenic, it has one more electron in its valence shell than silicon. When the bonds are formed in the solid, there will be an electron left over that's not used in bonding. The extra electrons supplied by the impurity can enter the conduction band and move through the solid under an applied voltage, and because the moving charge now consists of negatively charged electrons, the solid is said to be an **n-type semiconductor.**

Transistors are made from n- and p-type semiconductors and can be formed directly on the surface of a silicon chip, which has made possible the microcircuits in computers and calculators. Another interesting application is in solar batteries, which are easier to understand.

A silicon **solar battery** (also called a **solar cell**) is composed of a silicon wafer doped with arsenic (giving an n-type semiconductor) over which is placed a thin layer of silicon doped with boron (a p-type semiconductor). This is illustrated in Figure 9.40. In the dark there is an equilibrium between electrons and holes at the interface between the two layers, which is called a **p–n junction.** Some electrons from the n-type layer diffuse into the holes in the p-layer and are trapped. This leaves some positive holes in the n-layer. Equilibrium is achieved when the positive holes in the n-layer prevent further movement of electrons into the p-layer.

When light falls on the surface of the cell, the equilibrium is upset. Energy is absorbed, which permits electrons that were trapped in the p-layer to return to the n-layer. As these electrons move across the p–n junction into the n-layer, other electrons leave the n-layer through the wire, pass through the electrical circuit, and enter the p-layer. Thus an electric current flows when light falls on the cell and the external circuit is completed. This electric current can be used to run a motor, power a handheld calculator, or perform whatever other task we wish.

A hand-held solar calculator. The solar panel is located above the display.

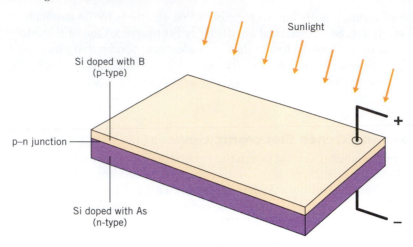

Sunlight

Si doped with B
(p-type)

p–n junction

+

Si doped with As
(n-type)

−

Figure 9.40 *Construction of a silicon solar battery.*

SUMMARY

Molecular Shapes and VSEPR Theory. The structures of most molecules can be described in terms of one or another of five basic geometries: **linear, planar triangular, tetrahedral, trigonal bipyramidal,** and **octahedral.** The **VSEPR theory** predicts molecular geometry by assuming that the electron pairs in the valence shell of an atom stay as far apart as possible so the repulsions between them are minimized. Figures 9.2 to 9.5 illustrate the structures obtained with different numbers of groups of electrons in the valence shell of the central atom in a molecule or ion and with different numbers of lone pairs and attached atoms. The correct shape of a molecule or polyatomic ion can usually be predicted from the Lewis structure.

Molecular Shape and Molecular Polarity. A molecule that contains identical atoms attached to a central atom will be nonpolar if there are no lone pairs of electrons in the central atom's valence shell. It will be polar if lone pairs are present, except in two cases: (1) when there are three lone pairs and two attached atoms and (2) when there are two lone pairs and four attached atoms. If all the atoms attached to the central atom are not alike, the molecule will usually be polar.

Valence Bond (VB) Theory. According to VB theory, a covalent bond is formed between two atoms when an atomic orbital on one atom **overlaps** with an atomic orbital on the other and a pair of electrons with paired spins is shared between the overlapping orbitals. In general, the better the overlap of the orbitals, the stronger is the bond. A given atomic orbital can only overlap with one other orbital on a different atom, so a given atomic orbital can only form one bond with an orbital on one other atom.

Hybrid Atomic Orbitals. Hybrid orbitals are formed by mixing pure s, p, and d orbitals. Hybrid orbitals overlap better with other orbitals than the pure atomic orbitals from which they are formed, so bonds formed by hybrid

orbitals are stronger than those formed by ordinary atomic orbitals. **Sigma bonds** (σ bonds) are formed by the following kinds of orbital overlap: $s-s$, $s-p$, end-to-end $p-p$, and overlap of hybrid orbitals. Sigma bonds allow free rotation around the bond axis. The side-by-side overlap of p orbitals produces a **pi bond** (π bond). Pi bonds do not permit free rotation around the bond axis because rotation around the axis of a π bond involves bond breaking. In complex molecules, the basic molecular framework is built with σ bonds. A double bond consists of one σ bond and one π bond. A triple bond consists of one σ bond and two π bonds.

Molecular Orbital (MO) Theory. The MO theory begins with the supposition that molecules are similar to atoms, except they have more than one positive center. They are treated as collections of nuclei and electrons, with the electrons of the molecule distributed among **molecular orbitals** of different energies. Molecular orbitals can spread over two or more nuclei, and they can be considered to be formed by the constructive and destructive interference of the overlapping electron waves corresponding to the atomic orbitals of the atoms in the molecule. **Bonding MOs** concentrate electron density between nuclei; **antibonding MOs** remove electron density from between nuclei. The rules for the filling of MOs are the same as those for atomic orbitals. The ability of MO theory to describe **delocalized orbitals** avoids the need for resonance theory. Delocalization of bonds leads to a lowering of the energy by an amount called the **delocalization energy** and produces more stable molecular structures.

Bonding in Solids. In solids, atomic orbitals of the atoms combine to yield **energy bands** that consist of many energy levels. The **valence band** is formed by orbitals of the valence shells of the atoms in the solid. A **conduction band** is a partially filled or empty band. In an **electrical conductor** the conduction band is either partially filled or is empty and overlaps a filled band. In an **insulator**, the **band gap** between the filled valence band and the empty conduction band is large, so no electrons populate the conduction band. In a **semiconductor,** the band gap between the filled valence band and the conduction band is small, and thermal energy can promote some electrons to the conduction

band. Silicon becomes a **p-type semiconductor,** in which the charge is carried by positive **"holes,"** if it is doped with a Group IIIA element such as boron. It becomes an **n-type semiconductor,** in which the charge is carried by electrons, if it is doped with a Group VA element such as arsenic.

Tools you have learned

The table below lists the tools you have learned in this chapter that are applicable to problem solving. Review them if necessary, and refer to them when working on the Thinking-It-Through problems and the Review Problems that follow.

Tool	Function
Basic molecular shapes (*page 368*)	Knowing how to draw them enables you to sketch the shapes of most molecules.
VSEPR theory (*page 370*)	This theory enables you to predict the shape of a molecule or polyatomic ion when its Lewis structure is known. It also enables you to determine the kind of hybrid orbitals used by the central atom in a molecule or polyatomic ion.
Molecular shape (*page 379*)	In this chapter you learned how we can use the shape of a molecule as a tool to determine whether the molecule is polar.
Orientations of hybrid orbitals (*pages 389 and 393*)	To allow us to explain the bonding in molecules based on their shapes.
Molecular orbital energy diagram (*pages 406 and 407*)	To allow us to explain the bonding in period 2 diatomic molecules.

THINKING IT THROUGH

For each of the following questions, list all the necessary information needed to obtain the answer and briefly describe what is to be done with it. Keep in mind that the goal here is *not* to find the answer itself; instead, you are just asked to assemble the available information needed to obtain the answer, state what additional data (if any) are needed, and describe how you would use the data to answer the question.

Level 1

1. The BrO_3^- ion is pyramidal. Without drawing the Lewis structure for the ion, explain how we know that there is at least one lone pair of electrons on the bromine atom.

2. The molecule SF_2 is nonlinear whereas the molecule XeF_2 is linear. Without drawing their Lewis structures, what differences could there be between the valence shells of S and Xe in these compounds?

3. The molecule SF_5Cl is octahedral. Is it polar or nonpolar? Explain.

4. Antimony forms a compound with hydrogen that is called stibine. Its formula is SbH_3, and the H—Sb—H bond angles are 91.3°. Which kinds of orbitals does Sb most likely use to form the Sb—H bonds, pure *p* orbitals or hybrid orbitals? Explain your reasoning.

5. Describe in detail how you would predict whether the SF_2 molecule is polar or nonpolar.

6. The sequence of energy levels in the molecular orbital energy level diagram for nitrogen monoxide, NO, is essen-

tially the same as for O_2. How would you determine the average bond order in the NO^+ ion?

7. All of the *s* orbitals in the 3*s* energy band of magnesium are filled by electrons. Nevertheless, magnesium is a good conductor of electricity. Explain this on the basis of the band theory of solids.

Level 2

8. The molecule XCl_3 is pyramidal. In which group in the periodic table is element X found? If the molecule were planar triangular, in which group would X be found? If the molecule were T-shaped, in which group would X be found? Why is it unlikely that element X is in Group VIA?

9. Consider the molecule $COCl_2$, whose Lewis structure is

$$\ddot{O}=C\overset{\ddot{\underset{\cdot\cdot}{C}l\cdot}}{\underset{\ddot{\underset{\cdot\cdot}{C}l\cdot}}{\big\langle}}$$

The VSEPR theory predicts that this is a planar triangular molecule with bond angles of 120°. However, in this molecule the O—C—Cl bond angles are not the same as the Cl—C—Cl bond angle. Suggest two reasons why all the bond angles are not the same.

10. Propadiene (allene), C_3H_4, and 1,3-butadiene, C_4H_6, each have two carbon–carbon double bonds. 1,3-butadiene also has a carbon–carbon single bond between the two center carbon atoms.
(a) Draw Lewis structures for both of these molecules. (In both of them, each carbon atom forms four covalent bonds.)

(b) What is the hybridization of the carbon atoms in each molecule?
(c) Is propadiene a planar molecule? Explain.
(d) One of these compounds can be drawn with both a polar and a nonpolar geometry. Show this with two sketches. Is the other compound polar or nonpolar? Explain.
(e) Does either molecule have its carbon atoms in a linear arrangement?

REVIEW QUESTIONS

Shapes of Molecules

9.1 Sketch the following molecular shapes and give the various bond angles in the structure: (a) planar triangular, (b) tetrahedral, (c) octahedral.

9.2 Sketch the following molecular shapes and give the bond angles in the structure: (a) linear, (b) trigonal bipyramidal.

VSEPR Theory

9.3 What is the underlying principle on which the VSEPR theory is based?

9.4 What arrangements of electron pairs are expected when the valence shell of the central atom contains (a) three pairs, (b) six pairs, (c) four pairs, or (d) five pairs of electrons?

9.5 What molecular shapes are expected when the valence shell of the central atom has (a) three bonding electron pairs and one lone pair, (b) four bonding electron pairs and two lone pairs, and (c) two bonding electron pairs and one lone pair? Give the name for each structure and also sketch its shape.

Predicting Molecular Polarity

9.6 Why is it useful to know the polarities of molecules?

9.7 How do we indicate a bond dipole when we draw the structure of a molecule?

9.8 What condition must be met if a molecule having polar bonds is to be nonpolar?

9.9 Use a drawing to show why the SO_2 molecule is polar.

Modern Bonding Theories

9.10 What is the theoretical basis of both valence bond (VB) theory and molecular orbital (MO) theory?

9.11 What shortcomings of Lewis structures and VSEPR theory do VB and MO theories attempt to overcome?

9.12 What is the main difference in the way VB and MO theories view the bonds in a molecule?

Valence Bond Theory

9.13 What is meant by *orbital overlap*?

9.14 What are the principal postulates of the valence bond theory?

9.15 Use sketches of orbitals to describe how the covalent bond in H_2 is formed.

9.16 Use sketches of orbitals to describe how VB theory would explain the formation of the H—Br bond in hydrogen bromide.

Hybrid Orbitals

9.17 What term is used to describe the mixing of atomic orbitals of the same atom?

9.18 Why do atoms usually prefer to use hybrid orbitals for bonding rather than plain atomic orbitals?

9.19 Sketch figures that illustrate the directional properties of the following hybrid orbitals: (a) sp, (b) sp^2, (c) sp^3, (d) sp^3d, (e) sp^3d^2.

9.20 Why do period 2 elements never use sp^3d or sp^3d^2 hybrid orbitals for bond formation?

9.21 What relationship is there, if any, between Lewis structures and the valence bond descriptions of molecules?

9.22 Using orbital diagrams, describe how sp^3 hybridization occurs in each atom: (a) carbon, (b) nitrogen, (c) oxygen. If these elements use sp^3 hybrid orbitals to form bonds, how many lone pairs of electrons would be found on each?

9.23 Sketch the way the orbitals overlap to form the bonds in each of the following: (a) CH_4, (b) NH_3, (c) H_2O. (Assume the central atom uses hybrid orbitals.)

9.24 We explained the bond angles of 107° in NH_3 by using sp^3 hybridization of the central nitrogen atom. Had the original unhybridized p orbitals of the nitrogen been used to overlap with $1s$ orbitals of each hydrogen, what would have been the H—N—H bond angles? Explain.

9.25 If the central oxygen in the water molecule did not use sp^3 hybridized orbitals (or orbitals of any other kind of hybridization), what would be the expected bond angle in H_2O (assuming no angle-spreading force)?

9.26 Using sketches of orbitals and orbital diagrams, describe sp^2 hybridization at (a) boron and (b) carbon.

Coordinate Covalent Bonds and VB Theory

9.27 The ammonia molecule, NH_3, can combine with a hydrogen ion, H^+ (which has an empty $1s$ orbital), to form the ammonium ion, NH_4^+. (This is how ammonia can neutralize acid and therefore function as a base.) Sketch the geometry of the ammonium ion, indicating the bond angles.

9.28 Use orbital diagrams to show the formation of a coordinate covalent bond in H_3O^+ when H_2O reacts with H^+.

Multiple Bonds and Hybrid Orbitals

9.29 How do σ and π bonds differ?

9.30 Why can free rotation occur easily around a σ-bond axis but not around a π-bond axis?

9.31 Using sketches, describe the bonds and bond angles in ethylene, C_2H_4.

9.32 Sketch the way the bonds form in acetylene, C_2H_2.

9.33 How does VB theory treat the benzene molecule? (Draw sketches describing the orbital overlaps and the bond angles.)

Molecular Orbital Theory

9.34 Why is the higher energy MO in H_2 called an anti-bonding orbital?

9.35 Using a sketch, describe the two lowest energy MOs of H_2 and their relationship to their parent atomic orbitals.

9.36 Explain why He_2 does not exist but H_2 does.

9.37 How does MO theory account for the paramagnetism of O_2?

9.38 On the basis of MO theory, explain why Li_2 molecules can exist but Be_2 molecules cannot. Could the ion $Be_2{}^+$ exist?

9.39 What are the bond orders in (a) $O_2{}^+$, (b) $O_2{}^-$, and (c) $C_2{}^+$?

9.40 What relationship is there between bond order and bond energy?

9.41 Sketch the shapes of the π_{2p_y} and $\pi_{2p_y}^*$ MOs.

9.42 What is a delocalized MO?

9.43 Use a Lewis-type structure to indicate delocalized bonding in the nitrate ion.

9.44 What problem encountered by VB theory does MO theory avoid by delocalized bonding?

9.45 Draw the representation of the benzene molecule that indicates its delocalized π system.

9.46 What effect does delocalization have on the stability of the electronic structure of a molecule?

9.47 What is delocalization energy? How is it related to resonance energy?

Bonding in Solids

9.48 On the basis of the band theory of solids, how do conductors, insulators, and semiconductors differ?

9.49 Define the terms (a) valence band and (b) conduction band.

9.50 Why does the electrical conductivity of a semiconductor increase with increasing temperature?

9.51 In calcium, why can't electrical conduction take place by movement of electrons through the $2s$ energy band? How does calcium conduct electricity?

9.52 What is a p-type semiconductor? What is an n-type semiconductor?

9.53 Name two elements that would make germanium a p-type semiconductor when added in small amounts.

9.54 Name two elements that would make silicon an n-type semiconductor when added in small amounts.

9.55 How does a solar cell work?

REVIEW PROBLEMS

Answers to problems whose numbers are printed in color are given in Appendix D.
More challenging problems are marked with asterisks.

VSEPR Theory

9.56 Predict the shapes of (a) $FCl_2{}^+$, (b) AsF_5, (c) AsF_3, (d) SbH_3, and (e) SeO_2.

9.57 Predict the shapes of (a) TeF_4, (b) $SbCl_6{}^-$, (c) $NO_2{}^-$, (d) $PCl_4{}^+$, and (e) $PO_4{}^{3-}$.

9.58 Predict the shapes of (a) $IO_4{}^-$, (b) $ICl_4{}^-$, (c) TeF_6, (d) $SiO_4{}^{4-}$, and (e) $ICl_2{}^-$.

9.59 Predict the shapes of (a) CS_2, (b) $BrF_4{}^-$, (c) ICl_3, (d) $ClO_3{}^-$, and (e) SeO_3.

9.60 Acetylene, a gas used in welding torches, has the Lewis structure $H—C\equiv C—H$. What would you expect the $H—C—C$ bond angle to be in this molecule?

9.61 Ethylene, a gas used to ripen tomatoes artificially, has the Lewis structure

$$\begin{array}{ccc} H & & H \\ | & & | \\ H—C & = & C—H \end{array}$$

What would you expect the $H—C—H$ and $H—C—C$ bond angles to be in this molecule? (Caution: Don't be fooled by the way the structure is drawn here.)

Predicting Molecular Polarity

9.62 Which of the following molecules would be expected to be polar? (a) HBr, (b) $POCl_3$, (c) CH_2O, (d) $SnCl_4$, (e) $SbCl_5$

9.63 Which of the following molecules would be expected to be polar? (a) PBr_3, (b) SeO_3, (c) $AsCl_3$, (d) ClF_3, (e) BCl_3

Valence Bond Theory

9.64 Hydrogen selenide is one of nature's most foul-smelling substances. Molecules of H_2Se have $H—Se—H$ bond angles very close to 90°. How would VB theory explain the bonding in H_2Se? Use sketches of orbitals to show how the bonds are formed. Illustrate with appropriate orbital diagrams as well.

9.65 Use sketches of orbitals to show how VB theory explains the bonding in the F_2 molecule. Illustrate with appropriate orbital diagrams as well.

Hybrid Orbitals

9.66 Use orbital diagrams to explain how the $BeCl_2$ molecule is formed. What kind of hybrid orbitals does beryllium use in $BeCl_2$?

9.67 Use orbital diagrams to describe the bonding in (a) $SnCl_4$ and (b) $SbCl_5$. Be sure to indicate hybrid orbital formation.

9.68 Draw Lewis structures for the following and use the geometry predicted by VSEPR theory for the electron pairs to determine what kind of hybrid orbitals the central atom uses in bond formation: (a) ClO_3^-, (b) SO_3, (c) OF_2.

9.69 Draw Lewis structures for the following and use the geometry predicted by VSEPR theory for the electron pairs to determine what kind of hybrid orbitals the central atom uses in bond formation: (a) $SbCl_6^-$, (b) $BrCl_3$, (c) XeF_4.

9.70 Use VSEPR theory to help you describe the bonding in the following molecules according to VB theory: (a) $AsCl_3$, (b) ClF_3.

9.71 Use VSEPR theory to help you describe the bonding in the following molecules according to VB theory: (a) $SbCl_5$, (b) $SeCl_2$.

Coordinate Covalent Bonds and VB Theory

9.72 Use orbital diagrams to show that the bonding in SbF_6^- involves the formation of a coordinate covalent bond.

9.73 What kind of hybrid orbitals are used by tin in $SnCl_6^{2-}$? Draw the orbital diagram for Sn in $SnCl_6^{2-}$. What is the geometry of $SnCl_6^{2-}$?

Multiple Bonding and Valence Bond Theory

9.74 A nitrogen atom can undergo sp^2 hybridization when it becomes part of a carbon–nitrogen double bond, as in $H_2C=NH$.
(a) Using a sketch, show the electron configuration of sp^2 hybridized nitrogen just before the overlapping occurs to make this double bond.
(b) Using sketches (and the analogy to the double bond in C_2H_4), describe the two bonds of the carbon–nitrogen double bond.
(c) Describe the geometry of $H_2C=NH$ (using a sketch that shows all expected bond angles).

9.75 A nitrogen atom can undergo sp hybridization and then become joined to carbon by a triple bond to give the structural unit $-C≡N:$. This triple bond consists of one σ bond and two π bonds.
(a) Write the orbital diagram for sp hybridized nitrogen as it would look before any bonds form.
(b) Using the carbon–carbon triple bond as the analogy, and drawing pictures to show which atomic orbitals overlap with which, show how the three bonds of the triple bond in $-C≡N:$ form.

(c) Again using sketches, describe all the bonds in hydrogen cyanide, $H-C≡N:$.
(d) What is the likeliest $H-C-N$ bond angle in HCN?

9.76 Tetrachloroethylene, a common dry-cleaning solvent, has the formula C_2Cl_4. Its structure is

Use the VSEPR and VB theories to describe the bonding in this molecule. What are the expected bond angles in the molecule?

9.77 Phosgene, $COCl_2$, was used as a war gas during World War I. It reacts with moisture in the lungs of its victims to form CO_2 and gaseous HCl, which cause the lungs to fill with fluid. Phosgene is a simple molecule having the structure

Describe the bonding in this molecule using VB theory.

9.78 What kind of hybrid orbitals do the numbered atoms use in the following molecule?

9.79 What kinds of bonds (σ or π) are found in the numbered bonds in the following molecule?

Molecular Orbital Theory

9.80 Use the MO energy diagram to predict which in each pair has the greater bond energy: (a) O_2 or O_2^+, (b) O_2 or O_2^-, (c) N_2 or N_2^+.

9.81 Assume that in the NO molecule the molecular orbital energy level sequence is similar to that for O_2. What happens to the NO bond length when an electron is removed from NO to give NO^+?

Bonding in Solids

9.82 Construct a diagram that illustrates the band structure of potassium.

9.83 Construct a diagram that illustrates the band structure of calcium.

ADDITIONAL EXERCISES

9.84 Formaldehyde has the Lewis structure

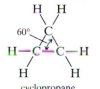

What would you predict its shape to be?

9.85 Describe the changes in molecular geometry that take place during the following reactions:
(a) $BF_3 + F^- \longrightarrow BF_4^-$
(b) $PCl_5 + Cl^- \longrightarrow PCl_6^-$
(c) $ICl_3 + Cl^- \longrightarrow ICl_4^-$
(d) $PCl_3 + Cl_2 \longrightarrow PCl_5$
(e) $C_2H_2 + H_2 \longrightarrow C_2H_4$

9.86 Cyclopropane is a triangular molecule with C—C—C bond angles of 60°. Explain why the σ bonds joining carbon atoms in cyclopropane are weaker than the carbon–carbon σ bonds in the noncyclic propane.

cyclopropane   propane

9.87 What facts about boron strongly suggested the need to consider sp^2 hybridization of its second shell atomic orbitals in BF_3?

9.88 Phosphorus trifluoride, PF_3, has F—P—F bond angles of 97.8°.
(a) How would VB theory use hybrid orbitals to explain this?
(b) How would VB theory use unhybridized orbitals to account for the bond angles?
(c) Do either of these models work very well?

9.89 A six-membered ring of carbons can hold a double bond but not a triple bond. Explain.

cyclohexene (exists)   cyclohexyne (unknown)

*9.90 There exists a hydrocarbon called butadiene, which has the molecular formula C_4H_6 and the structure

The C=C bond lengths are 134 pm (about what is expected for a carbon–carbon double bond), but the C—C bond length in this molecule is 147 pm, which is shorter than a normal C—C single bond. The molecule is planar (i.e., all the atoms lie in the same plane).
(a) What kind of hybrid orbitals do the carbon atoms use in this molecule to form the carbon–carbon bonds?

(b) Between which pairs of carbon atoms do we expect to find sideways overlap of p orbitals (i.e., π-type p–p overlap)?
(c) On the basis of your answer to part (b), do you expect to find localized or delocalized π bonding in the carbon chain in this molecule?
(d) Based on your answer to part (c), explain why the center carbon–carbon bond is shorter than a carbon–carbon single bond.

9.91 Draw the structure of the molecule CCl_2F_2 (carbon is the central atom). Is the molecule polar or nonpolar? Explain.

*9.92 *The more electronegative are the atoms bonded to the central atom, the less are the repulsions between the electron pairs in the bonds.* On the basis of this statement, predict the most probable structure for the molecule PCl_3F_2. Do we expect the molecule to be polar or nonpolar?

*9.93 *A lone pair of electrons in the valence shell of an atom has a larger effective volume than a bonding electron pair. Lone pairs therefore repel other electron pairs more strongly than do bonding pairs.* On the basis of these statements, describe how the bond angles in TeF_4 and BrF_4^- deviate from those found in a trigonal bipyramid and an octahedron, respectively. Sketch the molecular shapes of TeF_4 and BrF_4^- and indicate these deviations on your drawings.

*9.94 *The two electron pairs in a double bond repel other electron pairs more than the single pair of electrons in a single bond.* On the basis of this statement, which bond angles should be larger in SO_2Cl_2, the O—S—O bond angles or the Cl—S—Cl bond angles? (In the molecule, sulfur is bonded to two oxygen atoms and two chlorine atoms. *Hint:* Assign formal charges and work with the best Lewis structure for the molecule.)

*9.95 *A hybrid orbital does not distribute electron density symmetrically around the nucleus of an atom. Therefore, a lone pair in a hybrid orbital contributes to the overall polarity of a molecule.* On the basis of these statements and the fact that NH_3 is a very polar molecule and NF_3 is a nearly nonpolar molecule, justify the notion that the lone pair of electrons in each of these molecules is held in an sp^3 hybrid orbital.

9.96 In a certain molecule, a p orbital overlaps with a d orbital as shown below. What kind of bond is formed, σ or π? Explain your choice.

9.97 If we take the internuclear axis in a diatomic molecule to be the z axis, what kind of p orbital (p_x, p_y, or p_z) on one atom would have to overlap with a d_{xz} orbital on the other atom to give a pi bond?

Good Ozone—Where Is It?

5

Stratospheric Ozone Cycle

The *stratosphere* is an envelope of space between altitudes of 10 and 50 km (6 to 31 miles). The solar energy entering it is about 7% ultraviolet (UV) radiation whose wavelengths between 280 nm and 315 nm cause sunburn and skin cancer. So we are fortunate that nearly all UV radiation below 320 nm is removed in the *ozone layer* of the stratosphere by the *stratospheric ozone cycle,* a chemical chain reaction (Facets of Chemistry 13.1).

The ozone cycle is initiated by short UV wavelengths, those below 242 nm. These split O_2 molecules to electronically excited oxygen atoms, O^*.

$$O_2 \xrightarrow[\text{(λ = 242 nm or lower)}]{\text{UV radiation}} 2O^*$$

The cycle itself, thus initiated, has two steps that repeat over and over. In the first, ozone forms when O^* collides with O_2 at the surface of a neutral particle, M, which might be a molecule of N_2 or O_2.

$$O^* + O_2 + M \longrightarrow O_3 + M + \text{heat} \qquad (1)$$

The function of M is to absorb some of the energy of the collision. Otherwise the new O_3 molecule would have enough vibrational energy to split apart as soon as it formed. *What is desired, instead, is that UV energy split O_3.*

In the second step, ozone molecules absorb UV energy and break up.

$$O_3 + \text{UV energy} \xrightarrow[\text{(λ = 240–320 nm)}]{} O_2 + O^* \qquad (2)$$

The O^* that forms in Step 2 is a necessary *reactant* for Step 1, so when Step 2 occurs, it enables another Step 1 event. Thus, Steps 1 and 2 constitute a chemical chain reaction. Each "turn" of this cycle absorbs some UV energy; if we add Equations 1 and 2, the net result is

$$\text{UV energy} \longrightarrow \text{heat} \qquad (1 + 2)$$

Thus the conversion of UV energy into heat is the net effect of the ozone cycle. The cycle can repeat hundreds of times before other events, such as a chance combination of two atoms of O to give back O_2, terminate a chain.

The stratospheric ozone level varies with latitude, altitude, and month. Between 60° north and south of the equator, the level in 1974 varied between 260 and 360 Dobson units, reaching as high as 500 Dobson units. The *Dobson unit,* named for English scientist Gordon Dobson, equals 2.7×10^{16} molecules of O_3 in a column of air 1 cm^2 in area at its base (on Earth's surface) and extending through the stratosphere.

How the Ozone Cycle Is Broken. Any reaction that consumes O^*, besides its reaction with O_2 (Step 1), breaks the ozone cycle and reduces the supply of ozone. As the ozone level decreases, increasing amounts of solar UV radiation pass through to us, causing more skin cancer.

Atomic chlorine can combine with O^* and thus disrupt the cycle. Cl atoms have been furnished to the ozone layer for decades by our use of chlorofluorocarbons (CFCs), like CCl_3F and CCl_2F_2, as air conditioner fluids and aerosol gases. UV radiation breaks their C—Cl bonds, releasing Cl atoms. For example,

$$CCl_2F_2 \xrightarrow{\text{UV radiation}} CClF_2 + Cl$$

Atomic chlorine, thus formed, destroys ozone by a chain reaction.

$$Cl + O_3 \longrightarrow ClO + O_2 \qquad (3)$$

$$ClO + O \longrightarrow Cl + O_2 \qquad (4)$$

$$\text{Net:} \quad O_3 + O \longrightarrow 2O_2$$

The Cl atom from the breakup of only one CFC molecule can initiate the destruction of thousands of ozone molecules.

Mario Molina (USA/Mexico), F. S. Rowland (USA), and Paul Crutzen (Germany/Netherlands) shared the 1995 Nobel Prize in chemistry for their research into the effects of CFCs on the ozone layer. Their work led to a gradual, worldwide phasing out of the use of CFCs.

The Antarctic Ozone "Hole"

Since the 1970s, a decline in the ozone level has occurred over the Antarctic during each Antarctic spring, (fall in the Northern hemisphere), reaching its lowest

chemicals in our world

levels in October each year. By 1989, the loss was 70%. Thus, a continent-broad column of the atmosphere over the Antarctic, the *Antarctic ozone hole*, annually becomes more transparent to UV radiation. A significant recovery in the ozone level of the ozone hole occurs between November and March, *partly by the diffusion of stratospheric ozone from middle latitudes.* Such diffusion diminishes the ozone "cover" over populated areas, thus explaining how an ozone hole in the Antarctic threatens people elsewhere. Seasonally, the Arctic has a smaller, but growing "hole."

Why over the Antarctic, and Why in the Antarctic Spring? A circulating wind pattern called the *Antarctic vortex,* caused by Earth's rotation and the sharply uneven heating of the atmosphere, develops over the South Pole during the Antarctic winter, June to August. The reactions that destroy ozone occur in this vortex.

During the total darkness of the Antarctic winter, the stratospheric temperature drops below an important threshold, 195 K ($-78\ °C$). Now, vast, thin polar stratospheric clouds (PSCs) can form. They consist of crystals of ice and nitric acid trihydrate, the latter formed by the reaction of water with nitrogen dioxide. (NO_2 occurs both naturally and as a product of pollution.)

Molecular chlorine collects on the PSCs, being made from chlorine nitrate ($ClONO_2$) and hydrogen chloride.

$$HCl + ClONO_2 \longrightarrow Cl_2 + HONO_2$$

Chlorine nitrate arises when ClO, made during Antarctic sunshine by Equation 3, reacts with NO_2.

$$ClO + NO_2 \longrightarrow ClONO_2$$

Hydrogen chloride forms when Cl, available from the ozone-destroying cycle (Equations 3–4), reacts with methane (CH_4) to give CH_3 and HCl. Methane is a pollutant produced both naturally and by pipeline leaks.

When the sun reappears, chlorine molecules are released from the PSCs and are soon broken apart by UV radiation ($Cl_2 + UV \rightarrow 2Cl$). Therefore, *just as sunlight returns with the Antarctic spring, a large supply of chlorine atoms is released into the stratosphere.* They quickly destroy ozone by the cycle of Equations 3–4, and the Antarctic ozone hole once again appears. In October, 1998, the ozone level in the ozone hole dropped to 92 Dobson units; its lowest value, 88 Dobson units, was in late September, 1993. The area of ozone depletion over the Antarctic in October, 1998, $26 \times 10^6\ km^2$, was the largest ever observed (see Figure 5*a*).

The Greenhouse Effect and the Ozone Hole. The greenhouse gases that warm our lower atmosphere (see

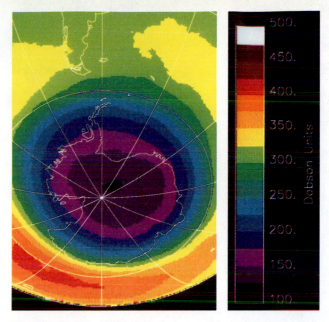

Figure 5a The 1998 ozone hole over the Antarctic.

Chemicals in Our World 4) lessen the transfer of Earth's heat to the stratosphere, causing still lower polar stratospheric temperatures during winter. This worsens the ozone hole problem, particularly in the Arctic where high altitude temperatures otherwise seldom fall below the 195 K threshold for the formation of polar stratospheric clouds. It thus appears that increases in levels of CO_2 and other greenhouse gases indirectly contribute to ozone holes over both poles. Hence, despite successful efforts to reduce releases of the CFCs, ozone hole problems are expected to increase into the 2020s.

Questions

1. Write the equations by means of which ozone cycles are initiated.

2. What two equations represent the ozone cycle?

3. What happens to the UV radiation as a result of the ozone cycle?

4. By means of equations, explain how CFCs can reduce the stratospheric ozone level.

5. Polar stratospheric clouds provide reservoirs for the subsequent release of chlorine atoms. How is this done? (Include equations.)

6. How does the reappearance of the sun at the onset of the Antarctic spring initiate the decline in ozone levels in the ozone hole?

The high temperature of molten rock can cause a subterranean buildup of extreme pressure great enough to blow glowing ash, gases, and lava sky high. We'll study the relationships among pressure, temperature, and volume of gases in this chapter.

Properties of Gases

This Chapter in Context We have just studied the *kinds* of matter—elements, compounds, and certain mixtures. We've also looked at the *structure* of matter and the characteristic reactions of certain of its *families*. In this chapter we begin a systematic study of the physical *states* of matter, namely, gases, liquids, and solids. We study gases first, because they are the easiest to understand, and their behavior will help to explain some of the properties of liquids and solids in the next chapter. Many gases are common substances, like the oxygen and nitrogen in air. When we have looked at the general facts about all gases, we'll see how the kinetic theory of matter, introduced in Section 6.2, helps us to understand them.

10.1 Properties Common to All Gases

Perhaps the most remarkable fact about gases is that despite wide differences in *chemical* properties, they all more or less obey the same set of *physical* laws, the *gas laws*. One question that we raise, in fact, is why are there *general* laws for gases but not for liquids and solids?

You already know some of the properties of gases. Whenever you inflate a balloon, for example, you have an experience with gas pressure, and the "feel" of a balloon suggests that the pressure acts equally in all directions.

The warning, "Do Not Incinerate," is printed on aerosol cans because a gas's temperature affects its pressure. A sealed can, if made too hot, is in danger of exploding from the increased pressure.[1]

[1]An amusing newspaper report, "The Awful Day of the Ham," told on herself by Donna Tabbert Long, described how a peaceful Sunday was punctuated by a huge explosion that shook the apartment building. When her husband asked her how the ham dinner was coming, she reported that when she last looked in the oven the ham (in its unopened tin) was "puffing up nicely." Her husband reached the kitchen door when the explosion occurred and soon reported, in a quiet voice tinged with awe, "There's ham hanging all over in there." (*Minneapolis Tribune,* August 2, 1980, page 5B.)

Air is roughly 21% O_2 and 79% N_2, but it has traces of several other gases.

423

To study the effects of pressure and temperature on all gases, we need to learn how four properties relate to each other, namely, pressure (P), volume (V), temperature (T), and amount or number of moles of gas (n).

10.2 Pressure. Its Measurement and Units

Pressure, as we learned in Section 6.4, is force *per unit area,* calculated by dividing the force by the area on which the force acts.

$$\text{Pressure} = \frac{\text{force}}{\text{area}}$$

Earth exerts a gravitational force on everything with mass on it or near it. What we call the *weight* of an object, like a book, is simply our measure of the gravitational force acting on it. Hence, we use "weight" for "force" in the equation for pressure when the gravitational force is involved.

The distinction between force and pressure is important. The weight of the books in a backpack causes less shoulder pain, for example, when the pack has wide straps rather than narrow straps of string. The force (the weight) is the same, but the pressure, or force *per unit area,* on your shoulders is much less when the weight is distributed over wide straps. The *ratio* of force to area, the pressure, is less, and you can feel the difference.

Atmospheric Pressure and the Barometer

When Earth's gravitational force acts on the air mass of the atmosphere, it creates an opposing force, namely, that of the air pushing on Earth's surface, which we also mentioned in Section 6.4. The ratio of this pushing force to Earth's surface area at any one place is the *atmospheric pressure* at that spot. You can see atmospheric pressure at work by pumping the air from any collapsible container (see Figure 10.1). Before some air is removed, the walls of the can experience atmospheric pressure identically inside and out. When some air is pumped out, however, the pressure inside decreases, making the atmospheric pressure outside of the can greater than the pressure inside. The net inward pressure is sufficiently great to make the can crumple.

To measure atmospheric pressure we use a **barometer.** The simplest type is the *Torricelli barometer* (see Figure 10.2), which consists of a tube sealed at one end, 80 cm or more in length, and filled with mercury. When the tube is inverted over a dish of mercury, some mercury runs out, but not all.[2] Atmospheric pressure, pushing on the surface of the mercury in the dish, holds some of the mercury in the tube. Opposing the atmosphere is the downward pressure caused by the weight of the mercury still inside the tube. When the two pressures become equal, no more mercury can run out, but a space inside the tube above the mercury level has been created having essentially no atmosphere; it's a *vacuum.*

Mercury, a shiny, metallic element, is a liquid above $-39\,^{\circ}$C.

The height of the mercury column, measured from the surface of the mercury in the dish, is directly proportional to atmospheric pressure. On days when the atmospheric pressure is high, more mercury is forced from the dish into the tube and the height of the column increases. When the atmospheric pressure drops, during an approaching storm, for example, some mercury flows out of the tube and the height of the column decreases. Most people live where this height fluctuates between 730 and 760 mm.

[2]Today, mercury is kept as much as possible in closed containers. Although atoms of mercury do not readily escape into the gaseous state, mercury vapor is a dangerous poison.

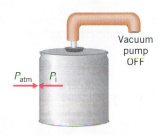

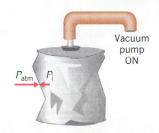

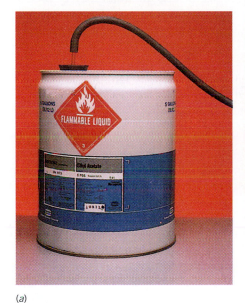

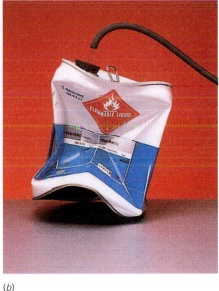

(a) (b)

Figure 10.1 *The effect of an unbalanced pressure.* (a) The pressure inside the can, P_i, is the same as the atmospheric pressure outside, P_{atm}. The pressures are balanced; $P_i = P_{atm}$. (b) When a vacuum pump reduces the pressure inside the can, P_i is made less than P_{atm}, and the unbalanced outside pressure quickly and violently makes the can collapse.

Units of Pressure

The height of the mercury column in a barometer varies with altitude. On top of Mount Everest, Earth's highest peak, for example, the air is so "thin" that the atmospheric pressure can maintain a height of only about 250 mm of mercury. At sea level, the height fluctuates around a value of 760 mm. Some days it is a little higher, some a little lower, depending on the weather. The average pressure at sea level has long been used by scientists as a unit of pressure and is called the **standard atmosphere** of pressure, abbreviated **atm.**[3] In Chapter 6 we said that 1 atm equals approximately 14.7 lb in.$^{-2}$.

In the SI, the unit of pressure is the **pascal,** symbolized **Pa**. In SI units, the pascal is the ratio of the force in *newtons* to the area in meters squared. It's a very small amount; 1 Pa is approximately the pressure exerted by the weight of a lemon spread over an area of 1 m^2.

To bring the standard atmosphere unit in line with other SI units, it has been redefined in terms of the pascal as follows.

$$1 \text{ atm} = 101{,}325 \text{ Pa} \qquad \text{(exactly)}$$

In Canada, atmospheric pressures in weather reports are given in kilopascals (kPa); thus, 1 atm is about 101 kPa.

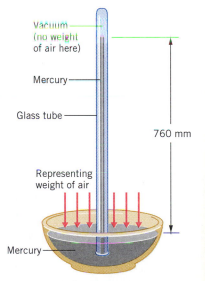

Figure 10.2 *The mercurial or Torricelli barometer.*

[3]Because any metal, including mercury, expands or contracts as the temperature increases or decreases, the height of the mercury column in the barometer varies a little with temperature. Therefore, the standard atmosphere was originally defined as the pressure able to support a column of mercury 760 mm high, measured at 0 °C.

For ordinary laboratory work, the *atmosphere* is an inconveniently large unit, and chemists often use a smaller unit, the **torr** (named after Torricelli). The torr is defined as 1/760th of 1 atm.

$$1 \text{ torr} = \tfrac{1}{760} \text{ atm}$$

$$1 \text{ atm} = 760 \text{ torr} \quad \text{(exactly)}$$

Thus, on most days near sea level, the torr is very close to the pressure that is able to support a column of mercury 1 mm high. In fact, the *millimeter of mercury* (abbreviated *mm Hg*) is often itself used as a pressure unit. Except when the most exacting measurements are being made, it is safe to use the relationship

$$1 \text{ torr} = 1 \text{ mm Hg}$$

The objection to "mm Hg" as a *pressure* unit is that a unit of length is not a good way to describe "force per unit area."

The torr and atm are the units that we will use throughout this book, but those going into medicine in the United States will often see "mm Hg" used in connection with the gases of respiration or of anesthesia.

Manometers

Gases used as reactants or formed as products in chemical reactions are kept from escaping by employing closed glassware. To measure pressures inside such vessels, a **manometer** is used. Two types are common, open-end and closed-end manometers.

Advantages of mercury over other liquids are its low reactivity, its low melting point, and particularly its very high density, which permits short manometer tubes.

The **open-end manometer** consists of a U-tube partly filled with a liquid, usually mercury (see Figure 10.3). One end of the U-tube's arm is open to the atmosphere. The other end is exposed to the entrapped gas. When the pressure of the gas, P_{gas}, equals atmospheric pressure, P_{atm}, the two mercury levels are equal (Figure 10.3a). When the gas pressure exceeds that of the atmosphere (Figure 10.3b), the gas pushes the mercury to make its level in the open tube stand higher than its level nearest the gas. The difference in the two heights, P_{Hg}, represents the millimeters of mercury pressure (or number of torr) over and above the

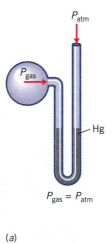 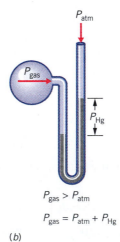 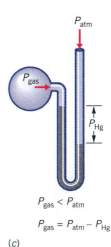

Figure 10.3 *An open-end manometer.* (a) The pressure of the entrapped gas, P_{gas}, equals P_{atm}, the atmospheric pressure. (Note that the arrows representing each pressure are here of *equal* length). (b) The entrapped gas pressure exceeds that of the atmosphere by an amount equal to the value of P_{Hg} when expressed in the unit of millimeters of mercury or mm Hg (but remember that 1 mm Hg = 1 torr). In other words, $P_{gas} = P_{atm} + P_{Hg}$. (c) The gas pressure is less than that of the atmosphere by an amount equal to P_{Hg}, so $P_{gas} = P_{atm} - P_{Hg}$.

atmospheric pressure that day. Thus, the pressure of the gas in torr when the gas pressure *exceeds* atmospheric pressure is found by *adding* P_{Hg} to P_{atm}.

$$P_{gas} = P_{atm} + P_{Hg} \quad \text{(used when } P_{gas} > P_{atm}) \quad (10.1)$$

The value of P_{atm} in torr is measured by a laboratory barometer at the time when P_{Hg} is determined. If the difference in the mercury heights[4] in the two arms is, for example, 120 mm on a day when P_{atm} is 752 torr, the total gas pressure is 872 torr, by Equation 10.1.

$$P_{gas} = 752 \text{ torr} + 120 \text{ torr} = 872 \text{ torr}$$

When the pressure on the entrapped gas is *less* than atmospheric pressure, the higher outside pressure is able to force mercury in the direction of the gas bulb (see Figure 10.3c). Now to find the gas pressure we must do a *subtraction*.

$$P_{gas} = P_{atm} - P_{Hg} \quad \text{(used when } P_{gas} < P_{atm}) \quad (10.2)$$

The level near the gas sample might, for example, become 200 mm *higher* than the level in the open arm, giving a pressure difference of 200 torr. If the atmospheric pressure is 752 torr, for example, then the pressure in the bulb is 552 torr, by Equation 10.2.

$$P_{gas} = 752 \text{ torr} - 200 \text{ torr} = 552 \text{ torr}$$

In a problem or exercise, if you can't remember whether to add or subtract, draw a picture of the manometer and roughly indicate the levels according to the data given (and always remember to ask whether the *size* of the answer is reasonable).

The **closed-end manometer** is particularly convenient when the gas pressures to be measured are less than atmospheric pressure (see Figure 10.4). When a closed-end manometer is manufactured, the end of the arm farthest from the gas sample is sealed, not left open. The tube is then pumped free of air, and mercury is allowed to enter and fill the tube. Then the system is opened to atmospheric pressure, and some mercury drains out until the pressure of the atmosphere is able to retain the remaining mercury at a natural level in the arm nearest the gas container. No atmospheric pressure now exists above the mercury in the sealed arm.

When a closed-end manometer is connected to a gas sample having a pressure *less* than atmospheric, the level in the closed arm cannot be maintained, and it falls. Some mercury is pulled into the arm nearest the gas sample, opening a space above the mercury in the closed arm. This space is now a vacuum just as in a Torricelli barometer. Therefore, the difference in heights of the two mercury levels, P_{Hg}, corresponds *directly* to the actual pressure of the gas sample. One advantage of a closed-end mercury manometer is that a separate measurement of the atmospheric pressure is not required as it is for the open-end manometer.

Liquids Other than Mercury in Manometers and Barometers

Mercury is so dense (13.6 g mL^{-1}) that when a gas pressure differs very little from atmospheric pressure, the difference in mercury levels is correspondingly small. The precision of measurement therefore suffers. By using a liquid less dense than mercury, the difference in liquid levels is larger. If water ($d = 1.0$ g mL^{-1}) is used, for example, a pressure difference of 1 mm Hg would show up as a

---

[4]In this chapter, when we have trailing zeros *before* the decimal point, for example, the 0 in 120, we intend them to be counted as significant. Thus in 120 torr there are three significant figures. In 760 torr, there are likewise three.

Use Equation 10.1 when P_{gas} exceeds P_{atm}.

Use Equation 10.2 when P_{gas} is less than P_{atm}.

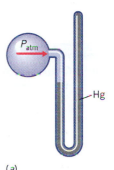

(a)

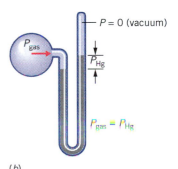

(b)

Figure 10.4 *A closed-end manometer for measuring gas pressures less than 1 atm or 760 torr. (a) When constructed, the tube is fully evacuated and then mercury is allowed to enter the tube so that the closed arm fills completely. (b) When the tube is connected to a bulb containing a gas at low pressure, the gas pressure is insufficient to keep the mercury level at the top of the closed arm, and mercury flows out of that arm and into the other arm. The system comes to rest when the difference in mercury levels, P_{Hg}, represents the pressure of the confined gas;* $P_{gas} = P_{Hg}$.

difference in heights of liquid levels 13.6 times as much or 13.6 mm (of water). In other words, a pressure that can sustain a 1 mm difference in height when mercury is the liquid can hold a 13.6 mm difference in height when the much less dense water is the liquid. A simple relationship exists between the two heights, h_A and h_B, when liquids A and B, having densities of d_A and d_B, are compared in a manometer or barometer measuring the same actual gas pressure.

$$h_B \times d_B = h_A \times d_A$$

Dividing by d_B gives

$$h_B = h_A \times \frac{d_A}{d_B} \tag{10.3}$$

EXAMPLE 10.1

Using a Fluid Other than Mercury in a Manometer

A liquid with a density of 0.854 g mL^{-1} is used in an open-end manometer to measure a gas pressure slightly greater than atmospheric pressure, which is 745.0 mm Hg at the time of measurement. The liquid level is 17.5 mm higher in the open arm than in the arm nearest the gas sample (see Figure 10.3b). What is the gas pressure in torr?

Analysis: The unit "torr" is a millimeter of *mercury* unit, so we need to calculate what the given height would be if mercury had been the liquid instead. Only then can we calculate the gas pressure in *torr*. Equation 10.3 is used, where liquid B is mercury and liquid A is the one with a density of 0.854 g mL^{-1}.

Solution:

$$h_{\text{Hg}} = 17.5 \text{ mm} \times \frac{0.854 \text{ g mL}^{-1}}{13.6 \text{ g mL}^{-1}} = 1.10 \text{ mm Hg}$$

The height of the corresponding column of mercury, 1.10 mm, translates into 1.10 torr, so the pressure of the gas in torr is

$$P_{\text{gas}} = 745.0 \text{ torr} + 1.10 \text{ torr}$$

$$= 746.1 \text{ torr}$$

Is the Answer Reasonable?
Remember that we're after an answer in torr, which means a property of *mercury,* namely, mm Hg. Mercury is over 13 times as dense as water, so mercury is even more dense than the given liquid. To put the answer in "mercury terms" the pressure of 17.5 mm of the given liquid must be divided by a number somewhat greater than 13. The result of the division is about 1 mm Hg (1 torr) higher than the starting pressure, and 746.1 torr is roughly that much more than 745 torr.

Practice Exercise 1

If water were used instead of mercury to construct a Torricelli barometer (Figure 10.3), the tube would have to be at least how long in millimeters and in feet? ◆

10.3 Pressure–Volume–Temperature Relationships for a Fixed Amount of Gas

Now that we have all of the physical units needed to describe gases, we can study how fixed masses of gases respond to changes in pressure, volume, and temperature.

Boyle's Law

Robert Boyle, an English scientist (1627–1691), learned by experiment how the volume of a fixed amount of gas varies with the gas pressure at a constant temperature (see Figure 10.5). His discovery, now called **Boyle's law** or the **pressure–volume law,** is that *the volume of a given amount of gas held at constant temperature varies inversely with the applied pressure.*

$$V \propto \frac{1}{P} \quad \text{(T and mass are constant)}$$

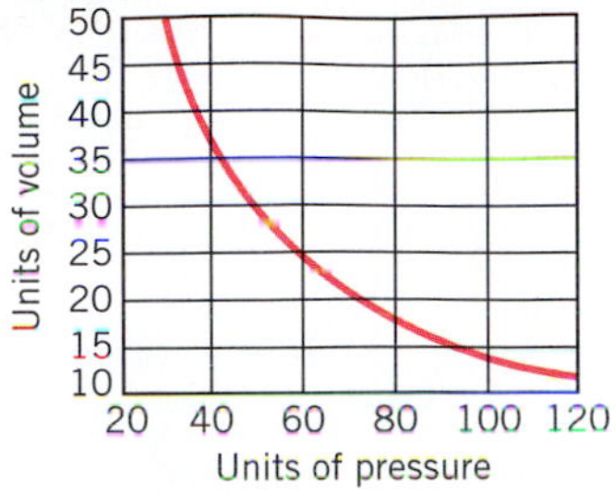

What is remarkable and is actually the heart of Boyle's discovery is that *this relationship is essentially the same for all gases at ordinary temperatures and pressures.* Neither liquids nor solids have a comparably general law.

We can remove the proportionality sign, $\propto$, by introducing a proportionality constant, C.

$$V = \frac{1}{P} \times C$$

Rearranging gives

$$PV = C$$

This equation tells us, for example, that if at constant temperature the gas pressure is doubled the gas volume is cut in half. (The curve in Figure 10.5 is characteristic of an inverse relationship such as Boyle's law. (See also Appendix A.3.)

Figure 10.5 *The relationship of volume to pressure, Boyle's law. The unit of pressure here is the same unit that Boyle used, inches of mercury. The volume varies reciprocally with the pressure. Thus, the gas volume is reduced by half, for example, when the pressure doubles from 30 to 60 units.*

The Concept of an *Ideal Gas*

When very precise measurements are made, Boyle's law doesn't quite work. No gas obeys Boyle's law exactly over a wide range of pressures, but most gases come very close to it at ordinary temperatures and pressures. It's when a gas is under high pressure (10 atm or more) or at a low temperature (below 200 K) or some combination of these *and is therefore on the verge of changing to a liquid that the gas most poorly fits Boyle's law. We'll see why later in the chapter.*

Although real gases do not *exactly* obey Boyle's law or any of the other gas laws that we'll study, it is often useful to imagine a hypothetical gas that would. We call such a hypothetical gas an *ideal gas.* An **ideal gas** obeys the gas laws exactly over all temperatures and pressures. A real gas behaves more and more like an ideal gas as its pressure decreases and its temperature increases.

By "ordinary" we mean near room temperature and atmospheric pressure.

Charles' Law

Jacques Charles discovered how the volume of a fixed amount of gas varies as the Kelvin temperature is changed at constant pressure (see Figure 10.6). The gas used for obtaining the data in Figure 10.6 is one that happens to liquefy at − 100 °C. Each plot is a straight line, so it is easy to extrapolate (meaning, *reasonably extend*) the lines to temperatures below − 100 °C. Such extrapolations, shown by the dashed lines in Figure 10.6, are simply reasonable predictions of how the gas would behave if it never changed to a liquid. Straight-line plots signify a *direct* proportionality, and other gases give these plots as well. Such results give us another general gas law, called **Charles' law** or the **temperature–volume law:** *the volume of a fixed amount of gas is directly proportional to its Kelvin temperature when the gas pressure is held constant.* In other words,

$$V \propto T \quad \text{(P and mass are constant)}$$

Jacques Alexandre César Charles (1746–1823), a French scientist and more a mathematician than a chemist, had a keen interest in hot air balloons. He was the first to inflate a balloon with hydrogen.

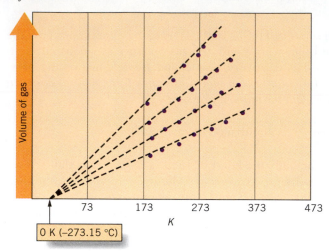

Figure 10.6 *Charles' law plots.* Each line shows how the gas volume changes with the Kelvin temperature for a different mass of the same gas.

Using a constant of proportionality, C', we can write

$$V = C'T \qquad (P \text{ and mass are constant})$$

When two variables are *directly* proportional, we say that they have a *linear* relationship, which the straight-line plots of Figure 10.6 display. *The temperature must be expressed in kelvins if the Charles' law relationship is to be expressible in this simple form.* Thus, in calculations involving gas temperatures, we always translate degrees Celsius into kelvins.

> The value of C' depends on the size and pressure of the gas sample.

The extrapolated lines in Figure 10.6 all converge on the same value, namely, $-273.15\ °C$. The lines would all reach this temperature if the corresponding gas volumes became zero. For a real gas, however, a zero volume is impossible. Yet this extrapolation is how scientists recognized that $-273.15\ °C$ must represent nature's coldest temperature. A lower temperature would imply a gas with a negative volume. So $-273.15\ °C$ came to be called **absolute zero** and was designated zero on the Kelvin scale of temperatures.

Gay-Lussac's Law

> Joseph Louis Gay-Lussac (1778–1850), a French scientist, was a codiscoverer of the element boron.

Although observations of Jacques Charles helped, Joseph Gay-Lussac is usually credited with discovering how the pressure and temperature of a fixed amount of gas at constant volume are related. The specified conditions exist, for example, when a gas is confined in a vessel with rigid walls, like an aerosol can. The relationship is called **Gay-Lussac's law** or the **pressure–temperature law:** *the pressure of a fixed amount of gas held at constant volume is directly proportional to the Kelvin temperature.* Thus,

$$P \propto T \qquad (V \text{ and mass are constant})$$

> We're using different symbols for the various gas law constants because they are different for each law.

Using still another constant of proportionality, Gay-Lussac's law becomes

$$P = C''T \qquad (V \text{ and mass are constant})$$

The Combined Gas Law

In many experiments with a fixed amount of gas, no effort need be made to keep either the temperature or the pressure constant. This is because there is a more general law that relates all three variables, namely, the **combined gas law:** *the ratio PV/T is a constant for a fixed amount of gas.*

$$\frac{PV}{T} = \text{constant} \qquad \text{(for a fixed amount of gas)}$$

If an initial set of conditions for a gas sample is given the label "1," then

$$\frac{P_1 V_1}{T_1} = C'''$$

where C''' is the constant. If any or all of the conditions are changed to create a new set, which we may label "2," then

$$\frac{P_2 V_2}{T_2} = C'''$$

Because both *PV/T* expressions equal the same constant, C''', the combined gas law may be written as

$$\frac{P_1 V_1}{T_1} = \frac{P_2 V_2}{T_2} \qquad (10.4)$$

Tools

ICombined gas law

Although temperatures must be in kelvins, the values of pressure and volume can be in any units *as long as they are used consistently for both sides.*

Equation 10.4 contains each of the other gas laws as special cases. Boyle's law, for example, requires a constant temperature, so T_1 equals T_2. When we cancel them in Equation 10.4, we're left with

Equation 10.4 is the combined gas law in the mathematical form most useful for calculations.

$$P_1 V_1 = P_2 V_2 \qquad \text{(when } T_1 \text{ cancels } T_2\text{)}$$

which is one way to write Boyle's law. Similarly, under constant pressure, P_1 cancels P_2, and Equation 10.4 reduces to

$$V_1/T_1 = V_2/T_2 \qquad \text{(when } P_1 \text{ cancels } P_2\text{)}$$

This, of course, is another way of writing Charles' law. Under constant volume, Gay-Lussac's condition, V_1 cancels V_2, and Equation 10.4 reduces to

$$P_1/T_1 = P_2/T_2 \qquad \text{(when } V_1 \text{ cancels } V_2\text{)}$$

This signifies that at constant volume, the ratio *P/T* is a constant (Gay-Lussac's law). You can see why the gas law represented by Equation 10.4 is called the *combined* gas law.

A sample of argon is trapped in a gas bulb at a pressure of 760 torr when the volume is 100 mL and the temperature is 35.0 °C. What must its temperature be if its pressure becomes 720 torr and its volume 200 mL?

EXAMPLE 10.2

Using the Combined Gas Law

Analysis: Notice that the mass of the sample is *constant.* This is our clue that the combined gas law equation applies.

Solution: It's a good practice in gas law calculations to collect the data first, remembering to convert degrees Celsius into kelvins.

Argon, a noble gas, is used for the low pressure atmosphere inside lightbulbs because it won't chemically corrode the hot, glowing filament of wire in the bulb.

$$V_1 = 100 \text{ mL} \qquad P_1 = 760 \text{ torr} \qquad T_1 = 308 \text{ K} \quad (35.0 \,°\text{C} + 273 \,°\text{C}) \times \frac{1 \text{ K}}{°\text{C}}$$

$$V_2 = 200 \text{ mL} \qquad P_2 = 720 \text{ torr} \qquad T_2 = \text{?}$$

The combined gas law equation,

$$\frac{P_1 V_1}{T_1} = \frac{P_2 V_2}{T_2}$$

The combined gas law equation rearranges to

$$T_2 = T_1 \times \frac{V_2}{V_1} \times \frac{P_2}{P_1}$$

is used, and after inserting the known values and rearranging we have

$$T_2 = 308 \text{ K} \times \frac{200 \text{ mL}}{100 \text{ mL}} \times \frac{720 \text{ torr}}{760 \text{ torr}}$$

$$= 584 \text{ K, or } 311 \text{ °C} \quad \left[\text{from } (584 \text{ K} - 273 \text{ K}) \times \frac{\text{°C}}{\text{K}} \right]$$

Is the Answer Reasonable?
The net effect is that a large *increase* in temperature must occur. Is this reasonable? Yes; notice that the volume is doubled (200 mL/100 mL), and Charles' law tells us that the temperature would have to be *increased* to achieve this. In fact, the kelvin temperature would have to be doubled (to 616 K) had it not been for the pressure change. The pressure decreases, and Gay-Lussac's law tells us that the temperature must be *decreased* to have this happen. But notice that the change in pressure is relatively small (720 torr/760 torr), not enough to offset the large increase in temperature mandated by Charles' law. The temperature, therefore, is not quite doubled.

Practice Exercise 2

What will be the final pressure of a sample of nitrogen with a volume of 950 m³ at 745 torr and 25.0 °C if it is heated to 60.0 °C and given a final volume of 1150 m³? ◆

As we said, the combined gas law can be used even when one of the three variables, pressure, volume, or temperature, is held constant. The next worked example illustrates this important fact and shows why separate equations for the laws of Boyle, Charles, and Gay-Lussac are really not necessary for gas law calculations.

EXAMPLE 10.3

Using the Combined Gas Law

Anesthetic gas is normally given to a patient when the room temperature is 20.0 °C and the patient's body temperature is 37.0 °C. What would this temperature change do to 1.60 L of gas if the pressure and mass stay constant?

Analysis: This is a simple application of Equation 10.4. Because the pressure is constant, we can cancel P_1 and P_2 from Equation 10.4 to leave

$$\frac{V_1}{T_1} = \frac{V_2}{T_2}$$

Solution: Let's collect the data, changing degrees Celsius to kelvins:

$$V_1 = 1.60 \text{ L} \qquad T_1 = 293 \text{ K} \quad (20.0 \text{ °C} + 273 \text{ °C}) \times \frac{1 \text{ K}}{\text{°C}}$$

$$V_2 = ? \qquad T_2 = 310 \text{ K} \quad (37.0 \text{ °C} + 273 \text{ °C}) \times \frac{1 \text{ K}}{\text{°C}}$$

All that's left is the arithmetic. Inserting the data into the equation gives

$$\frac{1.60 \text{ L}}{293 \text{ K}} = \frac{V_2}{310 \text{ K}}$$

Solving for V_2 gives

$$V_2 = 1.60 \text{ L} \times \frac{310 \text{ K}}{293 \text{ K}}$$

$$= 1.69 \text{ L}$$

The volume increases by 0.09 L, about 90 mL. Although this is only a 6% change, it is a factor that anesthesiologists have to consider in administering anesthetic gases.

Is the Answer Reasonable?
The answer is reasonable because the volume must increase with the temperature, but not by much because the volume-raising ratio of kelvin temperatures, only 310 K/293 K, is small.

Practice Exercise 3

A sample of nitrogen has a volume of 880 mL and a pressure of 740 torr. What pressure will change the volume to 870 mL at the same temperature? ◆

10.4 Ideal Gas Law

Relationships of Gas Volumes in Gas Phase Reactions

From our study of stoichiometry in Chapter 3, we know that the *mole* relationships among reactants and products in a reaction can be expressed simply. We merely use the whole-number coefficients of the balanced equations. For reactions in which reactants and products are gases, the *volume* relationships are equally simple, provided that comparisons are made under identical conditions of temperature and pressure. For example, hydrogen gas reacts with chlorine gas to give gaseous hydrogen chloride. Beneath the names in the following equation are the measured *volumes* with which these gases interact when they are at the same pressure and temperature.

Hydrogen + chlorine ⟶ hydrogen chloride
 1 vol   1 vol   2 vol

The discovery of the *general* nature of the relationships of gas volumes, expressible in simple, whole-number ratios, is usually attributed to Gay-Lussac.

Similarly simple, whole-number ratios by volume are observed when hydrogen combines with oxygen to give water, which is a gas above 100 °C.

Hydrogen + oxygen ⟶ water (gaseous state)
 2 vol   1 vol   2 vol

Notice that the reacting volumes, measured under identical temperatures and pressures, are in ratios of simple, whole numbers.[5]

Avogadro's Principle

When Amedeo Avogadro thought about how the volume ratios in the reactions of gases are in *whole* numbers, like the mole ratios, he was driven to conclude that equal volumes of gases must have identical numbers of molecules (at the same T and P). Today, we know that "equal numbers of *molecules*" is the same

Amedeo Avogadro (1776–1856), an Italian scientist, helped to put chemistry on a quantitative basis.

[5]The great French chemist, Antoine Laurent Lavoisier (1743–1794), was the first to observe the volume relationships of this particular reaction. In his 1789 textbook, *Elements of Chemistry,* he wrote that the formation of water from hydrogen and oxygen requires that two volumes of hydrogen be used for every volume of oxygen. Lavoisier was unable to extend the study of this behavior of hydrogen and oxygen to other gas reactions because he was beheaded during the French Revolution. (See Michael Laing, *The Journal of Chemical Education,* February 1998, page 177.)

	He (at. mass, 4.003)	Ne (at. mass, 20.18)	Ar (at. mass, 39.95)
Number of grams	4.003 g	20.18 g	39.95 g
Number of moles	1.000 mol	1.000 mol	1.000 mol
Number of atoms of gas	6.022×10^{23} atoms	6.022×10^{23} atoms	6.022×10^{23} atoms
Pressure	1.00 atm	1.00 atm	1.00 atm
Temperature	273 K	273 K	273 K
Volume	22.4 L	22.4 L	22.4 L
Density	0.179 g L^{-1}	0.901 g L^{-1}	1.78 g L^{-1}

Figure 10.7 *Avogadro's principle.* When measured at the same temperature and pressure, equal volumes of gases contain equal numbers of moles. Shown here are data for three Group VIIIA elements, helium, neon, and argon, all under identical temperatures and pressures. Notice that 1 mol of each gas, or Avogadro's number of gaseous atoms, occupies the same volume. The reason why the numbers of grams and densities differ is that the masses of the individual atoms are not the same.

as "equal numbers of *moles,*" so Avogadro's insight, now called **Avogadro's principle,** is expressed as follows. *When measured at the same temperature and pressure, equal volumes of gases contain equal numbers of moles* (see Figure 10.7).

Avogadro's principle means that in reactions of gases, their volumes and their numbers of moles must be in the same ratio. In the equation for the reaction of hydrogen and chlorine, for example, the coefficients are all in the same 1:1:2 ratio as are the numbers of volumes and the numbers of moles.

$$H_2(g) + Cl_2(g) \longrightarrow 2HCl(g)$$

Coefficients	1	1	2
Volumes	1 vol	1 vol	2 vol (experimental)
Molecules (or moles)	1	1	2 (Avogadro's principle)

A corollary to Avogadro's principle is that *the volume of a gas is directly proportional to its number of moles, n.*

$$V \propto n \qquad \text{(at constant } T \text{ and } P\text{)}$$

Avogadro's principle was a remarkable advance in our understanding of gases. His insight enabled chemists for the first time to determine the formulas of gaseous elements.[6]

Standard Molar Volume

Avogadro's principle implies that the volume occupied by one mole of *any* gas—its *molar volume*—must be identical for all gases under the same pressure and temperature. To put experimental molar volumes of different gases on the same basis for comparison, scientists have agreed to use 1 atm and 273.15 K

---

[6]Suppose that hydrogen chloride, for example, is correctly formulated as HCl, not as H_2Cl_2 or H_3Cl_3 or higher, and certainly not as $H_{0.5}Cl_{0.5}$. Then the only way that *two* volumes of hydrogen chloride could come from just *one* volume of hydrogen and *one* of chlorine is if each particle of hydrogen and each of chlorine were to consist of *two* atoms of H and Cl, respectively, H_2 and Cl_2. If these particles were single-atom particles, H and Cl, then one volume of H and one volume of Cl could give only *one* volume of HCl, not two. Of course, if the initial assumption were incorrect so that hydrogen chloride is, say, H_2Cl_2 instead of HCl, then hydrogen would be H_4 and chlorine would be Cl_4. The extension to larger subscripts works in the same way.

Table 10.1 Molar Volumes of Some
Gases at STP

Gas	Formula	Standard Molar Volume (L)
Helium	He	22.398
Argon	Ar	22.401
Hydrogen	H_2	22.410
Nitrogen	N_2	22.413
Oxygen	O_2	22.414
Carbon dioxide	CO_2	22.414

(0 °C) as the **standard conditions of temperature and pressure,** or **STP,** for short. Under STP, the volume of one mole of a gas is 22.4 L, or very close to this. Table 10.1 gives the experimental data supporting this value, now called the **standard molar volume** of a gas.

Formulation of the Ideal Gas Law

The constant C''' in the combined gas law, $PV/T = C'''$, is a constant only for a fixed number of moles of gas. C''', in fact, is directly proportional to n. To create an equation even more general than the combined gas law, we split C''' into the product of the number of moles, n, and still another constant.

$$\frac{PV}{T} = C''' = n \times \text{new constant}$$

This "new constant" is given the symbol R and is called the **universal gas constant.** We can now write the combined gas law in a still more general form called the **ideal gas law.**

In some references, it is called the universal gas law.

$$\frac{PV}{T} = nR$$

An ideal gas would obey this law exactly over all ranges of the gas variables. The equation, sometimes called the **equation of state for an ideal gas,** is usually rearranged and written as follows.

Ideal Gas Law (Equation of State for an Ideal Gas)

$$PV = nRT \qquad (10.5)$$

Ideal gas law

Equation 10.5 tells us how the four important variables for a gas, P, V, n, and T, are related. If we know the values of three, we can calculate the fourth. In fact, Equation 10.5 tells us that if three of the four variables are fixed for a given gas, *the fourth can only have one value.* We can define the *state* of a given gas simply by specifying any three of the four variables.

If n, P, and T in Equation 10.5 are known, for example, then V can have only one value.

To use the ideal gas law, we have to know the value of the universal gas constant, R. To calculate it, we use the standard conditions of pressure and temperature, and we use the standard molar volume, which sets n equal to 1 mol, the number of moles of the sample. We still have to decide what units to use for pressure and volume, so the value of R differs with these choices. Our choices for this chapter are to express volumes in *liters* and pressures in *atmospheres*. Thus for one molar volume at STP, $n = 1$ mol, $V = 22.4$ L, $P = 1.00$ atm, and $T = 273$ K. Using these values lets us calculate R as follows.

$$R = \frac{PV}{nT} = \frac{(1.00 \text{ atm})(22.4 \text{ L})}{(1.00 \text{ mol})(273 \text{ K})}$$

$$= 0.0821 \frac{\text{atm L}}{\text{mol K}}$$

Or, arranging the units in a commonly used order,

$$R = 0.0821 \text{ L atm mol}^{-1} \text{ K}^{-1}$$

When a problem gives the values of pressure, volume, or temperature in any units other than, respectively, atmospheres, liters, or kelvins, unit conversions must be carried out before using the value of R in the ideal gas law.

EXAMPLE 10.4
Using the Ideal Gas Law

A sample of oxygen at 24.0 °C and 745 torr was found to have a volume of 455 mL. How many grams of O_2 were in the sample?

Analysis: In this question we are given the pressure, volume, and temperature of a gas and then asked to determine the number of *grams* of gas. The ideal gas law equation, $PV = nRT$, lets us calculate the number of *moles, n,* from P, V, and T, but an easy moles-to-grams conversion for O_2, using its molecular mass, will give us the mass of O_2 in grams.

Solution: To use $R = 0.0821 \text{ L atm mol}^{-1} \text{ K}^{-1}$, we must have V in liters, P in atmospheres, and T in kelvins. Gathering the data and making the necessary unit conversions as we go, we have

$P = 0.980 \text{ atm}$ From 745 torr $\times \dfrac{1 \text{ atm}}{760 \text{ torr}}$

$V = 0.455 \text{ L}$ From 455 mL

$T = 297 \text{ K}$ From $(24.0 \text{ °C} + 273 \text{ °C}) \times \dfrac{1 \text{ K}}{\text{°C}}$

Solving the ideal gas law for n gives us

$$n = \frac{PV}{RT}$$

Substituting the proper values of P, V, R, and T into this equation gives

$$n = \frac{(0.980 \text{ atm})(0.455 \text{ L})}{(0.0821 \text{ L atm mol}^{-1} \text{ K}^{-1})(297 \text{ K})}$$

$$= 0.0183 \text{ mol of } O_2$$

The molecular mass of O_2 is 32.00, so 1 mol O_2 = 32.00 g O_2. A conversion factor from this relationship lets us convert "mol O_2" into "g O_2."

$$0.0183 \text{ mol } O_2 \times \frac{32.00 \text{ g } O_2}{1 \text{ mol } O_2} = 0.586 \text{ g } O_2$$

Thus, the sample of oxygen had a mass of 0.586 g.

Is the Answer Reasonable?
Doing a "head check" is tricky, but let's deal with very roughly rounded data. Let's ignore the values of temperature and pressure; they aren't all that far from the values of STP. Let's round 455 mL, the given volume, to 500 mL or half a liter. Because of the known standard molar volume for all gases, we know that 22 L of oxygen at STP is close to one mole and must have a mass of 32 g (the formula mass of

O_2). So divide 32 g by 22 L—the answer will be between 1 and 2 g/L—and again by 2 to give an answer between 0 and 1 g/half liter, which is what we got, an answer between 0 and 1 g per half liter of oxygen. This may seem far-fetched, given all of the rounding, but we're at least reassured that the decimal point in the answer isn't off.

Practice Exercise 4

What volume in milliliters does a sample of nitrogen with a mass of 0.245 g occupy at 21 °C and 750 torr? ◆

Using the Gas Laws—A Summary

"Which gas law should I use?" is a common question. The stock answer is, "It depends on the data given." But first apply some reasoning to the data in the statement of the problem. Is the amount of the gas sample stated either in moles or in a mass unit (which is convertible to moles)? Or does the problem call for the calculation of an amount of gas (in moles or in a mass unit)? There is only one gas law that includes moles, namely, the ideal gas law. So that's the law to use. If the size of the sample either in moles or in mass units is not given or not called for, the combined gas law most likely can be applied. This law simplifies, as we've said, if the temperature is constant (to Boyle's law), the pressure is constant (to Charles' law), or the volume is constant (to Gay-Lussac's law). Figure 10.8 summarizes these options.

Determining the Molecular Mass of a Gas

When a chemist makes a new compound, its molecular mass is usually determined to help establish its chemical identity. When the compound is a gas, its molecular mass can be found using experimental values of pressure, volume, temperature, and sample mass together with the ideal gas law.

Suppose, for example, that a gas has been trapped in a sealed bulb at known values of P, V, T, and mass. The ideal gas law equation now lets us use the values of P, V, and T to compute the number of *moles* in the sample by the equation

$$n = \frac{PV}{RT}$$

Then we calculate the ratio of *grams to moles* to find the molecular mass.

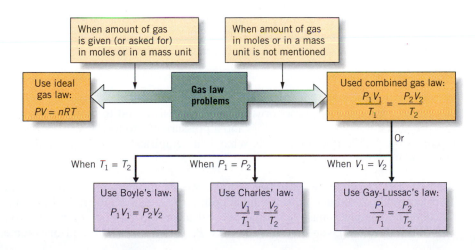

Figure 10.8 *Strategies for gas law problems.* The first decision depends on whether the data include the size of the sample in moles (or in some mass unit that can be converted to moles).

EXAMPLE 10.5

Determining the Molecular Mass of a Gas

A student collected a sample of a gas in a 0.220 L gas bulb until its pressure reached 0.757 atm at a temperature of 25.0 °C. The sample weighed 0.299 g. What is the molecular mass of the gas?

Analysis: To find the molecular mass, we calculate the ratio of the number of grams of the sample to the number of moles. We've been given the number of grams, and we can use the ideal gas equation to compute the number of moles.

Solution: Pressure is already in atmospheres and the volume is in liters, but we must convert degrees Celsius into kelvins. Gathering our data, we have

$$P = 0.757 \text{ atm} \qquad V = 0.220 \text{ L} \qquad T = (25.0 \text{ °C} + 273 \text{ °C}) \times \frac{1 \text{ K}}{\text{°C}} = 298 \text{ K}$$

We substitute the data into the ideal gas law equation.

$$PV = nRT$$

$$(0.757 \text{ atm})(0.220 \text{ L}) = (n)(0.0821 \text{ L atm mol}^{-1} \text{ K}^{-1})(298 \text{ K})$$

Rearranging to solve for n gives

$$n = \frac{(0.757 \text{ atm})(0.220 \text{ L})}{(0.0821 \text{ L atm mol}^{-1} \text{ K}^{-1})(298 \text{ K})}$$

$$= 6.81 \times 10^{-3} \text{ mol}$$

The ratio of grams to moles, therefore, is

$$\frac{0.299 \text{ g}}{6.81 \times 10^{-3} \text{ mol}} = 43.9 \text{ g mol}^{-1}$$

When the molecular mass of a substance is expressed in units of grams per mole, the quantity is sometimes called the **molar mass** and has the units g mol^{-1}.

Because there are 43.9 grams per mole, the molecular mass must be 43.9. (The gas used for this example is CO_2, with a molecular mass of 44.0.)

Is the Answer Reasonable?
Notice that the *volume* of the gas sample (0.220 L) is close to 1/100th of the standard molar volume (22.4 L). So the given mass of the sample (0.299 g) must be about 1/100th of the molecular mass of the gas. When we multiply 0.299 by 100, the result is about 30. Although that's not very close to 43.9, the calculated answer, we can say that at least the decimal place in 43.9 is right.

Practice Exercise 5

The label on a cylinder of a noble gas became illegible, so a student allowed some of the gas to flow into a 300 mL gas bulb until the pressure was 685 torr. The sample now weighed 1.45 g; its temperature was 27.0 °C. What is the molecular mass of this gas? Which of the Group VIIIA gases (the noble gases) was it? ◆

EXAMPLE 10.6

Calculating the Molecular Mass from Gas Density

A gaseous compound of phosphorus and fluorine was found to have a density of 3.50 g L^{-1} at a temperature of 25.0 °C and a pressure of 740 torr. It was also found to consist of 35.2% P and 64.8% F. What is its empirical formula, its molecular mass, and its molecular formula?

Analysis: To calculate the *empirical formula,* we use the percentage composition. To calculate the *molecular mass,* we need both the number of moles, n, of the sample and the sample's mass in grams. We calculate the number of moles from $PV = nRT$. But the given data include only P, T, and not V. Instead of volume, we're given the gas density, 3.50 g L^{-1}. So we'll pick the volume to be 1.00 L. At this volume the sample must have a mass of 3.50 g, because the gas density is 3.50 g *per liter*. So when we compute *n*, we can calculate the molecular mass by simply computing the following ratio.

$$\text{Molecular mass} = \frac{3.50 \text{ g}}{n}$$

To determine the *molecular formula,* we compare the molecular mass just found with that calculated from the empirical formula and make any needed multiplication of subscripts in the empirical formula. Because this is a complicated calculation, let's summarize the steps.

Step 1. Calculate the *empirical formula* from the percentage composition.

Step 2. Calculate the *molecular mass* by relating the grams per liter (density) data to the number of grams per mole, using the ideal gas law to find the number of moles in a 1 L sample.

Step 3. Calculate the *molecular formula* by comparing the experimental molecular mass (step 2) to that calculated from the empirical formula (step 1).

Solution: *Step 1.* To calculate the empirical formula from the percentages by mass of P (35.2%) and F (64.8%), we assume a sample with a mass of 100.0 g, which therefore holds the following number of moles of P and F.

$$35.2 \text{ g P} \times \frac{1 \text{ mol P}}{30.97 \text{ g P}} = 1.14 \text{ mol P} \quad \text{and} \quad 64.8 \text{ g F} \times \frac{1 \text{ mol F}}{19.00 \text{ g F}} = 3.41 \text{ mol F}$$

The formula, then, could be written as $P_{1.14}F_{3.41}$, but we divide both of the subscripts by the smaller, 1.14. Because $3.41 \div 1.14 = 2.99$, close enough to 3, we can write the empirical formula as PF_3.

Step 2. To calculate the number of moles of gas in 1.00 L, we use the ideal gas law, making unit conversions as needed.

$$T = 298 \text{ K} \qquad \text{From } (25.0 \text{ °C} + 273 \text{ °C}) \times \frac{1 \text{ K}}{\text{°C}}$$

$$P = 0.974 \text{ atm} \qquad \text{From } 740 \text{ torr} \times \frac{1 \text{ atm}}{760 \text{ torr}}$$

$$V = 1.00 \text{ L}$$

$$n = \frac{PV}{RT} = \frac{(0.974 \text{ atm})(1.00 \text{ L})}{(0.0821 \text{ L atm mol}^{-1} \text{ K}^{-1})(298 \text{ K})}$$

$$= 3.98 \times 10^{-2} \text{ mol}$$

Because the mass of the gas in 1.00 L is 3.50 g (from the density), the ratio of grams to moles is

$$\frac{3.50 \text{ g}}{3.98 \times 10^{-2} \text{ mol}} = 87.9 \text{ g mol}^{-1}$$

The molecular mass of the compound is thus 87.9.

Step 3. The molecular mass that we calculate from the empirical formula, PF_3, is 87.97. It is very close to that found in Step 2, so PF_3 is the molecular formula of the compound as well as its empirical formula.

Are the Answers Reasonable?
For a problem like this, with so many steps, a "head check" is very difficult. At least we found simple, whole numbers for the subscripts of the empirical formula. Our calculation that gave the molecular mass to be that of this formula gives us further confidence in our solutions.

A gaseous compound of phosphorus and fluorine with an empirical formula of PF_2 has a density of 5.60 g L^{-1} at 23.0 °C and 750 torr. Calculate its molecular mass and its molecular formula. ◆

10.5 Stoichiometry of Reactions between Gases

For reactions involving gases, Avogadro's principle lets us use a new kind of stoichiometric equivalency, one between *volumes* of gases. Earlier, for example, we noted the following reaction and its gas volume relationships.

$$2H_2(g) + O_2(g) \longrightarrow 2H_2O(g)$$
$$\text{2 vol} \qquad \text{1 vol} \qquad \text{2 vol}$$

We can now write the following stoichiometric equivalencies, *provided we understand that the pressure and temperature are constant.*

The recognition that equivalencies in gas *volumes* are numerically the same as those for numbers of moles of gas in reactions involving gases simplifies many calculations.

2 vol $H_2(g)$ ⇔ 1 vol $O_2(g)$	just as 2 mol H_2 ⇔ 1 mol O_2
2 vol $H_2(g)$ ⇔ 2 vol $H_2O(g)$	just as 2 mol H_2 ⇔ 2 mol H_2O
1 vol $O_2(g)$ ⇔ 2 vol $H_2O(g)$	just as 1 mol O_2 ⇔ 2 mol H_2O

EXAMPLE 10.7

Stoichiometry of Reactions of Gases

How many liters of oxygen at STP are needed to combine exactly with 1.50 L of hydrogen at STP?

Analysis: We need the chemical equation

$$2H_2(g) + O_2(g) \longrightarrow 2H_2O(g)$$

and we need Avogadro's principle, which tells us that for this reaction of gases, 2 vol $H_2(g)$ ⇔ 1 vol $O_2(g)$. We can write this equivalency because the *same* conditions of temperature and pressure, STP, are stipulated.

Solution: We multiply the given, 1.50 L H_2, by a conversion factor made from the equivalency 2 vol $H_2(g)$ ⇔ 1 vol $O_2(g)$. Using liters as the volume unit,

$$\text{Volume of } O_2 = 1.50 \text{ L } H_2 \times \frac{1 \text{ L } O_2}{2 \text{ L } H_2}$$

$$= 0.750 \text{ L } O_2$$

Is the Answer Reasonable?
The calculated fact that 0.750 L of O_2 combines with 1.50 L of H_2 is a reasonable answer because the coefficients and Avogadro's principle tell us that we need half as much O_2 by volume as H_2.

Methane burns according to the following equation.

$$CH_4(g) + 2O_2(g) \longrightarrow CO_2(g) + 2H_2O(g)$$

The combustion of 4.50 L of CH_4 consumes how many liters of O_2, both volumes measured at STP? ◆

In experimental work involving the generation of gases, questions arise such as, *What size flask* is needed to trap the gas that is going to be generated by a reaction? Or, *How many grams of some starting material* can generate enough gas to fill a given gas bulb at some value of pressure and temperature. The stoichiometric equivalencies between gas volumes and moles help us find the answers to such questions.

EXAMPLE 10.8

Calculating the Volume of a Gas Product Using Stoichiometric Equivalencies

A student needs to prepare some CO_2 and intends to use the following reaction in which $CaCO_3$ is heated strongly.

$$CaCO_3(s) \longrightarrow CO_2(g) + CaO(s)$$

The question concerns the size of the flask needed to accept the gas. How large should the flask meant to hold the CO_2 be if 1.25 g of $CaCO_3$ is used? The pressure of the CO_2 is to be 740 torr, and its final temperature is to be 25.0 °C.

Analysis: The balanced equation tells us

$$1 \text{ mol } CaCO_3 \Leftrightarrow 1 \text{ mol } CO_2$$

Once we calculate the *moles* of $CaCO_3$, we can use this value for n in the ideal gas law equation to find the gas volume.

Solution: The formula mass of $CaCO_3$ is 100.1, so

$$\text{Moles of } CaCO_3 = 1.25 \text{ g } CaCO_3 \times \frac{1 \text{ mol } CaCO_3}{100.1 \text{ g } CaCO_3}$$

$$= 1.25 \times 10^{-2} \text{ mol } CaCO_3$$

Because 1 mol $CaCO_3 \Leftrightarrow 1$ mol CO_2,

$$n = 1.25 \times 10^{-2} \text{ mol } CO_2$$

Before we use *n* in the ideal gas law equation, we must convert the given pressure and temperature into the units required by R.

$$P = 740 \text{ torr} \times \frac{1 \text{ atm}}{760 \text{ torr}} = 0.974 \text{ atm} \quad T = (25.0 \text{ °C} + 273 \text{ °C}) \times \frac{1 \text{ K}}{\text{°C}} = 298 \text{ K}$$

By rearranging the ideal gas law equation we obtain

$$V = \frac{nRT}{P}$$

$$= \frac{(1.25 \times 10^{-2} \text{ mol})(0.0821 \text{ L atm mol}^{-1} \text{ K}^{-1})(298 \text{ K})}{0.974 \text{ atm}}$$

$$= 0.314 \text{ L} = 314 \text{ mL}$$

The container must therefore have a volume of at least 314 mL.

Is the Answer Reasonable?
The reaction generates 25% more than 0.0100 mol. At STP, 0.0100 mol would occupy 1/100th of the standard molar volume, or 0.224 L. So we'll increase this by 25% by adding roughly 0.06 L to it, giving about 0.28 L. But the conditions are not STP; they involve a *lower* pressure (thus requiring more space for the gas than 0.28 L) and a *higher* temperature (also requiring more space than 0.28 L). So 0.28 L, while not a bad fit to 0.314 L, is understandably low.

In one lab, the gas collecting apparatus used a gas bulb with a volume of 250 mL. How many grams of $Na_2CO_3(s)$ would be needed to prepare enough $CO_2(g)$ to fill this bulb to a pressure of 738 torr at a temperature of 23 °C? The equation is

$$Na_2CO_3(s) + 2HCl(aq) \longrightarrow 2NaCl(aq) + CO_2(g) + H_2O \quad \blacklozenge$$

10.6 Dalton's Law of Partial Pressures

The volume of a gas equals the volume of its container, because all gases expand to fill whatever volume is available to them. This even applies to mixtures. Thus, in a 1.00 L flask of air, the volume of *each* gas present (N_2, O_2, and others) is the same, namely, 1.00 L.

In many applications, mixtures of nonreacting gases are used, and our question in this section is, How do the gas laws apply to gas mixtures? John Dalton discovered the answer.

In a mixture of nonreacting gases, each gas contributes to the total pressure in proportion to the fraction (by moles) in which it is present. This contribution to the total pressure is called the **partial pressure** of the gas. It is the pressure that the gas would exert all by itself if it were the only gas in the same container at the same temperature. The general symbol for the partial pressure of gas a is P_a. For a particular gas, the formula of the gas may be put into the subscript, as in P_{O_2}. What Dalton discovered about partial pressures is now called **Dalton's law of partial pressures:** *The total pressure of a mixture of nonreacting gases is the sum of their individual partial pressures.* In equation form, the law is

$$P_{total} = P_a + P_b + P_c + \cdots \tag{10.6}$$

Tools

Dalton's law of partial pressures

In dry CO_2-free air at STP, for example, P_{O_2} is 159.12 torr, P_{N_2} is 593.44 torr, and P_{Ar} is 7.10 torr. These partial pressures add up to 759.66 torr, just 0.34 torr less than 760 torr or 1.00 atm. The remaining 0.34 torr is contributed by several trace gases, including other noble gases.

EXAMPLE 10.9

Using Dalton's Law of Partial Pressures

The air we breathe is a common mixture of nonreacting gases. At least under ordinary temperatures and pressures, nitrogen and oxygen do not react.

Suppose you want to fill a pressurized tank having a volume of 4.00 L with oxygen-enriched air for use in diving, and you want the tank to contain 50.0 g of O_2 and 150 g of N_2. What will the total gas pressure have to be at 25 °C?

Analysis: If we calculate the pressure that each gas would have if it were *alone* in the tank and then add the partial pressures, we'll have the answer. To find the individual partial pressures, we apply the ideal gas law using a temperature of 298 K (for 25 °C) and a volume of 4.00 L. To use this law, however, we must first convert the amounts of the gases in grams to numbers of moles, a routine grams-to-moles conversion.

Solution:

For moles of oxygen, $n = 50.0 \text{ g } O_2 \times \dfrac{1 \text{ mol } O_2}{32.00 \text{ g } O_2} = 1.56 \text{ mol } O_2$

For moles of nitrogen, $n = 150 \text{ g } N_2 \times \dfrac{1 \text{ mol } N_2}{28.00 \text{ g } N_2} = 5.36 \text{ mol } N_2$

Next, we calculate the partial pressures. For each, $P_a = \dfrac{nRT}{V}$, where V is the volume of the tank.

$$P_{O_2} = \frac{1.56 \text{ mol} \times 0.0821 \text{ L atm K}^{-1} \text{ mol}^{-1} \times 298 \text{ K}}{4.00 \text{ L}}$$

$$= 9.54 \text{ atm}$$

$$P_{N_2} = \frac{5.36 \text{ mol} \times 0.0821 \text{ L atm K}^{-1} \text{ mol}^{-1} \times 298 \text{ K}}{4.00 \text{ L}}$$

$$= 32.8 \text{ atm}$$

For the total pressure, using Dalton's law, we have

$$P_{\text{total}} = 9.54 \text{ atm} + 32.8 \text{ atm}$$

$$= 42.3 \text{ atm}$$

Is the Answer Reasonable?

Once you've seen that the *total* number of moles of gases is about 7 mol, you'll know that this many moles would occupy 7 mol × 22.4 L/mol or close to 160 L at STP. Compressing a gas from 160 L to 4 L, or 1/40th as much, would require increasing the pressure from 1 atm to 40 times as much, which is rather close to our answer.

Practice Exercise 9

How many grams of oxygen are present at 25 °C in a 5.00 L tank of oxygen-enriched air under a total pressure of 30.0 atm when the only other gas is nitrogen at a partial pressure of 15.0 atm? ◆

Collecting Gases over Water

When gases that do not react with water are prepared in the laboratory, they can be trapped over water by an apparatus like that shown in Figure 10.9. A gas collected this way, in fact, is saturated with water vapor. Water vapor in a mixture of gases has a partial pressure like that of any other gas.

The vapor present in the space above *any* liquid always contains some of the liquid's vapor, which exerts its own pressure called the liquid's **vapor pressure.** Its value for any given substance depends only on the temperature. The vapor pressures of water at different temperatures, for example, are given in Table 10.2.

When we collect a (wet) gas over water, we usually want to know how much *dry* gas this corresponds to, and we can use data for the wet gas to find out. From Dalton's law we can write

> Even the mercury in a barometer has a tiny vapor pressure—0.0012 torr at 20 °C which is much too small to affect readings of barometers and the manometers studied in this chapter.

$$P_{\text{total}} = P_{\text{gas}} + P_{\text{water}}$$

The pressure of the gas is then, by rearrangement,

$$P_{\text{gas}} = P_{\text{total}} - P_{\text{water}}$$

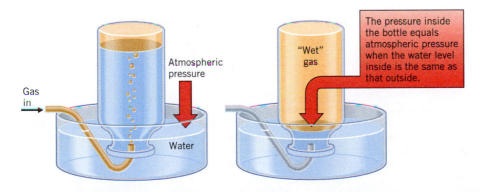

Gas in

Atmospheric pressure

"Wet" gas

The pressure inside the bottle equals atmospheric pressure when the water level inside is the same as that outside.

Water

Figure 10.9 *Collecting a gas over water.* As the gas bubbles through the water, water vapor goes into the gas, so the total pressure inside the bottle includes the partial pressure of the water vapor at the temperature of the water.

Table 10.2 **Vapor Pressure of Water at Various Temperatures**

Temperature (°C)	Vapor Pressure (torr)	Temperature (°C)	Vapor Pressure (torr)
0	4.579	50	92.51
5	6.543	55	118.0
10	9.209	60	149.4
15	12.79	65	187.5
20	17.54	70	233.7
25	23.76	75	289.1
30	31.82	80	355.1
35	42.18	85	433.6
37[a]	47.07	90	527.8
40	55.32	95	633.0
45	71.88	100	760.0

[a]Human body temperature.

We find values for P_{water} at various temperatures from Table 10.2. P_{total} is obtained from the laboratory barometer, provided that the water level inside the gas collecting bottle is the same as the water level outside. We thus calculate P_{gas}, and this pressure is what the gas would exert if it were dry and inside the same volume that was used to collect it.[7]

EXAMPLE 10.10

Calculation Using Data from the Collection of a Gas over Water

A sample of oxygen is collected over water at 20.0 °C and a pressure of 738 torr. Its volume is 310 mL. (a) What is the partial pressure of the oxygen? (b) What would be its volume when dry at STP?

Analysis: The initial pressure of 738 torr has to be corrected by a Dalton's law calculation, using the vapor pressure of water at 20 °C, which is 17.5 torr (rounded from 17.54 torr in Table 10.2). The result of this correction will be the answer to the first question. We will then have *initial* values of P (here meaning P_{O_2}), V, and T.

The second part of the problem asks for a final value of V at new values of P and T representing standard temperature and pressure. In other words, after answering the first question we will have values for P_1, V_1, and T_1, as well as values for P_2 and T_2. We're asked to calculate V_2. The combined gas law tells us how these values are all related.

Solution: First, we find the partial pressure of the oxygen, using Dalton's law.

$$P_{O_2} = P_{total} - P_{water}$$

$$= 738 \text{ torr} - 17.5 \text{ torr} = 720 \text{ torr}$$

---

[7]If the water levels are not the same inside the flask and outside, a correction has to be calculated and applied to the room pressure to obtain the true pressure in the flask. For example, if the water level is higher inside the flask than outside, the pressure in the flask is lower than atmospheric pressure. The difference in levels is in millimeters of *water,* so this has to be converted to the equivalent in millimeters of mercury, using the approach of Example 10.1, before the room pressure is corrected.

Second, we assemble the data.

$P_1 = 720$ torr (now P_{O_2}) $P_2 = 760$ torr (standard pressure)

$V_1 = 310$ mL $V_2 = ?$

$T_1 = (20.0\ °C + 273\ °C) \times \dfrac{1\ K}{°C} = 293\ K$ $T_2 = 273\ K$ (standard temperature)

We use these in the combined gas law equation:

$$\frac{P_1 V_1}{T_1} = \frac{P_2 V_2}{T_2}$$

to get

$$\frac{(720\ \text{torr})(310\ \text{mL})}{293\ K} = \frac{(760\ \text{torr})(V_2)}{273\ K}$$

Solving for V_2 gives us

$$V_2 = \frac{(720\ \text{torr})(310\ \text{mL})(273\ K)}{(760\ \text{torr})(293\ K)}$$

$$= 274\ \text{mL}$$

Thus, when the water vapor is removed from the gas sample, the dry oxygen will occupy a volume of 274 mL at STP.

Are the Answers Reasonable?

The dry pressure (720 torr) must be *less* than that of the wet sample (738 torr). And the dry volume (274 mL) must be *less* than the wet volume (310 mL), but not much less, so the sizes of the answers are reasonable.

Practice Exercise 10

Suppose you prepared a sample of nitrogen and collected it over water at 15 °C at a total pressure of 745 torr and a volume of 310 mL. Find the partial pressure of the nitrogen and the volume it would occupy at STP. ◆

Mole Fractions and Mole Percents

One of the useful ways of describing the composition of a mixture of gases is in terms of the *mole fractions* of the components. The **mole fraction** of a component of a mixture is the ratio of the number of its moles to the total number of moles of all components. Expressed mathematically, the mole fraction of substance A in a mixture of $A, B, C, \ldots, Z$ substances is

The concept of mole fraction applies to any uniform mixture in any physical state, gas, liquid, or solid.

$$X_A = \frac{n_A}{n_A + n_B + n_C + n_D + \cdots + n_Z} \tag{10.7}$$

Tools

◄ Mole fractions

where X_A is the mole fraction of component A, and $n_A, n_B, n_C, \ldots, n_Z$ are the numbers of moles of each component, $A, B, C, \ldots, Z$, respectively. The sum of all mole fractions for a mixture must always equal 1.

You can see in Equation 10.7 that the units (moles) must cancel and so a value of mole fraction has no units. Nevertheless, always remember the definition: a mole fraction stands for the ratio of *moles* of one component to the total number of *moles* of all components.

Sometimes a mole fraction is multiplied by 100 to give the **mole percent** of the component, symbolized **mol%.**

Mole Fractions of Gases from Partial Pressures

Partial pressure data can be used to calculate the mole fractions of individual gases in a gas mixture because the number of *moles* of each gas is directly proportional to its partial pressure. We can demonstrate this as follows. The partial pressure, P_A, for any one gas, A, in a gas mixture with a total volume V at a temperature T is found by the ideal gas law equation, $PV = nRT$. So to calculate the number of moles of A present, we have

$$n_A = \frac{P_A V}{RT}$$

For any particular gas mixture at a given temperature, the values of V, R, and T are all constants, making the ratio V/RT a constant, too. We can therefore simplify the previous equation by using C to stand for V/RT. In other words, we can write

$$n_A = P_A C$$

The result is the same as saying that *the number of moles of a gas in a mixture of gases is directly proportional to the partial pressure of the gas.* The constant C is the same for all gases in the mixture. So by using different letters to identify individual gases, we can let $P_B C$ stand for n_B, $P_C C$ stand for n_C, and so on in Equation 10.7. Thus,

$$X_A = \frac{P_A C}{P_A C + P_B C + P_C C + \cdots + P_Z C}$$

The constants, C, can be factored out and canceled, so

$$X_A = \frac{P_A}{P_A + P_B + P_C + \cdots + P_Z}$$

The denominator is the sum of the partial pressures of all the gases in the mixture, but this sum equals the total pressure of the mixture (Dalton's law of partial pressures). Therefore, the previous equation simplifies to

$$X_A = \frac{P_A}{P_{\text{total}}} \tag{10.8}$$

Thus, the mole fraction of a gas in a gas mixture is simply the ratio of its partial pressure to the total pressure. Equation 10.8 also gives us a simple way to calculate the partial pressure of a gas in a gas mixture when we know its mole fraction.

EXAMPLE 10.11

Using Partial Pressure Data to Calculate Mole Fractions

What are the mole fractions and mole percents of nitrogen and oxygen in air when their partial pressures are 160 torr for oxygen and 600 torr for nitrogen? Assume no other gases are present.

Analysis: This calls for Equation 10.8, which gives us the mole fraction of a gas from its partial pressure and the total pressure. But we must first calculate P_{total} which is just the sum of the partial pressures (Equation 10.6). When we have P_{total}, we can apply Equation 10.8 to each gas.

Solution:

$$P_{\text{total}} = 600 \text{ torr} + 160 \text{ torr} = 760 \text{ torr}$$

We use Equation 10.8 first for the mole fraction of N_2.

$$X_{N_2} = \frac{P_{N_2}}{P_{total}} = \frac{600 \text{ torr}}{760 \text{ torr}}$$

$$= 0.789, \text{ or } 78.9 \text{ mole percent of } N_2$$

We do the same for oxygen.

$$X_{O_2} = \frac{P_{O_2}}{P_{total}} = \frac{160 \text{ torr}}{760 \text{ torr}}$$

$$= 0.211, \text{ or } 21.1 \text{ mole percent of } O_2$$

Are the Answers Reasonable?
The two mole percents must add up to 100%, and they do.

<hr>

Practice Exercise 11

What is the mole fraction and the mole percent of oxygen in exhaled air if P_{O_2} is 116 torr and P_{total} is 760 torr? ◆

Partial Pressures from Mole Fractions

As we said, if we can calculate mole fractions from partial pressures, it must be equally simple to calculate partial pressures from mole fractions. A mountain climber, for example, might want to do this kind of calculation before working at high altitudes, wondering uneasily about the partial pressure of oxygen. We breathe easily at or near sea level because our lungs and the hemoglobin concentration of our blood are adapted to the partial pressure of oxygen in air, roughly 160 torr, at sea level. At the summit of Mount Everest (elevation 8848 m), however, the total atmospheric pressure is only about 250 torr. *The mole fractions of oxygen and nitrogen in air are essentially constant at all altitudes in the lower atmosphere,* so the mole fraction of oxygen, 0.211, is the same at sea level as at the top of Mount Everest. For any gas in air, what changes with altitude is not the mole fraction but the partial pressure, because proportionately fewer numbers of moles of the gas are present in a given volume. Thus, the value of P_{O_2} at any altitude is some fraction, the mole fraction, of the atmospheric pressure there. So for oxygen on top of Mount Everest, the value of P_{O_2} must be the mole fraction of O_2 times the atmospheric pressure there.

$$P_{O_2} = 0.211 \times 250 \text{ torr} = 52.8 \text{ torr (at Everest's summit)}$$

> ▶**Chemistry in Practice**◀ A value of P_{O_2} of only about 50 torr is too low for the survival of nearly all people, but Nepalese porters (Sherpas) and several other well-adapted people have cardiovascular systems that work at high altitudes. When the partial pressure of oxygen is too low, the lungs automatically (involuntarily) compensate by making breathing come faster and faster, deeper and deeper. The result is that more waste carbon dioxide is blown out of their systems than ought to be removed for the safe and efficient exchange of oxygen and carbon dioxide. The acid–base balance of the blood becomes upset, and the uptake of oxygen is done less and less well. The situation is extremely critical, but we have to know more about acids and bases to explain. (We'll return to this in *Chemicals in Our World 9.*) ◆

10.7 Graham's Law of Effusion

We'll begin by distinguishing between *effusion* and *diffusion*. **Diffusion** is the complete spreading out and intermingling of the molecules of one gas, like those of perfume or cologne, into and among those of another gas, like the air in a room (see Figure 10.10*a*). Helium-filled rubber balloons go limp overnight because helium atoms diffuse through invisible pores in the balloon and into the surrounding air. Air-filled balloons go limp more slowly because the ability to diffuse is reduced as the gas molecules become larger.

Effusion is the movement of gas molecules through an extremely tiny opening into a *vacuum,* not into another gas (see Figure 10.10*b*). Thomas Graham studied the relationship between effusion rates and molecular masses for a series of gases. He wanted to minimize as much as possible the "jostling" effects of crowding, namely, the collision events that slow molecules down or bump them aside or even to the rear. To accomplish this he made the effusion hole as small as possible in a material machined as thin as possible, thus minimizing the population of molecules in the hole. By letting the gas effuse into a vacuum, Graham was able to minimize backflow. Under these conditions, Graham found what we now call **Graham's law:** *The rates of effusion of gases are inversely proportional to the square roots of their densities, d, when compared at identical pressures and temperatures.*

Thomas Graham (1805–1869) was a Scottish scientist.

$$\text{Effusion rate} \propto \frac{1}{\sqrt{d}} \qquad (\text{constant } P \text{ and } T)$$

This relationship is changed to an equation in the usual way, using a proportionality constant, k.

$$\text{Effusion rate} \times \sqrt{d} = k \qquad (\text{constant } P \text{ and } T)$$

What is remarkable about k is that *it is virtually identical for all gases.* Thus if we have two gases, A and B, we can write

$$\text{Effusion rate } (A) \times \sqrt{d_A} = \text{effusion rate } (B) \times \sqrt{d_B} = k$$

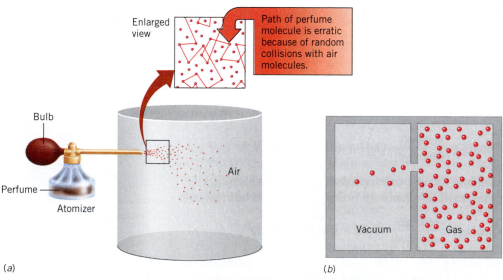

(a)   (b)

Figure 10.10 *Spontaneous movements of gases.* (*a*) Diffusion. (*b*) Effusion.

By rearrangement, we obtain

$$\frac{\text{effusion rate } (A)}{\text{effusion rate } (B)} = \frac{\sqrt{d_B}}{\sqrt{d_A}} = \sqrt{\frac{d_B}{d_A}} \qquad (10.9)$$

It can be shown[8] that the density of a gas is directly proportional to its molecular mass. Thus, we can re-express Equation 10.9 as follows (canceling proportionality constants):

$$\frac{\text{effusion rate } (A)}{\text{effusion rate } (B)} = \sqrt{\frac{d_B}{d_A}} = \sqrt{\frac{M_B}{M_A}} \qquad (10.10)$$

Tools

Graham's law
of effusion

where M_A and M_B are the molecular masses of gases A and B.

EXAMPLE 10.12
Using Graham's Law

Under all of the Graham's law conditions, which effuses more rapidly and by what factor, ammonia or hydrogen chloride?

Analysis: A gas effusion problem requires the use of Graham's law, which says that the gas with the smaller molecular mass will effuse more rapidly. So we need to find the molecular masses. Then the relative rates of effusion are found by using these in Equation 10.10.

Solution: The molecular masses are 17.03 for NH_3 and 36.46 for HCl, so we know that NH_3, with its smaller value, effuses more rapidly than HCl. The ratio of the effusion rates is given by

$$\frac{\text{effusion rate } (NH_3)}{\text{effusion rate } (HCl)} = \sqrt{\frac{36.46}{17.03}} = 1.463$$

Ammonia effuses 1.463 times more rapidly than HCl under the same conditions.

Is the Answer Reasonable?
The only quick check is to be sure that the answer ought to be greater than 1, as calculated. Ammonia molecules, having lower mass, should effuse faster than those of hydrogen chloride.

Practice Exercise 12

The hydrogen halide gases all have the same general formula, HX, where X can be Cl, Br, or I. If HCl(g) effuses 1.88 times more rapidly than one of the others, which hydrogen halide is the other, HBr or HI? ◆

---

[8]Density = mass/volume, so mass = (density)(volume).

Number of moles = n = mass/formula mass, so mass = (n)(formula mass).

Therefore, (density)(volume) = (n)(formula mass), from which it can be seen that (density) = (formula mass)(n)/(volume). So for fixed values of volume and number of moles (n), the density of a gas is proportional to its formula mass.

10.8 Kinetic Theory and the Gas Laws

Back in the nineteenth century, scientists who already knew the gas laws couldn't help but ask, "How do gases 'work'?" They wondered what had to be true about all gases to explain their conformity to a common set of gas laws. The **kinetic theory of gases** was the answer. We introduced some of its ideas in Chapter 6.

The strategy for answering the question about the nature of a gas began with postulating a *model* or picture of an ideal gas. Then the laws of physics and statistics were applied to see whether the observed gas laws could be predicted from the model. The results were splendidly successful.

> **Postulates of the Kinetic Theory of Gases**
> **1.** A gas consists of an extremely large number of very tiny particles that are in constant, random motion.
> **2.** The gas particles themselves occupy a net volume so small in relation to the volume of their container that their contribution to the total volume can be ignored.
> **3.** The particles often collide in perfectly elastic collisions[9] with themselves and with the walls of the container, and they move in straight lines between collisions neither attracting nor repelling each other.

The particles are assumed to be so small that they have no dimensions at all.

Billiard players master the applications of the laws of mechanics, if not their mathematical forms.

Physics, the study of matter, energy, and physical changes, has a branch called *mechanics* that deals with the laws of behavior of moving objects. These laws would have to apply to gases if they were, in fact, like supersmall, constantly moving billiard balls that constantly bounce off each other and the walls of their container, and so exert a net pressure on the walls (see Figure 10.11). The gas particles are assumed to be so small that their individual volumes could be ignored.

Just 1 mL of a gas at STP has over 2.5×10^{19} molecules.

The theoretical calculations based on the model gave results that agreed splendidly with the known behavior of gases, as described by the gas laws. Scientists, therefore, concluded that the model itself must be very close to

Figure 10.11 *Gas pressure as understood by the kinetic theory of gases.* The pressure of the gas in the balloon, *a*, is caused by collisions of the gas molecules with the balloon's walls, *b*.

(a)

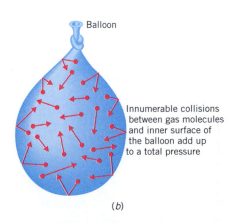

Balloon

Innumerable collisions between gas molecules and inner surface of the balloon add up to a total pressure

(b)

[9]In *perfectly elastic* collisions, no energy is lost by friction as the colliding objects deform momentarily.

the truth about gases. When the behavior of real gases departs somewhat from the ideal, the model even helps to explain this, as we'll see later in this section.

The Gas Laws Explained by the Kinetic Theory

According to the model, gases are mostly empty space. This explains why gases, unlike liquids and solids, can be compressed so much (squeezed to smaller volumes). It also explains why we have gas laws for gases, and *the same laws for all gases,* but not comparable laws for liquids or solids. The chemical identity of the gas does not matter, because gas molecules do not touch each other except when they collide, and there are extremely weak interactions, if any, between them.

We cannot go over the mathematical details, but we can describe some of the ways in which the model of an ideal gas used in the kinetic theory helps us understand the gas laws.

Kinetic Theory and Gas Temperature Revisited

The greatest triumph of the kinetic theory came with its explanation of gas temperature, which we discussed in Section 6.2. What the calculations showed was that the product of gas pressure and volume, PV, is proportional to the average molecular kinetic energy of the gas molecules.

$$PV \propto \text{average molecular KE}$$

But from the experimental study of gases, culminating in the equation of state for an ideal gas, we have another term to which PV is proportional, namely, the Kelvin temperature of the gas.

$$PV \propto T$$

(We know what the proportionality constant here is, namely, nR, because by the ideal gas law, PV equals nRT.) With PV proportional *both* to T and to "average molecular KE," then it must be true that the temperature of a gas is proportional to the average molecular KE of its particles.

$$T \propto \text{average molecular KE}$$

Kinetic Theory and the Pressure–Volume Law (Boyle's Law)

Using the model of an ideal gas, physicists were able to show that gas pressure is the net effect of innumerable collisions made by gas particles on the walls of the container. Let's imagine that one wall of a gas container is a movable piston that we can push in (or pull out) and so change the gas volume (see Figure 10.12). If we make the volume smaller without changing the temperature, no change occurs in the average molecular kinetic energy of the molecules. But now there would be more gas particles per unit volume. If we reduce the volume by one-half, for example, we double the number of molecules per unit volume. This would, therefore, double the number of collisions per second with a unit area and so double the pressure. Thus, cutting the volume in half requires that we double the pressure, not triple it or quadruple it, which is exactly what Boyle discovered:

$$P \propto \frac{1}{V} \quad \text{or} \quad V \propto \frac{1}{P}$$

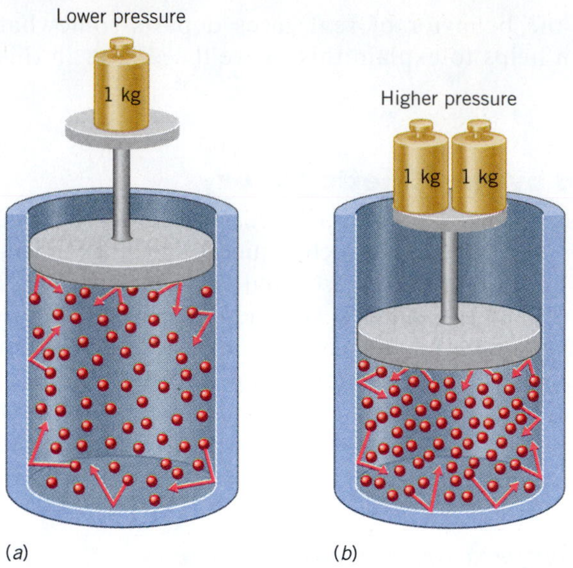

Figure 10.12 *The kinetic theory and the pressure–volume law (temperature held constant).* When the gas volume is made smaller in going from (*a*) to (*b*), the frequency of the collisions per unit area of the container's walls increases. Therefore, the pressure increases.

Lower pressure

1 kg

Higher pressure

1 kg | 1 kg

(*a*)

(*b*)

Kinetic Theory and the Pressure–Temperature Law (Gay-Lussac's Law)

The kinetic theory, as we have said, tells us that an increase in gas temperature increases the average velocity of gas particles. At higher velocities, the particles must strike the container's walls more frequently and with greater force. But at constant volume, the *area* being struck is still the same, so the force per unit area—the pressure—must therefore increase (see Figure 10.13). In this way the kinetic theory explains how gas pressure is proportional to gas temperature (under constant volume and mass), which is the pressure–temperature law of Gay-Lussac.

Figure 10.13 *The kinetic theory and the pressure–temperature law (volume held constant).* (*a*) The gas exerts only a small pressure, P_1, over and above the atmospheric pressure. (*b*) Raising the gas temperature raises the pressure to P_2, with the extra mercury being added to keep the gas from expanding its volume. At the higher temperature the gas particles move with greater average velocity (as suggested by the longer motion arrows), so normally they would take up more space. The extra mercury is a measure of the increased pressure needed to prevent this.

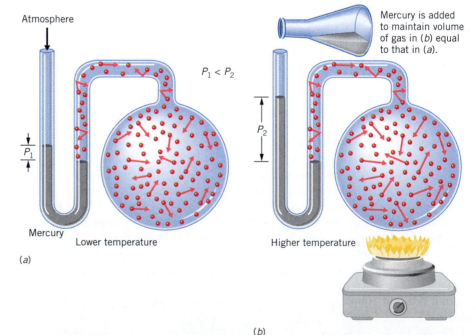

Atmosphere

$P_1 < P_2$

Mercury is added to maintain volume of gas in (*b*) equal to that in (*a*).

P_1

P_2

Mercury

Lower temperature

(*a*)

Higher temperature

(*b*)

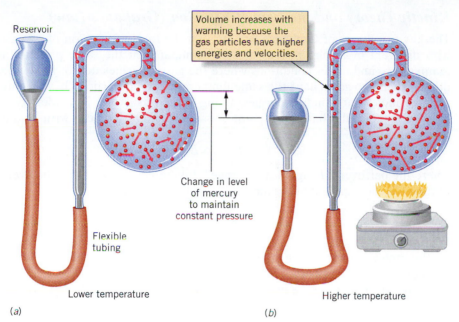

Volume increases with warming because the gas particles have higher energies and velocities.

Reservoir

Change in level of mercury to maintain constant pressure

Flexible tubing

Lower temperature

(a)

Higher temperature

(b)

Figure 10.14 *The kinetic theory and the temperature–volume law (pressure held constant).* The pressure is the same in both (a) and (b), as indicated by the mercury levels. However, at the lower temperature (a), the gas takes up less volume. At the higher temperature (b), the gas particles have a higher average energy and velocity. To hold the pressure constant (a condition of Charles' law), the gas has to be given more room by letting some mercury flow out of the tube and back into the reservoir.

Kinetic Theory and the Temperature–Volume Law (Charles' Law)

The kinetic theory tells us that increasing the temperature of a gas increases the average molecular kinetic energy of its particles, which tends to increase the pressure of the gas. The only way we can keep P constant is to allow the gas to expand (see Figure 10.14). Therefore, a gas expands with increasing T in order to keep P constant, which is another way of saying that V is proportional to T at constant P. Thus, the kinetic theory explains Charles' law.

Kinetic Theory and Dalton's Law of Partial Pressures

The law of partial pressures is actually evidence for that part of the third postulate in the kinetic theory that pictures gas particles moving in straight lines between collisions, neither attracting nor repelling each other. They act *independently,* in other words (see Figure 10.15). Only if the particles of each gas do act independently can the partial pressures of the gases add up in a simple way to give the total pressure.

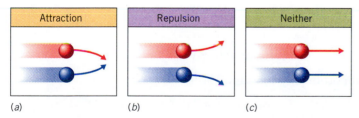

| Attraction | Repulsion | Neither |

(a)   (b)   (c)

Figure 10.15 *Gas molecules act independently when they neither attract nor repel each other.* Gas molecules would not travel in straight lines if they attracted each other as in (a) or repelled each other as in (b). This would influence the length of time between collisions with the walls and would therefore affect the collision frequency with the walls. In turn, this would affect the pressure each gas would exert. Only if the molecules traveled in straight lines with no attractions or repulsions, as in (c), would their individual pressures not be influenced by near misses or by collisions between the molecules.

Kinetic Theory and the Law of Effusion (Graham's Law)

The key conditions of Graham's law are that the rates of effusion of two gases with different molecular masses must be compared at the same pressure and temperature and under conditions where the gas molecules do not hinder each other. When two gases have the same temperature, their particles have identical average molecular kinetic energies. Using subscripts 1 and 2 to identify two gases with molecules having different masses m_1 and m_2, we can write that at a given temperature

$$\overline{KE_1} = \overline{KE_2}$$

where the bar over KE signifies "average." Taking $\overline{v_1^2}$ and $\overline{v_2^2}$ to be the average of the velocities squared of the molecules of the gases, we have

$$\tfrac{1}{2}m_1\overline{v_1^2} = \tfrac{1}{2}m_2\overline{v_2^2}$$

We have not extended the "average" notation (the bar) over the masses because all the molecules of a given substance have the same mass. Thus, the average mass of molecules of gas 1 is just m_1.

Now let's rearrange the previous equation to get the ratio of $\overline{v^2}$ terms.

$$\frac{\overline{v_1^2}}{\overline{v_2^2}} = \frac{m_2}{m_1}$$

Next, we'll take the square root of both sides. When we do this we obtain a ratio of quantities called the **root mean square** (abbreviated **rms**) speeds, which we will represent as $\overline{v_1}$ and $\overline{v_2}$.

$$\frac{\overline{v_1}}{\overline{v_2}} = \sqrt{\frac{m_2}{m_1}}$$

Suppose we have two molecules with velocities of 6 and 10 m s^{-1}. The average speed is $\tfrac{1}{2}(6 + 10) = 8$ m s^{-1}. The rms speed is obtained by squaring each speed, averaging the squared values, and then taking the square root of the result. Thus $\overline{v} = \sqrt{\tfrac{1}{2}(6^2 + 10^2)} = 8.2$ m s^{-1}.

The rms speed, $\overline{v_{rms}}$, is not actually the same as the average speed of the gas molecules, but instead represents the speed of a molecule that would have the average kinetic energy. (The difference is subtle, and the two averages do not differ by much, as noted in the margin.) For any substance, the individual mass of a molecule is proportional to the molecular mass. Representing a molecular mass of a gas by M, we can restate this as $m \propto M$. The proportionality constant is the same for all gases. (It's in grams per atomic mass unit when we express m in atomic mass units.) When we take a ratio of two molecular masses, the constant cancels anyway, so we can write

$$\frac{\overline{v_1}}{\overline{v_2}} = \sqrt{\frac{m_2}{m_1}} = \sqrt{\frac{M_2}{M_1}}$$

The rate of effusion of a gas, of course, has to be proportional to the average speed of its molecules.

$$\text{Effusion rate} \propto \overline{v}$$

Let's use k as the proportionality constant. This gives

$$\text{Effusion rate} = k\overline{v}$$

Taking a ratio of speeds causes k to cancel, so we can now write

$$\frac{\text{effusion rate gas 1}}{\text{effusion rate gas 2}} = \sqrt{\frac{M_2}{M_1}}$$

This is the way we expressed Graham's law in Equation 10.10. Thus still another gas law supports the model of an ideal gas.

Kinetic Theory and Absolute Zero

The kinetic theory found that gas temperature is proportional to the average molecular kinetic energy of the gas molecules.

$$T \propto \text{average molecular KE} \propto \tfrac{1}{2}m(\overline{v^2})$$

If the average molecular KE becomes zero, the temperature must also become zero. But mass (m) cannot become zero, so the only way that the average molecular KE can be zero is if v goes to zero. A gas particle cannot move any slower than it does at a dead standstill, so if the gas molecules stop moving entirely, the gas is as cold as anything can get. It's at absolute zero.[10]

10.9 Real Gases: Deviations from the Ideal Gas Law

The combined gas law states that the ratio PV/T is a constant. According to the ideal gas law, PV/T equals a product of two constants, nR. But, experimentally, for real gases PV/T is actually not quite a constant. When we use experimental values of P, V, and T for a real gas, like O_2, to plot actual values of PV/T as a function of P, we get the curve shown in Figure 10.16. The *horizontal* plot in Figure 10.16 is what we should see if PV/T were truly constant over all values of P, as it would be for an ideal gas.

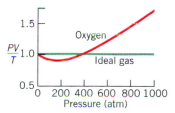

Figure 10.16 *A graph of PV/T versus P for an ideal gas is like plotting a constant versus P. The graph must be a straight line, as shown, because the ideal gas law equation tells us that PV/T = nR (a product of constants). The same plot for oxygen, however, is not a straight line, showing that O2 is not "ideal."*

A real gas, like oxygen, deviates from ideal behavior for two important reasons. First, the model of an ideal gas assumes that gas molecules individually have no volume, but of course they do. (If all of the kinetic motions of the gas molecules ceased and the molecules settled, you could imagine the net space that the molecules would occupy in and of themselves.) Therefore, the actual space left for the kinetic motions of the molecules of the real gas is not as large as the volume V of the container. Fortunately, the volume taken up by the molecules themselves is a very tiny fraction of the container's volume at ordinary pressures. This is why real gases fit the gas laws so well at ordinary pressures.

The fraction of space taken up by the molecules themselves would increase if we forced more and more molecules into the container and so increased the pressure. It would be like jamming more and more people into a small room. The fraction of space available for each person to move around diminishes even though the volume of the room is unchanged. The increasing fraction of space taken up by the individual molecular volumes is one reason why real gases deviate more and more from ideal behavior at increasing pressures. At the higher values of P in the plot for a real gas in Figure 10.16, the volume occupied by the molecules themselves becomes too large in relationship to the container volume, V. Thus, the value of V used in a gas law calculation is larger than warranted, which tends to make the ratio PV/T greater for a real gas than that for the ideal gas, particularly at high pressures.

The second fact concerning real gases is that their particles do attract each other somewhat, unlike one postulate about an ideal gas. Any tendency for the gas particles to cling together, however slight, and so strike the container's walls less often would cause smaller values of P than expected from the model of an ideal gas. Thus, the ratio PV/T is *less* than that for an ideal gas, particularly

---

[10]Actually, even at 0 K, there must be some slight motion. It's required by the Heisenberg uncertainty principle, which says (in one form) that it's impossible to know precisely both the speed and the location of a particle simultaneously. (If one knows the speed, then there's uncertainty in the location, for example.) If the molecules were actually dead still at absolute zero, there would be no uncertainty in their speed. But then the uncertainty in their *position* would be infinitely great. We would not know where they were! But we do know; they're in this or that container. Thus some uncertainty in speed must exist to have less uncertainty in position and so locate the sample!

Table 10.3 Van der Waals Constants

Substance	a $(L^2 \, atm \, mol^{-2})$	b $(L \, mol^{-1})$
Noble Gases		
Helium, He	0.03421	0.02370
Neon, Ne	0.2107	0.01709
Argon, Ar	1.345	0.03219
Krypton, Kr	2.318	0.03978
Xenon, Xe	4.194	0.05105
Other Gases		
Hydrogen, H_2	0.02444	0.02661
Oxygen, O_2	1.360	0.03183
Nitrogen, N_2	1.390	0.03913
Methane, CH_4	2.253	0.04278
Carbon dioxide, CO_2	3.592	0.04267
Ammonia, NH_3	4.170	0.03707
Water, H_2O	5.464	0.03049
Ethyl alcohol, C_2H_5OH	12.02	0.08407

where the problem of particle volume is least, at lower pressures. The curve of Figure 10.16, therefore, dips at lower pressures.

van der Waals Equation

J. D. van der Waals (1837–1923), a Dutch scientist, won the 1910 Nobel Prize in physics.

Many attempts have been made to modify the equation of state of an ideal gas to get an equation that better fits the experimental data for individual real gases. One of the more successful efforts was that of J. D. van der Waals. He found ways to correct the measured values of P and V to give better fits of the data to the general gas law equation. The result of his derivation is called the *van der Waals equation of state for a real gas*.

$$\left(P + \frac{n^2a}{V^2}\right)(V - nb) = nRT \tag{10.11}$$

The constants a and b are called the *van der Waals constants* (see Table 10.3). They are determined for each real gas by carefully measuring P, V, and T under different conditions. Then trial calculations are made to figure out what values of the constants give the best matches between the observed data and the van der Waals equation.

Notice that the constant a involves a correction to the pressure term of the ideal gas law, so the size of a would indicate something about attractions between molecules. The constant b helps to correct for the volume term, so the size of b would indicate something about the sizes of particles in the gas. Larger values of a mean stronger attractive forces between molecules; larger values of b mean larger molecular sizes. Thus, the most easily liquefied substances, like water and ethyl alcohol, have the largest values of the van der Waals constant a, suggesting attractive forces between their molecules. In the next chapter we'll continue the study of factors that control the physical state of a substance, particularly attractive forces and their origins.

SUMMARY

Gas Laws. An **ideal gas** is a hypothetical gas that obeys the gas laws exactly over all ranges of pressure and temperature. Real gases exhibit ideal gas behavior most closely at low pressures and high temperatures, which are conditions remote from those that liquefy a gas.

Boyle's Law (Pressure–Volume Law). Volume varies inversely with pressure at constant temperature and mass. $V \propto 1/P$.

Charles' Law (Temperature–Volume Law). Volume varies directly with the Kelvin temperature at constant pressure and mass. $V \propto T$.

Gay-Lussac's Law (Temperature–Pressure Law). Pressure varies directly with Kelvin temperature at constant volume and mass. $P \propto T$.

Avogadro's Principle. Equal volumes of gases contain equal numbers of moles when compared at the same temperature and pressure.

Combined Gas Law. PV divided by T for a given gas sample is a constant. $PV/T = C$.

Ideal Gas Law. $PV = nRT$. When P is in atm and V is in L, the value of R is 0.0821 L atm mol^{-1} K^{-1} (T being, as usual, in kelvins).

Dalton's Law of Partial Pressures. The total pressure of a mixture of gases is the sum of the partial pressures of the individual gases.

$$P_{\text{total}} = P_a + P_b + P_c + \cdots$$

Graham's Law of Effusion. The rate of effusion of a gas varies inversely with the square root of its density (or the square root of its molecular mass) at constant pressure and temperature.

Kinetic Theory of Gases. An ideal gas consists of innumerable very hard particles, with individual volumes so small they can be ignored, between which no forces of at-

traction or repulsion act. The particles are in constant, chaotic, random motion. When the laws of physics and statistics are applied to this model, and the results compared with the ideal gas law, the Kelvin temperature of a gas is proportional to the average kinetic energy of the gas particles. Pressure is the result of forces of collisions of the particles with the container's walls.

Real Gases. Because individual gas particles do have real volumes and because small forces of attraction do exist between them, real gases do not exactly obey the gas laws. The van der Waals equation of state for a real gas makes corrections for the volume of the gas molecules and for the attractive force between gas molecules. The measure of the attractive force is given by a, a van der Waals constant, and the other constant, b, is a measure of the relative size of the gas molecules.

Reaction Stoichiometry: A Summary. With the study in this chapter of the stoichiometry of reactions involving gases, we have completed our study of the tools needed for the calculations of all variations of reaction stoichiometry. The fact of central importance is that all such calculations must funnel through *moles.* Whether we start with grams of some compound in a reaction, or the molarity of its solution plus a volume, or P–V–T data for a gas in the reaction, *we must get the essential calculation into moles.* The equation's coefficients then give us the stoichiometric equivalencies needed to convert from the number of moles of one substance into the numbers of moles of any of the others in the reaction. Then we can move back to any other kind of unit we wish. The following flowchart summarizes what we have been doing.

The labels on the arrows of the flowchart suggest the basic tools. Formula masses or molecular masses get us from grams to moles or from moles to grams. Molarity and volume data move us from concentrations to moles or back. With P–V–T data we can find moles, or, knowing moles of a gas, we can find any one of P, V, or T, given the other two.

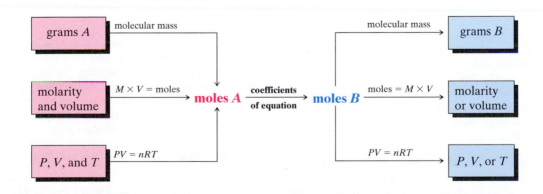

Tools you have learned

| Mo | Tc | Ru | Rh | Pd | Ag 07.8682 | Cd 112.411 | In 114.82 | Sn 118.710 | Sb 121.75 | Te 127.60 |
| 72 | 73 | 74 | 75 | 76 | 77 | 78 | 79 | 80 | 81 | 82 | 83 | 84 |

The table below lists the tools that you have learned in this chapter that are applicable to problem solving. Review them if necessary, and refer to them when working on the Thinking-It-Through problems and the Review Problems that follow.

Tool	Function
Combined gas law (*page 431*)	To be used to calculate a particular value of P, V, or T, given other values and the quantity of gas as fixed.
	To be used in applications of Boyle's, Charles', or Gay-Lussac's law.
Ideal gas law (*page 435*)	To be used when any three of the four variables of the physical state of a gas, P, V, T, or n, are known to calculate the value of the fourth.
Dalton's law of partial pressures (*page 442*)	To calculate the partial pressure of one gas in a mixture of gases from the total pressure and either the partial pressures of the other gases or their mole fractions.
	Given the volume and temperature of a gas collected over water and the atmospheric pressure, to calculate the volume the gas would have when dry.
Mole fractions (*page 445*)	To calculate the mole fraction or mole percent of one component of a mixture from other data about the mixture.
	To calculate the partial pressure of one gas in a mixture from its mole fraction.
Graham's law of effusion (*page 449*)	To calculate relative rates of effusion of gases.
	To calculate molecular masses of gases from relative rates of effusion.

THINKING IT THROUGH

Remember, the goal in the following problems is *not* to find the answer itself; instead, assemble the needed information and describe *how* you would use the data to answer the question. (For some problems, you may find your explanations easier if you actually perform some of the intermediate calculations.) Where problems can be solved by the factor-label method, be sure to set up the solution to the problem by arranging the conversion factors so the units cancel correctly to give the desired units of the answer. Obtain data for atomic masses from the inside front cover of the book.

The problems are divided into two groups. Those in Level 2 are more challenging than those in Level 1 and provide an opportunity to really hone your problem solving skills. Detailed answers to these problems can be found at our Web site: http://www.wiley.com/college/brady.

Level 1 Problems

1. The plunger of a bicycle tire pump is to be pushed in so that the pressure of the entrapped volume of air changes from 734 torr to 2.50 atm. The air hose has been closed off; no air can escape. No temperature change occurs. If the initial volume of air is 175 mL, what will its final volume be? (Set up the calculations.)

2. At 70 °C a sample of a gas has a volume of 550 mL. If you wished to reduce its volume to 500 mL, to what temperature must you cool it? Assume no change in pressure. (Set up the calculation.)

3. A sample of (dry) methane, CH_4, was collected at an atmospheric pressure of 1.00 atm and a temperature of 25 °C in a sealed glass flask which the manufacturer said could withstand an internal pressure of 1.50 atm. Assuming the claim to be accurate, describe how you would determine whether this flask could be safely heated to 50 °C.

4. A glass vessel with a volume of 1.50 L was filled with 1.68 g of dry nitrogen at 24 °C and attached to a closed-end mercury manometer. The mercury in the arm of the manometer attached to the glass vessel was measured to be 18 mm higher than that in the arm open to the atmosphere. What was the atmospheric pressure? (Explain the calculations required to answer the question.)

5. Hydrochloric acid was added to a sample of magnesium. The reaction

$$2HCl(aq) + Mg(s) \longrightarrow MgCl_2(aq) + H_2(g)$$

produced hydrogen gas which had a volume of 37.6 mL and a pressure of 746 torr when it was collected over water at 20 °C. How many grams of magnesium were consumed in the reaction? (Describe the calculations in detail.)

6. A gas sample weighing 1.67 g has a volume of 276 mL at 589 torr and 28 °C. What is the molecular mass of the gas?

Level 2 Problems

7. A 1.56 g sample of gas was collected over water at 25 °C and a pressure of 745 torr in a 275 mL container. What is the density of a dry sample of this gas at 45 °C and 770 torr? (Describe the calculations in detail.)

8. A small amount of hydrogen gas was burned in a 1.000 L container initially filled with 0.2500 g of dry O_2. Some O_2 remained after the reaction was complete. The container and its contents were returned to 25°C and the pressure in the container was measured to be 57.48 torr. Describe in detail how you would calculate how many grams of hydrogen had been burned. (*Hint:* Some of the water produced is liquid and some is gas.)

9. A 20.0 mL sample of dilute, aqueous hydrogen peroxide was placed in a 1.00 L sealed vessel. The pressure inside the flask was determined to be 740 torr. The system was maintained at 25 °C as the hydrogen peroxide decomposed according the following equation.

$$2H_2O_2(aq) \longrightarrow O_2(g) + 2H_2O(l)$$

After all of the hydrogen peroxide had thus decomposed, the pressure was measured again and found to be 907 torr. Set up the calculations you would use to determine the number of grams of H_2O_2 present in the initial sample. (Ignore the very small *change* in the gas space available in the flask caused by the formation of some additional water. Assume that oxygen is insoluble in water.) Is it necessary to do a Dalton's law calculation to correct for the vapor pressure of water? Explain.

10. A cylinder with a movable piston contains helium gas over water at 35 °C. The piston is lowered until the volume of the gas is halved. The pressure inside the cylinder rose to 840 torr. What was the initial pressure inside the cylinder before the piston was moved? (Describe the necessary calculations in detail.)

11. In a chemical analysis, the nitrogen in a 0.465 g sample of a compound containing C, H, N, and O was converted to N_2 gas. The gas had a volume of 76.6 mL at 25 °C and 752 torr. In a separate analysis, the compound was determined to also contain 32.0% C and 6.67% H. What is the empirical formula of the compound? (Explain the calculations in detail.)

12. Both CO_2 and SF_6 are colorless and odorless gases that are chemically quite stable in the atmosphere. If you planned to manufacture tennis balls and wished them to retain their "bounce" for the longest time, which gas would you use to give internal pressure to the balls? Explain. Assume that costs are not what matters the most, only quality.

13. Use kinetic molecular theory to explain the increase in size of a weather balloon as it rises from ground level into the upper atmosphere, keeping in mind that the pressure and temperature both decrease with increasing altitude. Consider separately the effects upon volume of changing the pressure and temperature. Which effect is dominant, and how do you know this?

14. A 1 liter bulb is filled with equal parts of hydrogen gas and oxygen gas. Hydrogen atoms are depicted as light gray spheres and oxygen atoms as red spheres. Which of the following figures best represents the contents of the bulb. Explain why you selected one figure and why you rejected each of the others.

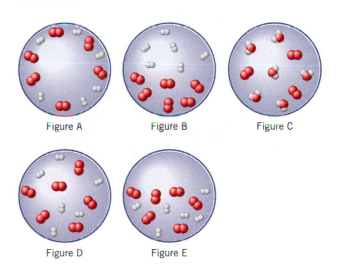

Figure A   Figure B   Figure C

Figure D   Figure E

15. At 25 °C the distribution of molecular speeds for argon atoms is as shown in the figure. Note that the figure is not symmetric about its maximum value but is steeper on the low-speed side.

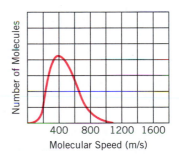

(a) What does the area under the curve represent?
(b) Sketch two additional curves to represent the distribution of molecular speeds for argon, one at a lower temperature and one at a higher temperature. Be sure to maintain in your other curves the property identified in part (a).

16. Samples of NO gas and O_2 gas each at 1.0 atm pressure are separated by a closed stopcock as shown in the figure. Both NO and O_2 are colorless gases. At 25 °C these gases react rapidly according to the equation: $2NO + O_2 \rightarrow 2NO_2$. The product of this reaction is a brown gas.

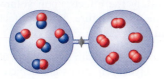

(a) What would you expect to see when the stopcock is turned to the open position? (Both bulbs do not instantaneously become brown.)
(b) What concept does the development of color in this example illustrate?
(c) Make a sketch to show the system once changes in color cease.

REVIEW QUESTIONS

Concept of Pressure; Manometers and Barometers

10.1 Define pressure. Explain why the height of the mercury in a barometer is independent of the cross-sectional area of the tube.

10.2 How are the following related?
(a) force and pressure   (c) torr and atm
(b) torr and mm Hg   (d) torr and pascal

10.3 At 20 °C the density of mercury is 13.6 g mL^{-1} and that of water is 1.00 g mL^{-1}. At 20 °C, the vapor pressure of mercury is 0.0012 torr and that of water is 18 torr. Give and explain two reasons why water would be an inconvenient fluid to use in a Torricelli barometer.

10.4 What is the advantage of using a closed-end manometer, rather than an open-end one, when measuring the pressure of a trapped gas?

P, V, T Relationships for a Fixed Amount of Gas

10.5 Express the following gas laws in equation form: (a) temperature–volume law, (b) temperature–pressure law, (c) pressure–volume law, (d) combined gas law.

10.6 Which of the four important variables in the study of the physical properties of gases are assumed to be held constant in each of the following laws? (a) Boyle's law, (b) Charles' law, (c) Gay-Lussac's law, (d) combined gas law

10.7 What is meant by an *ideal gas*?

Ideal Gas Law

10.8 State the ideal gas law in the form of an equation. What is the value of the gas constant in units of L atm mol^{-1} K^{-1}?

10.9 Using oxygen as an example, carefully distinguish between 1 *molecule,* 1 *mole,* 1 *molar mass,* and 1 *molar volume.*

10.10 At STP how many molecules of H_2 are in 22.4 L?

Dalton's Law and Graham's Law

10.11 State Dalton's law in the form of an equation.

10.12 How is the partial pressure of a gas related to the total pressure of a gas mixture?

10.13 State Graham's law in the form of an equation.

10.14 At a given temperature, how is the rate of effusion of a gas related to its molecular mass?

Kinetic Theory of Gases

10.15 What model of a gas was proposed by the kinetic theory of gases?

10.16 To what properties of an ideal gas is its temperature proportional?
(a) According to the kinetic theory.
(b) According to the combined gas law.

10.17 What aspects of the kinetic theory of gases explain why two gases, given the freedom to mix, will always mix *entirely?*

10.18 If the molecules of a gas at constant volume are somehow given a lower average kinetic energy, what two measurable properties of the gas will change and in what direction?

10.19 Explain *how* raising the temperature of a gas causes it to expand at constant pressure. (Describe how the model of an ideal gas connects the increase in temperature to the gas expansion.)

10.20 Explain in terms of the kinetic theory *how* raising the temperature of a confined gas makes its pressure increase.

10.21 How does the kinetic theory explain the existence of a minimum temperature, 0 K?

10.22 How does the kinetic theory explain the greater effusion rate of a gas with a low molecular mass compared to one with a higher molecular mass?

10.23 In what way does Dalton's law of partial pressures provide evidence that the molecules in a gas move and behave independently of each other?

10.24 What postulates of the kinetic theory are not strictly true, and why?

10.25 If a given gas *A* has a larger value of the van der Waals constant *b* than gas *B*, what does this suggest about gas *A*?

10.26 A small value for the van der Waals constant *a* suggests something about the molecules of the gas. What?

10.27 What equation does van der Waals' equation become more and more like when the volume of a gas becomes larger and larger without any change in temperature or in the number of moles?

REVIEW PROBLEMS

Answers to problems whose numbers are printed in color are given in Appendix D.
More challenging problems are marked with asterisks.

Pressure Unit Conversions

10.28 Carry out the following unit conversions: (a) 1.26 atm to torr, (b) 740 torr to atm, (c) 738 torr to mm Hg, (d) 1.45×10^3 Pa to torr.

10.29 Carry out the following unit conversions: (a) 0.625 atm to torr, (b) 825 torr to atm, (c) 62 mm Hg to torr, (d) 1.22 kPa to torr.

10.30 What is the pressure in torr of each of the following? (a) 0.329 atm (summit of Mt. Everest, the world's highest mountain) (b) 0.460 atm (summit of Mt. Denali, the highest mountain in the United States)

10.31 What is the pressure in atm of each of the following? (These are the values of the pressures exerted individually by N_2, O_2, and CO_2, respectively, in typical inhaled air.) (a) 595 torr   (b) 160 torr   (c) 0.300 torr

Manometers and Barometers

10.32 An open-end manometer containing mercury was connected to a vessel holding a gas at a pressure of 720 torr. The atmospheric pressure was 765 torr. Sketch a diagram of the apparatus showing the relative heights of the mercury in the two arms of the manometer. What is the difference in the heights of the mercury expressed in centimeters?

10.33 An open-end manometer containing mercury was connected to a vessel holding a gas at a pressure of 820 torr. The atmospheric pressure was 750 torr. Sketch a diagram of the apparatus showing the relative heights of the mercury in the two arms of the manometer. What is the difference in the heights of the mercury expressed in centimeters?

10.34 An open-end mercury manometer was connected to a flask containing a gas at an unknown pressure. The mercury in the arm open to the atmosphere was 65 mm higher than the mercury in the arm connected to the flask. The atmospheric pressure was 748 torr. What was the pressure of the gas in the flask (in torr)?

10.35 An open-end mercury manometer was connected to a flask containing a gas at an unknown pressure. The mercury in the arm open to the atmosphere was 82 mm lower than the mercury in the arm connected to the flask. The atmospheric pressure was 752 torr. What was the pressure of the gas in the flask (in torr)?

10.36 A student carried out a reaction using a bulb connected to an open-end mercury manometer on a day when the pressure was 746 torr. Before the reaction, the level of the mercury in both arms was at a height of 12.50 cm, measured by a meter stick placed between the two arms of the U-shaped tube. After the reaction, the level in the arm nearest the bulb was at 8.50 cm. Did the reaction produce or consume gases? What was the final pressure in the bulb (in torr)?

10.37 In another experiment, using the same apparatus as described in the previous problem, the mercury level in the arm nearest the bulb stood at 10.20 cm after the reaction. Before the reaction both levels were at 12.50 cm. The pressure in the lab that day was 741 torr. Did the reaction produce or consume gases? What was the final pressure in the bulb (in torr)?

10.38 Suppose that in a closed-end manometer the mercury in the closed arm was 12.5 cm higher than the mercury in the arm connected to a vessel containing a gas. What is the pressure of the gas expressed in torr?

10.39 Suppose a gas is in a vessel connected to both an open-end and a closed-end manometer. The difference in heights of the mercury in the closed-end manometer was 236 mm, while in the open-end manometer the mercury level in the arm open to the atmosphere was 512 mm below the level in the arm connected to the vessel. Calculate the atmospheric pressure. (It may help to sketch the apparatus.)

10.40 Suppose you were to construct a barometer using a fluid with a density of 1.22 g mL^{-1}. How high would the liquid level be in this barometer if the atmospheric pressure was 755 torr? (Mercury has a density of 13.6 g mL^{-1}.)

10.41 A suction pump works by creating a vacuum into which some fluid will rise. If the pump were able to create a perfect vacuum (i.e., giving a pressure equal to 0 torr), what is the maximum height, in meters, that water could be raised if the atmospheric pressure is 765 torr? Suppose that you work for a construction company in Florida and are told to use a suction pump to remove water from a pit. The water level is over 35 ft below the site where you must place the pump. Can *any* suction pump be expected to raise water to this height at sea level? (The density of water is 1.00 g mL^{-1}; that of mercury is 13.6 g mL^{-1}.)

Gas Laws for a Fixed Amount of Gas

10.42 A gas has a volume of 255 mL at 725 torr. What volume will the gas occupy at 365 torr if the temperature of the gas doesn't change?

10.43 A bicycle pump has a barrel that is 75.0 cm long (about 30 in.). If air is drawn into the pump at a pressure of 1.00 atm during the upstroke, how long must the downstroke be, in centimeters, to raise the pressure of the air to 5.50 atm (approximately the pressure in the tire of a 10-speed bike)? Assume no change in the temperature of the air.

10.44 A gas has a volume of 3.86 L at a temperature of 45 °C. What will the volume of the gas be if its temperature is raised to 80 °C while its pressure is kept constant?

10.45 A balloon has a volume of 2.50 L indoors at a temperature of 22 °C. If the balloon is taken outdoors on a cold day when the air temperature is −15 °C (5 °F), what will its volume be in liters? Assume constant air pressure within the balloon.

10.46 A sample of a gas has a pressure of 850 torr at 285 °C. To what Celsius temperature must the gas be heated to double its pressure if there is no change in the volume of the gas?

10.47 Before taking a trip, you check the air in a tire of your automobile and find it has a pressure of 45 lb in.$^{-2}$ on a day when the air temperature is 10 °C (50 °F). After traveling some distance, you find that the temperature of the air in the tire has risen to 43 °C (approximately 110 °F). What is the air pressure in the tire at this higher temperature, expressed in units of lb in.$^{-2}$?

10.48 A sample of helium at a pressure of 740 torr and in a volume of 2.58 L was heated from 24.0 to 75.0 °C. The volume of the container expanded to 2.81 L. What was the final pressure (in torr) of the helium?

10.49 When a sample of neon with a volume of 648 mL and a pressure of 0.985 atm was heated from 16.0 to 63.0 °C, its volume became 689 mL. What was its final pressure (in atm)?

10.50 What must be the new volume of a sample of nitrogen (in L) if 2.68 L at 745 torr and 24.0 °C is heated to 375.0 °C under conditions that let the pressure change to 760 torr?

10.51 When 280 mL of oxygen at 741 torr and 18.0 °C was warmed to 33.0 °C, the pressure became 760 torr. What was the final volume (in mL)?

10.52 A sample of argon with a volume of 6.18 L, a pressure of 761 torr, and a temperature of 20.0 °C expanded to a volume of 9.45 L and a pressure of 373 torr. What was its final temperature in °C?

10.53 A sample of a refrigeration gas in a volume of 455 mL, at a pressure of 1.51 atm, and at a temperature of 25.0 °C was compressed into a volume of 220 mL with a pressure of 2.00 atm. To what temperature (in °C) did it have to change?

Ideal Gas Law
10.54 What would be the value of the gas constant R in units of *mL torr mol^{-1} K^{-1}*?

10.55 The SI generally uses its base units to compute constants involving derived units. The SI unit for volume, for example, is the cubic meter, called the *stere*, because the meter is the base unit of length. We learned about the SI unit of pressure, the pascal, in this chapter. The temperature unit is the kelvin. Calculate the value of the gas constant in the SI units *m^3 Pa mol^{-1} K^{-1}*.

10.56 What volume in liters does 0.136 g of O_2 occupy at 20.0 °C and 748 torr?

10.57 What volume in liters does 1.67 g of N_2 occupy at 22.0 °C and 756 torr?

10.58 What pressure (in torr) is exerted by 10.0 g of O_2 in a 2.50 L container at a temperature of 27 °C?

10.59 If 12.0 g of water is converted to steam in a 3.60 L pressure cooker held at a temperature of 108 °C, what pressure would be produced?

10.60 A sample of carbon dioxide has a volume of 26.5 mL at 20.0 °C and 624 torr. How many grams of CO_2 are in the sample?

10.61 Methane is formed in landfills by the action of certain bacteria on buried organic matter. If a sample of methane collected from a landfill has a volume of 250 mL at 750 torr and 27 °C, how many grams of methane are in the sample?

10.62 A sample of 4.18 mol of H_2 at 18.0 °C occupies a volume of 24.0 L. Under what pressure, in atmospheres, is this sample?

10.63 If a steel cylinder with a volume of 1.60 L contains 10.0 mol of oxygen, under what pressure (in atm) is the oxygen if the temperature is 25.0 °C?

10.64 To three significant figures, calculate the density in g L^{-1} of the following gases at STP: (a) C_2H_6 (ethane), (b) N_2, (c) Cl_2, (d) Ar.

10.65 To three significant figures, calculate the density in g L^{-1} of the following gases at STP: (a) Ne, (b) O_2, (c) CH_4 (methane), (d) CF_4.

10.66 What density (in g L^{-1}) does oxygen have at 24.0 °C and 742 torr?

10.67 At 748.0 torr and 20.65 °C, what is the density of argon (in g L^{-1})?

10.68 A chemist isolated a gas in a glass bulb with a volume of 255 mL at a temperature of 25.0 °C and a pressure (in the bulb) of 10.0 torr. The gas weighed 12.1 mg. What is the molecular mass of this gas?

10.69 Boron forms a variety of unusual compounds with hydrogen. A chemist isolated 6.3 mg of one of the boron hydrides in a glass bulb with a volume of 385 mL at 25.0 °C and a bulb pressure of 11 torr.
(a) What is the molecular mass of this hydride?
(b) Which of the following is likely to be its molecular formula, BH_3, B_2H_6, or B_4H_{10}?

10.70 At 22.0 °C and a pressure of 755 torr, a gas was found to have a density of 1.13 g L^{-1}. Calculate its molecular mass.

10.71 A gas was found to have a density of 0.08747 mg mL^{-1} at 17.0 °C and a pressure of 760 torr. What is its molecular mass? Can you tell what the gas most likely is?

Stoichiometry of Reactions of Gases
10.72 How many liters of F_2 at STP are needed to react with 4.00 L of H_2, also at STP, in the following reaction?

$$H_2(g) + F_2(g) \longrightarrow 2HF(g)$$

10.73 In the Haber process for the synthesis of ammonia,

$$N_2(g) + 3H_2(g) \longrightarrow 2NH_3(g)$$

how many liters of N_2 are needed to react completely with 45.0 L of H_2, if the volumes of both gases are measured at STP?

10.74 How many milliliters of oxygen are required to react completely with 175 mL of C_4H_{10} if the volumes of both gases are measured at the same temperature and pressure? The reaction is

$$2C_4H_{10}(g) + 13O_2(g) \longrightarrow 8CO_2(g) + 10H_2O(g)$$

10.75 How many milliliters of O_2 are consumed in the complete combustion of a sample of hexane, C_6H_{14}, if the reaction produces 855 mL of CO_2? Assume all gas volumes are measured at the same temperature and pressure. The reaction is

$$2C_6H_{14}(g) + 19O_2(g) \longrightarrow 12CO_2(g) + 14H_2O(g)$$

10.76 Propylene, C_3H_6, reacts with hydrogen under pressure to give propane, C_3H_8:

$$C_3H_6(g) + H_2(g) \longrightarrow C_3H_8(g)$$

How many liters of hydrogen (at 740 torr and 24 °C) react with 18.0 g of propylene?

10.77 Nitric acid is formed when NO_2 is dissolved in water.

$$3NO_2(g) + H_2O(l) \longrightarrow 2HNO_3(aq) + NO(g)$$

How many milliliters of NO_2 at 25 °C and 752 torr are needed to form 12.0 g of HNO_3?

10.78 How many milliliters of O_2 measured at 27 °C and 654 torr are needed to react completely with 16.8 mL of CH_4 measured at 35 °C and 725 torr?

10.79 How many milliliters of H_2O vapor, measured at 318 °C and 735 torr, are formed when 33.6 mL of NH_3 at 825 torr and 127 °C react with oxygen according to the following equation?

$$4NH_3(g) + 3O_2(g) \longrightarrow 2N_2(g) + 6H_2O(g)$$

10.80 Calculate the maximum number of milliliters of CO_2, at 745 torr and 27 °C, that could be formed in the combustion of carbon monoxide if 300 mL of CO at 683 torr and 25 °C is mixed with 150 mL of O_2 at 715 torr and 125 °C.

10.81 A mixture of ammonia and oxygen is prepared by combining 300 mL of N_2 (measured at 750 torr and 28 °C) with 220 mL of O_2 (measured at 780 torr and 50 °C). How many milliliters of N_2 (measured at 740 torr and 100 °C) could be formed if the following reaction occurs?

$$4NH_3(g) + 3O_2(g) \longrightarrow 2N_2(g) + 6H_2O(g)$$

Dalton's Law of Partial Pressures

10.82 A 1.00 L container was filled by pumping into it 1.00 L of N_2 at 200 torr, 1.00 L of O_2 at 150 torr, and 1.00 L of He at 300 torr. All volumes and pressures were measured at the same temperature. What was the total pressure inside the container after the mixture was made?

10.83 A mixture of N_2, O_2, and CO_2 has a total pressure of 740 torr. In this mixture the partial pressure of N_2 is 120 torr and the partial pressure of O_2 is 400 torr. What is the partial pressure of the CO_2?

10.84 A 22.4 L container at 0 °C contains 0.30 mol N_2, 0.20 mol O_2, 0.40 mol He, and 0.10 mol CO_2. What are the partial pressures of each of the gases?

10.85 A 0.200 mol sample of a mixture of N_2 and CO_2 with a total pressure of 840 torr is exposed to solid CaO which reacts with CO_2 according to the equation

$$CaO(s) + CO_2(g) \longrightarrow CaCO_3(s)$$

After reaction was complete, the pressure of the gas had dropped to 320 torr. How many moles of CO_2 were in the original mixture?

10.86 A sample of carbon monoxide was prepared and collected over water at a temperature of 20 °C and a total pressure of 754 torr. It occupied a volume of 268 mL. Calculate the partial pressure of the CO in torr as well as its dry volume (in mL) under a pressure of 1.00 atm and a temperature of 20 °C.

10.87 A sample of hydrogen was prepared and collected over water at 25 °C and a total pressure of 742 torr. It occupied a volume of 288 mL. Calculate its partial pressure (in torr) and what its dry volume would be (in mL) under a pressure of 1.00 atm.

10.88 What volume of "wet" methane would you have to collect at 20.0 °C and 742 torr to be sure that the sample contains 244 mL of dry methane (also at 742 torr)?

10.89 What volume of "wet" oxygen would you have to collect if you need the equivalent of 275 mL of dry oxygen at 1.00 atm? (The atmospheric pressure in the lab is 746 torr.) The oxygen is to be collected over water at 15.0 °C.

10.90 What are the mole percents of the components of air inside the lungs when they have the following partial pressures? For N_2, 570 torr; O_2, 103 torr; CO_2, 40 torr; and water vapor, 47 torr.

10.91 Assuming no other components are present, calculate the mole percents of oxygen and nitrogen in the air at the following places.
(a) At the top of Mt. Everest (elevation 8.8 km) on a day when $P_{N_2} = 197$ torr and $P_{O_2} = 53$ torr.
(b) At sea level and 1 atm pressure, with $P_{N_2} = 600$ torr and $P_{O_2} = 160$ torr.
(c) Comparing the answers calculated for parts (a) and (b), what has to be the reason why it is hard to breathe without supplemental oxygen at high altitudes?

Graham's Law

10.92 Under conditions in which the density of CO_2 is 1.96 g L^{-1} and that of N_2 is 1.25 g L^{-1}, which gas will effuse more rapidly? What will be the ratio of the rates of effusion of N_2 to CO_2?

10.93 At 25 °C, which gas diffuses more rapidly, N_2, CH_4, or CO_2?

10.94 An unknown gas X effuses 1.65 times faster than C_3H_8. What is the molecular mass of gas X?

10.95 Uranium hexafluoride is a white solid that readily passes directly into the vapor state. (Its vapor pressure at 20.0 °C is 120 torr.) A trace of the uranium in this compound—about 0.7%—is uranium-235, which can be used in a nuclear power plant. The rest of the uranium is essentially uranium-238, and its presence interferes with these applications for uranium-235. Gas effusion of UF_6 can be used to separate the fluoride made from uranium-235 and the fluoride made from uranium-238. Which hexafluoride effuses more rapidly? By how much? (The very small difference that you will find is actually enough, although repeated effusions are necessary.)

ADDITIONAL EXERCISES

10.96 One of the oldest units for atmospheric pressure is lb in.$^{-2}$. Calculate the numerical value of the standard atmosphere in these units to three significant figures. Calculate the mass in pounds of a uniform column of water 33.9 ft high having an area of 1.00 in.^{2} at its base. (Use the following data: density of mercury = 13.6 g mL^{-1}; density of water = 1.00 g mL^{-1}; 1 mL = 1 cm^3; 1 lb = 454 g; 1 in. = 2.54 cm.)

*__**10.97**__ A typical automobile has a weight of approximately 3500 lb. If the vehicle is to be equipped with tires, each of which will contact the pavement with a "footprint" that is 6.0 in. wide by 3.2 in. long, what must the gauge pressure of the air be in each tire? (Gauge pressure is the amount that the gas pressure exceeds atmospheric pressure. Assume that atmospheric pressure is 14.7 lb in.$^{-2}$.)

10.98 Suppose you were planning to move a house by transporting it on a large trailer. The house has an estimated weight of 45.6 tons (1 ton = 2000 lb). The trailer is expected to weigh 8.3 tons. Each wheel of the trailer will have tires inflated to a gauge pressure of 85 psi (which is actually 85 psi above atmospheric pressure). If the area of contact between a tire and the pavement can be no larger than 100 in.^{2} (10 in. × 10 in.), what is the minimum number of wheels the trailer must have? (Remember, tires are mounted in multiples of two on a trailer. Assume that atmospheric pressure is 14.7 psi.)

*__**10.99**__ The motion picture *Titanic* described the tragedy of the collision of the ocean liner of the same name with an iceberg in the North Atlantic. The ship sank soon after the collision on April 14, 1912, and now rests on the seafloor at a depth of 12,468 ft. Recently, the wreck was explored by the research vessel *Nautile,* which has successfully recovered a variety of items from the debris field surrounding the sunken ship. Calculate the pressure in atmospheres and pounds per square inch exerted on the hull of the *Nautile* as it explores the seabed surrounding the *Titanic.* (Seawater has a density of approximately 1.025 g mL^{-1}; mercury has a density of 13.6 g mL^{-1}; 1 atm = 14.7 lb in.$^{-2}$.)

*__**10.100**__ Two flasks (which we will refer to as Flask 1 and Flask 2) are connected to each other by a U-shaped tube filled with an oil having a density of 0.826 g mL^{-1}. The oil level in the arm connected to Flask 2 is 16.24 cm higher than in the arm connected to Flask 1. Flask 1 is also connected to an open-end mercury manometer. The mercury level in the arm open to the atmosphere is 12.26 cm higher than the level in the arm connected to Flask 1. The atmospheric pressure is 0.827 atm. What is the pressure of the gas in Flask 2 expressed in torr?

*__**10.101**__ A bubble of air escaping from a diver's mask rises from a depth of 100 ft to the surface where the pressure is 1.00 atm. Initially, the bubble has a volume of 10.0 mL. Assuming none of the air dissolves in the water, how many times larger is the bubble just as it reaches the surface? Use your answer to explain why scuba divers constantly exhale as they slowly rise from a deep dive. (The density of seawater is approximately 1.025 g mL^{-1}; the density of mercury is 13.6 g mL^{-1}.)

*__**10.102**__ In a diesel engine, the fuel is ignited when it is injected into hot compressed air, heated by the compression itself. In a typical high-speed diesel engine, the chamber in the cylinder has a diameter of 10.7 cm and a length of 13.4 cm. On compression, the length of the chamber is shortened by 12.7 cm (a "5-inch stroke"). The compression of the air changes its pressure from 1.00 to 34.0 atm. The temperature of the air before compression is 364 K. As a result of the compression, what will be the final air temperature (in K and °C) just before the fuel injection?

*__**10.103**__ Early one cool (60.0 °F) morning you start on a bike ride with the atmospheric pressure at 14.7 lb in.$^{-2}$ and the tire gauge pressure at 50.0 lb in.$^{-2}$. (Gauge pressure is the amount that the pressure exceeds atmospheric pressure.) By late afternoon, the air had warmed up considerably, and this plus the heat generated by tire friction sent the temperature inside the tire to 104 °F. What will the tire gauge now read, assuming that the volume of the air in the tire and the atmospheric pressure have not changed?

10.104 The range of temperatures over which an automobile tire must be able to withstand pressure changes is roughly − 50 to 120 °F. If a tire is filled to 35 lb in.$^{-2}$ at − 50 °F (on a cold day in Alaska, for example), what will be the pressure in the tire (in the same pressure units) on a hot day in Death Valley when the temperature is 120 °F? (Assume that the volume of the tire does not change.)

10.105 Chlorine reacts with sulfite ion to give sulfate ion and chloride ion. How many milliliters of Cl_2 gas measured at 25 °C and 734 torr are required to react with all the SO_3^{2-} in 50.0 mL of 0.200 M Na_2SO_3 solution?

10.106 Ammonia effuses at a rate that is 2.93 times larger than that of an unknown gas. What is the molecular mass of the unknown?

10.107 A common laboratory preparation of hydrogen on a small scale uses the reaction of zinc with hydrochloric acid. Zinc chloride is the other product.
(a) Write the balanced equation for the reaction.
(b) If 12.0 L of H_2 at 760 torr and 20.0 °C is wanted, how many grams of zinc are needed, in theory?
(c) If the acid is available as 8.00 *M* HCl, what is the minimum volume of this solution (in milliliters) required to produce the amount of H_2 described in part (b)?

*****10.108** In an experiment designed to prepare a small amount of hydrogen by the method described in the preceding problem, a student was limited to using a gas-collecting bottle with a maximum capacity of 335 mL. The method involved collecting the hydrogen over water. What are the minimum number of grams of Zn and the minimum number of milliliters of 6.00 *M* HCl needed to produce the *wet* hydrogen that can exactly fit this collecting bottle at 740 torr and 25.0 °C?

10.109 Carbon dioxide can be made in the lab by the reaction of hydrochloric acid with calcium carbonate.

$$CaCO_3(s) + 2HCl(aq) \longrightarrow CaCl_2(aq) + H_2O(l) + CO_2(g)$$

How many milliliters of dry CO_2 at 20.0 °C and 745 torr can be prepared from a mixture of 12.3 g of $CaCO_3$ and 185 mL of 0.250 *M* HCl?

10.110 A mixture was prepared in a 500 mL reaction vessel from 300 mL of O_2 (measured at 25 °C and 740 torr) and 400 mL of H_2 (measured at 45 °C and 1250 torr). The mixture was ignited and the H_2 and O_2 reacted to form water. What was the final pressure inside the reaction vessel after the reaction was over if the temperature was held at 120 °C?

10.111 A student collected 18.45 mL of H_2 over water at 24 °C. The water level inside the collection apparatus was 8.5 cm higher than the water level outside. The barometric pressure was 746 torr. How many grams of zinc had to react with HCl(aq) to produce the H_2 that was collected?

10.112 A mixture of gases is prepared from 87.5 g of O_2 and 12.7 g of H_2. After the reaction of O_2 and H_2 is complete, what is the total pressure of the mixture if its temperature is 160 °C and its volume is 12.0 L? What are the partial pressures of the gases remaining in the mixture?

10.113 In many countries, the fertilizer ammonia is made using methane as the source of hydrogen (and energy). The overall process—and there are several steps—is by the following equation, where N_4O is used as an approximate formula for air, another raw material.

$$7CH_4 + 10H_2O + 4N_4O \longrightarrow 16NH_3 + 7CO_2$$

Methane is sold in units of *tcf,* where 1 tcf = 1×10^3 ft^3 = 28.3×10^3 L at STP.

(a) What volume of NH_3 (in liters at STP) can be made by this process from 1.00 tcf of methane (also at STP)?
(b) One thousand cubic feet of methane at STP represents how many kilograms of ammonia?

10.114 A sample of an unknown gas with a mass of 3.620 g was made to decompose into 2.172 g of O_2 and 1.448 g of S. Prior to the decomposition, this sample occupied a volume of 1120 mL at 750 torr and 25.0 °C.
(a) What is the percentage composition of the elements in this gas?
(b) What is the empirical formula of the gas?
(c) What is its molecular formula?

*****10.115** A sample of a new antimalarial drug with a mass of 0.2394 g was made to undergo a series of reactions that changed all of the nitrogen in the compound into N_2. This gas had a volume of 18.90 mL when collected over water at 23.80 °C and a pressure of 746.0 torr. At 23.80 °C, the vapor pressure of water is 22.110 torr.
(a) Calculate the percentage of nitrogen in the sample.
(b) When 6.478 mg of the compound was burned in pure oxygen, 17.57 mg of CO_2 and 4.319 mg of H_2O were obtained. What are the percentages of C and H in this compound? Assuming that any undetermined element is oxygen, write an empirical formula for the compound.
(c) The molecular mass of the compound was found to be 324. What is its molecular formula?

*****10.116** In one analytical procedure for determining the percentage of nitrogen in unknown compounds, weighed samples are made to decompose to N_2, which is collected over water at known temperatures and pressures. The volumes of N_2 are then translated into grams and then into percentages.
(a) Show that the following equation can be used to calculate the percentage of nitrogen in a sample having a mass of W grams when the N_2 has a volume of V mL and is collected over water at t_C °C at a total pressure of P torr. The vapor pressure of water occurs in the equation as $P^\circ_{H_2O}$.

$$\text{Percentage N} = 0.04489 \times \frac{V(P - P^\circ_{H_2O})}{W(273 + t_C)}$$

(b) Use this equation to calculate the percentage of nitrogen in the sample described in the preceding problem.

10.117 The "rotten-egg" odor is caused by hydrogen sulfide, H_2S. Most people can detect it at a concentration of 0.15 ppb (parts per billion), meaning 0.15 L of H_2S in 10^9 L of space. A typical student lab is 40 × 20 × 8 ft.
(a) At STP, how many liters of H_2S could be present in a typical lab to have a concentration of 0.15 ppb?
(b) How many milliliters of 0.100 *M* Na_2S would be needed to generate the amount of H_2S in part (a) by the following reaction with hydrochloric acid?

$$Na_2S(aq) + 2HCl(aq) \longrightarrow H_2S(g) + 2NaCl(aq)$$

Children soon learn that moist sand sticks together so they can build sand castles, but dry sand or sand under water will not. Surface tension is the property that's responsible for this, and it is one of the properties of liquids we discuss in this chapter.

Intermolecular Attractions and the Properties of Liquids and Solids

This Chapter in Context The *physical* properties of substances often concern us more on a day-to-day basis than *chemical* properties. This is especially true when substances are in their liquid or solid states (their *condensed states*). One of the goals of chemistry is to provide an understanding of how chemical composition and molecular structure determine the physical properties of matter. By studying the relationships involved, we have acquired a more complete knowledge of the structure of matter and an ability (at least to some degree) to design materials that have the kinds of physical properties we want. Examples of success include materials for transistors and computer chips and the many plastics that have come to replace traditional materials such as wood, steel, and natural rubber.

We began our study of the physical properties of substances in the last chapter where we discussed the properties of gases. In this chapter we will investigate the physical properties of the other two states of matter—liquids and solids. We shall also see what happens when substances change from one state to another.

11.1 Why Gases Differ from Liquids and Solids

As you learned in Chapter 10, virtually all gases, regardless of their chemical composition, obey the same set of gas laws rather well. A useful consequence of this, for example, is that we can use the ideal gas law to calculate molecular masses of gases. Such calculations rely on the fact that one mole of any gas, regardless of its composition, occupies very nearly the same volume at a given temperature and pressure.

There are two reasons why all gases behave so nearly alike. One is that the molecules themselves occupy an extremely small volume compared to the total volume of the gas, so a gas is almost entirely empty space. The other is that the forces of attraction between the gas molecules are uniformly so weak that they have hardly any influence on gas properties. We don't find similar "liquid laws" or "solid laws" because these two factors can't be ignored in liquids and solids; they play a very important role in determining the properties of these "condensed" states of matter.

467

In liquids and solids, the molecules are packed together very tightly and there is little empty space between. As a result, these states of matter are nearly *incompressible,* which means their volumes change very little when they are subjected to high pressures. Gases, on the other hand, are easily compressed, so this is one of the most obvious differences between liquids and gases and between solids and gases.

Most of the physical properties of gases, liquids, and solids are actually controlled by the strengths of **intermolecular attractions,** which are the attractive forces that exist between neighboring particles. In liquids and solids, these forces are much stronger than in gases. But why?

If you have ever played with magnets you know that the attractive force between them rapidly becomes weaker as the distance between the magnets increases. Thus, when two magnets are near each other, the attraction can be quite strong, but if they are far apart hardly any attraction is felt at all. This same phenomenon applies to the attractions between molecules as well.

As we will see, chemical composition influences intermolecular attractions to a large extent. In gases, though, the molecules are so far apart that the attractive forces are almost negligible, so any differences between them hardly matter at all. As a result, chemical composition has little effect on the properties of a gas. But in a liquid or a solid, the molecules are close together and the attractions are strong. Differences among these attractions caused by differences in chemical makeup are amplified, so the properties of liquids and solids depend quite heavily on chemical composition.

> The closer two molecules are, the more strongly they attract each other.

11.2 Intermolecular Attractions

As we noted in the preceding discussion, most of the *physical* properties of substances are controlled by the strengths of intermolecular attractions. Before we study how these forces arise, it is important to understand that the attractions *between* molecules (the *inter*molecular forces) are always much weaker than the attractions between atoms *within* molecules (*intra*molecular forces, which are the chemical bonds that hold molecules together). In a molecule of HCl, for example, the H and Cl atoms are held very tightly to each other by a covalent bond, and it is the strength of this bond that affects the *chemical properties* of HCl. The strength of the chemical bond also keeps the molecule intact as it moves about. When a particular chlorine atom moves, the hydrogen atom bonded to it is forced to follow along (see Figure 11.1). Attractions between neighboring HCl molecules, in contrast, are much weaker. These weaker attractions are what determine the *physical properties* of HCl.

From the preceding discussion, we can see that to understand and explain the physical properties of liquids and solids, we need to learn about the kinds and relative strengths of intermolecular attractions. These forces all arise from attractions between opposite charges and are classified according to how they originate.

> The force between two electrical charges is called the *electrostatic force* or *coulombic force.* Other forces in nature are the magnetic, gravitational, nuclear strong, and nuclear weak forces.

Tools

Intermolecular attractions— dipole–dipole forces

δ+ δ−
H—Cl

Dipole–Dipole Attractions

In Chapter 8 you saw that the HCl molecule is an example of a *polar molecule,* one with a partial positive charge at one end and a partial negative charge at the other. Because unlike charges attract, polar molecules tend to line up so that the positive end of one dipole is near the negative end of another. Thermal energy (molecular kinetic energy), however, causes the molecules to collide and become disoriented, so the alignment isn't perfect. Nevertheless, there is still a net attrac-

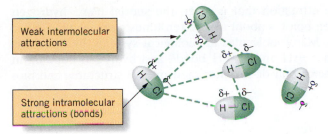

Weak intermolecular attractions

Strong intramolecular attractions (bonds)

Figure 11.1 *Attractions within and between hydrogen chloride molecules.* Strong *intramolecular* attractions (chemical bonds) exist between H and Cl atoms within HCl molecules. These attractions control the chemical properties of HCl. Weaker *intermolecular* attractions exist between neighboring HCl molecules. The intermolecular attractions control the physical properties of this substance.

tion between polar molecules (see Figure 11.2). We call this kind of intermolecular attractive force a **dipole–dipole attraction.** It is generally a much weaker force than a covalent bond, being only about 1% as strong. Dipole–dipole attractions fall off rapidly with distance, with the energy required to separate a pair of dipoles being proportional to $1/d^3$, where d is the distance between the dipoles.

There are two principal reasons for the relative weakness of dipole–dipole attractions. One is that the charges associated with dipoles are only *partial* charges, not full charges. The other reason, which we have already discussed, is that at ordinary temperatures (e.g., room temperature) collisions between molecules cause the dipoles to be somewhat misaligned, thereby reducing the effectiveness of the attractions.

Hydrogen Bonds

A particularly important kind of dipole–dipole attraction occurs when hydrogen is covalently bonded to a very small, highly electronegative atom, principally fluorine, oxygen, and nitrogen. In this situation, unusually strong dipole–dipole attractions are often observed, for which there are two reasons. First, F—H, O—H, and N—H bonds are very polar. Because of the large electronegativity differences, the ends of the bond dipoles carry substantial amounts of positive and negative charge. Second, the charges are highly concentrated because of the small sizes of the atoms involved. And third, the positive end of one dipole can get quite close to the negative end of another, also because of the small sizes of the atoms. All of these factors combine to produce an exception-

Tools

Intermolecular attractions— hydrogen bonding

Relative Electronegativities

F	4.1
O	3.5
N	3.1
H	2.1

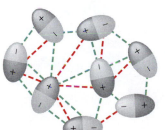

Attractions (– –) are greater than repulsions (– –), so the molecules feel a net attraction to each other.

Figure 11.2 *Dipole–dipole attractions.* Attractions between polar molecules occur because the molecules tend to align themselves so that opposite charges are near each other and like charges are as far apart as possible. The alignment is not perfect because the molecules are constantly moving and colliding.

Hydrogen bonding between water molecules in ice causes its molecules to be farther apart in the solid than in liquid water. This makes ice less dense than liquid water, which is why icebergs like this one float.

ally strong dipole–dipole attraction that is given the special name **hydrogen bond.** Typically, a hydrogen bond is about five to ten times stronger than other dipole–dipole attractions. Many molecules in biological systems, like proteins and nucleic acids, contain N—H and O—H bonds, and hydrogen bonding in these substances is one factor that determines their overall structures and biological functions.

Hydrogen Bonds in Water

The strength of the hydrogen bond gives water some very unusual and special properties. Most liquids become more dense when they change to their solid forms. Not so with water. In liquid water, the molecules experience hydrogen bonds that constantly break and re-form as the molecules move around (see Figure 11.3*b*). As water freezes, however, the molecules become locked in place, and each water molecule participates in four hydrogen bonds (Figure 11.3*c*). The resulting structure occupies a larger volume than the same amount of liquid water, so ice is less dense than the liquid. Because of this, ice cubes and icebergs float in the more dense liquid (much to the distress of the captain of the *Titanic*).

Tools

Intermolecular attractions— London forces

London Forces

The attractions between polar molecules such as HCl and SO_2 are fairly easy to understand. But attractive forces occur even between the particles of nonpolar substances, such as the atoms of the noble gases and the nonpolar molecules of Cl_2 and CH_4. These nonpolar substances can also be condensed to liquids and even to solids if they are cooled to low enough temperatures. Therefore, attractions between their particles, although weak, must exist to cause them to cling together. Electronegativity differences cannot be the origin of these attractions, so what is?

In 1930 Fritz London, a German physicist, explained how the particles in even nonpolar substances can experience intermolecular attractions. He noted that in any atom or molecule the electrons are constantly moving. If we could examine such motions in two neighboring particles, we would find that the movement of electrons in one influences the movement of electrons in the

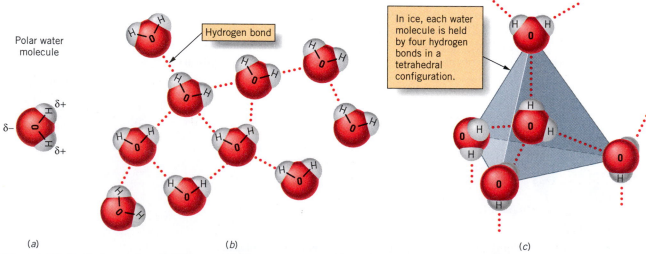

(a)   *(b)*   *(c)*

Figure 11.3 *Hydrogen bonding in water.* (*a*) The polar water molecule. (*b*) Hydrogen bonding produces strong attractions between water molecules in the liquid.

(*c*) Hydrogen bonding (dotted lines) between water molecules in ice, where each water molecule is held by four hydrogen bonds in a tetrahedral configuration.

other. This is because electrons repel each other and tend to stay as far apart as they can. Therefore, as an electron of one particle gets near the other particle, electrons on the second particle are pushed away. This happens continually as the electrons move around, so to some extent, the electron density in both particles flickers back and forth in a synchronous fashion. This is illustrated in Figure 11.4, which depicts a series of instantaneous "frozen" views of the electron density. Notice that *at any given moment the electron density of a particle can be unsymmetrical,* with more negative charge on one side than on the other. For that particular instant, the particle is a dipole, and we call it a momentary dipole or **instantaneous dipole.**

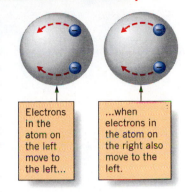

Electrons in the atom on the left move to the left...

...when electrons in the atom on the right also move to the left.

London forces are also called **dispersion forces.**

As an instantaneous dipole forms in one particle, it causes the electron density in its neighbor to become unsymmetrical, too. As a result, this second particle also becomes a dipole. We call it an **induced dipole** because it is caused by, or *induced* by, the formation of the first dipole. Because of the way the dipoles are formed, they always have the positive end of one near the negative end of the other, so there is a dipole–dipole attraction between them. It is a very short-lived attraction, however, because the electrons keep moving; the dipoles vanish as quickly as they form. But, in another moment, the dipoles will reappear in a different orientation and there will be another brief dipole–dipole attraction. In this way the short-lived dipoles cause momentary tugs between the particles. When averaged over a period of time, there is a net, overall attraction. It tends to be relatively weak, however, because the attractive forces are only "turned on" part of the time.

The momentary dipole–dipole attractions that we've just discussed are called *instantaneous dipole–induced dipole attractions,* or **London forces,** to distinguish them from the attractions that exist continuously between permanent dipoles in polar substances like HCl.

London forces exist between all molecules and ions. Although they are the only kind of attraction possible between nonpolar molecules, London forces also contribute significantly to the total intermolecular attraction between polar molecules, where they are present in addition to the regular dipole–dipole attractions. London forces even occur between oppositely charged ions, but their effects are relatively weak compared to ionic attractions. London forces contribute little to the net overall attractions between ions and are often ignored.

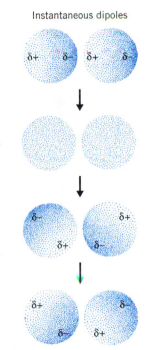

Instantaneous dipoles

Figure 11.4 *Instantaneous "frozen" views of the electron density in two neighboring particles.* Attractions occur between the instantaneous dipoles while they exist.

The Strengths of London Forces

To compare the strengths of intermolecular attractions, a property we can use is boiling point. As we will explain in more detail later in this chapter, the higher the boiling point, the stronger are the attractions between molecules in the liquid.

The strengths of London forces are found to depend chiefly on three factors. One is the **polarizability** of the electron cloud of a particle, which is a measure of the ease with which the electron cloud is distorted, and thus is a measure of the ease with which the instantaneous and induced dipoles can form. In general, as the size of the electron cloud increases, its polarizability also increases. When an electron cloud is large, the outer electrons are generally not held very tightly by the nucleus (or nuclei, if the particle is a molecule). This makes the electron cloud "mushy" and rather easily deformed (see Figure 11.5). Particles with large electron clouds therefore experience stronger London forces than do similar particles with small electron clouds.

The effects of size can be seen if we compare the boiling points of the halogens or the noble gases (see Table 11.1). Going down Group VIIIA, for example, the noble gas atoms become larger and the boiling points increase, reflecting

London forces decrease very rapidly as the distance between particles increases. The energy required to separate particles held by London forces varies as $1/d^6$, where d is the distance between the particles.

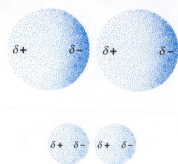

Table 11.1 Boiling Points of the Halogens and Noble Gases

Group VIIA	Boiling Point (°C)	Group VIIIA	Boiling Point (°C)
F_2	−188.1	He	−268.6
Cl_2	−34.6	Ne	−245.9
Br_2	58.8	Ar	−185.7
I_2	184.4	Kr	−152.3
		Xe	−107.1
		Rn	−61.8

Figure 11.5 *Effect of molecular size on the strengths of London forces. A large electron cloud is more easily deformed than a small one, so in a large molecule the charges on opposite ends of an instantaneous dipole are larger than in a small molecule. Large molecules therefore experience stronger London forces than small molecules.*

The effect of large numbers of atoms on the total strengths of London forces can be compared to the bond between loop and hook layers of the familiar product Velcro. Each loop-to-hook attachment is fairly weak, but when large numbers of them are involved, the overall bond between Velcro layers is quite strong.

stronger intermolecular attractions (stronger London forces) with the particles having larger electron clouds. A similar trend is found among the nonpolar halogen molecules in Group VIIA.

Another factor that affects the strengths of London forces is the number of atoms in a molecule. For molecules containing the same elements, London forces increase with the number of atoms, as illustrated by the hydrocarbons (see Table 11.2). Their molecules consist of chains of carbon atoms with hydrogen atoms bonded to them all along the chain. If we compare two hydrocarbon molecules of different chain length—for example, C_3H_8 and C_6H_{14}—the one having the longer chain also has the higher boiling point. This means that the molecule with the longer chain length experiences the stronger intermolecular attractive forces. They are stronger between the longer molecules because there are more places along their lengths where instantaneous dipoles can develop and lead to London attractions to other molecules (see Figure 11.6). Even if the strength of attraction at each location is about the same, the *total* attraction experienced between the longer C_6H_{14} molecules is greater than that felt between shorter C_3H_8 molecules. In Table 11.2 you can see that some of the hydrocarbons have boiling points that are quite high, which shows that the cumulative effects of London forces can be very strong attractions.

The third factor that affects the strengths of London forces is molecular shape. Even with molecules that have the same number of the same kinds of atoms, the shapes of the molecules can have an influence. For example, compare the molecules shown in Figure 11.7. Both have the same molecular

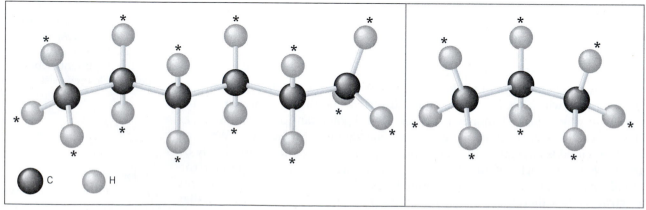

Figure 11.6 *The number of atoms in a molecule affects London forces. The C_6H_{14} molecule (left) has more sites (indicated by asterisks, *) along its chain where it can be attracted to other molecules nearby than does the shorter* C_3H_8 molecule (*right*). As a result, the boiling point of C_6H_{14} (hexane, 68.7 °C) is higher than that of C_3H_8 (propane, −42.1 °C).

Table 11.2 Boiling Points of Some Hydrocarbons[a]

Molecular Formula	Boiling Point at 1 atm (°C)
CH_4	− 161.5
C_2H_6	− 88.6
C_3H_8	− 42.1
C_4H_{10}	− 0.5
C_5H_{12}	36.1
C_6H_{14}	68.7
⋮	⋮
$C_{10}H_{22}$	174.1
⋮	⋮
$C_{22}H_{46}$	327

[a]The molecules of each hydrocarbon in this table have carbon chains of the type C—C—C—C—etc.; that is, one carbon follows another.

formula, C_5H_{12}. However, the more compact neopentane molecule, $(CH_3)_4C$, has a lower boiling point than the long chainlike *n*-pentane molecule, $CH_3CH_2CH_2CH_2CH_3$. Presumably, because of the compact shape of the $(CH_3)_4C$ molecule, the individual hydrogens on neighboring molecules cannot interact with each other as well as those on the chainlike molecule.

London Forces and Dipole–Dipole Forces Compared

At first thought, we might expect that London forces are quite weak compared to dipole–dipole attractions. However, for many substances the London forces are as strong as or stronger than any dipole–dipole forces also present. For example, consider HCl (bp − 84.9 °C) and HBr (bp − 67.0 °C). Molecules of HCl are more polar than those of HBr because Cl is more electronegative than Br.

neopentane, $(CH_3)_4C$
bp = 9.5 °C

n-pentane, $CH_3CH_2CH_2CH_2CH_3$
bp = 36.1 °C

Figure 11.7 *Two molecules with the formula C_5H_{12}.* Not all hydrogen atoms can be seen in these space-filling models. The neopentane molecule, $(CH_3)_4C$, has a more compact shape than the *n*-pentane molecule, $CH_3CH_2CH_2CH_2CH_3$. In the more compact structure, the H atoms cannot interact with those on neighboring molecules as well as the H atoms in the long chainlike structure, so the intermolecular attractions are weaker in the more compact molecule.

This means that the dipole–dipole attractions in HCl(*l*) are stronger than those in HBr(*l*). But their boiling points tell us that the intermolecular attractions in HBr(*l*) are stronger than those in HCl(*l*). This could only be true if the London forces in HBr are a lot stronger than in HCl, so that they can compensate for the weaker dipole–dipole attractions.

London forces are expected to increase going from HCl to HBr because Br is larger than Cl and is more polarizable. The fact that the increased strengths of the London forces are able to dominate over the changes in dipole–dipole attractions reveals that London forces contribute significantly to the overall intermolecular attractions in these substances. In fact, one source suggests that in HCl(*l*), the dipole–dipole attractions account for only about 20% of the total intermolecular attractions and that London forces account for about 80%.

Ion–Dipole and Ion–Induced Dipole Forces of Attraction

Ions are able to interact with the charged ends of polar molecules to give **ion–dipole attractions.** This occurs in water, for example, when ionic compounds dissolve to give hydrated ions. Cations become surrounded by water molecules that are oriented with the negative ends of their dipoles pointing toward the cation. Similarly, anions attract the positive ends of water dipoles.

The creation of a dipole in a neighbor is not restricted to instantaneous irregularities in the electron densities of neutral atoms and molecules. Ions, with full positive and negative charges, also induce dipoles in molecules nearby (like those of a solvent), leading to **ion–induced dipole attractions.** These can be quite strong because the charge on the ion does not flicker on and off like the instantaneous charges responsible for ordinary London forces.

11.3 Some General Properties of Liquids and Solids

We'll begin our study of the general physical properties of liquids and solids by examining two properties that depend mostly on how tightly packed molecules are, namely, *compressibility* and *diffusion*. Other properties depend much more on the strengths of intermolecular attractive forces, properties such as *retention of volume or shape, surface tension,* the ability of a liquid to *wet* a surface, the *viscosity* of a liquid, and a solid's or liquid's *tendency to evaporate.*

Properties That Depend Primarily on Tightness of Packing

Compressibility

The **compressibility** of a substance is a measure of the ability of the substance to be forced into a smaller volume. Gases are highly compressible because the molecules are far apart and the gas is mostly empty space. In Section 11.1, we noted that in a liquid or solid most of the space is taken up by the molecules, and that there is very little empty space into which to crowd other molecules. As a result, it is very difficult to compress liquids or solids to a smaller volume by applying pressure, so we say that these states of matter are nearly **incompressible.** This is a property that we often make use of. When you "step on the brakes" of a car, for example, you rely on the incompressibility of the brake fluid to transmit the pressure you apply with your foot to the brake shoes on the wheels. If some air should get into the brake lines, you're in big trouble, because when you apply pressure to the brakes it simply compresses the air and doesn't force the brake

Hydraulic machinery such as this earthmover uses the incompressibility of liquids to transmit forces that accomplish work.

shoes to stop the car. A similar application of the incompressibility of liquids is the foundation of the engineering science of *hydraulics,* which uses fluids to transmit forces that lift or move heavy objects.

Diffusion

As we indicated in Sections 6.2 and 10.7, diffusion occurs much more rapidly in gases than in liquids, and hardly at all in solids. In gases, molecules diffuse rapidly because they travel relatively long distances between collisions, as illustrated in Figure 11.8. In liquids, however, a given molecule suffers many collisions as it moves about, so it takes longer to move from place to place and diffusion is much slower. Diffusion in solids is almost nonexistent at room temperature because the particles of a solid are held tightly in place. At high temperatures, though, the particles of a solid sometimes have enough kinetic energy to jiggle their way past each other, and diffusion can occur slowly. Such high temperature solid-state diffusion is used to make electronic devices, like transistors (described in Section 9.9). Solid-state diffusion allows for small, carefully controlled amounts of some impurity—arsenic, for example—to be added to a semiconductor such as silicon or germanium. This process, called "doping," allows the conductivity of these materials to be modified in desirable ways.

Properties That Depend Primarily on the Strengths of Intermolecular Attractions

Retention of Volume and Shape

Volume and shape are obvious physical properties. You know that a solid such as an ice cube keeps both its shape and volume when transferred from one container to another. The same amount of water as a *liquid,* however, retains only its volume, not its shape, when transferred to a container with another shape. A gas does not necessarily retain either its volume or shape when transferred; a gas spontaneously adapts its shape and volume to any container. The underlying reason for the differences are simply intermolecular attractive forces.

In gases, intermolecular attractive forces are too weak to prevent the molecules from moving apart to fill the entire container. In liquids and solids, however, the attractions are much stronger and are able to hold the particles closely together, thereby preventing liquids and solids from expanding in volume in the manner of a gas. As a result, liquids and solids keep the same volume regardless of the size of their container. In a solid, the attractions are even stronger than in a liquid. They hold the particles more or less rigidly in place, so a solid retains its shape when moved from one container to another.

Surface Tension

A property that is especially evident for liquids is the phenomenon of **surface tension,** a property related to the tendency of a liquid to seek a shape that yields the minimum surface area. For a given volume, the shape with a minimum surface area is a sphere—it's a principle of solid geometry.

The tendency of a liquid to spontaneously assume a form with a minimum surface area explains many common observations. Surface tension, for example, is responsible for the tendency of raindrops to be little spheres. Surface tension also causes the sharp edges of glass tubing to become rounded when the glass is softened in a flame, an operation called "fire polishing." Surface tension is also what allows us to fill a water glass above the rim, giving the surface a rounded appearance (see Figure 11.9). The surface behaves as if it had a thin, invisible "skin" that lets the water in

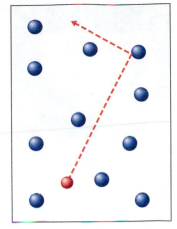
(a)

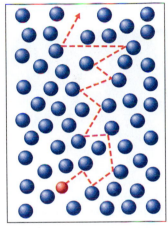
(b)

Figure 11.8 *Diffusion in a gas and a liquid viewed at the molecular level.* (a) Diffusion in a gas is rapid because relatively few collisions occur between widely spaced molecules. (b) Diffusion in a liquid is slow because of many collisions between closely spaced particles.

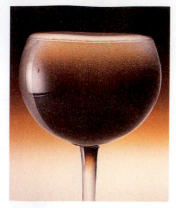

Figure 11.9 *Surface tension in a liquid.* Surface tension allows a glass to be filled with water above the rim.

An insect called a *waterstrider,* shown here, is able to walk on water because of the liquid's surface tension, which causes water to behave as though it has a skin that resists piercing by the insect's legs.

the glass pile up, trying to assume a spherical shape. Gravity, of course, works in opposition, tending to pull the water down. If too much water is added to the glass, the gravitational force finally wins and the skin breaks; the water overflows.

To understand surface tension, we need to examine why molecules would prefer to be within a liquid rather than at its surface. Let's begin by examining the environment surrounding molecules in these two conditions (see Figure 11.10). We see that a molecule that's *within* the liquid is surrounded by densely packed molecules on all sides, whereas one at the *surface* has neighbors beside and below it, but none above. As a result, a surface molecule is attracted to fewer neighbors than one within the liquid. With this in mind, let's imagine how we might change an interior molecule to one at the surface. To accomplish this, we would have to pull away some of the surrounding molecules. Because there are intermolecular attractions, removing neighbors requires work; so there's an increase in potential energy involved. This leads to the conclusion that *a molecule at the surface has a higher potential energy than a molecule in the bulk of the liquid.*

What does potential energy have to do with surface tension? Understanding the answer requires that you know how potential energy is related to stability. In general, a system becomes more stable when its potential energy decreases. That's why a yardstick standing on end falls over; by doing so its potential energy drops and it reaches a lower energy, more stable position. For a liquid, reducing its surface area (and thereby reducing the number of molecules at the surface) lowers its potential energy, and the lowest energy (most stable state) is achieved when the liquid has the smallest surface area possible. As noted above, for a given volume, the shape with the smallest surface area is a sphere, so liquids spontaneously tend to assume spherical shapes if they can.

Another consequence of surface tension is that it takes work to expand the surface area of a liquid. If you push on the surface, it resists expansion and pushes back, so the surface appears to behave as though it were a "skin" that resists penetration. This skin is what enables certain insects to "walk on water," as illustrated in the photo at left.

Figure 11.10 *Surface tension and intermolecular attractions.* In water, as in other liquids, molecules at the surface are surrounded by fewer molecules than those below the surface. As a result, surface molecules experience fewer attractions than molecules within the liquid.

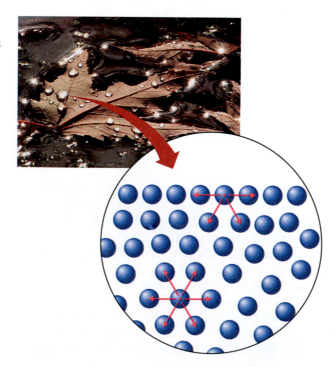

Liquids with strong intermolecular attractive forces will have large differences in potential energy between their interior and surface molecules. This produces a very strong tendency to reduce the number of surface molecules, so the liquid will have a large surface tension. The generalization, then, is that *liquids with strong intermolecular attractions have large surface tensions.* Not surprisingly, water's surface tension is among the highest known (comparisons being at the same temperature); its intermolecular forces are hydrogen bonds, the strongest kind of dipole–dipole attraction. In fact, the surface tension of water is roughly three times that of gasoline, which consists of relatively nonpolar hydrocarbon molecules able to experience only London forces.

In more accurate terms, the surface tension of a liquid is proportional to the energy needed to expand the liquid's surface area.

Wetting of a Surface by a Liquid

A property we associate with liquids, especially water, is their ability to wet things. **Wetting** is the spreading of a liquid across a surface to form a thin film. Water wets clean glass, such as the windshield of a car, by forming a thin film over the surface of the glass (see Figure 11.11*a*). Water won't wet a greasy windshield, however. Instead, on greasy glass water forms tiny beads (see Figure 11.11*b*).

Because of strong hydrogen bonding, the surface tension of water is roughly two to three times larger than the surface tension of any common organic solvent.

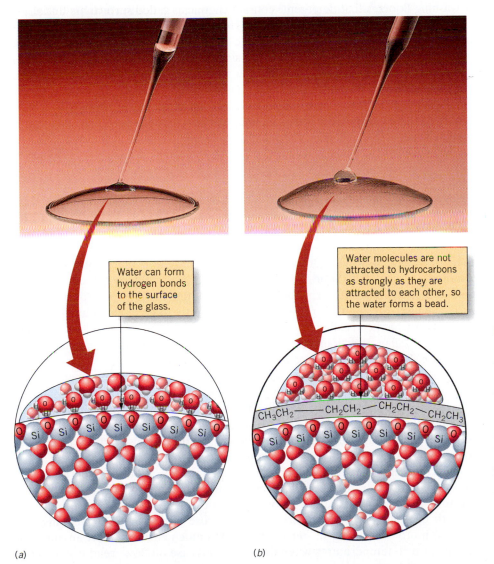

Water can form hydrogen bonds to the surface of the glass.

Water molecules are not attracted to hydrocarbons as strongly as they are attracted to each other, so the water forms a bead.

CH_3CH_2 — CH_2CH_2 — CH_2CH_2 — CH_2CH_3

(a)

(b)

Figure 11.11

Intermolecular attractions affect the ability of water to wet a surface. (a) Water wets a clean glass surface because the surface contains many oxygen atoms to which water can form hydrogen bonds. (*b*) If the surface has a layer of grease, to which water molecules are only weakly attracted, the water doesn't wet it. The water resists spreading and forms a bead instead.

For wetting to occur, the intermolecular attractive forces between the liquid and the surface must be of about the same strength as the forces within the liquid itself. Such a rough equality of forces exists when water touches clean glass. Water, despite its high surface tension and tendency to form beads, is yet able to wet clean glass because the glass surface contains lots of oxygen atoms. Water molecules can therefore form hydrogen bonds to the glass nearly as well as they can to each other. Thus, part of the energy needed to expand the water's surface area as wetting occurs is recovered by the formation of hydrogen bonds to the glass surface.

When the glass is coated by a film of oil or grease, however, the conditions are quite different (Figure 11.11*b*). The surface now is actually no longer glass, but oil and grease instead. Their molecules are relatively nonpolar, so their attractions to other molecules (including water) are largely limited to London forces. These are weak compared with hydrogen bonds, so the attractions *within* liquid water are much stronger than the attractions *between* water molecules and the greasy surface. The weak water-to-grease attractions can't overcome the attractions within water that give it surface tension, so the water doesn't spread out; it forms beads instead.

One of the reasons why detergents are used for such chores as doing laundry or washing floors is that detergents contain chemicals called **surfactants** that drastically lower the surface tension of water. This makes the water "wetter," which allows the detergent solution to spread more easily across the surface to be cleaned.

When a liquid has a low surface tension, like gasoline, we know that it has weak intermolecular attractions, and such a liquid easily wets solid surfaces. The weak attractions between molecules in gasoline, for example, are readily overcome by attractions to almost any surface, so gasoline easily spreads to a thin film. If you've ever spilled a little gasoline, you have experienced firsthand that it doesn't bead.

Glass is characterized by a vast network of silicon–oxygen bonds.

Viscosity

As everybody knows, syrup flows less readily or is more resistant to flow than water (both at the same temperature). Flowing is a change in the *form* of the liquid, and such resistance to a change in form is called the liquid's **viscosity.** We say that syrup is more *viscous* than water. The concept of viscosity is not confined to liquids, however, although it is with liquids that the property is most commonly associated. Solid things, even rock, also yield to forces acting to change their shapes, but normally do so only gradually and imperceptibly. Gases also have viscosity, but they respond almost instantly to form-changing forces.

Viscosity has been called the "internal friction" of a material. It is caused largely by intermolecular attractions; the stronger such attractions are, the more viscous the material is. Both permanent dipole–dipole and London forces affect viscosity. The ability of molecules to tangle with each other is also a factor. The long, floppy, entangling molecules in heavy machine oil (almost entirely a mixture of long chain, nonpolar hydrocarbons), plus the London forces in the material, give it a viscosity roughly 600 times that of water at 15 °C. Vegetable oils, like the olive oil or corn oil used to prepare salad dressings, consist of molecules that are also large but generally nonpolar. Olive oil is roughly 100 times more viscous than water.

Viscosity also depends on temperature, increasing as the temperature decreases. When water, for example, is cooled from its boiling point to room temperature, its viscosity increases by over a factor of three. The increase in viscosity with cooling is why operators of vehicles use a "light," thin (meaning less viscous) motor oil during subzero weather. It's much easier to start an automobile at subzero temperatures when the cold crankcase oil flows readily and is not thick and syrupy. During very hot months, a "heavy" (more viscous) motor oil is used so that it cannot slip as readily between cylinders and pistons and so be lost even as it is doing its lubricating work.

Evaporation and Sublimation

Finally, we come to one of the most important physical properties of liquids and solids—their tendency to undergo a change of state from liquid to gas or from solid to gas. For liquids, the change is called **evaporation,** and everyone has seen liquids evaporate; it's what happens to water, for example, when streets, wet from a rain shower, gradually dry. Solids can also change directly to the gaseous state by evaporation without going through the liquid state, except that for a solid-to-gas change of state we use a special term, **sublimation.** Solid carbon dioxide, commonly called *dry ice,* is an example. It is "dry" ice because it doesn't melt; instead, at atmospheric pressure it *sublimes;* it changes directly to gaseous CO_2. Naphthalene, the ingredient in some brands of moth flakes, is a substance that can sublime and seemingly disappear.

A change of state is also called a *phase change.*

To understand evaporation and sublimation, we have to examine the motions of molecules in liquids and solids. Within these two states, the molecules are not motionless. They bounce around against their neighbors, and at a given temperature, there is exactly the same distribution of kinetic energies in a liquid or a solid as there is in a gas. This means that Figure 6.1 on page 235 applies to liquids and solids as well as gases. Some molecules have low kinetic energies and move slowly, while others with higher kinetic energies move faster. A small fraction of the molecules have very large kinetic energies and therefore very high velocities. If one of these high velocity molecules is at the surface and is moving outward fast enough, it can escape the attractions of its neighbors and enter the vapor state. When this happens, we say the molecule has left by evaporation (or sublimation, if the substance is a solid).

Naphthalene sublimes when heated, and the vapor condenses directly to a solid when it encounters a cool surface. Here beautiful flaky naphthalene crystals have been formed on the bottom of a flask filled with ice water.

When a canteen of water is covered with a water-absorbing fabric that is kept wet, the evaporation of the water helps to keep the water in the canteen cool.

Evaporation and Cooling

One of the things we notice about the evaporation of a liquid is that it produces a cooling effect. Have you ever come out of the water after swimming and been chilled by a breeze? The evaporation of water from your body produced this effect. In fact, our bodies use the evaporation of water to maintain a constant body temperature. During warm weather or vigorous exercise we perspire, and the evaporation of our perspiration cools our skin. Evaporative cooling also occurs within the lungs, which puts water vapor into position to be exhaled, carrying heat out of the body with it. You've probably also used the cooling effect caused by evaporation by blowing gently on the surface of a hot bowl of soup or a cup of coffee. The stream of air stirs the liquid, bringing more hot liquid to the surface to be cooled as it evaporates.

We can see why liquids become cool during evaporation by examining Figure 11.12, which illustrates the kinetic energy distribution in a liquid at a particular temperature. A marker along the horizontal axis shows the minimum kinetic energy needed by a molecule to escape the attractions of its neighbors. Only molecules with kinetic energies equal to or greater than the minimum can leave the liquid. Others with less kinetic energy may begin to leave, but before they can escape they slow to a stop and then fall back. The situation is somewhat like attempting to launch a spaceship using a cannon. If the velocity of the projectile is not very great, it will not rise very far before it slows to a halt and falls to Earth. However, if it is going fast enough (if it has its "escape velocity"), it can overcome Earth's gravity and leave.

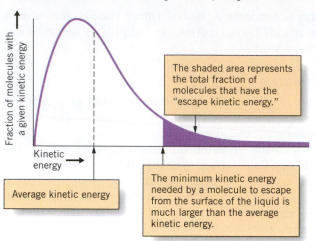

Figure 11.12 *Cooling of a liquid by evaporation.* Molecules that are able to escape from the liquid have kinetic energies larger than the average. When they leave, the average kinetic energy of the molecules left behind is less, so the temperature is lower.

Figure 11.12 also allows us to understand why liquids cool as they evaporate. Notice in Figure 11.12 that the minimum kinetic energy needed to escape is much larger than the average kinetic energy, which means that when molecules evaporate they carry with them large amounts of kinetic energy. As a result, the average kinetic energy of the molecules left behind decreases. (You might think of this as being similar to removing people taller than 6 ft from a large class of students. When this is done, the average height of those who are left is less.) Because the Kelvin temperature of the remaining liquid is directly proportional to the now lower average kinetic energy, the temperature is lower; in other words, evaporation causes the liquid that remains to be cooler.

Factors That Control the Rate of Evaporation

One of the important things we are going to be concerned about later in this chapter is the *rate of evaporation* of a liquid. There are several factors that control this. You are probably already aware of one of them—the surface area of the liquid. Because evaporation occurs from the liquid's surface and not from within, it makes sense that as the surface area is increased, more molecules are able to escape and the liquid evaporates more quickly. For liquids having the same surface area, the rate of evaporation depends on two factors, namely, temperature and the strengths of intermolecular attractions. Let's examine each of them separately.

The influence of temperature on evaporation rate is no surprise; you already know that hot water evaporates faster than cold water. The underlying reason is brought out in Figure 11.13, which shows the kinetic energy distributions for the *same* liquid at two temperatures. Notice two important features of the figure. First, the same minimum kinetic energy is needed for the escape of molecules at both temperatures. This minimum is determined by the kinds of attractive forces between the molecules, and is independent of temperature. Second, the shaded area of the curve represents the total fraction of molecules having kinetic energies equal to or greater than the minimum. At the higher temperature, the total fraction is larger, which means that at the higher temperature a greater total fraction has the ability to evaporate. As you might expect, when more molecules have the needed energy, more evaporate in a unit of time, and the rate of evaporation per unit surface area of a given liquid is greater at a higher temperature.

The effect of intermolecular attractions on evaporation rate can be seen by studying Figure 11.14. Here we have kinetic energy distributions for two *different* liquids—call them *A* and *B*—both at the same temperature. In liquid *A*, the

To understand how temperature and intermolecular forces affect the rate of evaporation, we must compare evaporation rates from the same size surface area. In this discussion, therefore, "rate of evaporation" means "rate of evaporation per unit surface area."

Tools

Effect of temperature on the rate of evaporation

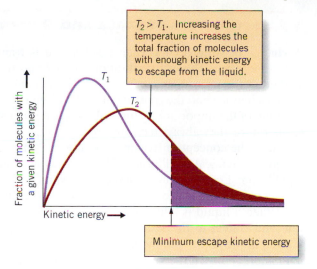

T_2 > T_1. Increasing the temperature increases the total fraction of molecules with enough kinetic energy to escape from the liquid.

Minimum escape kinetic energy

Figure 11.13 *Effect of increasing the temperature on the rate of evaporation of a liquid.* At the higher temperature, the total fraction of molecules with enough kinetic energy to escape is larger, so the rate of evaporation is larger.

attractive forces are weak; they might be of the London type, for example. As we see, the minimum kinetic energy needed by A molecules to escape is not very large because they are not attracted very strongly to each other. In liquid B, the intermolecular attractive forces are much stronger; they might be hydrogen bonds, for instance. Molecules of B, therefore, are held more tightly to each other at the liquid's surface and must have a higher kinetic energy to evaporate. As you can see from the figure, the total fraction of molecules with enough energy to evaporate is greater for A than for B, which means that A evaporates faster than B. In general, then, *the weaker the intermolecular attractive forces, the faster is the rate of evaporation at a given temperature.* You are probably also aware of this phenomenon. At room temperature, for example, nail polish remover [acetone, $(CH_3)_2CO$], whose molecules experience weak dipole–dipole and London forces of attraction, evaporates faster than water, whose molecules feel the effects of much stronger hydrogen bonds.

Tools

Intermolecular forces and the rate of evaporation

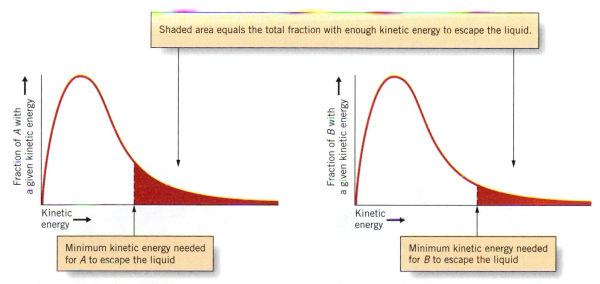

Shaded area equals the total fraction with enough kinetic energy to escape the liquid.

Minimum kinetic energy needed for A to escape the liquid

Minimum kinetic energy needed for B to escape the liquid

Figure 11.14 *Kinetic energy distribution in two different liquids, A and B, at the same temperature.* The minimum kinetic energy required by molecules of A to escape is less than for B because the intermolecular attractions in A are weaker than in B. This causes A to evaporate faster than B.

11.4 Changes of State and Dynamic Equilibrium

A **change of state** occurs when a substance is transformed from one physical state to another. The evaporation of a liquid and the sublimation of a solid, described in the preceding section, are two examples. Others are the melting of a solid such as ice and the freezing of a liquid such as water.

One of the important features about changes of state is that, at any particular temperature, they always tend toward a condition of *dynamic equilibrium*. We introduced the concept of dynamic equilibrium on page 168 with an example of a system at chemical equilibrium. The same general principles apply to a physical equilibrium. Let's examine in some detail what happens when a liquid evaporates in a sealed container and what is involved when equilibrium is reached.

In chemistry, when we use the term equilibrium, we always mean dynamic equilibrium unless otherwise indicated.

When a liquid is added to the empty container, molecules begin to evaporate and collect in the space above the liquid (see Figure 11.15*a*). As they fly around in the vapor, the molecules collide with each other, with the walls of the container, and with the surface of the liquid itself. Those that strike the liquid's surface tend to stick because their kinetic energies become scattered among the surface molecules. It is somewhat like throwing a ping-pong ball into a large box of ping-pong balls. The incoming ball knocks others around, and by giving them kinetic energy it loses some of its own. Because its kinetic energy has been reduced, there is a high probability that the incoming ball won't bounce out.

The change of a vapor to its liquid state is called **condensation.** We say that the vapor *condenses* when it changes to a liquid. The rate at which vapor molecules collide with a liquid's surface and condense depends on the concentration of molecules in the vapor. When there are only a few in a given volume, the number of collisions per second with a unit area of the liquid's surface is small, and the rate of condensation is slow. When there are many molecules per unit of volume in the vapor, many such collisions occur each second with a unit of surface area, so the rate of condensation is higher.

Consider now a liquid as it is first introduced into an empty sealed chamber. With virtually none of its molecules yet in the vapor state, the evaporation rate from the liquid is much larger than the condensation rate from the vapor. But as mole-

Figure 11.15 *Evaporation of a liquid into a sealed container.* (*a*) The liquid has just begun to evaporate into the container. The rate of evaporation is greater than the rate of condensation. (*b*) A dynamic equilibrium is reached when the rate of evaporation equals the rate of condensation. In a given time period, the number of molecules entering the vapor equals the number that leave, so there is no net change in the number of gaseous molecules.

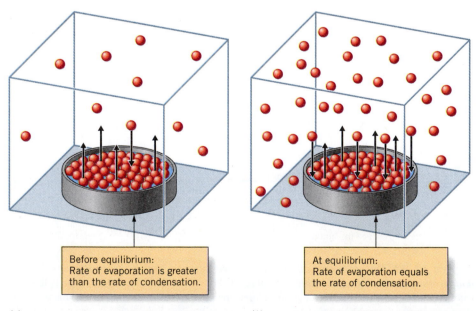

Before equilibrium:
Rate of evaporation is greater than the rate of condensation.

At equilibrium:
Rate of evaporation equals the rate of condensation.

(*a*)

(*b*)

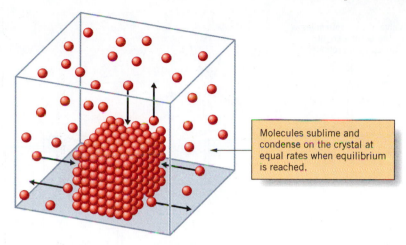

Molecules sublime and condense on the crystal at equal rates when equilibrium is reached.

Figure 11.16 *A solid–vapor equilibrium.*

cules accumulate in the vapor, the condensation rate increases and continues to increase until the rates of condensation and evaporation are the same (Figure 11.15b). From that moment on, the number of molecules in the vapor will remain constant, because over a given period of time the number that enters the vapor is the same as the number that leaves. At this point we have a condition of *dynamic equilibrium,* one in which two opposing effects, evaporation and condensation, are occurring at equal rates and their effects cancel. It is an *equilibrium* because there is no apparent change; the *number* of vapor molecules remains constant as does the number of liquid molecules. It is a *dynamic* equilibrium because activity has not ceased. Molecules continue to evaporate and condense; they just do so at equal rates.

Similar equilibria are also reached in melting and sublimation. When a solid is heated to make it melt, a temperature is reached, called the **melting point** of the solid, at which dynamic equilibrium exists between molecules in the solid and those in the liquid state. Molecules leave the solid and enter the liquid at the same rate as molecules leave the liquid and join the solid. As long as no heat is added or removed from a solid–liquid equilibrium mixture, melting and freezing occur at equal rates. For sublimation, the situation is exactly the same as in the evaporation of a liquid into a sealed container (see Figure 11.16). After a few moments, the rates of sublimation and condensation become the same and equilibrium is established. All these various equilibria have very profound effects on the properties of solids and liquids, as we shall soon see.

11.5 Vapor Pressures of Liquids and Solids

When a liquid evaporates, the molecules that enter the vapor exert a pressure called the **vapor pressure.** From the very moment a liquid begins to evaporate into the vapor space above it, there is a vapor pressure. If the evaporation is taking place inside a sealed container, this pressure grows until finally equilibrium is reached. Once the rates of evaporation and condensation become equal, the concentration of molecules in the vapor remains constant and the vapor exerts a constant pressure. This final pressure is called the **equilibrium vapor pressure of the liquid.** In general, when we refer to the vapor pressure, we really mean the equilibrium vapor pressure.

We can measure the vapor pressure of a liquid with an apparatus like that shown in Figure 11.17. Initially, both sides of the apparatus are open to the atmosphere, so the pressure in both arms of the manometer is the same. A small amount of liquid is then added to the flask from the funnel and the left side of

The vapor–liquid equilibrium is possible only in a closed container. When the container is open, vapor molecules drift away and the liquid might completely evaporate.

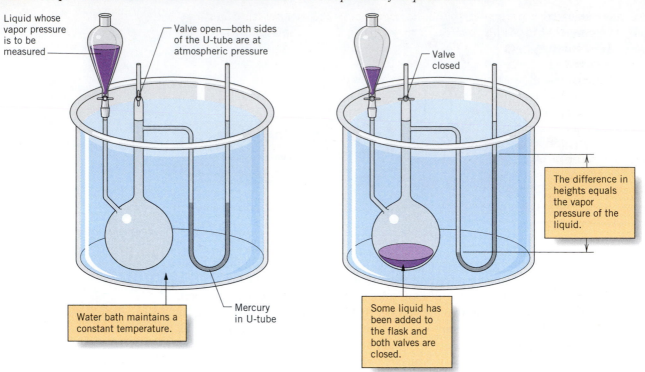

Liquid whose vapor pressure is to be measured

Valve open—both sides of the U-tube are at atmospheric pressure

Valve closed

The difference in heights equals the vapor pressure of the liquid.

Water bath maintains a constant temperature.

Mercury in U-tube

Some liquid has been added to the flask and both valves are closed.

Figure 11.17 *Measuring the vapor pressure of a liquid.*

the apparatus is sealed immediately. As some of the liquid evaporates, the pressure in the flask rises, forcing the fluid in the left side of the manometer downward. The increase in pressure, as measured by the difference in the levels in the manometer, is equal to the pressure exerted by the vapor, namely, the vapor pressure of the liquid.

Factors that affect the vapor pressure

Factors That Affect Equilibrium Vapor Pressure

Figure 11.18 shows plots of equilibrium vapor pressure versus temperature for a few liquids. From these graphs we see that both a liquid's temperature and its chemical composition are the major factors affecting its vapor pressure. Once

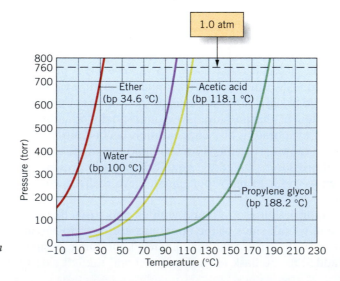

Figure 11.18 *Variation of vapor pressure with temperature for some common liquids.*

we have selected a particular liquid, however, only the temperature matters. The reason is that the vapor pressure of a given liquid is a function solely of its rate of evaporation *per unit area of the liquid's surface*. When this rate is large, a large concentration of molecules in the vapor state is necessary to establish equilibrium, which is another way of saying that the vapor pressure is relatively high when the evaporation rate is high. As the temperature of a given liquid increases, so does its rate of evaporation and so does its vapor pressure.

As chemical composition changes in going from one liquid to another, the strengths of intermolecular attractions change. If the attractions increase, the rates of evaporation at a given temperature decrease, and the vapor pressures decrease. These data on relative vapor pressures tell us that, of the four liquids in Figure 11.18, intermolecular attractions are strongest in propylene glycol, next strongest in acetic acid, third strongest in water, and weakest in ether. Thus, *we can use vapor pressures as indications of relative strengths of the attractive forces in liquids.*

A liquid with a high vapor pressure at a given temperature is said to be *volatile*.

Factors That Do Not Affect the Vapor Pressure

An important fact about vapor pressure is that *its magnitude doesn't depend on the total surface area of the liquid, nor on the volume of the liquid in the flask, nor on the volume of the flask itself, just as long as some liquid remains when equilibrium is reached.* The reason is because none of these factors affects the rate of evaporation *per unit surface area*.

Increasing the *total* surface area does increase the *total* rate of evaporation, but the larger area is also available for condensation; so the rate at which molecules return to the liquid also increases. The rates of both evaporation and condensation are thus affected equally, and no change occurs to the equilibrium vapor pressure.

Adding more liquid to the container can't affect the equilibrium either because evaporation occurs from the *surface*. Having more molecules in the bulk of the liquid does not change what is going on at the surface.

To understand why the vapor pressure doesn't depend on the *size* of the vapor space, consider a liquid in equilibrium with its vapor in a cylinder with a movable piston, as illustrated in Figure 11.19*a*. Withdrawing the piston increases the volume of the vapor space; as the vapor expands, the pressure it exerts becomes less, so there's a momentary drop in the pressure. The molecules of the vapor, being more spread out now, no longer strike the surface as frequently, so the rate of condensation has also decreased. The rate of evaporation hasn't changed, however, so for a moment the system is not at equilibrium and the substance is evaporating faster than it is condensing (Figure 11.19*b*). This condition prevails, changing more liquid into vapor, until the concentration of molecules in the vapor has risen enough to make the condensation rate again equal to the evaporation rate (Figure 11.19*c*). At this point the vapor pressure has returned to its original value. Therefore, the net result of expanding the space above the liquid is to change more liquid into vapor, but it does not affect the equilibrium vapor pressure. Similarly, we expect that reducing the volume of the vapor space above the liquid will also not affect the equilibrium vapor pressure.

Humidity is a measure of how nearly saturated the air is with moisture. At 100% humidity, the partial pressure of water vapor is equal to the equilibrium vapor pressure of water at the temperature of the air.

Recall that when the volume of a gas increases at constant temperature, its pressure decreases.

Practice Exercise 1

Suppose a liquid is in equilibrium with its vapor in a piston–cylinder apparatus like that just described. If the piston is withdrawn a short way and the system is allowed to return to equilibrium, what will have happened to the *total number* of molecules in both the liquid and the vapor? ◆

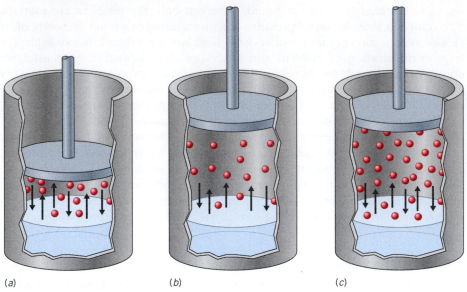

(a)   (b)   (c)

Figure 11.19 *Effect of a volume change on the vapor pressure of a liquid.*
(*a*) Equilibrium exists between liquid and vapor. (*b*) The volume is increased, which upsets the equilibrium and causes the pressure to drop. Because the concentration of molecules in the vapor has dropped, the rate of condensation is now less than the rate of evaporation, which hasn't changed. (*c*) After more liquid has evaporated, equilibrium is restored; the rates of condensation and evaporation are again equal, and the vapor pressure has returned to its initial value.

In many solids, such as NaCl, the attractive forces are so strong that virtually no particles have enough kinetic energy to escape at room temperature, so essentially no evaporation occurs. Their vapor pressures at room temperature are virtually zero.

We know that the water in this pot of vegetables is boiling because we can see bubbles of steam rising to the surface. Because the temperature of boiling water cannot rise above its boiling point, foods cooked in water can't become too hot and burn.

Vapor Pressures of Solids

Solids have vapor pressures just as liquids do. In a crystal, the particles are not stationary, as we have said. They vibrate back and forth about their equilibrium positions. At a given temperature there is a distribution of kinetic energies, so some particles vibrate slowly while others vibrate with a great deal of molecular kinetic energy. Some particles at the surface have large enough kinetic energies to break away from their neighbors and enter the vapor state. When particles in the vapor collide with the crystal, they can be recaptured, so condensation can occur too. Eventually, the concentration of particles in the vapor reaches a point where the rate of sublimation equals the rate of condensation, and a dynamic equilibrium is established. The pressure of the vapor that is in equilibrium with the solid is called the **equilibrium vapor pressure of the solid.** As with liquids, this equilibrium vapor pressure is usually referred to simply as the vapor pressure. Like that of a liquid, the vapor pressure of a solid is determined by the strengths of the attractive forces between the particles and by the temperature.

11.6 Boiling Points of Liquids

If you were asked to check whether a pot of water was boiling, what would you look for? The answer, of course, is *bubbles.* When a liquid boils, large bubbles usually form at many places on the inner surface of the container and rise to the top. If you were to place a thermometer into the boiling water, you would find that the temperature remains constant, regardless of how you adjust the flame under the pot. A hotter flame just makes the water bubble faster, but it doesn't raise the temperature. *Any pure liquid remains at a constant temperature while it is boiling,* a temperature called the liquid's **boiling point.**

If you measure the boiling point of water in Philadelphia, New York, or any place else that is nearly at sea level, your thermometer will read 100 °C or very close to it. However, if you try this experiment in Denver, Colorado, you will find that water boils at about 95 °C. Denver, at a mile above sea level, has a lower atmospheric pressure, so we find that the boiling point depends on the atmospheric pressure.

These observations raise some interesting questions. Why do liquids boil? And why does the boiling point depend on the pressure of the atmosphere? The answers become apparent when we realize that inside the bubbles of a boiling liquid is the *liquid's vapor,* not air. When water boils, the bubbles contain water vapor (steam); when alcohol boils, the bubbles contain alcohol vapor. As a bubble grows, liquid evaporates into it, and the pressure of the vapor pushes the liquid aside, making the size of the bubble increase (see Figure 11.20). Opposing the bubble's internal vapor pressure, however, is the pressure of the atmosphere pushing down on the top of the liquid, attempting to collapse the bubble. The only way the bubble can exist and grow is for the vapor pressure within it to equal (maybe just slightly exceed) the pressure exerted by the atmosphere. In other words, bubbles of vapor cannot even form until the temperature of the liquid rises to a point at which the liquid's vapor pressure equals the atmospheric pressure. Thus, in scientific terms, the **boiling point** is defined as *the temperature at which the vapor pressure of the liquid is equal to the prevailing atmospheric pressure.*

Now we can easily understand why water boils at a lower temperature in Denver than it does in New York City. Because the atmospheric pressure is lower in Denver, the water there doesn't have to be heated to as high a temperature to make its vapor pressure equal to the atmospheric pressure. The lower temperature of boiling water at places with high altitudes, like Denver, makes it necessary to cook foods longer. At the other extreme, a pressure cooker is a device that increases the pressure over the boiling water and thereby raises the boiling point. At the higher temperature, foods cook more quickly.

To make it possible to compare the boiling points of different liquids, chemists have chosen 1 atm as the reference pressure. The boiling point of a liquid at 1 atm is called its **normal boiling point.** (If a boiling point is reported without also mentioning the pressure at which it was measured, we assume it to be the normal boiling point.) Notice in Figure 11.18, page 484, that we can find the normal boiling

On the top of Mt. Everest, the world's tallest peak, water boils at only 69 °C.

The tiny bubbles that sometimes form on the surface of a pot before the water boils contain air driven from solution by the rising temperature.

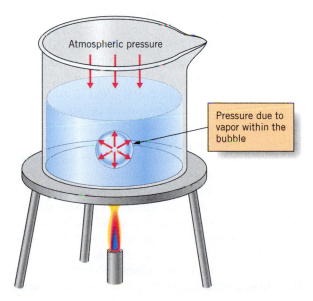

Figure 11.20 *A liquid at its boiling point.* The pressure of the vapor within a bubble in a boiling liquid pushes the liquid aside against the opposing pressure of the atmosphere. Bubbles can't form unless the vapor pressure of the liquid is at least equal to the pressure of the atmosphere.

Atmospheric pressure

Pressure due to vapor within the bubble

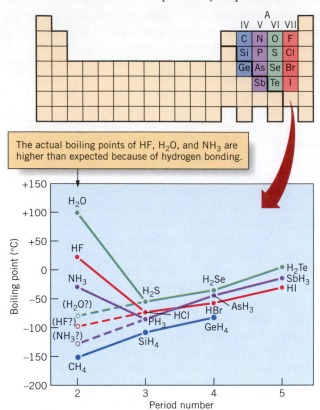

Figure 11.21 *Boiling points of the hydrogen compounds of the elements of Groups IVA, VA, VIA, and VIIA of the periodic table.*

Boiling points of substances

points of ether, water, acetic acid, and propylene glycol by noting the temperatures at which their vapor pressure curves cross the 1 atm pressure line.

Earlier we mentioned that the boiling point is a property whose value depends on the strengths of the intermolecular attractions in a liquid. When the attractive forces are strong, the liquid has a low vapor pressure at a given temperature, so it must be heated to a high temperature to bring its vapor pressure up to atmospheric pressure. High boiling points therefore result from strong intermolecular attractions, so we often use normal boiling point data to assess relative intermolecular attractions among different liquids.

The effects of intermolecular attractions on boiling point are easily seen by examining Figure 11.21, which gives plots of the boiling points versus period numbers for some families of binary hydrogen compounds. Notice, first, the gradual increase in boiling point for the hydrogen compounds of the Group IVA elements (CH_4 through GeH_4). These compounds are composed of nonpolar tetrahedral molecules. The boiling points increase from CH_4 to GeH_4 simply because the molecules become larger and their electron clouds become more polarizable, which leads to an increase in the strengths of the London forces.

When we look at the hydrogen compounds of the other nonmetals, we find the same trend from period 3 through period 5. Thus, for three compounds of the Group VA series, PH_3, AsH_3, and SbH_3, there is a gradual increase in boiling point, corresponding again to the increasing strengths of London forces. Similar increases occur for the three Group VIA compounds (H_2S, H_2Se, and H_2Te) and for the three Group VIIA compounds (HCl, HBr, and HI). Significantly, however, the period 2 members of each of these series (NH_3, H_2O, and HF) have much higher boiling points than might otherwise be expected. The reason is that each is involved in hydrogen bonding, which is a much stronger attraction than London forces.

One of the most interesting and far-reaching consequences of hydrogen bonding is that it causes water to be a liquid, rather than a gas, at temperatures near 25 °C. If it were not for hydrogen bonding, water would have a boiling point somewhere near −80 °C and could not exist as a liquid except at still lower temperatures. At such low temperatures it is unlikely that life as we know it could have developed.

Practice Exercise 2

The atmospheric pressure at the top of Mt. McKinley in Alaska, 3.85 miles above sea level, is 330 torr. Use Figure 11.18 to estimate the boiling point of water at the top of this mountain. ◆

11.7 Energy Changes during Changes of State

When a liquid or solid evaporates or a solid melts, there are increases in the distances between the particles of the substance. Particles that normally attract each other are forced apart, increasing the potential energies of the systems. Such energy changes affect our daily lives in many ways, especially the energy changes associated with the changes in state of water, changes which even control the weather on our planet. To study these energy changes, let's begin by examining how the temperature of a substance varies as it is heated.

Heating Curves and Cooling Curves

Figure 11.22*a* illustrates the way the temperature of a substance changes as we add heat to it *at a constant rate,* starting with the solid and finishing with the gaseous state of the substance. The graph is sometimes called a **heating curve** for the substance.

Notice that as heat is added to the solid, the temperature increases until the solid begins to melt. The temperature then remains constant as long as both solid and liquid phases coexist. When all the solid has melted, the continued addition of heat to what is now a liquid causes the temperature to climb once again. This continues until the boiling point is reached, at which point the temperature levels off again. As we noted, the temperature of the boiling liquid stays the same until all of it has boiled away and changed to a gas. Then, finally,

A heating curve can be used to measure accurately the melting point and boiling point of a liquid.

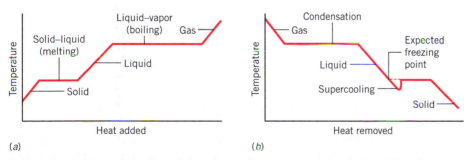

(a)   (b)

Figure 11.22 *Heating and cooling curves.* (*a*) A heating curve in which heat is added to a substance at a constant rate. The temperatures corresponding to the flat portions of the curve occur at the melting point and boiling point. (*b*) A cooling curve in which heat is removed from a substance at a constant rate. Condensation of vapor to a liquid occurs at the same temperature as the liquid boils. Supercooling is seen here as the temperature of the liquid dips below its freezing point (the same temperature as its melting point). Once a tiny crystal forms, the temperature rises to the freezing point.

more heat just raises the temperature of the gas. Let's analyze these changes in terms of changes in the kinetic and potential energies of the particles.

First, let's look at the portions of the graph that slope upward. These occur where we are increasing the temperature of the solid, the liquid, or the gas phases. Because temperature is related to average kinetic energy, nearly all of the heat we add in these regions of the heating curve goes to increasing the average kinetic energies of the particles. In other words, the added heat makes the particles go faster and collide with each other with more force.

In those portions of the heating curve where the temperature remains constant, the average kinetic energy of the particles is not changing. This means that all the heat being added must go to increase the *potential energies* of the particles. During melting, the particles held rigidly in the solid begin to separate slightly as they form the mobile liquid phase. This separation of the particles gives rise to a potential energy increase, which is what we measure by following the quantity of heat input during the melting process. During boiling, an even greater increase occurs in the distances between the molecules. Here they go from the relatively tight packing in the liquid to the widely spaced distribution of molecules in the gas. This gives rise to an even larger increase in the potential energy, which we see as a longer flat region on the heating curve during the boiling of the liquid.

The opposite of a heating curve is a **cooling curve** (see Figure 11.22*b*). Here we start with a gas and gradually cool it—remove heat from it at a constant rate—until we have reached a solid. The cooling curve looks very much like the opposite of a heating curve, except for what often happens when the temperature of the liquid approaches the freezing point. We might expect that when the freezing point is reached, removal of more heat would cause the immediate formation of some solid. Often, however, the liquid continues to cool *below its freezing point,* a phenomenon called **supercooling.**

Supercooling can happen when the freezing point is reached because the molecules of the liquid are still in a jumbled disordered arrangement, not in the ordered structure characteristic of a solid (as we'll soon describe). While the molecules move about, waiting for a few to form the beginnings of a crystal, the temperature continues to drop because heat is continually being removed. Finally, a tiny crystal forms and other molecules quickly join it, losing potential energy which is changed to kinetic energy. The sudden increase in molecular kinetic energy raises the temperature to the freezing point, and then freezing continues normally as more heat is taken away.

Molar Heats of Fusion, Vaporization, and Sublimation

Because phase changes occur at constant temperature and pressure, the potential energy changes associated with melting and vaporization can be expressed as enthalpy changes. Usually, enthalpy changes are expressed on a "per mole" basis and are given special names to identify the kind of change involved. For example, using the word **fusion** instead of "melting," the **molar heat of fusion, ΔH_{fusion},** is the heat absorbed by one mole of a solid when it melts to give a liquid at the same temperature and pressure. Similarly, the **molar heat of vaporization, $\Delta H_{vaporization}$,** is the heat absorbed when one mole of a liquid is changed to one mole of vapor at a constant temperature and pressure. Finally, the **molar heat of sublimation, $\Delta H_{sublimation}$,** is the heat absorbed by one mole of a solid when it sublimes to give one mole of vapor, once again at a constant temperature and pressure. The values of ΔH for fusion, vaporization, and sublimation are all positive because the phase change in each case is accompanied by a net increase in potential energy.

When a solid or liquid is heated, the volume expands only slightly, so there are only small changes in the average distance between the particles. This means that very small changes in potential energy take place, so almost all the heat added goes to increasing the kinetic energy.

Supercooling of a vapor is also possible. Condensation of supercooled water vapor onto solid surfaces leads to dew in warm weather and frost in freezing weather.

The concept of enthalpy was introduced in Chapter 6.

Tools

Enthalpy changes in phase changes

These are also called enthalpies of fusion, vaporization, and sublimation.

Facets of Chemistry 11.1

Refrigeration and Air Conditioning

The fact that the evaporation of a liquid is endothermic and the condensation of a gas is exothermic forms the basis for the operation of refrigerators and air conditioners. These devices use a gas that can be liquefied at room temperature—ammonia or a halogenated hydrocarbon—to pump heat from one place to another.

In a refrigerator, a compressor squeezes the gas into a small volume, which causes the pressure to increase. At the same time, the work done compressing the gas becomes stored as heat—the gas becomes warm. The warm gas is then circulated through cooling coils that are usually located on the back of the refrigerator. (Did you ever notice that these coils are warm?) When the gas cools under pressure, it liquefies when its temperature drops below its critical temperature, and this releases the substance's heat of vaporization, which is also given off to the surroundings. After being cooled, the pressurized liquid moves back into the refrigerator where it passes through a nozzle into a region of low pressure. At this low pressure, the liquid evaporates, and heat equal to the heat of vaporization is absorbed, which cools the inside of the refrigerator. Then the gas moves back to the compressor, and the cycle is repeated.

Through condensation and evaporation, the compressor uses the gas as the fluid in a heat pump that absorbs heat from inside the refrigerator and dumps it outside. An air conditioner works in essentially the same way, except that a room becomes the equivalent of the inside of a refrigerator, and the heat pump dumps the heat into the air outside.

You have probably made use of these enthalpy changes at one time or another. For example, you've added ice to a drink to keep it cool because as the ice melts, it absorbs heat (its heat of fusion). Your body uses the heat of vaporization of water to cool itself through the evaporation of perspiration. During the summer, ice cream trucks carry dry ice because the sublimation of CO_2 absorbs its heat of sublimation and keeps the ice cream cold. Refrigerators and air conditioners work on the basis of making the unwanted heat in one space provide the heat of vaporization for the system's operating fluid (see Facets of Chemistry 11.1).

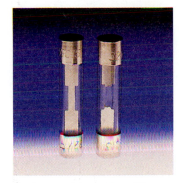

Heats of Vaporization and Vapor Pressures

The increase of vapor pressure with temperature noted in Section 11.5 is related to the heat of vaporization of the substance. The relationship, however, is not a simple proportionality. It involves *natural logarithms,* which are logarithms to the base *e*, a nonrepeating decimal number (an irrational number) having the value 2.71828. . . . The natural logarithm of a quantity *x*, symbolized as ln *x*, is the power to which *e* must be raised to give *x*. (If necessary, refer to Appendix A for discussions of logarithms and antilogarithms.)

The equation relating vapor pressure, heat of vaporization (ΔH_{vap}), and temperature is called the **Clausius–Clapeyron equation**[1]:

$$\ln P = \frac{-\Delta H_{vap}}{RT} + C \qquad (11.1)$$

The quantity ln *P* is the natural logarithm of the vapor pressure, *R* is the gas constant expressed in energy units ($R = 8.314$ J mol^{-1} K^{-1}), *T* is the absolute temperature, and *C* is a constant.

Fusion means melting. The thin metal band in an electrical fuse becomes hot as electricity passes through it. It protects a circuit by melting if too much current is drawn. On the right we see a fuse that has done its job.

Tools

Clausius–Clapeyron equation

The Clausius–Clapeyron equation can be derived from the principles of thermodynamics, a subject that is discussed in Chapter 18.

[1]Rudolf Clausius (1822–1888) was a German physicist. Benoit Clapeyron (1799–1864) was a French engineer. Both made significant contributions to thermodynamics.

The Clausius–Clapeyron equation provides a convenient graphical method for determining heats of vaporization from experimentally measured vapor pressure–temperature data. To see this, let's rewrite Equation 11.1 as follows.

$$\ln P = \left(\frac{-\Delta H_{vap}}{R}\right)\frac{1}{T} + C$$

Recall from algebra that a straight line is represented by the general equation

$$y = mx + b$$

where x and y are variables, m is the slope, and b is the intercept of the line with the y axis. In this case, we can make the substitutions

$$y = \ln P \qquad x = \frac{1}{T} \qquad m = \left(\frac{-\Delta H_{vap}}{R}\right) \qquad b = C$$

Therefore, we have

$$\ln P = \left(\frac{-\Delta H_{vap}}{R}\right)\frac{1}{T} + C$$

$$\updownarrow \qquad\qquad \updownarrow \quad\ \updownarrow \quad \updownarrow$$

$$y \ = \qquad\quad m \quad\ x \ + \ b$$

Thus, a graph of $\ln P$ versus $1/T$ should give a straight line that has a slope equal to $-\Delta H_{vap}/R$. Such straight line relationships are illustrated in Figure 11.23 in which experimental data are plotted for water, acetone, and ethanol.

Sometimes we wish to calculate the vapor pressure at some temperature given the vapor pressure at another temperature. Provided the value of ΔH_{vap} is known, we can use the Clausius–Clapeyron equation.

Suppose P_1 is the vapor pressure of the substance at a temperature T_1 and P_2 is the vapor pressure at temperature T_2. Then, we can write

$$\ln P_1 = \left(\frac{-\Delta H_{vap}}{R}\right)\frac{1}{T_1} + C$$

$$\ln P_2 = \left(\frac{-\Delta H_{vap}}{R}\right)\frac{1}{T_2} + C$$

Subtracting the second equation from the first gives

$$\ln P_1 - \ln P_2 = \left(\frac{-\Delta H_{vap}}{R}\right)\frac{1}{T_1} - \left(\frac{-\Delta H_{vap}}{R}\right)\frac{1}{T_2} + C - C$$

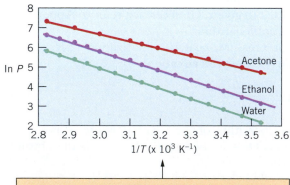

The numbers along the horizontal axis are equal to $1/T$ values multiplied by 1000 to make the axis easier to label.

Figure 11.23 *A graph showing plots of ln P versus 1/T for three liquids.*

Combining logarithm terms on the left and factoring out $\Delta H_{vap}/R$ on the right yields

For logarithms,

$\log a - \log b = \log (a/b)$

$\ln a - \ln b = \ln (a/b)$

$$\ln \frac{P_1}{P_2} = \frac{\Delta H_{vap}}{R} \left(\frac{1}{T_2} - \frac{1}{T_1} \right) \qquad (11.2)$$

Example 11.1 illustrates how we can use this "two point" form of the Clausius–Clapeyron equation.

EXAMPLE 11.1

Using the Clausius–Clapeyron Equation

Hexane, C_6H_{14}, a component of paint thinners, has $\Delta H_{vap} = 30.1$ kJ/mol. At room temperature (25 °C), the vapor pressure of hexane is 148 torr. What is the vapor pressure of hexane at 50 °C?

Analysis: The problem is straightforward; we just have to substitute quantities into Equation 11.2. To avoid errors, let's extract the pressure–temperature data from the problem and arrange it in a table. Arbitrarily taking "1" to be the data for the lower temperature, we have

	Vapor pressure	Temperature
1	148 torr	25 °C = 298 K
2	P_2	50 °C = 323 K

The unknown quantity, P_2, is in the logarithm term. To obtain it, we will first solve for the ratio, P_1/P_2. Then we can substitute the value of P_1 and solve for P_2.

Solution: Let's begin by substituting values into the right side of Equation 11.2. We must be careful about the energy units, however. Because R has energy units of joules, we must change the units of ΔH_{vap} from kilojoules to joules; thus, the value of ΔH_{vap} must be expressed as 30.1×10^3 J/mol.

$$\ln \frac{P_1}{P_2} = \frac{30.1 \times 10^3 \text{ J mol}^{-1}}{8.314 \text{ J mol}^{-1} \text{ K}^{-1}} \left(\frac{1}{323 \text{ K}} - \frac{1}{298 \text{ K}} \right)$$

$$= (3.62 \times 10^3 \text{ K}) \times (-2.60 \times 10^{-4} \text{ K}^{-1})$$

$$= -0.941$$

To obtain the ratio P_1/P_2 we have to take the antilogarithm of -0.941. This is done by raising e to the -0.941 power.

$$\frac{P_1}{P_2} = e^{-0.941} = 0.390$$

Solving the pressure ratio for P_2 gives

$$P_2 = \frac{P_1}{0.390} = \frac{148 \text{ torr}}{0.390}$$

$$= 379 \text{ torr}$$

Thus, the vapor pressure of hexane at 50 °C is calculated to be 379 torr.

Is the Answer Reasonable?
There is no simple way to estimate the answer, but there are a few checks we can make to see whether we've done the calculation correctly. First, we can check to be sure we've expressed the temperature in kelvins, which we have. We also see that the calculated vapor pressure, 379 torr, is higher at the higher temperature, which it should be. Therefore, the answer seems to be reasonable.

Practice Exercise 3

Use the data in the preceding example to calculate the vapor pressure of hexane at 60 °C. ◆

Intermolecular attractions and heats of vaporization and sublimation!

Energy Changes and Intermolecular Attractions

For a given substance, $\Delta H_{\text{vaporization}}$ is larger than ΔH_{fusion}. This is because the particles separate to greater distances during vaporization than during melting, so the changes in potential energy are greater. For the same substance, $\Delta H_{\text{sublimation}}$ is even larger than $\Delta H_{\text{vaporization}}$ because the attractive forces in the solid are larger than those in the liquid. As you would expect, overcoming these stronger forces requires more energy.

When a liquid evaporates or a solid sublimes, the particles go from a situation in which the attractive forces are very strong to one in which the attractive forces are so small they can almost be ignored. Therefore, the values of $\Delta H_{\text{vaporization}}$ and $\Delta H_{\text{sublimation}}$ give us directly the energy needed to separate molecules from each other. We can examine such values to obtain reliable comparisons of the strengths of intermolecular attractions.

In Table 11.3, notice that the heats of vaporization of water and ammonia are very large, which is just what we would expect for hydrogen-bonded substances. By comparison, CH_4, a nonpolar substance composed of atoms of similar size, has a very small heat of vaporization. Note also that polar substances such as HCl and SO_2 have fairly large heats of vaporization compared with nonpolar substances. For example, compare HCl with Cl_2. Even though Cl_2 contains two relatively large atoms, and therefore would be expected to have larger London forces than HCl, the latter substance has the larger $\Delta H_{\text{vaporization}}$. This must be due to dipole–dipole attractions between polar HCl molecules—attractions that are absent in nonpolar Cl_2.

Heats of vaporization also reflect the factors that control the strengths of London forces. For example, the data in Table 11.3 show the effect of chain length on the intermolecular attractions between hydrocarbons; as the chain length increases from one carbon in CH_4 to six carbons in C_6H_{14}, the heat of vaporization also increases. This is in agreement with the expected increase in the total strengths of the London forces. In our discussion of London forces, we also men-

The stronger the attractions, the more the potential energy will increase when the molecules become separated, and the larger will be the value of ΔH.

The Lewis structure of SO_2 tells us that the molecule is nonlinear, and this means that it must be polar.

Table 11.3 Some Typical Heats of Vaporization

Substance	$\Delta H_{\text{vaporization}}$ (kJ/mol)	Type of Attractive Force
H_2O	+ 43.9	Hydrogen bonding and London
NH_3	+ 21.7	Hydrogen bonding and London
HCl	+ 15.6	Dipole–dipole and London
SO_2	+ 24.3	Dipole–dipole and London
F_2	+ 5.9	London
Cl_2	+ 10.0	London
Br_2	+ 15.0	London
I_2	+ 22.0	London
CH_4	+ 8.16	London
C_2H_6	+ 15.1	London
C_3H_8	+ 16.9	London
C_6H_{14}	+ 30.1	London

tioned that their strengths increase as the electron clouds of the particles become larger, and we used the boiling points of the halogens to illustrate this. The heats of vaporization of the halogens in Table 11.3 are in agreement with this, too.

11.8 Dynamic Equilibrium and Le Châtelier's Principle

Throughout this chapter, we have studied various dynamic equilibria. One example was the equilibrium that exists between a liquid and its vapor in a closed container. You learned that when the temperature of a liquid is increased in this system, its vapor pressure also increases. Let's briefly review why this occurs.

Initially, the liquid is in equilibrium with its vapor, which exerts a certain pressure. When the temperature is increased, equilibrium no longer exists because evaporation occurs more rapidly than condensation. Eventually, as the concentration of molecules in the vapor increases, *the system reaches a new equilibrium* in which there is more vapor and a little less liquid. The greater concentration of the vapor causes a larger pressure.

The way a liquid's equilibrium vapor pressure responds to a temperature change is an example of a general phenomenon. *Whenever a dynamic equilibrium is upset by some disturbance, the system changes in a way that will, if possible, bring the system back to equilibrium again.* It's also important to understand that in the process of regaining equilibrium, the system undergoes a net change. Thus, when the temperature of a liquid is raised, there is some net conversion of liquid into vapor as the system returns to equilibrium. When the new equilibrium is reached, the amounts of liquid and vapor are not the same as they were before.

Throughout the remainder of this book, we will deal with many kinds of equilibria, both chemical and physical. It would be very time-consuming and sometimes very difficult to carry out a detailed analysis each time we wish to know the effects of some disturbance on an equilibrium system. Fortunately, there is a relatively simple and fast method for predicting the effect of a disturbance, one based on a principle proposed in 1888 by a brilliant French chemist, Henry Le Châtelier (1850–1936).

Although in the new equilibrium there is less liquid and more vapor, there is equilibrium *nonetheless.*

> **Le Châtelier's Principle**
> When a dynamic equilibrium in a system is upset by a disturbance, the system responds in a *direction* that tends to counteract the disturbance and, if possible, restore equilibrium.

Tools

Le Châtelier's principle

Let's see how we can apply Le Châtelier's principle to a liquid–vapor equilibrium that is subjected to a temperature increase. We cannot increase a temperature, of course, without adding heat. Thus, *the addition of heat is really the disturbing influence when a temperature is increased.* So let's incorporate "heat" as a member of the equation used to represent the liquid–vapor equilibrium.

$$\text{heat} + \text{liquid} \rightleftharpoons \text{vapor} \qquad (11.3)$$

Recall that double arrows, $\rightleftharpoons$, constitute the way we indicate a dynamic equilibrium in an equation. They imply opposing changes happening at equal rates. Evaporation is endothermic, so the heat is placed on the left side of Equation 11.3 to show that heat is absorbed by the liquid when it changes to the vapor, and that heat is released when the vapor condenses to a liquid.

Le Châtelier's principle tells us that when the temperature of the liquid–vapor equilibrium system is raised by adding heat, the system will try to adjust in a way that absorbs some of the added heat. This can happen if some liquid evaporates, because vaporization is endothermic. When liquid evaporates, the amount of vapor increases and the pressure rises. Thus, we have reached the correct conclusion in a very simple way, namely, that heating a liquid must increase its vapor pressure.

Recall that we often use the term **position of equilibrium** to refer to the relative amounts of the substances on opposite sides of the double arrows in an equilibrium expression such as Equation 11.3. Thus, we can think of how a disturbance affects the position of equilibrium. For example, increasing the temperature increases the amount of vapor and decreases the amount of liquid, and we say *the position of equilibrium has shifted;* in this case, it has shifted in the direction of the vapor, or it has *shifted to the right*. In using Le Châtelier's principle, it is often convenient to think of a disturbance as "shifting the position of equilibrium" in one direction or another in the equilibrium equation.

Practice Exercise 4

Use Le Châtelier's principle to predict how a temperature increase will affect the vapor pressure of a solid. (*Hint:* solid + heat ⇌ vapor.) ◆

11.9 Phase Diagrams

Tools

Phase diagramPhase diagram

Sometimes it is useful to know under what combinations of temperature and pressure a substance will be a liquid, a solid, or a gas, or the conditions of temperature and pressure that produce an equilibrium between any two phases. A simple way to determine this is to use a **phase diagram**—a graphical representation of the pressure–temperature relationships that apply to the equilibria between the phases of the substance.

Figure 11.24 is the phase diagram for water. On it, there are three lines that intersect at a common point. These lines give temperatures and pressures at which equilibria between phases can exist. For example, line *AB* is the vapor pressure curve for the solid (ice). Every point on this line gives a temperature and a pressure at which ice and its vapor are able to coexist in equilibrium. For instance, at − 10 °C ice has an equilibrium vapor pressure of 2.15 torr. This kind of information would be very useful to know if you wanted to design a system for making freeze-dried coffee (see Facets of Chemistry 11.2). It tells you that ice will sublime at − 10 °C if a vacuum pump can reduce the pressure of water vapor above the ice to less than 2.15 torr. The lower pressure disturbs the ice–vapor equilibrium by removing water vapor, and ice sublimes in a vain attempt to replace the water vapor and reestablish the equilibrium.

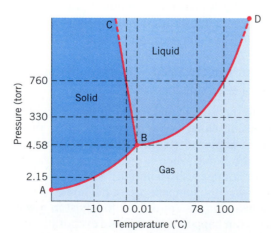

Figure 11.24 *The phase diagram for water, distorted to emphasize certain features.* Temperatures and pressures corresponding to the dashed lines on the diagram are referred to in the text discussion.

Facets of Chemistry 11.2

Freeze-drying

Anyone who has watched much television has certainly seen advertisements that praise the rich taste of freeze-dried instant coffee. Campers routinely carry freeze-dried foods because they are lightweight and require no refrigeration because bacteria cannot grow and reproduce in the complete absence of moisture (see Figure 1). Therefore, freeze-drying is clearly a useful way of preserving foods. It is accomplished by first freezing the food (or brewed coffee) and then placing it in a chamber that is connected to high capacity vacuum pumps. These pumps lower the pressure in the chamber to below the vapor pressure of ice, which causes the ice crystals to sublime. Drying foods in this way has an important advantage over other methods—the delicate molecules responsible for the flavor of foods are not destroyed, as they would be if the food were heated. Freeze-dried foods may also be easily reconstituted simply by adding water.

Freeze-drying is also used by biologists to preserve tissue cultures and bacteria. Removing moisture from tissue and storing the tissue at low temperatures allow the

Figure 1 Freeze-dried foods are often among the supplies carried by campers.

cells to survive in what amounts to a state of suspended animation for periods of at least several years.

Line *BD* is the vapor pressure curve for liquid water. It gives the temperatures and pressures at which the liquid and vapor are able to coexist in equilibrium. Notice that when the temperature is 100 °C, the vapor pressure is 760 torr. Therefore, this diagram also tells us that water boils at 100 °C when the pressure is 1 atm (760 torr), because that is the temperature at which the vapor pressure equals 1 atm. At the top of Mt. McKinley, in Alaska, the atmospheric pressure is only about 330 torr. The phase diagram tells us that the boiling point of water there will be 78 °C.

The melting point and boiling point can be read directly from the phase diagram.

The solid–vapor equilibrium line, *AB*, and the liquid–vapor line, *BD*, intersect at a common point, *B*. Because this point is on both lines, there is equilibrium between all three phases at the same time.

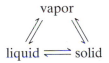

vapor

liquid ⇌ solid

The temperature and pressure at which this triple equilibrium occurs define the **triple point** of the substance. For water, the triple point occurs at 0.01 °C and 4.58 torr. Every known chemical substance except helium has its own characteristic triple point, which is controlled by the balance of intermolecular forces in the solid, liquid, and vapor.

In the SI, the triple point of water is used to define the Kelvin temperature of 273.16 K.

Line *BC*, which extends upward from the triple point, is the solid–liquid equilibrium line or *melting point line*. It gives temperatures and pressures at which the solid and the liquid are able to be in equilibrium. At the triple point, the melting of ice occurs at +0.01 °C (and 4.58 torr); at 760 torr, melting occurs very slightly lower, at 0 °C. Thus, we can tell that *increasing the pressure on ice lowers its melting point.*

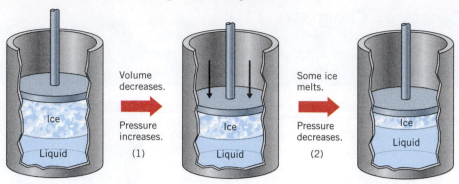

Figure 11.25   *The effect of pressure on the equilibrium* $H_2O(solid) \rightleftharpoons H_2O(liquid)$.
(1) Pushing down on the piston decreases the volume of both the ice and liquid water by a small amount and increases the pressure. (2) Some of the ice melts, producing the more dense liquid. As the total volume of ice and liquid water decreases, the pressure drops in accordance with Le Châtelier's principle.

The decrease in the melting point of ice that occurs when there is an increase in pressure can be predicted using Le Châtelier's principle and the knowledge that liquid water is more dense than ice. Let's reflect again on the equilibrium between ice and liquid water at 0 °C and 1 atm.

$$H_2O(s) \rightleftharpoons H_2O(l)$$

Imagine that the equilibrium is established in an apparatus like that shown in Figure 11.25 (and, again, keep in mind that ice is less dense than liquid water). If the piston is pushed in slightly, the pressure increases and the equilibrium is disturbed. According to Le Châtelier's principle, the system should respond in a way that tends to reduce the pressure. This can happen if some of the ice melts, so that the ice–liquid mixture won't require as much space. Then the molecules won't push as hard against each other, and the pressure will drop. Thus, a pressure increasing disturbance to the system favors a volume decreasing change, which corresponds to the melting of some ice.

Next, suppose we have ice at a pressure just below the solid–liquid line, *BC*. If, *at constant temperature,* we raise the pressure to a point just above the line, the ice will melt and become a liquid. This could only happen if the melting point becomes lower as the pressure is raised.

Water is very unusual. Almost all other substances have melting points that increase with increasing pressure. Consider, for example, the phase diagram for carbon dioxide (see Figure 11.26). For CO_2 the solid–liquid line slants to the

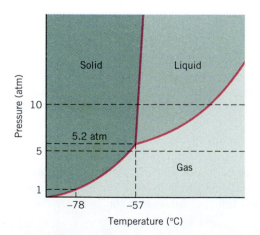

Figure 11.26   *The phase diagram for carbon dioxide.*

right (it slanted to the left for water). Also notice that carbon dioxide has a triple point that's above 1 atm. At atmospheric pressure, the only equilibrium that can be established is between solid carbon dioxide and its vapor. At a pressure of 1 atm this equilibrium occurs at a temperature of $-78\ °C$. This is the temperature of dry ice, which sublimes at atmospheric pressure at $-78\ °C$.

Besides specifying phase equilibria, the three intersecting lines on a phase diagram serve to define regions of temperature and pressure at which only a single phase can exist. For example, between lines *BC* and *BD* in Figure 11.24 are temperatures and pressures at which water exists as a liquid without being in equilibrium with either vapor or ice. At 760 torr, water is a liquid anywhere between $0\ °C$ and $100\ °C$. For instance, we are told by the diagram that we can't have ice with a temperature of $25\ °C$ if the pressure is 760 torr (which, of course, you already knew; ice never has a temperature of $25\ °C$). The diagram also says that we can't have water vapor with a pressure of 760 torr when the tempera-ture is $25\ °C$ (which, again, you already knew; the temperature has to be taken to $100\ °C$ for the vapor pressure to reach 760 torr). Instead, we are told by the phase diagram that the *only* phase for pure water at $25\ °C$ and 1 atm is the liquid. Below $0\ °C$ at 760 torr, water is a solid; above $100\ °C$ at 760 torr, water is a vapor. On the phase diagram for water, the phases that can exist in the different temperature–pressure regions are marked.

EXAMPLE 11.2

Interpreting a Phase Diagram

What phase would we expect for water at $0\ °C$ and 4.58 torr?

Analysis: The words, "What phase . . . ," as well as the specified temperature and pressure suggest that we refer to the phase diagram of water (Figure 11.24).

Solution: First we find $0\ °C$ on the temperature axis of the phase diagram of water. Then we move upward until we intersect a line corresponding to 4.58 torr. This intersection occurs in the "solid" region of the diagram. At $0\ °C$ and 4.58 torr, then, water exists as a solid.

EXAMPLE 11.3

Interpreting a Phase Diagram

What phase changes occur if water at $0\ °C$ is gradually compressed from a pressure of 2.15 torr to 800 torr?

Analysis: Asking about "what phase changes occur" suggests once again that we use the phase diagram of water (Figure 11.24).

Solution: At $0\ °C$ and 2.15 torr, water exists as a gas (water vapor). As the vapor is compressed, we move upward along the $0\ °C$ line until we encounter the solid–vapor line. Here, an equilibrium will exist as compression gradually transforms the gas into solid ice. Once all the vapor has frozen, further compression raises the pressure and we continue the climb along the $0\ °C$ line until we next encounter the solid–liquid line at 760 torr. As further compression takes place, the solid will gradually melt. After all the ice has melted, the pressure will continue to climb while the water remains a liquid. At 800 torr and $0\ °C$, the water will be liquid.

Practice Exercise 5

What phase changes will occur if water at $-20\ °C$ and 2.15 torr is heated to $50\ °C$ under constant pressure? ◆

Practice Exercise 6

What phase will water be in if it is at a pressure of 330 torr and a temperature of 50 °C? ◆

There is one final aspect of the phase diagram that has to be mentioned. The vapor pressure line for the liquid, which begins at point *B*, terminates at point *D*, which is known as the **critical point.** The temperature and pressure at *D* are called the **critical temperature, T_c,** and **critical pressure, P_c.** Above the critical temperature, a distinct liquid phase cannot exist, *regardless of the pressure.*

Figure 11.27 illustrates what happens to a substance as it approaches its critical point. In Figure 11.27*a*, we see a liquid in a container with some vapor above it. We can distinguish between the two phases because they have different densities, which causes them to bend light differently. This allows us to see the interface, or surface, between the more dense liquid and the less dense vapor. If this liquid is now heated, two things happen. First, more liquid evaporates. This causes an increase in the number of molecules per cubic centimeter of vapor which, in turn, causes the density of the vapor to increase. Second, the liquid expands (just like mercury does in a thermometer). This means that a given mass of liquid occupies more volume, so its density decreases. As the temperature of the liquid and vapor continue to increase, the vapor density rises and the liquid density falls; they approach each other. Eventually these densities become equal, and a separate liquid phase no longer exists; everything is the same (see Figure 11.27*b*). The highest temperature at which a liquid phase still exists is the critical temperature, and the pressure of the vapor at this temperature is the critical pressure. A substance that has a temperature above its critical temperature and a density near its liquid density is described as a **supercritical fluid.**

The values of the critical temperature and critical pressure are unique for every chemical substance. As with the other physical properties that we've discussed, the values T_c and P_c are controlled by the intermolecular attractions. Liquids with strong intermolecular attractions, like water, tend to have high critical temperatures. Substances with weak intermolecular attractions tend to have low critical temperatures.

Above the critical temperature, a substance cannot be liquefied by the application of pressure.

This interface between liquid and gas is called the meniscus.

Supercritical CO_2 has solvent properties that allow it to dissolve and remove caffeine from coffee beans. Some producers of decaffeinated coffee have switched from using potentially toxic solvents to the use of supercritical CO_2 because of the nontoxic nature of any traces of CO_2 that might remain in their decaffeinated product.

11.10 ○ Crystalline Solids

Solids have unique properties that are determined in large measure by the orderly packing of the particles within them. This order is revealed even when we examine the external features of crystals. For example, Figure 11.28 is a photo-

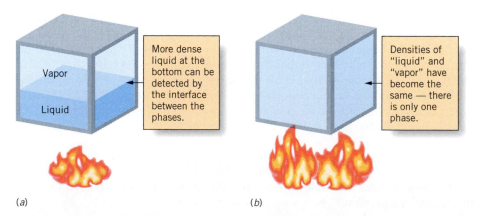

Figure 11.27 *Changes that are observed when a liquid is heated in a sealed container.* (*a*) Below the critical temperature. (*b*) Above the critical temperature.

(*a*)   (*b*)

graph of crystals of one of our most familiar chemicals, sodium chloride—ordinary table salt. Notice that each particle is very nearly a perfect little cube. You might think that the manufacturers went to a lot of trouble and expense to make such uniformly shaped crystals. Actually, they could hardly avoid it. Whenever a solution of NaCl is evaporated, the crystals that form have edges that intersect at 90° angles. Cubes, then, are the norm, not the exception, for NaCl.

When most substances freeze, or when they separate as a solid from a solution that is being evaporated, they normally form crystals that have highly regular features. The crystals have flat surfaces, for example, that meet at angles which are characteristic for the substance. The regularity of these surface features reflects the high degree of order among the particles that lie within the crystal. This is true whether the particles are atoms, molecules, or ions.

Figure 11.28 *Crystals of table salt.* The size of the tiny cubic sodium chloride crystals can be seen in comparison with a penny.

Crystal Lattices and Unit Cells

The particles in crystals are arranged in patterns that repeat over and over again in all directions. The resulting overall pattern is called a **crystal lattice.** Its high degree of regularity is the principal feature that makes solids different from liquids. A liquid lacks this long-range repetition of structure because the particles in a liquid are jumbled and disorganized as they move about.

Because there are millions of chemical compounds, it might seem that an enormous number of different kinds of lattices are possible. If this were true, studying the crystal lattices of solids would be hopelessly complex. Fortunately, however, the number of *kinds* of lattices that are mathematically possible is quite limited. This fact has allowed for a great deal of progress in understanding the structures of solids.

To describe the structure of a crystal it is convenient to view it as being composed of a huge number of simple, basic units called **unit cells.** By repeating this simple structural unit up and down, back and forth, in all directions, we can build the entire lattice. Figure 11.29, for example, shows the simplest and most symmetrical unit cell of all, called the **simple cubic unit cell.** This unit cell is a cube having points at each of its eight corners. Stacking these unit cells gives a *simple cubic lattice.* Figure 11.29c shows the packing of atoms in a substance that crystallizes in

The regular, symmetrical shapes of snowflakes are caused by the highly organized packing of water molecules within crystals of ice.

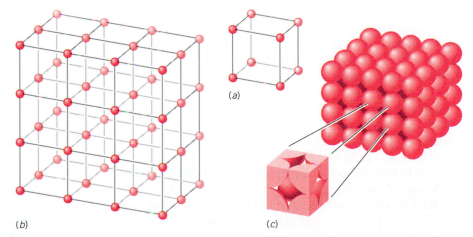

(a)

(b)

(c)

Figure 11.29 *A simple cubic unit cell.* (*a*) Locations of the lattice points in a simple cubic unit cell. (*b*) A portion of a simple cubic lattice built by stacking simple cubic unit cells. (*c*) A substance that forms crystals with a simple cubic lattice. In the unit cell for this substance, atoms have their nuclei at the lattice points. Notice that only a portion of each atom lies within this particular unit cell. The rest of each atom is located in adjacent unit cells.

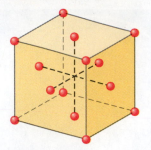

Figure 11.30 *A face-centered cubic unit cell.* Lattice points are found at each of the eight corners and in the center of each face.

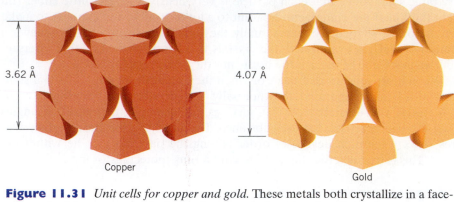

Copper

Gold

Figure 11.31 *Unit cells for copper and gold.* These metals both crystallize in a face-centered cubic structure with similar face-centered cubic unit cells. The atoms are arranged in the same way, but their unit cells have edges of different lengths because the atoms are of different sizes. (1 Å = 1×10^{-10} m.)

a simple cubic lattice and the unit cell for that substance. When the unit cell is "carved out" of the crystal we find only part of an atom (1/8th of an atom, actually) at each corner. The rest of each atom resides in adjacent unit cells.

Two other cubic unit cells are also possible: face-centered cubic and body-centered cubic. The **face-centered cubic** (abbreviated **fcc**) **unit cell** has identical particles at each of the corners plus another in the center of each face, as shown in Figure 11.30. Many common metals—copper, silver, gold, aluminum, and lead, for example—form crystals that have face-centered cubic lattices. Each of these metals has the same *kind* of lattice, but the *sizes* of their unit cells differ because the sizes of the atoms differ (see Figure 11.31).

The **body-centered cubic (bcc) unit cell** has identical particles at each corner plus one in the center of the cell, as illustrated in Figure 11.32. The body-centered cubic lattice is also common among a number of metals—examples are chromium, iron, and platinum. Again, these are substances with the same *kind* of lattice, but the dimensions of the lattices reflect the *different sizes* of the particular atoms.

Not all unit cells are cubic. Some have edges of different lengths or edges that intersect at angles other than 90°. Although you should be aware of the existence of other unit cells and lattices, we will limit the remainder of our discussion to cubic lattices and their unit cells.

We have seen that a number of metals have fcc or bcc lattices. The same is true for many compounds. Figure 11.33, for example, is a view of a portion of a sodium chloride crystal. The Cl^- ions (green) are located at the lattice points that correspond to a face-centered cubic unit cell. The smaller gray spheres represent Na^+ ions. Notice that they fill the spaces between the Cl^- ions. Sodium chloride is said to have a face-centered cubic lattice, and the cubic shape of this lattice is behind the cubic shape of a sodium chloride crystal.

Many of the alkali halides (Group IA–VIIA compounds), such as NaBr and KCl, crystallize with fcc lattices that have the same arrangement of ions as found in NaCl. Because sodium bromide and potassium chloride both have the same kind of lattice as sodium chloride, Figure 11.33 also could be used to describe the unit cells of NaBr and KCl. The *sizes* of their unit cells are different, however, because K^+ is a larger ion than Na^+ and Br^- is larger than Cl^-.

A major point to be learned from the preceding discussion is that *a single lattice type can be used to describe the structures of many different crystals.* For

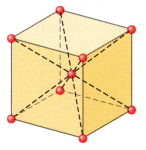

Figure 11.32 *A body-centered cubic unit cell.* Lattice points are located at each of the eight corners and in the center of the unit cell.

The way the ions are packed in NaCl is common to many compounds and is called the **rock salt structure.** (Rock salt is a common name for sodium chloride.)

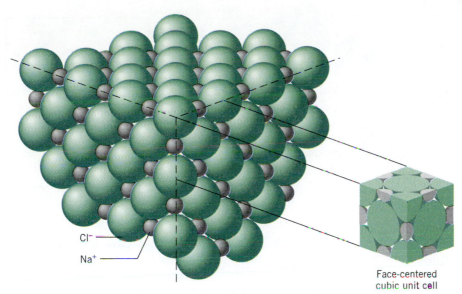

Cl⁻
Na⁺
Face-centered
cubic unit cell

Figure 11.33 *The packing of ions in a sodium chloride crystal.* A face-centered cubic unit cell can be seen in the structure of NaCl.

this reason, only a handful of different lattice types is all we need to describe the crystal structure of every possible element or compound.

11.11 X-Ray Diffraction

When atoms are bathed in X rays, they absorb some of the radiation and then emit it again in all directions. In effect, each atom becomes a tiny X-ray source. If we look at radiation from two such atoms (see Figure 11.34), we find that the X rays emitted are in phase in some directions but out of phase in others. In Chapter 7 you learned that constructive (in-phase) and destructive (out-of-phase) interferences create a phenomenon called *diffraction*. X-ray diffraction by crystals has enabled many scientists to win Nobel Prizes by determining the structures of extremely complex compounds in a particularly elegant way.

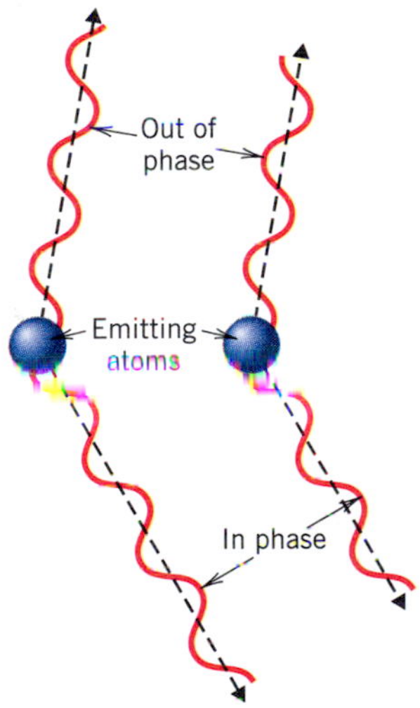

In a crystal, there are enormous numbers of atoms, but *they are evenly spaced throughout the lattice*. When the crystal is bathed in X rays, intense beams are diffracted because of constructive interference, and they appear only in specific directions. In other directions, no X rays appear because of destructive interference. When the X rays coming from the crystal fall on photographic film, the diffracted beams form a **diffraction pattern** (see Figure 11.35). The film is darkened only where the X rays strike.[2]

In 1913, the British physicist William Henry Bragg and his son William Lawrence Bragg discovered that just a few variables control the appearance of an X-ray diffraction pattern. These are shown in Figure 11.36, which illustrates the conditions necessary to obtain constructive interference of the X rays from successive layers of atoms (planes of atoms) in a crystal. A beam of X rays having a wavelength λ strikes the layers at an angle θ. Constructive interference causes an intense diffracted beam to emerge at the same angle θ. The Braggs derived an equation, now called the **Bragg equation,** relating λ, θ, and the distance between the planes of atoms, d,

$$n\lambda = 2d \sin \theta \qquad (11.4)$$

Out of phase

Emitting atoms

In phase

Figure 11.34 *Diffraction of X rays from atoms in a crystal.* X rays emitted from atoms are in phase in some directions and out of phase in other directions.

The two Braggs shared the 1915 Nobel Prize in physics.

[2]Modern X-ray diffraction instruments use electronic devices to detect and measure the intensities of the diffracted X rays.

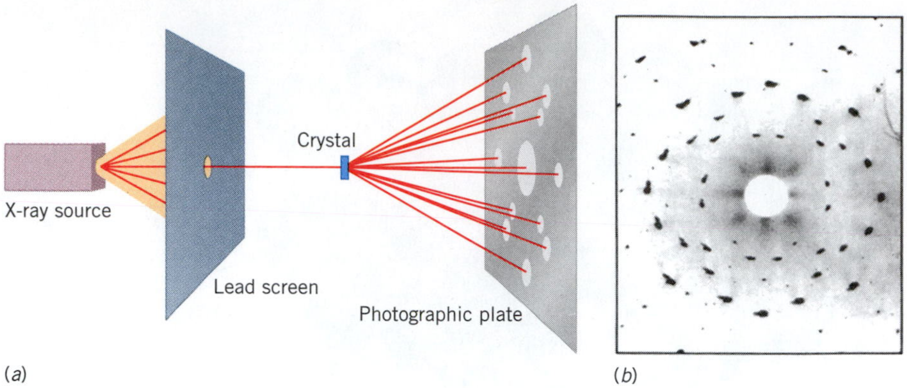

Crystal

X-ray source

Lead screen

Photographic plate

Figure 11.35 *X-ray diffraction.* (*a*) The production of an X-ray diffraction pattern. (*b*) An X-ray diffraction pattern produced by sodium chloride.

(*a*)

(*b*)

where *n* is a whole number. The Bragg equation is the basic tool used by scientists in the study of solid structures. Let's briefly see how they use it.

In any crystal, many different sets of planes can be passed through the atoms. Figure 11.37 illustrates this idea in two dimensions for a simple pattern of points. When a crystal produces an X-ray diffraction pattern, many spots are observed because of the diffraction from the many sets of planes. The physical geometry of the apparatus used to record the diffraction pattern allows the measurement of the angles at which the diffracted beams emerge from each distinct set of planes. Knowing the wavelength of the X rays, the values of *n* (which can be determined), and the measured angles, *θ*, the distances between planes of atoms, *d*, can be computed. The next step is to use the calculated interplanar distances to work backward to deduce where the atoms in the crystal must be located so that layers of atoms are indeed separated by these distances. If this sounds like a difficult task, it is! Some sophisticated mathematics as well as computers are needed to accomplish it. The efforts, however, are well rewarded because the calculations give the locations of atoms within the unit cell and the distances between them. This information, plus a lot of chemical "common sense," is used by chemists to arrive at the shapes and sizes of the molecules in the crystal. The general shape of DNA molecules, the chemicals of genes, was deduced by using X-ray diffraction. Today, X-ray diffraction is an important tool used by biochemists to determine structures of complex proteins and enzymes.

Figure 11.36 *Diffraction of X rays from successive layers of atoms in a crystal.* The layers of atoms are separated by a distance *d*. The X rays of wavelength λ enter and emerge at an angle *θ* relative to the layers of the atoms. For the emerging beam of X rays to have any intensity, the condition $n\lambda = 2d \sin \theta$ must be fulfilled, where *n* is a whole number.

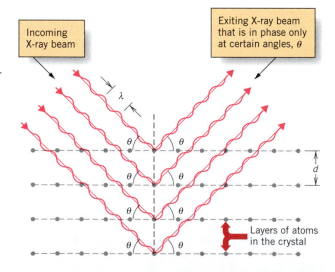

Incoming X-ray beam

Exiting X-ray beam that is in phase only at certain angles, *θ*

Layers of atoms in the crystal

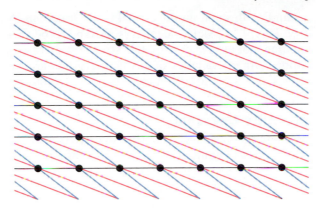

Figure 11.37 *A two-dimensional pattern of points with many possible sets of parallel lines.* Each set of lines is identified by a different color. In a three-dimensional crystal lattice there are many sets of parallel planes.

11.12 Physical Properties and Crystal Types

You know from personal experience that solids exhibit a wide range of properties. Some solids, like diamond, are very hard. Others, such as naphthalene (moth flakes) or ice, are soft by comparison, and are easily crushed. Some solids, like salt crystals or iron, have high melting points, whereas others, such as candle wax, melt at low temperatures. Some solids conduct electricity well, but others are nonconducting.

The physical properties just described depend on the kinds of particles in the solid as well as on the strengths of attractive forces holding the solid together. Even though we can't make exact predictions about such properties, some generalizations do exist. In discussing them, it is convenient to divide crystals into four types: ionic, molecular, covalent, and metallic.

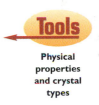

Tools

Physical properties and crystal types

Ionic Crystals

We already discussed some of the properties of **ionic crystals** in Chapter 2. They are relatively hard, have high melting points, and are brittle, properties reflecting strong attractive forces between ions of opposite charge as well as repulsions that occur when ions of like charge are near each other. You also learned that ionic compounds do not conduct electricity in their solid states but in their molten states conduct electricity well. This behavior is consistent with the mobility of ions in the liquid state but not in the solid state.

Molecular Crystals

Molecular crystals are solids in which the lattice sites are occupied either by atoms (as in solid argon or krypton) or by molecules (as in solid CO_2, SO_2, or H_2O). Such solids tend to be soft and have low melting points because the particles in the solid experience relatively weak intermolecular attractions. In crystals of argon, for example, the attractive forces are exclusively London forces. In SO_2, which is composed of polar molecules, there are dipole–dipole attractions as well as London forces. And in water crystals (ice) the molecules are held in place primarily by strong hydrogen bonds.

Molecular crystals are soft because little effort is needed to separate the particles or cause them to move past each other.

Covalent Crystals

Covalent crystals are solids in which lattice positions are occupied by atoms that are covalently bonded to other atoms at neighboring lattice sites. The result is a crystal that is essentially one gigantic molecule. These solids are sometimes called *network solids* because of the interlocking network of covalent bonds

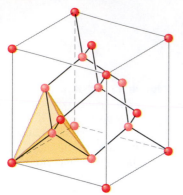

Figure 11.38 *The unit cell of diamond.* Notice that each carbon atom is covalently bonded to four others at the corners of a tetrahedron. This structure extends throughout an entire diamond crystal. (In diamond, of course, all the atoms are identical. They are different shades of gray here to make it easier to visualize the structure.)

In terms of the band theory, the electrons in the "electron sea" model correspond to the electrons in the conduction band.

The elements of Group IIA are also harder than their neighbors in Group IA.

extending throughout the crystal in all directions. A typical example is diamond (see Figure 11.38). Covalent crystals tend to be very hard and to have very high melting points because of the strong attractions between covalently bonded atoms. Other examples of covalent crystals are quartz (SiO_2, found in typical grains of sand) and silicon carbide (SiC, a common abrasive used in sandpaper).

Metallic Crystals

Metallic crystals have properties that are quite different from those of the other three types. Metallic crystals conduct heat and electricity well, and they have the luster characteristically associated with metals. A number of different models have been developed to explain metallic crystals, and in Chapter 9 the bonding in metals was discussed in terms of the *band theory of solids.* A simpler model, however, views the lattice positions of a metallic crystal as being occupied by *positive ions* surrounded by electrons in a cloud that spreads throughout the entire solid (see Figure 11.39). The electrons in this cloud belong to no single positive ion, but rather to the crystal as a whole. Because the electrons aren't localized on any atom, they are free to move easily, which accounts for the high electrical conductivity of metals. By their movement, the electrons can also transmit kinetic energy rapidly through the solid, so metals are also good conductors of heat. This model explains the luster of metals, too. When light shines on the metal, the loosely held electrons vibrate easily and readily reemit the light with essentially the same frequency and intensity.

It is not possible to make many simple generalizations about the melting points of metals. We've seen before that some, like tungsten, have very high melting points, whereas others, such as mercury, have quite low melting points. To some extent, the melting point depends on the charge of the positive ions in the metallic crystal. The atoms of the Group IA metals tend to exist as cations with a 1+ charge, and they are only weakly attracted to the "electron sea" that surrounds them. Atoms of the Group IIA metals, however, each lose two electrons to the metallic lattice and form ions with a 2+ charge. These more highly charged ions are attracted more strongly to the surrounding electron sea, so the Group IIA metals have higher melting points than their neighbors in Group IA. For example, magnesium melts at 650 °C whereas sodium, also in period 3, melts at only 98 °C. Those metals with very high melting points, like tungsten, must have very strong attractions between their atoms, which suggests that there probably is some covalent bonding between them as well.

The different ways of classifying crystals and a summary of their general properties are given in Table 11.4.

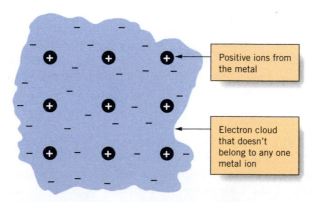

Figure 11.39 *The "electron sea" model of a metallic crystal.* In this highly simplified view of a metallic solid, metal atoms lose valence electrons to the solid as a whole and exist as positive ions surrounded by a mobile "sea" of electrons.

Positive ions from the metal

Electron cloud that doesn't belong to any one metal ion

Table 11.4 Types of Crystals

Crystal Type	Particles Occupying Lattice Sites	Type of Attractive Force	Typical Examples	Typical Properties
Ionic	Positive and negative ions	Attractions between ions of opposite charge	NaCl, CaCl$_2$, NaNO$_3$	Relatively hard; brittle; high melting points; nonconductors of electricity as solids, but conduct when melted
Molecular	Atoms or molecules	Dipole–dipole attractions, London forces, hydrogen bonding	HCl, SO$_2$, N$_2$, Ar, CH$_4$, H$_2$O	Soft; low melting points; nonconductors of electricity in both solid and liquid states
Covalent (network)	Atoms	Covalent bonds between atoms	Diamond, SiC (silicon carbide), SiO$_2$ (sand, quartz)	Very hard; very high melting points; nonconductors of electricity
Metallic	Positive ions	Attractions between positive ions and an electron cloud that extends throughout the crystal	Cu, Ag, Fe, Na, Hg	Range from very hard to very soft; melting points range from high to low; conduct electricity in both solid and liquid states; have characteristic luster

EXAMPLE 11.4

Identifying Crystal Types from Physical Properties

The metal osmium, Os, forms an oxide with the formula OsO$_4$. The soft crystals of OsO$_4$ melt at 40 °C, and the resulting liquid does not conduct electricity. To which crystal type does solid OsO$_4$ probably belong?

Analysis: You might be tempted to suggest that the compound is ionic simply because it is formed from a metal and a nonmetal. However, this is not always the case, especially when the metal has a large positive charge. Therefore, we must examine the properties of the compound and find the kind of solid that fits the properties.

Solution: The characteristics of the OsO$_4$ crystals—softness and low melting point—suggest that solid OsO$_4$ exists as molecular crystals that contain molecules of OsO$_4$. This is further supported by the fact that liquid OsO$_4$ does not conduct electricity, which is evidence for the lack of ions in the liquid.

Practice Exercise 7

Boron nitride, which has the empirical formula BN, melts under pressure at 3000 °C and is as hard as a diamond. What is the probable crystal type for this compound? ◆

Practice Exercise 8

Crystals of elemental sulfur are easily crushed and melt at 113 °C to give a clear yellow liquid that does not conduct electricity. What is the probable crystal type for solid sulfur? ◆

11.13 Noncrystalline Solids

If a cubic salt crystal is broken, the pieces still have flat faces that intersect at 90° angles. If you shatter a piece of glass, on the other hand, the pieces often have surfaces that are not flat. Instead, they tend to be smooth and curved (see Figure 11.40). This behavior illustrates a major difference between crystalline solids, like NaCl, and noncrystalline solids, or **amorphous solids,** such as glass.

The word *amorphous* is derived from the Greek word *amorphos,* which means "without form." Amorphous solids do not have the kinds of long-range repetitive internal structures that are found in crystals. In some ways their structures, being jumbled, are more like liquids than solids. Examples of amorphous solids are ordinary glass and many plastics. In fact, the word **glass** is often used as a general term to refer to any amorphous solid.

As suggested in Figure 11.40, substances that form amorphous solids often consist of long, chainlike molecules that are intertwined in the liquid state somewhat like long strands of cooked spaghetti. To form a crystal from the melted material, these long molecules would have to become untangled and line up in specific patterns. But as the liquid cools, the molecules slow down. Unless the liquid is cooled extremely slowly, the molecular motion decreases too rapidly for the untangling to take place, and the substance solidifies with the molecules still intertwined.

Compared with substances that produce amorphous solids, those that form crystalline solids behave quite differently when they are cooled. First, substances that form crystals solidify at a constant temperature, as you saw in Figure 11.22*b* on page 489. You also learned that sometimes the liquid can be cooled below its freezing point, and supercooling takes place. For most liquids, though, crystallization usually begins before long. But with substances that form amorphous solids, this never happens. The molecules cannot become untangled before they are frozen in place at a low temperature. Therefore, amorphous solids are sometimes described as **supercooled liquids,** a term suggesting the kind of structural disorder found in liquids. The term also suggests that the material's constituent molecules retain some residual ability at least to flex their chains if not to diffuse throughout the material and, over a long period of time, achieve an improved degree of crystalline orderliness. But the rate at which such change occurs is typically so small that under ordinary conditions it cannot be observed and the material is fully rigid. Glass, for example, which is a typical amorphous solid, over a very long time will develop regions of higher and higher order, as revealed by X-ray diffraction patterns of old glass. Figure 11.41 shows a typical cooling curve for an amorphous solid. No sharp breaks occur in the curve as the material is cooled until it becomes rigid.

Amorphous solids also soften gradually when heated. This is the reason you can heat glass tubing in a flame to soften it so that you can bend it. By contrast, if you warm an ice cube (crystalline water), it won't gradually become soft; instead, at 0 °C it will suddenly melt and drip all over you!

Within the solid, long molecules become tangled. Solid lacks the long-range order found in crystals.

Figure 11.40 *Glass is a noncrystalline solid.* When glass breaks, the pieces have sharp edges, but their surfaces are not flat planes. This is because in an amorphous solid like glass, long molecules are tangled and disorganized, so there is no long-range order characteristic of a crystal.

Some scientists prefer to reserve the term *solid* for crystalline substances, so they refer to glass as a *supercooled liquid.*

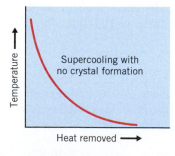

Supercooling with no crystal formation

Temperature →

Heat removed ⟶

Figure 11.41 *A cooling curve for a liquid that forms an amorphous solid.* There is no sharply defined freezing point.

SUMMARY

Physical Properties: Gases, Liquids, and Solids. Most physical properties depend primarily on intermolecular attractions. In gases, these attractions are weak because the molecules are so far apart. They are much stronger in liquids and solids, whose particles are packed together tightly.

Intermolecular Attractions. Polar molecules attract each other by **dipole–dipole attractions,** which arise because the positive end of one dipole attracts the negative end of another. Nonpolar molecules are attracted to each other by **London forces,** which are **instantaneous dipole–induced dipole attractions.** London forces are present between all particles, including atoms, polar and nonpolar molecules, and ions. London forces increase with increasing size of the particle's electron cloud; they also increase with increasing chain length among molecules such as the hydrocarbons. Compact molecules experience weaker London forces than similar long chain molecules. **Hydrogen bonding,** a special case of dipole–dipole attractions, occurs between molecules in which hydrogen is covalently bonded to a small, very electronegative atom—principally, nitrogen, oxygen, or fluorine. Hydrogen bonding is much stronger than the other types of intermolecular attractions. **Ion–dipole attractions** occur when ions interact with polar substances. **Ion–induced dipole attractions** result when an ion creates a dipole in a neighboring molecule or ion.

General Properties of Liquids and Solids. Properties that depend mostly on closeness of packing of particles are **compressibility** (or the opposite, incompressibility) and **diffusion.** Diffusion is slow in liquids and almost nonexistent in solids at room temperature. Properties that depend mostly on the strengths of intermolecular attractions are **retention of volume and shape, surface tension,** and **ease of evaporation.** Both solids and liquids retain volume; solids retain shape when transferred from one vessel to another. Surface tension is related to the energy needed to expand a liquid's surface area. Liquids can **wet** a surface if its molecules are attracted to the surface about as strongly as they are attracted to each other. Evaporation of liquids and solids is endothermic and produces a cooling effect. The overall rate of evaporation increases with increasing surface area. The rate of evaporation from a given surface area of a liquid increases with increasing temperature, and with decreasing intermolecular attractions. Evaporation of a solid is called **sublimation.**

Changes of State. Changes from one physical state to another, such as melting, vaporization, or sublimation, can occur as dynamic equilibria. In a **dynamic equilibrium,** opposite processes occur at equal rates, so there is no apparent change in the composition of the system. For liquids and solids, equilibria are established when vaporization occurs in a sealed container. A solid is in equilibrium with its liquid at the melting point.

Vapor Pressures. When the rates of evaporation and condensation of a liquid are equal, the vapor exerts a pressure called the **equilibrium vapor pressure** (or, more commonly, just the **vapor pressure**). The vapor pressure is controlled by the rate of evaporation *per unit surface area*. When the intermolecular attractive forces are large, the rate of evaporation is small and the vapor pressure is small. Vapor pressure increases with increasing temperature because the rate of evaporation increases as the temperature rises. The vapor pressure is independent of the *total* surface area of the liquid because as the surface area increases, the rates of evaporation and condensation are both increased equally; there is no change in the concentration of molecules in the vapor, so there is no change in the vapor pressure. Solids have vapor pressures just as liquids do.

Boiling Point. A substance boils when its vapor pressure equals the prevailing atmospheric pressure. The **normal boiling point** of a liquid is the temperature at which its vapor pressure equals 1 atm. Substances with high boiling points have strong intermolecular attractions.

Energy Changes Associated with Changes of State. On a heating curve, flat portions correspond to phase changes in which the heat added changes the potential energies of the particles without changing their average kinetic energy. On a cooling curve, **supercooling** occurs sometimes when the temperature of the liquid drops below the freezing point of the substance. The enthalpy changes for melting, vaporization of a liquid, and sublimation are the **molar heat of fusion,** ΔH_{fusion}, the **molar heat of vaporization,** $\Delta H_{\text{vaporization}}$, and the **molar heat of sublimation,** $\Delta H_{\text{sublimation}}$, respectively. They are all endothermic and are related in size as follows: $\Delta H_{\text{fusion}} < \Delta H_{\text{vaporization}} < \Delta H_{\text{sublimation}}$. The sizes of these enthalpy changes are large for substances with strong intermolecular attractive forces. The heat of vaporization is related to the vapor pressure by the **Clausius–Clapeyron equation.**

Le Châtelier's Principle. When the equilibrium in a system is upset by a disturbance, the system changes in a direction that minimizes the disturbance and, if possible, brings the system back to equilibrium. By this principle, we find that raising the temperature favors an endothermic change. Decreasing the volume favors a change toward a more dense phase.

Phase Diagrams. Temperatures and pressures at which equilibria can exist between phases are given graphically in a **phase diagram.** The three equilibrium lines intersect at the **triple point.** The liquid–vapor line terminates at the **critical point.** At the **critical temperature,** a liquid has a vapor pressure equal to its **critical pressure.** Above the critical temperature a liquid phase cannot be formed; the single phase that exists is called a **supercritical fluid.** The equilibrium lines also divide a phase diagram into temperature–pressure regions in which a substance can exist in just a single phase. Water is different from most substances in that its melting point decreases with increasing pressure.

Crystalline Solids. Crystalline solids have very regular features that are determined by the highly ordered arrangements of particles within their solids, which can be described in terms of repeating three-dimensional arrays called **lattices.** The simplest portion of the lattice is the **unit cell.** Three cubic unit cells are possible—**simple cubic, face-centered cubic,** and **body-centered cubic.** Many different substances can have the same kind of lattice. Information about crystal structures are obtained experimentally from X-ray diffraction patterns produced when crystals are bathed in X rays. Distances between planes of atoms can

be calculated by the Bragg equation, $n\lambda = 2d \sin \theta$, where n is a whole number, λ is the wavelength of the X rays, d is the distance between planes of atoms producing the diffracted beam, and θ is the angle at which the diffracted X-ray beam emerges relative to the planes of atoms producing the diffracted beam.

Crystal Types. Crystals can be divided into four general types: **ionic, molecular, covalent,** and **metallic.** Their properties depend on the kinds of particles within the lattice and on the attractions between the particles, as summarized in Table 11.4. A noncrystalline or **amorphous solid** is formed when a liquid is cooled without crystallization occurring. Such a solid has no sharply defined melting point and is called a **supercooled liquid.**

The table below lists the tools that you have learned in this chapter that are applicable to problem solving. Review them if necessary, and refer to them when working on the Thinking-It-Through problems and Review Problems that follow.

Tool	Function
Relationship between intermolecular forces and molecular structure *(pages 468, 469, and 470)*	To predict and compare the strengths of intermolecular attractions in substances based on their molecular structures and molecular shapes. You should be able to predict the types of intermolecular forces (dipole–dipole, London, and hydrogen bonding) experienced by substances based on molecular structures.
Factors that affect rates of evaporation and vapor pressure *(pages 480, 481, and 484)*	To compare the relative rates of evaporation based on the strengths of intermolecular forces in substances. You should also be able to compare strengths of intermolecular forces based on the relative magnitudes of vapor pressures at a given temperature.
Boiling points of substances *(page 488)*	To compare the strengths of intermolecular forces in substances based on their boiling points.
Enthalpy changes during phase changes *(page 490)*	To compare the strengths of intermolecular forces in substances based on relative values of $\Delta H_{\text{vaporization}}$ and $\Delta H_{\text{sublimation}}$.
Clausius–Clapeyron equation *(page 491)*	To determine the heat of vaporization from vapor pressure versus temperature data; to determine the vapor pressure at some temperature if $\Delta H_{\text{vaporization}}$ and the vapor pressure at another temperature are known.
Le Châtelier's principle *(page 495)*	To predict the direction in which the position of equilibrium is shifted when a dynamic equilibrium is upset by a disturbance. You should be able to predict how the position of equilibrium between phases is affected by temperature and pressure changes.
Phase diagram *(page 496)*	To identify temperatures and pressures at which equilibrium can exist between phases of a substance, and to identify conditions under which only a single phase can exist.
Physical properties of solids *(page 505)*	To judge the type of crystalline solid formed by a substance.

THINKING IT THROUGH

Remember, the goal for each of the following problems is to assemble the available information needed to obtain the answer, state what additional data (if any) are needed, and describe how you would use the data to answer the question. For problems involving unit conversions, list the relationships among the units that are needed to carry out the conversions. Construct the conversion factors that can be formed from these relationships. Then set up the solution to the problem by arranging the conversion factors so the units cancel correctly to give the desired units of the answer.

The problems are divided into two groups. Those in Level 2 are more challenging than those in Level 1 and provide an opportunity to really hone your problem solving skills. Detailed answers to these problems can be found at our Web site: http://www.wiley.com/college/brady

Level 1 Problems

1. When hot coffee in an insulated cup gradually cools, the cooler liquid at the surface circulates to the bottom as fresh hot liquid rises to the top. Why?

2. Vigorous exercise on a hot muggy day produces more discomfort than the same exercise on an equally hot but dry day. Why?

3. If you pour out half the contents of a filled can of gasoline and then seal the container, you will find that when the container is reopened sometime later there is a noticeable hiss, as gases escape from the can. Why is there an increase in the pressure inside the can even if its temperature hasn't changed?

4. London forces are generally weaker than hydrogen bonds, yet water boils at 100 °C whereas $C_{20}H_{42}$ (a component of candle wax) boils at 343 °C. Why?

5. At a given temperature, which of the liquids below is likely to have the greater vapor pressure?

$$CH_3CH_2{-}O{-}CH_2CH_3 \qquad CH_3CH_2{-}\overset{\overset{\displaystyle H}{|}}{N}{-}CH_2CH_3$$

6. A flask is partially filled with water and fitted with a stopcock. With the stopcock open, the water is brought to a boil. After a few minutes heating is stopped and the stopcock is closed. The flask is allowed to cool for a short time and then ice water is poured over the top of it. This causes the water to boil vigorously! Why?

7. Why are indoor relative humidities generally very low in Minnesota in winter, even when outside the humidity is 100% and it's snowing?

8. Why do clouds form when the humid air of a weather system known as a *warm front* encounters the cool, relatively dry air of a *cold front?*

9. When a liquid is spread over a larger surface area, it evaporates more rapidly. Why, then, does the surface area of a liquid in a sealed container not influence the equilibrium vapor pressure of the liquid?

10. Some liquid water is added to a large copper vessel. After waiting for a time period sufficient for the water to have reached equilibrium with its vapor it was discovered that the vapor pressure of the water was less than the equilibrium vapor pressure. What is the most probable reason for this?

11. When one mole of a liquid freezes, less energy is given off than when one mole of it condenses from a vapor to a liquid. Why?

12. Which of the following is likely to have the larger molar heat of vaporization?

$$CH_3CH_2CH_2{-}OH \qquad CH_3CH_2CH_2{-}SH$$

13. Which of the following is likely to have the larger molar heat of vaporization?

$$CH_3CH_2CH_2CH_2CH_2CH_2CH_2CH_2{-}Br$$
$$CH_3CH_2CH_2CH_2{-}SH$$

14. Room temperature is above hydrogen's critical temperature. If H_2 at a pressure of 1 atm is compressed at room temperature, will a pressure eventually be reached at which it will condense to a liquid?

Level 2 Problems

15. At a given temperature, which of the liquids below is likely to have the higher vapor pressure?

$$H{-}\overset{\overset{\displaystyle Cl}{|}}{\underset{\underset{\displaystyle H}{|}}{C}}{-}Cl \qquad H{-}\overset{\overset{\displaystyle Cl}{|}}{\underset{\underset{\displaystyle Cl}{|}}{C}}{-}Cl$$

16. If a gas such as Freon, CCl_2F_2, is allowed to expand freely into a vacuum, it cools slightly. Why does this cooling prove there are attractive forces between the CCl_2F_2 molecules?

17. Calcium oxide crystallizes with the same structure as NaCl. The radius of a Ca^{2+} ion is approximately 99 pm and that of an O^{2-} ion is approximately 140 pm. Using these values, estimate the density of CaO.

18. Melting point is sometimes used as an indication of the extent of covalent bonding in a compound—the higher the melting point, the more ionic the substance. On this basis, oxides of metals seem to become less ionic as the charge on the metal ion increases. Thus, Cr_2O_3 has a melting point of 2266 °C whereas CrO_3 has a melting point of only 196 °C. The explanation often given is similar in some respects to explanations of the variations in the strengths of certain intermolecular attractions given in this chapter. Provide an explanation for the greater degree of electron sharing in CrO_3 as compared with Cr_2O_3.

19. The apparatus shown in the figure below can be used to demonstrate that the equilibrium between a liquid and its vapor pressure is a *dynamic equilibrium*. The well of the left tube is filled with normal water, H_2O, and the well of the right tube with heavy water, D_2O. (The symbol D stands for deuterium, ^2H, an isotope of ordinary hydrogen.) The water samples are frozen and the air is pumped out of both sides of the apparatus. The water samples are allowed to melt, and equilibria are established between the two liquid samples and their respective vapors, at which point the pressures in both sides reach steady values. The large-bore stopcock between the two sides is then opened. From time to time, samples of liquid are drained from the bottom taps for analysis. Predict what periodic analyses of the two liquids will show and explain how these results will show that there is a dynamic equilibrium between the liquids and the vapors above them.

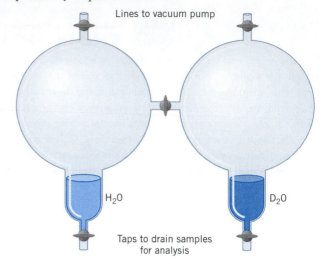

Lines to vacuum pump

H_2O

D_2O

Taps to drain samples
for analysis

REVIEW QUESTIONS

Comparisons among the States of Matter

11.1 List four items in your room whose physical properties account for their use in a particular application.

11.2 Under what conditions would we expect gases to obey the gas laws best?

11.3 Why is the behavior of a gas affected very little by its chemical composition?

11.4 Why are the intermolecular attractive forces stronger in liquids and solids than they are in gases?

Intermolecular Attractions

11.5 Describe *dipole–dipole attractions*.

11.6 What are *London forces?* How are they affected by molecular size?

11.7 What are *hydrogen bonds?*

11.8 Which nonmetal atoms in molecules besides hydrogen, are most often involved in hydrogen bonding? Why these and not others?

11.9 Which is expected to have the higher boiling point, C_8H_{18} or C_4H_{10}? Explain your choice.

11.10 Ethanol and dimethyl ether have the same molecular formula, C_2H_6O. Ethanol boils at 78.4 °C, whereas dimethyl ether boils at − 23.7 °C. Their structural formulas are

$$CH_3CH_2OH \qquad CH_3OCH_3$$
ethanol dimethyl ether

Explain why the boiling point of the ether is so much lower than the boiling point of ethanol.

11.11 How do the strengths of covalent bonds and dipole–dipole attractions compare? How do the strengths of ordinary dipole–dipole attractions compare with the strengths of hydrogen bonds?

11.12 Explain why London forces are called instantaneous dipole–induced dipole forces. What effect does the polarizability of atoms have on the strengths of London forces?

11.13 What are *ion–induced dipole attractions?* How are the strengths of these forces affected by the amount of charge on the cation, assuming the cations to be of equal size?

11.14 In which compound are the ion–induced dipole attractions stronger, CaO or CaS?

General Properties of Liquids and Solids

11.15 Name two physical properties of liquids and solids that are controlled primarily by how tightly packed the particles are. Name three that are controlled mostly by the strengths of the intermolecular attractions.

11.16 Why does diffusion occur more slowly in liquids than in gases? Why does diffusion occur extremely slowly in solids?

11.17 Compare how shape and volume change when (a) a gas, (b) a liquid, and (c) a solid is transferred from one container to another.

11.18 Why are liquids and solids so difficult to compress?

11.19 On the basis of kinetic theory, would you expect the rate of diffusion in a liquid to increase or decrease as the temperature is increased? Explain your answer.

11.20 What is *surface tension?* Why do molecules at the surface of a liquid behave differently from those within the interior?

11.21 What kinds of observable effects are produced by the surface tension of a liquid?

11.22 What relationship is there between surface tension and the intermolecular attractions in the liquid?

11.23 Which liquid is expected to have the larger surface tension at a given temperature, CCl_4 or H_2O? Explain your answer.

11.24 What does *wetting* of a surface mean? What is a *surfactant?* What is its purpose and how does it function?

11.25 Polyethylene plastic consists of long chains of carbon atoms, each of which is also bonded to hydrogens, as shown below:

$$\cdots-\underset{\underset{H}{|}}{\overset{\overset{H}{|}}{C}}-\underset{\underset{H}{|}}{\overset{\overset{H}{|}}{C}}-\underset{\underset{H}{|}}{\overset{\overset{H}{|}}{C}}-\underset{\underset{H}{|}}{\overset{\overset{H}{|}}{C}}-\underset{\underset{H}{|}}{\overset{\overset{H}{|}}{C}}-\underset{\underset{H}{|}}{\overset{\overset{H}{|}}{C}}-\underset{\underset{H}{|}}{\overset{\overset{H}{|}}{C}}-\underset{\underset{H}{|}}{\overset{\overset{H}{|}}{C}}-\cdots$$

Water forms beads when placed on a polyethylene surface. Why?

11.26 The structural formula for glycerol is

$$H-\underset{\underset{OH}{|}}{\overset{\overset{H}{|}}{C}}-\underset{\underset{OH}{|}}{\overset{\overset{H}{|}}{C}}-\underset{\underset{OH}{|}}{\overset{\overset{H}{|}}{C}}-H$$

Would you expect this liquid to wet glass surfaces? Explain your answer.

11.27 On the basis of what happens on a molecular level, why does evaporation lower the temperature of a liquid?

11.28 On the basis of the distribution of kinetic energies of the molecules of a liquid, explain why increasing the liquid's temperature increases the rate of evaporation.

11.29 How is the rate of evaporation of a liquid affected by increasing the surface area of the liquid? How is the rate of evaporation affected by the strengths of intermolecular attractive forces?

11.30 During the cold winter months, snow often disappears gradually without melting. How is this possible? What is the name of the process responsible for this phenomenon?

11.31 How is freeze-drying accomplished? What are its advantages?

11.32 Why do puddles of water evaporate more quickly in the summer than in the winter?

Changes of State and Equilibrium

11.33 What terms do we use to describe the following changes of state: (a) solid → gas, (b) liquid → gas, (c) gas → liquid, (d) solid → liquid, (e) liquid → solid?

11.34 Why do molecules of a vapor that collide with the surface of a liquid tend to be captured by the liquid?

11.35 When an equilibrium is established in the evaporation of a liquid into a sealed container, we refer to it as a *dynamic equilibrium*. Why?

11.36 At a molecular level, what is happening when a dynamic equilibrium is established between the liquid and solid forms of a substance? What is the temperature called at which there is an equilibrium between a liquid and a solid?

11.37 Is it possible to establish an equilibrium between a solid and its vapor?

Vapor Pressure

11.38 Define *equilibrium vapor pressure.*

11.39 Explain why changing the volume of a container in which there is a liquid–vapor equilibrium has no effect on the vapor pressure.

11.40 Why doesn't a change in the surface area of a liquid cause a change in the equilibrium vapor pressure?

11.41 What effect does increasing the temperature have on the vapor pressure of a liquid? Why?

11.42 Which compound should have the higher vapor pressure at 25 °C, butanol or diethyl ether? Their structures are

$$\underset{\text{butanol}}{CH_3CH_2CH_2CH_2-OH} \qquad \underset{\text{diethyl ether}}{CH_3CH_2-O-CH_2CH_3}$$

11.43 Why does humid air at 30 °C contain more moisture per cubic meter than humid air at 25 °C?

11.44 Why does rain often form when humid air is forced to rise over a mountain range?

11.45 Why does moisture condense on the outside of a cool glass of water in the summertime?

11.46 Why do we feel more uncomfortable in humid air at 90 °F than in dry air at 90 °F?

11.47 Why does the air in a heated building in the winter have such a low humidity?

Boiling Points of Liquids

11.48 Define *boiling point* and *normal boiling point.*

11.49 Why does the boiling point vary with atmospheric pressure?

11.50 Mt. Kilimanjaro in Tanzania is the tallest peak in Africa (19,340 ft). The normal barometric pressure at the top of this mountain is about 345 torr. At what Celsius temperature would water be expected to boil there? (See Figure 11.18.)

11.51 Why is the boiling point a useful property with which to identify liquids?

11.52 Explain how a pressure cooker works.

11.53 The radiator cap of an automobile engine is designed to maintain a pressure of approximately 15 lb/in.^{2} above normal atmospheric pressure. How does this help prevent the engine from "boiling over" in hot weather?

11.54 Butane, C_4H_{10}, has a boiling point of −0.5 °C (which is 31 °F). Despite this, liquid butane can be seen sloshing about inside a typical butane lighter, even at room temperature. Why isn't the butane boiling inside the lighter at room temperature?

11.55 Why does H_2S have a lower boiling point than H_2Se? Why does H_2O have a much higher boiling point than H_2S?

11.56 An H—F bond is more polar than an O—H bond, so HF forms stronger hydrogen bonds than H_2O. Nevertheless, HF has a lower boiling point than H_2O. Explain why this is so.

Energy Changes That Accompany Changes of State

11.57 Below is a cooling curve for one mole of a substance.

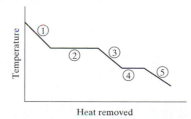

(a) On which portions of this graph do we find the average kinetic energy of the molecules of the substance changing?

(b) On which portions of this graph is the amount of heat removed related primarily to a lowering of the potential energy of the molecules?

(c) Which portion of the graph corresponds to the release of the heat of vaporization?

(d) Which portion of the graph corresponds to the release of the heat of fusion?

(e) Which is larger, the heat of fusion or the heat of vaporization?

(f) On the graph, indicate the melting point of the solid.

(g) On the graph, indicate the boiling point of the liquid.

(h) On the drawing, indicate how supercooling of the liquid would affect the graph.

11.58 What are the name and symbol given to the enthalpy change involved in (a) the conversion of one mole of liquid to one mole of vapor, (b) the conversion of one mole of solid to one mole of vapor, and (c) the conversion of one mole of solid to one mole of liquid?

11.59 Why is $\Delta H_{vaporization}$ larger than ΔH_{fusion}? How does $\Delta H_{sublimation}$ compare with $\Delta H_{vaporization}$?

11.60 Would the "heat of condensation," $\Delta H_{condensation}$, be exothermic or endothermic?

11.61 Which would be expected to have a higher molar heat of vaporization, water (bp 100 °C) or ethyl alcohol (bp 78.4 °C)? Explain your answer.

11.62 Hurricanes can travel for thousands of miles over warm water, but they rapidly lose their strength when they move over a large land mass or over cold water. Why?

11.63 Rain forms when water vapor condenses in clouds. Explain why this is the source of energy for producing winds in storms.

11.64 Ethanol (grain alcohol) has a molar heat of vaporization of 39.3 kJ/mol. Ethyl acetate, a common solvent, has a molar heat of vaporization of 32.5 kJ/mol. Which of these substances has the larger intermolecular attractions?

11.65 A burn caused by steam is much more serious than one caused by the same amount of boiling water. Why?

11.66 Arrange the following substances in order of their increasing values of $\Delta H_{vaporization}$: (a) HF, (b) CH_4, (c) CF_4, (d) HCl.

Le Châtelier's Principle

11.67 State Le Châtelier's principle in your own words.

11.68 What do we mean by the *position of equilibrium?*

11.69 Use Le Châtelier's principle to predict the effect of adding heat in the equilibrium: solid + heat ⇌ liquid.

Phase Diagrams

11.70 For most substances, the solid is more dense than the liquid. Use Le Châtelier's principle to explain why the melting point of such substances should *increase* with increasing pressure.

11.71 Define *critical temperature* and *critical pressure*.

11.72 What phases of a substance are in equilibrium at the triple point?

11.73 Why doesn't CO_2 have a normal boiling point?

11.74 At room temperature, hydrogen can be compressed to very high pressures without liquefying. On the other hand, butane becomes a liquid at high pressure (at room temperature). What does this tell us about the critical temperatures of hydrogen and butane?

Crystalline Solids and X-Ray Diffraction

11.75 What surface features do crystals have that suggest a high degree of order among the particles within them?

11.76 What is a *crystal lattice?* What is a *unit cell?* What relationship is there between a unit cell and a crystal lattice?

11.77 Describe simple cubic, face-centered cubic, and body-centered cubic unit cells. (Make a sketch of each.)

11.78 Make a sketch of a layer of sodium ions and chloride ions in a NaCl crystal. Indicate how the ions are arranged in a face-centered cubic pattern.

11.79 How do the crystal structures of copper and gold differ? In what way are they similar? On the basis of the location of the elements in the periodic table, what kind of crystal structure would you expect for silver?

11.80 Only 14 different kinds of crystal lattices are possible. How can this be true, considering the fact that there are millions of different chemical compounds that are able to form crystals?

11.81 Write the Bragg equation and define the symbols.

11.82 Explain, in general terms, how an X-ray diffraction pattern of a crystal and the Bragg equation provide information that allows chemists to figure out the structures of molecules.

***11.83** Why can't $CaCl_2$ or $AlCl_3$ form crystals with the same structure as NaCl? (*Hint:* Count the number of ions of Na^+ and Cl^- in the unit cell of NaCl.)

Crystal Types

11.84 What kinds of particles are located at the lattice sites in a metallic crystal?

11.85 What kinds of attractive forces exist between particles in (a) molecular crystals, (b) ionic crystals, and (c) covalent crystals?

11.86 Why are covalent crystals sometimes called network solids?

11.87 Magnesium chloride is ionic. What general physical properties are expected of its crystals?

Amorphous Solids

11.88 What does the word *amorphous* mean?

11.89 What is an amorphous solid? What happens when a substance that forms an amorphous solid is cooled from the liquid state to the solid state?

11.90 Why is glass sometimes called a supercooled liquid? Compare what happens when glass is heated in a flame with what occurs when a crystal, such as ice, is heated.

REVIEW PROBLEMS

Answers to problems whose numbers are printed in color are given in Appendix D. More challenging problems are marked with asterisks.

Intermolecular Attractions and Molecular Structure

11.91 Which liquid evaporates faster at 25 °C, diethyl ether (an anesthetic) or butanol (a solvent used in the preparation of shellac and varnishes)? Both have the molecular formula $C_4H_{10}O$, but their structural formulas are different, as shown below.

$$CH_3CH_2CH_2CH_2OH \qquad CH_3CH_2—O—CH_2CH_3$$
butanol   diethyl ether

11.92 Which should have the higher boiling point, butanol or diethyl ether?

11.93 What kinds of intermolecular attractive forces (dipole–dipole, London, hydrogen bonding) are present in the following substances?
(a) HF   (b) PCl_3 (c) SF_6 (d) SO_2

11.94 What kinds of intermolecular attractive forces are present in the following substances?

(a) $CH_3—\overset{\overset{\textstyle O}{\|}}{C}—OH$ (b) H_2S (c) SO_3 (d) CH_3NH_2

11.95 Which should have the higher boiling point, ethanol (CH_3CH_2OH, found in alcoholic beverages) or ethanethiol (CH_3CH_2SH, a foul-smelling liquid found in the urine of rabbits that have feasted on cabbage)?

11.96 How do the strengths of London forces compare in $CO_2(l)$ and $CS_2(l)$? Which of these is expected to have the higher boiling point? (Check your answer by referring to the *Handbook of Chemistry and Physics,* which is available in your school library.)

11.97 Below are the vapor pressures of some relatively common chemicals measured at 20 °C. Arrange these substances in order of increasing intermolecular attractive forces.

Benzene, C_6H_6	80 torr
Acetic acid, $HC_2H_3O_2$	11.7 torr
Acetone, C_3H_6O	184.8 torr
Diethyl ether, $C_4H_{10}O$	442.2 torr
Water	17.5 torr

11.98 The boiling points of some common substances are given here. Arrange these substances in order of increasing strengths of intermolecular attractions.

Ethanol, C_2H_5OH	78.4 °C
Ethylene glycol, $C_2H_4(OH)_2$	197.2 °C
Water	100 °C
Diethyl ether, $C_4H_{10}O$	34.5 °C

Energy Changes That Accompany Changes of State

11.99 The molar heat of vaporization of water at 25 °C is + 43.9 kJ/mol. How many kilojoules of heat would be required to vaporize 125 mL (0.125 kg) of water?

11.100 The molar heat of vaporization of acetone, C_3H_6O, is 30.3 kJ/mol at its boiling point. How many kilojoules of heat would be liberated by the condensation of 5.00 g of acetone?

*11.101 Suppose 45.0 g of water at 85 °C is added to 105.0 g of ice at 0 °C. The molar heat of fusion of water is 6.01 kJ/mol, and the specific heat of water is 4.18 J/g °C. On the basis of these data, (a) what will be the final temperature of the mixture, and (b) how many grams of ice will melt?

11.102 A cube of solid benzene (C_6H_6) at its melting point of 5.45 °C and weighing 10.0 g is placed in 10.0 g of water at 30 °C. Given that the heat of fusion of benzene is 9.92 kJ/mol, to what temperature will the water have cooled by the time all of the benzene has melted?

Clausius–Clapeyron Equation

11.103 Ethyl acetate, a solvent for certain plastics, has a vapor pressure of 72.8 torr at 20 °C and a vapor pressure of 186.2 torr at 40 °C. Estimate the value of ΔH_{vap} for this solvent in units of kilojoules per mole.

11.104 Isopropyl alcohol, used in rubbing alcohol, has a heat of vaporization of 42.09 kJ/mol and a vapor pressure of 31.6 torr at 20 °C. What is the vapor pressure of this alcohol at 60 °C?

11.105 The vapor pressure of liquid nitrogen is 28.1 torr at − 216 °C and 47.0 torr at − 213 °C. Calculate the heat of vaporization of N_2 in units of kJ/mol. Estimate the normal boiling point of liquid nitrogen.

11.106 Diethyl ether, a liquid that has been used as an anesthetic, has a vapor pressure of 185 torr at 0 °C. At 10 °C its vapor pressure is 0.384 atm. What is the heat of vaporization of diethyl ether in kJ/mol?

11.107 The heat of vaporization of toluene (used to make trinitrotoluene, TNT) is 38.0 kJ/mol. The vapor pressure of toluene is 20.0 torr at 18.4 °C. What will its vapor pressure be at 25 °C?

11.108 Referring to the data in the preceding problem, determine the temperature at which toluene will have a vapor pressure of 40 torr.

Phase Diagrams

11.109 Sketch the phase diagram for a substance that has a triple point at −15.0 °C and 0.30 atm, melts at −10.0 °C at 1 atm, and has a normal boiling point of 90 °C.

11.110 Based on the phase diagram of the preceding problem, below what pressure will the substance undergo sublimation? How does the density of the liquid compare with the density of the solid?

11.111 According to Figure 11.26, what phase(s) should exist for CO_2 at (a) $-60\,°C$ and 6 atm, (b) $-60\,°C$ and 2 atm, (c) $-40\,°C$ and 10 atm, and (d) $-57\,°C$ and 5.2 atm?

11.112 Looking at the phase diagram for CO_2 (Figure 11.26), how can we tell that solid CO_2 is more dense than liquid CO_2?

11.113 Describe what happens when the pressure on CO_2 is gradually raised from 1 atm to 20 atm at a constant temperature of $-56\,°C$. What happens if the pressure is increased from 1 atm to 20 atm at a constant temperature of $-58\,°C$?

11.114 Describe what happens when CO_2 at 2 atm is warmed from $-80\,°C$ to $0\,°C$ at constant pressure.

Crystalline Solids and X-Ray Diffraction

11.115 How many atoms are contained within a simple cubic unit cell? (*Hint:* To answer the question, add up the parts of atoms shown in Figure 11.29c.)

11.116 How many copper atoms are within the face-centered cubic unit cell of copper? (*Hint:* See Figure 11.31, and add up all the *parts* of atoms in the fcc unit cell.)

11.117 Copper crystallizes with a face-centered cubic unit cell. The length of the edge of a unit cell is 362 pm. Sketch the face of a unit cell, showing the nuclei of the copper atoms at the lattice points. The atoms are in contact along the diagonal from one corner to another. The length of this diagonal is four times the radius of a copper atom. What is the atomic radius of copper?

11.118 Silver forms face-centered cubic crystals. The atomic radius of a silver atom is 144 pm. Draw the face of a unit cell with the nuclei of the silver atoms at the lattice points. The atoms are in contact along the diagonal. Calculate the length of an edge of this unit cell.

*11.119 Potassium ions have a radius of 133 pm, and bromide ions have a radius of 195 pm. The crystal structure of potassium bromide is the same as for sodium chloride. Estimate the length of the edge of the unit cell in potassium bromide.

*11.120 The unit cell edge in sodium chloride has a length of 564.0 pm. The sodium ion has a radius of 95 pm. What is the *diameter* of a chloride ion?

Crystal Types

11.121 Tin(IV) chloride, $SnCl_4$, has soft crystals with a melting point of $-30.2\,°C$. The liquid is nonconducting. What type of crystal is formed by $SnCl_4$?

11.122 Elemental boron is a semiconductor, is very hard, and has a melting point of about $2250\,°C$. What type of crystal is formed by boron?

11.123 Gallium crystals are shiny and conduct electricity. Gallium melts at $29.8\,°C$. What type of crystal is formed by gallium?

11.124 Titanium(IV) bromide forms soft orange-yellow crystals that melt at $39\,°C$ to give a liquid that doesn't conduct electricity. The liquid boils at $230\,°C$. What type of crystals does $TiBr_4$ form?

11.125 Columbium is another name for one of the elements. This element is shiny, soft, and ductile. It melts at $2468\,°C$, and the solid conducts electricity. What kind of solid does columbium form?

11.126 Elemental phosphorus consists of soft white "waxy" crystals that are easily crushed and melt at $44\,°C$. The solid does not conduct electricity. What type of crystal does phosphorus form?

ADDITIONAL EXERCISES

11.127 Make a list of *all* of the attractive forces that exist in solid Na_2SO_3.

11.128 Should acetone molecules be attracted to water molecules more strongly than to other acetone molecules? Explain your answer. The structure of acetone is shown below in problem 11.129.

*11.129 The intermolecular forces of attraction in ethylene glycol are much stronger than in liquid acetone. Therefore, more energy is absorbed when a given amount of the glycol is converted to a vapor than when the same amount of acetone is changed to a vapor. Yet, if you spill acetone on your hand it produces a much greater cooling effect than if you spill an equivalent amount of the glycol on your hand. Why?

11.130 Acetic acid has a heat of fusion of 10.8 kJ/mol and a heat of vaporization of 24.3 kJ/mol. Use Hess's law to estimate the value for the heat of sublimation of acetic acid, in kilojoules per mole.

*11.131 When warm moist air sweeps in from the ocean and rises over a mountain range, it expands and cools. Explain how this cooling is related to the attractive forces between gas molecules. Why does this cause rain to form? When the air drops down the far side of the range, its pressure rises as it is compressed. Explain why this causes the air temperature to rise. How does the humidity of this air compare with the air that originally came in off the ocean? Now, explain why the coast of California is lush farmland, whereas valleys (such as Death Valley) that lie to the east of the tall Sierra Nevada mountains are arid and dry.

*11.132 If you shake a CO_2 fire extinguisher on a cool day, you can feel the liquid CO_2 sloshing around inside. However, if you shake the same fire extinguisher on a very hot summer day, you will not feel the liquid sloshing around. What is responsible for this difference?

ethylene glycol   acetone

11.133 The critical temperature of acetone is 236 °C, whereas the critical temperature of propane, C_3H_8, is 97 °C. Why are they so different?

11.134 Isopropyl alcohol (rubbing alcohol) has a heat of vaporization of 42.09 kJ/mol and a vapor pressure of 31.6 torr at 20 °C. Estimate the normal boiling point of isopropyl alcohol.

11.135 Propane, the fuel used in gas barbecues, has the following vapor pressures:

Temperature (°C)	Vapor pressure (torr)
− 92.4	40
− 87.0	60
− 79.6	100
− 61.2	300
− 47.3	600

Graphically determine the value of ΔH_{vap} for propane in units of kJ/mol.

*11.136** At 25 °C, liquid A has a vapor pressure of 100 torr while liquid B has a vapor pressure of 200 torr. The heat of vaporization of liquid A is 32.0 kJ/mol and that of liquid B is 18.0 kJ/mol. At what Celsius temperature will A and B have the same vapor pressure?

*11.137** Gold crystallizes in a face-centered cubic lattice. The edge of the unit cell has a length of 407.86 pm. The density of gold is 19.31 g/cm^3. Use these data and the atomic mass of gold to calculate the value of Avogadro's number.

Ozone, O_3, is a powerful oxidizing agent able to split molecules apart that have carbon–carbon double bonds. Because such double bonds occur widely in all living things, ozone is dangerous to plants and animals. The U.S. National Ambient Air Quality Standard for ozone is a daily maximum 1-hour average ozone concentration of only 0.12 ppm (120 ppb).[1] Dozens of U.S. urban areas exceed this at least once per year. The U.S. Environmental Protection Agency proposed in 1998 to reduce this standard to 0.08 ppm over an 8-hour average.

Ozone Formation. How does ozone originate in the air where we live? The *direct* and only source of ozone in the lower atmosphere is the combination of oxygen atoms with oxygen molecules when they collide at the surface of some particle, *M*.

$$O + O_2 + M \longrightarrow O_3 + M \qquad (1)$$

M is any molecule, like N_2 or O_2, that can absorb some of the kinetic energy involved in the collision. Thus the earlier question, "How does ozone originate?" becomes a new question, "How are oxygen *atoms* generated?"

Oxides of Nitrogen. Once oxygen atoms are generated, there will be ozone. The chief source of oxygen atoms is the breakup of molecules of nitrogen dioxide, an air pollutant, a breakup enabled by solar energy.

$$NO_2 + \text{solar energy} \longrightarrow NO + O \qquad (2)$$

The NO_2 for Equation 2 forms in air from nitrogen monoxide, NO, which is produced inside vehicle engine cylinders where the high temperature and pressure force the direct combination of nitrogen and oxygen.

$$N_2 + O_2 \xrightarrow[\text{pressure}]{\text{high temperature}} 2NO \qquad (3)$$

[1] *Ambient* means "all surrounding, all encompassing." The symbol *ppm* stands for "parts per million," and *ppb* means "parts per billion." Thus 0.12 ppm means 0.12 mL ozone in 10^6 mL of air.

As soon as newly made NO, now in the exhaust gas, hits the cooler outside air, it reacts further with oxygen to form nitrogen dioxide, NO_2, which gives smog its reddish-brown color (see Figure 6a).

$$2NO + O_2 \longrightarrow 2NO_2 \qquad (4)$$

Thus, NO_2 forms in air made smoggy by vehicle exhaust. But Reaction 2 converts NO_2 back to nitrogen monoxide. Because Reactions 2 and 4 do not occur at identical rates, some NO is always present in smoggy air, and *nitrogen monoxide is able to destroy ozone.*

$$NO + O_3 \longrightarrow NO_2 + O_2 \qquad (5)$$

The reactions of Equations 1, 2, and 5, when added together, give us *no net chemical effect!* (Try it.) How then does atmospheric ozone develop at all?

The reactions of Equations 1–5 proceed at their own rates. The rates are unequal, however, so the reactions actually do not exactly cancel each other. The net effect is that a small, steady concentration of ozone develops even in clean air. The ozone level in clean air, however, is very low, between 20 and 50 ppb. In the more polluted urban areas, levels as high as 400 ppb commonly occur for brief periods each year.

Unburned Hydrocarbons and the Ozone in Smog. Equation 5 represents the reaction that can make most ozone disappear, *but other substances are able to remove the NO needed for this reaction before the NO is used to destroy ozone.* It is the removal of NO other than by Reaction 5 that enables a buildup in the ozone level of the lower atmosphere.

Net destroyers of nitrogen monoxide are themselves present in vehicle exhaust, namely, the unburned or partially oxidized hydrocarbons remaining from the incomplete combustion of the fuel. In sunlight and oxygen, some unburned hydrocarbon molecules are changed into organic derivatives of hydrogen peroxide called *peroxy radicals,* usually symbolized as R—O—O (or simply RO_2).[2] Peroxy radicals originate chiefly by the

[2] The term "radical" is used for any species with an unpaired electron. Sometimes an electron dot is added to the formula of a radical, as in $RO_2·$. The chlorine *atom,* for example, is a radical and sometimes symbolized as Cl·. It has seven valence electrons, six occurring as three pairs (not shown) plus a "lone" unpaired electron.

chemicals in our world

Figure 6a The effect of smog on visibility. (Top) A clear day. (Bottom) A day of heavy smog.

reaction of unburned hydrocarbons with hydroxyl radicals, HO. These arise by a number of mechanisms in polluted air, but we will not go into the details of how HO radicals form. Suffice it to say, HO radicals and hydro-

carbons lead to ROO radicals, and these destroy NO molecules as follows.

$$ROO + NO \longrightarrow RO + NO_2$$

This reaction reduces the supply of ozone-destroying NO.

The development of ozone in smog thus depends on the formation of ROO radicals whose formation, in turn, depends on hydrocarbon emissions. Controlling such emissions has been a priority since 1975. Auto makers now install catalytic converters between the engine and the tailpipe that help to oxidize the hydrocarbons and carbon monoxide in exhaust gases to carbon dioxide and water. Other catalysts are able to aid in the reduction of nitrogen oxides to nitrogen. The catalysts are deactivated ("poisoned") by lead compounds, so "no-lead" gasoline must now be used.

Questions

1. What chemical reaction inside the cylinder of a vehicle engine launches the production of ozone in smog?

2. How does the NO_2 in smog form? (Write an equation.)

3. How is NO_2 involved in the production of ozone in smog? (Write equations.)

4. What chemical property of ozone makes it dangerous?

5. What is the connection between the presence of unburned hydrocarbons in smog and the ozone in smog? (Use equations as part of the answer.)

When a solution has a beautiful color, it's a bonus to a chemist. But it's the inner beauty of what goes on at the molecular level of a solution, both when it is forming and when it is made, that is more fascinating.

Solutions

This Chapter in Context We have studied two of the three *kinds* of matter, namely, elements and compounds (Chapters 2, 7, 8, and 9). In this chapter we'll look at the third kind, *mixtures,* of which *solutions* are simply special types. We have delayed their study because concepts surrounding intermolecular attractive forces (Chapters 10 and 11) as well as certain principles of thermochemistry (Chapter 6) illuminate the properties of solutions. One goal in this chapter is to learn how chemical composition makes some substances very soluble in water and others insoluble.

We've already introduced the most common terms used for solutions, for example, *solute, solvent,* and *solubility* (Section 3.9). If you aren't sure of them, it's important to review them now.

12.1 Formation of Solutions

A **solution** is a homogeneous mixture in which all of the particles are very small, typically on the order of atoms, ions, and small molecules, those with average diameters in the range of 0.05 to 0.25 nm. As solute particles become larger than those of this range, mixtures take on the special properties of colloidal dispersions (Section 12.10) and suspensions.

A Tendency toward Randomness, One Driving Force behind the Formation of a Solution

To understand the effect of solute and solvent on solubility, we need a detailed study of the *process* by which solutions form. We'll begin by taking a closer look at how gases spontaneously mix with or dissolve in each other.

Imagine a container holding two unmixed gases separated by a removable panel (see Figure 12.1*a*). When the panel is slid away (Figure 12.1*b*) the process of mixing begins *spontaneously* because of the random motions of the molecules. The gases thus mingle with no outside help. Once they have formed a homogeneous solution, their molecules will never spontaneously separate from

Homogeneous means "everywhere alike."

521

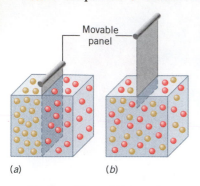

Figure 12.1 *Mixing of gases.* When two gases, initially in separate compartments (*a*), suddenly find themselves in the same container (*b*), they mix spontaneously.

one another and re-form the original, unmixed state. We say "never" because the unmixed state in the same container is, statistically, so overwhelmingly unlikely.

The spontaneous mixing of gases illustrates one of nature's strong "driving forces" for change, namely, *the tendency of a system, left to itself, to become increasingly disordered.*[1] At the instant the panel is removed (Figure 12.1), we have two gases in the same container, but they are separated and on opposite sides. This actually represents considerable *order,* like lines of boys and girls before a sixth grade dance. Because of the natural motions of the molecules, the ordered state suddenly becomes highly improbable as soon as the panel is removed. Now the vastly more probable distribution is one in which the molecules are thoroughly mixed.

Attractions between Solute and Solvent—Another Driving Force behind the Formation of a Solution

The only factor we need to understand how *gaseous* solutions form is nature's driving force for disorder. Another possible factor, attractive forces between particles, is negligible because such forces are very weak in gases.

Attractive forces are very important, however, to the formation of liquid solutions because such forces hold solute and solvent particles close together (see Figure 12.2). This works against getting them to intermingle. Yet ions and polar molecules must separate from each other if intermingling is to occur and a solution is to form. Success in preparing a nongaseous solution, therefore, depends not only on the drive toward disorder but also on the strengths of the intermolecular attractive forces in *both solvent and solute.*

Solutions of Liquids in Liquids

If we mix ethyl alcohol (beverage alcohol) and water they completely blend or dissolve in any proportion we choose; we say that the two are completely *miscible.* Benzene, C_6H_6, on the other hand, is a liquid that is virtually insoluble in water; the two are *immiscible.* To understand this different behavior, let's look closely at what happens in each case when the liquids are combined. Overall, we know that the molecules of solute and solvent must be pushed apart to make room for those of the other if a solution is to form.

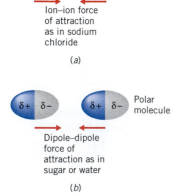

(a)

(b)

Figure 12.2 *Opposite charges attract.* (*a*) Interionic force of attraction. (*b*) Intermolecular force of attraction.

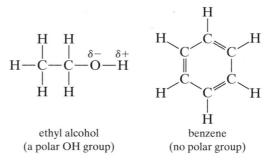

ethyl alcohol
(a polar OH group)

benzene
(no polar group)

Ethyl alcohol has a molecular structure that includes a polar O—H group. Its molecules, therefore, can form hydrogen bonds with water molecules, which also have O—H groups (see Figure 12.3). Water molecules can therefore attract alcohol molecules almost as strongly as they attract each other. Thus, the forces that must be overcome when water molecules are pushed apart to make room for alcohol molecules are compensated by similar attractions to the incoming alcohol molecules. The alcohol molecules are therefore able to move into the water and form the solution, and water molecules can similarly move into the alcohol. These two liquids, therefore, dissolve in each other. Because forces

[1]In Chapter 18 this driving force will be called the *entropy effect.*

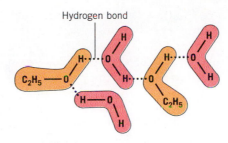

Figure 12.3 *Hydrogen bonds in aqueous ethyl alcohol.* Ethyl alcohol molecules experience hydrogen bonding ($\cdots$) to water molecules.

of attraction are easily accommodated in this system, nature's strong tendency toward the greater disorder of a solution is free to work.

Quite a different situation occurs if we try to dissolve benzene in water. At room temperature and pressure, the two liquids do not mix. Benzene molecules have no O—H groups and are otherwise nonpolar. Between them are only relatively weak London forces of attraction. Benzene molecules are therefore not attracted to water molecules very strongly. To accommodate the benzene molecules in a true solution, the water molecules would have to move apart to make room for them. This would require overcoming strong intermolecular attractions, which cannot be balanced by attractions to the incoming benzene molecules. As a result, nature's tendency toward greater disorder is prevented from having its way.[2]

It is also easy to see why a solution of water in benzene cannot be made. Suppose that we did manage to disperse water molecules in benzene. As they move about, the water molecules would occasionally encounter each other. Because they attract each other so much more strongly than they attract benzene molecules, water molecules would stick together at each such encounter by hydrogen bonds. This would continue to happen until all the water was in a separate phase. A solution of water in benzene would thus not be stable, and it would not form spontaneously.

Although benzene cannot dissolve in water, it does dissolve quite well in nonpolar liquids, like carbon tetrachloride, CCl_4. The forces of attraction between CCl_4 molecules are about as weak as those between benzene molecules. Therefore CCl_4 molecules can easily leave their own kind and mingle with the benzene molecules. Nature's drive toward disorder now easily overcomes the small resistance, and the solution readily forms.

In summary, we see that when the strengths of intermolecular attractions are *similar* in solute and solvent, solutions can form as illustrated by the easy formation of alcohol–water and benzene–carbon tetrachloride solutions. This is what is behind a rule of thumb that chemists widely use, namely the **"like dissolves like" rule:** when solute and solvent have molecules "like" each other in polarity, they tend to form a solution; when their molecules are quite different in polarity, solutions of any appreciable concentration do not form. The rule has long enabled chemists to use chemical composition and molecular structure to predict the likelihood of two substances dissolving in each other.

Tools

"Like dissolves like" rule

Solutions of Solids in Liquids

In moving to solutions of solids in liquids the basic principles remain the same. We'll look first at what happens when sodium chloride, a crystalline salt, dissolves in water.

Figure 12.4 depicts a section of a crystal of NaCl in contact with water. The dipoles of water molecules orient themselves so that the negative ends of some

[2]The full explanation for the immiscibility of benzene (or oil) and water is complicated. It requires more background in thermodynamics (Chapter 18) than we have thus far been able to provide. For a discussion with leading references, however, see T. P. Silverstein, *Journal of Chemical Education*, January 1998, page 116.

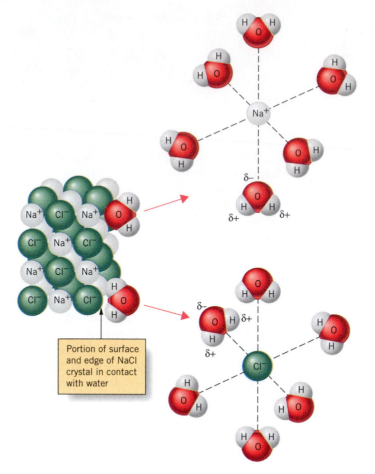

Figure 12.4 *Hydration of ions.* Hydration involves a complex redirection of forces of attraction and repulsion. Before this solution forms, water molecules are attracted only to each other, and Na$^+$ and Cl$^-$ ions have only each other in the crystal to be attracted to. In the solution, the ions have water molecules to take the places of their oppositely charged counterparts; in addition, water molecules find ions more attractive than even other water molecules.

Portion of surface and edge of NaCl crystal in contact with water

Water molecules collide everywhere along the crystal surface, but *successful* collisions—those that dislodge ions—are more likely to occur at corners and edges.

We're clearly at the borderline between chemical and physical changes when we place the binding of water molecules to sodium or chloride ions in the realm of physical changes.

point toward Na$^+$ ions and the positive ends of others point at Cl$^-$ ions. In other words, *ion–dipole* attractions occur that tend to tug and pull ions from the crystal. At the corners and edges of the crystal, ions are held by fewer neighbors within the solid and so are more readily dislodged than those elsewhere on the crystal's surface. As water molecules dislodge these ions, new corners and edges are exposed, and the dissolution of the crystal continues.

As they become free, the ions become completely surrounded by water molecules (also shown in Figure 12.4). The phenomenon is called the **hydration** of ions. The *general* term for the surrounding of a solute particle by solvent molecules is **solvation,** so hydration is just a special case of solvation. Ionic compounds are able to dissolve in water when the attractions between water dipoles and ions overcome the attractions of the ions for each other within the crystal.

Similar events explain why solids composed of polar molecules, like those of sugar, dissolve in water (see Figure 12.5). Attractions between the solvent and solute dipoles help to dislodge molecules from the crystal and bring them into solution. Again we see that "like dissolves like"; a polar solute dissolves in a polar solvent.

The same reasoning explains why nonpolar solids like wax are soluble in nonpolar solvents such as benzene. Wax is a solid mixture of long chain hydrocarbons. Their molecules attract each other weakly by London forces, so the wax molecules easily slip away from the solid even though the attractions between the molecules of the solvent (benzene) and the solute (wax) are weak themselves.

When intermolecular attractive forces within solute and solvent are sufficiently different, the two do not form a solution. For example, ionic solids or very polar molecular solids (like sugar) are insoluble in nonpolar solvents such as benzene, gasoline, or lighter fluid. The molecules of these solvents, all hydrocarbons, are unable to attract ions or very polar molecules with enough force to overcome the much stronger attractions that the ions or polar molecules experience within their own crystals. Nature's drive toward disorder can thus be stymied by interionic or intermolecular forces.

12.2 Heats of Solution

Because intermolecular attractive forces are important when liquids and solids are involved, the formation of a solution is inevitably associated with energy or enthalpy exchanges. The enthalpy exchanged between the system and its surroundings when one mole of a solute dissolves in a solvent at constant pressure to make a dilute solution is called the *molar enthalpy of solution,* or usually just the **heat of solution, ΔH_{soln}**.

It costs energy, that is, it is *endothermic,* to separate the particles of solute and also those of the solvent and make them spread out to make room for each other. This change would *increase* the potential energy of the system (because things that naturally attract each other are being pulled apart). But once the spread-apart particles come back together as a *solution,* the attractive forces between approaching solute and solvent particles yield a decrease in the system's potential energy, and this is an *exothermic* change. The heat of solution, ΔH_{soln}, is simply the net result of these two opposing enthalpy contributions. To look at this more closely, let us follow the potential energy changes that occur when a solid dissolves in a liquid.

Lattice Energies and Solvation Energies

In Chapter 6 you learned that enthalpy is a *state function.* So the magnitude of a *change* in enthalpy, like ΔH_{soln}, does not depend on *how* the system goes from one state to another. We may devise any route or path we please for the *process* of forming a solution as long as we go from the same initial state—separated samples of solute and solvent—to the same final state, the solution.

To take advantage of available energy data, we will devise a hypothetical two-step path and use an enthalpy diagram (see Figure 12.6). Overall, the two steps will take us from the solid solute and liquid solvent to the final solution, but you'll see that they do not represent the way a solution is actually made in the lab. In the first step, we imagine that the solid separates into its individual particles; in effect, we imagine that the solid is vaporized. In the second step, we imagine that the gaseous solute particles enter the solvent where they become solvated. As we said, we imagine this particular process to let us use existing experimental data, being permitted to do so only because the heat of solution is a state function.

As you can see from Figure 12.6, the first step is endothermic and *increases* the potential energy of the system. The particles in the solid attract each other, so energy must be supplied to separate them, energy called the *lattice energy.*[3] By receiving the requisite lattice energy, the solid is converted to its gaseous state.

In the second step, gaseous solute particles enter the solvent and are solvated. The potential energy of the system now *decreases,* because solute and solvent particles attract each other. This step is therefore exothermic, and its energy is called the **solvation energy,** the general term. The special term **hydration energy** is used for aqueous solutions.

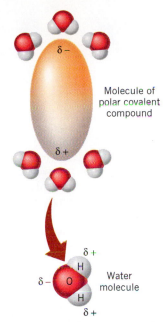

Figure 12.5 *Hydration of a polar molecule.* A polar molecule of a molecular compound can trade the forces of attraction it experiences for other molecules of its own kind for forces of attraction to molecules of water in an aqueous solution.

$$\Delta H_{soln} = H_{soln} - H_{components}$$

ΔH is positive for an endothermic change, so the arrow for Step 1 in Figure 12.6 points upward, the positive direction.

---

[3] The concept of lattice energy applies to any kind of crystalline solid, whether it is ionic or molecular. In Section 8.1 we discussed lattice energies as contributing to the stabilities of ionic compounds.

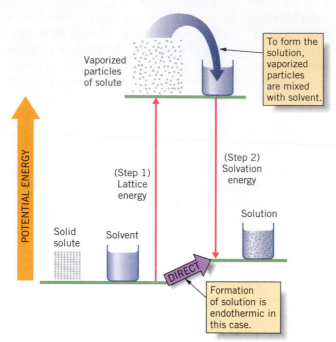

Figure 12.6 *Enthalpy diagram for a solid dissolving in a liquid.* In the real world, the solution is formed directly as indicated by the purple arrow. We can analyze the energy change by imagining the two separate steps, because enthalpy changes are functions of state and are independent of path. The energy change along the direct path is the algebraic sum of Step 1 and Step 2.

The magnitude of ΔH_{soln} depends somewhat on the final concentration of the solution being made.

The *heat of solution,* the net energy change, is the difference between the energy required for Step 1 and the energy released in Step 2. When the energy required for Step 1 exceeds the energy released in Step 2, the solution forms endothermically. But when more energy is released in Step 2 than is needed for Step 1, the solution forms exothermically.

We can test this analysis by comparing the experimental heats of solution of some salts in water with those calculated from lattice and hydration energies (Table 12.1). The calculations are described in Figures 12.7 and 12.8, using enthalpy diagrams for the formation of aqueous solutions of two salts, KI and NaBr.

The agreement between calculated and measured values in Table 12.1 is not particularly impressive in an absolute sense. This is partly because lattice and hy-

Table 12.1 Lattice Energies, Hydration Energies, and Heats of Solution for Some Group IA Metal Halides

Compound	Lattice Energy (kJ mol^{-1})^a	Hydration Energy (kJ mol^{-1})	ΔH_{soln}^{b} Calculated ΔH_{soln} (kJ mol^{-1})	Measured ΔH_{soln} (kJ mol^{-1})
LiCl	+ 833	− 883	− 50	− 37.0
NaCl	+ 766	− 770	− 4	+ 3.9
KCl	+ 690	− 686	+ 4	+ 17.2
LiBr	+ 787	− 854	− 67	− 49.0
NaBr	+ 728	− 741	− 13	− 0.602
KBr	+ 665	− 657	+ 8	+ 19.9
KI	+ 632	− 619	+ 13	+ 20.33

^{a}These values are the lattice energies given opposite signs. True lattice energies have negative values, because they relate to the exothermic *formation* of crystalline lattices from their constituent, gaseous ions (or molecules).

^{b}Heats of solution refer to the formation of extremely dilute solutions.

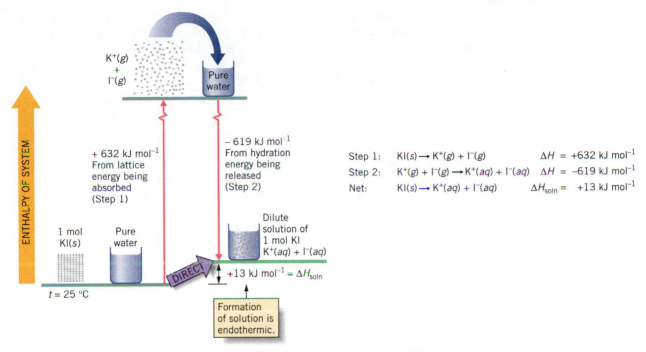

Figure 12.7 *Enthalpy of solution.* The formation of aqueous potassium iodide.

dration energies are not precisely known and partly because the model used in our analysis is evidently too simple. (The model, for example, postulates nothing about how nature's tendency toward randomness influences the formation of the solution.) Notice, however, that when "theory" predicts relatively large heats of solution, the experimental values are also relatively large, and that both val-

Small percentage errors in very large numbers can cause huge percentage changes in the *differences* between such numbers.

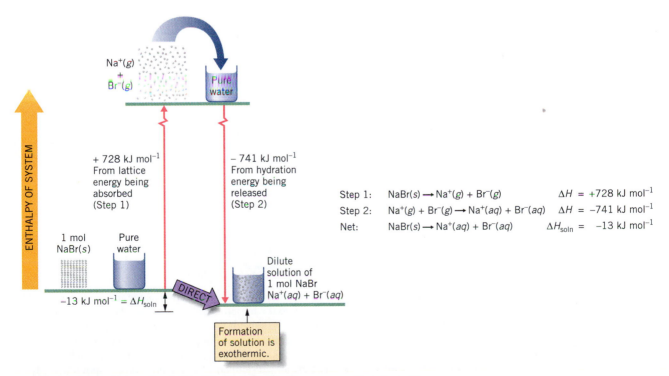

Figure 12.8 *Enthalpy of solution.* The formation of aqueous sodium bromide.

ues have the same sign (except for NaCl). Notice also that the changes in values show the same trends when we compare the three chloride salts—LiCl, NaCl, and KCl—or the three bromide salts—LiBr, NaBr, and KBr.

Solutions of Liquids in Liquids

To consider heats of solution when liquids dissolve in liquids, we will work with a three-step path to go from the initial to the final state (see Figure 12.9). First, we imagine that the molecules of one liquid are moved apart just far enough to make room for molecules of the other liquid. (We will designate one liquid as the solute.) Because we have to overcome forces of attraction, this step increases the system's potential energy and so is *endothermic*.

The second step is like the first, but is done to the other liquid (solvent). On an enthalpy diagram (Figure 12.10) we have climbed two energy steps and have both the solvent and the solute in their slightly "expanded" conditions.

The third step lets nature's drive for randomness take its course, as the molecules of expanded solvent and solute come together and intermingle. Because the molecules of the two liquids now experience mutual forces of attraction, the system's potential energy *decreases,* and Step 3 is exothermic. The value of ΔH_{soln} will, again, be the net energy change for these steps.

Ideal Solutions

The enthalpy diagram in Figure 12.10 shows the case when the sum of the energy inputs for Steps 1 and 2 is equal to the energy released in Step 3, so the overall value of ΔH_{soln} is zero. This is very nearly the case when we make a solution of benzene and carbon tetrachloride. Attractive forces between molecules of benzene are almost exactly the same as those between molecules of CCl_4, or between molecules of benzene and those of CCl_4. If all such intermolecular

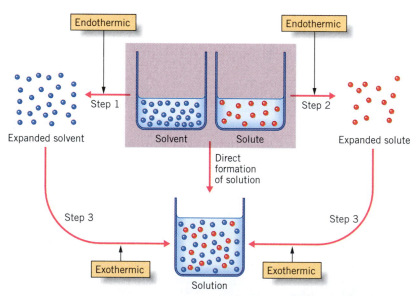

Figure 12.9 *Enthalpy of solution.* To analyze the enthalpy change for the formation of a solution of two liquids, we can imagine the hypothetical steps shown here. **Step 1.** The molecules of the liquid designated as the solvent move apart slightly to make room for the solute molecules, which is an endothermic process. **Step 2.** The molecules of the solute are made to take up a larger volume to make room for the solvent molecules, which is also an endothermic change. **Step 3.** The expanded samples of solute and solvent spontaneously intermingle, their molecules also attracting each other making the step exothermic.

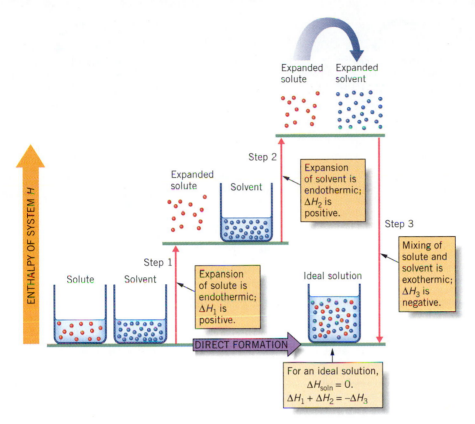

Figure 12.10 *Enthalpy changes in the formation of an ideal solution.* The three-step and the direct-formation paths both start and end at the same place with the same enthalpy outcome. The sum of the positive ΔH values for the two endothermic steps, 1 and 2, numerically equals the negative ΔH value for the exothermic step, 3. The net ΔH for the formation of an ideal solution is therefore zero.

forces were identical, the net ΔH_{soln} would be exactly zero, and the resulting solution would be called an **ideal solution.** Be sure to notice the difference between an *ideal solution* and an *ideal gas.* In an ideal gas, there are no attractive forces. In an ideal solution, there are attractive forces, but they are all the same.

Acetone and water form a solution exothermically (see Figure 12.11*a*). With these liquids, the third step releases more energy than the sum of the first two chiefly because molecules of water and acetone attract each other more strongly than acetone molecules attract each other.

Ethyl alcohol and hexane form a solution endothermically (see Figure 12.11*b*). In this case, the release of energy in the third step is not enough to compensate for the energy demands of Steps 1 and 2, and the solution becomes cool as it forms. The problem is chiefly that ethyl alcohol molecules attract each other more strongly than they can attract hexane molecules. Hexane molecules cannot push their way in and among those of ethyl alcohol without breaking up some of the hydrogen bonding in the alcohol.

Solutions of Gases in Liquids

The formation of a solution of a gas in a liquid is usually exothermic when water is the solvent. The gas is already expanded, so there is no energy cost associated with "expanding the solute." The solvent, however, must be expanded slightly to accommodate the molecules of the gas, and this does require a small energy input—small because the attractive forces do not change very much when the solvent molecules are forced apart just a little bit. When the gas molecules finally fill the spaces thus made for them, the net ΔH_{soln} depends on how strongly the gas molecules become attracted to those of the solvent. If this attraction is strong enough, there is enough of a lowering of the potential energy as the solution forms to make the overall formation of the solution exothermic. This is because

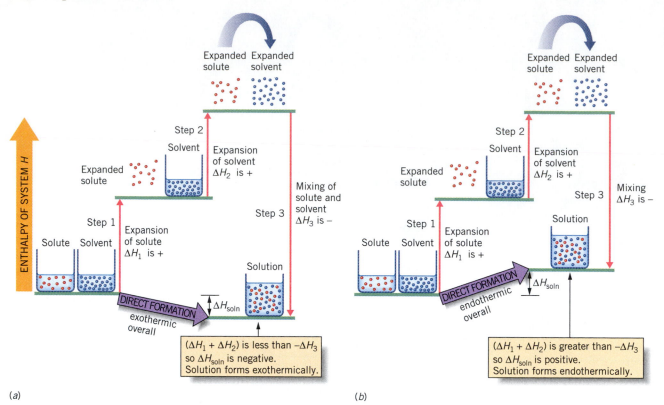

Figure 12.11 *Enthalpy changes when a real solution forms.* Because ΔH_{soln} is a state function, the multistep paths and the direct-formation paths have identical enthalpy outcomes. (*a*) The process is exothermic. (*b*) The process is endothermic.

the gas molecules change from a situation in which the attractive forces are almost zero to a situation in which there are significant forces of attraction. The net energy change associated with the two steps of expanding the solvent slightly and then filling the spaces with solute molecules is simply the solvation energy.

12.3 Solubility and the Effect of Temperature

By "solubility" we mean the mass of solute that forms a *saturated* solution with a given mass of solvent at a specified temperature. The units often are grams of solute per 100 g of solvent. Solubility, notice, refers to a *saturated* solution. If we try to make more solute dissolve, the extra will remain as a separate phase in contact with the solution. There will be a coming and going of solute particles between the dissolved and undissolved phases, of course, because in a saturated solution we have equilibrium between them (see Figure 12.12). We can write this equilibrium as follows.

$$solute_{undissolved} \rightleftharpoons solute_{dissolved}$$
(Solute contacts the   (Solute is in a
saturated solution.)   saturated solution.)

Effect of Temperature on the Solubility of Solids and Liquids in Other Liquids

As long as we keep the temperature constant, the equilibrium in a saturated solution holds. Only by subjecting the system to the specific stresses of heating or

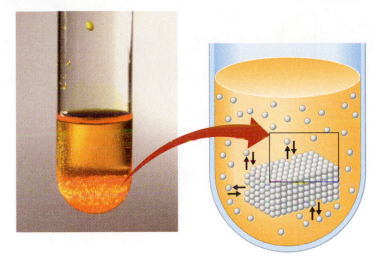

Figure 12.12 *A saturated solution.* In a saturated solution dynamic equilibrium exists between the undissolved solute and the solute in the solution.

cooling it through changing its temperature can we shift its equilibrium. The equilibrium shifts in whichever direction most directly absorbs the stress (Le Châtelier's principle).

Let's consider the more common situation, one in which adding heat causes more solute to dissolve into a solution that is initially saturated. We may even place "heat" within the equilibrium expression, putting it on the left side, the undissolved solute side, because heat is absorbed when solute dissolves.

$$solute_{undissolved} + heat \rightleftharpoons solute_{dissolved}$$
(Solute is in contact with a saturated solution.)

The stress of additional heat causes the equilibrium to shift to the *right,* a shift that "uses up the heat" and thus absorbs the stress *but also causes more solute to dissolve.* The individual solutes and solvents determine how much more dissolves, and there are large variations. Compare in Figure 12.13, for example, the widely different responses to increasing temperature in the solubilities of ammonium nitrate and sodium chloride in water.

Some solutes become *less* soluble with increasing temperature, for example, cerium(III) sulfate, $Ce_2(SO_4)_3$ (Figure 12.13). Heat must be *removed* from a saturated solution of cerium(III) sulfate to make more solute dissolve. For its equilibrium equation, we must show "heat" on the right side because heat is liberated when more of the solute dissolves into a saturated solution.

$$solute_{undissolved} \rightleftharpoons solute_{dissolved} + heat$$

When we *increase* the temperature of this system, the equilibrium absorbs the stress, the added heat, by shifting to the left, and some of the dissolved solute comes out of solution. Such systems are not common.

Effect of Temperature on the Solubility of a Gas in a Liquid

Table 12.2 gives data for the solubilities of several common gases in water at different temperatures, but all under 1 atm of pressure. You can see that at constant pressure, the solubilities of gases in water always decrease with increasing temperature. (This generalization should not be extended to nonaqueous solvents, where the situation is not so simple.)

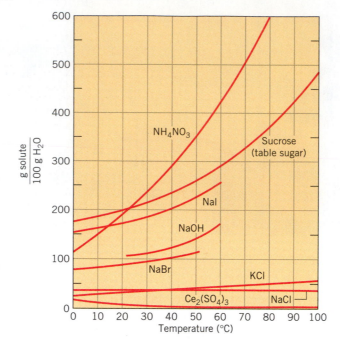

Figure 12.13 *Solubility in water versus temperature for several substances.* Most substances become more soluble when the temperature of the solution is increased, but the amount of this increased solubility varies considerably.

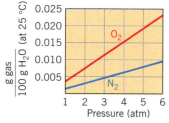

Figure 12.14 *Solubility in water versus pressure for two gases.* The amount of gas that dissolves increases as the pressure is raised.

12.4 Effect of Pressure on the Solubilities of Gases

The solubility of a gas in a liquid is affected not only by temperature but also by pressure. The gases in air, for example, chiefly oxygen and nitrogen, are not very soluble in water under ordinary pressures, but at elevated pressures both become increasingly soluble (see Figure 12.14). Facets of Chemistry 12.1 describes the dangerous implications of these changes to those who must work under higher than normal air pressures.

The relevant equilibrium is

$$\text{Gas} + \text{solvent} \rightleftharpoons \text{solution} \qquad (12.1)$$

Suppose we now increase the pressure on the system and so reduce the volume available to the gas. Le Châtelier's principle says that the system will change in a way to counteract this "stress," that is, the system will respond in a pressure-absorbing (pressure-reducing) way. In order to respond this way and reduce the

Table 12.2 Solubilities of Common Gases in Water[a]

Gas	Temperature			
	0 °C	20 °C	50 °C	100 °C
Nitrogen, N_2	0.0029	0.0019	0.0012	0
Oxygen, O_2	0.0069	0.0043	0.0027	0
Carbon dioxide, CO_2	0.335	0.169	0.076	0
Sulfur dioxide, SO_2	22.8	10.6	4.3	1.8[b]
Ammonia, NH_3	89.9	51.8	28.4	7.4[c]

[a]Solubilities are in grams of solute per 100 g of water when the gaseous space over the liquid is saturated with the gas and the total pressure is 1 atm.
[b]Solubility at 90 °C.
[c]Solubility at 96 °C.

Facets of Chemistry 12.1

The Bends—Decompression Sickness

When people, like deep-sea divers, work where the air pressure is much above normal, they have to be careful to return slowly to normal atmospheric pressure (see photo). Otherwise, they face the danger of the "bends"—severe pains in joints and muscles, fainting, possible deafness, paralysis, and death. Each 10 m of water depth adds roughly 1 atm to the pressure. Also at risk are workers who dig deep tunnels, where higher than normal air pressure is maintained to keep water out.

The bends can develop because the solubilities of both nitrogen and oxygen in blood are higher under higher pressure, as Henry's law says. Once the blood is enriched in these gases it must not be allowed to lose them suddenly. If microbubbles of nitrogen or oxygen appear at blood capillaries, they will block the flow of blood. Such a loss is particularly painful at joints, and any reduction in blood flow to the brain can be extremely serious.

For each atmosphere of pressure above the normal, about 20 minutes of careful decompression is usually recommended. This allows time for the respiratory system to

gather and expel excess nitrogen. Excess oxygen is mostly used up by normal metabolism.

Figure 1 If deep sea divers return too rapidly to atmospheric pressure they will suffer the "bends."

---

stress requires a reduction in the amount of *gas*, so some gas molecules must leave the gas phase and dissolve in the solution. Thus, increasing the pressure of the gas shifts Equilibrium 12.1 to the right, and the solubility of the gas increases. Conversely, if we make the pressure on the solution lower, the equilibrium must shift to the left, and some dissolved gas leaves the solution. Every time you pop open a carbonated beverage, dissolved carbon dioxide gas fizzes out in response to the sudden lowering of pressure.

To see more clearly the dynamics of how the effect of pressure on gas solubility works, imagine a closed container with a movable wall and partly filled with a solution of some gas in a liquid (see Figure 12.15). On the left, we begin

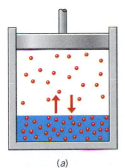
(a)

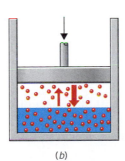
(b)

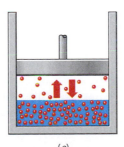
(c)

Figure 12.15 *How pressure increases the solubility of a gas in a liquid.* (*a*) At some specific pressure, equilibrium exists between the vapor phase and the solution. (*b*) An increase in pressure puts stress on the equilibrium. More gas molecules are dissolving than are leaving the solution. (*c*) More gas has dissolved and equilibrium is restored.

with equilibrium; gas molecules come and go at equal rates between the dissolved and undissolved states (Figure 12.15*a*). The rate at which gas molecules enter the solution is proportional to the frequency with which they collide with the surface of the solution. This frequency increases when we increase the pressure of the gas (Figure 12.15*b*) because the gas molecules are squeezed closer together. More of them now collide with a given surface area of the solution in each second of time. Because liquids and liquid solutions are incompressible, the increased pressure alone has no effect on the frequency with which gas molecules can leave the solution. Therefore the increased pressure increases the rate of the forward reaction in our system (Equation 12.1) over the rate of the reverse reaction.

As more gas dissolves, however, the forward rate steadily slows and the reverse rate increases. It increases because at a higher concentration of dissolved gas, there are simply more gas molecules in solution at each unit of surface area. Their frequency of escape is proportional to this concentration. But the forward gas-dissolving reaction dominates until enough additional gas has dissolved to cause the opposing rates to equalize. We again have equilibrium (Figure 12.15*c*); in terms of what species are where, it is not the original equilibrium system but a new one because more gas molecules are in solution.

William Henry (1775–1836) was an English chemist and physician.

For gases *that do not react with the solvent,* there is a simple relationship between gas pressure and gas solubility—the **pressure–solubility law,** usually called **Henry's law,** after William Henry.

Henry's law

> **Henry's Law (Pressure–Solubility Law)**
> The concentration of a gas in a liquid at any given temperature is directly proportional to the partial pressure of the gas on the solution.
>
> $$C_g = k_g P_g \qquad (T \text{ is constant})$$

In the equation, C_g is the concentration of the gas and P_g is the partial pressure of the gas above the solution. The proportionality constant, k_g, called the Henry's law constant, is unique to each gas. (Notice that constant temperature is assumed.) The equation is true only at low concentrations and pressures and, as we said, for gases that do not react with the solvent.

An alternate (and commonly used) expression of Henry's law is

Equation 12.2 is true because the two ratios, C_1/P_1 and C_2/P_2, equal the same Henry's law constant, k_g.

$$\frac{C_1}{P_1} = \frac{C_2}{P_2} \qquad\qquad (12.2)$$

where C_1 and P_1 refer to initial conditions and C_2 and P_2 to final conditions.

EXAMPLE 12.1

Using Henry's Law

At 20 °C the solubility of N_2 in water is 0.0150 g L^{-1} when the partial pressure of nitrogen is 580 torr. What will be the solubility of N_2 in water at 20 °C when its partial pressure is 800 torr?

Analysis: This problem deals with the effect of gas pressure on gas solubility, so Henry's law applies. We use this law in its form given by Equation 12.2 because it lets us avoid having to know or calculate the Henry's law constant.

Solution: Let's gather the data first.

$$C_1 = 0.0150 \text{ g } L^{-1} \qquad C_2 = ?$$

$$P_1 = 580 \text{ torr} \qquad P_2 = 800 \text{ torr}$$

Using Equation 12.2, we have

$$\frac{0.0150 \text{ g L}^{-1}}{580 \text{ torr}} = \frac{C_2}{800 \text{ torr}}$$

Solving for C_2,

$$C_2 = 0.0207 \text{ g L}^{-1}$$

The solubility under the higher pressure is 0.0207 g L^{-1}.

Is the Answer Reasonable?
In relationship to the initial concentration, the size of the answer makes sense because Henry's law tells us to expect a greater solubility at the higher pressure.

Practice Exercise 1

How many grams of nitrogen and oxygen are dissolved in 100 g of water at 20 °C when the water is saturated with air? At 1 atm pressure, the solubility of oxygen in water is 0.00430 g O_2/100 g H_2O, and the solubility of nitrogen in water is 0.00190 g N_2/100 g H_2O. In pure, dry air, P_{N_2} equals 593 torr and P_{O_2} equals 159 torr. ◆

Solubilities of Gases That Are Strongly Hydrated

The gases sulfur dioxide, ammonia, and, to a lesser extent, carbon dioxide are far more soluble in water than are oxygen or nitrogen (see Table 12.2). Part of the reason is that SO_2, NH_3, and CO_2 molecules have polar bonds and sites of partial charge that attract water molecules, forming hydrogen bonds to help hold the gases in solution. Ammonia molecules, in addition, not only can accept hydrogen bonds from water (O—H$\cdots$N) but also can donate them through their N—H bonds (N—H$\cdots$O).

The more soluble gases also react with water to some extent as the following chemical equilibria form.

$$CO_2(aq) + H_2O \rightleftharpoons H_2CO_3(aq) \rightleftharpoons H^+(aq) + HCO_3^-(aq)$$

$$SO_2(aq) + H_2O \rightleftharpoons H^+(aq) + HSO_3^-(aq)$$

$$NH_3(aq) + H_2O \rightleftharpoons NH_4^+(aq) + OH^-(aq)$$

The forward reactions contribute to the higher concentrations of the gases in solution, as compared to gases such as O_2 and N_2 that do not react with water at all. Gaseous sulfur trioxide is very soluble in water because it reacts quantitatively with water to form sulfuric acid.[4]

$$SO_3(g) + H_2O \longrightarrow H_2SO_4(aq)$$

---

[4]The "concentrated sulfuric acid" of commerce has a concentration of 93 to 98% H_2SO_4, or roughly 18 M. This solution takes up water avidly and very exothermically, removing moisture even out of humid air itself (and so must be kept in stoppered bottles). Because it is dense (d about 1.8 g mL^{-1}), oily, sticky, and highly corrosive, it must be handled with extreme care. Safety goggles and gloves must be worn when dispensing concentrated sulfuric acid. When making a more dilute solution, *always pour (slowly with stirring) the concentrated acid into the water.* If (less dense) water is poured onto concentrated sulfuric acid, it can layer on the acid's surface. At the interface, such intense heat can be generated that the steam thereby created could explode from the container, spattering acid around.

12.5 Temperature-Independent Concentration Units

In experimental work, solutions are used for such a variety of purposes that it shouldn't be surprising that no single concentration expression could serve all needs. For the stoichiometry of chemical reactions in solution, *molar concentration* or *molarity,* mol/L, provides by far the best set of units. Molarity, however, varies with temperature because a solution's volume (but not its number of moles of solute) varies somewhat with temperature. Most solutions expand when heated and contract when cooled.

To describe and compare the *physical properties* of solutions, temperature-insensitive expressions for concentration work best. We have already studied one, the *mole fraction,* or its near relative, the *mole percent* (see Section 10.6). Although it was introduced for mixtures of gases, the concept of mole fraction applies to any homogeneous mixture. We'll next study weight fractions (and weight percents) and molal concentrations, which are other temperature-insensitive concentration expressions.

> In a mixture, the mole fraction of component *A*, X_A, is given by
>
> $$X_A = \frac{\text{no. of moles of } A}{\left(\begin{array}{c}\text{total no. of moles} \\ \text{of all components}\end{array}\right)}$$

Weight Fraction and Weight Percent[5]

When we compute the ratio of the mass of one component to the total mass of the mixture or solution, we find the **weight fraction, $w_{\text{component}}$**. For a solution, we have

Tools

Weight percents

$$w_{\text{component}} = \frac{\text{mass of component}}{\text{mass of solution}}$$

For example, the weight fraction of NaCl in a solution consisting of 12.5 g of NaCl and 75.0 g of H_2O is

$$w_{\text{NaCl}} = \frac{12.5 \text{ g}}{(12.5 \text{ g} + 75.0 \text{ g})} = 0.143$$

Notice that the units cancel; weight fraction has no units. The weight fraction of water in this solution can be found in the same way, or, because the fractions must add up to 1.000, we could find the weight fraction of water by difference: $(1.000 - 0.143) = 0.857$.

When a weight fraction is multiplied by 100, we have the *weight percent* or **percent by weight.** When we know the percent by weight of one solute in a solution, we know the *number of grams of solute per 100 grams of solution.* A solution labeled "0.9% NaCl," for example, is one in which the ratio of solute to solution is 0.9 g of NaCl to 100 g of NaCl solution.

There are other ways to define *percent concentration,* but in most laboratories, when no specification of the kind of percentage is given, a *percent* concentration means the percent *by weight.* Often percent by weight is indicated in data by the symbol (w/w). Other expressions of concentration are *parts per million* (ppm) and *parts per billion* (ppb), where 1 ppm equals 1 g of component in 10^6 g of the mixture and 1 ppb equals 1 g of component in 10^9 g of the mixture.

Practice Exercise 2

How many grams of NaOH are needed to prepare 250 g of 1.00% NaOH solution in water? How many grams of water are needed? How many milliliters are needed, given that the density of water at room temperature is 0.988 g mL^{-1}? ◆

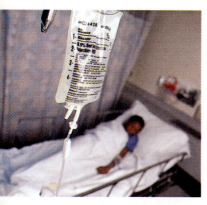

Water transferred into a patient by intravenous drip contains sodium chloride at a concentration of 0.9%.

[5]Strictly speaking, these terms should be *mass fraction* and *mass percentage,* but we bow to very widely accepted practice.

Practice Exercise 3

Hydrochloric acid can be purchased from chemical supply houses as a solution that is 37% HCl. What mass of this solution contains 7.5 g of HCl? ◆

Molal Concentration

The number of moles of solute per kilogram of *solvent* is called the **molal concentration** or the **molality** of a solution. The usual symbol for molality is *m*.

Tools

Molal
concentration

$$\text{Molality} = m = \frac{\text{mol of solute}}{\text{kg of solvent}}$$

For example, if we dissolve 0.5000 mol of sugar in 1000 g (1.000 kg) of water, we have a 0.5000 *m* solution of sugar. We would not even need a volumetric flask to prepare the solution, because we weigh the solvent. Some important physical properties of a solution are related in a simple way to its molality, as we'll soon see.

We can, of course, obtain any mass we need by taking the equivalent in volume, using the solvent's density to calculate the volume.

It is very important that you learn the distinction between *molarity* and *molality.*

$$\text{Molality} = m = \frac{\text{mol of solute}}{\text{kg of } solvent} \qquad \text{Molarity} = M = \frac{\text{mol of solute}}{\text{L of } solution}$$

Be sure to notice that molality is defined per kilogram of *solvent,* not kilogram of solution.

Molarity is useful in determining the number of moles of solute obtained when we pour a specified volume of solution *at the temperature for which the measuring device has been calibrated.*[6] The molarity of a solution, however, changes with temperature, as we pointed out, but molality does not.

The difference between molarity and molality is large when the solvent has a density considerably different from 1 g mL^{-1}, water's density. Consider bromine, for example; it is so corrosive that it is often dispensed as a dilute solution in carbon tetrachloride, a solvent of high density. If we had a 1 *molar* solution of Br_2 in CCl_4, the *volume* containing 1 mol of Br_2 would be 1000 mL of solution. But if we had a 1 *molal* Br_2 solution in CCl_4, the volume containing 1 mol of Br_2 would be about 680 mL. Because of its considerable density, we obtain 1000 *grams* of CCl_4 with much less volume than 1000 mL.

For CCl_4 at 25 °C,
$$d = 1.589 \text{ g mL}^{-1}$$

It should also be pointed out that when water is the solvent, a solution's molarity approaches its molality as the solution becomes more dilute. In very dilute solutions, 1 L of *solution* is nearly 1 L of water, which has a mass close to 1 kg. Now the ratio of moles/liter (molarity) is very nearly the same as the ratio of moles/kilogram (molality).

EXAMPLE 12.2

Calculation to Prepare a Solution of a Given Molality

An experiment calls for a 0.150 *m* solution of sodium chloride in water. How many grams of NaCl would have to be dissolved in 500 g of water to prepare a solution of this molality?

Analysis: We work with the basic definition; *molality* means moles of solute per kilogram of solvent. This is a *ratio,* so 0.150 *m* signifies the following ratio, where we substitute 1000 g for 1 kg.

$$\frac{0.150 \text{ mol NaCl}}{1000 \text{ g H}_2\text{O}}$$

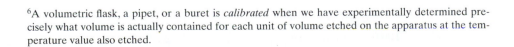

[6]A volumetric flask, a pipet, or a buret is *calibrated* when we have experimentally determined precisely what volume is actually contained for each unit of volume etched on the apparatus at the temperature value also etched.

To calculate the number of moles of NaCl we need for 500 g of H_2O, we use this ratio as a conversion factor.

Solution:

$$500 \text{ g } H_2O \times \frac{0.150 \text{ mol NaCl}}{1000 \text{ g } H_2O} = 0.0750 \text{ mol NaCl}$$

This gives us the number of *moles* of NaCl needed. We next use the formula mass of NaCl, 58.44, in the usual way as a tool to convert 0.0750 mol of NaCl to grams of NaCl.

$$0.0750 \text{ mol NaCl} \times \frac{58.44 \text{ g NaCl}}{1 \text{ mol NaCl}} = 4.38 \text{ g NaCl}$$

Thus, when 4.38 g of NaCl is dissolved in 500 g of H_2O, the concentration is 0.150 *m* NaCl.

Is the Answer Reasonable?
We'll round the formula mass of NaCl to 60, so 0.1 mol is 6 g and 0.2 mol is 12 g. We also notice that 0.150 *m* is halfway between 0.1 and 0.2 *m*. So for a whole kilogram of water, we'd need halfway between 6 g (0.1 mol) and 12 g (0.2 mol) of NaCl. For half as much water, or 500 g of water, we cut these limits in two, so we'd need halfway between 3 and 6 g of NaCl. Our answer is in this range, so it's probably right.

Practice Exercise 4

Water freezes at a lower temperature when it contains solutes. To study the effect of methyl alcohol on the freezing point of water, we might begin by preparing a series of solutions of known molalities. Calculate the number of grams of methyl alcohol (CH_3OH) needed to prepare a 0.250 *m* solution, using 2000 g of water. ◆

Practice Exercise 5

If you prepare a solution by dissolving 4.00 g of NaOH in 250 g of water, what is the molality of the solution? ◆

Conversions among Concentration Units

Sometimes, in working with a solution whose weight percent is known, we run into a situation where we would like to know its molality. We'll show how to handle a problem like this to demonstrate how *the basic definitions of various concentration expressions are finally all that we need to work our way to the answer.*

EXAMPLE 12.3

Finding Molality from Weight Percent

What is the molality of 10.0% aqueous NaCl?

Analysis: Working from the basic definitions of molality and percent by weight, we translate the heart of the question into a form that displays them. In other words, the question asks us to go from

$$\frac{10.0 \text{ grams of NaCl}}{100.0 \text{ grams of NaCl soln}} \quad \text{to} \quad \frac{? \text{ moles of NaCl}}{1 \text{ kg of water}}$$

The first ratio applies the definition of a weight percent, and the second uses that of molality. We can now see that we must carry out a grams-to-moles conversion for 10.0 g of NaCl. Then we must calculate the number of grams of *water* in 100.0 grams of NaCl *solution* and change the result to kilograms. Finally, to obtain the molality we seek, we take the ratio of the number of moles of NaCl to the kilograms of water.

Solution: We calculate the number of moles of NaCl in 100 g of the solution in the usual way by a grams-to-mole conversion using the formula mass of NaCl, 58.44.

$$10.0 \text{ g NaCl} \times \frac{1 \text{ mol NaCl}}{58.44 \text{ g NaCl}} = 0.171 \text{ mol NaCl}$$

To calculate the number of kilograms of *water* in 100 g of the *solution,* we must first subtract out the mass of the solute in 100 g of the solution and change the result to kilograms.

$$100.0 \text{ g soln} - 10.0 \text{ g NaCl} = 90.0 \text{ g or } 0.0900 \text{ kg H}_2\text{O}$$

Now we compute the molality by taking the following ratio.

$$\text{Molality} = \frac{0.171 \text{ mol NaCl}}{0.0900 \text{ kg H}_2\text{O}}$$

$$= 1.90 \text{ } m \text{ NaCl}$$

Thus a 10.0% NaCl solution is also 1.90 molal.

Is the Answer Reasonable?
If you're satisfied that you've calculated the moles of NaCl correctly (about 1/6th mol or about 0.17 mol) then notice that this is dissolved in 10% *less* than 0.1 kg water (10% of 0.1 kg is 0.09 kg). The ratio of 0.17 mol to 0.1 kg is the same as the ratio of 1.7 mol to 1 kg, which aligns the ratio with the definition of molality. So the molality should be about 10% *more* than 1.7, which the answer is.

Practice Exercise 6

A certain sample of concentrated hydrochloric acid is 37.0% HCl. Calculate the molality of this solution. ♥

Another kind of calculation is to find the molarity of a solution from its weight percent. This cannot be done without the density of the solution, as the next example shows.

EXAMPLE 12.4
Finding Molarity from Weight Percent

A certain supply of concentrated hydrochloric acid has a concentration of 36.0% HCl. The density of the solution is 1.19 g mL^{-1}. Calculate the molar concentration of HCl.

Analysis: Working again from basic definitions, here those of percent by weight and molarity, we see that our task is the conversion of

$$\frac{36.0 \text{ grams HCl}}{100 \text{ grams HCl soln}} \quad \text{to} \quad \frac{? \text{ moles HCl}}{1 \text{ liter HCl soln}}$$

The numerators show that we need a grams-to-moles conversion for the solute, HCl. The denominators tell us that we must carry out a grams-to-liters conversion for the HCl solution, which is why we need the solution's density. *Density is the tool that connects mass to volume.*

Solution: First, we find the number of moles of HCl in 36.0 g of HCl, which is the mass of solute in 100 g of 36.0% HCl solution. We do this in the usual way, using the molecular mass of HCl (36.46).

$$36.0 \text{ g HCl} \times \frac{1 \text{ mol HCl}}{36.46 \text{ g HCl}} = 0.987 \text{ mol HCl}$$

To find the number of liters of HCl solution in 100 g of HCl solution we use the density, 1.19 g mL^{-1}, as our tool. Because of the way we define density, *both* units in the given density refer to the *solution*. There are 1.19 g HCl *solution* per milliliter of HCl *solution*. The given density, for example, gives us the following conversion factors. (We *must* carry the complete units into the calculation if we intend to use the factor-label method.)

> When the density of a *solution* is given, "g/mL" always means the ratio of grams of *solution* to milliliters of *solution*.

$$\frac{1 \text{ mL HCl soln}}{1.19 \text{ g HCl soln}} \quad \text{and} \quad \frac{1.19 \text{ g HCl soln}}{1 \text{ mL HCl soln}}$$

So to calculate the number of milliliters (and then, liters) of HCl solution in 100 g of HCl solution, we simply use the first conversion factor.

$$100 \text{ g HCl soln} \times \frac{1 \text{ mL HCl soln}}{1.19 \text{ g HCl soln}} = 84.0 \text{ mL or } 0.0840 \text{ L HCl soln}$$

Thus, the volume occupied by 100 g of 36.0% HCl is 0.0840 L. The molarity of the solution is then found by the ratio of the number of moles of HCl to liters of HCl solution:

$$\text{Molarity} = \frac{0.987 \text{ mol HCl}}{0.0840 \text{ L HCl soln}} = 11.8 \text{ } M \text{ HCl}$$

Thus 36.0% HCl is also 11.8 *M* HCl.

Is the Answer Reasonable?
Doing rounding in the usual madcap manner, we see that there must be roughly 1 mol HCl in the 100 g of HCl solution. But because of the density, the volume of that solution has to be nearly 20% less than 100 mL, or a bit more than 80 mL. (The density, about 1.2 g/mL, is about 20% more than 1 g/mL.) If there's 1 mol of HCl in 100 g (or 80 mL) of the solution, then there has to be 10 mol (or 20% more than 10 mol) of HCl in ten times as much solution. This (10 *M* HCl) puts us in the ballpark of the answer, 11.8 *M* HCl.

Practice Exercise 7

Hydrobromic acid can be purchased as 40.0% HBr. The density of this solution is 1.38 g mL^{-1}. What is the molar concentration of HBr in this solution? ◆

12.6 Effects of Solutes on Vapor Pressures of Solutions

Colligative Properties

> After the Greek *kolligativ,* depending on number and not on nature.

The physical properties of solutions to be studied in this and succeeding sections are called **colligative properties,** because they depend mostly on the relative *populations* of particles in mixtures, not on their chemical identities. In this section, we examine the effects of solutes on the vapor pressures of solvents in liquid solutions.

Solutions of a Nonvolatile Solute

All liquid solutions of nonvolatile solutes have lower vapor pressures than their pure solvents. When the solution is dilute (and, always remember, the solute is nonvolatile), a simple law, **Raoult's law,** says that the vapor pressure of the solution, P_{soln}, equals the product of the mole fraction of the solvent, $X_{solvent}$, and its vapor pressure when pure, $P^\circ_{solvent}$. In equation form, Raoult's law is expressed as follows.

Francois Marie Raoult (1830–1901) was a French scientist.

Raoult's Law Equation

$$P_{solution} = X_{solvent}P^\circ_{solvent}$$

Because of the form of this equation, a plot of $P_{solution}$ versus $X_{solvent}$ should be *linear* at all concentrations when the system obeys Raoult's law (see Figure 12.16).

Usually we want to focus on how much the solute directly lowers the vapor pressure, that is, by how much does the *solute's* mole fraction concentration, X_{solute}, *change* the vapor pressure. We can show that the change in vapor pressure, or ΔP, is directly proportional to the mole fraction of solute, X_{solute}, as follows.

First, the value of ΔP is simply the following difference.

$$\Delta P = (P^\circ_{solvent} - P_{solution})$$

The mole fractions for our two-component system, $X_{solvent}$ and X_{solute}, must add up to 1; it's the nature of mole fractions.

$$X_{solvent} = 1 - X_{solute}$$

We now insert this expression for $X_{solvent}$ into the Raoult's law equation.

$$P_{solution} = X_{solvent}P^\circ_{solvent}$$

$$P_{solution} = (1 - X_{solute})P^\circ_{solvent} = P^\circ_{solvent} - X_{solute}P^\circ_{solvent}$$

So, by rearranging terms,

$$X_{solute}P^\circ_{solvent} - (P^\circ_{solvent} - P_{solution}) = \Delta P$$

The result is usually arranged as follows.

$$\Delta P = X_{solute}P^\circ_{solvent} \qquad (12.3)$$

Thus, the change in vapor pressure equals the mole fraction of the solute times the solvent's vapor pressure when pure.

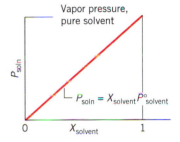

Figure 12.16 *A Raoult's law plot.* When the vapor pressure of a solution is plotted against the mole fraction of the solvent, the result is a straight line.

Tools

Raoult's law

EXAMPLE 12.5

Calculating with Raoult's Law

Carbon tetrachloride has a vapor pressure of 100 torr at 23 °C. This solvent can dissolve candle wax, which is essentially nonvolatile. Although candle wax is a mixture, we can take its molecular formula to be $C_{22}H_{46}$ (molecular mass 311). What is the vapor pressure at 23 °C of a solution prepared by dissolving 10.0 g of wax in 40.0 g of CCl_4 (molecular mass 154)?

Analysis: This problem involves the vapor pressure of a solution and so we need Equation 12.3. Its use requires us to calculate the mole fraction of the *solute*. For this, we must first calculate the total number of moles of solute and solvent.

Solution:

For CCl_4 $40.0 \text{ g } CCl_4 \times \dfrac{1 \text{ mol } CCl_4}{154 \text{ g } CCl_4} = 0.260 \text{ mol } CCl_4$

For $C_{22}H_{46}$ $10.0 \text{ g } C_{22}H_{46} \times \dfrac{1 \text{ mol } C_{22}H_{46}}{311 \text{ g } C_{22}H_{46}} = 0.0322 \text{ mol } C_{22}H_{46}$

The total number of moles = 0.292 mol

Now, we can calculate the mole fraction of the solute, $C_{22}H_{46}$.

$$X_{C_{22}H_{46}} = \frac{0.0322 \text{ mol}}{0.292 \text{ mol}} = 0.110$$

The amount of the lowering of the vapor pressure of the solute, ΔP, will be this particular mole fraction, 0.110, of the vapor pressure of pure CCl_4 (100 torr), calculated by Equation 12.3.

$$\Delta P = 0.110 \times 100 \text{ torr} = 11.0 \text{ torr}$$

The presence of the wax in the CCl_4 lowers the vapor pressure of the CCl_4 by 11.0 torr, that is, from 100 to 89.0 torr.

Is the Answer Reasonable?
Let's start by realizing that the vapor pressure of the *solution,* because of Raoult's law, *simply must be less than that of the pure solvent.* And it is, so we've not made some kind of really stupid mistake. Now let's apply the Raoult's law equation directly, that is, let's multiply the mole fraction of the *solvent* by the vapor pressure of the solvent. The mole fraction of the solvent must be 1 minus 0.11 (the other mole fraction) or 0.89. Taking 0.89 times 100 torr gives 89 torr, which is pleasantly close to the 89 torr we calculated.

Practice Exercise 8

Dibutyl phthalate, $C_{16}H_{22}O_4$ (molecular mass 278), is an oil sometimes used to soften plastic articles. It has a negligible vapor pressure, having to be heated to 148 °C before its vapor pressure is even 1 torr. What is the vapor pressure, at 20 °C, of a solution of 20.0 g of dibutyl phthalate in 50.0 g of octane, C_8H_{18} (molecular mass 114)? The vapor pressure of pure octane at 20 °C is 10.5 torr. ◆

The reason a nonvolatile solute lowers the vapor pressure of the solvent is complex, but perhaps, with the help of Figure 12.17, this could be understood as follows. For evaporation of the solvent to take place, molecules at or near the surface must have a kinetic energy equal to or greater than some minimum amount (Figure 12.17*a*). Suppose that at a particular temperature, 1.0% of the solvent molecules have this KE. Only these are able to evaporate and set up a dynamic equilibrium between the solution and the vapor of its solvent. Now let's suppose a portion of the molecules in the solution, say 20%, are those of a nonvolatile solute (Figure 12.17*b*). This means that out of the 1.0% of the molecules in the solution that have enough KE required for their escape into the vapor state, only 80% of them are actually solvent molecules. Therefore, 0.8%, not 1.0%, of the solvent molecules in the solution have enough energy to leave by evaporation. The remaining 0.2% are molecules of the nonvolatile solute, which by definition are incapable of leaving. Because the fraction of solvent molecules with enough energy to escape is lower in the solution, the rate of evaporation is

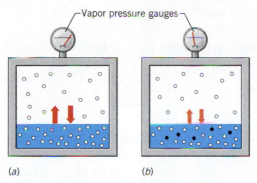

(a)   (b)

Figure 12.17 *Effect of a nonvolatile solute on the vapor pressure of a solvent.* (*a*) Equilibrium between a pure solvent and its vapor. In the liquid, molecules indicated by circles represent those with kinetic energies equal to or larger than the minimum required for evaporation. (*b*) In the solution, the same number of molecules have the energy needed by the *solvent* to escape, but some of those molecules (black circles) are not solvent, they are solute. Therefore, fewer solvent molecules have the energy to evaporate from the solution. The result is that equilibrium is achieved with fewer molecules in the vapor and the vapor pressure of the solution is less than that of the pure solvent.

also lower. Therefore, the concentration in the vapor required for equilibrium is less. It thus follows that the vapor will exert a lower pressure over the solution than over the pure solvent. And the amount of vapor pressure lowering corresponds to what we would calculate using Raoult's law.

Solutions That Contain Two or More *Volatile* Components

When two (or more) components of a liquid solution can evaporate, the vapor contains molecules of each. Each volatile component contributes its own partial pressure to the total pressure. By Raoult's law, the partial pressure of a particular component is directly proportional to the component's mole fraction in the solution. By Dalton's law of partial pressures, the total vapor pressure will be the sum of the partial pressures. To calculate these partial pressures, we use the Raoult's law equation for each component. When component A is present in a mole fraction of X_A, its partial pressure (P_A) is this fraction of its vapor pressure when pure, namely, P_A°.

$$P_A = X_A P_A^{\circ}$$

And, by the same argument, P_B, the partial pressure of component B, is

$$P_B = X_B P_B^{\circ}$$

Remember, P_A and P_B here are the *partial* pressures as calculated by Raoult's law.

The total pressure of the solution of liquids A and B is then, by Dalton's law of partial pressures, the sum of P_A and P_B.

$$P_{\text{total}} = X_A P_A^{\circ} + X_B P_B^{\circ}$$

Notice that this equation contains the Raoult's law equation as a special case. Thus if one component, say, component B, is nonvolatile, it has no vapor pressure (P_B° is zero) so the $X_B P_B^{\circ}$ term drops out, leaving the Raoult's law equation.

Acetone is a solvent for both water and molecular liquids that do not dissolve in water, like benzene. At 20 °C, acetone has a vapor pressure of 162 torr. The vapor pressure of water at 20 °C is 17.5 torr. Assuming that the mixture obeys Raoult's law, what would be the vapor pressure of a solution of acetone and water with 50.0 mol% of each?

EXAMPLE 12.6

Calculating the Vapor Pressure of a Solution of Two Liquids

Analysis: To find P_{total} we need to calculate the individual partial pressures and then add them.

Solution: A concentration of 50.0 mol% corresponds to a mole fraction of 0.500, so

$$P_{acetone} = 0.500 \times 162 \text{ torr} = 81.0 \text{ torr}$$

$$P_{water} = 0.500 \times 17.5 \text{ torr} = \underline{8.75 \text{ torr}}$$

$$P_{total} = 89.8 \text{ torr}$$

Is the Answer Reasonable?
The vapor pressure of the solution (89.8 torr) has to be much higher than that of pure water (17.5 torr), because of the volatile acetone, but much less than that of pure acetone (162 torr), because of the high mole fraction of water.

Practice Exercise 9

At 20 °C, the vapor pressure of cyclohexane, a hydrocarbon solvent, is 66.9 torr and that of toluene (another solvent) is 21.1 torr. What is the vapor pressure of a solution of the two at 20 °C when each is present at a mole fraction of 0.500? ◆

Ideal Solutions and Raoult's Law

As we described in Section 12.2, an *ideal solution* is one for which ΔH_{soln} is zero, and therefore one in which forces of attraction between all molecules are identical. Only an ideal solution would obey Raoult's law exactly, as illustrated in Figure 12.18 for an ideal, two-component system, both components being volatile liquids. The figure describes how the total vapor pressure of the solution is related to the mole fractions of the two components.

The two steeper lower lines show how the *partial* vapor pressures of the components change as their mole fractions change. These are straight lines, as required by the linear relationship of the Raoult's law equation. The top line in Figure 12.18 represents the resultant total pressure at a particular pair of values of mole fractions. Each point on the top line is the sum of the partial pressures given by points on the lines just below it.

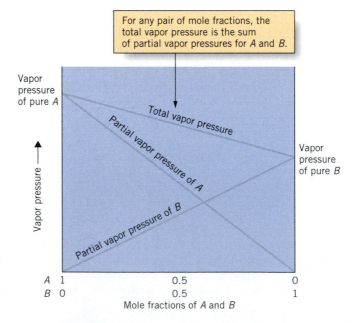

Figure 12.18 *The vapor pressure of an ideal, two-component solution of volatile compounds.*

Deviations from Raoult's Law

Very few real solutions made of two volatile components come close to being ideal. Figure 12.19 shows two typical ways in which some such solutions can behave. When Raoult's law is not obeyed for each component, the individual partial pressures are not linear with mole fraction. Hence, the sums of the partial pressures cannot be linear either. The sums, which are the values of total vapor pressure for the solutions of varying mole fraction composition, are represented by the upper plots in Figure 12.19. They display either upward or *positive deviations,* or they show downward or *negative deviations* from Raoult's law behavior. What is behind these deviations?

Negative deviations occur when solutions form exothermically. To see why, let's call the solute *A* and the solvent *B*. We have already explained that when a solution forms exothermically, the attractions between molecules of *A* and *B* in the *solution* are stronger than those between them when pure. *A* or *B* molecules, therefore, are held more strongly within the solution than as pure *A* or pure *B*. Hence they are less able to escape the solution than from their pure forms. As a result, their partial pressures above the solution are *lower* than calculated by Raoult's law, so the solution as a whole has a lower than expected vapor pressure. Negative deviations, in other words, imply stronger intermolecular forces between solute and solvent than exist in the pure components.

Just the opposite occurs when the intermolecular attractions within the solution of *A* and *B* are *weaker* than within their pure forms. It is thus easier for an *A* or *B* molecule to escape from the solution than from their pure liquids, so their partial pressures above the solution are *higher* than calculated by Raoult's law. Such solutions form endothermically and show positive overall deviations from Raoult's law.

Table 12.3 summarizes the relationships between Raoult's law predictions and the kinds of deviations just discussed.

Acetone–water solutions show negative deviations.

Ethyl alcohol–hexane solutions show positive deviations.

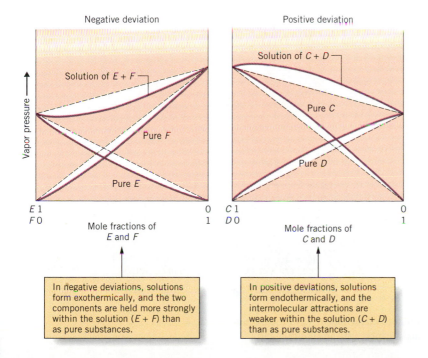

In negative deviations, solutions form exothermically, and the two components are held more strongly within the solution (*E* + *F*) than as pure substances.

In positive deviations, solutions form endothermically, and the intermolecular attractions are weaker within the solution (*C* + *D*) than as pure substances.

Figure 12.19 *Deviations from Raoult's law.* Typical deviations from ideal behavior of the total vapor pressures of real, two-component solutions of volatile substances.

Table 12.3 **Deviations from Raoult's Law**

Relative Attractive Forces	ΔH_{soln}	Raoult's Law Plots
A to A or B to B forces are greater than those between A and B	Positive (endothermic)	Positive deviations
A to A and B to B forces are all equal and equal to those between A and B	Zero	Straight lines
A to A or B to B forces are less than those of A to B	Negative (exothermic)	Negative deviations

12.7 Effects of Solutes on Freezing and Boiling Points of Solutions

Solutes influence more than the vapor pressures of solutions. They also affect boiling and freezing points. The boiling point of a solution, for example, is higher than that of the solvent itself, and the freezing point of a solution is lower than the pure solvent's freezing point. For aqueous solutions, we can understand why by returning to the phase diagram for water, first studied in Section 11.9.

Effect of a Solute on the Phase Diagram of Water

Figure 12.20a shows the phase diagram for pure water by the curves in blue. Remember that the curves are plots of the values of pressures and temperatures at which the various phases are able to be *in equilibrium*. The *triple point* occurs where all three phases are able to exist in equilibrium simultaneously. Two verti-

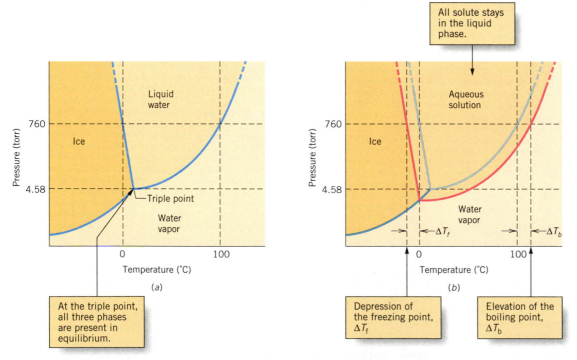

Figure 12.20 *Phase diagrams for water and an aqueous solution.* (*a*) Phase diagram for pure water. (*b*) Phase diagram for an aqueous solution of a nonvolatile solute.

cal dashed lines in Figure 12.20*a* extend from particular points on the solid–liquid and the vapor–liquid curves, points that correspond to a pressure of 760 torr (1 atm). These vertical lines thus intersect the temperature axis at the *normal* freezing point and the *normal* boiling point of pure water.

Now let's replace the pure water in the phase diagram by an aqueous solution (see Figure 12.20*b*). *The solute stays entirely in the liquid phase.* None of its molecules are in the vapor phase, which is pure water vapor. None are in the solid (ice) phase, because the structure of the ice crystals does not admit alien molecules. Let's see now how all this affects the curves that separate the phases in the phase diagram.

First, the curved line that separates ice from water vapor is unaffected because neither ice nor water vapor contains solute molecules. However, the solute makes the vapor pressure of the *solution* less than that of the pure solvent (water) at every temperature. Therefore, the corresponding curved line (in red) between the solution and vapor phases *is everywhere lower* than the line (light blue) for pure water. The triple point is thus forced to a lower value. The line defining the equilibrium between the solid and liquid phases, which slants upward from the triple point, is therefore pushed to the left. The two new curves (in red) for an aqueous solution in Figure 12.20*b* are everywhere *below* the curves for pure water.

The word "normal" in *normal* boiling or freezing point means that the system is under 1 atm of pressure when boiling or freezing.

One technology for obtaining salt-free water from the ocean is to freeze seawater, separate the ice crystals (pure water, when washed free of brine), and let the ice melt.

Freezing Point Depression and Boiling Point Elevation

Find the points where the new (red) curves in Figure 12.20*b* and those for pure water (soft blue) intersect at the horizontal dashed line set at a pressure of 760 torr. The two intersections of the red curves with this horizontal line represent, left to right, the freezing point and the normal boiling point of the *solution*. As you can see, the freezing point of the solution is *lower* than that of pure water by an amount equal to ΔT_f, and the boiling point is *higher* by ΔT_b. The first phenomenon is called **freezing point depression** by a solute, and the latter is **boiling point elevation.** We say that the solute *depresses* the freezing point and *elevates* the boiling point.

Freezing point depression and boiling point elevation are both *colligative properties*. The magnitudes of ΔT_f and ΔT_b are proportional to the relative populations of solute and solvent molecules. Molality (*m*) is the preferred concentration expression (rather than mole fraction or percent by weight) because of the resulting simplicity of the equations relating ΔT to concentration. The following equations work reasonably well only for dilute solutions, however.

$$\Delta T_f = K_f m \tag{12.4}$$

$$\Delta T_b = K_b m \tag{12.5}$$

K_f and K_b are the proportionality constants and are called, respectively, the **molal freezing point depression constant** and the **molal boiling point elevation constant.** The values of both K_f and K_b are characteristic of each solvent (see Table 12.4). The units of each constant are $°C\ m^{-1}$. For example, the value of K_f for a given solvent corresponds to the number of degrees of freezing point lowering for each molal unit of concentration. The K_f for water is 1.86 $°C\ m^{-1}$, and water freezes normally at 0 °C. Therefore, a 1.00 *m* solution in water freezes at 1.86 °C *below* the normal freezing point, or at −1.86 °C. A 2.00 *m* solution should freeze at −3.72 °C. (We say "should," but systems are seldom this ideal.) Similarly, because K_b for water is 0.51 $°C\ m^{-1}$, a 1.00 *m* aqueous solution at 1 atm pressure boils at (100 + 0.51) °C or 100.51 °C, and a 2.00 *m* solution should boil at 101.2 °C.

Tools

Freezing point depression; boiling point elevation

Table 12.4 Molal Boiling Point Elevation and Freezing Point Depression Constants

Solvent	BP (°C)	K_b (°C m^{-1})	MP (°C)	K_f (°C m^{-1})
Water	100	0.51	0	1.86
Acetic acid	118.3	3.07	16.6	3.57
Benzene	80.2	2.53	5.45	5.07
Chloroform	61.2	3.63	—	—
Camphor	—	—	178.4	37.7
Cyclohexane	80.7	2.69	6.5	20.0

When water seeps underneath the surfaces of roads and bridges and freezes, cracks and chuck holes begin to form.

▶**Chemistry in Practice**◀ Water expands when it freezes, enough to crack boulders, concrete, or vehicle radiators. To prevent the later, motorists dissolve commercial, nonvolatile, permanent-type antifreezes in water to use as the radiator fluid. The "antifreeze" concentration is made sufficiently high to have a freezing point substantially below that of pure water. In the most northerly tier of inland states as well as in Canada, motorists typically ask for protection to − 40 °C, which is approximately achieved with a 50:50 solution, by volume, of the antifreeze.

A sufficiently concentrated solution of methyl alcohol (CH_3OH, "wood alcohol") in water also prevents radiator fluid from freezing. But methyl alcohol (bp 65 °C) is volatile and slowly evaporates, leaving the radiator unprotected. That's why methyl alcohol is called "temporary antifreeze," but it's still used by watchful people because it's inexpensive. ◆

EXAMPLE 12.7

Estimating a Freezing Point using a Colligative Property

Estimate the freezing point of a permanent-type antifreeze solution made up of 100.0 g of ethylene glycol, $C_2H_6O_2$ (molecular mass 62.07), and 100.0 g of water (molecular mass 18.0). (The most we can ask for is an estimate because of the uncertainty over the precision of Equation 12.4 at this high concentration.)

Analysis: Equation 12.4 is clearly appropriate because the problem concerns freezing point depression. To use the equation, we first calculate the molality of the solution, that is, the number of moles of ethylene glycol per kilogram of water.

Solution: For the number of moles of ethylene glycol,

$$100.0 \text{ g } C_2H_6O_2 \times \frac{1 \text{ mol } C_2H_6O_2}{62.07 \text{ g } C_2H_6O_2} = 1.611 \text{ mol } C_2H_6O_2$$

$HOCH_2CH_2OH$
ethylene glycol
(bp 198 °C)

This 1.611 mol of $C_2H_6O_2$ is contained in 100.0 g or 0.1000 kg of water, so the molality is found by taking the following ratio.

$$\text{Molality} = \frac{1.611 \text{ mol}}{0.1000 \text{ kg}} = 16.11 \text{ } m$$

For our estimate we must next use Equation 12.4 although it is most reliably used only for much more dilute solutions. Table 12.4 says that K_f for water is 1.86 $°C\ m^{-1}$.

$$\Delta T_f = K_f m = (1.86\ °C\ m^{-1})(16.11\ m)$$

$$= 30.0\ °C$$

The solution, therefore, is estimated to freeze at 30.0 °C below 0 °C, or at −30.0 °C.

Is the Answer Reasonable?

Once we're satisfied that the molality is a bit more than 15 m, we can easily figure out that for each unit of this molality the freezing point must be depressed by about 2 °C, for a total depression of about 30 °C.

Practice Exercise 10

At what temperature will a 10% aqueous solution of sugar ($C_{12}H_{22}O_{11}$) boil? ◆

Determination of Molecular Masses

We have described the properties of freezing point depression and boiling point elevation as *colligative;* that is, they depend on the relative *numbers* of particles, not on their kinds. Because the effects are proportional to molal concentrations, experimental values of ΔT_f or ΔT_b should be useful for calculating the molecular masses of unknown solutes. We'll see how this works in the next example.

EXAMPLE 12.8

Calculating a Molecular Mass from Freezing Point Depression Data

A solution made by dissolving 5.65 g of an unknown molecular compound in 110.0 g of benzene froze at 4.39 °C. What is the molecular mass of the solute?

Analysis: We can use Equation 12.4 to calculate the number of moles of solute in the given solution. Then we take the ratio of grams of solute to number of moles of solute to find the molecular mass. Our first task, therefore, is to calculate the number of moles of solute in 5.65 g of the unknown.

Solution: Table 12.4 tells us that the melting point of benzene is 5.45 °C and that the value of K_f for benzene is 5.07 $°C\ m^{-1}$. The amount of freezing point depression is

$$\Delta T_f = 5.45\ °C - 4.39\ °C = 1.06\ °C$$

We now use Equation 12.4 to find the molality of the solution

$$\Delta T_f = K_f m$$

$$\text{Molality} = \frac{\Delta T_f}{K_f}$$

$$= \frac{1.06\ °C}{5.07\ °C\ m^{-1}} = 0.209\ m$$

This result means that there is a ratio of 0.209 mol of solute per 1 kg of benzene. But we do not have this much benzene, only 110.0 g or 0.1100 kg. So the actual number of moles of solute in the given solution is found by

$$0.1100\ \text{kg benzene} \times \frac{0.209\ \text{mol solute}}{1\ \text{kg benzene}} = 0.0230\ \text{mol solute}$$

With this result we know that 5.65 g of solute is the same as 0.0230 mol of solute, so the ratio of grams to moles is

$$\frac{5.65 \text{ g}}{0.0230 \text{ mol}} = 246 \text{ g mol}^{-1}$$

Thus, the molecular mass of the solute is 246.

Is the Answer Reasonable?

Once we're satisfied that the amount of freezing point depression is about 1 °C, we can concentrate on the implication of the molal freezing point depression constant for benzene; we'll round it to 5 °C per molal unit. So a depression that is 1/5th as much must mean that the molal concentration must also be (roughly) 1/5th as much, or about 0.2 mol per kilogram of solvent. In about 1/10th as much solvent, 0.1 kg benzene, there would be 1/10th as much solute, or 0.02 mol. But this amount comes from a mass of solute between 5 and 6 g. The ratio of grams to moles (the molecular mass) is then 5 or 6 divided by 0.02, or between 250 and 300 g mol^{-1}. The answer is close to that range, not within it because our estimate of the number of moles of solute is low by at least 10% (0.02 versus 0.023).

Practice Exercise 11

A solution made by dissolving 3.46 g of an unknown compound in 85.0 g of benzene froze at 4.13 °C. What is the molecular mass of the compound? ◆

12.8 Dialysis and Osmosis. Osmotic Pressure

In living things, membranes of various kinds keep mixtures and solutions organized and separated. Yet some substances have to be able to pass through membranes in order that nutrients and products of chemical work can be distributed correctly. These membranes, in other words, must have a selective *permeability*. They must keep some substances from going through while letting others pass. Such membranes are said to be *semipermeable*.

Latin permeare, to go through.

The degree of permeability varies with the kind of membrane. Cellophane, for example, is permeable to water and small solute particles—ions or molecules—but impermeable to very large molecules, like those of starch or proteins. Special membranes can even be prepared that are permeable only to water, not to any solutes.

Depending on the kind of membrane separating solutions of different concentration, two similar phenomena, *dialysis* and *osmosis*, can be observed. Both are functions of the relative populations of the particles of the dissolved materials, so they are colligative properties.

When a membrane is able to let both water and *small* solute particles through, such as the membranes in living systems, the process is called **dialysis**, and the membrane is called a *dialyzing membrane*. It does not permit huge molecules through, like those of proteins and starch.

> ▶**Chemistry in Practice**◀ Artificial kidney machines use dialyzing membranes to help remove the smaller molecules of wastes from the blood while letting the blood retain its large protein molecules. ◆

Osmosis

When a semipermeable membrane will let only solvent molecules get through, this movement is called *osmosis,* and the special membrane needed to observe it is called an *osmotic membrane.* Such membranes are rare, but they can be made.

The *direction* in which osmosis occurs, namely, the net movement of solvent from the more dilute solution (or pure solvent) into the more concentrated solution, may be explained with the help of Figure 12.21 as follows. An osmotic membrane separates pure water, *A*, from an aqueous solution, *B*. (We don't have to suppose that water is the solvent; it could be any solvent.) The solvent in each solution has its own "escaping tendency," which is just like vapor pressure. So the "escaping tendency" of the water from *A* is greater than that from *B* just as the vapor pressure of *A* is greater than that of *B*. The unequal escaping tendencies results in a net transfer of solvent *from* the less concentrated side (pure water) into the more concentrated solution. The phenomenon is **osmosis,** a net shift of solvent (only) through the membrane from the side less concentrated in *solute* to the side more concentrated.

> The direction of osmosis is here traditionally defined with reference to the concentration of the *solute,* not the solvent.

Osmotic Pressure

What can stop osmosis is the development of a back pressure. As illustrated in Figure 12.21*b*, a column of solution rises as solvent moves into the more concentrated solution. The extra weight of the liquid in the rising column creates a back pressure. We could actually prevent osmosis, or even force it to reverse, by applying an opposing pressure, as illustrated in Figure 12.21*c*. The exact back pressure needed to prevent any osmotic flow *when one of the liquids is pure solvent* is called the **osmotic pressure** of the solution.

Be careful to notice that the term "osmotic pressure" uses the word *pressure* in a novel way. A solution does not "have" a special pressure called osmotic pressure apart from osmosis. What the solution has is a *concentration* that can generate the occurrence of osmosis and the associated osmotic pressure in a special circumstance. Then, in proportion to the solution's concentration, it requires a specific back pressure to prevent osmosis. By *exceeding* this back

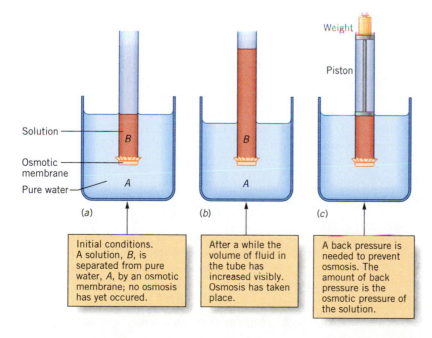

Weight

Piston

Solution

B   *B*

Osmotic membrane

Pure water   *A*   *A*

(a)   (b)   (c)

Initial conditions. A solution, *B*, is separated from pure water, *A*, by an osmotic membrane; no osmosis has yet occured.	After a while the volume of fluid in the tube has increased visibly. Osmosis has taken place.	A back pressure is needed to prevent osmosis. The amount of back pressure is the osmotic pressure of the solution.

Figure 12.21 *Osmosis and osmotic pressure.*

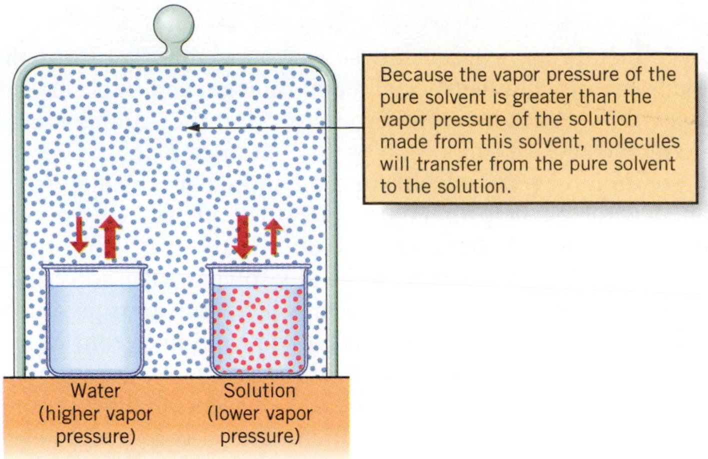

Because the vapor pressure of the pure solvent is greater than the vapor pressure of the solution made from this solvent, molecules will transfer from the pure solvent to the solution.

Water (higher vapor pressure)   Solution (lower vapor pressure)

Figure 12.22 *Principles at work in osmosis.* Here, beakers instead of a semipermeable membrane separate the pure solvent from its solution with a nonvolatile solute.

pressure, osmosis can be reversed. *Reverse osmosis* is used widely to purify seawater.

It might be helpful in understanding osmosis to refer to Figure 12.22, where we have a beaker of water in an enclosed space *already saturated in water vapor* and we have just added a beaker of an aqueous solution of a nonvolatile solute. This upsets equilibria. Water molecules tend to escape into the vapor phase from each beaker, but their rate of the escape from the pure water (larger upward arrow) is greater than their rate of escape from the solution (smaller upward arrow). This must be so because the vapor pressure of pure water is greater than that of the solution. The rate at which water molecules return to the solution from the saturated vapor state, however, is not diminished by the presence of the solute. Thus, at the solution–vapor interface, more water molecules return to the solution than leave it, and this takes water from the vapor state. To replace this water and thus keep the vapor phase saturated, more water must evaporate from the pure water. The net effect is that water moves from the beaker of pure water into the vapor space and then down into the solution. In osmosis, an osmotic membrane instead of beakers separates the liquids, but the same principles are at work.

It is not hard to imagine that the higher the concentration of the *solution* is, the more water will transfer to it (referring to Figure 12.22). In other words, the osmotic pressure of a solution is proportional to the relative concentrations of its solute and solvent. Although *molality* would, in principle, serve better to express this proportionality by an equation, in any *dilute* aqueous solution molality and molarity are virtually identical, as we noted earlier. It simplifies matters to use *molarity*.

The symbol for osmotic pressure is the Greek capital pi, Π. In a dilute aqueous solution, Π is proportional to both temperature, T, and molar concentration, M. Therefore, we can write

$$\Pi \propto MT$$

The proportionality constant turns out to be the gas constant, R, so for a *dilute* aqueous solution we can write

$$\Pi = MRT \qquad (12.6)$$

Of course, M equals mol/L, which equals n/V, where n is the number of moles and V is the volume in liters. We can therefore substitute n/V for M in Equation 12.6. After rearranging terms, we have an equation for osmotic pressure identical in form to the ideal gas law.

$$\Pi V = nRT \qquad (12.7)$$

Tools

Osmotic pressure

Equation 12.7 is the *van't Hoff equation for osmotic pressure.*

Osmotic pressures can be measured by an instrument called an *osmometer*, illustrated and explained in Figure 12.23. Osmotic pressures can be very high, even in dilute solutions, as Example 12.9 shows.

EXAMPLE 12.9
Calculating an Osmotic Pressure

A very dilute solution, 0.0010 *M* sugar in water, is separated from pure water by an osmotic membrane. What osmotic pressure in torr develops at 25 °C or 298 K?

Analysis: Equation 12.6 applies; recall that $R = 0.0821$ L atm mol^{-1} K^{-1}.

Solution:

$$\Pi = (0.0010 \text{ mol L}^{-1})(0.0821 \text{ L atm mol}^{-1} \text{ K}^{-1})(298 \text{ K})$$

$$= 0.024 \text{ atm}$$

The osmotic pressure is 0.024 atm. In torr we have

$$P = 0.024 \text{ atm} \times \frac{760 \text{ torr}}{1 \text{ atm}} = 18 \text{ torr}$$

The osmotic pressure of 0.0010 *M* sugar in water is 18 torr.

Is the Answer Reasonable?
About all that you can do with this problem is double-check your use of a calculator.

The fact that the density of water is on the order of 1 gram per milliliter is what makes molality and molarity very nearly the same for an *aqueous* solution when it is dilute.

Practice Exercise 12

What is the osmotic pressure in torr of a 0.0115 *M* glucose solution when the temperature is 25 °C? ◆

In the preceding example, you saw that a 0.0010 *M* solution of sugar has an osmotic pressure of 18 torr. This pressure would support a column of the solution roughly 25 cm or 10 in. high. If the solution had been 100 times as concentrated, 0.100 *M* sugar—still relatively dilute—the height of the column supported would be roughly 25 m or over 80 ft!

▶**Chemistry in Practice**◀ Osmotic pressure is a factor (not the only factor) in the rise of sap in trees. Water, of course, can move into a plant from the ground only if the ground water is more dilute in solutes than the plant sap. If this is not true, as in the overzealous application of fertilizer or the accidental spillage of salt onto the soil, the plant does not get the water it needs. It may even lose water to the soil, wilt, and die.

When agricultural fields are heavily irrigated but poorly drained, minerals in the soil tend to migrate toward the surface and stay there. They can make the growing zone of the soil too salty for crops. Much acreage of agricultural land has been lost to soil salination caused by irrigation without effective drainage. About a quarter of the land in Iraq with the potential for irrigation, for example, in what was once called the "fertile crescent," contains too much salt for crops. Some historians believe that the weakening and collapse of the Sumerian civilization, located in what is now southern Iraq, may have been hastened not just by attacks of outsiders but also by soil salination and subsequent famines. ◆

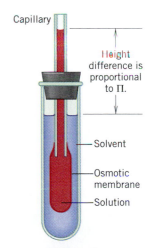

Figure 12.23 *Simple osmometer.* When solvent moves into the solution by osmosis, the level of the solution in the capillary rises. The height reached can be related to the osmotic pressure of the solution.

Molecular Mass Determinations
Using Osmotic Pressure

An osmotic pressure measurement taken of a dilute solution can be used to determine the molar concentration (see Equation 12.6) and thus also the molecular mass of a molecular solute. The next example shows how.

EXAMPLE 12.10

Calculating Molecular Mass from Osmotic Pressure

An aqueous solution with a volume of 100 mL and containing 0.122 g of an unknown molecular compound has an osmotic pressure of 16.0 torr at 20.0 °C. What is the molecular mass of the solute?

Analysis: A molecular mass is the ratio of grams to moles. We're given the number of *grams* of the solute, so we need to find the number of *moles* equivalent to this mass in order to compute the ratio.

To find the number of *moles* of solute we must first use Equation 12.6 to find the moles-per-liter concentration of the solution, its *molarity*. Remember that to use the gas constant ($R = 0.0821$ L atm mol^{-1} K^{-1}), we need P in atm and T in kelvins. Then we can use the given volume of the solution (in liters) and the molarity to calculate the number of moles of solute that are represented by 0.122 g of solute. Finally, knowing both the grams and the number of moles of solute, we take the ratio of the two to find the molecular mass.

$$16.0 \text{ torr} \times \frac{1 \text{ atm}}{760 \text{ torr}}$$
$$= 0.0211 \text{ atm}$$

Solution: First, the solution's molarity, M, is calculated using the osmotic pressure. This pressure (Π), 16.0 torr, corresponds to 0.0211 atm. The temperature, 20.0 °C, corresponds to 293 K (273 + 20.0). Using Equation 12.6,

$$\Pi = MRT$$

$$0.0211 \text{ atm} = (M)(0.0821 \text{ L atm mol}^{-1} \text{ K}^{-1})(293 \text{ K})$$

$$M = 8.77 \times 10^{-4} \frac{\text{mol solute}}{\text{L soln}}$$

The 0.122 g of solute is in 100 mL or 0.100 L of solution, not in a whole liter. So the number of moles of solute corresponding to 0.122 g is found by

$$\text{Moles of solute} = (0.100 \text{ L soln})\left(8.77 \times 10^{-4} \frac{\text{mol solute}}{\text{L soln}}\right)$$

$$= 8.77 \times 10^{-5} \text{ mol}$$

We have thus shown that 0.122 g of solute is the same as 8.77×10^{-5} mol. The ratio of grams to moles is calculated as follows.

$$\text{Molar mass} = \frac{0.122 \text{ g}}{8.77 \times 10^{-5} \text{ mol}} = 1.39 \times 10^3 \text{ g mol}^{-1}$$

Thus the molecular mass of the compound is 1.39×10^3.

Is the Answer Reasonable?
First, about all you can do is double-check the calculation of the number of moles, which turns out to be a little less than 10×10^{-5} mole, or 10^{-4} mol. Dividing the mass, roughly 0.12 g, by 10^{-4} mol gives 0.12×10^4 g per mol or 1.2×10^3 g per mol, which is very roughly what we found.

Practice Exercise 13

A solution of a carbohydrate prepared by dissolving 72.4 mg in 100 mL of solution has an osmotic pressure of 25.0 torr at 25.0 °C. What is the molecular mass of the compound? ◆

12.9 Colligative Properties of Solutions of Electrolytes

Expected Effect of the Dissociation of a Solute

The molal freezing point depression constant for water is 1.86 °C m^{-1}. So you might think that a 1.00 m solution of NaCl would freeze at −1.86 °C. Instead, it freezes at −3.37 °C. This greater depression of the freezing point by the salt, which is almost twice 1.86 °C, is not hard to understand if we remember that colligative properties depend on the concentrations of *particles*. We know that NaCl(s) dissociates into ions in water.

$$NaCl(s) \longrightarrow Na^+(aq) + Cl^-(aq)$$

If the ions are truly separated, 1.00 m NaCl actually has a concentration in dissolved solute particles of 2.00 m, twice the given molal concentration. Theoretically, then, 1.00 m NaCl should freeze at $2 \times (−1.86$ °C) or −3.72 °C. (Why it actually freezes a little higher than this, at −3.37 °C, will be discussed shortly.)

If we made up a solution of 1.00 m $(NH_4)_2SO_4$, we would have to consider the following dissociation.

$$(NH_4)_2SO_4(s) \longrightarrow 2NH_4^+(aq) + SO_4^{2-}(aq)$$

Thus, 1 mol of $(NH_4)_2SO_4$ can give a total of 3 mol of ions (2 mol of NH_4^+ and 1 mol of SO_4^{2-}). A 1 m solution of $(NH_4)_2SO_4$, therefore, has a calculated freezing point of $3 \times (−1.86$ °C) = −5.58 °C. It actually freezes at −3.6 °C. The relatively poor agreement between the experimental and calculated values suggests that the theory behind the calculation is flawed (see later).

When we want to roughly *estimate* a colligative property of a solution of an electrolyte, we recalculate the solution's molality using an assumption about the way the solute dissociates or ionizes.

We say "estimate" because the calculations can only give rough values when the simple assumption of 100% dissociation is used.

EXAMPLE 12.11

Estimating the Freezing Point of a Salt Solution

Estimate the freezing point of aqueous 0.106 m $MgCl_2$, assuming that it dissociates completely.

Analysis: The equation that relates a change in freezing temperature to molality is Equation 12.4.

$$\Delta T_f = K_f m$$

Its use here requires a value of K_f from Table 12.4—it's 1.86 °C m^{-1} for water—as well as a recalculation of m to take into account the assumption that the solute dissociates entirely.

Solution: When $MgCl_2$ dissolves in water it breaks up as follows.

$$MgCl_2(s) \longrightarrow Mg^{2+}(aq) + 2Cl^-(aq)$$

Because 1 mol of $MgCl_2$ gives 3 mol of ions, the effective (assumed) molality of what we have called 0.106 m $MgCl_2$ is three times as great.

$$\text{Effective molality} = (3)(0.106\ m) = 0.318\ m$$

Now we can use Equation 12.4.

$$\Delta T_f = (1.86\ °C\ m^{-1})(0.318\ m)$$
$$= 0.591\ °C$$

Thus, the freezing point is depressed below 0.000 °C by 0.591 °C, so we calculate that this solution freezes at −0.591 °C. (It actually freezes at −0.517 °C.)

Is the Answer Reasonable?
The molality, as recalculated, is roughly 0.3, so 3/10th of 1.86 (call it 2) is 0.6, which, after we add the unit, °C, and subtract from 0 °C gives −0.6 °C, which is close to the answer.

Practice Exercise 14

Calculate the freezing point of aqueous 0.237 *m* LiCl on the assumption that it is 100% dissociated. Calculate what its freezing point would be if the percent dissociation were 0%. ◆

Effects of Interionic Attractions

"Association" is the opposite of dissociation. It is the coming together of particles to form larger particles.

Neither the 1.00 *m* NaCl nor the 1.00 *m* $(NH_4)_2SO_4$ solution described earlier in this section freezes quite as low as calculated. In Example 12.11 we saw a similar discrepancy. What is wrong, of course, is our assumption that an electrolyte separates *completely* into its ions. This assumption is not quite correct. The ions are not really 100% free of each other. Some oppositely charged ions exist as very closely associated pairs, called *ion pairs,* which behave as single "molecules." Clusters larger than two ions probably also exist. The actual *particle* concentration in a 1.00 *m* NaCl solution, therefore, is not quite double the molality. It is less than 2.00 *m* because a small fraction of the Na^+ and Cl^- ions are associated in the solution. As a result, the lowering of the freezing point of 1.00 *m* NaCl is not quite as great as calculated on the basis of 100% dissociation.

As solutions of electrolytes are made more and more *dilute,* the observed and calculated freezing points come closer and closer together. At greater dilutions, the association of ions is less and less a complication because the ions can be farther apart. So the solutes behave more and more as if they were 100% separated into their ions at ever higher dilutions.

Jacobus van't Hoff (1852–1911), a Dutch scientist, won the first Nobel Prize in chemistry (1901).

Chemists compare the degrees of dissociation of electrolytes at different dilutions by a quantity called the **van't Hoff factor,** *i.* It is the ratio of the observed freezing point depression to the value calculated on the assumption that the solute dissolves as a nonelectrolyte.

$$i = \frac{(\Delta T_f)_{\text{measured}}}{(\Delta T_f)_{\text{calcd as a nonelectrolyte}}}$$

The hypothetical van't Hoff factor, *i,* is 2 for NaCl, KCl, and $MgSO_4$, which break up into two ions on 100% dissociation. For K_2SO_4, the theoretical value of *i* is 3 because one K_2SO_4 unit gives 3 ions. The actual van't Hoff factors for several electrolytes at different dilutions are given in Table 12.5. Notice that with decreasing concentration (that is, at higher dilutions) the experimental van't Hoff factors agree better with their corresponding hypothetical van't Hoff factors.

The increase in the percentage of complete dissociation that comes with greater dilution is not the same for all salts. In going from concentrations of 0.1 to 0.001 *m,* the increase in percentage dissociation of KCl, as measured by the change in *i,* is only about 7%. But for K_2SO_4, the increase for the same dilution is about 22%, a difference caused by the anion, SO_4^{2-}. It has twice the charge as the anion in KCl and so the SO_4^{2-} ion attracts K^+ more strongly than can Cl^-. Hence, letting an ion of 2− charge and an ion of 1+ charge get farther apart by

Table 12.5 Van't Hoff Factors versus Concentration

	Van't Hoff Factor, i			
	Molal Concentration (mol salt/kg water)			If 100% Dissociation Occurred
Salt	0.1	0.01	0.001	
NaCl	1.87	1.94	1.97	2.00
KCl	1.85	1.94	1.98	2.00
K_2SO_4	2.32	2.70	2.84	3.00
$MgSO_4$	1.21	1.53	1.82	2.00

dilution has a greater effect on their acting independently than giving ions of 1− and 1+ charge more room. When *both* cation and anion are doubly charged, the improvement in percent dissociation with dilution is even greater. Thus, as seen in Table 12.5, there is an almost 50% increase in the value of i for $MgSO_4$ as we go from a 0.1 to a 0.001 m solution.

Estimation of Percent Dissociation from Freezing Point Depression Data

In 1.00 m aqueous acetic acid, the following equilibrium exists.

$$HC_2H_3O_2(aq) \rightleftharpoons H^+(aq) + C_2H_3O_2^-(aq)$$

Equilibria of this type are discussed more fully in Chapters 4 and 16.

The solution freezes at −1.90 °C, which is only a little lower than it would be expected to freeze (−1.86 °C) if no ionization occurred. Evidently some ionization, but not much, has happened. We can estimate the percent ionization by the following calculation.

We first use the data to calculate the *apparent molality* of the solution, that is, the molality of all dissolved species, $HC_2H_3O_2 + C_2H_3O_2^- + H^+$. We again use Equation 12.4 ($\Delta T_f = K_f m$) only now letting m stand for the apparent molality; K_f is 1.86 °C m^{-1}.

$$\text{Apparent molality} = \frac{\Delta T_f}{K_f} = \frac{1.90\ °C}{1.86\ °C\ m^{-1}}$$

$$= 1.02\ m$$

If there is 1.02 mol of all solute species in 1 kg of solvent, we have 0.02 mol of solute species more than we started with, because we began with 1.00 mol of acetic acid. Since we get just one *extra* particle for each acetic acid molecule that ionizes, the extra 0.02 mol of particles must have come from the ionization of 0.02 mol of acetic acid. So the percent ionization is

$$\text{Percent ionization} = \frac{\text{mol of acid ionized}}{\text{mol of acid available}} \times 100\%$$

$$= \frac{0.02}{1.00} \times 100\% = 2\%$$

In other words, the percent ionization in 1.00 m acetic acid is estimated to be 2% by this procedure. (Other kinds of measurements offer better precision, and the percent ionization of acetic acid at this concentration is actually less than 1%.)

Association of Solute Molecules and Colligative Properties

Some molecular solutes produce *weaker* colligative effects than their molal concentrations would lead us to predict. This is not often seen, but it points to another phenomenon that we should briefly note, namely, the phenomenon of the **association** of solute particles. For example, when dissolved in benzene, benzoic acid molecules associate as *dimers*. They are held together by hydrogen bonds, indicated by the dotted lines in the following.

Di- signifies two, so a dimer is the result of the combination of two single molecules.

C_6H_5— means

$$2\ C_6H_5 - \overset{\overset{\textstyle O}{\|}}{C} - O - H \rightleftharpoons C_6H_5 - C \overset{O \cdots H - O}{\underset{O - H \cdots O}{\Big<}} \!\! \Big> C - C_6H_5$$

benzoic acid   benzoic acid dimer

Because of association, the depression of the freezing point of a 1.00 *m* solution of benzoic acid in benzene is only about one-half the calculated value. By forming a dimer, benzoic acid has an effective molecular mass that is twice as much as normally calculated. The larger effective molecular mass reduces the molal concentration of particles by half, and the effect on the freezing point depression is reduced by one-half.

12.10 Colloidal Dispersions

Importance of Particle Size: Solutions, Colloidal Dispersions, and Suspensions

As we mentioned at the beginning of this chapter, the *sizes* of particles, quite apart from their relative numbers or identities, affect the physical properties of homogeneous mixtures. If we mix powdered sand with water, for example, and shake the mixture constantly, it continues to be somewhat homogeneous, but the sand particles soon settle when we stop shaking it. Nothing like this occurs when we dissolve salt or sugar in water. Small ions and molecules do not settle out of solution. Their weights are too small for gravity to overcome the mixing influence of their kinetic motions.

The shaken mixture of sand and water is an example of a **suspension.** To form suspensions, particles must have at least one dimension larger than about 1000 nm (1 μm). Suspensions that don't settle at once, like powdered sand in water, can nearly always be separated by filtration using ordinary filter paper or an ordinary laboratory centrifuge. The high-speed rotor of a centrifuge creates a strong gravitation-like force sufficient to pull the suspended material down through the fluid.

Properties and Types of Colloidal Dispersions

A **colloidal dispersion**—often simply called a **colloid**—is a mixture in which the dispersed particles have at least one dimension in the range of 1 to 1000 nm (see Table 12.6). Colloidally dispersed particles are generally too small to be trapped by ordinary filter paper.

The tendency of colloidal dispersions in a fluid state not to separate is aided by the collisions that the dispersed particles experience from the constantly moving molecules of the "solvent." The erratic movement of colloidally dispersed particles caused by such uneven buffeting is called **Brownian movement** after a Scottish botanist, Robert Brown (1773–1858). He first

Table 12.6 Colloidal Systems

Type	Dispersed Phase[a]	Dispersing Medium[b]	Common Examples
Foam	Gas	Liquid	Soap suds, whipped cream
Solid foam	Gas	Solid	Pumice, marshmallow
Liquid aerosol	Liquid	Gas	Mist, fog, clouds, liquid air pollutants
Emulsion	Liquid	Liquid	Cream, mayonnaise, milk
Solid emulsion	Liquid	Solid	Butter, cheese
Smoke	Solid	Gas	Dust, particulates in smog, aerogels
Sol	Solid	Liquid	Starch in water, paint, jellies[c]
Solid sol	Solid	Solid	Alloys, pearls, opals

[a]The colloidal particles constitute the *dispersed phase*.
[b]The continuous matter into which the colloidal particles are dispersed is called the *dispersing medium*.
[c]Sols that become semisolid or semirigid, like gelatin dessert and fruit jellies, are called *gels*.

noticed it when he used a microscope to study pollen grains in water. Particles in suspension, of course, experience the same bumping around, but they are too large to be kept in suspension without outside help, such as by stirring or shaking the mixture.

Colloidal dispersions that do eventually separate are those in which the dispersed particles, over time, grow too large. For example, you can make a colloidal dispersion of olive oil in vinegar—a salad dressing—by vigorously shaking the mixture. Shaking breaks at least some of the oil into microdroplets having sizes in the right range for colloidal dispersions. The microdroplets, however, eventually coalesce, and the oil separates. Evidently, to prepare a *stable* colloidal dispersion, we must not only make the dispersed particles initially small enough but must also keep them from joining together.

The dispersed particles will not coalesce if they carry the same kind of electrical charge, either all positive or all negative (see Figure 12.24a). (Ions of the opposite charge are in the solvent, keeping the whole system electrically neutral.) Some of the most stable dispersions form when the surfaces of their colloidal particles have preferentially attracted ions of just one kind of charge from a dissolved salt. The dispersed particles of most **sols,** which are colloidal dispersions of solids in a fluid, are examples. Alternatively, dispersions may form when extremely large, like-charged ions, such as those of proteins, are involved (see Figure 12.24b).

Colloidal dispersions of one liquid in another are called **emulsions.** They are often relatively stable provided that a third component called an **emulsifying agent** is also present. Mayonnaise, for example, is an oil-in-water emulsion (see Figure 12.25) of an edible oil in a dilute solution of an edible organic acid. The

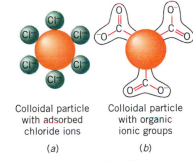

Colloidal particle with adsorbed chloride ions

Colloidal particle with organic ionic groups

(a)   (b)

Figure 12.24 *Stabilizing colloidal dispersions.* Colloidal particles often bear electrical charges that stabilize the dispersion, preventing the particles from joining together. (*a*) The colloidal particle has attracted chloride ions to itself. (*b*) A particle whose extremely large molecules carry negatively charged groups. In both cases, these colloidal particles cannot join together, grow larger, and eventually separate from the dispersing medium.

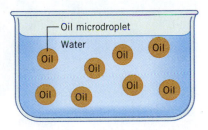

Figure 12.25 *An oil-in-water emulsion.*

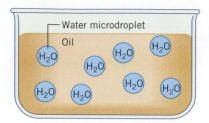

Figure 12.26 *A water-in-oil emulsion.*

emulsifying agent is lecithin, a component of egg yolk. Its molecules act to give an electrically charged surface to each microdroplet of the oil, which keeps the microdroplets from coalescing. In homogenized milk, microparticles of butterfat are coated with the milk protein casein. Water-in-oil emulsions are also possible (Figure 12.26); margarine is an example in which a soybean product is the emulsifying agent.

Starch has very large molecules and forms a colloidal dispersion in water. Usually, colloidal starch mixtures have a milky look or can even be opaque, because the colloidal particles are large enough to reflect and scatter visible light. Even when a beam of light is focused on a starch dispersion so dilute as to look as clear as water, the path of the beam is revealed by the light scattered to the side (see Figure 12.27). Light scattering by colloidal dispersions is called the **Tyndall effect,** after John Tyndall (1820–1893), a British scientist. Solutes in true solutions, however, involve species too small to scatter light, so solutions do not give the Tyndall effect.

> ▶**Chemistry in Practice**◀ The Tyndall effect is quite common. Whenever you see sunlight "streaming" through openings in a forest canopy or between clouds, you are observing the Tyndall effect. Sunlight is scattered by colloidally dispersed particles of smoke or microdroplets of liquids exuded by trees. Not all wavelengths of visible light are equally well scattered. As a result, when the sun is near the horizon at sunrise or sunset and the air is polluted by smoke, brilliant sky colors are seen (see Facets of Chemistry 12.2). ◆

Figure 12.27 *The Tyndall effect, light scattering by colloidal dispersions.* The tube on the left contains a colloidal starch dispersion, and the tube on the right has a colloidal dispersion of Fe_2O_3 in water. The middle tube has a solution of Na_2CrO_4, a colored solute. The thin red laser light is partly scattered in the two colloidal dispersions so it can be seen, but it passes through the middle solution unchanged.

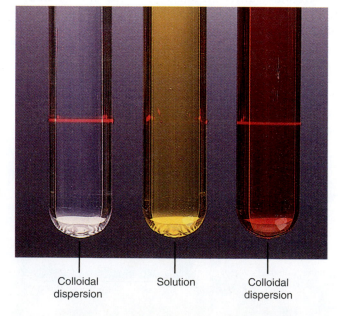

Colloidal dispersion   Solution   Colloidal dispersion

Facets of Chemistry 12.2

Sky Colors

When we look at a part of the sky well away from the sun on a reasonably clear day, we see the sky as blue. Why isn't it black? The answer is the occurrence of colloidal sized particles in the air. No part of our sky is totally free from them, and they scatter sunlight back toward our eyes (Tyndall effect). Then why isn't the sky white, like the sun? Why is it blue? And why is the western sky after sunset often orange or red?

The answers lie in the fact that not all frequencies of light are scattered by colloidal particles with the same intensity. White light from the sun is a mixture of the frequencies of light from all colors of the spectrum—from the lowest visible frequency, red, through orange, yellow, green, blue, and indigo, to the highest frequency, violet.

And the intensity of *scattered* light varies with the *fourth power of the frequency*. Thus, blue to violet frequencies are the most intensely scattered. So when the midday sunlight streams across the sky, these frequencies are preferentially scattered toward our vision, and we see a blue sky. When we're enjoying the midday sun, those far to the east or west of us are seeing either a sunset or a sunrise. And as we see the scattered blue light, they see more of what isn't scattered so well—the oranges and reds (see Figure 1).

When the sky has a lot of colloidal particles—maybe from a forest fire or from heavy smog—so much blue light is scattered as the setting sunlight streams toward us that the light which reaches us is left vivid in mostly reds and oranges.

Figure 1 Light scattering colors the sky at sunset.

How Soaps and Detergents Work

The "like dissolves like" rule helps explain how soaps and detergents work. Oily and greasy matter constitute the "glue" that binds soil to fabrics or skin. Hot water alone does a poor job of removing such dirt-binding materials, but even a small amount of detergent makes a cleansing difference.

All detergents have a common feature, whether they are ionic or molecular. Their structures include a long, nonpolar, hydrocarbon "tail" holding a very polar or ionic "head." A typical system in ordinary soap, for example, has the following anion. (The charge-balancing cation is Na$^+$ in most soaps.)

$$CH_3CH_2CH_2CH_2CH_2CH_2CH_2CH_2CH_2CH_2CH_2CH_2CH_2CH_2CH_2CH_2CH_2C\overset{O}{\overset{\|}{}}O^-$$

hydrocarbon tail   anionic head

In synthetic detergents, the hydrophilic group is not

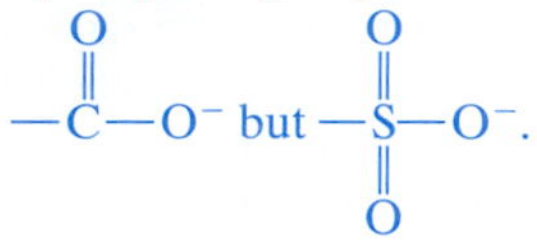

$$-\overset{O}{\overset{\|}{C}}-O^- \text{ but } -\overset{O}{\underset{\|}{\overset{\|}{S}}}-O^-.$$

The tail of this species, like all hydrocarbons, is nonpolar and therefore has a strong tendency to dissolve in oil or grease but not in water. Its natural tendency is to avoid water; we say that the tail is **hydrophobic** (water fearing). The entire ion is dragged into solution in hot water only because of its ionic head, which can be hydrated. The ionic head is said to be **hydrophilic** (water loving). To minimize contacts between the hydrophobic tails and water and to maximize contacts between the hydrophilic heads and water, the soap's anions gather into colloidal sized groups called soap **micelles** (see Figure 12.28).

When soap micelles encounter a greasy film, their anions find something else that can accommodate hydrophobic tails (see Figure 12.29). The tails of innumerable anions of soap work their way into the film; the tails, in a sense, dissolve in the grease or oil. But the hydrophilic heads of the anions will not be dragged into the grease; the heads stay out in the water. With a little agitating help from the washing machine, the greasy film breaks up into countless tiny droplets *each pincushioned by projecting, negatively charged heads*. Because the droplets, now colloidally dispersed, all have the same kind of charge, they repel each other and cannot again coalesce. For all practical purposes, the greasy film has been dissolved, and we send it on its way to be destroyed by bacteria at wastewater treatment plants or in the ground.

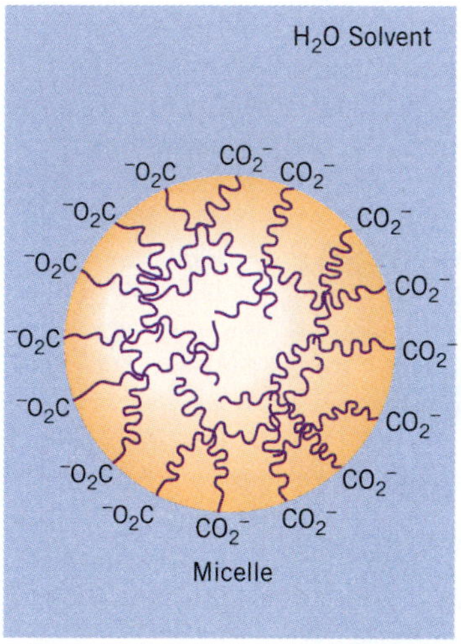

Figure 12.28 *Soap micelles.* The formation of soap micelles is driven by the water-avoiding properties of the hydrophobic tails of soap units and the water-attracting properties of their hydrophilic heads.

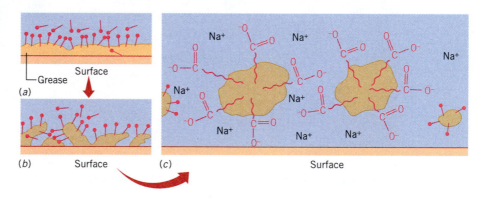

Figure 12.29 *How soap loosens grease that might be coating a cloth surface.* (*a*) The hydrophobic tails of soap anions work their way into the grease layer. The hydrophilic heads (represented by the solid dots) stay exposed to water. (*b*) The grease layer breaks up. (*c*) In an enlarged view, microdroplets of grease form, each pincushioned and made like-charged by soap anions. The microdroplets are now easily poured out with the wash water.

SUMMARY

Solutions. In **solutions,** the dissolved particles have dimensions on the order of 0.05 to 0.25 nm (sometimes up to 1 nm). Solutions are fully homogeneous. They cannot be separated by filtration, and they remain physically stable indefinitely. (They do not display the *Tyndall effect* or *Brownian movement,* either.)

When liquid solutions form by physical means, generally polar or ionic solutes dissolve well in polar solvents, like water. Nonpolar, molecular solutes dissolve well in nonpolar solvents. These observations are behind the **"like dissolves like" rule.** Nature's driving force for establishing disorder plus intermolecular attractions are the major factors in the formation of a solution. When both solute and solvent are nonpolar, nature's tendency toward more randomness dominates because intermolecular attractive forces are weak. Ion–dipole attractions and the **solvation** or **hydration** of dissolved species are major factors in forming solutions of ionic or polar solutes in something polar, like water.

Heats of Solution. The molar enthalpy of solution, the **heat of solution,** is the net of the **lattice energy** and the **solvation energy** (or, when water is the solvent, the **hydration energy**). The lattice energy is the increase in the potential energy of the system required to separate the molecules or ions of the solute from each other. The solvation energy corresponds to the potential energy lowering that occurs in the system from the subsequent attractions of these particles to the solvent molecules as intermingling occurs.

When liquids dissolve in liquids, an **ideal solution** forms if the potential energy increase needed to separate the molecules equals the energy lowering as the separated molecules come together. Usually, some net energy exchange occurs, however, and few solutions are ideal.

When a gas dissolves in a gas, the energy cost to separate the particles is virtually zero, because they're already separated. They remain separated in the gas mixture, so the net heat of solution is very small if not zero.

Pressure and Gas Solubility. At pressures not too much different from atmospheric pressure, the solubility of a gas in a liquid is directly proportional to the partial pressure of the gas—**Henry's law.**

Concentration Expressions. To study and use colligative properties, a solution's concentration ideally is expressed either in mole fractions or as a **molal concentration,** or **molality.** Such expressions provide a temperature-independent description of a concentration that, in the final analysis, gives a ratio of solute to solvent particles. The **molality** of a solution, *m,* is the ratio of the number of moles of solute to the kilograms of *solvent* (not solution, but solvent).

Weight fractions and **weight percents** are concentration expressions used often for reagents when direct information about number of moles of solutes is unimportant.

Colligative Properties. **Colligative properties** are those that depend on the ratio of the particles of the solute to the molecules of the solvent. These properties include the **lowering of the vapor pressure,** the **depression of a freezing point,** the **elevation of a boiling point,** and the **osmotic pressure** of a solution.

According to **Raoult's law,** the vapor pressure of a solution of a nonvolatile (molecular) solute is the vapor pressure of the pure solvent times the mole fraction of the solvent. An alternative expression of this law is that the change in vapor pressure caused by the solute equals the mole fraction of the solute times the pure solvent's vapor pressure. When the components of a solution are volatile liquids, then Raoult's law calculations find the *partial pressures* of the vapors of the individual liquids. The sum of the partial pressures equals the total vapor pressure of the solution.

An **ideal solution** is one that would obey Raoult's law exactly and in which all attractions between molecules are equal. An ideal solution has $\Delta H_{soln} = 0$. Solutions of liquids that form exothermically usually display negative deviations from Raoult's law. Those that form endothermically have positive deviations.

In proportion to its molal concentration, a solute causes a **freezing point depression** and a **boiling point elevation.** The proportionality constants are the *freezing point depression constant* and the *boiling point elevation constant,* and they differ from solvent to solvent. Freezing point and boiling point data from a solution made from known masses of both solute and solvent can be used to calculate the molecular mass of a solute.

When a solution is separated from the pure solvent (or from a less concentrated solution) by an *osmotic membrane,* **osmosis** occurs, which is the net flow of solvent into the more concentrated solution. The back pressure required to prevent osmosis is called the solution's **osmotic pressure,** which, in a dilute aqueous solution, is proportional to the product of the Kelvin temperature and the molar concentration. The proportionality constant is *R,* the ideal gas constant, and the equation relating these variables is $\Pi = mRT$.

When the membrane separating two different solutions is a *dialyzing membrane,* it permits **dialysis,** the passage not only of solvent molecules but also of small solute molecules or ions. Only very large molecules are denied passage.

Colligative Properties of Electrolytes. Because an electrolyte releases more ions in solution than indicated by the molal (or molar) concentration, a solution of an electrolyte has more pronounced colligative properties than a solution of a molecular compound at the same molality. The dissociation of a strong electrolyte approaches 100% in very dilute solutions, particularly when both ions are singly charged. Weak electrolytes provide solutions more like those of nonelectrolytes.

The ratio of the value of a particular colligative property, like ΔT_f for a freezing point depression, to the value expected under no dissociation is the **van't Hoff factor, *i*.** A solute whose formula unit breaks into two ions, like NaCl or $MgSO_4$, would have a van't Hoff factor of 2 if it were 100% dissociated. Observed van't Hoff factors approach those corresponding to 100% dissociation only as the solutions are made more and more dilute.

Other Homogeneous Mixtures. Relative particle sizes and homogeneity distinguish among the three simplest kinds of mixtures—**suspensions, colloidal dispersions,** and **solutions.** Suspensions have particles with at least one di-mension larger than 1000 nm; they are not homogeneous; and they settle out spontaneously. They can be separated by filtration.

Colloidal Dispersions. In colloidal dispersions, the dispersed particles have one or more dimensions in the range of 1 to 1000 nm. They are considered to be homogeneous, but we're at a borderline. They separate more slowly than suspensions in response to gravity, but many can be kept indefinitely if the dispersed particles bear like charges or are coated with an emulsifying agent. Colloidal particles can scatter light **(Tyndall effect),** can display **Brownian movement,** and cannot ordinarily be separated by filtration.

Tools *you have learned*

Mo	Tc	Ru	Rh	Pd	Ag 107.8682	Cd 112.411	In 114.82	Sn 118.710	Sb 121.75	Te 127.60
72	73	74	75	76	77	78	79	80	81	82	83	84

The table below lists the tools that you have learned in this chapter that are applicable to problem solving. Review them if necessary, and refer to them when working on the Thinking-It-Through problems and the Review Problems that follow.

Tool	Function
"Like dissolves like" rule *(page 523)*	To use chemical composition and structure to predict whether two substances can form a solution.
Henry's law *(page 534)*	To calculate the solubility of a gas at a given pressure from its solubility at another pressure.
Weight fraction; weight percent *(page 536)*	To calculate the mass of a solution that delivers a given mass of solute.
Molal concentration *(page 537)*	To have a temperature-independent concentration expression for use with colligative properties.
Raoult's law *(page 541)*	To calculate the effect of a solute on the vapor pressure of a solution.
Equations for freezing point depression and boiling point elevation *(page 547)*	To estimate freezing and boiling points or molecular masses. To estimate percent dissociation of a weak electrolyte *(page 557)*.
Equation for osmotic pressure *(page 552)*	To estimate the osmotic pressure of a solution. To calculate molecular masses from osmotic pressure and concentration data *(page 554)*.

THINKING IT THROUGH

Remember, the goal for each of the following problems is to assemble the available information needed to obtain the answer, state what additional data (if any) are needed, and describe how you would use the data to answer the question. For problems involving unit conversions, list the relationships among the units that are needed to carry out the conversions. Construct the conversion factors that can be formed from these relationships. Then set up the solution to the problem by arranging the conversion factors so the units cancel correctly to give the desired units of the answer.

The problems are divided into two groups. Those in Level 2 are more challenging than those in Level 1 and provide an opportunity to really hone your problem solving skills. Detailed answers to these problems can be found at our Web site: http://www.wiley.com/college/brady

Level 1 Problems

1. The solubility of a gas in water at 20 °C is 0.0176 g L^{-1} at 681 torr. How would you calculate its solubility at 0.989 atm at 20 °C?

2. A solution of sulfuric acid is 10.0% H_2SO_4. How many grams of this solution are needed to provide 0.100 mol of H_2SO_4? Explain the calculation. If the question had asked for the number of *milliliters* of the solution instead of the number of grams, what additional data about the solution would be needed, and how would they be used?

3. For an experiment involving osmosis, you need to know the molality of a 12.5% solution of sugar ($C_{12}H_{22}O_{11}$) in water. How can this be calculated?

4. The mole fraction of NaCl in an aqueous solution (with no other component) is 0.0100. How would you calculate the weight percent of NaCl in the solution?

5. A solution of KNO_3 has a concentration that is 0.750 m. How would you calculate the mole fraction of KNO_3 in the solution?

6. The stockroom has a solution that is 0.125 m NH_4Cl. Describe how you would calculate the amount of this solution you would need to supply 0.575 mol of NH_4Cl. If you want to measure the amount of the solution in milliliters, what additional data would you need?

7. A 50.00% H_2SO_4 solution has a specific gravity at 20 °C of 1.3977. Describe how you would calculate the molarity of the solution.

8. A solution of glucose ($C_6H_{12}O_6$, a nonvolatile solute) in 100 g of water has a vapor pressure of 20.134 torr at a temperature at which the vapor pressure of pure water is 23.756 torr. Describe how you would calculate the number of grams of glucose in the solution.

9. The cooling system of an automobile usually contains a solution of antifreeze prepared by mixing equal volumes of ethylene glycol, $C_2H_4(OH)_2$, and water. The density of ethylene glycol is 1.113 g mL^{-1}. Describe how you would calculate the freezing point of the mixture. How would you calculate the boiling point of the solution?

10. Describe how you would calculate the vapor pressure of a solution of tetrachloroethylene in methyl acetate at 40.0 °C in which the mole fraction of tetrachloroethylene is 0.450. At 40 °C, the vapor pressure of tetrachloroethylene is 40.0 torr and that of methyl acetate is 400 torr.

11. What is the approximate osmotic pressure of a 0.0125 M solution of $CaCl_2$ in water at 20 °C? Explain the calculation.

Level 2 Problems

12. If you add 75.0 g of water to 150 g of a 10.0% solution of NaCl, what is the percent concentration of the new solution?

13. A solution of $AlCl_3$ in water has a freezing point of −4.50 °C. Assuming complete dissociation of the solute and ideal solution behavior, describe how you would calculate the vapor pressure of this solution at 35 °C. What data, not given in the problem, would you need to obtain the answer?

14. A solution of glycerol, a nonvolatile solute, in 150 g of water has a vapor pressure of 26.36 torr at 30 °C. At this temperature, pure water has a vapor pressure of 31.82 torr. Explain how you would calculate the amount of water that would have to be added to this solution to raise the vapor pressure to 29.00 torr at 30 °C.

15. A solution of a nonvolatile solute in 150 g of water has a freezing point of −0.835 °C. An additional 1.20 g of the solute was added to the solution, and the freezing point dropped to −1.045 °C. How many grams of the solute had been dissolved in the original solution? Explain the calculation in detail.

16. The figure on the left shows a sealed vessel containing liquid water in equilibrium with water vapor. Liquid water is shown as pale blue with water vapor represented by individual water molecules. In the vessel on the right, green spheres represent a nonvolatile solute. (Only solute molecules on the surface layer of water are shown.) The vapor pressure in the vessel on the right, which contains the nonvolatile solute, is observed to be less than for pure water in the vessel on the left.

(a) Assume that the solute molecules on the surface of the solution on the right occupy 10% of the surface area. If the water in the vessel on the left was transferred to a container with 10% less cross sectional area, would the vapor pressure be reduced to the same value as the vapor pressure in the vessel on the right? Explain.

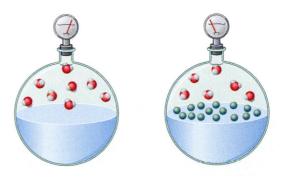

(b) Use these models to explain why the vapor pressure in the vessel on the right is less than the vapor pressure of the pure water in the vessel shown on the left. Consider the relative rates of evaporation and condensation in the two vessels and the concept of dynamic equilibria.

17. Acetic acid found in vinegar, CH_3CO_2H, is a weak acid which is partially dissociated in aqueous solution:

$$CH_3CO_2H(aq) \rightleftharpoons CH_3CO_2^-(aq) + H^+(aq).$$

(a) Based upon the partial dissociation of a weak acid, would you expect the van't Hoff factor for acetic acid to equal 1.0, 3.0, or be some number larger than 1.0 but much smaller than 2.0? Explain.

(b) From a freezing point depression experiment using a nonpolar solvent, the van't Hoff factor for acetic acid is found to be very close to 0.5. Explain this observation. Use Lewis structures to justify your explanation.

18. Make a sketch to represent solid NaCl and another to show NaCl(*aq*). Have your sketch of the solid show the 3-D structure of NaCl. In both sketches show the fundamental species that comprise NaCl. In the sketch for NaCl(*aq*) use water molecules to distinguish the fundamental species in aqueous solution from those in the solid. What intermediate step between your two sketches should you draw if you wished to analyze the role of lattice energy and hydration energy in determining the heat of solution of NaCl? Draw a sketch for the species involved in this intermediate step and use the three sketches to explain how lattice and hydration energies determine the heat of solution.

REVIEW QUESTIONS

Why Solutions Form

12.1 Name three kinds of *homogeneous* mixtures. In which are the particle sizes the smallest?

12.2 What is the "driving force" behind the spontaneous dissolving of two gases in each other?

12.3 When substances form liquid solutions, what two factors are involved in determining the solubility of the solute in the solvent?

12.4 Methyl alcohol, CH_3—O—H, and water are miscible in all proportions. What does this mean? Explain how the O—H unit in methyl alcohol contributes to this.

12.5 Gasoline and water are immiscible. What does this mean? Explain why they are immiscible, in terms of structural features of their molecules and forces of attraction between them.

12.6 Explain how ion–dipole forces help to bring potassium chloride into solution in water.

12.7 What does hydration refer to, and what kinds of substances, electrolytes or nonelectrolytes, does it tend to assist most in dissolving in water?

12.8 What kinds of substances are more soluble in water, hydrophilic compounds or hydrophobic compounds?

12.9 Explain why potassium chloride will not dissolve in carbon tetrachloride, CCl_4.

12.10 Iodine, I_2, is very slightly soluble in water but dissolves readily in carbon tetrachloride. Why won't it dissolve very much in water, and what makes so easy the formation of the solution of iodine in carbon tetrachloride?

12.11 Iodine dissolves far better in ethyl alcohol (to form "tincture of iodine," an old antiseptic) than in water. What does this tell us about the alcohol molecules, as compared with water molecules?

Heats of Solution

12.12 How do we define *heat of solution?* Give both a narrative definition and an equation.

12.13 No solution actually forms by the two-step process that we used as a model for the formation of a solution of some ionic compound in water. When we are interested in the molar enthalpy of solution, however, we can get away with this model. Explain. What makes attractive the particular two steps we used for the purpose of estimating ΔH_{soln}?

12.14 The value of ΔH_{soln} for a soluble compound is, say, $+26$ kJ mol^{-1}, and a nearly saturated solution is prepared in an insulated container (e.g., a coffee cup calorimeter). Will the system's temperature increase or decrease as the solute dissolves?

12.15 Referring to the preceding question, which value for this compound would be numerically larger, its lattice energy or its hydration energy?

12.16 Which would be expected to have the larger hydration energy, Al^{3+} or Li^+? Why? (Both ions are about the same size.)

12.17 Suggest a reason why the value of ΔH_{soln} for a gas such as CO_2, dissolving in water, is negative.

12.18 The value of ΔH_{soln} for the formation of an acetone–water solution is negative. Explain this in general terms that discuss intermolecular forces of attraction.

12.19 The value of ΔH_{soln} for the formation of an ethyl alcohol–hexane solution is positive. Explain this in general terms that involve intermolecular forces of attraction.

12.20 How can Le Châtelier's principle be applied to explain how the addition of heat causes undissolved solute to go into solution (in the cases of most saturated solutions)?

Temperature and Solubility

12.21 If a saturated solution of NH_4NO_3 at 70 °C is cooled to 10 °C, how many grams of solute will separate if the quantity of the solvent is 100 g? (Use data in Figure 12.13.)

12.22 A hot concentrated solution in which equal masses of NaI and NaBr are dissolved is slowly cooled from 50 °C. Which salt precipitates first? (Use data in Figure 12.13.)

Pressure and Solubility

12.23 What is Henry's law?

12.24 What makes ammonia or carbon dioxide so much more soluble in water than nitrogen?

12.25 Sulfur dioxide is an air pollutant wherever sulfur-containing fuels have been used. Rainfall that washes sulfur dioxide out of the air and onto the land is called acid rain. How does sulfur dioxide contribute to acid rain? Write an equation as part of your answer.

Expressions of Concentration

12.26 Write the definition for each of the following concentration units: mole fraction, mole percent, molality, percent by weight.

12.27 How does the molality of a solution vary with increasing temperature? How does the molarity of a solution vary with increasing temperature?

12.28 Suppose a 1.0 *m* solution of a solute is made using a solvent with a density of 1.15 g/mL. Will the molarity of this solution be numerically larger or smaller than 1.0? Explain.

Lowering of the Vapor Pressure

12.29 What specific fact about a physical property of a solution must be true to call it a colligative property?

12.30 Viewed at a molecular level, what causes a solution with a nonvolatile solute to have a lower vapor pressure than the solvent at the same temperature?

12.31 How is Raoult's law used when we want to calculate the total vapor pressure of a liquid solution made up of two *volatile* liquids?

12.32 What kinds of data would have to be obtained to find out if a binary solution of two miscible liquids is almost exactly an ideal solution?

12.33 An aqueous solution showed a *positive deviation* from the expected relationship between the vapor pressure of an ideal binary solution and its mole fraction composition. What does this mean? Explain the positive deviation in terms of intermolecular attractions.

Freezing Point Depression and Boiling Point Elevation

12.34 When an aqueous solution of sodium chloride starts to freeze, why don't the ice crystals contain ions of the salt?

12.35 Explain why a nonvolatile solute dissolved in water makes the system have (a) a higher boiling point than water, and (b) a lower freezing point than water.

Dialysis and Osmosis

12.36 Why do we call dialyzing and osmotic membranes *semipermeable?* What is the opposite of *permeable?*

12.37 What is the key difference between dialyzing and osmotic membranes?

12.38 What is meant by *dialysis?*

12.39 At a molecular level, explain why in osmosis there is a net migration of solvent from the side of the membrane less concentrated in solute to the side more concentrated in solute.

12.40 Two glucose solutions of unequal molarity are separated by an osmotic membrane. Which solution will *lose* water, the one with the higher or the lower molarity?

12.41 Which aqueous solution has the higher osmotic pressure, 10% glucose, $C_6H_{12}O_6$, or 10% sucrose, $C_{12}H_{22}O_{11}$? (Both are molecular compounds.)

12.42 When a solid is *associated* in a solution, what does this mean? What difference does it make to expected colligative properties?

Colligative Properties of Electrolytes

12.43 Why are colligative properties of solutions of ionic compounds usually more pronounced than those of solutions of molecular compounds of the same molalities?

12.44 What is the van't Hoff factor? What is its expected value for all nondissociating molecular solutes? If its measured value is slightly larger than 1.0, what does this suggest about the solute? What is suggested by a van't Hoff factor of approximately 0.5?

12.45 Which aqueous solution, if either, is likely to have the lower freezing point, 10% NaCl or 10% NaI?

12.46 Which aqueous solution, if either, is likely to have the higher boiling point, 0.50 *m* NaI or 0.50 *m* Na_2CO_3?

Kinds of Mixtures

12.47 What feature about mixtures distinguishes them from elements and compounds?

12.48 What single feature most distinguishes suspensions, colloidal dispersions, and solutions from each other?

Colloidal Dispersions

12.49 List three factors that can contribute to the stability of a colloidal dispersion.

12.50 What general name is given to a colloidal dispersion of one liquid in another?

12.51 What is an emulsifying agent? Give an example.

12.52 What is the Tyndall effect, and why do colloidal dispersions but not solutions show it?

12.53 What causes Brownian movement in fluid colloidal dispersions?

12.54 What is a sol, and how can one often be stabilized?

12.55 What simple test could be used to tell if a clear, aqueous fluid contains colloidally dispersed particles?

12.56 Soap, as a solute in water, spontaneously forms micelles. What are micelles, and what is the driving force for their formation?

REVIEW PROBLEMS

Answers to problems whose numbers are printed in color are given in Appendix D.
The more challenging problems are marked with asterisks.

Heat of Solution

12.57 Consider the formation of a solution of aqueous potassium chloride. Write the thermochemical equations for (a) the conversion of solid KCl into its gaseous ions and (b) the subsequent formation of the solution by hydration of the ions. The lattice energy of KCl is -690 kJ mol^{-1}, and the hydration energy of the ions is -686 kJ mol^{-1}. Calculate the heat of solution of KCl in kJ mol^{-1}.

12.58 Referring to Table 12.1, suppose the lattice energy of KCl were just 2% greater than the value given. What then would be the calculated ΔH_{soln} (in kJ mol^{-1})? How would this new value compare with that given in the table? By what percentage is the new value different from the value in the table? (The point of this calculation is to show that small percentage errors in two large numbers, such as lattice energy and hydration energy, can cause large percentage changes in the absolute difference between them.)

Henry's Law

12.59 The solubility of methane, the chief component of natural gas, in water at 20 °C and 1.0 atm pressure is 0.025 g L^{-1}. What is its solubility in water at 1.5 atm and 20 °C?

12.60 At 740 torr and 20 °C, nitrogen has a solubility in water of 0.018 g L^{-1}. At 620 torr and 20 °C, its solubility is 0.015 g L^{-1}. Show that nitrogen obeys Henry's law.

Expressions of Concentration

12.61 What is the molal concentration of glucose, $C_6H_{12}O_6$, a sugar found in many fruits, in a solution made by dissolving 24.0 g of glucose in 1.00 kg of water? What is the mole fraction of glucose in the solution? What is the weight percent of glucose in the solution?

12.62 If you dissolved 11.5 g of NaCl in 1.00 kg of water, what would be its molal concentration? What are the weight percent NaCl and the mole percent NaCl in the solution? The volume of this solution is virtually identical to the original volume of the 1.00 kg of water. Therefore, what is the molar concentration of NaCl in this solution? What would have to be true about any solvent for one of its dilute solutions to have essentially the same molar and molal concentrations?

12.63 A solution of ethyl alcohol, CH_3CH_2OH, in water has

a concentration of 1.25 m. Calculate the weight percent of ethyl alcohol.

12.64 A solution of NaCl in water has a concentration of 19.5%. Calculate the molality of the solution.

12.65 A solution of NH_3 in water is at a concentration of 5.00% by weight. Calculate the mole percent NH_3 in the solution. What is the molal concentration of the NH_3?

12.66 An aqueous solution of isopropyl alcohol, C_3H_8O, rubbing alcohol, has a mole fraction of alcohol equal to 0.250. What is the percent by mass of alcohol in the solution? What is the molality of the alcohol?

12.67 Sodium nitrate, $NaNO_3$, is sometimes added to tobacco to improve its burning characteristics. An aqueous solution of $NaNO_3$ has a concentration of 0.363 m. Its density is 1.0185 g mL^{-1}. Calculate the molar concentration of $NaNO_3$ and the weight percent of $NaNO_3$ in the solution. What is the mole fraction of $NaNO_3$ in the solution?

12.68 In an aqueous solution of sulfuric acid, the concentration is 1.89 mol% of acid. The density of the solution is 1.0645 g mL^{-1}. Calculate the following: (a) the molal concentration of H_2SO_4, (b) the weight percent of the acid, and (c) the molarity of the solution.

Raoult's Law

12.69 At 25 °C, the vapor pressure of water is 23.8 torr. What is the vapor pressure of a solution prepared by dissolving 65.0 g of $C_6H_{12}O_6$ (a nonvolatile solute) in 150 g of water? (Assume the solution is ideal.)

12.70 The vapor pressure of water at 20 °C is 17.5 torr. A 20% solution of the nonvolatile solute ethylene glycol, $C_2H_4(OH)_2$, in water is prepared. Estimate the vapor pressure of the solution.

12.71 At 25 °C the vapor pressures of benzene (C_6H_6) and toluene (C_7H_8) are 93.4 and 26.9 torr, respectively. A solution made by mixing 60.0 g of benzene and 40.0 g of toluene is prepared. At what applied pressure, in torr, will this solution boil?

12.72 Pentane (C_5H_{12}) and heptane (C_7H_{16}) are two hydrocarbon liquids present in gasoline. At 20 °C, the vapor pressure of pentane is 420 torr and the vapor pressure of heptane is 36.0 torr. What will be the total vapor pressure (in torr) of a solution prepared by mixing equal masses of the two liquids?

* **12.73** Benzene and toluene help get good engine performance from lead-free gasoline. At 40 °C, the vapor pressure of benzene is 180 torr and that of toluene is 60 torr. Suppose you wished to prepare a solution of these liquids that will have a total vapor pressure of 96 torr at 40 °C. What must be the mole percent concentrations of each in the solution?

* **12.74** The vapor pressure of pure methyl alcohol, CH_3OH, at 30 °C is 160 torr. How many grams of the nonvolatile solute glycerol, $C_3H_5(OH)_3$, must be added to 100 g of methyl alcohol to obtain a solution with a vapor pressure of 140 torr?

12.75 A solution containing 8.3 g of a nonvolatile nondissociating substance dissolved in 1 mol of chloroform, $CHCl_3$, has a vapor pressure of 511 torr. The vapor pressure of pure $CHCl_3$ at the same temperature is 526 torr. Calculate (a) the mole fraction of the solute, (b) the number of moles of solute in the solution, and (c) the molecular mass of the solute.

12.76 At 21.0 °C, a solution of 18.26 g of a nonvolatile, non-polar compound in 33.25 g of ethyl bromide, C_2H_5Br, had a vapor pressure of 336.0 torr. The vapor pressure of pure ethyl bromide at this temperature is 400.0 torr. What is the molecular mass of the compound?

Freezing Point Depression and Boiling Point Elevation

12.77 Calculate what would be the estimated boiling point of 2.00 m sugar in water. What would be its estimated freezing point? (It's largely the sugar in ice cream that makes it difficult to keep ice cream frozen hard.)

12.78 Glycerol, $C_3H_5(OH)_3$ (molecular mass 92), is essentially a nonvolatile liquid that is very soluble in water. A solution is made by dissolving 46.0 g of glycerol in 250 g of water. Calculate the approximate freezing point of the solution and the boiling point of the solution (at 1 atm).

12.79 How many grams of sucrose ($C_{12}H_{22}O_{11}$) are needed to lower the freezing point of 100 g of water by 3.00 °C?

12.80 What will be the boiling point of the solution described in the preceding problem?

12.81 A solution of 12.00 g of an unknown nondissociating compound dissolved in 200.0 g of benzene freezes at 3.45 °C. Calculate the molecular mass of the unknown.

12.82 A solution of 14 g of a nonvolatile, nondissociating compound in 1.0 kg of benzene boils at 81.7 °C. Calculate the molecular mass of the unknown.

12.83 What are the molecular mass and molecular formula of a nondissociating molecular compound whose empirical formula is C_4H_2N if 3.84 g of the compound in 500 g of benzene gives a freezing point depression of 0.307 °C?

12.84 Benzene reacts with hot concentrated nitric acid dissolved in sulfuric acid to give chiefly nitrobenzene, $C_6H_5NO_2$. A by-product is often obtained, which consists of 42.86% C, 2.40% H, and 16.67% N (by mass). The boiling point of a solution of 5.5 g of the by-product in 45 g of benzene was 1.84 °C higher than that of benzene. (a) Calculate the empirical formula of the by-product. (b) Calculate a molecular mass of the by-product and determine its molecular formula.

Osmotic Pressure

12.85 (a) Show that the following equation is true.

$$\text{Molar mass of solute} = \frac{(\text{grams of solute})RT}{\Pi V}$$

(b) An aqueous solution of a compound with a very high molecular mass was prepared in a concentration of 2.0 g L^{-1} at 25 °C. Its osmotic pressure was 0.021 torr. Calculate the molecular mass of the compound.

12.86 A saturated solution of 0.400 g of a polypeptide in 1.00 L of an aqueous solution has an osmotic pressure of 3.74 torr at 27 °C. What is the approximate molecular mass of the polypeptide?

Colligative Properties of Electrolyte Solutions

12.87 The vapor pressure of water at 20 °C is 17.5 torr. What would be the vapor pressure at 20 °C of a solution made by dissolving 10.0 g of NaCl in 100 g of water? (Assume complete dissociation of the solute and an ideal solution.)

12.88 How many grams of $AlCl_3$ would have to be dissolved in 150 mL of water to give a solution that has a vapor pressure of 38.7 torr at 35 °C? Assume complete dissociation of the solute and ideal solution behavior. (At 35 °C, the vapor pressure of pure water is 42.2 torr.)

12.89 What is the osmotic pressure, in torr, of a 2.0% solution of NaCl in water when the temperature of the solution is 25 °C?

12.90 Below are the concentrations of the most abundant ions in seawater.

Ion	Molality
Chloride	0.566
Sodium	0.486
Magnesium	0.055
Sulfate	0.029
Calcium	0.011
Potassium	0.011
Bicarbonate	0.002

Use these data to estimate the osmotic pressure of seawater at 25 °C in units of atm. What is the minimum pressure in atm needed to desalinate seawater by reverse osmosis?

12.91 What is the expected freezing point of a 0.20 m solution of $CaCl_2$? (Assume complete dissociation.)

12.92 The freezing point of a 0.10 m solution of mercury(I) nitrate is approximately −0.27 °C. Show that these data suggest that the formula of the mercury(I) ion is Hg_2^{2+}.

12.93 A 1.00 m aqueous solution of HF freezes at −1.91 °C. According to these data, what is the percent ionization of HF in the solution?

12.94 An aqueous solution of a weak electrolyte, HX, with a concentration of 0.125 m has a freezing point of −0.261 °C. What is the percent ionization of the compound (to two significant figures)?

Interionic Attractions and Colligative Properties

12.95 The van't Hoff factor for the solute in 0.100 m $NiSO_4$ is 1.19. What would this factor be if the solution behaved as if it were 100% dissociated?

12.96 What is the expected van't Hoff factor for K_2SO_4 in an aqueous solution, assuming 100% dissociation?

12.97 A 0.118 m solution of LiCl has a freezing point of −0.415 °C. What is the van't Hoff factor for this solute at this concentration?

12.98 What is the approximate osmotic pressure of a 0.118 m solution of LiCl at 10 °C? Express the answer in torr. (Use the data in the preceding problem.)

ADDITIONAL EXERCISES

****12.99** As discussed in Facets of Chemistry 12.1, the "bends" is a medical emergency that a diver faces if he or she rises too quickly to the surface from a deep dive. The origin of the problem is seen in the calculations in this problem. At 37 °C (normal body temperature), the solubility of N_2 in water is 0.015 g L^{-1} when its pressure over the solution is 1 atm. Air is approximately 78 mol% N_2. How many moles of N_2 are dissolved per liter of blood (essentially an aqueous solution) when a diver inhales air at a pressure of 1 atm? How many moles of N_2 dissolve per liter of blood when the diver is submerged to a depth of approximately 100 ft, where the total pressure of the air being breathed is 4 atm? If the diver suddenly surfaces, how many milliliters of N_2 gas, in the form of tiny bubbles, are released into the bloodstream from each liter of blood (at 37 °C and 1 atm)?

12.100 The vapor pressure of a mixture of 400 g of carbon tetrachloride and 43.3 g of an unknown compound is 137 torr at 30 °C. The vapor pressure of pure carbon tetrachloride at 30 °C is 143 torr, while that of the pure unknown is 85 torr. What is the approximate molecular mass of the unknown?

****12.101** An experiment calls for the use of the dichromate ion, $Cr_2O_7^{2-}$, in sulfuric acid as an oxidizing agent for isopropyl alcohol, C_3H_8O. The chief product is acetone, C_3H_6O, which forms according to the following equation.

$$3C_3H_8O + Na_2Cr_2O_7 + 4H_2SO_4 \longrightarrow$$
$$3C_3H_6O + Cr_2(SO_4)_3 + Na_2SO_4 + 7H_2O$$

(a) The oxidizing agent is available only as sodium dichromate dihydrate. What is the minimum number of grams of sodium dichromate dihydrate needed to oxidize 21.4 g of isopropyl alcohol according to the balanced equation?

(b) The amount of acetone actually isolated was 12.4 g. Calculate the percentage yield of acetone.

(c) The reaction produces a volatile by-product. When a sample of it with a mass of 8.654 mg was burned in oxygen, it was converted into 22.368 mg of carbon dioxide and 10.655 mg of water, the sole products. (Assume that any unaccounted for element is oxygen.) Calculate the percentage composition of the by-product and determine its empirical formula.

(d) A solution prepared by dissolving 1.338 g of the by-product in 115.0 g of benzene had a freezing point of 4.87 °C. Calculate the molecular mass of the by-product and write its molecular formula.

12.102 What is the osmotic pressure in torr of a 0.010 M aqueous solution of a molecular compound at 25 °C?

****12.103** Ethylene glycol, $C_2H_6O_2$, is used in some antifreeze mixtures. Protection against freezing to as low as −40 °F is sought.

(a) How many moles of solute are needed per kilogram of water to ensure this protection?

(b) The density of ethylene glycol is 1.11 g mL^{-1}. To how many milliliters of solute does your answer to part (a) correspond?

(c) Calculate the number of quarts of ethylene glycol that should be mixed with each quart of water to get the desired protection.

12.104 The osmotic pressure of a dilute solution of a slightly soluble polymer in water was measured using the osmometer in Figure 12.23. The difference in the heights of the liquid levels was determined to be 1.26 cm at 25 °C. Assume the solution has a density of 1.00 g mL^{-1}. (a) What is the osmotic pressure of the solution in torr? (b) What is the molarity of the solution? (c) At what temperature would the solution be expected to freeze? (d) On the basis of the results of these calculations, explain why freezing point depression cannot be used to determine the molecular masses of compounds composed of very large molecules.

12.105 A solution of ethyl alcohol, C_2H_5OH, in water has a concentration of 4.613 mol L^{-1}. At 20 °C, its density is 0.9677 g mL^{-1}. Calculate the following: (a) the molality of the solution and (b) the percent concentration of the alcohol. The density of ethyl alcohol is 0.7893 g mL^{-1} and the density of water is 0.9982 g mL^{-1} at 20 °C.

12.106 Consider an aqueous 1.00 m solution of Na_3PO_4, a compound with useful detergent properties.

(a) Calculate the boiling point of the solution on the assumption that it does not ionize at all in solution.
(b) Do the same calculation assuming that the van't Hoff factor for Na_3PO_4 reflects 100% dissociation into its ions.
(c) The 1.00 m solution boils at 100.183 °C at 1 atm. Calculate the van't Hoff factor for the solute in this solution.

12.107 In an aqueous solution of KNO_3, the concentration is 0.9159 mol% of the salt. The solution's density is 1.0489 g mL^{-1}. Calculate (a) the molal concentration of KNO_3, (b) the percent (w/w) of KNO_3, and (c) the molarity of the KNO_3 in the solution.

Experienced mountain trekkers have discovered that mosquitoes become inactive below about 14 °C (58 °F), which is no help for this deer. In this chapter you will learn about factors besides temperature that affect rates of reaction.

Kinetics: The Study of Rates of Reaction

This Chapter in Context One of the noteworthy characteristics of chemical reactions is that they occur at widely different speeds. Take, for example, some familiar reactions of oxygen, those with iron, carbon, and nitrogen. The rusting of iron is slow, but the combustion of coal dust (mostly carbon) is explosively fast. Air is about 20% oxygen and 80% nitrogen. Fortunately, nitrogen is unreactive to oxygen in air under ordinary conditions. But under the high temperature and pressure found in a working vehicle cylinder, these two elements in air do react somewhat, giving gases that contribute both to smog and to the greenhouse effect.

The oxygen-requiring reactions of metabolism accelerate when your body has a fever, stepping up the body's demand for oxygen. This puts a strain on the heart as it must now pump faster to deliver the oxygen and to remove waste carbon dioxide. Chemists in a wide variety of areas are thus interested in the factors that affect speeds of reactions and in how such speeds can be controlled.

13.1 Speeds at Which Reactions Occur

The **rate of reaction** for a given chemical change is the speed with which its reactants disappear and its products form.

Rate information often has practical value. For instance, knowing the rates of reactions and the factors that control them allows the manufacturers of chemical products to adjust the conditions of reactions efficiently so as to improve productivity and hold down operating costs.

At a more fundamental level, a study of the rate of a reaction often gives information about *how* its reactants actually change into its products. A balanced equation generally describes only a net overall change. Usually, however, the net change is the result of a series of simple reactions that are not at all evident from the equation. Consider, for example, the combustion of propane, C_3H_8.

$$C_3H_8(g) + 5O_2(g) \longrightarrow 3CO_2(g) + 4H_2O(g)$$

573

Anyone who has ever played billiards knows that this reaction simply cannot occur in a single, simultaneous collision between one propane molecule and five oxygen molecules. Just getting only three balls to come together with but one "click" on a flat, two-dimensional surface is extremely improbable. How unlikely it must be, then, for the *simultaneous* collision in three-dimensional space of six reactant molecules. Instead, the combustion of propane proceeds very rapidly by a series of much more probable steps, involving colliding chemical species of fleeting existence. *The series of individual steps that add up to the overall observed reaction is called the* **mechanism of the reaction.** Information about reaction mechanisms is one of the dividends paid by the study of rates, a study that is called **kinetics.**

Kinetics tells us how fast a reaction will occur. The study of *chemical equilibria* (next chapter) concerns the relative amounts of reactants and products that remain when nothing further happens.

13.2 Factors That Affect Reaction Rates

Before we take up the quantitative aspects of reaction rates, let's make an overall survey in qualitative language of the factors that govern them. There are chiefly five.

1. Chemical nature of the reactants.
2. Ability of the reactants to come in contact with each other.
3. Concentrations of the reactants.
4. Temperature.
5. Availability of rate-accelerating agents called *catalysts*.

Chemical Nature of the Reactants

Bonds break and new bonds form during reactions. The most fundamental differences among reaction rates, therefore, lie in the reactants themselves, in the inherent tendencies of their atoms, molecules, or ions to undergo changes in chemical bonds. Some reactions are fast by nature and others are slow. Because sodium atoms by nature lose electrons so easily, for example, a freshly exposed surface of metallic sodium tarnishes almost instantly when exposed to air and moisture. Under identical conditions, iron also reacts with air and moisture to form rust, but the reaction is much slower because iron atoms do not lose electrons as easily as do those of sodium.

Ability of the Reactants to Meet

Most reactions involve two or more reactants whose particles (atoms, ions, or molecules) must collide with each other for the reaction to occur. This is why reactions are so often carried out in liquid solutions or in the gas phase, states in which the particles are able to intermingle and collide with each other easily.

Consider, for example, that although *liquid* gasoline does burn rapidly, gasoline *vapor* explodes when mixed with air in the right proportions before ignition. (An *explosion* is an extremely rapid reaction that quickly generates hot expanding gases.) The combustion of vaporized gasoline illustrates a **homogeneous reaction,** because all of the reactants are in the same phase. Another example is the neutralization of sodium hydroxide, dissolved in water, by aqueous hydrochloric acid.

When the reactants are present in different phases—for example, when one is a gas and the other a liquid or a solid—the reaction is called a **heterogeneous reaction.** In a heterogeneous reaction, the reactants are able to meet only at the interface between the phases, so *the area of contact between the phases determines the rate of the reaction.* This area is controlled by the sizes of the particles of the reactants. By pulverizing a solid, the total surface area can be hugely increased (see Figure 13.1). This maximizes contacts between the atoms, ions, or molecules in the solid state with those in a different phase.

> ►**Chemistry in Practice**◄ If you have ever tried to light a campfire, you've learned that holding a match to finely divided kindling gets the fire going far more rapidly than trying to ignite a log directly. The kindling, with its greater surface area, allows for more contact with oxygen and a faster reaction.
>
> When grain dust particles are extremely small, dry, and mixed with air, grain elevators have been known to explode (see Figure 13.2). Although the reaction is heterogeneous, the contact between the two reactants, the dust and oxygen, is at a maximum, so an accidental spark sets off an explosion that destroys the elevator.
>
> The same situation is deliberately created in the operation of a coal-fueled power plant, where the coal is first pulverized and then blown with air into the combustion chamber. The combustion is extremely rapid and virtually 100% complete, thus delivering the maximum amount of the chemical energy in coal to the production of electricity. ◆

Although heterogeneous reactions such as those just illustrated are obviously important, they are very complex and difficult to analyze. In this chapter, therefore, we'll focus mostly on homogeneous systems.

Concentrations of the Reactants

The rates of both homogeneous and heterogeneous reactions are affected by the concentrations of the reactants. For example, wood burns relatively quickly in air but extremely rapidly in pure oxygen. Someone has estimated that if air were 30% oxygen instead of 21%, it would not be possible to put out forest fires. Even red hot steel wool, which only sputters and glows in air, bursts into flame when thrust into pure oxygen (see Figure 13.3).

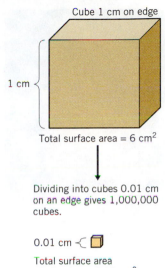

Cube 1 cm on edge

1 cm

Total surface area = 6 cm^2

Dividing into cubes 0.01 cm on an edge gives 1,000,000 cubes.

0.01 cm

Total surface area of all cubes = 600 cm^2

Figure 13.1 *Effect of crushing a solid.* When a single solid is subdivided into much smaller pieces, the total surface area on all of the pieces becomes very large.

Similar explosions caused by coal dust in underground coal mines have killed many miners.

Figure 13.2 *Particle size and reaction rate.* An explosion in this grain elevator in New Orleans, Louisiana, killed 35 people in December 1977.

Temperature of the System

Almost all chemical reactions occur faster at higher temperatures than they do at lower temperatures. You may have noticed, for example, that insects move more slowly when the air is cool. An insect is a cold-blooded creature, which means that its body temperature is determined by the temperature of its surroundings. As the air cools, insects cool, and so the rates of their chemical metabolism slow down, making insects sluggish.

Presence of Catalysts

Catalysts are substances that increase the rates of chemical reactions without being used up. It's amazing, but *catalysts affect every moment of our lives.* That's because the enzymes that direct our body chemistry are all catalysts. So are many of the substances used by the chemical industry to make gasoline, plastics, fertilizers, and other products that have become virtual necessities in our lives. We will discuss how catalysts affect reaction rates later in Section 13.9.

With this qualitative overview, let's now delve into the quantitative aspects of kinetics.

Figure 13.3 *Effect of concentration on rate.* Steel wool, after being heated to redness in a flame, burns spectacularly when dropped into pure oxygen.

13.3 Measuring a Rate of Reaction

A **rate** in any field of study is always expressed as a ratio in which a unit of time is in the denominator. Suppose, for example, that you have a job with a pay rate of ten dollars per hour. Because *per* can be translated as *divided by,* your pay rate can be written as a fraction (abbreviating hour as hr).

$$\text{Rate of pay} = \frac{10 \text{ dollars}}{1 \text{ hr}}$$

The fraction $\frac{1}{\text{hr}}$ can also be written as hr^{-1}, so your pay rate can also be given as

$$\text{Rate of pay} = 10 \text{ dollars hr}^{-1}$$

In general, $\frac{1}{x} = x^{-1}$.

Rates of Chemical Reactions

When chemical reactions occur, the concentrations of reactants decrease as they are used up, while the concentrations of the products increase as they form. So one way to describe a reaction's rate is to pick one species in the reaction's equation and describe its change in concentration per unit of time. The result is the rate of the reaction *with respect to that species.* Remembering that we always take "final minus initial," the rate of reaction with respect, say, to species X is

$$\text{Rate with respect to } X = \frac{(\text{concn of } X \text{ at time } t_2 - \text{concn of } X \text{ at time } t_1)}{(t_2 - t_1)}$$

$$= \frac{\Delta(\text{concn of } X)}{\Delta t}$$

Once again, we use the symbol Δ to mean a change.

Molarity (mol/L) is normally the concentration unit, and the second (s) is the most often used unit of time. Therefore, the units for reaction rates are most frequently the following.

$$\frac{\text{mol/L}}{\text{s}}$$

Because 1/L and 1/s can also be written as L^{-1} and s^{-1}, the units for a reaction rate can be expressed as **mol L^{-1} s^{-1}.** For instance, if the concentration of one product of a reaction increases by 0.50 mol/L each second, the rate of its formation is 0.50 mol L^{-1} s^{-1}. Similarly, if the concentration of a reactant decreases by 0.20 mol/L per second, its rate of reaction is 0.20 mol L^{-1} s^{-1}. Notice that the numerical value of a reaction rate is, by custom, given as a positive value whether something increases or decreases in concentration.

Rates and Coefficients

When we know the value of a reaction rate with respect to one species, the coefficients of the reaction's balanced equation may be used to find the rates with respect to the other species. For example, in the combustion of propane described earlier,

$$C_3H_8(g) + 5O_2(g) \longrightarrow 3CO_2(g) + 4H_2O(g)$$

five moles of O_2 *must* be consumed per unit of time for each mole of C_3H_8 used in the same time. Therefore, in this reaction oxygen *must* react five times faster than propane in units of mol L^{-1} s^{-1}. Similarly, CO_2 forms three times faster than C_3H_8 reacts and H_2O four times faster. The magnitudes of the rates relative to each other are thus in the same relationship as the coefficients in the balanced equation.

EXAMPLE 13.1

Relationships of Rates within a Reaction

Butane is lighter fluid.

Butane, C_4H_{10}, burns in oxygen to give CO_2 and H_2O according to the equation

$$2C_4H_{10}(g) + 13O_2(g) \longrightarrow 8CO_2(g) + 10H_2O(g)$$

If at a certain moment the butane concentration is decreasing at a rate of 0.20 mol L^{-1} s^{-1}, what is the rate at which the oxygen concentration is decreasing, and what are the rates at which the product concentrations are increasing?

Analysis: The question asks about the rates *relative to the given rate* of the reaction of butane, so it's basically a simple stoichiometry problem. This, as we said, is because the magnitudes of the rates relative to each other are in the same relationship as the coefficients in the balanced equation.

Solution: For oxygen

$$\frac{0.20 \text{ mol } C_4H_{10}}{L\ s} \times \frac{13 \text{ mol } O_2}{2 \text{ mol } C_4H_{10}} = \frac{1.3 \text{ mol } O_2}{L\ s}$$

Oxygen is reacting at a rate of 1.3 mol L^{-1} s^{-1}. For CO_2 and H_2O, we have similar calculations.

$$\frac{0.20 \text{ mol } C_4H_{10}}{L\ s} \times \frac{8 \text{ mol } CO_2}{2 \text{ mol } C_4H_{10}} = \frac{0.80 \text{ mol } CO_2}{L\ s}$$

$$\frac{0.20 \text{ mol } C_4H_{10}}{L\ s} \times \frac{10 \text{ mol } H_2O}{2 \text{ mol } C_4H_{10}} = \frac{1.0 \text{ mol } H_2O}{L\ s}$$

Therefore,

$$\text{Rate of formation of } CO_2 = 0.80 \text{ mol } L^{-1}\ s^{-1}$$

$$\text{Rate of formation of } H_2O = 1.0 \text{ mol } L^{-1}\ s^{-1}$$

Are the Answers Reasonable?
If what we've calculated is correct, then the ratio of the numerical values of the last two rates, namely 0.80 to 1.0, should check out to be the same as the ratio of the corresponding coefficients in the chemical equation, namely, 8 to 10 (the same as 0.8 to 1.0).

Hydrogen sulfide burns in oxygen to form sulfur dioxide and water.

$$2H_2S(g) + 3O_2(g) \longrightarrow 2SO_2(g) + 2H_2O(g)$$

If sulfur dioxide is being formed at a rate of 0.30 mol L^{-1} s^{-1}, what are the rates of disappearance of hydrogen sulfide and oxygen? ◆

Because the rates of reaction of reactants and products are all related, it doesn't matter which species we pick to follow concentration changes over time. For example, to study the decomposition of hydrogen iodide, HI, into H_2 and I_2,

$$2HI(g) \longrightarrow H_2(g) + I_2(g)$$

it is easiest to monitor the I_2 concentration because it is the only colored substance in the reaction. As the reaction proceeds, purple iodine vapor forms, and there are instruments that allow us to relate the intensity of the color to the iodine concentration. Then, once we know the rate of formation of iodine, we also know the rate of formation of hydrogen. It's the same because the coefficients of H_2 and I_2 are the same. And the rate of disappearance of HI, which has a coefficient of 2 in the equation, is twice as fast as the rate of formation of I_2.

Change of Reaction Rate with Time

A reaction rate is generally not constant throughout the reaction but commonly changes as the reactants are used up. This is because the rate usually depends on the concentrations of the reactants, and these change as the reaction proceeds. For example, Table 13.1 contains data for the decomposition of hydrogen iodide at a temperature of 508 °C. The data, which show the changes in molar HI concentration over time, are plotted in Figure 13.4. Notice that the molar HI concentration drops fairly rapidly during the first 50 seconds of the reaction, which means that the initial rate is relatively fast. However, later, in the interval between 300 s and 350 s, the concentration changes by only a small amount, so the rate has slowed considerably. Thus, the steepness of the curve at any moment reflects the rate of the reaction; the steeper the curve, the higher is the rate.

The rate at which the HI is being consumed at any particular moment can be determined from the slope (or tangent) of the curve measured at the time we have chosen. The slope, which can be read off the graph, is the ratio (expressed positively) of the change in concentration to the change in time. In Figure 13.4, for example, the rate of the decomposition of hydrogen iodide is determined for a time 100 seconds from the start of the reaction. After the tangent to the curve is drawn, we measure a concentration change (a decrease of 0.027 mol/L) and the time change (110 s) from the graph. Because the rate is based on a *decreasing* concentration, we use a minus sign for the *equation* that describes this rate so that the rate itself will be a positive quantity. We use square brackets to signify concentrations specifically in moles per liter; [HI] thus means the molar concentration of HI.

Appendix A has a discussion of slope.

Table 13.1 Data, at 508 °C, for the Reaction $2HI(g) \rightarrow H_2(g) + I_2(g)$

Concentration of HI (mol/L)	Time (s)
0.100	0
0.0716	50
0.0558	100
0.0457	150
0.0387	200
0.0336	250
0.0296	300
0.0265	350

$$\text{Rate}_{\text{with respect to HI}} = -\left(\frac{[\text{HI}]_{\text{final}} - [\text{HI}]_{\text{initial}}}{t_{\text{final}} - t_{\text{initial}}} \right) = -\left(\frac{-0.027 \text{ mol/L}}{110 \text{ s}} \right)$$

$$\text{Rate}_{\text{with respect to HI}} = 2.5 \times 10^{-4} \text{ mol } L^{-1} s^{-1}$$

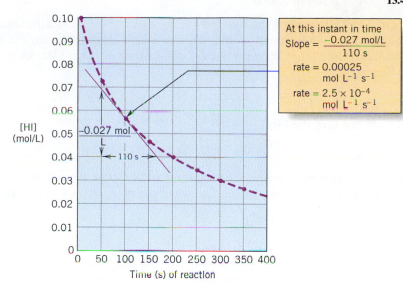

At this instant in time

Slope = $\dfrac{-0.027 \text{ mol/L}}{110 \text{ s}}$

rate = 0.00025 mol L^{-1} s^{-1}

rate = 2.5×10^{-4} mol L^{-1} s^{-1}

Figure 13.4 *Effect of time on concentration.* The data for this plot of the change in the concentration of HI with time for the reaction

$$2HI(g) \longrightarrow H_2(g) + I_2(g)$$

at 508 °C are taken from Table 13.1. The slope is negative because we're measuring the *disappearance* of HI. But when its value is used as a rate of reaction, we express the rate as positive, as we do all rates of reaction.

Thus, at this moment in the reaction, the rate with respect to HI is 2.5×10^{-4} mol L^{-1} s^{-1}. By working Practice Exercise 2, you will see that the rate with respect to HI later on is smaller.

Practice Exercise 2

Use the graph in Figure 13.4 to estimate the reaction rate with respect to HI 250 seconds after the start of the reaction. ◆

13.4 Concentration and Rate

Thus far we have focused on a rate with respect to *one* component of a reaction. We'll now broaden our focus to consider a rate expression that includes all reactants.

The rate of a homogeneous reaction at any instant is proportional to the product of the molar concentrations of the reactants, each molarity raised to some power or exponent that has to be found by experiment. Let's consider a chemical reaction with an equation of the following form.

$$A + B \longrightarrow \text{products}$$

Its rate of reaction can be expressed as follows.

$$\text{Rate} \propto [A]^m [B]^n \tag{13.1}$$

Tools

Rate law of a reaction

As we said, the values of the exponents n and m are found by experiment, which we'll go into shortly.

Rate Laws

The proportionality symbol, $\propto$, in Equation 13.1 can be replaced by an equals sign if we introduce a proportionality constant, k, which is called the **rate constant** for the reaction. This gives Equation 13.2.

$$\text{Rate} = k [A]^m [B]^n \tag{13.2}$$

The value of k depends on the particular reaction being studied as well as the temperature at which the reaction occurs.

Equation 13.2 is called the **rate law** for the reaction of A with B. Once we have found values for k, n, and m, the rate law allows us to calculate the rate of the reaction at any set of known values of concentrations. Consider, for example, the following reaction.

$$H_2SeO_3 + 6I^- + 4H^+ \longrightarrow Se + 2I_3^- + 3H_2O \qquad (13.3)$$

Its rate law is of the form

$$\text{Rate} = k\,[H_2SeO_3]^x\,[I^-]^y\,[H^+]^z$$

The units of the rate constant are generally such that the calculated rate will have the units $\text{mol L}^{-1}\text{s}^{-1}$.

The exponents have been found experimentally to be the following for the initial rate of this reaction (i.e., the rate when the reactants are first combined).

$$x = 1, \qquad y = 3, \qquad \text{and} \qquad z = 2$$

At 0 °C, k equals $5.0 \times 10^5 \text{ L}^5 \text{ mol}^{-5}\text{ s}^{-1}$. (We have to specify the temperature because k varies with it.) Substituting the exponents and the value of k into the rate law equation gives the rate law for the reaction of Equation 13.3.

When the exponent is equal to 1, as it is for $[H_2SeO_3]$, *it is usually omitted.*

$$\text{Rate} = (5.0 \times 10^5 \text{ L}^5 \text{ mol}^{-5}\text{ s}^{-1})\,[H_2SeO_3]\,[I^-]^3\,[H^+]^2 \qquad \text{(at 0 °C)} \quad (13.4)$$

Equation 13.4 will allow us to calculate the initial rate of the reaction at 0 °C for any set of concentrations of H_2SeO_3, I^-, and H^+.

EXAMPLE 13.2

Calculating Reaction Rate from the Rate Law

What is the rate of Reaction 13.3 at 0 °C when the reactant concentrations are the following: $[H_2SeO_3] = 2.0 \times 10^{-2}\,M$, $[I^-] = 2.0 \times 10^{-3}\,M$, and $[H^+] = 1.0 \times 10^{-3}\,M$?

Analysis: Because we already know the rate law (Equation 13.4), the answer to this question is merely a matter of substituting the given molar concentrations into this law.

Solution: To see how the units work out, let's write all of the concentration values as well as the rate constant's units in fraction form.

$$\text{Rate} = \frac{5.0 \times 10^5 \text{ L}^5}{\text{mol}^5 \text{ s}} \times \left(\frac{2.0 \times 10^{-2} \text{ mol}}{L}\right) \times \left(\frac{2.0 \times 10^{-3} \text{ mol}}{L}\right)^3$$

$$\times \left(\frac{1.0 \times 10^{-3} \text{ mol}}{L}\right)^2$$

To perform the arithmetic, we first raise the concentrations and their units to the appropriate powers.

$$\text{Rate} = \frac{5.0 \times 10^5 \text{ L}^5}{\text{mol}^5 \text{ s}} \times \left(\frac{2.0 \times 10^{-2} \text{ mol}}{L}\right) \times \left(\frac{8.0 \times 10^{-9} \text{ mol}^3}{L^3}\right)$$

$$\times \left(\frac{1.0 \times 10^{-6} \text{ mol}^2}{L^2}\right)$$

$$\text{Rate} = \frac{8.0 \times 10^{-11} \text{ mol}}{L\,s} = 8.0 \times 10^{-11} \text{ mol L}^{-1}\text{ s}^{-1}$$

Is the Answer Reasonable?
There's obviously no simple check. At least the answer has the correct units for a reaction rate.

Practice Exercise 3

The rate law for the decomposition of HI to I_2 and H_2 is

$$\text{Rate} = k\,[\text{HI}]^2$$

At 508 °C, the rate of the reaction of HI was found to be 2.5×10^{-4} mol L^{-1} s^{-1} when the HI concentration was 0.0558 M (see Figure 13.4). (a) What is the value of k? (b) What are the units of k? ◆

Exponents in the Rate Law

Although a rate law's exponents are generally unrelated to the chemical equation's coefficients, they sometimes are the same by coincidence, as is the case in the decomposition of hydrogen iodide.

$$2\text{HI}(g) \longrightarrow \text{H}_2(g) + \text{I}_2(g)$$

The rate law, as we've said, is

$$\text{Rate} = k\,[\text{HI}]^2$$

The exponent of [HI] in the rate law, namely 2, happens to match the coefficient of HI in the overall chemical equation, but *there is no way we could have predicted this match without experimental data*. Therefore, *never* simply assume the exponents and the coefficients are the same; it's a trap that many students fall into.

An exponent in a rate law is called the **order of the reaction**[1] with respect to the corresponding reactant. For instance, the decomposition of gaseous N_2O_5 into NO_2 and O_2,

$$2\text{N}_2\text{O}_5 \longrightarrow 4\text{NO}_2 + \text{O}_2$$

has the rate law

$$\text{Rate} = k\,[\text{N}_2\text{O}_5]$$

The exponent of $[\text{N}_2\text{O}_5]$ is 1, so the reaction rate is said to be *first order* in N_2O_5. The rate law for the decomposition of HI has an exponent of 2 for the HI concentration, so its reaction rate is *second order* in HI. The rate law in Equation 13.4 tells us that Reaction 13.3 is first order with respect to H_2SeO_3, third order with respect to I^-, and second order with respect to H^+.

The **overall order of a reaction** is the sum of the orders with respect to each reactant in the rate law. The decomposition of N_2O_5 is thus a first-order reaction, and that of HI is a second-order reaction. Reaction 13.3 is a sixth-order reaction, overall.

The exponents in a rate law are usually small whole numbers, but fractional and negative exponents are occasionally found. A negative exponent means that the concentration term really belongs in the denominator, which means that as the concentration of the species increases, the rate of reaction decreases. There are even zero-order reactions. They have reaction rates that are independent of the concentration of any reactant. An example is the elimination of ethyl alcohol by the body. Regardless of the blood alcohol level, the rate of alcohol removal by the body is constant. Another zero-order reaction is the decomposition of gaseous ammonia into H_2 and N_2 on a hot platinum surface. The rate at which ammonia decomposes is the same, regardless of its concentration in the gas.

---

[1] The reason for describing the *order* of a reaction is to take advantage of a great convenience, namely, the mathematics involved in the treatment of the data is the same for all reactions having the same order. We will not go into this very deeply, but you should be familiar with this terminology; it's often used to describe the effects of concentration on reaction rates.

The following reaction

$$BrO_3^- + 3SO_3^{2-} \longrightarrow Br^- + 3SO_4^{2-}$$

has the rate law

$$\text{Rate} = k\,[BrO_3^-]\,[SO_3^{2-}]$$

What is the order of the reaction with respect to each reactant? What is the overall order of the reaction? ◆

Determining the Exponents in a Rate Law

We've mentioned several times that the exponents in the rate law of an overall reaction must be determined experimentally. *This is the only way for us to know for sure what the exponents are.* To determine the exponents, we study how changes in concentration affect the rate of the reaction. For example, consider again the following hypothetical reaction.

$$A + B \longrightarrow \text{products}$$

Suppose, further, that the data in Table 13.2 have been obtained in a series of five experiments. We know the form of the rate law for the reaction will be

$$\text{Rate} = k\,[A]^n\,[B]^m$$

Notice in Table 13.2 that the concentration of B is constant for the first three sets of data. Changes in the rate are therefore caused only by changes in the concentration of A. Now all we have to do is figure out what the order of the reaction must be with respect to A to give the observed changes in the rate.

In Table 13.2, we find that when $[A]$ is doubled, the rate doubles; when $[A]$ is tripled, the rate triples. What must be the exponent, n, on the value of $[A]$ for this to be true? The answer is 1. We can see this if we try a few values for $[A]$. When $[A] = 0.10$, $[A]^1 = 0.10$, and when $[A] = 0.20$, $[A]^1 = 0.20$. Notice that when we double the concentration of A, we double the value of $[A]^1$, which means that the rate would double. If the coefficient n were 2 instead of 1, doubling the value of $[A]$ would multiply the rate by 2^2 or 4.

In the final three sets of data, the concentration of B changes while the concentration of A is held constant. This time it is the concentration of B that affects the rate. Now we see that when $[B]$ is doubled, the rate increases by a factor of 4 (which equals 2^2), and when $[B]$ is tripled, the rate does increase by a factor of 9 (which equals 3^2). The only way that B can affect the rate this way is if its concentration is squared in the rate law, so the exponent m must equal 2. This also can be shown using some trial values for $[B]$. If $[B] = 0.3$, then $[B]^2$ equals 0.09; if the con-

We'll display the exponent of $[A]$, 1, for emphasis.

Table 13.2 **Concentration–Rate Data for the Hypothetical Reaction** $A + B \rightarrow$ **products**

Initial Concentrations		Initial Rate of Formation
$[A]$	$[B]$	of Products (mol L^{-1} s^{-1})
0.10	0.10	0.20
0.20	0.10	0.40
0.30	0.10	0.60
0.30	0.20	2.40
0.30	0.30	5.40

centration is doubled, so that $[B] = 0.6$, then $[B]^2$ equals 0.36. Notice that 0.36 is four times as large as 0.09. Doubling the concentration therefore increases $[B]^2$ by a factor of 4, and that increases the rate by a factor of 4.

Having determined the exponents for the concentration terms, we now know that the rate law for the reaction must be

$$\text{Rate} = k\,[A]^1\,[B]^2$$

To calculate the value of k, we need only substitute rate and concentration data into the rate law for any one of the sets of data.

$$k = \frac{\text{rate}}{[A]^1\,[B]^2}$$

Using the data from the first set in Table 13.2,

$$k = \frac{0.20 \text{ mol L}^{-1}\text{s}^{-1}}{(0.10 \text{ mol L}^{-1})(0.10 \text{ mol L}^{-1})^2}$$

$$= \frac{0.20 \text{ mol L}^{-1}\text{s}^{-1}}{0.0010 \text{ mol}^3 \text{ L}^{-3}}$$

After canceling such units as we can, the value of k with the net units is

$$k = 2.0 \times 10^2 \text{ L}^2 \text{ mol}^{-2}\text{s}^{-1}$$

Practice Exercise 5

Use the data from the other four experiments (Table 13.2) to calculate k for this reaction. What do you notice about the values of k? ◆

Table 13.3 summarizes the reasoning used to determine the order with respect to each reactant from experimental data.

Table 13.3 Relationship between the Order of
a Reaction and Changes in Concentration and Rate

Factor by Which the Concentration Is Changed	Factor by Which the Rate Changes	Exponent on the Concentration Term in the Rate Law
2	Rate	0
3	is	0
4	unchanged	0
2	$2 = 2^1$	1
3	$3 = 3^1$	1
4	$4 = 4^1$	1
2	$4 = 2^2$	2
3	$9 = 3^2$	2
4	$16 = 4^2$	2
2	$8 = 2^3$	3
3	$27 = 3^3$	3
4	$64 = 4^3$	3

EXAMPLE 13.3

**Determining
the Exponents
of a Rate Law**

Sulfuryl chloride (bp 69.3 °C) is a dense liquid (1.67 g mL^{-1}) with a pungent odor and is corrosive to the skin and lungs.

Sulfuryl chloride, SO_2Cl_2, is used to manufacture the antiseptic chlorophenol. The following data were collected on the decomposition of SO_2Cl_2, at a certain temperature.

$$SO_2Cl_2(g) \longrightarrow SO_2(g) + Cl_2(g)$$

Initial Concentration of SO_2Cl_2 (mol L^{-1})	Initial Rate of Formation of SO_2 (mol L^{-1} s^{-1})
0.100	2.2×10^{-6}
0.200	4.4×10^{-6}
0.300	6.6×10^{-6}

What are the rate law and the value of the rate constant for this reaction?

Analysis: The first step is to write the general form of the expected rate law so we can see which exponents have to be determined. Then we study the data to see how the rate changes when the concentration is changed by a certain factor.

Solution: We expect the rate law to have the form

$$\text{Rate} = k\,[SO_2Cl_2]^x$$

We could also use Experiments 2 and 3. From the second to the third, the rate increases by the same factor, 1.5, as the concentration, so by these data, too, the reaction must be first order.

Let's examine the data from the first two experiments. Notice that when we double the concentration from 0.100 M to 0.200 M, the initial rate doubles (from 2.2×10^{-6} mol L^{-1} s^{-1} to 4.4×10^{-6} mol L^{-1} s^{-1}). If we look at the first and third, we see that when the concentration triples (from 0.100 M to 0.300 M), the rate also triples (from 2.2×10^{-6} mol L^{-1} s^{-1} to 6.6×10^{-6} mol L^{-1} s^{-1}). This behavior tells us that the reaction must be first order in the SO_2Cl_2 concentration. The rate law is therefore

$$\text{Rate} = k\,[SO_2Cl_2]^1$$

To evaluate k, we can use any of the three sets of data. Choosing the first,

$$k = \frac{\text{rate}}{[SO_2Cl_2]^1}$$

$$= \frac{2.2 \times 10^{-6}\ \text{mol L}^{-1}\,\text{s}^{-1}}{0.100\ \text{mol L}^{-1}}$$

$$= 2.2 \times 10^{-5}\ \text{s}^{-1}$$

Is the Answer Reasonable?
We should get the same value of k by picking any other pair of values. Thus, arbitrarily picking the last pair of data, at an initial molar concentration of SO_2Cl_2 of 0.300 mol L^{-1} and an initial rate of 6.6×10^{-6} mol L^{-1} s^{-1}, k again calculates to be 2.2×10^{-5} s^{-1}.

EXAMPLE 13.4

**Determining the
Exponents of a Rate Law**

Isoprene is used industrially to make a product identical with natural rubber.

Isoprene forms a dimer called dipentene. (A *dimer* is a compound formed by joining two simpler molecules.)

What is the rate law for this reaction, given the following data for the initial rate of formation of dipentene?

Initial Concentration of Isoprene (mol L^{-1})	Initial Rate of Formation of Dipentene (mol L^{-1} s^{-1})
0.50	1.98
1.50	17.8

Analysis: With only one reactant, we expect the rate law to be of the form

$$\text{Rate} = k \, [\text{isoprene}]^x$$

To discover x, we must compare the results of the two experiments.

Solution: From the first to the second experiment, we see that the isoprene concentration increases by a factor of 3. The *rate* of the reaction, however, increases by a factor of $(17.8/1.98) = 8.99$. This is very nearly a factor of 9, which is 3^2. Therefore, the value of x is 2 and the rate law is

$$\text{Rate} = k \, [\text{isoprene}]^2$$

Is the Answer Reasonable?
Given so few data, we don't have a way to try another calculation.

EXAMPLE 13.5

Determining the Exponents of a Rate Law

The following data were measured for the reduction of nitric oxide with hydrogen.

$$2NO(g) + 2H_2(g) \longrightarrow N_2(g) + 2H_2O(g)$$

Initial Concentrations (mol L^{-1})		Initial Rate of Formation of H$_2$O (mol L^{-1} s^{-1})
[NO]	[H$_2$]	
0.10	0.10	1.23×10^{-3}
0.10	0.20	2.46×10^{-3}
0.20	0.10	4.92×10^{-3}

What is the rate law for the reaction?

Nitric oxide is what forms to some extent from the air's nitrogen and oxygen in operating vehicle engine cylinders where the temperature and pressure are both high.

Analysis: This time we have two reactants. To see how their concentrations affect the rate we must vary only one concentration at a time. Therefore, we choose two experiments in which the concentration of one reactant doesn't change and examine the effect of a change in the concentration of the other reactant. Then we repeat the procedure for the second reactant.

Solution: We expect the rate law to have the form

$$\text{Rate} = k \, [\text{NO}]^n \, [\text{H}_2]^m$$

Let's look at the first two experiments. Here the concentration of NO remains the same, so the rate is being affected by the change in the H$_2$ concentration. When we double the H$_2$ concentration, the rate doubles, so the reaction is first order with respect to H$_2$. This means $m = 1$.

Next, we need to pick two experiments in which the H$_2$ concentration doesn't change. Working with the first and third, we see that [NO] doubles and the rate in-

When the NO in vehicle exhaust reaches the outside air, it quickly oxidizes to nitrogen dioxide, NO$_2$, a poison and the chief cause of the reddish color of smog.

creases by a factor of $4.92/1.23 = 4.00$. When doubling the concentration of a species quadruples the rate, the reaction is second order in that species, so $n = 2$.

Therefore, the rate law for the reaction is

$$\text{Rate} = k\,[\text{NO}]^2\,[\text{H}_2]$$

Is the Answer Reasonable?

The only data that we haven't used as a pair are the data for the second and third reactions. The value of [NO] increases by 2 in going from the second to the third set of data, so this should multiply the rate by 2^2 or 4, if we've found the right exponents. But the value for [H_2] halves at the same time, so this should take a rate that is otherwise four times as large and cut it by a factor of $(1/2)^1$ or in half. The net effect, then, is to make the rate of the third reaction two times as large as that of the second reaction, which is the observed rate change.

Practice Exercise 6

Ordinary sucrose (table sugar) reacts with water in an acidic solution to produce two simpler sugars, glucose and fructose, that have the same molecular formulas.

$$\text{C}_{12}\text{H}_{22}\text{O}_{11} + \text{H}_2\text{O} \longrightarrow \text{C}_6\text{H}_{12}\text{O}_6 + \text{C}_6\text{H}_{12}\text{O}_6$$
$$\text{sucrose} \qquad\qquad \text{glucose} \qquad \text{fructose}$$

In a series of experiments, the following data were obtained.

Initial Sucrose Concentration (mol L^{-1})	Rate of Formation of Glucose (mol L^{-1} s^{-1})
0.10	6.17×10^{-5}
0.20	1.23×10^{-4}
0.50	3.09×10^{-5}

What is the order of the reaction with respect to sucrose? ◆

Practice Exercise 7

A certain reaction has the following equation: $A + B \longrightarrow C + D$. Experiments yielded the following results.

Initial Concentrations (mol L^{-1})		Initial Rate of Formation of C (mol L^{-1} s^{-1})
[A]	[B]	
0.40	0.30	1.0×10^{-4}
0.80	0.30	4.0×10^{-4}
0.80	0.60	1.6×10^{-3}

What is the rate law for the reaction? ◆

13.5 Concentration and Time

The rate law tells us how the speed of a reaction varies with the concentrations of the reactants. Often, however, we wish to have more information about the status of a reaction than how fast it is going. For instance, if we were manufacturing some chemical in a large reaction vessel, we might want to know what the concentrations of the reactants and products are at some specified time after the reaction has started, so we could decide whether it is time to harvest the products. We might like to know how long it would take for the reactant concentrations to drop below some minimum optimum values, so we could replenish them. Obtaining information of this kind requires some sort of expression that relates concentration to time.

The relationship between the concentration of a reactant and time can be derived from a rate law using calculus, but we'll look only at the results of such derivations. Knowing the order of the reaction helps, because the mathematics is always the same for a reaction of a given order. Even so, the mathematical expressions that relate concentration and time in complex reactions can be complicated, so we will consider only very simple cases just to give you a taste of the subject.

> A rate law is actually a differential equation from which the relationship between concentration and time is obtained by an operation called *integration*. The equations we discuss in this section are therefore often called *integrated rate equations*.

Concentration versus Time for First-Order Reactions

A reaction that is, overall, first order has a rate law of the type

$$\text{Rate} = k\,[A]$$

It can be shown that the following equation is the relationship between the concentration of A and time.

$$\ln \frac{[A]_0}{[A]_t} = kt \tag{13.5}$$

Here, $[A]_0$ is the initial concentration of A ($t = 0$), and $[A]_t$ is the molar concentration of A at some time t after the start of the reaction. The equation involves a natural logarithm, ln (see Section 11.7, page 491, and Appendix A). Thus, the natural logarithm of a particular ratio of concentrations equals the product of the rate constant and time. Equation 13.5 can be used directly in calculations, as illustrated by the following examples.

Tools

Rate law, first-order reactions

> Logarithms, both common (base 10) and natural (base e), are discussed in Appendix A. Natural logarithms (ln) are related as follows to common logarithms (log).
>
> $$\ln x = 2.303 \log x$$

EXAMPLE 13.6

Concentration– Time Calculations for First-Order Reactions

Dinitrogen pentoxide, N_2O_5, is not very stable. In the gas phase or dissolved in a nonaqueous solvent, like carbon tetrachloride, it decomposes by a first-order reaction into N_2O_4 and O_2.

$$2N_2O_5 \longrightarrow 2N_2O_4 + O_2$$

The rate law is

$$\text{Rate} = k\,[N_2O_5]$$

At 45 °C, the rate constant for the reaction in carbon tetrachloride is $6.22 \times 10^{-4}\ \text{s}^{-1}$. If the initial concentration of N_2O_5 in the solution is 0.100 M, how many minutes will it take for the concentration to drop to 0.0100 M?

> N_2O_5, an acid anhydride, is changed to nitric acid by a reaction with water.
>
> $$N_2O_5 + H_2O \longrightarrow 2HNO_3$$

Analysis: To solve the problem, we simply substitute quantities into Equation 13.5 and solve for t.

Solution: First, let's assemble the data.

$$[N_2O_5]_0 = 0.100\ M \qquad\qquad [N_2O_5]_t = 0.0100\ M$$
$$k = 6.22 \times 10^{-4}\ \text{s}^{-1} \qquad\qquad t = ?$$

Substituting,

$$\ln\left(\frac{0.100\ M}{0.0100\ M}\right) = (6.22 \times 10^{-4}\ \text{s}^{-1})t$$

$$\ln(10.0) = (6.22 \times 10^{-4}\ \text{s}^{-1})t$$

Using a scientific calculator to find the natural logarithm[2] of 10.0,

$$2.303 = (6.22 \times 10^{-4} \text{ s}^{-1})t$$

Solving for t,

$$t = \frac{2.303}{6.22 \times 10^{-4} \text{ s}^{-1}}$$

$$= 3.70 \times 10^3 \text{ s}$$

This is the time in seconds. Dividing by 60 gives the time in minutes, 61.7 min.

Is the Answer Reasonable?
There's no simple check other than to redo the calculation.

EXAMPLE 13.7

Concentration–Time Calculations for First-Order Reactions

If the initial concentration of N_2O_5 in a carbon tetrachloride solution at 45 °C is 0.500 M (see Example 13.6), what will its concentration be after exactly one hour?

Analysis: This time we have to solve for an unknown concentration, which is within the logarithm expression. The easiest way to do this is to first solve for the *ratio* of the concentrations. Once this ratio is known, we can then substitute the known concentration and calculate the unknown one. Remember that the unit of k involves seconds, not hours, so we must convert the given 1 hr into seconds (1 hr = 3600 s).

Solution: Let's begin by listing the data.

$$[N_2O_5]_0 = 0.500 \ M \qquad\qquad [N_2O_5]_t = ? \ M$$
$$k = 6.22 \times 10^{-4} \text{ s}^{-1} \qquad\qquad t = 3600 \text{ s}$$

Now we solve for the concentration ratio

$$\ln\left(\frac{[N_2O_5]_0}{[N_2O_5]_t}\right) = (6.22 \times 10^{-4} \text{ s}^{-1}) \times 3600 \text{ s}$$

$$= 2.24$$

To take the antilogarithm (antiln), we use a pocket calculator to find e raised to the 2.24 power.

$$\text{antiln}\left[\ln\left(\frac{[N_2O_5]_0}{[N_2O_5]_t}\right)\right] = \frac{[N_2O_5]_0}{[N_2O_5]_t}$$

$$\text{antiln}(2.24) = e^{2.24}$$

$$= 9.4$$

This means that

$$\frac{[N_2O_5]_0}{[N_2O_5]_t} = 9.4$$

---

[2]There are special rules for significant figures for logarithms. When taking a logarithm of a quantity, the number of digits written *after the decimal point* equals the number of significant figures in the quantity. Here, 10.0 has three significant figures, so the natural logarithm of 10.0, which is 2.303, has three digits after the decimal.

Now we can substitute the known concentration, $[N_2O_5]_0 = 0.500\ M$. This gives

$$\frac{0.500\ M}{[N_2O_5]_t} = 9.4$$

Solving for $[N_2O_5]_t$ gives

$$[N_2O_5]_t = \frac{0.500\ M}{9.4}$$

$$= 0.053\ M$$

Thus, after one hour, the concentration of N_2O_5 will have dropped to 0.053 M.

Is the Answer Reasonable?

At least the final concentration of N_2O_5 is *less* than its initial concentration. You'd know that you made a huge mistake if the final concentration were greater; reactants are used up by reactions. Other than this, there is no simple check except to redo the calculations.

Practice Exercise 8

In Practice Exercise 6, the reaction of sucrose with water in an acidic solution was described.

$$\underset{\text{sucrose}}{C_{12}H_{22}O_{11}} + H_2O \longrightarrow \underset{\text{glucose}}{C_6H_{12}O_6} + \underset{\text{fructose}}{C_6H_{12}O_6}$$

The reaction is first order with a rate constant of $6.2 \times 10^{-5}\ s^{-1}$ at 35 °C, when the H^+ concentration is 0.10 M. Suppose, in an experiment, the initial sucrose concentration was 0.40 M.

(a) What will its concentration be after exactly 2 hours? (b) How many minutes will it take for the concentration of sucrose to drop to 0.30 M? ◆

The logarithmic relationship between concentration and time for a first-order reaction (Equation 13.5) provides an elegant way to determine accurately the rate constant by graphical means. Using the properties of logarithms,[3] Equation 13.5 can be rewritten in a form that corresponds to the equation for a straight line.

$$\ln [A]_t = -kt + \ln [A]_0$$
$$\updownarrow \qquad \updownarrow\!\updownarrow \qquad \updownarrow$$
$$y = mx + b$$

The equation for a straight line is usually written

$$y = mx + b$$

where x and y are variables, m is the slope, and b is the intercept of the line with the y axis.

Thus, a plot of the values of $\ln [A]_t$ (vertical axis) versus values of t (horizontal axis) should give a straight line that has a slope equal to $-k$. Such a plot is illustrated in Figure 13.5 for the decomposition of N_2O_5 into N_2O_4 and O_2 in the solvent carbon tetrachloride.

Concentration versus Time for Second-Order Reactions

For the sake of simplicity, we will only consider a second-order reaction of the kind that has a rate law of the following type.

$$\text{Rate} = k [B]^2$$

---

[3] The logarithm of a quotient, $\ln \dfrac{a}{b}$, can be written as the difference, $\ln a - \ln b$.

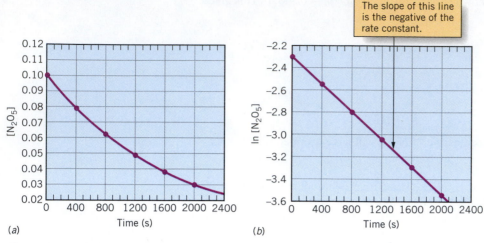

Figure 13.5 *The decomposition of N_2O_5. (a)* A graph of concentration versus time for the decomposition at 45 °C. *(b)* A straight line is obtained if the logarithm of the concentration is plotted versus time. The slope of this line equals the negative of the rate constant for the reaction.

Tools

Rate law,
second-order
reactions

The relationship between concentration and time for a reaction with such a rate law is given by Equation 13.6, an equation that is quite different from that for a first-order reaction.

$$\frac{1}{[B]_t} - \frac{1}{[B]_0} = kt \qquad (13.6)$$

$[B]_0$ is the initial concentration of B and $[B]_t$ is the concentration at time t. The next example illustrates how Equation 13.6 is applied to calculations.

EXAMPLE 13.8

Concentration–Time Calculations for Second-Order Reactions

NOCl is a corrosive, reddish-yellow gas that is intensely irritating to the eyes and skin.

Nitrosyl chloride, NOCl, decomposes slowly to NO and Cl_2.

$$2NOCl \longrightarrow 2NO + Cl_2$$

The rate law shows that the rate is second order in NOCl.

$$Rate = k\,[NOCl]^2$$

The rate constant k equals 0.020 L mol^{-1} s^{-1} at a certain temperature. If the initial concentration of NOCl in a closed reaction vessel is 0.050 M, what will the concentration be after 30 minutes?

Analysis: We're given a rate law and so can see that it is for a second-order reaction and has the simple form to which our study is limited. Thus, we conclude that we must calculate $[NOCl]_t$, the molar concentration of NOCl, after 30 minutes (1800 s) by Equation 13.6.

Solution: Let's begin by tabulating the data.

$$[NOCl]_0 = 0.050\ M \qquad\qquad [NOCl]_t = ?\ M$$

$$k = 0.020\ \text{L mol}^{-1}\,\text{s}^{-1} \qquad\qquad t = 1800\ \text{s}$$

The equation we wish to substitute into is

$$\frac{1}{[NOCl]_t} - \frac{1}{[NOCl]_0} = kt$$

Making the substitutions gives

$$\frac{1}{[NOCl]_t} - \frac{1}{0.050 \text{ mol L}^{-1}} = (0.020 \text{ L mol}^{-1} \text{ s}^{-1}) \times (1800 \text{ s})$$

Solving for $1/[NOCl]_t$ gives

$$\frac{1}{[NOCl]_t} - 20 \text{ L mol}^{-1} = 36 \text{ L mol}^{-1}$$

$$\frac{1}{[NOCl]_t} = 56 \text{ L mol}^{-1}$$

Taking the reciprocals of both sides gives us the value of $[NOCl]$.

$$[NOCl]_t = \frac{1}{56 \text{ L mol}^{-1}} = 0.018 \text{ mol L}^{-1}$$

$$= 0.018 \ M$$

Thus, the molar concentration of NOCl has decreased from 0.050 M to 0.018 M after 30 minutes.

Is the Answer Reasonable?
The concentration of NOCl has decreased, as it must, but there is, again, no simple check other than to redo the calculation.

Practice Exercise 9

For the reaction in the preceding example, determine how many minutes it would take for the NOCl concentration to drop from 0.040 M to 0.010 M. ◆

The rate constant k for a second-order reaction, one with a rate following Equation 13.6, can be determined graphically by a method similar to that used for a first-order reaction. Thus, Equation 13.6 can be rearranged as follows to correspond to an equation for a straight line.

$$\underset{\updownarrow}{\frac{1}{[B]_t}} = \underset{\updownarrow}{kt} + \underset{\updownarrow}{\frac{1}{[B]_0}}$$

$$y = mx + b$$

Therefore, when a reaction is second order (of the type we're studying), a plot of $1/[B]_t$ versus t should yield a straight line having a slope k. This is illustrated in Figure 13.6 for the decomposition of HI, using data in Table 13.1.

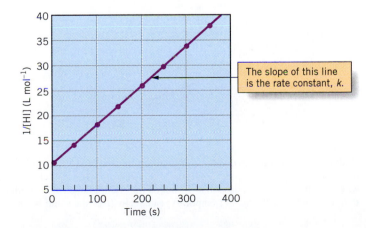

The slope of this line is the rate constant, k.

Figure 13.6 *Second order kinetics.* A graph of $1/[HI]$ versus time for the data in Table 13.1.

Half-lives for First- and Second-Order Reactions

Fast reactions are characterized by large values of k and small values of $t_{1/2}$.

The *half-life* of a reactant is a convenient way to describe how fast it reacts, particularly for an overall first-order process. A reactant's **half-life, $t_{1/2}$,** is the amount of time required for half of the reactant to disappear. A rapid reaction thus has a short half-life because half of the reactant disappears quickly. The equations for half-lives depend on the order of the reaction.

Half-lives of First-Order Reactions

When a reaction, overall, is first order, the half-life of the reactant can be obtained from Equation 13.5 (page 587) by setting $[A]_t$ equal to one-half of $[A]_0$.

$$[A]_t = \frac{1}{2}[A]_0$$

Tools

Half-lives

Substituting $(1/2)[A]_0$ for $[A]_t$ into Equation 13.5 and solving for t, which becomes $t_{1/2}$, gives

$$t_{1/2} = \frac{\ln 2}{k} \qquad (13.7)$$

Because $\ln$ 2 equals 0.693, Equation 13.7 is sometimes written

$$t_{1/2} = \frac{0.693}{k}$$

Because k is a constant for a given reaction, the half-life is also a constant for any particular first-order reaction (at any given temperature). Remarkably, in other words, *the half-life of a first-order reaction is not affected by the initial concentration of the reactant.* This can be illustrated by one of the most common first-order events in nature, the change that radioactive isotopes undergo during radioactive "decay." In fact, you have probably heard the term *half-life* used in reference to the life spans of radioactive substances.

Iodine-131, an unstable, radioactive isotope of iodine, undergoes a nuclear reaction whereby it emits a type of radiation and changes into a stable isotope of xenon.[4] The intensity of the radiation decreases, or *decays,* with time (see Figure 13.7). Notice that the time it takes for the first half of the ^{131}I to disappear is 8 days. Then, during the next 8 days half of the remaining ^{131}I disappears, and so on. Regardless of the initial amount, it takes 8 days for half of that amount of ^{131}I to disappear, which means that the half-life of ^{131}I is a constant.

Half-lives of Second-Order Reactions

The half-life of a second-order reaction *does* depend on initial reactant concentrations. We can see this by examining Figure 13.4 (page 579), which follows the

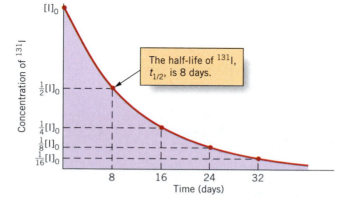

Figure 13.7 *First-order radioactive decay of iodine-131.* The initial concentration of the isotope is represented by $[I]_0$.

The half-life of ^{131}I, $t_{1/2}$, is 8 days.

[4]Iodine-131 is used in the diagnosis of thyroid disorders. The thyroid gland is a small organ located just below the "Adam's apple" and astride the windpipe. It uses iodide ion to make a hormone, so when a patient is given a dose of [131]I$^-$ mixed with nonradioactive I$^-$, both ions are taken up by the thyroid gland. The change in (temporary) radioactivity of the gland is a measure of thyroid activity.

decomposition of gaseous HI, a second-order reaction. The reaction begins with a hydrogen iodide concentration of 0.10 M. After 125 seconds, the concentration of HI drops to 0.050 M, so 125 s is the observed half-life when the initial concentration of HI is 0.10 M. If we then take 0.050 M as the next "initial" concentration, we find that it takes 250 seconds (at a *total* elapsed time of 375 seconds) to drop to 0.025 M. Thus, halving the initial concentration, from 0.10 M to 0.05 M, causes a doubling of the half-life, from 125 to 250 seconds.

It can be shown for a second-order reaction of the type we're studying, like the decomposition of HI, that the half-life is inversely proportional to the initial concentration of the reactant. The half-life is related to the rate constant by Equation 13.8.

$$t_{1/2} = \frac{1}{k \times (\text{initial concentration of reactant})} \qquad (13.8)$$

This equation comes from Equation 13.6 by letting $[B]_t = \frac{1}{2}[B]_0$.

EXAMPLE 13.9

Half-life Calculations

The half-life of radioactive iodine-131 is 8.0 days. What fraction of the initial iodine-131 would be present in a patient after 24 days if none of it were eliminated through natural body processes?

Analysis: Having learned that iodine-131 is a radioactive isotope and so decays by a first-order process, we know that the half-life is constant and does not depend on the ^{131}I concentration.

Solution: A period of 24 days is exactly three half-lives. If we take the fraction initially present to be 1, we can set up a table

Half-life	0	1	2	3
Fraction	1	$\frac{1}{2}$	$\frac{1}{4}$	$\frac{1}{8}$

Half of the iodine-131 is lost in the first half-life, half of that disappears in the second half-life, and so on. Therefore, the fraction remaining after three half-lives is $\frac{1}{8}$.

The fraction remaining after n half-lives is $\left(\frac{1}{2}\right)^n$, or simply $\frac{1}{2^n}$.

EXAMPLE 13.10

Half-life Calculations

The reaction $2HI(g) \rightarrow H_2(g) + I_2(g)$ has the rate law, Rate $= k\,[HI]^2$, with $k = 0.079$ L mol^{-1} s^{-1} at 508 °C. What is the half-life for this reaction at this temperature when the initial HI concentration is 0.050 M?

Analysis: The rate law tells us that the reaction is second order. To calculate the half-life, we need to use Equation 13.8.

Solution: The initial concentration is 0.050 mol L^{-1}; $k = 0.079$ L mol^{-1} s^{-1}. Substituting these values into Equation 13.8 gives

$$t_{1/2} = \frac{1}{(0.079 \text{ L mol}^{-1}\text{ s}^{-1})(0.050 \text{ mol L}^{-1})}$$

$$= 250 \text{ s} \qquad \text{(rounded)}$$

Practice Exercise 10

In Practice Exercise 6, the reaction of sucrose with water was found to be first order with respect to sucrose. The rate constant under the conditions of the experiments was 6.17×10^{-4} s^{-1}. Calculate the value of $t_{1/2}$ for this reaction in minutes. How many minutes would it take for three-quarters of the sucrose to react? (*Hint:* What fraction of the sucrose remains?) ◆

13.6 Theories about Reaction Rates

In Section 13.2 we mentioned that nearly all reactions proceed faster at higher temperatures. As a rule, the reaction rate increases by a factor of about 2 or 3 for each 10 °C increase in temperature, although the actual amount of increase differs from one reaction to another. Thus temperature has a very significant effect on rate, and, to understand why, we have to imagine what actually happens to the molecules in a reaction system. Stated another way, we need to develop some theoretical models that explain our observations. One of the simplest models is called *collision theory*.

Collision Theory

The kinetic theory provides insights for reaction rate theory.

The basic postulate of **collision theory** is that the rate of a reaction is proportional to the number of *effective* collisions per second among the reactant molecules. An *effective collision* is one that actually gives product molecules. Anything that can increase the frequency of effective collisions should, therefore, increase the rate.

One of the several factors that influences the number of effective collisions per second is *concentration*. As reactant concentrations increase, the number of collisions per second of all types, including effective collisions, cannot help but increase. We'll return to the significance of concentration in Section 13.8.

At the start of the reaction described by Figure 13.4, only about one of every billion billion (10^{18}) collisions leads to a net chemical reaction. In each of the other collisions, the reactant molecules just bounce off each other.

Not *every* collision between reactant molecules actually results in a chemical change. We know this because the reactant atoms or molecules in a gas or a liquid undergo an enormous number of collisions per second with each other. If each collision were effective, all reactions would be over in an instantaneous explosion. Therefore, for most reactions to occur at nonexplosive rates, as they do, *only a very small fraction of all the collisions can really lead to a net change.* Why is this so?

The Importance of Molecular Orientation

In most reactions, when two reactant molecules collide they must be oriented correctly for a reaction to occur. For example, the reaction represented by the following equation

$$2NO_2Cl \longrightarrow 2NO_2 + Cl_2$$

appears to proceed by a two-step mechanism. One step involves the collision of an NO_2Cl molecule with a chlorine atom.

$$NO_2Cl + Cl \longrightarrow NO_2 + Cl_2$$

The orientation of the NO_2Cl molecule when hit by the Cl atom is important (see Figure 13.8). The poor orientation shown in Figure 13.8a cannot result in the formation of Cl_2 because the two Cl atoms are not being brought close enough together for a new Cl—Cl bond to form as an N—Cl bond breaks. Figure 13.8b shows the necessary orientation if the collision of NO_2Cl and Cl is to effectively lead to products.

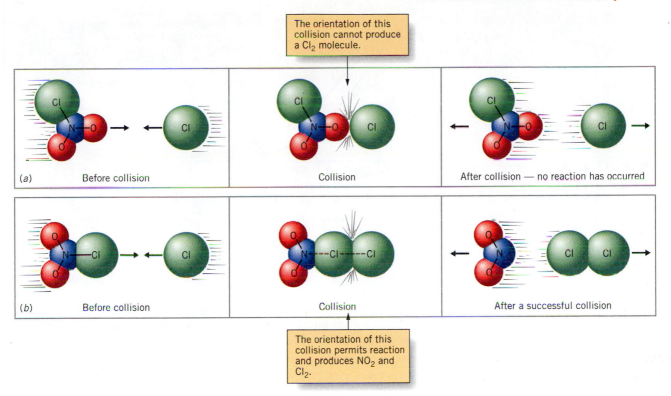

The orientation of this collision cannot produce a Cl_2 molecule.

(a)   Before collision   Collision   After collision — no reaction has occurred

(b)   Before collision   Collision   After a successful collision

The orientation of this collision permits reaction and produces NO_2 and Cl_2.

Figure 13.8 *The importance of molecular orientation during a collision in a reaction.* The key step in the decomposition of NO_2Cl to NO_2 and Cl_2 is the collision of a Cl atom with a NO_2Cl molecule. (*a*) A poorly oriented collision. (*b*) An effectively oriented collision.

The Importance of Molecular Kinetic Energy

Not all collisions, even those correctly oriented, are energetic enough to result in products, and this is the major reason that only a small percentage of all collisions actually lead to chemical change. The colliding particles must carry into the collision a certain minimum combined molecular kinetic energy, called the **activation energy, E_a**. In a successful collision, activation energy changes over to potential energy as the particles hit each other and chemical bonds become reorganized into those of the products. For most chemical reactions, the activation energy is quite large, and only a small fraction of all well-oriented, colliding molecules have it.

We can understand the existence of activation energy by studying in detail what actually takes place during a collision. For old bonds to break and new bonds to form, the atomic nuclei within the colliding particles must get close enough together. The molecules on a collision course must, therefore, be moving with a combined kinetic energy great enough to overcome the natural repulsions between electron clouds. Otherwise, the molecules simply veer away or bounce apart. Only fast molecules with large kinetic energies can collide with enough collision energy to enable their nuclei and electrons to overcome repulsions and thereby reach the positions required for the bond breaking and bond making that the chemical change demands.

How Temperature Greatly Affects Rates

With the concept of activation energy, we can now explain why the rate of a reaction increases so much with increasing temperature. We'll use the two plots in

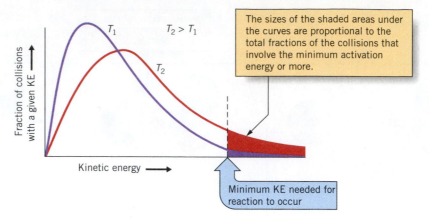

The sizes of the shaded areas under the curves are proportional to the total fractions of the collisions that involve the minimum activation energy or more.

Minimum KE needed for reaction to occur

Figure 13.9 *Kinetic energy distributions for a reaction mixture at two different temperatures.*

Figure 13.9, each plot corresponding to a different temperature for the same mixture of reactants. Each curve is a plot of the different *fractions* of all collisions (vertical axis) that have particular values of kinetic energy of collision (horizontal axis). (The total area under a curve then represents the total number of collisions, because all of the fractions must add up to this total.) Notice what happens to the plots when the temperature is increased; the maximum point shifts to the right and the curve flattens somewhat. However, *a modest increase in temperature generally does not affect the reaction's activation energy.* Within reason, the activation energy of a reaction is not affected by a change in temperature. In other words, as the curve flattens and shifts to the right with an increase in temperature, the value of E_a stays the same.

The shaded areas under the curves in Figure 13.9 represent the sum of all those fractions of the total collisions that equal or exceed the activation energy. This sum—we could call it the *reacting fraction*—is relatively much greater at the higher temperature than at the lower temperature because a significant fraction of the curve shifts beyond the activation energy in even a modest change to a higher temperature. In other words, at the higher temperature, a much greater fraction of the collisions occurring each second results in a chemical change, so the reactants disappear faster at the higher temperature.

Transition State Theory

Transition state theory is used to explain in detail what happens when reactant molecules come together in a collision. Most often, those in a head-on collision slow down, stop, and then fly apart unchanged. When a collision does cause a reaction, the particles that separate are those of the products. Regardless of what happens to them, however, as the molecules on a collision course slow down, their total kinetic energy decreases as it changes into potential energy (PE). It's like the momentary disappearance of the kinetic energy of a tennis ball when it hits the racket. In the deformed racket and ball, this energy becomes potential energy, which soon changes back to kinetic energy as the ball takes off in a new direction.

The law of conservation of energy requires that the total energy, the sum of KE and PE, be constant in a collision.

To visualize the relationship between activation energy and the development of total potential energy we sometimes use a *potential-energy diagram* (see Figure 13.10). The vertical axis represents changes in *potential* energy as the kinetic energy of the colliding particles changes over to this form. The horizontal axis is called the **reaction coordinate,** and it represents the extent to which the reactants have changed to the products. It helps us follow the path taken by the reaction as reactant molecules come together and change into product molecules. Activation energy

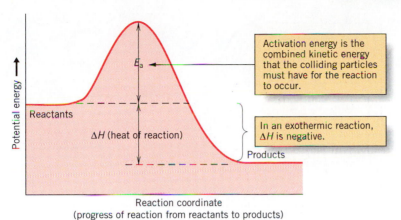

Activation energy is the combined kinetic energy that the colliding particles must have for the reaction to occur.

In an exothermic reaction, ΔH is negative.

Figure 13.10 *Potential-energy diagram for an exothermic reaction.*

thus appears as a potential-energy "hill" between the reactants and products. Only colliding molecules, properly oriented, that can deliver kinetic energy into potential energy at least as large as E_a are able to climb over the hill and produce products.

We can use a potential-energy diagram to follow the progress of both an unsuccessful and a successful collision (see Figure 13.11). As two reactant molecules collide, we say that they begin to climb the potential-energy barrier as they slow down and experience the conversion of their kinetic energy into potential energy. But if their combined initial kinetic energies are equivalent to a potential energy that is less than E_a, the molecules are unable to reach the top of the hill (Figure 13.11*a*). Instead, they fall back toward the reactants. They bounce apart chemically unchanged with their original total kinetic energy; no net reaction has occurred. On the other hand, if the combined kinetic energy of the colliding molecules equals or exceeds E_a, and if the molecules are oriented properly, they are able to pass over the activation energy barrier and form product molecules (Figure 13.11*b*).

Heats of Reaction, Energies of Activation, and Potential-Energy Diagrams

A reaction's potential-energy diagram, such as that of Figure 13.10, helps us to visualize the *heat of reaction,* ΔH, a concept introduced in Section 6.4. It's the difference between the potential energy of the products and the potential energy of the reactants. Figure 13.10 is for an *exothermic* reaction because the products have a *lower* potential energy than the reactants. In such a system, the net decrease in potential energy appears as an increase in the molecular kinetic

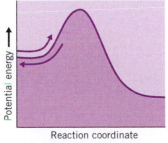

(*a*) An unsuccessful collision; the colliding molecules separate unchanged.

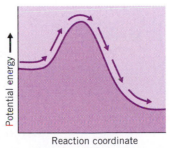

(*b*) A successful collision; the activation energy barrier is crossed and the products are formed.

Figure 13.11 *The difference between an unsuccessful and a successful collision.*

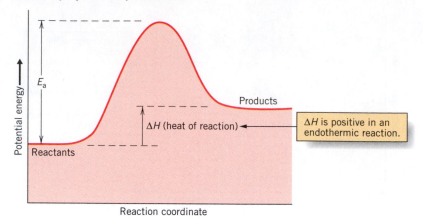

Figure 13.12 *A potential-energy diagram for an endothermic reaction.*

energy of the emerging product molecules. Thus, the temperature of the system increases during an exother-mic reaction because the average molecular kinetic energy of the system increases.

A potential-energy diagram for an endothermic reaction is shown in Figure 13.12. Now the products have a *higher* potential energy than the reactants and, in terms of the heat of reaction, a net input of energy is needed to form the products. Endothermic reactions produce a cooling effect as they proceed because there is a net conversion of molecular kinetic energy to potential energy. As the *total* molecular kinetic energy decreases, the *average* molecular kinetic energy decreases as well, and the temperature drops.

Notice that E_a for an endothermic process is invariably greater than (or it might be equal to) the heat of reaction. Thus if ΔH is both positive and *high*, E_a must also be high, making such reactions very slow. However, for an exothermic reaction (ΔH is negative) we cannot tell from ΔH how large E_a is. It could be high, making for a slow reaction despite its being exothermic. If E_a is low, the reaction would be rapid and all its heat would appear quickly.

In Chapter 6 we saw that when the direction of a reaction is reversed, the sign given to the enthalpy change, ΔH, is reversed. In other words, a reaction that is exothermic in the forward direction *must* be endothermic in the reverse direction, and vice versa. This might seem to suggest that reactions are generally reversible. Many are, but if we look again at the energy diagram for a reaction that is exothermic in the forward direction (Figure 13.10), it is obvious that in the opposite direction the reaction is endothermic *and must have a signifi-*

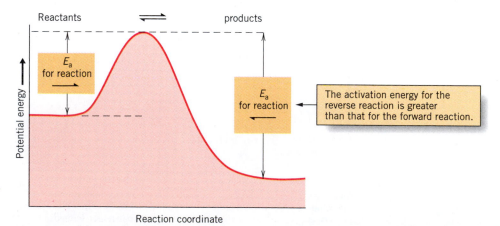

Figure 13.13 *Activation energy barrier for the forward and reverse reactions.*

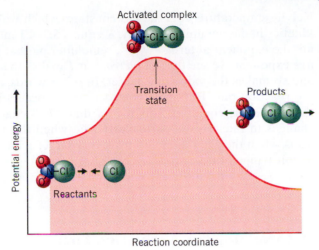

Figure 13.14 *Transition state and the activated complex.* Formation of an activated complex in the reaction between NO_2Cl and Cl.

$$NO_2Cl + Cl \longrightarrow NO_2 + Cl_2$$

cantly higher activation energy than the forward reaction. What differs most for the forward and reverse directions is the relative height of the activation energy barrier (see Figure 13.13).

One of the main reasons for studying activation energies is that they provide information about what actually occurs during an effective collision. For example, in Figure 13.8*b* on page 595, we described a way that NO_2Cl could react successfully with a Cl atom during a collision. During this collision, there is a moment when the N—Cl bond is partially broken and the new Cl—Cl bond is partially formed. This brief moment during a successful collision is called the reaction's **transition state.** The potential energy of the transition state corresponds to the high point on the potential-energy diagram (see Figure 13.14). The unstable chemical species that momentarily exists at this instant, O_2N----Cl----Cl, with its partially formed and partially broken bonds, is called the **activated complex.**

The size of the activation energy tells us about the relative importance of bond breaking and bond making during the formation of the activated complex. A very high activation energy suggests, for instance, that bond *breaking* contributes very heavily to the formation of the activated complex because bond breaking is an energy-absorbing process. On the other hand, a small activation energy may mean that bonds of about equal strength are being both broken and formed simultaneously.

13.7 Measuring the Activation Energy

As we have said, changing the temperature alters the rate of a reaction and so changes the value of the rate constant, k. The effect can be large, but how much k changes with temperature depends on the magnitude of the activation energy.

The activation energy is related to the rate constant by a relationship discovered in 1889 by Svante Arrhenius, whose name you may recall from our discussion of electrolytes and acids and bases in Chapter 4. The usual form of the **Arrhenius equation** is

$$k = A\,e^{-E_a/RT} \qquad (13.9)$$

where k is the rate constant, A is a proportionality constant sometimes called the **frequency factor**, e is the base of the natural logarithm system, and T is the

Tools

Arrhenius equation

Kelvin temperature. R is the gas constant, which we'll express in our study of kinetics in energy units,[5] namely, R equals 8.314 J mol^{-1} K^{-1}. When we evaluate the exponential term, $e^{-E_a/RT}$, remembering that E_a/RT appears as a *negative* exponent, we see that an *increase* in T makes the entire $e^{-E_a/RT}$ term *larger* and so makes the value of k larger. In other words, an increase in T increases the reaction rate. But again consider that T occurs in an *exponential* term in the Arrhenius equation. This means that small changes in T can mean *large* changes in rate. This is particularly true when the value of E_a is itself large. Thus, when the activation energy is very large, even a small increase in the reaction temperature causes a substantial percentage increase in the number of molecules having sufficient energy to react, so a large change in the rate is observed.

> Use your scientific calculator to determine some changes in the value of $e^{-E_a/RT}$ at different values of T. You might set E_a at 5.0×10^3 kJ mol^{-1} and see how $e^{-E_a/RT}$ changes between 298 K and 308 K, only a 10 degree difference.

Determining the Activation Energy Graphically

Equation 13.9 is normally used in its logarithmic form. If we take the natural logarithm of both sides, we obtain

$$\ln k = \ln A - E_a/RT$$

Let's rewrite the equation as

Arrhenius equation, alternate form

$$\ln k = \ln A - (E_a/R) \times (1/T) \qquad (13.10)$$

We know that the rate constant k varies with the temperature T, which also means that the quantity $\ln k$ varies with the quantity $(1/T)$. These two quantities, namely, $\ln k$ and $1/T$, are variables, so Equation 13.10 is in the form of an equation for a straight line.

$$\ln k = \ln A + (-E_a/R) \times (1/T)$$
$$\updownarrow \qquad \updownarrow \qquad \updownarrow \qquad \updownarrow$$
$$y = \quad b \ + \quad m \qquad x$$

Thus, to determine the activation energy, we can make a graph of $\ln k$ versus $1/T$, measure the slope of the line, and then use the relationship

$$\text{Slope} = -E_a/R$$

to calculate E_a. Example 13.11 illustrates how this is done.

---

[5]The units of R given here are actually SI units, namely, the joule (J), the mole (n), and the kelvin (K). To calculate R in these units we need to go back to the defining equation for the universal gas law, rearranging terms.

$$R = PV/nT$$

In Chapter 10 we learned that the standard conditions of pressure and temperature are 1 atm and 273.15 K; we expressed the standard molar volume in liters, namely, 22.414 L. But 1 atm equals 1.01325×10^5 N m^{-2}, where N is the SI unit of force, the newton, and m is the meter, the SI unit of length. So m^2 is area given in SI units. From Chapter 10, the ratio of force to area given by N m^{-2} is called the pascal, Pa, and that force times distance or N m defines one unit of energy in the SI and is called the joule, J. In the SI, volume must be expressed as m^3, to employ the SI unit of length to define volume, and 1 L equals 10^{-3} m^3. So now we can calculate R in SI units.

$$R = \frac{(1.01325 \times 10^5 \text{ N m}^{-2}) \times (22.414 \times 10^{-3} \text{ m}^3)}{(1 \text{ mol} \times 273.15 \text{ K})}$$

$$= 8.314 \text{ N m mol}^{-1} \text{ K}^{-1} = 8.314 \text{ J mol}^{-1} \text{ K}^{-1}$$

EXAMPLE 13.11

**Determining
Energy of Activation
Graphically**

Consider again the decomposition of NO_2 into NO and O_2. The equation is

$$2NO_2(g) \longrightarrow 2NO(g) + O_2(g)$$

The following data were collected for the reaction.

Rate Constant, k $(L\ mol^{-1}\ s^{-1})$	Temperature $(^\circ C)$
7.8	400
10	410
14	420
18	430
24	440

Determine the activation energy for the reaction in kilojoules per mole.

Analysis: Equation 13.10 applies, but the use of rate data to determine the activation energy graphically requires that we plot ln k, not k, versus the *reciprocal* of the *Kelvin* temperature, so we have to convert the given data into ln k and $1/T$ before we can make the graph.

Solution: To illustrate, using the first set of data, the conversions are

$$\ln k = \ln (7.8) = 2.05$$

$$\frac{1}{T} = \frac{1}{(400 + 273)\ K} = \frac{1}{673\ K}$$

$$= 1.486 \times 10^{-3}\ K^{-1}$$

We are carrying extra "significant figures" for the purpose of graphing the data. The remaining conversions give the table in the margin. Then we plot ln k versus $1/T$ as shown in the figure below.

ln k	$1/T\ (K^{-1})$
2.05	1.486×10^{-3}
2.30	1.464×10^{-3}
2.64	1.443×10^{-3}
2.89	1.422×10^{-3}
3.18	1.403×10^{-3}

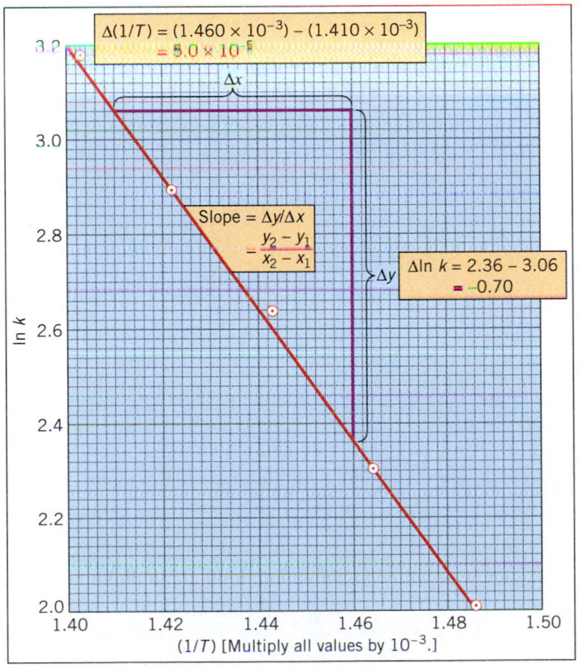

The slope of the curve is obtained as the ratio

There is a statistical method for computing the slope of the straight line that best fits the data. Ordinarily, this is what chemists would use.

$$\text{Slope} = \frac{\Delta(\ln k)}{\Delta(1/T)}$$

$$= \frac{-0.70}{5.0 \times 10^{-5} \text{ K}^{-1}}$$

$$= -1.4 \times 10^4 \text{ K} = -E_a/R$$

After changing signs and solving for E_a we have

$$E_a = (8.314 \text{ J mol}^{-1} \text{ K}^{-1})(1.4 \times 10^4 \text{ K})$$

$$= 1.2 \times 10^5 \text{ J mol}^{-1}$$

$$= 1.2 \times 10^2 \text{ kJ mol}^{-1}$$

Is the Answer Reasonable?
There's no simple check.

Calculating the Activation Energy

In obtaining an activation energy, sometimes we may not wish to go to the bother of graphing data, or we may have only two rate constants measured at two temperatures. In such cases the activation energy can be obtained algebraically. The following relationship can be derived from Equation 13.9.

$$\ln\left(\frac{k_2}{k_1}\right) = \frac{-E_a}{R}\left(\frac{1}{T_2} - \frac{1}{T_1}\right) \tag{13.11}$$

The following example illustrates the use of this equation.

EXAMPLE 13.12

Calculating an Energy of Activation from Two Rate Constants

The decomposition of HI has rate constants $k = 0.079$ L mol^{-1} s^{-1} at 508 °C and $k = 0.24$ L mol^{-1} s^{-1} at 540 °C. What is the activation energy of this reaction in kJ mol^{-1}?

Analysis: Where there are only two data sets, the simplest way to estimate E_a is to use Equation 13.11 and thus to find E_a algebraically.

Solution: Let's begin by organizing the data; a small table will be helpful. Choose one of the rate constants as k_1 (it doesn't matter which one) and then fill in the table.

	k (L mol^{-1} s^{-1})	T (K)
1	0.079	508 + 273 = 781
2	0.24	540 + 273 = 813

We also have $R = 8.314$ J mol^{-1} K^{-1}. Substituting into Equation 13.11 gives

$$\ln\left(\frac{0.24 \text{ L mol}^{-1} \text{ s}^{-1}}{0.079 \text{ L mol}^{-1} \text{ s}^{-1}}\right) = \frac{-E_a}{8.314 \text{ J mol}^{-1} \text{ K}^{-1}}\left(\frac{1}{813 \text{ K}} - \frac{1}{781 \text{ K}}\right)$$

$$\ln(3.0) = \frac{-E_a}{8.314 \text{ J mol}^{-1} \text{ K}^{-1}}(0.00123 \text{ K}^{-1} - 0.00128 \text{ K}^{-1})$$

$$1.10 = \frac{-E_a}{8.314 \text{ J mol}^{-1}}(-0.000050)$$

Multiplying both sides by 8.314 J mol^{-1} gives

$$9.15 \text{ J mol}^{-1} = E_a (0.000050)$$

Solving for E_a, we have

$$E_a = 1.8 \times 10^5 \text{ J mol}^{-1}$$

Converting to kilojoules,

$$E_a = 1.8 \times 10^2 \text{ kJ mol}^{-1}$$

Is the Answer Reasonable?
There's no simple check.

EXAMPLE 13.13

**Calculating
the Rate Constant
at a Particular
Temperature**

The reaction $2NO_2 \rightarrow 2NO + O_2$ has an activation energy of 111 kJ/mol. At 400 °C, $k = 7.8$ L mol^{-1} s^{-1}. What is the value of k at 430 °C?

Analysis: This time we know the activation energy and k at one temperature. We will need to use Equation 13.11 to obtain k at the other temperature. Since the logarithm term contains the ratio of the rate constants, we will solve for the value of this ratio, substitute the known value of k, and then solve for the unknown k.

Solution: Let's begin by writing Equation 13.11.

$$\ln\left(\frac{k_2}{k_1}\right) = \frac{-E_a}{R}\left(\frac{1}{T_2} - \frac{1}{T_1}\right)$$

Organizing the data gives us the following table.

	k (L mol^{-1} s^{-1})	T (K)
1	7.8	400 + 273 = 673 K
2	?	430 + 273 = 703 K

We must use $R = 8.314$ J mol^{-1} K^{-1} and express E_a in joules ($E_a = 1.11 \times 10^5$ J mol^{-1}). Next, we substitute values into the right side of the equation and solve for $\ln(k_2/k_1)$.

$$\ln\left(\frac{k_2}{k_1}\right) = \frac{-1.11 \times 10^5 \text{ J mol}^{-1}}{8.314 \text{ J mol}^{-1} \text{ K}^{-1}}\left(\frac{1}{703 \text{ K}} - \frac{1}{673 \text{ K}}\right)$$

$$= (-1.34 \times 10^4 \text{ K})(-6.34 \times 10^{-5} \text{ K}^{-1})$$

Therefore,

$$\ln\left(\frac{k_2}{k_1}\right) = 0.850$$

Taking the antilog gives the ratio of k_2 to k_1.

$$\frac{k_2}{k_1} = e^{0.850} = 2.34$$

Solving for k_2,

$$k_2 = 2.34\, k_1$$

Substituting the value of k_1 from the data table gives

$$k_2 = 2.34\,(7.8 \text{ L mol}^{-1} \text{ s}^{-1})$$

$$= 18 \text{ L mol}^{-1} \text{ s}^{-1}$$

Is the Answer Reasonable?

Although no simple check exists, we have at least found that the value of k for the higher temperature is greater than it is for the lower temperature, as it should be.

Practice Exercise 12

The reaction $CH_3I + HI \rightarrow CH_4 + I_2$ was observed to have rate constants $k = 3.2 \ L \ mol^{-1} \ s^{-1}$ at 350 °C and $k = 23 \ L \ mol^{-1} \ s^{-1}$ at 400 °C. (a) What is the value of E_a expressed in kJ/mol? (b) What would be the rate constant at 300 °C? ◆

13.8 Collision Theory and Reaction Mechanisms

At the beginning of this chapter we mentioned that most reactions do not occur in a single step. Instead, the net overall reaction is the result of a series of simple reactions, each of which is called an *elementary process*. An **elementary process** is a reaction whose rate law can be written from its own chemical equation, using its coefficients as exponents for the concentration terms without requiring experiments to determine the exponents. The entire series of elementary processes is called the reaction's **mechanism.** For most reactions, the individual elementary processes cannot actually be observed; instead, we only see the net reaction. Therefore, the mechanism a chemist writes is really a *theory* about what occurs step by step as the reactants are changed to the products.

What makes an elementary process "elementary?" Remember that the exponents of the concentration terms for the rate law of an *overall* reaction bear no *necessary* relationship to the coefficients in the overall balanced equation. As we keep emphasizing, the exponents in the *overall* rate law must be determined experimentally. So what makes an elementary process "elementary" is that a simple relationship between coefficients and exponents *does exist,* as we are about to see.

Because the individual steps in a mechanism usually cannot be observed, arriving at a chemically sensible set of elementary processes for a reaction is not simple. The task is disciplined, however, by a major test of a mechanism. *The overall rate law derived from the mechanism must agree with the observed rate law for the overall reaction.*

Writing mechanisms is challenging because making reasonable postulates about their elementary processes often depends on both "scientific intuition" and chemical knowledge gained only after considerable study. However, we'll look at some mechanisms to see how overall equations and rate laws can provide clues to them. First, we'll see how the exponents of the concentration terms in a rate law for an elementary process can be devised. Then we can put elementary processes together into a mechanism.

Predicting the Rate Law for an Elementary Process

In Section 13.6 you learned that the basic postulate of the collision theory is that the rate of a reaction is proportional to the number of effective collisions per second between the reactants. You also learned that the number of effective collisions is only a small fraction of the total. Nevertheless, if the *total* number of collisions were in some way doubled, the number of *effective* collisions would be doubled too. With this in mind, let's see how we can predict the rate law from postulated elementary processes.

Let's suppose that we know for a fact that a certain collision process takes place during a particular reaction. In Example 13.8 we introduced the decomposition of nitrosyl chloride into nitrogen monoxide, NO, and chlorine.

$$2NOCl \longrightarrow 2NO + Cl_2$$

Suppose, as chemists believe, that this reaction involves more than one step, one of them being a collision between a molecule of NOCl and an *atom* of Cl.

$$NOCl + Cl \longrightarrow NO + Cl_2 \qquad (13.12)$$

This equation represents an example of an elementary process. As with any reaction, we anticipate that its rate law will have the form

$$Rate = k[NOCl]^x[Cl]^y$$

Let's see if we can figure out what x and y are, based on the given information that this reaction occurs directly by collisions between NOCl molecules and Cl atoms. We will imagine that we are varying the concentration of each reactant of Equation 13.12 while that of the other is held constant. Then we'll reason how these changes ought to affect the rate of reaction.

To start, suppose we double the number of NOCl molecules in the container (thereby doubling the NOCl concentration) while keeping the number of Cl atoms the same. There now would be twice as many NOCl molecules with which each Cl could collide, so the number of NOCl-to-Cl collisions should double and the rate of the reaction should double. Our conclusion, then, is that doubling the concentration of NOCl should double the rate of this elementary process. This means that in the rate law for this elementary process, [NOCl] must appear with an exponent of 1 (i.e., $x = 1$).

A similar argument applies to the doubling of the Cl concentration in the container while keeping the number of NOCl molecules the same. Each molecule of NOCl would now have twice as many Cl atoms with which to collide, so the NOCl-to-Cl collision frequency should double and the rate should double. Thus, doubling the concentration of Cl doubles the rate, so in the rate law [Cl] should also have an exponent of 1 (i.e., $y = 1$). The expected rate law for the elementary process is therefore

> Recall from Section 13.4 that when the rate is doubled by doubling the reactant concentration, the exponent on the concentration term is 1.

$$Rate = k[NOCl]^1[Cl]^1$$

Ordinarily, of course, we don't write exponents that are equal to one. However, we've done so here to point out that the exponents in the rate law of an *elementary process* are the *same* as the coefficients of the reactants in the chemical equation for the process.

Let's now do an example of an elementary process that involves collisions between two identical molecules leading directly to the products shown.

$$2NO_2 \longrightarrow NO_3 + NO \qquad (13.13)$$

$$Rate = k[NO_2]^x$$

What can we say about the value of the exponent x? We reason as follows. If the NO_2 concentration were doubled, there would be *twice* as many individual NO_2 molecules and *each* would have *twice* as many neighbors with which to collide. The number of NO_2-to-NO_2 collisions per second would therefore be increased by a factor of 4, resulting in an increase in the rate by a factor of 4, which is 2^2. Earlier we saw that when the doubling of a concentration leads to a fourfold increase in the rate, the concentration of that reactant is raised to the second power in the rate law. Thus, if Equation 13.13 represents an elementary process, its rate law should be

> There are twice as many NO_2 molecules, each of which collides twice as often, so the total number of collisions per second doubly doubles, i.e., it increases by a factor of 4.

$$Rate = k[NO_2]^2$$

Once again the exponent in the rate law for the *elementary process* is the same as the coefficient in the chemical equation. Thus, our two exercises have illustrated that *the rate law for an elementary process can be predicted from the chemical equation for that process.*

> The exponents in the rate law for an elementary process are equal to the coefficients of the reactants in the chemical equation for that elementary process.

Let us emphasize again that this rule applies only to *elementary processes.* If all we know is the balanced equation for the overall reaction, the only way we can find the exponents of the rate equation is by doing experiments.

Predicting Reaction Mechanisms

How does the ability to predict the rate law of an elementary process help chemists predict reaction mechanisms? To answer this question, let's look at two reactions and what is believed to be their mechanisms. (There are many other, more complicated systems, and Facets of Chemistry 13.1 describes one type, the free radical chain reaction, that is particularly important.)

First, consider the gaseous reaction

$$2NO_2Cl \longrightarrow 2NO_2 + Cl_2 \tag{13.14}$$

Experimentally, the rate is first order in NO_2Cl, so the rate law is

$$\text{Rate} = k\,[NO_2Cl]$$

The first question we might ask is, Could the overall reaction (Equation 13.14) occur in a single step by the collision of two NO_2Cl molecules? The answer is no, because then it would be an elementary process and the rate law predicted for it would include a squared term, $[NO_2Cl]^2$. But the experimental rate law is first order in NO_2Cl. So the predicted and experimental rate laws don't agree, and we must look further to find the mechanism of the reaction.

On the basis of chemical intuition and other information that we won't discuss here, chemists believe the actual mechanism of the reaction in Equation 13.14 is the following two-step sequence of elementary processes.

$$NO_2Cl \longrightarrow NO_2 + Cl$$

$$NO_2Cl + Cl \longrightarrow NO_2 + Cl_2$$

The Cl atom formed here is called a reactive intermediate. We never actually observe the Cl because it reacts so quickly.

Notice that when the two reactions are added the intermediate, Cl, drops out and we obtain the net overall reaction given in Equation 13.14. *Being able to add the elementary processes and thus to obtain the overall reaction is another major test of a mechanism.*

In any multistep mechanism one step is usually much slower than the others. In this mechanism, for example, it is believed that the first step is slow and that once a Cl atom forms, it reacts very rapidly with another NO_2Cl molecule to give the final products.

The final products of a multistep reaction cannot appear faster than the products of the slow step, so the slow step in a mechanism is called the **rate-determining step** or the **rate-limiting step.** In the two-step mechanism above, then, the first reaction is the rate-determining step because the final products can't be formed faster than the rate at which Cl atoms form.

The rate-determining step is similar to a slow worker on an assembly line. Regardless of how fast the other workers are, the production rate depends on how quickly the slow worker does his or her job. The factors that control the speed of the rate-determining step therefore also control the overall rate of the

Facets of Chemistry 13.1

Free Radicals, Explosions, Octane Ratings, and Aging

A **free radical** is a very reactive species that contains one or more unpaired electrons. Examples are chlorine atoms formed when a Cl_2 molecule absorbs a photon (light) of the appropriate energy:

$$Cl_2 + \text{light energy } (h\nu) \longrightarrow 2Cl\cdot$$

(A dot placed next to the symbol of an atom or molecule represents an unpaired electron and indicates that the particle is a free radical.) The reason free radicals are so reactive is because of the tendency of electrons to become paired through the formation of either ions or covalent bonds.

Free radicals are important in many gaseous reactions, including those responsible for the production of photochemical smog in urban areas. Reactions involving free radicals have useful applications, too. Many plastics are made by polymerization reactions that take place by mechanisms that involve free radicals. In addition, free radicals play a part in one of the most important processes in the petroleum industry, *thermal cracking*. This reaction is used to break C—C and C—H bonds in long chain hydrocarbons to produce the smaller molecules that give gasoline a higher octane rating. An example is the formation of free radicals in the thermal cracking reaction of butane. When butane is heated to 700–800 °C, one of the major reactions that occurs is

$$CH_3-CH_2:CH_2-CH_3 \xrightarrow{\text{heat}} CH_3CH_2\cdot + CH_3CH_2\cdot$$

The central C—C bond of butane is shown here as a pair of dots, :, rather than the usual dash. When the bond is broken, the electron pair is divided between the two free radicals that are formed. This reaction produces two ethyl radicals, $CH_3CH_2\cdot$.

Free radical reactions tend to have high initial activation energies because chemical bonds must be broken to form the radicals. Once the free radicals are formed, reactions in which they are involved tend to be very rapid.

Free Radical Chain Reactions

In many cases, a free radical reacts with a reactant molecule to give a product molecule plus another free radical. Reactions that involve such a step are called **chain reactions.**

Many explosive reactions are chain reactions involving free radical mechanisms. One of the most studied reactions of this type is the formation of water from hydrogen and oxygen. The elementary processes involved can be described according to their roles in the mechanism.

The reaction begins with an **initiation step** that gives free radicals.

$$H_2 + O_2 \xrightarrow{\text{hot surface}} 2OH\cdot \quad \text{(initiation)}$$

The chain continues with a **propagation step,** which produces the product plus another free radical.

$$OH\cdot + H_2 \longrightarrow H_2O + H\cdot \quad \text{(propagation)}$$

The reaction of H_2 and O_2 is explosive because the mechanism also contains **branching steps.**

$$\left.\begin{array}{l} H\cdot + O_2 \longrightarrow OH\cdot + O\cdot \\ O\cdot + H_2 \longrightarrow OH\cdot + H\cdot \end{array}\right] \quad \text{(branching)}$$

Thus, the reaction of one $H\cdot$ with O_2 leads to the net production of two $OH\cdot$ plus an $O\cdot$. Every time an $H\cdot$ reacts with oxygen, there is an increase in the number of free radicals in the system. The free radical concentration grows rapidly, and the reaction rate becomes explosively fast.

Chain mechanisms also contain termination steps, which remove free radicals from the system. In the reaction of H_2 and O_2, the wall of the reaction vessel serves to remove $H\cdot$, which tends to halt the chain process.

$$2H\cdot \xrightarrow{\text{wall}} H_2$$

Free Radicals and Aging

Direct experimental evidence also exists for the presence of free radicals in functioning biological systems. These

Figure 1 People exposed to sunlight over long periods, like this woman from Nepal (a small country between India and Tibet), tend to develop wrinkles because ultraviolet radiation causes changes in their skin. Wrinkling is particularly evident when compared to the smooth child's skin.

highly reactive species play many roles, but one of the most interesting is their apparent involvement in the aging process. One theory suggests that free radicals attack protein molecules in collagen. Collagen is composed of long strands of fibers of proteins and is found throughout the body, especially in the flexible tissues of the lungs, skin, muscles, and blood vessels. Attack by free radicals seems to lead to cross-linking between these fibers, which stiffens them and makes them less flexible. The most readily observable result of this is the stiffening and hardening of the skin that accompanies aging or too much sunbathing (see Figure 1).

Free radicals also seem to affect fats (lipids) within the body by promoting the oxidation and deactivation of enzymes that are normally involved in the metabolism of lipids. Over a long period, the accumulated damage gradually reduces the efficiency with which the cell carries out its activities. Interestingly, vitamin E appears to be a natural free radical inhibitor. Diets that are deficient in vitamin E produce effects resembling radiation damage and aging.

Still another theory of aging suggests that free radicals attack the DNA in the nuclei of cells. DNA is the substance that is responsible for directing the chemical activities required for the successful functioning of the cell. Reactions with free radicals cause a gradual accumulation of errors in the DNA, which reduce the efficiency of the cell and can lead, ultimately, to malfunction and cell death.

reaction. This means that *the rate law for the rate-determining step is directly related to the rate law for the overall reaction.*

Because the rate-determining step is an elementary process, we can predict its rate law from the coefficients of its reactants. The coefficient of NO_2Cl in its relatively slow breakdown to NO_2 and Cl is 1. Therefore, the rate law predicted for the first step is

$$\text{Rate} = k\,[NO_2Cl]$$

> For a proposed mechanism to be acceptable, the rate law predicted by the rate-determining step should be the same as the rate law determined experimentally for the overall reaction.

Notice that the predicted rate law derived for the two-step mechanism agrees with the experimentally measured rate law. Although this doesn't *prove* that the mechanism is correct, it does provide considerable support for it. From the standpoint of kinetics, therefore, the mechanism is a reasonable one.

The second reaction mechanism that we will study is that of the following gas phase reaction.

$$2NO + 2H_2 \longrightarrow N_2 + 2H_2O \tag{13.15}$$

The experimentally determined rate law is

$$\text{Rate} = k\,[NO]^2\,[H_2] \quad (\textit{experimental})$$

We can quickly tell from this rate equation that Equation 13.15 could *not* itself be an elementary process. If it were, the exponent on $[H_2]$ would have to be 2. Obviously, a mechanism involving two or more steps must be involved.

A chemically reasonable mechanism that yields the correct form for the rate law consists of the following two steps.

$$2NO + H_2 \longrightarrow N_2O + H_2O \quad (\text{slow})$$

$$N_2O + H_2 \longrightarrow N_2 + H_2O \quad (\text{fast})$$

One test of the mechanism, as we said, is that the two equations must add to give the correct overall equation; and they do. Further, the chemistry of the second step has actually been observed in separate experiments. N_2O is a known compound, and it does react with H_2 to give N_2 and H_2O. Another test of the mechanism involves the coefficients of NO and H_2 in the predicted rate law for the first step, the supposed rate-determining step.

> Dinitrogen monoxide or nitrous oxide is the "laughing gas" used as an anesthetic in medicine and dentistry.

$$\text{Rate} = k\,[NO]^2\,[H_2] \quad (\textit{predicted})$$

This rate equation does match the experimental rate law, but there is still a serious flaw in the proposed mechanism. If the postulated slow step does actually describe an elementary process, it would involve the simultaneous collision between three molecules, two NO and one H_2. Such a collision is such an unlikely event that if it were really involved in the mechanism, the overall reaction would be extremely slow. Therefore, when chemists propose mechanisms for reactions, they avoid whenever possible elementary processes involving more than two-body or **bimolecular collisions.**

Chemists believe the reaction in Equation 13.15 proceeds by the following three-step sequence of bimolecular elementary processes.

$$2NO \rightleftharpoons N_2O_2 \qquad \text{(fast)}$$

$$N_2O_2 + H_2 \longrightarrow N_2O + H_2O \qquad \text{(slow)}$$

$$N_2O + H_2 \longrightarrow N_2 + H_2O \qquad \text{(fast)}$$

In this mechanism the first step is proposed to be a rapidly established equilibrium in which the unstable intermediate N_2O_2 forms in the forward reaction and then quickly decomposes into NO by the reverse reaction. The rate-determining step is the reaction of N_2O_2 with H_2 to give N_2O and a water molecule. The third step is the reaction mentioned above. Once again, notice that the three steps add to give the net overall change.

Since the second step is rate determining, the rate law for the reaction should match the rate law for this step. We predict this to be

$$\text{Rate} = k\,[N_2O_2]\,[H_2] \qquad (13.16)$$

However, the experimental rate law does not contain the species N_2O_2. Therefore, we must find a way to express the concentration of N_2O_2 in terms of the reactants in the overall reaction. To do this, let's look closely at the first step of the mechanism, which we view as a reversible reaction.

The rate in the forward direction, in which NO is the reactant, is

$$\text{Rate (forward)} = k_f\,[NO]^2$$

The rate of the reverse reaction, in which N_2O_2 is the reactant, is

$$\text{Rate (reverse)} = k_r\,[N_2O_2]$$

If we view this as a dynamic equilibrium, then the rate of the forward and reverse reactions are equal, which means that

$$k_f\,[NO]^2 = k_r\,[N_2O_2] \qquad (13.17)$$

Since we would like to eliminate N_2O_2 from the rate law in Equation 13.16, let's solve Equation 13.17 for $[N_2O_2]$.

$$[N_2O_2] = \frac{k_f}{k_r}\,[NO]^2$$

Substituting into the rate law in Equation 13.16 yields

$$\text{Rate} = k\left(\frac{k_f}{k_r}\right)[NO]^2\,[H_2]$$

Combining all the constants into one (k') gives

$$\text{Rate} = k'\,[NO]^2\,[H_2]$$

Now the rate law derived from the mechanism matches the rate law obtained experimentally. We are able to say, therefore, that the three-step mechanism does appear to be reasonable on the basis of kinetics.

The procedure we have worked through here applies to many reactions that proceed by mechanisms which involve sequential steps. Steps that precede the

Recall that in a dynamic equilibrium forward and reverse reactions occur at equal rates.

The occurrence of N_2O_2 in the proposed mechanism can only be surmised. The compound is never present at a detectable concentration because, as supposed, it's too unstable.

rate-determining step are considered to be rapidly established equilibria involving unstable intermediates.

Summary

Although chemists may devise other experiments to help prove or disprove the correctness of a mechanism, one of the strongest pieces of evidence is the experimentally measured rate law for the overall reaction. No matter how reasonable a particular mechanism may appear, if its elementary processes cannot yield a predicted rate law that matches the experimental one, the mechanism is wrong and must be discarded.

Practice Exercise 13

Ozone, O_3, reacts with nitric oxide, NO, to form nitrogen dioxide and oxygen.

$$NO + O_3 \longrightarrow NO_2 + O_2$$

This is one of the reactions involved in smog and is believed to occur by a one-step mechanism (the reaction above). If this is so, what is the expected rate law for the reaction? ◆

Practice Exercise 14

The mechanism for the decomposition of NO_2Cl is

$$NO_2Cl \longrightarrow NO_2 + Cl$$
$$NO_2Cl + Cl \longrightarrow NO_2 + Cl_2$$

What would the predicted rate law be if the second step in the mechanism were the rate-determining step? ◆

13.9 Catalysts

A **catalyst** is a substance that changes the rate of a chemical reaction without itself being used up. In other words, all of the catalyst added at the start of a reaction is present chemically unchanged after the reaction has gone to completion. The action caused by a catalyst is called **catalysis.** Broadly speaking, there are two kinds of catalysts. *Positive catalysts* speed up reactions, and *negative catalysts,* usually called *inhibitors,* slow reactions down. After this, when we use "catalyst" we'll mean positive catalyst, the usual connotation.

Although the catalyst is not part of the overall reaction, it does participate by changing the mechanism of the reaction. The catalyst provides a path to the products that has a rate-determining step with a lower activation energy than that of the uncatalyzed reaction (see Figure 13.15). Because the activation energy along this new route is smaller, a greater fraction of the collisions of the reactant molecules have the minimum energy needed to react, so the reaction proceeds faster.

Catalysts can be divided into two groups—**homogeneous catalysts,** which exist in the same phase as the reactants, and **heterogeneous catalysts,** which exist in a separate phase.

Homogeneous Catalysts

An example of homogeneous catalysis is found in the *lead chamber process* for manufacturing sulfuric acid. To make sulfuric acid by this process, sulfur is burned to give SO_2, which is then oxidized to SO_3. The SO_3 is dissolved in water as it forms to give H_2SO_4.

In the modern process for making sulfuric acid, the contact process, vanadium(V) oxide, V_2O_5, is the catalyst for the oxidation of sulfur dioxide to sulfur trioxide.

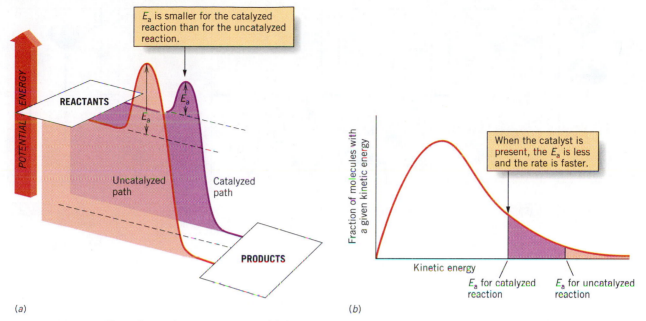

Figure 13.15 *Effect of a catalyst on a reaction.* (*a*) The catalyst provides an alternative, low-energy path from the reactants to the products. (*b*) A larger fraction of molecules have sufficient energy to react when the catalyzed path is available.

$$S + O_2 \longrightarrow SO_2$$

$$SO_2 + \tfrac{1}{2}O_2 \longrightarrow SO_3$$

$$SO_3 + H_2O \longrightarrow H_2SO_4$$

Unassisted, the second reaction, oxidation of SO_2 to SO_3, occurs slowly. In the lead chamber process, the SO_2 is combined with a mixture of NO, NO_2, air, and steam in large lead-lined reaction chambers. The NO_2 readily oxidizes the SO_2 to give NO and SO_3. The NO is then reoxidized to NO_2 by oxygen.

$$NO_2 + SO_2 \longrightarrow NO + SO_3$$

$$NO + \tfrac{1}{2}O_2 \longrightarrow NO_2$$

The NO_2 serves as a catalyst by being an oxygen carrier and by providing a low-energy path for the oxidation of SO_2 to SO_3. Notice, as must be true for any catalyst, the NO_2 is regenerated; it has not been permanently changed.

The NO_2 is regenerated in the second reaction and so is recycled over and over. Thus, only small amounts of it are needed in the reaction mixture to do an effective catalytic job.

Heterogeneous Catalysts

A heterogeneous catalyst is commonly a solid, and it usually functions by promoting a reaction on its surface. One or more of the reactant molecules are adsorbed onto the surface of the catalyst where an interaction with the surface increases their reactivity. An example is the synthesis of ammonia from hydrogen and nitrogen by the Haber process.

$$3H_2 + N_2 \longrightarrow 2NH_3$$

Adsorption occurs when molecules bind to a surface.

The reaction takes place on the surface of an iron catalyst that contains traces of aluminum and potassium oxides. It is thought that hydrogen molecules and nitrogen molecules dissociate while being held on the catalytic surface. The hydrogen atoms then combine with the nitrogen atoms to form ammonia.

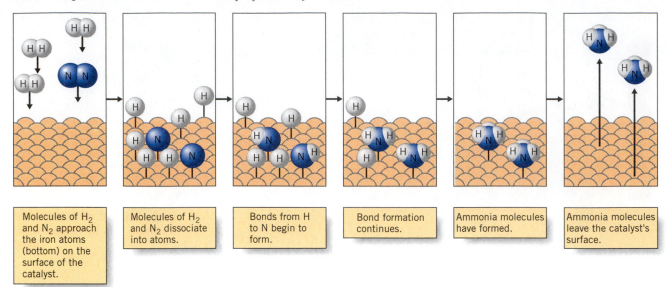

| Molecules of H_2 and N_2 approach the iron atoms (bottom) on the surface of the catalyst. | Molecules of H_2 and N_2 dissociate into atoms. | Bonds from H to N begin to form. | Bond formation continues. | Ammonia molecules have formed. | Ammonia molecules leave the catalyst's surface. |

Figure 13.16 *The Haber process.* Catalytic formation of ammonia molecules from hydrogen and nitrogen on the surface of a catalyst.

Finally, the completed ammonia molecule breaks away, freeing the surface of the catalyst for further reaction. This sequence of steps is illustrated in Figure 13.16.

Heterogeneous catalysts are used in many important commercial processes. The petroleum industry uses heterogeneous catalysts to crack hydrocarbons into smaller fragments and then re-form them into the useful components of gasoline (see Figure 13.17). The availability of such catalysts allows refineries to produce gasoline, jet fuel, or heating oil from crude oil in any ratio necessary to meet the demands of the marketplace.

A vehicle that uses unleaded gasoline is equipped with a catalytic converter (Figure 13.18) designed to lower the concentrations of exhaust gas pollutants, such as carbon monoxide, unburned hydrocarbons, and nitrogen oxides. Air is let into the exhaust stream that then passes over a catalyst that adsorbs CO, NO, and O_2. The NO dissociates into N and O atoms, and the O_2 also dissociates into atoms. Pairing of nitrogen atoms then produces N_2, and oxidation of CO by oxy-

Figure 13.17 *Catalysts are very important in the petroleum industry.* (*a*) Catalytic cracking towers at a Standard Oil refinery. (*b*) A variety of catalysts are available as beads, powders, or in other forms for various refinery operations.

(*a*)

(*b*)

gen atoms produces CO_2. Unburned hydrocarbons are also oxidized to CO_2 and H_2O. The catalysts of catalytic converters are deactivated or "poisoned" by lead-based octane boosters like tetraethyl lead $[Pb(C_2H_5)_4]$, so "leaded" gasoline cannot be legally used in vehicles anymore.

The poisoning of catalysts is also a major problem in many industrial processes. Methyl alcohol (methanol, CH_3OH), for example, is a promising fuel that can be made from coal and steam by the reaction

$$C \text{ (from coal)} + H_2O \longrightarrow CO + H_2$$

followed by

$$CO + 2H_2 \longrightarrow CH_3OH$$

A catalyst for the second step is copper(I) ion held in solid solution with zinc oxide. However, traces of sulfur, a contaminant in coal, must be avoided because sulfur reacts with the catalyst and destroys its catalytic activity.

In living systems, catalysts are called *enzymes,* and virtually all belong to the protein family of compounds. The B vitamins are required in the diet because the body uses them to make important enzymes. (We'll study enzymes further in Section 23.10.) Biological poisons often work as enzyme inactivators. Nerve poisons, for example, react with certain enzymes and thereby block the enzyme sites required for the operation of the nervous system.

Figure 13.18 *A modern catalytic converter of the type used in about 80% of new cars.* Part of the converter has been cut away to reveal the porous ceramic material that serves as the support for the catalyst.

SUMMARY

Reaction Rates. The speeds or **rates** of reaction are controlled by five factors: (1) the nature of the reactants, (2) the ability of reactants to meet, (3) the concentrations of the reactants, (4) the temperature, and (5) the presence of catalysts. The rates of **heterogeneous reactions** are determined largely by the area of contact between the phases; the rates of **homogeneous reactions** are determined by the concentrations of the reactants. The rate is measured by monitoring the change in reactant or product concentrations with time.

$$\text{Rate} = \Delta(\text{concentration})/\Delta(\text{time})$$

In any chemical reaction, the rates of formation of products and rates of disappearance of reactants are related by the coefficients of the balanced overall chemical equation.

Rate Laws. The **rate law** for a reaction relates the reaction rate to the molar concentrations of the reactants. The rate is proportional to the product of the molar concentrations of the reactants, each raised to an appropriate power. These exponents must be determined by experiments in which the concentrations are varied and the effects of the variations on the rate are measured. The proportionality constant, k, is called the **rate constant.** Its value depends on temperature but not on the concentrations of the reactants. The sum of the exponents in the rate law is the **order** (or overall order) of the reaction.

Concentration and Time. Equations exist that relate the concentration of a reactant at a given time t to the initial concentration and the rate constant. The time required for half of a reactant to disappear is the **half-life, $t_{1/2}$.** For a first-order reaction, the half-life is a constant that depends only on the rate constant for the reaction; it is independent of the initial concentration. The half-life for a second-order reaction is inversely proportional both to the initial concentration of the reactant and to the rate constant.

Theories of Reaction Rate. According to **collision theory,** the rate of a reaction depends on the number of **effective collisions** per second of the reactant particles, which is only an extremely small fraction of the total number of collisions per second. This fraction is small partly because collisions usually do not result in products unless the reactant molecules are suitably oriented. The major reason, however, why the fraction of successful collisions is small is that the colliding molecules must jointly possess a minimum molecular kinetic energy called the **activation energy, E_a.** As the temperature increases, a larger fraction of the collisions has this necessary energy, making more collisions effective each second and the reaction faster.

Transition state theory visualizes how the energies of molecules and the orientations of their nuclei interact as they collide. In this theory, the energy of activation is viewed as an energy barrier on the reaction's potential-energy diagram. The *heat of reaction* is the net potential-energy difference between the reactants and the products. In reversible reactions, the values of E_a for both the forward and reverse reactions can be identified on an energy diagram. The species at the high point on an energy diagram is the **activated complex** and is said to be in the **transition state.**

Determining the Activation Energy. The **Arrhenius equation** lets us see how changes in activation energy and temperature affect a rate constant. The Arrhenius equation also lets us determine E_a either graphically or by a calculation using the appropriate form of the Arrhenius equation. The calculation requires two rate constants determined at two temperatures. The graphical method uses more values of rate constants at more temperatures and thus usually yields more accurate results. The activation energy and the rate constant at one temperature can be used to calculate the rate constant at another temperature.

Reaction Mechanisms. The detailed sequence of elementary processes that lead to the net chemical change is the **mechanism** of the reaction. Since intermediates usually cannot be detected, the mechanism is a theory. Support for a mechanism comes from matching the predicted rate law for the mechanism with the rate law obtained from experimental data. For the **rate-determining step** or for any **elementary process** the corresponding rate law has exponents equal to the coefficients in the balanced equation for the elementary process.

Catalysis. **Catalysts** are substances that change a reaction rate but are not consumed by the reaction. Negative catalysts inhibit reactions. Positive catalysts provide alternative paths for reactions for which at least one step has a smaller activation energy than the uncatalyzed reaction. **Homogeneous catalysts** are in the same phase as the reactants. **Heterogeneous catalysts** provide a path of lower activation energy by having a surface on which the reactants are adsorbed and react. Catalysts in living systems are called enzymes.

Tools *you have learned*

Mo	Tc	Ru	Rh	Pd	Ag	Cd	In	Sn	Sb	Te
					07.8682	112.411	114.82	118.710	121.75	127.60
72	73	74	75	76	77	78	79	80	81	82	83	84

The table below lists the tools that you have learned in this chapter that are applicable to problem solving. Review them if necessary, and refer to them when working on the Thinking-It-Through problems and the Review Problems that follow.

Tool	**Function**
Rate law of a reaction *(page 579)*	To use data on how concentration affects rate to evaluate the exponents of the concentration terms in a rate law, determine the overall order of a reaction from the exponents of the concentration terms in its rate law, and calculate a rate constant.
Half-lives *(page 592)*	To calculate a half-life from experimental data and reaction order, or to calculate the rate of disappearance of a reactant.
Arrhenius equation *(pages 599 and 600)*	To determine an activation energy graphically and to interrelate rate constants, activation energy, and temperature.

THINKING IT THROUGH

Remember, the goal for each of the following problems is to assemble the available information needed to obtain the answer, state what additional data (if any) are needed, and describe how you would use the data to answer the question. For problems involving unit conversions, list the relationships among the units that are needed to carry out the conversions. Construct the conversion factors that can be formed from these relationships. Then set up the solution to the problem by arranging the conversion factors so the units cancel correctly to give the desired units of the answer.

The problems are divided into two groups. Those in Level 2 are more challenging than those in Level 1 and provide an opportunity to really hone your problem solving skills. Detailed answers to these problems can be found at our Web site: http://www.wiley.com/college/brady

Level 1 Problems

1. Consider the following reaction

$$C_3H_8(g) + 5O_2(g) \longrightarrow 3CO_2(g) + 4H_2O(g)$$

If at a given moment C_3H_8 is reacting at a rate of 0.400 mol L^{-1} s^{-1}, what are the rates of formation of CO_2 and H_2O? What is the rate at which O_2 is reacting? (Explain how you would obtain the answers.)

2. Suppose the following data were collected for the reaction

$$2A \longrightarrow B + 2C$$

Concentration of A (mol L^{-1})	Time (minutes)
0.2000	0
0.1902	10
0.1721	20
0.1482	30
0.1213	40
0.0945	50
0.0700	60

How would you determine the rate at which A is reacting at $t = 25$ minutes? Explain how you would determine the rates at which B and C are being formed at $t = 25$ min.

3. The reaction $I^-(aq) + OCl^-(aq) \rightarrow IO^-(aq) + Cl^-(aq)$ occurs in a basic solution. When the $[OH^-] = 1.00\ M$, the reaction is first order with respect to each reactant and has a rate constant of $60.0\ L\ mol^{-1}\ s^{-1}$. If $[I^-] = 0.0200\ M$ and $[OCl^-] = 0.0300\ M$ at a certain time, what is the rate at which OI^- is being formed? (Describe how you would find the answer. Provide any necessary equations.)

4. Carbon-14 is a radioactive isotope that decays by the nuclear reaction

$$^{14}_{6}C \longrightarrow {}^{14}_{7}N + {}^{0}_{-1}e$$

It is a first-order process with a half-life of 5770 years. What is the rate constant for the decay? Explain how you would determine the number of years that it will take for 1/100th of the ^{14}C in a sample of naturally occurring carbon (a mixture of carbon isotopes, including ^{14}C) to decay.

5. The reaction $2NO_2 \rightarrow 2NO + O_2$ has the rate constant $k = 0.63\ L\ mol^{-1}\ s^{-1}$ at 600 K. If the initial concentration of NO_2 in a reaction vessel is $0.050\ M$, how would you determine how long it will take for the concentration to drop to $0.020\ M$?

6. For the reaction described in the preceding question, how can you calculate the number of molecules of NO_2 that react each second when the NO_2 concentration is $3.0 \times 10^{-3}\ M$?

Level 2 Problems

7. The following data were collected at a certain temperature for the reaction

$$2N_2O_5(g) \longrightarrow 4NO_2(g) + O_2(g)$$

Experiment	$[N_2O_5]$ (mol L^{-1})	Rate of Formation of O_2 (mol L^{-1} s^{-1})
1	0.050	0.084
2	0.075	0.126

What is the order of the reaction with respect to N_2O_5? How would you obtain the value of the rate constant at this temperature?

8. For the reaction

$$2N_2O_5(g) \longrightarrow 4NO_2(g) + O_2(g)$$

it was observed at a particular moment that the pressure in a container of these gases was increasing at the rate of 20.0 torr per second. What additional information would you need to calculate the rate at which the N_2O_5 was reacting in units of mol L^{-1} s^{-1}? Explain how you would perform the calculation.

9. At high temperatures, collisions between argon atoms and oxygen molecules can lead to the formation of oxygen atoms.

$$Ar + O_2 \longrightarrow Ar + O + O$$

At 5000 K, the rate constant for this reaction is 5.49×10^6 L mol^{-1} s^{-1}. At 10,000 K, $k = 9.86 \times 10^8$ L mol^{-1} s^{-1}. Describe the calculation required to estimate the activation energy for the reaction.

10. The rate law for the following reaction,

$$F_2(g) + 2NO_2(g) \longrightarrow 2NO_2F(g)$$

as found by experiment, is

$$Rate = k\,[NO_2]\,[F_2]$$

The mechanism of the reaction is believed to consist of the following elementary processes.

$$F_2 + NO_2 \longrightarrow NO_2F + F \qquad (step\ 1)$$

$$F + NO_2 \longrightarrow NO_2F \qquad (step\ 2)$$

If this really *is* the mechanism, which is the rate-determining step? If the rate-determining step were the second step, derive the rate law that would be expected.

11. The five figures represent relative concentrations of the species A (red spheres), B (blue spheres) and AB (red–blue joined spheres) which form reactants and products for the reaction: $A + B \rightarrow AB$.

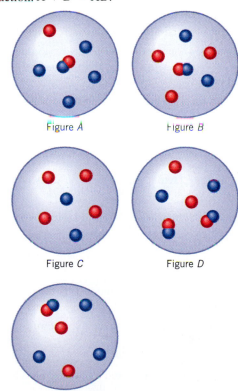

Figure A

Figure B

Figure C

Figure D

Figure E

For each of the following rate laws identify the figure with the fastest rate and the slowest rate.

(a) Rate = k $[A]^2$
(b) Rate = k $[A][B]$
(c) Rate = k $[A][B]^2$

12. The figure below shows the potential energy-reaction coordinate diagram for a chemical reaction which can occur via two reaction pathways.
(a) What fundamental kinetic concept does this diagram illustrate?
(b) What is the equation for the overall reaction?
(c) What is the mechanism for this reaction using the higher energy pathway?
(d) What is the mechanism for this reaction using the lower energy pathway?
(e) On the figure above, identify 1) the activation energy for the reaction for the higher energy path, 2) the activation energy for the reaction for the lower energy path, and 3) the ΔH for the reaction.

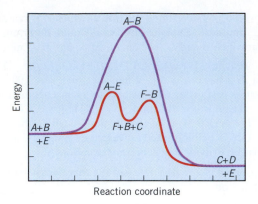

(f) Identify each of the species A to F as a reactant, intermediate, product, or catalyst.

REVIEW QUESTIONS

Factors That Affect Reaction Rate

13.1 What does *rate of reaction* mean in qualitative terms?

13.2 Give an example from everyday experience of (a) a very fast reaction, (b) a moderately fast reaction, and (c) a slow reaction.

13.3 In terms of reaction rates, what is an explosion?

13.4 From an economic point of view, why would industrial corporations want to know about the factors that affect the rate of a reaction?

13.5 Suppose we compared two reactions, one requiring the simultaneous collision of three molecules and the other requiring a collision between two molecules. From the standpoint of statistics, and all other factors being equal, which reaction should be faster? Explain your answer.

13.6 List the five factors that affect the rates of chemical reactions.

13.7 What is a *homogeneous reaction?* Give an example.

13.8 What is a *heterogeneous reaction?* Give an example.

13.9 Why are chemical reactions usually carried out in solution?

13.10 What is the major factor that affects the rate of a heterogeneous reaction?

13.11 How does particle size affect the rate of a heterogeneous reaction? Why?

13.12 The rate of hardening of epoxy glue depends on the amount of hardener that is mixed into the glue. What factor affecting reaction rates does this illustrate?

13.13 A Polaroid instant photograph develops faster if it's kept warm than if it is exposed to cold. Why?

13.14 What is a catalyst?

13.15 In cool weather, the number of chirps per minute from crickets diminishes. How can this be explained?

13.16 On the basis of what you learned in Chapter 11, why do foods cook faster in a pressure cooker than in an open pot of boiling water?

13.17 Persons who have been submerged in very cold water and who are believed to have drowned sometimes can be revived. On the other hand, persons who have been submerged in warmer water for the same length of time have died. Explain this in terms of factors that affect the rates of chemical reactions.

Concentration and Rate; Rate Laws

13.18 What are the units of reaction rate?

13.19 What is a *rate law?* What is the proportionality constant called?

13.20 What is meant by the *order* of a reaction?

13.21 What are the units of the rate constant for (a) a first-order reaction, (b) a second-order reaction, and (c) a third-order reaction?

13.22 How must the exponents in a rate law be determined?

13.23 Is there any way of using the coefficients in the balanced overall equation for a reaction to predict with certainty what the exponents are in the rate law?

13.24 If the concentration of a reactant is doubled and the reaction rate doubles, what must be the order of the reaction with respect to that reactant?

13.25 If the concentration of a reactant is doubled, by what factor will the rate increase if the reaction is second order with respect to that reactant?

13.26 In an experiment, the concentration of a reactant was tripled. The rate increased by a factor of 27. What is the order of the reaction with respect to that reactant?

13.27 Biological reactions usually involve the interaction of an enzyme with a *substrate,* the substance that actually undergoes the chemical change. In many cases, the rate of reaction depends on the concentration of the enzyme but is independent of the substrate concentration. What is the order of the reaction with respect to the substrate in such instances?

Concentration and Time, Half-lives

13.28 Give the equations that relate concentration to time for (a) a first-order reaction and (b) a second-order reaction.

13.29 What is meant by the term *half-life?*

13.30 How is the half-life of a first-order reaction affected by the initial concentration of the reactant?

13.31 How is the half-life of a second-order reaction affected by the initial reactant concentration?

13.32 Derive the equations for $t_{1/2}$ for first- and second-order reactions from Equations 13.5 and 13.6, respectively.

Effect of Temperature on Rate

13.33 What is the basic postulate of collision theory?

13.34 What two factors influence the effectiveness of molecular collisions in producing chemical change?

13.35 In terms of the kinetic theory, why does an increase in temperature increase the reaction rate?

13.36 Draw the potential-energy diagram for an endothermic reaction. Indicate on the diagram the activation energy for both the forward and reverse reactions. Also indicate the heat of reaction.

13.37 Explain, in terms of the law of conservation of energy, why an endothermic reaction leads to a cooling of the reaction mixture (provided heat cannot enter from outside the system).

13.38 Define the terms *transition state* and *activated complex.*

13.39 Draw a potential-energy diagram for an exothermic reaction and indicate on the diagram the location of the transition state.

13.40 Suppose a certain slow reaction is found to have a very small activation energy. What does this suggest about the importance of molecular orientation in the formation of the activated complex?

13.41 The decomposition of carbon dioxide,

$$CO_2 \longrightarrow CO + O$$

has a very large activation energy of approximately 460 kJ/mol. Explain why this is consistent with a mechanism that involves the breaking of a C=O bond.

13.42 State the Arrhenius equation (which relates the rate constant to temperature and the activation energy) and define the symbols.

Reaction Mechanisms

13.43 What is the definition of an *elementary process?* How are elementary processes related to the mechanism of a reaction?

13.44 What is a *rate-determining step?*

13.45 In what way is the rate law for a reaction related to the rate-determining step?

13.46 A reaction has the following mechanism.

$$2NO \longrightarrow N_2O_2$$

$$N_2O_2 + H_2 \longrightarrow N_2O + H_2O$$

$$N_2O + H_2 \longrightarrow N_2 + H_2O$$

What is the net overall change that occurs in this reaction?

13.47 If the reaction $NO_2 + CO \rightarrow NO + CO_2$ occurs by a one-step collision process, what would be the expected rate law for the reaction? The actual rate law is Rate = k $[NO_2]^2$. Could the reaction actually occur by a one-step collision between NO_2 and CO? Explain.

13.48 Oxidation of NO to NO_2—one of the reactions in the production of smog—appears to involve carbon monoxide. A possible mechanism is

$$CO + \cdot OH \longrightarrow CO_2 + H \cdot$$

$$H \cdot + O_2 \longrightarrow HOO \cdot$$

$$HOO \cdot + NO \longrightarrow \cdot OH + NO_2$$

(The formulas with dots represent extremely reactive species with unpaired electrons and are called *free radicals.*) Write the net chemical equation for the reaction.

13.49 Show that the following two mechanisms give the same net overall reaction.

Mechanism 1

$$OCl^- + H_2O \longrightarrow HOCl + OH^-$$

$$HOCl + I^- \longrightarrow HOI + Cl^-$$

$$HOI + OH^- \longrightarrow H_2O + OI^-$$

Mechanism 2

$$OCl^- + H_2O \longrightarrow HOCl + OH^-$$

$$I^- + HOCl \longrightarrow ICl + OH^-$$

$$ICl + 2OH^- \longrightarrow OI^- + Cl^- + H_2O$$

13.50 The experimental rate law for the reaction $NO_2 + CO \rightarrow CO_2 + NO$ is Rate = k $[NO_2]^2$. If the mechanism is

$$2NO_2 \longrightarrow NO_3 + NO \qquad \text{(slow)}$$

$$NO_3 + CO \longrightarrow NO_2 + CO_2 \qquad \text{(fast)}$$

show that the predicted rate law is the same as the experimental rate law.

Catalysts

13.51 How does a catalyst increase the rate of a chemical reaction?

13.52 What is a *homogeneous catalyst?* How does it function, in general terms?

13.53 What is a *heterogeneous catalyst?* How does it function?

13.54 What is the difference in meaning between the terms *adsorption* and *absorption?* (If necessary, use a dictionary.) Which one applies to heterogeneous catalysts?

13.55 What does the catalytic converter do in the exhaust system of an automobile? Why should leaded gasoline not be used in cars equipped with catalytic converters?

REVIEW PROBLEMS

Answers to problems whose numbers are printed in color are given in Appendix D. More challenging problems are marked with asterisks.

Measuring Rates of Reaction

13.56 The following data were collected at a certain temperature for the decomposition of sulfuryl chloride, SO_2Cl_2, a chemical used in a variety of organic syntheses.

$$SO_2Cl_2 \longrightarrow SO_2 + Cl_2$$

Time (min)	$[SO_2Cl_2]$ (mol L^{-1})
0	0.1000
100	0.0876
200	0.0768
300	0.0673
400	0.0590
500	0.0517
600	0.0453
700	0.0397
800	0.0348
900	0.0305
1000	0.0267
1100	0.0234

Make a graph of concentration versus time and determine the rate of formation of SO_2 at $t = 200$ minutes and $t = 600$ minutes.

13.57 The following data were collected for the decomposition of acetaldehyde, CH_3CHO (used in the manufacture of a variety of chemicals including perfumes, dyes, and plastics), into methane and carbon monoxide. The data were collected at a temperature of 530 °C.

$$CH_3CHO \longrightarrow CH_4 + CO$$

$[CH_3CHO]$ (mol L^{-1})	Time (s)
0.200	0
0.153	20
0.124	40
0.104	60
0.090	80
0.079	100
0.070	120
0.063	140
0.058	160
0.053	180
0.049	200

Make a graph of concentration versus time and determine the rate of reaction of CH_3CHO after 60 seconds and after 120 seconds.

13.58 In the reaction $3H_2 + N_2 \rightarrow 2NH_3$, how does the rate of disappearance of hydrogen compare to the rate of disappearance of nitrogen? How does the rate of appearance of NH_3 compare to the rate of disappearance of nitrogen?

13.59 For the reaction $2A + B \rightarrow 3C$, it was found that the rate of disappearance of B was 0.30 mol L^{-1} s^{-1}. What were the rate of disappearance of A and the rate of appearance of C?

13.60 In the combustion of hexane (a low-boiling component of gasoline),

$$2C_6H_{14}(g) + 19O_2(g) \longrightarrow 12CO_2(g) + 14H_2O(g)$$

it was found that the rate of reaction of C_6H_{14} was 1.20 mol L^{-1} s^{-1}.

(a) What was the rate of reaction of O_2?
(b) What was the rate of formation of CO_2?
(c) What was the rate of formation of H_2O?

13.61 At a certain moment in the reaction

$$2N_2O_5 \longrightarrow 4NO_2 + O_2$$

N_2O_5 is decomposing at a rate of 2.5×10^{-6} mol L^{-1} s^{-1}. What are the rates of formation of NO_2 and O_2?

Rate Laws for Reactions

13.62 The oxidation of NO (released in small amounts in the exhaust of automobiles) produces the brownish-red gas NO_2, which is a component of urban air pollution.

$$2NO + O_2 \longrightarrow 2NO_2$$

The rate law for the reaction is Rate $= k\,[NO]_2\,[O_2]$. At 25 °C, $k = 7.1 \times 10^9$ L^2 mol^{-2} s^{-1}. What would be the rate of the reaction if $[NO] = 0.0010$ mol L^{-1} and $[O_2] = 0.034$ mol L^{-1}?

13.63 The rate law for the decomposition of N_2O_5 is

$$\text{Rate} = k\,[N_2O_5]$$

If $k = 1.0 \times 10^{-5}$ s^{-1}, what is the reaction rate when the N_2O_5 concentration is 0.0010 mol L^{-1}?

13.64 For the reaction

$$2HCrO_4^- + 3HSO_3^- + 5H^+ \to 2Cr^{3+} + 3SO_4^{2-} + 5H_2O$$

the rate law is Rate $= k\,[HCrO_4^-]\,[HSO_3^-]^2\,[H^+]$.
(a) What is the order of the reaction with respect to each reactant?
(b) What is the overall order of the reaction?

13.65 For the reaction

$$NO + O_3 \longrightarrow NO_2 + O_2$$

the rate law is Rate $= k\,[NO]\,[O_3]$.
What is the order with respect to each reactant, and what is the overall order of the reaction?

13.66 The following data were collected for the reaction

$$M + N \longrightarrow P + Q$$

Initial Concentrations (mol L^{-1})		Initial Rate of Reaction
$[M]$	$[N]$	(mol L^{-1} s^{-1})
0.010	0.010	2.5×10^{-3}
0.020	0.010	5.0×10^{-3}
0.020	0.030	4.5×10^{-2}

What is the rate law for the reaction? What is the value of the rate constant (with correct units)?

13.67 Cyclopropane, C_3H_6, is a gas used as a general anesthetic. It undergoes a slow molecular rearrangement to propylene.

cyclopropane   propylene

At a certain temperature, the following data were obtained relating concentration and rate.

Initial Concentration of C_3H_6 (mol L^{-1})	Rate of Formation of Propylene (mol L^{-1} s^{-1})
0.050	2.95×10^{-5}
0.100	5.90×10^{-5}
0.150	8.85×10^{-5}

What is the rate law for the reaction? What is the value of the rate constant, with correct units?

13.68 The reaction of iodide ion with hypochlorite ion, OCl^- (the active ingredient in a "chlorine bleach" such as Clorox), follows the equation $OCl^- + I^- \to OI^- + Cl^-$. It is a rapid reaction that gives the following rate data.

Initial Concentrations (mol L^{-1})		Rate of Formation of
$[OCl^-]$	$[I^-]$	Cl^- (mol L^{-1} s^{-1})
1.7×10^{-3}	1.7×10^{-3}	1.75×10^4
3.4×10^{-3}	1.7×10^{-3}	3.50×10^4
1.7×10^{-3}	3.4×10^{-3}	3.50×10^4

What is the rate law for the reaction? Determine the value of the rate constant with its correct units.

13.69 The formation of small amounts of nitric oxide, NO, in automobile engines is the first step in the formation of smog. As noted earlier, nitric oxide is readily oxidized to nitrogen dioxide by the reaction

$$2NO(g) + O_2(g) \longrightarrow 2NO_2(g)$$

The following data were collected in a study of the rate of this reaction.

Initial Concentrations (mol L^{-1})		Rate of Formation of
$[O_2]$	$[NO]$	NO_2 (mol L^{-1} s^{-1})
0.0010	0.0010	7.10
0.0040	0.0010	28.4
0.0040	0.0030	255.6

What is the rate law for the reaction? What is the rate constant with its correct units?

13.70 At a certain temperature the following data were collected for the reaction $2ICl + H_2 \rightarrow I_2 + 2HCl$.

Initial Concentrations (mol L^{-1})		Initial Rate of Formation of I$_2$ (mol L^{-1} s^{-1})
[ICl]	[H$_2$]	
0.10	0.10	0.0015
0.20	0.10	0.0030
0.10	0.0500	0.00075

Determine the rate law and the rate constant (with correct units) for the reaction.

13.71 The following data were obtained for the reaction of $(CH_3)_3CBr$ with hydroxide ion at 55 °C.

$$(CH_3)_3CBr + OH^- \longrightarrow (CH_3)_3COH + Br^-$$

Initial Concentrations (mol L^{-1})		Initial Rate of Formation of (CH$_3$)$_3$COH (mol L^{-1} s^{-1})
[(CH$_3$)$_3$CBr]	[OH$^-$]	
0.10	0.10	1.0×10^{-3}
0.20	0.10	2.0×10^{-3}
0.30	0.10	3.0×10^{-3}
0.10	0.20	1.0×10^{-3}
0.10	0.30	1.0×10^{-3}

What is the rate law for the reaction? What is the value of the rate constant (with correct units) at this temperature?

Concentration and Time

13.72 Data for the decomposition of SO_2Cl_2 according to the equation $SO_2Cl_2(g) \rightarrow SO_2(g) + Cl_2(g)$ were given in Problem 13.56. Show graphically that these data fit a first-order rate law. Graphically determine the rate constant for the reaction.

13.73 For the data in Problem 13.57, decide graphically whether the reaction is first or second order. Determine the rate constant for the reaction described in that problem.

13.74 The decomposition of SO_2Cl_2 described in Problem 13.72 has a first-order rate constant $k = 2.2 \times 10^{-5}$ s^{-1} at 320 °C. If the initial SO_2Cl_2 concentration in a container is 0.0040 M, what will its concentration be (a) after 1.00 hour, (b) after 1.00 day?

13.75 If it takes 75.0 min for the concentration of a reactant to drop to 20% of its initial value in a first-order reaction, what is the rate constant for the reaction in the units min^{-1}?

13.76 The concentration of a drug in the body is often expressed in units of milligrams per kilogram of body weight. The initial dose of a drug in an animal was 25.0 mg/kg body weight. After 2.00 hours, this concentration had dropped to 15.0 mg/kg body weight. If the drug is eliminated metabolically by a first-order process, what is the rate constant for the process in units of min^{-1}?

13.77 In the preceding problem, what must the initial dose of the drug be in order for the drug concentration 3.00 hours afterward to be 5.0 mg/kg body weight?

13.78 The decomposition of hydrogen iodide follows the equation $2HI(g) \rightarrow H_2(g) + I_2(g)$. The reaction is second order and has a rate constant equal to 1.6×10^{-3} L mol^{-1} s^{-1} at 700 °C. If the initial concentration of HI in a container is 3.4×10^{-2} M, how many minutes will it take for the concentration to be reduced to 8.0×10^{-4} M?

13.79 The second-order rate constant for the decomposition of HI at 700 °C was given in the preceding problem. At 2.5×10^3 minutes after a particular experiment had begun, the HI concentration was equal to 4.5×10^{-4} mol L^{-1}. What was the initial molar concentration of HI in the reaction vessel?

Half-lives

13.80 The half-life of a certain first-order reaction is 15 minutes. What fraction of the original reactant concentration will remain after 2.0 hours?

13.81 Strontium-90 has a half-life of 28 years. How long will it take for all of the strontium-90 presently on Earth to be reduced to 1/32nd of its present amount?

13.82 Using the graph from Problem 13.56, determine the time required for the SO_2Cl_2 concentration to drop from 0.100 mol L^{-1} to 0.050 mol L^{-1}. How long does it take for the concentration to drop from 0.050 mol L^{-1} to 0.025 mol L^{-1}? What is the order of this reaction? (*Hint:* How is the half-life related to concentration?)

13.83 Using the graph from Problem 13.57, determine how long it takes for the CH_3CHO concentration to decrease from 0.200 mol L^{-1} to 0.100 mol L^{-1}. How long does it take the concentration to drop from 0.100 mol L^{-1} to 0.050 mol L^{-1}? What is the order of this reaction? (*Hint:* How is the half-life related to concentration?)

13.84 A certain first-order reaction has a rate constant $k = 1.6 \times 10^{-3}$ s^{-1}. What is the half-life for this reaction?

13.85 The decomposition of NOCl, the compound that gives a yellow-orange color to aqua regia (a mixture of concentrated HCl and HNO_3 that's able to dissolve gold and platinum), follows the reaction $2NOCl \rightarrow 2NO + Cl_2$. It is a second-order reaction with $k = 6.7 \times 10^{-4}$ L mol^{-1} s^{-1} at 400 K. What is the half-life of this reaction if the initial concentration of NOCl is 0.20 mol L^{-1}?

Calculations Involving the Activation Energy

13.86 The following data were collected for a reaction.

Rate Constant (L mol^{-1} s^{-1})	Temperature (°C)
2.88×10^{-4}	320
4.87×10^{-4}	340
7.96×10^{-4}	360
1.26×10^{-3}	380
1.94×10^{-3}	400

Determine the activation energy for the reaction in kJ/mol both graphically and by calculation using Equation 13.11. For the calculation of E_a, use the first and last sets of data in the table above.

13.87 Rate constants were measured at various temperatures for the reaction $HI(g) + CH_3I(g) \rightarrow CH_4(g) + I_2(g)$. The following data were obtained.

Rate Constant ($L\ mol^{-1}\ s^{-1}$)	Temperature (°C)
1.91×10^{-2}	205
2.74×10^{-2}	210
3.90×10^{-2}	215
5.51×10^{-2}	220
7.73×10^{-2}	225
1.08×10^{-1}	230

Determine the activation energy in kJ/mol both graphically and by calculation using Equation 13.11. For the calculation of E_a, use the first and last sets of data in the table above.

13.88 The decomposition of NOCl, $2NOCl \rightarrow 2NO + Cl_2$, has $k = 9.3 \times 10^{-5}\ L\ mol^{-1}\ s^{-1}$ at 100 °C and $k = 1.0 \times 10^{-3}\ L\ mol^{-1}\ s^{-1}$ at 130 °C. What is E_a for this reaction in $kJ\ mol^{-1}$? Use the data at 100 °C to calculate the frequency factor.

13.89 The conversion of cyclopropane, an anesthetic, to propylene (see Problem 13.67) has a rate constant $k = 1.3 \times 10^{-6}\ s^{-1}$ at 400 °C and $k = 1.1 \times 10^{-5}\ s^{-1}$ at 430 °C.
(a) What is the activation energy in kJ/mol?
(b) What is the value of A for this reaction, calculated using Equation 13.9?
(c) What is the rate constant for the reaction at 350 °C?

13.90 The reaction of CO_2 with water to form carbonic acid $CO_2(aq) + H_2O \rightarrow H_2CO_3(aq)$ has $k = 3.75 \times 10^{-2}\ s^{-1}$ at 25 °C and $k = 2.1 \times 10^{-3}\ s^{-1}$ at 0 °C. What is the activation energy for this reaction in kJ/mol?

13.91 If a reaction has $k = 3.0 \times 10^{-4}\ s^{-1}$ at 25 °C and an activation energy of 100.0 kJ/mol, what will the value of k be at 50 °C?

13.92 The decomposition of N_2O_5 has an activation energy of 103 kJ/mol and a frequency factor of $4.3 \times 10^{13}\ s^{-1}$. What is the rate constant for this decomposition at (a) 20 °C and (b) 100 °C?

13.93 At 35 °C, the rate constant for the reaction

$$C_{12}H_{22}O_{11} + H_2O \longrightarrow C_6H_{12}O_6 + C_6H_{12}O_6$$
$$\text{sucrose} \qquad\qquad \text{glucose} \qquad \text{fructose}$$

is $k = 6.2 \times 10^{-5}\ s^{-1}$. The activation energy for the reaction is $108\ kJ\ mol^{-1}$. What is the rate constant for the reaction at 45 °C?

ADDITIONAL EXERCISES

13.94 For the reaction and data given in Problem 13.57, make a graph of concentration versus time for the *formation* of CH_4. What are the rates of formation of CH_4 at $t = 40$ s and $t = 100$ s?

13.95 The age of wine can be determined by measuring the trace amount of radioactive tritium, ^3H, present in a sample. Tritium is formed from hydrogen in water vapor in the upper atmosphere by cosmic bombardment, so all naturally occurring water contains a small amount of this isotope. Once the water is in a bottle of wine, however, the formation of additional tritium from the water is negligible, so the tritium initially present gradually diminishes by a first-order radioactive decay with a half-life of 12.5 years. If a bottle of wine is found to have a tritium concentration that is 0.100 that of freshly bottled wine (i.e., $[^3H]_t = 0.100\ [^3H]_0$), what is the age of the wine?

13.96 One of the reactions that occurs in polluted air in urban areas is $2NO_2(g) + O_3(g) \rightarrow N_2O_5(g) + O_2(g)$. It is believed that a species with the formula NO_3 is involved in the mechanism for the reaction, and the observed rate law for the overall reaction is Rate $= k\ [NO_2]\ [O_3]$. Propose a mechanism for this reaction that includes the species NO_3 and is consistent with the observed rate law.

***13.97** Suppose a reaction occurs with the mechanism

(1) $\qquad\qquad 2A \rightleftharpoons A_2 \qquad\qquad$ (fast)

(2) $\qquad\qquad A_2 + E \longrightarrow B + C \qquad$ (slow)

in which the first step is a very rapid reversible reaction that can be considered to be essentially an equilibrium (forward and reverse reactions occurring at the same rate) and the second is a slow step.
(a) Write the rate law for the forward reaction in step (1).
(b) Write the rate law for the reverse reaction in step (1).
(c) Write the rate law for the rate-determining step.
(d) What is the chemical equation for the net reaction that occurs in this chemical change?
(e) Use the results of parts (a) and (b) to rewrite the rate law of the rate-determining step in terms of the concentrations of the reactants in the overall balanced equation for the reaction.

***13.98** A reaction that has the stoichiometry

$$2A + B \longrightarrow \text{products}$$

was found to yield the following data.

Initial Concentrations (mol L^{-1})		Initial Rate of Formation of A (mol L^{-1} s^{-1})
[A]	[B]	
0.020	0.030	0.0150
0.025	0.030	0.0188
0.025	0.040	0.0334

(a) What is the rate law for the reaction?
(b) What is the rate constant for the reaction with its correct units?

13.99 The decomposition of urea, $(NH_2)_2CO$, in 0.10 M HCl follows the equation

$$(NH_2)_2CO(aq) + 2H^+(aq) + H_2O \longrightarrow 2NH_4^+(aq) + CO_2(g)$$

At 60 °C, $k = 5.84 \times 10^{-6}$ min^{-1} and at 70 °C, $k = 2.25 \times 10^{-5}$ min^{-1}. If this reaction is run at 80 °C starting with a urea concentration of 0.0020 M, how many minutes will it take for the urea concentration to drop to 0.0012 M?

13.100 Show that for a reaction that obeys the general rate law

$$\text{Rate} = k\,[A]^n$$

a graph of log (rate) versus log [A] should yield a straight line with a slope equal to the order of the reaction. For the reaction in Review Problem 13.56, measure the rate of the reaction at $t = 150$, 300, 450, and 600 s. Then graph log (rate) versus log [SO$_2$Cl$_2$] and determine the order of the reaction with respect to SO$_2$Cl$_2$.

13.101 The bonding in O$_2$ was discussed in Section 9.7 as was the bonding in N$_2$. Molecular nitrogen is very unreactive, whereas molecular oxygen is very reactive. On the basis of what you have learned in this chapter, what fact about O$_2$ is responsible, at least in part, for its reactivity? What might account for the low degree of reactivity of molecular nitrogen?

13.102 It was mentioned that the rates of many reactions approximately double for each 10 °C rise in temperature. Assuming a starting temperature of 25 °C, what would the activation energy be, in kJ mol^{-1}, if the rate of a reaction were to be twice as large at 35 °C?

13.103 The development of a photographic image on film is a process controlled by the kinetics of the reduction of silver halide by a developer. The time required for development at a particular temperature is inversely proportional to the rate constant for the process. Below are published data on development times for Kodak's Tri-X film using Kodak D-76 developer. From these data, estimate the activation energy (in units of kJ mol^{-1}) for the development process. Also estimate the development time at 15 °C.

Temperature (°C)	Development Time (minutes)
18	10
20	9
21	8
22	7
24	6

13.104 The rate at which crickets chirp depends on the ambient temperature, because crickets are cold-blooded insects whose body temperature follows the temperature of their environment. It has been found that the Celsius temperature can be estimated by counting the number of chirps in 8 seconds and then adding 4. In other words, t_C = (number of chirps in 8 seconds) + 4.
(a) Calculate the number of chirps in 8 seconds for temperatures of 20, 25, 30, and 35 °C.
(b) The number of chirps per unit of time is directly proportional to the rate constant for a biochemical reaction involved in the cricket's chirp. On the basis of this assumption, make a graph of ln (chirps in 8 s) versus (1/T). Calculate the activation energy for the biochemical reaction involved.
(c) How many chirps would a cricket make in 8 seconds at a temperature of 120 °C?

*****13.105** The cooking of an egg involves the denaturation of a protein called albumin, and the time required to achieve a particular degree of denaturation is inversely proportional to the rate constant for the process. This reaction has a high activation energy, E_a = 418 kJ mol^{-1}. Calculate how long it would take to cook a traditional three-minute egg on top of Mt. McKinley in Alaska on a day when the atmospheric pressure there is 355 torr.

*****13.106** In the upper atmosphere, a layer of ozone, O$_3$, shields Earth from harmful ultraviolet radiation. The ozone is generated by the reactions

$$O_2 \xrightarrow{h\nu} 2O$$

$$O + O_2 \longrightarrow O_3$$

The migration of chlorofluorocarbon aerosol propellants and refrigerant fluids such as Freon 12 (CCl$_2$F$_2$) into the stratosphere would at least partially destroy the ozone shield by the reactions

$$CCl_2F_2 \longrightarrow CClF_2 + Cl$$

$$Cl + O_3 \longrightarrow ClO + O_2$$

$$ClO + O \longrightarrow Cl + O_2$$

Read Facets of Chemistry 13.1 and then explain how the second and third of these reactions constitute a chain reaction that could destroy huge numbers of O_3 molecules as well as O atoms that might otherwise react with O_2 to replenish the ozone.

*13.107 The following question is based on Facets of Chemistry 13.1. The reaction of hydrogen and bromine appears to follow the mechanism.

$$Br_2 \xrightarrow{hv} 2Br\cdot$$

$$Br\cdot + H_2 \longrightarrow HBr + H\cdot$$

$$H\cdot + Br_2 \longrightarrow HBr + Br\cdot$$

$$2Br\cdot \longrightarrow Br_2$$

(a) Identify the initiation step in the mechanism.
(b) Identify any propagation steps.
(c) Identify the termination step.
(d) The mechanism also contains the reaction

$$H\cdot + HBr \longrightarrow H_2 + Br\cdot$$

How does this reaction affect the rate of formation of HBr?

Acid Rain

Normally, water from even the purest of natural sources is slightly acidic because of its dissolved carbon dioxide. Rain, however, is often more acidic than normal, and then it is called **acid rain.** Recorded extremes are rain as acidic as lemon juice, observed in 1964 in the northeastern section of the United States, and rain as acidic as vinegar, which fell at Pitlochry, Scotland, in 1974.

When rainwater becomes more acidic by falling through polluted air or by leaching the deposits of certain pollutants from soil, the waters of nearby lakes and streams also become more acidic, sometimes enough so to render them lethal habitats for the hatchlings of game fish. Moreover, acid rain leaches aluminum ions from soil, ions that also damage fish and forest life. Whole forests in parts of Europe have thus been severely harmed (see Figure 7a).

Certain building materials are corroded by acid rain. Limestone and marble are particularly sensitive, because they are chiefly calcium carbonate, and all carbonates are dissolved by acids.

$$CaCO_3(s) + 2H^+(aq) \rightarrow Ca^{2+}(aq) + CO_2(g) + H_2O$$

Several major cathedrals in Europe, built of limestone and marble, have been harmed by acid rain. Acid is also corrosive to exposed metals such as railroad rails, vehicles, and machinery.

Dust particles settling on buildings, metals, and soil also carry acidic materials as solids, so specialists prefer the term **acid deposition** to *acid rain* to encompass all means whereby acids enter surface features of Earth. Southern parts of the Scandinavian peninsula receive acid deposition from Germany's Ruhr valley, the English Midlands, and countries of eastern Europe. Parts of southern Canada and the northern United States receive acid deposition from the industrial belt curving from Boston to Chicago.

Industrial areas of eastern Europe have released into their atmospheres enormous loads of the oxides of sulfur and nitrogen, the chief causes of acid rain. Let's see how human activities produce these gases and what has been done about them.

Oxides of Sulfur. During the combustion of coal and oil, their sulfur impurities are oxidized to sulfur dioxide. Although relatively small quantities of sulfur are in the fossil fuels, often much less than 3%, such enormous quantities of the fuels are burned, that hundreds of millions of tons of SO_2 are generated worldwide each year.

Sulfur dioxide dissolves in water by forming hydrates, $SO_2 \cdot nH_2O$, where n varies with the SO_2 concentration and the temperature.[1] The hydrates, which we could also represent as $SO_2(aq)$, are in equilibrium with hydrogen ion and hydrogen sulfite ion, HSO_3^-.

$$SO_2(aq) + H_2O \rightleftharpoons H^+(aq) + HSO_3^-(aq)$$

Although the percentage to which the forward reaction occurs is much less than 100%, when falling rain picks up gaseous SO_2 from the air, the rainwater reaching the ground is more acidic than normal. To make matters worse, both oxygen and the ozone (O_3) in smog convert some SO_2 to sulfur trioxide, particularly in sunlight when fine dust is present. When SO_3 reacts with water, sulfuric acid, a strong acid (one with a high percentage ionization), forms. Thus SO_3 contributes to the acidity of rain where air pollution occurs.

Volcanic eruptions also inject huge quantities of sulfur oxides into the atmosphere. However, they are *relatively* small sources, and we can do nothing to prevent such natural events. Volcanic SO_2 releases have been estimated to be roughly 13×10^6 tons per year, which is 5 to 10% of the quantity released by human activities.

Nitrogen Dioxide. Another contributor to acid rain, nitrogen dioxide (NO_2), forms from the nitrogen monoxide in vehicle exhausts (see *Chemicals in Our World 6*).

---

[1]Traditionally, water with dissolved sulfur dioxide is said to contain sulfurous acid, H_2SO_3, but evidence for the presence of such a molecule in water does not exist. Nevertheless, it's convenient to represent the formation of hydrogen ions in aqueous sulfur dioxide as the ionization of H_2SO_3.

chemicals in our world

Figure 7a Acid rain has destroyed the forest on the Krusne Hory mountains near Teplice, Czech Republic.

In water, NO_2 reacts to give nitric acid (HNO_3), a strong acid, and nitrous acid (HNO_2), a weak acid.

$$2NO_2(g) + H_2O \longrightarrow HNO_3(aq) + HNO_2(aq)$$

Meeting the Acid Rain Problem. The acid rain problem would largely be solved if no coal or oil were used for the generation of electrical power. But nuclear power offers its own problems, like waste disposal and the possibility of leaks or of more dramatic discharges of radioactive materials. Neither solar power nor wind power is yet economically feasible for large scale systems. Water power could not come close to meeting the demands. Coal and oil are necessary for the foreseeable future, so the prevention of the release of sulfur oxides when these fuels are used is essential. And since the 1970s, the releases of sulfur oxides have been cut in half. It's quite a success story, and the annual cost (about $1 billion) has been far less than earlier predicted (about $10 billion or more). How has this happened?

One factor has been political. The U.S. Congress in its various Clean Air Acts and Amendments has mandated *how much* the annual reductions must be, but not *how to do it.* Power companies and power plant operators were given the flexibility to take advantage of local opportunities and the freedom to decide among strategies according to marketplace forces. Should we burn high sulfur coal from Appalachian mines in Ohio, West Virginia, and Pennsylvania, or more expensive low-sulfur coal from mines in Wyoming and Montana? Or should we use a mixture of these coals to reduce overall emissions? Should we install smokestack scrubbers to remove the sulfur oxides? The sulfur dioxide in smoke-

stack gases, for example, reacts with wet calcium hydroxide according to the following equation and so is removed as a solid.

$$SO_2(g) + Ca(OH)_2(s) \longrightarrow CaSO_3(s) + H_2O$$

The companies were also given emissions marketing rights. "Rights-to-pollute" coupons were issued that power plants could buy, sell, or save. A plant that became unusually successful in lowering its emissions of sulfur oxides might recover some of the costs by selling its rights-to-pollute coupons to plants not yet as efficient, provided that total annual emissions were capped according to the Clean Air Act requirements. The target for 2010 is a total release of 9 million tons of sulfur oxides per year. Without controls, the expected release would be an estimated 19 million tons. Much control has been achieved without economy destroying efforts.

Mercury Pollution

Mercury is another pollutant that can get into the natural water supply, and, once present, it remains for a long time. Some mercury arises from natural sources, and some is released by burning coal.

Biological processes in aquatic bacteria convert mercury to derivatives of the methylmercury ion, CH_3Hg^+, whose compounds are somewhat soluble in fat and muscle tissue. This ion binds to proteins in cells and renders them incapable of carrying out their functions. Fish in mercury-polluted waters are able to accumulate so much mercury that humans eating the fish endanger their brain and nerve tissue.

Questions

1. What are the chief gases that are responsible for acid rain, and how do they get into Earth's atmosphere? (Write equations.) Why is *acid deposition* a better term?

2. How do sulfur dioxide and nitrogen dioxide contribute to an increase in the concentration of hydrogen ion in water? (Write the equilibria.)

3. Briefly describe some of the problems caused by acid rain in living systems and materials.

4. What form of mercury (give a formula) is a threat to humans?

5. In general terms, explain how mercury persists for so long in an aquatic environment.

A herd of sheep reacts to the actions of a sheep dog. In some respects, chemical equilibria behave in a similar way when subjected to a stress. Le Châtelier's principle, introduced in Chapter 12, will be used in this chapter to predict the way chemical equilibria behave in response to various disturbances.

Chemical Equilibrium— General Concepts

This Chapter in Context In Chapter 13 you learned that most chemical reactions slow down as the reactants are consumed and the products form. The focus there was learning about the factors that affect how *fast* the reactant and product concentrations change. In this and the following three chapters we will turn our attention to the ultimate fate of chemical systems, *dynamic chemical equilibrium*—the situation that exists when the concentrations cease to change.

You have already encountered the concept of a dynamic equilibrium several times earlier in this book. Recall that such an equilibrium is established when two opposing processes occur at equal rates. In a liquid–vapor equilibrium, for example, the processes are evaporation and condensation. In a chemical equilibrium, such as the ionization of a weak acid, they are the forward and reverse reactions represented by the chemical equation.

In this chapter we will study the equilibrium condition both qualitatively and quantitatively. By knowing the factors that affect the composition of an equilibrium system we have some control over the outcome of a reaction.

14.1 Dynamic Equilibrium in Chemical Systems

Equilibrium is the situation that exists when a reaction appears to have stopped and the concentrations of the reactants and products no longer change. For instance, you learned that acetic acid ($HC_2H_3O_2$) rapidly establishes the equilibrium

$$HC_2H_3O_2(aq) + H_2O \rightleftharpoons H_3O^+(aq) + C_2H_3O_2^-(aq)$$

in which $HC_2H_3O_2$ and H_2O are reacting at the same rate as H_3O^+ and $C_2H_3O_2^-$. Let's review how this chemical equilibrium comes about.

When acetic acid is added to the water, the forward reaction (ionization) occurs rapidly at first as molecules of $HC_2H_3O_2$ and H_2O collide, but then it slows as the number of $HC_2H_3O_2$ molecules decreases. Just the opposite effect is found for the

*The reaction read from left to right is the **forward reaction.** The reaction read from right to left is the **reverse reaction.***

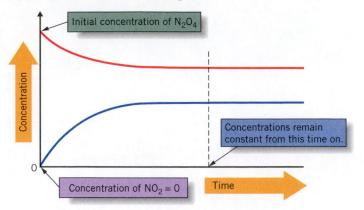

Figure 14.1 *The approach to equilibrium.* In the decomposition of $N_2O_4(g)$ into $NO_2(g)$,

$$N_2O_4(g) \longrightarrow 2NO_2(g)$$

the concentrations of N_2O_4 and NO_2 change relatively fast at first. As time passes, the concentrations change more and more slowly. When equilibrium is reached, the concentrations of N_2O_4 and NO_2 no longer change with time; they remain constant.

reverse reaction. The ions encounter each other more frequently as their concentrations increase, and the reverse reaction speeds up. Eventually, the rates of the forward and reverse reactions become equal, so the concentrations of the ions cease to change. At this point the system has reached a state of dynamic equilibrium.

It is an equilibrium because the concentrations don't change. It is dynamic because the opposing reactions never cease.

The reaction of acetic acid with water is just one example of a general phenomenon. For almost any reaction, the concentrations change relatively fast at first and then approach steady equilibrium values, as shown in Figure 14.1 for the decomposition of $N_2O_4(g)$ into $NO_2(g)$. If the system is not disturbed, these concentrations will never change. Keep in mind, however, that even after equilibrium has been reached, both the forward and reverse reactions continue to occur. This is what's implied by the double arrows, $\rightleftharpoons$, in the equilibrium equation.

Sometimes an equal sign ($=$) is used in place of the double arrows.

Almost all chemical systems eventually reach a state of dynamic equilibrium, although sometimes this equilibrium is extremely difficult (or even impossible) to detect. This is because in some reactions the amounts of either the reactants or products present at equilibrium are virtually zero. For instance, when a strong acid such as HCl is dissolved in water, it reacts so completely to produce H_3O^+ and Cl^- that no detectable amount of HCl molecules remains; we say that the HCl is completely ionized. Similarly, in water vapor at room temperature, there are no detectable amounts of either H_2 or O_2 from the equilibrium

$$2H_2O(g) \rightleftharpoons 2H_2(g) + O_2(g)$$

The water molecules are so stable that we can't detect whether any of them decompose. Even in cases such as these, however, it is often *convenient* to presume that an equilibrium does exist, and where it suits us, we will do so.

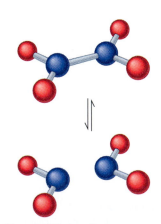

The equilibrium between N_2O_4 and NO_2.

14.2 Reaction Reversibility

The ionization of acetic acid is an example of a reaction that is able to proceed in either direction. Acetic acid molecules are able to react with water molecules to form the ions, and the ions are able to react to form the molecules. For most chemical reactions the composition of the equilibrium mixture is found to be in-

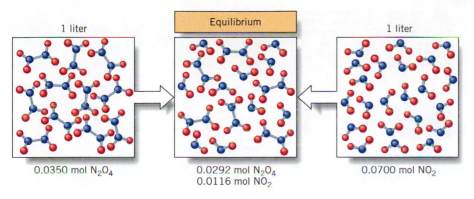

1 liter   Equilibrium   1 liter

0.0350 mol N_2O_4 0.0292 mol N_2O_4 0.0116 mol NO_2 0.0700 mol NO_2

Figure 14.2 *Reaction reversibility.* The same equilibrium composition is reached from either the forward or reverse direction, provided the *overall* system composition is the same.

dependent of whether we begin the reaction from the "reactant side" or the "product side." To illustrate this, let's consider some experiments that we might perform at 25 °C on the gaseous decomposition of N_2O_4 into NO_2 mentioned in Figure 14.1.

$$N_2O_4(g) \rightleftharpoons 2NO_2(g)$$

Suppose we set up the two experiments shown in Figure 14.2. In the first 1 liter flask we place 0.0350 mol N_2O_4. Since no NO_2 is present, some N_2O_4 must decompose for the mixture to reach equilibrium, so the reaction will proceed in the forward direction (i.e., from left to right). When equilibrium is reached, we find the concentration of N_2O_4 has dropped to 0.0292 mol L^{-1} and the concentration of NO_2 has become 0.0116 mol L^{-1}.

In the 1 liter flask on the right we place 0.0700 mol of NO_2 (*precisely* the amount of NO_2 that would form if 0.0350 mol of N_2O_4—the amount placed in the first flask—decomposed completely). In this second flask there is no N_2O_4 present initially, so NO_2 molecules must combine, following the reverse reaction (right to left) shown above, to give enough N_2O_4 for equilibrium. When we measure the concentrations at equilibrium, we find, once again, 0.0292 mol L^{-1} of N_2O_4 and 0.0116 mol L^{-1} of NO_2.

Stoichiometrically, 0.0700 mol of NO_2 could be formed from 0.0350 mol of N_2O_4.

We see here that the same equilibrium composition is reached whether we begin with pure NO_2 or pure N_2O_4, just as long as the *total* amount of nitrogen and oxygen to be divided between these two substances is the same. Similar observations apply to other chemical systems as well, which leads to the following generalization.

> For a given *overall* system composition, we always reach the same equilibrium concentrations whether equilibrium is approached from the forward or reverse direction.

It is in this sense that chemical reactions are said to be reversible.

14.3 The Equilibrium Law for a Reaction

For any chemical system at equilibrium there exists a simple relationship among the molar concentrations of the reactants and products. To show you this we will use the gaseous reaction of hydrogen with iodine to form hydrogen iodide. The equation for the resulting equilibrium is

$$H_2(g) + I_2(g) \rightleftharpoons 2HI(g)$$

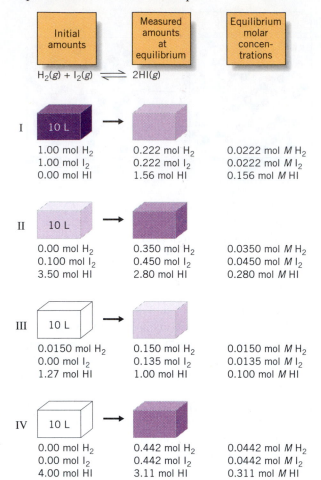

Initial amounts	Measured amounts at equilibrium	Equilibrium molar concentrations

$H_2(g) + I_2(g) \rightleftharpoons 2HI(g)$

I 10 L

1.00 mol H_2 0.222 mol H_2 0.0222 mol M H_2
1.00 mol I_2 0.222 mol I_2 0.0222 mol M I_2
0.00 mol HI   1.56 mol HI   0.156 mol M HI

II 10 L

0.00 mol H_2 0.350 mol H_2 0.0350 mol M H_2
0.100 mol I_2 0.450 mol I_2 0.0450 mol M I_2
3.50 mol HI   2.80 mol HI   0.280 mol M HI

III 10 L

0.0150 mol H_2 0.150 mol H_2 0.0150 mol M H_2
0.00 mol I_2 0.135 mol I_2 0.0135 mol M I_2
1.27 mol HI   1.00 mol HI   0.100 mol M HI

IV 10 L

0.00 mol H_2 0.442 mol H_2 0.0442 mol M H_2
0.00 mol I_2 0.442 mol I_2 0.0442 mol M I_2
4.00 mol HI   3.11 mol HI   0.311 mol M HI

Figure 14.3 *Four experiments to study the equilibrium among H_2, I_2, and HI gases.* Different amounts of the reactants and product are placed in a 10.0 L reaction vessel at 440 °C where the gases establish the equilibrium

$$H_2(g) + I_2(g) \rightleftharpoons 2HI(g)$$

When equilibrium is reached, different amounts of reactants and products remain in each experiment, which gives different equilibrium concentrations.

The molar concentrations are obtained by dividing the number of moles of each substance by the volume, 10.0 L.

Let's see what happens if we set up several experiments in which we start with different amounts of the reactants and/or product in a 10.0 L reaction vessel, as illustrated in Figure 14.3. Notice that when equilibrium is reached the amounts of H_2, I_2, and HI are different for each experiment, as are their molar concentrations. This isn't particularly surprising, but what is amazing is that the relationship among the concentrations is very simple and can actually be predicted (once we've learned how) from the balanced equation for the reaction.

For each experiment in Figure 14.3, if we square the molar concentration of HI at equilibrium and then divide this by the product of the equilibrium molar concentrations of H_2 and I_2, we obtain the same numerical value. This is shown in Table 14.1 where we have once again used square brackets around formulas as symbols for molar concentrations.

The fraction used to calculate the values in the last column of Table 14.1,

$$\frac{[HI]^2}{[H_2][I_2]}$$

is called the **mass action expression** for this equilibrium. The origin of this term isn't important; just consider it a name we use to refer to this kind of fraction. The numerical value of the mass action expression is called the **reaction quotient,** and is often symbolized by the letter Q. In Table 14.1, notice that when H_2, I_2, and HI are in dynamic equilibrium at 440 °C, the reaction quotient is equal to essentially the same constant value of 49.5. In fact, if we repeated the

Table 14.1 Equilibrium Concentrations and the Mass Action Expression

Experiment	Equilibrium Concentrations (mol L^{-1})			$\dfrac{[HI]^2}{[H_2][I_2]}$
	$[H_2]$	$[I_2]$	$[HI]$	
I	0.0222	0.0222	0.156	$(0.156)^2/(0.0222)(0.0222) = 49.4$
II	0.0350	0.0450	0.280	$(0.280)^2/(0.0350)(0.0450) = 49.8$
III	0.0150	0.0135	0.100	$(0.100)^2/(0.0150)(0.0135) = 49.4$
IV	0.0442	0.0442	0.311	$(0.311)^2/(0.0442)(0.0442) = 49.5$
				Average = 49.5

experiments in Figure 14.3 over and over again starting with different amounts of H_2, I_2, and HI, we would still obtain the same reaction quotient, provided the systems had reached equilibrium and the temperature was 440 °C. Therefore, for this reaction at equilibrium we can write

$$\frac{[HI]^2}{[H_2][I_2]} = 49.5 \qquad \text{(at 440 °C)} \qquad (14.1)$$

This relationship is called the **equilibrium law** for the system. Significantly, it tells us that for a mixture of these three gases to be at equilibrium at 440 °C, the value of the mass action expression (the reaction quotient) must equal 49.5. If the reaction quotient has any other value, then the gases are not in equilibrium and they must react further. The constant 49.5, which characterizes this equilibrium system, is called the **equilibrium constant.** The equilibrium constant is usually symbolized by K_c (the subscript c because we write the mass action expression using molar concentrations). Thus, we can state the equilibrium law as follows:

$$\frac{[HI]^2}{[H_2][I_2]} = K_c = 49.5 \qquad \text{(at 440 °C)} \qquad (14.2)$$

It is often useful to think of an equilibrium law such as Equation 14.2 as a *condition* that must be met for equilibrium to exist.

> For chemical equilibrium to exist in a reaction mixture, the reaction quotient Q must be equal to the equilibrium constant, K_c.

As you've probably noticed, we have repeatedly mentioned the temperature when referring to the value of K_c. This is because the value of the equilibrium constant changes when the temperature changes. Thus, if we had performed the experiments in Figure 14.3 at a temperature other than 440 °C, we would have obtained a different value for K_c.

Equation 14.2 can be called the *equilibrium condition* for the reaction of H_2, I_2, and HI at 440 °C.

In general, it is necessary to specify the temperature when giving a value of K_c, because K_c changes when the temperature changes. For example, for the reaction

$$CH_4(g) + H_2O(g) \rightleftharpoons$$
$$CO(g) + 3H_2(g)$$

$$K_c = 1.78 \times 10^{-3} \text{ at 800 °C}$$

$$K_c = 4.68 \times 10^{-2} \text{ at 1000 °C}$$

$$K_c = 5.67 \text{ at 1500 °C}$$

Predicting the Equilibrium Law

An important fact about the mass action expression and the equilibrium law is that it can *always* be predicted from the balanced chemical equation for the reaction. For example, for the general chemical equation

$$dD + eE \rightleftharpoons fF + gG$$

where D, E, F, and G represent chemical formulas and d, e, f, and g are their coefficients, the mass action expression is

$$\frac{[F]^f[G]^g}{[D]^d[E]^e}$$

Tools

Constructing the equilibrium law

The exponents in the mass action expression are the same as the stoichiometric co-efficients in the balanced equation.

The condition for equilibrium in this reaction is given by the equation

$$\frac{[F]^f [G]^g}{[D]^d [E]^e} = K_c$$

where the only concentrations that satisfy the equation are *equilibrium concentrations.*

Notice that in writing the mass action expression the molar concentrations of the products are always placed in the numerator and those of the reactants appear in the denominator. Also note that after being raised to appropriate powers the concentration terms are *multiplied,* not added.

Notice that even though we cannot predict the *rate law* from the balanced overall equation, we can predict the *equilibrium law.*

EXAMPLE 14.1

Writing the Equilibrium Law

Write the equilibrium law for the reaction

$$N_2(g) + 3H_2(g) \rightleftharpoons 2NH_3(g)$$

which is used to synthesize ammonia industrially.

Solution: The equilibrium law sets the mass action expression equal to the equilibrium constant. To form the mass action expression, we place the concentrations of the products in the numerator and the concentrations of the reactants in the denominator. The coefficients in the equation become exponents on the concentrations. Therefore,

$$\frac{[NH_3]^2}{[N_2][H_2]^3} = K_c$$

Notice that we omit writing the exponent when it is equal to one.

Practice Exercise 1

Write the equilibrium law for each of the following:
(a) $2H_2(g) + O_2(g) \rightleftharpoons 2H_2O(g)$
(b) $NH_3(aq) + H_2O(l) \rightleftharpoons NH_4^+(aq) + OH^-(aq)$ ◆

The rule that we always write the concentrations of the products in the numerator of the mass action expression and the concentrations of the reactants in the denominator is not required by nature. It is simply a convention chemists have agreed on. Certainly, if the mass action expression is equal to a constant,

$$\frac{[HI]^2}{[H_2][I_2]} = K_c$$

its reciprocal is also equal to a constant (let's call it K_c'),

$$\frac{[H_2][I_2]}{[HI]^2} = \frac{1}{K_c} = K_c'$$

The reason that chemists choose to stick to a set pattern—always placing the concentrations of the products in the numerator—is to avoid having to specify mass action expressions along with tabulated values of equilibrium constants. If we have the chemical equation for the equilibrium, we can *always* construct the correct mass action expression from it. For example, suppose we're told that at a particular temperature $K_c = 10.0$ for the reaction

$$2NO_2(g) \rightleftharpoons N_2O_4(g)$$

From the chemical equation we can write the correct mass action expression and the correct equilibrium law.

$$K_c = \frac{[N_2O_4]}{[NO_2]^2} = 10.0$$

The balanced chemical equation contains all the information we need to write the equilibrium law.

Manipulating Equations for Chemical Equilibria

Tools

Manipulating
equilibrium
equations

Sometimes it is useful to be able to combine chemical equilibria to obtain the equation for some other reaction of interest. In doing this, we perform various operations such as reversing an equation, multiplying the coefficients by some factor, and adding the equations to give the desired equation. In our discussion of thermochemistry, you learned how such manipulations affect ΔH values. Some different rules apply to changes in the mass action expressions and equilibrium constants.

Changing the Direction of an Equilibrium

When the direction of an equation is reversed, the new equilibrium constant is the reciprocal of the original. You have just seen this in the discussion above. As another example, when we reverse the equilibrium

$$PCl_3 + Cl_2 \rightleftharpoons PCl_5 \qquad K_c = \frac{[PCl_5]}{[PCl_3]\,[Cl_2]}$$

we obtain

$$PCl_5 \rightleftharpoons PCl_3 + Cl_2 \qquad K'_c = \frac{[PCl_3]\,[Cl_2]}{[PCl_5]}$$

The mass action expression for the second reaction is the reciprocal of that for the first, so K'_c equals $1/K_c$.

Multiplying the Coefficients by a Factor

When the coefficients in an equation are multiplied by a factor, the equilibrium constant is raised to a power equal to that factor. For example, suppose we multiply the coefficients of the equation

$$PCl_3 + Cl_2 \rightleftharpoons PCl_5 \qquad K_c = \frac{[PCl_5]}{[PCl_3]\,[Cl_2]}$$

by 2. This gives

$$2PCl_3 + 2Cl_2 \rightleftharpoons 2PCl_5 \qquad K''_c = \frac{[PCl_5]^2}{[PCl_3]^2[Cl_2]^2}$$

Comparing mass action expressions, we see that $K''_c = K_c^2$.

Adding Chemical Equilibria

When chemical equilibria are added, their equilibrium constants are multiplied. For example, suppose we add the following two equations.

We have numbered the equilibrium constants just to distinguish one from the other.

$$2N_2 + O_2 \rightleftharpoons 2N_2O \qquad K_{c1} = \frac{[N_2O]^2}{[N_2]^2[O_2]}$$

$$2N_2O + 3O_2 \rightleftharpoons 4NO_2 \qquad K_{c2} = \frac{[NO_2]^4}{[N_2O]^2[O_2]^3}$$

$$2N_2 + 4O_2 \rightleftharpoons 4NO_2 \qquad K_{c3} = \frac{[NO_2]^4}{[N_2]^2[O_2]^4}$$

If we multiply the mass action expression for K_{c1} by that for K_{c2}, we obtain the mass action expression for K_{c3}.

$$\frac{[\cancel{N_2O}]^2}{[N_2]^2[O_2]} \times \frac{[NO_2]^4}{[\cancel{N_2O}]^2[O_2]^3} = \frac{[NO_2]^4}{[N_2]^2[O_2]^4}$$

Therefore, $K_{c1} \times K_{c2} = K_{c3}$.

Practice Exercise 2

At 25 °C, $K_c = 7.0 \times 10^{25}$ for the reaction: $2SO_2(g) + O_2(g) \rightleftharpoons 2SO_3(g)$. What is the value of K_c for the reaction: $SO_3(g) \rightleftharpoons SO_2(g) + \frac{1}{2}O_2(g)$? ◆

Practice Exercise 3

At 25 °C, the following reactions have the equilibrium constants noted to the right of their equations.

$$2CO(g) + O_2(g) \rightleftharpoons 2CO_2(g) \qquad K_c = 3.3 \times 10^{91}$$

$$2H_2(g) + O_2(g) \rightleftharpoons 2H_2O(g) \qquad K_c = 9.1 \times 10^{80}$$

Use these data to calculate K_c for the reaction

$$H_2O(g) + CO(g) \rightleftharpoons CO_2(g) + H_2(g) ◆$$

14.4 Equilibrium Laws for Gaseous Reactions

When all the reactants and products are gases, we can formulate mass action expressions in terms of partial pressures as well as molar concentrations. This is possible because the molar concentration of a gas is proportional to its partial pressure. This comes from the ideal gas law,

$$PV = nRT$$

Solving for P gives

$$P = \left(\frac{n}{V}\right)RT$$

If you double the molar concentration of a gas without changing its temperature or volume, you double its pressure.

The quantity n/V has units of mol/L and is simply the molar concentration. Therefore, we can write

$$P = (\text{molar concentration}) \times RT \qquad (14.3)$$

This equation applies whether the gas is by itself in a container or part of a mixture. In the case of a gas mixture, P is the partial pressure of the gas.

The relationship expressed in Equation 14.3 lets us write the mass action expression for reactions between gases in terms either of molarities or partial pressures. However, when we make a switch we can't expect the numerical values of the equilibrium constants to be the same, so we use two different symbols for K. When molar concentrations are used, we use the symbol K_c. When partial pressures are used, then K_p is the symbol. For example, the equilibrium law for the reaction of nitrogen with hydrogen to form ammonia

$$N_2(g) + 3H_2(g) \rightleftharpoons 2NH_3(g)$$

can be written in either of the following two ways

$$\frac{[NH_3]^2}{[N_2][H_2]^3} = K_c \qquad \left(\begin{array}{l}\text{because molar concentrations} \\ \text{are used in the mass action} \\ \text{expression}\end{array}\right)$$

$$\frac{P_{NH_3}^2}{P_{N_2}P_{H_2}^3} = K_p \qquad \left(\begin{array}{l}\text{because partial pressures are} \\ \text{used in the mass action} \\ \text{expression}\end{array}\right)$$

The equilibrium molar concentrations can be used to calculate K_c, whereas the equilibrium partial pressures can be used to calculate K_p. We will discuss how to convert between K_c and K_p in Section 14.6.

Write the expression for K_p for the reaction

$$N_2O_4(g) \rightleftharpoons 2NO_2(g)$$

Solution: For K_p we use partial pressures in the mass action expression.

$$K_p = \frac{P_{NO_2}^2}{P_{N_2O_4}}$$

Practice Exercise 4

Using partial pressures, write the equilibrium law for the reaction

$$H_2(g) + I_2(g) \rightleftharpoons 2HI(g) \; \blacklozenge$$

EXAMPLE 14.2

Writing Expressions for K_p

14.5 The Significance of the Magnitude of K

Whether we work with K_p or K_c, a bonus of always writing the mass action expression with the product concentrations in the numerator is that the size of the equilibrium constant gives us a measure of how far the reaction proceeds toward completion when equilibrium is reached. For example, the reaction

$$2H_2(g) + O_2(g) \rightleftharpoons 2H_2O(g)$$

has $K_c = 9.1 \times 10^{80}$ at 25 °C. This means that when there is an equilibrium between these gases,

$$K_c = \frac{[H_2O]^2}{[H_2]^2[O_2]} = \frac{9.1 \times 10^{80}}{1}$$

By writing K_c as a fraction, $(9.1 \times 10^{80})/1$, we see that the numerator of the mass action expression must be enormous compared with the denominator, which means that the concentration of H_2O has to be enormous in comparison to the concentrations of H_2 and O_2. At equilibrium, therefore, most of the hydrogen and oxygen atoms in the system are found in the H_2O and very few are present in H_2 and O_2. Thus, the large value of K_c tells us that the reaction between H_2 and O_2 goes essentially to completion.

The reaction between N_2 and O_2 to give NO

$$N_2(g) + O_2(g) \rightleftharpoons 2NO(g)$$

Actually, you would need about 200,000 L of water vapor at 25 °C just to find one molecule of O_2 and two molecules of H_2.

has a very small equilibrium constant; $K_c = 4.8 \times 10^{-31}$ at 25 °C. The equilibrium law for this reaction is

$$\frac{[NO]^2}{[N_2][O_2]} = 4.8 \times 10^{-31}$$

Since $10^{-31} = 1/10^{31}$, we can write this as

$$\frac{[NO]^2}{[N_2][O_2]} = \frac{4.8}{10^{31}}$$

In air at 25 °C, the equilibrium concentration of NO should *be about 10^{-17} mol/L. It is usually higher because NO is formed in various reactions, such as those responsible for air pollution caused by automobiles.*

Here the denominator is huge compared with the numerator, so the concentrations of N_2 and O_2 must be very much larger than the concentration of NO. This means that in a mixture of N_2 and O_2 at this temperature, very little NO is formed. The reaction hardly proceeds at all toward completion before equilibrium is reached.

The relationship between the equilibrium constant and the position of equilibrium can thus be summarized as follows:

Tools

Significance of the magnitude of K

When K is very large	The reaction proceeds far toward completion. The position of equilibrium lies far toward the products.
When $K \approx 1$	The concentrations of reactants and products are nearly the same at equilibrium. The position of equilibrium lies approximately midway between reactants and products.
When K is very small	Extremely small amounts of products are formed. The position of equilibrium lies far toward the reactants.

Notice that we have omitted the subscript for K in this summary. The same qualitative predictions about the extent of reaction apply whether we use K_p or K_c.

One of the ways that we can use equilibrium constants is to compare the extents to which two or more reactions proceed to completion. Take care in making such comparisons, however, because unless the K's are greatly different, the comparison is valid only if both reactions have the same number of reactant and product molecules appearing in their balanced chemical equations.

Practice Exercise 5

Which of the following reactions will tend to proceed farthest toward completion?
(a) $H_2(g) + Br_2(g) \rightleftharpoons 2HBr(g)$ $K_c = 7.9 \times 10^{18}$
(b) $2NO(g) \rightleftharpoons N_2(g) + O_2(g)$ $K_c = 2.1 \times 10^{30}$
(c) $2BrCl \rightleftharpoons Br_2 + Cl_2$ (in CCl_4 solution) $K_c = 0.145$ ◆

14.6 The Relationship between K_p and K_c

For some reactions K_p is equal to K_c, but for many others the two constants have different values. It is therefore desirable to have a way to calculate one from the other. Converting between K_p and K_c uses the relationship between partial pressure and molarity described in Section 14.4. Equation 14.3 can be used to change K_p to K_c by substituting

$PV = nRT$

Solving for the concentration of the gas, n/V, gives

$$\frac{n}{V} = \frac{P}{RT}$$

(molar concentration) $\times RT$

for the partial pressure of each gas in the mass action expression for K_p. Similarly, K_c can be changed to K_p by solving Equation 14.3 for the molar concentrations, and then substituting the result, P/RT, into the appropriate expression for K_c. This sounds like a lot of work, and it is. Fortunately, there is a general equation, which can be derived from these relationships, that we can use to make these conversions simply.

$$K_p = K_c(RT)^{\Delta n_g} \qquad (14.4)$$

Tools

Converting between K_p and K_c

In this equation, the value of Δn_g is equal to the change in the *number of moles of gas* in going from the reactants to the products.

$$\Delta n_g = \text{(moles of } gaseous \text{ products)} - \text{(moles of } gaseous \text{ reactants)}$$

We use the coefficients of the balanced equation for the reaction to calculate the numerical value of Δn_g. For example, the equation

$$N_2(g) + 3H_2(g) \rightleftharpoons 2NH_3(g) \qquad (14.5)$$

tells us that two moles of NH_3 are formed when one mole of N_2 and three moles of H_2 react. In other words, two moles of gaseous product are formed from a total of four moles of gaseous reactants. That's a decrease of two moles of gas, so Δn_g for this reaction equals -2.

Δn_g is calculated from the coefficients of the equation, taking them to stand for moles.

For some reactions, the value of Δn_g is equal to zero. An example is the decomposition of HI.

$$2HI(g) \rightleftharpoons H_2(g) + I_2(g)$$

Notice that if we take the coefficients to mean moles, there are two moles of gas on each side of the equation. This means that $\Delta n_g = 0$. Since (RT) raised to the zero power is equal to 1, $K_p = K_c$.

EXAMPLE 14.3

Converting between K_p and K_c

At 500 °C, the reaction between N_2 and H_2 to form ammonia

$$N_2(g) + 3H_2(g) \rightleftharpoons 2NH_3(g)$$

has $K_c = 6.0 \times 10^{-2}$. What is the numerical value of K_p for this reaction?

Solution: The equation that we wish to use is

$$K_p = K_c(RT)^{\Delta n_g}$$

In the discussion above, we saw that $\Delta n_g = -2$ for this reaction. All we need now are appropriate values of R and T. The temperature, T, must be expressed in kelvins. (When used to stand for temperature, a capital letter T in an equation always means the absolute temperature.) Next we must choose an appropriate value for R. Referring back to Equation 14.3, if the partial pressures are expressed in atm and the concentration in mol L^{-1}, the only value of R that is consistent with these units is $R = 0.0821$ L atm mol^{-1} K^{-1}, and this is the *only* value of R that can be used in Equation 14.4.[1] Assembling the data, then, we have

$$K_c = 6.0 \times 10^{-2} \qquad\qquad \Delta n_g = -2$$

$$T = (500 + 273)\ \text{K} = 773\ \text{K} \qquad R = 0.0821\ \text{L atm mol}^{-1}\ \text{K}^{-1}$$

---

[1] The branch of science called thermodynamics, which is discussed in Chapter 18, defines equilibrium constants as unitless quantities. It treats each of the terms in the mass action expression as a ratio. For K_c, these terms are concentration ratios in which the numerator is the actual effective molar concentration (with units of mol L^{-1}) and the denominator is 1.0 M. For K_p, the mass action expression contains terms that are pressure ratios in which the numerators are the actual pressures (in atm) and the denominators equal 1.0 atm. Because the same units appear in the numerator and denominator of each term, they cancel, so K_c and K_p have no units. However, to convert between K_c and K_p, we need to use $R = 0.0821$ L atm mol^{-1} K^{-1} because of the nature of the units used to form the terms in mass action expressions.

Substituting these into the equation for K_p gives

$$K_p = (6.0 \times 10^{-2}) \times [(0.0821) \times (773)]^{-2}$$

$$= (6.0 \times 10^{-2}) \times (63.5)^{-2}$$

$$= 1.5 \times 10^{-5}$$

In this case, K_p has a numerical value quite different from that of K_c.

EXAMPLE 14.4

Converting between K_p and K_c

At 25 °C, K_p for the reaction $N_2O_4(g) \rightleftharpoons 2NO_2(g)$ has a value of 0.140. Calculate the value of K_c.

Solution: Once again, the equation we need is

$$K_p = K_c(RT)^{\Delta n_g}$$

This time, $\Delta n_g = 2 - 1 = +1$. Now let's tabulate the data.

$$K_p = 0.140 \qquad \Delta n_g = +1$$

$$T = 298 \text{ K} \qquad R = 0.0821 \text{ L atm mol}^{-1} \text{ K}^{-1}$$

Solving the equation for K_c gives

$$K_c = \frac{K_p}{(RT)^{\Delta n_g}}$$

Substituting values into this equation yields

$$K_c = \frac{0.140}{[(0.0821) \times (298)]^1}$$

$$= 5.72 \times 10^{-3}$$

Once again, there is a substantial difference between the values of K_p and K_c.

Things to Check
In working these problems, check to be sure you have used the correct value for R and that the temperature is expressed in kelvins. Also, examining Equation 14.4, we see that when Δn_g is positive, the value of K_p will be *larger* than K_c [because K_c is multiplied by (RT) raised to a positive power]. The opposite will be true, of course, if Δn_g is negative. Thus, in Example 14.4, Δn_g is positive and the given value of K_p (0.140) is larger than the calculated value of K_c (5.72×10^{-3}). On the other hand, in Example 14.3 where $\Delta n_g = -2$, the K_p value is smaller than the K_c value.

Practice Exercise 6

Methanol, CH_3OH, is a promising fuel that can be synthesized from carbon monoxide and hydrogen according to the reaction

$$CO(g) + 2H_2(g) \rightleftharpoons CH_3OH(g)$$

For this reaction at 200 °C, $K_p = 3.8 \times 10^{-2}$. Do you expect K_p to be larger or smaller than K_c? Calculate the value of K_c at this temperature. ◆

Practice Exercise 7

Nitrous oxide, N_2O, is a gas used as an anesthetic; it is sometimes called "laughing gas." This compound has a strong tendency to decompose into nitrogen and oxygen following the equation

$$2N_2O(g) \rightleftharpoons 2N_2(g) + O_2(g)$$

but the reaction is so slow that the gas appears to be stable at room temperature (25 °C). The decomposition reaction has $K_c = 7.3 \times 10^{34}$. What is the value of K_p for this reaction at 25 °C? ◆

14.7 Heterogeneous Equilibria

In a **homogeneous reaction**—or a **homogeneous equilibrium**—all of the reactants and products are in the same phase. Equilibria among gases are homogeneous because all gases mix freely with each other, so a single phase exists. There are also many equilibria in which reactants and products are dissolved in the same liquid phase.

When more than one phase exists in a reaction mixture, we call it a **heterogeneous reaction.** A common example is the combustion of wood, in which a solid fuel reacts with gaseous oxygen. Another is the thermal decomposition of sodium bicarbonate (baking soda), which occurs when the compound is sprinkled on a fire.

$$2NaHCO_3(s) \longrightarrow Na_2CO_3(s) + H_2O(g) + CO_2(g)$$

Safety-minded cooks keep a box of baking soda nearby because this reaction makes it an excellent fire extinguisher for burning fats or oil. The fire is smothered by the products of the reaction.

Heterogeneous reactions are able to reach equilibrium just as homogeneous reactions can. If $NaHCO_3$ is placed in a sealed container so that no CO_2 or H_2O can escape, the gases and solids come to a heterogeneous equilibrium.

$$2NaHCO_3(s) \rightleftharpoons Na_2CO_3(s) + H_2O(g) + CO_2(g)$$

Following our usual procedure, we can write the equilibrium law for this reaction as

$$\frac{[Na_2CO_3(s)][H_2O(g)][CO_2(g)]}{[NaHCO_3(s)]^2} = K$$

However, the equilibrium law for reactions involving pure liquids and solids can be written in an even simpler form. This is because the concentration of a pure liquid or solid is unchangeable; that is, *for any pure liquid or solid, the ratio of amount of substance to volume of substance is a constant.* For example, if we had a 1 mole crystal of $NaHCO_3$, it would occupy a volume of 38.9 cm³. Two moles of $NaHCO_3$ would occupy twice this volume, 77.8 cm³ (Figure 14.4), but the *ratio* of moles to liters (i.e., the molar concentration) remains the same. For $NaHCO_3$, the concentration of the substance in the solid is

$$\frac{1 \text{ mol}}{0.0389 \text{ L}} = \frac{2 \text{ mol}}{0.0778 \text{ L}} = 25.7 \text{ mol L}^{-1}$$

We include the physical states in the mass action expression here because we have a heterogeneous system.

1 mol NaHCO₃
38.9 cm³

Molarity = $\frac{1 \text{ mol NaHCO}_3}{0.0389 \text{ L}}$
= 25.7 mol/L

2 mol NaHCO₃
77.8 cm³

Molarity = $\frac{2 \text{ mol NaHCO}_3}{0.0778 \text{ L}}$
= 25.7 mol/L

Figure 14.4 *The concentration of a substance in the solid state is a constant.* Doubling the number of moles also doubles the volume, but the *ratio* of moles to volume remains the same.

This is the concentration of $NaHCO_3$ in the solid, regardless of the size of the solid sample. In other words, the concentration of $NaHCO_3$ is constant, provided that some of it is present in the reaction mixture.

Similar reasoning shows that the concentration of Na_2CO_3 in pure solid Na_2CO_3 is a constant, too. This means that the equilibrium law now has three constants, K plus two of the concentration terms. It makes sense to combine all of the numerical constants together.

$$[H_2O(g)][CO_2(g)] = \frac{K[NaHCO_3(s)]^2}{[Na_2CO_3(s)]} = K_c$$

The equilibrium law for a heterogeneous reaction is written without concentration terms for pure solids or liquids. Equilibrium constants that are given in tables represent all of the constants combined.[2]

EXAMPLE 14.5

Writing the Equilibrium Law for a Heterogeneous Reaction

The air pollutant sulfur dioxide can be removed from a gas mixture by passing the gases over calcium oxide. The equation is

$$CaO(s) + SO_2(g) \rightleftharpoons CaSO_3(s)$$

Write the equilibrium law for this reaction.

Solution: The concentrations of the two solids, CaO and $CaSO_3$, are incorporated into the equilibrium constant K_c for the reaction. The only concentration term that should appear in the mass action expression is that of SO_2. Therefore, the equilibrium law is simply

$$\frac{1}{[SO_2(g)]} = K_c$$

Practice Exercise 8

Write the equilibrium law for the following heterogeneous reactions.
(a) $2Hg(l) + Cl_2(g) \rightleftharpoons Hg_2Cl_2(s)$
(b) $NH_3(g) + HCl(g) \rightleftharpoons NH_4Cl(s)$
(c) Dissolving solid Ag_2CrO_4 in water: $Ag_2CrO_4(s) \rightleftharpoons 2Ag^+(aq) + CrO_4{}^{2-}(aq)$ ◆

14.8 Le Châtelier's Principle and Chemical Equilibria

In Section 14.9 you will see that it is possible to perform calculations that tell us what the composition of an equilibrium system is. However, many times we really don't need to know exactly what the equilibrium concentrations are. Instead, we may want to know what actions we should take to control the relative amounts of the reactants or products at equilibrium. For instance, if we were designing gasoline engines, we would like to know what could be done to minimize the formation of nitrogen oxide pollutants (see Facets of Chemistry 14.1). Or, if we were preparing ammonia, NH_3, by the reaction of N_2 with H_2, we might want to know how to maximize the yield of NH_3.

Le Châtelier's principle, introduced in Chapter 11, provides us with the means for making qualitative predictions about changes in chemical equilibria.

---

[2]Thermodynamics handles heterogeneous equilibria in a more elegant way by expressing the mass action expression in terms of "effective concentrations," or **activities.** In doing this, thermodynamics *defines* the *activity* of any pure liquid or solid as equal to 1, which means terms involving such substances drop out of the mass action expression.

Facets of Chemistry 14.1

Air Pollution and Le Châtelier's Principle

As everyone knows, one of the most serious causes of air pollution is the automobile. All sorts of obnoxious chemicals are present in the exhaust gases leaving the engine, and various methods have been devised to control the amounts of these emissions that enter the atmosphere. For instance, almost all cars today are equipped with catalytic converters (see Chapter 13). These devices mix air with the exhaust gases and promote oxidation of unburned fuel and carbon monoxide to carbon dioxide. They also cause the decomposition of nitrogen oxides to elemental nitrogen and oxygen. Catalytic converters are expensive, however, and other methods to accomplish the same goals have also been studied.

When air is drawn into a car's engine, both N_2 and O_2 are present. During combustion of the gasoline, oxygen reacts with the hydrocarbons in the fuel to produce CO_2, CO, and H_2O. However, N_2 and O_2 can also form NO.

$$N_2(g) + O_2(g) \rightleftharpoons 2NO(g)$$

At room temperature, K_c for this reaction is 4.8×10^{-31}. Its small value tells us that the equilibrium concentration of NO should be very small. Therefore, we don't find N_2 reacting with O_2 under ordinary conditions, so the atmosphere is quite stable.

The reaction of N_2 and O_2 to form NO is endothermic. Le Châtelier's principle tells us that at high temperatures, such as those found in the cylinders of a gasoline or diesel engine during combustion, this equilibrium should be shifted to the right, so at high temperatures some NO does form. Unfortunately, when the exhaust leaves the engine, it cools so rapidly that the NO can't decompose. The reaction rate at the lower temperature becomes too slow. The result is that some NO is present among the exhaust gases. Once in the atmosphere, this NO becomes oxidized to NO_2, which is responsible for the brownish haze often associated with severe air pollution.

One way to reduce the amount of NO pollution of the atmosphere is to reduce the amount of NO that's formed in automobile engines. Since the extent to which the reaction proceeds toward the formation of NO increases as the temperature is raised, the amount of NO that's formed can be reduced by simply running the com-

At times, air pollution from auto exhaust emissions poses a health hazard. This can be particularly severe in large cities like Los Angeles where heavy traffic and atmospheric conditions can combine to produce dangerous levels of smog.

bustion reaction at a lower temperature. One method that has been used to accomplish this is to lower the compression ratio of the engine. This is the ratio of the volume of the cylinder when the piston is at the bottom of its stroke divided by the volume after the piston has compressed the air–fuel mixture. At high compression ratios, the air–fuel mixture is heated to a high temperature before it's ignited. After combustion, the gases are very hot, which favors the production of NO. Lowering the compression ratio lowers the maximum combustion temperature which decreases the tendency for NO to be formed. Unfortunately, lowering the compression ratio also lowers the efficiency of the engine, which makes for poorer fuel economy.

Another method for controlling NO emissions that has been experimented with involves mixing water with the air–fuel mixture. Some of the heat from the combustion is absorbed by the water vapor, so the mixture of exhaust gases doesn't get as hot as it would otherwise. At these lower temperatures, the concentration of NO in the exhaust is greatly reduced.

It does this in much the same way that it allows us to predict the effects of outside influences on equilibria that involve physical changes, such as liquid–vapor equilibria. Recall that **Le Châtelier's principle** states that *if an outside influence upsets an equilibrium, the system undergoes a change in a direction that counteracts the disturbing influence and, if possible, returns the system to equilibrium.* Let's examine what kinds of "outside influences" can affect chemical equilibria.

Tools

Le Châtelier's principle

Adding or Removing a Reactant or Product

If we add or remove a reactant or product, we change one of the concentrations in the system. This changes the value of Q so that it is no longer equal to K_c. The result is that the equilibrium is upset. For the system to return to equilibrium, the concentrations must change in a way that causes Q to equal K_c again. This is brought about by the chemical reaction proceeding either to the right or left, and Le Châtelier's principle lets us predict which way the reaction goes.

$Cu(H_2O)_4{}^{2+}$ and $CuCl_4{}^{2-}$ are called **complex ions.** Some of the interesting properties and applications of complex ions of metals are discussed in Chapter 20.

> An equilibrium will shift in a direction that will partially consume a reactant or product that is added, or partially replace a reactant or product that is removed.

As an example, let's study the equilibrium between two ions of copper.

$$Cu(H_2O)_4{}^{2+}(aq) + 4Cl^-(aq) \rightleftharpoons CuCl_4{}^{2-}(aq) + 2H_2O$$
$$\text{blue} \qquad\qquad\qquad\qquad \text{yellow}$$

As noted, $Cu(H_2O)_4{}^{2+}$ is blue and $CuCl_4{}^{2-}$ is yellow. Mixtures of the two have an intermediate color and therefore appear blue-green, as illustrated in Figure 14.5, center.

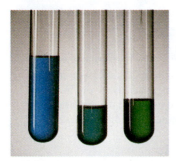

Figure 14.5 *The effect of concentration changes on the position of equilibrium.* The solution in the center contains a mixture of blue $Cu(H_2O)_4{}^{2+}$ and yellow $CuCl_4{}^{2-}$, so it has a blue-green color. At the right is some of the same solution after the addition of concentrated HCl. It has a more pronounced green color because the equilibrium is shifted toward $CuCl_4{}^{2-}$. At the left is some of the original solution after the addition of water. It is blue because the equilibrium has shifted toward $Cu(H_2O)_4{}^{2+}$.

If we add chloride ion to an equilibrium mixture of these copper ions, the system can get rid of some of it by reaction with $Cu(H_2O)_4{}^{2+}$. This gives more $CuCl_4{}^{2-}$ (Figure 14.5, right), and we say that the equilibrium has "shifted to the right." In this new position of equilibrium, there is less $Cu(H_2O)_4{}^{2+}$ and more $CuCl_4{}^{2-}$ and uncombined H_2O. There is also more Cl^-, because not all that we added reacts. In other words, in this new position of equilibrium, *all* the concentrations have changed in a way that causes Q to become equal to K_c. Similarly, the position of equilibrium is shifted to the left when we add water to the mixture (Figure 14.5, left). The system is able to get rid of some of the H_2O by reaction with $CuCl_4{}^{2-}$, so more of the blue $Cu(H_2O)_4{}^{2+}$ is formed.

If we were able to remove a reactant or product, the position of equilibrium would also be changed. For example, if we add Ag^+ to a solution that contains both copper ions in equilibrium, we see an enhancement of the blue color. As the Ag^+ reacts with Cl^- to form insoluble AgCl, the equilibrium shifts to the left to replace some of the Cl^-. In this case, the system replaces part of the substance that is removed.

The reaction shifts in a direction that will remove a substance that's been added or replace a substance that's been removed.

Changing the Volume in Gaseous Reactions

Changing the volume of a system composed of gaseous reactants and products changes their molar concentrations. It also changes the pressure, which is how we analyze the effect of this disturbing influence.

Let's consider the equilibrium

$$3H_2(g) + N_2(g) \rightleftharpoons 2NH_3(g)$$

If we reduce the volume of the reaction mixture, we expect the pressure to increase. The system can oppose the pressure change if it is able to reduce the number of molecules of gas, because fewer molecules of gas exert a lower pressure. If the reaction proceeds to the right, two NH_3 molecules appear when four molecules (one N_2 and three H_2) disappear. Therefore, this equilibrium is shifted to the right when the volume of the reaction mixture is reduced.

Now let's look at the equilibrium

$$H_2(g) + I_2(g) \rightleftharpoons 2HI(g)$$

If this reaction proceeds in either direction, there is no change in the number of molecules of gas. This reaction, then, cannot respond to pressure changes, so changing the volume of the reaction vessel has virtually no effect on the equilibrium.

The simplest way to analyze the effects of a volume change on an equilibrium system is to count the number of molecules of gaseous substances on both sides of the equation.

> Reducing the volume of a gaseous reaction mixture shifts the equilibrium in whichever direction will, if possible, decrease the number of molecules of gas.

As a final note here, moderate pressure changes have essentially no effect on reactions involving only liquids or solids. Substances in these states are virtually incompressible, and reactions involving them have no way to counteract pressure changes.

Changing the Temperature

To determine how a chemical equilibrium responds to a temperature change, we must know how the energy of the system changes when the reaction occurs. For instance, the reaction to produce NH_3 from N_2 and H_2 is exothermic; heat is evolved when NH_3 molecules are formed ($\Delta H_f^\circ = -46.19$ kJ/mol from Table 6.2). If we include heat as a product in the equilibrium equation

$$3H_2(g) + N_2(g) \rightleftharpoons 2NH_3(g) + \text{heat}$$

analyzing the effects of temperature changes becomes simple.

To raise the temperature, we add heat from some external source, such as a Bunsen burner. The equilibrium above can counter the temperature increase by absorbing some of the added heat. It does this by proceeding from right to left because the decomposition of some NH_3 to give more N_2 and H_2 is endothermic and consumes some of the heat that we've added. Thus, raising the temperature shifts this equilibrium to the left. In general,

> Increasing the temperature shifts an equilibrium in a direction that produces an endothermic change.

The effect of temperature on the equilibrium involving the complex ions of copper described earlier is demonstrated in Figure 14.6. Here we have used our observations to determine that the reaction is endothermic in the forward direction.

An important point to note in these examples is that when we change the temperature, the concentrations change *even though the volume stays the same and no chemical substances have been added or removed.* Thus, the system comes to a new position of equilibrium in which the value of the mass action expression has changed, which means that K has changed. We can predict how K changes with temperature from the sign of the heat of reaction. For instance, let's look once again at the equilibrium law for the formation of ammonia, an exothermic reaction.

$$\frac{[NH_3]^2}{[N_2][H_2]^3} = K_c$$

As we've noted, when the temperature increases, the concentration of NH_3 decreases while the concentrations of N_2 and H_2 increase. Therefore, the numera-

You may recall that this same kind of analysis was used in Chapter 12 to predict how solubility changes with temperature.

Keep in mind that when heat is added to an equilibrium mixture, it is added to all of the substances present (reactants *and* products). As the system returns to equilibrium, the net reaction that occurs is the one that is endothermic.

The only factor that can change K_c or K_p for a given equilibrium is a change in temperature.

Figure 14.6 *The effect of temperature on the position of equilibrium.* The reaction is

$$Cu(H_2O)_4^{2+} + 4Cl^- \rightleftharpoons CuCl_4^{2-} + 4H_2O$$

blue   yellow

In the center, the tube contains an equilibrium mixture of the two ions of copper. When the solution is cooled in ice *(left)*, the equilibrium shifts toward the blue $Cu(H_2O)_4^{2+}$. When heated in boiling water *(right),* the equilibrium shifts toward yellow $CuCl_4^{2-}$. This behavior indicates that the reaction is endothermic in the forward direction.

tor of the mass action expression becomes *smaller* and the denominator becomes *larger.* This gives a smaller reaction quotient and therefore a smaller value of K_c. Thus,

> If the forward reaction in an equilibrium is exothermic, raising the temperature causes the equilibrium constant to become smaller.

Of course, just the opposite change in K occurs if the forward reaction is endothermic .

Agents That Do Not Affect the Position of Equilibrium

There are some things that we can do to an equilibrium system that do not alter the position of equilibrium.

Adding a Catalyst

Recall that catalysts are substances that affect the speeds of chemical reactions without actually being used up. However, catalysts do not affect the position of equilibrium in a system. The reason is that a catalyst affects both the forward and reverse reactions equally. Both are speeded up to the same degree, so adding a catalyst to a system has no net effect on the system's equilibrium composition. The catalyst's only effect is to bring the system to equilibrium faster.

Adding an Inert Gas at Constant Volume

A change in volume is not the only way to change the pressure in an equilibrium system of gaseous reactants and products. The pressure can also be changed by keeping the volume the same and adding another gas. If this gas cannot react with any of the gases already present (i.e., if the added gas is *inert* toward the substances in equilibrium), the concentrations of the reactants and products won't change. The concentrations will continue to satisfy the equilibrium law and the reaction quotients will continue to equal K_c, so there will be no change in the position of equilibrium.

EXAMPLE 14.6

Application of Le Châtelier's Principle

The reaction $N_2O_4(g) \rightleftharpoons 2NO_2(g)$ is endothermic, with $\Delta H° = +56.9$ kJ. How will the amount of NO_2 at equilibrium be affected by (a) adding N_2O_4, (b) lowering the pressure by increasing the volume of the container, (c) raising the temperature, and (d) adding a catalyst to the system? Which of these changes will alter the value of K_c?

Solution: (a) Adding N_2O_4 will cause the equilibrium to shift to the right—in a direction that will consume some of the added N_2O_4. The amount of NO_2 will increase.
(b) When the pressure in the system drops, the system responds by producing more molecules of gas, which will tend to raise the pressure and partially offset the change. Since more gas molecules are formed if some N_2O_4 decomposes, the amount of NO_2 at equilibrium will increase.
(c) Because the reaction is endothermic in the forward direction, we write the equation showing heat as a reactant

$$\text{Heat} + N_2O_4(g) \rightleftharpoons 2NO_2(g)$$

Raising the temperature is accomplished by adding heat, so the system will respond by absorbing heat. This means that the equilibrium will shift to the right and the amount of NO_2 at equilibrium will increase.
(d) A catalyst causes a reaction to reach equilibrium more quickly, but it has no effect on the position of chemical equilibrium. Therefore, the amount of NO_2 at equilibrium will not be affected.

Finally, the *only* change that alters K is the temperature change. Raising the temperature (adding heat) will increase K_c because the forward reaction is endothermic.

Practice Exercise 9

Consider the equilibrium $PCl_3(g) + Cl_2(g) \rightleftharpoons PCl_5(g)$, for which $\Delta H° = -88$ kJ. How will the amount of Cl_2 at equilibrium be affected by (a) adding PCl_3, (b) adding PCl_5, (c) raising the temperature, and (d) decreasing the volume of the container? How (if at all) will each of these changes affect K_p for the reaction? ◆

14.9 Equilibrium Calculations

You have seen that the magnitude of an equilibrium constant gives us some feel for the extent to which the reaction proceeds at equilibrium. Sometimes, however, it is necessary to have more than merely a qualitative knowledge of equilibrium concentrations. This requires that we be able to use the equilibrium law for purposes of calculation.

Equilibrium calculations for gaseous reactions can be performed using either K_p or K_c, but for reactions in solution we must use K_c. Whether we deal with concentrations or partial pressures, however, the same basic principles apply.

Overall, we can divide equilibrium calculations into two main categories:

1. Calculating equilibrium constants from known equilibrium concentrations or partial pressures.
2. Calculating one or more equilibrium concentrations or partial pressures using the known value of K_c or K_p.

Calculating K_c from Equilibrium Concentrations

One way to determine the value of K_c is to carry out the reaction, measure the concentrations of reactants and products after equilibrium has been reached, and then use the equilibrium values in the equilibrium law to compute K_c. As an example, let's look again at the decomposition of N_2O_4.

$$N_2O_4(g) \rightleftharpoons 2NO_2(g)$$

In Section 14.2, we saw that if 0.0350 mol of N_2O_4 is placed into a 1 liter flask at 25 °C, the concentrations of N_2O_4 and NO_2 at equilibrium are $[N_2O_4]$ = 0.0292 mol/L and $[NO_2]$ = 0.0116 mol/L. To calculate K_c for this reaction, we substitute the equilibrium concentrations into the mass action expression of the equilibrium law.

$$\frac{[NO_2]^2}{[N_2O_4]} = K_c = \frac{(0.0116)^2}{(0.0292)}$$

Performing the arithmetic gives

$$K_c = 4.61 \times 10^{-3}$$

Although calculating an equilibrium constant this way is easy, sometimes we have to do a little work to figure out what all the concentrations are, as shown in Example 14.7.

EXAMPLE 14.7

Calculating K_c from Equilibrium Concentrations

At a certain temperature, a mixture of H_2 and I_2 was prepared by placing 0.200 mol of H_2 and 0.200 mol of I_2 into a 2.00 liter flask. After a period of time the equilibrium

$$H_2(g) + I_2(g) \rightleftharpoons 2HI(g)$$

was established. The purple color of the I_2 vapor was used to monitor the reaction, and from the decreased intensity of the purple color it was determined that, at equilibrium, the I_2 concentration had dropped to 0.020 mol L^{-1}. What is the value of K_c for this reaction at this temperature?

Analysis: The first step in any equilibrium problem is to write the balanced chemical equation and the related equilibrium law. The equation is already given, and the equilibrium law corresponding to it is

$$\frac{[HI]^2}{[H_2][I_2]} = K_c$$

To calculate the value of K_c we must substitute the *equilibrium concentrations* of H_2, I_2, and HI into the mass action expression. But what are they? We have been given only one directly, the value of $[I_2]$. To obtain the others, we have to do some reasoning based on the chemical equation.

The system starts with a set of initial concentrations, and to reach equilibrium a chemical change occurs. There is no HI present initially, so the reaction above must proceed to the right, because there has to be some HI present for the system to be at equilibrium. This change will increase the HI concentration and decrease the concentrations of H_2 and I_2. If we can determine what the *changes* are, we can figure out the equilibrium concentrations. *In fact, the key to solving almost all equilibrium problems is determining how the concentrations change as the system comes to equilibrium.*

To help us in our analysis we will construct a **concentration table** beneath the chemical equation. We will do this for most of the equilibrium problems we encounter from now on, so let's examine the general form of the table.

In the first row we write the initial *molar concentrations* of the reactants and products. (All entries in the table will be *molar concentrations, not moles.* These are the only quantities that satisfy the equilibrium law.) Then, in the second row we enter the changes in concentration, using a positive sign if the concentration is increasing and a negative sign if it is decreasing. Finally, adding the changes to the initial concentrations gives the equilibrium concentrations.

$$\begin{pmatrix} \text{Equilibrium} \\ \text{concentration} \end{pmatrix} = \begin{pmatrix} \text{Initial} \\ \text{concentration} \end{pmatrix} + \begin{pmatrix} \text{Change in} \\ \text{concentration} \end{pmatrix}$$

To set up the concentration table, it is essential that you realize that **the changes in concentration must be in the same ratio as the coefficients of the balanced equation.** This is because the changes are caused by the chemical reaction proceeding in one direction or the other. In this problem, the coefficients of H_2 and I_2 are the same, so as the reaction proceeds to the right, the concentrations of H_2 and I_2 must decrease by the *same* amount. Similarly, because the coefficient of HI is twice that of H_2 or I_2, the concentration of HI must increase by *twice* as much. Now let's look at the finished concentration table, how the individual entries are obtained, and the solution of the problem.

Solution: To construct the concentration table, we begin by entering the data given in the statement of the problem. These are shown in colored type. The values in regular type are then derived as described below.

	$H_2(g)$	+	$I_2(g) \rightleftharpoons$	$2HI(g)$
Initial concentrations (M)	0.100		0.100	0.000
Changes in concentrations (M)	−0.080		−0.080	+2(0.080)
Equilibrium concentrations (M)	0.020		0.020	0.160

Initial Concentrations
Notice that we have calculated the ratio of moles to liters for both the H_2 and I_2, (0.200 mol/2.00 L) = 0.100 M. Because no HI was placed into the reaction mixture, its initial concentration has been set to zero.

In any system, the initial concentrations are controlled by the person doing the experiment.

Changes in Concentrations
We have been given both the initial and equilibrium concentrations of I_2, so by difference we can calculate the change for I_2 (−0.080 M). The other changes are then calculated from the mole ratios specified in the chemical equation. As noted above, because H_2 and I_2 have the same coefficients (i.e., 1), their changes are equal. The coefficient of HI is 2, so its change must be twice that of I_2. In addition, notice that the changes for both reactants have the same sign, which is opposite that of the product. *This relationship among the signs of the changes is always true and can serve as a useful check when you construct a concentration table.*

The changes in concentrations are controlled by the stoichiometry of the reaction.

If we can find one of the changes, we can calculate the others from it.

Equilibrium Concentrations
For H_2 and HI, we just algebraically add the change to the initial value.

Now we can substitute the equilibrium concentrations into the mass action expression and calculate K_c.

$$K_c = \frac{(0.160)^2}{(0.020)(0.020)}$$

$$= 64$$

Concentration
table

The Concentration Table—A Summary

Let's review some key points that apply not only to this problem, but to others that deal with equilibrium calculations.

1. The only values that we can substitute into the mass action expression in the equilibrium law are equilibrium concentrations—the values that appear in the last row of the table.

2. When we enter initial concentrations into the table, they should be in units of moles per liter (mol L^{-1}). The initial concentrations are those present in the reaction mixture when it's prepared; we imagine that no reaction occurs until everything is mixed.

By expressing concentrations in moles per liter (mol L^{-1}), we follow what happens to reactants and products in each liter of the reaction mixture.

3. The changes in concentrations always occur in the same ratio as the coefficients in the balanced equation. For example, if we were dealing with the equilibrium

$$3H_2(g) + N_2(g) \rightleftharpoons 2NH_3(g)$$

and found that the $N_2(g)$ concentration decreases by 0.10 M during the approach to equilibrium, the entries in the "change" row would be as follows:

$3H_2(g)$	$+$	$N_2(g)$	$\rightleftharpoons$	$2NH_3(g)$
Change in concentration: $-3 \times (0.10\ M)$		$-1 \times (0.10\ M)$		$+2 \times (0.10\ M)$
$\downarrow$		$\downarrow$		$\downarrow$
$-0.30\ M$		$-0.10\ M$		$+0.20\ M$

If the initial concentration of a reactant is zero, its change must be positive (an increase) because the final concentration can't be negative.

4. In constructing the "change" row, be sure the reactant concentrations all change in the same direction, and that the product concentrations all change in the opposite direction. If the concentrations of the reactants decrease, all the entries for the reactants in the "change" row should have a minus sign, and all the entries for the products should be positive.

Keep these ideas in mind as we construct the concentration tables for other equilibrium problems.

Practice Exercise 10

An equilibrium was established for the reaction

$$CO(g) + H_2O(g) \rightleftharpoons CO_2(g) + H_2(g)$$

at 500 °C. (This is an industrially important reaction for the preparation of hydrogen.) At equilibrium, the following concentrations were found in the reaction vessel: $[CO] = 0.180\ M$, $[H_2O] = 0.0411\ M$, $[CO_2] = 0.150\ M$, and $[H_2] = 0.200\ M$. What is the value of K_c for this reaction? ◆

Practice Exercise 11

In a particular experiment, it was found that when $O_2(g)$ and $CO(g)$ were mixed and reacted according to the equation

$$2CO(g) + O_2(g) \rightleftharpoons 2CO_2(g)$$

the O_2 concentration decreased by 0.030 mol L^{-1} when the reaction reached equilibrium. How had the concentrations of CO and CO_2 changed? ◆

Practice Exercise 12

A student placed 0.200 mol of $PCl_3(g)$ and 0.100 mol of $Cl_2(g)$ into a 1.00 liter container at 250 °C. After the reaction

$$PCl_3(g) + Cl_2(g) \rightleftharpoons PCl_5(g)$$

came to equilibrium it was found that the flask contained 0.120 mol of PCl_3.
(a) What were the initial concentrations of the reactants and product?
(b) By how much had the concentrations changed when the reaction reached equilibrium?
(c) What were the equilibrium concentrations?
(d) What is the value of K_c for this reaction at this temperature? ◆

Calculating Equilibrium Concentrations Using K_c

In the simplest calculation of this type, all but one of the equilibrium concentrations are known, as illustrated in Example 14.8.

EXAMPLE 14.8

Using K_c to Calculate Concentrations at Equilibrium

The reversible reaction

$$CH_4(g) + H_2O(g) \rightleftharpoons CO(g) + 3H_2(g)$$

has been used as a commercial source of hydrogen. At 1500 °C, an equilibrium mixture of these gases was found to have the following concentrations: [CO] = 0.300 M, [H_2] = 0.800 M, and [CH_4] = 0.400 M. At 1500 °C, K_c = 5.67 for this reaction. What was the equilibrium concentration of $H_2O(g)$ in this mixture?

Solution: The first step, once we have the chemical equation for the equilibrium, is to write the equilibrium law for the reaction.

$$K_c = \frac{[CO][H_2]^3}{[CH_4][H_2O]}$$

The equilibrium constant and all of the equilibrium concentrations except that for H_2O are known, so we substitute these values into the equilibrium law and solve for the unknown quantity.

$$5.67 = \frac{(0.300)(0.800)^3}{(0.400)[H_2O]}$$

Solving for [H_2O] gives

$$[H_2O] = \frac{(0.300)(0.800)^3}{(0.400)(5.67)} = \frac{0.154}{2.27}$$

$$= 0.0678 \ M$$

Practice Exercise 13

Ethyl acetate, $CH_3CO_2C_2H_5$, is an important solvent used in lacquers, adhesives, the manufacture of plastics, and even as a food flavoring. It is produced from acetic acid and ethanol by the reaction

$$CH_3CO_2H(l) + C_2H_5OH(l) \rightleftharpoons CH_3CO_2C_2H_5(l) + H_2O(l)$$
$$\text{acetic acid} \qquad \text{ethanol}$$

At 25 °C, K_c = 4.10 for this reaction. In a reaction mixture, the following equilibrium concentrations were observed: [CH_3CO_2H] = 0.210 M, [H_2O] = 0.00850 M, and [$CH_3CO_2C_2H_5$] = 0.910 M. What was the concentration of C_2H_5OH in the mixture? ◆

Acetic acid has the structure

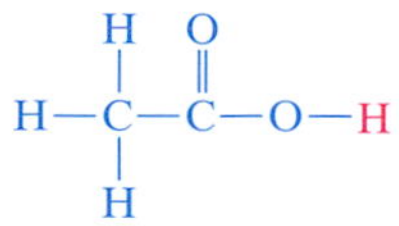

where the hydrogen released by ionization is indicated in red. The formula of the acid is written either as CH_3CO_2H (which emphasizes its molecular structure) or as $HC_2H_3O_2$ (which emphasizes that the acid is monoprotic).

Calculating Equilibrium Concentrations Using K_c and Initial Concentrations

A more complex type of calculation involves the use of initial concentrations and K_c to compute equilibrium concentrations. Although some of these problems can be so complicated that a computer is needed to solve them, we can learn the general principles involved by working on simple calculations. Even these, however, require a little applied algebra. This is where the concentration table can be very helpful.

EXAMPLE 14.9

Using K_c to Calculate Equilibrium Concentrations

The reaction

$$CO(g) + H_2O(g) \rightleftharpoons CO_2(g) + H_2(g)$$

has $K_c = 4.06$ at 500 °C. If 0.100 mol of CO and 0.100 mol of $H_2O(g)$ are placed in a 1.00 liter reaction vessel at this temperature, what are the concentrations of the reactants and products when the system reaches equilibrium?

Analysis: The key to solving this kind of problem is recognizing that at equilibrium the mass action expression must equal K_c.

$$\frac{[CO_2][H_2]}{[CO][H_2O]} = 4.06 \quad \text{(at equilibrium)}$$

We must find values for the concentrations that satisfy this condition.

Because we don't know what these concentrations are, we represent them algebraically as unknowns. This is where the concentration table is very helpful. To build the table, we need quantities to enter into the "initial concentrations," "changes in concentrations," and "equilibrium concentrations" rows.

Initial Concentrations

The initial concentrations of CO and H_2O are each 0.100 mol/1.00 L = 0.100 M. Since no CO_2 or H_2 are initially placed into the reaction vessel, their initial concentrations both are zero. (See below.)

Changes in Concentrations

Some CO_2 and H_2 must form for the reaction to reach equilibrium. This also means that some CO and H_2O must react. But how much? If we knew the answer, we could calculate the equilibrium concentrations. Therefore, the changes in concentration are our unknown quantities.

We could just as easily have chosen x to be the number of mol/L of H_2O that reacts or the number of mol/L of CO_2 or H_2 that forms. There's nothing special about having chosen CO to define x.

Let us allow x to be equal to the number of moles per liter of CO that react. The change in the concentration of CO is then $-x$ (it is negative because the change decreases the CO concentration). Because CO and H_2O react in a 1:1 mole ratio, the change in the H_2O concentration is also $-x$. Since one mole each of CO_2 and H_2 is formed from one mole of CO, the CO_2 and H_2O concentrations each increase by x (their changes are $+x$). We enter these quantities in the "Change" row.

Equilibrium Concentrations

We obtain the equilibrium concentrations as

$$\begin{pmatrix} \text{Equilibrium} \\ \text{concentration} \end{pmatrix} = \begin{pmatrix} \text{Initial} \\ \text{concentration} \end{pmatrix} + \begin{pmatrix} \text{Change in} \\ \text{concentration} \end{pmatrix}$$

The coefficients of x can be the same as the coefficients in the balanced chemical equation. This would assure us that the changes are in the same ratio as the coefficients in the equation.

Solution: Here is the completed concentration table.

	$CO(g)$ +	$H_2O(g) \rightleftharpoons$	$CO_2(g)$ +	$H_2(g)$
Initial concentrations (M)	0.100	0.100	0.0	0.0
Changes in concentrations (M)	$-x$	$-x$	$+x$	$+x$
Equilibrium concentrations (M)	$0.100 - x$	$0.100 - x$	x	x

Note that the last line in the table tells us that the equilibrium CO and H_2O concentrations are equal to the number of moles per liter that were present initially minus the number of moles per liter that react. The equilibrium concentrations of CO_2 and H_2 equal the number of moles per liter of each that forms, since no CO_2 or H_2 are present initially.

Next, we substitute the quantities from the "equilibrium concentrations" row into the mass action expression and solve for x.

The equilibrium concentration values must satisfy the equation given by the equilibrium law.

$$\frac{(x)(x)}{(0.100 - x)(0.100 - x)} = 4.06$$

which we can write as

$$\frac{x^2}{(0.100 - x)^2} = 4.06$$

In this problem we can solve the equation for x most easily by taking the square root of both sides.

$$\frac{x}{(0.100 - x)} = \sqrt{4.06} = 2.01$$

Clearing fractions gives

$$x = 2.01(0.100 - x)$$

$$x = 0.201 - 2.01x$$

Collecting terms in x gives

$$x + 2.01x = 0.201$$

$$3.01x = 0.201$$

$$x = 0.0668$$

Now that we know the value of x, we can calculate the equilibrium concentrations from the last row of the table.

$$[CO] = 0.100 - x = 0.100 - 0.0668 = 0.033 \ M$$

$$[H_2O] = 0.100 - x = 0.100 - 0.0668 = 0.033 \ M$$

$$[CO_2] = x = 0.0668 \ M$$

$$[H_2] = x = 0.0668 \ M$$

Checking the Answers

One way to check the answers is to substitute the equilibrium concentrations we've found into the mass action expression and evaluate the reaction quotient. If our answers are correct, Q should equal K_c. Let's do this.

$$Q = \frac{(0.0668)^2}{(0.033)^2} = 4.1$$

Rounding K_c to two significant figures gives 4.1, so the calculated concentrations satisfy the equilibrium law.

EXAMPLE 14.10

Using K_c to Calculate Equilibrium Concentrations

In the preceding example it was stated that the reaction

$$CO(g) + H_2O(g) \rightleftharpoons CO_2(g) + H_2(g)$$

has $K_c = 4.06$ at 500 °C. Suppose 0.0600 mol each of CO and H_2O are mixed with 0.100 mol each of CO_2 and H_2 in a 1.00 L reaction vessel. What will the concentrations of all the substances be when the mixture reaches equilibrium at this temperature?

Analysis: We will proceed in much the same way as in the preceding example. However, this time determining the algebraic signs of x will not be quite so simple because none of the initial concentrations is zero. The best way to determine the algebraic signs is to use the initial concentrations to calculate the initial reaction quotient. Then we can compare Q to K_c and by reasoning we will figure out which way the reaction must proceed to make Q equal to K_c.

Solution: The equilibrium law for the reaction is

$$\frac{[CO_2][H_2]}{[CO][H_2O]} = K_c$$

Let's use the initial concentrations, shown in the first row of the concentration table below, to determine the initial value of the reaction quotient.

$$Q_{initial} = \frac{(0.100)(0.100)}{(0.0600)(0.0600)} = 2.78 < K_c$$

As indicated, $Q_{initial}$ is less than K_c, so the system is not at equilibrium. To reach equilibrium Q must become larger, which requires an increase in the concentrations of CO_2 and H_2 as the reaction proceeds. This means that for CO_2 and H_2, x must be positive, and, therefore, for CO and H_2O, x must be negative.

Here is the completed concentration table.

	CO(g)	+	H₂O(g)	⇌	CO₂(g)	+	H₂(g)
Initial concentrations (*M*)	0.0600		0.0600		0.100		0.100
Changes in concentrations (*M*)	$-x$		$-x$		$+x$		$+x$
Equilibrium concentrations (*M*)	$0.0600 - x$		$0.0600 - x$		$0.100 + x$		$0.100 + x$

Notice that we have chosen the coefficients of x to be the same as the coefficients in the balanced chemical equation.

Substituting equilibrium quantities into the mass action expression in the equilibrium law gives us

$$\frac{(0.100 + x)^2}{(0.0600 - x)^2} = 4.06$$

Taking the square root of both sides yields

$$\frac{0.100 + x}{0.0600 - x} = 2.01$$

To solve for x we first multiply each side by $0.0600 - x$ to obtain

$$0.100 + x = 2.01(0.0600 - x)$$

$$0.100 + x = 0.121 - 2.01x$$

Collecting terms in x to one side and the constants to the other gives

$$x + 2.01x = 0.121 - 0.100$$

$$3.01x = 0.021$$

$$x = 0.0070$$

Now we can calculate the equilibrium concentrations:

$$[CO] = [H_2O] = (0.0600 - x) = 0.0600 - 0.0070 = 0.0530 \ M$$

$$[CO_2] = [H_2] = (0.100 + x) = 0.100 + 0.0070 = 0.107 \ M$$

Checking the Answers

As a check, let's evaluate the reaction quotient using the calculated equilibrium concentrations.

$$Q = \frac{(0.107)^2}{(0.0530)^2} = 4.08$$

This is acceptably close to the value of K_c. (That it is not *exactly* equal to K_c is because of the rounding of answers during the calculations.)

In each of the preceding two examples, we were able to simplify the solution by taking the square root of both sides of the algebraic equation obtained by substituting equilibrium concentrations into the mass action expression. Such simplifications are not always possible, however, as illustrated in the next example.

EXAMPLE 14.11

Using K_c to Calculate Equilibrium Concentrations

At a certain temperature $K_c = 4.50$ for the reaction

$$N_2O_4(g) \rightleftharpoons 2NO_2(g)$$

If 0.300 mol of N_2O_4 is placed into a 2.00 L container at this temperature, what will be the equilibrium concentrations of both gases?

Analysis: As in the preceding example, at equilibrium the mass action expression must be equal to K_c.

$$\frac{[NO_2]^2}{[N_2O_4]} = 4.50$$

We will need to find algebraic expressions for the equilibrium concentrations and substitute them into the mass action expression. To obtain these, we set up the concentration table for the reaction.

Initial Concentrations

The initial concentration of N_2O_4 is 0.300 mol/2.00 L $= 0.150 \ M$. Since no NO_2 was placed in the reaction vessel, its initial concentration is 0.000 M.

Changes in Concentrations

There is no NO_2 in the reaction mixture, so we know its concentration must increase. This means the N_2O_4 concentration must decrease as some of the NO_2 is formed. Let's allow x to be the number of moles per liter of N_2O_4 that reacts, so the change in the N_2O_4 concentration is $-x$. Because of the stoichiometry of the reaction, the NO_2 concentration must increase by $2x$, so its change in concentration is $+2x$.

Equilibrium Concentrations

As before, we add the change to the initial concentration in each column to obtain expressions for the equilibrium concentrations.

Solution: Here is the concentration table.

	$N_2O_4(g) \rightleftharpoons$	$2NO_2(g)$
Initial concentrations (M)	0.150	0.000
Changes in concentrations (M)	$-x$	$+2x$
Equilibrium concentrations (M)	$0.150 - x$	$2x$

Notice that we have chosen the coefficients of x to be the same as the coefficients in the balanced chemical equation.

Now we substitute the equilibrium quantities into the mass action expression:

$$\frac{(2x)^2}{(0.150 - x)} = 4.50$$

or

$$\frac{4x^2}{(0.150 - x)} = 4.50 \qquad (14.6)$$

This time the left side of the equation is not a perfect square, so we cannot just take the square root of both sides as in Example 14.10. However, because the equation involves terms in x^2, x, and a constant, we can use the quadratic formula. Recall that for a quadratic equation of the form

$$ax^2 + bx + c = 0$$

$$x = \frac{-b \pm \sqrt{b^2 - 4ac}}{2a}$$

Expanding Equation 14.6 above gives

$$4x^2 = 4.50(0.150 - x)$$

$$= 0.675 - 4.50x$$

Arranging terms in the standard order gives

$$4x^2 + 4.50x - 0.675 = 0$$

Therefore, the quantities we will substitute into the quadratic formula are as follows: $a = 4$, $b = 4.50$, and $c = -0.675$. Making these substitutions gives

$$x = \frac{-4.50 \pm \sqrt{(4.50)^2 - 4(4)(-0.675)}}{2(4)}$$

$$= \frac{-4.50 \pm \sqrt{31.05}}{8}$$

$$= \frac{-4.50 \pm 5.57}{8}$$

Because of the $\pm$ term, there are two values of x that satisfy the equation, $x = 0.134$ and $x = -1.26$. However, only the first value, $x = 0.134$, makes any sense chemically. Using this value, the equilibrium concentrations are

$$[N_2O_4] = 0.150 - 0.134 = 0.016 \ M$$

$$[NO_2] = 2(0.134) = 0.268 \ M$$

Notice that if we had used the negative root, -1.26, the equilibrium concentration of NO_2 would be negative. Negative concentrations are impossible, so $x = -1.26$ is not acceptable *for chemical reasons*. In general, whenever you use the quadratic formula in a chemical calculation, one root will be satisfactory and the other will lead to answers that are nonsense.

Checking the Answers

Once again, we can evaluate the reaction quotient using the calculated equilibrium values. When we do this, we obtain $Q = 4.49$, which is acceptably close to the value of K_c given.

Equilibrium problems can be even more complex than the one we have just discussed; however, you will not have to deal with them. In fact, in most of the problems you encounter there will be ways of simplifying the algebra. Only rarely will it even be necessary to resort to the quadratic formula.

_____ **Practice Exercise 14** _____

During an experiment, 0.200 mol of H_2 and 0.200 mol of I_2 were placed into a 1.00 liter vessel where the reaction

$$H_2(g) + I_2(g) \rightleftharpoons 2HI(g)$$

came to equilibrium. For this reaction, $K_c = 49.5$ at the temperature of the experiment. What were the equilibrium concentrations of H_2, I_2, and HI? ◆

Equilibrium Calculations when K_c Is Very Small

Many chemical reactions have equilibrium constants that are either very large or very small. For example, most weak acids have very small values for K_c. Therefore, only very tiny amounts of products form when these weak acids react with water.

When the K_c for a reaction is very small the equilibrium calculations can usually be simplified considerably, as shown in the next example.

EXAMPLE 14.12
Simplifying Equilibrium Calculations for Reactions with Small K_c

Hydrogen, a potential fuel, is found in great abundance in water. Before the hydrogen can be used as a fuel, however, it must be separated from the oxygen; the water must be split into H_2 and O_2. One possibility is thermal decomposition, but this requires very high temperatures. Even at 1000 °C, $K_c = 7.3 \times 10^{-18}$ for the reaction

$$2H_2O(g) \rightleftharpoons 2H_2(g) + O_2(g)$$

If at 1000 °C the H_2O concentration in a reaction vessel is set initially at 0.100 M, what will the H_2 concentration be when the reaction reaches equilibrium?

Solution: We know that at equilibrium

$$\frac{[H_2]^2[O_2]}{[H_2O]^2} = 7.3 \times 10^{-18}$$

We now set up the concentration table.

Initial Concentrations

The initial concentration of H_2O is 0.100 M; those of H_2 and O_2 are both 0.0 M.

Changes in Concentrations

We know the changes must be in the same ratio as the coefficients in the balanced equation, so we place x's in this row with coefficients equal to those in the chemical equation. Because there are no products present initially, their changes must be positive and the change for the water must be negative. This gives

	$2H_2O(g) \rightleftharpoons 2H_2(g) + O_2(g)$		
Initial concentrations (M)	0.100	0.0	0.0
Changes in concentrations (M)	$-2x$	$+2x$	$+x$
Equilibrium concentrations (M)	$0.100 - 2x$	$2x$	x

Notice that we have chosen the coefficients of x to be the same as the coefficients in the balanced chemical equation.

When we substitute the equilibrium quantities into the mass action expression we get

$$\frac{(2x)^2 x}{(0.100 - 2x)^2} = 7.3 \times 10^{-18}$$

or

$$(2x)^2 x = (4x^2)x = 4x^3$$

$$\frac{4x^3}{(0.100 - 2x)^2} = 7.3 \times 10^{-18}$$

This is a *cubic equation* (one term involves x^3) and can be rather difficult to solve unless we can simplify it. In this instance we are able to do so because the very small value of K_c tells us that hardly any of the H_2O will decompose. Whatever the actual value of x, we know that it is going to be very small. This means that $2x$ will also be small, so when this tiny value is subtracted from 0.100, the result will still be very, very close to 0.100. We will make the assumption, then, that the term in the denominator will be essentially unchanged from 0.100 by subtracting $2x$ from it; that is, we will assume that $(0.100 - 2x) \approx 0.100$.

Even before we solve the problem, we know that hardly any H_2 and O_2 will be formed, because K_c is so small.

This assumption greatly simplifies the math. We now have

$$\frac{4x^3}{(0.100)^2} = 7.3 \times 10^{-18}$$

$$4x^3 = (0.0100)(7.3 \times 10^{-18}) = 7.3 \times 10^{-20}$$

$$x^3 = 1.8 \times 10^{-20}$$

$$x = 2.6 \times 10^{-7}$$

Notice that the value of x that we've obtained is indeed very small. If we double it and subtract the answer from 0.100, we still get 0.100 when we round to the correct number of significant figures (the third decimal place).

Always be sure to check your assumptions when solving a problem of this kind.

$$0.100 - 2x = 0.100 - 2(2.6 \times 10^{-7}) = 0.09999948$$

$$= 0.100 \qquad \text{(rounded correctly)}$$

This check verifies that our assumption was valid. Finally, we have to obtain the H_2 concentration. Our table gives

$$[H_2] = 2x$$

Therefore,

$$[H_2] = 2(2.6 \times 10^{-7}) = 5.2 \times 10^{-7} \, M$$

Tools

Simplifying
assumptions
when
K is very small

The simplifying assumption made in the preceding example is valid because a very small number is subtracted from a much larger one. We could also have neglected x (or $2x$) if it were a very small number that was being added to a much larger one. Remember that you can only neglect an x that's *added* or *subtracted;* you can never drop an x that occurs as a multiplying or dividing factor.

Some examples are

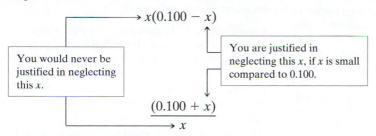

$$x(0.100 - x)$$

You would never be justified in neglecting this x.

You are justified in neglecting this x, if x is small compared to 0.100.

$$(0.100 + x)$$

$$x$$

As a rule of thumb, you can expect that these simplifying assumptions will be valid if the concentration from which x is subtracted, or to which x is added, is at least 1000 times greater than K. For instance, in the preceding example, $2x$ was subtracted from 0.100. Since 0.100 is much larger than $1000 \times (7.3 \times 10^{-18})$ we expect the assumption $0.100 - 2x \approx 0.100$ to be valid. However, even though the simplifying assumption is expected to be valid, always check to see if it really is after finishing the calculation. If the assumption proves invalid, then some other way to solve the algebra must be found.

Practice Exercise 15

In air at 25 °C and 1 atm, the N_2 concentration is 0.033 M and the O_2 concentration is 0.00810 M. The reaction $N_2(g) + O_2(g) \rightleftharpoons 2NO(g)$ has $K_c = 4.8 \times 10^{-31}$ at 25 °C. Taking the N_2 and O_2 concentrations given above as initial values, calculate the equilibrium NO concentration that should exist in our atmosphere from this reaction at 25 °C. ◆

SUMMARY

Dynamic Equilibrium. When the forward and reverse reactions in a chemical system occur at equal rates, a dynamic equilibrium exists and the concentrations of the reactants and products remain constant. For a given overall chemical composition, the amounts of reactants and products that are present at equilibrium are the same regardless of whether the equilibrium is approached from the direction of pure "reactants," pure "products," or any mixture of them. (In a chemical equilibrium, the terms *reactants* and *products* do not have the usual significance because the reaction is proceeding in both directions simultaneously. Instead, we use *reactants* and *products* simply to identify the substances on the left- and right-hand sides of the equation for the equilibrium.)

The Equilibrium Law. The **mass action expression** is a fraction. The concentrations of the products, raised to powers equal to their coefficients in the chemical equation, are multiplied together in the numerator. The denominator is constructed in the same way from the concentrations of the reactants raised to powers equal to their coefficients. The numerical value of the mass action expression is the **reaction quotient, Q.** At equilibrium, the reaction quotient is equal to the **equilibrium constant, K_c.** If partial pressures of gases are used in the mass action expression, K_p is obtained. The magnitude of the equilibrium

constant is roughly proportional to the extent to which the reaction proceeds to completion when equilibrium is reached.

Manipulating Equilibrium Equations. When an equation is multiplied by a factor n to obtain a new equation, we raise its K to the power n to obtain the K for the new equation. When two equations are added, we multiply their K's to obtain the new K. When an equation is reversed, we take the reciprocal of its K to obtain the new K.

Relating K_p to K_c. The values of K_p and K_c are only equal if the same number of moles of gas are represented on both sides of the chemical equation. When the numbers of moles of gas are different, K_p is related to K_c by the equation $K_p = K_c(RT)^{\Delta n_g}$. Remember to use $R = 0.0821$ L atm mol^{-1} K^{-1} and $T =$ absolute temperature. Also, be careful to calculate Δn_g as the difference between the number of moles of *gaseous* products and the number of moles of *gaseous* reactants in the balanced equation.

Heterogeneous Equilibria. An equilibrium involving substances in more than one phase is a **heterogeneous equilibrium.** The mass action expression for a heterogeneous equilibrium omits concentration terms for pure liquids and/or pure solids.

Le Châtelier's Principle. This principle states that *when an equilibrium is upset, a chemical change occurs in a direction that opposes the disturbing influence and brings the system to equilibrium again.* Adding a reactant or a product causes a reaction to occur that uses up part of what has been added. Removing a reactant or a product causes a reaction that replaces part of what has been removed. Increasing the pressure (by reducing the volume) drives a reaction in the direction of the fewer number of moles of gas. Pressure changes have virtually no effect on equilibria involving only solids and liquids. Raising the temperature causes an equilibrium to shift in an endothermic direction. The value of K increases with increasing temperature for reactions that are endothermic in the forward direction. A change in temperature is the only factor that changes K. Addition of a catalyst has no effect on an equilibrium.

Equilibrium Calculations. The initial concentrations in a chemical system are controlled by the person who combines the chemicals at the start of the reaction. The changes in concentration are determined by the stoichiometry of the reaction. Only equilibrium concentrations satisfy the equilibrium law. When these are used, the mass action expression is equal to K_c. When a change in concentration is expected to be very small compared to the initial concentration, the change may be neglected and the algebraic equation derived from the equilibrium law can be simplified. In general, this simplification is valid if the initial concentration is at least 1000 times larger than K.

In this chapter you have learned the following tools that we apply to various aspects of solving problems dealing with chemical equilibria.

Tool	Function
Mass action expression and the equilibrium law *(page 631)*	To define the relationship that exists between the equilibrium concentrations or partial pressures of the reactants and products in a chemical reaction. Be sure you know how to form the mass action expression from the balanced chemical equation. Writing the equilibrium law is one of the first steps in solving equilibrium problems.
Manipulation of equilibrium equations *(page 633)*	To permit you to determine K for a particular reaction by manipulating or combining one or more different chemical equations and their associated equilibrium constants. Be sure you know the rules that apply for the various operations.
Magnitude of the equilibrium constant *(page 636)*	To ascertain qualitatively the extent to which a reaction proceeds toward completion when equilibrium is reached. The size of K determines whether you are able to apply simplifying assumptions in solving problems.
$K_p = K_c(RT)^{\Delta n_g}$ *(page 637)*	To convert between K_p and K_c. Remember that Δn_g is the change in the number of moles of gas, which is computed using the coefficients of gaseous products and reactants.
Le Châtelier's principle *(page 641)*	To analyze the effects of disturbances on the position of equilibrium.
Concentration table *(page 648)*	To organize the approach to solving problems dealing with chemical equilibria. It incorporates initial concentrations, changes in concentrations, and equilibrium concentration expressions.
Simplifying assumptions *(page 656)*	To simplify the algebra when working equilibrium problems for reactions for which K is very small.

THINKING IT THROUGH

Remember, the goal for each of the following problems is to assemble the available information needed to obtain the answer, state what additional data (if any) are needed, and describe how you would use the data to answer the question. For problems involving unit conversions, list the relationships among the units that are needed to carry out the conversions. Construct the conversion factors that can be formed from these relationships. Then set up the solution to the problem by arranging the conversion factors so the units cancel correctly to give the desired units of the answer.

The problems are divided into two groups. Those in Level 2 are more challenging than those in Level 1 and provide an opportunity to really hone your problem solving skills. Detailed answers to these problems can be found at our Web site: http://www.wiley.com/college/brady.

Level 1 Problems

1. Suppose the following data were collected at 25 °C for the system $NO_2(g) + NO(g) \rightleftharpoons N_2O_3(g)$

Equilibrium Concentrations		
[NO$_2$]	[NO]	[N$_2$O$_3$]
0.020 M	0.033 M	8.6×10^{-3} M
0.040 M	0.0085 M	4.4×10^{-3} M
0.015 M	0.012 M	2.3×10^{-3} M

Explain in detail how you would use these data to show that they obey the principles of the equilibrium law.

2. The reaction $NO_2(g) + NO(g) \rightleftharpoons N_2O(g) + O_2(g)$ has $K_c = 0.914$ at a certain temperature. What manipulations are necessary to determine, at this same temperature, the value of K_c for the following reaction?

$$2N_2O(g) + 2O_2(g) \rightleftharpoons 2NO_2(g) + 2NO(g)$$

3. In the preceding question, is there sufficient information to calculate the value of K_p for the reaction below? If not, what additional data are needed?

$$NO_2(g) + NO(g) \rightleftharpoons N_2O(g) + O_2(g)$$

4. The reaction $NO_2(g) + NO(g) \rightleftharpoons N_2O(g) + O_2(g)$ has $\Delta H° = -42.9$ kJ. List the kinds of changes that would alter the amount of NO_2 present in the system at equilibrium. What could be done to this system *without* changing the amount of NO_2 at equilibrium?

5. At 100 °C, $K_c = 0.135$ for the reaction

$$3H_2(g) + N_2(g) \rightleftharpoons 2NH_3(g)$$

In a reaction mixture at equilibrium at this temperature, [NH$_3$] = 0.030 M and [N$_2$] = 0.50 M. Explain how you can calculate the molar concentration of $H_2(g)$. (Set up the calculation.)

6. In the reaction mixture described in the preceding question, explain how you can calculate the partial pressures of each of the gases in the mixture. (Set up the calculation.)

7. How would you use the results obtained in the preceding question to calculate the value of K_p for the following equilibrium?

$$3H_2(g) + N_2(g) \rightleftharpoons 2NH_3(g)$$

8. The reaction $N_2O_3(g) \rightleftharpoons NO(g) + NO_2(g)$ is to be studied at a certain temperature. A mixture was prepared in which the initial concentrations of NO and NO$_2$ were 0.40 M and 0.60 M, respectively. When equilibrium was reached, the concentration of N_2O_3 was measured to be 0.13 M. What were the equilibrium concentrations of NO and NO$_2$? (Explain each step in the solution to the problem.)

Level 2 Problems

9. One function of the catalysts in an automotive catalytic converter is to promote the decomposition of nitrogen oxides into nitrogen and oxygen. Given that the reaction of $N_2(g)$ with $O_2(g)$ to form $NO(g)$ is endothermic, explain how the functioning of a catalytic converter is related to chemical equilibrium and Le Châtelier's principle.

10. The following are equilibrium constants calculated for the reaction $3H_2(g) + N_2(g) \rightleftharpoons 2NH_3(g)$

t (°C)	K_p	K_c
25	7.11×10^5	4.26×10^8
300	4.61×10^{-9}	1.02×10^{-5}
400	2.62×10^{-10}	8.00×10^{-7}

Explain how you can determine in a simple way how the amount of NH$_3$ at equilibrium varies as the temperature of the reaction mixture is increased.

11. Suppose we begin the reaction

$$2H_2(g) + O_2(g) \rightleftharpoons 2H_2O(g) \qquad K_c = 1.4 \times 10^{17}$$

with the following initial concentrations: [H$_2$] = 0.020 M, [O$_2$] = 0.030 M, and [H$_2$O] = 0.00 M. If we let x equal the number of moles per liter of O$_2$ that react, why are we not justified in neglecting x in expressing the equilibrium concentrations of H$_2$ and O$_2$?

12. In the preceding question, how could the initial concentrations be recalculated so as to make the simplifying assumption valid when the calculation is performed? (*Hint:* The position of equilibrium does not depend on the direction from which equilibrium is approached.)

13. For the equilibrium $3NO_2(g) \rightleftharpoons N_2O_5(g) + NO(g)$, $K_c = 1.0 \times 10^{-11}$. If a 4.00 L container initially holds 0.20 mol of NO_2, how many moles of N_2O_5 will be present when this system reaches equilibrium? (Describe each step in the solution to the problem.)

14. To study the following reaction at 20°C,

$$NO(g) + NO_2(g) + H_2O(g) \rightleftharpoons 2HNO_2(g)$$

a mixture of $NO(g)$, $NO_2(g)$, and $H_2O(g)$ was prepared. For NO, NO_2, and HNO_2, the initial concentrations were as follows: $[NO] = [NO_2] = 2.59 \times 10^{-3}\ M$ and $[HNO_2] = 0.0\ M$. The initial partial pressure of $H_2O(g)$ was 17.5 torr. When equilibrium was reached, the HNO_2 concentration was $4.0 \times 10^{-4}\ M$. Explain in detail how to calculate the equilibrium constant, K_c, for this reaction.

15. In CCl_4 solution Cl_2 reacts with Br_2 to form BrCl according to the reaction:

$$Br_2 + Cl_2 \rightleftharpoons 2BrCl \qquad K_c = 2$$

If the initial concentrations were $[Br_2] = 0.6$ M, $[Cl_2] = 0.4$ M, and $[BrCl] = 0.0$ M, which of the following concentration versus time graphs could represent this reaction? Explain why you rejected each of the other four graphs.

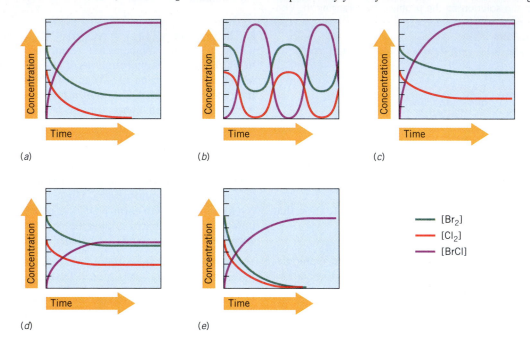

(a)   (b)   (c)

(d)   (e)

— $[Br_2]$
— $[Cl_2]$
— $[BrCl]$

16. Two 1.00 L bulbs are filled with 0.500 atm of $F_2(g)$ and $PF_3(g)$, as illustrated in the figure below. At a particular temperature $K_p = 4.0$ for the reaction of these gases to form $PF_5(g)$

$$F_2(g) + PF_3(g) \rightleftharpoons PF_5(g)$$

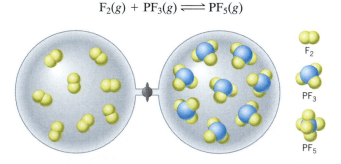

(a) The stopcock between the bulbs is opened and the pressure is observed to fall. Make a sketch of this apparatus showing the gas composition once the pressure is stable.
(b) List any chemical reactions that continue to occur once the pressure is stable.
(c) Suppose all the gas in the left bulb is forced into the bulb on the right and then the stopcock is closed. Make a second sketch to show the composition of the gas mixture after equilibrium is reached. Comment on how the composition for the one-bulb system differs from that for the two-bulb system.

REVIEW QUESTIONS

General
14.1 Sketch a graph showing how the concentrations of the reactants and products of a typical chemical reaction vary with time during the course of the reaction. Assume no products are present at the start of the reaction.

14.2 What is meant when we say that chemical reactions are *reversible*?

Mass Action Expression, K_p, and K_c
14.3 What is an *equilibrium law*?

14.4 How is the term *reaction quotient* defined?

14.5 Under what conditions does the reaction quotient equal K_c?

14.6 When a chemical equation and its equilibrium constant are given, why is it not necessary to also specify the form of the mass action expression?

14.7 At 225 °C, $K_p = 6.3 \times 10^{-3}$ for the reaction

$$CO(g) + 2H_2(g) \rightleftharpoons CH_3OH(g)$$

Would we expect this reaction to go nearly to completion?

14.8 Here are some reactions and their equilibrium constants.
(a) $2CH_4(g) \rightleftharpoons C_2H_6(g) + H_2(g)$ $K_c = 9.5 \times 10^{-13}$
(b) $CH_3OH(g) + H_2(g) \rightleftharpoons CH_4(g) + H_2O(g)$
                                     $K_c = 3.6 \times 10^{20}$
(c) $H_2(g) + Br_2(g) \rightleftharpoons 2HBr(g)$ $K_c = 2.0 \times 10^9$
Arrange these reactions in order of their increasing tendency to go toward completion.

Converting between K_p and K_c

14.9 State the equation relating K_p to K_c and define all terms. Which is the only value of R that can be properly used in this equation?

14.10 Use the ideal gas law to show that the partial pressure of a gas is directly proportional to its molar concentration. What is the proportionality constant?

Heterogeneous Equilibria

14.11 What is the difference between a *heterogeneous equilibrium* and a *homogeneous equilibrium*?

14.12 Why do we omit the concentrations of pure liquids and solids from the mass action expressions of heterogeneous reactions?

Le Châtelier's Principle

14.13 State Le Châtelier's principle in your own words.

14.14 How will the equilibrium

$$Heat + CH_4(g) + 2H_2S(g) \rightleftharpoons CS_2(g) + 4H_2(g)$$

be affected by the following?

(a) The addition of $CH_4(g)$.
(b) The addition of $H_2(g)$.
(c) The removal of $CS_2(g)$.
(d) A decrease in the volume of the container.
(e) An increase in temperature.

14.15 The reaction $CO(g) + 2H_2(g) \rightleftharpoons CH_3OH(g)$ has $\Delta H° = -18$ kJ. How will the amount of CH_3OH present at equilibrium be affected by the following?

(a) Adding $CO(g)$.
(b) Removing $H_2(g)$.
(c) Decreasing the volume of the container.
(d) Adding a catalyst.
(e) Increasing the temperature.

14.16 Consider the equilibrium

$$N_2O(g) + NO_2(g) \rightleftharpoons 3NO(g) \qquad \Delta H° = +155.7 \text{ kJ}$$

In which direction will this equilibrium be shifted by the following changes?
(a) Addition of N_2O.
(b) Removal of NO_2.
(c) Addition of NO.
(d) Increasing the temperature of the reaction mixture.
(e) Addition of helium gas to the reaction mixture at constant volume.
(f) Decreasing the volume of the container at constant temperature.

14.17 In Review Questions 14.15 and 14.16, which change(s) will alter the value of K_c?

14.18 Consider the equilibrium $2NO(g) + Cl_2(g) \rightleftharpoons 2NOCl(g)$ for which $\Delta H° = -77.07$ kJ. How will the amount of Cl_2 at equilibrium be affected by the following?
(a) Removal of $NO(g)$.
(b) Addition of $NOCl(g)$.
(c) Raising the temperature.
(d) Decreasing the volume of the container.

REVIEW PROBLEMS

Answers to problems whose numbers are printed in color are given in Appendix D.
More challenging problems are marked with asterisks.

Equilibrium Laws for K_p and K_c

14.19 Write the equilibrium law for each of the following reactions in terms of molar concentrations:
(a) $2PCl_3(g) + O_2(g) \rightleftharpoons 2POCl_3(g)$
(b) $2SO_3(g) \rightleftharpoons 2SO_2(g) + O_2(g)$
(c) $N_2H_4(g) + 2O_2(g) \rightleftharpoons 2NO(g) + 2H_2O(g)$
(d) $N_2H_4(g) + 6H_2O_2(g) \rightleftharpoons 2NO_2(g) + 8H_2O(g)$
(e) $SOCl_2(g) + H_2O(g) \rightleftharpoons SO_2(g) + 2HCl(g)$

14.20 Write the equilibrium law for each of the following gaseous reactions in terms of molar concentrations.

(a) $3Cl_2(g) + NH_3(g) \rightleftharpoons NCl_3(g) + 3HCl(g)$
(b) $PCl_3(g) + PBr_3(g) \rightleftharpoons PCl_2Br(g) + PClBr_2(g)$
(c) $NO(g) + NO_2(g) + H_2O(g) \rightleftharpoons 2HNO_2(g)$
(d) $H_2O(g) + Cl_2O(g) \rightleftharpoons 2HOCl(g)$
(e) $Br_2(g) + 5F_2(g) \rightleftharpoons 2BrF_5(g)$

14.21 Write the equilibrium law for the reactions in Problem 14.19 in terms of partial pressures.

14.22 Write the equilibrium law for the reactions in Problem 14.20 in terms of partial pressures.

14.23 Write the equilibrium law for each of the following reactions in aqueous solution.
(a) $Ag^+(aq) + 2NH_3(aq) \rightleftharpoons Ag(NH_3)_2^+(aq)$
(b) $Cd^{2+}(aq) + 4SCN^-(aq) \rightleftharpoons Cd(SCN)_4^{2-}(aq)$

14.24 Write the equilibrium law for each of the following reactions in aqueous solution.
(a) $HClO(aq) + H_2O \rightleftharpoons H_3O^+(aq) + ClO^-(aq)$
(b) $CO_3^{2-}(aq) + HSO_4^-(aq) \rightleftharpoons HCO_3^-(aq) + SO_4^{2-}(aq)$

Manipulating Equilibrium Equations
14.25 At 25 °C, $K_c = 1 \times 10^{-85}$ for the reaction

$$7IO_3^- + 9H_2O + 7H^+ \rightleftharpoons I_2 + 5H_5IO_6$$

What is the value of K_c for the following reaction?

$$I_2 + 5H_5IO_6 \rightleftharpoons 7IO_3^- + 9H_2O + 7H^+$$

14.26 Use the following equilibria

$$2CH_4(g) \rightleftharpoons C_2H_6(g) + H_2(g) \qquad K_c = 9.5 \times 10^{-13}$$

$$CH_4(g) + H_2O(g) \rightleftharpoons CH_3OH(g) + H_2(g)$$

$$K_c = 2.8 \times 10^{-21}$$

to calculate K_c for the reaction

$$2CH_3OH(g) + H_2(g) \rightleftharpoons C_2H_6(g) + 2H_2O(g)$$

14.27 Write the equilibrium law for each of the following reactions in terms of molar concentrations.
(a) $H_2(g) + Cl_2(g) \rightleftharpoons 2HCl(g)$
(b) $\frac{1}{2}H_2(g) + \frac{1}{2}Cl_2(g) \rightleftharpoons HCl(g)$
How does K_c for reaction (a) compare with K_c for reaction (b)?

14.28 Write the equilibrium law for the reaction

$$2HCl(g) \rightleftharpoons H_2(g) + Cl_2(g)$$

How does K_c for this reaction compare with K_c for reaction (a) in the preceding problem?

Converting between K_p and K_c
14.29 A 345 mL container holds NH_3 at a pressure of 745 torr and a temperature of 45 °C. What is the molar concentration of ammonia in the container?

14.30 In a certain container at 145 °C the concentration of water vapor is 0.0200 M. What is the partial pressure of H_2O in the container?

14.31 For which of the following reactions does $K_p = K_c$?
(a) $2H_2(g) + C_2H_2(g) \rightleftharpoons C_2H_6(g)$
(b) $N_2(g) + O_2(g) \rightleftharpoons 2NO(g)$
(c) $2NO(g) + O_2(g) \rightleftharpoons 2NO_2(g)$

14.32 For which of the following reactions does $K_p = K_c$?
(a) $CO_2(g) + H_2(g) \rightleftharpoons CO(g) + H_2O(g)$
(b) $PCl_3(g) + Cl_2(g) \rightleftharpoons PCl_5(g)$
(c) $N_2O_4(g) \rightleftharpoons 2NO_2(g)$

14.33 The reaction $CO(g) + 2H_2(g) \rightleftharpoons CH_3OH(g)$ has $K_p = 6.3 \times 10^{-3}$ at 225 °C. What is the value of K_c at this temperature?

14.34 The reaction $HCHO_2(g) \rightleftharpoons CO(g) + H_2O(g)$ has $K_p = 1.6 \times 10^6$ at 400 °C. What is the value of K_c for this reaction at this temperature?

14.35 The reaction $N_2O(g) + NO_2(g) \rightleftharpoons 3NO(g)$ has $K_c = 4.2 \times 10^{-4}$ at 500 °C. What is the value of K_p at this temperature?

14.36 One possible way of removing NO from the exhaust of a gasoline engine is to cause it to react with CO in the presence of a suitable catalyst.

$$2NO(g) + 2CO(g) \rightleftharpoons N_2(g) + 2CO_2(g)$$

At 300 °C, this reaction has $K_c = 2.2 \times 10^{59}$. What is K_p at 300 °C?

14.37 At 773 °C the reaction

$$CO(g) + 2H_2(g) \rightleftharpoons CH_3OH(g)$$

has $K_c = 0.40$. What is K_p at this temperature?

14.38 The reaction $COCl_2(g) \rightleftharpoons CO(g) + Cl_2(g)$ has $K_p = 4.6 \times 10^{-2}$ at 395 °C. What is K_c at this temperature?

Heterogeneous Equilibria
14.39 Calculate the molar concentration of water in (a) 18.0 mL of H_2O, (b) 100.0 mL of H_2O, and (c) 1.00 L of H_2O. Assume that the density of water is 1.00 g/mL.

14.40 The density of sodium chloride is 2.164 g cm^{-3}. What is the molar concentration of NaCl in a 12.0 cm^3 sample of pure NaCl? What is the molar concentration of NaCl in a 25.0 g sample of pure NaCl?

14.41 Write the equilibrium law corresponding to K_c for each of the following heterogeneous reactions.
(a) $2C(s) + O_2(g) \rightleftharpoons 2CO(g)$
(b) $2NaHSO_3(s) \rightleftharpoons Na_2SO_3(s) + H_2O(g) + SO_2(g)$
(c) $2C(s) + 2H_2O(g) \rightleftharpoons CH_4(g) + CO_2(g)$
(d) $CaCO_3(s) + 2HF(g) \rightleftharpoons CaF_2(s) + H_2O(g) + CO_2(g)$
(e) $CuSO_4 \cdot 5H_2O(s) \rightleftharpoons CuSO_4(s) + 5H_2O(g)$

14.42 Write the equilibrium law corresponding to K_c for each of the following heterogeneous reactions.
(a) $CaCO_3(s) + SO_2(g) \rightleftharpoons CaSO_3(s) + CO_2(g)$
(b) $AgCl(s) + Br^-(aq) \rightleftharpoons AgBr(s) + Cl^-(aq)$
(c) $Cu(OH)_2(s) \rightleftharpoons Cu^{2+}(aq) + 2OH^-(aq)$
(d) $Mg(OH)_2(s) \rightleftharpoons MgO(s) + H_2O(g)$
(e) $3CuO(s) + 2NH_3(g) \rightleftharpoons 3Cu(s) + N_2(g) + 3H_2O(g)$

14.43 The heterogeneous reaction $2HCl(g) + I_2(s) \rightleftharpoons 2HI(g) + Cl_2(g)$ has $K_c = 1.6 \times 10^{-34}$ at 25 °C. Suppose 0.100 mol of HCl and solid I_2 is placed in a 1.00 L container. What will be the equilibrium concentrations of HI and Cl_2 in the container?

14.44 At 25 °C, $K_c = 360$ for the reaction

$$AgCl(s) + Br^-(aq) \rightleftharpoons AgBr(s) + Cl^-(aq)$$

If solid AgCl is added to a solution containing 0.10 M Br$^-$, what will be the equilibrium concentrations of Br$^-$ and Cl$^-$?

Equilibrium Calculations

14.45 At a certain temperature, $K_c = 0.18$ for the equilibrium $PCl_3(g) + Cl_2(g) \rightleftharpoons PCl_5(g)$. Suppose a reaction vessel at this temperature contained these three gases at the following concentrations: $[PCl_3] = 0.0420\ M$, $[Cl_2] = 0.0240\ M$, $[PCl_5] = 0.00500\ M$.
(a) Is the system in a state of equilibrium?
(b) If not, which direction will the reaction have to proceed to get to equilibrium?

14.46 At 460 °C, the reaction

$$SO_2(g) + NO_2(g) \rightleftharpoons NO(g) + SO_3(g)$$

has $K_c = 85.0$. A reaction flask at 460 °C contains these gases at the following concentrations: $[SO_2] = 0.00250\ M$, $[NO_2] = 0.00350\ M$, $[NO] = 0.0250\ M$, $[SO_3] = 0.0400\ M$.
(a) Is the reaction at equilibrium?
(b) If not, which way will the reaction have to proceed to arrive at equilibrium?

14.47 At a certain temperature, the reaction

$$CO(g) + 2H_2(g) \rightleftharpoons CH_3OH(g)$$

has $K_c = 0.500$. If a reaction mixture at equilibrium contains 0.180 M CO and 0.220 M H$_2$, what is the concentration of CH$_3$OH?

14.48 $K_c = 64$ for the reaction

$$N_2(g) + 3H_2(g) \rightleftharpoons 2NH_3(g)$$

at a certain temperature. Suppose it was found that an equilibrium mixture of these gases contained 0.360 M NH$_3$ and 0.0192 M N$_2$. What was the concentration of H$_2$ in the mixture?

14.49 At 773 °C, a reaction mixture of CO(g), H$_2$(g), and CH$_3$OH(g) was allowed to come to equilibrium. The following equilibrium concentrations were then measured: $[CO] = 0.105\ M$, $[H_2] = 0.250\ M$, $[CH_3OH] = 0.00261\ M$. Calculate K_c for the reaction

$$CO(g) + 2H_2(g) \rightleftharpoons CH_3OH(g)$$

14.50 Ethylene, C$_2$H$_4$, and water react under appropriate conditions to give ethanol. The reaction is

$$C_2H_4(g) + H_2O(g) \rightleftharpoons C_2H_5OH(g)$$

An equilibrium mixture of these gases at a certain temperature had the following concentrations: $[C_2H_4] = 0.0148\ M$, $[H_2O] = 0.0336\ M$, $[C_2H_5OH] = 0.180\ M$. What is the value of K_c?

14.51 At high temperature, 2.00 mol of HBr was placed in a 4.00 L container where it decomposed to give the equilibrium $2HBr(g) \rightleftharpoons H_2(g) + Br_2(g)$. At equilibrium the concentration of Br$_2$ was measured to be 0.0955 M. What is K_c for this reaction at this temperature?

14.52 A 0.050 mol sample of formaldehyde vapor, CH$_2$O, was placed in a heated 500 mL vessel and some of it decomposed. The reaction is $CH_2O(g) \rightleftharpoons H_2(g) + CO(g)$. At equilibrium, the CH$_2$O($g$) concentration was 0.066 mol L^{-1}. Calculate the value of K_c for this reaction.

14.53 The reaction $NO_2(g) + NO(g) \rightleftharpoons N_2O(g) + O_2(g)$ reached equilibrium at a certain high temperature. Originally, the reaction vessel contained the following initial concentrations: $[N_2O] = 0.184\ M$, $[O_2] = 0.377\ M$, $[NO_2] = 0.0560\ M$, and $[NO] = 0.294\ M$. The concentration of NO$_2$, the only colored gas in the mixture, was monitored by following the intensity of the color. At equilibrium, the NO$_2$ concentration had become 0.118 M. What is the value of K_c for this reaction at this temperature?

14.54 At 25 °C, 0.0560 mol O$_2$ and 0.020 mol N$_2$O were placed in a 1.00 L container where the following equilibrium was established.

$$2N_2O(g) + 3O_2(g) \rightleftharpoons 4NO_2(g)$$

At equilibrium, the NO$_2$ concentration was 0.020 M. What is the value of K_c for this reaction?

14.55 At 25 °C, $K_c = 0.145$ for the following reaction in the solvent CCl$_4$: $2BrCl \rightleftharpoons Br_2 + Cl_2$. If the initial concentration of BrCl in the solution is 0.050 M, what will be the equilibrium concentrations of Br$_2$ and Cl$_2$ be?

14.56 At 25 °C, $K_c = 0.145$ for the following reaction in the solvent CCl$_4$: $2BrCl \rightleftharpoons Br_2 + Cl_2$. If the initial concentrations of Br$_2$ and Cl$_2$ are each 0.0250 M, what will their equilibrium concentrations be?

14.57 The equilibrium constant, K_c, for the reaction

$$SO_3(g) + NO(g) \rightleftharpoons NO_2(g) + SO_2(g)$$

was found to be 0.500 at a certain temperature. If 0.240 mol of SO$_3$ and 0.240 mol of NO are placed in a 2.00 L container and allowed to react, what will be the equilibrium concentration of each gas?

14.58 For the reaction in the preceding problem, a reaction mixture is prepared in which 0.120 mol NO$_2$ and 0.120 mol SO$_2$ are placed in a 1.00 L vessel. After the system reaches equilibrium, what will be the equilibrium concentrations of all four gases? How do these equilibrium values compare to those calculated in Problem 14.57? Account for your observation.

14.59 At a certain temperature the reaction

$$CO(g) + H_2O(g) \rightleftharpoons CO_2(g) + H_2(g)$$

has $K_c = 0.400$. Exactly 1.00 mol of each gas was placed in a 100 L vessel and the mixture underwent reaction. What was the equilibrium concentration of each gas?

14.60 At 25 °C, $K_c = 0.145$ for the following reaction in the solvent CCl_4: $2BrCl \rightleftharpoons Br_2 + Cl_2$. If the initial concentration of each substance in a solution is 0.0400 M, what will their equilibrium concentrations be?

14.61 The reaction $2HCl(g) \rightleftharpoons H_2(g) + Cl_2(g)$ has $K_c = 3.2 \times 10^{-34}$ at 25 °C. If a reaction vessel contains initially 0.0500 mol L^{-1} of HCl and then reacts to reach equilibrium, what will be the concentrations of H_2 and Cl_2?

14.62 At 200 °C, $K_c = 1.4 \times 10^{-10}$ for the reaction

$$N_2O(g) + 2NO_2(g) \rightleftharpoons 3NO(g)$$

If 0.200 mol of N_2O and 0.400 mol NO_2 are placed in a 4.00 L container, what would the NO concentration be if this equilibrium were established?

14.63 At 2000 °C, the decomposition of CO_2,

$$2CO_2(g) \rightleftharpoons 2CO(g) + O_2(g)$$

has $K_c = 6.4 \times 10^{-7}$. If a 1.00 L container holding 1.0×10^{-2} mol of CO_2 is heated to 2000 °C, what will be the concentration of CO at equilibrium?

14.64 At 500 °C, the decomposition of water into hydrogen and oxygen, $2H_2O(g) \rightleftharpoons 2H_2(g) + O_2(g)$, has $K_c = 6.0 \times 10^{-28}$. How many moles of H_2 and O_2 are present at equilibrium in a 5.00 L reaction vessel at this temperature if the container originally held 0.015 mol H_2O?

*14.65 At a certain temperature, $K_c = 0.18$ for the equilibrium $PCl_3(g) + Cl_2(g) \rightleftharpoons PCl_5(g)$. If 0.026 mol of PCl_5 is placed in a 2.00 L vessel at this temperature, what will the concentration of PCl_3 be at equilibrium?

*14.66 At 460 °C, the reaction

$$SO_2(g) + NO_2(g) \rightleftharpoons NO(g) + SO_3(g)$$

has $K_c = 85.0$. Suppose 0.100 mol of SO_2, 0.0600 mol of NO_2, 0.0800 mol of NO, and 0.120 mol of SO_3 are placed in a 10.0 L container at this temperature. What will the concentrations of all the gases be when the system reaches equilibrium?

*14.67 At a certain temperature, $K_c = 0.500$ for the reaction

$$SO_3(g) + NO(g) \rightleftharpoons NO_2(g) + SO_2(g)$$

If 0.100 mol SO_3 and 0.200 mol NO are placed in a 2.00 L container and allowed to come to equilibrium, what will the NO_2 and SO_2 concentrations be?

*14.68 At 25 °C, $K_c = 0.145$ for the following reaction in the solvent CCl_4: $2BrCl \rightleftharpoons Br_2 + Cl_2$. A solution was prepared with the following initial concentrations: $[BrCl] = 0.0400 M$, $[Br_2] = 0.0300 M$, and $[Cl_2] = 0.0200 M$. What will their equilibrium concentrations be?

*14.69 At a certain temperature, $K_c = 4.3 \times 10^5$ for the reaction

$$HCHO_2(g) \rightleftharpoons CO(g) + H_2O(g)$$

If 0.200 mol of $HCHO_2$ is placed in a 1.00 L vessel, what will be the concentrations of CO and H_2O when the system reaches equilibrium?

*14.70 The reaction $H_2(g) + Br_2(g) \rightleftharpoons 2HBr(g)$ has $K_c = 7.9 \times 10^{18}$ at 25 °C. If 0.100 mol of H_2 and 0.200 mol of Br_2 are placed in a 10.0 L container, what will all the equilibrium concentrations be at 25 °C?

ADDITIONAL EXERCISES

14.71 The reaction $N_2O_4(g) \rightleftharpoons 2NO_2(g)$ has $K_p = 0.140$ at 25 °C. In a reaction vessel containing these gases in equilibrium at this temperature, the partial pressure of N_2O_4 was 0.250 atm.
(a) What was the partial pressure of the NO_2 in the reaction mixture?
(b) What was the total pressure of the mixture of gases?

14.72 The following reaction in aqueous solution has $K_c = 1 \times 10^{-85}$ at a temperature of 25 °C.

$$7IO_3^- + 9H_2O + 7H^+ \rightleftharpoons I_2 + 5H_5IO_6$$

What is the equilibrium law for this reaction?

*14.73 At a certain temperature, $K_c = 0.914$ for the reaction

$$NO_2(g) + NO(g) \rightleftharpoons N_2O(g) + O_2(g)$$

A mixture was prepared containing 0.200 mol NO_2, 0.300 mol NO, 0.150 mol N_2O, and 0.250 mol O_2 in a 4.00 L container. What will be the equilibrium concentrations of each gas?

*14.74 At 400 °C, $K_c = 2.9 \times 10^4$ for the reaction

$$HCHO_2(g) \rightleftharpoons CO(g) + H_2O(g)$$

A mixture was prepared with the following initial concentrations: $[CO] = 0.20 M$, $[H_2O] = 0.30 M$. No formic acid, $HCHO_2$, was initially present. What was the equilibrium concentration of $HCHO_2$?

*14.75 At 27 °C, $K_p = 1.5 \times 10^{18}$ for the reaction

$$3NO(g) \rightleftharpoons N_2O(g) + NO_2(g)$$

If 0.030 mol of NO were placed in a 1.00 L vessel and this equilibrium were established, what would be the equilibrium concentrations of NO, N_2O, and NO_2?

14.76 Consider the equilibrium

$$2NaHSO_3(s) \rightleftharpoons Na_2SO_3(s) + H_2O(g) + SO_2(g)$$

How will the position of equilibrium be affected by the following?

(a) Adding $NaHSO_3$ to the reaction vessel.
(b) Removing Na_2SO_3 from the reaction vessel.
(c) Adding H_2O to the reaction vessel.
(d) Increasing the volume of the reaction vessel.

*14.77 For the reaction below, $K_p = 1.6 \times 10^6$ at 400 °C.

$$HCHO_2(g) \rightleftharpoons CO(g) + H_2O(g)$$

A mixture of $CO(g)$ and $H_2O(l)$ was prepared in a 2.00 L reaction vessel at 25 °C in which the pressure of CO was 0.177 atm. The mixture also contained 0.391 g of H_2O. The vessel was sealed and heated to 400 °C. When equilibrium was reached, what was the partial pressure of the $HCHO_2$?

*14.78 At a certain temperature, $K_c = 0.914$ for the reaction

$$NO_2(g) + NO(g) \rightleftharpoons N_2O(g) + O_2(g)$$

A mixture was prepared containing 0.200 mol of each gas in a 5.00 L container. What will be the equilibrium concentration of each gas? How will the concentrations then change if 0.050 mol of NO_2 is added to the equilibrium mixture?

14.79 At a certain temperature, $K_c = 0.914$ for the reaction

$$NO_2(g) + NO(g) \rightleftharpoons N_2O(g) + O_2(g)$$

Equal amounts of NO and NO_2 are to be placed in a 5.00 L container until the N_2O concentration at equilibrium is 0.050 M. How many moles of NO and NO_2 must be placed in the container?

Those who value game fish, whether bears or people, have a stake in the acidity of lakes, rivers, and streams. If the waters are ever so slightly too acidic, trout and salmon hatchlings cannot survive.

Acids and Bases: A Second Look

This Chapter in Context The Arrhenius view of acids and bases, discussed in Chapter 4, is that *acids* give hydrogen ions and *bases* give hydroxide ions in water (or are other substances that can neutralize hydrogen ions). There are broader and more useful views, however, so we now return to acid–base chemistry to learn about them. We'll also study how trends in the strengths of acids and bases correlate with the periodic table.

15.1 Brønsted Acids and Bases

An acid–base neutralization, according to Arrhenius, is a reaction in which an acid and a base combine to produce water and a salt. However, many reactions resemble neutralizations without involving H_3O^+, OH^-, or even H_2O. For example, a white cloud of ammonium chloride crystals forms when ammonia and hydrogen chloride gases are allowed to mix in air (see Figure 15.1).

$$NH_3(g) + HCl(g) \longrightarrow NH_4Cl(s)$$

Here, an acid and an acid neutralizer combine to give a salt, so the reaction "looks" like a neutralization, yet all that happens is that H^+ moves from HCl to NH_3. The newly formed ions, NH_4^+ and Cl^-, combine to give the salt, ammonium chloride. Water is not involved.

You'll recall that another name for H^+ is *proton*, because H^+, strictly speaking, is a hydrogen atom minus its electron. Thus H^+ is simply the *nucleus* of a hydrogen atom, a lone proton. So a *proton* transfer occurs when molecules of ammonia and hydrogen chloride mix in air. As ammonium ions and chloride ions form, they attract each other, gather, and settle into places in crystals of ammonium chloride.

proton transfers
from Cl to N

The first curved arrow signifies that the NH_3 molecule will remove H from HCl, using the electron pair on N for the new bond. The second curved arrow denotes the breaking of the H—Cl bond with the bonding electron pair going to the Cl^- ion.

Brønsted Concept

Johannes Brønsted[1] is credited with being the first to see clearly that the important event in most acid–base reactions is simply the transfer of a proton from one species to another. Brønsted proposed, therefore, that *acids* be redefined as species that donate protons and *bases* as species that accept protons. The heart of the *Brønsted concept of acids and bases* is that *acid–base reactions are proton transfer reactions.* Brønsted's definitions are therefore very simple.

> **Brønsted's Definitions of Acids and Bases**
> An **acid** is a proton donor.
> A **base** is a proton acceptor.

Accordingly, hydrogen chloride is an acid because when it reacts with ammonia, HCl molecules donate protons to NH_3 molecules. Similarly, ammonia is a base because NH_3 molecules accept protons.

Even when water is the solvent, chemists use the Brønsted definitions more often than those of Arrhenius. Thus, the reaction between hydrogen chloride and water to form hydronium ion (H_3O^+) and chloride ion (Cl^-), which is another proton transfer reaction, is clearly a Brønsted acid–base reaction. Molecules of HCl are the acid in this reaction, and water molecules are the base. HCl molecules collide with water molecules and during the collisions protons transfer.

Figure 15.1 *The reaction of gaseous HCl with gaseous NH_3.* As each gas escapes from its concentrated aqueous solution and mingles with the other, a cloud of microcrystals of NH_4Cl forms above the bottles.

$$H_2O \;+\; HCl(g) \longrightarrow H_3O^+(aq) \;+\; Cl^-(aq)$$
base   acid

or

water   hydrogen   collision "complex"   hydronium   chloride
         chloride   of proper orientation   ion   ion
                        and energy for
                        proton transfer

Conjugate Acids and Bases

Johannes Brønsted (1879–1947) was a Danish chemist.

Under the Brønsted view, it is useful to consider any acid–base reaction as a chemical equilibrium, having both a forward and a reverse reaction. We first encountered a chemical equilibrium in a discussion of weak acids on page 168. Let's examine it again in the light of Brønsted's definitions. Formic acid, $HCHO_2$, is a weak acid, so we represent its ionization as a chemical equilibrium in which water is not just a solvent but also a chemical reactant, a proton acceptor.[2]

---

[1]Thomas Lowry (1874–1936), a British scientist working independently of Brønsted, developed many of the same ideas but did not carry them as far. Some references call this view of acids and bases the Brønsted–Lowry theory.

[2]Just as we have represented acetic acid as $HC_2H_3O_2$, placing the ionizable or acidic hydrogen first (as we do in HCl or HNO_3), so we give the formula of formic acid as $HCHO_2$ and the formula of the formate ion as CHO_2^-. As the margin comment shows, however, the acidic hydrogen in formic acid resides on O, not on C. Many chemists prefer to write formic acid as CHO_2H.

$$HCHO_2(aq) + H_2O \rightleftharpoons H_3O^+(aq) + CHO_2^-(aq)$$

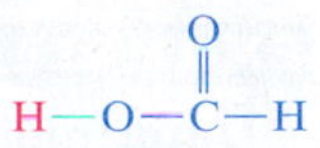

In the forward reaction, a formic acid molecule donates a proton to the water molecule and changes to a formate ion, CHO_2^- (see Figure 15.2a). Thus $HCHO_2$ behaves as a Brønsted acid, a proton donor. Because water accepts this proton from $HCHO_2$, water behaves as a Brønsted base, a proton acceptor.

Now let's look at the reverse reaction (see Figure 15.2b). In it, H_3O^+ behaves as a Brønsted acid because it donates a proton to the CHO_2^- ion. The CHO_2^- ion behaves as a Brønsted base by accepting the proton.

The equilibrium involving $HCHO_2$, H_2O, H_3O^+, and CHO_2^- is typical of proton transfer equilibria in general in that we can identify *two* acids (e.g., $HCHO_2$ and H_3O^+) and *two* bases (e.g., H_2O and CHO_2^-). Notice that in the aqueous formic acid equilibrium, the acid on the right of the arrows (H_3O^+) is formed from the base on the left (H_2O), and the base on the right (CHO_2^-) is formed from the acid on the left ($HCHO_2$).

Two substances, like H_3O^+ and H_2O, that differ from each other by only *one* proton are referred to as a **conjugate acid–base pair.** H_3O^+ and H_2O are thus such a pair. One member of the pair is called the **conjugate acid** because it is the proton donor of the two. The other member is the **conjugate base,** because it is the pair's proton acceptor. We say that H_3O^+ is the conjugate acid of H_2O, and H_2O is the conjugate base of H_3O^+. Notice that the acid member of the pair has one more H^+ than the base member.

The pair $HCHO_2$ and CHO_2^- is the other conjugate acid–base pair in the aqueous formic acid equilibrium. $HCHO_2$ has one more H^+ than CHO_2^-, so the conjugate acid of CHO_2^- is $HCHO_2$; the conjugate base of $HCHO_2$ is CHO_2^-. One way to highlight the two members of a conjugate acid–base pair in an equilibrium equation is to connect them by a line.

conjugate pair

$$HCHO_2 + H_2O \rightleftharpoons H_3O^+ + CHO_2^-$$

acid   base   acid   base

conjugate pair

In any Brønsted acid–base equilibrium, there are invariably *two* conjugate acid–base pairs. We must learn how to pick them out of an equation by inspection and to write them from formulas.

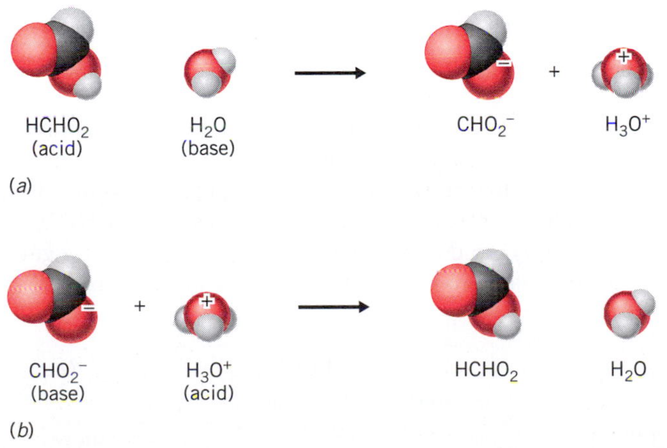

Chemists often use the terms *proton* and *hydrogen ion* interchangeably in discussions of acid–base chemistry.

Notice how the conjugate acid always has one more H^+ than the conjugate base.

$$\begin{matrix} & O \\ & \| \\ H-O-&C-H \end{matrix}$$
formic acid ($HCHO_2$)
Only the H in red is available
in an acid–base reaction.

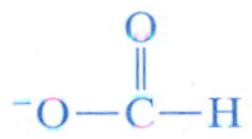

$$\begin{matrix} & O \\ & \| \\ {}^-O-&C-H \end{matrix}$$
formate ion (CHO_2^-)

Figure 15.2 *Brønsted acids and bases in aqueous formic acid.* (a) Formic acid transfers a proton to a water molecule. $HCHO_2$ is the acid and H_2O is the base. (b) When hydronium ion transfers a proton to the CHO_2^- ion, H_3O^+ is the acid and CHO_2^- is the base.

$HCHO_2$ (acid) H_2O (base) CHO_2^- H_3O^+

(a)

CHO_2^- (base) H_3O^+ (acid) $HCHO_2$ H_2O

(b)

EXAMPLE 15.1

Determining the Conjugate Bases of Brønsted Acids

What are the conjugate bases of nitric acid, HNO_3, and the hydrogen sulfate ion, HSO_4^-?

Analysis: A conjugate base is always found by removing one H^+ from a given acid and adjusting the charge on the remainder.

Solution: Removing one H^+ (both the atom and the charge) from HNO_3 leaves NO_3^-. The nitrate ion, NO_3^-, is thus the conjugate base of HNO_3. Deleting an H^+ from HSO_4^- leaves its conjugate base, SO_4^{2-}. (Notice that the charge goes from 1− to 2− because we've removed the positively charged H^+.)

Practice Exercise 1

Write the formula of the conjugate base for each of the following Brønsted acids. (a) H_2O (b) HI (c) HNO_2 (d) H_3PO_4 (e) $H_2PO_4^-$ (f) HPO_4^{2-} (g) H_2 (h) NH_4^+ ◆

EXAMPLE 15.2

Determining the Conjugate Acids of Brønsted Bases

What are the formulas of the conjugate acids of OH^- and PO_4^{3-}?

Analysis: To find a conjugate acid we add one H^+ to the formula of the given base and adjust the charge.

Solution: Adding H^+ to OH^- gives H_2O, the conjugate acid of the hydroxide ion. Adding H^+ to the formula PO_4^{3-} gives HPO_4^{2-}, the conjugate acid of PO_4^{3-}. Always be sure that the electrical charge is correctly shown.

Practice Exercise 2

Write the formula of the conjugate acid for each of the following Brønsted bases. (a) HO_2^- (b) SO_4^{2-} (c) CO_3^{2-} (d) CN^- (e) NH_2^- (f) NH_3 (g) $H_2PO_4^-$ (h) HPO_4^{2-} ◆

EXAMPLE 15.3

Identifying Conjugate Acid–Base Pairs in a Brønsted Acid–Base Reaction

Sodium hydrogen sulfate is used in the manufacture of certain kinds of cement and to clean oxide coatings from metals.

The anion of sodium hydrogen sulfate, HSO_4^-, reacts as follows with the phosphate ion, PO_4^{3-}.

$$HSO_4^-(aq) + PO_4^{3-}(aq) \longrightarrow SO_4^{2-}(aq) + HPO_4^{2-}(aq)$$

Identify the two conjugate acid–base pairs.

Analysis: We look for pairs that differ by only one H^+. *The members of each pair must be on opposite sides of the arrow.*

Solution: Two of the formulas in the equation contain "PO_4," so they must belong to the same conjugate pair. The one with the greater number of hydrogens, HPO_4^{2-}, must be the Brønsted acid, and the other, PO_4^{3-}, must be the Brønsted base. Therefore, one conjugate acid–base pair is HPO_4^{2-} and PO_4^{3-}. The other two ions, HSO_4^- and SO_4^{2-}, belong to the second conjugate acid–base pair; HSO_4^- is the conjugate acid and SO_4^{2-} is the conjugate base.

conjugate pair

$$\underset{\text{acid}}{HSO_4^-(aq)} + \underset{\text{base}}{PO_4^{3-}(aq)} \longrightarrow \underset{\text{base}}{SO_4^{2-}(aq)} + \underset{\text{acid}}{HPO_4^{2-}(aq)}$$

conjugate pair

Is the Answer Reasonable?
We have satisfied the requirements that each conjugate pair has one member on one side of the arrow and the other member on the opposite side of the arrow and that the members of each pair obviously differ from each other by one H^+.

_____ **Practice Exercise 3** _____

One kind of baking powder contains sodium bicarbonate and calcium dihydrogen phosphate. When water is added, a reaction occurs by the following net ionic equation.

$$HCO_3^-(aq) + H_2PO_4^-(aq) \longrightarrow H_2CO_3(aq) + HPO_4^{2-}(aq)$$

Identify the two Brønsted acids and the two Brønsted bases in this reaction. (The H_2CO_3 decomposes to release CO_2, which causes the cake batter to rise.) ◆

_____ **Practice Exercise 4** _____

When some of the strong cleaning agent "trisodium phosphate" is mixed with household vinegar, which contains acetic acid, the following equilibrium is one of the many that are established. (The products are favored.) Identify the pairs of conjugate acids and bases.

$$PO_4^{3-}(aq) + HC_2H_3O_2(aq) \rightleftharpoons HPO_4^{2-}(aq) + C_2H_3O_2^-(aq)$$ ◆

Amphoteric Substances

Some molecules or ions are able to function either as an acid or as a base, depending on the kind of substance mixed with them. For example, in its reaction with hydrogen chloride, water behaves as a *base* because it *accepts* a proton from the HCl molecule.

$$\underset{\text{base}}{H_2O} + \underset{\text{acid}}{HCl(g)} \longrightarrow H_3O^+(aq) + Cl^-(aq)$$

On the other hand, water behaves as an *acid* when it reacts with the weak base ammonia.

$$\underset{\text{acid}}{H_2O} + \underset{\text{base}}{NH_3(aq)} \rightleftharpoons NH_4^+(aq) + OH^-(aq)$$

Here, H_2O *donates* a proton to NH_3 in the forward reaction.

Substances that can be either acids or bases depending on the other substance present are said to be **amphoteric**. Another term is **amphiprotic** to stress that *proton* donating or accepting ability is of central concern.

Amphoteric or amphiprotic substances may be either molecules or ions. For example, anions of acid salts, such as the bicarbonate ion of baking soda, are amphoteric. The HCO_3^- ion can either donate a proton to a base or accept a proton from an acid. Thus, toward the hydroxide ion, the bicarbonate ion is an acid; it donates its proton to OH^-.

$$\underset{\text{acid}}{HCO_3^-(aq)} + \underset{\text{base}}{OH^-(aq)} \longrightarrow CO_3^{2-}(aq) + H_2O$$

Toward hydronium ion, however, HCO_3^- is a base; it accepts a proton from H_3O^+.

$$\underset{\text{base}}{HCO_3^-(aq)} + \underset{\text{acid}}{H_3O^+(aq)} \longrightarrow H_2CO_3(aq) + H_2O$$

From the Greek *amphoteros,* "partly one and partly the other."

Acid salts (Section 2.11), like $NaHCO_3$, $NaHSO_3$, and NaH_2PO_4, are so named because they can neutralize OH^-.

[Recall that H_2CO_3 (carbonic acid) almost entirely decomposes to $CO_2(g)$ and water as it forms.]

<hr>

Practice Exercise 5

The anion of sodium monohydrogen phosphate, Na_2HPO_4, is amphoteric. Using H_3O^+ and OH^-, write net ionic equations that illustrate this property. ◆

<hr>

15.2 Strengths of Brønsted Acids and Bases and Periodic Trends

The relative strengths of Brønsted *acids* are assigned according to their abilities to donate protons to one particular acceptor, usually water. Acids that are strong electrolytes in water are also described as *strong acids* because they are strong donors of protons to *water*. Acids that are weak electrolytes are also *weak acids;* they are poor donors of protons to water.

In similar manner, the relative strengths of Brønsted *bases* are assigned according to their abilities to accept and bind protons. The hydroxide ion is a *strong* Brønsted base, for example, because it will strongly bind a proton (in the form of a water molecule) once a donor has given it one. A *weak* Brønsted base is a weak proton binder. The water molecule, for example, is a weak base because the product of its accepting a proton, namely, the hydronium ion, holds H^+ weakly.

Comparing the Acid–Base Strengths of Conjugate Pairs

As we said, in the equilibrium present in an aqueous solution of any weak acid, the equilibrium equation actually shows two acids. One is stronger than the other, and the position of the equilibrium, which lies to the left, tells us which acid is stronger. Let's see how this works. We'll again use the familiar acetic acid equilibrium.

$$\underset{\text{acid}}{HC_2H_3O_2(aq)} + \underset{\text{base}}{H_2O} \rightleftharpoons \underset{\text{acid}}{H_3O^+(aq)} + \underset{\text{base}}{C_2H_3O_2^-(aq)}$$

It is the *chemical nature* of acetic acid not to ionize much.

The two Brønsted acids in this equilibrium are $HC_2H_3O_2$ and H_3O^+; and it's helpful to think of them as competing with each other in donating protons to acceptors. The fact that nearly all potential protons stay on the $HC_2H_3O_2$ molecules, and only a relative few spend their time on the H_3O^+ ions, means that the *hydronium ion is a better proton donor than the acetic acid molecule.* Thus, the hydronium ion is a stronger Brønsted acid than acetic acid, and we inferred this relative acidity from the position of equilibrium.

The acetic acid equilibrium also has two bases, namely, $C_2H_3O_2^-$ and H_2O. Both compete for any available protons. But at equilibrium, most of the protons originally carried by acetic acid are still found on $HC_2H_3O_2$ molecules; relatively few are joined to H_2O in the form of H_3O^+ ions. This means that *acetate ions must be more effective than water molecules at obtaining and holding protons from proton donors.* This is the same as saying that the acetate ion is a stronger base than the water molecule. So this illustrates a relative basicity that we are able to infer from the position of the acetic acid equilibrium.

Notice what our discussion of the two conjugate pairs in the acetic acid equilibrium has brought out.

The position of an acid–base equilibrium favors the weaker acid and base.

$$HC_2H_3O_2(aq) \; + \; H_2O \; \rightleftharpoons \; H_3O^+(aq) \; + \; C_2H_3O_2^-(aq)$$

weaker acid   weaker base   stronger acid   stronger base

The concepts we've just developed allow us now to understand better one of the driving forces for metathesis reactions discussed in Chapter 4, namely, the formation of a weak electrolyte from a solution of strong electrolytes.

Consider what happens when solutions of hydrochloric acid and sodium acetate are mixed. At the moment of mixing, the solution contains high concentrations of H_3O^+ and $C_2H_3O_2^-$ ions. But we've just seen that the hydronium ion (a strong acid) will donate a proton to the acetate ion to make acetic acid (a weak acid). Writing the net ionic equation as an equilibrium (leaving out the spectator ions, Na^+ and Cl^-) gives

When aqueous solutions of hydrochloric acid and sodium acetate are mixed, the vinegary odor of acetic acid can be soon noticed.

$$H_3O^+(aq) + C_2H_3O_2^-(aq) \rightleftharpoons HC_2H_3O_2(aq) + H_2O$$

stronger acid   stronger base   weaker acid   weaker base

We know that the position of equilibrium favors the weaker acid and base, so to reach equilibrium there must be a *net reaction* of the ions on the left to form the molecules on the right.

This analysis provides us with another way of stating the previous generalization.

> *Stronger acids and bases tend to react with each other to produce their weaker conjugates.*

You can see that once we know relative strengths of acids and bases, we can predict reactions.

Reciprocal Relationships within Conjugate Acid–Base Pairs

One aid in predicting relative strengths of acids and bases is the existence of a reciprocal relationship.

> The stronger a Brønsted acid is, the weaker is its conjugate base.

To illustrate, recall that $HCl(g)$ is a very strong Brønsted acid; it's 100% ionized in a dilute aqueous solution.

$$HCl(g) + H_2O \xrightarrow{100\%} H_3O^+(aq) + Cl^-(aq)$$

As we explained in Section 4.5, we don't write double equilibrium arrows for the ionization of a strong acid. By not doing so with $HCl(g)$ is another way of saying that the chloride ion, the conjugate base of $HCl(g)$, must be a very weak Brønsted base. Even in the presence of H_3O^+, a very strong proton donor, chloride ions aren't able to win protons. So HCl, the strong acid, has a particularly weak conjugate base, Cl^-.

There's a matching reciprocal relationship.

The hydronium ion is the strongest proton donor that can exist in water. Any stronger donor, like $HCl(g)$, gives up a proton to a water molecule to form H_3O^+.

> The weaker a Brønsted acid is, the stronger is its conjugate base.

Consider, for example, the conjugate pair, OH^- and O^{2-}. The hydroxide ion is the conjugate acid, and the oxide ion is the conjugate base. But the hydroxide ion must be an *extremely* weak Brønsted acid; in fact, we've known it so far only

The hydroxide ion is the strongest proton acceptor that can exist in water. Any stronger acceptor takes a proton from a water molecule to generate OH^-.

as a base. Given the extraordinary weakness of OH^- as an acid, its conjugate base, the oxide ion, must be an exceptionally strong base. In water, oxide ions are, in fact, such powerful proton acceptors that they are able to win protons quantitatively (100%) even from H_2O molecules, which are very weak proton donors, as we know. (Again, we don't write double equilibrium arrows.)

$$O^{2-} + H_2O \xrightarrow{100\%} 2OH^-$$
$$\text{base} \qquad \text{acid}$$

EXAMPLE 15.4

Using Conjugate Relationships to Predict Equilibrium Positions

Will the substances to the left of the arrows or those to the right be favored in the following equilibrium, given the fact (which may be new to you) that acetic acid is known to be a stronger acid than the hydrogen sulfite ion?

$$HSO_3^-(aq) + C_2H_3O_2^-(aq) \rightleftharpoons HC_2H_3O_2(aq) + SO_3^{2-}(aq)$$

Analysis: We just learned that the position of an acid–base equilibrium favors the weaker acid and base. So all we have to do is identify which acid and base make up the weaker set. When we do this, we'll have discovered which make up the stronger set.

Solution: We'll rewrite the equilibrium equation and use the given fact about acetic acid versus the hydrogen sulfite ion to start the writing of labels.

$$HSO_3^-(aq) + C_2H_3O_2^-(aq) \rightleftharpoons HC_2H_3O_2(aq) + SO_3^{2-}(aq)$$
$$\text{weaker acid} \qquad\qquad\qquad\qquad \text{stronger acid}$$

Now we'll use the reciprocal relationships to label the two bases. The stronger acid must have the weaker conjugate base; the weaker acid must have the stronger conjugate base.

$$HSO_3^-(aq) + C_2H_3O_2^-(aq) \rightleftharpoons HC_2H_3O_2(aq) + SO_3^{2-}(aq)$$
$$\text{weaker acid} \quad \text{weaker base} \qquad \text{stronger acid} \quad \text{stronger base}$$

Finally, the answer flows from the fact that the position of an acid–base equilibrium favors the weaker acid and base. So here the position of the equilibrium lies to the left.

Practice Exercise 6

Given the fact that HSO_4^- is a stronger acid than HPO_4^{2-}, which are favored in the following equilibrium, the substances on the left of the arrows or those on the right?

$$HSO_4^-(aq) + PO_4^{3-}(aq) \rightleftharpoons SO_4^{2-}(aq) + HPO_4^{2-}(aq) \; \blacklozenge$$

Considerable variations in weakness exist among Brønsted acids or bases. We'll next study how we may use the periodic table to help us assign relative strengths of acids (and to remember the assignments). If we know how to do this for *acids*, we'll have also done this for conjugate bases, because of the reciprocal relationships.

Tools

Periodic table and strengths of binary acids

Periodic Trends in the Strengths of Binary Acids

Many of the binary compounds between hydrogen and nonmetals, which we may represent by HX, H_2X, H_3X, etc., are acidic and are called **binary acids.** HCl is a common example, but Table 15.1 lists the others that are acids in water. The strong acids are marked by asterisks. *The names and formulas of the strong binary acids must now be learned* if you've not already done so.

Table 15.1 **Acidic Binary Compounds of Hydrogen and Nonmetals**[a]

Group VIA		Group VIIA	
(H_2O)		HF	Hydrofluoric acid
H_2S	Hydrosulfuric acid	*HCl	Hydrochloric acid
H_2Se	Hydroselenic acid	*HBr	Hydrobromic acid
H_2Te	Hydrotelluric acid	*HI	Hydriodic acid

[a]The *names* are for the aqueous solutions of these compounds. Strong acids are marked with asterisks.

The relative strengths of binary acids correlate with the periodic table in two ways.

> The strengths of the binary acids increase from left to right within the same period.

As we go left to right within a period, the increase in electronegativities causes corresponding H—X bonds to become more polar, making the partial positive charge on H greater. The larger that the $\delta+$ on H is in H—X, the nearer is the H to separating as H^+ and the stronger is the acid. Thus, the electronegativity increases going left to right from S to Cl in period 3, and HCl is a stronger acid than H_2S. A similar increase in electronegativity occurs going left to right in period 2, from O to F, and HF is a stronger acid than H_2O.

The second correlation is as follows:

Electronegativity was discussed in Section 8.5.

> The strengths of binary acids increase from top to bottom within the same group.

Among the binary acids of the halogens, for example, the following is the order of relative acidity.[3]

$$HF < HCl < HBr < HI$$

Thus, HF is the weakest acid in the series, and HI is the strongest.

The identical trend occurs in the series of the binary acids of Group VIA elements, having formulas of the general type H_2X. The farther X is from the top of the group, the stronger H_2X is as an acid. Thus, S lies below O, and H_2S is a stronger acid than H_2O.

These trends in acidity are opposite what we would expect on the basis of trends in electronegativities, which tell us that the H—F bond is more polar than the H—I bond and that the O—H bond is more polar than the H—S bond. Evidently, the relative electronegativity of X is just one factor affecting the acidity of an H—X bond. Another is the strength of the H—X bond.

In analyzing the changes associated with the donation of a proton by an acid, one of the most important factors to be considered, besides the polarity of the bond, is the strength of the H—X bond. Breaking of this bond is essential for the hydrogen to become separated as an H^+ ion, so anything that contributes to variations in bond strength will impact variations in acid strength.

[3]When we compare acid strengths, we compare the abilities of different acids to protonate a particular base. For strong acids such as HCl, HBr, and HI, water is too strong a proton acceptor to permit us to see differences among their proton-donating abilities. All three of these acids are completely ionized in water and so appear to be of equal strength, a phenomenon called the *leveling effect* (the differences are obscured or leveled out). To compare the acidities of these acids, a solvent that is a weaker proton acceptor than water has to be used.

In general, the bond strength is determined to a significant degree by the sizes of the atoms involved. Small atoms form strong covalent bonds whereas large atoms form much weaker bonds. Moving horizontally within a period, atomic size varies relatively little, so the strengths of the H—X bonds are nearly the same. As a result, the most significant influence on acid strength is variations in the polarity of H—X bonds. Within a group, however, there is a relatively large increase in atomic size from one element to the next, which means there is a rapid decrease in the strength of the H—X bonds as we descend a group. Apparently, this decrease in bond strength is enough to more than compensate for the decrease in the polarity of the H—X bonds, and the molecules become better able to release protons as we go down a group. The net effect of the two opposing factors, therefore, is an increase in the strengths of the binary acids as we go from top to bottom in the group.

Practice Exercise 7

Using *only* the periodic table, choose the stronger acid of each pair: (a) H_2Se or HBr, (b) H_2Se or H_2Te, (c) CH_3OH or CH_3SH. ◆

Oxoacids

Strong acids

Acids made of hydrogen, oxygen, and some other element are called **oxoacids** (see Table 15.2). Those that are strong acids in water are marked in the table by asterisks.

A feature common to the structures of all oxoacids is the presence of O—H groups bonded to some central atom. For example, the structures of two oxoacids of the Group VIA elements are

$$H-O-\underset{\underset{O}{\|}}{\overset{\overset{O}{\|}}{S}}-O-H \qquad H-O-\underset{\underset{O}{\|}}{\overset{\overset{O}{\|}}{Se}}-O-H$$

$$\begin{array}{cc} H_2SO_4 & H_2SeO_4 \\ \text{sulfuric acid} & \text{selenic acid} \end{array}$$

Table 15.2 Some Oxoacids of Nonmetals and Metalloids[a]

Group IVA		Group VA		Group VIA		Group VIIA	
H_2CO_3	Carbonic acid	*HNO_3	Nitric acid			HFO	Hypofluorous acid
		HNO_2	Nitrous acid				
		H_3PO_4	Phosphoric acid	*H_2SO_4	Sulfuric acid	*$HClO_4$	Perchloric acid
		H_3PO_3	Phosphorous acid[b]	H_2SO_3	Sulfurous acid	*$HClO_3$	Chloric acid
						$HClO_2$	Chlorous acid
						HClO	Hypochlorous acid
		H_3AsO_4	Arsenic acid	*H_2SeO_4	Selenic acid	*$HBrO_4$	Perbromic acid[c]
		H_3AsO_3	Arsenous acid	H_2SeO_3	Selenous acid	*$HBrO_3$	Bromic acid
						HIO_4 (H_5IO_6)[d]	Periodic acid
						HIO_3	Iodic acid

[a]Strong acids are marked with asterisks.
[b]Phosphorous acid, despite its formula, is only a diprotic acid.
[c]Pure perbromic acid is unstable; a dihydrate is known.
[d]H_5IO_6 is formed from $HIO_4 + 2H_2O$.

When an oxoacid ionizes, the hydrogen that's lost as an H^+ comes from the same kind of bond in every instance, specifically, an O—H bond. The "acidity" of such a hydrogen, meaning the ease with which it's released as H^+, is determined by how the group of atoms attached to the oxygen affects the polarity of the O—H bond. If this group of atoms makes the O—H bond more polar, it will cause the H to come off more easily as H^+ and thereby increase the acidity of the molecule.

The largest that $\delta+$ could be is a full $1+$, which would mean a detached hydrogen ion.

$$\overset{\delta-\quad\;\;\delta+}{G-O-H}$$

If the group of atoms, G, attached to the O—H group is able to draw electron density from the O atom, the O will pull electron density from the O—H bond, thereby making the bond more polar.

It turns out that there are two principal factors that determine how the polarity of the O—H bond is affected. One is the electronegativity of the central atom in the oxoacid; the other is the number of oxygens attached to the central atom.

The Effect of the Electronegativity of the Central Atom

To study the effects of the electronegativity of the central atom, we must compare oxoacids having the same number of oxygens. When we do this, we find that as the electronegativity of the central atom increases, the oxoacid becomes a better proton donor (i.e., a stronger acid). The following diagram illustrates the effect.

$$\overset{\delta-\quad\;\;\delta+}{-X-O-H}$$

As the electronegativity of X increases, electron density is drawn away from O, which draws electron density away from the O—H bond. This makes the bond more polar and makes the molecule a better proton donor.

Because electronegativity increases from bottom to top within a group and from left to right within a period, we can make the following generalization.

> When the central atoms of oxoacids hold the same number of oxygen atoms, the acid strength increases from bottom to top within the group and from left to right within a period.

Tools

Periodic table and strengths of oxoacids

In Group VIA, for example, H_2SO_4 is a stronger acid than H_2SeO_4 because sulfur is more electronegative than selenium. Similarly, among the halogens, acid strength increases for acids with the formula HXO_4 as follows:

$$HIO_4 < HBrO_4 < HClO_4$$

Oxoacid strength
Increases

Increases

Practice Exercise 8

Which is the stronger acid, $HClO_3$ or $HBrO_3$? ◆

The Effect of the Number of Oxygens on the Acidity of Oxoacids

To examine how the number of oxygens attached to the central atom affects the acidity of a molecule, we must compare acids that have the same central atom. When we do this, we find that as the number of *lone oxygens* increases, the oxoacid becomes a better proton donor. Thus, comparing HNO_3 with HNO_2, we find that HNO_3 is the stronger acid. To understand why, let's look at their molecular structures.

nitrous acid   <   nitric acid

Oxygen, as you know, is a very electronegative element and has a strong tendency to pull electron density away from any atom to which it is attached. In an oxoacid, lone oxygens pull electron density away from the central atom, which increases the central atom's ability to draw electron density away from the O—H bond. It's as though the lone oxygens make the central atom more electronegative. It makes sense, therefore, that the more lone oxygens there are attached to a central atom, the more polar will be the O—H bonds of the acid and the stronger will be the acid. Thus, in HNO_3 the two lone oxygens produce a greater effect than the one lone oxygen in HNO_2, and HNO_3 is the stronger acid.

Similar effects are seen among other oxoacids, as well. For example, H_2SO_4 is a stronger acid than H_2SO_3 because there are two lone oxygens attached to the sulfur in H_2SO_4 while there's only one in H_2SO_3.

The HSO_4^- ion is also a stronger acid than the HSO_3^- ion.

H_2SO_4
sulfuric acid   > H_2SO_3
sulfurous acid

Among the halogens, we find the same trend in acid strengths. For the oxoacids of chlorine, for instance, we find this trend.

$$HClO < HClO_2 < HClO_3 < HClO_4$$

Comparing their structures, we have

HClO   HClO₂   HClO₃   HClO₄

This leads to another generalization.

> For a given central atom, the acid strength of an oxoacid increases with the number of oxygens held by the central atom.

The ability of lone oxygens to affect acid strength extends to organic compounds as well. For example, compare the molecules below.

$$
\begin{array}{cc}
\underset{\displaystyle H}{\overset{\displaystyle H\quad H}{\underset{|\ \ \ |}{H-C-C-O-H}}} & \underset{\displaystyle H}{\overset{\displaystyle H\quad O}{\underset{|\ \ \ \parallel}{H-C-C-O-H}}}
\end{array}
$$

The molecule on the left is ethanol, also called ethyl alcohol. In water it is not acidic at all. Replacing the two hydrogens on the carbon adjacent to the OH group with an oxygen, however, yields acetic acid, the molecule on the right. The greater ability of oxygen to pull electron density from the carbon produces a greater polarity of the O—H bond, which is one factor that causes acetic acid to be a better proton donor than ethanol.

Oxoacids and the Effect of the Dispersal of Negative Charge to the Lone Oxygens

In the ionization of an acid, HA, the strength of the acid is analyzed in terms of the position of equilibrium in the reaction

$$ HA + H_2O \rightleftharpoons H_3O^+ + A^- $$

The stronger the acid, the farther to the right is the position of equilibrium. This position of equilibrium is determined by two factors. One is the ability of the acid to push protons onto water molecules (the strength of HA as a proton donor), and the other is the willingness of A^- to accept protons from H_3O^+ (the strength of A^- as a proton acceptor).

For oxoacids, the lone oxygens play a part in determining the basicity of the anion formed in the ionization reaction. Consider the acids H_2SO_4 and H_3PO_4.

$$
\begin{array}{cc}
\underset{\displaystyle O}{\overset{\displaystyle O}{\underset{\parallel}{\overset{\parallel}{H-O-S-O-H}}}} & \underset{\displaystyle O-H}{\overset{\displaystyle O}{\underset{|}{\overset{\parallel}{H-O-P-O-H}}}} \\
H_2SO_4 & H_3PO_4 \\
\text{sulfuric acid} & \text{phosphoric acid}
\end{array}
$$

The percentage ionization of the solute in 1 *M* H_3PO_4 is far less than in 1 *M* H_2SO_4.

As you can see in their structures, H_2SO_4 has two lone oxygens and H_3PO_4 has only one. Following the ionizations of their first protons, three lone oxygens are present in the resulting HSO_4^- anion but only two are in the $H_2PO_4^-$ anion.

$$
\left[\underset{\displaystyle O}{\overset{\displaystyle O}{\underset{\parallel}{\overset{\parallel}{H-O-S-O}}}}\right]^- \quad \left[\underset{\displaystyle O-H}{\overset{\displaystyle O}{\underset{|}{\overset{\parallel}{H-O-P-O}}}}\right]^-
$$

$$ HSO_4^- \qquad\qquad H_2PO_4^- $$

In oxoanions such as these, the *lone* oxygens carry most of the negative charge, not the oxygens still bonded to hydrogens. This charge actually spreads over the lone oxygens. In HSO_4^-, therefore, each oxygen carries a charge of about $\frac{1}{3}-$. This is not a *formal* charge obtained by the rules on page 350. We're only saying that the charge of $1-$ is spread more or less evenly over three oxygens to give each a charge of $\frac{1}{3}-$. By the same reasoning, in $H_2PO_4^-$ each of the two lone oxygens carries a charge of about $\frac{1}{2}-$. The smaller negative charge on the lone oxygens in HSO_4^- make this ion less able to attract H^+ ions from H_3O^+, so HSO_4^- is a weaker base than $H_2PO_4^-$. In other words, HSO_4^- cannot become H_2SO_4 again as readily as $H_2PO_4^-$ can become H_3PO_4. There are thus *two* factors that make H_2SO_4 more fully ionized than H_3PO_4 in water. One is

The HSO_4^- ion is also a stronger acid than the $H_2PO_4^-$ ion.

the greater tendency of H_2SO_4 to lose H^+, and the other is the lesser tendency of HSO_4^- to recapture H^+. So if we compare solutions of H_2SO_4 and H_3PO_4 of equal molar concentrations, there will be a larger percentage of HSO_4^- ions than of $H_2PO_4^-$ ions. Sulfuric acid, to put it another way, will be more fully ionized than phosphoric acid. Of course, this is how we define acid strength—in terms of the *percentage* of the acid molecules that become ionized in solution. So sulfuric acid is a stronger acid than phosphoric acid.

Practice Exercise 9

In each pair, select the stronger acid: (a) HIO_3 or HIO_4, (b) H_2SeO_3 or H_2SeO_4, (c) H_3AsO_3 or H_3AsO_4. ◆

15.3 Lewis Acids and Bases

When a proton transfers to a Brønsted base, the latter provides the pair of electrons needed to form the covalent bond to H. The proton accepts this pair as the bond forms. The general nature of these facts was noticed by G. N. Lewis, after whom Lewis symbols are named (Section 8.2), so he proposed very broad definitions of acids and bases, definitions that would include both the Arrhenius and Brønsted kinds but would embrace other systems as well. **Lewis acids and bases** are defined as follows.

> **Lewis Definitions of Acids and Bases**
> **1.** An **acid** is any ionic or molecular species that can accept a pair of electrons in the formation of a coordinate covalent bond.
> **2.** A **base** is any ionic or molecular species that can donate a pair of electrons in the formation of a coordinate covalent bond.
> **3.** **Neutralization** is the formation of a coordinate covalent bond between the donor (base) and the acceptor (acid).

Examples of Lewis Acid–Base Reactions

The reaction between BF_3 and NH_3 illustrates a Lewis acid–base neutralization. The reaction is exothermic because a bond is formed between N and B, the nitrogen donating an electron pair and the boron accepting it.

Compounds like BF_3NH_3, which are formed by simply joining two smaller molecules, are called **addition compounds.**

The ammonia molecule thus acts as a *Lewis base*. The boron atom in BF_3, having only six electrons in its valence shell and needing two more to achieve an octet, accepts the pair of electrons from the ammonia molecule. Hence, BF_3 is functioning as a *Lewis acid*.

The Lewis concept includes the Arrhenius and Brønsted acids and bases as special cases. When hydronium ion reacts with hydroxide ion, for example, a

In other words, hydrated metal ions tend to be proton donors in water. Let's see why. The reason will be, by now, a familiar argument.

The positive charge on the central metal ion attracts the water molecule, holds it by its O atom, and draws electron density from the O—H bonds. This intensifies the partial positive charge on H and weakens the O—H bond with respect to the transfer of H^+ to another water molecule in the formation of a hydronium ion. (See below.)

Electron density is reduced in the O—H bonds of water (indicated by the curved arrows) by the positive charge of the metal ion, thereby increasing the partial positive charge on the hydrogens. This promotes the transfer of H^+ to a water molecule.

The degree to which metal ions produce acidic solutions depends on two principal factors. One is the amount of charge on the cation, and the other is the cation's size. As you might expect, increasing the charge on the cation increases the metal ion's ability to draw electron density toward itself and away from the O—H bond and thereby favors the release of H^+. This means that highly charged metal ions ought to produce more acidic solutions than ions of low charge, and that is generally the case.

The reason the size of the cation also affects its acidity is that when the cation is small, the positive charge is highly concentrated. A highly concentrated positive charge is better able to pull electrons from an O—H bond than a positive charge that is more spread out. Therefore, for a given positive charge, the smaller the cation the more acidic are its solutions.

Both size and amount of charge can be considered together by referring to a metal ion's *positive charge density,* the ratio of the positive charge to the volume of the cation (its ionic volume).

$$\text{Charge density} = \frac{\text{ionic charge}}{\text{ionic volume}}$$

The higher the positive charge density is, the more effective the metal ion is at drawing electron density from the O—H bond and the more acidic is the hydrated cation.

Very small cations with large positive charges have large positive charge densities and tend to be quite acidic. An example is the hydrated aluminum ion, Al^{3+}. The hexahydrate, $Al(H_2O)_6^{3+}$, is one of several of this cation's hydrated forms that are present in an aqueous solution of an aluminum salt. This ion is acidic in water because of the equilibrium

$$[Al(H_2O)_6]^{3+}(aq) + H_2O \rightleftharpoons [Al(H_2O)_5(OH)]^{2+}(aq) + H_3O^+(aq)$$

The equilibrium, while not actually *strongly* favoring the products, does produce enough hydronium ion so that a 0.1 M solution of $AlCl_3$ in water has about the same concentration of hydronium ions as a 0.1 M solution of acetic acid, roughly 1×10^{-3} M.

octahedral $Al(H_2O)_6^{3+}$ ion

Periodic Trends in the Acidity of Metal Ions

Within the periodic table, atomic size increases down a group and decreases from left to right in a period. Cation sizes follow these same trends, so within a given group, the cation of the metal at the top of the group has the smallest volume and the largest charge density. Therefore, hydrated metal ions at the top of a group in the periodic table are the most acidic within the group.

The cations of the Group IA metals (Li^+, Na^+, K^+, Rb^+, or Cs^+), with charges of just 1+, have little tendency to increase the H_3O^+ concentration in an aqueous solution.

Within Group IIA, the Be^{2+} cation is very small and has sufficient charge density to cause the hydrated ion to be a weak acid. The other cations of Group IIA, Mg^{2+}, Ca^{2+}, Sr^{2+}, or Ba^{2+}, have charge densities that become progressively smaller as we go down the group. Although their hydrated ions all generate some hydronium ion in water, the amount is negligible.

Some transition metal ions are also acidic, especially those with charges of 3+. For example, solutions containing salts of Fe^{3+} and Cr^{3+} tend to be acidic because their ions in solution exist as $Fe(H_2O)_6^{3+}$ and $Cr(H_2O)_6^{3+}$, respectively, and undergo the same ionization reaction as does the $Al(H_2O)_6^{3+}$ ion discussed above.

Nonmetal Oxides as Acids

Recall from Chapter 4 that not all nonmetal oxides react with water; e.g., CO.

Nonmetal oxides are usually *acidic anhydrides,* those that react with water give acidic solutions. Typical examples of the formation of acids from nonmetal oxides are the following reactions.

$$SO_3(g) + H_2O \longrightarrow H_2SO_4(aq) \qquad \text{sulfuric acid}$$

$$N_2O_5(g) + H_2O \longrightarrow 2HNO_3(aq) \qquad \text{nitric acid}$$

$$CO_2(g) + H_2O \longrightarrow H_2CO_3(aq) \qquad \text{carbonic acid}$$

15.5 Ionization of Water and the pH Concept

There are literally thousands of weak acids and bases, and they vary widely in how weak they are. Both the acetic acid in vinegar and the carbonic acid in pressurized soda water, for example, are classified as weak. Yet carbonic acid is only about 3% as strong an acid as acetic acid. To study such differences quantitatively, we need to explore acid–base equilibria in greater depth. This requires that we first discuss a particularly important equilibrium that exists in *all* aqueous solutions, namely, the ionization of water itself.

Under the right voltage, pure water displays a slight ability to conduct an electrical current, indicating the presence of trace concentrations of ions. They arise from the very slight self-ionization or *autoionization* of water itself, represented by the following equilibrium equation.

We omit the usual (aq) following the symbols for ions in water for the sake of simplicity.

$$H_2O + H_2O \rightleftharpoons H_3O^+ + OH^-$$

The forward reaction requires a collision of two H_2O molecules.

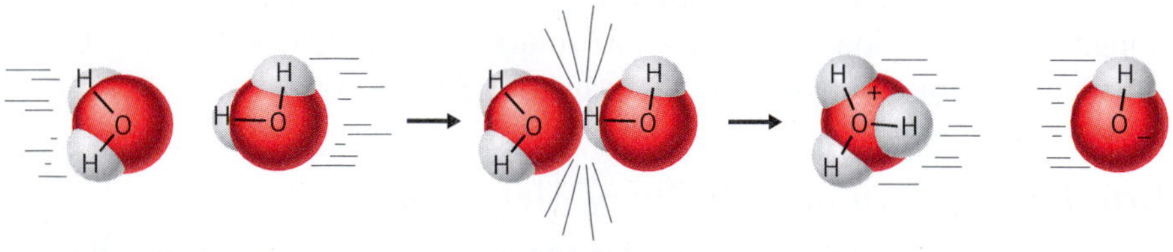

Its equilibrium law, following the procedures developed in Chapter 14, is

$$\frac{[H_3O^+][OH^-]}{[H_2O]^2} = K_c$$

In pure water, and even in dilute aqueous solutions, the molar concentration of water has essentially a constant value of 55.6 M. Therefore, the term $[H_2O]^2$ in the denominator is a constant which we can combine with K_c.

$$[H_3O^+][OH^-] = K_c \times [H_2O]^2$$

The product of the two constants, $K_c \times [H_2O]^2$, must also be a constant. Because of the importance of the autoionization equilibrium, it is given the special symbol K_w and is called the **ion product constant of water.**

$$[H_3O^+][OH^-] = K_w$$

Often, for convenience, we omit the water molecule that carries the hydrogen ion and write H^+ in place of H_3O^+. The equilibrium equation for the autoionization of water then simplifies as follows.

$$H_2O \rightleftharpoons H^+ + OH^-$$

The equation for K_w based on this is likewise simplified.

$$[H^+][OH^-] = K_w \qquad (15.1)$$

In pure water, the concentrations of H^+ and OH^- produced by autoionization are equal, having the following values at 25 °C.

$$[H^+] = [OH^-] = 1.0 \times 10^{-7}\,mol\,L^{-1}$$

Therefore, at 25 °C,

$$K_w = (1.0 \times 10^{-7})(1.0 \times 10^{-7})$$

$$K_w = 1.0 \times 10^{-14} \qquad (\text{at } 25\,°C) \qquad (15.2)$$

As with other equilibrium constants, the value of K_w varies with temperature (see Table 15.3). But, for simplicity, we will generally deal with systems at 25 °C, so we'll usually not specify the temperature each time. The value of K_w at 25 °C is so important that it should be learned.

The Autoionization of Water when Solutes Are Present

Water's autoionization takes place in *any* aqueous solution, but, because of the effects of other solutes, the molar concentrations of H^+ and OH^- may not be equal. Nevertheless, their product, K_w, is the same. Thus, although Equations 15.1 and 15.2 were derived for pure water, they also apply to dilute aqueous solutions. The significance of this must be emphasized. *In any aqueous solution, the product of $[H^+]$ and $[OH^-]$ equals K_w, although these two molar concentrations may not actually equal each other.*

Criteria for Acidic, Basic, and Neutral Solutions

A *neutral solution* is one that is neither acidic nor basic; in other words, the molar concentrations of H_3O^+ and OH^- are equal. An *acidic solution* is one in which some solute has made the molar concentration of H_3O^+ greater than that of OH^-. On the other hand, a *basic solution* exists when the molar concentration of OH^- exceeds that of H_3O^+. We therefore define acidic and basic solutions in terms of the *relative* molarities of H_3O^+ and OH^-.

The density of water is 1.00 g mL^{-1} or 1.00×10^3 g L^{-1}, so the molar concentration of water (at 18.0 g mol^{-1}) is 55.6 mol L^{-1}.

Tools

Ion product constant of water

Because of the autoionization of water, even the most acidic solution has some OH$^-$ and even the most basic solution has some H$_3$O$^+$.

Table 15.3 K_w at Various Temperatures

Temperature (°C)	K_w
0	1.5×10^{-15}
10	3.0×10^{-15}
20	6.8×10^{-15}
25	1.0×10^{-14}
30	1.5×10^{-14}
40	3.0×10^{-14}
50	5.5×10^{-14}
60	9.5×10^{-14}

Neutral solution	$[H_3O^+] = [OH^-]$
Acidic solution	$[H_3O^+] > [OH^-]$
Basic solution	$[H_3O^+] < [OH^-]$

EXAMPLE 15.5
Finding [H⁺]
from [OH⁻] or
Finding [OH⁻]
from [H⁺]

In a sample of blood at 25 °C, $[H^+] = 4.6 \times 10^{-8}$ *M*. Find the molar concentration of OH^-, and decide if the sample is acidic, basic, or neutral.

Analysis: The values of $[H^+]$ and $[OH^-]$ are *always* related to each other and to K_w at 25 °C as follows.

$$1.0 \times 10^{-14} = [H^+][OH^-]$$

Solution: We substitute the given value of $[H^+]$ into this equation and solve for $[OH^-]$.

$$1.0 \times 10^{-14} = (4.6 \times 10^{-8})\,[OH^-]$$

Solving for $[OH^-]$, and remembering that its units are mol L^{-1} or *M*,

$$[OH^-] = \frac{1.0 \times 10^{-14}}{4.6 \times 10^{-8}}\,M$$

$$= 2.2 \times 10^{-7}\,M$$

When we compare $[H^+]$ equaling 4.6×10^{-8} *M* with $[OH^-]$ equaling 2.2×10^{-7} *M*, we see that $[OH^-] > [H^+]$, so the blood is basic, although very slightly so.

Are the Answers Reasonable?
We know $K_w = 1.0 \times 10^{-14} = [H^+][OH^-]$. So one check is to note that if $[H^+]$ is slightly *less* than 1×10^{-7}, $[OH^-]$ will have to be slightly *more* than 1×10^{-7}, as it is. The calculation is so easy with a handheld calculator, however, that the best check is to redo the calculation.

Practice Exercise 11

An aqueous solution of sodium bicarbonate, $NaHCO_3$, has a molar concentration of hydroxide ion of 7.8×10^{-6} *M*. What is the molar concentration of hydrogen ion? Is the solution acidic, basic, or neutral? ◆

The pH Concept

In most solutions of weak acids and bases, the molar concentrations of H^+ and OH^- are very small, like those in Example 15.5. When you tried to compare the two values in that example, namely, 4.6×10^{-8} *M* and 2.2×10^{-7} *M*, you had to look in four places, two before the times signs and the two negative exponents. There's an easier approach involving only one number. A Danish chemist, S. P. L. Sørenson (1868–1939), suggested it.

To make comparisons of small values of $[H^+]$ easier, Sørenson defined a quantity that he called the **pH** of the solution as follows.

Tools
pH equation

$$pH = -\log [H^+] \tag{15.3}[4]$$

[4]In this equation and similar ones, like Equation 15.5, the logarithm of only the numerical part of the bracketed term is taken. The physical units must be mol L^{-1}, but they are set aside for the calculation.

The properties of logarithms let us rearrange Equation 15.3 as follows.

$$[H^+] = 10^{-pH} \qquad (15.4)$$

Equation 15.3 can be used to calculate the pH of a solution if its molar concentration of H^+ is known. On the other hand, if the pH is known and we wish to calculate the molar concentration of hydrogen ion, we apply Equation 15.4. We will illustrate these calculations with examples shortly.

The logarithmic definition of pH, Equation 15.3, has proved to be so useful that it has been adapted to quantities other than $[H^+]$. Thus, for any quantity X, we may define a term, pX, in the following way.

$$pX = -\log X \qquad (15.5)$$

Tools

Defining equation for pX

For example, to express small concentrations of hydroxide ion, we can define the **pOH** of a solution as follows.

$$pOH = -\log [OH^-]$$

Similarly, for K_w we can define **pK_w** as follows.

$$pK_w = -\log K_w$$

The numerical value of pK_w at 25 °C equals $-\log (1.0 \times 10^{-14})$ or $-(-14.00)$. Thus,

$$pK_w = 14.00 \qquad \text{(at 25 °C)}$$

A useful relationship among pH, pOH, and pK_w can be derived from Equation 15.1 that defines K_w.

$$[H^+][OH^-] = K_w \qquad \text{(Equation 15.1)}$$

By the properties of logarithms,

$$\log [H^+] + \log [OH^-] = \log K_w$$

We next multiply the terms on both sides by -1.

$$-\log [H^+] + -\log [OH^-] = -\log K_w$$

But each of these terms is in the form of pX (Equation 15.5), so

$$pH + pOH = pK_w = 14.00 \qquad \text{(at 25 °C)} \qquad (15.6)$$

Tools

Relationship of pH and pOH

This tells us that in an aqueous solution of any solute at 25 °C, the sum of pH and pOH is 14.00.

Definitions of Acidic, Basic, and Neutral Solutions According to pH

One meaning attached to pH is that it is a measure of the acidity of a solution. Hence, we may define *acidity, basicity,* and *neutral* in terms of pH values. In pure water, or in any solution that is *neutral,*

$$[H^+] = [OH^-] = 1.0 \times 10^{-7} \, M$$

Therefore, by Equation 15.3, *the pH of a neutral solution at 25 °C is 7.00.*[5]

An *acidic solution* is one in which $[H^+]$ is larger than $10^{-7} \, M$ and so has a pH *less* than 7.00. Thus, *as a solution's acidity increases, its pH decreases.*

[5]The rule for significant figures in logarithms is that *the number of decimal places in the logarithm of a number equals the number of significant figures in the number.* For example, 3.2×10^{-5} has just two significant figures. The logarithm of this number, displayed on a pocket calculator, is -4.494850022. We write the logarithm, when correctly rounded, as -4.49.

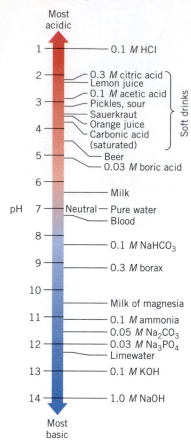

Figure 15.3 *The pH scale.*

A *basic solution* is one in which the value of $[H^+]$ is less than 10^{-7} M and so has a pH that is *greater* than 7.00. *As a solution's acidity decreases, its pH increases.* These simple relationships may be summarized as follows. At 25 °C:

pH < 7.00	**Acidic solution**
pH > 7.00	**Basic solution**
pH = 7.00	**Neutral solution**

The pH values for some common substances are given on a pH scale in Figure 15.3.

One of the deceptive features of pH, particularly to those just learning the concept, is that a hydrogen ion concentration multiplies by ten for each decrease in pH by only one unit. For example, by Equation 15.3, at pH 6 the hydrogen ion concentration is 1×10^{-6} mol L^{-1}. At pH 5, it's 1×10^{-5} mol L^{-1}, or ten times greater. That's a major difference.

▶**Chemistry in Practice**◀ The pH concept is used not only in chemistry but also throughout the life sciences, because the processes of life are very sensitive to the pH of the fluids associated with them. Small changes in pH matter. The fingerlings of game fish, for example, generally cannot live in water with a pH less than about 5.5. So when sulfur oxides or nitrogen oxides wash into lakes from acid rain, fish habitats can be ruined. (We'll discuss acid rain in *Chemicals in Our World 7* after this chapter.)

If the pH of your blood is not right, your ability to breathe can be fatally impaired. If the pH of your blood goes lower than 7.2, for example, or higher than 7.7, you are in a serious medical emergency. In untreated diabetes, the blood pH is on a downward move, from about 7.35 toward 7.2; in altitude sickness, it is moving upward, from 7.35 to 7.9. Unchecked hysterics, diarrhea, or certain drug overdoses also upset the pH of the blood. And consider that you can change the pH of a liter of pure water from 7 to 4 just by adding only 1 drop of concentrated hydrochloric acid to it. That's not enough to make the water taste sour. Fortunately, the blood has solutes that help to prevent such huge swings in its pH. More about this is given in *Chemicals in Our World 9*. ◆

The pH concept is almost never used when a pH value would be a negative number, namely, with solutions having hydrogen ion concentrations greater than 1 M. For example, in a solution where the value of $[H^+]$ is 2 M, the "2" is a simple enough number; we can't simplify it further by using its corresponding pH value. When $[H^+]$ is 2.00 M, the pH, calculated by Equation 15.3, would be $-\log$ (2.00) or -0.301. There is nothing wrong with negative pH values, but they offer no advantage over the actual value of $[H^+]$.

Measuring pH

One of the remarkable things about pH is that it can be easily measured using an instrument called a pH meter (Figure 15.4). An electrode system sensitive to the hydrogen ion concentration in a solution is first dipped into a solution of known pH to calibrate the instrument. Once calibrated, the apparatus can then be used to measure the pH of any other solution simply by immersing the electrodes into it. Most modern pH meters are able to determine pH values to

Table 15.4 Common Acid–Base Indicators

Indicator	Approximate pH Range	Color Change (lower to higher pH)
Methyl green	0.2–1.8	Yellow to blue
Thymol blue	1.2–2.8	Yellow to blue
Methyl orange	3.2–4.4	Red to yellow
Ethyl red	4.0–5.8	Colorless to red
Methyl purple	4.8–5.4	Purple to green
Bromocresol purple	5.2–6.8	Yellow to purple
Bromothymol blue	6.0–7.6	Yellow to blue
Litmus	4.7–8.3	Red to blue
Cresol red	7.0–8.8	Yellow to red
Thymol blue	8.0–9.6	Yellow to blue
Phenolphthalein	8.2–10.0	Colorless to pink
Thymolphthalein	9.4–10.6	Colorless to blue
Alizarin yellow R	10.1–12.0	Yellow to red
Clayton yellow	12.2–13.2	Yellow to amber

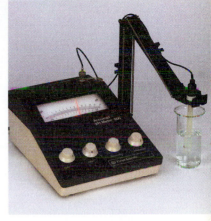

Figure 15.4 *A pH meter.*

within ±0.01 pH units, and research-grade instruments are capable of even greater precision in the pH range of 0.0 to 14.0.

Another less precise method of obtaining the pH uses acid–base indicators. As you learned in Chapter 4, these are dyes whose colors in aqueous solution depend on the acidity of the solution. Table 15.4 gives several examples. To obtain a rough idea of the pH, a drop of the solution to be tested is touched to a paper strip impregnated with one or more indicator dyes and the resulting color is compared with a color code. Some commercial test papers (e.g., Hydrion) are impregnated with several dyes, with their vials carrying the color code (see Figure 15.5).

Litmus paper, commonly found in chemistry labs, is often used qualitatively to test whether a solution is acidic or basic. It consists of porous paper impregnated with litmus dye, made either red or blue by exposure to acid or base, and then dried. Below pH 4.7, litmus is red and above pH 8.3 it is blue. The transition for the color change occurs over a pH range of 4.7 to 8.3 with the center of the change at about 6.5, very nearly a neutral pH. To test whether a solution is acidic, a drop of it is placed on blue litmus paper. If the dye turns pink, the solution is acidic. Similarly, to test if the solution is basic, a drop is touched to red litmus paper. If the dye turns blue, the solution is basic.

Figure 15.5 *Using a pH test paper.* The color of this Hydrion test strip changed to orange when it was dipped into the solution in the beaker. According to the color code, the pH of the solution is closer to 3 than to 4.

pH Calculations

Let's now look at some examples of pH calculations using Equations 15.3 and 15.4. The first example calculates a pH from a value either of $[H^+]$ or of $[OH^-]$. Another finds the value of $[H^+]$ or $[OH^-]$ given a pH.

EXAMPLE 15.6
Calculating pH and pOH from [H⁺]

Because rain washes pollutants out of the air, the lakes in many parts of the world have undergone pH changes. In a New England state, the water in one lake was found to have a hydrogen ion concentration of 3.2×10^{-5} mol L^{-1}. What are the calculated pH and pOH values of the lake's water? Is the water acidic or basic?

Analysis: We're given a value of [H$^+$], so we can use Equation 15.3 directly.

Solution:

$$pH = -\log [H^+] = -\log (3.2 \times 10^{-5})$$

$$= -(-4.49) = 4.49$$

The pH is less than 7.00, so the lake's water is acidic, too acidic for game fish to survive. Because the sum of pH and pOH equals 14.00, the pOH of the lake is

$$pOH = 14.00 - 4.49$$

$$= 9.51$$

Be sure to notice that *common* logs (base 10 logs) are used in pH calculations. Don't use the *natural* log function on your scientific calculator by mistake.

Are the Answers Reasonable?
The value of [H$^+$] is between 1×10^{-5} and 1×10^{-4}, so the pH has to be between 5 and 4, which it is. Therefore, the pOH must be between 9 and 10, as it is.

Practice Exercise 12

A carbonated beverage was found to have a hydrogen ion concentration of 3.67×10^{-4} mol L^{-1}. What are the calculated pH and pOH values of this beverage? Is it acidic, basic, or neutral? ◆

EXAMPLE 15.7
Calculating pH from [OH$^-$]

What is the calculated pH of a sodium hydroxide solution at 25 °C in which the hydroxide ion concentration equals 0.0026 *M*?

Analysis: The easiest way to calculate the pH when working with a basic solution is to calculate the pOH first, and then subtract it from 14.00.

Solution: Because [OH$^-$] equals 0.0026 *M*, we enter this value into the equation for pOH.

$$pOH = -\log [OH^-] = -(\log 0.0026)$$

$$= 2.59$$

We now subtract this pOH value from 14.00 to find the pH.

$$pH = 14.00 - 2.59$$

$$= 11.41$$

Notice that the pH in this basic solution is well above 7.

Is the Answer Reasonable?
The molar concentration of hydroxide ion is between 0.001 (or 10^{-3}) and 0.01 (or 10^{-2}), so the pOH must be between 3 and 2, which it is. Thus the pH ought to be between 11 and 12.

Practice Exercise 13

The molarity of OH$^-$ in the water in which a soil sample has soaked overnight is 1.47×10^{-9} mol L^{-1}. What is the calculated pH? ◆

EXAMPLE 15.8
Calculating $[H^+]$ from pH

"Calcareous soil" is soil rich in calcium carbonate. The pH of the moisture in such soil generally ranges from just over 7 to as high as 8.3. After one particular soil sample was soaked in water, the pH of the water was measured to be 8.14. What value of $[H^+]$ corresponds to a pH of 8.14? Is the soil acidic or basic?

Analysis: To calculate $[H^+]$ from pH, we use Equation 15.4.

$$[H^+] = 10^{-pH}$$

Solution: We substitute the pH value, 8.14, into the equation,

$$[H^+] = 10^{-8.14}$$

Because the pH has two digits following the decimal point, we obtain two significant figures in the hydrogen ion concentration. Therefore, the answer, correctly rounded, is

$$[H^+] = 7.2 \times 10^{-9}\ M$$

Because the pH > 7 and $[H^+] < 1 \times 10^{-7}\ M$, the soil is basic.

Is the Answer Reasonable?
The given value of 8.14 is between 8 and 9, so at a pH of 8.14, the hydrogen ion concentration must lie between 10^{-8} and 10^{-9} mol L^{-1}. And that's what we found.

Practice Exercise 14

Find the values of $[H^+]$ that correspond to each of the following values of pH. State whether each solution is acidic or basic.
(a) 2.90 (the approximate pH of lemon juice)
(b) 3.85 (the approximate pH of sauerkraut)
(c) 10.81 (the pH of milk of magnesia, a laxative)
(d) 4.11 (the pH of orange juice, on the average)
(e) 11.61 (the pH of dilute, household ammonia) ◆

15.6 Solutions of Strong Acids and Bases

pH of Dilute Solutions of Strong Acids and Bases

By now you know that strong acids and bases are considered to be 100% dissociated in an aqueous solution. This makes calculating the concentrations of H^+ and OH^- in their solutions a relatively simple task.

When the solute in a solution is a strong monoprotic acid, such as HCl or HNO_3, we expect to obtain one mole of H^+ for every mole of the acid in the solution. Thus, a 0.010 M solution of HCl contains 0.010 mol L^{-1} of H^+, and a 0.0020 M solution of HNO_3 contains 0.0020 mol L^{-1} of H^+. (In this chapter we will not consider strong diprotic acids, such as H_2SO_4, because only the first step in their ionization is complete. In solutions of H_2SO_4, for example, only about 10% of the HSO_4^- ions are further ionized to SO_4^{2-} ions and hydrogen ions.)

To calculate the pH of a solution of a strong monoprotic acid, we use the molarity of the H^+ obtained from the molar concentration of the acid. Thus, the 0.010 M HCl solution mentioned above has $[H^+] = 0.010\ M$, from which we calculate the pH to be equal 2.00.

For strong bases, calculation of the OH^- concentration is similarly straightforward. A 0.050 M solution of NaOH contains 0.050 mol L^{-1} of OH^- because the base is fully dissociated and each mole of NaOH releases one mole of OH^-

when it dissociates. For bases such as $Ba(OH)_2$ we have to recognize that two moles of OH^- are released by each mole of the base.

$$Ba(OH)_2(s) \longrightarrow Ba^{2+}(aq) + 2OH^-(aq)$$

Therefore, if a solution contained 0.010 mol $Ba(OH)_2$ per liter, the concentration of OH^- would be 0.020 M. Of course, once we know the OH^- concentration we can calculate pOH, from which we can calculate the pH. The following example illustrates the kinds of calculations we've just described.

EXAMPLE 15.9
Calculation of pH, pOH, and $[OH^-]$ in Solutions of Strong Acids or Strong Bases

Calculate the values of pH, pOH, and $[OH^-]$ in the following solutions: (a) 0.020 M HCl and (b) 0.00035 M $Ba(OH)_2$.

Analysis: Both HCl and $Ba(OH)_2$ (a Group IIA hydroxide) are strong electrolytes, so we assume 100% ionization for each. From each mole of HCl, we expect one mole of H^+, and from each mole of $Ba(OH)_2$, there are *two* moles of OH^- liberated.

Solution:
(a) In 0.020 M HCl, $[H^+] = 0.020$ M. Therefore, pH $= -\log(0.020) = 1.70$. Thus in 0.020 M HCl, the pH is 1.70. The pOH is $(14.00 - 1.70) = 12.30$. To find $[OH^-]$, we can use this value of pOH.

$$[OH^-] = 10^{-pOH} = 10^{-12.30}$$
$$= 5.0 \times 10^{-13} M$$

Notice how much smaller $[OH^-]$ is in this acidic solution than it is in pure water.
(b) As noted, one mole of $Ba(OH)_2$ dissolved per liter produces two moles of OH^- per liter. Therefore,

$$[OH^-] = 2 \times 0.00035 M = 0.00070 M$$
$$pOH = -\log(0.00070)$$
$$= 3.15$$

Thus the pOH of this solution is 3.15, and the pH $= (14.00 - 3.15) = 10.85$.

Are the Answers Reasonable?
In (a), the molarity of H^+ is numerically between 0.01 M and 0.1 M, or between 10^{-2} M and 10^{-1} M. So the pH must be between 1 and 2, as we found. In (b), the molarity of OH^- is between 0.0001 and 0.001, or between 10^{-4} and 10^{-3}. Hence, the pOH must be between 3 and 4, which it is.

Practice Exercise 15

Calculate $[H^+]$ and the pH in 0.0050 M NaOH. ◆

Suppression of the Ionization of Water by Acids or Bases

In the preceding calculations, we have made a critical and correct assumption, namely, that the autoionization of water contributes negligibly to the total $[H^+]$ in a solution of an acid and to the total $[OH^-]$ in a solution of a base.[6] Let's take a closer look at this.

---

[6] The autoionization of water cannot be neglected when working with a very dilute solution.

In a solution of an acid, there are actually *two* sources of H^+. One is from the ionization of the acid solute itself and the other is from the autoionization of water. Thus,

$$[H^+]_{total} = [H^+]_{from\ solute} + [H^+]_{from\ H_2O}$$

Except in very dilute solutions of acids, the second term, the contribution from the water, is small compared to the contribution from the solute. For instance, in Example 15.9, we saw that in 0.020 *M* HCl the molarity of OH^- was 5.0×10^{-13} *M*. The only source of OH^- in this acidic solution is from the autoionization of water, and the amounts of OH^- and H^+ *formed by the autoionization of water must be equal.* Therefore, $[H^+]_{from\ H_2O}$ also equals 5.0×10^{-13} *M*. If we now look at the total $[H^+]$ for this solution, we have

$$[H^+]_{total} = 0.020\ M + 5.0 \times 10^{-13}\ M$$
$$\text{(from HCl)} \qquad \text{(from H}_2\text{O)}$$

$$= 0.020\ M \text{ (rounded correctly)}$$

In any solution of an acid, the autoionization of water is suppressed by the H^+ provided by the solute. Therefore, except for very dilute solutions (10^{-6} *M* or less), *we'll assume that all the H^+ in the solution of an acid comes from the solute.* Similarly, *we'll assume that in a solution of a base, all the OH^- comes from the dissociation of the solute.*

A similar equation applies to the total OH^- concentration in a solution of a base.

$$[OH^-]_{total} = [OH^-]_{from\ solute} + [OH^-]_{from\ H_2O}$$

Only in very dilute solutions of acids or bases (10^{-6} *M* or less) does this assumption fail.

SUMMARY

Brønsted Acids and Bases. A **Brønsted acid** is a proton donor; a **Brønsted base** is a proton acceptor. Acid–base neutralization in the Brønsted theory is a proton transfer event. In an aqueous equilibrium involving a Brønsted acid or base, there are two **conjugate acid–base pairs.** The members of any given pair differ from each other by only one H^+. A substance that can be either an acid or a base, depending on the nature of the other reactant, is **amphoteric** or, with proton transfer reactions, **amphiprotic.**

Lewis Acids and Bases. A **Lewis acid** accepts a pair of electrons from a **Lewis base** in the formation of a coordinate covalent bond. Lewis bases often have filled valence shells and must have at least one unshared electron pair. Lewis acids have an incomplete valence shell that can accept an electron pair, or they have double bonds that allow electron pairs to be moved to make room for an incoming electron pair from a Lewis base.

We may summarize the three views of acids and bases as follows.

Arrhenius view	Acids give H^+ in water; bases provide OH^- in water.
Brønsted view	Acids are proton donors; bases are proton acceptors.
Lewis view	Acids are electron pair acceptors; bases are electron pair donors.

Relative Acidities and the Periodic Table. Binary acids contain only hydrogen and another nonmetal. Their strengths increase from top to bottom within a group and left to right across a period. **Oxoacids,** which contain oxygen atoms in addition to hydrogen and another element, increase in strength as the number of oxygen atoms on the same central atom increases. Oxoacids having the same number of oxygens increase in strength as the central atom moves from bottom to top within a group and from left to right across a period.

Acid–Base Properties of the Elements and Their Oxides. Oxides of most metals are basic anhydrides. Those of the Groups IA and IIA metals neutralize acids and tend to react with water to form metal hydroxides. The hydrates of metal ions tend to be proton donors when the positive charge density of the metal ion is itself sufficiently high, as it is when the ion has a 3+ charge. The small beryllium ion, Be^{2+}, forms a weakly acidic hydrated ion.

The Autoionization of Water and the pH Concept. Water reacts with itself to produce small amounts of H^+ and OH^- ions. The concentrations of these ions both in pure water *and* in dilute aqueous solutions are related by the expression

$$[H^+][OH^-] = K_w = 1.0 \times 10^{-14} \qquad \text{(at 25 °C)}$$

K_w is the **ion product constant of water.** In pure water,

$$[H^+] = [OH^-] = 1.0 \times 10^{-7}\ M$$

The **pH** of the solution is a measure of the acidity of a solution and is normally obtained by using a pH meter. As the pH decreases, the acidity increases. Sørenson's original defining equation for pH is $pH = -\log[H^+]$. In exponential form, this relationship between $[H^+]$ and pH is given by $[H^+] = 10^{-pH}$.

Comparable expressions can be used to describe low OH^- ion concentrations in terms of **pOH** values. Thus, $pOH = -\log[OH^-]$ and $[OH^-] = 10^{-pOH}$. At 25 °C, $pH + pOH = 14.00$. A solution is acidic if its pH is less than 7.00 and basic if its pH is greater than 7.00.

Solutions of Strong Acids and Strong Bases. In calculating the pH of solutions of strong acids and bases, we assume that they are 100% ionized.

The following table includes important equations needed to solve problems dealing with acid–base equilibria. You must have all of them at your fingertips when you begin to analyze a prob-lem. The key to success is in choosing the correct tools to apply to the solution.

Tool	Function
Periodic table and strengths of binary acids *(page 674)*	To help to understand and to recall the relative strengths of binary acids.
List of strong acids *(page 676)*	To judge if a given acid is weak under the assumption that any acid not on the list of strong acids is weak.
Periodic table and strengths of oxoacids *(page 677)*	To help to understand and to recall the relative strengths of oxoacids.
Ion product constant of water *(page 685)* $[H^+][OH^-] = K_w$	To calculate either the $[H^+]$ or $[OH^-]$ when one is known.
pH equation *(page 686)* $pH = -\log[H^+]$	To calculate pH values from hydrogen ion concentrations and to calculate $[H^+]$ from the measured pH of a solution.
Defining equation for pX *(page 687)* $pX = -\log X$	The equation is the model for functions for calculating pH, pOH, and pK_w.
Relationship of pH to pOH *(page 687)* $pH + pOH = 14.00$ **(at 25 °C)**	To calculate pH from pOH, and vice versa.

THINKING IT THROUGH

Remember that the goal in the following problems is *not* to find the answer itself; instead, you are asked to assemble the available information needed to obtain the answer, state what additional data (if any) are needed, and describe *how* you would use the data to answer the question.

Level 1 Problems

1. How do you obtain the formula for the conjugate base of CH_4? How can you tell whether this base is strong or weak?

2. How can you tell whether the following are amphoteric? $H_3O^+, H_2O, HSO_4^-, NO_3^-$

3. Can the ammonia molecule serve as a Lewis acid? Can it be a Lewis base? Explain the basis for your answer.

4. The hydration of a metal ion when it is placed into water can be viewed as a Lewis acid–base reaction. Explain using Al^{3+} as an example.

Level 2 Problems

5. Consider a metal ion, M^{n+}, in contact with n hydroxide ions. When the charge on the metal ion is small, the com-pound is basic. When the charge on the metal ion is large, the compound is acidic. From what you have learned in this chapter, explain this observation.

6. The reaction

$$N_2H_5^+ + NH_3 \longrightarrow N_2H_4 + NH_4^+$$

can be viewed as one in which one Lewis base displaces an-other from a compound. Among the reactants, which is the Lewis base? Which is the Lewis acid?

7. A 0.035 M solution of an acid has a pH of 3.5. Is the solute a strong acid or not? How can you tell?

8. In our calculations of the H^+ concentration in a solution of a strong acid, we ignored the small contribution made by the ionization of water. Suppose a solution of HCl had a concentration of 1.0×10^{-6} M. Show how you would calculate the molarity of H^+ in this solution taking into account the autoionization of water.

9. The pH of a 0.50 M solution of formic acid, $HCHO_2$, is 2.02. Explain how you would use this information to calculate the percentage of the formic acid that has ionized in the solution.

REVIEW QUESTIONS

Brønsted Acids and Bases

15.1 How is a Brønsted acid defined? How is a Brønsted base defined?

15.2 How are the formulas of the members of a conjugate acid–base pair related to each other? Within the pair, how can you tell which is the acid?

15.3 What is meant by the term amphoteric? Give two chemical equations that illustrate the amphoteric nature of water.

Trends in Acid–Base Strengths

15.4 Within the periodic table, how do the strengths of the binary acids vary from left to right across a period? How do they vary from top to bottom within a group?

15.5 Astatine, atomic number 85, is radioactive and does not occur in appreciable amounts in nature. On the basis of what you have learned in this chapter, answer the following.
(a) How would the acidity of HAt compare to HI?
(b) What might be a formula for an oxoacid of At?

15.6 Explain why nitric acid is stronger than nitrous acid.

$$HO\text{—}NO_2 \qquad HO\text{—}NO$$
nitric acid $\qquad\quad$ nitrous acid

15.7 Explain why H_2S is a stronger acid than H_2O.

15.8 Explain why $HClO_4$ is a stronger acid than H_2SeO_4.

15.9 Explain why the NO_2^- ion is a stronger base than the SO_3^- ion.

15.10 The position of equilibrium in the equation below lies far to the left. Identify the conjugate acid–base pairs. Which of the two acids is stronger?

$$HOCl(aq) + H_2O \rightleftharpoons H_3O^+(aq) + OCl^-(aq)$$

15.11 Consider the following: CO_3^{2-} is a weaker base than hydroxide ion, and HCO_3^- is a stronger acid than water. In the equation below, would the position of equilibrium lie to the left or to the right? Justify your answer.

$$CO_3^{2-}(aq) + H_2O \rightleftharpoons HCO_3^-(aq) + OH^-(aq)$$

15.12 Acetic acid, $HC_2H_3O_2$, is a weaker acid than nitrous acid, HNO_2. How do the strengths of the bases $C_2H_3O_2^-$ and NO_2^- compare?

15.13 Nitric acid, HNO_3, is a very strong acid. It is 100% ionized in water. In the reaction below, would the position of equilibrium lie to the left or to the right?

$$NO_3^-(aq) + H_2O \rightleftharpoons HNO_3(aq) + OH^-(aq)$$

15.14 $HClO_4$ is a stronger proton donor than HNO_3, but in water both acids appear to be of equal strength; they are both 100% ionized. Why is this so? What solvent property would be necessary in order to distinguish between the acidities of these two Brønsted acids?

15.15 Formic acid, $HCHO_2$, and acetic acid, $HC_2H_3O_2$, are classified as weak acids, but in water $HCHO_2$ is more fully ionized than $HC_2H_3O_2$. However, if we use liquid ammonia as a solvent for these acids, they both appear to be of equal strengths; both are 100% ionized in liquid ammonia. Explain why this is so.

15.16 Which of the molecules below is expected to be the stronger Brønsted acid? Why?

15.17 In which of the molecules below is the hydrogen printed in color the more acidic? Explain your choice.

Lewis Acids and Bases

15.18 Define Lewis acid and Lewis base.

15.19 Explain why the addition of a proton to a water molecule to give H_3O^+ is a Lewis acid–base reaction.

15.20 Methylamine has the formula CH_3NH_2 and the structure

Use Lewis structures to illustrate the reaction of methylamine with boron trifluoride, BF_3.

15.21 Use Lewis structures to show the Lewis acid–base reaction between CO_2 and H_2O to give H_2CO_3. Identify the Lewis acid and the Lewis base in the reaction.

15.22 Explain why the oxide ion, O^{2-}, can function as a Lewis base but not as a Lewis acid.

Acid–Base Properties of the Elements and Their Oxides

15.23 Suppose that a new element was discovered. Based on the discussions in this chapter, what properties (both physical and chemical) might be used to classify the element as a metal or a nonmetal?

15.24 If the oxide of an element dissolves in water to give an acidic solution, is the element more likely to be a metal or a nonmetal?

15.25 Many chromium salts crystallize as hydrates containing the ion $Cr(H_2O)_6^{3+}$. Solutions of these salts tend to be acidic. Explain why.

15.26 Which ion is expected to give the more acidic solution, Fe^{2+} or Fe^{3+}? Why?

15.27 Ions of the alkali metals have little effect on the acidity of a solution. Why?

15.28 What acid is formed when the following oxides react with water? (a) SO_3, (b) CO_2, (c) P_4O_{10}

Ionization of Water and the pH Concept

15.29 Write the chemical equation for the autoionization of water and the equilibrium law for K_w.

15.30 How are acidic, basic, and neutral solutions in water defined (a) in terms of $[H^+]$ and $[OH^-]$ and (b) in terms of pH?

REVIEW PROBLEMS

Answers to problems whose numbers are printed in color are given in Appendix D.
More challenging problems are marked with asterisks.

Brønsted Acids and Bases

15.31 Give the conjugate acids of the following.
(a) F^- (c) C_5H_5N (e) $HCrO_4^-$
(b) N_2H_4 (d) O_2^{2-}

15.32 Give the conjugate bases of the following.
(a) NH_3 (c) HCN (e) HNO_3
(b) HCO_3^- (d) H_5IO_6

15.33 Identify the conjugate acid–base pairs in the following reactions.
(a) $HNO_3 + N_2H_4 \rightleftharpoons NO_3^- + N_2H_5^+$
(b) $NH_3 + N_2H_5^+ \rightleftharpoons NH_4^+ + N_2H_4$
(c) $H_2PO_4^- + CO_3^{2-} \rightleftharpoons HPO_4^{2-} + HCO_3^-$
(d) $HIO_3 + HC_2O_4^- \rightleftharpoons IO_3^- + H_2C_2O_4$

15.34 Identify the conjugate acid–base pairs in the following reactions.
(a) $HSO_4^- + SO_3^{2-} \rightleftharpoons HSO_3^- + SO_4^{2-}$
(b) $S^{2-} + H_2O \rightleftharpoons HS^- + OH^-$
(c) $CN^- + H_3O^+ \rightleftharpoons HCN + H_2O$
(d) $H_2Se + H_2O \rightleftharpoons HSe^- + H_3O^+$

Trends in Acid–Base Strengths

15.35 Choose the stronger acid: (a) H_2S or H_2Se; (b) H_2Te or HI; (c) PH_3 or NH_3. Give your reasons.

15.36 Choose the stronger acid: (a) HBr or HCl; (b) H_2O or HF; (c) H_2S or HBr. Give your reasons.

15.37 Choose the stronger acid: (a) HIO_3 or HIO_4; (b) H_3AsO_4 or H_3AsO_3. Give your reasons.

15.38 Choose the stronger acid: (a) $HOCl$ or $HClO_2$; (b) H_2SeO_4 or H_2SeO_3. Give your reasons.

15.39 Choose the stronger acid: (a) H_3AsO_4 or H_3PO_4, (b) H_2CO_3 or HNO_3; (c) H_2SeO_4 or $HClO_4$. Give your reasons.

15.40 Choose the stronger acid: (a) $HClO_3$ or HIO_3; (b) HIO_2 or $HClO_3$; (c) H_2SeO_3 or $HBrO_4$. Give your reasons.

Lewis Acids and Bases

15.41 Use Lewis symbols to diagram the reaction

$$NH_2^- + H^+ \longrightarrow NH_3$$

Identify the Lewis acid and Lewis base in the reaction.

15.42 Use Lewis symbols to diagram the reaction

$$BF_3 + F^- \longrightarrow BF_4^-$$

Identify the Lewis acid and Lewis base in the reaction.

15.43 Aluminum chloride, $AlCl_3$, forms molecules with itself with the formula Al_2Cl_6. Its structure is

Use Lewis structures to show how the reaction $2AlCl_3 \rightarrow Al_2Cl_6$ is a Lewis acid–base reaction.

15.44 Beryllium chloride, $BeCl_2$, exists in the solid as a *polymer*, consisting of long chains of $BeCl_2$ units arranged as indicated below. The formula of the chain can be represented as $(BeCl_2)_n$, where n is a large number. Use Lewis structures to show how the reaction $nBeCl_2 \rightarrow (BeCl_2)_n$ is a Lewis acid–base reaction.

15.45 Use Lewis structures to diagram the reaction

$$CO_2 + O^{2-} \longrightarrow CO_3^{2-}$$

Identify the Lewis acid and Lewis base in this reaction.

15.46 Use Lewis structures to diagram the reaction

$$CO_2 + H_2O \longrightarrow H_2CO_3$$

Identify the Lewis acid and Lewis base in this reaction.

Autoionization of Water and pH

15.47 At the temperature of the human body, 37 °C, the value of K_w is 2.4×10^{-14}. Calculate $[H^+]$, $[OH^-]$, pH, and pOH of pure water at this temperature. What is the relation between pH, pOH, and K_w at this temperature? Is water neutral at this temperature?

15.48 Deuterium oxide, D_2O, ionizes like water. At 20 °C its K_w or ion product constant, analogous to that of water, is 8.9×10^{-16}. Calculate $[D^+]$ and $[OD^-]$ in deuterium oxide at 20 °C. Calculate also the pD and the pOD.

15.49 Calculate the H^+ concentration in each of the following solutions in which the hydroxide ion concentrations are
(a) 0.0024 M   (c) 5.6×10^{-9} M
(b) 1.4×10^{-5} M   (d) 4.2×10^{-13} M

15.50 Calculate the OH^- concentration in each of the following solutions in which the H^+ ion concentrations are
(a) 3.5×10^{-8} M   (c) 2.5×10^{-13} M
(b) 0.0065 M   (d) 7.5×10^{-5} M

15.51 Calculate the pH of each solution in Problem 15.49.

15.52 Calculate the pH of each solution in Problem 15.50.

15.53 A certain brand of beer had an H^+ ion concentration equal to 1.9×10^{-5} mol L^{-1}. What is the pH of the beer?

15.54 A soft drink was put on the market with $[H^+] = 1.4 \times 10^{-5}$ mol L^{-1}. What is its pH?

15.55 Calculate the molar concentrations of H^+ and OH^- in solutions that have the following pH values.
(a) 3.14   (b) 2.78   (c) 9.25   (d) 13.24   (e) 5.70

15.56 Calculate the molar concentrations of H^+ and OH^- in solutions that have the following pOH values.
(a) 8.26   (b) 10.25   (c) 4.65   (d) 6.18   (e) 9.70

Solutions of Strong Acids and Bases

15.57 What is the concentration of H^+ in 0.010 M HCl? What is the pH of the solution?

15.58 What is the concentration of H^+ in 0.0050 M HNO_3? What is the pH of the solution?

15.59 A sodium hydroxide solution is prepared by dissolving 6.0 g NaOH in 1.00 L of solution. What is the molar concentration of OH^- in the solution? What is the pOH and the pH of the solution?

15.60 A solution was made by dissolving 0.837 g $Ba(OH)_2$ in 100 mL final volume. What is the molar concentration of OH^- in the solution? What are the pOH and the pH?

15.61 A solution of $Ca(OH)_2$ has a measured pH of 11.60. What is the molar concentration of the $Ca(OH)_2$ in the solution?

15.62 A solution of HCl has a pH of 2.50. How many grams of HCl are there in 250 mL of this solution?

ADDITIONAL EXERCISES

15.63 What is the formula of the conjugate acid of dimethylamine, $(CH_3)_2NH$? What is the formula of its conjugate base?

15.64 Write equations that illustrate the amphiprotic nature of the bicarbonate ion.

15.65 Hydrogen peroxide is a stronger Brønsted acid than water. (a) Explain why this is so. (b) Is an aqueous solution of hydrogen peroxide acidic or basic?

*15.66 How would you expect the degree of ionization of $HClO_3$ to compare in the solvents $H_2O(l)$ and $HF(l)$? The reactions are

$$HClO_3 + H_2O \rightleftharpoons H_3O^+ + ClO_3^-$$

$$HClO_3 + HF \rightleftharpoons H_2F^+ + ClO_3^-$$

Justify your answer.

15.67 Hydrazine, N_2H_4, is a weaker Brønsted base than ammonia. In the following reaction, would the position of equilibrium lie to the left or to the right? Justify your answer.

$$N_2H_5^+ + NH_3 \rightleftharpoons N_2H_4 + NH_4^+$$

15.68 Identify the two Brønsted acids and two Brønsted bases in the reaction

$$NH_2OH + CH_3NH_3^+ \rightleftharpoons NH_3OH^+ + CH_3NH_2$$

15.69 In the reaction in the preceding exercise, the position of equilibrium lies to left. Identify the stronger acid in each of the conjugate pairs in the reaction.

*15.70 Suppose 38.0 mL of 0.00200 M HCl is added to 40.0 mL of 0.00180 M NaOH. What will be the pH of the final mixture?

*15.71 What is the pH of a 1.0×10^{-7} M solution of HCl?

15.72 Milk of magnesia is a suspension of magnesium hydroxide in water. Although $Mg(OH)_2$ is relatively insoluble, a small amount does dissolve in the water, which makes the mixture slightly basic and gives it a pH of 10.08. How many grams of $Mg(OH)_2$ are actually dissolved in 100 mL of milk of magnesia?

*15.73 A 1.0 M solution of acetic acid has a pH of 2.37. What percentage of the acetic acid is ionized in the solution?

Ceramics and Superconductors

8

Energy Loss from Electrical Resistance. Electrons in motion in an electrical current either bump into jiggling metal atoms or encounter defects in the metal's structure as they flow in a conductor. Therefore, at ordinary temperatures, metals like copper have a small but significant resistance to the flow of electricity, converting a percentage of the electrical energy to heat. To compensate, very high and energy wasting voltages in power cables must be used for the long-distance transmission of electricity. This is one reason why scientists and engineers have long been studying the resistance-free phenomenon of superconductivity.

The Superconducting State. A *superconductor* is a material in a state in which it offers no resistance to electricity. In addition, a superconductor repels a surrounding magnetic field. Prior to 1987, only a few materials, nearly all of them metals and metal alloys, could be put into such a state and then only if their temperatures were lowered to just a few degrees above absolute zero.

H. Kamerlingh-Onnes, a Dutch physicist, won the 1913 Nobel Prize in physics for his discovery that mercury is superconducting at 4.1 K in liquid helium (boiling point 4.2 K). The temperature at and below which a material is superconducting is given the symbol T_c. An intermetallic compound of niobium and germanium, Nb_3Ge, was discovered in 1975 for which T_c equals 23.2 K.

In 1986, K. A. Müller and J. G. Bednorz of IBM in Zürich, Switzerland, discovered that a previously known ceramic material,[1] a mixed oxide of lanthanum, barium, and copper with the formula $LaBa_2CuO_4$, has a value

[1]Ceramics make up a large number of diverse inorganic substances defined more by how they are made useful, namely, by very high temperature processing, than by their constituents. They can involve any number of the oxides of metallic elements as well as metalloids, like silicon. Industrial ceramics include clay products, like bricks, and any kind of tile. Whitewares used for stoneware or fine china are ceramics. So are porcelains. Glass is a ceramic substance.

of T_c equal to 30 K. For their work, Müller and Bednorz shared the 1987 Nobel Prize in physics. Keeping the material in a superconducting state at 30 K, however, still requires refrigeration technology too expensive and too cumbersome to permit any important practical applications.

High Temperature Superconductivity. To make superconductivity practical, materials had to be discovered that could pass over to the superconducting state at temperatures much higher than the boiling point of liquid helium. Specifically, a material with T_c above the boiling point of liquid nitrogen, 77.4 K, was sought. The technology for producing and handling liquid nitrogen is well known and relatively simple, and liquid nitrogen costs about as much as milk or beer. Today, materials that have values of T_c above the boiling point of nitrogen are called *high temperature superconductors.*

With the opening made by Müller and Bednorz toward ceramiclike substances, the search for better superconductors continued among mixed metal oxides. Chemists Maw-Kuen Wu and J. Ashburn (University of Alabama) were the first to synthesize a material superconducting above the boiling point of nitrogen. They substituted yttrium for lanthanum in the Müller–Bednorz compound and prepared a mixed oxide, $YBa_2Cu_3O_7$, with T_c equal to 93 K. Chemist Paul Chu (University of Houston) was on the same yttrium track, and the Chu and Wu teams made a joint announcement in early 1987. $YBa_2Cu_3O_7$ is sometimes called the "1-2-3 compound," because of the $1:2:3$ ratio of Y, Ba, and Cu. The significant structural feature of this substance is the layering of copper and oxygen atoms (see Figure 8*a*). The 1-2-3 compound is just one of a small family of the mixed oxides of these elements that are superconducting above 77 K.

Experimentally, the ceramic superconductors are easy to make. All it takes are the parent oxides and a high temperature furnace. Teams around the world worked arduously to find materials that pushed T_c to still higher values. A mixed oxide of thallium, calcium, barium, and copper—discovered in 1988 by A. M. Hermann and Z. Sheng (University of Arkansas)—was found that had zero electrical resistance at 107 K. An IBM team (San Jose, CA), working with the same elements, found a material for which $T_c = 122$ K. Since

chemicals in our world

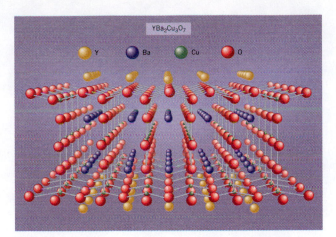

Figure 8a The 1-2-3 superconductor. (Adapted from A. W. Sleight, "Chemistry of High-Temperature Superconductors," *Science,* December 16, 1988, page 1521. Used by permission.)

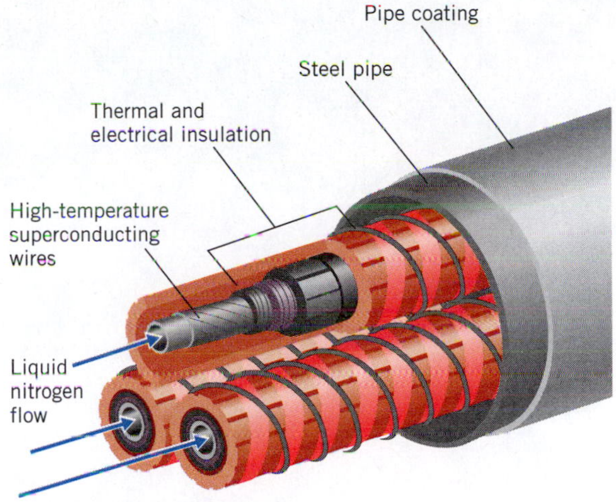

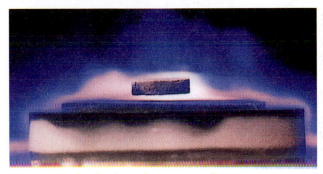

Figure 8b Superconducting electrical cables can be made from a bismuth–strontium–calcium–copper–oxygen ceramic, doped with a trace of lead. This ceramic flexes somewhat like thin glass filaments. The "wires" are protected by a silver alloy. (Source, American Superconductor/ Jared Schneidman.)

then, materials with values of T_c above 130 K (at atmospheric pressure) have been found.

Applications. As you can imagine, the chief problem with ceramic superconductors is that they are brittle. But during the 1990s scientists and engineers at American Superconductor Corp. (Westborough, MA) successfully made wirelike materials that could be used to build electrical cables (see Figure 8b). Just as the new millennium starts, Detroit Edison Company is using these cables to replace copper cables at its Frisbie power substation. The new cables carry up to three times the amount of electricity as the old ones. This enables a large expansion in electrical service without having to lay down more copper cables.

Another application of superconducting cables is the development of superconducting coils for the storage of standby electricity. In the absence of electrical resistance in superconducting cables, the electrons would just keep on circulating (in theory, forever). The electrical energy could then be tapped to smooth out blips in electrical transmission that mess up sensitive applications, like computer uses.

One of the most fascinating properties of materials in their superconducting states is that they can be levitated by a magnetic field.[2] Besides zero electrical resistance, a substance in its superconducting state permits no magnetic field within itself. A weak magnetic field is actually *repelled* by a superconductor, as we said, and this is what makes the levitation of superconductors

[2]*Levitation* means "floating on air"; the verb is "to levitate."

Figure 8c Levitation of a magnet above a supercooled sample of the 1-2-3 compound.

possible (Figure 8c). Someday we might levitate entire trains and send them at high speed with no more frictional resistance than offered to airplanes by the surrounding air. Prototypes of such "mag-lev" trains are currently being tested by Germany and Japan.

Questions

1. What two properties must a material possess to be called a superconductor?

2. What is the economic significance of finding a superconducting material with a value of T_c greater than 75 K?

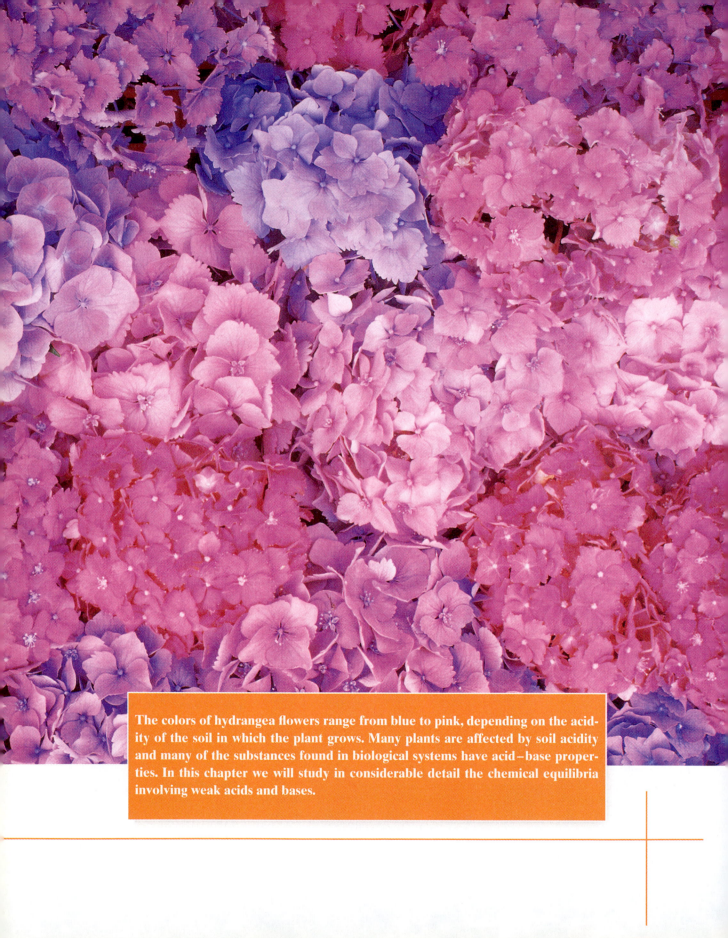

The colors of hydrangea flowers range from blue to pink, depending on the acidity of the soil in which the plant grows. Many plants are affected by soil acidity and many of the substances found in biological systems have acid–base properties. In this chapter we will study in considerable detail the chemical equilibria involving weak acids and bases.

Equilibria in Solutions of Weak Acids and Bases

This Chapter in Context In the preceding chapters we've discussed the general principles of chemical equilibrium and how to deal with calculating the hydrogen and hydroxide ion concentrations and pH of solutions of strong acids and bases. In this chapter we bring these concepts together as we examine the chemical equilibria that exist in solutions of weak acids or bases. There are literally thousands of such substances, many of which are found in biological systems as a consequence of the natural acidity or basicity of biological molecules. The principles we develop here have applications, therefore, not only in traditional chemistry labs but also in labs that focus on biochemically related problems.

16.1 Ionization Constants for Weak Acids and Bases

As you've learned, weak acids and bases are incompletely ionized in water and exist in solution as molecules that are in equilibria with the ions formed by their reactions with water. To deal quantitatively with these equilibria it is essential that you be able to write correct chemical equations for the equilibrium reactions, from which you can then obtain the corresponding correct equilibrium laws.

Reaction of a Weak Acid with Water

In aqueous solutions, all weak acids behave the same way. They are Brønsted acids and therefore proton donors. Some examples are $HC_2H_3O_2$, HSO_4^-, and NH_4^+. In water, these participate in the following equilibria.

$$HC_2H_3O_2 + H_2O \rightleftharpoons H_3O^+ + C_2H_3O_2^-$$

$$HSO_4^- + H_2O \rightleftharpoons H_3O^+ + SO_4^{2-}$$

$$NH_4^+ + H_2O \rightleftharpoons H_3O^+ + NH_3$$

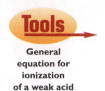

Tools

General equation for ionization of a weak acid

Notice that in each case, the acid reacts with water to give H_3O^+ and the corresponding conjugate base. We can represent these reactions in a general way using HA to stand for the formula of the acid.

$$HA + H_2O \rightleftharpoons H_3O^+ + A^- \tag{16.1}$$

As you can see from the equations above, HA does not have to be electrically neutral; it can be a molecule such as $HC_2H_3O_2$, a negative ion such as HSO_4^-, or a positive ion such as NH_4^+. (Of course, the actual charge on the conjugate base will then depend on the charge on the parent acid.)

Following the procedure developed in Chapter 14, we can also write a general equation for the equilibrium law for Reaction 16.1, using K_c' to stand for the equilibrium constant.

$$K_c' = \frac{[H_3O^+][A^-]}{[HA][H_2O]}$$

In our discussion of the autoionization of water in Chapter 15, we noted that in dilute aqueous solutions, $[H_2O]$ can be considered a constant, so it can be combined with K_c' to give a new equilibrium constant. Doing this gives

$$K_c' \times [H_2O] = \frac{[H_3O^+][A^-]}{[HA]} = K_a$$

The new constant K_a is called an **acid ionization constant.** Abbreviating H_3O^+ as H^+, the equation for the ionization of the acid can be simplified as

Some references call K_a the acid *dissociation* constant.

$$HA \rightleftharpoons H^+ + A^-$$

from which the expression for K_a is obtained directly.

$$K_a = \frac{[H^+][A^-]}{[HA]} \tag{16.2}$$

You should learn how to write the chemical equation for the ionization of a weak acid and be able to write the equilibrium law corresponding to its K_a. This is illustrated in Example 16.1.

EXAMPLE 16.1

Writing the K_a Expression for a Weak Acid

The meat products shown here contain nitrite ion as a preservative.

Nitrous acid, HNO_2, is a weak acid that's formed in the stomach when nitrite food preservatives encounter stomach acid. There has been some concern that this acid may form carcinogenic products by reacting with proteins. Write the chemical equation for the equilibrium ionization of HNO_2 in water and the appropriate K_a expression.

Solution: The conjugate base of HNO_2 is NO_2^-, so the equation for the ionization reaction is

$$HNO_2 + H_2O \rightleftharpoons H_3O^+ + NO_2^-$$

In the K_a expression, we leave out the H_2O that appears on the left and, for simplicity, represent H_3O^+ as H^+.

$$K_a = \frac{[H^+][NO_2^-]}{[HNO_2]}$$

Alternatively, we could have written the simplified equation for the ionization reaction, from which the K_a expression above is obtained directly.

$$HNO_2 \rightleftharpoons H^+ + NO_2^-$$

Practice Exercise 1

For each of the following acids, write the equation for its ionization in water and the appropriate expression for K_a: (a) $HCHO_2$, (b) $(CH_3)_2NH_2^+$, (c) $H_2PO_4^-$. ◆

For weak acids, values of K_a are usually quite small and can be conveniently represented in a logarithmic form similar to pH. Thus, we can define the **pK_a** of an acid as

$$pK_a = -\log K_a$$

Also, $K_a = 10^{-pK_a}$.

The strength of a weak acid is determined by its value of K_a; the larger the K_a, the stronger and more fully ionized the acid. Because of the negative sign in the defining equation for pK_a, the stronger the acid, the *smaller* is its value of pK_a. The values of K_a and pK_a for some typical weak acids are given in Table 16.1. A more complete list is located in Appendix E.

The values of K_a for strong acids are very large and are not tabulated. For many strong acids, K_a values have not been measured.

Practice Exercise 2

Two acids, HA and HB, have pK_a values of 3.16 and 4.14, respectively. Which is the stronger acid? What are the K_a values for these acids? ◆

Reaction of a Weak Base with Water

As with weak acids, all weak bases behave in a similar manner in water. They are weak Brønsted bases and are therefore proton acceptors. Examples are ammonia, NH_3, and acetate ion, $C_2H_3O_2^-$. Their reactions with water are

$$NH_3 + H_2O \rightleftharpoons NH_4^+ + OH^-$$

$$C_2H_3O_2^- + H_2O \rightleftharpoons HC_2H_3O_2 + OH^-$$

Table 16.1 K_a and pK_a Values for Weak Monoprotic Acids at 25 °C

Name of Acid	Formula	K_a	pK_a
Iodic acid	HIO_3	1.7×10^{-1}	0.23
Chloroacetic acid	$HC_2H_2O_2Cl$	1.36×10^{-3}	2.87
Nitrous acid	HNO_2	7.1×10^{-4}	3.15
Hydrofluoric acid	HF	6.8×10^{-4}	3.17
Cyanic acid	$HOCN$	3.5×10^{-4}	3.46
Formic acid	$HCHO_2$	1.8×10^{-4}	3.74
Barbituric acid	$HC_4H_3N_2O_3$	9.8×10^{-5}	4.01
Acetic acid	$HC_2H_3O_2$	1.8×10^{-5}	4.74
Hydrazoic acid	HN_3	1.8×10^{-5}	4.74
Butanoic acid	$HC_4H_7O_2$	1.52×10^{-5}	4.82
Propanoic acid	$HC_3H_5O_2$	1.34×10^{-5}	4.87
Hypochlorous acid	$HOCl$	3.0×10^{-8}	7.52
Hydrogen cyanide (*aq*)	HCN	6.2×10^{-10}	9.21
Phenol	HC_6H_5O	1.3×10^{-10}	9.89
Hydrogen peroxide	H_2O_2	1.8×10^{-12}	11.74

We see that in each instance, the base reacts with water to give OH^- and the corresponding conjugate acid. We can also represent these reactions by a general equation. If we represent the base by the symbol B, the reaction is

Tools

General equation for ionization of a weak base

$$B + H_2O \rightleftharpoons BH^+ + OH^- \qquad (16.3)$$

This yields the equilibrium law

$$K'_c = \frac{[BH^+][OH^-]}{[B][H_2O]}$$

As with solutions of weak acids, the quantity $[H_2O]$ in the denominator is effectively a constant that can be combined with K'_c to give a new constant that we call the **base ionization constant, K_b**.

$$K_b = \frac{[BH^+][OH^-]}{[B]} \qquad (16.4)$$

EXAMPLE 16.2

Writing the K_b Expression for a Weak Base

Hydrazine is a rocket fuel and is used in the manufacture of semiconductor devices.

Hydrazine, N_2H_4, is a weak base. It is a poisonous substance that's sometimes formed when chlorine bleach is added to an aqueous solution of ammonia. Write the equation for the reaction of hydrazine with water and write the expression for its K_b.

Solution: The conjugate acid of N_2H_4 is $N_2H_5^+$. Therefore, the equation for the ionization reaction in water is[1]

$$N_2H_4 + H_2O \rightleftharpoons N_2H_5^+ + OH^-$$

In the expression for K_b we omit the H_2O, so the equilibrium law is

$$K_b = \frac{[N_2H_5^+][OH^-]}{[N_2H_4]}$$

[1]When a proton is accepted by a weak base that contains a nitrogen atom, such as NH_3, N_2H_4, or $(CH_3)_2NH$, the proton binds to a lone pair of electrons on a nitrogen atom of the base. For reaction with water, we can diagram the reactions of these bases as

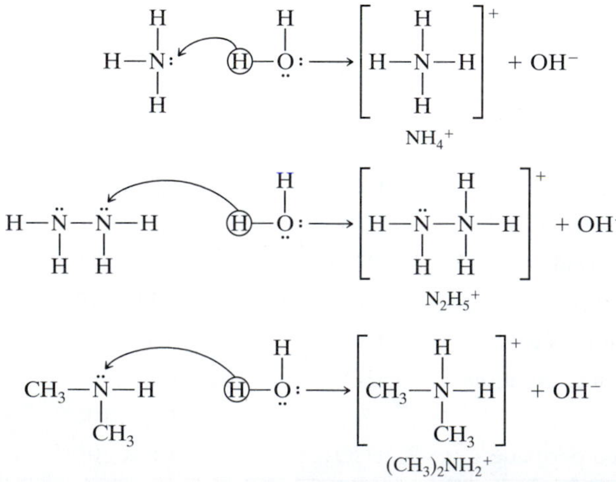

EXAMPLE 16.3
Writing the K_b Expression for a Weak Base

Solutions of chlorine bleach such as Clorox contain the hypochlorite ion, OCl^-, which is a weak base. Write the chemical equation for the reaction of OCl^- with water and the appropriate expression for K_b for this anion.

Solution: The reaction of hypochlorite ion as a base will yield its conjugate acid plus OH^-. We obtain the formula for the conjugate acid of OCl^- by adding one H^+ to the anion, which gives $HOCl$. Therefore, the equation for the equilibrium is

$$OCl^- + H_2O \rightleftharpoons HOCl + OH^-$$

As before, we omit H_2O from the expression for K_b.

$$K_b = \frac{[HOCl][OH^-]}{[OCl^-]}$$

Practice Exercise 3

For each of the following bases, write the equation for its ionization in water and the appropriate expression for K_b. (*Hint:* See footnote 1.) (a) $(CH_3)_3N$ (trimethylamine), (b) SO_3^{2-} (sulfite ion), (c) NH_2OH (hydroxylamine). ◆

hydroxylamine, NH_2OH

Because the K_b values for weak bases are usually small numbers, the same kind of logarithmic notation is often used to represent their equilibrium constants. Thus, **pK_b** is defined as

$$pK_b = -\log K_b$$

Also $K_b = 10^{-pK_b}$.

Table 16.2 lists some molecular bases and their corresponding values of K_b and pK_b. A more complete list is located in Appendix E.

Conjugate Acid–Base Pairs and Their Values of K_a and K_b

Formic acid, $HCHO_2$ (a substance partly responsible for the sting of a fire ant), is a typical weak acid that ionizes according to the equation

Formica is Latin for "ant."

$$HCHO_2 + H_2O \rightleftharpoons H_3O^+ + CHO_2^-$$

As you've seen, we write its K_a expression as

$$K_a = \frac{[H^+][CHO_2^-]}{[HCHO_2]}$$

Table 16.2 K_b and pK_b Values for Weak Molecular Bases at 25 °C

Name of Base	Formula	K_b	pK_b
Butylamine	$C_4H_9NH_2$	5.9×10^{-4}	3.23
Methylamine	CH_3NH_2	4.4×10^{-4}	3.36
Ammonia	NH_3	1.8×10^{-5}	4.74
Hydrazine	N_2H_4	1.7×10^{-6}	5.77
Strychnine	$C_{21}H_{22}N_2O_2$	1.0×10^{-6}	6.00
Morphine	$C_{17}H_{19}NO_3$	7.5×10^{-7}	6.13
Hydroxylamine	$HONH_2$	6.6×10^{-9}	8.18
Pyridine	C_5H_5N	1.7×10^{-9}	8.76
Aniline	$C_6H_5NH_2$	4.4×10^{-10}	9.36

$$
\underset{\text{formic acid, } HCHO_2}{H-\overset{\overset{\textstyle O}{\|}}{C}-O-H}
$$

$$
\underset{\text{formate ion, } CHO_2^-}{H-\overset{\overset{\textstyle O}{\|}}{C}-O^-}
$$

The conjugate base of formic acid is the formate ion, CHO_2^-, and when a solute that contains this ion (e.g., $NaCHO_2$) is dissolved in water, the solution is slightly basic. In other words, the formate ion is a weak base in water and participates in the equilibrium

$$
CHO_2^- + H_2O \rightleftharpoons HCHO_2 + OH^-
$$

The K_b expression for this base is

$$
K_b = \frac{[HCHO_2]\,[OH^-]}{[CHO_2^-]}
$$

There is an important relationship between the equilibrium constants for this acid–base pair: the product of K_a times K_b equals K_w. We can see this by multiplying the mass action expressions.

$$
K_a \times K_b = \frac{[H^+]\,[\cancel{CHO_2^-}]}{[\cancel{HCHO_2}]} \times \frac{[\cancel{HCHO_2}]\,[OH^-]}{[\cancel{CHO_2^-}]} = [H^+]\,[OH^-] = K_w
$$

In fact, this same relationship exists for *any* acid–base conjugate pair.

For *any* acid–base conjugate pair: $K_a \times K_b = K_w$ (16.5)

Another useful relationship, which can be derived by taking the negative logarithm of each side of Equation 16.5, is

$$
pK_a + pK_b = pK_w = 14.00 \qquad (\text{at } 25\,^\circ C) \qquad (16.6)
$$

There are some important consequences of the relationship expressed in Equation 16.5. One is that it is not necessary to tabulate both K_a and K_b for the members of an acid–base pair; if one K is known, the other can be calculated. For example, the K_a for $HCHO_2$ and the K_b for NH_3 will be found in most tables of acid–base equilibrium constants, but these tables usually will not contain the equilibrium constants for the ions that are the conjugates. Thus, tables usually will not contain the K_b for CHO_2^- or the K_a for NH_4^+.

Tools

Relationship between K_a and K_b

$-\log (K_a \times K_b) = -\log K_w$

$(-\log K_a) + (-\log K_b)$
$\qquad\qquad = -\log K_w$

$pK_a + pK_b = pK_w$

Tables of ionization constants usually give values for only the molecular member of an acid–base pair.

Practice Exercise 4

The value of K_a for $HCHO_2$ is 1.8×10^{-4}. What is K_b for the CHO_2^- ion? ◆

Another interesting and useful observation is that *there is an inverse relationship between the strengths of the acid and base members of a conjugate pair.* This is illustrated graphically in Figure 16.1. Because the product of K_a and K_b is a constant, the larger the value of K_a, the smaller is the value of K_b. In other words, *the stronger the conjugate acid, the weaker is its conjugate base* (a fact that we noted in Chapter 15 in our discussion of the strengths of Brønsted acids and bases). We will say more about this relationship in Section 16.3.

16.2 Equilibrium Calculations

Our goal in this section is to develop a general strategy for dealing quantitatively with the equilibria of weak acids and bases in water. Generally, these calculations fall into two categories: (1) calculating the value of K_a or K_b from the initial concentration of the acid or base and the measured pH of the solution (or some other data about the equilibrium composition of the solution) and (2) calculating equilibrium concentrations given K_a or K_b and initial concentrations.

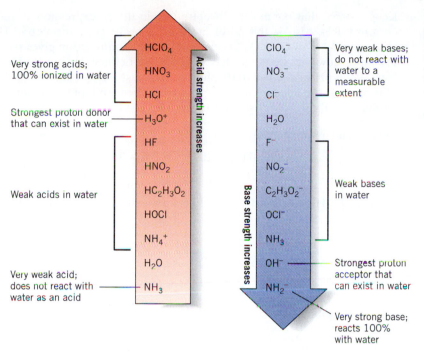

Figure 16.1 *The relative strengths of conjugate acid–base pairs.* The stronger the acid is, the weaker is its conjugate base. The weaker the acid is, the stronger is its conjugate base. Very strong acids are 100% ionized, and their conjugate bases do not react with water to any measurable extent.

Calculating K_a or K_b from Initial Concentrations and Equilibrium Data

In problems of this type, our goal is to obtain the equilibrium concentrations that are needed to evaluate the reaction quotients in the expressions for K_a or K_b. (The reason, of course, is that at equilibrium the reaction quotient, Q, equals the equilibrium constant.) We are usually given the molar concentration of the acid or base as it would appear on the label of a bottle containing the solution. Also provided is information from which we can obtain directly at least one of the equilibrium concentrations. This might be the measured pH of the solution, which provides an estimate of the equilibrium concentration of H^+. Alternatively, we might be given the **percentage ionization** of the acid or base, which we can define as follows:

$$\text{Percent ionization} = \frac{\text{moles ionized per liter}}{\text{moles available per liter}} \times 100\% \qquad (16.7)$$

Let's look at some examples that illustrate how to determine K_a and K_b from the kind of data mentioned.

Recall that the reaction quotient, Q, is the numerical value of the mass action expression for the reaction mixture.

Percent ionization

Lactic acid ($HC_3H_5O_3$), which is present in sour milk, also gives sauerkraut its tartness. It is a monoprotic acid. In a 0.100 *M* solution of lactic acid, the pH is 2.44 at 25 °C. Calculate the K_a and pK_a for lactic acid at this temperature.

EXAMPLE 16.4

Calculating K_a and pK_a from pH

Analysis: In any problem like this we begin by writing the chemical equation for the equilibrium so that we can set up the expression for K_a. We know the solute is an acid, and we know the general equation for the ionization of an acid, which we apply to the solute in question.

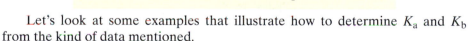

$$HC_3H_5O_3 \rightleftharpoons H^+ + C_3H_5O_3^- \qquad K_a = \frac{[H^+][C_3H_5O_3^-]}{[HC_3H_5O_3]}$$

Sauerkraut contains lactic acid. For some, nothing goes better on a hot dog than sauerkraut and mustard. Lactic acid gives the sauerkraut its sour taste.

Because the ionization reaction reaches equilibrium so quickly, we assume the measured pH *always* gives the equilibrium concentration of H^+.

Keep in mind that the molar concentrations in the K_a expression all refer to *equilibrium* values. Also, in this problem the equilibrium concentration of H^+ must be identical to that of $C_3H_5O_3^-$ because the ionization reaction gives these ions in a 1:1 ratio. However, we do not have the values of either $[H^+]$ or $[C_3H_5O_3^-]$; all we have is the pH of the 0.100 *M* solution. We must start by finding $[H^+]$ from pH. This gives us the *equilibrium* value of $[H^+]$. Then we'll put together a concentration table of the kind we used in Chapter 14.

Solution: First we convert pH into $[H^+]$.

$$[H^+] = 10^{-2.44} = 3.6 \times 10^{-3} \ M$$

$$= 0.0036 \ M$$

Because $[C_3H_5O_3^-] = [H^+]$, we now know that $[C_3H_5O_3^-] = 0.0036 \ M$. With these data we can now set up the concentration table.

	$HC_3H_5O_3$	$\rightleftharpoons$	H^+	$+$	$C_3H_5O_3^-$
Initial concentrations (*M*)	0.100		0		0 (Note 1)
Changes in concentrations caused by the ionization (*M*)	−0.0036		+0.0036		+0.0036 (Note 2)
Final concentrations at equilibrium (*M*)	(0.100 − 0.0036) = 0.096 (correctly rounded)		0.0036		0.0036

Note 1. The *initial* concentration of the acid is what the label tells us, 0.100 *M*; we take it to be the concentration before any ionization occurs. The *initial* values of $[H^+]$ and $[C_3H_5O_3^-]$ in the table correspond to the concentrations of the two ions *produced by solutes that are strong electrolytes.* Because no strong electrolytes are present in this solution, we set $[H^+]$ and $[C_3H_5O_3^-]$ equal to zero. (As we noted earlier, the *extremely* small amount of H^+ contributed by the self-ionization of water can be safely ignored. We will ignore it in all calculations involving acids in this chapter.)

Note 2. From the pH, we have obtained the equilibrium concentrations of H^+ and $C_3H_5O_3^-$. The *changes* for the ions are obtained by difference. For every H^+ ion that forms by the ionization, there is one less molecule of the initial acid. The minus sign in −0.0036 in the column for $HC_3H_5O_3$ is used because the concentration of this species is *decreased* by the ionization.

The last row of data gives us the equilibrium concentrations that we now use to calculate K_a.

$$K_a = \frac{(3.6 \times 10^{-3})(3.6 \times 10^{-3})}{0.096}$$

$$= 1.4 \times 10^{-4}$$

Thus the acid ionization constant for lactic acid is 1.4×10^{-4}. To find pK_a, we take the negative logarithm of K_a.

$$pK_a = -\log K_a = -\log (1.4 \times 10^{-4})$$

$$pK_a = 3.85$$

EXAMPLE 16.5

Calculating K_b and pK_b from pH

Methylamine, CH_3NH_2, is a weak base and one of several substances that give herring brine its pungent odor. In 0.100 *M* CH_3NH_2, only 6.4% of the base has undergone ionization. What are K_b and pK_b of methylamine?

Analysis: In this problem we've been given the percentage ionization of the base. We will use this to calculate the equilibrium concentrations of the ions in the solution. After we know these, solving the problem follows the same path as in the preceding example.

Solution: The first step is to write the chemical equation for the equilibrium and the equilibrium law. To write the chemical equation we need the formula for the conjugate acid of CH_3NH_2, which is $CH_3NH_3^+$ (we've added one H^+ to the nitrogen atom of the base to obtain the conjugate acid). Therefore, applying the general equation for the ionization of a weak base given on page 704, the chemical equation and equilibrium law are

$$CH_3NH_2(aq) + H_2O \rightleftharpoons CH_3NH_3^+(aq) + OH^-(aq)$$

$$K_b = \frac{[CH_3NH_3^+][OH^-]}{[CH_3NH_2]}$$

The percentage ionization tells us that 6.4% of the CH_3NH_2 has reacted. Therefore, the number of moles per liter of the base that has ionized at equilibrium in this solution is:

$$\text{Moles per liter of } CH_3NH_2 \text{ ionized} = 0.064 \times 0.100\ M = 0.0064\ M$$

This value represents the *decrease* in the concentration of CH_3NH_2, which we record in the "change" row of the concentration table as -0.0064. We can now use the change in $[CH_3NH_2]$ to determine the concentrations of the other species at equilibrium.

Methylamine. The fishy aroma you experience when preparing to consume pickled herring is due in part to the weak base methylamine.

methylamine

We take 6.4% of 0.100 M to find the moles per liter of the base that has ionized.

	$H_2O + CH_3NH_2 \rightleftharpoons$	$CH_3NH_3^+$ +	OH^-
Initial concentrations (M)	0.10	0	0 (Note 1)
Changes in concentrations caused by the ionization (M)	−0.0064	+0.0064	+0.0064 (Note 2)
Final concentrations at equilibrium (M)	(0.100 − 0.0064) = 0.094 (rounded)	0.0064	0.0064

Notice that because we know *both* the initial concentration of CH_3NH_2 and the change in this value, we need make no assumptions in computing the final concentration.

Note 1. Once again, we record the concentrations of solute species, assuming no initial ionization. We can safely ignore the extremely small contribution to $[OH^-]$ produced by the self-ionization of water.

Note 2. The concentration of CH_3NH_2 decreases by 0.0064 M, so the concentrations of the ions each increase by this amount. (Remember: However the reactants change, the products change in the opposite direction.)

Now we can calculate K_b.

$$K_b = \frac{(0.0064)(0.0064)}{(0.094)}$$

$$= 4.4 \times 10^{-4}$$

And we can calculate pK_b.

$$pK_b = -\log(4.4 \times 10^{-4})$$

$$= 3.36$$

Practice Exercise 5

When butter turns rancid, its foul odor is mostly that of butyric acid, a weak monoprotic acid similar to acetic acid in structure. In a 0.0100 M solution of butyric acid at 20 °C, the acid is 4.0% ionized. Calculate the K_a and pK_a of butyric acid at this temperature. (You don't need to know the formula for butyric acid; you can use any symbol you wish for the acid and its conjugate base—for example, HBu and Bu^-.) ◆

Practice Exercise 6

Few substances are more effective in relieving intense pain than morphine. Morphine is an alkaloid—an alkali-like compound obtained from plants—and alkaloids are all weak bases. In 0.010 M morphine, the pH is 10.10. Calculate the K_b and pK_b for morphine. ◆

Calculating Equilibrium Concentrations from K_a (or K_b) and Initial Concentrations

Acid–base equilibrium problems of this type often present a confusing picture to students. Usually, the difficulty is getting started on the right foot. Therefore, before we begin, let's take an overview of the "landscape" to develop a strategy for selecting the correct attack on the problem.

Almost any problem in which you are given K_a or K_b falls into one of three categories: (1) the solution contains a weak acid as its *only* solute, (2) the solution contains a weak base as its *only* solute, or (3) the solution contains *both* a weak acid and its conjugate base. The approach we take for each of these conditions is described below and is summarized in Figure 16.2.

1. If the solution contains only a weak acid as the solute, then the problem must be solved using K_a. This means that the correct chemical equation for the problem is the ionization of the weak acid. It also means that if you are given the K_b for the acid's conjugate base, you will have to calculate the K_a in order to solve the problem.

2. If the solution contains only a weak base, the problem must be solved using K_b. The correct chemical equation is for the ionization of the weak base. If the statement of the problem provides you with the K_a for the base's conjugate acid, then you must calculate the K_b to solve the problem.

Conditions 1 and 2 apply to solutions of weak molecular acids and bases and to solutions of salts that contain an ion that is an acid or a base. Condition 3 applies to solutions called buffers, which are discussed in Section 16.5.

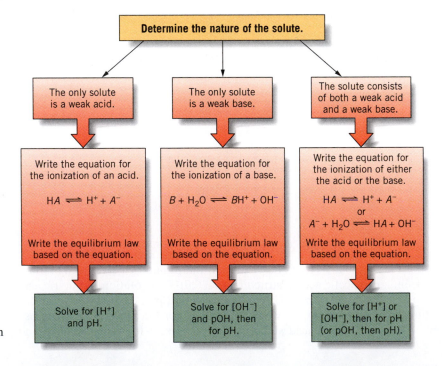

Figure 16.2 *Determining how to proceed in acid–base equilibrium problems.* The nature of the solute species determines how the problem is approached. Following this diagram will get you started in the right direction.

3. Solutions that contain two solutes, one a weak acid and the other its conjugate base, have some special properties that we will discuss later. To work problems for these kinds of mixtures, we can use *either* K_a or K_b—it doesn't matter which we use, because we will obtain the same answers either way. However, if we elect to use K_a, then the chemical equation we use must be for the ionization of the acid. If we decide to use K_b, then the correct chemical equation is for the ionization of the base. Usually, the choice of whether to use K_a or K_b is made on the basis of which constant is most readily available.

When you begin the solution of a problem, decide which of the three conditions above applies. If necessary, keep Figure 16.2 handy to guide you in your decision, so that you select the correct path to follow.

Simplifications in Acid–Base Equilibrium Calculations

In Chapter 14 you learned that when the equilibrium constant is small, it is frequently possible to make simplifying assumptions that greatly reduce the algebraic effort in obtaining equilibrium concentrations. For many acid–base equilibrium problems such simplifications are particularly useful.

Let's consider a solution of 1.0 M acetic acid, $HC_2H_3O_2$, for which $K_a = 1.8 \times 10^{-5}$. What is involved in determining the equilibrium concentrations in the solution?

To answer this question, we begin with the chemical equation for the equilibrium. Because the only solute in the solution is the weak acid, we must use K_a and therefore write the equation for the ionization of the acid. Let's use the simplified version.

$$HC_2H_3O_2 \rightleftharpoons H^+ + C_2H_3O_2^-$$

The equilibrium law is

$$K_a = \frac{[H^+][C_2H_3O_2^-]}{[HC_2H_3O_2]} = 1.8 \times 10^{-5}$$

We will take the given concentration, 1.0 M, to be the initial concentration of $HC_2H_3O_2$ (i.e., the concentration of the acid before any ionization has occurred). This concentration will drop slightly as the acid ionizes and the ions form. If we let x be the amount of acetic acid that ionizes per liter, we can construct the following concentration table.

As you know by now, acetic acid is what gives vinegar its sour taste.

acetic acid

The initial concentrations of the ions are set equal to zero because none of them have been supplied by a solute.

	$HC_2H_3O_2$	$\rightleftharpoons$	H^+	$+$	$C_2H_3O_2^-$
Initial concentrations (M)	1.0		0		0
Changes in concentrations caused by the ionization (M)	$-x$		$+x$		$+x$
Concentrations at equilibrium (M)	$1.0 - x$		x		x

The equilibrium constant, 1.8×10^{-5}, is quite small, so we can anticipate that very little of the acetic acid will be ionized at equilibrium. This means x will be very small, so we make the approximation $(1.0 - x) \approx 1.0$. Substituting values into the K_a expression gives

$$\frac{(x)(x)}{1.0} = 1.8 \times 10^{-5}$$

$$\frac{x^2}{1.0} = 1.8 \times 10^{-5}$$

Solving for x gives $x = 0.0042 \ M$. We see that the value of x is indeed negligible compared to $1.0 \ M$ (i.e., if we subtract $0.0042 \ M$ from $1.0 \ M$ and round correctly, we obtain $1.0 \ M$).

Notice that when we make this approximation, *the initial concentration of the acid is used as if it were the equilibrium concentration.* The approximation is valid when the equilibrium constant is small and the concentrations of the solutes are reasonably high—conditions that will apply to situations you will encounter in all but Section 16.4. In Section 16.4 we will discuss the conditions under which the approximation is not valid.

Let's look at some examples that illustrate typical acid–base equilibrium problems.

> As we will discuss later, if the solute concentration is at least 400 times the value of K, we can use initial concentrations as though they were equilibrium values.

EXAMPLE 16.6

Calculating the Values of [H⁺] and pH for a Solution of a Weak Acid from Its K_a Value

The calcium salt of propionic acid, calcium propionate, is used in baked products as a preservative.

A student planned an experiment that would use $0.10 \ M$ propionic acid, $HC_3H_5O_2$. Calculate the values of $[H^+]$ and pH for this solution. For propionic acid, $K_a = 1.4 \times 10^{-5}$.

Analysis: First, we note that the only solute in the solution is a weak acid, so we know we will have to use K_a and write the equation for the ionization of the acid.

$$HC_3H_5O_2 \rightleftharpoons H^+ + C_3H_5O_2^-$$

$$K_a = \frac{[H^+][C_3H_5O_2^-]}{[HC_3H_5O_2]}$$

The initial concentration of the acid is $0.10 \ M$, and the initial concentrations of the ions are both $0 \ M$. Then we construct the concentration table.

Solution: All concentrations in the table are in moles per liter.

	$HC_3H_5O_2$	$\rightleftharpoons$	H^+	$+$	$C_3H_5O_2^-$
Initial concentrations (M)	0.10		0		0
Changes in concentrations caused by the ionization (M)	$-x$		$+x$		$+x$
Final concentrations at equilibrium (M)	$(0.10 - x)$ ≈ 0.10		x		x

Notice that the equilibrium concentrations of H^+ and $C_3H_5O_2^-$ are the same; they are represented by x. Anticipating that x will be very small, we make the simplifying approximation $(0.10 - x) \approx 0.10$, so we take the equilibrium concentration of $HC_3H_5O_2$ to be $0.10 \ M$. Substituting these quantities into the K_a expression gives

$$K_a = \frac{[H^+][C_3H_5O_2^-]}{[HC_3H_5O_2]} = \frac{(x)(x)}{(0.10)} = 1.4 \times 10^{-5}$$

Solving for x yields

$$x = 1.2 \times 10^{-3}$$

Because $x = [H^+]$,

$$[H^+] = 1.2 \times 10^{-3} \ M$$

Finally, we calculate the pH

$$pH = -\log(1.2 \times 10^{-3})$$

$$= 2.92$$

Is the Answer Reasonable?
First, we see that the calculated pH is less than 7. This tells us that the solution is acidic, which it should be for a solution of an acid. If we wish to check the accuracy

of the calculation, we can substitute the calculated equilibrium concentrations into the mass action expression. If the calculated quantities are correct, the reaction quotient should equal K_a. Let's do the calculation.

$$\frac{[H^+][C_3H_5O_2^-]}{[HC_3H_5O_2]} = \frac{(x)(x)}{(0.10)} = \frac{(1.2 \times 10^{-3})^2}{0.10} = 1.4 \times 10^{-5} = K_a$$

The check works, so we know we have done the calculation correctly.

Practice Exercise 7

Nicotinic acid, $HC_2H_6NO_2$, is a B vitamin. It is also a weak acid with $K_a = 1.4 \times 10^{-5}$. Calculate $[H^+]$ and the pH of a 0.050 M solution of $HC_2H_6NO_2$. ◆

A solution of hydrazine, N_2H_4, has a concentration of 0.25 M. What is the pH of the solution, and what is the percentage ionization of the hydrazine? Hydrazine has $K_b = 1.7 \times 10^{-6}$.

EXAMPLE 16.7

Calculating the pH of a Solution and the Percentage Ionization of the Solute

Analysis: Hydrazine must be a weak base, because we have its value of K_b (the "b" in K_b tells us this is a *base* ionization constant). Since hydrazine is the only solute in the solution, we will have to write the equation for the ionization of a weak base and then set up the K_b expression.

Solution: Let's write the chemical equation and the K_b expression.

$$N_2H_4 + H_2O \rightleftharpoons N_2H_5^+ + OH^-$$

$$K_b = \frac{[N_2H_5^+][OH^-]}{[N_2H_4]}$$

Let's once again construct the concentration table. The only sources of $N_2H_5^+$ and OH^- are the ionization of the N_2H_4, so their initial concentrations are both zero. They will form in equal amounts, so we let their changes in concentration equal $+x$. The concentration of N_2H_4 will decrease by x, so its change is $-x$.

	N_2H_4	$+$	H_2O	$\rightleftharpoons$	$N_2H_5^+$	$+$	OH^-
Initial concentrations (M)	0.25				0		0
Changes in concentrations caused by the ionization (M)	$-x$				$+x$		$+x$
Final concentrations at equilibrium (M)	$(0.25 - x)$ ≈ 0.25				x		x

Because K_b is so small, $[N_2H_4] \approx 0.25$ M. (As before, we assume the initial concentration will be effectively the same as the equilibrium concentration, an approximation that we expect to be valid.) Substituting into the K_b expression,

$$\frac{(x)(x)}{0.25} = 1.7 \times 10^{-6}$$

Solving for x gives $x = 6.5 \times 10^{-4}$. This value represents the hydroxide ion concentration, from which we can calculate the pOH.

$$pOH = -\log(6.5 \times 10^{-4})$$

$$= 3.19$$

The pH of the solution can then be obtained from the relationship

$$pH + pOH = 14.00$$

Thus,

$$pH = 14.00 - 3.19$$

$$= 10.81$$

To calculate the percentage ionization, we need to know the number of moles per liter of the base that has ionized. From the concentration table, we see that this value is also equal to x, the amount of N_2H_4 that ionizes per liter (i.e., the change in the N_2H_4 concentration). Therefore,

$$\text{Percentage ionization} = \frac{6.5 \times 10^{-4}\ M}{0.25\ M} \times 100\%$$

$$= 0.26\%$$

The base is 0.26% ionized.

Are the Answers Reasonable?

First, we can quickly check to see whether the value of x and our assumed equilibrium concentration of N_2H_4 (0.25 M) satisfy the equilibrium law by substituting them into the mass action expression and comparing the result with K_b. Doing this, the value of the mass action expression we obtain is 1.7×10^{-6}, which is the same as K_b, so the value of x is correct. Because x is so small (i.e., because so little of the base is ionized at equilibrium), we expect a small percentage ionization, which agrees with the value we calculated.

Practice Exercise 8

Pyridine, C_5H_5N, is a bad-smelling liquid for which $K_b = 1.7 \times 10^{-9}$. What is the pH of a 0.010 M aqueous solution of pyridine? ◆

Practice Exercise 9

Phenol is an acidic organic compound for which $K_a = 1.3 \times 10^{-10}$. What is the pH of a 0.15 M solution of phenol in water? ◆

16.3 Solutions of Salts: Ions as Weak Acids and Bases

In Section 16.1 you saw that weak acids and bases are not limited to molecular substances. For instance, on page 701 NH_4^+ was given as an example of a weak acid, and on page 703 $C_2H_3O_2^-$ was cited as an example of a weak base. To prepare solutions of these ions, however, we cannot simply add them to water. Ions always come to us in compounds in which there is both a cation *and* an anion. Therefore, to place NH_4^+ in water, we need a salt such as NH_4Cl, and to place $C_2H_3O_2^-$ in water, we need a salt such as $NaC_2H_3O_2$.

Because a salt contains two ions, the pH of its solution can potentially be affected by either the cation or the anion, or perhaps even both. Therefore, we have to consider *both* ions as we assess the effect of a salt on the pH of a solution.

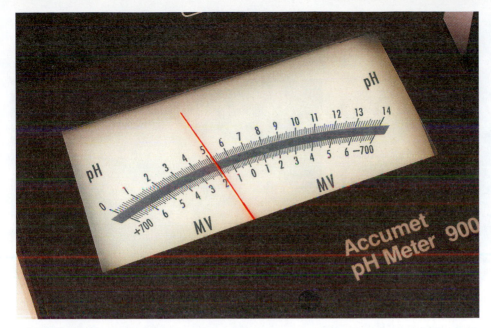

Ammonium ion as a weak acid. The measured pH of a solution of the salt NH_4NO_3 is less than 7, which indicates the solution is acidic.

Identification of acidic cations

Cations as Acids

Conjugate Acids of Molecular Bases Are Weak Acids

NH_4^+ is the conjugate acid of the molecular base, NH_3. We have already learned that NH_4^+, supplied for example by NH_4Cl, is a weak acid. Its equilibrium and K_a expression are

$$NH_4^+(aq) \rightleftharpoons NH_3(aq) + H^+(aq) \qquad K_a = \frac{[NH_3][H^+]}{[NH_4^+]}$$

Because the K_a values for ions are seldom tabulated, we would usually expect to calculate the K_a value using the relationship: $K_a \times K_b = K_w$. In Table 16.2 the K_b for NH_3 is given as 1.8×10^{-5}. Therefore,

$$K_a = \frac{K_w}{K_b} = \frac{1.0 \times 10^{-14}}{1.8 \times 10^{-5}} = 5.6 \times 10^{-10}$$

Another example is the hydrazinium ion, $N_2H_5^+$, which is also a weak acid.

$$N_2H_5^+(aq) \rightleftharpoons N_2H_4(aq) + H^+(aq) \qquad K_a = \frac{[N_2H_4][H^+]}{[N_2H_5^+]}$$

The tabulated K_b for N_2H_4 is 1.7×10^{-6}. From this, the calculated value of K_a equals 5.9×10^{-9}.

These examples illustrate that *the conjugate acids of molecular bases tend to be acidic*. Therefore,

> Salts that contain cations that are the conjugate acids of weak molecular bases can affect the pH of a solution. These cations are weak acids.

Metal Ions with High Charge Densities Are Weak Acids

In Section 15.4 you learned that small, highly charged cations such as Al^{3+} are acidic in water because the water molecules that surround the metal ion are able

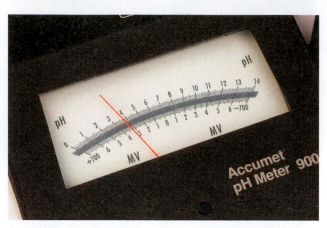

Acidity of chromium(III) ion in water. The pH of a solution of chromium(III) nitrate, which contains the blue-violet $Cr(H_2O)_6^{3+}$ ion, is distinctly acidic. This ion undergoes the same kind of equilibrium ionization as does the $Al(H_2O)_6^{3+}$ ion.

to release H^+ ions rather easily. For the hydrated aluminum ion, which can be represented as $Al(H_2O)_6^{3+}$, the equilibrium can be expressed as

$$Al(H_2O)_6^{3+}(aq) + H_2O \rightleftharpoons Al(H_2O)_5(OH)^{2+}(aq) + H_3O^+(aq)$$

Aluminum ion is just one example. Many other cations with charges of 3+, such as Cr^{3+} and Fe^{3+}, also yield aqueous solutions that are acidic. The equilibria in such solutions can be treated by the same procedures that we've used for other weak acids, but we will not discuss them further in this book. (Salts that may be of concern to us in this chapter will not contain these ions.)

> *It's the charge density (charge per unit volume) that matters, not just the size of the charge, so if the cation is very small, like Be^{2+}, the ratio of its smaller 2+ charge to its small volume can yet be large enough to make its hydrated ion a weak acid.*

Metal Ions with Small Charges Are "Nonacids"

None of the singly charged cations of the Group IA metals—Li^+, Na^+, K^+, Rb^+, or Cs^+—directly affects the pH of an aqueous solution. Except for Be^{2+}, neither do any of the doubly charged cations of the Group IIA metals—Mg^{2+}, Ca^{2+}, Sr^{2+}, or Ba^{2+}. In none of these is the ratio of charge to size apparently large enough.

Identification of basic anions

Acid	Base
HCl	Cl^-
$HC_2H_3O_2$	$C_2H_3O_2^-$

Anions as Bases

When a Brønsted acid loses a proton, its conjugate base is formed. Thus Cl^- is the conjugate base of HCl, and $C_2H_3O_2^-$ is the conjugate base of $HC_2H_3O_2$. Although both Cl^- and $C_2H_3O_2^-$ are bases, only the latter affects the pH of an aqueous solution. Why?

Earlier you learned that there is an inverse relationship between the strength of an acid and its conjugate base—the stronger the acid, the weaker is the conjugate base. Therefore, when an acid is *extremely strong,* as in the case of HCl or other "strong" acids that are 100% ionized (such as HNO_3), the conjugate base is *extremely weak*—too weak to affect in a measurable way the pH of a solution. Consequently, we have the following generalization.

> The anion of a strong acid is too weak a base to influence the pH of a solution.

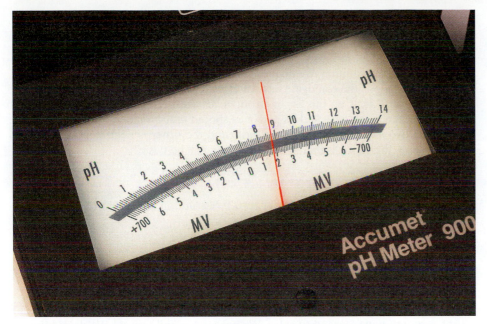

Acetate ion as a weak base in water. The pH of a solution of $KC_2H_3O_2$ is greater than 7, which shows that the solution is basic. Acetate ion reacts with water to yield small amounts of hydroxide ion.

$$C_2H_3O_2^- + H_2O \rightleftharpoons$$
$$HC_2H_3O_2 + OH^-$$

Acetic acid is much weaker than HCl, as evidenced by its value of K_a (1.8×10^{-5}). Because acetic acid is a weak acid, its conjugate base is much stronger than Cl^-. We can calculate the value of K_b for $C_2H_3O_2^-$ from the K_a for $HC_2H_3O_2$, also with the equation $K_a \times K_b = K_w$.

$$K_b = \frac{1.0 \times 10^{-14}}{1.8 \times 10^{-5}} = 5.6 \times 10^{-10}$$

This leads to another conclusion:

> The anion of a weak acid is a weak base and can influence the pH of a solution. It will tend to make the solution basic.

Predicting the Acid–Base Properties of a Salt

To decide if any given salt will affect the pH of an aqueous solution, we examine each of its ions and see what it alone might do. There are four possibilities:

1. If neither the cation nor the anion can affect the pH, the solution should be neutral.
2. If only the cation of the salt is acidic, the solution will be acidic.
3. If only the anion of the salt is basic, the solution will be basic.
4. If a salt has a cation that is acidic and an anion that is basic, the pH of the solution is determined by the *relative* strengths of the acid and base.

Let's work some examples to show how to use these generalizations.

EXAMPLE 16.8

Predicting the Effect of a Salt on the pH of a Solution

Sodium hypochlorite, NaOCl, is an ingredient in many common bleaching and disinfecting agents. Will a NaOCl solution be acidic, basic, or neutral?

Analysis: This is a salt, so we assume it to be 100% dissociated in water.

$$NaOCl(s) \xrightarrow{H_2O} Na^+(aq) + OCl^-(aq)$$

To answer the question, we take each ion, in turn, and examine its effect on the acidity of the solution.

Solution: The Na^+ ion is of a Group IA metal and is *not* acidic, so it does not affect the pH of the solution. We might say it has a "neutral" effect on the pH.

The OCl^- ion is the conjugate base of HOCl, which you should realize is a weak acid (because it isn't on the list of strong acids you're expected to know). Therefore, OCl^- is a weak base and its presence should tend to make the solution basic.

Of the two ions in the salt, one is basic and the other is neutral. The answer to the question, therefore, is that this salt solution will be basic.

Practice Exercise 10

For each compound, predict whether its 0.1 M solution in water will be acidic, basic, or neutral: (a) $NaNO_2$, (b) KCl, (c) NH_4Br. ◆

EXAMPLE 16.9
Calculating the pH of a Salt Solution

What is the pH of a 0.10 M solution of NaOCl? For HOCl, $K_a = 3.0 \times 10^{-8}$.

Analysis: Problems such as this are just like the other acid–base equilibrium problems you have learned to solve. We proceed as follows: (1) We determine the nature of the solute—is it a weak acid, a weak base, or are both a weak acid and weak base present? (2) We write the appropriate chemical equation and equilibrium law. (3) We proceed with the solution.

The solute is a salt, which is dissociated into the ions Na^+ and OCl^-. Only the latter can affect the pH. It is the conjugate *base* of HOCl, so for the purposes of problem solving, the active solute species is a weak base. On page 710 you learned that when the *only* solute is a weak base, we must use the K_b expression to solve the problem. This is where we begin the solution to the problem.

Solution: We write the chemical equation for the equilibrium ionization of the base, OCl^-, and its K_b expression.

$$OCl^- + H_2O \rightleftharpoons HOCl + OH^- \qquad K_b = \frac{[HOCl][OH^-]}{[OCl^-]}$$

The data provided in the problem is the K_a for HOCl. But we can easily calculate K_b because $K_a \times K_b = K_w$.

$$K_b = \frac{K_w}{K_a} = \frac{1.0 \times 10^{-14}}{3.0 \times 10^{-8}} = 3.3 \times 10^{-7}$$

Now let's set up the concentration table. The only sources of HOCl and OH^- are the reaction of the OCl^-, so the initial concentrations of the ions are zero and will each increase by x while the concentration of OCl^- will decrease by x.

	OCl^-	+	H_2O	$\rightleftharpoons$	HOCl	+	OH^-
Initial concentrations (M)	0.10				0		0
Changes in concentrations caused by the ionization (M)	$-x$				$+x$		$+x$
Final concentrations at equilibrium (M)	$(0.10 - x)$ ≈ 0.10				x		x

At equilibrium, the concentrations of HOCl and OH^- are the same, x.

$$[HOCl] = [OH^-] = x$$

Because K_b is so small, $[OCl^-] \approx 0.10\ M$. (Once again, x is so small we are able to use the initial concentration of the base as if it were the equilibrium concentration.) Substituting into the K_b expression gives

$$\frac{(x)(x)}{0.10} = 3.3 \times 10^{-7}$$

$$x = 1.8 \times 10^{-4}\ M$$

This value of x represents the OH^- concentration, from which we can calculate the pOH and then the pH.

$$pOH = -\log\,(1.8 \times 10^{-4})$$

$$= 3.74$$

$$pH = 14.00 - pOH = 14.00 - 3.74$$

$$= 10.26$$

The pH of this solution is 10.26.

Are the Answers Reasonable?

From the nature of the salt, we expect the solution to be basic. The calculated pH corresponds to a basic solution, so the answer seems to be reasonable. You've also seen that we can check the accuracy of the answer by substituting the calculated equilibrium concentrations into the mass action expression.

$$\frac{[HOCl]\,[OH^-]}{[OCl^-]} = \frac{(x)(x)}{0.10} = \frac{(1.8 \times 10^{-4})^2}{0.10} = 3.2 \times 10^{-7}$$

The result is quite close to the K_b calculated from the given K_a, so our answer is correct.

EXAMPLE 16.10

Calculating the pH of a Salt Solution

What is the pH of a 0.20 M solution of hydrazinium chloride, N_2H_5Cl? Hydrazine, N_2H_4, is a weak base with $K_b = 1.7 \times 10^{-6}$.

Analysis: This problem is quite similar to the preceding one. Looking over the statement of the problem, we should realize that N_2H_5Cl is a salt composed of $N_2H_5^+$ (the conjugate acid of N_2H_4) and Cl^-. Since N_2H_4 is a weak base, we expect the $N_2H_5^+$ ion to be a weak acid and thereby affect the pH of the solution. On the other hand, Cl^- is the conjugate base of HCl (a strong acid) and is too weak to influence the pH. Therefore, the only active solute species is the acid, $N_2H_5^+$, which means that to solve the problem we write the equation for the ionization of the acid and use the K_a expression.

Solution: We begin with the chemical equation for the equilibrium and write the K_a expression.

$$N_2H_5^+ \rightleftharpoons H^+ + N_2H_4 \qquad K_a = \frac{[H^+]\,[N_2H_4]}{[N_2H_5^+]}$$

The problem has given us K_b for N_2H_4, but we need K_a for $N_2H_5^+$. We obtain this by solving the equation $K_a \times K_b = K_w$ for K_a.

$$K_a = \frac{K_w}{K_b} = \frac{1.0 \times 10^{-14}}{1.7 \times 10^{-6}} = 5.9 \times 10^{-9}$$

Now we set up the concentration table. The initial concentrations of H^+ and N_2H_4 are both set equal to zero; there is no strong acid to give H^+ in the solution, and no

N_2H_4 is present before the reaction of water with $N_2H_5^+$. Next, we indicate that the concentration of $N_2H_5^+$ decreases by x and the concentrations of H^+ and N_2H_4 both increase by x.

	$N_2H_5^+$	$\rightleftharpoons$	H^+	+	N_2H_4
Initial concentrations (M)	0.20		0		0
Changes in concentrations caused by the ionization (M)	$-x$		$+x$		$+x$
Final concentrations at equilibrium (M)	$(0.20 - x)$ ≈ 0.20		x		x

At equilibrium, equal amounts of H^+ and N_2H_4 are present, and their equilibrium concentrations are each equal to x.

$$[H^+] = [N_2H_4] = x$$

We also assume that $[N_2H_5^+] \approx 0.20\ M$ and then substitute quantities into the mass action expression.

$$\frac{(x)(x)}{0.20} = 5.9 \times 10^{-9}$$

$$x = 3.4 \times 10^{-5}\ M$$

Since $x = [H^+]$, the pH of the solution is

$$pH = -\log(3.4 \times 10^{-5})$$

$$= 4.47$$

Is the Answer Reasonable?

The active solute species in the solution is a weak acid and the calculated pH is less than 7, so the answer seems reasonable. Check the accuracy yourself by substituting equilibrium concentrations into the mass action expression.

Practice Exercise 11

What is the pH of a 0.10 M solution of $NaNO_2$? ◆

Practice Exercise 12

What is the pH of a 0.10 M solution of NH_4Br? ◆

Solutions That Contain the Salt of a Weak Acid and a Weak Base

There are many salts whose ions are both able to affect the pH of the salt solution. Whether or not the salt has a net effect on the pH now depends on the relative strengths of its ions in functioning one as an acid and the other as a base. If they are matched in their respective strengths, the salt has no net effect on pH. In ammonium acetate, for example, the ammonium ion is an acidic cation and the acetate ion is a basic anion. However, the K_a of NH_4^+ is 5.6×10^{-10} and the K_b of $C_2H_3O_2^-$ just happens to be the same, 5.6×10^{-10}. The cation tends to produce H^+ ions to the same extent that the anion tends to produce OH^-. So in aqueous ammonium acetate, $[H^+] = [OH^-]$, and the solution has a pH of 7.

Consider, now, ammonium formate, NH_4CHO_2. The formate ion, CHO_2^-, is the conjugate base of the weak acid, formic acid, so it is a Brønsted base. Its K_b

is 5.6×10^{-11}. Comparing this value to the (slightly larger) K_a of the ammonium ion, 5.6×10^{-10}, we see that NH_4^+ is slightly stronger as an acid than the formate ion is as a base. As a result, a solution of ammonium formate is slightly acidic. We are not concerned here about calculating a pH, only in predicting if the solution is acidic, basic, or neutral.

EXAMPLE 16.11

Predicting How a Salt Affects the pH of Its Solution

Will an aqueous solution that is 0.20 M NH_4F be acidic, basic, or neutral?

Analysis: This is a salt in which the cation is a weak acid (it's the conjugate acid of a weak base) and the anion is a weak base (it's the conjugate base of a weak acid). The question, then, is, "How do the two ions compare in their abilities to affect the pH of the solution?"

Solution: The K_a of NH_4^+ (calculated from the K_b for NH_3) is 5.6×10^{-10}. Similarly, the K_b of F^- is 1.5×10^{-11} (calculated from the K_a for HF, 6.8×10^{-4}). Comparing the two equilibrium constants, we see that the acid (NH_4^+) is stronger than the base (F^-), so we expect the solution to be slightly acidic.

Practice Exercise 13

Will an aqueous solution of ammonium cyanide, NH_4CN, be acidic, basic, or neutral? ◆

16.4 Equilibrium Calculations when Simplifications Fail

In the preceding two sections we used initial concentrations of solutes as though they were equilibrium concentrations when we performed calculations. This is only an approximation, as we discussed on page 712, but it is one that works for many situations. Unfortunately, it does not work in all cases, so now that you have learned the basic approach to solving equilibrium problems, we will examine those conditions under which simplifying approximations do and do not work. We will also study how to solve problems when the approximations cannot be used.

When Simplifying Assumptions Fail

When a weak acid, HA, ionizes in water, its concentration is reduced as the ions form. If we let x represent the amount of acid that ionizes per liter, the equilibrium concentration becomes

$$[HA]_{\text{equilib}} = [HA]_{\text{initial}} - x$$

For reasons beyond the scope of this text,[2] the accuracy of acid–base equilibrium calculations is limited to about ±5%. So as long as x does not exceed ±5% of $[HA]_{\text{initial}}$, we assume its value is negligible and say that

$$[HA]_{\text{equilib}} \approx [HA]_{\text{initial}}$$

[2]To properly perform equilibrium calculations, we should use quantities called *activities* instead of molar concentrations. The activity of a substance is its effective concentration, which is affected by its electrical charge and the concentrations of other charged species in the solution. By using molar concentrations we introduce errors in the calculations that limit the accuracy we are able to obtain, so we generally carry only two significant figures. Using molar concentrations, however, makes the problems much easier to cope with mathematically.

It is not difficult to show that for x to be less than or equal to ±5% of $[HA]_{initial}$, $[HA]_{initial}$ must be greater than or equal to 400 times the value of K_a.

Tools

When simplifications work

> Simplifications work when $[HA]_{initial} \geq 400 \times K_a$

For the equilibrium problems in the preceding sections, this condition was fulfilled, so our simplifications were valid. We now want to study what to do when $[HA]_{initial} < 400 \times K_a$ (i.e., when we cannot justify simplifying the algebra).

The Quadratic Solution

When the algebraic equation obtained by substituting quantities into the equilibrium law is a quadratic equation, we can use the quadratic formula to obtain the solution. This is illustrated by the following example.

EXAMPLE 16.12

Using the Quadratic Formula in Equilibrium Problems

Chloroacetic acid, $HC_2H_2O_2Cl$, is used as an herbicide and in the manufacture of dyes and other organic chemicals. It is a weak acid with $K_a = 1.4 \times 10^{-3}$. What is the pH of a 0.010 M solution of $HC_2H_2O_2Cl$?

Analysis: Before we begin the solution, we check to see whether we can use our usual simplifying approximation. We do this by calculating $400 \times K_a$ and comparing the result to the initial concentration of the acid.

$$400 \times K_a = 400(1.4 \times 10^{-3}) = 0.56$$

The initial concentration of $HC_2H_2O_2Cl$ is *less than* 0.56, so we know the simplification does not work.

chloroacetic acid

Solution: Let's begin by writing the chemical equation and the K_a expression. (We know we must use K_a because the only solute is the acid.)

$$HC_2H_2O_2Cl \rightleftharpoons H^+ + C_2H_2O_2Cl^-$$

$$K_a = \frac{[H^+][C_2H_2O_2Cl^-]}{[HC_2H_2O_2Cl]} = 1.4 \times 10^{-3}$$

The initial concentration of the acid will be reduced by an amount x as it undergoes ionization to form the ions. From this, let's build the concentration table.

	$HC_2H_2O_2Cl$	$\rightleftharpoons$	H^+	$+$	$C_2H_2O_2Cl^-$
Initial concentrations (M)	0.10		0		0
Changes in concentrations (M)	$-x$		$+x$		$+x$
Equilibrium concentrations (M)	$(0.010 - x)$		x		x

Substituting equilibrium concentrations into the equilibrium law gives

$$\frac{(x)(x)}{(0.010 - x)} = 1.4 \times 10^{-3}$$

This time we cannot neglect x. To solve the problem, we first clear fractions by multiplying both sides by $(0.010 - x)$. This gives

$$x^2 = (0.010 - x)1.4 \times 10^{-3}$$

$$x^2 = (1.4 \times 10^{-5}) - (1.4 \times 10^{-3})x$$

The general quadratic equation has the form

$$ax^2 + bx + c = 0$$

The values of x that make this equation true are related to the coefficients a, b, and c by the quadratic formula:

$$x = \frac{-b \pm \sqrt{b^2 - 4ac}}{2a}$$

Rearranging our equation to follow the general form gives

$$x^2 + (1.4 \times 10^{-3})x - (1.4 \times 10^{-5}) = 0$$

so we make the substitutions $a = 1$, $b = 1.4 \times 10^{-3}$, and $c = -1.4 \times 10^{-5}$. Entering these into the quadratic formula gives

$$x = \frac{-1.4 \times 10^{-3} \pm \sqrt{(1.4 \times 10^{-3})^2 - 4(1)(-1.4 \times 10^{-5})}}{2(1)}$$

$$= \frac{-1.4 \times 10^{-3} \pm \sqrt{2.0 \times 10^{-6} + 5.6 \times 10^{-5}}}{2}$$

$$= \frac{-1.4 \times 10^{-3} \pm \sqrt{5.8 \times 10^{-5}}}{2}$$

$$= \frac{-1.4 \times 10^{-3} \pm 7.6 \times 10^{-3}}{2}$$

Because of the $\pm$ sign we obtain two values for x, but as you saw in Chapter 15, only one of them makes any sense. Here are the two values:

$$x = 3.1 \times 10^{-3} \, M \quad \text{and} \quad x = -4.5 \times 10^{-3} \, M$$

We know that x cannot be negative because that would give negative concentrations for the ions (which is impossible), so we must choose the first value as the correct one. This yields the following equilibrium concentrations.

$$[H^+] = 3.1 \times 10^{-3} \, M$$

$$[C_2H_2O_3Cl^-] = 3.1 \times 10^{-3} \, M$$

$$[HC_2H_2O_2Cl] = 0.010 - 0.0031$$

$$= 0.007 \, M$$

Finally, we calculate the pH of the solution.

$$pH = -\log (3.1 \times 10^{-3})$$

$$= 2.51$$

Notice that, indeed, x is not negligible compared to the initial concentration, so the simplifying approximation would not have been valid.

Checking the Answer
As before, a quick check can be performed by substituting the calculated equilibrium concentrations into the mass action expression.

$$\frac{(3.1 \times 10^{-3})^2}{0.007} = 1.4 \times 10^{-3}$$

The value we obtain equals K_a, so the equilibrium concentrations are correct.

Practice Exercise 14

Calculate the pH of a 0.0010 M solution of dimethylamine, $(CH_3)_2NH$, for which $K_b = 9.6 \times 10^{-4}$. ◆

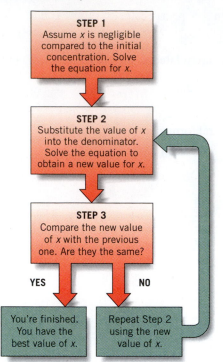

Figure 16.3 *The method of successive approximations.* Following the steps outlined here often leads to a rapid solution of equilibrium problems when the usual simplifications fail.

Solving by Successive Approximations

If you worked your way through the discussion of the use of the quadratic formula, you'll surely be better able to appreciate the *method of successive approximations.* When using a calculator without a built-in quadratic function, it is usually much faster and just as accurate. The procedure is outlined in Figure 16.3; follow the diagram as we apply the method below.

If we were to attempt to use the simplifying approximation in the preceding example, we would obtain the following:

$$\frac{x^2}{0.010} = 1.4 \times 10^{-3}$$

for which we obtain the solution $x = 3.7 \times 10^{-3}$. We will call this our *first approximation* to a solution to the equation

$$\frac{(x)(x)}{(0.010 - x)} = 1.4 \times 10^{-3}$$

Let's put $x = 3.7 \times 10^{-3}$ into the term in the denominator, $(0.100 - x)$, and recalculate x. This gives us

$$\frac{x^2}{(0.010 - 0.0037)} = 1.4 \times 10^{-3}$$

or

$$x^2 = (0.006) \times 1.4 \times 10^{-3}$$

Taking only the positive root,

$$x = 2.9 \times 10^{-3}$$

Notice that this value of x is much closer to the value calculated using the quadratic formula (which gave $x = 3.1 \times 10^{-3}$). Now we'll call $x = 2.9 \times 10^{-3}$ our *second approximation*, and repeat the process. This gives

$$\frac{x^2}{(0.010 - 0.029)} = 1.4 \times 10^{-3}$$

$$x^2 = (0.007) \times 1.4 \times 10^{-3}$$

$$x = 3.1 \times 10^{-3}$$

This is the identical value obtained using the quadratic formula. Notice also that the *change* in x between the two approximations grew smaller. The second approximation gave a smaller correction than the first. Each succeeding approximation differs from the preceding one by smaller and smaller amounts. We stop the calculation when the difference between two approximations is less than ten percent of the final approximation. Try this approach by reworking Practice Exercise 14 and solving for $[OH^-]$ in the $0.0010\ M\ (CH_3)_2NH$ solution using the method of successive approximations.[3]

16.5 Buffers: The Control of pH

We all know that acids cause things made of metals, like cars or battery terminals, to corrode faster. Chefs know that adding just a little lemon juice to milk makes it curdle. If the pH of your blood were to change from what it should be, namely, 7.35 (measured at room temperature), either to 7.00 or to 8.00, you would die. Thus, a change in pH can cause chemical reactions, sometimes entirely unwanted. Fortunately, there are ways to protect systems against large changes in pH.

   By a careful choice of solutes, a solution can be prepared that will experience no more than a small change in pH even if strong acid or strong base is added or is produced by some reaction. Such mixtures of solutes are called **buffers**, because they do just that—they buffer the system against a change in pH. The solution itself is said to be *buffered* or is described as a *buffer solution*. There are few areas of experimental work in chemistry, or in its applications in other fields, where the concept of buffers is not important. Buffers introduce no new concepts to what we have already studied, just a new application.

Every life-form is extremely sensitive to slight changes in pH.

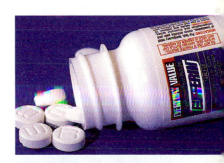

A buffered aspirin product. The buffering ingredients in this product are chosen to control stomach acidity caused by the main ingredient, aspirin (acetylsalicylic acid).

Components of Buffers

Usually, a buffer consists of two solutes. One provides a weak Brønsted acid and the other provides a weak Brønsted base. Usually, the acid and base represent a conjugate pair. If the acid is molecular, then the conjugate base is *supplied by a soluble salt of the acid*. For example, a common buffer system consists of acetic acid plus sodium acetate, with the salt's acetate ion serving as the Brønsted base. In your blood, carbonic acid (H_2CO_3, a weak diprotic acid) and the bicarbonate ion (HCO_3^-, its conjugate base) serve as one of the buffer systems used to

---

[3]Many modern handheld calculators have "solver" functions such as the quadratic formula that enable the user to solve equilibrium problems such as those discussed here without having to make approximations. You might wish to check the instruction manual for your calculator to see if it has these capabilities.

maintain a remarkably constant pH in the face of the body's production of organic acids by metabolism. Another common buffer consists of the weakly acidic cation, NH_4^+, supplied by a salt like NH_4Cl, and its conjugate base, NH_3.

One important point about buffers is the distinction between keeping a solution at a particular pH and keeping it neutral—at a pH of 7. Although it is certainly possible to prepare a buffer to work at pH 7, buffers can be made that will work around any pH value throughout the pH scale. One topic we must study, therefore, is how to pick the weak acid and its salt—and their mole ratio—that would make a buffer good for some chosen pH.

Buffer systems do *not* protect a solution against a tide of strong acid or base, just relatively small amounts. So another topic we have to consider is the *capacity* of a buffer, which is the amount of strong acid or base it can absorb without undergoing more than a specified change in pH.

How a Buffer Works

Tools

Reactions in a buffer when H^+ or OH^- is added

For simplicity, we will first confine our discussion to the HA/A^- type of buffer system, like the acetic acid/sodium acetate buffer. Let's first see *how* this system can buffer a solution.

To work, a buffer must be able to neutralize either a strong acid or strong base that is added. This is precisely what the weak base and weak acid components of the buffer do. If we add extra H^+ to the buffer (from a strong acid) the weak conjugate base can react with it as follows.

$$H^+(aq) + A^-(aq) \longrightarrow HA(aq)$$

Thus, the added H^+ changes some of the buffer's Brønsted base, A^-, to its conjugate (weak) acid, HA. This reaction prevents a large buildup of H^+ that would otherwise be caused by the addition of the strong acid.

A similar response occurs when a strong base is added to the buffer. The OH^- from the strong base will react with some HA.

$$HA(aq) + OH^-(aq) \longrightarrow A^-(aq) + H_2O$$

Here the added OH^- changes some of the buffer's Brønsted acid, HA, into its conjugate base, A^-. This prevents a buildup of OH^-, which would otherwise cause a large change in the pH. Thus, one member of a buffer team neutralizes H^+ that might get into the solution, and the other member neutralizes OH^-.

Figure 16.4 *A practical application of a buffer.* Baking soda, which is sodium bicarbonate, is sometimes added to swimming pools to control the pH of the water.

▶**Chemistry in Practice**◀ Although most buffer systems consist of two separate species that react with H^+ or OH^-, the bicarbonate ion is an example of a single ion that is able to serve both functions. The reactions are

$$HCO_3^-(aq) + H^+(aq) \longrightarrow H_2CO_3(aq)$$

$$HCO_3^-(aq) + OH^-(aq) \longrightarrow H_2O + CO_3^{2-}(aq)$$

Because sodium bicarbonate is nontoxic and because the pH of a solution of the salt is close to 7.0, there are many practical applications of the HCO_3^- buffer. One is in controlling the pH of a swimming pool (Figure 16.4). The chemicals that are added to a swimming pool to retard the growth of bacteria can also affect the pH of the water, making it unpleasant for swimmers. Adding nontoxic $NaHCO_3$ is an inexpensive and effective way to maintain the pool's pH at an acceptable value. ◆

Calculating the pH of a Buffer Solution

The Acetic Acid–Acetate Ion Buffer System

The acetate buffer is a solution of both acetic acid and sodium acetate, which provides the acetate ion in solution. The $HC_2H_3O_2$ in the buffer neutralizes OH^- as follows:

$$HC_2H_3O_2(aq) + OH^-(aq) \longrightarrow C_2H_3O_2^-(aq) + H_2O$$

and its acetate ion neutralizes H^+ as follows.

$$C_2H_3O_2^-(aq) + H^+(aq) \longrightarrow HC_2H_3O_2(aq)$$

The following example illustrates how we can calculate the pH of such a buffer mixture.

A mixture of $HC_2H_3O_2$ and $C_2H_3O_2^-$ is called the acetate buffer.

EXAMPLE 16.13

Calculating the pH of a Buffer

To study the effects of a weakly acidic medium on the rate of corrosion of a metal alloy, a student prepared a solution by making it both 0.11 M $NaC_2H_3O_2$ and also 0.090 M $HC_2H_3O_2$. What is the pH of this solution?

Analysis: The buffer solution contains both the weak acid $HC_2H_3O_2$ and its conjugate base $C_2H_3O_2^-$. Earlier we noted that when both solute species are present we can use *either* K_a or K_b to perform calculations, whichever is handy. In our tables we find $K_a = 1.8 \times 10^{-5}$ for $HC_2H_3O_2$, so the simplest approach is to use the equation for the ionization of the acid. We will also be able to use the simplifying approximations developed earlier; these always work for buffers, so *we will be able to use the initial concentrations as though they are equilibrium values.*

Solution: We begin with the chemical equation and the expression for K_a.

$$HC_2H_3O_2 \rightleftharpoons H^+ + C_2H_3O_2^- \qquad K_a = \frac{[H^+][C_2H_3O_2^-]}{[HC_2H_3O_2]} = 1.8 \times 10^{-5}$$

Let's set up the concentration table this time so we can proceed carefully. We will take the initial concentrations of $HC_2H_3O_2$ and $C_2H_3O_2$ to be the values given in the problem. There's no H^+ present from a strong acid, so we set this concentration equal to zero. If the initial concentration of H^+ is zero, its concentration must increase on the way to equilibrium, so under H^+ in the change row we enter $+x$. The other changes follow from that. Here's the completed table.

	$HC_2H_3O_2$	$\rightleftharpoons$	H^+	$+$	$C_2H_3O_2^-$
Initial concentrations (M)	0.090		0		0.11
Changes in concentrations (M)	$-x$		$+x$		$+x$
Equilibrium concentrations (M)	$(0.090 - x) \approx 0.090$		x		$(0.11 + x) \approx 0.11$

For buffer solutions the quantity x will be very small, so it is safe to make the simplifying approximations. What remains, then, is to substitute the quantities from the last row of the table into the K_a expression.

$$1.8 \times 10^{-5} = \frac{(x)(0.11)}{(0.090)}$$

Solving for x gives us

$$x = \frac{(0.090) \times (1.8 \times 10^{-5})}{(0.11)}$$

$$= 1.5 \times 10^{-5}$$

Because x equals $[H^+]$, we now have $[H^+] = 1.5 \times 10^{-5}\ M$. Then we calculate pH.

$$pH = -\log(1.5 \times 10^{-5})$$

$$= 4.82$$

Thus the pH of the buffer is 4.82.

Notice how small x is compared to the initial concentrations. The simplification was valid.

Checking the Answer

We can check the answer in the usual way by substituting our calculated equilibrium values into the mass expression. Let's do it.

$$\frac{[H^+][C_2H_3O_2^-]}{[HC_2H_3O_2]} = \frac{(1.5 \times 10^{-5})(0.11)}{(0.090)} = 1.8 \times 10^{-5}$$

The reaction quotient equals K_a, so the values we've obtained are correct equilibrium concentrations.

Practice Exercise 15

Calculate the pH of the buffer solution in the preceding example by using the K_b for $C_2H_3O_2^-$. (Be sure to write the chemical equation for the equilibrium as the reaction of $C_2H_3O_2^-$ with water. Then use the chemical equation as a guide in setting up the equilibrium expression for K_b. If you work the problem correctly, you should obtain the same answer as above.) ◆

Permissible Simplifications in Buffer Calculations

There are two useful simplifications that we can use in working buffer calculations. The first is the one we made in Example 16.13:

There are some buffer systems for which these simplifications might not work. However, you will not encounter them in this text.

> We will be justified in using the *initial* concentrations of both the weak acid and its conjugate base as though they were equilibrium values.

There is a further simplification that can be made because the mass action expression contains the ratio of the molar concentrations (in units of moles per liter) of the acid and conjugate base. Let's enter these units for the acid and its conjugate base into the mass action expression. For an acid HA,

$$K_a = \frac{[H^+][A^-]}{[HA]} = \frac{[H^+](\text{mol } A^-\ L^{-1})}{(\text{mol } HA\ L^{-1})} = \frac{[H^+](\text{mol } A^-)}{(\text{mol } HA)} \tag{16.8}$$

Notice that the units L^{-1} cancel from the numerator and denominator. This means that for a given acid–base pair, the $[H^+]$ is determined by the *mole* ratio of conjugate base to conjugate acid; we don't *have* to use molar concentrations.

> *For buffer solutions **only,*** we can use either molar concentrations or moles in the K_a (or K_b) expression to express the amounts of the members of the conjugate acid–base pair (but we must use the same units for each member of the pair).

A further consequence of the relationship derived above is that the pH of a buffer should not change if the buffer is diluted. Dilution changes the volume of a solution but it does not change the number of moles of the solutes, so their mole *ratio* remains constant and so does [H$^+$].

The Ammonia–Ammonium Ion Buffer

A solution of ammonium chloride in aqueous ammonia also provides a weak acid, NH$_4^+$, and its conjugate base, NH$_3$, and so it is a buffer. It is able to neutralize OH$^-$ by the following reaction.

$$NH_4^+(aq) + OH^-(aq) \longrightarrow NH_3(aq) + H_2O$$

The conjugate base can neutralize H$^+$.

$$NH_3(aq) + H^+(aq) \longrightarrow NH_4^+(aq)$$

EXAMPLE 16.14

Calculating the pH of an Ammonia/Ammonium Ion Buffer

To study the influence of an alkaline medium on the rate of a reaction, a student prepared a buffer solution by dissolving 0.12 mol of NH$_3$ and 0.095 mol of NH$_4$Cl in water. What is the pH of the buffer?

Analysis: The pH of the buffer is determined by the mole ratio of the members of the acid–base pair, so to calculate the pH we will be able to use the moles of NH$_3$ and NH$_4^+$ directly in the mass action expression. To set up the equilibrium law we can use either the K_a for NH$_4^+$ or the K_b for NH$_3$. Since K_b is tabulated, we will use it and write the equation for the ionization of the base.

Solution: We begin with the chemical equation and the K_b expression.

$$NH_3 + H_2O \rightleftharpoons NH_4^+ + OH^- \qquad K_b = \frac{[NH_4^+][OH^-]}{[NH_3]} = 1.8 \times 10^{-5}$$

The solution contains 0.12 mol of NH$_3$ and 0.095 mol of NH$_4^+$ from the complete dissociation of the salt NH$_4$Cl. We can enter these quantities into the mass action expression and solve for [OH$^-$].

The solution also contains 0.095 mol of Cl$^-$, of course, but this ion is not involved in the equilibrium.

$$1.8 \times 10^{-5} = \frac{(0.095)[OH^-]}{0.12}$$

solving for [OH$^-$] gives

$$[OH^-] = 2.3 \times 10^{-5}$$

To calculate the pH, we obtain pOH and subtract it from 14.00.

$$pOH = -\log(2.3 \times 10^{-5})$$

$$= 4.64$$

$$pH = 14.00 - 4.64$$

$$= 9.36$$

Checking the Answer

By now you know that you can check the answer by substituting equilibrium concentrations into the mass action expression. Try it.

Practice Exercise 16

Determine the pH of the buffer in the preceding example using the K_a for NH$_4^+$, which you can calculate from K_b for NH$_3$. (If you work the problem correctly, you should obtain the same answer as above.) ◆

Preparation of a Buffer with a Given pH

Biologists also must be concerned with possible toxic side effects produced by the components of a buffer.

In experimental work, a chemist or biologist usually first decides the pH at which a particular system should be buffered and then chooses the buffer components that will best deliver this pH. This choice requires careful attention to what most affects the pH of a buffer.

The Factors That Govern the pH of a Buffer Solution

We can more clearly see the two factors that dominate the pH of a buffered solution by rearranging Equation 16.8 to solve for $[H^+]$.

$$[H^+] = K_a \frac{[HA]}{[A^-]} \tag{16.9}$$

or

$$[H^+] = K_a \frac{\text{mol } HA}{\text{mol } A^-} \tag{16.10}$$

The first factor to affect $[H^+]$ (and therefore pH) is the K_a of the weak acid. The second is the *ratio* of the molarities (in Equation 16.9) or the ratio of moles (in Equation 16.10). The quantities we use in these two equations, of course, are the "initial" values—either concentrations or moles. In other words, we assume their values don't change as a result of ionization of the weak acid. To emphasize this, let's rewrite them as follows[4]

$$[H^+] = K_a \frac{[HA]_{\text{initial}}}{[A^-]_{\text{initial}}} \tag{16.11}$$

$$[H^+] = K_a \frac{(\text{mol } HA)_{\text{initial}}}{(\text{mol } A^-)_{\text{initial}}} \tag{16.12}$$

Notice particularly what happens if we prepare a buffer so as to make the concentrations of the two buffer components identical. Then the ratio $[HA]_{\text{initial}}/[A^-]_{\text{initial}}$ in Equation 16.11 equals 1, which means that $[H^+] = K_a$ and therefore pH = pK_a.

Selecting the Weak Acid for the Preparation of a Buffer Solution

When $[H^+] = K_a$, then $-\log[H^+] = -\log K_a$. So pH $= pK_a$.

Usually, buffers are made so that the ratio $[HA]_{\text{initial}}/[A^-]_{\text{initial}}$ is not greatly different from 1. Consequently, *what mostly determines where, on the pH scale, a buffer can work best is the pK_a of the weak acid.* Thus, to prepare a specific buffer for use at a prechosen pH, we first select a weak acid whose pK_a is near

---

[4]If you take a biology course, you're likely to run into a logarithmic form of Equation 16.11 called the *Henderson–Hasselbalch equation*. This is obtained by taking the negative logarithm of both sides of Equation 16.11 and rearranging the term involving the concentrations.

$$\text{pH} = pK_a + \log \frac{[A^-]_{\text{initial}}}{[HA]_{\text{initial}}}$$

In most buffers used in the life sciences, the anion A^- comes from a salt in which the cation has a charge of 1+, such as NaA, and the acid is monoprotic. With these as conditions, the equation is sometimes written (where the quantities in parentheses represent initial concentrations)

$$\text{pH} = pK_a + \log \frac{[\text{salt}]}{[\text{acid}]}$$

For practice, you may wish to apply this equation to buffer problems at the end of the chapter.

the pH we desire. Almost never can an acid be found, however, whose pK_a *exactly* equals the pH we want. So we try for an acid whose pK_a is *close* to this pH. Then, by experimentally adjusting the ratio $[HA]_{initial}/[A^-]_{initial}$, we can make a final adjustment to get the desired pH.

The desirable range for the ratio $[HA]_{initial}/[A^-]_{initial}$ is from $\frac{10}{1}$ to $\frac{1}{10}$. Outside this range we can run into problems with the solubilities of the solutes, or we can have a buffer of low capacity. (The question of buffer *capacity* will be studied soon.) According to Equation 16.11, keeping the ratio $[HA]_{initial}/[A^-]_{initial}$ within these limits means that the $[H^+]$ will range from $10 \times K_a$ to $0.1 \times K_a$. Taking the negative logarithm of both sides, this translates into the desirable range of pH values for a buffer normally being

$$pH = pK_a \pm 1 \qquad (16.13)$$

EXAMPLE 16.15

Preparing a Buffer Solution to Have a Predetermined pH

A solution buffered at a pH of 5.00 is needed in an experiment. Can we use acetic acid and sodium acetate to make it? If so, what mole ratio of acetic acid to acetate ion is needed?

Analysis: We first check the pK_a of acetic acid to see if it's in the desired range of $pH = pK_a \pm 1$. If so, then we proceed to calculate the required mole ratio.

Solution: Because we want the pH to be 5.00, the pK_a of the selected acid should be 5.00 ± 1, meaning the range of pK_a between 4.00 and 6.00. Because $K_a = 1.8 \times 10^{-5}$ for acetic acid, $pK_a = 4.74$. So the pK_a of acetic acid falls in the desired range, and acetic acid can be used together with the acetate ion to make the buffer.

We will use Equation 16.12 to answer the second question: What mole ratio of solutes is needed?

$$[H^+] = K_a \frac{(\text{mol } HC_2H_3O_2)_{initial}}{(\text{mol } C_2H_3O_2^-)_{initial}} \qquad (16.14)$$

Solving for the mole ratio gives

$$\frac{(\text{mol } HC_2H_3O_2)_{initial}}{(\text{mol } C_2H_3O_2^-)_{initial}} = \frac{[H^+]}{K_a}$$

The desired pH = 5.00, so $[H^+] = 1.0 \times 10^{-5}$; also $K_a = 1.8 \times 10^{-5}$. Substituting gives

$$\frac{(\text{mol } HC_2H_3O_2)_{initial}}{(\text{mol } C_2H_3O_2^-)_{initial}} = \frac{1.0 \times 10^{-5}}{1.8 \times 10^{-5}} = 0.56$$

This is the *mole* ratio of the buffer components we want. So we have to prepare the solution such that

$$\text{moles of } HC_2H_3O_2 = 0.56 \times (\text{moles of } C_2H_3O_2^-)$$

Suppose that we want 1.0 L of the solution, and that we indeed use sodium acetate as the source of the acetate ion. We could, for example, dissolve 0.10 mol of $NaC_2H_3O_2$ in water and then add 0.056 mol of $HC_2H_3O_2$, making the final volume equal to 1.0 L. We could have chosen quantities at one-tenth this, for example 0.010 mol of $NaC_2H_3O_2$ and 0.0056 mol of $HC_2H_3O_2$. The *ratio* would still be the same. The solution would still be buffered at a pH of 5.00.

Practice Exercise 17

A student needed an aqueous buffer for a pH of 3.90. Would formic acid and its salt, sodium formate, make a good pair for this purpose? If so, what mole ratio of the acid, $HCHO_2$, to the anion of this salt, CHO_2^-, is needed? ◆

Buffer Capacity

There is a limit to a buffer's *capacity*—how much strong acid or strong base the buffer is able to absorb before its buffering ability is essentially destroyed. For example, if enough strong acid is added to react with all of the base component of the buffer, the mixture will no longer be able to neutralize more strong acid if it enters the solution. Similarly, if enough strong base is added to neutralize all of the acid component of the buffer, the mixture will no longer be able to react with additional strong base. In either case, the buffering ability of the mixture will have been exhausted.

A buffer's *capacity* is determined by the sizes of the actual molarities of its components. So we must decide before making the buffer solution what outer limits to the change in pH can be tolerated—0.1 unit, 0.2 unit, or something else. Then, consistent with the necessary *ratio* of base to acid, we calculate the actual number of moles of each that we will dissolve in the solution. Let's do a calculation to show how effective a buffer is at resisting pH changes produced by a relatively small amount of extra acid (or base).

A buffer of low capacity might be needed if its ions and molecules could, if too concentrated, interfere with another use of the solution.

EXAMPLE 16.16
The Capacity of a Buffer to Resist Changes in pH

A student had available 1.00 L of an acetic acid/sodium acetate buffer that contained 1.00 M $HC_2H_3O_2$ and 1.00 M $NaC_2H_3O_2$. This entire solution was intended for an experiment in which 0.15 mole of H^+ was expected to be generated by a reaction without changing the volume of the solution. By how much will the pH of the buffer change by the addition of this much H^+? Could the buffer be used in this experiment if the maximum allowable pH change could be no larger than 0.2 unit?

Analysis: To answer the question, we need to know the initial pH of the buffer and the pH after the addition of the strong acid. We use the given concentrations of the $HC_2H_3O_2$ and $C_2H_3O_2^-$ to obtain the initial pH. Next, we determine how the concentrations of the $HC_2H_3O_2$ and $C_2H_3O_2^-$ change when the system absorbs the strong acid. We then use these new concentrations to calculate the new pH of the buffer. The difference between the new and original pH values tells us how much the pH changes.

Solution: First, we calculate the pH of the buffer before addition of the H^+. For acetic acid, we have seen that

$$K_a = \frac{[H^+][C_2H_3O_2^-]}{[HC_2H_3O_2]} = 1.8 \times 10^{-5}$$

The given data tell us that initially, $[HC_2H_3O_2]_{initial} = 1.00$ M and $[C_2H_3O_2^-]_{initial} = 1.00$ M. We will assume these initial values are effectively the same as the equilibrium concentrations. Substituting them into the mass action expression lets us calculate the initial pH of the buffer.

$$\frac{[H^+](1.00)}{1.00} = 1.8 \times 10^{-5}$$

$$[H^+] = 1.8 \times 10^{-5}$$

$$pH = 4.74$$

Next, we consider the reaction that takes place as H^+ enters the solution. Recall that the conjugate base of the buffer neutralizes H^+.

The equilibrium constant for this reaction is 5.6×10^4, which is the reciprocal of K_a for acetic acid. Its large value tells us the reaction proceeds very far toward completion.

$$H^+ + C_2H_3O_2^- \longrightarrow HC_2H_3O_2$$

The reaction is nearly complete, in the sense that virtually all the added H^+ is consumed. As an approximation, we assume complete reaction, so for each mole of H^+

added, a mole of $C_2H_3O_2^-$ is changed to a mole of $HC_2H_3O_2$. Since 0.15 mol L^{-1} of H^+ is added, the concentration of $HC_2H_3O_2$ increases by this amount and the concentration of $C_2H_3O_2^-$ decreases by this amount. Applying these changes gives the final concentrations.

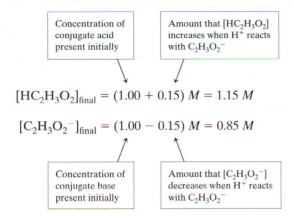

We now substitute these values into the mass action expression to calculate the final $[H^+]$ in the solution.

$$\frac{[H^+](0.85)}{(1.15)} = 1.8 \times 10^{-5}$$

$$[H^+] = 2.4 \times 10^{-5} \text{ mol } L^{-1}$$

From this we calculate the final pH,

$$pH = 4.62$$

By adding 0.15 mol of H^+ to a liter of the buffer, its pH has gone from 4.74 to 4.62, a change of only 0.12 pH unit (which is smaller than the limit of ±0.2 pH unit). The buffer solution, in other words, was able to absorb 0.15 mole of H^+ without exceeding the limit, even though the added H^+ consumed 15% of the available base, $C_2H_3O_2^-$.

Are the Answers Reasonable?

There's no simple way to estimate how *much* the pH will change when H^+ or OH^- is added to the buffer; however, we can anticipate the *direction* of the change. Adding H^+ to a buffer will lower the pH somewhat and adding OH^- will raise it. In this example, the buffer is absorbing H^+, so its pH should drop, and our calculations agree. If our calculated pH had been higher than the original, we might expect to find errors in the calculation of the final values of $[HC_2H_3O_2]$ and $[C_2H_3O_2^-]$.

Practice Exercise 18

Suppose that the buffer of the preceding example is used in an experiment in which 0.11 mol of OH^- ion will be generated (with no volume change). Can the buffer handle this without having the pH change by more than 0.2 unit? Calculate the new pH. ◆

Example 16.16 and Practice Exercise 18 demonstrate how effective a buffer can be at absorbing H^+ and OH^- with only minimal changes in pH. In many experiments it's necessary to limit the pH change to ranges smaller than ±0.2 pH unit (e.g., to ±0.1 pH unit or less). Generally, the mole quantities of buffer required to do this must be upwards of 10 times greater than the expected influx of strong acid or base. Maintaining the pH within such narrow ranges is particularly important in work on biological systems where a change of even 0.1 pH unit in some internal fluid is usually lethal to a living system.

The Effectiveness of a Buffer System

Although buffers do not have unlimited *capacities,* they still operate remarkably well in preventing wide swings in pH. The question of *capacity* is related to the question of buffer *effectiveness.* How well does a buffer system work? How well does it actually hold the pH? In Example 16.16, we added 0.15 mole of strong acid to 1.00 L of a buffer and the pH changed by 0.12 pH unit. Just think of how much the pH would change had we added this much acid to 1.00 L of pure water. The solution would now have

$$[H^+] = 0.15 \text{ mol L}^{-1}$$

and a pH of

$$pH = -\log (0.15)$$
$$= 0.82$$

With the buffer, the pH changes from 4.74 to 4.62; without the buffer the pH (starting with pure water) changes from 7.00 to 0.82. This is an enormous difference. Buffers, unless overwhelmed by the addition of excessive amounts of strong acid or base, truly prevent wide swings of pH.

16.6 Ionization of Polyprotic Acids

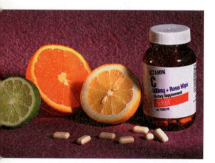

Vitamin C, found in many fruits and vegetables, is a weak diprotic acid with the formula $H_2C_6H_6O_6$.

Until now our discussion of weak acids has focused entirely on equilibria involving monoprotic acids. There are, of course, many acids capable of supplying more than one H^+ per molecule. Recall that these are called polyprotic acids. Examples include sulfuric acid, H_2SO_4, carbonic acid, H_2CO_3, and phosphoric acid, H_3PO_4. These acids undergo ionization in a series of steps, each of which releases one proton. For weak polyprotic acids, such as H_2CO_3 and H_3PO_4, each step is an equilibrium. Even sulfuric acid, which we consider a strong acid, is not completely ionized. Loss of the first proton to yield the HSO_4^- ion is complete, but the loss of the second proton is incomplete and involves an equilibrium. In this section, we will focus our attention on weak polyprotic acids.

Let's begin with the weak diprotic acid, H_2CO_3. In water, the acid ionizes in two steps, each of which is an equilibrium that transfers an H^+ ion to a water molecule.

$$H_2CO_3 + H_2O \rightleftharpoons H_3O^+ + HCO_3^-$$
$$HCO_3^- + H_2O \rightleftharpoons H_3O^+ + CO_3^{2-}$$

As usual, we can use H^+ in place of H_3O^+ and simplify these equations to give

$$H_2CO_3 \rightleftharpoons H^+ + HCO_3^-$$
$$HCO_3^- \rightleftharpoons H^+ + CO_3^{2-}$$

Each step has its own ionization constant, K_a, which we identify as K_{a_1} for the first step and K_{a_2} for the second. For carbonic acid,

$$K_{a_1} = \frac{[H^+][HCO_3^-]}{[H_2CO_3]} = 4.3 \times 10^{-7}$$

$$K_{a_2} = \frac{[H^+][CO_3^{2-}]}{[HCO_3^-]} = 5.6 \times 10^{-11}$$

Table 16.3 Acid Ionization Constants for Polyprotic Acids

Name	Formula	Acid Ionization Constant for Successive Ionizations (25 °C)		
		First	Second	Third
Carbonic acid	H_2CO_3	4.3×10^{-7}	5.6×10^{-11}	
Hydrogen sulfide	H_2S (aq)	9.5×10^{-8}	1×10^{-19}	
Phosphoric acid	H_3PO_4	7.1×10^{-3}	6.3×10^{-8}	4.5×10^{-13}
Arsenic acid	H_3AsO_4	5.6×10^{-3}	1.7×10^{-7}	4.0×10^{-12}
Sulfuric acid	H_2SO_4	Large	1.0×10^{-2}	
Selenic acid	H_2SeO_4	Large	1.2×10^{-2}	
Telluric acid	H_6TeO_6	2×10^{-8}	1×10^{-11}	
Sulfurous acid	H_2SO_3	1.2×10^{-2}	6.6×10^{-8}	
Selenous acid	H_2SeO_3	3.5×10^{-3}	5×10^{-8}	
Tellurous acid	H_2TeO_3	3×10^{-3}	2×10^{-8}	
Ascorbic acid	$H_2C_6H_6O_6$	6.7×10^{-5}	2.7×10^{-12}	
Oxalic acid	$H_2C_2O_4$	6.5×10^{-2}	6.1×10^{-5}	
Citric acid	$H_3C_6H_5O_7$	7.4×10^{-4}	1.7×10^{-5}	4.0×10^{-7}

Notice that each ionization makes a contribution to the total molar concentration of H^+, and one of our goals here is to relate the K_a values and the concentration of the acid to this molarity. At first glance, this seems to be a formidable task, but certain simplifications are justified that make the problem relatively simple to solve.

Simplifications in Calculations Involving Polyprotic Acids

The principal factor that simplifies calculations involving many polyprotic acids is the large differences between successive ionization constants. Notice that for H_2CO_3 K_{a_1} is much larger than K_{a_2} (they differ by a factor of nearly 10,000). Similar differences between K_{a_1} and K_{a_2} are observed for many diprotic acids. One of the reasons is that an H^+ is lost much more easily from the neutral H_2A molecule than from the HA^- ion. The stronger attraction of the opposite charges inhibits the second ionization. Typically, K_{a_1} is greater than K_{a_2} by a factor of between 10^4 and 10^5, as the data in Table 16.3 show. For a triprotic acid, the second acid ionization constant is similarly greater than the third, as illustrated in Table 16.3 by phosphoric acid, H_3PO_4.

Because K_{a_1} is so much larger than K_{a_2}, virtually all the H^+ in a solution of the acid comes from the first step in the ionization. In other words,

Additional K_a values of polyprotic acids are located in Appendix E.

$$[H^+]_{equilib} = [H^+]_{first\ step} + [H^+]_{second\ step}$$

$$[H^+]_{first\ step} \gg [H^+]_{second\ step}$$

Therefore, we make the approximation that

$$[H^+]_{equilib} \approx [H^+]_{first\ step}$$

This means that as far as calculating the H^+ concentration is concerned, we can treat the acid as though it were a monoprotic acid and ignore the second step in the ionization. Example 16.17 illustrates how this is applied.

Calculate the concentrations of all the species produced in the ionization of 0.040 M H_2CO_3 as well as the pH of the solution.

Analysis: As in any equilibrium problem, we begin with the chemical equations and the K_a expressions

$$H_2CO_3 \rightleftharpoons H^+ + HCO_3^- \qquad K_{a_1} = \frac{[H^+][HCO_3^-]}{[H_2CO_3]} = 4.3 \times 10^{-7}$$

$$HCO_3^- \rightleftharpoons H^+ + CO_3^{2-} \qquad K_{a_2} = \frac{[H^+][CO_3^{2-}]}{[HCO_3^-]} = 5.6 \times 10^{-11}$$

We will want to calculate the concentrations of H^+, HCO_3^-, and CO_3^{2-}, and the concentration of H_2CO_3 that remains at equilibrium. As noted in the preceding discussion, the problem is simplified considerably by assuming that the second reaction occurs to a negligible extent compared to the first. This assumption (which we will justify later in the problem) allows us to calculate the H^+ and HCO_3^- concentrations as though the solute were a monoprotic acid. This type of problem is one we've worked on in Section 16.2.

To obtain $[CO_3^{2-}]$, we will have to use the second step in the ionization. Once again, the large difference between K_{a_1} and K_{a_2} permit some simplifications. The first involves $[H^+]$, which we have already examined.

$$[H^+]_{\text{equilib}} \approx [H^+]_{\text{first step}}$$

The second involves $[HCO_3^-]$, for which we can write

$$[HCO_3^-]_{\text{equilib}} = [HCO_3^-]_{\substack{\text{formed in} \\ \text{first step}}} - [HCO_3^-]_{\substack{\text{lost in} \\ \text{second step}}}$$

However, because $K_{a_1} \gg K_{a_2}$, we expect that

$$[HCO_3^-]_{\substack{\text{formed in} \\ \text{first step}}} \gg [HCO_3^-]_{\substack{\text{lost in} \\ \text{second step}}}$$

Therefore, we can make the approximation that

$$[HCO_3^-]_{\text{equilib}} \approx [HCO_3^-]_{\substack{\text{formed in} \\ \text{first step}}}$$

Solution: Treating H_2CO_3 as though it were a monoprotic acid yields a problem similar to many others we worked earlier in this chapter. We know some of the H_2CO_3 ionizes; let's call this amount x. If x moles per liter of H_2CO_3 ionizes, then x moles per liter of H^+ and HCO_3^- are formed, and the concentration of H_2CO_3 is reduced by x.

	H_2CO_3	$\rightleftharpoons$	H^+	$+$	HCO_3^-
Initial concentrations (M)	0.040		0		0
Changes in concentrations (M)	$-x$		$+x$		$+x$
Equilibrium concentrations (M)	$0.040 - x$		x		x

$$[H^+] = [HCO_3^-] = x$$

$$[H_2CO_3] = 0.040 - x$$

Substituting into the expression for K_{a_1} gives

$$\frac{(x)(x)}{(0.040 - x)} = 4.3 \times 10^{-7}$$

Because K_{a_1} is so small, we expect that $(0.040 - x) \approx 0.040$. (This is the usual approximation we made in solving problems of this kind.) The algebra then reduces to

$$\frac{x^2}{0.040} = 4.3 \times 10^{-7}$$

Solving for x gives $x = 1.3 \times 10^{-4}$, so $[H^+] = [HCO_3^-] = 1.3 \times 10^{-4}\ M$. From this we can now calculate the pH.

$$pH = -\log{(1.3 \times 10^{-4})} = 3.89$$

Now let's calculate the concentration of CO_3^{2-}. We obtain this from the expression for K_{a_2}.

$$K_{a_2} = \frac{[H^+][CO_3^{2-}]}{[HCO_3^{2-}]} = 5.6 \times 10^{-11}$$

In our analysis, we concluded that we can use the approximations

$$[H^+]_{equilib} \approx [H^+]_{first\ step}$$

$$[HCO_3^-]_{equilib} \approx [HCO_3^-]_{first\ step}$$

Substituting the values obtained above gives us

$$\frac{(1.3 \times 10^{-4})[CO_3^{2-}]}{1.3 \times 10^{-4}} = 5.6 \times 10^{-11}$$

$$[CO_3^{2-}] = 5.6 \times 10^{-11} = K_{a_2}$$

Let's summarize the results. At equilibrium, we have

$$[H_2CO_3] = 0.040\ M$$

$$[H^+] = [HCO_3^-] = 1.3 \times 10^{-4}\ M\ (and\ pH = 3.89)$$

$$[CO_3^{2-}] = 5.6 \times 10^{-11}\ M$$

Checking the Simplifying Assumptions
In the second step of the ionization, H^+ and CO_3^{2-} are formed in equal amounts from HCO_3^-. This means that the contribution to $[H^+]$ by *the second step equals* $[CO_3^{2-}]$. Therefore, $[H^+]_{second\ step} = 5.6 \times 10^{-11}\ M$. Comparing this value to the $[H^+]$ computed using K_{a_1}, we see that the second step of the ionization contributes a negligible amount to the total hydrogen ion concentration in the solution.

$$[H^+]_{equilib} = [H^+]_{first\ step} + [H^+]_{second\ step}$$

$$= (1.3 \times 10^{-4}\ M) + (5.6 \times 10^{-11}\ M)$$

$$= 1.3 \times 10^{-4}\ M\ (correctly\ rounded)$$

Thus, we were safe in treating H_2CO_3 as though it were a monoprotic acid when we computed the H^+ concentration.

Our second approximation,

$$[HCO_3^-]_{equilib} \approx [HCO_3^-]_{formed\ in\ first\ step}$$

was also valid. The concentration of HCO_3^- formed in the first step equals $1.3 \times 10^{-4}\ M$. This value is decreased by 5.6×10^{-11} mol L^{-1} when the H^+ and CO_3^{2-} form in the second step. Because 5.6×10^{-11} is negligible compared to 1.3×10^{-4}, the equilibrium concentration of HCO_3^- equals the concentration computed using K_{a_1}.

Notice that $0.040 > 400 \times K_{a_1}$, so it is safe to use the initial concentration, $0.040\ M$, as though it were the equilibrium concentration.

*Keep in mind that for a given solution, the values of H^+ used in the expressions for K_{a_1} and K_{a_2} are identical. At equilibrium there is only *one* equilibrium H^+ concentration.*

Checking the Accuracy of the Answers
By now you know that you can always perform a quick check of the answers by substituting them into the appropriate mass action expressions.

Practice Exercise 19

Ascorbic acid (vitamin C) is a diprotic acid, $H_2C_6H_6O_6$. See Table 16.3. Calculate $[H^+]$, pH, and $[C_6H_6O_6^{2-}]$ in a 0.10 M solution of ascorbic acid. ◆

One of the most interesting observations we can derive from the preceding example is that *in a solution that contains a polyprotic acid **as the only solute**, the molar concentration of the ion formed in the second step of the ionization numerically equals* K_{a_2}.

16.7 Solutions of Salts of Polyprotic Acids

You learned earlier that the pH of a salt solution is controlled by whether the salt's cation, anion, or both are able to react with water. For simplicity, the salts of polyprotic acids that we will discuss here will be limited to those containing *nonacidic* cations, such as sodium or potassium ion. In other words, we will study salts in which the anion alone is basic and thereby affects the pH of a solution.

A typical example of a salt of a polyprotic acid is sodium carbonate, Na_2CO_3. It is the salt of H_2CO_3, and the salt's carbonate ion is a Brønsted base that is responsible for *two* equilibria that furnish OH^- ion and so affect the pH of the solution. These equilibria and their corresponding expressions for K_b are as follows:

$$CO_3^{2-}(aq) + H_2O \rightleftharpoons HCO_3^-(aq) + OH^-(aq) \qquad (16.15)$$

$$K_{b_1} = \frac{[HCO_3^-][OH^-]}{[CO_3^{2-}]}$$

$$HCO_3^-(aq) + H_2O \rightleftharpoons H_2CO_3(aq) + OH^-(aq) \qquad (16.16)$$

$$K_{b_2} = \frac{[H_2CO_3][OH^-]}{[HCO_3^-]}$$

The calculations required to determine the concentrations of the species involved in these equilibria are very much like those involving weak diprotic acids. The difference is that the chemical reactions here are those of bases instead of acids. In fact, *the simplifying assumptions in our calculations will be almost identical to those for weak polyprotic acids.*

First, let's compare the K_b values for the successive equilibria. To obtain these constants, we will use the relationship that $K_a \times K_b = K_w$ along with the values of K_{a_1} and K_{a_2} for H_2CO_3.

Notice that in the equilibrium in which CO_3^{2-} is the base (Equation 16.15), the conjugate acid is HCO_3^-. Therefore, to calculate K_b for CO_3^{2-} (which we called K_{b_1}), we must use the K_a for HCO_3^-, which corresponds to K_{a_2} for carbonic acid. Thus,

$$K_{b_1} = \frac{K_w}{K_{a_2}} = \frac{1.0 \times 10^{-14}}{5.6 \times 10^{-11}} = 1.8 \times 10^{-4}$$

Similarly, in the equilibrium in which HCO_3^- is the base (Equation 16.16), the conjugate acid is H_2CO_3. Therefore, to calculate K_b for HCO_3^- (which we called K_{b_2}), we must use the K_a for H_2CO_3, which is K_{a_1} for carbonic acid.

The salt Na_3PO_4 is sold under the name trisodium phosphate or TSP. Solutions of the salt in water are quite basic and are used as an aid in cleaning grime from painted surfaces. Some states restrict the sale of TSP because the phosphate ion it contains presents a pollution problem by promoting the growth of algae in lakes.

In general, K_b values for anions of polyprotic acids are not tabulated. When needed, they are calculated from the appropriate K_a values for the acids.

$$K_{b_2} = \frac{K_w}{K_{a_1}} = \frac{1.0 \times 10^{-14}}{4.3 \times 10^{-7}} = 2.3 \times 10^{-8}$$

Now we can compare the two K_b values. For CO_3^{2-}, $K_{b_1} = 1.8 \times 10^{-4}$, and for the (much) weaker base, HCO_3^-, K_{b_2} is 2.3×10^{-8}. Thus CO_3^{2-} has a K_b nearly 10,000 times that of HCO_3^-. This means that the reaction of CO_3^{2-} with water (Equation 16.15) generates far more OH^- than the reaction of HCO_3^- (Equation 16.16). The contribution of the latter to the total pool of OH^- is relatively so small that we can safely ignore it.

$$[OH^-]_{equilib} = [OH^-]_{\substack{formed\ in \\ first\ step}} + [OH^-]_{\substack{formed\ in \\ second\ step}}$$

$$[OH^-]_{\substack{formed\ in \\ first\ step}} \gg [OH^-]_{\substack{formed\ in \\ second\ step}}$$

$$[OH^-]_{equilib} \approx [OH^-]_{\substack{formed\ in \\ first\ step}}$$

Thus, if we wish to calculate the pH of a solution of a basic anion of a polyprotic acid, *we may work exclusively with K_{b_1} and ignore any further reactions.* Let's study an example of how this works.

What is the pH of a 0.15 *M* solution of Na_2CO_3?

Analysis: We can ignore the reaction of HCO_3^- with water so the only relevant equilibrium is

$$CO_3^{2-} + H_2O \rightleftharpoons HCO_3^- + OH^- \qquad K_{b_1} = \frac{[HCO_3^-][OH^-]}{[CO_3^{2-}]} = 1.8 \times 10^{-4}$$

(The value of K_{b_1} was calculated in the discussion preceding this example.) Also, the initial concentration of CO_3^{2-} (0.15 *M*) is larger than $400 \times K_{b_1}$, so we can safely use 0.15 *M* as the equilibrium concentration of CO_3^{2-}.

Solution: Some of the CO_3^{2-} will react; we'll let this be x mol L^{-1}. The concentration table, then, is as follows.

	CO_3^{2-}	$+$	H_2O	$\rightleftharpoons$	HCO_3^-	$+$	OH^-
Initial concentrations (*M*)	0.15				0		0
Changes in concentrations caused by the ionization (*M*)	$-x$				$+x$		$+x$
Equilibrium concentrations (*M*)	$(0.15 - x) \approx 0.15$				x		x

Let's now insert the values in the last row of the table into the equilibrium expression for the carbonate ion.

$$\frac{[HCO_3^-][OH^-]}{[CO_3^{2-}]} = \frac{(x)(x)}{0.15} = 1.8 \times 10^{-4}$$

$$x = 5.2 \times 10^{-3}$$

Notice that the value of x really is sufficiently smaller than 0.15 to justify our using 0.15 as a good approximation of $(0.15 - x)$. Because $x = 5.2 \times 10^{-3}$, we have

$$[OH^-] = 5.2 \times 10^{-3} \text{ mol } L^{-1}$$

$$pOH = -\log(5.2 \times 10^{-3}) = 2.28$$

Some detergents that use sodium carbonate as their caustic ingredient. Detergents are best able to dissolve grease and grime if their solutions are basic. In these products, the basic ingredient is sodium carbonate. The salt Na_2CO_3 is relatively nontoxic, and the carbonate ion it contains is a relatively strong base. The carbonate ion also serves as a water softener by precipitating Ca^{2+} ions as insoluble $CaCO_3$.

Hence,

$$pH = 14.00 - 2.28$$

$$= 11.72$$

Thus, the pH of 0.15 M Na$_2$CO$_3$ is calculated to be 11.72. Once again we see that an aqueous solution of a salt composed of a basic anion and a neutral cation is basic.

Practice Exercise 20

What is the pH of a 0.20 M solution of Na$_2$SO$_3$ at 25 °C? For the diprotic acid, H$_2$SO$_3$, $K_{a_1} = 1.2 \times 10^{-2}$ and $K_{a_2} = 6.6 \times 10^{-8}$. ◆

Practice Exercise 21

Reasoning by analogy from what you learned about solutions of weak polyprotic acids, what is the molar concentration of H$_2$SO$_3$ in a 0.010 M solution of Na$_2$SO$_3$? ◆

16.8 Acid–Base Titrations Revisited

In Section 4.9, we studied the overall procedure for an acid–base titration, and we saw how titration data can be used in various stoichiometric calculations. In performing this procedure, the titration is halted at the **end point** when the change in color of an indicator occurs. Ideally, this end point should occur at the **equivalence point,** when stoichiometrically equivalent amounts of acid and base have combined. To obtain this ideal result, the selection of the specific indicator requires foresight. We will understand this better by studying how the pH of the solution being titrated changes with the addition of titrant.

The titrant is the solution being slowly added from a buret to a solution in the receiving flask.

When the pH of a solution at different stages of a titration is plotted against the volume of titrant added, we obtain a **titration curve.** The values of pH in these plots can be calculated by the procedures studied in this chapter, so this discussion will serve as a review.

Titration of a Strong Acid by a Strong Base

The titration of HCl(aq) with standard NaOH(aq) illustrates the titration of a strong acid by a strong base. The molecular and net ionic equations are

$$HCl(aq) + NaOH(aq) \longrightarrow NaCl(aq) + H_2O$$

$$H^+(aq) + OH^-(aq) \longrightarrow H_2O$$

Let's consider what happens to the pH of a solution, initially 25.00 mL of 0.2000 M HCl, as small amounts of the titrant, 0.2000 M NaOH, are added. We will calculate the pH of the resulting solution at various stages of the titration, retaining only two significant figures, and plot the values against the volume of titrant.

Before any titrant has been added, the receiving flask contains just 0.2000 M HCl. Because this is a strong acid, we know that

$$[H^+] = [HCl] = 0.2000 \ M$$

So

$$pH = -\log (0.2000)$$

$$= 0.70 \text{ (the initial pH)}$$

An acid–base titration using phenolphthalein as an indicator.

The amount of HCl initially present in 25.00 mL of 0.2000 M HCl is

$$25.00 \text{ mL HCl soln} \times \frac{0.2000 \text{ mol HCl}}{1000 \text{ mL HCl soln}} = 5.000 \times 10^{-3} \text{ mol HCl}$$

Now suppose we add 10.00 mL of 0.2000 M NaOH from the buret. The amount of NaOH added is

$$10.00 \text{ mL NaOH soln} \times \frac{0.2000 \text{ mol NaOH}}{1000 \text{ mL NaOH soln}} = 2.000 \times 10^{-3} \text{ mol NaOH}$$

This base neutralizes 2.000×10^{-3} mol of HCl, so the amount of HCl remaining is

$$(5.000 \times 10^{-3} - 2.000 \times 10^{-3}) \text{ mol HCl} = 3.000 \times 10^{-3} \text{ mol HCl remaining}$$

The total volume of the solution is now $(25.00 + 10.00)$ mL = 35.00 mL = 0.03500 L, so the $[H^+]$ is calculated as

$$[H^+] = \frac{3.000 \times 10^{-3} \text{ mol}}{0.03500 \text{ L}} = 8.571 \times 10^{-2} \text{ } M$$

The corresponding pH is 1.07 (rounded).

Table 16.4 shows the results of calculating the pH of the solution in the receiving flask after the addition of further small volumes of the titrant. After 25.00 mL of 0.2000 NaOH is added, all of the HCl is neutralized and we have reached the equivalence point. At this point the solution is a mixture of Na^+ and Cl^- ions, neither of which have an effect on pH, so the solution is neutral with a pH = 7.00.

Any further addition of titrant is simply the addition of more and more OH^- ion to an increasing volume of solution in the flask. So after the HCl has been neutralized by 25.00 mL of NaOH solution, the pH calculation reflects only a dilute solution of NaOH.

Figure 16.5 shows a plot of the pH of the solution in the receiving flask versus the volume of the 0.2000 M NaOH solution added—the *titration curve* for the titration of a strong acid with a strong base. Notice how slowly the pH of the solution increases until we are almost exactly at the equivalence point, pH = 7.00. With the addition of more base, the curve continues to rise very sharply and then, almost as suddenly, bends to reflect just a gradual increase in pH.

Despite the high precision assumed for the molarity and the volume of the titrant, we will retain only two significant figures in the calculated pH. Precision higher than this is hard to obtain and seldom sought in actual lab work.

Table 16.4 Titration of 25.00 mL of 0.2000 M HCl with 0.2000 M NaOH

Initial Volume of HCl (mL)	Initial Amount of HCl (mol)	Volume of NaOH Added (mL)	Amount of NaOH (mol)	Amount of Excess Reagent (mol)	Total Volume of Solution (mL)	Molarity of Ion in Excess (mol L^{-1})	pH
25.00	5.000×10^{-3}	0	0	5.000×10^{-3} (H^+)	25.00	0.2000 (H^+)	0.70
25.00	5.000×10^{-3}	10.00	2.000×10^{-3}	3.000×10^{-3} (H^+)	35.00	8.571×10^{-2} (H^+)	1.07
25.00	5.000×10^{-3}	20.00	4.000×10^{-3}	1.000×10^{-3} (H^+)	45.00	2.222×10^{-2} (H^+)	1.65
25.00	5.000×10^{-3}	24.00	4.800×10^{-3}	2.000×10^{-4} (H^+)	49.00	4.082×10^{-3} (H^+)	2.39
25.00	5.000×10^{-3}	24.90	4.980×10^{-3}	2.000×10^{-5} (H^+)	49.90	4.000×10^{-4} (H^+)	3.40
25.00	5.000×10^{-3}	24.99	4.998×10^{-3}	2.000×10^{-6} (H^+)	49.99	4.000×10^{-5} (H^+)	4.40
25.00	5.000×10^{-3}	25.00	5.000×10^{-3}	0	50.00	0	7.00
25.00	5.000×10^{-3}	25.01	5.002×10^{-3}	2.000×10^{-6} (OH^-)	50.01	3.999×10^{-5} (OH^-)	9.60
25.00	5.000×10^{-3}	25.10	5.020×10^{-3}	2.000×10^{-5} (OH^-)	50.10	3.992×10^{-4} (OH^-)	10.60
25.00	5.000×10^{-3}	26.00	5.200×10^{-3}	2.000×10^{-4} (OH^-)	51.00	3.922×10^{-3} (OH^-)	11.59
25.00	5.000×10^{-3}	50.00	1.000×10^{-2}	5.000×10^{-3} (OH^-)	75.00	6.667×10^{-2} (OH^-)	12.82

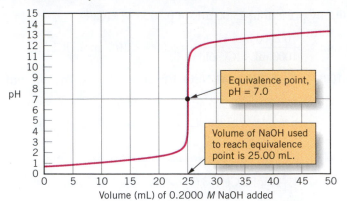

Figure 16.5 *Titration curve for titrating a strong acid with a strong base.* Here we follow how the pH changes during the titration of 0.2000 *M* HCl with 0.2000 *M* NaOH.

Notice also that the equivalence point occurs at a pH of 7.00, *which characterizes the titration of any strong monoprotic acid with any strong base.* The only ions at the equivalence point in our example, aside from the traces of those from the self-ionization of water, are Na^+ and Cl^-. Na^+ is not an acidic cation, and Cl^- is not a basic anion, so the solution at this point is neutral with a pH of 7.00.

Titration of a Weak Acid by a Strong Base

Let's do the calculations and draw a titration curve for the titration of 0.2000 *M* $HC_2H_3O_2$ with 0.2000 *M* NaOH. The molecular and net ionic equations are

$$HC_2H_3O_2(aq) + NaOH(aq) \longrightarrow NaC_2H_3O_2(aq) + H_2O$$

$$HC_2H_3O_2(aq) + OH^-(aq) \longrightarrow C_2H_3O_2^-(aq) + H_2O$$

Let us start with 25.00 mL of 0.2000 *M* $HC_2H_3O_2$ and use 0.2000 *M* NaOH as the titrant. Because the concentrations of both solutions are the same, we know that we will need exactly 25.00 mL of the base to reach the equivalence point. With this as background, let's look at what is involved in calculating the pH at various points along the titration curve.

1. **Before the titration begins.** The solution at this point is simply a solution of the weak acid, $HC_2H_3O_2$. We must use K_a to calculate the pH.

2. **During the titration but before the equivalence point.** As we add NaOH to the $HC_2H_3O_2$, the chemical reaction produces $C_2H_3O_2^-$, so the solution contains both $HC_2H_3O_2$ and $C_2H_3O_2^-$ (it is a buffer solution). We can use K_a to calculate the pH here as well.

3. **At the equivalence point.** Here all the $HC_2H_3O_2$ has reacted and the solution contains the salt $NaC_2H_3O_2$ (i.e., we have a mixture of the ions Na^+ and $C_2H_3O_2^-$). The anion is a weak base, so we use K_b to calculate the pOH and then the pH.

4. **After the equivalence point.** The OH^- introduced by the further addition of NaOH has nothing with which to react and causes the solution to become basic. The concentration of OH^- produced by the unreacted base is used to calculate the pOH and ultimately the pH.

We will look at each of these briefly.

Before the Titration Begins

The question here is, What is the pH of 0.2000 *M* $HC_2H_3O_2$? For acetic acid, $K_a = 1.8 \times 10^{-5}$. We let $x = [H^+] = [C_2H_3O_2^-]$, and use $[HC_2H_3O_2] = 0.2000$ *M*.

We worked a similar problem on page 712.

$$K_a = \frac{[H^+][C_2H_3O_2^-]}{[HC_2H_3O_2]} = \frac{(x)(x)}{(0.2000)} = 1.8 \times 10^{-5}$$

$$x^2 = 3.6 \times 10^{-6}$$

$$x = 1.9 \times 10^{-3}$$

So, $[H^+] = 1.9 \times 10^{-3}$ M, which yields a pH of 2.72. This is the pH before any NaOH is added.

During the Titration but Before the Equivalence Point

Once we add NaOH, the solution contains both $HC_2H_3O_2$ and $C_2H_3O_2^-$, so we now have a buffer system. Let's calculate the pH after 10.00 mL of the base has been added. The amount of NaOH added is

$$10.00 \text{ mL NaOH soln} \times \frac{0.2000 \text{ mol NaOH}}{1000 \text{ mL soln}} = 2.000 \times 10^{-3} \text{ mol NaOH}$$

This reacted with 2.000×10^{-3} mol of $HC_2H_3O_2$ to produce 2.000×10^{-3} mol of $C_2H_3O_2^-$. The amount of $HC_2H_3O_2$ remaining in the solution (realizing that we began with 5.000×10^{-3} mol), is simply

$$(5.000 \times 10^{-3} - 2.000 \times 10^{-3}) \text{ mol } HC_2H_3O_2 = 3.000 \times 10^{-3} \text{ mol } HC_2H_3O_2$$

The total volume is now (25.00 mL + 10.00 mL) = 35.00 mL = 0.03500 L. Therefore, the concentrations of acetate ion and acetic acid now are

$$[C_2H_3O_2]^- = \frac{2.000 \times 10^{-3} \text{ mol}}{0.03500 \text{ L}} = 5.714 \times 10^{-2} \text{ mol L}^{-1}$$

$$[HC_2H_3O_2] = \frac{3.000 \times 10^{-3} \text{ mol}}{0.03500 \text{ L}} = 8.571 \times 10^{-2} \text{ mol L}^{-1}$$

We can now substitute these values into the mass action expression and calculate $[H^+]$.

$$\frac{[H^+](5.714 \times 10^{-2})}{(8.571 \times 10^{-2})} = 1.8 \times 10^{-5}$$

From which we obtain $[H^+] = 2.7 \times 10^{-5}$ M. The pH is then calculated to be pH = 4.57.

As long as we have not reached the equivalence point—as long as unreacted acetic acid remains together with acetate ion—we can compute the pH of the solution in the receiving flask by this kind of buffer calculation.

At the Equivalence Point

At the equivalence point we've combined equal numbers of moles of $HC_2H_3O_2$ and NaOH. Their reaction yields $NaC_2H_3O_2$, so the solution is simply a dilute solution of sodium acetate. It no longer is a buffer. We now have to deal with the reaction of the acetate ion with water and the associated base ionization constant.

$$C_2H_3O_2^-(aq) + H_2O \rightleftharpoons HC_2H_3O_2(aq) + OH^-(aq)$$

$$K_b = \frac{[HC_2H_3O_2][OH^-]}{[C_2H_3O_2^-]} = \frac{K_w}{K_a} = 5.6 \times 10^{-10}$$

At the equivalence point, we have added 25.00 mL of 0.2000 M NaOH to the original 25.00 mL of 0.2000 M $HC_2H_3O_2$. The volume is now 50.00 mL = 0.05000 L. We began with 5.000×10^{-3} mol of $HC_2H_3O_2$, so we now have

5.000×10^{-3} mol of $C_2H_3O_2^-$ (in 0.05000 L). The concentration of $C_2H_3O_2^-$, therefore, is

$$[C_2H_3O_2^-] = \frac{5.000 \times 10^{-3} \, \text{mol}}{0.05000 \, \text{L}} = 0.1000 \, M$$

To solve the problem we realize that a small amount of the $C_2H_3O_2^-$ reacts to give equal amounts of OH^- and $HC_2H_3O_2$. We let the concentrations of these equal x, so $[OH^-] = [C_2H_3O_2^-] = x$. Substituting into the mass action expression gives

$$\frac{(x)(x)}{0.1000} = 5.6 \times 10^{-10}$$

$$x^2 = 5.6 \times 10^{-11}$$

$$x = 7.5 \times 10^{-6}$$

This means that $[OH^-] = 7.5 \times 10^{-6} \, M$, so the pOH is calculated to be 5.12. Finally, the pH is $(14.00 - 5.12)$ or 8.88. Thus the pH at the equivalence point in this titration is 8.88, somewhat basic because the solution contains the Brønsted base, $C_2H_3O_2^-$. *In the titration of any weak acid with a strong base, the pH at the equivalence point will be greater than 7.*

It is interesting that although we describe the overall reaction in the titration as neutralization, at the equivalence point the solution is slightly basic, not neutral.

After the Equivalence Point

What happens if we continue to add $NaOH(aq)$ to the solution beyond the equivalence point? Now the additional OH^- shifts the following equilibrium to the left.

$$C_2H_3O_2^-(aq) + H_2O \rightleftharpoons HC_2H_3O_2(aq) + OH^-(aq)$$

The generation of additional OH^- *by this route* is thus suppressed as we add more and more NaOH solution, so effectively the only source of OH^- to affect the pH is from the base added after the equivalence point. From here on, therefore, the pH calculations become identical to those for the last half of the titration curve for HCl and NaOH. Each point is just a matter of calculating the additional number of moles of NaOH, calculating the new final volume, taking their ratio (to obtain the molar concentration), and then calculating pOH and pH.

The Titration Curve

Table 16.5 summarizes the titration data, and the titration curve is given in Figure 16.6. Once again, notice how the change in pH occurs slowly at first, then

Table 16.5 Titration of 25.00 mL of 0.2000 M $HC_2H_3O_2$ with 0.2000 M NaOH

Volume of Base Added (mL)	Molar Concentration of Species in Parentheses	pH
None	1.9×10^{-3} (H^+)	2.72
10.00	2.7×10^{-5} (H^+)	4.57
24.90	7.2×10^{-8} (H^+)	7.14
24.99	7.1×10^{-9} (H^+)	8.14
25.00	7.5×10^{-6} (OH^-)	8.88
25.01	4.0×10^{-5} (OH^-)	9.60
25.10	4.0×10^{-4} (OH^-)	10.60
26.00	3.9×10^{-3} (OH^-)	11.59
35.00	3.3×10^{-2} (OH^-)	12.52

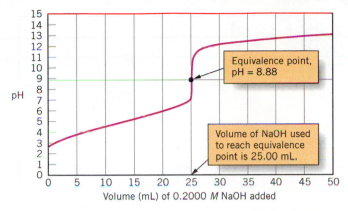

shoots up through the equivalence point, and finally looks just like the last half of the curve in Figure 16.5.

Practice Exercise 22

Suppose we titrate 20.0 mL of 0.100 *M* $HCHO_2$ with 0.100 *M* NaOH. Calculate the pH (a) before any base is added, (b) when half of the $HCHO_2$ has been neutralized, (c) after a total of 15.0 mL of base has been added, and (d) at the equivalence point. ◆

Practice Exercise 23

Suppose 30.0 mL of 0.15 *M* NaOH is added to 50.0 mL of 0.20 *M* $HCHO_2$. What will be the pH of the resulting solution? ◆

Titration of a Weak Base by a Strong Acid

The calculations involved here are nearly identical to those of the weak acid–strong base titration. We will review what is required but will not actually perform the calculations.

If we titrate 25.00 mL of 0.2000 *M* NH_3 with 0.2000 *M* HCl, the molecular and net ionic equations are

$$NH_3(aq) + HCl(aq) \longrightarrow NH_4Cl(aq)$$

$$NH_3(aq) + H^+(aq) \longrightarrow NH_4^+(aq)$$

1. **Before the titration begins.** The solution at this point is simply a solution of the weak base, NH_3. Since the only solute is a base, we must use K_b to calculate the pH.

2. **During the titration but before the equivalence point.** As we add HCl to the NH_3, the neutralization reaction produces NH_4^+, so the solution contains both NH_3 and NH_4^+ (it is a buffer solution). We can use K_b for NH_3 to calculate the pH here as well. (We could instead use K_a for NH_4^+ if it were available.)

3. **At the equivalence point.** Here all the NH_3 has reacted and the solution contains the salt NH_4Cl. The anion, Cl^-, comes from the strong acid HCl and is so weak that it cannot affect the pH. However, the cation NH_4^+ is a weak acid derived from the weak base NH_3, so we use its K_a to calculate the pH. In doing the calculation, it's important to use the total volume of the solution to calculate the concentration of the NH_4^+.

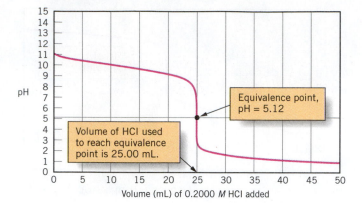

Figure 16.7 *Titration curve for the titration of a weak base with a strong acid.* Here we follow the pH as 0.2000 M NH_3 is titrated with 0.2000 M HCl.

4. **After the equivalence point.** The H^+ introduced by further addition of HCl has nothing with which to react, so it causes the solution to become more and more acidic. The concentration of H^+ produced by the unreacted HCl is used directly to calculate the pH.

Figure 16.7 illustrates the titration curve for this system.

Acid–Base Indicators

Most dyes that work as acid–base indicators are also weak acids. Therefore, let's represent an indicator by the formula H*In*. In its un-ionized state, H*In* has one color. Its conjugate base, *In*⁻, has a different color, the more strikingly different the better. In solution, the indicator is involved in a typical acid–base equilibrium:

$$HIn(aq) \rightleftharpoons H^+(aq) + In^-(aq)$$

acid form   base form
(one color)   (another color)

The corresponding acid ionization constant, K_{In}, is given by

$$K_{In} = \frac{[H^+][In^-]}{[HIn]}$$

In a strongly acidic solution, when the H^+ concentration is high, the equilibrium is shifted to the left and most of the indicator exists in its "acid form." Under these conditions, the color we observe is that of H*In*. If the solution is made basic, the H^+ concentration drops and the equilibrium shifts to the right, toward *In*⁻, and the color we observe is that of the "base form" of the indicator.

The *observed* change in color for an indicator actually occurs over a range of pH values. This is because of the human eye's limited ability to discern color changes. As discussed further in Facets of Chemistry 16.1, sometimes a change of as much as 2 pH units is needed for some indicators—more for litmus—before the eye notices the color change. This is why tables of acid–base indicators, such as Table 16.6, provide approximate pH ranges for the color changes.

How Acid–Base Indicators Work

In a typical acid–base titration, you've seen that as we pass the equivalence point, there is a sudden and large change in the pH. For example, in the titration of HCl with NaOH described earlier, the pH just one-half drop before the equivalence point (when 24.97 mL of the base has been added) is 3.92. Just one-half drop later (when 25.03 mL of base has been added) we have passed the

Facets of Chemistry 16.1

Color Ranges for Indicators

Suppose that the color of H*In* is red and the color of its anion, *In*⁻, is yellow. In terms of the eye's abilities, the ratio of [H*In*]/[*In*⁻] has to be roughly at least as large as 10 to 1 for the eye to see the color as definitely red. And this ratio has to be at least as small as 1 to 10 for the eye to see the color as definitely yellow. Therefore, the color transition that the eye actually sees at the end point occurs as the ratio of [H*In*] to [*In*⁻] changes from 10/1 to 1/10.

We can calculate the relationship of pH to pK_a for the indicator at the start of this range as follows. We know that

$$K_{In} = \frac{[H^+]\,[In^-]}{[HIn]}$$

Therefore, at the start of the range, we must have

$$K_{In} = [H^+] \times \frac{10}{1}$$

$$pK_{In} = pH - \log 10$$

Or, at the start of the transition we should have

$$pH = pK_{In} - 1$$

At the other end of the range, when [H*In*]/[*In*⁻] is 1/10, we have

$$K_{In} = [H^+] \times \frac{1}{10}$$

$$pK_{In} = pH - \log(10^{-1})$$

So at the end of the color transition, the pH should be

$$pH = pK_{In} + 1$$

Therefore, the *range* of pH over which a definite change in color occurs is

$$pH = pK_{In} \pm 1$$

A range of ±1 pH unit on either side of the value of pK_{In} corresponds to a span of 2 pH units over which the eye can definitely notice the color change.

equivalence point and the pH has risen to 10.08. This large swing in pH (from 3.92 to 10.08) causes a sudden shift in the position of equilibrium for the indicator, and we go from a condition where most of the indicator is in its acid form to a condition in which most is in the base form. This is observed visually as a change in color from that of the acid form to that of the base form.

Table 16.6 Some Common Acid–Base Indicators

Indicator	Color Change (Acidic to Basic)	pH Range in Which Color Change Occurs
Thymol blue	Red to yellow	1.2–2.8
Bromophenol blue	Yellow to blue	3.0–4.6
Congo red	Blue to red	3.0–5.0
Methyl orange	Red to yellow	3.2–4.4
Bromocresol green	Yellow to blue	3.8–5.4
Methyl purple	Purple to green	4.8–5.4
Methyl red	Red to yellow	4.8–6.0
Bromocresol purple	Yellow to purple	5.2–6.8
Bromothymol blue	Yellow to blue	6.0–7.6
Cresol red	Yellow to red	7.0–8.8
Thymol blue	Yellow to blue	8.0–9.6
Phenolphthalein	Colorless to pink	8.2–10.0
Alizarin yellow	Yellow to red	10.1–12.0

Selecting the Best Indicator for an Acid–Base Titration

The explanation in the preceding paragraph assumes that the midpoint of the color change of the indicator corresponds to the pH at the equivalence point. At this midpoint we expect that there will be equal amounts of both forms of the indicator, which means that $[In^-] = [HIn]$. If this condition is true, then

$$K_{In} = \frac{[H^+]\,[\cancel{In^-}]}{[\cancel{HIn}]}$$

so

$$K_{In} = [H^+]$$

Taking the negative logarithms of both sides gives us

$$-\log K_{In} = -\log [H^+]$$

or

$$pK_{In} = pH_{(\text{at the equivalence point})}$$

This important result tells us that once we know the pH that the solution should have at the equivalence point of a titration, we also know the pK_{In} that the indicator should have to function most effectively. Thus, we want an indicator whose pK_{In} equals (or is as close as possible to) the pH at the equivalence point. Phenolphthalein, for example, changes a solution from colorless to pink as the solution's pH changes over a range of 8.2 to 10.0. (See Figure 16.8.) *Phenolphthalein is therefore a*

pH 8.2   pH 10.0
Phenolphthalein

pH 3.2   pH 4.4
Methyl orange

pH 6.0   pH 7.6
Bromothymol blue

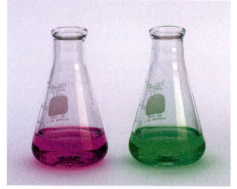

pH 4.8   pH 5.4
Methyl purple

Figure 16.8 *Colors of some common acid–base indicators.*

nearly perfect indicator for the titration of a weak acid by a strong base, where the pH at the equivalence point is on the basic side. (Refer back to Figure 16.6.)

Phenolphthalein also works very well for the titration of a strong acid by a strong base. As discussed earlier, just one drop of titrant can bring the pH from below 7 to nearly 10, which spans the pH range for this indicator.

The Importance of Color Intensity in an Indicator

Another factor involving indicators is that we want to use as little as possible. Because they are weak acids, indicators are also neutralized and consume titrant. We have to accept some loss of titrant this way, but no more than absolutely necessary. So the best indicators are those with the most intense colors. Then, even extremely small amounts can give striking colors and color changes.

SUMMARY

Acid and Base Ionization Constants. A weak acid HA ionizes according to the general equation

$$HA \rightleftharpoons H^+ + A^-$$

The equilibrium constant is called the **acid ionization constant, K_a**.

$$K_a = \frac{[H^+][A^-]}{[HA]}$$

A weak base B ionizes by the general equation

$$B + H_2O \rightleftharpoons BH^+ + OH^-$$

The equilibrium constant is called the **base ionization constant, K_b**.

$$K_b = \frac{[BH^+][OH^-]}{[B]}$$

The smaller the values of K_a (or K_b), the weaker are the substances as Brønsted acids (or bases).

Another way to compare the relative strengths of acids or bases is to use the negative logarithms of K_a and K_b, called **pK_a** and **pK_b**, respectively.

$$pK_a = -\log K_a \qquad pK_b = -\log K_b$$

The *smaller* the pK_a or pK_b, the *stronger* is the acid or base. For a conjugate acid–base pair,

$$K_a \times K_b = K_w$$

$$pK_a + pK_b = 14.00 \qquad \text{(at 25 °C)}$$

Equilibrium Calculations—Determining K_a and K_b. The values of K_a and K_b can be obtained from initial concentrations of the acid or base and either the pH of the solution or the percentage ionization of the acid or base. The measured pH gives the equilibrium value for $[H^+]$. The **percentage ionization** is defined as

$$\text{Percentage ionization} = \frac{\text{amount ionized}}{\text{amount available}} \times 100\%$$

Equilibrium Calculations—Determining Equilibrium Concentrations when K_a or K_b Is Known. Problems fall into one of three categories: (1) the only solute is a weak acid (we must use the K_a expression), (2) the only solute is a weak base (we must use the K_b expression), and (3) the solution contains both a weak acid and its conjugate base (we can use either K_a or K_b). When the initial concentration of the acid (or base) is larger than 400 times the value of K_a (or K_b), it is safe to use initial concentrations of acid or base as though they were equilibrium values in the mass action expression. When this approximation cannot be used, we use either the quadratic formula or the method of successive approximations (usually the latter is faster).

Ions as Acids or Bases. A solution of a salt is acidic if the cation is acidic but the anion is neutral. Metal ions with high charge density generally are acidic, but those of Groups IA and IIA (except Be^{2+}) are not. Cations such as NH_4^+, which are the conjugate acids of weak *molecular* bases, are themselves proton donors and are acidic.

When the anion of a salt is the conjugate base of a *weak* acid, the anion is a weak base. Anions of strong acids, such as Cl^- and NO_3^-, are such weak Brønsted bases that they cannot affect the pH of a solution.

If the salt is derived from a weak acid *and* a weak base, its net effect on pH has to be determined on a case by case basis by determining which of the two ions is the stronger.

Buffers. A solution that contains both a weak acid and a weak base (usually an acid–base conjugate pair) is called a **buffer,** because it is able to absorb $[H^+]$ from a strong acid or $[OH^-]$ from a strong base without suffering large changes in pH. For the general acid–base pair, HA and A^-, the following reactions neutralize H^+ and OH^-.

$$\text{When } H^+ \text{ is added:} \quad A^- + H^+ \longrightarrow HA$$

$$\text{When } OH^- \text{ is added:} \quad HA + OH^- \longrightarrow A^- + H_2O$$

The pH of a buffer is controlled by the ratio of weak acid to weak base, expressed either in terms of molarities or moles.

$$[H^+] = K_a \times \frac{[HA]}{[A^-]} = K_a \times \frac{\text{moles } HA}{\text{moles } A^-}$$

Because the $[H^+]$ is determined by the mole ratio of HA to A^-, dilution does not change the pH of a buffer.

In performing buffer calculations, the usually valid simplifications are

$$[HA]_{\text{equilibrium}} = [HA]_{\text{initial}}$$

$$[A^-]_{\text{equilibrium}} = [A^-]_{\text{initial}}$$

The value of $[A^-]_{\text{initial}}$ is found from the molarity of the *salt* of the weak acid in the buffer. Once $[H^+]$ is found by this calculation, the pH is calculated.

Buffer calculations can also be performed using K_b for the weak acid component. Similar equations apply, the principal difference being that it is the OH^- concentration that is calculated. Thus, using K_b we have

$$[OH^-] = K_b \times \frac{[\text{conjugate base}]}{[\text{conjugate acid}]}$$

$$= K_b \times \frac{\text{moles conjugate base}}{\text{moles conjugate acid}}$$

Buffers are most effective when the pK_a of the acid member of the buffer pair lies within ± 1 unit of the desired pH.

Although the molar or mole *ratio* determines the pH at which a buffer works, the *amounts* of the acid and its conjugate base give the buffer its *capacity* for taking on strong acid or base with little change in pH.

Equilibria in Solutions of Polyprotic Acids. A polyprotic acid has a K_a value for the ionization of each of its hydrogen ions. Successive values of K_a often differ by a factor of 10^4 to 10^5. This allows us to calculate the pH of a solution of a polyprotic acid by using just the value of K_{a_1}. If the polyprotic acid is the only solute, the anion formed in the second step of the ionization, A^{2-}, has a molar concentration numerically equal to K_{a_2}.

Solutions of Salts of Polyprotic Acids. The anions of weak polyprotic acids are bases that react with water in successive steps, the last of which has the molecular polyprotic acid as a product. For a diprotic acid, $K_{b_1} = K_w/K_{a_2}$ and $K_{b_2} = K_w/K_{a_1}$. Usually, $K_{b_1} \gg K_{b_2}$, so virtually all the OH^- produced in the solution comes from the first step. The pH of the solution can be calculated using just K_{b_1} and the reaction

$$A^{2-} + H_2O \rightleftharpoons HA^- + OH^-$$

Acid–Base Titrations and Indicators. The pH at the equivalence point depends on the ions of the salt made by the titration. In the titration of a strong acid by a strong base, the pH at the equivalence point is 7 because neither of the ions produced in the reaction affects the pH of an aqueous solution.

For the titration of a weak acid by a strong base, the approach toward calculation of the pH varies depending on which stage of the titration is considered. Before the titration begins, the only solute is the weak acid. After starting but before the equivalence point the solution is a buffer. At the equivalence point, the active solute is the conjugate base of the weak acid. After the equivalence point, excess base determines the pH of the mixture. At the equivalence point the pH is above 7. Similar calculations apply in the titration of a weak base by a strong acid. In such titrations, the pH at the equivalence point is below 7.

The indicator chosen for a titration should have the center of its color change near the pH at the equivalence point. For the titration of a strong or weak acid with a strong base, phenolphthalein is the best indicator. When a weak base is titrated with a strong acid, several indicators including methyl orange work satisfactorily.

The following table lists the important equations needed to solve problems dealing with acid–base equilibria. You must have all of them at your fingertips when you begin to analyze a problem.

The key to success is in choosing the correct tools to apply to the solution.

Tool	Function
General equation for the ionization of a weak acid *(page 702)*	To enable you to write the correct chemical equation for the ionization of a weak acid and the correct K_a expression.
General equation for the ionization of a of weak base *(page 704)*	To enable you to write the correct chemical equation for the ionization a weak base and the correct K_b expression.
$K_a \times K_b = K_w$ *(page 706)*	To calculate K_a given K_b, or vice versa.
$\text{Percent ionization} = \dfrac{\text{moles per liter ionized}}{\text{moles per liter available}} \times 100\%$ *(page 707)*	To calculate the percentage ionization from the initial concentration of acid (or base) and the change in the concentrations of the ions. If the percentage ionization or pH is known, the K_a (or K_b) can be calculated.

Identification of acidic cations and basic
is **anions** *(pages 715 and 716)*

To determine whether a salt solution is acidic, basic, or neutral. Also, it the first step in calculating the pH of a salt solution.

Criterion for applying simplifications in acid–base equilibrium calculations:
$$[X]_{\text{initial}} \geq 400 \times K \quad \textit{(page 722)}$$

To determine whether initial concentrations can be used as though they are equilibrium values in the mass action expression when working acid–base equilibrium problems.

Method of successive approximations
(page 724)

To provide a quick procedure for calculating equilibrium concentrations when the usual simplifications are inappropriate.

Reactions when H⁺ or OH⁻ are added to a buffer *(page 726)*

To determine the effect of strong acid or strong base on the pH of a buffer. Adding H^+ lowers $[A^-]$ and raises $[HA]$; adding OH^- lowers $[HA]$ and raises $[A^-]$.

THINKING IT THROUGH

Remember, the goal for each of the following problems is to assemble the available information needed to obtain the answer, state what additional data (if any) are needed, and describe how you would use the data to answer the question.

The problems are divided into two groups. Those in Level 2 are more challenging than those in Level 1 and provide an opportunity to really hone your problem solving skills. Detailed answers to these problems can be found at our Web site: http://www.wiley.com/college/brady.

Level 1 Problems

1. Propionic acid, $HC_3H_5O_2$, is a monoprotic acid. You have prepared a 0.100 M solution of this acid. What experimental information is required to determine the value of K_a for this acid? Explain how you would use this information to calculate K_a.

2. A 0.10 M solution of a weak base is prepared. The value of K_b for the base is 2.0×10^{-8}. Describe how you would calculate the pH of the solution. Explain any simplifications you would make and how you would justify their validity.

3. A 0.10 M solution of a weak acid with $K_a = 2.5 \times 10^{-6}$ has a pH of 3.28. Is the weak acid the only solute in the solution? (Describe in detail how you would answer this question.)

4. What information do you need to actually prepare, in the laboratory, a buffer that has a pH of 5.00 using acetic acid and potassium acetate? Describe any calculations that would be necessary.

5. How would you calculate the acetate ion concentration in a solution that contains 0.0010 M HCl and 0.10 M $HC_2H_3O_2$? Describe and justify any simplifications you would make in solving the problem.

6. A 0.035 M solution of an acid has a pH of 3.5. Is the solute a strong acid or not? How can you tell?

7. Acetic acid ($HC_2H_3O_2$) and propionic acid ($HC_3H_5O_2$) have nearly identical pK_a values. If you had an aqueous solution that is 0.10 M in acetic acid and also 0.10 M in the sodium salt of the other acid, propionic acid, would this solution function as a *buffer*? Explain. Write net ionic equations that describe its chemical behavior if a small amount of hydrochloric acid is added or if a small amount of sodium hydroxide is added.

8. Describe how you would calculate the value of K_{b_1} for the PO_4^{3-} ion.

9. In a solution of sodium sulfite, there are actually three sources of OH^- ion. Write chemical equations for the three reactions involved. Which source(s) of OH^- can be ignored, and why?

Level 2 Problems

10. A simple way to measure the pK_a of an acid is to mix equal numbers of moles of the acid and a base (such as NaOH) to form an aqueous solution and then measure the pH with a pH meter. Explain why this method avoids having to perform equilibrium calculations.

11. Suppose 25.0 mL of 0.10 M HCl is added to 50.0 mL of 0.10 M KNO_2. Explain how you would calculate the pH of the resulting mixture.

12. The value of K_w increases as the temperature of water is raised.
(a) Explain what effect increasing the temperature of pure water will have on its pH. (Will the pH increase, decrease, or stay the same?)
(b) As the temperature of pure water increases, will the water become more acidic, more basic, or remain neutral? Justify your answer.

13. A solution of dichloroacetic acid, $HC_2HO_2Cl_2$, has a concentration of 0.50 M. Describe how you would calculate the freezing point of this solution. (Since it is a dilute aqueous solution, assume that the molarity equals the molality.)

14. Suppose 250 mL of $NH_3(g)$ measured at 20 °C and 750 torr is dissolved in 100 mL of 0.050 M HCl. Describe how you would calculate the pH of the resulting solution.

15. How many milliliters of 0.0010 M HCl would have to be added to 100 mL of a 0.100 M solution of $HC_2H_3O_2$ to make the $C_2H_3O_2^-$ concentration equal 1.0×10^{-5} M? (Explain all the calculations that are necessary to obtain the answer.)

16. What is the concentration of H_2CO_3 in a 0.10 M solution of Na_2CO_3? (Explain how you would calculate the answer. In what way is this calculation similar to finding the CO_3^{2-} concentration in a solution of H_2CO_3?)

17. Show that the concentration of PO_4^{3-} in a 0.10 M solution of H_3PO_4 is not equal to K_{a_3} for this acid. Describe how you would calculate the PO_4^{3-} concentration in this solution.

18. Describe in detail how you would calculate the hydrogen ion concentration in 0.10 M H_2SO_4. (*Hint:* The H^+ concentration is *not* 0.20 M.)

19. Below is a diagram illustrating a mixture HF and F^- in an aqueous solution. Describe how the number of HF molecules and F^- ions will change after three OH^- ions are added. How will the number of HF and F^- change if two H^+ ions are added? Explain your answers by using chemical equations.

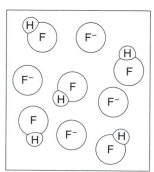

20. A student wrote the following equation for the equilibrium reaction of the weak base CH_3NH_2 with water. What, if anything, is wrong with it? Explain your answer.

$$CH_3NH_2(aq) + H_2O \rightleftharpoons CH_4NH_2^+(aq) + OH^-(aq)$$

21. Solid ammonium chloride was added to a milky suspension of insoluble $Mg(OH)_2$ and the mixture was shaken. Afterward, it was a clear, colorless solution. Use chemical equilibria to describe what happened in the mixture to cause the $Mg(OH)_2$ to dissolve.

22. $LiBr(aq)$ is a neutral solution and $KF(aq)$ is a basic solution. Both solutions, however, at equal concentrations show the identical freezing points. Discuss the implications of the acid–base properties of these solutions with respect to possible reactions of the ions with water and how the solutions could still show identical freezing point depressions. (Hint: Consider the number of solute species present from all possible sources.)

23. You wish to find the concentration of HCl in a jug of muriatic acid used to adjust the pH in your home swimming pool. You remove a 25.00 mL sample from the jug and titrate it with a standardized NaOH solution. As you proceed you remember that you forgot to add any phenolphthalein indicator. When you add a few drops of indicator the sample becomes dark pink. Before you take another sample of the acid and repeat the titration you notice that you do not have enough standardized NaOH for another titration. How could you still determine the concentration of HCl in the muriatic acid?

REVIEW QUESTIONS

Acid and Base Ionization Constants

16.1 Write the general equation for the ionization of a weak acid in water. Give the equilibrium law corresponding to K_a.

16.2 Write the chemical equation for the ionization of each of the following weak acids in water. (Some are polyprotic acids; for these write only the equation for the first step in the ionization.)
(a) HNO_2 (c) $HAsO_4^{2-}$
(b) H_3PO_4 (d) $(CH_3)_3NH^+$

16.3 For each of the acids in Review Question 16.2, write the appropriate K_a expression.

16.4 Write the general equation for the ionization of a weak base in water. Give the equilibrium law corresponding to K_b.

16.5 Write the chemical equation for the ionization of each of the following weak bases in water.
(a) $(CH_3)_3N$ (c) NO_2^-
(b) AsO_4^{3-} (d) $(CH_3)_2N_2H_2$

16.6 For each of the bases in Review Question 16.5, write the appropriate K_b expression.

16.7 Benzoic acid, $C_6H_5CO_2H$, is an organic acid whose sodium salt, $C_6H_5CO_2Na$, has long been used as a safe food additive to protect beverages and many foods against harmful yeasts and bacteria. The acid is monoprotic. Write the equation for its K_a.

16.8 Write the equation for the equilibrium that the benzoate ion, $C_6H_5CO_2^-$ (Review Question 16.7), would produce in water as it functions as a Brønsted base. Then write the expression for the K_b of the conjugate base of benzoic acid.

16.9 The pK_a of HCN is 9.21 and that of HF is 3.17. Which is the stronger Brønsted base: CN^- or F^-?

Acid–Base Properties of Salt Solutions

16.10 Aspirin is acetylsalicylic acid, a monoprotic acid whose K_a value is 3.27×10^{-4}. Does a solution of the sodium salt of aspirin in water test acidic, basic, or neutral? Explain.

16.11 The K_b value of the oxalate ion, $C_2O_4^{2-}$, is 1.9×10^{-10}. Is a solution of $K_2C_2O_4$ acidic, basic, or neutral? Explain.

Solubility and Simultaneous Equilibria

This Chapter in Context The principles of chemical equilibria developed in Chapter 14 and used to study weak acids and bases in Chapter 16 also apply to weakly soluble salts. In this chapter we'll go beyond the solubility rules and their indefinite categories of "soluble" and "insoluble" and study how else we might deal with the question, "How soluble?"

The solubility rules are least helpful when more than one salt is in the same solution. For example, $PbCl_2$ and $AgCl$ are both insoluble salts, according to the rules. So you might think that if we slowly add Cl^- ion (using dilute sodium chloride) to a dilute solution containing both lead(II) nitrate and silver nitrate both $PbCl_2$ and $AgCl$ would precipitate together, side by side, from the start. What actually happens is that nearly all of the Ag^+ ion precipitates as $AgCl$ *before* any Pb^{2+} ion separates as $PbCl_2$ (see Figure 17.1). This happens because $AgCl$ is much less soluble than $PbCl_2$. To study such differences in solubility, we need to examine solubility equilibria quantitatively.

Chapter Outline

17.1
Solubility Equilibria for Salts
17.2
Solubility Equilibria
of Metal Oxides and Sulfides
17.3
Separating Metal Ions
by Selective Precipitation
17.4
Complex Ions and Their Equilibria in
Aqueous Solutions

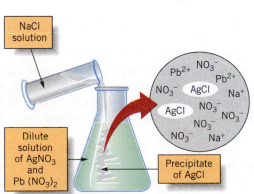

Figure 17.1 *Relative solubilities.* When dilute sodium chloride is added to a solution containing both Ag^+ ions and Pb^{2+} ions (both dissolved as their nitrate salts), the less soluble $AgCl$ precipitates before the more soluble $PbCl_2$.

759

> ▶**Chemistry in Practice**◀ Every heartbeat requires the contraction of heart muscle, and every contraction requires the momentary presence of calcium ion inside heart muscle cells. But also inside such cells is phosphate ion, and the two ions, Ca^{2+} and PO_4^{3-}, form a very insoluble precipitate. In the face of this fact, how does the heart muscle cell manage its need for calcium ion? If your career lies in biology, human physiology, medicine, or biochemistry, you'll eventually learn how. But the story begins with the fundamental principles that we'll study next. ◆

17.1 Solubility Equilibria for Salts

Solubility Product Constant, K_{sp}

Let's begin with the salt, AgCl, which the rules tell us is "insoluble." When we place crystalline AgCl in water, a very small amount dissolves, and the following equilibrium exists when the solution has become saturated.

$$AgCl(s) \rightleftharpoons Ag^+(aq) + Cl^-(aq)$$
(In a saturated solution of AgCl)

This is a heterogeneous equilibrium because it involves a solid reactant (AgCl) in equilibrium with ions in aqueous solution. Using the procedure developed in Section 14.7 (page 693), we write the equilibrium law as follows, omitting the solid from the mass action expression.

Solubility product expression

$$[Ag^+][Cl^-] = K_{sp} \qquad (17.1)$$

The equilibrium constant, K_{sp}, is called the **solubility product constant** (because the system is a *solubility* equilibrium and the constant equals a *product* of ion concentrations).

It is very important to make the distinction between solubility and solubility product. The *solubility* of a salt is the amount of the salt that dissolves in a given amount of solvent to give a saturated solution. *Solubility product* is the product of the molar concentrations of the ions in the saturated solution.

The solubilities of salts change with temperature, so a value of K_{sp} applies only at the temperature at which it was determined. Some typical K_{sp} values are in Table 17.1 and in Appendix E. (Insoluble oxides and sulfides make up a special group to be discussed in Section 17.2.)

Ion Products

Ion product of a salt

The term on the left in Equation 17.1, the product of the molar concentrations of the dissolved ions of the solute, is called the **ion product** of AgCl. At any dilution of the salt throughout the entire range of possibilities for an *unsaturated* solution, there will be varying values for the ion concentrations and, therefore, for the ion product. However, the ion product acquires a constant value, K_{sp}, in a *saturated* solution. Only then is the ion product called the *solubility product constant*. When a solution of AgCl is less than saturated, the value of the ion product is less than K_{sp}.

Many salts produce more than one of a given ion when they dissociate, and this introduces exponents into the ion product expression. Silver chromate, Ag_2CrO_4, for example, has the following solubility equilibrium.

$$Ag_2CrO_4(s) \rightleftharpoons 2Ag^+(aq) + CrO_4^{2-}(aq)$$

Table 17.1 Solubility Product Constants

Type	Salt	Ions of Salt	K_{sp} (25 °C)
Halides	$CaF_2 \rightleftharpoons Ca^{2+} + 2F^-$		3.9×10^{-11}
	$PbF_2 \rightleftharpoons Pb^{2+} + 2F^-$		3.6×10^{-8}
	$AgCl \rightleftharpoons Ag^+ + Cl^-$		1.8×10^{-10}
	$AgBr \rightleftharpoons Ag^+ + Br^-$		5.0×10^{-13}
	$AgI \rightleftharpoons Ag^+ + I^-$		8.3×10^{-17}
	$PbCl_2 \rightleftharpoons Pb^{2+} + 2Cl^-$		1.7×10^{-5}
	$PbBr_2 \rightleftharpoons Pb^{2+} + 2Br^-$		2.1×10^{-6}
	$PbI_2 \rightleftharpoons Pb^{2+} + 2I^-$		7.9×10^{-9}
Hydroxides	$Al(OH)_3 \rightleftharpoons Al^{3+} + 3OH^-$		3×10^{-34} (a)
	$Ca(OH)_2 \rightleftharpoons Ca^{2+} + 2OH^-$		6.5×10^{-6}
	$Fe(OH)_2 \rightleftharpoons Fe^{2+} + 2OH^-$		7.9×10^{-16}
	$Fe(OH)_3 \rightleftharpoons Fe^{3+} + 3OH^-$		1.6×10^{-39}
	$Mg(OH)_2 \rightleftharpoons Mg^{2+} + 2OH^-$		7.1×10^{-12}
	$Zn(OH)_2 \rightleftharpoons Zn^{2+} + 2OH^-$		3.0×10^{-16} (b)
Carbonates	$Ag_2CO_3 \rightleftharpoons 2Ag^+ + CO_3^{2-}$		8.1×10^{-12}
	$MgCO_3 \rightleftharpoons Mg^{2+} + CO_3^{2-}$		3.5×10^{-8}
	$CaCO_3 \rightleftharpoons Ca^{2+} + CO_3^{2-}$		4.5×10^{-9} (c)
	$SrCO_3 \rightleftharpoons Sr^{2+} + CO_3^{2-}$		9.3×10^{-10}
	$BaCO_3 \rightleftharpoons Ba^{2+} + CO_3^{2-}$		5.0×10^{-9}
	$CoCO_3 \rightleftharpoons Co^{2+} + CO_3^{2-}$		1.0×10^{-10}
	$NiCO_3 \rightleftharpoons Ni^{2+} + CO_3^{2-}$		1.3×10^{-7}
	$ZnCO_3 \rightleftharpoons Zn^{2+} + CO_3^{2-}$		1.0×10^{-10}
Chromates	$Ag_2CrO_4 \rightleftharpoons 2Ag^+ + CrO_4^{2-}$		1.2×10^{-12}
	$PbCrO_4 \rightleftharpoons Pb^{2+} + CrO_4^{2-}$		1.8×10^{-14} (d)
Sulfates	$CaSO_4 \rightleftharpoons Ca^{2+} + SO_4^{2-}$		2.4×10^{-5}
	$SrSO_4 \rightleftharpoons Sr^{2+} + SO_4^{2-}$		3.2×10^{-7}
	$BaSO_4 \rightleftharpoons Ba^{2+} + SO_4^{2-}$		1.1×10^{-10}
	$PbSO_4 \rightleftharpoons Pb^{2+} + SO_4^{2-}$		6.3×10^{-7}
Oxalates	$CaC_2O_4 \rightleftharpoons Ca^{2+} + C_2O_4^{2-}$		2.3×10^{-9}
	$MgC_2O_4 \rightleftharpoons Mg^{2+} + C_2O_4^{2-}$		8.6×10^{-5}
	$BaC_2O_4 \rightleftharpoons Ba^{2+} + C_2O_4^{2-}$		1.2×10^{-7}
	$FeC_2O_4 \rightleftharpoons Fe^{2+} + C_2O_4^{2-}$		2.1×10^{-7}
	$PbC_2O_4 \rightleftharpoons Pb^{2+} + C_2O_4^{2-}$		2.7×10^{-11}

^{a}Alpha form. ^{b}Amorphous form. ^{c}Calcite form. ^{d}At 10 °C.

When fully dissociated, each formula unit of Ag_2CrO_4 gives two Ag^+ ions and one CrO_4^{2-} ion. In the ion product of such a salt, each concentration term carries an exponent that equals the number of ions released per formula unit. So the ion product for silver chromate is $[Ag^+]^2[CrO_4^{2-}]$; the exponent of 2 appears because the formula of Ag_2CrO_4 has two Ag^+ ions per formula unit.

Practice Exercise 1

What are the ion product expressions for the following salts if they are fully dissociated in water? (a) $BaCrO_4$; (b) Ag_3PO_4 ◆

Silver chromate, Ag_2CrO_4, precipitates when sodium chromate is added to a solution of silver nitrate.

The Determination of K_{sp}

Using Molar Solubility to Determine K_{sp}

One way to determine the value of K_{sp} for a slightly soluble salt is to measure its **molar solubility,** that is, the number of moles of salt dissolved in one liter of its *saturated* solution. K_{sp} is then calculated under the assumption that all of the salt that dissolves is 100% dissociated into the ions implied in the salt's formula. This works reasonably well for slightly soluble salts made up of univalent ions, like silver bromide, which we use in the following example.

EXAMPLE 17.1

Calculating K_{sp} from Molar Solubility Data

Silver bromide, AgBr, is the light-sensitive compound used in nearly all photographic film. The molar solubility of AgBr is 7.1×10^{-7} mol L^{-1} at 25 °C. Calculate K_{sp} for AgBr at this temperature.

Analysis: As usual, we begin with the chemical equation for the equilibrium and the equilibrium law (here, the expression for K_{sp}).

$$AgBr(s) \rightleftharpoons Ag^+(aq) + Br^-(aq)$$

$$K_{sp} = [Ag^+][Br^-]$$

We have been given the moles of AgBr needed to make one liter of a *saturated* solution, 7.1×10^{-7} mol L^{-1}. We'll show how to use this information, following procedures outlined in Chapter 14, to construct a *concentration table*. (We'll be going into more detail than necessary for the simple calculation, but we want to describe a systematic approach that then works smoothly in more complicated situations.)

In the table, we first enter the initial concentrations of the ions shown in the expression for K_{sp}, those of any such ions that might already be in the solution before solid AgBr is added. But here, we don't have this complication; neither Ag$^+$ nor Br$^-$ has earlier been put into the solution from some other source (like AgNO$_3$ or NaBr).

Next, in the "change" row, we enter data on how the concentrations of the Ag$^+(aq)$ and Br$^-(aq)$ change when the AgBr dissolves. Because dissolving the salt always increases these concentrations, the quantities are entered as positive values.

As usual, we obtain the equilibrium values by adding the "initial concentrations" to the "changes."

There are no entries in the column under AgBr(s) because this substance does not appear in the K_{sp} expression.

	AgBr(s)	$\rightleftharpoons$	Ag$^+$(aq)	+	Br$^-$(aq)
Initial concentrations (M)	No entries in this column		0 (Note 1)		0
Changes in concentrations when AgBr dissolves (M)			$+7.1 \times 10^{-7}$ (Note 2)		$+7.1 \times 10^{-7}$
Equilibrium concentrations (M)			7.1×10^{-7} (Note 3)		7.1×10^{-7}

Note 1. When AgBr is placed in *pure* water, the *initial* concentrations of its ions are zero.

Note 2. When some AgBr dissolves, Ag$^+$ and Br$^-$ ions are released in a 1:1 ratio. So, assuming 100% dissociation, when 7.1×10^{-7} mol of AgBr dissolves (per liter), 7.1×10^{-7} mol of Ag$^+$ ion and 7.1×10^{-7} mol of Br$^-$ ion go into solution. The concentrations of these species thus change (increase) by these amounts.

Note 3. By adding the initial concentrations and the changes, we obtain the final (equilibrium) concentrations.

Solution: We now substitute the equilibrium concentrations into the K_{sp} expression.

$$K_{sp} = [Ag^+][Br^-]$$
$$= (7.1 \times 10^{-7})(7.1 \times 10^{-7})$$
$$= 5.0 \times 10^{-13}$$

The K_{sp} of AgBr is thus calculated to be 5.0×10^{-13} at 25 °C.

Is the Answer Reasonable?

The answer is the square of the molar solubility of the salt. This has to be the calculated K_{sp} for any salt whose ions are in a ratio of 1:1, assuming 100% dissociation.

Practice Exercise 2

The solubility of thallium(I) iodide, TlI, in water at 20 °C is 1.8×10^{-5} mol L^{-1}. Using this fact, calculate K_{sp} for TlI on the assumption that it is 100% dissociated in the solution. ◆

Silver chloride, silver bromide, and thallium(I) iodide are examples of 1:1 salts, meaning that the ratio of cation to anion is 1:1. The K_{sp} value of AgBr is simply the square of the salt's molar solubility, as we noted at the end of Example 17.1. When the assumption of 100% dissociation is valid, the K_{sp} value for any 1:1 salt in water having no other solute will always be the square of the molar solubility. When a salt's ions are not both singly charged, as they are in AgBr, the relationship between molar solubility and the solubility product constant is complicated by interionic attractions in the solution. We leave a description of this complication to Facets of Chemistry 17.1.

When the ions of a slightly soluble salt are not in a 1:1 ratio, the use of molar solubility data to calculate the salt's K_{sp} can be illustrated by the next examples.

> We assume that any soluble 1:1 salt made of univalent ions, like NaCl, is fully dissociated in water.

The molar solubility of silver chromate, Ag_2CrO_4, in water is 6.7×10^{-5} mol L^{-1} at 25 °C. What is K_{sp} for Ag_2CrO_4, assuming 100% dissociation?

EXAMPLE 17.2
Calculating K_{sp} from Molar Solubility Data

Analysis: The equilibrium equation and K_{sp} expression are

$$Ag_2CrO_4(s) \rightleftharpoons 2Ag^+(aq) + CrO_4^{2-}(aq) \qquad K_{sp} = [Ag^+]^2[CrO_4^{2-}]$$

The solute is pure water, so neither Ag^+ nor CrO_4^{2-} are present before the salt dissolves. The initial concentrations of Ag^+ and CrO_4^{2-} are therefore zero. When one liter of pure water dissolves 6.7×10^{-5} mol of Ag_2CrO_4, we can see by the stoichiometry of the equilibrium equation that 100% dissociation would produce 6.7×10^{-5} mol of CrO_4^{2-} and $2 \times (6.7 \times 10^{-5}$ mol) of Ag^+. With this information, we prepare the concentration table.

	$Ag_2CrO_4(s)$ $\rightleftharpoons$	$2Ag^+(aq)$	+	$CrO_4^{2-}(aq)$
Initial concentrations (M)		0		0
Changes in concentrations when Ag_2CrO_4 dissolves (M)		$+2 \times (6.7 \times 10^{-5}) =$ $+1.3 \times 10^{-4}$		$+6.7 \times 10^{-5}$
Equilibrium concentrations (M)		1.3×10^{-4}		6.7×10^{-5}

Facets of Chemistry 17.1

Ion Pairs

Like silver bromide, calcium sulfate, $CaSO_4$, has a 1:1 ratio of ions. How well does it work, then, to take the square of the molar solubility of $CaSO_4$ to calculate the value of this salt's K_{sp}? Let's see.

The equation for the solubility product constant of $CaSO_4$ is

$$K_{sp} = [Ca^{2+}][SO_4^{2-}]$$

We know by experiment that the actual molar solubility of $CaSO_4$ at 25 °C is 1.53×10^{-2} mol L^{-1}. The square of this molar solubility would give a value of 2.34×10^{-4} for the (calculated) K_{sp} of calcium sulfate. However, this is not the true value of the actual *ion product* in the saturated solution, namely, $[Ca^{2+}][SO_4^{2-}]$, an expression that requires that we use only the molarities of the *separated ions*. In a saturated solution of $CaSO_4$, the true value of K_{sp} is known to be closer to 2.4×10^{-5}, which is about one-tenth the value of K_{sp} calculated from the molar solubility. The true value of K_{sp} implies that the molarities of the separated ions are less than we assume from the molar solubility. What's wrong?

Formation of Ion Pairs

The ions released when calcium sulfate dissolves in water are divalent, not univalent. Nature, which is never "wrong," is unyieldingly insistent that "unlike charges attract," and a cation with a charge of 2+, like Ca^{2+}, has a substantially stronger attraction for an anion with a charge of 2−, like SO_4^{2-}, than is the attraction of Ag^+ for Cl^-.

Because of interionic attractions in the solution, calcium ions and sulfate ions are unable to leave each other completely alone. At least some tend to pair up, one on one, to form **ion pairs.** These remain dissolved, so we could represent them by the formula $CaSO_4(aq)$. Thus, calcium sulfate is not 100% dissociated in its saturated solution.

In a saturated solution, an ion pair is a moleculelike combination of two ions that are not bound in the undissolved crystals but move around in the dissolved state. About one-third of all of the calcium ions in a saturated solution of calcium sulfate are not separated from sulfate ions.

The formation of ion pairs means that using molar solubility data for salts of polyvalent ions to determine K_{sp} values can only give maximum values. We leave to other courses the full development of this complication.

Solution: Substituting the equilibrium concentrations into the mass action expression for K_{sp} gives

$$K_{sp} = (1.3 \times 10^{-4})^2(6.7 \times 10^{-5})$$
$$= 1.1 \times 10^{-12}$$

So the K_{sp} of Ag_2CrO_4 is calculated to be 1.1×10^{-12} at 25 °C.

Is the Answer Reasonable?

Being right depends on doing the calculation for the *change* in concentration of the Ag^+ ion, and this depends on correctly using its coefficient of 2 in the equilibrium equation. *Be sure you feel comfortable about this step.*

Now let's introduce another complication. Let's suppose that the aqueous system into which we're dissolving a slightly soluble salt is not pure water but already contains a solute.

EXAMPLE 17.3
Calculating K_{sp} from Molar Solubility Data

The molar solubility of $PbCl_2$ in a 0.10 *M* NaCl solution is 1.7×10^{-3} mol L^{-1} at 25 °C. Calculate the K_{sp} for $PbCl_2$.

Analysis: Again, we start by writing the equation for the equilibrium and the K_{sp} expression.

$$PbCl_2(s) \rightleftharpoons Pb^{2+}(aq) + 2Cl^-(aq) \qquad K_{sp} = [Pb^{2+}][Cl^-]^2$$

In this problem, the $PbCl_2$ is being dissolved not in pure water but in 0.10 M NaCl, which contains 0.10 M Cl^-. (It also contains 0.10 M Na^+, but that's not important here because Na^+ doesn't affect the equilibrium and so doesn't appear in the K_{sp} expression.) The initial concentration of Pb^{2+} is zero because none is in solution to begin with. The *initial* concentration of Cl^-, however, is 0.10 M. When the $PbCl_2$ dissolves in the NaCl solution, the Pb^{2+} concentration increases by 1.7×10^{-3} M and the Cl^- concentration increases by $2 \times (1.7 \times 10^{-3}$ $M)$. With these data we build the concentration table.

We are assuming that for each formula unit of $PbCl_2$ that dissolves, we get one Pb^{2+} ion and two Cl^- ions.

	$PbCl_2(s)$ $\rightleftharpoons$	$Pb^{2+}(aq)$	+	$2Cl^-(aq)$
Initial concentrations (M)		0		0.10
Changes in concentrations when $PbCl_2$ dissolves (M)		$+1.7 \times 10^{-3}$		$+2 \times (1.7 \times 10^{-3})$ $= +3.4 \times 10^{-3}$
Equilibrium concentrations (M)		1.7×10^{-3}		$0.10 + 0.0034 = 0.10$ (rounded to the second decimal place)

Solution: Substituting the equilibrium concentrations into the K_{sp} expression gives

$$K_{sp} = (1.7 \times 10^{-3})(0.10)^2$$
$$= 1.7 \times 10^{-5}$$

The K_{sp} of $PbCl_2$ is calculated to be 1.7×10^{-5} at 25 °C.

Is the Answer Reasonable?
As with the preceding worked example, being correct depends on using the coefficient of Cl^- from the equilibrium equation.

Practice Exercise 3

One liter of water is able to dissolve 2.15×10^{-3} mol of PbF_2 at 25 °C. If the solute is 100% dissociated, what is the K_{sp} of PbF_2? ◆

Practice Exercise 4

At 25 °C, the molar solubility of $CoCO_3$ in a 0.10 M Na_2CO_3 solution is 1.0×10^{-9} mol L^{-1}. If the solutes are 100% dissociated, what is the K_{sp} for $CoCO_3$? ◆

Practice Exercise 5

The molar solubility of PbF_2 in a 0.10 M $Pb(NO_3)_2$ solution at 25 °C is 3.1×10^{-4} mol L^{-1}. If the solutes are 100% dissociated, what is the K_{sp} for PbF_2? ◆

Calculating Molar Solubility from K_{sp}

Besides calculating K_{sp} from solubility information, we can also compute solubilities from values of K_{sp}. The following examples show how to make such calculations.[1]

---

[1] These calculations ignore the fact that not all of the salt that dissolves is 100% dissociated into the ions implied in the salt's formula. This is particularly a problem with the salts of polyvalent ions, so the solubilities of such salts, when calculated from their K_{sp} values, must be taken as rough estimates.

EXAMPLE 17.4

Calculating Molar Solubility from K_{sp}

What is the molar solubility of AgCl in pure water at 25 °C?

Analysis: To solve this problem, we need three things.

1. The equilibrium equation for a saturated solution of AgCl.
2. The K_{sp} equation, which we figure out from the equilibrium equation.
3. The value of K_{sp} for AgCl.

The relevant equations are

$$AgCl(s) \rightleftharpoons Ag^+(aq) + Cl^-(aq) \qquad K_{sp} = [Ag^+][Cl^-] = 1.8 \times 10^{-10}$$

The value of K_{sp} is from Table 17.1. Now we can build a concentration table. The solvent is pure water, so neither Ag^+ nor Cl^- ion is present in solution at the start; their *initial* concentrations are zero. Let's define x as the molar solubility of the salt—the number of moles of AgCl that dissolves in one liter. Since 1 mol of AgCl yields 1 mol Ag^+ and 1 mol Cl^-, the concentrations of each of these ions increase by x. Thus, our concentration table is as follows.

	AgCl(s)	$\rightleftharpoons$	$Ag^+(aq)$	+	$Cl^-(aq)$
Initial concentrations (M)			0		0
Changes in concentrations when AgCl dissolves (M)			$+x$		$+x$
Equilibrium concentrations (M)			x		x

> Notice that the coefficients of x, namely, 1, are the same as the coefficients of Ag^+ and Cl^- in the equation for the equilibrium.

Solution: We make the substitutions into the K_{sp} expression:

$$(x)(x) = 1.8 \times 10^{-10}$$
$$x = 1.3 \times 10^{-5}$$

The calculated molar solubility of AgCl in water at 25 °C is 1.3×10^{-5} mol L^{-1}.

Is the Answer Reasonable?
For a 1:1 salt, the calculated value of K_{sp} is the *square* of the salt's molar solubility. So the molar solubility of such a salt should be the square root of the K_{sp}, which is what we found.

EXAMPLE 17.5

Calculating Molar Solubility from K_{sp}

Calculate the molar solubility of lead iodide, PbI_2, from its K_{sp} in water at 25 °C.

Analysis: The relevant equations and K_{sp} data (from Table 17.1) are

$$PbI_2(s) \rightleftharpoons Pb^{2+}(aq) + 2I^-(aq) \qquad K_{sp} = [Pb^{2+}][I^-]^2 = 7.9 \times 10^{-9}$$

The solvent is water, so the initial concentrations of the ions are zero. As before, we will define x as the molar solubility of the salt. By doing this, the coefficients of x in the "change" row are made identical to the coefficients of the ions in the chemical equation, thereby assuring that the changes in the concentrations are in the correct mole ratio.

	$PbI_2(s)$	$\rightleftharpoons$	$Pb^{2+}(aq)$	$+$	$2I^-(aq)$
Initial concentrations (M)			0		0
Changes in concentrations when PbI_2 dissolves (M)			$+x$		$+2x$
Equilibrium concentrations (M)			x		$2x$

The precipitation of yellow PbI_2 occurs when the two colorless solutions, one with sodium iodide and the other of lead(II) nitrate, are mixed.

Solution: Substituting,

$$K_{sp} = (x)(2x)^2 = 4x^3 = 7.9 \times 10^{-9}$$
$$x^3 = 2.0 \times 10^{-9}$$
$$x = 1.3 \times 10^{-3}$$

Thus, the molar solubility of PbI_2 is calculated to be 1.3×10^{-3} mol L^{-1}.

Is the Answer Reasonable?

Setting x equal to the molarity of the Pb^{2+} ion *requires* us to use $2x$ for the molarity of Cl^-. If you're comfortable with this, you should feel right about the rest of the calculation.

Practice Exercise 6

What is the calculated molar solubility in water at 25 °C of each compound? (a) $AgBr$; (b) Ag_2CO_3 ◆

Common Ion Effect and Solubility

Suppose that we stir some calcium carbonate with water long enough to establish the following equilibrium.

$$CaCO_3(s) \rightleftharpoons Ca^{2+}(aq) + CO_3^{2-}(aq)$$

Then suppose we add to the solution a very soluble salt of calcium, like $CaCl_2$. This puts more $Ca^{2+}(aq)$ into solution. Now $[Ca^{2+}(aq)]\,[CO_3^{2-}(aq)]$, the ion product, momentarily *exceeds* K_{sp}. The effect is that the added Ca^{2+} has *upset the equilibrium.* Le Châtelier's principle tells us that the equilibrium must respond in a way that will remove as much as possible of the added Ca^{2+}. It can do this only by shifting to the left, *causing CO_3^{2-} to precipitate* as $CaCO_3$. Eventually, equilibrium is reestablished, but with a lower concentration of CO_3^{2-} in solution.

In the newly equilibrated system, there are two sources of Ca^{2+}, the added $CaCl_2$ and the $CaCO_3$ still in solution. An ion in a solution that has been supplied from more than one solute is called a **common ion.** The addition of the common ion lowers the solubility of $CaCO_3$; it becomes less soluble in the presence of $CaCl_2$ (or any other soluble calcium salt) than it is in pure water.

The lowering of the solubility of an ionic compound by the addition of a common ion is called the **common ion effect,** and it illustrates Le Châtelier's principle (page 640). Figure 17.2 shows how even a relatively soluble salt, NaCl, can be forced out of its saturated solution simply by adding concentrated hydrochloric acid, which serves as the source of a common ion, Cl^-. The common ion effect can dramatically lower the solubility of a salt, as the next example demonstrates.

Figure 17.2 *The common ion effect.* The test tube initially contained a saturated solution of sodium chloride in equilibrium with a trace amount of undissolved salt. The equilibrium is $NaCl(s) \rightleftharpoons Na^+(aq) + Cl^-(aq)$. We see in this photo the result of adding a few drops of concentrated hydrochloric acid, which contains a high concentration of Cl^-, an ion common to both the acid and the salt.

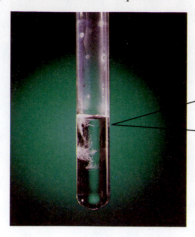

The addition of the common ion, Cl^-, has shifted the equilibrium to the left, and we can see some crystals of NaCl that were caused to precipitate from the solution.

EXAMPLE 17.6

Calculation Involving the Common Ion Effect

What is the molar solubility of PbI_2 in a 0.10 M NaI solution?

Analysis: As usual, we begin with the chemical equation for the equilibrium and the appropriate K_{sp} expression. The value of K_{sp} is obtained from Table 17.1.

$$PbI_2(s) \rightleftharpoons Pb^{2+}(aq) + 2I^-(aq) \qquad K_{sp} = [Pb^{2+}][I^-]^2 = 7.9 \times 10^{-9}$$

As before, we imagine that we are adding the PbI_2 to the NaI solution, so we need to know the concentrations of any of the ions of PbI_2 present in the NaI solution. These will be our initial concentrations.

The salt NaI does not contain any Pb^{2+}, but it does have I^-. Therefore, initially, no Pb^{2+} is in solution, but there is 0.10 mol of I^- per liter from the 0.10 mol L^{-1} of NaI.

Next, we let x be the molar solubility of PbI_2. When x mol of PbI_2 dissolves per liter, the concentration of Pb^{2+} changes by $+x$ and that of I^- by twice as much, or $+2x$. Finally, the equilibrium concentrations are obtained by summing the initial concentrations and the changes. Here is the concentration table.

	$PbI_2(s)$	$\rightleftharpoons$	$Pb^{2+}(aq)$	+	$2I^-(aq)$
Initial concentrations (M)			0		0.10
Changes in concentrations when $PbCl_2$ dissolves (M)			$+x$		$+2x$
Equilibrium concentrations (M)			x		$0.10 + 2x$

Solution: Substituting equilibrium values into the K_{sp} expression gives

$$K_{sp} = (x)(0.10 + 2x)^2 = 7.9 \times 10^{-9}$$

Just a brief inspection reveals that solving this expression for x will be difficult if we cannot simplify the math. Fortunately, a simplification is possible, because the small value of K_{sp} for PbI_2 tells us that the salt has a very low solubility. This means very little of the salt will dissolve, so x (or even $2x$) will be quite small. Let's assume that $2x$ will be much smaller than 0.10. If this is so, then

$$0.10 + 2x \approx 0.10$$

Substituting 0.10 M for the I^- concentration gives

$$K_{sp} = (x)(0.10)^2 = 7.9 \times 10^{-9}$$

$$x = \frac{7.9 \times 10^{-9}}{(0.10)^2}$$

$$= 7.9 \times 10^{-7} \, M$$

Thus, the molar solubility of PbI_2 in 0.10 M NaI solution is calculated to be $7.9 \times 10^{-2} \, M$.

Is the Answer Reasonable?
One check is to see if our simplifying assumption is valid. Notice that $2x$, which equals 1.6×10^{-6}, is indeed vastly smaller than 0.10, just as we anticipated.

A Mistake to Avoid

READ THESE TWO PARAGRAPHS!

The most common mistake that students make with problems like Example 17.6 is to use the coefficient of a formula in the solubility equilibrium at the wrong moment in the calculation. The coefficient of I^- in the PbI_2 equilibrium is 2. The mistake is to use this 2 to double the *initial* concentration of I^-. However, it can be used only to double the *change* in concentration. The *initial* concentration of I^- was provided not by PbI_2 but by NaI. The *change* in the concentration of I^- comes solely from the small solubility of PbI_2, so the coefficient of 2 enters into the calculation only to describe this change.

To avoid this mistake, always view the formation of the final solution as a two-step process. You begin with a solvent into which the "insoluble solid" will be placed. In some problems, the solvent may be pure water, in which case the initial concentrations of the ions will be zero. In other problems, like Example 17.6, the solvent will be a *solution* that contains a common ion. When this is so, first decide what the concentration of the common ion is and enter this value into the "initial concentration" row of the table. Next, imagine that the solid is added to the solvent and a little of it dissolves. *The amount that dissolves is what gives the values in the "change" row.* These entries must be in the same ratio as the coefficients in the equilibrium, which is accomplished if we let x be the molar solubility of the salt. Then the coefficients of x will be the same as the coefficients of the ions in the chemical equation for the equilibrium.

In Example 17.5 we found that the molar solubility of PbI_2 in *pure* water is $1.3 \times 10^{-3} \, M$. In water that contains 0.10 M NaI (Example 17.6), the solubility of PbI_2 is $7.9 \times 10^{-7} \, M$, well over a thousand times smaller. As we said, the common ion effect can cause huge reductions in solubilities of sparingly soluble compounds.

Practice Exercise 7
Calculate the molar solubility of AgI in 0.20 M NaI solution. Compare it to its calculated concentration in pure water. ◆

Practice Exercise 8
Calculate the molar solubility of $Fe(OH)_3$ in a solution where the OH^- concentration is initially 0.050 M and the dissociation of $Fe(OH)_3$ is 100%. ◆

Determining if a Precipitate Will Form in a Solution

Sometimes in setting up an experiment, the plans tentatively include an aqueous solution in which several ions are to be dissolved (or will form). Before proceeding, the chemist must ask, "Will any combination of the desired ions, at the planned concentrations, be of a salt that is too insoluble to be in solution?" If so, the plans for the experiment must be changed.

To use K_{sp} data and the initial values of ion molarities to predict if a precipitate will form, we first use the ion concentrations to compute ion products. Then these are compared to K_{sp} values. Remember that K_{sp} equals the ion product *for a saturated solution,* meaning one in which undissolved solute is present so that there is equilibrium between the dissolved and undissolved solutes. For example, 7.1×10^{-7} mol L^{-1} of AgBr is present in a *saturated* solution of AgBr, so $[Ag^+]$ and $[Br^-]$ each equal 7.1×10^{-7} M. The ion product, $[Ag^+][Br^-]$, therefore equals 5.0×10^{-13}, which is the K_{sp} of AgBr.

A substance will not precipitate from a solution if its concentration is such that the solution is either unsaturated or saturated.

But, as we have stressed, it is not necessary to have a *saturated* solution to compute an ion product. For example, in an *unsaturated* solution of AgBr containing just 7.1×10^{-8} mol L^{-1} of the solute, the molarities of Ag^+ and Cl^- each equal 7.1×10^{-8} mol L^{-1}. The computed ion product now is

$$[Ag^+][Br^-] = (7.1 \times 10^{-8})(7.1 \times 10^{-8}) = 5.0 \times 10^{-15}$$

This is less than the K_{sp} of AgBr (5.0×10^{-13}). No precipitate of AgBr will form in this solution because it is actually capable of dissolving more AgBr.

In a supersaturated solution, the concentrations of the ions are larger than in a saturated solution, so the ion product is larger than K_{sp}.

A *supersaturated* solution is one that contains more solute than is necessary for saturation. It is unstable, and there is a tendency for the extra solute to precipitate. In a supersaturated solution of AgBr, the computed ion product would be larger than K_{sp}.

When comparing a computed ion product with the K_{sp} for the salt, we can tell if a particular combination of ion concentrations will produce a precipitate or not. Here are the possibilities.

It is usually difficult to prevent the extra salt from precipitating out of a supersaturated solution. (Sodium acetate is a notable exception. Supersaturated solutions of this salt are easily made.)

Precipitate will form	Ion product $> K_{sp}$	(supersaturated)
No precipitate will form	$\begin{cases} \text{Ion product} = K_{sp} \\ \text{Ion product} < K_{sp} \end{cases}$	(saturated) (unsaturated)

Let's now consider these possibilities in some sample calculations.

EXAMPLE 17.7
Predicting whether a Precipitate Will Form

A student wished to prepare 1.0 L of a solution containing 0.015 mol of NaCl and 0.15 mol of $Pb(NO_3)_2$. Knowing from the solubility rules that the chloride of Pb^{2+} is "insoluble," there was concern that a precipitate of $PbCl_2$ might form. Will it?

Analysis: To answer this question, we will compute the ion product for $PbCl_2$ *using the concentrations of the ions in the solution to be prepared.* If the computed ion product is larger than K_{sp}, then a precipitate will form. To perform the calculation correctly, we need the correct form of the ion product. We can obtain this by writing the solubility equilibrium and the K_{sp} expression that applies to a saturated solution.

$$PbCl_2(s) \rightleftharpoons Pb^{2+}(aq) + 2Cl^-(aq) \qquad K_{sp} = [Pb^{2+}][Cl^-]^2$$

From Table 17.1, $K_{sp} = 1.7 \times 10^{-5}$.

Solution: The planned solution would have the following concentrations.

$$[Pb^{2+}] = 0.15 \text{ mol L}^{-1} = 0.15 \, M$$

$$[Cl^-] = 0.015 \text{ mol L}^{-1} = 0.015 \, M$$

We use these values to compute the ion product for $PbCl_2$ in such a solution.

$$[Pb^{2+}][Cl^-]^2 = (0.15)(0.015)^2 = 3.4 \times 10^{-5}$$

But this value of the ion product is larger than the K_{sp} of $PbCl_2$, 1.7×10^{-5}. This means that a precipitate of $PbCl_2$ will form if the preparation of this solution is attempted.

Is the Answer Reasonable?
Double-check that the molarities of the ions in the planned solution, as they are given, make sense. If they do, then there will be meaning to the rechecking of the calculation itself.

Practice Exercise 9

Will a precipitate of $CaSO_4$ form in a solution if the Ca^{2+} concentration is 0.0025 M and the SO_4^{2-} concentration is 0.030 M? ◆

Practice Exercise 10

Will a precipitate form in a solution containing 3.4×10^{-4} M CrO_4^{2-} and 4.8×10^{-5} M Ag^+? ◆

EXAMPLE 17.8

Predicting whether a Precipitate Will Form

What possible precipitate might form by mixing 50.0 mL of 1.0×10^{-4} M NaCl with 50.0 mL of 1.0×10^{-6} M $AgNO_3$? Will it form? (Assume that the volumes are additive.)

Analysis: In this problem we are being asked, in effect, whether a metathesis reaction will occur between NaCl and $AgNO_3$. We should be able to predict whether this *might* occur by using the solubility rules. If the solubility rules suggest a precipitate, we can then calculate the ion product for the compound using the concentrations of the ions in the final solution. If this ion product exceeds K_{sp} for the salt, then a precipitate will form.

To calculate the ion product correctly requires that we use the concentrations of the ions *after the solutions have been mixed*. Therefore, before computing the ion product, we must first take into account that mixing the solutions dilutes each of the solutes.

Solution: Let's begin by writing the equation for the potential metathesis reaction between NaCl and $AgNO_3$. We use the solubility rules to determine whether each product is soluble or insoluble.

$$NaCl(aq) + AgNO_3(aq) \longrightarrow AgCl(s) + NaNO_3(aq)$$

The solubility rules indicate that we expect a precipitate of AgCl. But are the concentrations of Ag^+ and Cl^- actually high enough?

In the original solutions, the 1.0×10^{-6} M $AgNO_3$ contains 1.0×10^{-6} M Ag^+ and the 1.0×10^{-4} M NaCl contains 1.0×10^{-4} M Cl^-. What are the concentrations of these ions after dilution? To determine these we use the equation that applies to all dilution problems involving molarity.

$$M_i V_i = M_f V_f \tag{17.2}$$

Solving for M_f gives

This equation was developed in Section 3.10 on page 135.

$$M_f = \frac{M_i V_i}{V_f} \tag{17.3}$$

The initial volumes of both solutions are 50.0 mL and when the two solutions are combined, the final total volume is 100 mL. Therefore,

The units for V_i and V_f can be any we please provided they are the same in both. The units for M, of course, are mol L^{-1}.

$$[Ag^+]_{final} = \frac{(50.0 \text{ mL})(1.0 \times 10^{-6} M)}{100.0 \text{ mL}} = 5.0 \times 10^{-7} M$$

$$[Cl^-]_{final} = \frac{(50.0 \text{ mL})(1.0 \times 10^{-4} M)}{100.0 \text{ mL}} = 5.0 \times 10^{-5} M$$

Now we use these to calculate the ion product for AgCl, which we can write from the dissociation reaction of the salt.

$$AgCl(s) \rightleftharpoons Ag^+(aq) + Cl^-(aq)$$

$$\text{Ion product for AgCl} = [Ag^+][Cl^-]$$

Substituting the concentrations computed above gives

$$\text{Ion product} = (5.0 \times 10^{-7})(5.0 \times 10^{-5}) = 2.5 \times 10^{-11}$$

In Table 17.1, the K_{sp} for AgCl is given as 1.8×10^{-10}. Notice that the ion product is *smaller* than K_{sp}, which means that a precipitate will not form when these solutions are poured together.

Practice Exercise 11

What precipitate might be expected if we pour together 100.0 mL of $1.0 \times 10^{-3} M$ $Pb(NO_3)_2$ and 100.0 mL of $2.0 \times 10^{-3} M$ $MgSO_4$? Will some form? (Assume that the volumes are additive.) ◆

Practice Exercise 12

What precipitate might be expected if we pour together 50.0 mL of 0.10 M $Pb(NO_3)_2$ and 20.0 mL of 0.040 M NaCl? Will some form? (Assume that the volumes are additive.) ◆

17.2 Solubility Equilibria of Metal Oxides and Sulfides

Most water-insoluble metal oxides, like Fe_2O_3, dissolve in acid. For example,

$$Fe_2O_3(s) + 6H^+(aq) \longrightarrow 2Fe^{3+}(aq) + 3H_2O$$

The acid neutralizes the oxide ion and thus releases the metal ion from the solid. Other metal oxides dissolve in water without the aid of an acid. They do so, however, by reacting with water instead of by a simple dissociation of ions that remain otherwise unchanged. Sodium oxide, for example, consists of Na^+ and O^{2-} ions, and it readily dissolves in water. The solution, however, does not contain the oxide ion, O^{2-}. Instead, the hydroxide ion forms. The equation for the reaction is

$$Na_2O(s) + H_2O \longrightarrow 2NaOH(aq)$$

This actually involves the reaction of oxide ions with water as the crystals of Na_2O break up.

$$O^{2-}(s) + H_2O \longrightarrow 2OH^-(aq)$$

The oxide ion is simply too powerful a base to exist in water at any concentration worthy of experimental note. We can understand why from the extraordinarily high (estimated) value of K_b for O^{2-}, 1×10^{22}. Thus there is no way to supply *oxide ions* to an aqueous solution in order to form an insoluble metal oxide directly. Oxide ions react with water, instead, to generate the hydroxide ion. When an insoluble metal *oxide* instead of an insoluble metal *hydroxide* does precipitate from a solution, it forms because the specific metal ion is able to react with OH^-, extract O^{2-}, and leave H^+ or a neutralized form in the solution. The silver ion, for example, precipitates as brown silver oxide, Ag_2O, when OH^- is added to aqueous silver salts.

$$2Ag^+(aq) + 2OH^-(aq) \longrightarrow Ag_2O(s) + H_2O$$

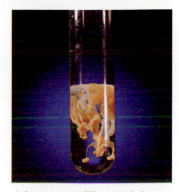

A brown, mudlike precipitate of silver oxide forms as sodium hydroxide solution is added to a solution of silver nitrate.

Hydrogen Sulfide as a Diprotic Acid

When we shift from oxygen to sulfur in Group VIA and consider metal sulfides, we find many similarities to oxides. One is that the sulfide ion, S^{2-}, like the oxide ion, is such a strong Brønsted base that it does not exist in any ordinary aqueous solution. The sulfide ion has not been detected in an aqueous solution even in the presence of 8 M NaOH where one might think that the reaction

$$OH^- + HS^- \longrightarrow H_2O + S^{2-}$$

could generate some detectable S^{2-}. An 8 M NaOH solution is at a concentration well outside the bounds of the "ordinary." Thus, Na_2S, like Na_2O, dissolves in water *by reacting with it,* not by releasing an otherwise unchanged divalent anion, S^{2-}.

$$Na_2S(s) + H_2O \longrightarrow 2Na^+(aq) + HS^-(aq) + OH^-(aq)$$

Just as some metal oxides form by a reaction between a metal ion and OH^-, many metal sulfides form when their metal ions are exposed to HS^-. Some metal ions even react with H_2S directly to form sulfides. Simply bubbling hydrogen sulfide gas into an aqueous solution of any one of a number of metal ions— Cu^{2+}, Pb^{2+}, and Ni^{2+}, for example—causes their sulfides to precipitate. Many of these have distinctive colors (Figure 17.3) that can be used to help identify which metal ion is in solution. A typical reaction is that of Cu^{2+} with H_2S.

$$Cu^{2+}(aq) + H_2S(aq) \rightleftharpoons CuS(s) + 2H^+(aq) \qquad K = 1.7 \times 10^{15}$$

The extremely large value of the equilibrium constant tells us that the forward reaction is essentially the only reaction; the reverse reaction is very

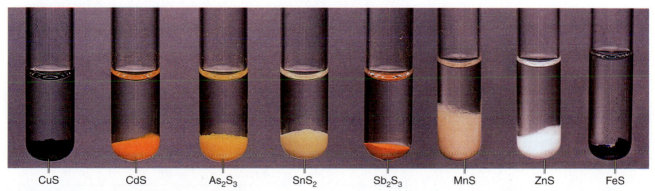

Figure 17.3 *The colors of some metal sulfides.*

CuS   CdS   As$_2$S$_3$ SnS$_2$ Sb$_2$S$_3$ MnS   ZnS   FeS unimportant. If we turn this equilibrium around, we would have something that looks very much like an equilibrium constant for defining a solubility product. Let's look at this possibility more closely.

K_{spa} Values for Metal Sulfides

CuS is extremely insoluble in aqueous acid, judging from the value of K for the reaction of Cu^{2+} with H_2S just described. However, if we were to write the equilibrium that describes the solubility of CuS in water, we would write

$$CuS(s) \rightleftharpoons Cu^{2+}(aq) + S^{2-}(aq)$$

This assumes that S^{2-} exists in water, and it doesn't, as we said. So we must instead write the equilibrium when CuS(s) dissolves in water as follows in order to show which anions are actually produced.

$$CuS(s) + H_2O \rightleftharpoons Cu^{2+}(aq) + HS^-(aq) + OH^-(aq) \qquad (17.4)$$

Now the ion product is $[Cu^{2+}][HS^-][OH^-]$, and the solubility product constant for CuS is expressed by the following equation.

$$K_{sp} = [Cu^{2+}][HS^-][OH^-]$$

Several values of K_{sp} of this form for a number of metal sulfides are given in the last column in Table 17.2. Notice particularly how much the K_{sp} values vary—from 2×10^{-53} to 3×10^{-11}, a spread by a factor of 10^{42}.

Acid-Insoluble Sulfides

Several metal sulfides, the *acid-insoluble sulfides,* have K_{sp} values so low that they do not dissolve in acid. The cations in this group can be precipitated from the other cations simply by bubbling hydrogen sulfide into a sufficiently acidic

Table 17.2 Metal Ions Separable by Selective Precipitation of Sulfides[a]

Metal Ion	Sulfide	K_{spa}	K_{sp}
Acid-Insoluble Sulfides			
Hg^{2+}	HgS (black form)	2×10^{-32}	2×10^{-53}
Ag^+	Ag_2S	6×10^{-30}	6×10^{-51}
Cu^{2+}	CuS	6×10^{-16}	6×10^{-37}
Cd^{2+}	CdS	3×10^{-7}	3×10^{-28}
Pb^{2+}	PbS	3×10^{-7}	3×10^{-28}
Sn^{2+}	SnS	1×10^{-5}	1×10^{-26}
Base-Insoluble Sulfides (Acid-Soluble Sulfides)			
Zn^{2+}	α-ZnS	3×10^{-4}	3×10^{-25}
	β-ZnS	3×10^{-2}	3×10^{-23}
Co^{2+}	CoS	5×10^{-1}	5×10^{-22}
Ni^{2+}	NiS	4×10^1	4×10^{-20}
Fe^{2+}	FeS	6×10^2	6×10^{-19}
Mn^{2+}	MnS (pink form)	3×10^{10}	3×10^{-11}
	MnS (green form)	3×10^7	3×10^{-14}

[a]Data are for 25 °C. See R. J. Meyers, *J. Chem. Ed.*, Vol. 63, 1986, p. 689.

solution that contains several metal ions. However, when the solution is acidic, we have to treat the solubility equilibria differently. In acid, HS^- and OH^- would be neutralized, leaving their conjugate acids, H_2S and H_2O. When we introduce acid ($2H^+$) into the left side of Equilibrium 17.4, for example, to neutralize the bases on the right side (HS^- and OH^-), we obtain

$$CuS(s) + H_2O + 2H^+(aq) \rightleftharpoons Cu^{2+}(aq) + H_2S(aq) + H_2O$$

After canceling the H_2O from each side, we have the net equation for the $CuS(s)$ solubility equilibrium *in dilute acid*.

$$CuS(s) + 2H^+(aq) \rightleftharpoons Cu^{2+}(aq) + H_2S(aq)$$

This changes the mass action expression for the solubility product equilibrium, which we will now call the **acid solubility product, K_{spa}**. The a in the subscript indicates that the medium is acidic.

$$K_{spa} = \frac{[Cu^{2+}][H_2S]}{[H^+]^2}$$

Tools

Acid solubility product

Table 17.2 also gives K_{spa} values for the metal sulfides. Notice that all K_{spa} values are about 10^{21} larger than the K_{sp} values. Metal sulfides are clearly vastly more soluble in dilute acid than in water. Yet several—the acid-insoluble sulfides—are so insoluble that even the most soluble of them, SnS, barely dissolves, even in moderately concentrated acid. So there are two families of sulfides, the *acid-insoluble* sulfides and the acid-soluble ones, otherwise known as the *base-insoluble* sulfides.

17.3 Separating Metal Ions by Selective Precipitation

Selective Precipitation of Metal Sulfides

The large differences in K_{spa} values between the acid-insoluble and the base-insoluble metal sulfides make it possible to separate the corresponding cations when they are in the same solution. The sulfides of the acid-insoluble cations are selectively precipitated by hydrogen sulfide from a solution kept at a pH that keeps the other cations in solution. A solution saturated in H_2S is used for which the molarity of H_2S is 0.1 M.

Let's work an example to show how we can calculate the pH needed to allow the selective precipitation of two metal cations as their sulfides. We will use Cu^{2+} and Ni^{2+} ions to represent a cation from each class.

EXAMPLE 17.9

Selective Precipitation of the Sulfides of Acid-Insoluble Cations from Base-Insoluble Cations

Over what range of hydrogen ion concentrations (and pH) is it possible to separate Cu^{2+} from Ni^{2+} when both metal ions are present in a solution at a concentration of 0.010 M and the solution is made saturated in H_2S (where $[H_2S] = 0.1$ M)?

Analysis: We must work with the chemical equilibria for the metal sulfides and their associated equilibrium expressions (the equations for their K_{spa}).

$$CuS(s) + 2H^+(aq) \rightleftharpoons Cu^{2+}(aq) + H_2S(aq) \qquad K_{spa} = \frac{[Cu^{2+}][H_2S]}{[H^+]^2} = 6 \times 10^{-16}$$

$$NiS(s) + 2H^+(aq) \rightleftharpoons Ni^{2+}(aq) + H_2S(aq) \qquad K_{spa} = \frac{[Ni^{2+}][H_2S]}{[H^+]^2} = 4 \times 10^{1}$$

The K_{spa} values (from Table 17.2) tell us that NiS is much more soluble in an acidic solution than CuS. Therefore, we want to make the H^+ concentration large enough to prevent NiS from precipitating, and small enough that CuS does precipitate.

The problem reduces to two questions. The first is, "What hydrogen ion concentration would be needed to keep the Cu^{2+} *in solution?*" (The answer to this question will give us the *upper limit* on $[H^+]$; we would really want a lower H^+ concentration so CuS *will* precipitate.) The second question is, "What is the hydrogen ion concentration just before NiS precipitates?" The answer to this will be the *lower limit* on $[H^+]$. At any lower H^+ concentration, NiS will precipitate, so we want an H^+ concentration equal to or larger than this value. Once we know these limits, we know that any hydrogen ion concentration in between them will permit CuS to precipitate but retain Ni^{2+} in solution.

Solution: We will find the upper limit first. If CuS *does not precipitate,* the Cu^{2+} concentration will be the given value, 0.010 *M*, so we substitute this along with the H_2S concentration (0.1 *M*) into the expression for K_{spa}.

$$K_{spa} = \frac{[Cu^{2+}][H_2S]}{[H^+]^2} = \frac{(0.010)(0.1)}{[H^+]^2} = 6 \times 10^{-16}$$

Now we solve for $[H^+]$.

$$[H^+]^2 = \frac{(0.010)(0.1)}{6 \times 10^{-16}} = 2 \times 10^{12}$$

$$[H^+] = 1 \times 10^6 \ M$$

If we could make $[H^+] = 1 \times 10^6$ *M*, we could prevent CuS from forming. However, it isn't possible to have 10^6 or a million moles of H^+ per liter! What the calculated $[H^+]$ tells us, therefore, is that *no matter how acidic the solution is, we cannot prevent CuS from precipitating when we saturate the solution with* H_2S. (You can now see why CuS is classed as an "acid-insoluble sulfide.")

To obtain the lower limit, we calculate the $[H^+]$ required to give an equilibrium concentration of Ni^{2+} equal to 0.010 *M*. If we keep the value of $[H^+]$ *equal to or larger* than this value, then NiS will be prevented from precipitating. The calculation is exactly like the one above. First, we substitute values into the K_{spa} expression.

$$K_{spa} = \frac{[Ni^{2+}][H_2S]}{[H^+]^2} = \frac{(0.010)(0.1)}{[H^+]^2} = 4 \times 10^1$$

Once again, we solve for $[H^+]$.

$$[H^+]^2 = \frac{(0.010)(0.1)}{4 \times 10^1}$$

$$[H^+] = 5 \times 10^{-3} \ M$$

The pH is therefore 2.3.

If we maintain the pH of the solution of 0.010 *M* Cu^{2+} and 0.010 *M* Ni^{2+} at 2.3 or lower (more acidic), as we make the solution saturated in H_2S, virtually all the Cu^{2+} will precipitate as CuS, but all the Ni^{2+} will stay in solution.

In actual experimental work involving the separation of acid-insoluble cations from base-insoluble cations, a solution more acidic than pH 2.3 (calculated in Example 17.9) is employed. A pH of about 0.5 is used, which corresponds to $[H^+] = 0.3$ *M*. We can also see why NiS can be classified as a "base-insoluble sulfide." If the solution is *basic* when it is made saturated in H_2S, the pH will surely be larger than 2.3 and NiS will precipitate.

Practice Exercise 13

Consider a solution containing Hg^{2+} and Fe^{2+}, both at molarities of 0.010 M. It is to be saturated with H_2S. Calculate the lower limit to the pH of this solution that must be maintained to keep Fe^{2+} in solution while Hg^{2+} precipitates as HgS. ◆

Selective Precipitation of Metal Carbonates

The principles of selective precipitation by the control of pH also apply to any system where the anion is that of a weak acid. The metal carbonates are examples, and many are quite insoluble in water (see Table 17.1). The dissociation of magnesium carbonate in water, for example, involves the following equilibrium and K_{sp} equations.

$$MgCO_3(s) \rightleftharpoons Mg^{2+}(aq) + CO_3{}^{2-}(aq) \qquad K_{sp} = 3.5 \times 10^{-8}$$

For strontium carbonate, the equations are

$$SrCO_3(s) \rightleftharpoons Sr^{2+}(aq) + CO_3{}^{2-}(aq) \qquad K_{sp} = 9.3 \times 10^{-10}$$

Would it be possible to separate the magnesium ion from the strontium ion by taking advantage of the difference in K_{sp} of their carbonates?

We can do so if we can control the carbonate ion concentration. Because the carbonate ion is a relatively strong Brønsted base, its control is available, indirectly, through adjusting the pH of the solution. This is because the hydrogen ion is one of the species in the following equilibria.[2]

The control of the concentration of the carbonate ion by pH is thus like the control of the concentration of hydrogen sulfide ion by pH.

$$H_2CO_3 \rightleftharpoons H^+ + HCO_3{}^- \qquad K_{a_1} = 4.3 \times 10^{-7}$$

$$HCO_3{}^- \rightleftharpoons H^+ + CO_3{}^{2-} \qquad K_{a_2} = 5.6 \times 10^{-11}$$

You can see that if we increase the hydrogen ion concentration, both equilibria will shift to the left, in accordance with Le Châtelier's principle, and this will reduce the concentration of the carbonate ion. The value of $[CO_3{}^{2-}]$ thus decreases with decreasing pH. On the other hand, if we decrease the hydrogen ion concentration, making the solution more basic, we will cause the two equilibria to shift to the right. The concentration of the carbonate ion thus increases with increasing pH.

Now let's see what kinds of calculations we must do to find the pH range within which the less soluble carbonate will precipitate while the more soluble carbonate remains dissolved.

It will be useful to combine the two hydrogen carbonate equilibria into one overall equation that relates the molar concentrations of carbonic acid, carbonate ion, and hydrogen ion. So we first add the two.

$$\begin{aligned} H_2CO_3 &\rightleftharpoons H^+ + HCO_3{}^- \\ \underline{HCO_3{}^- }&\rightleftharpoons \underline{H^+ + CO_3{}^{2-}} \\ H_2CO_3 &\rightleftharpoons 2H^+ + CO_3{}^{2-} \end{aligned} \qquad (17.5)$$

---

[2]The situation involving aqueous carbonic acid is complicated by the presence of dissolved CO_2, which we could represent in an equation as $CO_2(aq)$. In fact, this is how most of the dissolved CO_2 exists, namely, as $CO_2(aq)$, not as $H_2CO_3(aq)$. But we may use $H_2CO_3(aq)$ as a surrogate or stand-in for $CO_2(aq)$, because the latter changes smoothly to the former on demand. The following two successive equilibria involving carbonic acid exist in a solution of aqueous CO_2.

$$CO_2(aq) + H_2O \rightleftharpoons H_2CO_3(aq) \rightleftharpoons H^+(aq) + HCO_3{}^-(aq)$$

The value of K_{a_1} cited here for $H_2CO_3(aq)$ is really the product of the equilibrium constants of these two equilibria.

Recall from Chapter 16 that when we add two equilibria to get a third, the equilibrium constant of the later is the product of those of the two equilibria that are combined. Thus, for Equilibrium 17.5 the equilibrium constant, which we'll symbolize as K_a, is obtained as follows.

$$K_a = K_{a_1} \times K_{a_2} = \frac{[H^+]^2[CO_3{}^{2-}]}{[H_2CO_3]} \tag{17.6}$$

We can safely use Equation 17.6 *only* when two of the three concentrations are known.

$$= (4.3 \times 10^{-7}) \times (5.6 \times 10^{-11})$$

$$= 2.4 \times 10^{-17}$$

With this K_a value we may now study how to find the pH range within which Mg^{2+} and Sr^{2+} can be separated by taking advantage of their difference in K_{sp} values.

EXAMPLE 17.10

Separation of Metal Ions by the Selective Precipitation of Their Carbonates

A solution contains magnesium nitrate and strontium nitrate, each at a concentration of 0.10 M. Carbon dioxide is to be bubbled in to make the solution saturated in $CO_2(aq)$, approximately 0.030 M. What pH range would make it possible for the carbonate of one cation to precipitate but not that of the other?

Analysis: There are really two parts to this. First, what is the range in values of $[CO_3{}^{2-}]$ that allow one carbonate to precipitate but not the other? Second, given this range, what are the values of the solution's pH that correspond to this range in $[CO_3{}^{2-}]$ values?

Solution: The range in $[CO_3{}^{2-}]$ values is obtained by using the K_{sp} values of the two carbonates.

$$K_{sp} = [Mg^{2+}][CO_3{}^{2-}] = 3.5 \times 10^{-8}$$

$$K_{sp} = [Sr^{2+}][CO_3{}^{2-}] = 9.3 \times 10^{-10}$$

$MgCO_3$ is the more soluble of the two, so the $[CO_3{}^{2-}]$ value needed to give its saturated solution is found as follows.

$$[CO_3{}^{2-}] = \frac{K_{sp}}{[Mg^{2+}]} = \frac{3.5 \times 10^{-8}}{0.10}$$

$$= 3.5 \times 10^{-7} M$$

If the value of $[CO_3{}^{2-}]$ goes higher than 3.5×10^{-7} M, $MgCO_3$ will start to precipitate and thus will contaminate that of the less soluble carbonate, $SrCO_3$, that we're trying to use to selectively remove Sr^{2+} from the solution.

The value of $[CO_3{}^{2-}]$ that we must create in order to have Sr^{2+} precipitate corresponds to that of a saturated solution of $SrCO_3$. So the $[CO_3{}^{2-}]$ needed for saturation is found as follows.

$$[CO_3{}^{2-}] = \frac{K_{sp}}{[Sr^{2+}]} = \frac{9.3 \times 10^{-10}}{0.10}$$

$$= 9.3 \times 10^{-9} M$$

In other words, the value of $[CO_3{}^{2-}]$ must be adjusted, by adjusting the pH, so that it is *larger* than 9.3×10^{-9} M to make $SrCO_3$ precipitate. But, as found earlier, the $[CO_3{}^{2-}]$ value must not exceed 3.5×10^{-7} M or else magnesium ion will precipitate.

In summary, the range of carbonate ion molarities under which $SrCO_3$ but not $MgCO_3$ precipitates from the given solution are

$$[CO_3{}^{2-}] > 9.3 \times 10^{-9} \; M$$

$$[CO_3{}^{2-}] \leq 3.5 \times 10^{-7} \; M$$

The second phase of our calculation now asks what values of $[H^+]$ correspond to the calculated limits on $[CO_3{}^{2-}]$, in the light of Equilibrium 17.5 and the associated K_a equation, Equation 17.6? First we solve Equation 17.6 for the square of $[H^+]$, using the molarity of the dissolved CO_2, 0.030 M, as the value of $[H_2CO_3]$. This gives us

$$[H^+]^2 = K_a \times \frac{[H_2CO_3]}{[CO_3{}^{2-}]} = 2.4 \times 10^{-17} \times \frac{0.030}{[CO_3{}^{2-}]}$$

The value of $[H_2CO_3]$ is taken to be that of $[CO_2(aq)]$, namely, 0.030 M, and K_a, by Equation 17.6, is 2.4×10^{-17}.

We will use this equation for each of the two boundary values of $[CO_3{}^{2-}]$ to calculate the corresponding two values of $[H^+]^2$. Once we have them, the steps to values of $[H^+]$ and pH are easy.

For magnesium carbonate. To *prevent* the precipitation of $MgCO_3$, $[CO_3{}^{2-}]$ must be no higher than $3.5 \times 10^{-7} \; M$, so $[H^+]^2$ must not be less than

$$[H^+]^2 = 2.4 \times 10^{-17} \times \frac{0.030}{3.5 \times 10^{-7}}$$

$$= 2.1 \times 10^{-12}$$

Taking the square root,

$$[H^+] = 1.4 \times 10^{-6} \; M$$

This corresponds to a pH of 5.85. At a higher (more basic) pH, magnesium carbonate precipitates.

For strontium carbonate. To *cause* the precipitation of $SrCO_3$, $[CO_3{}^{2-}]$ must be at least as high as $9.3 \times 10^{-9} \; M$, as we calculated above. This corresponds to a value of $[H^+]^2$ found as follows.

$$[H^+]^2 = 2.4 \times 10^{-17} \times \frac{0.030}{9.3 \times 10^{-9}}$$

$$= 7.7 \times 10^{-11}$$

$$[H^+] = 8.8 \times 10^{-6} \; M$$

At a higher hydrogen ion concentration, $SrCO_3$ won't precipitate. In pH terms, the pH must be no less than 5.06.

In summary, when the pH of the given solution is kept between 5.06 and 5.85, Sr^{2+} will precipitate as $SrCO_3$ but Mg^{2+} will remain dissolved.

Is the Answer Reasonable?
The only way to check the answer is to go back over the reasoning and the calculations.

_____
Practice Exercise 14
_____

A solution contains calcium nitrate and nickel nitrate, each at a concentration of 0.10 M. Carbon dioxide is to be bubbled in to make its concentration equal 0.030 M. What pH range would make it possible for the carbonate of one cation to precipitate but not that of the other? ◆

17.4 Complex Ions and Their Equilibria in Aqueous Solutions

Complex Ions of Metals

In our previous discussions of metal-containing compounds, we left you with the impression that the only kinds of bonds in which metals are ever involved are ionic bonds. For some metals, like the alkali metals of Group IA, this is close enough to the truth to warrant no modifications. But for many other metal ions, especially those of the transition metals and the post-transition metals, it is not. This is because the ions of many of these metals are able to behave as Lewis acids (i.e., as electron pair acceptors in the formation of coordinate covalent bonds). Thus, by participating in Lewis acid–base reactions they become *covalently* bonded to other atoms. Copper(II) ion is a typical example.

In aqueous solutions of copper(II) salts, like $CuSO_4$ or $Cu(NO_3)_2$, the copper is not present as simple Cu^{2+} ions. Instead, each Cu^{2+} ion becomes bonded to four water molecules to give a pale blue ion with the formula $Cu(H_2O)_4^{2+}$ (see Figure 17.4). We call this species a **complex ion** because it is composed of a number of simpler species (i.e., it is *complex,* not simple). The chemical equation for the formation of the $Cu(H_2O)_4^{2+}$ ion is

$$Cu^{2+} + 4H_2O \longrightarrow Cu(H_2O)_4^{2+}$$

which can be diagrammed using Lewis structures as follows.

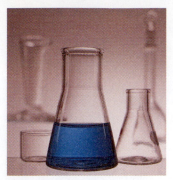

Figure 17.4 *The complex ion of Cu²⁺ and water.* A solution containing copper sulfate has a blue color because it contains the complex ion $Cu(H_2O)_4^{2+}$.

As you can see in this analysis, the Cu^{2+} ion accepts pairs of electrons from the water molecules, so Cu^{2+} is a Lewis acid and the water molecules are each Lewis bases.

The number of complex ions formed by metals, especially the transition metals, is enormous, and the study of the properties, reactions, structures, and bonding in complex ions like $Cu(H_2O)_4^{2+}$ has become an important specialty within chemistry. A part of learning about this specialty is becoming familiar with some of the terminology.

A Lewis base that attaches itself to a metal ion is called a **ligand.** Ligands can be neutral molecules with unshared pairs of electrons (like H_2O), or they can be anions (like Cl^- or OH^-). The atom in the ligand that actually provides the electron pair is called the **donor atom,** and the metal ion is the **acceptor.** The result of combining a metal ion with one or more ligands is a *complex ion,* or simply just a **complex.** The word "complex" avoids problems when the particle formed is electrically neutral, as sometimes happens. Compounds that contain complex ions are generally referred to as **coordination compounds** because the bonds in a complex ion can be viewed as coordinate covalent bonds. Sometimes the complex itself is called a **coordination complex.**

In an aqueous solution, the formation of a complex ion is really a reaction in which water molecules are replaced by other ligands. Thus, when NH_3 is added to a solution of copper ion, the water molecules in the $Cu(H_2O)_4^{2+}$ ion are re-

> Recall that a *Lewis base* is an electron pair donor in the formation of a coordinate covalent bond.

> "Ligand" is from the Latin *ligare,* meaning "to bind."

placed, one after another, by molecules of NH_3 until the deeply blue complex $Cu(NH_3)_4{}^{2+}$ is formed (Figure 17.5). Each successive reaction is an equilibrium, so the entire chemical system involves many species and is quite complicated. Fortunately, when the ligand concentration is *large* relative to that of the metal ion, the concentrations of the intermediate complexes are very small and we can work only with the *overall* reaction for the formation of the final complex. Our study of complex ion equilibria will be limited to these situations. The equilibrium equation for the formation of $Cu(NH_3)_4{}^{2+}$, therefore, can be written as though the complex forms in one step.

$$Cu(H_2O)_4{}^{2+}(aq) + 4NH_3(aq) \rightleftharpoons Cu(NH_3)_4{}^{2+}(aq) + 4H_2O$$

We will simplify this equation for the purposes of dealing quantitatively with the equilibria by omitting the water molecules. (It's safe to do this because the concentration of H_2O in aqueous solutions is taken to be effectively a constant and need not be included in mass action expressions.) In simplified form, we write the equilibrium above as follows:

$$Cu^{2+}(aq) + 4NH_3(aq) \rightleftharpoons Cu(NH_3)_4{}^{2+}(aq)$$

We have two goals here: to study such equilibria themselves and to learn how they can be used to influence the solubilities of metal ion salts.

According to Le Châtelier's principle, when the concentration of ammonia is high, the position of equilibrium in this reaction is shifted far to the right, so effectively all complex ions with water molecules are changed to those with ammonia molecules.

Formation Constants of Complex Ions

When the chemical equation for the equilibrium is written so that the complex ion is the product, the equilibrium constant for the reaction is called the **formation constant, K_{form}**. The equilibrium law for the formation of $Cu(NH_3)_4{}^{2+}$ in the presence of excess NH_3, for example, is

$$\frac{[Cu(NH_3)_4{}^{2+}]}{[Cu^{2+}][NH_3]^4} = K_{form}$$

Tools

Formation constants of complex ions

Sometimes this equilibrium constant is called the **stability constant.** The larger its value, the greater is the concentration of the complex at equilibrium, and so the more stable is the complex.

Table 17.3 provides several more examples of complex ion equilibria and their associated equilibrium constants. (Additional examples are in Appendix

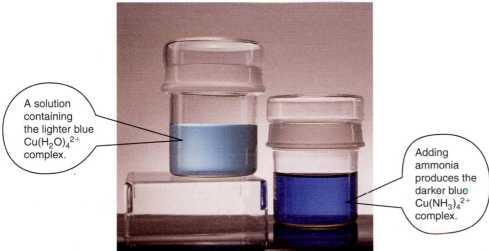

A solution containing the lighter blue $Cu(H_2O)_4{}^{2+}$ complex.

Adding ammonia produces the darker blue $Cu(NH_3)_4{}^{2+}$ complex.

Figure 17.5 *The complex ion of Cu^{2+} and ammonia.*

Table 17.3 Formation Constants and Instability Constants for Some Complex Ions

Ligand	Equilibrium	K_{form}	K_{inst}
NH_3	$Ag^+ + 2NH_3 \rightleftharpoons Ag(NH_3)_2^+$	1.6×10^7	6.3×10^{-8}
	$Co^{2+} + 6NH_3 \rightleftharpoons Co(NH_3)_6^{2+}$	5.0×10^4	2.0×10^{-5}
	$Co^{3+} + 6NH_3 \rightleftharpoons Co(NH_3)_6^{3+}$	4.6×10^{33}	2.2×10^{-34}
	$Cu^{2+} + 4NH_3 \rightleftharpoons Cu(NH_3)_4^{2+}$	1.1×10^{13}	9.1×10^{-14}
	$Hg^{2+} + 4NH_3 \rightleftharpoons Hg(NH_3)_4^{2+}$	1.8×10^{19}	5.6×10^{-20}
F^-	$Al^{3+} + 6F^- \rightleftharpoons AlF_6^{3-}$	1×10^{20}	1×10^{-20}
	$Sn^{4+} + 6F^- \rightleftharpoons SnF_6^{2-}$	1×10^{25}	1×10^{-25}
Cl^-	$Hg^{2+} + 4Cl^- \rightleftharpoons HgCl_4^{2-}$	5.0×10^{15}	2.0×10^{-16}
Br^-	$Hg^{2+} + 4Br^- \rightleftharpoons HgBr_4^{2-}$	1.0×10^{21}	1.0×10^{-21}
I^-	$Hg^{2+} + 4I^- \rightleftharpoons HgI_4^{2-}$	1.9×10^{30}	5.3×10^{-31}
CN^-	$Fe^{2+} + 6CN^- \rightleftharpoons Fe(CN)_6^{4-}$	1.0×10^{24}	1.0×10^{-24}
	$Fe^{3+} + 6CN^- \rightleftharpoons Fe(CN)_6^{3-}$	1.0×10^{31}	1.0×10^{-31}

E.) Notice that the most stable complex in the table, $Co(NH_3)_6^{3+}$, has, as you would expect, the largest value of K_{form}.

Instability Constants

Some chemists prefer to describe the relative stabilities of complex ions differently. The *inverses* of formation constants are cited and are called **instability constants, K_{inst}**. This approach focuses attention on the *breakdown* of the complex, not its formation. Therefore, the associated equilibrium equation is written as the reverse of the formation of the complex. The equilibrium for the copper–ammonia complex would be written as follows, for example:

$$Cu(NH_3)_4^{2+}(aq) \rightleftharpoons Cu^{2+}(aq) + 4NH_3(aq)$$

The equilibrium constant for this equilibrium is called the *instability constant*.

$$K_{inst} = \frac{[Cu^{2+}][NH_3]^4}{[Cu(NH_3)_4^{2+}]} = \frac{1}{K_{form}}$$

K_{inst} is called an *instability* constant because the larger its value is, the more *unstable* the complex is. The data in the last column of Table 17.3 show this. The least stable complex in the table, $Co(NH_3)_6^{2+}$, has the largest value of K_{inst}.

The Effect of Complex Ion Formation on the Solubilities of Salts

Tools

Complex ion formation and salt solubility

The silver halides are extremely insoluble salts, as we've learned. The K_{sp} of AgBr at 25 °C, for example, is only 5.0×10^{-13}. In a saturated solution, the concentration of each ion of AgBr is only 7.1×10^{-7} mol L^{-1}. Suppose that we start with a saturated solution in which undissolved AgBr is present and equilibrium exists. Now suppose that we start to add aqueous ammonia to the system. NH_3 molecules are strong ligands for silver ions, and they begin to form complexes, $Ag(NH_3)_2^+$, with the trace amount of Ag^+ ions initially in solution. The reaction is

Complex ion equilibrium: $\quad Ag^+(aq) + 2NH_3(aq) \rightleftharpoons Ag(NH_3)_2^+(aq)$

Because the forward reaction withdraws *uncomplexed* Ag^+ ions from solution, it upsets the solubility equilibrium:

Solubility equilibrium: $AgBr(s) \rightleftharpoons Ag^+(aq) + Br^-(aq)$

Ammonia, by withdrawing uncomplexed Ag^+ ions supplied by this solubility equilibrium, causes the equilibrium to shift to the right to generate more Ag^+ ions from $AgBr(s)$, another example of Le Châtelier's principle at work. In other words, ammonia induces more $AgBr(s)$ to dissolve. Our example illustrates a general phenomenon:

> The solubility of a slightly soluble salt increases when one of its ions can be changed to a soluble complex ion.

To analyze what happens, lets put the two equilibria together.

Complex ion equilibrium: $Ag^+(aq) + 2NH_3(aq) \rightleftharpoons Ag(NH_3)_2^+(aq)$

Solubility equilibrium: $AgBr(s) \rightleftharpoons Ag^+(aq) + Br^-(aq)$

Sum of equilibria: $AgBr(s) + 2NH_3(aq) \rightleftharpoons Ag(NH_3)_2^+(aq) + Br^-(aq)$

The equilibrium constant for the net overall reaction is written in the usual way. The term for $[AgBr(s)]$ is omitted because it refers to a solid and so has a constant value.

$$K_c = \frac{[Ag(NH_3)_2^+][Br^-]}{[NH_3]^2}$$

Because we have added two equations to obtain the third equilibrium equation, the K_c for this expression is the product of K_{form} and K_{sp}. We know the values for K_{form} and K_{sp} for the silver bromide–ammonia system, so by multiplying the two, we find the overall value of K_c.

Recall that when equilibria are added, their equilibrium constants are multiplied.

$$K_c = (1.6 \times 10^7)(5.0 \times 10^{-13})$$
$$= 8.0 \times 10^{-6}$$

This approach thus gives us a way to calculate the solubility of a sparingly soluble salt when one of its ligands is put into its solution. The next example shows how this works.

EXAMPLE 17.11

Calculating the Solubility of a Slightly Soluble Salt in the Presence of a Ligand

How many moles of AgBr can dissolve in 1.0 L of 1.0 M NH_3?

Analysis: A few preliminaries have to be done before we can take advantage of a concentration table. We need the overall equation and its associated equation for the equilibrium constant. The overall equilibrium is

$$AgBr(s) + 2NH_3(aq) \rightleftharpoons Ag(NH_3)_2^+(aq) + Br^-(aq)$$

The equation for K_c is

$$K_c = \frac{[Ag(NH_3)_2^+][Br^-]}{[NH_3]^2} = 8.0 \times 10^{-6} \text{ (as calculated earlier)}.$$

Now let's prepare the table.

$AgBr(s) + 2NH_3(aq) \rightleftharpoons Ag(NH_3)_2^+(aq) + Br^-(aq)$			
Initial concentrations (M)	1.0	0	0
Changes in concentrations caused by NH_3 (M)	$-2x$ (Note 1)	$+x$	$+x$
Equilibrium concentrations (M)	$(1.0 - 2x)$	x	x (Note 2)

Note 1. The 2 in $-2x$ signifies that each mole of AgBr that dissolves removes twice as many moles of NH_3.

Note 2. Letting the concentration of Br^- equal that of $Ag(NH_3)_2^+$ is valid because *and only because* K_{form} is such a large number. So essentially *all* Ag^+ ions that do dissolve from the insoluble AgBr are changed to the complex ion. There are relatively few uncomplexed Ag^+ ions in the solution.

Solution: Let's substitute the values in the last row of the concentration table into the equation for K_c

$$K_c = \frac{(x)(x)}{(1.0 - 2x)^2} = 8.0 \times 10^{-6}$$

This can be simplified by taking the square root of both sides. Then,

$$\frac{x}{(1.0 - 2x)} = \sqrt{8.0 \times 10^{-6}} = 2.8 \times 10^{-3}$$

Solving for x gives

$$x = 2.8 \times 10^{-3}$$

In other words, 2.8×10^{-3} mol of AgBr dissolves in 1.0 L of 1.0 M NH_3. This is not very much, of course, but in contrast, only 7.1×10^{-7} mol of AgBr dissolves in 1.0 L of pure water. Thus AgBr is nearly 4000 times more soluble in the 1.0 M NH_3 than in pure water.

Is the Answer Reasonable?

Everything depends on the reasoning used in preparing the concentration table, particularly letting $-2x$ represent the change in the concentration of NH_3 as a result of the presence of AgBr and the formation of the complex. If you are comfortable with this, then the calculation should be straightforward.

Practice Exercise 15

Calculate the solubility of silver chloride in 0.10 M NH_3 and compare it with its solubility in pure water. (Refer to Table 17.1 for the K_{sp} for AgCl.) ◆

Practice Exercise 16

How many moles of NH_3 have to be added to 1.0 L of water to dissolve 0.20 mol of AgCl? The complex ion $Ag(NH_3)_2^+$ forms. ◆

SUMMARY

Solubility Equilibria for Salts. The **ion product** of a sparingly soluble salt is the product of the molar concentrations of its ions, each concentration raised to the power equal to the subscript of the ion in the salt's formula. At a given temperature, this *ion product* in a *saturated* solution of the salt equals a constant called the **solubility product constant,** or K_{sp}.

The **common ion effect** is the ability of an ion of a soluble salt to suppress the solubility of a sparingly soluble compound that has the same (the "common") ion.

If soluble salts are mixed together in the same solution, a cation of one and an anion of another will precipitate if the ion product exceeds the solubility product constant of the salt formed from them.

When metal oxides and sulfides dissolve in water, neither O^{2-} nor S^{2-} ions exist in the solution. Instead, these anions react with water, and OH^- or HS^- ions form.

Selective Precipitation. Metal sulfides are vastly more soluble in an acidic solution than in water. To express their solubility equilibria in acid, the **acid solubility constant** or K_{spa} is used. Several metal sulfides have such low values of K_{spa} that they are insoluble even at low pH. These acid-insoluble cations can thus be selectively separated from the base-insoluble cations by making the solution both quite acidic as well as saturated in H_2S. By adjusting the pH, salts of other weak acids, like the carbonate salts, can also be selectively precipitated.

Complex Ions of Metals. **Coordination compounds** contain **complex ions** (also called **complexes** or **coordination complexes**), formed from a metal ion and a number of ligands. **Ligands** are Lewis bases and may be electrically neutral or negatively charged. Water and ammonia are common neutral ligands. The equilibrium constant for the formation of a complex in the presence of an excess of ligand is called the **formation constant** of the complex. Salts whose cations form stable complexes, like Ag^+ salts whose cation forms a stable complex with ammonia, $Ag(NH_3)_2^+$, are made more soluble when the ligand is present. **Instability constants** are the inverses of formation constants. The larger that the value of an instability constant is, the more unstable is the complex.

The following tools were introduced in this chapter. You will find them useful for solving certain problems.

Tool	Function
Solubility product expression, K_{sp} (*page 760*)	Calculate K_{sp} from the molar solubility of the salt. Calculate a solubility from the K_{sp}. Calculate a solubility when a common ion is present.
Ion product of a salt (*page 760*)	Can be used to predict whether a precipitate of the salt will form in a solution by comparing the ion product with K_{sp}. If ion product $> K_{sp}$, a precipitate forms. If ion product $= K_{sp}$ or ion product $< K_{sp}$, a precipitate doesn't form.
Acid solubility product constant, K_{spa} (*page 775*)	Use K_{spa} data to calculate the solubility of a metal sulfide at a given pH. Use K_{spa} data for two or more metal sulfides to calculate the pH at which one will selectively precipitate from a solution saturated in H_2S.
Formation constants (stability constants) of complexes (*page 781*)	Make judgments concerning the relative stabilities of complexes.
Complex ion formation and salt solubility (*page 782*)	Calculate how the solubility of a sparingly soluble salt changes when its cation is able to form a complex ion with a ligand added to the solution.

THINKING IT THROUGH

For each of the following, assemble the available information needed to obtain the answer, state what additional data (if any) are needed, and describe *how* you would use the data to answer the question. Keep in mind that the goal is not to find the answer itself, but instead to find the path that leads to the answer.

The problems are divided into two groups. Those in Level 2 are significantly more challenging than those in Level 1 and provide an opportunity to really hone your problem solving skills. Answers to these problems can be found on our Web site at http://www.wiley.com/college/brady.

Level 1 Problems

1. In water, the solubility of lead(II) chloride is 0.016 M. Explain how you would use this information to calculate the value of K_{sp} for $PbCl_2$.

2. Suppose 50.0 mL each of 0.0100 M solutions of NaBr and $Pb(NO_3)_2$ are poured together. How could you determine whether a precipitate would form? Describe how you would calculate the concentrations of the ions at equilibrium.

3. Suppose you wished to control the PO_4^{3-} concentration in a solution of phosphoric acid by controlling the pH of the solution. If you assume you know the H_3PO_4 concentration, what combined equation would be useful for this purpose?

4. What is the highest concentration of Pb^{2+} that can exist in a solution of 0.10 M HCl? (Explain how you would calculate the answer.)

5. Will lead(II) bromide be less soluble in 0.10 M $Pb(C_2H_3O_2)_2$ or 0.10 M NaBr? (Describe how you would perform the calculations necessary to answer the question.)

Level 2 Problems

6. How many milliliters of 0.10 M HCl would have to be added to 100 mL of a saturated solution of $PbCl_2$ solution to reduce the Pb^{2+} concentration to 0.0050 M? (Explain the calculations in detail. Don't forget to take into account the combined volumes of the two solutions.)

7. If a solution of 0.10 M Mn^{2+} and 0.10 M Cd^{2+} is gradually made basic, what will the concentration of Cd^{2+} be when $Mn(OH)_2$ just begins to precipitate? Assume no change in the volume of the solution. (Explain how you would approach the problem and state what data you need to answer the question.)

8. A solution of $MgBr_2$ can be changed to a solution of $MgCl_2$ by adding AgCl(s) and stirring the mixture well. In terms of the equilibria involved, explain how this happens.

9. After solid $CaCO_3$ was added to a slightly basic solution, the pH was measured to be 8.50. What was the molar solubility of $CaCO_3$ in this solution? (Explain the calculations involved. Take into account the reaction of CO_3^{2-} with water.)

10. Suppose that some dipositive cation, M^{2+}, is able to form a complex ion with a ligand, L, by the following equation.

$$M^{2+} + 2L \rightleftharpoons [M(L)_2]^{2+}$$

The cation also forms a sparingly soluble salt, MCl_2. In which of the following circumstances would a given quantity of ligand be more able to bring larger quantities of the salt into solution? Explain and justify the calculation involved.

(a) $K_{form} = 1 \times 10^2$ and $K_{sp} = 1 \times 10^{-15}$
(b) $K_{form} = 1 \times 10^{10}$ and $K_{sp} = 1 \times 10^{-20}$

11. Silver forms a sparingly soluble iodide salt, AgI, and it also forms a soluble iodide complex, AgI_2^-. How much potassium iodide must be added to 100 mL of water to dissolve 0.020 mol of AgI? (Explain in detail how you would obtain the answer. *Hint:* Be sure to include *all* the iodide that's added to the solution.)

12. You are given a sample containing NaCl(*aq*) and NaBr(*aq*), both with concentrations of 0.020 M. Some of the figures below represent a series of snapshots of what molecular level views of the sample would show as a 0.200 M $Pb(NO_3)_2(aq)$ solution is slowly added. In these figures, spectator ions are not shown, nor are ions whose concentrations are much less than the other ions. The K_{sp} constants for $PbCl_2$ and $PbBr_2$ are 1.7×10^{-5} and 2.1×10^{-6}, respectively.

(a) Arrange the figures in a time sequence to show what happens as the lead nitrate solution is added.
(b) Explain why you exclude any figures that do not belong in the observed time sequence.

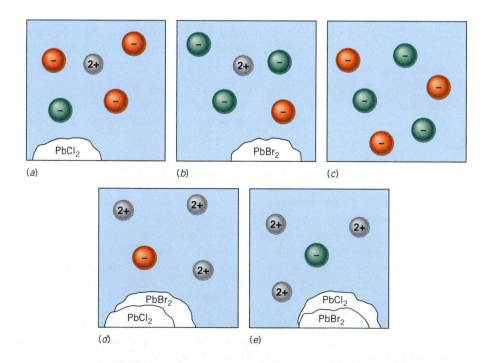

(a) (b) (c)

(d) (e)

REVIEW QUESTIONS

Solubility Products

17.1 What is the difference between an *ion product* and an *ion product constant?*

17.2 Use the equilibrium below to demonstrate why the K_{sp} expression does not include the concentration of $Ba_3(PO_4)_2$ in the denominator.

$$Ba_3(PO_4)_2(s) \rightleftharpoons 3Ba^{2+}(aq) + 2PO_4^{3-}(aq)$$

17.3 What is the common ion effect? How does Le Châtelier's principle explain it? Use the solubility equilibrium for AgCl to illustrate the common ion effect.

17.4 With respect to K_{sp}, what conditions must be met if a precipitate is going to form in a solution?

Oxides, Sulfides, and Selective Precipitations

17.5 Potassium oxide is readily soluble in water, but the resulting solution contains essentially no oxide ion. Explain, using an equation, what happens to the oxide ion.

17.6 Consider cobalt(II) sulfide.
(a) Write its solubility equilibrium and K_{sp} equation for a saturated solution in water.
(b) Write its solubility equilibrium and K_{spa} equation for a saturated solution in aqueous acid.

Complex Ion Equilibria

17.7 Using Le Châtelier's principle, explain how the addition of aqueous ammonia dissolves silver chloride. If HNO_3 is added after the AgCl has dissolved in the NH_3 solution, it causes AgCl to re-precipitate. Explain why.

17.8 For $PbCl_3^-$, $K_{form} = 2.5 \times 10^1$. If a solution containing this complex ion is diluted with water, $PbCl_2$ precipitates. Write the equations for the equilibria involved and use them together with Le Châtelier's principle to explain how this happens.

REVIEW PROBLEMS

Answers to problems whose numbers are printed in color are given in Appendix D.
More challenging problems are marked with asterisks.

Solubility Products

17.9 Write the K_{sp} expressions for each of the following compounds.
(a) CaF_2 (c) $PbSO_4$ (e) PbI_2
(b) Ag_2CO_3 (d) $Fe(OH)_3$ (f) $Cu(OH)_2$

17.10 Write the K_{sp} expressions for each of the following compounds.
(a) AgI (c) $PbCrO_4$ (e) $ZnCO_3$
(b) Ag_3PO_4 (d) $Al(OH)_3$ (f) $Zn(OH)_2$

Determining K_{sp}

17.11 Barium sulfate, $BaSO_4$, is so insoluble that it can be swallowed without significant danger, even though Ba^{2+} is toxic. At 25 °C, 1.00 L of water dissolves only 0.00245 g of $BaSO_4$. Calculate K_{sp} for $BaSO_4$.

17.12 A student found that a maximum of 0.800 g $AgC_2H_3O_2$ is able to dissolve in 100 mL of water. What is the molar solubility and the K_{sp} of this salt?

17.13 It was found that the molar solubility of $BaSO_3$ in 0.10 M $BaCl_2$ is 8.0×10^{-6} M. What is the value of K_{sp} for $BaSO_3$?

17.14 A student prepared a saturated solution of $CaCrO_4$ and found that when 156 mL of the solution was evaporated, 0.649 g of $CaCrO_4$ was left behind. What is the value of K_{sp} for this salt?

17.15 At 25 °C, the molar solubility of Ag_3PO_4 is 1.8×10^{-5} mol L^{-1}. Calculate K_{sp} for this salt.

17.16 The molar solubility of $Ba_3(PO_4)_2$ in water at 25 °C is 1.4×10^{-8} mol L^{-1}. What is the value of K_{sp} for this salt?

Using K_{sp} to Calculate Solubilities

17.17 What is the molar solubility of $PbBr_2$ in water?

17.18 What is the molar solubility of Ag_2CrO_4 in water?

17.19 Calculate the molar solubility of Ag_2CO_3 in water. (Ignore the reaction of the CO_3^{2-} ion with water.)

17.20 Calculate the molar solubility of PbI_2 in water.

17.21 At 25 °C, the value of K_{sp} for LiF is 1.7×10^{-3}, and that for BaF_2 is 1.7×10^{-6}. Which salt, LiF or BaF_2, has the larger molar solubility in water? Calculate the solubility of each in units of mol L^{-1}.

17.22 At 25 °C, the value of K_{sp} for AgCN is 2.2×10^{-16} and that for $Zn(CN)_2$ is 3×10^{-16}. In terms of grams per 100 mL of solution, which salt is the more soluble in water? (Do the calculations.)

17.23 A salt whose formula is of the form MX has a value of K_{sp} equal to 3.2×10^{-10}. Another sparingly soluble salt, MX_3, must have what value of K_{sp} if the molar solubilities of the two salts are to be identical?

17.24 A salt having a formula of the type M_2X_3 has $K_{sp} = 2.2 \times 10^{-20}$. Another salt, M_2X, has to have what K_{sp} value if M_2X has twice the molar solubility of M_2X_3?

17.25 Calcium sulfate is found in plaster. At 25 °C the value of K_{sp} for $CaSO_4$ is 2.4×10^{-5}. What is the calculated molar solubility of $CaSO_4$ in water?

17.26 Chalk is $CaCO_3$, and at 25 °C its $K_{sp} = 4.5 \times 10^{-9}$. What is the molar solubility of $CaCO_3$? How many grams of $CaCO_3$ dissolve in 100 mL of aqueous solution? (Ignore the reaction of CO_3^{2-} with water.)

Common Ion Effect

17.27 Copper(I) chloride has $K_{sp} = 1.9 \times 10^{-7}$. Calculate the molar solubility of CuCl in (a) pure water, (b) 0.0200 M HCl solution, (c) 0.200 M HCl solution, and (d) 0.150 M $CaCl_2$ solution.

17.28 Gold(III) chloride, $AuCl_3$, has $K_{sp} = 3.2 \times 10^{-25}$. Calculate the molar solubility of $AuCl_3$ in (a) pure water, (b) 0.010 M HCl solution, (c) 0.010 M $MgCl_2$ solution, and (d) 0.010 M $Au(NO_3)_3$ solution.

17.29 Calculate the molar solubility of Ag_2CrO_4 in 0.200 M $AgNO_3$ at 25 °C. For Ag_2CrO_4 at 25 °C, $K_{sp} = 1.2 \times 10^{-12}$.

17.30 What is the molar solubility of $Mg(OH)_2$ in 0.20 M NaOH? For $Mg(OH)_2$, $K_{sp} = 7.1 \times 10^{-12}$.

17.31 Calculate the molar solubility of $CaSO_4$ in 0.015 M $CaCl_2$.

17.32 Calculate the molar solubility of AgCl in 0.050 M $AlCl_3$.

Precipitation

17.33 Does a precipitate of $PbCl_2$ form when 0.0150 mol of $Pb(NO_3)_2$ and 0.0120 mol of NaCl are dissolved in 1.00 L of solution?

17.34 Silver acetate, $AgC_2H_3O_2$, has $K_{sp} = 2.3 \times 10^{-3}$. Does a precipitate form when 0.015 mol of $AgNO_3$ and 0.25 mol of $Ca(C_2H_3O_2)_2$ are dissolved in a total volume of 1.00 L of solution?

17.35 Does a precipitate of $PbBr_2$ form if 50.0 mL of 0.0100 M $Pb(NO_3)_2$ is mixed with (a) 50.0 mL of 0.0100 M KBr and (b) 50.0 mL of 0.100 M NaBr?

17.36 Would a precipitate of silver acetate form if 22.0 mL of 0.100 M $AgNO_3$ were added to 45.0 mL of 0.0260 M $NaC_2H_3O_2$? For $AgC_2H_3O_2$, $K_{sp} = 2.3 \times 10^{-3}$.

17.37 Both AgCl and AgI are very sparingly soluble salts, but the solubility of AgI is much less than that of AgCl, as can be seen by their K_{sp} values. Suppose that a solution contains both Cl^- and I^- with $[Cl^-] = 0.050$ M and $[I^-] = 0.050$ M. If solid $AgNO_3$ is added to 1.00 L of this mixture (so that no appreciable change in volume occurs), what is the value of $[I^-]$ when AgCl first begins to precipitate?

17.38 Suppose that Na_2SO_4 is added gradually to 100 mL of a solution that contains both Ca^{2+} ion (0.15 M) and Sr^{2+} ion (0.15 M). (a) What will the Sr^{2+} concentration be (in mol L^{-1}) when $CaSO_4$ just begins to precipitate? (b) What percentage of the strontium ion has precipitated when $CaSO_4$ just begins to precipitate?

Oxides, Sulfides, and Selective Precipitations

17.39 What value of $[H^+]$ and what pH permits the selective precipitation of the sulfide of just one of the two metal ions in a solution that has a concentration of 0.010 M Pb^{2+} and 0.010 M Co^{2+}?

17.40 Can a selective separation of Mn^{2+} from Sn^{2+} be accomplished by a suitable adjustment of the pH of a solution that is 0.010 M in Mn^{2+}, 0.010 M in Sn^{2+}, and saturated in H_2S? (Assume the green form of MnS in Table 17.2.)

Complex Ion Equilibria

17.41 Write the chemical equilibria and equilibrium laws that correspond to K_{form} for the following complexes. (a) $CuCl_4^{2-}$ (b) AgI_2^- (c) $Cr(NH_3)_6^{3+}$

17.42 Write the chemical equilibria and equilibrium laws that correspond to K_{inst} for the following complexes. (a) $Ag(S_2O_3)_2^{3-}$ (b) $Zn(NH_3)_4^{2+}$ (c) SnS_3^{2-}

17.43 Write equilibria that correspond to K_{inst} for each of the following complex ions and write the equations for K_{inst}. (a) $Co(NH_3)_6^{3+}$ (b) HgI_4^{2-} (c) $Fe(CN)_6^{4-}$

17.44 Write the equilibria that are associated with the equations for K_{form} for each of the following complex ions. Write also the equations for the K_{form} of each. (a) $Hg(NH_3)_4^{2+}$ (b) SnF_6^{2-} (c) $Fe(CN)_6^{3-}$

17.45 How many grams of solid NaCN have to be added to 1.2 L of water to dissolve 0.11 mol of $Fe(OH)_3$ in the form of $Fe(CN)_6^{3-}$? Use data as needed from Tables 17.1 and 17.3. (For simplicity, ignore the reaction of CN^- ion with water.)

17.46 Silver ion forms a complex with thiosulfate ion, $S_2O_3^{2-}$, that has the formula $Ag(S_2O_3)_2^{3-}$. This complex has $K_{form} = 2.0 \times 10^{13}$. How many grams of AgBr ($K_{sp} = 5.0 \times 10^{-13}$) will dissolve in 125 mL of 1.20 M $Na_2S_2O_3$ solution?

ADDITIONAL EXERCISES

17.47 Magnesium hydroxide, $Mg(OH)_2$, found in milk of magnesia, has a solubility of 7.05×10^{-3} g L^{-1} at 25 °C. Calculate K_{sp} for $Mg(OH)_2$.

17.48 Does iron(II) sulfide dissolve in 8 M HCl? Perform the calculations that prove your answer.

17.49 What is the pH of a saturated solution of $Mg(OH)_2$?

*17.50 Suppose that 50.0 mL of 0.12 M $AgNO_3$ is added to 50.0 mL of 0.048 M NaCl solution. (a) What mass of AgCl would form? (b) Calculate the final concentrations of all of the ions in the solution that is in contact with the precipitate. (c) What percentage of the Ag^+ ion has precipitated?

*17.51 A sample of hard water was found to have 278 ppm Ca^{2+} ion. Into 1.00 L of this water, 1.00 g of Na_2CO_3 was dissolved. What is the new concentration of Ca^{2+} in parts per million? (Assume that the addition of Na_2CO_3 does not change the volume, and assume that the density of the aqueous solutions involved are all 1.00 g mL^{-1}.)

*17.52 What value of $[H^+]$ and what pH would allow the selective separation of the carbonate of just one of the two metal ions in a solution that has a concentration of 0.010 M La^{3+} and 0.010 M Pb^{2+}? For $La_2(CO_3)_3$, $K_{sp} = 4.0 \times 10^{-34}$; for $PbCO_3$, $K_{sp} = 7.4 \times 10^{-14}$. A saturated solution of CO_2 in water has a concentration of H_2CO_3 equal to 3.3×10^{-2} M.

*17.53 When solid NH_4Cl is added to a suspension of $Mg(OH)_2(s)$, some of the $Mg(OH)_2$ dissolves.
(a) Write equations for *all* the chemical equilibria that exist in the solution after the addition of the NH_4Cl.
(b) Use Le Châtelier's principle to explain why adding NH_4Cl causes $Mg(OH)_2$ to dissolve.

(c) How many moles of NH_4Cl must be added to 1.0 L of a suspension of $Mg(OH)_2$ to dissolve 0.10 mol of $Mg(OH)_2$?
(d) What is the pH of the solution after the 0.10 mol of $Mg(OH)_2$ has dissolved in the solution containing the NH_4Cl?

*17.54 After solid $CaCO_3$ was added to a slightly basic solution, the pH was measured to be 8.50. What was the molar solubility of $CaCO_3$ in this solution? (See Thinking It Through, Problem 9 on page 786.)

17.55 What is the molar solubility of $Fe(OH)_3$ in water? (*Hint:* You have to take into account the self-ionization of water.)

17.56 The value of K_{inst} for $SnCl_4^{2-}$ is 5.6×10^{-3}.
(a) What is the value of K_{form}?
(b) Is this complex ion more or less stable than those in Table 17.3?

*17.57 Note 2 in Example 17.11 (page 784) says, "There are relatively few uncomplexed Ag^+ ions in the solution." Calculate the molar concentration of Ag^+ ion actually left after the complex forms as described in Example 17.11.

*17.58 What are the concentrations of Pb^{2+}, Br^-, and I^- in an aqueous solution that's in contact with both PbI_2 and $PbBr_2$?

The Carbonate Buffer and Mountaineering

9

In biochemistry, *respiration* refers to the transport of the blood gases, oxygen and carbon dioxide, and to their involvement in *metabolism*—the manifold of all chemical reactions in the body. Small changes in the pH of the blood have deadly consequences because H^+ ions participate in the equilibria of respiration in its earliest phases. Two such equilibria are the following, where H*Hb* represents hemoglobin, a protein in blood, and *Hb*O_2^- is oxyhemoglobin.[1]

$$HHb(aq) + O_2(aq) \rightleftharpoons H^+(aq) + HbO_2^-(aq) \quad (1)$$

$$HCO_3^-(aq) + H^+(aq) \rightleftharpoons H_2O + CO_2(aq) \quad (2)$$

Oxyhemoglobin of Equilibrium 1 is the form in which oxygen is carried in the blood. The bicarbonate ion and carbon dioxide of Equilibrium 2 are the chief forms in which carbon dioxide is transported in the blood. Very little carbonic acid is present in blood. However, $CO_2(aq)$ "stands in" for H_2CO_3 as the chief base neutralizer in blood because it is able to *directly* neutralize hydroxide ion by the following reaction.

$$CO_2(aq) + OH^-(aq) \longrightarrow HCO_3^-(aq) \quad (3)$$

Both Equilibria 1 and 2 shift to the right in the lungs and to the left in actively metabolizing cells. The shifts occur best when the pH of the blood is 7.35 ± 0.05. To hold the blood's pH within this range, the system relies on buffers. Without the blood buffers, we could not live.

The Blood's Carbonate Buffer. Together, $CO_2(aq)$ and $HCO_3^-(aq)$ constitute one of the important buffers in blood, the carbonate buffer. It neutralizes extra base by the reaction of Equation 3. The buffer neutralizes extra acid by using the bicarbonate ion.

$$HCO_3^-(aq) + H^+(aq) \longrightarrow CO_2(aq) + H_2O$$

[1] We have simplified the first equilibrium by showing only one O_2 instead of the four O_2 molecules known to be taken up by each H*Hb* molecule.

The relevant buffer equation, which can be derived from the defining equation for the first acid ionization constant of carbonic acid (see page 734), is as follows, only we use CO_2 for H_2CO_3.

$$pH = pK_a - \log\frac{[CO_2]}{[HCO_3^-]} \quad (4)$$

The value of pK_a for the first ionization of carbonic acid is 6.4, but *it cannot be used here* because it is for a solution at a temperature of 25 °C, not 37 °C, the temperature of the blood. Specialists in blood chemistry have agreed to an "apparent" pK_a value of 6.1, so we can write the following equation for the CO_2/HCO_3^- buffer in blood.

$$pH = 6.1 - \log\frac{[CO_2]}{[HCO_3^-]} \quad \text{(at 37 °C)}$$

Normally, in arterial blood, $[HCO_3^-]$ is 24 mmol L^{-1}, and $[CO_2]$ is 1.2 mmol L^{-1}. The pH of human arterial blood is then calculated to be 7.4.

$$pH = 6.1 - \log\frac{[1.2 \text{ mmol L}^{-1}]}{[24 \text{ mmol L}^{-1}]} = 7.4$$

Statistically, the average human arterial blood pH in health is 7.35, so the calculated and the observed values agree well.

The Response of the Blood to Excess Acid. To illustrate how the carbonate buffer works, let's suppose that the blood is challenged with a sudden influx of acid in the equivalent of 10 mmol of H^+ ion per liter of blood. (Acids are metabolically produced at higher rates in some conditions, like diabetes.) An extra 10 mmol per liter of H^+ would neutralize 10 mmol L^{-1} of HCO_3^- and so reduce $[HCO_3^-]$ from 24 mmol L^{-1} to 14 mmol L^{-1}. The HCO_3^- would change almost entirely to $CO_2(aq)$, so 10 mmol L^{-1} of *new* $CO_2(aq)$ would appear in the blood. If the new CO_2 could not be removed by breathing, the value of $[CO_2]$ would increase by 10 mmol L^{-1}, from 1.2 mmol L^{-1} to 11.2 mmol L^{-1}. This would be fatal because, by Equation 4, the resulting pH of the blood would be 6.2, much too low to permit life to continue. (Even a drop to a pH of 7.2 is serious.)

$$pH = 6.1 - \log\frac{[11.2 \text{ mmol L}^{-1}]}{[14 \text{ mmol L}^{-1}]} = 6.2$$

chemicals in our world

Figure 9a A heavy burden on the ability of the blood's carbonate buffer is caused by strenuous exertion at high altitude, as here on Mont Blanc, France's highest mountain. Overbreathing removes too much base-neutralizing CO_2 from the blood, and the pH of the blood increases.

However, essentially all the $CO_2(aq)$ leaves the blood in the lungs and is breathed out as $CO_2(g)$. Although the level of HCO_3^- stays the same, at 14 mmol L^{-1}, the level of $CO_2(aq)$ quickly drops back to 1.2 mmol L^{-1}. So the pH quickly bounces back up to 7.2.

$$pH = 6.1 - \log\frac{[1.2 \text{ mmol } L^{-1}]}{[14 \text{ mmol } L^{-1}]} = 7.2$$

A blood pH of 7.2 is still dangerously low for health, but it doesn't cause immediate death. The breathing mechanism provides further upward adjustment of the blood pH. The body, when healthy, responds quickly to a lowering of the blood pH by increasing the rate of breathing. This works to remove more CO_2. It's quite common for rapid breathing to pull the level of $CO_2(aq)$ in blood from 1.2 mmol L^{-1} down to 0.70 mmol L^{-1}, sometimes a bit lower. This would bring the pH of the blood in the example situation—the sudden influx of 10 mmol L^{-1} of H^+—back to 7.4.

$$pH = 6.1 - \log\frac{[0.7 \text{ mmol } L^{-1}]}{[14 \text{ mmol } L^{-1}]} = 7.4$$

Thus two mechanisms, buffering and breathing, have protected the system against the otherwise lethal influx of 10 mmol L^{-1} of strong acid. Neither of these two mechanisms could save the situation alone.

High Altitude Disease. Those who take themselves to high altitude (> 3000 m) without having allowed sufficient time for becoming acclimatized have the opposite problem, a *loss* of "acid" from the body, not a gain. They are in an environment with a partial pressure of oxygen lower than at sea level (Figure 9a). To compensate, a climber breathes at a higher rate. But overbreathing at high altitude may cause the loss of CO_2 to occur too rapidly, *and this has the same effect as a loss of acid.* (Remember that CO_2 is a base neutralizer.) The loss of acid is equivalent to a *gain* of base, so the pH of the blood *increases* with overbreathing at high altitude. One metabolic problem is now an impaired ability to *release* O_2 from HbO_2^- at tissues needing oxygen. This is because a lack of H^+ prevents Equilibrium 1 from shifting to the *left* to liberate oxygen as needed and where needed. The combination of this problem, the release of oxygen, and the lower availability of oxygen at high altitude makes the situation life threatening. The only remedy (other than supplemental oxygen) is a *prompt* return to a lower elevation.

Suggested Readings

C. S. Houston, "Mountain Sickness," *Scientific American,* October, 1992, page 4, 58.

H. Valtin and F. J. Gennari, *Acid–Base Disorders,* Little, Brown and Company, Boston, 1987.

M. P. Ward, J. S. Milledge, and J. B. West, *High Altitude Medicine and Physiology,* University of Pennsylvania Press, Philadelphia, 1989.

Questions

1. Suppose that the blood is suddenly made to accept 12 mmol L^{-1} of hydrochloric acid.
   (a) What is the resulting pH of the blood if no CO_2 is allowed to escape?
   (b) What is the resulting pH of the blood if 12 mmol L^{-1} of CO_2 can be quickly exhaled?
   (c) What additional change occurs to remove more CO_2?

2. In uncontrolled hysterics, the body involuntarily removes CO_2 at an excessive rate. How does this affect the pH of the blood, increase it or decrease it? Explain.

3. By taking an overdose of a narcotic, a person began to breathe in a shallow manner. How does this affect the pH of the blood, increase it or decrease it? Explain.

Mother nature's way of shuffling the deck. When a storm like this sweeps through a neighborhood, things are not the same afterward. In this chapter we will study the two principal factors that control what happens spontaneously in our world. One of them is nature's natural tendency toward disorder, which is given a chance to work when a tornado scatters everything in its path and leaves wreckage in its wake.

Thermodynamics

This Chapter in Context In our studies of chemical reactions, you've learned about the factors that control how fast reactions occur and you've learned how we can describe chemical systems at equilibrium. We are now ready to examine another fundamental aspect of chemical reactivity by seeking answers to the question, "What determines whether or not a chemical reaction is possible when substances are combined?" To answer this question we will expand on topics introduced in Chapter 6— energy changes in chemical reactions. As you will see, not only will we be able to determine the possibility of reaction, but we will find another explanation of chemical equilibrium and another way of finding equilibrium constants.

18.1 Introduction

The study of **thermodynamics** is concerned principally with energy changes and the flow of energy from one substance to another. Remarkably, a detailed study of energy transfers has led to an understanding of why certain changes are inevitable and why others are simply impossible. We can thus understand *why* rivers flow downhill instead of up, and *why* during the passage of time stone walls crumble instead of being formed from heaps of sand.

You were introduced to some of the principles of thermodynamics in Chapter 6 when you learned to calculate enthalpy changes (heats of reaction). You should recall that enthalpy changes deal with the exchange of heat between chemical systems and their surroundings. Such transfers of heat, however, represent only one aspect of thermodynamics. Another is nature's drive toward disorder (randomness) which we discussed briefly in Chapter 12. Both energy and disorder are prime topics in thermodynamics.

Thermodynamics is organized around three fundamental, experimentally derived laws of nature. For ease of reference, they are identified by number and are called the *first law,* the *second law,* and the *third law.* The first law of thermodynamics, which was discussed briefly in Chapter 6, deals with exchanges of energy between a system and its surroundings and is basically a statement of the law of

Thermo- implies heat, *-dynamics* implies movement.

If necessary, review in Chapter 6 the meanings of such terms as *system, surroundings, state, state function,* and *isothermal* and *adiabatic* changes.

793

conservation of energy. This law, you recall, serves as the foundation for Hess's law, which we used in our computations involving enthalpy changes in Chapter 6.

The First Law of Thermodynamics Revisited

To begin our study of thermodynamics, let's review some of the principles you learned in Chapter 6. As a start, recall that the **internal energy** of a system, which is given the symbol E, is the system's total energy—the sum of all the kinetic and potential energies of its particles.

$$E_{\text{system}} = (KE)_{\text{system}} + (PE)_{\text{system}}$$

Although we can't actually measure the system's total energy, we *can* measure energy changes, and the equation

$$\Delta E = E_{\text{final}} - E_{\text{initial}}$$

or, for a chemical reaction,

$$\Delta E = E_{\text{products}} - E_{\text{reactants}}$$

serves to relate the algebraic sign of ΔE to the direction of energy flow. Thus, ΔE is positive if energy flows into a system and negative if energy flows out.

The first law of thermodynamics considers two ways by which energy can be exchanged between a system and its surroundings. One is by the absorption or release of heat, which is given the symbol q and is called the **heat of reaction.** The other involves **work,** w. If *work is done on a system,* as in the compression of a gas, the system gains and stores energy. Conversely, if *the system does work on the surroundings,* as when a gas expands and pushes a piston, the system loses some energy by changing part of its potential energy to kinetic energy which is transferred to the surroundings. The **first law of thermodynamics** expresses the net change in energy mathematically by the equation

$\Delta E = (\text{heat input})$
$\quad + (\text{work input})$

$$\Delta E = q + w$$

In words, this equation states that the change in the internal energy equals the sum of the energy gained as heat and the energy gained by work being done on the system. This equation also defines the algebraic signs of q and w for energy flow. Thus, when . . .

q is $(+)$	Heat is absorbed by the system.
q is $(-)$	Heat is released by the system.
w is $(+)$	Work is done on the system.
w is $(-)$	Work is done by the system.

In Chapter 6 you learned that ΔE is a state function, which means that its value does not depend on how a change from one state to another is carried out. The opposite is true for q and w.

Practice Exercise 1

Which of the following changes is accompanied by the most negative value of ΔE? (a) A spring is compressed and heated. (b) A compressed spring expands and is cooled. (c) A spring is compressed and cooled. (d) A compressed spring expands and is heated. ◆

Work in Chemical Systems

There are two kinds of work that chemical systems can do (or have done on them) that are of concern to us. One is electrical work, which is examined in the next chapter and will be discussed there. The other is work associated with the

In a steam engine, such as the one powering this locomotive, work is derived from the expansion of steam, which is produced by supplying heat to boil water. The relationships between heat and work are important factors in the design of engines of all kinds.

expansion or contraction of a system under the influence of an external pressure. Such "pressure–volume" or P–V work was discussed in Chapter 6 where it was shown that this kind of work is given by the equation

$$w = -P\Delta V$$

where P is the *external pressure* on the system.

If P–V work is the only kind of work involved in a chemical change, the equation for ΔE takes the form

$$\Delta E = q + (-P\Delta V) = q - P\Delta V$$

When a reaction takes place in a container whose volume cannot change (so that $\Delta V = 0$), $\Delta E = q$, which means the entire energy change must appear as heat that's either absorbed or released. For this reason, ΔE is called the **heat of reaction at constant volume** (q_v).

$$\Delta E = q_v$$

Work being done on a gas. When a gas is *compressed* by an external pressure, work is done on the gas and w is positive. Because V decreases when the gas is compressed, the negative sign in the equation $w = -P\Delta V$ assures that w will be a positive quantity.

Practice Exercise 2

Molecules of an ideal gas have no intermolecular attractions and therefore undergo no change in potential energy on expansion of the gas. If the expansion is also at constant temperature, there is no change in the kinetic energy, so the isothermal (constant temperature) expansion of an ideal gas has $\Delta E = 0$. Suppose such a gas expands from a volume of 1.0 L to 12.0 L against a constant opposing pressure of 14.0 atm. In units of L atm, what are q and w for this change? ◆

Practice Exercise 3

If a gas is compressed under adiabatic conditions (allowing no transfer of heat to or from the surroundings) by application of an external pressure, the temperature of the gas increases. Why? ◆

Enthalpy and Enthalpy Changes

Rarely do we carry out reactions in containers of fixed volume. Usually, reactions take place in containers open to the atmosphere where they are exposed to a constant pressure. To study heats of reactions under constant pressure conditions, the concept of enthalpy was invented. Recall that the **enthalpy H** is defined by the equation

$$H = E + PV$$

At constant pressure, an **enthalpy change ΔH** is then

$$\Delta H = \Delta E + P\Delta V$$

Substituting for ΔE gives

$$\Delta H = (q - P\Delta V) + P\Delta V$$

$$= q_p$$

Diesel engines are used to power large trucks and other heavy equipment. In the cylinders of a diesel engine, air is compressed to very small volumes, raising the temperature to the point at which the fuel ignites spontaneously when injected into the cylinders.

where q_p is the **heat of reaction at constant pressure.** Thus, the value of ΔH for a system equals the heat of reaction at constant pressure.

Converting between ΔE and ΔH for a Chemical Reaction

In Chapter 6 we noted that ΔE and ΔH are not equal. They differ by the product $P\Delta V$. We can see this by rearranging the equation above.

$$\Delta H - \Delta E = P\Delta V$$

The ΔV values for reactions involving only solids and liquids are very tiny, so ΔE and ΔH for these reactions are nearly the same size.

The only time ΔE and ΔH differ by a significant amount is when gases are formed or consumed in a reaction, and even then they do not differ by much. To calculate ΔE from ΔH (or ΔH from ΔE), we must have a way of calculating $P\Delta V$.

If we assume the gases in the reaction behave as ideal gases, then we can use the ideal gas law. Solved for V this is

$$V = \frac{nRT}{P}$$

A volume change can therefore be expressed as

$$\Delta V = \Delta\left(\frac{nRT}{P}\right)$$

For a change at constant pressure and temperature we can rewrite this as

$$\Delta V = \Delta n\left(\frac{RT}{P}\right)$$

Thus, when the reaction occurs, the volume change is caused by a change in the number of moles of *gas*. Of course, in chemical reactions not all reactants and products need be gases, so to be sure we compute Δn correctly, let's express the change in the number of moles of gas as Δn_{gas}. It's defined as

Remember, in calculating Δn, we only count moles of *gases* among the reactants and products.

$$\Delta n_{gas} = (n_{gas})_{products} - (n_{gas})_{reactants}$$

The $P\Delta V$ product is therefore,

$$P\Delta V = P\cdot \Delta n_{gas}\left(\frac{RT}{P}\right)$$

$$= \Delta n_{gas}\,RT$$

Substituting into the equation for ΔH gives

Converting between ΔE and ΔH

$$\Delta H = \Delta E + \Delta n_{gas}\,RT \qquad (18.1)$$

The following example illustrates how small the difference is between ΔE and ΔH.

EXAMPLE 18.1

Conversion between ΔE and ΔH

The decomposition of calcium carbonate in limestone is used industrially to make carbon dioxide.

$$CaCO_3(s) \longrightarrow CaO(s) + CO_2(g)$$

The reaction is endothermic and has $\Delta H° = +571$ kJ. What is the value of $\Delta E°$ for this reaction? What is the percentage difference between $\Delta E°$ and $\Delta H°$?

Analysis: The problem asks us to convert between ΔE and ΔH, so we need to use Equation 18.1. The superscript $°$ on ΔH and ΔE tells us the temperature is 25 °C (standard temperature for measuring heats of reaction). It is also important to remember that in calculating Δn_{gas}, we count just the numbers of moles of gas.

We also need a value for R. Since we want to calculate the term $\Delta n_{gas}\,RT$ in units of kilojoules, we must have R in appropriate energy units. The value of R we will use in this and other calculations that involve energies is found in a table of constants like the one inside the rear cover of this book. It is: $R = 8.314$ J mol^{-1} K^{-1}.

Solution: The equation we need is

$$\Delta H = \Delta E + \Delta n_{gas}\, RT$$

Solving for ΔE and applying the superscript $°$ gives

$$\Delta E° = \Delta H° - \Delta n_{gas}\, RT$$

To calculate Δn_{gas} we take the coefficients in the equation to represent numbers of moles. Therefore, there is one mole of gas among the products and no moles of gas among the reactants, so

$$\Delta n_{gas} = 1 - 0 = 1$$

The temperature is 298 K, and $R = 8.314$ J mol^{-1} K^{-1}. Substituting,

$$\Delta E° = +571 \text{ kJ} - (1 \text{ mol})(8.314 \text{ J mol}^{-1} \text{ K}^{-1})(298 \text{ K})$$

$$= +571 \text{ kJ} - 2.48 \text{ kJ}$$

$$= +569 \text{ kJ}$$

The percentage difference is calculated as follows:

$$\% \text{ difference} = \frac{2.48 \text{ kJ}}{571 \text{ kJ}} \times 100\%$$

$$= 0.434\%$$

Notice how small the relative difference is between ΔE and ΔH.

Limestone is also an important building material. Here we see limestone blocks being cut at a quarry on the island of Malta, where it is the main building material for most of the island's cities.

Notice that for the second term we've changed joules to kilojoules before performing the subtraction.

Is the Answer Reasonable?

A gas is formed in the reaction, so the system must expand and push against the opposing pressure of the atmosphere when the reaction occurs at constant pressure. Energy must be supplied to accomplish this. At constant volume this expansion would not be necessary; the pressure would simply increase. Therefore, decomposing the $CaCO_3$ should require more energy at constant pressure ($\Delta H°$) than at constant volume ($\Delta E°$) by an amount equal to the work done pushing back the atmosphere, and we see that $\Delta H°$ is larger than $\Delta E°$, so the answer is reasonable.

Practice Exercise 4

The reaction $CaO(s) + 2HCl(g) \longrightarrow CaCl_2(s) + H_2O(g)$ has $\Delta H° = -217.1$ kJ. Calculate $\Delta E°$ for this reaction. What is the percentage difference between $\Delta E°$ and $\Delta H°$? ◆

18.2 Enthalpy Changes and Spontaneity

Now that we've reviewed the way thermodynamics treats energy changes, we turn our attention to one of our main goals in studying this subject—finding relationships among the factors that control whether events are spontaneous. By **spontaneous change** we mean one that occurs by itself, without continuous outside assistance. Examples are water flowing over a waterfall and the melting of ice cubes in a cold drink on a warm day. These are events that proceed on their own.

Some spontaneous changes occur very rapidly. An example is the set of biochemical reactions that take place when you accidentally touch something very hot—they cause you to jerk your hand away quickly. Other spontaneous events, such as the gradual erosion of a mountain, occur slowly and many years

We take many spontaneous events for granted. What would you think if you dropped a book and it rose to the ceiling?

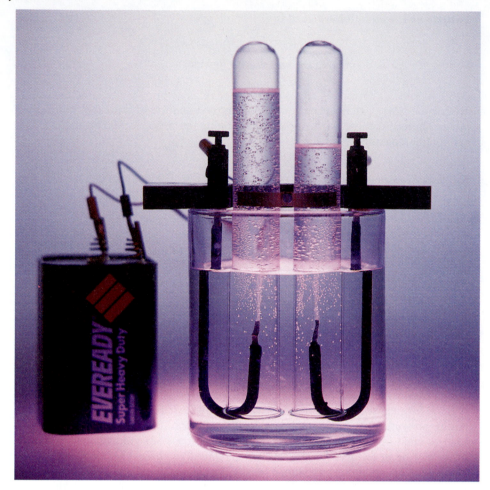

Figure 18.1 *A nonspontaneous chemical reaction.* From your own experience, you know that water does not spontaneously decompose into hydrogen and oxygen. Yet it can be made to decompose by a process called electrolysis, which uses electricity to cause the reaction $2H_2O(l) \longrightarrow 2H_2(g) + O_2(g)$ to proceed. (We will discuss electrolysis in detail in Chapter 19.) This nonspontaneous change will only continue as long as electricity is supplied.

may pass before a change is noticed. Still others occur at such an extremely slow rate under ordinary conditions that they appear not to be spontaneous at all. Gasoline–oxygen mixtures appear perfectly stable indefinitely at room temperature because under these conditions they react very slowly. However, if heated, their rate of reaction increases tremendously and they react explosively.

Each day we also witness events that are obviously *not* spontaneous. We may pass by a pile of bricks in the morning and later in the day find that they have become a brick wall. We know from experience that the wall didn't get there by itself. A pile of bricks becoming a brick wall is *not* spontaneous; it requires the intervention of a bricklayer. Similarly, the decomposition of water into hydrogen and oxygen is not spontaneous. We see water all the time and we know that it's stable. Nevertheless, we can cause water to decompose by passing an electric current through it in a process called *electrolysis* (Figure 18.1).

$$2H_2O(l) \xrightarrow{\text{electrolysis}} 2H_2(g) + O_2(g)$$

This decomposition will continue, however, only as long as the electric current is maintained. As soon as the supply of electricity is cut off, the decomposition ceases. This example demonstrates the difference between spontaneous and nonspontaneous changes. Once a spontaneous event begins, it has a tendency to continue until it is finished. A nonspontaneous event, on the other hand, can continue only as long as it receives some sort of outside assistance.

Nonspontaneous changes have another common characteristic. They are able to occur only when accompanied by some spontaneous change. For example, a bricklayer consumes food, and a series of spontaneous biochemical reactions then occur that supply the necessary muscle power to build a wall. Similarly, the nonspontaneous electrolysis of water requires some sort of spontaneous mechanical or chemical change to generate the needed electricity. In short, *all* nonspontaneous events occur at the expense of spontaneous ones. Everything that happens can be traced, either directly or indirectly, to spontaneous changes.

> The driving of nonspontaneous reactions to completion by linking them to spontaneous ones is an important principle in biochemistry.

Spontaneity and Potential Energy Changes

Because spontaneous changes are so important, it is necessary for us to understand the factors that favor spontaneity. Let's begin by examining some everyday events such as those depicted in Figure 18.2. One thing that each of these have in common is a lowering of the potential energy. Both the snowboard and the water lose potential energy as they move from a higher to a lower altitude. Similarly, the chemical substances in the gasoline–oxygen mixture lose chemical energy by evolving heat as the gasoline burns to produce CO_2 and H_2O.

Many (but not all) spontaneous events are accompanied by a lowering of the potential energy, so an energy lowering is *one* of the factors that work in *favor* of spontaneity. Since a change that lowers the potential energy of a system can be said to be exothermic, we can state this factor another way—*exothermic changes have a tendency to proceed spontaneously.*

> Most, but not all, chemical reactions that are exothermic occur spontaneously.

We have seen that thermodynamics looks to ΔE and ΔH for the potential energy change in reacting systems. You learned that it is ΔE that we observe when the reaction happens at constant volume, and it is ΔH that we measure when the reaction occurs at constant pressure. Since we are most interested in reactions at constant pressure, we will devote our attention to ΔH. This quantity, you recall, has a negative sign for an exothermic change, so we can say that reactions for which ΔH is negative *tend* to proceed spontaneously. The enthalpy

> If we study systems at constant temperature, the molecular kinetic energy remains constant and any energy changes we measure must be changes in potential energy.

Figure 18.2 *Spontaneous events.* Many common events are spontaneous. Here we see three of them—a person rides a "snowboard" downhill, water cascades over a waterfall, and fuel burns.

change thus serves as *one* of the thermodynamic factors that influence whether or not a given process can occur by itself.

18.3 Entropy and Spontaneous Change

In Chapter 12 we discussed the solution process, and you learned that one of the principal driving forces in the formation of a solution is the increase in disorder, or randomness, that occurs when particles of the solute mix with those of the solvent. In fact, this increase in disorder can be so important that it can outweigh energy effects. For example, the dissolving of NaI in water is endothermic ($\Delta H_{solution}$ is positive); nevertheless, NaI crystals will dissolve quite spontaneously in water, even though the energy effect would seemingly make the process nonspontaneous.

The dissolving of salts such as NaI in water is just one example of a change that occurs spontaneously even though it is endothermic. Other examples are the melting of ice when the weather becomes warm and the evaporation of water from a puddle or a pond. Both of these changes are also endothermic and spontaneous, and they take place for the same reason that NaI dissolves in water. Each is accompanied by an increase in the randomness of the distribution of particles in the system. When ice melts, the water molecules leave the highly organized crystalline state and become jumbled and disorganized in the liquid. When water evaporates, the molecules are no longer confined to the region of other water molecules; instead, they are free to roam throughout the entire atmosphere.

It is a universal phenomenon that something that brings about randomness is more likely to occur than something that brings about order. We can see the reason for this if we examine the close relationship between randomness and statistical probability. Suppose we try a simple experiment with a new deck of playing cards. When first unwrapped, they are separated according to suit and arranged numerically. Now, suppose we toss the deck in the air and let the cards fall to the floor. When we sweep them up and restack them, we will almost certainly find they have become rearranged. We expect this to occur because, once the deck is tossed in the air, all of the possible card sequences become possible. *There are many millions of ways for the cards to be arranged in a new sequence, but only one way for them to come together again in the original sequence.* As soon as the cards are tossed, therefore, a random sequence becomes much more probable than the ordered one with which we began. The spontaneous change from the ordered arrangement to a random (disordered) one takes place simply because of the laws of chance.

The same laws of chance that apply to the ordering of a deck of playing cards also apply to the ordering within distributions of particles in chemical and physical systems. Consider, for example, the melting of an ice cube (Figure 18.3). Of all the possible ways of placing water molecules into a container, the highly ordered arrangement within the crystalline ice cube has a very low probability, while the disordered arrangement of the molecules found in the liquid is more probable. We can see this more clearly, perhaps, by analogy. Just imagine tossing bricks in the air and observing how they land. The chance that they will fall one on the other to form a brick wall is certainly very small; the brick wall is an improbable result. A much more likely arrangement is a jumbled, disordered pile. Thus, the collapse of a brick wall and the melting of an ice cube have something in common. In each case, the system passes spontaneously from a state of low probability to one of higher probability.

The cards are highly ordered.

After they've been tossed they are disordered.

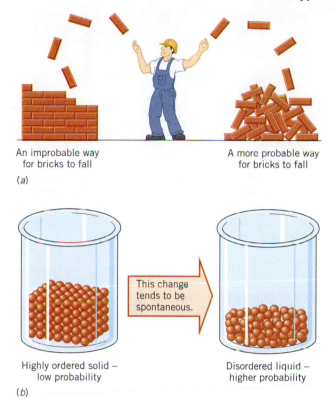

Figure 18.3 *Disorder has a higher probability than order.* (*a*) Tossing bricks into the air is more likely to give a pile of bricks than a brick wall. This is because there are more ways for the bricks to be jumbled than for them to form a wall. (*b*) If water molecules are dropped into a container, they would not be expected to land in just the right places to give a highly ordered solid. If the way they land is determined purely by chance, the disordered liquid is more likely to occur.

Entropy Changes and Spontaneity

Because statistical probability is so important in determining the outcome of chemical and physical events, thermodynamics defines a quantity, called **entropy** (symbol S), that describes the degree of randomness of a system. The larger the value of the entropy, the larger is the degree of randomness of the system and, therefore, the larger is its statistical probability.

The greater the statistical probability of a particular state, the greater is the entropy.

Like enthalpy, entropy is a state function. It depends only on the state of the system, so a **change in entropy, ΔS,** is independent of the path from start to finish. As with other thermodynamic quantities, ΔS is defined as "final minus initial" or "products minus reactants." Thus

$$\Delta S = S_{final} - S_{initial}$$

or, for a chemical system,

$$\Delta S = S_{products} - S_{reactants}$$

As you can see, when S_{final} is larger than $S_{initial}$ (or when $S_{products}$ is larger than $S_{reactants}$), the value of ΔS is positive. A positive value for ΔS means an increase in the randomness of the system during the change, and we have seen that this kind of change tends to be spontaneous. This leads to a general statement about entropy:

As you will learn soon, absolute values for S can be obtained. This is entirely different from E and H, where we can only determine differences (i.e., ΔE and ΔH).

An increase in entropy is a factor that favors spontaneity. Any event that is accompanied by an increase in the entropy of the system will have a *tendency* to occur spontaneously.

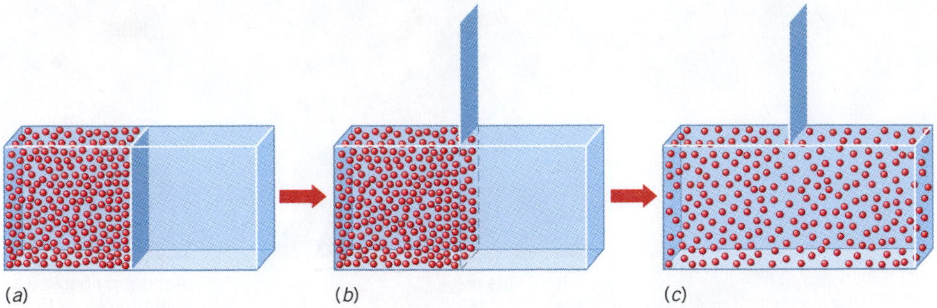

Figure 18.4 *The expansion of a gas into a vacuum.* (*a*) A gas in a container separated from a vacuum by a partition. (*b*) The gas at the moment the partition is removed. With the entire volume of the now-larger container available, this distribution of particles has a low probability. (*c*) The gas expands to achieve a more probable (higher entropy) particle distribution.

Tools

Predicting the
sign of Δ*S*

Predicting the Sign of Δ*S*

It is often possible to predict whether Δ*S* is positive or negative for a particular change. This is because several factors influence the magnitude of the entropy in predictable ways.

Volume

For gases, the entropy increases with increasing volume, as illustrated in Figure 18.4. At the left we see a gas confined to one side of a container, separated from a vacuum by a removable partition. Let's suppose the partition could be pulled away in an instant, as shown in Figure 18.4*b*. Now we find a situation in which all the molecules of the gas are at one end of a larger container. This is a very improbable distribution of the molecules, so the gas expands spontaneously to achieve a more probable (higher entropy) particle distribution.

Temperature

The entropy is also affected by the temperature; the higher the temperature, the larger is the entropy. For example, when a substance is a solid at absolute zero, its particles are essentially motionless. The atoms are at their equilibrium lattice positions, and there is a minimum in the amount of disorder (and entropy) of the particles (Figure 18.5*a*). If some heat is added to the solid, the kinetic energy of the particles increases along with the temperature. This causes the particles to move and vibrate within the crystal, so at a particular moment (pictured in Figure 18.5*b*) the particles are not found exactly at their lattice sites. Thus, we have introduced some disorder and increased the entropy. At a still higher temperature the instantaneous disorder is even greater and the solid has a still higher entropy (Figure 18.5*c*).

Physical State

One of the major factors that affects the entropy of a system is its physical state, which is demonstrated in Figure 18.6. We have already concluded that the distribution of particles in a liquid is more probable and of a higher entropy than that in a solid. A gas has an even more random, higher entropy particle distribution. In fact, a gas has such a large entropy compared with a liquid or solid that changes which produce gases from liquids or solids are almost always accompanied by increases in entropy.

$S_{solid} < S_{liquid} \ll S_{gas}$

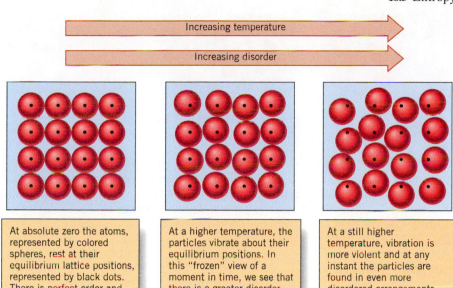

Figure 18.5 *Variation of entropy and disorder with temperature.* As the temperature rises, the system becomes more disordered and the entropy increases.

(a) At absolute zero the atoms, represented by colored spheres, rest at their equilibrium lattice positions, represented by black dots. There is perfect order and minimum entropy.

(b) At a higher temperature, the particles vibrate about their equilibrium positions. In this "frozen" view of a moment in time, we see that there is a greater disorder than at absolute zero.

(c) At a still higher temperature, vibration is more violent and at any instant the particles are found in even more disordered arrangements.

Entropy Changes in Chemical Reactions

When a chemical reaction produces or consumes gases, the sign of its entropy change is usually easy to predict. This is because the entropy of a gas is so much larger than that of either a liquid or solid. For example, the thermal decomposition of sodium bicarbonate produces two gases, CO_2 and H_2O.

$$2NaHCO_3(s) \xrightarrow{\text{heat}} Na_2CO_3(s) + CO_2(g) + H_2O(g)$$

Because the amount of gaseous products is larger than the amount of gaseous reactants, we can predict that the entropy change for the reaction is positive. On the other hand, the reaction

$$CaO(s) + SO_2(g) \longrightarrow CaSO_3(s)$$

(which can be used to remove sulfur dioxide from a gas mixture) has a negative entropy change.

For reactions that involve gases, we can simply calculate the change in the number of moles of gas, Δn_{gas}, on going from reactants to products. When Δn_{gas} is positive, so is the entropy change.

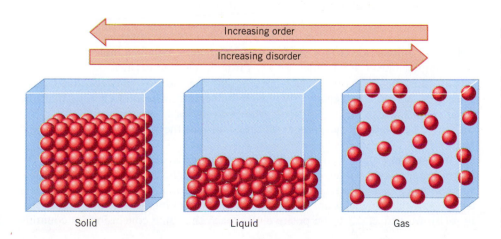

Increasing order

Increasing disorder

Solid   Liquid   Gas

Figure 18.6 *Comparison of the entropies of the solid, liquid, and gaseous states of a substance.* The crystalline solid is highly ordered and has a very low entropy. The liquid is less ordered than the solid and has a higher entropy, but all the particles are still found at one end of the container. The gas has the highest entropy because the particles are randomly distributed throughout the entire container.

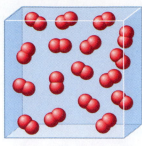 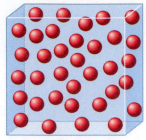

Lower entropy   Higher entropy

Figure 18.7 *Entropy is affected by molecular complexity.* If marbles were tossed into a weightless container in space, it is less likely they would arrange themselves in pairs than drift around individually. Similarly, hydrogen atoms grouped in pairs in H_2 is a less probable (lower entropy) particle distribution than individual H atoms.

Molecular Complexity

For chemical reactions, another major factor that controls the sign of ΔS is changes in the degree of complexity of the molecules. For example, Figure 18.7 illustrates two different ways for atoms of hydrogen to be distributed in a container. If there were no attractions between the atoms, which is the more likely way for the atoms to find themselves, as individual H atoms drifting in the container or as pairs of atoms in H_2? Clearly, the more *random* distribution is that of individual atoms, and this is also the one of higher entropy. Therefore, the reaction

For a given number of atoms, the more particles there are, the less complex each must be. Two moles of H has more particles than one mole of H_2; an atom of H is less complex than a molecule of H_2.

$$H_2(g) \longrightarrow 2H(g)$$

is accompanied by a positive ΔS. Notice that on forming the less complex particles, the total number of particles increases. We conclude, then, that *when a reaction occurs with a decrease in the degree of molecular complexity and an increase in the number of particles, there tends to be an increase in entropy.*

EXAMPLE 18.2

Predicting the Sign of ΔS

Predict the algebraic sign of ΔS for the reactions:

(a) $2NO_2(g) \longrightarrow N_2O_4(g)$

(b) $C_3H_8(g) + 5O_2(g) \longrightarrow 3CO_2(g) + 4H_2O(g)$

Analysis: As we examine the equations we look for changes in the number of moles of gas and changes in the number of particles on going from reactants to products.

Solution: In reaction (a) we are forming fewer, more complex molecules (N_2O_4) from simpler ones (NO_2). Since we are forming fewer molecules, molecular complexity must be increasing, which means that the entropy must be decreasing. Therefore, ΔS must be negative. We reach the same conclusion by noting that one mole of gaseous product is formed from two moles of gaseous reactant. When there is a decrease in the number of moles of gas, the reaction tends to have a negative ΔS.

For reaction (b), we can count the number of molecules on both sides. On the left of the equation we have six molecules; on the right there are seven. The only way to form more molecules from the same number of atoms is to make them less complex. If the molecules are less complex, they are of higher entropy, so for reaction (b), we expect ΔS to be positive. We reach the same conclusion by counting the number of moles of gas on both sides of the equation. On the left there are 6 moles of gas; on the right there are 7 moles of gas. Because the number of moles of gas is increasing, we expect ΔS to be positive.

Practice Exercise 5

Predict the sign of the entropy change for (a) the condensation of steam to liquid water and (b) the sublimation of a solid. ◆

Practice Exercise 6

Predict the sign of ΔS for the following reactions. Justify your answers.

(a) $2SO_2(g) + O_2(g) \longrightarrow 2SO_3(g)$
(b) $CO(g) + 2H_2(g) \longrightarrow CH_3OH(g)$
(c) $2H_2(g) + O_2(g) \longrightarrow 2H_2O(l)$
(d) $N_2(g) + 3H_2(g) \longrightarrow 2NH_3(g)$
(e) $Ca(OH)_2(s) \xrightarrow{H_2O} Ca^{2+}(aq) + 2OH^-(aq)$ ◆

18.4 The Second Law of Thermodynamics and the Gibbs Free Energy

We have studied two factors, enthalpy and entropy, that affect whether or not a physical or chemical event will be spontaneous. Sometimes these factors work together. For example, when a stone wall crumbles, its potential energy decreases and the stones become disordered. Therefore, the enthalpy decreases and entropy increases. Since both of these favor a spontaneous change, the two factors complement one another. In other situations, the effects of enthalpy and entropy are in opposition. Such is the case, as we have seen, in the melting of ice or the evaporation of water. The endothermic nature of these changes tends to make them nonspontaneous, while the increase in the randomness of molecules tends to make them spontaneous. In the reaction of H_2 with O_2 to form H_2O, the enthalpy and entropy changes are also in opposition. In this case, the exothermic nature of the reaction is sufficient to overcome the negative value of ΔS and cause the reaction to be spontaneous.

When enthalpy and entropy oppose one another, their relative importance in determining spontaneity is far from obvious. In addition, temperature becomes a third factor that can influence the direction in which a change is spontaneous. For example, if we attempt to raise the temperature of an ice–water slush to 25 °C, all the solid will melt. At 25 °C the change *solid → liquid* is spontaneous. On the other hand, if we attempt to cool an ice–water slush to −25 °C, freezing occurs, so at −25 °C the opposite change is spontaneous. *Thus, there are actually three factors that can influence spontaneity: the enthalpy change, the entropy change, and the temperature.* The balance between these factors come into focus through the second law of thermodynamics.

The Second Law of Thermodynamics

One of the most far-reaching observations in science is incorporated into the **second law of thermodynamics,** which states, in effect, that *whenever a spontaneous event takes place in our universe, the total entropy of the universe increases* ($\Delta S_{total} > 0$). Notice that the increase in entropy that's referred to here is for the *total* entropy of the *universe* (system *plus* surroundings), not just the system alone. This means that a system's entropy can decrease, just as long as there is a larger increase in the entropy of the surroundings so that the *overall* entropy change is positive. Because everything that happens relies on spontaneous changes of some sort, the entropy of the universe is constantly rising. As described in Facets of Chemistry 18.1, this has some serious implications regarding the safekeeping of our environment.

Now let's examine more closely the entropy change for the universe. As we've suggested, this quantity equals the sum of the entropy change for the system plus the entropy change for the surroundings.

$$\Delta S_{total} = \Delta S_{system} + \Delta S_{surroundings}$$

It can be shown that the entropy change for the surroundings is equal to the heat transferred *to* the surroundings *from* the system, $q_{surroundings}$, divided by the Kelvin temperature, T, at which it is transferred.[1]

$$\Delta S_{surroundings} = \frac{q_{surroundings}}{T}$$

The law of conservation of energy requires that the heat transferred to the surroundings equals the negative of the heat added to the system, so we can write

$$q_{surroundings} = -q_{system}$$

In our study of the first law of thermodynamics we saw that for changes at constant temperature and pressure, $q_{system} = \Delta H$ for the system. By substitutions, therefore, we arrive at the relationship

$$\Delta S_{surroundings} = \frac{-\Delta H_{system}}{T}$$

and the entropy change for the entire universe becomes

$$\Delta S_{total} = \Delta S_{system} - \frac{\Delta H_{system}}{T}$$

By rearranging the right side of this equation, we obtain

$$\Delta S_{total} = \frac{T\Delta S_{system} - \Delta H_{system}}{T}$$

Now let's multiply both sides of the equation by T to give

$$T\Delta S_{total} = T\Delta S_{system} - \Delta H_{system}$$

or

$$T\Delta S_{total} = -(\Delta H_{system} - T\Delta S_{system})$$

---

[1] Adding heat to either a system or its surroundings increases the temperature and therefore the entropy, so the entropy change is directly proportional to the amount of heat added. The *relative* magnitude of the entropy increase, however, is inversely proportional to the temperature at which the heat is added. If heat is added near absolute zero, there is a large and very noticeable increase in disorder (and therefore a large increase in S), but if the same amount of heat is added at high temperature, where the disorder is already large, the relative effect is much less (so the increase in S is smaller). Roughly, then, $\Delta S \propto q/T$. This is discussed in further detail in Facets of Chemistry 18.2.

Facets of Chemistry 18.1

Pollution and the Second Law of Thermodynamics

A problem we hear about often is pollution. Environmentalists and politicians talk frequently about how to avoid polluting the environment and how to clean up the mess that already exists. Such discussions raise some interesting questions. Can we entirely avoid polluting our environment? Why are pollutants so easily spread and so difficult to eliminate? Many of the answers can be found in the second law of thermodynamics.

According to the second law, any spontaneous activity is accompanied by an increase in the total entropy of the universe. This means that whenever we create order somewhere, our activities must generate an even greater disorder in our surroundings.

Much of our time and effort is spent in creating order in the world around us. We clean up our desks and put out the garbage, mow the lawn and rake leaves in the autumn. Overall our activities are spontaneous, because the biochemical reactions driven by the foods we eat and digest are spontaneous and allow us to do these things. According to the second law, however, the total entropy of the universe must be increased by our spontaneous activities. Therefore, the increased order that we create for ourselves has to be balanced by an even larger increase in disorder somewhere in our surroundings—the environment.

Pollution involves the scattering of undesirable substances through our surroundings and is accompanied by an enormous increase in entropy. It is also a direct result of our efforts to create an orderly world. For example, we vacuum the floors in our homes to keep them clean, but the combustion of fuels to generate the electricity to power the vacuum cleaner releases polluting gases such as SO_2 into the atmosphere. Once released, such pollutants cannot be recovered effectively because of the enormous expense that would be required.

The entropy effect on pollution is important to understand, particularly when we consider the consequences of releasing harmful substances into the environment. Such materials include the highly toxic elements released in trace amounts when coal or nuclear fuels are used, as well as many toxic chemicals such as DDT and dioxin. When these chemicals are released, their spread is unavoidable because of the large entropy increase that occurs as they are scattered about. Eliminating them once they have had an opportunity to disperse is an almost impossible task, because doing so requires an enormous expenditure of energy and must generate even more dis-

Pollution is a potential problem posed by landfills, which attempt to contain the refuse of modern society. Alternatives, such as burning rubbish, have their own drawbacks. For this reason, recycling of paper, glass, and plastic is an important way to minimize the spread of pollution and to extend the useful life of current landfills.

order around us. Only if a pollutant is extremely hazardous does it warrant the effort necessary to reduce its concentration in the environment, and even if this is done, we can never eliminate the pollutant entirely. We can only reduce our risk of injury.

The surest way to overcome pollution, of course, is not to create it in the first place. For the benefits of activities that pollute, we try to reduce the risks as much as we judge possible. Thus, if we must use coal to generate electricity, it is worth the effort to remove as much sulfur as we can from the coal *before* it is burned. Beyond that, we must simply face the fact that we can never avoid pollution entirely. It's a no-win situation and a trade-off between risks and benefits.

Because ΔS_{total} must be positive for a spontaneous change, the quantity in parentheses, $(\Delta H_{\text{system}} - T\Delta S_{\text{system}})$, must be negative.

> For a change to be spontaneous,
>
> $$\Delta H_{\text{system}} - T\Delta S_{\text{system}} < 0 \qquad (18.2)$$

Equation 18.2 gives us a way of examining the balance between ΔH, ΔS, and temperature in determining the spontaneity of an event, and it becomes convenient at this point to introduce another thermodynamic state function. It is called the **Gibbs free energy, G,** named to honor one of America's most important scientists, Josiah Willard Gibbs (1839–1903). (It's called *free energy* because it is related, as we will see later, to the maximum energy in a change that is "free" or "available" to do useful work.) The Gibbs free energy is defined as

$$G = H - TS \qquad (18.3)$$

so for changes at constant T and P, this equation becomes

$$\Delta G = \Delta H - T\Delta S \qquad (18.4)$$

Because G is defined entirely in terms of state functions, it is also a state function. This means that we can define the **Gibbs free energy change, ΔG,** as

$$\Delta G = G_{\text{final}} - G_{\text{initial}} \qquad (18.5)$$

By comparing Equations 18.2 and 18.5, we arrive at the special importance of the free energy change:

> *At constant temperature and pressure, a change can be spontaneous only if it is accompanied by a decrease in the free energy of the system.*

Tools ➜

ΔG as a predictor of spontaneity

Reactions that occur with a free energy decrease are sometimes said to be **exergonic.** Those that occur with a free energy increase are sometimes said to be **endergonic.**

In other words, for a change to be spontaneous, G_{final} must be less than G_{initial} and ΔG must be negative. With this in mind, we can now examine how ΔH, ΔS, and T are related in determining spontaneity.

When ΔH Is Negative and ΔS Is Positive

The collapse of a building (Figure 18.8) is accompanied by a large lowering of the potential energy as the debris falls and a large increase in entropy as the ordered structure becomes a jumbled pile of rubble. For this change, ΔH is negative and ΔS is positive, both of which favor spontaneity.

$$\Delta H \text{ is negative } (-)$$
$$\Delta S \text{ is positive } (+)$$
$$\Delta G = \Delta H - T\Delta S$$
$$= (-) - [T(+)]$$

Notice that regardless of the absolute temperature, which must be a positive number, ΔG will be negative. This means that regardless of the temperature, such a change must be spontaneous. In fact, the ultimate collapse of buildings and stone walls is inevitable because such changes are always spontaneous.

Figure 18.8 *The collapse of a building is spontaneous.* When a demolition crew brings down a building such as this, the collapse will always be spontaneous once begun because it is accompanied by a lowering of the potential energy and an increase in entropy.

When ΔH Is Positive and ΔS Is Negative

When a change is endothermic and is accompanied by a lowering of the entropy, both factors work against spontaneity.

$$\Delta H \text{ is positive } (+)$$
$$\Delta S \text{ is negative } (-)$$
$$\Delta G = \Delta H - T\Delta S$$
$$= (+) - [T(-)]$$

Now, no matter what the temperature is, ΔG will be positive and the change must be nonspontaneous. An example would be a building rising from a pile of rubble. If you saw such a thing in a film, common experience would tell you that the film must be being played backward. Buildings *do* collapse, but they don't rise spontaneously from a jumbled heap of wreckage.

When ΔH and ΔS Have the Same Sign

When ΔH and ΔS have the same algebraic sign, the temperature becomes the determining factor in controlling spontaneity. If ΔH and ΔS are both positive, then

$$\Delta G = (+) - [T(+)]$$

Thus, ΔG is the difference between two positive quantities, ΔH and $T\Delta S$. This difference will only be negative if the term $T\Delta S$ is larger in magnitude than ΔH, and this will only be true when the temperature is high. In other words, *when ΔH and ΔS are both positive, the change will be spontaneous at high temperature but not at low temperature.* A familiar example is the melting of ice.

$$H_2O(s) \longrightarrow H_2O(l)$$

This is a change that is endothermic and also accompanied by an increase in entropy. We know that at high temperature (above 0 °C) melting is spontaneous, but at low temperature (below 0 °C) it is not.

For similar reasons, when ΔH and ΔS are both negative, ΔG will be negative (and the change spontaneous) only when the temperature is low.

$$\Delta G = (-) - [T(-)]$$

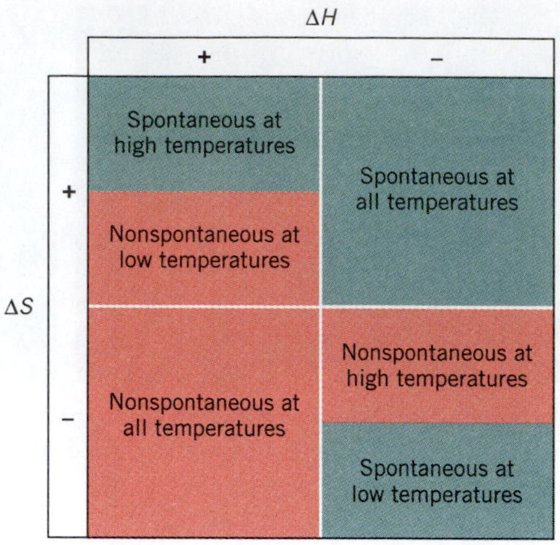

Figure 18.9 *Summary of the effects of the signs of* ΔH *and* ΔS *on spontaneity.* When ΔH and ΔS have the same algebraic sign, the temperature will determine whether a particular change is spontaneous or not.

Only when the negative value of ΔH is larger in magnitude than the negative value of $T\Delta S$ will ΔG be negative. Such a change is only spontaneous at low temperature. An example is the freezing of water.

$$H_2O(l) \longrightarrow H_2O(s)$$

This is an exothermic change that is accompanied by an decrease in entropy; it is only spontaneous at low temperatures (i.e., below 0 °C).

Figure 18.9 summarizes the effects of the signs of ΔH and ΔS on ΔG, and hence on the spontaneity of physical and chemical events.

18.5 The Third Law of Thermodynamics

Earlier we described how the entropy of a substance depends on temperature, and we noted that at absolute zero the order within a crystal is a maximum and the entropy is a minimum. The **third law of thermodynamics** goes one step further by stating: *At absolute zero the entropy of a perfectly ordered pure crystalline substance is zero.*

$$S = 0 \quad \text{at} \quad T = 0\,\text{K}$$

Because we know the point at which entropy has a value of zero, it is possible by *experimental measurement* and calculation to determine the total amount of entropy that a substance has at temperatures above 0 K. If the entropy of one mole of a substance is determined at a temperature of 298 K (25 °C) and a pressure of 1 atm, we call it the **standard entropy, $S°$**. Table 18.1 lists the standard entropies for a number of substances.[2] Notice that entropy has the dimensions of energy/temperature (i.e., joules per kelvin); the reason why is explained in Facets of Chemistry 18.2.

25 °C and 1 atm are the same standard conditions we used in our discussion of $\Delta H°$ in Chapter 6.

[2]In our earlier discussions of standard states (Chapter 6) we defined the *standard pressure* as 1 atm. This was the original pressure unit used by thermodynamicists. However, in the SI, the recognized unit of pressure is the pascal (Pa), not the atmosphere. After considerable discussion, the SI adopted the **bar** as the standard pressure for thermodynamic quantities: 1 bar = 10^5 Pa. One bar differs from one atmosphere by only 1.3%, and for thermodynamic quantities that we deal with in this text, their values at 1 atm and at 1 bar differ by an insignificant amount. Since you are more familiar with atmospheres than pascals, we shall continue to refer to the standard pressure as 1 atm.

Table 18.1 Standard Entropies of Some Typical Substances (25 °C, 1 atm)

Substance	$S°$ (J mol^{-1} K^{-1})	Substance	$S°$ (J mol^{-1} K^{-1})
$Ag(s)$	42.55	$H_2O(g)$	188.7
$AgCl(s)$	96.2	$H_2O(l)$	69.96
$Al(s)$	28.3	$HCl(g)$	186.7
$Al_2O_3(s)$	51.00	$HNO_3(l)$	155.6
$C(s, graphite)$	5.69	$H_2SO_4(l)$	157
$CO(g)$	197.9	$HC_2H_3O_2(l)$	160
$CO_2(g)$	213.6	$Hg(l)$	76.1
$CH_4(g)$	186.2	$Hg(g)$	175
$CH_3Cl(g)$	234.2	$K(s)$	64.18
$CH_3OH(l)$	126.8	$KCl(s)$	82.59
$CO(NH_2)_2(s)$	104.6	$K_2SO_4(s)$	176
$CO(NH_2)_2(aq)$	173.8	$N_2(g)$	191.5
$C_2H_2(g)$	200.8	$NH_3(g)$	192.5
$C_2H_4(g)$	219.8	$NH_4Cl(s)$	94.6
$C_2H_6(g)$	229.5	$NO(g)$	210.6
$C_8H_{18}(l)$	466.9	$NO_2(g)$	240.5
$C_2H_5OH(l)$	161	$N_2O(g)$	220.0
$Ca(s)$	154.8	$N_2O_4(g)$	304
$CaCO_3(s)$	92.9	$Na(s)$	51.0
$CaCl_2(s)$	114	$Na_2CO_3(s)$	136
$CaO(s)$	40	$NaHCO_3(s)$	102
$Ca(OH)_2(s)$	76.1	$NaCl(s)$	72.38
$CaSO_4(s)$	107	$NaOH(s)$	64.18
$CaSO_4 \cdot \frac{1}{2}H_2O(s)$	131	$Na_2SO_4(s)$	149.4
$CaSO_4 \cdot 2H_2O(s)$	194.0	$O_2(g)$	205.0
$Cl_2(g)$	223.0	$PbO(s)$	67.8
$Fe(s)$	27	$S(s)$	31.9
$Fe_2O_3(s)$	90.0	$SO_2(g)$	248.5
$H_2(g)$	130.6	$SO_3(g)$	256.2

Once we have the entropies of a variety of substances, we can calculate the **standard entropy change, $\Delta S°$**, for chemical reactions in much the same way as we calculated $\Delta H°$ in Chapter 6.

$$\Delta S° = (\text{sum of } S° \text{ of the products}) - (\text{sum of } S° \text{ of the reactants}) \quad (18.6)$$

If the reaction we are working with happens to correspond to the formation of 1 mol of a compound from its elements, then the $\Delta S°$ that we calculate can be referred to as the **standard entropy of formation, $\Delta S°_f$.** Values of $\Delta S°_f$, are not tabulated, however; if we need them for some purpose, we must calculate them from tabulated values of $S°$.

Tools

Standard entropies

This is simply a Hess's law type of calculation. Note, however, that elements have nonzero $S°$ values, which must be included in the bookkeeping.

Facets of Chemistry 18.2

Why the Units of Entropy Are Energy/Temperature

Entropy is a state function, just like enthalpy, so the value of ΔS doesn't depend on the "path" that is followed during a change. In other words, we can proceed from one state to another in any way we like and ΔS will be the same. If we choose a *reversible path* in which just a slight alteration in the system can change the direction of the process, we can measure ΔS directly. For example, at 25 °C an ice cube will melt and there is nothing we can do at that temperature to stop it. This change is *nonreversible* in the sense described above, so we couldn't use this path to measure ΔS. At 0 °C, however, it is simple to stop the melting process and reverse its direction. At 0 °C the melting of ice is a reversible process.

If we set up a change so that it is reversible, we can calculate the entropy change as $\Delta S = q/T$, where q is the heat added to the substance and T is the temperature at which the heat is added. We can understand this in the following way. In a system of molecules held together by attractive forces, disorder is introduced by adding heat,

which makes the molecules move more violently. In a crystal, for instance, the molecules vibrate about their equilibrium lattice positions, so that at any instant we really have a "not-quite-ordered" collection of particles. Adding heat increases molecular motion and leads to greater instantaneous disorder. In other words, as illustrated in Figure 18.5, if we could freeze the motion of the molecules the way a photographer's flash freezes action, we would see that after we have added some heat there is greater disorder and therefore greater entropy. This entropy increase is *directly proportional to the amount of heat added.* However, the added heat makes a more noticeable and significant difference if it is added at low temperature, where little disorder exists, rather than at high temperature where there is already substantial disorder. For a given quantity of heat the entropy change is therefore *inversely proportional to the temperature* at which the heat is added. Thus, $\Delta S = q/T$, and entropy has units of energy divided by temperature (e.g., $J\ K^{-1}$).

EXAMPLE 18.3

Calculating $\Delta S°$ from Standard Entropies

Substance	$S°$ (J/mol K)
$CO(NH_2)_2(aq)$	173.8
$H_2O(l)$	69.96
$CO_2(g)$	213.6
$NH_3(g)$	192.5

Notice that the unit *mol* cancels in each term, so the units of $\Delta S°$ are joules per kelvin.

Urea (a compound found in urine) is manufactured commercially from CO_2 and NH_3. One of its uses is as a fertilizer where it reacts slowly with water in the soil to produce ammonia and carbon dioxide. The ammonia provides a source of nitrogen for growing plants.

$$CO(NH_2)_2(aq) + H_2O(l) \longrightarrow CO_2(g) + 2NH_3(g)$$
$$\text{urea}$$

What is the standard entropy change when one mole of urea reacts with water?

Solution: The standard entropies of the reactants and products are found in Table 18.1. Let's first collect the data and assemble them as we have done in the margin.

Applying Equation 18.6, we have

$$\Delta S = [S°_{CO_2(g)} + 2S°_{NH_3(g)}] - [S°_{CO(NH_2)_2(aq)} + S°_{H_2O(l)}]$$

$$= \left[1\ mol \times \left(\frac{213.6\ J}{mol\ K}\right) + 2\ mol \times \left(\frac{192.5\ J}{mol\ K}\right)\right]$$

$$- \left[1\ mol \times \left(\frac{173.8\ J}{mol\ K}\right) + 1\ mol \times \left(\frac{69.96\ J}{mol\ K}\right)\right]$$

$$= (598.6\ J/K) - (243.8\ J/K)$$

$$= 354.8\ J/K$$

Thus, the standard entropy change for this reaction is +354.8 J/K (which we can also write as +354.8 $J\ K^{-1}$).

Is the Answer Reasonable?
In the reaction, gases are formed from liquid reactants. Since gases have much larger entropies than liquids, we expect $\Delta S°$ to be positive, which agrees with our answer.

Practice Exercise 7

Calculate the standard entropy change, $\Delta S°$, in $J\,K^{-1}$ for the following:

(a) $CaO(s) + 2HCl(g) \longrightarrow CaCl_2(s) + H_2O(l)$
(b) $C_2H_4(g) + H_2(g) \longrightarrow C_2H_6(g)$ ◆

18.6 Standard Free Energy Changes

When ΔG is determined at 25 °C (298 K) and 1 atm, we call it the **standard free energy change, $\Delta G°$**. There are several ways of obtaining $\Delta G°$ for a reaction. One of them is to compute $\Delta G°$ from $\Delta H°$ and $\Delta S°$.

$$\Delta G° = \Delta H° - (298\ \text{K})\Delta S°$$

Experimental measurement of $\Delta G°$ is also possible, but we will discuss how this is done later.

← **Tools**

Calculating
$\Delta G°$ from $\Delta H°$
and $\Delta S°$

Calculate $\Delta G°$ for the reaction of urea with water from values of $\Delta H°$ and $\Delta S°$.

$$CO(NH_2)_2(aq) + H_2O(l) \longrightarrow CO_2(g) + 2NH_3(g)$$

Analysis: We can calculate $\Delta G°$ with the equation

$$\Delta G° = \Delta H° - T\Delta S°$$

The data needed to calculate $\Delta H°$ come from Table 6.2 and require a Hess's law calculation. To obtain $\Delta S°$, we normally would need to do a similar calculation with data from Table 18.1. However, we already performed this calculation in Example 18.3.

Solution: First we calculate $\Delta H°$ from data in Table 6.2.

$$\Delta H° = [\Delta H°_{f\,CO_2(g)} + 2\Delta H°_{f\,NH_3(g)}] - [\Delta H°_{f\,CO(NH_2)_2(aq)} + \Delta H°_{f\,H_2O(l)}]$$

$$= \left[1\ \text{mol} \times \left(\frac{-393.5\ \text{kJ}}{\text{mol}}\right) + 2\ \text{mol} \times \left(\frac{-46.19\ \text{kJ}}{\text{mol}}\right)\right]$$

$$- \left[1\ \text{mol} \times \left(\frac{-319.2\ \text{kJ}}{\text{mol}}\right) + 1\ \text{mol} \times \left(\frac{-285.9\ \text{kJ}}{\text{mol}}\right)\right]$$

$$= (-485.9\ \text{kJ}) - (-605.1\ \text{kJ})$$

$$= +119.2\ \text{kJ}$$

In Example 18.3 we found $\Delta S°$ to be $+354.8\ J\,K^{-1}$. To calculate $\Delta G°$ we also need the Kelvin temperature, which we will express to at least four significant figures to match the number of significant figures in $\Delta S°$. Since standard temperature is *exactly* 25 °C, $T_K = (25.00 + 273.15)\ \text{K} = 298.15\ \text{K}$. Also, we must be careful to express $\Delta H°$ and $T\Delta S°$ in the same energy units, so we'll change the units of the

EXAMPLE 18.4

Calculating $\Delta G°$ from $\Delta H°$ and $\Delta S°$

Sometimes, the temperature is specified as a subscript in writing the symbol for the standard free energy change. For example, $\Delta G°$ can also be written $\Delta G°_{298}$. As you will see later, there are times when it is desirable to indicate the temperature explicitly.

To be exact:

$$T_K = t_C + 273.15$$

354.8 J K^{-1} = 0.3548 kJ K^{-1}

entropy change to give $\Delta S° = +0.3548$ kJ K^{-1}. Substituting into the equation for $\Delta G°$,

$$\Delta G° = +119.2 \text{ kJ} - (298.15 \text{ K})(0.3548 \text{ kJ K}^{-1})$$

$$= +119.2 \text{ kJ} - 105.8 \text{ kJ}$$

$$= +13.4 \text{ kJ}$$

Therefore, for this reaction, $\Delta G° = +13.4$ kJ.

Is the Answer Reasonable?

There's not much we can do to check the reasonableness of the answer. We just need to check to be sure we've done the calculations correctly.

Practice Exercise 8

Use the data in Table 6.2 and Table 18.1 to calculate $\Delta G°$ for the formation of iron(III) oxide (the iron oxide in rust). The equation for the reaction is

$$4\text{Fe}(s) + 3\text{O}_2(g) \longrightarrow 2\text{Fe}_2\text{O}_3(s) \quad \blacklozenge$$

In Section 6.8 you learned that it is useful to have tabulated standard heats of formation, $\Delta H_f°$, because they can be used with Hess's law to calculate $\Delta H°$ for many different reactions. **Standard free energies of formation, $\Delta G_f°$,** can be used in similar calculations to obtain $\Delta G°$.

Using $\Delta G_f°$ values

$$\Delta G° = (\text{sum of } \Delta G_f° \text{ of products}) - (\text{sum of } \Delta G_f° \text{ of reactants}) \quad (18.7)$$

The $\Delta G_f°$ values for some typical substances are found in Table 18.2. Example 18.5 shows how we can use them to calculate $\Delta G°$ for a reaction.

EXAMPLE 18.5

Calculating $\Delta G°$ from $\Delta G_f°$

Ethanol is blended with gasoline to produce a clean-burning fuel for automobiles.

Ethanol, C_2H_5OH (also called ethyl alcohol), is made from grain by fermentation and has been used as an additive to gasoline to produce a product called gasohol. What is $\Delta G°$ for the combustion of liquid ethanol to give $CO_2(g)$ and $H_2O(g)$?

Solution: First, we need the balanced equation for the reaction.

$$C_2H_5OH(l) + 3O_2(g) \longrightarrow 2CO_2(g) + 3H_2O(g)$$

Now we use Equation 18.7.

$$\Delta G° = [2\Delta G_f°{}_{CO_2(g)} + 3\Delta G_f°{}_{H_2O(g)}] - [\Delta G_f°{}_{C_2H_5OH(l)} + 3\Delta G_f°{}_{O_2(g)}]$$

As with $\Delta H_f°$, the $\Delta G_f°$ for any element in its standard state is zero. Therefore, using the data from Table 18.2,

$$\Delta G° = \left[2 \text{ mol} \times \left(\frac{-394.4 \text{ kJ}}{\text{mol}}\right) + 3 \text{ mol} \times \left(\frac{-228.6 \text{ kJ}}{\text{mol}}\right)\right]$$

$$- \left[1 \text{ mol} \times \left(\frac{-174.8 \text{ kJ}}{\text{mol}}\right) + 3 \text{ mol} \times \left(\frac{0 \text{ kJ}}{\text{mol}}\right)\right]$$

$$= (-1474.6 \text{ kJ}) - (-174.8 \text{ kJ})$$

$$= -1299.8 \text{ kJ}$$

The standard free energy change for the reaction equals -1299.8 kJ.

Checking the Answer for Reasonableness

As before, there's no easy way to estimate the answer. To check your answer, be sure you've got the correct algebraic sign for each term.

Practice Exercise 9

Calculate $\Delta G°$ in kilojoules for the following reactions using the data in Table 18.2.

(a) $2NO(g) + O_2(g) \longrightarrow 2NO_2(g)$
(b) $Ca(OH)_2(s) + 2HCl(g) \longrightarrow CaCl_2(s) + 2H_2O(g)$ ◆

Table 18.2 Standard Free Energies of Formation
of Typical Substances (25 °C, 1 atm)

Substance	$\Delta G_f^°$ (kJ mol^{-1})	Substance	$\Delta G_f^°$ (kJ mol^{-1})
Ag(s)	0	$H_2O(g)$	−228.6
AgCl(s)	−109.7	$H_2O(l)$	−237.2
Al(s)	0	HCl(g)	−95.27
$Al_2O_3(s)$	−1576.4	$HNO_3(l)$	−79.91
C(s, graphite)	0	$H_2SO_4(l)$	−689.9
CO(g)	−137.3	$HC_2H_3O_2(l)$	−392.5
$CO_2(g)$	−394.4	Hg(l)	0
$CH_4(g)$	−50.79	Hg(g)	+31.8
$CH_3Cl(g)$	−58.6	K(s)	0
$CH_3OH(l)$	−166.2	KCl(s)	−408.3
$CO(NH_2)_2(s)$	−197.2	$K_2SO_4(s)$	−1316.4
$CO(NH_2)_2(aq)$	−203.8	$N_2(g)$	0
$C_2H_2(g)$	+209	$NH_3(g)$	−16.7
$C_2H_4(g)$	+68.12	$NH_4Cl(s)$	−203.9
$C_2H_6(g)$	−32.9	NO(g)	+86.69
$C_8H_{18}(l)$	+17.3	$NO_2(g)$	+51.84
$C_2H_5OH(l)$	−174.8	$N_2O(g)$	+103.6
Ca(s)	0	$N_2O_4(g)$	+98.28
$CaCO_3(s)$	−1128.8	Na(s)	0
$CaCl_2(s)$	−750.2	$Na_2CO_3(s)$	−1048
CaO(s)	−604.2	$NaHCO_3(s)$	−851.9
$Ca(OH)_2(s)$	−896.76	NaCl(s)	−384.0
$CaSO_4(s)$	−1320.3	NaOH(s)	−382
$CaSO_4 \cdot \frac{1}{2}H_2O(s)$	−1435.2	$Na_2SO_4(s)$	−1266.8
$CaSO_4 \cdot 2H_2O(s)$	−1795.7	$O_2(g)$	0
$Cl_2(g)$	0	PbO(s)	−189.3
Fe(s)	0	S(s)	0
$Fe_2O_3(s)$	−741.0	$SO_2(g)$	−300.4
$H_2(g)$	0	$SO_3(g)$	−370.4

18.7 Free Energy and Maximum Work

In our society, one of the chief uses of spontaneous chemical reactions is the production of useful work. For example, fuels are burned in gasoline or diesel engines to power automobiles and heavy machinery, and chemical reactions in batteries start our autos and run all sorts of modern electronic gadgets, including cellular phones, beepers, and laptop computers.

When chemical reactions occur, however, their energy is not always harnessed to do work. For instance, if gasoline is burned in an open dish, the energy evolved is lost entirely as heat and no useful work is accomplished. Engineers, therefore, seek ways to capture as much energy as possible in the form of work. One of their primary goals is to maximize the efficiency with which chemical energy is converted to work and to minimize the amount of energy transferred unproductively to the environment as heat.

Scientists have discovered that the maximum conversion of chemical energy to work occurs if a reaction is carried out under conditions that are said to be reversible. A process is defined as **reversible** if its driving force is opposed by another force that is just the slightest bit weaker, so that the slightest increase in the opposing force will cause the direction of the change to be reversed. An example of a nearly reversible process is illustrated in Figure 18.10, where we have a compressed gas in a cylinder pushing against a piston that's held in place by liquid water above it. If a water molecule evaporates, the external pressure drops slightly and the gas can expand just a bit. Gradually, as one water molecule after another evaporates, the gas inside the cylinder slowly expands. At any time, however, the process can be reversed by the condensation of a water molecule.

Although we could obtain the maximum work by carrying out a change reversibly, a reversible process requires so many steps that it proceeds at an extremely slow speed. If the work cannot be done at a reasonable rate, it is of little value to us. Our goal, then, is to approach reversibility for maximum efficiency, but to carry out the change at a pace that will deliver work at acceptable rates.

A reversible process requires an infinite number of tiny steps. This takes forever to accomplish.

The relationship of useful work to reversibility was illustrated earlier (Section 6.4) in our discussion of the discharge of an automobile battery. Recall that when the battery is shorted with a heavy wrench, no work is done and all the en-

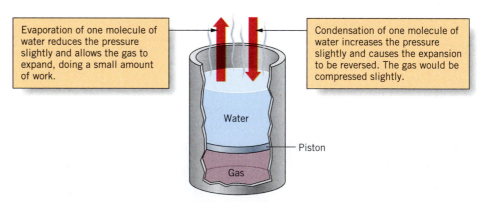

Figure 18.10 *The reversible expansion of a gas.* As water molecules in the liquid above the piston evaporate one at a time, the external pressure gradually decreases and the gas slowly expands. The expansion would be reversed if a molecule of water were to condense into the liquid. The ability of the expansion to be reversed by the slightest increase in the opposing pressure is what makes this a reversible process.

Therefore,

$$\Delta H = T\Delta S$$

and

$$\Delta S = \frac{\Delta H}{T} \qquad (18.8)$$

Thus, if we know ΔH for the phase change and the temperature at which the two phases coexist, we can calculate ΔS for the phase change. Another interesting relationship that we can obtain is

$$T = \frac{\Delta H}{\Delta S} \qquad (18.9)$$

Thus, if we know ΔH and ΔS, we can calculate the temperature at which equilibrium will occur.

EXAMPLE 18.7

Calculating the Equilibrium Temperature for a Phase Change

For the phase change $Br_2(l) \rightarrow Br_2(g)$, $\Delta H° = +31.0$ kJ mol^{-1} and $\Delta S° = 92.9$ J mol^{-1} K^{-1}. Assuming that ΔH and ΔS are nearly temperature independent, calculate the approximate Celsius temperature at which $Br_2(l)$ will be in equilibrium with $Br_2(g)$ at 1 atm (i.e., the normal boiling point of liquid Br_2).

Solution: The temperature at which equilibrium exists is given by Equation 18.9,

$$T = \frac{\Delta H}{\Delta S}$$

If ΔH and ΔS do not depend on temperature, then we can use $\Delta H°$ and $\Delta S°$ in this equation. That is,

$$T = \frac{\Delta H°}{\Delta S°}$$

Substituting the data given in the problem,

$$T = \frac{3.10 \times 10^4 \text{ J mol}^{-1}}{92.9 \text{ J mol}^{-1} \text{ K}^{-1}}$$

$$= 334 \text{ K}$$

The Celsius temperature is $334 - 273 = 61 °C$. Notice that we were careful to express $\Delta H°$ in joules, not kilojoules, so the units would cancel correctly. It is also interesting that the boiling point we calculated is quite close to the measured normal boiling point of 58.8 °C.

ΔH and ΔS do not change much with changes in temperature. This is because temperature changes affect the enthalpies and entropies of both the reactants and products by about the same amount, so the differences between reactants and products stay fairly constant.

Is the Answer Reasonable?
The $\Delta H°$ value equals 31,000 and the $\Delta S°$ value equals approximately 100, which means the temperature should be about 310 K. Our value, 334 K, is not far from that, so the answer is reasonable.

Practice Exercise 11

The heat of vaporization of mercury is 60.7 kJ/mol. For $Hg(l)$, $S° = 76.1$ J mol^{-1} K^{-1} and for $Hg(g)$, $S° = 175$ J mol^{-1} K^{-1}. Estimate the normal boiling point of liquid mercury. ◆

Free Energy Diagram for a Phase Change

We have said that in a phase change such as $H_2O(l) \rightarrow H_2O(s)$, equilibrium can exist for a given pressure only at one particular temperature; for water at a pressure of 1 atm, this temperature is 0 °C, the freezing point. At other temperatures, the phase change proceeds entirely to completion in one direction or another. One way to gain a better understanding of this is by studying **free energy diagrams,** which depict how the free energy changes as we proceed from the "reactants" to the "products."

Figure 18.11 illustrates three different free energy diagrams for water–ice mixtures. On the left of each diagram is indicated the free energy of pure liquid water, and on the right the free energy of the pure solid. Points along the horizontal axis represent mixtures of both phases. Thus, going from left to right across a graph we are able to see how the free energy varies as a system that consists entirely of $H_2O(l)$ changes ultimately to one that consists entirely of $H_2O(s)$.

Below 0 °C, we see that the free energy of the liquid is higher than that of the solid. You've learned that a spontaneous change will occur if the free energy can decrease. Therefore, the first diagram tells us that if we start with the liquid phase, or any mixture of liquid or solid, freezing will occur until only the solid is present. This is because the free energy decreases continually until all the liquid has frozen.

Above 0 °C, we have the opposite situation. The free energy decreases in the direction of $H_2O(s) \rightarrow H_2O(l)$, and it continues to drop until all the solid has melted. This means that if we have ice, or any mixture of ice and liquid water, it will continue to melt until only the liquid is present.

Above or below 0 °C the system is unable to establish an equilibrium mixture of liquid and solid. Melting or freezing occurs until only one phase is present. However, at 0 °C there is no change in free energy if either melting or freezing occurs, so there is no driving force for either change. Therefore, as long as a system of ice and liquid water is insulated from warmer or colder surroundings, any particular mixture of the two phases is stable and a state of equilibrium exists.

Free Energy Diagrams for Chemical Reactions

The free energy changes that occur in most chemical reactions are more complex than those in phase changes. As an example, let's study a reaction you've seen before — the decomposition of N_2O_4 into NO_2.

$$N_2O_4(g) \longrightarrow 2NO_2(g)$$

Figure 18.11 *Free energy diagrams for the conversion of $H_2O(l)$ to $H_2O(s)$. At the left of each diagram, the system consists entirely of $H_2O(l)$. At the right is $H_2O(s)$. The horizontal axis represents the extent of conversion from $H_2O(l)$ to $H_2O(s)$.*

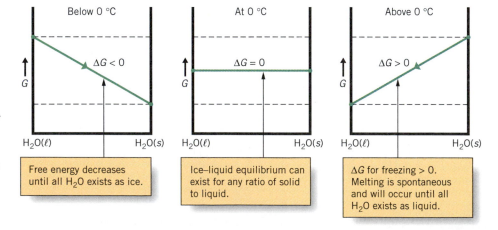

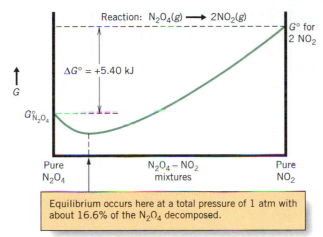

Figure 18.12 *Free energy diagram for the decomposition of $N_2O_4(g)$.* The minimum on the curve indicates the composition of the reaction mixture at equilibrium. Because $\Delta G°$ is positive, the position of equilibrium lies close to the reactant. Not much product will form by the time the system reaches equilibrium.

In our discussion of chemical equilibrium in Chapter 14 (page 629), we noted that equilibrium in this system can be approached from *either* direction, with the same equilibrium concentrations being achieved provided we begin with the same overall system composition.

Figure 18.12 shows the free energy diagram for the reaction. Notice that in going from reactant to product, the free energy has a minimum. It drops below that of either pure N_2O_4 or pure NO_2.

Any system will spontaneously seek the lowest point on its free energy curve. If we begin with pure $N_2O_4(g)$, the reaction will proceed from left to right and some $NO_2(g)$ will be formed, because proceeding in the direction of NO_2 leads to a lowering of the free energy. If we begin with pure $NO_2(g)$ a change also will occur. Going downhill on the free energy curve now takes place as the reverse reaction occurs [i.e., $2NO_2(g) \rightarrow N_2O_4(g)$]. Once the bottom of the "valley" is reached, the system has come to equilibrium. As you learned in Chapter 14, if the system isn't disturbed, the composition of the equilibrium mixture will remain constant. Now we see that the reason is because any change (moving either to the left or right) would require an uphill climb. Free energy increases are not spontaneous, so this doesn't happen.

An important thing to notice in Figure 18.12 is that some reaction takes place spontaneously in the forward direction even though $\Delta G°$ is positive. However, the reaction doesn't proceed far before equilibrium is reached. For comparison, Figure 18.13 shows the shape of the free energy curve for a reaction with a negative $\Delta G°$. We see here that at equilibrium there has been a much greater conversion of reactants to products. Thus, $\Delta G°$ *tells us where the position of equilibrium lies between pure reactants and pure products.* When $\Delta G°$ is positive, the position of equilibrium lies close to the reactants and little reaction occurs by the time equilibrium is reached. When $\Delta G°$ is negative, the position of equilibrium lies close to the products and a large amount of products will have formed by the time equilibrium is reached.

When a system moves "downhill" on its free energy curve, $G_{final} < G_{initial}$ and ΔG is negative. Changes with negative ΔG are spontaneous.

Tools

Using $\Delta G°$ to determine the position of equilibrium

Predicting the Outcome of a Chemical Reaction

In general, the value of $\Delta G°$ for most reactions is much larger numerically than the $\Delta G°$ for the $N_2O_4 \longrightarrow NO_2$ reaction. In addition, the extent to which a reaction proceeds is very sensitive to the size of $\Delta G°$. If the $\Delta G°$ value for a reaction is reasonably large—about 20 kJ or more—almost no observable reaction

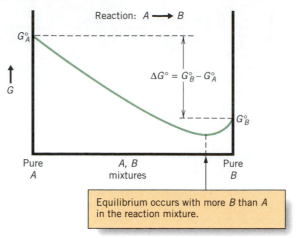

Reaction: $A \longrightarrow B$

$\Delta G° = G_B° - G_A°$

Pure A   A, B mixtures   Pure B

Equilibrium occurs with more *B* than *A* in the reaction mixture.

Figure 18.13 *Free energy curve for a reaction having a negative* $\Delta G°$. *Because* $G_B°$ *is less than* $G_A°$, $\Delta G°$ *is negative. This causes the position of equilibrium to lie far to the right, near the products. When the system reaches equilibrium, there will be a large amount of products present and little reactants.*

will occur when $\Delta G°$ is positive. On the other hand, the reaction will go almost to completion if $\Delta G°$ is both large and negative.[3] From a practical standpoint, then, *the size and sign of* $\Delta G°$ *serve as indicators of whether an observable spontaneous reaction will occur.*

EXAMPLE 18.8

Using $\Delta G°$ as a Predictor of the Outcome of a Reaction

Would we expect to be able to observe the following reaction at 25 °C?

$$NH_4Cl(s) \longrightarrow NH_3(g) + HCl(g)$$

Analysis: We will need to determine the magnitude and sign of $\Delta G°$ for the reaction. If $\Delta G°$ is reasonably large and positive, the reaction won't be observed. If it is reasonably large and negative, we can expect to see the reaction go nearly to completion.

Solution: First let's calculate $\Delta G°$ for the reaction using the data in Table 18.2. The procedure is the same as that discussed earlier.

$$\Delta G° = [\Delta G°_{f\,NH_3(g)} + \Delta G°_{f\,HCl(g)}] - [\Delta G°_{f\,NH_4Cl(s)}]$$

$$= [(-16.7 \text{ kJ}) + (-95.27 \text{ kJ})] - [-203.9 \text{ kJ}]$$

$$= +91.9 \text{ kJ}$$

Because $\Delta G°$ is large and positive, we only expect to observe the spontaneous formation of small amounts of products at this temperature.

Checking the Answer for Reasonableness
Be sure to check the algebraic signs of each term in the calculation.

Practice Exercise 12

Use the data in Table 18.2 to determine whether the reaction

$$2SO_2(g) + O_2(g) \longrightarrow 2SO_3(g)$$

should "occur spontaneously" at 25 °C. ◆

---

[3]To actually see a change take place, the speed of a spontaneous reaction must be reasonably fast. For example, the decomposition of the nitrogen oxides into N_2 and O_2 is thermodynamically spontaneous ($\Delta G°$ is negative), but their rates of decomposition are so slow that these substances appear to be stable and some are obnoxious air pollutants.

Practice Exercise 13

Use the data in Table 18.2 to determine whether we should expect to see the formation of $CaCO_3(s)$ in the following reaction at 25 °C.

$$CaCl_2(s) + H_2O(g) + CO_2(g) \longrightarrow CaCO_3(s) + 2HCl(g) \quad \blacklozenge$$

Equilibrium and Temperature

So far, we have confined our discussion of the relationship of free energy and equilibrium to a special case, 25 °C. But what about other temperatures? Equilibria certainly can exist at temperatures other than 25 °C, and in Chapter 14 you learned how to apply Le Châtelier's principle to predicting the way temperature affects the position of equilibrium. Now let's see how thermodynamics deals with this.

You've learned that at 25 °C the position of equilibrium is determined by the difference between the free energy of pure products and the free energy of pure reactants. This difference is given by ΔG°_{298}, where we have now used the subscript "298" to indicate the temperature, 298 K. ΔG°_{298} is defined as

$$\Delta G^{\circ}_{298} = (G^{\circ}_{products})_{298} - (G^{\circ}_{reactants})_{298}$$

At temperatures other than 25 °C, it is still the difference between the free energies of the products and reactants that determines the position of equilibrium. We might write this as ΔG°_T. Thus, at a temperature other than 25 °C (298 K), we have

$$\Delta G^{\circ}_T = (G^{\circ}_{products})_T - (G^{\circ}_{reactants})_T$$

where $(G^{\circ}_{products})_T$ and $(G^{\circ}_{reactants})_T$ are the total free energies of the pure products and reactants, respectively, at this other temperature.

Next, we must find a way to compute ΔG°_T. Earlier we saw that ΔG° can be obtained from the equation

$$\Delta G^{\circ} = \Delta H^{\circ} - (298 \text{ K})\Delta S^{\circ}$$

At a different temperature, T, the equation becomes

$$\Delta G^{\circ}_T = \Delta H^{\circ}_T - T\Delta S^{\circ}_T$$

Now we seem to be getting closer to our goal. If we can compute or estimate the values of ΔH°_T and ΔS°_T for a reaction, we have solved our problem.

The size of ΔG°_T obviously depends very strongly on the temperature—the equation above has temperature as one of its variables. However, the magnitudes of the ΔH and ΔS for a reaction are relatively insensitive to the temperature. This is because the enthalpies and entropies of *both* the reactants and products increase about equally with increasing temperature, so their differences, ΔH and ΔS, remain nearly the same. As a result, we can use ΔH°_{298} and ΔS°_{298} as reasonable approximations of ΔH°_T and ΔS°_T. This allows us to rewrite the equation for ΔG°_T as

$$\Delta G^{\circ}_T \approx \Delta H^{\circ}_{298} - T\Delta S^{\circ}_{298} \qquad (18.10)$$

The following examples illustrate how this equation is useful.

The magnitudes of ΔH and ΔS are relatively insensitive to temperature changes. However, ΔG is very temperature sensitive because

$$\Delta G = \Delta H - T\Delta S.$$

Tools

Calculating ΔG° at temperatures other than 25 °C

EXAMPLE 18.9

ΔG° at Temperatures Other than 25 °C

Earlier we saw that at 25 °C the ΔG° for the reaction

$$N_2O_4(g) \longrightarrow 2NO_2(g)$$

has a value of +5.40 kJ. What is the approximate value of ΔG°_T for this reaction at 100 °C?

Solution: To make use of Equation 18.10, we need values of $\Delta H°$ and $\Delta S°$. The $\Delta H°$ for the reaction can be calculated from the data in Table 6.2 on p. 260. There we find the following standard heats of formation.

$$N_2O_4(g) \qquad \Delta H_f° = +9.67 \text{ kJ/mol}$$

$$NO_2(g) \qquad \Delta H_f° = +33.8 \text{ kJ/mol}$$

We combine these by a Hess's law calculation to compute $\Delta H°$ for the reaction.

$$\Delta H° = [2\Delta H_{f\,NO_2(g)}°] - [\Delta H_{f\,N_2O_4(g)}°]$$

$$= \left[2 \text{ mol} \times \left(\frac{33.8 \text{ kJ}}{\text{mol}}\right)\right] - \left[1 \text{ mol} \times \left(\frac{9.67 \text{ kJ}}{\text{mol}}\right)\right]$$

$$= +57.9 \text{ kJ}$$

Next, we compute $\Delta S°$ for the reaction using data from Table 18.1.

$$N_2O_4(g) \qquad S° = 304 \text{ J/mol K}$$

$$NO_2(g) \qquad S° = 240.5 \text{ J/mol K}$$

This is also a Hess's law kind of calculation.

$$\Delta S° = \left[2 \text{ mol} \times \left(\frac{240.5 \text{ J}}{\text{mol K}}\right)\right] - \left[1 \text{ mol} \times \left(\frac{304 \text{ J}}{\text{mol K}}\right)\right]$$

$$= +177 \text{ J K}^{-1} \text{ or } 0.177 \text{ kJ K}^{-1}$$

The temperature is 100 °C, which is 373 K, so we can call the free energy change $\Delta G_{373}°$. Substituting into Equation 18.10 using $T = 373$ K, we get

$$\Delta G_{373}° \approx (+57.9 \text{ kJ}) - (373 \text{ K})(0.177 \text{ kJ K}^{-1})$$

$$\approx -8.1 \text{ kJ}$$

Notice that at this higher temperature, the sign of $\Delta G_T°$ has become negative.

Checking the Answer

Once again, check to be sure the signs are correct for the various terms in the calculation.

EXAMPLE 18.10

Position of Equilibrium as a Function of Temperature

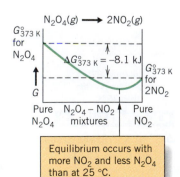

Equilibrium occurs with more NO_2 and less N_2O_4 than at 25 °C.

Using the results of Example 18.9, sketch a free energy diagram for the decomposition of N_2O_4 into NO_2 at 100 °C. For this reaction, how does the position of equilibrium change as the temperature is increased?

Solution: Since $\Delta G_T°$ is negative for the reaction at 100 °C, the $G°$ of the reactants must be higher than the $G°$ of the products. Therefore, when we draw the free energy curve, the minimum must lie closer to the products than to the reactants, as shown in the diagram in the margin. Comparing this diagram to the one in Figure 18.12, we see that at the higher temperature the minimum lies closer to the products. This means that at the higher temperature the decomposition will proceed further toward completion when equilibrium has been reached.

Is the Answer Reasonable?

The reaction is endothermic ($\Delta H° = +57.9$ kJ). Le Châtelier's principle tells us that the position of equilibrium in an endothermic reaction is shifted toward the products when the temperature is increased. This is in agreement with the conclusion we reached in this problem.

Practice Exercise 14

Use the data in Table 18.2 to determine ΔG_{298}° for the reaction

$$2NaHCO_3(s) \longrightarrow Na_2CO_3(s) + CO_2(g) + H_2O(g)$$

Then calculate the approximate value for ΔG° for the reaction at 200 °C using the data in Tables 6.2 and 18.1. How does the position of equilibrium for this reaction change as the temperature is increased? ◆

18.9 Calculating Equilibrium Constants from Thermodynamic Data

In the preceding discussion, you learned in a qualitative way that the position of equilibrium in a reaction is determined by the sign and magnitude of ΔG°. You also learned that the direction in which a reaction proceeds depends on where the system composition stands relative to the minimum on the free energy curve. Thus, the reaction will proceed spontaneously in the forward direction only if it will lead to a lowering of the free energy (i.e., if ΔG is negative).

Quantitatively, the relationship between ΔG (the predictor of spontaneity) and ΔG° is expressed by the following equation, which we will not attempt to justify.

$$\Delta G = \Delta G^{\circ} + RT \ln Q \qquad (18.11)$$

Here R is the gas constant in appropriate energy units (e.g., 8.314 J mol^{-1} K^{-1}), T is the Kelvin temperature, and $\ln Q$ is the natural logarithm of the reaction quotient. For gaseous reactions, Q is calculated using partial pressures expressed in atmospheres; for reactions in solution, Q is calculated from molar concentrations. Equation 18.11 allows us to predict the direction of the spontaneous change in a reaction mixture if we know ΔG° and the composition of the mixture, as illustrated in Example 18.11.

Tools

Relating the reaction quotient to ΔG

Recall that in Chapter 14 we defined Q as the *reaction quotient*—the numerical value of the mass action expression.

EXAMPLE 18.11

Determining the Direction of a Spontaneous Reaction

The reaction $2NO_2(g) \rightleftharpoons N_2O_4(g)$ has $\Delta G_{298}^{\circ} = -5.40$ kJ per mole of N_2O_4. In a reaction mixture, the partial pressure of NO_2 is 0.25 atm and the partial pressure of N_2O_4 is 0.60 atm. In which direction must this reaction proceed to reach equilibrium?

In Chapter 14, you learned that you can also predict the direction of a reaction by comparing Q with K (page 652).

Analysis: Since we know that reactions proceed spontaneously *toward* equilibrium, we are really being asked to determine whether the reaction will proceed spontaneously in the forward or reverse direction. We can use Equation 18.11 to calculate ΔG for the forward reaction. If ΔG is negative, then the forward reaction is spontaneous. However, if the calculated ΔG is positive, the forward reaction is nonspontaneous and it is really the reverse reaction that is spontaneous.

Solution: First, we need the correct form for the mass action expression so we can calculate Q correctly. Expressed in terms of partial pressures, the mass action expression is

$$\frac{P_{N_2O_4}}{P_{NO_2}^2}$$

Therefore, the equation we will use is

$$\Delta G = \Delta G^\circ + RT \ln \left(\frac{P_{N_2O_4}}{P_{NO_2}^2} \right)$$

Next, let's assemble the data:

$$\Delta G^\circ_{298} = -5.40 \text{ kJ mol}^{-1} \qquad\qquad T = 298 \text{ K}$$
$$= -5.40 \times 10^3 \text{ J mol}^{-1} \qquad\qquad P_{N_2O_4} = 0.60 \text{ atm}$$
$$R = 8.314 \text{ J mol}^{-1} \text{ K}^{-1} \qquad\qquad P_{NO_2} = 0.25 \text{ atm}$$

Notice we have changed the energy units of ΔG° to joules so they will be compatible with those calculated using R. Substituting quantities gives[4]

$$\Delta G = -5.40 \times 10^3 \text{ J mol}^{-1} + (8.314 \text{ J mol}^{-1} \text{ K}^{-1})(298 \text{ K}) \ln \left[\frac{0.60 \text{ atm}}{(0.25 \text{ atm})^2} \right]$$

$$= -5.40 \times 10^3 \text{ J mol}^{-1} + (8.314 \text{ J mol}^{-1} \text{ K}^{-1})(298 \text{ K})(2.26)$$

$$= -5.40 \times 10^3 \text{ J mol}^{-1} + 5.60 \times 10^3 \text{ J mol}^{-1}$$

$$= +2.0 \times 10^2 \text{ J mol}^{-1}$$

Since ΔG is positive, the forward reaction is nonspontaneous. The reverse reaction is the one that will occur, and to reach equilibrium, some N_2O_4 will have to decompose.

Is the Answer Reasonable?
There is no simple check for reasonableness in this example. However, we can check to be sure the energy units in both terms on the right are the same. Notice that here we have changed the units for ΔG° to joules to match those of R. Also notice that the temperature is expressed in kelvins, to match the temperature units in R.

<hr>

Practice Exercise 15

In which direction will the reaction described in the preceding example proceed to reach equilibrium if the partial pressure of NO_2 is 0.60 atm and the partial pressure of N_2O_4 is 0.25 atm? ◆

<hr>

Thermodynamic Equilibrium Constants

Earlier you learned that when a system reaches equilibrium the free energy of the products equals the free energy of the reactants and ΔG equals zero. We also know that at equilibrium the reaction quotient equals the equilibrium constant.

$$\text{At equilibrium} \qquad \begin{cases} \Delta G = 0 \\ Q = K \end{cases}$$

If we substitute these into Equation 18.11, we obtain

$$0 = \Delta G^\circ + RT \ln K$$

which can be rearranged to give

Tools

Determining
thermodynamic
equilibrium
constants

$$\Delta G^\circ = -RT \ln K \tag{18.12}$$

[4]As we noted in the footnote on page 588, when taking the logarithm of a quantity, the number of digits *after the decimal point* should equal the number of significant figures in the quantity. Since 0.60 atm and 0.25 atm both have two significant figures, the logarithm of the quantity in square brackets (2.26) is rounded to give two digits after the decimal point.

The equilibrium constant K calculated from this equation is often called the **thermodynamic equilibrium constant** and corresponds to K_p for reactions involving gases (with partial pressures expressed in atmospheres) and to K_c for reactions in solution (with concentrations expressed in mol L^{-1}).

Equation 18.12 is useful because it permits us to determine equilibrium constants from either measured or calculated values of $\Delta G°$. As you know, $\Delta G°$ can be determined by a Hess's law type of calculation from tabulated values of $\Delta G_f°$. It also allows us to obtain values of $\Delta G°$ from measured equilibrium constants.

EXAMPLE 18.12

Thermodynamic Equilibrium Constants

The brownish haze associated with air pollution is caused by nitrogen dioxide, NO_2, a red-brown gas. Nitric oxide, NO, is formed in auto engines, and some of it escapes into the air where it is oxidized to NO_2 by oxygen.

$$2NO(g) + O_2(g) \rightleftharpoons 2NO_2(g)$$

The value of K_p for this reaction is 1.7×10^{12} at 25.00 °C. What is $\Delta G°$ for the reaction, expressed in joules per mole? In kilojoules per mole?

Solution: We must substitute into the equation

$$\Delta G° = -RT \ln K_p$$

First, let's list the data. (For the temperature, we will use the equation $T_K = t_C + 273.15$ to be sure we have enough significant figures.)

$$R = 8.314 \text{ J mol}^{-1} \text{ K}^{-1}$$

$$T = 298.15 \text{ K}$$

$$K_p = 1.7 \times 10^{12}$$

Substituting values,

$$\Delta G° = -(8.314 \text{ J mol}^{-1} \text{ K}^{-1} \times 298.15 \text{ K}) \ln (1.7 \times 10^{12})$$

$$= -(8.314 \text{ J mol}^{-1} \text{ K}^{-1} \times 298.15 \text{ K}) \times (28.16)$$

$$= -6.980 \times 10^4 \text{ J mol}^{-1} \text{ (to four significant figures)}$$

Expressed in kilojoules, $\Delta G° = -69.80 \text{ kJ mol}^{-1}$.

Is the Answer Reasonable?

The value of K_p tells us that the position of equilibrium lies far to the right, which means $\Delta G°$ must be large and negative. Therefore, the answer, -69.80 kJ mol^{-1}, seems reasonable.

The units in this calculation give $\Delta G°$ on a per mole (mol^{-1}) basis. This reminds us that we are viewing the coefficients of the reactants and products as representing moles, rather than some other sized quantity.

EXAMPLE 18.13

Thermodynamic Equilibrium Constants

Sulfur dioxide, which is sometimes present in polluted air, reacts with oxygen when it passes over the catalyst in automobile catalytic converters. The product is the very acidic oxide SO_3.

$$2SO_2(g) + O_2(g) \rightleftharpoons 2SO_3(g)$$

For this reaction, $\Delta G° = -1.40 \times 10^2$ kJ mol^{-1} at 25 °C. What is the value of K_p?

Solution: Once again we use the equation

$$\Delta G° = -RT \ln K_p$$

To make the units cancel correctly, we have added the unit mol^{-1} to the value of $\Delta G°$. This just emphasizes that the amounts of reactants and products are specified in mole-sized quantities.

Our data are $R = 8.314$ J mol^{-1} K^{-1}, $T = 298$ K, and

$$\Delta G° = -1.40 \times 10^2 \text{ kJ mol}^{-1}$$

$$= -1.40 \times 10^5 \text{ J mol}^{-1}$$

To calculate K_p, let's first solve for $\ln K_p$.

$$\ln K_p = \frac{-\Delta G°}{RT}$$

Substituting values gives

$$\ln K_p = \frac{-(-1.40 \times 10^5 \text{ J mol}^{-1})}{(8.314 \text{ J mol}^{-1} \text{ K}^{-1})(298 \text{ K})}$$

$$= +56.5$$

To calculate K_p, we take the antilogarithm,

$$K_p = e^{56.5}$$

$$= 3 \times 10^{24}$$

Notice that we have expressed the answer to only one significant figure. As discussed earlier, when taking a logarithm, the number of digits written after the decimal place equals the number of significant figures in the number. Conversely, the number of significant figures in the antilogarithm equals the number of digits after the decimal in the logarithm.

If necessary, review the discussion of logarithms and antilogarithms in Appendix A.

Is the Answer Reasonable?
The value of $\Delta G°$ is large and negative, so the position of equilibrium should favor the products. The large value of K_p is therefore reasonable.

_____ **Practice Exercise 16** _____

The reaction $N_2(g) + 3H_2(g) \rightleftharpoons 2NH_3(g)$ has $K_p = 6.9 \times 10^5$ at 25.0 °C. Calculate $\Delta G°$ for this reaction in units of kilojoules. ◆

_____ **Practice Exercise 17** _____

The reaction $H_2(g) + I_2(g) \rightleftharpoons 2HI(g)$ has $\Delta G° = +3.3$ kJ mol^{-1} at 25.0 °C. What is the value of K_p at this temperature? ◆

Thermodynamic Equilibrium Constants at Temperatures Other than 25 °C

Earlier we discussed the way temperature affects the position of equilibrium in a reaction. At 25 °C, the relative proportions of reactants and products at equilibrium are determined by $\Delta G°_{298}$, and we have just seen that we can calculate the value of the equilibrium constant at 25 °C from $\Delta G°_{298}$. As the temperature moves away from 25 °C, the position of equilibrium also changes because of changes in the value of $\Delta G°_T$. Therefore, $\Delta G°_T$ can be used to calculate K at temperatures other than 25 °C by the same methods used in Examples 18.12 and 18.13.

Nitrous oxide, N_2O, is an anesthetic known as laughing gas because it sometimes relieves patients of their inhibitions. The decomposition of nitrous oxide has $K_p = 1.8 \times 10^{36}$ at 25 °C. The equation is

$$2N_2O(g) \rightleftharpoons 2N_2(g) + O_2(g)$$

For this reaction, $\Delta H° = -163$ kJ and $\Delta S° = +148$ J K^{-1}. What is the approximate value for K_p for this reaction at 40 °C?

Analysis: To apply Equation 18.12, we need to have the value of $\Delta G°$ at 40 °C (313 K) which we can represent as $\Delta G°_{313}$. We can estimate $\Delta G°_{313}$ using the values of $\Delta H°$ and $\Delta S°$ measured at 25 °C.

$$\Delta G°_{313} \approx \Delta H°_{298} - (313 \text{ K})\Delta S°_{298}$$

Recall that this is permitted because $\Delta H°$ and $\Delta S°$ do not change much with temperature.

Substituting the values of $\Delta H°$ and $\Delta S°$ provided in the problem gives

$$\Delta G°_{313} \approx -1.63 \times 10^5 \text{ J} - (313 \text{ K})(+148 \text{ J K}^{-1})$$

Notice that we've converted kilojoules to joules. Performing the arithmetic gives

$$\Delta G°_{313} \approx -2.09 \times 10^5 \text{ J (rounded)}$$

The next step is to use this value of $\Delta G°_{313}$ to compute K_p with Equation 18.12. First, let's solve for $\ln K_p$.

$$\ln K_p = \frac{-\Delta G°_{313}}{RT}$$

Substituting, with $R = 8.314$ J mol^{-1} K^{-1}, $T = 313$ K, and expressing $\Delta G°_{313}$ in joules on a per mole basis gives

$$\ln K_p = \frac{2.09 \times 10^5 \text{ J mol}^{-1}}{(8.314 \text{ J mol}^{-1} \text{ K}^{-1})(313 \text{ K})}$$

$$= +80.3$$

Taking the antilogarithm,

$$K_p = e^{80.3}$$

$$= 7 \times 10^{34}$$

Notice that at this higher temperature, N_2O is actually slightly more stable than at 25 °C, as reflected in the slightly smaller value for the equilibrium constant for its decomposition.

Is the Answer Reasonable?
In Chapter 14 you learned that Le Châtelier's principle predicts that when a reaction is exothermic, as this one is, an increase in temperature decreases the size of the equilibrium constant. The smaller value of K_p at the higher temperature here is consistent with Le Châtelier's principle, so the answer seems reasonable.

Practice Exercise 18

The reaction $N_2(g) + 3H_2(g) \rightleftharpoons 2NH_3(g)$ has a standard heat of reaction of -92.4 kJ and a standard entropy of reaction equal to -198.3 J K^{-1}. Estimate the value of K_p for this reaction at 50 °C. ◆

EXAMPLE 18.14

Calculating *K* at Temperatures Other than 25 °C

18.10 Bond Energies and Heats of Reaction

You have seen how thermodynamic data allow us to predict the spontaneity of chemical reactions as well as the nature of chemical systems at equilibrium. In addition to these useful and important benefits of the study of thermodynamics, there is a bonus. By studying heats of reaction, and heats of formation in particular, we can obtain fundamental information about the chemical bonds in the substances that react, because the origin of the energy changes in chemical reactions is changes in bond energies.

Bond strength relates to bond energy. A strong bond is one with a large bond energy.

Recall that the **bond energy** is the amount of energy needed to break a chemical bond to give electrically neutral fragments. It is a useful quantity to know in the study of chemical properties, because during chemical reactions, bonds within the reactants are broken and new ones are formed as the products appear. The first step—bond breaking—is one of the factors that controls the reactivity of substances. Elemental nitrogen, for example, has a very low degree of reactivity, which is generally attributed to the very strong triple bond in N_2. Reactions that involve the breaking of this bond in a single step simply do not occur. When N_2 does react, it is by a stepwise involvement of its three bonds, one at a time.

The heats of formation of the nitrogen oxides are endothermic because of the large bond energy of the N_2 molecule.

Measurement of Bond Energies

The bond energies of simple diatomic molecules such as H_2, O_2, and Cl_2 are usually measured *spectroscopically*. A flame or an electric spark is used to excite (energize) the molecules, causing them to emit light. An analysis of the spectrum of emitted light allows scientists to compute the amount of energy needed to break the bond.

The emission spectrum of a molecule gives information about the energy levels in the molecule, just as the emission spectrum of an atom gives information about atomic energy levels.

For more complex molecules, thermochemical data can be used to calculate bond energies in a Hess's law kind of calculation. We will use the standard heat of formation of methane to illustrate how this is accomplished. However, before we can attempt such a calculation, we must first define a thermochemical quantity that we will call the **atomization energy,** symbolized ΔH_{atom}. This is the amount of energy needed to rupture all the chemical bonds in one mole of gaseous molecules to give gaseous atoms as products. For example, the atomization of methane is

The kind of bond breaking described here divides the electrons of the bond equally between the two atoms. It could be symbolized as

$$A : B \longrightarrow A \cdot + \cdot B$$

$$CH_4(g) \longrightarrow C(g) + 4H(g)$$

and the enthalpy change for the process is ΔH_{atom}. For this particular molecule, ΔH_{atom} corresponds to the total amount of energy needed to break all the C—H bonds in one mole of CH_4; therefore, division of ΔH_{atom} by 4 would give the average C—H bond energy in methane, expressed in kJ mol^{-1}.

Atomization energies are always endothermic. The formation of a molecule from its *gaseous* atoms is always exothermic because bond formation is exothermic.

Figure 18.14 shows how we can use the standard heat of formation, ΔH_f°, to calculate the atomization energy. Across the bottom we have the chemical equation for the formation of CH_4 from its elements. The enthalpy change for this reaction, of course, is ΔH_f°. In this figure we also can see an alternative three-step path that leads to $CH_4(g)$. One step is the breaking of H—H bonds in the H_2 molecules to give gaseous hydrogen atoms, another is the vaporization of carbon to give gaseous carbon atoms, and the third is the combination of the gaseous atoms to form CH_4 molecules. These changes are labeled 1, 2, and 3 in the figure.

Since ΔH is a state function, the net enthalpy change from one state to another is the same regardless of the path that we follow. This means that the sum of the enthalpy changes along the upper path must be the same as the enthalpy change along the lower path, ΔH_f°. Perhaps this can be more easily seen in Hess's law terms if we write the changes along the upper path in the form of thermochemical equations.

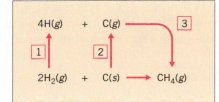

Figure 18.14 *Two paths leading to the formation of methane from its elements in their standard states.* As described in the text, steps 1, 2, and 3 of the upper path involve the formation of gaseous atoms of the elements and the formation of the bonds in CH_4. The lower path corresponds to the direct combination of the elements in their standard states to give CH_4. Because ΔH is a state function, the sum of the enthalpy changes along the upper path must equal the enthalpy change for the lower path (ΔH_f°).

Steps 1 and 2 have enthalpy changes that are called *standard heats of formation of gaseous atoms.* Values for these quantities have been measured for many of the elements, and some are given in Table 18.3. Step 3 is the opposite of atomization, and its enthalpy change will therefore be the negative of ΔH_{atom} (recall that if we reverse a reaction, we change the sign of its ΔH).

A more complete table of standard heats of formation of gaseous atoms is located in Appendix E.

(Step 1) $\qquad 2H_2(g) \longrightarrow 4H(g) \qquad \Delta H_1^\circ = 4\Delta H_{f\,H(g)}^\circ$

(Step 2) $\qquad\qquad C(s) \longrightarrow C(g) \qquad \Delta H_2^\circ = \Delta H_{f\,C(g)}^\circ$

(Step 3) $\quad 4H(g) + C(g) \longrightarrow CH_4(g) \qquad \Delta H_3^\circ = -\Delta H_{atom}$

$\qquad\qquad 2H_2(g) + C(s) \longrightarrow CH_4(g) \qquad \Delta H^\circ = \Delta H_{f\,CH_4(g)}^\circ$

Notice that by adding the first three equations, we get the equation for the formation of CH_4 from its elements in their standard states. This means that summing the ΔH° values of the first three equations should give ΔH_f° for CH_4.

When the equations are added, $C(g)$ and $4H(g)$ cancel.

$$\Delta H_1^\circ + \Delta H_2^\circ + \Delta H_3^\circ = \Delta H_{f\,CH_4(g)}^\circ$$

Let's substitute for ΔH_1°, ΔH_2°, and ΔH_3°, and then solve for ΔH_{atom}. First, we substitute for the numbered ΔH° quantities.

$$4\Delta H_{f\,H(g)}^\circ + \Delta H_{f\,C(g)}^\circ + (-\Delta H_{atom}) = \Delta H_{f\,CH_4(g)}^\circ$$

Next, we solve for $(-\Delta H_{atom})$.

$$-\Delta H_{atom} = \Delta H_{f\,CH_4(g)}^\circ - 4\Delta H_{f\,H(g)}^\circ - \Delta H_{f\,C(g)}^\circ$$

Changing signs and rearranging the right side of the equation just a bit gives

$$\Delta H_{atom} = 4\Delta H_{f\,H(g)}^\circ + \Delta H_{f\,C(g)}^\circ - \Delta H_{f\,CH_4(g)}^\circ$$

Now all we need are values for the ΔH_f°'s on the right side. From Table 18.3 we obtain $\Delta H_{f\,H(g)}^\circ$ and $\Delta H_{f\,C(g)}^\circ$, and the value of $\Delta H_{f\,CH_4(g)}^\circ$ is obtained from Table 6.2. We'll round these to the nearest 0.1 kJ/mol.

$$\Delta H_{f\,H(g)}^\circ = +217.9 \text{ kJ/mol}$$

$$\Delta H_{f\,C(g)}^\circ = +716.7 \text{ kJ/mol}$$

$$\Delta H_{f\,CH_4(g)}^\circ = -74.8 \text{ kJ/mol}$$

Substituting these values gives

$$\Delta H_{atom} = 1663.1 \text{ kJ/mol}$$

and division by 4 gives an estimate of the average C—H bond energy in this molecule.

$$\text{Bond energy} = \frac{1663.1 \text{ kJ/mol}}{4}$$

$$= 415.8 \text{ kJ/mol of C—H bonds}$$

Table 18.3 Standard Heats of Formation of Some Gaseous Atoms from the Elements in Their Standard States

Atom	ΔH_f° per mole of atoms (kJ mol^{-1})[a]
H	217.89
Li	161.5
Be	324.3
B	560
C	716.67
N	472.68
O	249.17
S	276.98
F	79.14
Si	450
P	332.2
Cl	121.47
Br	112.38
I	107.48

[a]All values are positive because forming the gaseous atoms from the elements involves bond breaking and is endothermic.

Table 18.4 Some Average
Bond Energies

Bond	Bond Energy (kJ mol^{-1})
C—C	348
C=C	612
C≡C	960
C—H	412
C—N	305
C=N	613
C≡N	890
C—O	360
C=O	743
C—F	484
C—Cl	338
C—Br	276
C—I	238
H—H	436
H—F	565
H—Cl	431
H—Br	366
H—I	299
H—N	388
H—O	463
H—S	338
H—Si	376

This value is quite close to the one in Table 18.4, which is an average of the C—H bond energies in many different compounds. The other bond energies in Table 18.4 are also based on thermochemical data and were obtained by similar calculations.

Uses of Bond Energies

An amazing thing about many covalent bond energies is that they are very nearly the same in many different compounds. This suggests, for example, that a C—H bond is very nearly the same in CH_4 as it is in a large number of other compounds that contain this kind of bond.

Because the bond energy doesn't vary much from compound to compound, we can use tabulated bond energies to estimate the heats of formation of substances. For example, let's calculate the standard heat of formation of methyl alcohol vapor, $CH_3OH(g)$. The structural formula for methanol is

$$
\begin{array}{c}
\quad\ \ \text{H} \\
\quad\ \ | \\
\text{H—C—O—H} \\
\quad\ \ | \\
\quad\ \ \text{H}
\end{array}
$$

To perform this calculation, we set up two paths from the elements to the compound, as shown in Figure 18.15. The lower path has an enthalpy change corresponding to $\Delta H^\circ_{f\,CH_3OH(g)}$, while the upper path takes us to the gaseous elements and then through the energy released when the bonds in the molecule are formed. This latter energy can be computed from the bond energies in Table 18.4. As before, the sum of the energy changes along the upper path must be the same as the energy change along the lower path, and this permits us to compute $\Delta H^\circ_{f\,CH_3OH(g)}$.

Steps 1, 2, and 3 involve the formation of the gaseous atoms from the elements, and their enthalpy changes are obtained from Table 18.3.

$$\Delta H^\circ_1 = \Delta H^\circ_{f\,C(g)} = 1 \text{ mol} \times (716.7 \text{ kJ/mol}) = 716.7 \text{ kJ}$$

$$\Delta H^\circ_2 = 4\Delta H^\circ_{f\,H(g)} = 4 \text{ mol} \times (217.9 \text{ kJ/mol}) = 871.6 \text{ kJ}$$

$$\Delta H^\circ_3 = \Delta H^\circ_{f\,O(g)} = 1 \text{ mol} \times (249.2 \text{ kJ/mol}) = 249.2 \text{ kJ}$$

The sum of these values, +1837.5 kJ, is the total energy input (the net ΔH°) for the first three steps.

The formation of the CH_3OH molecule from the gaseous atoms is exothermic because energy is always released when atoms become joined by covalent bonds. In this molecule we can count three C—H bonds, one C—O bond, and one O—H bond. Their formation releases energy equal to their bond energies, which we obtain from Table 18.4.

Figure 18.15 *Two paths for the formation of methyl alcohol vapor from its elements in their standard states. The numbered paths are referred to in the text discussion.*

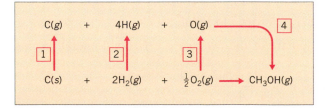

Bond	Energy (kJ)
3(C—H)	$3 \times (412 \text{ kJ/mol}) = 1236$
C—O	360
O—H	463

Adding these together gives a total of 2059 kJ. Therefore, ΔH_4° is −2059 kJ (because it is exothermic). Now we can compute the total enthalpy change for the upper path.

$$\Delta H^\circ = (+1837.5 \text{ kJ}) + (-2059 \text{ kJ})$$

$$= -222 \text{ kJ}$$

The value just calculated should be equal to ΔH_f° for $CH_3OH(g)$. For comparison, it has been found experimentally that ΔH_f° for this molecule (in the vapor state) is −201 kJ/mol. At first glance, the agreement doesn't seem very good, but on a relative basis the calculated value (−222 kJ) differs from the experimental one by only about 10%.

SUMMARY

First Law of Thermodynamics. The change in the **internal energy** of a system, ΔE, equals the sum of the heat absorbed by the system, q, and the work done on the system, w. ΔE is a state function, but q and w are not. The values of q and w depend on how the change takes place. For pressure–volume work, $w = -P\Delta V$, where $\Delta V = V_{\text{final}} - V_{\text{initial}}$. The heat of reaction at constant volume, q_v, is equal to ΔE, whereas the heat of reaction at constant pressure, q_p, is equal to ΔH. The value of ΔH differs from ΔE by the work expended in pushing back the atmosphere when the change occurs at constant atmospheric pressure. In general, the difference between ΔE and ΔH is quite small. For a chemical reaction, $\Delta H = \Delta E + \Delta n_{\text{gas}} RT$, where Δn_{gas} is the change in the number of moles of *gas* on going from reactants to products.

Spontaneity. A **spontaneous change** occurs without outside assistance, whereas a nonspontaneous change requires continuous help and can occur only if it is accompanied by and linked to some other spontaneous event.

The potential energy change is one factor that influences spontaneity. Exothermic changes, with their negative values of ΔH, tend to proceed spontaneously. Events that occur with an increase in randomness also tend to occur by themselves. Randomness is associated with statistical probability—the more random the distribution of a collection of particles, the greater is the probability of its occurrence. The thermodynamic quantity associated with randomness is **entropy, S**. An increase in entropy favors a spontaneous change. In general, gases have much higher entropies than liquids, which have somewhat higher entropies than solids.

Entropy increases with volume for a gas and with the temperature. During a chemical reaction, the entropy tends to increase if the complexity of the molecules decreases (e.g., forming diatomic molecules from triatomic molecules).

Second Law of Thermodynamics. This law states that the entropy of the universe increases whenever a spontaneous change occurs. All of our activities, being traced ultimately to spontaneous events, must increase the total entropy or disorder in the universe.

Gibbs Free Energy. The **Gibbs free energy change, ΔG,** allows us to determine the combined effects of temperature and of enthalpy and entropy changes on the spontaneity of a chemical or physical change. A change is spontaneous only if the free energy of the system decreases (ΔG is negative). When ΔH and ΔS have the same algebraic sign, the temperature becomes the critical factor in determining spontaneity.

Third Law of Thermodynamics. The entropy of a pure crystalline substance is equal to zero at absolute zero (0 K). Because we know where the zero point is on the entropy scale, it is possible to measure absolute amounts of entropy possessed by substances. **Standard entropies, S°,** are calculated for 25 °C and 1 atm (Table 18.1) and can be used to calculate ΔS° for chemical reactions.

$$\Delta S^\circ = (\text{sum of } S^\circ \text{ of products}) - (\text{sum of } S^\circ \text{ of reactants})$$

Standard Free Energy Changes. When ΔG is measured at 25 °C and 1 atm, it is the **standard free energy change, ΔG°.** As with enthalpy changes, the **standard free energies**

of formation, ΔG_f° (Table 18.2), can be used to obtain $\Delta G°$ for chemical reactions by a Hess's law type of calculation.

$$\Delta G° = (\text{sum of } \Delta G_f^\circ \text{ of products})$$
$$- (\text{sum of } \Delta G_f^\circ \text{ of reactants})$$

For any system, the value of ΔG is equal to the maximum amount of energy that can be obtained in the form of useful work. This maximum work can be obtained only if the change takes place reversibly. All real changes are irreversible and we always obtain less work than is theoretically available; the rest is lost as heat.

Free Energy and Equilibrium. When a system reaches equilibrium, $\Delta G = 0$ and no useful work can be obtained from it. At any particular pressure, an equilibrium between two phases of a substance (e.g., liquid $\rightleftharpoons$ solid, or solid $\rightleftharpoons$ vapor) can only occur at one temperature. The entropy change can be computed as $\Delta S = \Delta H/T$. The temperature at which the equilibrium occurs can be calculated from $T = \Delta H/\Delta S$.

In chemical reactions, a minimum on the free energy curve occurs partway between pure reactants and pure products. This minimum can be approached from either the reactants or products, and the composition of the equilibrium mixture is determined by where the minimum lies along the reactant $\rightarrow$ product axis; when it lies close to the products, the proportion of product to reactants is large and the reaction goes far toward completion.

When a reaction has a value of $\Delta G°$ that is both large and negative, it will appear to occur spontaneously because a lot of products will be formed by the time equilibrium is reached. If $\Delta G°$ is large and positive, it may be difficult to observe any reaction at all because only tiny amounts of products will be formed. The sign and magnitude of $\Delta G°$ can therefore be used to predict the *apparent* spontaneity of a chemical reaction.

At a temperature other than 25 °C, the value of ΔG_T° can be estimated using $\Delta H°$ and $\Delta S°$ as approximations of ΔH_T° and ΔS_T°. The equation is $\Delta G_T^\circ \approx \Delta H_{298\,K}^\circ - T\Delta S_{298\,K}^\circ$. Computing ΔG_T° allows us to see how temperature affects the position of equilibrium in a chemical reaction.

Thermodynamic Equilibrium Constants. The spontaneity of a reaction is determined by ΔG (how the free energy changes with a change in concentration). This is related to the standard free energy change, $\Delta G°$, by the equation $\Delta G = \Delta G° + RT \ln Q$, where Q is the reaction quotient for the system. At equilibrium, $\Delta G° = -RT \ln K$, where $K = K_p$ for gaseous reactions and $K = K_c$ for reactions in solution. For temperatures other than 25 °C, we can estimate ΔG_T° by the equation $\Delta G_T^\circ \approx \Delta H_{298}^\circ - T\Delta S_{298}^\circ$ and then use it to calculate K.

Bond Energies and Heats of Reaction. The bond energy equals the amount of energy needed to break a bond to give neutral fragments. The sum of all the bond energies in a molecule is the **atomization energy, ΔH_{atom}**, and, on a mole basis, it corresponds to the energy needed to break one mole of molecules into individual atoms. The heat of formation of a gaseous compound equals the sum of the energies needed to form atoms of the elements that are found in the substance plus the negative of the atomization energy.

Tools you have learned

The table below lists the tools that you have learned in this chapter. You will need them to answer chemistry questions. Review them if necessary, and refer to the tools when working on the Thinking-It-Through problems and the Review Problems that follow.

Tool	Function
$\Delta H = \Delta E + \Delta n_{gas} RT$ *(page 796)*	To convert between ΔE and ΔH for a reaction.
Predicting the sign of ΔS *(page 802)*	To determine whether the entropy change favors spontaneity.
Sign of ΔG *(page 808)*	To determine whether or not a change is spontaneous.
Standard entropies *(page 811)*	To calculate the value of $\Delta S°$ for a reaction.
$\Delta G° = \Delta H° - (298\ K)\Delta S°$ *(page 813)*	To calculate $\Delta G°$ from $\Delta H°$ and $\Delta S°$ values.

Standard free energies of formation (*page 814*)	To calculate $\Delta G°$ for a reaction from tabulated values of $\Delta G_f°$.
Value of $\Delta G°$ for a reaction (*page 821*)	To determine qualitatively whether or not a significant amount of products will form.
$\Delta G_T° \approx \Delta H_{298}° - T\Delta S_{298}°$ (*page 823*)	To estimate the value of $\Delta G°$ at temperatures other than 25 °C.
$\Delta G = \Delta G° + RT \ln Q$ (*page 825*)	To relate the reaction quotient to ΔG, which enables us to determine whether a reaction is at equilibrium, and, if not, the direction the reaction must proceed to reach equilibrium.
$\Delta G° = -RT \ln K$ (*page 826*)	To relate $\Delta G°$ to the equilibrium constant; K equals K_p for gaseous reactions and K_c for reactions in solution.

THINKING IT THROUGH

Remember, the goal for each of the following problems is to assemble the available information needed to obtain the answer, state what additional data (if any) are needed, and describe how you would use the data to answer the question. For problems involving unit conversions, list the relationships among the units that are needed to carry out the conversions. Construct the conversion factors that can be formed from these relationships. Then set up the solution to the problem by arranging the conversion factors so the units cancel correctly to give the desired units of the answer.

The problems are divided into two groups. Those in Level 2 are more challenging than those in Level 1 and provide an opportunity to really hone your problem solving skills. Detailed answers to these problems can be found at our Web site: http://www.wiley.com/college/brady.

Level 1 Problems

1. Describe how you would calculate the work done by a gas as it expands at constant temperature from a volume of 3.00 L and a pressure of 5.00 atm to a volume of 8.00 L. The external pressure against which the gas expands is 1.00 atm. Set up the calculation so the answer is in units of joules. (1 atm = 101,325 Pa.)

2. The reaction $2N_2O(g) \rightarrow 2N_2(g) + O_2(g)$ has $\Delta H° = -163.14$ kJ. What is the value of ΔE for the decomposition of 180 g of N_2O at 25 °C? If we assume that ΔH doesn't change appreciably with temperature, what is ΔE for this same reaction at 200 °C? (Set up the calculations.)

3. What factors must you consider to determine the sign of ΔS for the reaction $2N_2O(g) \rightarrow 2N_2(g) + O_2(g)$ if it occurs at constant temperature?

4. For the reaction in Question 3, what additional information do we need to determine whether this reaction should tend to proceed spontaneously in the forward direction?

5. What factors must you consider to determine the sign of ΔS for the following reaction?

$$2HI(g) \longrightarrow H_2(g) + I_2(s)$$

6. How would you calculate the value of $\Delta S_f°$ for HI? (Set up the calculation.)

7. The reaction

$$2C_4H_{10}(g) + 13O_2(g) \longrightarrow 8CO_2(g) + 10H_2O(g)$$

has $\Delta G° = -5407$ kJ. Show how you would determine the value of $\Delta G_f°$ for $C_4H_{10}(g)$. (Set up the calculation.)

8. Explain how you would calculate the value of $\Delta G_{500\ K}°$ for the reaction in Question 7. (Set up the calculation.)

9. A 10.0 L vessel at 20 °C contains butane, $C_4H_{10}(g)$, at a pressure of 2.00 atm. What is the maximum amount of work that can be obtained by the combustion of this butane if the gas is brought to a pressure of 1 atm and the temperature is brought to 25 °C? Assume the products are also returned to this same temperature and pressure. (Set up the calculation.)

10. Given the following reactions and their values of $\Delta G°$, explain in detail how you would calculate the value of $\Delta G_f°$ for $N_2O_5(g)$. Set up the calculation.

$$2H_2(g) + O_2(g) \longrightarrow 2H_2O(l) \qquad \Delta G° = -474.4 \text{ kJ}$$

$$N_2O_5(g) + H_2O(l) \longrightarrow 2HNO_3(l) \qquad \Delta G° = -37.6 \text{ kJ}$$

$$\tfrac{1}{2}N_2(g) + \tfrac{3}{2}O_2(g) + H_2(g) \longrightarrow HNO_3(l) \quad \Delta G° = -79.91 \text{ kJ}$$

11. How would you determine the atomization energy of the molecule whose structure is shown below?

$$
\begin{array}{ccccccc}
& H & & H & H & H & \\
& | & & | & | & | & \\
H- & C & = & C - & C - & N - & H \\
& & & & | & & \\
& & & & H & &
\end{array}
$$

12. Explain how you would use bond energies and the heats of formation of atoms of the elements to calculate the heat of formation of gaseous hydrogen cyanide, HCN.

Level 2 Problems

13. In the SI, pressure is expressed in pascals. One pascal is equal to one newton (N, the SI unit of force) per square meter: $1 \text{ Pa} = 1 \text{ N/m}^2$. Work is force times distance, and the joule is defined therefore as $1 \text{ J} = 1 \text{ N·m}$. Show that when the pressure is expressed in pascals and volume in cubic meters that the product $P\Delta V$ yields work in units of joules.

14. When an ideal gas expands or contracts at constant temperature, $\Delta E = 0$. In terms of the definition of an ideal gas and the kinetic theory interpretation of temperature, explain why this is true.

15. An ideal gas in a cylinder fitted with a piston expands at constant temperature from a pressure of 5.0 atm and a volume of 12 L to a final volume of 6.0 L against a constant opposing pressure of 2.0 atm. How much heat does the gas absorb, expressed in units of L·atm (liter × atm)? Set up the calculation. (*Hint:* Refer to the preceding question.)

16. The reaction in Question 5 has $\Delta H° = -53.2$ kJ. How can you determine how the sign and magnitude of $\Delta G_T°$ will be affected by a change in temperature?

17. At room temperature (25 °C), the gas ClNO is impure because it decomposes slightly according to the equation

$$2ClNO(g) \rightleftharpoons Cl_2(g) + 2NO(g)$$

The extent of decomposition is about 5%. What is the approximate value of $\Delta G_{298}°$ for this reaction at this temperature? (Explain each step in the solution to the problem.)

18. The reaction $N_2O(g) + O_2(g) \rightleftharpoons NO_2(g) + NO(g)$ has $\Delta H° = -42.9$ kJ and $\Delta S° = -26.1$ J/K. Suppose 0.100 mol of N_2O and 0.100 mol of O_2 were placed in a 10.0 L container at 500 °C and this equilibrium were established. Explain how you would determine what percentage of the N_2O would have reacted.

19. When a cylinder of NO(g) is opened to the air, a brown cloud of gas appears a short distance from the cylinder opening. The brown color is due to the formation of $NO_2(g)$ by the reaction:

$$2NO(g) + O_2(g) \longrightarrow 2NO_2(g).$$

Both NO and O_2 are colorless gases.
(a) Determine the signs for ΔG, ΔS and ΔH for this observed change. Explain your reasoning.
(b) If the sign of ΔG had been the same but the sign of ΔS had been reversed, would it have been possible to determine the sign of ΔH? Explain.

20. For two chemical reactions the equilibrium constants were determined at several temperatures. Plots of ln K versus $1/T$ gave the following results:

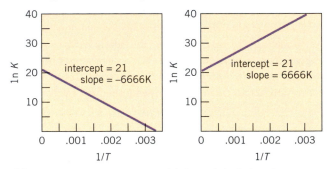

(a) Explain how the value of $\Delta S°$ and $\Delta H°$ for these reactions can be found from these plots. What assumptions are made in determining $\Delta S°$ and $\Delta H°$ this way?
(b) Are either $\Delta S°$ or $\Delta H°$ the same for these two reactions? Explain your reasoning.

REVIEW QUESTIONS

First Law of Thermodynamics

18.1 What is the origin of the name *thermodynamics?*

18.2 What kinds of energy contribute to the internal energy of a system? Why can't we measure or calculate a system's internal energy?

18.3 How is a change in the internal energy defined in terms of the initial and final internal energies?

18.4 What is the algebraic sign of ΔE for an endothermic change? Why?

18.5 State the first law of thermodynamics in words. What equation defines the change in the internal energy in terms of heat and work? Define the meaning of the symbols, including the significance of their algebraic signs.

18.6 Which quantities in the statement of the first law are state functions and which are not?

18.7 Which thermodynamic quantity corresponds to the heat of reaction at constant volume? Which corresponds to the heat of reaction at constant pressure?

18.8 What are the units of $P\Delta V$ if pressure is expressed in pascals and volume is expressed in cubic meters?

18.9 How are ΔE and ΔH related to each other?

18.10 If there is a decrease in the number of moles of gas during an exothermic chemical reaction, which is numerically larger, ΔE or ΔH? Why?

Spontaneous Change and Enthalpy

18.11 What is a *spontaneous change*?

18.12 List five changes that you have encountered recently that occurred spontaneously. List five changes that are non-spontaneous that you have caused to occur.

18.13 Which of the items that you listed in Question 18.12 are exothermic (leading to a lowering of the potential energy) and which are endothermic (accompanied by an increase in potential energy)?

18.14 At constant pressure, what role does the enthalpy change play in determining the spontaneity of an event?

Entropy

18.15 An instant cold pack purchased in a pharmacy contains a packet of solid ammonium nitrate surrounded by a pouch of water. When the packet of NH_4NO_3 is broken, the solid dissolves in water and a cooling of the mixture occurs because the solution process for NH_4NO_3 in water is endothermic. Explain, in terms of what happens to the molecules and ions, why this mixing occurs spontaneously.

18.16 What is *entropy*?

18.17 Will the entropy change for each of the following be positive or negative?
(a) Moisture condenses on the outside of a cold glass.
(b) Raindrops form in a cloud.
(c) Gasoline vaporizes in the carburetor of an automobile engine.
(d) Air is pumped into a tire.
(e) Frost forms on the windshield of your car.
(f) Sugar dissolves in coffee.

18.18 On the basis of our definition of entropy, suggest why entropy is a state function.

18.19 State the second law of thermodynamics.

18.20 In animated cartoons, visual effects are often created (for amusement) that show events that ordinarily don't occur in real life because they are accompanied by huge entropy decreases. Can you think of an example of this? Explain why there is an entropy decrease in your example.

18.21 How is entropy related to pollution? What are our chances of eliminating pollution from the environment?

Third Law of Thermodynamics and Standard Entropies

18.22 What is the third law of thermodynamics?

18.23 Would you expect the entropy of an alloy (a solution of two metals) to be zero at 0 K? Explain your answer.

18.24 Why does entropy increase with increasing temperature?

18.25 Does glass have $S = 0$ at 0 K? Explain.

Gibbs Free Energy

18.26 What is the equation expressing the change in the Gibbs free energy for a reaction occurring at constant temperature and pressure?

18.27 In terms of the algebraic signs of ΔH and ΔS, under what circumstances will a change be spontaneous:
(a) At all temperatures?
(b) At low temperatures but not at high temperatures?
(c) At high temperatures but not at low temperatures?

18.28 Under what circumstances will a change be nonspontaneous regardless of the temperature?

Free Energy and Work

18.29 How is free energy related to useful work?

18.30 What is a reversible process? How is the amount of work obtained from a change related to reversibility?

18.31 How is the *rate* at which energy is withdrawn from a system related to the amount of that energy which can appear as useful work?

18.32 When glucose is oxidized by the body to generate energy, part of the energy is used to make molecules of ATP (adenosine triphosphate). However, of the total energy released in the oxidation of glucose, only 38% actually goes to making ATP. What happens to the rest of the energy?

18.33 Why are real, observable changes not considered to be reversible processes?

Free Energy and Equilibrium

18.34 In what way is free energy related to equilibrium?

18.35 Considering the fact that the formation of a bond between two atoms is exothermic and is accompanied by an entropy decrease, explain why all chemical compounds decompose into individual atoms if heated to a high enough temperature.

18.36 When a warm object is placed in contact with a cold one, they both gradually come to the same temperature. On a molecular level, explain how this is related to entropy and spontaneity.

18.37 Sketch a graph to show how the free energy changes during a phase change such as the melting of a solid.

18.38 Sketch the shape of the free energy curve for a chemical reaction that has a positive $\Delta G°$. Indicate the composition of the reaction mixture corresponding to equilibrium.

18.39 Many reactions that have large, negative values of $\Delta G°$ are not actually observed to happen at 25 °C and 1 atm. Why?

Effect of Temperature on the Free Energy Change

18.40 Why is the value of ΔG for a change so dependent on temperature?

18.41 What equation can be used to obtain an approximate value for ΔG_T° for a reaction at a temperature other than 25 °C?

Thermodynamics and Equilibrium

18.42 Suppose a reaction has a negative ΔH° and a negative ΔS°. Will more or less product be present at equilibrium as the temperature is raised?

18.43 Write the equation that relates the free energy change to the value of the reaction quotient for a reaction.

18.44 How is the equilibrium constant related to the standard free energy change for a reaction? (Write the equation.)

18.45 What is a thermodynamic equilibrium constant?

18.46 What is the value of ΔG° for a reaction for which $K = 1$?

Bond Energies and Heats of Reaction

18.47 Define the term *atomization energy*.

18.48 Why are the heats of formation of gaseous atoms from their elements endothermic quantities?

18.49 The gaseous C_2 molecule has a bond energy of 602 kJ mol^{-1}. Why isn't the standard heat of formation of $C(g)$ equal to half this value?

REVIEW PROBLEMS

Answers to problems whose numbers are printed in color are given in Appendix D.
More challenging problems are marked with asterisks.

First Law of Thermodynamics

18.50 A certain system absorbs 300 J of heat and has 700 J of work performed on it. What is the value of ΔE for the change? Is the overall change exothermic or endothermic?

18.51 The value of ΔE for a certain change is -1455 J. During the change, the system absorbs 812 J of heat. Did the system do work, or was work done on the system? How much work, expressed in joules, was involved?

18.52 Suppose that you were pumping an automobile tire with a hand pump that pushed 24.0 in.^{3} of air into the tire on each stroke, and that during one such stroke the opposing pressure in the tire was 30.0 lb/in.^{2} above the normal atmospheric pressure of 14.7 lb/in.^{2}. Calculate the number of joules of work accomplished during each stroke. (1 L atm = 101.325 J.)

18.53 Consider the reaction between aqueous solutions of baking soda, $NaHCO_3$, and vinegar, $HC_2H_3O_2$.

$$NaHCO_3(aq) + HC_2H_3O_2(aq) \longrightarrow$$
$$NaC_2H_3O_2(aq) + H_2O(l) + CO_2(g)$$

If this reaction occurs at atmospheric pressure ($P = 1$ atm), how much work is done by the system in pushing back the atmosphere when 1.00 mol $NaHCO_3$ reacts at a temperature of 25 °C? (*Hint:* Review the gas laws.)

18.54 Calculate ΔH° and ΔE° for the following reactions at 25 °C. (If necessary, refer to the data in Table E.1 in Appendix E.)
(a) $3PbO(s) + 2NH_3(g) \rightarrow 3Pb(s) + N_2(g) + 3H_2O(g)$
(b) $NaOH(s) + HCl(g) \rightarrow NaCl(s) + H_2O(l)$
(c) $Al_2O_3(s) + 2Fe(s) \rightarrow Fe_2O_3(s) + 2Al(s)$
(d) $2CH_4(g) \rightarrow C_2H_6(g) + H_2(g)$

18.55 Calculate ΔH° and ΔE° for the following reactions at 25 °C. (If necessary, refer to the data in Table E.1 in Appendix E.)
(a) $2C_2H_2(g) + 5O_2(g) \rightarrow 4CO_2(g) + 2H_2O(g)$
(b) $C_2H_2(g) + 5N_2O(g) \rightarrow 2CO_2(g) + H_2O(g) + 5N_2(g)$
(c) $NH_4Cl(s) \rightarrow NH_3(g) + HCl(g)$
(d) $(CH_3)_2CO(l) + 4O_2(g) \rightarrow 3CO_2(g) + 3H_2O(g)$

Factors That Affect Spontaneity

18.56 Use the data from Table 6.2 to calculate ΔH° for the following reactions. On the basis of their values of ΔH°, which are favored to occur spontaneously?
(a) $CaO(s) + CO_2(g) \rightarrow CaCO_3(s)$
(b) $C_2H_2(g) + 2H_2(g) \rightarrow C_2H_6(g)$
(c) $3CaO(s) + 2Fe(s) \rightarrow 3Ca(s) + Fe_2O_3(s)$
(d) $Ca(OH)_2(s) \rightarrow CaO(s) + H_2O(l)$
(e) $2NaCl(s) + H_2SO_4(l) \rightarrow Na_2SO_4(s) + 2HCl(g)$

18.57 Use the data from Table 6.2 to calculate ΔH° for the following reactions. On the basis of their values of ΔH°, which are favored to occur spontaneously?
(a) $2C_2H_2(g) + 5O_2(g) \rightarrow 4CO_2(g) + 2H_2O(g)$
(b) $C_2H_2(g) + 5N_2O(g) \rightarrow 2CO_2(g) + H_2O(g) + 5N_2(g)$
(c) $Fe_2O_3(s) + 2Al(s) \rightarrow Al_2O_3(s) + 2Fe(s)$
(d) $NH_4Cl(s) \rightarrow NH_3(g) + HCl(g)$
(e) $Ag(s) + KCl(s) \rightarrow AgCl(s) + K(s)$

18.58 When a coin is tossed, there is a 50–50 chance of it landing heads or tails. Suppose that you tossed four coins. On the basis of the different possible outcomes, what is the probability of all four coins coming up heads? What is the probability of an even head–tails distribution? (Now work Problem 18.59.)

18.59 Suppose that you had two equal-volume containers sharing a common wall with a hole in it. Suppose there were four molecules in this system. What is the probability that all four would be in one container at the same time? What is the probability of finding an even distribution of molecules between the two containers? (*Hint:* If you haven't already done so, work Problem 18.58 first.) What do the results of Problems 18.58 and 18.59 suggest about why gases expand spontaneously?

18.60 Predict the algebraic sign of the entropy change for the following reactions.
(a) $PCl_3(g) + Cl_2(g) \rightarrow PCl_5(g)$
(b) $SO_2(g) + CaO(s) \rightarrow CaSO_3(s)$
(c) $CO_2(g) + H_2O(l) \rightarrow H_2CO_3(aq)$
(d) $Ni(s) + 2HCl(aq) \rightarrow H_2(g) + NiCl_2(aq)$

18.61 Predict the algebraic sign of the entropy change for the following reactions.
(a) $I_2(s) \rightarrow I_2(g)$
(b) $Br_2(g) + 3Cl_2(g) \rightarrow 2BrCl_3(g)$
(c) $NH_3(g) + HCl(g) \rightarrow NH_4Cl(s)$
(d) $CaO(s) + H_2O(l) \rightarrow Ca(OH)_2(s)$

Third Law of Thermodynamics

18.62 Calculate $\Delta S°$ for the following reactions in $J\ K^{-1}$ from the data in Table 18.1. On the basis of their values of $\Delta S°$, which of these reactions are favored to occur spontaneously?
(a) $N_2(g) + 3H_2(g) \rightarrow 2NH_3(g)$
(b) $CO(g) + 2H_2(g) \rightarrow CH_3OH(l)$
(c) $2C_2H_6(g) + 7O_2(g) \rightarrow 4CO_2(g) + 6H_2O(g)$
(d) $Ca(OH)_2(s) + H_2SO_4(l) \rightarrow CaSO_4(s) + 2H_2O(l)$
(e) $S(s) + 2N_2O(g) \rightarrow SO_2(g) + 2N_2(g)$

18.63 Calculate $\Delta S°$ for the following reactions in $J\ K^{-1}$, using the data in Table 18.1.
(a) $Ag(s) + \frac{1}{2}Cl_2(g) \rightarrow AgCl(s)$
(b) $H_2(g) + \frac{1}{2}O_2(g) \rightarrow H_2O(g)$
(c) $H_2(g) + \frac{1}{2}O_2(g) \rightarrow H_2O(l)$
(d) $CaCO_3(s) + H_2SO_4(l) \rightarrow$
$$CaSO_4(s) + H_2O(g) + CO_2(g)$$
(e) $NH_3(g) + HCl(g) \rightarrow NH_4Cl(s)$

18.64 Calculate $\Delta S°_f$ for the following compounds in units of $J\ mol^{-1}\ K^{-1}$.
(a) $C_2H_4(g)$ (c) $NaCl(s)$ (e) $HC_2H_3O_2(l)$
(b) $N_2O(g)$ (d) $CaSO_4 \cdot 2H_2O(s)$

18.65 Calculate $\Delta S°_f$ for the following compounds in units of $J\ mol^{-1}\ K^{-1}$.
(a) $Al_2O_3(s)$ (c) $N_2O_4(g)$ (e) $CaSO_4 \cdot \frac{1}{2}H_2O(s)$
(b) $CaCO_3(s)$ (d) $NH_4Cl(s)$

18.66 Nitrogen dioxide, NO_2, an air pollutant, dissolves in rainwater to form a dilute solution of nitric acid. The equation for the reaction is

$$3NO_2(g) + H_2O(l) \longrightarrow 2HNO_3(l) + NO(g)$$

Calculate $\Delta S°$ for this reaction in $J\ K^{-1}$.

18.67 Good wine will turn to vinegar if it is left exposed to air because the alcohol is oxidized to acetic acid. The equation for the reaction is

$$C_2H_5OH(l) + O_2(g) \longrightarrow HC_2H_3O_2(l) + H_2O(l)$$

Calculate $\Delta S°$ for this reaction in $J\ K^{-1}$.

Gibbs Free Energy

18.68 Phosgene, $COCl_2$, was used as a war gas during World War I. It reacts with the moisture in the lungs to produce HCl, which causes the lungs to fill with fluid, and CO_2. For $COCl_2(g)$, $S° = 284$ J/mol K and $\Delta H°_f = -223$ kJ/mol. Use this information and the data in Table 18.1 to calculate $\Delta G°_f$ for $COCl_2(g)$ in kJ mol^{-1}.

18.69 Aluminum oxidizes rather easily, but it forms a thin protective coating of Al_2O_3 that prevents further oxidation of the aluminum beneath. Use the data for $\Delta H°_f$ (Table 6.2) and $S°$ to calculate $\Delta G°_f$ for $Al_2O_3(s)$ in kJ mol^{-1}.

18.70 Compute $\Delta G°$ in kJ for the following reactions, using the data in Table 18.2.
(a) $SO_3(g) + H_2O(l) \rightarrow H_2SO_4(l)$
(b) $2NH_4Cl(s) + CaO(s) \rightarrow$
$$CaCl_2(s) + H_2O(l) + 2NH_3(g)$$
(c) $CaSO_4(s) + 2HCl(g) \rightarrow CaCl_2(s) + H_2SO_4(l)$
(d) $C_2H_4(g) + H_2O(g) \rightarrow C_2H_5OH(l)$
(e) $Ca(s) + 2H_2SO_4(l) \rightarrow CaSO_4(s) + SO_2(g) + 2H_2O(l)$

18.71 Compute $\Delta G°$ in kJ for the following reactions, using the data in Table 18.2.
(a) $2HCl(g) + CaO(s) \rightarrow CaCl_2(s) + H_2O(g)$
(b) $H_2SO_4(l) + 2NaCl(s) \rightarrow 2HCl(g) + Na_2SO_4(s)$
(c) $3NO_2(g) + H_2O(l) \rightarrow 2HNO_3(l) + NO(g)$
(d) $2AgCl(s) + Ca(s) \rightarrow CaCl_2(s) + 2Ag(s)$
(e) $NH_3(g) + HCl(g) \rightarrow NH_4Cl(s)$

18.72 Plaster of Paris, $CaSO_4 \cdot \frac{1}{2}H_2O(s)$, reacts with liquid water to form gypsum, $CaSO_4 \cdot 2H_2O(s)$. Write a chemical equation for the reaction and calculate $\Delta G°$ in kJ, using the data in Table 18.2.

18.73 When phosgene, the war gas described in Review Problem 18.68, reacts with water vapor, the products are $CO_2(g)$ and $HCl(g)$. Write an equation for the reaction and compute $\Delta G°$ in kJ. For $COCl_2(g)$, $\Delta G°_f$ is -210 kJ/mol.

18.74 Given the following,

$$4NO(g) \longrightarrow 2N_2O(g) + O_2(g) \qquad \Delta G° = -139.56\ kJ$$

$$2NO(g) + O_2(g) \longrightarrow 2NO_2(g) \qquad \Delta G° = -69.70\ kJ$$

calculate $\Delta G°$ for the reaction

$$2N_2O(g) + 3O_2(g) \longrightarrow 4NO_2(g)$$

18.75 Given these reactions and their $\Delta G°$ values,

$$COCl_2(g) + 4NH_3(g) \longrightarrow CO(NH_2)_2(s) + 2NH_4Cl(s)$$
$$\Delta G° = -332.0\ kJ$$

$$COCl_2(g) + H_2O(l) \longrightarrow CO_2(g) + 2HCl(g)$$
$$\Delta G° = -141.8 \text{ kJ}$$

$$NH_3(g) + HCl(g) \longrightarrow NH_4Cl(s) \qquad \Delta G° = -91.96 \text{ kJ}$$

calculate the value of $\Delta G°$ for the reaction

$$CO(NH_2)_2(s) + H_2O(l) \longrightarrow CO_2(g) + 2NH_3(g)$$

Free Energy and Work
18.76 Gasohol is a mixture of gasoline and ethanol (grain alcohol), C_2H_5OH. Calculate the maximum work that could be obtained at 25 °C and 1 atm by burning 1 mol of C_2H_5OH.

$$C_2H_5OH(l) + 3O_2(g) \longrightarrow 2CO_2(g) + 3H_2O(g)$$

18.77 What is the maximum amount of useful work that could possibly be obtained at 25 °C and 1 atm from the combustion of 48.0 g of natural gas, $CH_4(g)$, to give $CO_2(g)$ and $H_2O(g)$?

Free Energy and Equilibrium
18.78 Chloroform, formerly used as an anesthetic and now believed to be a carcinogen (cancer-causing agent), has a heat of vaporization $\Delta H_{\text{vaporization}} = 31.4 \text{ kJ mol}^{-1}$. The change, $CHCl_3(l) \rightarrow CHCl_3(g)$, has $\Delta S° = 94.2 \text{ J mol}^{-1} \text{ K}^{-1}$. At what temperature do we expect $CHCl_3$ to boil (i.e., at what temperature will liquid and vapor be in equilibrium at 1 atm pressure)?

18.79 For the melting of aluminum, $Al(s) \rightarrow Al(l)$, $\Delta H° = 10.0 \text{ kJ mol}^{-1}$ and $\Delta S° = 9.50 \text{ J/mol K}$. Calculate the melting point of Al. (The actual melting point is 660°C.)

18.80 Isooctane, an important constituent of gasoline, has a boiling point of 99.3 °C and a heat of vaporization of 37.7 kJ mol^{-1}. What is ΔS (in $\text{J mol}^{-1} \text{ K}^{-1}$) for the vaporization of 1 mol of isooctane?

18.81 Acetone (nail polish remover) has a boiling point of 56.2 °C. The change, $(CH_3)_2CO(l) \rightarrow (CH_3)_2CO(g)$, has $\Delta H° = 31.9 \text{ kJ mol}^{-1}$. What is $\Delta S°$ for this change?

Free Energy and Spontaneity of Chemical Reactions
18.82 Determine whether the following reaction will be spontaneous. (Do we expect appreciable amounts of products to form?)

$$C_2H_4(g) + 2HNO_3(l) \longrightarrow$$
$$HC_2H_3O_2(l) + H_2O(l) + NO(g) + NO_2(g)$$

18.83 Which of the following reactions (equations unbalanced) would be expected to be spontaneous at 25 °C and 1 atm?
(a) $PbO(s) + NH_3(g) \rightarrow Pb(s) + N_2(g) + H_2O(g)$
(b) $NaOH(s) + HCl(g) \rightarrow NaCl(s) + H_2O(l)$
(c) $Al_2O_3(s) + Fe(s) \rightarrow Fe_2O_3(s) + Al(s)$
(d) $2CH_4(g) \rightarrow C_2H_6(g) + H_2(g)$

18.84 Calculate the value of $\Delta G°_{373}$ in kJ for the following reaction at 100 °C, using data in Tables 6.2 and 18.1.

$$C_2H_4(g) + H_2(g) \longrightarrow C_2H_6(g)$$

18.85 Calculate the value of $\Delta G°_{373}$ in kJ for the following reaction at 100 °C, using data in Tables 6.2 and 18.1.

$$5SO_3(g) + 2NH_3(g) \longrightarrow 2NO(g) + 5SO_2(g) + 3H_2O(g)$$

Thermodynamic Equilibrium Constants
18.86 Calculate the value of the thermodynamic equilibrium constant for the following reactions at 25 °C. (Refer to the data in Appendix E.)
(a) $2PCl_3(g) + O_2(g) \rightleftharpoons 2POCl_3(g)$
(b) $2SO_3(g) \rightleftharpoons 2SO_2(g) + O_2(g)$

18.87 Calculate the value of the thermodynamic equilibrium constant for the following reactions at 25 °C. (Refer to the data in Appendix E.)
(a) $N_2H_4(g) + 2O_2(g) \rightleftharpoons 2NO(g) + 2H_2O(g)$
(b) $N_2H_4(g) + 6H_2O_2(g) \rightleftharpoons 2NO_2(g) + 8H_2O(g)$

18.88 The reaction $NO_2(g) + NO(g) \rightleftharpoons N_2O(g) + O_2(g)$ has $\Delta G°_{1273} = -9.67 \text{ kJ}$. A 1.00 L reaction vessel at 1000 °C contains 0.0200 mol NO_2, 0.040 mol NO, 0.015 mol N_2O, and 0.0350 mol O_2. Is the reaction at equilibrium? If not, in which direction will the reaction proceed to reach equilibrium?

18.89 The reaction $CO(g) + H_2O(g) \rightleftharpoons HCHO_2(g)$ has $\Delta G°_{673} = +79.8 \text{ kJ mol}^{-1}$. If a mixture at 400 °C contains 0.040 mol CO, 0.022 mol H_2O, and 3.8×10^{-3} mol $HCHO_2$ in a 2.50 L container, is the reaction at equilibrium? If not, in which direction will the reaction proceed spontaneously?

18.90 A reaction that can convert coal to methane (the chief component of natural gas) is

$$C(s) + 2H_2(g) \rightleftharpoons CH_4(g)$$

for which $\Delta G° = -50.79 \text{ kJ mol}^{-1}$. What is the value of K_p for this reaction at 25 °C? Does this value of K_p suggest that studying this reaction as a means of methane production is worth pursuing?

18.91 Adenosine triphosphate (ATP) is made by cells to provide chemical energy for biochemical processes. Its reaction with water, which does not occur in cells, is used as a way to measure how much chemical energy ATP can provide. The products are adenosine diphosphate (ADP) and free phosphate ion.

$$ATP + H_2O \rightleftharpoons ADP + PO_4^{3-}$$

The value of ΔG_T° for this reaction at 37 °C (normal human body temperature) is -33 kJ mol^{-1}. Calculate the value of the equilibrium constant for the reaction at this temperature.

18.92 What is the value of the equilibrium constant for a reaction for which $\Delta G^\circ = 0$? What will happen to the composition of the system if we begin the reaction with the pure products?

18.93 Methanol, a potential replacement for gasoline as an automotive fuel, can be made from H_2 and CO by the reaction $CO(g) + 2H_2(g) \rightleftharpoons CH_3OH(g)$. At 500 K, this reaction has $K_p = 6.25 \times 10^{-3}$. Calculate ΔG_{500}° for this reaction in units of kilojoules.

18.94 Refer to the data in Table 18.2 to determine ΔG° in kilojoules for the reaction $2NO(g) + 2CO(g) \rightleftharpoons N_2(g) + 2CO_2(g)$. What is the value of K_p for this reaction at 25 °C?

18.95 Use the data in Appendix E.1 to calculate ΔG_{773}° (at 500 °C) in kilojoules for the following reaction:

$$2NO(g) + 2CO(g) \rightleftharpoons N_2(g) + 2CO_2(g)$$

What is the value of K_p for the reaction at this temperature?

Bond Energies and Heats of Reaction

18.96 Use the data in Table 18.4 to compute the approximate atomization energy of NH_3.

18.97 Approximately how much energy would be released during the formation of the bonds in one mole of acetone molecules?

$$CH_3-\overset{\overset{\textstyle O}{\|}}{C}-CH_3$$

acetone

18.98 The standard heat of formation of ethanol vapor, $C_2H_5OH(g)$, is -235.3 kJ mol^{-1}. Use the data in Table 18.3 and the average bond energies for C—C, C—H, and O—H bonds to estimate the C—O bond energy in this molecule.

$$CH_3-CH_2-OH$$

ethanol

18.99 The standard heat of formation of ethylene, $C_2H_4(g)$, is $+52.284$ kJ mol^{-1}. Calculate the C=C bond energy in this molecule.

18.100 Carbon disulfide, CS_2, has the Lewis structure $\ddot{S}=C=\ddot{S}$, and for $CS_2(g)$, $\Delta H_f^\circ = +115.3$ kJ mol^{-1}. Use the data in Table 18.3 to calculate the average C=S bond energy in this molecule.

18.101 Gaseous H_2S has $\Delta H_f^\circ = -20.15$ kJ mol^{-1}. Use data in Table 18.3 to calculate the average S—H bond energy in this molecule.

18.102 For $SF_6(g)$, $\Delta H_f^\circ = -1096$ kJ mol^{-1}. Use Table 18.3 to calculate the average S—F bond energy in SF_6.

18.103 Use the results of the preceding problem and the data in Table 18.2 to calculate the standard heat of formation of $SF_4(g)$. The measured value of ΔH_f° for $SF_4(g)$ is -718.4 kJ mol^{-1}. What is the percentage difference between your calculated value of ΔH_f° and the experimentally determined value?

18.104 For the substance $SO_2F_2(g)$, $\Delta H_f^\circ = -858$ kJ mol^{-1}. The structure of the SO_2F_2 molecule is

$$\ddot{F}-\overset{\overset{\textstyle \ddot{O}}{\|}}{\underset{\underset{\textstyle \ddot{O}}{\|}}{S}}-\ddot{F}$$

Use the value of the S—F bond energy calculated in Problem 18.102 and the data in Table 18.3 to determine the average S=O bond energy in SO_2F_2 in units of kJ mol^{-1}.

18.105 Use the data in Tables 18.3 and 18.4 to estimate the standard heat of formation of acetylene, H—C≡C—H, in the gaseous state.

18.106 What would be the approximate heat of formation of CCl_4 vapor at 25°C and 1 atm?

18.107 Which substance should have the more exothermic heat of formation, CF_4 or CCl_4?

ADDITIONAL EXERCISES

***18.108** Would you expect the value of ΔH_f° for benzene, C_6H_6, computed from tabulated bond energies, to be very close to the experimentally measured value of ΔH_f°? Justify your answer.

18.109 For each of the reactions in Problem 18.86, calculate K_p at 300 °C.

18.110 If pressure is expressed in atmospheres and volume is expressed in liters, $P\Delta V$ has units of L atm (liters ×

atmospheres). In Chapter 10 you learned that 1 atm = 101,325 Pa, and in Chapter 1 you learned that 1 L = 1 dm^3. Use this information to determine the number of joules corresponding to 1 L atm.

*18.111 When an ideal gas expands at a constant temperature, $\Delta E = 0$ for the change. Why?

*18.112 When a real gas expands at a constant temperature, $\Delta E > 0$ for the change. Why?

18.113 How much work is accomplished by the following chemical reaction if it occurs inside a bomb calorimeter?

$$2C_4H_{10}(g) + 13O_2(g) \longrightarrow 8CO_2(g) + 10H_2O(g)$$

How much work is done if the reaction is carried out at 1 atm, with all reactants and products at 25 °C? [For $C_4H_{10}(g)$, $\Delta H_f^\circ = -126$ kJ mol^{-1}.]

*18.114 A cylinder fitted with a piston contains 5.00 L of a gas at a pressure of 4.00 atm. The entire apparatus is contained in a water bath to maintain a constant temperature of 25 °C. The piston is released and the gas expands until the pressure inside the cylinder equals the atmospheric pressure outside, which is 1 atm. Assume ideal gas behavior and calculate the amount of work done by the gas as it expands at constant temperature.

*18.115 The experiment described in the preceding exercise is repeated, but this time a weight, which exerts a pressure of 2 atm, is placed on the piston. When the gas expands, its pressure drops to this 2 atm pressure. Then the weight is removed and the gas is allowed to expand again to a final pressure of 1 atm. Throughout both expansions the temperature of the apparatus was held at a constant 25 °C. Calculate the amount of work performed by the gas in each step. How does the combined total amount of work in this two-step expansion compare to the amount of work done by the gas in the one-step expansion described in the preceding exercise?

18.116 What is the algebraic sign of ΔS for the following changes?
(a) $2Ag^+(aq) + CrO_4^{2-}(aq) \longrightarrow Ag_2CrO_4(s)$
(b) $NaCl(s) \rightarrow Na^+(aq) + Cl^-(aq)$
(c) $NH_3(g) \xrightarrow{H_2O} NH_3(aq)$
(d) naphthalene$(g) \rightarrow$ naphthalene(s)
(e) A gas is cooled from 40 °C to 25 °C at a constant pressure.
(f) A gas is compressed at constant temperature from 4.0 L to 2.0 L.

18.117 When potassium iodide spontaneously dissolves in water, the mixture becomes cool. For this change, which is of a larger magnitude, $T\Delta S$ or ΔH?

18.118 The reaction $Cl_2(g) + Br_2(g) \rightarrow 2BrCl(g)$ has a very small value for ΔS° (+11.6 J K^{-1}). Why?

18.119 For a phase change, why is there only one temperature at which there can be an equilibrium between the phases?

*18.120 The enthalpy of combustion, $\Delta H^\circ_{combustion}$, of oxalic acid, $H_2C_2O_4(s)$, is -246.05 kJ mol^{-1}. Consider the following data:

Substance	ΔH_f° (kJ mol^{-1})	S° (J mol^{-1} K^{-1})
C(s)	0	5.69
CO$_2$(g)	−393.5	213.6
H$_2$(g)	0	130.6
H$_2$O(l)	−285.8	69.96
O$_2$(g)	0	205.0
H$_2$C$_2$O$_4$(s)	?	120.1

(a) Write the balanced thermochemical equation that describes the combustion of one mole of oxalic acid.
(b) Write the balanced thermochemical equation that describes the formation of one mole of oxalic acid from its elements.
(c) Use the information in the table above and the equations in parts (a) and (b) to calculate ΔH_f° for oxalic acid.
(d) Calculate ΔS_f° for oxalic acid and ΔS° for the combustion of one mole of oxalic acid.
(e) Calculate ΔG_f° for oxalic acid and ΔG° for the combustion of one mole of oxalic acid.

18.121 Many biochemical reactions have positive values for ΔG° and seemingly should not be expected to be spontaneous. They occur, however, because they are chemically coupled with other reactions that have negative values of ΔG°. An example is the set of reactions that forms the beginning part of the sequence of reactions involved in the metabolism of glucose, a sugar. Given these reactions and their corresponding ΔG° values,

glucose + phosphate $\longrightarrow$ glucose-6-phosphate + H$_2$O

$$\Delta G^\circ = +13.13 \text{ kJ}$$

ATP + H$_2$O $\longrightarrow$ ADP + phosphate

$$\Delta G^\circ = -32.22 \text{ kJ}$$

calculate ΔG° for the coupled reaction

glucose + ATP $\longrightarrow$ glucose-6-phosphate + ADP

18.122 Cars, trucks, and other machines that use gas or diesel engines for power have cooling systems. In terms of thermodynamics, what makes these cooling systems necessary?

*18.123 Ethyl alcohol, C_2H_5OH, has been suggested as an alternative to gasoline as a fuel. In Example 18.5 we calculated ΔG° for combustion of 1 mol of C_2H_5OH; in Example 18.6 we calculated ΔG° for combustion of 1 mol of octane. Let's assume that gasoline has the same properties as octane (one of its constituents). The density of C_2H_5OH is 0.7893 g/mL; the density of octane, C_8H_{18}, is 0.7025 g/mL. Calculate the maximum work that could be obtained by burning 1 gallon (3.78 liters) each of C_2H_5OH and C_8H_{18}. On a *volume* basis, which is a better fuel?

18.124 Use the data in Table 18.3 to calculate the bond energy in the nitrogen molecule and the oxygen molecule.

18.125 The heat of vaporization of carbon tetrachloride, CCl_4, is 29.9 kJ mol^{-1}. Using this information and data in Tables 18.3 and 18.4, estimate the standard heat of formation of liquid CCl_4.

The availability of small, high-capacity batteries makes it possible for us to enjoy the conveniences of modern electronics, including the now-popular cell phone. Among the topics we will study in this chapter is how spontaneous redox reactions can be harnessed to produce electricity.

Electrochemistry

This Chapter in Context Oxidation and reduction (redox) reactions occur in many chemical systems. Examples include the rusting of iron, the action of bleach on stains, and the reactions of photosynthesis in the leaves of green plants. All of these changes involve the transfer of electrons from one chemical species to another. In this chapter we will study how it is possible to separate the oxidation and reduction steps in redox reactions and cause them to occur in different physical locations. When we are able to do this, we can use electricity to make nonspontaneous redox reactions happen and we can use spontaneous redox reactions to produce electricity.

Because electricity plays a role in these systems, the processes involved are described as *electrochemical changes*. The study of such changes is called *electrochemistry*.

The applications of electrochemistry are widespread. In industry, many important chemicals, including liquid bleach (sodium hypochlorite) and lye (sodium hydroxide), are manufactured by electrochemical reactions. Batteries, which produce electrical energy by means of chemical reactions, are used to power toys, flashlights, electronic calculators, laptop computers, heart pacemakers, video cameras, cellular phones, and even some automobiles. In the laboratory, electrical measurements enable us to monitor chemical reactions of all sorts, even those in systems as tiny as a living cell.

Besides the practical applications, we will also see how electrical measurements and the principles of thermodynamics combine to give fundamental information about chemical reactions, such as free energy changes and equilibrium constants.

19.1 Electrolysis

When electricity is passed through a molten (melted) ionic compound or through a solution of an electrolyte, a chemical reaction occurs that we call **electrolysis.** A typical electrolysis apparatus, called an **electrolysis cell** or

Sodium chloride melts at 801 °C.

electrolytic cell, is shown in Figure 19.1. This particular cell contains molten sodium chloride. (A substance undergoing electrolysis must be molten or in solution so its ions can move freely and conduction can occur.) *Inert electrodes*— electrodes that won't react with the molten NaCl—are dipped into the cell and then connected to a source of direct current (DC) electricity.

To study electrochemical changes, we must use direct current in which electrons move in only one direction, not in the oscillating, back and forth pattern of alternating current.

The DC source serves as an "electron pump," pulling electrons away from one electrode and pushing them through the external wiring onto the other electrode. The electrode from which electrons are removed becomes positively charged, while the other electrode becomes negatively charged. When electricity starts to flow, chemical changes begin to happen. At the positive electrode, the *anode,* oxidation occurs as electrons are pulled away from negatively charged chloride ions. The DC source pumps these electrons through the external electrical circuit to the negative electrode, the *cathode.* At the cathode, reduction takes place as the electrons are forced onto positively charged sodium ions.

The chemical changes that occur at the electrodes can be described by chemical equations.

At the melting point of NaCl, metallic sodium is a liquid.

$$Na^+(l) + e^- \longrightarrow Na(l) \qquad \text{(cathode)}$$

$$2Cl^-(l) \longrightarrow Cl_2(g) + 2e^- \qquad \text{(anode)}$$

You may recognize these as half-reactions of the type we used in Section 5.2 to balance equations by the ion–electron method. This is no accident. The half-reactions generated by the ion–electron method correspond to the chemical changes that occur at the electrodes during electrochemical changes.

By agreement among scientists, the names anode and cathode are assigned according to the nature of the reaction taking place at the electrode. If the reaction is oxidation, the electrode is called the *anode;* if it's reduction, the electrode is called the *cathode.*

The identification of an electrode as the anode or cathode depends on whether oxidation or reduction occurs there. In *any* apparatus in which an electrochemical reaction is taking place, the following definitions apply:

The **anode** is the electrode at which oxidation occurs.
The **cathode** is the electrode at which reduction occurs.

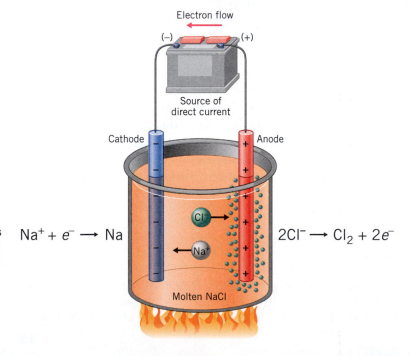

Figure 19.1 *Electrolysis of molten sodium chloride.* In this electrolysis cell, the passage of an electric current decomposes molten sodium chloride into metallic sodium and gaseous chlorine. Unless the products are kept apart, they react on contact to re-form NaCl.

Electron flow

(−)   (+)

Source of direct current

Cathode   Anode

$Na^+ + e^- \longrightarrow Na$ $2Cl^- \longrightarrow Cl_2 + 2e^-$

Cl^-

Na^+

Molten NaCl

Conduction of Charge in Electrochemical Cells

The conduction of electricity by a metal such as copper occurs by the movement of *electrons.* However, in a molten salt such as sodium chloride, or in a solution of an electrolyte (a solution that contains ions), *electrical charge is carried through the liquid by the movement of ions, not electrons.* The transport of electrical charge by ions is called **electrolytic conduction.** In an electrolytic cell it is able to continue *only* when chemical reactions take place at the electrodes.

When charged electrodes are dipped into molten NaCl they become surrounded by a layer of ions of the opposite charge. Let's look closely at what happens at one of the electrodes, say the anode (Figure 19.2). Here the positive charge of the electrode attracts negative Cl^- ions, which form a coating on the electrode's surface. The charge on the anode pulls electrons from the ions, causing them to be oxidized and changing them into neutral Cl atoms that join to become Cl_2 molecules. Because the molecules are neutral, they are not held by the electrode and so move away from the electrode's surface. Their places are quickly taken by negative ions from the surrounding liquid, which tends to leave the surrounding liquid positively charged. Other negative ions from farther away move toward the anode to keep the liquid there electrically neutral. In this way, negative ions gradually migrate toward the anode. By a similar process, positive ions diffuse through the liquid toward the negatively charged cathode where they become reduced.

> Within the electrolytic cell, charge is transported by the movement of ions. In the external electrical circuit, the electrical charge is transported by electrons that move through wires.

Cell Reactions

The overall reaction that takes place in the electrolysis cell is called the **cell reaction.** To obtain it, we add the individual electrode half-reactions together, but only after we make sure that the number of electrons gained in one half-reaction equals the number lost in the other, a requirement of every redox reaction. The procedure we use is the one described as the *ion–electron method* (Section 5.2). Thus, to obtain the cell reaction for the electrolysis of molten NaCl, we multiply the half-reaction for the reduction of sodium by 2 and then add the two half-reactions to obtain the net reaction. (Notice that $2e^-$ appear on each side, and so they cancel.)

> The terms cation and anion are derived from the directions the ions move during electrochemical reactions. Cations (positive ions) move toward the cathode and anions (negative ions) migrate toward the anode.

> Electrons are *transferred,* so no net gain or loss of electrons occurs.

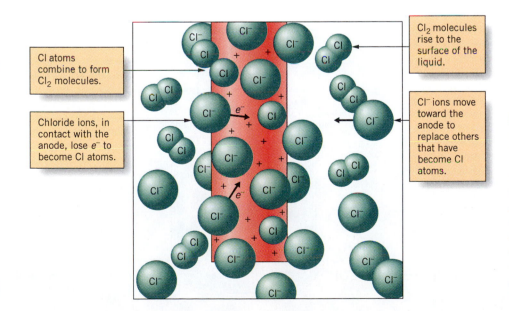

Cl atoms combine to form Cl_2 molecules.

Chloride ions, in contact with the anode, lose e^- to become Cl atoms.

Cl_2 molecules rise to the surface of the liquid.

Cl^- ions move toward the anode to replace others that have become Cl atoms.

Figure 19.2 *An atomic level view of changes at the anode in the electrolysis of molten NaCl.* The positive charge of the electrode attracts a coating of Cl^- ions. At the surface of the electrode, electrons are pulled from the ions, yielding neutral Cl atoms, which combine to form Cl_2 molecules that move away from the electrode and eventually rise to the surface as a gas.

$$2Na^+(l) + 2e^- \longrightarrow 2Na(l) \qquad \text{(cathode)}$$

$$2Cl^-(l) \longrightarrow Cl_2(g) + 2e^- \qquad \text{(anode)}$$

$$2Na^+(l) + 2Cl^-(l) + \cancel{2e^-} \longrightarrow 2Na(l) + Cl_2(g) + \cancel{2e^-} \qquad \text{(cell reaction)}$$

As you know, table salt is quite stable. It doesn't normally decompose because the reverse reaction, the reaction of sodium and chlorine to form sodium chloride, is highly spontaneous. Therefore, we often write the word *electrolysis* above the arrow in the equation to show that electricity is the driving force for this otherwise nonspontaneous reaction.

$$2Na^+(l) + 2Cl^-(l) \xrightarrow{\text{electrolysis}} 2Na(l) + Cl_2(g)$$

Electrolysis Reactions in Aqueous Solutions

When electrolysis is carried out in an aqueous solution, the electrode reactions are more difficult to predict because the oxidation and reduction of water can also occur. This happens, for example, in the electrolysis of a solution of potassium nitrate (Figure 19.3). The products are hydrogen and oxygen. At the cathode, water is reduced, not K^+.

$$2H_2O(l) + 2e^- \longrightarrow H_2(g) + 2OH^-(aq) \qquad \text{(cathode)}$$

At the anode, water is oxidized, not the nitrate ion.

$$2H_2O(l) \longrightarrow O_2(g) + 4H^+(aq) + 4e^- \qquad \text{(anode)}$$

When electrons appear as a reactant, the process is reduction; when they appear as a product, it is oxidation.

Color changes of an acid–base indicator dissolved in the solution confirm that the solution becomes basic around the cathode, where OH^- is formed, and acidic around the anode, where H^+ is formed (see Figure 19.4). In addition, the gases H_2 and O_2 can be separately collected.

The overall cell reaction can be obtained as before. Because the number of electrons lost has to equal the number gained, the cathode reaction must occur twice each time the anode reaction occurs once.

Figure 19.3 *Electrolysis of an aqueous solution of potassium nitrate.* The products of the electrolysis are the gases hydrogen and oxygen.

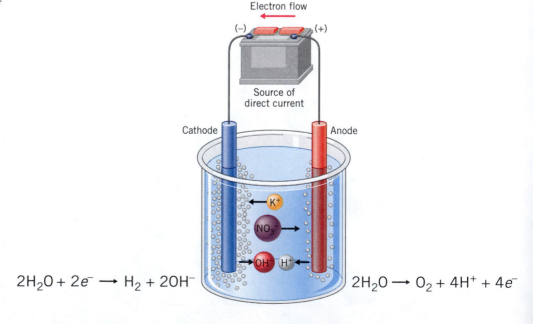

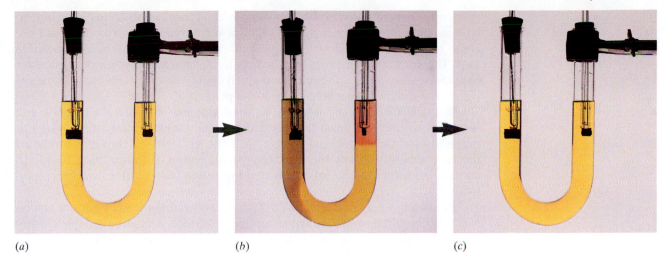

(a)

Before electrolysis begins. Yellow color of indicator shows solution is neutral.

(b)

During electrolysis, color changes show H^+ is formed at the anode (right) and OH^- is formed at the cathode (left).

(c)

After electrolysis is stopped and the solution is stirred, the yellow color shows that H^+ and OH^- have combined to give a neutral solution.

Figure 19.4 *Electrolysis of an aqueous solution of potassium nitrate in the presence of acid–base indicators.* (a) The initial yellow color indicates that the solution is neutral (neither acidic nor basic). (b) As the electrolysis proceeds, H^+ is produced at the anode (along with O_2) and causes the solution there to become pink. At the cathode, H_2 is evolved and OH^- ions are formed, which turns the solution around that electrode a bluish violet. (c) After the electrolysis is stopped and the solution is stirred, the color becomes yellow again as the H^+ and OH^- ions formed by the electrolysis neutralize each other.

$$4H_2O(l) + 4e^- \longrightarrow 2H_2(g) + 4OH^-(aq) \qquad \text{(cathode)}$$

$$2H_2O(l) \longrightarrow O_2(g) + 4H^+(aq) + 4e^- \qquad \text{(anode)}$$

After adding, we combine the coefficients for water and cancel the electrons from both sides to obtain the cell reaction.

$$6H_2O(l) \longrightarrow 2H_2(g) + O_2(g) + 4H^+(aq) + 4OH^-(aq)$$

Notice that hydrogen ions and hydroxide ions are produced in equal numbers. If the solution is stirred, they combine to form water.

$$6H_2O(l) \longrightarrow 2H_2(g) + O_2(g) + \underbrace{4H^+(aq) + 4OH^-(aq)}_{4H_2O}$$

The net change, then, is

$$2H_2O \xrightarrow{\text{electrolysis}} 2H_2(g) + O_2(g)$$

At this point you may have begun to wonder whether the potassium nitrate serves any function, since neither K^+ nor NO_3^- ions are changed by the reaction. Nevertheless, if the electrolysis is attempted with pure distilled water, nothing happens. There is no current flow, and no H_2 or O_2 forms. Apparently, the potassium nitrate must have some function.

The function of KNO_3 (or some other electrolyte) is to maintain electrical neutrality in the vicinity of the electrodes. If the KNO_3 were not present and the electrolysis were to occur anyway, the solution around the anode would become positively charged. It would become filled with H^+ ions, with no negative ions to balance their charge. Similarly, the solution surrounding the cathode would become negatively charged as it is filled with OH^- ions with no nearby positive ions. The formation of positively or negatively charged solutions just doesn't happen because doing so requires too much energy, so in the absence of an electrolyte the electrode reactions cannot take place.

When KNO_3 is in the solution, K^+ ions can move toward the cathode and mingle with the OH^- ions as they are formed. Similarly, the NO_3^- ions can move toward the anode and mingle with the H^+ ions as they are produced there. In this way, at any moment, each small region of the solution is able to contain the same number of positive and negative charges and thereby remain neutral.

Predicting the outcome of electrolysis reactions in aqueous solutions can be tricky because the reactions at the electrode surfaces are often complex, especially when they involve the evolution of hydrogen or oxygen. For example, the oxidation of H_2O to give O_2 is thermodynamically easier (in terms of the sign and magnitude of $\Delta G°$) than the oxidation of Cl^- to give Cl_2. The two competing half-reactions at the anode when aqueous sodium chloride is electrolyzed would be

$$2H_2O(l) \longrightarrow O_2(g) + 4H^+(aq) + 4e^-$$

$$2Cl^-(aq) \longrightarrow Cl_2(g) + 2e^-$$

As we said, oxygen should form by the first half-reaction, but because of the complexity of the electrode reactions, chlorine forms instead (provided that the chloride ion concentration is reasonably high). Thus, even though the evolution of O_2 is thermodynamically preferred, we don't see this happen because of complicating factors that are beyond the scope of this text.

Although it can be difficult to anticipate beforehand exactly what will happen in the electrolysis of aqueous solutions, we still can use what we learn experimentally about one electrolysis to predict what will happen in others. For instance, when a solution of $CuBr_2$ is electrolyzed, the cathode becomes coated with a deposit of copper and the solution around the anode acquires an orange-yellow color (see Figure 19.5). These observations tell us that copper ion is being reduced at the cathode and that bromide ion is being oxidized to bromine at the anode. The electrode half-reactions are

$$Cu^{2+}(aq) + 2e^- \longrightarrow Cu(s) \qquad \text{(cathode)}$$

$$2Br^-(aq) \longrightarrow Br_2(aq) + 2e^- \qquad \text{(anode)}$$

The net cell reaction is

$$Cu^{2+}(aq) + 2Br^-(aq) \xrightarrow{\text{electrolysis}} Cu(s) + Br_2(aq)$$

The behavior of a particular ion in an aqueous solution toward oxidation or reduction by electrolysis is the same regardless of the source of the ion. Therefore, once we know what happens in the electrolysis of $CuBr_2$, we can predict at least partially what will happen in the electrolysis of other salts that contain either of these ions. Solutions of $CuCl_2$, $CuSO_4$, $Cu(NO_3)_2$, and $Cu(C_2H_3O_2)_2$ all contain Cu^{2+} ion, and when they are electrolyzed we can expect that in each case Cu^{2+} will be reduced at the cathode. (However, we can't say at this point what will happen at the anode.) Similarly, solutions of KBr, NaBr, $CaBr_2$, and $FeBr_2$ all contain Br^-, so when solutions of these salts are electrolyzed we can expect that Br_2 will be formed at the anode.

Later in this chapter you will learn how to make reasonable predictions of what will happen in electrolysis reactions, even when there are competing reactions.

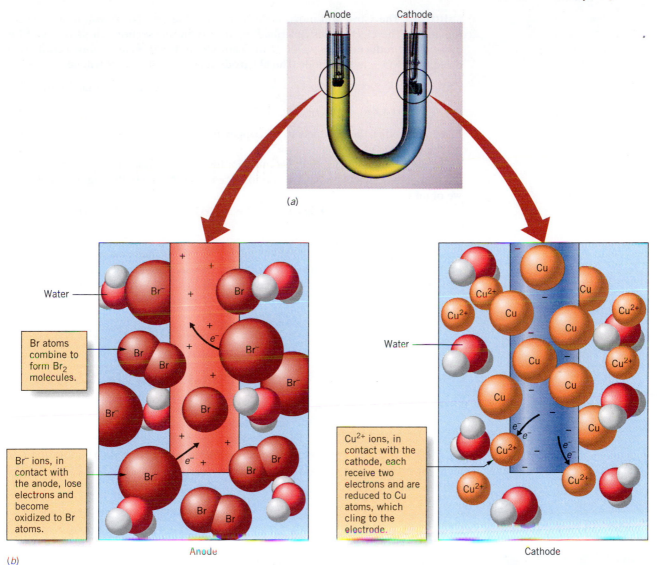

Figure 19.5 *Electrolysis of an aqueous solution of copper(II) bromide.* The blue color of the CuBr₂ solution is due to copper ion [actually, $Cu(H_2O)_4^{2+}$ ion]. At the anode on the left, bromide ions are oxidized to bromine atoms, which combine to form Br_2 molecules, imparting an orange-yellow color to the solution. At the cathode, copper ions are reduced. Here we see the copper as a black deposit building up on the electrode and flaking away to fall to the bottom of the apparatus. The diagrams below the photograph indicate what is happening at the atomic level at the electrodes. Once again, notice that electron transfer occurs at the surface of the electrodes and that electrons do not travel through the solution.

From the results of the electrolysis of solutions of KNO_3 and of $CuBr_2$, predict the products that will appear if a solution of $Cu(NO_3)_2$ is electrolyzed. Give the net cell reaction.

Analysis: A solution of $Cu(NO_3)_2$ contains Cu^{2+} and NO_3^- ions. To determine what will happen in this electrolysis, we need to recall what happens at the cathode when Cu^{2+} is the cation and what happens at the anode when NO_3^- is the anion.

EXAMPLE 19.1

Predicting the Outcome of an Electrolysis

Solution: The Cu^{2+} will move toward the cathode and NO_3^- will move toward the anode. The electrolyses described previously in this section tell us that Cu^{2+} is more easily reduced than water at the cathode and that water is more easily oxidized than NO_3^- at the anode. The electrode reactions, therefore, will be

$$Cu^{2+}(aq) + 2e^- \longrightarrow Cu(s) \qquad \text{(cathode: reduction of } Cu^{2+})$$

$$2H_2O(l) \longrightarrow O_2(g) + 4H^+(aq) + 4e^- \qquad \text{(anode: oxidation of } H_2O)$$

Thus, we predict the products to be copper metal (at the cathode) and oxygen gas (at the anode).

To obtain the cell reaction, we must multiply the cathode reaction by 2 so that the numbers of electrons gained and lost are equal. After combining half-reactions, we obtain

$$2Cu^{2+}(aq) + 2H_2O(l) \longrightarrow 2Cu(s) + O_2(g) + 4H^+(aq)$$

Practice Exercise 1

From the electrolysis reactions described in this section, predict the products that will form when a solution of KBr undergoes electrolysis. Write the equation for the net cell reaction. ◆

19.2 Stoichiometric Relationships in Electrolysis

Much of the early research in electrochemistry was performed by Michael Faraday. It was he who coined the terms anode, cathode, electrode, electrolyte, and electrolysis. In about 1833, Faraday discovered that the amount of chemical change that occurs during electrolysis is directly proportional to the amount of electrical charge that is passed through an electrolysis cell. For example, the reduction of copper ion at a cathode is given by the equation

$$Cu^{2+}(aq) + 2e^- \longrightarrow Cu(s)$$

The equation tells us that to deposit one mole of metallic copper requires two moles of electrons. Therefore, to deposit two moles of copper requires four moles of electrons, and that takes twice as much electricity. The half-reaction for an oxidation or reduction, therefore, relates the amount of chemical substance consumed or produced to the amount of electrons that the electric current must supply. To use this information, however, we must be able to relate it to electrical measurements that can be made in the laboratory.

The SI unit of electric current is the **ampere (A)** and the SI unit of charge is the **coulomb (C).** A coulomb is the amount of charge that passes by a given point in a wire when an electric current of one ampere flows for one second. This means that coulombs are the product of amperes of current multiplied by seconds. Thus

$$1 \text{ coulomb} = 1 \text{ ampere} \times 1 \text{ second}$$

$$1 \text{ C} = 1 \text{ A} \times \text{s}$$

For example, if a current of 4 A flows through a wire for 10 s, 40 C pass by a given point in the wire.

$$(4 \text{ A}) \times (10 \text{ s}) = 40 \text{ A} \times \text{s}$$

$$= 40 \text{ C}$$

Michael Faraday (1791–1867), a British scientist and both a chemist and a physicist, made key discoveries leading to electric motors, generators, and transformers.

Experimentally, it has been determined that 1 mol of electrons carries a charge of 96,485 C. We will use this number rounded to three significant figures.

$$1 \text{ mol } e^- \Leftrightarrow 9.65 \times 10^4 \text{ C} \quad \text{(to three significant figures)}$$

In electrochemistry, 1 mol of electrons is sometimes called 1 **faraday ($\mathscr{F}$)**, in honor of Michael Faraday. Thus,

$$1 \mathscr{F} = 9.65 \times 10^4 \text{ C/mol } e^-$$

The number of coulombs per mole of electrons is called the **Faraday constant.**

Now we have a way to relate laboratory measurements to the amount of chemical change that occurs during an electrolysis. Measuring the current in amperes and the time in seconds allows us to calculate the charge sent through the system in coulombs. From this we can get the amount of electrons (in moles), which we can then use to calculate the amount of chemical change produced.

Tools

Faraday constant

How many grams of copper are deposited on the cathode of an electrolytic cell if an electric current of 2.00 A is run through a solution of $CuSO_4$ for a period of 20.0 min?

EXAMPLE 19.2

Calculations Related to Electrolysis

Analysis: The balanced half-reaction serves as our tool for relating chemical change to amounts of electricity. The ion being reduced is Cu^{2+}, so the half-reaction is

$$Cu^{2+}(aq) + 2e^- \longrightarrow Cu(s)$$

Therefore,

$$1 \text{ mol Cu} \Leftrightarrow 2 \text{ mol } e^-$$

The product of current (in amperes) and time (in seconds) will give us charge (in coulombs). We can relate this to the number of moles of electrons by the Faraday constant. Then, from the number of moles of electrons we calculate moles of copper, from which we calculate the mass of copper by using the atomic mass.

$$\text{Faraday constant} = \frac{9.65 \times 10^4 \text{ C}}{\text{mole } e^-}$$

Solution: First we convert minutes to seconds; 20.0 min = 1.20×10^3 s. Then we multiply the current by the time to obtain the number of coulombs (1 A $\times$ s = 1 C).

$$20.0 \text{ min} \times \frac{60 \text{ s}}{1 \text{ min}} = 1.20 \times 10^3 \text{ s}$$

$$(1.20 \times 10^3 \text{ s}) \times (2.00 \text{ A}) = 2.40 \times 10^3 \text{ A} \times \text{s}$$

$$= 2.40 \times 10^3 \text{ C}$$

Because 1 mol $e^- \Leftrightarrow 9.65 \times 10^4$ C,

$$2.40 \times 10^3 \text{ C} \times \frac{1 \text{ mol } e^-}{9.65 \times 10^4 \text{ C}} = 0.0249 \text{ mol } e^-$$

Next, we use the relationship between mol e^- and mol Cu from the balanced half-reaction along with the atomic mass of copper.

$$0.0249 \text{ mol } e^- \times \left(\frac{1 \text{ mol Cu}}{2 \text{ mol } e^-} \right) \times \left(\frac{63.55 \text{ g Cu}}{1 \text{ mol Cu}} \right) = 0.791 \text{ g Cu}$$

The electrolysis will deposit 0.791 g of copper on the cathode.

We could have combined all these steps in a single calculation by stringing together the various conversion factors and using the factor-label method to cancel units.

$$2.00 \text{ A} \times 20.0 \text{ min} \times \frac{60 \text{ s}}{1 \text{ min}} \times \frac{1 \text{ mol } e^-}{9.65 \times 10^4 \text{ A s}} \times \frac{1 \text{ mol Cu}}{2 \text{ mol } e^-} \times \frac{63.55 \text{ g Cu}}{1 \text{ mol Cu}}$$

$$= 0.790 \text{ g Cu}$$

The small difference between the two answers is because of rounding off in the stepwise calculation.

Is the Answer Reasonable?
There is no simple check on the numerical answer, but the cancellation of units tells us we've set up the calculation correctly.

EXAMPLE 19.3

Calculations Related to Electrolysis

Electrolysis provides a useful way to deposit a thin metallic coating on an electrically conducting surface. The technique is called electroplating. How much time would it take in minutes to deposit 0.500 g of metallic nickel on a metal object using a current of 3.00 A? The nickel is reduced from the +2 oxidation state.

Analysis: We need an equation for the reduction. Because the nickel is reduced to the free metal from the +2 state, we can write

$$Ni^{2+}(aq) + 2e^- \longrightarrow Ni(s)$$

This gives the relationship

$$1 \text{ mol Ni} \Leftrightarrow 2 \text{ mol } e^-$$

We wish to deposit 0.500 g of Ni, which we can convert to moles. Then we can calculate the number of moles of electrons required, which in turn is used with the Faraday constant to determine the number of coulombs required. Because this is the product of amperes and seconds, we can calculate the time needed to deposit the metal.

Solution: First, we calculate the number of moles of electrons required.

$$0.500 \text{ g Ni} \times \left(\frac{1 \text{ mol Ni}}{58.69 \text{ g Ni}} \right) \times \left(\frac{2 \text{ mol } e^-}{1 \text{ mol Ni}} \right) = 0.0170 \text{ mol } e^-$$

Then we calculate the number of coulombs needed.

$$0.0170 \text{ mol } e^- \times \left(\frac{9.65 \times 10^4 \text{ C}}{1 \text{ mol } e^-} \right) = 1.64 \times 10^3 \text{ C}$$

$$= 1.64 \times 10^3 \text{ A} \times \text{s}$$

This tells us that the product of current multiplied by time equals $1.64 \times 10^3 \text{ A} \times \text{s}$. The current is 3.00 A. Dividing $1.64 \times 10^3 \text{ A} \times \text{s}$ by 3.00 A gives the time required in seconds, which we then convert to minutes.

$$\left(\frac{1.64 \times 10^3 \text{ A} \times \text{s}}{3.00 \text{ A}} \right) \times \left(\frac{1 \text{ min}}{60 \text{ s}} \right) = 9.11 \text{ min}$$

We could also have combined these calculations in a single string of conversion factors.

In this string of conversion factors, we have replaced 9.65×10^4 C by 9.65×10^4 A s. Remember, 1 C = 1 A × s.

$$0.500 \text{ g Ni} \times \frac{1 \text{ mol Ni}}{58.69 \text{ g Ni}} \times \frac{2 \text{ mol } e^-}{1 \text{ mol Ni}} \times \frac{9.65 \times 10^4 \text{ A s}}{1 \text{ mol } e^-} \times \frac{1}{3.00 \text{ A}} \times \frac{1 \text{ min}}{60 \text{ s}}$$

$$= 9.13 \text{ min}$$

As before, the small difference between the two answers is caused by rounding in the stepwise calculation.

Is the Answer Reasonable?
As in the preceding example, the units cancel correctly, so we can be confident we've set up the problem correctly.

EXAMPLE 19.4
Calculations Related to Electrolysis

What current is needed to deposit 0.500 g of chromium metal from a solution of Cr^{3+} in a period of 1.00 hr?

Solution: This problem is quite similar to Example 19.3. First we write the balanced half-reaction.

$$Cr^{3+}(aq) + 3e^- \longrightarrow Cr(s)$$

This gives the relationship 1 mol Cr ⇔ 3 mol e^-, which we use as before.

$$0.500 \text{ g Cr} \times \left(\frac{1 \text{ mol Cr}}{52.00 \text{ g Cr}}\right) \times \left(\frac{3 \text{ mol } e^-}{1 \text{ mol Cr}}\right) = 0.0288 \text{ mol } e^-$$

Then we find the number of coulombs.

$$0.0288 \text{ mol } e^- \times \left(\frac{9.65 \times 10^4 \text{ C}}{1 \text{ mol } e^-}\right) = 2.78 \times 10^3 \text{ C}$$

$$= 2.78 \times 10^3 \text{ A} \times \text{s}$$

Since we want the metal to be deposited in 1.00 hr (3600 s), the current is

$$\frac{2.78 \times 10^3 \text{ A} \times \text{s}}{3600 \text{ s}} = 0.772 \text{ A}$$

We can also set up the calculation as a string of conversion factors.

$$0.500 \text{ g Cr} \times \frac{1 \text{ mol Cr}}{52.00 \text{ g Cr}} \times \frac{3 \text{ mol } e^-}{1 \text{ mol Cr}} \times \frac{9.65 \times 10^4 \text{ A s}}{1 \text{ mol } e^-} \times \frac{1}{3600 \text{ s}} = 0.773 \text{ A}$$

Once again, rounding in the stepwise calculation causes a slight discrepancy between the two answers.

Is the Answer Reasonable?
Once again, the units cancel properly, so the problem is set up correctly.

Practice Exercise 2

How many moles of hydroxide ion will be produced at the cathode during the electrolysis of water with a current of 4.00 A for a period of 200 s? The cathode reaction is $2e^- + 2H_2O \longrightarrow H_2(g) + 2OH^-(aq)$. ◆

Practice Exercise 3

How many minutes will it take for a current of 10.0 A to deposit 3.00 g of gold from a solution of $AuCl_3$? ◆

Practice Exercise 4

What current must be supplied to deposit 3.00 g of gold from a solution of $AuCl_3$ in 20.0 min? ◆

19.3 Industrial Applications of Electrolysis

Besides being a useful tool in the chemistry laboratory, electrolysis has many important industrial applications. In this section we will briefly examine the chemistry of electroplating and the production of some of our most common chemicals.

Electroplating

Electroplating, which was mentioned in Example 19.3, is a procedure in which electrolysis is used to apply a thin (generally 0.03 to 0.05 mm thick) ornamental or protective coating of one metal over another. It is a common technique for improving the appearance and durability of metal objects. For instance, a thin, shiny coating of metallic chromium is applied over steel objects to make them attractive and to prevent rusting. Silver and gold plating are applied to jewelry made from less expensive metals, and silver plate is common on eating utensils (knives, forks, spoons, etc.).

Figure 19.6 illustrates a typical apparatus used for plating silver. Silver ion in the solution is reduced at the cathode where it is deposited as metallic silver on the object to be plated. At the anode, silver from the metal bar is oxidized, replenishing the supply of the silver ion in the solution. As time passes, silver is gradually transferred from the bar at the anode onto the object at the cathode.

The exact composition of the electroplating bath varies, depending on the metal to be deposited, and it can affect the appearance and durability of the finished surface. For example, silver deposited from a solution of silver nitrate ($AgNO_3$) does not stick to other metal surfaces very well. However, if it is deposited from a solution of silver cyanide containing $Ag(CN)_2^-$, the coating adheres well and is bright and shiny. Other metals that are electroplated from a cyanide bath are gold and cadmium. Nickel, which can also be applied as a protective coating, is plated from a nickel sulfate solution, and chromium is plated from a chromic acid (H_2CrO_4) solution.

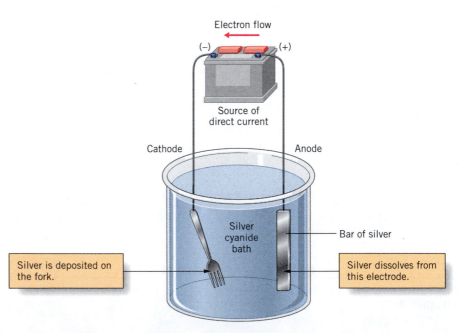

Figure 19.6 *Apparatus for electroplating silver.* Silver dissolves at the anode where it is oxidized to Ag^+. Silver is deposited on the cathode (the fork) where Ag^+ is reduced.

Production of Aluminum

Aluminum does not occur as the free metal in nature. It is even so reactive that the traditional methods for obtaining a metal from its ores do not work. Because of this, until the latter part of the nineteenth century, aluminum was so costly to isolate that only the very wealthy could afford aluminum products. Early efforts to produce aluminum by the electrolysis of a molten form of an aluminum compound were unfeasible. Its anhydrous halide salts (those with no water of hydration) are difficult to prepare and are volatile—they tend to evaporate rather than melt. On the other hand, its oxide, Al_2O_3, has such a high melting point (over 2000 °C) that no practical method of melting it could be found.

In 1886, a young student at Oberlin College, Charles M. Hall, discovered that Al_2O_3 dissolves in the molten form of a mineral called cryolite, Na_3AlF_6, to give a conducting mixture with a relatively low melting point from which aluminum could be produced electrolytically. The process was also discovered by Paul Héroult in France at nearly the same time, and today this method for producing aluminum is usually called the **Hall–Héroult process** (see Figure 19.7). Purified aluminum oxide, which is obtained from an ore called *bauxite,* is dissolved in molten cryolite in which the oxide dissociates to give Al^{3+} and O^{2-} ions. At the cathode, aluminum ions are reduced to produce the free metal, which forms as a layer of molten aluminum below the less dense solvent. At the carbon anodes, oxide ion is oxidized to give free O_2.

Aluminum is used today as a structural metal, in alloys, and in such products as aluminum foil, electrical wire, window frames, and kitchen pots and pans.

A large cell can produce as much as 900 lb of aluminum per day.

$$Al^{3+} + 3e^- \longrightarrow Al(l) \qquad \text{(cathode)}$$

$$2O^{2-} \longrightarrow O_2(g) + 4e^- \qquad \text{(anode)}$$

The net cell reaction is

$$4Al^{3+} + 6O^{2-} \longrightarrow 4Al(l) + 3O_2(g)$$

The oxygen formed at the anode attacks the carbon electrodes (producing CO_2), so the electrodes must be replaced frequently.

The production of aluminum consumes enormous amounts of electrical energy and is therefore very costly, not only in terms of dollars but also in terms of energy resources. For this reason, recycling of aluminum has a high priority as we seek to minimize our use of energy.

Production of aluminum by electrolysis. A row of electrolytic cells fills this aluminum production plant in Ghana (a country located on the west coast of Africa). In the foreground, a crane is about to lift a large ladle filled with molten aluminum.

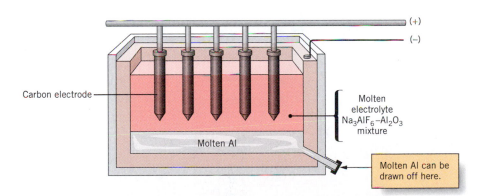

Carbon electrode

Molten electrolyte Na_3AlF_6–Al_2O_3 mixture

Molten Al

Molten Al can be drawn off here.

Figure 19.7 *Production of aluminum by electrolysis.* In the apparatus used to produce aluminum electrolytically by the Hall–Héroult process, Al_2O_3 is dissolved in molten cryolite, Na_3AlF_6. Al^{3+} is reduced to metallic Al and O^{2-} is oxidized to O_2, which reacts with the carbon anodes to give CO_2. Periodically, molten aluminum is drawn off at the bottom of the cell and additional Al_2O_3 is added to the cryolite. The carbon anodes also must be replaced from time to time as they are consumed by their reaction with O_2.

Production of Magnesium

Metallic magnesium has a number of structural uses because of its low density ("light weight"). Around the home, for example, you may have a magnesium alloy ladder.

The major source of magnesium is seawater in which, on a mole basis, Mg^{2+} is the third most abundant ion, exceeded only by Na^+ and Cl^-. To obtain metallic magnesium, seawater is made basic, which causes Mg^{2+} ions to precipitate as $Mg(OH)_2$. The precipitate is separated by filtration and dissolved in hydrochloric acid.

$$Mg(OH)_2 + 2HCl \longrightarrow MgCl_2 + 2H_2O$$

The resulting solution is evaporated to give solid $MgCl_2$, which is then melted and electrolyzed. Free magnesium is deposited at the cathode and chlorine gas is produced at the anode.

$$MgCl_2(l) \xrightarrow{\text{electrolysis}} Mg(l) + Cl_2(g)$$

Production of Sodium

Sodium is prepared by the electrolysis of molten sodium chloride (see Section 19.1). The metallic sodium and the chlorine gas that form must be kept apart or they will react violently and re-form NaCl. The **Downs cell** accomplishes this separation (see Figure 19.8).

Both sodium and chlorine are commercially important. Chlorine is used largely to manufacture plastics such as polyvinyl chloride (PVC), many solvents, and industrial chemicals. A small percentage of the annual chlorine production is used to chlorinate drinking water.

PVC is widely used to make pipes and conduits for water and sanitary systems.

Sodium has been used in the manufacture of tetraethyllead, an octane booster for gasoline that has been phased out in the United States but which is still used in many other countries. Sodium is also used as a coolant in certain nuclear reactors, and in the production of sodium vapor lamps. These lamps, with their familiar bright yellow color, have the advantage of giving off most of their energy in a portion of the spectrum that humans can see. In terms of useful light

Figure 19.8 *Commercial electrolysis of molten sodium chloride.* Cross section of the *Downs cell* used for the electrolysis of molten sodium chloride. The cathode is a circular ring that surrounds the anode. The electrodes are separated from each other by an iron screen. During the operation of the cell, molten sodium collects at the top of the cathode compartment, from which it is periodically drained. The chlorine gas bubbles out of the anode compartment and is collected.

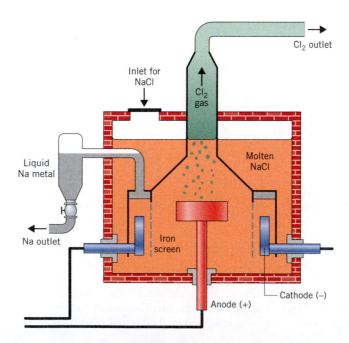

output versus energy input, they are about 15 times more efficient than ordinary incandescent lightbulbs and about 3 to 4 times more efficient than the bluish white mercury vapor lamps sometimes used for street lighting.

Refining of Copper

One of the most interesting and economically attractive applications of electrolysis is the purification or refining of metallic copper. When copper is first obtained from its ore, it is about 99% pure. The impurities—mostly silver, gold, platinum, iron, and zinc—decrease the electrical conductivity of the copper enough that even 99% pure copper must be further refined before it can be used in electrical wire.

The impure copper is used as the anode in an electrolysis cell that contains a solution of copper sulfate and sulfuric acid as the electrolyte (see Figure 19.9). The cathode is a thin sheet of very pure copper. When the cell is operated at the correct voltage, only copper and impurities more easily oxidized than copper (iron and zinc) dissolve at the anode. The less active metals simply fall off the electrode and settle to the bottom of the container. At the cathode, copper ions are reduced, but the zinc ions and iron ions remain in solution because they are more difficult to reduce than copper. Gradually, the impure copper anode dissolves and the copper cathode, about 99.96% pure, grows larger. The accumulating sludge—called anode mud—is removed periodically, and the value of the silver, gold, and platinum recovered from it virtually pays for the entire refining operation.

Copper refining. Copper cathodes, 99.96% pure, are pulled from the electrolytic refining tanks at Kennecott's Utah copper refinery. It takes about 28 days for the impure copper anodes to dissolve and deposit the pure metal on the cathodes.

Copper refining is one of the chief sources of gold in the United States.

Electrolysis of Brine

One of the most important commercial electrolysis reactions is the electrolysis of concentrated aqueous sodium chloride solutions called **brine.** An apparatus that could be used for this in the laboratory is shown in Figure 19.10. At the cathode, water is much more easily reduced than sodium ion, so H_2 forms.

$$2H_2O(l) + 2e^- \longrightarrow H_2(g) + 2OH^-(aq) \qquad \text{(cathode)}$$

As we noted earlier, even though water is more easily oxidized than chloride ion, complicating factors at the electrodes actually allow chloride ion to be oxidized instead. At the anode, therefore, we observe the formation of Cl_2.

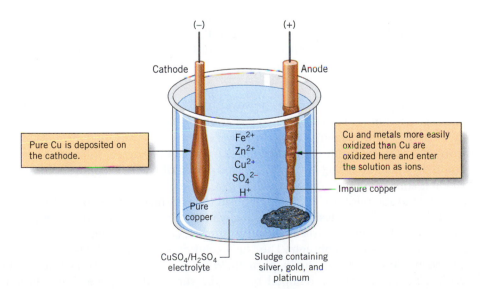

(−) Cathode   (+) Anode

Pure Cu is deposited on the cathode.

Fe^{2+}
Zn^{2+}
Cu^{2+}
SO_4^{2-}
H^+

Cu and metals more easily oxidized than Cu are oxidized here and enter the solution as ions.

Impure copper

Pure copper

$CuSO_4/H_2SO_4$ electrolyte

Sludge containing silver, gold, and platinum

Figure 19.9 *Purification of copper by electrolysis.* An impure copper anode dissolves and pure copper is deposited on the cathode. Metals less easily reduced than copper remain in solution, while metals less easily oxidized than copper settle to the bottom of the apparatus as "anode mud."

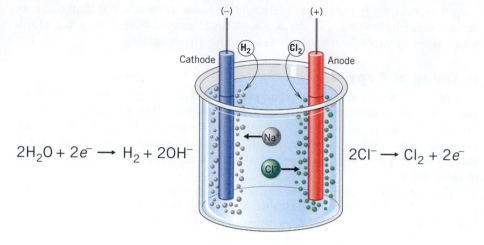

Figure 19.10 *The electrolysis of brine.* In the electrolysis of *brine*, a concentrated solution of sodium chloride, hydrogen gas is formed at the cathode and chlorine gas is formed at the anode. The solution becomes basic as the reaction progresses.

$$2H_2O + 2e^- \longrightarrow H_2 + 2OH^-$$

$$2Cl^- \longrightarrow Cl_2 + 2e^-$$

$$2Cl^-(aq) \longrightarrow Cl_2(g) + 2e^- \qquad \text{(anode)}$$

The net cell reaction is therefore

$$2Cl^-(aq) + 2H_2O(l) \longrightarrow H_2(g) + Cl_2(g) + 2OH^-(aq)$$

If we include the sodium ion, already in the solution as a spectator ion and not involved in the electrolysis directly, we can see why this is such an important reaction.

$$\underbrace{2Na^+(aq) + 2Cl^-(aq)}_{2NaCl(aq)} + 2H_2O \xrightarrow{\text{electrolysis}} H_2(g) + Cl_2(g) + \underbrace{2Na^+(aq) + 2OH^-(aq)}_{2NaOH(aq)}$$

Sodium hydroxide is commonly known as *lye* or as *caustic soda*.

Thus, the electrolysis converts inexpensive salt to valuable chemicals: H_2, Cl_2, and NaOH. The hydrogen is used to make other chemicals, including hydrogenated vegetable oils. The chlorine is used for the purposes mentioned earlier. Among the uses of sodium hydroxide, one of industry's most important bases, are the manufacture of soap and paper, the neutralization of acids in industrial reactions, and the purification of aluminum ores.

In the industrial electrolysis of brine to make pure NaOH, an apparatus as simple as that in Figure 19.10 cannot be used for three chief reasons. First, it is necessary to capture the H_2 and Cl_2 separately to prevent them from mixing and reacting (explosively). Second, the NaOH from the reaction is contaminated with unreacted NaCl. Third, if Cl_2 is left in the presence of NaOH, the solution becomes contaminated by hypochlorite ion (OCl^-), which forms by the reaction of Cl_2 with OH^-.

$$Cl_2 + 2OH^- \longrightarrow Cl^- + OCl^- + H_2O$$

In one manufacturing operation, however, the Cl_2 is not removed as it forms, and its reaction with hydroxide ion is used to manufacture aqueous sodium hypochlorite. For this purpose, the solution is stirred vigorously during the electrolysis so that very little Cl_2 escapes. As a result, a stirred solution of NaCl gradually changes during electrolysis to a solution of NaOCl, a dilute solution of which is sold as liquid laundry bleach (e.g., Clorox).

Most of the pure NaOH manufactured today is made in an apparatus called a **diaphragm cell.** The design varies somewhat, but Figure 19.11 illustrates its basic features. The cell consists of a steel wire mesh cathode that encloses a porous asbestos shell—the diaphragm. The NaCl solution is added to the top of the cell and seeps slowly through the diaphragm. When it contacts the iron cathode, hydrogen is evolved and is pumped out of the surrounding space. The solution,

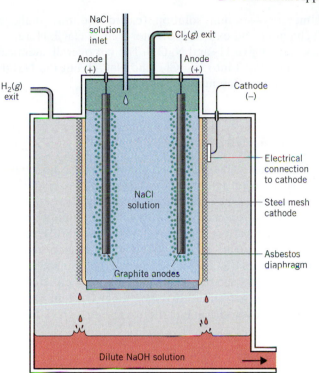

Figure 19.11 *A diaphragm cell used in the commercial production of NaOH by the electrolysis of aqueous NaCl.* This is a cross section of a cylindrical cell in which the NaCl solution is surrounded by an asbestos diaphragm supported by an iron mesh cathode. (From J. E. Brady and G. E. Humiston, *General Chemistry: Principles and Structure,* 4th ed. Copyright © 1986, John Wiley & Sons, New York. Used by permission.)

now containing dilute NaOH, drips off the cell into the reservoir below. Meanwhile, within the cell, chlorine is generated at the anodes dipping into the NaCl solution. Because there is no OH$^-$ in this solution, the Cl$_2$ can't react to form OCl$^-$ ion and simply bubbles out of the solution and is captured.

The NaOH obtained from the diaphragm cell, although free of OCl$^-$, is still contaminated by small amounts of NaCl. The production of pure NaOH is accomplished using a **mercury cell** (see Figure 19.12). In this apparatus, mercury serves as the cathode, and the cathode reaction actually involves the reduction of sodium ions to sodium atoms, which dissolve in the liquid mercury. The

The mercury cell, which gives pure NaOH, poses an environmental hazard because of the potential for mercury pollution. The waste products discharged from plants using mercury cells must be carefully monitored to prevent this.

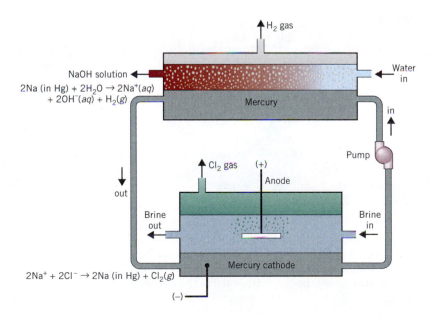

Figure 19.12 *Electrolysis of aqueous sodium chloride (brine) using a mercury cell.* At the anode, chloride ions are oxidized to chlorine, Cl$_2$. At the cathode, sodium ions are reduced to sodium atoms, which dissolve in the mercury. This mercury is pumped to a separate compartment, where it is exposed to water. As sodium atoms come to the surface of the mercury, they react with water to give H$_2$ and NaOH. (From J. E. Brady and G. E. Humiston, *General Chemistry: Principles and Structure,* 4th ed. Copyright © 1986, John Wiley & Sons, New York. Used by permission.)

mercury–sodium solution (called sodium amalgam) is pumped to another chamber and exposed to water free of NaCl. Here, the sodium atoms react with water to give H_2 and NaOH. The net overall chemical change is still the conversion of NaCl into H_2, Cl_2, and NaOH, but the NaOH produced in the mercury cell isn't contaminated by NaCl.

19.4 Galvanic Cells

If you have silver fillings in your teeth, you may have experienced a strange and perhaps even unpleasant sensation while accidentally biting on a piece of aluminum foil. The sensation is caused by a very mild electric shock produced by a voltage difference between your metal fillings and the aluminum. In effect, you created a battery. Batteries not too much different from this serve to power all sorts of gadgets, as mentioned at the beginning of this chapter. The energy for batteries comes from spontaneous redox reactions in which the electron transfer is forced to take place through a wire. Cells that provide electricity in this way are called **galvanic cells,** after Luigi Galvani (1737–1798), an Italian anatomist who discovered that electricity can cause the contraction of muscles.

> Galvanic cells are also called **voltaic cells,** after another Italian scientist Alessandro Volta (1745–1827), the inventor of the battery.

Setting Up a Galvanic Cell

If a shiny piece of metallic copper is placed into a solution of silver nitrate, a spontaneous reaction occurs. Gradually, a grayish white deposit forms on the copper and the solution itself becomes pale blue as hydrated Cu^{2+} ions enter the solution (see Figure 19.13). The equation is

$$2Ag^+(aq) + Cu(s) \longrightarrow Cu^{2+}(aq) + 2Ag(s)$$

No usable energy can be harnessed from this reaction, however, because all of the released energy appears as heat.

(a)

(b)

(c)

Figure 19.13 *Reaction of copper with a solution of silver nitrate.* (a) A coil of copper wire stands next to a beaker containing a silver nitrate solution. (b) When the copper wire is placed in the solution, copper dissolves, giving the solution its blue color, and metallic silver deposits as glittering crystals on the wire. (c) After a while, much of the copper has dissolved and nearly all of the silver has deposited as the free metal.

To produce *electrical* energy, the two half-reactions involved in the net reaction must be made to occur in separate containers or compartments called **half-cells.** An apparatus to accomplish this—a galvanic cell—is illustrated in Figure 19.14. On the left, a silver electrode dips into a solution of $AgNO_3$, and on the right, a copper electrode dips into a $Cu(NO_3)_2$ solution. The two solutions are connected by an external electrical circuit as well as by a *salt bridge,* the function of which will be described shortly. When the circuit is completed by closing the switch, the reduction of Ag^+ to Ag occurs spontaneously in the beaker on the left and oxidation of Cu to Cu^{2+} occurs spontaneously in the beaker on the right. Because of the nature of the reactions taking place, we can identify the silver electrode as the cathode and the copper electrode as the anode.

$$Ag^+(aq) + e^- \longrightarrow Ag(s) \qquad \text{(reduction—cathode)}$$

$$Cu(s) \longrightarrow Cu^{2+}(aq) + 2e^- \qquad \text{(oxidation—anode)}$$

When these reactions take place, electrons left behind by oxidation of the copper travel as an electric current through the external circuit to the cathode where they are picked up by the silver ions, which are thereby reduced.

For a galvanic cell to work, the solutions in both half-cells must remain electrically neutral. This requires that ions be permitted to enter or leave the solutions. For example, when copper is oxidized, the solution surrounding the electrode becomes filled with Cu^{2+} ions, so negative ions are needed to balance their charge. Similarly, when Ag^+ ions are reduced, NO_3^- ions are left behind in the solution and positive ions are needed to maintain neutrality. The salt bridge shown in Figure 19.14 serves these purposes. A **salt bridge** is a tube filled with an electrolyte solution, commonly KNO_3 or KCl, and fitted with porous plugs at each end. During operation of the cell, negative ions can diffuse from the salt bridge into the copper half-cell, or Cu^{2+} ions can leave the solution and enter the salt bridge. Both processes help keep this half-cell electrically neutral. At the other half-cell, positive ions from the salt bridge can enter or negative NO_3^- ions can leave to keep it electrically neutral.

Without the salt bridge, electrical neutrality couldn't be maintained and no electrical current could be produced by the cell. Therefore, *electrolytic contact*—

Remember, the anode is *always* the electrode at which oxidation takes place, and the cathode is where reduction takes place.

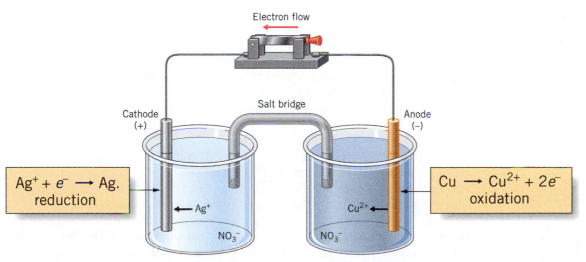

Figure 19.14 *A galvanic cell.* The cell consists of two half-cells where the oxidation and reduction half-reactions take place.

contact by means of a solution containing ions—must be maintained for the cell to function.

Charges on the Electrodes

In an *electrolysis* cell, an external DC source pushes electrons onto the cathode and pulls them from the anode. As a result, the cathode carries a negative charge and the anode carries a positive charge. In a *galvanic* cell, these charges are reversed. At the anode of the galvanic cell in Figure 19.14, copper atoms spontaneously leave the electrode and enter the solution as Cu^{2+} ions. The electrons that are left behind give the anode a slight negative charge. At the cathode, electrons spontaneously join Ag^+ ions to produce neutral atoms, but the effect is the same as if Ag^+ ions become part of the electrode, so the cathode acquires a slight positive charge. During the operation of the cell, the amount of positive and negative charge on the electrodes is kept small by the flow of electrons (an electric current) through the external circuit from the anode to the cathode when the circuit is complete. Remember, however, that it is the nature of the chemical change, not the electrical charge, that determines whether we label an electrode as a cathode or an anode.

> The small difference in charge between the electrodes is forced by the spontaneity of the overall reaction, that is, by the favorable free energy change. Nature's tendency toward electrical neutrality prevents a large buildup of charge on the electrodes and promotes the spontaneous flow of electricity through the external circuit.

Electrolytic Cell	**Galvanic Cell**
Cathode is negative (reduction).	Cathode is positive (reduction).
Anode is positive (oxidation).	Anode is negative (oxidation).

Even though the charges on the cathode and anode differ between electrolytic cells and galvanic cells, the ions in solution always move in the same direction. A cation is a positive ion that always moves away from the anode toward the cathode. In both types of cells, positive ions move toward the cathode. They are attracted there by the negative charge on the cathode in an electrolysis cell; they diffuse toward the cathode in our galvanic cell to balance the charge of negative ions left behind when the Ag^+ ions are reduced. Similarly, anions are negative ions that move away from the cathode and toward the anode. They are attracted to the positive anode in an electrolysis cell, and they diffuse toward the anode in our galvanic cell to balance the charge of the Cu^{2+} ions entering the solution.

> The events in an *electrolytic cell* are forced by electrical energy provided *from* the outside; those of the *galvanic cell* deliver electrical energy spontaneously *to* the outside.

Cell Notation for Galvanic Cells

As a matter of convenience, chemists have devised a shorthand way of describing the makeup of a galvanic cell. For example, the copper–silver cell that we have been using in our discussion is described as follows.

$$Cu(s) \,|\, Cu^{2+}(aq) \,\|\, Ag^+(aq) \,|\, Ag(s)$$

By convention, the anode half-cell is specified on the left, with the electrode material of the anode given first. In this case, the anode is copper metal, but in other galvanic cells an electrode could be an inert metal, like platinum, provided that the redox reaction in the half-cell is between two dissolved species (e.g., Fe^{2+} and Fe^{3+}). The single vertical bar represents a phase boundary—here, between the copper electrode and the solution that surrounds it. The double vertical bars represent the salt bridge, which separates the two half-cells. On the right, the cathode half-cell is described, with the material of the cathode given last. Thus, the electrodes themselves (copper and silver) are specified at opposite ends of the cell description.

> Also, notice that for each half-cell, the reactant in the redox change is given first. In the anode compartment, Cu is the reactant and is oxidized to Cu^{2+}, whereas in the cathode compartment, Ag^+ is the reactant and is reduced to Ag.

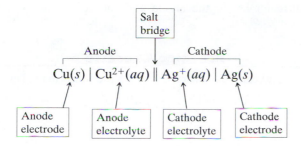

Anode electrode	Anode electrolyte	Cathode electrolyte	Cathode electrode

EXAMPLE 19.5
Describing Galvanic Cells

The following spontaneous reaction occurs when metallic zinc is dipped into a solution of copper sulfate.

$$Zn(s) + Cu^{2+}(aq) \longrightarrow Zn^{2+}(aq) + Cu(s)$$

Describe a galvanic cell that could take advantage of this reaction. What are the half-cell reactions? What is the abbreviated cell notation? Make a sketch of the cell and label the cathode and anode, the charges on each electrode, the direction of ion flow, and the direction of electron flow.

Analysis: Answering all these questions relies on identifying the anode and cathode from the equation for the cell reaction. The first step, therefore, is to divide the cell reaction into half-reactions. Then, the anode is where oxidation takes place and the cathode is where reduction takes place.

Solution: The half-reactions are

$$Zn(s) \longrightarrow Zn^{2+}(aq) + 2e^-$$
$$Cu^{2+}(aq) + 2e^- \longrightarrow Cu(s)$$

Zinc is oxidized, so it is the anode. The anode half-cell is therefore a zinc electrode dipping into a solution that contains Zn^{2+} [e.g., from dissolved $Zn(NO_3)_2$ or $ZnSO_4$]. Symbolically, the anode half-cell can be represented as

$$Zn(s) \mid Zn^{2+}(aq)$$

Copper ion is reduced, so the cathode half-cell consists of a copper electrode dipping into a solution continuing Cu^{2+} [e.g., from dissolved $Cu(NO_3)_2$ or $CuSO_4$]. This half-cell can be represented as

$$Cu^{2+}(aq) \mid Cu(s)$$

The cell notation places the zinc anode half-cell on the left and the copper cathode half-cell on the right separated by double bars that represent the salt bridge.

$$\underset{\text{anode}}{Zn(s) \mid Zn^{2+}(aq)} \parallel \underset{\text{cathode}}{Cu^{2+}(aq) \mid Cu(s)}$$

A sketch of the cell is shown in the margin. The anode always carries a negative charge in a galvanic cell, so the zinc electrode is negative and the copper electrode is positive. Electrons in the external circuit travel from the negative electrode to the positive electrode (i.e., from the Zn anode to the Cu cathode). Anions move toward the anode and cations move toward the cathode.

Practice Exercise 5

Sketch and label a galvanic cell that makes use of the following spontaneous redox reaction.

$$Mg(s) + Fe^{2+}(aq) \longrightarrow Mg^{2+}(aq) + Fe(s)$$

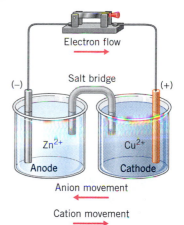

The zinc–copper cell. In the cell notation described in this Example, we indicate the anode on the left and the cathode on the right. In this drawing of the apparatus, the anode half-cell is also shown on the left, but it could just as easily be shown on the right, as in Figure 19.14. Be sure you understand that where we place the apparatus on the lab bench doesn't affect which half-cell is the anode and which is the cathode.

Write the half-reactions for the anode and cathode. Give the abbreviated cell notation. ◆

Practice Exercise 6

Write the anode and cathode half-reactions for the following galvanic cell.

$$Al(s) \mid Al^{3+}(aq) \parallel Pb^{2+}(aq) \mid Pb(s) \; \blacklozenge$$

19.5 Cell Potentials and Reduction Potentials

A galvanic cell has the ability to push electrons through an external circuit. The magnitude of this ability is expressed as a **potential.** Potential is expressed in an electrical unit called the **volt (V),** which is a measure of the amount of energy, in joules, that can be delivered per coulomb of charge as the current moves through the circuit. Thus, a current flowing under a potential of 1 volt can deliver 1 joule of energy per coulomb.

$$1 \, V = 1 \, J/C \tag{19.1}$$

> The potential generated by a galvanic cell has also been called an **electromotive force (emf),** suggesting that it can be thought of as the force with which an electric current is pushed through a wire. Electrical current and emf are often likened to the flow of water in a hose ("current") and the pressure of the water ("emf").

Cell Potentials of Galvanic Cells

The voltage or potential of a galvanic cell varies with the amount of current flowing through the circuit. The *maximum* potential that a given cell can generate is called its **cell potential, E_{cell},** and it depends on the composition of the electrodes, the concentrations of the ions in the half-cells, and the temperature. Therefore, to compare the potentials for different cells we use the **standard cell potential,** symbolized $E°_{cell}$. This is the potential of the cell when all of the ion concentrations are 1.00 M, the temperature is 25 °C, and any gases that are involved in the cell reaction are at a pressure of 1 atm.

> If current is drawn from a cell, some of the cell's voltage is lost overcoming its own internal resistance, and the measured voltage isn't E_{cell}.

Cell potentials are rarely larger than a few volts. For example, the standard cell potential for the galvanic cell constructed from silver and copper electrodes shown in Figure 19.15 is only 0.46 V, and one cell in an automobile battery produces only about 2 V. Batteries that generate higher voltages contain a number of cells arranged in series so that their potentials are additive.

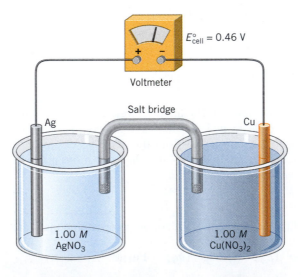

Figure 19.15 *A cell designed to generate the standard cell potential.* The concentrations of the ions in the half-cells are 1.00 M.

Reduction Potentials

It is useful to imagine that the measured overall cell potential arises from a competition between the two half-cells for electrons. Thus, we think of each half-cell as having a certain natural tendency to acquire electrons and proceed as a *reduction*. The magnitude of this tendency is expressed by the half-reaction's **reduction potential.** When measured under standard conditions, namely, 25 °C, concentrations of 1.00 M for all solutes, and a pressure of 1 atm, the reduction potential is called the **standard reduction potential.** To represent a standard reduction potential, we will add a subscript to the symbol $E°$ that identifies the substance undergoing reduction. Thus, the standard reduction potential for the half-reaction

$$Cu^{2+}(aq) + 2e^- \longrightarrow Cu(s)$$

is specified as $E°_{Cu^{2+}}$.

> Standard reduction potentials are also called **standard electrode potentials.**

When two half-cells are connected, the one with the larger reduction potential (the one with the greater tendency to undergo reduction) acquires electrons from the half-cell with the lower reduction potential, which is therefore forced to undergo oxidation. The measured cell potential actually represents the magnitude of the *difference* between the reduction potential of one half-cell and the reduction potential of the other. In general, therefore,

$$E°_{cell} = \left(\begin{array}{c}\text{standard reduction} \\ \text{potential of the} \\ \text{substance reduced}\end{array}\right) - \left(\begin{array}{c}\text{standard reduction} \\ \text{potential of the} \\ \text{substance oxidized}\end{array}\right) \quad (19.2)$$

Tools

Standard reduction potentials

As an example, let's look at the copper–silver cell. From the cell reaction,

$$2Ag^+(aq) + Cu(s) \longrightarrow 2Ag(s) + Cu^{2+}(aq)$$

we can see that silver ion is reduced and copper is oxidized. If we compare the two possible reduction half-reactions,

$$Ag^+(aq) + e^- \longrightarrow Ag(s)$$

$$Cu^{2+}(aq) + 2e^- \longrightarrow Cu(s)$$

the one for Ag^+ must have a greater tendency to proceed than the one for Cu^{2+}, because it is the silver ion that is actually reduced. This means that the standard reduction potential of Ag^+ must be algebraically larger than the standard reduction potential of Cu^{2+}. In other words, if we knew the values of $E°_{Ag^+}$ and $E°_{Cu^{2+}}$, we could calculate $E°_{cell}$ with Equation 19.2 by subtracting the smaller reduction potential from the larger one.

> Equation 19.2 will be used frequently, so be sure you know it.

$$E°_{cell} = E°_{Ag^+} - E°_{Cu^{2+}}$$

Assigning Standard Reduction Potentials

Unfortunately there is no way to measure the standard reduction potential of an isolated half-cell. All we can measure is the difference in potential produced when two half-cells are connected. Therefore, to assign values to the various standard reduction potentials, a reference electrode has been arbitrarily chosen and its standard reduction potential has been assigned a value of *exactly* 0 V. This reference electrode is called the **standard hydrogen electrode** (see Figure 19.16). Gaseous hydrogen at a pressure of 1 atm is bubbled over a platinum electrode coated with very finely divided platinum, which provides a large

> The large surface area provided by the fine platinum coating on the electrode enables the electrode reaction to occur rapidly.

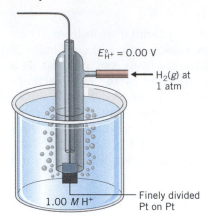

Figure 19.16 *The hydrogen electrode.* The half-reaction is $2H^+(aq) + 2e^- \rightleftharpoons H_2(g)$.

surface area on which the electrode reaction can occur. This electrode is surrounded by a solution whose temperature is 25 °C and in which the hydrogen ion concentration is 1.00 M. The half-cell reaction at the platinum surface, written as a reduction, is

$$2H^+(aq, 1.00\ M) + 2e^- \rightleftharpoons H_2(g, 1\ \text{atm}) \qquad E^\circ_{H^+} = 0\ \text{V}$$

The double arrows indicate only that the reaction is reversible, not that there is true equilibrium. Whether the half-reaction occurs as reduction or oxidation depends on the reduction potential of the half-cell with which it is paired.

Figure 19.17 illustrates the hydrogen electrode connected to a copper half-cell to form a galvanic cell. To obtain the cell reaction, we have to know what is oxidized and what is reduced, and we need to know which is the cathode and which is the anode. We can determine this by measuring the charges on the electrode, because we know that in a galvanic cell the cathode is the positive electrode and the anode is the negative electrode. When we use a voltmeter to measure the potential of the cell, we find that the copper electrode carries a positive charge and the hydrogen electrode a negative charge. Therefore, copper must be the cathode, and Cu^{2+} is reduced to Cu when the cell operates. Similarly, hydrogen must be the anode, and H_2 is oxidized to H^+. The half-reactions and cell reaction, therefore, are

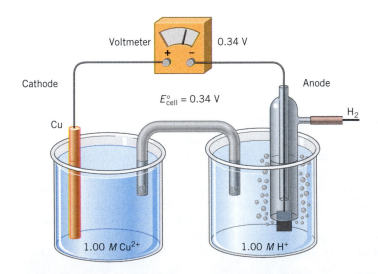

Figure 19.17 *A galvanic cell composed of copper and hydrogen half-cells.* The cell reaction is $Cu^{2+}(aq) + H_2(g) \rightarrow Cu(s) + 2H^+(aq)$.

$$Cu^{2+}(aq) + 2e^- \longrightarrow Cu(s) \qquad \text{(cathode)}$$

$$H_2(g) \longrightarrow 2H^+(aq) + 2e^- \qquad \text{(anode)}$$

$$Cu^{2+}(aq) + H_2(g) \longrightarrow Cu(s) + 2H^+(aq) \qquad \text{(cell reaction)}$$

Using Equation 19.2, we can express E°_{cell} in terms of $E^{\circ}_{Cu^{2+}}$ and $E^{\circ}_{H^+}$.

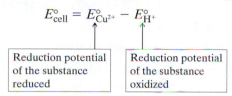

$$E^{\circ}_{cell} = E^{\circ}_{Cu^{2+}} - E^{\circ}_{H^+}$$

| Reduction potential of the substance reduced | Reduction potential of the substance oxidized |

The measured standard cell potential is 0.34 V and $E^{\circ}_{H^+}$ equals 0.00 V. Therefore,

$$0.34 \text{ V} = E^{\circ}_{Cu^{2+}} - 0.00 \text{ V}$$

Relative to the hydrogen electrode, then, the standard reduction potential of Cu^{2+} is +0.34 V. (We have written the value with a plus sign because some reduction potentials are negative, as we will see.)

In a galvanic cell, the measured *cell* potential is **always** taken to be a positive value. This is important to remember.

Now let's look at a galvanic cell set up between a zinc electrode and a hydrogen electrode (see Figure 19.18). This time we find that the hydrogen electrode is positive and the zinc electrode is negative; here, hydrogen is the cathode and zinc is the anode. This means that hydrogen ion is being reduced and zinc is being oxidized. The half-reactions and cell reaction are therefore

Remember, in a galvanic cell, the cathode is (+) and the anode is (−).

$$2H^+(aq) + 2e^- \longrightarrow H_2(g) \qquad \text{(cathode)}$$

$$Zn(s) \longrightarrow Zn^{2+}(aq) + 2e^- \qquad \text{(anode)}$$

$$2H^+(aq) + Zn(s) \longrightarrow H_2(g) + Zn^{2+}(aq) \qquad \text{(cell reaction)}$$

From Equation 19.2, the standard cell potential is

$$E^{\circ}_{cell} = E^{\circ}_{H^+} - E^{\circ}_{Zn^{2+}}$$

Substituting into this the measured standard cell potential of 0.76 V and $E^{\circ}_{H^+} = 0.00$ V, we have

$$0.76 \text{ V} = 0.00 \text{ V} - E^{\circ}_{Zn^{2+}}$$

which gives

$$E^{\circ}_{Zn^{2+}} = -0.76 \text{ V}$$

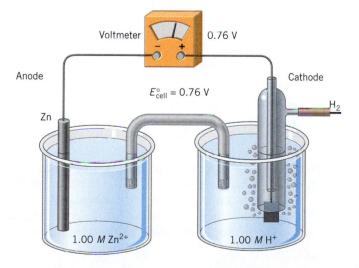

Figure 19.18 *A galvanic cell composed of zinc and hydrogen half-cells.* The cell reaction is $Zn(s) + 2H^+(aq) \rightarrow Zn^{2+}(aq) + H_2(g)$.

Table 19.1 Standard Reduction Potentials at 25 °C

Half-reaction	$E°$ (volts)
$F_2(g) + 2e^- \rightleftharpoons 2F^-(aq)$	+2.87
$S_2O_8^{2-}(aq) + 2e^- \rightleftharpoons 2SO_4^{2-}(aq)$	+2.01
$PbO_2(s) + HSO_4^-(aq) + 3H^+(aq) + 2e^- \rightleftharpoons PbSO_4(s) + 2H_2O$	+1.69
$2HOCl(aq) + 2H^+(aq) + 2e^- \rightleftharpoons Cl_2(g) + 2H_2O$	+1.63
$MnO_4^-(aq) + 8H^+(aq) + 5e^- \rightleftharpoons Mn^{2+}(aq) + 4H_2O$	+1.51
$PbO_2(s) + 4H^+(aq) + 2e^- \rightleftharpoons Pb^{2+}(aq) + 2H_2O$	+1.46
$BrO_3^-(aq) + 6H^+(aq) + 6e^- \rightleftharpoons Br^-(aq) + 3H_2O$	+1.44
$Au^{3+}(aq) + 3e^- \rightleftharpoons Au(s)$	+1.42
$Cl_2(g) + 2e^- \rightleftharpoons 2Cl^-(aq)$	+1.36
$O_2(g) + 4H^+(aq) + 4e^- \rightleftharpoons 2H_2O$	+1.23
$Br_2(aq) + 2e^- \rightleftharpoons 2Br^-(aq)$	+1.07
$NO_3^-(aq) + 4H^+(aq) + 3e^- \rightleftharpoons NO(g) + 2H_2O$	+0.96
$Ag^+(aq) + e^- \rightleftharpoons Ag(s)$	+0.80
$Fe^{3+}(aq) + e^- \rightleftharpoons Fe^{2+}(aq)$	+0.77
$I_2(s) + 2e^- \rightleftharpoons 2I^-(aq)$	+0.54
$NiO_2(s) + 2H_2O + 2e^- \rightleftharpoons Ni(OH)_2(s) + 2OH^-(aq)$	+0.49
$Cu^{2+}(aq) + 2e^- \rightleftharpoons Cu(s)$	+0.34
$SO_4^{2-}(aq) + 4H^+(aq) + 2e^- \rightleftharpoons H_2SO_3(aq) + H_2O$	+0.17
$2H^+(aq) + 2e^- \rightleftharpoons H_2(g)$	0
$Sn^{2+}(aq) + 2e^- \rightleftharpoons Sn(s)$	−0.14
$Ni^{2+}(aq) + 2e^- \rightleftharpoons Ni(s)$	−0.25
$Co^{2+}(aq) + 2e^- \rightleftharpoons Co(s)$	−0.28
$PbSO_4(s) + 2e^- \rightleftharpoons Pb(s) + SO_4^{2-}(aq)$	−0.36
$Cd^{2+}(aq) + 2e^- \rightleftharpoons Cd(s)$	−0.40
$Fe^{2+}(aq) + 2e^- \rightleftharpoons Fe(s)$	−0.44
$Cr^{3+}(aq) + 3e^- \rightleftharpoons Cr(s)$	−0.74
$Zn^{2+}(aq) + 2e^- \rightleftharpoons Zn(s)$	−0.76
$2H_2O + 2e^- \rightleftharpoons H_2(g) + 2OH^-(aq)$	−0.83
$Al^{3+}(aq) + 3e^- \rightleftharpoons Al(s)$	−1.66
$Mg^{2+}(aq) + 2e^- \rightleftharpoons Mg(s)$	−2.37
$Na^+(aq) + e^- \rightleftharpoons Na(s)$	−2.71
$Ca^{2+}(aq) + 2e^- \rightleftharpoons Ca(s)$	−2.76
$K^+(aq) + e^- \rightleftharpoons K(s)$	−2.92
$Li^+(aq) + e^- \rightleftharpoons Li(s)$	−3.05

Substances located to the left of the double arrows are *oxidizing agents*, because they become reduced when the reactions proceed in the forward direction. The best oxidizing agents are those most easily reduced, and they are located at the top of the table (e.g., F_2).

Substances located to the right of the double arrows are *reducing agents;* they become oxidized when the reactions proceed from right to left. The best reducing agents are those found at the bottom of the table (e.g., Li).

Notice that the reduction potential of zinc is negative. A negative reduction potential simply means that the substance is not as easily reduced as H^+. In this case, it tells us that Zn is oxidized when it is paired with the hydrogen electrode.

The standard reduction potentials of many half-reactions can be compared to that for the hydrogen electrode in the manner described above. Table 19.1

lists values obtained for some typical half-reactions. They are arranged in decreasing order—the half-reactions at the top have the greatest tendency to occur as reduction, while those at the bottom have the least tendency to occur as reduction.

EXAMPLE 19.6

Calculating Half-cell Potentials

We mentioned earlier that the standard cell potential of the silver–copper galvanic cell has a value of 0.46 V. The cell reaction is

$$2Ag^+(aq) + Cu(s) \longrightarrow 2Ag(s) + Cu^{2+}(aq)$$

and we have seen that the reduction potential of Cu^{2+}, $E^\circ_{Cu^{2+}}$, is +0.34 V. What is the value of $E^\circ_{Ag^+}$, the reduction potential of Ag^+?

Analysis: Since we know the potential of the cell and one of the two reduction potentials, we can use Equation 19.2 to calculate the unknown reduction potential. This requires that we identify the substance oxidized and the substance reduced.

Solution: Silver changes from Ag^+ to Ag, so it is reduced; copper is oxidized from Cu to Cu^{2+}. Therefore, according to Equation 19.2,

$$E^\circ_{cell} = E^\circ_{Ag^+} - E^\circ_{Cu^{2+}}$$

Substituting values for E°_{cell} and $E^\circ_{Cu^{2+}}$,

$$0.46 \text{ V} = E^\circ_{Ag^+} - 0.34 \text{ V}$$

$$E^\circ_{Ag^+} = 0.46 \text{ V} + 0.34 \text{ V}$$

$$= 0.80 \text{ V}$$

The standard reduction potential of silver ion is therefore +0.80 V.

Practice Exercise 7

The galvanic cell described in Practice Exercise 5 has a standard cell potential of 1.96 V. The standard reduction potential of Fe^{2+} corresponding to the half-reaction $Fe^{2+}(aq) + 2e^- \rightleftharpoons Fe(s)$ is −0.44 V. Calculate the standard reduction potential of magnesium. Check your answer by referring to Table 19.1. ↑

19.6 Using Standard Reduction Potentials

One of the goals of chemistry is to be able to predict reactions, and the half-reactions and standard reduction potentials of Table 19.1 serve this purpose particularly well for redox reactions. There are a number of questions we can address using the data in this table. Would a reaction occur, for example, between Cl_2 and Br^- to give Cl^- and Br_2? Or would Br_2 react with Cl^- to give Br^- and Cl_2? What reaction would happen if the oxidized and reduced form of one species, like Fe and a salt of Fe^{2+}, were mixed with a similar pair of another, like Ni and a salt of Ni^{2+}? Would Fe reduce Ni^{2+}, or would Ni reduce Fe^{2+}? How can we use the data in Table 19.1 to calculate the standard cell potential when a given redox reaction is set up as a galvanic cell? And finally, can we use the data in Table 19.1 to predict the outcome of electrolysis when there are competing reactions? Let's take up examples of these questions.

Facets of Chemistry 19.1

Corrosion of Iron and Cathodic Protection

A problem that has plagued humanity ever since the discovery of methods for obtaining iron and other metals from their ores has been corrosion—the reaction of a metal with substances in the environment. The rusting of iron in particular is a serious problem because iron and steel have so many uses.

The rusting of iron is a complex chemical reaction that involves both oxygen and moisture (see Figure 1). Iron won't rust in pure water that's oxygen free, and it won't rust in pure oxygen in the absence of moisture. The corrosion process is apparently electrochemical in nature, as shown in the accompanying diagram. At one place on the surface, iron becomes oxidized in the presence of water and enters solution as Fe^{2+}.

$$Fe(s) \longrightarrow Fe^{2+}(aq) + 2e^-$$

At this location the iron is acting as an anode.

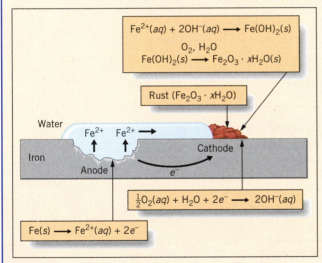

Figure 1 *Corrosion of iron.* Iron dissolves in anodic regions to give Fe^{2+}. Electrons travel through the metal to cathodic sites where oxygen is reduced, forming OH^-. The combination of the Fe^{2+} and OH^-, followed by air oxidation, gives rust.

The electrons that are released when the iron is oxidized travel through the metal to some other place where the iron is exposed to oxygen. This is where reduction takes place (it's a cathodic region on the metal surface), and oxygen is reduced to give hydroxide ion.

$$\tfrac{1}{2}O_2(aq) + H_2O + 2e^- \longrightarrow 2OH^-(aq)$$

The iron(II) ions that are formed at the anodic regions gradually diffuse through the water and eventually contact the hydroxide ions. This causes a precipitate of $Fe(OH)_2$ to form, which is very easily oxidized by O_2 to give $Fe(OH)_3$. This hydroxide readily loses water. In fact, complete dehydration gives the oxide,

$$2Fe(OH)_3 \longrightarrow Fe_2O_3 + 3H_2O$$

When partial dehydration of the $Fe(OH)_3$ occurs, *rust* is formed. It has a composition that lies between that of the hydroxide and that of the oxide, Fe_2O_3, and is usually referred to as a hydrated oxide. Its formula is generally represented as $Fe_2O_3 \cdot xH_2O$.

This mechanism for the rusting of iron explains one of the more interesting aspects of this damaging process. Perhaps you have noticed that when rusting occurs on the body of a car, the rust appears at and around a break (or a scratch) in the surface of the paint, but the damage extends under the painted surface for some distance. Apparently, the Fe^{2+} ions that are formed at the anode sites are able to diffuse rather long distances to the hole in the paint, where they finally react with air to form the rust.

Cathodic Protection

One way to prevent the rusting of iron is to coat it with another metal. This is done with "tin" cans, which are actu-

Figure 2 *Cathodic protection.* Before launching, a shiny new zinc anode disk is attached to the bronze rudder of this boat to provide cathodic protection. Over time, the zinc will corrode instead of the less reactive bronze. (The rudder is painted with a special blue paint to inhibit the growth of barnacles.)

ally steel cans that have been coated with a thin layer of tin. However, if the layer of tin is scratched and the iron beneath is exposed, the corrosion is accelerated because iron has a lower reduction potential than tin; the iron becomes the anode in an electrochemical cell and is easily oxidized.

Another way to prevent corrosion is called *cathodic protection*. It involves placing the iron in contact with a metal that is *more easily* oxidized. This causes iron to be a cathode and the other metal to be the anode. If corrosion occurs, iron is protected from oxidation because it is cathodic and the other metal reacts instead.

Zinc is most often used to provide cathodic protection to other metals. For example, zinc sacrificial anodes can be attached to the rudder of a boat (see Figure 2). When the rudder is submerged, the zinc will gradually corrode but the metal of the rudder will not. Periodically, the anodes are replaced to provide continued protection.

Steel objects that must withstand the weather are often coated with a layer of zinc, a process called galvanizing. You've seen this on chain-link fences and metal garbage pails. Even if the steel is exposed through a scratch, it is prevented from being oxidized because it is in contact with a metal that is more easily oxidized.

Predicting Spontaneous Redox Reactions

It's easy to predict the spontaneous reaction that occurs between the substances in two half-reactions, because we know that *the half-reaction with the more positive reduction potential always takes place as written (namely, as a reduction), while the other half-reaction is forced to run in reverse (as an oxidation).* This applies whether the reaction is taking place in the separate compartments of a galvanic cell or in a single container with all of the substances combined in the same reaction mixture.

EXAMPLE 19.7

Predicting a Spontaneous Reaction

What spontaneous reaction occurs if Cl_2 and Br_2 are added to a solution that contains both Cl^- and Br^-?

Analysis: We know that the more easily reduced substance *will be* reduced. If we can find reduction potentials for Cl_2 and Br_2, we can use their $E°$ values to determine this and proceed as before.

Solution: There are two possible reduction reactions.

$$Cl_2(g) + 2e^- \longrightarrow 2Cl^-(aq)$$

$$Br_2(aq) + 2e^- \longrightarrow 2Br^-(aq)$$

Referring to Table 19.1, we find that Cl_2 has a more positive reduction potential (1.36 V) than does Br_2 (1.07 V). This means Cl_2 will be reduced and the half-reaction for Br_2 will be reversed. Therefore, the spontaneous reaction has the following half-reactions.

$$Cl_2(g) + 2e^- \longrightarrow 2Cl^-(aq) \qquad \text{(a reduction)}$$

$$2Br^-(aq) \longrightarrow Br_2(aq) + 2e^- \qquad \text{(an oxidation)}$$

The net reaction will be

$$Cl_2(g) + 2Br^-(aq) \longrightarrow Br_2(aq) + 2Cl^-(aq)$$

Experimentally, chlorine does indeed oxidize bromide ion to bromine, a fact used to recover bromine from seawater and natural brine solutions.

Strictly speaking, the $E°$ values only tell us what to expect under standard conditions. However, only when $E°_{cell}$ is small can changes in the concentrations change the direction of the spontaneous reaction.

The reactants and products of *spontaneous* redox reactions are easy to spot when reduction potentials are listed in order of most positive to least positive (most negative), as in Table 19.1. This makes predicting possible redox reactions particularly easy. With *any* pair of half-reactions, the one higher up in the table has the more positive reduction potential and occurs as a reduction. The other half-reaction is reversed and occurs as an oxidation. Therefore, the *reactants* are found on the left side of the higher half-reaction and on the right side of the lower half-reaction.

<table>
<tr><td>

EXAMPLE 19.8

Predicting the Outcome of Redox Reactions

The activity series of the metals (Section 5.4), also used to predict reactions, is derived from Table 19.1.

</td><td>

Predict the reaction that will occur when Ni and Fe are added to a solution that contains both Ni^{2+} and Fe^{2+}.

Analysis: We are given a situation involving possible changes of ions to atoms or of atoms to ions. In other words, the system involves a possible redox reaction, and so we need Table 19.1. In this table, we see that Ni^{2+} has a more positive (less negative) reduction potential than Fe^{2+}, so Ni^{2+} is reduced and its half-cell reaction will be written just as in Table 19.1. The half-cell reaction for Fe^{2+} in Table 19.1, however, must be reversed; it is Fe that will be oxidized.

Solution: The half-cell reactions are

$$Ni^{2+}(aq) + 2e^- \longrightarrow Ni(s) \qquad \text{(reduction)}$$
$$\underline{Fe(s) \longrightarrow Fe^{2+}(aq) + 2e^- \qquad \text{(oxidation)}}$$
$$Ni^{2+}(aq) + Fe(s) \longrightarrow Ni(s) + Fe^{2+}(aq) \qquad \text{(net reaction)}$$

Practice Exercise 8

Use the positions of the half-reactions in Table 19.1 to predict the spontaneous reaction when Br^-, SO_4^{2-}, H_2SO_3, and Br_2 are mixed in an acidic solution. ◆

</td></tr>
</table>

Determining the Cell Reaction and Cell Potential of a Galvanic Cell

In any cell, we know that the substance most easily reduced (the one with the highest reduction potential) is the one that undergoes reduction. Example 19.9 shows how reduction potentials can be used to predict the standard cell potential and the overall cell reaction in a galvanic cell.

<table>
<tr><td>

EXAMPLE 19.9

Predicting the Cell Reaction and Cell Potential of a Galvanic Cell

</td><td>

What would be the cell reaction and the standard cell potential of a galvanic cell employing the following half-reactions?

$$Al^{3+}(aq) + 3e^- \rightleftharpoons Al(s) \qquad E^\circ_{Al^{3+}} = -1.66 \text{ V}$$
$$Cu^{2+}(aq) + 2e^- \rightleftharpoons Cu(s) \qquad E^\circ_{Cu^{2+}} = +0.34 \text{ V}$$

Which half-cell would be the anode?

Analysis: In the spontaneous cell reaction, the substance that is more easily reduced will be the one that undergoes reduction. This means that the half-reaction with the larger (more positive) reduction potential will take place as reduction while the other half-reaction will be reversed and occur as oxidation. The cell potential is simply the difference between the two reduction potentials, calculated using Equation 19.2.

Solution: The second half-reaction has the more positive reduction potential and will occur as a reduction; the other will occur as an oxidation. Thus, Cu^{2+} is reduced

</td></tr>
</table>

and Al is oxidized. To obtain the cell reaction, we add the two half-reactions, remembering that the electrons must cancel.

$$3[Cu^{2+}(aq) + 2e^- \longrightarrow Cu(s)] \qquad \text{(reduction)}$$

$$2[Al(s) \longrightarrow Al^{3+}(aq) + 3e^-] \qquad \text{(oxidation)}$$

$$\overline{3Cu^{2+}(aq) + 2Al(s) \longrightarrow 3Cu(s) + 2Al^{3+}(aq)} \qquad \text{(cell reaction)}$$

Remember that half-reactions are combined following the same procedure used in the ion–electron method of balancing redox reactions (Section 5.2).

The anode in the cell is aluminum because that is where oxidation takes place.

To obtain the cell potential, we substitute into Equation 19.2.

$$E^\circ_{cell} = (E^\circ \text{ of substance reduced}) - (E^\circ \text{ of substance oxidized})$$

$$E^\circ_{cell} = E^\circ_{Cu^{2+}} - E^\circ_{Al^{3+}}$$

$$= (0.34 \text{ V}) - (-1.66 \text{ V})$$

$$= 2.00 \text{ V}$$

An important point to notice here is that although we multiply the half-reactions by factors to make the electrons cancel, *we do not multiply the reduction potentials by these factors.*[1] To obtain the cell potential, we simply subtract one reduction potential from the other.

Practice Exercise 9

What are the overall cell reaction and the standard cell potential of a galvanic cell employing the following half-reactions?

These are the reactions in an Edison cell, a type of rechargeable storage battery.

$$NiO_2(s) + 2H_2O + 2e^- \rightleftharpoons Ni(OH)_2(s) + 2OH^-(aq) \qquad E^\circ_{NiO_2} = 0.49 \text{ V}$$

$$Fe(OH)_2(s) + 2e^- \rightleftharpoons Fe(s) + 2OH^-(aq) \qquad E^\circ_{Fe(OH)_2} = -0.88 \text{ V} \; \blacklozenge$$

Practice Exercise 10

What are the overall cell reaction and the standard cell potential of a galvanic cell employing the following half-reactions?

$$Cr^{3+}(aq) + 3e^- \rightleftharpoons Cr(s) \qquad E^\circ_{Cr^{3+}} = -0.74 \text{ V}$$

$$MnO_4^-(aq) + 8H^+(aq) + 5e^- \rightleftharpoons Mn^{2+}(aq) + 4H_2O \qquad E^\circ_{MnO_4^-} = +1.51 \text{ V} \; \blacklozenge$$

Determining whether a Reaction is Spontaneous from the Calculated Cell Potential

Since we can predict the spontaneous redox reaction that will take place among a mixture of reactants, it should be possible to predict whether or not a particular reaction, *as written,* can occur spontaneously. We can do this by calculating the cell potential that corresponds to the reaction in question and seeing if the potential is *positive.*

In a galvanic cell, the calculated cell potential for the spontaneous reaction is always positive. If the calculated cell potential is negative, the reaction is spontaneous in the reverse direction.

Tools

Standard cell potentials

[1]Reduction potentials are intensive quantities; they have the units volts, which are joules *per coulomb.* The same number of joules are available for each coulomb of charge regardless of the total number of electrons shown in the equation. Therefore, reduction potentials are never multiplied by factors before they are subtracted to give the cell potential.

These generalizations apply under standard conditions: 1 *M* concentrations of all ions, 1 atm pressure for gases, and 25 °C.

For example, to obtain the cell potential for a spontaneous reaction in our previous examples, we subtracted the reduction potentials in a way that gave a positive answer. Therefore, if we compute the cell potential for a particular reaction based on the way the equation is written and the potential comes out positive, we know the reaction is spontaneous. If the calculated cell potential comes out negative, however, the reaction is nonspontaneous. In fact, it is really spontaneous in the opposite direction.

EXAMPLE 19.10

Determining whether a Reaction Is Spontaneous by Using the Calculated Cell Potential

Determine whether the following reactions are spontaneous as written. If they are not, give the reaction that is spontaneous.

(1) $Cu(s) + 2H^+(aq) \longrightarrow Cu^{2+}(aq) + H_2(g)$
(2) $3Cu(s) + 2NO_3^-(aq) + 8H^+(aq) \longrightarrow 3Cu^{2+}(aq) + 2NO(g) + 4H_2O$

Analysis: If we can find the half-reactions for each of these and their corresponding reduction potentials, we should be able to calculate their values of E°_{cell} for the reactions as written. The signs of E°_{cell} will then tell us whether the reactions are spontaneous under standard conditions.

Solution: (1) The half-reactions involved in this reaction are

$$Cu(s) \longrightarrow Cu^{2+}(aq) + 2e^-$$

$$2H^+(aq) + 2e^- \longrightarrow H_2(g)$$

The H^+ is reduced and Cu is oxidized. Substituting values from Table 19.1 into Equation 19.2,

$$E^\circ_{cell} = E^\circ_{H^+} - E^\circ_{Cu^{2+}}$$

$$E^\circ_{cell} = (0.00\ V) - (0.34\ V)$$

$$= -0.34\ V$$

Copper doesn't dissolve in acids like HCl because the only oxidizing agent is H^+, which is a weaker oxidizing agent than Cu^{2+}.

The calculated cell potential is negative, so reaction 1 is not spontaneous in the forward direction. The spontaneous reaction is actually the reverse of 1.

$$Cu^{2+}(aq) + H_2(g) \longrightarrow Cu(s) + 2H^+(aq)$$

(2) The half-reactions involved in this equation are

$$Cu(s) \longrightarrow Cu^{2+}(aq) + 2e^-$$

$$NO_3^-(aq) + 4H^+(aq) + 3e^- \longrightarrow NO(g) + 2H_2O$$

The Cu is oxidized while the NO_3^- is reduced. According to Equation 19.2,

$$E^\circ_{cell} = E^\circ_{NO_3^-} - E^\circ_{Cu^{2+}}$$

Substituting values from Table 19.1 gives

$$E^\circ_{cell} = (0.96\ V) - (0.34\ V)$$

$$= +0.62\ V$$

Copper dissolves in HNO_3 because it contains the oxidizing agent NO_3^-.

Because the calculated cell potential is positive, reaction 2 is spontaneous in the forward direction, as written.

Practice Exercise 11

Which of the following reactions occur spontaneously in the forward direction?

(a) $Br_2(aq) + Cl_2(g) + 2H_2O \longrightarrow 2Br^-(aq) + 2HOCl(aq) + 2H^+(aq)$
(b) $3Zn(s) + 2Cr^{3+}(aq) \longrightarrow 3Zn^{2+}(aq) + 2Cr(s)$ ◆

Using Reduction Potentials to Predict Electrolysis Reactions

Tools

Standard reduction potentials

In our earlier discussion of electrolysis, we used the results of experiments to determine the nature of the electrode reactions in aqueous solutions. Although the value of experiments cannot be minimized, it is often possible to predict the outcome of electrolysis reactions using our table of reduction potentials. For example, in the electrolysis of copper(II) bromide, the competing reduction reactions at the cathode are

$$Cu^{2+}(aq) + 2e^- \rightleftharpoons Cu(s) \qquad\qquad E° = +0.34 \text{ V}$$

$$2H_2O + 2e^- \rightleftharpoons H_2(g) + 2OH^-(aq) \qquad E° = -0.83 \text{ V}$$

The fact that Cu^{2+} has a more positive $E°$ than water tells us that Cu^{2+} is the more easily reduced species and should be preferentially reduced at the cathode. As we saw earlier, this is exactly what we observe.

Reduction potentials also work in this case to predict the outcome at the anode. The competing oxidation reactions are

$$2H_2O \longrightarrow O_2(g) + 4H^+(aq) + 4e^-$$

$$2Br^-(aq) \longrightarrow Br_2(aq) + 2e^-$$

In Table 19.1 we find the products of these reactions on the left of the half-reactions corresponding to the reduction potentials.

$$O_2(g) + 4H^+(aq) + 4e^- \rightleftharpoons 2H_2O \qquad E° = +1.23 \text{ V}$$

$$Br_2(aq) + 2e^- \rightleftharpoons 2Br^-(aq) \qquad E° = +1.07 \text{ V}$$

These reduction potentials tell us that O_2 is more easily reduced than Br_2. But if O_2 is the *more easily reduced,* then the product, H_2O, must be *the less easily oxidized.* Stated another way, *the half-reaction with the smaller reduction potential is the more easily reversed as an oxidation.* The conclusion, therefore, is that Br^- is more easily oxidized than H_2O. It seems reasonable that when the electrolysis is carried out, the more easily oxidized substance should be the one oxidized, so we expect Br_2 to be the product formed at the anode. In the experiment described on page 850, we saw that the product formed at the anode in the electrolysis of $CuBr_2$ was indeed Br_2.

EXAMPLE 19.11

Predicting the Products in an Electrolysis Reaction

Electrolysis is planned for an aqueous solution that contains a mixture of 0.50 M $ZnSO_4$ and 0.50 M $NiSO_4$. On the basis of reduction potentials, what products are expected to be observed at the electrodes? What is the expected net cell reaction?

Analysis: We need to consider the competing reactions at the cathode and the anode. At the cathode, the half-reaction with the most positive (or least negative) reduction potential will be the one expected to occur. At the anode, the half-reaction with the *least positive* reduction potential is the one most easily reversed, and should occur as an oxidation.

Solution: At the cathode, the competing reduction reactions involve the two cations and water. The reactions and their reduction potentials are

$$Ni^{2+}(aq) + 2e^- \rightleftharpoons Ni(s) \qquad\qquad E° = -0.25 \text{ V}$$

$$Zn^{2+}(aq) + 2e^- \rightleftharpoons Zn(s) \qquad\qquad E° = -0.76 \text{ V}$$

$$2H_2O + 2e^- \rightleftharpoons H_2(g) + 2OH^-(aq) \qquad E° = -0.83 \text{ V}$$

The least negative reduction potential is that of Ni^{2+}, so we expect this ion to be reduced at the cathode and solid nickel to be formed.

At the anode, the competing oxidation reactions are for water and SO_4^{2-} ion. In Table 19.1, substances oxidized are found on the right side of the half-reactions. The two half-reactions having these as products are

$$S_2O_8^{2-}(aq) + 2e^- \rightleftharpoons 2SO_4^{2-}(aq) \qquad E° = +2.01 \text{ V}$$

$$O_2(g) + 4H^+ + 4e^- \rightleftharpoons 2H_2O \qquad E° = +1.23 \text{ V}$$

The half-reaction with the least positive $E°$ (the second one here) is most easily reversed as an oxidation, so we expect the oxidation half-reaction to be

$$2H_2O \rightleftharpoons O_2(g) + 4H^+ + 4e^-$$

At the anode, we expect O_2 to be formed.

The predicted net cell reaction is obtained by combining the two expected electrode half-reactions, making the electron loss equal to the electron gain.

$$2H_2O \longrightarrow O_2(g) + 4H^+(aq) + 4e^- \qquad \text{(anode)}$$

$$\underline{2 \times [Ni^{2+}(aq) + 2e^- \longrightarrow Ni(s)]} \qquad \text{(cathode)}$$

$$2H_2O + 2Ni^{2+}(aq) \longrightarrow O_2(g) + 4H^+(aq) + 2Ni(s) \qquad \text{(net cell reaction)}$$

Are the Answers Reasonable?

We can check the locations of the half-reactions in Table 19.1 to confirm our conclusions. For the reduction step, the higher up in the table a half-reaction is, the greater its tendency to occur as reduction. Among the competing half-reactions at the cathode, the one for Ni^{2+} is highest, so we expect that Ni^{2+} is the easiest to reduce and $Ni(s)$ should be formed at the cathode.

For the oxidation step, the lower down in the table a half-reaction is, the easier it is to reverse and cause to occur as oxidation. On this basis, the oxidation of water is easier than the oxidation of SO_4^{2-}, so we expect H_2O to be oxidized and O_2 to be formed at the anode.

Of course, we could also test our prediction by carrying out the electrolysis experimentally.

Practice Exercise 12

In the electrolysis of an aqueous solution containing both Cd^{2+} and Sn^{2+}, what product do we expect at the cathode? ◆

Although we can use reduction potentials to predict electrolysis reactions most of the time, there are occasions when standard reduction potentials do not successfully predict electrolysis products. Sometimes concentration effects or the formation of complex ions interfere. And sometimes the electrodes themselves are the culprits. For example, in the electrolysis of aqueous NaCl using inert platinum electrodes, we find experimentally that Cl_2 is formed at the anode. Is this what we would have expected? Let's examine the reduction potentials of O_2 and Cl_2 to find out.

$$Cl_2(aq) + 2e^- \rightleftharpoons 2Cl^-(aq) \qquad E° = +1.36 \text{ V}$$

$$O_2(g) + 4H^+(aq) + 4e^- \rightleftharpoons 2H_2O \qquad E° = +1.23 \text{ V}$$

Because of its less positive reduction potential, we would expect the oxygen half-reaction to be the easier to reverse (with water being oxidized to O_2). Thus, reduction potentials predict that O_2 should be formed, but experiment shows that Cl_2 is produced. The explanation for why this happens is beyond the scope of this text, but the unexpected result does teach us that we must be cautious in predicting products in electrolysis reactions solely on the basis of reduction potentials.

19.7 Cell Potentials and Thermodynamics

The fact that cell potentials allow us to predict the spontaneity of redox reactions is no coincidence. There is a relationship between the cell potential and the free energy change for a reaction. In Chapter 18 we saw that ΔG for a reaction is a measure of the maximum useful work that can be obtained from a chemical reaction. Specifically, the relationship is

$$-\Delta G = \text{maximum work} \tag{19.3}$$

Determining a Free Energy Change from a Cell Potential

In an electrical system, work is supplied by the electric current that is pushed along by the potential of the cell. It can be calculated from the equation

$$\text{maximum work} = nFE_{\text{cell}} \tag{19.4}$$

where n is the number of moles of electrons transferred, F is the Faraday constant (9.65×10^4 coulombs per mole of electrons), and E_{cell} is the potential of the cell in volts. To see that Equation 19.4 gives work (which has the units of energy) we can analyze the units. In Equation 19.1 you saw that 1 volt = 1 joule/coulomb. Therefore,

$$\text{maximum work} = (\text{mol } e^-) \times \left(\frac{\text{coulombs}}{\text{mol } e^-}\right) \times \left(\frac{\text{joules}}{\text{coulomb}}\right) = \text{joules}$$

$$\begin{array}{ccc} \updownarrow & \updownarrow & \updownarrow \\ n & F & E_{\text{cell}} \end{array}$$

Combining Equations 19.3 and 19.4 gives us

Note that when E_{cell} is positive, ΔG will be negative and the cell reaction will be spontaneous. We used this idea in the preceding section in predicting spontaneous cell reactions.

$$\Delta G = -nFE_{\text{cell}} \tag{19.5}$$

If we are dealing with the *standard* cell potential, we can calculate the *standard* free energy change.

$$\Delta G° = -nFE°_{\text{cell}} \tag{19.6}$$

Tools

Free energy
changes
from cell
potentials

EXAMPLE 19.12

Calculating the Standard Free Energy Change

Calculate $\Delta G°$ for the following reaction, given that its standard cell potential is 0.320 V at 25 °C.

$$NiO_2(s) + 2Cl^-(aq) + 4H^+(aq) \longrightarrow Cl_2(g) + Ni^{2+}(aq) + 2H_2O$$

Analysis: This is a straightforward application of Equation 19.6. Taking the coefficients in the equation to stand for *moles*, two moles of Cl^- are oxidized to Cl_2 and two moles of electrons are transferred ($n = 2$).

Solution: Using Equation 19.6, we have

$$\Delta G^\circ = -(2 \text{ mol } e^-) \times \left(\frac{9.65 \times 10^4 \text{ C}}{1 \text{ mol } e^-} \right) \times \left(\frac{0.320 \text{ J}}{\text{C}} \right)$$

$$= -6.18 \times 10^4 \text{ J}$$

$$= -61.8 \text{ kJ}$$

Practice Exercise 13

Calculate ΔG° for the reaction that takes place in the galvanic cell described in Practice Exercises 5 and 7. ◆

Determining Equilibrium Constants

One of the useful applications of electrochemistry is in the determination of equilibrium constants. In Chapter 18 you saw that ΔG° is related to the equilibrium constant by the expression

$$\Delta G^\circ = -RT \ln K_c$$

where we have used K_c for the equilibrium constant because electrochemical reactions occur in solution. We've seen in this chapter that ΔG° is also related to E°_{cell}:

$$\Delta G^\circ = -n\mathscr{F} E^\circ_{cell}$$

Therefore, E°_{cell} and the equilibrium constant are also related. Equating the right sides of the two equations above and solving for E°_{cell} gives[2]

Tools

Equilibrium constants from cell potentials

$$E^\circ_{cell} = \frac{RT}{n\mathscr{F}} \ln K_c \tag{19.7}$$

For the units to work out correctly, the value of R must be 8.314 J mol^{-1} K^{-1}, T must be the temperature in kelvins, $\mathscr{F}$ equals 9.65×10^4 C per mole of e^-, and n equals the number of moles of electrons transferred in the reaction.

EXAMPLE 19.13

Calculating Equilibrium Constants from E°_{cell}

Calculate K_c for the reaction in Example 19.12.

Solution: The problem simply involves substituting values into Equation 19.7. The reaction in Example 19.12 has $E^\circ_{cell} = 0.320$ V and $n = 2$. The temperature is 25 °C or 298 K. Let's solve Equation 19.7 for $\ln K_c$ and then substitute values.

$$\ln K_c = \frac{E^\circ_{cell} n\mathscr{F}}{RT}$$

[2]For historical reasons, this equation is sometimes expressed in terms of common logs (base 10 logarithms), and for reactions at 25 °C (298 K), all of the constants (R, T, and $\mathscr{F}$) can be combined to give 0.0592 joules/coulomb. Because joules/coulomb equals volts, Equation 19.7 reduces to

$$E^\circ_{cell} = \frac{0.0592 \text{ V}}{n} \log K_c$$

where n is the number of moles of electrons transferred in the cell reaction as it is written.

Substituting values and using the relationship that $1\text{ V} = 1\text{ J C}^{-1}$,

$$\ln K_c = \frac{0.320\text{ J C}^{-1} \times 2 \times 9.65 \times 10^4\text{ C mol}^{-1}}{8.314\text{ J mol}^{-1}\text{ K}^{-1} \times 298\text{ K}}$$

$$= 24.9$$

Taking the antilogarithm,

$$K_c = e^{24.9} = 7 \times 10^{10}$$

The huge value of K_c tells us that the reaction of Example 19.12 proceeds virtually to completion.

Practice Exercise 14

The calculated standard cell potential for the reaction

$$Cu^{2+}(aq) + 2Ag(s) \rightleftharpoons Cu(s) + 2Ag^+(aq)$$

is $E°_{cell} = -0.46$ V. Calculate K_c for this reaction. Does this reaction proceed very far toward completion? ◆

19.8 Effect of Concentration on Cell Potentials

When all of the ion concentrations in a cell are 1.00 M and when the partial pressures of any gases involved in the cell reaction are 1 atm, the cell potential is equal to the standard potential. When the concentrations or pressures change, however, so does the potential. For example, in an operating cell or battery, the potential gradually drops as the reactants are used up and as the cell reaction approaches its natural equilibrium status. When it reaches equilibrium, the potential has dropped to zero—the battery is dead.

The Nernst Equation: The Relationship of Cell Potential to Ion Concentrations

The effect of concentration on the cell potential can be obtained from thermodynamics. In Chapter 18, you learned that the free energy change is related to the reaction quotient Q by the equation

$$\Delta G = \Delta G° + RT \ln Q$$

Substituting for ΔG and $\Delta G°$ from Equations 19.5 and 19.6 gives

$$-n\mathscr{F}E_{cell} = -n\mathscr{F}E°_{cell} + RT \ln Q$$

Dividing both sides by $-n\mathscr{F}$ gives

$$E_{cell} = E°_{cell} - \frac{RT}{n\mathscr{F}} \ln Q \qquad (19.8)$$

This equation is commonly known as the **Nernst equation,** named after Walter Nernst.[3]

Walter Nernst (1864–1941), a German chemist and physicist, was also the discoverer of the third law of thermodynamics.

Tools

Nernst equation

---

[3]Using common logarithms instead of natural logarithms and calculating the constants for 25 °C gives another form of the Nernst equation that is sometimes used.

$$E_{cell} = E°_{cell} - \frac{0.0592\text{ V}}{n} \log Q$$

EXAMPLE 19.14
Calculating the Effect of Concentration on E_{cell}

A cell employs the following half-reactions.

$$Ni^{2+}(aq) + 2e^- \rightleftharpoons Ni(s) \qquad E^\circ_{Ni^{2+}} = -0.25 \text{ V}$$

$$Cr^{3+}(aq) + 3e^- \rightleftharpoons Cr(s) \qquad E^\circ_{Cr^{3+}} = -0.74 \text{ V}$$

Calculate the cell potential when $[Ni^{2+}] = 1.0 \times 10^{-4}$ M and $[Cr^{3+}] = 2.0 \times 10^{-3}$ M.

Analysis: Because the concentrations are not 1.00 M, we must use the Nernst equation, but first we need the cell reaction. We need it to determine the number of electrons transferred, n, and we need it to determine the correct form of the mass action expression from which we calculate Q. We must also note that the reacting system is heterogeneous; both solid metals and a liquid solution of their dissolved ions are involved. So we have to remember that a mass action expression does not contain concentration terms for solids, such as Ni and Cr.

Solution: Nickel has the more positive (less negative) reduction potential, so its half-reaction will occur as a reduction. This means that chromium will be oxidized. Making electron gain equal to electron loss, the cell reaction is found as follows.

$$3[Ni^{2+}(aq) + 2e^- \longrightarrow Ni(s)] \qquad \text{(reduction)}$$

$$\underline{2[Cr(s) \longrightarrow Cr^{3+}(aq) + 3e^-]} \qquad \text{(oxidation)}$$

$$3Ni^{2+}(aq) + 2Cr(s) \longrightarrow 3Ni(s) + 2Cr^{3+}(aq) \qquad \text{(cell reaction)}$$

The mass action expression, Q, for the cell reaction is

$$Q = \frac{[Cr^{3+}]^2}{[Ni^{2+}]^3}$$

A total of six electrons is transferred, which means $n = 6$. Now we can write the Nernst equation for the system.

$$E_{cell} = E^\circ_{cell} - \frac{RT}{n\mathscr{F}} \ln \frac{[Cr^{3+}]^2}{[Ni^{2+}]^3}$$

Notice that we've constructed the mass action expression, from which we will calculate the reaction quotient, using the concentrations of the ions raised to powers equal to their coefficients in the net cell reaction, and that we have not included concentration terms for the two solids. This is the procedure we followed for heterogeneous equilibria in Chapter 14.

Next we need E°_{cell}. Since Ni^{2+} is reduced,

$$E^\circ_{cell} = E^\circ_{Ni^{2+}} - E^\circ_{Cr^{3+}}$$

$$= (-0.25 \text{ V}) - (-0.74 \text{ V})$$

$$= 0.49 \text{ V}$$

Substituting the value for E°_{cell} and $R = 8.314$ J mol^{-1} K^{-1}, $T = 298$ K, $\mathscr{F} = 9.65 \times 10^4$ C mol^{-1}, $[Ni^{2+}] = 1.0 \times 10^{-4}$ M, and $[Cr^{3+}] = 2.0 \times 10^{-3}$ M into the equation, we get

$$E_{cell} = 0.49 \text{ V} - \frac{8.314 \text{ J mol}^{-1} \text{ K}^{-1} \times 298 \text{ K}}{6 \times 9.65 \times 10^4 \text{ C mol}^{-1}} \ln \frac{(2.0 \times 10^{-3})^2}{(1.0 \times 10^{-4})^3}$$

$$= 0.49 \text{ V} - (0.00428 \text{ V}) \ln (4.0 \times 10^6)$$

$$= 0.49 \text{ V} - (0.00428 \text{ V})(15.20) = 0.49 \text{ V} - 0.0651 \text{ V}$$

$$= 0.42 \text{ V}$$

The potential of the cell is expected to be 0.42 V, or about 14% less than the standard cell potential of 0.49 V.

Practice Exercise 15

In a certain zinc–copper cell,

$$Zn(s) + Cu^{2+}(aq) \longrightarrow Zn^{2+}(aq) + Cu(s)$$

the ion concentrations are $[Cu^{2+}] = 0.0100\ M$ and $[Zn^{2+}] = 1.0\ M$. What is the cell potential at 25 °C? ◆

Determining Concentrations from Experimental Cell Potentials

One of the principal uses of the relationship between concentration and cell potential is in the measurement of concentrations. Experimental determination of cell potentials combined with modern developments in electronics has provided a means of monitoring and analyzing the concentrations of all sorts of substances in solution, even some that are not themselves ionic and that are not involved directly in electrochemical changes. The basic concept behind all these approaches is illustrated in Example 19.15.

EXAMPLE 19.15

Using the Nernst Equation to Determine Concentrations

To measure the concentration of Cu^{2+} in a large number of samples of water in which the copper ion concentration is expected to be quite small, an apparatus was assembled that consists of a silver electrode, dipping into a 1.00 M solution of $AgNO_3$, connected by a salt bridge to a second half-cell containing a copper electrode that can be dipped successively into each water sample. In the analysis of one sample, the cell potential at 25 °C was measured to be 0.62 V, the copper electrode serving as the anode. What was the concentration of copper ion in this sample?

Analysis: In this problem, we have the cell potential, E_{cell}, and we can calculate E°_{cell}. The unknown quantity is one of the concentration terms in the Nernst equation.

Solution: The first step is to write the proper equation for the cell reaction, because we need it to compute E°_{cell} and to construct the mass action expression for use in the Nernst equation. Because copper is the anode, it is being oxidized. This also means that Ag^{+} is being reduced. Therefore, the equation for the cell reaction is

$$Cu(s) + 2Ag^{+}(aq) \longrightarrow Cu^{2+}(aq) + 2Ag(s)$$

Two moles of electrons are transferred, so $n = 2$ and the Nernst equation is

$$E_{cell} = E^{\circ}_{cell} - \frac{RT}{2\mathcal{F}} \ln \frac{[Cu^{2+}]}{[Ag^{+}]^2}$$

The mass action expression, Q, for this reaction is

$$Q = \frac{[Cu^{2+}]}{[Ag^{+}]^2}$$

The value of E°_{cell} can be obtained from the tabulated standard reduction potentials in Table 19.1. Following our usual procedure and recognizing that silver ion is reduced,

$$E^{\circ}_{cell} = E^{\circ}_{Ag^{+}} - E^{\circ}_{Cu^{2+}}$$

$$= (0.80\ V) - (0.34\ V)$$

$$= 0.46\ V$$

Facets of Chemistry 19.2

Measurement of pH

The Nernst equation tells us that the potential of a cell changes with the concentrations of the ions involved in the cell reaction. As noted in the conclusion to Example 19.15, one of the most useful results of this is that cell potential measurements provide a way indirectly to measure and monitor ion concentrations in aqueous solutions.

Scientists have developed a number of specialized electrodes that can be dipped into one solution after another and whose potential is affected in a reproducible way by the concentration of only one species in the solution. An example is the glass electrode of a pH meter for the measurement of pH (see Figure 1). The electrode is constructed from a hollow glass tube sealed with a special thin-walled glass membrane at the bottom. The tube is filled partway with a dilute solution of HCl, and dipping into this HCl solution is a silver wire coated with a layer of silver chloride. The potential of the electrode is controlled by the difference between the hydrogen ion concentrations inside and outside the thin glass membrane at the bottom. Since the H^+ concentration inside the electrode is constant, the electrode's potential varies only with the concentration of H^+ in the solution outside. In fact, this potential is proportional to the logarithm of the H^+ concentration and therefore to the pH of the outside solution.

A glass electrode is always used with another reference electrode whose potential is a constant, and this gives a galvanic cell whose potential depends on the pH of the solution into which the electrodes are immersed. The measurement of the potential of this galvanic cell and the translation of the potential into pH are the job of a pH meter such as the one shown in Figure 15.4 on page 689.

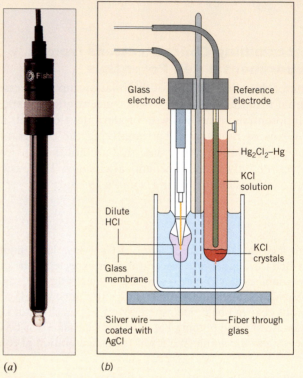

(a)   *(b)*

Figure 1 *Glass electrode.* (*a*) Photograph of a typical glass electrode. (*b*) Cutaway view of the construction of a glass electrode and a reference electrode.

Now we can substitute values into the Nernst equation and solve for the concentration ratio in the mass action expression.

$$0.62\ \text{V} = 0.46\ \text{V} - \frac{8.314\ \text{J mol}^{-1}\,\text{K}^{-1} \times 298\ \text{K}}{2 \times 9.65 \times 10^4\ \text{C mol}^{-1}}\ \ln \frac{[Cu^{2+}]}{[Ag^+]^2}$$

Solving for $\ln ([Cu^{2+}]/[Ag^+]^2)$ gives

$$\ln \frac{[Cu^{2+}]}{[Ag^+]^2} = -12$$

Taking the antilog gives us the value of the mass action expression.

$$\frac{[Cu^{2+}]}{[Ag^+]^2} = 6 \times 10^{-6}$$

Since we know that the concentration of Ag^+ is 1.00 *M*, we can now solve for the Cu^{2+} concentration.

$$\frac{[Cu^{2+}]}{(1.00)^2} = 6 \times 10^{-6}$$

$$[Cu^{2+}] = 6 \times 10^{-6} \, M$$

Notice that this is indeed a very small concentration, and it can be obtained very easily by simply measuring the potential generated by the electrochemical cell. Determining the concentrations in many samples is also very simple—just change the water sample and measure the potential again.

The ease of such operations and the fact that they lend themselves well to automation and computer analysis make electrochemical analyses especially attractive to scientists.

Practice Exercise 16

In the analysis of two other water samples by the procedure described in Example 19.15, cell potentials (E_{cell}) of 0.57 V and 0.82 V were obtained. Calculate the Cu^{2+} ion concentration in each of these samples. ◆

Practice Exercise 17

A galvanic cell was constructed by connecting a nickel electrode that was dipped into 1.20 M $NiSO_4$ solution to a chromium electrode that was dipping into a solution containing Cr^{3+} at an unknown concentration. The potential of the cell was measured to be 0.552 V, with the chromium serving as the anode. The standard cell potential for this system is 0.487 V. What was the concentration of Cr^{3+} in the solution of unknown concentration? ◆

19.9 Practical Applications of Galvanic Cells

One of the most familiar uses of galvanic cells is the generation of portable electrical energy. In this section we will briefly examine some of the more common present-day uses of galvanic cells, popularly called batteries, and some promising future applications.

Strictly speaking, a cell is a single electrochemical unit consisting of cathode and anode half-cells. A **battery** is a collection of cells connected in series.

The Lead Storage Battery

The common **lead storage battery** used to start an automobile is composed of a number of galvanic cells, each having a potential of about 2 V, that are connected in series so that their voltages are additive. Most automobile batteries contain six such cells and give about 12 V, but 6, 24, and 32 V batteries are also available.

A typical lead storage battery is shown in Figure 19.19. The anode of each cell is composed of a set of lead plates, the cathode consists of another set of plates that hold a coating of PbO_2, and the electrolyte is sulfuric acid. When the battery is discharging the electrode reactions are

$$PbO_2(s) + 4H^+(aq) + SO_4^{2-}(aq) + 2e^- \longrightarrow PbSO_4(s) + 2H_2O \qquad \text{(cathode)}$$

$$Pb(s) + SO_4^{2-}(aq) \longrightarrow PbSO_4(s) + 2e^- \qquad \text{(anode)}$$

The net reaction taking place in each cell is

$$PbO_2(s) + Pb(s) + \underbrace{4H^+(aq) + 2SO_4^{2-}(aq)}_{2H_2SO_4} \longrightarrow 2PbSO_4(s) + 2H_2O$$

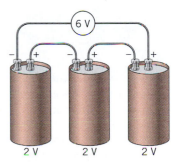

If three 2 volt cells are connected in series, their voltages are additive to provide a total of 6 volts.

As the cell discharges, the sulfuric acid concentration decreases, which provides a convenient means for checking the state of the battery. Because the density of a sulfuric acid solution decreases as its concentration drops, the concentration

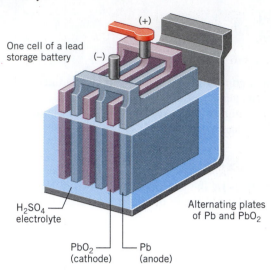

Figure 19.19 *Lead storage battery.* A 12 volt lead storage battery, such as those used in most automobiles, consists of six cells like the one shown here. Notice that the anode and cathode each consist of several plates connected together. This allows the cell to produce the large currents necessary to start a car.

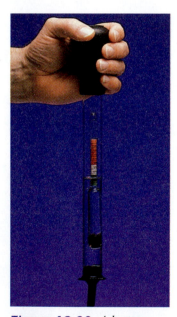

Figure 19.20 *A battery hydrometer.* Battery acid is drawn into the glass tube. The depth to which the float sinks is inversely proportional to the concentration of the acid and, therefore, to the state of charge of the battery.

can be determined very simply by measuring the density with a **hydrometer,** which consists of a rubber bulb that is used to draw the battery fluid into a glass tube containing a float (see Figure 19.20). The depth to which the float sinks is inversely proportional to the density of the liquid—the deeper the float sinks, the lower is the density and the weaker is the charge on the battery. The narrow neck of the float is usually marked to indicate the state of charge of the battery.

The principal advantage of the lead storage battery is that the cell reactions that occur spontaneously during discharge can be reversed by the application of a voltage from an external source. In other words, the battery can be recharged by electrolysis. The reaction for battery recharge is

$$2PbSO_4(s) + 2H_2O \xrightarrow{\text{electrolysis}} PbO_2(s) + Pb(s) + 4H^+(aq) + 2SO_4^{2-}(aq)$$

Disadvantages of the lead storage battery are that it is very heavy and that its corrosive sulfuric acid can spill.

The most modern lead storage batteries use a lead–calcium alloy as the anode. This reduces the need to have the individual cells vented, and the battery can be sealed, thereby preventing spillage of the electrolyte.

The Zinc–Carbon Dry Cell

The ordinary, relatively inexpensive, 1.5 V **zinc–carbon dry cell,** or Leclanché cell (named after its inventor George Leclanché), is used to power flashlights, tape recorders, and the like, but it is not really dry (see Figure 19.21). Its outer shell is made of zinc, which serves as the anode. The exposed outer surface at the bottom of the cell is the negative end of the battery. The cathode—the positive terminal of the battery—consists of a carbon (graphite) rod surrounded by a moist paste of graphite powder, manganese dioxide, and ammonium chloride.

The anode reaction is simply the oxidation of zinc.

$$Zn(s) \longrightarrow Zn^{2+}(aq) + 2e^- \qquad \text{(anode)}$$

The cathode reaction is complex, and a mixture of products is formed. One of the major reactions is

$$2MnO_2(s) + 2NH_4^+(aq) + 2e^- \longrightarrow Mn_2O_3(s) + 2NH_3(aq) + H_2O \quad \text{(cathode)}$$

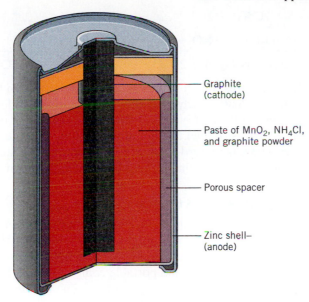

Graphite
(cathode)

Paste of MnO_2, NH_4Cl,
and graphite powder

Porous spacer

Zinc shell–
(anode)

Figure 19.21 *A cutaway view of a zinc–carbon dry cell.*

The ammonia that forms at the cathode reacts with some of the Zn^{2+} produced from the anode to form a complex ion, $Zn(NH_3)_4^{2+}$. Because of the complexity of the cathode half-cell reaction, no simple overall cell reaction can be written.

The major advantages of the dry cell are its relatively low price and the fact that it normally works without leaking. One disadvantage is that the cell loses its ability to function rather rapidly under heavy current drain, a loss that occurs because ions formed in the cell reaction cannot easily diffuse away from the electrodes. If unused for a while, however, these batteries spontaneously rejuvenate somewhat as the products diffuse away from the electrodes. Another disadvantage of the zinc–carbon dry cell is that it can't be recharged. Devices that claim otherwise simply drive the reaction products away from the electrodes, which allows the cell to function again. This doesn't work very many times, however. Before too long the zinc casing develops holes and the battery must be discarded.

A more popular version of the Leclanché battery uses a basic, or alkaline, electrolyte and is called an **alkaline battery** or **alkaline dry cell.** It too uses Zn and MnO_2 as reactants, but under basic conditions (Figure 19.22). The half-cell reactions are

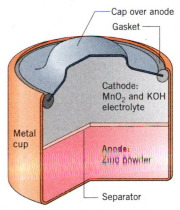

Cap over anode
Gasket

Cathode:
MnO_2 and KOH
electrolyte

Metal
cup

Anode:
Zinc powder

Separator

Figure 19.22 *A simplified diagram of an alkaline dry cell.*

$$Zn(s) + 2OH^-(aq) \longrightarrow ZnO(s) + H_2O + 2e \qquad \text{(anode)}$$

$$2MnO_2(s) + H_2O + 2e^- \longrightarrow Mn_2O_3(s) + 2OH^-(aq) \qquad \text{(cathode)}$$

$$Zn(s) + 2MnO_2(s) \longrightarrow ZnO(s) + Mn_2O_3(s) \qquad \text{(net cell reaction)}$$

and the voltage is about 1.54 V. It has a longer shelf-life and is able to deliver higher currents for longer periods than the less expensive zinc–carbon cell. The reason is that there are no ions produced around the electrodes that have to diffuse away to maintain the flow of current.

The Nickel–Cadmium Storage Cell

The **nickel–cadmium storage cell,** or **nicad battery,** powers rechargeable electronic calculators, electric shavers, and power tools. It produces a potential of about 1.4 V, which is slightly lower than that of the zinc–carbon cell. The electrode reactions in the cell are

$$Cd(s) + 2OH^-(aq) \longrightarrow Cd(OH)_2(s) + 2e^- \qquad \text{(anode)}$$

$$NiO_2(s) + 2H_2O + 2e^- \longrightarrow Ni(OH)_2(s) + 2OH^-(aq) \qquad \text{(cathode)}$$

$$Cd(s) + NiO_2(s) + 2H_2O \longrightarrow Cd(OH)_2(s) + Ni(OH)_2(s) \qquad \text{(net cell reaction)}$$

The nickel–cadmium battery is rechargeable and can be sealed to prevent leakage, which is particularly important in electronic devices. For the storage of a given amount of electrical energy, however, the nicad battery is considerably more expensive than the lead storage battery. Disposal of nicad batteries poses an environmental hazard because of the toxicity of cadmium. Efforts are underway to encourage the recycling of these batteries, rather than disposal in landfills.

Specialized Batteries

As everyone knows, the field of electronics has undergone fantastic changes in the past decade or so. Electronic calculators, digital watches, cellular telephones, electronically controlled cameras that focus automatically and adjust themselves for the correct exposure, and portable computers have been made possible by microelectronics centered around almost invisible circuits etched into the surfaces of silicon chips. This trend toward miniaturization has also been taken up by the makers of the batteries that power the electronics. Two of the common miniature batteries in use today are the "mercury battery" and the "silver oxide battery."

The Mercury Battery

This battery, one of the earliest small batteries to be developed for commercial use, first appeared in the early 1940s. It consists of a zinc anode and a mercury(II) oxide (HgO) cathode. The electrolyte is a concentrated solution of potassium hydroxide held in pads of absorbent material between the electrodes (see Figure 19.23). The reactions that take place at the electrodes are relatively simple.

$$Zn(s) + 2OH^-(aq) \longrightarrow ZnO(s) + H_2O + 2e^- \qquad \text{(anode)}$$

$$HgO(s) + H_2O + 2e^- \longrightarrow Hg(l) + 2OH^-(aq) \qquad \text{(cathode)}$$

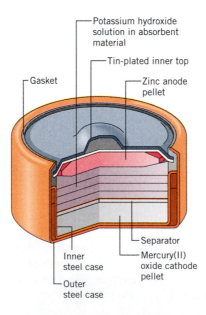

Potassium hydroxide solution in absorbent material

Tin-plated inner top

Gasket

Zinc anode pellet

Separator

Inner steel case

Mercury(II) oxide cathode pellet

Outer steel case

Figure 19.23 *A mercury battery.*

The net cell reaction is

$$Zn(s) + HgO(s) \longrightarrow ZnO(s) + Hg(l)$$

The mercury battery develops a potential of about 1.35 V. An especially useful feature of this cell is that its potential remains nearly constant throughout its life. By contrast, the potential of the ordinary dry cell drops off slowly but continuously as it is used. Like the nicad battery, however, the disposal of mercury batteries in landfills poses an environmental hazard because of the toxicity of mercury.

The Silver Oxide Battery

This battery is a more recent entry into the world of miniature batteries and is now used in many wristwatches, calculators, and autoexposure cameras. The anode material in the cell is zinc, and the substance that reacts at the cathode is silver oxide, Ag_2O (see Figure 19.24). The electrolyte is basic, and the half-reactions that occur at the electrodes are

$$Zn(s) + 2OH^-(aq) \longrightarrow Zn(OH)_2(s) + 2e^- \qquad \text{(anode)}$$

$$Ag_2O(s) + H_2O + 2e^- \longrightarrow 2Ag(s) + 2OH^-(aq) \qquad \text{(cathode)}$$

The overall net cell reaction is

$$Zn(s) + Ag_2O(s) + H_2O \longrightarrow Zn(OH)_2(s) + 2Ag(s)$$

As you might expect, these cells are quite expensive because they contain silver. An advantage that they have over the mercury battery, however, is that they generate a higher potential, about 1.5 V.

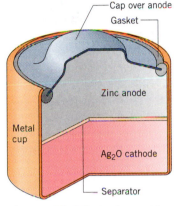

Figure 19.24 *A silver oxide battery.*

Fuel Cells

The galvanic cells we've discussed until now have a limited life span in use because the electrode reactants are eventually depleted. Fuel cells are different; they are electrochemical cells in which the electrode reactants are supplied continuously and are able to operate without theoretical limit as long as the supply of reactants is maintained. This makes fuel cells an attractive source of power where long-term generation of electrical energy is required. They are the subject of *Chemicals in Our World 10,* which follows this chapter, so we will not discuss them further here.

SUMMARY

Electrolysis. In an **electrolytic cell,** a flow of electricity causes an otherwise nonspontaneous reaction to occur. A negatively charged **cathode** causes reduction of one reactant and a positively charged **anode** causes oxidation of another. Ion movement instead of electron transport occurs in the electrolyte. The electrode reactions are determined by which species is most easily reduced and which is most easily oxidized, but in aqueous solutions complex surface effects at the electrodes can alter the expected order. In the electrolysis of water, an electrolyte must be present to maintain electrical neutrality around the electrodes.

Quantitative Aspects of Electrolysis. A **faraday** ($\mathscr{F}$) is the charge carried by 1 mol of electrons, 9.65×10^4 C. The product of current (**amperes**) and time (seconds) gives **coulombs.**

These relationships and the half-reactions that occur at the anode or cathode permit us to relate the amount of chemical change to measurements of current and time.

Applications of Electrolysis. Electroplating, the production of aluminum and magnesium, the refining of copper, and the electrolysis of molten and aqueous sodium chloride are examples of practical applications of electrolysis.

Galvanic Cells. In a **galvanic cell** a spontaneous redox reaction, in which the individual half-reactions occur in separated **half-cells,** causes the electron transfer to occur through an external electrical circuit. Reduction occurs at the positively charged cathode, and oxidation takes place at the negatively charged anode. The half-cells must be connected electrolytically to complete an electrical circuit. A **salt bridge** accomplishes this; it permits electrical neutrality to be maintained. The **potential** (expressed in volts) produced by a cell is equal to the **standard cell potential** when all ion concentrations are 1 *M* and the partial pressures of any gases involved equal 1 atm. The **cell potential** can be considered to be the difference between the **reduction potentials** of the half-cells. In the spontaneous reaction, the half-cell with the higher reduction potential undergoes reduction and forces the other to undergo oxidation. The reduction potentials of isolated half-cells can't be measured, but values are assigned by choosing the **hydrogen electrode** as a reference electrode; its standard reduction potential is assigned a value of exactly 0 V. Species more easily reduced than H^+ have positive reduction potentials; those less easily reduced have negative reduction potentials. Reduction potentials can be used to predict the cell reaction and to calculate E°_{cell}. They can also be used to predict spontaneous redox reactions not occurring in galvanic cells and to predict whether or not a given reaction is spontaneous. Often, they can be used to determine the products of electrolysis reactions.

Thermodynamics and Cell Potentials. The values of ΔG° and K_c for a reaction can be calculated from E°_{cell}. The Nernst equation relates the cell potential to the standard cell potential and the reaction quotient. It allows the cell potential to be calculated for ion concentrations other than 1.00 *M*. The important equations are

$$\Delta G^\circ = -n\mathscr{F}E^\circ_{cell}$$

$$E^\circ_{cell} = \frac{RT}{n\mathscr{F}} \ln K_c$$

$$E_{cell} = E^\circ_{cell} - \frac{RT}{n\mathscr{F}} \ln Q$$

Practical Galvanic Cells. The **lead storage battery** and the **nickel–cadmium battery** are rechargeable. The **zinc–carbon cell**—the common dry cell—is not. The common **alkaline battery** uses essentially the same reactions as the less expensive dry cell. The **zinc–mercury(II) oxide cell** and the **zinc–silver oxide cell** are commonly used in small electronic devices. **Fuel cells** are able to provide continuous power because they consume fuel that can be fed continuously.

Tools *you have learned*

The table below lists the tools you have learned in this chapter that are applicable to problem solving. Review them if necessary, and refer to them when working on the Thinking-It-Through problems and the Review Problems that follows.

Tool	Function
Faraday constant *(page 853)*	To relate coulombs (obtained from the product of current and time) to moles of chemical change in electrochemical reactions.
Standard reduction potentials	To calculate standard cell potentials *(page 867)*; to predict the outcome of electrolysis reactions *(page 877)*.
Standard cell potentials	To predict redox reactions *(page 875)*; to calculate free energy changes *(page 879)* and equilibrium constants *(page 880)*; and to relate concentration of species in galvanic cells to the cell potential *(page 881)*.
$\Delta G^\circ = -n\mathscr{F}E^\circ$ *(page 879)*	To calculate standard free energy changes from cell potentials, and vice versa.
$E^\circ = \dfrac{RT}{n\mathscr{F}} \ln K_c$ *(page 880)*	To calculate equilibrium constants from cell potentials.
Nernst equation *(page 881)*	To calculate the cell potential from E°_{cell} and concentration data; to calculate the concentration of a species in solution from E°_{cell} and a measured value for E_{cell}.

THINKING IT THROUGH

Remember, the goal for each of the following problems is to assemble the available information needed to obtain the answer, state what additional data (if any) are needed, and describe how you would use the data to answer the question. For problems involving unit conversions, list the relationships among the units that are needed to carry out the conversions. Construct the conversion factors that can be formed from these relationships. Then set up the solution to the problem by arranging the conversion factors so the units cancel correctly to give the desired units of the answer.

The problems are divided into two groups. Those in Level 2 are more challenging than those in Level 1 and provide an opportunity to really hone your problem solving skills. Detailed answers to these problems can be found at our Web site: http://www.wiley.com/college/brady.

Level 1 Problems

1. Explain how you would calculate the number of grams of Pb that will be deposited on the cathode of an electrolytic cell if 500 mL of a solution of $Pb(NO_3)_2$ is electrolyzed for a period of 30.00 min at a current of 4.00 A.

2. At a current of 2.00 A, describe in detail how you would calculate the number of minutes it would take to deposit a coating of silver on a metal object by electrolysis of $Ag(CN)_2^-$ if the silver coating is to be 0.020 mm thick and is to have an area of 25 cm^2. The density of silver is 10.50 g/cm^3.

3. After the electrolysis described in Question 1, show how you would calculate the pH of the solution.

4. How can you determine whether the reaction $Au + NO_3^- + 4H^+ \rightarrow Au^{3+} + NO + 2H_2O$ will occur spontaneously?

5. In a galvanic cell established between Fe(s) and $Fe^{2+}(aq)$ plus Co(s) and $Co^{2+}(aq)$, how can you determine which electrode will carry the positive charge?

6. Explain how you would use data in this chapter to calculate the value of the equilibrium constant at 25 °C for the following reaction.

$$2Br^-(aq) + I_2(s) \rightleftharpoons Br_2(aq) + 2I^-(aq)$$

Level 2 Problems

7. How many milliliters of H_2 gas at 25 °C and a total pressure of 755 torr will be generated by the electrolysis of water using a KNO_3 electrolyte if a current of 1.25 A is passed through the electrolysis cell for a period of 10.00 minutes? Describe the calculation in detail.

8. A current was passed through two electrolytic cells connected in series (so the same current passed through both of them). In one cell was a solution of $Cr_2(SO_4)_3$, and in the other, a solution of $AgNO_3$. If 1.45 g of Ag was deposited on the cathode in the second cell, how many grams of Cr were deposited on the cathode in the first cell? The temperature of the cell was 25 °C. Describe the calculations in detail.

9. The following cell was established:

$$Sn(s) \mid Sn^{2+}(? \ M) \parallel Cl^-(1.00 \times 10^{-2} \ M) \mid AgCl(s), Ag(s)$$

The half-reaction

$$AgCl(s) + e^- \rightleftharpoons Ag(s) + Cl^-(aq)$$

has $E^\circ_{AgCl} = 0.2223$ V, and $E^\circ_{Sn^{2+}} = -0.1375$ V. The tin electrode was dipped into a solution of Sn^{2+} and the potential of the cell was measured to be 0.3114 V. Describe how you would calculate the concentration of Sn^{2+} in the solution.

10. What is the value of K_c for the reaction in Question 9 at 35 °C?

11. Describe in detail how you would calculate the potential of a cell in which the cell reaction is

$$NiO_2(s) + 4H^+(aq) + 2Ag(s) \longrightarrow$$
$$Ni^{2+}(aq) + 2H_2O + 2Ag^+(aq)$$

if the pH is 2.50 and the concentrations of Ni^{2+} and Ag^+ are each 0.020 M.

12. Make a sketch of a galvanic cell that utilizes a standard hydrogen electrode and another electrode that would enable you to determine the K_{sp} for AgCl. Explain how you would find the K_{sp} value from the measured cell potential.

13. A galvanic cell utilizing a Cu electrode in 1.00 M $CuSO_4$ and metal Z in 1.00 M $Z(NO_3)_2$ is set up as shown below. In a second experiment, also shown below, a copper rod is placed in a beaker of 1.00 M $Z(NO_3)_2$ and a metal Z

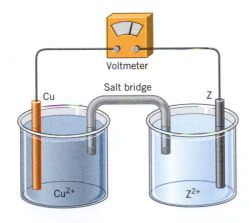

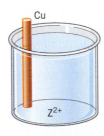

rod is placed in a 1.00 *M* $CuSO_4$ solution as shown in the figure. The colorless solution of 1.00 *M* $Z(NO_3)_2$ is observed to develop a blue color, while the blue color of the 1.00 *M* CuSO4 solution appears to remain constant.
(a) Write the chemical equation for the spontaneous cell reaction occurring in the galvanic cell.
(b) Which way do electrons flow in the wire connecting the Cu and metal *Z* electrodes?

(c) The salt bridge contains a saturated KNO_3 solution. Toward which electrodes do the ions in the salt bridge move?
(d) Does the color of the $CuSO_4$ solution remain constant, fade, or become darker?
(e) Over time, will the voltage shown on the voltmeter remain constant, fall, or increase?

REVIEW QUESTIONS

Electrolysis

19.1 What electrical charges do the anode and the cathode carry in an electrolytic cell? What does the term *inert electrode* mean?

19.2 How do *electrolytic conduction* and *metallic conduction* differ?

19.3 Why must electrolysis reactions occur at the electrodes in order for electrolytic conduction to continue?

19.4 What is the difference between a *direct current* and an *alternating current?*

19.5 Why must NaCl be melted before it is electrolyzed to give Na and Cl_2? Write the anode, cathode, and overall cell reactions for the electrolysis of molten NaCl.

19.6 Write half-reactions for the oxidation and the reduction of water.

19.7 What happens to the pH of the solution near the cathode and anode during the electrolysis of KNO_3? What function does KNO_3 serve in the electrolysis of a KNO_3 solution?

19.8 Write the anode reaction for the electrolysis of an aqueous solution that contains (a) NO_3^-, (b) Br^-, and (c) NO_3^- and Br^-.

19.9 Write the cathode reaction for the electrolysis of an aqueous solution that contains (a) K^+, (b) Cu^{2+}, and (c) K^+ and Cu^{2+}.

19.10 What products would we expect at the electrodes if a solution containing both KBr and $Cu(NO_3)_2$ were electrolyzed?

Stoichiometric Relationships in Electrolysis

19.11 What is a *faraday?* What relationships relate faradays to current and time measurements?

Applications of Electrolysis

19.12 What is *electroplating?* Sketch an apparatus to electroplate silver.

19.13 Describe the Hall–Héroult process for producing metallic aluminum. What half-reaction occurs at the anode? What half-reaction occurs at the cathode? What is the overall cell reaction?

19.14 In the Hall–Héroult process, why must the carbon anodes be replaced frequently?

19.15 Describe how magnesium is recovered from seawater. What electrochemical reaction is involved?

19.16 How is metallic sodium produced? What are some uses of metallic sodium? Write equations for the anode and cathode reactions.

19.17 Describe the electrolytic refining of copper. What economic advantages offset the cost of electricity for this process? What chemical reactions occur at (a) the anode and (b) the cathode?

19.18 Describe the electrolysis of aqueous sodium chloride. How do the products of the electrolysis compare for stirred and unstirred reactions? Write chemical equations for the reactions that occur at the electrodes.

19.19 What are the advantages and disadvantages in the electrolysis of aqueous NaCl of (a) the diaphragm cell and (b) the mercury cell?

Galvanic Cells

19.20 What is a galvanic cell? What is a half-cell?

19.21 What is the function of a *salt bridge?*

19.22 In the copper–silver cell, why must the Cu^{2+} and Ag^+ solutions be kept in separate containers?

19.23 Which redox processes take place at the anode and cathode in a galvanic cell? What electrical charges do the anode and cathode carry in a galvanic cell?

19.24 Explain how the movement of the ions relative to the electrodes is the same in both galvanic and electrolytic cells.

19.25 When magnesium metal is placed into a solution of copper sulfate, the magnesium dissolves to give Mg^{2+} and copper metal is formed. Write a net ionic equation for this reaction. Describe how you could use the reaction in a galvanic cell. Which metal, copper or magnesium, is the cathode?

19.26 Aluminum will displace tin from solution according to the equation $2Al(s) + 3Sn^{2+}(aq) \rightarrow 2Al^{3+}(aq) + 3Sn(s)$. What would be the individual half-cell reactions if this were the cell reaction in a galvanic cell? Which metal would be the anode and which the cathode?

19.27 At first glance, the following equation may appear to be balanced: $MnO_4^- + Sn^{2+} \rightarrow SnO_2 + MnO_2$. What is wrong with it?

Cell Potentials and Reduction Potentials

19.28 For a galvanic cell, what is the meaning of the term *potential?* What are its units?

19.29 What is the difference between a *cell potential* and a *standard cell potential?*

19.30 How are standard reduction potentials combined to give the standard cell potential for a spontaneous reaction?

19.31 What ratio of units gives volts? What are the units of amperes × volts × seconds?

19.32 Is it possible to measure the potential of an isolated half-cell? Explain your answer.

19.33 Describe the hydrogen electrode. What is the value of its standard reduction potential?

19.34 What do the positive and negative signs of reduction potentials tell us?

19.35 If $E°_{Cu^{2+}}$ had been chosen as the standard reference electrode and had been assigned a potential of 0.00 V, what would the reduction potential of the hydrogen electrode be relative to it?

19.36 Using a voltmeter, how can you tell which is the anode and which is the cathode in a galvanic cell?

Using Standard Reduction Potentials

19.37 Compare Table 5.2 with Table 19.1. What can you say about the basis for the activity series for metals?

19.38 Make a sketch of a galvanic cell for which the cell notation is

$$Ag(s) \mid Ag(aq) \parallel Fe^{3+}(aq) \mid Fe(s)$$

(a) Label the anode and the cathode.
(b) Indicate the charge on each electrode.

19.39 Make a sketch of a galvanic cell in which inert platinum electrodes are used in the half-cells for the system

$$Pt(s) \mid Fe^{2+}(aq), Fe^{3+}(aq) \parallel Br_2(aq), Br^-(aq) \mid Pt(s)$$

Label the diagram and indicate the composition of the electrolytes in the two cell compartments. Show the signs of the electrodes and label the anode and cathode.

Cell Potentials and Thermodynamics

19.40 Write the equation that relates the standard cell potential to the standard free energy change for a reaction.

19.41 What is the equation that relates the equilibrium constant to the cell potential?

19.42 What is the maximum amount of work, expressed in joules, that can be obtained from the discharge of a silver oxide battery (see p. 889) if 0.500 g of Ag_2O reacts? $E°$ for the cell reaction is 1.50 V.

The Effect of Concentration on Cell Potential

19.43 You have learned that the principles of thermodynamics allow the following equation to be derived, $\Delta G = \Delta G° + RT \ln Q$, where Q is the reaction quotient. Without referring to the text, use this equation and the relationship between ΔG and the cell potential to derive the Nernst equation.

19.44 The cell reaction during the discharge of a lead storage battery is

$$Pb(s) + PbO_2(s) + 4H^+(aq) + 2SO_4^{2-}(aq) \longrightarrow$$
$$2PbSO_4(s) + 2H_2O$$

The standard cell potential is 2.05 V. What is the correct form of the Nernst equation for this reaction at 25 °C?

Practical Galvanic Cells

19.45 What are the anode and cathode reactions during the discharge of a lead storage battery? How can a battery produce a potential of 12 V if the cell reaction has a standard potential of only 2 V?

19.46 What are the anode and cathode reactions during the charging of a lead storage battery? What are the reactions during discharge of the battery?

19.47 What reactions occur at the electrodes in the ordinary dry cell?

19.48 What chemical reactions take place at the electrodes in an alkaline dry cell?

19.49 Give the half-cell reactions and the cell reaction that take place in a nicad battery during discharge. What are the reactions that take place during the charging of the cell?

19.50 Describe the chemical reactions that take place in (a) a silver oxide battery and (b) a mercury battery.

19.51 What advantages do fuel cells offer over conventional means of obtaining electrical power by the combustion of fuels?

19.52 How is a hydrometer constructed? How does it measure density? Why can a hydrometer be used to check the state of charge of a lead storage battery?

19.53 What role does the element calcium play in the construction of a modern automobile battery?

REVIEW PROBLEMS

Answers to problems whose numbers are printed in color are given in Appendix D.
More challenging problems are marked with asterisks.

Quantitative Aspects of Electrolysis

19.54 How many coulombs are passed through an electrolysis cell by (a) a current of 4.00 A for 600 s? (b) A current of 10.0 A for 20.0 min? (c) A current of 1.50 A for 6.00 hr?

19.55 How many moles of electrons correspond to the answers to each part of Problem 19.54?

19.56 How many moles of electrons are required to (a) reduce 0.20 mol Fe^{2+} to Fe? (b) Oxidize 0.70 mol Cl^- to Cl_2?

(c) Reduce 1.50 mol Cr^{3+} to Cr? (d) Oxidize 1.0×10^{-2} mol Mn^{2+} to MnO_4^-?

19.57 How many moles of electrons are required to (a) produce 5.00 g Mg from molten $MgCl_3$? (b) Form 41.0 g Cu from a $CuSO_4$ solution?

19.58 How many moles of Cr^{3+} would be reduced to Cr by the same amount of electricity that produces 12.0 g Ag from a solution of $AgNO_3$?

19.59 In Problem 19.58, if a current of 4.00 A was used, how many minutes would the electrolysis take?

19.60 How many grams of $Fe(OH)_2$ are produced at an iron anode when a basic solution of NaOH undergoes electrolysis at a current of 8.00 A for 12.0 min?

19.61 How many grams of Cl_2 would be produced in the electrolysis of molten NaCl by a current of 4.25 A for 35.0 min?

19.62 How many hours would it take to produce 75.0 g of metallic chromium by the electrolytic reduction of Cr^{3+} with a current of 2.25 A?

19.63 How many hours would it take to generate 35.0 g of lead from $PbSO_4$ during the charging of a lead storage battery using a current of 1.50 A? The half-reaction is

$$PbSO_4 + 2e^- \longrightarrow Pb + SO_4^{2-}$$

19.64 How many amperes would be needed to produce 60.0 g of magnesium during the electrolysis of molten $MgCl_2$ in 2.00 hr?

19.65 A large electrolysis cell that produces metallic aluminum from Al_2O_3 by the Hall–Héroult process is capable of yielding 900 lb (409 kg) of aluminum in 24 hr. What current is required?

19.66 A solution of NaCl in water was electrolyzed with a current of 2.50 A for 15.0 min. How many milliliters of 0.100 M HCl would be required to neutralize the resulting solution?

19.67 How many milliliters of gaseous H_2, measured at STP, would be produced at the cathode in the electrolysis of water with a current of 0.750 A for 15.00 min?

Cell Notation

19.68 Write the half-reactions and the balanced cell reaction for the following galvanic cells.
(a) $Cd(s) \mid Cd^{2+}(aq) \parallel Au^{3+}(aq) \mid Au(s)$
(b) $Pb(s), PbSO_4(s) \mid SO_4^{2-}(aq) \parallel$
$\qquad H^+(aq), SO_4^{2-}(aq) \mid PbO_2(s), PbSO_4(s)$
(c) $Cr(s) \mid Cr^{3+}(aq) \parallel Cu^{2+}(aq) \mid Cu(s)$

19.69 Write the half-reactions and the balanced cell reaction for the following galvanic cells.
(a) $Zn(s) \mid Zn^{2+}(aq) \parallel Cr^{3+}(aq) \mid Cr(s)$
(b) $Fe(s) \mid Fe^{2+}(aq) \parallel Br_2(aq), Br^-(aq) \mid Pt(s)$
(c) $Mg(s) \mid Mg^{2+}(aq) \parallel Sn^{2+}(aq) \mid Sn(s)$

19.70 Write the cell notation for the following galvanic cells. For half-reactions in which all the reactants are in solution or are gases, assume the use of inert platinum electrodes.
(a) $Cd^{2+}(aq) + Fe(s) \rightarrow Cd(s) + Fe^{2+}(aq)$
(b) $Br_2(aq) + 2Cl^-(aq) \rightarrow Cl_2(g) + 2Br^-(aq)$
(c) $Au^{3+}(aq) + 3Ag(s) \rightarrow Au(s) + 3Ag^+(aq)$

19.71 Write the cell notation for the following galvanic cells. For half-reactions in which all the reactants are in solution or are gases, assume the use of inert platinum electrodes.
(a) $NO_3^-(aq) + 4H^+(aq) + 3Fe^{2+}(aq) \rightarrow$
$\qquad 3Fe^{3+}(aq) + NO(g) + 2H_2O$
(b) $NiO_2(s) + 4H^+(aq) + 2Ag(s) \rightarrow$
$\qquad Ni^{2+}(aq) + 2H_2O + 2Ag^+(aq)$
(c) $Mg(s) + Cd^{2+}(aq) \rightarrow Mg^{2+}(aq) + Cd(s)$

Using Reduction Potentials

19.72 For each pair of substances, use Table 19.1 to choose the better reducing agent.
(a) $Sn(s)$ or $Ag(s)$ (c) $Co(s)$ or $Zn(s)$
(b) $Cl^-(aq)$ or $Br^-(aq)$ (d) $I^-(aq)$ or $Au(s)$

19.73 For each pair of substances, use Table 19.1 to choose the better oxidizing agent.
(a) $NO_3^-(aq)$ or $MnO_4^-(aq)$ (c) $PbO_2(s)$ or $Cl_2(g)$
(b) $Au^{3+}(aq)$ or $Co^{2+}(aq)$ (d) $NiO_2(s)$ or $HOCl(aq)$

19.74 Use the data in Table 19.1 to calculate the standard cell potential for each of the following reactions:
(a) $Cd^{2+}(aq) + Fe(s) \rightarrow Cd(s) + Fe^{2+}(aq)$
(b) $Br_2(aq) + 2Cl^-(aq) \rightarrow Cl_2(g) + 2Br^-(aq)$
(c) $Au^{3+}(aq) + 3Ag(s) \rightarrow Au(s) + 3Ag^+(aq)$

19.75 Use the data in Table 19.1 to calculate the standard cell potential for each of the following reactions:
(a) $NO_3^-(aq) + 4H^+(aq) + 3Fe^{2+}(aq) \rightarrow$
$\qquad 3Fe^{3+}(aq) + NO(g) + 2H_2O$
(b) $NiO_2(s) + 4H^+(aq) + 2Ag(s) \rightarrow$
$\qquad Ni^{2+}(aq) + 2H_2O + 2Ag^+(aq)$
(c) $Mg(s) + Cd^{2+}(aq) \rightarrow Mg^{2+}(aq) + Cd(s)$

19.76 From the positions of the half-reactions in Table 19.1, determine whether the following reactions are spontaneous under standard state conditions.
(a) $2Au^{3+} + 6I^- \rightarrow 3I_2 + 2Au$
(b) $3Fe^{2+} + 2NO + 4H_2O \rightarrow 3Fe + 2NO_3^- + 8H^+$
(c) $3Ca + 2Cr^{3+} \rightarrow 2Cr + 3Ca^{2+}$

19.77 Use the data in Table 19.1 to determine which of the following reactions should occur spontaneously under standard state conditions.
(a) $Br_2 + 2Cl^- \rightarrow Cl_2 + 2Br^-$
(b) $Ni^{2+} + Fe \rightarrow Fe^{2+} + Ni$
(c) $H_2SO_3 + H_2O + Br_2 \rightarrow 4H^+ + SO_4^{2-} + 2Br^-$

19.78 Use the data in Table 19.1 to calculate the standard cell potential for the reaction

$$MnO_4^- + 8H^+ + 5Fe^{2+} \longrightarrow 5Fe^{3+} + Mn^{2+} + 4H_2O$$

19.79 Use the data in Table 19.1 to calculate the standard cell potential for the reaction

$$2Ag^+ + Fe \longrightarrow 2Ag + Fe^{2+}$$

19.80 Write the cell notation for the galvanic cells formed by using the following pairs of half-reactions. Calculate their standard cell potentials.
(a) $Co^{2+}(aq) + 2e^- \rightleftharpoons Co(s)$
   $Zn^{2+}(aq) + 2e^- \rightleftharpoons Zn(s)$
(b) $Ni^{2+}(aq) + 2e^- \rightleftharpoons Ni(s)$
   $Mg^{2+}(aq) + 2e^- \rightleftharpoons Mg(s)$
(c) $Au^{3+}(aq) + 3e^- \rightleftharpoons Au(s)$
   $Sn^{2+}(aq) + 2e^- \rightleftharpoons Sn(s)$

19.81 Write the cell notation for the galvanic cells formed by using the following pairs of half-reactions. For half-reactions in which all the reactants are in solution, assume the use of inert platinum electrodes. Calculate their standard cell potentials.
(a) $BrO_3^-(aq) + 6H^+(aq) + 6e^- \rightleftharpoons Br^-(aq) + 3H_2O$
   $Cu^{2+}(aq) + 2e^- \rightleftharpoons Cu(s)$
(b) $Fe^{3+}(aq) + e^- \rightleftharpoons Fe^{2+}(aq)$
   $Ag^+(aq) + e^- \rightleftharpoons Ag(s)$
(c) $NO_3^-(aq) + 4H^+(aq) + 3e^- \rightleftharpoons NO(g) + 2H_2O$
   $MnO_4^-(aq) + 8H^+(aq) + 5e^- \rightleftharpoons Mn^{2+}(aq) + 4H_2O$

19.82 From the half-reactions below, determine the cell reaction and standard cell potential.

$$BrO_3^- + 6H^+ + 6e^- \rightleftharpoons Br^- + 3H_2O \qquad E^\circ_{BrO_3^-} = 1.44 \text{ V}$$

$$I_2 + 2e^- \rightleftharpoons 2I^- \qquad E^\circ_{I_2} = 0.54 \text{ V}$$

19.83 What is the standard cell potential and the net reaction in a galvanic cell that has the following half-reactions?

$$MnO_2 + 4H^+ + 2e^- \rightleftharpoons Mn^{2+} + 2H_2O \quad E^\circ_{MnO_2} = 1.23 \text{ V}$$

$$PbCl_2 + 2e^- \rightleftharpoons Pb + 2Cl^- \quad E^\circ_{PbCl_2} = -0.27 \text{ V}$$

19.84 What will be the spontaneous reaction among H_2SO_3, $S_2O_3^{2-}$, HOCl, and Cl_2 under standard state conditions? The half-reactions involved are

$$2H_2SO_3 + 2H^+ + 4e^- \rightleftharpoons S_2O_3^{2-} + 3H_2O \; E^\circ_{H_2SO_3} = 0.40 \text{ V}$$

$$2HOCl + 2H^+ + 2e^- \rightleftharpoons Cl_2 + 2H_2O \qquad E^\circ_{HOCl} = 1.63 \text{ V}$$

19.85 What will be the spontaneous reaction among Br_2, I_2, Br^-, and I^-? (Assume 1.00 M ion concentrations.)

19.86 Will the following reaction occur spontaneously under standard state conditions?

$$SO_4^{2-} + 4H^+ + 2I^- \longrightarrow H_2SO_3 + I_2 + H_2O$$

Use the data in Table 19.1 to answer this question.

19.87 Use the data below to determine whether the reaction

$$S_2O_8^{2-} + Ni(OH)_2 + 2OH^- \rightarrow 2SO_4^{2-} + NiO_2 + 2H_2O$$

will occur spontaneously under standard state conditions.

$$NiO_2 + 2H_2O + 2e^- \rightleftharpoons Ni(OH)_2 + 2OH^- \; E^\circ_{NiO_2} = 0.49 \text{ V}$$

$$S_2O_8^{2-} + 2e^- \rightleftharpoons 2SO_4^{2-} \qquad E^\circ_{S_2O_8^{2-}} = 2.01 \text{ V}$$

19.88 If electrolysis is carried out on an aqueous solution of aluminum sulfate, what products are expected at the electrodes? Write the equation for the net cell reaction.

19.89 If electrolysis is carried out on an aqueous solution of cadmium iodide, what products are expected at the electrodes? Write the equation for the net cell reaction.

Cell Potentials and Thermodynamics
19.90 Calculate ΔG° for the following reaction *as written.*

$$2Br^- + I_2 \rightarrow 2I^- + Br_2$$

19.91 Calculate ΔG° for the reaction

$$2MnO_4^- + 6H^+ + 5HCHO_2 \rightarrow 2Mn^{2+} + 8H_2O + 5CO_2$$

for which $E^\circ_{cell} = 1.69$ V.

19.92 Given the following half-reactions and their standard reduction potentials,

$$2ClO_3^- + 12H^+ + 10e^- \rightleftharpoons Cl_2 + 6H_2O \quad E^\circ_{ClO_3^-} = 1.47 \text{ V}$$

$$S_2O_8^{2-} + 2e^- \rightleftharpoons 2SO_4^{2-} \qquad E^\circ_{S_2O_8^{2-}} = 2.01 \text{ V}$$

calculate (a) E°_{cell}, (b) ΔG° for the cell reaction, and (c) the value of K_c for the cell reaction.

19.93 Calculate K_c for the system, $Ni^{2+} + Co \rightleftharpoons Ni + Co^{2+}$. Use the data in Table 19.1. Assume $T = 298$ K.

19.94 The system $2AgI + Sn \rightleftharpoons Sn^{2+} + 2Ag + 2I^-$ has a calculated $E^\circ_{cell} = -0.015$ V. What is the value of K_c for this system?

19.95 Determine the value of K_c at 25 °C for the reaction

$$2H_2O + 2Cl_2 \rightleftharpoons 4H^+ + 4Cl^- + O_2$$

The Effect of Concentration on Cell Potential
19.96 The cell reaction

$$NiO_2(s) + 4H^+(aq) + 2Ag(s) \longrightarrow$$
$$Ni^{2+}(aq) + 2H_2O + 2Ag^+(aq)$$

has $E^\circ_{cell} = 2.48$ V. What will be the cell potential at a pH of 5.00 when the concentrations of Ni^{2+} and Ag^+ are each 0.010 M?

19.97 $E^\circ_{cell} = 0.135$ V for the reaction

$$3I_2(s) + 5Cr_2O_7^{2-}(aq) + 34H^+ \longrightarrow$$
$$6IO_3^-(aq) + 10Cr^{3+}(aq) + 17H_2O$$

What is E°_{cell} if $[Cr_2O_7^{2-}] = 0.010$ M, $[H^+] = 0.10$ M, $[IO_3^-] = 0.00010$ M, and $[Cr^{3+}] = 0.0010$ M?

19.98 A cell was set up having the following reaction.

$$Mg(s) + Cd^{2+}(aq) \rightarrow Mg^{2+}(aq) + Cd(s) \qquad E^\circ_{cell} = 1.97 \text{ V}$$

The magnesium electrode was dipping into a 1.00 M solution of $MgSO_4$ and the cadmium electrode was dipping into a solution of unknown Cd^{2+} concentration. The potential of the cell was measured to be 1.54 V. What was the unknown Cd^{2+} concentration?

19.99 A silver wire coated with AgCl is sensitive to the presence of chloride ion because of the half-cell reaction

$$AgCl(s) + e^- \rightleftharpoons Ag(s) + Cl^- \qquad E^\circ_{AgCl} = 0.2223 \text{ V}$$

A student, wishing to measure the chloride ion concentration in a number of water samples, constructed a galvanic cell using the AgCl electrode as one half-cell and a copper wire dipping into 1.00 M $CuSO_4$ solution as the other half-cell. In one analysis, the potential of the cell was measured to be 0.0925 V with the copper half-cell serving as the cathode. What was the chloride ion concentration in the water? (Take $E^\circ_{Cu^{2+}} = 0.3419$ V.)

*19.100 At 25 °C, a galvanic cell was set up having the following half-reactions.

$$Fe^{2+}(aq) + 2e^- \rightleftharpoons Fe(s) \qquad E^\circ_{Fe^{2+}} = -0.447 \text{ V}$$

$$Cu^{2+}(aq) + 2e^- \rightleftharpoons Cu(s) \qquad E^\circ_{Cu^{2+}} = +0.3419 \text{ V}$$

The copper half-cell contained 100 mL of 1.00 M $CuSO_4$. The iron half-cell contained 50.0 mL of 0.100 M $FeSO_4$. To the iron half-cell was added 50.0 mL of 0.500 M NaOH solution. The mixture was stirred and the cell potential was measured to be 1.175 V. Calculate the value of K_{sp} for $Fe(OH)_2$.

*19.101 Suppose a galvanic cell was constructed at 25 °C using a Cu/Cu^{2+} half-cell (in which the molar concentration of Cu^{2+} was 1.00 M) and a hydrogen electrode having a partial pressure of H_2 equal to 1 atm. The hydrogen electrode dips into a solution of unknown hydrogen ion concentration, and the two half-cells are connected by a salt bridge. The precise value of $E^\circ_{Cu^{2+}}$ is +0.3419 V.
(a) Derive an equation for the pH of the solution with the unknown hydrogen ion concentration, expressed in terms of E_{cell} and E°_{cell}.
(b) If the pH of the solution were 5.15, what would be the observed potential of the cell?
(c) If the potential of the cell were 0.645 V, what would be the pH of the solution?

ADDITIONAL EXERCISES

19.102 In the absence of unusual electrode effects, what should be expected to be observed at the cathode when electrolysis is performed on a solution that contains $NiSO_4$ and $CdSO_4$ at equal concentrations? (See Table 19.1.)

19.103 A solution containing vanadium (chemical symbol V) in an unknown oxidation state was electrolyzed with a current of 1.50 A for 30.0 min. It was found that 0.475 g of V was deposited on the cathode. What was the original oxidation state of the vanadium ion?

19.104 Consider the following half-reactions and their standard reduction potentials.

$$S_2O_8^{2-} + 2e^- \rightleftharpoons 2SO_4^{2-} \qquad E^\circ_{S_2O_8^{2-}} = +2.01 \text{ V}$$

$$O_2 + 4H^+ + 4e^- \rightleftharpoons 2H_2O \qquad E^\circ_{H_2O} = +1.23 \text{ V}$$

What is the most likely product at the anode in the electrolysis of Na_2SO_4?

19.105 The electrolysis of 250 mL of a brine solution (NaCl) was carried out for a period of 20.00 min with a current of 2.00 A. The resulting solution was titrated with 0.620 M HCl. How many milliliters of the HCl solution was required for the titration to achieve a neutral solution?

19.106 A solution of NaCl is neutral, with an expected pH of 7. If electrolysis were carried out on 500 mL of a NaCl solution with a current of 0.500 A, how many seconds would it take for the pH of the solution to rise to a value of 9.00?

*19.107 A watt is a unit of electrical power and is equal to one joule per second (1 watt = 1 J s^{-1}). How many hours can a calculator drawing 5×10^{-4} watt be operated by a mercury battery (see p. 888) having a cell potential equal to 1.34 V if a mass of 1.00 g of HgO is available at the cathode?

*19.108 Suppose that a galvanic cell were set up having the net cell reaction

$$Zn(s) + 2Ag^+(aq) \longrightarrow Zn^{2+}(aq) + 2Ag(s)$$

The Ag^+ and Zn^{2+} concentrations in their respective half-cells initially are 1.00 M, and each half-cell contains 100 mL of electrolyte solution. If this cell delivers current at a constant rate of 0.10 A, what will the cell potential be after 15.00 hr?

19.109 It was desired to determine the reduction potential of Pt^{2+} ion. A galvanic cell was set up in which one half-cell consisted of a Pt electrode dipping into a 0.0100 M solution of $Pt(NO_3)_2$ and the other was prepared by dipping a silver wire coated with AgCl into a 0.100 M solution of HCl. The potential of the cell was measured to be 0.778 V, and it was found that the Pt electrode carried a positive charge. Given the following half-reaction and its reduction potential,

$$AgCl(s) + e^- \rightleftharpoons Ag(s) + Cl^-(aq) \qquad E^\circ_{AgCl} = 0.2223 \text{ V}$$

calculate the standard reduction potential for the half-reaction $Pt^{2+}(aq) + 2e^- \rightleftharpoons Pt(s)$.

19.110 Consider the following half-reactions and their reduction potentials:

$$MnO_4^- + 8H^+ + 5e^- \longrightarrow Mn^{2+} + 4H_2O$$

$$E^\circ_{MnO_4^-} = +1.507 \text{ V}$$

$$ClO_3^- + 6H^+ + 6e^- \longrightarrow Cl^- + 3H_2O$$

$$E^\circ_{ClO_3^-} = +1.451 \text{ V}$$

(a) What is the value of E°_{cell}?
(b) What is the value of ΔG° for this reaction?
(c) What is the value of K_c for the reaction at 25 °C?

(d) Write the Nernst equation for this reaction.
(e) What is the potential of the cell when $[MnO_4^-] = 0.20$ M, $[Mn^{2+}] = 0.050$ M, $[Cl^-] = 0.0030$ M, $[ClO_3^-] = 0.110$ M, and the pH of the solution is 4.25?

19.111 Consider the following galvanic cell

$$Ag(s) \,|\, Ag^+(3.0 \times 10^{-4} \text{ } M) \,\|$$

$$Fe^{3+}(1.1 \times 10^{-3} \text{ } M), Fe^{2+}(0.040 \text{ } M) \,|\, Pt(s)$$

Calculate the cell potential. Determine the sign of the electrodes in the cell. Write the equation for the spontaneous cell reaction.

Hydrogen, Oxygen, and Fuel Cells

10

Some of the chemical energy available from the reaction of two simple chemicals, hydrogen and oxygen, becomes electricity in a hydrogen fuel cell. The reaction is very straightforward; hydrogen reacts with oxygen to give water. No pollutants are generated.

$$2H_2 + O_2 \longrightarrow 2H_2O + \text{energy}$$

As we studied in Chapter 18, fundamental principles of thermodynamics constrain the percentage of chemical energy that can be harnessed for useful work. Some of the chemical energy *must* appear as heat when fuels are burned. This is true even at maximum efficiency of conversion, attainable only if the reactions could take place reversibly. Typically, a power plant converts only 35 to 40% of the chemical energy of coal or oil to electricity. The engines of vehicles convert from 25 to 30% of the chemical energy in diesel oil or gasoline to useful work. The rest comes out as heat, which is why vehicles must have good engine-cooling systems. Hydrogen fuel cells are better. They have efficiencies of 40 to 50%, just one of many reasons why they are attractive.

Hydrogen fuel cells have been used for electricity on space missions beginning with NASA's Gemini spacecraft in the early 1960s. From onboard supplies of pure hydrogen and oxygen, the cells not only produce electricity but also replenish the water supplies of the astronauts. One problem with the cells of the Gemini and Apollo missions, however, is that they used aqueous KOH as an internal conducting material or electrolyte. That was acceptable for a space mission, but it would not work for vehicles on our roads. We cannot have systems capable of spraying caustic solutions about in the event of a collision. Moreover, aqueous hydroxides readily absorb and react with many pollutants in the air, like carbon dioxide, sulfur oxides, and nitrogen oxides.

The Hydrogen Fuel Cell and Electrolyte Substitutes. The fuel cells used in space missions, painstakingly assem-

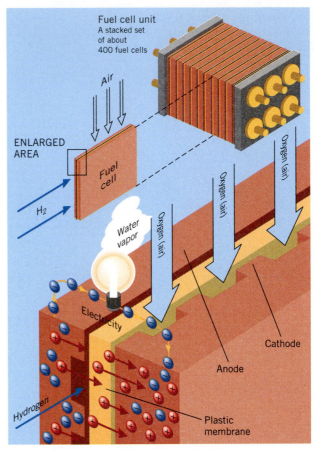

Figure 10a The hydrogen fuel cell. Hydrogen is broken up into electrons and protons at the surface of platinum, which is supported by a thin plastic film that permits protons to pass through. The electrons must get to the other side by going through an external circuit. On the other side, the protons and electrons combine with oxygen to produce water. (Courtesy of International Fuel Cells/Samuel Velasco, *Fortune,* March 30, 1998, page 122[F]).

chemicals in our world

Figure 10*b* The Daimler-Benz NEBUS, or new electric bus, which is powered by hydrogen fuel cells. The tanks of compressed hydrogen are housed on the top of the bus.

bled by hand, are both cumbersome and extremely expensive, costing between $300,000 and $400,000 per kilowatt of generating capacity. (Roughly 45 kilowatts of power must be supplied from a typical spacecraft's fuel cells.) The price of such cells is coming down, however, because the caustic electrolyte can be replaced by ultrathin films of fluorinated plastics coated with very thin layers of platinum catalyst. With its platinum, the film is a solid conductor, an *electrolyte sheet*. Gore-Tex, the material popular in tough, water-resistant outdoor garments, is one of the films used for supporting the platinum and for providing something of a "filter" that lets hydrogen ions through. Du Pont's Nafion is another such film.

An individual hydrogen fuel cell sandwiches the electrolyte sheet between pieces of graphite paper, which is permeable to both hydrogen and oxygen (see Figure 10*a*). One sheet is the anode (negative electrode) and the other is the cathode. This larger assembly is then sandwiched between two more plates machined with grooves for channeling gases. As seen in Figure 10*a*, hydrogen is channeled to the anode and oxygen (actually, air) is fed to the cathode. The platinum catalyzes the production of protons and electrons from the hydrogen, but the overall driving force is the powerful tendency of these two fundamental particles to combine with oxygen to form water. The film, however, will not allow electrons to pass through directly, only protons. So the electrons must take the long way around, through a circuit, to get to the other side and to participate in the chemical reaction. The movement of the electrons in the external circuit, of course, is electricity.

Each of the cells just described provides just under 1 volt. But connecting a large number of them in series is all that is needed to provide the volts for enough power for driving a car or a bus or for handling all of the electrical needs of a home. The energy not converted to electricity is heat, which can be used to warm the vehicle or surrounding space.

The costs of hydrogen fuel cells are still too high for the marketplace, but prototype vehicles and homes have been built. And anybody who has been into computers for even a few years knows how prices come down dramatically with improved materials, engineering, and the economies offered by mass production.

Vehicles. The power density—the kilowatts of power per cubic foot of apparatus volume—is a very important parameter when it comes to designing hydrogen fuel cell systems for vehicles or homes. Car engineers typically demand a power density of at least 30 kW/ft^3. A midsize car could then be designed that uses the volume of a large suitcase for the power unit. By the start of the twenty-first century, a power density of 50 kW/ft^3 will be available. The cost in 1990 was roughly $5000 per kW. By 1999, the cost had fallen to about $500/kW. Automobile designers want the cost closer to $25 to $30 per kW, but this will happen only with the economies available by markets of a quarter of a million vehicles a year. But progress is occurring. Daimler-Benz has already made and tested a 40-foot bus, the NEBUS (for "new electric bus"), which is nonpolluting and quiet (see Figure 10*b*).

Availability of Hydrogen. One serious problem with hydrogen fuel cells is that hydrogen isn't available at gas stations. Small hydrogen molecules leak too easily through the tiniest of pinholes, so hydrogen has to be handled with special equipment. To retrofit gas stations with such equipment would be very costly. It's easy for a bus to use hydrogen; it returns to its base each day where its hydrogen tanks can be refilled.

To get around this problem for automobiles and trucks, engineers are working to develop "on-board" liquid-fuel sources of hydrogen, like methanol or gasoline. It's possible to crack methanol or gasoline to strip off their hydrogen, so if either of these liquid fuels were in the tank, they would supply hydrogen on demand. Mercedes-Benz has already test driven a car that makes its own hydrogen from methanol.

Questions

1. On spacecraft, hydrogen fuel cells furnish electricity and what else?

2. What makes the early version of a hydrogen fuel cell impractical for use in vehicles?

3. What is "power density," and why is it important in the design of vehicles that use hydrogen fuel cell systems?

Whether you're a Rembrandt or an amateur artist, you depend on colored pigments to create your masterpiece. For centuries, artists have relied on pigments formed from colored metal compounds to create the paints they use. In this chapter, one of the topics we will explore is why some metal compounds are colored and why some are not.

Properties of Metals and Metal Complexes

This Chapter in Context In previous chapters you learned many of the concepts that chemists have applied in developing their understanding of how the elements react with each other and the kinds of compounds they produce. For example, in our discussions of chemical kinetics, you learned how various factors affect the rates of reactions, and in our study of thermodynamics you learned how enthalpy and entropy changes affect the possibility of observing chemical changes. Our focus, however, was primarily on the concepts themselves, with examples of chemical behavior being used to reinforce and justify them. With these concepts available to us now, we will direct our emphasis in the opposite direction and examine some of the physical and chemical properties of the elements and their compounds, beginning in this chapter with some properties of metals and their compounds.

20.1 Preparation of Metals from Compounds

When metals form compounds, they almost always lose electrons (become oxidized) and exist in positive oxidation states. Therefore, isolating metals from compounds generally involves reduction. In this section we will look at some of the ways this can be accomplished. Before we do this, however, let's take a brief look at where in nature we find metals and their compounds.

Occurrence of Metals

Most metals are reactive enough that they do not occur as free elements in nature. Instead, they are found combined with other elements in compounds. Where we find such compounds depends a great deal on their solubilities in water. For example, you learned that salts of the alkali metals (metals of Group IA) are soluble in water. It is no surprise, therefore, to find Na^+ and K^+ ions in the sea. In fact, the oceans provide a huge storehouse of many minerals, as illustrated by the table in the margin. Sodium and magnesium (as Na^+ and Mg^{2+}) are the metal ions in largest concentration in seawater, partly because their

Concentrations of Ions in Seawater

Ion	Molarity
Chloride	0.550
Sodium	0.468
Magnesium	0.055
Sulfate	0.029
Calcium	0.011
Potassium	0.011
Bicarbonate	0.002

901

natural abundance among the elements is large,[1] and also because their sulfates, halides, and carbonates are water soluble. Because the ions of Na and Mg are so abundant in seawater, and because seawater is free for the taking, the chief commercial source of both of these metals is the sea.

Oxygen is Earth's most abundant element, and because it is so reactive it is able to combine with nearly all metals. As you learned when studying the solubility rules in Chapter 4, most metal oxides are insoluble in water (exceptions are the oxides of the alkali metals and some of the oxides of the Group IIA metals). Because of these facts, many metals occur in nature as insoluble oxides found buried in the ground. An example is iron(III) oxide, which gives its rich rust-red color to iron ores and various clay minerals that contain Fe_2O_3 (Figure 20.1).

As you learned in Chapter 4, nearly all metal carbonates are also insoluble, and the ability of sea creatures, such as clams, oysters, and coral, to extract carbonate ions from seawater to construct their shells accounts for the large formations of carbonate minerals of calcium and magnesium present in many locations around the planet. Over eons of time, the calcium carbonate skeletons of these creatures have built up and then been moved by upheavals in Earth's crust. Such movements have often lifted them to heights far above sea level and at times subjected them to tremendous compressive forces that transformed them into limestone and marble.

Other anions yield insoluble compounds with metal ions as well, which accounts for deposits of metal sulfides, such as the copper sulfides (CuS and Cu_2S) and lead sulfide (PbS). Calcium phosphate, $Ca_3(PO_4)_2$, is the chief source of phosphate fertilizers, without which farmers could not hope to maintain the large crop harvests required to feed the world's growing population.

A few metals, most notably gold and platinum, do occur in the uncombined state in nature. These are metals with very low degrees of reactivity. Although they form many compounds, the metals are found free in nature because their compounds are not particularly stable and are easily decomposed by heat. Let's apply what we've learned in Chapter 18 to understand the factors that affect the ease with which such decompositions occur.

Thermal Stability of Metal Compounds: Thermodynamic Considerations

Thermal decomposition of a compound involves the conversion of the substance into its elements by heat. Some metal compounds are extremely resistant to such decomposition. For example, magnesium oxide is used to make bricks to line the walls of high-temperature ovens because it has a very high melting point and virtually no tendency to undergo decomposition. On the other hand, mercury(II) oxide decomposes at a relatively low temperature. Priestley, for example, in his experiments on oxygen, produced metallic mercury and oxygen from mercury(II) oxide by simply heating it with sunlight focused on the HgO by means of a magnifying glass. In this case, HgO decomposes quite spontaneously at elevated temperatures according to the equation

$$2HgO(s) \longrightarrow 2Hg(g) + O_2(g)$$

The practicality of using a thermal decomposition reaction of this type to produce a free metal depends on the extent to which the reaction proceeds to

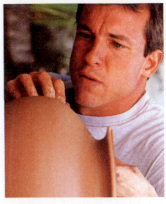

Figure 20.1 *The color of iron oxide.* Red Fe_2O_3 gives the clay used by this sculptor its rich color.

Sodium and magnesium are not the only substances obtained from the sea. The oceans are a source of many raw materials, including petroleum, found beneath the seafloor, and nonmetals such as bromine and iodine, which occur as anions in seawater itself.

The type of stone called marble is composed of calcium carbonate, which was deposited by sea creatures long ago and then transformed by heat and pressure caused by movements of the Earth's crust. How fortunate for us, because it enabled the great sculptor Michelangelo to create this masterpiece known as *La Pietà.*

---

[1]On Earth, sodium is the fourth most abundant element on an atom basis (approximately 2.6% of the atoms on Earth are Na); magnesium is the seventh most abundant (about 1.8%).

completion at a given temperature. In Chapter 18 you learned that at 25 °C the position of equilibrium in a reaction is governed by $\Delta G°$. Let's specify the temperature as a subscript by calling this $\Delta G°_{298}$. If we take $\Delta G°_T$ to be the equivalent of $\Delta G°_{298}$, but at some other temperature, we have the relationship,

$$\Delta G°_T = \Delta H°_T - T\Delta S°_T$$

where $\Delta H°_T$ and $\Delta S°_T$ are the enthalpy and entropy changes that accompany the reaction. For most systems ΔH and ΔS do not change much with temperature so that $\Delta H°_T$ and $\Delta S°_T$ can reasonably be approximated by $\Delta H°_{298}$ and $\Delta S°_{298}$.

A thermal decomposition reaction will be feasible when $\Delta G°_T$ is negative, since under these conditions an appreciable amount of product will be formed. We must now look at the magnitudes of $\Delta H°_T$ and $\Delta S°_T$, because they control the sign and magnitude of $\Delta G°_T$.

For the decomposition of an oxide, a gas (O_2) and sometimes the metal vapor is produced, so the reaction occurs with a sizable increase in entropy. This means that $\Delta S°_T$ will be positive, favoring the decomposition.

The enthalpy change for the decomposition, $\Delta H°_T$, is simply the negative of the heat of formation of the oxide and, since $\Delta H°_f$ is usually negative for the formation of metal oxides, $\Delta H°_T$ for the decomposition reaction will usually be positive. As a result, the sign of $\Delta G°_T$ is determined by the difference between two *positive* quantities, $\Delta H°_T$ and $T\Delta S°_T$.

If the metal oxide has a large negative heat of formation, a great deal of energy is evolved when the oxide is formed. This means that $\Delta H°_T$ for the decomposition will have a large positive value. As a result, the difference $\Delta H°_T - T\Delta S°_T$ can be negative *only* at very high temperatures, where $T\Delta S°_T$ is larger than $\Delta H°_T$. We express this by saying that the metal oxide is very stable with respect to thermal decomposition. On the other hand, if the $\Delta H°_f$ of the metal oxide is relatively small, as with HgO and certain other oxides (for example, Ag_2O, CuO, and Au_2O_3), then $\Delta H°_T$ for the decomposition reaction is a small positive quantity and $\Delta G°_T$ for the reaction can be negative at relatively low temperatures. These oxides, therefore, are said to have relatively low thermal stabilities. Let's look at some examples.

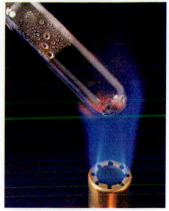

Decomposition of mercury(II) oxide. Heated to a temperature of only 400 °C, HgO turns black and releases oxygen. Drops of mercury begin to collect on the walls of the test tube as mercury vapor condenses on the cooler parts of the tube.

Above what temperature would the decomposition of Ag_2O be expected to proceed to an appreciable extent toward completion? At 25 °C, $\Delta H°_f$ for Ag_2O is -31.1 kJ mol^{-1}, and $\Delta S°_f = -66.1$ J mol^{-1} K^{-1}.

EXAMPLE 20.1

Calculating the Temperature Required for Thermal Decomposition

Analysis: The decomposition of Ag_2O,

$$Ag_2O(s) \longrightarrow 2Ag(s) + O_2(g)$$

is the reverse of its formation. Therefore, for this reaction at 25 °C we can write

$$\Delta H°_{298} = -\Delta H°_f = +31.1 \text{ kJ mol}^{-1}$$

$$\Delta S°_{298} = -\Delta S°_f = +66.1 \text{ J mol}^{-1} \text{ K}^{-1}$$

Let's calculate the temperature at which $\Delta G°_T = 0$, because above that temperature $\Delta G°_T$ will be negative and the decomposition reaction will proceed to a significant extent.

Solution: We will assume, as stated above, that $\Delta H°_T$ and $\Delta S°_T$ are approximately independent of temperature so that we can use $\Delta H°_{298}$ and $\Delta S°_{298}$ in the equation for $\Delta G°_T$.

$$\Delta G°_T = \Delta H°_{298} - T\Delta S°_{298}$$

When $\Delta G_T^\circ = 0$,

$$0 = \Delta H_{298}^\circ - T\Delta S_{298}^\circ$$

Solving for T gives

$$T = \frac{\Delta H_{298}^\circ}{\Delta S_{298}^\circ}$$

$$= \frac{31{,}100 \text{ J mol}^{-1}}{66.1 \text{ J mol}^{-1} \text{ K}^{-1}}$$

$$= 470 \text{ K}$$

This corresponds to a Celsius temperature of 197 °C.

Because ΔH_{298}° and ΔS_{298}° are both positive, ΔG_T° will become negative at temperatures above 197 °C. This means that above 197 °C the reaction should become feasible, with much of the Ag_2O undergoing decomposition.

Is the Answer Reasonable?

There is no simple check on this answer, although we can check to be sure that the units cancel properly, and they do. Notice that we were careful to change the units of ΔH_{298}° from kilojoules to joules. If we had not done that, the units would not have canceled properly.

EXAMPLE 20.2

Calculating the Temperature Required for Thermal Decomposition

Above what temperature would we expect gold(III) oxide to decompose? At 25°C, $\Delta H_f^\circ = +80.8$ kJ mol^{-1} for Au_2O_3. Also, for Au_2O_3, $S° = 125$ J mol^{-1} K^{-1}; for Au, $S° = 47.7$ J mol^{-1} K^{-1}; for O_2, $S° = 205$ J mol^{-1} K^{-1}.

Analysis: First we will calculate the values of ΔH_{298}° and ΔS_{298}° for the decomposition reaction. ΔH_{298}° is just the negative of ΔH_f°. We will use the entropy data provided to calculate directly the value of ΔS_{298}° for the decomposition. Once we have ΔH_{298}° and ΔS_{298}°, we can calculate the temperature at which ΔG_T° becomes negative for the decomposition. Above that temperature we expect substantial decomposition of the Au_2O_3.

Solution: The chemical equation for the decomposition is

$$Au_2O_3(s) \longrightarrow 2Au(s) + \tfrac{3}{2}O_2(g)$$

The ΔH_{298}° for the reaction is the negative of ΔH_f°, so $\Delta H_{298}^\circ = -80.8$ kJ mol^{-1}. To calculate ΔS_{298}° we use a Hess's law type of calculation.

$$\Delta S_{298}^\circ = (2S_{Au}^\circ + \tfrac{3}{2}S_{O_2}^\circ) - S_{Au_2O_3}^\circ$$

$$= \left(2 \text{ mol} \times \frac{47.7 \text{ J}}{\text{mol K}} + \frac{3}{2}\text{mol} \times \frac{205 \text{ J}}{\text{mol K}}\right) - 1 \text{ mol} \times \frac{125 \text{ J}}{\text{mol K}}$$

$$= +278 \text{ J K}^{-1} = +0.278 \text{ kJ K}^{-1}$$

Again, we obtain ΔG_T° by assuming that ΔH_T° and ΔS_T° are the same as ΔH_{298}° and ΔS_{298}°.

$$\Delta G_T^\circ = \Delta H_{298}^\circ - T\Delta S_{298}^\circ$$

Substituting,

$$\Delta G_T^\circ = -80.8 \text{ kJ} - T(+0.278 \text{ kJ K}^{-1})$$

Note that regardless of the temperature, ΔG_T° will be negative. This is because the absolute temperature is always a positive quantity. What this tells us is that Au_2O_3

is unstable with respect to decomposition at *any* temperature. It exists only because at low temperatures the rate of its decomposition is slow.

The instability of Au_2O_3 is such that over time any oxide that might form would spontaneously decompose, which explains why gold is always discovered as the free metal in nature.

If ΔH_f is positive for a substance, then decomposition will have a negative ΔH. Since decomposition also has a positive $T\Delta S$, ΔG will be negative at all temperatures and the substance will be unstable toward thermal decomposition regardless of the temperature.

Practice Exercise 1

Calculate the temperature required to observe a significant amount of thermal decomposition of MgO. At 25 °C, magnesium oxide has $\Delta H_f^\circ = -601.7$ kJ mol^{-1}; for Mg, $S^\circ = 32.5$ J mol^{-1} K^{-1}, and for O_2, $S^\circ = 205$ J mol^{-1} K^{-1}. For MgO, $S^\circ = 26.9$ J mol^{-1} K^{-1}. ◆

Reduction by Chemical Reducing Agents

Except in a few cases, thermal decomposition is not a practical way of producing a free metal. Instead, a compound is made to react with some substance that is a better reducing agent than the metal being sought. One of the most common agents used for the reduction of metal oxides is carbon. Tin and lead, for example, can be produced by heating their oxides with carbon.

$$2SnO + C \xrightarrow{\text{heat}} 2Sn + CO_2$$

$$2PbO + C \xrightarrow{\text{heat}} 2Pb + CO_2$$

Carbon is used in large quantities in commercial metallurgy because of its abundance and low cost.[2]

Hydrogen is another reducing agent that can be used to liberate metals of moderate chemical activity from their compounds. For instance tin and lead oxides are also reduced when heated under a stream of H_2.

$$SnO + H_2 \xrightarrow{\text{heat}} Sn + H_2O$$

$$PbO + H_2 \xrightarrow{\text{heat}} Pb + H_2O$$

The use of a more active metal to carry out the reduction is also possible. In Chapter 19 we saw that a galvanic cell could be established between two different metals, for example, Zn and Cu. In that cell the more active reducing agent, Zn, causes the Cu^{2+} to be reduced. Aluminum was first prepared in 1825 by the reaction of aluminum chloride with the more active metal, potassium.

$$AlCl_3 + 3K \longrightarrow 3KCl + Al$$

▶**Chemistry in Practice**◀ Using an active metal to reduce a compound of a less active metal has a number of practical applications. One that is handy around the home is using aluminum to remove the tarnish from silver. Generally, silver tarnishes by a gradual reaction with hydrogen sulfide, present in very small amounts in the air. The product of the reaction is silver sulfide, Ag_2S, which is black and forms a dull film over the bright metal. Polishing a silver object using a mild abrasive restores the shine, but it also gradually removes silver from the object as the silver sulfide is rubbed away. An alter-

---

[2]The metallurgy of iron and steel and the use of carbon in reducing iron ore are described in Chapter 3 of the supplement *Descriptive Chemistry of the Elements* which is available to accompany this book.

(*a*) (*b*) (*c*)

Using aluminum to remove tarnish from silver. (*a*) A badly tarnished silver vase stands next to a container of detergent (Soilax) dissolved in water. The bottom of the container is lined with aluminum foil. (*b*) The vase, shown partly immersed in the detergent, rests on the aluminum foil. (*c*) After a short time, the vase is removed, rinsed with water, and wiped with a soft cloth. Where the vase was immersed in the liquid, much of the tarnish has been reduced to metallic silver.

native method, and one that requires little effort, involves lining the bottom of a sink with aluminum foil, adding warm water and detergent (which acts as an electrolyte), and then submerging the tarnished silver object in the detergent–water mixture and placing it in contact with the aluminum. A galvanic cell is established in which the more active aluminum is the anode and the silver object is the cathode. In a short time, the silver sulfide is reduced, restoring the shine and depositing the freed silver metal on the object. As this happens, a small amount of the aluminum foil is oxidized and caused to dissolve. Besides requiring little physical effort, this silver polishing technique doesn't remove silver from the object being polished. ◆

As a practical source of metals, the reduction of compounds with other elements that are better reducing agents suffers from a serious limitation—specifically, the availability (and cost) of the reducing agent. Each possible reducing agent must then be generated by reacting one of its compounds with a still better reducing agent. Ultimately, of course, there must be some "best" reducing agent. How could this substance be prepared if there were no better reducing agent available that could be used to reduce its compounds?

The solution to this dilemma is electrolysis where, by applying a suitable potential to a molten salt, virtually any oxidation–reduction process can be brought about. Therefore, metals that themselves are very powerful reducing agents are nearly always prepared by electrolysis. Among the representative elements these include the very active elements in Groups IA and IIA as well as aluminum in Group IIIA. In Chapter 19, the use of electrolysis in the production of sodium, magnesium, and aluminum was described.

20.2 Covalent/Ionic Nature of Metal Compounds

Frequently we tend to think of metal compounds with nonmetals as essentially ionic in character, and in many instances this is true. However, we should remember that no bond is purely ionic and, most importantly, there is a gradual transition between ionic and covalent bonding. What we wish to examine in this section are some of the factors that allow us to predict trends in ionic (or covalent) bonding within a series of compounds.

Polarization of Cations and Anions

The basic concept that we will work with here is that the electron cloud surrounding an atom or ion is somewhat soft and "mushy" and can be distorted when in the presence of another ion. Consider, for example, the two ions in Figure 20.2. The positive ion here is depicted as smaller than the negative one, reflecting the fact that cations generally are smaller than anions. Because of its positive charge, the cation tends to draw the electron cloud of the anion toward itself. This gives an electron distribution somewhat like that in Figure 20.2b, in which the anion has been distorted so that in addition to being an ion, it is now something of a dipole as well. We say, therefore, that the electron cloud of the anion has been *polarized* by the cation.

When the anion becomes polarized, electron density is drawn into the region between the cation and anion. Because a covalent bond consists of electrons shared *between* nuclei, the polarization of the anion results in the partial formation of a covalent bond. The extent to which this polarization occurs determines the degree of covalent character in the ionic bond. The limit of this, of course, is where the electrons of the anion are pulled toward the cation so much that a nonpolar covalent bond is formed.

We would like to be able to compare the relative covalent character among metal compounds. To do this, we need to know what determines the extent to which a cation is able to polarize an anion. One factor is the amount of charge on the cation and another is the cation's size. As you might expect, the greater the positive charge, the more effective the cation will be at distorting the shape of the anion. In other words, we expect the amount of polarization of the anion to be directly proportional to the amount of charge on the cation.

The effect of cation size is a bit more subtle. For a given amount of charge, small cations are more effective than large ones at polarizing an anion. This is because the charge is more concentrated when the cation is small and also because the center of positive charge can get closer to the anion. Therefore, we expect the degree of polarization of an anion to be inversely proportional to the cation's size (i.e., the larger the cation, the less concentrated is the charge and the smaller is the degree of polarization).

Ionic Potential and the Polarization of Anions

To bring the effects of charge and size together, it is convenient to define a quantity called the **ionic potential,** given the symbol ϕ, which is the ratio of a cation's charge, q, to its radius, r.

$$\phi = \frac{q}{r}$$

A cation with a large ionic potential would be one with a large charge and a small radius, and such a cation would produce a large polarization of an anion. Therefore, we can state that *the larger the ionic potential of the cation, the greater will be the polarization of the anion and the more covalent will be the bond between the two ions.* We can see how these ideas apply if we examine some examples.

As we proceed down within a group, the cations have the same charge but become progressively larger, so their ionic potentials decrease. In Group IA, for instance, we find Li^+ to be very small, Na^+ somewhat larger, K^+ larger still, and

(a)

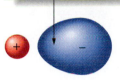

> Polarization of the anion causes electron density to be drawn into the region between the cation and anion.

(b)

Figure 20.2 *Polarization of an anion by the positive charge on the cation.* (a) Unpolarized anion and cation. (b) The electrical charge of the anion is distorted by the positive charge of the cation. Electron density is pulled toward the cation, placing additional electron density in the region between the two nuclei. The result is the partial formation of a covalent bond.

In Chapter 8 you learned that we can use electronegativity differences to predict relative degrees of ionic character in covalent bonds. Here we take the opposite view and examine how we can compare relative degrees of covalent character in ionic bonds. In most instances, the same conclusions would be reached regardless of which of the two methods we elect to use.

The polarizing effect of a cation is directly proportional to its charge and inversely proportional to its size.

so on. Therefore, because of its small size, Li^+ has the largest ionic potential of the cations of Group IA, and we expect it to form compounds with the greatest degree of covalent character. As it happens, among the alkali metals Li is the only element that forms compounds with organic molecules (called organolithium compounds) that have properties usually associated with covalently bonded substances, such as low melting points and appreciable solubilities in nonpolar solvents. In contrast, the corresponding compounds of the other alkali metals are predominantly ionic.

Moving to the right, to Group IIA, the ions have a charge of 2+ and they are also smaller. Therefore, their ionic potentials are larger than their neighbors in Group IA. We would expect their compounds to exhibit a greater degree of covalent character, and this is what is observed. Beryllium, for example, forms no compounds that are truly ionic, and its chloride ($BeCl_2$) is such a poor electrical conductor in the molten state that NaCl must be added to carry out electrolysis. Magnesium, just below beryllium in Group IIA, yields the Mg^{2+} ion that is larger than Be^{2+} and has a smaller ionic potential. Magnesium forms mostly ionic compounds, but it also forms organomagnesium compounds which, like organolithium compounds, exhibit properties characteristic of covalent substances. In Group IIA, as in the first group, we see that as we descend the group there is a trend toward decreasing covalent character in the bonds.

The effect of charge is also seen for compounds such as $SnCl_4$ and $PbCl_4$ in which the metals are in Group IVA. These, at first glance, appear to be salts of the cations Sn^{4+} and Pb^{4+}; however, each is quite covalent. For example, both are liquids with low melting points ($SnCl_4$, mp = -33 °C; $PbCl_4$, mp = -15 °C). In fact, we might compare them with the other Group IVA tetrahalides CCl_4 (mp = -23 °C), $SiCl_4$ (mp = -70 °C), and $GeCl_4$ (mp = -49.5 °C). The very high ionic potential of a 4+ ion thus appears to polarize an anion such as Cl^- to the extent that covalently bonded molecules result.

In addition to vertical trends in covalent character, we can also look at variations in the nature of the bonding as we move from left to right within a period. Here, as you know, the cations become progressively smaller and more highly charged, for example, Li^+, Be^{2+}, and B^{3+}, and so on. As a result, the ionic potential increases rapidly and the degree of covalent bonding in their compounds does so too. Thus LiCl is predominantly ionic, $BeCl_2$ is covalent, and so are the remaining halides as we continue across period 2 (i.e., BCl_3, CCl_4, etc.).

In period 3 the cations are larger than those in the preceding period, with the result that the covalent character of the bonds shows up later in the period. For example, NaCl and $MgCl_2$ are both essentially ionic, and it is not until Group IIIA, with $AlCl_3$, that the halides become largely covalent. In the

Al_2Cl_6 is said to be a dimer *of $AlCl_3$, which means it is formed from two $AlCl_3$ molecules.*

vapor state, pairs of $AlCl_3$ molecules combine to form covalently bonded Al_2Cl_6 molecules.

Formation of Al_2Cl_6 from two $AlCl_3$ molecules

The type of outer shell electron configuration possessed by the cation also affects its ability to polarize an anion. Cations with pseudonoble gas configurations of 18 electrons ($ns^2\,np^6\,nd^{10}$), such as those formed from the metals immediately following a transition series, appear to have higher ionic potentials than similarly sized ions having the same net charge but with only an octet of elec-

Table 20.1 Comparison of Melting Points of Chloride Salts of Pre- and Post-transition Metals in Which There Are Similar M—Cl Distances

Pre-transition Elements			Post-transition Elements		
Compound	M—Cl Distance (pm)	Melting Point (°C)	Compound	M—Cl Distance (pm)	Melting Point (°C)
NaCl	281	800	AgCl	277	455
RbCl	329	715	TlCl	331	430
$CaCl_2$	274	772	$CdCl_2$	276	568
$SrCl_2$	302	873	$PbCl_2$	302	501
			$HgCl_2$	291	276
$MgCl_2$	246	708	$ZnCl_2$	253	283
			$SnCl_2$	242	246

trons. This is because the d electrons that are added to the atom as the transition series is crossed do not completely shield the nuclear charge, which has also increased. The net effect is that a cation with a pseudonoble gas structure presents to a neighboring anion an effective positive charge that is actually somewhat greater than its net ionic charge. As a result, these cations behave as if their effective ionic potentials were higher than those calculated from their net ionic charges, so they are more effective at distorting the electron cloud of the anion.

Covalent Character and Melting Points of Metal Compounds

We often associate high melting points with a high degree of ionic bonding, where the attractive forces between particles within the solid are very strong. On the other hand, low melting points are often characteristic of substances that contain covalently bonded molecules in which the attractive forces between neighboring molecules are relatively weak. As a rough guide, then, we can use melting points of compounds of similar composition as a measure of their degree of covalent bonding, and we expect their melting points to decrease as the bonding becomes progressively more covalent.[3]

In Table 20.1 we compare salts having similar metal–chlorine bond distances. The salts on the left of the table have cations with an octet in their outer shell while those on the right contain cations with either a pseudonoble gas $ns^2 np^6 nd^{10}$ structure ($Ag^+, Zn^{2+}, Cd^{2+}, Hg^{2+}$) or a pseudonoble gas structure plus two electrons (Tl^+, Pb^{2+}, Sn^{2+}). The effectiveness of these post-transition metal ions at polarizing the chloride ion is revealed by the consistently lower melting points of their salts. Since on each line we have cations of essentially the same effective size (as shown by nearly the same M—Cl distances), the difference in covalence must be related to their effective ionic potentials, being less for elements that precede a transition series than for elements at the end of the series or for those in Groups IIIA and IVA that follow the transition elements.

---

[3]This generalization breaks down for polymeric covalent structures such as those found in quartz or diamond. In this case, because of the size of the giant molecules, very high melting points are found even though the bonding is covalent.

Some ionic compounds owe their colors to anions that are themselves colored. This solution contains sodium chromate, some of which is shown in solid form on the watch glass. Both the solution and the solid owe their color to the chromate ion, which is yellow.

Colors of Metal Compounds

The ions in most ionic compounds do not absorb light in the visible region of the spectrum. Therefore, they appear white or colorless when bathed in white light (e.g., NaCl is a colorless crystalline substance). The absorption that does take place with ionic compounds occurs in the shorter wavelength (higher frequency, higher energy) ultraviolet region. The energy provided by this high-energy radiation is used to shift an electron from the anion to the cation, as illustrated below

The absorption of a band of wavelengths from the ultraviolet "rainbow" produces a **charge transfer absorption band.**

As the bond between the metal and nonmetal becomes more covalent (that is, as electron density shifts away from the anion in the direction of the cation), less energy is required to produce the charge transfer. As a result, the absorption band shifts toward the lower energy visible region of the spectrum, where the removal of some colors from white light gives rise to reflected colors that represent the remainder of the visible spectrum. Therefore, depth of color in compounds (particularly those of elements with completed subshells) can often be taken to be a measure of the degree of covalent character in the metal–non-

(*a*)

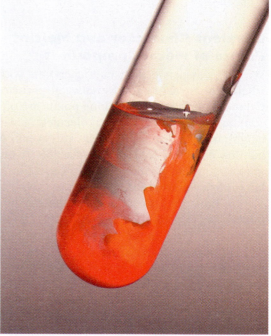

(*b*)

Charge transfer absorption as an indicator of covalent character. (*a*) Crystals of mercuric chloride and sodium iodide are colorless, showing that the ions themselves do not absorb visible light. (*b*) A few drops of sodium iodide were added to a colorless solution of mercuric chloride, yielding a precipitate of brightly colored mercuric iodide. The color of the HgI_2 precipitate, which is caused by a charge transfer absorption in the visible region of the spectrum, indicates that there is significant polarization of the iodide ions by mercury(II) ions.

Table 20.2 Colors of the Silver Halides

Compound	Color	Anion Radius (pm)
AgF	White	136
AgCl	White	181
AgBr	Cream	195
AgI	Yellow	216

metal bonds. Let's use this now to further our understanding of factors that affect the covalent character in metal–nonmetal bonds.

Influence of the Size and Charge of the Anion on Covalent Character

For a given cation, the bond becomes less ionic as the anion becomes more easily deformed, so let's examine trends in anion polarizability. In general, it is found that for a given charge, anions become more readily distorted as they become larger. In a large anion, the outer electrons are farther from the nucleus and are spread more thinly. Therefore, they are able to be influenced to a greater degree by a cation. Thus, among the halides, F^-, Cl^-, Br^-, and I^-, we expect to find the greatest amount of covalent character in compounds in which a metal is combined with I^-. Among the silver halides (Table 20.2) we see that as the anion grows in size, the compounds become progressively deeper in color, indicating a progressive rise in covalent character in the Ag—X bonds.[4]

We also find a similar relationship between ionic character and anion size if we compare metal oxides ($r_{O^{2-}} = 140$ pm) and sulfides ($r_{S^{2-}} = 184$ pm). Both of these anions are colorless as evidenced by the fact that both Na_2O and Na_2S are colorless. However with aluminum we find Al_2O_3 to be white while Al_2S_3 is yellow, suggesting a greater degree of ionic bonding in the oxide. This is further supported by comparing the melting points of these two compounds; Al_2O_3 melts at 2045 °C, while Al_2S_3 melts at a much lower temperature, 1100 °C. Additional examples are provided in Table 20.3.

For anions of a given size, polarization also increases with an increase in the charge on the anion. We can compare, for example, compounds containing chlo-

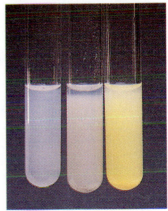

Colors of precipitates of the silver halides. Silver chloride is white, silver bromide has a cream color, and silver iodide is distinctly yellow. The trends in color indicate increasing covalent character in the silver–halide bonds as the halide ion becomes larger and more polarizable.

Table 20.3 Properties of Some Oxides and Sulfides

Oxides			Sulfides		
Compound	Color	Melting Point (°C)	Compound	Color	Melting Point (°C)
Al_2O_3	White	2045	Al_2S_3	Yellow	1100
Ga_2O_3	White	1900	Ga_2S_3	Yellow	1255
Sb_2O_3	White	656	Sb_2S_3	Yellow-red	550
Bi_2O_3	Yellow	860	Bi_2S_3	Brown-black	685 dec.
SnO_2	White	1127	SnS_2	Yellow	882
ZnO	White	1975	ZnS	White	1550

[4]The salts NaF, NaCl, NaBr, and NaI, which are predominantly ionic, are colorless. This indicates that the halide ions themselves are colorless. The color of the AgX compound is therefore a reflection of covalent bonding.

Table 20.4 Colors of Metal Chlorides and Sulfides

Chlorides ($r_{Cl^-} = 181$ pm)		Sulfides ($r_{S^{2-}} = 184$ pm)	
AgCl	White	Ag_2S	Black
CuCl	White	Cu_2S	Black
AuCl	Yellow	Au_2S	Brown-black
$CdCl_2$	White	CdS	Yellow
$HgCl_2$	White	HgS	Black or red depending on crystal structure
$PbCl_2$	White	PbS	Black
$SnCl_2$	White	SnS	Black
$AlCl_3$	White	Al_2S_3	Yellow
$GaCl_3$	White	Ga_2S_3	Yellow
$BiCl_3$	White	Bi_2S_3	Brown-black

ride ($r = 181$ pm) and sulfide ($r = 184$ pm) as shown in Table 20.4. In each case the sulfide has a deeper color than the corresponding chloride salt, indicating that the compounds containing the more highly charged sulfide ion are more covalent. We also see that, in general, most metal sulfides, except those of the alkali and alkaline earth metals, are deeply colored and possess a quite substantial degree of covalent bonding.

The presence of these colored compounds can often be seen about us. For instance, as discussed earlier, when silver tarnishes it reacts with traces of H_2S in the air to give a dull film of Ag_2S. Hydrogen sulfide also darkens lead-based paints, which were once used extensively. The paint pigment $Pb_3(OH)_2(CO_3)_2$, called *white lead,* reacts with H_2S to form black PbS. The fact that many of these compounds possess rather striking colors has also been put to use throughout history. For example, the brilliant yellow of natural CdS and the "vermilion" red of HgS have made them useful as pigments for the oil paints used by artists. Not long ago, several sticks of black PbS that had been used as a type of mascara were recovered from an ancient Egyptian burial ground.

Cadmium sulfide as a paint pigment. A solution of sodium sulfide is added to a solution of cadmium nitrate. Solutions of both reactants are colorless, but when they are combined they yield a yellow precipitate of cadmium sulfide, CdS. The artist's oil paint shown beside the flask is from a tube labeled cadmium yellow. The yellow pigment in the paint is cadmium sulfide.

20.3 Complex Ions of Metals; A More Detailed Look

In Section 17.4 (page 780), we introduced you to **complex ions** (or more simply, **complexes**) of metals. Recall that these are substances formed when molecules or anions become bonded covalently to metal ions to form more complex species. Two examples that were given were the pale blue $Cu(H_2O)_4^{2+}$ ion and the deep blue $Cu(NH_3)_4^{2+}$ ion. In Chapter 17 we discussed how the formation of complex ions can affect the solubilities of salts, but the importance of these substances reaches far beyond solubility equilibria. The number of complex ions formed by metals, especially the transition metals, is enormous, and the study of the properties, reactions, structures, and bonding in complexes like $Cu(H_2O)_4^{2+}$ has become an important specialty within chemistry. The study of metal complexes even extends into biochemistry. This is because nearly all the metals our bodies require ultimately become bound in complexes in order to perform their biochemical functions.

Before we proceed further, let's review some of the basic terminology we will use in our discussions. The molecules or ions that become attached to a metal ion [e.g., the NH_3 molecules in $Cu(NH_3)_4{}^{2+}$] are called **ligands.** Ligands are neutral molecules or anions that contain one or more atoms with at least one unshared electron pair which can be donated to the metal ion in the formation of a metal–ligand bond. The reaction of a ligand with a metal ion is therefore a Lewis acid–base reaction in which the ligand is the Lewis base (electron pair donor) and the metal ion is the Lewis acid (electron pair acceptor).

$$
\begin{array}{c}
\text{complex ion}
\end{array}
$$

$$
\underset{\substack{\text{silver ion}\\(\text{Lewis acid})}}{Ag^+} + \underset{\substack{\text{ammonia}\\\text{molecule}\\(\text{Lewis base})}}{\overset{\displaystyle H}{\underset{\displaystyle H}{:N-H}}} \longrightarrow \left[\overset{\displaystyle H}{\underset{\displaystyle H}{Ag \leftarrow N-H}} \right]^+
$$

Coordinate covalent bond between ammonia and the silver ion

A ligand atom that donates an electron pair to the metal is said to be a **donor atom** and the metal is the **acceptor.** Thus, in the example above, the nitrogen of the ammonia molecule is the donor atom and the silver ion is the acceptor. Because of the way the metal–ligand bond is formed, it can be considered to be a coordinate covalent bond, and compounds that contain metal complexes are often called **coordination compounds.**

A coordinate covalent bond is one in which both of the shared electrons originate on the same atom. Once formed, of course, a coordinate covalent bond is just like any other covalent bond.

Ligands

As we've noted, ligands may be either anions or neutral molecules. In either case, *ligands are Lewis bases* and, therefore, contain at least one atom with one or more lone pairs (unshared pairs) of electrons.

Anions that serve as ligands include many simple monatomic ions, such as the halide ions (F^-, Cl^-, Br^-, I^-) and the sulfide ion (S^{2-}). Common polyatomic anions that are ligands are nitrite ion ($NO_2{}^-$), cyanide ion (CN^-), hydroxide ion (OH^-), thiocyanate ion (SCN^-), and thiosulfate ion ($S_2O_3{}^{2-}$). (This is really only a small sampling, not a complete list.)

The most common neutral molecule that serves as a ligand is water, and most of the reactions of metal ions in aqueous solutions are actually reactions of their complex ions—ions in which the metal is attached to some number of water molecules. This number isn't always the same, however. Copper(II) ion, for example, forms the complex ion $Cu(H_2O)_4{}^{2+}$ (as we've noted earlier), but cobalt(II) combines with water molecules to form $Co(H_2O)_6{}^{2+}$. Another common neutral ligand is ammonia, NH_3, which has one lone pair of electrons on the nitrogen atom. If ammonia is added to an aqueous solution containing the $Ni(H_2O)_6{}^{2+}$ ion, for example, the color changes dramatically from green to blue as ammonia molecules displace water molecules (see Figure 20.3).

$$Ni(H_2O)_6{}^{2+}(aq) + 6NH_3(aq) \longrightarrow Ni(NH_3)_6{}^{2+}(aq) + 6H_2O$$
$$\underset{\text{(green)}}{} \qquad\qquad\qquad\qquad\qquad \underset{\text{(blue)}}{}$$

Each of the ligands that we have discussed so far is able to use just one atom to attach itself to a metal ion. Such ligands are called **monodentate ligands,** indicating that they have only "one tooth" with which to "bite" the metal ion.

There are also many ligands that have two or more donor atoms, and collectively they are referred to as **polydentate ligands.** The most common of these

Figure 20.3 *Complex ions of nickel.* (*Left*) A solution of nickel chloride, which contains the green $Ni(H_2O)_6{}^{2+}$ ion. (*Right*) Adding ammonia to the nickel chloride solution leads to formation of the blue $Ni(NH_3)_6{}^{2+}$ ion.

have two donor atoms, so they are called **bidentate ligands.** When they form complexes, *both* donor atoms become attached to the same metal ion. Oxalate ion and ethylenediamine (abbreviated *en* in writing the formula for a complex) are examples of bidentate ligands.

$$:O: \quad :O:$$

oxalate ion   ethylenediamine, en

When these ligands become attached to a metal ion, ring structures are formed as shown below. Complexes that contain such ring structures are called **chelates.**[5]

an oxalate complex   an ethylenediamine complex

Structures like these are important in "complex ion chemistry," as we shall see later in this chapter.

One of the most common polydentate ligands is a compound called ethylenediaminetetraacetic acid, mercifully abbreviated EDTA.

Sometimes the ligand EDTA is abbreviated using small letters, i.e., edta.

EDTA (sometimes, H_4EDTA)

The H atoms attached to the oxygen atoms are easily removed as protons, which gives an anion with a charge of 4−. The structure of the anion, $EDTA^{4-}$, is shown below with the donor atoms in color.

$EDTA^{4-}$

The structure of an EDTA complex. The nitrogen atoms are blue, oxygen is red, carbon is black, and hydrogen is white. The metal ion is in the center of the complex bonded to the two nitrogens and four oxygens.

As you can see, the $EDTA^{4-}$ ion has six donor atoms, and this permits it to wrap itself around a metal ion and form very stable complexes.

EDTA is a particularly useful and important ligand. It is relatively nontoxic, which allows it to be used in small amounts in foods to retard spoilage. If you look at the labels on bottles of salad dressings, for example, you often will find

---

[5]The term *chelate* comes from the Greek *chele,* meaning claw. These bidentate ligands grasp the metal ions with two "claws" (donor atoms) somewhat as a crab holds its prey. (Who says scientists have no imagination?)

EDTA as a food additive. Salad dressings such as these contain EDTA as a preservative.

that one of the ingredients is $CaNa_2EDTA$ (calcium disodium EDTA). The $EDTA^{4-}$ available from this salt forms soluble complex ions with any traces of metal ions that might otherwise promote reactions of the salad oils with oxygen, and thereby lead to spoilage.

Many shampoos contain Na_4EDTA to soften water. The $EDTA^{4-}$ binds to Ca^{2+}, Mg^{2+}, and Fe^{3+} ions, which removes them from the water and prevents them from interfering with the action of soaps in the shampoo.

EDTA is also sometimes added in small amounts to whole blood to prevent clotting. It ties up calcium ions, which the clotting process requires. EDTA has even been used as a treatment in cases of poisoning because it can help remove poisonous heavy metal ions, like Pb^{2+}, from the body when they have been accidentally ingested.

The *calcium* salt of EDTA is used because the $EDTA^{4-}$ ion would otherwise extract Ca^{2+} ions from bones, and that would be harmful.

Writing Formulas for Complexes

When we write the formula for a complex, we follow two rules:

1. The symbol for the metal ion is always given first, followed by the ligands.
2. The charge on the complex is the algebraic sum of the charge on the metal ion and the charges on the ligands.

For example, the formula of the complex ion of Cu^{2+} and H_2O, which we mentioned earlier, was written $Cu(H_2O)_4{}^{2+}$ with the Cu first followed by the ligands. The charge on the complex is $2+$ because the copper ion has a charge of $2+$ and the water molecules are neutral. Copper(II) ion also forms a complex ion with four cyanide ions, CN^-, $Cu(CN)_4{}^{2-}$. The metal ion contributes two $(+)$ charges and the four ligands contribute a total of four $(-)$ charges, one for each cyanide ion. The algebraic sum is therefore $2-$, so the complex ion has a charge of $2-$.

The formula for a complex ion is often placed within brackets, with the charge *outside*. The two complexes just mentioned would thus be written as $[Cu(H_2O)_4]^{2+}$ and $[Cu(CN)_4]^{2-}$. The brackets emphasize that the ligands are at-

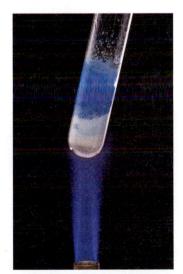

Crystals of $CuSO_4 \cdot 5H_2O$ lose water when heated. When copper sulfate pentahydrate is heated, the crystals lose water, which destroys the blue $[Cu(H_2O)_4]^{2+}$ complex and leaves the nearly white anhydrous $CuSO_4$.

Square brackets here do not mean molar concentration. It is usually clear from the context of a discussion whether we intend the brackets to mean molarity.

tached to the metal ion and are not free to roam about. One of the many complex ions formed by the chromium(III) ion contains five water molecules and one chloride ion as ligands. To indicate that all are attached to the Cr^{3+} ion, we use brackets and write the complex ion as $[CrCl(H_2O)_5]^{2+}$. When this complex is isolated as a chloride salt, the formula is written $[CrCl(H_2O)_5]Cl_2$, in which $[CrCl(H_2O)_5]^{2+}$ is the cation and so is written first. The formula $[CrCl(H_2O)_5]Cl_2$ clearly shows that five water molecules and a chloride ion are bonded to the chromium ion, and the other two chloride ions are present to provide electrical neutrality for the salt.

In Chapter 2 you learned about hydrates, and one was the beautiful blue hydrate of copper sulfate, $CuSO_4 \cdot 5H_2O$. It was much too early then to make the distinction, but the formula should have been written as $[Cu(H_2O)_4]SO_4 \cdot H_2O$ to show that four of the five water molecules are held in the crystal as part of the complex ion $[Cu(H_2O)_4]^{2+}$. The fifth water molecule is held in the crystal by being hydrogen bonded to the sulfate ion.

Many other hydrates of metal salts actually contain complex ions of the metals in which water is the ligand. Cobalt salts like cobalt(II) chloride, for example, crystallize from aqueous solutions as hexahydrates (meaning they contain six H_2O molecules per formula unit of the salt). The compound $CoCl_2 \cdot 6H_2O$ (Figure 20.4) actually is $[Co(H_2O)_6]Cl_2$, and it contains the pink complex $[Co(H_2O)_6]^{2+}$. This ion also gives solutions of cobalt(II) salts a pink color as can be seen in Figure 20.4. Although most hydrates of metal salts contain complex ions, the distinction is seldom made, and it's acceptable to write the formula for these hydrates in the usual fashion, e.g., $CuSO_4 \cdot 5H_2O$ instead of $[Cu(H_2O)_4]SO_4 \cdot H_2O$.

Figure 20.4 *Color of the cobalt(II) ion in hexahydrate salts.* The ion $[Co(H_2O)_6]^{2+}$ is pink and gives its color both to crystals and to an aqueous solution of $CoCl_2 \cdot 6H_2O$.

EXAMPLE 20.3

Writing the Formula for a Complex Ion

Write the formula for the complex ion formed by the metal ion Cr^{3+} and six NO_2^- ions as ligands. Decide whether the complex could be isolated as a chloride salt or a potassium salt, and write the formula for the appropriate salt.

Analysis: The net charge on the complex ion must first be determined. If it is a positive ion, it could be isolated as a chloride salt; if it is a negative ion, then a potassium salt could form.

Solution: Six NO_2^- ions contribute a total charge of $6-$; the metal contributes a charge of $3+$. The algebraic sum is $(6-) + (3+) = 3-$. The formula of the complex ion is therefore $[Cr(NO_2)_6]^{3-}$.

Because the complex is an anion, it requires a cation to form a neutral salt, so the complex could be isolated as a potassium salt, not as a chloride salt. For the salt to be electrically neutral, three K^+ ions are required for each complex ion, $[Cr(NO_2)_6]^{3-}$. The formula of the salt would therefore be $K_3[Cr(NO_2)_6]$.

Practice Exercise 2

Write the formula of the complex ion formed by Ag^+ and two thiosulfate ions, $S_2O_3^{2-}$. If we were able to isolate this complex ion as its ammonium salt, what would be the formula for the salt? ◆

Practice Exercise 3

Aluminum chloride crystallizes from aqueous solutions as a hexahydrate. Write the formula for the salt and suggest a formula for the complex ion formed by aluminum ion and water. ◆

The Chelate Effect

An interesting aspect of the complexes formed by ligands such as ethylenediamine and oxalate ion is their stabilities compared to similar complexes formed by monodentate ligands. For example, the complex $[Ni(en)_3]^{2+}$ is considerably more stable than $[Ni(NH_3)_6]^{2+}$, even though both complexes have six nitrogen atoms bound to a Ni^{2+} ion. We can compare them quantitatively by examining their formation constants.

$$Ni^{2+}(aq) + 6NH_3(aq) \rightleftharpoons [Ni(NH_3)_6]^{2+}(aq) \qquad K_{form} = 2.0 \times 10^8$$

$$Ni^{2+}(aq) + 3en(aq) \rightleftharpoons [Ni(en)_3]^{2+}(aq) \qquad K_{form} = 4.1 \times 10^{17}$$

The ethylenediamine complex is more stable than the ammonia complex by a factor of 2×10^9 (2 billion)! This exceptional stability of complexes formed with polydentate ligands is called the **chelate effect,** so named because it occurs in compounds that have these *chelate ring* structures.

There are two related reasons for the chelate effect, which we can understand best if we examine the ease with which the complexes undergo dissociation once formed. One reason appears to be associated with the probability of the ligand being removed from the vicinity of the metal ion when a donor atom becomes detached. If one end of a bidentate ligand comes loose from the metal, the donor atom cannot wander very far because the other end of the ligand is still attached. There is a high probability that the loose end will become reattached to the metal ion before the other end can let go, so overall the ligand appears to be bound tightly. With a monodentate ligand, however, there is nothing to hold the ligand near the metal ion if it becomes detached. The ligand can easily wander off into the surrounding solution and be lost. As a result, a monodentate ligand doesn't behave as if it is as firmly attached to the metal ion as a polydentate ligand.

The second reason is related to the entropy change for the dissociation. In Chapter 18 you learned that when there is an increase in the number of particles in a chemical reaction, the entropy change is positive. Dissociation of both complexes produces more particles, so the entropy change is positive for both of them. However, comparing the two reactions,

$$[Ni(NH_3)_6]^{2+}(aq) \rightleftharpoons Ni^{2+}(aq) + 6NH_3(aq) \qquad K_{inst} = 5.0 \times 10^{-9}$$

$$[Ni(en)_3]^{2+}(aq) \rightleftharpoons Ni^{2+}(aq) + 3en(aq) \qquad K_{inst} = 2.4 \times 10^{-18}$$

we see that dissociation of the ammonia complex gives a net increase of *six* in the number of particles, whereas dissociation of the ethylenediamine complex yields a net increase of *three*. This means the entropy change is more positive for the dissociation of the ammonia complex than for the ethylenediamine complex. The larger entropy change translates into a more favorable $\Delta G°$ for the decomposition, so at equilibrium the ammonia complex should be dissociated to a greater extent. This is, in fact, what is suggested by the values of their instability constants, K_{inst} (recall that $K_{inst} = 1/K_{form}$).

Ammonia is a monodentate ligand, and each NH_3 can supply one donor atom to a metal. Ethylenediamine (en) is a bidentate ligand, and each has two nitrogen donor atoms. Therefore, six ammonia molecules and three ethylenediamine molecules supply the same number of donor atoms.

Recall that $\Delta G°$ is related to the equilibrium constant. A more favorable $\Delta G°$ for dissociation should yield a larger equilibrium constant, and, indeed, we find the instability constant is larger for the ammonia complex.

20.4 **Nomenclature of Metal Complexes**

The naming of chemical compounds was introduced in Chapter 2 where we discussed the nomenclature system for simple inorganic compounds. This system, revised and kept up-to-date by the International Union of Pure and Applied Chemistry (IUPAC), has been extended to cover metal complexes. Below are some of the rules that have been developed to name coordination complexes. As you will see, some of the names arrived at following the rules are difficult to pronounce at first, and may even sound a little odd. However, the primary purpose of this and any other system of nomenclature is to provide a method that gives each unique compound its own unique name, and that permits us to write the formula of the compound given the name.

Rules of Nomenclature for Coordination Complexes

1. **Cationic species are named before anionic species.** This is the same rule that applies to other ionic compounds such as NaCl, where we name the cation first followed by the anion (i.e., sodium chloride).

2. **The names of anionic ligands always end in the suffix -*o*.**
   (a) Ligands whose names end in -*ide* have this suffix changed to -*o*.

Anion		Ligand
chloride	Cl^-	chloro-
bromide	Br^-	bromo-
cyanide	CN^-	cyano-
oxide	O^{2-}	oxo-

   (b) Ligands whose names end in -*ite* or -*ate* become -*ito* and -*ato*, respectively.

Anion		Ligand
carbonate	CO_3^{2-}	carbonato-
thiosulfate	$S_2O_3^{2-}$	thiosulfato-
thiocyanate	SCN^-	thiocyanato- (when bonded through sulfur)
		isothiocyanato- (when bonded through nitrogen)
oxalate	$C_2O_4^{2-}$	oxalato-
nitrite	NO_2^-	nitrito- (when bonded through oxygen; written ONO in formula for complex)[a]

[a]An exception to this is when the nitrogen of the NO_2^- ion is bonded to the metal, in which case the ligand is named nitro-.

3. **A neutral ligand is given the same name as the neutral molecule.** Thus the molecule ethylenediamine, when serving as a ligand, is called ethylenediamine in the name of the complex. Two very important exceptions to this, however, are water and ammonia. These are named as follows when they serve as ligands.

$$H_2O \quad \text{aqua} \qquad NH_3 \quad \text{ammine (note the double } m\text{)}$$

4. **When there is more than one of a particular ligand, their number is specified by the prefixes *di-* = 2, *tri-* = 3, *tetra-* = 4, *penta-* = 5, *hexa-* = 6, and so forth. When confusion might result by using these prefixes, the following are used instead: *bis-* = 2, *tris-* = 3, *tetrakis-* = 4.** Following this rule, the presence of two chloride ligands in a complex would be indicated as *dichloro-* (notice, too, the ending on the ligand name). However, if two ethylenedi-

amine ligands are present, use of the prefix *di-* might cause confusion. Someone reading the name might wonder whether diethylenediamine means two ethylenediamine molecules or one molecule of a substance called diethylenediamine. To avoid this problem we place the ligand name in parentheses preceded by *bis-*; that is, *bis(ethylenediamine)*.

Bis- is employed here so that if the name is used in verbal communication it is clear that the meaning is two ethylenediamine molecules.

5. **In the *formula* of a complex, the symbol for the metal is written first, followed by those of the ligands. Among the ligands, anionic ligands are written first (in alphabetical order), followed by neutral ligands (also in alphabetical order). In the *name* of the complex, the ligands are named first, in alphabetical order *without regard to charge,* followed by the name of the metal.** For example, suppose we had a complex composed of Co^{3+}, two Cl^- (chloro- ligands), one CN^- (cyano- ligand), and three NH_3 (ammine- ligands). The formula of this electrically neutral complex would be written $[CoCl_2(CN)(NH_3)_3]$. In the name of this complex, the ligands would be specified before the metal as *triamminedichlorocyano-* (*triammine-* for the three NH_3 ligands, *dichloro-* for the two Cl^- ligands, and *cyano-* for the CN^- ligand). Notice that in alphabetizing the names of the ligands, we ignore the prefixes *tri-* and *di-*. Thus *triammine-* is written before *dichloro-* because *ammine-* precedes *chloro-* alphabetically. For the same reason, *dichloro-* is written before *cyano-*.

In the *formula,* the metal ion appears first, followed by the ligands; in the *name,* the ligands are specified first, followed by the metal.

The ligands are alphabetized according to the first letter of the name of the ligand, not the first letter of the prefix.

6. **Negative (anionic) complex ions always end in the suffix *-ate.*** This suffix is appended to the English name of the metal atom in most cases. However, if the name of the metal ends in *-ium, -um,* or *-ese,* the ending is dropped and replaced by *-ate.*

Metal	Metal as Named in an Anionic Complex
Aluminum	aluminate
Chromium	chromate
Manganese	manganate
Nickel	nickelate
Cobalt	cobaltate
Zinc	zincate
Platinum	platinate
Vanadium	vanadate

For metals whose symbols are derived from their Latin names, the suffix *-ate* is appended to the Latin stem. (An exception, however, is mercury; in an anion it is named *mercurate.*)

Metal	Stem	Metal as Named in an Anionic Complex
Iron	ferr-	ferrate
Copper	cupr-	cuprate
Lead	plumb-	plumbate
Silver	argent-	argentate
Gold	aur-	aurate
Tin	stann-	stannate

For neutral or positively charged complexes, the metal is *always* specified with the English name for the element, *without any suffix.*

7. **The oxidation state of the metal in the complex is written in Roman numerals within parentheses following the name of the metal.** For example,

$$[Co(NH_3)_6]^{3+} \text{ is the hexaamminecobalt(III) ion}$$

$$[CuCl_4]^{2-} \text{ is the tetrachlorocuprate(II) ion}$$

No space

As you learned earlier, the charge on the complex is the algebraic sum of the charges on the ligands and the charge on the metal ion.

Notice that there are no spaces between the names of the ligands and the name of the metal, and that there is no space between the name of the metal and the parentheses that enclose the oxidation state expressed in Roman numerals.

The following are some additional examples. Study them carefully to see how the rules given above apply. Then try the practice exercises that follow.

Notice, once again, that the alphabetical order of the ligands is determined by the first letter in the name of the ligand, not the first letter in the prefix.

$[Ni(CN)_4]^{2-}$	tetracyanonickelate(II) ion
$K_3[CoCl_6]$	potassium hexachlorocobaltate(III)
$[CoCl_2(NH_3)_4]^+$	tetraamminedichlorocobalt(III) ion
$Na_3[Co(NO_2)_6]$	sodium hexanitrocobaltate(III)
$[Ag(NH_3)_2]^+$	diamminesilver(I) ion
$[Ag(S_2O_3)_2]^{3-}$	dithiosulfatoargentate(I) ion
$[Mn(en)_3]Cl_2$	tris(ethylenediamine)manganese(II) chloride
$[PtCl_2(NH_3)_2]$	diamminedichloroplatinum(II)

Practice Exercise 4

Name the following compounds: (a) $K_3[Fe(CN)_6]$ and (b) $[CrCl_2(en)_2]_2SO_4$. ◆

Practice Exercise 5

Write the formula for each of the following: (a) hexachlorostannate(IV) ion, (b) ammonium diaquatetracyanoferrate(II). ◆

20.5 Coordination Number and Structure

One of the most interesting aspects of the study of complexes is the kinds of structures that are formed. In many ways, this is related to the **coordination number** of the metal ion, which is *the number of donor atoms attached to the metal ion.* For example, in the complex $[Cu(H_2O)_4]^{2+}$ the copper is surrounded by the four oxygen atoms that belong to the water molecules, so the coordination number of Cu^{2+} in this complex is 4. Similarly, the coordination number of Cr^{3+} in the $[Cr(H_2O)_6]^{3+}$ ion is 6, and the coordination number of Ag^+ in $[Ag(NH_3)_2]^+$ is 2.

Sometimes the coordination number isn't immediately obvious from the formula of the complex. For example, you learned that there are many polydentate ligands which contain more than one donor atom that can bind simultane-

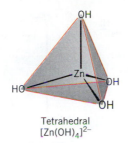

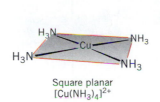

Tetrahedral
$[Zn(OH)_4]^{2-}$

Square planar
$[Cu(NH_3)_4]^{2+}$

Figure 20.5 *Tetrahedral and square planar geometries.* These are structures that occur for complexes in which the metal ion has a coordination number of 4. For the copper complex, we are viewing the square planar structure tilted back into the plane of the paper.

ously to a metal ion. Often, a metal is able to accommodate two or more polydentate ligands to give complexes with formulas such as $[Cr(H_2O)_2(en)_2]^{3+}$ and $[Cr(en)_3]^{3+}$. In each of these examples, the coordination number of the Cr^{3+} is 6. In the $[Cr(en)_3]^{3+}$ ion, there are three ethylenediamine ligands that each supply two donor atoms, for a total of 6, and in $[Cr(H_2O)_2(en)_2]^{3+}$, the two ethylenediamine ligands supply a total of 4 donor atoms and the two H_2O molecules supply another 2, so once again the total is 6.

Structures of Complexes

For metal complexes, there are certain geometries that are usually associated with particular coordination numbers.

Coordination Number 2. Examples are complexes such as $[Ag(NH_3)_2]^+$ and $[Ag(CN)_2]^-$. Usually, these complexes have a linear structure such as

$$[H_3N—Ag—NH_3]^+ \quad \text{and} \quad [NC—Ag—CN]^-$$

(Since the Ag^+ ion has a filled *d* subshell, it behaves as any of the representative elements as far as predicting geometry by VSEPR theory, so these structures are exactly what we would expect based on that theoretical model.)

Coordination Number 4. Two common geometries occur when four ligand atoms are bonded to a metal ion—tetrahedral and square planar. These are illustrated in Figure 20.5. The tetrahedral geometry is usually found with metal ions that have completely filled *d* subshells, such as Zn^{2+}. The complexes $[Zn(NH_3)_4]^{2+}$ and $[Zn(OH)_4]^{2-}$ are examples.

Square planar geometries are observed for complexes of Cu^{2+}, Ni^{2+}, and especially Pt^{2+}. Examples are $[Cu(NH_3)_4]^{2+}$, $[Ni(CN)_4]^{2-}$, and $[PtCl_4]^{2-}$. The most well-studied square planar complexes are those of Pt^{2+}, because they are considerably more stable than the others.

Coordination Number 6. The most common coordination number for complex ions is 6. Examples are $[Al(H_2O)_6]^{3+}$, $[Co(C_2O_4)_3]^{3-}$, $[Ni(en)_3]^{2+}$, and $[Co(EDTA)]^-$. With few exceptions, all complexes with a coordination number of 6 are octahedral. This holds true for those formed from both monodentate and bidentate ligands, as illustrated in Figure 20.6. In describing the shapes of octahedral complexes, most chemists use one of the simplified drawings of the octahedron shown in Figure 20.7.

Ordinarily, the VSEPR theory isn't used to predict the structures of transition metal complexes because it can't be relied on to give correct results when the metal has a partially filled *d* subshell.

Recall that EDTA has six donor atoms.

A description of how to sketch the structure of an octahedral molecule or ion was given in Chapter 9 on page 370.

Practice Exercise 6

What is the coordination number of the metal ion in (a) $[Cr(C_2O_4)_3]^{3-}$, (b) $[CoCl_2(C_2O_4)_2]^{3-}$, (c) $[Cr(C_2O_4)_2(en)]^-$, and (d) $[Co(EDTA)]^-$? ◆

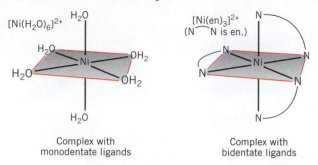

Figure 20.6 *Octahedral complexes.* Complexes with this geometry can be formed with either monodentate ligands such as water or with polydentate ligands such as ethylenediamine (en). To simplify the drawing of the ethylenediamine complex, the atoms joining the donor nitrogen atoms in the ligand, —CH_2—CH_2—, are represented as the curved line between the N atoms. Also notice that the nitrogen atoms of the bidentate ligand span adjacent positions within the octahedron. This is the case for all polydentate ligands that you will encounter in this book.

20.6 Isomers of Coordination Complexes

When you write the chemical formula for a particular compound, you might be tempted to think that you should also be able to predict exactly what the structure of the molecule or ion is. As you may realize by now, this just isn't possible in many cases. Sometimes we can use simple rules to make reasonable structural guesses, as in the discussion of the drawing of Lewis structures in Chapter 8, but

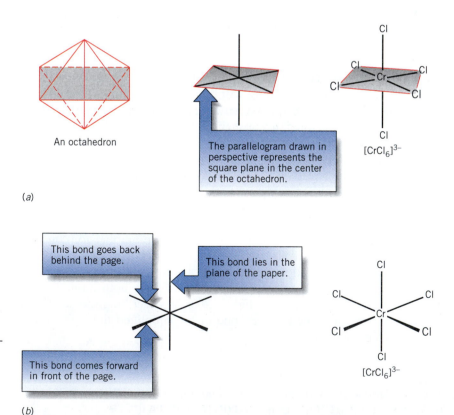

Figure 20.7 *Simplified representations of the octahedral complex, $[CrCl_6]^{3-}$.* (a) Drawings similar to those you learned to construct in Chapter 9. (b) An alternative method of representing the octahedron.

these rules apply only to simple molecules and ions. For more complex substances, there usually is no way of knowing for sure what the structure of the molecule or ion is without performing the necessary experiments to determine the structure.

One of the reasons that structures can't be predicted with certainty from chemical formulas alone is that there are usually many different ways that the atoms in the formula can be arranged. In fact, it is frequently possible to isolate two or more compounds that actually have the same chemical formula. For example, three different solids, each with its own characteristic color and other properties, can be isolated from a solution of chromium(III) chloride. All three have the same overall composition, and in the absence of other data, their formulas are written $CrCl_3 \cdot 6H_2O$. However, experiments have shown that these solids are actually the salts of three different complex ions. Their actual formulas (and colors) are

Chromium(III) chloride hexa-hydrate. The hydrated chromium(III) chloride purchased from chemical supply companies is actually $[CrCl_2(H_2O)_4]Cl \cdot 2H_2O$. Its green color in both the solid state and solution is due to the complex ion, $[CrCl_2(H_2O)_4]^+$.

$$[Cr(H_2O)_6]Cl_3 \qquad \text{purple}$$

$$[CrCl(H_2O)_5]Cl_2 \cdot H_2O \qquad \text{blue-green}$$

$$[CrCl_2(H_2O)_4]Cl \cdot 2H_2O \qquad \text{green}$$

Even though their overall compositions are the same, each of these substances is a distinct chemical compound with its own characteristic set of properties.

The existence of two or more compounds, each having the same chemical formula, is known as **isomerism.** In the example above, each salt is said to be an **isomer** of $CrCl_3 \cdot 6H_2O$. For coordination compounds, there are a variety of ways for isomerism to occur. In the example above, isomers exist because of the different possible ways that the water molecules and chloride ions can be held in the crystals. In one instance, all the water molecules serve as ligands, while in the other two, part of the water is present as water of hydration and some of the chloride is bonded to the metal ion. Another example, which is similar in some respects, is $Cr(NH_3)_5SO_4Br$. This "substance" can be isolated as two isomers.

$$[CrSO_4(NH_3)_5]Br \qquad \text{and} \qquad [CrBr(NH_3)_5]SO_4$$

They can be distinguished chemically by their differing abilities to react with Ag^+ and Ba^{2+}. The first isomer reacts in aqueous solution with Ag^+ to give a precipitate of AgBr, but it doesn't react with Ba^{2+}. This tells us that Br^- exists as a free ion in the solution. It also suggests the SO_4^{2-} is bound to the chromium and is unavailable to react with Ba^{2+} to give insoluble $BaSO_4$.

The second isomer, $[CrBr(NH_3)_5]SO_4$, reacts in solution with Ba^{2+} to give a precipitate of $BaSO_4$, which means there is free SO_4^{2-} in the solution. There is no reaction with Ag^+, however, because Br^- is bonded to the chromium and is not available as free Br^- in the solution. Thus, we see that even though both isomers have the same overall composition, they behave chemically in quite different ways and are therefore distinctly different compounds.

Ag_2SO_4 is soluble but $BaSO_4$ is insoluble. $BaBr_2$ is soluble but AgBr is insoluble.

Stereoisomerism

One of the most interesting kinds of isomerism found among coordination compounds is called **stereoisomerism,** which is defined as *differences among isomers that arise as a result of the various possible orientations of their atoms in space.* In other words, when stereoisomerism exists, we have compounds in which the same atoms are attached to each other, but they differ in the way those atoms are arranged in space relative to one another.

One form of stereoisomerism is called **geometric isomerism;** it is best understood by considering an example. Consider the square planar complexes having the formula $PtCl_2(NH_3)_2$. There are two ways to arrange the ligands around the platinum, as illustrated below. In one isomer, called the ***cis* isomer,** the chloride ions are *next to each other* and the ammonia molecules are also next to each other. In the other isomer, called the ***trans* isomer,** identical ligands are *opposite each other*. In identifying (and naming) isomers, *cis* means "on the same side," and *trans* means "on opposite sides."

The isomer on the left, *cis*-$PtCl_2(NH_3)_2$, is the anticancer drug known as *cisplatin.* It is interesting that only the *cis* isomer of this compound is clinically active against tumors. The *trans* isomer is totally ineffective.

cis isomer
cis-diamminedichloroplatinum(II)

trans isomer
trans-diamminedichloroplatinum(II)

Geometric isomerism also occurs for octahedral complexes. For example, consider the ions $[CrCl_2(H_2O)_4]^+$ and $[CrCl_2(en)_2]^+$. Both can be isolated as *cis* and *trans* isomers.

cis   *trans* $[CrCl_2(H_2O)_4]^+$

Curved lines connecting nitrogen atoms represent —CH_2—CH_2—, which links the nitrogen atoms in ethylenediamine.

cis   *trans*

$[CrCl_2(en)_2]^+$

N   N is en (ethylenediamine, $NH_2CH_2CH_2NH_2$).

In the *cis* isomers, the chloride ligands are both on the same side of the metal ion; in the *trans* isomers, the chloride ligands are on opposite ends of a line that passes through the center of the metal ion.

Chirality

There is a second kind of stereoisomerism that is much more subtle than geometric isomerism. This occurs when molecules are exactly the same except for one small difference—they differ from each other in the same way that your left hand differs from your right hand.

Although similar in appearance, your left and right hands are not exactly alike, as you discover if you try to fit your right hand into a left-hand glove (Figure 20.8). There is, however, a special relationship between the two hands, which you can see if you stand in front of a mirror. If you hold your left hand so that it is facing the mirror, and then look at its reflection, as illustrated in Figure 20.9, you will see that it looks exactly like your right hand. If it were possible to reach "through the looking glass," your right-hand glove would fit the reflection of your left hand perfectly.

In this analysis, we say our left and right hands are *mirror images of each other.* The thumbtacks in the photograph in the margin on p. 925 are also mirror

Figure 20.8 *One difference between left and right hands. A left-hand glove won't fit the right hand.*

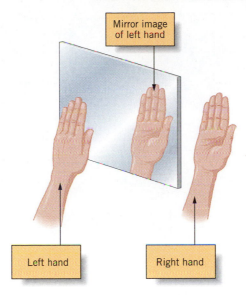

Figure 20.9 *Hands are mirror images of each other.* The image of the left hand reflected in the mirror appears the same as the right hand.

images of each other. The mirror image of one looks just like the other. Yet, there is a special difference between thumbtacks and hands, which is seen in a test for "handedness" that we call superimposability.

Superimposability is the ability of two objects to fit "one within the other" with no mismatch of parts. Your left and right hands lack this ability, and we say they are *nonsuperimposable*. This is why your right hand doesn't fit into your left-hand glove; your right hand does not match *exactly* the space that corresponds to your left hand. A pair of thumbtacks, on the other hand, are superimposable. If you imagine merging two thumbtacks so one goes inside the other, there is no mismatch of parts.

In the final analysis, *only if two objects pass the test of superimposability may we call them identical.* Your left and right hands are mirror images of each other, but they are not superimposable and are therefore not identical. A pair of thumbtacks are mirror images of each other and are superimposable, so they are identical.

The technical term for handedness is **chirality**.[6] Two plain thumbtacks do not have chirality, but your left and right hands do. Two gloves in a pair also have chirality. The test for chirality is the test for superimposability.

Thumbtacks and their mirror images are superimposable; they are not chiral and are identical.

> If one object is the mirror image of the other, and if they are *not* superimposable, then the objects are **chiral.**

In the realm of molecules and ions, we can apply the same test for handedness as before to determine whether two structures that are mirror images of each other are actually the same (and therefore not really two structures, but one) or whether they are different. Two structures that are mirror images, but not superimposable, are isomers with a special name of **enantiomers.**

The most common examples of chirality among coordination compounds occur with octahedral complexes that contain two or three bidentate ligands—

---

[6]The term **chiral** comes from the Greek *cheir,* meaning "hand."

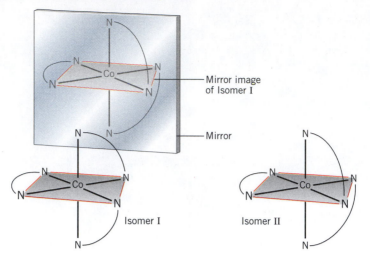

Figure 20.10 *The two iso-mers of [Co(en)₃]³⁺. Isomer II is constructed as the mirror image of isomer I. No matter how Isomer II is turned about, it is not superimposable on Isomer I.*

for instance, $[CoCl_2(en)_2]^+$ and $[Co(en)_3]^{3+}$. For the complex $[Co(en)_3]^{3+}$, the enantiomers are shown in Figure 20.10. For the complex $[CoCl_2(en)_2]^+$, only the *cis* isomer is chiral, as described in Figure 20.11.

As you can see, chiral isomers differ in only a very minor way from each other. This difference is so subtle that most of their properties are identical. They have identical melting points and boiling points, and in nearly all of their reactions, their behavior is exactly alike. The only way that the difference be-tween chiral molecules or ions manifests itself is in the way that they interact with physical or chemical "probes" that also have a handedness about them. For example, if two reactants are both chiral, then a given isomer of one of them will usually behave slightly differently toward each of the two isomers of the other

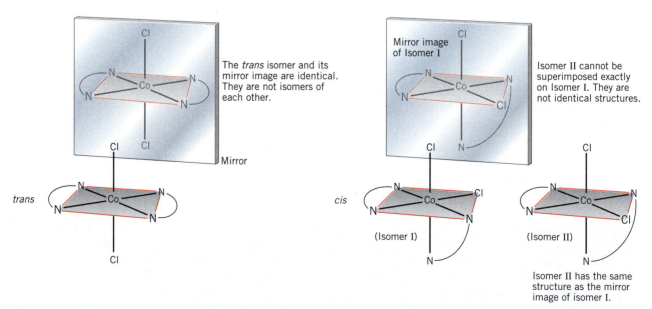

Figure 20.11 *Isomers of the [CoCl₂(en)₂]⁺ ion.* The mirror image of the *trans* isomer can be superimposed exactly on the original, so the *trans* isomer is not chiral. The *cis* iso-mer (Isomer I) is chiral, however, because its mirror image (Isomer II) cannot be super-imposed on the original.

reactant. As you will see in Chapter 23, this has very profound effects in bio-chemical reactions, in which nearly all of the molecules involved are chiral.

Chiral isomers like those described in Figures 20.10 and 20.11 also have a peculiar effect on polarized light, which is described in Facets of Chemistry 20.1. Because of this phenomenon, chiral isomers are said to be **optical isomers.**

20.7 Bonding in Transition Metal Complexes

Complexes of the transition metals differ from the complexes of other metals two special ways: (1) they are usually colored, whereas complexes of the representative metals are usually (but not always) colorless, and (2) their magnetic properties are often affected by the ligands attached to the metal ion. For example, it is not unusual for a given metal ion to form complexes with different ligands to give a rainbow of colors, as illustrated in Figure 20.12 for a series of complexes of cobalt. Also, because transition metal ions often have incompletely filled d subshells, we expect to find many of them with unpaired d electrons, and compounds that contain them should be paramagnetic. But for a given metal ion, the number of unpaired electrons is not always the same from one complex to another. For example, Fe^{2+} has four of its six $3d$ electrons unpaired in the $[Fe(H_2O)_6]^{2+}$ ion, but all of its electrons are paired in the $[Fe(CN)_6]^{4-}$ ion. As a result, the $[Fe(H_2O)_6]^{2+}$ ion is paramagnetic and the $[Fe(CN)_6]^{4-}$ ion is diamagnetic.

Crystal Field Theory

Any theory that attempts to explain the bonding in complex ions must also explain their colors and magnetic properties. One of the simplest theories that does this is the **crystal field theory.** The theory gets its name from its original use in explaining the behavior of transition metal ions in crystals. It was discovered later that the theory works well for transition metal complexes, too.

Crystal field theory ignores covalent bonding in complexes. It assumes that the primary stability comes from the electrostatic attractions between the positively charged metal ion and the negative charges of the ligand anions or dipoles. Crystal field theory's unique approach, though, is the way it examines how the negative charges on the ligands affect the energy of the complex by influencing the energies of the d orbitals of the metal ion, and this is what we will focus our attention on here. To understand the theory, therefore, it is essential that you know how the d orbitals are shaped and especially how they are ori-

More complete theories consider the covalent nature of metal–ligand bonding, but crystal field theory nevertheless provides a useful model for explaining the colors and magnetic properties of complexes.

Figure 20.12 *Colors of complex ions depend on the nature of the ligands.* Each of these brightly colored solutions contains a complex ion of Co^{3+}. The variety of colors arises because of the different ligands (molecules or anions) that are bonded to the cobalt ion in the complexes.

Facets of Chemistry 20.1

Optical Activity

The molecules of enantiomers have identical internal geometries (bond angles and lengths). They also have the same atoms or groups joined to the same central atoms. As a result, enantiomers have identical polarities, which also causes them to have identical boiling points, melting points, densities, and most other physical properties. One way enantiomers do differ, however, is in their interactions with plane-polarized light.

Light is *electromagnetic radiation* that possesses both electric and magnetic components which behave like vectors. These vectors oscillate in directions perpendicular to the direction in which the light wave is traveling (Figure 1a). In ordinary light, the oscillations of the electric and magnetic fields of the photons are oriented randomly around the direction of the light beam. In *plane-polarized light,* all the vibrations occur in the same plane (Figure 1b). Ordinary light can be polarized in several ways. One is to pass it through a special film of plastic, as in a pair of Polaroid sunglasses. This has the effect of filtering out all the vibrations except those that are in one plane (Figure 1b).

When plane-polarized light is passed through a nonchiral substance, the plane of the vibrations is unaffected. However, chiral substances have a peculiar effect on polarized light. A chiral isomer will rotate the plane of the polarized light as the light passes through it (Figure 2). Looking toward the light source, we find that one enan-

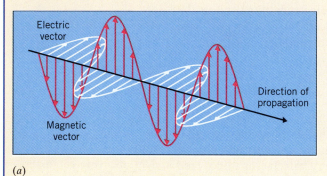

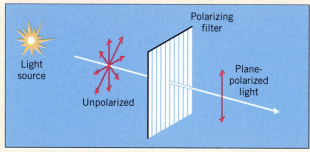

(a)   *(b)*

Figure 1 (*a*) Light possesses electric and magnetic vectors that oscillate perpendicular to the direction of propagation of the light wave. (*b*) In unpolarized light, the vectors are oriented randomly, but in *plane-polarized light,* all the vectors are in the same plane.

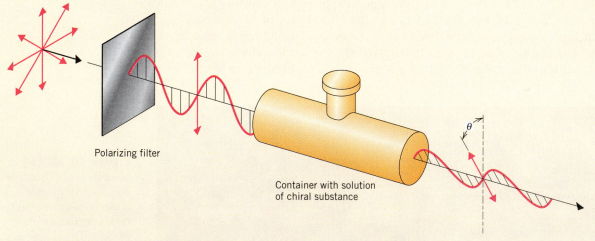

Figure 2 When plane-polarized light passes through a solution of a chiral substance, the plane of polarization is rotated either to the left or to the right. In this illustration, the plane of polarization of the light is rotated to the left (as seen facing the light source).

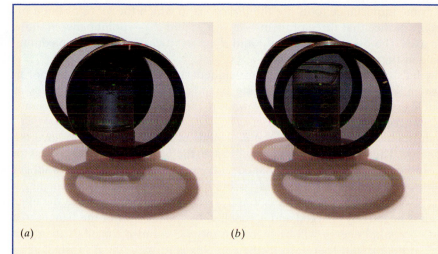

Figure 3 (*a*) Polarized light rotated by a sugar solution is not blocked completely by crossed polarizing filters. (*b*) Rotating the front polarizer to the right causes polarized light passing through the sugar solution to be blocked.

tiomer rotates the plane of polarization clockwise whereas the other enantiomer rotates it counterclockwise. Because of their effects on polarized light, chiral substances are said to be **optically active.**

The phenomenon of optical activity is not difficult to observe. In Figure 3*a* we see a pair of polarizing filters with a beaker containing a concentrated solution of sugar (a compound with chiral molecules) placed between them. The filters are oriented so that their planes of polarization are at an angle of 90°. As a result, the light waves that pass through the filter in back are blocked by the one in front. Notice, however, that some of the light that passes through the solution in the beaker is not blocked by the filter in front. This is because the sugar rotates the plane of polar-

ization as the light passes through the solution. Because the light from the first filter has been rotated by the solution, it is not blocked completely by the second.

In Figure 3*b*, we see the same apparatus, but the filter in front has been rotated to the right (clockwise) by about 70°. Where the light doesn't pass through the solution, it is no longer blocked completely. However, the light passing through the solution *is* blocked. Since we had to rotate the filter to the right to cause the light coming through the solution to be blocked, the enantiomer in the solution must rotate the plane of polarization to the right. A solution of the other enantiomer at the same concentration would produce a similar rotation, but in the opposite direction.

ented in space relative to each other. The *d* orbitals were described in Chapter 7, and they are illustrated again in Figure 20.13.

First, notice that four of the *d* orbitals have the same shape but point in different directions. These are the $d_{x^2-y^2}$, d_{xy}, d_{xz}, and d_{yz}. Each has four lobes of electron density. The fifth *d* orbital, labeled d_{z^2}, has two lobes that point in opposite directions along the *z* axis plus a small donut-shaped ring of electron density around the center that is concentrated in the *xy* plane.

The labels for the *d* orbitals come from the mathematics of quantum mechanics.

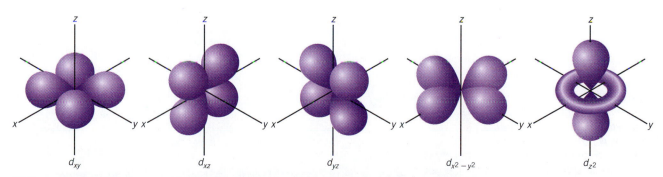

Figure 20.13 *The shapes and directional properties of the five d orbitals of a d subshell.*

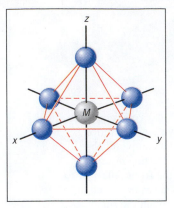

Figure 20.14 *An octahedral complex ion with ligands that lie along the x, y, and z axes.*

Of prime importance to us are the *directions* in which the lobes of the d orbitals point. Notice that three of them—d_{xy}, d_{xz}, and d_{yz}—point *between* the x, y, and z axes. The other two—the d_{z^2} and $d_{x^2-y^2}$ orbitals—have their maximum electron densities *along* the x, y, and z axes.

Now let's consider constructing an octahedral complex within this coordinate system. We can do this by bringing ligands in along each of the axes as shown in Figure 20.14. The question we want to answer is, "How do these ligands affect the energies of the d orbitals?"

In an isolated atom or ion, all the d orbitals of a given d subshell have the same energy. Therefore, an electron will have the same energy regardless of which d orbital it occupies. In an octahedral complex, however, this is no longer true. If the electron is in the $d_{x^2-y^2}$ or d_{z^2} orbital, it is forced to be nearer the negative charge of the ligands than if it is in a d_{xy}, d_{xz}, or d_{yz} orbital. Since the electron itself is negatively charged and is repelled by the charges of the ligands, the electron's potential energy will be higher in the $d_{x^2-y^2}$ and d_{z^2} orbitals than in a d_{xy}, d_{xz}, or d_{yz} orbital. Therefore, as the complex is formed, the d subshell actually splits into *two* new energy levels as shown in Figure 20.15. Here we see that regardless of which orbital the electron occupies, its energy increases because it is repelled by the negative charges of the approaching ligands. However, the electron is repelled *more* (and has a higher energy) if it is in an orbital that points directly at the ligands than if it occupies an orbital that points between them.

In an octahedral complex, the energy difference between the two sets of d-orbital energy levels is called the **crystal field splitting.** It is usually given the symbol Δ (delta), and its magnitude depends on the following factors:

The nature of the ligand. *Some ligands produce a larger splitting of the energies of the d orbitals than others.* For a given metal ion, for example, cyanide always gives a large value of Δ and F^- always gives a small value. We will have more to say about this later.

The oxidation state of the metal. *For a given metal and ligand, the size of Δ increases with an increase in the oxidation number of the metal.* As electrons are removed from a metal and the charge on the ion becomes more positive, the ion becomes smaller. This means that the ligands are attracted to the metal more strongly and they can approach the center of the complex more closely. As a result, they also approach the d orbitals along the x, y, and z axes more closely, and thereby cause a greater repulsion. This causes a greater splitting of the two d-orbital energy levels and a larger value of Δ.

The row in which the metal occurs in the periodic table. *For a given ligand and oxidation state, the size of Δ increases going down a group.* In other

Figure 20.15 *The changes in the energies of the d orbitals of a metal ion as an octahedral complex is formed.* As the ligands approach the metal ion, the d orbitals split into two new energy levels.

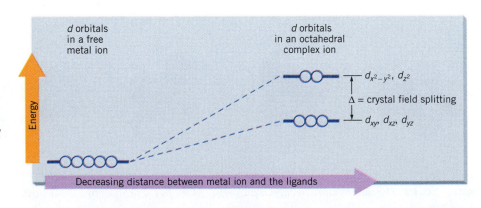

words, an ion of an element in the first row of transition elements has a smaller value of Δ than the ion of a heavier member of the same group. Thus, comparing complexes of Ni^{2+} and Pt^{2+} with the same ligand, we find that the platinum complex has the larger crystal field splitting. The explanation of this is that in the larger ion (e.g., Pt^{2+}), the d orbitals are larger and more diffuse and extend farther from the nucleus in the direction of the ligands. This produces a larger repulsion between the ligands and the orbitals that point at them.

The magnitude of Δ is very important in determining the properties of complexes, including the stabilities of oxidation states of the metal ions, the colors of complexes, and their magnetic properties. Let's look at some examples.

Relative Stabilities of Oxidation States

Comparing the cations formed by chromium, it is found that the Cr^{2+} ion is very easily oxidized to Cr^{3+}. This is easily explained by crystal field theory. In water, we expect these ions to exist as the complexes, $[Cr(H_2O)_6]^{2+}$ and $[Cr(H_2O)_6]^{3+}$, respectively. Let's examine the energies and electron populations of the d-orbital energy levels in each complex (Figure 20.16).

The element chromium has the electron configuration

$$Cr \quad [Ar]\, 3d^5 4s^1$$

Removing two electrons gives the Cr^{2+} ion, and removing three gives Cr^{3+}.

$$Cr^{2+} \quad [Ar]\, 3d^4$$

$$Cr^{3+} \quad [Ar]\, 3d^3$$

Next, we distribute the d electrons among the various d orbitals following Hund's rule, but for Cr^{2+} we have to make a choice. Should the electrons all be forced into the lower of the two energy levels, or should they be spread out? From the diagram we see that the fourth electron does not pair with one of the others in the lower energy d-orbital level. Instead, three electrons go into the lower level and the fourth is in the upper level. We will discuss *why* this happens later, but for now, let's use the two energy diagrams to explain why Cr^{2+} is so easy to oxidize.

There are actually two factors that favor the oxidation of Cr(II) to Cr(III). First, the electron that is removed from Cr^{2+} to give Cr^{3+} comes from the *higher* energy level, so oxidizing the chromium(II) *removes* a high-energy electron. The second factor is the effect caused by increasing the oxidation state of

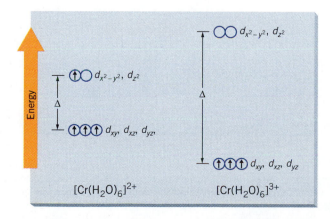

$[Cr(H_2O)_6]^{2+}$ $[Cr(H_2O)_6]^{3+}$

Figure 20.16 *Energy level diagrams for the $[Cr(H_2O)_6]^{2+}$ and $[Cr(H_2O)_6]^{3+}$ ions.* The magnitude of Δ is larger for the chromium(III) complex because the Cr^{3+} ion is smaller than the Cr^{2+} ion and the ligands are drawn closer to the metal ion, thereby increasing the repulsions felt by the $d_{x^2-y^2}$ and d_{z^2} orbitals.

the chromium. As we've pointed out, this increases the magnitude of Δ, and as you can see, the energy of the three electrons that remain is lowered. Thus, both the removal of a high-energy electron and the lowering of the energy of the electrons that are left behind help make the oxidation occur, and the $[Cr(H_2O)_6]^{2+}$ ion is very easily oxidized to $[Cr(H_2O)_6]^{3+}$.

Colors of Complex Ions

When light is absorbed by an atom, molecule, or ion, the energy of the photon raises an electron from one energy level to another. In many substances, such as sodium chloride, the energy difference between the highest energy populated level and the lowest energy unpopulated level is quite large, so the frequency of a photon that carries the necessary energy lies outside the visible region of the spectrum. The substance appears white because visible light is unaffected; it is reflected unchanged.

Remember, $E = h\nu$. The energy of the photon absorbed determines the frequency (and wavelength) of the absorbed light.

For complex ions of the transition metals, the energy difference between the *d*-orbital energy levels is not very large, and photons with frequencies in the visible region of the spectrum are able to raise an electron from the lower energy set of *d* orbitals to the higher energy set. This is shown in Figure 20.17 for the $[Cr(H_2O)_6]^{3+}$ ion.

As you know, white light contains photons of all the frequencies and colors in the visible spectrum. If we shine white light through a solution of a complex, the light that passes through has all the colors *except* those that have been absorbed. It is not difficult to determine what will be seen if we know what colors are being absorbed. All we need is a color wheel like the one shown in Figure 20.18. Across from each other on the color wheel are **complementary colors.** Green-blue is the complementary color to red, and yellow is the complementary color to violet-blue. If a substance absorbs a particular color when bathed in white light, the perceived color of the reflected or transmitted light is the complementary color. In the case of the $[Cr(H_2O)_6]^{3+}$ ion, the light absorbed when the electron is raised from one set of *d* orbitals to the other has a frequency of 5.22×10^{14} Hz, which is the color of yellow light. This is why a solution of this ion appears violet-blue.[7]

Because of the relationship between the energy and frequency of light, we see that the color of the light absorbed by a complex depends on the magnitude

Figure 20.17 *Absorption of a photon by the $[Cr(H_2O)_6]^{3+}$ complex. (a) The electron distribution in the ground state of the $[Cr(H_2O)_6]^{3+}$ ion. (b) Light energy raises an electron from the lower energy set of d orbitals to the higher energy set.*

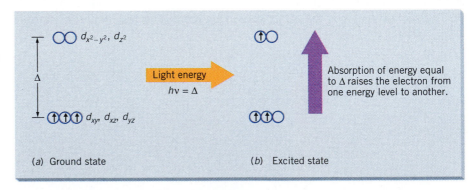

[7]The perception of color is actually somewhat more complex than this because of the varying sensitivity of the human eye to various wavelengths. For example, the eye is much more sensitive to green than to red. If a compound reflects both of these colors with equal intensity, it will appear greenish simply because the eye sees green better than it sees red.

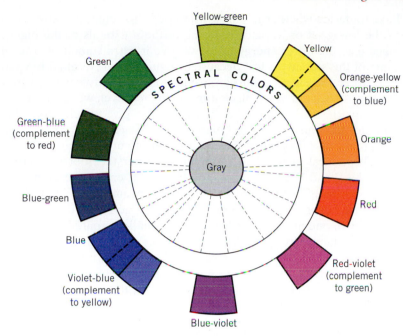

Figure 20.18 *A color wheel.* Colors that are across from each other are called complementary colors. When a substance absorbs a particular color, light that is reflected or transmitted has the color of its complement. Thus something that absorbs red light appears green-blue, and vice versa.

of Δ; the larger the size of Δ, the more energy the photon must have and the higher will be the frequency of the absorbed light. For a given metal in a given oxidation state, the size of Δ depends on the ligand. Some ligands give a large crystal field splitting, while others give a small splitting. For example, ammonia produces a larger splitting than water, so the complex $[Cr(NH_3)_6]^{3+}$ absorbs light of higher energy and higher frequency than $[Cr(H_2O)_6]^{3+}$. ($[Cr(NH_3)_6]^{3+}$ absorbs blue light and appears yellow.) Because changing the ligand changes Δ, the same metal ion is able to form a variety of complexes with a large range of colors.

A ligand that produces a large crystal field splitting with one metal ion also produces a large Δ in complexes with other metals. For example, cyanide ion is a very effective ligand and always gives a very large Δ, regardless of the metal to which it is bound. Ammonia is less effective than cyanide ion but more effective than water. Thus, ligands can be arranged in order of their effectiveness at producing a large crystal field splitting. This sequence is called the **spectrochemical series.** Such a series containing some common ligands arranged in order of their decreasing strength is

$$CN^- > NO_2^- > en > NH_3 > H_2O > C_2O_4^{2-} > OH^- > F^- > Cl^- > Br^- > I^-$$

For a given metal ion, cyanide ion produces the largest Δ and iodide ion produces the smallest.

The order of the ligands can be determined by measuring the frequencies of the light absorbed by complexes.

Magnetic Properties

Let's return to the question of the electron distribution among the *d* orbitals in Cr^{2+} complexes. As you saw above, this ion has four *d* electrons, and we noted that in placing these electrons in the *d* orbitals we had to make a decision about where to place the fourth electron. There's no question about the fate of the first three, of course. They just spread out across the three *d* orbitals in the lower level with their spins unpaired. In other words, we just follow Hund's rule, which we learned to apply in Chapter 7. But when we come to the fourth electron we have to decide whether to pair it with one of the electrons already in a *d* orbital of the lower set or to place it in one of the *d* orbitals of the higher set.[8] If we place it in the lower energy level, we give it extra stability (lower energy), but some of this stability is lost because it requires energy, called the **pairing energy,** *P*, to force the electron into an orbital that's already occupied by an electron. On the other hand, if we place it in the higher level, we are relieved of the burden of pairing the electron, but it also tends to give the electron a higher energy. Thus, for the fourth electron, "pairing" and "placement" work in opposite directions in the way they affect the energy of the complex.

The critical factor in determining whether the fourth electron enters the lower level and becomes paired, or whether it enters the higher level with the same spin as the other *d* electrons, is the magnitude of Δ. If Δ is larger than the pairing energy *P*, then greater stability is achieved if the fourth electron is paired with one in the lower level. If Δ is small compared to *P*, then greater stability is obtained by spreading the electrons out as much as possible. The complexes $[Cr(H_2O)_6]^{2+}$ and $[Cr(CN)_6]^{4-}$ illustrate this well.

Water is a ligand that does not produce a large Δ, so $P > Δ$, and minimum pairing of electrons takes place. This explains the energy level diagram for the $[Cr(H_2O)_6]^{2+}$ complex in Figure 20.19*a*. When cyanide is the ligand, however, a very large Δ is obtained, and this leads to pairing of the fourth electron with one in the lower set of *d* orbitals. This is shown in Figure 20.19*b*. It is interesting to note that it can be demonstrated experimentally that $[Cr(H_2O)_6]^{2+}$ has four unpaired electrons and the $[Cr(CN)_6]^{4-}$ ion has just two.

For octahedral chromium(II) complexes, there are two possibilities in terms of the number of unpaired electrons. They contain either four or two unpaired electrons, depending on the magnitude of Δ. When there is the maximum number of unpaired electrons, the complex is described as being **high-spin;** when there is the minimum number of unpaired electrons it is described as being **low-spin.** High- and low-spin octahedral complexes are possible when the metal has a d^4, d^5, d^6, or d^7 electron configuration. Let's look at another example—one containing the Fe^{2+} ion, which has the electron configuration

$$Fe^{2+} \quad [Ar]\, 3d^6$$

At the beginning of this section we mentioned that the $[Fe(H_2O)_6]^{2+}$ ion is paramagnetic and has four unpaired electrons, while the $[Fe(CN)_6]^{4-}$ ion is diamagnetic, meaning it has no unpaired electrons. Now we can see why, by referring to Figure 20.20. Water produces a weak splitting and a minimum amount of

Figure 20.19 *The effect of Δ on the electron distribution in a complex with four d electrons. (a) When Δ is small, the electrons remain unpaired. (b) When Δ is large, the lower energy level accepts all four electrons and two electrons become paired.*

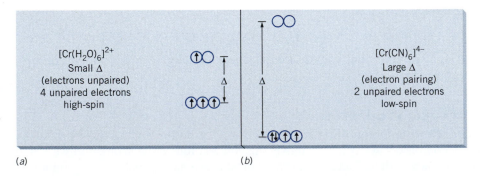

(a)   (b)

[Cr(H₂O)₆]²⁺
Small Δ
(electrons unpaired)
4 unpaired electrons
high-spin

[Cr(CN)₆]⁴⁻
Large Δ
(electron pairing)
2 unpaired electrons
low-spin

[8]We've never had to make this kind of decision before because the energy levels in atoms were always widely spaced. In complex ions, however, the spacing between the two *d*-orbital energy levels is fairly small.

pairing of electrons. When the six d electrons in the Fe^{2+} ion are distributed, one must be paired in the lower level after filling the upper level. The result is four unpaired d electrons, and a high-spin complex. Cyanide ion, however, produces a large splitting, so $\Delta > P$. This means that maximum pairing of electrons in the lower level takes place, and a low-spin complex is formed. Six electrons are just the right amount to completely fill all three of these d orbitals, and since all the electrons are paired, the complex is diamagnetic.

Crystal Field Theory and Other Geometries

The crystal field theory can be extended to geometries other than octahedral. The effect that changing the structure of the complex has on the energies of the d orbitals is to change the splitting pattern.

Square Planar Complexes

We can form a square planar complex from an octahedral one by removing the ligands that lie along the z axis. As this happens, the ligands in the xy plane are able to approach the metal a little closer because they are no longer being repelled by ligands along the z axis. The effect of these changes on the energies of the d orbitals is illustrated in Figure 20.21. The repulsions felt by the d orbitals

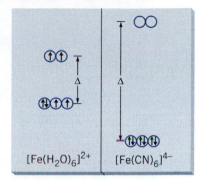

Figure 20.20 *The distribution of d electrons in* $[Fe(H_2O)_6]^{2+}$ *and* $[Fe(CN)_6]^{4-}$. *The magnitude of Δ for the cyanide complex is much larger than for the water complex. This produces a maximum pairing of electrons in the lower energy set of d orbitals in* $[Fe(CN)_6]^{4-}$.

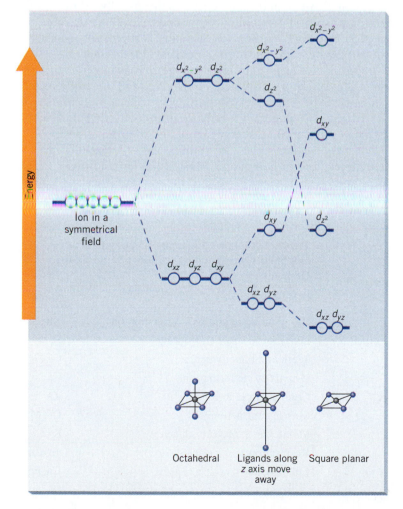

Figure 20.21 *Energies of d orbitals in complexes with various structures.* The splitting pattern of the d orbitals changes as the geometry of the complex changes.

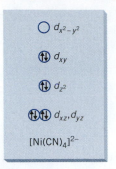

Figure 20.22 *Distribution of the electrons among the d orbitals of nickel in the diamagnetic [Ni(CN)$_4$]$^{2-}$ ion.*

that point in the z direction are reduced, so we find that the energies of the d_{z^2}, d_{xz}, and d_{yz} orbitals drop. At the same time, the energies of the orbitals in the xy plane feel greater repulsions, so the $d_{x^2-y^2}$ and d_{xy} rise in energy.

Nickel(II) ion (which has eight d electrons) forms a complex with cyanide ion that is square planar and diamagnetic. In this complex the strong field splitting produced by the cyanide ions yields a large energy separation between the d_{xy} and $d_{x^2-y^2}$ orbitals, so that a low-spin complex results. The electron distribution in this complex is illustrated in Figure 20.22.

Tetrahedral Complexes

The splitting pattern for the d orbitals in a tetrahedral complex is illustrated in Figure 20.23. Notice that the order of the energy levels is exactly opposite to that in an octahedral complex. In addition, the size of Δ is also much smaller for a tetrahedral complex than for an octahedral one. (Actually, $\Delta_{tet} \approx \frac{4}{9}\Delta_{oct}$ for the same metal ion with the same ligands.) This small Δ is always less than the pairing energy, so tetrahedral complexes are always high-spin complexes.

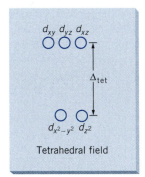

Figure 20.23 *Splitting pattern of the d orbitals for a tetrahedral complex. The crystal field splitting for the tetrahedral structure (Δ_{tet}) is smaller than that for the octahedral structure (Δ_{oct}). For complexes with the same ligands, $\Delta_{tet} \approx \frac{4}{9}\Delta_{oct}$.*

20.8 The Role of Metal Ions in Biological Systems

Most of the compounds in our bodies have structures based on carbon as the principal element, and their functions are usually related to the geometries assumed by carbon-containing compounds as well as the breaking and formation of carbon–carbon, carbon–oxygen, and carbon–nitrogen bonds. In Chapter 23 we will discuss some classes of biochemical molecules.

Our bodies also require certain metal ions in order to operate, and without them life cannot be sustained. For this reason, the study of metals in biological systems has become a very important branch of biochemistry, and a large number of research papers are published annually that deal with this topic. Table 20.5 lists some of the essential metals and their functions.

A few metals, such as sodium and potassium, are found as simple monatomic ions in body fluids. Most metal ions, however, become bound by ligands and do their work as part of metal complexes. As examples, we will look briefly at two metals—iron and cobalt.

Table 20.5 **Some Examples of Body Functions and Their Corresponding Essential Metals in the Human Body**

Body Function	Metal[a]
Blood pressure and blood coagulation	Na, Ca
Oxygen transport and storage	Fe
Teeth and bone structure	Ca
Urinary stone formation	Ca
Control of pH in blood	Zn
Muscle contraction	Ca, Mg
Maintenance of stomach acidity	K
Respiration	Fe, Cu
Cell division	Ca, Fe, Co

[a] The *ions* of these metals, not their *atoms*, are used by the body.

Iron is one of the essential elements required by our bodies. We obtain it in a variety of ways in our diets. Iron is involved in oxygen transport in our blood and in retaining oxygen in muscle tissue so that it's available when needed. The iron is present as Fe^{2+} held in a complex in which the basic ligand structure is

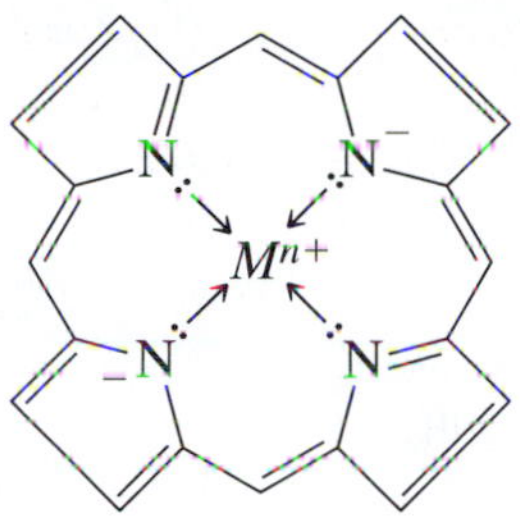

This ligand composition, with its square planar arrangement of nitrogen atoms that bind to a metal ion, is called a *porphyrin structure*. It is the ligand structure in a biologically active unit called heme. Heme is the oxygen-carrying component of the blood protein hemoglobin (which is described in greater detail in Chapter 23) and in myoglobin, which is found in muscle tissue.

In the lungs, O_2 molecules are absorbed by the blood and become bound to Fe^{2+} ions in the heme units of hemoglobin (Figure 20.24). Blood circulation then carries the O_2 to tissues where it is needed, at which time it is released by the Fe^{2+}. One of the important functions of the porphyrin ligand in this process is to prevent the Fe^{2+} from being oxidized by the O_2. (In fact, if the iron is oxidized to Fe^{3+}, it no longer is able to carry O_2.) In muscle tissue, heme units in the protein myoglobin take O_2 from hemoglobin and hold it until it's needed. In this way, muscle tissue is able to store O_2 so that plenty of it is available when the muscle must work hard.

Heme units are also present in proteins called cytochromes where the iron is involved in electron transfer reactions that employ the +2 and +3 oxidation states of iron.

The porphyrin structure is present in chlorophyll too, where the metal ion bound in the center is Mg^{2+}. Chlorophyll absorbs sunlight (solar energy) which is used by plants to convert carbon dioxide and water into glucose and oxygen.

The ability of transition metals to exist in different oxidation states is one reason they are used in biological systems. They can easily take part in many oxidation–reduction reactions.

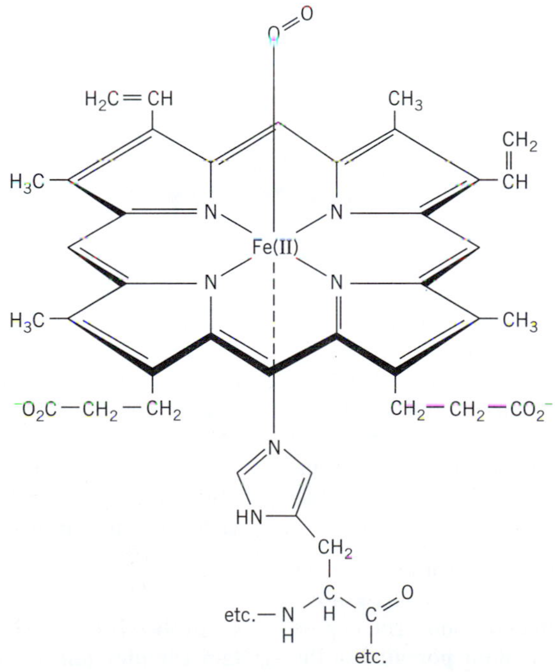

Figure 20.24 *Iron(II) bound to oxygen in heme.* The porphyrin ring ligand in heme surrounds an Fe^{2+} ion that binds an O_2 molecule. In its oxygenated form, the Fe^{2+} is octahedrally coordinated in the structure.

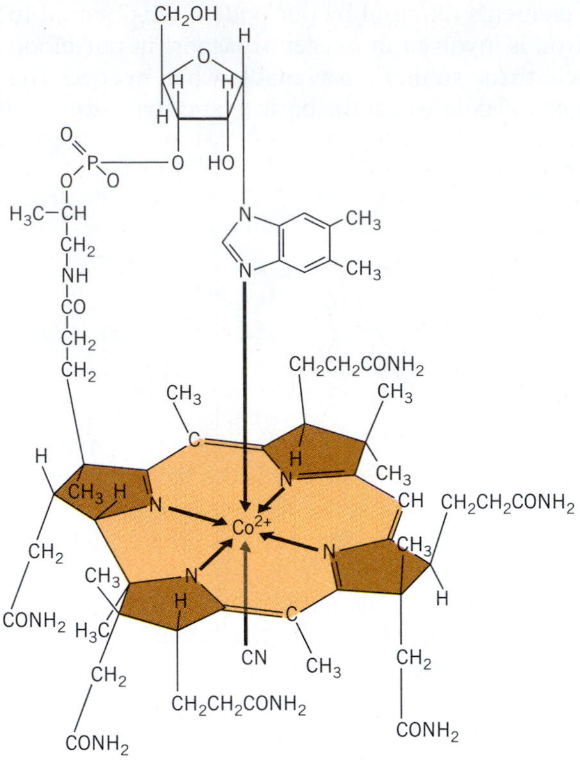

Figure 20.25 *The structure of cyanocobalamin.* Notice the cobalt ion in the center of the square planar arrangement of nitrogen atoms that are part of the ligand structure. Overall, the cobalt is surrounded octahedrally by donor atoms. For determining this structure, using X-ray diffraction, British Chemist Dorothy Crowfoot Hodgkin won the 1964 Nobel prize in Chemistry.

A structure similar to heme is that found in cyanocobalamin, the form of vitamin B_{12} found in vitamin pills (Figure 20.25). Here, a Co^{2+} ion is held in a square planar ligand structure (called a corrin ring) that is slightly different from that found in heme. Certain enzymes (biological catalysts) require cobalamins to function. Vitamin B_{12} is essential in our diets, and a deficiency in this vitamin leads to a disease called pernicious anemia.

We have illustrated here just a few examples of the important role that metal ions play in living systems. They are roles that cannot be fulfilled by other carbon-based compounds, some of which will be described in Chapter 23.

SUMMARY

Preparation of Metals. Most metals occur in compounds where they exist in positive oxidation states. Ions of sodium and potassium are found in seawater, along with magnesium. Insoluble carbonate deposits of Ca and Mg arise from the shells of marine animals. Many metals are found in the earth as oxides, sulfides, and phosphates. Compounds with small heats of formation tend to be thermally unstable and can be decomposed by heat. Carbon is an inexpensive chemical reducing agent. Very active metals must be obtained by electrolysis.

Covalent/Ionic Nature of Metal Compounds. Cations tend to polarize anions, which leads to some degree of covalent character in the metal–nonmetal bonds. The degree of polarization is proportional to the **ionic potential** of the cation, which is the ratio of the cation's charge to its radius. Anion polarization also increases with increasing size and charge of the anion. Melting points are a rough estimate of the degree of covalent character in metal compounds; the lower the melting point, the more covalent is the compound. Compounds formed from ions that are themselves colorless are sometimes colored. The color arises when photons in the visible region of the spectrum are able to cause an electron to shift from the anion to the cation. This yields a **charge transfer absorption band** in the visible region. In general, the more deeply colored the compound, the more covalent are the metal–nonmetal bonds.

Complex Ions of Metals. Coordination compounds contain **complex ions** (also called **complexes** or **coordination**

complexes), formed from a metal ion and a number of ligands. **Ligands** are Lewis bases and may be **monodentate, bidentate,** or, in general, **polydentate,** depending on the number of **donor atoms** that they contain. Water is the most common monodentate ligand. Polydentate ligands bind to a metal through two or more donor atoms and yield ring structures called **chelates.** Common bidentate ligands are oxalate ion and ethylenediamine (en); a common polydentate ligand is ethylenediaminetetraacetic acid (EDTA), which has six donor atoms.

In the formula of a complex, the metal is written first, followed by the formulas of the ligands. Brackets are often used to enclose the set of atoms that make up the complex, with the charge on the complex written outside the brackets. The charge on the complex is the algebraic sum of the charges on the metal ion and the charges on the ligands.

Complexes of polydentate ligands are more stable than similar complexes formed with monodentate ligands, partly because a polydentate ligand is less likely to be lost completely if one of its donor atoms becomes detached from the metal ion. The larger positive entropy change for dissociation of complexes with monodentate ligands also favors their dissociation compared with complexes of polydentate ligands. The phenomenon is called the **chelate effect.**

Nomenclature of Complexes. Complexes are named following a set of rules developed by the IUPAC. These are summarized on pages 918 to 920.

Coordination Number and Structure. The **coordination number** of a metal ion in a complex is the number of donor atoms attached to the metal ion. Polydentate ligands supply two or more donor atoms, which must be taken into account when determining the coordination number from the formula of the complex. Geometries associated with common coordination numbers are as follows: for coordination number 2, linear; for coordination number 4, tetrahedral and square planar (especially for Pt^{2+} complexes); and for coordination number 6, octahedral. The *porphyrin structure* is found in many biologically important metal complexes. Be sure you can draw octahedral complexes in the simplified ways described in Figure 20.7.

Isomers of Coordination Compounds. When two or more distinct compounds have the same chemical formula, they are **isomers** of each other. **Stereoisomers** have the same atoms attached to each other, but the atoms are arranged differently in space. In a *cis* **isomer,** attached groups of atoms are on the same side of some reference plane through the molecule. In a *trans* **isomer,** they are on opposite sides. *Cis* and *trans* isomerism is a form of **geometrical isomerism. Chiral** isomers are exactly the same in every way but one—they are not **superimposable** on their mirror images. These kinds of isomers exist for complexes of the type $M(AA)_3$, where M is a metal ion and AA is a bidentate ligand, and also for complexes of the type *cis*-$M(AA)_2a_2$, where a is a monodentate ligand. Chiral isomers that are related as object to mirror image are said to be **enantiomers** and are **optical isomers.**

Crystal Field Theory. In an octahedral complex, the ligands influence the energies of the d orbitals by splitting the d subshell into two energy sublevels. The lower one consists of the d_{xy}, d_{xz}, and d_{yz} orbitals; the higher energy level consists of the d_{z^2} and $d_{x^2-y^2}$ orbitals. The energy difference between the two new d sublevels is the **crystal field splitting,** Δ, and for a given ligand it increases with an increase in the oxidation state of the metal. For a given metal ion, Δ depends on the ligand, and it depends on the period number in which the metal is found.

In the **spectrochemical series,** ligands are arranged in order of their ability to cause a large Δ. Cyanide ion produces the largest crystal field splitting; iodide ion, the smallest. **Low-spin complexes** result when Δ is larger than the **pairing energy**—the energy needed to cause two electrons to become paired in the same orbital. **High-spin complexes** occur when Δ is smaller than the pairing energy. Light of energy equal to Δ is absorbed when an electron is raised from the lower energy set of d orbitals to the higher set, and the color of the complex is determined by the colors that remain in the transmitted light. Crystal field theory can also explain the relative stabilities of oxidation states, in many cases.

Different splitting patterns of the d orbitals occur for other geometries. The patterns for square planar and tetrahedral geometries are described in Figures 20.21 and 20.23, respectively. The value of Δ for a tetrahedral complex is only about 4/9 that of Δ for an octahedral complex.

Metals in Living Systems. Most metals required by living organisms perform their actions when bound as complex ions. Heme contains Fe^{2+} held in a square planar porphyrin ligand and binds to O_2 in hemoglobin and myoglobin. Heme is also found in cytochromes where it participates in redox reactions involving Fe^{2+} and Fe^{3+}. Vitamin B_{12}, required by the body to prevent the vitamin deficiency disease called pernicious anemia, contains Co^{2+} in a ring structure similar to the porphyrin ring.

REVIEW QUESTIONS

Occurrence of Metals and Recovery of Metals from Compounds

20.1 Which are the three most abundant metals in seawater? Why don't we find large amounts of silver ion in seawater?

20.2 What is the origin of limestone?

20.3 Why are many metals found as oxides and sulfides in the earth?

20.4 Why is carbon used so often as an industrial reducing agent?

20.5 Write the equation for the reduction of copper(II) oxide with hydrogen.

20.6 Why isn't sodium obtained by having sodium chloride react with a chemical reducing agent?

20.7 Why isn't thermal decomposition a practical method for obtaining metals such as sodium or magnesium?

20.8 In general, why do compounds tend to decompose at high temperatures but not at low temperatures?

20.9 In terms of thermodynamics, what must be true for us to be able to observe a substantial degree of decomposition of a compound?

20.10 The value of $\Delta G°$ for the decomposition of a metal oxide is negative if the heat of formation of the compound is positive. Why are we able to isolate such compounds at room temperature if they are unstable toward decomposition?

20.11 Many explosives have positive heats of formation. How does this explain why they tend to explode?

Covalent Character of Metal–Nonmetal Bonds

20.12 Describe how polarization of an anion by a cation produces a partial covalent character in the bond.

20.13 What two properties of a cation affect the degree to which an anion is polarized?

20.14 Define *ionic potential.*

20.15 The ionic radius of Be^{2+} is 31 pm, that of Mg is 65 pm, and that of Al^{3+} is 50 pm. Calculate the ionic potentials of these three cations. Comment on the statement that in many ways compounds of Be more closely resemble those of Al than those of Mg.

20.16 How does the melting point of a compound reflect the covalent character of its metal–nonmetal bonds?

20.17 What is a *charge transfer absorption band?*

20.18 Crystals of NaCl are white. What does this tell us about the energy required to cause an electron to transfer from a Cl^- ion to a Na^+ ion?

20.19 Why does the increasing covalent character of metal–nonmetal bonds shift the charge transfer absorption into the visible region of the spectrum?

20.20 Aqueous solutions of $Cd(NO_3)_2$ are colorless, as are solutions of Na_2S. Do the ions Cd^{2+}, NO_3^-, Na^+, and S^{2-} absorb visible light?

20.21 When solutions of $Cd(NO_3)_2$ and Na_2S are mixed, a yellow precipitate forms. What is the origin of the color of this solid?

Complex Ions of Metals

20.22 The formation of the complex ion $[Cu(H_2O)_4]^{2+}$ is described as a Lewis acid–base reaction.
(a) What are the formulas of the Lewis acid and the Lewis base in this reaction?
(b) What is the formula of the ligand?
(c) What is the name of the species that provides the donor atom?
(d) What atom is the donor atom, and why is it so designated?

(e) What is the name of the species that is the acceptor?

20.23 To be a ligand, a species should also be a Lewis base. Explain.

20.24 Why are substances that contain complex ions often called coordination compounds?

20.25 Give the names of two species we mentioned that are electrically neutral, monodentate ligands.

20.26 Give the formulas of four species that have 1− charges and are monatomic, monodentate ligands.

20.27 Use Lewis structures to diagram the formation of $[Cu(NH_3)_4]^{2+}$ and $[CuCl_4]^{2-}$ ions from their respective components.

20.28 What must be true about the structure of a ligand classified as bidentate?

20.29 What is a *chelate?* Use Lewis structures to diagram the way that the oxalate ion, $C_2O_4^{2-}$, functions as a chelating agent.

20.30 How many donor atoms does $EDTA^{4-}$ have?

20.31 Explain how a salt of $EDTA^{4-}$ can retard the spoilage of salad dressing.

20.32 How does a salt of $EDTA^{4-}$ in shampoo make the shampoo work better in hard water?

20.33 The cobalt(III) ion, Co^{3+}, forms a 1:1 complex with $EDTA^{4-}$. What is the net charge, if any, on this complex, and what would be a suitable formula for it (using the symbol EDTA)?

20.34 In which complex, $[Cu(H_2O)_4]^{2+}$ or $[Cu(NH_3)_4]^{2+}$, are the bonds from the acceptor to the donor atoms stronger? What evidence can you cite for this? Describe what one does in the lab and what one sees during a test for aqueous copper(II) ion that involves these species.

20.35 Which complex is more stable, $[Cr(NH_3)_6]^{3+}$ or $[Cr(en)_3]^{3+}$? Why?

Coordination Number and Structure

20.36 What is a *coordination number?* What structures are generally observed for complexes in which the central metal ion has a coordination number of 4?

20.37 What is the most common structure observed for coordination number 6?

20.38 Sketch the structure of an octahedral complex that contains only identical monodentate ligands.

20.39 Sketch the structure of the octahedral $[Co(EDTA)]^-$ ion. Remember that donor atoms in a polydentate ligand span adjacent positions in the octahedron. Use curved lines to represent parts of the ligand that connect donor atoms.

Isomers of Coordination Compounds

20.40 What are *isomers?*

20.41 Define *stereoisomerism, geometric isomerism, chiral isomers,* and *enantiomers.*

20.42 What are *cis* and *trans* isomers?

20.43 What condition must be fulfilled in order for a molecule or ion to be chiral?

20.44 What is *cisplatin?* Draw its structure.

20.45 What are *optical isomers?*

Bonding in Complexes

20.46 On appropriate coordinate axes, sketch and label the five *d* orbitals.

20.47 Which *d* orbitals point between the *x*, *y*, and *z* axes? Which point along the coordinate axes?

20.48 Explain why an electron in a $d_{x^2-y^2}$ or d_{z^2} orbital in an octahedral complex will experience greater repulsions because of the presence of the ligands than an electron in a d_{xy}, d_{xz}, or d_{yz} orbital.

20.49 Sketch the *d*-orbital energy level diagram for a typical octahedral complex.

20.50 Explain why octahedral cobalt(II) complexes are easily oxidized to cobalt(III) complexes. Sketch the *d*-orbital energy diagram and assume a large value of Δ when placing electrons in the *d* orbitals.

20.51 Explain how the same metal in the same oxidation state is able to form complexes of different colors.

20.52 If a complex appears red, what color light does it absorb? What color light is absorbed if the complex appears yellow?

20.53 What does the term *spectrochemical series* mean? How can the order of the ligands in the series be determined?

20.54 What do the terms *low-spin complex* and *high-spin complex* mean?

20.55 For which *d*-orbital electron configurations are both high-spin and low-spin complexes possible?

20.56 Indicate by means of a sketch what happens to the *d*-orbital electron configuration of the $[Fe(CN)_6]^{4-}$ ion when it absorbs a photon of visible light.

20.57 The complex $[Co(C_2O_4)_3]^{3-}$ is diamagnetic. Sketch the *d*-orbital energy level diagram for this complex and indicate the electron populations of the orbitals.

20.58 Consider the complex $[M(H_2O)_2Cl_4]^-$ illustrated below on the left. Suppose the structure of this complex is distorted to give the structure on the right, where the water molecules along the *z* axis have moved away from the metal somewhat and the four chloride ions along the *x* and *y* axes have moved closer. What effect will this distortion have on the energy level splitting pattern of the *d* orbitals? Use a sketch of the splitting pattern to illustrate your answer.

Before distortion   After distortion

Metals in Living Systems

20.59 In what ways is the porphyrin structure important in biological systems?

20.60 If a metal ion is held in the center of a porphyrin ring structure, what is its coordination number? (Assume the porphyrin is the only ligand.)

20.61 What function does heme serve in hemoglobin? What does it do in myoglobin?

20.62 How are the ligand ring structures similar in vitamin B_{12} and in heme? What metal is coordinated in cobalamin?

20.63 What are some of the roles played by calcium ion in the body?

REVIEW PROBLEMS

Answers to problems whose numbers are printed in color are given in Appendix D.

Thermal Stability of Metal Compounds

20.64 Calculate the temperature at which $\Delta G_T^\circ = 0$ for the decomposition of mercury(II) oxide according to the following equation.

$$2HgO(s) \longrightarrow 2Hg(g) + O_2(g)$$

For HgO, $\Delta H_f^\circ = -90.8$ kJ mol^{-1} and $S^\circ = 70.3$ J mol^{-1} K^{-1}; for Hg(g), $\Delta H_f^\circ = +61.3$ kJ mol^{-1} and $S^\circ = 175$ J mol^{-1} K^{-1}; for O_2, $S^\circ = 205$ J mol^{-1} K^{-1}.

20.65 From the data below, determine the temperature at which $K_p = 1$ for the reaction

$$CuO(s) \longrightarrow Cu(s) + \tfrac{1}{2}O_2(g)$$

For CuO, $\Delta H_f^\circ = -155$ kJ mol^{-1}. Absolute entropies:

CuO(s), 42.6 J mol^{-1} K^{-1}; Cu(s), 33.2 J mol^{-1} K^{-1}; and $O_2(g)$, 205 J mol^{-1} K^{-1}.

Covalent Character of Metal–Nonmetal Bonds

20.66 Predict which of the following exhibit the greater degree of covalent bonding.
(a) Bi_2O_3 or Bi_2O_5
(b) PbS or PbO
(c) $PbCl_2$ or PbI_2

20.67 Predict which of the following exhibit the greater degree of covalent bonding.
(a) CaO ($r_{Ca^{2+}} = 99$ pm) or SnO ($r_{Sn^{2+}} = 93$ pm)
(b) Na_2S or MgS
(c) LiCl ($r_{Cl^-} = 181$ pm) or Li_2S ($r_{S^{2-}} = 184$ pm)

20.68 Predict which of the following should be more ionic.
(a) SnO or SnS (b) SnS or SnS_2

20.69 Predict which of the following should be more ionic.
(c) SnS or PbS (d) BCl_3 or $BeCl_2$

20.70 Which of the following compounds should have the higher melting point?
(a) $CaCl_2$ or $SnCl_2$ ($r_{Ca^{2+}} = 99$ pm and $r_{Sn^{2+}} = 93$ pm)
(b) $BeCl_2$ or BeF_2

20.71 Which of the following compounds should have the higher melting point?
(a) $BeCl_2$ or $MgCl_2$ (b) $SnCl_2$ or $SnCl_4$

20.72 For each pair, predict which compound should be more deeply colored.
(a) $HgCl_2$ ($r_{Cl^-} = 181$ pm) or HgS ($r_{S^{2-}} = 184$ pm)
(b) Ag_2O or Ag_2S

20.73 For each pair, predict which compound should be more deeply colored.
(a) SnO or PbO
(b) SrS ($r_{Sr^{2+}} = 112$ pm) or PbS ($r_{Pb^{2+}} = 120$ pm)

20.74 Which compound in each pair in Problem 20.72 would have the lower melting point?

20.75 Which compound in each pair in Problem 20.73 would have the lower melting point?

Complex Ions
20.76 The iron(III) ion forms a complex with six cyanide ions, called the ferricyanide ion. What is the net charge on this complex ion, and what is its formula?

20.77 The silver ion forms a complex ion with two ammonia molecules. What is the formula of this ion? Can this complex ion exist as a salt with the sodium ion or with the chloride ion? Write the formula of the possible salt. (Use brackets and parentheses correctly.)

20.78 Write the formula, including its correct charge, for a complex that contains Co^{3+}, two Cl^-, and two ethylenediamine ligands.

20.79 Write the formula, including its correct charge, for a complex that contains Cr^{3+}, two NH_3 ligands, and four NO_2^- ligands.

Naming Complexes
20.80 How would the following molecules or ions be named as ligands when writing the name of a complex ion?
(a) $C_2O_4^{2-}$ (b) S^{2-} (c) Cl^-
(d) $(CH_3)_2NH$ (dimethylamine)

20.81 How would the following molecules or ions be named as ligands when writing the name of a complex ion?
(a) NH_3 (b) N^{3-} (c) SO_4^{2-} (d) $C_2H_3O_2^-$

20.82 Give IUPAC names for each of the following.
(a) $[Ni(NH_3)_6]^{2+}$ (d) $[Mn(CN)_4(NH_3)_2]^{2-}$
(b) $[CrCl_3(NH_3)_3]^-$ (e) $[Fe(C_2O_4)_3]^{3-}$
(c) $[Co(NO_2)_6]^{3-}$

20.83 Give IUPAC names for each of the following.
(a) $[AgI_2]^-$ (d) $[CrCl(NH_3)_5]SO_4$
(b) $[SnS_3]^{2-}$ (e) $K_3[Co(C_2O_4)_3]$
(c) $[Co(en)_2(H_2O)_4]_2(SO_4)_3$

20.84 Write chemical formulas for each of the following.
(a) tetraaquadicyanoiron(III) ion
(b) tetraammineoxalatonickel(II)
(c) diaquatetracyanoferrate(III) ion
(d) potassium hexathiocyanatomanganate(III)
(e) tetrachlorocuprate(II) ion

20.85 Write chemical formulas for each of the following.
(a) tetrachloroaurate(III) ion
(b) bis(ethylenediamine)dinitroiron(III) ion
(c) sodium tetraamminedicarbonatocobaltate(III)
(d) ethylenediaminetetraacetatoferrate(II) ion
(e) diamminedichloroplatinum(II)

Coordination Number and Structure
20.86 What is the coordination number of Fe in $[FeCl_2(H_2O)_2(en)]$? What is the oxidation number of iron in this complex?

20.87 What is the coordination number of nickel in $[Ni(C_2O_4)_2(NO_2)_2]^{4-}$?

20.88 NTA is the abbreviation for nitrilotriacetic acid, a substance that was used at one time in detergents. Its structure is

The four donor atoms of this ligand are shown in red. Sketch the structure of an octahedral complex containing this ligand. Assume that two water molecules are also attached to the metal ion and that each oxygen donor atom in the NTA is bonded to a position in the octahedron that is adjacent to the nitrogen of the NTA.

20.89 The compound shown below is called diethylenetriamine and is abbreviated "dien."

$$H_2\ddot{N}-CH_2-CH_2-\ddot{N}H-CH_2-CH_2-\ddot{N}H_2$$

When it bonds to a metal, it is a ligand with three donor atoms.
(a) Which are most likely the donor atoms?
(b) What is the coordination number of cobalt in the complex $Co(dien)_2^{3+}$?
(c) Sketch the structure of the complex $[Co(dien)_2]^{3+}$.
(d) Which complex would be expected to be more stable in aqueous solution, $[Co(dien)_2]^{3+}$ or $[Co(NH_3)_6]^{3+}$?
(e) What would be the structure of triethylenetetraamine?

Isomers of Coordination Compounds

20.90 Below are two structures drawn for a complex. Are they isomers, or are they identical? Explain.

Structure I   Structure II

20.91 Below is a structure for one of the isomers of the complex $[Co(H_2O)_3(dien)]^{3+}$. Are isomers of this complex chiral? Justify your answer.

N⌢N⌢N represents dien.

20.92 Sketch and label the isomers of the square planar complex $[Pt(NH_3)_2ClBr]$.

20.93 The complex $[Co(NH_3)_3Cl_3]$ can exist in two isomeric forms. Sketch them.

20.94 Sketch the chiral isomers of $[Co(C_2O_4)_3]^{3-}$.

20.95 Sketch the chiral isomers of $[Cr(en)_2Cl_2]^+$. Is there a nonchiral isomer of this complex?

Bonding in Complexes

20.96 In which complex do we expect to find the larger Δ?
(a) $[Cr(H_2O)_6]^{2+}$ or $[Cr(H_2O)_6]^{3+}$
(b) $[Cr(en)_3]^{3+}$ or $[CrCl_6]^{3-}$

20.97 Arrange the following complexes in order of increasing wavelength of the visible light absorbed by them:

$[Cr(H_2O)_6]^{3+}, [CrCl_6]^{3-}, [Cr(en)_3]^{3+}, [Cr(CN)_6]^{3-},$
$[Cr(NO_2)_6]^{3-}, [CrF_6]^{3-}, [Cr(NH_3)_6]^{3+}.$

20.98 Which complex should be expected to absorb light of the highest frequency, $[Cr(H_2O)_6]^{3+}$, $[Cr(en)_3]^{3+}$, or $[Cr(CN)_6]^{3-}$?

20.99 Which complex should absorb light at the longer wavelength?
(a) $[Fe(H_2O)_6]^{2+}$ or $[Fe(CN)_6]^{4-}$
(b) $[Mn(CN)_6]^{3-}$ or $[Mn(CN)_6]^{4-}$

20.100 In each pair below, which complex is expected to absorb light of the shortest wavelength? Justify your answers.
(a) $[Ru(NH_3)_5Cl]^{2+}$ or $[Fe(NH_3)_5Cl]^{2+}$
(b) $[Ru(NH_3)_6]^{2+}$ or $[Ru(NH_3)_6]^{3+}$

20.101 A complex $[CoA_6]^{3+}$ is red. The complex $[CoB_6]^{3+}$ is green. Which ligand, A or B, produces the larger crystal field splitting, Δ? Explain your answer.

20.102 Referring to the two ligands, A and B, described in the preceding problem, which complex would be expected to be more easily oxidized, $[CoA_6]^{2+}$ or $[CoB_6]^{2+}$? Explain your answer.

20.103 Referring to the complexes in the preceding two problems, would the color of $[CoA_6]^{2+}$ more likely be red or blue?

20.104 Would the complex $[CoF_6]^{4-}$ more likely be low-spin or high-spin? Could it be diamagnetic?

20.105 Sketch the d-orbital energy level diagrams for $[Fe(H_2O)_6]^{3+}$ and $[Fe(CN)_6]^{3-}$ and predict the number of unpaired electrons in each.

ADDITIONAL EXERCISES

*20.106 In the decomposition of HgO described in Problem 20.64, what are the equilibrium molar concentrations of $Hg(g)$ and $O_2(g)$ above solid HgO at the temperature at which $\Delta G_T^\circ = 0$?

20.107 Calculate K_p at 100, 500, and 1000 °C for the reaction $MoO_3(s) \rightleftharpoons Mo(s) + \frac{3}{2}O_2(g)$ given the following data:

	ΔH_f° (kJ mol^{-1})	S° (J mol^{-1} K^{-1})
$MoO_3(s)$	−754.3	78.2
$Mo(s)$	0.0	28.6
$O_2(g)$	0.0	205.0

20.108 Is the complex $[Co(EDTA)]^-$ chiral? Illustrate your answer with sketches.

20.109 The complex $[Pt(NH_3)_2Cl_2]$ can be obtained as two distinct isomeric forms. Make a model of a tetrahedron and show that if the complex were tetrahedral, two isomers would be impossible.

*20.110 A solution was prepared by dissolving 0.500 g of $CrCl_3 \cdot 6H_2O$ in 100 mL of water. A silver nitrate solution was added and gave a precipitate of AgCl that was filtered from the mixture, washed, dried, and weighed. The AgCl had a mass of 0.538 g.
(a) What is the formula of the complex ion of chromium in this compound?
(b) What is the correct formula for the compound?
(c) Sketch the structure of the complex ion in this compound.
(d) How many different isomers of the complex can be drawn?

For some, relaxing in the sun on a tropical isle is a dream come true. In recent times, however, we've learned that harmful ultraviolet rays from the sun can cause skin cancer. We are protected from these rays to a degree by a shield of ozone in the upper atmosphere, which some scientists feel is in jeopardy because of pollutants released by human activities. Ozone is a form of the element oxygen. In this chapter, one of the subjects we will explore is the variety of molecular structures formed by the elemental nonmetals.

Some Chemical Properties of the Nonmetals and Metalloids

This Chapter in Context In Chapter 20 you saw how we could apply chemical principles learned in earlier chapters to the study of the chemical properties of metals and metal compounds. In this chapter we continue with that theme as we turn our attention to the study of the nonmetals and metalloids (Figure 21.1). Our intent here is not to be encyclopedic; instead, we will focus our attention on several aspects of the chemistries of these elements. Our goal will be to study trends and similarities in chemical properties and reactions that extend over groups of elements in the periodic table. We will also study some common threads that exist in the structural relationships among classes of compounds, both simple and complex. For a more detailed discussion of the chemistry of the individual elements, we direct you to our supplement *Descriptive Chemistry of the Elements.*

When we wish to refer to nonmetals and metalloids together and in a way that distinguishes them from the metals, we will use the general term *nonmetallic element.*

21.1 Obtaining Metalloids and Nonmetals in Their Free States

In the last chapter we made the rather broad generalization that metals, in their compounds, usually exist only in positive oxidation states. This is not always true for the nonmetallic elements, so the procedures used to obtain them as free elements vary.

Metalloids

In most of their compounds, the metalloids are combined with nonmetals, either in a molecular structure such as SiO_2 or in an oxoanion such as those found in the silicates. In such combinations, the metalloid has a lower electronegativity than the nonmetal, so the metalloid exists in a positive oxidation state. To obtain the metalloid in its elemental state, therefore, a chemical reduction must be carried out. This is usually accomplished with either hydrogen or carbon as the reducing agent. For example, boron is obtained by passing a mixture of BCl_3

945

Figure 21.1 *Distribution of nonmetals and metalloids in the periodic table.*

vapor and hydrogen gas over a hot wire, on which the following reaction takes place.

$$2BCl_3(g) + 3H_2(g) \longrightarrow 2B(s) + 6HCl(g)$$

On the other hand, elemental silicon is produced by heating SiO_2 with carbon in an electric furnace, where the reaction

$$SiO_2(s) + C(s) \longrightarrow Si(s) + CO_2(g)$$

takes place once the temperature exceeds approximately 3000 °C (below this temperature the reverse reaction is actually favored).

The remaining metalloids may be obtained from their oxides by heating them with either carbon or hydrogen; for example,

$$GeO_2(s) + C(s) \longrightarrow Ge(s) + CO_2(g)$$

$$GeO_2(s) + 2H_2(g) \longrightarrow Ge(s) + 2H_2O(g)$$

Similarly, we have

$$2As_2O_3(s) + 3C(s) \longrightarrow 4As(s) + 3CO_2(g)$$

$$2Sb_2O_3(s) + 3C(s) \longrightarrow 4Sb(s) + 3CO_2(g)$$

and

$$As_2O_3(s) + 3H_2(g) \longrightarrow 2As(s) + 3H_2O(g)$$

$$Sb_2O_3(s) + 3H_2(g) \longrightarrow 2Sb(s) + 3H_2O(g)$$

As you may recall, in very pure form silicon and germanium have widespread applications in the electronics industry, where they are used in transistors and photoconduction devices. Sophisticated electronic circuits embedded in tiny wafers of silicon make possible a myriad of computer-controlled devices that we've come to take for granted, such as handheld computer games, computers that enable modern automobile engines to manage exhaust emissions, handheld calculators, and automatic-exposure cameras.

Uses of silicon as a semiconductor. A technician checks a silicon wafer during one stage in the manufacture of computer "chips." The silicon disk contains hundreds of individual chips, each consisting of thousands of electronic components. After all of the individual components have been added, the disk will be cut up to give separate chips that will be incorporated into parts for computer circuits.

There are few aspects of modern life that are not impacted significantly by computer circuits embedded in silicon or germanium.

Noble gas atoms have completed *s* and *p* subshells in their outer shells and very little tendency to form bonds to other atoms.

The Nonmetals

Unlike the metals and metalloids, it is difficult to make general statements about the preparation of the nonmetals. Some nonmetals, such as the noble gases, are always found uncombined in nature. Others, while present in many naturally occurring compounds, also are found extensively in the free state as well. For instance, our atmosphere is composed primarily of elemental nitrogen, N_2 (about 80%), and oxygen, O_2 (about 20%). Although nitrogen and oxygen are also found in a vast number of compounds, certainly their most economical source is simply the air itself. The atmosphere is also the major source of the noble gases, even though their concentrations in air are very small. Of the noble gases, only helium and radon are not obtained primarily from the atmosphere. Helium is found in gaseous deposits beneath Earth's crust where it has collected after being produced by the capture of electrons by alpha particles (helium nuclei, He^{2+}), which are formed during the radioactive decay of elements such as uranium. Radon itself is radioactive and is produced by the radioactive decay of still heavier elements. Since radon spontaneously decomposes into other elements, it occurs only in minute quantities in nature.

Sulfur and carbon are two other elements that occur naturally in both the combined and free states. There are, for instance, many naturally occurring sulfates (for example, $BaSO_4$ and $CaSO_4 \cdot 2H_2O$) and sulfides (FeS_2, CuS, HgS, PbS, and ZnS). In the free state, sulfur has been found in large underground deposits from which it is mined by forcing superheated water under pressure into the sulfur, causing the sulfur to melt. Once molten, the sulfur–water mixture is forced to the surface using compressed air.

Turning to carbon, we find that most of its naturally occurring compounds are carbonates, for example, limestone ($CaCO_3$). In the free state most carbon is found in either of its two principal forms, diamond or graphite.

Elemental sulfur in nature. Sulfur is one of the substances released from the earth during volcanic eruptions. Here we see molten sulfur pouring from a vent in the Kawah Ijen volcano in Indonesia.

Obtaining Nonmetals from Compounds

Since nonmetals combine with each other as well as with metals, no simple generalizations can be made concerning their recovery from compounds. When combined with a metal, the nonmetal is found in a negative oxidation state, so an oxidation must be brought about to generate the free element. For example, the halogens Cl_2, Br_2, and I_2 can be conveniently prepared in the laboratory by having one of their salts react with an oxidizing agent such MnO_2 in an acidic solution:

$$2X^- + MnO_2 + 4H^+ \longrightarrow X_2 + Mn^{2+} + 2H_2O$$

where X = Cl, Br, or I.

As you learned in Chapter 19, chlorine is a very important industrial chemical, and vast quantities (approximately 10 million tons annually) are produced by electrolysis of NaCl, both aqueous and molten. Chlorine is used in large amounts in water treatment, in the manufacture of pharmaceuticals, pesticides, and solvents, and in the production of vinyl chloride, which is used to manufacture vinyl plastics.

The halogens themselves can also serve as oxidizing agents in replacement reactions. Since the tendency to acquire electrons (electronegativity) decreases as we proceed downward in a group, the ability of the halogen to serve as an oxidizing agent decreases too. This is seen in their reduction potentials (Table 21.1), which decrease from fluorine to iodine. As a result, a given halogen is a better oxidizing agent than the other halogens below it in Group VIIA and is able to displace them from their binary compounds with metals. Thus F_2 will displace Cl^-, Br^-, and I^-, while Cl_2 will displace only Br^- and I^- but not F^-, and so on. This is illustrated by the following typical reactions.

For fluorine:

$$F_2 + \begin{Bmatrix} 2NaCl \\ 2NaBr \\ 2NaI \end{Bmatrix} \longrightarrow 2NaF + \begin{Bmatrix} Cl_2 \\ Br_2 \\ I_2 \end{Bmatrix}$$

Coal mining. Coal is largely carbon, mostly in a graphite form. Here we see it being mined underground in Australia. Similar mines throughout the world, both above and below ground, have provided an essential fuel for the growth of industry.

Table 21.1 Reduction Potentials of the Halogens

Reaction	$E°$ (V)
$F_2 + 2e^- \rightleftharpoons 2F^-$	2.87
$Cl_2 + 2e^- \rightleftharpoons 2Cl^-$	1.36
$Br_2 + 2e^- \rightleftharpoons 2Br^-$	1.09
$I_2 + 2e^- \rightleftharpoons 2I^-$	0.54

Remember, the larger the reduction potential, the greater is the tendency of the substance to acquire electrons and, therefore, to be an oxidizing agent.

For chlorine:

$$Cl_2 + \begin{Bmatrix} 2NaBr \\ 2NaI \end{Bmatrix} \longrightarrow 2NaCl + \begin{Bmatrix} Br_2 \\ I_2 \end{Bmatrix}$$

$$Cl_2 + NaF \longrightarrow \text{no reaction}$$

For bromine:

$$Br_2 + 2NaI \longrightarrow 2NaBr + I_2$$

$$Br_2 + \begin{Bmatrix} 2NaF \\ 2NaCl \end{Bmatrix} \longrightarrow \text{no reaction}$$

Iodine cannot replace any of the other halides from their compounds.

The relative oxidizing power of the halogens is used in the commercial preparation of Br_2. Bromine is isolated from seawater and brine solutions pumped from deep wells by passing Cl_2, followed by air, through the liquid. The Cl_2 oxidizes the Br^- to Br_2, and the air sweeps the volatile Br_2 from the solution.

Fluorine, because of its position as the most powerful chemical oxidizing agent, can only be obtained by electrolytic oxidation. This process must be carried out in the absence of water because water is more easily oxidized than the fluoride ion. In an aqueous solution, therefore, the oxidation

$$2H_2O(l) \longrightarrow O_2(g) + 4H^+(aq) + 4e^-$$

will occur in preference to

$$2F^-(aq) \longrightarrow F_2(g) + 2e^-$$

H^+ is more easily reduced than K^+, which is why H_2 appears at the cathode.

In practice, a molten mixture of KF and HF, which has a lower melting point than KF alone, is electrolyzed, producing H_2 at the cathode and F_2 at the anode.

Nonmetals can also be extracted from their compounds by reduction if the nonmetal happens to exist in a positive oxidation state. We have seen how this applies to the metalloids, and the general procedure employed there can also be used for the nonmetals. For instance, elemental phosphorus is produced from a phosphate such as $Ca_3(PO_4)_2$, where it exists in the +5 oxidation state. To obtain the phosphorus, the $Ca_3(PO_4)_2$ is heated with a mixture of carbon and SiO_2 (sand).[1]

$$Ca_3(PO_4)_2 + 3SiO_2 + 5C \longrightarrow 3CaSiO_3 + 5CO + 2P$$

In this reaction, the SiO_2 is present to combine with the calcium to form calcium silicate, $CaSiO_3$, a compound with a relatively low melting point. Because the reduction of SiO_2 by carbon requires *extremely* high temperatures, only the phosphorus is reduced to the element at the lower temperatures used in this reaction.

21.2 Molecular Structures of the Nonmetals and Metalloids

Only the noble gases exist in nature as single atoms. All the other elements are found in more complex forms—some as diatomic molecules and the rest in much more complex molecular structures. In this section we will study what

---

[1]Phosphorus was first produced in 1669 when alchemist Hennig Brandt heated a mixture of sand and dried urine. He condensed the vapors by passing them through water to give the new element. It was named phosphorus from the Greek *phosphoros* (light bringer) because the solid in a sealed bottle glowed in the dark. The glow is actually the result of slow oxidation of the phosphorus surface by residual oxygen in the container.

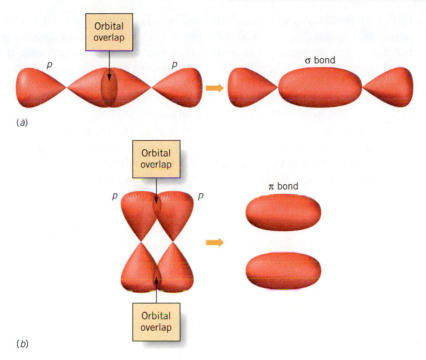

Figure 21.2 *Formation of sigma and pi bonds.* (*a*) Sigma bonds formed by the head-to-head overlap of *p* orbitals. (*b*) Pi bonds formed by the side-by-side overlap of *p* orbitals.

these structures are and use bonding theory to understand them. Let's begin by reviewing how atoms form covalent bonds.

You learned in Chapter 9 that atoms are able to share electrons in two basic ways. One is by the formation of σ bonds, which can involve the overlap of *s* orbitals or an end-to-end overlap of *p* orbitals or hybrid orbitals. The second kind of covalent bond is the π bond, which normally requires the sideways overlap of unhybridized *p* orbitals (Figure 21.2). (Pi bonds can also be formed by *d* orbitals, but we do not discuss these in this book.)

Not all atoms have the same tendency to form π bonds. The ability of an atom to form π bonds determines its ability to form multiple bonds, and this in turn greatly affects the kinds of molecular structures that the element produces. One of the most striking illustrations of this is the molecular structures of the elemental nonmetals and metalloids.

If we take a brief look at the electronic structures of the nonmetals and metalloids, we see that their atoms generally contain valence shells that are only partially filled. (Exceptions to this, of course, are the noble gases, which have electronically complete *s* and *p* subshells in their valence shells.) We have also seen that atoms tend to complete their valence shells by bond formation. Therefore, in their elemental states, most of the nonmetallic elements do not exist as single atoms, but instead form bonds to each other and produce more complex molecular species.

One of the controlling factors in determining the complexity of the molecular structures of the nonmetals and metalloids is their ability to form multiple bonds. Small atoms, such as those in period 2, are able to approach each other closely. As a result, effective sideways overlap of their *p* orbitals can occur, and these atoms form strong π bonds. Therefore, carbon, nitrogen, and oxygen are able to form multiple bonds about as easily as they are able to form single bonds. On the other hand, when the atoms are large—which is the case for atoms from periods 3, 4, and so on—π-type overlap between their *p* orbitals is relatively ineffective, so π bonds formed by large atoms are relatively weak compared to σ bonds. Therefore, rather than form a double bond consisting of one σ bond and one π bond, these elements prefer to use two separate σ bonds to bond their atoms together. This leads to a useful generalization: Elements in period 2 are able to form multiple bonds fairly readily, while elements below them in periods 3, 4, 5, and 6 have a tendency to prefer single bonds. Let's see how this affects the structures of the elemental nonmetallic elements.

Oxygen and nitrogen have six and five electrons, respectively, in their valence shells. This means that an oxygen atom needs two electrons to complete its valence shell, and a nitrogen atom needs three. Although a perfectly satisfactory Lewis structure for O_2 can't be drawn, experimental evidence suggests that the oxygen molecule does possess a double bond. The nitrogen molecule, which we discussed earlier, contains a triple bond. Atoms of oxygen and nitrogen, because of their small size, are capable of multiple bonding because they are able to form strong π bonds. This allows them to form a sufficient number of bonds with just a single neighbor to complete their valence shells, so they are able to form diatomic molecules.

> The molecular orbital theory, which is able to provide an excellent explanation of the bonding in O_2, also tells us that there is a double bond in the O_2 molecule.

Oxygen, in addition to forming the stable species O_2 (dioxygen), also can exist in another very reactive molecular form called **ozone,** which has the formula O_3. The structure of ozone can be represented as a resonance hybrid

$$\ddot{\underset{\cdot\cdot}{O}}=\overset{\displaystyle\ddot{O}}{}\diagdown_{\displaystyle\ddot{\underset{\cdot\cdot}{O}}} \longleftrightarrow \ddot{\underset{\cdot\cdot}{O}}\diagup^{\displaystyle\overset{\ddot{O}}{}}=\ddot{\underset{\cdot\cdot}{O}}$$

> ΔG_f° for O_3 is +163 kJ mol^{-1}, which indicates that the molecule has a strong tendency to decompose to give O_2.

This unstable molecule can be generated by the passage of an electric discharge through ordinary O_2, and the pungent odor of ozone can often be detected in the vicinity of high-voltage electrical equipment. It is also formed in limited amounts in the upper atmosphere by the action of ultraviolet radiation from the sun on O_2. The presence of ozone in the upper atmosphere shields Earth and its creatures from exposure to intense and harmful ultraviolet light. (See *Chemicals in Our World 5,* following Chapter 9.)

The existence of an element in more than one form, either as the result of differences in molecular structure as with O_2 and O_3, or as the result of differences in the packing of molecules in the solid, is a phenomenon called **allotropy.** The different forms of the element are called **allotropes.** Thus, O_2 is one allotrope of oxygen and O_3 is another. Allotropy is not limited to oxygen, as you will soon see.

Let's look now at another period 2 nonmetal, carbon. An atom of carbon has four electrons in its valence shell, so it must share four electrons to complete its octet. There is no way for carbon to form a quadruple bond, so a simple C_2 species is not stable under ordinary conditions. Instead, carbon completes its octet in other ways, leading to several allotropic forms of the element. One of these is diamond. In diamond, each carbon atom uses sp^3 hybrid orbitals to form covalent bonds to four other carbon atoms at the corners of a tetrahedron. Each of those atoms, in turn, is bonded through sp^3 hybrid orbitals to three more, and so on, as illustrated in Figure 21.3a. This produces a gigantic three-dimensional network of interconnected covalent bonds. A diamond crystal of gem quality, such as that in Figure 21.3b, therefore consists of a huge number of carbon atoms covalently bonded together in one enormous molecule. Since breaking a diamond crystal involves rupturing a very large number of covalent bonds, diamond is extremely hard.

A second allotrope of carbon is graphite. This is the form that probably is the most familiar to you. In graphite, the carbon atoms are arranged in the form of hexagonal rings connected together in large planar sheets, perhaps somewhat

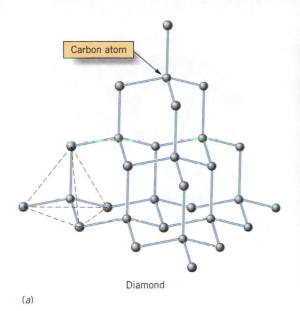

Carbon atom

Diamond

(a)

(b)

Figure 21.3 *Diamond.* (a) In the molecular structure of diamond, each carbon atom is covalently bonded to four other carbon atoms. (b) A gem-quality diamond.

reminiscent of chicken wire (Figure 21.4). Each carbon is surrounded by three nearest neighbors at angles of 120° from one another. This means that the molecular framework is based on σ bonds produced by the overlap of sp^2 hybrid orbitals on the carbon atoms. On each of the carbons throughout the entire structure, there remains an unhybridized p orbital containing an unpaired electron. These p orbitals are ideally situated to form π bonds with their neighbors, and one of the many resonance forms of a graphite sheet is illustrated in Figure 21.5.

In the total graphite structure, shown in Figure 21.6, the planes of carbon atoms are stacked in layers so that each carbon atom lies above another in every second layer. Within any given layer, adjacent carbon atoms are fairly close together (141 pm), while the spacing between successive planes is much greater (335 pm). The different planes of carbon atoms are not held together by covalent bonds, but instead by much weaker London forces. As a result, the layers are able to slide over one another with relative ease, and as you may know, graphite has many applications as a dry lubricant. As every schoolchild is taught, the lead in a lead pencil is actually graphite, and when one writes with a pencil, the layers of graphite slide off one another and onto the paper.

Graphite also has interesting electrical properties. Although the structure of a graphite layer is derived from covalent bonds, graphite is able to conduct electricity. The reason is that the overlap of p orbitals that are perpendicular to the

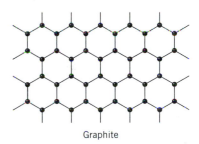

Graphite

Figure 21.4 *Sigma-bond framework of a graphite sheet.*

Unlike graphite, other nonmetals are electrical insulators.

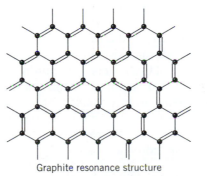

Graphite resonance structure

Figure 21.5 *One of the resonance structures for graphite.* The p orbitals of the carbon rings in graphite are oriented correctly to form a network of π bonds.

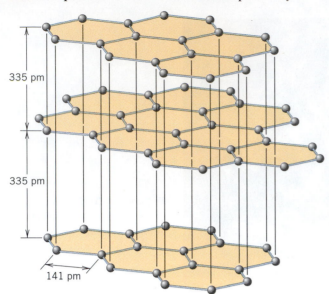

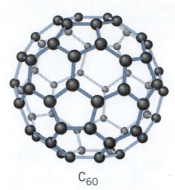

C₆₀

Figure 21.7 *The structure of buckminsterfullerene, C₆₀.* The shape of the molecule led to its nickname, *buckyball.* In some ways, the structure resembles a soccer ball, with the bonds between carbon atoms corresponding to the seams in the ball.

Figure 21.6 *The stacking of sheets of carbon atoms in the graphite structure.* The individual layers of carbon atoms are weakly attracted to each other and can slide over each other relatively easily, particularly when gas molecules become trapped between the layers.

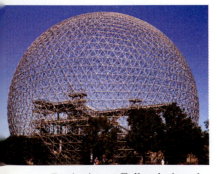

Buckminster Fuller designed the Geodesic Dome, which stands on the site of the 1967 World's Fair in Montreal, Quebec, Canada. It housed the United States pavillion.

carbon rings produces a delocalized π electron cloud that extends over the top and bottom surfaces of the entire layer. Within this π network, electrons are able to move relatively easily, so electrons can be injected into the π network at one end of a layer and pulled out of the other end, thus allowing electrical conduction. It is also interesting that the electrical conductivity is good parallel with the carbon layers, but very poor perpendicular to them. This is because electrons cannot easily move from one layer to another.

Another series of carbon allotropes called the **fullerenes** was discovered in the late 1980s. The simplest of them, illustrated in Figure 21.7, consist of a collection of five- and six-membered rings of carbon atoms arranged to give a soccer-ball-shaped molecule with the formula C_{60}. This allotrope was named **buckminsterfullerene** in honor of architect R. Buckminster Fuller who used similar ring arrangements in designing a type of structure called a Geodesic Dome—a spacious, lightweight, and strong building design. In some ways, the structure of a buckyball is like graphite in which a layer of carbon atoms has curved around on itself to form a ball-shaped molecule.

There is still one element of period 2 whose structure we have not considered, namely, boron. This element, found in Group IIIA, is quite unlike any of the others, since there is no simple way for it to complete its valence shell. The boron atom has just three electrons in its valence shell. Sharing electrons in such a way as to give three ordinary single bonds still would leave each boron atom with only six electrons around it. To achieve a stable electron configuration, therefore, elemental boron forms a rather complex molecular structure that cannot be described by simple Lewis structures.

Boron exists in several different crystalline forms, each of which is characterized by clusters of 12 boron atoms located at the vertices of an icosahedron (a 20-sided geometric figure) as shown in Figure 21.8. Each boron atom within a given cluster is equidistant from five others, and in the solid, each of these is also joined to yet another boron atom outside the cluster (Figure 21.8*b*). The electrons available for bonding are therefore delocalized to a large extent over many boron atoms.

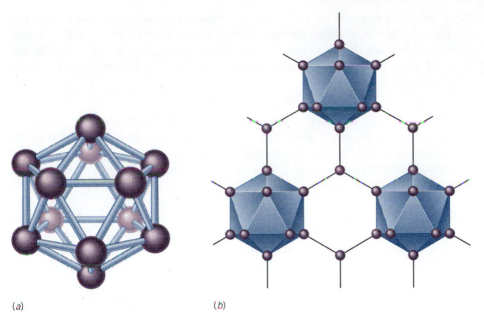

(a)   (b)

Figure 21.8 *The structure of elemental boron. (a)* The arrangement of 12 boron atoms in a B_{12} cluster. *(b)* The element boron consists of an interconnecting network of B_{12} clusters that produces a very hard and high-melting solid.

The linking together of B_{12} units produces a large three-dimensional covalent solid that is very difficult to break down. As a result, boron is very hard (it is the second hardest element) and has a very high melting point (about 2200 °C).

Nonmetallic Elements below Period 2

In graphite, carbon exhibits multiple bonding, as do nitrogen and oxygen in their molecular forms. As we noted earlier, their ability to do this reflects their ability to form strong π bonds—a requirement for the formation of a double or triple bond. When we move to the third and successive periods, a different state of affairs exists. Here, we have much larger atoms that are able to form relatively strong σ bonds but much weaker π bonds. Because their π bonds are so weak, these elements prefer to form only single bonds (σ bonds), and the molecular structures of the free elements reflect this.

Elements of Group VIIA

Each of the elements in Group VIIA is diatomic in the free state. This is because π bonding is not necessary in any of their molecular structures. Chlorine, for example, requires just one electron to complete its octet, so it only needs to form one covalent bond with another atom. Therefore, it forms one σ bond to another chlorine atom and is able to exist as diatomic Cl_2. Bromine and iodine form diatomic Br_2 and I_2 molecules for the same reason. The structures of the remaining nonmetals and metalloids are considerably more complex, however.

Elements of Group VIA

Below oxygen in Group VIA is sulfur, which has the Lewis symbol

$$:\overset{\cdot}{\underset{\cdot\cdot}{S}}\cdot$$

A sulfur atom requires two electrons to obtain an octet, so it must form two covalent bonds. But sulfur doesn't form π bonds well to other sulfur atoms; instead, it prefers to form two stronger single bonds to *different* sulfur atoms. Each of these also prefers to bond to two different sulfur atoms, and this gives rise to a

$$-\ddot{S}-\ddot{S}-\ddot{S}-\ddot{S}-\ddot{S}-$$

sequence. Actually, in sulfur's most stable form, called **orthorhombic sulfur,** the sulfur atoms are arranged in an eight-member ring to give a molecule with the formula S_8 (properly named cyclooctasulfur). The S_8 ring has a puckered crown-like shape, which is illustrated in Figure 21.9. Another allotrope is **monoclinic sulfur,** which also contains S_8 rings that are arranged in a slightly different crystal structure.

Selenium, below sulfur in Group VIA, also forms Se_8 rings in one of its allotropic forms. And selenium and tellurium also exist in a gray form in which there are long Se_x and Te_x chains (where the subscript x is a large number).

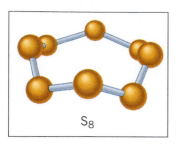

Figure 21.9 *The structure of the puckered S_8 ring.*

Elements of Group VA

Like nitrogen, the other elements in Group VA all have five valence electrons. Phosphorus is an example.

$$:\dot{P}\cdot$$

To achieve a noble gas structure, the phosphorus atom must acquire three more electrons. Because there is little tendency for phosphorus to form multiple bonds, as nitrogen does when it forms N_2, the octet is completed by the formation of three single bonds to three *different* phosphorus atoms.

The simplest elemental form of phosphorus is a waxy solid called **white phosphorus** because of its appearance. It consists of P_4 molecules in which each phosphorus atom lies at a corner of a tetrahedron, as illustrated in Figure 21.10. Notice that in this structure each phosphorus is bound to three others. This allotrope of phosphorus is very reactive, partly because of the very small P—P—P bond angle of 60°. At this small angle, the orbitals of the phosphorus atoms don't overlap very well, so the bonds are weak. As a result, breaking a P—P bond occurs easily. When a P_4 molecule reacts, this bond breaking is the first step, so P_4 molecules are readily attacked by other chemicals, especially oxygen. White phosphorus is so reactive toward oxygen that it ignites and burns spontaneously in air. For this reason, white phosphorus is used in military incendiary devices, and you've probably seen movies in which exploding phosphorus shells produce arching showers of smoking particles (Figure 21.11).

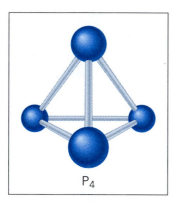

Figure 21.10 *The molecular structure of white phosphorus, P_4. The bond angles of 60° make the phosphorus–phosphorus bonds quite weak and cause the molecule to be very reactive.*

A second allotrope of phosphorus that is much less reactive is called **red phosphorus.** At the present time, its structure is unknown, although it has been suggested that it contains P_4 tetrahedra joined at the corners as shown in Figure 21.12. Red phosphorus is also used in explosives and fireworks, and it is mixed with fine sand and used on the striking surfaces of matchbooks. As a match is drawn across the surface, friction ignites the phosphorus, which then ignites the ingredients in the tip of the match.

A third allotrope of phosphorus is called **black phosphorus,** which is formed by heating white phosphorus at very high pressures. This variety has a layered structure in which each phosphorus atom in a layer is covalently bonded to three others in the same layer. As in graphite, these layers are stacked one atop another, with only weak forces between the layers. As you might expect, black phosphorus has many similarities to graphite.

Figure 21.11 *White phosphorus reacts vigorously with oxygen.* Streamers of white smoke arc through the air as a phosphorus bomb explodes. The burning particles are chunks of white phosphorus.

The elements arsenic and antimony, which are just below phosphorus in Group VA, are also able to form somewhat unstable yellow allotropic forms containing As_4 and Sb_4 molecules, but their most stable forms have a metallic appearance with structures similar to black phosphorus.

Elements of Group IVA

Finally, we look at the heavier nonmetallic elements in Group IVA, silicon and germanium. To complete their octets, each must form four covalent bonds. Unlike carbon, however, they have very little tendency to form multiple bonds, so they don't form allotropes that have a graphite structure. Instead, each of them forms a solid with a structure similar to diamond.

Summary

In this section we have examined the structures of the elemental forms of the nonmetallic elements. Some of them are simply diatomic molecules, but others are quite complex. This complexity can be traced to two factors. One is the number of electrons that the atom of the element needs to complete its octet. The second is the preference of the heavier (and larger) nonmetals and metalloids for single bonds. This preference prevents them from forming diatomic molecules and causes them to form more complex structures.

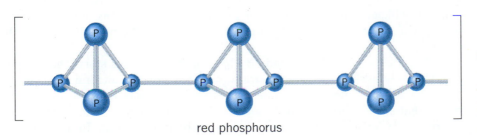

red phosphorus

Figure 21.12 *Proposed molecular structure of red phosphorus.* Red phosphorus is believed to be composed of long chains of P_4 tetrahedra connected at their corners.

21.3 Hydrogen Compounds of Nonmetals and Metalloids

Hydrogen is a quite special element, possessing only one electron in its $1s$ orbital and having no electrons in subshells beneath the outer shell. It can achieve the noble gas electron configuration of helium either by acquiring an additional electron to give the hydride ion, H^-, as it does when it reacts with the very active metals, or by forming a covalent bond. Because of its relatively high electronegativity, hydrogen forms covalent bonds with the nonmetals and metalloids. As a class, binary compounds with hydrogen are often called **hydrides,** whether the hydrogen exists in a positive or negative oxidation state.

Except for the very special case of boron, discussed below, and the hydrogen difluoride ion HF_2^-, in which the H atom lies equidistant between the two F atoms (that is, $[F\!-\!H\!-\!F]^-$), hydrogen is capable of binding to only one other atom by way of electron sharing, because it seeks only one additional electron. As a result, the structural chemistry of the simple hydrides is rather straightforward and, perhaps, even somewhat mundane. With the halogens, hydrogen forms compounds with the general formula HX (e.g., HF, HCl). Similarly, we find, quite expectedly, that the Group VIA elements form molecules of general formula H_2X (e.g., H_2O, H_2S), the elements of Group VA form H_3X (usually written XH_3, for example, NH_3, PH_3), and those in Group IVA form H_4X (or XH_4, for example, CH_4, SiH_4). The geometries of these molecules are all readily predicted by the VSEPR theory discussed in Section 9.2.

Boron Hydrides

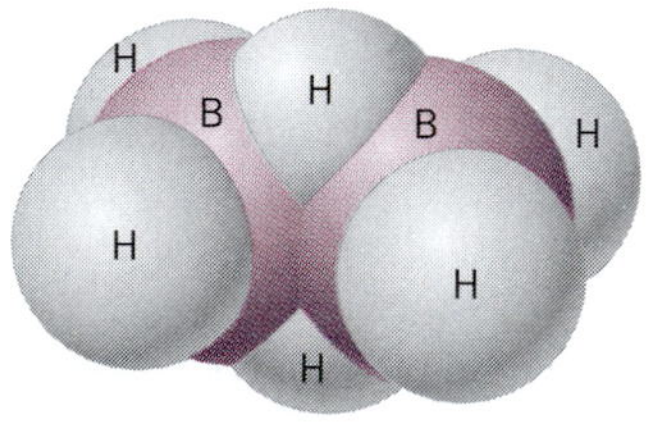

The fact that a boron atom has three valence electrons would suggest that its simplest hydride might have the formula BH_3. However, this compound does not exist; instead, the simplest compound formed between boron and hydrogen is **diborane,** which has the formula B_2H_6. Its structure is shown in Figure 21.13, where we see that each boron atom is surrounded by four hydrogens roughly at the vertices of a tetrahedron. One might therefore be tempted to write a Lewis structure of the molecule as follows.

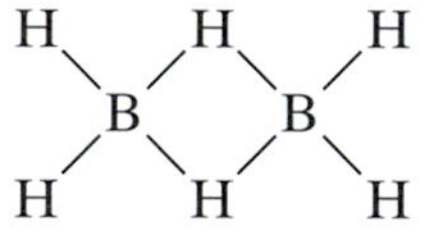

However, there are some things that are very wrong here. First, there are not enough valence electrons available in the molecule to form 8 bonds, as suggested by the structure. Each boron contributes three valence electrons and each hydrogen contributes one, so the total number of valence electrons in the molecule is 12. Forming eight ordinary covalent bonds would require 16 valence electrons. Furthermore, we know that the valence shell of hydrogen is the $1s$ subshell, which can accommodate a maximum of two electrons, not the four that are indicated for the bridging H atoms. How, then, are the bonds formed in this molecule?

Molecular orbital theory finds B_2H_6 as another success story. The bridging $B\!-\!H\!-\!B$ bonds are explained as follows. To have the tetrahedral arrangement around the boron atoms, each boron uses sp^3 hybrid orbitals. The four $B\!-\!H$ bonds that are not bridged are formed in the normal way—by the overlap of a hydrogen $1s$ orbital with an sp^3 hybrid orbital of boron. In each of the $B\!-\!H\!-\!B$ bridges, an sp^3 hybrid orbital from each of the boron atoms overlaps

Figure 21.13 *The molecular structure of diborane,* B_2H_6. This simplest hydrogen compound of boron is unusual because two of its hydrogen atoms are bonded to two different atoms simultaneously. (*a*) A ball-and-stick model. (*b*) A space-filling model.

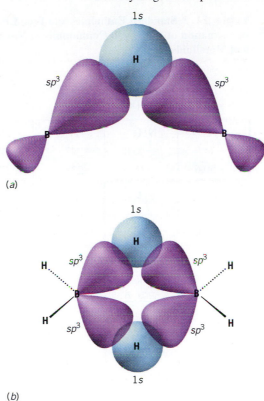

(a)

(b)

Figure 21.14 *Bonding in the hydrogen bridges of diborane, B_2H_6.* (a) The 1s orbital of a hydrogen atom overlaps simultaneously with two sp^3 hybrid orbitals, one from each of the boron atoms, to give a delocalized three-centered B—H—B bond. (b) The two three-center bonds in B_2H_6.

with the 1s orbital of a hydrogen atom, as shown in Figure 21.14. The result is a delocalized "three-center" bond that places about half of an electron pair between a bridging H atom and each of the boron atoms.

Boron forms quite a variety of hydrides, some of them fairly complex. We will not explore these further. However, one of their characteristic features is the existence of hydrogen atoms involved in multicentered bonds similar to those in B_2H_6.

Preparation of Hydrides

The hydrides of the nonmetallic elements are formed as products in many different chemical reactions; however, we will consider only two general methods of preparation here. One of these is by the direct combination of the elements, and the other is by the addition of protons (H^+) to the conjugate bases of the hydrides.

Hydrides Formed by Direct Combination of the Elements

Two simple examples of this kind of reaction occur when hydrogen is combined with chlorine or oxygen.

$$H_2(g) + Cl_2(g) \longrightarrow 2HCl(g)$$

$$2H_2(g) + O_2(g) \longrightarrow 2H_2O(g)$$

However, this method is not applicable to all of the hydrides, as we can see by examining some of their thermodynamic properties shown in Table 21.2. Here we see that only the hydrides of the more electronegative nonmetals possess

Table 21.2 Standard Enthalpies and Free Energies of Formation of Hydrogen Compounds of Nonmetals and Metalloids

$XH_n(g)$
ΔH_f° (kJ mol^{-1})
ΔG_f° (kJ mol^{-1})

CH_4	NH_3	H_2O	HF
−74.9	−46.0	−242	−271
−50.6	−16	−228	−273
SiH_4	PH_3	H_2S	HCl
+34	+5.4	−21	−92.5
+56.9	+13	−33	−95.4
GeH_4	AsH_3	H_2Se	HBr
+91	+66.5	+30	−36
+117	+69.0	+16	−53.6
	SbH_3	H_2Te	HI
	+145	+154	+26
	+148	+138	+2

negative free energies of formation. Those lying below the heavy line in the table have positive values for ΔG_f°, which means that the equilibrium constants for their formation from the elements are quite small. Therefore, from a practical standpoint, these hydrides cannot be made directly from the free elements. Instead an indirect procedure must be employed.

Even when thermodynamics favors the formation of the nonmetal hydride, the rate of formation can vary significantly. In period 2, for instance, fluorine reacts immediately with hydrogen when they are placed in contact. On the other hand, H_2 and O_2 mixtures are stable virtually indefinitely, unless the reaction is initiated in some way, for example, by applying heat or introducing a catalyst.

When chemists carry out reactions with oxygen-sensitive chemicals, they often fill their apparatus with N_2 gas. Because N_2 is so unreactive, there is little concern that this gas will participate in any reactions with the chemicals they are using.

Nitrogen, in many ways, appears to be unreactive compared with oxygen, not only toward hydrogen, but toward nearly all other chemical reagents as well. This is because of the high stability of the N_2 molecule, which is a result of its strong triple bond (the bond energy of N_2 is 946 kJ/mole, compared to 502 and 159 kJ/mole for O_2 and F_2, respectively). The large bond energy of N_2 becomes part of the activation energy for any reaction that requires splitting the N_2 molecule into nitrogen atoms.

The production of ammonia by the Haber process—a catalyzed reaction of N_2 with H_2—is undoubtedly one of the most important industrial chemical reactions. This is because nearly all useful nitrogen compounds, such as chemical fertilizers, can be prepared from NH_3 in one way or another. The Haber process, developed in Germany during World War I, employs the reaction

$$N_2(g) + 3H_2(g) \rightleftharpoons 2NH_3(g) \qquad \Delta H^\circ = -92 \text{ kJ}$$

As we can see, the reaction is exothermic. Applying Le Châtelier's principle, we can predict that the greatest yield of NH_3 would be achieved if the reaction is able to come to equilibrium at low temperature and high pressure. However, the reaction takes place very slowly at ordinary temperatures, even in the presence of a catalyst (iron containing a small amount of oxide serves as a heterogeneous catalyst). Therefore, a high temperature is used, even though the amount of NH_3 produced is somewhat lessened. The actual conditions that are employed are a pressure of approximately 1000 atm and a temperature of 400 to 500 °C.

Although CH_4 has a negative free energy of formation, the direct combination of carbon and hydrogen to form this compound is not easily accomplished. In fact, considerable effort has been expended by industrial chemists to find ways to combine elemental carbon and hydrogen to form hydrocarbon compounds, since doing so would provide a way of converting solid fuels such as coal into easily handled gaseous or liquid fuels.

Preparing Hydrides by Brønsted Acid–Base Reactions

Another method of preparing nonmetal hydrides involves the addition of protons, from a Brønsted acid, to the conjugate base of a nonmetal hydride, a reaction that we might depict as

$$X^{n-} + nHA \longrightarrow H_nX + nA^-$$

where X^{n-} is the conjugate base of the hydride H_nX and HA is the Brønsted acid. Let's look at some examples.

The hydrogen halides are commonly prepared in the laboratory by treating a halide salt with a nonvolatile acid such as sulfuric or phosphoric acid.

$$NaCl(s) + H_2SO_4(l) \longrightarrow HCl(g) + NaHSO_4(s)$$

$$NaCl(s) + H_3PO_4(l) \longrightarrow HCl(g) + NaH_2PO_4(s)$$

The reaction of sodium chloride with concentrated sulfuric acid. (*Left*) A drop of concentrated sulfuric acid is added to solid NaCl in the bottom of a flask. (*Right*) HCl gas leaving the mouth of the flask contacts a moist strip of blue litmus paper. The HCl dissolves in the water in the strip of litmus paper to give hydrochloric acid, which changes the litmus from blue to red.

The reaction of sodium iodide with concentrated sulfuric acid. (Left) Solid sodium iodide. (*Right*) A drop of concentrated sulfuric acid has been added, causing iodide ion to be oxidized to iodine, I_2. The molecular I_2 reacts with additional I^- to give triiodide ion, I_3^-, which has a red-brown color.

In these examples HCl is removed as a gas, which causes the reaction to proceed to completion. However, the reaction proceeds quite quickly and easily when H_2SO_4 is used, but the mixture must be heated to drive off HCl when H_3PO_4 is used.

With the heavier halogens, bromine and iodine, sulfuric acid cannot be used because it is a sufficiently strong oxidizing agent to oxidize the halide ion to the free halogen. For example, when treated with concentrated H_2SO_4, iodide ion reacts as follows:

$$2I^- + HSO_4^- + 3H^+ \longrightarrow I_2 + SO_2 + 2H_2O$$

Phosphoric acid, being a much weaker oxidizing agent than H_2SO_4, simply supplies protons to I^-, and HI can therefore be produced in a reaction analogous to the production of HCl above, that is,

$$NaI(s) + H_3PO_4(l) \longrightarrow HI(g) + NaH_2PO_4(s)$$

In reactions such as these, the nature of the Brønsted acid that is required to form the nonmetal hydride depends on the Brønsted base strength of the nonmetal anion—the stronger the base, the weaker need be the Brønsted acid. With this in mind, let's apply the trends in acid–base strengths that we discussed in Chapter 15 to an understanding of these reactions.

You learned in Chapter 15 that the strengths of binary acids (i.e., the nonmetal hydrides) increase from left to right across a period. Thus, for the hydrides CH_4, NH_3, H_2O, and HF, the trend in acid strength is as follows:

Weakest acid			Strongest acid
CH_4	NH_3	H_2O	HF

$$\longrightarrow$$

Acid strength increases

You also learned that there is a reciprocal relationship between the strength of a Brønsted acid and its conjugate base—the weaker the acid, the stronger is the conjugate base. This means that the strengths of the conjugate bases of the nonmetal hydrides increase from right to left, so comparing the bases CH_3^-, NH_2^-, OH^-, and F^-, the trend in base strength is

$$\begin{array}{cccc} \text{Strongest} & & & \text{Weakest} \\ \text{base} & & & \text{base} \\ CH_3^- & NH_2^- & OH^- & F^- \end{array}$$

$$\xleftarrow{\hspace{3cm}}$$
$$\text{Base strength} \\ \text{increases}$$

Removing additional protons from any of these bases yields ions that are even stronger bases, so there is a large increase in base strength from right to left among the anions of the nonmetals.

$$\begin{array}{cccc} \text{Very strong} & & & \text{Weak} \\ \text{base} & & & \text{base} \\ C^{4-} & N^{3-} & O^{2-} & F^- \end{array}$$

$$\xleftarrow{\hspace{3cm}}$$
$$\text{Base strength} \\ \text{increases}$$

Because HF is fairly acidic, the anion F^- is a weak base and the production of HF requires a strong acid such as H_2SO_4. Oxide ion, on the other hand, is a much stronger base than F^-, and when treated with even a relatively weak source of protons, oxide ion gobbles them up. Thus, oxide ion reacts with water to give OH^- in a reaction that is sometimes called *hydrolysis*. (In general, **hydrolysis** means *reaction with water*.)

$$O^{2-} + H_2O \longrightarrow 2OH^-$$

It also reacts with weak or strong acids to give water.

$$O^{2-} + 2H^+ \longrightarrow H_2O$$

As you learned in Chapter 4, even insoluble metal oxides react with weak or strong acids to give water and the metal salt. For example,

$$MgO(s) + 2HC_2H_3O_2(aq) \longrightarrow Mg^{2+}(aq) + 2C_2H_3O_2^-(aq) + H_2O$$

Nitride ion, N^{3-}, is expected to be an even stronger base than O^{2-}. Therefore, it is not surprising to find that Mg_3N_2 reacts with the very weak acid, H_2O, to produce NH_3.

$$Mg_3N_2(s) + 6H_2O \longrightarrow 3Mg(OH)_2(s) + 2NH_3(aq)$$

Metal carbides, which can be prepared by heating an active metal with carbon, react with water (hydrolyze) in the same fashion. Aluminum carbide, for instance, which contains C^{4-} ions, reacts according to the equation,

$$Al_4C_3(s) + 12H_2O \longrightarrow 4Al(OH)_3(s) + 3CH_4(g)$$

There are also carbides in which the anion is acetylide ion, C_2^{2-}. An example is calcium carbide, CaC_2. The C_2^{2-} ion has the structure

$$[:C\equiv C:]^{2-}$$

and on hydrolysis yields acetylene, C_2H_2, $H-C\equiv C-H$.

$$CaC_2(s) + 2H_2O \longrightarrow Ca(OH)_2(s) + C_2H_2(g)$$

A close-up view of two pieces of magnesium oxide, obtained by burning magnesium metal in air. The reaction gives mostly MgO, but some Mg_3N_2 also forms by reaction with nitrogen in the air. When a drop of water containing phenolphthalein indicator was added to the magnesium oxide at the top, the Mg_3N_2 in it reacted with the water to give insoluble $Mg(OH)_2$ and NH_3. Being a weak base, the NH_3 caused the phenolphthalein to turn pink.

Long ago, lamps on carriages and early automobiles used a flame of burning acetylene, formed by the slow drip of water on calcium carbide, to provide light. The acetylene used in welding torches is also made this way and then stored dissolved in acetone in pressurized cylinders.

This general method of preparation of hydrides also extends to the third, fourth, and fifth periods, with the same trends in the strength of the Brønsted acid required to liberate the hydride. In period 3 we have these anions:

Group	IVA	VA	VIA	VIIA
Anion	Si^{4-}	P^{3-}	S^{2-}	Cl^-

We again expect the anions to become increasingly basic as we move from right to left (from Cl^- to Si^{4-}). Therefore, the strength of the Brønsted acid needed to protonate the anion decreases. To form HCl from NaCl, a strong acid is required. Sulfide ion, on the other hand, is sufficiently basic to be highly hydrolyzed in aqueous solution (as we found in Chapter 17), and solutions containing a soluble sulfide such as Na_2S always have an odor of H_2S because of the reaction

$$HS^-(aq) + H_2O \rightleftharpoons H_2S(g) + OH^-(aq)$$

Many insoluble metal sulfides dissolve in acids with the evolution of H_2S.

Phosphides, like sulfides, also hydrolyze on contact with water. However, because the P^{3-} ion is more basic than the S^{2-} ion, the hydrolysis proceeds essentially to completion. Thus, aluminum phosphide, AlP, reacts with water to produce phosphine, PH_3.

$$AlP(s) + 3H_2O \longrightarrow Al(OH)_3(s) + PH_3(g)$$

Moving left to Group IVA, we again find that a hydrolysis reaction serves to prepare silicon hydrides. A metal silicide such as Mg_2Si (which can be formed by simply heating Mg and Si together) reacts with water to generate a mixture of silicon–hydrogen compounds called **silanes**: SiH_4, Si_2H_6, Si_3H_8, etc., up to Si_6H_{14}.

The heavier nonmetals behave in much the same fashion as those above them. Thus H_2Se and H_2Te, like H_2S, can be prepared by adding an acid to a metal selenide or telluride. Arsine, AsH_3, like phosphine, PH_3, is made by the hydrolysis of a metal arsenide such as Na_3As or AlAs, and the germanes GeH_4, Ge_2H_6, and Ge_3H_8 are produced by the action of dilute HCl on Mg_2Ge.

21.4 Catenation between Nonmetal Atoms

Catenation is the ability of atoms of the same element to become covalently bound to each other. A simple example is hydrogen peroxide, H_2O_2, which has the Lewis structure

$$\begin{array}{c} H \\ | \\ :\ddot{O}-\ddot{O}: \\ | \\ H \end{array}$$

Many such compounds are hydrides, and some additional examples are given in Table 21.3. Some other examples are[2]

---

[2]The halogens do not exhibit this property except in the simple elemental state such as F_2, Cl_2, etc., and in the triiodide ion, I_3^-.

Table 21.3 Catenation among Nonmetal Hydrides

Group IVA	CH_4	SiH_4	GeH_4	SnH_4
	C_2H_6	Si_2H_6	Ge_2H_6	Sn_2H_6
	C_3H_8	Si_3H_8	Ge_3H_8	
	$\vdots$	$\vdots$		
	C_nH_{2n+2} and many others	Si_6H_{14}		
Group VA	NH_3	PH_3	AsH_3	SbH_3
	N_2H_4	P_2H_4		
Group VIA	H_2O	H_2S	H_2Se	H_2Te
	H_2O_2	H_2S_2		
		$\vdots$		
		H_2S_6		

hydrazine   disilane   propane

The ability of the nonmetallic elements to form compounds in which they bond to other like atoms varies greatly. You will notice, for example, that in Group VIA only oxygen and sulfur form such compounds. In Group VA we find that both nitrogen and phosphorus catenate, but the chain length seems to be limited to two atoms. When we proceed to Group IVA, all of the elements, down to and including tin, exhibit this property, and here we find chains containing three, four, and even more atoms. We also see that the tendency toward catenation generally decreases downward in a group, as evidenced by the trend toward shorter chains demonstrated by the heavier elements in Group IVA, Ge and Sn.

Of all of the elements, carbon has the greatest capacity to form bonds to itself. In fact, the broad area of organic chemistry is concerned with hydrocarbons and compounds that are derived from them by substituting other elements for hydrogen. Organic compounds, then, are compounds in which the molecular framework consists primarily of carbon–carbon chains. The unique ability of carbon to form such diverse compounds containing these long, stable carbon chains may be the principal reason why life has evolved around the element carbon instead of around another element such as silicon. We will discuss organic chemistry in considerably greater detail in Chapter 23.

Catenation is not restricted to the nonmetal hydrides alone; it occurs in other compounds as well. Sulfur, for example, has a rather marked tendency to form bonds to other sulfur atoms, as we saw in our discussion of the free element. This carries over to its compounds too. For instance, when an aqueous solution containing sulfide ion, S^{2-}, is heated with elemental sulfur, a series of **polysulfide ions,** S_2^{2-}, S_3^{2-}, ... , S_x^{2-}, are formed. A typical example might be as follows:

In this example, an S_8 ring is attacked by several sulfide ions and is broken into fragments that become part of the various polysulfide ions. Addition of a strong acid to a solution of polysulfides produces the corresponding hydrides (H_2S_2, H_2S_3, . . . , H_2S_x).

Another similar reaction occurs when an aqueous solution containing sulfite ion is heated with sulfur. This reaction produces thiosulfate ion,[3] $S_2O_3^{2-}$.

$$\left[\begin{array}{c} :\ddot{O}: \\ \| \\ :\ddot{S}-S-\ddot{O}: \\ | \\ :\ddot{O}: \end{array} \right]^{2-}$$

In the Lewis structures for these oxoanions of sulfur, we have selected those with minimum formal charges.

Mild oxidation of thiosulfate ion produces a link between the sulfur atoms to give another catenated sulfur species, tetrathionate ion, $S_4O_6^{2-}$, which has the structure

$$\left[\begin{array}{c} :\ddot{O}: \qquad\quad :\ddot{O}: \\ \| \qquad\qquad \| \\ :\ddot{O}-S-\ddot{S}-\ddot{S}-S-\ddot{O}: \\ \| \qquad\qquad \| \\ :\ddot{O}: \qquad\quad :\ddot{O}: \end{array} \right]^{2-}$$

Tetrathionate ion is formed in the reaction of iodine with thiosulfate ion, a useful redox reaction in analytical chemistry.

The tetrathionate ion is one member of a family of **polythionates,** which contain sulfur–sulfur bonds. Some other oxoanions with linked sulfur atoms are dithionate ($S_2O_6^{2-}$), trithionate ($S_3O_6^{2-}$), pentathionate ($S_5O_6^{2-}$), and hexathionate ($S_6O_6^{2-}$). Typical structures are

$$\left[\begin{array}{c} :\ddot{O}: \quad :\ddot{O}: \\ \| \qquad \| \\ :\ddot{O}-S-\!-\!-S-\ddot{O}: \\ \| \qquad \| \\ :\ddot{O}: \quad :\ddot{O}: \end{array} \right]^{2-} \qquad \left[\begin{array}{c} :\ddot{O}: \qquad\qquad :\ddot{O}: \\ \| \qquad\qquad\qquad \| \\ :\ddot{O}-S-\ddot{S}-\ddot{S}-\ddot{S}-S-\ddot{O}: \\ \| \qquad\qquad\qquad \| \\ :\ddot{O}: \qquad\qquad :\ddot{O}: \end{array} \right]^{2-}$$

dithionate ion   pentathionate ion

In addition to sulfur, some other common nonmetals also form catenated species that are not hydrides. Some examples are the following:

1. Carbon. For example, oxalic acid:

$$\begin{array}{c} :\ddot{O}: \;\; :\ddot{O}: \\ \| \qquad \| \\ H\ddot{O}-C-C-\ddot{O}H \end{array}$$

2. Nitrogen. For example, azides containing the ion N_3^- (derived from hydrazoic acid, HN_3):

$$\left[\ddot{N}\!=\!N\!=\!\ddot{N} \right]^-$$

3. Oxygen. For example, peroxides such as peroxymonosulfate ion, SO_5^{2-}, and peroxydisulfate ion, $S_2O_8^{2-}$, which may be considered to be derived, at least in a formal way, from H_2O_2.

$$\left[\begin{array}{c} :\ddot{O}: \\ \| \\ :\ddot{O}-S-\ddot{O}-\ddot{O}: \\ \| \\ :\ddot{O}: \end{array} \right]^{2-} \qquad \left[\begin{array}{c} :\ddot{O}: \qquad\qquad :\ddot{O}: \\ \| \qquad\qquad\qquad \| \\ :\ddot{O}-S-\ddot{O}-\ddot{O}-S-\ddot{O}: \\ \| \qquad\qquad\qquad \| \\ :\ddot{O}: \qquad\qquad :\ddot{O}: \end{array} \right]^{2-}$$

peroxymonosulfate ion   peroxydisulfate ion

[3]In general, the prefix *thio-* implies substitution of sulfur for oxygen. Thus the thiosulfate ion ($S_2O_3^{2-}$) is a sulfate ion (SO_4^{2-}) in which an oxygen atom has been replaced by a sulfur atom. Similarly, thiocyanate ion is SCN^- whereas cyanate ion is OCN^-.

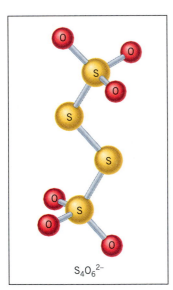

Structure of the tetrathionate ion, $S_4O_6^{2-}$.

21.5 Oxygen Compounds of Nonmetals and Metalloids

The oxides and those compounds derived from them, the oxoacids (acids containing oxygen) and salts of oxoanions (anions containing oxygen), are among the most important compounds of the nonmetals. Here we find many of the common reagents we use in the laboratory, for example, acids such as H_2SO_4 and HNO_3 and their salts, the sulfates and nitrates. In addition, most minerals are either oxides, such as SiO_2 (silica) and Fe_2O_3 (hematite), or contain oxoanions of nonmetals, for example, the carbonates (limestone) and silicates (asbestos, mica, and others).

We have already seen that nonmetal oxides exhibit acidic properties. In many cases, the oxide simply combines with water to form an oxoacid, as with CO_2 and SO_3.

$$CO_2(aq) + H_2O \longrightarrow H_2CO_3(aq)$$

$$SO_3(g) + H_2O \longrightarrow H_2SO_4(aq)$$

You also should recall that neutralization of these acids gives their corresponding anions.

Not all nonmetal oxides combine with water as illustrated above. Some oxides, like CO and NO, are inert toward water and do not yield acidic aqueous solutions. In some cases, such as SiO_2, the oxide is insoluble in water and cannot react with it. And sometimes the oxide does react with water but undergoes a redox reaction to give an acid in which the oxidation number of the nonmetal has changed. Nitrogen dioxide, for example, undergoes a reaction with water in which some of the NO_2 is oxidized while the rest is reduced.

$$3NO_2(g) + H_2O \longrightarrow 2HNO_3(aq) + NO(g)$$

This kind of reaction, in which the same reactant is both the oxidizing agent and reducing agent, is called a **disproportionation** reaction.

This reaction occurs when rainwater passes through air polluted with NO_2 and is one of the reactions responsible for acid rain.

Formation of Oxides of the Nonmetallic Elements

Table 21.4 contains a list of many of the oxides of the nonmetals and metalloids. As with the hydrides, there is more than one way such compounds can be made. Our goal here is to study some of them.

Table 21.4 Typical Oxides of the Nonmetals and Metalloids

Group IIIA	Group IVA	Group VA	Group VIA	Group VIIA
B_2O_3	CO, CO_2	N_2O, NO, N_2O_3, NO_2, N_2O_4, N_2O_5	O_2, O_3	OF_2, O_2F_2
	SiO_2	P_4O_6, P_4O_{10}	SO_2, SO_3	Cl_2O, ClO_2, Cl_2O_7
	GeO_2	As_4O_6, As_4O_{10}	SeO_2, SeO_3	Br_2O, BrO_2
		Sb_4O_6, Sb_4O_{10}	TeO_2, TeO_3	I_2O_5, I_2O_7

Table 21.5 Thermodynamic Properties of Some Nitrogen Oxides

Oxide	ΔH_f° (kJ mol^{-1})	ΔG_f° (kJ mol^{-1})
$N_2O(g)$	+81.5	+104
$NO(g)$	90.4	86.8
$NO_2(g)$	38	51.9
$N_2O_4(g)$	9.7	98.3
$N_2O_5(g)$	11	115

Combustion of sulfur. Sulfur burns with a pale blue flame to give sulfur dioxide, SO_2.

When ΔG° is positive, the equilibrium constant has a value smaller than 1.0 and the reaction does not proceed far toward completion at equilibrium.

Direct Combination of the Elements

One method that can be used in most cases is simply the direct union of the elements, as typified by the following reactions.

$$S(s) + O_2(g) \longrightarrow SO_2(g)$$

$$C(s) + \tfrac{1}{2}O_2(g) \longrightarrow CO(g)$$

$$C(s) + O_2(g) \longrightarrow CO_2(g)$$

Not all oxides can be prepared this way, however. For example, in Table 21.5 we see that many of the oxides of nitrogen have positive free energies of formation. This means the equilibrium constants for their formation from the elements are small, so making them by direct combination of the elements is impractical.[4] Obtaining them in substantial quantity, therefore, requires indirect procedures.

Oxidation of a Lower Oxide by Oxygen

In some cases an oxide can be prepared from a lower oxide by further reaction with oxygen. The synthesis of SO_3, for example, consists of catalytic oxidation of SO_2.

$$2SO_2(g) + O_2(g) \longrightarrow 2SO_3(g)$$

This very thermodynamically favorable reaction ($\Delta G^\circ = -140$ kJ) is slow under ordinary conditions but proceeds rapidly in the presence of a catalyst. Industrially, this oxidation is important because it is a critical step in the synthesis of sulfuric acid, the chemical with the largest annual industrial production worldwide. This reaction is also promoted in catalytic mufflers on cars and presents one of the "down sides" to this form of air pollution control. This is because the SO_3 that's formed can react with water to form the strong acid H_2SO_4, whereas SO_2 forms weakly acidic solutions of sulfurous acid. The result is a more acidic "acid rain."

In the late 1990s, the annual production of sulfuric acid topped 95 billion pounds (440 million moles)!

The combustion of CO is also an important industrial reaction because CO is often used as a fuel.

$$2CO(g) + O_2(g) \longrightarrow 2CO_2(g)$$

[4]Because their heats of formation are positive, more of the nitrogen oxides should be formed as the temperature is increased. This is predicted by Le Châtelier's principle. Environmentally, this is a problem because small amounts of NO are formed inside gasoline and diesel engines at the high temperatures involved in combustion of the fuel. Although thermodynamics suggests that NO ought to decompose when it reaches ordinary temperatures, the rate of decomposition is so slow that the NO is able to accumulate as an air pollutant. One of the functions of the catalytic converters on cars is to hasten the decomposition of NO back to N_2 and O_2 so it can't escape to pollute the air.

When white-hot carbon is treated with steam, an endothermic reaction takes place producing a mixture of CO and H_2, both of which can burn. This mixture is called water gas.

$$C(s) + H_2O(g) \longrightarrow CO(g) + H_2(g)$$

Reactions of Nonmetal Hydrides with Oxygen

Another technique that serves to produce oxides is the combustion of hydrides of the nonmetallic elements. As you learned earlier, methane (the chief constituent of natural gas) and other hydrocarbons burn to produce CO_2 and H_2O when an excess of O_2 is present.

$$CH_4(g) + 2O_2(g) \longrightarrow CO_2(g) + 2H_2O(l)$$

$$2C_8H_{18}(l) + 25O_2(g) \longrightarrow 16CO_2(g) + 18H_2O(l)$$
octane
(in gasoline)

Recall that when insufficient O_2 is available (as in an automobile engine), some CO may be produced along with CO_2.

A reaction of this general type, which is of great commercial importance, is the oxidation of ammonia. In this case a platinum catalyst is used and the reaction is

$$4NH_3(g) + 5O_2(g) \xrightarrow{\text{catalyst}} 4NO(g) + 6H_2O(g)$$

The NO formed in this reaction is readily oxidized further to produce NO_2.

$$2NO(g) + O_2(g) \longrightarrow 2NO_2(g)$$

As noted earlier, NO_2 disproportionates when dissolved in water to yield HNO_3 and NO.

$$3NO_2(g) + H_2O \longrightarrow 2HNO_3(aq) + NO(g)$$

The commercial application of this sequence of reactions accounts for the major source of nitric acid and nitrates used in the manufacture of explosives, fertilizers, plastics, and many other useful substances. In fact, the development of this process in Germany by Wilhelm Ostwald, accompanied by the successful preparation of NH_3 from N_2 and H_2 by Haber, are said to have prolonged World War I, because the Allied blockade of Germany was unable to halt the German manufacture of munitions which had depended, prior to these processes, on the importation of nitrates from other countries.

Other Oxidation–Reduction Reactions

Nonmetal oxides are formed in many redox reactions in which the oxidizing agent is an oxoacid. For instance, when nitric acid serves as an oxidizing agent, the nitrate ion is reduced, and depending on conditions, nitrogen in any oxidation state can be produced. When concentrated nitric acid is used, the reduction product is frequently NO_2, whereas dilute solutions of HNO_3 often yield NO as the reduction product.

$$4HNO_3(aq) + Cu(s) \longrightarrow Cu(NO_3)_2(aq) + 2NO_2(g) + 2H_2O \quad \text{(concentrated)}$$

$$8HNO_3(aq) + 3Cu(s) \longrightarrow 3Cu(NO_3)_2(aq) + 2NO(g) + 4H_2O \qquad \text{(dilute)}$$

Similarly, hot concentrated sulfuric acid is a fairly potent oxidizing agent, the reduction product usually being SO_2. For example,

$$Cu(s) + 2H_2SO_4(l) \longrightarrow CuSO_4(aq) + SO_2(g) + 2H_2O$$

Molecular Structures of the Oxides

The nature of the oxides of the nonmetallic elements is determined to a significant degree by the ability of the nonmetallic element to form multiple bonds, in this case to oxygen atoms. In our earlier discussion of the structures of the free elements, we saw that period 2 elements have a substantial tendency to enter into such bonding. This carries through to the oxides as well, and the oxides of the period 2 elements are simple monomeric species. Carbon, for example, forms CO and CO_2, whose Lewis structures are

$$:C\equiv O: \qquad :\ddot{O}=C=\ddot{O}:$$

The most common oxides of nitrogen are NO and NO_2. Both have an odd number of electrons and therefore are not able to satisfy the octet rule, since at least one atom must be left with an odd number of electrons. The structure of the NO molecule is sometimes represented as a resonance hybrid of the structures

$$\cdot\ddot{N}=\ddot{O}: \qquad \longleftrightarrow \qquad :\ddot{N}=\ddot{O}\cdot$$

Molecular orbital theory gives a better picture of this molecule, which we might compare to the species O_2^+. These two (NO and O_2^+) are **isoelectronic,** that is, they have the same number of electrons. Therefore, because N and O are neighbors in the periodic table, we might expect that electronically NO and O_2^+ would not be too different. The molecular orbital energy-level diagram for O_2^+, obtained by removing one of the antibonding electrons from O_2 (Table 9.1, page 408), is shown in Figure 21.15a. The corresponding energy diagram for NO is found in Figure 21.15b.

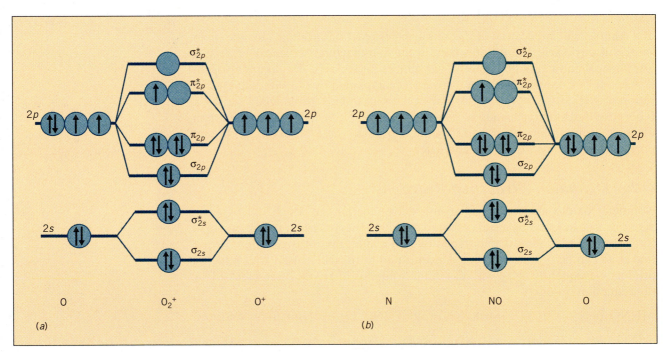

Figure 21.15 *Molecular orbital energy diagrams for O_2^+ and NO. (a) Energy diagram for O_2^+. (b) Energy diagram for NO. Notice that the oxygen atomic orbitals have a lower energy than the nitrogen atomic orbitals.*

The molecule NO_2 is another interesting species. Using Lewis formulas, we can represent its structure by two resonance forms, neither of which is capable of fully satisfying the octet rule.

In each of these structures an unpaired electron resides on the nitrogen atom, and in samples of NO_2 there is a dimerization equilibrium,

$$2NO_2(g) \rightleftharpoons N_2O_4(g)$$
brown   colorless

A **dimer** is a molecule formed by joining two identical smaller molecules.

in which colorless N_2O_4 (dinitrogen tetroxide) is formed from the brown NO_2. The N_2O_4 molecule has the structure

(one of four resonance structures)

In N_2O_4, the unpaired electrons of two NO_2 molecules have paired to form a co-valent bond between the two nitrogen atoms. Liquid N_2O_4 is an effective oxidizing agent and has been used as the oxidant in liquid-fuel rocket engines.

In period 3 there is considerably less tendency for the nonmetals to enter into multiple bonding. Sulfur, though, does appear able to do so, as evidenced by the oxides SO_2 and SO_3. The Lewis structures having minimum formal charges on the atoms are

With phosphorus there is even less tendency toward multiple bonding, and when phosphorus reacts with oxygen two oxides are formed, depending on reaction conditions.

$$P_4(s) + 3O_2(g) \longrightarrow P_4O_6(s)$$

$$P_4(s) + 5O_2(g) \longrightarrow P_4O_{10}(s)$$

The structures of these oxides bear a remarkably simple relationship to the elemental P_4 unit. As you learned earlier (Figure 21.10) the P_4 molecule is tetrahedral, with each phosphorus atom bonded to three others. In Figure 21.16a is

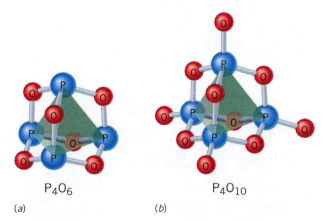

P_4O_6

(a)

P_4O_{10}

(b)

Figure 21.16 *Molecular structures of the oxides of phosphorus. (a) The structure of P_4O_6. (b) The structure of P_4O_{10}.*

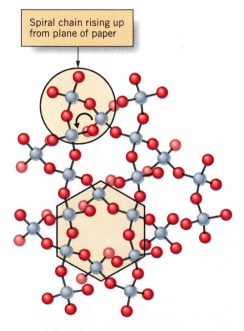

Figure 21.17 *The molecular structure of quartz, a crystalline form of SiO$_2$.* Each silicon atom is bonded to four oxygen atoms at the corners of a tetrahedron. The oxygen atoms are each bonded to another silicon to give a three-dimensional network.

Crystals of quartz. Quartz is composed of silicon dioxide, SiO$_2$.

shown the structure of P$_4$O$_6$ in which we find an oxygen atom inserted between phosphorus atoms along each of the six edges of the P$_4$ tetrahedron. In the P$_4$O$_{10}$ molecule, Figure 21.16*b*, there is an additional oxygen atom attached to each of the P atoms at the vertices of the tetrahedron.

Silicon, in Group IVA, forms only one oxide, SiO$_2$. Silicon dioxide, while having the same empirical formula as carbon dioxide, is nevertheless quite different structurally because silicon has virtually no tendency to form Si—O multiple bonds. In quartz, and other forms of SiO$_2$, each Si atom is surrounded tetrahedrally by four oxygen atoms, each of which is bound to another Si atom. The resulting structure, a portion of which is shown in Figure 21.17, is quite complex, extending in a network fashion in three dimensions. Overall, the structure has a hexagonal nature to it, as illustrated in Figure 21.18, and crystals of quartz are often hexagonal in cross section.

Figure 21.18 *Two features of the quartz structure.* The structure consists of spiral —Si—O—Si—O—Si—O— chains linked side by side. In this diagram, the spiral chains rise up from the plane of the paper. The overall hexagonal nature of the structure is also indicated.

One of the interesting features of quartz is that two kinds of crystals are formed which are nonsuperimposable mirror images of each other. The reason for this is that the quartz structure is built up of parallel —Si—O—Si—O—Si— spiral chains, one of which is outlined by the circle in Figure 21.18. These helical chains are linked together side by side to give the overall structure of the quartz crystal. Helices, like the threads on a screw, have a natural "handedness" to them, so that they either twist to the left or to the right (Figure 21.19). This makes the helices chiral, and within a crystal of quartz, all the spirals twist in the same direction, so the quartz crystal structure is chiral overall. The chirality gives crystals of quartz the ability to rotate the plane of polarized light—to the left for one kind of crystal and to the right for the other kind. It is interesting to note, however, that when quartz is melted, the chiral helices collapse and liquid SiO_2 is not optically active.

d-quartz   *l*-quartz

Figure 21.19 *Chiral nature of quartz crystals.* In quartz, the —O—Si—O—Si— helices can twist either clockwise or counterclockwise. This imparts a chiral character to a quartz crystal, and two different kinds of crystals (*d*-quartz and *l*-quartz) can be found that are nonsuperimposable mirror images of each other.

21.6 Oxoacids and Oxoanions of the Nonmetallic Elements

The oxoacids and their anions can be divided into two categories, simple monomeric acids and anions containing one atom of the nonmetal (Table 21.6), and complex polymeric acids and anions. We will study the simple species first, looking for trends and similarities in structure and properties where we can.

In general, the oxoacids consist of an atom of nonmetal or metalloid to which is bonded one or more —OH groups plus, perhaps, additional oxygen atoms not bonded to hydrogen, to give a generalized formula

$$XO_m(OH)_n$$

Some typical examples are

The ability of chiral substances to affect polarized light was discussed in Facets of Chemistry 20.1 on page 928.

$$PO(OH)_3$$
$$H_3PO_4$$

$$SO_2(OH)_2$$
$$H_2SO_4$$

$$NO_2(OH)$$
$$HNO_3$$

In Chapter 15 we saw that polarization of the O—H bond of the hydroxyl group by the central atom permits the H atom to be removed, more or less readily, as an H^+ ion. To reflect this, the formulas of the acids are written with the hydrogens first, followed by the formula of the anion that remains when the protons are removed by neutralization.

The geometric structures of the monomeric acids and anions, like those of the simple oxides, are readily predicted using the VSEPR theory. Rather than discuss them at length here, try Problems 21.96 and 21.97 at the end of the chapter.

Oxoacids of Period 2 Elements

Period 2 elements cannot have more than eight electrons in their valence shells, which limits the oxoacids they can form. Let's take a brief look at them going from right to left in the period.

For fluorine, only one oxoacid, HOF, is known. It is unstable and difficult to prepare. In part, this is related to the electronegativity of fluorine, which is

Table 21.6 Monomeric Oxoacids and Oxoanions of the Nonmetallic Elements

Group IIIA	Group IVA	Group VA	Group VIA	Group VIIA
H_3BO_3 (no simple borates)	H_2CO_3 (CO_3^{2-})	HNO_2 (NO_2^-) HNO_3 (NO_3^-)		HOF
	$H_4SiO_4{}^a$ (SiO_4^{4-})	H_3PO_3 (HPO_3^{2-}) H_3PO_4 (PO_4^{3-})	$SO_2(aq)^b$ (SO_3^{2-}) H_2SO_4 (SO_4^{2-})	HOCl (OCl^-) $HClO_2$ (ClO_2^-) $HClO_3$ (ClO_3^-) $HClO_4$ (ClO_4^-)
	$H_4GeO_4{}^a$ (GeO_4^{4-})	H_3AsO_3 (AsO_3^{3-}) H_3AsO_4 (AsO_4^{3-})	H_2SeO_3 (SeO_3^{2-}) H_2SeO_4 (SeO_4^{2-})	HOBr (OBr^-) $HBrO_2$ (BrO_2^-) $HBrO_3$ (BrO_3^-) $HBrO_4$ (BrO_4^-)
			H_2TeO_3 (TeO_3^{2-}) $Te(OH)_6$ $[TeO(OH)_5^-]$	HOI (OI^-) HIO_3 (IO_3^-) H_5IO_6 $(H_2IO_6^{3-})$ HIO_4 (IO_4^-)

^{a}Not observed to exist.
$^b H_2SO_3$ does not exist in aqueous solutions of SO_2.

greater than that of oxygen. If we assign oxidation numbers as usual, we are forced to give oxygen an oxidation number of zero instead of its normal value of -2 in other oxoacids.

$$\underset{+1}{H}\!-\!\underset{0}{\overset{..}{\underset{..}{O}}}\!-\!\underset{-1}{\overset{..}{\underset{..}{F}}}\!:$$

Moving to the left, the next element to form oxoacids is nitrogen. This element forms two oxoacids, nitric acid (HNO_3) and nitrous acid (HNO_2). Their structures have been discussed earlier (page 678), and as you know, HNO_3 is a strong acid whereas HNO_2 is weak.

Carbon forms just one oxoacid, H_2CO_3, by the reaction of CO_2 with water. As you probably recall, it is a weak diprotic acid and forms both bicarbonate,

HCO_3^-, and carbonate, CO_3^{2-}, salts. Bicarbonates are only isolated for mono-valent cations, however. Dipositive cations such as Ca^{2+} precipitate as carbon-ates by loss of H_2O and CO_2 when solutions of their bicarbonate salts are evaporated.

$$Ca^{2+}(aq) + 2HCO_3^-(aq) \xrightarrow{\text{evaporation}} CaCO_3(s) + H_2O(g) + CO_2(g)$$

This is what is responsible for formation of limestone deposits in caverns.

Boron, in Group IIIA, forms boric acid, $B(OH)_3$. However, this is not a Brønsted acid like the other oxoacids we've studied. Rather, it behaves as a Lewis acid. The chemical equation for the reaction is

$$B(OH)_3 + H_2O \rightleftharpoons B(OH)_4^- + H^+$$

which can be diagrammed using Lewis structures as

We see that loss of a proton by the water molecule bound to the boron makes the solution somewhat acidic.

When boric acid reacts with water, the OH groups attached to boron do not transfer H^+ ions to water molecules. Boric acid is different from other oxoacids in this respect.

Oxoacids of Elements in Periods 3 and Below

Once again, let's work from right to left in the periodic table, and begin by studying the oxoacids of the halogens. With the exception of fluorine, the halo-gens react with water in a **disproportionation** reaction—a reaction in which a portion of a reactant is oxidized while the rest is reduced.[5]

$$X_2 + H_2O \rightleftharpoons HOX + H^+ + X^-$$

Thus Cl_2 gives hypochlorous acid, HOCl, and hydrochloric acid, HCl; Br_2 yields HOBr and HBr, and so forth. In the presence of a base, we expect the corre-sponding acids to be neutralized and the oxoanions to be formed, thus driving these reactions toward completion. With chlorine, for example, we have

$$Cl_2 + 2OH^- \longrightarrow OCl^- + Cl^- + H_2O$$

and, in general, there is the reaction

$$X_2 + 2OH^- \longrightarrow OX^- + X^- + H_2O$$

However, here we have an interesting case of kinetics influencing the products of a chemical reaction.

The hypohalite ions themselves (OCl^-, OBr^-, and OI^-) have a tendency to disproportionate to give the halate (XO_3^-) and halide (X^-) ions.

$$3OX^- \longrightarrow XO_3^- + 2X^-$$

The equilibrium constants for this reaction with the different halogens are

$$3OCl^- \longrightarrow ClO_3^- + 2Cl^- \qquad K_c \approx 10^{27}$$

$$3OBr^- \longrightarrow BrO_3^- + 2Br^- \qquad K_c \approx 10^{15}$$

$$3OI^- \longrightarrow IO_3^- + 2I^- \qquad K_c \approx 10^{20}$$

[5]In this reaction, the halogen, X_2, is both the oxidizing and reducing agent.

The large values for these equilibrium constants suggest that all of the OX^- ions should be transformed rather completely to the corresponding XO_3^- ions. Solutions of hypochlorite ion, however, are reasonably stable when kept cool, although disproportionation to ClO_3^- does occur if a solution of hypochlorite is heated. Hypobromite ion, on the other hand, rapidly disproportionates at room temperature, while OI^- reacts so rapidly that it is never even observed at all when I_2 is dissolved in base. Thus, it is slow kinetics rather than thermodynamics that accounts for the "stability" of solutions of OCl^-. If all of the OX^- ions were to disproportionate as rapidly as OI^-, the reaction of the elemental halogens with base would lead directly to the production of the XO_3^- ions.

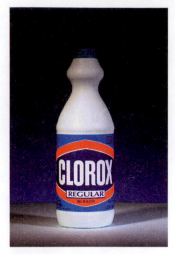

This familiar brand of chlorine bleach is a dilute aqueous solution of sodium hypochlorite, NaOCl.

> ►**Chemistry in Practice**◄ The kinetic stability of cold solutions containing OCl^- enables the marketing of chlorine bleach solutions such as Clorox, which consist of a 5% aqueous solution of NaOCl. ◆

Sulfur, in Group VIA, forms two oxides, SO_2 and SO_3. Earlier we noted that SO_3 reacts with water to give H_2SO_4, and neutralization yields the corresponding oxoanions HSO_4^- and SO_4^{2-}. Sulfur dioxide, although formally the acidic anhydride of H_2SO_3, doesn't actually form this acid in aqueous solution. Solutions of SO_2 are acidic from the reaction of water with dissolved SO_2 to give some H^+ and HSO_3^- ions.

$$SO_2(aq) + H_2O \rightleftharpoons H^+(aq) + HSO_3^-(aq)$$

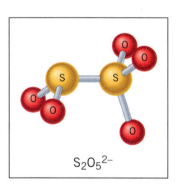

$S_2O_5^{2-}$

Partial neutralization of the acid gives a solution that contains the bisulfite ion, HSO_3^-. Salts containing this anion can be isolated if the cation is large and singly charged (e.g., Cs^+), but solid salts of smaller cations such as Na^+ contain the disulfite ion, $S_2O_5^{2-}$. When dissolved in water, salts such as $Na_2S_2O_5$ hydrolyze to give solutions of sodium bisulfite because of the equilibrium

$$S_2O_5^{2-} + H_2O \rightleftharpoons 2HSO_3^-$$

The other two nonmetallic elements of Group VIA, selenium and tellurium, also form oxoacids. Both selenium and tellurium form acids in the +4 oxidation state—selenous acid, H_2SeO_3, and tellurous acid, H_2TeO_3. They are both weak acids and are also easily reduced. In the +6 oxidation state, selenium forms selenic acid, H_2SeO_4, which is similar to sulfuric acid, but tellurium forms an acid with the formula $Te(OH)_6$ which is composed of octahedral molecules.

With phosphorus, in Group VA, hydrolysis of the oxides P_4O_6 and P_4O_{10} yields H_3PO_3 and H_3PO_4. Of these two, only H_3PO_4 is able to provide three protons for neutralization and thus produce salts containing the PO_4^{3-} ion. Complete neutralization of H_3PO_3, on the other hand, gives the ion HPO_3^{2-}. The differences in behavior between these two acids can be traced to their structures.

phosphoric acid   phosphorous acid

In each acid, phosphorus is bonded to four atoms. In H_3PO_3, however, one of the hydrogens is covalently bonded directly to the phosphorus atom and cannot easily be removed as a proton. As a result, H_3PO_3 is only a diprotic acid.

Arsenic also forms two simple oxoacids, H_3AsO_4 and H_3AsO_3. However, unlike phosphorus, both of these are triprotic acids because in each of them the hydrogens are bonded to oxygens that are then bonded to the arsenic. Complete neutralization gives the anions AsO_4^{3-} (arsenate ion) and AsO_3^{3-} (arsenite ion).

For silicon in Group IVA, no simple oxoacid can be isolated. The anion SiO_4^{4-}, called the orthosilicate ion, is in principle the anion of the acid H_4SiO_4 (orthosilicic acid).

$$
\left[\begin{array}{c} :\ddot{O}: \\ | \\ :\ddot{O}-Si-\ddot{O}: \\ | \\ :\ddot{O}: \end{array}\right]^{4-}
\qquad
\begin{array}{c} OH \\ | \\ HO-Si-OH \\ | \\ OH \end{array}
$$

orthosilicate ion   orthosilicic acid (cannot be isolated)

However, attempts to prepare this acid fail because of the tendency to form polymeric structures, as described below.

Polymeric Oxoacids and Oxoanions

A **polymer** is a substance that is formed by linking together two or more simpler units called **monomers.** Polymeric oxoacids and oxoanions consist of two or more nonmetal or metalloid atoms linked together in various ways by oxygen bridges. They occur primarily with the elements in the third and succeeding periods, although boron, a second-period element, also forms complex borates (which we shall not discuss here).

In the polymeric species we will examine, the basic structural unit is the XO_4 tetrahedron (Figure 21.20). Polymeric acids or ions are composed of two or more XO_4 tetrahedra linked together through commonly shared corners, resulting in structures that range from simple to complex. Fortunately, there are relatively simple structural relationships that exist between the polymeric acids and anions and the simple oxoacids discussed earlier in this section.

The oxoacids formed by the nonmetallic elements of period 3, in their highest oxidation states, can be represented by the formulas

Plastics like polyethylene and nylon are polymers. Such compounds are discussed in Chapter 23.

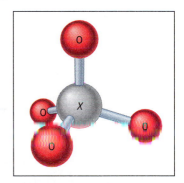

Figure 21.20 *The basic XO_4 tetrahedron found in polymeric oxoacids.*

$$
\begin{array}{c} OH \\ | \\ HO-Si-OH \\ | \\ OH \end{array}
\qquad
\begin{array}{c} O \\ \| \\ HO-P-OH \\ | \\ OH \end{array}
\qquad
\begin{array}{c} O \\ \| \\ HO-S-OH \\ \| \\ O \end{array}
\qquad
\begin{array}{c} O \\ \| \\ O=Cl-OH \\ \| \\ O \end{array}
$$

silicic acid (orthosilicic acid) (never isolated)   phosphoric acid (orthophosphoric acid)   sulfuric acid   perchloric acid

We will refer to these as the "ortho" acids, although in the names of the latter two the prefix *ortho-* is omitted.

As noted earlier, the hydrogen atoms of the —OH groups of oxoacids are acidic and may be neutralized by reaction with a base to yield anions. For the oxoacids above, we have the following anions formed by complete neutralization.

$$
\left[\begin{array}{c} :\ddot{O}: \\ | \\ :\ddot{O}-Si-\ddot{O}: \\ | \\ :\ddot{O}: \end{array}\right]^{4-}
\quad
\left[\begin{array}{c} :O: \\ \| \\ :\ddot{O}-P-\ddot{O}: \\ | \\ :\ddot{O}: \end{array}\right]^{3-}
\quad
\left[\begin{array}{c} :O: \\ \| \\ :\ddot{O}-S-\ddot{O}: \\ \| \\ :O: \end{array}\right]^{2-}
\quad
\left[\begin{array}{c} :O: \\ \| \\ \ddot{O}=Cl-\ddot{O}: \\ \| \\ :O: \end{array}\right]^{-}
$$

orthosilicate   orthophosphate   sulfate   perchlorate

Notice that the number of negative charges on the anion is equal, in each case, to the number of —OH groups attached to the central atom in the parent acid. In addition, in each of the anions the central atom is surrounded by four sets of electron pairs (in the X—O bonds) and each therefore has the tetrahedral structure shown in Figure 21.20.

The polymeric oxoacids and anions are related to these simple species in a rather direct way, because they are formed, if not always in practice, at least in principle, by joining these simple tetrahedral units together into more complex structures. We may view this as being accomplished by the removal of the constituents of a water molecule from two —OH groups on adjacent molecules of the acid. We can illustrate this in a general way as

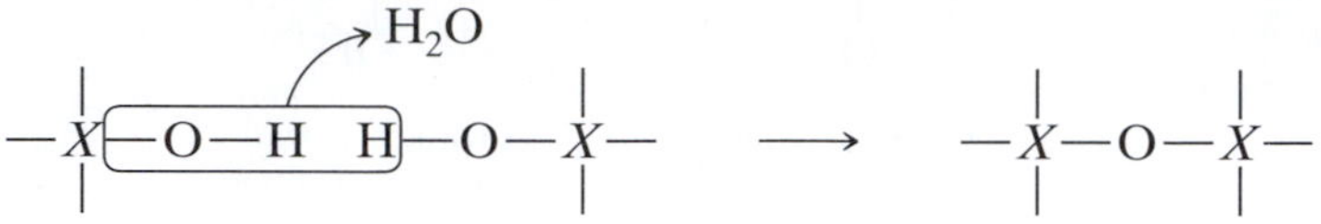

$$-X-O-H \quad H-O-X- \longrightarrow -X-O-X-$$

where the resulting product contains an oxygen bridge, which is characteristic of these polymeric species in general.

Another point to notice is that the bridging oxygen atom had its origin in an —OH group bonded to the central atom. Therefore, the maximum complexity of the polymeric acid (or anion) is determined by the number of —OH groups that are bound to the nonmetal in the simple monomeric acid. Thus, on the basis of the structures of the monomeric acids given above, we expect that silicon can form a maximum of four oxygen bridges to other Si atoms, phosphorus can form three, sulfur two, and chlorine only one.

Let us now begin to examine some of these polymers. When one molecule of water is eliminated from between two molecules of acid, species with the general formula $H_nX_2O_7$ are formed (Table 21.7). These constitute the "pyro" acids,[6] so named because in some cases they can be made by heating the free acid or an acid salt, driving out water. For example, heating a bisulfate salt such as $KHSO_4$ produces the pyrosulfate ion, $S_2O_7^{2-}$ (also called the disulfate ion).

$$2KHSO_4 \xrightarrow{\text{heat}} K_2S_2O_7 + H_2O$$

Table 21.7 "Pyro" Acids and Oxoanions

Silicon	Phosphorus	Sulfur	Chlorine
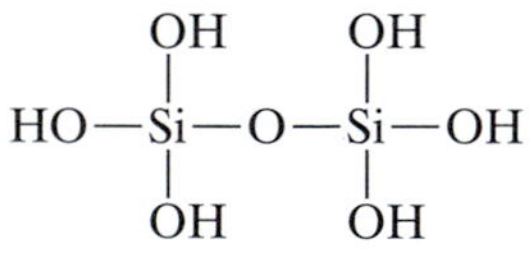 $H_6Si_2O_7$ pyrosilicic acid disilicic acid	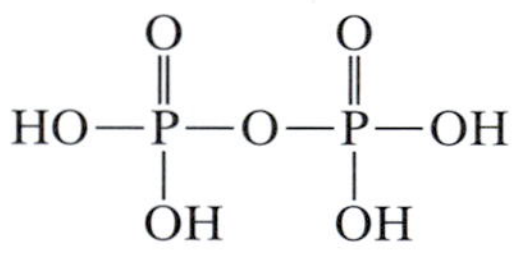 $H_4P_2O_7$ pyrophosphoric acid diphosphoric acid	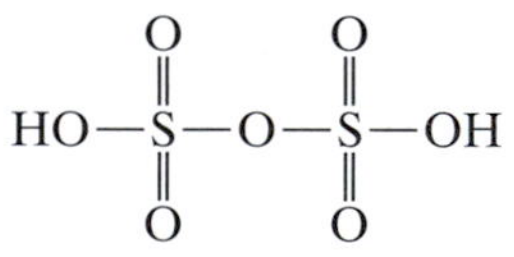 $H_2S_2O_7$ pyrosulfuric acid disulfuric acid	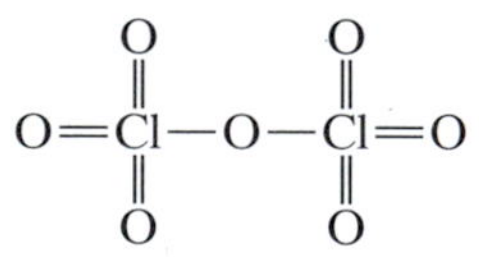 dichlorine heptaoxide
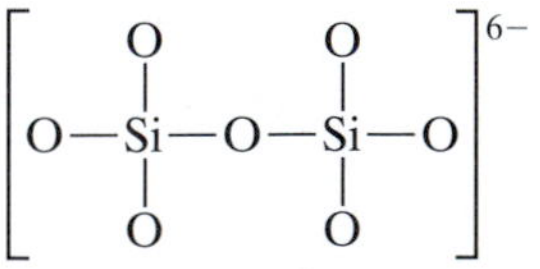 $Si_2O_7^{6-}$ pyrosilicate ion disilicate ion	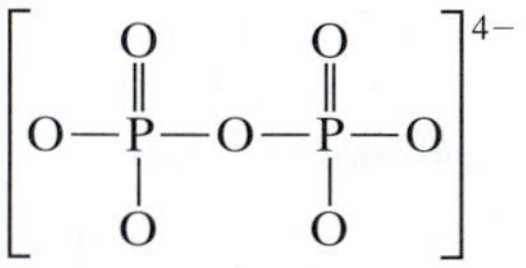 $P_2O_7^{-4}$ pyrophosphate ion diphosphate ion	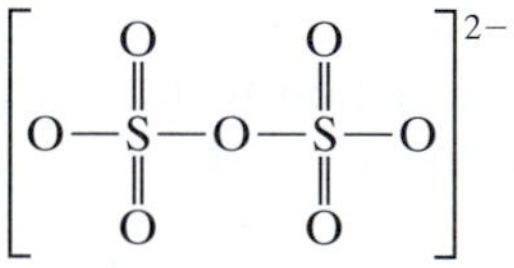 $S_2O_7^{2-}$ pyrosulfate ion disulfate ion	

[6]The prefix "pyro-" comes from the Greek *pyros,* meaning fire.

Similarly, pyrophosphoric acid (diphosphoric acid) is formed when H_3PO_4 is heated.

$$2H_3PO_4 \longrightarrow H_4P_2O_7 + H_2O$$

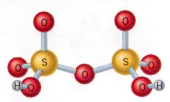

Not all pyro acids are formed this way, however. The free acid $H_2S_2O_7$ is made by dissolving SO_3 in concentrated sulfuric acid.

$$SO_3 + H_2SO_4 \longrightarrow H_2S_2O_7$$
disulfuric acid
pyrosulfuric acid

$H_2S_2O_7$

Removing the protons from the —OH groups of pyro acids gives the corresponding oxoanions. In Table 21.7 we have also included the oxide, Cl_2O_7, which is essentially structurally identical to the pyroanions (Figure 21.21). In fact, Cl_2O_7 is prepared by treating $HClO_4$ with a powerful dehydrating agent such as P_4O_{10}. In addition, Cl_2O_7 is the acid anhydride of $HClO_4$ and reacts with water to form this acid.

The elimination of water and the simultaneous formation of an oxygen bridge becomes progressively easier from chlorine to silicon. To prepare Cl_2O_7 from $HClO_4$ requires a very powerful dehydrating agent. As we move to the left, conditions become less severe until, with silicon, the acidification of a solution containing a soluble silicate leads to the immediate precipitation of hydrated silica, $SiO_2(H_2O)_x$, the end result of large-scale spontaneous dehydration–polymerization. With the exception of silicon, the dehydration process that leads to polymerization can be reversed by the addition of water to the polymeric species. Pyrosulfuric acid, for example, reacts with water to form H_2SO_4.

$$H_2S_2O_7 + H_2O \longrightarrow 2H_2SO_4$$

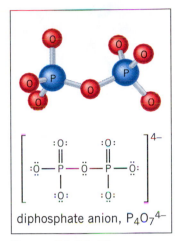

diphosphate anion, $P_4O_7{}^{4-}$

Figure 21.21 *The structure of the diphosphate anion, $P_2O_7{}^{4-}$. This structure is typical of all of the "pyro" anions.*

The pyrosilicate ion is found in certain minerals. However, the free acid, $H_6Si_2O_7$, cannot be isolated.

Structures with Two Bridging Oxygen Atoms

Except for chlorine, all of the dimeric "pyro" acids discussed above have additional —OH groups that can participate in further bridging, again by elimination of H_2O, to give more complex structures. When two oxygen bridges are formed, a chainlike structure consisting of a string of many XO_4 tetrahedra is produced. These constitute the "meta" series of acids and anions. The structure is illustrated below for phosphorus.

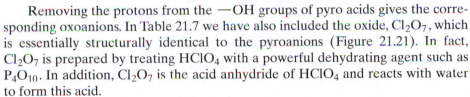

The acid formed is called metaphosphoric acid and has the empirical formula HPO_3. The corresponding anion, the metaphosphate ion, has the empirical formula $PO_3{}^-$.

If H_2SO_4 uses both —OH groups to form bridges, all the hydrogens are eliminated and a structure with the formula SO_3 is formed. In fact, one form of sulfur trioxide is a solid composed of hairlike crystals of SO_3 units linked in long chains.

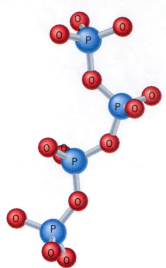

Structure of a typical metaphosphate anion.

The metasilicate ion is found in several minerals, including spodumene, $LiAl(SiO_3)_2$, a major source of the element lithium.

Silicon also forms metasilicates that contain the polymeric anion with the empirical formula SiO_3^{2-}. The anion carries a 2− charge on each SiO_3 unit because its parent acid, which cannot actually be isolated, is a polymer with the repeating unit H_2SiO_3.

$$-O\left(\begin{array}{c} OH \\ | \\ Si-O \\ | \\ OH \end{array}\right)_x \qquad -O\left(\begin{array}{c} O \\ | \\ Si-O \\ | \\ O \end{array}\right)_x^{2-}$$

metasilicic acid
(never isolated)   metasilicate ion

Structures with Three Bridging Oxygen Atoms

Only H_3PO_4 and H_4SiO_4 have sufficient OH groups to form three oxygen bridges. When phosphorus does this, all the hydrogens are lost and we have an oxide of phosphorus. In fact, the oxide is one we've discussed earlier—P_4O_{10}. Its structure is shown on page 969.

When silicon forms three oxygen bridges, each SiO_4 tetrahedron shares three corners with other SiO_4 tetrahedra. This gives rise to planar sheets of SiO_4 units as illustrated in Figure 21.22. The repeating unit in this silicate has the em-

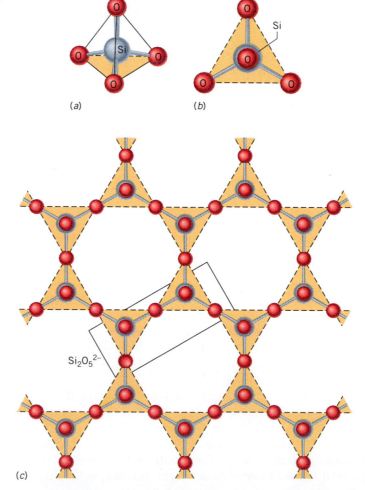

Figure 21.22 *Planar sheet silicate structure.* (*a*) An SiO_4 tetrahedron. (*b*) An SiO_4 tetrahedron viewed from above. (*c*) A sheet silicate structure in which SiO_4 tetrahedra (viewed from above) share three corners each to give a planar layered structure. The oxygens at the top of each tetrahedron carry a negative charge while the shared oxygens are neutral. The empirical formula of the repeating unit in the structure is $Si_2O_5^{2-}$.

pirical formula $Si_2O_5^{2-}$. Each silicon has one nonbridging oxygen that carries a negative charge, which accounts for the charge on the structure.

Several different minerals are known to contain $(Si_2O_5^{2-})_x$ sheets. They differ in the way the silicate layers are stacked, and in the nature of the cations and other anions that are also present in the structure. Nevertheless, they all bear certain similarities. Examples are talc (used in bath powder) and soapstone, in which the $(Si_2O_5^{2-})_x$ sheets are packed together with cations in such a way that there is a minimum of attractive forces between successive layers. These layers therefore slide over each other easily, and both of these minerals feel slippery.

In other related minerals there is substitution of another element for Si. In mica, for instance, every fourth Si atom is replaced by an Al^{3+} ion. The properties of mica are therefore different from talc; however, the layer structure is still apparent. A mica-type material with which you may be familiar is vermiculite, which is added to soil for use in propagating houseplants and as a cushioning filler in packaging items for shipment. This solid flakes into thin flat layers characteristic of mica.

Structures with Four Bridging Oxygen Atoms

Silicic acid, $Si(OH)_4$, is the only period 3 acid with sufficient —OH groups to form four oxygen bridges. When this happens, there are no free OH groups and we form the oxide, SiO_2, whose structure we discussed on page 970.

Summary

We have discussed a variety of fairly complex structures in the preceding pages, yet there is a unifying structural theme that unites them. They have in common the linking of XO_4 tetrahedra by the elimination of H_2O from —OH groups and the formation of oxygen bridges. For the polymeric acids, the number of hydrogens per atom X equals the number of —OH groups *not involved* in bridging between molecules of the monomeric acids. For the polymeric anions, the number of negative charges per atom X is equal to the number of hydrogens removed from the parent polymeric acid.

We have not treated the full range of complexity of structures in our study of the polymeric oxoanions. There are anions of phosphorus and silicon in which chains of tetrahedra circle and form rings, and among the silicates there are chainlike structures, such as those in the fibrous mineral asbestos, in which some SiO_4 tetrahedra share three corners while others share just two corners.

In concluding this section we note that there are many similarities of structure among the oxides, polymeric anions, and even the simple anions of the third-period nonmetallic elements. Some of these relationships are summarized in Table 21.8. In all of these species, the dominant structural feature is the simple tetrahedron, and even the most complex structures are simply obtained by combining tetrahedra in a variety of different ways.

21.7 Halogen Compounds of Nonmetals and Metalloids

The halogens (F, Cl, Br, and I) as a class form a large number of different compounds with the nonmetallic elements. In this section we will look at some of their binary compounds, focusing on the factors that control the kinds of compounds the halogens form as well as one aspect of their reactivity, their tendency to react with water.

Table 21.8 **Summary of Oxoanions of the Period 3 Nonmetallic Elements**

Anions	Silicon	Phosphorus	Sulfur	Chlorine
Simple anions (no X—O—X bridges)	SiO_4^{4-}	PO_4^{3-}	SO_4^{2-}	ClO_4^{-}
"Pyro" anions (one X—O—X bridge)	$Si_2O_7^{6-}$	$P_2O_7^{4-}$	$S_2O_7^{2-}$	Cl_2O_7
"Meta" anions (two X—O—X bridges)	$(SiO_3)_x^{2x-}$	$(PO_3)_x^{x-}$	$(SO_3)_x$	
Two-dimensional sheets (three X—O—X bridges)	$(Si_2O_5)_x^{2x-}$	$P_4O_{10}{}^a$		
Three-dimensional network (four X—O—X bridges)	$(SiO_2)_x$			

$^a P_4$ tetrahedron with oxygens inserted along the edges and an additional oxygen attached at each vertex of the tetrahedron.

Table 21.9 is a list (although not an exhaustive one) of many of the compounds that are formed between the halogens and the nonmetals and metalloids. The structures of the substances found in this table can, without exception, be predicted on the basis of the VSEPR theory.

In Table 21.9 we see that most of the nonmetals form more than one compound with a given halogen. The number of halogen atoms that can become bound to a particular nonmetal can be related to two factors. One is the outer shell electron configuration of the nonmetal, and the other has to do with the sizes of the atoms.

Influence of the Electronic Structures of the Atoms

Each halogen atom has seven electrons in its valence shell and requires only one more to achieve a noble gas configuration. This can be accomplished by forming a single covalent bond with another atom, and in most compounds the halogen

Table 21.9 **Halogen Compounds of the Nonmetals and Metalloids**

Group IIIA	BX_3 (X = F, Cl, Br, I) BF_4^{-}			
Group IVA	CX_4 (X = F, Cl, Br, I)	SiF_4 SiF_6^{2-} $SiCl_4$	GeF_4 GeF_6^{2-} $GeCl_4$	
Group VA	NX_3 (X = F, Cl, Br, I) N_2F_4	PX_3 PX_5 (X = F, Cl, Br, I)	AsX_3 (X = F, Cl, Br, I) AsF_5	SbX_3 (X = F, Cl, Br, I) SbF_5, $SbCl_5$
Group VIA	OF_2 (O_2F_2) OCl_2 OBr_2	SF_2, SF_4, SF_6 SCl_2, SCl_4	SeF_2, SeF_4, SeF_6 $SeCl_2$, $SeCl_4$ $SeBr_2$, $SeBr_4$	TeF_4, TeF_6 $TeCl_4$
Group VIIA		ClF, ClF_3, ClF_5	BrF, BrF_3, BrF_5 $BrCl$	IF_3, IF_5, IF_7 ICl, IBr

atom is singly bonded. There are compounds, however, in which the halogen atom serves as the central atom and exceeds an octet when bonded to other halogen atoms. This is possible when the central atom is chlorine, bromine, or iodine because their outer shells contain unoccupied d subshells. On this basis, then, we can divide the covalently bonded halogen compounds into two groups, those that obey the octet rule and those that do not.

The compositions of the compounds that obey the octet rule are determined by the number of electrons that the central atom requires to reach an octet, because each bonded halogen atom furnishes one electron to the central atom. For example, in Group VIIA (the halogens themselves) only one electron is needed and only one bond is formed, so the elemental halogens are diatomic. Substitution of one halogen for another is also possible, which gives substances called **interhalogen compounds.** Examples are ClF, BrF, BrCl, BrI, and ICl.

In Group VIA, each element requires two electrons to reach an octet, so two halogen atoms become bonded to give compounds such as OF_2 and SCl_2. Similarly, in Group VA, three electrons are given to the central atom by three halogens in molecules such as NF_3, PF_3, AsF_3, and so on. And in Group IVA, each element picks up four electrons from four halogen atoms in compounds such as CCl_4 and $SiCl_4$.

Boron, in Group IIIA, is once again a special case. Since the boron atom has only three valence electrons, it forms only three ordinary covalent bonds to the halogens. With fluorine, however, BF_3 can add on an additional F^- ion to form the BF_4^- (tetrafluoroborate) anion.

> Fluorine, being a period 2 element, cannot exceed an octet in its outer shell and only forms single covalent bonds to other atoms.

$$:\ddot{F}-\overset{\overset{\displaystyle :\ddot{F}:}{|}}{\underset{\underset{\displaystyle :\ddot{F}:}{|}}{B}} \;+\; \left[:\ddot{F}:\right]^- \;\longrightarrow\; \left[:\ddot{F}-\overset{\overset{\displaystyle :\ddot{F}:}{|}}{\underset{\underset{\displaystyle :\ddot{F}:}{|}}{B}}\leftarrow\ddot{F}:\right]^-$$

We see that the fourth B—F bond can be visualized as a coordinate covalent bond (although by now you know that we really cannot distinguish the source of the electrons once the bond has been formed). Since the boron halides have less than an octet of electrons around the central atom, they are all powerful Lewis acids, and the formation of the BF_4^- ion is a typical Lewis acid–base reaction, with the fluoride ion being the Lewis base.

The second category of halogen compounds consists of substances in which more than four pairs of electrons surround the central atom. These are limited to those nonmetals beyond the second period, because the second-period elements have a valence shell that can contain a maximum of only eight electrons, corresponding to the completion of the $2s$ and $2p$ subshells. The elements below the second period, however, also have a low-energy set of vacant d orbitals in their valence shells. These d orbitals may be used, through hybridization, to make additional electrons available for bonding, as illustrated below for sulfur in the molecule SF_6.

In its ground state, the electron configuration of sulfur can be represented as

S [Ne] ⇅   ⇅ ↑ ↑   ○○○○○
          $3s$ $3p$ $3d$

The use of two $3d$ orbitals by sulfur permits the formation of a set of sp^3d^2 hybrid orbitals that are each singly occupied.

S [Ne] ↑↑↑↑↑↑   ○○○
          sp^3d^2 unhybridized $3d$ orbitals

This makes available six unpaired electrons that can each pair with one electron of a fluorine atom to form SF_6.

S (in SF_6) [Ne]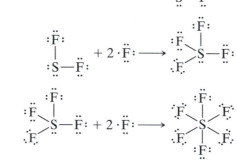
sp^3d^2 unhybridized $3d$ orbitals

(Colored arrrows represent fluorine electrons.)

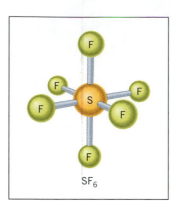

SF$_6$

Figure 21.23 *The structure of the SF$_6$ molecule.*

The structure of SF_6 is octahedral (Figure 21.23) as we would expect, both from the directional properties of the sp^3d^2 hybrid set as well as from the VSEPR theory, which says that when there are six electron pairs in the valence shell of an atom, they will situate themselves at the vertices of an octahedron.

When an atom expands its valence shell beyond an octet to accommodate bonding to additional halogen atoms, it must unpair electrons to do so. Each time such unpairing occurs, it yields *two* half-filled orbitals which are used to bond to *two* additional halogens. As a result, the formulas of the halogen compounds of a given nonmetal differ, one from the next, by two halogen atoms. Thus, sulfur forms SF_2, SF_4, and SF_6.

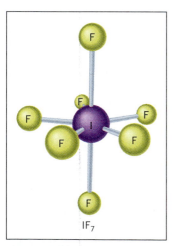

IF$_7$

Figure 21.24 *The structure of the IF$_7$ molecule.*

Similarly, iodine forms IF_3, IF_5, and IF_7. Notice that for a given nonmetal, *the total number of halogens that can be accommodated equals the number of valence electrons possessed by a neutral atom of the element.* A sulfur atom has six valence electrons and can bind to a maximum of six halogen atoms; iodine has seven valence electrons and can form bonds to seven halogen atoms.

The second factor that influences the number of halogen atoms that become bound to a nonmetal, at least to some degree, is the relative sizes of the different atoms. Notice in Table 21.9 that in Group VIIA, the compounds that contain the largest number of fluorine atoms are formed by the halogens found toward the bottom of the group. Thus, iodine forms IF_7 (Figure 21.24), but only five fluorines can be accommodated around chlorine and bromine in ClF_5 and BrF_5. This makes sense because iodine is a large atom and is therefore able to accommodate a larger number of bonded fluorine atoms with a minimum of crowding.

Reactivity of Halogen Compounds toward Water

Another interesting facet of the chemistry of the nonmetal halides is their reactivity toward compounds containing an —OH group, the most familiar of which is water. Here, once again, we find that both thermodynamics and kinetics are involved in determining the course of reactions.

The type of reaction we are interested in is the hydrolysis of the nonmetal halide to produce either the oxoacid, or an oxide, plus the corresponding hydrogen halide. Some examples are the reactions of PCl_5, $SiCl_4$, and SF_4 with water.

$$PCl_5 + 4H_2O \longrightarrow H_3PO_4 + 5HCl$$

$$SiCl_4 + 2H_2O \longrightarrow SiO_2 + 4HCl$$

$$SF_4 + 2H_2O \longrightarrow SO_2 + 4HF$$

These reactions occur very rapidly and proceed to completion with the evolution of considerable amounts of heat. In fact, it is quite common for many of the halogen compounds of the elements below period 2 to react very rapidly in this same way. For example, the tin(IV) and lead(IV) halides, which we saw in Chapter 20 were covalent, also hydrolyze in this manner with the formation of a mixture of species including complexes of Sn^{4+} with the halide ion.

A very important class of compounds called silicones are produced by hydrolysis of compounds such as $(CH_3)_2SiCl_2$. The formation of silicone polymers can be thought of as proceeding by formation of a hydroxy intermediate which then eliminates water in much the same way that $Si(OH)_4$ polymerizes.

The $-CH_3$ group is called a methyl group, and these polymers are called methylsilicone polymers.

Depending on the chain length, and the degree to which chains may be cross-linked to one another, the silicones may be oils, greases, or rubbery solids. They are useful in waterproofing fabrics. In low-temperature applications the oils remain fluid (hydrocarbon oils become very viscous), and the rubbers retain their elastic properties (ordinary rubber becomes brittle). Silicones are also unaffected by hydrocarbon solvents and greases that soften ordinary rubber, and they are not attacked by ozone in the air, which causes ordinary rubber to crack. They even have pharmaceutical applications; the anti-gas ingredient in some antacid products sold as remedies for indigestion goes by the name "simethicone" and is a liquid methylsilicone polymer.

There are also some nonmetal halogen compounds that are quite *unreactive* toward water, for example, CCl_4, SF_6, and NF_3. For these compounds, it is unfavorable kinetics, rather than thermodynamics, that prevents the hydrolysis from taking place. For example, we can compare the potential hydrolysis reactions of CCl_4 and $SiCl_4$, for which we have

$$SiCl_4(l) + 2H_2O(l) \longrightarrow SiO_2(s) + 4HCl(g) \qquad \Delta G° = -282 \text{ kJ}$$

$$CCl_4(l) + 2H_2O(l) \longrightarrow CO_2(g) + 4HCl(g) \qquad \Delta G° = -377 \text{ kJ}$$

Recall that the more negative the value of $\Delta G°$, the larger is the equilibrium constant for the reaction, so according to these data, the hydrolysis of CCl_4 (car-

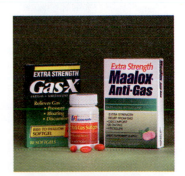

Some antacids that contain methylsilicone polymers as antigas ingredients.

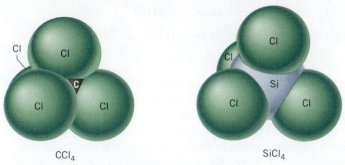

Figure 21.25 *Space-filling models of CCl$_4$ (left) and SiCl$_4$ (right) showing the relative sizes of carbon and silicon.* The large chlorine atoms leave very little access to the small carbon atom in CCl$_4$, so this substance cannot be attacked by molecules such as H$_2$O. Because of the larger size of the silicon atom, there is less crowding by the chlorine atoms in SiCl$_4$ and other molecules such as H$_2$O can reach the silicon more easily.

bon tetrachloride) should proceed even further toward completion than the hydrolysis of SiCl$_4$. Yet, although SiCl$_4$ reacts very rapidly with water, even moisture in the air, we find that CCl$_4$ is inert toward water. In fact, for many years CCl$_4$ was sold in retail stores for use as a dry cleaning solvent (before it was found to be very toxic). Why do these two substances behave so differently?

For a water molecule to initiate the hydrolysis reaction, it must be able to attack the central atom in the molecule. As you can see in Figure 21.25, in CCl$_4$ the small size of the carbon atom and the relatively large size of the chlorine atoms leave little room for a water molecule to approach the carbon atom. In SiCl$_4$, on the other hand, the larger Si atom provides a greater opportunity for attack, and in addition, the presence of low-energy 3d orbitals in the valence shell of the Si atom permits a temporary bonding of the water molecule to the Si atom prior to the expulsion of a molecule of HCl. The mechanism of this hydrolysis is believed to be

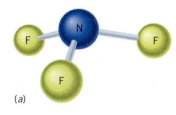

(a)

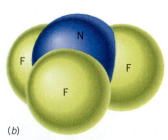

(b)

Figure 21.26 *The shape of the NF$_3$ molecule.* (a) Ball-and-stick model. (b) Space-filling model.

SF$_6$ is so stable it's used as an electrical insulator in high-voltage equipment.

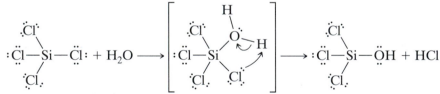

Repetition of this process eventually yields Si(OH)$_4$ (orthosilicic acid), which loses most of its water spontaneously to give a hydrated SiO$_2$.

The stability of SF$_6$ and NF$_3$ toward hydrolysis can also be attributed to the absence of a low-energy reaction mechanism. Like CCl$_4$, SF$_6$ should also undergo a hydrolysis reaction that proceeds essentially to completion, as indicated by its very large and negative $\Delta G°$.

$$SF_6(g) + 4H_2O(l) \longrightarrow H_2SO_4(aq) + 6HF(g) \qquad \Delta G° = -423 \text{ kJ}$$

However, crowding of the fluorine atoms around the sulfur atom apparently prevents attack by water (even up to 500 °C), as well as by most other reagents. This crowding is absent with SF$_4$, and its hydrolysis by water is instantaneous.

The resistance of NF$_3$ toward attack by water cannot be attributed to interference by the fluorine atoms as in SF$_6$, since the NF$_3$ molecule is pyramidal, with the nitrogen atom being quite openly exposed to an attacking water molecule (Figure 21.26). We might compare NF$_3$ with NCl$_3$, which does hydrolyze (if it doesn't explode first—NCl$_3$ is extremely unstable).

$$NCl_3 + 3H_2O \longrightarrow NH_3 + 3HOCl$$

The mechanism for the reaction appears to involve the initial formation of a hydrogen bond from water to the lone pair on the nitrogen atom, followed by expulsion of hypochlorous acid.

The failure of NF_3 to hydrolyze is believed to be related to the low basicity of the nitrogen of the NF_3 molecule. Because fluorine is so electronegative, the nitrogen carries a substantial positive charge, which prevents the hydrogen of water from attaching via a hydrogen bond. In effect, this makes a point of attachment of water unavailable in NF_3, which is a necessary step in the hydrolysis if the mechanism proposed for hydrolysis of NCl_3 is to be followed.

SUMMARY

Obtaining Nonmetallic Elements in Their Free States. Metalloids normally exist in positive oxidation states in compounds and are obtained by reduction. Common reducing agents are carbon and hydrogen. In nature, the noble gases are never found in compounds. Oxygen and nitrogen are obtained from air, and sulfur is found in natural deposits, often deep below Earth's surface. Carbon is found in coal and diamonds. Nonmetals are also found in naturally occurring compounds.

If a nonmetal is combined with a metal, oxidation is used to change it to the free element. Among the halogens, oxidizing strength decreases from fluorine to iodine. In Group VIIA, a given halogen (e.g., F_2) is able to oxidize the halide ion below it (e.g., Cl^-), thereby displacing it from its compounds. Fluorine, being the most difficult to prepare by oxidation of its anion, is obtained by electrolysis of molten mixtures of HF and KF. Elemental phosphorus is obtained from calcium phosphate by reaction with carbon and SiO_2.

Molecular Structures of the Nonmetals. Elements of period 2, because of their small size, form strong π bonds. As a result, these elements easily participate in multiple bonding between like atoms, which accounts for diatomic molecules of O_2 and N_2, and the π-bonded structure of graphite. Elements of periods 3, 4, and 5 are large and their p orbitals do not overlap well to form strong π bonds, so these elements prefer single σ bonds between like atoms, which leads to more complex molecular structures.

Different forms of the same element are called **allotropes.** Oxygen exists in two allotropic forms: dioxygen (O_2) and **ozone** (O_3). Carbon forms three allotropes: **diamond, graphite,** and C_{60} molecules called **buckminsterfullerene** (one member of the **fullerene** family of structures). Elemental boron consists of B_{12} clusters linked through other boron atoms to give an extremely hard solid. Sulfur forms S_8 molecules that can be arranged in two different allotropic forms. Phosphorus occurs as **white phosphorus** (P_4), **red phosphorus,** and **black phosphorus.** Silicon only forms a diamondlike structure.

Hydrogen Compounds of Nonmetals and Metalloids. Except for boron, the simple compounds of the nonmetallic elements with hydrogen contain the same number of hydrogen atoms as the number of electrons these elements require to reach an octet in their valence shells. The simplest boron hydride is diborane, B_2H_6, which is unusual because it contains hydrogen bridges between boron atoms. These hydrogen atoms participate in three-center bonds.

Two methods of preparing nonmetal hydrides discussed are direct combination of the elements and adding protons to the anions of the nonmetals. Direct combination is feasible when ΔG_f° for the compound is negative. Protonation of the anion becomes progressively easier moving from right to left within a period and from bottom to top within a group in the periodic table. This trend parallels the increasing basicity of the anions. When the anion is a poor base, a strong Brønsted acid is required. When the anion is very basic, it is able to capture protons from water to produce the hydride and a metal hydroxide (a process called **hydrolysis**).

Catenation among Nonmetal Atoms. Catenation occurs when atoms of the same element bond to produce a chain of like atoms. Carbon forms the largest number of such compounds. To a lesser extent, silicon and germanium form similar compounds that are analogous to the hydrocarbons. Hydrogen peroxide (H_2O_2) and hydrazine (N_2H_4)

are examples of catenated compounds of oxygen and nitrogen. Sulfur forms polysulfide anions, S_x^{2-}, and a few other anions that contain sulfur–sulfur bonds (thiosulfate ion and polythionate ions).

Oxygen Compounds of Nonmetallic Elements. Oxygen compounds can be made by direct combination of the elements, oxidation of a lower oxide by oxygen, reaction of nonmetal hydrides with oxygen, and as products of other redox reactions.

Oxides of carbon (CO and CO_2) and nitrogen (N_2O, NO, NO_2, N_2O_3, N_2O_4, and N_2O_5) are relatively simple molecules. NO is **isoelectronic** with O_2^+, and the bonding in this compound is explained well by molecular orbital theory. Dimerization of brown NO_2 yields colorless N_2O_4. Sulfur forms two oxides, SO_2 and SO_3. Phosphorus also forms two oxides, P_4O_6 and P_4O_{10}, whose molecular structures are related to the structure of the P_4 molecule. Silicon forms SiO_2 in which SiO_4 tetrahedra are linked at their corners to other SiO_4 tetrahedra. In quartz, linked spiral chains of SiO_4 tetrahedra yield two chiral crystal forms of the oxide.

Oxoacids and Oxoanions. Oxoacids contain atoms bonded to one or more OH groups that are able to release their hydrogens as H^+ ions. Loss of H^+ yields oxoanions. In period 2, the oxoacids are HOF (Group VIIA), HNO_2 and HNO_3 (Group VA), H_2CO_3 (Group IVA), and $B(OH)_3$ (Group IIIA). Boric acid is unusual because the hydrogens of the OH groups bonded to boron are not acidic. Instead, boric acid functions as a Lewis acid.

In period 3 and below, the oxoacids are as follows: among the halogens, acids with the general formulas HOX, HXO_2, HXO_3, and HXO_4; in Group VIA, H_2XO_3 and H_2XO_4 (except that H_2SO_3 is really just a solution of SO_2 in water); in Group VA, H_3XO_3 and H_3XO_4 (however, H_3PO_3 is a diprotic acid); and in Group IVA, H_4XO_4, except that the acid itself cannot be isolated.

Polymeric oxoacids and oxoanions discussed were those of the period 3 elements. They can be viewed as forming by the elimination of H_2O from —OH groups on two different oxoacid molecules and the subsequent formation of an oxygen bridge between two nonmetal atoms. In each of these oxoacids, the central atom is surrounded by a tetrahedral arrangement of oxygen atoms. Linking two such tetrahedra by an oxygen bridge yields acids with the formula $H_n X_2O_7$ and anions with the formula $X_2O_7^{n-}$. Further linking of tetrahedra produces polymeric species with the empirical formula $H_m XO_3$ (for the acid) and XO_3^{m-} (for the anion). Silicon is able to form anions in which SiO_4 tetrahedra are linked through three corners to give a planar sheet structure, which is found in minerals such as mica and talc. Linking all four corners of SiO_4 tetrahedra gives SiO_2, as in quartz.

Halogen Compounds of Nonmetals. When a halogen is bound to some central atom, it forms one covalent bond. The molecules produced in this way have structures easily derived by the VSEPR theory. In the simplest compounds, the number of halogens attached to a nonmetal is determined by the number of electrons needed to reach an octet in its valence shell. If a nonmetal bonds to additional halogen atoms, it does so two at a time. Among the interhalogen compounds, the number of fluorine atoms that can be bonded to the central atom is determined in part by the size of the central atom (the larger the central atom, the more fluorine atoms are able to fit around it).

The hydrolysis of halogen compounds of nonmetals to give the hydrogen halide plus either the oxoacid or oxide is favored by thermodynamics. Hydrolysis of compounds such as $(CH_3)_2SiCl_2$ leads to formation of silicone polymers. For some halogen compounds the hydrolysis reactions are prevented by the lack of a low-energy reaction mechanism, and the compounds appear to be stable in the presence of water.

REVIEW QUESTIONS

Recovery of Nonmetals and Metalloids from Compounds

21.1 Why are metalloids usually recovered from their compounds by chemical reduction rather than by oxidation?

21.2 Write chemical equations for
(a) the chemical reduction of BCl_3 with hydrogen.
(b) the production of Si from SiO_2 using carbon as a reducing agent.
(c) the reduction of As_2O_3 with hydrogen.

21.3 What is the major commercial source of N_2 and O_2?

21.4 Why is helium found in underground deposits? Why are only small quantities of radon observed in nature?

21.5 Chlorine, Cl_2, can be made in the lab by the reaction of HCl with $KMnO_4$, with $MnCl_2$ being among the products. Write a balanced net ionic equation for the reaction, keeping in mind that $KMnO_4$ is water soluble.

21.6 Complete the following chemical equations. If no reaction occurs, write N.R.
(a) $Cl_2 + KI \rightarrow$ (c) $I_2 + MgCl_2 \rightarrow$
(b) $Br_2 + CaF_2 \rightarrow$ (d) $F_2 + SrCl_2 \rightarrow$

21.7 How is Br_2 recovered from seawater?

21.8 Why can't fluorine be produced by the electrolysis of aqueous NaF? What products are formed at inert electrodes in the electrolysis of aqueous NaF?

21.9 Why is a molten mixture of KF and HF used in the production of F_2 by electrolysis rather than molten KF by itself?

21.10 Hydrogen fluoride is a gas that can be liquefied by cooling it to 19.6 °C (just slightly below room temperature). Why can't electrolysis be carried out on liquid HF to form F_2 and H_2?

21.11 Write the chemical equation for the production of elemental phosphorus from calcium phosphate and SiO_2. What is the reducing agent in the reaction? What is the function of the SiO_2? Why isn't SiO_2 reduced to Si in the reaction?

21.12 Under the proper conditions, iodide ion will react with H_2SO_4 to generate I_2 and H_2S. Write a balanced net ionic equation for this reaction.

Molecular Structures of the Nonmetals and Metalloids

21.13 Which of the nonmetals occur in nature in the form of isolated atoms?

21.14 Why are the period 2 elements able to form much stronger π bonds than the nonmetals of period 3? Why does a period 3 nonmetal prefer to form all σ bonds instead of one σ bond and several π bonds?

21.15 Even though the nonmetals of periods 3, 4, and 5 do not tend to form π bonds between like atoms, each of the halogens is able to form diatomic molecules (Cl_2, Br_2, I_2). Why?

21.16 What are *allotropes*? How do they differ from *isotopes*?

21.17 What are the two allotropes of oxygen?

21.18 Construct the molecular orbital diagram for O_2 and explain why it has two unpaired electrons. What is the net bond order in O_2?

21.19 Draw the Lewis structure for O_3. Is the molecule linear, based on the VSEPR theory? Assign formal charges to the atoms in the Lewis structure. Does this suggest the molecule is polar or nonpolar?

21.20 What beneficial function does ozone serve in Earth's upper atmosphere?

21.21 What are three allotropes of carbon?

21.22 Describe the structure of diamond. What kind of hybrid orbitals does carbon use to form bonds in diamond? What is the geometry around carbon in this structure?

21.23 Describe the structure of graphite. What kind of hybrid orbitals does carbon use in the formation of the molecular framework of graphite?

21.24 Why does graphite have lubricating properties?

21.25 Electrical conduction in graphite is good in a direction that's parallel with the carbon layers, but very poor in a direction that's perpendicular to them. Why?

21.26 Describe the C_{60} molecule. What is it called? What name is given to the series of similar substances?

21.27 In elemental boron, there are clusters of boron atoms linked through other boron atoms. What is the formula for the boron clusters? What is the shape of a cluster?

21.28 What is the molecular structure of sulfur in its most stable allotropic form?

21.29 Make a sketch that describes the molecular structure of white phosphorus.

21.30 What are the P—P—P bond angles in the P_4 molecule? If phosphorus uses p orbitals to form the phosphorus–phosphorus bonds, what bond angle would give the best orbital overlap? On the basis of your answers to these two questions, explain why P_4 is so chemically reactive.

21.31 What structure has been proposed for red phosphorus? How do the reactivities of red and white phosphorus compare?

21.32 What is the molecular structure of black phosphorus? In what way does the structure of black phosphorus resemble that of graphite?

21.33 What is the molecular structure of silicon? Suggest a reason why silicon doesn't form an allotrope that's similar in structure to graphite.

Hydrogen Compounds of Nonmetals and Metalloids

21.34 What two general methods were described in this chapter for the preparation of nonmetal hydrides?

21.35 Draw Lewis structures for the following hydrogen compounds. (a) SiH_4, (b) PH_3, (c) SeH_2, (d) HBr

21.36 What are the molecular shapes of each of the following? PH_3, H_2S, CH_4, H_2Se, Si_2H_6

21.37 Hydrogen forms ionic hydrides with the Group IA elements and also with the heavier elements in Group IIA. Write chemical equations for the reactions of Na and Ca with elemental H_2.

21.38 What is the Haber process? Why is the chemical reaction in the Haber process run at a high pressure and high temperature?

21.39 What is the simplest hydrogen compound formed by boron? What is it called? Describe the shape of the molecule.

21.40 Describe the boron–hydrogen bonds in diborane. What is the average bond order between boron and a bridging hydrogen in diborane?

21.41 If ΔG_f° is positive for a compound, what does this suggest about the feasibility of synthesizing the compound from its elements? Explain.

21.42 Which of the following nonmetal hydrides can be formed, at least in principle, by the direct combination of the elements? (a) NH_3, (b) HBr, (c) H_2Se, (d) H_2S, (e) PH_3

21.43 What accounts for the low degree of reactivity of elemental nitrogen?

21.44 Why do the anions of the nonmetals become easier to protonate as we move from right to left across a period (e.g., from Cl^- to Si^{4-})?

21.45 Write balanced chemical equations for the hydrolysis of each of the following:
(a) Ca_3P_2 (b) Na_2S (c) Al_4C_3 (d) NaF

21.46 How does the basicity of the anions of the nonmetals vary from left to right in a period of the periodic table? How does the strength of the Brønsted acid required to protonate these bases vary?

21.47 How does the basicity of the anions of the nonmetals vary from top to bottom in a group in the periodic table?

21.48 Which is a stronger base, NH_2^- or N^{3-}?

Catenation between Nonmetal Atoms

21.49 What does the term *catenation* mean? Which element demonstrates this property to the largest extent?

21.50 Give an example of a compound of (a) oxygen and (b) nitrogen in which catenation occurs.

21.51 Write an equation that describes the preparation of polysulfide ions.

21.52 Draw the Lewis structure for the hexathionate ion.

21.53 Would the polysulfide ions, S_x^{2-}, be expected to be linear? Explain your answer.

21.54 The peroxydiphosphate ion has the same general composition as the peroxydisulfate ion, $S_2O_8^{2-}$, with phosphorus substituted for sulfur. Draw the Lewis structure for the peroxydiphosphate ion and indicate the charge on the ion. (*Hint:* The charge is *not* 2−.)

Oxygen Compounds of the Nonmetallic Elements

21.55 Can the nitrogen oxides be synthesized in large percentage yields directly from the elements? Explain.

21.56 What oxides are formed by phosphorus? Describe their structures and relate them to the structure of the molecular form of phosphorus that occurs in white phosphorus.

21.57 Write balanced molecular equations for the complete reaction of each of the oxides of phosphorus with water.

21.58 What accounts for the ease with which NO_2 dimerizes to form N_2O_4?

21.59 On the basis of what you have learned about chemical bonding and bond energies, why is the dimerization of NO_2 to give N_2O_4 exothermic?

21.60 Carbon and silicon form oxides with the same empirical formula, XO_2, yet they are vastly different in their properties. Why?

21.61 What are the formulas of the oxides of arsenic and antimony?

21.62 Describe three methods that can be used to form oxides of nonmetals.

21.63 What products might form in the combustion of arsine, AsH_3?

21.64 What are the oxides of sulfur? Draw their Lewis structures and use VSEPR theory to predict their geometry.

21.65 Draw all four resonance structures of the N_2O_4 molecule.

21.66 Compounds can be prepared that contain the nitrosyl ion, NO^+. On the basis of the molecular orbital energy diagram for NO, explain why the NO^+ ion should be relatively stable compared with NO itself.

Oxoacids and Oxoanions

21.67 Complete and balance the following equations:

(a) $SO_3 + H_2O \rightarrow$ (c) $As_4O_{10} + H_2O \rightarrow$

(b) $P_4O_6 + H_2O \rightarrow$ (d) $Cl_2O_7 + H_2O \rightarrow$

21.68 Write equations for the ionization equilibria in a solution of (a) phosphoric acid and (b) phosphorus acid.

21.69 Write Lewis structures for H_3PO_4 and H_3PO_3. Write the formulas for all of the sodium salts that can be obtained by either partial or complete neutralization of these acids.

21.70 Fluorine forms the oxoacid HOF, but not HFO_3. Draw a Lewis structure for a hypothetical HFO_3 molecule and assign formal charges to the atoms. Explain why the existence of HFO_3 is unlikely based on the Lewis structure. (Remember, fluorine is a period 2 element and can have no more than eight electrons in its valence shell.)

21.71 Use Lewis structures to show how boric acid produces H_3O^+ in water.

21.72 What oxide of nitrogen is the acidic anhydride of (a) HNO_2 and (b) HNO_3?

21.73 What is a *disproportionation* reaction? Give two examples.

21.74 Write the chemical equation for the disproportionation reaction that occurs when NO_2 dissolves in water.

Polymeric Oxoacids and Oxoanions

21.75 Illustrate the types of structures that are obtained by joining XO_4 tetrahedra through the sharing of the following: (a) one corner, (b) two corners, (c) three corners, (d) four corners.

21.76 Use Lewis structures to show how $H_2S_2O_7$ would be formed by the removal of one molecule of H_2O from two molecules of H_2SO_4.

21.77 Write the structural formulas for (a) disulfuric acid, (b) trimetaphosphate ion, (c) pyrosilicate ion, (d) pyrophosphoric acid.

21.78 Pyrophosphoric acid may be obtained from its sodium salt in the following way. The sodium salt is dissolved in water and treated with lead nitrate to precipitate lead pyrophosphate which, when treated with H_2S, liberates the free acid. Write chemical equations for all of the reactions that take place.

21.79 Write generalized formulas for the "ortho" anions, "pyro" anions, and "meta" anions of the period 3 elements.

21.80 Use electron dot structures to show how hydrolysis of $(CH_3)_2SiCl_2$ gives rise to silicone polymers.

21.81 If some $(CH_3)_3SiCl$ is added to $(CH_3)_2SiCl_2$ before hydrolysis, shorter polymeric chains are produced than if pure $(CH_3)_2SiCl_2$ is hydrolyzed. Explain.

21.82 What is the basic structural unit in the structure of quartz? Why do two kinds of quartz crystals exist? What is the relationship between them?

Halogen Compounds of Nonmetallic Elements

21.83 Why are solutions of OCl^- stable while OI^- disproportionates rapidly to give I^- and IO_3^-?

21.84 In SF_6, sp^3d^2 hybrid orbitals are used to make all six valence electrons of sulfur available for bonding. What type of hybrid orbitals would be used in the bonding in SF_4? How does the resulting molecular structure compare with that predicted by the electron pair repulsion theory?

21.85 How does the size of the central atom influence the number of fluorine atoms that are able to bind to a given halogen?

21.86 Sulfur hexafluoride is so stable toward reaction with other reagents and toward decomposition that it is used as a gaseous insulator in high-voltage generators. What accounts for this inert behavior?

21.87 Why is $SiCl_4$ so susceptible to attack by water while CCl_4 is unreactive toward water?

REVIEW PROBLEMS

Answers to problems whose numbers are printed in color are given in Appendix D.

Hydrogen Compounds of Nonmetallic Elements

21.88 Complete and balance the following equations:
(a) $AlP(s) + H_2O \rightarrow$ (c) $FeS(s) + HCl(aq) \rightarrow$
(b) $Mg_2C(s) + H_2O \rightarrow$ (d) $MgSe(s) + H_2SO_4(aq) \rightarrow$

21.89 Complete and balance the following equations:
(a) $Mg_3N_2(s) + H_2O \rightarrow$
(b) $KBr(s) + H_2SO_4(\text{concd.}) \rightarrow$
(c) $Mg_2Si(s) + HCl(aq) \rightarrow$
(d) $CaC_2(s) + H_2O \rightarrow$

Oxygen Compounds of the Nonmetallic Elements

21.90 Complete and balance the following equations:
(a) $CO + O_2 \rightarrow$ (c) $P_4O_6 + O_2 \rightarrow$
(b) $C_2H_6 + O_2(\text{excess}) \rightarrow$ (d) $NH_3 + O_2 \rightarrow NO +$

21.91 Complete and balance the following equations:
(a) $PH_3 + O_2(\text{excess}) \rightarrow$ (c) $SiH_4 + O_2 \rightarrow$
(b) $H_2S + O_2 \rightarrow$ (d) $SbH_3 + O_2(\text{excess}) \rightarrow$

21.92 Use the data in Table 21.5 to calculate K_p for the reaction $2NO_2(g) \rightleftharpoons N_2(g) + 2O_2(g)$. Does your answer indicate that the position of equilibrium lies to the left or to the right? Suggest why NO_2 is able to exist at all.

21.93 In air, the partial pressure of N_2 is approximately 0.80 atm and the partial pressure of O_2 is approximately 0.20 atm. If the equilibrium described in the preceding problem were able to be established, what would be the partial pressure of NO_2 in air?

21.94 Calculate ΔH° for the reaction

$$2NO_2(g) \rightleftharpoons N_2O_4(g)$$

Explain why increasing the temperature leads to the dissociation of N_2O_4 into NO_2.

21.95 Calculate ΔH° for the following reactions:

$$CH_4(g) + 2O_2(g) \longrightarrow CO_2(g) + 2H_2O(g)$$

$$CH_4(g) + 4N_2O(g) \longrightarrow CO_2(g) + 2H_2O(g) + 4N_2(g)$$

On the basis of these calculations, explain why race drivers would use nitrous oxide (N_2O) as an oxidizer to power race car engines.

Oxoacids and Oxoanions

21.96 What are the molecular structures (shapes) of the following oxoacids of the halogens? (a) HIO_3, (b) $HClO_2$, (c) H_5IO_6, (d) $HClO_4$

21.97 What are the geometric structures of the following?
(a) PO_4^{3-} (c) NO_2^- (e) SO_3^{2-}
(b) ClO_2^- (d) $S_2O_3^{2-}$ (f) ClO_3^-

Halogen Compounds of Nonmetallic Elements

21.98 Draw Lewis structures and describe the molecular shapes of the following:
(a) PCl_3 (b) BrF_3 (c) SiF_6^{2-} (d) BBr_3 (e) BrF_4^-

21.99 Draw Lewis structures and describe the molecular shapes of the following:
(a) SF_6 (b) SF_4 (c) SCl_2 (d) I_3^- (e) ClF_5

ADDITIONAL EXERCISES

21.100 Using N_2, H_2, O_2, and H_2O as starting materials, write chemical equations for a series of reactions that would lead to the formation of HNO_3.

21.101 Using sulfur, oxygen, and water as starting materials, write chemical equations for a series of reactions that would lead to the formation of H_2SO_4.

21.102 Using Lewis structures, suggest the first step in a mechanism that would account for the rapid reaction of

PCl_5 with water. The ultimate products of the reaction are H_3PO_4 and HCl.

21.103 Using Lewis structures, suggest the first step in a mechanism that would account for the rapid reaction of SF_4 with water. The ultimate products are HF and SO_2.

Diamonds from Graphite

Cut and polished diamonds, sparkling brightly in the light, have fascinated people throughout history. Gem quality diamonds grace crowns and scepters, gowns, lapels, and fingers (see Figure 11a).

Because a diamond (of any quality) is unusually hard, it is able to put a scratch into the surface of anything else. So diamonds of inferior quality are used to make diamond-studded cutting and grinding tools that are used extensively by industry.

Imagine the amazement of scientists in 1797 when it was discovered that diamonds are simply a form of pure carbon. Another form, graphite, is "dirt cheap," so people naturally asked whether graphite could be converted into its more valuable cousin. At the atomic level, graphite consists of covalent *sheets* of carbon atoms (see Figure 11b). The sheets can slip over each other without much difficulty, making graphite a good, dry lubricant. Diamond, in contrast, is a covalent solid with an interlocking, *tetrahedral network* of atoms (see Figure 11c).

As you might expect, many attempted without success to make diamond from graphite. In 1938, a thermodynamic analysis of the problem was finally performed. Its results are summarized in Figure 11d in a graphical form, showing plots of $\Delta G/T$ versus the absolute temperature for the conversion of graphite to diamond.

$$C(s, \text{graphite}) \longrightarrow C(s, \text{diamond})$$

For the change to be *spontaneous*, ΔG must be negative, as we learned in Chapter 18. Because T is always positive, the ratio $\Delta G/T$ must therefore be negative for a spontaneous conversion of graphite to carbon. On the graph of Figure 11d the region where the ratio is negative is shaded.

Each line on the graph shows how $\Delta G/T$ varies with temperature for a given pressure. Thus, the line labeled

Figure 11a Diamond necklace, featuring a 68-carat stone.

1 atm shows that at a pressure of 1 atm, $\Delta G/T$ becomes less positive as the temperature increases. However, notice that $\Delta G/T$ never becomes negative at 1 atm. *It is thus thermodynamically impossible to convert graphite to diamond at atmospheric pressure.*

From the graph, we can also see that $\Delta G/T$ becomes less positive (more negative) with increasing pressure. At 20,000 atm, the ratio is negative at temperatures below 470 K (approximately 200 °C), and at higher pressures $\Delta G/T$ becomes negative at still higher temperatures. From the graph, therefore, scientists were able to establish the conditions of temperature and pressure at which the conversion from graphite to diamond is at least theoretically possible.

This was not the end of the problem, however, because suitable materials still had to be found that would allow the transition to occur at a reasonable rate. Success was elusive, but it finally came in the early 1950s. Pressures in excess of 100,000 atm and temperatures above 2800 °C were needed. The diamonds that were

chemicals in our world

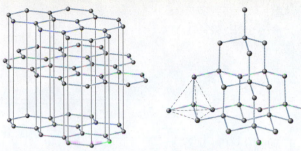

Figure 11b The structure of graphite.

Figure 11c The structure of diamond.

made were generally yellowish because of traces of nitrogen. But they were of industrial quality, and before long, over a million carats of industrial diamonds were being synthesized annually. Some of the synthetic diamonds are of substantial size — as large as 2 carats.

Diamond Coatings

When a mixture of 1% methane (CH_4) in hydrogen at a pressure of 50 millibars (38 torr) is passed over a glowing wire at 2200 °C, the methane molecules crack into atoms of carbon and hydrogen. When the carbon atoms deposit on a nearby surface, they gradually form a film of carbon *with a tetrahedral network structure*. The film, in other words, has the diamond structure, not that of graphite.

Some Applications of Diamond-Coated Surfaces. Many instruments, including computers, have components constantly endangered by temperature increases. A

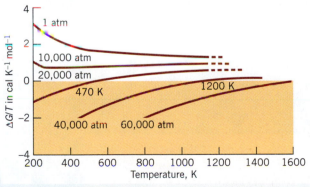

Figure 11d Thermodynamic analysis of the conversion of graphite to diamond. (From *Chemical and Engineering News,* April 5, 1971, p. 51. Used by permission.)

semiconductor laser diode, for example, must be in thermal contact with a material of high *thermal conductivity,* one able to function as a "heat sink" to carry heat away. Diamond has a higher thermal conductivity than all metals, being two to four times higher (depending on the impurities) than copper, silver, or gold. For several years, large, almost gem quality, yellow diamonds have been manufactured for making heat sinks. Diamond-coated surfaces might someday be made to work as well and at lower cost.

Diamond point-contact transistors in semiconducting devices are able to withstand operating temperatures up to five times higher than those made of silicon. This opens the way for the use of such devices in environments like jet engines or vehicle motors.

Another application of diamond films is to make wear-resistant cutting tools. A diamond coating of only 3 μm in thickness greatly prolongs the life of such tools and so reduces costs.

Speakers for high frequency sounds—"tweeters" —are improved by using diamond-coated diaphragms. With such a surface, sound distortions in the high frequency region are largely shifted to frequencies above those the human ear can discern. The reproduction of sound is thus of much higher fidelity.

Other applications being developed for diamond coatings are scratch-resistant coatings for lenses, sunglasses, and watch glasses.

Questions

1. For the conversion of graphite to diamond, what is the significance of the value of ΔG being *negative*?

2. Ultrahigh pressures tend to favor the conversion of graphite to diamond. Why?

3. What specific *particles* constitute the "building blocks" for the diamond structure when diamond films are made?

4. What physical property of diamond is behind the uses of diamond films for making computer parts?

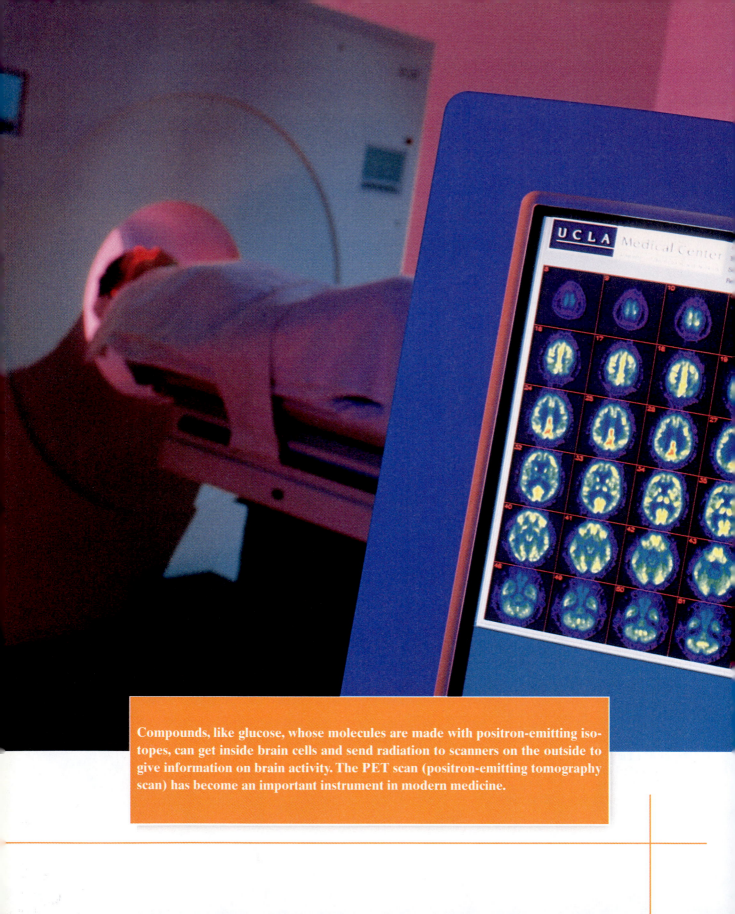

Compounds, like glucose, whose molecules are made with positron-emitting iso-topes, can get inside brain cells and send radiation to scanners on the outside to give information on brain activity. The PET scan (positron-emitting tomography scan) has become an important instrument in modern medicine.

Nuclear Reactions and Their Role in Chemistry

This Chapter in Context As in an ocean storm, nearly all of the action in chemistry occurs near the surface, among electron energy levels, and not in the depths among atomic nuclei. The nuclei of most isotopes are exceptionally stable and are completely unaffected by events high above. But there are exceptions. Many elements have one or more isotopes with unstable nuclei, resulting in properties both useful and dangerous. You will study them in this chapter.

22.1 Conservation of Mass–Energy

Isotopes with unstable atomic nuclei are **radioactive,** so called because they emit high-energy streams of particles or electromagnetic radiation, sometimes both at the same time. Often called **radionuclides,** radioactive isotopes undergo *nuclear reactions,* changes new to our study. Scientists exploit these reactions for many reasons, such as to study chemical changes, to conduct many kinds of analyses (including the dating of rocks and ancient objects), to diagnose or treat many diseases, and to obtain energy.

Interconvertibility of Mass and Energy

To begin our study of radionuclides we reexamine two physical laws that, until this chapter, have been assumed to be separate and independent, namely, the laws of conservation of energy and conservation of mass. They may be safely treated as distinct for chemical reactions but not for nuclear reactions. These two laws, however, are only different aspects of a deeper, more general law.

As atomic and nuclear physics developed in the early 1900s, physicists realized that the mass of a particle cannot be treated as a constant in all circumstances. The mass, m, of a particle depends on the particle's velocity, v, meaning the velocity relative to the observer. A particle's mass is related to this velocity, and to the velocity of light, c, by the following equation.

$$m = \frac{m_0}{\sqrt{1 - (v/c)^2}} \qquad (22.1)$$

$c = 3.00 \times 10^8$ m s^{-1}, the speed of light.

993

Notice what happens when v is zero and the particle has no velocity (relative to the observer). The ratio v/c is then zero, the whole denominator reduces to a value of 1, and Equation 22.1 becomes

$$m = m_0$$

This is why the symbol m_0 stands for the particle's *rest mass*.

Rest mass is what we measure in all lab operations, because any object, like a chemical sample, is either at rest (from our viewpoint) or is not moving extraordinarily rapidly. Only as the particle's velocity approaches the speed of light, c, does the v/c term in Equation 22.1 become important. As v approaches c, the ratio v/c approaches 1, and so $[1 - (v/c)^2]$ gets closer and closer to 0. The whole denominator, in other words, approaches a value of 0. If it actually reached 0, then m, which would be $(m_0 \div 0)$, would go to infinity. In other words, the mass, m, of the particle moving at the velocity of light would be infinitely great, a physical impossibility. This is why the speed of light is seen as an absolute upper limit on the speed that any particle can have.

At the velocities of everyday experience, the mass of anything calculated by Equation 22.1 equals the rest mass to several significant figures. The difference cannot be detected by weighing devices. Thus, in all of our normal work, mass appears to be conserved, and the law of conservation of mass functions this way in chemistry.

In a universe where mass changes with velocity, physicists also realized that a new understanding of energy is necessary. Mass and energy were seen to be interconvertible, just like potential and kinetic energies. What is conserved with respect to the energy of a system is *all* its forms of energy, including the system's mass calculated as an equivalent amount of energy. Alternatively, what is conserved about the mass of a system is *all* its forms, including the system's energy expressed as an equivalent amount of mass. The single, deeper law that summarizes these conclusions is now called the **law of conservation of mass–energy.**

> **Law of Conservation of Mass–Energy**
> The sum of all the energy in the universe and of all the mass (expressed as an equivalent in energy) is a constant.

Einstein Equation

Albert Einstein, one of the many famous physicists of the twentieth century, was able to show that when mass converts to energy, the amount of the change in energy, ΔE, is related to the change in rest mass, Δm_0, by the following equation, now called the **Einstein equation.**

$$\Delta E = \Delta m_0 c^2$$

Again, c is the velocity of light, 3.00×10^8 m s^{-1}.

Because the velocity of light is very large, even an enormous value of ΔE corresponds to an extremely small change in mass, Δm_0. The combustion of methane, for example, releases considerable heat per mole.

$$CH_4(g) + 2O_2(g) \longrightarrow CO_2(g) + 2H_2O(l) \qquad \Delta H^\circ = -890 \text{ kJ}$$

The source of the 890 kJ of heat energy is a loss of mass. When calculated by the Einstein equation, the 890 kJ originates from the conversion into energy of 9.89 ng of mass, about 1×10^{-7}% of the mass of 1 mol of CH_4 and 2 mol of O_2. Such a tiny change is not detectable by laboratory balances. Although the Ein-

Even at $v = 1000$ m s^{-1} (about 2250 mph), the denominator (unrounded) is 0.99999333, or within 7×10^{-4}% of 1.

Albert Einstein (1879–1955); Nobel Prize, 1921 (physics).

The Einstein equation is given as $E = mc^2$ in the popular press.

stein equation has no practical application in chemistry, its importance certainly became clear in physics when atomic fission was first observed in 1939.

22.2 Nuclear Binding Energies

The actual mass of an atomic nucleus is always a little smaller than the sum of the rest masses of all of its nucleons, its protons and neutrons. The mass difference, sometimes called the **mass defect,** represents mass that changed into energy as the nucleons gathered to form the nucleus, this energy being released from the system. The absorption of the same amount of energy by the nucleus would be required to break it apart into its nucleons again, so the energy is called the **binding energy** of the nucleus.

Nuclear binding energy is not energy actually possessed by the nucleus but is, instead, the energy the nucleus would have to absorb to break apart. Thus, the *higher* the binding energy, the *more stable* is the nucleus.

Nuclear Binding Energy of Helium

We can calculate nuclear binding energies using the Einstein equation. Helium-4, for example, has atomic number 2, so its nucleus consists of 4 nucleons, 2 protons and 2 neutrons. The rest mass of one helium-4 nucleus is known to be 4.001506 u. However, the sum of the rest masses of its four separated nucleons is slightly more, 4.031882748 u, which we can show as follows. The rest mass of an isolated proton is 1.007276470 u and that of a neutron is 1.008664904 u.

> The atomic mass unit, u, equals $1.6605402 \times 10^{-24}$ g.

$$\text{For 2 protons: } 2 \times 1.007276470 \text{ u} = 2.014552940 \text{ u}$$

$$\text{For 2 neutrons: } 2 \times 1.008664904 \text{ u} = \underline{2.017329808} \text{ u}$$

$$\text{Total rest mass of nucleons in } ^4\text{He} = 4.031882748 \text{ u}$$

The mass defect, the difference between the calculated and measured rest masses for the helium-4 nucleus, is 0.030377 u, found by

$$4.031882748 \text{ u} - 4.001506 \text{ u} = 0.030377 \text{ u}$$

Using Einstein's equation, the nuclear binding energy that is equivalent to the mass defect of 0.030377 u is found as follows, where we have to convert u (the atomic mass unit) to kilograms by the relationships $1 \text{ u} = 1.6605402 \times 10^{-24}$ g, 1 kg = 1000 g, and $1 \text{ J} = 1 \text{ kg m}^2 \text{ s}^{-2}$.

$$\Delta E = \Delta mc^2 = \underbrace{(0.030377 \text{ u}) \frac{(1.6605402 \times 10^{-24} \text{ g})}{1 \text{ u}} \frac{1 \text{ kg}}{1000 \text{ g}}}_{\Delta m \text{ (in kg)}} \underbrace{(3.00 \times 10^8 \text{ m s}^{-1})^2}_{c^2}$$

$$= 4.54 \times 10^{-12} \text{ kg m}^2 \text{ s}^{-2}$$

$$= 4.54 \times 10^{-12} \text{ J}$$

There are four nucleons in the helium-4 nucleus, so the binding energy per nucleon is $(4.54 \times 10^{-12} \text{ J})/4$ nucleons or 1.14×10^{-12} J/nucleon.

The formation of just one nucleus of helium-4 releases 4.54×10^{-12} J. If we could make Avogadro's number or 1 mol of helium-4 nuclei—the total mass would be only 4 grams—the net release of energy would be

$$(6.02 \times 10^{23} \text{ nuclei}) \times (4.54 \times 10^{-12} \text{ J/nucleus}) = 2.73 \times 10^{12} \text{ J}$$

This is a huge amount of energy from forming only 4 g of helium. It could keep a 100 watt lightbulb lit for nearly 900 years!

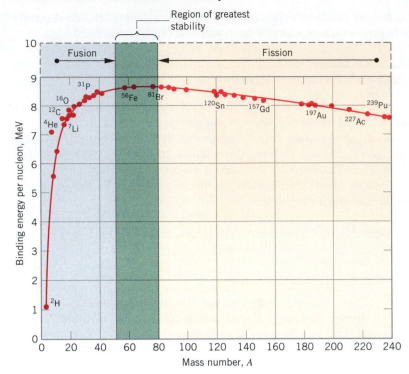

Figure 22.1 *Binding energies per nucleon.* The energy unit here is the megaelectron volt or MeV; 1 MeV = 10^6 eV = 1.602×10^{-13} J. (From D. Halliday and R. Resnick, *Fundamentals of Physics,* 2nd ed., revised, 1986, John Wiley & Sons, Inc. Used by permission.)

The formation of a nucleus from its nucleons is called **nuclear fusion,** and our calculation shows that the energy potentially available from nuclear fusion is enormous. The technological problems of achieving sustained fusion are immense, however, and practical power from fusion appears to be decades off.

Relationship between Binding Energy per Nucleon and Nuclear Stability

Figure 22.1 shows a plot of binding energies per nucleon versus mass numbers for most of the elements. The curve passes through a maximum at iron-56, which means that the nuclei of iron-56 atoms are the most stable of all. The plot in Figure 22.1, however, does not have a sharp maximum. Thus, a large number of elements with intermediate mass numbers in the broad center of the periodic table include the most stable isotopes in nature.

As we continue to follow the plot of Figure 22.1 to the highest mass numbers, the nuclei decrease in stability as the binding energies decrease. Among the heaviest atoms, therefore, we might expect to find isotopes that could change to more stable forms by breaking up, by undergoing nuclear fission. **Nuclear fission,** the subject of Section 22.8, is the spontaneous breaking apart of a nucleus to form isotopes of intermediate mass number.

Quite often you'll see the terms atomic fusion and atomic fission used for nuclear fusion and nuclear fission.

22.3 Radioactivity

Except for hydrogen, all atomic nuclei have more than one proton. Each proton bears a positive charge, and we have learned that like charges repel. So we might well ask how can *any* nucleus be stable? Electrostatic forces of attraction and repulsion, such as the kinds present among the ions in a crystal of sodium chloride, are not the only forces at work in the nucleus, however. Protons do, indeed, repel each other electrostatically, but another force, a force of attraction called the *nuclear strong force,* also acts in the nucleus. The nuclear strong force overcomes the electrostatic force of repulsion between protons, and it binds both protons and neutrons into a nuclear package. Moreover, the neutrons, by helping to keep the protons farther apart, also lessen repulsions between protons.

Still another nuclear force is called the *weak force,* and it is involved in beta decay.

Radioactive Decay

One consequence of the difference between the nuclear strong force and the electrostatic force occurs among nuclei carrying large numbers of protons but with too few intermingled neutrons to dilute the electrostatic repulsions between protons. Such nuclei are often unstable and occur among radionuclides. Their nuclei carry more energy than do other arrangements of the same nucleons. To achieve less energy and thus more stability, radionuclides undergo **radioactive decay,** meaning that they eject small nuclear fragments, and many simultaneously release high-energy electromagnetic radiation. This phenomenon is called **radioactivity.** About 50 of the approximately 350 naturally occurring isotopes are radioactive.

Adjacent neutrons experience no electrostatic repulsion between each other, only the strong force (of attraction).

Alpha Radiation

Naturally occurring atomic radiation consists principally of three kinds: alpha, beta, and gamma radiation.

Alpha radiation consists of a stream of the nuclei of helium atoms, called **alpha particles,** symbolized as ^4_2He, where 4 is the mass number and 2 is the atomic number. The alpha particle bears a charge of 2+, but the charge is omitted from the symbol.

Alpha particles are the most massive of any commonly emitted by radionuclides. When ejected (Figure 22.2), alpha particles move through the atom's electron orbitals reaching emerging speeds of up to one-tenth the speed of light. Their size, however, prevents them from going far. After traveling at most only a few centimeters in air, alpha particles collide with air molecules, lose kinetic energy, pick up electrons, and become neutral helium atoms. Alpha particles cannot penetrate the skin, although enough exposure causes a severe skin burn. If carried in air or on food into the soft tissues of the lungs or the intestinal tract, emitters of alpha particles can cause serious harm, including cancer.

Alpha-emitting radionuclides are common among the elements near the end of the periodic table.

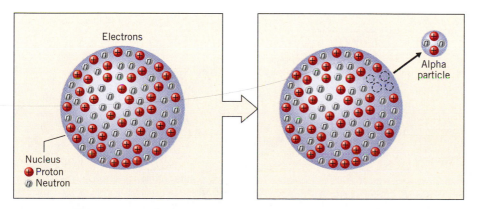

Figure 22.2 *Emission of an alpha particle from an atomic nucleus.*

Nuclear Equations

To symbolize the decay of a nucleus, we construct a **nuclear equation,** which we can illustrate by the alpha decay of uranium-238 to thorium-234.

$$^{238}_{92}\text{U} \longrightarrow {}^{234}_{90}\text{Th} + {}^{4}_{2}\text{He}$$

Unlike chemical reactions, nuclear reactions produce new isotopes, so we need separate rules for balancing nuclear equations. A nuclear equation is balanced when

1. The sums of the mass numbers on each side of the arrow are equal.
2. The sums of the atomic numbers on each side are equal.

In the nuclear equation for the decay of uranium-238, the atomic numbers balance ($90 + 2 = 92$), and the mass numbers balance ($234 + 4 = 238$). Notice that electrical charges are not included, even though they are there (initially). The alpha particle, for example, has a charge of $2+$. Initially, therefore, the thorium particle has as much opposite charge, $2-$. These charged particles, however, eventually pick up or lose electrons either from each other or from molecules in the matter through which they travel.

Beta Radiation

Naturally occurring **beta radiation** consists of a stream of electrons, which in this context are called **beta particles.** In a nuclear equation, the beta particle has the symbol $^{0}_{-1}e$, because the electron's mass number is 0 and its charge is $1-$. Hydrogen-3 (tritium) is a beta emitter that decays by the following equation.

Tritium is a synthetic radionuclide.

$$^{3}_{1}\text{H} \longrightarrow {}^{3}_{2}\text{He} + {}^{0}_{-1}e + \bar{\nu}$$

tritium   helium-3   beta particle   antineutrino
(electron)

Both the antineutrino (to be described shortly) and the beta particle come from the atom's nucleus, not its electron shells. We do not think of them as having a prior existence in the nucleus, any more than a photon exists before its emission from an excited atom (see Figure 22.3). Both the beta particle and the antineutrino are created during the decay process in which a neutron is transformed into a proton.

$$^{1}_{0}n \longrightarrow {}^{0}_{-1}e + {}^{1}_{1}p + \bar{\nu}$$

neutron   beta particle   proton   antineutrino
(in the nucleus)   (emitted)   (remains in
the nucleus)

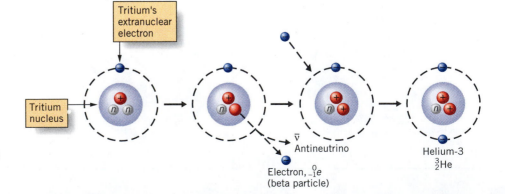

Figure 22.3 *Emission of a beta particle from a tritium nucleus.*

Unlike alpha particles, which are all emitted with the same discrete energy from a given radionuclide, beta particles emerge from a given beta emitter with a continuous spectrum of energies. Their energies vary from zero to some characteristic fixed upper limit for each radionuclide. This fact once gave nuclear physicists considerable trouble, partly because it was an apparent violation of energy conservation. To solve this problem, Wolfgang Pauli proposed in 1927 that beta emission is accompanied by yet another decay particle, this one electrically neutral and almost massless. Enrico Fermi suggested the name *neutrino* ("little neutral one"), but eventually it was named the *antineutrino*, symbolized by $\bar{\nu}$.

An electron is extremely small, so a beta particle is less likely to collide with the molecules of anything through which it travels. Depending on its initial kinetic energy, a beta particle can travel up to 300 cm in dry air, much farther than alpha particles. Only the highest energy beta particles can penetrate the skin, however.

Enrico Fermi (1901–1954); Nobel Prize, 1938 (physics).

Gamma Radiation

Gamma radiation, which often accompanies either alpha or beta radiation, consists of high-energy photons given the symbol $_0^0\gamma$ or, often, simply γ in equations. Gamma radiation is extremely penetrating and is effectively blocked only by very dense materials, like lead.

The emission of gamma radiation involves transitions between energy levels *within* the nucleus. Nuclei have energy levels of their own, much as atoms have orbital energy levels. When a nucleus emits an alpha or beta particle, it sometimes is left in an excited energy state. By the emission of a gamma-ray photon, the nucleus relaxes into a more stable state.

Electron Volt as an Energy Unit

The energy carried by a given radiation is usually described by an energy unit new to our study, the **electron volt,** abbreviated **eV**; 1 eV is the energy an electron receives when accelerated under the influence of 1 volt. It is related to the joule as follows.

$$1 \text{ eV} = 1.602 \times 10^{-19} \text{ J}$$

As you can see, the electron volt is an extremely small amount of energy, so multiples are commonly used, like the kilo-, mega-, and gigaelectron volt.

$$1 \text{ keV} = 10^3 \text{ eV}$$

$$1 \text{ MeV} = 10^6 \text{ eV}$$

$$1 \text{ GeV} = 10^9 \text{ eV}$$

In the interconversion of mass and energy,

$$1 \text{ eV} = 1.783 \times 10^{-36} \text{ kg}$$
$$1 \text{ GeV} = 1.783 \times 10^{-24} \text{ g}$$

An alpha particle radiated by radium-224 has an energy of 5 MeV. Hydrogen-3 (tritium) emits beta radiation at an energy of 0.05 to 1 MeV. The gamma radiation from cobalt-60, the radiation currently used to kill bacteria and other pests in certain foods, consists of photons with energies of 1.173 MeV and 1.332 MeV.

X Rays

X rays, like gamma rays, consist of high-energy electromagnetic radiation, but their energies are usually less than those of gamma radiation. Although X rays are emitted by some synthetic radionuclides, they normally are generated by special machines in which a high-energy electron beam is focused onto a metal target.

X rays used in diagnosis typically have energies of 100 keV or less.

EXAMPLE 22.1
Writing a Balanced Nuclear Equation

Ions of cesium, which is in the same family as sodium, travel in the body to many of the same sites where sodium ions go.

Marie Curie (1867–1934); Nobel Prizes, 1903 (physics) and 1911 (chemistry).

Cesium-137, $^{137}_{55}Cs$, one of the radioactive wastes from a nuclear power plant or an atomic bomb explosion, emits beta and gamma radiation. Write the nuclear equation for the decay of cesium-137.

Analysis: We start with an incomplete equation using the given information and then figure out any other data needed to complete the equation.

Solution: The incomplete nuclear equation is

$$^{137}_{55}Cs \longrightarrow ^{0}_{-1}e + ^{0}_{0}\gamma + \underline{\quad}$$

Mass number goes here.
Atomic symbol goes here.
Atomic number goes here.

The atomic symbol can be obtained from the table inside the front cover after we have determined the atomic number, Z. Z is found using the fact that the atomic number (55) on the left side of the equation must equal the sum of the atomic numbers on the right side.

$$55 = -1 + 0 + Z$$
$$Z = 56$$

The periodic table tells us that element 56 is Ba (barium). To determine which isotope of barium forms, we recall that the sums of the mass numbers on either side of the equation must also be equal. Letting A be the mass number of the barium isotope,

$$137 = 0 + 0 + A$$
$$A = 137$$

The balanced nuclear equation, therefore, is

$$^{137}_{55}Cs \longrightarrow ^{0}_{-1}e + ^{0}_{0}\gamma + ^{137}_{56}Ba$$

Is the Answer Reasonable?
Besides double-checking that element 56 is barium, the answer satisfies the requirements of a nuclear equation, namely, the sums of the mass numbers, 137, are the same on both sides as are the sums of the atomic numbers.

Practice Exercise 1

Radium-226, $^{226}_{88}Ra$, is an alpha and gamma emitter. Write a balanced nuclear equation for its decay. (Marie Curie earned one of her two Nobel Prizes for isolating the element radium, which soon became widely used to treat cancer.) ◆

Practice Exercise 2

Write the balanced nuclear equation for the decay of strontium-90, a beta emitter. (Strontium-90 is one of the many radionuclides present in the wastes of operating nuclear power plants.) ◆

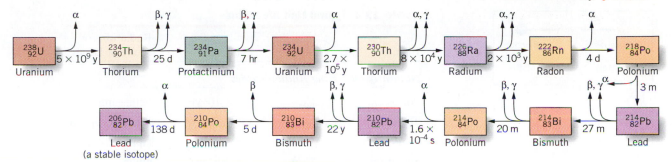

Figure 22.4 *The uranium-238 radioactive disintegration series.* The time given beneath each arrow is the half-life period of the preceding isotope (y = year, d = day, hr = hour, m = minute, and s = second).

Radioactive Disintegration Series

Sometimes one radionuclide does not decay to a stable isotope but decays instead to another unstable radionuclide. The decay of one radionuclide after another will continue until a stable isotope forms. The sequence of such successive nuclear reactions is called a **radioactive disintegration series.** Four series occur naturally. Uranium-238 is at the head of one (Figure 22.4).

We studied half-lives in Section 13.5. One **half-life** period in nuclear science is the time it takes for a given sample of a radionuclide to decay to one-half of its initial amount. Because radioactive decay is a first-order process, the period of time taken by one half-life is independent of the initial number of nuclei. The huge variations in the half-lives of several radionuclides are shown in Table 22.1.

Positron and Neutron Emissions; Electron Capture

Many synthetic isotopes emit positrons, particles with the mass of an electron but a positive instead of a negative charge. A **positron** is a positive beta particle, a positive electron, and its symbol is ^0_1e. It forms in the nucleus by the conversion of a proton to a neutron (Figure 22.5). Positron emission, like beta emission, is accompanied by a chargeless and virtually massless particle, a *neutrino* (v), the counterpart of the antineutrino ($\bar{v}$) in the realm of antimatter (defined below). Cobalt-54, for example, is a positron emitter and changes to a stable isotope of iron.

$$^{54}_{27}\text{Co} \longrightarrow {}^{54}_{26}\text{Fe} + {}^0_1e + v$$
$$\qquad\qquad\qquad\quad \text{positron}\quad \text{neutrino}$$

A positron, when emitted, eventually collides with an electron, and the two annihilate each other (Figure 22.6). Their masses change into the energy of two gamma-ray photons called *annihilation radiation photons,* each with an energy of 511 keV.

$$^0_{-1}e + {}^0_1e \longrightarrow 2\,{}^0_0\gamma$$

Because a positron destroys a particle of ordinary matter (an electron), it is called a particle of antimatter. To be classified as **antimatter,** a particle must have a counterpart among one of ordinary matter, and the two must annihilate each other when they collide. For example, a neutron destroys an antineutron.

Neutron emission, another kind of nuclear reaction, does not lead to an isotope of a different element. Krypton-87, for example, decays as follows to krypton-86.

$$^{87}_{36}\text{Kr} \longrightarrow {}^{86}_{36}\text{Kr} + {}^1_0n$$
$$\qquad\qquad\qquad\qquad \text{neutron}$$

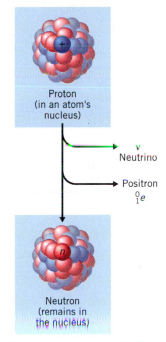

Figure 22.5 *The emission of a positron replaces a proton by a neutron.*

The symbol $^+\beta$ is sometimes used for the positron.

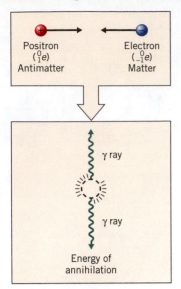

Figure 22.6 *Gamma radiation is produced when a positron and an electron collide.*

Table 22.1 Typical Half-life Periods

Element	Isotope	Half-life	Radiations or Mode of Decay
Naturally Occurring Radionuclides			
Potassium	$^{40}_{19}K$	1.3×10^9 yr	beta, gamma
Tellurium	$^{123}_{52}Te$	1.2×10^{13} yr	electron capture
Neodymium	$^{144}_{60}Nd$	5×10^{15} yr	alpha
Samarium	$^{149}_{62}Sm$	4×10^{14} yr	alpha
Rhenium	$^{187}_{75}Re$	7×10^{10} yr	beta
Radon	$^{222}_{86}Rn$	3.82 days	alpha
Radium	$^{226}_{88}Ra$	1590 yr	alpha, gamma
Thorium	$^{230}_{90}Th$	8×10^4 yr	alpha, gamma
Uranium	$^{238}_{92}U$	4.51×10^9 yr	alpha
Synthetic Radionuclides			
Tritium	$^{3}_{1}T$	12.26 yr	beta
Oxygen	$^{15}_{8}O$	124 s	positron
Phosphorus	$^{32}_{15}P$	14.3 days	beta
Technetium	$^{99m}_{43}Tc$	6.02 hr	gamma
Iodine	$^{131}_{53}I$	8.07 days	beta
Cesium	$^{137}_{55}Cs$	30 yr	beta
Strontium	$^{90}_{38}Sr$	28.1 yr	beta
Americium	$^{243}_{95}Am$	7.37×10^3 yr	alpha

Electron capture, yet another kind of nuclear reaction, is very rare among natural isotopes but common among synthetic radionuclides. Vanadium-50 nuclei, for example, can capture orbital K shell or L shell electrons, change to nuclei of stable atoms of titanium, and emit X rays and neutrinos.

The K shell is the shell with principal quantum number $n = 1$; the L shell is at $n = 2$.

$$^{50}_{23}V + {}^{0}_{-1}e \xrightarrow{\text{electron capture}} {}^{50}_{22}Ti + X \text{ rays} + \nu$$

The net effect of electron capture is the conversion of a proton into a neutron (Figure 22.7).

$$\underset{\substack{\text{proton} \\ \text{(in the} \\ \text{nucleus)}}}{^{1}_{1}p} + \underset{\substack{\text{electron} \\ \text{(captured} \\ \text{from the} \\ \text{K shell)}}}{^{0}_{-1}e} \longrightarrow \underset{\substack{\text{neutron} \\ \text{(in the} \\ \text{nucleus)}}}{^{1}_{0}n}$$

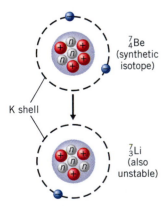

Figure 22.7 *Electron capture.* Electron capture is the collapse of an orbital electron into the nucleus, and this changes a proton into a neutron.

Electron capture does not change an atom's mass number, only its atomic number. It also leaves a hole in the K or L shell, and the atom emits photons of X rays as other orbital electrons drop down to fill the hole. Moreover, the nucleus that has just captured an orbital electron may be in an excited energy state and so can emit a gamma ray photon.

22.4 Band of Stability

When all known isotopes of each element, both stable and unstable, are arrayed on a plot according to numbers of protons and neutrons, an interesting zone can be defined (Figure 22.8). The two curved lines in the array of Figure 22.8 enclose this zone, called the **band of stability,** within which lie all stable nuclei. (No isotope above element 83, bismuth, is included in Figure 22.8 because none has a *stable* isotope.) Within the band of stability are also some unstable isotopes, because smooth lines cannot be drawn to exclude them.

Any isotope not represented anywhere on the array, inside or outside the band of stability, probably has a half-life too short to permit its detection. An isotope with 50 neutrons and 60 protons, for example, would be too unstable to justify the time and money for an attempt to make it.

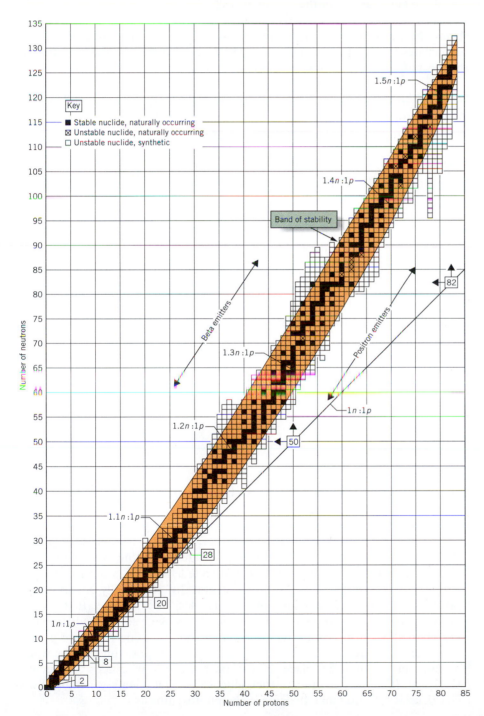

Figure 22.8 *The band of stability.*

Notice that the band curves slightly upward as the number of protons increases. The curvature means that the *ratio* of neutrons to protons gradually increases from 1:1, a ratio indicated by the straight line in Figure 22.8. The reason is easy to understand. More protons require more neutrons to provide a compensating nuclear strong force and to dilute electrostatic proton–proton repulsions.

Isotopes occurring above and to the left of the band of stability tend to be beta emitters. Isotopes lying below and to the right of the band are positron emitters. The isotopes with atomic numbers above 83 tend to be alpha emitters. Are there any reasons for these tendencies?

Alpha Emitters. The alpha emitters, as we said, occur mostly among the radionuclides above atomic number 83. Their nuclei have too many protons, and the most efficient way to lose protons is by the loss of an alpha particle.

Beta Emitters. Beta emitters are generally above the band of stability and so have neutron:proton ratios that evidently are too high. By beta decay a nucleus loses a neutron and gains a proton, thus decreasing the ratio.

$$\ _{0}^{1}n \longrightarrow\ _{1}^{1}p +\ _{-1}^{0}e + \bar{\nu}$$

The proton can also be given the symbol $_{1}^{1}H$ in nuclear equations.

For example, by beta decay fluorine-20 decreases its neutron:proton ratio from 11/9 to 10/10.

$$\ _{9}^{20}F \longrightarrow\ _{10}^{20}Ne +\ _{-1}^{0}e + \bar{\nu}$$

$$\frac{\text{neutron}}{\text{proton}} = \frac{11}{9} \qquad \frac{10}{10}$$

The surviving nucleus, that of neon-10, is closer to the center of the band of stability. Figure 22.9, an enlargement of the fluorine part of Figure 22.8, explains this change further. Figure 22.9 also shows how the beta decay of magnesium-27 to aluminum-27 also lowers the neutron:proton ratio.

Positron Emitters. In nuclei with too few neutrons to be stable, positron emission increases the neutron:proton ratio. A fluorine-17 nucleus, for example, increases its neutron:proton ratio, improves its stability, and moves into the band of stability by emitting a positron and a neutrino and changing to oxygen-17 (see Figure 22.9).

$$\ _{9}^{17}F \longrightarrow\ _{8}^{17}O +\ _{1}^{0}e + \nu$$

$$\frac{\text{neutron}}{\text{proton}} = \frac{8}{9} \qquad \frac{9}{8}$$

Positron decay by magnesium-23 to sodium-23 produces a similarly favorable shift (also shown in Figure 22.9).

▶**Chemistry in Practice**◀ Positron emitters are now being used in an important method for studying brain function called the PET scan, for *positron emission tomography*. The technique begins by chemically incorporating positron-emitting radionuclides into molecules, like glucose, that can be absorbed by the brain directly from the blood. It's like inserting radiation generators that act from within the brain rather than focusing X rays or gamma rays from the outside. Carbon-11, for example, is a positron emitter whose atoms can be used in place of carbon-12 atoms in glucose molecules, $C_6H_{12}O_6$. (One way to prepare such glucose is to let a leafy vegetable, Swiss chard, use $^{11}CO_2$ to make the glucose by photosynthesis.)

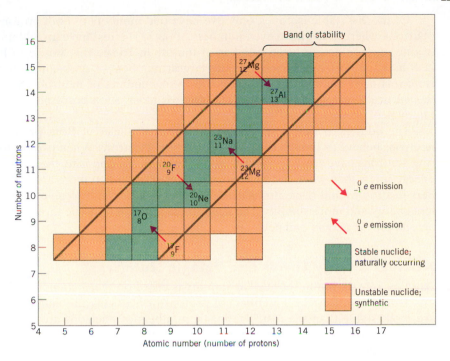

Figure 22.9 *An enlarged section of the band of stability.* Beta decay from magnesium-27 and fluorine-20 reduces their neutron/proton ratio and moves them closer to the band of stability. Positron decay from magnesium-23 and fluorine-17 increases this ratio and moves these nuclides closer to the band of stability, too.

A PET scan using carbon-11 glucose detects situations where glucose is not taken up normally, e.g., in manic depression, schizophrenia, and Alzheimer's disease. After the carbon-11 glucose is ingested, radiation detectors outside the body pick up the annihilation radiation of positron emission from specifically those brain sites that use glucose. PET scan technology, for example, showed that the uptake of glucose by the brains of smokers is less than that of nonsmokers (see Figure 22.10). ◆

Odd–Even Rule

Nature favors even numbers for protons and neutrons, as summarized by the **odd–even rule.**

Odd–Even Rule

When the numbers of neutrons and protons in a nucleus are both even, the isotope is far more likely to be stable than when both numbers are odd.

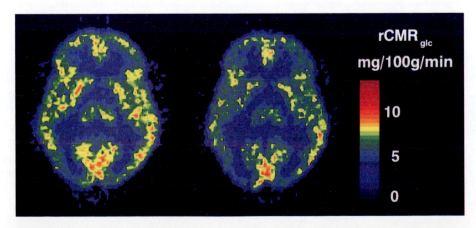

Figure 22.10 *Positron emission tomography (PET) in the study of brain activity.* (*Left*) Nonsmoker. (*Right*) Smoker. The PET scan reveals widespread reduction in the rate of glucose metabolism when nicotine is present. (Photos courtesy of E. D. London.)

Of the 264 stable isotopes, only five have odd numbers of both protons and neutrons, whereas 157 have even numbers of both. The rest have an odd number of one nucleon and an even number of the other. To see this, notice in Figure 22.8 how the horizontal lines with the largest numbers of dark squares (stable isotopes) most commonly correspond to even numbers of neutrons. Similarly, the vertical lines with the most dark squares most often correspond to even numbers of protons.

The isotopes $^2_1$H, $^6_3$Li, $^{10}_5$B, $^{14}_7$N, and $^{138}_{57}$La all have odd numbers of both protons and neutrons.

The odd–even rule is related to the spins of nucleons. Both protons and neutrons behave as though they spin, like orbital electrons. When two protons or two neutrons have paired spins, meaning the spins are opposite, their combined energy is less than when the spins are unpaired. Only when there are even numbers of protons and of neutrons can all spins be paired and so give the nucleus less energy and more stability. The least stable nuclei tend to be those with both an odd proton and an odd neutron.

Nuclear Energy Levels and "Magic Numbers"

Another rule of thumb for nuclear stability is based on *magic numbers* of nucleons. Isotopes with specific numbers of protons or neutrons, the **magic numbers,** are more stable than the rest. The magic numbers of nucleons are 2, 8, 20, 28, 50, 82, and 126, and where they fall is shown in Figure 22.8 (except for magic number 126).

When the numbers of both protons and neutrons are the same magic number, as they are in $^4_2$He, $^{16}_8$O, and $^{40}_{20}$Ca, the isotope is very stable. $^{100}_{50}$Sn also has two identical magic numbers. Although this isotope of tin is unstable, having a half-life of only several seconds, it is much more stable than nearby radionuclides, whose half-lives are in milliseconds. Thus, although tin-100 lies well outside the band of stability, it is stable enough to be observed. One stable isotope of lead, $^{208}_{82}$Pb, involves two different magic numbers, 82 protons and 126 neutrons.

Magic numbers do not cancel the need for a favorable neutron/proton ratio. An atom with 82 protons and 82 neutrons lies far outside the band of stability, yet 82 is a magic number.

The existence of magic numbers supports the hypothesis that a nucleus has a shell structure with energy levels analogous to electron energy levels. Electron levels, as you already know, are associated with their own special numbers, those that equal the maximum number of electrons allowed in a principal energy level: 2, 8, 18, 32, 50, 72, and 98 (for principal levels 1, 2, 3, 4, 5, 6, and 7, respectively). The total numbers of electrons in the atoms of the most chemically stable elements—the noble gases—also make up a special set: 2, 10, 18, 36, 54, and 86 electrons. Thus, special sets of numbers are not unique to nuclei.

22.5 Transmutation

The change of one isotope into another is called **transmutation,** and radioactive decay is only one cause. Transmutation can also be forced by the bombardment of nuclei with high-energy particles, such as alpha particles from natural emitters, neutrons from atomic reactors, or protons made by stripping electrons from hydrogen. To make them better bombarding missiles, protons and alpha particles can be accelerated in an electrical field (Figure 22.11). This gives them greater energy and enables them to sweep through the target atom's orbital electrons and become buried in its nucleus. Although beta particles can be accelerated, their disadvantage is that they are repelled by a target atom's own electrons.

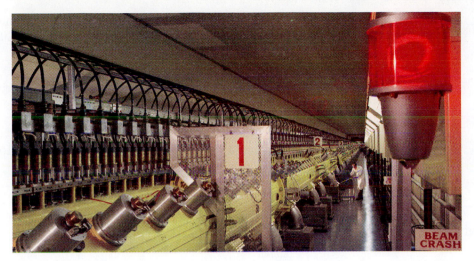

Figure 22.11 *Linear accelerator.* In this linear accelerator at Brookhaven National Laboratory on Long Island, New York, protons can be accelerated to just under the speed of light and given up to 33 GeV of energy before they strike their target. (1 GeV = 10^9 eV.)

Formation and Decay of Compound Nuclei

Both the energy and the mass of a bombarding particle enter the target nucleus at the moment of capture. The energy of the new nucleus, called a **compound nucleus,** quickly becomes distributed among all of the nucleons, but the nucleus is nevertheless rendered somewhat unstable. To get rid of the excess energy, a compound nucleus generally ejects something, a neutron, proton, or electron, and often emits gamma radiation as well. This leaves a new nucleus of an isotope different than the original target, so a transmutation has occurred overall.

Compound here refers only to the idea of *combination,* not to a chemical.

Ernest Rutherford observed the first example of artificial transmutation. When he let alpha particles pass through a chamber containing nitrogen atoms, an entirely new radiation was generated, one much more penetrating than alpha radiation. It proved to be a stream of protons (Figure 22.12). Rutherford was able to show that the protons came from the decay of the compound nuclei of fluorine-18, produced when nitrogen-14 nuclei captured bombarding alpha particles.

$$\ce{^4_2He} + \ce{^{14}_7N} \longrightarrow \ce{^{18}_9F*} \longrightarrow \ce{^{17}_8O} + \ce{^1_1p}$$

| alpha particle | nitrogen nucleus | fluorine (compound nucleus) | oxygen (a rare but stable isotope) | proton (high energy) |

The asterisk, *, symbolizes a high-energy nucleus, a *compound nucleus.*

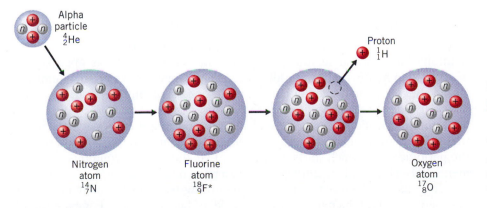

Figure 22.12 *Transmutation of nitrogen into oxygen.* When the nucleus of nitrogen-14 captures an alpha particle it becomes a compound nucleus of fluorine-18. This then expels a proton and becomes the nucleus of oxygen-17.

In the synthesis of alpha particles from lithium-7, protons are used as bombardment particles. The resulting compound nucleus, that of beryllium-8, splits in two.

$$\underset{\text{proton}}{^{1}_{1}p} + \underset{\text{lithium}}{^{7}_{3}\text{Li}} \longrightarrow \underset{\text{beryllium}}{^{8}_{4}\text{Be}^*} \longrightarrow \underset{\substack{\text{alpha} \\ \text{particles}}}{2\,^{4}_{2}\text{He}}$$

Decay Modes of Compound Nuclei

A given compound nucleus can be made in a variety of ways. Aluminum-27, for example, forms by any of the following routes.

$$^{4}_{2}\text{He} + {}^{23}_{11}\text{Na} \longrightarrow {}^{27}_{13}\text{Al}^*$$

$$^{1}_{1}p + {}^{26}_{12}\text{Mg} \longrightarrow {}^{27}_{13}\text{Al}^*$$

$$\underset{\text{deuteron}}{^{2}_{1}\text{H}} + {}^{25}_{12}\text{Mg} \longrightarrow {}^{27}_{13}\text{Al}^*$$

A deuteron is the nucleus of a deuterium atom, just as a proton is the nucleus of a protium (hydrogen) atom.

Once it forms, the compound nucleus, $^{27}_{13}\text{Al}^*$, has no memory of how it was made. It only knows how much energy it has. Depending on this energy, different paths of decay are available, and all of the following routes have been observed. They illustrate how the synthesis of such a large number of synthetic isotopes, some stable and others unstable, has been possible.

$$^{27}_{13}\text{Al}^* \xrightarrow{\substack{\text{The mode of decay is} \\ \text{a function of the} \\ \text{nuclear energy of} \\ \text{aluminum-27.}}}$$

$$\longrightarrow \underset{\substack{\text{aluminum-27} \\ \text{(now stable)}}}{^{27}_{13}\text{Al}} + \underset{\substack{\text{gamma} \\ \text{radiation}}}{^{0}_{0}\gamma}$$

$$\longrightarrow \underset{\substack{\text{magnesium-26} \\ \text{(stable)}}}{^{26}_{12}\text{Mg}} + \underset{\substack{\text{proton} \\ \text{radiation}}}{^{1}_{1}p}$$

$$\longrightarrow \underset{\substack{\text{aluminum-26} \\ \text{(unstable;} \\ t_{1/2} = 7.4 \times 10^5 \text{ yr)}}}{^{26}_{13}\text{Al}} + \underset{\substack{\text{neutron} \\ \text{radiation}}}{^{1}_{0}n}$$

$$\longrightarrow \underset{\substack{\text{magnesium-25} \\ \text{(stable)}}}{^{25}_{12}\text{Mg}} + \underset{\substack{\text{neutron} \\ \text{radiation}}}{^{1}_{0}n} + \underset{\substack{\text{proton} \\ \text{radiation}}}{^{1}_{1}p}$$

$$\longrightarrow \underset{\substack{\text{sodium-23} \\ \text{(stable)}}}{^{23}_{11}\text{Na}} + \underset{\substack{\text{alpha} \\ \text{radiation}}}{^{4}_{2}\text{He}}$$

Radionuclides above Atomic Number 83

Over a thousand isotopes have been made by transmutations. Most do not occur naturally; they appear in the band of stability (Figure 22.8) as open squares, nearly 900 in number. The naturally occurring isotopes above atomic number 83 all have very long half-lives. Others might have existed, but their half-lives probably were too short to permit them to last into our era. All of the elements from neptunium (atomic number 93) and up, the **transuranium elements,** are synthetic. Elements from atomic numbers 93 to 103 complete the actinide series of the periodic table, which starts with element 90, thorium. Beyond this series, elements 104–112, 114, 116, and 118 have also been made.

The study of the chemistry of the elements from 104 and higher must be done virtually one atom at a time because their half-lives are so short.

To make the heaviest elements, bombarding particles larger than neutrons are used, such as alpha particles or the nuclei of heavier atoms. For example, element 110—its temporary name is ununnilium (Uun)—was made when a neutron was ejected from the compound nucleus formed by the fusion of nickel-62 and lead-208.

$$^{62}_{28}\text{Ni} + ^{208}_{82}\text{Pb} \longrightarrow ^{270}_{110}\text{Uun} \longrightarrow ^{269}_{110}\text{Uun} + ^{1}_{0}n$$

Some atoms of element 111 formed when a neutron was lost from the compound nucleus made by bombarding bismuth-209 with nickel-64. Atoms of element 112 emerged from the loss of a neutron from a compound nucleus made by bombarding lead-208 with zinc-70. The synthesis of element 114 was done by Russian scientists by bombarding plutonium-242 with calcium-48. The loss of three neutrons from the compound nucleus gave the 287 isotope of element 114 having a half-life of about 5s. American scientists made element 118 by bombarding lead-208 with krypton-36. Element 116 was a decay product of 118.

Uun is the IUPAC symbol of element 110, named ununnilium, until a permanent name is adopted.

22.6 Detecting and Measuring Radiation

Atomic radiation is often described as **ionizing radiation** because it creates ions by knocking electrons from molecules in the matter through which the radiation travels. The generation of ions is behind some of the devices for detecting radiation.

Geiger Counter

The Geiger–Müller tube, one part of a *Geiger counter,* detects beta and gamma radiation having energy high enough to penetrate the tube's window. Inside the tube is a gas under low pressure in which ions form when radiation enters. The ions permit a pulse of electricity to flow, which activates a current amplifier, a pulse counter, and an audible click.

Scintillation Counters

A *scintillation counter* (Figure 22.13) has a surface coated with a special chemical that emits a tiny flash of light when struck by a particle of radiation. These flashes can be magnified electronically and automatically counted.

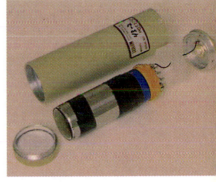

Figure 22.13 *Scintillation probe.* Energy received from radiations striking the phosphor at the end of this probe is processed by the photomultiplier unit and sent to an instrument (not shown) for recording.

Film Dosimeters

The darkening of a photographic film exposed to radiation over a period of time is proportional to the total quantity of radiation received. *Film dosimeters* work on this principle (Figure 22.14), and people who work near radiation sources use them. Each person keeps a log of total exposure, and if a predetermined limit is exceeded, the worker must be reassigned to an unexposed workplace.

Units for Radiation

The Curie and the Becquerel, Units of Activity

The SI unit of activity is the **becquerel (Bq),** and it equals one disintegration or other kind of nuclear transformation per second.

The **curie (Ci),** named after Marie Curie, the discoverer of radium, is an older unit, being equal to the activity of a 1.0 g sample of radium.

$$1 \text{ Ci} = 3.7 \times 10^{10} \text{ disintegrations s}^{-1} = 3.7 \times 10^{10} \text{ Bq}$$

The becquerel is named after Henri Becquerel (1852–1908), the discoverer of radioactivity, who won a Nobel Prize in 1903.

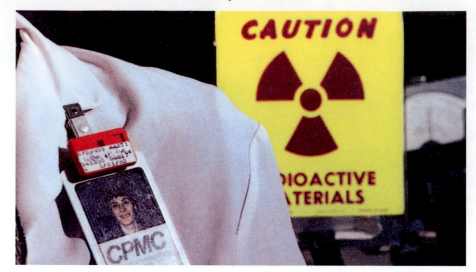

Figure 22.14 *Badge dosimeter.* The doses of various kinds of radiation received by the film in the dosimeter for predetermined periods of time are entered into the worker's log.

The Rad and the Gray, Units of Absorbed Dose

The *gray* is named after Harold Gray, a British radiologist.

The energy equivalent to the quantity of radiation absorbed by some material is defined in units of *absorbed dose* (or simply *dose*). The **gray (Gy)** is the SI unit of absorbed dose, and 1 gray corresponds to 1 joule of energy absorbed per kilogram of absorbing material. The **rad** is an older unit of absorbed dose, 1 rad being the absorption of 10^{-2} joule per kilogram of tissue. Thus, 1 Gy equals 100 rad. In terms of danger, if every individual in a large population received 450 rad (4.5 Gy), roughly half of the population would die in 60 days.

Rad stands for **r**adiation **a**b**s**orbed **d**ose.

The Sievert and the Rem, Units of Dose Equivalent

The gray is not a good basis for comparing the biological effects of radiation in tissue, because these effects depend not just on the energy absorbed but also on the kind of radiation and the tissue itself. The **sievert (Sv),** the SI unit of *dose equivalent,* was invented to meet this problem. If we let D stand for absorbed dose in grays, Q for *quality factor* (meaning relative effectiveness for causing harm in tissue), and N for any other modifying factors, then the dose equivalent, H, in sieverts is defined as the following product.

$$H = DQN$$

In other words, we multiply the absorbed dose (D) from some radiation by a factor (Q) that takes into account the biologically significant properties of the radiation and by any other factor (N) bearing on the net effect. The value of Q for alpha radiation is 20, but it is only 1.0 for beta and gamma radiation. The alpha particle has a large value of Q because of its relatively large size and charge, which enable it to damage molecules with considerable efficiency.

Rem stands for **r**oentgen **e**quivalent for **m**an, where the *roentgen* is a unit related to X ray and gamma radiation.

The **rem** is an older unit of dose equivalent, one still used in medicine. Its value is generally taken to equal 10^{-2} Sv. The U.S. government has set guidelines of 0.3 rem per week as the maximum exposure workers may receive. (For comparison, a chest X ray typically involves about 0.007 rem or 7 mrem.)

Radiation Sickness

A whole body dose of 25 rem (0.25 Sv) induces noticeable changes in human blood. A set of symptoms called *radiation sickness* develops at about 100 rem, becoming severe at 200 rem. Among the symptoms are nausea, vomiting, a drop in the white cell count, diarrhea, dehydration, prostration, hemorrhaging, and loss of hair. If each person in a large population absorbed 400 rem, half would die in 60 days. A 600 rem dose would kill everyone in the group in a week. Many workers received at least 400 rem in the moments following the steam explosion that tore apart one of the nuclear reactors at the Ukraine energy park near Chernobyl in 1986.

The results of radiation damage appear earliest in tissues with rapidly dividing cells, like bone marrow and the gastrointestinal tract.

Radiation-Produced Free Radicals

Even small absorbed doses can be biologically harmful. The danger does not lie in the heat energy associated with the dose, which is usually very small. Rather, the harm is in the ability of ionizing radiation to create unstable ions or neutral species with odd (unpaired) electrons, species that can set off other reactions. Water, for example, can interact as follows with ionizing radiation.

$$H—\ddot{\ddot{O}}—H \xrightarrow{\text{radiation}} \left[H—\dot{\ddot{O}}—H\right]^+ + {}_{-1}^{0}e$$

The new cation, $\left[H—\dot{O}—H\right]^+$, is unstable and one breakup path is

$$\left[H—\dot{\ddot{O}}—H\right]^+ \longrightarrow \underset{\text{proton}}{H^+} + \underset{\substack{\text{hydroxyl} \\ \text{radical}}}{:\dot{O}—H}$$

The proton might pick up a stray electron to become a hydrogen atom, H·.

Both the hydrogen atom and the hydroxyl radical are examples of **free radicals,** which are neutral or charged particles having one or more unpaired electrons and often called simply *radicals*. They are chemically very reactive. What they do next depends on the other chemical species nearby, but radicals can set off a series of totally undesirable chemical reactions inside a living cell. This is what makes the injury from absorbed radiation of far greater magnitude than the energy alone could inflict. A dose of 600 rem is a lethal dose in a human, but the same dose absorbed by pure water causes the ionization of only one water molecule in every 36 million.

Background Radiation

The presence of radionuclides in nature makes it impossible for us to be free from all exposure to ionizing radiation. High-energy cosmic rays shower on us from the sun. They interact with the air's nitrogen molecules to produce carbon-14, a beta emitter, which enters the food chain via the foods made as carbon dioxide is used in photosynthesis (see "Carbon-14 Dating" in Section 22.7). From soil and from building stone comes the radiation of radionuclides native to Earth's crust. The top 40 cm of Earth's soil holds an average of 1 gram of radium, an alpha emitter, per square kilometer. Naturally occurring potassium-40, a beta emitter, adds its radiation wherever potassium ions are found in the body. The presence of carbon-14 and potassium-40 together produce about 5×10^5 nuclear disintegrations per minute inside an adult human. Radon gas seeps into basements from underground formations (see Facets of Chemistry 22.1). In fact, a little over half of the radiation we experience, on average, is from radon-222 and its decay products.

Radium salts were once put into the paint used for the numerals on watch faces to make them glow in the dark.

Diagnostic X rays, both medical and dental, also expose us to ionizing radiation. All these sources produce a combined **background radiation** that averages 360 mrem per person annually in the United States. The averages are roughly 82% from natural radiation and 18% from medical sources.

Facets of Chemistry 22.1

Radon-222 in the Environment

Several years ago an engineer unintentionally but repeatedly set off radiation alarms at his workplace, a nuclear power plant in Pennsylvania. Such alarms are triggered when radiation levels surpass a predetermined value. The problem eventually was traced to his home, where the radiation level was 2700 picocuries per liter of air, compared with the level in the average house of 1 picocurie per liter. (The picocurie is 10^{-12} curie.) The chief culprit was radon-222, a natural intermediate in the uranium-238 disintegration series (Figure 22.4). (One picocurie per liter is equivalent to the decay of two radon atoms per minute.) The engineer had carried radon-222 and its decay products on his clothing from his home to the power plant.

One result of this episode was to send geologists scurrying to map the general area for concentrations of uranium-238. They found that the bedrock of the Reading Prong, a geological formation that cuts across Pennsylvania, New Jersey, New York, and up into the New England states, is relatively rich in uranium.

Radon-222 is a noble gas and is as *chemically* unreactive as argon, krypton, and the other noble gases. But this decay product of uranium-238 has an unstable nucleus. Its half-life is 4 days, and it emits both alpha and gamma radiation. Being a gas, radon-222 diffuses through rocks and soil and enters buildings. In fact, buildings can act much like chimneys and draw radon as a chimney draws smoke. Radon-222 in the air is breathed into the lungs. Because of the short half-life, some of it decays in the lungs before it is exhaled. The decay products are not gases, and several have short half-lives. When these are left in the lungs, their radiation can cause lung cancer. Scientists of the National Research Council estimate that roughly 12% of all lung cancer deaths, 15,000 to 22,000 per year (out of a total

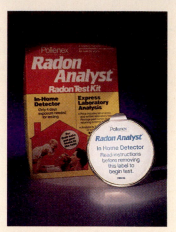

Figure 1 Test kits for measuring radon levels in rooms can be purchased by home owners.

157,000 deaths per year from lung cancer), are caused by exposure to radon. Almost 90% of these victims are also smokers, indicating that smoking and radon exposure make a particularly dangerous mix.

The Environmental Protection Agency (EPA) has recommended an upper limit of 4 picocuries per liter in the air of homes. Estimates released in the late 1980s by the EPA were that roughly 1% of the homes in the United States have radon levels that generate radiation of more than 10 picocuries per liter. Those who estimate risks present in day-to-day activities say that such a level poses less a risk to life than the risk of driving a car. Radon test kits for homeowners are available at home supply stores (see Figure 1).

Radiation Protection

Gamma radiation and X rays are so powerful that they are effectively shielded only by very dense materials, like lead. Otherwise, one should stay as far from a source as possible, because the intensity of radiation diminishes with the *square* of the distance. This relationship is the **inverse square law,** which can be written mathematically as follows, where d is the distance from the source.

$$\text{Radiation intensity} \propto \frac{1}{d^2}$$

When the intensity, I_1, is known at distance d_1, then the intensity I_2 at distance d_2 can be calculated by the following equation.

When the ratio is taken, the proportionality constant cancels, so we don't have to know what its value is.

$$\frac{I_1}{I_2} = \frac{d_2{}^2}{d_1{}^2}$$

This law applies only to a small source that radiates equally in all directions, with no intervening shields.

Practice Exercise 3

If an operator 10 m from a small source is exposed to 1.4 units of radiation, what will be the intensity of the radiation if he moves to 1.2 m from the source?

22.7 Applications of Radioactivity

The fact that the *chemical* properties of both the radioactive and stable isotopes of an element are the same is exploited in some uses of radionuclides. The chemical and physical properties enable scientists to get radionuclides into place in systems of interest. Then the radiation is exploited for medical or analytical purposes. Tracer analysis is an example.

Radioactive Tracers

In **tracer analysis,** the chemical form and properties of a radionuclide enable the system to distribute it to a particular location. The intensity of the radiation then tells something about how that site is working. In the form of the iodide ion, for example, iodine-131 is carried by the body to the thyroid gland, the only user of iodide ion in the body. The gland takes up the iodide ion to synthesize the hormone thyroxine. An underactive thyroid gland is unable to concentrate iodide ion normally and will emit less intense radiation under standard test conditions.

Tracer analyses are also used to pinpoint the locations of brain tumors, which are uniquely able to concentrate the pertechnetate ion, TcO_4^-, made from technetium-99m.[1] This strong gamma emitter, which resembles the chloride ion in some respects, is one of the most widely used radionuclides in medicine.

The tumor appears as a "hot spot" when subjected to radiological analysis.

Neutron Activation Analysis

A number of stable nuclei can be changed into emitters of gamma radiation by capturing neutrons, and this makes possible a procedure called **neutron activation analysis.** Neutron capture followed by gamma emission can be represented by the following equation (where A is a mass number and X is a hypothetical atomic symbol).

$$^{A}X \ + \ ^{1}_{0}n \ \longrightarrow \ ^{(A+1)}X^* \ \longrightarrow \ ^{(A+1)}X \ + \ ^{0}_{0}\gamma$$

^{A}X isotope of element X being analyzed	$^{1}_{0}n$ neutron	$^{(A+1)}X^*$ compound nucleus (unstable)	$^{(A+1)}X$ more stable form of a new isotope of X	$^{0}_{0}\gamma$ gamma radiation

A neutron-enriched compound nucleus emits gamma radiation at its own set of unique frequencies, and these sets of frequencies are now known for each isotope. (Not all isotopes, however, become gamma emitters by neutron capture.) The element can be identified by measuring the specific *frequencies* of gamma radiation emitted. The *concentration* of the element can be determined by measuring the *intensity* of the gamma radiation.

---

[1] The m in technetium-99m means that the isotope is in a metastable form. Its nucleus is at a higher energy level than the nucleus in technetium-99, to which technetium-99m decays.

Neutron activation analysis is so sensitive that concentrations as low as $10^{-9}\%$ can be determined. A museum might have a lock of hair of some famous but long dead person suspected of having been slowly murdered by arsenic poisoning. If so, some arsenic would be in the hair, and neutron activation analysis could find it without destroying the specimen of hair.

Radiological Dating

The determination of the age of a geological deposit or an archaeological find by the use of the radionuclides that are naturally present is called **radiological dating.** It is based partly on the premise that the half-lives of radionuclides have been constant throughout the entire geological period. This premise is supported by the finding that half-lives are insensitive to all environmental forces such as heat, pressure, magnetism, or electrical stresses.

In geological dating, a pair of isotopes is sought that are related as a "parent" to a "daughter" in a radioactive disintegration series, like the uranium-238 series (Figure 22.4). Uranium-238 (as "parent") and lead-206 (as "daughter") have thus been used as a radiological dating pair of isotopes. The half-life of uranium-238 is very long, a necessary criterion for geological dating. After the concentrations of uranium-238 and lead-206 are determined in a rock specimen, the *ratio* of the concentrations together with the half-life of uranium-238 is used to calculate the age of the rock.

For ^{238}U, $t_{1/2} = 4.51 \times 10^9$ yr.
For ^{40}K, $t_{1/2} = 1.3 \times 10^9$ yr.

Probably the most widely used isotopes for dating rock are the potassium-40/argon-40 pair. Potassium-40 is a naturally occurring radionuclide with a half-life nearly as long as that of uranium-238. One of its modes of decay is electron capture, and argon-40 forms.

$$^{40}_{19}\text{K} + ^{\ 0}_{-1}e \longrightarrow ^{40}_{18}\text{Ar}$$

The argon produced by this reaction remains trapped within the crystal lattices of the rock and is freed only when the rock sample is melted. How much has accumulated is measured with a mass spectrometer (page 59), and the observed ratio of argon-40 to potassium-40, together with the half-life of the parent, permits the age of the specimen to be estimated. Because the half-lives of uranium-238 and potassium-40 are so long, samples have to be at least 300,000 years old for either of the two parent–daughter isotope pairs to work.

Carbon-14 Dating

For dating organic remains, like wooden or bone objects from ancient tombs, carbon-14 analysis is used. The ratio of carbon-14 (a beta emitter) to stable carbon in the ancient sample is measured and compared to the ratio in recent organic remains. The data can then be used to calculate the age of the sample.

Extraordinary precautions are taken to ensure that specimens are not contaminated by more recent sources of carbon or carbon compounds.

There are two approaches to carbon-14 dating. The older method, introduced by its discoverer, Willard F. Libby (Nobel Prize in chemistry, 1960), measures the *radioactivity* of a sample taken from the specimen. The radioactivity is proportional to the concentration of carbon-14.

The newer and current method of carbon-14 dating relies on a device resembling a mass spectrometer that is able to separate the atoms of carbon-14 from the other isotopes of carbon (as well as from nitrogen-14) *and count all of them,* not just the carbon-14 atoms that decay. This approach permits the use of smaller samples (0.5–5 mg versus 1–20 g for the Libby method); it works at much higher efficiencies; and it gives more precise dates. Objects of up to 70,000 years old can be dated, but the highest accuracy involves systems no older than 7000 years.

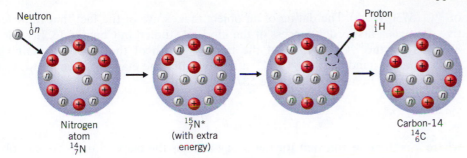

Figure 22.15 *Formation of carbon-14.* Carbon-14 forms in the atmosphere when neutrons in secondary cosmic radiation are captured by nitrogen-14 nuclei.

Carbon-14 has been produced in the atmosphere for many thousands of years by the action of cosmic rays on atmospheric nitrogen. Primary cosmic rays are streams of particles from the sun, mostly protons but also the nuclei of many elements (as high as atomic number 26). Their collisions with atmospheric atoms and molecules generate secondary cosmic rays, which include all subatomic particles. Secondary cosmic rays are what contribute to our background radiation, and people living at high altitudes receive more of this radiation than other people.

Neutrons of the proper energy in secondary cosmic rays can transmute nitrogen atoms into carbon-14 atoms according to the following equation (Figure 22.15).

$$\underset{\substack{\text{neutron} \\ \text{(in cosmic} \\ \text{ray)}}}{{}^{1}_{0}n} + \underset{\substack{\text{nitrogen} \\ \text{atom in the} \\ \text{atmosphere}}}{{}^{14}_{7}N} \longrightarrow \underset{\substack{\text{compound} \\ \text{nucleus}}}{{}^{15}_{7}N^*} \longrightarrow \underset{\substack{\text{carbon-14} \\ \text{atom (a beta} \\ \text{emitter)}}}{{}^{14}_{6}C} + \underset{\substack{\text{high-energy} \\ \text{proton}}}{{}^{1}_{1}p}$$

The newly formed carbon-14 diffuses into the lower atmosphere. It becomes oxidized to carbon dioxide and enters Earth's biosphere by means of photosynthesis. Carbon-14 thus becomes incorporated into plant substances and into the materials of animals that eat plants. As the carbon-14 decays, more is ingested by the living thing. The net effect is an overall equilibrium involving carbon-14 in the global system. As long as the plant or animal is alive, its ratio of carbon-14 atoms to carbon-12 atoms is constant.

At death, an organism's remains have as much carbon-14 as they can ever have, and they now slowly lose this carbon-14 by decay. The decay is a first-order process with a rate independent of the *number* of original carbon atoms. The ratio of carbon-14 to carbon-12, therefore, can be related to the years that have elapsed between the time of death and the time of the measurement. The critical assumption in carbon-14 dating is that the steady-state availability of carbon-14 from the atmosphere has remained largely unchanged over the period for which measurements are valid.[2]

In contemporary biological samples the ratio $^{14}C/^{12}C$ is about 1.2×10^{-12}. Thus, each fresh 1.0 g sample of biological carbon in equilibrium with the $^{14}CO_2$ of the atmosphere has a ratio of 5.8×10^{10} atoms of carbon-14 to 4.8×10^{22} atoms of carbon-12. The ratio decreases by a factor of 2 for each half-life period

[2]The available atmospheric pool of carbon-14 atoms fluctuates somewhat with the intensities of cosmic ray showers, with slow, long-term changes in Earth's magnetic field, and with the huge injections of carbon-12 into the atmosphere from the large-scale burning of coal and petroleum in the 1900s. To reduce the uncertainties in carbon-14 dating, results of the method have been corrected against dates made by tree-ring counting. For example, an uncorrected carbon-14 dating of a Viking site at L'anse aux Meadows, Newfoundland, gave a date of AD 895 ± 30. When corrected, the date of the settlement became AD 997, almost exactly the time indicated in Icelandic sagas for Leif Eriksson's landing at "Vinland," now believed to be the L'anse aux Meadows site.

of ^{14}C (5730 years). The dating of an object makes use of the fact that radioactive decay is a first order process of the kind we studied in Chapter 13. If we let r_0 stand for the ^{14}C/^{12}C ratio at the time of death of the carbon-containing species and r_t stand for the ^{14}C/^{12}C ratio now, after the elapse of t years, we can substitute into Equation 13.5 (page 587) to obtain

$$\ln \frac{r_0}{r_t} = kt \qquad (22.2)$$

where k is the rate constant for the decay and t is the elapsed time. We can obtain the rate constant from the half-life of ^{14}C using Equation 13.7

$$\ln 2 = kt_{1/2}$$

Substituting 5730 yr for $t_{1/2}$ and solving for k gives $k = 1.21 \times 10^{-4}$yr^{-1}. We can now substitute this value into Equation 22.2 to give

$$\ln \frac{r_0}{r_t} = (1.21 \times 10^{-4}\,\text{yr}^{-1})\,t \qquad (22.3)$$

Equation 22.3 can be used to calculate the age of a once-living object if its current ^{14}C/^{12}C ratio can be measured.

Suppose, now, that a sample of an ancient wooden object is found to have a ratio of ^{14}C to ^{12}C equal to 3.0×10^{-13}. Taking r_0 equal to 1.2×10^{-12} and substituting into Equation 22.3,

$$\ln\left(\frac{1.2 \times 10^{-12}}{3.0 \times 10^{-13}}\right) = (1.21 \times 10^{-4}\,\text{yr}^{-1})\,t$$

$$\ln 4.0 = (1.21 \times 10^{-4})\,t$$

Solving for t gives an age of 1.2×10^4 years.

22.8 Nuclear Fission

Because of their electrical neutrality, neutrons penetrate an atom's electron cloud relatively easily and so are able to enter the nucleus. Enrico Fermi discovered in the early 1930s that even slow-moving, *thermal neutrons* can be captured. (Thermal neutrons are those whose average kinetic energy puts them in thermal equilibrium with their surroundings at room temperature.)

Fission Reactions

When Fermi directed thermal neutrons at a uranium target, several different species of nuclei were produced, a result that Fermi could not explain. None had an atomic number above 82. He and other physicists did not follow up on a suggestion by Ida Noddack-Tacke, a chemist, that nuclei of much lower atomic number probably formed. The forces holding a nucleus together were considered too strong to permit such a result.

Discovery of Fission

Otto Hahn received the 1944 Nobel Prize in chemistry.

In 1939, two German chemists, Otto Hahn and Fritz Strassman, finally verified that uranium fission produces several isotopes much lighter than uranium, for example, barium-140 (atomic number 56). Shortly after the report of Hahn and Strassman, two German physicists, Lise Meitner and Otto Frisch, were able to show that a nucleus of one of the rare isotopes in natural uranium, uranium-235, splits into roughly two equal parts after it captures a slow neutron. They named this breakup nuclear **fission.** The general reaction can be represented as follows.

$$^{235}_{92}U + ^1_0n \longrightarrow X + Y + b\,^1_0n$$

X and Y can be a large variety of nuclei with intermediate atomic numbers. Over 30 have been identified. The coefficient b has an average value of 2.47, the average number of neutrons produced by fission events. A typical specific fission is

$$^{235}_{92}U + ^1_0n \longrightarrow ^{236}_{92}U^* \longrightarrow ^{94}_{36}Kr + ^{139}_{56}Ba + 3\,^1_0n$$

What actually undergoes fission is the compound nucleus of uranium-236. It has 144 neutrons and 92 protons for a neutron:proton ratio of roughly 1.6. Initially, the emerging krypton and barium isotopes have the same ratio, and this is much too high for them. The neutron:proton ratio for stable isotopes with 36 to 56 protons is nearer 1.2 to 1.3 (Figure 22.8). Therefore, the initially formed, neutron-rich krypton and barium nuclides promptly eject neutrons, called *secondary neutrons,* that generally have much higher energies than thermal neutrons.

An isotope that can undergo fission after neutron capture is called a **fissile isotope.** The naturally occurring fissile isotope of uranium used in reactors is uranium-235, whose abundance among the uranium isotopes today is only 0.72%. Two other fissile isotopes, uranium-233 and plutonium-239, can be made in nuclear reactors.

Nuclear Chain Reactions

The secondary neutrons released by fission become thermal neutrons as they are slowed by collisions with surrounding materials. They can now be captured by unchanged uranium-235 atoms. Because each fission event produces, on the average, more than two new neutrons, the potential exists for a **nuclear chain reaction** (Figure 22.16). A *chain reaction* is a self-sustaining process whereby products from one event cause one or more repetitions of the process.

If the sample of uranium-235 is small enough, the loss of neutrons to the surroundings is sufficiently rapid to prevent a chain reaction. However, at a certain *critical mass* of uranium-235, a few kilograms, this loss of neutrons is insufficient to prevent a sustained reaction. A virtually instantaneous fission of the sample ensues, in other words, an atomic bomb explosion. To trigger an atomic bomb, therefore, two subcritical masses of uranium-235 (or plutonium-239) are forced together to form a critical mass.

Energy Yield from Fission

The binding energy per nucleon (Figure 22.1) in uranium-235 (about 7.6 MeV) is lower than the binding energies of the new nuclides (about 8.5 MeV). The net change for a single fission event can be calculated as follows (where we intend only *two* significant figures in each result).

1 MeV = 1.602×10^{-13} J

Binding energy in krypton-94:

$$(8.5 \text{ MeV/nucleon}) \times 94 \text{ nucleons} = 800 \text{ MeV}$$

Binding energy in barium-139:

$$8.5 \text{ MeV/nucleon}) \times 139 \text{ nucleons} = \underline{1200 \text{ MeV}}$$

Total binding energy of products: 2000 MeV

Binding energy in uranium-235:

$$(7.6 \text{ MeV/nucleon}) \times 235 \text{ nucleons} = 1800 \text{ MeV}$$

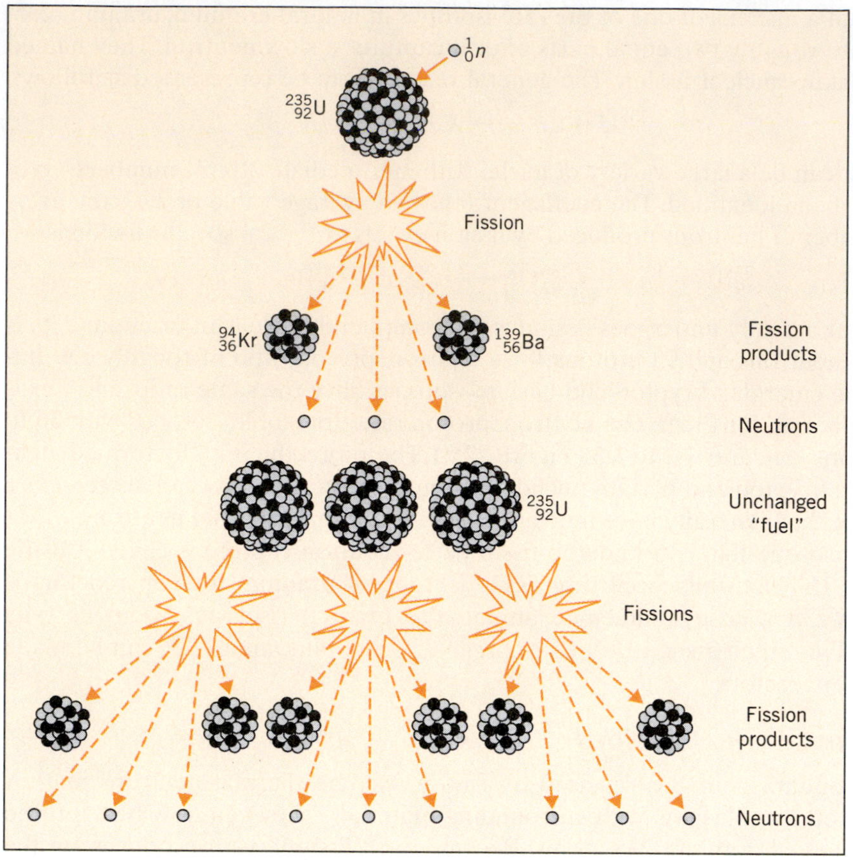

Figure 22.16 *Nuclear chain reaction.* Whenever the concentration of a fissile isotope is high enough (at or above the critical mass), the neutrons released by one fission can be captured by enough unchanged nuclei to cause more than one additional fission event. In civilian nuclear reactors, the fissile isotope is too dilute for this to get out of control. In addition, control rods of nonfissile materials that are able to capture excess neutrons can be inserted or withdrawn from the reactor core to make sure that the heat generated can be removed as fast as it forms.

The energy available from 1 kg of uranium-235 is equivalent to the energy of 3000 tons of soft coal or 13,200 barrels of oil.

The difference in total binding energy is (2000 MeV − 1800 MeV), or 200 MeV (3.2×10^{-11} J), which has to be taken as just a rough calculation. This is the energy released by each fission event going by the equation given. The energy produced by the fission of 1 kg (4.25 mol) of uranium-235 is calculated to be roughly 8×10^{13} J, enough to keep a 100 watt lightbulb in energy for 3000 years.

Electrical Energy from Fission

Virtually all civilian nuclear power plants throughout the world operate on the same general principles. The heat of fission is used either directly or indirectly to increase the pressure of some gas that then drives an electrical generator.

The heart of a nuclear power plant is the *reactor,* where fission takes place in the *fuel core.* The nuclear fuel is generally uranium oxide, enriched to 2–4% in uranium-235 and formed into glasslike pellets. These are housed in long, sealed metal tubes called *cladding.* Bunches of tubes are assembled in spacers that permit a coolant to circulate around the tubes. A reactor has several such assemblies in its fuel core. The coolant carries away the heat of fission.

There is no danger of a nuclear power plant undergoing an atomic bomb explosion. An atomic bomb requires pure uranium-235 or plutonium-239. The concentration of fissile isotopes in a reactor is in the range of 2 to 4%, and much of the remainder is the common, nonfissile uranium-238. However, if the coolant fails to carry away the heat of fission, the reactor core can melt down, and the molten mass might even go through the thick-walled containment vessel in which the reactor is kept. Or the high heat of the fusion might split molecules of coolant water into hydrogen and oxygen, which, on recombining, would produce

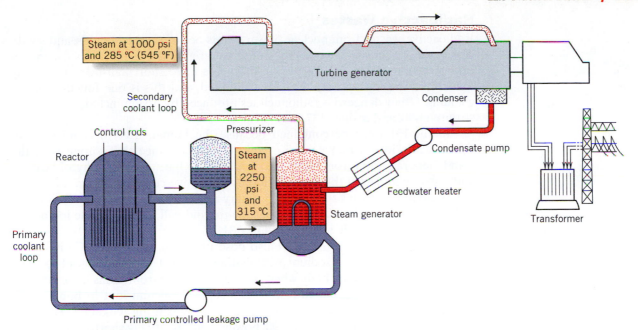

Figure 22.17 *Pressurized water reactor, the type used in most of the nuclear power plants in the United States.* Water in the primary coolant loop is pumped around and through the fuel elements in the core, and it carries away the heat of the nuclear chain reactions. The hot water delivers its heat to the cooler water in the secondary coolant loop, where steam is generated to drive the turbines. (Drawing from WASH-1261, U.S. Atomic Energy Commission, 1973.)

an immense explosion. This is what happened when the reactor at the Russian energy park at Chernobyl failed in 1986.

To convert secondary neutrons to thermal neutrons, the fuel core has a *moderator,* which is the coolant water itself in nearly all civilian reactors. Collisions between secondary neutrons and moderator molecules heats up the moderator. This heat energy eventually generates steam that enables an electric turbine to run. Ordinary water is a good moderator, but so are heavy water (D_2O) and graphite.

Two main types of reactors dominate civilian nuclear power, the *boiling water reactor* and the *pressurized water reactor.* Both use ordinary water as the moderator and so are sometimes called *light water reactors.* Roughly two-thirds of the reactors in the United States are the pressurized water type (Figure 22.17). Such a reactor has two loops, and water circulates in both. The primary loop moves water through the reactor core, where it picks up thermal energy from fission. The water is kept in the *liquid* state by being under high pressure (hence the name *pressurized* water reactor).

The hot water in the primary loop transfers thermal energy to the secondary loop at the steam generator (Figure 22.17). This makes steam at high temperature and pressure, which is piped to the turbine. As the steam drives the turbine, the steam pressure drops. The condenser at the end of the turbine, cooled by water circulating from a river or lake or from huge cooling towers, forces a maximum pressure drop within the turbine by condensing the steam to liquid water. The returned water is then recycled to high-pressure steam. (In the boiling water reactor, there is only one coolant loop. The water heated in the reactor itself is changed to the steam that drives the turbine.)

Radioactive Wastes

Radioactive wastes from nuclear power plants occur as gases, liquids, and solids. The gases are mostly radionuclides of krypton and xenon but, with the exception of xenon-85 ($t_{1/2} = 10.4$ years), the gases have short half-lives and decay quickly. During decay, they must be contained, and this is one function of the cladding. Other dangerous radionuclides produced by fission include iodine-131, strontium-90, and cesium-137.

Iodine-131 must be contained because the human thyroid gland concentrates iodide ion to make the hormone thyroxine. Once the isotope is in the gland, beta radiation from iodine-131 could cause harm, possibly thyroid cancer and possibly impaired thyroid function. An effective countermeasure to iodine-131 poisoning is to take doses of ordinary iodine (as sodium iodide). Statistically, this increases the likelihood that the thyroid will take up stable iodide ion rather than the unstable anion of iodine-131.

Cesium-137 has a half-life of 30 years; that of strontium-90 is 28.1 years. Both are beta emitters.

Cesium-137 and strontium-90 also pose problems to humans. Cesium is in Group IA together with sodium, so radiating cesium-137 cations travel to some of the same places in the body where sodium ions go. Strontium is in Group IIA with calcium, so strontium-90 cations can replace calcium ions in bone tissue, sending radiation into bone marrow and possibly causing leukemia. The half-lives of cesium-137 and strontium-90, however, are relatively short.

The entire reactor of a power plant becomes a solid waste problem when its working life is over.

Some radionuclides in wastes are so long-lived that solid reactor wastes must be kept away from all human contact for dozens of centuries, longer than any nation has ever yet endured. Probably the most intensively studied procedure for making solid radioactive wastes secure is to convert them to glasslike or rocklike solids, and bury them deeply within a rock stratum or in a mountain believed to be geologically stable with respect to earthquakes or volcanoes.

SUMMARY

The Einstein Equation. Mass and energy are interconvertible. The **Einstein equation,** $\Delta E = \Delta mc^2$ (where c is the speed of light), lets us calculate one from the other. The total of the energy in the universe and all the mass calculated as an equivalent of energy is a constant, which is the **law of conservation of mass–energy.**

Nuclear Binding Energies. When a nucleus forms from its nucleons, some mass changes into energy. This amount of energy, the **nuclear binding energy,** would be required to break up the nucleus again. The higher the binding energy per nucleon, the more stable is the nucleus.

Radioactivity. The *electrostatic force* by which protons repel each other is overcome in the nucleus by the *nuclear strong force.* The ratio of neutrons to protons is a factor in nuclear stability. By radioactivity, the naturally occurring **radionuclides** adjust their neutron/proton ratios, lower their energies, and so become more stable by emitting **alpha** or **beta radiation,** sometimes gamma radiation as well.

The loss of an **alpha particle** leaves a nucleus with four fewer units of mass number and two fewer of atomic number. Loss of a **beta particle** leaves a nucleus with the same mass number and an atomic number one unit higher.

Gamma radiation lets a nucleus lose some energy without a change in mass or atomic number.

Depending on the specific isotope, synthetic radionuclides emit alpha, beta, and gamma radiation. Some emit **positrons** (positive electrons) that produce gamma radiation by annihilation collisions with electrons. Other synthetic radionuclides decay by **electron capture** and emit X rays. Some radionuclides emit neutrons.

Nuclear equations are balanced when the mass numbers and atomic numbers on either side of the arrow respectively balance. The energies of emission are usually described in **electron volts** or multiples thereof.

A few very long-lived radionuclides in nature, like ^{238}U, are at the heads of **radioactive disintegration series,** which represent the successive decays of "daughter" radionuclides until a stable isotope forms.

Nuclear Stability. Stable nuclides generally fall within a curving band, called the **band of stability,** when all known nuclides are plotted according to their numbers of neutrons and protons. Radionuclides that have too high neutron/proton ratios eject beta particles to adjust their ratios downward. Those with neutron/proton ratios too low generally emit positrons to change their ratios upward.

Isotopes whose nuclei consist of even numbers of both neutrons and protons are generally much more stable than those with odd numbers of both; this is the **odd–even rule.** Having all neutrons paired and all protons paired is energetically better (more stable) than having any nucleon unpaired.

Isotopes with specific numbers of protons or neutrons, the **magic numbers** of 2, 8, 20, 28, 50, 82, and 126, are generally more stable than others.

Transmutation. When a bombardment particle—a proton, deuteron, alpha particle, or neutron—is captured, the resulting **compound nucleus** contains the energy of both the captured particle and its nucleons. The mode of decay of the compound nucleus is a function of its extra energy, not its extra mass. Many radionuclides have been made by these nuclear reactions, including all of the elements from 93 and higher.

Detecting and Measuring Radiations. Instruments to detect and measure radiation—Geiger counters, for example—take advantage of the ability of radiation to generate ions in air or other matter.

The electron volt describes how much energy a radiation carries; $1 \text{ eV} = 1.602 \times 10^{-19}$ J.

The **curie (Ci)** and the **becquerel (Bq),** the SI unit, describe how active a source is; $1 \text{ Ci} = 3.7 \times 10^{10}$ Bq where $1 \text{ Bq} = 1$ disintegration/s.

The SI unit of absorbed dose, the **gray (Gy),** is used to describe how much energy is absorbed by a unit mass of absorber; $1 \text{ Gy} = 1$ J/kg. An older unit, the **rad,** is equal to 0.01 Gy.

The **sievert (Sv),** an SI unit, and the **rem,** an older unit, are used to compare doses absorbed by different tissues and caused by different kinds of radiation. A 600 rem whole body dose is lethal.

The normal **background radiation** causes millirem exposures per year. Naturally occurring radon, cosmic rays, radionuclides in soil and stone building materials, medical X rays, and releases from nuclear tests or from nuclear power plants all contribute to this background.

Protection against radiation can be achieved by using dense shields (e.g., lead or thick concrete), by avoiding overuse of radionuclides or X rays in medicine, and by taking advantage of the **inverse square law.** This law tells us that the intensity of radiation decreases with the square of the distance from its source.

Dating. Radiological dating uses the known half-lives of naturally occurring radionuclides to date geological and archeological objects. During life, living systems take up carbon-14. Following the organism's death, the age of the remains can be determined from the ratio of carbon-14 to carbon-12.

Fission. Uranium-235, which occurs naturally, and plutonium-239, which can be made from uranium-238, are **fissile isotopes** that serve as the fuel in present-day reactors. When either isotope captures a thermal neutron, the isotope splits in one of several ways to give two smaller isotopes plus energy and more neutrons. The neutrons can generate additional fission events, enabling a nuclear chain reaction. If a critical mass of a fissile isotope is allowed to form, the **nuclear chain reaction** proceeds out of control, and the material detonates as an atomic bomb explosion.

Fission Reactors. The goal of controlling the chain reaction and keeping it from becoming supercritical has resulted in various designs for commercial reactors. *Pressurized water reactors* are most commonly used, and these have two loops of circulating fluids. In the primary loop, water circulates around the reactor core and absorbs the heat of fission. In the secondary loop, water accepts the heat and changes to high-pressure steam, which drives the electrical generator. If the circulation of fluid in the primary loop fails, the reactor must be shut down immediately to prevent a core meltdown and a possible breach of the reactor's containment vessel. One major problem with nuclear energy is the storage of radioactive wastes.

REVIEW QUESTIONS

Conservation of Mass–Energy

22.1 When a substance is described as *radioactive,* what do we know about it?

22.2 In chemical calculations involving chemical reactions we can regard the law of conservation of mass as a law independent of the law of conservation of energy despite Einstein's union of the two. What fact(s) makes this possible?

22.3 How can we know that the speed of light is the absolute upper limit on the speed of any object?

22.4 State the following.
(a) law of conservation of mass–energy
(b) Einstein equation

22.5 Why isn't the sum of the masses of all nucleons in one nucleus equal to the mass of the actual nucleus?

Radioactivity

22.6 Three kinds of radiation make up nearly all of the radiation observed from naturally occurring radionuclides. What are they?

22.7 Give the composition of each of the following.
(a) alpha particle   (c) positron
(b) beta particle   (d) deuteron

22.8 Why is the penetrating ability of alpha radiation less than that of beta or gamma radiation?

22.9 With respect to their formation, how do gamma rays and X rays differ?

22.10 How does electron capture generate X rays?

Nuclear Stability

22.11 What data are plotted and what criterion is used to identify the actual band in the band of stability?

22.12 Both barium-123 and barium-140 are radioactive, but which is more likely to have the *longer* half-life? Explain your answer.

22.13 Tin-112 is a stable nuclide but indium-112 is radioactive and has a very short half-life ($t_{1/2}$ = 14 min). What does tin-112 have that indium-112 does not to account for this difference in stability?

22.14 Lanthanum-139 is a stable nuclide but lanthanum-140 is unstable ($t_{1/2}$ = 40 hr). What rule of thumb concerning nuclear stability is involved?

22.15 As the atomic number increases, the neutron/proton ratio increases. What does this suggest is a factor in nuclear stability?

22.16 Radionuclides of high atomic number are far more likely to be alpha emitters than those of low atomic number. Offer an explanation for this phenomenon.

22.17 Although lead-164 has two magic numbers, 82 protons and 82 neutrons, it is unknown. Lead-208, however, is known and stable. What problem accounts for the nonexistence of lead-164?

22.18 What decay particle is emitted from a nucleus of low to intermediate atomic number but a relatively high neutron/proton ratio? How does the emission of this particle benefit the nucleus?

22.19 What decay particle is emitted from a nucleus of low to intermediate atomic number but a relatively low neutron/proton ratio? How does the emission of this particle benefit the nucleus?

22.20 What does electron capture do to the neutron/proton ratio in a nucleus, increase it, decrease it, or leave it alone? Which kinds of radionuclides are more likely to undergo this change, those above or those below the band of stability?

Transmutations

22.21 Compound nuclei form and decay almost at once. What accounts for the instability of a compound nucleus?

22.22 What explains the existence of several decay modes for the compound nucleus aluminum-27?

22.23 Rutherford theorized that a compound nucleus forms when helium nuclei hit nitrogen-14 nuclei. If this compound nucleus decayed by the loss of a neutron instead of a proton, what would be the other product?

Detecting and Measuring Radiations

22.24 What specific property of radiation is used by the Geiger counter?

22.25 Dangerous doses of radiation can actually involve very small quantities of energy. Explain.

22.26 What units, SI and common, are used to describe each of the following?
(a) the *activity* of a radioactive sample
(b) the *energy* of a particle or of a photon of radiation
(c) the quantity of energy absorbed by a given mass
(d) dose equivalents for comparing biological effects

22.27 A sample giving 3.7×10^{10} disintegrations/s has what activity in Ci and in Bq?

22.28 Explain the necessity in health sciences for the *sievert*.

Applications of Radionuclides

22.29 Why should a radionuclide used in diagnostic work have a short half-life? If the half-life is too short, what problem arises?

22.30 An alpha emitter is not used in diagnostic work. Why?

22.31 In general terms, explain how neutron activation analysis is used and how it works.

22.32 What is one assumption in the use of the uranium/lead ratio for dating ancient geologic formations?

22.33 If a sample used for carbon-14 dating is contaminated by air, there is a potentially serious problem with the method. What is it?

22.34 List some of the kinds of radiation that make up our background radiation.

Nuclear Energy

22.35 Why is it easier for a nucleus to capture a neutron than a proton?

22.36 What do each of the following terms mean?
(a) thermal neutron   (c) fissile isotope
(b) nuclear fission

22.37 Which fissile isotope occurs in nature?

22.38 What fact about the fission of uranium-235 makes it possible for a *chain reaction* to occur?

22.39 Explain in general terms why fission generates more neutrons than needed to initiate it.

22.40 Why would there be a *subcritical mass* of a fissile isotope? (Why isn't *any* mass of uranium-235 critical?)

22.41 What purpose is served by a *moderator* in a nuclear reactor?

22.42 Why is there no possibility of an atomic bomb explosion from a nuclear power plant?

REVIEW PROBLEMS

Answers to problems whose numbers are printed in color are given in Appendix D.

Conservation of Mass–Energy

22.43 Calculate the mass equivalent in grams of 1.00 kJ.

22.44 Calculate the mass in kilograms of a 1.00 kg object when its velocity, relative to us, is (a) 3.00×10^7 m s^{-1}, (b) 2.90×10^8 m s^{-1}, and (c) 2.99×10^8 m s^{-1}. (Notice the progression of these numbers toward the velocity of light, 3.00×10^8 m s^{-1}.)

22.45 Calculate the quantity of mass in nanograms that is changed into energy when one mole of liquid water forms by the complete combustion of hydrogen, all measurements being made at 1 atm and 25 °C.

22.46 Show that the mass equivalent to the energy released by the combustion of 1 mol of methane (890 kJ) is 9.89 ng.

Nuclear Binding Energies

22.47 Calculate the binding energy in J/nucleon of the deuterium nucleus whose mass is 2.0135 u.

22.48 Calculate the binding energy in J/nucleon of the tritium nucleus whose mass is 3.01550 u.

Radioactivity

22.49 Complete the following nuclear equations by writing the symbols of the missing particles.

(a) $^{211}_{82}\text{Pb} \longrightarrow \, ^{0}_{-1}e +$ _____

(b) $^{177}_{73}\text{Ta} \xrightarrow{\text{electron capture}}$ _____

(c) $^{220}_{86}\text{Rn} \longrightarrow \, ^{4}_{2}\text{He} +$ _____

(d) $^{19}_{10}\text{Ne} \longrightarrow \, ^{0}_{1}e +$ _____

22.50 Write the symbols of the missing particles to complete the following nuclear equations.

(a) $^{245}_{96}\text{Cm} \longrightarrow \, ^{4}_{2}\text{He} +$ _____

(b) $^{146}_{56}\text{Ba} \longrightarrow \, ^{0}_{-1}e +$ _____

(c) $^{58}_{29}\text{Cu} \longrightarrow \, ^{0}_{1}e +$ _____

(d) $^{68}_{32}\text{Ge} \xrightarrow{\text{electron capture}}$ _____

22.51 Write a balanced nuclear equation for each of the following changes.

(a) alpha emission from plutonium-242

(b) beta emission from magnesium-28

(c) positron emission from silicon-26

(d) electron capture by argon-37

22.52 Write the balanced nuclear equation for each of the following nuclear reactions.

(a) electron capture by iron-55

(b) beta emission by potassium-42

(c) positron emission by ruthenium-93

(d) alpha emission by californium-251

22.53 Write the symbols, including the atomic and mass numbers, for the radionuclides that would give each of the following products.

(a) fermium-257 by alpha emission

(b) bismuth-211 by beta emission

(c) neodymium-141 by positron emission

(d) tantalum-179 by electron capture

22.54 Each of the following nuclides forms by the decay mode described. Write the symbols of the parents, giving both atomic and mass numbers.

(a) rubidium-80 formed by electron capture

(b) antimony-121 formed by beta emission

(c) chromium-50 formed by positron emission

(d) californium-253 formed by alpha emission

22.55 Krypton-87 decays to krypton-86. What other particle forms?

22.56 Write the symbol of the nuclide that forms from cobalt-58 when it decays by electron capture.

Nuclear Stability

22.57 If an atom of potassium-38 had the option of decaying by positron emission or beta emission, which route would it likely take, and why? Write the nuclear equation.

22.58 Suppose that an atom of argon-37 could decay by either beta emission or electron capture. Which route would it likely take, and why? Write the nuclear equation.

22.59 If we begin with 3.00 mg of iodine-131 ($t_{1/2}$ = 8.07 hr), how much remains after 6 half-life periods?

22.60 A sample of technetium-99m with a mass of 9.00 ng will have decayed to how much of this radionuclide after 4 half-life periods (about 1 day)?

Transmutations

22.61 When vanadium-51 captures a deuteron ($^{2}_{1}\text{H}$), what compound nucleus forms? (Write its symbol.) This particle expels a proton ($^{1}_{1}p$). Write the nuclear equation for the overall change from vanadium-51.

22.62 The alpha-particle bombardment of fluorine-19 generates sodium-22 and neutrons. Write the nuclear equation, including the intermediate compound nucleus.

22.63 Gamma-ray bombardment of bromine-81 causes a transmutation in which neutrons are one product. Write the symbol of the other product.

22.64 Neutron bombardment of cadmium-115 results in neutron capture and the release of gamma radiation. Write the nuclear equation.

22.65 When manganese-55 is bombarded by protons, neutrons are released. What else forms? Write the nuclear equation.

22.66 Which nuclide forms when sodium-23 is bombarded by alpha particles and the compound nucleus emits a gamma ray photon?

22.67 Which isotope of element 112 will be formed by bombarding lead-208 with zinc-70 if the compound intermediate nucleus loses a neutron?

22.68 Write the symbol of the nuclide whose nuclei would be the target for bombardment by nickel-64 nuclei to produce nuclei of $^{272}_{111}$Uuu after the intermediate compound nucleus loses a neutron. (Uuu is the temporary symbol for element 111.)

Detecting and Measuring Radiations

22.69 Suppose that a radiologist who is 2.0 m from a small, unshielded source of radiation receives 2.8 units of radiation. To reduce the exposure to 0.28 units of radiation, to what distance from the source should the radiologist move?

22.70 By what percentage should a radiation specialist increase the distance from a small unshielded source to reduce the radiation intensity by 10.0%?

22.71 If exposure from a distance of 1.60 m gave a worker a dose of 8.4 rem, how far should the worker move away from the source to reduce the dose to 0.50 rem for the same period?

22.72 During work with a radioactive source, a worker was told that he would receive 50 mrem at a distance of 4.0 m during 30 min of work. What would be the received dose if the worker moved closer, to 0.50 m, for the same period?

Applications of Radionuclides

22.73 What percentage of cesium chloride made from cesium-137 ($t_{1/2} = 30$ yr; beta emitter) remains after 150 yr? What *chemical* product forms?

22.74 A sample of waste has a radioactivity, caused solely by strontium-90 (beta emitter, $t_{1/2} = 28.1$ yr), of 0.245 Ci g^{-1}. How many years will it take for its activity to decrease to 1.00×10^{-6} Ci g^{-1}?

22.75 A worker in a laboratory unknowingly became exposed to a sample of radiolabeled sodium iodide made from iodine-131 (beta emitter, $t_{1/2} = 8.07$ days). The mistake was realized 28.0 days after the accidental exposure at which time the activity of the sample was 25.6×10^{-5} Ci/g. The safety officer needed to know how active the sample was at the time of the exposure. Calculate this value in Ci/g.

22.76 Technetium-99*m* (gamma emitter, $t_{1/2} = 6.02$ hr) is widely used for diagnosis in medicine. A sample prepared in the early morning for use that day had an activity of 4.52×10^{-6} Ci. What will its activity be at the end of the day, that is, after 8.00 hr?

22.77 A 500 mg sample of rock was found to have 2.45×10^{-6} mol of potassium-40 ($t_{1/2} = 1.3 \times 10^9$ yr) and 2.45×10^{-6} mol of argon-40. How old was the rock? (What assumption is made about the origin of the argon-40?)

22.78 If a rock sample was found to contain 1.16×10^{-7} mol of argon-40, how much potassium-40 would also have to be present for the rock to be 1.3×10^9 years old?

22.79 A tree killed by being buried under volcanic ash was found to have a ratio of carbon-14 atoms to carbon-12 atoms of 4.8×10^{-14}. How long ago did the eruption occur?

22.80 A wooden door lintel from an excavated site in Mexico would be expected to have what ratio of carbon-14 to carbon-12 atoms if the lintel is 9.0×10^3 yr old?

Nuclear Energy

22.81 Complete the following nuclear equation by supplying the symbol for the other product of the fission.

$$^{235}_{92}U + {}^1_0n \longrightarrow {}^{94}_{38}Sr + \underline{\hspace{1cm}} + 2\,{}^1_0n$$

22.82 Both products of the fission in the previous problem are unstable. According to Figures 22.8 and 22.9, what is the most likely way for each of them to decay, by alpha emission, beta emission, or by positron emission? Explain. What are some of the possible fates of the extra neutrons produced by the fission shown in the previous problem?

ADDITIONAL EXERCISES

22.83 What is the nuclear equation for each of the following changes?
(a) beta emission from aluminum-30
(b) alpha emission from einsteinium-252
(c) electron capture by molybdenum-93
(d) positron emission by phosphorus-28

*22.84** Calculate the binding energy in J/nucleon of the nucleus of an atom of iron-56. The observed mass of one *atom* is 55.9349 u. What information lets us know that no isotope has a *larger* binding energy per nucleon?

*22.85** Calculate the binding energy in J/nucleon of uranium-235. The observed mass of one *atom* is 235.0439 u.

22.86 Give the nuclear equation for each of these changes.
(a) positron emission by carbon-10

(b) alpha emission by curium-243
(c) electron capture by vanadium-49
(d) beta emission by oxygen-20

*22.87** If a positron is to be emitted spontaneously, how much more *mass* (as a minimum) must an *atom* of the parent have than an *atom* of the daughter nuclide? Explain.

22.88 There is a gain in binding energy per nucleon when light nuclei fuse to form heavier nuclei. Yet a tritium atom and a deuterium atom, in a mixture of these isotopes, do not spontaneously fuse to give helium (and energy). Explain why not.

22.89 ^{214}Bi decays to isotope A by alpha emission; A then decays to B by beta emission, which decays to C by another beta emission. Element C decays to D by still another beta emission, and D decays by alpha emission to a stable isotope, *E*. What is the proper symbol of element *E*? (Contributed by Prof. W. J. Wysochansky, Georgia Southern University.)

22.90 ^{15}O decays by positron emission with a half-life of 124 s. (a) Give the proper symbol of the product of the decay. (b) How much of a 750 mg sample of ^{15}O remains after 5.0 min of decay? (Contributed by Prof. W. J. Wysochansky, Georgia Southern University.)

22.91 Alpha decay of ^{238}U forms ^{234}Th. What kind of decay from ^{234}Th produces ^{234}Ac? (Contributed by Prof. Mark Benvenuto, University of Detroit—Mercy.)

22.92 A sample of rock was found to contain 2.07×10^{-5} mol of ^{40}K and 1.15×10^{-5} mol of ^{40}Ar. If we assume that all of the ^{40}Ar came from decay of ^{40}K, what is the age of the rock in years ($t_{1/2} = 1.3 \times 10^9$ years for ^{40}K.)

22.93 The ^{14}C content of an ancient piece of wood was found to be one-eighth of that in living trees. How many years old is this piece of wood ($t_{1/2} = 5730$ years for ^{14}C)?

22.94 Dinitrogen trioxide, N_2O_3, is largely dissociated into NO and NO_2 in the gas phase where there exists the equilibrium, $N_2O_3 \rightleftharpoons NO + NO_2$. In an effort to determine the structure of N_2O_3, a mixture of NO and $*NO_2$ was prepared containing isotopically labeled N in the NO_2. After a period of time the mixture was analyzed and found to contain substantial amounts of both $*NO$ and $*NO_2$. Explain how this is consistent with the structure for N_2O_3 being ONONO.

22.95 The reaction $(CH_3)_2Hg + HgI_2 \rightarrow 2CH_3HgI$ is believed to occur through a transition state with the structure

$$
\begin{array}{ccc}
H_3C & & I \\
& Hg \quad Hg & \\
CH_3 & & I
\end{array}
$$

If this is so , what should be observed if CH_3HgI and $*HgI_2$ are mixed, where the asterisk denotes a radioactive isotope of Hg? Explain your answer.

22.96 *Racemization* is a chemical reaction in which an enantiomer is converted into its mirror image. One possible mechanism for the racemization of octahedral complex ions containing three bidentate ligands involves the temporary loss of one of the ligands

$$
d-[M(AA)_3] \rightleftharpoons \begin{bmatrix} [M(AA)_2] \\ + \\ AA \end{bmatrix} \rightleftharpoons l-[M(AA)_3]
$$

where *d* and *l* simply designate the two enantiomers of $[M(AA)_3]$. This change can be pictured as shown in Figure 22.18. Can you suggest a simple experiment, making use of radioisotopes, that would be able to confirm whether or not this mechanism is operative in the racemization of the $[Co(C_2O_4)_3]^{3-}$ ion?

*22.97 A large, complex piece of apparatus has built into it a cooling system containing an unknown volume of cooling liquid. It is desired to measure the volume of the coolant without draining the lines. To the coolant was added 10.0 mL of methanol whose molecules included atoms of ^{14}C and that had a specific activity of 580 counts per minute per gram (cpm/g), determined using a Geiger counter. The coolant was permitted to circulate to assure complete mixing before a sample was withdrawn that was found to have a specific activity of 29 cpm/g. Calculate the volume of coolant in the system in milliters. The density of methanol is 0.792 g/mL, and the density of the coolant is 0.884 g/mL.

*22.98 A complex ion of chromium(III) with oxalate ion was prepared from ^{51}Cr-labeled $K_2Cr_2O_7$. having a specific activity of 843 cpm/g (counts per minute per gram), and ^{14}C-labeled oxalic acid, $H_2C_2O_4$, having a specific activity of 345 cpm/g. Chromium-51 decays by electron capture with the emission of gamma radiation, whereas ^{14}C is a pure beta emitter. Because of the characteristics of the beta and gamma detectors, each of these isotopes may be counted independently. A sample of the complex ion was observed to give a gamma count of 165 cpm and a beta count of 83 cpm. From these data, determine the number of oxalate ions bound to each Cr(III) in the complex ion. (*Hint*: For the starting materials calculate the cpm per mole of Cr and oxalate, respectively.)

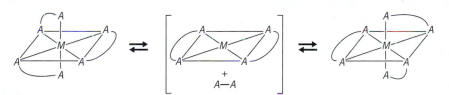

Figure 22.18 *Racemization.* A possible mechanism for the racemization of an octahedral $[M(AA)_3]$ complex. (AA = bidentate ligand.)

Advances in technology, like strong, light-weight plastics for equipment and clothing, have expanded the opportunities for adrenaline surges such as this skier is no doubt feeling.

Organic Compounds and Biochemicals

23

This Chapter in Context At the molecular level of life, nature uses compounds of carbon. The amazing variety of living systems down to the uniqueness of each individual is possible largely because of the properties of this element. Atoms of carbon are able to covalently combine with each other and with atoms of other nonmetals to form an enormous number of different molecules. We'll take a look at some of these in this chapter.

23.1 Nature of Organic Chemistry

Organic chemistry is the study of the preparation, properties, identification, and reactions of those compounds of carbon not classified as inorganic. The latter include the oxides of carbon, the bicarbonates and carbonates of metal ions, the metal cyanides, and a handful of other compounds. There are several million known carbon compounds, and all but a very few are organic.

Virtually all plastics, synthetic and natural fibers, dyes and drugs, insecticides and herbicides, ingredients in perfumes and flavoring agents, and all petroleum products are organic compounds. All the foods you eat consist chiefly of organic compounds in the families of the carbohydrates, fats and oils, proteins, and vitamins. The substances that make up furs and feathers, hides and skins, and all cell membranes are also organic.

The Uniqueness of the Element Carbon

What makes the existence of so many organic compounds possible is not just the multivalency of carbon—its atoms always have four bonds in organic compounds. Rather, carbon atoms are unique in their ability to form strong covalent bonds to each other *while at the same time bonding atoms of other nonmetals strongly.* The molecules in polyethylene, for example, have *carbon chains* that are thousands of carbon atoms long, with hydrogen atoms attached to each carbon.

Sulfur atoms can also form long chains, but they are unable to hold the atoms of any other element strongly at the same time.

$$H-\underset{\underset{H}{|}}{\overset{\overset{H}{|}}{C}}-\underset{\underset{H}{|}}{\overset{\overset{H}{|}}{C}}-\underset{\underset{H}{|}}{\overset{\overset{H}{|}}{C}}-\underset{\underset{H}{|}}{\overset{\overset{H}{|}}{C}}-\underset{\underset{H}{|}}{\overset{\overset{H}{|}}{C}}-\underset{\underset{H}{|}}{\overset{\overset{H}{|}}{C}}-\underset{\underset{H}{|}}{\overset{\overset{H}{|}}{C}}-\underset{\underset{H}{|}}{\overset{\overset{H}{|}}{C}}-\underset{\underset{H}{|}}{\overset{\overset{H}{|}}{C}}-\underset{\underset{H}{|}}{\overset{\overset{H}{|}}{C}}-\underset{\underset{H}{|}}{\overset{\overset{H}{|}}{C}}-\underset{\underset{H}{|}}{\overset{\overset{H}{|}}{C}}-\text{etc.}$$

polyethylene (small segment of one molecule)

The longest known sequence of atoms of other members of the carbon family, each also holding hydrogens, is eight for silicon, five for germanium, two for tin, and one for lead.

Few of the possible isomers of the compounds with large numbers of carbon atoms have actually been made, but there is nothing except too much crowding within the molecules of some of them to prevent their existence.

Isomerism among Organic Compounds

Another reason for the huge number of organic compounds is *isomerism,* introduced in Section 8.4 (page 338) where the isomers of C_4H_{10}, butane and isobutane, were discussed. *Isomers,* recall, are compounds with identical molecular formulas but whose molecules have different structures. Two other examples of isomers are ethyl alcohol and dimethyl ether (Figure 23.1).

butane
bp −0.5 °C

isobutane
bp −11.7 °C

ethyl alcohol
bp 78.5 °C

dimethyl ether
bp −23 °C

As the number of carbons per molecule increases, the number of possible isomers for any given formula becomes astronomic.

Formula	Number of Isomers
C_8H_{18}	18
$C_{10}H_{22}$	75
$C_{20}H_{42}$	366,319
$C_{40}H_{82}$	6.25×10^{13} (estimated)

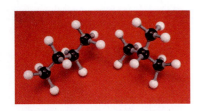
(a)

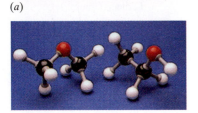
(b)

Figure 23.1 *Isomers.* (*a*) The isomers of C_4H_{10}. Butane is on the left and 2-methyl-propane (isobutane) is on the right. (*b*) The isomers of C_2H_6O. Dimethyl ether is on the left and ethanol is on the right.

Organic Families and Their Functional Groups

The study of the huge number of organic compounds is manageable because they can be sorted into *organic families* defined by *functional groups.* A major goal of this chapter is to show how they enable us to organize and understand the properties of organic compounds.

Functional groups are small structural units within molecules at which most of the compound's chemical reactions occur (Table 23.1). For example, as we learned in Section 8.4, all *alcohols* have the *alcohol group,* and a molecule of the simplest member of the alcohol family, methyl alcohol, has only one carbon. Ethyl alcohol molecules have two carbons. All members of the family of *carboxylic acids* have the *carboxyl group.* The carboxylic acid with two carbons per molecule is the familiar weak acid, acetic acid.

alcohol group

methyl alcohol

carboxyl group

acetic acid

Table 23.1 Some Important Families of Organic Compounds

Family	Characteristic Structural Feature[a]	Example
Hydrocarbons	Only C and H present	
	Families of Hydrocarbons:	
	Alkanes: Only single bonds	CH_3CH_3
	Alkenes: C=C	$CH_2{=}CH_2$
	Alkynes: C≡C	HC≡CH
	Aromatic: Benzene ring	
Alcohols	ROH	CH_3CH_2OH
Ethers	ROR′	CH_3OCH_3
Aldehydes	$\overset{O}{\overset{\|}{R}}CH$	$CH_3\overset{O}{\overset{\|}{C}}H$
Ketones	$R\overset{O}{\overset{\|}{C}}R'$	$CH_3\overset{O}{\overset{\|}{C}}CH_3$
Carboxylic acids	$R\overset{O}{\overset{\|}{C}}OH$	$CH_3\overset{O}{\overset{\|}{C}}OH$
Esters	$R\overset{O}{\overset{\|}{C}}OR'$	$CH_3\overset{O}{\overset{\|}{C}}OCH_3$
Amines	RNH_2, $RNHR'$, $RNR'R''$	CH_3NH_2 ; CH_3NHCH_3 ; $CH_3\overset{CH_3}{\overset{\|}{N}}CH_3$
Amides	$R\overset{O}{\overset{\|}{C}}{-}\overset{R''(H)}{\overset{\|}{N}}R'(H)$	$CH_3\overset{O}{\overset{\|}{C}}NH_2$

[a] R, R′, and R″ represent hydrocarbon groups—*alkyl groups*—defined in the text.

One family in Table 23.1, the *alkane* family, has no functional group, just C—C and C—H single bonds. These bonds are virtually nonpolar, because C and H are so alike in electronegativity. Therefore, alkane molecules are the least able of all organic molecules to attract ions or polar molecules. Hence, they do not react, at least at room temperature, with polar or ionic reactants such as strong acids and bases and common oxidizing agents, like the dichromate or the permanganate ion.

Alkanes do react with fluorine, chlorine, bromine, and hot nitric acid. They also burn, a reaction with oxygen.

Condensed Structures

The structural formulas in Table 23.1 are "condensed" because this saves both space and time in writing structures *without sacrificing any structural information.* In condensed structures, C—H bonds are usually "understood." When three H atoms are attached to carbon, they are set alongside the C, as in CH_3, but sometimes H_3C. When a carbon is bonded to two other H atoms, the condensed symbol is usually CH_2, sometimes H_2C. Thus the condensed structure of ethanol is CH_3CH_2OH, and that of dimethyl ether is CH_3OCH_3 or H_3COCH_3.

Functional Groups and Polar Reactants

When a polar group of atoms, like the OH group or even a halogen atom, is attached to carbon, the molecule has a polar site. It now can attract polar and ionic reactants and undergo chemical changes, but generally at or near just this functional group. This is why compounds of widely varying size but with the same functional group display very similar kinds of reactions.

The reactions of *amines*, for example, are similar from amine to amine. Therefore, we need only learn the handful of *kinds* of reactions exhibited by all amines and then adapt this knowledge to a specific example as needed. A special symbol makes this easy.

The Symbol R in Structural Formulas

The symbol R is from the German radikal.

Chemists use the symbol R in a structure to represent any and all purely alkane-like groups, those parts of molecules that do not react with a specific reactant. One type of amine, for example, is represented by the formula, $R{-}NH_2$. R may be CH_3, CH_3CH_2, $CH_3CH_2CH_2$, and so forth. Because the amines are compounds with an ammonialike group, they are Brønsted bases and, like ammonia, neutralize hydrochloric acid. We can, therefore, summarize in just one equation the reaction of hydrochloric acid with any amine of the $R{-}NH_2$ type, regardless of how large the hydrocarbon portion of the molecule is, by the following equation.

$$R{-}NH_2 + HCl \longrightarrow R{-}NH_3^+ + Cl^-$$

The study of organic chemistry is thus not the study of individual compounds, taking them one at a time. It is the study of the common properties of functional groups, how to change one into another, and how each group affects physical properties. The symbol R makes this study easier.

Straight chain

Branched chain

Isooctane, one of the many alkanes in gasoline, is the standard for the octane ratings of various types of gasoline. Pure isooctane is assigned an octane rating of 100.

Open-Chain and Ring Compounds

The continuous sequence of carbon atoms in polyethylene is called a **straight chain.** This means *only* that no carbon atom holds more than two other carbons. It means nothing about the *conformation* of the molecule. If we made a molecular model of polyethylene and coiled it into a spiral, we would still say that its carbon skeleton has a straight chain. It has no branches.

Branched chains are also very common. Isooctane, for example, has a *main chain* of five carbon atoms (in black) carrying three CH_3 *branches* (shown in red).

$$\overset{\displaystyle CH_3 \quad CH_3}{\underset{\displaystyle CH_3}{CH_3CCH_2CHCH_3}}$$

isooctane

Carbon **rings** are also common. Cyclohexane, for example, has a ring of six carbon atoms.

Carbon ring

$$\begin{array}{c} CH_2 \\ H_2C \qquad CH_2 \\ H_2C \qquad CH_2 \\ CH_2 \end{array}$$

cyclohexane   cyclohexane
(fully condensed)

Just about everything is "understood" in the very convenient, fully condensed structure of cyclohexane. When polygons, like the hexagon, are used to represent rings, the following conventions are used.

1. C occurs at each corner unless O or N (or another multivalent atom) is explicitly written at a corner.

2. A line connecting two corners is a covalent bond between adjacent ring atoms.

3. Remaining bonds, as required by the covalence of the atom at a corner, are understood to hold H atoms.

4. Double bonds are always explicitly shown.

The rule about double bonds will be modified with benzene and related compounds.

We can illustrate these rules with the following cyclic compounds.

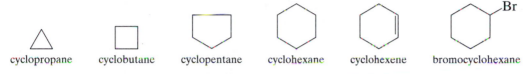

| cyclopropane | cyclobutane | cyclopentane | cyclohexane | cyclohexene | bromocyclohexane |

There is no theoretical upper limit to the size of a ring.

Many compounds have **heterocyclic** rings. Their molecules carry an atom other than carbon in a ring position, as in pyrrole, piperidine, and tetrahydropyran.

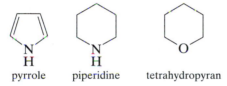

| pyrrole | piperidine | tetrahydropyran |

The tetrahydropyran ring occurs in molecules of sugar.

The atom other than carbon is called the *heteroatom,* and any atoms of H or another element bonded to it are not "understood" but always shown.

Structure and Physical Properties: General Principles

One goal of organic chemistry is to enable the prediction of properties from molecular structure. Functional groups and molecular size determine, for example, whether a compound is soluble in water.

Effects of Functional Groups on Solubility in Water

Piperidine and tetrahydropyran (shown above) are both freely soluble in water, whereas cyclohexane is insoluble. The cyclohexane molecule, having only non-polar C—C and C—H single bonds, is unable to form hydrogen bonds with water molecules (see page 522). Molecules of piperidine and tetrahydropyran, however, can form hydrogen bonds ($\cdots$) with water.

piperidine hydrogen bonded to
two water molecules

tetrahydropyran hydrogen
bonded to a water molecule

Thus organic compounds whose molecules have N and O atoms are more soluble in water than molecules of alkanes of about the same size.

Effects of Molecular Size on Physical Properties

Even with N or O atoms, a compound's solubility in water decreases as the hydrocarbon portion of its molecules becomes larger and larger. The cholesterol molecule, for example, has an OH group, but the group is overwhelmed by the huge hydrocarbon group. So cholesterol is practically insoluble in water.

cholesterol

Cholesterol, however, is correspondingly more soluble in such nonpolar organic solvents as benzene and chloroform. (Remember, "like dissolves like.")

The insolubility of cholesterol in water accounts for its separation from the bloodstream as heart disease slowly develops.

23.2 Hydrocarbons

In insufficient oxygen, some carbon monoxide and carbon also form when hydrocarbons burn.

The alkanes make up just one family of a multifamily group of compounds called the **hydrocarbons** whose molecules consist only of C and H atoms. Besides the alkanes, the hydrocarbons include *alkenes, alkynes,* and *aromatic hydrocarbons* (Table 23.1). All are insoluble in water. All burn, giving carbon dioxide and water if sufficient oxygen is available.

The alkanes are **saturated compounds,** compounds with only single bonds. Alkenes and alkynes are **unsaturated compounds,** because their molecules have double bonds (alkenes) or triple bonds (alkynes). The aromatic hydrocarbons are also unsaturated because the carbon atoms of their rings, when represented by simple Lewis structures, also have double bonds.

The special nature of the bonds in benzene was discussed in Section 9.8.

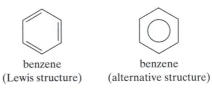

benzene
(Lewis structure)

benzene
(alternative structure)

Sources of Hydrocarbons

The *fossil fuels*—coal, petroleum, and natural gas—supply us with virtually all of our hydrocarbons. One of the operations in petroleum refining is to boil crude oil (petroleum freed of natural gas) and selectively condense the vapors between preselected temperature ranges. The liquid collected at each range is called a *fraction,* and the operation is *fractional distillation.* Each fraction is a complex mixture made up almost entirely of hydrocarbons, mostly alkanes. *Gasoline,* for example, is the fraction boiling roughly between 40 and 200 °C. The *kerosene* and *jet fuel* fraction overlaps this range, going from 175 to 325 °C.

Petroleum refinery. The towers contain catalysts that break up large molecules in hot crude oil to sizes suitable for vehicle engines.

The molecules of the alkanes in gasoline generally have 5 to 10 carbon atoms; those in kerosene, 12 to 18. *Paraffin wax* is part of the nonvolatile residue of petroleum refining, and it consists of alkanes with over 20 carbons per molecule. Low-boiling fractions of crude oil, those having molecules of 4 to 8 carbons, are used as nonpolar solvents.

Diesel fuel is another fraction of petroleum.

> ▶**Chemistry in Practice**◀ The physical and chemical properties of hydrocarbons help nature handle oil spills, although the short-term damage of such spills to wildlife, tourism, and coastal industries, like fishing, can be immense. Being less dense than water, most of the oil floats rather than sinks. The lower boiling components tend to evaporate. One or more species of over 30 different genera of oil-degrading bacteria and fungi gradually consume much of what remains. Of the 35,000 metric tons of oil that spilled from the *Exxon Valdez* into Prince William Sound, Alaska, in 1989, roughly 20% evaporated and oxidized in the sunlight, 50% degraded either in the water or on the beaches, 14% was recovered, and 13% sank. ◆

Alkanes

All open-chain **alkanes** (those without rings) have the general formula C_nH_{2n+2}, where n equals the number of carbon atoms. Table 23.2 gives the structures, names, and some properties of the first 10 unbranched open-chain alkanes. Their boiling points steadily increase with molecular mass, illustrating how London forces become greater with molecular size. The alkanes are generally less dense than water.

London forces are discussed in Section 11.2.

IUPAC System of Nomenclature

The IUPAC rules for naming organic compounds follow a regular pattern. The last syllable in an IUPAC name designates the family of the compound. The names of all saturated hydrocarbons, for example, end in *-ane*. For each family there is a rule for picking out and naming the *parent chain* or *parent ring* within a specific molecule. The compound is then regarded as having *substituents* at-

IUPAC stands for International Union of Pure and Applied Chemistry (page 65).

Table 23.2 Straight-Chain Alkanes

IUPAC Name	Molecular Formula	Structure	Boiling Point (°C)	Melting Point (°C)	Density (g mL^{-1}, 20 °C)
Methane	CH_4	CH_4	−161.5	−182.5	
Ethane	C_2H_6	CH_3CH_3	−88.6	−183.3	
Propane	C_3H_8	$CH_3CH_2CH_3$	−42.1	−189.7	
Butane	C_4H_{10}	$CH_3(CH_2)_2CH_3$	−0.5	−138.4	
Pentane	C_5H_{12}	$CH_3(CH_2)_3CH_3$	36.1	−129.7	0.626
Hexane	C_6H_{14}	$CH_3(CH_2)_4CH_3$	68.7	−95.3	0.659
Heptane	C_7H_{16}	$CH_3(CH_2)_5CH_3$	98.4	−90.6	0.684
Octane	C_8H_{18}	$CH_3(CH_2)_6CH_3$	125.7	−56.8	0.703
Nonane	C_9H_{20}	$CH_3(CH_2)_7CH_3$	150.8	−53.5	0.718
Decane	$C_{10}H_{22}$	$CH_3(CH_2)_8CH_3$	174.1	−29.7	0.730

tached to its parent chain or ring. Let's see how these principles work with naming alkanes.

IUPAC Rules for Naming the Alkanes

1. The name ending for all alkanes (and cycloalkanes) is *-ane* .
2. The *parent chain* is the longest continuous chain of carbons in the structure. For example, the branched-chain alkane

$$\begin{array}{c} CH_3 \\ | \\ CH_3CH_2CHCH_2CH_2CH_3 \end{array}$$

is regarded as being "made" from the following parent

$$CH_3CH_2CH_2CH_2CH_2CH_3$$

by replacing a hydrogen atom on the third carbon from the left with CH_3.

$$\begin{array}{c} CH_3 \\ \searrow\;\nearrow \\ \downarrow H \\ CH_3CH_2CHCH_2CH_2CH_3 \longrightarrow CH_3CH_2\overset{\displaystyle CH_3}{\overset{|}{C}}HCH_2CH_2CH_3 \end{array}$$

3. A prefix is attached to the name ending, *-ane,* to specify the number of carbon atoms *in the parent chain.* The prefixes through chain lengths of 10 carbons are as follows. The names in Table 23.2 show their use.

meth-	1 C	hex-	6 C
eth-	2 C	hept-	7 C
prop-	3 C	oct-	8 C
but-	4 C	non-	9 C
pent-	5 C	dec-	10 C

The parent chain of our example has six carbons, so the parent is named hexane—*hex* for six carbons and *ane* for being in the alkane family. Therefore, the alkane whose name we are devising is viewed as a derivative of this parent, *hexane.*

4. The carbon atoms of the parent chain are numbered starting from whichever end of the chain gives the location of the first branch the lower of two possible numbers. Thus the correct direction for numbering our example is from left to right, not right to left, because this locates the branch (CH_3) at position 3, not position 4.

$$\begin{array}{cc} \begin{array}{c} CH_3 \\ | \\ CH_3CH_2CHCH_2CH_2CH_3 \\ 1\quad2\quad3\quad4\quad5\quad6 \end{array} & \begin{array}{c} CH_3 \\ | \\ CH_3CH_2CHCH_2CH_2CH_3 \\ 6\quad5\quad4\quad3\quad2\quad1 \end{array} \\ \text{(correct direction of numbering)} & \text{(incorrect direction of numbering)} \end{array}$$

5. Each branch attached to the parent chain is named, so we must now learn the names of some of the alkanelike branches.

The Alkyl Groups

Any branch that consists only of carbon and hydrogen and that has only single bonds is called an **alkyl group,** and the names of all alkyl groups end in *-yl*. Think of an alkyl group as an alkane minus one of its hydrogen atoms. For example,

H—C—H $\xrightarrow{\text{remove one H}}$ H—C— or CH$_3$—

methane   methyl

H—C—C—H $\xrightarrow{\text{remove one H}}$ H—C—C— or CH$_3$CH$_2$—

ethane   ethyl

Two alkyl groups can be obtained from propane because the middle position in its chain of three is not equivalent to either of the end positions.

H—C—C—C—H $\xrightarrow{\text{remove one H}}$ H—C—C—C— or CH$_3$CH$_2$CH$_2$—

propane   propyl

H—C—C—C—H $\xrightarrow{\text{remove one H}}$ H—C—C—C—H or CH$_3$CHCH$_3$

propane   isopropyl

We will not need to know the IUPAC names for any alkyl groups with four or more carbon atoms.

6. The name of each alkyl group is attached to the name of the parent as a prefix, placing its chain location number in front and separating the number from the name by a hyphen. Thus, the named original example is 3-methylhexane.

CH$_3$

CH$_3$CH$_2$CHCH$_2$CH$_2$CH$_3$

3-methylhexane

7. When two or more groups are attached to the parent, each is named and located with a number. The names of alkyl substituents are assembled in alphabetical order. Always use *hyphens* to separate numbers from words. Here is an application.

CH$_3$CH$_2$ CH$_3$

CH$_3$CH$_2$CH$_2$CHCH$_2$CHCH$_3$

7   6   5   4   3   2   1

4-ethyl-2-methylheptane

8. When two or more substituents are identical, multiplier prefixes are used: di- (for 2), tri- (for 3), tetra- (for 4), and so forth. The location number of every group must occur in the final name. Always separate a number from another number in a name by a *comma*. For example,

CH$_3$ CH$_3$

CH$_3$CHCH$_2$CHCH$_2$CH$_3$

Correct name: 2,4-dimethylhexane

Incorrect names: 2,4-methylhexane   2-methyl-4-methylhexane
                3,5-dimethylhexane   2-4-dimethylhexane

Unsystematic or *common names* are still widely used for many compounds. For example, the common name of 2-methylpropane is *isobutane*.

9. When identical groups are on the *same* carbon, the number of this position is repeated in the name. For example,

$$CH_3$$
$$CH_3CCH_2CH_2CH_3$$
$$CH_3$$

Correct name:	2,2-dimethylpentane
Incorrect names:	2-dimethylpentane
	2,2-methylpentane
	4,4-dimethylpentane

These are not all of the IUPAC rules for alkanes, but they will handle all of our needs.

EXAMPLE 23.1

Using the IUPAC Rules to Name an Alkane

What is the IUPAC name for the following compound?

$$CH_3CH_2 \quad CH_3$$
$$CH_3CH_2CH_2CHCH_2CCH_3$$
$$CH_3$$

Analysis: The compound is an alkane because it is a hydrocarbon with only single bonds. We must therefore use the IUPAC rules for alkanes.

Solution: The ending to the name must be *-ane*. The longest chain is seven carbons long, so the name of the parent alkane is *heptane*. We have to number the chain from right to left in order to reach the first branch with the lower number.

At carbon 2 there are two one-carbon methyl groups. At carbon 4, there is a two-carbon ethyl group. Alphabetically, *ethyl* comes before *methyl,* so we must assemble these names as follows to make the final name. (Names of alkyl groups are alphabetized *before* any prefixes such as di- or tri- are affixed.)

4-ethyl-2,2-dimethylheptane

A hyphen separates a number from a word.	A comma separates two numbers.	There is no hyphen, no comma, and no space here.

Is the Answer Reasonable?
The most common mistake is to pick a shorter chain than the true "parent," a mistake we did not make. Another common mistake is to number the chain incorrectly. One overall check that some people use is to count the number of carbons implied by the name (in our example, $2 + 1 + 1 + 7 = 11$) and compare it to the count obtained directly from the structure. If the counts don't match, you know you can't be right.

Practice Exercise 1

Write the IUPAC names of the following compounds. In searching for the parent chain, be sure to look for the longest continuous chain of carbons *even if the chain twists and goes around corners.* ◆

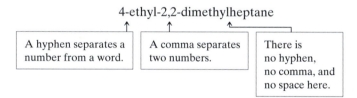

(a)
$$CH_3CH_2$$
$$CHCH_3$$
$$CH_2CH_2$$
$$CH_3$$

(b)
$$CH_3 \quad CH_2CH_2CH_3$$
$$CHCHCH_2CH_3$$
$$CH_3CH$$
$$CH_3$$

(c)
$$CH_3 \quad CH_3 \quad CH_3$$
$$CH_3CH_2CHCHCHCH_2CHCH_3$$
$$CH_3CH_2$$

Chemical Properties of Alkanes

Alkanes, as we have indicated, are generally stable at room temperature toward such different reactants as concentrated sulfuric acid (or any other common acid), concentrated aqueous bases (like NaOH), and even the most reactive metals. Fluorine attacks virtually all organic compounds, including the alkanes, to give mixtures of products. Like all hydrocarbons, the alkanes burn in air to give carbon dioxide and water. *Hot* nitric acid, chlorine, and bromine also react with alkanes. The chlorination of methane, for example, can be made to give the following compounds.

CH_3Cl CH_2Cl_2 $CHCl_3$ CCl_4

methyl   methylene   chloroform   carbon
chloride   chloride   tetrachloride

These are the common names for the chlorinated derivatives of methane, not the IUPAC names.

When heated at high temperatures in the absence of air, alkanes "crack," meaning that they break up into smaller molecules. The cracking of methane, for example, yields finely powdered carbon and hydrogen.

$$CH_4 \xrightarrow{\text{high temperatures}} C + 2H_2$$

The controlled cracking of ethane gives ethene, commonly called ethylene.

$$CH_3CH_3 \xrightarrow{\text{high temperatures}} CH_2{=}CH_2 + H_2$$

ethane   ethene

> ▶**Chemistry in Practice**◀ The finely divided carbon made by cracking methane is used as a filler in auto tires. The other product, hydrogen, is used either as a fuel or as a raw material for the chemical industry.
>
> Ethene ("ethylene"), from the cracking of ethane, is one of the most important raw materials in the organic chemicals industry. It is used to make polyethylene plastic items as well as ethyl alcohol and ethylene glycol (an antifreeze). ◆

$HOCH_2CH_2OH$
ethylene glycol

Alkenes and Alkynes

Hydrocarbons with one or more double bonds are members of the **alkene family.** Open-chain alkenes have the general formula C_nH_{2n}. Hydrocarbons with triple bonds are in the **alkyne family,** and have the general formula C_nH_{2n-2} (when open-chain).

Alkenes and alkynes, like all hydrocarbons, are insoluble in water and are flammable, as we've said. The most familiar alkenes are ethene and propene (commonly called propylene), the raw materials for polyethylene and polypropylene, respectively. Ethyne ("acetylene"), an important alkyne, is the fuel for oxyacetylene torches.

The molecular orbital structures of double and triple bonds are discussed in Section 9.6.

$CH_2{=}CH_2$ $CH_3CH{=}CH_2$ $HC{\equiv}CH$

ethene   propene   ethyne
(ethylene)   (propylene)   (acetylene)

Naming the Alkenes

The IUPAC rules for the names of alkenes are adaptations of those for alkanes, but with two important differences. First, the parent chain *must include the double bond* even if this means that the parent chain is shorter than another. Sec-

The IUPAC accepts both ethene and ethylene as the name of $CH_2{=}CH_2$. The common name of propene is propylene, and other simple alkenes have common names as well.

No number is needed to name $CH_2=CH_2$, ethene, or $CH_3CH=CH_2$, propene.

ond, the parent alkene chain must be *numbered from whichever end gives the first carbon of the double bond the lower of two possible numbers.* This (lower) number, followed by a hyphen, precedes the name of the parent chain, unless there is no ambiguity about where the double bond occurs. The number locations of branches are not a factor in numbering the chain. Otherwise, alkyl groups are named and located as before. Some examples of correctly named alkenes are as follows.

$$CH_3CH_2CH=CH_2 \qquad CH_3CH=CHCH_3$$

1-butene
(not 1,2-butene,
not 3-butene)

2-butene

$$CH_3CH_2\overset{\overset{\displaystyle CH_3}{|}}{C}HCH_2CH=\overset{\overset{\displaystyle CH_3}{|}}{C}CH_3$$

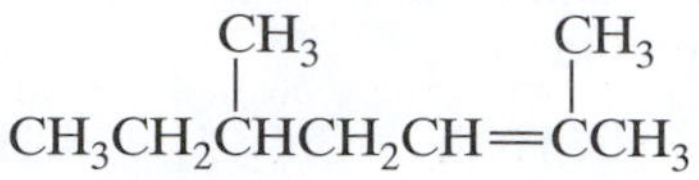

2,5-dimethyl-2-heptene
(not 3,6-dimethyl-5-heptene)

cyclohexene

Notice that only one number is used to locate the double bond, the number of the first carbon of the double bond to be reached as the chain is numbered.

Some alkenes have two double bonds and are called *dienes.* Some have three double bonds and are called *trienes,* and so forth. Each double bond has to be located by a number.

$$CH_2=CHCH=CHCH_3 \qquad CH_2=CHCH_2CH=CH_2 \qquad CH_2=CHCH=CHCH=CH_2$$

1,3-pentadiene

1,4-pentadiene

1,3,5-hexatriene

Geometric Isomerism among the Alkenes

As explained in Section 9.6, there is no free rotation at a carbon–carbon double bond (see page 400). Many alkenes, therefore, exhibit **geometric isomerism.** Thus *cis*-2-butene and *trans*-2-butene are **geometric isomers** of each other. They not only have the same molecular formula, C_4H_8, but also the same skeletons and the same organization of atoms and bonds, namely, $CH_3CH=CHCH_3$. The two 2-butenes differ in the *directions* taken by the two CH_3 groups attached at the double bond.

Because ring structures also lock out free rotation, geometric isomers of ring compounds are also possible. These two isomers of 1,2-dimethylcyclopropane are examples.

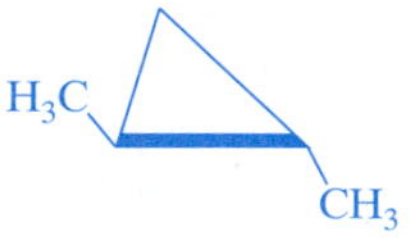

trans isomer
bp 28 °C

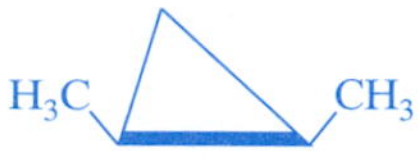

cis isomer
bp 37 °C

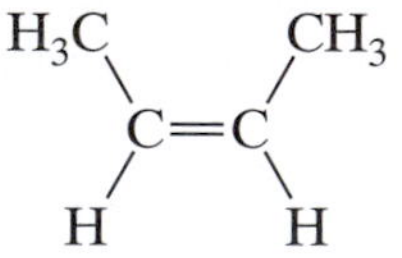

cis-2-butene
bp 3.72 °C

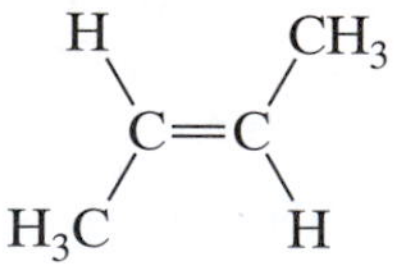

trans-2-butene
bp 0.88 °C

Cis means "on the same side"; *trans* means "on opposite sides." This difference in orientation gives the two geometric isomers of 2-butene measurable differences in physical properties, as their boiling points show. Because each has a double bond, however, the *chemical* properties of *cis*- and *trans*-2-butene are very similar.

Addition Reactions of Alkenes

Electron-seeking species are naturally attracted to the electron density at the π bond of the double bond of alkenes. Thus, alkenes react readily with protons provided by strong proton donors. Alkenes thus give **addition reactions,** reactions in which the pieces of a reactant become separately attached to the carbons of a double bond. Ethene, for example, readily reacts with hydrogen chloride as follows.

$$CH_2{=}CH_2 + H{-}Cl(g) \longrightarrow Cl{-}CH_2{-}CH_3$$

We say that the hydrogen chloride molecule *adds* across the double bond, one piece of HCl going to the carbon at one end and the other piece going to the other end. The pair of electrons of the π bond move out and take H^+ from HCl, which releases Cl^-. A (+) charge is left at one end of the original carbon–carbon double bond as a bond to H forms at the other end.

$$CH_2{=}CH_2 + H{-}Cl \longrightarrow {}^+CH_2{-}CH_2 \longrightarrow Cl{-}CH_2{-}CH_3$$

ethyl carbocation   1-chloroethane

The result of H^+ transfer is a very unstable cation, called a *carbocation,* an ion with a positive charge on carbon. This charged site then quickly attracts the Cl^- ion to give the product, 1-chloroethane.

2-Butene, like ethene, also adds HCl.

$$CH_3CH{=}CHCH_3 + HCl \longrightarrow CH_3CHCH_2CH_3$$

       |
       Cl

2-butene   2-chlorobutane

> A carbocation, having a carbon with only a sextet of valence electrons, not an octet, generally has only a fleeting existence.

Alkene double bonds also add hydrogen bromide, hydrogen iodide, and sulfuric acid ("hydrogen sulfate").

A water molecule adds to a double bond if an acid catalyst, like sulfuric acid, is present. The catalyst participates in the first step by donating a proton to one end of the double bond to create a carbocation. This unstable species preferentially attracts what is the most abundant electron-rich species present, namely, a water molecule,

> Alkynes give similar addition reactions.

$$CH_2{=}CH_2 + H{-}OSOH \xrightarrow{HSO_4^-} CH_2{-}^+CH_2 \longrightarrow CH_3{-}CH_2 \longrightarrow CH_3CH_2OH + H^+$$

ethene   sulfuric acid (trace; much H_2O present)   (H_2O attacks instead of HSO_4^-)   protonated form of ethanol   ethanol

This represents recovered catalyst.

Note that in the *overall* result, the pieces of the water molecule, H and OH, become attached at the different carbons of the double bond. 2-Butene (either *cis* or *trans*) gives a similar reaction.

Other inorganic compounds that add to an alkene double bond are chlorine, bromine, and hydrogen. Chlorine and bromine react rapidly at room temperature. Ethene, for example, reacts with bromine to give 1,2-dibromoethane.

$$CH_2{=}CH_2 + Br{-}Br \longrightarrow CH_2{-}CH_2$$

ethene   Br   Br

1,2-dibromoethane

The Addition of Hydrogen—Hydrogenation

The product of the addition of hydrogen to an alkene is an alkane, and the reaction is called *hydrogenation*. It requires a catalyst—powdered platinum, for example—and sometimes a higher temperature and pressure than available under an ordinary room atmosphere. The hydrogenation of 2-butene (*cis* or *trans*) gives butane.

$$CH_3CH{=}CHCH_3 + H{-}H \xrightarrow[\text{catalyst}]{\text{heat, pressure,}} CH_3CH{-}CHCH_3 \quad or \quad CH_3CH_2CH_2CH_3$$

2-butene
(*cis* or *trans*)

with H and H below the CH—CH

butane

Oxidation of a Double Bond

Ozone, by attacking the double bonds in the chlorophyll of green plants, is able to prevent photosynthesis and so kill the plants.

Ozone reacts with anything that has carbon–carbon double or triple bonds, and the reaction breaks the molecules into fragments at each such site, giving a variety of products. Because many important compounds in living systems have alkene double bonds, ozone is a very dangerous material in the wrong places (see *Chemicals in Our World 6*).

Aromatic Hydrocarbons
The Benzene Ring

Hydrocarbons and their oxygen or nitrogen derivatives that are not aromatic are called *aliphatic compounds.*

The most common **aromatic compounds** contain the *benzene ring,* a ring of six carbon atoms, each holding one H or one other atom or group. The benzene ring is represented as a hexagon either with alternating single and double bonds or with a circle. For example,

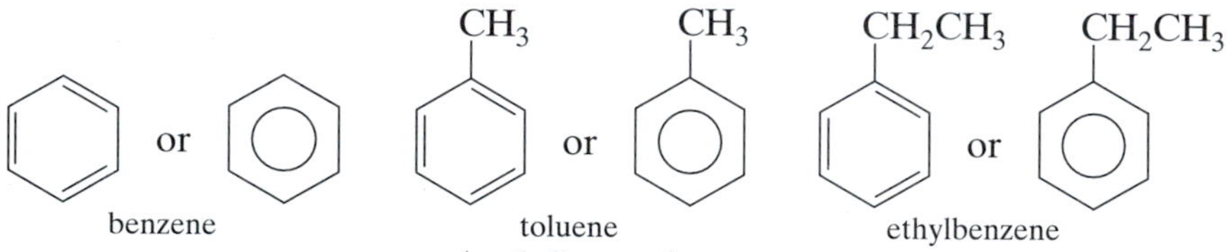

	CH_3		CH_2CH_3

benzene   or   toluene (methylbenzene)   or   ethylbenzene

The circle better suggests the delocalized bonds of the benzene ring, which are described in Section 9.8.

In the molecular orbital view of benzene, discussed in Section 9.8, the delocalization of the ring's π electrons strongly stabilizes the ring. This explains why benzene does not easily give addition reactions; they would interfere with the delocalization of electron density. Instead, the benzene ring most commonly undergoes **substitution reactions,** those in which one of the ring H atoms is replaced by another atom or group. For example, benzene reacts with chlorine in the presence of iron(III) chloride to give chlorobenzene, instead of a 1,2-dichloro compound. To dramatize this point, we must use a resonance structure for benzene (see page 356).

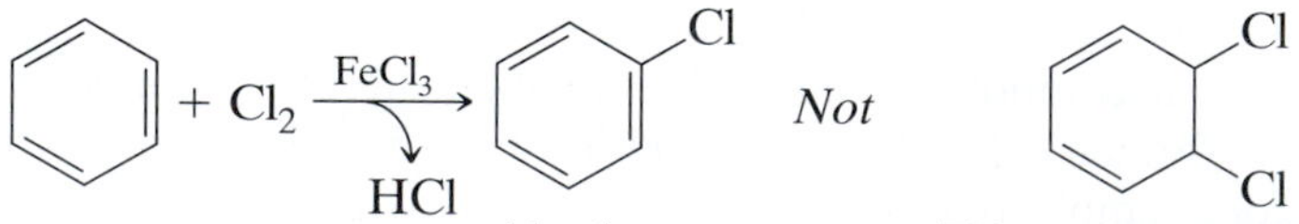

$$\xrightarrow[\text{HCl}]{FeCl_3}$$

+ Cl_2 → chlorobenzene   *Not*   (This would form if chlorine *added* to the double bond.)

You can infer that *substitution,* but not addition, leaves intact the closed-circuit, delocalized, and very stable π electron network of the benzene ring.

Provided that a suitable catalyst is present, benzene reacts by substitution with chlorine, bromine, and nitric acid as well as with sulfuric acid. (Recall that Cl_2 and Br_2 readily *add* to alkene double bonds.)

$$C_6H_6 + Br_2 \xrightarrow{\text{FeBr}_3 \text{ catalyst}} C_6H_5-Br + HBr$$
bromobenzene

$$C_6H_6 + HNO_3 \xrightarrow{\text{H}_2\text{SO}_4 \text{ catalyst}} C_6H_5-NO_2 + H_2O$$
nitrobenzene

$$C_6H_6 + H_2SO_4 \longrightarrow C_6H_5-SO_3H + H_2O$$
benzenesulfonic acid

The sulfonic acids are roughly as strong acids as sulfuric acid.

23.3 Alcohols and Ethers, Organic Derivatives of Water

Our study of functional groups will be limited largely to those of greatest importance to biochemistry. The alcohol group, for example, occurs in or is chemically important to all major foodstuffs, like carbohydrates, lipids (fats and oils), and proteins.

An **alcohol** is any compound with an OH group attached to a carbon with three other groups also attached by *single bonds*. Using the symbol R to represent any alkyl group, alcohols have ROH as their general structure.

The four structurally simplest alcohols are the following. (Their common names are in parentheses below their IUPAC names.)

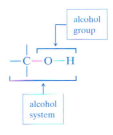

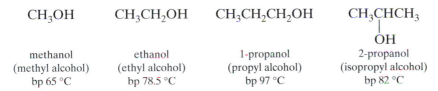

<table>
<tr><td>CH₃OH</td><td>CH₃CH₂OH</td><td>CH₃CH₂CH₂OH</td><td>CH₃CHCH₃
|
OH</td></tr>
<tr><td>methanol
(methyl alcohol)
bp 65 °C</td><td>ethanol
(ethyl alcohol)
bp 78.5 °C</td><td>1-propanol
(propyl alcohol)
bp 97 °C</td><td>2-propanol
(isopropyl alcohol)
bp 82 °C</td></tr>
</table>

Ethanol is the alcohol in beverages and is also added to gasoline to make "gasohol."

In the body, methanol causes blindness and often death, and ethanol causes loss of coordination and inhibitions. When abused, ethanol ruins the liver and the brain.

IUPAC Names of Alcohols

The name ending for an alcohol is *-ol*. It replaces the *-e* ending of the name of the hydrocarbon that corresponds to the parent. The parent chain of an alcohol must be the longest *that includes the carbon holding the OH group*. The chain is numbered to give the site of the OH group the lower number regardless of where alkyl substituents occur.

Ethers

Molecules of **ethers** contain two alkyl groups joined to one oxygen, the two R groups being alike or different. We give only the common names for the following examples.

<table>
<tr><td>CH₃OCH₃</td><td>CH₃CH₂OCH₂CH₃</td><td>CH₃OCH₂CH₃</td><td>R—O—R′</td></tr>
<tr><td>dimethyl ether
bp −23 °C</td><td>diethyl ether
bp 34.5 °C</td><td>methyl ethyl ether
bp 11 °C</td><td>ethers
(general structure)</td></tr>
</table>

Diethyl ether was the "ether" once widely used as an anesthetic in surgery.

Pentane (bp 36.1 °C), which is like diethyl ether in having no OH group but has nearly the identical molecular mass, has about the same boiling point as diethyl ether.

The contrasting boiling points of alcohols and ethers illustrate the influence of hydrogen bonding. Hydrogen bonds cannot exist between molecules of ethers, and the simple ethers have very low boiling points, being substantially lower than those of alcohols of comparable molecular masses. For example, 1-butanol, $CH_3CH_2CH_2CH_2OH$ (bp 117 °C), boils 83 degrees higher than its structural isomer, diethyl ether (bp 34.5 °C).

Major Reactions of Alcohols and Ethers

Ethers are almost as chemically inert as alkanes. They burn (as do alkanes), and they are split apart when boiled in concentrated acids.

The alcohols, in contrast, have a rich chemistry. We'll look at their oxidation and dehydration reactions as well as some substitution reactions.

Oxidation Reactions of Alcohols

When the carbon atom of the alcohol system, the *alcohol carbon atom,* also holds at least one H, this H can be removed by an oxidizing agent. The H atom of the OH group also leaves, and the two H's become part of a water molecule. We may think of the oxidizing agent as providing the O atom for H_2O. The number of H atoms left on the original alcohol carbon then determines the *family* of the product.

The oxidation of an alcohol of the RCH_2OH type, with 2 H atoms on the alcohol carbon, produces first an aldehyde, which is further oxidized to a carboxylic acid.

The aldehyde group, CH=O, is one of the most easily oxidized, and the carboxyl group, CO_2H, is one of the most oxidation resistant of the functional groups.

$$RCH_2OH \xrightarrow{\text{oxidation}} \underset{\text{aldehyde}}{\overset{\overset{\displaystyle O}{\parallel}}{RCH}} \xrightarrow[\text{oxidation}]{\text{further}} \underset{\text{carboxylic acid}}{\overset{\overset{\displaystyle O}{\parallel}}{RCOH}}$$

The net ionic equation for the formation of the aldehyde when dichromate ion is used is

$$3RCH_2OH + Cr_2O_7{}^{2-} + 8H^+ \longrightarrow 3RCH{=}O + 2Cr^{3+} + 7H_2O$$

Aldehydes are much more easily oxidized than alcohols, so unless the aldehyde is removed from the solution as it forms, it will consume oxidizing agent that has not yet reacted and be changed to the corresponding carboxylic acid.

The oxidation of an alcohol of the R_2CHOH type produces a ketone. For example, the oxidation of 2-propanol gives propanone.

$$\underset{\text{2-propanol}}{3CH_3\overset{\overset{\displaystyle OH}{|}}{C}HCH_3} + Cr_2O_7{}^{2-} + 8H^+ \longrightarrow \underset{\substack{\text{propanone}\\\text{(acetone)}}}{3CH_3\overset{\overset{\displaystyle O}{\parallel}}{C}CH_3} + 2Cr^{3+} + 7H_2O$$

Ketones strongly resist oxidation, so a ketone does not have to be separated from the oxidizing agent as it forms.

Alcohols of the type R_3COH have no removable H atom on the alcohol carbon, so they cannot be oxidized in a similar manner.

EXAMPLE 23.2

Alcohol Oxidation Products

What organic product can be made by the oxidation of 2-butanol with dichromate ion? If no oxidation can occur, state so.

Analysis: We first have to look at the structure of 2-butanol.

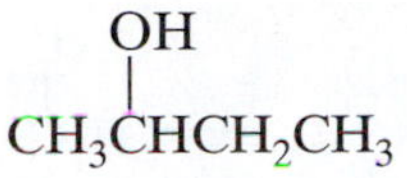

$$\underset{\text{2-butanol}}{CH_3\overset{\displaystyle OH}{\overset{|}{C}}HCH_2CH_3}$$

2-Butanol can be oxidized because it has an H atom on the alcohol carbon (carbon 2). The oxidation results in the detachment of both this H atom and the one joined to the O atom, leaving a double bond to O.

Solution: We carry out the changes that the analysis found. We simply erase the two H atoms that we identified and insert a double bond from C to O. The product is 2-butanone.

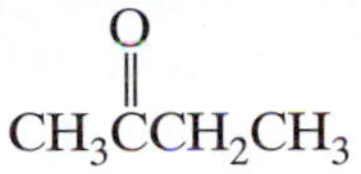

$$\underset{\text{2-butanone (a ketone)}}{CH_3\overset{\displaystyle O}{\overset{\|}{C}}CH_2CH_3}$$

The starting material had two alkyl groups, CH_3 and CH_2CH_3. It doesn't matter what these alkyl groups are, however, because the reaction takes the same course in all cases. All alcohols of the R_2CHOH type can be oxidized to ketones in this way.

Is the Answer Reasonable?
The "skeleton" of heavy atoms, C's and O, does not change in this kind of oxidation, and 2-butanone has the same skeleton as 2-butanol. The oxidation also produces a double bond from C to O, as we've shown.

Practice Exercise 2

What are the structures of the products that form by the oxidation of the following alcohols? (a) ethanol   (b) 3-pentanol ◆

Dehydration Reactions of Alcohols

In the presence of a strong acid, like concentrated sulfuric acid, an alcohol molecule can undergo the loss of a water molecule leaving behind a carbon–carbon double bond. This reaction, called **dehydration,** is one example of an **elimination reaction.** For example,

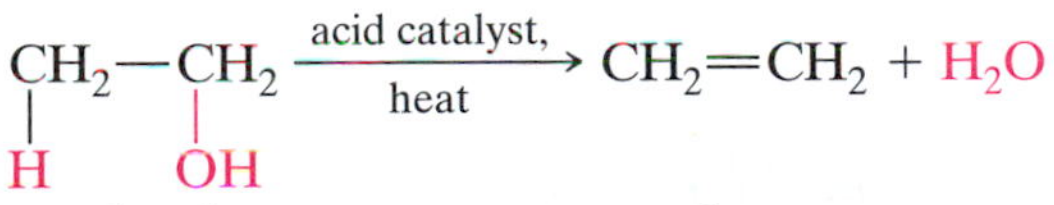

$$\underset{\substack{\\ \text{ethanol}}}{CH_2\!\!-\!\!CH_2} \xrightarrow[\text{heat}]{\text{acid catalyst,}} CH_2\!\!=\!\!CH_2 + H_2O$$
$$\underset{\text{ethanol}}{\overset{\displaystyle ||}{HOH}}\underset{\text{ethene}}{}$$

$$\underset{\substack{\\ \text{1-propanol}}}{CH_3CH\!\!-\!\!CH_2} \xrightarrow[\text{heat}]{\text{acid catalyst,}} CH_3CH\!\!=\!\!CH_2 + H_2O$$
$$\underset{\text{1-propanol}}{\overset{\displaystyle ||}{HOH}}\underset{\text{propene}}{}$$

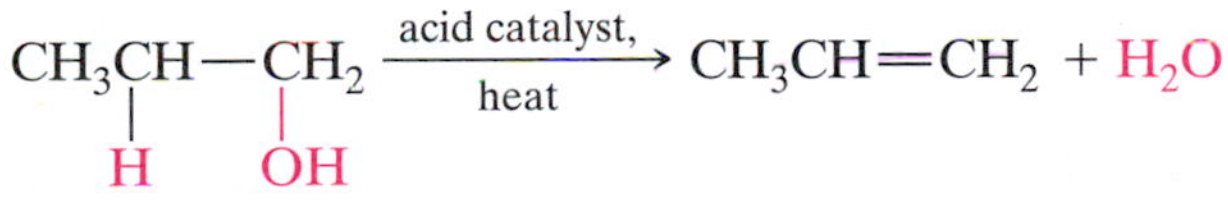

What makes the elimination of water possible is the proton-accepting ability of the O atom of the OH group. Alcohols resemble water in that they react like Brønsted bases toward concentrated strong acids to give an equilibrium mixture involving a protonated form. Ethanol, for example, reacts with (and dissolves in) concentrated sulfuric acid by the following reaction. (H_2SO_4 is written as $H\!-\!OSO_3H$.)[1]

---

[1]The first curved arrow at the top of the next page illustrates how an atom attached to one molecule, like the O of the HO group, uses an unshared pair of electrons to pick up an atom of another molecule, like the H of the catalyst. The second curved arrow shows that *both* electrons of the bond from H to O in sulfuric acid remain in the hydrogen sulfate ion that forms. Thus H transfers as H^+.

This is like the reaction of H_2SO_4 with H_2O to give the H_3O^+ and HSO_4^- ions.

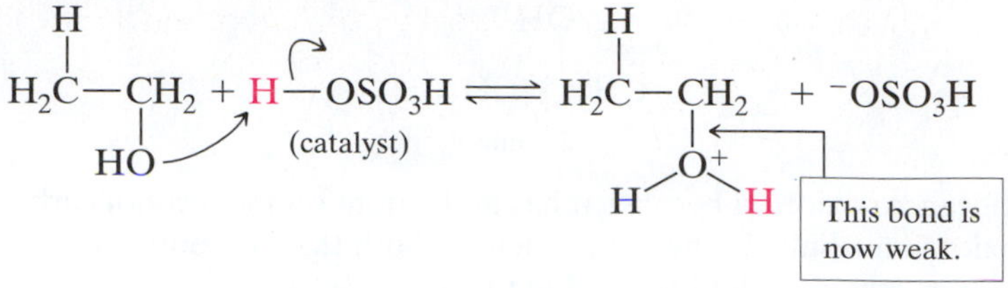

$$H_2C{-}CH_2 + H{-}OSO_3H \rightleftharpoons H_2C{-}CH_2 + {^-}OSO_3H$$

(catalyst)

This bond is now weak.

The organic cation is nothing more than the ethyl derivative of the hydronium ion, $CH_3CH_2OH_2^+$, and *all three bonds to oxygen in this cation are weak,* just like all three bonds to oxygen in H_3O^+.

A water molecule now leaves, taking with it the electron pair that held it to the CH_2 group. The remaining organic species is a carbocation.

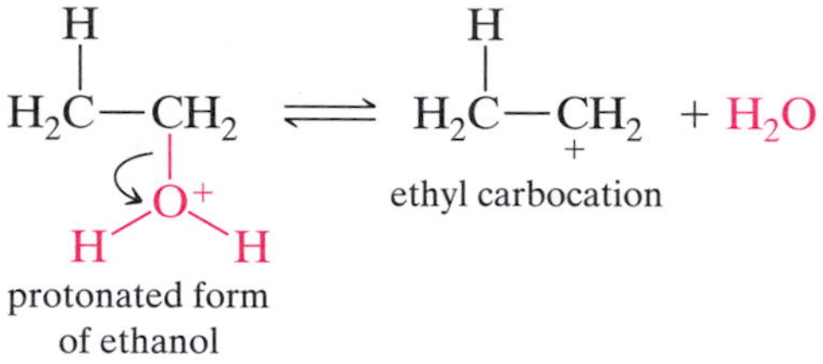

$$H_2C{-}CH_2 \rightleftharpoons H_2C{-}CH_2 + H_2O$$

protonated form
of ethanol

ethyl carbocation

Carbocations, as we said, are unstable. The ethyl carbocation, $CH_3CH_2^+$, to become more stable, loses a proton, donating it to some proton acceptor in the medium. The electron pair holding the departing proton stays behind to become the second bond of the new double bond in the product. All carbon atoms now have outer octets. We'll use the HSO_4^- ion (written here as HSO_3O^-) as the proton acceptor.

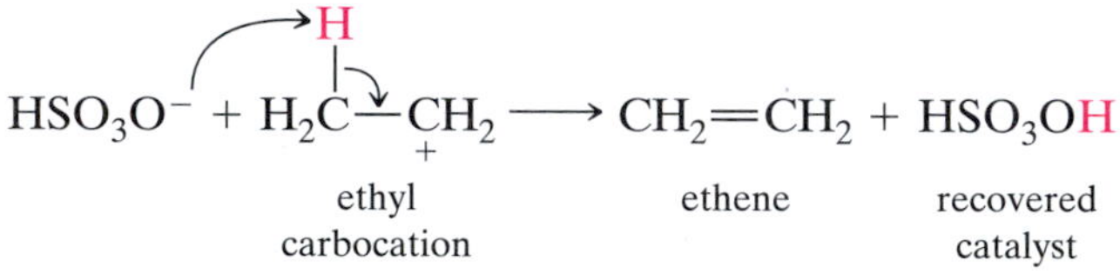

$$HSO_3O^- + H_2C{-}CH_2 \longrightarrow CH_2{=}CH_2 + HSO_3OH$$

ethyl
carbocation

ethene

recovered
catalyst

The last few steps in the dehydration of ethanol—the separation of H_2O, loss of the proton, formation of the double bond, and recovery of the catalyst—probably occur simultaneously. Notice two things about the catalyst, H_2SO_4. First, it works to convert a species with strong bonds, the alcohol, to one with weak bonds strategically located, the protonated alcohol. Second, the catalyst is recovered.

Substitution Reactions of Alcohols

Under acidic conditions, the OH group of an alcohol can be replaced by a halogen atom, using a concentrated hydrohalogen acid. For example,

$$CH_3CH_2OH + HI \text{ (concd.)} \xrightarrow{\text{heat}} CH_3CH_2I + H_2O$$

ethanol

iodoethane
(ethyl iodide)

$$CH_3CH_2CH_2OH + HBr \text{ (concd.)} \xrightarrow{\text{heat}} CH_3CH_2CH_2Br + H_2O$$

1-propanol

1-bromopropane
(propyl bromide)

These reactions, like the earlier reaction between chlorine and benzene, are **substitution reactions.** The first step in each is the transfer of H^+ to the OH of the alcohol to give the protonated form of the OH group.

$$R{-}OH + H^+ \longrightarrow R{-}OH_2^+$$

Once again, the acid catalyst works to weaken an important bond. Given the high concentration of halide ion, X^-, it is this species that successfully interacts with $R-OH_2^+$ to displace OH_2 and give $R-X$.

23.4 Amines, Organic Derivatives of Ammonia

Amines are organic derivatives of ammonia in which one, two, or three hydrocarbon groups have replaced hydrogens. Examples, together with their common (not IUPAC) names, are

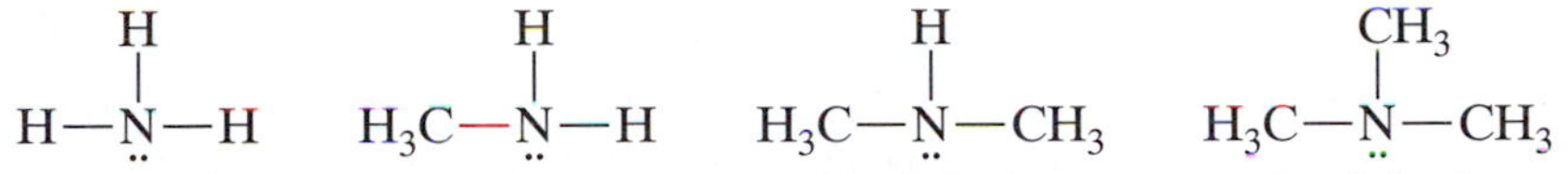

ammonia
bp −33.4 °C

methylamine
bp −8 °C

dimethylamine
bp 8 °C

trimethylamine
bp 3 °C

The hydrocarbon groups, which are all methyl groups in these structures, do not have to be identical.

The N—H bond is not as polar as the O—H bond, so amines boil at lower temperatures than alcohols of comparable molecular masses. Amines of low molecular mass are soluble in water. Hydrogen bonding between molecules of water and the amine facilitates this.

CH_3CH_2OH,   bp 78.5 °C
$CH_3CH_2NH_2$,   bp 17 °C
$CH_3CH_2CH_3$,   bp −42 °C

Basicity and Reactions of Amines

As we said earlier, the amines are Brønsted bases. Behaving like ammonia, amines that do dissolve in water establish an equilibrium in which a low concentration of hydroxide ion exists. For example,

ethylmethylamine

ethylmethylammonium ion

As a result, aqueous solutions of amines test basic to litmus and have pH values above 7.

When an amine is mixed with a stronger proton donor than water, for example, the hydronium ion in hydrochloric acid, the amine and acid react almost quantitatively. The amine accepts a proton and changes almost 100% into its protonated form. For example,

ethylmethylamine

ethylmethylammonium ion
(a protonated amine)

Even water-insoluble amines give this reaction. By thus changing into ions that can be hydrated, amines that are otherwise insoluble in water become much more soluble.

▶**Chemistry in Practice**◀ Many important medicinal chemicals, like quinine, are amines, but they are usually supplied to patients in protonated forms so that the drug can be administered as an aqueous solution, not as a solid. This strategy is particularly important for medicinals that must be given by intravenous drip. ◆

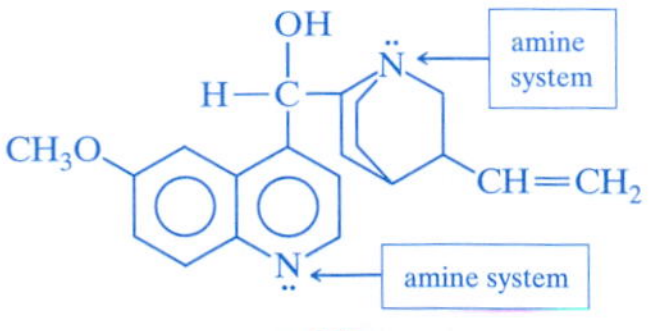

quinine
(an antimalarial drug)

Acidity of Protonated Amines

A protonated amine is a substituted ammonium ion. Like the ammonium ion itself, protonated amines are weak Brønsted acids. They can neutralize strong base. For example,

$$CH_3NH_3^+(aq) \quad + OH^-(aq) \longrightarrow CH_3NH_2(aq) + H_2O$$

methylammonium ion   methylamine

This reverses the protonation of an amine and releases the uncharged amine molecule.

23.5 Organic Compounds with Carbonyl Groups

We introduced the carbon–oxygen double bond, the **carbonyl group,** in Section 8.4 (page 339). It occurs in several organic families, however, so it uniquely defines none. What is attached to the carbon atom in C=O determines the specific family.

Types and Names of Carbonyl Compounds

AL-de-hide
KEY-tone

When the carbonyl group binds an H atom plus a hydrocarbon group (or a second H), the compound is an **aldehyde.** When C=O holds two hydrocarbon groups at C, the compound is a **ketone.** A **carboxylic acid** carries an OH on the carbon of the carbonyl group.

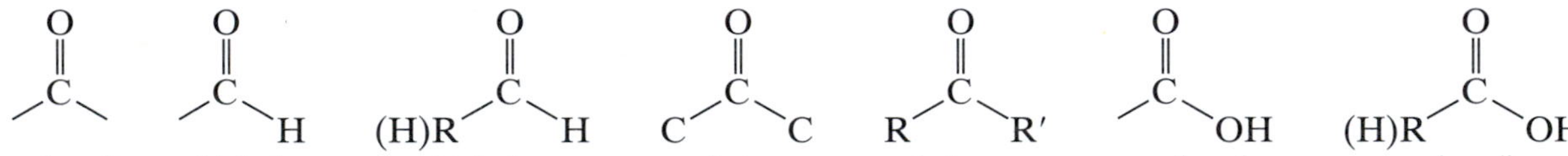

| carbonyl group | aldehyde group | aldehydes | keto group | ketones | carboxyl group | carboxylic acids |

The *aldehyde group* is often condensed to CHO, with the double bond of the carbonyl group being "understood." The *ketone group* is sometimes condensed to CO. The *carboxyl group* is often written as CO_2H.

The carbonyl group occurs widely. As an aldehyde group, it's in the molecules of most sugars, like glucose. Another common sugar, fructose, has the keto group. The carboxyl group is present in all of the building blocks of proteins, the amino acids.

The carbonyl group is a polar group, and it helps to make compounds having it much more soluble in water than hydrocarbons of roughly the same molecular mass.

Aldehydes and Ketones
Nomenclature

The IUPAC name ending for an aldehyde is -al. The parent chain is the longest chain *that includes the aldehyde group.* Thus, the three-carbon aldehyde is named *propanal,* because "propane" is the name of the three-carbon alkane and the *-e* in propane is replaced by *-al.* The numbering of the chain always starts by assigning the carbon of the aldehyde group position 1. This rule, therefore, makes it unnecessary to include the number locating the aldehyde group in the name, as illustrated by the name 2-methylpropanal.

$$\underset{\substack{\text{methanal} \\ \text{bp} -21\ °C}}{\overset{\overset{\textstyle O}{\|}}{HCH}} \qquad \underset{\substack{\text{ethanal} \\ \text{bp } 21\ °C}}{\overset{\overset{\textstyle O}{\|}}{CH_3CH}} \qquad \underset{\substack{\text{propanal} \\ \text{bp } 49\ °C}}{\overset{\overset{\textstyle O}{\|}}{CH_3CH_2CH}} \qquad \underset{\substack{\text{2-methylpropanal} \\ \text{(not 2-methyl-1-propanal)} \\ \text{bp } 64\ °C}}{\overset{\overset{\textstyle CH_3}{|}}{CH_3CH}\!-\!\overset{\overset{\textstyle O}{\|}}{CH}}$$

Aldehydes cannot form hydrogen bonds between their own molecules, so they boil at lower temperatures than alcohols of comparable molecular masses.

The name ending for the IUPAC names of ketones is *-one*. The parent chain must include the carbonyl group and be numbered from whichever end reaches the carbonyl carbon first. The number of the ketone group's location must be part of the name whenever there would otherwise be uncertainty.

CH_3OH (molecular mass 32), bp 65 °C

$CH_2{=}O$ (molecular mass, 30), bp −21 °C

$$\underset{\substack{\text{propanone} \\ \text{(acetone)} \\ \text{bp } 56.5\ °C}}{\overset{\overset{\textstyle O}{\|}}{CH_3CCH_3}} \qquad \underset{\substack{\text{3-pentanone} \\ \text{bp } 101.5\ °C}}{\overset{\overset{\textstyle O}{\|}}{CH_3CH_2CCH_2CH_3}} \qquad \underset{\substack{\text{5-methyl-2-hexanone} \\ \text{(not 2-methyl-5-hexanone)} \\ \text{bp } 145\ °C}}{\overset{\overset{\textstyle CH_3}{|}}{CH_3CHCH_2CH_2}\overset{\overset{\textstyle O}{\|}}{CCH_3}}$$

We need not write "2-propanone," because if the carbonyl carbon is anywhere else in a three-carbon chain, the compound is the *aldehyde*, propanal.

Hydrogenation of Aldehydes and Ketones

The double bond in the carbonyl group of both aldehydes and ketones adds hydrogen. The reaction is just like the addition of hydrogen to an alkene double bond and under roughly the same conditions, namely, a metal catalyst, heat, and pressure. The reaction is called either *hydrogenation* or *reduction*. For example,

$$\underset{\text{ethanal}}{\overset{\overset{\textstyle O}{\|}}{CH_3CH}} + H\!-\!H \xrightarrow[\text{catalyst}]{\text{heat, pressure,}} \underset{\substack{| \\ H}}{\overset{\overset{\textstyle O-H}{|}}{CH_3CH}} \quad \text{or} \quad \underset{\text{ethanol}}{CH_3CH_2OH}$$

$$\underset{\substack{\text{propanone} \\ \text{(acetone)}}}{\overset{\overset{\textstyle O}{\|}}{CH_3CCH_3}} + H\!-\!H \xrightarrow[\text{catalyst}]{\text{heat, pressure,}} \underset{\substack{| \\ H}}{\overset{\overset{\textstyle O-H}{|}}{CH_3CCH_3}} \quad \text{or} \quad \underset{\text{2-propanol}}{\overset{\overset{\textstyle OH}{|}}{CH_3CHCH_3}}$$

Propanone (acetone) is commonly used as a fingernail polish remover.

The H atoms take up positions at opposite ends of the carbonyl group's double bond, which then becomes a single bond holding an OH group.

Oxidation of Aldehydes

Aldehydes and ketones are in separate families because of their remarkably different behavior toward oxidizing agents. As we noted on page 1042, aldehydes are easily oxidized, but ketones strongly resist oxidation. Even in storage in a bottle, aldehydes are slowly oxidized by the oxygen of the air trapped in the bottle.

Carboxylic Acids and Their Derivatives
IUPAC Names of Carboxylic Acids

The name ending of the IUPAC names of carboxylic acids is *-oic acid*. The parent chain must be the longest that includes the carboxyl carbon, which is numbered as position 1. The name of the hydrocarbon with the same number of carbons as the parent is then changed by replacing the terminal *-e* with *-oic acid*. For example,

Because carboxylic acids have both a lone oxygen and an OH group, their molecules strongly hydrogen bond to each other. Their high boiling points, relative to alcohols of comparable molecular mass, reflect this.

$$HCO_2H \qquad CH_3CO_2H \qquad \overset{\displaystyle CH_3}{\overset{|}{CH_3CHCH_2CO_2H}}$$

methanoic acid   ethanoic acid   3-methylbutanoic acid
bp 101 °C   bp 118 °C   bp 176 °C

The carboxyl group is a weakly acidic group, but all carboxylic acids neutralize such bases as the hydroxide, bicarbonate, and carbonate ions. The general equation for the reaction with OH^- is

RCO_2Na is a salt consisting of Na^+ and RCO_2^-, the *carboxylate ion*.

$$RCO_2H + NaOH \xrightarrow{H_2O} RCO_2Na + H_2O$$

Esters of Carboxylic Acids

A *derivative* of a carboxylic acid is a compound that can be made from the acid, or which can be changed to the acid by hydrolysis.

Carboxylic acids are used to synthesize two important derivatives of the acids, *esters* and *amides*. In **esters**, the OH of the carboxyl group is replaced by OR.

$$(H)R\overset{\displaystyle O}{\overset{\|}{C}}OR' \text{ or } (H)RCO_2R', \text{ for example: } CH_3\overset{\displaystyle O}{\overset{\|}{C}}O(CH_2)_7CH_3$$

esters   octyl ethanoate
(fragrance of oranges)

The IUPAC name of an ester begins with the name of the alkyl group attached to the O atom. This is followed by a separate word, one taken from the name of the parent carboxylic acid but altered by changing *-ic acid* to *-ate*. For example,

$$HCO_2CH_3 \qquad CH_3CO_2CH_2CH_3 \qquad \overset{\displaystyle CH_3 \qquad\quad CH_3}{\overset{|\qquad\qquad\;\; |}{CH_3CHCH_2CO_2CHCH_3}}$$

methyl methanoate   ethyl ethanoate   isopropyl 3-methylbutanoate
bp 31.5 °C   bp 77 °C   bp 142 °C

Formation of Esters

One way to prepare an ester is to heat a solution of the parent carboxylic acid and the alcohol in the presence of an acid catalyst. (We'll not go into the details of how this happens.) The following kind of equilibrium forms, but a substantial stoichiometric excess of the alcohol (usually the less expensive reactant) is commonly used to drive the position of the equilibrium toward the ester.

$$R\overset{\displaystyle O}{\overset{\|}{C}}OH + HOR' \underset{heat}{\overset{H^+ \text{ catalyst,}}{\rightleftharpoons}} R\overset{\displaystyle O}{\overset{\|}{C}}OR' + H_2O$$

carboxylic acid   alcohol   ester

Esters are responsible for many pleasant fragrances in nature.

For example,

$$CH_3CH_2CH_2\overset{\displaystyle O}{\overset{\|}{C}}OH + HOCH_2CH_3 \underset{heat}{\overset{H^+ \text{ catalyst,}}{\rightleftharpoons}} CH_3CH_2CH_2\overset{\displaystyle O}{\overset{\|}{C}}OCH_2CH_3 + H_2O$$

butanoic acid   ethanol   ethyl butanoate
bp 166 °C   bp 78.5 °C   (fragrance of pineapple)
bp 120 °C

Ester	Aroma
$HCO_2CH_2CH_3$	rum
$HCO_2CH_2CH(CH_3)_2$	raspberries
$CH_3CO_2(CH_2)_4CH_3$	bananas
$CH_3CO_2(CH_2)_2CH(CH_3)_2$	pears
$CH_3CO_2(CH_2)_7CH_3$	oranges
$CH_3(CH_2)_2CO_2CH_2CH_3$	pineapples
$CH_3(CH_2)_2CO_2(CH_2)_4CH_3$	apricots

Hydrolysis of Esters

An ester is hydrolyzed to its parent acid and alcohol when the ester is heated together with a stoichiometric excess of water (plus an acid catalyst). The identical equilibrium as shown above forms, but in ester hydrolysis water is in excess, so the equilibrium shifts to the left to favor the carboxylic acid and alcohol, another illustration of Le Châtelier's principle.

Esters are also split apart by the action of aqueous base, only now the carboxylic acid emerges not as the free acid but as its anion. The reaction is called ester *saponification*. We may illustrate it by the action of aqueous sodium hydroxide on a simple ester, ethyl acetate.

$$CH_3\overset{\overset{\displaystyle O}{\|}}{C}OCH_2CH_3(aq) + NaOH(aq) \xrightarrow{\text{heat}} CH_3\overset{\overset{\displaystyle O}{\|}}{C}O^-(aq) + Na^+(aq) + HOCH_2CH_3(aq)$$

| ethyl ethanoate (ethyl acetate) | ethanoate ion (acetate ion) | ethanol |

> ▶**Chemistry in Practice**◀ Ester groups abound among the molecules of the fats and oils in our diets. The hydrolysis of their ester groups occurs when we digest them, with an enzyme as the catalyst, not a strong acid. Because this digestion occurs in a region of the intestinal tract where the fluids are slightly basic, the anions of carboxylic acids form. ◆

Amides of Carboxylic Acids

Carboxylic acids can also be converted to *amides,* a functional group found in proteins. In **amides,** the OH of the carboxyl group is replaced by trivalent nitrogen, which may also hold any combination of H atoms or hydrocarbon groups.

$$(H)R\overset{\overset{\displaystyle O}{\|}}{C}NH_2 \text{ or } (H)RCONH_2, \text{ for example: } CH_3\overset{\overset{\displaystyle O}{\|}}{C}NH_2$$

| simple amides | ethanamide (acetamide) |

The *simple amides* are those in which the nitrogen bears no hydrocarbon groups, only 2 H atoms. In the place of either or both of these H atoms, however, there can be a hydrocarbon group, and the resulting substance is still in the amide family.

The IUPAC names of the simple amides are devised by first writing the name of the parent carboxylic acid. Then its ending, *-oic acid,* is replaced by *-amide.* For example,

$CH_3CH_2CONH_2$	$CH_3CH_2CH_2CH_2CONH_2$	$CH_3\overset{\overset{\displaystyle CH_3}{\|}}{C}HCH_2CH_2CONH_2$
propanamide	pentanamide	4-methylpentanamide

One of the ways to prepare simple amides parallels the synthesis of esters, that is, by heating a mixture of the carboxylic acid and an excess of ammonia.

In general:
$$R\overset{\overset{\displaystyle O}{\|}}{C}OH + H-NH_2 \xrightarrow{\text{heat}} R\overset{\overset{\displaystyle O}{\|}}{C}NH_2 + H_2O$$

| carboxylic acid | ammonia | simple amide |

An example:
$$CH_3\overset{\overset{\displaystyle O}{\|}}{C}OH + H-NH_2 \xrightarrow{\text{heat}} CH_3\overset{\overset{\displaystyle O}{\|}}{C}NH_2 + H_2O$$

| ethanoic acid | ammonia | ethanamide |

Amides, like esters, can be hydrolyzed. When simple amides are heated with water, they change back to their parent carboxylic acids and ammonia. Both strong acids and strong bases promote the reaction. As the following equations show, the reaction is the reverse of the formation of an amide.

Urea is an important nitrogen fertilizer because it reacts with soil moisture to release ammonia (and carbon dioxide).

$$NH_2\overset{\overset{\displaystyle O}{\|}}{C}NH_2$$
urea

In general:

$$R\overset{\overset{\displaystyle O}{\|}}{C}NH_2 + H-OH \xrightarrow{\text{heat}} R\overset{\overset{\displaystyle O}{\|}}{C}OH + NH_3$$
simple amide   carboxylic acid

An example:

$$CH_3\overset{\overset{\displaystyle O}{\|}}{C}NH_2 + H-OH \xrightarrow{\text{heat}} CH_3\overset{\overset{\displaystyle O}{\|}}{C}OH + NH_3$$
ethanamide   ethanoic acid

Nonbasicity of Amides

Despite the presence of the NH_2 group in simple amides, the amides are not Brønsted bases like amines or ammonia. The carbonyl group is the cause. Because of its O atom, the entire carbonyl group is electronegative and therefore draws electron density toward itself. This "tightens" the unshared pair of electrons on N enough to prevent the N atom from accepting a proton in dilute acid. Amides, therefore, are neutral compounds in an acid–base sense.

23.6 Organic Polymers

Types of Polymers

Nearly all of the organic compounds that we have studied so far have relatively low molecular masses. Both in nature and in the world of synthetics, however, many substances consist of **macromolecules** made up of hundreds or even thousands of atoms. Synthetics made of macromolecules are examples of how chemists have been able to take very ordinary substances in nature, like coal, oil, air, and water, and make new materials, never seen before, with useful applications. Recording tapes, skis, composites in recreational vehicles and their tires, backpacking gear, and all sorts of other recreational equipment derive strength from macromolecules (Figure 23.2). Anything useful for making thread and cloth consists of macromolecules. As paints cure, some of their molecules change into other molecules of enormous sizes.

In nature, substances with macromolecules are almost everywhere you look. Trees and anything made of wood, for example, derive their strength from lignins and cellulose, both consisting of enormous molecules that overlap and intertwine.

Figure 23.2 *Synthetics in recreation.* Lightweight synthetic fabrics have made it easier to pack strong tents into remote regions (here, the Wasatch Mountains of Utah).

From the Greek *poly-*, many, + *meros,* parts.

Polymers

Some macromolecular substances have more structural order than others. A **polymer,** for example, is a macromolecular substance all of whose molecules have a small characteristic structural feature that repeats itself over and over. The molecules of polypropylene, for example, have the following system.

$$\text{etc.}-CH_2\overset{\overset{\displaystyle CH_3}{|}}{C}HCH_2\overset{\overset{\displaystyle CH_3}{|}}{C}HCH_2\overset{\overset{\displaystyle CH_3}{|}}{C}HCH_2\overset{\overset{\displaystyle CH_3}{|}}{C}HCH_2\overset{\overset{\displaystyle CH_3}{|}}{C}HCH_2\overset{\overset{\displaystyle CH_3}{|}}{C}HCH_2\overset{\overset{\displaystyle CH_3}{|}}{C}HCH_2\overset{\overset{\displaystyle CH_3}{|}}{C}H-\text{etc.}$$
polypropylene

If you study this structure, you can see that one structural unit occurs repeatedly (actually thousands of times). In fact, the structure of a polymer is usually represented by the use of only its repeating unit, enclosed in parentheses, with a subscript n standing for several thousand units.

$$CH_2=CH \atop |\ CH_3 \qquad -CH_2CH- \atop |\ CH_3 \qquad (CH_2CH)_n \atop |\ CH_3$$

propylene   repeating unit in polypropylene   polypropylene

The value of n is not a constant for every molecule in a given polymer sample. A polymer, therefore, does not consist of molecules identical in *size,* just identical in *kind* but having the same repeating unit. Notice that despite the *-ene* ending to "polypropylene," the substance has no double bonds. The polymer is named after its starting material.

The repeating unit of a polymer is contributed by a chemical raw material called a **monomer.** Thus propylene is the monomer for polypropylene. The reaction that makes a polymer out of a monomer is called **polymerization,** and the verb is "to polymerize."

Hundreds of alkenes and their halogen derivatives have been tested as monomers, and Table 23.3 gives some examples of those that have been particularly successful.

Teflon, a polymer of $F_2C=CF_2$, contains only carbon and fluorine atoms. Chemically, Teflon is so inert that only molten sodium and potassium attack it.

Polystyrene is a polymer of styrene, $C_6H_5-CH=CH_2$, and its structure is like that of polypropylene but with phenyl groups, C_6H_5-, instead of methyl groups attached to every other carbon. Sometimes a gas, like carbon dioxide, is blown through molten polystyrene as it is molded into articles. As the hot liquid congeals, tiny pockets of gas are trapped, and the product is a foamed plastic, like the familiar polystyrene cups or insulation materials (Figure 23.3).

Teflon-coated frying pan.

Natural Rubber

Dienes, alkenes with two double bonds, are also important monomers. Natural rubber is a polymer of isoprene.

Figure 23.3 *Styrofoam.* Made of polystyrene, Styrofoam is widely used as an insulation.

Table 23.3 Polymers of Substituted Alkenes

Polymer	Monomer	Uses
Polyvinyl chloride (PVC)	$CH_2{=}CHCl$	Insulation, credit cards, bottles, plastic pipes
Saran	$CH_2{=}CCl_2$ and $CH_2{=}CHCl$	Packaging film, fibers, tubing
Teflon	$F_2C{=}CF_2$	Nonstick surfaces, valves
Orlon	$CH_2{=}CHC{\equiv}N$	Fabrics
Polystyrene	⟨○⟩$-CH{=}CH_2$	Foamed items, insulation
Lucite	$CH_2{=}\overset{\overset{\textstyle CH_3}{\textstyle \vert}}{C}CO_2CH_3$	Windows, coatings, molded items

Much of the plastic pipe used in sewers is made of polyvinyl chloride, or PVC.

$$CH_2{=}\overset{\overset{\textstyle CH_3}{\textstyle \vert}}{C}-CH{=}CH_2$$

isoprene
(2-methyl-1,3-butadiene)

natural rubber
(Note the repeating units, furnished by isoprene, between the dashed lines.)

Notice that the polymer molecule retains several alkene groups, which is why rubber is easily attacked by ozone. Chemists have largely solved this problem by finding antioxidant compounds to add to the rubber when rubber products are made.

Copolymers

Butyl rubber is a copolymer of isoprene and 2-methylpropene (isobutylene).

Sometimes two monomers are mixed together and polymerized. The operation is called *copolymerization,* and the product is a **copolymer.** Saran, for example, is a copolymer of $CH_2{=}CCl_2$ and $CH_2{=}CHCl$, and its molecules have the following units in their structure.

$$-CH_2CCl_2-CH_2CHCl-$$

some features of the structure of Saran

The Saran structure is actually not this uniform because the units provided by the monomers do not always alternate.

Effects of Molecular Size and Geometry on Polymers

For polymers to be useful, they must be as chemically inert as possible under the conditions in which they are employed. They must not be readily attacked by air or any of its pollutants; they must be stable toward water and microorganisms; and they must sometimes be stable even at elevated temperatures, like Teflon on a nonstick frying pan. Polymers like polyethylene and polypropylene, which are entirely alkanelike, are as chemically inert as almost any polymer can be, making them particularly useful for storing food material. Fibers made of polypropylene are used to manufacture indoor–outdoor carpets and artificial turf (see Figure 23.4).

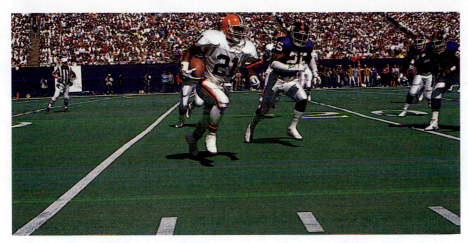

Figure 23.4 *Artificial turf on a football field.*

> ▶**Chemistry in Practice**◀ Their stability toward water (and soil microorganisms) makes most polymers a major problem in landfill operations. Such waste-disposal problems have led fast-food restaurants and grocery stores to change how they package food items. Some polymers can be melted down and reused, but for this purpose it is very important that polymeric materials be sorted by polymer type. It doesn't work to try to make recycled objects from a jumble of polymer types. ◆

Landfill problems have prompted increasing numbers of communities to use incineration and recycling for waste disposal and energy recovery.

Polymers and Physical Properties

Beyond chemical stability, physical properties are the features of polymers most sought after. Desirable properties of Teflon, for example, are its chemical inertness and its slipperiness toward just about anything. Nylon (a polyamide) isn't eaten by moths—a chemical property, in fact—but its superior strength and its ability to be made into fibers and fabrics of great beauty are what make nylon valuable. Dacron (a polyester) does not mildew, like cotton, and when made into fibers for a sail, it is superior to cotton (Figure 23.5). Dacron has greater strength with lower mass than cotton, and Dacron fibers do not stretch as much.

The best fiber-forming polymers have molecules that are not only very long but also have shapes that let the molecules align side by side, overlap end to end, and twist into cables of molecules, which further overlap and twist into fibers (Figure 23.6). Because the polymer molecules are large, substantial London forces of attraction occur between them.

Polyesters and Polyamides

Polyesters and polyamides are copolymers made from difunctional monomers. For a polyester, one monomer is a dicarboxylic acid and the other is a diol (an organic compound with two alcohol groups). Water molecules split out during

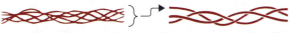

Polymer molecules overlap. Overlapping molecules twist together to form cables. Cables intertwine to form fibers.

Figure 23.6 *Synthetic fibers.* When molecules of polymers are very long and symmetrical about their axes, they can overlap and twist into cablelike systems that, in turn, can intertwine to give strong fibers.

Figure 23.5 *Dacron.* The exceptional wet strength and lightness of Dacron are put to good use in sailing boats.

the copolymerization, and a succession of ester groups forms as the polymer chain grows. We'll let *A* and *B* represent those parts of the "parents" that hold the functional groups.

$$HO-\overset{\overset{\displaystyle O}{\|}}{C}-A-\overset{\overset{\displaystyle O}{\|}}{C}-OH + H-O-B-O-H + \text{etc.} \longrightarrow$$

dicarboxylic acid   diol

The polymer is called Dacron when it is made into fibers and Mylar when used to make thin films, like the backing for cassette tapes.

$$\left(\overset{\overset{\displaystyle O}{\|}}{C}-A-\overset{\overset{\displaystyle O}{\|}}{C}-O-B-O\right)_n + nH_2O$$

a polyester

In Dacron, *A* is a benzene ring with the two carboxyl groups attached at opposite ends, and *B* is CH_2CH_2.

Nylon

A number of polymers are polyamides, and they make up a large family called the *nylon family*. Probably the most common member is nylon-66. We can visualize its relationship to its monomers by the following equation.

$$-\overset{\overset{\displaystyle O}{\|}}{C}OH + H-NH(CH_2)_6NH-H + HO\overset{\overset{\displaystyle O}{\|}}{C}(CH_2)_4\overset{\overset{\displaystyle O}{\|}}{C}OH + H-NH(CH_2)_6NH-H$$

1,6-diaminohexane   1,6-hexane-dicarboxylic acid

$$\downarrow -nH_2O$$

$$\text{etc.}\left(NH(CH_2)_6NH\overset{\overset{\displaystyle O}{\|}}{C}(CH_2)_4\overset{\overset{\displaystyle O}{\|}}{C}\right)_n\text{etc.}$$

nylon-66

Nylon molecules are very long and are very symmetrical around their long axes, and they have *regularly* spaced NH and C=O groups. These two groups are able to attract each other by a hydrogen bond, so as nylon molecules aggregate side by side, innumerable hydrogen bonds form *between* the chains. This makes possible very strong fibers.

23.7 Major Types of Biochemicals

The world of living things includes natural polymers of major importance, such as proteins, starch, cellulose, and the chemicals of genes.

Biochemistry is the systematic study of the chemicals of living systems, their organization into cells, and their chemical interactions. Biochemicals have no life in themselves as isolated compounds, yet life has a molecular basis. Only when chemicals are organized into cells in tissues can interactions occur that enable tissue repair, cell reproduction, the generation of energy, the removal of wastes, and the management of a number of other functions.

Three Important Requirements for Life

Starch, table sugar, and cotton are all *carbohydrates*.

A living system requires materials, energy, and information or "blueprints." Our focus in this chapter will be on the substances that supply them—carbohydrates, lipids, proteins, and nucleic acids—the basic materials whose molecules, together with water and a few kinds of ions, make up cells and tissues.

Our brief survey of biochemistry will focus mainly on the *structures* of selected biochemical materials. This is where any study of biochemistry must begin, because the chemical, physical, and biological properties of biochemicals are determined by structure. Where appropriate, we'll describe some of their reactions, particularly with water.

A variety of compounds are required for cells to work. The membranes that enclose all of the cells of your body, for example, are made up mostly of lipid molecules, although molecules of proteins and carbohydrates are also incorporated. Most hormones are in either the lipid or the protein family. Essentially all cellular catalysts—enzymes—are proteins, but many enzyme molecules cannot function without the presence of relatively small molecules of vitamins or certain metal ions.

Lipids and carbohydrates are also our major sources of the chemical energy we need to function. In times of fasting or starvation, however, the body is also able to draw on its proteins for energy.

The information needed to operate a living system is borne by molecules of nucleic acids. Their structures carry the *genetic code* that instructs cells how to make its proteins, including its enzymes. Hundreds of diseases, like cystic fibrosis and sickle-cell anemia, are caused by defects in the molecular structures of nucleic acids. Viruses, like the AIDS virus, work by taking over the genetic machinery of a cell.

Lipids include the fats and oils in our diet, like butter, margarine, salad oils, and baking shortenings (e.g., lard).

Meat and egg albumen are particularly rich in *proteins*.

Individual genes are actually sections of molecules of a nucleic acid called DNA for short.

23.8 Carbohydrates

Carbohydrates are naturally occurring polyhydroxyaldehydes or polyhydroxyketones, or else they are compounds that react with water to give these. The carbohydrates include table sugar (sucrose) as well as starch and cellulose.

Monosaccharides

Carbohydrates that do not react with water are called **monosaccharides.** The most common is glucose, a pentahydroxyaldehyde and probably the most widely occurring structural unit in the entire living world. Glucose is the chief carbohydrate in blood, and it provides the building unit for such important polysaccharides as cellulose and starch. Fructose, a pentahydroxyketone, is produced together with glucose when we digest table sugar. Honey is also rich in fructose.

$$\underset{\overset{|}{OH}\ \overset{|}{OH}\ \ \overset{|}{OH}\ \ \overset{|}{OH}\ \ \overset{|}{OH}}{CH_2CH-CH-CH-CH-CH} \qquad \qquad \underset{\overset{|}{OH}\ \overset{|}{OH}\ \ \overset{|}{OH}\ \ \overset{|}{OH}\ \overset{|}{OH}}{CH_2CH-CH-CH-\overset{\overset{\textstyle O}{\|}}{C}HCCH_2}$$

glucose (open-chain form),   fructose (open-chain form),
a polyhydroxyaldehyde   a polyhydroxyketone

Cyclic Forms of Monosaccharides

When dissolved in water, the molecules of most carbohydrates exist in a mobile equilibrium involving more than one structure. Glucose, for example, exists as two cyclic forms and one open-chain form in equilibrium in water (see Figure 23.7). The open-chain form, the only one with a free aldehyde group, is represented by less than 0.1% of the solute molecules. Yet the solute in an aqueous glucose solution still gives the reactions of a polyhydroxyaldehyde. This is possible because the equilibrium between this form and the two cyclic forms shifts to

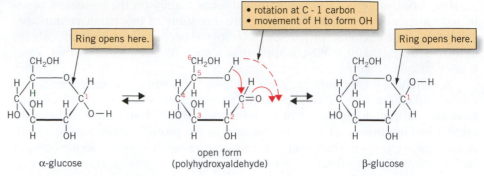

Figure 23.7 *Structures of glucose.* Three forms of glucose are in equilibrium in an aqueous solution. The curved arrows in the open form show how bonds become reoriented as the molecule closes into a cyclic form. Depending on how the CH=O group is turned at the moment of ring closure, the new OH group at C-1 takes up one of two possible orientations, α or β.

supply more of any of its members when a specific reaction occurs to just one (in accordance with Le Châtelier's principle).

Disaccharides

Carbohydrates whose molecules are split into two monosaccharide molecules by reacting with water are called **disaccharides.** *Sucrose* (table sugar, cane sugar, or beet sugar) is an example, and its hydrolysis gives glucose and fructose.

The six-membered rings of the cyclic forms of monosaccharides are not actually *flat* rings.

Human blood has roughly 100 mg of glucose per 100 mL. The brain normally relies on the chemical energy of glucose for its function.

To simplify, let's represent sucrose as Glu—O—Fru, where Glu is a glucose unit and Fru is a fructose unit, both joined by an oxygen bridge, —O—. The hydrolysis of sucrose, the chemical reaction by which we digest it, can thus be represented as follows.[2]

$$\text{Glu—O—Fru} + H_2O \xrightarrow[\text{(hydrolysis)}]{\text{digestion}} \text{glucose} + \text{fructose}$$
sucrose

Lactose (milk sugar) hydrolyzes to glucose and galactose (Gal), an isomer of glucose, and this is the reaction by which we digest lactose.

$$\text{Gal—O—Glu} + H_2O \xrightarrow[\text{(hydrolysis)}]{\text{digestion}} \text{galactose} + \text{glucose}$$
lactose

[2]Considerable detail is lost with the simplified structure of sucrose (and other carbohydrates like it to come). We leave the details, however, to other books because we seek a broader view.

Polysaccharides

Some of the most important naturally occurring polymers are the **polysaccharides,** carbohydrates whose molecules involve thousands of monosaccharide units linked to each other by oxygen bridges. They include starch, glycogen, and cellulose. The complete hydrolyses of all three yield only glucose, so their structural differences involve details of the oxygen bridges.

Starch

Plants store glucose units for energy needs in molecules of starch, found often in seeds and tubers (e.g., potatoes). Starch consists of two kinds of glucose polymers. The structurally simpler kind is *amylose,* which makes up roughly 20% of starch. We may represent its structure as follows, where O is the oxygen bridge linking glucose units.

$$\text{Glu} \left(\text{O} - \text{Glu} \right)_n \text{OH}$$

amylose (*n* is very large.)

The average amylose molecule has over 1000 glucose units linked together by oxygen bridges. These are the sites that are attacked and broken when water reacts with amylose during digestion. Molecules of glucose are released and are eventually delivered into circulation in the bloodstream.

$$\text{Amylose} + n\text{H}_2\text{O} \xrightarrow[\text{(hydrolysis)}]{\text{digestion}} n \text{ glucose}$$

The bulk of starch is made up of *amylopectin,* whose molecules are even larger than those of amylose. The amylopectin molecule consists of several amylose molecules linked by oxygen bridges from the end of one amylose unit to a site somewhere along the "chain" of another amylose unit.

$$
\begin{array}{c}
\text{etc.} \\
|\\
\text{Glu} \left(\text{O} - \text{Glu} \right)_m \text{O} \\
|\\
\text{Glu} \left(\text{O} - \text{Glu} \right)_n \text{O} \\
|\\
\text{Glu} \left(\text{O} - \text{Glu} \right)_o \text{O} \\
|\\
\text{etc.}
\end{array}
$$

amylopectin (*m, n,* and *o* are large numbers.)

Molecular masses ranging from 50,000 to several million are observed for amylopectin samples from the starches of different plant species. (A molecular mass of 1 million corresponds to about 6000 glucose units.)

Glycogen

Animals store glucose units for energy as glycogen, a polysaccharide with a molecular structure very similar to that of amylopectin. When we eat starchy foods and deliver glucose molecules into the bloodstream, any excess glucose not needed to maintain a healthy concentration in the blood is removed from circulation by particular tissues, like the liver and muscles. Liver and muscle cells convert glucose to glycogen. Later, during periods of high energy demand or fasting, glucose units are released from the glycogen reserves so that the concentration of glucose in the blood stays high enough for the needs of the brain and other tissues.

Cellulose

Cellulose is the chief material in a plant cell wall, and it makes up about 100% of cotton.

Cellulose is a polymer of glucose, much like amylose, but with the oxygen bridges oriented with different geometries. We lack the enzyme needed to hydrolyze its oxygen bridges, so we are unable to use cellulose materials like lettuce for food, only for fiber. Animals that eat grass and leaves, however, have bacteria living in their digestive tracts that convert cellulose into small molecules, which the host organism then appropriates for its own use.

23.9 Lipids

Lipids are natural products that tend to dissolve in such nonpolar solvents as diethyl ether and benzene but do not dissolve in water. Molecules of lipids are relatively nonpolar with large segments that are entirely hydrocarbonlike.

The lipid family is huge and diverse, because the only structural requirement is that large portions of lipid molecules be so hydrocarbonlike that they have poor solubility in water. Thus, the lipid family includes cholesterol as well as sex hormones, like estradiol and testosterone. You can see from their structures how largely hydrocarbonlike they are.

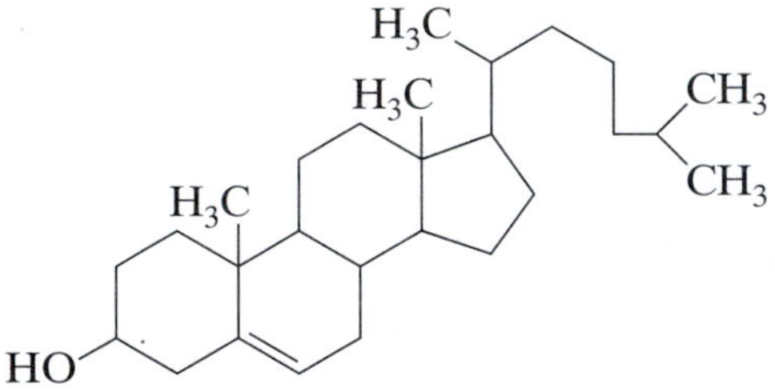

cholesterol 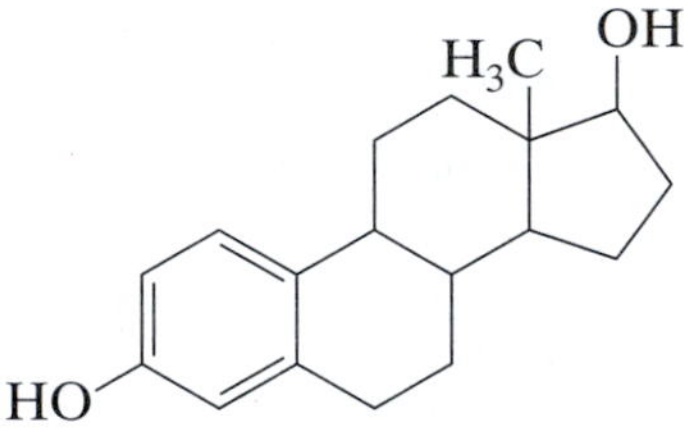   estradiol (a female sex hormone) 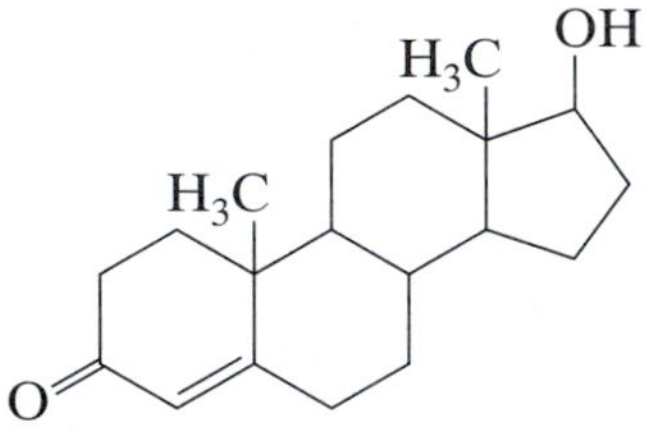   testosterone (a male sex hormone)

Triacylglycerols

$$CH_2OH$$
$$CHOH$$
$$CH_2OH$$
glycerol

The lipid family also includes the edible fats and oils in our diets—substances such as olive oil, corn oil, peanut oil, butterfat, lard, and tallow. These are **triacylglycerols**, that is, esters between glycerol, an alcohol with three OH groups, and any three of several long-chain carboxylic acids.

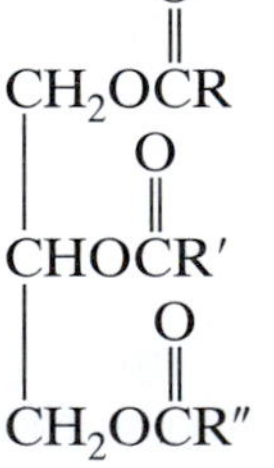

triacylglycerols (general structural features) 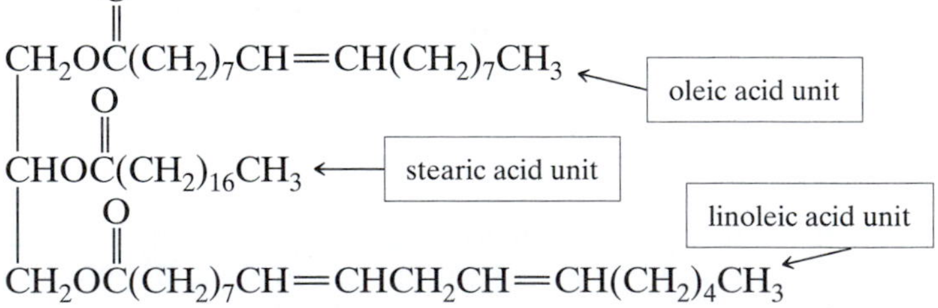   triacylglycerol (typical molecule present in a vegetable oil)

Fatty Acids

The carboxylic acids used to make triacylglycerols, the **fatty acids,** generally have just one carboxyl group on an unbranched chain with an even number of carbon atoms (Table 23.4). Their long hydrocarbon chains make triacylglycerols

Table 23.4 Common Fatty Acids

Fatty Acid	Number of Carbon Atoms	Structure	Melting Point (°C)
Myristic acid	14	$CH_3(CH_2)_{12}CO_2H$	54
Palmitic acid	16	$CH_3(CH_2)_{14}CO_2H$	63
Stearic acid	18	$CH_3(CH_2)_{16}CO_2H$	70
Oleic acid	18	$CH_3(CH_2)_7CH{=}CH(CH_2)_7CO_2H$	4
Linoleic acid	18	$CH_3(CH_2)_4CH{=}CHCH_2CH{=}CH(CH_2)_7CO_2H$	−5
Linolenic acid	18	$CH_3CH_2CH{=}CHCH_2CH{=}CHCH_2CH{=}CH(CH_2)_7CO_2H$	−11

mostly like hydrocarbons in physical properties, including insolubility in water. Many fatty acids have alkene groups.

Triacylglycerols obtained from vegetable sources, like olive oil, corn oil, and peanut oil, are called *vegetable oils* and are liquids at room temperature. Triacylglycerols from animal sources, like lard and tallow, are called *animal fats* and are solids at room temperature. The vegetable oils generally have more alkene double bonds per molecule than animal fats, and so are said to be *polyunsaturated*. The double bonds are usually *cis,* and so the molecules are kinked, making it more difficult for them to nestle close together, experience London forces, and so be in the solid state.

Lard is hog fat. Tallow is beef fat.

Cooking oils are advertised as *polyunsaturated* because of their several alkene groups per molecule.

Digestion of Triacylglycerols

We digest the triacylglycerols by hydrolysis, meaning by their reaction with water. Our digestive juices in the upper intestinal tract have enzymes called *lipases* that catalyze these reactions. For example, the complete digestion of the triacylglycerol shown above occurs by the following reaction.

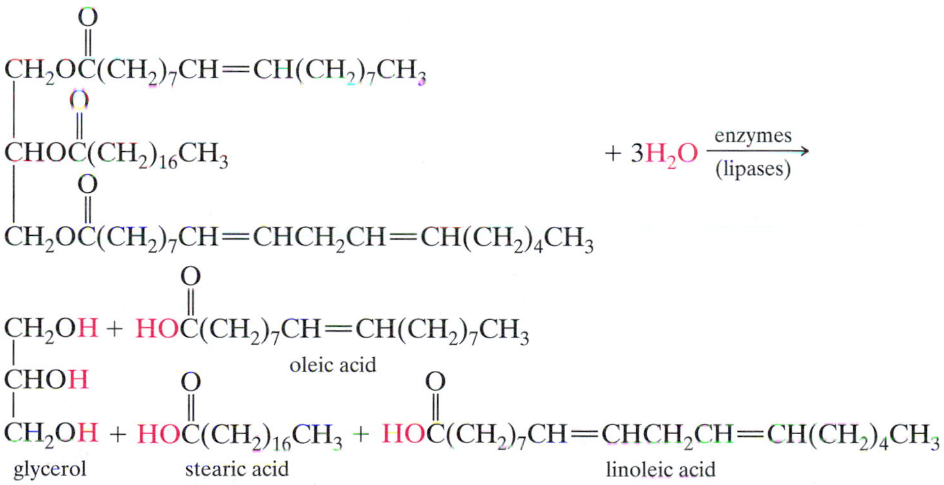

$$
\begin{array}{l}
CH_2O\overset{O}{\overset{\|}{C}}(CH_2)_7CH{=}CH(CH_2)_7CH_3 \\
| \\
CHO\overset{O}{\overset{\|}{C}}(CH_2)_{16}CH_3 \qquad\qquad + 3H_2O \xrightarrow[\text{(lipases)}]{\text{enzymes}} \\
| \\
CH_2O\overset{O}{\overset{\|}{C}}(CH_2)_7CH{=}CHCH_2CH{=}CH(CH_2)_4CH_3
\end{array}
$$

$$
\begin{array}{l}
CH_2OH + HO\overset{O}{\overset{\|}{C}}(CH_2)_7CH{=}CH(CH_2)_7CH_3 \\
| \qquad\qquad\qquad\qquad \text{oleic acid} \\
CHOH \qquad\quad O \qquad\qquad\qquad O \\
| \qquad\qquad\quad \| \qquad\qquad\qquad\quad \| \\
CH_2OH + HO\overset{}{C}(CH_2)_{16}CH_3 + HO\overset{}{C}(CH_2)_7CH{=}CHCH_2CH{=}CH(CH_2)_4CH_3 \\
\text{glycerol} \qquad \text{stearic acid} \qquad\qquad\qquad \text{linoleic acid}
\end{array}
$$

Actually, the *anions* of the acids form, because the medium in which lipid digestion occurs is basic.

> ▶**Chemistry in Practice**◀ When the hydrolysis of a triacylglycerol is carried out in the presence of sufficient base so as to release the fatty acids as their anions, the reaction is called **saponification.** The mixture of the salts of long-

Surface-active agents, like detergents and soap, lower the surface tension of water and so help to make oil-in-water emulsions.

chain fatty acids is what makes up ordinary *soap*. We discussed how soaps and detergents work in Section 12.10. In pioneer times, it was common to make soap by stirring pork fat (lard) or beef fat with hot, aqueous lye in a cast-iron kettle. Standard recipes existed that helped to ensure that the product had a minimum amount of leftover reactants, either lye or fat. ◆

Hydrogenation of Vegetable Oils

Vegetable oils are generally less expensive to produce than butterfat, but because the oils are liquids, few people care to use them as bread spreads. Remember that animal fats, like butterfat, are solids at room temperature, and vegetable oils differ from animal fats only in the number of carbon–carbon double bonds per molecule. Simply adding hydrogen to the double bonds of a vegetable oil, therefore, changes the lipid from a liquid to a solid. If only one of the three double bonds in the triacylglycerol molecule shown above were hydrogenated, the product (with two double bonds) would melt above room temperature.

▶**Chemistry in Practice**◀ The *limited* hydrogenation of a vegetable oil changes it into a solid used as the chief component of margarine. The better brands melt on the tongue, which is one reason for butter's popularity. The peanut oil in "old fashioned" peanut butter is likewise partially hydrogenated to make a homogeneous spread. ◆

The Lipids of Cell Membranes

The lipids involved in the structures of cell membranes in animals are not triacylglycerols. Some are diacylglycerols with the third site on the glycerol unit taken up by an attachment to a phosphate unit. This, in turn, is joined to an amino alcohol unit by an esterlike network. The phosphate unit carries one negative charge, and the amino unit has a positive charge. These lipids are called *glycerophospholipids*. Lecithin is one example.

The amino alcohol unit in lecithin is contributed by choline, a cation:

$$HOCH_2CH_2\overset{+}{N}(CH_3)_3$$

Ethanolamine (in its protonated form),

$$HOCH_2CH_2NH_3{}^+,$$

is another amino alcohol that occurs in phospholipids.

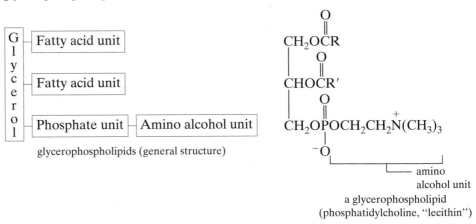

glycerophospholipids (general structure)

a glycerophospholipid
(phosphatidylcholine, "lecithin")

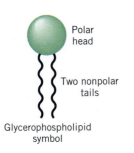

Glycerophospholipid
symbol

The glycerophospholipids illustrate that it is possible for lipid molecules to carry polar, even ionic sites and still not be very soluble in water. The combination of both nonpolar and polar or ionic units within the same molecule enables the glycerophospholipids and similar substances to be major building units for the membranes of animal cells.

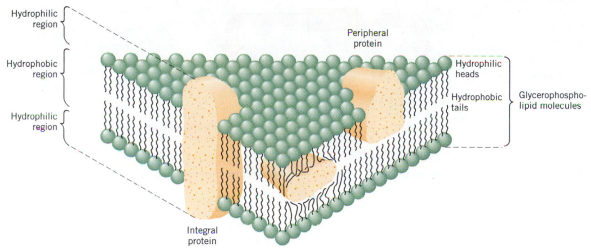

Figure 23.8 *Animal cell membrane.* The lipid molecules of an animal cell membrane are organized as a bilayer.

The Lipid Bilayer of Cell Membranes

The purely hydrocarbonlike portions of a glycerophospholipid molecule, the long R groups contributed by the fatty acid units, are *hydrophobic.* The portions bearing the electrical charges are *hydrophilic.* In an aqueous medium, therefore, the molecules of a glycerophospholipid aggregate in a way that minimizes the exposure of the hydrophobic side chains to water and maximizes contact between the hydrophilic sites and water. These interactions are roughly what take place when glycerophospholipid molecules aggregate to form the *lipid bilayer* membrane of an animal cell (Figure 23.8). The hydrophobic side chains intermingle in the center of the layer where water molecules do not occur. The hydrophilic groups are exposed to the aqueous medium inside and outside of the cell. Not shown in Figure 23.8 are cholesterol and cholesterol ester molecules, which help to stiffen the membranes. Thus cholesterol is essential to the cell membranes of animals.

If a pin were stuck through a cell membrane and then withdrawn, the membrane would automatically close up. It is remarkable that a unit as vital to life as a cell depends for its structural integrity on the sum of many weak forces, the forces involved in water-attracting and water-avoiding dipole–dipole interactions, hydrogen bonds, and London forces.

Cell membranes also include protein units, which provide several services. Some are molecular recognition sites for molecules such as hormones and neurotransmitters. Other proteins provide channels for the movements of ions, like Na^+, K^+, Ca^+, Cl^-, HCO_3^-, and others, into or out of the cell. Some are channels for the transfer of small organic molecules, like glucose.

Hydrophobic and *hydrophilic* were introduced on page 562.

Neurotransmitters are small molecules that travel between the end of one nerve cell and the surface of the next to transmit the nerve impulse.

23.10 Proteins

The proteins are a huge family of substances that make up about half of the human body's dry weight. They are found in all cells and in virtually all parts of cells. They are the stress-bearing constituents of skin, muscles, and tendons. In teeth and bones, proteins function like steel rods in concrete, namely, as reinforcing members without which the minerals would crumble. Proteins serve as

Proteins truly merit the name, taken from the Greek *proteios,* meaning "of the first rank."

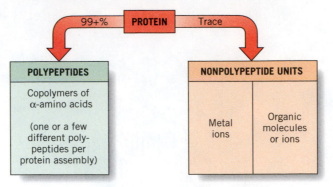

Figure 23.9 *Components of proteins.* Some proteins consist exclusively of polypeptide molecules, but most also have structures that include traces of nonpolypeptide units such as small organic molecules or metal ions, or both.

enzymes, hormones, and neurotransmitters. They carry oxygen in the bloodstream as well as some of the waste products of metabolism. No other group of compounds has such a variety of functions in living systems.

The dominant structural units of **proteins** are macromolecules called **polypeptides,** which are made from a set of monomers called **α-amino acids.** Most protein molecules include, besides their polypeptide units, small organic molecules or metal ions, and the whole protein lacks its characteristic biological function without these species (Figure 23.9).

Amino Acids

The monomer units for polypeptides are a group of about 20 α-amino acids all of which share the following structural features. R here stands for a structural group, an *amino acid side chain.* Some examples of the set of 20 amino acids used to make proteins are given below. All are known by their common names. Each also has a three-letter symbol.

$$\overset{\alpha\text{-position} \longrightarrow}{^+NH_3\overset{\overset{O}{\overset{\|}{}}}{CH}CO^-}$$
$$| \\ R$$

α-amino acids (general structural features)

R = H, glycine (Gly)

= CH_3, alanine (Ala)

= CH_2—⬡, phenylalanine (Phe)

= $CH_2CH_2CO_2H$, glutamic acid (Glu)

= $CH_2CH_2CH_2CH_2NH_2$, lysine (Lys)

= CH_2SH, cysteine (Cys)

The simplest amino acid is aminoacetic acid, or glycine, for which the "side chain" is H.

Glycine, like all of the amino acids in their pure states, exists as a *dipolar ion.* Such an ion results from an internal self-neutralization, by the transfer of a proton from the proton-donating carboxyl group to the proton-accepting amino group.

$$NH_2CH_2CO_2H \longrightarrow {}^+NH_3CH_2CO_2{}^-$$

aminoacetic acid (glycine) glycine, dipolar ionic form

Polypeptides

Polypeptides are copolymers of the amino acids. The carboxyl group of one amino acid becomes joined to the amino group of another by means of the same kind of carbonyl–nitrogen bond found in amides, but here called the **peptide**

bond. Let's see how two amino acids, glycine and alanine, can become linked by a (multistep) splitting out of water.

$$^+NH_3CH_2\overset{\overset{O}{\|}}{C}-O^- + H-\overset{\overset{H}{|}}{\underset{\underset{H}{|}}{N}}\overset{\overset{O}{\|}}{\underset{\underset{CH_3}{|}}{C}}HCO^- \xrightarrow[\text{by several steps}]{\text{in the body,}} {}^+NH_3CH_2\overset{\overset{O}{\|}}{C}-NH\overset{\overset{O}{\|}}{\underset{\underset{CH_3}{|}}{C}}HCO^- + H_2O$$

peptide bond

glycine (Gly)   alanine (Ala)   glycylalanine (Gly-Ala)

The product of this reaction, glycylalanine, is an example of a *dipeptide*. (Notice how the three-letter symbols for the amino acids make up what biochemists sometimes use as the *structural* formulas of such products.)

We could have taken glycine and alanine in different roles and written the equation for the formation of a different dipeptide, Ala-Gly.

$$^+NH_3\overset{\overset{O}{\|}}{\underset{\underset{CH_3}{|}}{C}}HC-O^- + H-\overset{\overset{H}{|}}{\underset{\underset{H}{|}}{N}}CH_2\overset{\overset{O}{\|}}{C}-O^- \xrightarrow[\text{by several steps}]{\text{in the body,}} {}^+NH_3\overset{\overset{O}{\|}}{\underset{\underset{CH_3}{|}}{C}}HC-NHCH_2\overset{\overset{O}{\|}}{C}O^- + H_2O$$

peptide bond

alanine (Ala)   glycine (Gly)   alanylglycine (Ala-Gly)

We can think of the formation of these two dipeptides as being like the formation of two 2-letter words from the letters N and O. Taken in one order we get NO; in the other, ON. They have entirely different meanings, yet are made of the same pieces.

Note that each dipeptide has one CO_2^- at one end of the chain and one NH_3^+ group at the other end. Each end of a dipeptide molecule, therefore, could become involved in the formation of yet another peptide bond involving any of the 20 amino acids. For example, if glycylalanine were to combine with phenylalanine as follows, a *tripeptide* would form with the sequence Gly-Ala-Phe.

peptide bonds

$$^+NH_3CH_2\overset{\overset{O}{\|}}{C}-NH\overset{\overset{O}{\|}}{\underset{\underset{CH_3}{|}}{C}}HC-O^- + H-\overset{\overset{H}{|}}{\underset{\underset{CH_2C_6H_5}{|}}{N}}\overset{\overset{O}{\|}}{C}HCO^- \longrightarrow {}^+NH_3CH_2\overset{\overset{O}{\|}}{C}-NH\overset{\overset{O}{\|}}{\underset{\underset{CH_3}{|}}{C}}HC-NH\overset{\overset{O}{\|}}{\underset{\underset{CH_2C_6H_5}{|}}{C}}HCO^- + H_2O$$

glycylalanine (Gly-Ala)   phenylalanine (Phe)   glycylalanylphenylalanine (Gly-Ala-Phe)

The tripeptide still has one CO_2^- and one NH_3^+ group at opposite ends of the chain, so it could react at either end with still another amino acid to make a tetrapeptide. Each additional unit still leaves the product with one CO_2^- and one NH_3^+ group. You can see how a very long sequence of amino acid units might be joined together. Note how the sequence of side chains could be in any order. We took only one sequence, Gly-Ala-Phe, for purposes of illustration, but five other tripeptides are possible using Gly, Ala, and Phe. The six tripeptides differ only in the sequence in which the three side chains, H, CH_3, and $CH_2C_6H_5$, are located at the α-carbon atoms.

Proteins

Many proteins consist of a single polypeptide. Most proteins, however, involve assemblies of two or more polypeptides. These are identical in some proteins,

The artificial sweetener aspartame (NutraSweet) is the methyl ester of a dipeptide.

$$\begin{array}{l} \overset{\overset{O}{\|}}{C}OCH_3 \quad \text{(methyl ester group)}\\ | \\ CHCH_2C_6H_5 \\ | \\ NH \quad \text{(peptide bond)}\\ | \\ C=O \\ | \\ CHCH_2CO_2^- \\ | \\ NH_3^+ \end{array}$$

aspartame

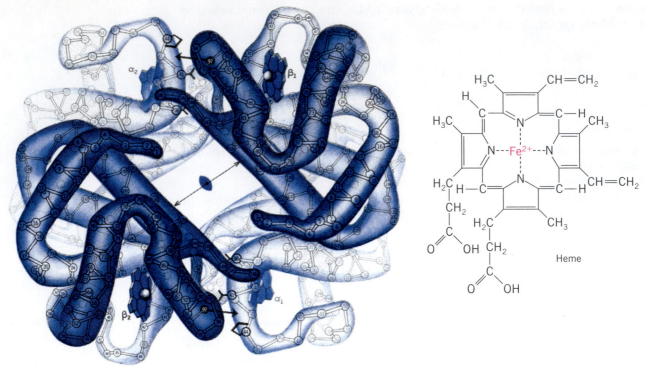

Figure 23.10 *Hemoglobin.* Its four polypeptide chains occupy the twisted tubelike forms. Two units are alike, the α-chains; the two β-chains are also alike. The flat disks are the heme molecules. (From R. E. Dickerson and E. I. Geis, *The Structure and Action of Proteins.* © 1969, W. A. Benjamin, Inc., Menlow Park, CA. All rights reserved. Used by permission.)

Hemoglobin is the oxygen carrier in blood.

but in others the aggregating polypeptides are different. Moreover, a relatively small organic molecule may be included in the aggregation, and a metal ion is sometimes present as well. Thus, the terms "protein" and "polypeptide" are not synonyms. Hemoglobin, for example, has all of the features just described (Figure 23.10). It is made of four polypeptides—two similar pairs—and one molecule of heme, the organic compound that causes the red color of blood. Heme, in turn, holds an iron(II) ion. The *entire* package is the protein, hemoglobin. If one piece is missing or altered in any way—for example, if iron occurs as Fe^{3+} instead of Fe^{2+}—the substance is not hemoglobin, and it does not transport oxygen in the blood.

The Importance of Protein Shape

Notice in Figure 23.10 how the strands of each polypeptide unit in hemoglobin are coiled and that the coils are kinked and twisted. Such shapes of polypeptides are determined by the amino acid sequence, because the side chains are of different sizes and some are hydrophilic and others are hydrophobic. Polypeptide molecules become twisted and coiled in whatever way minimizes the contact of hydrophobic groups with the surrounding water and maximizes the contacts of hydrophilic groups with water molecules.

The final shape of a protein, called its *native form,* is as critical to its ability to function as anything else about its molecular architecture. For example, just the exchange of one side-chain R group by another changes the shape of hemoglobin and causes a debilitating condition known as sickle-cell anemia.

Physical agents, like heat, and chemicals, like poisons and certain solvents, can distort the native form of a protein and render it biologically useless without breaking any covalent bond. When this happens, we say that the protein has been *denatured,* and the change is nearly always irreversible.

The shapes of polypeptides change even when H^+ ions are donated to accepting sites or are released from donating sites. Thus any change in the pH of the protein's surrounding medium can have disastrous consequences for the ability of the protein to function. This is why a living system must exercise extremely tight control over the pH of its fluids by means of its buffers (and, incidentally, why the study of medicine requires extensive grounding in acid–base chemistry).

Glutamic acid has a proton-donating side chain.

$$-CH_2CH_2CO_2H$$

Lysine has a proton-accepting side chain.

$$-CH_2CH_2CH_2CH_2NH_2$$

Enzymes

The catalysts in living cells are called **enzymes,** and virtually all are proteins. Some enzymes require metal ions, such as Mn^{2+}, Co^{2+}, Cu^{2+}, and Zn^{2+}, all of which are on the list of the *trace elements* that must be in a good diet. Some enzymes also require molecules of the B vitamins to be complete enzymes.

Some of our most dangerous poisons work by deactivating enzymes, often those needed for the transmission of nerve signals. For example, the botulinum toxin that causes botulism, a deadly form of food poisoning, deactivates an enzyme in the nervous system. Heavy metal ions, like Hg^{2+} or Pb^{2+}, are poisons because they deactivate enzymes.

Heavy metal ions bond to the HS groups of cysteine side chains in polypeptides.

23.11 Nucleic Acids

The enzymes of an organism are made under the chemical direction of a family of compounds called the *nucleic acids.* Both the similarities and the uniqueness of every species as well as every individual member of a species depend on structural features of these compounds.

DNA and RNA

The **nucleic acids** occur as two broad types, namely, **RNA,** or ribonucleic acids, and **DNA,** or deoxyribonucleic acids. DNA is the actual chemical of a gene, the individual unit of heredity and the chemical basis through which we inherit all of our characteristics.

The main chains or "backbones" of DNA molecules consist of alternating units contributed by phosphoric acid and a monosaccharide (see Figure 23.11). In RNA the monosaccharide is ribose (hence the R in RNA). In DNA, the monosaccharide is deoxyribose (*deoxy* means "lacking an oxygen unit"). Thus, both DNA and RNA have the following system, where *G* stands for *group,* each *G* unit representing a unique nucleic acid side chain or *base.*

$$
\begin{array}{ccccccc}
G^1 & & G^2 & & G^3 & \\
| & & | & & | & \\
\text{phosphate}-\text{sugar}-\text{phosphate}-\text{sugar}-\text{phosphate}-\text{sugar}-\text{etc.}
\end{array}
$$

backbone system in all nucleic acids—many thousands of repeating units long
(In DNA, the sugar is deoxyribose. In RNA, the sugar is ribose.)

ribose

deoxyribose

Side Chain Bases

The side chains, *G,* are all heterocyclic amines whose molecular shapes have much to do with their function. Being amines, they are referred to as the *bases*

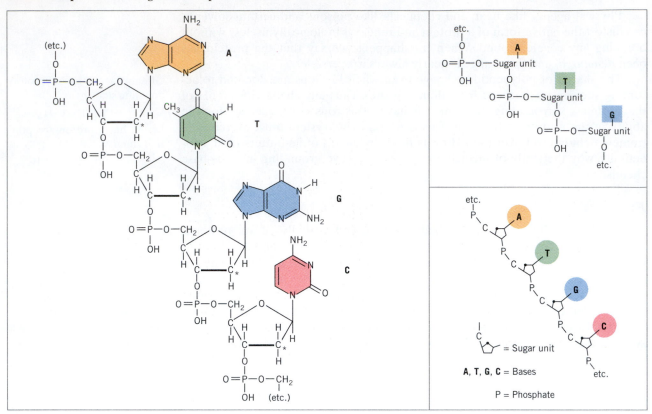

Figure 23.11 *Nucleic acids.* A segment of a DNA chain featuring each of the four DNA bases. When the sites marked by asterisks each carry an OH group, the main "backbone" would be that of RNA. In RNA U would replace T. The insets show how simplified versions of a DNA strand can be drawn.

of the nucleic acids and are represented by single letters—A for adenine, T for thymine, U for uracil, G for guanine, and C for cytosine.

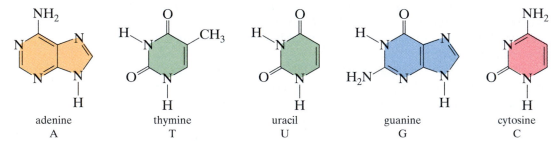

| adenine | thymine | uracil | guanine | cytosine |
| A | T | U | G | C |

A, T, G, and C occur in DNA. A, U, G, and C are in RNA. These few bases are the "letters" of the genetic alphabet. The messages of all genes are composed with just four letters, A, T, G, and C.

The DNA Double Helix

Crick and Watson shared the 1962 Nobel Prize in physiology or medicine with Maurice Wilkins, using X-ray data from Rosalind Franklin to deduce the helical structure of DNA.

In 1953, F. H. C. Crick of England and J. D. Watson, an American, deduced that DNA occurs in cells as two intertwined, oppositely running strands of molecules coiled like a spiral staircase and called the **DNA double helix** (Figure 23.12). Hydrogen bonds help to hold the two strands side by side, but other factors are also involved.

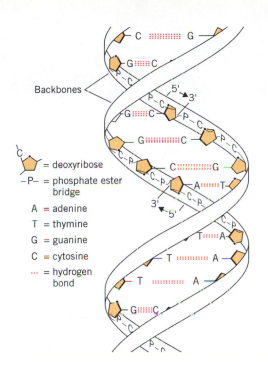

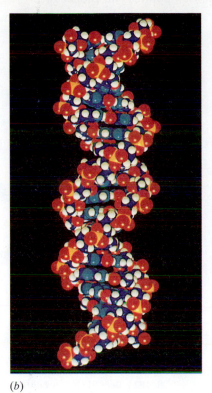

Figure 23.12 *The DNA double helix.* (*a*) A schematic drawing in which the hydrogen bonds between the two strands are indicated by dotted lines (*b*) A model of a short section of a DNA double helix. The "backbones" are in blue.

(*a*)

(*b*)

The bases have N—H and O=C groups, and hydrogen bonds (···) can exist between them.

$$\underset{\delta+}{N}—\underset{\delta-}{H}···O=C$$

However, the bases with the best "matching" molecular geometries for maximum hydrogen bonding occur only as particular *pairs* of bases. The functional groups of each pair are in exactly the right locations in their molecules to allow hydrogen bonds between pair members. Adenine (A) pairs with thymine (T), and cytosine (C) pairs with guanine (G) (see Figure 23.13). A pairs only with T in DNA, never with G or C. C pairs only with G, never with A or T. Thus opposite every G on one strand in a DNA double helix, a C occurs on the other. Opposite every A on one strand in DNA, a T is found on the other.

Adenine (A) can also pair with uracil (U), but U occurs in RNA. The A-to-U pairing is an important factor in the work of RNA.

The Replication of DNA

Prior to cell division, the cell produces duplicate copies of its DNA so that each daughter cell will have a complete set. Such reproductive duplication is called DNA **replication.**

The accuracy of DNA replication results from the limitations of the base pairings: A only with T and C only with G (Figure 23.14). The unattached letters, A, T, G, and C, in Figure 23.14 here represent not just the bases but the whole monomer molecules of DNA. These are called *nucleotides* and are molecules made of a phosphate–sugar unit holding one particular base. The nucleotides are made by the cell and are present in the cellular "soup."

cytosine unit

a typical nucleotide, one using cytosine as its side chain

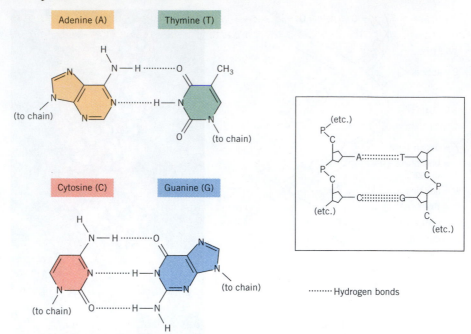

Figure 23.13 *Base pairing in DNA.* The hydrogen bonds are indicated by dotted lines.

An enzyme catalyzes each step of replication. As replication occurs, the two strands of the parent DNA double helix separate, and the monomers for new strands assemble along the exposed single strands. Their order of assembling is determined altogether by the specificity of base pairing. For example, a base such as T on a parent strand can accept only a nucleotide with the base A. Two daughter double helices result that are identical to the parent, and each carries one strand from the parent. One new double helix goes to one daughter cell and the second goes to the other.

Genes

The DNA in one human cell nucleus has an estimated 3 billion base pairs.

A single human gene has between 1000 and 3000 bases, but they do not occur continuously on a DNA molecule. In multicelled organisms, a gene is neither a whole DNA molecule nor a continuous sequence in one. A single gene consists of the totality of particular *segments* of a DNA chain that, taken together, carry the necessary genetic instruction for a particular polypeptide. The separated sec-

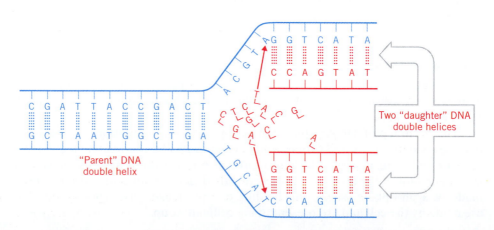

Figure 23.14 *Base pairing and the replication of DNA.*

tions making up a gene are called **exons**—a unit that helps to *express* a message. The sections of a DNA strand between exons are called **introns**—units that *in*-terrupt the gene.

DNA-Directed Synthesis of Polypeptides

Each polypeptide in a cell is made under the direction of its own gene. In broad outline, the steps between a gene and an enzyme occur as follows. (We momentarily ignore both the introns as well as how the cell ignores them.)

$$\underset{\text{gene}}{\text{DNA}} \xrightarrow{\text{transcription}} \text{RNA} \xrightarrow{\text{translation}} \text{polypeptide}$$

(The genetic message is read off in the cell nucleus and transferred to RNA.) (The genetic message, now on RNA outside the nucleus, is used to direct the synthesis of a polypeptide.)

The step labeled **transcription** accomplishes just that—the genetic message, represented by a sequence of bases on DNA, is transcribed into a *complementary* sequence of bases on RNA, but U, not T, is on the RNA. **Translation** means the conversion of the base sequence on RNA into a side-chain sequence on a new polypeptide. It is like translating from one language (the DNA/RNA base sequences) to another language (the polypeptide side-chain sequence).

The cell tells which amino acid unit must come next in a growing polypeptide by means of a code that relates unique groups of nucleic acid bases to specific amino acids. The code, in other words, enables the cell to translate from the four-letter RNA alphabet (A, U, G, and C) to the 20-letter alphabet of amino acids (the amino acid side chains). To see how the cell does this, we have to learn more about RNA.

Something like 140,000 genes are needed to account for all of the kinds of proteins in a human body.

The RNA of a Cell

Four types of RNA are involved in the connection between gene and polypeptide. One is called *ribosomal* RNA, or rRNA, and it is packaged together with enzymes in small granules called *ribosomes*. Ribosomes are manufacturing stations for polypeptides.

Another RNA is called *messenger* RNA, or mRNA, and it brings the blueprints for a specific polypeptide from the cell nucleus to the manufacturing station (a ribosome). mRNA is the carrier of the genetic message to the polypeptide assembly site.

mRNA is made from another kind of RNA called *heterogeneous nuclear* RNA, or hnRNA. This is the RNA first made at the direction of a DNA unit, so it contains sections that were specified by both exons and introns (hence, *heterogeneous*).

The last type of RNA is called *transfer* RNA, or tRNA. It has responsibility for picking up the prefabricated parts, the amino acids, and getting them from the cell "soup" to the ribosome.

Little agreement exists on the name *heterogeneous nuclear RNA* among biochemistry references. Some call it pre RNA; others call it *primary transcript RNA,* or ptRNA.

Polypeptide Synthesis

When some chemical signal tells a cell to make a particular polypeptide, the cell nucleus makes the hnRNA that corresponds to the gene. As we indicated, the base sequence in hnRNA is an exact complement to a base sequence in the DNA being transcribed—both the exon sections of the DNA and the introns. Reactions and enzymes in the cell nucleus then remove those parts of the new hnRNA that were caused by introns. It is as if the message, now on hnRNA, is edited to remove nonsense sections. The result of this editing is messenger RNA.

The mRNA now moves outside the cell nucleus and becomes joined to a ribosome where the enzymes for polypeptide synthesis are found. The polypeptide manufacturing site now awaits the arrival of tRNA molecules bearing amino acid units. All is in readiness for polypeptide synthesis—for *translation*.

The Genetic Code

To continue with our use of the language of coding, we can describe the genetic message as being written in code words made of three letters—three bases. For example, when the bases G, G, and U occur side by side on an mRNA chain, they together specify the amino acid glycine. Although three other triplets also mean glycine, one of the genetic codes for glycine is GGU. Similarly, GCU is a code "word" for alanine.

Each triplet of bases *as it occurs on mRNA* is called a **codon,** and each codon corresponds to a specific amino acid. Which amino acid goes with which codon constitutes the **genetic code.** One of the most remarkable features of the code is that it is essentially universal. The codon specifying alanine in humans, for example, also specifies alanine in the genetic machinery of bacteria, aardvarks, camels, rabbits, and stinkbugs. Our chemical kinship with the entire living world is profound (even humbling).

A triplet of bases complementary to a codon occurs on tRNA and is called an **anticodon.**

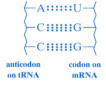

anticodon   codon on
on tRNA   mRNA

A tRNA molecule, bearing the amino acid corresponding to its anticodon, can line up at a strand of mRNA only where the anticodon "fits" by hydrogen bonding to the matching codon. If the anticodon is ACC, then the A on the anticodon must find a U on the mRNA at the same time that its neighboring two C bases find neighboring G bases on the mRNA strand. Thus the anticodon and codon line up as shown in the margin, where the dotted lines indicate hydrogen bonds.

mRNA codons determine the sequence in which tRNA units can line up, and this sets the sequence in which amino acid units become joined in a polypeptide. Figure 23.15 carries this description further, where we enter the process soon after special steps (not shown) have launched it.

In Figure 23.15*a*, a tRNA–glycine unit arrives at mRNA where it finds two situations, namely, a dipeptide already started and a codon to which the tRNA–glycine can fit. In Figure 23.15*b* the tRNA–glycine unit is in place. It is the only such unit that could be in this place. Base pairing between the two triplets—codon to anticodon—ensures this.

In Figure 23.15*c*, the whole dipeptide unit moves over and is joined to the glycine unit by a newly formed peptide bond. This leaves a tripeptide attached via a tRNA unit to the mRNA.

Because the next codon, GCU, is coded for the amino acid alanine, only a tRNA–alanine unit can come into the next site. So alanine, with its methyl group as the side chain, is the only amino acid that enters the polypeptide chain at this point. These steps repeat over and over until a special codon is reached that is designed to stop the process. The polypeptide is now complete. It has a *unique* sequence of side chains that has been initially specified by the molecular structure of a gene.

Genetic Defects

Knowledge of the locations and structures of genes will help to cure at least some genetic defects, and the *Human Genome Project* is a major scientific effort to "map" all of the genes of the body by locating where they are on the various chromosomes.

About 2000 diseases are attributed to various kinds of defects in the genetic machinery of cells. With so many steps from gene to polypeptide, we should expect many opportunities for things to go wrong. Suppose, for example, that just one base in a gene were wrong. Then one base on an mRNA strand would be the wrong base—"wrong" with respect to getting the polypeptide we want. This

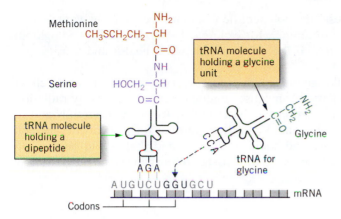

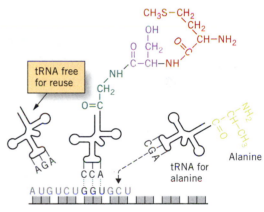

(a) A growing polypeptide is shown here just two amino acid units long. A third amino acid, glycine, carried by its own tRNA, is moving into place at the codon GGU. (Behind the assemblage shown here is a large particle called a ribosome, not shown, that stabilizes the assemblage and catalyzes the reactions.)

(b) The tRNA–glycine unit is in place, and now the dipeptide will transfer to the glycine, form another peptide bond, and make a tripeptide unit.

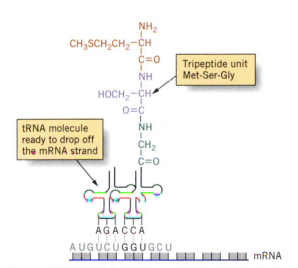

(c) The tripeptide unit has formed, and the next series of reactions that will add another amino acid unit can now start.

(d) One tRNA leaves to be reused. Another tRNA carrying a fourth amino acid unit, alanine, is moving into place along the mRNA strand to give an assemblage like that shown in part (b). Then the tripeptide unit will transfer to alanine as another peptide bond is made and a tetrapeptide unit forms. A cycle of events thus occurs as the polypeptide forms.

Figure 23.15 *Polypeptide synthesis at an mRNA strand*

would change every remaining codon. If the next three codons on mRNA, for example, were UCU—GGU—GCU—U—etc., and the first G were deleted, then the sequence would become UCU—GUG—CUU—etc. All remaining triplets change! You can imagine how this could lead to an entirely different polypeptide than what we want.

Or suppose that the GGU codon were replaced by, say, a GCU codon so that the sequence

$$\text{UCU—GGU—GCU—etc.}$$

becomes

$$\text{UCU—GCU—GCU—etc.}$$

In one of the subunits of hemoglobin in persons with sickle-cell anemia, an isopropyl group is a side chain where a $CH_2CH_2CO_2H$ side chain should be.

One amino acid in the resulting polypeptide would not be what we want. Something like this makes the difference, for example, between normal hemoglobin and sickle-cell hemoglobin and the difference between health and a tragic genetic disease.

Atomic radiation, like gamma rays, beta rays, or even X rays, as well as chemicals that mimic radiation can cause genetic defects, too. A stray radiation hitting a gene might cause two opposite bases to fuse together chemically and so make the cell reproductively dead. When it came time for it to divide, the cell could not replicate its genes and could not divide, and that would be the end of it. If this happens to enough cells in a tissue, the organism might be fatally affected.

Viruses

The protein of a virus is a protective overcoat for its nucleic acid, and it provides an enzyme to catalyze a breakthrough by the virus into a host cell.

Viruses are packages of chemicals usually consisting of nucleic acid and protein. Their nucleic acids are able to take over the genetic machinery of the cells of particular host tissues, manufacture more virus particles, and multiply enough to burst the host cell. Because cancer cells divide irregularly, and usually more rapidly than normal cells, some viruses are thought to be among the agents that can cause cancer.

Genetic Engineering

Many Nobel Prizes have been won for the fundamental research that made genetic engineering possible.

Insulin is a polypeptide hormone made by the pancreas, and we require insulin to manage the concentration of glucose in the blood. People who do not make insulin have insulin-dependent diabetes. In severe diabetes, the individual must take insulin by injection (to avoid the protein-digesting enzymes of the digestive tract). People who once relied on the insulin isolated from the pancreases of animals—cows, pigs, and sheep, for example—sometimes experienced allergic responses because the animal insulin molecules are not identical to those of human insulin. *Human* insulin, however, is now being manufactured by recombinant DNA technology.

Recombinant DNA

The DNA of certain strains of bacteria can be modified so the bacteria have genes that they normally do not have. But now let's back up a bit. Bacteria make their own polypeptides using the same genetic code as humans. However, unlike humans, bacteria carry DNA not only in chromosomes but also in large, circular, supercoiled DNA molecules called *plasmids.* Each plasmid carries just a few genes, but several copies of a plasmid can exist in one bacterial cell. Each plasmid replicates independently of the chromosome.

Plasmids can be removed from bacteria and snipped open by special enzymes, exposing open ends of plasmid DNA molecules. To these are next attached new DNA, and then the plasmids are reclosed. The new DNA might have the gene for directing the synthesis of a polypeptide completely foreign to the bacteria, like the subunits of human insulin. The changed plasmid DNA is called **recombinant DNA,** and the technology associated with this procedure is called **genetic engineering.**

The remarkable feature of bacterial plasmids that hold recombinant DNA is that when the bacteria multiply, they produce still more of the *altered* plasmids. Because bacteria can reproduce rapidly, the number of altered bacteria can become large very quickly. Between their cell divisions, the altered bacteria manufacture the proteins for which they are genetically programmed, including the

Facets of Chemistry 23.1

DNA in Court

Because all people are *humans*, nearly all DNA is the same in all of us. However, parts of DNA molecules differ from person to person. So, unless you have an identical sibling, your DNA is also special. No one else has ever had DNA *identical* to yours; no one else ever will. Moreover, every cell in your body, except sperm, ova, or hair shafts, has the entire set of the DNA molecules unique to you.

Short Tandem Repeats

In the human genome, the entire set of human genes, there are many regions of DNA molecules that have short sequences of side-chain bases repeated in tandem. They're called STRs, for *short tandem repeats*. These regions all have a common core sequence, but both the number and the lengths of the repeated sequences vary from person to person. The variations in STRs between individuals form the basis of the most important advance in forensic science (the science of crime evidence) since the development of fingerprints. The method is variously called DNA profiling, DNA typing, or DNA fingerprinting.

The sometimes fierce courtroom controversies over the use of DNA profiling have nothing to do with the basic science of DNA itself or of the uniqueness of an individual's genes but, rather, with the quality of the evidence gathering and the control procedures used. When these are done so well as to be above dispute, the power of DNA profiling to distinguish among individuals is extremely high.

The technology of DNA sequencing is changing rapidly, so what we describe here may well be no longer the exact method in use when the reader sees these words. But the general principles will be the same, namely, to isolate DNA from the crime scene sample, from possible culprits, from the victim, and from people known to be innocent; then, see how the various DNAs match up.

Restriction Fragments

With the help of Figure 1 we'll study a technique that has been used when a sample from a crime scene is relatively large, for example, a spot of blood the size of a dime. The sample is isolated and its DNA is extracted. In Step 1, one of a repertoire of special enzymes called *restriction en-*

zymes is used to cut the DNA into pieces called *restriction fragments*. Restriction enzymes catalyze the hydrolytic splitting of DNA molecules only at places that are adjacent to specific, short sequences of bases. For example, one commonly used restriction enzyme (about 100 are known) cleaves only at sites adjacent to a series of four bases,

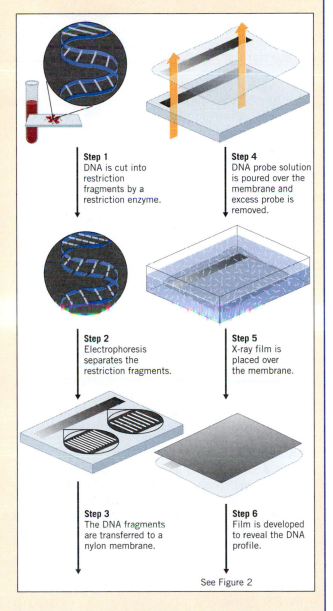

Step 1
DNA is cut into restriction fragments by a restriction enzyme.

Step 2
Electrophoresis separates the restriction fragments.

Step 3
The DNA fragments are transferred to a nylon membrane.

Step 4
DNA probe solution is poured over the membrane and excess probe is removed.

Step 5
X-ray film is placed over the membrane.

Step 6
Film is developed to reveal the DNA profile.

See Figure 2

Figure 1 *Obtaining a DNA profile.* Several steps are used to create a profile of the RFLPs from the DNA obtained from a crime scene sample or a defendant. (Courtesy of Cellmark Diagnostics, Germantown, Maryland.)

AATT paired on the opposite strand of the DNA double helix to TTAA.

$$\text{etc.}-G\overset{\downarrow}{}A-A-T-T-C-\text{etc.}$$
$$\text{etc.}-C-T-T-A-A\overset{}{\underset{\uparrow}{}}G-\text{etc.}$$

section of a DNA double helix
(Arrows are sites of hydrolysis,
catalyzed by a restriction enzyme.)

Notice that one restriction fragment (top strand) will end on its left with AATTC. A fragment from the bottom ends on its right with CTTAA.

RFLPs

The remainders of each of the restriction fragments vary widely in length from individual to individual. In other words, humans have restriction fragments that, in length, are polymorphic (from the Greek *poly-morphism* for "multiform"). In short (!), humans display *restriction-fragment length polymorphisms,* which, mercifully, may be called RFLPs. A person's unique genetic profile or "DNA fingerprint" consists in the variations in length and relative amounts of the RFLPs.

Separation of RFLPs

After chopping up the person's DNA, the technique continues with separating the RFLPs and displaying them. Being polymorphic in length means that the RFLPs are polymorphic in molecular mass and so are separable by *gel electrophoresis.* "Gel" refers to a supporting semisolid, semiliquid material through which ions can migrate. "Electrophoresis," from the Greek *phoresis,* "being carried," refers to the movement of ions under the impulse of an electrical force.

In Step 2 (Figure 1), a solution of RFLPs is spread in a narrow band at one end of the gel-coated surface, a current is turned on, and the ions move out, *lighter ions moving more rapidly than heavier ions,* assuming that the charges on the ions are equal. Several samples can be run side by side at the same time on the same gel-coated plate and so under identical conditions. The RFLPs are thus separated and now exist in narrow, parallel zones in the gel. What remains is to locate the zones and make them visible.

DNA Probes

A blotting technique transfers the (invisible) RFLPs from the gel to a nylon membrane (Step 3), which is then immersed in a solution containing a *DNA probe.* One kind of probe is a chemical that renders DNA phosphorescent wherever its molecules are located. Another kind of probe uses short-length, radiolabeled molecules of DNA (Step 4 in Figure 1). One or more atoms in the probe molecules are from a beta-emitting isotope.

By base pairing, the short DNA probe molecules bind to the RFLPs wherever they are on the nylon surface, and

any excess probe is then washed off. The binding of probe to restriction fragments makes each zone of the RFLPs a beta-emitting zone, but the zones are still invisible. To find them, a sheet of X-ray film is placed over the nylon surface (Step 5), and the beta radiation acts on the film. After several days, the X-ray film is developed (Step 6), and dark bands appear wherever bands of separated RFLPs are present.

"Bar Code" Profiles

The final result, the *DNA profile,* looks much like a store's bar code, being a series of parallel lines on a piece of paper that differ from person to person in number, thickness, and spatial separation (see Figure 2). When the profile of a suspect matches that of a crime scene sample of blood or semen, it's similar to having a match of fingerprints. Without a match, the suspect is innocent. In London, England, about 20% of rape suspects have been cleared by failures of DNA profile matches. When a match occurs, however, the suspect is almost certainly guilty (many experts say, *certainly*), assuming that all the standards of handling evidence and doing the profiling have been followed.

When the crime scene specimen is very small, then the DNA must first be "amplified" or cloned so that the amount of DNA fragments is large enough for the rest of the analysis. A reaction called the *polymerase chain reaction,* catalyzed by a polymerase enzyme, is used for the cloning.

In 1998, the U.S. Federal Bureau of Investigation established a national database using the databases of the 50 states. It is now possible to compare the DNA profile of a suspect or the profile of DNA obtained from a crime scene with all of the profiles in the system.

Additional Reading

"Blood and DNA Evidence," Chapter 9 in B. Robertson and G. A. Vignaux, *Interpreting Evidence,* John Wiley & Sons, Inc., New York, 1995.

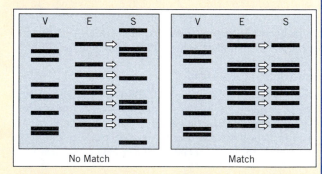

Figure 2 *DNA analysis in crime lab testing.* The DNA profiles of the victim (V) and the suspect (S) are compared to the evidence (E) obtained at the crime scene. On the left, there is no match between S and E, but on the right, the match occurs. In neither can the evidence be attributed to the victim. (Courtesy of Cellmark Diagnostics, Germantown, Maryland.)

proteins specified by the recombinant DNA. In this way, bacteria can be induced to make *human* insulin. The tech-nology is not limited to bacteria; yeast cells can also be used for genetic engineering.

One of the hopes of specialists in genetic engineering is to find ways to correct genetic defects. Considerable research is being done, for example, to alter the nucleic acids of viruses to enable them to "piggyback" into human cells the DNA that will cure cystic fibrosis. Hemophiliacs, those who bleed extensively from even the slightest cut because they lack a factor for blood clotting, may be helped by the genetic engineering of this clotting factor. Many other applications of genetic engineering are being investigated.

Genetic engineering carries over to the use of DNA as a forensic (courtroom) tool. Facets of Chemistry 23.1 discusses how DNA profiling (DNA "fingerprinting") has become an important weapon both for convicting the guilty and for freeing the innocent.

Medical diagnosis technology using "biochips" is another recent and very significant development involving DNA. Biochips are further described in *Chemicals in Our World 12* following this chapter.

SUMMARY

Organic Compounds. **Organic chemistry** is the study of the covalent compounds of carbon, except for its oxides, carbonates, bicarbonates, and a few others. **Functional groups** attached to nonfunctional and hydrocarbonlike groups are the basis for the classification of organic compounds. Members of a family have the same kinds of chemical properties. Their physical properties are a function of the relative proportions of functional and nonfunctional groups. Opportunities for hydrogen bonding strongly influence the boiling points and water solubilities of compounds with OH or NH groups.

Saturated Hydrocarbons. The **alkanes**—saturated compounds found chiefly in petroleum—are the least polar and the least reactive of the organic families. Their carbon skeletons can exist as **straight chains, branched chains,** or **rings.** When they are open chain, there is free rotation about single bonds. Rings can include heteroatoms, but **heterocyclic compounds** are not hydrocarbons.

By the IUPAC rules, the names of alkanes end in *-ane*. A prefix denotes the carbon content of the parent chain, the longest continuous chain, which is numbered from whichever end is nearest a branch. The **alkyl groups** are alkanes minus a hydrogen.

Alkanes in general do not react with strong acids or bases or strong redox reactants. At high temperatures, their molecules crack to give H_2 and unsaturated compounds.

Unsaturated Hydrocarbons. The lack of free rotation at a carbon–carbon double bond makes **geometric isomerism** possible. The pair of electrons of a π bond makes alkenes act as Brønsted bases, enabling alkenes to undergo **addition reactions** with hydrogen chloride and water (in the presence of an acid catalyst). Alkenes also add hydrogen (in the presence of a metal catalyst). Bromine and chlorine add to alkenes under mild, uncatalyzed conditions. Strong oxidizing agents, like ozone, readily attack alkenes and break their molecules apart.

Aromatic hydrocarbons, like benzene, do not give the addition reactions shown by alkenes. Instead, they undergo **substitution reactions** that leave the energy lowering, pi electron network intact. In the presence of suitable catalysts, benzene reacts with chlorine, bromine, nitric acid, and sulfuric acid.

Alcohols and Ethers. The **alcohols,** ROH, and the **ethers,** ROR′, are alkyl derivatives of water. Ethers have almost as few chemical properties as alkanes. Alcohols undergo an **elimination reaction** that splits out H_2O and changes them to alkenes. Concentrated hydrohalogen acids, like HI, react with alcohols by a **substitution reaction,** which replaces the OH group by halogen. Oxidizing agents convert alcohols of the type RCH_2OH first into aldehydes and then into carboxylic acids. Oxidizing agents convert alcohols of the type R_2CHOH into ketones. Alcohols form esters with carboxylic acids.

Amines. The simple **amines,** RNH_2, as well as the more substituted relatives, RNHR′ and RNR′R″, are organic derivatives of ammonia and thus are weak bases that can neutralize strong acids. The conjugate acids of amines are good proton donors and can neutralize strong bases.

Carbonyl Compounds. **Aldehydes,** RCH=O, among the most easily oxidized organic compounds, are oxidized to carboxylic acids. **Ketones,** RCOR′, strongly resist oxidation. Aldehydes and ketones add hydrogen to make alcohols.

The **carboxylic acids,** RCO_2H, are weak acids but neutralize strong bases. Their anions are Brønsted bases. Carboxylic acids can also be changed to esters by heating them with alcohols.

Esters, RCO_2R', react with water to give back their parent acids and alcohols. When aqueous base is used for hydrolysis, the products are the salts of the parent carboxylic acids as well as alcohols.

The simple **amides,** $RCONH_2$, can be made from acids and ammonia, and they can be hydrolyzed back to their parent acids and trivalent nitrogen compounds. The amides are not basic.

IUPAC Nomenclature of Compounds with Functional Groups. Each family of compounds has a characteristic name ending:

-ane	alkanes	-al	aldehydes
-ene	alkenes	-one	ketones
-yne	alkynes	-oic acid	carboxylic acids
-ol	alcohols	-oate	esters

Each member of a family has a defined "parent," which is the longest chain that includes the functional group. How the chain is numbered varies with the family, but generally the numbering starts from whichever end of the parent chain reaches the functional group with the lower number. The names of side chains, such as the alkyl groups, are included in the name ahead of the name of the parent, and their locations are designated by the numbers of the carbons at which they are attached to the parent.

Polymers. Macromolecules with more or less repeating structural units are called **polymers.** They are made from one or more **monomers** by **polymerization.**

The polymers most useful for making filaments and fibers are those with very long, symmetrical molecules. If regularly spaced polar groups occur that can attract each other, either by ordinary dipole–dipole attractions or by hydrogen bonds, the fibers are particularly strong.

Carbohydrates. The glucose unit is present in starch, glycogen, cellulose, sucrose, lactose, and simply as a **monosaccharide** in honey. It's the chief "sugar" in blood. As a monosaccharide, glucose is a pentahydroxyaldehyde in one form but also exists in two cyclic forms. The latter are joined by oxygen bridges in the **disaccharides** and **polysaccharides.** In lactose (milk sugar), a glucose and a galactose unit are joined. In sucrose (table sugar), glucose and fructose units are linked.

The major polysaccharides—starch (amylose and amylopectin), glycogen, and cellulose—are all polymers of glucose but with different patterns and geometries of branching. The digestion of the disaccharides and starch requires the hydrolyses of the oxygen bridges to give monosaccharides.

Lipids. Natural products in living systems that are relatively soluble in nonpolar solvents, but not in water, are all in a large group of compounds called **lipids.** They include sex hormones and cholesterol, the triacylglycerols of nutritional importance, and the glycerophospholipids and other phospholipids needed for cell membranes.

The **triacylglycerols** are triesters between glycerol and three of a number of long-chain, often unsaturated **fatty acids.** In digestion of triacylglycerols, hydrolysis of the ester groups occurs, and anions of the fatty acids form (together with glycerol).

Molecules of glycerophospholipids have large segments that are hydrophobic and others that are hydrophilic. Glycerophospholipid molecules are the major components of cell membranes, where they are arranged in a lipid bilayer.

Amino Acids, Polypeptides, and Proteins. The **amino acids** in nature are mostly compounds with NH_3^+ and CO_2^- groups joined to the same carbon atom (called the alpha position of the amino acid). About 20 amino acids are important to the structures of polypeptides and proteins.

When two amino acids are linked by a **peptide bond** (an amide bond), the compound is a dipeptide. In **polypeptides,** several (sometimes thousands) of regularly spaced peptide bonds occur, involving as many amino acid units. The uniqueness of a polypeptide lies in its chain length and in the sequence of the side-chain groups.

Some **proteins** consist of only a polypeptide, but most proteins involve two or more (sometimes different) polypeptides and often a nonpolypeptide component (an organic molecule) and a metal ion.

Nearly all **enzymes** are proteins. Major poisons work by deactivating enzymes.

Nucleic Acids. The formation of a polypeptide with the correct amino acid sequence is under the control of **nucleic acids.** These are polymers whose backbones are made of a repeating sequence of pentose sugar units. On each sugar unit is a **base,** a heterocyclic amine such as adenine (A), thymine (T), guanine (G), cytosine (C), or uracil (U). In **DNA,** the sugar is deoxyribose and the bases are A, T, G, and C. In **RNA,** the sugar is ribose and the bases are A, U, G, and C.

DNA exists in cell nuclei as **double helices.** The two strands are associated side by side, with hydrogen bonds occurring between particular pairs of bases. In DNA A is always paired to T; C is always paired to G. The base U replaces T in RNA, and A can also pair with U. The bases, A, U, G, and C, are the four letters of the genetic alphabet, and the specific combination that codes for one amino acid in polypeptide synthesis is a set of three bases, side by side, on a strand. An individual gene in higher organisms consists of sections of a DNA chain, called **exons,** separated by other sections called **introns.** The **replication** of DNA is the synthesis by the cell of exact copies of the original two strands of a DNA double helix.

Polypeptide Synthesis. The DNA that carries the gene for a particular polypeptide is used to direct the synthesis of a molecule of heterogeneous nuclear RNA. The hnRNA is then modified by the removal of those segments corresponding to the introns of the DNA. This leaves messenger

RNA, the carrier of the message of the gene to the ribosomes. The ribosomes contain rRNA (ribosomal RNA). The overall change from gene to mRNA is called **translation.** mRNA is like an assembly line at a factory (ribosomes) awaiting the arrival of parts—the amino acids. These are borne on transfer RNA molecules, tRNA.

A unit of three bases on mRNA constitutes a **codon** for one amino acid. A tRNA–amino acid unit has an **anticodon** that can pair to a codon on mRNA. In this way a tRNA–amino acid unit "recognizes" only one place on mRNA for delivering the amino acid to the polypeptide developing at the mRNA assembly line. The key that re-

lates codons to specific amino acids is the **genetic code,** and it is essentially the same code for all organisms in nature. The involvement of mRNA in directing the synthesis of a polypeptide is called **translation.**

Even small changes or deletions in the sequences of bases on DNA can have large consequences in the forms of genetic diseases. Viruses are able to use their own nucleic acids to take over the genetic apparatus of the host tissue. In **genetic engineering,** the genetic machinery of microorganisms is taken over, forced to take **recombinant DNA,** and made to manufacture polypeptides not otherwise synthesized by the organism.

REVIEW QUESTIONS

Structural Formulas

23.1 Which of the following compounds are inorganic?
(a) CH_3OH (c) CCl_4 (e) K_2CO_3
(b) CO (d) $NaHCO_3$

23.2 In general terms, what makes possible so many organic compounds?

23.3 What is the condensed structure of R if R—CH_3 is ethane?

23.4 Which of the following structures are possible, given the numbers of bonds that various atoms can form?
(a) $CH_2CH_2CH_3$ (c) $CH_3CH=CHCH_2CH_3$
(b) $CH_3=CHCH_2CH_3$

23.5 Write neat, condensed structures of the following.

(a)

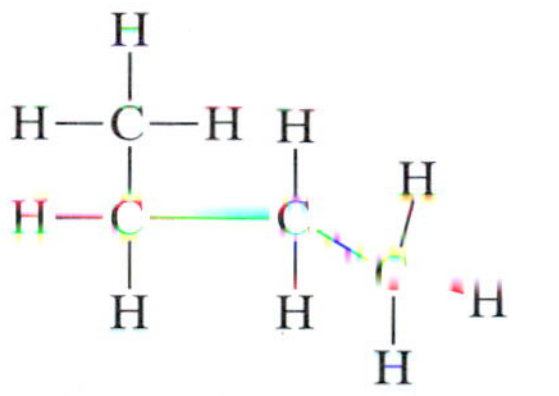

(b)

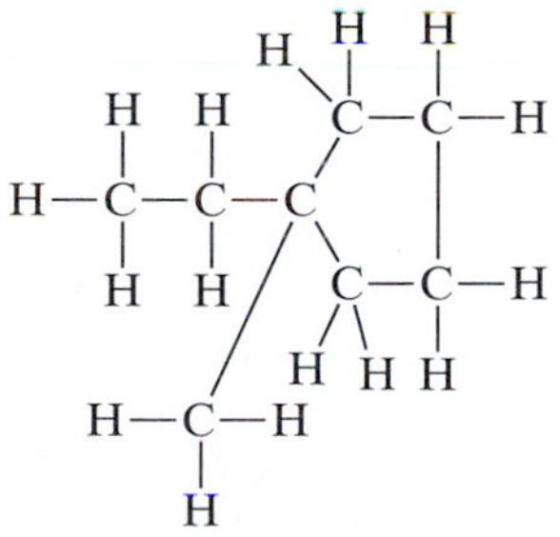

(c)

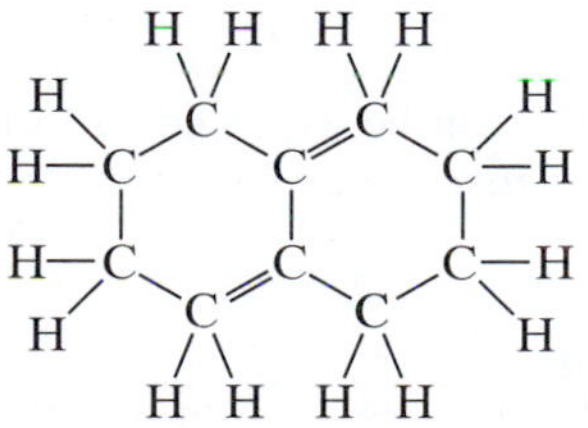

Families of Organic Compounds

23.6 In $CH_3CH_2NH_2$, the NH_2 group is called the *functional group.* In *general terms,* why is it called this?

23.7 In general terms, why do functional groups impart more chemical reactivity to families that have them? Why don't the alkanes display as many reactions as, say, the amines?

23.8 How are *macromolecules* and *polymers* alike? How are they different?

Isomers

23.9 What must be true about two substances if they are to be called isomers of each other?

23.10 Are H_2O and D_2O isomers? If not, why not?

23.11 What prevents free rotation about a double bond and so makes geometric isomers possible?

Properties and Structure

23.12 Explain why $CH_3CH_2CH_2OH$ is more soluble in water than $CH_3CH_2CH_2CH_3$.

23.13 Of the two compounds in Question 23.12, which has the higher boiling point? Explain.

23.14 Examine the structures of the following compounds.

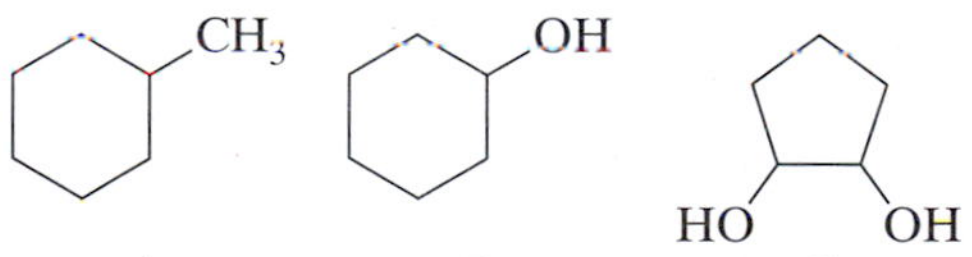

A   B   C

(a) Which has the highest boiling point? Explain.
(b) Which has the lowest boiling point? Explain.

23.15 Which of the following compounds is more soluble in water? Explain.

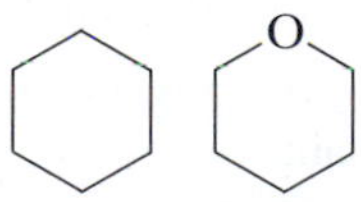

A   B

23.16 Which of the following compounds has the higher boiling point? Explain.

$$CH_3CH_2\underset{\underset{CH_3}{|}}{CH}CH_2\underset{\underset{CH_3}{|}}{CH}CH_3 \quad CH_3CH_2CH_2CH_2CH_3$$

A   B

23.17 Why do aldehydes and ketones have boiling points that are lower than those of their corresponding alcohols?

23.18 Acetic acid boils at 118 °C, higher even than 1-propanol, which boils at 97 °C. How can the higher boiling point of acetic acid be explained?

23.19 Methyl ethanoate has a higher molecular mass than its parent acid, ethanoic acid. Yet methyl ethanoate (bp 59 °C) boils much lower than ethanoic acid (bp 118 °C). How can this be explained?

23.20 Ethanamide is a solid at room temperature. 1-Aminobutane, which has about the same molecular mass, is a liquid. Explain.

23.21 Amines, RNH_2, do not have as high boiling points as alcohols of comparable molecular masses. Why?

Names
23.22 Write the condensed structures of the following compounds.
(a) 2,2-dimethyloctane
(b) 1,3-dimethylcyclopentane
(c) 1,1-diethylcyclohexane
(d) 6-ethyl-5-isopropyl-7-methyl-1-octene
(e) *cis*-2-pentene

23.23 Write condensed structures of the following compounds.
(a) 3-methylbutanal   (d) isopropyl ethanoate
(b) 4-methyl-2-octanone   (e) 2-methylbutanamide
(c) 2-chloropropanoic acid

23.24 Write condensed structures of the following compounds.
(a) 2,3-butanedione
(b) butanedicarboxylic acid
(c) 2-aminopropanal
(d) cyclohexyl 2-methylpropanoate
(e) sodium 2,3-dimethylbutanoate

23.25 "3-Butanol" is not a proper name, but a structure could still be written for it. What is this structure, and what IUPAC name should be used?

Chemical Properties of Organic Compounds
23.26 What are the products of the complete combustion of any hydrocarbon, assuming an abundant supply of oxygen?

23.27 Brønsted acids are known to react with alkenes but not with alkanes. Explain.

23.28 Propene is known to react with concentrated sulfuric acid as follows.

$$CH_3CH{=}CH_2 + H_2SO_4 \longrightarrow CH_3\underset{\underset{OSO_3H}{|}}{CH}CH_3$$

The reaction occurs in two steps. Write the equation for the first.

23.29 2-Methylpropene reacts with hydrogen chloride as follows.

$$CH_3\underset{\underset{CH_3}{|}}{C}{=}CH_2 + H{-}Cl \longrightarrow CH_3\underset{\underset{Cl}{|}}{\overset{\overset{CH_3}{|}}{C}}CH_3$$

The reaction occurs in two steps. Write the equation of each step.

23.30 Briefly explain how the C—O bond in isopropyl alcohol is weakened when a strong acid is present.

23.31 Which isomer of 1-butanol cannot be oxidized by dichromate ion? Write its structure and IUPAC name.

23.32 A monofunctional organic oxygen compound dissolves in aqueous base but not in aqueous acid. The compound is in which of the families of organic compounds that we studied? Explain.

23.33 A monofunctional organic nitrogen compound dissolves in aqueous hydrochloric acid but not in aqueous sodium hydroxide. What kind of organic compound is it?

23.34 Why are amides neutral while amines are basic?

23.35 Hydrazine is a Brønsted base but urea is not. Offer an explanation.

$$NH_2NH_2 \qquad NH_2\overset{\overset{O}{\|}}{C}NH_2$$

hydrazine   urea

23.36 Write the equilibrium that is present when $CH_3CH_2CH_2NH_2$ is in an aqueous solution.

23.37 An organic compound dissolves in both aqueous base and in aqueous acid, but it is insoluble in water. Offer an explanation.

23.38 Predict which of the following species can neutralize the hydronium ion in dilute, aqueous HCl at room temperature.

(a) $CH_3\overset{\overset{O}{\|}}{C}O^-$ (c) HO^- (e) $CH_3CH_2CH_2CH_3$

(b) CH_3NH_2 (d) $CH_3\overset{\overset{O}{\|}}{C}NH_2$ (f) a pyrrolidine ring with NH

23.39 Write the products that can be expected to form in the following situations. If no reaction occurs, write "no reaction."
(a) $CH_3CH_2CH_2NH_2 + HBr(aq) \longrightarrow$
(b) $CH_3CH_2CH_2CH_3 + HI(aq) \longrightarrow$
(c) $CH_3CH_2CH_2NH_3^+ + H_3O^+ \longrightarrow$
(d) $CH_3CH_2CH_2NH_3^+ + OH^- \longrightarrow$

23.40 Write the equation for the equilibrium that is present in a solution of propanoic acid and methanol with a trace of strong acid.

Biochemistry

23.41 What is *biochemistry?*

23.42 What are the three fundamental needs for sustaining life, and what are the general names for the substances that supply these needs?

Carbohydrates

23.43 How are *carbohydrates* defined?

23.44 What monosaccharide forms when the following polysaccharides are completely hydrolyzed? (a) starch, (b) glycogen, (c) cellulose

23.45 Glucose exists in three forms only one of which is a polyhydroxyaldehyde. How then can an aqueous solution of any form of glucose give the reactions of an aldehyde?

23.46 Name the compounds that form when sucrose is digested.

23.47 The digestion of lactose gives what compounds? Name them.

23.48 The complete hydrolysis of the following compounds gives what product(s)? (Give the names only.)
(a) amylose   (b) amylopectin

23.49 Describe the relationships among amylose, amylopectin, and starch.

23.50 Why are humans unable to use cellulose as a source of glucose?

23.51 What function is served by glycogen in the body?

Lipids

23.52 How are *lipids* defined?

23.53 Why are lipids more soluble than carbohydrates in nonpolar solvents?

23.54 Cholesterol is not an ester, yet it is defined as a lipid. Why?

23.55 A product such as corn oil is advertised as "polyunsaturated," but all nutritionally important lipids have unsaturated C=O groups. What does "polyunsaturated" refer to?

23.56 Is it likely that the following compound could be obtained by the hydrolysis of a naturally occurring lipid? Explain.

$CH_3(CH_2)_2CH=CHCH_2CH=CHCH_2CH=CH(CH_2)_7CO_2H$

Amino Acids, Polypeptides, and Proteins

23.57 Describe the specific ways in which the monomers for all polypeptides are (a) alike and (b) different.

23.58 What is the *peptide bond?*

23.59 Describe the structural way in which two *isomeric* polypeptides would be different.

23.60 Describe the structural ways in which two different polypeptides can differ.

23.61 Why is a distinction made between the terms *polypeptide* and *protein?*

23.62 In general terms, what forces are at work that determine the final *shape* of a polypeptide strand?

23.63 Explain in general terms how a change in pH might alter the shape of a polypeptide.

23.64 What kind of substance makes up most enzymes?

23.65 The most lethal chemical poisons act on what kinds of substances? Give examples.

Nucleic Acids and Heredity

23.66 In general terms only, how does the body solve the problem of getting a particular amino acid sequence into a polypeptide rather than a mixture of randomly organized sequences?

23.67 In what specific structural way does DNA carry genetic messages?

23.68 What kind of force occurs between the two DNA strands in a double helix?

23.69 How are the two DNA strands in a double helix structurally related?

23.70 In what ways do DNA and RNA differ structurally?

23.71 Which base pairs with (a) A in DNA, (b) A in RNA, and (c) C in DNA or RNA?

23.72 A specific gene in the DNA of a higher organism is said to be *segmented* or *interrupted.* Explain.

23.73 On what kind of RNA are codons found?

23.74 On what kind of RNA are anticodons found?

23.75 What function does each of the following have in polypeptide synthesis?
(a) a ribosome   (b) mRNA   (c) tRNA

23.76 What kind of RNA forms directly when a cell uses DNA at the start of the synthesis of a particular enzyme? What kind of RNA is made from it?

23.77 The process of *transcription* begins with which nucleic acid and ends with which one?

23.78 The process of *translation* begins with which nucleic acid? What is the end result of translation?

23.79 What is a virus, and (in general terms only) how does it infect a tissue?

23.80 Genetic engineering is able (in general terms) to accomplish what?

23.81 What is *recombinant DNA?*

REVIEW PROBLEMS

Answers to problems whose numbers are printed in color are given in Appendix D.

Structural Formulas

*23.82 Write full (expanded) structures for each of the following molecular formulas. Remember how many bonds the various kinds of atoms must have. In some you will have to use double or triple bonds. (*Hint:* A trial-and-error approach will have to be used.)
(a) CH_5N (c) $CHCl_3$ (e) C_2H_2
(b) CH_2Br_2 (d) NH_3O (f) N_2H_4

*23.83 Write full (expanded) structures for each of the following molecular formulas. Follow the guidelines of Problem 23.82.
(a) C_2H_6 (c) CH_2O (e) C_3H_3N
(b) CH_2O_2 (d) HCN (f) CH_4O

Families of Organic Compounds

23.84 Name the family to which each compound belongs.
(a) $CH_3CH{=}CH_2$
(b) CH_3CH_2OH
(c) $CH_3CH_2\overset{\displaystyle O}{\overset{\|}{C}}OCH_3$
(d) $CH_3CH_2CH_2\overset{\displaystyle O}{\overset{\|}{C}}OH$
(e) $CH_3CH_2CH_2NH_2$
(f) $HOCH_2CH_2CH_3$

23.85 To what organic family does each compound belong?
(a) $CH_3C{\equiv}CH$
(b) $CH_3CH_2\overset{\displaystyle O}{\overset{\|}{C}}H$
(c) $CH_3\overset{\displaystyle O}{\overset{\|}{C}}CH_2CH_2CH_3$
(d) $CH_3{-}O{-}CH_2CH_3$
(e) $CH_3CH_2NH_2$
(f) $CH_3CH_2\overset{\displaystyle O}{\overset{\|}{C}}NH_2$

23.86 Identify by letter the compounds in Problem 23.84 that are saturated compounds.

23.87 Which parts of Problem 23.85 give the structures of saturated compounds?

23.88 Classify the following compounds as amines or amides. If another functional group occurs, name it too.
(a) $CH_3CH_2NH_2$
(b) $CH_3CH_2NHCH_3$
(c) $CH_3CH_2\overset{\displaystyle O}{\overset{\|}{C}}NH_2$
(d) $CH_3\overset{\displaystyle O}{\overset{\|}{C}}CH_2NH_2$

23.89 Name the functional groups present in each of the following structures.

(a) 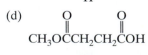 O̶ ⌐NH

(b) $CH_3O{-}\bigcirc{-}N{-}\overset{\displaystyle O}{\overset{\|}{C}}CH_3$

(c) O⌐ ...C O H

(d) $CH_3O\overset{\displaystyle O}{\overset{\|}{C}}CH_2CH_2\overset{\displaystyle O}{\overset{\|}{C}}OH$

Isomers

23.90 Decide whether the members of each pair are identical, are isomers, or are unrelated.
(a) $CH_3{-}CH_3$ and $CH_3{-}\overset{\displaystyle}{\underset{\displaystyle CH_3}{C}H}$ ... CH_3

(b) $CH_3\ \underset{CH_2}{\overset{}{}}\ \underset{CH_3}{}$ and $\overset{CH_2}{\underset{CH_3 \ CH_3}{}}$

(c) CH_3CH_2OH and $CH_3CH_2CH_2OH$

(d) $CH_3CH{=}CH_2$ and $CH_2{-}{-}{-}CH_2$ ... CH_2

(e) $H{-}\overset{\displaystyle O}{\overset{\|}{C}}{-}CH_3$ and $CH_3{-}\overset{\displaystyle O}{\overset{\|}{C}}{-}H$

(f) $CH_3\overset{\displaystyle CH_3}{\overset{\|}{C}H}CH_3$ and $CH_3\overset{\displaystyle CH_3}{\overset{\|}{C}H}$ ... CH_3

(g) $CH_3CH_2CH_2NH_2$ and $CH_3CH_2{-}NH{-}CH_3$

23.91 Examine each pair and decide if the two are identical, are isomers, or are unrelated.

(a) $CH_3CH_2\overset{\displaystyle O}{\overset{\|}{C}}OH$ and $HO\overset{\displaystyle O}{\overset{\|}{C}}CH_2CH_3$

(b) $H\overset{\displaystyle O}{\overset{\|}{C}}OCH_2CH_3$ and $CH_3CH_2\overset{\displaystyle O}{\overset{\|}{C}}OH$

(c) $H\overset{\displaystyle O}{\overset{\|}{C}}OCH_2CH_2OH$ and $HOCH_2CH_2\overset{\displaystyle O}{\overset{\|}{C}}OH$

(d) $CH_3\overset{\displaystyle O}{\overset{\|}{C}}CH_2CH_3$ and $CH_3CH_2\overset{\displaystyle O}{\overset{\|}{C}}CH_3$

(e) $CH_3{-}CH{-}CH_3$
$\quad\ \ CH_2{-}CH_2\ \ CH_3\ \ CH_3$
$\quad\ \ \ \ \ \ \ \ CH_2{-}C{-}CH$
$\quad\ \ \ \ \ \ \ \ \ \ \ \ CH_3\ \ CH_3$
and
$CH_3{-}\overset{CH_3}{\overset{\|}{C}H}{-}CH_2{-}CH_2{-}CH_2{-}\overset{CH_3}{\underset{CH_3}{\overset{\|}{C}}}{-}\overset{CH_3}{\overset{\|}{C}H}{-}CH_3$

(f) $CH_3{-}NH{-}\overset{\displaystyle O}{\overset{\|}{C}}{-}CH_3$ and $CH_3CH_2\overset{\displaystyle O}{\overset{\|}{C}}NH_2$

(g) H—O—O—H and H—O—H

Names of Hydrocarbons

23.92 Write the IUPAC names of the following hydrocarbons.
(a) $CH_3CH_2CH_2CH_2CH_3$
(b) $CH_3CH_2CH_2CHCH_3$
$\qquad\qquad\qquad\qquad |$
$\qquad\qquad\qquad\quad CH_3$

(c) $\qquad CH_3$
$\qquad\qquad |$
$CH_3CHCH_2CHCHCH_3$
$\qquad\qquad\qquad |$
$\qquad\qquad\quad CH_3$

23.93 Write the IUPAC names of the following compounds.
(a) $\qquad\quad CH_3$
$\qquad\qquad\quad |$
$CH_3CH_2CHCH_2CHCH_3$
$\qquad\qquad\qquad |$
$\qquad\qquad\quad CH_3$

(c) $\qquad CH_3$
$\qquad\qquad |$
$CH_3CHCH=CHCH_3$

(b) $CH_3CH_2CH=CHCH_2CH_3$

Geometric Isomerism

23.94 Write the structures of the cis and trans isomers, *if any*, for the following.
(a) $CH_2=CHBr$
(b) $CH_3CH=CHCH_2CH_3$
(c) $\qquad Br$
$\qquad\qquad |$
$CH_3C=CHCl$

*23.95 Write the structures of the cis and trans isomers, if any, for the following compounds.
(a) $\qquad CH_3$
$\qquad\qquad |$
$CH_3C=CHCH_3$

(c) $\qquad\qquad Cl \quad Cl$
$\qquad\qquad\quad |\qquad |$
$CH_3CH_2C=CCH_2CH_3$

(b) $\qquad\qquad CH_3$
$\qquad\qquad\qquad |$
$CH_3CH=CCH_2CH_3$

Chemical Reactions of Alkenes

23.96 Write the structures of the products that form when ethylene reacts with each of the following substances by an addition reaction. (Assume that needed catalysts or other conditions are provided.)
(a) H_2
(b) Cl_2
(c) Br_2
(d) $HCl(g)$
(e) $HBr(g)$
(f) H_2O (in acid)

23.97 The isopropyl cation is more stable than the propyl cation, so the former forms almost exclusively when propene undergoes addition reactions with proton donating species. On the basis of this fact, predict the final products of the reaction of propene with each of the following reagents.
(a) hydrogen chloride
(b) hydrogen iodide
(c) water in the presence of an acid catalyst

23.98 Repeat Problem 23.96 using 2-butene.

23.99 Repeat Problem 23.96 using cyclohexene.

23.100 In general terms, why doesn't benzene give the same kinds of addition reactions as cyclohexene?

23.101 If benzene were to *add* one mole of Br_2, what would form? What forms, instead, when Br_2, benzene, and an $FeBr_3$ catalyst are heated together? (Write the structures.)

Structures and Properties of Other Functional Groups

23.102 Write condensed structures and the IUPAC names for all of the saturated alcohols with three or fewer carbon atoms per molecule.

23.103 Write the condensed structures and the IUPAC names for all of the possible alcohols with the general formula $C_4H_{10}O$.

23.104 Write the condensed structures of all of the possible ethers with the general formula $C_4H_{10}O$. Following the pattern for the common names of ethers given in the chapter, what are the likely common names of these ethers?

23.105 Write the structures of the isomeric alcohols of the formula $C_4H_{10}O$ that could be oxidized to aldehydes. Write the structure of the isomer that could be oxidized to a ketone.

23.106 Write the structures of the products of the acid-catalyzed dehydrations of the following compounds
(a) OH

(b) $\qquad\qquad$ OH
$\qquad\qquad\qquad |$
$\qquad\qquad$—CHCH$_3$

(c) $\qquad$—CH$_2$CH$_2$OH

23.107 Write the structures of the substitution products that form when the alcohols of Problem 23.106 are heated with concentrated hydriodic acid.

23.108 Write the structures of the aldehydes or ketones that can be prepared by the oxidation of the compounds given in Problem 23.106.

*23.109 Write the structure of the product of the acid-catalyzed dehydration of 2-propanol. Write the mechanism of the reaction.

*23.110 When ethanol is heated in the presence of an acid catalyst, ethene and water form; an elimination reaction occurs. When 2-butanol is heated under similar conditions, *two* alkenes form (although not in equal quantities). Write the structures and IUPAC names of these two alkenes.

*23.111 If the formation of the two alkenes in Problem 23.110 were determined purely by statistics, then the mole ratio of the two alkenes should be in the ratio of what whole numbers? Why?

23.112 Which of the following two compounds could be easily oxidized? Write the structure of the product of the oxidation.

$\qquad\qquad O \qquad\qquad\qquad\quad O$
$\qquad\qquad \|\qquad\qquad\qquad\quad \|$
$CH_3CH_2CH \quad CH_3CH_2CCHCH_3$
$\qquad\qquad\qquad\qquad\qquad\qquad\quad |$
$\qquad\qquad\qquad\qquad\qquad\qquad CH_3$

$\qquad\qquad\quad$ **A** $\qquad\qquad\qquad\qquad$ **B**

23.113 Which of the following compounds has the chemical property given below? Write the structure of the product of the reaction.

$$\underset{\textbf{A}}{CH_3CH_2CH_2OH} \quad \underset{\textbf{B}}{CH_3CH_2\overset{\displaystyle O}{\overset{\|}{C}}OH} \quad \underset{\textbf{C}}{CH_3CH_2\overset{\displaystyle O}{\overset{\|}{C}}H}$$

(a) is easily oxidized
(b) neutralizes NaOH
(c) forms an ester with methyl alcohol
(d) can be oxidized to an aldehyde
(e) can be dehydrated to an alkene

23.114 Write the structures of the products that form in each of the following situations. If no reaction occurs, write "no reaction."
(a) $CH_3CH_2CO_2^- + HCl(aq) \longrightarrow$
(b) $CH_3CH_2CO_2CH_3 + H_2O \xrightarrow{\text{heat}}$
(c) $CH_3CH_2CH_2CO_2H + NaOH(aq) \longrightarrow$

23.115 Write the structures of the products that form in each of the following situations. If no reaction occurs, write "no reaction."
(a) $CH_3CH_2CO_2H + NH_3 \xrightarrow{\text{heat}}$
(b) $CH_3CH_2CH_2CONH_2 + H_2O \xrightarrow{\text{heat}}$
(c) $CH_3CH_2CO_2CH_3 + NaOH(aq) \xrightarrow{\text{heat}}$

*23.116 What organic products with smaller molecules form when the following compound is heated with water?

$$\underset{\quad\;\;CH_3CH_2}{CH_3CH_2\overset{\displaystyle O}{\overset{\|}{\underset{|}{N}}}CCH_3}$$

23.117 Which of the following species give the reaction specified? Write the structures of the products that form.

$$CH_3CH_2\overset{\displaystyle O}{\overset{\|}{C}}NH_2 \quad CH_3CH_2CH_2NH_2 \quad CH_3CH_2NH_3^+$$

(a) neutralizes dilute hydrochloric acid
(b) hydrolyzes
(c) neutralizes dilute sodium hydroxide

Polymers
23.118 Polyvinyl chloride, or PVC, is widely used to manufacture drainage pipes. It is a polymer of vinyl chloride, $CH_2{=}CHCl$. Write the structure of PVC, first showing a molecular portion of four successive monomer units and then the condensed structure.

*23.119 If the following two compounds polymerized in the same way that Dacron forms, what would be the repeating unit of their polymer?

$$HO\overset{\displaystyle O}{\overset{\|}{C}}CH_2CH_2\overset{\displaystyle O}{\overset{\|}{C}}OH \quad HOCH_2CH_2OH$$

*23.120 Kodel, a polyester made from the following monomers, is used to make fibers for weaving crease-resistant fabrics.

$$HOCH_2{-}\langle\rangle{-}CH_2OH \quad HO\overset{\displaystyle O}{\overset{\|}{C}}{-}\langle\rangle{-}\overset{\displaystyle O}{\overset{\|}{C}}OH$$

monomers for Kodel

What is the structure of the repeating unit in Kodel?

*23.121 If the following two compounds polymerized in the same way that nylon forms, what would be the repeating unit in the polymer?

$$HO\overset{\displaystyle O}{\overset{\|}{C}}{-}\langle\rangle{-}\overset{\displaystyle O}{\overset{\|}{C}}OH \quad NH_2CH_2CH_2CH_2CH_2NH_2$$

Lipids
*23.122 Write the structure of a triacylglycerol that could be made from palmitic acid, oleic acid, and linoleic acid.

*23.123 Write the structures of the products of the complete hydrolysis of the following triacylglycerol.

$$\begin{array}{l} CH_2O\overset{\displaystyle O}{\overset{\|}{C}}(CH_2)_7CH{=}CHCH_2CH{=}CH(CH_2)_4CH_3 \\ |\quad\;\; \overset{\displaystyle O}{\overset{\|}{}} \\ CHO\overset{\displaystyle O}{\overset{\|}{C}}(CH_2)_{12}CH_3 \\ |\quad\;\; \overset{\displaystyle O}{\overset{\|}{}} \\ CH_2O\overset{\displaystyle O}{\overset{\|}{C}}(CH_2)_7CH{=}CH(CH_2)_7CH_3 \end{array}$$

*23.124 Write the structure of the triacylglycerol that would result from the complete hydrogenation of the lipid whose structure is given in Problem 23.123.

*23.125 If the compound in Problem 23.123 is saponified, what are the products? Give their names and structures.

23.126 What parts of glycerophospholipid molecules provide hydrophobic sites? Hydrophilic sites?

23.127 In general terms, describe the structure of the membrane of an animal cell. What kinds of molecules are present? How are they arranged? What forces keep the membrane intact? What functions are served by the protein components?

Amino Acids and Polypeptides
23.128 Write the structure of the dipeptide that could be hydrolyzed to give two molecules of glycine.

23.129 What is the structure of the tripeptide that could be hydrolyzed to give three molecules of alanine?

*23.130 What are the structures of the two dipeptides that could be hydrolyzed to give glycine and phenylalanine?

*23.131 Write the structures of the dipolar ionic forms of the amino acids that are produced by the complete digestion of the following compound.

$$\underset{\qquad CH_3 \quad\;\; CH_2SH \;\; CH_2C_6H_5}{{}^+NH_3\underset{|}{CH}C\overset{\displaystyle O}{\overset{\|}{}}NH\underset{|}{CH}C\overset{\displaystyle O}{\overset{\|}{}}NH\underset{|}{CH}C\overset{\displaystyle O}{\overset{\|}{}}O^-}$$

ADDITIONAL EXERCISES

23.132 The H_2SO_4-catalyzed addition of water to 2-methyl-propene could give two isomeric alcohols.
(a) Write their condensed structures.
(b) Actually, only one forms. It is not possible to oxidize this alcohol to an aldehyde or to a ketone having four carbons. Which alcohol, therefore, is the product?
(c) Write the structures of the two possible carbocations that could form when sulfuric acid donates a proton to 2-methylpropene.
(d) One of the two carbocations of part (c) is more stable than the other. Which one is it, and how can you tell?

23.133 An unknown alcohol was either **A** or **B**. When the unknown was oxidized with an *excess* of strong oxidizing agent, sodium dichromate in acid, the *organic* product isolated was able to neutralize sodium hydroxide.

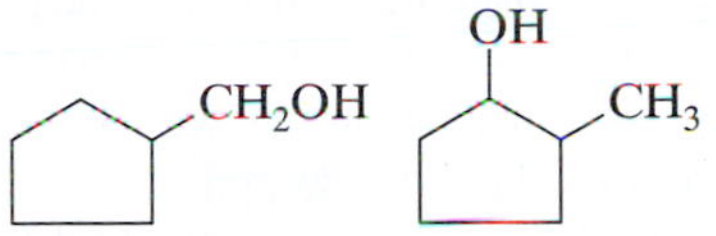

(a) Which was the alcohol? Give the reasons for this choice.
(b) Write the balanced net ionic equation for the oxidation.

*__23.134__ Suggest a reason why trimethylamine, $(CH_3)_3N$, has a *lower* boiling point (bp 3 °C) than dimethylamine, $(CH_3)_2NH$ (bp 8 °C), despite having a larger formula mass.

*__23.135__ Write the structures of the organic products that form in each situation. If no reaction occurs, write "no reaction."

(a)

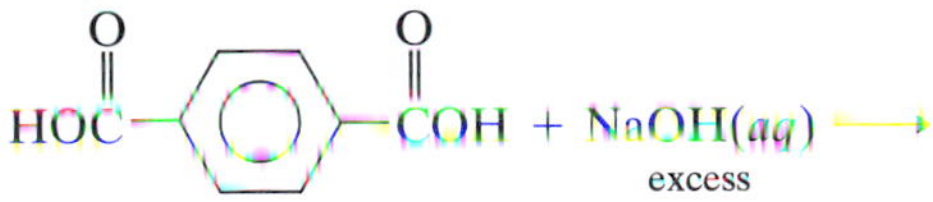

$$HOC\text{-}\langle O \rangle\text{-}COH + NaOH(aq) \longrightarrow$$
excess

(b)

$$CH_3CCH_2CH_2CH_3 + H_2 \xrightarrow[\text{catalyst}]{\text{heat, pressure,}}$$

(c) $CH_3\overset{+}{N}H_2CH_2CH_2CH_3 + OH^-(aq) \longrightarrow$

(d)

$$CH_3CH_2OCH_2CH_2COCH_3 + H_2O \xrightarrow{\text{heat}}$$

(e)

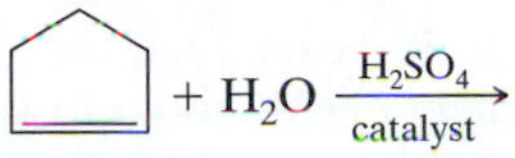

$$\langle\;\rangle + H_2O \xrightarrow[\text{catalyst}]{H_2SO_4}$$

(f)

$$\langle O \rangle\text{-}CH=CH\text{-}\langle O \rangle + HBr \longrightarrow$$

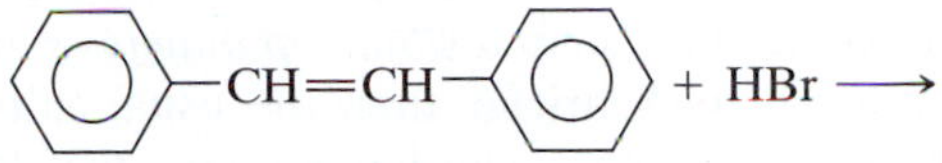

*__23.136__ How many tripeptides can be made from three different amino acids? If the three are glycine (Gly), alanine (Ala), and phenylalanine (Phe), what are the structures of the possible tripeptides? (*Hint:* Gly-Ala-Phe is one structure.)

*__23.137__ A 0.5574 mg sample of an organic acid, when burned in pure oxygen, gave 1.181 mg of CO_2 and 0.3653 mg H_2O. Calculate the empirical formula of the acid.

*__23.138__ When 0.2081 g of an organic acid was dissolved in 50.00 mL of 0.1016 M NaOH, it took 23.78 mL of 0.1182 M HCl to neutralize the NaOH that was not used up by the sample of the organic acid. Calculate the molecular mass from the data given. Is the answer necessarily the *molecular mass* of the organic acid? Explain.

Let's dream ahead some 10 to 30 years. Suppose that you have a sore throat and finally decide to go to your clinic; enough is enough. The nurse takes a throat swab and sends it to the clinic's lab for *biochip diagnosis*. In a very short time—much briefer than the usual culture plate technology—the physician has a report that tells not only which specific microorganism is responsible but also whether it is in a mutant form. If a bacterium is responsible, the same technology would tell if the one bothering you is resistant to certain antibiotics but not to others. Thus, you would be given the specific antibiotic for dealing with the problem, and you'd not waste time or money on others.

Genetic information on the same swab could also tell you whether you've inherited a defective gene that disposes you toward breast cancer, prostate cancer, or other maladies. Simple changes in lifestyle might then help you significantly forestall the threat. Otherwise, an aggressive treatment that will work best can be planned, not as firing a shotgun in the dark at a noise, hoping to hit an enemy without harming friends, but as precise sniping at an enemy clearly seen.

Biochip diagnosis could tell someone who is HIV positive if his or her particular viral strain is resistant to the medications commonly used. The exact nature of the mutation could then be used to select the treatment.

A Specific DNA Probe in Each Zone of a Biochip. Actually, none of the above is dreaming. The technology using the biochip is exploding. Several new companies have formed to develop biochips, and every pharmaceutical concern is working in the field. A biochip is like a computer chip, only it carries information not as sequences of zeros and ones but as fragments of DNA molecules, in other words, as sequences of nucleotide units A, T, G, and C.

Imagine a glass or silicon plate; it might be as large as 10 cm^2 or as small as 20 μm^2. The surface can be divided into tiny squared zones, hundreds, thousands, even millions in number. All but one zone is covered to enable the uncovered zone to receive identical copies of the molecules having one particular DNA segment. Each such molecule is called a *probe,* and each probe is generally on the order of 20 nucleotides long. Roughly 100 million molecules of a given probe are loaded onto one zone.

The development of automated machines for synthesizing any desired nucleotide sequence has made possible the preparation of an unlimited variety of probes. Which sequences are actually desired is learned from genome projects. The sequences are those of either a healthy or a mutant human gene or are those of the genes of any microorganism chosen. The probe chosen, for example, might be a telltale sequence of nucleotide units belonging to part of one gene, such as a sequence signifying a potential cancer danger.

The probe molecules become attached to the surface of their assigned zone. Then another zone is exposed and given a batch of identical molecules but of another DNA probe, one with a different sequence of nucleotides. (The methods of attaching the probes to the plates differ with the various corporations; we'll not describe any details.) Zone by zone, different sets of probe molecules, each set with its unique DNA sequence, are loaded onto the chip until all zones are filled. One chip might focus entirely on breast cancer in all of its mutant varieties; another chip, on HIV; and so on. A chip might, by using all of its zones, carry one single gene in its entirety, short segment by short segment, one in each zone. (The Human Genome Project is rapidly providing the nucleotide sequences of human genes.)

When a sample of genetic material, say, from the throat swab, is to be tested, its DNA molecules are chemically joined to a fluorescent marker molecule. The chip is then treated with a solution of the test material, and base pairing—A to T and G to C—leads gene molecules to "matching" zones (see Figure 12a). Which zone has accepted test material shows up by intense fluorescence. Sometimes, couplings happen between test

chemicals in our world

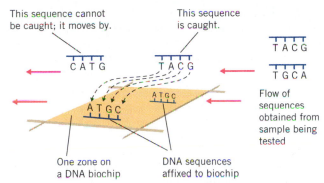

This sequence cannot be caught; it moves by.

This sequence is caught.

C A T G

T A C G

T A C G

T G C A

Flow of sequences obtained from sample being tested

A T G C

A T G C

One zone on a DNA biochip

DNA sequences affixed to biochip

Figure 12a DNA sequences affixed to one zone of a DNA biochip are able to "catch" matching or complementary sequences in a test solution as it flows over the biochip.

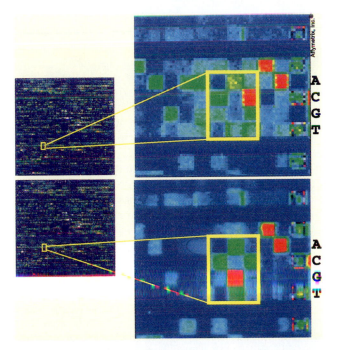

Figure 12b GeneChip® p53 Probe Array Image. The top image shows the pattern for a normal *p53* gene. The The bottom image is that of a mutant *p53* gene. The area of the mutation is enlarged. (Source, Affymetrix, Inc., Santa Clara, CA.)

material molecules and those of a probe that do not exactly match. The binding is then not as tight, and the fluorescence is less intense.

Biochips, Breast Cancer, and the No-Name Protein. When something goes chemically haywire with a cell's DNA and the cell cannot repair the damage, it's supposed to self-destruct. (Cell suicide is called *apoptosis,* and it involves several steps.) By cellular self-destruction, the dangers of the proliferation of such a cell into a lethal tumor are avoided.

What the cell mobilizes when its DNA has been damaged are extra quantities of a protein not yet fully named, the p53 protein ("p" for protein and "53" for its formula mass of 53,000). The gene that directs its synthesis is the *p53* gene (note the italics, which indicate a gene). The protein, p53, activates another gene that makes a protein that arrests the cell cycle so that DNA repair might occur. If this fails, the cell dies.

It has been estimated that about 6.5 million of the cancer cases that arise each year throughout the world involve tumors having a mutant form of the *p53* gene. Roughly half of all lung, colon, and rectal cancers in the United States as well as about 40% of lymphomas, stomach, and pancreatic cancers involve *p53* mutations. In many human tumors, all that is wrong with the molecules of the p53 protein is that they are improperly folded!

Mutations of the *p53* gene also occur in breast cancer. Certain of them cause cancerous breast cells to be resistant to a widely used chemotherapeutic agent, doxorubicin. Biochip technology applied to an individual's *p53* gene (Figure 12b) will thus enable a breast cancer specialist to select medications (or radiation therapy) according to that individual's situation.

Questions

1. What kind of chemistry makes it possible for the DNA of materials being tested to lodge at only one kind of probe molecule?

2. What is normally the connection between damage to a cell and the p53 protein?

3. From the perspective of society as a whole, why are mutations of the *p53* gene so threatening?

Review of Mathematics

A.1 Exponentials

Very large and very small numbers are often expressed as powers of ten. This is often called **exponential notation.** When the quantity is expressed as a number between 1 and 10 multiplied by 10 raised to a power (e.g., 3.2×10^5), we also call it **scientific notation.** Some examples are given in Table A.1.

If you have a scientific calculator, it almost certainly is able to express numbers in scientific notation and perform arithmetic on numbers expressed this way. In fact, scientific calculators are discussed later in this Appendix. However, even though you have a calculator that works with numbers in scientific notation, it is important that you understand how to write these numbers and how to perform arithmetic with them. If you are unsure of yourself, work through the following review.

Table A.1

Number	Exponential Form
1	1×10^0
10	1×10^1
100	1×10^2
1000	1×10^3
10,000	1×10^4
100,000	1×10^5
1,000,000	1×10^6
0.1	1×10^{-1}
0.01	1×10^{-2}
0.001	1×10^{-3}
0.0001	1×10^{-4}
0.00001	1×10^{-5}
0.000001	1×10^{-6}
0.0000001	1×10^{-7}

Positive Exponents Suppose you come across a number such as 6.4×10^4. This is just an alternative way of writing 64,000. In other words, the number 6.4 is multiplied by 10 four times.

$$6.4 \times 10^4 = 6.4 \times 10 \times 10 \times 10 \times 10 = 64,000$$

Instead of actually writing out all of these 10s, notice that the decimal point is simply moved four places to the right in going from 6.4×10^4 to 64,000. Here are a few other examples. Study them to see how the decimal point changes.

$$53.476 \times 10^2 = 5347.6$$

$$0.0016 \times 10^5 = 160$$

$$0.000056 \times 10^3 = 0.056$$

In some calculations it is helpful to reexpress large numbers in scientific notation. To do this, count the number of places you would have to move the decimal point to the left to put it just after the first digit in the number. For example, you would have to move the decimal point four places left in the following number.

$$\underset{4\ \ 3\ \ 2\ \ 1}{6\,0\,5\,3\,0}$$

Therefore, we can rewrite 60530 as 6.0530×10^4. Thus the number of moves equals the exponent.

Sometimes it is necessary to change a number from one exponential form to another. For example, suppose we wish to reexpress the number 6.0530×10^4 as something times 10^2.

$$6.0530 \times 10^4 = \boxed{} \times 10^2$$

What number goes in the box? Although there are rules that you could learn, it is better to think it through as follows. The number 6.0530×10^4 is a product of two parts, 6.0530 and 10^4. Together they equal 60530. We want to change the exponential part to 10^2, which means the exponential part of the product is becoming *smaller*. The only way the product can remain equal to 60530 is if the decimal part becomes *larger*. Since we are decreasing the exponential part by a factor of 100 (i.e., 10^2), we must increase the decimal part by a factor of 100, which means we change it from 6.0530 to 605.30. Therefore,

$$6.0530 \times 10^4 = \boxed{605.30} \times 10^2$$

Negative Exponents A number such as 1.4×10^{-4} is an alternative way of writing

$$\frac{1.4}{10 \times 10 \times 10 \times 10}$$

A-1

Thus a negative exponent tells us how many times to divide by ten. Dividing by ten, of course, can be done by simply moving the decimal point. Thus, 1.4×10^{-4} can be reexpressed as 0.00014. Notice that the decimal point is four places to the *left* of the 1. Here are some examples of numbers expressed as negative exponentials and their equivalents.

$$3567.9 \times 10^{-3} = 3.5679$$

$$0.01456 \times 10^{-2} = 0.0001456$$

$$45691 \times 10^{-3} = 45.691$$

To express a number smaller than 1 in scientific notation, we count the number of places the decimal has to be moved to put it to the right of the first digit. The number that we count equals the negative exponent. For example, you would have to move the decimal point five places to the right to put it to the right of the first digit in

$$0.000068901$$

This number can be reexpressed as 6.8901×10^{-5}.

Multiplying Numbers Written in Scientific Notation The real value of exponential forms of numbers comes in carrying out multiplications and divisions of large and small numbers. Suppose we want to multiply 1000 by 10000. Doing this by a long process fills the page with zeros (and some 1s). But notice the following relationships.

Numbers	1000	×	10000		
	$[10 \times 10 \times 10]$	×	$[10 \times 10 \times 10 \times 10]$	=	10,000,000
Exponential form	10^3	×	10^4	=	10^7
Exponents	3	+	4	=	7

If we *add* 3 and 4, the exponents of the exponential forms of 1000 and 10000, we get the exponent for the answer. Suppose we want to multiply 4160 by 20000. If we reexpress each number in scientific notation we have

$$(4.16 \times 10^3) \times (2 \times 10^4)$$

This now simplifies to

$$(4.16) \times (2) \times [10^3 \times 10^4]$$

The answer can be easily worked without losing the location of the decimal point. The exponents are added to give the exponent on 10 in the answer, and the numbers 4.16 and 2 are multiplied. The product is 8.32×10^7. Thus there are 2 steps in multiplying numbers that are written in scientific notation.

1. Multiply the decimal parts of the numbers (the parts that precede the 10s).

2. Add the exponents of the 10s algebraically to obtain the exponent on the 10 in the answer.

Dividing Numbers Written in Scientific Notation Suppose we want to divide 10,000,000 by 1000. Doing this by long division would be primitive to say the least. Notice the following relationships.

$$\frac{10,000,000}{1000} = \frac{10^7}{10^3} = 10^4 = 10,000$$

The exponent in the result is $(7 - 3)$. So, to divide numbers in exponential form, we subtract exponents.

To extend this operation one step, let's divide 0.00468 by 0.0000400. The answer can be obtained as follows, expressing the numbers in scientific notation.

$$\frac{4.68 \times 10^{-3}}{4.00 \times 10^{-5}} = \frac{4.68}{4.00} \times 10^{[(-3)-(-5)]}$$

$$= 1.17 \times 10^2$$

Notice that the division proceeds in two steps.

1. Carry out the division of the numbers standing before the 10s.

2. Algebraically subtract the exponent in the denominator from the exponent in the numerator.

The Pocket Calculator and Exponentials In scientific work the concept of an exponential is very important because it is much easier to handle either very large or very small numbers in exponential form. The review we've provided in this appendix is intended to help you refresh your memory from secondary school about these kinds of numbers. As noted earlier, almost any scientific calculator gives you the ability to multiply and divide large or small numbers when they are reexpressed as exponentials. We encourage you to study the Instruction Manual for your pocket calculator and read Section A.4 of this appendix so that you can do these kinds of calculations easily. However, we caution you that doing this without a basic understanding of exponentials will leave many traps in your path to skill in performing chemistry calculations.

A.2 Logarithms

A **logarithm** is simply an exponent. For example, if we have an expression

$$N = a^b$$

we see that to obtain N we must raise a to the b power. The exponent b is said to be the logarithm of N; it is the exponent to which we must raise a, which we call the *base*, to obtain the number N.

$$\log_a N = b$$

We can define a logarithm system for any base we want. For example, if we choose the base 2, then

$$\log_2 8 = 3$$

which means that to obtain the number 8, we must raise the base (2) to the third power.

$$8 = 2^3$$

The most frequently encountered logarithm systems in the sciences are *common logarithms* and *natural logarithms,* which are discussed separately below. However, for any system of logarithms there are some useful relationships that evolve from the behavior of exponents in arithmetic operations. We encounter these at various times throughout the sciences, so it is important that you are familiar with them.

For Multiplication:

If $N = A \times B$

$$\log N = \log A + \log B$$

For Division:

If $N = \dfrac{A}{B}$

$$\log N = \log A - \log B$$

For Exponentials:

If $N = A^b$

$$\log N = b \log A$$

Common Logarithms The **common logarithm** or **log** of a number N is the exponent to which 10 must be raised to equal the number N. Thus, if

$$N = 10^x$$

then the common logarithm of N, or **log N,** is simply x.

$$\log N = x$$

As mentioned earlier, there is another kind of logarithm called a **natural logarithm** that we'll take up later. However, when we use the terms "logarithm" or "log N" without specifying the base, we always mean common logarithms related to the base 10.

Some examples of common logarithms are as follows.

$N = 10$	$\log N = 1$	because	$10 = 10^1$
$= 100$	$= 2$		$100 = 10^2$
$= 1000$	$= 3$		$1000 = 10^3$
$= 0.1$	$= -1$		$0.1 = 10^{-1}$
$= 0.00001$	$= -5$		$0.00001 = 10^{-5}$

The logarithms of these numbers were easy to figure out by just using the definition of logarithms, but usually it isn't this simple. For example, what is the value of x in the following equations?

$$4.23 = 10^x$$

$$\log 4.23 = x$$

In other words, to what power x must we raise 10 to obtain 4.23? Obviously, the exponent is not a whole number. Therefore, to determine the value of x we have to use a table of logarithms or a hand-held calculator with the capability of using logs. If you are using a calculator for calculations involving logarithms, you should refer to the instruction booklet for directions.

Antilogarithms There are times when we know what the logarithm of a number is and we seek to find the number. The procedure is called taking the **antilogarithm** or **antilog.** Thus, if

$$N = 10^x$$

and

$$\log N = x$$

then,

$$N = \text{antilog of } x$$

Finding an antilog is also discussed in Section A.4.

Natural Logarithms In the sciences there are many phenomena in which observed quantities are related to each other logarithmically, which means that there is an exponential relationship between them. However, this exponential relationship does not involve 10 raised to a power. Instead it involves powers of a number symbolized by the letter e, which is said to be the base of the system of natural logarithms.

The quantity e is an irrational number, just like another quantity you are familiar with, π. This means that e is a nonrepeating decimal. To eight significant figures its value is

$$e = 2.7182818 \ldots$$

To truly appreciate the significance of this number and the system of logarithms based on it, you must have a good foundation in calculus. A discussion of calculus is beyond the scope of this text, so we will not attempt to explain the origin of the value of e further. Nevertheless, we can still use the results and discuss the system of **natural logarithms** that have the number e as its base.

Natural logarithms use the symbol **ln** instead of log. Thus, if the number N equals e^x,

$$N = e^x$$

then

$$\log_e N = \ln N = x$$

Scientific calculators have the ability to calculate natural logarithms and their corresponding antilogarithms. This is discussed in Section A.4 of this appendix. Sometimes it is useful to know the simple relationship between common and natural logarithms. To four significant figures,

$$\ln N = 2.303 \log N$$

In other words, to find $\ln N$, simply find the common log of N and multiply it by 2.303.

A.3 Graphing

One of the most useful ways of describing the relationship between two quantities is by means of a graph. It allows us to obtain an overall view of how one quantity changes when the other changes. Constructing graphs from experimental data, or from data calculated from an equation, is therefore a common operation in the sciences, so it is important that you understand how to present data graphically. Let's look at an example.

Suppose that we want to know how the volume of a given amount of gas varies as we change its pressure. In Chapter 10 we find an equation that gives the pressure-volume relationship for a fixed amount of gas at a constant temperature,

$$PV = \text{constant} \qquad (A.1)$$

where P = pressure and V = volume.

Suppose, now, that we want to see graphically how the pressure and volume of a gas are related as the pressure is raised from 1.0 to 10.0 atm in steps of 1.0 atm. Let's also suppose that for this sample of gas the value of the constant in our equation is 0.25 L atm. To construct the graph we first must decide how to label and number the axes.

Usually, the horizontal axis, called the **abscissa,** is chosen to correspond to the **independent variable**—the variable whose values are chosen first and from which the values of the **dependent variable** are determined. In our example, we are choosing values of pressure (1.0 atm, 2.0 atm, etc.), so we will label the abscissa "pressure." We will also mark off the axis evenly from 1.0 to 10.0 atm, as shown in Figure A.1. Next we put the label "volume" on the vertical axis, called the **ordinate.** Before we can number this axis, however, we need to know over what range the volume will vary. Using Equation A.1 we can calculate the data in Table A.2, and we see that the volume ranges from a low of 0.025 L to a high of 0.25 L. We can therefore mark off the ordinate evenly in increments of 0.025 L, starting at the bottom with 0.025 L. Notice that in labeling the axes we have indicated the units of P and V in parentheses.

Table A.2

Pressure (atm)	Volume (L)
1.0	0.25
2.0	0.125
3.0	0.083
4.0	0.062
5.0	0.050
6.0	0.042
7.0	0.036
8.0	0.031
9.0	0.028
10.0	0.025

Next we plot the data as shown in Figure A.2. To clearly show each plotted point, we use a small dot that has its center located at the coordinates of the point. Then we draw a *smooth* curve through the points.

Sometimes, when plotting experimentally measured data, the points do not fall exactly on a smooth line (Figure A.3). However, nature generally is not irregular even though that's what the data appear to suggest. The fluctuations are usually due to experimental error of some sort. Therefore, rather than draw an irregular line connecting all the data points (the dashed line), we draw a smooth curve that passes as close as possible to all of the points, even though it may not actually pass through any of them.

Slope One of the properties of curves and lines on a graph is their *slope* or steepness. Consider Figure A.4, which is a straight line drawn on a set of *xy* coordinate axes. The slope of the line is defined as the change in y, Δy, divided by the change in x, Δx.

$$\text{slope} = \frac{\Delta y}{\Delta x} = \frac{y_2 - y_1}{x_2 - x_1}$$

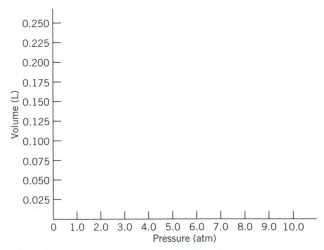

Figure A.1 Choosing and labeling the axes of a graph.

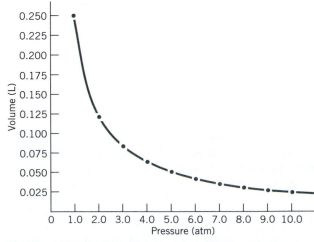

Figure A.2 Plotting points and drawing the curve.

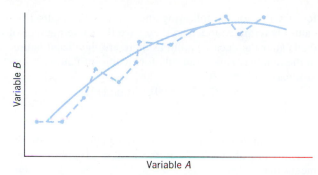

Figure A.3 Experimental data do not all fall on a smooth curve when they are plotted because they contain experimental error. A smooth curve is drawn as close to all of the data points as possible, rather than the jerky dashed line.

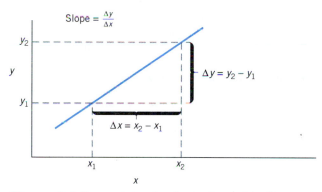

Figure A.4 Determining the slope of a straight line.

Curved lines have slopes too, but the slope changes from point to point on the curve. To obtain the slope graphically, we can draw a line that is tangent to the curve at the point where we want to know the slope. The slope of this tangent is the same as the slope of the curve at this point. In Figure A.5 we see a curve for which the slope is determined at two different points. At point M the curve is rising steeply, and $\Delta y/\Delta x$ for the tangent is large. At point N the curve is rising less steeply, and $\Delta y/\Delta x$ for the tangent at N is small. Therefore, the slope at M is larger than the slope at N.

A.4 Electronic Calculators

Performing arithmetic calculations is commonplace in all of the sciences, and electronic calculators have made these operations simple and routine. So-called **scientific calculators** are inexpensive (most cost less than this book) and are able to perform complex functions at the touch of a button. But these electronic wizards are only as versatile as the person who presses the keys.

This section is not intended to teach you how to use your particular calculator—each calculator comes with its own instruction booklet. Instead, our goal is to point out common errors that are easy to make with some kinds of calculators, to explain how to speed up certain types of cal-

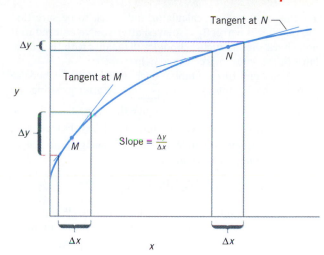

Figure A.5 For a given Δx, Δy is much larger at M than it is at N. Therefore, the slope at M is larger than at N.

culations, and to explain how certain kinds of calculations are done on the calculator.

General Operation Most scientific calculators use an *algebraic entry system,* in which keys are pressed in the order that the quantities and mathematical operations appear in the computation. Thus

$$2 \times 3 = 6$$

is entered by pressing keys in this order: 2, ×, 3, =. We will indicate this by using boxes to represent the keys.

$$\boxed{2}\ \boxed{\times}\ \boxed{3}\ \boxed{=}$$

When the keys are pressed in this sequence, the answer, 6, appears on the display. In explaining how to use the calculator for certain calculations, we will assume an algebraic entry system.

Scientific Notation Earlier in this appendix and in Section 1.5 we saw that numbers expressed as

$$2.3 \times 10^5$$

$$6.7 \times 10^{-12}$$

are said to be in **exponential notation** or **scientific notation.** Scientific calculators are able to do arithmetic with numbers such as these, provided they are entered correctly.

Many calculators have a key labeled EXP or EE that is used to enter the exponential portion of numbers that are expressed in scientific notation. If we read the number 2.3×10^5 as "two point three times ten to the fifth," the EXP (or EE) key should be read as ". . . times ten to the. . . ." Thus the sequence of keys that enter the number 2.3×10^5 is

$$\boxed{2}\ \boxed{.}\ \boxed{3}\ \boxed{\text{EXP}}\ \boxed{5}$$

Try this on your own calculator and check to be sure that the number is correctly displayed after you've entered it. If not, check the instruction booklet that came with the calculator or ask your instructor for help.

When entering a number such as 6.7×10^{-12}, be sure to press the "change sign" or $\boxed{+/-}$ key after pressing the $\boxed{\text{EXP}}$ key. Calculators will not give the correct entry if the $\boxed{-}$ key is pressed instead of the $\boxed{+/-}$ key. For eample, clear the display and then try the following sequence of keystrokes on your calculator.

$$\boxed{6}\,\boxed{.}\,\boxed{7}\,\boxed{\text{EXP}}\,\boxed{+/-}\,\boxed{1}\,\boxed{2}$$

If you wish to multiply 2.3×10^5 and 6.7×10^{-12}, enter the first one, press the $\boxed{\times}$ key, then enter the second number, and finally press the $\boxed{=}$ key. The answer displayed on the calculator should be 1.541×10^{-6}.

$$(2.3 \times 10^5) \times (6.7 \times 10^{-12}) = 1.541 \times 10^{-6}$$

If your display does not show an exponential, your calculator's instruction manual will tell you how to proceed.

Chain Calculations It is not uncommon to have to calculate the value of a fraction in which there is a string of numbers that must be multiplied together in the numerator and another string of numbers to be multiplied together in the denominator. For example,

$$\frac{5.0 \times 7.3 \times 8.5}{6.2 \times 2.5 \times 3.9}$$

Many students will compute the value of the numerator and write down the answer, then compute the value of the denominator and write it down, and finally divide the value of the numerator by the value of the denominator (whew!). Although this gives the correct answer, there is a simpler way to do the arithmetic that doesn't require writing down any intermediate values. The procedure is as follows:

1. Enter the first value from the numerator (5.0).

2. Any of the other values from the numerator are entered by *multiplication* and any of the values from the denominator are entered by *division*.

The following is one sequence that gives the answer.

$$5.0 \times 7.3 \times 8.5 \div 6.2 \div 2.5 \div 3.9 = 5.13234 \ldots$$

Notice that each value from the denominator is entered by division. It also doesn't matter in what sequence the numbers are entered. The following gives the same answer.

$$5.0 \div 6.2 \times 7.3 \div 2.5 \div 3.9 \times 8.5 = 5.13234 \ldots$$

Logarithms and Antilogarithms Your scientific calculator works with both common logarithms and natural logarithms. To find the common log of a number, enter the number in the calculator and press the $\boxed{\log}$ or $\boxed{\text{LOG}}$ but-

ton. Try this for the following and check your results. (The values here are rounded so they have the same number of digits after the decimal point as there are significant figures in the number. This is the rule for significant figures in logarithms.)

$$\log 12.35 = 1.0917 \text{ (rounded)}$$

$$\log (3.70 \times 10^{-4}) = -3.432 \text{ (rounded)}$$

Obtaining the antilogarithm is equally easy. For the first number above, 12.35 is the antilog of 1.0917, which means that

$$12.35 = 10^{1.0917}$$

Some calculators have a key labeled $\boxed{10^x}$; others use a combination of an inverse function key, usually labeled $\boxed{\text{INV}}$, and the key used to obtain the logarithm. Thus, entering the value 1.0917 and pressing either $\boxed{10^x}$ or the sequence $\boxed{\text{INV}}\,\boxed{\log}$ yields the value 12.35. Try it.

To obtain natural logarithms of a number we use a key labeled either $\boxed{\ln x}$ or $\boxed{\text{LN}}$ or sometimes $\boxed{\ln}$. Thus the natural logarithm of 12.35 is obtained by entering the number and pressing the $\boxed{\ln x}$ key. Try it.

$$\ln 12.35 = 2.5137 \text{ (rounded)}$$

The value 12.35 is the antiln of 2.5137, which means

$$12.35 = e^{2.5137}$$

Therefore, to obtain the antiln of 2.5137, we enter the number into the calculator and press either the key labeled $\boxed{e^x}$ or the sequence $\boxed{\text{INV}}\,\boxed{\ln x}$. Try it.

$$\text{antiln of } 2.5137 = 12.35$$

Powers and Roots Squares and square roots are handled easily with keys labeled $\boxed{x^2}$ and $\boxed{\sqrt{x}}$. For higher powers and higher roots (e.g., cube roots or fourth roots) we use the key that on most scientific calculators is labeled $\boxed{x^y}$. To use this function, we must enter two values—the number and its exponent. For example, to raise the number 2.6 to the 5th power,

$$(2.6)^5$$

we enter the number 2.6, press the $\boxed{x^y}$ key, enter the value of the exponent (5), and then press either the "equals" key, or the next function key desired if we are performing a chain calculation. Thus, the sequence

$$\boxed{2}\,\boxed{.}\,\boxed{6}\,\boxed{x^y}\,\boxed{5}\,\boxed{=}$$

produces the result 118.81376 on the display of the calculator. Therefore,

$$(2.6)^5 = 118.81376$$

To take a root, the procedure is quite similar, because we make use of the relationship

$$\sqrt[n]{(\text{number})} = (\text{number})^{1/n}$$

Thus, the cube root of a number such as 56.4 is obtained as

$$\sqrt[3]{56.4} = (56.4)^{1/3} = (56.4)^{0.333333} \ldots$$

In other words, we raise 56.4 to the one-third power, for which the decimal equivalent is 0.3333333 . . . (with as many threes as will fit on the display). Let's try it. Clear the display and enter the number 56.4. Next, press the $\boxed{x^y}$ key.

Then enter 0.33333 with as many threes after the decimal point as you can. Finally, press the "equals" key. The answer should be 3.834949. . . . If we had wanted to take the 5th root of 56.4, we would have raised the number to the 1/5 power, or 0.20. Try it. You should get 2.24004 . . . as the answer. (Your calculator may handle x^y functions differently. If so, consult the manual.)

Electron Configurations of the Elements

Atomic Number			Atomic Number			Atomic Number		
1	H	$1s^1$	31	Ga	$[Ar]\,4s^2\,3d^{10}\,4p^1$	61	Pm	$[Xe]\,6s^2\,4f^5$
2	He	$1s^2$	32	Ge	$[Ar]\,4s^2\,3d^{10}\,4p^2$	62	Sm	$[Xe]\,6s^2\,4f^6$
3	Li	$[He]\,2s^1$	33	As	$[Ar]\,4s^2\,3d^{10}\,4p^3$	63	Eu	$[Xe]\,6s^2\,4f^7$
4	Be	$[He]\,2s^2$	34	Se	$[Ar]\,4s^2\,3d^{10}\,4p^4$	64	Gd	$[Xe]\,6s^2\,4f^7\,5d^1$
5	B	$[He]\,2s^2\,2p^1$	35	Br	$[Ar]\,4s^2\,3d^{10}\,4p^5$	65	Tb	$[Xe]\,6s^2\,4f^9$
6	C	$[He]\,2s^2\,2p^2$	36	Kr	$[Ar]\,4s^2\,3d^{10}\,4p^6$	66	Dy	$[Xe]\,6s^2\,4f^{10}$
7	N	$[He]\,2s^2\,2p^3$	37	Rb	$[Kr]\,5s^1$	67	Ho	$[Xe]\,6s^2\,4f^{11}$
8	O	$[He]\,2s^2\,2p^4$	38	Sr	$[Kr]\,5s^2$	68	Er	$[Xe]\,6s^2\,4f^{12}$
9	F	$[He]\,2s^2\,2p^5$	39	Y	$[Kr]\,5s^2\,4d^1$	69	Tm	$[Xe]\,6s^2\,4f^{13}$
10	Ne	$[He]\,2s^2\,2p^6$	40	Zr	$[Kr]\,5s^2\,4d^2$	70	Yb	$[Xe]\,6s^2\,4f^{14}$
11	Na	$[Ne]\,3s^1$	41	Nb	$[Kr]\,5s^1\,4d^4$	71	Lu	$[Xe]\,6s^2\,4f^{14}\,5d^1$
12	Mg	$[Ne]\,3s^2$	42	Mo	$[Kr]\,5s^1\,4d^5$	72	Hf	$[Xe]\,6s^2\,4f^{14}\,5d^2$
13	Al	$[Ne]\,3s^2\,3p^1$	43	Tc	$[Kr]\,5s^2\,4d^5$	73	Ta	$[Xe]\,6s^2\,4f^{14}\,5d^3$
14	Si	$[Ne]\,3s^2\,3p^2$	44	Ru	$[Kr]\,5s^1\,4d^7$	74	W	$[Xe]\,6s^2\,4f^{14}\,5d^4$
15	P	$[Ne]\,3s^2\,3p^3$	45	Rh	$[Kr]\,5s^1\,4d^8$	75	Re	$[Xe]\,6s^2\,4f^{14}\,5d^5$
16	S	$[Ne]\,3s^2\,3p^4$	46	Pd	$[Kr]\,4d^{10}$	76	Os	$[Xe]\,6s^2\,4f^{14}\,5d^6$
17	Cl	$[Ne]\,3s^2\,3p^5$	47	Ag	$[Kr]\,5s^1\,4d^{10}$	77	Ir	$[Xe]\,6s^2\,4f^{14}\,5d^7$
18	Ar	$[Ne]\,3s^2\,3p^6$	48	Cd	$[Kr]\,5s^2\,4d^{10}$	78	Pt	$[Xe]\,6s^1\,4f^{14}\,5d^9$
19	K	$[Ar]\,4s^1$	49	In	$[Kr]\,5s^2\,4d^{10}\,5p^1$	79	Au	$[Xe]\,6s^1\,4f^{14}\,5d^{10}$
20	Ca	$[Ar]\,4s^2$	50	Sn	$[Kr]\,5s^2\,4d^{10}\,5p^2$	80	Hg	$[Xe]\,6s^2\,4f^{14}\,5d^{10}$
21	Sc	$[Ar]\,4s^2\,3d^1$	51	Sb	$[Kr]\,5s^2\,4d^{10}\,5p^3$	81	Tl	$[Xe]\,6s^2\,4f^{14}\,5d^{10}\,6p^1$
22	Ti	$[Ar]\,4s^2\,3d^2$	52	Te	$[Kr]\,5s^2\,4d^{10}\,5p^4$	82	Pb	$[Xe]\,6s^2\,4f^{14}\,5d^{10}\,6p^2$
23	V	$[Ar]\,4s^2\,3d^3$	53	I	$[Kr]\,5s^2\,4d^{10}\,5p^5$	83	Bi	$[Xe]\,6s^2\,4f^{14}\,5d^{10}\,6p^3$
24	Cr	$[Ar]\,4s^1\,3d^5$	54	Xe	$[Kr]\,5s^2\,4d^{10}\,5p^6$	84	Po	$[Xe]\,6s^2\,4f^{14}\,5d^{10}\,6p^4$
25	Mn	$[Ar]\,4s^2\,3d^5$	55	Cs	$[Xe]\,6s^1$	85	At	$[Xe]\,6s^2\,4f^{14}\,5d^{10}\,6p^5$
26	Fe	$[Ar]\,4s^2\,3d^6$	56	Ba	$[Xe]\,6s^2$	86	Rn	$[Xe]\,6s^2\,4f^{14}\,5d^{10}\,6p^6$
27	Co	$[Ar]\,4s^2\,3d^7$	57	La	$[Xe]\,6s^2\,5d^1$	87	Fr	$[Rn]\,7s^1$
28	Ni	$[Ar]\,4s^2\,3d^8$	58	Ce	$[Xe]\,6s^2\,4f^1\,5d^1$	88	Ra	$[Rn]\,7s^2$
29	Cu	$[Ar]\,4s^1\,3d^{10}$	59	Pr	$[Xe]\,6s^2\,4f^3$	89	Ac	$[Rn]\,7s^2\,6d^1$
30	Zn	$[Ar]\,4s^2\,3d^{10}$	60	Nd	$[Xe]\,6s^2\,4f^4$	90	Th	$[Rn]\,7s^2\,6d^2$

Atomic Number			Atomic Number			Atomic Number		
91	Pa	$[Rn]\, 7s^2\, 5f^2\, 6d^1$	100	Fm	$[Rn]\, 7s^2\, 5f^{12}$	108	Hs	$[Rn]\, 7s^2\, 5f^{14}\, 6d^6$
92	U	$[Rn]\, 7s^2\, 5f^3\, 6d^1$	101	Md	$[Rn]\, 7s^2\, 5f^{13}$	109	Mt	$[Rn]\, 7s^2\, 5f^{14}\, 6d^7$
93	Np	$[Rn]\, 7s^2\, 5f^4\, 6d^1$	102	No	$[Rn]\, 7s^2\, 5f^{14}$	110	Uun	$[Rn]\, 7s^2\, 5f^{14}\, 6d^8$
94	Pu	$[Rn]\, 7s^2\, 5f^6$	103	Lr	$[Rn]\, 7s^2\, 5f^{14}\, 6d^1$	111	Uuu	$[Rn]\, 7s^2\, 5f^{14}\, 6d^9$
95	Am	$[Rn]\, 7s^2\, 5f^7$	104	Rf	$[Rn]\, 7s^2\, 5f^{14}\, 6d^2$	112	Uub	$[Rn]\, 7s^2\, 5f^{14}\, 6d^{10}$
96	Cm	$[Rn]\, 7s^2\, 5f^7\, 6d^1$	105	Db	$[Rn]\, 7s^2\, 5f^{14}\, 6d^3$	114	Uuq	$[Rn]\, 7s^2\, 5f^{14}\, 6d^{10}\, 7p^2$
97	Bk	$[Rn]\, 7s^2\, 5f^9$	106	Sg	$[Rn]\, 7s^2\, 5f^{14}\, 6d^4$	116	Uuh	$[Rn]\, 7s^2\, 5f^{14}\, 6d^{10}\, 7p^4$
98	Cf	$[Rn]\, 7s^2\, 5f^{10}$	107	Bh	$[Rn]\, 7s^2\, 5f^{14}\, 6d^5$	118	Uuo	$[Rn]\, 7s^2\, 5f^{14}\, 6d^{10}\, 7p^6$
99	Es	$[Rn]\, 7s^2\, 5f^{11}$						

Equivalent Weights and Normality

C.1 The Equivalent in Acid–Base Reactions

For certain kinds of reactions, it is possible to define chemical quantities that always combine in a one-to-one ratio, regardless of the coefficients in the balanced equation. One is acid–base reactions and the other is redox reactions.

In an acid–base neutralization or proton-transfer reaction, the quantity of acid that provides one mole of H^+ will *always* neutralize the quantity of base that provides one mole of OH^-. These two always combine in a 1:1 ratio.

$$H^+ + OH^- \longrightarrow H_2O$$

Suppose that 36 g of some acid is able to furnish exactly one mole of H^+, and 40 g of a base is able to furnish exactly 1 mol of OH^-. Because 1 mol of H^+ reacts with 1 mol of OH^-, we know that the 36 g of the acid is just enough to react with the 40 g of base. We can make this statement without even knowing what the acid or base is; we only have to know how much of the acid gives 1 mol H^+ and how much of the base gives 1 mol OH^-.

This reasoning forms the basis of the definition of another kind of chemical quantity called the **equivalent,** abbreviated **eq.** The exact definition depends on the kind of reaction, acid–base or redox but, in both, the *equivalent* is always defined so that equivalents of reactants always react in a one-to-one ratio. This is the key concept to remember. If there are two reactants, *A* and *B*,

> 1 eq of reactant *A* reacts with exactly 1 eq of reactant *B*.

For acids and bases, the definitions of the equivalent that meet this standard are the following.

> **One equivalent of an acid** is the amount of the acid that is able to furnish 1 mol of hydrogen ion.
> **One equivalent of a base** is the amount of the base that is able to neutralize 1 mol of hydrogen ion.

From the chemical equation for the neutralization reaction, we see that these definitions assure a one-to-one combining ratio for equivalents of acids and bases.

C.2 The Equivalent in Redox Reactions

The definitions of equivalents for a redox reaction parallel those for acid–base reactions. They retain the central principle of a 1:1 ratio between an equivalent of an oxidizing agent and an equivalent of a reducing agent. If one particle of an oxidizing agent accepts one electron, something else must donate that electron. Thus the *equivalents* for redox reactions are defined as follows.

> **One equivalent of an oxidizing agent** is the amount of the substance that is able to accept 1 mol of electrons.
> **One equivalent of a reducing agent** is the amount of the substance that is able to donate 1 mol of electrons.

C.3 Equivalent Weights

Let's now apply these principles to real chemicals and see how the equivalent is related to the more familiar chemical unit, the *mole*. The *equivalent* is not an SI unit. Many chemists have moved away from using it. In certain fields, however, including many scientific fields outside of chemistry, the concept of the equivalent is widely employed.

Equivalents of Acids and Bases Consider the acids HCl and H_2SO_4. One mole of HCl is enough acid to supply 1 mol of H^+, so 1 mole of HCl must be the same as 1 eq HCl.

$$1 \text{ mol HCl} \Leftrightarrow 1 \text{ mol } H^+$$

So,

$$1 \text{ mol HCl} = 1 \text{ eq HCl}$$

On the other hand, 1 mol of H_2SO_4 is enough acid to supply 2 mol of H^+, provided the H_2SO_4 is completely neutralized. For complete neutralization, then, 1 mol of H_2SO_4 must be equal to 2 eq H_2SO_4.

$$1 \text{ mol } H_2SO_4 \Leftrightarrow 2 \text{ mol } H^+$$

So,

$$1 \text{ mol } H_2SO_4 = 2 \text{ eq } H_2SO_4 \text{ (for complete reaction)}$$

Thus, there is a simple relationship between moles and equivalents for acids. *The number of equivalents in 1 mol of an acid is equal to the number of hydrogen ions that are neutralized when one molecule of the acid reacts.*

Phosphoric acid, H_3PO_4, is a triprotic acid when its *complete* neutralization entails the reaction of three H^+ ions per molecule. Thus, 1 mol of H_3PO_4 equals 3 equivalents of this acid

$$1 \text{ mol } H_3PO_4 \Leftrightarrow 3 \text{ mol } H^+$$

1 mol H_3PO_4 = 3 eq H_3PO_4 (for complete neutralization)

You can see the pattern, which we can summarize as follows.

If an acid is monoprotic

1 mol of acid = 1 eq of acid

If an acid is diprotic

1 mol of acid = 2 eq of acid

If an acid is triprotic

1 mol of acid = 3 eq of acid

The size of an equivalent does depend on the reaction. We emphasized for the case of phosphoric acid that we were referring to its *complete* neutralization. But suppose that the following reaction were being used.

$$H_3PO_4 + 2NaOH \longrightarrow Na_2HPO_4 + 2H_2O$$

Here the H_3PO_4 molecule is donating only two H^+. In this reaction, we have to write:

$$1 \text{ mol } H_3PO_4 \Leftrightarrow 2 \text{ mol } H^+$$

So for this partial neutralization,

$$1 \text{ mol } H_3PO_4 = 2 \text{ eq } H_3PO_4$$

This must caution us about the use of the concept of the equivalent.

Determining the number of equivalents per mole of base is as simple as for an acid. Consider the two bases NaOH and $Ba(OH)_2$. One mole of NaOH is enough base to supply 1 mol of OH^-, so an equivalent of NaOH must be the same as one mole of NaOH.

$$1 \text{ mol } NaOH \Leftrightarrow 1 \text{ mol } OH^-$$

$$1 \text{ mol } NaOH = 1 \text{ eq } NaOH$$

One mole of $Ba(OH)_2$ is able to supply 2 mol of OH^- when completely neutralized, so 1 mol of $Ba(OH)_2$ must then be 2 eq base.

$$1 \text{ mol } Ba(OH)_2 \Leftrightarrow 2 \text{ mol } OH^-$$

$$1 \text{ mol } Ba(OH)_2 = 2 \text{ eq } Ba(OH)_2$$

Thus, for a metal hydroxide, the number of equivalents in 1 mol of the base is equal to the number of hydroxides in one formula unit of the base. Thus, for $Al(OH)_3$,

$$1 \text{ mol } Al(OH)_3 = 3 \text{ eq } Al(OH)_3$$

When the base is not one that supplies OH^- ions, the same principles apply. For example, one HCO_3^- ion can neutralize one H^+ ion.

$$HCO_3^- + H^+ \longrightarrow CO_2 + H_2O$$

Thus,

$$1 \text{ mol } NaHCO_3 \Leftrightarrow 1 \text{ mol of neutralized } H^+$$

$$1 \text{ mol } NaHCO_3 = 1 \text{ eq } NaHCO_3$$

The carbonate ion, however, can neutralize two H^+ ions.

$$CO_3^{2-} + 2H^+ \longrightarrow CO_2 + H_2O$$

When this is the stoichiometry of the neutralization reaction, then we can write

$$1 \text{ mol } Na_2CO_3 \Leftrightarrow 2 \text{ mol of neutralized } H^+$$

$$1 \text{ mol } Na_2CO_3 = 2 \text{ eq } Na_2CO_3$$

Equivalent Weights of Acids and Bases In working problems using equivalents, it is often useful to know the mass of an equivalent of each of the reactants. Earlier, for example, we described a case in which 36 g of an acid gave 1 mol of H^+ and 40 g of a base gave 1 mol of OH^-. These quantities provide 1 eq each of the acid and the base. Knowing how much an equivalent of each of them weighs establishes a mass relationship between the two reactants that can be used in stoichiometric calculations. Thus, if we had a sample of 20 g of the base (0.50 eq of base), we know that we would only need 18 g of the acid (0.50 eq of acid).

The mass in grams of one equivalent is called the **equivalent weight.**[1] An equivalent weight, in other words, is the number of grams per equivalent, the ratio of the number of grams in a mole to the number of equivalents in a mole.

$$\text{Equivalent weight} = \frac{\text{no. g per mol}}{\text{no. eq per mol}}$$

We know the following two facts about sulfuric acid, for example. First, when it is neutralized completely,

$$1 \text{ mol } H_2SO_4 = 2 \text{ eq } H_2SO_4$$

Second,

$$1 \text{ mol } H_2SO_4 = 98.07 \text{ g } H_2SO_4$$

Therefore, we can write

$$98.07 \text{ g } H_2SO_4 = 2 \text{ eq } H_2SO_4$$

---

[1]Equivalent *weight* is the term almost universally used, although equivalent *mass* would be correct. We will follow here the usual custom.

The equivalent weight, therefore, is found by taking the ratio of grams to equivalents:

$$\frac{98.07 \text{ g } H_2SO_4}{2 \text{ eq } H_2SO_4} = 49.04 \text{ g eq}^{-1} \text{ for } H_2SO_4$$

The same kind of calculations apply to any base.

Equivalents of Oxidizing Agents and Reducing Agents Careful attention to stoichiometry is also essential in calculating the equivalent weights of oxidizing and reducing agents. Consider, for example, the reaction of potassium permanganate as an oxidizing agent. We have learned in Chapter 5 that when a reaction using $KMnO_4$ is carried out in an *acidic* solution, the permanganate ion is reduced to the manganese(II) ion by the following half-reaction.

$$\underset{+7}{MnO_4^-}(aq) + 8H^+(aq) + 5e^- \longrightarrow \underset{+2}{Mn^{2+}}(aq) + 4H_2O$$

Mn goes from the +7 to the +2 state by a gain of five electrons. We do not, in fact, need the half-reaction to know this. All we need is just the change in oxidation numbers themselves to figure out how many electrons are gained (or lost) by some reactant. Because each mole of MnO_4^- gains 5 mol of electrons, each mole of this oxidizing agent must correspond to 5 eq. In other words,

$$1 \text{ mol } KMnO_4 \Leftrightarrow 5 \text{ mol } e^- \quad \text{(acidic medium)}$$

$$1 \text{ mol } KMnO_4 = 5 \text{ eq } KMnO_4 \quad \text{(acidic medium)}$$

In a neutral or slightly basic medium, however, MnO_4^- is reduced to MnO_2.

$$\underset{+7}{MnO_4^-} \longrightarrow \underset{+4}{MnO_2}$$

Now, Mn changes from its +7 oxidation state to the +4 state, which means a three-electron change. Thus a gain of only 3 mol of electrons by each mole of MnO_4^- occurs as MnO_2 forms, so there are 3 eq/mol of $KMnO_4$ when it is used in a neutral or basic medium.

$$1 \text{ mol } KMnO_4 \Leftrightarrow 3 \text{ mol of } e^- \quad \text{(neutral or basic medium)}$$

$$1 \text{ mol } KMnO_4 = 3 \text{ eq } KMnO_4 \quad \text{(neutral or basic medium)}$$

Thus *the number of equivalents per mole of an oxidizing agent equals the number of moles of electrons accepted by 1 mol during a redox reaction.* A similar statement applies to reducing agents. *The number of equivalents per mole of a reducing agent equals the number of moles of electrons donated by 1 mol during a redox reaction.* To determine the equivalents per mole for an oxidizing agent or a reducing agent, we use the oxidation states of the reactants and products.

Consider sodium dichromate, $Na_2Cr_2O_7$, which contains $Cr_2O_7^{2-}$, a strong oxidizing agent. In an acidic medium, the chromium in the dichromate ion changes from a +6 oxidation state to a +3 state.

$$\underset{+6}{Cr_2O_7^{2-}} \longrightarrow \underset{+3}{2Cr^{3+}}$$

This means that *each* Cr in $Cr_2O_7^{2-}$ gains three electrons, but there are two Cr in $Cr_2O_7^{2-}$. So each whole $Cr_2O_7^{2-}$ ion gains six electrons. In other words,

$$1 \text{ mol } Cr_2O_7^{2-} \Leftrightarrow 6 \text{ mol of } e^-$$

$$1 \text{ mol } Na_2Cr_2O_7 = 6 \text{ eq } Na_2Cr_2O_7$$

Thus, we must consider both the change in the oxidation number *and* the number of atoms in the actual reactant that undergo the change.

Consider another example, that of oxalic acid, $H_2C_2O_4$. When it acts as a reducing agent, it becomes oxidized to CO_2.

$$\underset{+3}{H_2C_2O_4} \longrightarrow \underset{+4}{2CO_2}$$

Carbon is oxidized from a +3 state in $H_2C_2O_4$ to a +4 state in CO_2, but *two carbons* in each molecule of the reactant, $H_2C_2O_4$, make this change. It is a $1e^-$ change *per carbon atom* in $H_2C_2O_4$. Each whole molecule of $H_2C_2O_4$ must, therefore, give up two electrons. Therefore,

$$1 \text{ mol } H_2C_2O_4 \Leftrightarrow 2 \text{ mol of } e^-$$

$$1 \text{ mol } H_2C_2O_4 = 2 \text{ eq } H_2C_2O_4$$

The reduction of sodium chromate to chromium (III) oxide provides another example that relates moles to equivalents. The change is

$$\underset{+6}{Na_2CrO_4} \longrightarrow \underset{+3}{Cr_2O_3}$$

Thus each Cr in Na_2CrO_4 changes from the +6 to the +3 state. This corresponds to a three-electron change for each Cr in Na_2CrO_4. We are concerned about equivalents of *reactants,* so only the number of Cr atoms in the *reactant* concerns us, and each one changes by three electrons. Therefore,

$$1 \text{ mol } Na_2CrO_4 \Leftrightarrow 3 \text{ mol of } e^-$$

$$1 \text{ mol } Na_2CrO_4 = 3 \text{ eq } Na_2CrO_4$$

Equivalent Weights of Oxidizing Agents and Reducing Agents Calculating the equivalent weight of an oxidizing agent or a reducing agent is exactly like that for acids and bases. For $Na_2Cr_2O_7$, for example, we know the following.

$$1 \text{ mol } Na_2Cr_2O_7 = 6 \text{ eq } Na_2Cr_2O_7$$

$$1 \text{ mol } Na_2Cr_2O_7 = 261.97 \text{ g } Na_2Cr_2O_7$$

Therefore,

$$261.97 \text{ g } Na_2Cr_2O_7 = 6 \text{ eq } Na_2Cr_2O_7$$

Because *an equivalent weight is simply the ratio of grams per equivalent,* we set up the ratio and carry through the calculation.

$$\frac{261.97 \text{ g Na}_2\text{Cr}_2\text{O}_7}{6 \text{ eq Na}_2\text{Cr}_2\text{O}_7} = 43.66 \text{ g eq}^{-1} \text{ for Na}_2\text{Cr}_2\text{O}_7$$

Retaining four significant figures, the equivalent weight of $Na_2Cr_2O_7$ is 43.66 g; 43.66 g of $Na_2Cr_2O_7$ takes up 1 mol of electrons in any redox reaction in which Cr changes from the +6 to the +3 state.

You must always remember that the number of equivalents per mole varies not only from compound to compound but also according to the experimental conditions (acidic, basic or neutral solution) and the nature of the other reactant. The equivalent weight of a compound is not a constant under all conditions.

C.4 Normality

In working with equivalents when the reactants are in solution, it is convenient to define a quantity similar to molarity, but in terms of equivalents instead of moles. This quantity is called the *normality* of a solution. The **normality** of a solution is a concentration unit that expresses the number of *equivalents* of solute per liter of solution.

$$\text{Normality} = \frac{\text{equivalents of solute}}{1 \text{ L of solution}}$$

Thus, a solution that contains one equivalent of solute per liter of solution has a concentration that we specify as "one normal," abbreviated 1 N. Like molarity, it provides a conversion factor that relates equivalents and volume. For example, a solution labeled 2.00 N H_2SO_4 contains 2.00 eq of H_2SO_4 per liter (1000 mL), and provides the two conversion factors

$$\frac{2.00 \text{ eq H}_2\text{SO}_4}{1 \text{ L soln}} \quad \text{and} \quad \frac{1 \text{ L soln}}{2.00 \text{ eq H}_2\text{SO}_4}$$

One of the kinds of problems that often comes up in the lab when a stock solution's concentration is given in the units of molarity is to convert this to the units of normality. There is a useful rule of thumb for this calculation.

> The normality of a solution equals its molarity multiplied by the number of equivalents per mole.
>
> normality = molarity × no. of eq per mol

This holds true both for acid–base systems and redox systems.

Titrations Using Normality as a Concentration Unit One of the benefits of defining equivalents and normality is that it makes calculations for titrations very simple. Consider the units that we obtain if we multiply the normality of a solution by the volume used in a particular experiment.

$$\underbrace{\frac{\text{eq solute}}{\text{L soln}}}_{N} \times \underbrace{\text{L soln}}_{V} = \text{eq solute}$$

Thus, normality (N) times volume in liters (V) gives equivalents. In any reaction, the number of equivalents of one reactant exactly equals the number of equivalents of another; it's the way equivalents are defined. So we can write the following relationship that applies to any reaction between two substances, A and B, in which the concentrations are in eq/L, the units of normality.

$$N_A V_A = N_B V_B \qquad \text{(C.1)}[2]$$

Equation C.1 applies both to acid–base reactions and to redox reactions. This is the beauty of the concept of equivalents. *By definition*, we can rely on the 1:1 ratio of equivalents that makes Equation C.1 possible, regardless of the system.

---

[2]Don't try to use this equation when the concentrations are in units of molarity. It won't necessarily work because we cannot say that there is *always* a 1:1 ratio of the *moles* of one reactant to the *moles* of another.

Answers to Practice Exercises and Selected Review Problems

Chapter 1

Practice Exercises

1. (a) mg (b) μm (c) ps
(a) 1×10^{-9} m (b) 1×10^{-2} m (c) 1×10^{-12} m
(a) cg (b) Mm (c) ns

2. 50 °F = 10 °C, 68 °F = 293 K

3. (a) 108 in. (b) 1.25×10^5 cm (c) 0.0107 ft
(d) 8.58 km/L

4. 2.70 g/mL

5. (a) 0.272 cm^3 (b) 171 g

6. specific gravity = 2.70, density = 169 lb/ft^3

7. density = 0.902 g/mL, 7.53 lb/gal

Review Problems

1.43 (a) 1×10^{-2} m = 0.01 m (b) 1×10^3 m = 1000 m
(c) 1×10^{12} pm (d) 1×10^{-1} m = 0.1 m
(e) 1×10^{-3} kg = 0.001 kg (f) 1×10^{-2} g = 0.01 g

1.45 (a) 140 °F (b) 68 °F (c) 7.5 °C (d) 15 °C
(e) 313 K (f) 253 K

1.47 99.43 °F; person has a fever.

1.49 -269 °C, -452 °F

1.51 77 K

1.53 (a) 4 (b) 5 (c) 4 (d) 2 (e) 4 (f) 1

1.55 (a) 0.69 (b) 83.24 (c) 6×10^{-3} (d) 22.84 (e) 775.4

1.57 (a) 2.34×10^3 (b) 3.10×10^7 (c) 2.87×10^{-4}
(d) 4.50×10^4 (e) 4.00×10^{-6} (f) 3.24×10^5

1.59 (a) 210,000 (b) 0.00000335 (5 zeros after decimal point) (c) 3,800 (d) 0.0000000000046 (11 zeros after decimal point) (e) 0.00000346 (5 zeros after decimal point)
(f) 850,000,000

1.61 (a) 2.0×10^4 (b) 8.0×10^7 (c) 1.0×10^3
(d) 2.4×10^5 (e) 2.0×10^{18}

1.63 (a) 3.20×10^{-3} km (b) 8.2×10^3 μg (c) 7.53×10^{-5} kg
(d) 0.1375 L (e) 25 mL (f) 3.42×10^{-9} dm

1.65 (a) 91 cm (b) 2.3 kg (c) 2.8×10^3 mL or 2800 mL
(d) 200 mL (e) 88 km/hr (f) 80.4 km

1.67 360 mL

1.69 2205 lb

1.71 188 cm (rounds to 190 cm)

1.73 (a) 5.2 m^2 (b) 3.1×10^3 mm^2 or 3100 mm^2
(c) 1.0×10^2 L

1.75 2452 km/hr

1.77 0.798 g/mL

1.79 31.6 mL

1.81 276 g

1.83 11 g/mL

1.85 0.715

1.87 1470 g

1.89 1.20×10^3 lb

Additional Exercises

1.93 (a) 9.556 mL (b) 1.477 g/mL

1.95 345 francs

1.97 1.331×10^5 ft^3

1.101 Yes, the unknown liquid is ethylene glycol.

1.104 1.2 mi/hr

Chapter 2

Practice Exercises

1. (a) 1 N, 2 Cl (b) 1 Fe, 1 S, 4 O (c) 3 Ca, 2 P, 8 O
(d) 1 Co, 2 N, 12 O, 12 H

2. left side: 1 Mg, 2 O, 4 H, 2 Cl; right side: the same

3. $Mg(OH)_2(s) + 2HCl(aq) \rightarrow MgCl_2(aq) + 2H_2O(l)$

4. 26.9814 u

5. 5.2955 times heavier

6. 10.8 u

7. $^{240}_{94}Pu$

8. 17 protons, 18 neutrons, 17 electrons

9. (a) K, Ar, Al (b) Cl (c) Ba (d) Ne (e) Li (f) Ce

10. (a) NaF (b) Na$_2$O (c) MgF$_2$ (d) Al$_4$C$_3$

11. (a) Cr^{2+} : CrCl$_2$ and CrO; Cr^{3+} : CrCl$_3$ and Cr$_2$O$_3$
(b) Cu^+ : CuCl and Cu$_2$O; Cu^{2+} : CuCl$_2$ and CuO

12. (a) Na_2CO_3 (b) $(NH_4)_2SO_4$ (c) $KC_2H_3O_2$
(d) $Sr(NO_3)_2$ (e) $Fe(C_2H_3O_2)_3$

13. potassium sulfide, magnesium phosphide, nickel(II) chloride, iron(III) oxide

14. (a) Al_2S_3 (b) SrF_2 (c) TiO_2 (d) $CrBr_2$

15. phosphorus trichloride, sulfur dioxide, dichlorine heptaoxide

16. $HF(aq)$: hydrofluoric acid;
   $HBr(aq)$: hydrobromic acid.
   NaF: sodium fluoride;
   NaBr: sodium bromide

17. sodium arsenate

18. $NaHSO_3$, sodium hydrogen sulfite

19. (a) zinc (II) nitrate (b) arsenic trifluoride

Review Problems

2.117 (c)

2.119 29.3 g of N

2.121 5.54 g of NH_3

2.123 2.286 g of O

2.125 $1.9926482 \times 10^{-23}$ g

2.127 1.008 g

2.129 2.0158 u

2.131 63.55 u

2.133

	neutrons	protons	electrons
(a)	138	88	88
(b)	8	6	6
(c)	124	82	82
(d)	12	11	11

2.135

	electrons	protons	neutrons
(a)	36	35	46
(b)	23	26	32
(c)	27	29	34
(d)	36	37	50

Additional Exercises

2.138 35.6 μ

2.142 NaH_2PO_4 and Na_2HPO_4

2.146 191,600 gallons

Chapter 3

Practice Exercises

1. 3.44 mol of N

2. 1.11 mol of S

3. 28.4 g of Ag

4. 2.906×10^{12} atoms of Pb

5. 105.99 g

6. 13.2 g of Na_2CO_3

7. 0.467 mol of H_2SO_4

8. 59.5 g of Fe

9. 25.94% N, 74.06% O; there are no other elements present.

10. 36.85% N, 63.15% O

11. 10.5 g of Fe

12. NO

13. N_2O_5

14. Na_2SO_4

15. CH_2O

16. N_2H_4

17. $3CaCl_2(aq) + 2K_3PO_4(aq) \rightarrow Ca_3(PO_4)_2(s) + 6KCl(aq)$

18. $2Fe(s) + 3Cl_2(g) \rightarrow 2FeCl_3(s)$

19. 0.183 mol H_2SO_4

20. 0.958 mol of O_2

21. 78.5 g Al_2O_3

22. 30.01 g NO

23. theoretical yield = 30.9 g $HC_2H_3O_2$, % yield = 86.1%

24. 0.2499 M Na_2SO_4

25. 333 mL KCl solution

26. 3.47 g $CaCl_2$

27. Dilute 25.0 mL of 0.500 M H_2SO_4 solution to 100 mL.

28. 26.8 mL of NaOH solution

Review Problems

3.17 1 N : 2 O

3.19 2.59×10^{-3} mol Na atoms

3.21 (a) 6 C : 11 H (b) 12 C : 11 O (c) 2 H : 1 O
(d) 2 H : 1 O

3.23 1.05 mol Al

3.25 4.32 mol Al

3.27 (a) 2 mol Al/3 mol S or 3 mol S/2 mol Al
(b) 3 mol S/1 mol $Al_2(SO_4)_3$ or 1 mol $Al_2(SO_4)_3$/3 mol S
(c) 0.600 mol Al (d) 3.48 mol S

3.29 0.0725 mol N_2 and 0.218 mol H_2

3.31 0.833 mol UF_6

3.33 7.07×10^{23} carbon atoms

3.35 3.76×10^{24} atoms

3.37 3.01×10^{23} atoms

3.39 (a) 23.0 g (b) 32.1 g (c) 35.5 g

3.41 (a) 75.4 g Fe (b) 392 g O (c) 35.1 g Ca

3.43 0.319 mol Ni

3.45 (a) 84.0 μ (b) 294.2 μ (c) 96.1 μ (d) 342.2 μ
(e) 249.7 μ

3.47 (a) 388 g (b) 151 g (c) 34.9 g (d) 139 g

3.49 (a) 0.215 mol (b) 0.0916 mol (c) 0.0794 mol
(d) 4.31×10^{-8} mol = 0.0431 μmol

3.51 0.0750 mol Ca, 3.01 g Ca

3.53 1.30 mol N, 62.5 g $(NH_4)_2CO_3$

3.55
(a) 19.2% Na, 1.68% H, 25.8% P, 53.3% O
(b) 12.2% N, 5.26% H, 26.9% P, 55.6% O
(c) 62.0% C, 10.4% H, 27.6% O
(d) 29.4% Ca, 23.6% S, 47.0% O
(e) 23.3% Ca, 18.6% S, 55.7% O, 2.34% H

3.57 22.9% P, 77.1% Cl

3.59 Theoretical data (83.89% C, 10.35% H, 5.76% N) are consistent with experimental values.

3.61 0.474 g O

3.63 (a) SCl (b) CH_2O (c) NH_3 (d) AsO_3 (e) HO

3.65 $NaTcO_4$

3.67 CCl_2

3.69 $Na_2B_4O_7$

3.71 C_2H_6O

3.73 $C_{19}H_{30}O_2$

3.75 (a) $Na_2S_4O_6$ (b) $C_6H_4Cl_2$ (c) $C_6H_3Cl_3$

3.77 $C_{19}H_{30}O_2$

3.79 empirical formula HgBr, molecular formula Hg_2Br_2

3.81 empirical formula CHNO, molecular formula $C_3H_3N_3O_3$

3.83 14 mol Fe

3.85 $4Fe + 3O_2 \rightarrow 2Fe_2O_3$

3.87
(a) $Ca(OH)_2 + 2HCl \rightarrow CaCl_2 + 2H_2O$
(b) $2AgNO_3 + CaCl_2 \rightarrow Ca(NO_3)_2 + 2AgCl$
(c) $2Fe_2O_3 + 3C \rightarrow 4Fe + 3CO_2$
(d) $2NaHCO_3 + H_2SO_4 \rightarrow Na_2SO_4 + 2H_2O + 2CO_2$
(e) $2C_4H_{10} + 13O_2 \rightarrow 8CO_2 + 10H_2O$

3.89
(a) $Mg(OH)_2 + 2HBr \rightarrow MgBr_2 + 2H_2O$
(b) $2HCl + Ca(OH)_2 \rightarrow CaCl_2 + 2H_2O$
(c) $Al_2O_3 + 3H_2SO_4 \rightarrow Al_2(SO_4)_3 + 3H_2O$
(d) $2KHCO_3 + H_3PO_4 \rightarrow K_2HPO_4 + 2H_2O + 2CO_2$
(e) $C_9H_{20} + 14O_2 \rightarrow 9CO_2 + 10H_2O$

3.91 (a) 0.030 mol $Na_2S_2O_3$ (b) 0.24 mol HCl
(c) 0.15 mol H_2O (d) 0.15 mol H_2O

3.93 (a) $4P + 5O_2 \rightarrow P_4O_{10}$ (b) 8.85 g O_2 (c) 14.2 g P_4O_{10}
(d) 3.26 g P

3.95 30.31 g HNO_3

3.97 (a) Limiting reactant is Fe_2O_3. (b) 195 g Fe is formed.

3.99 26.7 g of $FeCl_3$ are left over.

3.101 theoretical yield = 66.98 g $BaSO_4$, % yield = 94.73%

3.103 % yield = 88.74%

3.105 9.2 g C_7H_8

3.107 0.25 mol HCl/1 liter HCl solution or
1 liter HCl solution/0.25 mol of HCl

3.109 (a) 1.00 M NaOH (b) 0.577 M $CaCl_2$
(c) 3.33 M KOH (d) 0.150 M $H_2C_2O_4$

3.111 (a) 1.46 g NaCl (b) 16.2 g $C_6H_{12}O_6$
(c) 6.12 g H_2SO_4

3.113 0.11 M H_2SO_4

3.115 300 mL

3.117 230 mL

3.119 0.113 M KOH

3.121 12 mL $NiCl_2$ solution, 0.36 g $NiCO_3$

3.123 188 mL NaOH

Additional Exercises

3.125 382,000 years

3.127 empirical formula $CaC_2N_2S_2$, molecular formula $CaC_2N_2S_2$

3.129 12.9 g

3.131 56,041 lb liquid hydrogen is left over.

3.133 50.0 mL of 0.10 M HCl solution must be added.

Chapter 4

Practice Exericses

1.
(a) $MgCl_2(s) \rightarrow Mg^{2+}(aq) + 2Cl^-(aq)$
(b) $Al(NO_3)_3(s) \rightarrow Al^{3+}(aq) + 3NO_3^-(aq)$
(c) $Na_2CO_3(s) \rightarrow 2Na^+(aq) + CO_3^{2-}(aq)$
(d) $(NH_4)_2SO_4(s) \rightarrow 2NH_4^+(aq) + SO_4^{2-}(aq)$

2.
molecular: $CdCl_2(aq) + Na_2S(aq) \rightarrow CdS(s) + 2NaCl(aq)$
ionic: $Cd^{2+}(aq) + 2Cl^-(aq) + 2Na^+(aq) + S^{2-}(aq) \rightarrow$
$$CdS(s) + 2Na^+(aq) + 2Cl^-(aq)$$
net ionic: $Cd^{2+}(aq) + S^{2-}(aq) \rightarrow CdS(s)$

3.
(a) *molecular:*
$AgNO_3(aq) + NH_4Cl(aq) \rightarrow AgCl(s) + NH_4NO_3(aq)$
ionic:
$Ag^+(aq) + NO_3^-(aq) + NH_4^+(aq) + Cl^-(aq) \rightarrow$
$$AgCl(s) + NH_4^+(aq) + NO_3^-(aq)$$
net ionic: $Ag^+(aq) + Cl^-(aq) \rightarrow AgCl(s)$
(b) *molecular:*
$Na_2S(aq) + Pb(C_2H_3O_2)_2(aq) \rightarrow PbS(s) + 2NaC_2H_3O_2(aq)$
ionic:
$2Na^+(aq) + S^{2-}(aq) + Pb^{2+}(aq) + 2C_2H_3O_2^-(aq) \rightarrow$
$$PbS(s) + 2Na^+(aq) + 2C_2H_3O_2^-(aq)$$
net ionic: $Pb^{2+}(aq) + S^{2-}(aq) \rightarrow PbS(s)$

4.
(a) $HCHO_2(aq) + H_2O \rightarrow H_3O^+(aq) + CHO_2^-(aq)$
(b) $H_3PO_4(aq) + H_2O \rightarrow H_2PO_4^-(aq) + H_3O^+(aq)$
$H_2PO_4^-(aq) + H_2O \rightarrow HPO_4^{2-}(aq) + H_3O^+(aq)$
$HPO_4^{2-}(aq) + H_2O \rightarrow PO_4^{3-}(aq) + H_3O^+(aq)$

5. $HNO_2(aq) + H_2O \rightleftharpoons H_3O^+(aq) + NO_2^-(aq)$

6. *molecular:*
$2HCl(aq) + Ca(OH)_2(aq) \rightarrow CaCl_2(aq) + 2H_2O(l)$
ionic: $2H^+(aq) + 2Cl^-(aq) + Ca^{2+}(aq) + 2OH^-(aq) \rightarrow$
$$Ca^{2+}(aq) + 2Cl^-(aq) + 2H_2O(l)$$
net ionic: $H^+(aq) + OH^-(aq) \rightarrow H_2O(l)$

7.
(a) *molecular:*
$HCl(aq) + KOH(aq) \rightarrow H_2O(l) + KCl(aq)$
ionic: $H^+(aq) + Cl^-(aq) + K^+(aq) + OH^-(aq) \rightarrow$
$$H_2O(l) + K^+(aq) + Cl^-(aq)$$
net ionic: $H^+(aq) + OH^-(aq) \rightarrow H_2O(l)$

(b) *molecular:*
$HCHO_2(aq) + LiOH(aq) \rightarrow H_2O(l) + LiCHO_2(aq)$
ionic: $HCHO_2(aq) + Li^+(aq) + OH^-(aq) \rightarrow$
$$H_2O(l) + Li^+(aq) + CHO_2^-(aq)$$
net ionic: $HCHO_2(aq) + OH^-(aq) \rightarrow H_2O(l) + CHO_2^-(aq)$
(c) *molecular:*
$N_2H_4(aq) + HCl(aq) \rightarrow N_2H_5Cl(aq)$
ionic:
$N_2H_4(aq) + H^+(aq) + Cl^-(aq) \rightarrow N_2H_5^+(aq) + Cl^-(aq)$
net ionic: $N_2H_4(aq) + H^+(aq) \rightarrow N_2H_5^+(aq)$

8. *molecular:*
$Al(OH)_3(s) + 3HCl(aq) \rightarrow AlCl_3(aq) + 3H_2O(l)$
ionic: $Al(OH)_3(s) + 3H^+(aq) + 3Cl^-(aq) \rightarrow$
$$Al^{3+}(aq) + 3Cl^-(aq) + 3H_2O(l)$$
net ionic: $Al(OH)_3(s) + 3H^+(aq) \rightarrow Al^{3+}(aq) + 3H_2O(l)$

9.
(a) *molecular:*
$KCHO_2(aq) + HCl(aq) \rightarrow KCl(aq) + HCHO_2(aq)$
ionic: $K^+(aq) + CHO_2^-(aq) + H^+(aq) + Cl^-(aq) \rightarrow$
$$K^+(aq) + Cl^-(aq) + HCHO_2(aq)$$
net ionic: $CHO_2^-(aq) + H^+(aq) \rightarrow HCHO_2(aq)$
(b) *molecular:* $CuCO_3(s) + 2HC_2H_3O_2(aq) \rightarrow$
$$Cu(C_2H_3O_2)_2(aq) + H_2O(l) + CO_2(g)$$
ionic and net ionic:
$CuCO_3(s) + 2HC_2H_3O_2(aq) \rightarrow$
$$Cu^{2+}(aq) + 2C_2H_3O_2^-(aq) + H_2O(l) + CO_2(g)$$
(c) No net reaction occurs.
molecular: $Ca(C_2H_3O_2)_2(aq) + 2AgNO_3(aq) \rightarrow$
$$2AgC_2H_3O_2(aq) + Ca(NO_3)_2(aq)$$
ionic:
$Ca^{2+}(aq) + 2C_2H_3O_2^-(aq) + 2Ag^+(aq) + 2NO_3^-(aq) \rightarrow$
$$2Ag^+(aq) + 2C_2H_3O_2^-(aq) + Ca^{2+}(aq) + 2NO_3^-(aq)$$
(d) *molecular:*
$2NaOH(aq) + NiCl_2(aq) \rightarrow 2NaCl(aq) + Ni(OH)_2(s)$
ionic: $2Na^+(aq) + 2OH^-(aq) + Ni^{2+}(aq) + 2Cl^-(aq) \rightarrow$
$$2Na^+(aq) + 2Cl^-(aq) + Ni(OH)_2(s)$$
net ionic: $2OH^-(aq) + Ni^{2+}(aq) \rightarrow Ni(OH)_2(s)$

10. 0.40 M Fe^{3+} and 1.2 M Cl^-

11. 0.750 M Na^+

12. 60.0 mL KOH

13. 0.0120 mol $BaSO_4$; 0.480 M Cl^-, 0.300 M Mg^{2+}, 0.060 M SO_4^{2-}

14. (a) 5.41×10^{-3} mol Ca^{2+} (b) 5.41×10^{-3} mol Ca^{2+}, same as (a)   (c) 5.41×10^{-3} mol $CaCl_2$, same as (a)
(d) 0.600 g $CaCl_2$ (e) 30% $CaCl_2$

15. 0.178 M H_2SO_4

16. 0.0220 M HCl, % (w/w) HCl = 0.079%

Review Problems

4.25
(a) $LiCl(s) \rightarrow Li^+(aq) + Cl^-(aq)$
(b) $BaCl_2(s) \rightarrow Ba^{2+}(aq) + 2Cl^-(aq)$
(c) $Al(C_2H_3O_2)_3(s) \rightarrow Al^{3+}(aq) + 3C_2H_3O_2^-(aq)$
(d) $(NH_4)_2CO_3(s) \rightarrow 2NH_4^+(aq) + CO_3^{2-}(aq)$
(e) $FeCl_3(s) \rightarrow Fe^{3+}(aq) + 3Cl^-(aq)$

4.27
(a) *ionic:*
$2NH_4^+(aq) + CO_3^{2-}(aq) + Mg^{2+}(aq) + 2Cl^-(aq) \rightarrow$
$$2NH_4^+(aq) + 2Cl^-(aq) + MgCO_3(s)$$
net ionic: $CO_3^{2-}(aq) + Mg^{2+}(aq) \rightarrow MgCO_3(s)$
(b) *ionic:*
$Cu^{2+}(aq) + 2Cl^-(aq) + 2Na^+(aq) + 2OH^-(aq) \rightarrow$
$$Cu(OH)_2(s) + 2Na^+(aq) + 2Cl^-(aq)$$
net ionic: $Cu^{2+}(aq) + 2OH^-(aq) \rightarrow Cu(OH)_2(s)$
(c) *ionic:*
$3Fe^{2+}(aq) + 3SO_4^{2-}(aq) + 6Na^+(aq) + 2PO_4^{3-}(aq) \rightarrow$
$$Fe_3(PO_4)_2(s) + 6Na^+(aq) + 3SO_4^{2-}(aq)$$
net ionic: $3Fe^{2+}(aq) + 2PO_4^{3-}(aq) \rightarrow Fe_3(PO_4)_2(s)$
(d) *ionic:*
$2Ag^+(aq) + 2C_2H_3O_2^-(aq) + Ni^{2+}(aq) + 2Cl^-(aq) \rightarrow$
$$2AgCl(s) + Ni^{2+}(aq) + 2C_2H_3O_2^-(aq)$$
net ionic: $Ag^+(aq) + Cl^-(aq) \rightarrow AgCl(s)$

4.29 $S^{2-}(aq) + Cu^{2+}(aq) \rightarrow CuS(s)$

4.31 *molecular:*
$AgNO_3(aq) + NaBr(aq) \rightarrow AgBr(s) + NaNO_3(aq)$
ionic: $Ag^+(aq) + NO_3^-(aq) + Na^+(aq) + Br^-(aq) \rightarrow$
$$AgBr(s) + Na^+(aq) + NO_3^-(aq)$$
net ionic: $Ag^+(aq) + Br^-(aq) \rightarrow AgBr(s)$

4.33 $HClO_4(aq) + H_2O(l) \rightarrow H_3O^+(aq) + ClO_4^-(aq)$

4.35 $N_2H_4(aq) + H_2O(l) \rightleftharpoons N_2H_5^+(aq) + OH^-(aq)$

4.37 $HNO_2(aq) + H_2O(l) \rightleftharpoons H_3O^+(aq) + NO_2^-(aq)$

4.39
$H_2CO_3(aq) + H_2O(l) \rightleftharpoons H_3O^+(aq) + HCO_3^-(aq)$
$HCO_3^-(aq) + H_2O(l) \rightleftharpoons H_3O^+(aq) + CO_3^{2-}(aq)$

4.41
(a) *molecular:*
$Ca(OH)_2(aq) + 2HNO_3(aq) \rightarrow 2H_2O(l) + Ca(NO_3)_2(aq)$
ionic: $Ca^{2+}(aq) + 2OH^-(aq) + 2H^+(aq) + 2NO_3^-(aq) \rightarrow$
$$2H_2O(l) + Ca^{2+}(aq) + 2NO_3^-(aq)$$
net ionic: $2OH^-(aq) + 2H^+(aq) \rightarrow$
$$2H_2O(l) \quad \text{or} \quad OH^-(aq) + H^+(aq) \rightarrow H_2O(l)$$
(b) *molecular:*
$Al_2O_3(s) + 6HCl(aq) \rightarrow 2AlCl_3(aq) + 3H_2O(l)$
ionic: $Al_2O_3(s) + 6H^+(aq) + 6Cl^-(aq) \rightarrow$
$$2Al^{3+}(aq) + 6Cl^-(aq) + 3H_2O(l)$$
net ionic: $Al_2O_3(s) + 6H^+(aq) \rightarrow 2Al^{3+}(aq) + 3H_2O(l)$
(c) *molecular:*
$Zn(OH)_2(s) + H_2SO_4(aq) \rightarrow ZnSO_4(aq) + 2H_2O(l)$
ionic: $Zn(OH)_2(s) + 2H^+(aq) + SO_4^{2-}(aq) \rightarrow$
$$Zn^{2+} + SO_4^{2-}(aq) + 2H_2O(l)$$
net ionic: $Zn(OH)_2(s) + 2H^+(aq) \rightarrow Zn^{2+} + 2H_2O(l)$

4.43
(a) $2H^+(aq) + CO_3^{2-}(aq) \rightarrow H_2O(l) + CO_2(g)$
(b) $OH^-(aq) + NH_4^+(aq) \rightarrow NH_3(aq) + H_2O(l)$

4.45
(a) Reaction takes place because $Cr(OH)_3(s)$ is formed.
(b) Reaction takes place because water is formed.

4.47 (a) soluble   (b) soluble   (c) insoluble   (d) soluble
(e) insoluble   (f) insoluble

4.49 (a) insoluble   (b) soluble   (c) soluble   (d) insoluble
(e) soluble   (f) insoluble

4.51

(a) *molecular:*
$3HNO_3(aq) + Cr(OH)_3(s) \rightarrow Cr(NO_3)_3(aq) + 3H_2O(l)$
ionic: $3H^+(aq) + 3NO_3^-(aq) + Cr(OH)_3(s) \rightarrow$
$$Cr^{3+}(aq) + 3NO_3^-(aq) + 3H_2O(l)$$
net ionic: $3H^+(aq) + Cr(OH)_3(s) \rightarrow Cr^{3+}(aq) + 3H_2O(l)$

(b) *molecular:*
$HClO_4(aq) + NaOH(aq) \rightarrow NaClO_4(aq) + H_2O(l)$
ionic: $H^+(aq) + ClO_4^-(aq) + Na^+(aq) + OH^-(aq) \rightarrow$
$$Na^+(aq) + ClO_4^-(aq) + H_2O(l)$$
net ionic: $H^+(aq) + OH^-(aq) \rightarrow H_2O(l)$

(c) *molecular:* $Cu(OH)_2(s) + 2HC_2H_3O_2(aq) \rightarrow$
$$Cu(C_2H_3O_2)_2(aq) + 2H_2O(l)$$
ionic and net ionic:
$Cu(OH)_2(s) + 2HC_2H_3O_2(aq) \rightarrow$
$$Cu^{2+}(aq) + 2C_2H_3O_2^-(aq) + 2H_2O(l)$$

(d) *molecular:*
$ZnO(s) + H_2SO_4(aq) \rightarrow ZnSO_4(aq) + H_2O(l)$
ionic: $ZnO(s) + 2H^+(aq) + SO_4^{2-}(aq) \rightarrow$
$$Zn^{2+}(aq) + SO_4^{2-}(aq) + H_2O(l)$$
net ionic: $ZnO(s) + 2H^+(aq) \rightarrow Zn^{2+}(aq) + H_2O(l)$

4.53

(a) *molecular:*
$Na_2SO_3(aq) + Ba(NO_3)_2(aq) \rightarrow 2NaNO_3(aq) + BaSO_3(s)$
ionic: $2Na^+(aq) + SO_3^{2-}(aq) + Ba^{2+}(aq) + 2NO_3^-(aq) \rightarrow$
$$2Na^+(aq) + 2NO_3^-(aq) + BaSO_3(s)$$
net ionic: $Ba^{2+}(aq) + SO_3^{2-}(aq) \rightarrow BaSO_3(s)$

(b) *molecular:* $2HCHO_2(aq) + K_2CO_3(aq) \rightarrow$
$$2KCHO_2(aq) + CO_2(g) + H_2O$$
ionic: $2HCHO_2(aq) + 2K^+(aq) + CO_3^{2-}(aq) \rightarrow$
$$2K^+(aq) + 2CHO_2^-(aq) + CO_2(g) + H_2O$$
net ionic: $2HCHO_2(aq) + CO_3^{2-}(aq) \rightarrow$
$$2CHO_2^-(aq) + CO_2(g) + H_2O$$

(c) *molecular:* $2NH_4Br(aq) + Pb(C_2H_3O_2)_2(aq) \rightarrow$
$$2NH_4C_2H_3O_2(aq) + PbBr_2(s)$$
ionic:
$2NH_4^+(aq) + 2Br^-(aq) + Pb^{2+}(aq) + 2C_2H_3O_2^-(aq) \rightarrow$
$$2NH_4^+(aq) + 2C_2H_3O_2^-(aq) + PbBr_2(s)$$
net ionic: $Pb^{2+}(aq) + 2Br^-(aq) \rightarrow PbBr_2(s)$

(d) *molecular:* $2NH_4ClO_4(aq) + Cu(NO_3)_2(aq) \rightarrow$
$$2NH_4NO_3(aq) + Cu(ClO_4)_2(aq)$$
ionic:
$2NH_4^+(aq) + 2ClO_4^-(aq) + Cu^{2+}(aq) + 2NO_3^-(aq) \rightarrow$
$$2NH_4^+(aq) + 2NO_3^-(aq) + Cu^{2+}(aq) + 2ClO_4^-(aq)$$
net ionic: N.R.

4.55 There are numerous possible answers. For instance:

(a) $HCl(aq) + NaHCO_3(aq) \rightarrow$
$$NaCl(aq) + H_2O(l) + CO_2(g)$$
(b) $Fe(NO_3)_2(aq) + 2NaOH(aq) \rightarrow$
$$Fe(OH)_2(s) + 2NaNO_3(aq)$$
(c) $BaBr_2(aq) + Na_2SO_3(aq) \rightarrow BaSO_3(s) + 2NaBr(aq)$
(d) $2AgNO_3(aq) + (NH_4)_2S(aq) \rightarrow Ag_2S(s) + 2NH_4NO_3(aq)$
(e) $ZnO(s) + 2HCl(aq) \rightarrow ZnCl_2(aq) + H_2O(l)$

4.57 (a) 0.0438 mol K^+, 0.0438 mol OH^- (b) 0.015 mol Ca^{2+}, 0.030 mol Cl^- (c) 0.020 mol Al^{3+}, 0.060 mol Cl^-

4.59 (a) Cr^{2+} $0.25\ M$, NO_3^- $0.50\ M$ (b) Cu^{2+} $0.10\ M$, SO_4^{2-} $0.10\ M$ (c) Na^+ $0.48\ M$, PO_4^{3-} $0.16\ M$ (d) Al^{3+} $0.15\ M$, SO_4^{2-} $0.23\ M$

4.61 $0.070\ M$ Na_3PO_4

4.63 1.0 g $Al_2(SO_4)_3$

4.65 $0.0565\ M$ $Ba(OH)_2$

4.67 6.0 mL of $NiCl_2$ solution, 0.43 g $AgCl$

4.69 188 mL NaOH solution

4.71 13.3 mL of $AlCl_3$ solution

4.73 $0.167\ M$ Fe^{3+}, 3.67 g Fe_2O_3 unreacted

4.75 (a) 8.00×10^{-3} mol AgCl (b) Na^+ $0.220\ M$, Cl^- $0.0600\ M$, Ag^+ zero, NO_3^- $0.160\ M$

4.77 48.40% lead

4.79 $0.114\ M$ HCl

4.81 2.67×10^{-3} mol $HC_3H_5O_3$

4.83 $MgSO_4 \cdot 7H_2O$

4.85 58.60% NaCl

Additional Exercises

4.87 (a) strong electrolyte (b) nonelectrolyte (c) strong electrolyte (d) nonelectrolyte (e) weak electrolyte (f) nonelectrolyte (g) strong electrolyte (h) weak electrolyte

4.90 100% of the sample was NaCl.

4.93 (a) C 57.84%, H 3.64%, O 38.5% (b) $C_4H_3O_2$ (c) $C_8H_6O_4$ (d) diprotic acid

Chapter 5

Practice Exercises

1. Aluminum is oxidized and is the reducing agent. Chlorine is reduced and is the oxidizing agent.

2. (a) Ni $+2$, Cl -1 (b) Mg $+2$, Ti $+4$, O -2 (c) K $+1$, Cr $+6$, O -2 (d) H $+1$, P $+5$, O -2 (e) V $+3$, C 0, H $+1$, O -2

3. $+8/3$ or $+2.67$

4. $Cl^- + SnCl_3^- + 2HgCl_2 \rightarrow SnCl_6^{2-} + Hg_2Cl_2$

5. $4Cu + 2NO_3^- + 10H^+ \rightarrow 4Cu^{2+} + N_2O + 5H_2O$

6. $2MnO_4^- + 3C_2O_4^{2-} + 4OH^- \rightarrow 2MnO_2 + 6CO_3^{2-} + 2H_2O$

7. (a) *molecular:* $Mg(s) + 2HCl(aq) + MgCl_2(aq) + H_2(g)$
ionic: $Mg(s) + 2H^+(aq) + 2Cl^-(aq) \rightarrow$
$$Mg^{2+}(aq) + 2Cl^-(aq) + H_2(g)$$
net ionic: $Mg(s) + 2H^+(aq) \rightarrow Mg^{2+}(aq) + H_2(g)$
(b) *molecular:* $2Al(s) + 6HCl(aq) \rightarrow 2AlCl_3(aq) + 3H_2(g)$
ionic: $2Al(s) + 6H^+(aq) + 6Cl^-(aq) \rightarrow$
$$2Al^{3+}(aq) + 6Cl^-(aq) + 3H_2(g)$$
net ionic: $2Al(s) + 6H^+(aq) \rightarrow 2Al^{3+}(aq) + 3H_2(g)$

8. (a) $2Al(s) + 3CuCl_2(aq) \rightarrow 2AlCl_3(aq) + 3Cu(s)$
(b) no reaction

9. $2C_4H_{10} + 13O_2 \rightarrow 8CO_2 + 10H_2O$

10. $C_2H_5OH + 3O_2 \rightarrow 2CO_2 + 3H_2O$

11. $4Fe(s) + 3O_2(g) \rightarrow 2Fe_2O_3(s)$

12. $P_4(s) + 5O_2(g) \rightarrow P_4O_{10}(s)$

13. 2.37 g $Na_2S_2O_3$

14. (a) $5Sn^{2+} + 16H^+ + 2MnO_4^- \rightarrow 5Sn^{4+} + 2Mn^{2+} + 8H_2O$
(b) 0.120 g Sn (c) 40.0% Sn (d) 50.7% SnO_2

Review Problems

5.24 (a) H_3AsO_3 is oxidized and is the reducing agent. HNO_3 is reduced and is the oxidizing agent. (b) NaI is oxidized and is the reducing agent. HOCl is reduced and is the oxidizing agent. (c) $H_2C_2O_4$ is oxidized and is the reducing agent. $KMnO_4$ is reduced and is the oxidizing agent. (d) Al is oxidized and is the reducing agent. H_2SO_4 is reduced and is the oxidizing agent.

5.26 (a) -2 (b) $+4$ (c) 0 (d) -3

5.28 (a) Na $+1$, H $+1$, P $+5$, O -2 (b) Ba $+2$, Mn $+6$, O -2 (c) Na $+1$, S $+10/4 = +2.5$, O -2 (d) Cl $+3$, F -1

5.30 (a) $+2$ (b) $+5$ (c) -1 (d) $+4$ (e) -2

5.32 (a) Na $+1$, O -2, Cl $+1$ (b) Na $+1$, O -2, Cl $+3$ (c) Na $+1$, O -2, Cl $+5$ (d) Na $+1$, O -2, Cl $+7$

5.34 (a) Pb $+2$, S -2 (b) Ti $+4$, Cl -1 (c) Sr $+2$, O -2, I $+5$ (d) Cr $+3$, S -2

5.36
(a) $6H^+ + BiO_3^- + 2e^- \rightarrow Bi^{3+} + 3H_2O$ reduction
(b) $2H_2O + Pb^{2+} \rightarrow PbO_2 + 4H^+ + 2e^-$ oxidation

5.38
(a) $Fe + 2OH^- \rightarrow Fe(OH)_2 + 2e^-$ oxidation
(b) $SO_2Cl_2 + 2OH^- + 2e^- \rightarrow SO_3^{2-} + 2Cl^- + H_2O$
                                              reduction

5.40
(a) $2S_2O_3^{2-} + OCl^- + 2H^+ \rightarrow S_4O_6^{2-} + Cl^- + H_2O$
(b) $2NO_3^- + 4H^+ + Cu \rightarrow 2NO_2 + 2H_2O + Cu^{2+}$
(c) $IO_3^- + 3AsO_3^{3-} \rightarrow I^- + 3AsO_4^{3-}$
(d) $Zn + SO_4^{2-} + 4H^+ \rightarrow Zn^{2+} + SO_2 + 2H_2O$
(e) $NO_3^- + 4Zn + 10H^+ \rightarrow NH_4^+ + 4Zn^{2+} + 3H_2O$
(f) $2Cr^{3+} + 3BiO_3^- + 4H^+ \rightarrow Cr_2O_7^{2-} + 3Bi^{3+} + 2H_2O$
(g) $I_2 + 5OCl^- + H_2O \rightarrow 2IO_3^- + 5Cl^- + 2H^+$
(h) $2Mn^{2+} + 5BiO_3^- + 14H^+ \rightarrow 2MnO_4^- + 5Bi^{3+} + 7H_2O$
(i) $3H_3AsO_3 + Cr_2O_7^{2-} + 8H^+ \rightarrow$
                             $3H_3AsO_4 + 2Cr^{3+} + 4H_2O$
(j) $2I^- + HSO_4^- + 3H^+ \rightarrow I_2 + SO_2 + 2H_2O$

5.42
(a) $2CrO_4^{2-} + 3S^{2-} + 4H_2O \rightarrow 2CrO_2^- + 3S + 8OH^-$
(b) $2MnO_4^- + 3C_2O_4^{2-} + 4H_2O \rightarrow$
                          $2MnO_2 + 6CO_2 + 8OH^-$
(c) $4ClO_3^- + 3N_2H_4 \rightarrow 4Cl^- + 6NO + 6H_2O$
(d) $NiO_2 + 2Mn(OH)_2 \rightarrow Ni(OH)_2 + Mn_2O_3 + H_2O$
(e) $3SO_3^{2-} + 2MnO_4^- + H_2O \rightarrow$
                          $3SO_4^{2-} + 2MnO_2 + 2OH^-$

5.44 $S_2O_3^{2-} + 4OCl^- + H_2O \rightarrow 2SO_4^{2-} + 4Cl^- + 2H^+$

5.46
(a) $Mn(s) + 2HCl(aq) \rightarrow MnCl_2(aq) + H_2(g)$
$Mn(s) + 2H^+(aq) + 2Cl^-(aq) \rightarrow Mn^{2+}(aq) + 2Cl^-(aq) + H_2(g)$; $Mn(s) + 2H^+(aq) \rightarrow Mn^{2+}(aq) + H_2(g)$
(b) $Cd(s) + 2HCl(aq) \rightarrow CdCl_2(aq) + H_2(g)$
$Cd(s) + 2H^+(aq) + 2Cl^-(aq) \rightarrow Cd^{2+}(aq) + 2Cl^-(aq) + H_2(g)$; $Cd(s) + 2H^+(aq) \rightarrow Cd^{2+}(aq) + H_2(g)$
(c) $Sn(s) + 2HCl(aq) \rightarrow SnCl_2(aq) + H_2(g)$
$Sn(s) + 2H^+(aq) + 2Cl^-(aq) \rightarrow Sn^{2+}(aq) + 2Cl^-(aq) + H_2(g)$; $Sn(s) + 2H^+(aq) \rightarrow Sn^{2+}(aq) + H_2(g)$
(d) $Ni(s) + 2HCl(aq) \rightarrow NiCl_2(aq) + H_2(g)$
$Ni(s) + 2H^+(aq) + 2Cl^-(aq) \rightarrow Ni^{2+}(aq) + 2Cl^-(aq) + H_2(g)$; $Ni(s) + 2H^+(aq) \rightarrow Ni^{2+}(aq) + H_2(g)$

(e) $2Cr(s) + 6HCl(aq) \rightarrow 2CrCl_3(aq) + 3H_2(g)$
$2Cr(s) + 6H^+(aq) + 6Cl^-(aq) \rightarrow 2Cr^{3+}(aq) + 6Cl^-(aq) + 3H_2(g)$; $2Cr(s) + 6H^+(aq) \rightarrow 2Cr^{3+}(aq) + 3H_2(g)$

5.48
(a) $3Ag(s) + 4HNO_3(aq) \rightarrow 3AgNO_3(aq) + NO(g) + 2H_2O$
(b) $Ag(s) + 2HNO_3(aq) \rightarrow AgNO_3(aq) + NO_2(g) + H_2O$

5.50
(a) N.R.
(b) $2Cr(s) + 3Pb^{2+}(aq) \rightarrow 2Cr^{3+}(aq) + 3Pb(s)$
(c) $2Fe(s) + 2Ag^+(aq) \rightarrow Fe^{2+}(aq) + 2Ag(s)$
(d) $3Ag(s) + Au^{3+}(aq) \rightarrow 3Ag^+(aq) + Au(s)$

5.52
(a) $2C_6H_6 + 15O_2 \rightarrow 12CO_2 + 6H_2O$
(b) $C_3H_8 + 5O_2 \rightarrow 3CO_2 + 4H_2O$
(c) $C_{21}H_{44} + 32O_2 \rightarrow 21CO_2 + 22H_2O$

5.54
(a) $2C_6H_6 + 9O_2 \rightarrow 12CO + 6H_2O$; $2C_3H_8 + 7O_2 \rightarrow 6CO + 8H_2O$; $2C_{21}H_{44} + 43O_2 \rightarrow 42CO + 44H_2O$
(b) $2C_6H_6 + 3O_2 \rightarrow 12C + 6H_2O$; $C_3H_8 + 2O_2 \rightarrow 3C + 4H_2O$; $C_{21}H_{44} + 11O_2 \rightarrow 21C + 22H_2O$

5.56 $2CH_3OH + 3O_2 \rightarrow 2CO_2 + 4H_2O$

5.58
(a) $2Mn^{2+} + 5BiO_3^- + 14H^+ \rightarrow 2MnO_4^- + 5Bi^{3+} + 7H_2O$
(b) 72.3 g $NaBiO_3$

5.60 3.53 g Cu

5.62
(a) $5Sn^{2+} + 2MnO_4^- + 16H^+ \rightarrow 2Mn^{2+} + 5Sn^{4+} + 8H_2O$
(b) 17.4 mL $KMnO_4$ solution

5.64 (a) 9.463% (b) 18.40%

5.66 (a) 0.02993 g H_2O_2 reacted (b) 2.993%

5.68 (a) $2CrO_4^{2-} + 3SO_3^{2-} + H_2O \rightarrow 2CrO_2^- + 3SO_4^{2-} + 2OH^-$ (b) 0.875 g Cr (c) 25.4%

5.70 (a) 5.405×10^{-3} mol $C_2O_4^{2-}$
(b) 0.5998 g $CaCl_2$
(c) 24.35%

Additional Exercises

5.73 Cu < Pb < Fe < Al

5.77
(a) $Zn(s) + Sn^{2+}(aq) \rightarrow Sn(s) + Zn^{2+}(aq)$
(b) $2Cr(s) + 6H^+(aq) \rightarrow 3H_2(g) + 2Cr^{3+}(aq)$
(c) N.R.
(d) $Mn(s) + Pb^{2+}(aq) \rightarrow Pb(s) + Mn^{2+}(aq)$
(e) $Zn(s) + Co^{2+}(aq) \rightarrow Co(s) + Zn^{2+}(aq)$

5.79
(a) $2Zn(s) + O_2(g) \rightarrow 2ZnO(s)$
(b) $4Al(s) + 3O_2(g) \rightarrow 2Al_2O_3(s)$
(c) $2Mg(s) + O_2(g) \rightarrow 2MgO(s)$
(d) $2Fe(s) + O_2(g) \rightarrow 2FeO(s)$ or $4Fe(s) + 3O_2(g) \rightarrow 2Fe_2O_3(s)$
(e) $2Ca(s) + O_2(g) \rightarrow 2CaO(s)$

5.83 11.33 g Cu remained unreacted. Total mass Cu bar + silver = 14.78 g.

5.85 16.7 mL NaOH solution

Chapter 6

Practice Exercises

1. 5200 J or 5.2 kJ, 1200 cal or 1.2 kcal

2. 3.68 kJ, 73.7 kJ/mol

3. $2.5H_2(g) + 1.25O_2(g) \rightarrow 2.5H_2O(g)$ $\Delta H° = -647.2$ kJ

4. $2Cu(s) + O_2(g)$

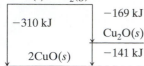

5. $\Delta H° = -44.0$ kJ

6. $Na(s) + \frac{1}{2}H_2(g) + C(s) + \frac{3}{2}O_2(g) \rightarrow NaHCO_3(s)$
$\Delta H° = -947.7$ kJ/mol

7. $\Delta H° = -113.4$ kJ   (b) $\Delta H° = -177.8$ kJ

Review Problems:

6.34 5.30×10^2 kJ

6.36 (a) 109 kcal   (b) 531 kJ

6.38 -17 J

6.40 7.32 kJ

6.42 135 J

6.44 (a) 1670 J   (b) 1670 J   (c) 23.2 J/°C   (d) 0.464 J/g °C

6.46 (a) 3.5×10^3 kcal   (b) 56 mi

6.48 25.12 J/mol °C

6.50 30.4 kJ

6.52 Heat evolved $= 3.8 \times 10^3$ J. Heat evolved per mole $HNO_3 = 53$ kJ/mol.

6.54 (a) 22,200 J   (b) -222 kJ/mol

6.56 (a) $2CO(g) + O_2(g) \rightarrow 2CO_2(g)$ $\Delta H° = -566$ kJ
(b) -283 kJ

6.58 $4Al(s) + 2Fe_2O_3(s) \rightarrow 2Al_2O_3(s) + 4Fe(s)$
$\Delta H° = -1708$ kJ

6.60 162 kJ

6.62 $Ge(s) + O_2(g)$

	-255 kJ
-534.7 kJ	GeO(s)
GeO$_2$(s)	-280 kJ

6.64 Reverse the first equation:

$2NO_2 \rightarrow N_2O_4$ $\Delta H = -57.93$ kJ
Add the first equation to the second:

$\dfrac{2NO + O_2 \rightarrow 2NO_2}{2NO + O_2 \rightarrow N_2O_4}$ $\dfrac{\Delta H = -113.14 \text{ kJ}}{\Delta H = -171.07 \text{ kJ}}$

6.66 -78.60 kJ

6.68 -127.8 kJ

6.70 -53.0 kJ

6.72 -1268 kJ

6.74 -188 kJ mol^{-1}

6.76 only (b)

6.78
(a) $2C(graphite) + 2H_2(g) + O_2(g) \rightarrow HC_2H_3O_2(l)$
$\Delta H_f° = -487.0$ kJ/mol
(b) $Na(s) + \frac{1}{2}H_2(g) + C(graphite) + \frac{3}{2}O_2(g) \rightarrow NaHCO_3(s)$
$\Delta H_f° = -947.7$ kJ/mol
(c) $Ca(s) + \frac{1}{8}S_8(s) + 3O_2(g) + 2H_2(g) \rightarrow CaSO_4 \cdot 2H_2O(s)$
$\Delta H_f° = -2021.1$ kJ/mol

6.80 (a) -196.6 kJ   (b) -177.8 kJ

6.82
(a) $\frac{1}{2}H_2(g) + \frac{1}{2}Cl_2(g) \rightarrow HCl(g)$ $\Delta H_f° = -92.30$ kJ
(b) $\frac{1}{2}N_2(g) + 2H_2(g) + \frac{1}{2}Cl_2(g) \rightarrow NH_4Cl(s)$
$\Delta H_f° = -315.4$ kJ

6.84 $C_{12}H_{22}O_{11}(s) + 12O_2(g) \rightarrow 12CO_2(g) + 11H_2O(g)$
$\Delta H_{combustion}° = -5.65 \times 10^3$ kJ/mol;
for $C_{12}H_{22}O_{11}(s)$, $\Delta H_f° = -2.22 \times 10^3$ kJ/mol

Additional Exercises

6.87 28.83 °C

6.89 $+463.9$ kJ, endothermic

6.90 -17 kJ

6.91 -266 kJ

6.94 $+364$ kJ, right to left

6.96 -43 kJ

6.100 $HCHO_2(l) + H_2(g) \rightarrow CH_3OH(l) + \frac{1}{2}O_2(g)$
$\Delta H° = +125$ kJ

6.102 $+25$ kJ

Chapter 7

Practice Exercises

1. 5.45×10^{14} Hz

2. 2.874 m

3. 656.5 nm, red

4. for $n = 3$, $\ell = 0$ (s), 1 (p), 2 (d); for $n = 4$, $\ell = 0$ (s), 1 (p), 2 (d), 3 (f)

5. 9 orbitals

6.
(a) Mg $1s^2 2s^2 2p^6 3s^2$
(b) Ge $1s^2 2s^2 2p^6 3s^2 3p^6 3d^{10} 4s^2 4p^2$
(c) Cd $1s^2 2s^2 2p^6 3s^2 3p^6 3d^{10} 4s^2 4p^6 4d^{10} 5s^2$
(d) Gd $1s^2 2s^2 2p^6 3s^2 3p^6 3d^{10} 4s^2 4p^6 4d^{10} 4f^7 5s^2 5p^6 5d^1 6s^2$

7.
(a) Na ⇅ ⇅ ⇅⇅⇅ ↑
$1s^2$ $2s^2$ $2p^6$ $3s^1$

(b) S ⇅ ⇅ ⇅⇅⇅ ⇅ ↑↑↑
$1s^2$ $2s^2$ $2p^6$ $3s^2$ $3p^4$

(c) Fe
⇅ ⇅ ⇅⇅⇅ ⇅ ⇅⇅⇅ ⇅↑↑↑↑ ⇅
$1s^2$ $2s^2$ $2p^6$ $3s^2$ $3p^6$ $3d^6$ $4s^2$

8.
(a) P [Ne] ⇅ ↑↑↑   3 unpaired electrons
$3s^2$ $3p^3$

(b) Sn [Kr] ⇅⇅⇅⇅⇅ ⇅ ↑↑○
$4d^{10}$ $5s^2$ $5p^2$
2 unpaired electrons

9. (a) $4s^24p^4$ (b) $5s^25p^2$ (c) $5s^25p^5$

10. (a) Sn (b) Ga (c) Cr (d) S^{2-}

11. (a) Be (b) C

Review Problems

7.66 6.98×10^{14} Hz

7.68 1.02×10^{15} Hz

7.70 2.97 m

7.72 5.0×10^6 m $= 5.0 \times 10^3$ km

7.74 2.6×10^{-19} J; for 1 mol, 1.6×10^5 J $= 1.6 \times 10^2$ kJ

7.76 (a) violet (b) 7.31×10^{14} Hz
(c) 4.85×10^{-19} J

7.78 1094 nm; no, not in visible spectrum

7.80 17.37 nm, infrared

7.82 (a) p (b) f

7.84 (a) $n = 3, \ell = 0$ (b) $n = 5, \ell = 2$

7.86 $\ell = 0, 1, 2, 3, 4, 5$

7.88 (a) $m_\ell = -1, 0, +1$
(b) $m_\ell = -3, -2, -1, 0, +1, +2, +3$

7.90 $\ell = 4, n = 5$

7.92

n	ℓ	m_ℓ	m_s
2	1	-1	$-\frac{1}{2}$
2	1	-1	$+\frac{1}{2}$
2	1	0	$-\frac{1}{2}$
2	1	0	$+\frac{1}{2}$
2	1	$+1$	$-\frac{1}{2}$
2	1	$+1$	$+\frac{1}{2}$

7.94 21 electrons have $\ell = 1$ and 20 electrons have $\ell = 2$.

7.96
(a) S $1s^22s^22p^63s^23p^4$
(b) K $1s^22s^22p^63s^23p^64s^1$
(c) Ti $1s^22s^22p^63s^23p^63d^24s^2$
(d) Sn $1s^22s^22p^63s^23p^63d^{10}4s^24p^64d^{10}5s^25p^2$

7.98 (a) paramagnetic (b) paramagnetic (c) paramagnetic (d) not paramagnetic (diamagnetic) (e) not paramagnetic (diamagnetic)

7.100 (a) 0 (b) 3 (c) 3

7.102 (a) Ni [Ar] $3d^84s^2$ (b) Cs [Xe] $6s^1$
(c) Ge [Ar] $3d^{10}4s^24p^2$ (d) Br [Ar] $3d^{10}4s^24p^5$
(e) Bi [Xe] $4f^{14}5d^{10}6s^26p^3$

7.104 (a)
Mg ⇅ ⇅ ⇅⇅⇅ ⇅
    $1s^2$ $2s^2$ $2p^6$ $3s^2$

(b)
Ti ⇅ ⇅ ⇅⇅⇅ ⇅ ⇅⇅⇅ ↑↑○○○ ⇅
    $1s^2$ $2s^2$ $2p^6$ $3s^2$ $3p^6$ $3d^2$ $4s^2$

7.106
(a) Ni [Ar] ⇅⇅⇅↑↑ ⇅
         $3d^8$ $4s^2$

(b) Cs [Xe] ↑
         $6s^1$

(c) Ge [Ar] ⇅⇅⇅⇅⇅ ⇅ ↑↑○
         $3d^{10}$ $4s^2$ $4p^2$

(d) Br [Ar] ⇅⇅⇅⇅⇅ ⇅ ⇅⇅↑
         $3d^{10}$ $4s^2$ $4p^5$

7.108 (a) 5 (b) 4 (c) 4 (d) 6

7.110 (a) Na $3s^1$ (b) Al $3s^23p^1$ (c) Ge $4s^24p^2$
(d) P $3s^23p^3$

7.112
(a) Na ↑   (b) Al ⇅ ↑○○
        $3s^1$ $3s^2$ $3p^1$

(c) Ge ⇅ ↑↑○ (d) P ⇅ ↑↑↑
        $4s^2$ $4p^2$ $3s^2$ $3p^3$

7.114 (a) 1+ (b) 6+ (c) 7+

7.116 (a) Na (b) Sb

7.118 Sb

7.120 (a) Na (b) Co^{2+} (c) Cl^-

7.122 (a) C (b) O (c) Cl

7.124 (a) Cl (b) Br

7.126 Mg

Additional Exercises

7.128 6.31×10^{22} photons

7.130 91.2 nm

7.132 2.754×10^6 m/s

7.134 365 nm

7.136 (a) $2s$ subshell should get filled before the $2p$.
(b) [Kr] implies that the $3d$ and the $4s$ subshells are filled already. (c) Nothing is wrong. (d) Configuration should be [Xe] $4f^{14}5d^76s^2$ because the $6s$ subshell becomes filled completely before the $5d$, except when transferring an electron from the $6s$ to $5d$ subshell results in a half-filled ($5d^5$) or filled ($5d^{10}$) $5d$ subshell, which is not the case here.

7.138 14 electrons

7.142 (a) $+328$ kJ, endothermic (b) $+141$ kJ, endothermic
(c) -844 kJ, exothermic

Chapter 8

Practice Exercises

1. (a) Cr^{2+} [Ar] $3d^4$ (b) Cr^{3+} [Ar] $3d^3$ (c) Cr^{6+} [Ar]

2. (a) $\cdot \ddot{Se} :$ (b) $: \ddot{I} \cdot$ (c) $\cdot Ca \cdot$

3. $\cdot Mg \cdot + \cdot \ddot{O} : \longrightarrow Mg^{2+} + [: \ddot{O} :]^{2-}$

4. (a) Br (b) Cl (c) Cl

5. SO_2 O S O

                              O
NO_3^- O N O

HClO$_3$ O

Cl   O   H

O

H$_3$PO$_4$ O

H   O   P   O   H

O

H

6. SO$_2$ 18e^- PO$_4^{3-}$ 32e^- NO$^+$ 10e^-

7. OF$_2$ $:\!\ddot{F}\!-\!\ddot{O}\!-\!\ddot{F}\!:$ NH$_4^+$ $\left[\begin{array}{c} H \\ | \\ H\!-\!N\!-\!H \\ | \\ H \end{array}\right]^+$

SO$_2$ $:\!\ddot{O}\!-\!\ddot{S}\!=\!\ddot{O}\!:$ NO$_3^-$ $\left[\begin{array}{c} :\!\ddot{O}\!: \\ \| \\ N \\ \diagup \diagdown \\ :\!\ddot{O}\quad\ddot{O}\!: \end{array}\right]^-$

ClF$_3$ $\begin{array}{c} :\!\ddot{F}\!: \\ | \\ \ddot{C}l \\ \diagup \diagdown \\ :\!\ddot{F}\quad\ddot{F}\!: \end{array}$ HClO$_4$ $H\!-\!\ddot{O}\!-\!\overset{\displaystyle :\!\ddot{O}\!:}{\underset{\displaystyle :\!\ddot{O}\!:}{C}l}\!-\!\ddot{O}\!:$

8. (a) $\overset{2-}{:\!\ddot{N}}\!-\!\overset{\oplus}{N}\!\equiv\!\overset{\oplus}{O}\!:$ (b) $[\ddot{S}\!=\!C\!=\!\ddot{N}]^{\ominus}$

9. (a) SO$_2$ $\ddot{O}\overset{\displaystyle\ddot{S}}{\diagup\diagdown}\ddot{O}$ (b) HClO$_3$ $\ddot{O}\overset{\displaystyle\overset{:\!\ddot{O}\!:}{\|}{C}l}{\diagup\diagdown}\underset{H}{\ddot{O}}$

(c) H$_3$PO$_4$ $H\!-\!\ddot{O}\!-\!\overset{\displaystyle :\!O:}{\underset{\displaystyle :\!\ddot{O}\!-\!H}{P}}\!-\!\ddot{O}\!-\!H$

10. (a)

$\left[:\!\ddot{O}\overset{\ddot{N}}{\diagup\diagdown}\ddot{O}:\right]^- \longleftrightarrow \left[:\!\ddot{O}\overset{\ddot{N}}{\diagup\diagdown}\ddot{O}:\right]^-$

(b)

$\left[\begin{array}{c} :\!\ddot{O}: \\ | \\ :\!\ddot{O}\!-\!P\!-\!\ddot{O}: \\ | \\ :\!\ddot{O}: \end{array}\right]^- \longleftrightarrow \left[\begin{array}{c} :\!\ddot{O}: \\ \| \\ :\!\ddot{O}\!-\!P\!=\!\ddot{O} \\ | \\ :\!\ddot{O}: \end{array}\right]^- \longleftrightarrow$

$\left[\begin{array}{c} :\!\ddot{O}^+ \\ | \\ :\!\ddot{O}\!-\!P\!-\!\ddot{O}: \\ \| \\ :\!O\!: \end{array}\right]^- \longleftrightarrow \left[\begin{array}{c} :\!\ddot{O}: \\ \| \\ \ddot{O}\!=\!P\!-\!\ddot{O}: \\ | \\ :\!O\!: \end{array}\right]^-$

11. $[:\!\ddot{O}\!-\!H]^- + H^+ \longrightarrow \underset{H\diagup\quad\diagdown H}{\ddot{O}}$

Review Problems

8.57 Magnesium loses 2 electrons:
Mg [Ne] $3s^2 \rightarrow$ Mg^{2+} [Ne] $+ 2e^-$
Each bromine gains one electron:
Br [Ar] $3d^{10}4s^24p^5 + e^- \rightarrow$ Br$^-$ [Kr]

8.59 Pb^{2+} [Xe] $6s^25d^{10}4f^{14}$ Pb^{4+}[Xe] $5d^{10}4f^{14}$

8.61 Mn^{3+} [Ar] $3d^4$; ion has 4 unpaired electrons.

8.63 (a) $\cdot\!\dot{S}i\cdot$ (b) $\cdot\!\dot{\ddot{S}}b\cdot$ (c) $\cdot\!Ba\cdot$ (d) $\cdot\!\dot{A}l\cdot$ (e) $:\!\dot{\ddot{S}}\cdot$

8.65 (a) [K]$^+$ (b) [Al]$^{3+}$ (c) $\left[:\!\ddot{S}\!:\right]^{2-}$ (d) $\left[:\!\dot{\ddot{S}}i\!:\right]^{4-}$

(e) [Mg]$^{2+}$

8.67 (a) $:\!\ddot{B}r\!\cdot\overset{\frown}{}\!Ca\overset{\frown}{}\!\cdot\ddot{B}r\!: \longrightarrow$ [Ca]$^{2+}$ + $2\left[:\!\ddot{B}r\!:\right]^-$

(b) $:\!\ddot{O}\overset{\frown}{}\!\dot{A}l\overset{\frown}{}\!\ddot{O}\overset{\frown}{}\!\dot{A}l\overset{\curvearrowright}{}\!\ddot{O}: \longrightarrow$ 2[Al]$^{3+}$ + $3\left[:\!\ddot{O}\!:\right]^{2-}$

(c) $K\overset{\frown}{}\!\cdot\dot{\ddot{S}}\cdot\overset{\frown}{}\!K \longrightarrow$ 2[K]$^+$ + $\left[:\!\ddot{S}\!:\right]^{2-}$

8.69 1.4×10^3 g

8.71 3.44×10^{-7} m, ultraviolet

8.73

(a) $:\!\ddot{B}r\!\cdot + \cdot\ddot{B}r\!: \longrightarrow :\!\ddot{B}r\!-\!\ddot{B}r\!:$

(b) $2H\cdot + \cdot\ddot{O}\!: \longrightarrow H\!-\!\ddot{O}\!-\!H$

(c) $3H\cdot + \cdot\ddot{N}\cdot \longrightarrow \underset{\displaystyle H}{H\!-\!\overset{|}{N}\!-\!H}$

8.75 (a) H$_2$Se   (b) H$_3$As   (c) SiH$_4$

8.77 (a) S   (b) Si   (c) Br   (d) C

8.79 The N—S bond is the most polar.

8.81

(a)   Cl
Cl   Si   Cl
           Cl

(b)   F   P   F
            F

(c)   H   P   H
            H

(d)   Cl   S   Cl

8.83 (a) 32   (b) 26   (c) 8   (d) 20

8.85 (a) $\left[\begin{array}{c} :\!\ddot{C}l\!: \\ | \\ :\!\ddot{C}l\!-\!As\!-\!\ddot{C}l\!: \\ | \\ :\!\ddot{C}l\!: \end{array}\right]^+$ (b) $\left[:\!\ddot{O}\!-\!\ddot{C}l\!-\!\ddot{O}\!:\right]^-$

(c) $H\!-\!\ddot{O}\!-\!\ddot{N}\!=\!\ddot{O}$ (d) $:\!\ddot{F}\!-\!\ddot{X}e\!-\!\ddot{F}\!:$

8.87 (a) $:\!\ddot{C}l\!-\!\overset{\displaystyle :\!\ddot{C}l\!:}{\underset{\displaystyle :\!\ddot{C}l\!:}{S}i}\!-\!\ddot{C}l\!:$ (b) $:\!\ddot{F}\!-\!\overset{\displaystyle :\!\ddot{F}\!:}{P}\!-\!\ddot{F}\!:$ (c) $\underset{H\diagup\quad|\quad\diagdown H}{\overset{\displaystyle P}{\underset{\displaystyle H}{}}}$

(d) $\ddot{C}l\overset{\displaystyle\ddot{S}}{\diagup\diagdown}\ddot{C}l$

8.89 (a) :S̈=C=S̈: (b) [:C≡N:]$^-$

8.91 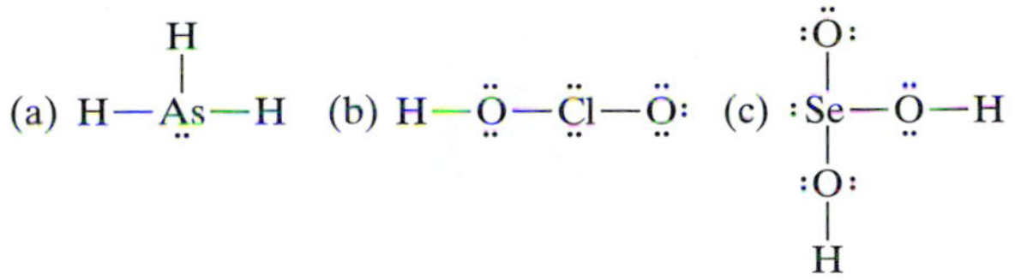

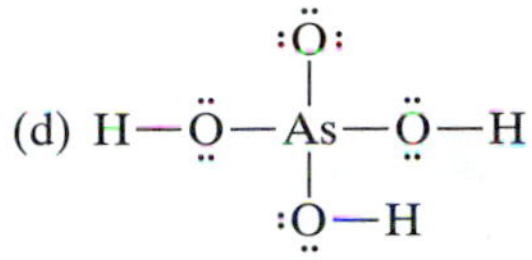

8.93 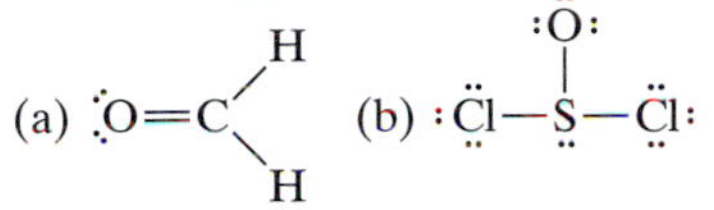

8.95 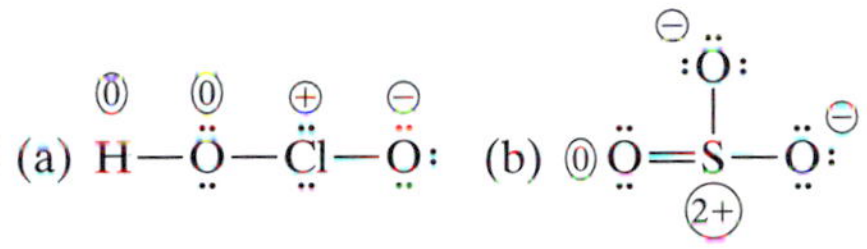

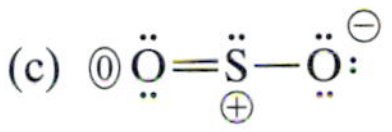

8.97 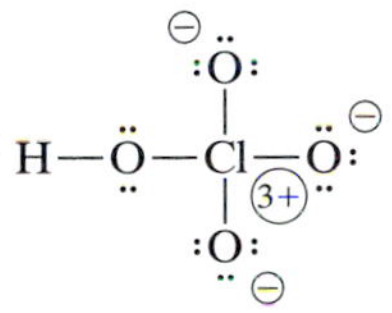

preferred structure is 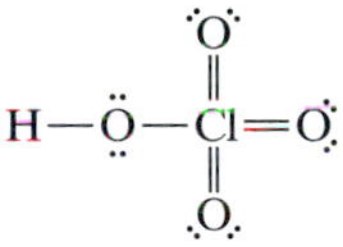

8.99 :C̈l—Be—C̈l: is preferred because all formal charges are zero

8.101

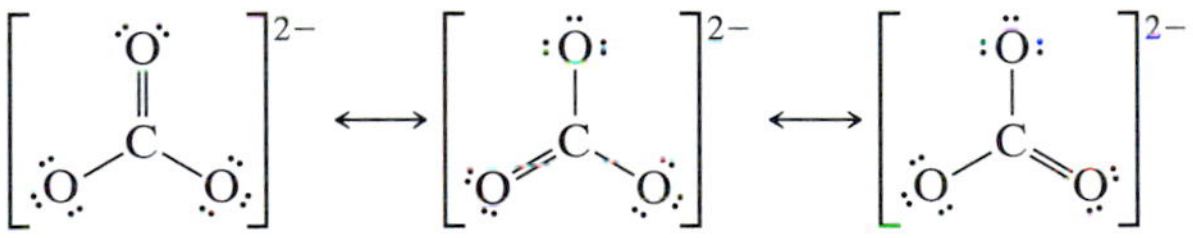

average bond order = 4/3 or 1.33

8.103 NO_3^- has longer bonds than NO_2^-.

8.105 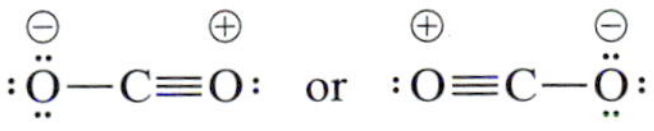

These structures have formal charges and are less stable than the preferred structure, in which all atoms have a formal charge of zero.

8.107 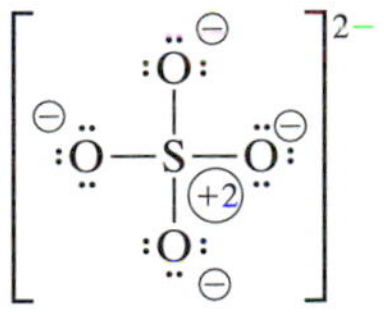

preferred structure:

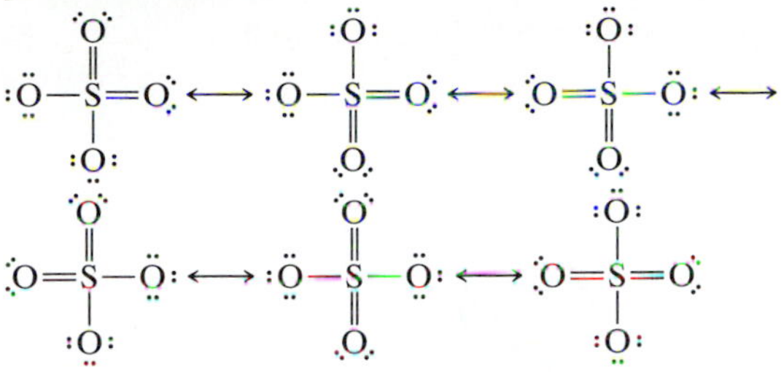

average bond order = 1.5

8.109 $H_2O + H^+ \longrightarrow H_3O^+$

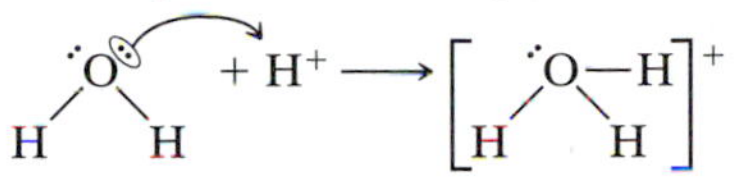

Additional Exercises

8.111 +148 kJ, 4363 kJ; 29.4 times larger

8.113 −2269 kJ mol^{-1}

8.118 (a) There are 2 extra electrons ($26e^-$ instead of $24e^-$) and unfavorable formal charges. (b) There are unfavorable formal charges. (c) C cannot accommodate 10 electrons.

8.120 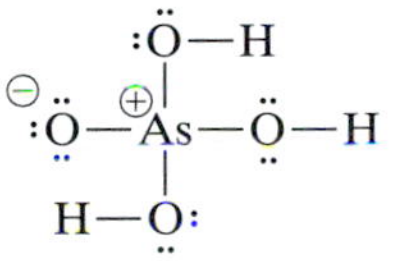

lower energy Lewis structure:

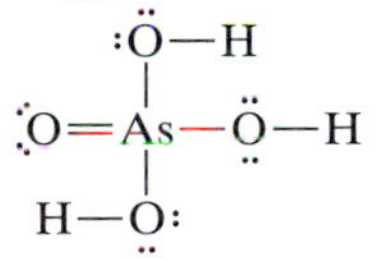

8.123 :C̈l—S̈—S̈—C̈l:

Chapter 9

Practice Exercises

1. trigonal bipyramidal
2. SO_3^{2-} trigonal pyramidal, XeO_4 tetrahedral, OF_2 bent or nonlinear
3. planar triangular
4. polar molecules: (b), (c), (d), (e)
5. The half-filled 1*s* atomic orbital of hydrogen overlaps with the half-filled 3*p* orbital of chlorine.

Cl (in HCl) (⇅) (⇅)(⇅)(↑x) where *x* = hydrogen electron
3*s* 3*p*

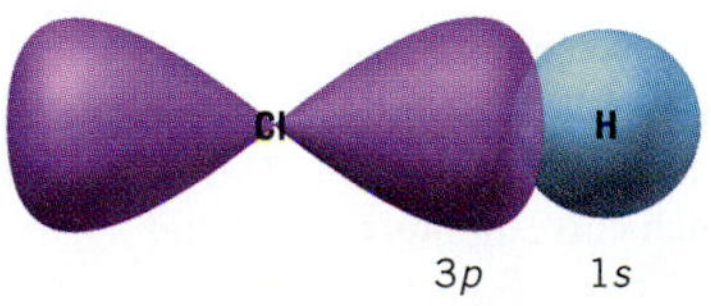

6. P (in PH$_3$)

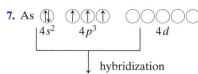 where x = hydrogen electron

Se in H$_2$Se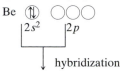

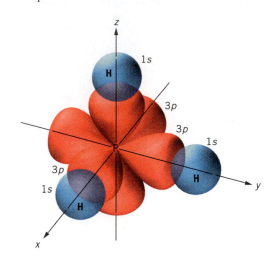

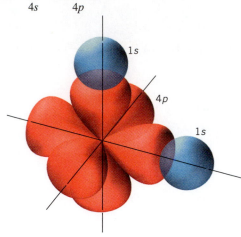

9.66 Beryllium uses sp hybrid orbitals in BeCl$_2$.

Be ⬆⬇ ◯◯◯
$|2s^2$ $|2p$

↓ hybridization

Be in BeCl$_2$ (↑x)(↑x) ◯◯
sp 2 unhybridized p orbitals

9.68 (a) sp^3 (b) sp^2 (c) sp^3

9.70 (a) sp^3 (b) sp^3d

9.72 sp^3d^2 octahedral
Sb in SbF$_6^-$ (↑x)(↑x)(↑x)(↑x)(↑x)(xx) ◯◯◯
sp^3d^2 3 unhybridized p orbitals

7. As ⬆⬇ ↑↑↑ ◯◯◯◯◯
$|4s^2$ $4p^3$ $4d$

↓ hybridization

As in AsCl$_5$ (↑x)(↑x)(↑x)(↑x)(↑x) ◯◯◯◯
sp^3d 4 unhybridized d orbitals

8. (a) sp^3 (b) sp^3d

9. (a) sp^3 (b) sp^3d

10. sp^3d^2
P in PCl$_6^-$ (↑x)(↑x)(↑x)(↑x)(↑x)(xx) ◯◯◯
sp^3d^2 3 unhybridized p orbitals
PCl$_6^-$ is octahedral.

11. $(\sigma_{2s})^2(\sigma_{2s}^*)^2(\pi_{2py})^2(\pi_{2px})^2(\sigma_{2pz})^2(\pi_{2py}^*)^1(\pi_{2px}^*)^0(\sigma_{2pz}^*)^0$, bond order = 2.5

9.74 (a) N in H$_2$C=NH ⬆⬇↑↑↑ ↑
sp^2 unhybridized p orbital

(b)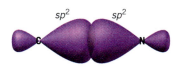

Overlap of 2 sp^2 hybridized orbitals forms σ bond.

Unhybridized p orbitals form π bond.

Review Problems

9.56
(a) bent, nonlinear (b) trigonal bipyramidal
(c) trigonal pyramidal (d) trigonal pyramidal
(e) bent, nonlinear

9.58 (a) tetrahedral (b) square planar (c) octahedral
(d) tetrahedral (e) linear

9.60 180°

9.62 Polar molecules: (a), (b), and (c)

9.64 The half-filled p orbitals of Se overlap with the half-filled $1s$ orbitals of the hydrogen atoms. The angle between p orbitals is 90°, which corresponds to the bond angles of the molecule.

(c)
$$\text{H} \underset{120°}{\overset{120°}{\underset{\text{H}}{\text{C}}}}=\underset{120°}{\ddot{\text{N}}}-\text{H}$$

9.76 C uses sp^2 hybrid orbitals in bonding; expected bond angles = 120°

9.78 1 sp^3, 2 sp, 3 sp^2, 4 sp^2

9.80 (a) O$_2^+$ (b) O$_2$ (c) N$_2$

9.82

Additional Exercises

9.84 planar triangular

9.86 The normal bond angle of an sp^3 hybridized carbon is 109.5°. In cyclopropane, the bond angles of the carbons joining each other are 60°, which are very far from normal, so the cyclopropane ring is under much strain.

9.88
(a) The half-filled sp^3 hybridized orbitals of P overlap with the half-filled p orbitals of F (expected bond angle = about 109.5°).
(b) The half-filled p orbitals of P overlap with the half-filled p orbitals of F (expected bond angle = about 90°).
(c) The observed bond angle (97.8°) fits neither of the models perfectly; it is in between the bond angles predicted by the models.

9.90
(a) sp^2
(b) Since butadiene is a system of alternating double and single bonds, all the carbons have sideways overlaps of p orbitals. (In general, we expect sideways overlap of p orbitals between carbons that are connected by double or triple bonds.)
(c) delocalized
(d) The central C—C bond is shorter than a C—C single bond because of the delocalized π bonding in the carbon chain; the delocalization energy stabilizes the bond.

9.92

nonpolar

9.94 The O—S—O bond angles are larger.

9.96 A π bond is formed because the p and d orbitals overlap above and below the imaginary axis between the two nuclei.

Chapter 10

Practice Exercises

1. 1.03×10^4 mm, 33.8 ft
2. 688 torr
3. 750 torr
4. 214 mL
5. 132 g/mol, Xe
6. 138 g/mol, molecular formula P_2F_4
7. 9.00 L O_2
8. 1.06 g Na_2CO_3
9. 98.2 g O_2
10. $P_{N_2} = 732$ torr, $V_{STP} = 283$ mL
11. 0.153, 15.3 mol% O_2
12. HI

Review Problems

10.28 (a) 958 torr
(b) 0.974 atm
(c) 738 mm Hg
(d) 10.9 torr

10.30 (a) 250 torr
(b) 350 torr

10.32 4.5 cm Hg

10.34 813 torr

10.36 826 torr; the reaction produced gases.

10.38 125 torr

10.40 8.42×10^3 mm

10.42 507 mL

10.44 4.28 L

10.46 843 °C

10.48 796 torr

10.50 5.73 L

10.52 −53 °C

10.54 6.24×10^4 mL torr mol^{-1} K^{-1}

10.56 0.104 L

10.58 2340 torr

10.60 0.0398 g CO_2

10.62 4.16 atm

10.64 (a) 1.34 g/L (b) 1.25 g/L (c) 3.17 g/L (d) 1.78 g/L

10.66 1.28 g/L

10.68 88.2 g/mol

10.70 27.5 g/mol

10.72 4.00 L F_2

10.74 1.14×10^3 mL O_2

10.76 10.7 L H_2

10.78 36.3 mL O_2

10.80 217 mL CO_2

10.82 650 torr

10.84 N_2 228 torr, O_2 152 torr, He 304 torr, CO_2 76 torr

10.86 $P_{CO} = 736$ torr, $V = 260$ mL

10.88 250 mL

10.90 N_2 75.0%, O_2 13.6%, CO_2 5.3%, H_2O 6.2%

10.92 N_2 effuses 1.25 times faster than CO_2.

10.94 16.2 g/mol

Additional Exercises

10.96 14.7 lb/in.^{2}

10.98 12 tires

10.99 378 atm or 5.55×10^3 lb/in.^{2}

10.100 741 torr

10.102 646 K, 373 °C

10.104 50 lb/in.^{2}

10.105 253 mL Cl_2

10.108 0.844 g Zn, 4.30 mL HCl

10.111 0.0467 g Zn

Chapter 11

Practice Exercises

1. In liquid, number of molecules decreases, and in vapor, number of molecules increases; sum of molecules remains the same.

2. 75 °C

3. 532 torr

4. Vapor pressure will increase with temperature increase.

5. solid becomes gas (sublimation).

6. liquid

7. covalent crystal

8. molecular crystal

Review Problems

11.91 diethyl ether

11.93 (a) London, hydrogen bonding
(b) London, dipole–dipole
(c) London
(d) London, hydrogen bonding

11.95 ethanol

11.97 diethyl ether < acetone < benzene < water < acetic acid

11.99 305 kJ

11.101 (a) 0 °C   (b) 48.0 g ice melts

11.103 35.8 kJ/mol

11.105 4.88 kJ/mol, bp −189 °C, 84 K

11.107 28.3 torr

11.109

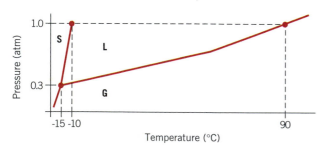

11.111 (a) solid   (b) gas   (c) liquid   (d) solid and liquid and gas

11.113 At −56 °C, gas becomes liquid, then solid; at −58 °C, gas becomes solid.

11.115 1 atom per unit cell

11.117 128 pm

11.119 656 pm

11.121 molecular

11.123 metallic

11.125 metallic

Chapter 12

Practice Exercises

1. 0.00430 g O_2, 0.00190 g N_2

2. 2.50 g NaOH, 248 g, 251 mL H_2O

3. 20 g HCl solution

4. 16.0 g CH_3OH

5. 0.400 m NaOH

6. 16.1 m HCl

7. 6.81 mol/L HBr

8. 9.02 torr

9. 44.0 torr

10. 100.16 °C

11. 157 g/mol

12. 214 torr

13. 538 g/mol

14. For 100% dissociation, $T_f = -0.882$ °C, for 0% dissociation, $T_f = -0.441$ °C

Review Problems

12.57
(a) $KCl(s) \rightarrow K^+(g) + Cl^-(g)$ $\Delta H = +690$ kJ/mol
(b) $K^+(g) + Cl^-(g) \rightarrow$
                    $K^+(aq) + Cl^-(aq)$ $\Delta H = -686$ kJ/mol
    $KCl(s) \rightarrow K^+(aq) + Cl^-(aq)$ $\Delta H = +4$ kJ/mol

12.59 0.038 g/L

12.61 0.133 m $C_6H_{12}O_6$, 2.39×10^{-3}, 2.34%

12.63 5.44%

12.65 5.28%, 3.09 m NH_3

12.67 0.359 M $NaNO_3$, 3.00%, 6.49×10^{-3}

12.69 22.8 torr

12.71 69.4 torr

12.73 benzene 30%, toluene 70%

12.75 (a) 0.029
(b) 0.030
(c) 2.8×10^2 g/mol

12.77 bp 101.0 °C, fp −3.72 °C

12.79 55.1 g $C_{12}H_{22}O_{11}$

12.81 152 g/mol

12.83 127 g/mol, $C_8H_4N_2$

12.85
(a) Begin with $\Pi V = nRT$ and $n =$ mass/molar mass. Substitute for n,

$\Pi V = $ mass/(molar mass $\times RT$)

Rearrange; multiply by molar mass and divide by ΠV:

molar mass $\times \Pi V/\Pi V = $ mass/molar mass $\times RT \times$ molar mass/ΠV

After canceling,

molar mass $=$ mass $\times RT/\Pi V$

(b) 1.8×10^6 g/mol

12.87 16.5 torr

12.89 1.3×10^4 torr

12.91 -1.1 °C

12.93 3% ionization

12.95 2

12.97 1.89

Additional Problems

12.99 N_2 at 1 atm, 5.3×10^{-4} mol N_2; air at 4 atm, 1.7×10^{-3} mol N_2; 30 mL N_2

12.101

(a) 35.4 g $Na_2Cr_2O_7$

(b) 59.9%

(c) oxygen 15.70%, carbon 70.52%, hydrogen 13.77%; $C_6H_{14}O$

(d) 102 g/mol, $C_6H_{14}O$

12.104 (a) 0.926 torr   (b) $4.98 \times 10^{-5}\ M$

(c) -9.25×10^{-5} °C   (d) It is impossible to measure a Δt this small.

12.106 (a) 100.51 °C

(b) 102.04 °C

(c) 0.36

Chapter 13

Practice Exercises

1. rate of disappearance of $H_2S = 0.30$ mol L^{-1} s^{-1}, rate of disappearance of $O_2 = 0.45$ mol L^{-1} s^{-1}

2. 1.0×10^{-4} mol L^{-1} s^{-1}

3. (a) 0.080   (b) L mol^{-1} s^{-1}

4. BrO_3^- first order, SO_3^{2-} first order, overall second order

5. $k = 2.0 \times 10^2$ L^2 mol^{-2} s^{-1}; values of k are the same.

6. first order

7. rate $= k\,[A]^2\,[B]^2$

8. (a) 0.26 M $C_{12}H_{22}O_{11}$ (b) 77 min

9. 63 min

10. $t_{1/2} = 18.7$ min; it takes 37.4 min for three-quarters to react.

11. first order

12. (a) $E_a = 1.4 \times 10^2$ kJ mol^{-1}

(b) 0.30 L mol^{-1} s^{-1}

13. rate $= k\,[NO]\,[O_3]$

14. rate $= k\dfrac{[NO_2Cl]^2}{[NO_2]}$ (assuming the first step is an equilibrium)

Review Problems

13.56

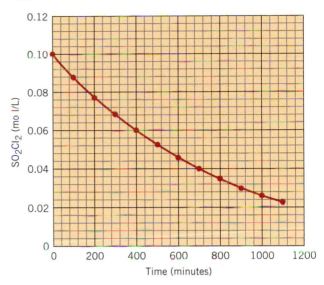

At $t = 200$ min, rate $= 1 \times 10^{-4}$ mol L^{-1} min^{-1}, at $t = 600$ min, rate $= 6 \times 10^{-5}$ mol L^{-1} min^{-1}.

13.58 Rate of disappearance of H_2 is three times the rate of disappearance of N_2; NH_3 appears two times as fast as N_2 disappears.

13.60 (a) -11.4 mol L^{-1} s^{-1} (b) 7.20 mol L^{-1} s^{-1}

(c) 8.40 mol L^{-1} s^{-1}

13.62 2.4×10^2 mol L^{-1} s^{-1}

13.64 (a) $HCrO_4^-$ first order, HSO_3^- second order, H^+ first order   (b) fourth order

13.66 rate $= k\,[M]\,[N]^2$, $k = 2.5 \times 10^3$ L^2 mol^{-2} s^{-1}

13.68 rate $= k\,[OCl^-][I^-]$, $k = 6.1 \times 10^0$ L mol^{-1} s^{-1}

13.70 rate $= k\,[ICl][H_2]$, $k = 0.13$ L mol^{-1} s^{-1}

13.72

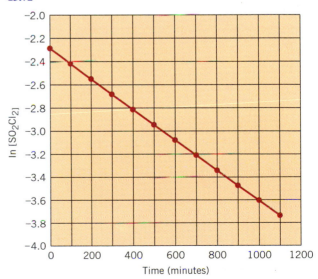

slope $= -0.0013$, $k = 1.3 \times 10^{-3}$ min^{-1}

13.74 (a) $3.7 \times 10^{-3}\ M$ (b) $6.0 \times 10^{-4}\ M$

13.76 $4.26 = 10^{-3}\ \text{min}^{-1}$

13.78 $1.3 \times 10^{4}\ \text{min}$

13.80 $1/2^8$ or $1/256$

13.82 It takes approximately 500 min to decrease concentration to half of its original value; this is a first-order reaction.

13.84 $t_{1/2} = 4.3 \times 10^{2}\ \text{s}$ or 7.2 min

13.86 79 kJ/mol

13.88 $E_a = 98.9\ \text{kJ/mol},\ A = 6.6 \times 10^{9}$

13.90 $E_a = 78\ \text{kJ/mol}$

13.92 (a) $1.9 \times 10^{-5}\ \text{s}^{-1}$ (b) $0.16\ \text{s}^{-1}$

Additional Exercises

13.95 41.5 years

13.97 (a) rate $= k_{1(\text{forward})} [A]^2$ (b) rate $= k_{1(\text{reverse})} [A_2]$ (c) rate $= k_2 [A_2] [E]$ (d) $2A + E \rightarrow B + C$ (e) rate $= k_2(k_{1(\text{forward})}/k_{1(\text{reverse})}) [A]^2 [E]$ or   rate $= k' [A]^2 [E]$

13.99 $6.36 \times 10^{3}\ \text{min}$

13.102 52.9 kJ/mol

13.104 (a) If $t = 20\ °C$, 16 chirps; if $t = 25\ °C$, 21 chirps; if $t = 30\ °C$, 26 chirps; if $t = 35\ °C$, 31 chirps

(b)

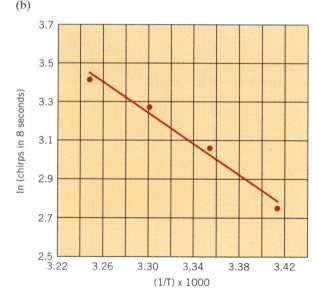

$E_a = 33\ \text{kJ/mol}$ (c) none, it would die

13.107 (a) $\text{Br}_2 \rightarrow 2\text{Br·}$ (b) $\text{Br·} + \text{H}_2 \rightarrow \text{HBr} + \text{H·}$ $\text{H·} + \text{Br}_2 \rightarrow \text{HBr} + \text{Br·}$ (c) $2\text{Br·} \rightarrow \text{Br}_2$ (d) It decreases the concentration of HBr.

Chapter 14

Practice Exercises

1. (a) $K_c = \dfrac{[H_2O]^2}{[H_2]^2 [O_2]}$ (b) $K_c = \dfrac{[NH_4^+] [OH^-]}{[NH_3] [H_2O]}$

2. 1.2×10^{-13}

3. 1.9×10^{5}

4. $K_p = \dfrac{P_{HI}^2}{P_{H_2} P_{I_2}}$

5. (b)

6. K_p should be smaller than K_c, $K_c = 57$

7. $K_p = 1.8 \times 10^{36}$

8. (a) $K_c = \dfrac{1}{[Cl_2(g)]}$ (b) $K_c = \dfrac{1}{[NH_3(g)] [HCl(g)]}$ (c) $K_c = [Ag^+]^2[CrO_4^{2-}]$

9. (a) decrease; will not affect K_p (b) increase; will not affect K_p (c) increase; K_p will decrease   (d) decrease; K_p will increase

10. $K_c = 4.06$

11. [CO] is decreased by 0.060 mol/L, [CO_2] is increased by 0.060 mol/L.

12. (a) [PCl_3] $= 0.200\ M$, [Cl_2] $= 0.100\ M$, [PCl_5] $=$ zero (b) by $0.080\ M$ (c) [PCl_3] $= 0.120\ M$, [Cl_2] $= 0.020\ M$, [PCl_5] $= 0.080\ M$ (d) $K_c = 33$

13. $8.98 \times 10^{-3}\ M$

14. [HI] $= 0.312\ M$, [H_2] $= 0.044\ M$, [I_2] $= 0.044\ M$

15. [NO] $= 1.1 \times 10^{-17}\ M$

Review Problems

14.19

(a) $K_c = \dfrac{[POCl_3]^2}{[PCl_3]^2 [O_2]}$ (b) $K_c = \dfrac{[SO_2]^2 [O_2]}{[SO_3]^2}$

(c) $K_c = \dfrac{[NO]^2 [H_2O]^2}{[N_2H_4] [O_2]^2}$ (d) $K_c = \dfrac{[NO_2]^2 [H_2O]^8}{[N_2H_4] [H_2O_2]^6}$

(e) $K_c = \dfrac{[SO_2] [HCl]^2}{[SOCl_2] [H_2O]}$

14.21

(a) $K_p = \dfrac{P_{POCl_3}^2}{P_{PCl_3}^2 P_{O_2}}$ (b) $K_p = \dfrac{P_{SO_2}^2 P_{O_2}}{P_{SO_3}^2}$

(c) $K_p = \dfrac{P_{NO}^2 P_{H_2O}^2}{P_{N_2H_4} P_{O_2}^2}$ (d) $K_p = \dfrac{P_{NO_2}^2 P_{H_2O}^8}{P_{N_2H_4} P_{H_2O_2}^6}$

(e) $K_p = \dfrac{P_{SO_2} P_{HCl}^2}{P_{SOCl_2} P_{H_2O}}$

14.23

(a) $K_c = \dfrac{[Ag(NH_3)_2^+]}{[Ag^+][NH_3]^2}$ (b) $K_c = \dfrac{[Cd(SCN)_4^{2-}]}{[Cd^{2+}][SCN^-]^4}$

14.25 $K_c = 1 \times 10^{85}$

14.27

(a) $K_c = \dfrac{[HCl]^2}{[H_2] [Cl_2]}$ (b) $K_c = \dfrac{[HCl]}{[H_2]^{1/2} [Cl_2]^{1/2}}$

K_c for equation (b) is the square root of K_c for equation (a).

14.29 $0.0375\ M$ NH₃

14.31 (b)

14.33 $K_c = 11$

14.35 $K_p = 2.7 \times 10^{-2}$

14.37 $K_p = 5.4 \times 10^{-5}$

14.39 (a) 55.5 M, (b) 55.5 M, (c) 55.5 M

14.41

(a) $K_c = \dfrac{[CO]^2}{[O_2]}$ (b) $K_c = [H_2O][SO_2]$

(c) $K_c = \dfrac{[CH_4][CO_2]}{[H_2O]^2}$ (d) $K_c = \dfrac{[H_2O][CO_2]}{[HF]^2}$

(e) $K_c = [H_2O]^5$

14.43 $[HI] = 1.47 \times 10^{-12} M, [Cl_2] = 7.37 \times 10^{-13} M$

14.45 (a) no   (b) to the left

14.47 $4.36 \times 10^{-3} M$ CH_3OH

14.49 $K_c = 0.398$

14.51 $K_c = 0.0955$

14.53 $K_c = 0.915$

14.55 $[Cl_2] = 0.011 M, [Br_2] = 0.011 M$

14.57 $[SO_3] = 0.0703 M, [NO] = 0.0703 M,$
$[NO_2] = 0.0497 M, [SO_2] = 0.0497 M$

14.59 $[CO] = 0.0123 M, [H_2O] = 0.0123 M,$
$[CO_2] = 0.0077 M, [H_2] = 0.0077 M$

14.61 $[H_2] = 8.9 \times 10^{-19} M, [Cl_2] = 8.9 \times 10^{-19} M$

14.63 $[CO] = 5.0 \times 10^{-4} M$

14.65 $[PCl_3] = 3.0 \times 10^{-5} M$

14.67 $[NO_2] = 0.0281 M, [SO_2] = 0.0281 M$

14.69 $[CO] = 0.20 M, [H_2O] = 0.20 M$

Additional Exercises

14.71 (a) 0.187 atm   (b) 0.437 atm

14.73 $[NO_2] = 0.0449 M, [NO] = 0.0700 M, [N_2O] = 0.0426 M, [O_2] = 0.0676 M$

14.75 $[NO] = 1.4 \times 10^{-8} M, [N_2O] = 0.010 M, [NO_2] = 0.010 M$

14.79 0.261 mol NO, 0.261 mol NO_2

Chapter 15

Practice Exercises

1. (a) OH^- (b) I^- (c) NO_2^- (d) $H_2PO_4^-$ (e) HPO_4^{2-}
(f) PO_4^{3-} (g) H^- (h) NH_3

2. (a) H_2O_2 (b) HSO_4^- (c) HCO_3^- (d) HCN
(e) NH_3 (f) NH_4^+ (g) H_3PO_4 (h) $H_2PO_4^-$

3. $HCO_3^- + H_2PO_4^- \rightleftharpoons H_2CO_3 + HPO_4^{2-}$
     Base 1   Acid 2   Acid 1   Base 2

4. $PO_4^{3-} + HC_2H_3O_2 \rightleftharpoons HPO_4^{2-} + C_2H_3O_2^-$
     Base 1   Acid 2   Acid 1   Base 2

5.
$HPO_4^{2-} + H_3O^+ \rightarrow H_2PO_4^- + H_2O$ HPO_4^{2-} acts as a base
$HPO_4^{2-} + OH^- \rightarrow PO_4^{3-} + H_2O$ HPO_4^{2-} acts as an acid

6. substances on the right

7. (a) HBr (b) H_2Te (c) CH_3SH

8. $HClO_3$

9. (a) HIO_4 (b) H_2SeO_4 (c) H_3AsO_4

10.
(a) The fluoride ion is more likely to behave as a Lewis base because it has a filled octet of electrons (lone electron pairs) that it could donate; fluoride would not be able to accept more electrons because it cannot have more than an octet.
(b) $BeCl_2$ is more likely to behave as a Lewis acid since beryllium does not have an octet and therefore is able to accept more electrons.

11. $[H^+] = 1.3 \times 10^{-9}$, basic

12. pH $= 3.44$, pOH $= 10.56$, acidic

13. pH $= 5.17$

14. (a) $[H^+] = 1.3 \times 10^{-3}$, acidic   (b) $[H^+] = 1.4 \times 10^{-4}$, acidic   (c) $[H^+] = 1.5 \times 10^{-11}$, basic   (d) $[H^+] = 7.8 \times 10^{-5}$, acidic   (e) $[H^+] = 2.5 \times 10^{-12}$, basic

15. $[H^+] = 2.0 \times 10^{-12}$, pH $= 11.70$

Review Problems

15.31 (a) HF (b) $N_2H_5^+$ (c) $C_5H_5NH^+$ (d) HO_2^-
(e) H_2CrO_4

15.33. (a) $HNO_3 + N_2H_4 \rightleftharpoons NO_3^- + N_2H_5^+$
         Acid 1   Base 2   Base 1   Acid 2

(b) $NH_3 + N_2H_5^+ \rightleftharpoons NH_4^+ + N_2H_4$
    Base 1   Acid 2   Acid 1   Base 2

(c) $H_2PO_4^- + CO_3^{2-} \rightleftharpoons HPO_4^{2-} + HCO_3^-$
    Acid 1   Base 2   Base 1   Acid 2

(d) $HIO_3 + HC_2O_4^- \rightleftharpoons IO_3^- + H_2C_2O_4$
    Acid 1   Base 2   Base 1   Acid 2

15.35 (a) H_2Se, larger central atom   (b) III, more electronegative atom   (c) PH_3, larger central atom

15.37 (a) HIO_4, central atom holds more oxygen atoms
(b) H_3AsO_4, central atom holds more oxygen atoms

15.39
(a) H_3PO_4, more electronegative P
(b) HNO_3, more electronegative N
(c) $HClO_4$, more electronegative Cl

15.41

15.43

15.45

15.47 $[H^+] = 1.5 \times 10^{-7}, [OH^-] = 1.5 \times 10^{-7}, pH = 6.82$, $pOH = 6.82; pH + pOH = 13.64$; yes, water is neutral because $[H^+] = [OH^-]$.

15.49 (a) $4.2 \times 10^{-12}\,M$ (b) $7.1 \times 10^{-10}\,M$
(c) $1.8 \times 10^{-6}\,M$ (d) $0.024\,M$

15.51 (a) 11.38 (b) 9.15 (c) 5.74 (d) 1.62

15.53 4.72

15.55
(a) $[H^+] = 7.2 \times 10^{-4}\,M, [OH^-] = 1.4 \times 10^{-11}\,M$
(b) $[H^+] = 1.7 \times 10^{-3}\,M, [OH^-] = 6.0 \times 10^{-12}\,M$
(c) $[H^+] = 5.6 \times 10^{-10}\,M, [OH^-] = 1.8 \times 10^{-5}\,M$
(d) $[H^+] = 5.8 \times 10^{-14}\,M, [OH^-] = 0.17\,M$
(e) $[H^+] = 2.0 \times 10^{-6}\,M, [OH^-] = 5.0 \times 10^{-9}\,M$

15.57 $[H^+] = 0.010\,M, pH = 2.00$

15.59 $[OH^-] = 0.15\,M; pOH = 0.82$,
$pH = 13.18$

15.61 $2.0 \times 10^{-3}\,M$

Additional Exercises

15.63 conjugate acid $(CH_3)_2NH_2^+$, conjugate base $(CH_3)_2N^-$

15.65 (a) H_2O_2 has more oxygen atoms than H_2O, and the oxygen–hydrogen bonds are more polar. (b) acidic

15.67 Equilibrium lies to the right, favoring the weaker acid and base.

15.70 $pH = 4.29$

15.71 $pH = 6.80$

Chapter 16

Practice Exercises

1.
(a) $HCHO_2 + H_2O \rightleftharpoons H_3O^+ + CHO_2^-$
$$K_a = \frac{[H_3O^+][CHO_2^-]}{[HCHO_2]}$$
(b) $(CH_3)_2NH_2^+ + H_2O \rightleftharpoons H_3O^+ + (CH_3)_2NH$
$$K_a = \frac{[H_3O^+][(CH_3)_2NH]}{[(CH_3)_2NH_2^+]}$$
(c) $H_2PO_4^- + H_2O \rightleftharpoons H_3O^+ + HPO_4^{2-}$
$$K_a = \frac{[H_3O^+][HPO_4^{2-}]}{[H_2PO_4^-]}$$

2. If $pK_a = 3.16$ then $K_a = 6.9 \times 10^{-4}$, stronger acid; if $pK_a = 4.14$ then $K_a = 7.2 \times 10^{-5}$, weaker acid.

3.
(a) $(CH_3)_3N + H_2O \rightleftharpoons OH^- + (CH_3)_3NH^+$
$$K_b = \frac{[OH^-][(CH_3)_3NH^+]}{[(CH_3)_3N]}$$
(b) $SO_3^{2-} + H_2O \rightleftharpoons OH^- + HSO_3^-$
$$K_b = \frac{[OH^-][HSO_3^-]}{[SO_3^{2-}]}$$
(c) $NH_2OH + H_2O \rightleftharpoons OH^- + NH_3OH^+$
$$K_b = \frac{[OH^-][NH_3OH^+]}{[NH_2OH]}$$

4. 5.6×10^{-11}

5. $K_a = 1.7 \times 10^{-5}, pK_a = 4.78$

6. $K_b = 1.6 \times 10^{-6}, pK_b = 5.79$

7. 3.08

8. 8.59

9. 5.35

10. (a) basic (b) neutral (c) acidic

11. 8.07

12. 5.13

13. basic

14. 10.79

15. 4.82

16. 9.36

17. moles of CHO_2^- : moles of $HCHO_2 = 1.43 : 1.00$, or $1.00 : 0.70$

18. 4.84; yes, the buffer can handle this much OH^- with less than 0.20 pH change.

19. $[H^+] = 2.6 \times 10^{-3}\,M, pH = 2.59$, $[C_6H_6O_6^{2-}] = 2.7 \times 10^{-12}\,M$

20. 10.24

21. $[H_2SO_3] = 8.3 \times 10^{-3}\,M$

22. (a) 2.37 (b) 3.74 (c) 4.23 (d) 8.22

23. 3.66

Review Problems

16.32 1.5×10^{-11}

16.34 5.6×10^{-3}

16.36 7.1×10^{-11}

16.38 $K_a = 0.02, pK_a = 1.7$

16.40 $K_b = 6 \times 10^{-4}, pK_b = 3.2$

16.42 $[HC_3H_5O_2] = 0.15\,M, [H^+] = 4.6 \times 10^{-3}\,M$, $[C_3H_5O_2^-] = 4.6 \times 10^{-3}\,M, pH = 2.34$

16.44 2.78

16.46 10.26

16.48 $0.46\,M$

16.50 $[HCN] = 1.8 \times 10^{-3}\,M, pH = 11.26$

16.52 5.73

16.54 4.29

16.56 7.0×10^{-10}

16.58 0.77%

16.60 10.67

16.62 % ionization = 6.7%

16.64 10.44

16.66 3.22

16.68 4.97

16.70 9.00

16.72 pH decreases by 0.23 units

16.74 pH will decrease by 0.06 units

16.76 $[NH_4Cl]/[NH_3] = 1$

16.78 22 g

16.80 0.96 g

16.82 $pH_{initial} = 4.79$, $pH_{final} = 4.76$; if the HCl were added to pure water, pH = 1.78

16.84 $[H^+] = 3.2 \times 10^{-3} M$, $[OH^-] = 3.1 \times 10^{-12} M$, pH = 2.50, $[C_6H_6O_6{}^{2-}] = 2.7 = 10^{-12} M$, $[HC_6H_6O_6{}^-] = 3.1 \times 10^{-3} M$, $[H_2C_6H_6O_6] = 0.15 M$

16.86 $[H^+] = [H_2PO_4{}^-] = 0.14 M$, $[HPO_4{}^{2-}] = 6.3 \times 10^{-8} M$, $[PO_4{}^{3-}] = 2.0 \times 10^{-19} M$, $[H_3PO_4] = 2.9 M$, pH = 0.85

16.88 $[H^+] = [HC_4H_4O_6{}^-] = 0.021 M$, $[C_4H_4O_6{}^{2-}] = 4.3 \times 10^{-5} M$, $[H_2C_4H_4O_6] = 0.48 M$

16.90 $[H^+] = [H_2PO_3{}^-] = 0.095 M$, $[HPO_3{}^{2-}] = 2.6 \times 10^{-7} M$, pH = 1.02

16.92 $[OH^-] = [HSO_3{}^-] = 1.3 \times 10^{-4} M$, pH = 10.13

16.94 25 g

16.96 pH = 9.70

16.98 $[PO_4{}^{3-}] = 0.41 M$, $[HPO_4{}^{2-}] = 9.4 \times 10^{-2} M$, $[H_2PO_4{}^-] = 1.6 \times 10^{-7} M$, $[H_3PO_4] = 2.4 \times 10^{-18} M$, pH = 12.97

16.100 8.22; indicator cresol red or thymol blue

16.102 12.85

16.104 (a) 2.87 (b) 4.58 (c) 4.74 (d) 8.72

Additional Exercises

16.107 9.56

16.110
(a) 3.8 g $HC_2H_3O_2$, 18 g $NaC_2H_3O_2 \cdot 2H_2O$
(b) $[HC_2H_3O_2] = 0.25 M$, $[NaC_2H_3O_2] = 0.60 M$
(c) 12.60
(d) 1.40

16.112 pH > 7, because K_b of CN^- is larger than K_a of $NH_4{}^+$

16.114 630 mL

16.116 2.52

16.118 $[H_3PO_4] = 0.26 M$, $[H_2PO_4{}^-] = [H^+] = 0.043 M$, $[HPO_4{}^{2-}] = 6.3 \times 10^{-8} M$, $[PO_4{}^{3-}] = 6.6 \times 10^{-19} M$

Chapter 17

Practice Exercises

1. (a) $K_{sp} = [Ba^{2+}][CrO_4{}^{2-}]$ (b) $K_{sp} = [Ag^+]^3[PO_4{}^{3-}]$

2. 3.2×10^{-10}

3. 3.98×10^{-8}

4. 1.0×10^{-10}

5. 3.9×10^{-8}

6. (a) 7.1×10^{-7} mol L^{-1} (b) 1.3×10^{-4} mol L^{-1}

7. 4.2×10^{-16} mol L^{-1}

8. 1.3×10^{-35} mol L^{-1}

9. Precipitate will form.

10. Precipitate will not form.

11. Precipitate will not form.

12. Precipitate will not form.

13. if pH is less than or equal to 2.9

14. If pH is above 5.40 and lower than 6.13 $CaCO_3$ will precipitate but not $NiCO_3$.

15. 4.9×10^{-3} mol AgCl dissolves in 1.0 L of 0.10 M NH_3; 1.3×10^{-5} mol AgCl dissolves in 1.0 L of H_2O (~380 times less).

16. 4.1 mol NH_3

Review Problems

17.9 (a) $K_{sp} = [Ca^{2+}][F^-]^2$ (b) $K_{sp} = [Ag^+]^2[CO_3{}^{2-}]$
(c) $K_{sp} = [Pb^{2+}][SO_4{}^{2-}]$ (d) $K_{sp} = [Fe^{3+}][OH^-]^3$
(e) $K_{sp} = [Pb^{2+}][I^-]^2$ (f) $K_{sp} = [Cu^{2+}][OH^-]^2$

17.11 1.10×10^{-10}

17.13 8.0×10^{-7}

17.15 2.8×10^{-18}

17.17 8.1×10^{-3} mol L^{-1}

17.19 1.3×10^{-4} mol L^{-1}

17.21 $[LiF] = 0.041$ mol L^{-1} (more soluble), $[BaF_2] = 7.5 \times 10^{-3}$ mol L^{-1}

17.23 2.8×10^{-18} mol L^{-1}

17.25 4.9×10^{-3} mol L^{-1}

17.27 (a) 4.4×10^{-4} mol L^{-1} (b) 9.5×10^{-6} mol L^{-1}
(c) 9.5×10^{-7} mol L^{-1} (d) 6.3×10^{-7} mol L^{-1}

17.29 3.0×10^{-11} mol L^{-1}

17.31 1.5×10^{-3} mol L^{-1}

17.33 Precipitate will not form.

17.35 (a) Precipitate will not form.
(b) Precipitate will form.

17.37 2.3×10^{-8} mol L^{-1}

17.39 pH ≤ 1.35

17.41 (a) $K_{form} = \dfrac{[CuCl_4{}^{2-}]}{[Cu^{2+}][Cl^-]^4}$ (b) $K_{form} = \dfrac{[AgI_2{}^-]}{[Ag^+][I^-]^2}$
(c) $K_{form} = \dfrac{[Cr(NH_3)_6{}^{3+}]}{[Cr^{3+}][NH_3]^6}$

17.43 (a) $K_{inst} = \dfrac{[Co^{3+}][NH_3]^6}{[Co(NH_3)_6{}^{3+}]}$ (b) $K_{inst} = \dfrac{[Hg^{2+}][I^-]^4}{[HgI_4{}^{2-}]}$
(c) $K_{inst} = \dfrac{[Fe^{2+}][CN^-]^6}{[Fe(CN)_6{}^{4-}]}$

17.45 450 g

Additional Exercises

17.48 FeS is soluble in 8 M HCl : 3.92 mol FeS dissolves per liter of 8 M HCl.

17.50 (a) 0.34 g AgCl (b) $[Na^+] = 0.024 M$, $[NO_3{}^-] = 0.060 M$, $[Ag^+] = 0.036 M$, $[Cl^-] = 5.0 \times 10^{-9} M$
(c) 40%

17.52 If pH ≤ 3.49, neither carbonate will precipitate; if 3.49 < pH ≤ 4.15, $PbCO_3$ will precipitate but $La_2(CO_3)_3$ will not; if pH > 4.15, both carbonates precipitate.

17.54 $5.0 \times 10^{-4} M$

17.57 $1.8 \times 10^{-10} M$

Chapter 18

Practice Exercises

1. (b)

2. $w = -154$ L atm $q = +154$ L atm

3. When the gas is compressed work is done on the system, and the energy the system gains is converted to heat energy. Since there is no exchange of heat with the surroundings the temperature of the gas will increase.

4. -214.6 kJ, % difference $= 1.14\%$

5. (a) negative (b) positive

6. (a) negative (b) negative (c) negative (d) negative (e) positive

7. (a) -229 kJ (b) $\Delta S° = -120.9$ J K^{-1}, $\Delta G° = -1482$ J K^{-1}

9. (a) -69.7 kJ (b) -120.1 kJ

10. A maximum of 788.2 kJ of work can be obtained per mol Al.

11. 341 °C or 614 K

12. Yes, the reaction will occur spontaneously.

13. No, the reaction is not spontaneous.

14. $\Delta G° = +33$ kJ, $\Delta G°_{473} = -29.0$ kJ

15. The reaction is spontaneous in the forward direction.

16. -33.3 kJ

17. 0.26

18. 3.84×10^4

Review Problems

18.50 $+1000$ J, endothermic

18.52 $+121$ J

18.54
(a) $\Delta H° = +24.58$ kJ, $\Delta E = +19.6$ kJ
(b) $\Delta H° = -178$ kJ, $\Delta E = -175$ kJ
(c) $\Delta H = \Delta E = +847.6$ kJ
(d) $\Delta H = \Delta E = +65.029$ kJ

18.56 (a) -178 kJ, favored (b) -311 kJ, favored (c) $+1084.3$ kJ, not favored (d) $+65.2$ kJ, not favored (e) $+64.2$ kJ, not favored

18.58 probability of all heads is $1/16 = 0.0625$; probability of an even head-tail distribution is $6/16 = 0.375$.

18.60 (a) negative (b) negative (c) negative (d) positive

18.62 (a) -198.3 J K^{-1}, not favored (b) -332.3 J K^{-1}, not favored (c) $+92.6$ J K^{-1}, favored (d) $+14$ J K^{-1}, favored (e) $+159.6$ J K^{-1}, favored

18.64
(a) -52.8 J mol^{-1} K^{-1}
(b) -74.0 J mol^{-1} K^{-1}
(c) -90.1 J mol^{-1} K^{-1}
(d) -868.9 J mol^{-1} K^{-1}
(e) -318 J mol^{-1} K^{-1}

18.66 -269.7 J mol^{-1} K^{-1}

18.68 -209 kJ

18.70 (a) -82.3 kJ (b) -8.8 kJ (c) $+70.7$ kJ (d) -14.3 kJ (e) -715 kJ

18.72 -4.7 kJ

18.74 $+0.16$ kJ

18.76 maximum work 1299.8 kJ

18.78 333 K or 60 °C

18.80 $+101$ J mol^{-1} K^{-1}

18.82 The reaction will be spontaneous.

18.84 -91.8 kJ

18.86 (a) $K_p = 10^{263}$ (b) $K_p = 2.88 \times 10^{-25}$

18.88 The system is not at equilibrium, it will shift to the right.

18.90 $K_p = 8.000 \times 10^8$

18.92 $K_p = 1$; the equilibrium will shift toward the reactants.

18.94 $K_p = 10^{121}$

18.96 1.16×10^3 kJ mol^{-1}

18.98 C—O bond energy: 354 kJ

18.100 C=S bond energy: 577.7 kJ

18.102 S—F bond energy: 308.0 kJ

18.104 S=O bond energy: 587.8 kJ

18.106 $\Delta H = -149$ kJ

Additional Exercises

18.113 At constant pressure $w = -7.43$ kJ, in bomb calorimeter $w = $ zero.

18.114 -15.0 L atm

18.117 $T\Delta S$ must be larger than ΔH for ΔG to be negative.

18.120 (a) $H_2C_2O_4(s) + \frac{1}{2} O_2(g) \rightarrow 2CO_2(g) + H_2O(l)$ (b) $H_2 + 2C + 2O_2 \rightarrow H_2C_2O_4$ (c) -826.8 kJ (d) $\Delta S°_{formation} = +274.6$ J mol^{-1} K^{-1}, $\Delta S°_{combustion} = -431.9$ J mol^{-1} K^{-1} (e) $\Delta G°_{formation} = -908.6$ kJ mol^{-1}, $\Delta G°_{combustion} = -117.3$ kJ

18.123 $w_{ethanol} = 8.44 \times 10^4$ kJ, $w_{octane} = 1.23 \times 10^5$ kJ; octane is a better fuel.

18.125 -179 kJ/mol

Chapter 19

Practice Exercises

1.

anode	$2Br^-(aq) \rightarrow Br_2(aq) + 2e^-$
cathode	$2H_2O(l) + 2e^- \rightarrow H_2(g) + 2OH^-(aq)$
net	$2H_2O(l) + 2Br^-(aq) \rightarrow$ $Br_2(aq) + H_2(g) + 2OH^-(aq)$

2. 8.29×10^{-3} mol OH$^-$

3. 7.35 min

4. 3.67 A

5.
anode $\quad$ $Mg(s) \rightarrow Mg^{2+}(aq) + 2e^-$
cathode $\quad$ $Fe^{2+}(aq) + 2e^- \rightarrow Fe(s)$
$Mg(s)|Mg^{2+}(aq)\|Fe^{2+}(aq)|Fe(s)$

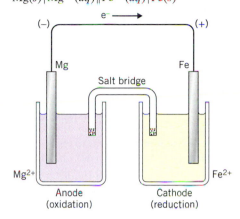

6.
anode $\quad$ $Al(s) \rightarrow Al^{3+}(aq) + 3e^-$
cathode $\quad$ $Pb^{2+}(aq) + 2e^- \rightarrow Pb(s)$

7. -2.40 V

8. $Br_2 + H_2SO_3 + H_2O \rightarrow 2Br^- + SO_4^{2-} + 4H^+$

9. $NiO_2(s) + 2H_2O(l) + Fe(s) \rightarrow$
$$Ni(OH)_2(s) + Fe(OH)_2(s), 1.37 \text{ V}$$

10. $3MnO_4^-(aq) + 24H^+(aq) + 5Cr(s) \rightarrow$
$$3Mn^{2+}(aq) + 12H_2O + 5Cr^{3+}(aq), 2.25 \text{ V}$$

11. (a) not spontaneous $\quad$ (b) spontaneous

12. $Sn(s)$

13. -3.78×10^2 kJ

14. 2.9×10^{-16}

15. 1.04 V

16. If $E_{cell} = 0.57$ V then $[Cu^{2+}] = 1.9 \times 10^{-4}$ M; if $E_{cell} = 0.82$ V then $[Cu^{2+}] = 6.9 \times 10^{-13}$ M.

17. $[Cr^{3+}] = 6.7 \times 10^{-4}$ M

Review Problems

19.54 (a) 2.40×10^3 C $\quad$ (b) 1.2×10^4 C $\quad$ (c) 3.24×10^4 C

19.56 (a) 0.40 mol e^- $\quad$ (b) 0.70 mol e^- $\quad$ (c) 4.50 mol e^-
(d) 5.0×10^{-2} mol e^-

19.58 0.0371 mol Cr^{3+}

19.60 2.68 g $Fe(OH)_2$

19.62 51.5 hr

19.64 66.2 A

19.66 233 mL

19.68
(a) anode $\quad$ $Cd(s) \rightarrow Cd^{2+}(aq) + 2e^-$
$\quad$ cathode $\quad$ $Au^{3+}(aq) + 3e^- \rightarrow Au(s)$
$\quad$ cell $\quad$ $3Cd(s) + 2Au^{3+}(aq) \rightarrow 3Cd^{3+}(aq) + 2Au(s)$
(b) anode $\quad$ $Pb(s) + SO_4^{2-}(aq) \rightarrow PbSO_4(s) + 2e^-$
$\quad$ cathode $\quad$ $PbO_2(s) + SO_4^{2-}(aq) + 4H^+(aq) + 2e^- \rightarrow$
$$PbSO_4(s) + 2H_2O(l)$$

cell $\quad$ $Pb(s) + PbO_2(s) + 2SO_4^{2-}(aq) + 4H^+(aq) \rightarrow$
$$2PbSO_4(s) + 2H_2O(l)$$
(c) anode $\quad$ $Cr(s) \rightarrow Cr^{3+}(aq) + 3e^-$
$\quad$ cathode $\quad$ $Cu^{2+}(aq) + 2e^- \rightarrow Cu(s)$
$\quad$ cell $\quad$ $2Cr(s) + 3Cu^{2+}(aq) \rightarrow 2Cr^{3+}(aq) + 3Cu(s)$

19.70
(a) $Fe(s)|Fe^{2+}(aq)\|Cd^{2+}(aq)|Cd(s)$
(b) $Pt(s), Cl^-(aq)|Cl_2(g)\|Br_2(aq)|Br^-(aq), Pt(s)$
(c) $Ag(s)|Ag^+(aq)\|Au^{3+}(aq)\|Au(s)$

19.72 (a) $Sn(s)$ $\quad$ (b) $Br^-(aq)$ $\quad$ (c) $Zn(s)$ $\quad$ (d) $I^-(aq)$

19.76 (a) spontaneous $\quad$ (b) not spontaneous
(c) spontaneous

19.78 0.74 V

19.80
(a) $Zn(s)|Zn^{2+}(aq)\|Co^{2+}(aq)|Co(s), 0.48$ V
(b) $Mg(s)|Mg^{2+}(aq)\|Ni^{2+}(aq)|Ni(s), 2.12$ V
(c) $Sn(s)|Sn^{2+}(aq)\|Au^{3+}(aq)|Au(s) 1.56$ V

19.82 $BrO_3^- + 6H^+ + 6I^- \rightarrow Br^- + 3H_2O + 3I_2, 0.90$ V

19.84 $4HOCl + 2H^+ + S_2O_3^{2-} \rightarrow 2Cl_2 + H_2O + 2H_2SO_3$

19.86 not spontaneous

19.88 $2H_2O \rightarrow 2H_2 + O_2, -2.06$ V

19.90 1.0×10^2 kJ

19.92 (a) 0.54 V $\quad$ (b) -5.2×10^2 kJ $\quad$ (c) 1.7×10^{91}

19.94 $K_c = 0.31$

19.96 2.07 V

19.98 2.97×10^{-15} M

19.100 2.03×10^{-15}

Additional Exercises

19.103 V^{3+}

19.106 0.965 s

19.107 663 hr

19.108 1.54 V

Chapter 20

Practice Problems

1. 5566 K or 5293 °C

2. $[Ag(S_2O_3)_2]^{3-}, (NH_4)_3[Ag(S_2O_3)_2]$

3. $AlCl_3 \cdot 6H_2O, [Al(H_2O)_6]^{3+}$

4. (a) potassium hexacyanoferrate(III)
(b) dichlorobis(ethylenediamine)chromium(III) sulfate

5. (a) $[SnCl_6]^{2-}$ $\quad$ (b) $(NH_4)_2[Fe(CN)_4(H_2O)_2]$

6. (a) 6 $\quad$ (b) 6 $\quad$ (c) 6 $\quad$ (d) 6

Review Problems

20.64 735 K $\quad$ or $\quad$ 462 °C

20.66 (a) Bi_2O_5 $\quad$ (b) PbS $\quad$ (c) PbI_2

20.68 (a) SnO $\quad$ (b) SnS

20.70 (a) $CaCl_2$ $\quad$ (b) BeF_2

20.72 (a) HgS $\quad$ (b) Ag_2S

20.74 (a) HgS $\quad$ (b) Ag_2S

20.76 $[Fe(CN)_6]^{3-}$

20.78 $[CoCl_2(en)_2]^+$

20.80 (a) oxalato (b) sulfido (c) chloro
(d) dimethylamine

20.82
(a) hexaamminenickel(II) ion
(b) triamminetrichlorochromate(II) ion
(c) hexanitrocobaltate(III) ion
(d) diamminetetracyanomanganate(II) ion
(e) trioxalatoferrate(III) ion

20.84 (a) $[Fe(CN)_2(H_2O)_4]^+$ (b) $Ni(C_2O_4)(NH_3)_4$
(c) $[Fe(CN)_4(H_2O)_2]^-$ (d) $K_3[Mn(SCN)_6]$ (e) $[CuCl_4]^{2-}$

20.86 coordination number 6, oxidation number +2

20.88

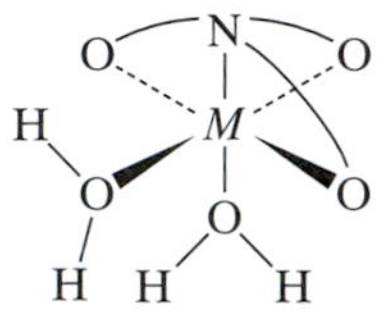

20.90 The two structures are identical, because they are both
cis isomers.

20.92

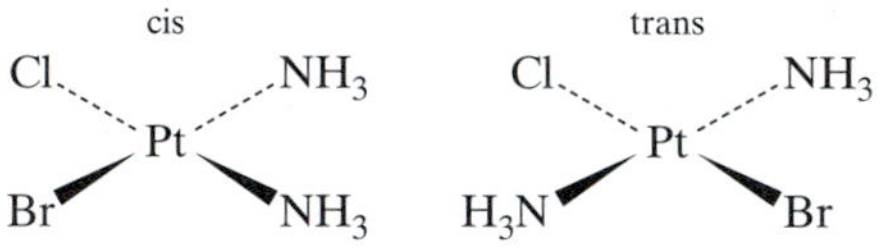

20.94 The atoms joining the donor oxygen atoms
(—C—C—) are represented by a curved line.

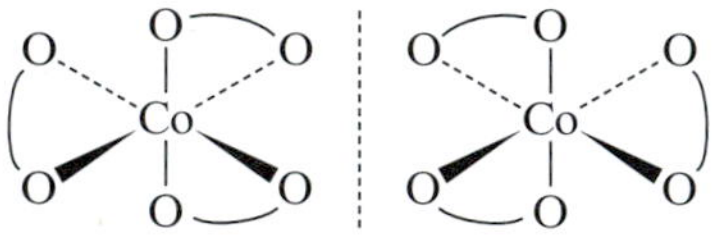

20.96 (a) $[Cr(H_2O)_6]^{3+}$ (b) $[Cr(en)_3]^{3+}$

20.98 $[Cr(CN)_6]^{3-}$

20.100 The shorter wavelength of light has higher frequency;
the frequency absorbed is proportional to the size of Δ.
(a) $[RuCl(NH_3)_5]^{2+}$; for a given ligand and oxidation state, the
size of Δ increases going down the group. (b) $[Ru(NH_3)_6]^{3+}$;
at higher oxidation state the magnitude of Δ is also higher.

20.102 $[CoA_6]^{2+}$; oxidation will remove a higher energy
electron.

20.104 paramagnetic, high spin

Additional Exercises

20.110 (a) $[CrCl(H_2O)_5]^{2+}$ (b) $[CrCl(H_2O)_5]Cl_2 \cdot H_2O$

(c)

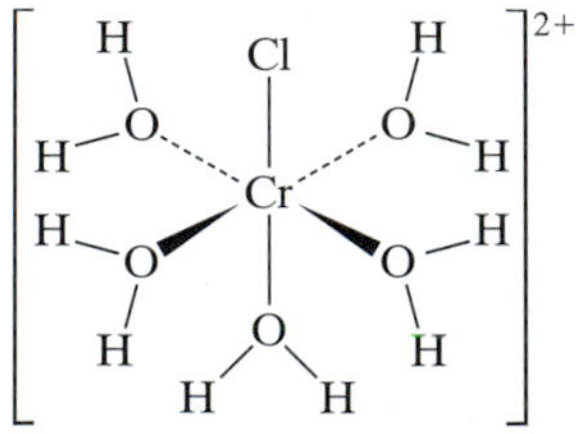

(d) only one

Chapter 21

Review Problems

21.88
(a) $AlP + 3H_2O \rightarrow PH_3 + Al(OH)_3$
(b) $Mg_2C + 4H_2O \rightarrow 2Mg(OH)_2 + CH_4$
(c) $FeS + 2HCl \rightarrow FeCl_2 + H_2S$
(d) $MgSe + H_2SO_4 \rightarrow MgSO_4 + H_2Se$

21.90
(a) $2CO + O_2 \rightarrow 2CO_2$
(b) $2C_2H_6 + 7O_2 \rightarrow 4CO_2 + 6H_2O$
(c) $P_4O_6 + 2O_2 \rightarrow P_4O_{10}$
(d) $4NH_3 + 5O_2 \rightarrow 4NO + 6H_2O$

21.92 $K_p = 1.57 \times 10^{18}$; the equilibrium lies far to the right;
NO_2 is able to exist because its rate of decomposition is very
slow.

21.94 $\Delta H° = -57.9$ kJ/mol; since the reaction is exothermic
(heat is generated), increasing the temperature will shift the
reaction to the left.

21.96 (a) trigonal pyramidal (b) nonlinear, bent
(c) octahedral (d) tetrahedral

21.98 (a) trigonal pyramidal

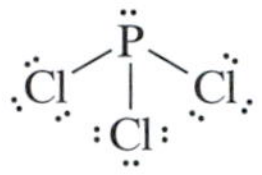

(b) T-shaped

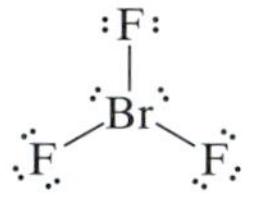

(c) octahedral

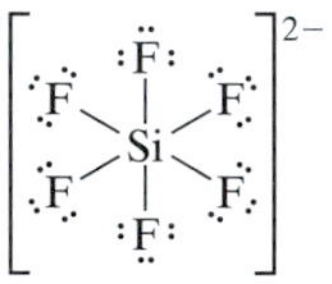

(d) planar triangular

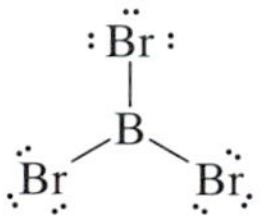

(e) square planar

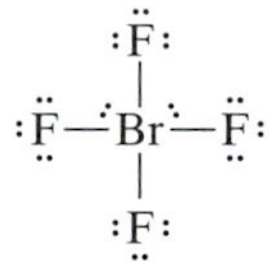

Additional Exercises

21.100
$N_2 + 3H_2 \rightleftharpoons 2NH_3$

$4NH_3 + 5O_2 \xrightarrow{\text{catalyst}} 4NO + 6H_2O$
$2NO + O_2 \rightarrow 2NO_2$
$3NO_2 + H_2O \rightarrow 2HNO_3 + NO$

21.102

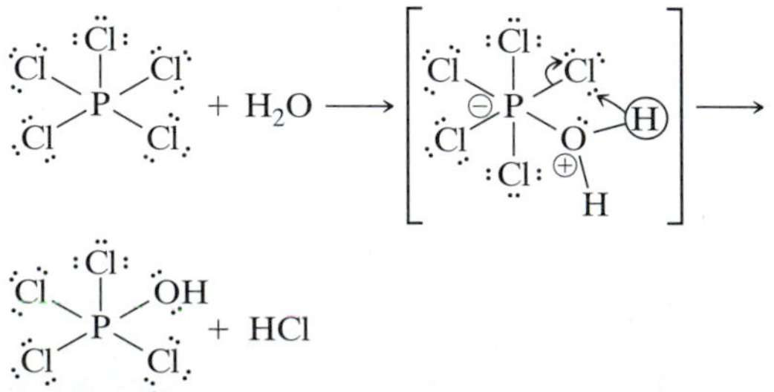

Chapter 22

Practice Exercises

1. $^{226}_{88}\text{Ra} \rightarrow \,^{222}_{86}\text{Rn} + \,^{4}_{2}\text{He} + \,^{0}_{0}\gamma$

2. $^{90}_{38}\text{Sr} \rightarrow \,^{0}_{-1}e + \,^{90}_{39}\text{Y}$

3. 97 units

Review Problems

22.43 1.11×10^{-11} g

22.45 -3.18 ng

22.47 1.8×10^{-13} J/nucleon

22.49 (a) $^{211}_{83}\text{Bi}$ (b) $^{177}_{72}\text{Hf}$ (c) $^{216}_{84}\text{Po}$ (d) $^{19}_{9}\text{F}$

22.51 (a) $^{242}_{94}\text{Pu} \rightarrow \,^{4}_{2}\text{He} + \,^{238}_{92}\text{U}$

(b) $^{28}_{12}\text{Mg} \rightarrow \,^{0}_{-1}e + \,^{28}_{13}\text{Al}$

(c) $^{26}_{14}\text{Si} \rightarrow \,^{0}_{1}e + \,^{26}_{13}\text{Al}$ (d) $^{37}_{18}\text{Ar} + \,^{0}_{-1}e \rightarrow \,^{37}_{17}\text{Cl}$

22.53 (a) $^{261}_{102}\text{No}$ (b) $^{211}_{82}\text{Pb}$ (c) $^{141}_{61}\text{Pm}$ (d) $^{179}_{74}\text{W}$

22.55 $^{1}_{0}n$, a neutron forms

22.57 positron emission, because that would move it toward the band of stability; $^{38}_{19}\text{K} \rightarrow \,^{0}_{1}e + \,^{38}_{18}\text{Ar}$

22.59 0.0469 mg

22.61 $^{53}_{24}\text{Cr}$, $^{51}_{23}\text{V} + \,^{2}_{1}\text{H} \rightarrow \,^{53}_{24}\text{Cr}^* \rightarrow \,^{1}_{1}p + \,^{52}_{23}\text{V}$

22.63 $^{80}_{35}\text{Br}$

22.65 $^{55}_{26}\text{Fe}$, $^{55}_{25}\text{Mn} + \,^{1}_{1}p \rightarrow \,^{1}_{0}n + \,^{55}_{26}\text{Fe}$

22.67 $^{277}_{112}\text{Uub}$

22.69 6.3 m

22.71 6.6 m

22.73 3.1% cesium chloride remains; $BaCl_2$ forms.

22.75 2.84×10^{-3} Ci/g

22.77 1.3×10^9 years old; we made the assumption that all the ^{40}Ar came from the decay of ^{40}K.

22.79 2.7×10^1 years ago

22.81 $^{140}_{54}\text{Xe}$

Additional Exercises

22.92 8.3×10^8 years

22.94 Once ONO*NO forms from NO and radioactive *NO_2, it has an equal chance to dissociate either into NO and O*NO (*NO_2) or into ONO (NO_2) and *NO.

22.97 170 mL

Chapter 23

Practice Exercises

1. (a) 3-methylhexane
(b) 4-ethyl-2,3-dimethylheptane
(c) 5-ethyl-2,4,6-trimethyloctane

2. (a) $CH_3\overset{\displaystyle O}{\overset{\|}{C}}\text{—H}$ or $CH_3\overset{\displaystyle O}{\overset{\|}{C}}\text{—OH}$ (depending on the strength of oxidizing agent)

(b) $CH_3CH_2\overset{\displaystyle }{\underset{\displaystyle \overset{\|}{O}}{C}}CH_2CH_3$

Review Problems

23.82

(a) H—C—N with H's (amine structure)

(b) H—C—Br with Br on top, H below

(c) Cl—C—H with Cl top and bottom

(d) N—O—H with H's on N

(e) H—C≡C—H

(f) N—N with H's

23.84 (a) alkene (b) alcohol (c) ester
(d) carboxylic acid (e) amine (f) alcohol

23.86 (b), (e), (f)

23.88 (a) amine (b) amine (c) amide (d) amine, ketone (carbonyl group is also present)

23.90 (a) identical (b) identical (c) unrelated
(d) isomers (e) identical (f) identical (g) isomers

23.92 (a) pentane (b) 2-methylpentane
(c) 2,4-dimethylhexane

23.94 (a) no cis or trans isomerism

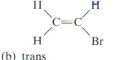

(b) trans   cis

(c) trans   cis

23.96

(a) CH_3CH_3 (b) $\overset{CH_2-CH_2}{\underset{\overset{|}{Cl}\quad\overset{|}{Cl}}{}}$ (c) $\overset{CH_2-CH_2}{\underset{\overset{|}{Br}\quad\overset{|}{Br}}{}}$

(d) $\overset{CH_3-CH_2}{\underset{\quad\overset{|}{Cl}}{}}$ (e) $\overset{CH_3-CH_2}{\underset{\quad\overset{|}{Br}}{}}$ (f) $\overset{CH_3-CH_2}{\underset{\quad\overset{|}{OH}}{}}$

23.98

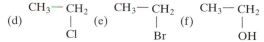

(a) $CH_3CH_2CH_2CH_3$ (b) $CH_3\underset{\overset{|}{Cl}}{CH}-\underset{\overset{|}{Cl}}{CH}CH_3$

(c) $CH_3CH-CHCH_3$ (d) $CH_3CH_2CHCH_3$
    | |
    Br Br   Cl

(e) $CH_3CH_2CHCH_3$ (f) $CH_3CH_2CHCH_3$
      | |
     Br   OH

23.100 An addition reaction would disrupt the delocalized π electrons which give the extra stability to the benzene; consequently, benzene would lose its aromatic character.

23.102 CH_3OH methanol
        CH_3CH_2OH ethanol
        $CH_3CH_2CH_2OH$ 1–propanol
        CH_3CHCH_3 2–propanol
          |
          OH

23.104 $CH_3CH_2CH_2-O-CH_3$ methyl *n*-propyl ether
        $CH_3CH_2-O-CH_2CH_3$ diethyl ether
        $CH_3CH-O-CH_3$ isopropyl methyl ether
           |
         CH_3

23.106 (a) 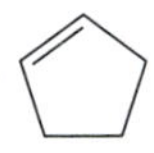 (b) 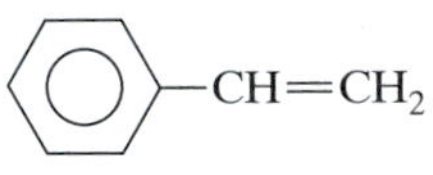 $-CH=CH_2$

(c) 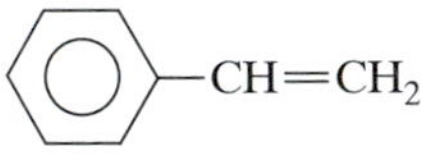 $-CH=CH_2$

23.108 (a)  (b) 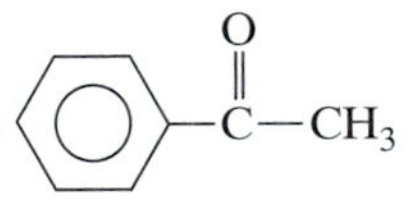 $-\overset{\overset{\displaystyle O}{||}}{C}-CH_3$

(c) $-CH_2-\overset{\overset{\displaystyle O}{||}}{C}-H$

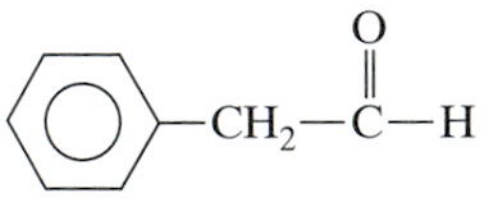

23.110 $CH_3CH=CHCH_3$ is 2-butene and $CH_3CH_2CH=CH_2$ is 1-butene.

23.112 Compound **A** could be easily oxidized; the product is CH_3CH_2COOH.

23.114 (a) $CH_3CH_2COOH + Cl^-$
(b) $CH_3CH_2COOH + CH_3OH$
(c) $CH_3CH_2CH_2COO^-Na^+ + H_2O$

23.116 $CH_3CH_2NHCH_2CH_3 + CH_3COOH$

23.118 $-(CH_2CHCl)_n-$ or
$-CH_2-CH-CH_2-CH-CH_2-CH-CH_2-CH-$
       | | | |
      Cl   Cl   Cl   Cl

23.120

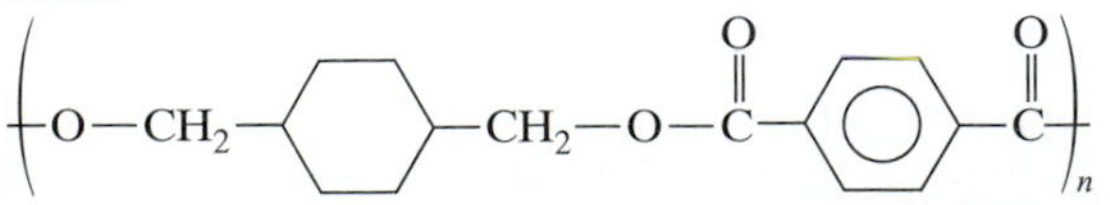

23.122

23.124

23.126 hydrophobic site, fatty acid chains; hydrophilic site, phosphate group (charged units)

23.128 $^+NH_3CH_2\overset{\overset{\displaystyle O}{||}}{C}NHCH_2COO^-$

23.130 $^+NH_3CH_2\overset{\overset{\displaystyle O}{||}}{C}NHCHCOO^-$
                      |
                      $CH_2C_6H_5$

or

$^+NH_3CHC NHCH_2COO^-$
   |
  $CH_2C_6H_5$

Additional Exercises

23.133 (a) Compound **A** was the alcohol, because it will be oxidized into a carboxylic acid which neutralizes NaOH.

(b) 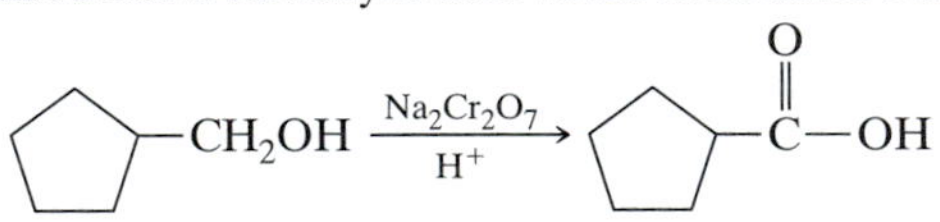 $-CH_2OH$ $\xrightarrow[\text{H}^+]{\text{Na}_2\text{Cr}_2\text{O}_7}$ $-\overset{\overset{\displaystyle O}{||}}{C}-OH$

23.135 (a) 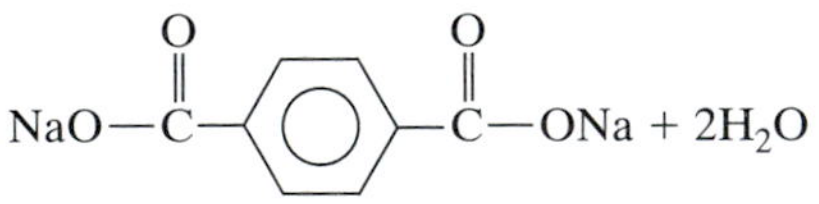 $NaO-\overset{\overset{\displaystyle O}{||}}{C}-$ $-\overset{\overset{\displaystyle O}{||}}{C}-ONa + 2H_2O$

(b)   OH
       |
   $CH_3CHCH_2CH_2CH_3$
(c) $CH_3NHCH_2CH_2CH_3 + H_2O$
(d) $CH_3CH_2OCH_2CH_2COOH + CH_3OH$

(e) 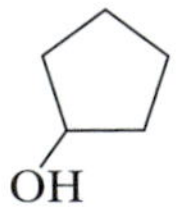
     OH

(f) 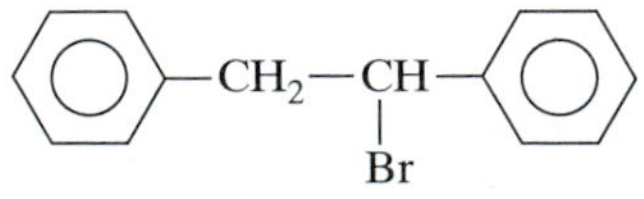 $-CH_2-CH-$
                 |
                 Br

23.138 83.04 g/mol; this is the molecular mass only if the acid is monoprotic.

Tables of Selected Data

Table E.1 Thermodynamic Data for Selected Elements,
Compounds, and Ions (at 25 °C)

Substance	ΔH_f° (kJ/mol)	S° (J/mol K)	ΔG_f° (kJ/mol)
Aluminum			
Al(s)	0	28.3	0
Al^{3+}(aq)	−524.7		−481.2
$AlCl_3$(s)	−704	110.7	−629
Al_2O_3(s)	−1676	51.0	−1576.4
$Al_2(SO_4)_3$(s)	−3441	239	−3100
Arsenic			
As(s)	0	35.1	0
AsH_3(g)	+66.4	223	+68.9
As_4O_6(s)	−1314	214	−1153
As_2O_5(s)	−925	105	−782
H_3AsO_3(aq)	−742.2		
H_3AsO_4(aq)	−902.5		
Barium			
Ba(s)	0	66.9	0
Ba^{2+}(aq)	−537.6	9.6	−560.8
$BaCO_3$(s)	−1219	112	−1139
$BaCrO_4$(s)	−1428.0		
$BaCl_2$(s)	−860.2	125	−810.8
BaO(s)	−553.5	70.4	−525.1
$Ba(OH)_2$(s)	−998.22	112	−875.3
$Ba(NO_3)_2$(s)	−992	214	−795
$BaSO_4$(s)	−1465	132	−1353
Beryllium			
Be(s)	0	9.50	0
$BeCl_2$(s)	−468.6	89.9	−426.3
BeO(s)	−611	14	−582
Bismuth			
Bi(s)	0	56.9	0
$BiCl_3$(s)	−379	177	−315
Bi_2O_3(s)	−576	151	−497

(continued)

Table E.I **Thermodynamic Data for Selected Elements, Compounds, and Ions (at 25 °C)** *(Continued)*

Substance	ΔH°_f (kJ/mol)	S° (J/mol K)	ΔG°_f (kJ/mol)
Boron			
B (s)	0	5.87	0
BCl_3 (g)	−404	290	−389
B_2H_6 (g)	+36	232	+87
B_2O_3 (s)	−1273	53.8	−1194
$B(OH)_3$ (s)	−1094	88.8	−969
Bromine			
Br_2 (l)	0	152.2	0
Br_2 (g)	+30.9	245.4	+3.11
HBr (g)	−36	198.5	+53.1
Br^- (aq)	−121.55	82.4	−103.96
Cadmium			
Cd (s)	0	51.8	0
Cd^{2+} (aq)	−75.90	−73.2	−77.61
$CdCl_2$ (s)	−392	115	−344
CdO (s)	−258.2	54.8	−228.4
CdS (s)	−162	64.9	−156
$CdSO_4$ (s)	−933.5	123	−822.6
Calcium			
Ca (s)	0	41.4	0
Ca^{2+} (aq)	−542.83	−53.1	−553.58
$CaCO_3$ (s)	−1207	92.9	−1128.8
CaF_2 (s)	−741	80.3	−1166
$CaCl_2$ (s)	−795.8	114	−750.2
$CaBr_2$ (s)	−682.8	130	−663.6
CaI_2 (s)	−535.9	143	
CaO (s)	−635.5	40	−604.2
$Ca(OH)_2$ (s)	−986.6	76.1	−896.76
$Ca_3(PO_4)_2$ (s)	−4119	241	−3852
$CaSO_3$ (s)	−1156		
$CaSO_4$ (s)	−1433	107	−1320.3
$CaSO_4 \cdot \frac{1}{2}H_2O$ (s)	−1573	131	−1435.2
$CaSO_4 \cdot 2H_2O$ (s)	−2020	194.0	−1795.7
Carbon			
C (s, graphite)	0	5.69	0
C (s, diamond)	+1.88	2.4	+2.9
CCl_4 (l)	−134	214.4	−65.3
CO (g)	−110	197.9	−137.3
CO_2 (g)	−394	213.6	−394.4
CO_2 (aq)	−413.8	117.6	−385.98

Table E.1 Thermodynamic Data for Selected Elements, Compounds, and Ions (at 25 °C) *(Continued)*

Substance	ΔH_f° (kJ/mol)	S° (J/mol K)	ΔG_f° (kJ/mol)
Carbon			
$H_2CO_3(aq)$	−699.65	187.4	−623.08
$HCO_3^-(aq)$	−691.99	91.2	−586.77
$CO_3^{2-}(aq)$	−677.14	−56.9	−527.81
$CS_2(l)$	+89.5	151.3	+65.3
$CS_2(g)$	+117	237.7	+67.2
$HCN(g)$	+135.1	201.7	+124.7
$CN^-(aq)$	+150.6	94.1	+172.4
$CH_4(g)$	−74.9	186.2	−50.79
$C_2H_2(g)$	+227	200.8	+209
$C_2H_4(g)$	+51.9	219.8	+68.12
$C_2H_6(g)$	−84.5	229.5	−32.9
$C_3H_8(g)$	−104	269.9	−23
$C_4H_{10}(g)$	−126	310.2	−17.0
$C_6H_6(l)$	+49.0	173.3	+124.3
$CH_3OH(l)$	−238	126.8	−166.2
$C_2H_5OH(l)$	−278	161	−174.8
$HCHO_2(g)$	−363	251	+335
$HC_2H_3O_2(l)$	−487.0	160	−392.5
$CH_2O(g)$	−108.6	218.8	−102.5
$CH_3CHO(g)$	−167	250	−129
$(CH_3)_2CO(l)$	−248.1	200.4	−155.4
$C_6H_5CO_2H(s)$	−385.1	167.6	−245.3
$CO(NH_2)_2(s)$	−333.5	104.6	−197.2
$CO(NH_2)_2(aq)$	−319.2	173.8	−203.8
$CH_2(NH_2)CO_2H(s)$	−532.9	103.5	−373.4
Chlorine			
$Cl_2(g)$	0	223.0	0
$Cl^-(aq)$	−167.2	56.5	−131.2
$HCl(g)$	−92.5	186.7	−95.27
$HCl(aq)$	−167.2	56.5	−131.2
$HClO(aq)$	−131.3	106.8	−80.21
Chromium			
$Cr(s)$	0	23.8	0
$Cr^{3+}(aq)$	−232		
$CrCl_2(s)$	−326	115	−282
$CrCl_3(s)$	−563.2	126	−493.7
$Cr_2O_3(s)$	−1141	81.2	−1059
$CrO_3(s)$	−585.8	72.0	−506.2
$(NH_4)_2Cr_2O_7(s)$	−1807		
$K_2Cr_2O_7(s)$	−2033.01		

(continued)

Table E.1 Thermodynamic Data for Selected Elements, Compounds, and Ions (at 25 °C) *(Continued)*

Substance	ΔH_f° (kJ/mol)	S° (J/mol K)	ΔG_f° (kJ/mol)
Cobalt			
Co (s)	0	30.0	0
Co^{2+} (aq)	−59.4	−110	−53.6
CoCl$_2$ (s)	−325.5	106	−282.4
Co(NO$_3$)$_2$ (s)	−422.2	192	−230.5
CoO (s)	−237.9	53.0	−214.2
CoS (s)	−80.8	67.4	−82.8
Copper			
Cu (s)	0	33.15	0
Cu^{2+} (aq)	+64.77	−99.6	+65.49
CuCl (s)	−137.2	86.2	−119.87
CuCl$_2$ (s)	−172	119	−131
Cu$_2$O (s)	−168.6	93.1	−146.0
CuO (s)	−155	42.6	−127
Cu$_2$S (s)	−79.5	121	−86.2
CuS (s)	−53.1	66.5	−53.6
CuSO$_4$ (s)	−771.4	109	−661.8
CuSO$_4 \cdot 5H_2O$ (s)	−2279.7	300.4	−1879.7
Fluorine			
F$_2$ (g)	0	202.7	0
F$^-$ (aq)	−332.6	−13.8	−278.8
HF (g)	−271	173.5	−273
Gold			
Au (s)	0	47.7	0
Au$_2$O$_3$ (s)	+80.8	125	+163
AuCl$_3$ (s)	−118	148	−48.5
Hydrogen			
H$_2$ (g)	0	130.6	0
H$_2$O (l)	−285.9	69.96	−237.2
H$_2$O (g)	−241.8	188.7	−228.6
H$_2$O$_2$ (l)	−187.8	109.6	−120.3
H$_2$O$_2$ (g)	−136.3	233	−105.6
H$_2$Se (g)	+76	219	+62.3
H$_2$Te (g)	+154	234	+138
Iodine			
I$_2$ (s)	0	116.1	0
I$_2$ (g)	+62.4	260.7	+19.3
HI (g)	+26	206	+1.30

Table E.1 Thermodynamic Data for Selected Elements, Compounds, and Ions (at 25 °C) *(Continued)*

Substance	ΔH_f° (kJ/mol)	S° (J/mol K)	ΔG_f° (kJ/mol)
Iron			
$Fe\,(s)$	0	27	0
$Fe^{2+}\,(aq)$	−89.1	−137.7	−78.9
$Fe^{3+}\,(aq)$	−48.5	−315.9	−4.7
$Fe_2O_3\,(s)$	−822.2	90.0	−741.0
$Fe_3O_4\,(s)$	−1118.4	146.4	−1015.4
$FeS\,(s)$	−100.0	60.3	−100.4
$FeS_2\,(s)$	−178.2	52.9	−166.9
Lead			
$Pb\,(s)$	0	64.8	0
$Pb^{2+}\,(aq)$	−1.7	10.5	−24.4
$PbCl_2\,(s)$	−359.4	136	−314.1
$PbO\,(s)$	−217.3	68.7	−187.9
$PbO_2\,(s)$	−277	68.6	−219
$Pb(OH)_2\,(s)$	−515.9	88	−420.9
$PbS\,(s)$	−100	91.2	−98.7
$PbSO_4\,(s)$	−920.1	149	−811.3
Lithium			
$Li\,(s)$	0	28.4	0
$Li^+\,(aq)$	−278.6	10.3	
$LiF\,(s)$	−611.7	35.7	−583.3
$LiCl\,(s)$	−408	59.29	−383.7
$LiBr\,(s)$	−350.3	66.9	−338.87
$Li_2O\,(s)$	−596.5	37.9	−560.5
$Li_3N\,(s)$	−199	37.7	−155.4
Magnesium			
$Mg\,(s)$	0	32.5	0
$Mg^{2+}\,(aq)$	−466.9	−138.1	−454.8
$MgCO_3\,(s)$	−1113	65.7	−1029
$MgF_2\,(s)$	−1124	79.9	−1056
$MgCl_2\,(s)$	−641.8	89.5	−592.5
$MgCl_2 \cdot 2H_2O\,(s)$	−1280	180	−1118
$Mg_3N_2\,(s)$	−463.2	87.9	−411
$MgO\,(s)$	−601.7	26.9	−569.4
$Mg(OH)_2\,(s)$	−924.7	63.1	−833.9
Manganese			
$Mn\,(s)$	0	32.0	0
$Mn^{2+}\,(aq)$	−223	−74.9	−228

(continued)

Table E.1 **Thermodynamic Data for Selected Elements, Compounds, and Ions (at 25 °C)** *(Continued)*

Substance	ΔH_f° (kJ/mol)	S° (J/mol K)	ΔG_f° (kJ/mol)
Manganese			
MnO_4^- (aq)	−542.7	191	−449.4
$KMnO_4$ (s)	−813.4	171.71	−713.8
MnO (s)	−385	60.2	−363
Mn_2O_3 (s)	−959.8	110	−882.0
MnO_2 (s)	−520.9	53.1	−466.1
Mn_3O_4 (s)	−1387	149	−1280
$MnSO_4$ (s)	−1064	112	−956
Mercury			
Hg (l)	0	76.1	0
Hg (g)	+61.32	175	+31.8
Hg_2Cl_2 (s)	−265.2	192.5	−210.8
$HgCl_2$ (s)	−224.3	146.0	−178.6
HgO (s)	−90.83	70.3	−58.54
HgS (s, red)	−58.2	82.4	−50.6
Nickel			
Ni (s)	0	30	0
$NiCl_2$ (s)	−305	97.5	−259
NiO (s)	−244	38	−216
NiO_2 (s)			−199
$NiSO_4$ (s)	−891.2	77.8	−773.6
$NiCO_3$ (s)	−664.0	91.6	−615.0
$Ni(CO)_4$ (g)	−220	399	−567.4
Nitrogen			
N_2 (g)	0	191.5	0
NH_3 (g)	−46.0	192.5	−16.7
NH_4^+ (aq)	−132.5	113	−79.37
N_2H_4 (g)	+95.40	238.4	+159.3
N_2H_4 (l)	+50.6	121.2	+149.4
NH_4Cl (s)	−314.4	94.6	−203.9
NO (g)	+90.4	210.6	+86.69
NO_2 (g)	+34	240.5	+51.84
N_2O (g)	+81.5	220.0	+103.6
N_2O_4 (g)	+9.16	304	+98.28
N_2O_5 (g)	+11	356	+115
HNO_3 (l)	−174.1	155.6	−79.9
NO_3^- (aq)	−205.0	146.4	−108.74

Table E.1 Thermodynamic Data for Selected Elements, Compounds, and Ions (at 25 °C) *(Continued)*

Substance	ΔH_f° (kJ/mol)	S° (J/mol K)	ΔG_f° (kJ/mol)
Oxygen			
$O_2(g)$	0	205.0	0
$O_3(g)$	+143	238.8	+163
$OH^-(aq)$	−230.0	−10.75	−157.24
Phosphorus			
$P(s, \text{white})$	0	41.09	0
$P_4(g)$	+314.6	163.2	+278.3
$PCl_3(g)$	−287.0	311.8	−267.8
$PCl_5(g)$	−374.9	364.6	−305.0
$PH_3(g)$	+5.4	210.2	+12.9
$P_4O_6(s)$	−1640		
$POCl_3(g)$	−1109.7	646.5	−1019
$POCl_3(l)$	−1186	26.36	−1035
$P_4O_{10}(s)$	−2984	228.9	−2698
$H_3PO_4(s)$	−1279	110.5	−1119
Potassium			
$K(s)$	0	64.18	0
$K^+(aq)$	−252.4	102.5	−283.3
$KF(s)$	−567.3	66.6	−537.8
$KCl(s)$	−436.8	82.59	−408.3
$KBr(s)$	−393.8	95.9	−380.7
$KI(s)$	−327.9	106.3	−324.9
$KOH(s)$	−424.8	78.9	−379.1
$K_2O(s)$	−361	98.3	−322
$K_2SO_4(s)$	−1433.7	176	−1316.4
Silicon			
$Si(s)$	0	19	0
$SiH_4(g)$	+33	205	+52.3
$SiO_2(s, \text{alpha})$	−910.0	41.8	−856
Silver			
$Ag(s)$	0	42.55	0
$Ag^+(aq)$	+105.58	72.68	+77.11
$AgCl(s)$	−127.1	96.2	−109.8
$AgBr(s)$	−100.4	107.1	−96.9
$AgNO_3(s)$	−124	141	−32
$Ag_2O(s)$	−31.1	121.3	−11.2

(continued)

Table E. I Thermodynamic Data for Selected Elements, Compounds, and Ions (at 25 °C) *(Continued)*

Substance	ΔH_f° (kJ/mol)	S° (J/mol K)	ΔG_f° (kJ/mol)
Sodium			
$Na(s)$	0	51.0	0
$Na^+(aq)$	−240.12	59.0	−261.91
$NaF(s)$	−571	51.5	−545
$NaCl(s)$	−413	72.38	−384.0
$NaBr(s)$	−360	83.7	−349
$NaI(s)$	−288	91.2	−286
$NaHCO_3(s)$	−947.7	102	−851.9
$Na_2CO_3(s)$	−1131	136	−1048
$Na_2O_2(s)$	−510.9	94.6	−447.7
$Na_2O(s)$	−510	72.8	−376
$NaOH(s)$	−426.8	64.18	−382
$Na_2SO_4(s)$	−1384.49	149.49	−1266.83
Sulfur			
$S(s, \text{rhombic})$	0	31.8	0
$SO_2(g)$	−297	248	−300
$SO_3(g)$	−396	256	−370
$H_2S(g)$	−20.6	206	−33.6
$H_2SO_4(l)$	−813.8	157	−689.9
$H_2SO_4(aq)$	−909.3	20.1	−744.5
$SF_6(g)$	−1209	292	−1105
Tin			
$Sn(s, \text{white})$	0	51.6	0
$Sn^{2+}(aq)$	−8.8	−17	−27.2
$SnCl_4(l)$	−511.3	258.6	−440.2
$SnO(s)$	−285.8	56.5	−256.9
$SnO_2(s)$	−580.7	52.3	−519.6
Zinc			
$Zn(s)$	0	41.6	0
$Zn^{2+}(aq)$	−153.9	−112.1	−147.06
$ZnCl_2(s)$	−415.1	111	−369.4
$ZnO(s)$	−348.3	43.6	−318.3
$ZnS(s)$	−205.6	57.7	−201.3
$ZnSO_4(s)$	−982.8	120	−874.5

Table E.2 Heats of Formation of Gaseous Atoms from Elements
in Their Standard Sites

Element	$\Delta H^\circ_f(\text{kJ mol}^{-1})^a$	Element	$\Delta H^\circ_f(\text{kJ mol}^{-1})^a$
Group IA		*Group IVA*	
H	217.89	C	716.67
Li	161.5	Si	450
Na	108.2		
K	89.62	*Group VA*	
Rb	82.0	N	472.68
Cs	78.2	P	332.2
Group IIA		*Group VIA*	
Be	324.3	O	249.17
Mg	146.4	S	276.98
Ca	178.2		
Sr	163.6	*Group VIIA*	
Ba	177.8	F	79.14
		Cl	121.47
Group IIIA		Br	112.38
B	560	I	107.48
Al	329.7		

^{a}All values in this table are positive because forming the gaseous atoms from the elements is endothermic; it involves bond breaking.

Table E.3 Average Bond Energies

Bond	Bond Energy (kJ mol^{-1})
C—C	348
C=C	612
C≡C	960
C—H	412
C—N	305
C=N	613
C≡N	890
C—O	360
C=O	743
C—F	484
C—Cl	338
C—Br	276
C—I	238
H—H	436
H—F	565
H—Cl	431
H—Br	366
H—I	299
H—N	388
H—O	463
H—S	338
H—Si	376

Table E.4 **Vapor Pressure of Water as a Function of Temperature**

Temp (°C)	Vapor Pressure (torr)	Temp (°C)	Vapor Pressure (torr)	Temp (°C)	Vapor Pressure (torr)	Temp (°C)	Vapor Pressure (torr)
0	4.58	26	25.2	52	102.1	78	327.3
1	4.93	27	26.7	53	107.2	79	341.0
2	5.29	28	28.3	54	112.5	80	355.1
3	5.68	29	30.0	55	118.0	81	369.7
4	6.10	30	31.8	56	123.8	82	384.9
5	6.54	31	33.7	57	129.8	83	400.6
6	7.01	32	35.7	58	136.1	84	416.8
7	7.51	33	37.7	59	142.6	85	433.6
8	8.04	34	39.9	60	149.4	86	450.9
9	8.61	35	41.2	61	156.4	87	468.7
10	9.21	36	44.6	62	163.8	88	487.1
11	9.84	37	47.1	63	171.4	89	506.1
12	10.5	38	49.7	64	179.3	90	525.8
13	11.2	39	52.4	65	187.5	91	546.0
14	12.0	40	55.3	66	196.1	92	567.0
15	12.8	41	58.3	67	205.0	93	588.6
16	13.6	42	61.5	68	214.2	94	610.9
17	14.5	43	64.8	69	223.7	95	633.9
18	15.5	44	68.3	70	233.7	96	657.6
19	16.5	45	71.9	71	243.9	97	682.1
20	17.5	46	75.6	72	254.6	98	707.3
21	18.6	47	79.6	73	265.7	99	733.2
22	19.8	48	83.7	74	277.2	100	760.0
23	21.1	49	88.0	75	289.1		
24	22.4	50	92.5	76	301.4		
25	23.8	51	97.2	77	314.1		

Table E.5 Solubility Product Constants

Salt	Solubility Equilibrium	K_{sp}
Fluorides		
MgF_2	$MgF_2(s) \rightleftharpoons Mg^{2+}(aq) + 2F^-(aq)$	6.6×10^{-9}
CaF_2	$CaF_2(s) \rightleftharpoons Ca^{2+}(aq) + 2F^-(aq)$	3.9×10^{-11}
SrF_2	$SrF_2(s) \rightleftharpoons Sr^{2+}(aq) + 2F^-(aq)$	2.9×10^{-9}
BaF_2	$BaF_2(s) \rightleftharpoons Ba^{2+}(aq) + 2F^-(aq)$	1.7×10^{-6}
LiF	$LiF(s) \rightleftharpoons Li^+(aq) + F^-(aq)$	1.7×10^{-3}
PbF_2	$PbF_2(s) \rightleftharpoons Pb^{2+}(aq) + 2F^-(aq)$	3.6×10^{-8}
Chlorides		
$CuCl$	$CuCl(s) \rightleftharpoons Cu^+(aq) + Cl^-(aq)$	1.9×10^{-7}
$AgCl$	$AgCl(s) \rightleftharpoons Ag^+(aq) + Cl^-(aq)$	1.8×10^{-10}
Hg_2Cl_2	$Hg_2Cl_2(s) \rightleftharpoons Hg_2^{2+}(aq) + 2Cl^-(aq)$	1.2×10^{-18}
$TlCl$	$TlCl(s) \rightleftharpoons Tl^+(aq) + Cl^-(aq)$	1.8×10^{-4}
$PbCl_2$	$PbCl_2(s) \rightleftharpoons Pb^{2+}(aq) + 2Cl^-(aq)$	1.7×10^{-5}
$AuCl_3$	$AuCl_3(s) \rightleftharpoons Au^{3+}(aq) + 3Cl^-(aq)$	3.2×10^{-25}
Bromides		
$CuBr$	$CuBr(s) \rightleftharpoons Cu^+(aq) + Br^-(aq)$	5×10^{-9}
$AgBr$	$AgBr(s) \rightleftharpoons Ag^+(aq) + Br^-(aq)$	5.0×10^{-13}
Hg_2Br_2	$Hg_2Br_2(s) \rightleftharpoons Hg_2^{2+}(aq) + 2Br^-(aq)$	5.6×10^{-23}
$HgBr_2$	$HgBr_2(s) \rightleftharpoons Hg^{2+}(aq) + 2Br^-(aq)$	1.3×10^{-19}
$PbBr_2$	$PbBr_2(s) \rightleftharpoons Pb^{2+}(aq) + 2Br^-(aq)$	2.1×10^{-6}
Iodides		
CuI	$CuI(s) \rightleftharpoons Cu^+(aq) + I^-(aq)$	1×10^{-12}
AgI	$AgI(s) \rightleftharpoons Ag^+(aq) + I^-(aq)$	8.3×10^{-17}
Hg_2I_2	$Hg_2I_2(s) \rightleftharpoons Hg_2^{2+}(aq) + 2I^-(aq)$	4.7×10^{-29}
HgI_2	$HgI_2(s) \rightleftharpoons Hg^{2+}(aq) + 2I^-(aq)$	1.1×10^{-28}
PbI_2	$PbI_2(s) \rightleftharpoons Pb^{2+}(aq) + 2I^-(aq)$	7.9×10^{-9}
Hydroxides		
$Mg(OH)_2$	$Mg(OH)_2(s) \rightleftharpoons Mg^{2+}(aq) + 2OH^-(aq)$	7.1×10^{-12}
$Ca(OH)_2$	$Ca(OH)_2(s) \rightleftharpoons Ca^{2+}(aq) + 2OH^-(aq)$	6.5×10^{-6}
$Mn(OH)_2$	$Mn(OH)_2(s) \rightleftharpoons Mn^{2+}(aq) + 2OH^-(aq)$	1.6×10^{-13}
$Fe(OH)_2$	$Fe(OH)_2(s) \rightleftharpoons Fe^{2+}(aq) + 2OH^-(aq)$	7.9×10^{-16}
$Fe(OH)_3$	$Fe(OH)_3(s) \rightleftharpoons Fe^{3+}(aq) + 3OH^-(aq)$	1.6×10^{-39}
$Co(OH)_2$	$Co(OH)_2(s) \rightleftharpoons Co^{2+}(aq) + 2OH^-(aq)$	1×10^{-15}
$Co(OH)_3$	$Co(OH)_3(s) \rightleftharpoons Co^{3+}(aq) + 3OH^-(aq)$	3×10^{-45}
$Ni(OH)_2$	$Ni(OH)_2(s) \rightleftharpoons Ni^{2+}(aq) + 2OH^-(aq)$	6×10^{-16}
$Cu(OH)_2$	$Cu(OH)_2(s) \rightleftharpoons Cu^{2+}(aq) + 2OH^-(aq)$	4.8×10^{-20}
$V(OH)_3$	$V(OH)_3(s) \rightleftharpoons V^{3+}(aq) + 3OH^-(aq)$	4×10^{-35}
$Cr(OH)_3$	$Cr(OH)_3(s) \rightleftharpoons Cr^{3+}(aq) + 3OH^-(aq)$	2×10^{-30}
Ag_2O	$Ag_2O(s) + H_2O \rightleftharpoons 2Ag^+(aq) + 2OH^-(aq)$	1.9×10^{-8}
$Zn(OH)_2$	$Zn(OH)_2(s) \rightleftharpoons Zn^{2+}(aq) + 2OH^-(aq)$	3.0×10^{-16}

(continued)

Table E.5 Solubility Product Constants (*Continued*)

Salt	Solubility Equilibrium	K_{sp}
Hydroxides		
$Cd(OH)_2$	$Cd(OH)_2(s) \rightleftharpoons Cd^{2+}(aq) + 2OH^-(aq)$	5.0×10^{-15}
$Al(OH)_3$ (alpha form)	$Al(OH)_3(s) \rightleftharpoons Al^{3+}(aq) + 3OH^-(aq)$	3×10^{-34}
Cyanides		
$AgCN$	$AgCN(s) \rightleftharpoons Ag^+(aq) + CN^-(aq)$	2.2×10^{-16}
$Zn(CN)_2$	$Zn(CN)_2(s) \rightleftharpoons Zn^{2+}(aq) + 2CN^-(aq)$	3×10^{-16}
Sulfites		
$CaSO_3$	$CaSO_3(s) \rightleftharpoons Ca^{2+}(aq) + SO_3^{2-}(aq)$	3×10^{-7}
Ag_2SO_3	$Ag_2SO_3(s) \rightleftharpoons 2Ag^+(aq) + SO_3^{2-}(aq)$	1.5×10^{-14}
$BaSO_3$	$BaSO_3(s) \rightleftharpoons Ba^{2+}(aq) + SO_3^{2-}(aq)$	8×10^{-7}
Sulfates		
$CaSO_4$	$CaSO_4(s) \rightleftharpoons Ca^{2+}(aq) + SO_4^{2-}(aq)$	2.4×10^{-5}
$SrSO_4$	$SrSO_4(s) \rightleftharpoons Sr^{2+}(aq) + SO_4^{2-}(aq)$	3.2×10^{-7}
$BaSO_4$	$BaSO_4(s) \rightleftharpoons Ba^{2+}(aq) + SO_4^{2-}(aq)$	1.1×10^{-10}
$RaSO_4$	$RaSO_4(s) \rightleftharpoons Ra^{2+}(aq) + SO_4^{2-}(aq)$	4.3×10^{-11}
Ag_2SO_4	$Ag_2SO_4(s) \rightleftharpoons 2Ag^+(aq) + SO_4^{2-}(aq)$	1.5×10^{-5}
Hg_2SO_4	$Hg_2SO_4(s) \rightleftharpoons Hg_2^{2+}(aq) + SO_4^{2-}(aq)$	7.4×10^{-7}
$PbSO_4$	$PbSO_4(s) \rightleftharpoons Pb^{2+}(aq) + SO_4^{2-}(aq)$	6.3×10^{-7}
Chromates		
$BaCrO_4$	$BaCrO_4(s) \rightleftharpoons Ba^{2+}(aq) + CrO_4^{2-}(aq)$	2.1×10^{-10}
$CuCrO_4$	$CuCrO_4(s) \rightleftharpoons Cu^{2+}(aq) + CrO_4^{2-}(aq)$	3.6×10^{-6}
Ag_2CrO_4	$Ag_2CrO_4(s) \rightleftharpoons 2Ag^+(aq) + CrO_4^{2-}(aq)$	1.2×10^{-12}
Hg_2CrO_4	$Hg_2CrO_4(s) \rightleftharpoons Hg_2^{2+}(aq) + CrO_4^{2-}(aq)$	2.0×10^{-9}
$CaCrO_4$	$CaCrO_4(s) \rightleftharpoons Ca^{2+}(aq) + CrO_4^{2-}(aq)$	7.1×10^{-4}
$PbCrO_4$	$PbCrO_4(s) \rightleftharpoons Pb^{2+}(aq) + CrO_4^{2-}(aq)$	1.8×10^{-14}
Carbonates		
$MgCO_3$	$MgCO_3(s) \rightleftharpoons Mg^{2+}(aq) + CO_3^{2-}(aq)$	3.5×10^{-8}
$CaCO_3$	$CaCO_3(s) \rightleftharpoons Ca^{2+}(aq) + CO_3^{2-}(aq)$	4.5×10^{-9}
$SrCO_3$	$SrCO_3(s) \rightleftharpoons Sr^{2+}(aq) + CO_3^{2-}(aq)$	9.3×10^{-10}
$BaCO_3$	$BaCO_3(s) \rightleftharpoons Ba^{2+}(aq) + CO_3^{2-}(aq)$	5.0×10^{-9}
$MnCO_3$	$MnCO_3(s) \rightleftharpoons Mn^{2+}(aq) + CO_3^{2-}(aq)$	5.0×10^{-10}
$FeCO_3$	$FeCO_3(s) \rightleftharpoons Fe^{2+}(aq) + CO_3^{2-}(aq)$	2.1×10^{-11}
$CoCO_3$	$CoCO_3(s) \rightleftharpoons Co^{2+}(aq) + CO_3^{2-}(aq)$	1.0×10^{-10}
$NiCO_3$	$NiCO_3(s) \rightleftharpoons Ni^{2+}(aq) + CO_3^{2-}(aq)$	1.3×10^{-7}
$CuCO_3$	$CuCO_3(s) \rightleftharpoons Cu^{2+}(aq) + CO_3^{2-}(aq)$	2.3×10^{-10}
Ag_2CO_3	$Ag_2CO_3(s) \rightleftharpoons 2Ag^+(aq) + CO_3^{2-}(aq)$	8.1×10^{-12}
Hg_2CO_3	$Hg_2CO_3(s) \rightleftharpoons Hg_2^{2+}(aq) + CO_3^{2-}(aq)$	8.9×10^{-17}

Table E.5 Solubility Product Constants *(Continued)*

Salt	Solubility Equilibrium	K_{sp}
Carbonates		
$ZnCO_3$	$ZnCO_3(s) \rightleftharpoons Zn^{2+}(aq) + CO_3^{2-}(aq)$	1.0×10^{-10}
$CdCO_3$	$CdCO_3(s) \rightleftharpoons Cd^{2+}(aq) + CO_3^{2-}(aq)$	1.8×10^{-14}
$PbCO_3$	$PbCO_3(s) \rightleftharpoons Pb^{2+}(aq) + CO_3^{2-}(aq)$	7.4×10^{-14}
Phosphates		
$Ca_3(PO_4)_2$	$Ca_3(PO_4)_2(s) \rightleftharpoons 3Ca^{2+}(aq) + 2PO_4^{3-}(aq)$	2.0×10^{-29}
$Mg_3(PO_4)_2$	$Mg_3(PO_4)_2(s) \rightleftharpoons 3Mg^{2+}(aq) + 2PO_4^{3-}(aq)$	6.3×10^{-26}
$SrHPO_4$	$SrHPO_4(s) \rightleftharpoons Sr^{2+}(aq) + HPO_4^{2-}(aq)$	1.2×10^{-7}
$BaHPO_4$	$BaHPO_4(s) \rightleftharpoons Ba^{2+}(aq) + HPO_4^{2-}(aq)$	4.0×10^{-8}
$LaPO_4$	$LaPO_4(s) \rightleftharpoons La^{3+}(aq) + PO_4^{3-}(aq)$	3.7×10^{-23}
$Fe_3(PO_4)_2$	$Fe_3(PO_4)_2(s) \rightleftharpoons 3Fe^{2+}(aq) + 2PO_4^{3-}(aq)$	1×10^{-36}
Ag_3PO_4	$Ag_3PO_4(s) \rightleftharpoons 3Ag^{+}(aq) + PO_4^{3-}(aq)$	2.8×10^{-18}
$FePO_4$	$FePO_4(s) \rightleftharpoons Fe^{3+}(aq) + PO_4^{3-}(aq)$	4.0×10^{-27}
$Zn_3(PO_4)_2$	$Zn_3(PO_4)_2(s) \rightleftharpoons 3Zn^{2+}(aq) + 2PO_4^{3-}(aq)$	5×10^{-36}
$Pb_3(PO_4)_2$	$Pb_3(PO_4)_2(s) \rightleftharpoons 3Pb^{2+}(aq) + 2PO_4^{3-}(aq)$	3.0×10^{-44}
$Ba_3(PO_4)_2$	$Ba(PO_4)_2(s) \rightleftharpoons 3Ba^{2+}(aq) + 2PO_4^{3-}(aq)$	5.8×10^{-38}
Ferrocyanides		
$Zn_2[Fe(CN)_6]$	$Zn_2[Fe(CN)_6](s) \rightleftharpoons 2Zn^{2+}(aq) + Fe(CN)_6^{4-}(aq)$	2.1×10^{-16}
$Cd_2[Fe(CN)_6]$	$Cd_2[Fe(CN)_6](s) \rightleftharpoons 2Cd^{2+}(aq) + Fe(CN)_6^{4-}(aq)$	4.2×10^{-18}
$Pb_2[Fe(CN)_6]$	$Pb_2[Fe(CN)_6](s) \rightleftharpoons 2Pb^{2+}(aq) + Fe(CN)_6^{4-}(aq)$	9.5×10^{-19}

Table E.6 Formation Constants of Complexes (25°C)

Complex Ion Equilibrium	K_f	Complex Ion Equilibrium	K_f
Halide Complexes		$Cu^{+} + 2I^{-} \rightleftharpoons [CuI_2]^{-}$	8×10^{8}
$Al^{3+} + 6F^{-} \rightleftharpoons [AlF_6]^{3-}$	2.5×10^{4}	$Ag^{+} + 2I^{-} \rightleftharpoons [AgI_2]^{-}$	1×10^{11}
$Al^{3+} + 4F^{-} \rightleftharpoons [AlF_4]^{-}$	2.0×10^{8}	$Pb^{2+} + 4I^{-} \rightleftharpoons [PbI_4]^{2-}$	3×10^{4}
$Be^{2+} + 4F^{-} \rightleftharpoons [BeF_4]^{2-}$	1.3×10^{13}	$Hg^{2+} + 4I^{-} \rightleftharpoons [HgI_4]^{2-}$	1.9×10^{30}
$Sn^{4+} + 6F^{-} \rightleftharpoons [SnF_6]^{2-}$	1×10^{25}	**Ammonia Complexes**	
$Cu^{+} + 2Cl^{-} \rightleftharpoons [CuCl_2]^{-}$	3×10^{5}	$Ag^{+} + 2NH_3 \rightleftharpoons [Ag(NH_3)_2]^{+}$	1.6×10^{7}
$Ag^{+} + 2Cl^{-} \rightleftharpoons [AgCl_2]^{-}$	1.8×10^{5}	$Zn^{2+} + 4NH_3 \rightleftharpoons [Zn(NH_3)_4]^{2+}$	7.8×10^{8}
$Pb^{2+} + 4Cl^{-} \rightleftharpoons [PbCl_4]^{2-}$	2.5×10^{15}	$Cu^{2+} + 4NH_3 \rightleftharpoons [Cu(NH_3)_4]^{2+}$	1.1×10^{13}
$Zn^{2+} + 4Cl^{-} \rightleftharpoons [ZnCl_4]^{2-}$	1.6	$Hg^{2+} + 4NH_3 \rightleftharpoons [Hg(NH_3)_4]^{2+}$	1.8×10^{19}
$Hg^{2+} + 4Cl^{-} \rightleftharpoons [HgCl_4]^{2-}$	5.0×10^{15}	$Co^{2+} + 6NH_3 \rightleftharpoons [Co(NH_3)_6]^{2+}$	5.0×10^{4}
$Cu^{+} + 2Br^{-} \rightleftharpoons [CuBr_2]^{-}$	8×10^{5}	$Co^{3+} + 6NH_3 \rightleftharpoons [Co(NH_3)_6]^{3+}$	4.6×10^{33}
$Ag^{+} + 2Br^{-} \rightleftharpoons [AgBr_2]^{-}$	1.7×10^{7}	$Cd^{2+} + 6NH_3 \rightleftharpoons [Cd(NH_3)_6]^{2+}$	2.6×10^{5}
$Hg^{2+} + 4Br^{-} \rightleftharpoons [HgBr_4]^{2-}$	1×10^{21}	$Ni^{2+} + 6NH_3 \rightleftharpoons [Ni(NH_3)_6]^{2+}$	2.0×10^{8}

(continued)

Table E.6 Formation Constants of Complexes (25°C)

Complex Ion Equilibrium	K_f	Complex Ion Equilibrium	K_f
Cyanide Complexes		$Co^{2+} + 3\,en \rightleftharpoons [Co(en)_3]^{2+}$	1.3×10^{14}
		$Co^{3+} + 3\,en \rightleftharpoons [Co(en)_3]^{3+}$	4.8×10^{48}
$Fe^{2+} + 6CN^- \rightleftharpoons [Fe(CN)_6]^{4-}$	1.0×10^{24}	$Ni^{2+} + 3\,en \rightleftharpoons [Ni(en)_3]^{2+}$	4.1×10^{17}
$Fe^{3+} + 6CN^- \rightleftharpoons [Fe(CN)_6]^{3-}$	1.0×10^{31}	$Cu^{2+} + 2\,en \rightleftharpoons [Cu(en)_2]^{2+}$	3.5×10^{19}
$Ag^+ + 2CN^- \rightleftharpoons [Ag(CN)_2]^-$	5.3×10^{18}	$Mn^{2+} + 3\,bipy \rightleftharpoons [Mn(bipy)_3]^{2+}$	1×10^6
$Cu^+ + 2CN^- \rightleftharpoons [Cu(CN)_2]^-$	1.0×10^{16}	$Fe^{2+} + 3\,bipy \rightleftharpoons [Fe(bipy)_3]^{2+}$	1.6×10^{17}
$Cd^{2+} + 4CN^- \rightleftharpoons [Cd(CN)_4]^{2-}$	7.7×10^{16}	$Ni^{2+} + 3\,bipy \rightleftharpoons [Ni(bipy)_3]^{2+}$	3.0×10^{20}
$Au^+ + 2CN^- \rightleftharpoons [Au(CN)_2]^-$	2×10^{38}	$Co^{2+} + 3\,bipy \rightleftharpoons [Co(bipy)_3]^{2+}$	8×10^{15}
Complexes with Other Monodentate Ligands		$Mn^{2+} + 3\,phen \rightleftharpoons [Mn(phen)_3]^{2+}$	2×10^{10}
		$Fe^{2+} + 3\,phen \rightleftharpoons [Fe(phen)_3]^{2+}$	1×10^{21}
Methylamine (CH_3NH_2)		$Co^{2+} + 3\,phen \rightleftharpoons [Co(phen)_3]^{2+}$	6×10^{19}
$Ag^+ + 2CH_3NH_2 \rightleftharpoons [Ag(CH_3NH_2)_2]^+$	7.8×10^6	$Ni^{2+} + 3\,phen \rightleftharpoons [Ni(phen)_3]^{2+}$	2×10^{24}
Thiocyanate ion (SCN^-)		$Co^{2+} + 3C_2O_4^{2-} \rightleftharpoons [Co(C_2O_4)_3]^{4-}$	4.5×10^6
$Cd^{2+} + 4SCN^- \rightleftharpoons [Cd(SCN)_4]^{2-}$	1×10^3	$Fe^{3+} + 3C_2O_4^{2-} \rightleftharpoons [Fe(C_2O_4)_3]^{3-}$	3.3×10^{20}
$Cu^{2+} + 2SCN^- \rightleftharpoons [Cu(SCN)_2]$	5.6×10^3		
$Fe^{3+} + 3SCN^- \rightleftharpoons [Fe(SCN)_3]$	2×10^6	**Complexes of Other Ligands**[a]	
$Hg^{2+} + 4SCN^- \rightleftharpoons [Hg(SCN)_4]^{2-}$	5.0×10^{21}	$Zn^{2+} + EDTA^{4-} \rightleftharpoons [Zn(EDTA)]^{2-}$	3.8×10^{16}
		$Mg^{2+} + 2NTA^{3-} \rightleftharpoons [Mg(NTA)_2]^{4-}$	1.6×10^{10}
Hydroxide Ion		$Ca^{2+} + 2NTA^{3-} \rightleftharpoons [Ca(NTA)_2]^{4-}$	3.2×10^{11}
$Cu^{2+} + 4OH^- \rightleftharpoons [Cu(OH)_4]^{2-}$	1.3×10^{16}		
$Zn^{2+} + 4OH^- \rightleftharpoons [Zn(OH)_4]^{2-}$	2×10^{20}		
Complexes with Bidentate Ligands[a]			
$Mn^{2+} + 3\,en \rightleftharpoons [Mn(en)_3]^{2+}$	6.5×10^5		
$Fe^{2+} + 3\,en \rightleftharpoons [Fe(en)_3]^{2+}$	5.2×10^9		

[a] en, ethylenediamine; bipy, bipyridyl; phen, 1,10-phenanthroline; EDTA^{4-}, ethylendiaminetetraacetate ion; and NTA^{3-}, nitrilotriacetate ion.

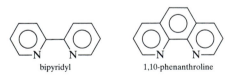

bipyridyl   1,10-phenanthroline

Table E.7 Ionization Constants of Weak Acids and Bases (Alternative Formulas in Brackets)

Monoprotic Acids	Name	K_a
$HC_2O_2Cl_3$ [Cl_3CCO_2H]	trichloroacetic acid	2.2×10^{-1}
HIO_3	iodic acid	1.69×10^{-1}
$HC_2HO_2Cl_2$ [Cl_2CHCO_2H]	dichloroacetic acid	5.0×10^{-2}
$HC_2H_2O_2Cl$ [$ClCH_2CO_2H$]	chloroacetic acid	1.36×10^{-3}
HNO_2	nitrous acid	7.1×10^{-4}
HF	hydrofluoric acid	6.8×10^{-4}
$HOCN$	cyanic acid	3.5×10^{-4}
$HCHO_2$ [HCO_2H]	formic acid	1.8×10^{-4}
$HC_3H_5O_3$ [$CH_3CH(OH)CO_2H$]	lactic acid	1.38×10^{-4}

Table E.7 Ionization Constants of Weak Acids and Bases (Alternative Formulas in Brackets) *(Continued)*

Monoprotic Acids	Name	K_a
$HC_4H_3N_2O_3$	barbituric acid	9.8×10^{-5}
$HC_7H_5O_2$ [$C_6H_5CO_2H$]	benzoic acid	6.28×10^{-5}
$HC_4H_7O_2$ [$CH_3CH_2CH_2CO_2H$]	butanoic acid	1.52×10^{-5}
HN_3	hydrazoic acid	1.8×10^{-5}
$HC_2H_3O_2$ [CH_3CO_2H]	acetic acid	1.8×10^{-5}
$HC_3H_5O_2$ [$CH_3CH_2CO_2H$]	propanoic acid	1.34×10^{-5}
$HOCl$	hypochlorous acid	3.0×10^{-8}
$HOBr$	hypobromous acid	2.1×10^{-9}
HCN	hydrocyanic acid	6.2×10^{-10}
HC_6H_5O	phenol	1.3×10^{-10}
HOI	hypoiodous acid	2.3×10^{-11}
H_2O_2	hydrogen peroxide	1.8×10^{-12}

Polyprotic Acids	Name	K_{a_1}	K_{a_2}	K_{a_3}
H_2SO_4	sulfuric acid	Large	1.0×10^{-2}	
H_2CrO_4	chromic acid	5.0	1.5×10^{-6}	
$H_2C_2O_4$	oxalic acid	5.6×10^{-2}	5.4×10^{-5}	
H_3PO_3	phosphorous acid	3×10^{-2}	1.6×10^{-7}	
H_2SO_3	sulfurous acid	1.2×10^{-2}	6.6×10^{-8}	
H_2SeO_3	selenous acid	4.5×10^{-3}	1.1×10^{-8}	
H_2TeO_3	tellurous acid	3.3×10^{-3}	2.0×10^{-8}	
$H_2C_3H_2O_4$ [$HO_2CCH_2CO_2H$]	malonic acid	1.4×10^{-3}	2.0×10^{-6}	
$H_2C_8H_4O_4$	phthalic acid	1.1×10^{-3}	3.9×10^{-6}	
$H_2C_4H_4O_6$	tartaric acid	9.2×10^{-4}	4.3×10^{-5}	
$H_2C_6H_6O_6$	ascorbic acid	6.8×10^{-5}	2.7×10^{-12}	
H_2CO_3	carbonic acid	4.5×10^{-7}	4.7×10^{-11}	
H_3PO_4	phosphoric acid	7.1×10^{-3}	6.3×10^{-8}	4.5×10^{-13}
H_3AsO_4	arsenic acid	5.6×10^{-3}	1.7×10^{-7}	4.0×10^{-12}
$H_3C_6H_5O_7$	citric acid	7.1×10^{-4}	1.7×10^{-5}	6.3×10^{-6}

Weak Bases	Name	K_b
$(CH_3)_2NH$	dimethylamine	9.6×10^{-4}
CH_3NH_2	methylamine	4.4×10^{-4}
$CH_3CH_2NH_2$	ethylamine	4.3×10^{-4}
$(CH_3)_3N$	trimethylamine	7.4×10^{-5}
NH_3	ammonia	1.8×10^{-5}
N_2H_4	hydrazine	9.6×10^{-7}
NH_2OH	hydroxylamine	6.6×10^{-9}
C_5H_5N	pyridine	1.5×10^{-9}
$C_6H_5NH_2$	aniline	4.1×10^{-10}
PH_3	phosphine	1×10^{-28}

Table E.8 Standard Reduction Potentials (25 °C)

$E°$ (Volts)	Half-Cell Reaction
+2.87	$F_2(g) + 2e^- \rightleftharpoons 2F^-(aq)$
+2.08	$O_3(g) + 2H^+(aq) + 2e^- \rightleftharpoons O_2(g) + H_2O$
+2.05	$S_2O_8^{2-}(aq) + 2e^- \rightleftharpoons 2SO_4^{2-}(aq)$
+1.82	$Co^{3+}(aq) + e^- \rightleftharpoons Co^{2+}(aq)$
+1.77	$H_2O_2(aq) + 2H^+(aq) + 2e^- \rightleftharpoons 2H_2O$
+1.69	$PbO_2(s) + SO_4^{2-}(aq) + 4H^+(aq) + 2e^- \rightleftharpoons PbSO_4(s) + 2H_2O$
+1.63	$2HOCl(aq) + 2H^+(aq) + 2e^- \rightleftharpoons Cl_2(g) + 2H_2O$
+1.51	$Mn^{3+}(aq) + e^- \rightleftharpoons Mn^{2+}(aq)$
+1.507	$MnO_4^-(aq) + 4H^+(aq) + 3e^- \rightleftharpoons MnO_2(s) + 2H_2O$
+1.49	$MnO_4^-(aq) + 8H^+(aq) + 5e^- \rightleftharpoons Mn^{2+}(aq) + 4H_2O$
+1.46	$PbO_2(s) + 4H^+(aq) + 2e^- \rightleftharpoons Pb^{2+}(aq) + 2H_2O$
+1.44	$BrO_3^-(aq) + 6H^+(aq) + 6e^- \rightleftharpoons Br^-(aq) + 3H_2O$
+1.42	$Au^{3+}(aq) + 3e^- \rightleftharpoons Au(s)$
+1.36	$Cl_2(g) + 2e^- \rightleftharpoons 2Cl^-(aq)$
+1.33	$Cr_2O_7^{2-}(aq) + 14H^+(aq) + 6e^- \rightleftharpoons 2Cr^{3+}(aq) + 7H_2O$
+1.24	$O_3(g) + H_2O + 2e^- \rightleftharpoons O_2(g) + 2OH^-(aq)$
+1.23	$MnO_2(s) + 4H^+(aq) + 2e^- \rightleftharpoons Mn^{2+}(aq) + 2H_2O$
+1.23	$O_2(g) + 4H^+(aq) + 4e^- \rightleftharpoons 2H_2O$
+1.20	$Pt^{2+}(aq) + 2e^- \rightleftharpoons Pt(s)$
+1.07	$Br_2(aq) + 2e^- \rightleftharpoons 2Br^-(aq)$
+0.96	$NO_3^-(aq) + 4H^+(aq) + 3e^- \rightleftharpoons NO(g) + 2H_2O$
+0.94	$NO_3^-(aq) + 3H^+(aq) + 2e^- \rightleftharpoons HNO_2(aq) + H_2O$
+0.91	$2Hg^{2+}(aq) + 2e^- \rightleftharpoons Hg_2^{2+}(aq)$
+0.87	$HO_2^-(aq) + H_2O + 2e^- \rightleftharpoons 3OH^-(aq)$
+0.80	$NO_3^-(aq) + 4H^+(aq) + 2e^- \rightleftharpoons 2NO_2(g) + 2H_2O$
+0.80	$Ag^+(aq) + e^- \rightleftharpoons Ag(s)$
+0.77	$Fe^{3+}(aq) + e^- \rightleftharpoons Fe^{2+}(aq)$
+0.69	$O_2(g) + 2H^+(aq) + 2e^- \rightleftharpoons H_2O_2(aq)$
+0.54	$I_2(s) + 2e^- \rightleftharpoons 2I^-(aq)$
+0.49	$NiO_2(s) + 2H_2O + 2e^- \rightleftharpoons Ni(OH)_2(s) + 2OH^-(aq)$
+0.45	$SO_2(aq) + 4H^+(aq) + 4e^- \rightleftharpoons S(s) + 2H_2O$
+0.401	$O_2(g) + 2H_2O + 4e^- \rightleftharpoons 4OH^-(aq)$
+0.34	$Cu^{2+}(aq) + 2e^- \rightleftharpoons Cu(s)$
+0.27	$Hg_2Cl_2(s) + 2e^- \rightleftharpoons 2Hg(l) + 2Cl^-(aq)$
+0.25	$PbO_2(s) + H_2O + 2e^- \rightleftharpoons PbO(s) + 2OH^-(aq)$
+0.2223	$AgCl(s) + e^- \rightleftharpoons Ag(s) + Cl^-(aq)$
+0.172	$SO_4^{2-}(aq) + 4H^+(aq) + 2e^- \rightleftharpoons H_2SO_3(aq) + H_2O$
+0.169	$S_4O_6^{2-}(aq) + 2e^- \rightleftharpoons 2S_2O_3^{2-}(aq)$
+0.16	$Cu^{2+}(aq) + e^- \rightleftharpoons Cu^+(aq)$
+0.15	$Sn^{4+}(aq) + 2e^- \rightleftharpoons Sn^{2+}(aq)$
+0.14	$S(s) + 2H^+(aq) + 2e^- \rightleftharpoons H_2S(g)$
+0.07	$AgBr(s) + e^- \rightleftharpoons Ag(s) + Br^-(aq)$

Table E.8 Standard Reduction Potentials (25 °C) *(Continued)*

$E°$ (Volts)	Half-Cell Reaction
0.00	$2H^+(aq) + 2e^- \rightleftharpoons H_2(g)$
−0.13	$Pb^{2+}(aq) + 2e^- \rightleftharpoons Pb(s)$
−0.14	$Sn^{2+}(aq) + 2e^- \rightleftharpoons Sn(s)$
−0.15	$AgI(s) + e^- \rightleftharpoons Ag(s) + I^-(aq)$
−0.25	$Ni^{2+}(aq) + 2e^- \rightleftharpoons Ni(s)$
−0.28	$Co^{2+}(aq) + 2e^- \rightleftharpoons Co(s)$
−0.34	$In^{3+}(aq) + 3e^- \rightleftharpoons In(s)$
−0.34	$Tl^+(aq) + e^- \rightleftharpoons Tl(s)$
−0.36	$PbSO_4(s) + 2e^- \rightleftharpoons Pb(s) + SO_4^{2-}(aq)$
−0.40	$Cd^{2+}(aq) + 2e^- \rightleftharpoons Cd(s)$
−0.44	$Fe^{2+}(aq) + 2e^- \rightleftharpoons Fe(s)$
−0.56	$Ga^{3+}(aq) + 3e^- \rightleftharpoons Ga(s)$
−0.58	$PbO(s) + H_2O + 2e^- \rightleftharpoons Pb(s) + 2OH^-(aq)$
−0.74	$Cr^{3+}(aq) + 3e^- \rightleftharpoons Cr(s)$
−0.76	$Zn^{2+}(aq) + 2e^- \rightleftharpoons Zn(s)$
−0.81	$Cd(OH)_2(s) + 2e^- \rightleftharpoons Cd(s) + 2OH^-(aq)$
−0.83	$2H_2O + 2e^- \rightleftharpoons H_2(g) + 2OH^-(aq)$
−0.88	$Fe(OH)_2(s) + 2e^- \rightleftharpoons Fe(s) + 2OH^-(aq)$
−0.91	$Cr^{2+}(aq) + 2e^- \rightleftharpoons Cr(s)$
−1.16	$N_2(g) + 4H_2O + 4e^- \rightleftharpoons N_2H_4(aq) + 4OH^-(aq)$
−1.18	$V^{2+}(aq) + 2e^- \rightleftharpoons V(s)$
−1.216	$ZnO_2^{-2}(aq) + 2H_2O + 2e^- \rightleftharpoons Zn(s) + 4OH^-(aq)$
−1.63	$Ti^{2+}(aq) + 2e^- \rightleftharpoons Ti(s)$
−1.66	$Al^{3+}(aq) + 3e^- \rightleftharpoons Al(s)$
−1.79	$U^{3+}(aq) + 3e^- \rightleftharpoons U(s)$
−2.02	$Sc^{3+}(aq) + 3e^- \rightleftharpoons Sc(s)$
−2.36	$La^{3+}(aq) + 3e^- \rightleftharpoons La(s)$
−2.37	$Y^{3+}(aq) + 3e^- \rightleftharpoons Y(s)$
−2.37	$Mg^{2+}(aq) + 2e^- \rightleftharpoons Mg(s)$
−2.71	$Na^+(aq) + e^- \rightleftharpoons Na(s)$
−2.76	$Ca^{2+}(aq) + 2e^- \rightleftharpoons Ca(s)$
−2.89	$Sr^{2+}(aq) + 2e^- \rightleftharpoons Sr(s)$
−2.90	$Ba^{2+}(aq) + 2e^- \rightleftharpoons Ba(s)$
−2.92	$Cs^+(aq) + e^- \rightleftharpoons Cs(s)$
−2.92	$K^+(aq) + e^- \rightleftharpoons K(s)$
−2.93	$Rb^+(aq) + e^- \rightleftharpoons Rb(s)$
−3.05	$Li^+(aq) + e^- \rightleftharpoons Li(s)$

Glossary

A

Abscissa The horizontal axis. (Appendix A.3)

Absolute Zero 0 K. Nature's lowest temperature. (1.4, 10.3)

Acceptor Ion The central metal ion in a complex ion. (17.4, 20.3)

Accuracy Freedom from error. The closeness of a measurement to the true value. (1.5)

Acid *Arrhenius theory*—a substance that produces hydronium ions (hydrogen ions) in water. (2.11, 4.4)

Brønsted theory—a proton donor. (15.1)

Lewis theory—an electron-pair acceptor. (15.3)

Acid–Base Indicator A dye with one color in acid and another color in base. (4.9, 15.5, 16.8)

Acid–Base Neutralization The reaction of an acid with a base. (2.11, 4.4, 15.3)

Acid Deposition The settling of acidic materials, whether they are gases, liquids or solids, onto surfaces; acid rain. (*Chemicals in Our World 7*)

Acidic Anhydride An oxide that reacts with water to make the solution acidic. (4.4, 15.4)

Acidic Solution An aqueous solution in which $[H^+] > [OH^-]$. (15.5)

Acid Ionization Constant (K_a) $K_a = \dfrac{[H^+][A^-]}{[HA]}$ for the equilibrium,

$$HA \rightleftharpoons H^+ + A^- \quad (16.1)$$

Acid Rain Rain made acidic by dissolved sulfur and nitrogen oxides. (*Chemicals in Our World 10*)

Acid Salt A salt of a partially neutralized polyprotic acid, for example, $NaHSO_4$ or $NaHCO_3$. (2.11)

Acid Solubility Product The special solubility product expression for metal sulfides in dilute acid and related to the equation for their dissolving. For a divalent metal sulfide, MS,

$$MS(s) + 2H^+(aq) \longrightarrow$$
$$M^{2+}(aq) + H_2S(aq)$$

$$K_{spa} = \frac{[M^{2+}][H_2S]}{[H^+]^2} \quad (17.2)$$

Actinide Elements (Actinide Series) Elements 90–103. (2.6)

Activated Complex The chemical species that exists with partly broken and partly formed bonds in the transition state. (13.6)

Activation Energy The minimum kinetic energy that must be possessed by the reactants in order to give an effective collision (one that produces products). (13.6)

Activities Effective concentrations which properly should be substituted into a mass action expression to satisfy the equilibrium law. The activity of a solid is defined as having a value of 1. (14.8)

Activity Series A list of metals in order of their reactivity as reducing agents. (5.4)

Actual Yield See *Yield, Actual.*

Addition Compound A molecule formed by the joining of two simpler molecules through formation of a covalent bond (usually a coordinate covalent bond). (8.10)

Addition Reaction The addition of a molecule to a double or triple bond. (23.2)

Adiabatic Change A change within a system during which no heat energy enters or leaves the system. (6.4)

Alcohol An organic compound whose molecules have the OH group attached to tetrahedral carbon. (23.3)

Aldehyde An organic compound whose molecules have the group $-CH=O$. (23.5)

Alkali Metals The Group IA elements (except hydrogen)—lithium, sodium, potassium, rubidium, and cesium. (2.6)

Alkaline Battery (Alkaline Dry Cell) A zinc–manganese dioxide galvanic cell of 1.54 V used commonly in flashlight batteries. (19.9)

Alkaline Earth Metals The Group IIA elements—beryllium, magnesium, calcium, strontium, barium, and radium. (2.6)

Alkalis The alkali metals. (2.6)

Alkane A hydrocarbon whose molecules have only single bonds. (2.10, 23.2)

Alkene A hydrocarbon whose molecules have one or more double bonds. (23.2)

Alkyl Group An organic group of carbon and hydrogen atoms related to an alkane but with one less hydrogen atom (e.g., CH_3, methyl; CH_3CH_2, ethyl). (23.2)

Alkyne A hydrocarbon whose molecules have one or more triple bonds. (23.2)

Allotrope Alternate form of an element. For example, O_2 and O_3 are allotropes of oxygen. (21.2)

Allotropy Existence of an element in more than one form. (21.2)

Alpha Particle (^4_2He) The nucleus of a helium atom. (22.3)

Alpha Radiation A high-velocity stream of alpha particles produced by radioactive decay. (22.3)

Amide An organic compound whose molecules have any one of the following groups. (23.5)

$$\underset{-CNH_2}{\overset{O}{\underset{\|}{}}} \qquad \underset{-CNHR}{\overset{O}{\underset{\|}{}}} \qquad \underset{-CNR_2}{\overset{O}{\underset{\|}{}}}$$

Amine An organic compound whose molecules contain the group NH_2, NHR, or NR_2. (23.4)

α-Amino Acid One of 20 monomers of polypeptides. (23.10)

Amorphous Solid A noncrystalline solid. A glass. (11.13)

Ampere (A) The SI unit for electric current; one coulomb per second. (19.2)

Amphiprotic Compound A compound that can act either as a proton donor or a proton acceptor; an amphoteric compound. (15.1)

Amphoteric Compound A compound that can react as either an acid or a base. (15.1)

Amplitude The height of a wave, which is a measure of the wave's intensity. (7.1)

Angstrom (Å) 1 Å = 10^{-10} m = 100 pm = 0.1 nm. (7.9)

Anhydrous Without water. (2.3)

Anion A negatively charged ion. (2.9)

Anode The positive electrode in a gas discharge tube. The electrode at which oxidation occurs during an electrochemical change. (19.1)

Antibonding Molecular Orbital A molecular orbital that denies electron density to the space between nuclei and destabilizes a molecule when occupied by electrons. (9.7)

Anticodon A triplet of bases on a tRNA molecule that pairs to a matching triplet—a codon—on an mRNA molecule during mRNA-directed polypeptide synthesis. (23.11)

Antilogarithm When $N = a^b$, N is the antilog of b. (Appendix A.2)

Antimatter Any particle annihilated by a particle of ordinary matter. (22.3)

Aqueous Solution A solution that has water as the solvent. (4.1)

Aromatic Compound An organic compound whose molecules have the benzene ring system. (23.2)

Arrhenius Acid See *Acid.*

Arrhenius Base See *Base.*

Arrhenius Equation An equation that relates the rate constant of a reaction to the reaction's activation energy. (13.7)

Association The joining together of molecules by intermolecular attractions such as hydrogen bonds. (12.9)

Atmosphere, Standard (atm) 101,325 Pa. The pressure that supports a column of mercury 760 mm high at 0 °C; 760 torr. (6.4, 10.2)

Atmospheric Pressure The weight per unit area of Earth's surface that is caused by the gases of the atmosphere. (6.4, 10.2)

Atom A neutral particle having one nucleus; the smallest representative sample of an element. (1.2)

Atomic Mass The average mass (in u) of the atoms of the isotopes of a given element as they occur naturally. (2.4)

Atomic Mass Unit (u) $1.6605402 \times 10^{-24}$ g; 1/12th the mass of one atom of carbon-12. Also called a dalton. Formerly known as the amu. (2.4)

Atomic Number (Z) The number of protons in a nucleus. (2.5)

Atomic Spectrum The line spectrum produced when energized or excited atoms emit electromagnetic radiation. (7.2)

Atomic Weight See *Atomic Mass.*

Atomization Energy (ΔH_{atom}) The energy needed to rupture all of the bonds in one mole of a substance in the gas state and produce its atoms, also in the gas state. (18.10)

Aufbau Principle A set of rules enabling the construction of an electron structure of an atom from its atomic number. (7.5)

Avogadro's Number 6.0221367×10^{23}; the number of particles or formula units in one mole. (3.1)

Avogadro's Principle Equal volumes of gases contain equal numbers of molecules when they are at identical temperatures and pressures. (10.4)

Axial Bond A covalent bond oriented parallel to the vertical axis in a trigonal bipyramidal molecule. (9.1)

Azimuthal Quantum Number The quantum number ℓ. (See also: *secondary quantum number.*) (7.3)

B

Background Radiation The atomic radiation from the natural radionuclides in the environment and from cosmic radiation. (22.6)

Balance An apparatus for measuring mass. (1.4)

Balanced Equation A chemical equation that has on opposite sides of the arrow the same number of each atom and the same net charge. (2.3, 3.5)

Band Gap The energy separation between the filled valence shell and the nearest conduction band (empty band). (9.9)

Band of Stability The envelope that encloses just the stable nuclides in a plot of all nuclides constructed according to their numbers of protons versus their numbers of neutrons. (22.4)

Band Theory of Solids An energy band of closely spaced energy levels exists in solids, some of which levels make up a valence band and others of which make up the (partially filled or empty) conduction band. (9.9)

Bar The standard pressure for thermodynamic quantities; 1 bar = 10^5 pascals. (18.5)

Barometer An apparatus for measuring atmospheric pressure. (10.2)

Base *Arrhenius theory*—a substance that releases OH^- ions in water. (2.11, 4.4)

Brønsted theory—a proton acceptor. (15.1)

Lewis theory—an electron pair donor. (15.3)

Base Ionization Constant, K_b $K_b = \dfrac{[BH^+][OH^-]}{[B]}$ for the equilibrium,

$$B + H_2O \rightleftharpoons BH^+ + OH^- \quad (16.1)$$

Base Units The units of the fundamental measurements of the SI. (1.4)

Basic Anhydride An oxide that can neutralize acid or that reacts with water to give OH^-. (4.4, 15.4)

Basic Solution An aqueous solution in which $[H^+] < [OH^-]$. (15.5)

Battery One or more galvanic cells arranged to serve as a practical source of electricity. (19.9)

Becquerel (Bq) 1 disintegration s^{-1}. The SI unit for the activity of a radioactive source. (22.6)

Bent Molecule A molecule that is nonlinear. (9.2)

Beta Particle ($_{-1}^{0}e$) An electron emitted by radioactive decay. (22.3)

Beta Radiation A stream of electrons produced by radioactive decay. (22.3)

Bidentate Ligand A ligand that has two atoms that can become simultaneously attached to the same metal ion. (20.3)

Bimolecular Collision A collision of two molecules. (13.8)

Binary Acid An acid with the general formula H_nX, where X is a nonmetal. (2.11, 15.2)

Binary Compound A compound composed of two different elements. (2.9)

Binding Energy, Nuclear The energy equivalent of the difference in mass between an atomic nucleus and the sum of the masses of its nucleons. (22.2)

Biochemistry The study of the organic substances in organisms. (23.7)

Black Phosphorus An allotrope of phosphorus with a graphitelike structure. (21.2)

Body-Centered Cubic (bcc) Unit Cell A unit cell having identical atoms, molecule, or ions at the corners of a cube plus one more particle in the center of the cube. (11.10)

Boiling Point The temperature at which the vapor pressure of the liquid equals the atmospheric pressure. (11.6)

Boiling Point Elevation A colligative property of a solution by which the solution's boiling point is higher than that of the pure solvent. (12.7)

Bond Angle The angle formed by two bonds that extend from the same atom. (9.1)

Bond Dipole A dipole within a molecule associated with a specific bond. (9.3)

Bond Dissociation Energy See *Bond Energy.*

Bond Distance See *Bond Length.*

Bond Energy The energy needed to break one mole of a particular bond to give electrically neutral fragments. (8.3, 8.8, 9.7, 18.10)

Bonding Molecular Orbital A molecular orbital that introduces a buildup of electron density between nuclei and stabilizes a molecule when occupied by electrons. (9.7)

Bond Length The distance between two nuclei that are held by a chemical bond. (8.3, 8.8)

Bond Order The *net* number of pairs of bonding electrons. Bond order = $1/2 \times$ (no. of bonding e^- − no. of antibonding e^-). (8.8, 9.8)

Boundary That which divides a system from its surroundings. (6.4)

Boyle's Law See *Pressure–Volume Law.*

Bragg Equation $n\lambda = 2d \sin \theta$. The equation used to convert X-ray diffraction data into crystal structure. (11.11)

Branched-Chain Compound An organic compound in whose molecules the carbon atoms do not all occur one after another in a continuous sequence. (23.1)

Branching Step A step in a chain reaction that produces more chain-propagating species than it consumes. (Facets of Chemistry 13.1)

Brine An aqueous solution of sodium chloride, often with other salts. (19.3)

Brønsted Acid See *Acid.*

Brønsted Base See *Base.*

Brownian Movement The random, erratic motions of colloidally dispersed particles in a fluid. (12.10)

Buckminsterfullerene An allotrope of carbon with a fullerence structure and the formula C_{60}. (21.2)

Buffer (a) A pair of solutes that can keep the pH of a solution almost constant if either acid or base is added. (b) A solution containing such a pair of solutes. (16.5)

Buret A long tube of glass usually marked in mL and 0.1 mL units and equipped with a stopcock for the controlled addition of a liquid to a receiving flask. (4.9)

By-product The substances formed by side reactions. (3.8)

C

Calorie (cal) 4.184 J. The energy that will raise the temperature of 1.00 g of water from 14.5 to 15.5 °C. 1 cal = 4.184J. (In popular books on foods, the term *Calorie,* with a capital C, means 1000 cal or 1 kcal.) (6.1, 6.5)

Calorimeter An apparatus used in the determination of the heat of a reaction. (6.4)

Calorimetry The science of measuring the quantities of heat that are involved in chemical or physical changes. (6.5)

Carbohydrates Polyhydroxyaldehydes or polyhydroxyketones or substances that yield these by hydrolysis and that are obtained from plants or animals. (23.8)

Carbonyl Group An organic functional group consisting of a carbon atom joined to an oxygen atom by a double bond; C=O. (23.5)

Carboxylic Acid An organic compound whose molecules have the carboxyl group, CO_2H. (23.5)

Catalysis Rate enhancement caused by a catalyst. (13.9)

Catalyst A substance that in relatively small proportion accelerates the rate of a reaction without being permanently chemically changed. (13.2, 13.9)

Catenation The ability of atoms of the same element to become covalently bonded to one another. (21.4)

Cathode The negative electrode in a gas discharge tube. The electrode at which reduction occurs during an electrochemical change. (19.1)

Cathode Ray A stream of electrons ejected from a hot metal and accelerated toward a positively charged site in a vacuum tube. (Facets of Chemistry 2.1)

Cation A positively charged ion. (2.9)

Cell Potential, E_{cell} The emf of a galvanic cell when no current is drawn from the cell. (19.5)

Cell Reaction The overall chemical change that takes place in an electrolytic cell or a galvanic cell. (19.1)

Celsius Scale A temperature scale on which water freezes at 0 °C and boils at 100 °C (at 1 atm) and that has 100 divisions called Celsius degrees between these two points. (1.4)

Centimeter (cm) 0.01 m. (1.4)

Chain Reaction A self-sustaining change in which the products of one event cause one or more new events. (Facets of Chemistry 13.1, 22.8)

Change of State Transformation of matter from one physical state to another. In thermochemistry, any change in a variable used to define the state of a particular system—a change in composition, pressure, volume, or temperature. (11.4)

Charge Transfer Absorption Band An absorption band that arises from the absorption of photons, each of which supplies energy to transfer an electron from an anion to a cation. (20.2)

Charles' Law See *Temperature–Volume Law.*

Chelate A complex ion containing rings formed by polydentate ligands. (20.3)

Chelate Effect The extra stability found in complexes that contain chelate rings. (20.3)

Chemical Bond The force of electrical attraction that holds atoms together in compounds. (2.8, 7.1)

Chemical Change See *Chemical Reaction*

Chemical Energy The potential energy of chemicals that is transferred during chemical reactions. (1.3, 6.1)

Chemical Equation A before-and-after description that uses formulas and coefficients to represent a chemical reaction. (2.3, 3.5)

Chemical Equilibrium See *Dynamic Equilibrium.*

Chemical Formula See *Formula.*

Chemical Property The ability of a substance, either by itself or with other substances, to undergo a change into new substances. (1.7)

Chemical Reaction A change in which new substances (products) form from starting materials (reactants). (1.1, 1.3)

Chemical Symbol A formula for a substance. (2.1)

Chemistry The study of the compositions of substances and the ways by which their properties are related to their compositions. (1.1)

Chiral See *Chirality*. (20.6)

Chirality The "handedness" of an object; the property of an object (like a molecule) that makes it unable to be superimposed onto a model of its own mirror image. (20.6)

Cis Isomer A stereoisomer whose uniqueness is in having two groups on the same side of some reference plane. (20.6, 23.2)

Clausius–Clapeyron Equation The relationship between the vapor pressure, the temperature, and the molar heat of vaporization of a substance (where C is a constant).

$$\ln P = \frac{\Delta H_{vap}}{RT} + C \quad (11.7)$$

Closed-End Manometer See *Manometer.*

Codon An individual unit of hereditary instruction that consists of three side-by-side side chains on a molecule of mRNA. (23.11)

Coefficients Numbers in front of formulas in chemical equations. (2.3)

Colligative Property A property such as vapor pressure lowering, boiling point elevation, freezing point depression, and osmotic pressure whose physical value depends only on the ratio of the numbers of moles of solute and solvent particles and not on their chemical identities. (12.6)

Collision Theory The rate of a reaction is proportional to the number of collisions that occur each second between the reactants. (13.6)

Colloidal Dispersion (Colloid) A homogeneous mixture in which the particles of one or more components have at least one dimension in the range of 1 to 1000 nm—larger than those in a solution but smaller than those in a suspension. (12.10)

Combined Gas Law See *Gas Law, Combined.*

Combustion A rapid reaction with oxygen accompanied by a flame and the evolution of heat and light. (5.5)

Common Ion The ion in a mixture of ionic substances that is common to the formulas of at least two. (17.1)

Common Ion Effect The solubility of one salt is reduced by the presence of another having a common ion. (17.1)

Common Logarithm See *Logarithm.*

Competing Reaction A reaction that reduces the yield of the main product by forming by-products. (3.8)

Complementary Color The color of the reflected or transmitted light when one component of white light is removed by absorption. (20.7)

Complex Ion (or simply a *Complex*) The combination of one or more anions or neutral molecules (ligands) with a metal ion. (17.4, 20.3)

Compound A substance consisting of chemically combined atoms from two or more elements and present in a definite ratio. (2.1)

Compound Nucleus An atomic nucleus carrying excess energy following its capture of some bombarding particle. (22.5)

Compressibility Capability to undergo a reduction in volume under increasing pressure. (11.3)

Concentrated Solution A solution that has a large ratio of the amounts of solute to solvent. (3.9)

Concentration The ratio of the quantity of solute to the quantity of solution (or the quantity of solvent). See *Molal Concentration, Molar Concentration, Mole Fraction, Normality, Percentage Concentration.* (3.9)

Concentration Table A part of the strategy for organizing data needed to make certain calculations, particularly any involving equilibria. (14.9)

Condensation The change of a vapor to its liquid state. (11.4)

Conduction Band A vacant or partially filled but uninterrupted band in the valence band region of a solid. (9.9)

Conformation A particular relative orientation or geometric form of a flexible molecule. (9.5)

Conjugate Acid The species in a conjugate acid–base pair that has the greater number of H^+ units. (15.1, 16.3)

Conjugate Acid–Base Pair Two substances (ions or molecules) whose formulas differ by only one H^+ unit. (15.1, 16.1)

Conjugate Base The species in a conjugate acid–base pair that has the fewer number of H^+ units. (15.1, 16.1)

Conservation of Energy, Law of See *Law of Conservation of Energy.*

Conservation of Mass–Energy, Law of See *Law of Conservation of Mass–Energy.*

Continuous Spectrum The electromagnetic spectrum corresponding to the mixture of frequencies present in white light. (7.2)

Contributing Structure One of a set of two or more Lewis structures used in applying the theory of resonance to the structure of a compound. A resonance structure. (8.9)

Conversion Factor A ratio constructed from the relationship between two units such as 2.54 cm/1 in., from 1 in. = 2.54 cm. (1.6)

Cooling Curve A graph showing how the temperature of a substance changes as heat is removed from it at a constant rate as the substance undergoes changes in its physical state. (11.7)

Coordinate Covalent Bond A covalent bond in which both electrons originated from one of the joined atoms, but otherwise like a covalent bond in all respects. (8.10)

Coordination Complex See *Coordination Compound.*

Coordination Compound A complex or its salt. (17.4, 20.3)

Coordination Number The number of donor atoms that surround a metal ion. (20.5)

Copolymer A polymer made from two or more different monomers. (23.6)

Core Electrons The inner electrons of an atom that are not exposed to the electrons of other atoms when chemical bonds form. (7.6)

Corrosion The slow oxidation of metals exposed to air or water. (5.5)

Coulomb (C) The SI unit of electrical charge; the charge on 6.25×10^{18} electrons; the amount of charge that passes a fixed point of a wire conductor when a current of 1 A flows for 1 s. (19.2)

Covalent Bond A chemical bond that results when atoms share electron pairs. (8.3)

Covalent Crystal A crystal in which the lattice positions are occupied by atoms that are covalently bonded to the atoms at adjacent lattice sites. (11.12)

Critical Mass The mass of a fissile isotope above which a self-sustaining chain reaction occurs. (22.8)

Critical Point The point at the end of a vapor pressure versus temperature curve for a liquid and that corresponds to the critical pressure and the critical temperature. (11.9)

Critical Pressure (P_c) The vapor pressure of a substance at its critical temperature. (11.9)

Critical Temperature (T_c) The temperature above which a substance cannot exist as a liquid regardless of the pressure. (11.9)

Crystal Field Splitting (Δ) The difference in energy between sets of d orbitals in a complex ion. (20.7)

Crystal Field Theory A theory that considers the effects of the polarities or the charges of the ligands in a complex ion on the energies of the d orbitals of the central metal ion. (20.7)

Crystal Lattice The repeating symmetrical pattern of atoms, molecules, or ions that occurs in a crystal. (11.10)

Cubic Meter (m^3) The SI derived unit of volume. (1.4)

Curie (Ci) A unit of activity for radioactive samples, equal to 3.7×10^{10} disintegrations s^{-1}. (22.6)

D

Dalton One atomic mass unit, u. (2.4)

Dalton's Atomic Theory Matter consists of tiny, indestructible particles called atoms. All atoms of one element are identical. The atoms of different elements have different masses. Atoms combine in definite ratios by atoms when they form compounds. (2.2)

Dalton's Law of Partial Pressures See *Partial Pressures, Law of.*

Data The information (often in the form of physical quantities) obtained in an experiment or other experience or from references. (1.2)

Decimal Multipliers Factors—exponentials of 10 or decimals—that are used to define larger or smaller SI units. (1.4)

Decomposition A chemical reaction that changes one substance into two or more other substances. (2.1)

Dehydration An elimination reaction in which water is a product. (23.3)

Delocalization Energy The difference between the energy a substance would have if its molecules had no delocalized molecular orbitals and the energy it has because of such orbitals. (9.8)

Delocalized Molecular Orbital A molecular orbital that spreads over more than two nuclei. (9.8)

Density The ratio of an object's mass to its volume. (1.7)

Dependent Variable The experimental variable of a pair of variables whose value is determined by the other, the independent variable. (Appendix A.3)

Derived Unit Any unit defined solely in terms of base units. (1.4)

Dialysis The passage of small molecules and ions, but not species of a colloidal size, through a semipermeable membrane. (12.8)

Diamagnetism The condition of not capable of being attracted to a magnet. (7.4)

Diaphragm Cell An electrolytic cell used to manufacture sodium hydroxide by the electrolysis of aqueous sodium chloride. (19.3)

Diatomic Molecule A molecule composed of two atoms each (2.3)

Diatomic Substance A substance made from the atoms of only two elements. (2.9)

Diborane A compound formed from boron and hydrogen with the formula B_2H_6. (21.3)

Diffraction Constructive and destructive interference by waves. (7.3)

Diffraction Pattern The image formed on a screen or a photographic film caused by the diffraction of electromagnetic radiation such as visible light or X rays. (11.11)

Diffusion The spontaneous intermingling of one substance with another. (10.7, 11.3)

Dilute Solution A solution in which the ratio of the quantities of solute to solvent is small. (3.9)

Dimensional Analysis See *Factor-Label Method.*

Dimer A molecule formed by joining two identical smaller molecules. (21.5)

Dipole Partial positive and partial negative charges separated by a distance. (8.5)

Dipole–Dipole Attractions Attractions between molecules that are dipoles. (11.2)

Dipole Moment The product of the sizes of the partial charges in a dipole multiplied by the distance between them; a measure of the polarity of a molecule. (8.5)

Diprotic Acid An acid that can furnish two H^+ per molecule. (2.11, 4.4)

Disaccharide A carbohydrate whose molecules can be hydrolyzed to two monosaccharides. (23.8)

Dispersion Forces Another term for London forces. (11.2)

Disproportionation A reaction in which the same reactant is both the oxidizing agent and the reducing agent. (21.5, 21.6)

Dissociation The separation of preexisting ions when an ionic compound dissolves or melts. (4.1)

Distorted Tetrahedron A description of a molecule in which the central atom is surrounded by 5 electron pairs, one of which is a lone pair of electrons. The central atom is bonded to four other atoms. The structure is also said to have a see-saw shape. (9.2)

DNA Deoxyribonucleic acid; a nucleic acid that hydrolyzes to deoxyribose, phosphate ion, adenine, thymine, guanine, and cytosine, and that is the carrier of genes. (23.11)

DNA Double Helix Two oppositely running strands of DNA held in a helical configuration by interstrand hydrogen bonds. (23.11)

Donor Atom The atom on a ligand that makes an electron pair available in the formation of a complex. (17.4, 20.3)

Doped Semiconductor A semiconductor chip (silicon) to which a trace amount of an element (like boron or arsenic) has been added to alter the conducting properties of the semiconductor. (9.9)

Double Bond A covalent bond consisting of one sigma bond and one pi bond. (8.3, 9.6)

Double Replacement Reaction (Metathesis Reaction) A reaction of two salts in which cations and anions exchange partners (e.g., $AgNO_3 + NaCl \rightarrow AgCl + NaNO_3$). (4.3)

Downs Cell An electrolytic cell for the industrial production of sodium. (19.3)

Ductility A metal's ability to be drawn (or stretched) into wire. (2.7)

Dynamic Equilibrium A condition in which two opposing processes are occurring at equal rates. (4.5, 14.1)

E

Effective Nuclear Charge The net positive charge an outer electron experiences as a result of the partial screening of the full nuclear charge by core electrons. (7.9)

Effusion The movement of a gas through a very tiny opening into a region of lower pressure. (10.7)

Effusion, Law of (Graham's Law) The rates of effusion of gases are inversely proportional to the square roots of their densities when compared at identical pressures and temperatures:

$$\text{Effusion rate} \propto \frac{1}{\sqrt{d}} \ (\text{constant } P, T)$$

where d is the gas density. (10.7)

Einstein Equation $\Delta E = \Delta m_0 c^2$ where ΔE is the energy obtained when a quantity of rest mass, Δm_0, is destroyed, or the energy lost when this quantity of mass is created. (22.1)

Elastomers Polymers with elastic properties. (21.6)

Electrochemical Change A chemical change that is caused by or that produces electricity. (19.1)

Electrochemistry The study of electrochemical changes. (19.1)

Electrolysis The production of a chemical change by the passage of electricity through a solution that contains ions or through a molten ionic compound. (19.1)

Electrolysis Cell An apparatus for electrolysis. (19.1, 19.4)

Electrolyte A compound that conducts electricity either in solution or in the molten state. (4.1)

Electrolytic Cell See *Electrolysis Cell.*

Electrolytic Conduction The transport of electrical charge by ions. (19.1)

Electrolyze To pass electricity through an electrolyte and cause a chemical change. (19.1)

Electromagnetic Energy Energy transmitted by wavelike oscillations in the strengths of electrical and magnetic fields; light energy. (7.1)

Electromagnetic Radiation The successive series of oscillations in the strengths of electrical and magnetic fields associated with light, microwaves, gamma rays, ultraviolet rays, infrared rays, and the like. (7.1)

Electromagnetic Spectrum The distribution of frequencies of electromagnetic radiation among various types of such radiation—microwave, infrared, visible, ultraviolet, X, and gamma rays. (7.1)

Electromotive Force (emf) The voltage produced by a galvanic cell and that can make electrons move in a conductor. (19.5)

Electron (e^- or $_{-1}^{0}e$) (a) A subatomic particle with a charge of $1-$ and mass of 0.0005485712 u ($9.1093897 \times 10^{-28}$ g) and that occurs outside an atomic nucleus. The particle that moves when an electric current flows. (2.5) (b) A beta particle. (22.3)

Electron Affinity The energy change (usually expressed in kJ mol^{-1}) that occurs when an electron adds to an isolated gaseous atom or ion. (7.9)

Electron Capture The capture by a nucleus of an orbital electron and that changes a proton into a neutron in the nucleus. (22.3)

Electron Cloud Because of its wave properties, an electron's influence spreads out like a cloud around the nucleus. (7.8)

Electron Configuration The distribution of electrons in an atom's orbitals. (7.5)

Electron Density The concentration of the electron's charge within a given volume. (7.8)

Electronegativity The relative ability of an atom to attract electron density toward itself when joined to another atom by a covalent bond. (8.5)

Electronic Structure The distribution of electrons in an atom's orbitals. (7.1, 7.5)

Electron Pair Bond A covalent bond. (8.3)

Electron Spin The spinning of an electron about its axis that is believed to occur because the electron behaves as a tiny magnet. (7.4)

Electron Volt (eV) The energy an electron receives when it is accelerated under the influence of 1 V and equal to 1.6×10^{-19} J. (22.3)

Electroplating Depositing a thin metallic coating on an object by electrolysis. (19.3)

Element A substance in which all of the atoms have the same atomic number. A substance that cannot be broken down by chemical reactions into anything that is both stable and simpler. (2.1)

Elementary Process One of the individual steps in the mechanism of a reaction. (13.8)

Elimination Reaction The loss of a small molecule from a larger molecule as in the elimination of water from an alcohol. (23.3)

Emission Spectrum See *Atomic Spectrum.*

Empirical Facts Facts discovered by performing experiments. (1.2)

Empirical Formula A chemical formula that uses the smallest whole-number subscripts to give the proportions by atoms of the different elements present. (3.4)

Emulsifying Agent A substance that stabilizes an emulsion. (12.10)

Emulsion A colloidal dispersion of one liquid in another. (12.10)

Enantiomers Stereoisomers whose molecular structures are related as an object to its mirror image but that cannot be superimposed. (20.6)

Endergonic Descriptive of a change in which a system's free energy increases. (18.4)

Endothermic Descriptive of a change in which a system's internal energy increases. (6.3, 6.4)

End Point The moment in a titration when the indicator changes color and the titration is ended. (4.9, 16.8)

Energy Something that matter possesses by virtue of an ability to do work. (1.3, 6.1)

Energy Band A large number of closely spaced energy levels in a solid formed by combining atomic orbitals of similar energy from each of the atoms in the solid. (9.9)

Energy Level A particular energy an electron can have in an atom or a molecule. (7.2)

Enthalpy (H) The heat content of a system. (6.4, 13.1)

Enthalpy Change (ΔH) The difference in enthalpy between the initial state and the final state for some change. (6.4, 13.1)

Enthalpy of Solution See *Heat of Solution.*

Entropy (S) The thermodynamic quantity that describes the degree of randomness of a system. The greater the disorder or randomness, the higher is the statistical probability of the state and the higher is the entropy. (18.3)

Entropy Change (ΔS) For a change, the sum of the values of S for the products minus the sum of the values of S for the reactants. (18.3)

Enzyme A catalyst in a living system. (23.10)

Equation of State for an Ideal Gas See *Gas, Law, Ideal.*

Equatorial Bond A covalent bond located in the plane perpendicular to the long axis of a trigonal bipyramidal molecule. (9.1)

Equilibrium See *Dynamic Equilibrium.*

Equilibrium Constant The value that the mass action expression has when the system is at equilibrium. (14.3)

Equilibrium Law The mathematical equation for a particular equilibrium system that sets the mass action expression equal to the equilibrium constant. (14.3)

Equilibrium Vapor Pressure of a Liquid The pressure exerted by a vapor in equilibrium with its liquid state. (11.5)

Equilibrium Vapor Pressure of a Solid The pressure exerted by a vapor in equilibrium with its solid state. (11.5)

Equivalence See *Stoichiometric Equivalence.* (3.1)

Equivalence Point The moment in a titration when the number of equivalents of the reactant added from a buret equals the number of equivalents of another reactant in the receiving flask. (16.8)

Equivalent (eq) (a) The amount of an acid (or base) that provides 1 mol of H^+ (or 1 mol of OH^-). (b) The amount of an oxidizing agent (or a reducing agent) that can accept (or provide) 1 mol of electrons in a particular redox reaction. (Appendix C)

Equivalent Weight The mass in grams of one equivalent. (Appendix C)

Ester An organic compound whose molecules have the ester group. (23.5)

$$\overset{\displaystyle O}{\underset{\text{ester group}}{-C-O-C}}$$

Ether An organic compound in whose molecules two hydrocarbon groups are joined to an oxygen. (23.3)

Evaporate To change from a liquid to a vapor. (11.3)

Exact Number A number obtained by a direct count or that results by a definition and that is considered to have an infinite number of significant figures. (1.5)

Exergonic Descriptive of a change in a system in which there is a free energy decrease. (18.4)

Exon One of a set of sections of a DNA molecule (separated by introns) that, taken together, constitute a gene. (23.11)

Exothermic Descriptive of a change in which energy leaves a system and enters the surroundings. (6.3, 6.4)

Exponential Notation See *Scientific Notation.*

Extensive Property A property of an object that is described by a physical quantity whose magnitude is proportional to the size or amount of the object (e.g., mass or volume). (1.7)

F

Face-Centered Cubic (fcc) Unit Cell A unit cell having identical atoms, molecules, or ions at the corners of a cube and also in the center of each face of the cube. (11.10)

Factor-Label Method A problem-solving technique that uses the correct cancellation of the units of physical quantities as a guide for the correct setting up of the solution to the problem. (1.6)

Fahrenheit Scale A temperature scale on which water freezes at 32 °F and boils at 212 °F (at 1 atm) and between which points there are 180 degree divisions called Fahrenheit degrees. (1.4)

Family of Elements See *Group.*

Faraday (𝓕) One mole of electrons; 9.65×10^4 coulombs. (19.2)

Faraday Constant (𝓕) 9.65×10^4 coulombs/mol e^-. (19.2)

Fatty Acid One of several long-chain carboxylic acids produced by the hydrolysis (digestion) of a lipid. (23.9)

First Law of Thermodynamics A formal statement of the law of conservation of energy. See *Law of Conservation of Energy.* (6.4, 18.1)

First-Order Reaction A reaction with a rate law in which rate $= k[A]^1$, where A is a reactant. (13.4)

Fissile Isotope An isotope capable of undergoing fission following neutron capture. (22.8)

Fission The breaking apart of atomic nuclei into smaller nuclei accompanied by the release of energy, and the source of energy in nuclear reactors. (22.1, 22.8)

Force Anything that can cause an object to change its motion or direction. (10.2)

Formal Charge The apparent charge on an atom in a molecule or polyatomic ion as calculated by a set of rules that generally assign a bonding pair of electrons to the more electronegative of the two atoms held by the bond. (8.8)

Formation Constant (K_{form}) The equilibrium constant for an equilibrium involving the formation of a complex

ion. Also called the stability constant. (17.4)

Formula A representation of the composition of a substance that uses the chemical symbols of the elements present and subscripts that describe the proportions by atoms. (2.3, 2.8)

Formula Mass The sum of the atomic masses (in u) of all of the atoms represented in the chemical formula of an ionic compound. (3.2)

Formula Unit A particle that has the composition given by the chemical formula. (2.8)

Fossil Fuels Coal, oil, and natural gas. (*Chemicals in Our World 8*)

Free Energy See *Gibbs Free Energy.*

Free Energy Diagram A plot of the changes in free energy for a multicomponent system versus the composition. (18.9)

Free Radical An atom, molecule, or ion that has one or more unpaired electrons. (Facets of Chemistry 13.1, *Chemicals in Our World 11,* 22.6)

Freezing Point Depression A colligative property of a liquid solution by which the freezing point of the solution is lower than that of the pure solvent. (12.7)

Frequency (ν) The number of cycles per second of electromagnetic radiation. (7.1)

Frequency Factor The proportionality constant, A, in the Arrhenius equation. (13.7)

Fuel Cell An electrochemical cell in which electricity is generated from the reactions that accompany the burning of a fuel. (*Chemicals in Our World 10*)

Fullerene An allotrope of carbon made of an extended joining together of five- and six-membered rings of carbon atoms. (21.2)

Functional Group The group of atoms of an organic molecule that enters into a characteristic set of reactions that are independent of the rest of the molecule. (23.1)

Fusion (a) Melting. (11.7) (b) The formation of atomic nuclei by the joining together of the nuclei of lighter atoms. (22.2)

G

G See *Gibbs Free Energy.*

ΔG See *Gibbs Free Energy Change.*

ΔG° See *Standard Free Energy Change.*

Galvanic Cell An electrochemical cell in which a spontaneous redox reaction produces electricity. (19.4)

Gamma Radiation Electromagnetic radiation with wavelengths in the range of 1 Å or less (the shortest wavelengths of the spectrum). (22.3)

Gas Constant, Universal (R) $R = 0.0821$ liter atm $mol^{-1}\ K^{-1}$ (10.4)

Gas Law, Combined For a given mass of gas, the product of its pressure and volume divided by its Kelvin temperature is a constant.

$$PV/T = \text{a constant} \qquad (10.3)$$

Gas Law, Ideal $PV = nRT$ (10.4)

Gay-Lussac's Law See *Pressure–Temperature Law.*

Genetic Code The correlation of codons with amino acids. (23.11)

Genetic Engineering The use of enzymes, nucleic acids, and genes to modify an organism's genes for some purpose. (23.11)

Geometric Isomer One of a set of isomers that differ only in geometry. (20.6, 23.2)

Geometric Isomerism The existence of isomers whose molecules have identical atomic organizations but different geometries; cis/trans isomers. (20.6, 23.2)

Gibbs Free Energy (G) A thermodynamic quantity that relates enthalpy (H), entropy (S), and temperature (T) by the equation

$$G = H - TS \qquad (18.4)$$

Gibbs Free Energy Change (ΔG) The difference in free energy between the initial and final states for some reaction or physical change. (18.4)

Glass Any amorphous solid. (11.13, *Chemicals in Our World 9*)

Graham's Law See *Effusion, Law of.*

Gram (g) 0.001 kg. (1.4)

Gray (Gy) The SI unit of radiation absorbed dose. 1 Gy = 1 J kg^{-1} (22.6)

Greenhouse Effect The retention of solar energy made possible by the ability of the greenhouse gases (e.g., CO_2, CH_4, H_2O, and the chlorofluorocarbons) to absorb outgoing radiation and re-radiate some of it back to Earth. (*Chemicals in Our World 5*)

Ground State The lowest energy state of an atom or molecule. (7.2, 7.3)

Group A vertical column of elements in the periodic table. (2.6)

H

ΔH See *Enthalpy Change.*

ΔH_{atom} See *Atomization Energy.*

$\Delta H°$ See *Standard Heat of Reaction.*

$\Delta H_f°$ See *Standard Heat of Formation.*

ΔH_{fusion} See *Molar Heat of Fusion.*

$\Delta H_{sublimation}$ See *Molar Heat of Sublimation.*

$\Delta H_{vaporization}$ See *Molar Heat of Vaporization.*

Haber Process An industrial synthesis of ammonia from nitrogen and hydrogen under pressure and heat in the presence of a catalyst. (13.9)

Half-cell That part of a galvanic cell in which either oxidation or reduction takes place. (19.4)

Half-life ($t_{1/2}$) The time required for a reactant concentration or the mass of a radionuclide to be reduced by half. (13.5, 22.3)

Half-reaction A hypothetical reaction that constitutes exclusively either the oxidation or the reduction half of a redox reaction and in whose equation the correct formulas for all species taking part in the change are written together with enough electrons to give the correct electrical balance. (5.2)

Halogen Family Group VIIA in the periodic table—fluorine, chlorine, bromine, iodine, and astatine. (2.6)

Hall–Héroult Process A method for manufacturing aluminum by the electrolysis of aluminum oxide in molten cryolite. (19.3)

Hard Water Water with dissolved Mg^{2+}, Ca^{2+}, Fe^{2+}, or Fe^{3+} ions at a concentration high enough (above 25 mg L^{-1}) to interfere with the use of soap. (Facets of Chemistry 4.1)

Heat A form of energy. Also called thermal energy. (1.3, 6.2)

Heat Capacity The quantity of heat needed to raise the temperature of an object by 1 °C. (6.5)

Heating Curve A plot of the temperature change in a sample versus the quantity of heat added. (11.7)

Heat of Formation, Standard See *Standard Heat of Formation.*

Heat of Reaction The heat exchanged between a system and its surroundings when a chemical change occurs in the system. (6.4)

Heat of Reaction at Constant Pressure The heat of a reaction in an open system. (6.4)

Heat of Reaction at Constant Volume The heat of a reaction in a sealed vessel, like a bomb calorimeter. (6.4, 18.1)

Heat of Reaction, Standard See *Standard Heat of Reaction.*

Heat of Solution (ΔH_{soln}) The energy exchanged between the system and its surroundings when one mole of a solute dissolves in a solvent to make a dilute solution. (12.2)

Henry's Law See *Pressure–Solubility Law.*

Hertz (Hz) 1 cycle s^{-1}; the SI unit of frequency. (7.1)

Hess's Law For any reaction that can be written in steps, the standard heat of reaction is the same as the sum of the standard heats of reaction for the steps. (6.7)

Hess's Law Equation For the change,

$$aA + bB + \ldots \longrightarrow$$
$$nN + mM + \ldots$$

$$\Delta H° = \left(\begin{array}{c} \text{sum of } \Delta H_f° \text{ of all} \\ \text{of the reactants} \end{array}\right) -$$
$$\left(\begin{array}{c} \text{sum of } \Delta H_f° \text{ of all} \\ \text{of the products} \end{array}\right) \quad (6.8)$$

Heterocyclic Compound A compound whose molecules have rings that include one or more multivalent atoms other than carbon. (23.1)

Heterogeneous Catalyst A catalyst that is in a different phase than the reactants and onto whose surface the reactant molecules are adsorbed and where they react. (13.9)

Heterogeneous Equilibrium An equilibrium involving more than one phase. (14.7)

Heterogeneous Mixture A mixture that has two or more phases with different properties. (2.1)

Heterogeneous Reaction A reaction in which not all of the chemical species are in the same phase. (13.2, 14.7)

High-Spin Complex A complex ion or coordination compound in which there is the maximum number of unpaired electrons. (20.7)

Homogeneous Catalyst A catalyst that is in the same phase as the reactants. (13.9)

Homogeneous Equilibrium An equilibrium system in which all components are in the same phase. (14.7)

Homogeneous Mixture A mixture that has only one phase and that has uniform properties throughout. (2.1)

Homogeneous Reaction A reaction in which all of the chemical species are in the same phase. (13.2, 14.7)

Hund's Rule Electrons that occupy orbitals of equal energy are distributed with unpaired spins as much as possible among all such orbitals. (7.5)

Hybrid Atomic Orbitals Orbitals formed by mixing two or more of the basic atomic orbitals of an atom and that make possible more effective overlaps with the orbitals of adjacent atoms than do ordinary atomic orbitals. (9.5)

Hydrate A compound that contains molecules of water in a definite ratio to other components. (2.1)

Hydration The development in an aqueous solution of a cage of water molecules about ions or polar molecules of the solute. (12.1)

Hydration Energy The enthalpy change associated with the hydration of gaseous ions or molecules as they dissolve in water. (12.2)

Hydride A binary compound formed between hydrogen and another element. (21.3)

Hydrocarbon An organic compound whose molecules consist entirely of carbon and hydrogen atoms. (2.10, 23.2)

Hydrogen Bond An extrastrong dipole–dipole attraction between a hydrogen bound covalently to nitrogen, oxygen, or fluorine, and another nitrogen, oxygen, or fluorine atom. (11.2)

Hydrogen Electrode The standard of comparison for reduction potentials and for which $E°_{H^+}$ has a value of 0.00 V (25 °C, 1 atm) when $[H^+] = 1\ M$ in the reversible half-cell reaction:

$$2H^+(aq) + 2e^- \rightleftharpoons H_2(g) \quad (19.6)$$

Hydrolysis A reaction with water. (21.3)

Hydrometer A device for measuring specific gravity. (19.9)

Hydrophilic Group A polar molecular unit capable of having dipole–dipole attractions or hydrogen bonds with water molecules. (12.10)

Hydrophobic Group A nonpolar molecular unit with no affinity for the molecules of a polar solvent, like water. (12.10)

Hypothesis A tentative explanation of the results of experiments. (1.2)

I

Ideal Gas A hypothetical gas that obeys the gas laws exactly. (10.3)

Ideal Gas Law $PV = nRT$ (10.4)

Ideal Solution A hypothetical solution that would obey the vapor pressure–concentration law (Raoult's law) exactly. (12.2, 12.6)

Immiscible Insoluble. (12.1)

Incompressible Incapable of losing volume under increasing pressure. (11.3)

Independent Variable The experimental variable of a pair of variables whose value is first selected and from which the value of the dependent variable then results. (Appendix A.3)

Indicator A chemical put in a solution being titrated and whose change in color signals the end point. (4.9, 16.8)

Induced Dipole A dipole created when the electron cloud of an atom or a molecule is distorted by a neighboring dipole or by an ion. (11.2)

Inert Gases See *Noble Gases*.

Initiation Step The step in a chain reaction that produces reactive species that can start chain propagation steps. (Facets of Chemistry 13.1)

Inner Transition Elements Members of the two long rows of elements below the main body of the periodic table—elements 58–71 and elements 90–103. (2.6)

Inorganic Compound A compound made from any elements except those compounds of carbon classified as organic compounds. (2.11)

Instability Constant (K_{inst}) The reciprocal of the formation constant for an equilibrium in which a complex ion forms. (17.4)

Instantaneous Dipole A momentary dipole in an atom, ion, or molecule caused by the erratic movement of electrons. (11.2)

Insulated System A system with boundaries that permit no energy to pass through. (6.4)

Intensive Property A property whose physical quantity is independent of the size of the sample, such as density or temperature. (1.7)

Interhalogen Compound A compound whose molecules are made from two different halogens. (21.7)

Intermolecular Attractions Attractions *between* neighboring molecules. (11.1)

Internal Energy (E) The sum of all of the kinetic energies and potential energies of the particles within a system. (18.1)

International System of Units (SI) The successor to the metric system of measurements that retains most of the units of the metric system and their decimal relationships but employs new reference standards. (1.4)

Intron One of a set of sections of a DNA molecule that separate the exon sections of a gene from each other. (23.11)

Inverse Square Law The intensity of a radiation is inversely proportional to the square of the distance from its source. (22.6)

Ion An electrically charged particle on the atomic or molecular scale of size. (2.8)

Ion–Dipole Attraction The attraction between an ion and the charged end of a polar molecule. (11.2)

Ion–Electron Method A method for balancing redox reactions that uses half-reactions. (5.2)

Ionic Bond The attractions between ions that hold them together in ionic compounds. (8.1)

Ionic Character The extent to which a covalent bond has a dipole moment and is polarized. (8.5)

Ionic Compound A compound consisting of positive and negative ions. (2.8)

Ionic Crystal A crystal that has ions located at the lattice points. (11.12)

Ionic Equation A chemical equation in which soluble strong electrolytes are written in dissociated or ionized form. (4.2)

Ionic Potential The ratio of an ions charge to its radius ($\phi = \dfrac{q}{r}$) (20.2)

Ionic Reaction A chemical reaction in which ions are involved. (4.2)

Ion–Induced Dipole Attraction An electrostatic attraction to a dipole induced in a nearby molecule (or ion) by an ion. (11.2)

Ionization Energy (IE) The energy needed to remove an electron from an isolated, gaseous atom, ion, or molecule (usually given in units of kJ mol^{-1}). (7.9)

Ionization Reaction A reaction of chemical particles that produces ions. (4.4)

Ionizing Radiation Any high-energy radiation—X rays, gamma rays, or radiation from radionuclides—that generates ions as it passes through matter. (22.6)

Ion Product The mass action expression for the solubility equilibrium involving the ions of a salt and equal to the product of the molar concentrations of the ions, each concentration raised to a power that equals the number of ions obtained from one formula unit of the salt. (17.1)

Ion Product Constant of Water (K_w) $K_w = [H^+][OH^-]$ (15.5)

Isoelectronic Having the same number of electrons (e.g., NO and O_2^+ are isoelectronic). (21.5)

Isomer One of a set of compounds that have identical molecular formulas but different structures. (8.4, 20.6, 23.2)

Isomerism The existence of sets of isomers. (8.4, 20.6, 23.2)

Isotopes Constituents of the same element but whose atoms have different mass numbers. (2.4, 2.5)

IUPAC Rules The formal rules for naming substances as developed by the International Union of Pure and Applied Chemistry. (21.2)

J

Joule (J) The SI unit of energy.

$$1 \text{ J} = 1 \text{ kg m}^2 \text{ s}^{-2}$$
$$1 \text{ J} = 4.184 \text{ cal (exactly)} \quad (6.1)$$

K

K See *Kelvin.*

K_a See *Acid Ionization Constant.*

K_b See *Base Ionization Constant.*

K_{sp} See *Solubility Product Constant.*

K_{spa} See *Acid Solubility Product.*

K_w See *Ion Product Constant of Water.*

K-Capture See *Electron Capture.*

Kelvin (K) One degree on the Kelvin scale of temperature and identical in size to the Celsius degree. (1.4)

Kelvin Scale The temperature scale on which water freezes at 273.15 K and boils at 373.15 K and that has 100 degree divisions called kelvins between these points. K = °C + 273.15. (1.4)

Ketone An organic compound whose molecules have the carbonyl group (C=O) flanked by hydrocarbon groups. (23.5)

Kilocalorie (kcal) 1000 cal.(6.1)

Kilogram (kg) The base unit for mass in the SI and equal to the mass of a cylinder of platinum–iridium alloy kept by the International Bureau of Weights and Measures at Sèvres, France. 1 kg = 1000 g. (1.4)

Kilojoule (kJ) 1000 J. (6.1)

Kinetic Energy (KE) Energy of motion. KE = $(1/2)mv^2$. (1.3, 6.1)

Kinetics The study of the factors that govern how rapidly reactions occur. (13.1)

Kinetic Theory of Gases A set of postulates used to explain the gas laws. A gas consists of an extremely large number of very tiny, very hard particles in constant, random motion. They have negligible volume and, between collisions, experience no forces between themselves. (10.8)

Kinetic Theory of Matter The particles of substances (atoms, ions, molecules) are in a state of constant agitation or motion, except at absolute zero (0 K). (6.2)

L

Lanthanide Elements Elements 58–71. (2.6)

Lattice Energy Energy released by the imaginary process in which isolated ions come together to form a crystal of an ionic compound. (8.1)

Law A description of behavior (and not an *explanation* of behavior) based on the results of many experiments. (1.2)

Law of Conservation of Energy The energy of the universe is constant; it can be neither created not destroyed but only transferred and transformed. (1.3)

Law of Conservation of Mass No detectable gain or loss in mass occurs in chemical reactions. Mass is conserved. (2.2)

Law of Conservation of Mass–Energy The sum of all the mass in the universe and of all of the energy, expressed as an equivalent in mass (calculated by the Einstein equation), is a constant. (22.1)

Law of Definite Proportions In a given chemical compound, the elements are always combined in the same proportion by mass. (2.2)

Law of Gas Effusion See *Effusion, Law of.*

Law of Heat Summation (Hess's Law) For any reaction that can be written in steps, the standard heat of reaction is the sum of the standard heats of reaction for the steps. (6.7)

Law of Multiple Proportions Whenever two elements form more than one compound, the different masses of one element that combine with the same mass of the other are in a ratio of small whole numbers. (2.2)

Law of Partial Pressures See *Partial Pressures, Law of.*

Lead Storage Battery A galvanic cell of about 2 V involving lead and lead(IV) oxide in sulfuric acid. (19.9)

Le Châtelier's Principle When a system that is in dynamic equilibrium is subjected to a disturbance that upsets the equilibrium, the system undergoes a change that counteracts the disturbance and, if possible, restores the equilibrium. (11.8, 14.8)

Lewis Acid An electron-pair acceptor. (15.3)

Lewis Base An electron-pair donor. (15.3)

Lewis Structure (Lewis Formula) A structural formula drawn with Lewis symbols and that uses dots and dashes to show the valence electrons and shared pairs of electrons. (8.3, 8.7, 8.8)

Lewis Symbol The symbol of an element that includes dots to represent the valence electrons of an atom of the element. (8.2)

Ligand A molecule or an anion that can bind to a metal ion to form a complex. (17.4, 20.3)

Like Dissolves Like Rule Strongly polar and ionic solutes tend to dissolve in polar solvents, and nonpolar solutes tend to dissolve in nonpolar solvents. (12.1)

Limiting Reactant The reactant that determines how much product can form when nonstoichiometric amounts of reactants are used. (3.7)

Line Spectrum An atomic spectrum. So named because the light emitted by an atom and focused through a narrow slit yields a series of lines when projected on a screen. (7.2)

Linear Molecule A molecule all of whose atoms lie on a straight line. (9.1, 9.2)

Lipid Any substance found in plants or animals that can be dissolved in nonpolar solvents. (23.9)

Liter (L) 1 dm^3. 1 L = 1000 mL = 1000 cm^3 (1.4)

Ln *x* The natural logarithm (to the base *e*) of *x*. (Appendix A.2)

Localized Bond A covalent bond in which the pair of electrons is localized between two nuclei. (9.8)

Logarithm In general, the exponent, *b*, in $N = a^b$. In *common* logarithms, or *logs,* the exponent, *x*, in $N = 10^x$. In *natural* logarithms, symbolized *ln*, the exponent, *x*, in $N = e^x$, where $e = 2.7182818. . . .$ (Appendix A.2)

London Forces Weak attractive forces caused by instantaneous dipole–induced dipole attractions. (11.2)

Lone Pair A pair of electrons in the valence shell of an atom that is not shared with another atom. An unshared pair of electrons. (9.2)

Low-Spin Complex A coordination compound or a complex ion with electrons paired as much as possible in the lower energy set of *d* orbitals. (20.7)

M

Macromolecule A molecule whose molecular mass is very large. (23.6)

Magic Numbers The numbers 2, 8, 20, 28, 50, 82, and 126, numbers whose significance in nuclear science is that a nuclide in which the number of protons or neutrons equals a magic number has nuclei that are relatively more stable than those of other nuclides nearby in the band of stability. (22.4)

Magnetic Quantum Number (m_ℓ) A quantum number that can have values from $-\ell$ to $+\ell$. (7.3)

Main Group Elements See *Representative Elements.*

Main Reaction The desired reaction between the reactants as opposed to competing reactions that give by-products. (3.8)

Malleability A metal's ability to be hammered or rolled into thin sheets. (2.7)

Manometer A device for measuring the pressure within a closed system. The two types—*closed end* and *open end*—differ according to whether the operating fluid (e.g., mercury) is exposed at one end to the atmosphere or not. (10.2)

Mass A measure of the amount of matter that there is in a given sample. (1.3)

Mass Action Expression A fraction in which the numerator is the product of the molar concentrations of the products, each raised to a power equal to its coefficient in the equilibrium equation, and the denominator is the product of the molar concentrations of the reactants, each also raised to a power that equals its coefficient in the equation. (For gaseous reactions, partial pressures can be used in place of molar concentrations.) (14.3)

Mass Defect The mass that changed into energy as the nucleons gathered to form a nucleus. (22.2)

Mass Number (*A*) The numerical sum of the protons and neutrons in an atom of a given isotope. (2.5)

Matter Anything that has mass and occupies space. (1.3)

Maximum Work The maximum amount of work that by theory could be obtained from a given change. It is equal to the value of ΔG for the change. (18.7)

Mechanism of a Reaction The series of individual steps (called elementary processes) in a chemical reaction that gives the net, overall change. (13.1, 13.8)

Melting Point The temperature at which a substance melts; the temperature at which a solid is in equilibrium with its liquid state. (11.4)

Meniscus The interface between a liquid and a gas. (11.9)

Mercury Battery A galvanic cell of about 1.35 V involving zinc and mercury(II) oxide. (19.9)

Mercury Cell A device for the manufacture of sodium hydroxide. (19.3)

Metal An element or an alloy that is a good conductor of electricity, that has a shiny surface, and that is malleable and ductile; an element that normally forms positive ions and has an oxide that is basic. (2.7)

Metallic Crystal A solid having positive ions at the lattice positions that are attracted to a "sea of electrons" that extends throughout the entire crystal. (11.12)

Metalloids Elements with properties that lie between those of metals and nonmetals, and that are found in the periodic table around the diagonal line running from boron (B) to astatine (At). (2.7)

Metathesis Reaction See *Double Replacement Reaction.*

Meter (m) The SI base unit for length. (1.4)

Micelle A colloidal-sized group of ions, such as the anions of a detergent, that have clustered to maximize the contact between the ions' hydrophilic heads and water and to minimize the contact between the ions' hydrophobic parts and water. (12.10)

Milliliter (mL) 0.001 L. 1000 mL = 1 L. (1.4)

Millimeter (mm) 0.001 m. 1000 mm = 1 m. (1.4)

Millimeter of Mercury (mm Hg) A unit of pressure equal to 1/760 atm. 760 mm Hg = 1 atm. 1 mm Hg = 1 torr. (10.2)

Miscible Mutually soluble. (12.1)

Mixture Any matter consisting of two or more substances physically combined in no particular proportion by mass. (2.1)

Model, Scientific A picture or a mental construction derived from a set of ideas and assumptions that are imagined to be true because they can be used to explain certain observations and measurements, for example, the model of an ideal gas. Also called theoretical model. (1.2, 10.8)

Molal Boiling Point Elevation Constant (K_b) The number of degrees (°C) per unit of molal concentration that the boiling point of a solution is higher than that of the pure solvent. (12.7)

Molal Concentration (*m*) The number of moles of solute in 1000 g of solvent. (12.5)

Molal Freezing Point Depression Constant (K_f) The number of degrees (°C) per unit of molal concentration that the freezing point of a solution is lower than that of the pure solvent. (12.7)

Molality The molal concentration. (12.5)

Molar Concentration (M) The number of moles of solute per liter of solution. The molarity of a solution. (3.10)

Molar Heat Capacity The heat that can raise the temperature of 1 mol of a substance by 1 °C; the heat capacity per mole. (6.5)

Molar Heat of Fusion The heat absorbed when 1 mol of a solid melts to give 1 mol of the liquid at constant temperature and pressure. (11.7)

Molar Heat of Sublimation The heat absorbed when 1 mol of a solid sublimes to give 1 mol of its vapor at constant temperature and pressure. (11.7)

Molar Heat of Vaporization The heat absorbed when 1 mol of a liquid changes to 1 mol of its vapor at constant temperature and pressure. (11.7)

Molarity See *Molar Concentration.*

Molar Mass See *Formula Mass.*

Molar Solubility The number of moles of solute required to give 1 L of a saturated solution of the solute. (17.1)

Molar Volume, Standard The volume of 1 mol of a gas under standard conditions of temperature and pressure; 22.4/1383 L. (10.4)

Mole (mol) The SI unit for amount of substance; the formula mass in grams of an element or compound; a quantity of chemical substance that contains Avogadro's number of formula units (6.02×10^{23}). (3.1)

Molecular Compound A compound consisting of neutral (but often polar) molecules. (2.8)

Molecular Crystal A crystal that has molecules or individual atoms at the lattice points. (11.12)

Molecular Equation A chemical equation that gives the full formulas of all of the reactants and products and that is used to plan an actual experiment. (4.2)

Molecular Formula A chemical formula that gives the actual composition of one molecule. (2.8, 3.4)

Molecular Kinetic Energy The kinetic energy of a molecule. (6.2)

Molecular Mass The sum of the atomic masses of all of the atoms present in one molecule of a molecular compound. (3.2)

Molecular Orbital (MO) An orbital that extends over two or more atomic nuclei. (9.7)

Molecular Orbital Theory (MO Theory) A theory about covalent bonds that views a molecule as a collection of positive nuclei surrounded by electrons distributed among a set of bonding and antibonding orbitals of different energies. (9.4, 9.7)

Molecular Weight See *Molecular Mass.*

Molecule A neutral particle composed of two or more atoms combined in a definite ratio of whole numbers. (1.2, 2.2, 8.3)

Mole Fraction The ratio of the number of moles of one component of a mixture to the total number of moles of all components. (10.6)

Mole Percent The mole fraction of a component expressed as a percentage. (10.6)

Monoclinic Sulfur An allotrope of sulfur, S_8. (21.2)

Monodentate Ligand A ligand that can attach itself to a metal ion by only one atom. (20.3)

Monomer A substance of relatively low formula mass that is used to make a polymer. (23.6)

Monoprotic Acid An acid that can furnish one H^+ per molecule. (2.11, 4.4)

Monosaccharide A carbohydrate that cannot be hydrolyzed. (23.8)

N

Natural Logarithm See *Logarithm.*

Negative Charge A type of electrical charge possessed by certain particles such as the electron. A negative charge is attracted by a positive charge and is repelled by another negative charge. (2.5)

Nernst Equation (19.8)

$$E_{\text{cell}} = E_{\text{cell}}^{\circ} - \frac{RT}{n\mathcal{F}} \ln Q$$

Net Ionic Equation An ionic equation from which spectator ions have been omitted. It is balanced when both atoms and electrical charge balance. (4.2)

Neutralization, Acid–Base The destruction of an acid by a base or of a base by an acid. (2.11, 4.4, 15.3)

Neutral Solution A solution in which $[H^+] = [OH^-]$. (15.5)

Neutron $(n, {}_{0}^{1}n)$ A subatomic particle with a charge of zero, a mass of 1.008665 u $(1.674954 \times 10^{-24}$ g), and that exists in all atomic nuclei except those of the hydrogen-1 isotope. (2.5)

Neutron Activation Analysis A technique to analyze for trace impurities in a sample by studying the frequencies and intensities of the gamma radiation they emit after they have been rendered radioactive by neutron bombardment of the sample. (22.7)

Neutron Emission A nuclear reaction in which a neutron is ejected. (22.3)

Nickel–Cadmium Storage Cell (Nicad Battery) A galvanic cell of about 1.4 V involving cadmium and nickel(IV) oxide. (19.9)

Noble Gases Group VIIIA in the periodic table—helium, neon, argon, krypton, xenon, and radon. (2.6)

Node A place where the amplitude or intensity of a wave is zero. (7.3)

Nomenclature The names of substances and the rules for devising names. (2.11, 21.2)

Nonelectrolyte A compound that in its molten state or in solution cannot conduct electricity. (4.1)

Nonlinear Molecule A molecule in which the atoms do not lie in a straight line. (9.2)

Nonmetal (Nonmetallic Element) A nonductile, nonmalleable, nonconducting element that tends to form negative ions (if it forms them at all) far more readily than positive ions and whose oxide is likely to show acidic properties. (2.7)

Nonoxidizing Acid An acid in which the anion is a poorer oxidizing agent than the hydrogen ion (e.g., HCl, H_2SO_4, H_3PO_4). (5.3)

Nonpolar Covalent Bond An electron pair bond at the ends of which are atoms of equal or very nearly equal electronegativity. (8.5)

Nonvolatile Descriptive of a substance with a high boiling point, a low vapor pressure, and that does not evaporate. (12.6)

Normal Boiling Point The temperature at which the vapor pressure of a liquid equals 1 atm. (11.6)

Normality (N) The number of equivalents of solute per liter of solution. (Appendix C.4)

n-Type Semiconductor A semiconductor doped with an impurity that causes the moving charge to consist of negatively charged electrons. (9.9)

Nuclear Chain Reaction See *Chain Reaction*.

Nuclear Equation A description of a nuclear reaction that uses the special symbols of isotopes, that describes some kind of nuclear transformation or disintegration, and that is balanced when the sums the atomic numbers on either side of the arrow are equal and the sums of the mass numbers are also equal. (22.3)

Nuclear Fission See *Fission*.

Nuclear Fusion See *Fusion*.

Nuclear Reaction A change in the composition or energy of the nuclei of isotopes accompanied by one or more events such as the radiation of nuclear particles or electromagnetic energy, transmutation, fission, or fusion. (22.1)

Nucleic Acids Polymers in living cells that store and translate genetic information and whose molecules hydrolyze to give a sugar unit (ribose from ribonucleic acid, RNA, or deoxyribose from deoxyribonucleic acid, DNA), a phosphate, and a set of four or five nitrogen-containing, heterocyclic bases (adenine, thymine, guanine, cytosine, and uracil). (23.11)

Nucleon A proton or a neutron. (2.5)

Nucleus The hard, dense core of an atom that holds the atom's protons and neutrons. (2.5)

O

Octahedral Molecule A molecule in which the planar surfaces of an octahedron are created by connecting adjacent nuclei with imaginary lines. (9.1)

Octahedron An eight-sided figure that can be envisioned as two square pyramids sharing the common square base. (9.1)

Octet (of Electrons) Eight electrons in the valence shell of an atom. (8.1)

Octet Rule An atom tends to gain or lose electrons until its outer shell has eight electrons. (8.1, 8.3)

Odd–Even Rule When the numbers of protons and neutrons in an atomic nucleus are both even, the isotope is more likely to be stable than when both numbers are odd. (22.4)

Open-End Manometer See *Manometer*.

Open System A system open to the atmosphere. (6.4)

Optical Isomers Stereoisomers other than geometric (cis/trans) isomers and that include substances that can rotate the plane of plane-polarized light. (20.6)

Optically Active A property of being able to rotate the plane of plane polarized light. (20.6)

Orbital An electron waveform with a particular energy and a unique set of values for the quantum numbers n, ℓ, and m_ℓ. (7.3)

Orbital Diagram A diagram showing an atom's orbitals in which the electrons are represented by arrows to indicate paired and unpaired spins. (7.5)

Order (of a Reaction) The sum of the exponents in the rate law is the *overall* order. Each exponent gives the order of the reaction with respect to a specific reactant. (13.4)

Ordinate The vertical axis. (Appendix A.3)

Organic Chemistry The study of the compounds of carbon that are not classified as inorganic. (2.10, 23.1)

Organic Compound Any compound of carbon other than a carbonate, bicarbonate, cyanide, cyanate, carbide, or gaseous oxide. (8.4, 23.1)

Orthorhombic Sulfur An allotrope of sulfur, S_8, having a puckered eight-member ring structure. (21.2)

Osmosis The passage of solvent molecules, but not those of solutes, through a semipermeable membrane; the limiting case of dialysis. (12.8)

Osmotic Pressure The back pressure that would have to be applied to prevent osmosis; one of the colligative properties. (12.8)

Outer Electrons The electrons in the occupied shell with the largest principal quantum number. An atom's electrons in its valence shell. (7.6)

Outer Shell The occupied shell in an atom having the highest principal quantum number (n). (7.6)

Overlap of Orbitals A portion of two orbitals from different atoms that share the same space in a molecule. (9.4)

Oxidation A change in which an oxidation number increases (becomes more positive). A loss of electrons. (5.1)

Oxidation Number The charge that an atom in a molecule or ion would have if all of the electrons in its bonds belonged entirely to the more electronegative atoms; the oxidation state of an atom. (5.1)

Oxidation–Reduction Reaction A chemical reaction in which changes in oxidation numbers occur. (5.1, 5.3)

Oxidation State See *Oxidation Number*.

Oxidizing Acid An acid in which the anion is a stronger oxidizing agent than H^+ (e.g., $HClO_4$, HNO_3). (5.3)

Oxidizing Agent The substance that causes oxidation and that is itself reduced. (5.1)

Oxoacid An acid that contains oxygen besides hydrogen and another element (e.g., HNO_3, H_3PO_4, H_2SO_4). (2.11, 15.2)

Ozone A very reactive allotrope of oxygen having the formula O_3. (21.2)

P

Pairing Energy The energy required to force two electrons to become paired and occupy the same orbital. (20.7)

Paramagnetism The weak magnetism of a substance whose atoms, molecules, or ions have unpaired electrons. (7.4)

Partial Charge Charges at opposite ends of a dipole that are fractions of full 1+ or 1− charges. (8.5)

Partial Pressure The pressure contributed by an individual gas to the total pressure of a gas mixture. (10.6)

Partial Pressures, Law of (Dalton's Law of Partial Pressures) The total pressure of a mixture of gases equals the sum of their partial pressures. (10.6)

Parts per Billion (ppb) The number of parts of one component of a mixture to 10^9 total parts. (Facets of Chemistry 12.2)

Parts per Million (ppm) The number of parts of one component of a mixture to 10^6 total parts. (Facets of Chemistry 12.2)

Pascal (Pa) The SI unit of pressure equal to 1 newton m^{-2};

$$133.3224 \text{ Pa} = 1 \text{ torr.} \quad (10.2)$$

Pauli Exclusion Principle No two electrons in an atom can have the same values for all four of their quantum numbers. (7.4)

Peptide Bond The amide linkage in molecules of polypeptides. (23.10)

Percent by Mass (Percent by Weight) (a) The number of grams of an element combined in 100 g of a compound. (3.3) (b) The number of grams of a substance in 100 g of a mixture or solution. (12.5, Facets of Chemistry 12.2)

Percent Composition A list of the percentages by weight of the elements in a compound. (3.3)

Percentage Concentration A ratio of the amount of solute to the amount of solution expressed as a percent. (3.9)

Weight/weight The grams of solute in 100 g of solution. (12.5)

Volume/volume The volumes of solute in 100 volumes of solution. (Facets of Chemistry 12.2)

Percentage Ionization (16.2)

$$\frac{[\text{Amt of species ionized}]}{[\text{Initial amt of species}]} \times 100\%.$$

Percentage Yield The ratio (taken as a percent) of the mass of product obtained to the mass calculated from the reaction's stoichiometry. (3.8)

Period A horizontal row of elements in the periodic table. (2.6)

Periodic Law The properties of the elements are a periodic function of their atomic numbers. (2.6)

Periodic Table A table in which the symbols for the elements are listed in order of increasing atomic number and arranged so that elements with similar properties lie in the same column (group). (2.6)

PET See *Positron Emission Tomography.*

pH $-\log[\text{H}^+]$. (15.5)

Phase A homogeneous region within a sample. (2.1)

Phase Diagram A pressure–temperature graph on which are plotted the temperatures and the pressures at which equilibrium exists between the states of a substance. It defines regions of T and P in which the solid, liquid, and gaseous states of the substance can exist. (11.9)

Photon A unit of energy in electromagnetic radiation equal to $h\nu$, where ν is the frequency of the radiation and h is Planck's constant. (7.1)

Physical Change Any change in which substances do not change into other substances. (1.7, 2.1)

Physical Property A property that can be specified without reference to another substance and that can be measured without causing a chemical change. (1.7)

Physical State The condition of aggregation of a substance's formula units, whether as a solid, a liquid, or a gas. (1.3)

Pi Bond (π Bond) A bond formed by the sideways overlap of a pair of p orbitals and that concentrates electron density into two separate regions that lie on opposite sides of a plane that contains an imaginary line joining the nuclei. (9.6)

pK_a $-\log K_a$ (16.1)

pK_b $-\log K_b$ (16.1)

pK_w $-\log K_w$ (15.5)

Planar Triangular Molecule A molecule in which a central atom holds three other atoms located at the corners of an equilateral triangle and that includes the central atom at its center. (9.1)

Planck's Constant (h) The ratio of the energy of a photon to its frequency; $6.6260755 \times 10^{-34}$ J Hz^{-1}. (7.1)

p–n Junction The interface between n- and p-type semiconductors in a transistor. (9.9)

pOH $-\log[\text{OH}^-]$ (15.5)

Polar Covalent Bond (Polar Bond) A covalent bond in which more than half of the bond's negative charge is concentrated around one of the two atoms. (8.5)

Polar Molecule A molecule in which individual bond polarities do not cancel and in which, therefore, the centers of density of negative and positive charges do not coincide. (8.5)

Polyatomic Ion An ion composed of two or more atoms. (2.9)

Polydentate Ligand A ligand that has two or more atoms that can become simultaneously attached to a metal ion. (20.3)

Polymer A substance formed by linking together many simpler units called monomers. (23.6)

Polymerization A chemical reaction that converts a monomer into a polymer. (23.6)

Polypeptide A polymer of α-amino acids that makes up all or most of a protein. (23.10)

Polyprotic Acid An acid that can furnish more than one H$^+$ per molecule. (2.11, 4.4)

Polysaccharide A carbohydrate whose molecules can be hydrolyzed to hundreds of monosaccharide molecules. (23.8)

Polysulfide Ion An ion with the general formula S_x^{2-}. (21.4)

Position of Equilibrium The relative amounts of the substances on both sides of the double arrows in the equation for an equilibrium. (4.3, 11.8)

Positive Charge A type of electrical charge possessed by certain particles such as the proton. A positive charge is attracted by a negative charge and is repelled by another positive charge. (2.5)

Positron ($_0^1e$) A positively charged particle with the mass of an electron. (22.3)

Positron Emission Tomography (PET) A technique to obtain an X-ray-like image of part of the body by letting the body take in positron-emitting isotopes and then studying the gamma emission. (22.4)

Post-transition Metal A metal that occurs in the periodic table immediately to the right of a row of transition elements. (2.9)

Potential See *Electromotive Force.*

Potential Energy Stored energy. (1.3, 6.1)

Precipitate A solid that separates from a solution usually as the result of a chemical reaction. (3.9, 4.3)

Precipitation In chemistry, the formation of a precipitate. (4.3)

Precision How reproducible measurements are; the fineness of a measurement as indicated by the number of significant figures reported in the physical quantity. (1.5)

Pressure Force per unit area. (6.4, 10.2)

Pressure–Concentration Law See *Vapor Pressure–Concentration Law.*

Pressure–Solubility Law (Henry's Law) The concentration of a gas dissolved in a liquid at any given temperature is directly proportional to the partial pressure of this gas above the solution. (12.4)

Pressure–Temperature Law (Gay-Lussac's Law) The pressure of a given mass of gas is directly proportional to its Kelvin temperature if the volume is kept constant. $P \propto T$. (10.3)

Pressure–Volume Law (Boyle's Law) The volume of a given mass of a gas is inversely proportional to its pressure if the temperature is kept constant. $V \propto 1/P$. (10.3)

Pressure–Volume Work The energy transferred as work when a system expands or contracts against the pressure exerted by the surroundings. At constant pressure, work $= -P\Delta V$. (6.4)

Principal Quantum Number (n) The quantum number that defines the principal energy levels and that can have values of $1, 2, 3, \ldots, \infty$. (7.3)

Products The substances produced by a chemical reaction and whose formulas follow the arrows in chemical equations. (2.3)

Propagation Step A step in a chain reaction for which one product must serve in a succeeding propagation step as a reactant and for which another (final) product accumulates with each repetition of the step. (Facets of Chemistry 13.1)

Property A characteristic of matter. (1.7)

Protein A macromolecular substance found in cells that consists wholly or mostly of one or more polypeptides that often are combined with an organic molecule or a metal ion. (23.10)

Proton ($_1^1p$ or $_1^1H^+$) (a) A subatomic particle, with a charge of 1+ and a mass of 1.00727252 u (1.6726430 $\times$ 10^{-24}g), that is found in atomic nuclei. (2.5) (2) The name often used for the hydrogen ion and symbolized as H^+. (2.11)

Proton Acceptor A Brønsted base. (15.1)

Proton Donor A Brønsted acid. (15.1)

p-Type Semiconductor A semiconductor doped with an impurity that enables the charge to move as positively charged holes. (9.9)

Pure Substance An element or a compound. (2.1)

Q

Qualitative Analysis The use of experimental procedures to determine what elements are present in a substance. (3.3, 4.9)

Qualitative Observation Observations that do not involve numerical information. (1.4)

Quantitative Analysis The use of experimental procedures to determine the percentage composition of a compound or the percentage of a component of a mixture. (3.3, 4.9)

Quantitative Observation An observation involving a measurement and numerical information. (1.4)

Quantized Descriptive of a discrete, definite amount as of *quantized energy.* (7.2)

Quantum The energy of one photon. (7.1)

Quantum Mechanics See *Wave Mechanics.* (7.3)

Quantum Number A number related to the energy, shape, or orientation of an orbital, or to the spin of an electron. (7.2)

R

R See *Gas Constant, Universal.*

Rad A unit of radiation absorbed dose equal to 10^{-5} J g^{-1} or 10^{-2} Gy. (22.6)

Radiation The emission of electromagnetic energy or nuclear particles. (22.1)

Radical See *Free Radical.*

Radioactive The ability to emit various types of atomic radiation or gamma rays. (22.1)

Radioactive Decay The change of a nucleus into another nucleus (or into a more stable form of the same nucleus) by the loss of a small particle or a gamma ray photon. (22.3)

Radioactive Disintegration Series A sequence of nuclear reactions beginning with a very long-lived radionuclide and ending with a stable isotope of lower atomic number. (22.3)

Radioactivity The emission of one or more kinds of radiation from an isotope with unstable nuclei. (22.3)

Radiological Dating A technique for measuring the age of a geologic formation or an ancient artifact by determining the ratio of the concentrations of two isotopes, one radioactive and the other a stable decay product. (22.7)

Radionuclide A radioactive isotope. (22.1)

Raoult's Law The vapor pressure of one component above a mixture of molecular compounds equals the product of its vapor pressure when pure and its mole fraction. (12.6)

Rare Earth Metals The lanthanides. (2.6)

Rate A ratio in which a unit of time

appears in the denominator, for example, 40 miles hr^{-1} or 3.0 mol L^{-1} s^{-1}. (13.3)

Rate Constant The proportionality constant in the rate law; the rate of reaction when all reactant concentrations are 1 M. (13.4)

Rate-Determining Step (Rate-Limiting Step) The slowest step in a reaction mechanism. (13.8)

Rate Law An equation that relates the rate of a reaction to the molar concentrations of the reactants raised to powers. (13.4)

Rate of Reaction How quickly the reactants disappear and the products form and usually expressed in units of mol L^{-1} s^{-1}. (13.1)

Reactant, Limiting See *Limiting Reactant.*

Reactants The substances brought together to react and whose formulas appear before the arrow in a chemical equation. (2.3)

Reaction Coordinate The horizontal axis of a potential energy diagram of a reaction. (13.6)

Reaction Quotient (Q) The numerical value of the mass action expression. See *Mass Action Expression.* (14.3)

Reagent A chemical commonly kept on hand in a laboratory to be used in chemical reactions. (4.4)

Recombinant DNA DNA in a bacterial plasmid altered by the insertion of DNA from another organism. (23.11)

Redox Reaction An oxidation–reduction reaction. (5.1)

Red Phosphorus An allotrope of phosphorus believed to be formed by the linking of P_4 tetrahedra at their corners. (21.2)

Reducing Agent A substance that causes reduction and is itself oxidized. (5.1)

Reduction A change in which an oxidization number decreases (becomes less positive and more negative). A gain of electrons. (5.1)

Reduction Potential A measure of the tendency of a given half-reaction to occur as a reduction. (19.5)

Rem A dose in rads multiplied by a

factor that takes into account the variations that different kinds of radiation have in their damage-causing abilities in tissue. (22.6)

Replication In nucleic acid chemistry, the reproductive duplication of DNA double helices prior to cell division. (23.11)

Representative Element An element in one of the A groups in the periodic table. Also called main group element. (2.6)

Resonance A concept in which the actual structure of a molecule or polyatomic ion is represented as a composite or average of two or more Lewis structures, which are called the resonance or contributing structures (and none of which has real existence). (8.9)

Resonance Energy The difference in energy between a substance and its principal resonance (contributing) structure. (8.9)

Resonance Hybrid The actual structure of a molecule or polyatomic ion taken as a composite or average of the resonance or contributing structures. (8.9)

Resonance Structure A Lewis structure that contributes to the hybrid structure in resonance-stabilized systems; a contributing structure. (8.9)

Reverse Osmosis The use of pressure to force the migration of solvent from a solution through a semipermeable membrane. (Facets of Chemistry 12.3)

Reversible Process A process that occurs by an infinite number of steps during which the driving force for the change is just barely greater than the force that resists the change. (18.7)

Ring, Carbon A closed-chain sequence of carbon atoms. (23.1)

RNA Ribonucleic acid; a nucleic acid that gives ribose, phosphate ion, adenine, uracil, guanine, and cytosine when hydrolyzed. (23.11)

Rock Salt Structure The face-centered cubic structure observed for sodium chloride, which is also possessed by crystals of many other compounds. (11.10)

Roentgen Unit of exposure intensity for X rays or gamma radiation. (22.6)

Rydberg Equation An equation used

to calculate the wavelengths of all the spectral lines of hydrogen. (7.2)

S

Salt An ionic compound in which the anion is not OH^- or O^{2-} and the cation is not H^+. (2.11, 4.6)

Salt Bridge A tube that contains an electrolyte that connects the two half-cells of a galvanic cell. (19.4)

Saponification The reaction of an organic ester with a strong base to give an alcohol and the salt of the organic acid. (23.9)

Saturated Organic Compound A compound whose molecules have only single bonds. (23.2)

Saturated Solution A solution that holds as much solute as it can at a given temperature. A solution in which there is an equilibrium between the dissolved and the undissolved states of the solute. (3.9)

Scientific Law See *Law.*

Scientific Method The observation, explanation, and testing of an explanation by additional experiments. (1.2)

Scientific Notation The representation of a quantity as a decimal number between 1 and 10 multiplied by 10 raised to a power (e.g., 6.02×10^{23}). (Appendices A.1, A.4)

Secondary Quantum Number (ℓ) The quantum number whose values can be $0, 1, 2, \ldots, (n-1)$, where n is the principal quantum number. (7.3)

Second Law of Thermodynamics Whenever a spontaneous event takes place, it is accompanied by an increase in the entropy of the universe. (18.4)

Second-Order Reaction A reaction with a rate law of the type rate = $k[A]^2$ or rate = $k[A][B]$, where A and B are reactants. (13.4)

Seesaw Structure A description given to a molecule in which the central atom has 5 electron pairs in its valence shell, one of which is a lone pair and the others are used in bonds to other atoms. See also: Distorted tetrahedron. (9.2)

Semiconductor A substance that conducts electricity weakly. (2.7)

Shell All of the orbitals associated with a given value of n (the principal quantum number). (7.3)

SI (International System of Units) The modified metric system adopted in 1960 by the General Conference on Weights and Measures. (1.4)

Side Reaction A reaction that occurs simultaneously with another reaction (the main reaction) in the same mixture to produce by-products. (3.8)

Sievert (Sv) The SI unit for dose equivalent. (22.6)

Sigma Bond (σ Bond) A bond formed by the head-to-head overlap of two atomic orbitals and in which electron density becomes concentrated along and around the imaginary line joining the two nuclei. (9.6)

Significant Figures The number of digits in a physical quantity that are known to be certain plus one more. (1.5)

Silane A compound of silicon and hydrogen (e.g., SiH_4, Si_2H_6, Si_3H_8, etc.) (21.3)

Silver Oxide Battery A galvanic cell of about 1.5 V involving zinc and silver oxide and used when miniature batteries are needed. (19.9)

Simple Cubic Unit Cell A cell or unit of crystal structure that has atoms, molecules, or ions only at the corners of a cube. (11.10)

Single Bond A covalent bond in which a single pair of electrons is shared. (8.3)

Single Replacement Reaction A reaction in which one element replaces another in a compound; usually a redox reaction. (5.4)

Skeleton Equation An unbalanced equation showing only the formulas of reactants and products. (5.2)

Soap A salt of a fatty acid whose anions form micelles and which is able to hold oil and grease particles in suspension in water. (12.10)

Sol The colloidal dispersion of a solid in a fluid. (12.10)

Solar Battery (Solar Cell) A silicon wafer doped with arsenic and placed over a silicon wafer doped with boron to give a system that conducts electricity when light falls on it. (9.9)

Solubility The ratio of the quantity of solute to the quantity of solvent in a saturated solution and that is usually expressed in units of (g solute)/(100 g solvent) at a specified temperature. (3.9, 4.3)

Solubility Product Constant (K_{sp}) The equilibrium constant for the solubility of a salt and that, for a saturated solution, is equal to the product of the molar concentrations of the ions, each raised to a power equal to the number of its ions in one formula unit of the salt. (17.1)

Solute Something dissolved in a solvent to make a solution. (3.9)

Solution A homogeneous mixture in which all particles are of the size of atoms, small molecules, or small ions. (2.1, 3.9, 12.1)

Solvation The development of a cage-like network of a solution's solvent molecules about a molecule or ion of the solute. (12.1)

Solvation Energy The enthalpy of the interaction of gaseous molecules or ions of solute with solvent molecules during the formation of a solution. (12.2)

Solvent A medium, usually a liquid, into which something (a solute) is dissolved to make a solution. (3.9)

sp Hybrid Orbital A hybrid orbital formed by mixing one s and one p atomic orbital. The angle between a pair of sp hybrid orbitals is 180°. (9.5)

sp^2 Hybrid Orbital A hybrid orbital formed by mixing one s and two p atomic orbitals. The angle between two sp^2 hybrid orbitals is 120°. (9.5)

sp^3 Hybrid Orbital A hybrid orbital formed by mixing one s and three p atomic orbitals. The angle between two sp^3 hybrid orbitals is 109.5° (9.5)

sp^3d Hybrid Orbital A hybrid orbital formed by mixing one s, three p, and one d atomic orbital. sp^3d hybrids point to the corners of a trigonal bipyramid. (9.5)

sp^3d^2 Hybrid Orbital A hybrid orbital formed by mixing one s, three p, and two d atomic orbitals. sp^3d^2 hybrids point to the corners of an octahedron. (9.5)

Specific Gravity The ratio of the density of a substance to the density of water. (1.7)

Specific Heat The quantity of heat that will raise the temperature of 1 g of a substance by 1 °C, usually in units of cal g^{-1} °C^{-1} or J g^{-1} °C^{-1}. (6.5)

Spectator Ion An ion whose formula appears in an ionic equation identically on both sides of the arrow, that does not participate in the reaction, and that is excluded from the net ionic equation. (4.2)

Spectrochemical Series A listing of ligands in order of their ability to produce a large crystal field splitting. (20.7)

Spin Quantum Number (m_s) The quantum number associated with the spin of a subatomic particle and which for the electron can have a value of $+\frac{1}{2}$ or $-\frac{1}{2}$. (7.4)

Spontaneous Change A change that occurs by itself without outside assistance. (18.2)

Square Planar Molecule A molecule with a central atom having four bonds that point to the corners of a square. (9.2)

Square Pyramid A pyramid with four triangular sides and a square base. (9.2)

Stability Constant See *Formation Constant.*

Stabilization Energy See *Resonance Energy.*

Standard Atmosphere See *Atmosphere, Standard.*

Standard Cell Potential (E°_{cell}) The potential of a galvanic cell at 25 °C and when all ion concentrations are exactly 1 M and the partial pressures of all gases are 1 atm. (19.5)

Standard Conditions of Temperature and Pressure (STP) 273.15 K (0 °C) and 1 atm (760 torr). (10.4)

Standard Electrode Potential See *Standard Reduction Potential.*

Standard Enthalpy Change ($\Delta H°$) See *Standard Heat of Reaction.*

Standard Enthalpy of Formation (ΔH°_f) See *Standard Heat of Formation.*

Standard Entropy ($S°$) The entropy of 1 mol of a substance at 25 °C and 1 atm. (18.5)

Standard Entropy Change ($\Delta S°$) The entropy change of a reaction when determined with reactants and products at 25 °C and 1 atm and on the scale of the mole quantities given by the coefficients of the balanced equation. (18.5)

Standard Entropy of Formation ($\Delta S_f°$) The value of $\Delta S°$ for the formation of one mole of a substance from its elements in their standard states. (18.5)

Standard Free Energy Change ($\Delta G°$) $\Delta G° = \Delta H° - (298 \text{ K})\Delta S°$ (18.6)

Standard Free Energy of Formation ($\Delta G_f°$) The value of $\Delta G°$ for the formation of *one* mole of a compound from its elements in their standard states. (18.6)

Standard Heat of Formation ($\Delta H_f°$) The amount of heat absorbed or evolved when one mole of the compound is formed from its elements in their standard states. (6.8)

Standard Heat of Reaction ($\Delta H°$) The enthalpy change of a reaction when determined with reactants and products at 25 °C and 1 atm and on the scale of the mole quantities given by the coefficients of the balanced equation. (6.6)

Standard Molar Volume See *Molar Volume, Standard.*

Standard Reduction Potential The reduction potential of a half-reaction at 25 °C when all ion concentrations are 1 M and the partial pressures of all gases are 1 atm. (19.5)

Standard Solution Any solution whose concentration is accurately known. (4.9)

Standard State The condition in which a substance is in its most stable forms at 25 °C and 1 atm. (6.8)

Standing Wave A wave whose peaks and nodes do not change position. (7.3)

State Function A function or variable whose value depends only on the initial and final states of the system and not on the path taken by the system to get from the initial to the final state. (P, V, T, H, S, and G are all state functions.) (6.4)

State of Matter A physical state of a substance: solid, liquid, or gas. (1.3. See also *Standard State.*)

State of a System The set of specific values of the physical properties of a system—its composition, physical form, concentration, temperature, pressure, and volume. (6.4)

Stereoisomerism The existence of isomers whose structures differ only in spatial orientations (e.g., geometric isomers and optical isomers). (20.6)

Stock System A system of nomenclature that uses Roman numerals to specify oxidation states. (2.11)

Stoichiometric Equivalence The ratio by moles between two elements in a formula or two substances in a chemical reaction. (3.1)

Stoichiometry A description of the relative quantities by moles of the reactants and products in a reaction as given by the coefficients in the balanced equation. (3.1)

Stopcock A device on a buret that is used to control the flow of titrant. (4.9)

Stored Energy See *Potential Energy.*

STP See *Standard Conditions of Temperature and Pressure.*

Straight-Chain Compound An organic compound in whose molecules the carbon atoms are joined in one continuous open-chain sequence. (23.1)

Strong Acid An acid that is essentially 100% ionized in water. A good proton donor. An acid with a large value of K_a. (4.5, 15.6)

Strong Base Any powerful proton acceptor. A base with a large value of K_b. A metal hydroxide that dissociates essentially 100% in water. (4.5, 15.6)

Strong Electrolyte Any substance that ionizes or dissociates in water to essentially 100%. (4.1)

Structural Formula A chemical formula that shows how the atoms of a molecule or polyatomic ion are arranged, to which other atoms they are bonded, and the kinds of bonds (single, double, or triple). (8.3)

Subatomic Particles Electrons, protons, neutrons, and atomic nuclei. (2.3)

Sublimation The conversion of a solid directly into a gas without passing through the liquid state. (11.3)

Subshell All of the orbitals of a given shell that have the same value of the secondary quantum number, ℓ. (7.3)

Substance See *Pure Substance.*

Substitution Reaction The replacement of an atom or group on a molecule by another atom or group. (23.2, 23.3)

Supercooled The condition of a substance in its liquid state below its freezing point. (11.7)

Supercooled Liquid A liquid at a temperature below its freezing point. An amorphous solid. (11.7, 11.13)

Supercritical Fluid A substance at a temperature above its critical temperature. (11.9)

Superimposability A test of structural chirality in which a model of one structure and a model of its mirror image are compared to see if the two could be made to blend perfectly, with every part of one coinciding simultaneously with the parts of the other. (20.6)

Supersaturated Solution A solution whose concentration of solute exceeds the equilibrium concentration. (3.9)

Surface Tension A measure of the amount of energy needed to expand the surface area of a liquid. (11.3)

Surfactant A substance that lowers the surface tension of a liquid and promotes wetting. (11.3)

Surroundings That part of the universe other than the system being studied and separated from the system by a real or an imaginary boundary. (6.4)

Suspension A homogeneous mixture in which the particles of at least one component are larger than colloidal particles (>1000 nm in at least one dimension). (12.10)

System That part of the universe under study and separated from the surroundings by a real or an imaginary boundary. (6.4)

T

$t_{1/2}$ See *Half-life.*

Temperature A measure of the hotness or coldness of something and proportional to the average molecular kinetic energy of the atoms, molecules, or ions present. (1.3, 6.2)

Temperature–Volume Law (Charles' Law) The volume of a given mass of a gas is directly proportional to its Kelvin temperature if the pressure is kept constant. $V \propto T$. (10.3)

Tetrahedral Molecule A molecule with a central atom bonded to four other atoms located at the corners of an imaginary tetrahedron. (9.1)

Tetrahedron A four-sided figure with four triangular faces and shaped like a pyramid. (9.1)

Theoretical Model See *Model, Scientific.*

Theoretical Yield The yield of a product calculated from the reaction's stoichiometry. (3.8)

Theory A tested explanation of the results of many experiments. (1.2)

Thermal Decomposition (20.1)

Thermal Energy See *Heat.*

Thermal Property A physical property, like heat capacity or heat of fusion, that concerns a substance's ability to absorb heat without changing chemically. (6.5)

Thermochemical Equation A balanced chemical equation accompanied by the value of $\Delta H°$ that corresponds to the mole quantities specified by the coefficients. (6.6)

Thermochemistry The study of the energy changes of chemical reactions. (6.1)

Thermodynamic Equilibrium Constant (K) The equilibrium constant that is calculated from $\Delta G°$ (the standard free energy change) for a reaction at T (in K) by the equation, $\Delta G° = -RT \ln K$. (18.9)

Thermodynamics The study of the laws that govern the energy and entropy changes of physical and chemical events. (6.4, 18.1)

Third Law of Thermodynamics For a pure crystalline substance at 0 K, $S = 0$. (18.5)

Titrant The solution added from a buret during a titration. (4.9)

Titration An analytical procedure in which a solution of unknown concentration is combined slowly and carefully with a standard solution until a color change of some indicator or some other signal shows that equivalent quantities have reacted. Either solution can be the titrant in a buret with the other solution being in a receiving flask. (4.9, 16.8)

Torr A unit of pressure equal to 1/760 atm. 1 mm Hg. (10.2)

Total Energy The sum of the kinetic and potential energies of a system. (6.1)

Tracer Analysis The use of small amounts of a radioisotope to follow (trace) the course of a chemical or biological change. (22.7)

Transcription The synthesis of mRNA at the direction of DNA. (23.11)

Trans Isomer A stereoisomer whose uniqueness lies in having two groups that project on opposite sides of a reference plane. (20.6, 23.2)

Transition Elements The elements located between Groups IIA and IIIA in the periodic table. (2.6)

Transition Metals The transition elements. (2.6)

Transition State The brief moment during an elementary process in a reaction mechanism when the species involved have acquired the minimum amount of potential energy needed for a successful reaction, an amount of energy that corresponds to the high point on a potential energy diagram of the reaction. (13.6)

Transition State Theory A theory about the formation and breakup of activated complexes. (13.6)

Translation The synthesis of a polypeptide at the direction of a molecule of mRNA. (23.11)

Transmutation The conversion of one isotope into another. (22.5)

Transuranium Elements Elements 93–103. (22.5)

Traveling Wave A wave whose peaks and nodes move. (7.3)

Triacylglycerol An ester of glycerol and three fatty acids. (23.9)

Trigonal Bipyramid A six-sided figure made of two three-sided pyramids that share a common face. (9.1)

Trigonal Bipyramidal Molecule A molecule with a central atom holding five other atoms that are located at the corners of a trigonal bipyramid. (9.1, 9.2)

Trigonal Pyramidal Molecule A molecular structure that consists of an atom, situated at the top of a three sided pyramid, which is bonded to three other atoms located at the corners of the base of the pyramid. (9.2)

Triple Bond A covalent bond in which three pairs of electrons are shared. (8.3, 9.6)

Triple Point The temperature and pressure at which the liquid, solid, and vapor states of a substance can coexist in equilibrium. (11.9)

Triprotic Acid An acid that can furnish three H^+ ions per molecule. (4.4)

T-Shaped Molecule A molecule having 5 electron pairs in its valence shell, two of which are lone pairs. The other three are used in bonds to other atoms. The molecule has the shape of the letter T, with the central atom located at the intersection of the two crossing lines. (9.2)

Tyndall Effect The scattering of light by colloidally dispersed particles that gives a milky appearance to the mixture. (12.10)

U

Uncertainty Principle There is a limit to our ability to measure a particle's speed and position simultaneously. (7.8)

Unit Cell The smallest portion of a crystal that can be repeated over and over in all directions to give the crystal lattice. (11.10)

Universal Gas Constant (R) The ratio of PV to nT for gases, $R = 0.0821$ L atm mol^{-1} K^{-1} (10.4)

Unsaturated Compound A compound whose molecules have one or more double or triple bonds. (23.2)

Unsaturated Solution Any solution with a concentration less than that of a saturated solution of the same solute and solvent. (3.9)

V

Vacuum An enclosed space containing no matter whatsoever. A *partial vacuum* is an enclosed space containing a gas at a very low pressure. (10.2)

Valence Band The vacant or partially filled band of outer shell orbitals in a solid. (9.9)

Valence Bond Theory (VB Theory) A theory of covalent bonding that views a bond as being formed by the sharing of one pair of electrons between two overlapping atomic or hybrid orbitals. (9.4)

Valence Electrons The electrons of an atom in its valence shell that participate in the formation of chemical bonds. (7.6)

Valence Shell The electron shell with the highest principal quantum number, n, that is occupied by electrons. (7.6)

Valence Shell Electron Pair Repulsion Theory (VSEPR Theory) The geometry of a molecule at any given central atom is determined by the repulsions of both bonding and nonbonding (lone pair) electrons in the valence shell. (9.2)

Van der Waals Constants Empirical constants that make the van der Waals equation conform to the gas law behavior of a real gas. (10.9)

Van der Waals Equation An equation of state for a real gas that corrects the ideal gas law for the excluded volume of the gas and for intermolecular attractions. (10.9)

Van't Hoff Factor The ratio of the observed freezing point depression to the value calculated on the assumption that the solute dissolves as un-ionized molecules. (12.9)

Vapor Pressure The pressure exerted by the vapor above a liquid (usually referring to the *equilibrium* vapor pressure when the vapor and liquid are in equilibrium with each other). (10.6, 11.5)

Vapor Pressure–Concentration Law (Raoult's Law) The vapor pressure of one component above a mixture of molecular compounds equals the product of its vapor pressure when pure and its mole fraction. (12.6)

VB Theory See *Valence Bond Theory.*

Viscosity The resistance of a fluid to a change in its form. (11.3)

Visible Spectrum That region of the electromagnetic spectrum whose frequencies can be detected by the human eye. (7.1)

Volatile Descriptive of a liquid that has a low boiling point, has a high vapor pressure at room temperature, and therefore evaporates easily. (3.8, 12.6)

Volt (V) The SI unit of electromotive force or emf in joules per coulomb. $1 \text{ V} = 1 \text{ J C}^{-1}$. (19.5)

Voltaic Cell See *Galvanic Cell.*

VSEPR Theory See *Valence Shell Electron Pair Repulsion Theory.*

V-Shaped Molecule See *Nonlinear Molecule.*

W

Wavelength (λ) The distance between crests in the wavelike oscillations of electromagnetic radiations. (7.1).

Wave Function (ψ) A mathematical function that describes the intensity of an electron wave at a specified location in an atom. The square of the wave function at a particular location specifies the probability of finding an electron there. (7.3)

Wave Mechanics A theory of atomic structure based on the wave properties of matter. (7.3)

Weak Acid An acid with a low percentage ionization in solution; a poor proton donor; an acid with a low value of K_a. (4.5, 16.2)

Weak Base A base with a low percentage ionization in solution; a poor proton acceptor; a base with a low value of K_b. (4.5, 15.6)

Weak Electrolyte A substance that has a low percentage ionization or dissociation in solution. (4.5)

Weighing The operation of measuring the mass of something. (1.3)

Weight The force with which something is attracted to Earth by gravity. (1.3)

Weight Fraction The ratio of the mass of one component of a mixture to the total mass. (12.5)

Wetting The spreading of a liquid across a solid surface. (11.3)

White Phosphorus An allotrope of phosphorus, P_4. (21.2)

Work The energy expended in moving an opposing force through some particular distance. Work has units of *force* $\times$ *distance*. (6.4, 18.1)

X

X Ray A stream of very high energy photons emitted by substances when they are bombarded by high-energy beams of electrons or emitted by radionuclides that have undergone K-electron capture. (22.3)

Y

Yield, Actual The quantity of product isolated from a chemical reaction. (3.8)

Yield, Percentage The ratio, given as a percent, of the quantity of product actually obtained in a reaction to the theoretical yield. (3.8)

Yield, Theoretical The quantity of a product calculated by the stoichiometry of the reaction. (3.8)

Z

Zinc–Carbon Dry Cell A galvanic cell of about 1.5 V involving zinc and manganese dioxide under mildly acidic conditions. (19.9)

Photo Credits

Chapter 1 *Opener:* ©Corbis. *Page 2:* ©Index Stock. *Page 4:* Gerard Lacz/Peter Arnold, Inc. *Page 5:* Stuart Westmorland/ Tony Stone Images/New York, Inc. *Pages 6, 9, and 14 (top left and bottom):* Michael Watson. *Page 8 (left):* West Stock Photography. *Page 8 (right):* Michael Schumann/SABA. *Page 10:* Courtesy Bureau International des Poids et Mesures. *Page 13 (top):* Andy Washnik. *Page 13 (cup of water):* Andy Washnik. *Page 13 (football player):* Jim Cummins/FPG International. *Page 13 (paper clip and penny):* Coco McCoy/Rainbow. *Page 14 (top and in center):* Courtesy Central Scientific Co. *Page 14 (top right):* Courtesy Cole-Parmer Instrument Co. *Page 24:* Jean-Yves Ruszniew/Corbis-Bettmann. *Page 26:* Andy Washnik. *Page 31 (top):* M. Claye/Jacana/Photo Researchers. *Page 31 (bottom):* Sinclair Stammers/Science Photo /Photo Researchers.

Chemicals in Our World 1 *Page 41:* Christel Rosenfeld/ Tony Stone Images/New York, Inc.

Chapter 2 *Opener:* Ian O'Leary/Tony Stone Images/New York, Inc. *Page 45 (left):* Thomas Brase/Tony Stone Images/ New York, Inc. *Pages 45 (center and right) and 46:* Andy Washnik. *Page 47 (top):* Michael Watson. *Page 47 (bottom):* Breck Kent. *Page 49:* Peter Lerman. *Page 51 (top):* James Brady. *Page 51 (bottom left):* Richard Megna/Fundamental Photographs. *Page 51 (bottom right):* Michael Watson. *Page 52:* Courtesy NASA. *Page 53:* Robert Capece. *Page 62:* Courtesy E.F. Smith Collection, Rare Book and Manuscript Library, University of Pennsylvania. *Page 67:* ©Index Stock. *Page 68 (top):* ©Tony Stone Images/New York, Inc. *Page 68 (bottom):* Courtesy General Cable Corporation. *Page 69 (diamonds):* Gerald Zanetti/The Stock Market. *Page 69 (Mercury):* OPC, Inc. *Page 69 (silicon wafer):* Courtesy Sematech. *Page 69 (Earth):* Courtesy NASA. *Page 69 (non-metallic elements):* Michael Watson. *Page 77:* Andy Washnik. *Page 79:* ©Bachmann/Photo Researchers. *Page 82:* Michael Watson. *Page 85:* Robert Capece.

Chapter 3 *Opener:* ©Index Stock. *Page 99:* ©Photri, Inc. *Page 101 (top):* Michael Watson. *Page 101 (bottom):* David Parker/Photo Researchers. *Page 102:* Tom McHugh/Photo Researchers. *Page 104:* Michael Watson. *Page 111:* Paul Silverman/ Fundamental Photographs. *Page 117:* Richard Megna/Fundametal Photographs. *Page 123:* Courtesy Orgo-Thermit. *Page 124:* Richard Elliot/Tony Stone Images/New York, Inc. *Page 127:* Courtesy Miles Laboratories. *Page 128:* Richard Megna/Fundamental Photographs. *Page 129:* Andy Washnik. *Page 130:* Lawrence Migdale/Photo Researchers. *Pages 132 and 134:* Michael Watson. *Page 133:* John Eastcott/Photo Researchers. *Page 136:* OPC, Inc.

Chemicals in Our World 2 *Page 148:* Courtesy of Sandia National Laboratories' Intelligent Micromachine Initiative; www.mdl.sandia.gov/Micromachine.

Chapter 4 *Opener:* Martin Barraud/Tony Stone Images/ New York, Inc. *Pages 152, 167, 172 (top), and 185:* Michael Watson. *Pages 155–157, 171, 173, and 180 (top):* Andy Washnik. *Page 160:* Courtesy of Betz Company. *Pages 162 (top) and 180 (bottom):* Robert Capece. *Page 162 (bottom):* Ken Karp. *Page 168:* Bill Stormont/The Stock Market. *Page 172 (bottom):* ©The Denver Post. Reproduced with permission.

Chapter 5 *Opener:* Robert E. Daemmrich/Tony Stone Images/New York, Inc. *Page 194:* Tony Wiles/Tony Stone Images/New York, Inc. *Page 195:* OPC, Inc. *Page 197 (top):* Courtesy Loctite Corporation. *Pages 197 (bottom) and 215 (bottom):* Robert Capece. *Page 198:* ©Corbis-Bettmann. *Page 201:* Peter Lerman. *Pages 206, 215 (top), 216 (bottom), and 218:* Andy Washnik. *Pages 208, 210, and 211:* Michael Watson. *Page 212:* OPC, Inc. *Page 214:* Sandra Baker/Gamma Liaison. *Page 216 (top):* Lynn Johnson.

Chemicals in Our World 3 *Page 229:* ©New York Times Pictures. Reproduced with permission.

Chapter 6 *Opener:* Michael T. Sedam/Corbis-Bettmann. *Page 232:* Andy Washnik. *Page 237:* ©Index Stock. *Page 248:* Ken Wardius/The Picture Cube. *Page 253:* Courtesy Robert Williams, Space Telescope Science Institute and NASA. *Page 254:* Archive Photos/The Image Bank.

Chapter 7 *Opener:* Barbara Filet/Tony Stone Images/New York, Inc. *Page 276 (left):* Peter Arnold/Peter Arnold, Inc. *Page 276 (center):* IPA/The Image Works. *Page 276 (right):* Edward Degginger/The Image Works. *Page 279:* From *The Gift of Color,* Eastman Kodak Company. *Page 280:* Gary Braasch/ Tony Stone Images/New York, Inc. *Page 283:* Wayne Eastep/ Tony Stone Images/New York, Inc. *Page 288:* Phil Savoie/ Bruce Coleman, Inc. *Page 289 (top):* John Paul Endress/The Stock Market. *Page 289 (bottom):* ©Index Stock. *Page 294:* Ray Pfortner/Peter Arnold, Inc.

Chapter 8 *Opener:* Oli Tennent/Tony Stone Images/New York, Inc. *Page 332:* ©Lawrence Berkeley National Laboratory/Photo Researchers. *Page 339:* Andy Washnik. *Page 348:* E. R. Degginger/Photo Researchers. *Page 354:* J. H. Robinson/Photo Researchers. *Page 356:* Courtesy Polystyrene Packaging Council.

Chapter 9 *Opener:* ©VCG 1997/FPG International. *Page 413:* Andy Washnik.

Chemicals in Our World 5 Page 421: Courtesy R. Wooldridge.

Chapter 10 *Opener:* Frank Whitney/The Image Bank. *Page 425:* OPC, Inc. *Page 450:* Andy Washnik.

Chapter 11 *Opener:* ©Tony Stone Images/New York, Inc. *Page 470:* John Beatty/Tony Stone Images/New York, Inc. *Page 474:* IPA/Peter Arnold, Inc. *Pages 476 (top), 477, 479, and 491:* Michael Watson. *Page 476 (center):* Hermann Eisenbeiss/Photo Researchers. *Page 476 (bottom):* Pat O'Hara/Tony Stone Images/New York, Inc. *Page 486:* Courtesy Corning Consumer Products Company. *Page 497:* Courtesy Oregon Freeze Dry, Inc. *Pages 501 (top) and 508:* Robert Capece. *Page 501 (bottom):* Eric Grave/Science Source/Photo Researchers.

Chemicals in Our World 6 *Page 519:* Jim Mendenhall.

Chapter 12 *Opener:* ©Telegraph Colour Library/FPG International. *Page 531:* Andy Washnik. *Pages 533 and 561:* Jeff Hunter/The Image Bank. *Page 536:* Yoav Levy/Phototake. *Page 560:* OPC, Inc.

Chapter 13 *Opener:* Raymond Gehman/Corbis-Bettmann. *Page 575:* Courtesy USDA. *Page 576:* OPC, Inc. *Page 607:* Leonard Lee Rue III/Photo Researchers. *Page 612 (left):* Courtesy American Petroleum Institute. *Page 612 (right):* Courtesy Englehard Corporation. *Page 613:* Courtesy AC Spark Plug.

Chemicals in Our World 7 *Page 625:* Simon Fraser/Science Photo Library/Photo Researchers.

Chapter 14 *Opener:* Dewitt Jones/Corbis-Bettmann. *Page 641:* Pete Seaward/Tony Stone Images/New York, Inc. *Pages 642 and 644:* Michael Watson.

Chapter 15 *Opener:* Paul A. Souders/Corbis-Bettmann. *Pages 668–689:* Andy Washnik.

Chemicals in Our World 8 *Page 699:* David Perker/IMI/University of Birmingham TC Consortium/Science Photo Library/Photo Researchers.

Chapter 16 *Opener:* ©Tony Stone Images/New York, Inc. *Pages 702, 709, 711, 712, 715–717, 734, and 748:* Andy Washnik. *Pages 708, 725, and 738:* Coco McCoy/Rainbow. *Page 726:* Courtesy Chris Stocker. *Page 739:* Paul Silverman/Fundamental Photographs. *Page 740:* Peter Lerman.

Chapter 17 *Opener:* Peter Aitken/Photo Researchers. *Page 759:* Andy Washnik. *Pages 761, 767, 768, and 773 (top):* Michael Watson. *Page 773 (bottom):* OPC, Inc. *Pages 780 and 781:* Andy Washnik.

Chemicals in Our World 9 *Page 791:* Pascal Tournaire/Agence Vandystadt/Allsport.

Chapter 18 *Opener:* Alan R. Moller/Tony Stone Images/New York, Inc. *Page 794:* Richard A. Cooke III/Tony Stone Images/New York, Inc. *Page 795 (top):* Jenny Woodcock/Corbis-Bettmann. *Page 795 (bottom):* Peter Vadnai/The Stock Market. *Page 797:* Bob Krist/Corbis-Bettmann. *Page 798:* Charles D. Winters/Photo Researchers. *Page 799 (left):* David Stoecklein/The Stock Market. *Page 799 (center):* Adam Jones/Photo Researchers. *Page 799 (right):* Lowell J. Georgia/Photo Researchers. *Page 807:* David Woodfall/Tony Stone Images/New York, Inc. *Page 809:* Jim Zipp/Photo Researchers. *Page 814:* Courtesy American Coalition for Ethanol. *Page 817:* Courtesy Lockheed Missile and Space Co.

Chapter 19 *Opener:* Peter Turnley/Corbis-Bettmann. *Pages 849, 851, and 862:* Michael Watson. *Page 852:* Courtesy Edgar Fahs Smith Collection, University of Pennsylvania. *Page 857:* Courtesy Kaiser Aluminum. *Page 859:* Courtesy ASARCO, Inc. *Page 872:* Courtesy James Brady. *Page 881:* Science Photo Library/Photo Researchers. *Page 884:* Courtesy Fisher Scientific. *Page 886:* OPC, Inc.

Chemicals in Our World 10 *Page 899:* Courtesy Mercedes Benz of North America.

Chapter 20 *Opener:* Donna Disario/The Stock Market. *Page 902 (top):* Dan Boler/Tony Stone Images/New York, Inc. *Page 902 (bottom):* Art Resource. *Page 903:* Richard Megna/Fundamental Photographs. *Pages 906, 910, 912, 913, 915 (top), and 925:* Andy Washnik. *Page 911:* Jerry Mason/Photo Researchers. *Page 915 (bottom), 916, 923, 924, 929. and 929:* Michael Watson.

Chapter 21 *Opener:* Paul Steel/The Stock Market. *Page 946:* David Parker/Photo Researchers. *Page 947 (top):* ©API/Explorer/Photo Researchers. *Page 947 (bottom):* ©Telegram Colour Library/FPG International. *Page 951:* William Rivelli/The Image Bank. *Page 953:* Lee Snider/Corbis-Bettmann. *Page 955:* Larry Burrows, Life Magazine, ©1966, Time, Inc. *Pages 959, 960, 961, 974, and 983:* Andy Washnik. *Page 966:* Richard Megna/Fundamental Photographs. *Page 970:* Roberto deGugliemo/Photo Researchers.

Chemicals in Our World 11 *Page 990:* A. J. Copley/Visuals Unlimited.

Chapter 22 *Opener:* Mark Harmel/Tony Stone Images/New York, Inc. *Page 994:* Philippe Halsman Studio. *Page 999:* Courtesy Los Alamos Scientific Laboratory, University of California. *Page 1000:* Courtesy AIP Neils Bohr Library. *Page 1005:* Courtesy of E. D. London, National Institute on Drug Abuse. *Page 1007:* Courtesy Brookhaven National Laboratory. *Page 1009:* Courtesy Ludlum Measurements, Inc. *Page 1010:* Yoav Levy/Phototake. *Page 1012:* Charles D. Winters/Photo Researchers.

Chapter 23 *Opener:* Jess Stock/Tony Stone Images/New York, Inc. *Pages 1028 (top), 1035, and 1038:* Robert Capece. *Page 1028 (bottom):* Michael Watson. *Page 1032:* Martin Bond/Photo Researchers. *Page 1050:* Richard Price/FPG International. *Page 1051 (top):* Courtesy Regal Ware. *Page 1051 (bottom):* Michael Ventura/Bruce Coleman, Inc. *Page 1053 (top):* Robert Tringali Jr./Sports Chrome, Inc. *Page 1053 (center):* Peter Holden/Visuals Unlimited. *Page 1053 (bottom):* ©Telegraph Colour Library/FPG International. *Page 1067:* Nelson Max/Peter Arnold, Inc.

Chemicals in Our World 12 *Page 1085:* Image courtesy Affymetrix, Inc., Santa Clara, CA.

Index

Relationships Among Units

(Values in boldface are exact.)

Length

1 in. = **2.54** cm
1 ft = **30.48** cm
1 yd = **91.44** cm
1 mi = **5280** ft
1 ft = **12** in.
1 yd = **36** in.

Volume

1 liq. oz = **29.57353** mL
1 qt = **946.352946** mL
1 gallon = **3.785411784** L
1 gallon = **4** qt = **8** pt
1 qt = **2** pt= **32** liq. oz

Mass

1 oz = **28.349523125** g
1 lb = **453.59237** g
1 lb = **16** oz

Pressure

1 atm = **760** torr
1 atm = **101,325** Pa
1 atm = 14.696 psi (lb/in.^{2})
1 atm = 29.921 in. Hg

Energy

1 cal = **4.184** J
1 eV = 1.6022×10^{-19} J
1 eV/molecule = 96.49 kJ/mol
1 eV/molecule = 23.06 kcal/mol
1 J = 1 kg m^2 s^{-2} = 10^7 erg

Physical Constants

Rest mass of electron	$m_e = 5.48579903 \times 10^{-4}$ u ($9.1093897 \times 10^{-28}$ g)
Rest mass of proton	$m_p = 1.007276470$ u ($1.6726231 \times 10^{-24}$ g)
Rest mass of neutron	$m_n = 1.008664904$ u ($1.6749286 \times 10^{-24}$ g)
Electronic charge	$e = 1.60217733 \times 10^{-19}$ C
Atomic mass unit	u = $1.6605402 \times 10^{-24}$ g
Gas constant	R = 0.082057 L atm mol^{-1} K^{-1}
	= 8.314510 J mol^{-1} K^{-1}
	= 1.9872 cal mol^{-1} K^{-1}
Molar volume, ideal gas	= 22.41383 L (at STP)
Avogadro's number	= 6.0221367×10^{23} things/mol
Speed of light in a vacuum	= 2.99792458×10^8 m s^{-1}
Plank's constant	= $6.6260755 \times 10^{-34}$ J s
Faraday constant	= 9.6485309×10^4 C mol^{-1}

Laboratory Reagents

Values are for the average concentrated reagents available commercially.

Reagent	Percent (wt/wt)	Mole Solute Liter Solution	Gram Solute 100 ml Solution
NH_3	29	15	26
$HC_2H_3O_2$	99.7	17	105
HCl	37	12	44
HNO_3	71	16	101
H_3PO_4	85	15	144
H_2SO_4	96	18	177